International Conference on Mechatronics, Manufacturing and Materials Engineering 2016 (MMME 2016)

MATEC Web of Conferences Volume 63 (2016)

Hong Kong, China
11-12 June 2016

Editors:

Wen-Pei Sung

Jimmy Chih-Ming Kao

ISBN: 978-1-5108-2677-9

Printed from e-media with permission by:

Curran Associates, Inc.
57 Morehouse Lane
Red Hook, NY 12571

Some format issues inherent in the e-media version may also appear in this print version.

Printed by Curran Associates, Inc. (2016)

For additional information, please contact EDP Sciences – Web of Conferences
at the address below.

EDP Sciences – Web of Conferences
17, Avenue du Hoggar
Parc d'Activité de Courtabœuf
BP 112
F-91944 Les Ulis Cedex A
France

Phone: +33 (0) 1 69 18 75 75
Fax: +33 (0) 1 69 28 84 91

contact-edps@webofconferences.org

Additional copies of this publication are available from:

Curran Associates, Inc.
57 Morehouse Lane
Red Hook, NY 12571 USA
Phone: 845-758-0400
Fax: 845-758-2633
Email: curran@proceedings.com
Web: www.proceedings.com

TABLE OF CONTENTS

MATEC Web of Conferences
Open access conference proceedings in Materials science, Engineering and Chemistry
A flexible, fast and efficient publication service, for an open access recording of your event and a wide dissemination of your proceedings
www.matec-conferences.org
edp open Open Access Publishing from EDP Sciences

ISSN (electronic edition): 2261-236X

MMME 2016 Committee

Conference Chairman

Prof. Jimmy C.M. Kao, National Sun Yat-Sen University
Prof. Wen-Pei Sung, National Chin-Yi University of Technology

Program Committee

Andrey Nikolaevich Belousov, Prof. Andrey Belousov, Head Laboratory of Applied Nanothechnology, Ukraine
Ghamgeen Izat Rashed, Wuhan University, China
Sajjad Jafari, Semnan University, Iran
Krupa Ranjeet Rasane, Department of Electronic Communications Engineering KLE Society's College of Engineering, India
Ahmed N. Abdalla, Universiti Malaysia Pahang, Malaysia
Yan Wang, The University of Nottingham, U.K.
Prof. Yu-Kuang Zhao, National Chin-Yi University of Technology, Taiwan
Yi-Ying Chang, National Chin-Yi University of Technology, Taiwan
Viranjay M.Srivastava, Jaypee University of Information Technology, Solan, H.P. India
Chenggui Zhao, Yunnan Normal University, China
Hsiang-Chuan Liu, Asia University
Hao-En Chueh Yuanpei University
Zhou Liang, Donghua University, China
Liu Yunan, University of Michigan, USA
Wang Liying, Institute of Water Conservancy and Hydroelectric Power, China
Rahim Jamian, Universiti Kuala Lumpur Malaysian Spanish Institute, Malaysia
Lixin Guo, Northeastern University, China
Wen-Sheng Ou, National Chin-Yi University of Technology
Mostafa Shokshok, National University of Malaysia, Malaysia
Ramezan ali Mahdavinejad, University of Tehran, Iran
Anita Kovač Kralj, University of Maribor, Slovenia
Tjamme Wiegers, Delft University of Technology, Netherlands
Bhagavathi Tarigoppula, Bradley University, USA

Co-sponsor

International Frontiers of science and technology Research Association
HongKong Control Engineering and Information Science Research Association
National Chin-Yi University of Technology

Preface

2016 International Conference on Mechatronic, Manufacturing and Materials Engineering (MMME 2016) was held in Hong Kong, during June 11-12, 2016. The conference is an international forum for the presentation of technological advances and research results in the fields of Mechatronic, Manufacturing, Information and Materials Engineering. The conference brought together leading researchers, engineers and scientists in the domain of interest from around the world to share the valuable experiences with the scientist and scholars around the world.

Today Machinery and Electronics industry is very important in our economic life and it is involved in man's foods, clothes, housing, and transportation and so on. In this conference, we discussed the Mechatronic, Manufacturing, Information and its application in many fields. We received more than 500 submissions from email and electronic submission system, which were reviewed by international experts, and about 200 papers have been selected for presentation, representing 11 national and international organizations. It will be the most comprehensive Conference focused on the Mechatronic, Manufacturing, Information and Materials Engineering.

We would like to thank the conference chairs, organization staff, the authors and the members of International Technological Committees for their hard work.

May, 2016

Wen-Pei Sung
National Chin-Yi University of Technology

Statement of Peer review

In submitting conference proceedings to MATEC Web of Conferences, I certify to the Publisher that I adhere to the **Policy on Publishing Integrity** of the journal in order to safeguard good scientific practice in publishing.

1. All articles have been subjected to peer review administered by the proceedings editors.
2. Reviews have been conducted by expert referees, who have been requested to provide unbiased and constructive comments aimed, whenever possible, at improving the work.
3. Proceedings editors have taken all reasonable steps to ensure the quality of the materials they publish and their decision to accept or reject a paper for publication has been based only on the merits of the work and the relevance to the journal.

Title, date and place of the conference

2016 International Conference on Mechatronics, Manufacturing and Materials Engineering (MMME2016)
June 11-12, 2016
Hong Kong

Proceedings editor(s):

Wen-Pei Sung, National Chin-Yi University of Technology
Jimmy Chih-Ming Kao, National Sun Yat-Sen University

Date and editor's signature

MATEC Web of Conferences **63**, 01001 (2016)
MMME 2016
DOI: 10.1051/matecconf/20166301001

Radon Exhalation Patterns in A Dead-end Tunnel

Shu Guang WANG[a], Xiao Bing NIU, Qing CHEN, Wei YANG

Northwest Institute of Nuclear Technology, Shaanxi, Xi'an, 710024, China

Abstract. The regularity of radon exhalation rate in the over-broken granite tunnel is susceptible to weather conditions and ventilation styles. Based on the calculation model of radon exhalation in tunnel, some experiments have been carried out to analyze the variations of radon exhalation in cases of natural ventilation, blowing ventilation and exhaust ventilation separately. The results show that there is a linear relation between the radon exhalation and the natural ventilation quantity, and also between the radon exhalation and the ambient temperature; the radon exhalation in the case of exhaust ventilation is 63% higher than that in the blowing case under the condition of the same ventilation quantity and ambient temperature. Therefore, it is suggested that operation in the tunnel in high temperature be avoided in summer, and the blowing ventilation be adopted as an effective way for ventilation.

Introduction

In addition to the physical properties of rock and the exposed area of surrounding rock, the main factors affecting the radon exhalation rate in tunnel include the weather conditions and ventilation styles. As early as 1951, Smith, O. F. discovered the effects of weather conditions on the soil radon exhalation; in 1963, Schroeder, G. L. began to test the control of radon exhalation during the tunnel construction in pressurized ventilation[1]. F. Perrier[2] et al, through on-site monitoring of the tunnel with the length of 38m and slope of 8% in Pulchoki Mount of Nepal, achieved the following conclusions: obvious natural wind only existed in this tunnel in winter (from October to next March, the temperature outside the tunnel is lower than that inside) and the radon concentration in the tunnel changed with the ventilation quantity. Perrier[2] pointed out that the change in radon concentration was very sensitive to airflow exchange, but he did not make a quantitative description of the changing relationship between them. Mechanical ventilation had been widely used as a common way of radon reduction in underground construction[1], but the relationship between mechanical ventilation and radon exhalation rate was still unclear.

Through experiments, this paper studies the regularity of radon exhalation respectively in natural ventilation and mechanical ventilation in an over-broken granite dead-end tunnel in Northwest China, and analyzes the effects of outside temperature on the radon exhalation in tunnel in different ventilation cases. The pollution of radon in tunnel in summer is the most serious, so this paper just focuses on the situation in summer.

[a] Corresponding author: hitwsg@163.com

1 Formula of Radon Exhalation Rate in Ventilated Tunnel

The calculation model of change of concentration of radon in the ventilated tunnel is shown in Fig.1. Take the micro unit of dx wide (right in Fig.1), then at a certain time, the equilibrium equation of radon in the micro unit radon is:

$$C_1Q_1 + \delta s dx = C_2Q_2 + \lambda A dx C_1 \quad (1)$$

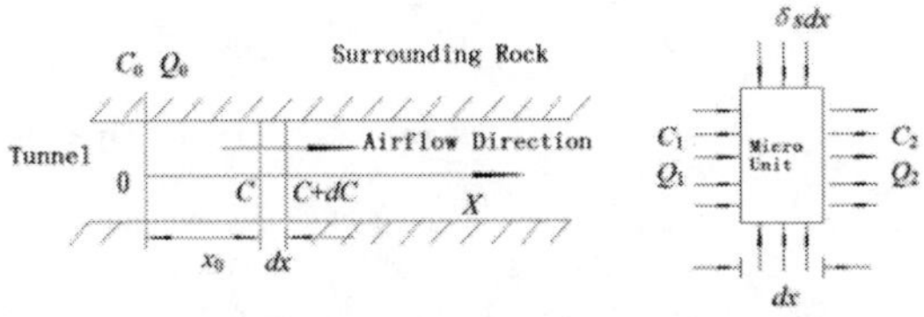

a. Ventilated Tunnel b. Micro Unit

Fig.1 Calculation model of radon concentration in the ventilated tunnel

Where: C_1 is the concentration of radon flowing into the micro-unit airflow, Bq/m^3; Q_1 is the volume of air flowing into the micro unit, m^3/s; δ is the radon exhalation rate on the surrounding rock surface, Bq/(s·m^2); s is the sectional perimeter of tunnel, m; A is the cross-sectional area of tunnel, m^2; dx is the width of micro unit, m; C_2 is the concentration of radon flowing out of micro unit, Bq/m^3; Q_2 is the volume of air flowing out of micro unit, m^3/s; λ is the radon decay constant, s^{-1}.

The physical meaning of Eq.1: at a certain time, the sum of the amount of radon C_1Q_1 flowing into the micro unit by the tunnel airflow and the amount of radon δsdx exhaled from the surrounding rock surface to the micro unit is equal to the sum of the amount of radon C_2Q_2 flowing out of the micro unit and the decay amount of radon $\lambda AdxC_1$ in the micro unit.

Also, C_1 is equal to C shown at the left of Fig.,Q_1 is equal to Q_0+qx_0, C_2 is equal to $C+dC$, Q_2 is equal to $Q_0+q(x_0+dx)$. Therefore, the Eq.1 can be rewritten as

$$\delta sdx+C(Q_0+qx_0)=[Q_0+q(x_0+dx)](C+dC)+\lambda ACdx \tag{2}$$

Where, C is the radon concentration at the coordinates x_0, Bq/m3; q is the natural ventilation quantity of tunnel with per unit length, $m^3/(m\cdot s)$; Q_0 is the ventilation quantity of tunnel at the origin of coordinates, (m^3/s); δ is radon exhalation rate of surrounding rock, $Bq/(s\cdot m^2)$; s is the sectional perimeter of tunnel, taking the average of 12m; A is cross-sectional area of tunnel, taking the design value of $9m^2$.

Omit the second-order infinitesimal and take the radon exhalation rate δ as a constant, then Eq.2 is changed as

$$\frac{dC}{\delta s-\lambda AC-qC}=\frac{dx}{Q_0+qx_0}$$

Substitute the boundary condition $C\,|_{x=0}=C_0$, then the solution of the equation is:

$$C=\frac{\delta s}{\lambda A+q}+\left(C_0-\frac{\delta s}{\lambda A+q}\right)\left(\frac{Q_0}{Q_0+qx_0}\right)^{\frac{\lambda A+q}{q}}$$

After deformation, the above equation is:

$$\delta=\frac{(\lambda A+q)[C-C_0(\frac{Q_0}{Q_0+qx_0})^{\frac{\lambda A+q}{q}}]}{s[1-(\frac{Q_0}{Q_0+qx_0})^{\frac{\lambda A+q}{q}}]} \tag{3}$$

The actual measured q of the tunnel is approximately $10^{-4}m^3/(s\cdot m)$, and λA is about $10^{-5}m^2/s$. Therefore it can be approximately considered that $\lambda A+q\approx q$, then the Eq.3 can be simplified as:

$$\delta=\frac{C(qx_0+Q_0)-C_0Q_0}{sx_0} \tag{4}$$

Where, (qx_0+Q_0) is the sectional ventilation quantity at the coordinate x_0.

The Eq.4 is the formula of radon exhalation rate. As s and x_0 are dimension parameters of tunnel and are the known quantities in the formula, the radon exhalation rate can be calculated by just measuring the radon concentrations (C and C_0) and ventilation quantities (q and Q_0) of two sections.

2 Impact of Natural Ventilation on Radon Exhalation Rate

During the measurement of radon exhalation rate, according to the air state of the measured environment, the measurement methods can be divided into two kinds, namely dynamic and static methods [3, 4]. The so-called dynamic method is to measure the radon exhalation rate in the case of air flow, while the static method is to do so in the still air. There is always natural ventilation in the experimented tunnel. Therefore, only the dynamic method is used for the measurement of radon exhalation rate.

2.1 Experimental methods and measures

The quantity of natural ventilation in tunnel is usually small and the wind speed is small, so it is not easy to conduct the all-section measurement. Meanwhile, the distribution of natural ventilation flow field is and the wind directions above and below the section are opposite (Fig. 2). In order to avoid the effects of external weather conditions on the measurement results and improve the reliability of the measurement data, an air door is set up 100m away from the entry of the tunnel (Fig. 3) as the natural ventilation's section of wind speed measurement. The size of inside section is adjusted by adjusting the number of inserted door planks. The natural ventilation quantity can be calculated according to sectional wind speed and vent area. During measurement, adjust the width of air door to 13cm, set up 3 measurement points on the measurement section (Fig. 3 (b)) and average the sectional wind speeds. Calculate the ventilation quantity according to the sectional wind speed and ventilation area[5, 6].

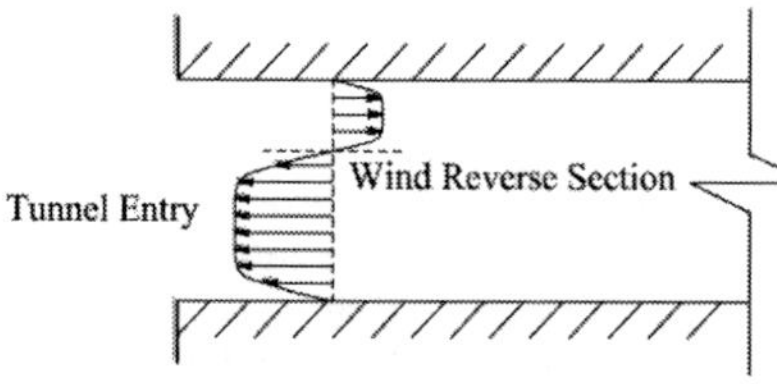

(a) Summer Conditions

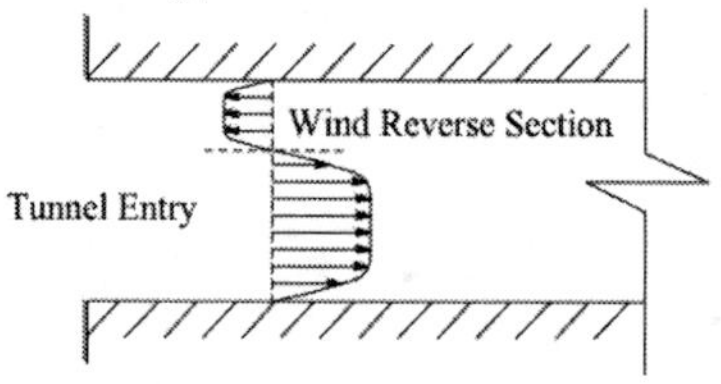

(b) Winter Conditions

Fig. 2 Seasonal change law of natural ventilation in tunnel

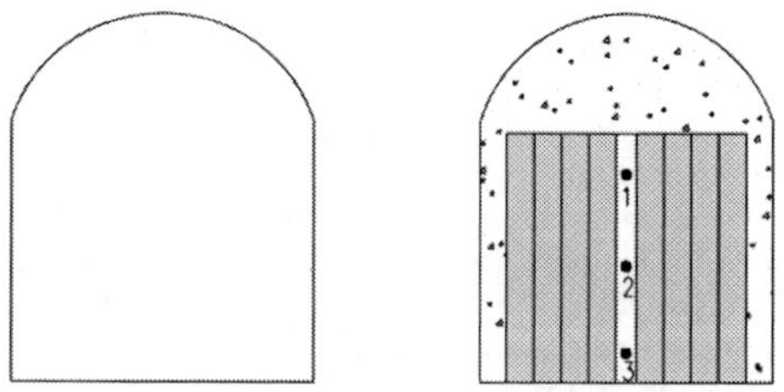

(a) Section of Original Tunnel (b) Rectification Section and Measurement Points

Fig. 3 Layout of recitation section and measurement point of tunnel ventilation

DOI: 10.1051/matecconf/20166301001

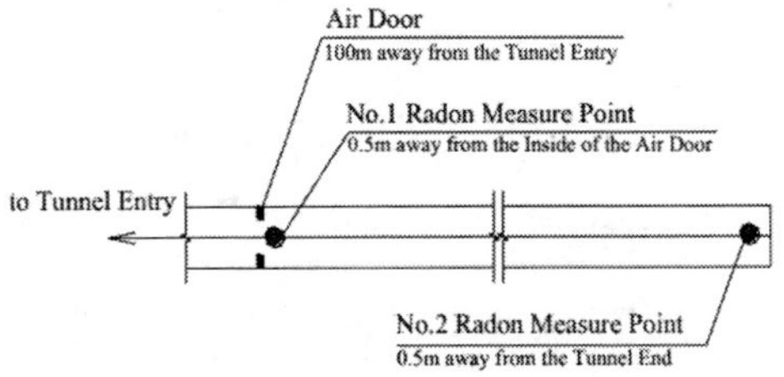

Fig. 4 Overall layout of radon exhalation ratemeasurement in natural ventilation case

The overall layout of the experiment is shown in Fig. 4. The distance from the measured tunnel section to the end of tunnel is about 600m. Study the regularity of radon exhalation according to the measurement principle with the dynamic method in the natural ventilation as well as Eq.4. Not only the natural ventilation quantity, surface area of tunnel and its length shall be measured, but it is necessary to monitor the concentration of radon in tunnel as well as the outside temperature.

(1) Radon concentration measurement Under the normal conditions in summer, there is a large amount of radon accumulated in the tunnel. The transfer of the radon is mainly controlled by the convection and diffusion, so that the radon concentration reaches relative balanced state in a certain time. Experimental results show that, in the absence of mechanical ventilation, the radon concentrations in the tunnel are basically the same everywhere, but the radon concentration in the rock wall is slightly higher with the presence of cracks (or rock bolt orifice). Therefore, if the effect of surrounding rock nonuniformity is ignored, and the radon concentration is measured in natural state theoretically, the radon detector can be placed in any position. The measuring method is as follows: set up two radon concentration monitoring points (Fig. 4) in the tunnel, both located in the center of the cross section of the tunnel, and average the radon concentrations at two measurement points. The main instrument used is RAD7 continuous electronic radon detector, with the accuracy of ±5% (measuring range is from 3.7Bq/m^3 to 7.4×10^{4}Bq/m^3).

(2) Monitor temperature Set up a thermometer screen outside the tunnel and place the metrological sensor in it to continuously monitor the outside temperature. The main component used is HCD1019-5P multi- parameter metrological sensor, with the temperature measurement accuracy of <±0.3℃ (measurement range is from -40 to +123.8℃).

2.2 Data analysis

2.2.1 Natural ventilation quantity and radon exhalation rate

Radon exhalation rate is calculated according to Eq.4. The relationship between the natural ventilation quantity and radon exhalation rate is shown in Fig.5.

After Linear regression analysis, the regression equation of radon exhalation rate and natural ventilation quantity in the natural state is:

$$\delta = 9.0765Q - 0.4854 \tag{5}$$

It can be seen that the radon exhalation rate is positively correlated with the natural ventilation quantity.

In theory, in the case of natural ventilation: Q_0=0, $q\neq 0$. Substitute this condition into Eq.3 and it is obtained as follows after transformation:

$$\delta = \frac{C}{S}\cdot Q + \frac{\lambda VC}{S} \tag{6}$$

Where, Q is the natural ventilation quantity, m^3/s; V is the tunnel volume, m^3; S is the surface area of the tunnel, S=sx, m^2.

It can be seen, among the physical quantities in the Eq.6, the rest are the constants except for the natural ventilation quantity Q as a variable. Namely, this equation is like the style of a unitary linear function: $y=ax+b$, indicating linear relationship between radon exhalation rate δ and the natural ventilation quantity Q. This conclusion is consistent with the analysis results of the above measured data.

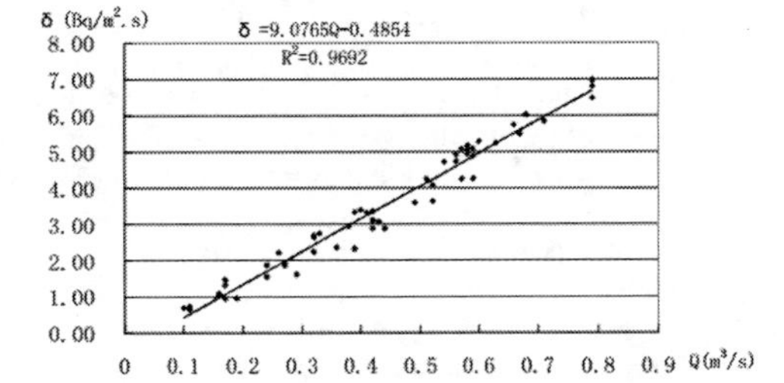

Fig. 5 Relationship between natural ventilation quantity and radon exhalation rate

2.2.2 Correlation analysis of radon exhalation rate and temperature

(1) Regression equation Based on the mechanism of radon exhalation, the key factor affecting the radon exhalation rate is the pressure difference between the inside and outside of the tunnel and the pressure difference is due to the temperature change. Therefore, the study of relationship between the environmental temperature and radon exhalation rate shall be emphasized. Effective data of a total of 56 groups is obtained in the experiment. The scatter diagram of temperature and radon exhalation rate is shown in Fig. 6.

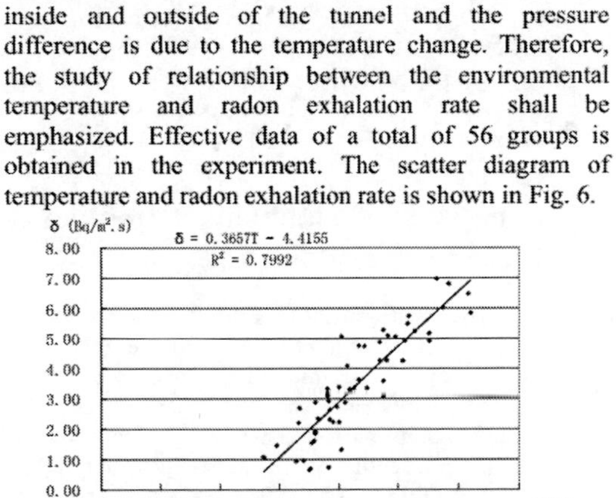

Fig. 6 The outside temperature VS radon exhalation rate in natural ventilation case

Fig.6 shows that there is an obvious linear relationship between temperature and radon exhalation rate. After the linear regression analysis, the linear regression equation of radon exhalation rate and temperature is:

$$\delta = 0.3657T - 4.4155 \tag{7}$$

Where, T is the temperature outside the tunnel, ℃.

(2) Test of linear regression significance Test the significance of linear regression with the F testing method:

$$F = (n-2)\frac{R^2}{1-R^2} = 214.92$$

Look up the table of F distribution critical values and obtain: when α=0.05, the degree of freedom is (1, n-2), namely the critical value λ of (1, 54) is $4.00 < \lambda_{\alpha=0.05} < 4.08$, then $F > \lambda_{\alpha=0.05}$, namely the linear regression is significant. Look up the table of F distribution critical values and obtain: when α=0.01, the degree of freedom is (1, n-2), namely the critical value λ of (1, 54) is $7.08 < \lambda_{\alpha=0.01} < 7.31$, then $F > \lambda_{\alpha=0.01}$, namely, the linear regression is highly significant.

(3) Prediction interval According to mathematical statistical theory, the approximate prediction interval with the confidence coefficient 1-α=0.95 of the predicted value δ_0 of δ at $T=T_0$ based on the regression Eq.7 is (δ_0-2σ, δ_0+2σ). The approximate prediction interval with the confidence coefficient 1-α=0.99 of the predicted value δ_0 of δ at $T=T_0$ is (δ_0-3σ, δ_0+3σ). Where, σ is the variance of the predicted values.

$$\sigma^2 = [\sum_{i=1}^{n}(\delta_i - \bar{\delta})^2 - a\sum_{i=1}^{n}(\delta_i - \bar{\delta})(T_i - \bar{T})]/(n-2) = 0.59$$

Namely, σ=0.77. In the above equation, $\bar{\delta}$ is the average value of the measured radon exhalation rates; δ_i is the ith value of the measured radon exhalation rate; $\bar{T}$ is the average value of the measured temperature; Ti is the ith value of the measured temperature; a is the monomial coefficient of the regression equation, 0.3657; n is the sample size, 56.

Whereby, the approximate prediction interval with the confidence coefficient 1-α=0.95 of the predicted value δ_0 is (δ_0-1.54, δ_0+1.54), while the approximate prediction interval with the confidence coefficient 1-α=0.99 of the predicted value δ_0 is (δ_0-2.31, δ_0+2.31).

3 Impact of Mechanical Ventilation on Radon Exhalation Rate

The measurement of radon exhalation rate of tunnel in the mechanical ventilation with the dynamic method should meet the following conditions. (1) Radon exhalation in tunnel should be uniform and the radon concentrations on the same section are equal. (2) The mechanical ventilation quantity is constant. To meet the above conditions, a fixed fan is selected in the experiment for the mechanical ventilation in the tunnel, and the measurement of various data in the tunnel is implemented after the stable operation of ventilation system for 24 hours.

3.1 Experimental methods and measures

Select the studied tunnel of about 120m long. The overall measurement layout is shown in Fig. 7. According to the measurement principle of dynamic method as well as the calculation formulas of radon exhalation rate (4), not only the surface area and length of the tunnel shall be measured, but it is necessary to measure the ventilation quantities and radon concentrations in the two sections. In order to ensure the reliability of measurement results, the following methods and measures are applied during measurement.

(1) Tunnel selection In order to reduce the effects of changes in outer weather conditions on the measured data in the tunnel, the experimental section is located 100m away from the tunnel entry, with a total length of 100m, namely the shaded area in Fig. 7.

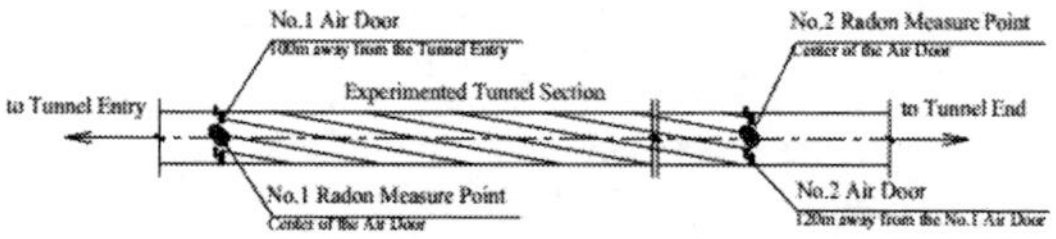

Fig. 7 Overall measurement layout of radon exhalation rate in mechanical ventilation case

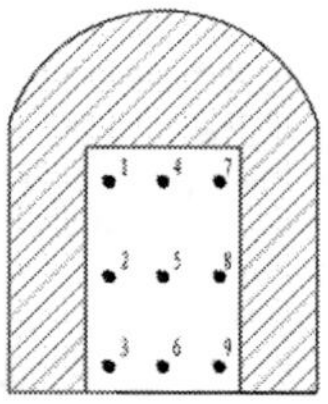

Fig. 8 Positions of wind speed measurement points in the mechanical ventilation section

(2) Ventilation system design and air distribution The ventilation system in tunnel is equipped with the single-wind-path rigid duct. On the duct, two fans of the same type are connected in parallel: one of which can achieve the blowing ventilation, namely the airflow direction in tunnel is outward from inward area when this fan operates; the other of which can achieve the exhaust ventilation, namely the airflow direction in tunnel is inward from outward area when this fan operates. Therefore, the research on the radon exhalation can be conducted separately in the blowing ventilation and exhaust ventilation.

The ventilation duct extends out of tunnel entry approximately 30m and the duct mouth is in an open space. The impact of air backflow in the mechanical ventilation on the radon concentration in tunnel can be ignored. The duct in the tunnel is throughout the entire selected section, and its end is about 200m away from the air door. The impact of duct mouth on the flow field in the tunnel can be ignored.

(3) Air volume measurement Considering that the natural ventilation quantity in the mechanical ventilation may result in the difference of total ventilation quantity in

the sections of two ends of selected tunnel, it is necessary to separately measure the air volume in two sections. It is difficult to calculate the area of the cross section through all-section measurement due to the irregularity of the real tunnel section. In order to facilitate measurement and calculation, two air doors are separately set up at both ends of the selected tunnel (Fig. 7) as the measurement sections of wind speeds. A total of 9 measurement points are set up on the section (Fig. 8), and the average value of results form the 9 points is taken as the sectional wind speed. Finally, the total ventilation quantity shall be calculated according to the sectional wind speed in the section and the ventilation area.

(4) Measurement of radon concentration After mechanical ventilation, a large amount of radon flow out of tunnel due to the exhaust effect of ventilation and the distribution of radon concentration changes due to the dilution effect of ventilation. The experimental results show that, in the mechanical ventilation, the radon concentration in tunnel gradually increases along the flow direction and the maximum radon concentration will be at the end of airflow [7]. There are some similar conclusions in literature [8].

The air door changes the tunnel shape and affects the airflow in tunnel. A vortex will be formed when the air flows through the air door [9], seriously affecting the uniformity of radon distribution in tunnel.

According to the above situation, in the mechanical ventilation, it is necessary to separately measure the radon concentrations at both ends of tunnel; the radon concentration in the center of air door can be considered to represent the concentration level of radon in the air flowing through the air door. The layout of radon detector in mechanical ventilation is shown in Fig. 7.

(5) Monitor of temperature outside tunnel

(6) Calculate the radon exhalation rate based on equation (4) according to the radon concentration and air volume at the air door.

3.2 Data processing and analysis

3.2.1 Calculation of radon exhalation rate

According to Eq.4, radon exhalation rate is calculated in accordance with the following equation:

$$\delta = \frac{C(qx_0 + Q) - C_0 Q}{sx_0} \quad (8)$$

Where, C is the concentration of radon in the air flowing through the air door, Bq/m^3, where the external air door is referred to in the case of blowing ventilation and External air door in the case of exhaust ventilation; C_0 is the initial concentration of radon in the air flowing through the air door, Bq/m^3, where the External air door is referred to in the case of blowing ventilation and external air door in the case of exhaust ventilation; Q is mechanical ventilation quantity through the air door, m^3/s, where the External air door is referred to in the case of blowing ventilation and external air door in the case of exhaust ventilation; (qx_0+Q) is the total air volume through the air door, m^3/s, where the external air door is referred to in the case of blowing ventilation and External air door in the case of exhaust ventilation; x_0 is the length of the measured tunnel, taking 120m; s is the cross-section perimeter of tunnel, taking the average value of 12m.

3.2.2 Data Analysis in blowing ventilation

(1) Regression analysis Obtain 109 groups of effective data in blowing ventilation. The scatter diagram of temperature and radon exhalation rate is shown in Fig. 9, which indicates an obvious linear relationship. Therefore, the linear fitting is conducted.

After linear regression analysis, the linear regression equation of radon exhalation rate is

$$\delta = 1.7814T - 28.94 \quad (9)$$

Where, δ is the radon exhalation rate, Bq/(m^2.s); T is the temperate outside tunnel, $15℃ \le T \le 34℃$

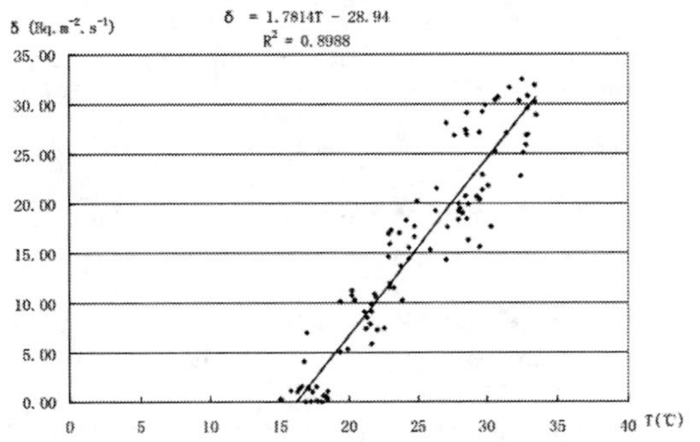

Fig. 9 Radon exhalation rate VS the temperature outside tunnel in blowing ventilation case

(2) Significance test Test the significance of the linear regression with the F testing method. The test result shows that the linear regression is highly significant.

(3) Prediction interval The approximate prediction interval with the confidence coefficient $1\text{-}\alpha=0.95$ of the predicted value δ_0 is $(\delta_0\text{-}6.74, \delta_0\text{+}6.74)$, while the approximate prediction interval with the confidence coefficient $1\text{-}\alpha=0.99$ of the predicted value δ_0 is $(\delta_0\text{-}10.11, \delta_0\text{+}10.11)$.

3.2.3 Data analysis in exhaust ventilation

(1) Regression Analysis Obtain 89 groups of effective measurement data in exhaust ventilation. The scatter diagram of temperature and radon exhalation rate is shown in Fig. 10, which indicates an obvious linear relationship. Therefore, the linear fitting is conducted. After linear regression analysis, the linear regression equation of radon exhalation rate is

$$\delta = 2.8979T - 45.168 \quad (10)$$

Where, $\delta$is the radon exhalation rate, Bq/(m^2.s); T is the temperate outside tunnel, $15℃ \le T \le 25℃$.

 DOI: 10.1051/matecconf/20166301001

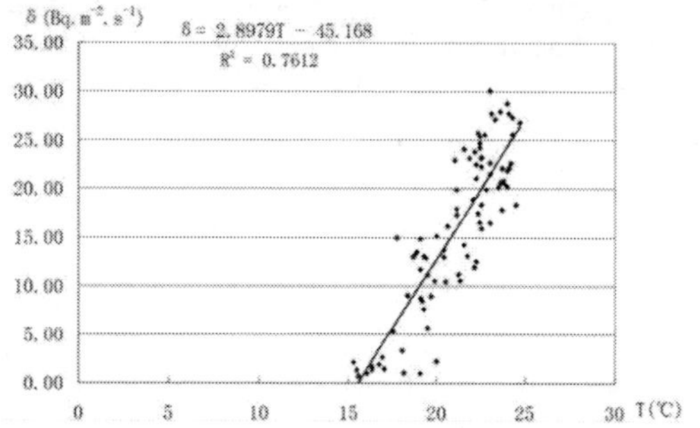

Fig. 10 Radon exhalation rate VS the temperature outside tunnel in exhaust ventilation case

(2) Test of linear regression significance After test with the *F* testing method, the linear regression is proved to be highly significant.

(3) Prediction interval The approximate prediction interval with the confidence coefficient of 0.95 of the predicted value δ_0 is (δ_0-8.22, δ_0+8.22), while the approximate prediction interval with the confidence coefficient of 0.99 of the predicted value δ_0 is (δ_0-12.33, δ_0+12.33).

3.2.4 Analysis of effects of two ventilation styles on radon exhalation rate

It can be seen from the data analysis that, with the same ambient temperature, the radon exhalation rate in exhaust ventilation is higher than that in blowing ventilation. According to the above regression equation, if k_1 and k_2 are respectively assumed as the slopes of Eq.9 and Eq.10, then $(k_2- k_1)/k_1 \approx 0.63$, meaning that the radon exhalation rate in exhaust ventilation is 63% higher than that in blowing ventilation. This result is slightly higher than that of research on some tunnel in literature [10]: the radon exhalation space density in exhaust ventilation is 40%-60% higher than that in blowing ventilation. The reason for this difference may be the higher permeability of over-broken rock in the experimented tunnel.

Conclusions and Recommendations

Based on the on-site experiments, this paper separately studies the regularity of radon exhalation in summer natural ventilation and two mechanical ventilations and achieves the following conclusions:

(1) In natural ventilation, the correlation between the radon exhalation rate in tunnel and the natural ventilation quantity as well as the external ambient temperature is linear. In the summer natural state, the radon exhalation rate is positively correlated with the ambient temperature, namely, the radon exhalation rate increases with the ambient temperature. The higher ambient temperature will cause the higher radon concentration in tunnel.

(2) In mechanical ventilation, the correlation between the radon exhalation rate in tunnel and the ambient temperature outside tunnel is linear. At the same ambient temperature, the radon exhalation rate in exhaust ventilation is 63% higher than that in blowing ventilation. The results show that under the same conditions, the exhaust ventilation can better accelerate the radon exhalation in rock than the blowing ventilation. Therefore, the blowing ventilation in tunnel is more conducive to radon control.

Based on the above conclusions, it is recommended to avoid working in over-broken granite tunnel and other underground constrictions when there is radon pollution during high temperature period in summer. If available, it is suggested to operate in early spring, late autumn or winter. When the mechanical ventilation is needed in the underground work space, it is necessary to select the blowing ventilation as the main ventilation style.

References

1. Z.Si-ming. Uranium and Radon Ventilation. *Uranium Mining and Metallurgy*, 4 (1992)
2. Frédeéric Perrier, Patrick Richon, Umesh Gautam, et al. Seasonal variations of natural ventilation and radon-222exhalation in a slightly rising dead-end tunnel. *Journal of Environmental Radioactivity*, 97 (2007)
3. LI Ren-jie. Determination of radon reduction rate and exploration on its influence factors. *Uranium Mining and Metallurgy*, 19 (2000)
4. Wu Hui-shan. Radon Measurement and Practical Data. Atomic Energy Press, (2001)
5. Zhao Yi-hui. Mine ventilation and air conditioning.: China University of Mining and Technology press, (1990)
6. Du Wei-xin. The Mine Ventilation Technology. China University of Mining and Technology press, (2009)
7. YE Yong-jun, DING De-xin, ZHOU Xing-huo, et al. Study on the maximal driving length of dummy drift in uranium mines. *Chinese Journal of Nuclear Science and Engineering*, 29 (2009)
8. André Unger, Stefan Finsterle, Gudmundur S. Bodvarsson. Estimating large-scale fractured rock properties from radon data collected in a ventilated tunnel. *PROCEEDINGS TOUGH Symposium*. Lawrence Berkeley Lab, Berkeley, Calif. .(2003)
9. Cai Zeng-ji, Long Tian-yu. Fluid Mechanical Pump and Fan (The Fifth Edition) .China Architecture & Building Press. (2009)
10. Zhu Si-ming, Calculation Method of Foreign Uranium Tunnel Air. China Nuclear Society Symposium on Radiation Environment of Mine, 1 (1994)
11. LIU Yong, ZHANG Xin-hua. Permeation and control of radon and its radiation products in underground mining of uranium deposits. *China Mining Magazine*, 13.(2004)
12. LI Ming, ZHANG Xin-hua, LIU Yong. Research on the regularity of emission and dispersion of radon in tunnel construction. *Modern Tunneling Technology*, 42 (2005)

MATEC Web of Conferences **63**, 01002 (2016)
MMME 2016

DOI: 10.1051/matecconf/20166301002

Design of EMCCD Driving System for Underwater Low-Light Camera

Yuchen Hou[1, a], Houde Wu[1] and Wenhai Xu[1]

[1] *Institute of Optoelectronic Information Engineering, Dalian Maritime University, 116026 Dalian, China*

Abstract. For the core of underwater low-light camera, i.e. EMCCD, the whole solution of driving system is presented in this paper. The timing signals which meet the driving requirements of EMCCD are produced from FPGA. The high-speed integrated driver chips convert the timing signals into general power driving signals with the amplitude under 12V; a Class A push-pull amplifier circuit built in discrete components transforms the timing signals into high-voltage multiplying signals. Besides, the impedance matching for the circuit optimizes the driving system. The experimental results indicate that the driving system generates the high-voltage multiplying signals, with their high level adjusted from 30V to 48V and frequency raised up to 10MHz. The whole driving system satisfies the requirements of EMCCD, and has the capability to ensure the normal operation of EMCCD.

1 Introduction

Determining the damage degree of shipwrecks quickly and precisely is the precondition for drawing up and implementing the scheme of underwater salvage. At present, the exploration of the damage degree of shipwrecks is usually realized by recording images with general high-resolution camera. However, secondary light source need to be utilized in the low-light underwater conditions. In this case, the performance of general high-resolution cameras reduces greatly because of the backscattering of water-body, so that high-quality images cannot be obtained. For the purpose of overcoming the problem, the underwater low-light camera with the ability to detect weak light emerged at the right moment.

With its favourable characteristics such as high quantum efficiency, high sensitivity, high readout rate, ultra-low readout noise and adjustable gain [1], EMCCD (Electron-Multiplying Charge Coupled Device) can take high-definition images in the deep sea with hardly any illumination. Therefore, it is utilized as the core of the underwater low-light camera. In the whole design of the driving system, it is difficult to generate the high-voltage multiplying signal, which is the most important part of the application of EMCCD [2]. In addition, the scheme design for the driving system and the establishment of the system which makes the EMCCD work well based on the design lay the foundation for the overall system design of the underwater low-light camera.

2 CCD97 overview

Produced by E2V in the United Kingdom, CCD97 is a type of frame transfer area EMCCD [3]. CCD97 utilizes the technology of gaining on-chip. While the readout frequency is 11MHz, the output noise can be inferior to the level of an electron [4]. As a consequence, CCD97 is chosen as the core of underwater low-light camera.

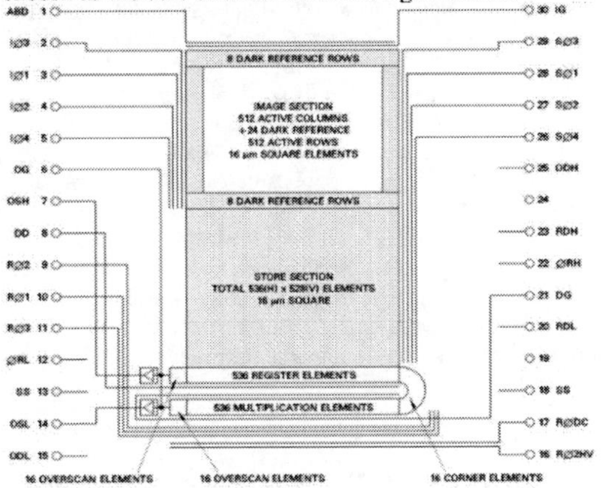

Figure 1. Structure chart of CCD97.

As shown in Figure 1, the active image area of CCD97 is 8.192×8.192 mm^2, and the image pixel size is 16×16um^2. The number of image section pixels is 528(H) ×536(V). In practical application, there are 8 dark pixels at the beginning and end of each line, and 12 at the beginning and end of each row [5]. These dark pixels covered completely by aluminium shells without outside light source is used as the reference of dark noise. As a result, the number of active pixels in the image section is 512(H) ×512(V). The corresponding store section has 528(H) ×536(V) elements to store the charge produced in the image section. When reading out the image signals

[a] Dlmuhyc@dlmu.edu.cn

from OSL (multiplication output channel), the charge will go through 536 register elements and 16 corner elements in the lower right corner, then the charge enter into 536 multiplication elements. Before entering the sense amplifier, the charge is multiplied in order to reduce the influence caused by the sense amplifier [6].

To give full play to the performance of CCD97, the low-noise driving circuits and bias circuits are necessary [7]. Except for RΦ2HV, the amplitudes of other driving signals are under 12V. RΦ2HV is the high-voltage multiplication signal, with an amplitude up to 50V which is adjustable. The frequency of driving signals that matches the output signal is 10MHz. Nevertheless, the phases of signals should be adjustable, and the drive capability neither too strong to cause a large overshoot voltage nor too weak to influence the operation of CCD97. Consequently, the design of driving system should consider amplitude, frequency, phase and driving capability [8].

3 Overall structure of underwater low-light camera

As presented in Figure 2, the overall structure of underwater low-light camera mainly contains three parts: EMCCD module, AFE module and FPGA module. FPGA module, as the control centre of the whole system, receives the commands from the computer through the RS422 interface and generates the driving signals utilized as the input of the peripheral driving circuitry of CCD97 on the EMCCD module. Through the peripheral driving circuitry, the driving signals are converted into the power driving signals that meet the requirements of CCD97, which then converts the light signals into analogue signals. Through the coaxial cable, analogue signals enter into the AFE module with the functions such as correlated couple sample, black level clamp, programmable gain amplifier and analog-to-digital converter. The digital signals converted by the AFE module are processed by the FPGA module and finally transmitted to the computer through LVDS driving module.

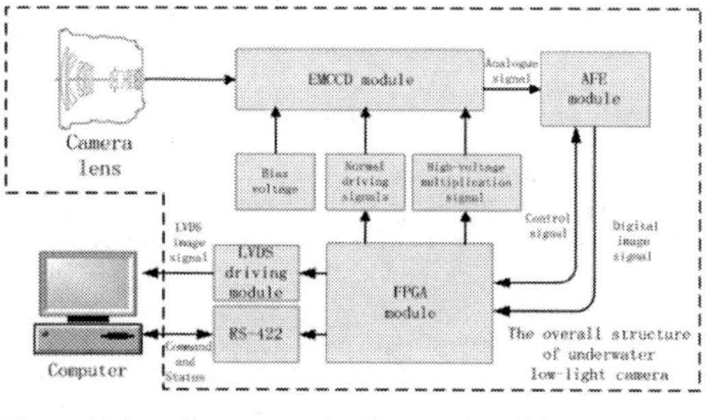

Figure 2. Overall structure of underwater low-light camera

4 Design of EMCCD driving system

The design of EMCCD driving system involves the hardware driving circuit of EMCCD and production of timing signals by FPGA.

4.1 Hardware part

For the normal operation of CCD97, 14 driving signals are necessary. According to the requirement of driving CCD97, the driving signals are divided into the general driving signals and the high-voltage multiplication driving signals. The former takes on the characteristics of low amplitude, high frequency and large load, in which case, the high-speed integrated driving chip meets the requirements of design. For the latter——RΦ2HV, the amplitude reaches up to 50V, which cannot be achieved by the high-speed integrated driving chip. Given this, the driving circuit is made up of discrete components. The amplitudes of driving signals are shown in Table 1.

Table 1. Amplitudes of driving signals.

Name		PULSE AMPLITUDE (V)		
		MIN	TYPICAL	MAX
IΦ1,2,3,4	HIGH	+5	+7	+9
	LOW	-6	+5	-4
SΦ1,2,3,4	HIGH	+5	+7	+9
	LOW	-6	+5	-4
RΦ1,2,3	HIGH	+8	+12	+13
	LOW	-	0	-
ΦRL, ΦRH	HIGH	0	+10	-
	LOW	0	0	-
RΦ2HV	HIGH	+20	+40	+50
	LOW	0	+4	-

4.1.1 Design of general driving circuit

In this paper, EL7457 produced by Intersil in the USA is applied to achieve the conversions of output signals.

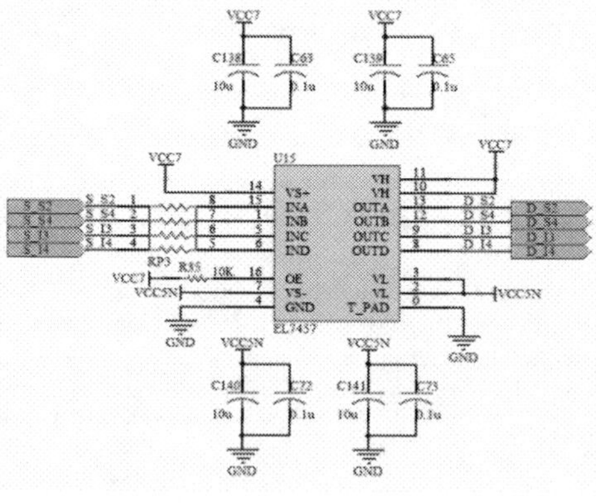

Figure 3. General driving circuit

EL7457 is featured by 4 channels, 8V ≥ VLOW ≥ -5V, -2V ≤ VHIGH ≤ 16.5V, 2A peak drive and clocking that speeds up to 40MHz, which completely meet the requirements of normal driving signals.

As shown in Figure 3, S2, S4, I3 and I4 are four vertical transferring signals, and the chip converts 4 timing signals into power driving signals with a high level of +7V and a low level of d -5V at the same time.

4.1.2 High-voltage multiplication driving circuit

The high-level voltages of the high-voltage multiplication driving signal is adjustable and divided into 32 levels ranging from 20V to 48V in the design. The schematic of high-voltage power supply circuit is shown in Figure 4.

A voltage supply of 50V is supplied to the LT3012, i.e. a high-voltage LDO (Low Dropout Regulator), and the commands from FPGA control the resistance of MAX5435 (Digital Potentiometer), and thus change the output voltage of LT3012. For the low-voltage power supply circuit, LT1118 is adopted, which has functions of current source and current sink, in order to keep the low voltage stable. In the pulse shaping circuit, the comparer MAX999 converts the TTL from FPGA into the square signal with a peak-peak value of 5V, which subsequently improves the driving power through SN74LVC2T45, and finally controls the open or close of the push-pull MOSFET to achieve the electric multiplying function of EMCCD.

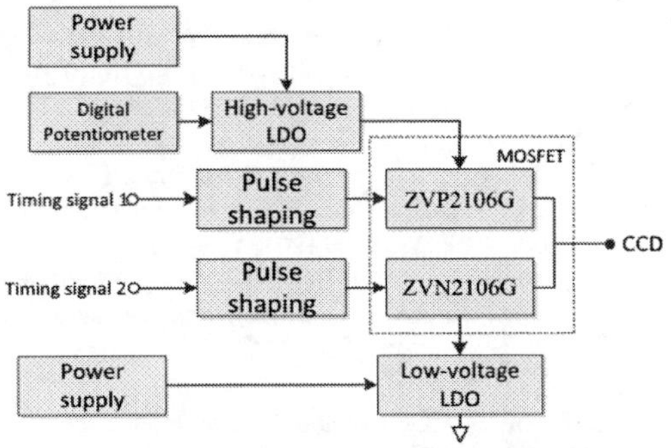

Figure 4. High-voltage multiplication driving circuit

The Class A push-pull driving amplifier circuit is shown in Figure 5. ZVP2016G and ZVN2106G, a pair of matched MOSFET (Metal Oxide Semiconductor Field Effect Transistor), are utilized to establish the circuit. The ZVP2106G has the following parameters: V_{DS}= -60V ,I_D=-450mA and $V_{GS(th)}$>0.8V; as for ZVN2106G, V_{DS}=60V, I_D=710mA and $V_{GS(th)}$<-1.5V. When no signal is input into HV_A, the voltage is about 1.4V on 1SS226 and approximately 0.7V on 1SS193. Hence, V_{GS} is about 0.7V [10] and the P-type MOSFET is pinch-off. When the signal enters into HV_A, the voltage of high level is clipped by 1SS193; the low level makes the voltage of source electrode lower than V_{GS}, so that P-type MOSFET is on and the HV_OUT outputs high level. Similarly, the signal of HV_B can control the output of low level. If to control HV_OUT of high level and low level is to generate alternation, the driving circuit would output the high-voltage multiplication driving signal.

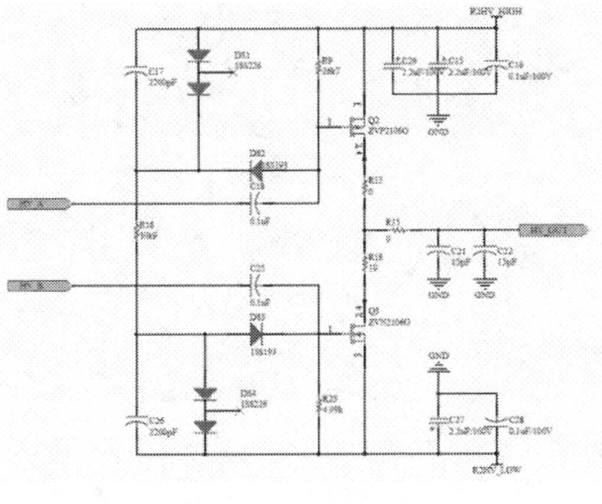

Figure 5. Class A push-pull driving amplifier circuit

4.2 Software part

The timing signals are generated from FPGA, the code of which is programmed in Verilog on the platform of XILINX ISE 14.7. The frequency of system clocking is 200MHz, and the cycle 5ns. Under this condition, the timing signals can adjust phase position and duty ratio in 5ns, before the signals go through the circuit practically. The simulation waveforms of timing signals are shown in Figure 6.

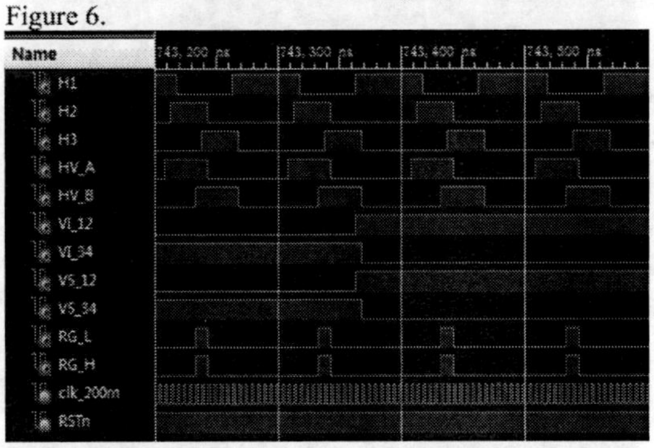

Figure 6. Simulation waveforms of timing signals.

HV_A and HV_B are the control signals of high-voltage multiplication driving circuit. If the pair of MOSFETs are on at same time, the low impedance path forms to cause the great loss of current, and even lead to a burning. In consequence, the phase relationship of HV_A and HV_B should be controlled strictly. In order to prevent the pair of MOSFETs from turning on simultaneously, the duty ratio of HV_A and HV_B should be adjusted to reduce the turn-on time of MOSFETs, which decreases the power dissipation and improves the stability of system.

5 Result and conclusion

From Figure 7, it can be seen that the high-voltage multiplication driving signal generated by the driving system has a voltage amplitude of 46V and a frequency of 10MHz, which satisfies the driving requirements of CCD97 according the design scheme.

DOI: 10.1051/matecconf/20166301002

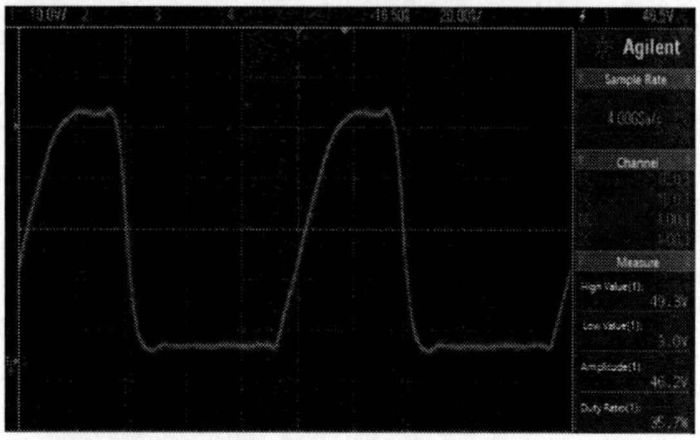

Figure 7. High-voltage multiplication driving signal

Figures 8-10 are the images shot by the camera when the high-level voltages of high-voltage multiplication driving signals are 30V, 40V and 48V respectively. The image signals which have gained correct multiplication constitute the distinct images, the details of which are clearly visible.

Figure 8. Obtained image at a multiplying high voltage of 38V

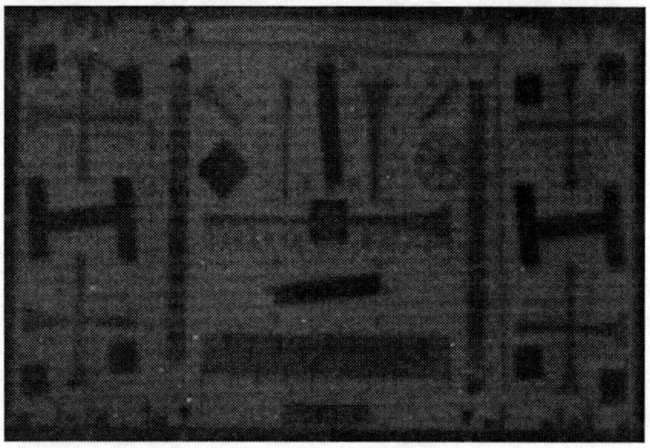

Figure 9. Obtained image at a multiplying high voltage of 40V

Figure 10. Obtained image at a multiplying high voltage of 43V

In the low-light conditions of 10^{-3} lux which requires an output clocking of 10MHz and a frame rate of 25fps, the underwater low-light camera captures the distinct pictures with a low noise. In consequence, the design of the whole driving system is successful, which meets the requirements on underwater low-light camera capturing images.

References

1. C.L. Zhang, Q. Chen, B.B. Zhou, Recent progress toward ultra-sensitivity EMCCD, *Infrared Technology*, **29**,4,192-195(2007)
2. Z.B. Xie, T. Zhang, J.J Zhang, High-voltage Clock Design of Space EMCCD Imaging Detection System, *Science Technology and Engineering*, **11**, 17, 3945-3950(2011)
3. Y.D. Bai, S.C. Zhang, Driving Circuit Design of Electric Multiplying CCD, *Electro-Optic Technology Application*, **28**, 3, 51-54 (2013).
4. E2V Technology LTD, CCD97-00 Back Illuminated Eletronic Multiplying CCD Sensor, www.e2v.com, 2004.
5. J. Sun, B.P. Zhang, R.X. Cao, Driving Circuit Design of EMCCD, *Modern Electronics Technique*, **34**, 4, 150-154 (2011)
6. J.W. He, X. He, Z.H. Wei, Design of EMCCD Cameras for Star Detection, Optics and Precision Engineering, **18**,6, 1396-1403 (2010)
7. Q.Q. Huang, Design and Optimization of EMCCD Driving Circuit, *Nanjing University of Science and Technology*, 2014.
8. J.B Lai, G.Q. Ren, M.F. Wang, Y.F. Liao, Design of High Frequency Asymmetric EMCCD Multiplication Driving Circuit, *Modern Electronics Technique*, **38**, 22, 111-114 (2015)
9. P.F. Cheng, M.J. Gu, M.C. Wang, Design of Array CCD of Frame Transfer Driving Circuit Based on FPGA, *Infrared Technology*, **28**, 9, 519-522 (2006)
10. J.H. Jin, B.H. Li, X.M. Chen, Analysis of EMCCD Electronic Multiplying Driving Circuit and Design of PCB, Astronomical Research & Technology, **9**, 3, 316-322 (2012)

MATEC Web of Conferences **63**, 01003 (2016) DOI: 10.1051/matecconf/20166301003
MMME 2016

Transient Dynamics Analysis of The Reachstacker Speader Based On ANSYS

Yu Feng SHU[1],Si Li FAN[1], Hai Ying ZHANG[1]

[1]*DongGuan Polytechnic, 523808, Dongguan, China.*

Abstract: Reachstacker is an indispensable handling machinery, it will inevitably lead to unbalanced force at the job site. This paper does transient dynamics analysis for the spreader mechanism, which is one of the most significance key components. We get dynamic response of the spreader in lifting instant, results not only provide a reference for designers to understand the mechanical characteristics of spreader comprehensively, but also bedding for the future research.

1 Introduction

Reachstacker is widely used in container yard and marina. As one of the most significance key components, spreader is often in an unbalanced state during working,in particular the moment of the container has just been lifted[1].Imbalance effect is more obvious with the unbalanced force, along with which the cargo and container will produce an acceleration, so the spreader will inevitably suffer large dynamic load. In this way, the maximum stress and displacement point of the mechanism may be constantly changing over time. In summary, in order to fully understand the mechanical properties of the spreader mechanism under dynamic load, static analysis is certainly not comprehensive. Spreader is work with telescopic, lateral displacement, rotation,etc, which can be applied to 20 and 40-feet standard container. This paper take the 45T reachstacker for example, rated load is 450KN and total weight is about 7500KG[2-3]. we select time integration method to calculate the transient response of spreader mechanism. Taking into account the effects between accuracy and solving speed, this paper sets the integration time step for 0.2s, the minimum time step is 0.05s and the maximum value is 0.5s.

2 Transient Dynamics Analysis

2.1 Finite element model

This paper selects the 10 nodes solid187 unit,which has strong adaptability to irregular structure[4].Weighing the effect between accuracy and model size in meshing, we can set individual size for different parts in the element size dialog box[2]. So the spreader modelis being divided into total of 340,033 nodes and 171,269 units. The spreader's main parts are manufactured by high-strength steel,therefore, itsmechanical properties are as follows:(Elastic modulus) $E = 211GPa$,(Poisson's ratio) $\mu = 0.3$,(Density) $\rho = 7850kg/m^3$, (Constant mass damper) $\alpha = 0.1693$, (Constant stiffness damper) $\beta = 0.00046$. Figure 1 shows the finite element model of spreader.

2.2 Load Step

As we know, reachstacker spreader contains four locking pins, and load increasing of each locking pin is always a certain process. We assume that locking pin came into contact with container surfacefrom the moment t_0=0, field tests show that container is about to leave ground from the moment t_1=1.2s, in other words, total weight of cargo and containers equals the force of four locking pins. We select"Ramped Load"in the "Full Transient Options"dialog box for this process, which intended load is gradually increasing. This paper selects 1.5s foranalysis process, and we change"Ramped Load" into "Stepped Load"in the second load step setting options, which meansload is constant in this period. Load time history of the single locking pin surface is shown in Figure 2.

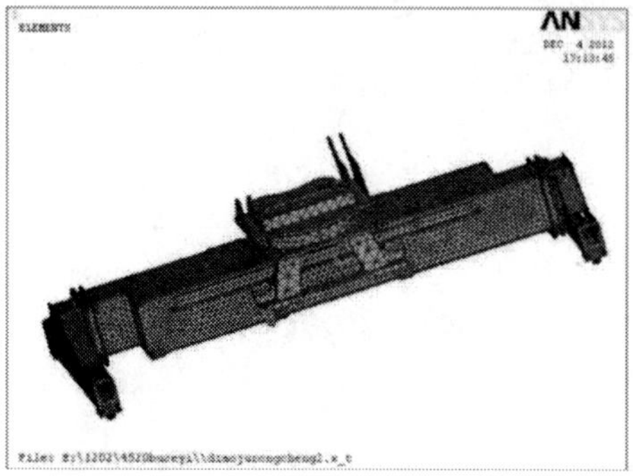

Figure1. Finite element model

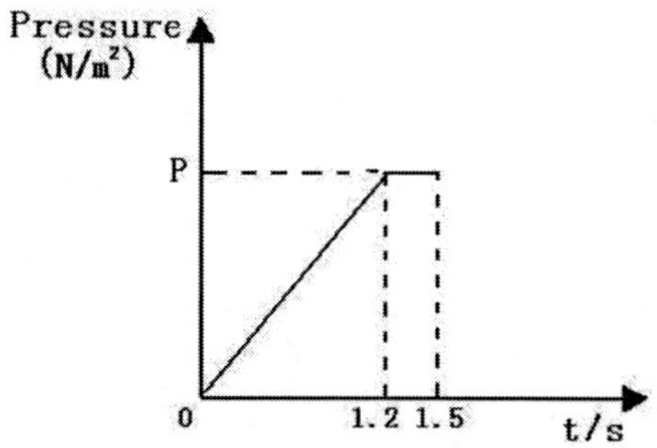

Figure 2. Load time history

2.3 Analysis Cases

There is no doubt that spreader force under40 feet container conditions is even worse than 20 feet[5]. In order to examine the further dynamic response of spreader mechanism and explain the body harm caused by partial load in quantifiable form,taking into account that left and right of spreader is 800mm, front and rear is 100mm, this paper does transient dynamic analysis for spreader mechanism under four typical conditions withrated load, as shown in Table 1.

Table 1. Analysis cases

	Load /T	Container specification /feet	Rotatio n/°	Left and right sideswa y/m	Front and rear sideswa y/m
1	45	40	0	0	0
2	45	40	0	800	0
3	45	40	0	0	100
4	45	40	0	800	100

2.4 Load Determination

It is necessary to consider the gravity due to huge structure. Firstly, we should define gravitational acceleration in –Zdirection; Secondly, the force of each locking pin will be converted into pressure, which should be applied to the plane of locking pin. Four locking pins are numbered in this paper, and force of the locking pin No.1 is F1, pressure is P1, force of the locking pin No.2 is F2, pressure is P2, and so on[6-7]. Number of locking pin is shown in Figure 3.

Figure 3.locking pin number

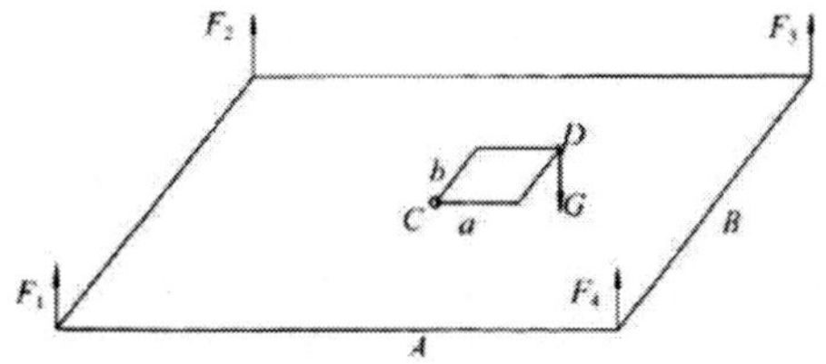

Figure 4. Computing model

Allcases are partial load condition except 1, sowe need to calculate the force on each locking pin, following relationship can be easily obtained according to the knowledge of mechanical balance and computing model shown in Fig.4:

$$G = mg \tag{1}$$

$$F_i = P_i S \tag{2}$$

$$F_1 + F_2 + F_3 + F_4 = G \tag{3}$$

$$(F_1 + F_2)(\frac{A}{2} + a) = (F_3 + F_4)(\frac{A}{2} - a) \tag{4}$$

$$(F_1 + F_4)(\frac{B}{2} + b) = (F_2 + F_3)(\frac{B}{2} - b) \tag{5}$$

G : Total weight of cargo and container.

F_i : Force of the locking pin numbered i.

P_i : Pressure of the locking pin numbered i.

S : Bearing area of the locking pin.

A : Length of40 feet container.

B : Width of 40 feet container.

a : Displacement of lateral partial load.

b : Displacement of longitudinal partial load[8].

Among the parameters above:

$m = 45t$， $S = 2832.69mm^2$

$A = 11985mm$， $B = 2259mm$.

a and b values of all cases are listed in Table 2.

Table 2.a and b values of all cases

	Case 1(mm)	Case2(mm)	Case3(mm)	Case4(mm)
a	0	800	0	800
b	0	0	100	100

We can get the pressure values of each locking pin via solving these equation, as shown in Table 3.

Table 3. Pressure values of each locking pin in different cases

	$P_1(N/m^2)$	$P_2(N/m^2)$	$P_3(N/m^2)$	$P_4(N/m^2)$
1	38920602	38920602	38920602	38920602
2	38920602	28474378	49366826	38920602
3	38920602	38920601	45803553	32037651
4	38920602	28474378	56249777	32037651

3 Solution and Analysis

X is perpendicular to the direction of beam end face, Y is perpendicular to the direction of beam cross-section, Z is the direction of gravity[9]. This paper study the 134194 node which lies in the vicinity of the maximum displacement, we can get X、Y、Z displacement variation history of 134194 node under all cases. As a result of space limitations, we only showpart of the calculation results.

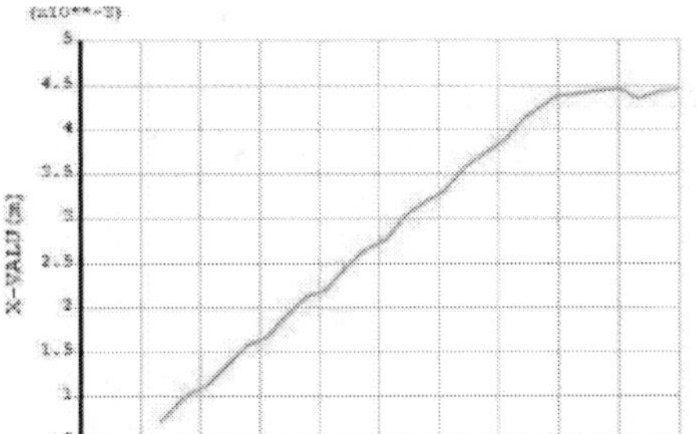

（a）X displacements variation history of case 1

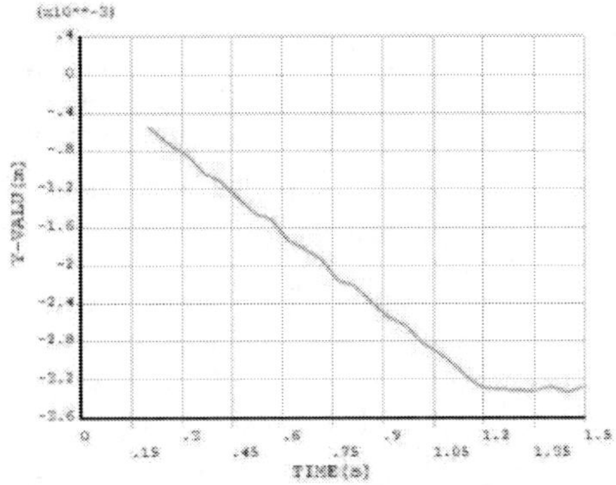

（b）Y displacements variation history of case 1

Figure 5. X、Y displacements variation history of case 1

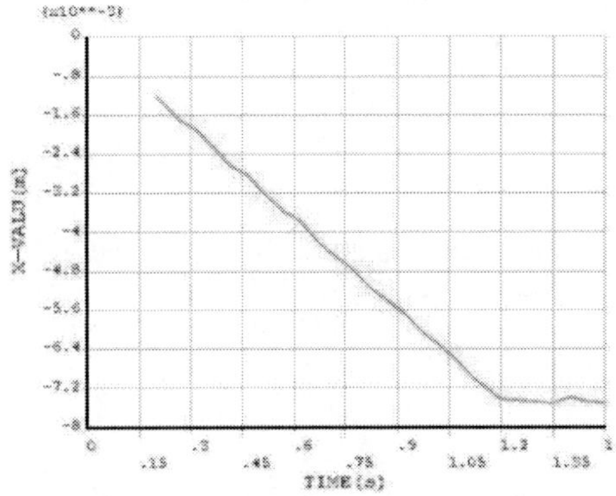

（a）X displacements variation history of case 4

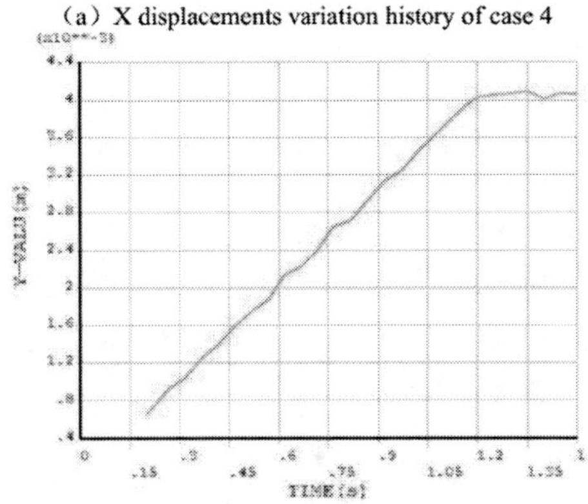

（b）Y displacements variation history of case 4

Figure 6.X、Y displacements variation history of case 4

Seen from the calculation results,the maximum stress under all cases are less than the material yield limit 600Mpa. Apart from a few high-stress areas, overall stress of spreader box skeleton is low (<180MPa), which fully reflects the excellent mechanical properties of high strength steel[5]. The maximum stress are basically the same position under all cases, lying on the head ribs of beams body. Furthermore, stress concentration occurs in the vicinity of maintenance windows because of structural mutation.

Forcase 1, with the increase of ramped loading, X

displacement is increasing, Y and Z displacements are constantly decreasing, with the increase ofstepped loading, they have tended to be a certain value; with the ramp load applied in Partial load conditions, Y displacement is increasing, X and Z displacements areconstantly decreasing, they have tended to be a certain valuewith the increase of stepped loading.Table 4showsthe gradient of three maximum stress values and displacementswhich lie in partial load conditionbyquantifiable form with respect to case 1.

Table 4.Gradient of the maximum stress and displacement

Case	σ_{max} (MPa)	Stress gradient	D_{max} (m)	Incremental value
2	540	19.5%↑	0.039311	16.3%↑
3	516	14.2%↑	0.042642	26.1%↑
4	507	12.2%↑	0.048142	42.4%↑

4 Summary

The maximum stress are basically the same position under all cases, lying on the head ribs of beams body, there are large stress concentration in the vicinity of maintenance window.

X displacement contributes to the maximum deformation is larger than Y and Z under No partial load conditions.

Y displacement contributes to the maximum deformation is larger than X and Z under partial load conditions.

To reduce the maximum deformation of spreader mechanism under partial load conditions , designers should try to increase the stiffness which perpendicular to the direction of telescopic beam cross-section.

References

1. Li Zhang,Jianlong Lin,Huiyu Xiang. Modal analysis and Experiment(Tsinghua UniversityPress, Beijing 2011), p,56-69.
2. Baz,A,Ruzzene,M. Dynamic stability of Periodic shells with moving loads[J]. Journal of Sound and Vibration, 2006,830-844.
3.Diqing Wu, Yanlin Lu, Wanbin Chen. Modal analysis and optimization of forklift frame based on finite element: submitted to Journal of Light Industry Machinery (2008), p, 48-50.
4.Zhu Rensheng, Zheng Zhenzhen, Liu Yongmei.Finite Element Analysis for MW Wind Turbine Tower[J]. Applied Mechanics and Materials, 2011:130-134.
5.Wei Zhou.Strength calculation and modal analysis of subway stainless steel body: submitted to Journal of Urban Rapid Rail Transit (2007), p, 47-49.
6.Shenbo Yan. Optimization Design of Reachstackers's boom: Master,submitted to Daqing Petroleum Institute (2006), p, 52-54.
7.Altintas,Y,Y.Cao.Virtual design and optimization of machine tool spindles. Annals-Manufacturing Technology, 2005,54(1):379-382.
8. Liqun Wu, Wanrong Wu. Application of Pro / E in dynamic performance analysis of the hydraulic excavator: submitted to Mechanical Research and Application (2007), p, 83-84.
9. YE Meixin, HUANG Qiong, WU Qinqin. Analysis of steel-concrete composite structure with overlap slab of Xingguang bridge, 2007:121-124.

MATEC Web of Conferences **63**, 01004 (2016) DOI: 10.1051/matecconf/20166301004
MMME 2016

High Temperature Exposure of HPC – Experimental Analysis of Residual Properties and Thermal Response

Zbyšek Pavlík[1, a], Jan Fořt[1], Milena Pavlíková[1], Jaroslav Pokorný[1], Anton Trník[1], Jiří Studnička[1], David Čítek[2], Jiří Kolísko[2] and Robert Černý[1]

[1]*Department of Materials Engineering and Chemistry, Faculty of Civil Engineering, CTU Prague, Thákurova 7, 166 29 Prague, Czech Republic*
[2]*Department of Experimental methods, Klokner Institute, CTU Prague, Šolínova 7, 166 08 Prague, Czech Republic*

Abstract. The effect of high temperature exposure on properties of a newly designed High Performance Concrete (HPC) is studied in the paper. The HPC samples are exposed to the temperatures of 200, 400, 600, 800, and 1000°C respectively. Among the basic physical properties, bulk density, matrix density and total open porosity are measured. The mechanical resistivity against disruptive temperature action is characterised by compressive strength, flexural strength and dynamic modulus of elasticity. To study the chemical and physical processes in HPC during its high-temperature exposure, Simultaneous Thermal Analysis (STA) is performed. Linear thermal expansion coefficient is determined as function of temperature using thermodilatometry (TDA). In order to describe the changes in microstructure of HPC induced by high temperature loading, MIP measurement of pore size distribution is done. Increase of the total open porosity and connected decrease of the mechanical parameters for temperatures higher than 200 °C were identified.

1 Introduction

The population boom and building grounds limitations is reflected in construction of high-rise buildings. Hence, normal strength concrete is often replaced by the high performance concrete (HPC) or ultra-high performance concrete (UHPC). The use of HPC for high rise building application creates tasks for the assessment its load carrying capacity and behaviour during exposure to high temperatures when damage of chemical and structural changes takes place. In order to uplift HPC mechanical parameters, decrease of the water cement ratio and application of superplasticizers are common procedure. Due to the lower dosage of batch water, the pore volume decreases what results in improved mechanical strength of concrete. Despite of these improvements, decrease of the pore volume leads to the decrease of the concrete fire resistance.

High temperature exposure, especially rapid temperature rise connected to the fire, can cause serious damage in micro and meso structure of concrete [1]. Concrete spalling and consequent bar exposure decrease durability of such structures. The fire exposure generates increase of the compressive stress parallel to the exposed surface causing tensile stress upright to the heated surface due to the thermal expansion of the water inside the pores [2-3]. Hence, the increase of the temperature builds up the water vapour pressure in the concrete pores and tensile stress in concrete structure. The different mechanisms of spalling were described by Sanchayan and Foster [4] and many parameters influencing spalling were discussed. However, prediction of the spalling represents complex task for further evaluation. A spalling is usually sorted into following categories: aggregate spalling, explosive spalling, surface spalling and corner spalling [5-6]. The first three types of spalling occur during first 20-30 minutes of fire exposure. The corner spalling arises after 30-60 minutes of fire [7].

Properties such a compressive strength, flexure strength and modulus of elasticity are usually studied in relation with the high temperature exposure. In the study of Hertz [8], free water content, moisture gradient and its influence in case of elevated temperature exposal was examined and found as a reason of spalling. Studies [2, 5] focused on an estimation of the critical amount of water in concrete concluded that moisture about less than 3 mass% will not cause spalling. The range of 3-4 mass% of free water generates risk of explosive spalling in dependence on the concrete microstructure [4]. Unfortunately, Fu and Li [9] found that for microstructure of the HPC or UHPC with very small pores and permeability, moisture content slightly higher than zero can cause spalling.

In case of fire, the release of the bound water from the calcium silicate hydrates (CSHs) takes place at about 110-150 °C [10]. Increase of the tensile stresses in the

[a] Corresponding author: pavlikz@fsv.cvut.cz

concrete structure is caused by the thermal expansion of the aggregates from 300 °C, where some cracks appearance was noted [11]. This explains the fact that only negligible decrease of the concrete mechanical parameters under 300 °C is usually observed. The decomposition of the $Ca(OH)_2$ can reduce the durability of the concrete in case of rehydration of the lime as a consequence of the cooling of the concrete after high temperature exposure. Here, surface cracking and spalling may arise. The C–S–H decomposition starts at about 550 °C but became significant only above 600 °C. The decomposition rate increases dramatically with temperature above 600 °C and it is a main reason for the strength loss of concrete exposed to temperatures above 600 °C [12-13] The main products of the C-S-H decomposition are C_3S and β-C_2S (decomposes at about 650 °C). Significant loss of concrete durability, strength and other structural changes occur at temperatures higher than 600 °C [14]. Temperature about 800 °C is responsible for the concrete crumbling and temperature above 1150 °C turn minerals into the glassy phase.

2 Experimental

Newly developed HPC composed of Portland cement CEM I 42.5, silica sand with maximum size of 2 mm, mineral microfillers, and water was examined. In order to decrease the amount of the batch water, the superplasticizer based on polycarboxylate ether was used for reaching the proper workability of fresh mixture and for decrease of concrete porosity. The detailed composition of studied concrete is shown in Table 1.

Table 1. Composition of researched HPC.

Material	Amount [kg]
Silica sand 0 – 2 mm	1 160
CEM I 42.5 R	700
Mineral fillers	250
Superplasticizer	30
Water	160

2.1 Sampling and curing

HPC prisms having dimensions of 40 × 40 × 160 mm were casted, cured for 24 hours in highly humid environment of approx. 98 % and after removing from forms placed for 27 days in water having temperature of approx. 20 °C.

The cured samples were before the particular tests dried for 72 hours at 105 °C. In order to evaluate the effect of high temperature exposure on HPC properties, the samples were exposed 2 hours to the temperatures of 200, 400, 600, 800, and 1 000 °C respectively. The applied heating rate was 1 °C/min.

2.2 Basic physical properties

Among the basic physical properties, bulk density, matrix density, and total open porosity were measured. The bulk density measurement was performed by weighting the sample dry mass and measurement of its volume by the digital caliper. The matrix density was accessed by a helium pycnometer (Pycnomatic ATC, Thermo Scientific) and the total open porosity was calculated on the basis of knowledge of the bulk and matrix density according to the formula

$$\phi = 1 - \frac{\rho_{bulk}}{\rho_{matrix}}, \qquad (1)$$

where ϕ [-] is the total open porosity, ρ_{bulk} [kg/m^3] the bulk density and ρ_{mat} [kg/m^3] the matrix density [15]. The relative expanded uncertainty of applied testing method was 5 %.

2.3 Mechanical parameters

Mechanical resistivity of tested HPC was characterized by flexural strength, compressive strength and dynamic modulus of elasticity measurement. The flexural strength was determined according to the procedure described in the technical standard ČSN EN 12390-5 - Testing hardened concrete - Part 5: Flexural strength of test specimens. For the measurement, prismatic samples having dimensions of 40 × 40 × 160 mm were used. The compressive strength was measured according to the standard ČSN EN 12390-3 - Testing hardened concrete – Part 3: Compressive strength of test specimens. The experiments were done on the halves of broken prisms in the flexural strength test; the loading area was 40 × 40 mm. The uncertainty of both testing methods is 1.4 %. The dynamic Young's modulus was measured by the pulse ultrasonic method using a DIO 562 device working on the frequency of 50 kHz [16].

2.4 Thermodilatometry (TDA)

Thermal stability of the studied HPC was investigated by a horizontal thermodilatometer allowing monitoring of the length changes of a studied material during a negligible loading as a function of temperature and time. The samples with the dimensions of 20 × 20 × 160 mm were examined in the temperature interval from 25 to 1000 °C with a heating rate 1 °C/min in a static air atmosphere. From the measured temperature induced length changes, linear thermal expansion coefficient was accessed as function of temperature $\alpha(T)$ according to the following equation

$$\alpha = \frac{d(l/l_0)}{dT} = \frac{d\varepsilon}{dT}, \qquad (2)$$

where α [1/°C] is the linear thermal expansion coefficient, l [m] the sample length at chosen temperature, l_0 [m] the initial sample length, ε [-] the thermal strain.

2.5 Simultaneous Thermal Analysis (STA)

This analysis was performed using apparatus Labsys Evo (Setaram) that allows simultaneous DSC-TG measurements [17-18]. For our measurements, finely ground samples were prepared and placed into an alumina crucible with a volume of 100 mm^3. The mass of the samples was about 80 mg. The STA was realized in the temperature range from 25 to 1200 °C with a heating rate of 5 °C/min in an argon atmosphere (flow rate 40 ml/min).

2.6 Mercury intrusion porosimetry (MIP)

The effect of elevated temperatures on changes in porous structure of studied HPC was observed using MIP analysis. For the measurement, the combination of porosimeters Pascal 140 and Pascal 440 (Thermo Scientific) was used. At the evaluation of the measured data, the circular cross section of capillaries was assumed. The mercury contact angle was assumed to be 130°.

3 Results and discussion

Basic physical and mechanical properties of the tested HPC are given in Tables 2-3.

Table 2. Basic physical properties of HPC.

Temperature [°C]	Bulk density [kg/m^3]	Matrix density [kg/m^3]	Total open porosity [%]
reference	2 349	2 554	7.8
200	2 363	2 561	7.3
400	2 156	2 667	16.2
600	2 130	2 670	17.3
800	2 111	2 924	25.9
1000	2 075	2 920	29.4

Table 3. Mechanical parameters of HPC.

Temperature [°C]	Flexural strength [MPa]	Compressive strength [MPa]	Young's Modulus [GPa]
reference	11.1	113.8	45.9
200	13.7	123.8	47.2
400	6.3	105.2	39.2
600	3.7	75.8	30.2
800	2.4	50.6	23.6
1000	1.6	20.5	13.1

Here, the presented results were calculated on the basis of five independent measurements on different samples. According to the ČSN EN 12390-5, the flexural strength data is calculated from the three independent measurements and compressive strength from the six.

For samples exposed to the maximum temperature of 200 °C, decrease of the total open porosity value and related increase in mechanical strength was observed. Up to this temperature, only minor changes in concrete structure were monitored and the thermal strain led to the limited improvement of the mechanical parameters.

Decrease of the bulk density and consequent increase of the total open porosity are distinct in relation to the temperatures higher than 200 °C. These can be attributed to the structural changes in concrete matrix due to the high temperature action. One can also observe decrease in the strength characteristics of the same trend as changes in the porosity. The significant decrease of the compressive strength was visible at 600 °C, where approx. 30% drop occurred.

The above given findings are similar to the conclusions of the study published by Netinger et al. [19], where only slight mechanical changes were observed up to 500 °C. Starting from 600 °C, the mechanical characteristics decreased proportionally with increasing temperature. The temperature of 1000 °C caused an almost complete loss of mechanical parameters of HPC.

Data obtained by the MIP given in Figure 1 confirmed that the temperature loading influenced the total open porosity as well as diameter of pores. While differences between the pore size distributions of reference sample and sample exposed to 200 °C were almost negligible, the temperature exposure of 400 °C shift distinctly the pore size distribution curve due to the increase of pore volume. Concurrently, the increased total open porosity has negative effect on the mechanical parameters. Similar finding was observed for every increase of temperature up to 1000 °C, which significantly increased the total open porosity as well as pore diameter.

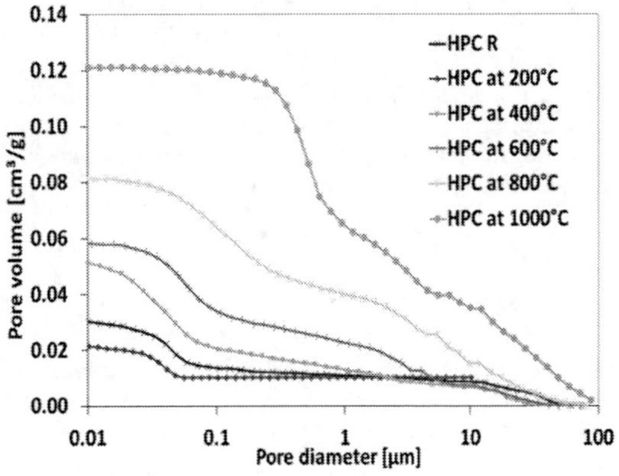

Figure 1. Cumulative pore volume

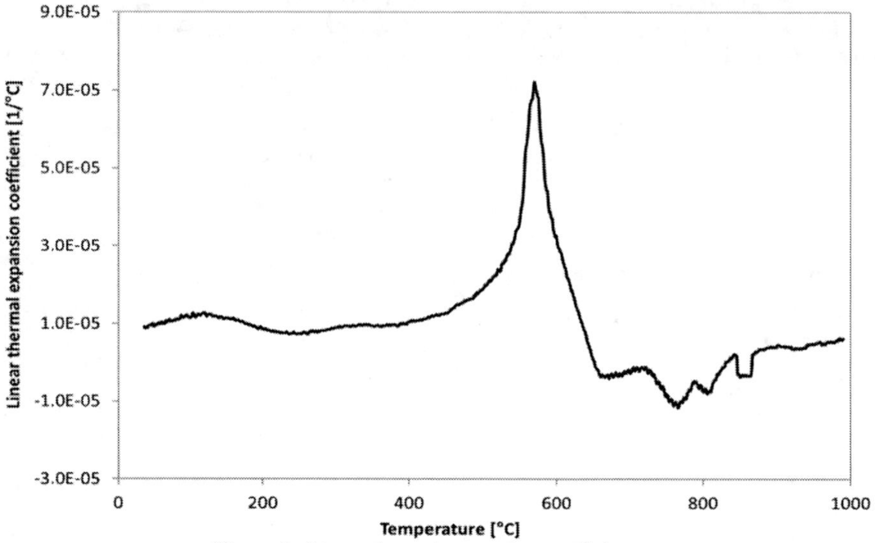

Figure 2. Linear thermal expansion coefficient.

The temperature dependent linear thermal expansion coefficient is presented in Figure 2. The distinct thermal strain we assign to the two main changes in the concrete matrix. The first is the decomposition of portlandite usually connected with the cracks formation. The change in the crystal phase of the aggregate, i.e. the transformation of the alpha quartz to the beta quartz modification, is the second phenomena responsible for the linear thermal expansion of the studied material.

In the temperature interval from 573 to ~ 650 °C, no significant expansion was observed and the studied material even underwent limited thermal shrinkage. The shrinkage recorded at the temperature of about 700-750 °C was found by Estevez [20] and can be assigned to the formation of the new silica compound.

The thermal strain data is shown in Figure 3. It reveals the increase of the strain in the dependency on the increasing temperature up to 600 °C. The subsequent decrease of the strain, visible from 650 to 850 °C is the result of the reaction of the silica microfillers to the high temperature exposure as was mentioned above.

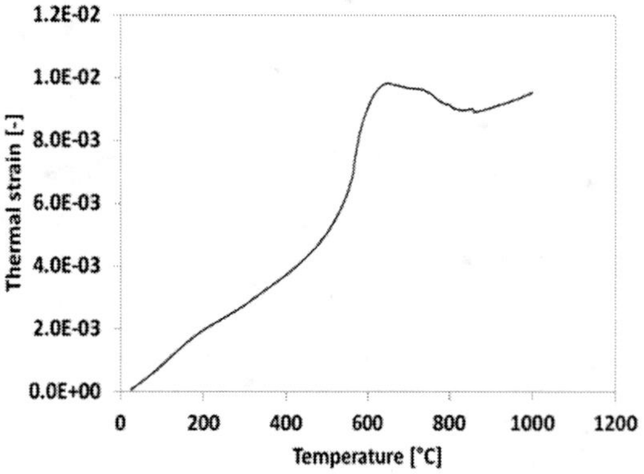

Figure 3. Thermal strain.

The data obtained by STA is shown in Figure 4. Here, information about decomposition and crystallization in the studied HPC during the high temperature exposure are accessed. The release of the bound water from pores due to the dehydration and loss of water from C-S-H gels took place in the temperature interval from 20 to 200 °C. The decomposition of the ettringite was recorded at the same temperature interval. The portlandite $Ca(OH)_2$ disintegration was found at 445 °C. The temperature about 573 °C is typical for transformation of the quartz from the alpha to the beta modification. This reaction is connected with volume expansion, sharp endothermic heat flow peak and no mass change. In the temperature interval 580-800 °C, the calcite decomposition took place. At the temperature above 800 °C, there was found one significant exothermal peak corresponding to a crystallization process. In our study, this peak corresponds to the crystallization of wollastonite [21].

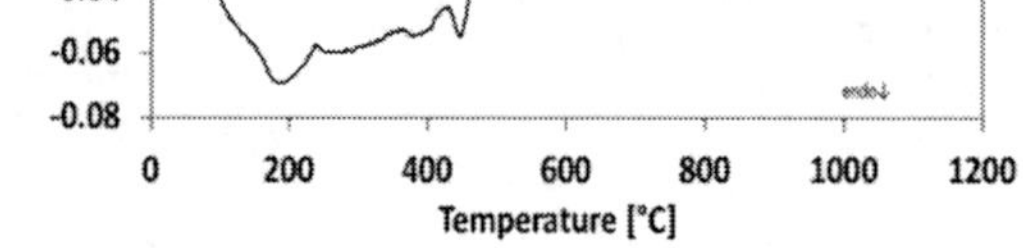

Figure 4. DSC data of the studied HPC.

4 Conclusions

The residual properties of HPC after high temperature exposure and its performance during the high temperature action were studied. Increase of the total open porosity and connected decrease of the mechanical parameters for temperatures higher than 200 °C were identified. This concrete performance was explained by STA data that allowed identification of transformation and decomposition of the hydrated products in the HPC matrix. The temperature of the dehydratation of the physically bonded water and C-S-H gels, portlandite decomposition, ettringite dehydration, trasformation of the alpha quartz to the beta modification, calcite disintegration, wollastonite crystallization and microsilica transformation were revealed and documented.

Obtained results will be employed for further research aimed at the formulation of the computational model for prediction of the HPC structures behavior exposed to the fire and high temperature action.

Acknowledgment

This research has been supported by the Czech Science Foundation, under project No 15-05791S.

References

1. J. Zhao, J.J. Zheng, G.F. Peng, K. van Breugel, Cement Concrete Res., **64**, 64 (2014)
2. L.T. Phan, N.J. Carino, ACI Mater. J., **99**, 54 (2002)
3. F.J. Ulm, P. Acker, M. Lévy, J. Eng. Mech., **125**, 283 (1999)
4. S. Sanchayan, S.J. Foster, Mater. Struc., **49**, 769 (2016)
5. K. Sakr, E. El-Hakim, Cement Concrete Res., **35**, 590 (2005)
6. O. Arioz, Fire Saf. J., **42**, 516 (2007)
7. J. Xiao, G. Konig, Fire Saf. J., **39**, 89 (2004)
8. K.D. Hertz, Fire Saf. J. **38**, 103 (2003)
9. Y. Fu, L. Li, Mater. Structur., **44**, 361 (2011)
10. G.A. Khoury, Prog. Struct. Eng. Mater., **2**, 256 (2000)
11. A.H. Akca, N.O. Zihnioglu, Constr. Build Mater., **44**, 317 (2013)
12. G.F. Peng, Z.S. Huang, Constr. Build. Mater., **36**, 593 (2008)
13. Q. Ma, R. Guo, Z. Zhao, Z. Lin, K. He, Constr. Build. Mater., **93**, 371 (2015)
14. W.M. Lin, T.D. Lin, L.J. Power-Couche, ACI Mater., **93**, 239 (1996)
15. M. Benes, R. Stefan, Int. J. Heat. Mass. Tran., **85**, 110 (2015)
16. Z. Pavlík, M. Keppert, M. Pavlíková, J. Žumár, J. Fořt, R. Černý, Cement Wapno Beton, **2**, 67, (2014)
17. Z. Pavlík, A. Trník, M. Keppert, M. Pavlíková, J. Žumár, R. Černý, Int. J. Thermophys., **35**, 767, (2014)
18. Z. Pavlík, J. Fořt, M. Záleská, M. Pavlíková, A. Trník, I. Medved', M. Keppert, P.G. Koutsoukos, R. Černý, J. Clean. Prod., **112**, 409 (2016)
19. I. Netinger, I, Kesegic, I. Guljas, Fire Saf. J., **46**, 425 (2011)
20. L.P: Estevez, Thermochemica Acta, **518**, 27 (2011)
21. A. Yazdani, H.R. Rezaie, H. Ghassai, J. Ceram. Process Res., **11**, 348 (2010)

MATEC Web of Conferences **63**, 01005 (2016) DOI: 10.1051/matecconf/20166301005
MMME 2016

Cherry Picking Robot Vision Recognition System Based on OpenCV

Qi Rong ZHANG [a], Pei PENG and Yan Mei JIN

Information Engineering, Guizhou University of Engineering Science, Bijie 551700, Guizhou , China

Abstract. Through OpenCV function, the cherry in a natural environment image after image preprocessing, color recognition, threshold segmentation, morphological filtering, edge detection, circle Hough transform, you can draw the cherry's center and circular contour, to carry out the purpose of the machine picking. The system is simple and effective.

Keywords. OpenCV; Cherry picking robot; Vision recognition system

1 Introduction

Cherry picking operation is the most time-consuming and laborious in cherry production chain of a link. At the same time, picking operation quality directly affects the follow-up and storage products. How to get high quality products at low cost is must pay attention to and consider problems in fruit production. Picking operation due to strong seasonal, labour intensity, high cost, thus ensuring timely harvest fruit harvesting operations and reducing costs are some of the important ways of agricultural income. Because of the complexity of the picking work, picking automation level is still low. At present, the domestic fruit picking jobs are basically carried out manually, which costs about 50% -70% of the total cost, and the time is more concentrated. Because our country totally dependent on manual picking and cherry mature period is short, pick large cost and low efficiency. Cherry often waste caused by not timely harvest, which limits the size of the cherry planting. Development of mechanical harvesting has become the urgent needs of our industry and the inevitable trend of cherries. Cherry is called "spring after the first fruit". It doesn't take good field soil. Planting technology and fruit trees management difficulty is not large, and economic efficiency is high. Cherry picking reduces the cost of picking robot. It brings huge profits to give farmers more planting cherry enthusiasm, and reduce the cost of lower prices. Local cherries are more competitive and market. Cherry picking robots can effective and efficient harvest, the key is the robot vision recognition system for image processing.

2 OpenCV introduction

Using OpenCV function can be achieved Cherry recognition technology. It does not require additional code in the c + +, basically. It does not require additional code in the c + +, basically.

3 Recognition Technology

The technology uses Windows 7 as a development platform, using Visual Studio 2010 C ++ and OpenCV 2.4.9 open source library as a development tool for programming cherry picking robot vision recognition system. Figure 1 is a step of cherry picking robot vision recognition system.

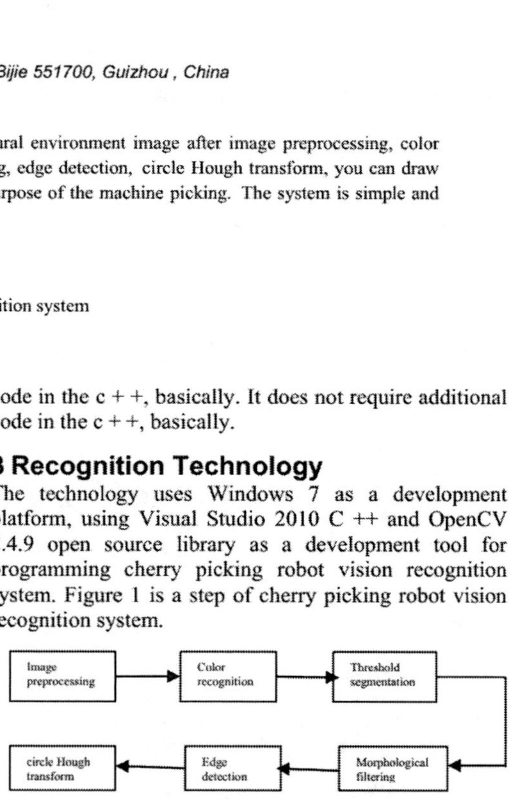

Figure 1. Step of cherry picking robot vision recognition system

First of all, we have a cherry image for image pre-processing, with the median filter, and then use the R-1 / 2G-1 / 2B color recognition, followed by thresholding. We then split a good image before morphological closing operation, then the morphological opening operation. Then we have the image edge detection, and finally the image circle Hough transform. Finally it draws a circle and circular contour of cherry.

3.1. Image preprocessing (median filter)

Cherry picking robot in the natural environment due to hardware quality objective of the natural environment and shooting equipment and shooting angle, distance and other factors, Cherry captured images typically contain noise. These noises to extract the cherry feature recognition and extraction of the impact is bigger, which

[a] Qi Rong ZHANG: zqr_hk@163.com

can make the image distortion, the extraction results cannot achieve the result of practical application, so we usually need to get rid of the noise of image. This technology is using median filter, the input of a cherry images using OpenCV function library of medianBlur () function to realize the function. The function prototype is void medianBlur (InputArray src, OutputArray dst, int Ksize). The function of the third parameter is the linear size of aperture, it must be an odd number greater than 1, this technology uses a linear aperture size is seven.

3.2 Color recognition

Because of the cherry red color with branches and leaves of green has bigger difference, so the recognition with cherry color features is effective method. Due to collect images cherry hardware equipment use RGB model more, so this technology is also used the model. In this model, all the color of the image that is a red, green, and blue, a combination of these three basic colors. Through OpenCV split () function can realize the RGB model of channel separation, separated channels.at (0) blue, channels.at (1) green, channels.at (2) red. But red component separated from the experimental results show that the effect not beautiful. Some scholars use 2 * red - green - blue, for some fruits better recognition [1]. It was found by using a combination of red -1/2 green -1/2 blue component, which can process image than 2 * Red - Green - Blue good color effect [2], so that this technology uses a combination of component.

3.3 Threshold segmentation

Threshold segmentation method is to determine an image gray value range of gray threshold, then the image of each pixel gray value compared with the threshold value, and according to the comparison results corresponding pixels can be divided into two categories. These two types of pixel generally belong to the image of two types of areas, so as to achieve the goal of segmentation. A greater difference between the segmentation methods for gray value range image segmentation effect is good. We can use fixed threshold segmentation, the use of OpenCV threshold () function to implement this function. The function prototype is double threshold (InputArray, OutputArray DST, double thresh, double very much, int type) with double thresh parameter values from 0 to 254, how many values according to the specific circumstances of the image. Parameter type of the value is 0, 1,2,3,4 respectively. 0 represents binary threshold, 1 representing the anti-binary threshold, 2 representing truncated threshold, 3 representing the anti-threshold value to 0, 4 representing the threshold value to 0. Specific what the threshold type will be according to the actual circumstances of the cherry image. Therefore, the specific values of thresh and type parameters determine the effectiveness of cherries image fixed threshold segmentation process.

3.4 Morphological filtering

After threshold segmentation, cherry image due to factors such as the shade of leaves, illumination, the cherry fruit image will appear a few holes, outside the fruit images are some of the small noise, the noise and holes will affect the subsequent processing of the images, so first for closing operation to remove cherry inside the hole, then to open operation to remove cherry external image noise. Closing operations use OpenCV morphological morphologyEx () function and getStructuringElement () function, morphologyEx () function in the identifier selection MORPH_CLOSE. GetStructuringElement () function of the first parameter, there are three kinds of shapes to choose from, MORPH_RECT said rectangle, MORPH_CROSS said cross shape, MORPH_ELLIPSE said oval, the size of the second parameter indicates the kernel function. According to the different images of cherries, in order to optimal effect, we can be on different images with different kernel size for closed operation. This technology uses oval kernel and the kernel size 20 for processing. Open operation in morphologyEx () function in the identifier choose MORPH_OPEN, according to the different images of cherries, in order to optimal effect, we can be on different images using different kernel size to open operation. This technology uses elliptic kernel and the kernel size 70 for processing.

3.5 Edge detection

OpenCV provide four kinds of edge detection methods, which respectively is a Canny operator, Sobel operator, Laplacian operator, Scharr filter. We found through experiment contrast using Canny operator to extract the contour which is thinner and cherry fruit good connectivity. Sobel edge operator to extract the contour is slightly rough and uneven thickness. Laplacian operator to extract the edges of cherry is coarse and connectivity is better. Scharr filter to extract the contour edges have fuzzy horizontal line, so the result is bad. The result of edge detection can be shown in figure 2.

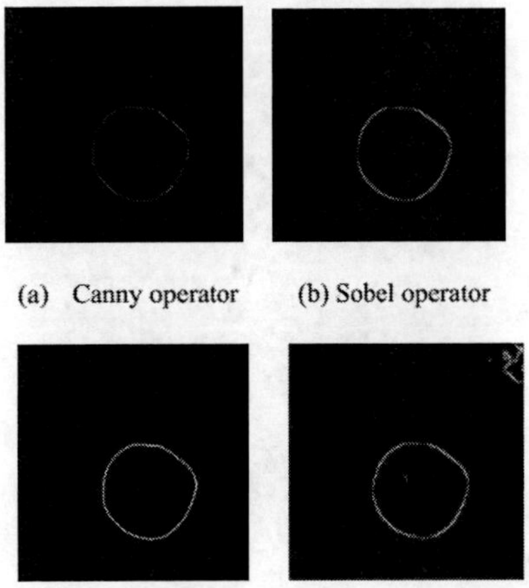

(a) Canny operator (b) Sobel operator

(c) Laplacian operator (d) Scharr filter

Figure 2. Edge detection image processing results

DOI: 10.1051/matecconf/20166301005

This technology uses Canny operator to extract the edge. OpenCV use the Canny () function to achieve this function; you need to pay attention to the former first to use this function in addition to Gaussian noise.

3.6 Circle Hough transform

Filmed in the actual cameras in the picture, there are different degrees of cherry branches, leaves and cherry occlusion problem between each other. Cherry as approximation is circular, using circle Hough transform to detect the boundary of the cherry. Using OpenCV HoughCircles () function of circle Hough transform, we can find the picture of cherry's center and radius, then Circle () function draws a circle and rounded profile, so the robot can pick operation.

4 The cherry picking robot vision recognition system of the experimental results and analysis

In order to test the function of this system, we input a cherry image to this system for processing the whole process. Test results can be shown in Figure 3.

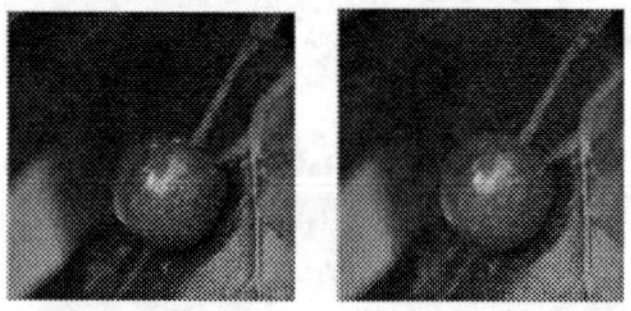

(a)Original image (b) Image preprocessing

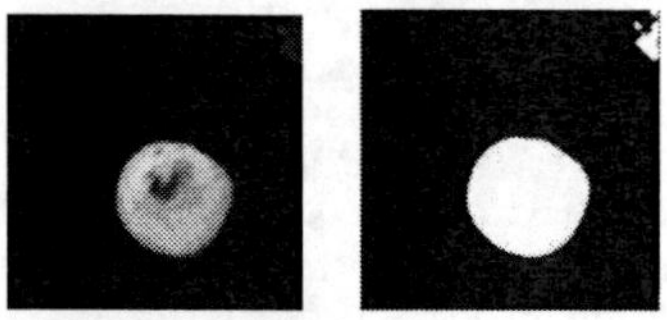

(c)Color recognition (d) Thresholding

(e)Morphological closing operation

(f)Morphological open operation

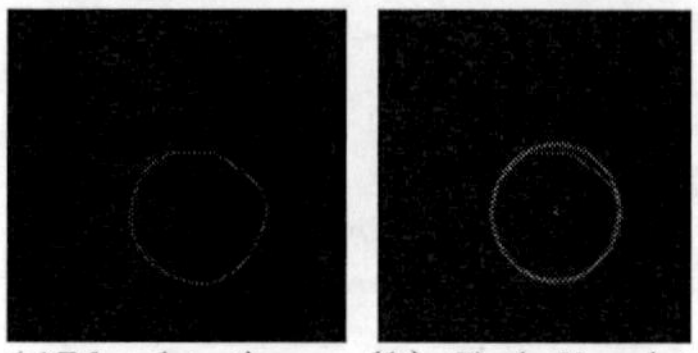

(g)Edge detection (h) Circle Hough transform

Figure 3. The processing of cherry picking robot vision recognition system

From figure c, you can see cherry has successfully through the color recognition extracted from the leaves and twigs in the background. You can see from figure d, threshold selection of threshold segmentation can remove part of the holes on the picture. Figure d and e results were not significantly different, because if there is no small noise in the cherry image, the morphological closing operation of the treatment effect cannot see. From the figure f, you can clearly see the upper right corner of the figure g without the noise. Figure h shows that draw the outline of the cherry and the center of the circle.

5 Conclusion

This technology for cherry block adjacent to each other and overlap each other more serious problem cannot be effectively solved directly. To solve the above problem is only to shoot a cherry, picking a cherry finish after taking the next cherry images. It is best to deal with no shade of cherries. But the system is simple and effective, which does not need to how much programming basis. It can be run directly calling OpenCV function. For occlusion, due to the direct call OpenCV circle Hough transform function processing results are poor. The solution is to re-write the code through improved Hough transform to solve.

Acknowledgments

This work was supported by 2015 Annual Science and Technology Department of Guizhou Province, Bijie City Science and Technology Bureau, Guizhou University of Engineering Science Joint Science and Technology Fund Project, Cherry picking robot vision recognition system research, Guizhou Branch of LH word [2015] No. 7599; the grants from 2014 Science and Technology Department of Guizhou Province, Bijie City Science and Technology Bureau, Bijie University of Science and Technology Joint Fund Program, Guizhou Branch of LH word [2014] No. 7537

References

1. X.J.Zhou, Y.L.li, 2008. Based on machine vision pomelo to identify the natural scenarios mature orange Transactions of the Chinese Society of Agricultural Engineering, 24(1):175-178

DOI: 10.1051/matecconf/20166301005

2. X.Qin. Research and Implementation of a natural scene Apple Detection System [D]。Xinyang: North West Agriculture and Forestry University，2011：1-48

Full-automatic Special Drill Hydraulic System and PLC Control

Xue Jun TIAN

Lingnan Normal University Electromechanical Engineering Research Institution Zhanjiang GuangdongChina

Abstract: A hydraulic-driven and PLC full-automatic special drill is introduced, working principle of the hydraulic system and PLC control system are analyzed and designed, this equipment has the advantages of high efficiency, superior quality and low cost etc.

Keywords: Full automatic special machine tool; Hydraulic system; PLC

Introduction

A full-automatic special drill consists of pedestal, workbench, spindle box, gearbox, cooling pump, rocker arm, drill and hopper etc. It can fulfill the one-off process from feeding, clamping to drilling as well as volume production. The whole operation process is described as follows: place the workpiece in the hopper, press the start button and the drill will work automatically; then take out the workpiece a[1]fter the drilling process is finished, the process of which not only greatly improves the work efficiency and the drilling quality but also relieves workers out of fatigue resulting from repeatedly carrying the parts. Compared with common drills, full-automatic special drill can realize process automation. The whole machining process is stable, secure and efficient because the hydraulic system features strong clamp force.

1 Full-automatic special drill hydraulic system

Working station structure of full-automatic special drill is shown in Fig. 1.

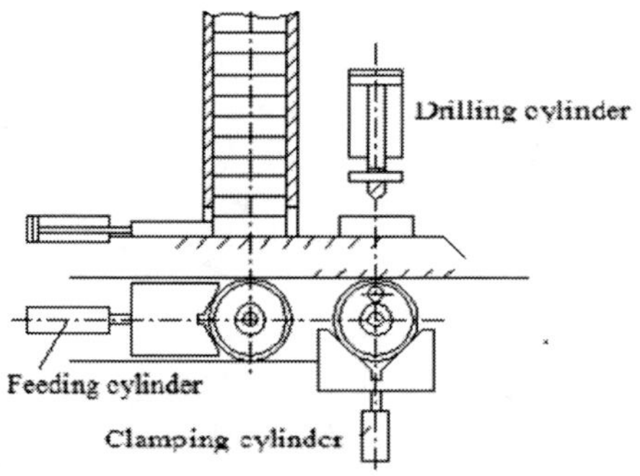

Fig. 1 Working station structure chart of full-automatic special drill

The working process is as follows: press the button→ the feeding cylinder advances→ the feeding cylinder starts to retreats→ The feeding cylinder resets and the clamping cylinder advances→ the drilling cylinder advances quickly→ the drilling cylinder advances→ the drilling cylinder retreats quickly→ the clamping cylinder retreats → the feeding cylinder advances, repeatedly.

[2] According to the above working process, selection of the reversing valve is analyzed below: the feeding cylinder is non-interfering with other cylinders, a two-position four-way solenoid valve can meet the feeding requirements; the clamping cylinder maintains certain pressure in the working process; if a two-position four-way magnetic exchange valve is selected, the pressure cannot be maintained when the drilling cylinder advances quickly and retreats quickly, so a

DOI: 10.1051/matecconf/20166301006

three-position four-way solenoid valve is used; a three-position four-way solenoid valve is used for reversing of the drilling cylinder. A two-position two-way solenoid valve instead of the medium function of the reversing valve can used for unloading due to mutual interference between cylinders. Besides, To ensure stable operation, hydraulic cylinder metering-out control loop is used due to vertical installation of the drilling cylinder. Sequential action of three cylinders is controlled by travel switch and solenoid valve[1]. Full-automatic special drill hydraulic system loop is shown in Fig. 2, electric control analysis in Fig. 3 and electric control loop in Fig. 4.

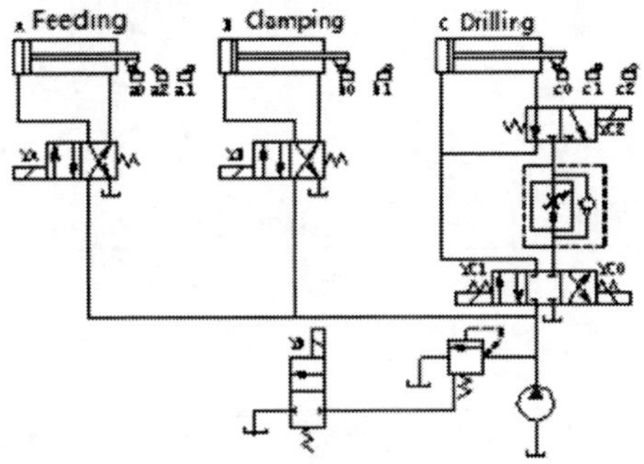

Fig. 2 Full-automatic special drill hydraulic system loop

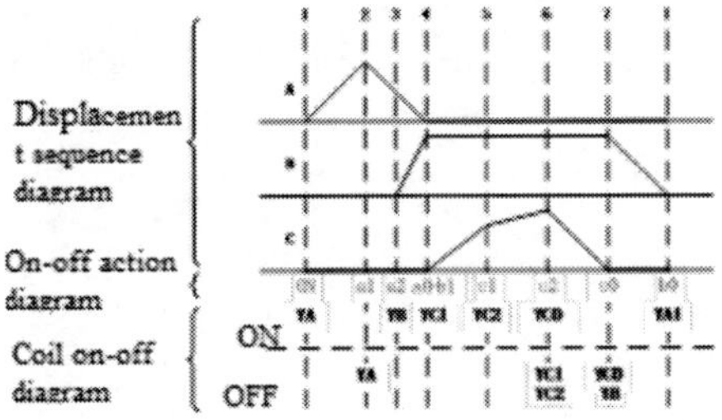

Fig. 3 Full-automatic special drill electric control analysis chart

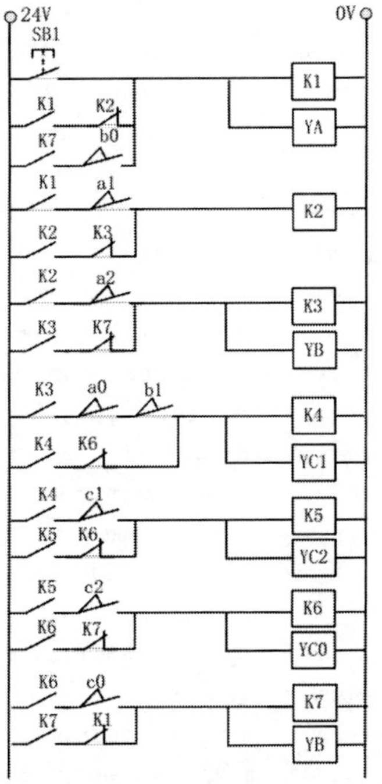

Fig. 4 Full-automatic special drill electric control loop diagram

① The feeding cylinder advances:

Press the switch and the feeding cylinder starts to work.

Inlet oil line: oil tank →oil hydraulic pump→ (YA electromagnet is powered on, the reversing valve changes the direction)→ left cavity of the feeding cylinder;

Return oil line: left cavity of the feeding cylinder → reversing valve→ oil tank.

② Movement of the clamping cylinder:

When the feeding cylinder piston rod moves to a2, YB solenoid valve is powered on and the clamping cylinder stays still; when the piston rod moves to a1 and the clamping cylinder extends, the feeding cylinder withdraws.

Inlet oil line of the feeding cylinder: oil tank →oil hydraulic pump→ (YA electromagnet is powered off, the reversing valve changes the direction)→left cavity of the feeding cylinder;

Return oil line of the feeding cylinder: left cavity of the feeding cylinder → reversing valve→ oil tank;

Inlet oil line of the clamping cylinder: oil tank →oil hydraulic pump→ (YA electromagnet is powered on, the reversing valve changes the direction)→left cavity of the clamping cylinder;

Return oil line of the clamping cylinder: left cavity of the clamping cylinder → reversing valve→ oil tank.

③ Movement of the drilling cylinder:

When the feeding cylinder piston rod returns to a0 and the clamping cylinder moves to b1, YC1 is powered on and the drilling cylinder extends; then the drilling cylinder is under differential connection and moves quickly.

Inlet oil line of the drilling cylinder: oil tank→ oil hydraulic pump→ (YC1 electromagnet is powered on, the reversing valve changes the direction) →overflow valve→ left cavity of the drilling cylinder;

Return oil of the drilling cylinder: right cavity of the drilling cylinder→ left cavity of the drilling cylinder (the drilling cylinder is under differential connection);

Movement of the feeding cylinder and the clamping cylinder remains unchanged.

④ Movement of the drilling cylinder at c1

When the drilling cylinder piston rod moves to the travel switch c1, YC 2 is powered on and the piston rod slows down.

Inlet oil line of the drilling cylinder: oil tank→ oil hydraulic pump→ (YC1 electromagnet is powered on, the reversing valve changes the direction) →overflow valve→ left cavity of the drilling cylinder;

Return oil line of the clamping cylinder: left cavity of the clamping cylinder → (YC2 electromagnet is powered on, the reversing valve changes the direction)→compensated flow control valve →oil tank.

Movement of the feeding cylinder and the clamping cylinder remains unchanged.

⑤ Movement of the drilling cylinder at c2

When the drilling cylinder piston rod moves to the travel switch c2, YC1 is powered off, YC0 is powered on, YC2 is also powered on and the drilling cylinder starts to retreat.

Inlet oil line of the drilling cylinder: oil tank→ oil hydraulic pump→ (YC0 is powered on and the reversing valve changes the direction)→throttle governing valve→(YC2 is powered on and the reversing valve changes the direction)→ right cavity of the drilling cylinder;

Return oil line of the drilling cylinder; left cavity of the drilling cylinder →overflow valve→ three-position four-way solenoid valve→ oil tank.

2 Full-automatic special drill PLC system

1) Hardware structure of PLC control system

PLC control system hardware consists of input circuit, output circuit and PLC basic unit[2]. Input circuit consists of travel switch, start button and protection switch. To satisfy the above control requirements of the full-automatic special drill, Mitsubishi FX2N-16MR-001 (8 unput/8 point relay output) type PLC is selected.

2) PLC control program under automatic working condition

In case of PLC control, plan PLC input and output ports first; PLC hardware wiring is shown in Fig. 5:

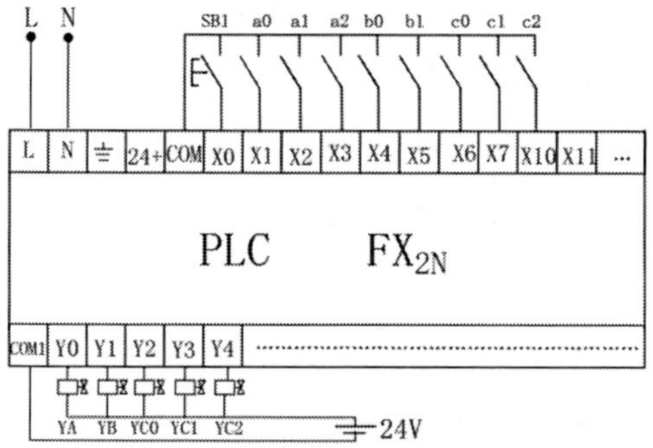

Fig. 5 Full-automatic special drill PLC wiring diagram

According to IO port stated above, PLC ladder diagram is designed as shown in Fig. 6:

DOI: 10.1051/matecconf/20166301006

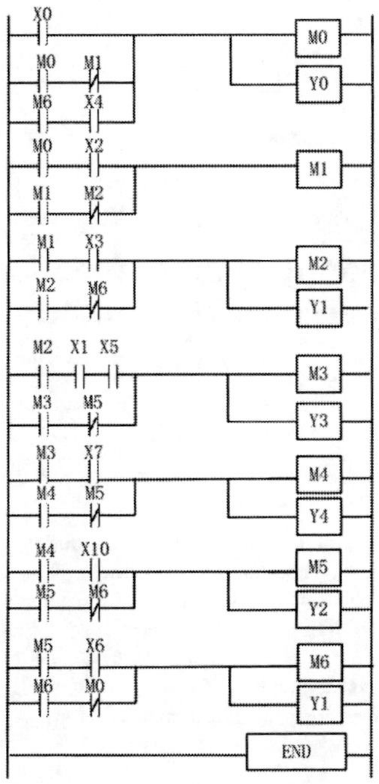

Fig.6 Full-automatic special drill PLC ladder diagram

3 Conclusions

Stable performance and good operation of full-automatic special drill demonstrates reasonable and feasible design of the hydraulic system. Full-automatic special drill containing PLC control system greatly lowers the fault rate, improves reliability and automation degree of equipment operation, reduces labor intensity and reaches the expected effect.

References:

[1] Y. X. Lu, *Hydropneumatics Manual* [M]. Beijing: China Machine Press. (2002. 1.)

[2] J. W. Luo, *Hydraulic Drive and Control* [M]. Chongqing: Chongqing University Press. (2006. 7.)

MATEC Web of Conferences **63**, 01007 (2016) DOI: 10.1051/matecconf/20166301007
MMME 2016

Design and Implementation of Electric Steering Gear Inspection System for Unmanned Aerial Vehicles Based on Virtual Instruments

Xing ZHENG [1,a], He YANG [2] and Shao Jun LIANG [1]

[1] *Wuhan Mechanical Technology College, Wuhan, 430075, China*
[2] *Hubei University of Education, Wuhan, 430075, China*

Abstract. A kind of UAV electric servo detection system based on Virtual Instrument is designed in this paper, including the hardware platform based on PC-DAQ virtual instrument architecture and the software platform based on LabVIEW function, structure and system implementation methods. The function, structure and system implementation method of software platform is also described. The gear limits checking, zero testing, time domain characteristics test results showed that the system achieves testing requirements well, and can complete detection of electric steering gear automatically, fast, easy and accurate.
Keywords. Virtual Instruments, Unmanned Aerial Vehicles, Electric Steering Gear, Inspection

1 Introduction

As an operational component of flight control system for an unmanned aerial vehicle, steering gear generates corresponding force and torque according to command of a flight control computer for manipulating steering panel or thrust guide mechanism, in order to control flight attitude and trajectory. Therefore, performances of the steering gear directly impacts performance and flight safety of the flight control system. In examining, debugging and maintaining the steering gear, it is necessary to comprehensively test the performances. To ensure that the unmanned aerial vehicle can smoothly and safely fulfill its flight task, performance of the steering gear shall be also tested before the vehicle takes off.

At present, electric steering gears of unmanned aerial vehicles are mostly manually inspected. In other words, in inspecting a steering gear, an inspector shall regulate position of potentiometer and switching state to generate various input signals such as analog and switch quantity signals necessary for exciting the steering gear. All input signals of the steering gear are indicated by an analog meter or manually observed. During testing, various performance parameters, including startup current, operating current, startup voltage and times of magnification, shall be manually read and recorded by the inspector. Such manual inspection has some problems like difficulty for testing, low efficiency and poor precision. To solve difficulties for inspecting an electric steering gear, it is necessary to develop an electric steering gear inspection system for the unmanned aerial vehicle.

2 System Development Plan

Based on full demonstration, a detailed development plan is made. Built up with virtual instruments and technologies, the system is made up two major parts, including hardware and software platforms. Virtual instruments belong to a system of computer instruments designed and designed by users with a virtual panel on a computer-based hardware platform and test functions realized by testing software. In essence, they simulate interfaces of traditional instruments by computer display, to output measurement results in multiple forms. Data of signals are calculated, analyzed and processed with computer software. Meanwhile, signals are acquired, measured and conditioned by I/O equipment, in order to make availability of all testing functions.

Being fundamental for virtual instruments, hardware platform is mainly responsible for signal acquisition, amplification, analog-to-digital conversion and display and so on. With a PC-DAQ virtual software architecture, system hardware is composed of multi-functional data acquisition card, signal conditioning board and power supply.

Being critical for virtual instruments, software platform is primarily used for calculating, analyzing and processing data of tested signals. System software is developed by LabVIEW graphic programming language. Data are managed and conditioned by Microsoft SQL Server, including tested items, parameter processing, test case generation, system management, databases and other functional modules.

3 Development Plan for Hardware Platform

[a] Xing ZHENG: zhengxcn@foxmail.com

DOI: 10.1051/matecconf/20166301007

In consideration of reliability, openness, scalability, portability, standardization and requirements for speed and functions, system hardware is designed with PC-DAQ virtual instrument architecture pursuant to practical testing requirements, made up of portable computer, data acquisition card, signal conditioning board and power supply. The components of system hardware are shown in Fig 1 as follows.

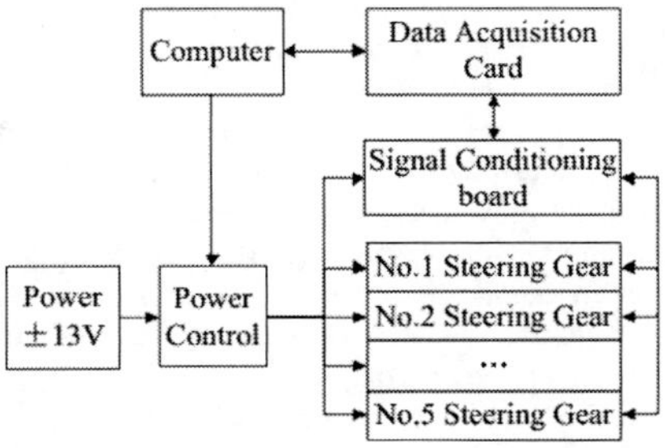

Figure 1. Components of Hardware Platform

3.1. Portable computer

In view that the system shall be portable and integrated, a portable computer is used as computer platform of the system. Required to be designed with current common basic configuration, this system has a socket for PCMCIA bus.

3.2 Data acquisition card

NI's DAQCard-6024E data acquisition card is used. As one of E-series plug-and-play multi-functional data acquisition cards based on PCMCIA bus, it has 16 channels of analog inputs with a sampling frequency of 200KS/s, 2 channels of analog outputs with a sampling frequency of 10KS/s, 2 24-bit timer/counter and 8 digital I/O ports that support TTL.

3.3 Signal conditioning board

As an intermediary module that connects computer, power supply and steering gear, it mainly includes sampling/follower module, current sampling module and power control module for sampling output current signals and controlling working power of steering gear.

3.4 Power supply

Power is supplied by constant-voltage supply and dual-redundant power supply of nickel-cadmium battery pack, on the other hand, to provide multi-channel DC power supply for the signal conditioning board, on the other hand, to make high-power working power available to the steering gear.

4 Development Plan for Software Platform

Developed by LabVIEW graphic programming language, software platform is made up of several functional modules, including tested items, parameter processing, test case generation, system management and databases. From Fig 2, it may be discovered that database is essential for the software of that system, where data are organized and managed through databases. To meet the requirements for different tested items, software modules "Tested Items and Parameter Processing" are set during software design. Tested items and relevant parameters are set by a tester according to different functional requirements. Set content is stored in the database of the tested system, in order that test cases may be directly transferred from the database for subsequent tests. In this way, it is unnecessary to set tested items and parameters every time. It is convenient to uniformly manage data in a database, and acquired data may be stored in hard disk drive in the form of data files.

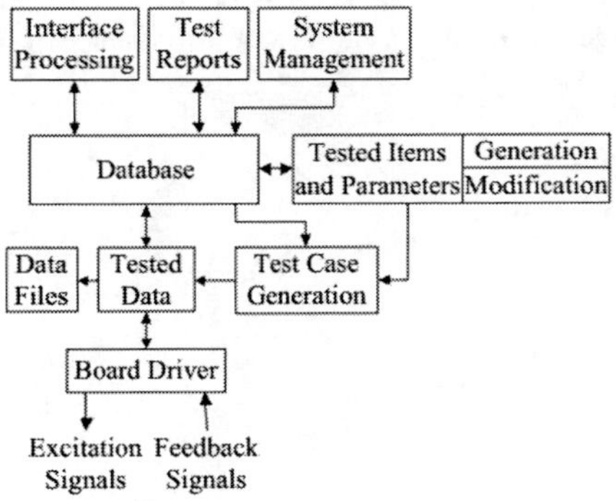

Figure 2. Components of Software Platform

Database is an indispensable part of data processing system in a computer. In organizing numerous data, they may be efficiently managed by database technologies. Waveform files or arrays are available from LabVIEW for storing acquired data. Frequency of excitation and feedback signals isn't quite high in the tested steering gear, so data are stored in the form of a database for convenient management. The database is operated in LabVIEW by ADO database interface technologies through ActiveX.

5 Specific Implementation

The system testing process is shown in Fig 3 as follows. Fig 3 shows the main program flowchart of tested system software, from which it may be discovered that the whole tested system is designed into multiple modules in terms of software, including modules of initialization and test case generation, in order to be in line with basic design principles of software engineering. A test may be performed once the testing program is started up, and test results may be replayed. To test software, created test cases may be used or directly manually tested, while basic or combinational test cases may be stored in a database. Test results may be stored in databases and data files. In the mean time, they are displayed.

In performing a test, test cases are needed at first. Type of steering gear and distribution of I/O ports are concerned about identification of tested data. To choose type of a controller, provided that type of the steering gear is given and I/O is defined, corresponding control panel and program interface will be available from the program. Then, users can use existed test cases in the database and make modifications based on that, or input the test cases by hands. Once the test cases are judged to be legitimate, they will be added to the database and initialized (including database and board of the test).

The test may be performed once the test cases are determined. It is divided into manual and automatic tests. During testing, sampling waveforms may be checked on a real-time basis (occasionally time-delayed). It is possible to change excitation signals during testing and ensure the test is performed on a real-time basis. A control panel will be available from the program to change inputs of a controller by hands on a real-time basis. For an automatic comprehensive test, a test case editor will be provided. Once the test is started, data will be automatically tested and then collected pursuant to cases.

In this paper, data of I/O ports are divided into analog quantity, switching value and digital value (serial codes or pulse sequence). For input signals of the controller, excitation signals are synchronously provided by analog quantities of each channel, and switching values of all channels are in sync with the level translation. Sampling signals are sampled at certain frequency according to characteristics of separate channel signals, and generally synchronously sampled. To save space, different signals are organized in distinct ways. For instance, for analog signals, it is only necessary to record values of all sampling points after recording starts or ends, because the sampling frequency is constant in each channel. As regards switching value, only initial level and time of level changes are recorded.

To display data, oscilloscope, data sheets and analog display models of panels will be provided. For switching value and analog quantity, waveforms may be displayed in an oscilloscope, while digital value is marked at the response time.

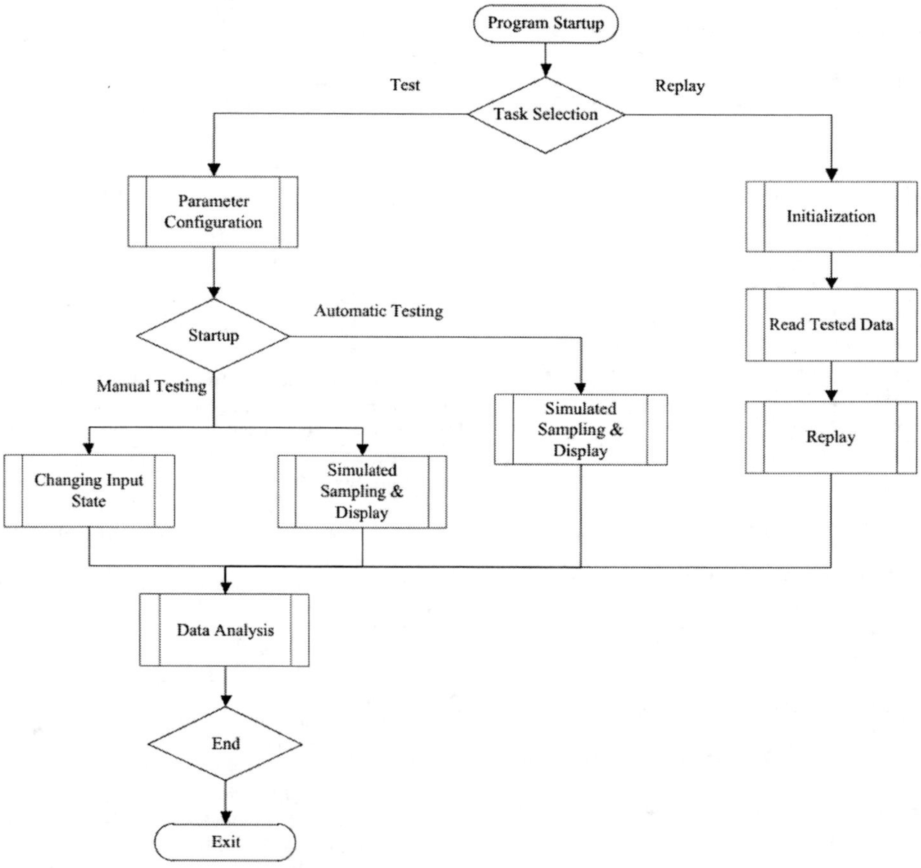

Figure 3. Flowchart of Tested Main Program

6 Results

Generally, electric steering gear of an unmanned aerial vehicle stably controls flight by following. According to requirements for static and dynamic performances of the steering gear, tested items are confirmed to mainly include limit check, zero-bit test, zero-modulation test, following, time-domain and frequency-domain characteristics. The electric steering gear inspection system for unmanned aerial vehicle designed by above methods is used for inspecting major tested items. Fig 4, 5 and 6 respectively show polarity test results, zero-bit test results and analysis of time-domain characteristics. All results suggest that inspection requirements are reasonably met and inspection purposes are fulfilled.

References

1. Li Ying, Yu Jianguo, Shao Kaifeng, Automatic Test System of Electric Helm Gear Based on Single Chip Microcomputer & PC, Computer Measurement & Control, Vol.16, 955(2008)
2. Su Na, Development and application of virtual instrument LabVIEW, Computer CD Software and Applications, Vol.23, 131(2013)
3. Yan Gang, Zhou Jun, Design and Application of Steering Gear Angle Detection System, Electronics World, Vol.16, 141(2013)
4. Lu Xiaoli, Research on Automatic Test Method of Linear Actuator, Journal of Mudanjiang College of Education, Vol.4, (2007)
5. Jiang Yang, Cong Shuquan, Zhu Wenlai, Design and Implementation of Intelligence Test System for Electric Actuator Based on LabVIEW, Journal of Astronautic Metrology and Measurement, Vol.31, 65(2011)
6. Zeng Man, He Weiguo, Systematic Implementation of Steering Engine Comprehensive Performance Test Platform, Computer Measurement & Control, Vol.18, 1518(2010)

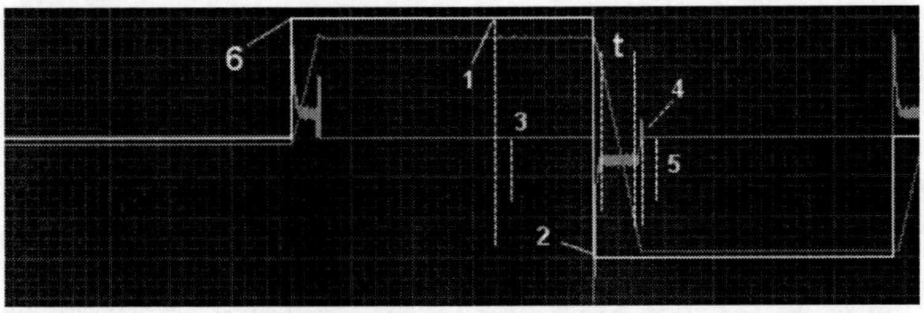

Figure 4. Curves of Polarity Tests (White: Excitation; Red: Feedback; Green: Current)

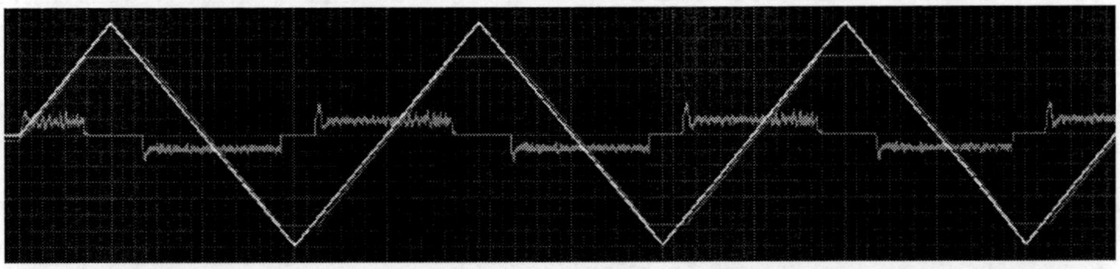

Figure 5. Curves of Zero-bit Tests (White: Excitation; Red: Feedback; Green: Current)

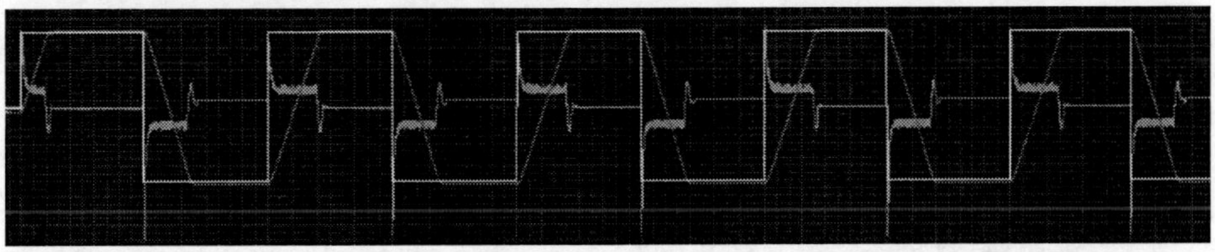

Figure 6. Waveforms for Analyzing Time-domain Characteristics (White: Excitation; Red: Feedback; Green: Current)

MATEC Web of Conferences **63**, 01008 (2016) DOI: 10.1051/matecconf/20166301008
MMME 2016

Drilling azimuth gamma embedded design

Yi Ren ZHOU[a], Xiao Lin QIU, Zhi Qiang GUO, Jian Hui ZHANG
Nanchang Institute of Technology, No.901 Heroe Ave, Qingshanhu ,Nanchang ,Jiangxi, 330044.China

Abstract: Embedded drilling azimuth gamma design, the use of radioactive measuring principle embedded gamma measurement while drilling a short section analysis. Monte Carlo method, in response to the density of horizontal well logging numerical simulation of 16 orientation, the orientation of horizontal well analysed, calliper, bed boundary location, space, different formation density, formation thickness, and other factors inclined strata dip the impact by simulating 137Cs sources under different formation conditions of the gamma distribution, to determine the orientation of drilling density tool can detect window size and space, draw depth of the logging methods. The data 360 ° azimuth imaging, image processing method to obtain graph, display density of the formation, dip and strata thickness and other parameters, the logging methods obtain real-time geo-steering. To establish a theoretical basis for the orientation density logging while drilling method implementation and application of numerical simulation in-depth study of the MWD azimuth and density log response factors of horizontal wells.

Keywords: Embedded; LWD; azimuth gamma; numerical simulation; Monte Carlo method

1. Introduction

The design of embedded MWD gamma position to msp430 embedded MWD gamma position consider to be the carrier, industrial-oriented research applications can accurately detect the formation and reserve evaluation of the success of intelligent systems, this design makes a great contribution to solve the problem of the introduction of the high cost of foreign advanced LWD system.

From the 20th century of 60s,it has developed for 80 years, along with the birth of the drilling position with thedensity logging tool and logging technology which had bring a new impetus, they think of the resistivity logging while drilling technology, and nuclear porosity parameters LWD technology, acoustic parameters LWD technology, nuclear magnetic resonance logging-while-drilling technologies make a progress in drilling industry. However, the development is going through some bottlenecks, with foreign logging-while-drilling density of a Monte Carlo simulation (Monte Carlo) appear that takes hope to break through these bottlenecks . But we still fall behind in comparison with foreign transmission technology, and we still as a freshman at the research work in the azimuth gamma density in this area , Monte Carlo model can be effectively in recording photon radialand longitudinal formations. Monte Carlo model Research would make a progress in the drilling orientation density logging, the instrument has been designed for detecting and playing a important role in the promotion, and to promote the orientation density loggingwhile drilling application and development of the oil well logging industry.

LWD with wired and wireless are two types of transmission. There are cable transmission, special drill pipe transmission and optical transmission[1]. The transmission rate is about 2Mbit/s[2]; wireless transmission isbased on using mud pulse, electromagnetic and acoustic waves[3], etc. ,Such as Baker Hughes who launched overseas drilling focused electrical imaging logging tool[1][3], the logging tool is adapted to different salinity of the mud,and the application get a big success; but there is a big gap in the domestic research in this area."Embedded-based msp430 drilling position Gamma" willdistinguish the reservoirs in different directions, and the interpretation of gamma intensity in different directions of information storage layer boundary judgment, judgment horizon, interpretation, etc. It provide a piercing direction Technical solutions.

2. Embedded drilling azimuth gamma hardware integration

Embedded drilling azimuth gamma instrument is consist of the sector detection module, gamma counting module, the gamma sensor module.

The sector detection module by the trysail accelerometer or 3-axis fluxgate and a control circuit. It will produce centrifugal force during rotation, it will cause a bit of impact on the three-axis plus table, so I chose trysail fluxgate sector detection as the sample

[a] Corresponding author: renyiz@126.com
Jiangxi Education Fund projects, Numbering: GJJ151174

instrument. MSP430 processor control the circuit ,it is shown in Figure 3.1, the peripheral circuit includes a serial Flash, ferroelectric memory, 485 communication port, RTC (real-time controller always) circuit. The control circuit, by connecting the SPI with the trysail flux gate, to complete configuration and data read controlcircuit to do the relevant calculation, the current high-side position, gamma counting module through port 485 sent. The main function of the sector detection module is the drilling plane which is separated several sectors, It is divided into four sectors up and down at least, and no more than 16 or 32 sectors.

Gamma counting module uses MSP430 as the processor (as Figure 1), 8MHz frequency, surrounding with serial Flash 4M, story data; ferroelectric memory, data is temporarily stored; 485 communicated with the sector detection module and the MWD module; it also equippedwith vibration and temperature sensors to detect underground anomalies. Azimuth sensor is packaged by alloy (usually tungsten alloy) , only a window was left in one direction , the current gamma design, all came from the window which is in the direction of where the ground; and while in which direction the current window, you need to provided by the sector detection module.

Gamma Sensor: Use CBG-DGA gamma sensor with 75 ° window, gamma range 0-500API, working temperature 0-150 ℃.

Figure 1 MSP430 processor

3. LWD azimuth gamma software architecture

Software hierarchy shown in Figure 2

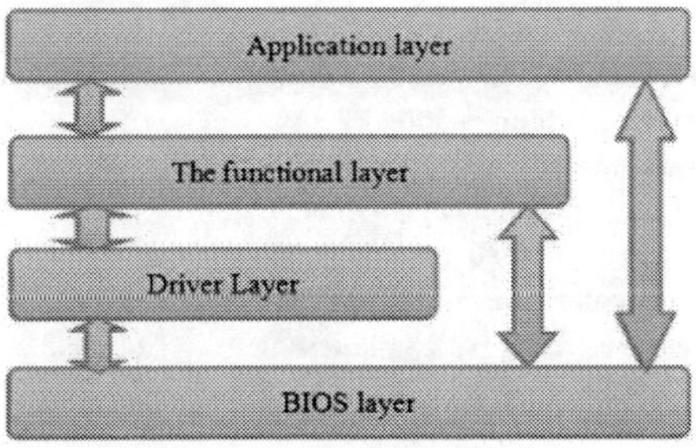

Figure 2 Software Hierarchy

BIOS layer: a package of basic hardware operation, the operation of the register of a transparent, provide a common call interface. In accordance with the functional components it was classified; all features BIOS level debugging is complete, generate a library; the applicationin the formof libraries to link calls.

To use the BIOS module:

- UART
- SPI
- IO
- WatchDog
- Timer

ii. Driver layer: for packaging features provide module-specific features function call entry, driver layer for the control chip is through BIOS calls to achieve, function driver layer may be collected and summarized and logical form comparison of standard call interface, after successful commissioning alone generate library functions, application layer drive functionality in the form of library functions to call. Use Table 1 to the drive module

Table 1 using a driver layer to the drive module

Number	Drive	Function Module
1	Flash	S Flash Module
2	E2PROM	E2PROM Module，To be completed
3	FRAM	Ferroelectric memory module

iii.Functional layer: logical functions closely classify the instrument, at this level it is achieved by the application of layer to call. The functional layer is designed in accordance with the logic function will be closely linked logically functional design as a class, the functional layer is closely related to the specific application. In addition to some special design features can be fixed, but most of it need separately design alone and debugging; functional layer in the system is in the form of source code management and debugging; function organize the same functions in a source file; name the functional layer function to do a specific function prefix, easy to understand and recall. Table 2 functional modules used.

Table 2 functional layer used in function block

Number	Features	Function Module	Prefix function
1	Protocol processing	Statute for data analysis and packaging	Cmd_
2	azimuth gamma	azimuth gamma all algorithms and processing	AZG_
3	System Configuration	Associated with the system configuration storage and readout	SysCfg_
4	system	Processing system, Operation, Control, etc.	ASys_
5	Monitoring and recording	Monitoring and recording system health	Supv_

iv.Application Layer: The application management, in accordance with certain features designed to call the function layer, to achieve the management of the entire system. The main function is defined as a main (), to achieve the main process control, calling each function module, complete application functionality; logically classifying the operation to define and call functions using ASYS_.

4. Experimental Method

I .Using gamma logging LWD azimuth to response for the simulation program experiments, numerical simulation for various of factors, improve various parameters of the logging methods, the factors into position, calliper clearance and ground conditions, by simulating the responding of the probe formation parameters, finding out the law, drawing response formula, better improve LWD azimuth detection parameter gamma logging method, lay the foundation for data processing. Using the principle of gamma photon interaction with the formation, the use of Monte Carlo simulation method, the LWD azimuth gamma logging tool window can be selected to achieve the desired effect can be the best to reflect the formation of information.

Ii .In both density quite different formations, respectively, from the detector and 137Cs radioactive source was simulated to obtain the optimum parameters of the length of the source pitch.

iii. Gamma logging LWD to establish azimuth the model using Monte Carlo methods, 137Cs radioactive source in a different density of the formation, the gamma field different litho strata distribution of numerical simulation, the 137Cs radioactive sources "back" shield, and between the radiation source and the detector, and thedistance between the probe are used shielding material is isolated, so 137Cs radioactive source emits gamma photons can be detected in the main instrument in accordance with the direction release, and it emits gammaphoton does not directly reach the detector from the borehole and received by the detector, and the impact detector count, and further information on the formation of judgment, to give this kind of distribution of radioactive sources, by setting the shield can achieve the best detection effect.

Iv .Borehole measurement conditions change, such as changing the hole diameter, changing the distance between the logging tool and the borehole wall, which gap (Standoff), etc., using Monte Carlo simulation method to study the condition of the wellbore LWD azimuth gamma Effect of horse logging methods; changing density, litho logy, thickness, interface parameters such as the formation, by the Monte Carlo simulation method to study gamma logging LWD azimuth respond to different formation conditions.

V .The resulting application response relationship for dipping strata numerical simulation data by MATLAB, Origin8 and other computer languages, the resulting process simulation, the final azimuth of the LWD density to achieve imaging gamma logging; tilted to get under formation conditions Figure gamma imaging processing, determining formation dip and strata thickness, and use C #, C ++ formation processing software.

5. Conclusion

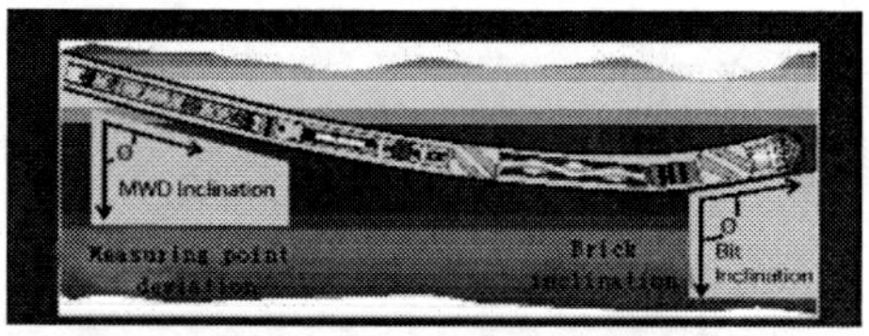

Figure 3 Embedded drilling azimuth gamma

Embedded gamma logging LWD azimuth LWD system in the process, as shown in Figure 3. Detecting the formation azimuth 16, to obtain information on the data layer of 360 ° , using effective means of information and data imaging visual display of the detected formation. In the Monte Carlo model design, it will be divided into the wellbore surrounding the 16 sectors, the detector close to the wall, 22.5 ° rotates once every stratum detect simulated to give gamma formations surrounding a borehole 16 in the direction of Ma photon counting rate, to give the corresponding information data formation processing. Then change to adifferent detection conditions calliper , litho logy, stratathickness, research azimuth gamma logging LWD a lot of log response .

Verification of embedded systems technology gamma LWD parameters are as follows

- Measuring range: 0-500API
- Gamma measurement accuracy: ±2%
- Operating temperature: 0-150℃
- Sectorisation: 4-16 Sector
- Real-time transmission: On gamma、Under gamma
- Accuracy sector: 0.2° (Static state) , 2°(180RPM)
- Anti-vibration: 15g
- Impact resistance: 500g

Embedded systems LWD Gamma has a very significant azimuth horizon interpretation advantages for thin reservoirs and complex formation, can greatly

DOI: 10.1051/matecconf/20166301008

improve the drilling rate to enhance oil production. See Figure 4.

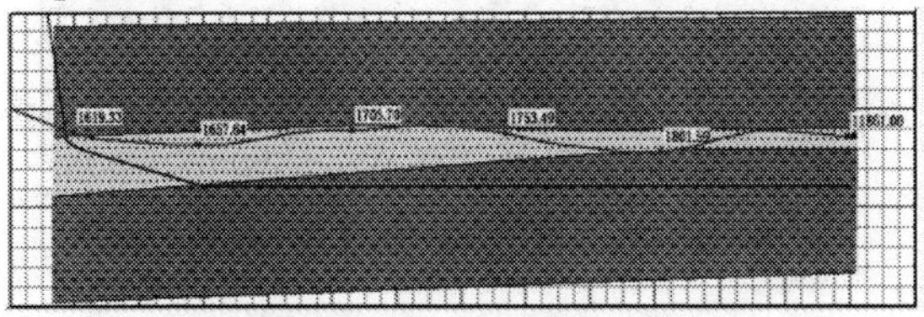

Figure 4 Embedded systems used in LWD oil sheet Gamma

After several tests, this design parameters and function were verified. Reflects the embedded system by LWD gamma vibration is small, high precision, to ensure that the measured data in a composite drilling reliability.

References

[1] Liu Xinping, etc. LWD status quo and prospect data transmission technology [J]. Well Logging Technology, 2014, 32 (2) : 249-253.

[2] Paul Lurie, Philip Head, Jacke E S. Smart Drilling with Electric Drill String[c].SPE 79886, 2013.

[3]David Patrick Murphy .Advances in MWD and Formation valuation[J].World Oil, 2013, 226(3).

About the Author:

Zhou Yiren (1958-), male, Jiangxi Yichun, Nanchang Institute of Technology solid scientific research center, senior engineer, the main research directions: embedded systems. (Nanchang 330044)

Qiu Xiaolin, 1962.7, male, Jiangxi, Nanchang Institute of Technology Dean, Ph.D., professor, research direction: Materials Science (Nanchang 330044).

Guo Zhiqiang, 1991.9, male, Shanxi Datong, Nanchang Institute of Technology research laboratory, research direction: electromagnetic theory (Nanchang 330044)

MATEC Web of Conferences **63**, 01009 (2016) DOI: 10.1051/matecconf/20166301009
MMME 2016

A New Cold Rotary Forging Technology for Automotive Starter Guiding Cylinder with Internal Helical Involute Spline

Ping WANG [1,a], Xiao Fei DONG [1]

[1] *School of Mechanical Engineering, Yancheng Institute of Technology, 224051 Yancheng Jiangsu, China*

Abstract. A new cold rotary forging technology of the internal helical involute spline was presented based on an analysis of the structure of automotive starter guide cylinder. 3D rigid-plastic finite element model was employed. Billet deformation, Billet equivalent stress and forming load were investigated under the DEFORM 3D software environment, then the forming process parameters were applied in the forming trials, and the simulation results are conformed with the experimental results. The validity of 3D finite element simulation model has been verified. The research results show that the proposed cold rotary forging technology can be efficient in handling of the forming manufacturing problems of automobile starter guide cylinder with internal helical involute spline.

1 Introduction

Automotive starter guiding cylinder is the critical part of automotive starter transmission system, it works in a complex condition, and it has to meet the requirements for the excellent mechanical properties. At present, most automotive starter guiding cylinders are manufactured by metal cutting, in which metal streamlines are cut off, thus lowering the mechanical properties. With the rapid increase of automotive starter guiding cylinders production, to explore the advanced manufacturing method of automotive starter guiding cylinders is of great significance to the development of automobile parts enterprises.

Cold rotary forging is an advanced and innovative metal forming technology that, based on the principle of continuous uniaxial pressure and using rotational movements in the dies [1], offers the following advantages: the reduction of the forging load, uniform quality, the strong effect on the mechanical properties, close tolerance and savings in materials cost etc. Many scholars have made some researches on the rotary forging technology. Montoya et al [1] made a kinematic and sensitivity analysis of rotary forging process by means of a simulation model. Han et al [2-4] clarified the difference between the cold rotary forging and conventional forging in the forming process. Qin [5] revealed the contact force response in cold rotary forging through the modeling and simulation. M.J. Roy, D.M. Maijer [6-7] investigated the spinning of a common aluminium automotive casting alloy A356 at elevated temperatures. Wang et al [8-12] developed a 3D rigid–plastic FE method to analyze the cold rotary forging process. Zhou et al [13-18] studied the cold rotary forging process by using analytical and experimental methods. All of these research results provided useful guidelines for the studies on the cold rotary forging process. However, the cold rotary forging technology for internal helical involute spline is not involved.

In this study, Base on the analysis of the structure of automotive starter guiding cylinder with internal helical involute spline, a new cold rotary forging technology was proposed. The cold rotary forging process is simulated using the DEFORM 3D software. The forging experiment was performed under the same process conditions used in the FE simulations. Comparison of the simulation and experimental results shows that the cold rotary forging technology is able to successfully find a defect-free cold precision forging operation for internal helical involute spline.

2 The description of cold rotary forging technology of internal helical involute spline

The cold rotary forging process of internal helical involute spline is illustrated in Figure. 1.

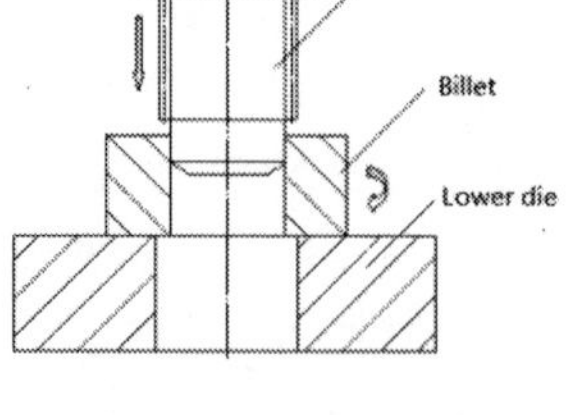

[a] Corresponding author: wp_ycit@163.com

Figure 1. Schematic diagram of cold rotary forging of internal helical involute spline.

The structure characteristics of the punch is mainly composed of the lower guiding part and the upper forming part, the lower guiding part is used to avoid material flow to the center, and the upper forming part is used to ensure the billet flow in a helical angle so as to form the internal helical involute spline in the process of extrusion. As the punch moves down, the billet rotates around its own axis under the action of the punch. When the rotary forging process is completed, the punch leaves the lower die under the action of returned material device.

3 Forming process and die set design

3.1 Forming process design

According to the final shape of automotive guide cylinder with helical involute spline, the forming process can be divided into ten stages: making billet, annealing, shot blasting, the saponification of phosphorus, forward extrusion, upsetting, backward extrusion cavity, machining guiding hole, the saponification of phosphorus, cold rotary forging internal helical involute spline. The schematic process of these stages was shown in Figure. 2.

(a) (b) (c) (d) (e)

Figure. 2 The forming process of automotive guide cylinder: (a) making billet, (b) forward extrusion, and upsetting, (c) backward extrusion cavity, (d)making guiding hole,(e) cold rotary forging internal helical involute spline.

3.2 Die set design

The die set is mainly composed of upper die and lower die. As shown in Figure.3.

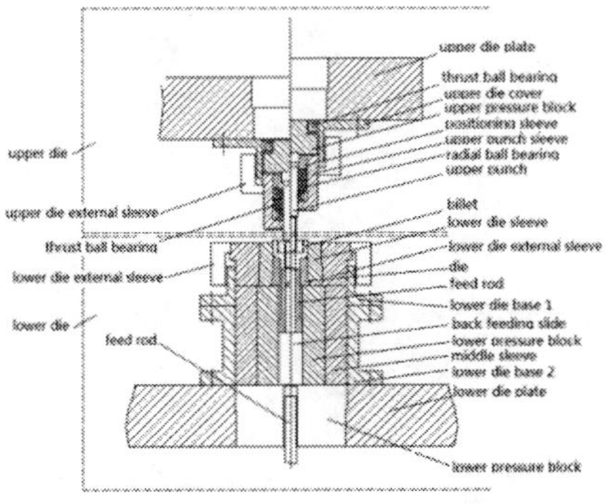

Figure. 3 Schematics of the die set.

When die set works, the upper die drives punch down, and all parts of the punch extrusion in turn in the inner hole of billet, from the beginning of punch extrusion billet to the end of the forming of internal helical involute spline, the punch will extrusion billet along the direction of a spiral angle, and the extrusion process is in stable state. After finishing the extrusion of internal helical involute spline, the punch rotates out of the billet, and the punch withdraws from the lower die under the action of feeding device, the whole rotary extrusion process is complete.

4 Finite element analysis

4.1 Finite element model

In order to evaluate the feasibility of the proposed cold rotary forging technology, a 3D finite element simulation was carried out using commercial finite element software DEFORM 3D. Considering the complexity of the internal helical involute spline, dies geometrical model was built under the NX7.5 software. The models were subsequently transformed into the STL standard format files and imported into DEFORM 3D analysis software by its preprocessor. A 3D rigid-plastic FE model was established under the DEFORM 3D environment, as shown in Figure. 4.

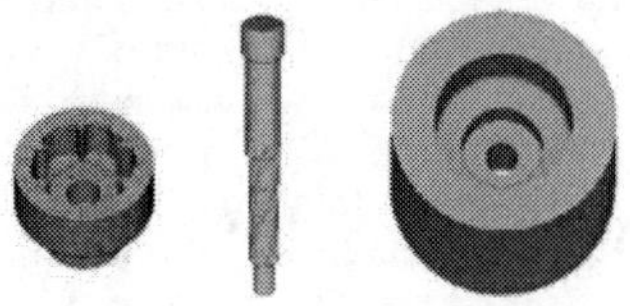

(a) billet (b) punch (c) die

Figure. 4 Finite element model: (a) billet, (b) punch, (c) die.

The billet material for used simulations and experiments was AISI5120, the material model was considered as rigid plastic with Von Mises yield criterion and isotropic hardening model. The mesh was divided into three-dimensional tetrahedral, the gridding number is 197410. Billet temperature is set to 20℃. The punch and die were assumed to be rigid bodies, so their deformations can be neglected when setting up FEM model. The punch speed is set to 12mm/s, and billet is set to turn around its own axis. The solving method of sparse matrix is set to Newton Raphson. The type of friction between billet and punch was assumed as shear friction, and the coefficient of friction is set to 0.12.

4.2 Forming Load Analysis

Forming load has important influences on the accuracy of the forged parts and the die life. The forming load result of the rotary forging process is shown in Figure.5.

DOI: 10.1051/matecconf/20166301009

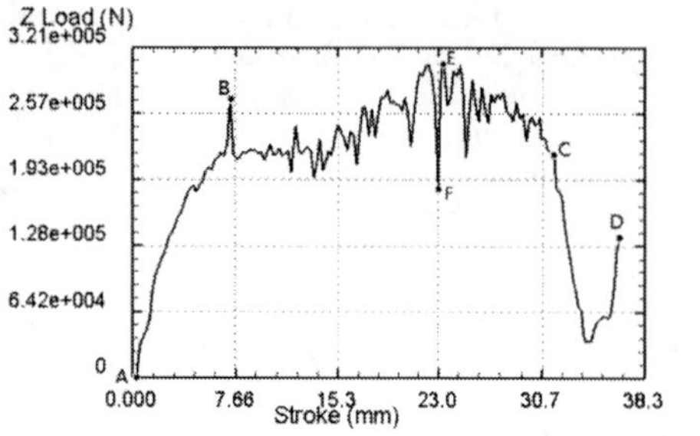

Figure. 5 The forming load result of the rotary forging process.

For the convenience of the analysis of the forming load, and the forming load curve is divided into three stages: the initial forming stage, the middle forming stage and the final forming stage. In the initial stage, the curve of the forming loads started from point A to point B. Due to the frictional resistance between the punch and the billet and the billet deformation resistance, the forming load increased quickly from 0 to 272.92KN. The punch was squeezed into the billet to form the initial shape of the internal helical involute spline. In the middle forming stage, the curve of the forming loads started from point B to point C, Along with the increase of the punch stroke, the deformation of the billet became more and more complex, the forming loads began to fluctuate, the maximum value of the forming load occured in point E (305.57KN), and the minimum value of the forming loads occured in Point F (219.02KN), the average value of the forming loads is 246.35KN. At this stage, the main part of the internal helical involute spline has been formed. On the whole, then the forming process of the internal spiral involute spline is in a relatively stable state. In the final forming stage, the curve of the forming loads started from point C to point D. The general trend of the forming loads is decreased at this stage. The forming temperature generated by extrusion concentrated in the forming area, which made the plastic flow of the billet increasing, the forming loads decreased. When the forming process closed to end, the forming loads increased slightly, as the metal flow was hampered by the die.

4.3 Billet Deformation Analysis

The billet deformation analysis is shown in Figure. 6. In the initial forming stage, as shown in Figure. 6(a), the punch moves down and begins to extrusion billet, and then the material is in the elastic deformation stage. With the increase of metal deformation, extrusion stress increases gradually, when the stress exceeds the material yield limit, the material begins to flow. Due to the limitation of the punch guiding parts, the material can flow only along the circumferential and axial. In the extrusion process of the internal helical involute spline, the friction force between the punch and the billet will drive billet counterclockwise rotating. In the middle forming step, as shown in Figure. 6(b), the forming is in stable state, and the metal flowed smoothly. In the final forming stage, as shown in Figure. 6(c), the internal helical involute spline was formed, and the vector changes of the each node flow speed was uniform, the forming performance of the internal helical involute spline is good, and no defects occurs.

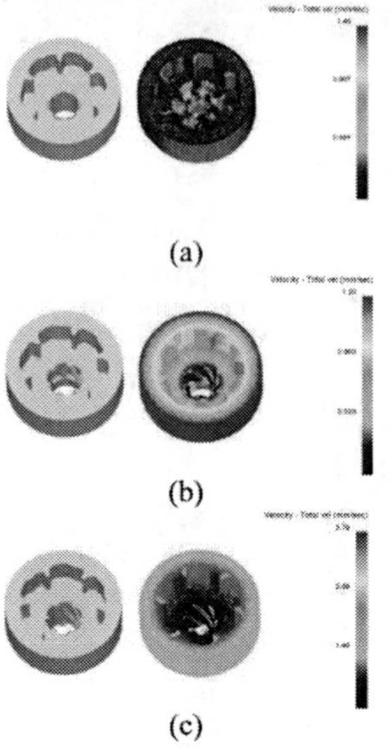

Figure.6 Billet deformation and metal flow results of rotary forging: (a) the initial forming stage, (b) the middle forming stage,(c) the final forming stage.

4.4 Billet Equivalent Stress Analysis

The equivalent stress distribution of the billet is shown in Figure. 7. It can be seen that the maximum stress is located in the contacting area of the punch forming part bottom and the billet in the initial forming stages, as shown in Figure. 7(a).

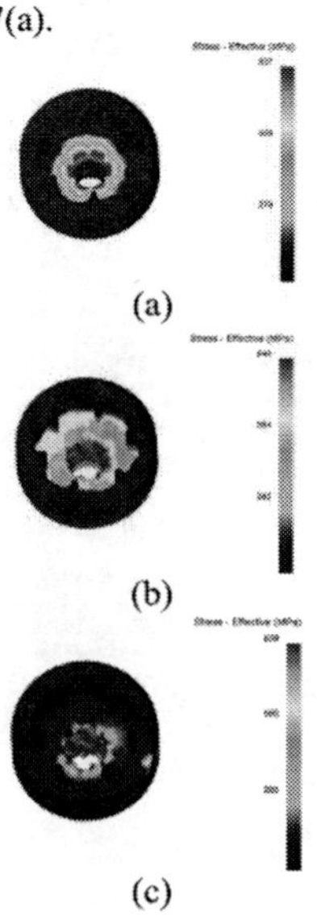

Figure.7 Billet equivalent stress distribution of the rotary forging: (a) the initial forming stage, (b) the middle forming stage, (c) the final forming stage.

5 Experimental verification

The primary aim of the experiment was to verify the feasibility of the proposed rotary forging technology. This experiment of the cold rotary forging process of automobile starter guide cylinder with the internal helical involute spline was conducted using YJ61-160A hydraulic testing press, which is an industrial metal extrusion hydraulic press made by Jiangdong machinery co., LTD. China, and a forging die sets similar to the simulative process were then fabricated, as shown in Figure. 8. The rest of the experimental conditions were the same with the simulation conditions.

Figure. 8 Tools used for the experiment.

A well-shaped the internal helical involute spline was achieved by the proposed rotary forging process, as shown in Figure. 9(b), and no forming defects occurred. The requirements of mechanical properties of the part are fully met. For comparison, Figure. 9(a) shows the simulated model of the internal helical involute spline. It is evident that a very good agreement was obtained between the simulation results and experimental results.

(a) (b)

Figure. 9 Models comparison: (a) the simulated model, (b) the experimental model.

6 Conclusions

A new cold rotary forging technology for automotive starter guiding cylinder with the internal helical involute spline was proposed. The finite element analysis based the software of Deform-3D and the experiments were performed to investigate metal flow, stress distributions and the forming load. As a result of this study, the following conclusions can be drawn:

(1)The structure of the internal helical involute spline of automotive starter guiding cylinder can be formed through the cold rotary forging technology.

(2)The finite element model built in this paper can predict the forming defects in the rotary forging forming process of automotive starter guiding cylinder with the internal helical involute spline. The model can be used to analyze and optimize the rotary forging forming process.

(3) The friction condition and the velocity of the punch are the main factors of affecting the rotary forging forming. Reasonable friction condition and velocity of the punch can effectively reduce the forming load and improve the full rate of the forming of the internal helical involute spline.

(4)The validity and the feasibility of cold rotary forging process scheme were verified by finite element simulation and actual production.

Acknowledgment

The project was supported by joint innovation project of Jiangsu Province under Grant BY2015057-28.

References

1. I. Montoya, M.T. Santos, I. Pérez, B. González, J.F. Puigjaner, Kinematic and sensitivity analysis of rotary forging process by means of a simulation model, Int J Mater Form Suppl. 1, 383−386(2008)
2. X. H. Han and H. Lin, Comparison between cold rotary forging and conventional forging, Journal of Mechanical Science and Technology. 23, 2668−2678(2009)
3. X. H. Han and H. Lin, Process design and control in cold rotary forging of non-rotary gear parts , J. Mater. Process. Technol. 214, 2402−2416(2014)
4. X. H. Han and H. Lin, 3D FE modelling of contact pressure response in cold rotary forging, Tribology International. 57, 115−123(2013)
5. X. P. Qin, Modeling and simulation of contact force in cold rotary forging, J. Cent. South Univ. 21,35−42(2014)
6. M.J. Roy, D.M. Maijer, Analysis and modeling of a rotary forming process for cast aluminium alloy A356, J. Mater. Process. Technol. 226,188−204(2015)
7. M.J. Roy, D.M. Maijer, Response of A356 to warm rotary forming and subsequent T6 heat treatment, Materials Science and Engineering. A611, 223−233(2014)
8. G. C. Wang and G. Q. Zhao, A three-dimensional rigid–plastic FEM analysis of rotary forging deformation of a ring workpiece [J], J. Mater. Process. Technol 95, 112−115(1999)
9. G. C. Wang and G. Q. Zhao, Simulation and analysis of rotary forging a ring workpiece using finite element method[J], Finite. Elem. Anal. Des 38, 1151−1164(2002)
10. G. Liu, S. J. Yuan, Z. R. Wang and D. C. Zhou, Explanation of the mushroom effect in the rotary forging of a cylinder[J], J. Mater. Process. Technol 151 , 178−182(2004)
11. X. H. Han and H. Lin, Effect of process parameters on wear in cold rotary forging by using 3D FE numerical simulation [J]. Iron making and Steelmaking 40,50−60(2013)
12. H.K. Moon, M.C. Lee, M.S. Joun, An approximate efficient finite element approach to simulating a rotary forming process and its application to a wheel-

bearing assembly [J]. Finite Elements in Analysis and Design 44:17–23(2007)
13. D. C. Zhou, Y. D. Han and Z. R. Wang, Research on rotary forging and its distribution of deformation[J], J. Mater. Process. Technol 31, 161–168(1992)
14. H. K. Oh and S. Choi, A study on center thinning in the rotary forging of a circular plate[J], J. Mater.Process. Technol 66, 101–106(1997)
15. G. C. Wang, J. Guan and G. Q. Zhao, A photo plastic experimental study on deformation of rotary forging a ring workpiece [J], J. Mater. Process. Technol 169, 108–114(2005)
16. X. B. Deng, L. Hua, X. H. Han, Y.L. SONG, Numerical and experimental investigation of cold rotary forging of a 20CrMnTi alloy spur bevel gear [J]. Material and Design 32,1376–1389(2011).
17. Jun-Song Jin, Lei Deng, Xin-Yun,Wang, Ju-Chen Xi , A new rotary forming process for rim thickening of a disc-like sheet metal part[J]. J. Mater. Process. Technol 212, 2247–2254(2012)
18. C.C.Wong, A.Danno, K.K. Tong, M.S.Yong, Cold rotary forming of thin-wall component from flat-disc blank [J]. J. Mater. Process. Technol 208,53–62(2008)

MATEC Web of Conferences **63**, 01010 (2016) DOI: 10.1051/matecconf/20166301010
MMME 2016

Research on current sharing of paralleled IGBTs in different DC breaker circuit topologies

Ying CHEN[1,a] and Fang ZHUO[1]

[1]*School of electrical engineering, Xi'an Jiaotong University, 710049 Xi'an, China*

Abstract. IGBT modules used in series and parallel to satisfy the requirement in high-power DC circuit breakers are often prone to large-current destruction due to current unbalance between paralleled IGBTs. It is of great importance to identify the current unbalance causes and to find a method optimizing the current sharing of paralleled IGBTs. In this paper the current-sharing influencing factors are discussed and verified by simulation. Two possible circuit topologies used in DC circuit breakers are proposed and simulated to see their performance in current sharing. The results show that one of them can provide us with a simple and effective method to achieve good current balancing in the DC circuit breaker application.

1 Introduction

DC breakers are increasingly being used in many power electronic applications such as Flexible AC transmission systems (FACTS) and high-voltage direct current (HVDC) transmission systems [1]. Currently, these applications are often operated at progressively higher power levels and in order to fulfil this demand, power semiconductor devices used in these applications need to reach a high power level. Among all the semiconductor devices used in such circumstances the insulated-gate bipolar transistor (IGBT) is a major representative with advantages like easy driving, low power dissipation and snubberless operation [2][3]. During the past few decades, developments in semiconductor technology have led to fast advance in high-power IGBTs [4]-[11].

Although The blocking voltages of these IGBTs rise up to 6.5 kV and current ratings are steadily increasing [12][13], they cannot fully meet the high power requirement and are often connected in parallel due to their limited power capacity[14][15]. When the power and voltage ratings are within the limit, it is also advantageous to use several IGBTs with lower ratings in series and parallel to cut the cost. But unfortunately the current (voltage) imbalance between parallel (series)-connected IGBTs can lead to imbalance temperature and losses, as well as over voltage or current destruction, so attention should be paid during the IGBT module design [3]. The factors causing the imbalance are similar for both the voltage and the current sharing, and here we mainly focus on the later. The current balance influencing factors falls into 3 categories: the device characteristics, the gate-drive circuit and the main circuit design [2]-[11]. Many efforts have been made in order to achieve equal current sharing in the parallel-connected IGBT modules [16]-[19]. However, these methods are rather wasting or complicated.

In this paper we discuss the current balancing performances of 2 alternative topologies to see if any of them can provide us with a simple and effective method to achieve good current balancing in the DC circuit breaker application. In the first part we review the electrical characteristics of the IGBTs and discuss the factors influencing the turn-on, turn-off, and the steady state. In the second part we illustrate the effect of each major influencing factor with the Saber software. Then we provide the 2 possible circuit topologies used in a dc circuit breaker and analyze their performances.

2 Electrical characteristics of IGBT and influencing factors

The IGBT symbol and the equivalent circuit is shown in Fig. 1. The device can be seen as a voltage-controlled current source and an equivalent resistance r, together with parasitic capacitances, inductances and resistances [20]. The IGBT switching process can be divided into the steady state and the transient. The main factor influencing the steady state is the equivalent resistance, and this is related to the saturation voltage and the collector current. For the transient, the key factor is the gate current and this is decided by the gate-drive voltage, the internal gate resister r_g, the external gate resister R_g, and the capacitances C_{GE} and C_{CG}. To explain this, the turn on and turn off behavior of the device should be first introduced.

[a] Corresponding author: yingchen@stu.xjtu.edu.cn

DOI: 10.1051/matecconf/20166301010

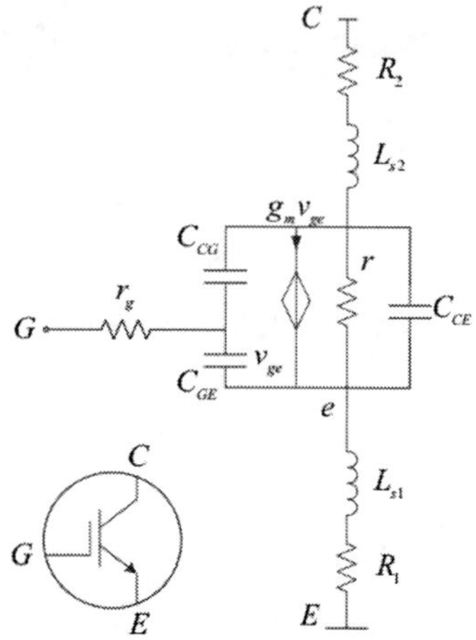

Figure 1. Equivalent circuit of an IGBT.

As can be seen Fig. 2, the processes of turning on and off can be each divided into 3 phases. During the turn-on process the gate voltage starts rising right after the driving voltage is given, and when it reaches the threshold V_{th} the collector current begins to rise. When the current stops rising the gate voltage experience a short break at V_M, and during this break the collector-emitter voltage starts to fall until the gate voltage begin to rise again. The turn-off process is the reverse of the turn-on process.

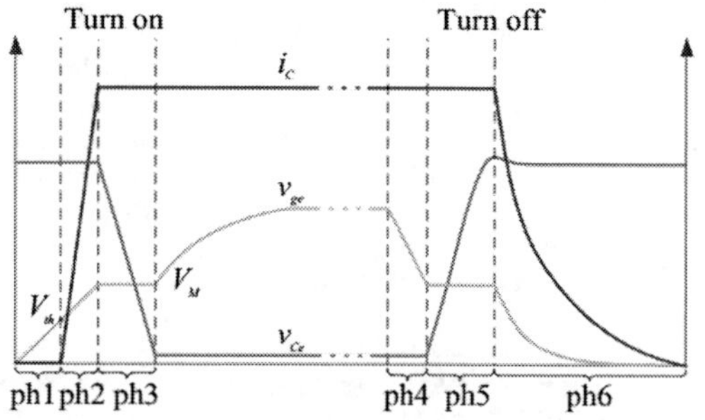

Figure 2. Turn-on and turn-off waveforms of an IGBT.

To get a better understanding of the waveforms and to find out the influential factors of the waveform features, it is necessary to see what happens inside the IGBT during the turn-on and turn-off process. The equivalent gate circuit of an IGBT is illustrated in Fig. 3.

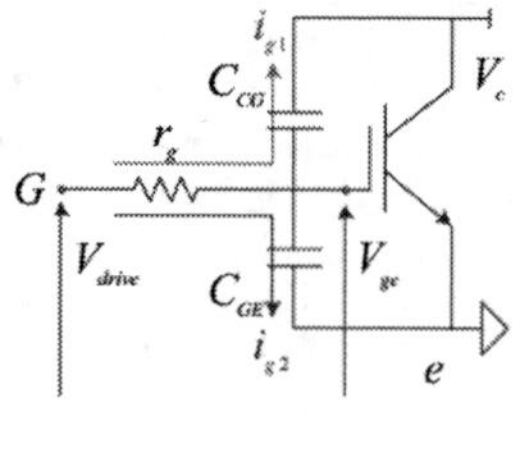

Figure 3. The equivalent gate circuit of an IGBT.

The increase of the gate voltage indicates that the gate current is charging the capacitor C_{GE} and when the voltage across the capacitor is higher than the threshold V_{th} the collector current starts to rise. Later when voltage of C_{GE} comes to the Miller plateau the current stops rising. During the Miller plateau the gate current charges C_{CG} and when it's fully charged the gate current goes back to C_{GE} again. So this process indicate that the charging speed is determined by the gate current magnitude which further depends on the gate-drive voltage, r_g, and R_g, while the Miller plateau is related to C_{GE} and C_{CG}. The turn-off process is the reverse of that of the turn on.

According to Fig. 3 the following relationships can be obtained.

$$i_g = \frac{V_{drive} - V_{ge}}{r_g} \tag{1}$$

$$i_g = i_{g1} + i_{g2} \tag{2}$$

$$i_{g2} = C_{GE}\frac{dV_{ge}}{dt} \tag{3}$$

$$i_{g1} = C_{CG}\frac{d\left(V_c - V_{ge}\right)}{dt} \tag{4}$$

If we assume that V_c is fixed, which is true in most cases, the time of each phases in the turn-on process in Fig. 2 can be obtained as

$$t_{ph1} = \left(R_g + r_g\right)\left(C_{CG} + C_{GE}\right)\ln\left[1/\left(1 - \frac{V_{th}}{V_{GD}}\right)\right] \tag{5}$$

$$t_{ph2} = \left(R_g + r_g\right)\left(C_{CG} + C_{GE}\right)\ln\left[1/\left(1 - \frac{V_M}{V_{GD}}\right)\right] - t_{ph1} \tag{6}$$

$$t_{ph3} = \frac{\left(V_c - V_F\right)\left(R_g + r_g\right)C_{CG}}{V_{GD} - V_M} \tag{7}$$

where V_M is the gate Miller plateau voltage, V_{GD} refers to the applied gate-drive voltage, R_g is the external gate resistance, and V_F means the on-state voltage drop. From the analysis above we can figure out that the influential factors of the transient are the threshold voltage, capacitances, gate resistances, stray impedances in the branches and the time delay. For the steady state, the influential factor is the equivalent resistances in each IGBT branch.

3 Simulation verification of the effect of influencing factors

To further study the effect of the influential factors mentioned above one by one, the software Saber is used for the simplicity in separating the effect of each factor. Two IGBT modules are connected in parallel with different gate drives, as shown in Fig. 4.

The simulation result is shown in Fig. 5. The IGBT on state resistance and the line resistances, as well as the external and internal gate resistances are similar factors so only one of the two in each pair is shown. As can be seen from the figure, the gate-emitter capacitance, the time delay and the gate resistance have important impact on the current sharing of two paralleled IGBTs. The line resistance affect the current sharing greatly in steady state.

The less important factor is the threshold voltage and the least is the line stray inductances. The simulation results coherent with the calculated results.

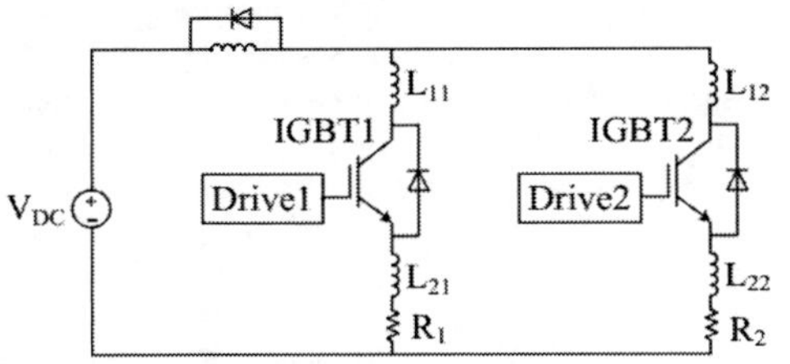

Figure 4. The simulation circuit.

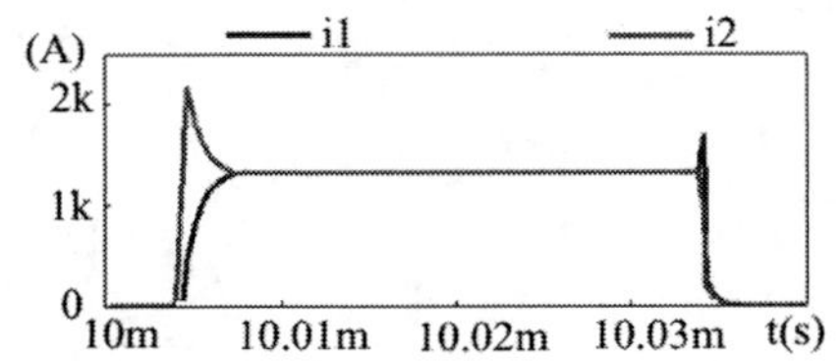

(a) Cge1=130%Cge2.

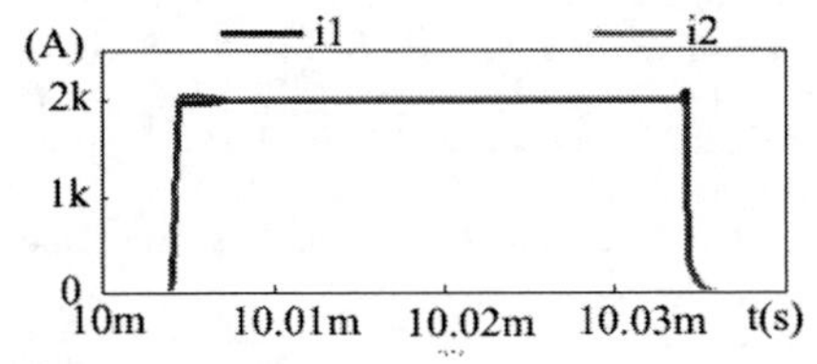

(b) Vth1=130%Vth2.

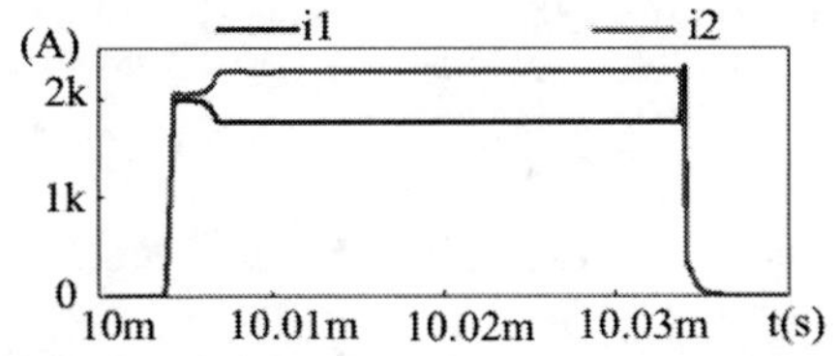

(c) R1=130%R2.

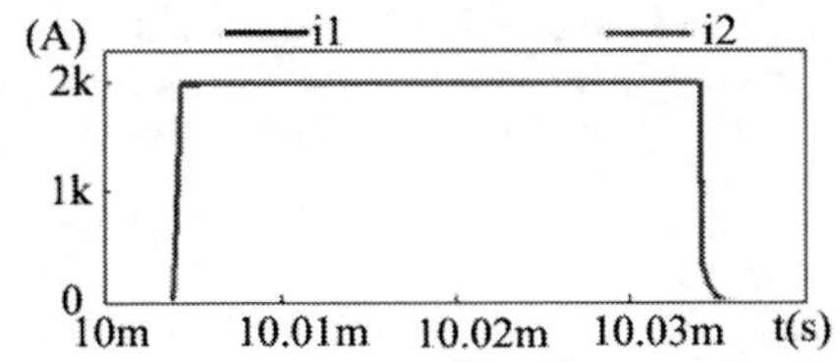

(d) L11=130%L12.

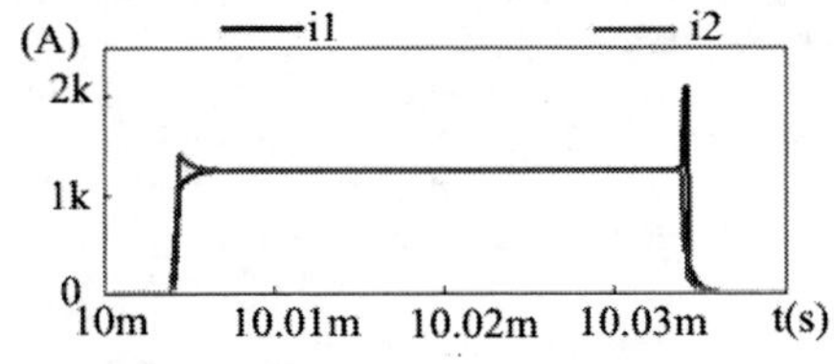

(e) IGBT1 delay 100ns.

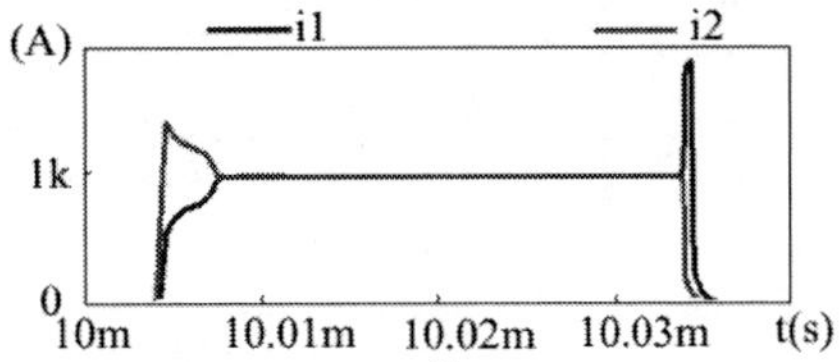

(f) Rg1=130%Rg2.

Figure 5. The simulation result.

4 Two alternative topologies and their performance analysis

In the DC circuit breaker application, two types of series and parallel connection of the IGBTs are possible, as illustrated in Fig. 6.

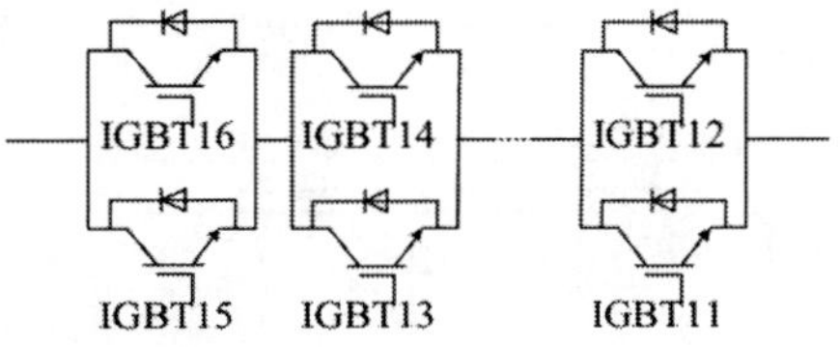

(a) Topology 1.

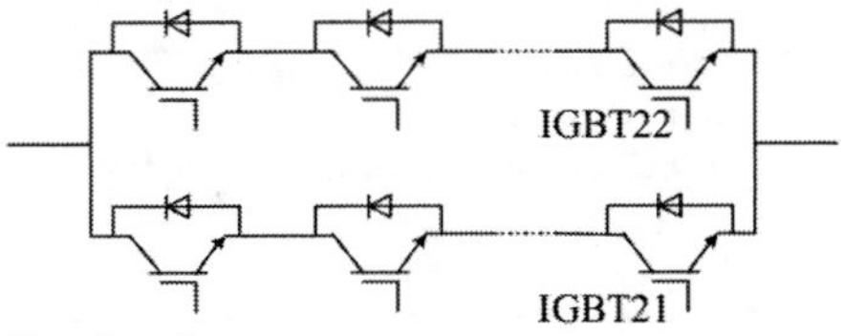

(b) Topology 2.

Figure 6. The alternative topologies.

The two topologies are set under simulation to see their performances. Three pairs of IGBTs are presented in the simulation topology and parameters of IGBT11 and IGBT21 are changed in each case. The currents in the related branches, i.e. i_{11}, i_{12}, i_{21}, i_{22}, are detected and the results are shown in Fig. 7. It is necessary to note that topology 1 can be seen as three identical sub-systems each formed by two paralleled IGBTs (e.g. IGBT11 and IGBT12), and the characteristics of every sub-system are the same as described in part 3. Consequently, the current is re-shared in each sub-system in topology 1. As can be seen from Fig. 7, the unbalance degree of topology 2 is much smaller than that of topology 1, and when the number of IGBTs increases, the unbalance degree of currents in parallel further decreases. This is because in topology 1 all the IGBTs other than IGBT11 have the same parameters in purpose of comparing with topology 2, so the current unbalance is just subjected to the aimed sub-system (i.e. IGBT11 and IGBT12) while in topology 2 the unbalance is shared by the other IGBTs. So a conclusion can be drawn that when certain degree of unbalance in current is allowed, it is better to use the

DOI: 10.1051/matecconf/20166301010

second topology and choose IGBTs of smaller rated current and voltage to increase the number of devices.

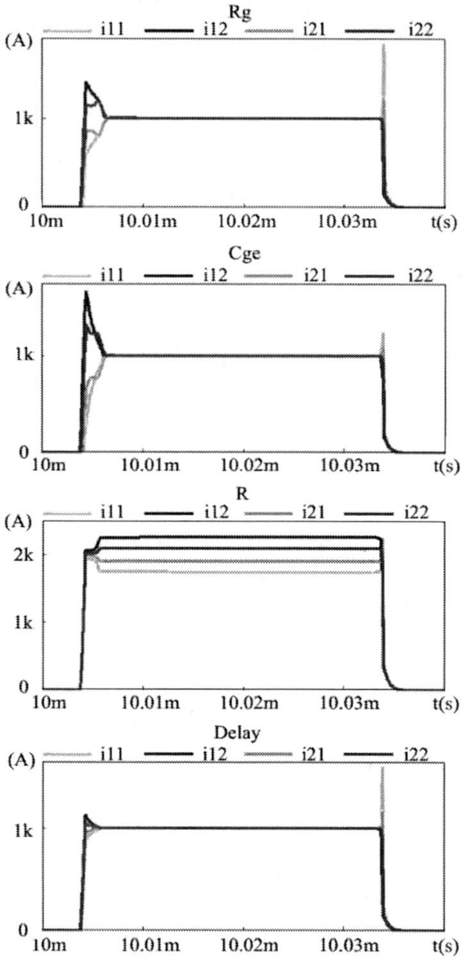

Figure 7. The alternative connecting topologies.

5 Conclusion

In this paper, the electrical characteristics of the IGBTs are reviewed and the current-sharing influencing factors of parallel-connected IGBTs are discussed. The effect of each major influencing factor is illustrated with the help of software Saber. Two possible circuit topologies used in a DC circuit breaker are proposed and simulated to see their current balancing performances in order to find out if any of them can provide us with a simple and effective method to achieve good current balancing in the DC circuit breaker application. It is shown that one of the alternative topologies has certain self-balancing effect and performs better under the same conditions. The result can help with better design of the DC circuit breakers especially in high-power applications when the devices have to be connected in series and parallel. Future work will involve the experimental verification of the performance as well as the reliability analysis.

Acknowledgment

This work was supported by the National Key Basic Research Program of China (973 Program), No. 2015CB251004. The authors are with the State Key Laboratory of Electrical Insulation and Power Equipment, School of Electrical Engineering, Xian Jiaotong University, Xi'an, Shaanxi 710049, China.

References

1. N.Y.A. Shammas, R. Withanage, D. Chamund, Microelectron. J. **39**, 899-907 (2008)
2. N. Mohan, T. M. Undeland, W. P. Robbins, *Power Electronics: Converters, Applications, Design, 2nd ed.* (Wiley, New York, 1995)
3. B. J. Baliga, *Modern Power Devices* (Wiley, New York, 1987)
4. M. Rahimo, A. Kopta, R. Schnell, U. Schlapbach, S. Zehringer, *Proc. 25th Int. PCIM Conf*, 314-319 (2004)
5. F. Blaabjerg, J. K. Pedersen, U. Jaeger, *Conf. Rec. IEEE IAS Annu. Meeting*, 997-1005 (1995)
6. A. Kopta, M. Rahimo, S. Eicher, U. Schlapbach, *Proc. IEEE ISPSD*, 1-4 (2006)
7. M. Rahimo, A. Kopta, S. Eicher, U. Schlapbach, S. Linder, *Proc. IEEE ISPSD*, 67-70 (2005)
8. A. Kopta, M. Rahimo, U. Schlabach, D. Schneider, E. Carroll, S. Linder, *Conf. Rec. IEEE IAS Annu. Meeting*, 794-798 (2005)
9. F. Chimento, W. Hermansson, T. Jonsson, IEEE Trans. Ind. Appl. **48**, 1046-1053 (2012)
10. A. A. Jaecklin, *Proc. Bipolar/BiCMOS Circuits and Technology*, 61-66 (1998)
11. L. Kong, J. Zhang, Z. J. Qiu, Z. L. Meng, *Proc. ICEMS*, 1851-1856(2013)
12. A. Blinov, D. Vinnikov, V. Ivakhno, *Proc. CPE*, 360-364 (2011)
13. N. Chen, F. Chimento, M. Nawaz, L. W. Wang, IEEE Trans. Ind. Appl. **51**, 539-546 (2015)
14. U. Schlapbach, *Proc. 6th Int. CIPS*, 1-7 (2010)
15. F. Chimento, F. Nicosia, S. Musumeci, A. Raciti, M. Melito, G. Sorrentino, Proc. IEEE ISIE, 931-936 (2007)
16. R. Chokhawala, J. Catt, B. R. Pelly, IEEE Trans. Ind. Appl. **31**, 603-611 (1995)
17. S. Pendharkar, M. Trivedi, K. Shenai, IEEE Trans. Electron Dev. **45**, 2222-2231 (1998)
18. G. Belverde, M. Melito, S. Musumeci, A. Raciti, *Conf. Rec. EPE*, 1-10(2001)
19. S. Musumeci, R. Pagano, A. Raciti, M. Melito, F. Frisina, *Proc. IEEE IECON*, 555-560 (2002)
20. B.Jayant Baliga, *Fundamentals of Power Semiconductor Devices*. (Springer Verlag, New York, 2008.)

MATEC Web of Conferences **63**, 01011 (2016) DOI: 10.1051/matecconf/20166301011
MMME 2016

A study with Paperless Electronic Medical Advice

Sheng Li HU[1, a], Jin Qing CHI[2]

[1]*Central Hospital of Cangzhou, Information Center, Hebei Province, China*
[2]*Central Hospital of Cangzhou, Bidding office, Hebei Province, China*

Abstract. The purpose of this study is to drive the EMR's further application and rapid expansion clinically based on the electronic medical advice as the core thereby realizing the paperless EMR. This paper has analyzed the implementation effects of electronic medical advice, followed by discussion of frequently asked questions (FAQ) and development direction, and finally analyzed the solution of paperless EMR. It indicates that electronic signatures and unitrust time stamp are keys to achieve the paperless EMR. The laws and regulations, new ideas, paperless consultation sheets, all data integration and sharing, virtual printing technology, application of wireless and mobile ward-round trolley, etc. within the health care field shall be proposed to be improved. It is significantly important to save the medical costs, reduce the medical negligence, achieve the paperless EMR and build a high-quality digital hospital.

1 The advantages of Paperless Electronic Medical Advice

1.1 Background

It is an important content to deepen the public hospital reform by establishing and completing the hospital information system (HIS) based on the Electronic Medical Record (EMR) as the core. It has important meanings in promotion of medical service equalization; driving the medical service management to be developed towards scientification, specialization and informatization; enhancing the hospital's work efficiency and protecting the medical safety. Electronic medical advice is a core of EMR, so the construction and development of electronic medical advice system are extremely important.

1.2 Effect analysis of electronic medical advice implementation

Electronic medical advice is an important part of EMR and a core business of hospitals. It is not only a proof of treatment settlement, but also legal evidence. From the implementation effects we can see, the electronic order system can optimize the clinical working process thereby reducing the workload of medical staff, improving work efficiency, and guaranteeing the medical safety [1-12].

For doctors, electronic medical advice can make the medical records clear, neat, standard thereby avoiding the medical hidden dangers due to poor handwriting [1]. According to different symptoms, the medical advice is maintained differently. The medical advice is automatically linked to the corresponding charges. The contents of those charges consist of standard charging codes and charging names that are ticked in the saved medical advice database when using. It can directly enter into the medical advice interface. This can save plenty of time in making the medical advice; therefore, doctors will have more time to communicate with patients.

For nurses, the medical risks made by copying the medical advice to the treatment book, injection book, oral administration book, inspection book could be reduced. Automatic charging begins after medic advice received. Those fees ac-cord with the commodity price and no need to add the costs again, therefore, nurses can get more time to deliver the humanized warm care to patients and patient's satisfaction will be improved.

For mangers, the medical advice can be improved. In this system, the trade name or common name can be used for searching. However, finally the drug common name, patent medicine name of new active compounds and compound preparation name that meet the prescription regulations will be submitted. This will avoid a wrong trade name occurred in the medical advice. When submitting the medical advice, the usage, dosage and departments should be checked throughout in order to guarantee the accuracy of medical advice, as well as avoid the potential risks resulting from signature missing.

2 Frequently Asked Questions and Countermeasures in Electronic Medical Advice

2.1 Frequently asked questions

[a] Corresponding author: hushenglihsl@139.com

In the implementation of electronic medical advice, it may have all kinds of mistakes either man made or technical defects. More common problems include injection dosage, specification, unit mistakes, confused long-term and temporary medical advice, inconsistent performers and signatures, and discrepancy between stopping time and actual time resulting in overcharge and charging missing, etc.

2.2 Countermeasure analysis

First from the side of management, on the one hand, medical staff should enhance the training in laws and regulations and reinforce the legal awareness; on the other hand, the management system of electronic medical advice should be completed and effective rewards and punishments should be set out. A communication coordination mechanism should be established among doctors, nurses, pharmacists, laboratories' and inspectors in order to ensure a smooth communication and avoid unnecessary medical disputes. When performing rescuing tasks, relevant medical advice and expenses should be additional recording without delay in order to ensure the accuracy of medical advice and expenses. Nurses have responsibilities on supervision of doctor's diagnosis and treatment scheme and the medical advice should be implemented after seriously checked. Meanwhile, the advanced methods of other hospital's medical advice should be studied in order to use the echoing check method making nurses mutual supervision, improving the mistake detection rate, and preventing occurrence of errors. To establish a special treatment book and nurse hand-over book; faithfully implement checking system; be standard in charging; and avoid overcharge and missing charge[2]. After nurses received and processed the medical advice, if they have to leave for something, the job number must be withdrawn in time in case unnecessary troubles.

Secondary from the view of technical means, to complete the electronic medical advice system and avoid the some mistakes from technical level, such as missing signatures, non-received and unprocessed orders, and the discharge notification and ward transfer won't be dealt with. Only if those medical orders are processed, the next operation will take place.

3 Development trend of electronic medical advice

3.1. Paperless EMR is based on the electronic medical advice as the core

By studying the published documents and development trend of medical informationization in recent years, we have found that electronic medical advice system is combined with data mining and artificial intelligence technology developing towards paperless, intelligent and standard.

It is in a starting stage. Paperless management will be achieved in long-term and temporary medical advice, progress notes, various kinds of special format records, all kinds of tests and examination application forms.

3.2 Intelligent electronic medical advice

When making new medical advice, stopping medical advice and cancelling medical advice, there is an automatic dialog box popped up at nurse station, intelligently reminding nurses to receive and perform. When medical advice is stopping, it is also stopping charging, which will avoid extra charge caused by nurses forget stopping the medical advice.

3.3 Electronic medical advice standardization

Standardization for medical advice, especially, the standardization management is used for medical non-medication and examination advice. For example, discharged next day and discharge tomorrow, etc. shall be normalized as discharge tomorrow. This is convenient for quality control of electronic medical advice. If there is in no medical advice of discharge tomorrow, the discharge notice cannot be processed. Through medical advice standardization, the medical advice quality can be improved via technological means.

4 Paperless emr with electronic medical advice as the core

4.1. Issues paperless EMR faced

At present, most of hospitals are still in a completely paper medical record stage or paper and paperless medical record coexisting administrative stage. Due to paperless EMR has many advantages of compared with paper medical record and it is also an important symbol of digital hospital and is an inevitable trend of modern hospital management. We will discuss about issues that paperless EMR is faced, as well as solutions.

From article we can see, paperless EMR is facing many problems, including imperfectly related laws and regulations, incapably guaranteed safety of electronic signatures, without establishing certificate authority (CA) within health care field, without choosing qualified unitrust Time Stamp Authority (TSA) etc. The hysteretic laws and regulations restrict the development of paperless EMR. Moreover, the ideas of hospital managers falling behind and the costs of medical informationization difficult to shoulder are also factors to barrier the paperless EMR.

4.2. Electronic signatures and unitrust time stamp

When patients print the long-term and temporary medical advice after discharge, for purpose of medical advice is accordance with norm of law, printed paper medical advice sheet must be signed in handwriting. Medical advices are involving in numerous medical staff; review signatures waste time and energy and easily go wrong.

Therefore, using electronic signatures and unitrust time stamp replacing hand-written signatures is a key to achieve the paperless EMR. Apart from using the electronic signatures in the paperless EMR, medical staff must retain all previous amendment trace and time in the EMR when amending the medical records. It is predicted that in the near future, along with improvement of electronic signatures, unitrust time stamp and related laws and regulations, paperless EMR will enter into an integrated development period.

4.3. Update ideas

To further improve various rules and regulations within the health care field, make the EMR be in accordance with laws and regulations, and solve the lagging legal system resulting from application of new technology. In addition, hospital managers shall also update the new ideas, advance with times, and properly increase the input in medical informationization.

4.4. Consultation issues

Doctors will sign their names after they make a medical advice order, the consultation form in the order will be delivered to a relevant department, medical staff in those related departments will confirm after seeing. The relevant consultation notes will be filled after consultation termination. Experts who take part in consultation will process the electronic signatures [3]. The entire process can realize the paperless management.

4.5. Integration and sharing of all system data

Conclusions EMR should achieve the data interface with HIS, PACS, LIS, operation anesthesia, physical examinations, electronic application form, etc. The information sharing is achieved based on the data integration. All information related to patients in the EMR, including various test and examination results, ultrasound, CT imaging, operation anesthesia records and videos, etc. At this time, EMR is a multimedia EMR, including text, sound and images, etc. With respect to system integration scheme, HL7 interface mode is recommended to use for effective integration. The Drug Consultation and Drug Safety Detection System shall be implanted into the EMR. The overdose, contraindication and incompatibility will be remained intelligently to guide the rational medications for doctors when recording the medical advice in order to prevent the phenomenon of irrational drug use [4].

4.6. Virtual print technology

Take advantages of virtual printing technology can directly converse all kinds of data information in the EMR into to PDF, JPG, TIF, BMP, PCX, PNG format. Then those format data files will be saved in the hard disks or light disks by binary forms after they are encrypted. Other application programs cannot open those medical records, which guarantees medical data unable to be amended thereby realizing EMR data off-line storage and working out the problems of system performance decline caused by infinitely data increasing in the database as well as an important step of achievement of paperless EMR filing.

4.7. Applications of wireless mobile ward-around trolley

Doctors can carried out the ward-around by wireless computer ward-around trolley in each ward via deploying the WLAN and using the barcode technology. They can directly make medical advice, check various kinds of images such as CT, CR image device and all test results. Nurses can directly view medical advice information and execution as well as type-in patient's indicators. The application of wireless mobile ward-around trolley is an important milestone of realization of paperless EMR.

4.8. All self-service devices shall be placed in a notable place of hospitals

After implementing EMR, for the sake of maximum protection of patient's right to know, hospitals shall place various kinds of self-service devices at a notable place. This will let patients check the expense details and conduct a satisfaction evaluation for the entire treatment. This is very important in easing the current tension of doctor-patient relationship and creating a harmonious hospital.

4.9. Paper documents

At the present, EMR is unable to be really paperless, namely there are still some paper medical records, such as consent for operation that is needed relatives of patients signing the documents, some independent devices such as cerebral electrical device that provides the paper reports due to unable to access the network because of interface problems.

4.10. Set out an emergency response for paperless medical record

For some problems occurred during the paperless medical record, an emergency response shall be set out, such as network paralysis caused the system unable to be normal operation, frequent electronic signatures lead to the KEY often changing, the temporary KEY unable to be made up by the third party certificate authority due to weekends or nights, all of which may cause medical staff unable to login resulting in patients waiting for long time.

5 Conclusion

Paperless EMR is still in starting stage; many aspects are still imperfect and need us to constantly make great efforts thereby striving for early implementation of real paperless EMR and construction of E-hospitals.

The research work was supported by Cangzhou Science and Technology Research and Development Program guidance. In the process of paper writing, my friend Zhi Guo gave the vigorous help and put forward much valuable and impartial advice, and at the same time, expresses my heartfelt thanks to the experts who revised my paper.

References

1. Cao Yang, Wang Yan. Optimization of Doctor's Order Inputs inter-face and functions of Anamnesis. Chinese Journal of Medical instrumentation, 34, 5, (2010).
2. Cao Yang, Ding Aimin, Wang Yan. New Model of Doctor-Nurse Communication Based on Electronic Medical Advice Platform. Chinese Journal of Medical instrumentation, 36, 1, (2012).
3. Chen Xiaohua, Wang Li, Chen Qun. Electronic check prescription application submission work. Journal of Nursing Science, 26, 7, (2011).
4. Gu hui. Defect analysis and countermeasures in the process of executing electronic doctor's advice by nurses. Journal of Clinical Medicine in practice, 15, 22, (2011).
5. Hu Shengli, Bai Hongmin. Research on electronic signatures application in electronic medical records. Chinese Hospital, 15, 11, (2011).
6. Li Wei, Li Fengsheng, Wang Qisheng.Analysis based on digital trans-mission system of hospital construction under doctor's orders, Hospital administration Journal of Chinese People's liberation Army,17,10, (2010).
7. Tao Hongyan. ICU nurses prevent the implementation of response measures electronic prescription errors. National Medical Frontiers of China, 16, (2011).
8. Xie Haiying, Li Ni, Yan Chengmei etc. Effectiveness of Printing and Bedside Use of Medical Orders in Nursing Work. Nursing Journal of Chinese People's Liberation Army, 25,4B, (2008).
9. Yang Li, Sun Sunchun.Cooperation of doctor and nurse solving faculty electronic medical order input question. China Medicine Guides, 18, 25, (2014).
10. Yue LI, Zhang Xiaoling, Guo Lipingl. New doctor checking method in the electronic Prescription. Yiyao Qianyan, 12, 13, (2012).
11. Zhang Danjing. PDA application experience in clinical nursing. Chinese Nurse Management9, 3, (2009).
12. Zheng Xichuan, Tan Shensheng, Wu Yunzhen. Implementation the computerized Provider order Entry through Optimizing Clinical Workflow. Chinese Digital Medicine, 7, 3, (2011).

Acoustical Semi-blind Deconvolution for Bearing Defect Detection

Yu Wang[1,a], Wen Zhong[1], Yang Yin[1], Xiaoyin Liu[1] and Chunhua Song[1]

[1]*Xihua University, School of Mechanical Engineering, Chengdu, China*

Abstract. Acoustical machine monitoring is frequently complicated by noisy environments at a production site. This paper presents a semi-blind deconvolution algorithm to extract only one desired acoustic source signal from different sources which are convoluted and mixed by mechanical systems before being measured. The method is based on blind model transformation, robust independent component analysis, reference signal and spectral distance. The new algorithm is tested on simulation and experimental cases. Results demonstrate that blind separation of acoustic signals is possible even when measurements are distanced from vibration exciting sources of faulty bearings. Furthermore, the method can eliminate the effect of structural resonances and large reverberation time of mixtures, which often causes severe problems in classical acoustical diagnostic methods of rolling element bearings.

1 Introduction

Machine faults modify machine sounds characteristically, therefore observing machine sounds can be a useful mean for condition monitoring and fault diagnosis [1]. Acoustical diagnosis utilizes acoustic signals which are detected by microphones, so non-contact detection and easy handling are excellent features of this diagnosis [2]. However, it is hard to apply the acoustic diagnosis to the precise diagnosis, because the machine sound from the diagnosed part is usually corrupted by other interfering sources and non-directive noise, which is often much louder. Therefore, signal processing techniques such a blind source separation (BSS) should be employed as a preprocessor to increase the portion of relevant information in the measured signal [3]. Due to time-delayed superposition of the sources and reflections from the walls, acoustical mixtures have to be modeled with convolutive mixtures [4]. Existing BSS methods are not well geared to handle such situations, because the modifying paths of vibrations (dynamics of the system) and propagation paths of sound (in air) are neither controllable nor known [5,6]. We show a modified blind deconvolution method, enabling to tackle problems of both multiple convoluted sources and monitoring measurements, as it uses an analytically derived spectral distance in the optimization process of the cost function used. As a consequence, this method possesses a good behaviour when the reverberation time of mixtures is sufficiently large and the machine is operating in a noisy environment.

The paper is organized as following: The basic problem is described in Section 2. Section 3 presents the proposed method. Section 4 and Section 5 show the numerical simulation, experimental setup and results of the separation procedure. Conclusions are presented in Section 6.

2 Problem formulation

In the following, all signals are assumed of stochastic, zero-mean and finite-power. At a discrete time index t, a convolutive mixing process is usually described as a linear Multiple-Input Multiple-Output (MIMO) system:

$$x_i(t)=\sum_{j=1}^{n}\sum_{\tau=0}^{K_{ij}}h_{ij}(\tau)s_j(t-\tau),\ i=1,\cdots,m\ \ (m\geq n) \quad (1)$$

where $\mathbf{x}(t)=[x_1(t),\cdots,x_m(t)]^{\mathrm{T}}$ denotes observed signals from an array of m microphones, $\mathbf{s}(t)=[s_1(t),\cdots,s_n(t)]^{\mathrm{T}}$ denotes n unknown original source signals, $[\bullet]^{\mathrm{T}}$ denotes a transposition of a vector or a matrix, h_{ij} is impulse responses of a system (mixing filters), and K_{ij} denotes the corresponding length of h_{ij} which may change in time.

For brevity, the convolutive mixing model Eq.(1) can be written in a following matrix form:

$$\mathbf{x}(t)=\sum_{\tau=0}^{K}\mathbf{H}_{\tau}\mathbf{s}(t-\tau) \quad (2)$$

where $\mathbf{H}_{\tau}$ is an $m\times n$ mixing matrix which contains the τ -th filter coefficients.

Assuming that all source signals arrive at sensors or microphones at the same time without being filtered, the convolutive mixture model Eq.(2) simplifies to a instantaneous mixture model:

$$\mathbf{x}(t)=\mathbf{H}\mathbf{s}(t) \quad (3)$$

[a] Corresponding author: redtu160@163.com

where $\mathbf{H}$ is an $m\times n$ matrix containing mixing coefficients.

The goal of the classical BSS is to find an $m\times n$ unmixing matrix $\mathbf{W}$ to recover all sources:

$$\mathbf{y}(t)=\mathbf{Wx}(t)=\mathbf{WHs}(t)=\mathbf{PDs}(t) \quad (4)$$

where $\mathbf{P}\in\mathbf{R}^{m\times m}$ is a permutation matrix, and $\mathbf{D}\in\mathbf{R}^{m\times m}$ is a diagonal scaling matrix.

In contrast to most classical BSS algorithms, the constrained independent component analysis (cICA) algorithm does not need recover all sources. Furthermore, its goal is to recover only one source. For extracting one independent source signal, a reliable and flexible contrast function $J(y)$ of the cICA is defined by:

$$\max\ J(y)\approx \rho[\mathrm{E}\{G(y)\}-\mathrm{E}\{G(v)\}]^2 \quad (5)$$

$$\text{s.t.}\quad g(\mathbf{w})=\varepsilon(y,r)-\xi\le 0,\ \ h(\mathbf{w})=\mathrm{E}\{y^2\}-1=0 \quad (6)$$

where ρ is a positive constant, v is a Gaussian variable having zero mean and unity variance, $G(\bullet)$ is a nonquadratic function. The equality constraint $h(\mathbf{w})$ ensures that the contrast function $J(y)$ and the weight vector $\mathbf{w}$ are bounded. $\varepsilon(y,r)$ is the closeness measure between the extracted signal y and the reference signal r, ξ is a threshold, and $\mathrm{E}\{\bullet\}$ denotes a sample mean operator. From the contrast function Eq.(5) and Eq.(6), Lu and Rajapakse derived the cICA algorithm [7,8]:

$$\mathbf{w}_{k+1}=\mathbf{w}_k-\eta\mathbf{R}_{\mathbf{xx}}^{-1}L'_{\mathbf{w}_k}/\delta(\mathbf{w}_k) \quad (7)$$

$$L'_{\mathbf{w}_k}=\bar{\rho}\mathrm{E}\{\mathbf{x}G'_y(y)\}-0.5\mu\mathrm{E}\{\mathbf{x}g'_y(\mathbf{w}_k)\}-\lambda\mathrm{E}\{\mathbf{x}y\} \quad (8)$$

$$\delta(\mathbf{w}_k)=\bar{\rho}\mathrm{E}\{\mathbf{x}G''_{y^2}(y)\}-0.5\mu\mathrm{E}\{g''_{y^2}(\mathbf{w}_k)\}-\lambda \quad (9)$$

where k denotes the iteration index, η is the learning rate, $\mathbf{R}_{\mathbf{xx}}$ is the covariance matrix of the observed signals $\mathbf{x}$, $\bar{\rho}=\rho\bullet sign(E\{G(y)\}-E\{G(v)\})$, $G'_y(y)$ and $G''_{y^2}(y)$ are the first and the second derivatives of $G(y)$ with respect to y, and $g'_y(\mathbf{w}_k)$ and $g''_{y^2}(\mathbf{w}_k)$ are the first and the second derivatives of $g(\mathbf{w}_k)$ with respect to y. The optimum multipliers μ and λ are found by iteratively updating them based on a gradient-ascent method: $\mathbf{w}_0 r(t)\mathbf{w}_{k+1}=\mathbf{w}^{+}_{k+1}/\|\mathbf{w}^{+}_{k+1}\|$

$$\mu_{k+1}=\max\{0,\mu_k+\gamma g(\mathbf{w}_k)\} \quad (10)$$

$$\lambda_{k+1}=\lambda_k+\gamma h(\mathbf{w}_k) \quad (11)$$

3 Method development

Unfortunately, there are two important issues that should be and solved when the cICA algorithm comes to mixed acoustic signals in a shop floor. One issue is that the cICA based on instantaneous mixture can not be applied to convolutive mixtures of sources, and another one is the design of the reference signal.

Hence, instead of modeling individual samples at a discrete time index *t*, *N* successive samples of observation column vector are stacked to construct a new block. The equations for such a block can be written as follows:

$$\begin{aligned} x(t)&=h(0)s(t)+\cdots+h(K)s(t\text{-}K),\\ x(t\text{-}1)&=h(0)s(t\text{-}1)+\cdots+h(K)s(t\text{-}K\text{-}1),\cdots,\\ x(t\text{-}N)&=h(0)s(t\text{-}N)+\cdots+h(K)s(t\text{-}K\text{-}N)\end{aligned} \quad (12)$$

The m-dimensional observation sequence can be written as an mN-dimensional column vector $\hat{\mathbf{x}}(t)$ in Eq.(13), and the n-dimensional source sequence can be written as an $n(N+K)$-dimensional vector in Eq.(14).

$$\hat{\mathbf{x}}(t)=[x^{\mathrm{T}}(t),x^{\mathrm{T}}(t\text{-}1),\cdots,x^{\mathrm{T}}(t\text{-}N\text{+}1)]^{\mathrm{T}} \quad (13)$$

$$\hat{\mathbf{s}}(t)=[s^{\mathrm{T}}(t),s^{\mathrm{T}}(t\text{-}1),\cdots,s^{\mathrm{T}}(t\text{-}N\text{-}K\text{+}1)]^{\mathrm{T}} \quad (14)$$

where $x^{\mathrm{T}}(t)=[x_1(t),\cdots,x_m(t)]^{\mathrm{T}}$, $s^{\mathrm{T}}(t)=[s_1(t),\cdots,s_n(t)]^{\mathrm{T}}$.

Then, the convolutive mixing model Eq.(1) and Eq.(3) can be reformulated as [9]:

$$\hat{\mathbf{x}}(t)=\hat{\mathbf{H}}\hat{\mathbf{s}}(t) \quad (15)$$

where $\hat{\mathbf{H}}$ is a block-Toeplitz matrix and has dimensions $mN\times n(N+K)$:

$$\hat{\mathbf{H}}=\begin{bmatrix} h(0) & h(1) & \cdots & h(K) & 0 & \cdots & \cdots & 0\\ 0 & h(0) & h(1) & \cdots & h(K) & 0 & \cdots & 0\\ \vdots & & & & & & & \vdots\\ 0 & \cdots & \cdots & 0 & h(0) & h(1) & \cdots & h(K)\end{bmatrix} \quad (16)$$

This equation is very interesting. It tells us Eq.(15) has the same structure as the instantaneous mixture model given in Eq.(3), and the task of blind convolutive model can be solved by the cICA algorithm based on instantaneous model.

However, the second issue arises accordingly, i.e. the design of the reference signal. James and Gibson pointed out that the shape, the phase and the occurrence time of the reference signal may influence the output result [10]. This cICA method requires that the desired source signal is strong enough to be observed in a sensor signal, which cannot be always satisfied. Consequently, in contrast to the conventional measure distance in the standard cICA algorithm, the closeness measure $\varepsilon(y,r)$ is newly defined as:

$$\varepsilon(y,r)=\frac{1}{2N}\sum_{f=0}^{N}\left[\frac{Y(f)}{R(f)}+\frac{R(f)}{Y(f)}\right]-1,\quad f=1,2,\cdots,N \quad (17)$$

where $Y(f)$ and $R(f)$ are power spectra of $y(t)$ and $r(t)$ respectively.

Next, let us now show a valid approach to find the reference signal, whose phase and waveform do not have to be well matched to the desired source signal.

The simulation of the simplified signal model of rolling element bearing can be expressed as:

$$s(t)=e^{-B(t-\Delta t)}\cos(2\pi f_n t+\varphi) \quad (18)$$

Here, the original phase φ is assumed to be zero, and B is an appropriate value. Assuming that there is a priori information about frequency available, a reference signal $r(t)$ can be easily obtained based on Eq.(18):

$$r(t) = sign(\sin(\pi f_c t)) \qquad (19)$$

where $sign(\bullet)$ means sign operation, f_c is a certain characteristic frequency.

Finally, the block diagram of the proposed algorithm is shown in Figure 1.

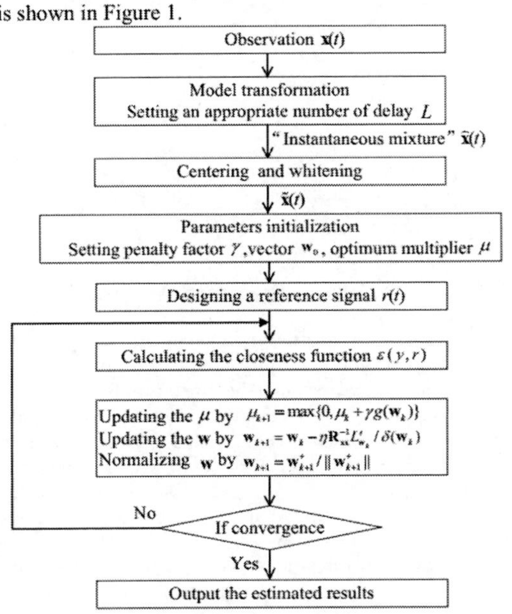

Figure 1. Algorithm flow chart for single source estimation

4 Numerical simulation

To verify the validity of the proposed algorithm, a computer simulation was carried out, in which two simulated signals of rolling bearings with outrace fault were created. First, three simulated source signals were generated as follows:

$$s_1(t) = e^{-580(t-0.25)}\cos(2\pi \bullet 2570 \bullet t)$$

$$s_1(t) = e^{-580(t-0.25)}\cos(2\pi \bullet 2570 \bullet t)$$

$$s_3(t) = \sin(2\pi \bullet 345 \bullet t) + \sin(2\pi \bullet 2530 \bullet t) + \cos(2\pi \bullet 2990 \bullet t)$$

Here, the sample rate fs=8192 Hz and the sample points n=8192 (intervals between adjacent pulses: $s_1(t)$:100 sample points, $s_2(t)$:176 sample points). Second, each source signal was normalized to unit variance and had zero mean. Third, the convolutive mixing process is defined by Eq.(1) and the convolutive mixing matrix $h_{ij}(\tau)$ was generated randomly with length 128 in order to simulate the case where the time-delay is sufficiently large.

Figure 2. shows the source signals and their amplitude spectra. It can be seen that the first and second signals are typical of structural systems excited by impacts, i.e., periodic trains of decaying oscillations. These types of signals are often considered as models of faulty bearing vibrations. The third one is a periodic signal. Artificially, the mixtures were corrupted by additive white Gaussian noise, and signal to noise ratio (SNR) SNR1=7.5, SNR2=5.6 and SNR3=8.3 approximately. The waveforms and spectra of the mixed signals are shown in Figure 3. Intuitively, both the waveforms and the spectra of the mixed signals provide little information of the periodic impulses.

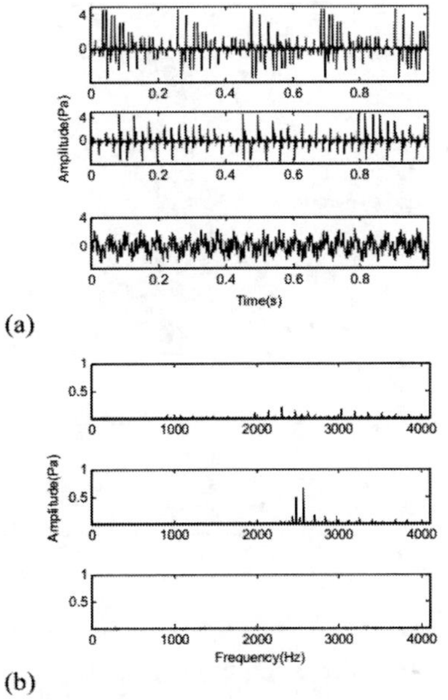

Figure 2. Simulated source signals (a) Time domain waveforms (b) Amplitude spectra

Next, the proposed algorithm was applied to recover the three source signals. Some information about the algorithm parameters is first given: L=40 and $G(u) = u^2$. The reference signal and estimated result are shown in Figure 4. and Figure 5. respectively. It can be seen that our method successfully separated the transient impulse signals and provides a very good result. This simulation demonstrates the effectiveness of the proposed algorithm. Nevertheless, we found that it could enhance the separation quality by increasing the filter length, but the computational complexity of the algorithm increases dramatically with the factor (without regard to sample points). It seems that further investigations are necessary concerning the effects of the parameter.

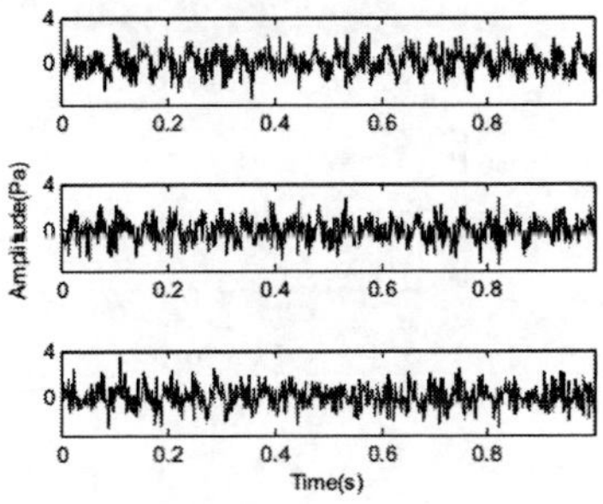

(a)

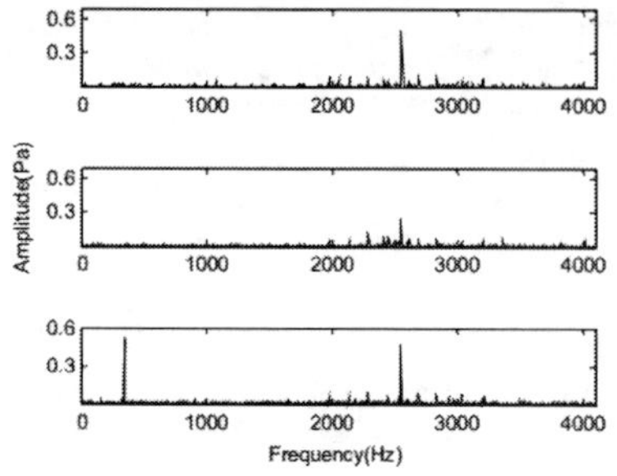

(b)

Figure 3. Simulated mixed signals (a) Time domain waveforms (b) Amplitude spectra

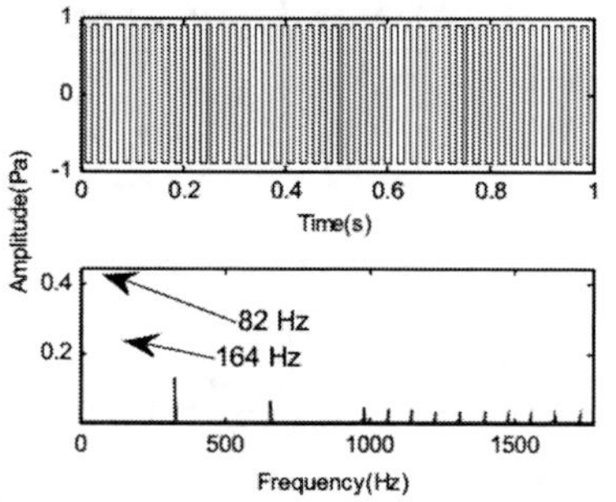

(a)

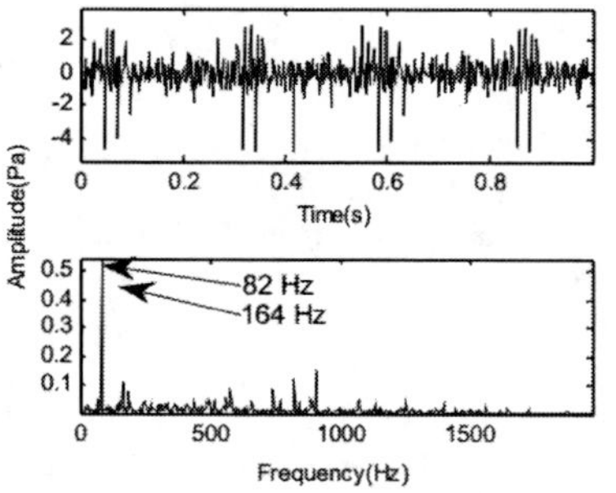

(b)

Figure 4. Time domain waveforms and envelope spectra (a) Simulated reference signal (b) Estimated signal (Characteristic frequency 82 Hz)

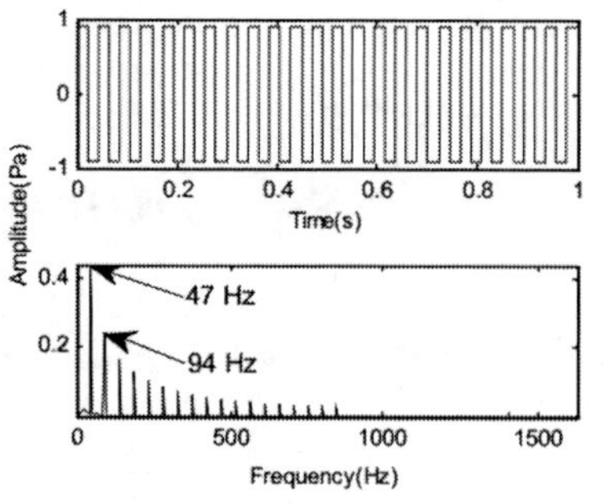

(a)

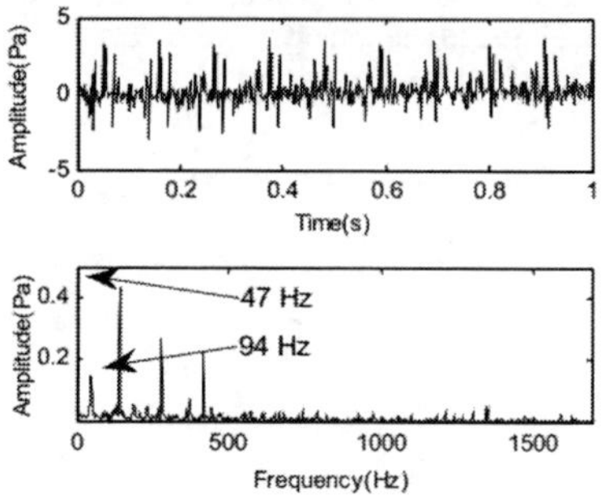

(b)

Figure 5. Time domain waveforms and envelope spectra (a) Simulated reference signal (b) Estimated signal (Characteristic frequency 47 Hz)

5 Experimental setup and results

To further investigate the effectiveness of the presented algorithm for the recovery of fault bearing acoustic signals, a test bed in our laboratory was built. The test bed was equipped with an electric motor with changeable rotation speed, which was connected with the main shaft of the system through a belt pulley. Two rolling bearings were used to mount the shaft. One rolling element bearing with an artificial defect in the outer race was tested. All acoustic signals were collected by two microphones in different positions, which are illustrated in Figure 6. Moreover, a small drilling machine was used to be an interference source and its sound is shown in Figure 7. The experiment was executed at 13.33 Hz rotation rate of shaft. The sound signals were captured at the sampling frequency of 8192 Hz and the sample points were 8192.

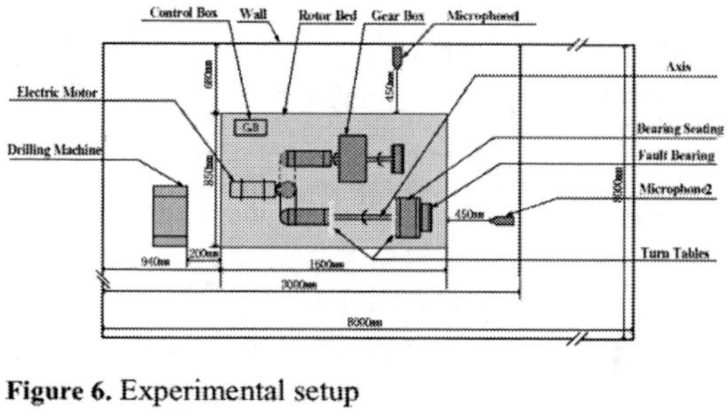

Figure 6. Experimental setup

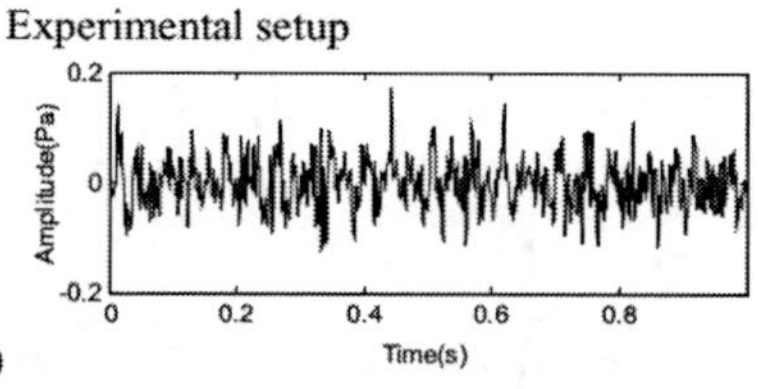

(a)

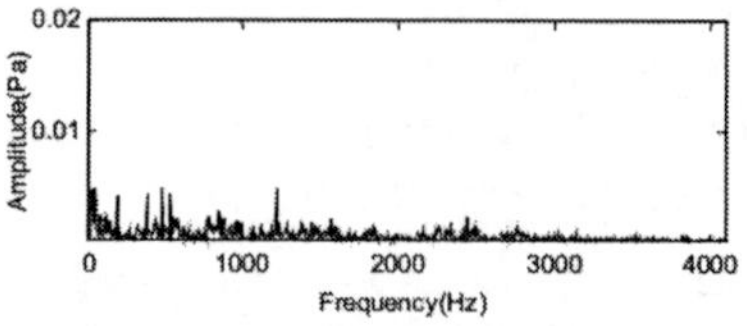

(b)

Figure 7. Acoustic signal of the drilling machine (a) Time domain waveform (b) Amplitude spectrum

DOI: 10.1051/matecconf/20166301012

The signals in Figure 8. represent two raw sensor signals without any processing. It is seen that no impacts can be located. Fault related signatures are usually modulated by the signals at high resonance frequencies of the bearing and surrounding structures, thus a 1000Hz cut-off frequency is selected in this work to eliminate the effects of low-frequency noisy components. In Figure 9., filtered signals using a high-pass filter (high pass frequency f=1000Hz) are presented. Figure 10. shows the envelope spectra of the filtered signals. The signature of the rolling element fault is not inspected through the filtered signals and their corresponding envelope spectra for and the actual outer race fault. Hence, the raw signals and filtered signals were heavily interfered by other undesired signals and can not to be utilized for bearing defect detection directly.

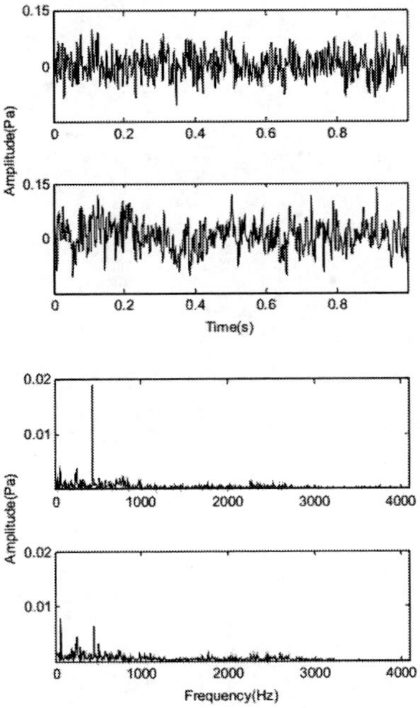

(a)

(b)

Figure 8. Sensor signals (a) Time domain waveforms (b) Amplitude spectra

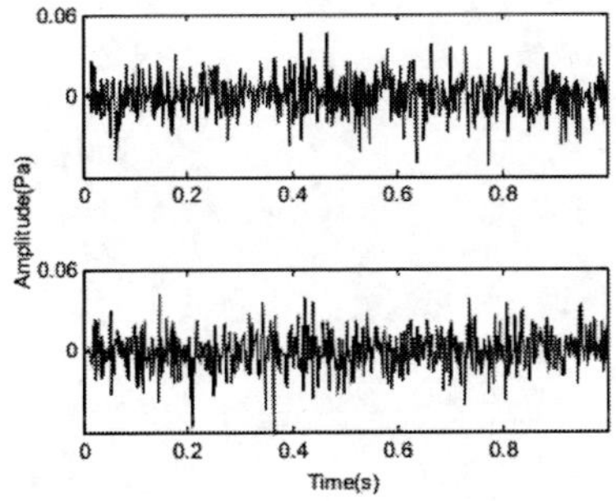

(a)

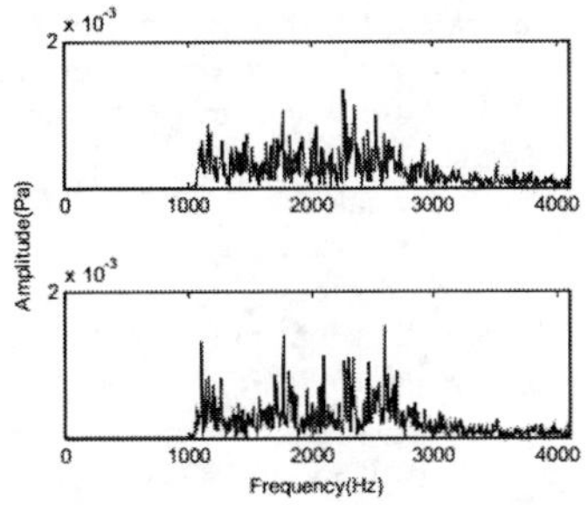

(b)

Figure 9. Filtered signals using a high-pass filter (cut-off frequency f=1000Hz) (a) Time domain waveforms (b) Amplitude spectra

Then, the proposed algorithm was utilized herein to extract the source signal from sensor signals. The algorithm parameters are first given as follows: L=30 and $G(u)=u^2$. The estimated signal in time and frequency domains is presented in Figure 10. The kurtosis of the signal increased dramatically to 38.55 compared to -0.40 for the raw signal and 1.25 for the filtered signal. Intuitively, the recovered signal may come from the faulty bearing according to the periodic impulses. This idea will be confirmed based on the knowledge about the diagnosed bearing and its rotation speed (about 800 rpm). The parameters of the bearing are given as follows: Pitch diameter: D=39mm, Ball diameter: d=7.5mm, Number of rolling elements: Z=12, Contact angular: 0.

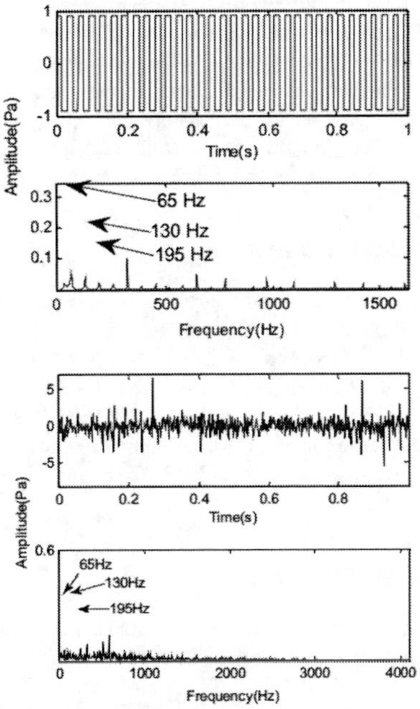

(a)

(b)

Figure 10. Time domain waveforms and envelope spectra (a) Reference signal (b) Estimated signal (Characteristic frequency 64.6 Hz)

It can be calculated that the characteristic frequency of defected outer race is 64.6 Hz theoretically. As can be viewed, the ball pass frequency of the outer race and its harmonics are all visible, and the analysis result (65 Hz) is very close to the calculated result (64.6 Hz). Hence, it is concluded that the impacts are confirmed to be generated by the defective bearing and the estimated signal in Figure 10. belongs to the defected bearing. The results of the experiment can be evaluated as satisfactory and demonstrate that the proposed algorithm can be implemented to extract acoustical impulse signals of faulty bearings.

It is also worth mentioning that the scaling ambiguity, permutation ambiguity, and delay ambiguity exist in the recovered results, but these ambiguities are not essential, since the most relevant information about the faults is contained in the waveforms or spectra of the source signals and not in their amplitudes, order or time-delays.

6 Conclusions

In this paper, we proposed a blind deconvolution method for detection the incipient fault of rolling bearings by acoustic analysis. The new approach has several advantages listed as below. First, the proposed algorithm can perform successfully in complicated acoustic situations with low computational cost even when the time-delay of mixtures is very large. Second, it also does not require the knowledge of the number of sources like in classical BSP (the number of sensors is greater than the number of the sources) and it works well in the presence of multiple additive noise. Experiments with simulated signals and real machine sound from a test bed and drill machine in a room are conducted to demonstrate the effectiveness of this algorithm. The results show that this algorithm can be efficiently applied to rolling element bearing fault detection in practice, and is very promising in acoustic-based machine diagnosis.

Acknowledgements

This work is supported by the National Natural Science Foundation of China (Approved Grant: 51305357) and Scientific Research Fund of Sichuan Provincial Education Department (Approved Grant: 13ZB0018). The authors are also grateful to the Open Research Fund of Key Laboratory of Manufacturing and Automation, Xihua University (SZJJ2012-024) for funding this discovery project.

References

1. M. Knaak, M. Kunter, D. Filberi, *14th International Conference on Digital Signal Processing*, **1** (2002)
2. R. M. Vilela, J.C. MetrGlho, J.C. Cardoso, *Proceedings of the 12th IEEE Mediterranean,* **1** (2004)
3. W. Li, F. Gu, A. D. Ball, C. E. Phipps, Mechanical Systems and Signal Processing, **15** (2001)
4. A. Ypma, A. Leshem, R. P.W. Duin, Neurocomputing , **49** (2002)
5. G. Gelle, M. Colas, G. Delaunay, Mechanical Systems and Signal Processing, **14** (2000)
6. J. Antoni, Mechanical Systems and Signal Processing, **19** (2005)
7. W. Lu, J.C. Rajapakse, IEEE Trans. Neural Networks, **16** (2005)
8. W. Lu, J.C. Rajapakse, Neurocomputing **69** (2006)
9. S. Choi, A. Cichocki, Electronics Letters, **37** (2001)
10. C.J. James, O.J. Gibson, IEEE Trans. Biomed. Eng. **50** (2003)
11. R.B. Randall, J. Antoni, S. Chobsaard, Mechanical Systems and Signal Processing, **15** (2001)

The influence of Ac parameters in the process of micro-arc oxidation film electric breakdown

Jin MA
Institute of Mechanical and Electronic Engineering, Wuhan University of Technology,Wuhan, China, 430070

Abstract. This paper studies the electric breakdown discharge process of micro-arc oxidation film on the surface of aluminum alloy. Based on the analysis of the AC parameters variation in the micro-arc oxidation process, the following conclusions can be drawn: The growth of oxide film can be divided into three stages, and Oxide film breakdown discharge occurs twice in the micro-arc oxidation process. The first stage is the formation and disruptive discharge of amorphous oxide film, producing the ceramic oxide granules, which belong to solid dielectric breakdown. In this stage the membrane voltage of the oxide film plays a key role; the second stage is the formation of ceramic oxide film, the ceramic oxide granules turns into porous structure oxide film in this stage; the third stage is the growth of ceramic oxide film, the gas film that forms in the oxide film's porous structure is electric broken-down, which is the second breakdown discharge process, the current density on the oxide film surface could affect the breakdown process significantly.

Keywords: micro-arc oxidation; AC parameter; solid dielectric; gas film; break down

1 Introduction

Micro-arc oxidation, also called micro-plasma oxidation, is a technique to grow ceramic oxide films in micro areas with high temperature environment through complex electrochemical, plasma-chemical and thermo-chemical processes caused by the micro plasma discharging on the light alloy surface. The ceramic films have some unique characteristics such as strong adhesive strength with substrate, good electrical insulation, corrosion resistance and high hardness, which can largely enhance the properties of alloy surface [1]. In order to study the influence of AC parameters on the oxide film breakdown process systematically and quantitatively, the test of micro-arc oxidation on the aluminum alloy is carried out by applying AC power in weak alkaline electrolyte in this paper.

2 Material and Equipment

The alloy material for experimentation is 2A12 aluminum alloy which has been made into cuboid samples with the size of 14×8×2 mm. The top 4mm of the sample is the wire connection area which is covered by insulating material. The area of sample surface contacting with the electrolyte aqueous solution is 2.16 cm^2.

The micro-arc oxidation device is homemade with an AC output voltage ranging from 0 to 700 V (step-less voltage regulation) with 50Hz frequency. As a result of AC power, the two AC electrodes sample electrode (a) and stainless steel electrode (c) are cathode and anode mutually. An auxiliary inert silver electrode (b) is fastened 1mm away from the sample surface. The electrolyte solution is alkalescency electrolyte solution. Homemade cooling device is installed in the cooling bay to keep the temperature of electrolyte aqueous solution under 25℃. The electrolyte can keep flowing by using a water pump.

[a] Corresponding author: matinjin@whut.edu.cn

3 Method and Process

Samples with the same specification are disposed by micro-arc oxidation test, using AC power source in the electrolyte solution which has the same ratio of solubility. The initial AC current is at 1 A. Experimental results are measured and plotted as shown in Fig 1 and Fig 2.

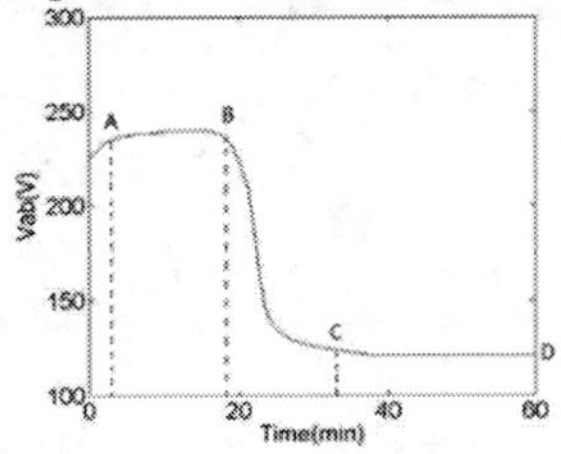

Figure 1. Variation curve of Vab in micro-arc oxidation process

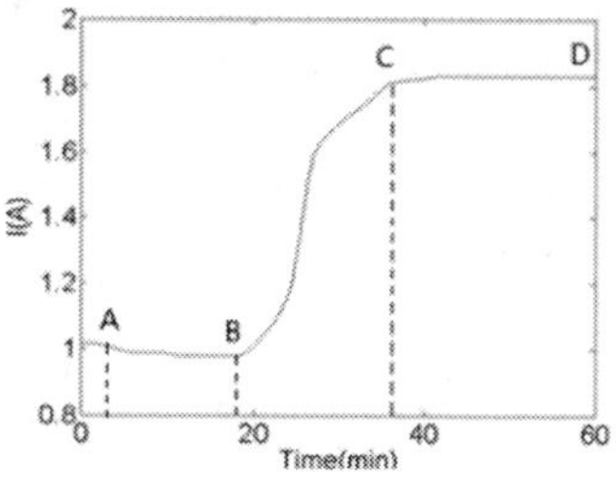

Figure 2. Variation curve of I in micro-arc oxidation process

According to the Fig.1 and Fig.2, it can be judged that the point A and B is the important time nodes in the process of micro-arc oxidation.

The disposing time for samples is 10 sec, 5 min, 20 min and 40 min respectively. After the micro-arc oxidation treatment, measure the thickness of samples surface oxide film, analyze the morphology of oxide films.

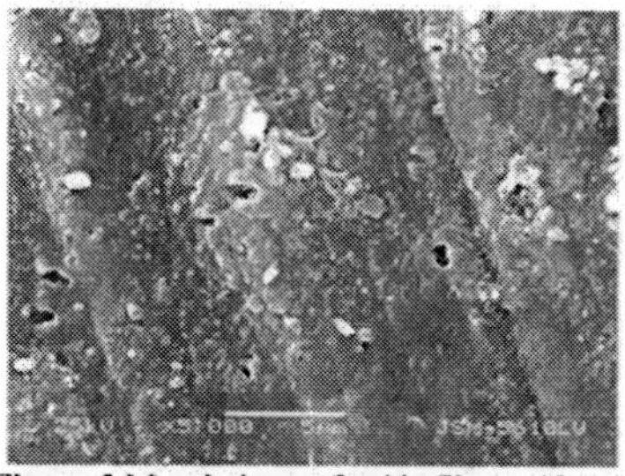

Figure 3. Morphology of oxide film at 10 sec

Figure 4. Morphology of oxide film at 5 min

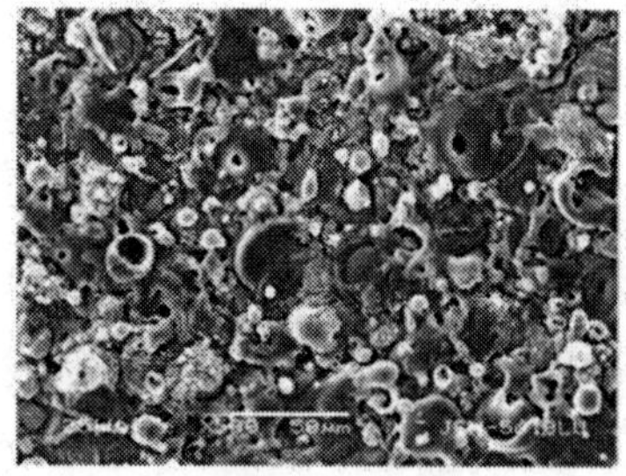

Figure 5. Morphology of oxide film at 20 min

Figure 6. Morphology of oxide film at 40 min

It can be seen from Fig.3 that a thin oxide film has formed at 10 sec of the micro-arc oxidation process with separate oxide granules in size of appr 1 um preliminary occurred on it. According to Fig.4, when oxidation time extended to 5 min, the number of oxide granules with the average size appr. 5um gradually increased and started mutual cross-linking to cover the sample surface. After 20 min, as is shown in Fig.5, the size of granules was appr. 10 um, with granules becoming mutual superposed and forming a porous film structure. As can be seen from Fig.6, granules fused mutually and no obvious boundaries remained at 40 min, crater shape hole was formed with the size of appr.10 um.

4 Results and Discussion

According to the experimental phenomenon and the variation of electrical parameters in the process of micro-arc oxidation, the growth mechanism of micro-arc oxidation film can be divided into three stages.

4.1 Stage 1 Forming of amorphous oxide film

On the sample surface, the Al element got rapid electrochemical reactions when the sample was put into electrolyte and start charging, a thin amorphous oxide film with the thickness less than 1um got rapidly formed[2]. After the film was formed, charge accumulates on both sides of the film under the effect of electric field, thus forming a capacitor [3]. In the electrolyte side of the amorphous oxide film, a certain concentration gradient of charge is accumulated under the effect of electric field [4]; with the ignorance of the concentration between auxiliary electrode and the oxidation film, V_{ab} can be regarded as the voltage between amorphous oxide film. When the charge accumulates till V_{ab} reaches the breakdown voltage of amorphous oxide film, the weakest part would be punctured so the micro-arc discharging phenomenon emerges, which produces high temperature and high pressure, oxidizing Al into ceramic oxide granules (as is shown in Fig.3). This is the first breakdown discharge process. Due to micro-arc discharging, the accumulated electric quantity on the surface of amorphous oxide film decreased. Under the effect of electric field, the charged ions in the electrolyte migrate as a supplementary of micro-arc electric consuming. When the electric quantity of consumption and supplementary reaches a dynamic equilibrium, V_{ab} remains relatively stable.

When the first electric breakdown occurs, V_{ab} values tend to be the limit value 183 v, which means the critical breakdown voltage of oxide film is 183V. According to the formula of critical breakdown field strength [5], oxide film critical breakdown field strength can be estimated by Eq. (1):

$$E_b = \sqrt{2}V/b \qquad (1)$$

Where E_b = 2588KV/cm and b is thickness of oxide film.

As the value of solid dielectric electric breakdown field strength is about 10^4-10^6V/cm orders of magnitude, and the critical breakdown field strength of amorphous oxide film is within the scope of this order of magnitude, so the breakdown of amorphous oxide film should belong to solid dielectric breakdown.

In the process of oxidation film breakdown for the first time, membrane voltage of oxide film plays a key role. Specifically, membrane voltage must be higher than the critical breakdown voltage of oxide film so that the oxide film can be breakdown.

4.2 Stage 2 Forming of the ceramic oxide film.

The ceramic oxide granules formed by micro-arc discharging strengthened the weak part of initial oxide film. As a result, accumulation in both sides of amorphous oxide film continually increases, punctured and strengthened the weak part of film by ceramic oxide granule constantly. As a consequence, V_{ab} increased gradually at the forming stage of ceramic oxide film (in the period A to B as shown in Fig.1).With the proceeding of micro-arc oxidation, the granules increased and cross-linked gradually (as shown in Fig.4), forming porous structure of ceramic oxide film by cover the sample surface. The thickness of film is about 5-10 um at this stage.

4.3 Stage 3 Growth of ceramic oxide film.

The ceramic oxide granules gradually formed porous structure of ceramic oxide film. Due to the electrolytic reaction, the gas film in the pore space is formed when oxygen precipitated from anode and emerges from the porous structure of ceramic oxide film, as shown in Fig.7.

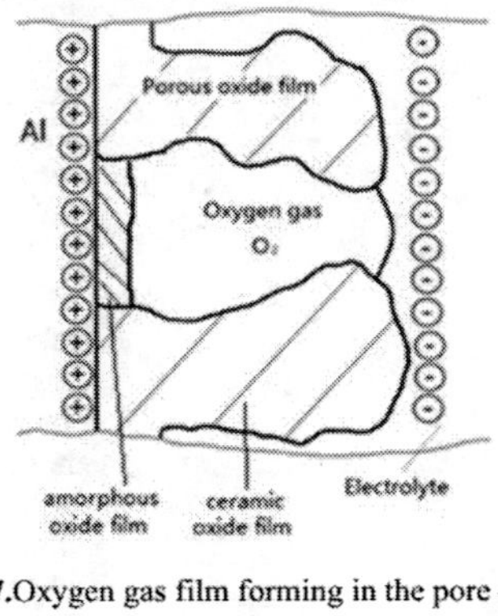

Figure 7. Oxygen gas film forming in the pore of porous structure

The breakdown field intensity of oxygen is about 30KV/cm [5] and the thickness of oxide film is about 10um. Therefore the gas film would be broken down when the voltage between both sides is above 30V, and then micro-arc discharges. This is the second breakdown discharge process. The gaseous dielectric breakdown is easier other than the solid dielectric breakdown at the first stage. Under the joule heating effect, micro-arc discharge released a lot of heat in the pore, forming a high temperature and high pressure micro area to oxidate Al into high temperature molten oxides. The molten oxides expanded and blend with each other with continuous growth under high pressure, as is shown in Fig.5. At this stage, the growth rate of oxide film increased rapidly.

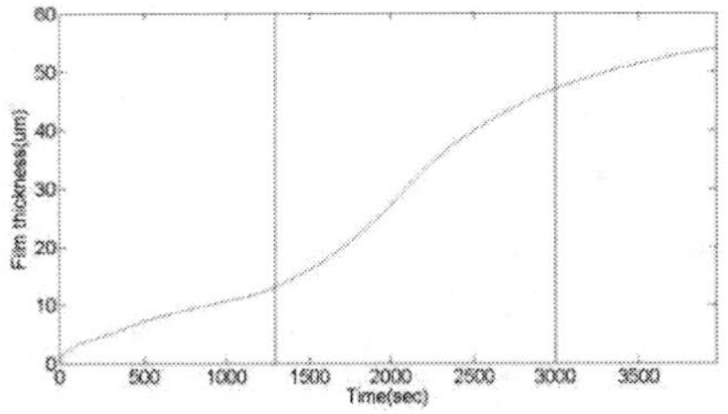

Figure 8. The thickness curve of oxide film in micro-arc oxidation process

According to Fig 8, at stage 1 and stage 2, oxide film growth is very slowly. At stage 3 the growth of oxide film becomes quickly.

Because of rapid discharging of gas film breakdown, supplementary electric should be obtained from electrolyte outside the pores. However, the migration speed of charged ions in electrolyte cannot compensate the consuming electric quantity due to the limitation of the electrolyte conductivity. As a consequence, Vab value decreased rapidly and current value increased rapidly (in the period B to C as is shown in Fig.1 and Fig.2.).

Along with the oxidation film growing, the thickness of film is increasing while electric field intensity between film sides is gradually weakening. When electric field intensity is unable to breakdown the gas film, the micro-arc stopped discharging and the film's thickness reached a max-value.

By keeping the voltage between electrode (a) and electrode (c) at 600V, different initial current densities (10A/dm2、20A/dm2、30A/dm2、40A/dm2、50A/dm2、60A/dm2) are employed to carry on micro-arc oxidation experiments. Variation of current is measured in the process of micro-arc oxidation test and drawn as shown in Fig 9.

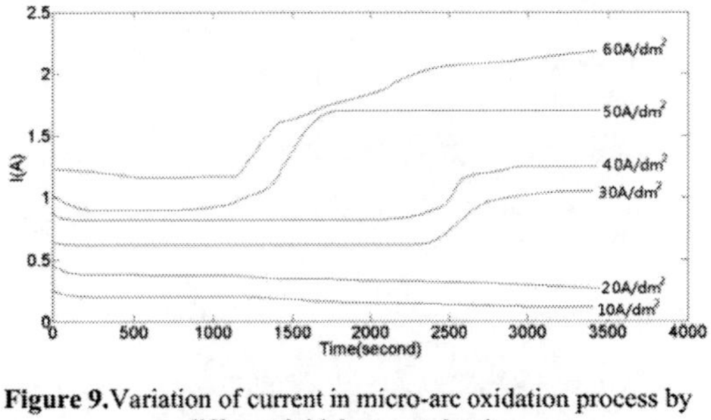

Figure 9. Variation of current in micro-arc oxidation process by different initial current density

From the figure 8, which shows the process of micro-arc oxidation, when the initial current density is less than 30 A/dm^2, the second breakdown does not occur. This is

 DOI: 10.1051/matecconf/20166301013

due to the liquid - gas interface formation between the electrolyte and the gas, charged particles must have sufficient kinetic energy to pass through the liquid - gas interface. With enough energy, it rushes into gas film and generates gas ionization that cause gas film electric breakdown. With too low current density, the charged particles in the electrolyte move too slow, which would decrease the gas film breakdown probability. So the second breakdown cannot occur on the sample surface, and the growth of the ceramic oxide film is in the stagnation.

5 Conclusions

Based on the analysis and discussion, the following conclusions can be drawn:

1)The growth of ceramic oxide film can be divided into three stages: the formation and breakdown stage for amorphous oxide film; the formation stage for porous ceramic oxide film and the growth stage for ceramic oxide film.

2) In the first stage, amorphous oxide film forms quickly at the test beginning, the ceramic oxide granules is produced as a result of oxide film's electric breakdown discharge. This is the first breakdown discharge process which belongs to solid dielectric breakdown. Membrane voltage must be higher than the critical breakdown voltage of oxide film to enable the oxide film to be breakdown.

3) In the second stage, ceramic oxide granules gradually grow up into porous structure of ceramic oxide film.

4) In the third stage, oxygen gas film is formed in the pores of ceramic oxide film. Due to the gas film's breakdown, the micro-arc discharge release large amounts of heat to form micro area of high temperature and high pressure, which provide the condition for ceramic oxide film's continuous growth. This is the second breakdown discharge process which belongs to gaseous dielectric breakdown. In this stage, initial current density must be higher than $30A/dm^2$ to make the second breakdown occur, and the ceramic oxide film can growth rapidly.

References

1. Hu.Zhenqian, Ma.Jin, Heat Treatment of Metals, **28**, 23-26 (2003)
2. Ma.Jin, Hu.Zhenqian, Material&Heat Treatment, **41**, 142-144 (2012)
3. Jia.Zheng, Dai.Changsong,Chen.Ling, *Measurement Methods of Electrochemistry, Chemical Industry Press,*(2006)
4. W. Plieth, *Electrochemistry for Materials Science, Science Press*, (2008)
5. Yin. Zhiwen, *Dielectric Physics, Science Press,* (2006)
6. Ma.Jin, ICMME 2014, *Applied Mechanics and Materials,* **697**, 7-11(2015)
7. Vangolu. Y, Alsaran. A, Yildirim.O.S, Wear, **271**, 2322-2327(2011)
8. Jacob.Roshan, Srinivasan. S.A, Sivaprasad. K, Muthupandi. V, Materials Science Forum, **765**,634-638(2013)
9. Shin. Sowoon, Choi. Yong-Wook, Choi.Jinsub, Materials Letters, **105**, 117-119(2013)
10. Tao.Hong, Ho.Soking, Li.echo, Zhang. Xuezhu, Advanced Materials Research, **905**, 146-150(2014)
11. Dey, Arjun, Rani. R.Uma, Thota. Hari Krishna, **39**, 3313-3320(2013)
12. Urbahs. A, Urbaha. M, Zujevs. V, Savkovs, K, Rijkuris, G, Transport Means - Proceedings of the International Conference, 101-105(2013)
13. Durdu.Salih, Aytaç.Aylin, Usta.Metin, Journal of Alloys and Compounds, **509**, 8601-8606 (2011)
14. Yavuz.H.G, Gunyuz.M, Ozkara.I.M, Baydogan. M, Cimenoglu.H, Defect and Diffusion Forum, **312-315**, 877-88(2011)
15. Ma.Jin, Journal of Materials Protection, **47**, 41-43(2014)

MATEC Web of Conferences **63**, 01014 (2016) DOI: 10.1051/matecconf/20166301014
MMME 2016

Research of barometric altitude measurement technology

Ye ZHAN[1,a], Li Min CHANG[1] and Jun LI[1]

[1] *Aircraft control Department, Air Force Aviation University, Changchun, China*

Abstract. The atmospheric pressure measurement is closed related to the aircraft flight and it is the key factor directly affecting the aviation flight safety. With the development of aircraft, the barometric altitude measurement technology is more and more prominent and important. According to the barometric altitude measurement principle and the general barometric altitude measurement methods, four kinds of measurement methods of bellows, vibration cylinder, silicon resonant and silicon piezoresistance are analyzed with the principle and compared from the application.

1 Introduction

In the aerospace field, the atmospheric pressure measurement is closed related to the aircraft flight. From the aviation perspective, the height of aircraft is indicated by the pressure altimeter in the flight. The pressure altimeter is the instrument indicating the flight altitude through the atmospheric pressure. Measuring and choosing the flight altitude correctly is greatly important for playing the aircraft performance sufficiently, reducing the fuel consumption, saving the flight time and ensuring the flight safety. The pressure altimeter is installed on all modern kinds of aircraft. However, this kind of altimeter is with a certain hidden trouble because of the natural principle characteristics used by barometric altimeter.

2 Barometric altitude measuring principle

Assuming that the atmosphere is relatively still to the earth, there is no horizontal and vertical movement of atmosphere. Then, a thin layer of air column is got at any height, which is shown in Figure 1. The air column height is dH and the cross section area is dF. The air density is ρ and the downward orientation is the positive direction. And the static equilibrium condition for the air column is as followed.

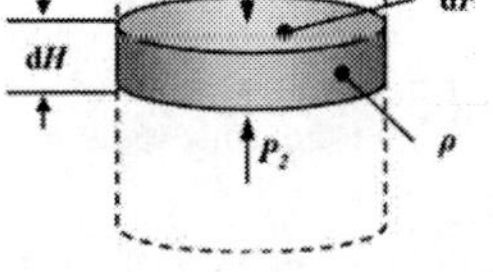

Figure 1. Force analysis of thin layer air column

$$dP\cdot dF+\rho g\cdot dH\cdot dF=0 \quad dP=P_1+P_2$$

Due to $P=\rho gRT$, the equation can be transformed to

$$\frac{dP}{P}=-\frac{1}{RT}dH \tag{1}$$

Due to the changing of atmospheric temperature vertical change rate τ with the change of time and place, the temperature at one height can be sensed accurately if the average temperature vertical change rate τ=-6.5K/*km* of troposphere is used in practical application. Therefore, the standard barometric altitude formula is taken place by the Laplace barometric altitude formula on approximation.

T is taken place by T_m of temperature arithmetic mean and the equation (2) is integrated as follow.

$$\int_{P_b}^{P_H}\frac{dP}{P}=\int_{H_b}^{H}\frac{1}{RT_m}dH \tag{2}$$

And the following equation can be got.

$$H=RT_m\ln\frac{P_b}{P_H} \tag{3}$$

The algorithms are different in different levels. In the troposphere below 11000m, Tm of average temperature is calculated by the following equation (4).

$$T_m=\frac{T_b+T_H}{2}=\frac{T_b+(T_b+\tau H)}{2}=T_b+\frac{\tau H}{2} \tag{4}$$

In the flight altitude range of 11000~20000m, the average temperature T_m is calculated by the following equation (5).

$$T_m=\frac{\frac{T_b+T_{11}}{2}H_{11}+T_{11}(H-H_{11})}{H}=\frac{(T_b-T_H)H_{11}}{2H}+T_{11} \tag{5}$$

The altitude exist the one-to-one relationship with the barometric pressure. It has been found that the altitude calculated by the approximate formula is more close to the actual barometric altitude if the datum temperature (T_b) and the high altitude temperature (T_H and T_{11}) used

[a] Corresponding author: zhanye32@aliyun.com

the actual measured values in actual atmosphere. Therefore, the barometric altitude approximate formula is used in some air data system.

3 Barometric altitude measuring technology

3.1 Bellows barometric altimeter

The bellows barometric altimeter is designed according to the standard atmospheric conditions. The air static pressure outside will change when the barometric altitude changes. And the pressure sensed by the vacuum bellows changes so the bellows center will output the tiny linear displacement. The displacement is transmitted and magnified through the connecting gear and the indicating section will indicate the barometric altitude after changing. Its core component is the vacuum bellows, which is shown in Figure 2.

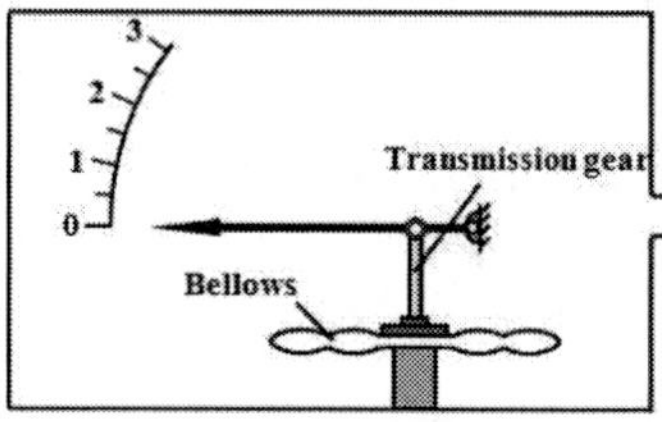

Figure 2. Principle diagram of barometric altimeter

The sensing component of bellows barometric altimeter is the vacuum bellows. The bellows inside is vacuum and the outside senses the atmosphere static pressure. Its pressure-displacement characteristic curve is similar with the atmospheric pressure-altitude characteristic curve. It is shown in Figure 3.

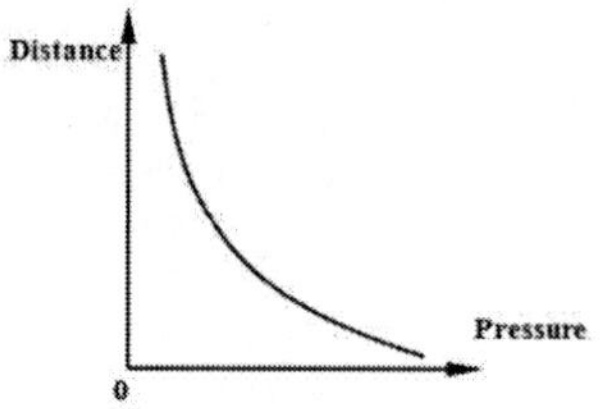

Figure 3. Characteristic curve of vacuum bellows

The early atmosphere data measurement was done though the bellows instrument. Although the structure of bellows instrument is simple and the operation is reliable, the precision is low.

3.2 Vibration cylinder pressure altimeter

The vibration cylinder pressure sensor component is mainly composed of the static pressure sensor and the total pressure sensor. The structure of vibration cylinder sensor is shown in Figure 4. It consists of the vibration cylinder, the single coil excitement, the vibration pickup coil, the foundation bed and the outer protective barrel. The vibration cylinder is the core component, whose top is closed as free and the other end is fixed on the foundation bed. The space is a vacuum between the vibration cylinder and the protective barrel, which is as a reference for the measured pressure. The pressure sensed by the vibration cylinder is the absolute pressure when the measured pressure went into the space between vibration cylinder and coil.

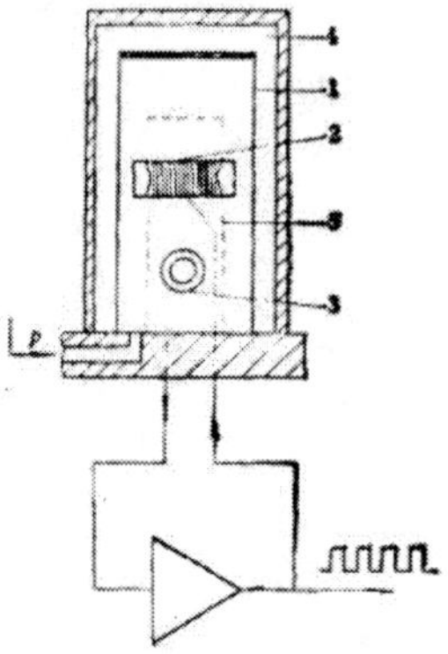

Figure 4. Principle block diagram of vibration cylinder sensor

The vibration cylinder pressure sensor component is the sensitive part of altimeter. It receives the static pressure and the total pressure from the pitot tube of aircraft and the pressure are converted into the frequency signal which is with a one-to-one correspondence relationship with pressure.

When the power is not on, the vibration cylinder is the stationary state. When turning on the DC power supply, the natural noise of amplifier products a weak random pulse in the excitation coil. The pulse signal make the magnetic field change through the excitation coil, which forming the pulse power. The pulse power cause the deformation of vibration cylinder wall and the vibration cylinder vibrates with the resonance frequency of low amplitude.

When the measured pressure is zero, the natural vibration frequency of vibration cylinder is f_0. When the measured pressure is changed, the vibration cylinder stiffness change and its resonant frequency also change accordingly. The vibration pick-up coil sense the vibration frequency changing with pressure on one hand and keep the vibration cylinder maintain the persistent oscillation through the positive feedback on the other hand. And the vibration frequency is $f = f_0\sqrt{1+AP}$. P is the measured pressure, A is the vibration cylinder constant. Due to the amplitude limiting in the circuit, the maximum deformation of vibration cylinder is limited to a fixed value to ensure that the frequency pulse signal with a certain range and the corresponding relation with pressure is output.

Compared with the traditional bellows pressure altimeter, the vibration cylinder pressure altimeter is with

DOI: 10.1051/matecconf/20166301014

the advantages of high precision, simple structure, small volume and light quality.

3.3 Silicon resonant pressure altimeter

The silicon resonant pressure sensor consists of two parts. One part is the pressure membrane and the other part is the oscillator. The two parts are closely combined and about 2 microns is only kept in the middle to make the oscillator produce the movement. The pressure membrane is connected with the measured air source at work and the resonant beam is installed on the sensor. There is a pressure difference between the measured air source and the pressure membrane. The pressure membrane deformation is made by the pressure difference, where a deformation force exists. The force is used on the resonant beam leading the natural frequency of resonant beam changed. The natural frequency has the linear function relation with the measured air source in the scope of work pressure. Therefore, the pressure can be got by measuring the sensor frequency.

When being working, the resonant beam frequency is got by adding the vibration signals on resonant beam to make it produce the vibration. The sensor resonant beam is shown in Figure5.

Figure 5. Structure diagram of resonant beam

Two resistances are respectively put on the silicon beam and the resistance in the middle of silicon beam is the vibration resistance. The thermal excitation voltage V_1 uses the constant-current source for power and the vibration pick-up signal is V_0. The sensor structure is shown in Figure6.

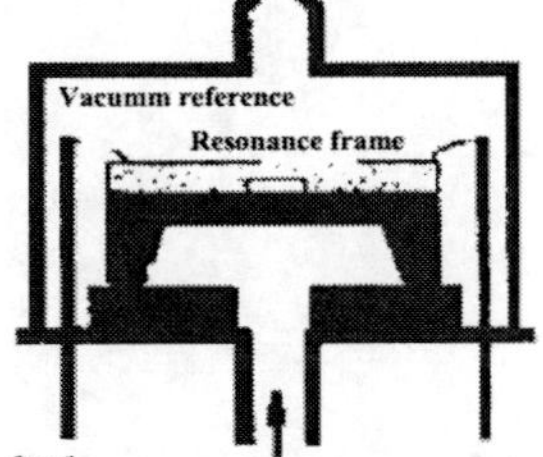

Figure 6. Structure diagram of silicon resonant pressure sensor

Due to the silicon resonant pressure sensor advantages of small volume, light weight, high precision and simple output signal processing, it can be got an intelligent pressure test system with the single-chip computer or dedicated computer. The digital signals of total pressure, static pressure and total temperature are sent to the computer directly at the static pressure tube exit and the needed are calculated by the computer. This way can reduce the delay error of hydrostatic pressure, improve the dynamic characteristics of flight parameters, reduce the member of airborne computer and play the computer potential abundantly.

3.4 Silicon piezoresistive pressure altimeter

When the pressure is made on a semiconduct along one axial, the semiconduct will produce the deformation and its resistivity will change significantly. And the phenomenon is called the. The sensor made of the piezoresistive effect of semiconduct materials is called the piezoresistive sensor.

The structure of silicon piezoresistive pressure altimeter is shown in Figure7. The system detects the air static pressure through the silicon resistance pressure sensor and completes the sampling for sensor analog signal through the $\sum-\Delta ADC$ converter. The temperature sensors are directly integrated within the pressure sensor, which can insure that the temperature detected by sensor is accordance with the actual temperature of silicon piezoresistive pressure sensor.

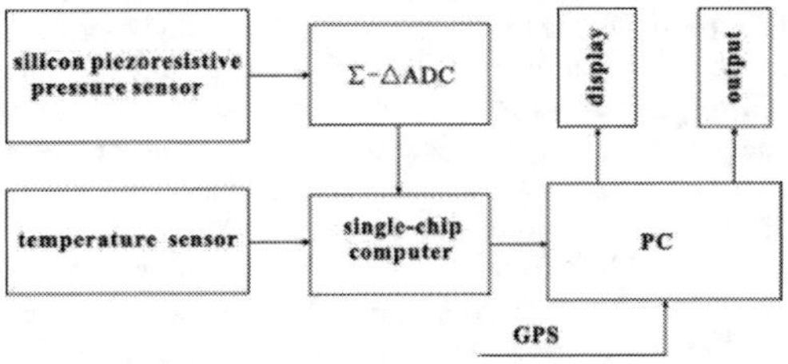

Figure 7. Structure diagram of silicon piezoresistive pressure altimeter

When the aircraft speed is lower than 1 Mach, the aircraft geopotential height is measured indirectly through the air static pressure by the barometric altimeter based on equation (6) in international standard atmosphere environment.

$$\begin{cases} H = \frac{T_b}{U}[(\frac{p_H}{p_b})^{-U_{R/g}} - 1] \\ H = 11000 + (\frac{RT_b}{g})ln(\frac{p_b}{p_H}) \end{cases} \quad (6)$$

T_H is the atmospheric temperature lower limit value and P_H is the pressure lower limit value of international standard atmosphere in appropriate level.

U is the vertical temperature gradient of international standard atmosphere in appropriate level.

T_b is the standard atmosphere sea-level temperature and T_0=288. 15 K (0~11000m).

P_b is the standard atmosphere sea-level static pressure and P_0=101325Pa.

R is the air special gas constant and $R = 287.05287 m^2 / (K^o s^2)$.

g is the gravity acceleration of standard sea level and $g = 9.80665m / s^2$.

It should be pointed out that the geopotential height measured by barometric altimeter has the same datum with the altitude (geometry) but two data are not consistent. With the height increasing, the difference is more and more. The conversion can be made according the following equation (7). In the equation, r is the earth radius.

$$H = rH / (r + h) \qquad (7)$$

The silicon piezoresistive pressure sensor is made of the silicon piezoresistive effect of semiconductor material. The silicon piezoresistive pressure altimeter has the characteristics of high sensitivity, quick dynamic response, high measure precision, good stability, wide operating temperature range, miniaturization, mass production. But the silicon piezoresistive pressure sensor outputs are susceptible to a variety of factors, such as temperature, humidity and power fluctuations. And the cross sensitivity of pressure and temperature is the main influencing factor.

4 Conclusion

The pressure measurement technology development is concentrated in two aspects. One is using the micro mechanical processing technology and very nature integrated technology to make the pressure sensor volume as small as possible and to meet the need of special occasions or make the pressure sensor be easy to connect with the computer to develop the intelligent pressure sensor. The second is using the emerging semiconductor materials, such as SiC and POR-Si or improving the Wheatstone circuit, such as Double Wheatstone bridge circuit and eight arms bridge circuit.

References

1. Wei Dang, Shengguo Huang, *Design of a digital atmospheric pressure altimeter*. (Measurement and control technology, **23(12)**, 2004)
2. Zhenhai Hao, Shengguo Huang, *Development of high precision barometric altimeter*. (Journal of Nanjing University of Aeronautics & Astronautics, **41(1)**, 2009)
3. Limin Chang, Xinglong Qi, Qingwei Dong, *Realization of air data computer system based on silicon resonant pressure sensor*. (Computer Measurement & Control, **16(12)**, 2008)
4. Lei Gao, Yinhui Ma, Wenqing Wang, *Design and implementation of silicon piezoresistive barometric altimeter*. (Microcomputer & Applications, **33(20)**, 2014)
5. Kai Chen, Chaohui Hu, Yuechun Wei, Jie Yan, *Design of air data system of an aviation standby instrument system*. (Chinese Journal of Sensor and Actuators, **19(4)**, 2006)
6. Yuhua Wang, Xiaonan Wu, Jinyong Yu, *Design of barometric altimeter based on single chip microcomputer*. (Instrument Technology, **8**, 2013)

MATEC Web of Conferences **63**, 01015 (2016) DOI: 10.1051/matecconf/20166301015
MMME 2016

The research on the consistency of the testing accuracy for smoke measuring equipment based on ionization chamber principle

Ting Ting WANG[1,a], Pei ZHANG[1], Hong ZHANG[1], Lei Lei XIE[1], Ya Jun YAN[1], Miao Miao DAI[1], and Ya Yun WANG[1]

[1] *Shanghai Institute of Quality Inspection and Technical Research, 201114 Shanghai, China*

Abstract. A convenient method is proposed here to determine the relative accuracy of measuring ionization chamber (MIC). A non-dimensional quantity is presented here to characterize the relative accuracy of the testing sample to the reference sample. There are two key points of the program. One is comparison by pairs. The other is eliminating the uncertainty in the program by exchanging the position and control unit in each testing group.

1 Introduction

As we known, there are two kinds of widely used smoke measuring equipment, which are obscuration meter and measuring ionization chamber (MIC) [1]. Smoke density can be measured by different techniques [2-6]. However, for reference measurements of smoke density in connection with testing of ionization smoke detectors for automatic fire detection systems, it is advantageous to use a measuring instrument based on the ionization chamber principle. This is because of the complex nature of smoke, which means that measuring errors are likely to occur if the reference measuring instrument reacts to other properties of the smoke than the detector under test.

When the measuring ionization chamber (MIC) [7-9] is used in the smoke tunnel measurements and fire sensitivity tests, the high measuring accuracy is very important for the overall measuring accuracy of the whole system is limited by the measuring accuracy of the measuring ionization chamber (MIC). The radioactive source in the MIC makes the certification of MIC very expensive and complicated. A convenient method is proposed here to determine the relative accuracy of measuring ionization chamber (MIC) which may be widely used by testing laboratory and manufactory.

2 Measuring ionization chamber (MIC)

The response threshold of detectors using ionization is characterized by a non-dimensional quantity y or X which are derived from the relative change of the current flowing in a measuring ionization chamber, and which are related to the particle concentration of the test aerosol, measured in the proximity of the detector, at the moment that it generates an alarm signal.

According to BS EN54-7:2001 "Fire detection and fire alarm systems" and other relative standard, MIC consists of a measuring chamber, an electronic amplifier and a method of continuously sucking in a sample of the aerosol or smoke to be measured.

The measuring chamber contains a measuring volume a suitable means by which the sampled air is sucked in and passes the measuring volume in such a way that the aerosol/smoke particles diffuse into the volume. This diffusion is such that the flow of ions within the measuring volume is not disturbed by air movements. The air within the measuring volume is ionized by alpha radiation from an americium radioactive source, such that there is bipolar flow of ions when an electrical voltage is applied between the electrodes. This flow of ions is affected by the aerosol or smoke particles in a known manner. The relative variation in the current of ions is used as a measurement of the aerosol or smoke concentration.

The measured smoke densities is expressed in terms of the dimensionless quantity defined by the equation:

$$X = \frac{I_0 - I}{I_0}, \quad 0 \le X \le 1 \qquad (1)$$

Where I_0 is the quiescent ionization chamber current in clean air and I is the ionization chamber current in presence of smoke.

3 Testing program

3.1. Basis of the program

According to the chapter 4.4 of CNAS-CL 06:2014 "The traceability requirements of the measured results", a program is designed to get the relative accuracy of the measuring ionization chamber (MIC). According to

[a] Corresponding author: wangtt@sqi.org.cn

CNAS-CL 06:2014, several methods of measurement traceability have been pointed.

When measuring results can not be traced back to the international system of units (SI) or is not related to SI, the measurement should be traced back to the RM, method/standard agreed by conventional measurement angecy, or through the way such as comparison between laboratory, to prove the consistency of the measurement result with other laboratories.

3.2 Key points of the program

A non-dimensional quantity R_{ns} is used here to characterize the relative accuracy of the test sample named n and reference sample named S.

There are two key points of the program. One is comparison by pairs which is tested as a group. The other is eliminating the uncertainty in the program as much as possible.

3.2.1 Comparison by pairs

A specified measuring ionization chamber (MIC) is chosen as reference sample S. Other testing measuring ionization chamber (MIC) are marked as No. 1, 2,...N. Compare the testing sample n and the reference sample S respectively. Take a testing sample named n and place it with the reference sample S in the same smoke environment. Test the non-dimensional quantity X of both samples and compare them. If you have many testing samples, you should make the testing pair by pair. Schematic diagram is shown in figure 1.

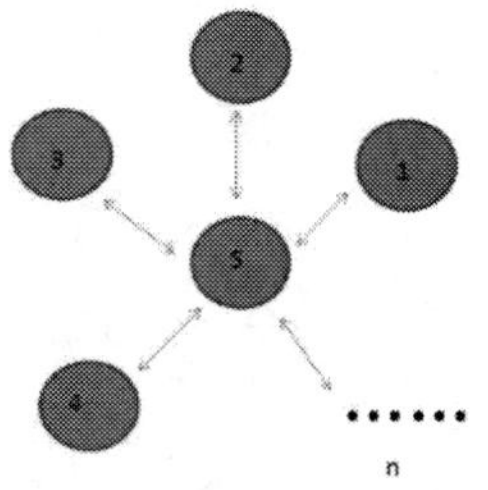

Figure 1. Schematic diagram of Comparison by pairs.

3.2.2 Eliminating the uncertainty

There are two factors that affect the measurement uncertainty, which are the position of the samples and the selection of the control unit.

Position of the samples:

Sample n and S are placed in the vertical direction of the smoke flow. In order to fully eliminate the effects of inhomogeneous smoke in the tunnel, the testing must be carried out twice when comparing the sample n and S by exchanging the position. Swap positions can effectively reduce the measurement uncertainty. Schematic diagram is shown in figure 2.

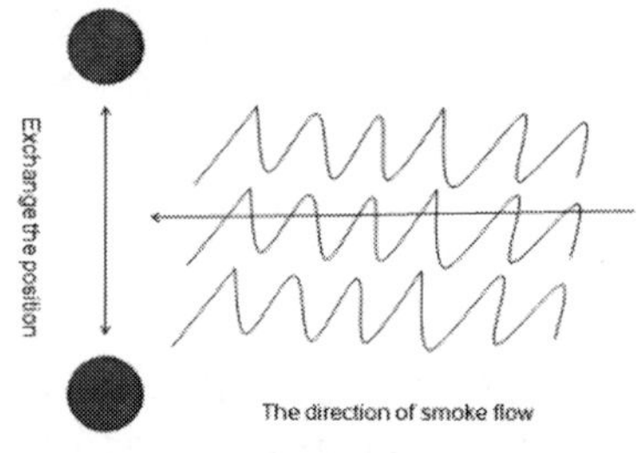

Figure 2. Schematic diagram of the position of the samples.

Selecting the measuring ionization chamber (MIC) control unit:

For the testing must be done pair by pair, two measuring ionization chamber (MIC) control units are needed here. In order to eliminate the measurement uncertainty, the same type measuring ionization chamber (MIC) control units come from the same manufacturer should be selected. Also, swap the control unit can eliminate the tiny difference between the control units.

3.3 Testing procedures

Here is the detail procedure:

1. Clean and alignment the smoke tunnel and measuring ionization chambers (MICs) to ensure they work property.
2. Install the samples in the correct position and connect the control units respectively.
3. Preheat the equipment for 15 minutes.
4. Check the performance of the measuring ionization chamber (MIC). Adjust the chamber voltage VDC.
5. Adjust the aerosol generator. The measuring ionization chambers (MICs) (sample n and S) are exposed to a slowly increasing smoke density produced by a paraffin oil aerosol generator. The rate of increase of aerosol density is $0.05 min^{-1}$. The flow rate of aerosol is $20\ l\ min^{-1}$.
6. Record the values X of the sample n and S. Record as Xn and Xs.
7. Denote Xs and Xn abscissa and ordinate respectively. Carry on the linear numerical fitting to get the slope $R_{ns-1.}$
8. Swap the position and the control units of sample n and S and record the values X also.
9. Repeat the step 7 to get the slope $R_{ns-2,}$ $R_{ns-3,}$ $R_{ns-4.}$
10. The relative accuracy of sample n is R_{ns}, which is $(R_{ns-1}+R_{ns-2}+R_{ns-3}+R_{ns-4})/4.$
11. The relative error of the sample n is $|1-R_{ns}|\times 100\%$. If the relative error is less than 5%, the relative accuracy of the testing sample n to the relative S is good. This tolerance 5% comes from chapter 5.1.4 of BS EN54-7:2001.

Note: Must empty the smoke tunnel before each testing.

DOI: 10.1051/matecconf/20166301015

4 Example

Here, an example is presented to testify our program.

Testing environment: Air temperature is 20.6℃ and air humidity is 45.4%. Air velocity is 0.2m/s.
Condition: The rate of increase of aerosol density is $0.05 min^{-1}$. The flow rate of aerosol is $20 l min^{-1}$. The flow rate of vacuum pump is $30 l min^{-1}$.
Test equipment: measuring ionization chamber (MIC) : Delta MIC EC-912, smoke tunnel: AWT 2800.

Mark the testing sample as a. Based on the choice of location and control unit of the reference sample S, There is four situations of each testing group.

Situation 1:

Table 1. The choice of location and control unit for situation 1.

Location1, control unit 1	●
Location1, control unit 2	
Location2, control unit 1	
Location2, control unit 2	

Table 2 shows the values for under testing sample a and reference sample S in situation 1. Figure 3 shows the values X of sample a to S.

Table 2. Values for under testing sample a and reference sample S in situation 1.

Xa (VDC:18.8V)	Xs (VDC:19.2V)
0.00	0.00
0.05	0.05
0.11	0.10
0.16	0.15
0.20	0.20
0.26	0.25
0.32	0.30
0.36	0.35
0.41	0.40
0.46	0.45
0.50	0.50

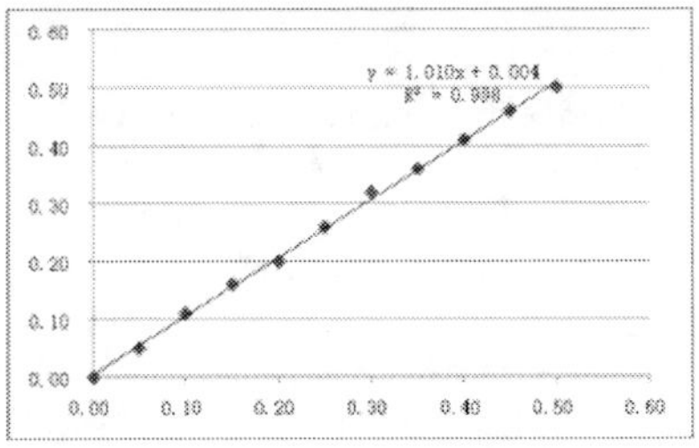

Figure 3. Values X of sample a to S.

From figure 3, it can be derived that the slope R_{ns-1} is 1.010, and the correlation coefficient is 0.998.

Situation 2:

Table 3. The choice of location and control unit for situation 2.

Location1, control unit 1	
Location1, control unit 2	●
Location2, control unit 1	
Location2, control unit 2	

Table 4 shows the values for under testing sample a and reference sample S in situation 2. Figure 4 shows the values X of sample a to S.

Table 4. Values for under testing sample a and reference sample S in situation 2.

Xa (VDC:18.4V)	Xs (VDC:18.9V)
0.00	0.00
0.05	0.06
0.10	0.09
0.15	0.15
0.20	0.20
0.25	0.25
0.30	0.30
0.35	0.35
0.40	0.39
0.45	0.44
0.50	0.49

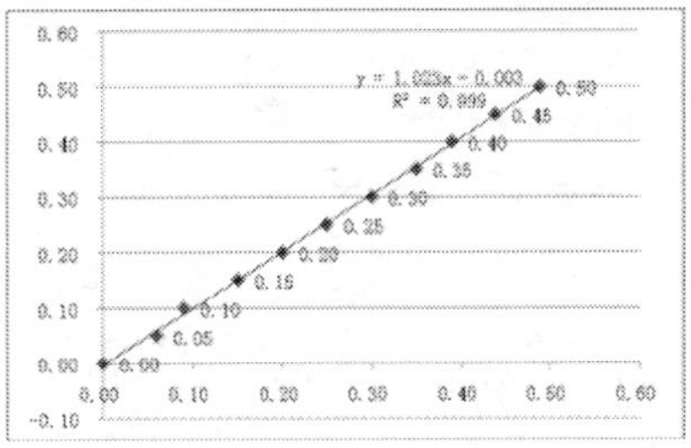

Figure 4. Values X of sample a to S.

From figure 4, it can be derived that the slope R_{ns-2} is 1.023, and the correlation coefficient is 0.999.

Situation 3:

Table 5. The choice of location and control unit for situation 3.

Location1, control unit 1	
Location1, control unit 2	
Location2, control unit 1	●
Location2, control unit 2	

Table 6 shows the values for under testing sample a and reference sample S in situation 3. Figure 5 shows the values X of sample a to S.

Table 6. Values for under testing sample a and reference sample S for situation 3.

Xa (VDC:18.8V)	Xs (VDC:19.4V)
0.00	0.00
0.05	0.06
0.10	0.11
0.15	0.16
0.20	0.21
0.25	0.25
0.30	0.30

0.35	0.36
0.40	0.41
0.45	0.46
0.50	0.50

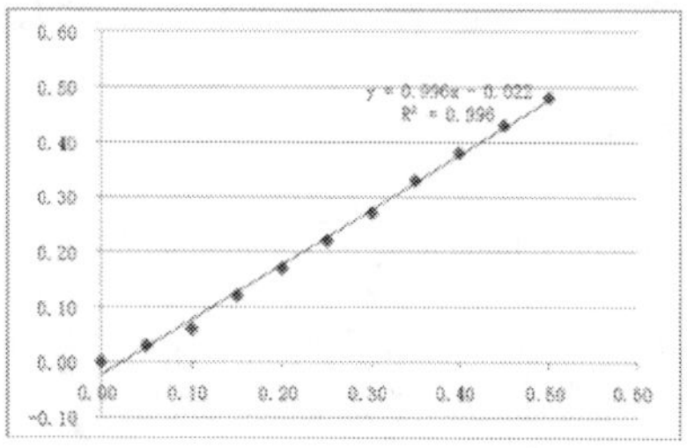

Figure 5. Values X of sample a to S.

From figure 5, it can be derived that the slope R_{ns-3} is 0.996, and the correlation coefficient is 0.996.

Situation 4:

Table 7. The choice of location and control unit for situation 4.

Location1, control unit 1	
Location1, control unit 2	
Location2, control unit 1	
Location2, control unit 2	●

Table 8 shows the values for under testing sample a and reference sample S in situation 4. Figure 6 shows the values X of sample a to S.

Table 8. Values for under testing sample a and reference sample S in situation 4.

Xa (VDC:19.0V)	Xs (VDC:19.8V)
0.00	0.00
0.05	0.06
0.10	0.11
0.15	0.16
0.20	0.21
0.25	0.25
0.30	0.30
0.35	0.36
0.40	0.41
0.45	0.46
0.50	0.50

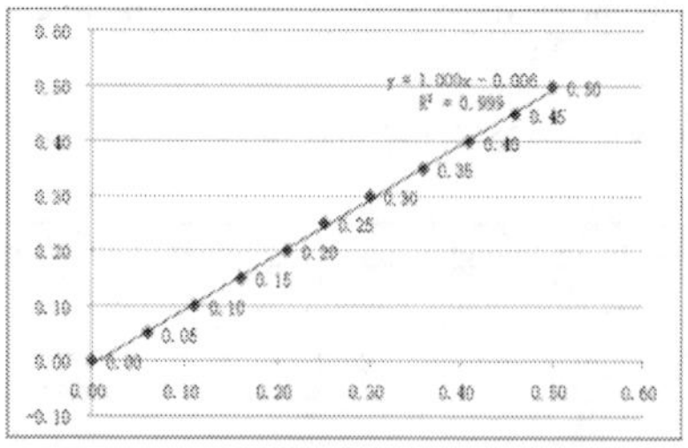

Figure 6. Values X of sample a to S.

From figure 6, it can be derived that the slope R_{ns-4} is 1.000, and the correlation coefficient is 0.999.

Then the relative accuracy R_{ns} of sample a to sample S is 1.00725 and the relative error of the sample a to sample S is 0.725% . The relative error is smaller than the tolerance. So the testing performance of sample a and s has no obvious difference and they have a good consistency.

5 Conclusion

A convenient method is proposed in this paper to determine the relative accuracy of measuring ionization chamber (MIC). A non-dimensional quantity R_{ns} is used here to characterize the relative accuracy of the testing sample named n and reference sample named S. There are two key points of the program. One is comparison by pairs. The other is eliminating the uncertainty in the program by exchanging the position and control unit in each testing group. This method may be widely used by testing laboratory and manufactory.

This paper is funded by the project "14DZ2294100 The platform construction project of professional testing technical service for digital electronic product" by shanghai science and technology committee.

References

1. S. Aggarwal, V. Motevalf, Fire safety J. **29**, 99(1997).
2. S. J. Chen, D. C. Hovde, K. A. Peterson, A. W. Marshall, Fire safety J. **42**, 507(2007).
3. M. A. Jackson, I. Robins, Fire safety J. **22**, 185(1994).
4. S. Shibata, T. Higashino, A. Sawada, T. Oyabu, Y. Takei, H. Nabto, K. Toko, IEEJ Trans. on Sensors and Micromachines **130**, 38(2010).
5. W. Krüll, R. Tobera, I. Willms, H. Essen, N. Wahl, Procedia Engin. 45, 584(2012).
6. G. Heskeatad, J. S. Newman, Fire safety J. **18**, 355(1992).
7. N. G. Bernigau, H. O. Luck, J. Aerosol Sci. 17, 511(1986).
8. A. Scheidweiler, Fire Tech. **12**, 113(1976).
9. C. Helsper, H. Fissan, J. Muggli, A. Scheidweiler, Fire Tech. **19**, 14(1983).

Study of UHF Frequency Response Characteristics of the Semicircle Dipole Inner Sensor for Partial Discharge Detection in GIS

Wen SHI [1], Jing Gang YANG [2,a], Le GUO [1], Lei CHEN [1], Ming Jie JIANG [1]
[1] *Guodian nanjing automation co., LTD ,Nanjing 211106,Jiangsu, China*
[2] *Jiangsu Electric Power Company Research Institute, Nanjing 211103, Jiangsu, China*

Abstract. The paper designed a semicircle dipole inner sensor for UHF partial discharge detection in GIS. By comparing the symmetric semicircle dipole sensor and asymmetric semicircle dipole sensor characteristics, the symmetric semicircle dipole sensor's coupling property is better than that of asymmetric semicircle dipole sensor. At the same time, by changing the symmetric semicircle dipole transducer radius, the position of the feeder pole , the height of the feeder pole, analyze its influence on sensor features.

1 The introduction

In the mid - 1960 - s, the birth of GIS, makes the high voltage power equipment qualitative leap. In recent decades, the GIS with its compact structure, space and building cover an small area, the advantages of the high reliability and easy installation, obtained the rapid development [1]. At the same time, the safe and reliable operation for a long time also become the focus of attention. Studies have shown that [2-4], GIS has internal SF6 gas leakage, switch fault, internal discharge hidden trouble. In all kinds of fault, large of the failure caused by the insulation defect [5, 6]. Therefore, on-line insulation monitoring of GIS is very necessary.

UHF is more effective for monitoring GIS partial discharge, with strong anti-interference ability, high sensitivity of monitoring [7-9]. Depending on the location of the sensor installation, UHF sensor is divided into inner sensor and external sensor. Inner sensor installed in the body of the GIS, can effectively resist the outside electromagnetic interference, and has higher efficiency of sensor. As the special inner sensor installation position, it determines that the design and installation is better done when manufactured the GIS . If it really needs to install the inner sensor on equipments which have been put into operation, it only chooses at GIS hand hole, has certain limitations. For the GIS equipment has been put into operation, can install the external sensors, it is usually installed on the GIS basin insulator, through testing the leaked electromagnetic wave signal of the insulator to realize the partial discharge detection. The shortcoming of this method is the sensor works in the complex electromagnetic environment , vulnerable to the effects of electromagnetic radiation around. The anti-interference ability is poor. Visible, the inner sensor installed on GIS shell can obtain high detection sensitivity, and it has the characteristics of flexible installation, good safety.

[a] Corresponding author: huzi_yang@163.com

2 Measure of sensor performance metrics

Uhf inner sensor is a kind of uhf antenna which used to receive electromagnetic waves that produced by internal partial discharge of GIS . Therefore, we can use the antenna performance indicators to judging the performance of the sensor. For antenna input impedance, standing wave ratio, gain are the important parameters, so these parameters for evaluating the performance of the inner sensor is of great significance. However, from the perspective of monitoring partial discharge signals, we pay more attention to the sensor output voltage amplitude which is the sensor's output characteristic in the same field environment, Here, we introduce the sensitivity [10] as the main evaluation index of the sensor performance. To sum up, the input impedance and the sensitivity are the mainly measure of the sensor.

2.1 The input impedance

The sensor receive signals as a receiving antenna. The signals which coupling to the partial discharge that can be converted into voltage signal output, as the power of the peripheral receiving circuit, based on thevenin theorem, the circuit can be equivalent to the ideal voltage source and the resistance in series, a simplified circuit resistance can be understood as the input impedance of receiving antenna (sensor) looking from peripheral circuit terminal [11].

When the sensor input impedance matched with the impedance of peripheral circuit, it can achieve the maximum output power. Starting from the equipment, only when the sensor and the transmission lines realize the impedance matching, the peripheral circuit can receive the partial discharge signals that coupled by the sensor for maximum efficiency. From macro available, Impedance matching can be measured by reflection coefficient or standing wave ratio.

In radio frequency circuit, on a point of the transmission line, the reflection coefficient (voltage) is defined as the reflected wave voltage ratio the incident wave voltage, such as type 1.

$$\eta = \frac{U_2}{U_1} \qquad (1)$$

U_2 is the reflected wave voltage, U_1 is the incident wave voltage.

Standing wave coefficient (voltage) is defined as the signal antinode point voltage and the ratio of the node voltage, such as type 2.

$$\rho = \frac{|U|_{max}}{|U|_{min}} = \frac{1+|\eta|}{1-|\eta|} \qquad (2)$$

2.2 The sensitivity

Sensitivity is a kind of transfer function of the coupling characteristics of the sensor, it is to point to in a certain frequency range, when the incident electric field for the 1vm-1, the ability of the sensor output voltage [10]. Visible, the sensitivity is the most key performance indicators measure uhf inner sensor.

As a receiving antenna, sensor receives electromagnetic signals generated by partial discharge, converted to voltage signal and output. Thus, according to the practical significance of the sensitivity, the sensitivity of the sensor can be defined as: the amplitude frequency response of the sensor H_e as the sensor output voltage $U_{(f)}$ the ratio of the incident field strength $E_{(f)}$, the unit is m [10]. Such as type 3.

$$H_e = \frac{U(f)}{E(f)} \qquad (3)$$

H_e characterization of the ability of the sensor that converts the incident electric field to output voltage signal, Its amplitude is proportional to the sensor detection sensitivity. In the same frequency band, when H_e the larger, shows that the sensitivity of the sensor is the better, the sensor performance is fine.

3 Performance analysis of semicircle dipole inner sensor

Semicircle dipole sensor uses semicircle board instead of symmetric oscillator. Semicircle dipole inner sensor can be divided into symmetric semicircle dipole and asymmetric semicircle dipole according to the structural characteristics .As shown in figure 1, figure 2. Figure 1 shows simulation model of symmetrical semicircle dipole .Figure 2 is asymmetric semicircle dipole, its radius ratio of 1.4:1.

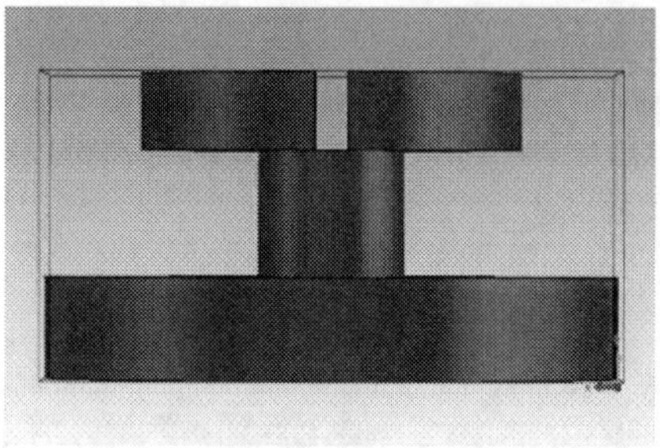

Figure 1. Symmetric Semicircle Dipole Inner Sensor

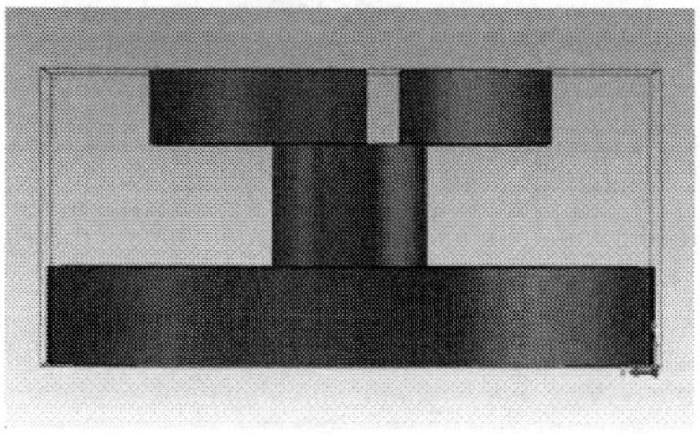

Figure 2. Asymmetric Semicircle Dipole Inner Sensor

Compared the sensitivity characteristics and standing wave ratio of the symmetric semicircle dipole and asymmetric semicircle dipole semicircle inner sensors, figure 3 shows the sensitivity graph of the two sensors, figure 4 is the standing wave ratio characteristic of the two sensors. It can be seen from the figure 3 that the coupling characteristics and coupling frequencies of the symmetrical semicircle dipole sensor are more accord with the GIS internal wave spectrum characteristics of partial discharge. From figure 4,it can be seen that the start frequency which the value less than 2 of the standing wave ratio curve of the symmetrical dipole is lower than that of the asymmetric dipole.

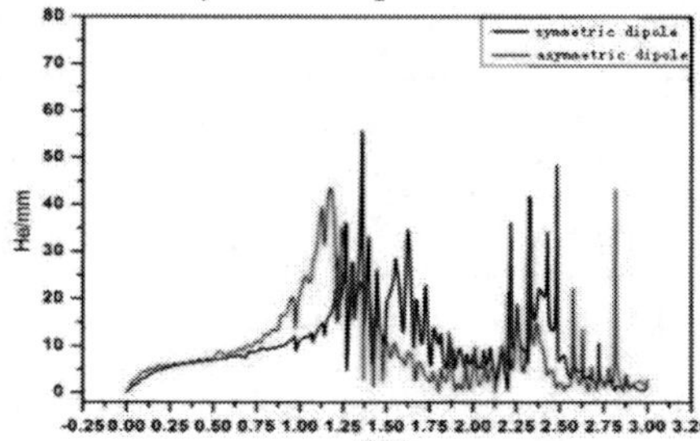

Figure 3. Contrast Sensitivity

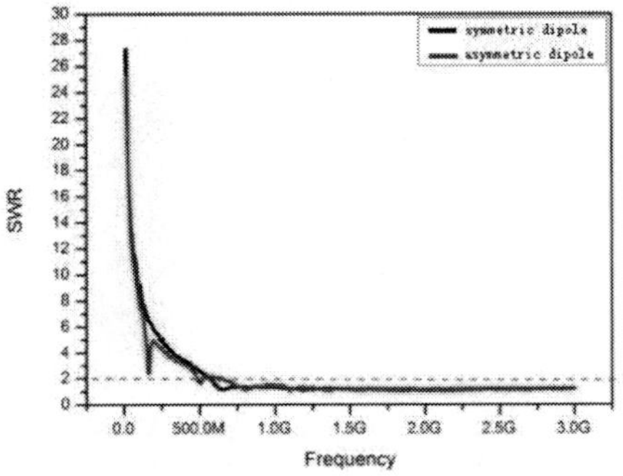

Figure 4. Contrast Standing Wave Ratio

4 Characteristics analysis of symmetrical semicircle dipole transducer

The factors that affect the semicircle dipole transducer characteristics are the radius of the round plate, the feeder pole position , The location of the semicircular shell from the GIS (the height of the feeder pole), etc. By simulation analysing the influence of these factors on the sensor performance.

4.1 Analysis the influence of radius of semi-circle plate of sensor characteristics

Compared the sensitivity of the symmetric semicircle dipole sensor when the semi-circle plate radius are 50 mm , 75 mm and 90 mm, shown in figure 5.

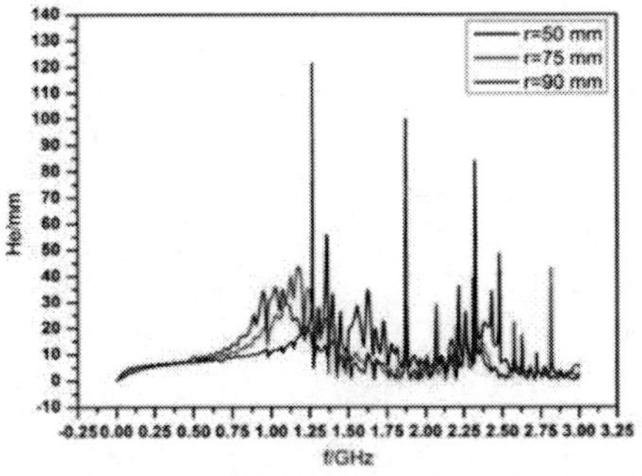

Figure 5. Effecting Sensitivity by Semi-circle Plate Radius

From figure 5 , along with the increase of the radius of semi-circle plate sensor, the peak of the sensitivity moves to the lower frequency band, In combination with the frequency characteristic of GIS internal partial discharge signals,it can be seen that when the radius of semicircle dipole is 90 mm, conform to the principle of design.

4.2 Analysis the influence of the feeder pole location of sensor characteristics

The feeder pole position of sensor sensitivity has a certain impact. Compared the sensor sensitivity when feeding point in the center of the symmetric semicircle dipole and 10 mm from the center. The sensitivity curve as shown in figure 6. From the figure, for symmetrical dipole senor, when center feeding, its sensitivity curve peak is almost four times higher than that of noncentral feeding. Visible, using the center feeding can more improve the sensitivity of the sensor.

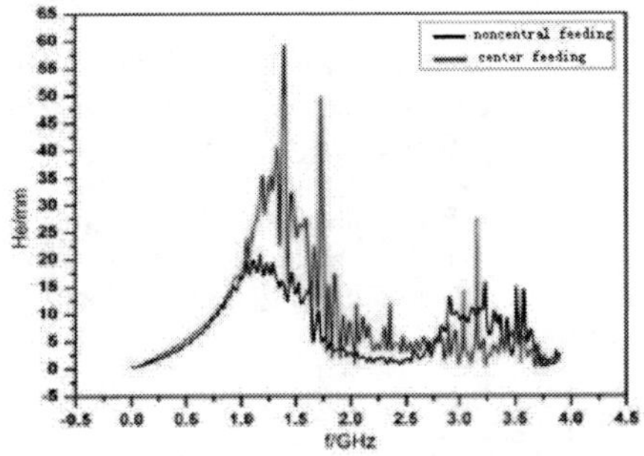

Figure 6. Effecting Sensitivity by feeding point location

4.3 Analysis the influence of the feeder pole height of sensor features

The distance of the symmetric semicircular dipole sensor from the GIS shell (the height of the feeder pole) has great effects on the sensitivity of the sensor. Compared the sensitivity influence of the symmetric semicircle dipole sensor when the height of the feeder pole are10 mm, 20 mm, 30 mm and 40 mm. Shown as figure 7.

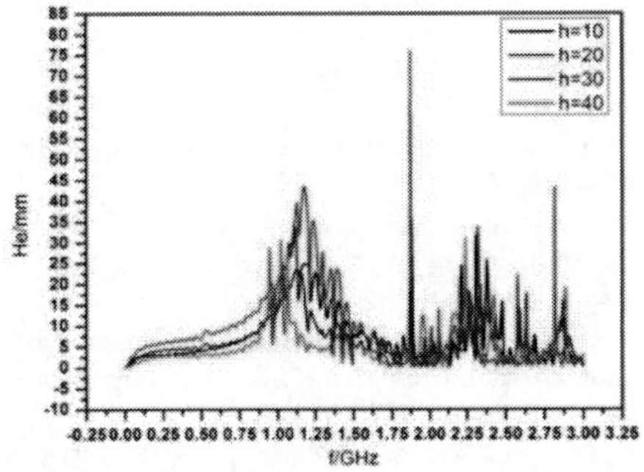

Figure 7. Effecting Sensitivity by feeder pole height

From the above, when the height of the feeder pole is 20 mm, the sensitivity characteristic of the sensor amplitude is the highest, the coupling characteristics is the best.

5 Summary

In this paper, compared the symmetry semicircle dipole and the asymmetric half inner sensors. At the same time, the influence sensitivity factor of symmetrical semicircle dipole inner sensor was analyzed, as the following conclusion:

The symmetric semicircle dipole sensor's coupling property is better than that of asymmetric semicircle dipole sensor.

The symmetric semicircle dipole sensor 's radius, the feeder pole position and the height of the feeder pole are affect the sensitivity of the sensor characteristics.

With the increase of radius of semi-circle plate, the peak of the sensor sensitivity moves to the lower frequency band.

For symmetrical dipole senor, when center feeding, its sensitivity curve peak is almost four times higher than that of noncentral feeding.

Visible, for the symmetrical semicircle dipole inner sensor, when the radius of semi-circle plate for 90 mm , using center feeding and when the height of the feeder pole is 20 mm, the sensitivity of the sensor characteristics is the best.

References

1. Qiu Yuchang.Beijing: water conservancy electric power publishing,34-89(1994).
2. Barnett R.Paul,Nelson Jeffrey H.2010 International Conference on High Voltage Engineering and Application New Orleans, 676-679(2010).
3. You LinHui. Mechanical and electrical information, **30,** 67-68 (2010).
4. Zheng-Guo zhu,Jian-Wu jiang,Jian-Ling zhong,et. High voltage electric equipment, 46-48， **5**, 98-102(2010).
5. Braun J.M.,Fujimoto N.,Rizzetto S. Proc 3 Int Conf Prop Appl Dielectr Mater:1136-1139（1991）.
6. Li-jun jin, Wei-dong liu. Journal of China power, 35 (3) : 52-55（2002）.
7. Schichler U.,Gorablenkow J.,Diessner A . Eighth International Conference on Dielectric Materials,Measurements and Applications:139-144（2000）.
8. Lee Seung-Chul,Jung Seung-Yong,Koo Ja-Yoon,et al. Transactions of the Korean Institute of Electrical Engineers, 57(2):252-255（2008）.
9. Meijer Sander,Judd Martin D., Proceedings of 2008 International Conference on Condition Monitoring and D iagnosis:1031-1035（2007）.
10. M.D.Judd,O.Farish,J.S.Pearson.IEE Proc.-Sci, Meas, Technol.,Vol.144,No.3,May 1997.
11. Jian-sheng wang,Yu-chang qiu,Xiang-hua wu,et. Proceedings of the csee, 20 (8) : 42 to 45（2000）.

MATEC Web of Conferences 63, 01017 (2016) DOI: 10.1051/matecconf/20166301017
MMME 2016

Analysis of partial discharge Ultrasonic wave characteristic on typical Defects in GIS

Jing Gang YANG[1,a], Wen SHI[2], Hong Tao LI, Bin GONG [1], Wei Yi JIANG [1]

[1] *Jiangsu Electric Power Corporation, Nanjing 211103, Jiangsu , China*
[2]*Guodian nanjing automation co., LTD ,Nanjing 211106,Jiangsu, China*

Abstract. For analyzing partial discharge ultrasonic wave characteristic on typical defects in SF6 gas insulation switchgear (GIS), a 110 kV GIS was taken as an object to design and simulate the typical partial discharge models in GIS. There are point electrode model on bus and cavity of GIS, floating electrode model, air interstice model in insulator, metal particle and iron core electromagnetic shock models. PD signals on different defects were thus detected with the ultrasonic method. The experimental results indicated that the ultrasonic signals of different defects showed different characteristics in waveform, amplitude and phase resolved partial discharge chart, which is helpful in PD pattern recognition for GIS.

1 Introduction

Please submit sources files directly to the conference organizer. If the conference editors chose to provide print-ready PDF documents to the publisher, you have to submit high-resolution PDF file with all fonts embedded (see PDF guidelines) instead of the sources files. In this case, please remember that no final corrections will be made by the publisher.

Gas insulation switchgear (GIS) as an important equipment in the power system operation, in the event of failure, will cause huge economic losses and adverse social impact. Therefore, the operation reliability of the GIS is very important. The GIS internal insulation defect of the vast majority comes from the development of partial discharge, so the detection of partial discharge is an effective means of preventing GIS insulation fault[1-3].

Ultrasound is one of the hot methods of domestic and foreign research of GIS partial discharge [4], this method diagnosis of GIS by detecting partial discharge for ultrasonic signal, , so as to realize the accurate extraction of partial discharge signals, can be more comprehensive research intrinsic characteristics of partial discharge[5].

2 Ultrasonic detection of partial discharge

The final printed area will be 172 x 252 mm. In partial discharge in GIS can be induced in charge neutralization process, and the corresponding produce steep pulse current, the effect of electric current pulse will make the local area which occurs the partial discharge expanding and instantaneous heat, forming a similar explosion, the expansion of the heated region back to the original size after discharge, the density of the medium in an instant change caused by the partial discharge which produced a higher shrinkage volume change, form the ultrasound, in the form of spherical wave from the partial discharge point spread to all around, so when partial discharge occurs is also accompanied by ultrasound. On the other hand, ultrasound can also detect mechanical wave produced by the particle motion, on general engineering application of the ultrasonic to distinguish the motion of particles.

Combines both the ultrasonic method and digital measurement, this paper aims to detect partial discharge model typical discharge or free particles to beat the ultrasonic signal, after digital processing, spectra and statistics can be obtained, for further study of different defect types of partial discharge characteristics to provide data and test basis.

3 Typical flaws with the design of the detection system

GIS internal reason of the failure of tend to be various, in the manufacturing and assembly process, GIS will inevitably leave some minor defects. According to the relevant literature and the experience of field test, Produced high-voltage bus for a metal spike, wall bulging metal spikes, structure as shown in figure 1.

[a] Corresponding author: 15105168828@163.com

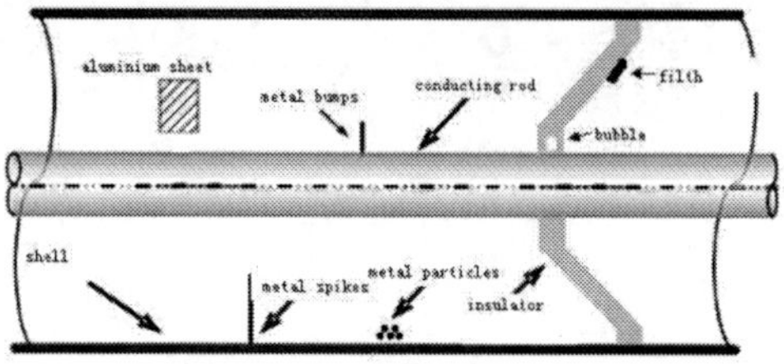

Figure 1. GIS common defects schematic

Five kinds of common defects of GIS model designed, including a high-voltage bus metal spike, GIS wall metal spike， floating potential, insulator within the air gap and free metal particles etc. The metal spike model is a steel needle which is 53 mm length, the equivalent radius of curvature of the cutting edge is 100 μ m; Metal suspended potential defects by polyethylene insulation screw to fix the metal aluminium on the high voltage bus bar, the two are separated by a distance (test respectively use 1 mm, 1.5 mm, 2 mm and 4 mm gap distance), insulating screw can be used to adjust the clearance between guide rod and suspended aluminum distance.

Typical defects of GIS experiment and testing system of ultrasonic signal main structure schematic is shown in figure 2. The system is mainly composed of the maximum output voltage is 250 kV AC testing transformer which without partial discharge, protection resistance Rz (water resistance), coupling capacitance Ck (85 pF), detection impedance (50 Ω), C1 and C0 divider, 110 kV gis test section which built-in defect can be replaced (Cx), AIA ultrasonic signal detector and pulse current method of detection system. Pulse current method as an auxiliary means, is used to compare and confirm collection for discharge signals.

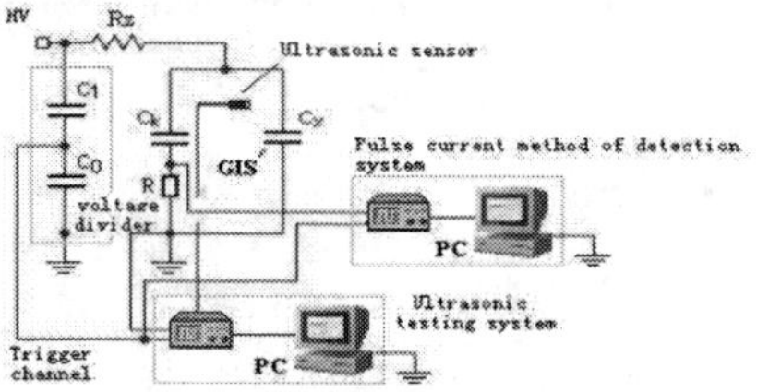

Figure 2. Test system structure diagram

Install the defects in GIS cavity in the body and the SF6 gas charging 0.45 MPa when test.

4 Partial discharge characteristics of defects

4.1 Busbar metal spike defects

Due to reasons such as processing or installation, the GIS may form metal protrusions on high voltage busbar, spike part of its easy to form a strong field intensity area that can make the surrounding gas ionization, then producing partial discharge, its long-term effects can destroy GIS insulation condition[6].

When test, the voltage from low to high, until stable the measured ultrasonic signal, record its initial discharge and strong discharge PRPD (phase - resolved partial discharge, partial discharge signals phase map) spectra,as shown in figure 3.

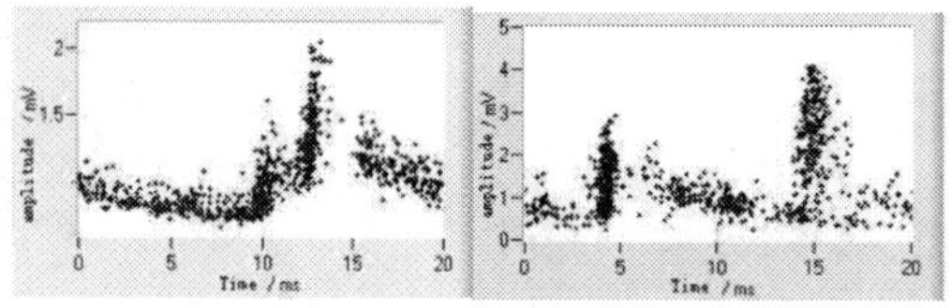

(a) starting discharge stage (b) strong discharge stage

Figure 3. Initial discharge and strong discharge PRPD spectra

From PRPD spectra can be found in this type of discharge in two stages: only negative half cycle of power frequency appear a small amount of discharge pulse and the positive half cycle basic not discharge during low voltage, the cause of this phenomenon is that the positive polarity, tip quality of positive ions ionized to larger then moving slowly ,but electrons to cutting-edge mobile quickly. At high voltage, gas ionization is very intense when the positive half cycle power frequency, a lot of positive ions to be near plate equivalent reduced the plate spacing, Continue to ionization and form electron avalanche discharge and lingers, which may arise in the positive half cycle breakdown produces large amplitude of discharge pulses. Partial discharge level increased, the positive half cycle also appeared a lot of discharge pulse, and the positive and negative half cycle of discharge pulse in power frequency phase peak near the positive and negative.

4.2 GIS wall metal spike defects

GIS shell wall connected to the earth, but due to the presence of metal spikes and make the electric field distortion, the spikes in the electric field makes the SF6 gas ionization occurred around, causing partial discharge. When the experiment, the applied voltage from low to high, recording the initial discharge and strong discharge PRPD spectra as shown in figure 4.

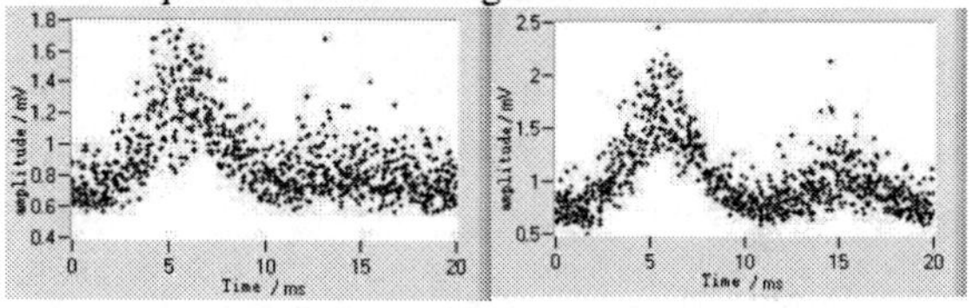

(a) starting discharge stage (b) strong discharge stage

Figure 4. Initial discharge and discharge strong PRPD spectra

It can be see that signal mainly concentrated in the positive half exemption value nearby when the initial discharge, as voltage continues to rise, discharge and discharge frequency continues to increase, and the negative half cycle also appeared some discharge pulse, phase concentration near the 270 °.

When the metal spikes, therefore, the location is not at the same time, the cause of the partial discharge PRPD spectra will present different characteristics due to the polarity effect, in the low voltage, phase spectra show unimodal pattern, when the voltage is higher, two phase spectrum of discharge of the graph is bimodal pattern, but the amplitude and density of the signal distribution difference.

4.3 GIS built-in suspension potential defects

Found in the process of experiment, the size of the suspended electrode gap distance on the initial discharge voltage and discharge repetition rate has a significant impact, but little impact spectra characteristics of discharge, so this article only to the discharge gap distance to the defect of 2 mm model data for analysis.

Suspended potential defects caused by partial discharge is in the nature of the local breakdown. Because floating potential and high voltage bus form a capacitor, when both ends of the equivalent capacitance voltage exceeds the withstand voltage, high voltage busbar and gaps between floating electrode was breakdown, partial discharge. When the applied voltage is higher, defects in a power frequency cycle may occur many times charge discharge process, appear many times breakdown phenomenon. Records of initial discharge and strong discharge PRPD spectrum diagram as shown in figure 5 .

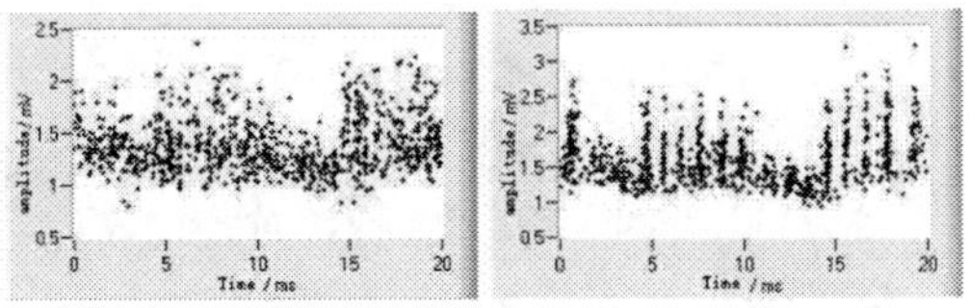

(a) starting discharge stage (b) strong discharge stage

Figure 5. Iinitial discharge and discharge strong PRPD spectra

4.4 The basin insulator internal air bubble defect

Solid dielectric in high voltage electrical equipment for insulation effect and mechanical strength of equipment, solid insulation gas gap (bubble) is the most common type of defect, when solid insulation appear dielectric defects high voltage equipment insulation will breakdown, causing serious accidents.

The experiment results show that the bubble size has a certain influence on the initial discharge voltage and discharge repeat rate, but has less effect on the discharge signal characteristics, so this article choose size 2 mm (thickness) x 1 mm (diameter) of the bubble experiment results to analyzing.

Considering the dispersivity of the discharge and the discharge mode of time accumulated, experiments on samples can step up to voltage until the first discharge signals collected, then the stability of this voltage applied of 10 minutes, and then gradually increase the voltage, each level voltages remain for 10 minutes or so on the discharge phenomenon. Record the starting discharge and strong discharge PRPD spectrum diagram as shown in figure 6.

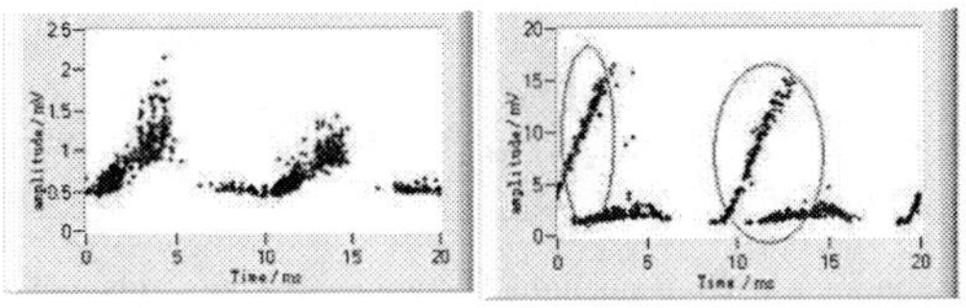

(a) starting discharge stage (b) strong discharge stage

Figure 6. Initial discharge and discharge strong PRPD spectra

PRPD spectra showed obvious phenomenon of "rabbit ear" (in part). Due to near zero air gap applied electric field in applied voltage polarity reversal, and the air gap within the dual pole field intensity in the same direction, the superposition of two field intensity in the air gap within the field strength increase, and discharge, so the "rabbit ear" signal spectra.

With the increase of applied voltage, "rabbit ear" part becomes more obvious, the spectra of the air gap discharge mode change is not big, generally considered as the partial discharge, air gap of defects are happening gas molecules within the discharge decomposition, produce the electronegativity molecules, such as CO_2, H_2O, on the one hand these molecules influence the gases in the air gap discharge effect, on the other hand, with medium produces chemical reaction, generate crystalline objects attached in the surface of the medium, change the conductivity on the surface of the medium. Constant bombardment due to discharge of molecular and electronic media surface, the medium surface roughness increases, the influence on the discharge space charge accumulation, lead to the change of discharge spectra. Comparing the spectra of different voltages can also be found that with the increase of applied voltage and the phase of "rabbit ear" followed wide.

4.5 Free metal particles defect

Free particle impact cavity and partial discharge occurs in flight will produce ultrasonic, this article only on the particles of a study on ultrasonic features of partial discharge.

The experiment results show that the material, the size of the metal particles will affect the intensity of particles to beat, the smaller size, The lighter material, the more light beat the strong, but in the event of beating, the ultrasonic signal features are basically identical, so this article choose the experiment results of aluminum particles which diameter is 2 mm to analyze.

Recorded the initial discharge PRPD spectra and flight figure as shown in figure 7.

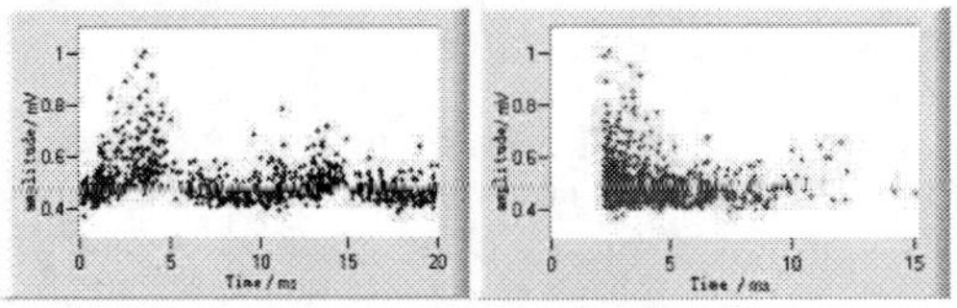

(a) PRPD spectra (b) flight figure

Figure 7. start discharging the PRPD spectra, flight figure

Initial discharge, the signal is concentrated in the power supply voltage is near the peak, the signal amplitude is smaller, particle flight jitter is not active, discharge and discharge time interval between the statistics are not relevant, no obvious characteristics on the drawing in the flight. This is because when the voltage is not high, the metal particles are not violent beating, a model is equivalent to a wall in metal bumps defect model, at this point the first discharge signals appeared in the positive peak of the power supply.

With the applied voltage rise gradually, the particles gradually occur beating, discharge more and more intense, recorded the PRPD spectra and flight diagram of the starting discharge, as shown in figure 8.

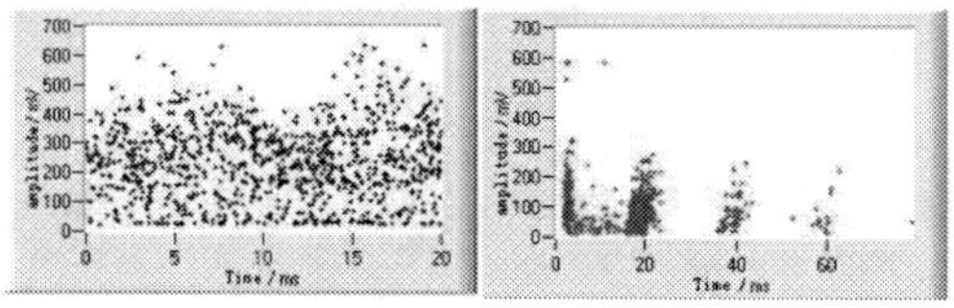

(a) PRPD spectra (b) flight figure

Figure 8. metal particle movement PRPD spectra, flight figure

Ultrasonic pulse signal when the free particle free beats more evenly distributed in 0 ~ 360 ° phase cycle, more than 90% of the measurement results are similar , so it can be said that the free movement of particles in the free beats of large signal amplitude and no obvious phase correlation is one big characteristic. Although particle collision shell moment has nothing to do with the power frequency phase, but ultrasonic signal amplitude, time of flight are showing strong cyclical change, the flying time is about 20 ms.

This is because when particles collide with the metal case, Immediately get to have a new charge up charge force upward movement, if ignore flight discharge lead to loss of charge occurs, an odd number of times after half cycle power frequency particles will receive a charge in the opposite direction, when the maximum charge force and gravity in the same direction (pointing to the electrode), the impact has great momentum to detect ultrasonic signal also reached maximum, so the flight figure carrying flight time and signal amplitude comparison of regular periodic variation characteristics.

5 Discussion

Through this study, can get the following typical defects of GIS partial discharge of ultrasonic signal features:

1) metal spike defects : When the metal spikes in different places, its cause partial discharge PRPD spectra will present different characteristics due to the polarity effect, in the low voltage, phase spectra show unimodal pattern, the high voltage, phase spectrum of two kinds of discharge the graph is bimodal pattern, but the signal amplitude and the signal density distribution difference;

2) suspension potential defects: partial discharge is mainly focused on the falling edge of the power supply voltage, as the voltage continues to rise, phase appears to have the phenomenon of zero drift, and appear the phenomenon of multiple discharge - discharge extinction - discharge;

3) Insulator internal air bubble defect: the biggest feature of the air bubble defect is in positive and negative half cycle rise along their "rabbit ear" shape distribution, and with the increase of voltage "rabbit ear" amplitude increases, the trend of "rabbit ear" starting phase decreases;

4) Free metal particles defect: particles without movement, the equivalent of the spikes defect on the grounding body, partial discharge signals show its discharge characteristics of thorns; Discharge occurs When the particle movement, defect discharge occurs in the whole power frequency cycle, power frequency ultrasonic signal amplitude, time of flight but showed strong cyclical changes;

GIS partial discharge is a complicated process, encountered in the process of ultrasonic wave in the spread of disorder will produce a series of reflection and refraction, so using ultrasonic wave method to detect internal defects of GIS still needs further research and exploration.

References

1. Schei A, Kyrkjeeide S, Larsen V. Asia Pacific. IEEE, Yokohama, 919–924 (2002).
2. Li Da-jian,Liang Ji-zhong,Bu Ke-wei,Yang Jing-gang,Li Yan-ming.High Voltage Apparatus,45(1):72-75 (2009).
3. Jin Yi, Liu Wei, Zha Xian-guang, et.Jiangsu Electrical Engineering, 31(2):12-15 (2012).
4. Lundgaard LE. Partial discharge XIII. IEEE Electrical Insulation Magazine,8 (4): 10.1109/1157.145095 (1992).
5. Wang Jianming, LI Pen.Jiangsu Electrical Engineering, 31(6): 62-65 (2013).
6. Zhou Yun-feng, Zhang Zi-yang, Yang Jing-gang,et.Jiangsu Electrical Engineering,31(4):3-5 (2012).

MATEC Web of Conferences 63, 01018 (2016) DOI: 10.1051/matecconf/20166301018
MMME 2016

Design of disk type PM synchronous generator based on halbach

Chuan ZHANG[1], Shu Qin LIU[1,a]

[1]School of Electrical Engineering, Shandong University, Ji'nan 250061, Shandong Province, China;

Abstract. This paper expounds the application merits of disk type PM synchronous generator in small-scale wind power generation .The characteristics of the disk type PM synchronous generator were analyzed and 8P, 200W disk type generator with 90 degree Halbach magnet structure was designed. Simulation and calculation are carried out by using finite element method. By comparing the 90 degree Halbach magnet structure and the traditional magnet structure of the disk type generator, the results shows that the 90 degree Halbach magnet structure can improve the air gap magnetic density, reduce the leakage of main magnetic circuit and the loss.

1 Introduction

Disc type permanent magnet synchronous motor with axial flux characteristics plays an irreplaceable role in many fields. Various studies have been made about the axial permanent magnet motor in the application of high speed motor and generator [1-4], also it has a wide application in electric vehicle [5-6], flywheel energy storage [7], driven ship elevator [8], and ship drive[9]. A coreless axial-flux machine with Halbach structure in direct-drive wind power generation field has the following advantages: (1) small axial size, high power density, simple structure; (2) coreless structure has no cogging effect which can effectively reduce the motor vibration and noise caused by the cogging; (3) the magnetic pole has a special Halbach structure which hold better sine degree of air gap magnetic density, meanwhile, avoiding the additional loss caused by harmonic. Thus, the study on the coreless disc generator with Halbach structure in the direct drive wind power generation field has important practical significance.

The main contributions of this paper can be summarized as follows. Regarding to the Halbach array structure of coreless double disc rotor permanent magnet generator, the structure and corresponding design features were analyzed. And, the three dimensional electromagnetic finite element calculation of the designed 8P and 200W disk motor is carried out ; The characteristics of magnetic field distribution of motors are studied through the analysis of the finite element field plots results and numerical results; The influence of different Halbach magnetic pole structure and magnetic pole thickness on the air magnetic flux and peak value is studied as well.

2 Structural design of coreless, double external rotor disc type permanent magnet generator with Halbach array

2.1. Structure of magnetic circuit Based on Halbach

Coreless double exterior rotor disc type generator is composed of the pole, the back iron , armature windings, bearings and other structures. Magnet is made of special structural configuration of Halbach array and it is attached to the back of a certain thickness of ductile iron that has a strong magnetic permeability. Armature winding consists of multi-strand wire wrapped package in accordance with a certain pitch. The structure of disc type generator with Halbach array shown in figure 1. As shown in figure.1 （a）, each magnetic pole is composed of axial magnetization pole a and auxiliary magnetic pole b which is perpendicular to a. Every four different polarization direction of the magnetic poles formed a set of loop. The material of back iron is made of nodular cast iron which can ensure the strength of the rotor as well as lower the cost of machine. Windings were wrapped by using a tool specially designed for this machine. After the windings were made we encapsulated the windings into epoxy resin which can leave out the iron core and solve the problem of cogging torque when the speed of wind power generate is low.

[a] Corresponding author: lshuqin@sdu.edu.cn

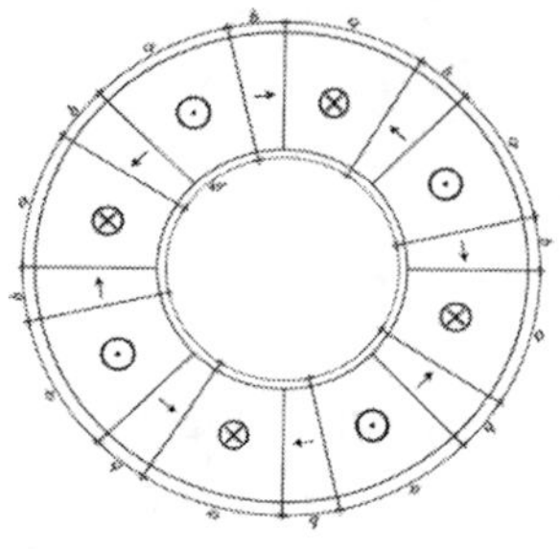

(a)schematic diagram of magnet with Halbach array

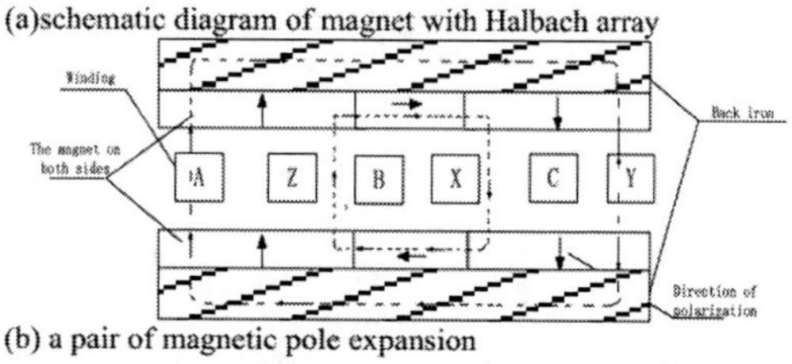

(b) a pair of magnetic pole expansion

Figure 1.the expand structure of disc type generator with 90° Halbach array

As shown in figure 1(b),the path of main magnetic flux are marked with dotted lines. In figure 1 (b), there are a pair of upper and lower rotor NS poles respectively. It is shown in the figure that the main magnetic circuit starts from the N pole on the right side of the rotor to the S pole in lower rotor through the winding, and then gets to the N pole on the left side of the rotor. Then it gets through the wingding again into the S pole in upper rotor, and finally returns to the N pole on the right rotor to complete a closed magnetic circuit.

2.2. Determination of inner and outer diameter of double external rotor disc type generator

The direction of magnetic field line through stator winding is axial direction and the conductor which parallel to the pole is the effective part. The position of a single conductor in the plane is represented by r and θ as shown in figure 2.

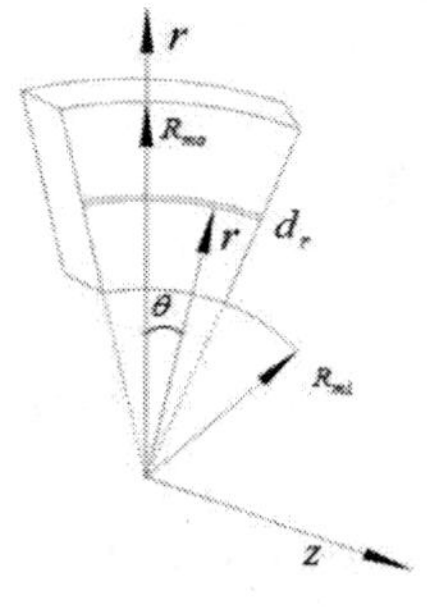

Figure 2. electromagnetic calculation diagram

Air gap magnetic density at the mean radius of generator is represented by $B_\delta(\theta)$. The mechanical angle is represented by Ω .The average electromotive force of each conductor can be expressed by[4] :

$$\frac{p}{\pi}\int_0^{\pi/p} e\,d\theta = \frac{1}{8}\Omega(D_o^2 - D_i^2)\frac{p}{\pi}\int_0^{\pi/p} B_\delta(\theta)d\theta = \frac{1}{8}\Omega B_{\delta av}(D_o^2 - D_i^2) \quad (1)$$

I_Φ is the effective value of phase current and E_Φ is effective value of phase voltage. Phase number is represented by m . Air gap flux density is represented by B_δ .Average air gap magnetic density in each pole distance is $B_{\delta av} = \alpha_i B_\delta$. The power of the generator p can be expressed by[4]:

$$p = mE_\Phi I_\Phi = \frac{\pi}{2\sqrt{2}} mE_{\Phi av} I_\Phi \quad (2)$$

The electrical load A_{av} at mean radius of the generator can be expressed by[4]:

$$A_{av} = \frac{4mNI_\Phi}{a\pi(D_0 + D_i)} \quad (3)$$

Put formula (3) into formula (2) :

$$p = \frac{\pi^3}{960\sqrt{2}} n\alpha_i k_w B_\delta A_{av}(D_o^2 - D_i^2)(D_o + D_i) \quad (4)$$

Set $D_o = \gamma D_i$

$$p = \frac{\pi^3}{960\sqrt{2}} n\alpha_i k_w B_\delta A_{av}(1 - \frac{1}{\gamma^2})(1 + \frac{1}{\gamma})D_o^3 \quad (5)$$

Set D_o, A_{av} and B_δ as constant and γ as the variable. Taking the derivative of the last formula, it can be seen that the output is maximized when the Ratio of inner and outer radius is set to$\gamma = D_o / D_i = \sqrt{3}$ [10]。Also the motor inertia, the magnetic flux leakage factor, efficiency and other factors should be considered synthetically. Generally speaking, the ratio of inner and outer diameter of the axial flux disc type permanent magnet generator is between 1.5-2.2 [11].

3 Magnetic field simulation

3.1. Basic parameters of generator

The rated speed of the disc type generator is 500 rpm, and alternating current with 50Hz is desired when the speed is between 350 rpm and 500 rpm. According to formula $p = \frac{60f}{n}$, the values of p range from 6 to 8, and it set to 8 in this article. The basic parameters of double exterior rotor disc type generator is shown as Table 1.

Table 1. The basic parameters of double exterior rotor disc type generator.

Item	Value
Rated Power/ W	200
Rated Voltage/ V	50
Rated Speed/ rpm	500
Poles	8
Rated Frequency/ Hz	50
Rotor Inner Diameter/ mm	110
Rotor External Diameter/ mm	230
Rotor Back Iron Thickness/ mm	6
Halbach Degree	90°
Magnetic Materials	NdFeB N50

3.2. Simulation and analysis of magnetic field of disk type generator

The 3-D model of disc type generator can be build based on the parameters given in Table 1 and the simulative calculation for electromagnetic field can be made. The finite element model of generator is shown in figure3. Halbach magnet structure is shown in figure3 (a) and traditional pole structure is shown in figure3 (b). In order to observe the air gap flux density only the lower Halbach pole group and back iron are left in figure3.

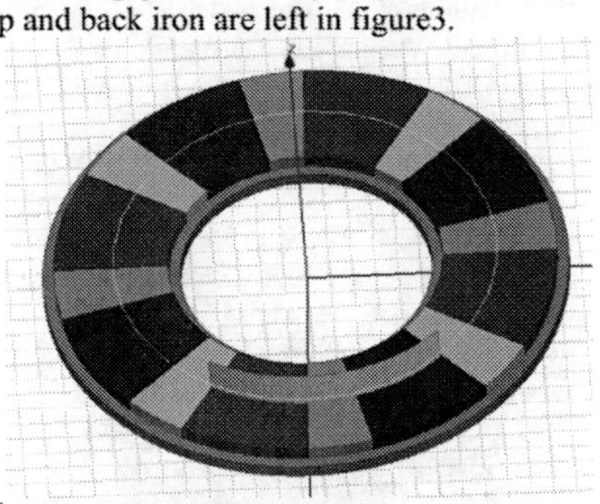

(a) Halbach magnet structure

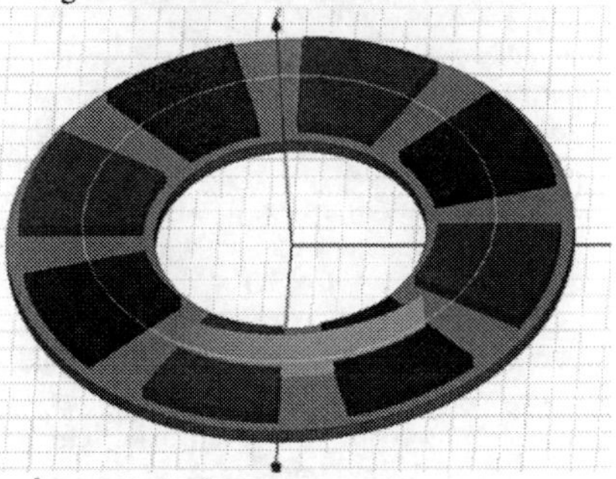

(b) traditional magnet structure

Figure 3. finite element model of disk type generator

The section of the center line in Figure 3 is the position of the magnetic field analysis. After the calculation, the scalar diagram corresponding to the section in figure 3 is shown in figure 4. The air gap magnetic density of the 90 degree Halbach magnet structure and the traditional magnet structure all have the characteristics of uniform distribution. In the case of the Halbach magnet structure, the maximum magnetic density of the back iron is about 1.8T. In the case of the traditional magnet structure, the maximum magnetic density of the back iron is about 2.2T which presents the saturation phenomenon. Because of the magnetic flux is shared by the he Halbach magnetic pole structure, the magnetic density in the back iron part is small in the case of Halbach magnet structure.

The section of the center line in Figure 3 is the position of the magnetic field analysis. After the calculation, the scalar diagram corresponding to the section in figure 3 is shown in figure 4. The air gap magnetic density of the 90 degree Halbach magnet structure and the traditional magnet structure all have the characteristics of uniform distribution. In the case of the Halbach magnet structure, the maximum magnetic density of the back iron is about 1.8T. In the case of the traditional magnet structure, the maximum magnetic density of the back iron is about 2.2T which presents the saturation phenomenon. Because of the magnetic flux is shared by the he Halbach magnetic pole structure, the magnetic density in the back iron part is small in the case of Halbach magnet structure.

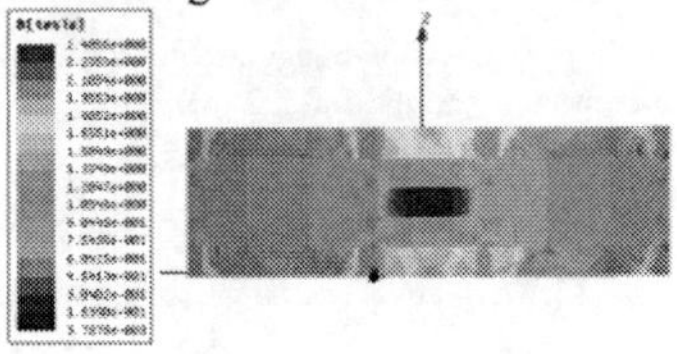

(a) the magnetic flux density corresponding to the section in figure3(a)

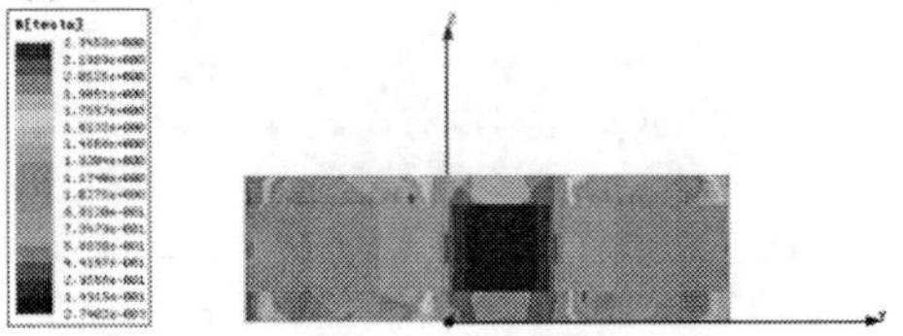

(b) the magnetic flux density corresponding to the section in figure3(b)

Figure 4. magnetic density distribution map of the disk type generator with Halbach magnet structure or traditional magnet structure

3.3. Disk type generator air gap magnetic density of Halbach magnet structure

For easy quantitative analyzing the air gap magnetic density, it draw a center line of the air gap that direction is perpendicular to the axial in the middle of the air gap as shown in figure 3. The pole-arc coefficient of the magnet is 30°/45°.In other words , the axial magnetization magnetic pole angle is 30 degrees. Axial magnetic density of the air gap on the center line of the curve is shown in figure 5 computed by the magnetic field. As can be seen from figure 5,axial magnetic density vector curve has four peaks and four low ebb. It meets the characteristics of the 4 S and 4 N poles of the disc type generator. In the case of the Halbach magnet structure which is corresponding to figure 3(a), The range of the magnetic density peak is 0.80T-0.85T. In the case of the traditional magnet structure which is corresponding to figure 3(b), The range of the magnetic density peak is 0.72T-0.76T. The width of the magnet on the air gap center is 43mm. In the case of the Halbach magnet structure, the peak width of air gap magnetic density is about 41mm. In the case of the traditional magnet structure, the peak width of air gap magnetic density is about 26mm.

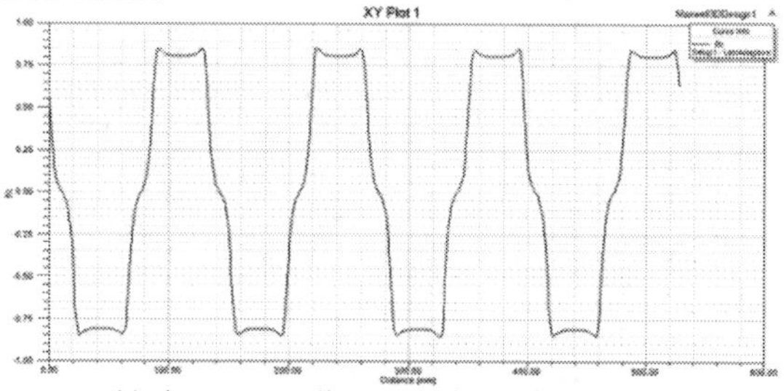

(a) air gap center line magnetic density curve corresponding to the figure3 (a)

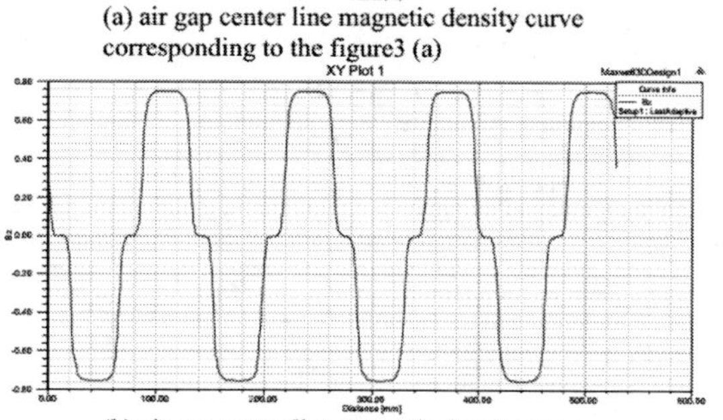

(b) air gap center line magnetic density curve corresponding to the figure3 (b)

Figure5. magnetic density curve when the axial magnetization magnetic pole angle is 30 degrees

4 Conclusion

In this paper, the design of the disc type generator is studied from the aspects of the structure and application of the small direct drive wind power generator. The calculation formula of the stator winding of the disc motor is induced, and Halbach magnet structure stator magnetic circuit is designed. Using finite element method to analyze the magnetic field of disk type generator with 90 degree Halbach magnet structure and traditional magnet structure, we can get the following conclusions:

1. In the case of the Halbach magnet structure, the maximum magnetic density of the back iron is about 1.8T while in the case of the traditional magnet structure, the maximum magnetic density of the back iron is about 2.2T. Obviously, using Halbach magnet structure can effectively reduce the degree of magnetic saturation of the back iron as well as the magnetic loss.

2. In the case of the Halbach magnet structure the range of the magnetic density peak is 0.80T-0.85T while in the case of the traditional magnet structure the range of the magnetic density peak is 0.72T-0.76T.Thus,using Halbach magnetic pole structure can effectively improve the air gap density.

3. In the case of the Halbach magnet structure, the peak width of air gap magnetic density is about 41mm which is account for 95% of the width of the magnetic pole while in the case of the traditional magnet structure, the peak width of air gap magnetic density is about 26mm which is account for 95% of the width of the magnetic pole. Thus, the magnetic flux leakage of the main magnetic circuit can be reduced effectively by using the Halbach magnet structure.

Supported by The Fundamental Research Funds of Shandong University : 2015JC042. Natural Science Foundation of Shandong province: ZR2012EEM031.

References

1. Sadeghierad M, LesaniH,Monsef H, et al. High-speed axial-fluxpermanent-magnet generator with coreless stato[J].Canadian Journal Electrical and Computer Engineering,2009,34(2):63-67.
2. R. L. Ficheux, F. Caricchi, F. Crescimbini and O. Honorati, "Axial-flux permanent-magnet motor for direct-drive elevator systems without machine room," in IEEE Transactions on Industry Applications, vol. 37, no. 6, pp. 1693-1701, Nov/Dec 2001.
3. B. J. Chalmers, W. Wu and E. Spooner, "An axial-flux permanent-magnet generator for a gearless wind energy system,"Power Electronics, Drives and Energy Systems for Industrial Growth, 1996., Proceedings of the 1996 International Conference on, New Delhi, 1996, pp. 610-616 vol. 1.
4. Wen-Shan Zhang.Research on Axial Flux Generator for Small-Scale Vertical-Axis Wind Power Application [D].2012
5. JavadiS ,Mirsalim M. A coreless axial-flux permanent-magnet generator for automotive applications[J].IEEETransationson Magnetics, 2008 44(12): 4591-4598.
6. Li Lina-Research on the axial-radial flux compound-structure permanent-magnet synchronous machine (C SPM SM) used for HEV [D].Harbin Institute of Technology, 2008
7. Wu Qian. Research on an axial / radial-axial flux compound-structure permanent-magnet synchronous machine used for hybrid electric vehicle[D].Harbin Institute of Technology,2008
8. Yang Zhiyi. Research on key techniques of energy storage of flywheel battery [D].Hefei University of Technology, 2002
9. H.Ficheux,F. Caricchi, et al. Axial-flux permanent-magnet motor for direct-drive elevator systems without machine room[J]. Transaction on Energy Conversion, 2000, 10(2): 190-197.

DOI: 10.1051/matecconf/20166301018

10. F Caricchi, F Crescimbini, O Honorati. Modular axial flux permanent magnet motor for ship propulsion drives[j].IEEE Transactions on Energy Conversion,Vol. 14, No. 3, September 1999:673-679.
11. Tang Ren Yuan.Modern permanent magnet motor theory and design[M].Beijing: Mechanical Industry Press, 1997.

MATEC Web of Conferences **63**, 01019 (2016) DOI: 10.1051/matecconf/20166301019
MMME 2016

The Acoustic Characteristic Investigation of Micro Plasma Shock Wave Obtained by Femtosecond Laser Ablation of Silicon Wafer Surface

Dong ZHONG[1,a], Zhong Ming LI[1], Chun CHEN[2], Ruo Lin RUAN[3]

[1] *College of electronic and information, Hubei University of Science and Technology, Xianning, China*

[2]*College of Physical Education, Hubei University of Science and Technology, Xianning, China*

[3]*College of biomedical engineering, Hubei University of Science and Technology, Xianning, China*

Abstract. We experimentally investigate the acoustic characteristic of micro plasma shock wave obtained by femtosecond laser ablation of silicon wafer surface using F-P optical fiber acoustic sensor with anti electromagnetic interference. The experimental results show that each of the femtosecond laser pulse can only produce a plasma shock wave. Under the different femtosecond laser energy, the acoustic attenuation signal of single shock wave has been analyzed by using wavelet transform, the results indicated that the energy change of femtosecond laser has no effect on the frequency of the acoustic signal of the micro plasma shock wave and have only influence on the amplitude, meanwhile, the main frequency of the acoustic attenuation signal are distributed within range of 0-100KHz, According to these frequency components, we can conclude that the decay process of micro plasma shock wave is the fast transduction of the ultrasonic signal into the acoustic signal. This study can provide the experimental basis for femtosecond laser ablation mechanism and femtosecond laser plasma control application research.

1 Introduction

Research on femtosecond laser ablation phenomenon has a strong effect on comprehending mechanism of the interaction between femtosecond laser and materials, improving the efficiency of femtosecond laser micromachining, minimizing laser damage, and so on. When interacting with materials, non-linear phenomena will be induced, such as multi-photon effects, self focusing and avalanche ionization. Due to femtosecond laser has high peak power and intensity, the use of ultrashort lasers, i.e., fs lasers, has a reduced thermal effect on the ablated material compared to ns LA(laser ablation) and an ablation depth with improved precision can be obtained[1,2]. These special characteristics of fs LA have

[a] Corresponding author: zhongdong0129@163.com

initiated new research paths for more detailed understanding of ultrafast laser-material interactions, plasma excited species expansion dynamics, and mechanisms of material removal[3]. Regardless of laser pulse duration (fs-ns), laser produced-plasmas (LPPs) are highly transient in both space and time, so studying plasma expansion dynamics and emitting species' kinetic properties is essential for most applications [4].

According to the existed experimental phenomena and results, people put forward different theories when dealing with different material, mainly including two-temperature model[5], micro explosion theory[6], theory of Coulomb explosion[7,8] and threshold theory[9]. Although the ablation process of different materials basically coincides with the threshold theory, still there is no clear understanding of the plasma dynamics characteristics and the interaction of femtosecond laser ablating different materials so far, and it is not even clear of the mechanism of its origin. This field needs more theoretical and experimental research [10].

Based on this, in recent years, people have carried out the research of dynamic characteristics and detection method of the femtosecond laser plasma. Some researchers developed an ultrafast time-resolved imaging technique for the plasma dynamics characteristics utilizing the optical polarigraphy technique and a chirped supercontinum[11]. Including laser induced breakdown spectroscopy [12],broadband ultrasonic technique[13], optimization of laser-induced breakdown spectroscopy [14], and so on, these methods also had been used widely. For diagnosis of the plasma wave, researchers employed a technique based on the modification of the spectrum of an intense femtosecond pulse generating the plasma wave, and an effective diagnostic method of laser-induced plasma based on measuring of the scattered microwave radiation has been proposed[]. The latest new method has been proposed to investigate the optical emission features of plasmas by using time of flight emission spectroscopy, but the decay processes occur on a nanosecond time scale and are functions of plasma parameters such as electron density and temperature. Measuring those parameters is difficult due to fast relaxation (1 ns), relatively low electron density.

The main difficulty lies in that femtosecond laser pulse time is so short that creating a pulse with high energy would produce plasma with small size. The existing technical means can only observe the early emerging, progress and ionization of plasma by over speed spectroscopy technique, and the emission wavelength shift and spectral intensity of spectral to analyze the expansion regularity of plasma, considering the atmospheric environmental impact around the aspects. The plasma energy status and decay process of target ablation stripping cannot be reflected. Considering the plasma is charged particles, the existing electrical probe in situ detection can be affected by the electromagnetic interference of plasma charged particles, thus affecting the accuracy of measurement. In this paper, the optical fiber F-P probe was used in situ measurement the acoustic signal process of micro plasma shock wave produced by femtosecond laser ablation of silicon wafer. We make a research on characteristics and the decay process of plasma obtained by femtosecond laser ablation target material, providing experimental basis for femtosecond laser ablation mechanism and plasma control.

2 Experimental setup

The schematic of experimental setup is shown in Fig.1. A commercial Ti: sapphire femtosecond laser with 180 fs laser pulses and 780 nm central wavelength, 500 Hz repetition rate was used in the system (Japanese Cyber Laser Company; LS-IF-FW-C-401). The pulse energy could be continuously varied by a variable neutral density filter and a mechanical shutter could turn the laser on and off. The laser beam is focused normally onto a sample mounted vertically on a translation stage. Before femtosecond laser irradiation, the silicon wafer is cleaned by ultrasonic bath for 15 min in acetone and rinsed in deionized water. For the F-P optical fiber sensing system, a tunable semiconductor laser with 1545 nm wavelength and 0.2 mw Power was used as the light source (Santec Company). The multi-channel data acquisition card of NI Company was used for data collection. The highly efficient signal processing circuit developed independently by our laboratory was used for photoelectric conversion. High speed CCD used in the experiment is a type of PCO.dimax HD high speed digital camera produced by PCO Company of Germany, the exposure time is 1.5us-40ms.

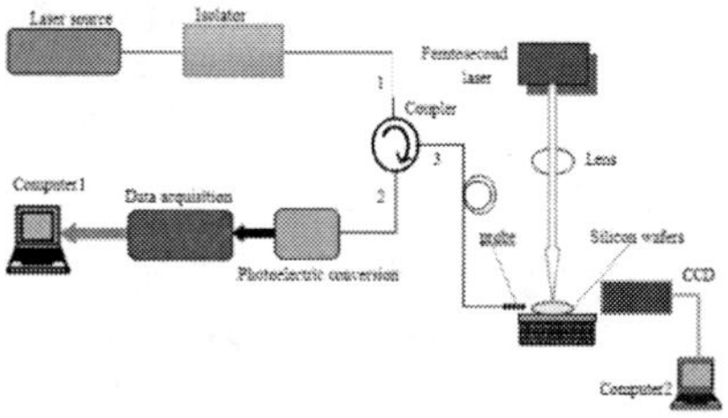

Fig.1 Schematic layout of the experimental setup

The ablation was carried out in air by irradiating the silicon wafer with different numbers of laser pulses and different laser energy. The numbers of the femtosecond laser pulses was set to 500Hz. The energy of the femtosecond laser was set to 110mw. The laser micro plasma shock wave could be formed when the laser beam irradiated on the silicon wafer. The acoustic signals of laser micro plasmas shock wave were detected by the optical fiber sensing probe and collected by data acquisition module. Then we can analyze the acoustic characteristic of laser micro plasmas shock wave by wavelet transform.

Meanwhile, CCD is connected with the computer by the 1394 interface and finished shooting photo storage through the MATLAB software programming to call CCD own function package, the high speed CCD camera had been applied to observe the expansion process of laser micro plasma. In comparison, the frequency and amplitude of the acoustic emission spectrum has been analyzed, the dynamic characteristic of the femtosecond laser micro plasma expansion has been researched.

3 Results and discussion

The acoustic signal of micro plasma shock wave with femtosecond laser pulse at 500Hz was studied. To determine the relation between the number of laser pulse and the number of micro plasma shock wave acoustic signal, we measured the acoustic signal of micro plasma shock wave with laser pulse at the number 500Hz. As showed in Fig.2, five acoustic signal of micro plasma shock wave can be obtained in 10 ms.

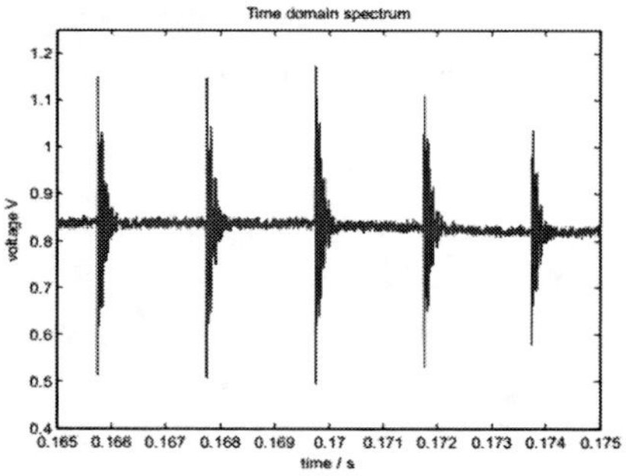

Fig.2 The acoustic signal of micro plasma shock wave by femtosecond laser ablation of silicon wafer surface under the condition of 500 Hz laser pulses

The micro plasma shock wave we have detected could be explained by the theory of micro-explosions. Multiphoton ionization and the ionization by electron impact (avalanche ionization) play the major role in femtosecond dielectric breakdown. The interplay of these two processs turns the material under irradiation into plasmoid. Because the pulse duration is much shorter than the electron-phonon energy transfer time, the electrons are heated while the ions stay cold. The micro plasma formed would produce a pressure which can be measured by the effective deposited energy inside the material.

When the pressure exceeds the Young modulus of the material, an explosion would appear thus triggers a shock wave. As for the energy deposition, the pressure outstripping the Young modulus of the material takes place with the subsequent shockwave at the end of the laser pulse. The shock wave decreases with the distance rapidly for the damping in the air.

Due to the wider signal bandwidth of femtosecond laser radiation, the composition of frequency is quite abundance and we can analyze the frequency spectrum of acoustic signal to investigate the acoustic characteristic of laser micro plasmas shock wave by wavelet transform. The acoustic signal is decomposed into different frequency channels, the characteristics of different frequency components can be obtained for the signal in different frequency bands.

In the study, the acoustic signals of single micro plasma shock wave were decomposed using db5 wavelet from which the spectrum features were extracted. The acoustic signal characteristic of single micro plasma shock wave was analyzed by db5 wavelet and the change of spectrum curve at different laser energy was obtained. The energy of the femtosecond laser was set to 110 mw. Figure 3 shows the acoustic signal spectrum of single micro plasma shock wave by db5 wavelet with femtosecond laser energy 110mw. In fig.3, a and d marking in the figure respectively stands for the low and the high frequency signals after wavelet transformation, the numbers after a and d indicate the progression of wavelet decomposition. As is shown from the Fig. 3, the acoustic emission signal frequency component of micro plasma shock wave is abundant, mainly focusing among the range of 0-100KHz. The amplitude of the signal is mainly affected by the change of laser energy which has little effect on the whole frequency range. The wavelet decomposition shows that level a_6 and d_1 is the low-frequency and high-frequency of acoustic emission signals derived from the micro plasma shock wave, mainly in 0~10KHz and 70~80KHz.

Through wavelet transform we can clearly determine the frequency component of the signal, it includes the frequencies of 0-100KHz. We can see that the energy mainly concentrate in the a6 which is the lowest frequency part of the signal. The proportion of other states' energy takes a decline trend from d6 to d1. The low frequency components (LF) take the major part of the signal. We can infer that the high frequency components (HF) attenuate quickly in the propagation of the micro plasma shock wave. In this way the dominant frequency of the signal shows a decreasing trend.

4 Conclusions

In this work, the direct femtosecond laser structuring technique developed previously by us for metals is extended to silicon that allows producing black silicon by structuring silicon surface with a grating of equally spaced parallel nanostructure-textured microgrooves.

Acknowledgment

The authors gratefully acknowledge the financial support for this work provided by the Dr. Start-up fund of Hubei University of science and technology No. BK1524 and the science research project in Hubei province department of education, Grant No. B2015077, the National Natural Science Foundation of China (NSFC) under the Grant No. 61575148, No. 61271256 and No. 51171061. the colleges and universities of Hubei Province innovation and entrepreneurship training plan, No. 201510927017, No. 201510927018, No. 201410927025 and 201410927029, the Team

DOI: 10.1051/matecconf/20166301019

Plans Program of the Outstanding Young Science and Technology Innovation of Colleges and Universities in Hubei Province, Grant No.T201513, the Program of the Natural Science Foundation of Hubei Province, Grant No. 2014BAA315 and No. 2015CFB452, the teaching reform program of Hubei University of science and technology, No. 2015-XA-007, No. 2014-XA-007 and 2014-XC-025.

References

[1] F. Beffa, H. Jackel, M. Achtenhagen, C. Harder, and D. Erni, Appl. Phys.Lett. 77, 2301 (2000).

[2] D. Grojo, J. Hermann, and A. Perrone, J. Appl. Phys. 97, 063306 (2005).

[3] R. Stoian, D. Ashkenasi, A. Rosenfeld, and E. E. B. Campbell, Phys. Rev. B 62, 13167 (2000).

[4] S. S. Harilal, A. Hassanein, and M. Polek, J. Appl. Phys. 110, 053301(2011)

[5] Schmidt V, Husinsky W, and Betz G. Appl. Surf. Sci. 197: 145(2015).

[6] Glezer E N and Mazur E. Appl. Phys. Lett., 71: 882(1997).

[7] Henyk M, Mitzner R, and Wolfframm D et al. Appl. Surf. Sci. ,154-155: 249(2000).

[8] Henyk M, Wolfframm D, and Reif J. Appl. Surf. Sci. , 168(2014)

[9] Juodkazis S, Matsuo S, and Misawa H et al. Appl. Surf. Sci., 197-198: 705(2002).

[10] Lihe Yan, Xiaofang Wang, Jinhai Si, Shigeki Matsuo, Tao Chen, Wenjiang Tan, Feng Chen, and Xun Hou . APPLIED PHYSICS LETTERS 100, 111107 (2012).

[11] M. Kotzagianni, S. Couris. APPLIED PHYSICS LETTERS 100, 264104 (2012).

[12] S. I. Kudryashov, A. A. Ionin, S. V. Makarov, N. N. Melnil. International Symposium on High Power Laser Ablation 2012 AIP Conf. Proc. 1464, 244-255 (2012).

[13] Mostafa Alshershby, Yu Ren, Jiang Qin, ZuoqiangHao,and Jingquan Lina). APPLIED PHYSICS LETTERS 102, 204101 (2013).

[14] Go Odachi, Ryosuke Sakamoto, Kento Hara, Takashi Yagi, Applied Surface Science. (2013).

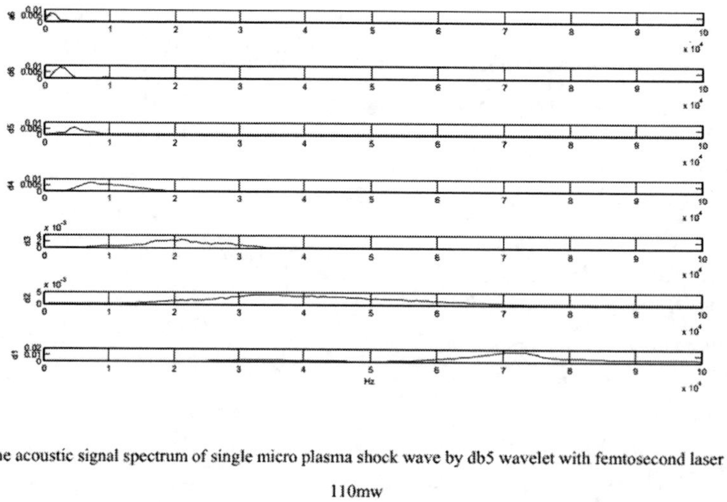

Fig.3 The acoustic signal spectrum of single micro plasma shock wave by db5 wavelet with femtosecond laser energy 110mw

MATEC Web of Conferences **63**, 01020 (2016) DOI: 10.1051/matecconf/20166301020
MMME 2016

Easily magnetic anomalies earthquake prediction

Min JIANG
Ningbo Polytechnic ; Zhejiang Ningbo 315800

Abstract: Low power consumption long time offset magnetic field detector (earthquake prediction) .The design of the hardware circuit of the magnetic field detector seismic geomagnetic acquisition and pre processing module mainly includes. Electronic compass, compass. monitoring device while the magnetic azimuth for monitoring and analyzing the object, GSM, but it can also be applied to other seismic precursor information analysis, such as earthquake precursory infrasound abnormality, only need infrasound abnormality intelligent sensor replace geomagnetic anomaly intelligent sensor, and modify the relevant parameters can be.

Key words: earthquake prediction; GSM circuit; LED; Electronic compass

1 INTRODUCTION

Earthquake is a common natural disaster, and the prediction of earthquake is a worldwide problem. Earthquake prediction is a worldwide famous puzzle and it is a necessity of scientists to do research on this significant thesis. In order to avoid the heavy losses caused by the earthquake, to find an effective method of imminent earthquake early warning is the hard direction of the earthquake researchers. In recent years, scientists have found that monitoring of the geomagnetic anomaly signal of imminent earthquake is one of effective methods of impending earthquake early warning. China suffers natural disasters, especially earthquake calamity, all the time from ancient times. If accurate prediction is possible, Chinese people may avoid that great loss of lives or fortune and they would not fear destructive earthquakes. We discussed the research background and status of magnetic measurements earthquake, proposed the design of earthquake early warning system on the basis of explore seismomagnetic relations. The working principle of the system is real time monitoring the earth's magnetic field vertical component Z and magnetic north point a. To give an accurate earthquake prediction is very difficult since that there is too much information relating to earthquake and scientists could hardly distinguish precursor from artificial interference or background noise. As a result, author choose magnetic methods as a breakthrough point after reading lots of inferences. It's possible to record geomagnetic data per second benefiting from the use of new instruments. In this paper, many methods are used to eliminate interference and then first difference method is used to analyze geomagnetic data. The frequency of anomalies gives more information than the amplitude of first difference of geomagnetic data and such operation could avoid unknown artificial inference that may existed and give a comparison with natural background.

Low power consumption long time offset magnetic field detector (earthquake prediction) consists of electronic compass, compass. Is based on such a fact: Beichuan middle school in 2008 May 12, afternoon 2:15 class time is before the Wenchuan earthquake 13 minutes. According to the recollections of grade two (1) class students, physics teacher of this class is "Electromagnetic" he brought the electromagnetic demonstrator and compass for teaching. At the beginning of the class, the electromagnetic demonstrator is not open, the teacher found on the table of the compass appears chaotic pendulum. Sitting in the middle of a row of students recalled that the teacher's compass is a bit like a bamboo dragonfly, a large pointer, with the vertical support vertical, the teacher found the compass is wrong, has been staring at the compass". "He took his fingers, the compass and turned again, the teacher again with a finger to stop the compass, to open the finger compass or chaos." Students recall, the teacher began to write on the blackboard, from time to time with the benefit of hindsight compass, students on the compass turn indiscriminately whispered together to discuss. The students, and so on for a long time did not stop to make everyone laugh. Is based on such a fact:

2 THEORETICAL BASIS

Overview of the earth's magnetic field: the earth's magnetic field is called the geomagnetic field. Geomagnetic field approximation to the one placed in the centre of the earth's magnetic dipole magnetic field, the magnetic dipole that geocentric magnetic dipole or geocentric dipole, magnetic lines of force distribution as shown in the fig.1. The earth's magnetic field is a stable magnetic field, the strength of about 0.5~0.6 Gauss, because the magnetic field is a variety of different sources of magnetic field is the result of the superposition, so in terms of its nature, is divided into two parts: one is the main source of stable magnetic field within the solid earth, covers an area of magnetic field of more than 95%; the second is main source to the variation of the magnetic field outside of the solid earth, the biggest change covers the magnetic field strength of 2% ~ 4%.

Fig.1 Schematic diagram of magnetic dipole magnetic field

In addition, the New York Times reported in 1858 November 11, before the earthquake in Setubal, Portugal, local only appeared in the compass abnormal phenomenon, even the mariner's compass pointer occur irregular circles of the fact. Before and after the two similar impending earthquake anomalies difference of 150 years, can explain the strong geomagnetic anomalies before the earthquake is not accidental.

The geomagnetic field is closely related to the structure of the earth's interior. A large number of observations and studies have found that the magnetic azimuth angle of the epicenter of the earthquake occurred in the area of the epicenter of the earthquake occurred a few minutes to a few days before the magnitude 5 or above. So in pre earthquake on the research of the magnetic azimuth anomaly information, not only to deepen knowledge and understanding of seismomagnetic effect, and probably become one of the important way of earthquake early warning.

In the determination of space electric field at the same time, choose to use the low power magnetic deviation detector consists of electronic compass, a compass. Using the special timing device opened 30 seconds per minute. Before the earthquake electromagnetic anomaly magnetoelectric effect is the main factor that caused by the earthquake, the earthquake electromagnetic anomaly before confirmed by a large number of examples. Use of the space electric field and the earth magnetic field changes in the design of earthquake prediction, the electric field, magnetic field parameters measurement is not necessary to enter the interior of the earth, technically easy to implement, low power consumption long time offset a major advantage of the detector is not the earth magnetic field debugging can work normally. The static when the power circuit is only 0.1~1 μ A, through to 30 seconds per minute, while working current general < 10 mA, so it is very save electricity, two new general 2 number No.5 cell.

 DOI: 10.1051/matecconf/20166301020

3 HARDWARE

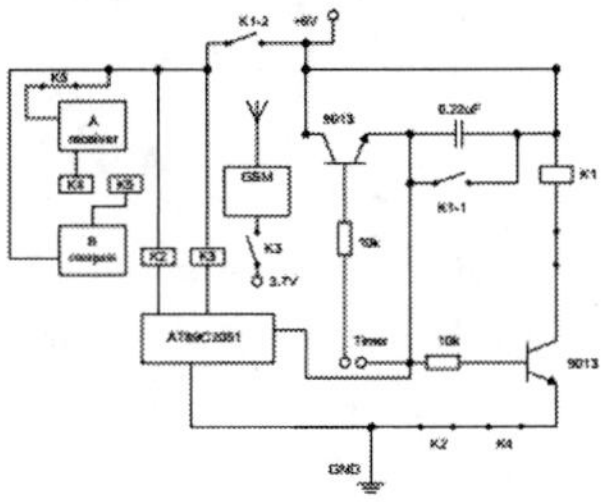

Fig.2 Earthquake prediction device hardware

When the remote alarm circuit in the normal electromagnetic GSM, remote automatic alarm when abnormal electromagnetic, alarm self-locking want to disarm by remote control after launch "output ff" command. A charge can alarm more than 200 times, is very convenient. GPRS remote monitoring device can also analysis level and the exact time of the earthquake epicentre of the earthquake. Most of the devastating earthquake in China. This device can know a few days to a few hours in advance before the earthquake electromagnetic anomaly can avoid many of the terrible disasters. Fig.2 Earthquake prediction device hardware.

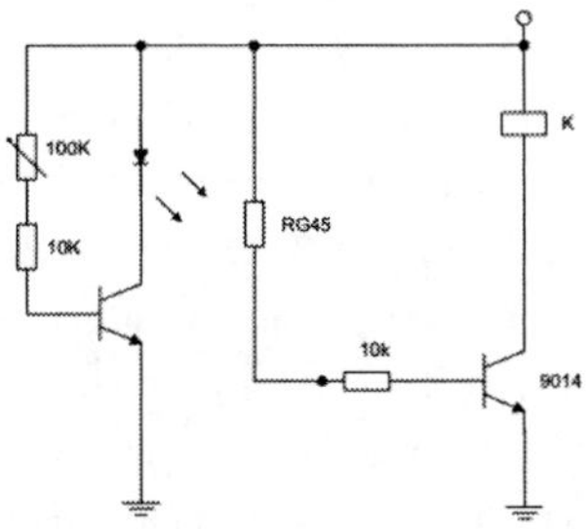

Fig.3 Earthquake prediction device LED Circuit

Used to increase the earth's magnetic field, is the switching element, open collector output contactless, long life. Fast switching speed and frequency band (DC to 100kHz). Because the earth's magnetic field is very weak so about two ferrite rod enhanced, and achieved good results. DN6851 Hall components through T2, T1 amplification direct drive music integrated circuit, the circuit is connected 4 times per minute, time of about 10 seconds. Alarm when the magnetic field changes before the earthquake. The music integrated circuit is actually a large-scale CMOS (complementary symmetry metal oxide semiconductor integrated circuits English abbreviation) circuit, transistor best the power triode in 300MW silicon NPN type electrode dissipation power larger than that of the, 9013, 8050, 3DG12, 3DK4 and 3DX201 type requires current amplification factor of 100. LED is the best choice of 5mm red light emitting diodes. The 8 Ω ,0.25w small caliber horn loudspeakers. The power supply is connected in series with two 5 numbers (to be equipped with a plastic cell rack), and the voltage is 3V.

4 SOFTWARE

The discrete wavelet transform (DWT) is a mathematical tool that transforms a data vector of length M into a different vector of the same length. It is based on recursive sums and differences of the vector components; the sum can be compared with the low frequency amplitude in the Fourier transform, and the differences with the high frequency amplitudes. It is similar to the discrete Fourier transform concerning the properties of orthogonality and invertibility, and the wavelets represent the basis vectors of the transform, like the sine and cosine functions represent the basis functions of the Fourier transform. One of the basic advantages of wavelets is that an event can be simultaneously described in the frequency domain as well as in the time domain, unlike the usual Fourier transform where an event is accurately described either in the frequency or in the time domain. This feature permits a multiresolution analysis of data with different behaviour on different scales. Fig.4 is Earthquake prediction MCU software. Signal sampling and wavelet transform are used to achieve single-chip software. In the work, the software is combined with the hardware circuit to realize the discrete wavelet transform.

```
            ORG         0000H
            AJMP   START1
            ORG         0030H
SAME        EQU         4EH
START1:     MOV         P1,#00H
START:      MOV         A,P3
            ANL         A,#1FH
            CJNE   A,#1FH,SHIY
            AJMP   START1
SHIY:       MOV         SAME,A
            LCALL  YS
            MOV         A,P3
            ANL         A,#1FH
            CJNE   A,#1FH,SHIY1
            AJMP   START1
SHIY1:      CJNE   A,SAME,START1
            CJNE   A,#17H,NEXT1
            AJMP   LEFT
NEXT1:      CJNE   A,#0FH,NEXT2
            AJMP   RIGHT
NEXT2:      CJNE   A,#1DH,NEXT3
            AJMP   EARGE
NEXT3:      CJNE   A,#1EH,NEXT4
            AJMP   BRAKE
```

Fig.4 Earthquake prediction MCU software

The ordinary compass warning circuit and its working principle: watch movement is mainly used for circuit four times per minute is switched on, the red LED light per minute 30 seconds were detected, the MG45 (20k) photosensitive resistance for receiving the magnetic offset, if found abnormal driving a music integrated circuit work, two No.5 dry batteries are available for about one year time. The entire production cost is lower. Components and models see third parts. Welding should be paid special attention to, because the music integrated circuit CMOS circuit, so electric iron shell must to have a good grounding, also can dial to the power plug of the electric soldering iron using waste heat of the electric iron welding so that it can be avoid external field induced breakdown in a music integrated circuit, and cause permanent damage. Welding the electric iron power should not be more than 30W, and stay on the circuit board time should be as short as possible, each solder joint time generally do not exceed the 2sec. Do not use flux soldering paste or paste, if necessary, after use must be welded oil clean.

Warning circuit of Electronic Compass: is mainly composed of DN6851 Hall element and now use it to abnormalities in the monitoring of the earth's magnetic field, the basic principle is using the switch type Hall sensor (Hall switch) is a new non contact integrated circuit switch, DN6851 has simple structure single, plastic shell, small volume. The combined use of two ferrite rod. The Hall effect is the production of a voltage difference (the Hall voltage) across an electrical conductor, transverse to an electric current in the conductor and a magnetic field perpendicular to the current. It was discovered by Edwin Hall in 1879.The Hall coefficient is defined as the ratio of the induced electric field to the product of the current density and the applied magnetic field. It is a characteristic of the material from which the conductor is made, since its value depends on the type, number, and properties of the charge carriers that constitute the current.

5 SUMMARY

Geomagnetic field is closely related to the interior structure of Earth. large number of observations and researches show that the angel of magnetic azimuth in epicentral area would be obvious anomaly within a few minutes to several days before an earthquake of magnitude 5 or over. Therefore, a research in abnormal information of the angel at an impending earthquake not only deepens the knowledge and understanding of seismomagnetic effect, but also probably become one of the important ways of earthquake early warning. Low power consumption long time offset magnetic field detector (earthquake prediction) .The design of the hardware circuit of the magnetic field detector seismic geomagnetic acquisition and pre processing module mainly includes. Electronic compass, compass. monitoring device while the magnetic azimuth for monitoring and analyzing the object, GSM, but it can also be applied to other seismic precursor information analysis, such as earthquake precursory infrasound abnormality, only need infrasound abnormality intelligent sensor replace geomagnetic anomaly intelligent sensor, and modify the relevant parameters can be.

REFERENCES

1. Min Jiang,2013.11 How to weave the

electromagnetic spider web to predict earthquakes , 2013 3rd International Conference on Education and Education Management(EEM 2013), p546-551

2. Min Jiang,2014.1 Electromagnetic spider web Application in earthquake prediction, International Core Journal of Scientific Research & Engineering Index,p111-114.
3. Min Jiang,2014.11 The Group of Electromagnetic spider web, Economic, Business Management and Education Innovation(EBMEI 2014),p287-291.
4. Min Jiang,2014.12 Low Power Consumption Electromagnetic spider web, 2014 4th International Conference on Education and Education Management(EEM 2014), p74-78.
5. Min Jiang, Ping Yao Wang 2015.9 Simple space electric field and earth magnetic anomaly remote alarm system,2015 International Conference on Frontiers of Manufacturing Science and Measuring Technology, p452-458.
6. Min Jiang ,2016.4 Application of 3 kinds of practical electromagnetic spiders in electromagnetic spider web. ICEICE 2016 02083. MATEC Web of Conferences 44.

MATEC Web of Conferences **63**, 01021 (2016) DOI: 10.1051/matecconf/20166301021
MMME 2016

Illegally additives determination by thin layer chromatography

Xin LIANG [1], Jie ZHANG [1], Ting Ting LIU [1], Li LI [1,a]

[1]*Department of Pharmaceutics, Qiqihar Medical University, Qiqihar, China*

Abstract. To find out a simple TLC method, which could identify m-phenyenediamine and o-phenylenediamine in several colors of oxidation hair coloring products simultaneously. The colors includes black, yellow, brown and red. A small amount of the sample was placed on a silica gel GF_{254} plate by sample applicator and run using a solvent containing ethyl acetate- butanol -water (37.5:37.5:25 vol). This study established TLC conditions which could let phenylenediamines better separaed and be detected in oxidation hair coloring products simultaneously.

1 Introduction

Currently phenylenediamines had been detected in many hair coloring products, the detect method has been establish ever before. But there is not any study to expode phenylenediamines may be detected in what kind of color. It is well reported that these dye products are of health concern due to risks of allergic dermatitis, nephrotoxic and carcinogenic effects [1-5]. In Chinese hygienic standard for cosmetics, OPD, MPD are forbidden to be added [6].

Thin layer chromatography (TLC) is a widely employed laboratory technique and is similar to paper chromatography. However, instead of using a stationary phase of paper, it involves a stationary phase of a thin layer of adsorbent like silica gel, alumina, or cellulose on a flat, inert substrate. Compared to paper, it has the advantage of faster runs, better separations, and the choice between different adsorbents. For even better resolution and to allow for quantification, high-performance TLC can be used. An older popular use had been to differentiate chromosomes by observing distance in gel (separation of was a separate step).[7]

Thin-layer chromatography can be used to monitor the progress of a reaction, identify compounds present in a given mixture, and determine the purity of a substance. Specific examples of these applications include: analyzing ceramides and fatty acids, detection of pesticides or insecticides in food and water, analyzing the dye composition of fibers in forensics, assaying the radiochemical purity of radiopharmaceuticals, or identification of medicinal plants and their constituents. [8]

A number of enhancements can be made to the original method to automate the different steps, to increase the resolution achieved with TLC and to allow more accurate quantitative analysis. This method is referred to as HPTLC, or "high-performance TLC".

The norms for determining oxidation type dye in hair dye is high performance liquid chromatography. But the new methods are more convenient and fast.

2 Experiment

2.1 Materials

N-butanol was purchased from Tianjin tianli Chemical Reagent Co., Ltd. Ethyl acetate and anhydrous ethanol was purchased from Tianjin fuchen Chemical Reagent Factory. All reagents were analytical purity. m-phenyenediamine(MPD), o-phenylenediamine(OPD) and p-Phenylenediamine(PPD) from Tianjin guangfu institute of fine chemicals were used as reference substances. Oxidation hair coloring products was purchased from the same company, but different in colors.

3 Method

3.1. Reference solutions

Weigh about 50 mg of the reference substance accurately in a standard 10 ml flask. Add 5 ml of 96% ethanol and 250 mg of ascorbic acid. Make the solution alkaline by addition of the ammonia solution to give an apparent pH of 10 (test with indicator paper). Make up to 10 ml with 96% ethanol and mix[9].

3.2 Treatment of test samples

Discard the first 2 or 3 cm of cream extruded from the tube. Put the following into a centrifuge tube previously

[a] Corresponding author: 836009379@qq.com
Project supported by Qiqihar Medical University (Grant No. QY2016Z-22)

flushed out: 300 mg ascorbic acid with 3 g cream. Add 25 % ammonia drop-wise until the pH is 10. Make up to 10 ml with 96 % ethanol. stopper and then centrifuge at 4 000 r/min. For 10 minutes.

3.3 Developing solvents

In the EU's governing cosmetic products rules, benzene - 2-butanol - water (50:25:25) was used in identification of certain oxidation colorants in hair dyes.

Considering of the toxicity of benzene and the non applicability to quick identification, the developing solvents were replaced and optimized[10]. According to Zhang Jingbao's classification of common solvents[11], benzene as electron acceptor, then according to the boiling point, viscosity, volatile, toxic, cost and other factors, preliminary determination of ethyl acetate as a substitute. Because the restrictions of the actual reagent used, using n-butanol instead of butyl alcohol. Similarly, ethyl acetate substitute for benzene. Change the proportion of developing solvents to determine the best ratio of percentage by graphic method, combined with the developing time, select the best developing solvent system.

3.4 System suitability test

3.4.1 Detection sensitivity

Preparation of 5.0mg/mL reference solutions, and were diluted to 0.5mg/mL, 1.0mg/mL, 2.0mg/mL, 4.0mg/mL. Then apply at points 1, 2 and 3 respectively 2 μL, using the best developing solvent system to run .View in the UV lamp 254nm.olvent system.

3.4.2 Rf value

Apply 2 μL of each of the reference solutions, develop, written down the distance of baseline to spread advanced. View in the UV light at the wavelength of 254nm, draw the main spot of OPD, MPD, PPD, write down the baseline to the spot center distance.

3.4.3 Resolution

Pipette with an equal amount of OPD, MPD, PPD reference solution mixed. Apply 2 μL at first point and apply 2 μL reference solution at following points, develop, write down the baseline to spread advanced distance. Under the 254 nm UV lamp examination, draw the main spot location, radius of OPD, MPD, PPD. Writing down the distance of baseline to spot center, and calculating the separating degree R.

3.4.4 Sample quantity choice

Using the supernatant, apply at points 1, 2, 3 and 4 respectively 2 μL, 4 μL, 6 μL, 8 μL of the test solution samples. Using the best developing solvent system. Observation under the UV lamp 254nm.

3.4.5 Identification

Compare the R_f value and the colour obtained from the sample with those of the reference substances chromatographed.

4 Results

4.1 Optimization of developing solvents

Change the ratio of ethyl acetate and n-butanol. The R_f values and development time of OPD, MPD, PPD as shown in table 1.

Table 1. Different ratio of TLC chromatogram development using ethyl acetate - n-butanol – water

	Developing solvents (%,v/v)		
groups	ethyl acetate	n-butanol	water
1	60.0	15.0	25.0
2	45.0	35.0	25.0
3	37.5	37.5	25.0
4	30.0	45.0	25.0
5	15.0	60.0	25.0

When the percentage of the ethyl acetate is 37.5 %, namely, ethyl acetate: n-butanol: water = 37.5: 37.5: 25, adjacent spots maximum separation degree, the separation of the components that get ideal. (Table.2)

Table 2. R_f values of TLC chromatogram development using ethyl acetate - n-butanol – water

	R_f values		
groups	R_f OPD	R_f MPD	R_f PPD
1	0.83	0.82	0.73
2	0.82	0.71	0.72
3	0.85	0.78	0.73
4	0.78	0.74	0.67
5	0.88	0.78	0.77

4.2 Detection sensitivity

0.5mg/mL, 1.0mg/mL, 2.0mg/mL, 4.0mg/mL of reference solutions were detected. Apply at points 1, 2 and 3 respectively 2 μL, using the best developing solvent system to run .View in the UV lamp 254nm.olvent system.

DOI: 10.1051/matecconf/20166301021

The concentration that produced the spot with the least observable spot was noted. The sensitivity of o-phenylendiamine is 0.5 mg/mL. The sensitivity of m-phenylendiamine is 0.5 mg/mL. The sensitivity of p-phenylendiamine is 0.5 mg/mL (Figure1).

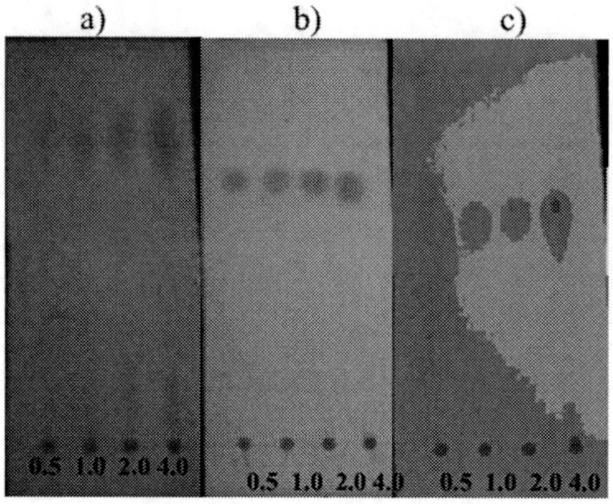

Figure 1. Uv spectrophotometry showing the observable spots of different concentrations. (a) OPD (0.5 - 4.0 mg/ml applied), (b) MPD (0.5 - 4.0 mg/ml applied), and PPD (0.5 - 4.0 mg/ml applied). The assay was carried a silica gel GF_{254} plate which had been eluted with ethyl acetate- butanol -water (37.5:37.5:25 vol).

4.3 R_f values

2 μL of each of the reference solutions was applied, developed, written down the distance of baseline to spread advanced. View in the UV light at the wavelength of 254nm, draw the main spot of OPD, MPD, PPD, write down the baseline to the spot center distance. Measure the corresponding distances and calculate R_f values(Table 2).

4.4 Resolution

Under the 254 nm UV lamp examination, draw the main spot location, radius of OPD, MPD, PPD. Writing down the distance of baseline to spot center, and calculating the separating degree.

Calculate the degree of separation between adjacent spots, the resolution of o-phenylendiamine and m-phenylenediamine is 1.7, the resolution of m-phenylenediamine and p-phenylenediamine is 1.2.

The resolution are both greater than 1, so been eluted with ethyl acetate- butanol -water (37.5:37.5:25), the separating degree of spots is good (Figure2).

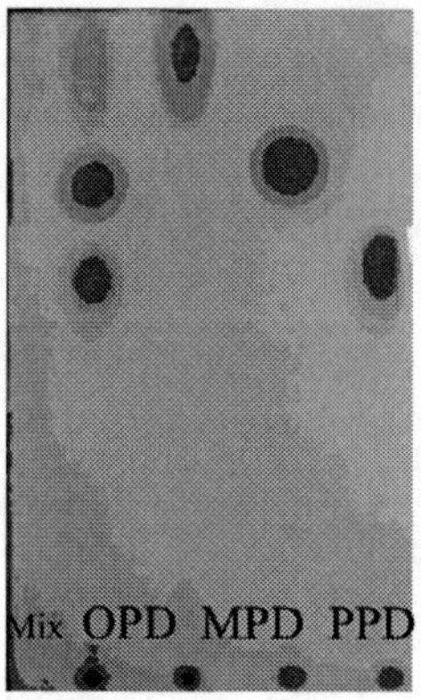

Figure 2. Uv spectrophotometry showing the observable spots of the resolution of three reference substance. The assay was carried a silica gel GF254 plate which had been eluted with ethyl acetate- butanol -water (37.5:37.5:25 vol).

4.5 Sample quantity choice

2 μL, 4 μL, 6 μL, 8 μL of the test solution samples was applied. The assay was carried a silica gel GF254 plate which had been eluted with ethyl acetate- butanol -water (37.5:37.5:25 vol). Observation under the UV lamp 254nm.

Four colors of commercial hair dyes, 6μL spots were clear, separation effect is good, is the best point sample.

4.6 Identification

The R_f value was compared with the colour obtained from the sample with those of the reference substances chromatographed.

Four kinds color of commercial hair dyes failed to detect the illegal additives OPD, MPD. But the yellow dye hair cream Color was detected with PPD(Figure3).

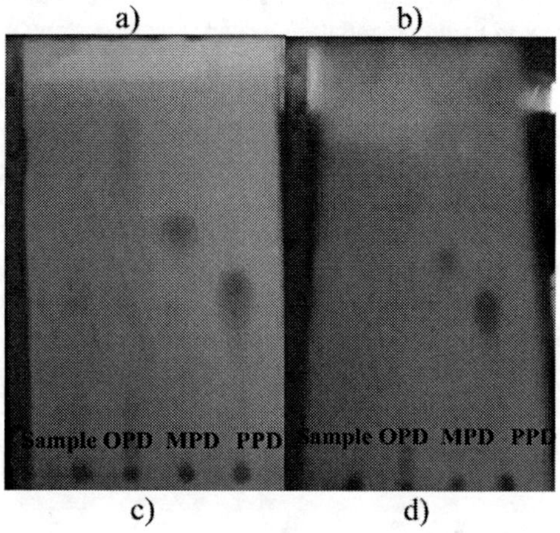

DOI: 10.1051/matecconf/20166301021

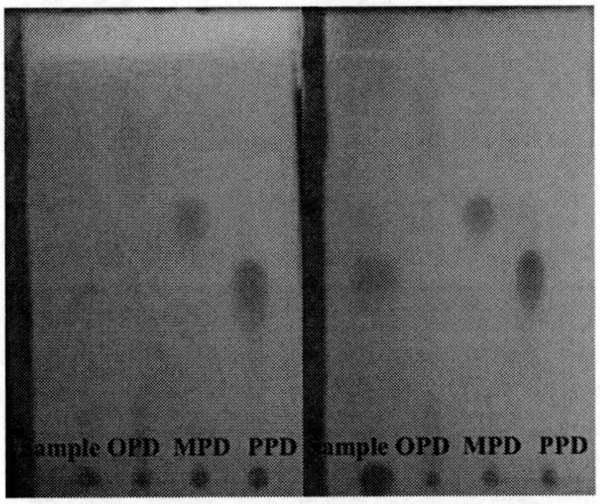

Figure 3 UV spectrophotometry showing the observable spots of four commercial hair dyes.(a) red dye hair cream, (b) black dye hair cream, (c) brown dye hair cream ,(d) yellow dye hair cream. The assay was carried a silica gel GF_{254} plate which had been eluted with ethyl acetate- butanol -water (37.5:37.5:25 vol).

5 Discussion

The test choice was a non-toxic ethyl acetate - n-butanol - water (2:1:1) system, but the separation effect is not so well, and too slow, the Rf Value was not ideal. After elevate the ratio of n-butanol, finally determine to take ethyl acetate- butanol -water (37.5:37.5:25 vol) as the developing solvent. The separation effect was good, OPD, MPD, PPD can be well separated, the spot is round,tailing phenomena was not obvious.

TLC method is simple, good reproducibility, high sensitivity, can be used as an identification method of check illegal additives in the hair dye. By thin layer chromatography (TLC) to identify the hair dye illegally added ingredients, should have the standard samples and test samples at the same time for thin layer chromatographic analysis. After developing, the spots of standard and test samples have same shape, Rf value and color, can be a preliminary conclusion is that the same compounds. There was no illegal ingredient added in the four kinds of color commercial permanent colorants, But the yellow dye hair cream Color was detected with PPD.

The results showing that yellow color commercial permanent colorants tends to be added with illegal ingredients, but there are still no evidences indicating that yellow color commercial permanent colorants contain OPD,MPD.

Acknowledgement

This paper is supported by Qiqihar Medical University (Grant No. QY2016Z-22).

References

1. B.Desmedt, P. Courselle, J.O. De Beer, V. Rogiers, E.Deconinck, K. De Paepe. Illegal cosmetics on the EU market: a threat for human health? Archives of Toxicology **Volume 88**, Issue 9, 1765-1766 (2014)

2.Desmedt B, Rogiers V, Courselle P, De Beer J, De Paepe K, Deconinck E. Development and validation of a fast chromatographic method for screening and quantification of legal and illegal skin whitening agents. JPharm Biomed Anal **83**:82-88(2013)

3.Desmedt B, Van Hoeck E, Rogiers V, Courselle P, De Beer J, De Paepe K, Deconinck E. Characterization of suspected illegal skin whitening cosmetics. J Pharm Biomed Anal **90**:85-91(2014)

4.Karamagi C, Owino E, Katabira E. Hydroquinone neuropathy following use of skin bleaching creams:case report. East Afr Med J 78(4):223-224. (2001)

5.Olumide YM, Akinkugbe AO, Altraide D, Mohammed T, Ahamefule N, Ayanlowo S, Onyekonwu C, Essen N. Complications of chronic use of skin lightening cosmetics. Int J Dermatol **47**(4):344-353(2008)

6.The ministry of health of the People's Republic of China. Hygienic Standard for Cosmetics. (Beijing: military medical science press, 2007)

7.Harry W. Lewis and Christopher J. Moody. Experimental Organic Chemistry: Principles and Practice (Illustrated ed.). WileyBlackwell. 159–173(1989)

8.A.I. Vogel, A.R. Tatchell, B.S. Furnis, A.J. Hannaford, and P.W.G. Smith. Vogel's Textbook of Practical Organic Chemistry (5th ed.).

9.The rules governing cosmetic products in the European Union Cosmetic products Methods of analysis. European Commission. Enterprise Directorate-General Pharmaceuticals and cosmetics, 35-41(1999).

10.Liu Ziyao, Sun Shuwen, Xiao Haiying, Leng YiPing. TLC is analysed in the choice of new agents and alternative method. Journal of China modern distance education of Chinese medicine. **9** (10): 222-223(2011)

11.Zhang Jingbao, Ban Yundong. Commonly used solvents of FUZZY clustering analysis. Journal of chromatography. **7** (5): 256-261(1989).

MATEC Web of Conferences 63, 01022 (2016) DOI: 10.1051/matecconf/20166301022
MMME 2016

Brushless DC Motor Control System Design Based on DSP2812

Min WEI[1, a], Juan WEI[2]

[1]*School of Electronic Engineering, Xi'an Shiyou University, 710065 xi'an, Peoples R China*
[2]*School of Telecommunications Engineering, Xidian University, 710071 xi'an, Peoples R China*

Abstract. By comparison various control methods currently for permanent magnet brushless DC motor, on the basis of motor principle analysis, a current smallest and most real-time all-digital rare earth permanent magnet brushless DC motor control system is designed. The high-speed digital signal processor DSP2812 is applied as the main control unit. The fuzzy PID control algorithm is used to control rectifier regulator and speed, which the speed and current is double closed loop in the system. The principle of control system, control strategy and software is analyzed in this paper. The system has some features such as less overshoot, rapid response speed, good performance of anti-jamming, simple structure, high control precision, flexible in changing control policies and so on. Validity of the design is verified by prototype test.

1 Introduction

Permanent magnet brushless DC motor is a superior speed performance, simple structure, reliable operation, easy maintenance electromechanical integration motor. It is now widely used in servo control, medical equipment, instrumentation, robotics, household appliances, computers and military and other fields [1].

In general, there are three kinds of permanent magnet brushless DC motor control method currently such as the utilize of ASIC controller chip, on the basis of the series single-chip microcomputer control and applying the high-performance digital signal processor control. The control system constituted by ASIC controller chip is simple, low cost. Bur since the IC control chip can complete all the work from the position detection decoding to the PWM modulation output in the hardware, the system functions can not be further expanded. In most cases SCM control system is applied. However, from the two aspects: the working speed and accuracy, brushless DC motor control system based on the general SCM could not be compared with high-performance digital signal processor. Rather than to obtain approximate results by look-up table method, rapid DSP computing capability enables the digital control system to calculate in real time. The internal structure of control system can achieve more complex control algorithms, so that the speed loop and current loop can be implemented in digital form, and a full digital brushless DC motor control system is achieved.

In this paper, the control scheme based on the DSP2812 brushless DC motor is presented Current and speed double closed-loop control strategy is used to achieve precise control for whole system.

This paper is organized as follows. Section 2 describes the constitute of the driving system. In Section 3 and 4, we describe the motor mathematical model and electric running control analysis respectively. Section 5 describes software implementation. This leads to a concrete parameter selection which can guarantee normal and safety operation of the system. We conclude the section by a detailed prototype test.

2 Constitute of the driving system

The system block diagram is shown in Figure1.The DSP is mainly responsible for receiving circuit feedback signal from current sensors and position sensors, and sending control commands to produce motor drive signals [2].The pulse signal of rotor position sensor is captured by capturing unit to determine the rotor position, And then appropriate driver logic level is outputted to drive control chip, and the motor is driven to rotate by the MOSFET. After the pulse signal from position sensor is captured by the controller, the current speed of the motor is calculated, and the motor speed is controlled to follow the set speed by control program. The motor current is gathered by controller with A/D and current detection circuit. Motor current is control by PID operation. Once having abnormal conditions such as overload, low voltage drive timing and so on, fault protection can be realized by driver protection circuit.

[a] Corresponding author: weimin_pb@163.com

DOI: 10.1051/matecconf/20166301022

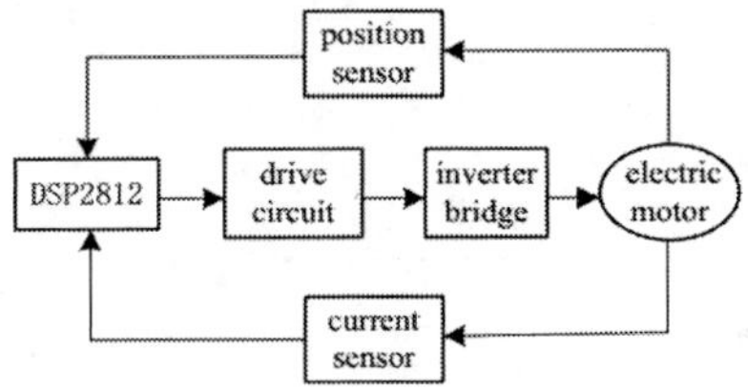

Figure 1. System Block Diagram

3 Motor mathematical model

Suppose that the motor phase winding is completely symmetrical, and that the air-gap magnetic field EMF waveform is greater than 120 degrees trapezoidal wave, the self and mutual inductances of the motor windings can be considered constant, and the motor equivalent circuit can be described as Figure 2. Where u_A, u_B, u_C are A, B, C phase winding terminal voltage respectively; e_A, e_B, e_C are A, B, C phase winding back potential respectively; R is a phase resistance; And L-M is a phase winding equivalent inductance.

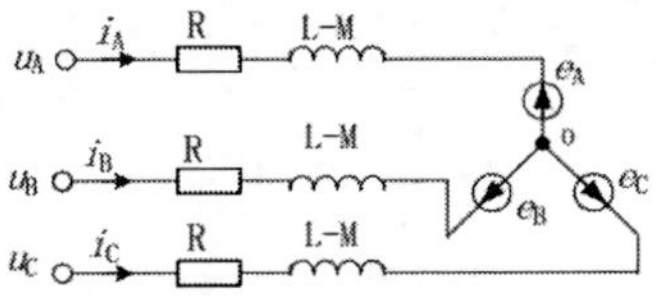

Figure 2. Equivalent circuit of motor

The voltage balance equation of the motor equivalent circuit can be expressed as

$$\begin{bmatrix} u_A \\ u_B \\ u_C \end{bmatrix} = \begin{bmatrix} R & 0 & 0 \\ 0 & R & 0 \\ 0 & 0 & R \end{bmatrix} \begin{bmatrix} i_A \\ i_B \\ i_C \end{bmatrix} + \begin{bmatrix} L-M & 0 & 0 \\ 0 & L-M & 0 \\ 0 & 0 & L-M \end{bmatrix} p \begin{bmatrix} i_A \\ i_B \\ i_C \end{bmatrix} + \begin{bmatrix} e_A \\ e_B \\ e_C \end{bmatrix} \tag{1}$$

And

$$i_A + i_B + i_C = 0 \tag{2}$$

Electromagnetic torque is

$$T = \frac{1}{\omega}(e_A i_A + e_B i_B + e_C i_C) \tag{3}$$

p is a differential operator (d/dt) in the formula (1). ω is the rotor mechanical angular velocity in the formula (3).

4 Electric running control analysis

The speed and current double closed-loop PID control is achieved by software system. The PID control algorithm is the incremental control algorithm. The adjustment process is a new current pulse width modulation (PWM) signal generating process, and adjusting the width of the PWM signal can adjust the average current.

Two-phase conduction six states control method is used in electric running inverter bridge circuit. At any time, only upper and lower two power switch is turned on which are belong to different bridge arms. Speed is controlled through the lower bridge arm power switch working on PWM state. Control signal waveform of six power tubes is shown in Figure 3.

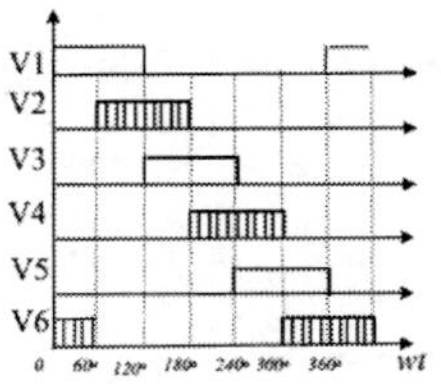

Figure 3. The control signal wave form of electromotor

For example, when the power switch V1 in the inverter bridge circuit upper bridge arm and the power switch V2 in the lower bridge arm are on conductive magnetic state, working circuit is positive battery terminal→ V1 → A phase winding → C phase winding → V2 →negative battery terminal. Ignoring the pressure drop of conducting pipe, the electric running equivalent circuit of the magnetic state is shown in Figure4, according to Kirchhoff's voltage law, combined with the formula (1), the loop voltage equation can be obtained

$$u_A - u_C = (L-M)\frac{d}{dt}(i_A - i_C) + R((i_A - i_C) + (e_A - e_C) \tag{4}$$

Let

$i = i_A = -i_C$;

$u_A - u_C = U$ is battery voltage;

$e_A - e_C = E$ is the phase potential when trapezoidal wave gap magnetic field waveform flat top width is greater than 120°.

According to the three elements method, we can obtain

$$i_{(\infty)} = \frac{U - 2E}{2R} \tag{5}$$

$$\tau = \frac{L-M}{R} \tag{6}$$

Let $i_{(0+)} = i_0$, with formula (5) and (6), circuit current can be obtained

$$i = \frac{U-2E}{2R} + (i_0 - \frac{U-2E}{2R})e^{-tR/L-M} \tag{7}$$

Electromagnetic torque of the motor is

$$T = \frac{1}{\omega}(e_A i_A + e_C i_C) = \frac{2Ei}{w} \tag{8}$$

PWM can be conducted by V2, then average current can be controlled, and thus the average electromagnetic torque can be changed. The magnetic state of electrical angle of 60° is analyzed here, and the other same five magnetic states can be also analyzed in the same way.

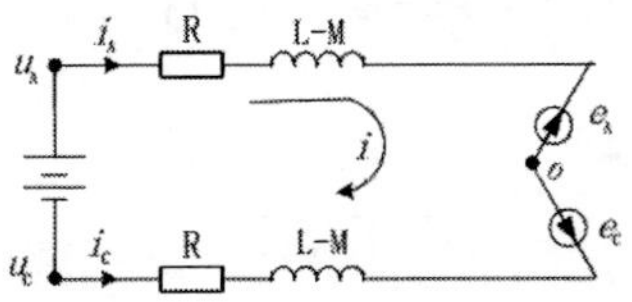

Figure 4. Equivalent circuit of electric running

DOI: 10.1051/matecconf/20166301022

5 Software implementation

System software consists of two parts which is the main program and interrupt routines. The main program includes system initialization, capture the current position and set the interrupt flag. Program flow chart of main program is shown in Figure 5. Subroutines are mainly position capture interrupt subroutine, A/D interrupt service subroutine, the timer interrupt subroutine, fault protection interrupt subroutine. The motor commutation and current loop adjustment is the core components of software.

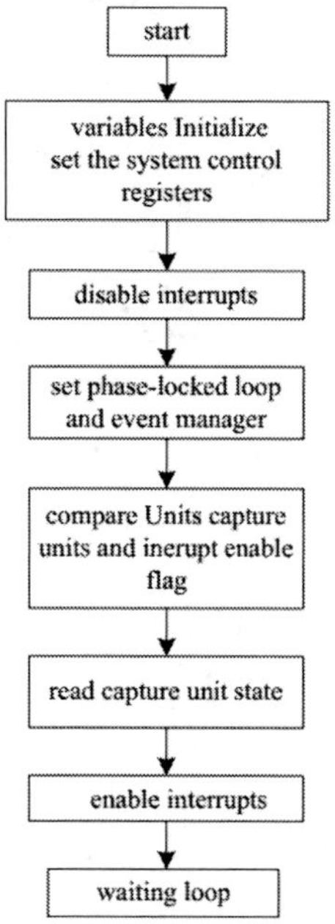

Figure 5. Program flow chart of main program

5.1 Position capture interrupt subroutine

The rotor position signal is required to timely, accurate detected by control systems. And then the corresponding output state of the PWM circuit is determined. It can ensure that the motor stator windings commutate in real time and precise in real time and accurately. Therefore, using the capture interrupt mode, we can get the position signal. Program flow chart of position capture interrupt subroutine is shown in Figure 6.

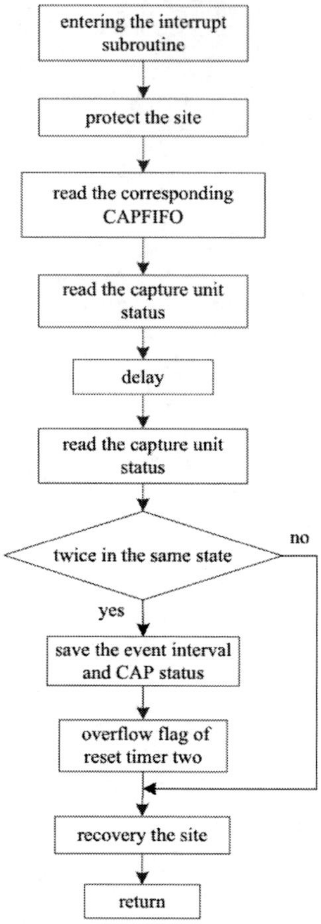

Figure 6. Program flow chart of position capture interrupt subroutine

5.2 A/D interrupt subroutine

In the A/D converter interrupt service routine, the first step is to adjust subroutine using the speed loop PI. A new current reference value by speed adjustment is used to adjust the current ring. Program flow char of A /D interrupt subroutine is shown in Figure 7.

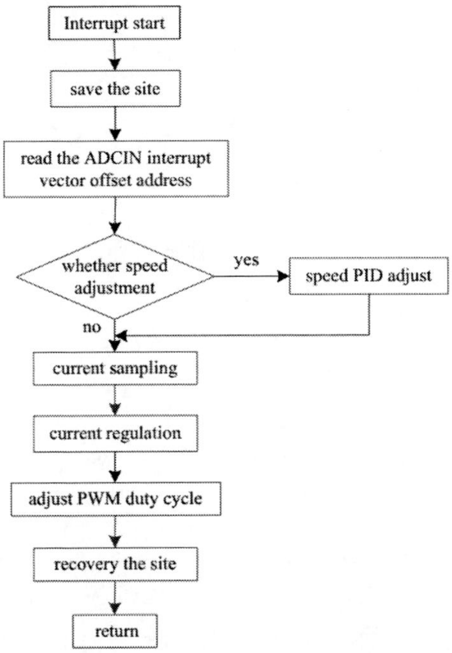

Figure 7. Program flow char of A/D interrupt subroutine

6 Results

The experiment of whole system is conducted, which is used the brushless DC motor controller designed in this paper combined with rare earth permanent magnet motor body. The measured results are as follows:
Rated voltage: 220V (AC)
Rated power: 200W
Rated torque: 2Nm
Overload multiples: 2.0
Base speed: 900r / min
Speed ratio: 1:10
Maximum system efficiency: 86% (motor and controller)

7 Conclusion

The brushless DC motor control system based on DSP2812 have some features such as simple structure, small in size, high precision control, flexible in changing control policies. Feasibility of the design is verified by prototype test.

Acknowledgments

This study is supported by grants from the Xi 'an science and technology plan projects (CXY1515(1)), Science and technology fund in shaanxi province department of education (2013JK1047), Shaanxi science and technology research and development projects (2012GY2-41), and the Natural Science Foundation of Shaanxi Province (2014JM2-6116)

References

1. F. Chai. *Permanent magnet brushless motor and drive technology* (China Machine Press, Beijing, 2013)
2. K. F. Su, Q. lv, etc., TMS320F2812 *principle and Development* (Electronic Industry Press, Beijing, 2005)
3. C. L. Xia, H. W. Fang, Electrical Technology, 27 3(2012)
4. S. Zhang, *Brushless DC the motor Principles and Applications*(China Machine Press, Beijing,1996)

Reseach on the reduction of rocket motor jet noise by water injection

Hao ZOU[a], Yi Hua XU and Yu Lin LI
School of Aircraft Engineering, Nanchang Hangkong University, 330063, China

Abstract: Injecting water in the mixing layer of rocket motor jets is a means to reduce jet noise. The calculation of the sound pressure signals at the prescribed receivers was performed by FW-H acoustics model under the condition of water injection and without water injection. The calculation results show that the jet noise is with obvious directivity. The total sound pressure levels are obviously much higher in 10°to 30°direction than that in other direction. The sound pressure levels at the condition of water injected are lower than that of without water injection at the all receiver points, which indicates that water injection can reduce jet noise effectively.

1 Introduction

When Rocket engine is at work, Propellant combusting in chamber generates high temperature and high pressure combustion gas. And then the gas inject into the surrounding environment through the laval nozzle at a supersonic speed. The high speed jet mix sharply with relatively static medium around, which will form the violent turbulence pulsation in the jet boundary layer, and spread to all around in the form of compressing wave. It will produce violent jet noise to the surrounding. The jet noise not only do harm people at the auditory organs and the upper respiratory tract around the rocket emission but also damage to the load and structure of the aircraft and ground facilities.

The research on jet noise reduction has made great progress so far. Such as microjet noise reduction technology[1], it makes microjet and engine plume generated strong coupling action. The coupling action makes the jet flow field, especially the structure of the turbulent mixing region, change obviously, which change the structure of the jet noise field and achieve the purpose of noise reduction. Based on the sound source, jet noise can be reduced by two methods[2], reducing the jet velocity and changing the structure of the jet field. At present, noise is reduced mainly by lobe nozzle and water injection. Lobed nozzle reduces noise by increasing the contact area of the wake flow with the outside atmosphere, so as to enhance the gas mixture to achieve noise reduction target. In this way, the mixing of airflow is enhanced, the goal of noise reduction is achieved. However, this method will cause the negative effects like the loss of engine's thrust, the increasing resistance, the increasing mass of aircraft and so on while reducing noise. It is proved by practice[3-5] that injecting water in the jet flow field of engine is an effective way to reduce noise. Low temperature water rapidly vaporizes and absorbs noise when it is contacted with high temperature gas. For example, in order to minimize effect of jet noise resulted from the rocket ignition on the aircraft structure, a lot of water would be used to suppress overpressure when the rocket ignition. Krothapalli[6] inject the fine atomization water droplets directly into the gas jet shear layer, with the help of particle image velocimetry, draw a conclusion that water injection reduces the turbulence intensity and shear stress in the jet. NASA and the European space agency claimed that it reduced the jet noise by 8 to 12 db to use the water spray at the plume when rocket was launching. Xu Yue, etc[7]. applied the modified finite volume method to study the effect of water injection on the jet noise. They obtained that the influence of water mass flow rate on the effective gas parameters.

Above all, a lot of researches on jet noise reduction have been carried out. Of them, experimental and numerical study on the water injection to reduce noise are also carried out. But the numerical calculation is based on the equivalent jet parameter method to estimate the size of the noise. In order to eliminate the error caused by the method of equivalent parameter. In this paper, CFD method will be applied to calculate the near field flow field and sound source, combined with the method of solving the wave equation and the far field acoustic field distribution, to research rocket engine jet flow and gas dynamic field noise characteristics, and calculate the

[a] Hao ZOU :erebus1991@gmail.com

effect of water injection on jet flow field and sound field. It provides numerical reference for water injection to reduce noise of the rocket engine.

2 Computing method

Lighthill[8] modified navier-stokes equation into the following form:

$$\frac{\partial^2 \rho}{\partial t^2} - c_0^2 \nabla^2 \rho = \frac{\partial \tau_{ij}}{\partial x_i \partial x_j} \tag{1}$$

The left side of the equation is a classic acoustic wave operator, the right is a source term associated with turbulence mostly, known as Lighthill stress tensor. Lighthill equation of sound source is quadrupole source and also the source of turbulence noise, which is unknown and needed to get through computing the flow field. This equation is mainly suitable for the condition that high speed and turbulence as the main noise sources, such as high speed jet. On the basis of the Lighthill equation, Ffcows Williams and Hawkings[9] develop FW-H equation.

$$\frac{\partial^2 p'}{c_0^2 \partial t^2} - \nabla^2 p' = \frac{\partial^2}{\partial x_i \partial x_j}\{T_{ij} H(f)\}$$
$$-\frac{\partial}{\partial x_i}\{[\tau_{ij} n_j + \rho u_i (u_n - v_n)]\delta(f)\}$$
$$+\frac{\partial}{\partial_t}\{[\rho_0 \upsilon_n + \rho(u_n - v_n)]\delta(f)\} \tag{2}$$

Where, T_{ij} is the Lighthill stress tensor, defined as

$$T_{ij} = \rho u_i u_j + P_{ij} - a_0^2(\rho - \rho_0)\delta_{ij} \tag{3}$$

τ_{ij} is the compressive stress tensor, u_i is the fluid velocity component in the x_i direction, u_n is the fluid velocity component normal to the surface ($f = 0$), v_i is the surface velocity components in the x_i direction, v_n is the surface velocity component normal to the surface, $\delta(f)$ is Dirac delta function, and $H(f)$ is Heaviside function.

FW-H equation, containing a monopole, dipole and quadrupole, can describe all of the aerodynamic noise problem in theory. The complete solution consists of surface integrals and volume integrals. The surface integrals represent the contributions from monopole and dipole acoustic sources and partially from quadrupole sources, whereas the volume integrals represent quadrupole (volume) sources in the region outside the source surface. This paper adopts the FW-H equation to compute the noise flow field.

For Lagrange discrete phase model, fluid phase is treated as the continuous phase and the discrete phase is dealt by computing a large number of particles and bubbles or droplet movement in the flow field. Between dispersed phase and fluid phase, exchange can be exist in momentum, quality and energy, The requirements of volume ratio for the second phase as dispersed phase is very low. The volume rate of injection water droplet in this paper is less than 10%, therefor a discrete phase model can be used.

3 Computing model

A small experimental engine was the object of study. Its nozzle is an axisymmetric laval nozzle. The export is of 22 mm in diameter, and the throat is of 10 mm in diameter. The structure is shown in figure 1.

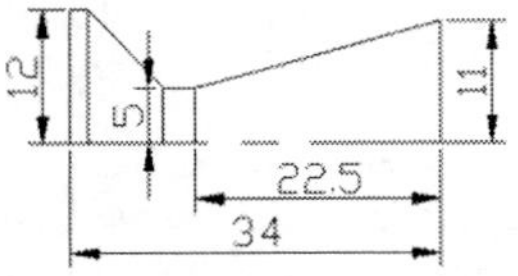

Fig. 1 Rocket engine nozzle structure

The engine combustion Chamber is of 3600K in total temperature and 10Mpa in total pressure. The pressure ratio of outlet to inlet is 0.01013, which is lower than the critical pressure ratio of the jet gas (the critical pressure ratio of air is 0.582). Tthe jet expand completely and can be simply described by the inlet pressure of jet, outlet pressure of jet is atmospheric pressure by default.

In this paper, computational domain selected is of 1000 mm at axial direction and of 220 mm at radial direction, as shown in Figure2. Nozzle entry is set as pressure entrance type of boundary. Nozzle wall is set as wall type of boundary. The rest boundary of the computational domain are set as export pressure. The divergent surfaces starting from the outlet of nozzle were took as sound source of FW-H integral surface. The sound source surface were divided into two parts. The first part starts with the nozzle exit, and 6D in axial length. The second part is of 3D for the initial diameter, 6D for the end diameter, 10D for axial length, where D is the nozzle diameter. A noise monitoring point is uniformly distributed at every 10 degrees in a circle at the center of the nozzle exit with 50d as the radius of circle from 10 ° to 90 °, as shown in figure 3.

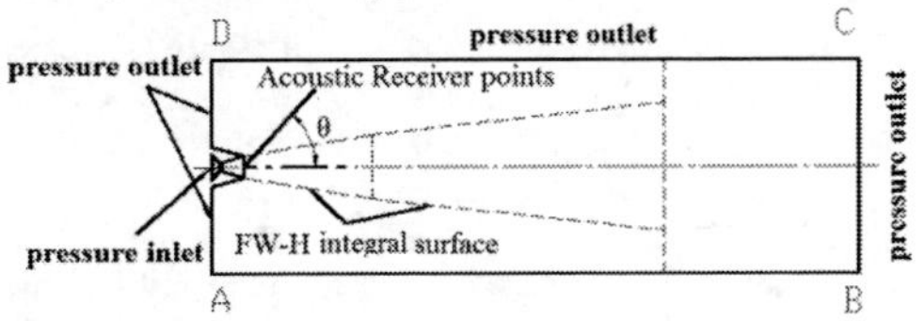

Fig. 2 Computational domain and the distribution of noise monitoring

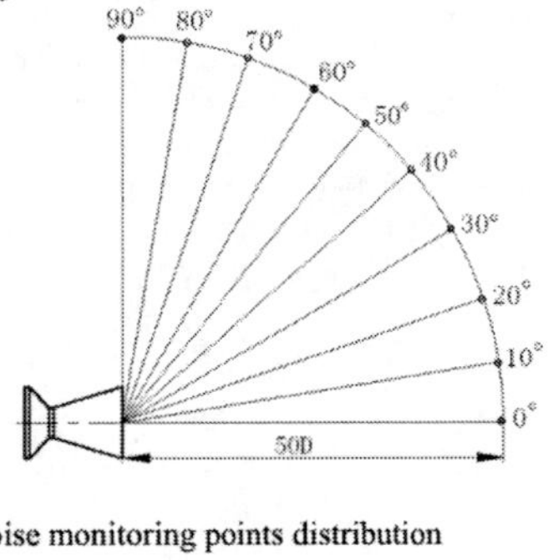

Fig.3 noise monitoring points distribution

DOI: 10.1051/matecconf/20166301023

To guarantee the whole calculation domain to be meshed structural grid, the whole region is divided into several sub-domains for meshing respectively. The nozzle and the sound source surface internal grid is refined, the number of grid is about 2 millions. The configuration of grid is shown in figure 4.

Fig.4 The axial cross-section grid configuration

Standard k-epsilon model, which most commonly used in engineering, is chosen as turbulence model. Without adding water, the medium attribute is set to the ideal gas, The medium is the mixture of air and water vapour after water injection. Water is injected into the flow field in the form of water droplets at a certain speed, then mixing with gas, and vaporized rapidly.

4 Calculation result and analysis

4.1 Flow field without water injection

In order to give a good initial boundary condition for the unsteady calculation, steady state is computed before the unsteady calculation. Under the condition of the steady state flow field, general calculation is exhanged to the condition of the unsteady, the time step is set as $\Delta t = 1 \times 10^{-5}$s. FW-H noise model is open. In order to reduce the influence of initial conditions on the unsteady calculation results, the unsteady calculation takes a long time until the flow field is stable and convergent and the noise of the total sound pressure level changes with time no longer.

Figure 5, figure 6 and figure 7 show the pressure, temperature and Mach number distribution at the axial across section without water spray respectively. From these figures, it is known that the complex expansion shock wave system exist in supersonic jet flow field of rocket engine. As the nozzle exit pressure is greater than the ambient pressure, gas expands at the exit, which makes the pressure and temperature of gas flow decreasing, and the Mach number increasing at the central region. When the air flow reaches the free boundary, the pressure is reduced to the ambient pressure. The gas flow is compressed under the action of the free boundary. The pressure and temperature at the center of the flow increase, and the Mach number decreases gradually until intersection with the free boundary again. And it will generates a new expansion wave system. The supersonic jet flow field of rocket engine is the alternating flow field of the expansion wave and the compression wave until the shock wave vanishes under the action of viscous dissipation.

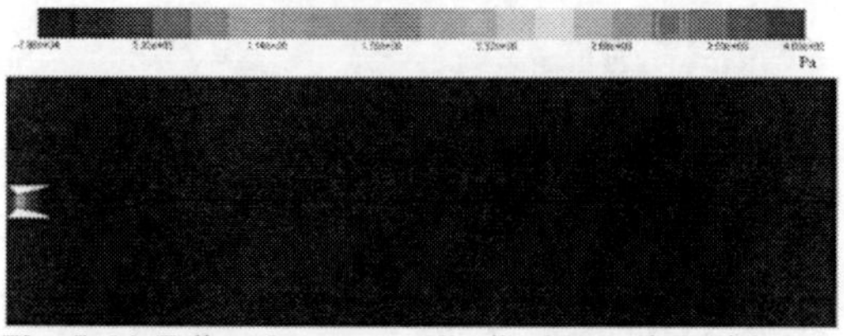

Fig. 5 (a) Full scale pressure distribution at the axial cross section

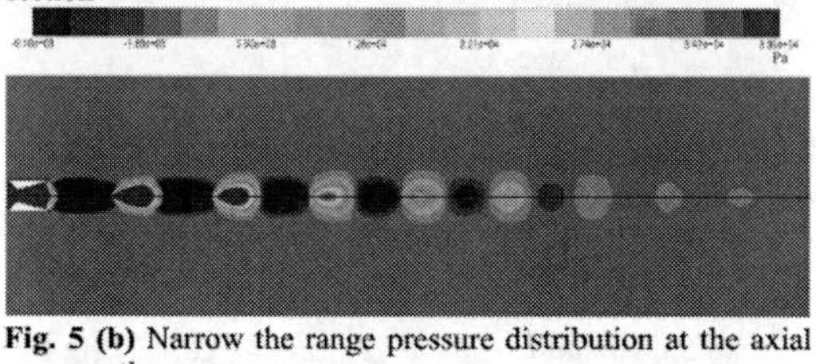

Fig. 5 (b) Narrow the range pressure distribution at the axial cross section

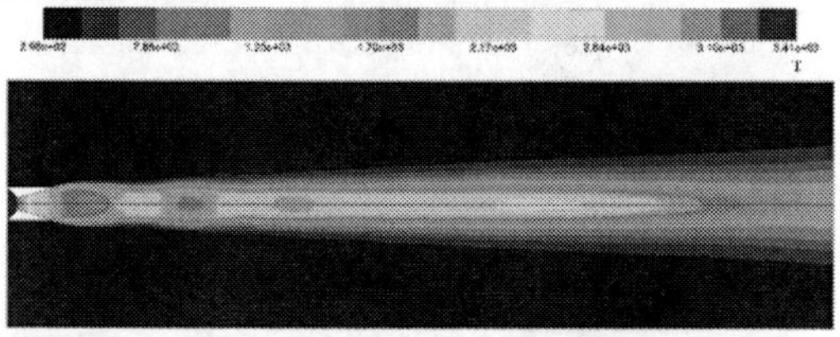

Fig.6 The axial cross section temperature distribution temperature distribution at the axial cross section

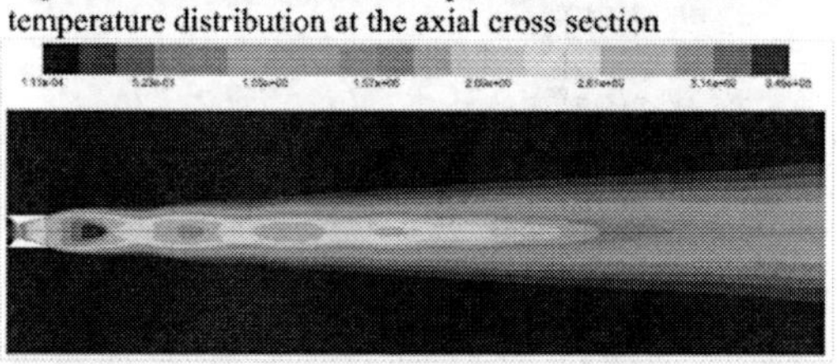

Fig.7 Mach number distribution at Tthe axial cross section speed of Mach number distribution

4.2 fluid Flow field with water injection

Water spray to the jet flow with the angle of 45°to the axis, and the spray point is at the position of 2D far from the nozzle outlet port (where, D is nozzle outlet port diameter). The mass flow rate of the water is 0.88kg/s. The calculation results of flow are shown in figure 8 to 11. These figures show the H_2O mass fraction distribution, Water particles concentration distribution, temperature distribution and Mach number distribution after water injected respectively.

From these figures, it is known that water evaporates after being fixed with high temperature combustion gas, the maximum H_2O mass fraction reaches to 0.466. The maximum concentration of water particles is only 10kg/m^3, the most of water sprayed into the jet flow was vaporized, which make high temperature region of jet flow became small. Due to the evaporation of liquid water, and energy exchange with gas flow, the temperature of flow field at downstream of the nozzle exit decrease obviously, the highest temperature of flow field at downstream of the nozzle exit is less than 1300 k. And the Mach number of flow was decreased. To compare figure 7 and figure 11, it would be found that the Mach number after the water spray was 0.04 smaller than that of without water, and the high Mach region became small too, only a small part of the area, the highest Mach number is up to 3.45 Ma . the supersonic flow does not exist in the downstream flow field area.

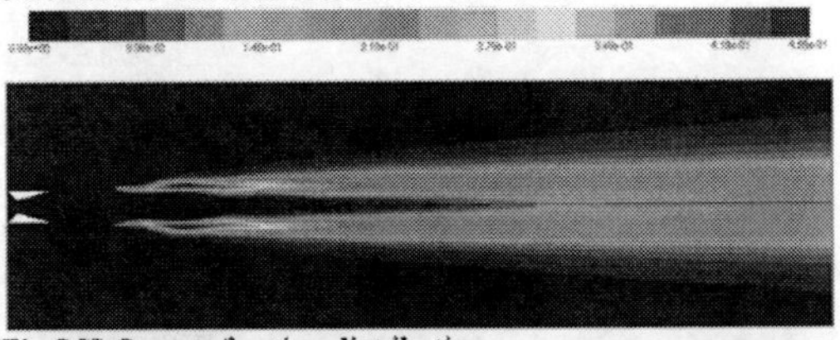

Fig.8 H_2O mass fraction distribution

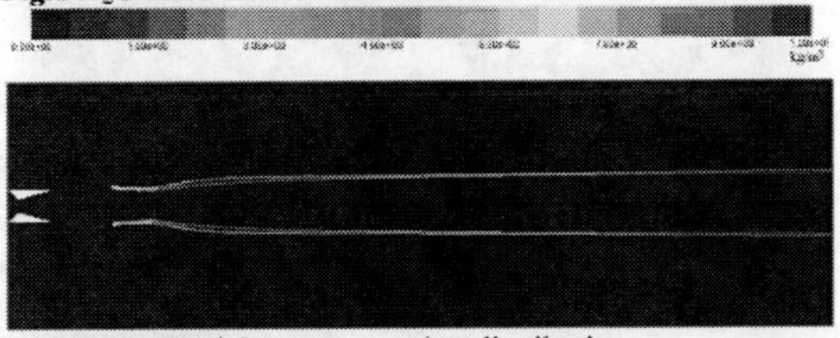

Fig.9 Water particles concentration distribution

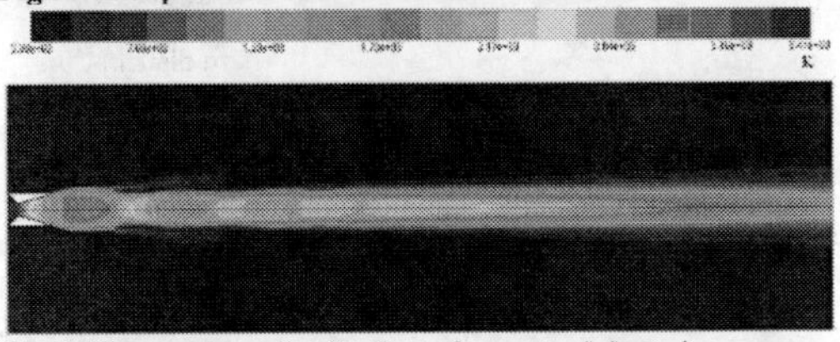

Fig.10 Temperature distribution after water injected

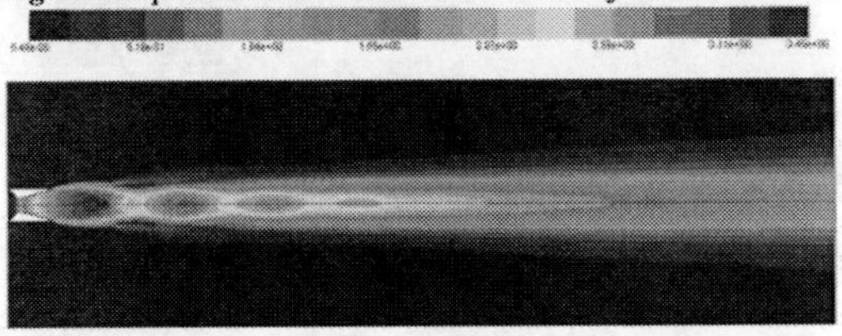

Fig.11 Mach number distribution after water injected

4.3 Analysis of water spray influence on noise reduction

The calculation of the sound pressure signals at the prescried receivers was performed by FW-H acoustics model. And the sound pressure signals were transformed with the Fourier transformation. The total sound pressure level of the monitoring points were obtained. The results are expressed as the function of the angle for the monitoring point, as shown in figure 11.

In figure 12, the solid curve shows the total sound pressure level change with the azimuth angle induced by jet flow field at the condition without water injection, while the dotted line shows that induced by jet flow field with water injection.

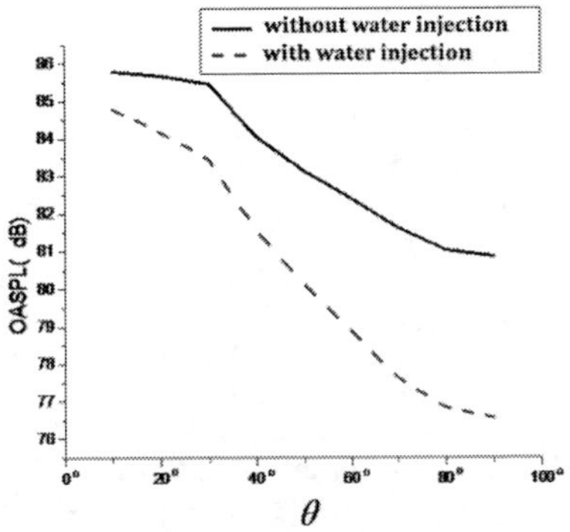

Fig.12 Change of the total sound pressure level with the azimuth angle

From the figure 12, jet noise is not evenly distributed in the circumference direction, but of obvious directivity. The total sound pressure levels are obviously much higher in 10°to 30°direction than that in other direction. The maximum total sound pressure level is 85.78db. According to the overall trend of the curve, from upstream direction to downstream direction total sound pressure level gradually decreases. This is because the rocket jet noise is dominated by the turbulent jet flow away from the nozzle exit. Throughout the 10° ~ 90°direction, within the scope of the dashed curve are located in below the solid curve, and the closer to the 90° direction, the greater the gap is, which indicate that the overall sound pressure levels at the condition of water injected are lower than that of without water injection. The maximum total sound pressure level is 84.78db after water is injected. Water injection at the nozzle exit have effect of noise reduction, the closer to upstream direction, the more effective noise reduction is.

5 Conclusion

1) Under the same initial conditions, the highest temperature of flow field at downstream of the nozzle exit is less than 1300 k after water was injected. Meanwhile, the Mach number with water injection was less than that of without water. And the construction of flow field was changed obviously.

2) The jet noise is with obvious directivity. The total sound pressure levels are obviously much higher in 10°to 30°direction than that in other direction.

3) The overall sound pressure levels at the condition of water injected are lower than that of without water injection, which indicates that water injection can reduce jet noise effectively.

Acknowledgement

This work was financially sponsored by national natural science foundation of China (No.51266013).

References

[1] Xu Yue Research Progress on Aeroengine jet noise reduction by microjet, *Aero. Sci. & Tech.*, (2): 52-54 (2011)

[2] Qiao Weiyang, *Aeroacoustics of Flight Vehicle*. Beihang University Press, Beijing(2010).

[3] Eric Zoppellari, Daniel Juve. Reduction of jet noise by water injection. AIAA-97-1622 (1997).

[4] Thomas D Norum. Reduction in multi- component jet noise by water injection. AIAA-04-2976, (2004).

[5] S.Sankaran, Jopaul K. Ignatius etc. Suppression of high mach number rocket jet noise by water injection. *Journal of spacecraft and rockets*, (6):1164-1170 (2009).

[6] Anjaneyulu Krothapalli, BRenton Greska etc, High speed jet noise reduction using microjets[R]. AIAA-02-2450 (2002).

[7] Xu yue, Zhouxu etc, Reduction of rocket engine jet noise by water injection. *Journal of Aerospace Power*, **25**(04): 816-820 (2010).

[8] Lighthill.M.J. On sound generated aerody- namically: I. General theory. *Proceeding of the Royal Society of London, Series A: Mathematical and Physical Sciences*, **211**(1107): 564-587 (1952).

[9] J. E. Ffowcs Williams, D. L. Hawkings. Sound generation by turbulence and surfaces in arbitrary motion. *Proceeding of the Royal Society of London, Series A: Mathematical and Physical Sciences*, **264** (1151): 321-342 (1969).

Research on the Interior Sound Quality in Hybrid Electric Vehicle

Lian Ying LIAO[1,2,a], Yan Yan ZUO[1] and Hao Dong MENG[2]

[1] *Institute of Noise and Vibration, Jiangsu University, Zhenjiang, 212013, China*
[2] *Changzhou Institute of Technology, Changzhou, 213002, China*

Abstract. Even the overall level of vehicle interior noise of hybrid electric vehicle (HEV) reduced to a certain degree, the vibration and noise generated by the engine, motor, generator and power split have made greater effect on the vehicle interior sound quality in HEV. In order to research the feature of vehicle interior sound quality in HEV, the HEV is used to be the research object, the binaural noise sample of the driver when playing different kinds of music in the vehicle with the speed of sixty kilometers per hour is collected. ArtemiS is used to conduct frequency division processing, so as to obtain the relative weight of each frequency band and the overall noise. The tone, roughness and sharpness of sound quality subjective evaluation parameters are quantified, the SPSS is used to establish the linear regression model of the sample, and the best masking music tracks are found out. Then, the sound samples that contains the best music tracks and the simple vehicle interior noise are re-collected, the regression model and ArtemiS are used to predict the subjective evaluation value. The research results show that when adding the music, the tone degree rises and the lowering degree decreases, thus the disturbing degree reduces, which significantly improves the sound quality in the HEV.

1 Introduction

The strength of music in vehicle is usually higher than the noise level. At the same time, music spectrum is completely different with the low frequency noise in vehicle. Besides, tone is also an important characteristic. These factors can effectively distribute the attention of people on the low frequency noise in vehicle, so as to improve the vehicle interior sound quality [1,2,3]. In order to analyze the improvement degree of music on the vehicle interior sound quality, 24 tracks that are selected in accordance with the provisions are played in vehicle with the uniform speed condition of 60 km/h of HEV [4], and the 24 sound samples contain tracks and vehicle interior low frequency noise collected by SQuadriga I and binaural microphone are studied. Finally, combined with the subjective evaluation criteria and psychoacoustic acoustics parameters, the masking effect of music is subjectively analyzed and predicted. The study results provide theoretical guidance for improving the vehicle interior sound quality in HEV.

2 The subjective evaluation model of sound quality

The masking effect of music means that uses the method of playing music in vehicle to weaken the interference of low frequency noise in vehicle on ears, so as to achieve the purpose of improving the vehicle interior sound quality. At present, the subjective evaluation method is usually used to measure the improvement degree of vehicle interior sound quality. Since the final evaluation index of sound quality is the subjective feeling of people, thus the psychoacoustic acoustics parameters, such as language articulation, tone, loudness, sharpness, roughness ect. are usually used to assist the subjective evaluation and analysis.

In the music, there are many single-tone components. The tone is just the main description on the proportion of single frequency component in noise. Its calculation model is shown in formula (1).

$$K = C \cdot W_N^{0.79} \cdot \left(\sqrt{\sum_{i=1}^{N} [W_1(\Delta z_i) W_2(f_i) W_3(\Delta L_i)]^2} \right)^{0.29} \tag{1}$$

In the formula, C is the normalized constant; WN means the relationship between loudness of no-single frequency component NN and loudness of single frequency component N; W1(Δzi) means the relationship between ith single frequency component and the critical frequency band difference; W2(fi) means the relationship between ith single frequency component and the frequency; W3(ΔLi) means the sound level surplus effect of ith single frequency component. the values are as formula （2）, formula （3）, formula （4） and formula （5）.

$$W_N = 1 - N_N / N \tag{2}$$

$$W_1(\Delta z_i) = [0.13/(\Delta z + 0.13)]^{1/0.29} \tag{3}$$

[a] Corresponding author: liaolianying@163.com

$$W_2(f_i)=1/\sqrt{1+0.2(f_i/700+700/f_i)^2} \tag{4}$$

$$W_3(\Delta L_i)=(1-e^{-\Delta L_i/15}) \tag{5}$$

The loudness difference value of left and right ear in high frequency band is very large, which needs to be corrected, so as to correctly describe the subjective feelings in ears [5], the correction model is shown in formula (6).

$$N_{EQ}=N_M+0.04\times(e^{ILD/4}-1) \tag{6}$$

In the formula, N_{EQ} is the perception value of loudness in ears; N_M is the arithmetic average value of loudness in ears; *ILD* is the sound pressure level difference in ears.

The roughness describes the change relationship between loudness and time in the range of 20~300 Hz of modulation frequency, which gives people a kind of rough psychological feeling, the roughness after music masking is shown in formula (7).

$$R=0.3f_{\mathrm{mod}}\int_0^{24Bark}\Delta L dz \tag{7}$$

In the formula, fmod is the modulation frequency; ΔL is the masking depth.

When the modulation frequency is 0.25~20 Hz, the change of loudness with time is more slowly, which gives people a kind of jitter feeling, that is fluctuation, the value is shown in formula (8).

$$F=0.36\int_0^{24Bark}\log(N'_{\max}/N'_{\min})dz/(T/0.25+0.25/T) \tag{8}$$

In the formula, N'_{max} and $N'_{\min}$ are the greatest feature loudness and the smallest feature loudness; T means the short time difference of two consecutive loudness.

Since the evaluation personnel does not have professional training in advance, at the same time, each sample may only has subtle difference, therefore, the paired comparison method is selected to be used as the subjective evaluation criterion [6]. Make ranking of N samples with two in a group, there are N! kinds of combinations in total. Make comparison playing and let the evaluation team select relatively sweet samples, finally, make quantization process on the results, as shown in table 1.

Table 1. Subjective evaluation test table of paired comparison method.

Group	A is sweeter than B	A is as sweet as B	B is sweeter than A
N_1			
N_2			
…			

As shown in table 1, in the same group, if A is sweeter than B, A gets 2 scores, B gets 0 score. If A is as sweet as B, A and B get 1 score. If B is sweeter than A, B gets 2 scores. When the evaluation is completed, make the unity and normalize the results [7].

3 Collection and pretreatment of sound samples

The driver wears binaural microphone, vice-driver holds HEAD portable sound analyzer SQuadriga I. According to GB/T18697, selecting the high speed road with few vehicles and the sampling frequency is 44.1 kHz, the sound signals of 26 music with different styles played in vehicle with the speed of 60 km/h are collected. Through the subjective comparison, remove 2 acoustic signals that are seriously interfered, and then get 24 sound samples with vehicle interior low frequency noise.

3.1 Equal loudness process

The collected sound samples are concentrated in the range of 40dB~60dB, different loudness makes great effect on the subjective feeling of people. Refer to the sound sample with the loudness value of 20song, make equal loudness process on 24 sound samples. Table 2 lists the change circumstances of all the parameters of sound sample 2 and sound sample 6 before and after equal loudness process.

Table 2. All the parameters of sample 2 and sample 6 before and after equal loudness process.

Objective parameters	2#	2# amendment	6#	6# amendment
Loudness	17.8	19.7	25.95	20.15
Sharpness	1.08	1.115	2.035	1.895
Roughness	1.735	1.84	1.715	1.495
Fluctuation	0.047	0.049	0.059	0.053
Tone	0.079	0.078	0.266	0.274
A sound pressure level	62.9	64.4	66.75	62.95

Before the amendment, N2#<20song, N6#>20song, after the amendment, the values are about 20song. This can effectively avoid the subjective evaluation error caused by the loudness. Through the comparison, the sharpness, roughness, jitter degree, tone and other parameters basically maintain the positive relationship with loudness. In essence, equal loudness process is equivalent to a weight function. It does not change the differences relationship between all the parameters while amending the loudness.

3.2 Filtering in different frequency band

In order to study the effect of different frequency bands on the overall sound quality, at the same time considering too much number of samples will cause the fatigue of evaluators, so as to effect the accuracy of the results, divide the sound samples into 4 frequency bands: mid-low frequency (20~160Hz), mid frequency (160~1280Hz), mid-high frequency (1280~4000Hz) and high frequency (higher than 4000Hz). Conduct division frequency band filtering on the 24 samples after equal loudness process and then get 96 division frequency band sample signals.

4 Subjective evaluation analysis on sound quality

4.1 Subjective evaluation process

As for most of the psychological evaluation tests, 20 evaluators can get more accurate results [8]. 24 students and teachers that have driving experience and are more familiar with the vehicle interior noise environment are selected as the main evaluators. The proportion of men and women in evaluators is selected as 3:1. Before the subjective evaluation, conduct sound listening training on the evaluators, so as to unify the evaluation standard of subjective evaluation.

Sum up all the evaluation scores of each evaluators on each sound sample, thus get the preference evaluation value of each evaluators on the sound sample. Use SPSS software to conduct the correlation analysis on the evaluation result of each evaluators and the evaluation results of other evaluators. After the analysis, remove the evaluation results whose correlation coefficients are less than 0.7. Finally, conduct normalization on the evaluation result according to formula (9).

$$X = 0.8(X_i - X_{min})/(X_{max} - X_{min}) + 0.1 \tag{9}$$

X_i is the random evaluation score; X_{min} and X_{max} are the minimum value and maximum value in the same group. The scatter diagram and linear relationship in the whole frequency band and division frequency bands are shown in Fig.1.

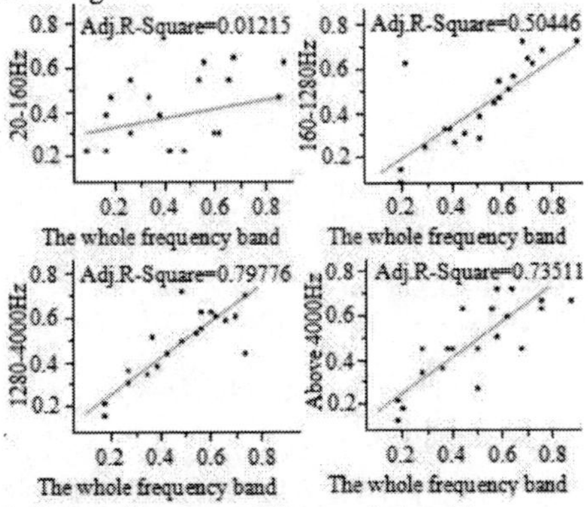

Fig.1. Scatter diagram and linear relationship

Fig.1 shows that, the adjustment determination coefficients Adj.R-Square of the whole frequency band, mid-low frequency band, mid frequency band, mid-high frequency band and high frequency band are 0.01215, 0.50446, 0.79779 and 0.73511. The larger the Adj.R-Square is, the determination coefficient R is larger, and the correlation degree is higher. Therefore, the effect of 1280~4000 Hz frequency band on the subjective evaluation of the whole frequency band is the largest. That frequency band mainly reflects the tone of samples. Secondly, the relationship of high frequency band with more than 4000 Hz is most closed to the language articulation and intelligibility. But if the frequency band has too many ingredients, the sharpness will be too large, which generates auditory fatigue on people. Low frequency samples are mainly come from the vehicle interior low frequency noise, while the music has very little low frequency component. Therefore, the auditory feelings of 20~160 Hz are almost the same, which have smallest effect on the whole frequency band. In the following linear regression model, the sound samples of 20~160 Hz are neglected.

4.2 Subjective evaluation and linear regression analysis

Assume y_k is the subjective evaluation value of samples, among them, k=1, 2, 3, 4 separately represent the entire frequency band, 160~1280 Hz, 1280~4000 Hz and more than 4000Hz frequency band. x_{ki} represents the subjective evaluation parameter of k frequency band. Among them, i=1~8 separately represent the loudness, sharpness, roughness, fluctuation, tone, AI index, A sound pressure level, linear sound pressure level of k frequency band. The subjective evaluation models of the four frequency bands are expressed and shown in formula (10).

$$[y] = [\beta] \cdot [x] + [\varepsilon] \tag{10}$$

In the formula, $[y]$ is the subjective evaluation value matrix, $[\beta]$ is the regression coefficient matrix, $[x]$ is the subjective parameter matrix, $[\varepsilon]$ is the residual matrix.

Use SPSS software to establish the multiple linear regression model of samples, the fitting degree inspection is shown in table 3.

Table 3. Fitting degree inspection of multiple linear regression model.

Model	R	Error ε	Durbin-Watson
The whole frequency band	0.863	0.1385	2.157
160~1280Hz	0.932	0.1031	1.768
1280~4000Hz	0.906	0.1103	1.419
Over 4000Hz	0.889	0.13996	2.366

The R value of each frequency band is very close to 1 and the fitting degree is very good. At the same time, the statistic of Durbin-Watson is about 2, which shows the residual error obeys the normal distribution. Using the method of linear regression analysis to establish the subjective evaluation model is effective. The regression coefficient matrix and residual matrix are shown in formula (11) and formula (12).

$$[\beta] = \begin{bmatrix} 0.089 & -0.048 & 1.283 & 16.075 & 5.012 & -1.531 & -0.09 & 0.089 \\ -0.033 & 2.331 & -1.503 & 11.818 & 0.322 & -0.593 & -0.149 & 0.207 \\ -0.006 & 0.084 & -0.585 & 13.127 & 0.438 & 3.176 & 0.126 & -0.076 \\ 0.191 & -0.044 & 0.614 & -2.21 & 1.058 & -3.494 & -0.158 & 0.125 \end{bmatrix} \tag{11}$$

$$[\varepsilon] = \{-5.574, -4.171, -5.171, 4.74\}^T \tag{12}$$

 DOI: 10.1051/matecconf/20166301024

5 The effect of music on vehicle interior sound quality

Through the subjective evaluation analysis before, the preference character of sample 20 and sample 23 are found to be good, while the subjective evaluation result of sample 12 is worst. Therefore, sample 20 and sample 23 are more suitable to mask the low frequency noise in the vehicle. In order to quantify the effect of music masking effect on vehicle interior noise, re-collect the vehicle interior noise sample of No. 20 and the simple vehicle interior noise sample according to the previous acquisition process. After the process of HEAD ArtemiS, the differences of the two samples in tone, roughness and sharpness are found to be the largest. The results are shown in Fig.2.

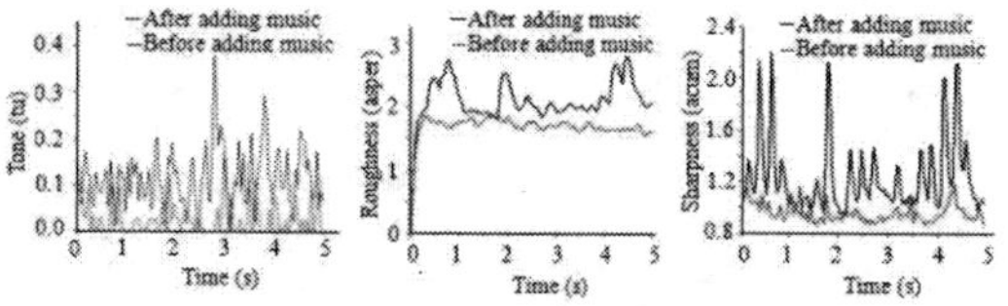

Fig.2. Subjective evaluation parameters change before and after adding music

The tone, roughness and sharpness in the original low frequency noise in vehicle are milder, which is a kind of more low-pitched acoustic environment. In general, when staying in this kind of environment for a long time, people will feel more oppressive and the attention is easily to be distracted, which effects the driving safety. After adding the music, the tone, roughness and sharpness of samples rise obviously. After the sense of rhythm in vehicle being strengthened, the fatigue feelings generated in low-pitched acoustic environment in vehicle can be avoid effectively, which brings pleasure for people in physical and mental.

While compared with the sound quality of the two noise samples before and after adding music, the annoyance level reduced for about 3 levels. In the same site and environment, the original 24 evaluators make the re-evaluation on the sound samples after adding music according to the same division level [9]. The evaluation results are shown in figure 3.

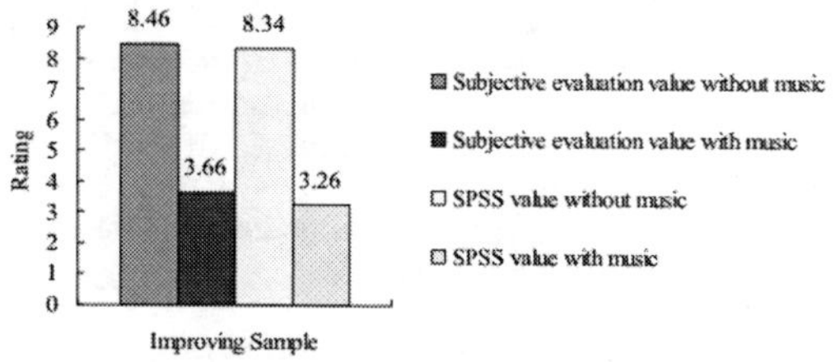

Fig.3. Subjective and objective evaluation value of samples before and after being improved

The subjective evaluation values of annoyance level before and after adding music have made great improvement. The subjective evaluation values of annoyance level increase from 8.46 scores to 3.66 scores, which increase for 9 rankings. At the same time, the consistency of prediction model imitated by SPSS and the values evaluated by the actual listening team is very high. Thus explains using music to improve the vehicle interior sound quality is feasible and effective.

6 Conclusions

The effect of 1280~4000 Hz frequency band on the subjective evaluation of the whole frequency band is the largest. The second one is the high frequency band of more than 4000 Hz. The last one is the low frequency band of 20~160 Hz, which has the smallest effect on the whole frequency band, when regression model is established, the low frequency band can be neglected.

When the samples are under equal loudness process, the samples can be equivalent to a recorded weight function, which will not only make the loudness of each sample be coincident, but also do not change the difference relationship of the remaining variables.

The essence of using music to improve the vehicle interior noise is to raise the tone of vehicle interior noise and reduce the low-pitched degree, so as to reduce the annoyance level and improve the vehicle interior sound quality.

Acknowledgements

This work was financially supported by both National Natural Science Foundation of China (51575238) and the Jiangsu Colleges and Universities Natural Science Research Project (15KJB580001).

References

1. X.P. Chen, *Acoustics and Psychoacoustics*(Posts & Telecom, Peiking, 2010)
2. Y. Pan, Z.H. Meng, ACTA ACUSTICA **38**, 2(2013)
3. D.X. Mao, Y. Wang, Techl Acous **25**,6 (2006)
4. Y.H. Gao, *Pure Electric Vehicle Interior Sound Quality Analysis and Evaluation System*(JLU, Jilin, 2013)
5. J. Zhang, D.X. Mao, Sci. China **40**,4 (2010)
6. D.X. Mao, W.Z. Yu and Z.M. Wang, ACTA ACUSTICA **30** (2005)
7. J.H. Yoon, I.H. Yang, J.E. Jeong, App. Acous **73**(2012)
8. M.X. Shen, S.G. Zuo, Tech Acous **28**,3 (2009)
9. H.Y. Liu, J.P. Gu, Z.H. Meng, Tech Acous **27**, 5(2008)

Application research of solar energy optoelectronic system in the design of culture square

Qi Feng YAN

Bozhou college Anhui,Bozhou 236800

Abstract. In recent years,as people worsen recognition of revival environment and world energy issue,they have started caring about solar energy,which is convenient to achieve,cheap and renewable clean energy,called the 21 century golden energy.Now our project each of designer need worth thinking is how to apply clean and energy-saving solar energy optoelectronic system to square landscape design.To survey and study about solar energy optoelectronic technology in the culture square design has vital influence from the point of economic and social sustainable development and earth ecology.

Key words:solar energy optoelectronic system;conventional energy;ecologicalization;culture square design.

1 Introduction

As our country is advancing further Reform and opening up policy and rising living standard,people whether in countryside or in city, develop their demand to culture square function from original space to daily multi-function reception room,which presents more demands for design.The culture square design doesn't pursue natural landscape,but also pay more attention to design and construction of human landscape so that we can close nature more,feel more comfortable,enjoy more complete facilities lighting,heating,refrigeration equipment),more environmental materials etc.However,statistical data shows if exploitation quantity of energy in the world remains the present level,petroleum on earth will dry up 42 years later,natural gas and coal will use up 66 years later and 150 years later respectively.Thus,in terms of energy utilization , enough construction of culture square will use up more non-renewable energy,meanwhile worsen the environment which completely obey our original intention to construct squares.Solar energy is considered as a convenient and clean renewable resource.Project we worth studying is to apply solar energy optoelectronic system to integrated design of cultural square.

[a] Corresponding author: 1060198680@qq.com

2 Research background and meaning of solar energy optpelectronic technology in square landscape design

2.1 Research background

It is known that developed countries like USA,Japan,Germany,France,Italy are giant countries on energy consumption which get achievement in application technology of solar energy construction .It is French famous heating of thermal storage walls that belongs to invention of French Felix Trombe .Meanwhile,recent years occur zero-consumption house in the world ,it is completed with solar energy belonging to energy consumption and it's really clean and pollution-free which represents development trend of the 21st century solar energy construction.As a giant country with great energy consumption,America not only invest lots of expenditure in exploitation and utilization of solar energy ,but the government also implements tax reduction for consumers.Thus,in recent years,solar energy gains fast development and application as well as promotion,because it possesses leading position in either its study about use technology of solar energy ,study of design optimization or study about new materials,product development,improvement and promotion of construction parts structure.As it were,now America has formed solar energy construction industry integration.

Our country has a vast territory,possesses abundant solar energy.Except that parts of areas (like Zichuan,Guiyang province) are not fit for utilization,most of our country suitably utilizes solar energy.It is estimated that each year amounts of solar energy radiation inland receives is equal to 240 billion tons standard coals.Annual average radiation amount is up to 5,900 megajoule(MJ)/Sq.Our optoelectronic system 'application research began in the late 1970s, we have experienced constant exploration and practice for 40 years,eventually achieve partial results.The development of solar energy systems equipment has a large number of social practice life, and has gained good social and economic benefits.Therefore, effective application for the system in the cultural square design can not only save a lot of conventional energy, and the prevention and control of environmental pollution has effects that cannot ignore.So we advocate "ecological square design, clean e energy application" .In fact,it is the problem of how to utilize well solar energy system.Certainly,the progress of science and technology, will also increase utilization efficiency of the system, decrease construction cost which expands increasingly and promote its application scope.

2.2 research meaning

Square construction is unnecessary process in the modernization city and new-type countryside construction.To study square landscape design application in the solar energy optoelectronic technology has the following three reality meanings: Advocating low-carbon life, improving people' life quality

In recent years, with continuous development of our country's gross national product (GNP),matched lighting, sound, refrigeration, heating and water-consuming system in the square design has become the basic requirement for urban and rural culture square. Culture square construction introduces solar energy optoelectronic system into the lighting facilities, air conditioners, stereos, solar water heaters, air purifiers, water purifiers and so on.It not only saves conventional energy but it is also safe and reliable, healthy and environmental. What's more is a good way to effectively solve the urban and rural culture square construction Promotion.

2.3 Optimizing energy structure, replacing the conventional energy resource

Main energy our country is consuming is coal, oil and natural gas.However,conventional energy resources in China are quite limited. the more and more decreasing resource will dry up one day.Therefore, vigorous development for renewable energy sources such as solar photovoltaic resource not only can replace and save conventional energy, but also it is a really feasible sustainable development road.Square landscape design promotes use of solar photovoltaic technology,which not only saves space and excavation engineering, but also will have important significance on controlling the conventional energy consumption, optimizing energy structure.

3 Characteristic analysis of utilization about china's solar energy photovoltaic resources in square landscape design

As a kind of new energy, solar energy photoelectricity compares with conventional energy such as fossil fuels (coal, oil, natural gas) and nuclear fuel, it has the following several characteristics:(a)

3.1 solar photoelectricity is inexhaustible, inexhaustible

Each year, according to the existing data,it is estimated that solar energy radiation the Chinese land received is the equivalent of 2.403 trillion tons of standard coal, an annual radiation amount is about 5.9 billion joules per square meter.China has a vast territory,solar energy resources is available ,abundant and inexhaustible .For the remote and inconvenient areas like the western China shows its superiority.As long as the one-time investment construction is in place,usual maintenance costs are much smaller compared to the other energy.Of course, in addition to the local area (such as sichuan, guizhou and other places) are not suitable for solar energy utilization, most areas in China is fit for use of solar energy.It is estimated that each year solar radiation amount China inland receives is equal to 2.4 trillion tons of standard coal, an annual radiation amount is about 5.9 billion joules per square meter.

During"five-year" period, our relative research will divide the solar energy resources into five categories areas.The first,second,third type of regions' annual sunshine duration is more than 2000 h, the amount of radiation is higher than 5000 MJ / m². A is rich solar energy resources, the area is larger, accounting for about two-thirds of the nation's total,which has the good condition of utilizing solar energy.The fourth or fifth class area has worse solar energy resources condition, but still has a certain use value.

Solar energy is the most abundant energy human discovered up to now.The sun constantly happens to nuclear reaction and radiates energy outward,and its quality has a loss.It is estimated that the sun lifetime is about 60 billion years,yet the earth life is about 5 billion years. Therefore, we can say the solar energy is inexhaustible, relative to a few thousand years of civilization of human society and history of human for a shorter time to use fossil fuels, solar energy has more perpetuity.

3.2 Solar energy is a green, clean energy

Conventional energy will release a large number of harmful gas in the use ,nuclear fuel will discharge radioactive waste, which can pollute environment.But the use of solar energy can greatly reduce environmental pollution, so the solar energy is clean energy worthy of promoting in the culture square design.

4 Solar photovoltaic dispersion

While solar radiation spreads the global,the incident power per unit area is small. So,to get greater efficiency must be required larger lighting area, which will involve the equipment, material, structure,the occupation of land and other cost problems, yet cultural square just can satisfy the above all kinds of related conditions, so the introduction of solar photovoltaic system in the culture square is a very wise choice

Ⅳ.The application principle of photovoltaic system in the culture square design Square.

As an important part of public open space, shows as a continuous, holistic, unified streets background,the vision shows the richness of landscape elements, such as trees, fountains, sculpture, seat and modern lighting system, air purification systems, water heating system and so on .Square design is the most typical modern city landscape design.Designing excellent square has strengthened the cohesion of the city itself and external attraction, forms urban public space while construct city guard space against disaster, at the same time, improve the service functions of the city, such as multiple roles.Therefore, plan and design application of solar photovoltaic technology in square landscape design is particularly important to improve the city's image, enhance the attraction of the city .

4.1 the application principle of "people-oriented"

The ultimate goal of cultural square design applies arts, science, technology and other means to meet the demand of existence and development in a social setting.The use of city square should fully embody care for people, previous square is small green space without air purification system and high-tech lighting system, taking firm ground or construction as the principal thing;Modern plaza appears a large green space, air purification system, high-tech lighting system, water treatment system, audio system, air-conditioning system etc to collocate and transform through the clever solar energy photovoltaic technology facilities,so as to realize the "functionality", "technological" and "environmental protection" of square ,strengthen the awareness that cultural square is considered as a public center "living room".Planning design of modern culture square considers "person" as the main body, indicates"human nature ", and it is further close to people's lives.

Square should have enough firm ground to supply people for activities, and shall ensure that not less than 25% green area,irrigation,weeding and other kinds of security of green land need photovoltaic system to guarantee so that people to relieve summer heat, and enrich color and stratification of landscape level.

Square need have seats,water fountain , public toilets, telephone booth, small kiosks and other service facilities,and also have some sculptures, sketch to enrich the content to make the square have more cultural connotation and artistic appeal.The normal operation need photovoltaic system as guarantee, only make design novel, layout rational and environment beautiful, functions complete ,which can fully meet the general public' different needs from high art appreciation to fitness, entertainment and leisure.

Improve the practicability of water space, water landscape form, sound, color are the three elements in the design.Both still water and flowing water will bring the ripples of heart, give a person infinite association.But it needs solar energy photovoltaic system to provide security.Whether it is a man-made waterfall, fountain or not will take passion and satisfies human visual enjoyment.

4.2 highlighting the design principles of lighting landscape

The modern people has increasingly rich nightlife,brightening of city square has become a key problem in the construction of city.It is a lighting technology to shape city night scene with lights to beautify the city, show the city individuality, enhance the quality of life, improve the investment environment, promote the development of tourism industry which plays an active role.[3] Only blend deeply solar energy photovoltaic technologyin in the design to adjust mutual relationship suitable to local conditions, integrating various elements inside the space,together with getting good effect.

4.3 ombining with local light condition,blending reasonably the principle of architecture design concept.

Firstly, use cultural square solar energy photovoltaic system constructs must consider local illumination factor in the design, because the light in different region and season has a big difference, integrating cultural square solar energy photovoltaic system designs should make full use of sunlight conditions.At the same time, cultural square using solar energy lighting system construction must meet the aesthetic needs for cultural square, but in terms of the current domestic architectural design, above two principles are comprehensively designed into a cultural square which need make great efforts.This needs reasonably arranging relations of design and construction when doing design ,but the prospects are good too.

4.4 Efficiency balance principle

Significance appears increasingly that modern urban comprehensively utilize square urban space and comprehensively solve the problem of environmental problems.Therefore, planning design of city square have thoughts of "ecology first" humanities and environment coordinating development.Solar energy photovoltaic system as the application of green resources in construction industry is a fairly new concept.In modern society, solar energy resources are mainly used in solar power, solar heating, solar water heaters and solar energy air conditioning and so on. The application of solar technology only is confined to smaller areas,and it does not need spend more cost.At the same time, the equipment of the heating system in solar energy photovoltaic system played a key role, so in the culture square design, the process of the use of solar photovoltaic systems need pay equal attention to principle of economic benefit, social benefit and environment benefit .

4.5 Reasonably using solar energy,realizing principles of energy ecologicalization

The design of square uses ecological energy and material as far as possible,recycles the materials as far as possible,maximizes the potential of energy and materials.[4] Solar energy photovoltaic system is the most common technical means in the daily life.To this end, when we use it in the cultural plaza design,firstly we should note the solar photovoltaic technology.Only comprehensively consider parameters of the solar energy photovoltaic system ,the materials application as well as the various building environment to make the design of solar energy photovoltaic system to do the best.

5 Conclusion

In a word, cultural square is leading actor among the urban culture square and the new countryside culture square design, reflecting the diversity of modern urban and rural areas.In addition, the application of solar photovoltaic system is the lifeblood concerning the social and economic development,but the development of traditional energy causes more and more serious crisis problem .Therefore ,in the modern cultural square design, we shall consciously strengthen the popularization and application of solar photovoltaic system.As a good designer, we not only require artistry and scientific nature of design works, more attention should be paid to the ecology, energy conservation, create vibrant urban and rural cultural square,so that propose and utilize the square.At the same time,the consciousness of sustainable development should be understood,know new energy technology and application specification as far as possible, in order to advocate and actively use solar photoelectricity and other renewable resources in the design ,eventually design a comfortable, beautiful, green and environmental urban and rural culture square.

These guidelines, written in the style of a submission to *MATEC Web of Conferences*, explain how to prepare your paper using Microsoft Word.

Please submit sources files directly to the conference organizer. If the conference editors chose to provide print-ready PDF documents to the publisher, you have to submit high-resolution PDF file with all fonts embedded (see PDF guidelines) instead of the sources files. In this case, please remember that no final corrections will be made by the publisher.

Here are some examples:

1. Huang Yicheng. Energy encyclopedia [M]. Beijing: China encyclopedia publishing house, 1997:600-611
2. Gao Feng, Sun chengquan , Liu Quangen. Some ideas about the situation of the exploitation and thoughts about utilization of solar energy in our country [J]. Energy Engineering, 2000 (5) : 8-11.
3. (USA) Kevin lynch, Xiang Bingren. City impression [M]. Beijing: China building industry press, 1990:42-53
4. Chen Lishun, Dai Dashuang. model study about ecological supply of Urban energy [J].Development and economics of science and technology ,the fifth Phrase, 2006:152-153

MATEC Web of Conferences **63**, 01026 (2016) DOI: 10.1051/matecconf/20166301026
MMME 2016

The switch rail detection system based on laser sensor

Ji Ming SA [1,2], Yu Jun GU [1], Ai Cheng SUN [1], Yong Feng PAN [1]

[1] *School of Information Engineering , Wuhan University of Technology Wuhan,Hubei*

[2] *Key Laboratory of Fiber Optic Sensing Technology and Information Processing 430070, Wuhan,Hubei*

Abstract: As a carrier, turnout is an extremely important part of transport, when railways is operating. So the detection of turnout should meet the requirement. However, the detection effort is mainly completed manually at the present stage, which is low accuracy. Thus the study of the switch rail detection system based on laser sensor is necessary. In this paper, we discuss the scheme of the switch rail detection by using Gocator 2030 laser sensor and SIMENS 840D numerical control system. We study the algorithm and data collection of the switch rail detection based on laser sensor. This detection system provides accurate data collection and information display for the enterprise of producing turnout. After the test, the system has a faster detection speed and higher detection accuracy.

Key words: Laser sensor, Switch rail detection, High-precision, Automation

Introduction

In order to meet the rapid development of railway, the safety of railway transport has gradually aroused public's attention. The accuracy of turnout matching improves steps by steps. The existing turnout detection methods in our country are basically artificial measurement, and the test results will be affected by many artificial factors. The error of measurement is about 1mm, so the artificial measurement can't match the accuracy requirements of high-precision processing equipment. The switch rail detection system based on laser sensor has advantages of small workload, fast measurement and high accuracy. It also can give clear and concise result. So the study of the switch rail detection system based on laser sensor is very important.

In this paper, we take the Gocator2030 laser sensor and the 840D SIEMENS numerical control system as an example, study the specified data measurement of the switch section by laser sensor. The system achieves a high precision measurement and it has important practical significance.

1 The design of detection system

1.1 The principle of detection system

This system measures the switch rail profile by laser sensor. Each detection system has two laser sensors. The data collected by sensor transmits to SIEMENS 840D CNC system through Ethernet. The CNC system achieves these functions, including data processing, information display and interface printing. We put two laser sensors on both sides to realize the detection of the switch rail contour. As shown in Figure 1.

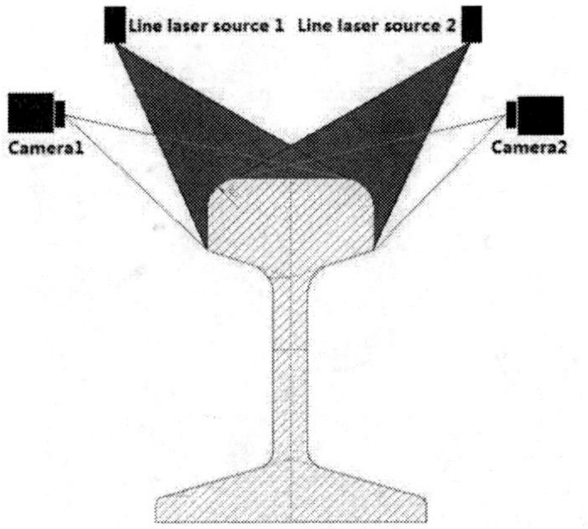

Figure 1 The switch rail profile measurement

The red area is the area of laser irradiation, blue line segment represents the sensor view.

1.2 The profile of Siemens 840D

The switch rail detection system is installed on the Siemens 840D system.

Siemens 840D system (referred to as 840D) is a CNC system based on Personal Computer. Its MMC system uses NT Windows or XP operating system, has all kinds of interface. 840D has better human-computer interaction and integration of the upper application system which facilitates the user to customize the system function and parameter adjustment.

According to customer's request for data reading of machine tool, SIEMENS developed the software for data acquisition, such as SinCOM software and MCIS software. 840D has an open man-machine interface and a communication interface, which uses Ethernet TCP/IP protocol for communication and has flexible reading and writing functions. 840D is suitable for the development of CNC machine.

1.3 Gocator 2030 Laser Sensor

Gocator 2030, which combines laser source with camera together perfectly, has many advantages such as easy to use, strong independence, good development and portable design. The sensor is shown in Figure 2.

(1)Easy to use: The inside Web server of Gocator 2030 can configure parameters and measurement methods by any computer. So gocator 2030 can be very easy to update and configuration without installing additional software.

(2) Independence and Development: The Gocator 2030 system can easily make extensive use of LMI multi sensor hubs, master the power distribution and has a safety interlock device of laser. Gocator 2030 has a Microsecond synchronizer, an encoder and an input array processor.

(3) Portable Design: Gocator 2030 has small volume and its mass is less than 1.5kg. So it is very suitable to set up the Gocator 2030 around the robotic arms and in a tight space.

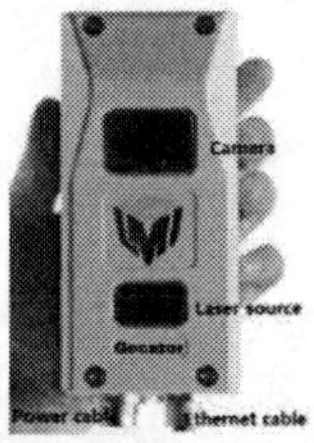

Figure 2 Gocator 2030 laser sensor

2 The design of mechanical structure

The Switch processing is to process the railhead. So we need to extract the complete railhead processing surface. Now we take 4305 switch rail as an example and the standardized process is shown in figure 3. The 4305 switch rail has 8 detection sections. The range of its height is 127.6mm to 150.6mm, which of width id 0 to 71mm. The detection system need to extract effective data from 8 sections. The two sensors' corresponding height and angle to rail will be adjusted to the most suitable condition. We need a high intensity and flexible mechanical structure.

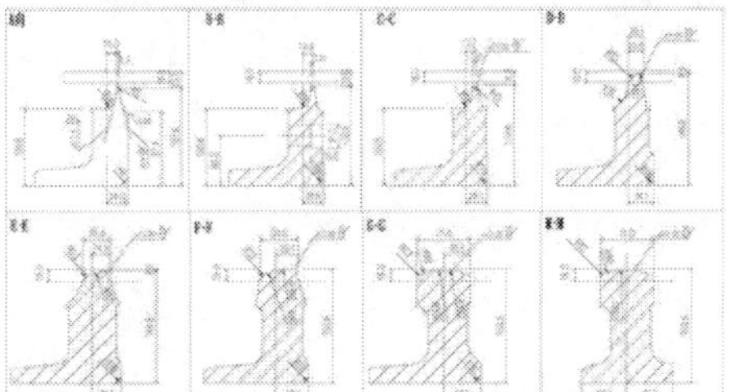

Figure 3 the 4305 switch rail standardized process

The mechanical structure of the system is divided into three parts, which are fixed plate, aluminum profile and sensor connecting plate. The overall configuration is shown as figure 4.

(1)The fixed plate is used to adjust the height and angle of

 DOI: 10.1051/matecconf/20166301026

the two aluminum profiles.

(2) The aluminum profile is used to connect the fixed plate and the sensor connected to plate and to adjust the height of the sensor connecting plate.

(3)The sensor connecting plate is used to change the rolling angle and the rotation angle of the sensor, which ensures the two laser line segments superposed together.

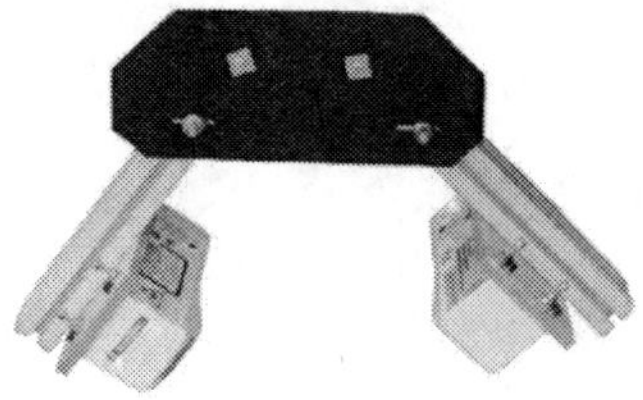

Figure 4 the overall configuration

3.Data processing and algorithm analysis

The contour of the image is converted into N data points by Gocator 2030. Then these data points are stored as CSV file format in the SIEMENS 840D CNC system.

The image data points of the rail have three problems, that are different camera visual angles, some data points lost and burr contour.

In this chapter, we need to design an algorithm, which can deal with the three problems and get the detection data which we need from the processed data.

3.1 Correcting the contour data

In order to correct the contour distortion caused by different camera visual angles and curve the two sections into a complete contour, we calibrate a cuboid standard block, which is 150mm high, 100mm long and 20mm wide, to get the fitting data. The standard block is shown as figure 5. The two groups of curve data can be convert into a same world view by translation and rotation based on the fitting data. The conversion process is shown as figure 6.

The angle of line segment B and line segment C reflect the tilt angle of rail. In fact, line segment B and line segment C are overlap. The expression of line segment B and line segment C is y=ax+b. It is shown in formula (1) and Formula (2) that computing the value of parameters a and b.

The slope of line segment B and line segment C can be obtained as ab and ac based on Least squares. Then the contour data can be rotated in the coordinate system that makes the slope of line segment B and line segment C are both equal to 0. The rotation of the contour data is shown as figure 6(b).

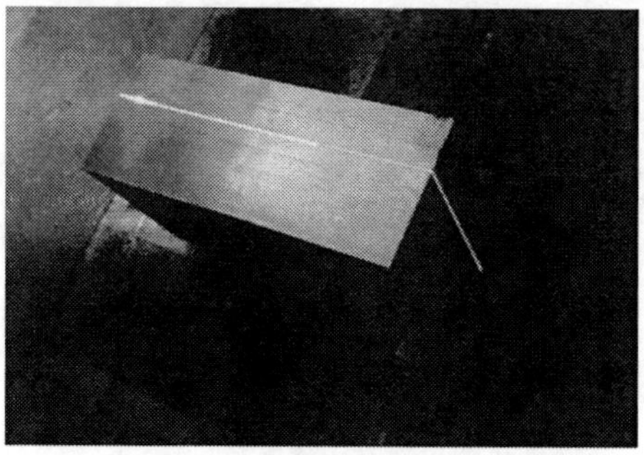

Figure 5 the standard block

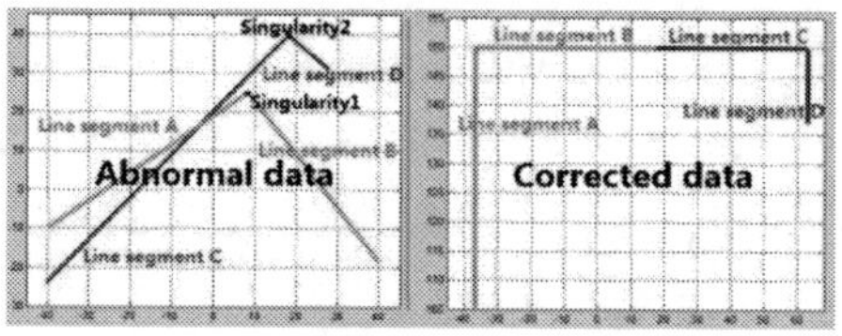

Figure 6 deformity correction

$$a=\frac{\sum_{i=1}^{N}X_i\sum_{i=1}^{N}Y_i-N\sum_{i=1}^{N}X_iY_i}{\sum_{i=1}^{N}X_i\sum_{i=1}^{N}X_i-N\sum_{i=1}^{N}X_iX_i} \quad \text{..............................(1)}$$

$$b=\frac{\sum_{i=1}^{N}Y_i-a\sum_{i=1}^{N}X_i}{N} \quad \text{..(2)}$$

The rotation of the contour data is shown as figure 7. The process of converting original point (Xsp,Ysp) to new point (Xtp,Ytp) needs three steps:

(1) The original coordinate system origin point (Xtp,Ytp) shifts to new coordinate system origin point (0,0).

(2) The original X axis rotates θx and the original Y axis rotates θy.

(3) The original X axis expand ratio of Kx and Y axis stretches ratio of Ky.

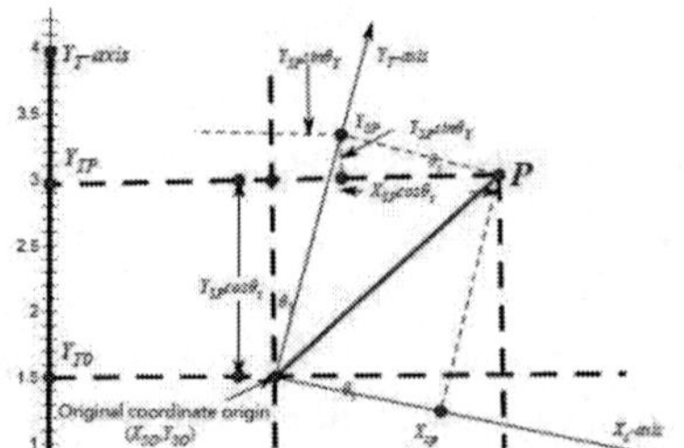

Figure 7 coordinate system conversion

The converted coordinate system is:

$$X_{tp} = X_{so} + X_{sp}K_x \cos\theta_x - Y_{sp}K_y \sin\theta_y \quad \text{..................(3)}$$

$$Y_{tp} = Y_{so} + X_{sp}K_x \sin\theta_x + Y_{sp}K_y \cos\theta_y \quad \text{..................(4)}$$

The proportion coefficient k in formula (3) and formula (4) is 1, so the two rotation angles are:

$$\theta_{ab} = \arctan a_b \text{...(5)}$$

$$\theta_{cd} = \arctan a_c \text{...(6)}$$

θ_{ab} is the rotation angle of green contour data in figure 6(a). θ_{cd} is the rotation angle of red contour data in figure 6(a). ab is the slope of line segment B in figure 6(a). ac is the slope of line segment C in figure 6(a).

The following part introduces how to translate the contour data. Line segment A and Line segment D are parallel. Line segment B and line segment C are an overlapping line. The standard block is 100mm height and 150mm width. The two values Xso,Yso of the two contour data can be computed based on formula (3) and formula (4).

The contour data which has been rotated and translated is shown as figure 6(b). Rotation angle value, x axis translation value and z axis translation value of green contour data and Rotation angle value, x axis translation value and z axis translation value of red contour data has been computed. These 6 contour correcting values can correct the original contour data. The corrected contour is shown as figure 8.

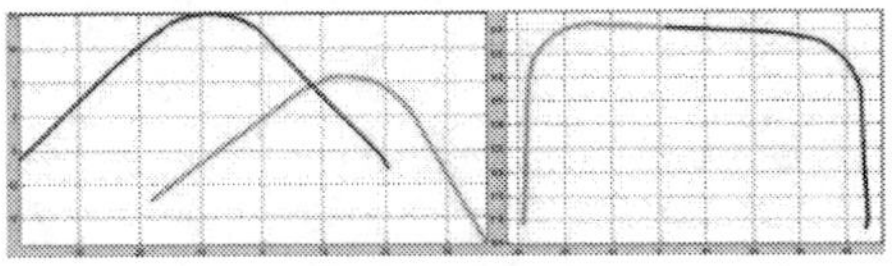

Figure 8 the correction of contour data

3.2 Contour data deburring

The burr of contour data, which is zigzag, is generated by light interference. The burr of contour data is shown as figure 9(a). The burr of contour data can be suppressed or eliminated by using Kalman filtering and sliding filter based on Hamming function.

Kalman filter, which is using the linear state equation, is an optimal estimation algorithm by observing the system data of input and output. The contour data can be processed by using formula(7). The contour data which has been processed by kalman filter is shown in figure9(b). The red line is original contour data, and the blue line is processed contour data by using kalman filter. This method has small/subtle inhibitory effect on the burr of the contour data.

x(k)=Ax(k-1)+Bu(k)+w(k)..(7)

Sliding filter based on Hamming is a weighted smoothing filter. It is a spatial domain filtering technique with low frequency enhancement. The neighborhood size of the sliding filter is directly related to the smoothing effect. The larger the neighborhood is, the better the effect is. However, if the neighborhood is too large, the edge information loss is very serious. So it is very important to choose the value of neighborhood reasonably.

Choosing the value of neighborhood reasonably is a weighted method. Reasonable weighting factor g(r) can be computed by Hamming function W(r).

The Hamming function is:

W(r)=0.54+0.46cos(r* π /m)...........................(8)

The smoothing formula of the five point method is:

t_i=0.04(T_{i-2}+T_{i+2})+0.24(T_{i-1}+T_{i+1})+0.44T_i........................(9)

It is shown in figure 9(c) that the impact of burr on the contour can be ignored.

3.3 Extracting the required values from 8 sets of rail contour data

In the paper, the switch rail line 4305 is selected as sample. The height and the width of 8 sets of the rail contour data needs to be calculated. In order to improve the accuracy of calculation, the following measures can be taken:

1. The height of the rail is the maximum value in the contour data. In order to prevent the interference from the jumping point, we take the maximum 7 values in the average.
2. The width of rail is the width of the rail height decreased by 16mm. In order to make the data more accurate, we select 5 adjacent points on both sides which are decreased by 16mm of the rail height.

DOI: 10.1051/matecconf/20166301026

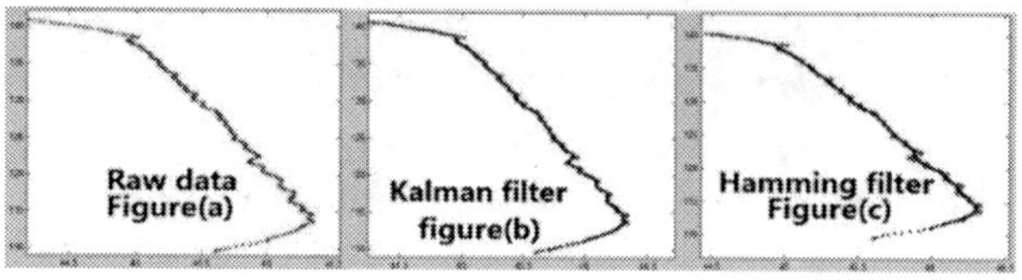

Figure 9 Kalman filter and Hamming filter

4. The detection result

After the contour data acquisition, the smoothed and completed rail contour can be obtained through the data correction and filtering. The required values can be obtained by using some processing algorithms.

In the paper, we select 150mm high, 100mm long, 20mm wide rectangular as the standard block. Than we measure the standard block 5 times. The detection result is shown as figure 10. The values conform to the size of the standard block. The Numerical errors of the standard block values is less than 0.01mm.

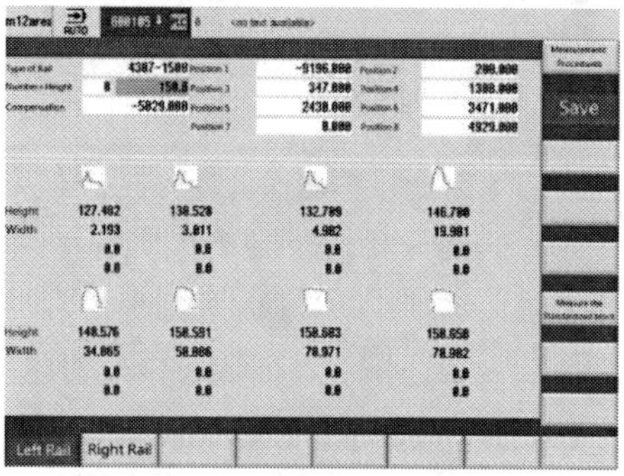

Figure 10 the result of standard block

The detection result of the switch rail line 4305 is shown in figure 11.

The Numerical errors of the rail line 4305 is less than 0.15mm. Comparing with artificial measurement, the detection accuracy of the switch rail detection system has a great improvement.

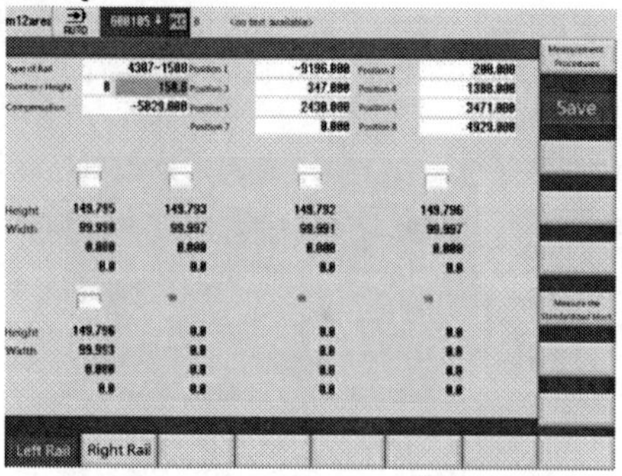

Figure 11 the result of the rail line 4305

5. Summary

With the development of railway, turnout contour detection has a demand of high precision, high efficiency and intelligence . At present, the main use of the turnout contour detection is manual detection. The accuracy of the results depends on the experience of the workers. Besides the manual detection has problems of low detection efficiency and poor data reliability. In the paper, we propose a switch point detection system based on laser sensor, to improve the detection accuracy and detection efficiency.

Based on the principle of coordinate system transformation, we propose a contour section splicing model. It combines the 2 contour sections into a complete contour. The burrs on the contour can be removed by hamming smoothing filter. Finally, the contour data is displayed through the monitor. So the switch rail detection system based on laser sensor with high precision, high efficiency and intelligent meets the requirements of the turnout processing enterprise.

References

1. Weijie Gao, study on rail contour detection system based on machine vision [D], Beijing Jiaotong University, 2012.

2. Gang Dai, study on detection and information processing technology of the cigarette box structure [D], Nanjing University of Science and Technology, 2013

3. Shuding Wu, Study on applying digital detection and multi-channel visualization technology in detection of high-speed railway turnouts [D], Southwest Jiaotong University,2012.

4. Hongyong Mao, Duanwei Shi, An Algorithm to Recognize the Target Object Contour Base on 2D Point Clouds by Laser-CCD-Scanning[D], Wuhan University, 2015.

5. Jiwu Wang, Weijie Gao, Yisong Wang, Study on the height measurement based on the image processing

DOI: 10.1051/matecconf/20166301026

technique[J]. Journal of AROB 16th ,Japan January 27-29, 2011.

6. Wen Li Rail contour detection system based on laser contour sensor [D], China Academy of Railway Sciences, 2014.

MATEC Web of Conferences 63, 01027 (2016) DOI: 10.1051/matecconf/20166301027
MMME 2016

Study on Prediction of Top Oil Temperature for Transformers Based on Bayesian Network Model

Ran LI [a], Ya Yuan YI

School of Electrical Engineering, South West Jiaotong University, Chengdu 610031, China

Abstract. The top oil temperature for transformer has a great influence on transformer's operational life and load capacity, therefore,it is important to predict the top oil temperature.On the basis of analyzing and summarizing the main impacts on the top oil temperature,an idea is proposed to predict the top oil temperature by means of Bayesian network,and Bayesian network model is established.The model takes active power,reactive power,load current, ambient temperature and previous time oil temperature as its quantitative indicators, and trains the sample data to find out the probability distribution between various factors.The model is verified according to data collected from the transformer of SSZ11-50kV/220.The results show that the relative error between predictive value and measured value is small,which can be accepted completely in engineering.Therefore,Bayesian network is reasonable and can be widely applied to forecast the top oil temperature.

1 Introduction

With the rapid development of science and technology, electricity that an emerging green energy has been widely used in various fields which range from the national defense science and technology to the people's livelihood. The large demand for electricity lead to the power system developing along the direction[1] of large capacity, high voltage, large grid , power grid,and smart grid,and the development is more and more rapid, which requires power supply and reliability of the power system to achieve a higher level.In power systems,the transformer plays an very important role[2-3] in economic transmission, flexible allocation, safely use of electricity.Winding hot spot temperature is an important performance index of the transformer, and it is closely related to the operation life and the load capacity of transformers.Once transformer's windings get overheated[3],the internal insulation will decreaseand ultimatelyaffect the normal operation of the transformer.However,it is difficult to measure windings hot spot temperature[4].In practice, top oil temperature are usually used insteading of windings hot spot temperature to scale whether the transformer operate normally. Therefore, predicting the top oil temperature timely and taking protective measures in advance notonly to ensure the stable operation of the transformer,but also extendedlife. Which has an important significance in the filed of saving resource and protecting environment.

With constantly update of modern mathematical theory, there are many approaches[5-10]are used to predict transformer's top oil temperature currently,such as thedirect measurement, numerical analysis, artificial predic- tion and so on.IEEE C57.91-1995 standard [5] has pro-posed transformer top oil temperature rise exponential model, but the model does not fully reflect the impact of ambient temperature.Qing He[6]have taken the load current and ambient temperature into consideration,and establishpredictive models.Comparing withregression model,the predict model is practical.However,but the accuracy remain to be improved.Considering the dynamic changes of ambient temperature around the transformer, B.C.Lesieutre have established an improved top oil temperature prediction model.Model parameters are identified by using the least squares method, and from the predictive results, the model is acceptable in some degree[7].The neural network[8-10] has been applied to predict the top oil temperature according to simulate the relation between the inputs and outputs.

As a branch of mechanical learning,Bayesian network possess the characteristic of inference and visualization when using Bayesian networks to deal with uncertainty and interdependence problem between the multi-factor in complex systems.The generation of oil temperature for transformeris such a problem.There are many factors[11] affecting the top oil temperature,and both of those are interactive.Therefore,Bayesian networks can be used as a effective method of analyzing and forecasting the top oil temperature.After careful analysis and consideration,the paper choose 5 factors,namely active power,reactive power,load current, ambient temperature around transformer and the previous time oil temperature,and build Bayesian

[a] LI Ran:1061479141@qq.com

network model according to the 5 parameters.The model has been tested,and obtain predictive data.When predictive datais compared withthemonitor data,it is found that the predictive data is close to monitor data.And when Bayesian network model is compared withBP model,the Bayesian network model outperforms obviously theBP model consistently in relative error.Therefore,Bayesian network model is appropriate to predict transformer's top oil temperature.

2 Descriptionof Bayesian Network

2.1 The Bayesian theory

The Bayesian theory has been proposed by Thomas Bayes in the early 18th century.It is based on Bayesian formula,which is expressed as follows:

$$P(B_i|A)=\frac{P(B_i)P(A|B_i)}{\sum_{j=1}^{n}P(B_j)P(A|B_j)} \quad (1)$$

In the formula, B_1, B_2,……,B_n are pairwise mutually exclusive events,and $P(B_i)>0$,i=1,2,3,…,n.Occurrenceof an event "A" is always with the simultaneous of one of the B_1, B_2,…,B_n.Bayesian formula is to find the probability of each event "B_i" lead to "A" under the condition that the event "A" has happened.The top oil temperature is affected by many factors.In the processing of predicting the top oil temperature,the probabilities of the events that affect the top oil temperature is calculated by Bayesian formula,and then to forecast transformer's oil temperature by means of the probabilities.

2.2 The Bayesian network

Bayesian network has been proposed by Judea Pearl in 1988. Its essence is an inference network,which is based on uncertainty probability. Bayesian networkprovides an effective method with regard to the causal information presentation.In addition, it is also one of the most effective theoretical models in the areas of expression and reasoning of uncertain knowledge currently[12].Bayesian network is a directed acyclic graph (DAG), that is made up of nodes and directed edges.Nodes represent random variables,and directed edges between nodes represent the relationship between the nodes (points to his son by the parent node). The intensity of the relationship between nodes is expressed by the conditional probability,If there is no parent nodes,it is expressed by the prior probability.

The following Figure1 is a simple Bayesian network.

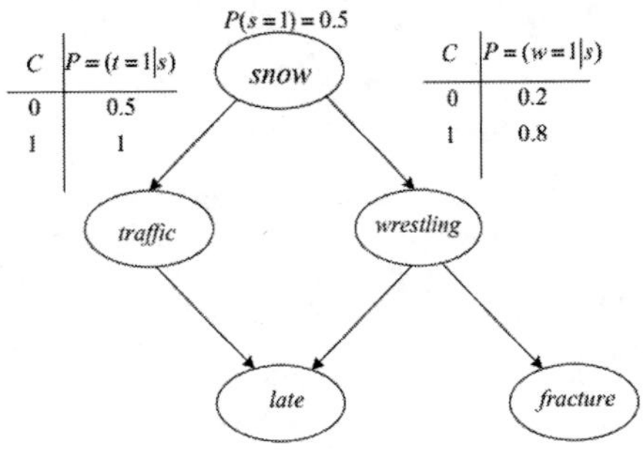

Figure1.a simple Bayesian network

In the Figure1,there are 5 random variables,namely snow, traffic jam, late,wrestling and fracture,and they are briefly represented by "s", "t", "l", "w"and "f".Variables of 1 or 0, is on behalf of the event is true or false.For example, the probability of snowing is 0.5,which can be expressed by P(s=1)=0.5.And P(t=1|s) represents that the probability of traffic jam is 1 if snowing.According to causality,the s is a parent node of w,and f is a child node of w.

2.3 establishment of Bayesian network model

In order to predict the top oil temperature by the Bayesian network,the first step is to establish Bayesian network model.The process of establishing model is divided into 3 steps:

1) choosing the variables nodes
2) determining the network structure
3) deciding the probability distribution of variables

In order to predict the top oil temperature precisely,it is important to take the key factors that influence the top oil temperature into account.However,it is impossible to think over allfactors.The paper divides elements into 3 sorts:environmental parameters,characteristicparameters, and running parameters.Meanwhile,those parameters are interactive.For example,the relationships exiting in the power,voltage,and currentcan be describe as follows:

$$P=UI\cos\varphi \quad (2)$$

$$Q=UI\sin\varphi \quad (3)$$

Where P is active power,Q is reactive power,U is voltage,and I is load current.φ is phase angle.

After consideration,active power,reactive power,load current, ambient temperature around the transformer and historical oil temperature are taken as the variables nodes.

Generally speaking,there are 3 ways to build network structure[13].The first method is that the experts in this field determine network nodes and structure, and provide distribution parameters of the node variables .The second one is that network nodes and structure are determined by experts,butacquire distribution parameters fromsample data by machine learning.The third one is that experts only identify network's nodes, and its structure and parameters are acquired by machine learning.The first method relies entirely on expert knowledge, and expert cognitive is limited, so prediction results will appear large deviation due to the limitations.The first method is just suitable for some simple questions. The second method is fit for the prediction problem where relationship between variables is clear and there are large capacity of sample data.The efficiency of forecast analysis can be improved through parameters learning.The third way is completely dependent on data-driven,so it is apt to address the questions that have larger variable quantity, more complex relationship and adequate learning samples.The prediction of top oil temperature belongs to the problems are described by the above third method.So the network structure can be decide easily.

After determining the network structure,the next step is to calculate the conditional probability of variables.In Bayesian networks,the probability distribution can be obtained by training the sample data.But the sample data must have a large number.The one of the most principal steps is network inference[17],by which the predictive results can be obtained.

3Bayesian probability distribution calculation

3.1 compute of condition probability

Theconditional probabilitieswhichreflectthe dependent relationships of variousfactors，it plays very important roles in the emergence of top oil temperature.Only find out the conditional probabilities,can predict the top oil temperature.At present,there are a lot of algorithms[14-15]applied to compute conditional probabilities.such as the least squares(LS),moment method(MM),and maximumlikelihood estimate (MLE).This paper selects MLE.

Assume sample is $O=\{O_1,O_2,\ldots\ldots O_m\}$,the likelihood function [16]can be expressed by:

$$L=\sum_{i=1}^{n}\sum_{m=1}^{M}\log P(X_i|P_a(X_i),O_m) \quad (4)$$

When the network node variable is discrete,there will be $\theta_{i,j,k}=P(X_i=k|P_a(X_i)=j)$,so the formula(4) can rewrite as

$$L=\sum_{i}\sum_{m}\log\prod_{j,k}\theta_{i,j,k}{}^{I_{i,j,k,m}}=\sum_{i}\sum_{m}\sum_{j,k}I_{i,j,k,m}\log\theta_{i,j,k}= \\ \sum_{i,j,k}I(X_i=k,P_a(X_i)=j|O_m)\log\theta_{i,j,k} \quad (5)$$

$$I_{i,j,k,m}=\begin{cases}0, & X_i=k,P_a(X_i)=j\\ 1, & others\end{cases} \quad (6)$$

Where, $\sum_{m}I(X_i=k,P_a(X_i)=j|O_m)$ recordsthe numberof $(X_i=k,P_a(X_i)=j)$ .By Lagrange multiplier method,the maximum likelihood estimate of $\theta_{i,j,k}$ can be calculated, the result is

$$\hat{\theta}_{i,j,k}=\frac{\sum_{m}I(X_i=k,P_a(X_i)=j|O_m)}{\sum_{m}\sum_{k}I(X_i=k,P_a(X_i)=j|O_m)} \quad (7)$$

3.2 Bayesian network inference

Bayesian networkinference is to calculate probability of nodes variable by means of Bayesian network structure and conditional probability table after giving sample data . Probability inference is one of the most key steps in the process of forecasting top oil temperature.In the past 20 years,the Bayesian network has developed some effective inference algorithms,that consisting of accurate algorithm and approximatealgorithm.The accurate algorithm makes conclusions in full accordance with the calculation formula,so it is just suitable to addressing simple questions.But approximate algorithm is widely used in large-scale network structure to solve the problem,that is complex and have a large amount of calculation.The prediction of the top oil temperature involves a lot of variables,and the variables are interactive,so it is an extremely complicated question. Therefore it is necessary to adopt approximate algorithm.The commonly used approximate algorithm in network inferenceinclude stochasticsamplingalgorithm, search-basedalgorithm,model simplification algorithm, and loopy belief propagation algorithm.The paper makes use of stochastic sampling algorithm.

4 Prediction of top oil temperature by Bayesian network

The paper uses Bayesian Network Tool(BNT) in Matlab to realize the establishment of network structure, network variable parameter learning and prediction of top oil temperature.Matlab is a strong tool in numeral calculations, It can not only reduce computational expense greatly,but also improve the prediction efficiency in some degree.

4.1 sample data

The data used in the paper is provided by a certain grid company in Chengdu.Sampling a transformer of SSZ11-50kV/220 every five minutes from 0:00 to 24:00 on May 3 and 4,we can easily obtain top oil temperature data.And ambient temperature around the transformer,load current, active power and reactive power of the corresponding moment are also supplied. The sample data is divided into training data and test data according to the ratio of 9:1.So there are 520 sets of data used to training Bayesian network,56 sets of data used to test the correctness of model.

4.2 training and learning of Bayesian network

Bayesian network structure can be acquired from learning sample data.There are mainly two types of commonly use structure learning method, namely based on dependency test study and based on search score learning.The latter is widely used,and its principle is to establish Bayesian network structure according to certain search strategy and score criterion in all nodes structure space.This paper has adopted the based on search score learning.Constructing Bayesian network structure consists of two steps:model selection and model optimization.K2 algorithm is applied to learn Bayesian network structure in the paper.It uses posteriori probability as scoring function. According to the pre-determined network node sequence, the posteriori probability associated with the node is obtained by using the greedy search,and decide the node of largest posterior probability as its parent.After allparent nodes determined, the construction of Bayes networkstructure is completed finally.Then the following step is to calculate conditional probability.Above 2.1,maximum likelihood estimate has been introduced in detail.

4.3 prediction results and analysis

At present,there are many indicators are used in most of literature to represent the deviation of the predictive values and actual values,such as squared error and mean absolute error, mean relative error and meansquare error. The paper select relative error to evaluate the prediction accuracy.

56 sets of data are used to test Bayesian network,and 56 predictive top oil temperature are generated,and the comparison with actual data is expressed in the Figure2 BP neural network also apply to prediction of transformer top oil temperature,and the predicted values are compared with the actual values.The comparison is described in the following Figure3:

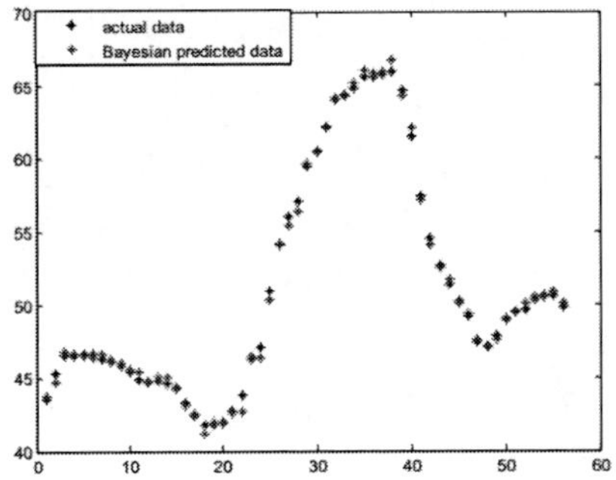

Figure 2. The waveform of Bayesian predicted value and actual value

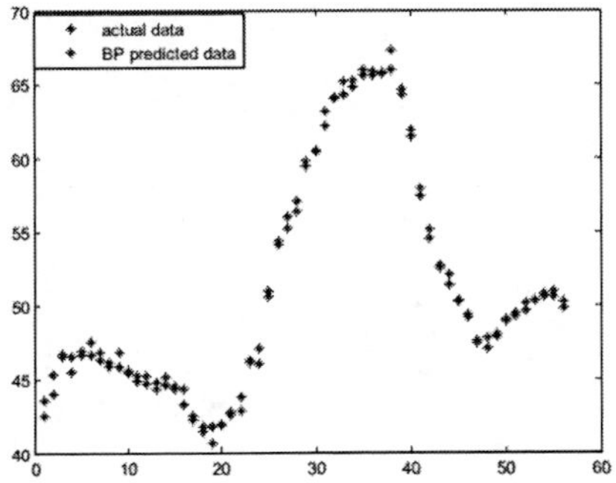

Figure 3. The waveform of BP predicted value and actual value

From Figure2 and 3,we can easily have a simple cognition that Bayesian forecasting result is closer to actual value.

In order to observe results clearly,the relative error between predictive values and measured values have been computed,and I paint the results in a picture,which is represented in Figure4.

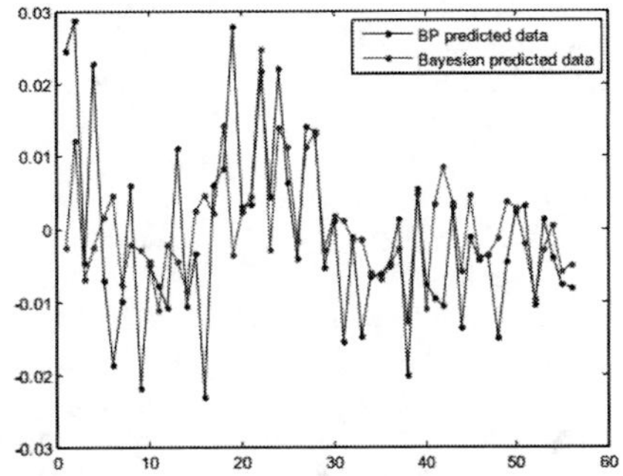

Figure 4. The waveform of relative between predicted value and actual value

As we can see,the relative error of Bayesian network fluctuate around"zero",and the fluctuationranges small. When the Bayesian network is compared with the BP model, the Bayesian network is superiorto theBP model consistently in relative error part.The Bayesian network's maximum relative error is 1.82%,which can be accepted absolutely in engineering.The predicted data can be used to assist decision-maker to guide transformer operation normal.

5 Conclusions

This paper analyzes some factors affecting transformer top oil temperature, and determines the five main factors as Bayesian networks' random variables.On the basis of the analysis of Bayesian network, sampling information active power, reactive power, ambient temperature, load current and the last time the top oil temperature are used as the inputs to train Bayesian network by means of K2 algorithm.And the optimal network structure is determine as a result of employing greedy search.And finally,taking advantage of stochastic sampling algorithm to deduce the transformer top oil temperature.In the end,the prediction of top oil temperature for transformer is realized.Due to many factors that affect the oil temperature, and this article only selects five variables, so deviation exist in forecast results.But the error is acceptable to some extent. Therefore,proposing Bayesian network to predict top oil temperature for transformer can be applied to practice.

References

1. WANG Yongqiang, YUE Guoliang, HE Jie, et al. Studyon Prediction of TopOil Temperaturefor PowerTransformerBasedon Kalman Filter Algorithm [J]. High Voltage Apparatus,**50(8)**,74-79 (2014)
2. XIONG Hao, CHEN Weigen, DU Lin, et al. Study on prediction of Top-oil Temperature for Power Transformer Based on T-S Model[J]. Proceedings of the CSEE,**27(30)**,15-19 (2007)

 DOI: 10.1051/matecconf/20166301027

3. CHEN Jinming, WU Yi, ZHU Haibing, et al. Study and application of Transformer Top-oil predicting[J]. Electrotechnical Application, **33(22)**,89-93 (2015)
4. LIU Hui,SONG Guobing,LUO Junming.Dynamic modeling based on the transformer top oil temperature[J].TRANSFORMER,**39** (**9**),5-9(2002)
5. IEEE Standard C.57.91 — 1995 IEEE guide for loading mineral-oil-immersed power transformer[S](1995)
6. Qing H, Si J, Tylavsky J D. Prediction of top-oil temperaturefor transformers using neural networks [J]. IEEE Transactions on Power Delivery, **15(4)**, 1205-1211 (2000)
7. Lesieutre B C, Hagman W H, Kirtley Jr J L. An improved transformer top oil temperature model for use inan on-line monitoring an diagnostic system [J]. IEEE Transactions on Power Delivery,**2(1)**, 249-256 (1997)
8. Galdi V,Ippolito L, PiccoloA, et al.Neural diagnostic system for transformer thermal overload protection[J]. Proc. Inst. Elect. Eng., Elect. Power Application, **147(5)**,415-421 (2000)
9. Pradhan M K, Ramu T S. On-line monitoring of temperature in power transformers using optimal linear combination of ANNs[C]. IEEE International Symposium on Electrical Insulation, Indianapolis, USA, 70-74 (2004)
10. CHEN Weigen, XI Hongjuan, Su Xiaoping, et al. Application of Generalized Regression Neural Network to Transformer Winding Hot Spot Temperature Forecasting[J]. High Voltage Engineering, **38(1)**,16-21 (2012)
11. CHEN Weigen, SU Xiaoping, CHEN Xi, et al. Influence Factor Analysis and Improvement of the Thermal Model for Predicting Transformer Top Oil Temperature[J].High Voltage Engineering , **37(6)** , 1329-1335(2011)
12. HU Yusheng,TU Xuyan,CUI Xiaoyu,et al. Ainferential Method of Uncertain Knowledge Based onBayes-Network[J].ComputerIntegrated manufacturing System,**7(12)**, 65-68 (2001)
13. ZHOUZhiyuan, SHEN Guchao,ZHU Xiaolong.et al. Application of Bayesian Network in Intelligence Prediction[J].Information Science,**32(10)**,3-8(2014)
14. Wang YansongYu Jilai. Joint conditions probability forecastmethod for wind speed and wind power[J]. Proceedings of the CSEE, **31(7)**,7-14 (2011)
15. REN Jia, GAO Xiaoguang.*Bayesian network parameter learning anddecisionsupport forUAVs*[M]. Beijing: National Defence Industry Publisher, 70-100 (2012)
16. JANGFeng,GAOWen,YAO Hongxu.The *inference andlearning ofBayesian networks*[C]// Proceedings of national networkand informationsecuritytechnology conference in 2005.268-275(2005)
17. LI Haitao,JIN Guang,ZHOU Jinglun,et al.Survey of Bayesian network inference algorithms[J].Systems Engineering and Electronics,**30(5)**,935-939 (2008)

MATEC Web of Conferences **63**, 01028 (2016) DOI: 10.1051/matecconf/20166301028
MMME 2016

Based on Full Life Cycle Battery Green Recycling System Research of Mechanical and Electronic Products

Peng XIE [1,a], Hong Mei ZHANG[1] and Li AN [1]

[1]*College of Science, Air Force Engineering University, Xi'an, China*

Abstract. The mechanical and electronic products especially the battery bring the great convenience to our lives, but also consume a large amount of resource, and pollute environment .In this paper we analysis the problems in battery recycling and utilization in China, and point out the shortcomings as well as propose the construction of battery full life cycle green recycling system.

1. INTRODUCTION

In September 2013, the chairman Xi Jinping points out: we both keep ecological environment protection, and keep economic growth. We prefer to keep ecological environment protection, rather than economic growth, and in reality ecological environment protection is treasure. Vividly expressed government attitude and determination on ecological civilization construction.

The mechanical and electronic products especially the battery bring the great convenience to our lives, but also consume a large amount of resource, and pollute environment . Therefore construction the battery full life cycle green recycling has very important significance.

The full life cycle recycling system of mechanical and electronic products has qualities of economic and environment protection. From cycle economic view, minimizing pollution caused by discarded batteries,supplying new energy or refresh to participate the new life cycle, and forming life cycle recycling to gain maximal profits.

According to related statistics, The batteries of mechanical and electronic products in China replace 1,000,000 sets every year, if small mechanical and electrical product of test terminals including, it can reach about 2,000,000 sets. The discarded batteries contains toxic substances ,such as lead, cadmium, mercury and so on, but driven by short-term economic interests, most of discarded batteries processed by irregularly and simply ,then inevitably make negative impacts on environments.

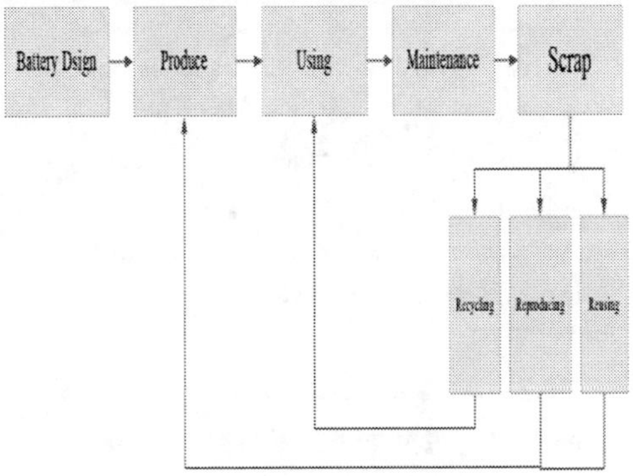

Figure 1 The Diagram of Battery Full Life Cycle

[a] Corresponding author:xpf68@163.com

2. THE PROBLEMS AND REASONS IN DISCARDED BATTERIES RECYCLING

2.1 the problems in discarded batteries recycling

2.1.1 Recollecting rate is low

According to related statistics, The batteries of mechanical and electronic products in China recollecting rate is 10%.

2.1.2 The processing channels are not clear

Some discarded batteries can not find factories to process and stack for long time, which result in environment pollution risk, the other discarded batteries process difficultly due to economic, remote location,or transportation. In short, the discarded batteries hasn't process timely and efficiently, in turn affected initiative on recycling activities, and generated a vicious circle.

2.2 The negative reasons in discarded batteries recycling

2.2.1 Advertisement

On the market, The discarded batteries of mechanical and electronic products divided two types: disposable dry cell which contains toxic metals mercury , and rechargeable battery which included lead-acid battery and nicked cadmium and lithium ion battery and so on. All batteries contains toxic hazardous substances, that can enter our body in various ways, and damage the nervous system and bone after long-term accumulation.

Lacking of related risk knowledge,result in weak in recycling consciousness.

2.2.2 Not yet established a completed and effective recycling system

With inadequate recognition and weak recycling consciousness, the discarded batteries recycling still at small scale. Although few non-governmental environmental protection organizations and individuals engaged in recycling, it is also small quantity and small scale. Comparatively speaking, the recycling quantity is a drop in the bucket, so far it hasn't form a effectively recycling networks.

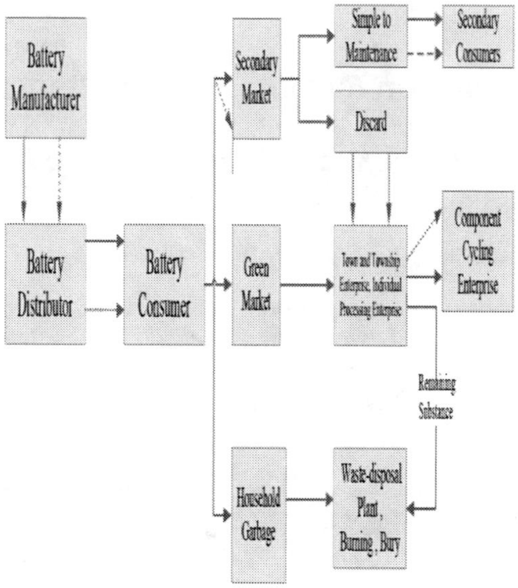

Figure 2 The Status of Waste Battery Cycling and Dispose

2.2.3 High technology, low profits, difficult to form the industrialization production

It is different to recycle discarded batteries, because of recycling process technology requirements is higher than manufacturing requirements. At present, the processing technology abroad is completed. Due to low return, long cycle time, little government subsidies, and less investors, it is unable to form industrial scale, which is one important reason of recycling difficulties.

3. THE BATTERY GREEN RECYCLING SYSTEM CONSTRUCTION OF MECHANICAL AND ELECTRONIC PRODUCTS BASED ON FULL LIFE CYCLE

DOI: 10.1051/matecconf/20166301028

3.1 Strengthen the awareness

Strengthening the ideological quality education, setting up environmental protection consciousness, and widen publicized the laws and regulation of discarded batteries by various ways such as education book, public service ads, leaflets, the mechanical and electronic enterprises must fully recognize the significant of discarded TV recycling project and recycling economic development, and carry out the expand responsibility activities, advocate cleaning production, improve discarded batteries recycling rate.

3.2 Economic and law do both

There are details about battery recycling, subsidies, provisions and penalties in American battery law and resource regeneration (RCRA). With America related laws, we could legislated relevant laws as soon as possible in China , which must be go by and observed, form virtuous cycle. The central and local governments roll out related encourage policies, which contain cleaning new resource, green economic, increasing state subsidies, supporting new emerging industry.

3.3 Strengthen discarded recycling reproducing technology and application research

Reproducing technology is a general term of technical measure and engineering activity, which guided by products full life cycle design and management, aimed for high quality, high efficiency, energy saving and environmental protection products, by the means of advanced technology and industrialized production, to repair and modify discarded batteries. By lengthen the mechanical and electrical products life, life, reduced charge time, increased recharge cycles, simply battery renew technology and other ways to increase battery service time. To Strength the battery reproducing technology research and application is increasingly important, broaden usage ways, raise battery recycling rate, and provide supply for the development of the mechanical and electrical products recycle economic.

3.4 The implementation of mechanical and electrical products recycle design

To develop mechanical and electrical products recycle economic, on one hand, raising battery recycling rate, and on other hand reducing discarded batteries generation, which required consideration on every stage of whole period from design to usage termination. By minimizing demanding of natural resource and environment pollution, maximizing the function, to meet mechanical and electrical products optimized design. The combination of modern design and computer technology, to implement battery green design such DFR, DRD,DFE and other green design concept and method facing to usage and maintainer,

Together with economic benefit, social benefit and environment, to improve every stage of mechanical and electrical products battery full life cycle and resource utilization rate.

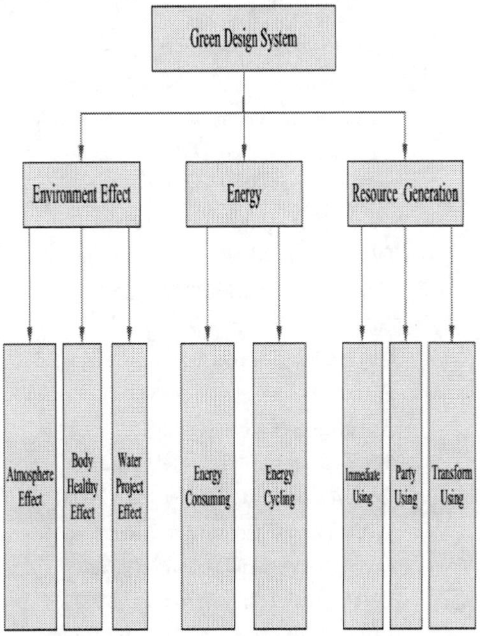

Figure 3 Green Design System

4. CONCLUSION

The recycle economic development bring social benefit and economic benefit newly, discarded battery of Mechanical and Electronic products green cycling as an important tool, it can reduce environmental destruction, increase resource utilization rate, and it is also inevitable demanding of construction of economic society and environment-friendly society .Based on full life Cycle battery green Recycling system of Mechanical and Electronic products not only promote discarded batteries recycling and reuse, but also based on the theory of cycling economic, and from controlling waste products generation in Mechanical and Electronic products generation and processing, coordinate development of economic ,society and environment.

REFERENCES

1. Charles Petzold . Microsoft C# Windows Programming (Part I; Party II). Beijing: Peking University Press(2012)
2. Jeremy Greenwood, Boyan Jovanovic. Financial Development , Growth, and The Distribution of Income[J]. Journal Economy,2013,98:1076-1107.
3. Oded Golor, Joseph Zeira. Income Distribution and Macroeconomics[J]. The Reviews of Economic Sduies,2012, 60(1):35-52.
4. Clark, George, Lixin Colin Xu, Heng Zou. Finance and Income in Inequality: Test of Alternative Theories[C]. 2013, NO.2984.
5. The measures of life cycle[J]. Industry and Environment (Chinese version). 2003(Z1).
6. WANG Zhen. The ways of romoting life cycle development[J]. Industry and Environment (Chinese version). 2003(Z1).
7. Life Cycle Assessment [J]. Toothpaste Industry. 1997(04).
8. HU Zhiyuan, ZHANG Cheng, PU Geng qiang, WANG Chengtao. Oxygen oil life cycle emission assessment[J]. Internal Combustion Engine Project.
9. XUE Feng The methods of extending life cycle [J]. Industry and Environment (Chinese version). 2003(Z1).
10. ZHANG Yaping, DENG Nan Environment policy research of products orientation.[J].Environment science and technology .2003 (04).
11. LIIU Hongmei Food life cycle carbon footprint assessment research[J]. JING CHU institute and technology assessment academic journal. 2012(09).
12. PENG Xiaoyan. ISO14040 Environment management——Life cycle assessment—— Principle and frame [J].world standard and quality management 1998(04).
13. ZENG Guisheng, GUO Qin, HU Chang'an. The technology of recovered valuable metals from discarded lithium-ion [J]. Hydrometallurgy 2008(04.)
14. Li Jian, ZHAO Qian, CUI Hongxiang. The benifit analysis of discarded lithium-ion recovery[J]. China Resources Comprehensive Utilization. 2007(05) .
15. JIN Yujian, MEI Gguangjun,LI Shuyuan. The research of recovered cobalt salt from lithium ion battery positive pole in salting out method[J].Journal of environmental sciences(SCI) 2006(07.)
16. WU fang. Recovery of the cobalt and lithium from discarded lithium-ion [J] Nonferrous metals academic journal 2004(04).
17. WEN Junjie LIJian Discarded lithium-ion recycling valuable metals crafts research[J] Environment protection 2001(12).

AUTHOR: XIE Peng, Male, Born in Zhumadian City, Henan Province. Master degree, University

DOI: 10.1051/matecconf/20166301028

Lecturer. Electronic science teaching and research department of science college, Air Force Engineering University.

Research Fields: Network and date integration.

An innovative research on autonomous learning ability's cultivation for the college students majoring in art design

Xin Guang REN[1], Yong Min CUI[1,a]

[1] *Rongcheng College,Harbin University of Science and Technology,Rongcheng;Shandong,China*

Abstract. The autonomous learning ability plays an important role on the process of improving college or university students' comprehensive qualities. For the college students majoring in Art Design, this kind of ability can help them master the professional skills of art and design effectively and contribute to the improvement of comprehensive quality and professional attainment. This paper mainly puts forward some effective measures to improve the college students' autonomous learning ability through two ways: first, analyzing the problems during the process of forming the autonomous learning ability, which exists among the students majoring in Art Design; second, acknowledging the important meanings of cultivating this ability. These measures are conductive to the development of college students who study this major and their cultivation, and meeting the social demand for practical talents in this field.

With the rocketing development of the Internet science technology, the information from mass media such as WeChat, Microblog and so on updates more quickly than the information in textbooks. So Internet becomes an important approach when people obtain the new information day by day. It also makes up for the textbooks' deficiencies and is convenient for teachers to do a better teaching. For the college students majoring in Art Design, the cultivation of autonomous learning ability and its improvement have a vital effect on improving themselves comprehensive qualities. Therefore, the students majoring in Art Design should change their learning styles from the traditional infusion to autonomous learning and adapt to the new learning style. Students should make full use of the huge Internet resources and put the terminal such as Internet and mobile phone as a kind of learning media, besides to the school learning. College and university is not only the place to learn the professional knowledge but also a bridge from college to society. This period is the critical point for them to improve the self-restraint ability and the communicative ability. So for the college students majoring in Art Design, just studying in college is not enough. The knowledge acquiring from the traditional class and this kind of teachers are hard to meet the demand of society in this field. Then the cultivation of autonomous learning ability is of great importance for their study.

1 The Problem of Autonomous Learning Ability Existing Among the Students Majoring in Art Design

With the quickly development of economy and technology in China, design can meet people's aesthetic demand in their common life and then the art design industry has a rapid growth in the society. The college students majoring in Art Design can adopt flexible forms of employment and do various jobs related to the art design. Therefore, the high requirement of the art design personnel is made in the sustainable development aspect. The college and university should pay more attention on strengthening the cultivation of Art the Design majors' innovative abilities, especially their autonomous learning ability's cultivation and improvement.

1.1 The Different Foundations of the College Students Whose Major are Art Design

The Art Design Major is the combination of modern scientific technology and art. It has a high demand to the Art Design majors' comprehensive quality and needs to cultivate the overall design personnel who has good and professional aesthetic knowledge and a strong sense of art expressive ability. The difference of students' basic capabilities and study is a big obstacles which exist in the Art Design major's teaching process and the cultivation of autonomous learning ability. With the popularity of Chinese higher education and its growth, many high school students do badly in study respect and have little interest in the art, but they still chose the Art Design major as a shortcut to install college or university. In the result, some phenomena appear among the college students of this major, for example the different basic capabilities and study, the different learning attitude as well as the difficulty of students' management. In

[a] Yong Min Cui:renxinguang12@163.com

addition, owing to the difference of students, many rural students are limited by their living standards and then they may lack of the experience of modern art design and the design foundation. Consequently, during the teaching practical activity of the Art Design major, the college or university should face the differences of professional foundations and backgrounds. They can do some organized activity to improve students' professional skills and knowledge' foundation in the art design aspect through building an integrated teaching system.

1.2 The Serious Insufficient of Autonomous Learning Ability among the College Art Design Majors

The college students of China accept the examination-oriented education in a long run and are accustomed to the passive learning style, especially the students whose major are Art Design. They lack of the autonomous learning ability through the learning style in high school and the mode of examination-oriented education. With the flourishing and growth of Chinese economy and society, the new skills, materials and the thoughts emerge as an endless stream. Being a modern art design worker, if autonomous learning ability is insufficient, that is to say, just learning through the passive learning style may lead to the loss of innovative and creative ability, it's hard to catch the pace of art design practice at last. The Art Design major is related to aesthetics, design, management, architecture, dress, media fields and so on. If the college students only attend to the teaching class, it is difficult to have the autonomous learning sense and build their own career planning accurately by themselves. Then they cannot have a comprehensive understanding related to the company and industry etc. and the problem of employment occurs with this problem during the graduation period.

1.3 The Confusion during the Cultivation Process of Art Design Majors' Autonomous Learning Ability

With the development and popularization of the Internet technology, the Internet has become an important way to accept new things for the ordinary people. Besides the textbooks and other books, there are several other cyber or practical ways for the college students whose major are Art Design to study in the autonomous learning style. For example, students need to design software, downloading can be used on the Internet; students need to design the material, searching on the Internet is a good way; students need to design originality things or thoughts, some reference on the Internet can be used etc... The network and computer bring students the sufficient resources and high work efficiency. However, at the same time, the students also can copying others ideas to finish their own works, which makes extensive learning style become a normality. In addition, the college students depend on the Internet deeply and it results in the some phenomena appearing universally, for instance, a serious shortage of professional art and design books reading quantity, professional learning not enough in depth and breadth and the lack of professional practice opportunities. Actually it makes students' autonomous learning channel single and is no use in training and improving their autonomous learning ability.

2 The Important Meaning of Cultivating the Art Design Majors' Autonomous Learning Ability

2.1 Autonomous Learning Contributing to Inspiring Design Inspiration of the Art Design Major Students

Design comes from life and in return it serves life. As to the college students majoring in Art Design, learning the professional knowledge is not only existing on their classroom, but also learning in anytime and anywhere. The traditional learning style cannot totally inspire students' inspiration in their professional field. The cultivation of design inspiration requires students to master a lot of information during daily study and common life by several ways, for example, reading books, searching on the Internet and so on. And thus students need to form a good habit of autonomous learning. The autonomous learning means they not only need to study the professional knowledge but also learning some knowledge related to humanities, art, science and technology etc. so that they can broaden their eyes and inspire the design inspiration.

Therefore, learning the professional course of art and design does not mean to meet the demand of the college students' comprehensive development in art and design aspects. The students must do the autonomous learning after class and be good at observing things, experience life, enhance the knowledge reserve and independent study ability so as to better stimulate their inspiration, create better works of art and design.

2.2 Autonomous Learning Contributing to Developing the Art Design Majors' Creative Thinking

During the design process of the Art Design majors, they need to expand the creative thinking of design through autonomous learning. Exploration, discovery, innovation are the strong power to the development of all disciplines. So, for the Art Design major in college and university, innovation is the soul of design creation. Because art covers many respects of people's common life, continuous innovation in the field of art and design makes all fields' communication possible in art and culture aspects. People like fresh things. So they are in the creation process of art and design. The static ideas tend to drain the origin of creation. So only through autonomous learning to dabble in different knowledge and absorb the useful things used for design from other fields can they continuously create the innovative works of art and design. Therefore, it is of great importance for the Art Design majors to cultivate autonomous learning ability.

3 The Main Approach to Improve Art Design Majors' Autonomous Learning Ability

3.1 A Good Autonomous Learning Environment Created on Campus

The environment of school can reflect the explicit or implicit features in the campus culture. It is a kind of cultural accumulation under some certain rules. The library in colleges or universities carries its cultural connotation and reveals the culture of the school in some extent. It is an important place for students to learn autonomously and plays an indispensable role on improving Art Design majors' autonomous learning ability. According to the needs related to teaching art and design professional knowledge, the library in colleges and universities should constantly update the professional books and resources, form the specialized characteristics database which can reflect the artistic design and pay more attention on the training of the information literacy among the students majoring in Art Design. Opening the art stacks in the library is convenient for the Art Design majors to look up and reference, and gives students a learning atmosphere to enrich themselves and keep their learning enthusiasm. Students can get abundant resources and services as well as study in a good learning atmosphere. The library truly has the effect of educating students to some extent.

Combined with the characteristics of the Art Design majors, the colleges or universities should make some regulations and rules in their management. Only making the scientific and overall systems can they avoid students' arbitrariness and subjectivity and use the rules to manage students' daily behaviors and learning attitudes so that making the students form a good learning style. Under the influence of good learning styles in class, even in the colleges or universities, the good habit of autonomous learning can be formed among the Art Design majors. Under the managing system in colleges, it is easy to neglect the communication with students under equality and freedom conditions and ignore the humanistic care for students. Consequently, administrators and teachers in colleges or universities should strengthen the communication with the students, respect the college students' personality characteristics, make efforts to find the human-centered factor in the process of the education and improve the students' autonomous learning ability.

Due to the emphasis on creative and practical ability of the Art Design major's characteristics, the college or university should make a relevant incentive mechanism to active and encourage the students' innovation and make efforts to launch various activities combined with the professional knowledge and specialized practice. For example, to establish a professional mentors studio related to visual communication design, environmental design and digital media art, to undertake some works based on the social actual projects, contributing to increasing the professional students' autonomous learning enthusiasm after class and guiding and helping students to run their own business; encouraging students to participate in the high level professional competition and its related professional and academic exchanges actively, making full use of social practice during their holiday and stimulating and training their autonomous learning ability in the practice. Colleges and universities establish the enterprise cooperation model with the enterprises to transmit Art Design majors to enterprises Dinggang internship, to practice their professional theory in the enterprises, to consolidate professional knowledge and to continuously test students' autonomous learning ability.

3.2 Strengthening the Comprehensive Quality of Teachers and Helping to Guide Students' Autonomous Learning

Improving the teachers' comprehensive quality and professional quality has a positive effect on the improvement of Art Design majors' autonomous learning. High level and high quality professional teachers of Art Design major tend to have a broad mind and wide vision, which does good to giving students a reasonable guide in the career planning. In order to improve the comprehensive qualities of teachers, teachers are required to innovate their teaching mode during the teaching process, so as to ensure to carrying out the teaching process efficiently, to cultivate the college students' autonomous learning ability and the ability of exploration actively. On the one hand, teachers should fully consider students' professional foundation and learning efficiency when they teach the students. With the aid of teaching materials and teaching equipment, teachers use lectures, exercises, questions and answers, to organize the discussion and guide the students to find the approaches of autonomous learning, and helping students master the professional knowledge. On the other hand, teachers should make good use of the network information, bring the new technology, new method and new ideas to their curriculum design and encourage students to study through searching professional material on the Internet. These behaviors can help students to keep up with the professional frontier, learn the new material and new concept etc. and also can stimulate the Art Design majors' learning motivation and interest and help students learning autonomously with the aid of the network.

In addition, the Art Design majors' cultural foundation is relatively weak in colleges and universities. They go their own way without discipline and have little attention on the learning planning and assessment. Besides guiding students in the class, teachers help them master the professional knowledge about art and design through the most effective way, and consciously improve students' autonomous learning ability through all kinds of approaches, so as to enable students to constantly improve the creative design ability, master the professional technology on the frontier as well as have the ability to adapt to the social needs and development. Therefore, teachers must enhance their life-long learning ability, constantly improve their own educational philosophy and improve the knowledge structure in order to make the professional education continue to meet to the social development demands in the new period.

3.3 Stimulating Learning Motivation and Strengthening Students' Self-Monitoring Ability

In order to make the Art Design majors cultivate autonomous learning ability, college administrators and teachers must inspire students' intrinsic professional learning motivation at first. In psychology, students are affected by many factors in their studying. However, the decisive factors is the students' learning motivation. Other inseparable factors contain the students' social value, learning interest, ability and consciousness etc... During their learning process, it is inevitably to encounter a variety of learning difficulties and then the learning motivation would gradually weaken. At this time it is particularly critical for the students to continue to adhere to the study by will. So, besides teaching knowledge in the classroom, teachers should pay more attention on guiding students, letting them understand the meanings and values of the professional courses and improving their teaching methods so as to leading the students to keep good curiosity and thirst for the professional knowledge which would help Art Design majors cultivating autonomous learning ability.

On the process of learning, the students not only need to have learning motivation and learning will but also have the scientific and rational planning during the autonomous learning. They need to have some self-monitoring ability to overcome the inertia and avoid the blindly learning. Above all, the college students majoring in Art Design should do self-monitoring to develop a practical learning plan and objectives. On the same time the plan should consider their own actual situation. They need to adjust their plan in time by taking appropriate self-monitoring, continuous review and observing autonomous learning situation. Being Art Design majors in colleges and universities, upholding the long-term autonomous study plan can increase their autonomous learning ability and improve their professional skills. It can let them skilled use their knowledge and design skills and enhance their own employment power.

In short, the rapid development of the current knowledge and information society affects people's life style and the way of thinking. For the Art Design major's education and teaching work in Colleges and universities, it is necessary to break the traditional teaching style and innovate the teaching concept to deal with the great potential of the Art Design major's development in the information society. The cultivation of Art Design majors' autonomous learning ability can not only improve the students' comprehensive quality and professional quality, but also enhance their competitiveness of the future. Therefore, when the colleges and universities innovate teaching ideas in the Art Design major, they also need to enhance the students' autonomous learning ability.

References

[1] Xiaoqing Li and Jianguo Li, The Improving Strategies of the Art Design Majors in Colleges or Universities [J], The Home of Drama, 2014. (18)
[2] Jiajia Xue. On the Improvement of the Fine Arts Majors' Autonomous Learning Ability in Colleges or Universities[J], Statistics and management, 2016. 1
[3] Min Yang and Hongjun Zhang, Analysis on How to Improve the Ability of Autonomous Learning Ability in the Art Voluntary School [J], The Age of Teaching, 2013. (23)
[4] Cheng Ji and Quan Wen,On the Cultivation and Innovation of the Autonomous Learning Ability [J], Literary Theory, 2015. (18)
[5] Xiaoquan Zeng, The Cultivation and Research on the Art Design Majors' Autonomous Learning Ability [J], The Higher Education Forum, 2010. 2
[6] Shifu Xu, A Study on the Environmental Art Design Major's Autonomous Learning Approaches under the Background of Internet [J], Educational Research on the Art, 2015. 9
[7] Zhi Liu, The Exploration on the Cultivation mode of the Art Design Majors' Autonomous innovative Ability [J],China Higher Education, 2014. (17)

Simulation of coal low-temperature oxidation heating process in gob with "U+L" ventilation

Pei Ling ZHOU [1,2], Ying Hua ZHANG [1,2], Xu Jie SHEN [1,2], Zhi An HUANG [1,2,a], Yu Kun GAO [1,2]

[1] *State Key Laboratory of High-Efficient Mining and Safety of Metal Mines(University of Science and Technology Beijing), Ministry of Education, Beijing 100083, China*

[2] *Mine Emergency Technology Research Center(University of Science and Technology Beijing), State Administration of production safety supervision and administration, Beijing 100083, China*

Abstract. In a gob with U + L ventilation, a tail roadway exists, which has important effects on the oxidation heating process and gas concentration in gob areas. Research on the heating process and gas concentration in the "U+L" ventilation can provide the basis for the prevention of spontaneous combustion, thus, the regularities of the oxidation heating process and gas concentration in gob areas were researched by simulation. Results showed that compared with U ventilation, U + L ventilation caused the high temperature zone and high temperature points in the gob areas to increase in depth and width and to be influenced by the distance between the crossheading of the tail roadway and workface. The heating rate of the high-temperature point in the gob with tail roadway was 1.5 times of that in gob without tail roadway, but was unaffected by the location of the tail roadway. Tail roadway had diversion effects on the airflow, especially near return side and the maximum reduction of gas concentration can be 0.36%.

1 INTRODUCTION

Spontaneous combustion in a gob is one of the major sources of coal mine disasters. Under a U + L ventilation condition, the effect of tail roadway on spontaneous combustion cannot be ignored. In research regarding spontaneous combustion "three-zone," many domestic and foreign scholars have focused on the gob areas of U-type ventilation; they have used theoretical analysis, experimental research, and numerical simulation to conduct research on the distribution of "three zones" in an air-flowing field and an oxygen concentration field, which reflects an actual situation [1-3] at a certain extent. However, research on the spontaneous combustion in the temperature field is limited.

On the other hand, Su Quan-zhi [4] conducted research on the spontaneous combustion in the pressure field and air-flowing field in a gob with U + L ventilation [5,6]. The result showed that the existence of tail roadway made the pressure field change in a gob, thus leading to the offset of spontaneous combustion zones to tail roadways in the air-flowing field and increasing the risk of spontaneous combustion. Hence, the effect of tail roadways on spontaneous combustion in a gob cannot be

[a] Corresponding author:huang_za@qq.com

DOI: 10.1051/matecconf/20166301030

ignored. However, studying a pressure field or an air-flowing field alone cannot fundamentally reflect the spontaneous combustion characteristics in a gob. As an indicator of spontaneous combustion, temperature is a significant parameter to be studied. Therefore, from the perspective of multi-field coupling, conducting research on the spontaneous combustion "three-zone" in a gob with tail roadway and combining the pressure field, air-flow field, and temperature field will yield results that are close to the actual situation and have important implications for production.

2 MODELING OF MULTI-FIELD COUPLING

2.1 Mathematical model of multi-field coupling

A gob area is a porous medium system consisting of caving rock mass and residual coal. Spontaneous combustion is a result of the interaction and mutual coupling of internal pressure, air flow, oxygen concentration, and temperature. Spontaneous combustion is a result of the interaction and mutual coupling of internal pressure, air flow, oxygen concentration, and temperature [7-10]. In a gob with U + L ventilation, a mathematical model of multi-field coupling can be established by using the simultaneous governing equations of temperature field, pressure field, and oxygen concentration field. We can obtain a 3D mathematical model of the low-temperature oxidation of coal by using the following equation:

$$\begin{cases} \frac{\partial \overline{Q}_x}{\partial x}+\frac{\partial \overline{Q}_y}{\partial y}+\frac{\partial \overline{Q}_z}{\partial z}=0 \\ \frac{\partial C}{\partial \tau}+\overline{Q}_x\cdot\frac{\partial C}{\partial x}+\overline{Q}_y\cdot\frac{\partial C}{\partial y}+\overline{Q}_z\cdot\frac{\partial C}{\partial z}=D_e\left(\frac{\partial^2 C}{\partial x^2}+\frac{\partial^2 C}{\partial y^2}+\frac{\partial^2 C}{\partial z^2}\right)-V(T) \\ \rho_e c_e\frac{\partial T}{\partial \tau}=q(T)+\lambda_e\left(\frac{\partial^2 T}{\partial x^2}+\frac{\partial^2 T}{\partial y^2}+\frac{\partial^2 T}{\partial z^2}\right)-\rho_g c_g\left(\overline{Q}_x\cdot\frac{\partial T}{\partial x}+\overline{Q}_y\cdot\frac{\partial T}{\partial y}+\overline{Q}_z\cdot\frac{\partial T}{\partial z}\right) \\ \frac{\partial}{\partial x}\left(K_x\frac{\partial H}{\partial x}\right)+\frac{\partial}{\partial y}\left(K_y\frac{\partial H}{\partial y}\right)+\frac{\partial}{\partial z}\left(K_z\frac{\partial H}{\partial z}\right)=0 \end{cases} \quad (1)$$

In multivariate equations, x, y, and z are the coordinates of 3D space; $\overline{Q}$ is the leakage flow rate; K is the absolute permeability; C is the concentration of O2 (the value is 24%); D is the oxygen diffusion coefficient in coal; V (T) is the actual oxygen uptake rate; H is the partial pressure of heat. These equations are convection diffusion governing equations, chemical reaction governing equations, heat transfer governing equations, and momentum transfer governing equations.

A calculation model was established with a tail roadway. On the basis of the coal–oxygen reaction mechanism, a UDF computation program was written and imported to FLUENT.

2.2 Modeling gob of U + L ventilation

A certain mine was utilized as a prototype, and a geometric model of a gob was built. The gob trend was researched in 100 m. The working surface length was 100 m; the height was 10 m; the ventilation airways were 4.4 m wide, 2.6 m high, and 10 m long; the tail roadway was 3.5 m wide, 2.2 m high, and 5 m long.

From the working surface to the distance, the caving zone in gob was divided into three zones: the natural accumulation area (10 m), load influence zone (30 m), and compaction stability region (60 m); these zones were named a, b, and c, respectively. Accoding to the working surface data and calculating formula, the bulking coefficients are K_{pa} = 1.94, K_{pb} = 1.87, and K_{pc} = 1.5. The relationship between porosity and bulking coefficient is $\varepsilon=1-\frac{1}{K}$; thus, the porosity of each district is 0.61, 0.58, and 0.46; D_{ma} = 0.35, $1/\alpha_a$ = 2850, and C_a = 45.1; D_{mb} = 0.2, $1/\alpha_b$ = 10700, and C_b = 93.1; D_{mc} = 0.1, $1/\alpha_c$ = 181000, and C_c = 632, D_m is the average particle diameter, $1/\alpha$ is the viscous drag coefficient, and C is the inertial resistance coefficient.

3 MODEL VERIFICATION

Owing to the limitations in production and cost, field verification cannot be conducted. The simulation experiments were conducted in a laboratory to validate the model. Given that the temperature distribution in a gob is the simulation result of multi-field states and physical parameters, the multi-field states, physical parameters, model will be correct if the temperature rising law is correct. In this case, the simulation results after changing the simulation materials are also correct. Considering that the oxidation heating rate of coal is slow, spontaneous combustion phenomena will occur only in accumulation and good thermal conditions. Accordingly, we selected pyrite as material, which had a fast oxidation heating rate.

On the basis of the similarity criteria for dimensional analysis, combining the actual situation at the scene, and according to a similarity ratio of 1:200, the model was designed with stainless steel housing and within 30 mm-thick extruded polystyrene board insulation materials. The content was 520 mm × 500 mm × 50 mm, and the size of both intake and return gates was 50 mm × 50 mm × 50 mm.

The model verification was conducted as follows. First, a numerical simulation was performed. The model and multi-field parameters above were used. The volume was as large as the volume of the similar model. The material was pyrite. Second, a similar simulation experiment was conducted, and the temperature distribution of pyrite was simulated. The physical map of the model and the temperature monitoring points are shown in Figure 1. Third, the simulation results were validated with the test data. If the simulation results were in good agreement when the simulation material was pyrite; thus, the model was correct. In this case, the simulation would also be correct when the simulation material was coal. Thus, the correctness of the simulation model could be verified.

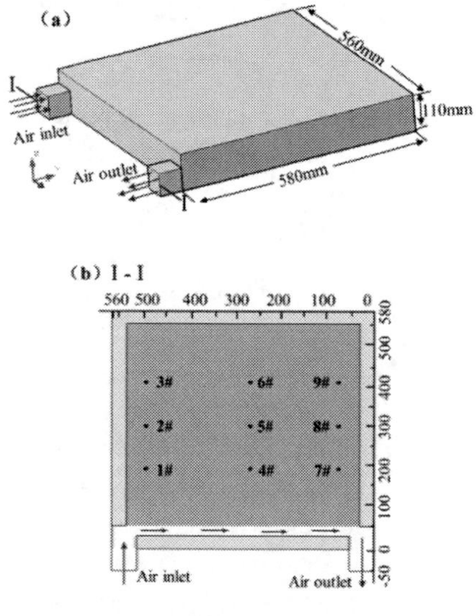

Figure 1. Model map and measurement point position

In Fig.1, the #1, #2, and #3 thermocouples were in the inlet side, whereas the #7, #8, and #9 thermocouples were in the outlet side. The validation was divided into two steps. Temperature distribution of the reaction after 10h was verified. The comparison results are shown in Figure 2. From Figure 2, the results of similar experiments were consistent with the results of numerical simulation; thus, the correctness of the model was verified.

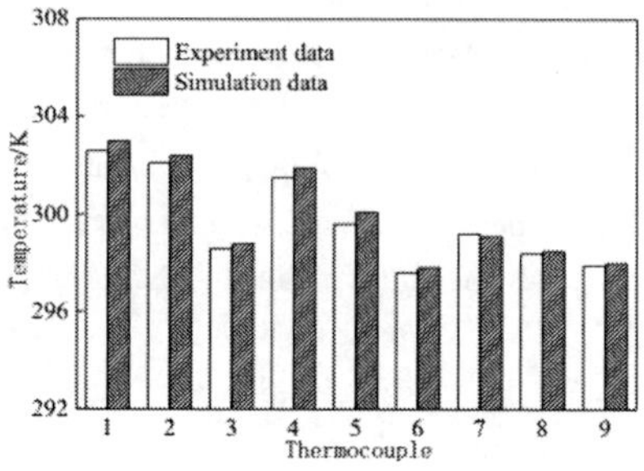

Figure 2. Temperature after 9 hours of simulation and experiment

4 RESULTS AND ANALYSIS

4.1 Low-temperature oxidation heating process in gob with "U+L" ventilation

The temperature field was gained because of the heat storing in gob areas caused by the oxidation reactions between coal and oxygen. The oxidation heating process of temperature filed in gob was shown in Figure 3. The temperature field and changes on day 1, 4, 7 and 10 were shown in (a), (b), (c) (d). The average temperature in gob areas increased as the oxidation of coal proceeded, and the high-temperature areas basically occurred on day 4 with a depth of 60 cm.

Owing to the outflow of air in tail roadway, the width of the spontaneous combustion zone in the return side was small, the spontaneous combustion zone became wide and deep in the oxygen concentration field, the frequency factor of deep coal for reaction was increased. Consequently, the heat release of reaction increased, the temperature increased, and the high-temperature belt became wide and deep.

Given that the tail roadway lied in the porous media, heat flowed out from it; thus, the temperature increased significantly near the tail roadway. In the return side areas, the temperature was highest near the tail roadway, approximately 3.3 K higher than the ambient temperature.

The temperature curve of high temperature point changing with time was shown in Figure 4. The temperature of high temperature point increased exponentially. The temperature was increased to 312.8 K from the initial value on day 10 with a increasing heating rate and the increased risk of spontaneous combustion.

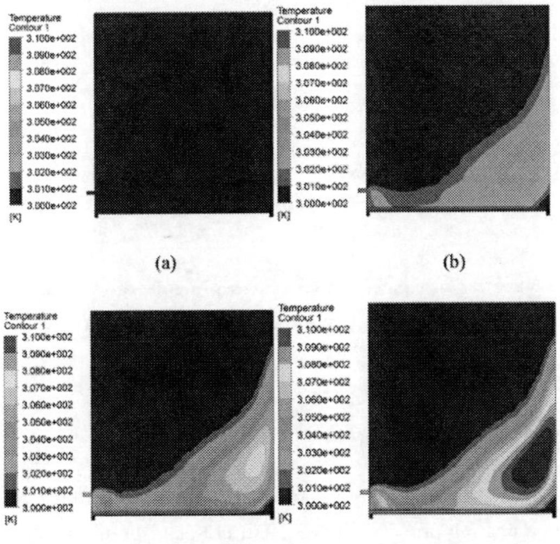

Figure 3. Oxidation heating process of temperature filed in gob. (a) 1d;(b)4d; (c)7d; (d) 10d

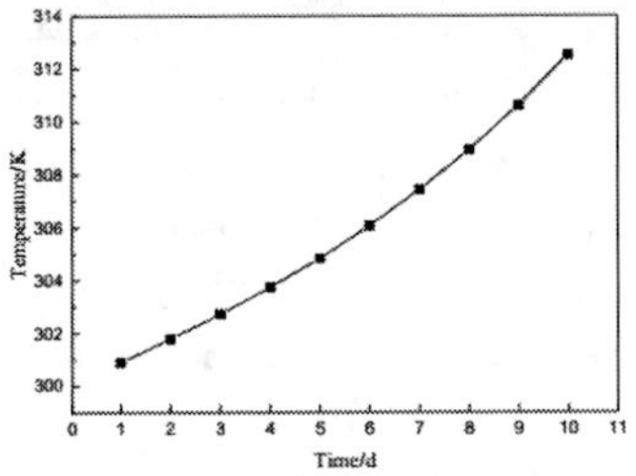

Figure 4. Temperature of high temperature point changing with time

The preceding results indicated that the existence of tail roadway affected oxidation heating process, thus, temperature in gob was affected. The location of the tail roadway also affected the oxidation heating process and temperature field.

4.2 Influence of the position of tail roadway on the temperature field in gob areas

The distribution law of the temperature field within 10 days was calculated to analyze the influence of tail roadway on the temperature field in gob mainly from the two aspects of the high-temperature point. The depth and temperature of high temperature point was shown as Figure 5.

(1) Comparison of the U and U + L ventilations indicated that the tail roadway had a great influence on the location and the temperature of the high-temperature points. The heating rate of the high-temperature point was 0.8 (K/d) when no tail roadway existed, but it can be 1.24 (K/d) when tail roadway existed, heating rate of the high-temperature point in the gob of U + L ventilation was 1.5 times of that in U ventilation.

(2) The location of tail roadway had an no influence on the high-temperature point. neither the depth and temperature of high temperature point none changed with the different distance. When the distances between the crossheading and working surface were 5, 10, 15, 20, and 30 m, the heating rates were 1.09, 1.11, 1.13, 1.15, and 1.24 (K/d), respectively.

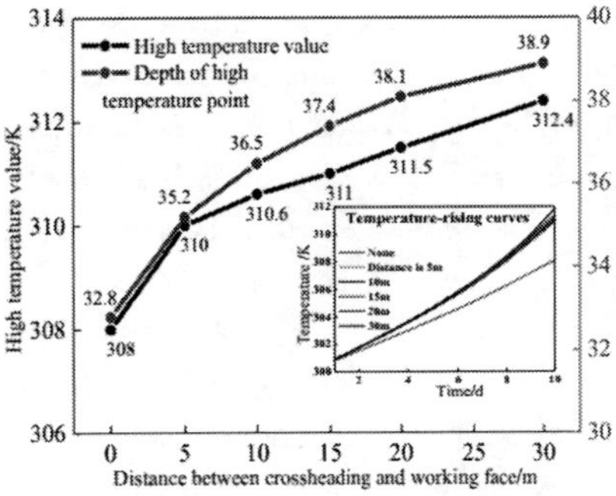

Figure 5. Effects of the distance between crossheading and workface on the high temperature points in gob

4.3 Flowrate of tail roadway and the distribution of gas concentration

The use of tail roadway had an effect on oxidation heating process in gob areas while the tail roadway was mainly used for gas drainage. The Figure 6 show that flowrate of roadway when differential pressure of tail roadway and outlet was different, while the distance between crossheading and workface was 20m, differential pressure of inlet and outlet was 200pa, and differential pressure of outlet and tail roadway was 0Pa，20Pa，50Pa，100Pa，200Pa.

With the increasing of differential pressure of outlet and tail roadway, flowrate of outlet decreasing, but flowrate of tail roadway increasing. The flowrate of tail roadway became greater than outlet when the differential

DOI: 10.1051/matecconf/20166301030

pressure was more than 150pa, this is the results of that a large part of the airflow flows out from the tail roadway.

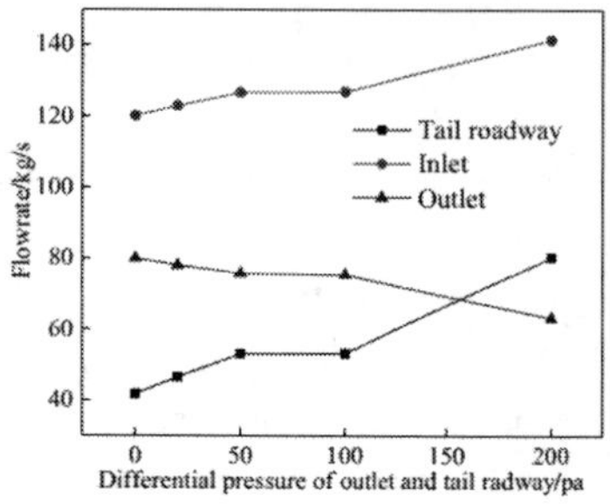

Figure 6. Flowrate of roadway when differential pressure of tail roadway and outlet was different

In the gob with "U+L" ventilation, due to the diversion effects of the tail roadway on the airflow, the gas concentration was affected too. It can be seen from Figure 7 that the maximum reduction of gas concentration was about 0.36% in the position 10m away from return side, and the maximum reduction of gas concentration was about o.32% in the position 100m away from return side. The gas concentration 50 m from workface shown in (a) and (b) of Figure 7 in areas 10 m and 100 m from return side decreased by 0.09% and 0.02%. The gas concentration 250 m from workface in areas 10 m and 100 m from return side decreased by 0.36% and 0.32%. In the areas nearer from the outlet, the differences of gas concentration in gob between with and without tail roadway were larger and it could be concluded that the gas drainage was best in the gob with U and L ventilation.

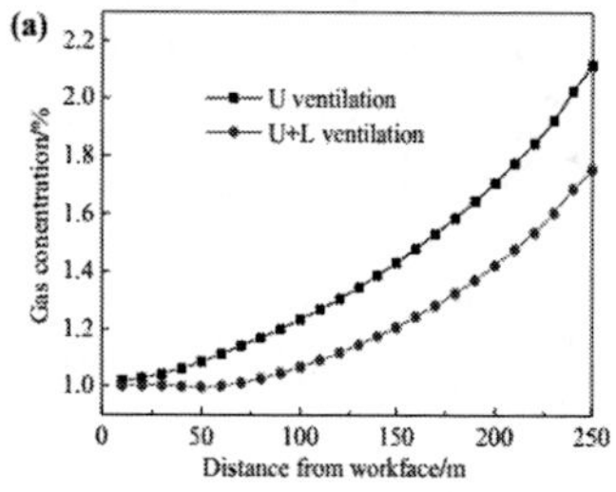

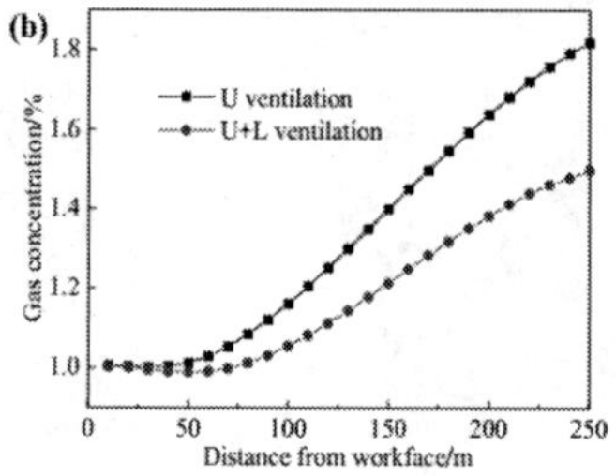

Figure 7. Gas concentration contrast in gob with "U" and "U+L" ventilation. (a) 10m from return side; (b)100m from return side

5 CONCLUSION

(1) The existence of tail roadway had a great influence on the heating process, the temperature near the tail roadway, approximately 3.3 K higher than the ambient temperature in the return side areas.

(2) Compared the gob with U and U + L ventilation systems, in the gob with a U + L ventilation, the location of high-temperature points was deeper and the temperature was higher, heating rate of the high-temperature point in the gob of U + L ventilation was 1.5 times of that in U ventilation .

(3) The distance between the crossheading of tail roadway and working surface had no effects on high temperature in the gob.

(4) Tail roadway had diversion effects on the airflow, and the gas drainage effect was the best near return side, the maximum reduction of gas concentration can be 0.36%.

References

[1] J. Huang, S. Q. Yang, X. T. Chu, Q. Xu. Numerical Simulation on Air leakage flow field in spontaneous combustion three zone of goaf. Coal Sci. Techol. 37(2009)

[2] L. M. Yuan, C. A. Smith, Numerical study on effects of coal properties on spontaneous heating in longwall gob areas. Fuel. 87(2008)

[3] K. Brooks, D. A. Glasser. simplified model of spontaneous combustion in coal stockpiles. Fuel. 65(1986)

[4] Q. Z. Su. Study on variation law of spontaneous combustion "three-zone "in gob areas of fully mechanized coal face. PhD thesis, Taiyuan University of Technology,

Taiyuan, China.

[5] K. Wang, S. G. Jiang, W. Q. Zhang, Z. Y. Wu. Numerical simulation of tail roadway change to gas flow field in goaf. J. Min. Saf. Eng. 29(2012)

[6] S. Q. Yang, T. Cheng, Q. Xu, W. L. Lv, J. Huang. Theoretical analysis and numerical simulation of influence of the change of negative pressure and air volume of inner interlocked tail road on coal spontaneous combustion. J. Chin. Coal Soc. 36(2011)

[7] W. Liu. (2014) Multi-physics coupling mechanism and three-dimensional numerical simulation of spontaneous combustion of gob areas. PhD thesis, Chin. Univ. Min. Techno. Beijing, China.

[8] Q. L. He, D. M. Wang. Numerical simulation of spontaneous combustion process in goaf areas by fully-mechanized caving roof coal. J. Chin. Univ. Min. Technol. 33(2004)

[9] H. Y. Niu, X. H. Zhang. Research on the coal self - ignition and prevention techniques classification. Industrial Saf. Environ. protection. 33(2007)

[10] M. G. Yu, X. H. Chang, H. L. Jia. L. X. Lu. analysis of spontaneous combustion three zone in goaf based on Matlab. J. Chin. Coal Soc. 35(2010)

MATEC Web of Conferences **63**, 01031 (2016) DOI: 10.1051/matecconf/20166301031
MMME 2016

Stereo Matching Based On Election Campaign Algorithm

Qing Hua XIE[1,a], Xiang Wei ZHANG[1], Wen Ge LV[1] and Si Yuan CHENG[1]

[1] *School of Electromechanical Engineering, Guangdong University of Technology, Guangzhou china*

Abstract. Stereo matching is one of the significant problems in the study of the computer vision. By getting the distance information through pixels, it is possible to reproduce a three-dimensional stereo. In this paper, the edges are the primitives for matching, the grey values of the edges and the magnitude and direction of the edge gradient were figured out as the properties of the edge feature points, according to the constraints for stereo matching, the energy function was built for finding the route minimizing by election campaign optimization algorithm during the process of stereo matching was applied to this problem the energy function. Experiment results show that this algorithm is more stable and it can get the matching result with better accuracy.

Keywords: Stereo Matching, Energy Function, Election Campaign Algorithm, Optimization.

Introduction

In binocular stereo vision, stereo matching get the image of the depth map from finding the scene with a point on the left and the right images.

According to the matching primitives, the matching to generate the disparity data distribution is different, the stereo matching can be based on region, feature, phase matching. According to constrains of the search strategy, stereo matching can be local and global matching. Noise, grey value difference, occlusion, perspective distortion will affect the match result. The method based on feature matching does not depend directly on the pixel gray, it has strong anti-interference ability, and fast calculation. Phase matching method lose effectiveness when the presence assumption of local structure does not exist. Local matching with low computational complexity, running fast, but is sensitive to occlusion and local ambiguity region. Global matching use the global optimal search strategy, which has better accuracy, can reduce the influence of occlusion and local ambiguity area such as texture consistency.

In this paper, the edge points are used as matching primitives, and the energy function with the constraint conditions of the stereo matching is established. In order to find the optimal solution path of energy function, election campaign algorithm is used to realize the optimal matching of feature points.

Stereo Matching Disparity Map

Edge points on the right and left two parallel binocular image were extracted, matching point set of each images is formed by the gray value and gradient. Matching energy function is established based on the constraint conditions, then election campaign algorithm is used to search the optimum, matched conjugate points and the parallax information.

Get the Edge Points constraint conditions

In this paper, Sobel Operator is used to extract edge points.

$$S_x = -f(x-1,y-1)+f(x-1,y+1)-2f(x,y-1)+2f(x,y+1)-f(x+1,y-1)+f(x+1,y+1) \quad (1)$$

$$S_y = f(x-1,y-1)+2f(x-1,y)+f(x-1,y+1)-f(x+1,y-1)-2f(x+1,y)-f(x+1,y+1) \quad (2)$$

[a] Qing Hua XIE: xqhcool@163.com

$f(x,y)$ is the grey value of point (x,y), the gradient and direction angle at the point (x,y) is defined as the formula below.

$$|\nabla f| = |S_x| + |S_y| \quad (3)$$

$$\theta(x,y) = \arctan\left|\frac{S_y}{S_x}\right| \quad (4)$$

Energy function is defined as

$$\nabla^2 f = f(x-1,y-1)+f(x-1,y)+f(x-1,y+1)+f(x-1,y+1)+f(x,y-1)+ f(x,y+1)+f(x+1,y-1)+f(x+1,y)+f(x+1,y+1)-8+f(x,y) \quad (5)$$

Constraint Conditions

Outer pole line constraint. In two images, a conjugate point in the corresponding outer pole line.

Uniqueness constraint. The point in the left image only has a conjugate point in the right image.

Consistency constraint. The matching feature points in the images should have similar properties.

Continuity constraint. Two points very close on the object, they should not have a big difference of depth.

Order preserving constraint. The sequence of two points in the left image will have same sequence of the corresponding matching points in the right image.

Election Campaign Algorithm

Election campaign algorithm (ECA) is a new optimization algorithm simulating election process. Election candidates always pursue the maximum support from voters by means of various election behaviors. They get the overview of their support from the voters according to the sample survey, and decide what to do on the next move. Solution space is imagined as voters and current solutions are imagined as candidates in ECA. The function value of a feasible solution is named as the prestige of a voter and the function value of a current solution is named as the prestige of a candidate.

In ECA, the candidates influence the voters round them, the voter's support is in proportion to the distance between the candidates and the voters. The voter will vote to the candidate which he pefer the most. The sum of location coordinates of every voters supported the candidate powered by its contribution is a new location coordinates, which is named support focus, it is the next position of the candidate. Such computational cycle is done continually until a candidate finds the position of the highest support, which is the global solution of the optimization problems.

Framework of the election campaign algorithm is described as the follows:

Set the Algorithm Parameters. Main parameters of Election campaign algorithm are:

- The number of candidates.
- The number of voters.
- The number of floating voters.
- Target accuracy.

Generate the Candidates and Calculate the Prestige of Them. Generated the define number of candidates in feasible solution field on the uniform distribution. Use the objective function to calculate the prestige of each candidate.

Generate the Voters. The uniform distribution is employed to generate the voters in feasible solution field.

Compute the Investigate Mean Square of Candidates. Higher prestige of a candidate, smaller the mean square deviation of local voters, so that ECA is able to converge to local optimization solution rapidly and steadily. The following formula is used to describe the relation of the prestige and the mean square deviation of a candidate.

$$\sigma_{C_i} = \frac{(P_{Max} - P_{C_i})}{(P_{Max} - P_{Min})} \quad (6)$$

Where σ_{C_i} represents the mean square deviation of candidate C_i; P_{Max} and P_{Min} are the maximum and minimum prestige of candidates.

Next step, voters are generated around each candidate on the normal distribution.

$$N_{C_i} = \frac{(P_{Max} - P_{C_i})}{(P_{Max} - P_{Min})}(N_V - N_{FV}) \quad (7)$$

N_{C_i} is the voter number of candidate C_i, N_V is the number of all voters, N_{FV} is the number of floating voters. Floating voters are generated on the uniform distribution. Here, it is supposed that candidate's effect on voters declined linearly.

Calculate the Supported of the Voters. Candidates can influence the voters within their effect range, the effect on voter V_j from candidate C_i is

$$F_{C_iV_j} = \frac{D_{C_iMax} - D_{C_iV_j}}{D_{C_iMax}} P_{C_i} \quad (8)$$

$F_{C_iV_j}$ is the effect on voter V_j form candidate C_i, V_j is the global or local voter. Here, it is supposed that candidate's effect on voters declined linearly.

Then Compute the prestige of voters by means of objective function.

The support of a voter is proportional to his prestige, and then the proportional constant will be reduced, so the prestige of a voter can used to denote the support of a voter directly. A voter may be influenced by several candidates; the voter should distribute his support to candidates proportionally on the magnitude of effect from candidate to voter. The support from voter V_j to candidate C_i is

$$F_{V_j} = \sum_{i=1}^{m} \max(F_{C_1V_j}, F_{C_2V_j}, \text{L}, F_{C_iV_j}, \text{L}\ F_{C_mV_j}) \quad (9)$$

$$S_{C_iV_j} = \frac{F_{C_iV_j}}{F_{V_j}} P_{V_j} \quad (10)$$

$S_{C_iV_j}$ represents the support from the voter V_j to candidate C_i.

Find the Support Focus of Candidates. A new position coordinate will achieve by means of summing the products of the support from the voters to the candidate and the position coordinate of the voters. It is named the support focus.

$$S_{C_i} = \sum_{j=1}^{n} \max(S_{C_iV_1}, S_{C_iV_2}, \text{L}, S_{C_iV_j}, \text{L}\ S_{C_iV_n}) \quad (11)$$

$$x^{*}_{C_i} = \frac{\sum_{j=1}^{n} S_{C_iV_j} x_{V_j}}{S_{C_i}} \quad (12)$$

$x^{*}_{C_i}$ is the support focus of the candidate C_i. The support focus of a candidate is obtained by investigating, which depends on those voters whose distances to the candidate are nearer and the prestige is higher relatively. The next post of the candidate should be the support focus, where the candidate will have the higher support.

Calculate the Prestige of the Candidates and Compare the Prestige of the Voters with the Candidates. In order to jump out of local optimization solution and increase search rate, the prestige of candidates are compared to that of voters, if the prestige of a voter is higher than that of a candidate, the voter with higher prestige will substitute for the candidate and the that candidate of lower prestige will be eliminated in election.

Check Whether the Condition is Reach. Check whether the condition is reach, otherwise return to step B to execute the period. Here, the condition could be the number of select candidates in the target accuracy or max cycle. Do that circularly until the highest support is found.

$$| P^{t}_{C_i} - P^{t-1}_{C_i} | \le \xi \quad (13)$$

$P^{t}_{C_i}$ represents the prestige of candidate C_i in cycle t. When the absolute value of the difference between the prestige of the candidates in the present cycle and the previous cycle less than or equal to the target accuracy ξ, the computation stop.

Do that circularly until the highest support is found.

Flow scheme of the election campaign algorithm is shown in Fig. 1.

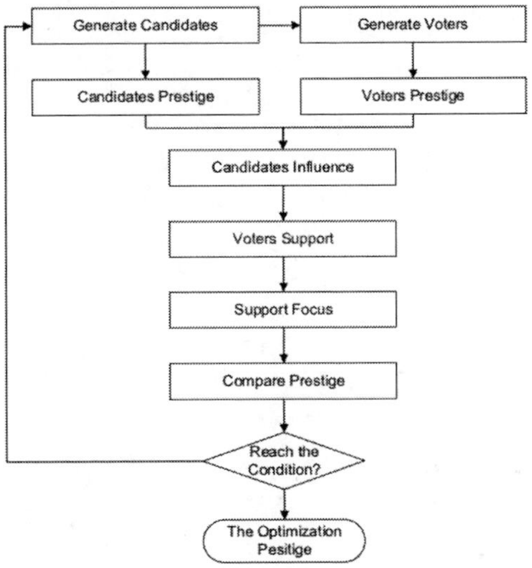

Fig. 1 Flow scheme of election campaign algorithm

Stereo Matching Based On Election Campaign Algorithm

The energy function is used as the object function of Election Campaign Algorithm. The matching relationship between edge feature point sets in the same corresponding line on the left and right images, $L\{l_1, l_2, \text{L}\ , l_s\}$ and $R\{r_1, r_2, \text{L}\ , r_t\}$, will find by the optimization algorithm.

Using Election campaign algorithm, the number of candidates define as 10, the number of voters is 50, the number of floating voters is 10, and the target accuracy is set as 1%. The experiment get the disparity map of two images, as is show in Fig. 2, it prove the feasibility of the algorithm.

Fig. 2 Stereo matching experiment

Conclusions

With the analysis of the stereo matching algorithm and constraint conditions, this paper use edge feature points as matching primitives, and election campaign algorithm as the energy function optimal solution. The disparity map obtained from the results can show that algorithm is feasible.

Election campaign algorithm has good performance to avoid the solution trapped in local optima. This is particularly important when dealing with highly multimodal functions such as the matching of edge feature point sets. For future work, another feature and different energy function will be used in the matching of point sets.

References

[1] X. Su and T.M. Khoshgoftaar. A progressive edge-based stereo correspondence method. In Proc. of the Int. Symposium on Visual Compute. (ISVC), volume 1, pages 248 257, 2007.

[2] K. Kyung-Yoo, K. Sung-Yeol, and H. Yo-Sung. Stereo matching using global and local segmentation. In Proc. of the Int. Technical Conf. on Circus/Systems, Comp. and Communist. (ITC-CSCC 07), pages 1225 1226, 2007.

[3] X. Shenghua, L. Jiping, P. Jianping, and W. Yong. Straight line extraction for stereo image feature matching. In IEEE Proc. of the Congress on Image and Signal Processing, 2008.

[4] G. Nieradka. Method for edge detection in images using fuzzy relation. In Proceedings of the SPIE, volume 6937, pages 69373E 69373E 8, 2008.

[5] Grzegorz Nieradka, Bohdan Butkiewicz, Features stereo matching based on fuzzy logic, IFSA-EUSFLAT 2009 Proceedings, 36, 1 (2009) 227-249.

Reference to a book:

[6] SU Yin-jie, Stereo matching based on ant colony optimization algorithm, Micro computer information, 24, 8-3 (2008) 227-249.

MATEC Web of Conferences 63, 01032 (2016) DOI: 10.1051/matecconf/20166301032
MMME 2016

Synchronization Between Two Different Switched Chaotic Systems By Switching Control

Li Ming DU[1,a], Feng Ying WANG[1], Jie DONG[1] and Zheng Yu LI[1]

[1] *Faculty of Information & Control Engineering, Shenyang Jianzhu University, Shenyang 110168*

Abstract. This paper is concerned with the synchronization problem of two different switched chaotic systems, considering the general case that the master-slave switched chaotic systems have uncertainties. Two basic problems are considered: one is projective synchronization of switched chaotic systems under arbitrary switching; the other is projective synchronization of switched chaotic systems by design of switching when synchronization cannot achieved by using any subsystems alone. For the two problems, common Lyapunov function method and multiple Lyapunov function method are used respectively, an adaptive control scheme has been presented, some sufficient synchronization conditions are attainted, and the switching signal is designed . Finally, the numerical simulation is provide to show the effectiveness of our method.

1 Introduction

Chaos is very interesting nonlinear phenomenon and has been intensively studied in the last three decades [1-3]. In particular, chaos synchronization has become one of the hot issues since Pecora and Carroll introduced a method to synchronize two identical chaotic systems with different initial conditions[4]. For it's potential applications in secure communication, chemical and biological systems, information science and many other fields, lots of researches on the synchronization of coupled chaotic dynamical systems have been done and many theoretical and experimental results have been obtained[5-19]. Such as feedback control methods[5-7], backstepping design technique[8], impulsive control method[9], fuzzy sliding mode control method [10], adaptive method[11,12] and other kind of synchronization[13-19].

Most studies are assumed that the master and the slave systems are single accurate mathematical model of the chaotic system with same dynamical structures. However, it is difficult to find two exactly identical chaotic systems. On the other hand, the master and slave chaotic systems may change very quickly–even by jumping or switching. To the best of the author's knowledge, no synchronization criteria have been reported for this kind of studies. Obviously, synchronization between two different switched chaotic systems are more interesting and worth researching. For example, it can improves safety features of chaos synchronization secret communication while applying in secret communication because of the richness of selection in switched chaotic systems which achieves the variety of transmisson signal.

[a] Corresponding author: duliboy@163.com

This paper studies the projective synchronization problems between different switched chaotic systems with uncertainty. By utilizing switching control theory and adaptive control technology, an adaptive control scheme has been presented, some sufficient synchronization conditions are attainted, and the switching signal is designed based on multiple Lyapunov function method.

2 Problem Formulation

Consider the master-slave switched chaotic systems is described as follow

$$\dot{x} = A_\sigma x + B_\sigma f_\sigma(t,x,\theta), \tag{1}$$

$$\dot{y} = A_\sigma y + B_\sigma (g_i(t,y,\tilde{\theta}) + u_\sigma(t)), \tag{2}$$

where $t \in R^+, x \in R^n$ is the state variable of master system, $y \in R^n$ is the state variable of the slave system, $\sigma : R^+ \to M = \{1,2,...,m\}$ is the switching signal, $A_i, B_i, i \in M$ denote known matrices with appropriate dimensions of the ith subsystem, $\theta, \tilde{\theta}$ is constant matrix which may be unknown, $f_i(t,x,\theta) : J \times R^n \times R^{n\times n} \to R^n$ with $f_i(t,0,\theta) \equiv 0$ and $g_i(t,x,\tilde{\theta}) : J \times R^n \times R^{n\times n} \to R^n$ with $g_i(t,0,\tilde{\theta}) \equiv 0$ are continuous vector-value function, $u_i(t)$, $i \in M$ is the controller of the ith subsystem. It is easy to see that many chaotic systems can be described as the subsystem form of (1) and (2), such as the Lorenz system, Chen's system, Liu system, and Chua's circuit.

Definition 1. For the given master system $\dot{x}_i = f_i(x_i)$, and slave system $\dot{y}_i = g_i(y_i)$, where $i = 1,2,...,n$, $x_i \in R$, $y_i \in R$, We called the master and slave systems achieve projective synchronization if there exist scaling factors h, which satisfied $\lim_{t\to\infty}\|x_i - hy_i\| = 0$.

The error dynamical system between Eqs. (1) and (2) can be expressed as

$$\dot{e} = A_\sigma e + B_\sigma(g_\sigma(t,y,\tilde{\theta}) - hf_\sigma(t,x,\theta) + u_\sigma(t)), \quad (3)$$

where $e = y - hx$ is the synchronization error. So the synchronization problem is now replaced by the problem of stabilizing the system (3).

Assumption 1: The nonlinear part of the system (1) and (2) $f : J \times R^n \times R^{n\times n} \to R^n$ satisfies the following condition

$$\|g_i(t,y,\tilde{\theta}) - hf_i(t,x,\theta)\| \le l\|y - hx\|, \quad i \in M, \quad (4)$$

It is easy to see that the synchronization problem turn into the stabilization problem of the system (3).

3 Common Lyapunov function method

In this section, we will discuss the projective synchronization between different switched chaotic systems by common Lyapunov function method. First, we design the controllers as the form of (5).

$$u_i = K_i e - \frac{1}{2}\lambda r B_i^T Pe, \ \lambda \in R_+, K_i \in R^{n\times n}, i \in M, \quad (5)$$

Where r represents the dynamical estimate of l^2 in assumption (4), and The estimate $r(t)$ is updated according to the following algorithm:

$$\dot{r} = \frac{1}{2}\lambda\|e\|^2, \ \lambda \in R^+, \|e\| = (\sum_{i=1}^n e^2)^{\frac{1}{2}}. \quad (6)$$

Under control law (5) and the adaptation algorithm (6), the ith subsystem of (3) can be characterized as follow form:

$$\dot{e} = (A_i + B_i K_i - \frac{1}{2}\lambda r B_i B_i^T P)e + B_i(g_i(t,y,\tilde{\theta}) - hf_i(t,x,\theta)),$$

$$\dot{r} = \frac{1}{2}\lambda\|e\|^2, \quad (7)$$

where $e = y - hx$, then we discuss the asymptotic stabilization of the system (7).

Theorem 1. consider the error system (7) of master system (1) and the controlled slave system (2) satisfying Assumption 1. If there exist a positive definite matrix P, a matrix K_i and a positive constant λ such that the following LMI

$$(A_i + B_i K_i)^T P + P(A_i + B_i K_i) + \lambda^{-1} I < 0, \ i \in M, \quad (8)$$

is satisfied. then the error system (7) is global asymptotic stable with the adaptive controller(5) under arbitrary switching , that is to say, the master-slave systems achieve projective synchronization.

Proof. Construct a common Lyapunov function in the form of

$$V = e^T Pe + \frac{1}{\lambda}(l^2 - r)^2, \quad (9)$$

The derivative of Lyapunov function along trajectory of each subsystem in system (7) is

$$\begin{aligned}
\dot{V} &= \dot{e}^T Pe + e^T P\dot{e} - \frac{2}{\lambda}(l^2 - r)\dot{r} \\
&= e^T(P(A_i + B_i K_i) + (A_i + B_i K_i)^T P)e - \lambda r e^T P B_i B_i^T Pe \\
&\quad + 2e^T PB_i(g_i(t,y,\tilde{\theta}) - hf_i(t,x,\theta)) - \frac{2}{\lambda}(l^2 - r)\dot{r} \\
&\le e^T(P(A_i + B_i K_i) + (A_i + B_i K_i)^T P)e - \lambda r\|B_i^T Pe\|^2 \\
&\quad + 2\|B_i^T Pe\|\|g_i(y) - hf_i(x)\| - \frac{2}{\lambda}(l^2 - r)\dot{r} \\
&\le e^T(P(A_i + B_i K_i) + (A_i + B_i K_i)^T P)e - \lambda r\|B_i^T Pe\|^2 \\
&\quad + 2l\|B_i^T Pe\|\|e\| - \frac{2}{\lambda}(l^2 - r)\dot{r} \\
&\le e^T(P(A_i + B_i K_i) + (A_i + B_i K_i)^T P)e - \lambda r\|B_i^T Pe\|^2 \\
&\quad + \lambda l^2\|B_i^T Pe\|^2 + \frac{1}{\lambda}\|e\|^2 - \frac{2}{\lambda}(l^2 - r)\dot{r} .
\end{aligned} \quad (10)$$

By substituting Eq. (6) into Eq. (10), we have

$$\dot{V} \le e^T((A_i + B_i K_i)P + (A_i + B_i K_i)^T P + \lambda^{-1} I)e < 0. \quad (11)$$

According to the stability theory of switching system, error system (7) is asymptotically stable with arbitrary switching conditions, i.e. $\|e(t)\| = \|y(t) - hx(t)\| \to 0$ as $t \to \infty$. It is obviously that that the master-slave systems achieve projective synchronization, which complete the proof.

4 Multiple Lyapunov function method

In this section, we will study the case that none of subsystems can achieve synchronization. For this kind of condition, the common Lyapunov function could not get, so we discuss how to realize synchronization by the suitable design of switching ruler between sub chaotic systems and give a multiple Lyapunov function methos.

Theorem 2. Under the assumption 1, If there exist $\delta_{ij} \ge 0$ ($\delta_{ij} \le 0$) and positive defined matrix P_i such that LIMs

$$(A_i + B_i K_i)^T P_i + P_i(A_i + B_i K_i) + \lambda^{-1} I + \sum_{j=1}^m \delta_{ij}(P_j - P_i) < 0,$$
$$i = 1,2,\ldots,m \quad (12)$$

hold, then synchronization between master systems (1) and slave systems(2) is achieved under the switching controller (5) with switching signal

$$\sigma(t) = i, \text{ if } e(t) \in \bar{\Omega}_i , \quad i \in M , \qquad (13)$$

where

$$\Omega_i = \{e \in R^n \mid e^T(P_j - P_i)e \geq 0, \forall j \in M\} \qquad (14)$$

and $\bar{\Psi} = \Psi, \cdots, \bar{\Omega}_i = \Omega_i - \bigcup_{j=1}^{i-1} \bar{\Omega}_j, \cdots, \bar{\Psi}_m = \Psi_m - \bigcup_{j=1}^{m-1} \bar{\Psi}_j$.

Proof. Let $\delta_{ij} \geq 0$,then for any $e \in R^n \setminus \{0\}$, exist a $i \in M$ to make $e^T(P_j - P_i)e \geq 0$, " jÎ M , according to (12), we get

$$(A_i + B_i K_i)^T P_i + P_i(A_i + B_i K_i) + \lambda^{-1} I < 0, \ i \in M, \qquad (15)$$

Then, for any $i \in M$, let construct Lyapunov function as the form of

$$V(t) = e^T P_i e + \frac{1}{\lambda}(l^2 - r)^2, \qquad (16)$$

The derivative of ith Lyapunov function along the trajectory of (7) is

$$\dot{V}(t) = \dot{e}^T P_i e + e^T P_i \dot{e} - \frac{2}{\lambda}(l^2 - r)\dot{r}$$
$$= e^T(P_i(A_i + B_i K_i) + (A_i + B_i K_i)^T P_i)e - \lambda \hat{r} e^T P_i B_i B_i^T P_i e$$
$$+ 2e^T P_i B_i (g_i(t, y, \tilde{\theta}) - hf_i(t, x, \theta)) - \frac{2}{\lambda}(l^2 - r)\dot{r}$$
$$\leq e^T(P_i(A_i + B_i K_i) + (A_i + B_i K_i)^T P_i)e - \lambda r \left\|B_i^T P_i e\right\|^2$$
$$+ 2l \left\|B_i^T P_i e\right\| \left\|e\right\| - \frac{2}{\lambda}(l^2 - r)\dot{r}$$
$$\leq e^T(P_i(A_i + B_i K_i) + (A_i + B_i K_i)^T P_i)e - \lambda r \left\|B_i^T P_i e\right\|^2$$
$$+ \lambda l^2 \left\|B_i^T P_i e\right\|^2 + \frac{1}{\lambda}\left\|e\right\|^2 - \frac{2}{\lambda}(l^2 - r)\dot{r} .$$
$$\leq e^T((A_i + B_i K_i)P_i + (A_i + B_i K_i)^T P_i + \lambda^{-1} I)e.$$

From (15), it is easy to see that error system (7) is asymptotically stable under switching ruler(13), and the theorem 2 is proved.

5 Numerical Simulation

In this section we give numerical experiments on the problems of projective synchronization of master system(1) and slave system(2) to illustrate the validity of Theorems.

Let take Lorzen and Genesio system as the subsystems of master switched chaotic system and Liu and Rossler system as the subsystems of slave switched chaotic sysem. The Lorenz system is a very typical chaotic system, it is presented as:

$$\begin{cases} \dot{x}_1 = a_1(x_2 - x_1) \\ \dot{x}_2 = a_2 x_1 - x_2 - x_1 x_3 \\ \dot{x}_3 = -a_3 x_3 + x_1 x_2 \end{cases} , \qquad (17)$$

where x is the state variable of master system, a_1, a_2, a_3 are the parameters of system, If we choose $a_1 = 10, \ a_2 = 28, \ a_3 = 8/3$, the system stays in chaotic state. As we assume that the initial condition, $x_1(0) = 0.1$, $x_2(0) = 0.5$, $x_3(0) = 0.3$, chaotic attractor is shown in Figure 1.

Genesio system is described as:

$$\begin{cases} \dot{x}_1 = x_2 \\ \dot{x}_2 = x_3 \\ \dot{x}_3 = -\hat{a}_1 x_1 - \hat{a}_2 x_2 - \hat{a}_3 x_3 + x_1^2 \end{cases} , \qquad (18)$$

where $\hat{a}_1 = 6, \hat{a}_2 = 2.92, \hat{a}_3 = 1.2$, and the initial point of system $x_1(0) = 0.1$, $x_2(0) = 0.5$, $x_3(0) = 0.3$, the chaotic attractor of Genesio system is displayed as Figure 2.

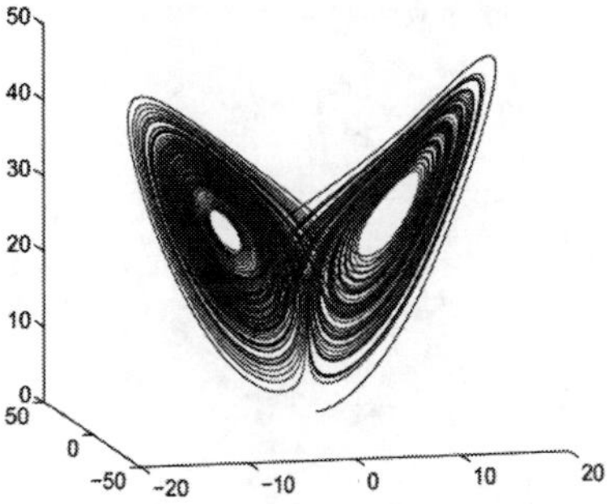

Figure 1. A chaotic attractor of Lorenz system

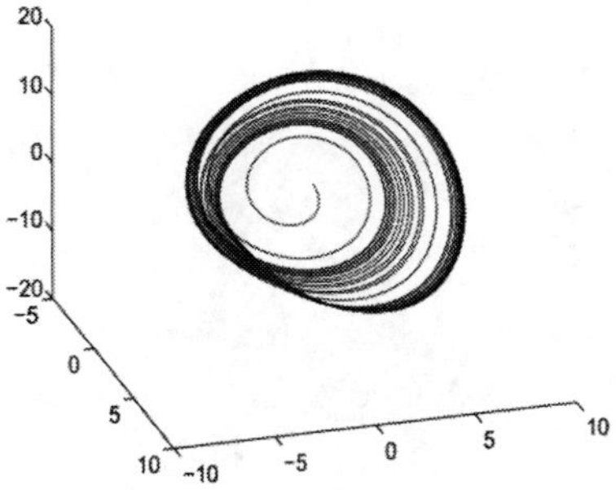

Figure 2. A chaotic attractor of Genesio system

Liu system is another typical chaotic system model, which is described as:

$$\begin{cases} \dot{y}_1 = b_1(y_2 - y_1) \\ \dot{y}_2 = b_2 y_1 - b_4 y_1 y_3 \\ \dot{y}_3 = -b_3 y_3 + b_5 y_1^2 \end{cases} , \qquad (19)$$

where y is the state variables of slave system, b_1, b_2, b_3, b_4, b_5 denote the parameters of system, if $b_1 = 10$, $b_2 = 28$, , $b_4 = 1, b_5 = 4$ $b_3 = 8/3$, the system stands in chaos, as the initial point of system $y_1(0) = 5$, $y_2(0) = -6$, $y_3(0) = 4$, its chaotic attractor is displayed as Figure 3.

Rossler system has plentiful chaotic features and is described by:

$$\begin{cases} \dot{y}_1 = -y_2 - y_3 \\ \dot{y}_2 = y_1 + \hat{b}_2 y_2 \\ \dot{y}_3 = \hat{b}_1 + y_3(y_1 - \hat{b}_3) \end{cases}, \tag{20}$$

where $\hat{b}_1, \hat{b}_2, \hat{b}_3$, denote the parameters of system, if $\hat{b}_1 = 0.2$, $\hat{b}_2 = 0.2, \hat{b}_3 = 5.7$, the system is chaotic, as the initial point of system $y_1(0) = 5$, $y_2(0) = -6$, $y_3(0) = 4$, its chaotic attractor is displayed as Figure 4.

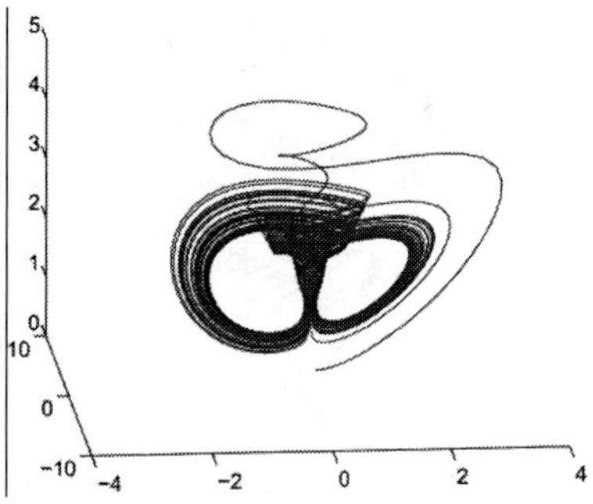

Figure 3. A chaotic attractor of Liu system

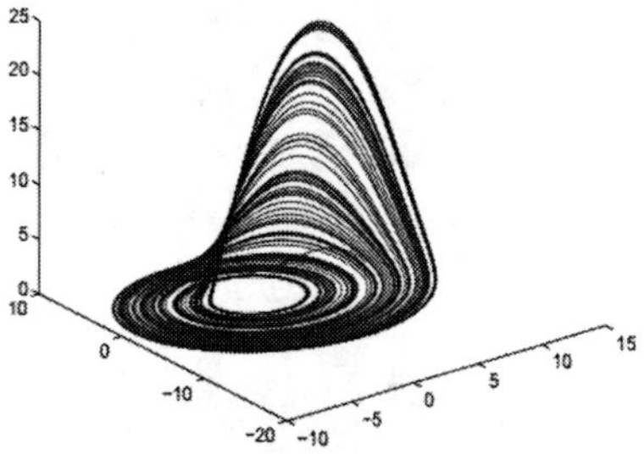

Figure 4. A chaotic attractor of Rossler system

It is easy to see that the system (17) (18)and system (19) (20)can be rewrited as the form of Eqs. (1) and (2), if we take

$$A_1 = \begin{bmatrix} -10 & 10 & 0 \\ 28 & 0 & 0 \\ 0 & 0 & -8/3 \end{bmatrix}, \quad A_2 = \begin{bmatrix} 0 & 1 & 0 \\ 0 & 0 & 1 \\ -6 & -2.92 & -1.2 \end{bmatrix},$$

$$B_1 = B_2 = \begin{bmatrix} 1 & 0 & 0 \\ 0 & 1 & 0 \\ 0 & 0 & 1 \end{bmatrix}, \quad K_1 = \begin{bmatrix} -3 & 0 & 0 \\ 0 & -2 & 0 \\ 0 & 0 & -5 \end{bmatrix},$$

$$K_2 = \begin{bmatrix} -4 & 0 & 0 \\ 0 & -3 & 0 \\ 0 & 0 & -2 \end{bmatrix}, \quad h = 3, \quad \lambda = 2,$$

then we can get the positive matrix

$$P = \begin{bmatrix} 0.8780 & -0.3505 & 0 \\ -0.0505 & 0.1812 & 0 \\ 0 & 0 & 3.5787 \end{bmatrix},$$

which satisfied the inequality (8). If the initial state of system $x(0) = (0.1 \;\; 0.3 \;\; 0.5)$, $y(0) = (5 \;\; -6 \;\; 4)$, $r(0) = 8$, and put in control as $t > 5$,the simulation result is displayed as Figure 5.

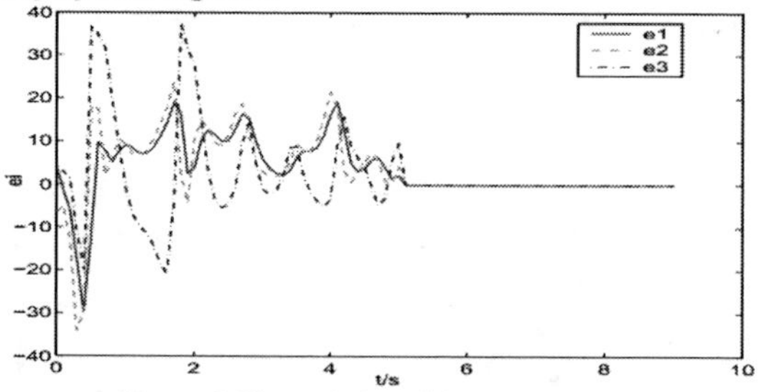

Figure 5. The evolution of the error vector

From Figure 5, we can clearly see that the error state of system reaches zero quickly with adaptive controller under arbitrary switching, which demonstrates the validity of our results.

Remark 1. In the paper, if we take $h = 1$, then the two different switched chaotic systems will get complete synchronization with the adaptive controller under arbitrary switching, and if we take $h = -1$, the master-slave switched chaotic systems will get antisynchronization under arbitrary switching.

6 CONCLUSIONS

In the paper, we present a novel method to solve the problem of projective synchronization between different switched chaotic systems, By adopting common Lyapunov function method and multiple Lyapunov function method, an adaptive controller is achieved, which realize the projective synchronization between the master-slave system with different switched chaotic systems under arbitrary switching and designed switching. Numerical simulation indicates the accuracy of the results.

Acknowledgement

This work was supported by the National Natural Science Foundation of China under Grants 61272253 and Shenyang Jianzhu University Subject Culvert Project XKHY2-72.

References

1. A.L. Fradkov , R. J. Evans, Annual Reviews in Control , **29(1)**, 33–56, (2005).
2. Y.Chang, G.Chen, Chaos, Solitons & Fractals, **27(1)**, 75-86, (2006).
3. E.Ott, C.Grebogi, J.A.Yorke, Phys Rev Lett, **64(11)**, 1196–1199, (1990).
4. L.M.Pecora , T.L.Carroll, Phys Rev Lett , **64(8)**, 821–824, (1990).
5. J.H.Park , O.M.Kwon, Chaos, Solitons & Fractals, **23(2)**,495–501,(2005).
6. F.Wang, C.Liu, Physics Letters A , **360(2)**, 274-278, (2006).
7. A.E.Matouk, Communications in Nonlinear Science and Numerical Simulation, **16(2)**, 975-986, (2011).
8. X.Q.Wu, J.A.Lu , Chaos, Solitons & Fractals, **18(4)**,721–729,(2003).
9. A.Khadra, X.Liu, X.Shen, IEEE Trans. On Circuits and Systems, **50(3)**,341-350, (2003).
10. H.T.Yau. Mechanical Systems and Signal Processing, **22(2)**, 408-418, (2008).
11. J.Lu , X.Wu , X,Han, J.Lû, Phys Lett A , **329**,327–333, (2004).
12. J.H.Park. Int J Nonlin Sci Numer Simulat , **6**,201–206, (2005).
13. J.Meng ,X.Y.Wang, Phys. Lett. A, 369(4): 294-298,2007.
14. C.Li, X.F.Liao,W. Kwokwo , Physica D,**194(3)**,187-202,(2004).
15. C.Masoller, H.Zanette, **Physica A**, **300**,359-366,(2001).
16. Y.W.Wang, Z.H.Guan, Chaos, Solitons & Fractals,**27**,97–101,(2006).
17. N.F.Rulkov, M.M.Sushchik, L.S.Tsimring, H.D.I. Abarbanel, Phys Rev E,**51**,980 - 94, (1995).
18. M. Hu, Z.Xu, Nonlinear Analysis: Real World Applications, **9(3)**:1253-1260, (2008).
19. D.Ghosh, S. Banerjee, A. R. Chowdhury, Physics Letters A, **374(21)**,2143-2149, (2010).

Adsorption/oxidation of sulfur-containing gases on nitrogen-doped activated carbon

Qiang LIU[1], Ming KE[1], Pei YU[1], Hai Qiang HU[1], Xi Ming YAN[1]

[1]*College of Science, China University of Petroleum-Beijing, 102249 Beijing, China*

Abstract. Coconut shell-based activated carbon (CAC) was used for the removal of methyl mercaptan (MM). CAC was modified by urea impregnation and calcined at 450°C and 950°C. The desulfurization activity was determined in a fixed bed reactor under room temperature. The results showed that the methyl mercaptan adsorption/oxidation capacity of modified carbon caicined at 950°C is more than 3 times the capacity of original samples. On the other hand, the modified carbon caicined at 950°C also has a high capacity for the simultaneous adsorption/oxidation of methyl mercaptan and hydrogen sulfide.The introduce of basic nitrogen groups siginificantly increases the desulfurization since it can facilitate the electron transfer process between sulfur and oxygen. The structure and chemical properties are characterized using Boehm titration, N_2 adsorption-desorption method, thermal analysis and elemental analysis. The results showed that the major oxidation products were dimethyl disulfide and methanesulfonic acid which adsorbed in the activated carbon.

1 Introduction

Methyl mercaptan (CH_3SH) and hydrogen sulfide (H_2S), which exit in the chemical gas from crude oil, petrolum, coal, natural gas and many other industrial gases, are the major compositions of sulfur compounds [1-3]. Release of the sulfur-containing gases into the atmosphere will cause environmental problems since they can be oxidated into sulfate and sulfonic species. However, a small amount of methyl mercaptan and hydrogen sulfide can poison the catalyst in many industrial processes such as the synthesis of ammonia and Fischer-Tropsch. In few decades, there are many efforts for the deep removal of hydrogen sulfide and methyl mercaptan. The main processes for the removal of methyl mercaptan and hydrogen sulfide include adsorption, biological degradation, incineration and adsorption/oxidation, et al [4-5]. Among the above technologies, adsorption/oxidation is recognized as a promising method due to the high efficiency, low investment and mild operating conditions.

Activated carbons are widely used in the gas purification process due to the well-devoloped porous structure, large removal capacity and easy regeneration property. The removal capacity for methyl mercaptan and hydrogen sulfide also depends on the surface chemistry, such as nitrogen groups and metal oxides. The purpose of the paper is to study the introduction of nitrogen through the impregnation and calcination method and its effect on the adsorption/oxidation of methyl mercaptan in the existence of hydrogen sulfide. In order to study the reaction mechnisam, the original and modified carbons were determined by Boehm titration, N_2 adsorption-desorption method, thermal analysis and elemental analysis.

2 Experimental

2.1 Materials

The original coconut shell-based activated carbon was obtained from Beijing Kecheng Guanghua New Technology Co, LTD. The size of the original activated carbon was 40-60 mesh. The activated carbon was washed with distilled water and dried at 115°C for 12h. This coconut shell-based activated carbon was designed as CAC.

Firstly, CAC was oxidated using 30 wt.% HNO_3 solution at 60°C and stirred for 2h. Then, the oxidative activated carbon was washed using distilled water until neutral pH and dried at 115°C for 12h. Secondly, 3g oxidated sample was mixed with a saturated urea solution and stirred for 24h at 25°C. Then the mixture was heated to accelerate the evaporation process and the carbon was dried at 115°C. Thirdly, the impregnation activated carbon was heated in nitrogen from room temperature to 450°C or 950°C at 10°C /min and kept at this temperature for half an hour. Then boiling water was used to wash the carbon. At last, the samples were heated at 115 °C for 12h. The samples were designated as CAC-450(450°C) and CAC-950 (950°C). A supernumerary letter E was added to represent the exhausted samples after the desulfurization test.

[a] Corresponding author: keming@cup.edu.cn

2.2 Methods

2.2.1 Breakthrough capacity

Dynamic tests were performed to evaluate the removal capacity of CH_3SH under atmospheric pressure. Before the test, moist air was used moisten the adsorbents and the time was 1h. Then the moist gas mixture was used to test the desulfurization activity.The inlet concentration of CH_3SH was 0.3% in volume. The reactor condition was as follows: reaction temperature 25°C, relative humidity 80% (25°C), overall gas hourly space velocity (GHSV) 5000h^{-1}. The test was stopped when the outlet concentration of CH_3SH was 50ppm. To study the stimultaneous removal capacity of CH_3SH and H_2S, the mixture air containing 0.3% CH_3SH and 0.03% H_2S was also used. The concentrations of CH_3SH and H_2S were analysed using GC-6890 gas chromatography with a PFPD detector. The adsorption capacity was calculated by integration of the area above the breakthrough curves.

2.2.2 pH of carbon surface

0.4g carbon power was mixed with 20 ml water and the mixture was kept at room temperature for 12h. Then the pH of the suspension liquid was denoted as pH of the carbon surface. For the exhausted samples, the pH was desigated as pHE.

2.2.3 Sorption of nitrogen

Nitrogen isotherms were measured at 77K using a NOVA2000e (Quantachrome instruments) surface area analyzer. The samples were heated at 120°C and outgassed at the temperature for 12h. The specific surface area was determined using Braunuer-Emmett-Teller measurements (BET) method. Nonlocal Density Functional Theory (NLDFT) was used to calculate the micropore volume (V_{mic}) and pore size distribution. The total pore volume (V_t) was obtained by the amount of N_2 adsorped when p/p_0=0.99.

2.2.4 Boehm titration

The method of Boehm was used to determine the acidic and basic surface groups. The numbers of acidic sites and basic sites were calculated according to the corresponding amounts of NaOH and HCl.

2.2.5 Thermal analysis

Thermal analysis was performed using a TGA/DSC1 (Mettler Toledo) instruments thermal analyzer. The setting were as follows : heating rate, 10°C /min, argon atmosphere with 50ml/min.

2.2.6 Elemental analysis

The content of carbon, hydrogen and nitrogen was determined using Vario EL elementary analyzer.

3 Results and discussion

Elemental analysis results are shown in Table 1. The nitrogen content was siginificantly increased after modifying with urea and calcining at certain temperature. Since the calcination temperature is important for the dissociation of nitogen-containing species, the nitrogen content in CAC-450 is highest due to the incomplete decomposition of nitrogen-containing compounds. When the calcination temperature reaches at 950°C, many nitrogen-containing groups decompose thus resulting in a decrease in the nitrogen content. After the calcination at 950°C, the majority of nitrogen incorporated into the sample matrix. The nitrogen mainly existed in the form of pyridine-like or quaternary configuration.

The surface chemistry of the activated carbon was studied and the results were shown in Table 2. The pH of the sample surface increased after introduction of nitrogen. The Boehm titration results were also shown in Table2. Since the nitrogen-containing species and oxygen-based species can have similar pKa, so the results include only acidic and basic groups. The amount of basic groups increased after modification with urea. The results were in aggrement with the changes in the surface pH.

Table1. Elemental analysis of the carbon samples

Sample	C	H	N
CAC	81.15	0.75	0.13
CAC-450	87.38	1.38	2.25
CAC-950	90.92	0.91	1.16

Table2. pH and surface chemistry from Boehm titration (number of groups in mmol/g)

Sample	pH	Acidic	Basic	Basic/Total
CAC	7.51	0.35	0.43	0.55
CAC-450	8.47	0.39	0.57	0.59
CAC-950	9.43	0.31	0.65	0.68

The surface area and pore structure are important in the removal process. N_2 adsorption isotherms and micropore size distribution curves are shown in Figure1 and Figure2. The textural parameters were calculated based on the nitrogen isotherms and the results were shown in Table3. The pore < 2nm is called micropore based on the IUPAC classification. For the different samples, the adsorption isotherms were all similar and the results show that they were micropore materials. For the sample treated at 450°C, the surface area and pore volume decreased compared with the originanl sample. Otherwise, for the sample treated at 950°C, the surface area and pore volume slightly increased. For the exhaustion samples, the decrease in the total pore volume might be due to the reaction products adsorbed on the carbon surface. For the

DOI: 10.1051/matecconf/20166301033

exhaustion sample CAC-950E, the relative large decrease in the pore volume, especially the micropore volume, suggested that the existence of more strongly adsorbed productions after the breakthrough test. This will be discussed in the later part of the paper.

Table3. Structural parameters of carbon samples from sorption of nitrogen

Sample	S_{BET}(m^2/g)	V_{mic}(cm^3/g)	V_t(cm^3/g)	V_{mic}/V_t
CAC	772	0.279	0.390	0.72
CAC-450	738	0.258	0.419	0.62
CAC-950	783	0.279	0.421	0.66
CAC-450E	717	0.256	0.415	0.62
CAC-950E	695	0.242	0.407	0.59

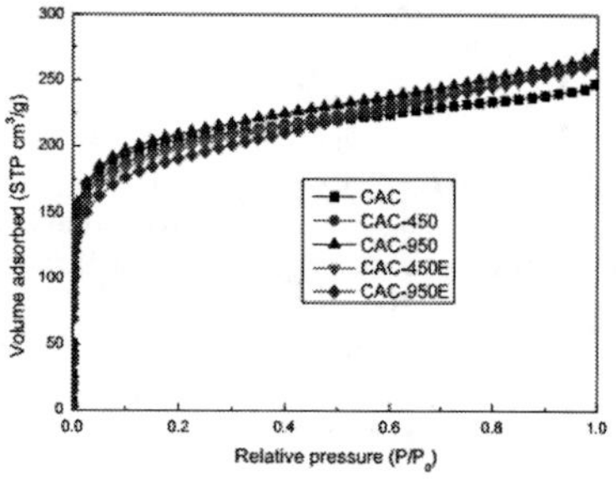

Figure1.Nitrogen adsorption isotherms of the initial and exhausted samples

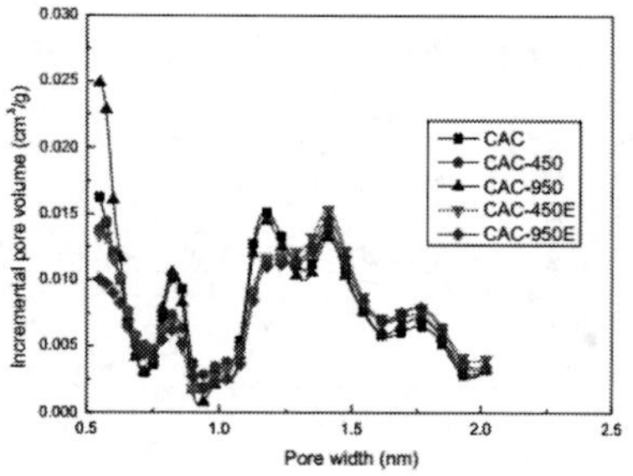

Figure2. Micropore size distributions of the initial and exhausted samples

The breakthrough results were shown in Figure3 and Table4. After modification with urea, the CH_3SH capacity increased. For the carbon treated at 950°C, the CH_3SH capacity exceeds more than 3 times of the CAC sample. The reason maybe that the nitrogen incorporated into the carbon martix can catalysis the oxidation reaction through improving the electron transfer rates from sulfur to oxygen. The intersting behavior is that the sample treated at 950°C shows good simultaneous adsorption/oxidation of methyl mercaptan and hydrogen sulfide. Since the content of hydrogen sulfide is much lower than methyl mercaptan, hydrogen sulfide can not be detected in the outlet gas. The results show that the modified sample can be used for the simultaneous adsorption/oxidation of methyl mercaptan and hydrogen sulfide. For all the samples, pH of the carbon surface decreased after the breakthrough test. For the modified samples, the large decrease in the pH indicated that some acid species were part of the oxidation products. The results can be explained in the later DTG results.

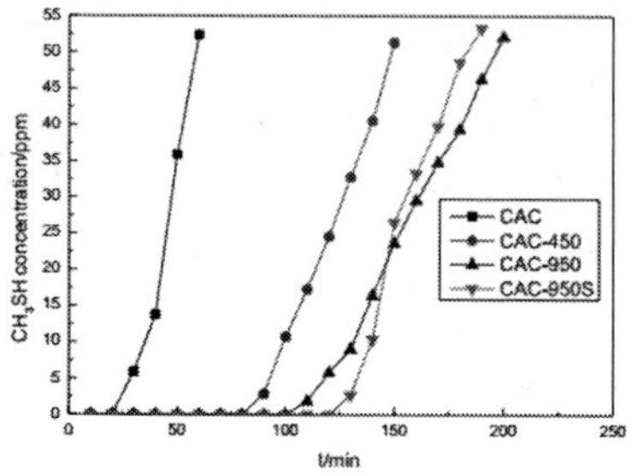

Figure3. CH_3SH breakthrough curves

Table4. pH of exhausted samples, their CH_3SH breakthrough capacity and the amount of water preadsorbed on the carbon surface

Sample	Water preadsorbed (mg/g)	CH_3SH breakthrough capacity (mg/g)	pHE
CACE	81.2	71.8	6.25
CAC-450E	89.5	169.5	7.38
CAC-950E	78.6	232.6	4.85
CAC-950SE	78.6	230.2	4.71

Thermal analysis (TA) can be used to evaluate the adsorbed species on the carbon surface. The differential thermogravimetry (DTG) results are shown in Fig4. For all the samples, there are two main peaks. The first peak centered at about 80 °C which means desorption of water. According to the results reported by many researchers, the second peak which centered beween 80°C and 300°C may represent desorption of dimethyl disulfide [6]. An interseting behavior was found for the sample treated at 950°C. There is another peak centered between 300°C to 350°C. According to the previous research, the peak centered between 300°C to 350°C may represent the strongly adsorbed species [7]. Based on the pH decrease and the literature, the peak centered between 300°C to 350°C may represent strongly adsorbed sulfonic acid species. The DTG results also were in accordance with the breakthrough tests.

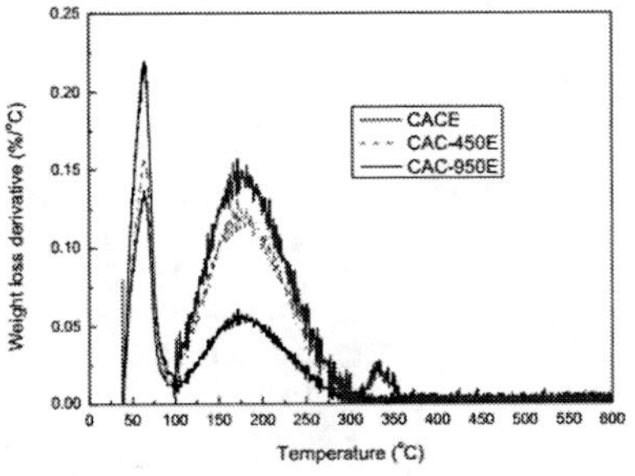

Figure4. DTG curves for the exhausted samples

4 Conclusions

The coconut shell-based activated carbon can be used as good materies for the adsorption/oxidation of sulfur-containing gases in many industrial processes. The sample modified by urea and calcined at 950°C indicates high adsorption/oxidation capacity for methyl mercaptan and the adsorption capacity is more than 3 times the capacity of original sample. On the other hand, the modified sample also shows high efficiency for the simultaneous adsorption/oxidation of methyl mercaptan and hydrogen sulfide. The basic nitrogen groups are highly dispersed on the carbon surface and act as catalytic sites for the oxidation of sulfur-containing gases. Dimethyl disulfide and sulfonic acid species are the major oxidation productions.

References

1. A. Tangerman, E.G. Winkel. *Phosphorus Sulfur* **188**(2013), 396-402.
2. A. Bagreev, J.A. Menendez, I. Dukhno, Y. Tarasenko, T.J. Bandosz, *Carbon*, **43**(2005), 208-210.
3. S. Bashkova, A. Bagreev, T.J. Bandosz, *Langmuir*, **19** (2003), 6115-6121.
4. C. Cammarano, E. Huguet, R. Cadours, C. Leroi, *Appl. Catal. B Environ*, **156** (2014),128-133.
5. M.G. Conti-Ramsden, K. Nkrumah-Amoako, N.W. Brown, *Adsorption* **19**(2013), 989-996.
6. C.A. Cellier, V. Vromman, V. Ruaux, E.M. Gaigneaux, *J. Phys. Chem. B*, **108**(2004), 9989-10001.
7. T. Zeng, W.A. Arnold, B.M. Toner, *Environ. Sci. Technol*, **47** (2013), 1287-1296.

MATEC Web of Conferences **63**, 01034 (2016) DOI: 10.1051/matecconf/20166301034
MMME 2016

Research on automatic current sharing control methods for control power supply

Xian Bin DAI

Electrical Engineering Department, Shenyang Institute of Engineering, China

Abstract. High-power switching devices in control power supply have different saturated forward voltage drops and the inconsistency of turning on/off times and they lead to the inconsistency in the external characteristics of the inverter modules in parallel operation. Modules with good performance in external characteristics undertake more currents and lead to overloading status and modules with bad performance in external characteristics stay in light-loading status, which increases the thermal stress of module undertaking more currents and influences the service life of high-power switching devices. Based on the simulation analysis of the small-signal module using control power supply automatic current sharing method, it is able to find out the characteristics of current-sharing loop control, namely, slow response speed of the current-sharing loop, which is beneficial for improving the stability of the entire control power supply system.

1 Preface

Control power supply mainly provides power supply for secondary system automatic equipment such as smart power substation control, monitoring, protection, signal and adjustment, etc. and it enjoys the reputation as "heart" of the power operation system. If abnormal operation or malfunction occurs to the control power supply, computer monitoring equipment, network communication equipment, intelligent protection and adjustment equipment in the substation will also suffer from abnormal operation or power failure and the power grid will be running in the status free from protection and surveillance. Once accidents happen to the power grid, relay protection will refuse to act or no selective act will be made, which will cause the widespread blackout and major economic losses. Therefore, control power supply is related to the safe and stable operation of the smart substation and the smart power grid and it is the basic equipment for the entire power operation system and its reliability must be guaranteed.

Generally speaking, control power supply is composed of AC input unit, rectifier module, surveillance unit, DC output and relevant auxiliary units. [1-2] AC input unit is input to supply power for the rectifier module through one loop in the AC switching & control circuit. Rectifier module is the key device in the power supply system. In general, high-power and high-frequency switching power supply will be used as the rectifier module. As its core circuit -- "inverter bridge" is composed of high-power switching devices. Surveillance unit mainly conducts surveillance and control of the entire operating power supply system and it contains MCU, measurement module, control module, human-machine interactivity module and telecommunication module. Currently, control power supply used in smart substations is being gradually replaced by high-frequency, modularized and intellectualized switching power supply to enable the control power supply to display characteristics such as small volume, high efficiency, small ripple coefficient, rapid dynamic response, high control precision and stack output of module, etc. Nevertheless, some problems also exist such as current sharing/voltage sharing problems among rectifier modules in parallel connection. [3-4]

2 Equations Current Sharing of Rectifier Module

Generally speaking, control power supply usually uses high-power, intellectualized and high-frequency switching power supply as its rectifier module and as its core circuit -- "inverter bridge" is composed of high-power switching devices. In order to elevate the output power of the rectifier module, it usually uses multiple inverter bridges in serial connection in operation, which

[a] Corresponding author:dxbcy@126.com

DOI: 10.1051/matecconf/20166301034

leads to the uneven current/current among devices in parallel connection and it also influences the service life of the high-power switching devices. The high-power switching devices used to form the inverter bridges in parallel connection have different saturated forward voltage drops and the inconsistency of turning on/off times and they lead to the inconsistency in the external characteristics of the inverter modules in parallel operation. Modules with good performance in external characteristics undertake more currents and lead to overloading status and modules with bad performance in external characteristics stay in under-loading status, which increases the thermal stress of module undertaking more currents and influences the service life of high-power switching devices. [5]Therefore, in the design for control power supply, it is necessary to adopt current-sharing technology in order to guarantee the even distribution of power supply stress and thermal stress among the modules to keep one or multiple modules from running in the extreme current status.

General current sharing methods include output impedance method, master-slave control method, automatic current sharing method based on average current value, external current sharing controller method and automatic current sharing method. Though there are many current sharing methods, they often display deficiencies in practical application. [6] Specifically speaking: current sharing precision is too low in output impedance method; it is unable to realize redundancy technique and it fails to well guarantee the reliability of the parallel power supply module system through adopting master-slave control method and automatic current sharing method based on average current value; external current sharing controller method makes system become more complicated and it is not beneficial for converting the technique to practical product. As to automatic current sharing method, it has characteristics such as high current sharing precision, good dynamic response and the possibility of realizing redundancy technique, etc, which makes it become more and more popular. Automatic current sharing method -- in n modules in parallel connection, it takes the module with largest output current as the master module and the others are considered as slave modules. No module is preset manually as the main module and they are sequenced according to the size of current value, namely, module with larger current value automatically becomes the master module. Therefore it is called "automatic current sharing method".

3 Rectifier Module's Hardware Circuit Design

Rectifier Module's Hardware Circuit Block Design is as illustrated in Figure 1. It is mainly composed of EMI filter, AC/DC circuit with three-phase passive PFC technique, DC/DC convertor, input/output detection and protection circuit and auxiliary power supply, etc. Based on the aforesaid hardware circuit scheme, the design prototype's parameters are as follows: input -- AC380V, power factor -- $\cos\varphi \geq 0.98$; output DC voltage range -- 0～250V, output DC -- 10A.

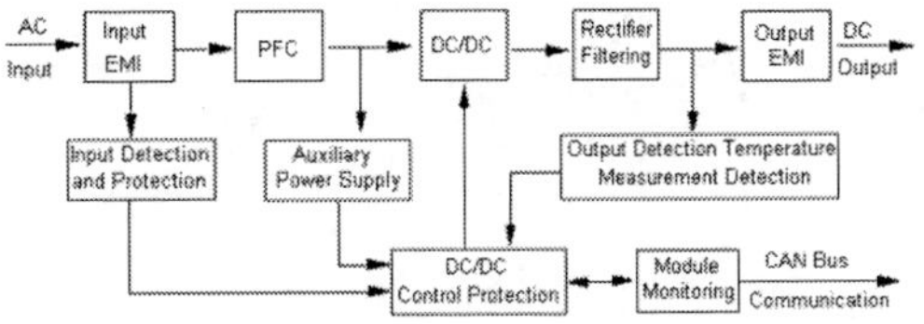

Figure 1. Rectifier Module's Hardware Circuit Block Design

Three-phase Passive PFC Circuit Design -- PFC technique refers to the technique used to restrain harmonic pollution and improve power factor. Passive PFC technique uses passive components to improve power factor and reduce harmonic wave so as to meet with the rectifier power supply module's DC output standard. PFC circuit's overall design principle diagram is as illustrated in Figure 2.

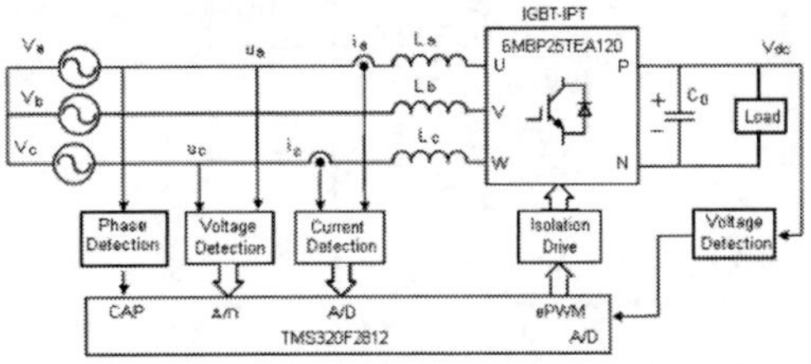

Figure 2. Filter Circuit

PFC Circuit is composed of three-phase rectifier main circuit and digital control circuit. The rectifier main circuit's structure mainly uses Boost type of six-switch & three-phase PFC circuit and its switch and protection circuit uses IGBT-IPM intelligent module with model number of 6MBP25TEA1200 manufactured by Fuji Company.

The digital control circuit mainly includes electric signal sampling circuit at DC/AC sides, sampling conditioning circuit, control processor (TMS320F2812 fixed-point 32-bit DSP chip manufactured by TI Company) and IMP drive and isolation .Sampling quantity at DC/AC sides mainly includes phase-a/phase-b current and voltage at AC side, phase position of phase-a voltage and output voltage at DC side. Sampling quality is finished by Hall current, voltage transformer and voltage sensor manufactured by LEM Company

4 Modeling in Automatic Current Sharing Method

Current-sharing controller's circuit diagram in automatic current sharing method is as illustrated as Figure 3. In the

Figure, U0 is follower for output voltage detection and it isolates load from influencing the voltage detection; U1 is current sharing loop amplifier; U2 is voltage loop amplifier; U3 is current loop amplifier.

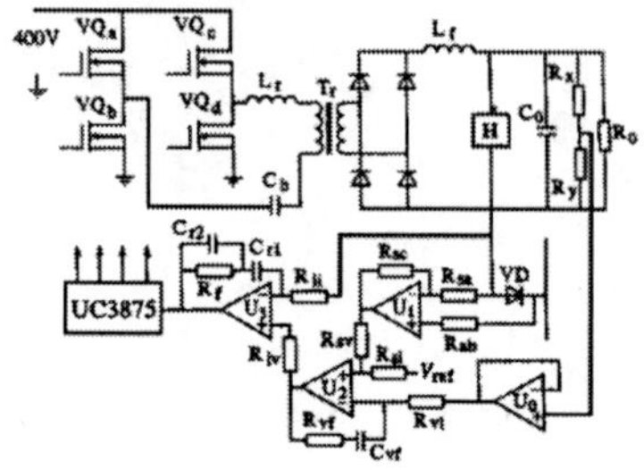

Figure 3. Current Sharing Controller's Circuit Diagram

Control block diagram based on small-signal automatic current sharing method is as illustrated as Figure 4.

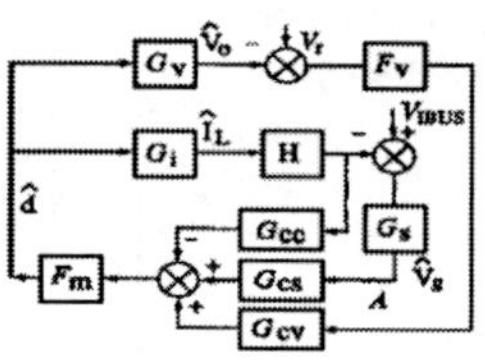

Figure4. Control Block Diagram of Small-signal Automatic Current Sharing method

$$T_S = F_m G_{cs} G_s H G_i \quad (1)$$

$$G_{cs} = 1 + W_t \frac{S + W_z}{S(S + W_p)}) \quad (2)$$

$$G_s = 40 \frac{R_{st}}{R_{vs} + R_{vt}} (1 + \frac{SR_{vf}C_{vf} + 1}{SR_{vt}C_{vf}}) \quad (3)$$

$$G_i = \frac{NV_{in}}{Z_f + R_d} \quad (4)$$

In formula (1) to (4), H is the transfer function from detection current to voltage signal, Fm is pulse width modulation transfer function, Gv is transfer function of output voltage, Gi is transfer function of inductive current, Gs is transfer function of current sharing amplifier, Gcc is current loop's current amplifier transfer function, Gcs is current sharing loop's current amplifier transfer function and Gcv is voltage loop's current amplifier transfer function.

5 Simulation and Analysis of Automatic Current Sharing Method

In this simulation, it uses six modules in parallel connection for experiment. In circuit in Figure 3, parameters of all devices are as follows: input - 380V, output-220V, switching frequency fs - 50kHz, equivalent ratio of main transformer N - 10/20, primary resonant inductor Lr - 20μH, output filtering inductor Lf - 160μH, output filtering capacity Co -1200μH, current measurement conversion ratio - 0.012 (V/A), current sampling gain H - 20, Fm - 0.20;

Simulation is conducted through MATLAB and bold plots obtained for the current sharing loop's open-loop gain are as illustrated as Figure 5 and 6. Based on the analysis of bold plots of current sharing loop's transfer function, crossover frequency basically occurs at 500kHz(rad/s) with slow dynamic response speed.

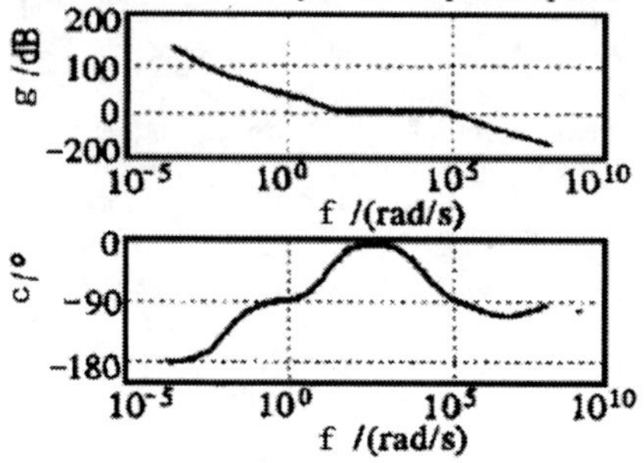

Figure 5. Bolt Plots in TS Full Loading

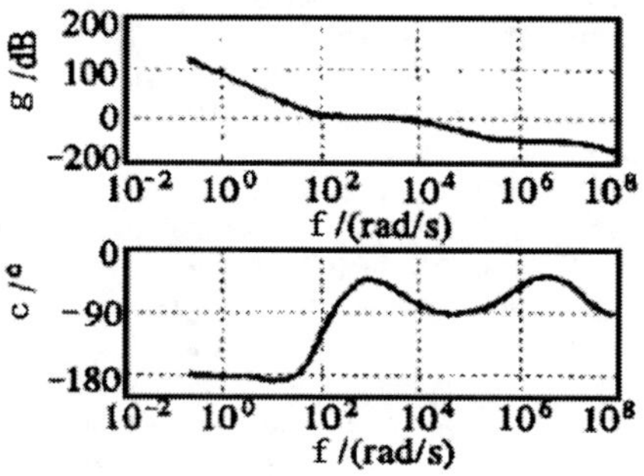

Figure 6. Bolt Plots in TS Light Loading

6 Conclusion

Based on analysis of small-single module in automatic current sharing method, it is able to find out current sharing loop control and slow response speed, which is beneficial for controlling the stability of the entire control power supply. The continuous development of electric and electronic techniques will enable current sharing techniques to become more and more mature and perfect and it will inevitably promote control power supply to develop towards larger output power.

Acknowledgements

The paper is financially aided by "General Scientific Research Project (L2014515) among Colleges and Universities in Liaoning Province" Fund.

References

1. Zijing Yang, Qiwei Lu, Duolun Wang, Cong Wang, DSP-based Digital Control Electric Operation Power Supply Development, *Power Electronics*, 2012(7): 73-75.
2. HIRAKI E, TANAKA T, NAKAOKA M. Zero-voltageand zero-current soft-switching PWM Inverter.EPE 2012 Dresden, ISBN:90-7 58115-08-5.
3. Wei Wei, Jun Li, Lanjie Wei, Research on A Soft Switching Electric Operation Power Supply, *Power Electronics*,2013(11): 47-49.
4. T.-F. Wu, H.-M. Hsieh, Y.-E. Wu, et al, Parallel-inverter system with failure isolation and hot-swap features, *IEEE Trans. Industry Applications*, 2007, 43 (5): 1329 ~ 1340.
5. Chen LiangLiang,XiaoLan,Yan Yanguang. DC LooP-Current detecting and Restraining Methods For Parallel Inverters System. *Trans.of NJUA&A*, 2004, 21(l):l-6.
6. Bo Zhang, Parallel Connection and Current Sharing Technique for High-frequency Switching Power Supply in Power Operating System, *The World of Power Supply*, 2012(3): 26-29.

MATEC Web of Conferences 63, 01035 (2016) DOI: 10.1051/matecconf/20166301035
MMME 2016

The Development of Tanks War Mobile Game based on Android System

Yi Ping SHI [1,a]

[1] *College of Electrical & Electronic Engineering, Shanghai University of Engineering Science, Shanghai, China*

Abstract. Along with the rapid development of mobile games and the popularity of Google Android operating system, the Android mobile game designed for recreation, exercise thinking will become mainstream. This paper mainly introduces the development process of mobile game and analyzes the key technologies in combination with development of tank war game based on Android system.

1 Introduction

With the popularity of the Android operating system and the large increase in the number of the terminal equipments of Android system, the application development based on the Android application system have attracted more and more developers. The mobile game development has always been the most important part of mobile development. The development of Android application layer is based on Java development, so it has the advantage of transplantation games by J2ME development quickly. But transplantation of game can't play unique characteristics of the Android operating system. The market based on the Android game development is not enough mature. It is very necessary to develop a tanks war mobile game which adapts to the Android completely on the basis of understanding of the Android application development.

2 Introduction of the Android System[1-3]

Android is the open source operating system of intelligence mobile phone based on Linux platform developed by Google. It is the first open operating system for the mobile terminal. The Android SDK provides the necessary tools and API for Android application using the Java language on the Android platform.

Android system architecture is divided into four layers. They are application layer, application framework layer, system runtime library layer and Linux kernel layer.

2.1 Application layer.

The application layer contains many core applications within the Android platform, such as Email client, browser etc. These applications are to use Java language to programming. The program running on the virtual machine and all application software belong to the application layer.

2.2 Application framework layer.

Application framework layer uses API framework to develop the core application. Developers can also use the framework to write their own applications.

2.3 System runtime library.

System runtime library layer: when using the Android application framework, the Android system will use C/C++ library to the individual components and make it better service for developers.

2.4 *Linux kernel layer:.*

The core service of Android system is based on Linux2.6 kernel. Such as security, memory management, process management, the network protocol stack, and driver model all rely on Linux2.6 kernel.

Usually a complete Android application includes four components: activity, intent receiver, service and content providers. The Activity component is the user interface. An application usually contains more than one Activity. They are stored in the Activity stack following the principle of "advanced" to switch to each other. When decided to use the above which components to build Android applications, it needs to be registered in the AndroidManifest.xml. This is an XML configuration file.It can declare the application components and their features and requirements in this file. This configuration file is necessary for each Android application.

3 The Design of Tanks War Mobile Game

[a] Yi Ping SHI :syp@sues.edu.cn

3.1 The game plan

Tanks war games which is a mobile entertainment games can bring joy to people's life. The rules of the game is simple. When start the game, players control their own tanks and use the direction key to control the tank movement. When the enemy tanks began to appear, Player can press the "OK" button to launch the bullet. After all enemy tanks is destroyed, the player wins and the game is over. Enemy tanks have been firing bullets to the player's tank. When the player's tank was hit by a bullet or our home was destroyed by a bullet, the game over.

3.2 The Design of System Module.

According to the function requirement analysis, tanks war game needs to complete the following functions: the layout of the game interface, the control function and other auxiliary function.

The overall module structure of tanks war game includes main interface module and control module. The main interface module mainly can give the player the first part of the senses after entering the game. It includes the interface of game graphics area, the display of players and enemy tanks number, the update interface of score display and initialization display of obstacles and base camp.

The game uses "OK" key to start. After started the game there will enter an initial interface. There are player tanks, the walls and the warehouse in the interface. The tanks can't be controlled. When press the "OK" button, player tanks begin to accept control the movement by up, down , left and right key. Press the "OK" button to launch the bullet. When the player is launched by enemy bullets or home is destroyed, the game is out of control. At the end of the game, it will automatically pop up a dialog box. The dialog content is " oh! You're lost !"

3.3 The key technologies

3.3.1 The building the game's interface

In order to build the game's interface, it includes the drawing of our tanks and the enemy tanks, the bullet, the retaining wall and the warehouse. In tanks war game GamePanel class is inherits from the View class. The View class is mainly used to display the View, decorate the canvas, provide drawings button touch screen events and functions, etc. Gamepanel class is responsible for rendering initial main interface and accepting the collision events of the bullet. According to target of bullet impact, it will call corresponding functions of Tank class, Shells class or Wall class respectively.

3.3.2 The movement of the tanks.

In order to move the tanks, it can use the following two ways.One is to use a thread to implement multiple tanks movement. This method has the advantage of stable performance, but some operating more trouble. It has a certain understanding for thread. Another is to use Iterator to implement multiple tanks movement. It uses Iterator to access the various elements in a container object and does not need to expose the internals of the object. In this paper, it uses the second method.

The movement of the tank using the up and down or so key control. View class encapsulates the touchscreen monitor function onTouchEvent (MotionEvent event)。 The event preserves the touch screen motion. Through the getAction function it can get press or mobile operation, which can control the movement of the tank.

3.3.3 Operation mechanism

Every game needs an engine to promote the operation of the game. Tanks war game's engine uses a thread to the Iterator container elements to perform the corresponding action, until our tanks are destroyed or our base camp is destroyed. Through continued remove and draw of obstacles, tanks and bullets, it maintains the update contents in the main screen displays until the player tanks or home was destroyed.

4 The Implementation of Tanks War Mobile Game based on the Android[4-6]

4.1. System Development Environment

Android Virtual Device is to debug their Android applications for developers. It can run on the PC. It provides the mobile phone user with the interface and operation way as real shape. It is just different from underlying hardware drivers. Developers can use the mouse instead of mobile phone on the AVD points.

To realize all this, the developers need to set up an Android development environment. The development environment of tanks war game is using Microsoft Windows 7 operating system + JDK + Eclipse + Android SDK development environment. Android game development environment includes the following three parts:

4.1.1 JDK.

The JDK (Java Development Kit) is Sun Microsystems products for Java developer. The JDK is the core of the Java. It includes the Java runtime environment, Java tools and Java class library. The JDK is mainly used develop Java applications on mobile devices and embedded devices. It contains some basic components and common packages.

4.1.2 Eclipse.

Eclipse is an open source and extensible development platform based on Java. It is just a framework and a set of services. It can build the development environment

DOI: 10.1051/matecconf/20166301035

through the plug-in components. Eclipse comes with a standard set of plug-ins, including the Java Development tools (Java Development Kit). Eclipse has main characteristics as following: open, and extensible integrated development environment, successful graphical interface application interface(SWT/JFace), powerful plugin loaded, full of Chinese help files and nature of open source.

4.1.3 Android SDK Tool.

The Android SDK tool is a software development kit. It provides a set of application software development tools for software development engineers to establish a particular software package, software framework, hardware platform, operating system. It is exclusive of the Android software development kit. Android SDK tools include Android emulator, Android development tools plugin for the Eclipse IDE, debug monitor service, Android debug bridge, Android asset packing tool, Android interface description language, sqlite database, Android virtual devices etc.

4.2 Class Module Design.

Developed source directory of tanks war mobile game is shown in Fig.1.

Fig.1 the source directory

GameFactory.java is to create bullets class for tanks fire. GamePanel.java is the class of the main interface and control class in the game. Point.java is the coordinate class. Shells.java is bullets class which contains the bullet movement method. Tank.java is tank class which includes tanks image reading method and moving method. Wall.java is obstacles class which contains the drawing method of the obstacles.

4.3 The Implementation of the game interface

The main window is the human-computer interaction. In the program it can create a window which inherits from the View class.

4.3.1 The creation of the game main interface

Creating GamePanel class is used to initialize, add and store the various components in tanks war games. Map[] [] is used to record the current state at every point. It uses initial Map() function to initialize the walls and the base camp.

4.3.2 The draw of the bullet method

Creating Shells class with public void drawShells() method to realize the bullet drawing.

4.3.3 The drawing of a single wall

A unit of wall is made up of a blue rectangle and a white border. It calls drawWall() method to draw a single wall.

4.3.4 The drawing of tanks

It uses the external material pictures. Drawing is first to read the external image. It uses tank overloading function to draw the initial player tanks and the enemy's tanks in the process of the game.

4.4 The Implementation of the control Module.

In the main window the game can input and output data. Tanks need to be get four direction movement of up, down, left and right. The "OK" key controls tanks to take fire.

4.4.1 Tanks move

The game must support multiple tanks movement at the same time. It uses Iterator and ArrayList method to operate each element in the container. It implements the enemy tanks movement randomly, and the behavior of judgment when the bullet touch the tanks.

4.4.2 Tanks to generate

Player tanks are generated at the start of the game. The enemy tanks are randomly generated in the certain coordinates range in the process of game. The enemy is only three tanks.

4.4.3 The realization of the bullets function

When a player presses the "OK" button, it will launch a bullet from a tank or when the enemy triggers tanks random events. The bullet will move along the emission direction beyond borders or touch.

4.4.4 Initialization of the map drawing

The game uses a map to express the battle scene. The point of map in the game has four states. It respectively is the idle state, occupied by a tank, occupied by a obstacle and occupied by a stronghold. It records the status using two-dimensional array.

4.5 The game test

In the process of programming, the programmer can directly click the Run Project icon on Eclipse toolbar and call the Android simulator to take simulated test on the game. The game interface is shown in Fig.2.

DOI: 10.1051/matecconf/20166301035

Test process is as follows: after loading the tanks war game software the player can use the mouse to click startup icon of tanks war on the virtual machine. When the game page is started, the player can press the "OK" button to start the game. He can control their tanks. When the enemy tanks begin to appear, the player can use the virtual keyboard direction key or the keyboard direction key to control the direction and press the "OK" button to launch the bullet. When the player's tank was hit by a bullet or our home was destroyed by a bullet, game is over.

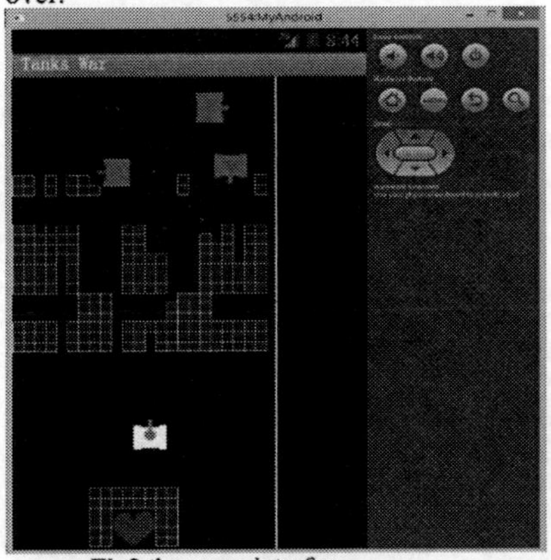

Fig2.the game interface

The game software function test includes the start of the game, exit, handling, the bullet trigger, statistical scores and other functions as well as the response to the mouse click events and key events. Through above test it get the following conclusion: the tanks war game software can realize the function of the tanks war games, friendly interface, easy to operate.After finishing the game development, the programmer can convert them to the APK document. He can install and execute them on Android emulator or an Android phone.

5. Conclusion

With the rapid development of the Android platform, all kinds of mobile games based on Android development emerge in endlessly. The development of mobile phone game based on Android system will focus on the construction of the Android development environment, the use of the Android game development technology, game plan, the design and implementation of game software. The software test result of tanks war mobile game shows that the software can realize the function of the system correctly and can successfully run on Android emulator and Android mobile phone.

References

1. Rogers, *The Android Application Development* , (Beijing: People's Posts and Telecommunications Press,2010)
2. Komatineni, *Mobile Development Series Pro Android 3*, (Beijing: People's Posts and Telecommunications Press,2011)
3. HuaiYu Fan, *Essence of Android Development* . (Beijing: The Mechanical Industrial Press,2012)
4. Hongcheng Huang, *The Design and Development of Android Mobile Application* , (Beijing: People's Posts and Telecommunications Press,2012)
5. Yulong Hao, *The Basis of Android Programming*,(Beijing: Tsinghua university Press,2012)
6. Yafeng Wu, *Real Treasure of Android Game Development* , (Beijing: People's Posts and Telecommunications Press, 2010)

The Effect of General Corrosion on the Guided Wave Inspection of the Pipeline

Jin Heng ZHANG [1,a], Chang De LI[1,b], Yang YANG[1,c]

[1] *School of electronics and information, Nanchang Institute of Technology, Nanchang*

Abstract. The guided wave method can inspect pipelines very quickly and widely. For instance, it can inspect the overall pipelines by digging several detection pits or removing part of coating material to set the array ring. However, it will make the guided wave attenuate more seriously and make the signals hard to identify when setting the array ring on the general corrosion. In this study, the wave propagation will be discussed when the general corrosion is under the array ring and the severe localized corrosion is inside the general corrosion via experiment and finite element method. The results showed that the excitation energy will be lower when the array ring set on the pipe surface with the general corrosion. By two-dimensional Fourier transform analysis, its non-uniform contact surface will increase asymmetric modal and mix signals. The energy attenuation will increase when the corrosion depth is deepened or the inspection frequency is risen. For example, the 2 mm deep general corrosion will attenuate -1.09 dB/m at 20 kHz and attenuate -3.01 dB/m at 40 kHz; the 4 mm deep general corrosion will attenuation -5.76 dB/m at 20 kHz and attenuation -23.19 dB/m at 40 kHz. However, the coherent signals which were caused by the general corrosion will decay with increasing frequency. For example, the coherent signals of 2 mm deep general corrosion are -23.67 dB at 20 kHz and -35.44 dB at 40 kHz; then, the 20 mm long and 3.5 mm deep localized corrosion which signal is -26.34 dB at 20 kHz and -26.94 dB at 40 kHz will be detected easily at high frequency. It can provide detectors to understand the impact when the array ring set on the area of general corrosion and the way to distinguish the localized corrosion which is inside the area of general corrosion.

Keywords: Guided wave, Finite element method, Two-dimensional Fourier transform, General corrosion, Localized corrosion

1 Introduction

In petrochemical plants, the use of pipelines to transport all kinds of fluids are very common. In response to different environmental conditions will be derived from different pipes, such as coated pipelines. Although the pipelines can a hold temperature, fire and stabilization process, but the water vapor will easily penetrate the gap between the clad material and the pipeline. It is easy to form a wide range of general corrosion and possibly mixed with severe localized corrosion, such as Fig 1. And because of the coating material, it will not be easy to perceive the whole pipeline situation, if there have corrosion on the pipeline. In recent years, the guided wave method can inspect pipelines very quickly and widely by digging several detection pits or removing part of coating material to set the array ring.However, it will make the guided wave attenuate more seriously and make the signals hard to identify when setting the array ring on the general corrosion.

For guided wave detection of general corrosion and localized corrosion pipelines. In 2005 ASTM Standard G46 [1] was defined the measurement and evaluation method of corrosion, uniform corrosion on the pipeline and in accordance with the number of holes, the average depth and size of the holes per unit area and made different levels of distinction. In 2009 Su, for the research of coating pipelines with general corrosion, to understand the causes of corrosion and its risk parameters and propose different detection methods to prevent it. In 2010, Carandente et al [2] also discuss the relationship between the defect and the defect echo, but for a more detailed appearance defects do research, and found that defects tapered recess which case the reflected signal a trend similar to the defect. In 2011, Løvstad and Cawley [3] pierce hole circumferential distribution and axial distribution of the case, the analysis of the reflection coefficient of torque hole guided wave effects and effective use of the single-hole signal superimposed way to predict the random distribution of holes echo situation.

[a] Corresponding author: 416244710@qq.com

DOI: 10.1051/matecconf/20166301036

In 2012, Cobb et al [4] conducted research for the uniform corrosion, and they changes temperature, different covering parameters such as attenuation investigate the case, as the determination of the severity of uniform corrosion. In2013 EPRI's test report [5], also pointed out that the general corrosion on the pipeline will cause attenuation of guided wave energy and cause other problems caused by signal enhancement judgment on background signals.

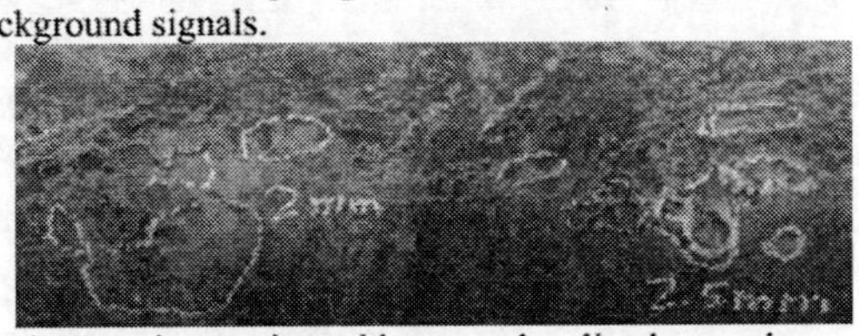

Fig 1 general corrosion with severe localized corrosion

2 Research approach

2.1 Dispersion curve

Dispersion curves is the relation diagram between group velocities or phase velocities and frequency. As shown in Fig.2, it is the group velocity dispersion curve for a 6-inch pipe, it can see that the used in this research, there is a non-dispersion mode T(0,1).

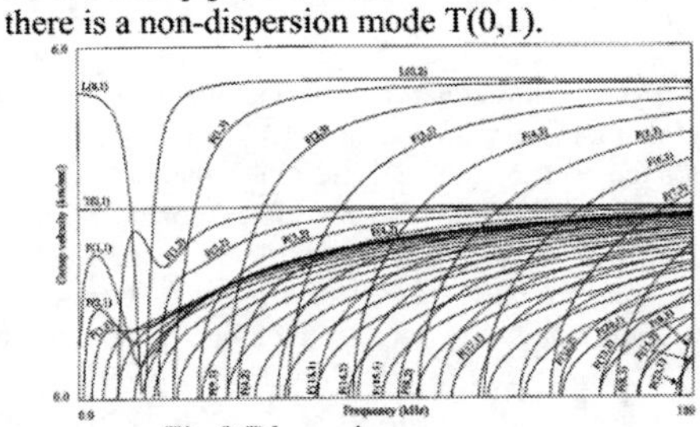

Fig 2 Dispersion curves

2.2 The Setting of the general corrosion model

General corrosion is the case of the entire metal surface corrosion on the pipelines and with a consistent average corrosion depth.The simulation will construct numerous round holes to approximate the case of general corrosion when the ratio of the maximum depth and the average depth of the holes is 1 can be expressed as general corrosion. The general corrosion pipeline model will set a single aperture of 15mm, depth of 1mm round hole and construct on different numbers and distribution.The paint must first fraction prior to the construction of the pipeline corrosion, thereby setting coordinates (X,Y), X represents the direction of the longitudinal axis of the pipeline, Y represents the circumferential location of the pipeline which the circumferential tube every 6 degrees (about 9mm arc length)-divided into units, each tube axial 9mm divided into a unit.And then, use Matlab software to draw the coordinate points, based on the random nonce selected in the circumferential and axial positions and the coordinate values established hole thereon, as shown in Fig 3.

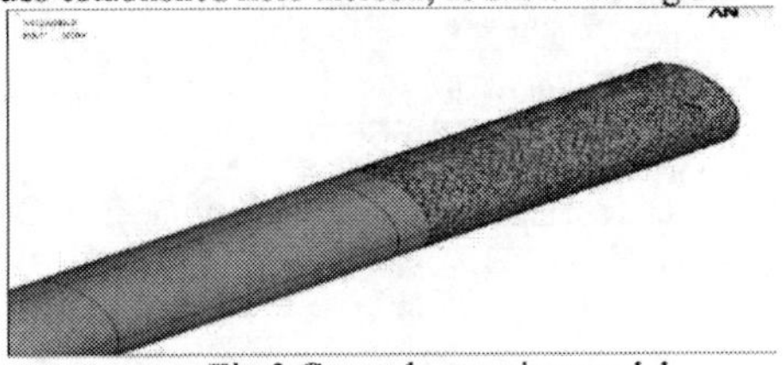

Fig 3 General corrosion model

2.3 Two-dimensional fourier transform

After the solution of finite element, the time-domain signal can be retrieved at any point in the structure. Through the two-dimensional fourier transform method can further analyze the relationship between the frequency and wave number, to investigate the wave propagation mode. It isassume the harmonic waves, which propagating along with elastic, homogeneous, and isotropic material, the particle displacement can be expressed as

$$u(x,t) = A(\omega)e^{i(\omega t - kx - \theta)} \qquad (1)$$

Wherein x is position, t is the time, $A(\omega)$ is the amplitude dependent with frequency, k is the wave number, ω is the angular frequency, θ is the phase.The two-dimensional fourier transform method, including a function of time and position, the conversion equation can be expressed as

$$H(k,f) = \int_{-\infty}^{+\infty}\int_{-\infty}^{+\infty} u(x,t)e^{-i(kx+\omega t)}dxdt \qquad (2)$$

3 The simulation set and analysis

3.1. The simulation of wave propagation

In this paper, 6-inch steel pipelines were construct which is 3.6 meters long, outer radius of 84 mm, 7 mm thickness and the general corrosion zone is 0.5 m, such as Fig 4.Then construct a distributed evenly full circle tube surface corrosion depth 1, 2 and 4 mm in the model simulations, and using frequency of 20, 30 and 40 kHz for testing.Fig 5(a)-(f) can found that the corrosion area is formed of two main echo signal wave packet. The main reflection signals are caused by the edges of corrosion. The corrosion signal will decrease with the increasing frequency. It exhibit that uniform corrosion at high frequencies will have less obvious reflections. When the uniform corrosion depth was 4 mm, it will make more apparent reflection signals because of changing cross-sectional area increase and cause greater reflection signals.And the increasing corrosion depth also caused more penetrable and reduce the reflected energy.

DOI: 10.1051/matecconf/20166301036

In order to define the extent of corrosion decay, we will measure its incident wave amplitude (I) and the transmission wave amplitude to analysis the attenuation. The result are shown in Fig 6. General corrosion attenuation with 1 mm and 2 mm depth does not have much downward trend with the enhanc frequency, but the corrosion depth. However, when etching depth of 4 mm its attenuated will be significantly decreased. For example, the 2 mm deep general corrosion will attenuate -1.09 dB/m at 20 kHz and attenuate -3.01 dB/m at 40 kHz; the 4 mm deep general corrosion will attenuation -5.76 dB/m at 20 kHz and attenuation -23.19 dB/m at 40 kHz.

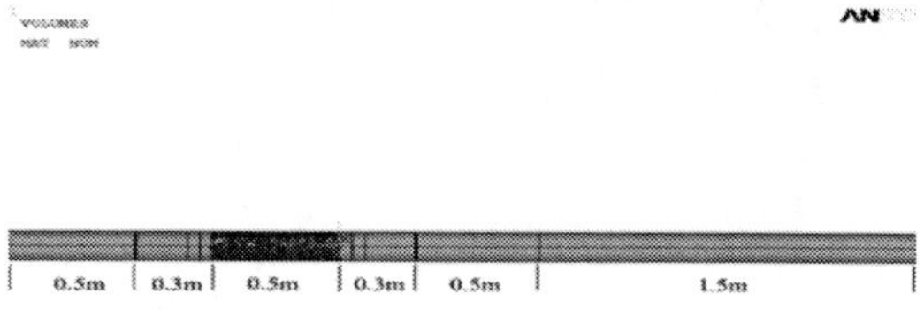

Fig 4 The size and characteristics of the pipeline

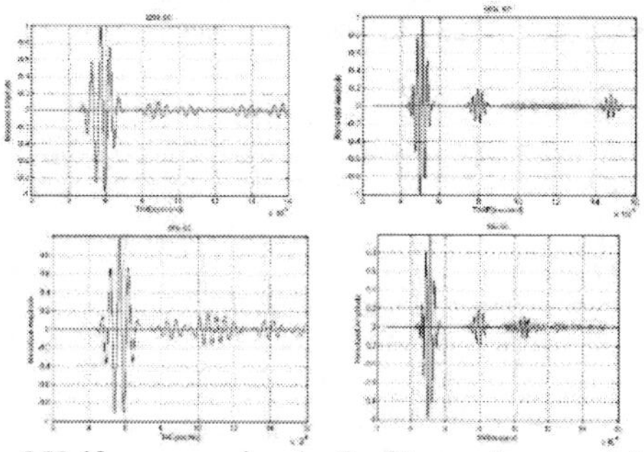

Fig 5 Uniform corrosion depth of 1 mm, 2 mm and 4 mm of the time domain signal

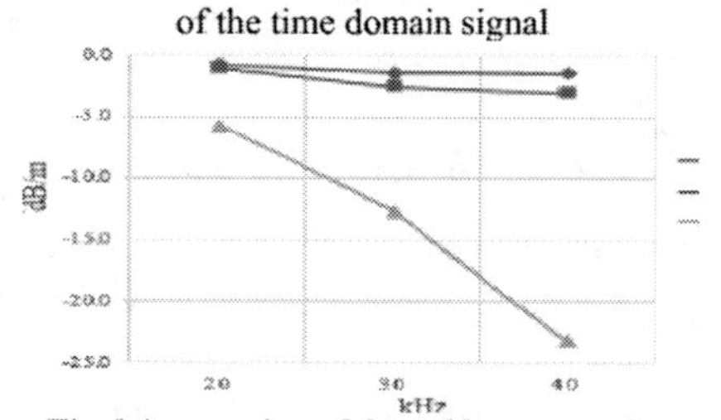

Fig 6 Attenuation of the uniform corrosion

3.2 Local defects results and discussion

As stated previously, the same 6-inch steel pipeline was used and constructed a defect at two meters away from the pipe bottom. Then use a center frequency of 20, 30 and 40 kHz 5 cycles of modulation signal to excite the torque mode. First, set a local defects D1 length of 20 mm, and then, set a local defects D2 length of 40 mm, both of the two defects depths are 3.5 mm and the circumferential are 1/10 distribution perimeter. Using the center frequency of 40kHz, a wavelength of 81.5 mm of the T(0,1) guided wave to detect D1, its L / λ = 0.245, the reflection coefficient is 0.0635 local defects. At theFrequency of 30 kHz, the wavelength of 100.8 mm of T(0,1) guided wave testing, its L / λ = 0.184, partial defect reflection coefficient is 0.0629; At theFrequency of 20 kHz, the wavelength of 162 mm of T(0,1) guided wave testing, its L / λ = 0.123, partial defect reflection coefficient is 0.0565;To detect defect D2, At theFrequency of 40 kHz, the wavelength of 81.5 mm of T(0,1) guided wave testing, its L / λ = 0.490, defect reflection coefficient is 0.151. At theFrequency of 30 kHz, the wavelength of 100.8 mm of T(0,1) guided wave testing, its L / λ = 0.368, defect reflection coefficient is 0.0352. At theFrequency of 20 kHz, the wavelength of 162 mm of T(0,1) guided wave testing, its L / λ = 0.245, defect reflection coefficient is 0.0635. The result are shown in Fig7. The reflection coefficient of local defect D1 will also rise with increasing frequency, and the reflection coefficient of local defect D2 will decrease with increasing frequency.

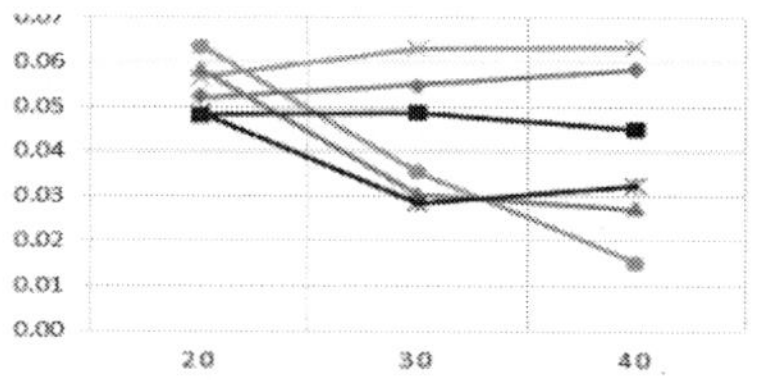

Fig 7 Defects reflection coefficient

3.3 The discussion of uniform corrosion with localize defects

In order to understand whether localize defects signals will interfere with the uniform corrosion. It will continued the setting in 3.1 section and added the localized defects D1, D2 in the uniform corrosion. The result are shown in Fig8-9.When the local defects D1 in 1mm and 2mm uniform corrosion, the increasing frequency will make the localized defects signal more obvious. Because the uniform corrosion signal will decrease with the increasing frequency and the local defects D1 will increase with the increasing frequency. However, the local defects D2 signal will mix with uniform corrosion signal at all frequencies. For example, the coherent signals of 2 mm deep general corrosion are -23.67 dB at 20 kHz and -35.44 dB at 40 kHz; then, the localized corrosion D1 which signal is -26.34 dB at 20 kHz and -26.94 dB at 40 kHz will be detected easily at high frequency. It can provide detectors to understand the way to distinguish the localized corrosion which is inside the area of general corrosion.

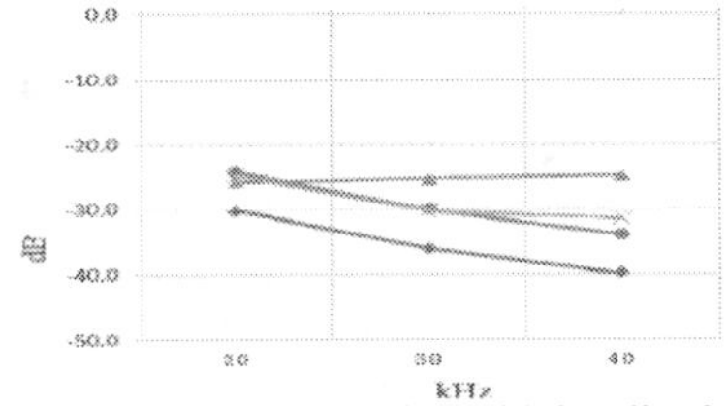

Fig 8 1-mm uniform corrosion with localized defects

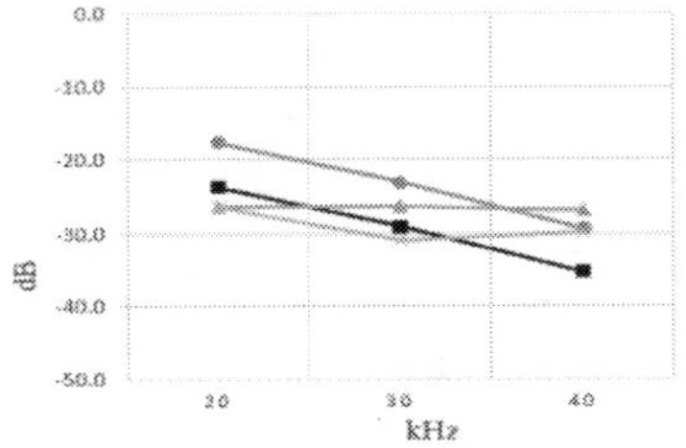

Fig 9 2-mm uniform corrosion with localized defects

4 Experimental setting and measurement results

4.1 Instrument

The experimental measurement instrument is used in the United Kingdom GUL (Guided Ultrasonics Ltd.). The equipment including circular piezoelectric transducers, the signal processor Wavemaker G3 and application software Wave Pro. T(0,1) is used in the study. The excitation in the frequency regime of 6 inch pipeline is 0.0-3.0, while the operation corresponding to the frequency of 21.39-33.90 kHz.

4.2 Experiment Settings

In order to understand the results of uniform corrosion detection signal, we use a pipeline with uniform corrosion. The pipeline is 6-inch pipe, radius of 84 mm, thickness 7.1 mm, where the length of uniform corrosion zone is 1.83 m. Next, set the probe on the pipe with clean surface to detect the uniform corrosion. Then, produce two artificial local defects D1 and D2 one after another, such as Fig 10. The defect D1 and D2 depth are all 2.47 mm, perimeter are all 30 mm, axial length are 30 mm and 60 mm. Finally do the detect again.

4.3 The results and discussion

Uniform corrosion signals are clumps of continuous signals, because the actual uniform corrosion area was formed by many corrosion section. Its distribution are not surrounded the full circle of the pipe but on one side of the pipeline. With different corrosion section, it different lengths, and cross-sectional area will vary so the reflected signal will follow the change. By the experiment, it can be found that the D1 signal has a significant value and exceed the measurable limit from low to high frequency. This artificial defects D1 axis length of 30 mm with a frequency range of the wavelength ratio between 0.18-0.28. Though the signal was a slight downward at high frequencies, the signals still have more obvious peak as the simulation do and the coherent signals which were caused by the general corrosion will decay with increasing frequency. It will be beneficial to highlight the local defects signal at the high frequency.Then, processing the artificial defects D2 again make it into a 60 mm shaft and do the same test. This artificial defects D2 axis length of 60 mm with a frequency range of the wavelength ratio between 0.36-0.56.Comparing with simulation results showed that the signal will be decreased and it can also be seen in the experiment. Such of this defect was easily mixed together with the uniform corrosion at low and high frequency such as Fig 11.

Integration of the above result, the coherent signals which were caused by the general corrosion will decay with increasing frequency, and the local defects signals which was inside the general corrosion will be affect by itself shaft length and the detection frequency.

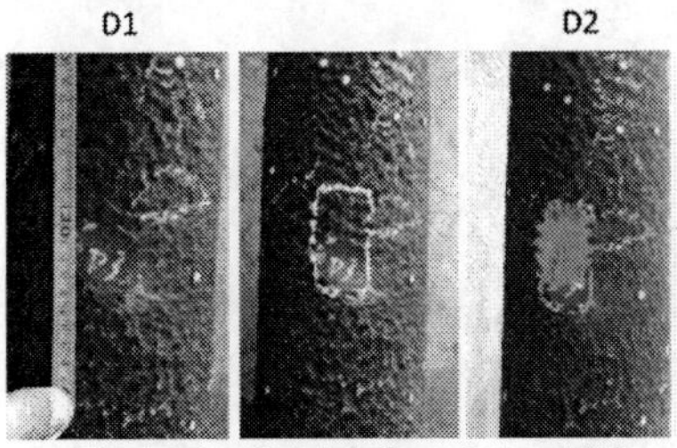

Fig 10 Artificial defects D1 and D2

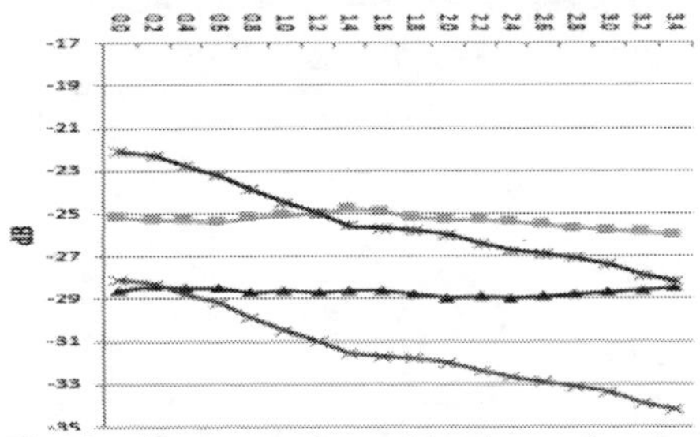

Fig 11 uniform corrosion and local corrosion signal

Conclusions

The excitation energy will be lower when the array ring set on the pipe surface with the general corrosion. By two-dimensional Fourier transform analysis, its non-uniform contact surface will increase asymmetric modal and mix signals. The energy attenuation will increase when the corrosion depth is deepened or the inspection frequency is risen. For example, the 2 mm deep general corrosion will attenuate -1.09 dB/m at 20 kHz and attenuate -3.01 dB/m at 40 kHz; the 4 mm deep general corrosion will attenuation -5.76 dB/m at 20 kHz and attenuation -23.19 dB/m at 40 kHz. However, the coherent signals which were caused by the general corrosion will decay with increasing frequency. For example, the coherent signals of 2 mm deep general corrosion are -23.67 dB at 20 kHz and -35.44 dB at 40 kHz; then, the 20 mm long and 3.5 mm deep localized corrosion which signal is -26.34 dB at 20 kHz and -26.94

dB at 40 kHz will be detected easily at high frequency. It can provide detectors to understand the impact when the array ring set on the area of general corrosion and the way to distinguish the localized corrosion which is inside the area of general corrosion.

References

1. ASTM Standard G46, "Standard Guide for Examination and Evaluation of Pitting Corrosion," ASTM International, West Conshohocken, PA, 2005.
2. R. Carandente, J. Ma and P. Cawley, "The Scattering of the Fundamental Torsional Mode from Axi-symmetric Defects with Varying Depth Profile in Pipes," Journal of the Acoustical Society of America, Vol. 127, pp. 3340-3348, 2010.
3. A. Løvstad and P. Cawley, "The Reflection of the Fundamental Torsional Guided Wave from Multiple Circular Holes in Pipes," NDT&E International, Vol. 44, pp. 553-562, 2011.
4. A. C. Cobb, H. Kwun, L. Caseres and G. Janega, "Torsional Guided Wave Attenuation in Piping from Coating, Temperature, and Large-area Corrosion," NDT&E International, Vol. 47, pp. 163-170, 2012.
5. Obtaining Credit for Guided Wave as a Buried Pipe Direct Examination. EPRI, Palo Alto, CA: 2013.

MATEC Web of Conferences **63**, 01037 (2016) DOI: 10.1051/matecconf/20166301037
MMME 2016

Summary on Sensorless permanent magnet Brushless DC Motor Control Strategies

Hai Xia LI[1,2],Yang CAO[2]

[1]*Guangxi Scientific Experiment Center of Mining, Metallurgy and Environment,Guilin University of Technology, Gunag xi guilin 541004,China*

[2]*Guilin university of technology ,college of Mechanical and Control Engineering,Guang xi, guilin,541000,china*

Abstract: This paper aims at discussing the development process and application of permanent magnet brushless DC motor. By referring to the related literatures, this thesis gives an overview of several common non-position sensor detection technologies, analyzing their strengths and weaknesses as well as a number of new and improved methods in practical applications. Besides, The application situation of the electric door with sensorless permanent magnet brushless DC motor was illustrated.

1 The introduction

The permanent magnet brushless DC motor (DC Motor Brushless, referred to asBLDCM) , is a multivariable and nonlinearcontrol object, integrated with modern electronic technology, controltheory and motor technology.Thefirst permanent magnet brushless DCmotor is manufactured in 1962, which marks the permanent magnet brushless DC motor to enter a new stage of development. Harrison and others , in the mid-1950s , for the first time ,to apply for a patent inapplication , which the motor`sbody tube commutationreplaces the mechanical commutation , thus the modern permanent magnet brushless DC motor prototype appeared. Germany's MANNESMANN company ,in 1978, launched the MAC permanent magnet brushless DC motor and its driving device, which makes permanent magnet brushless DC motor officially entered the practical stage of the mass production[1].

Permanent magnetic motor was widely used in vehicles, ships, industrial power, medical equipment and household electrical appliance etc, because of its characteristics of the little electromagnetic interference , the reliable operation, the convenient maintenance, economy and diversification etc.Traditional permanent magnetic motor are mostly based on the Holzer-position sensor, but the sensor accuracy and installation location greatly limits its dynamic performance and adjust the performance, as well , the cost of its production is too high. Therefore, with the no-position sensorbeginning to be used, many control methods of permanent magnet brushless DC motor rised .In this paper,the advantages and disadvantages of various methods and their improved technology will be summarized and discussed in detail.

2 The technology of Non-position

Sensor Detection

2.1Based on the position of the back emf position detectiontechnology

(1)Based on the counter electromotive force passing zero detection

Based on the counter electromotive force passing zero detection is the simplest method. Its principle is prior to ignore the influence of armature reaction ; by disconnectedthe zero's opposite electromotive force ,to measuresix key position signal, as shown in Figure 1. Literature [2,3] rotor through measure each phase permanent magnet brushless DC motor winding voltage, followed by the software calculated back EMF zero crossing, but they use the hardware detection circuit, and a phase compensation method differ. The calculation method of literature [4] is measuring the motor's winding-line voltage, get zero position of the back EMF . It is function (1) of stator winding three-phase instantaneous voltage.

$$u_{as} = r_s i_{as} + L_s \frac{di_{as}}{dt} + e_{as}$$
$$u_{bs} = r_s i_{bs} + L_s \frac{di_{bs}}{dt} + e_{as} \text{-----------------(1)}$$
$$u_{cs} = r_s i_{cs} + L_s \frac{di_{cs}}{dt} + e_{cs}$$

r_s ----------------phase resistance

L_S ----------------Phase Inductance

e_{as}, e_{bs}, e_{cs} ---Three-phase winding open-circuit voltage

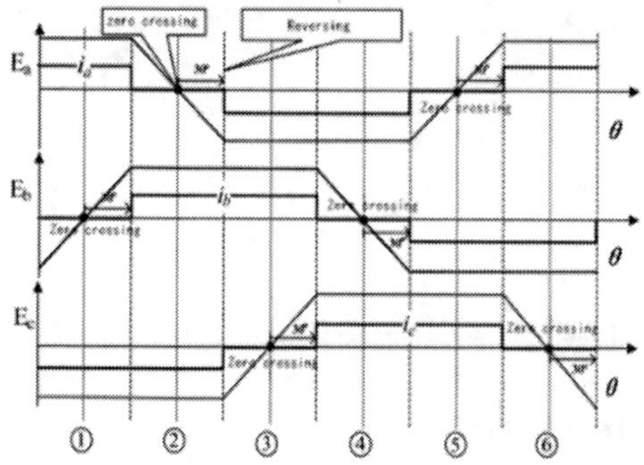

Fig.1Bemf zero and in phase diagram

Thistest technology is simple ,but some obvious shortcomings also exist. First of all, it is very difficult and vulnerable that measuring the zero position. The current solution is to use the depth filter circuit to filter.Second, when the motor's speed is slow ,the potential amplitude value of amplitude is too small, and difficult to get the accurate value. Third, in theory,trapezoidal wave is the sole criterion for judging back EMF waveform.Butin actual measurement it's difficult to get the trapezoidal wave .

To avoid the defects, A method of directly detecting back EDF , without the assumed neutral point , has been proposed, which could achieve full duty cycle operation .[5]At the same timethe method of direct detection zero ,with many limitations, has been proposed.in China.[6] In order to compensate the position error in real time, a new automatic correction test methodof rotor position phase , which using non conduction phase current deviation , is proposed [7].

(2)Fly-wheel diode method

Ogasawara et al ,in 1991 ,was first to proposethe method of continued flow diode, also known as the three-phase conduction method. The biggest advantage of the continuous flow diode detection method is to expand the speed range of the system ,in particular the low speed . But the method needs to open the loop for start ,withthe specific PWM modulation mode and six way detection circuit, which greatly increases the complexity of the hardware circuit. In addition,

DOI: 10.1051/matecconf/20166301037

in this method, six mutually isolated ways make the hardware design difficulty and cost too much. Therefore, at home and abroad, few people use this method, so the technology is not relatively mature and the relevant literature is relatively little.

(3)The electric potential integral method

Back EMF integral method is presented in literature [8], the basic idea of this method is: in the use of integral circuit to realize the detection of rotor magnetic pole position, with comparing resulting of the integrator output ad the reference voltage, in reversal signal. As shown in Figure 2 Schematic diagram shown

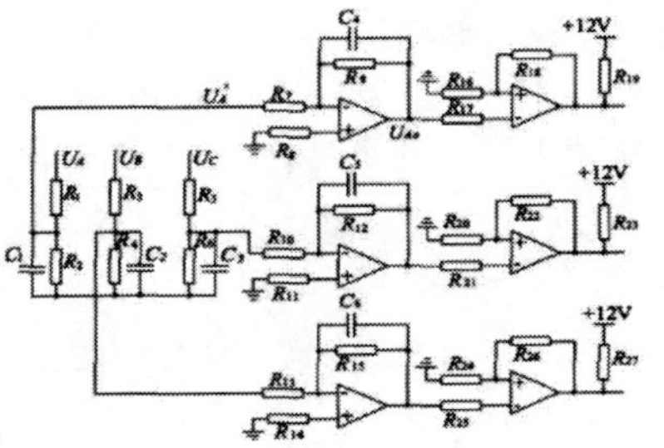

in.

Fig.2 The principle diagram of the back emf integral method

Inliterature [9],the method has used in the airconditioningcompressorapplication.

compared with the traditional back EMF method, back EMF integral method is relatively stable.

Another different principle of back EMF integral method: integral method of back EMF logic level,[10]was carried out. The results indicate that only level detection in the detection of polar could greatly improve detection performance of the back EMF vibration amplitude at start and low speed.

(4) third harmonic detection method

Using Fourier series decomposition method to decompose the phase windings ofBLDCM.The integral value of third harmonic component and is the motor commutation point zero, as shown in figure

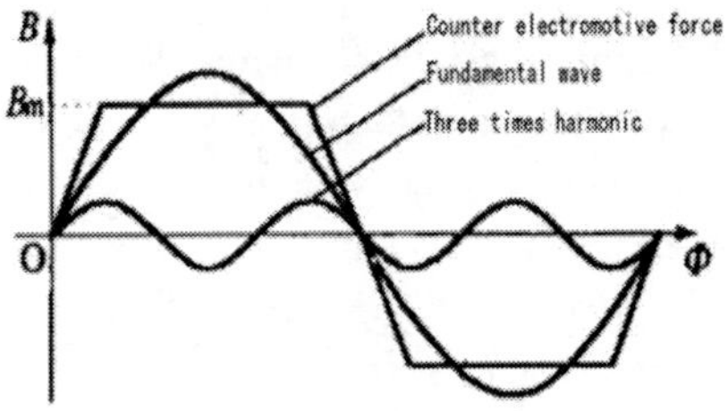

3.

Fig.3 Counter electromotive force, harmonic relationship

In literature[11] ,using TMS320F240 DSP chip to Proved the correctness of this method .Experimental results show the method is not affected by fly-wheel diode conduction effect, don't need a depth filter, and can detect the rotor position under high-speed operation.But in practical application, there are many obvious defects in assembly and low-speed operation. In order to improve the rotor position detection accuracy of the motor ,in high speed region, the method of the software phase locked loop of the literature[12] is achieved.

2.2 based on observer method

The main idea of the state observer method is to estimate the rotor position of the motor mathematical model by measuring the input of the actual system in real time.

In the literature [13], the motor control was carried out by using the extended Calman filter method, which was not affected by the back EMF waveformandthe change of system parameters, as well，the low speed performance was good.[14]Experimental results show it can completely replace the application of position sensor in fans, pumps and other occasions. Synovial observer to estimate thewaveform of the motor back EMF , in order to obtain commutation signals.[15] The results show that the observer combination can enhance the system disturbance and parameter changes of robustness. But this calculation method is very large, need to calculate the powerful DSP to solve, so its application is not very extensive.

2.3 Based on flux linkage function of rotor position detection

Due to the commutation time flux function value is very sensitive, so this method does not need the depth filter, and the flux function value and speed is independent, and low-speed range can extension. But the exact value of voltage and current ,and the accuracy of the parameters of the motor greatly reduces the robustness and anti-interference ability of the method. In order to make better correction of flux and position value, the control structure of double current loop is given in literature [16]. This method can be realized in full speed range.

2.4 based on the inductance of the position detection

(1) the current rate method

The main principle of current rate method is that calculate the slope of the detected current changes,next to get the position. In every electromotive force rate ,commutation positionis zero. This detection method need high-precision. In general, the sensor can obtain the value of the current signal. But ,the waveform with a small change is extremely difficult to detect,so few applications, in engineering ,were used.

(2) Method inductance

The inductance method has two forms: one is used in the salient pole permanent magnet brushless DC motor, the principle is to judge the change of the inductance of the motor winding during the starting process,with applying a specific detection voltage to the motor. However, this method is more difficult and can only be applied to the salient pole permanent magnet brushless DC motor, therefore, the present application is less. For strengthening the terminal voltage detection process more quick to get the position of the motor rotor, in literature [22], making a number of improvements, which using injection of weak high frequency AC signal in the stator windings. The other inductance method is using the method of applying the square wave voltage pulse to detect the current amplitude in the winding, through the inductance`s difference to get the rotor position signal, and the experimental results show better effect [17]. But the method also has a lot of disadvantages. It depends on the highly accurate current detection, and the ability to resist interference is poor. Ifdeliberately magnifying inductance characteristic of motor rotor, the detection will be more accurate. So it will become a very mature control method.

2.5 the artificial intelligence test

(1) fuzzy control

In the literature[18],at the frist Zadeh proposed the fuzzy control theory in 1960s. The fuzzy control method works realized by the fuzzy controller`s algorithm rules .

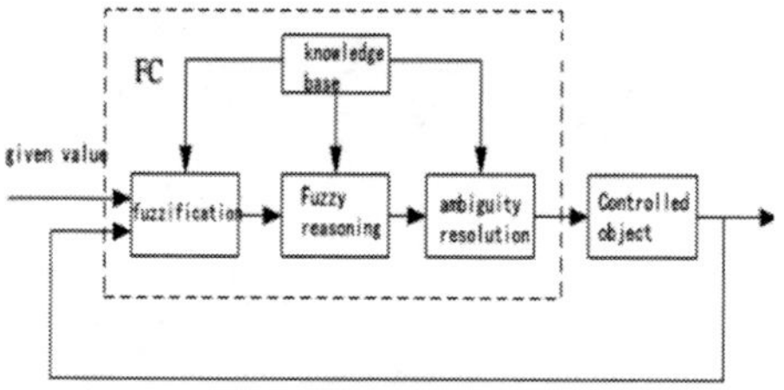

Fig.4 The fuzzy controller

Fuzzy control does not need to set up a dynamic analysis model of the object ;and if the design of fuzzy rules is proper, the whole system will have good robustness. In addition, the fuzzy controller has the advantages of simple design, good stability etc. In the literature[19] ,of this method,debugging time was limited, and operating environment is complex and changeable, but the complexity of the fuzzy rules beyond a certain limit. To solve these problems, using the genetic algorithm to make fuzzy control system to autonomous learning, freegroupandself-correction.

(2) neural network

The most common training method of neural network is BP (error propagation back) algorithm, which is proposed in 1986 by Rumelh, Hinton and Williams, as shown in figure 5.

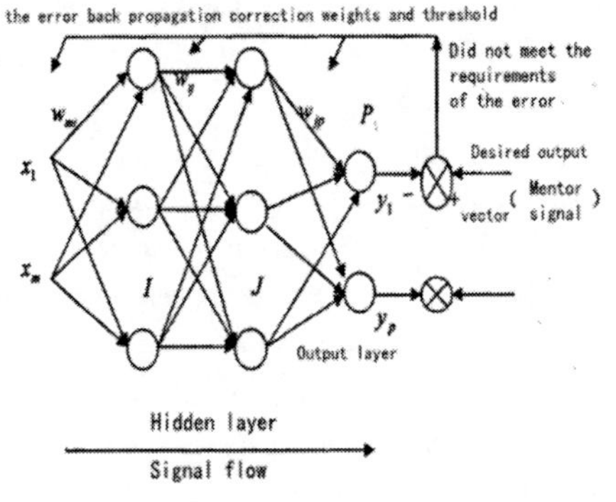

Fig. 5 BP network structure

BP neural network has simple algorithm structure and real-time control, but BP network also exist many shortcomings.If the system is so complex, the practicality will be not strong (the convergence speed is slow, and sometimes does not converge at global optimum). BP algorithm of reference [21] additional momentum and adaptive BP algorithm is used to improve these shortcomings.

(3) genetic algorithm

In 1969, the United States Michigan university professor Holland first proposed this method, later the DeJong, Goldberg and others summed up the formation of a new global optimization search algorithm [22]. Thegenes and the genetic algorithm is a bionic simulation of long-term evolution. In an iterative process, in each iteration must retain a set of candidate solution, according to the result and a number of indicators, then using genetic operator operation to obtain a new generation of candidate solutions, repeat the steps until the condition is met.

Genetic algorithm is efficient heuristic search without any auxiliary information help features, this is what the other traditional optimization method does not have.It is suitable for large-scale parallel computing.

3、in the application of electric door

The above principle of sensorless control methods and introduction ofheir advantages and disadvantages is a comprehensive, without specific application scenario.

For example , the applicationdedicated to electric door motor, because electric door working environment is more special, some specific requirements will puts forward some specific requirements to use the working characteristic of motor.The traditional motor can't satisfy the requirements of the automatic door , in general electric motor is embodied in the following features:

(1)Electric door motor requires high controllability, and the good regulation widthand dynamic range.

(2)Electric door motor must possess simple structure, high reliability, good stability, strong adaptability .It should frequent start work at low speed under the bad environment, for a long time.

With the update of control technology, it is expected in the future of new control method will be more and more applied on the electric door ; these methods will also become the research hot spot in the future.

4、summary

This paper summarizes the basic principle, advantages, disadvantages and improving ways of several kinds of permanent magnet brushless dc motor sensorless control method 。Sensorless motor control technology not only avoids weaknesses of the position sensor motor.but also makes full use of the software control technology.

With the development of the research,Artificial intelligence control method, wll be more applied to the electric door`s motor control category.

Literature References

[1] jia-qing li, brushless dc motor control applications, Beijing: Beijing university of aeronautics and astronautics press.(2014)

[2] wen-hua gao, brushless dc motor bemf zero crossing detection research, machinery and electronic,2008, 6.

[3] zhu. Key position sensorless brushless dc motor control technology research. Central south university,control science and engineering.(2014)

[4] article analyzes, Cheng Shan beauty, Cai Kai etc. Bemf zero crossing detection new method of brushless dc motor rotor position. Journal of electrotechnics, 2009, 24 (7) : 52 ~ 58.

[5] Jianwen Shao. Further Improvement of Direct Back EMF Detection for Sensorless Brushless DC Motor (BLDC) motors Drives [C]. 20 th Annual IEEE Applied Power Electronics Conferenceand Exposition,APEC 2005:933-937.

[6] Huang, Yushui. An Improved BEMF Detection Method for Sensor - less BLDC Motors Motors [C]. IEEE ICIT 2008 - Conference Proceedings, 2008:1-4.

[7] officer two yong, ping-gang song, Yu Xiao yu. Brushless dc motor based on fly-wheel diode test. Micro &special motor, 2005 (1) : 24 ~ 28.

[8] Zhang jie, the research of air conditioning compressor motor drive system, Harbin institute of technology, power electronics and power transmission.(2015)

[9] Zhou bo, wei jia, xin-hua Mu, etc. The logic level of integral comparison method to implement the sensorless brushless dc motor control [J]. Journal of electrical engineering technology, 2000 (8) : 4-9.

[10] Wu ding hui, zhi-cheng Ji, yan-xia Shen, etc. Three times in the stator harmonic method in the application of the brushless dc motor. The micro motors, 2003 (3) : 13 ~ 16.

[11] Faeq Maher, Ishak Dahaman. A New Scheme Sensorless Control of BLDC Motor motors Using Software PLL and Third Harmonic Back EMF - [C], 2009 ieee Symposium on Industrial Electronicsand Applications, ISIEA 2009 - Proceedings, 2009:861-865.

[12] Zhang Xianghu, based on the kalman filter of sensorless brushless dc motor control China university of mining, electrical engineering.(2015)

[13] Lin Yiping. Gain correction based on EFK aspect of speed sensorless BLDCM control [J]. Electric drive, 2007, 5 (6) : 12-15.

[14] Gao Zheng, all-digital fuzzy synovial observer based on DSP, the design and implementation of hangzhou university of electronic science and technology, computer system structure.(2015)

[15] Chen Yidong Yang Yulin, Wang Liqiao, governing the country, Wan Cheng wide, asymmetry of power grid failure all the wind generator power converter control strategy of electric power system automation, 2011 (7).

[16] Zhou Yangzhong Lin Qixing, Ma Junting, salient pole permanent magnet brushless dc motor without position sensor type instantaneous torque observation, proceedings of the csee, 2013 (18).

[17] Paul peng wang, square wave and the characteristics of partial discharge pulse voltage on the motor insulation life influence mechanism research, southwest jiaotong university, the electric power system and its automation.(2013)

[18] SheZhiTing, speed sensorless DTC of induction motor parameter identification and control method of research. Hunan university, control science and engineering.(2011)

[19] Hou Chunjie, brushless dc motor based on genetic algorithm for adaptive fuzzy control, henan university of science and technology.(2013)

[20] xu kun, guo-li li, wei-dong jiang. Brushless dc motor based on neural network [J]. Control and detection, 2006 (6) : 53-56.

[21] Wang Xiangchen hong-xing wu. Intelligent control in the application of brushless dc motor control [J]. Journal of micromotor, 2008 9 (1) : 62-65.

[22] lian-hong Wang, artificial immune optimization and the research of classification algorithm and its application of hunan university, control theory and control engineering.(2009)

Design of comprehensive general maintenance service system of aerial reconnaissance camera

Xu LI[1,a], Xiang Bin YU[1], Ye ZHAN[1] and Yan ZANG

[1] *Aircraft control Department, Air Force Aviation University, Changchun, China*

Abstract. Aiming at the problem of lack of security equipment for airborne reconnaissance camera and universal difference between internal and external field and model, the design scheme of comprehensive universal system based on PC-104 bus architecture and ARM wireless test module is proposed is proposed using the ATE design. The scheme uses the "embedded" technology to design the system, which meets the requirements of the system. By using the technique of classified switching, the hardware resources are reasonably extended, and the general protection of the various types of aerial reconnaissance cameras is realized. Using the concept of "wireless test", the test interface is extended to realize the comprehensive protection of the aerial reconnaissance camera and the field. The application proves that the security system works stably, has good generality, practicability, and has broad application prospect.

1 Introduction

The aerial photographic reconnaissance is an important guarantee to win the local war in new era so our army has improved and developed a series of reconnaissance aircraft. However, due to the reasons of different manufacturers of airborne reconnaissance camera, vary types and models, different protection demands of internal and external and serious technical barriers, a comprehensive and universal security equipment has not been formed. Therefore, based on the working principle, control process and interface definition of various types of reconnaissance camera and the protection demands of internal and external, a design scheme of a general security system based on PC-104 bus architecture and ARM wireless test module is proposed with ATE design ideas and the wireless test concept.

2 Demand analysis

The optical imaging camera is mainly used for the aerial reconnaissance camera. At present, there are six kinds of middle-low altitude frame camera, high altitude frame camera, middle-low altitude panoramic camera, middle-low altitude CCD camera, middle-high altitude CCD camera, infrared line scanner. Although these cameras are with different kinds, different manufacturers and different technical standards, the basic principle, the camera interface and the work flow are so similar.

2.1 Basic principle analysis

From the composition structure, these cameras can be divided into 3 parts of optical imaging, electronic control and data storage. The optical imaging part is the basic core of the camera, which use the optical devices in the focal plane to form an image. The camera electronic control is the control core for camera, which achieves the automatically adjusting, calculation and controlling for lens, focal length, roll compensation, heating and task parameters and determines camera working status through receiving the operation instructions of aircraft reconnaissance management system, and flight parameters. At the same time, it feed-back the real-time status and self-test results to airborne reconnaissance mission processing system through the data communication interface. Data storage is used to store the image data, which is stored in different storage modes, such as film storage, hard disk storage, data recorder and so on.

2.2 Camera interface analysis

The camera interface consists of two types of power interface and data communication interface. 28V DC power is mainly used for camera working and heating and three phase 115V/400Hz AC power is also needed to provide with some cameras at the same time. The data communication interface mainly completes the interaction work with the aircraft reconnaissance and management system and the inertial navigation system and the cameras all use ANRIC 429 serial communication. Due to the large current consumption of camera, the power interface and the data communication

[a] Corresponding author: dlixut@aliyun.com

interface are connected with the independent cable to avoid interference.

2.3 Work flow analysis

The camera work process can be divided into 5 stages: temperature control or work waiting, starting self-test, re-ready, being ready, taking pictures. When entering the " temperature control or work waiting" state after power on automatically, the camera configures the equipping and hanging scheme and set the parameters of camera photosensitive and loading amount. After receiving starting self-test instruction, the camera enters the "starting self-test" state. And it reports the self-test results and returns to the "temperature control or work waiting "state at the end of self-test. After receiving the instruction, the camera enters the "re-ready" state. At this point, the flight parameter information received by camera is in the effective range and the camera is ready and enters the "being ready" state. At this time, the camera can take pictures after receiving the instructions. When the camera receives the instruction of stopping taking pictures, it will be back to the "being ready" state and returns to the "temperature control or work waiting" state. The camera operating process can be controlled by the serial communication interface of ANRIC 429.

2.4 Maintenance demand analysis

In the past, the various types of aircraft reconnaissance camera equipment mainly use the working way of out-position. This way needs the maintenance personnel first remove the camera, and then sent to the dedicated security center for the maintain security. The demolition and transport will consume a lot of time, which can't meet the time requirements of rapid take-off and continuous operation on the field. At present, the field support of aerial reconnaissance camera is used the equipment after equipped to test. But because of being no complete simulation of the camera work conditions for the self-test equipment, many failures can't be found in time, which gives the flight training and operational risk. In order to completely eliminate such problems, the protection of equipment must support in situ.

3 Hardware structure design

On the basis of in-depth analysis and summary of all kinds of aerial reconnaissance cameras, we put forward the overall design objectives of "portable design, multi model general, internal and external field", which is based on the in-depth communication with the maintenance personnel and the comprehensive understanding of the protection needs.

3.1 Portable design

In the past, most of the aviation reconnaissance camera support equipment is based on industrial control mechanism, the equipment volume, the weight is large and the carrying capacity is inconvenient. So we are seriously on the PC-104, CAN, VXI, PXI and other bus architecture based on the combination of security needs, development costs and other considerations, the final selection is that the PC-104 bus architecture build a security system. Using PC-104 bus stacking design greatly reduces the test equipment volume. At the same time, the technology and method of the high density multilayer signal conditioning board, the built-in touch screen, and the small ICA package are used for the first time to realize the design of the 4U box. Compared to the previous airborne reconnaissance camera inspection instrument, the size is about 8 times smaller and easy to carry, which improves the mobility.

The integrated security system for aerial reconnaissance camera is a portable box structure, which can be divided into two parts, the interface panel and the control resource. The layout of the various parts and sub modules are shown in Figure 1. The interface panel is mainly the machine-machine interface, man-machine interface of the security system, which consisted of the ICA connector, the built-in touch monitor, USB interface, LAN interface, WLAN antenna, system power switch, external power insurance and other parts. ICA connector is used for the wired connection of security system and measured camera when checking the camera in infield. The built-in touch monitor is used to display the test software, test data, receive the user's operating instructions and provide the man-machine interface for the testing process.

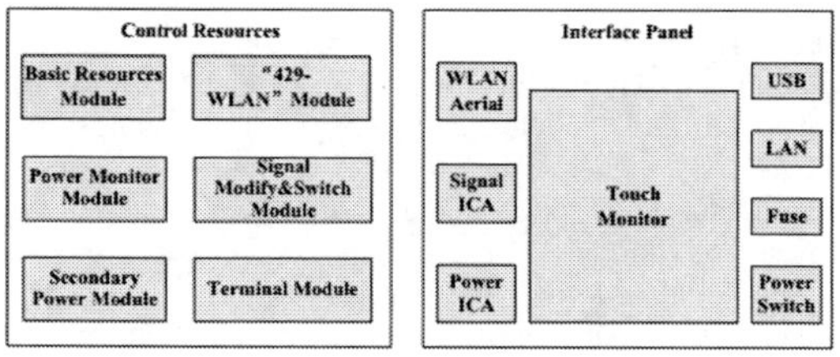

Figure 1. System Layout Diagram

The control resource is the core of the system, which is based on the PC-104 bus construction and consisted of the basic resources module, 429-WLAN conversion module, signal conditioning control module, power monitoring module, power supply module and terminal module two module and other functional modules. The basic resources module includes PC-104 computer, C429 communication card and multi-function data acquisition card, which constitutes the basic test resource of the system. They mainly complete the ANRIC 429 data transceiver and analog quantity, the amount of the collection and transmission. They use the PC-104 stacking characteristics to stack together through the installation of the screw from the bottom up, greatly reducing the volume. The 429-WLAN conversion module can realize the bidirectional conversion between ANRIC 429 bus data and 802.11B data, which provides the conditions for the field in situ test. The signal conditioning control module mainly completes the testing signal processing and the switching control work of basic

DOI: 10.1051/matecconf/20166301038

resources, which consists CamTester circuit board, DC contactor and AC contactor. CamTester circuit board includes the rolling door control circuit, the internal and external ANRIC 429 switching circuit, the computer power supply circuit, AC and DC voltage conditioning circuit and other functions. The power monitoring module is used in the infield test to monitor the camera power supply voltage and phase sequence and the current and appear abnormal promptly to cut off power and give the alarm prompt, which is mainly consists of CamTesterPow circuit board, AC and DC voltage sensor and AC DC current sensor. The CamTesterPow on the circuit board comprises the circuit of AC DC voltage monitoring circuit, AC and DC current monitoring circuit, rolling shutter door consumed current measuring circuit, 115V / 400Hz three-phase alternating current phase sequence measuring circuit function. It also equipped with different specifications of the fuse which control of the device is not less than the consumption of the allowable value.

3.2 Universal design

The "classified switching" technology makes the hardware resources of the system are divided into three levels: resource establishment, resource expansion and resource management. Firstly, the hardware resource based on PC-104 bus is built by using PC-104 bus equipment series and standardization. The resource expansion is carried out by using the terminal module and the relay matrix, and the ICA interface and the special cable are connected with the camera. The resource management is responsible for the system software. System software is responsible for data analysis, processing and related calculation, output, display, etc. The signal format conversion of the camera and the receiving channel is completed in the signal processing control module. In this way, different types of cameras can be connected to the hardware resource through different connecting cables, and all the test resources are available to it. Different control software is run according to different objects and the control software makes the distribution according to the needs which achieve general safeguard is for multi class multi camera.

3.3 Integrated design

Learning from the "wireless test" concept, a special field working module is developed. The external field working module can be used in in situ docking with the external field aerial camera, and the camera's ANRIC 429 data is converted to the data of the wireless local area network. Therefore, the real-time data is achieved exchanging between the integrated security platform located in the infield and the located in the field of aviation camera through the construction of wireless local area network, which meet the design requirements of the inside and outside the field of comprehensive test.

The external field module is constructed using the ARM chip with the internet communication interface, which is mainly consisted of ARM main processing chip circuit, relay circuit, power module, camera interface and other functional modules. The principle and information interaction are shown in Figure 2. The ARM main processing chip completes the conversion of ANRIC 429 and the wireless LAN format data, through the DO channel, the relay circuit controlling the communication interface of the camera. The external field working module is connected with the camera by the communication interface of the camera, and the power supply of the camera is provided by the aircraft power supply.

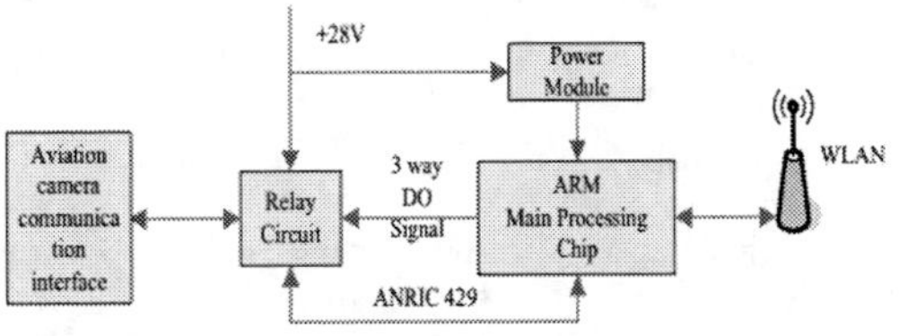

Figure 2. Field Work Module Principle Structure Diagram

4 Software sturcture design

Because the control logic and working process of the cameras are similar, the “virtual system” technology and the multi-level packaging technology are introduced to make the hierarchical, modular and interface design for the system measurement and control resources in order to save the software development cycle and improve the robustness and maintainability of the software system. The “virtual system” is based on the hardware driver, the general algorithm and the general function module. The system forms the hierarchical measurement and control resource in the multi-level package. Each package is actually a specific function, which provides the interface definition for the development of the.h file format. The developers only need to test the needs of the corresponding interface file to achieve the required resources for the control. At the same time, as long as the function of the module is modified, the module and above all calls are achieved. Using multi-level packaging technology to hide the complexity in the entity and providing the simple, refined, diverse access external interface improve the system's maintainability and scalability.

The virtual system is divided into three levels. The top level is the system monitoring program module, which is used to allocate, initialize and monitor the whole device function and hardware resource. The second level is the main control program and the function sub module for each test item for realizing the function of each module, including the conversion of the communication protocol, the collection of the power signal, the relay control, data preservation, etc. The underlying program includes the hardware board driver, the camera work condition data packets and other independent public process algorithm. Each level contains a variety of functional modules and sub modules. The software architecture is shown in Figure 3.

DOI: 10.1051/matecconf/20166301038

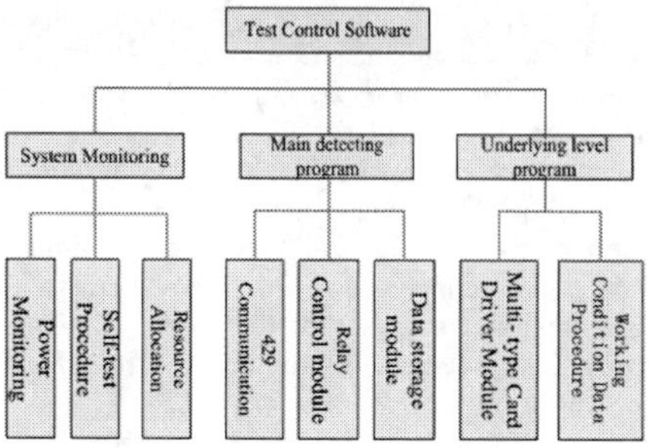

Figure 3. Characteristic curve of vacuum bellows

5 Conclusion

In this paper, a comprehensive general security device for aerial reconnaissance camera is designed based on PC-104 bus architecture. The use of hardware resources are divided into three levels of resource establishment, resource expansion and resource management by using the "classified switching" technology. Through the scheduling of the system software, the general protection of various types of aircraft reconnaissance cameras is realized. Using the concept of "wireless test", the special field working module is developed, which can realize the comprehensive protection of the aerial reconnaissance camera and the field. Using the "virtual system" technology, the measurement and control resources are formed and a comprehensive control of the test system is achieved through the multi-level packaging of the basic resources. At present, the security platform has been applied in some aviation troops, the effect is good.

References

1. Zhong Li-ping. General ATE Frame Design of Weapon Support System [J]. Automatic Measurement and Control. 2007,Vol.26,No.3: 84- 85.
2. SUN Yan-jun, YAO Qin, DU Xin-min, Xia Rui. The Design Projeet of Aero-Device ' s Wireless Test System[J]. Journal of Projectiles,Rockets,Missiles and Guidance. 2004,Vol.24,No.3: 184- 186.
3. Li Xu, Zhan Ye, Chang Limin, Zang Yan. Design of Comprehensive General Support System for Airborne Inertial Platform. 2013 International Conference on Mechatronic Sciences, Electric Engineering and Computer（MEC 2013）. 223-226.
4. FANG Wei,WU Jie,GAO Guo-zhao. Unified Virtual Systems Management Model Based on CIM Modeling [J]. Computer Engineering. 2011 Vol.37,No.16: 270- 272.
5. FANG Li,CHEN Xiang-ping,WU Liang-kun. Research on Frame of Interface Test System Software Based on Components [J]. Instrument Technique and Sensor. 2011, No.8: 50- 54.

Analysis of Limiting Measures of Three-phase Short-circuit Current of 500kV Intensive Receiving-end Power Grid in the Early Stage of UHV Construction

Xiao Hang ZHANG [1,a], Si Qing SHENG [1], Fu Qiang LI [2], Yu Ou HU [2], Wen Chao ZHANG [3] and Yan PAN [3]

[1]*College of Electrical and Electronic Engineering, North China Electric Power University, Baoding, Hebei, China*
[2]*North China Grid Company Limited, Beijing, China*
[3]*NARI Group Corporation Beijing Monitoring Technology Center, Beijing, China*

Abstract. This paper took the intensive receiving-end power grid as the research object and analysed the problems of three-phase short-circuit current over-limited of 500kV intensive power grid in the early stage of UHV construction. Firstly, this paper analysed the influence factors of the short-circuit current based on self-impedance analysis model of two-port network theory and the equivalent model of UHV power grid. Then, the applicable current-limiting measures were put forward. Finally, this paper proposed one scheme to solve the problems of three-phase short-circuit current over-limited of 500kV buses in one practical receiving-end power grid and verified the feasibility of this scheme from three different aspects. Aiming at the over-limited influential factors, the applicable current-limiting measures can be taken as follows: separating the 500kV bus at UHV substation, lapping the lines outside the substation and decentralizing the power supplies. The proposed measures can effectively solve the problems of short-circuit current over-limited in the early stage of UHV construction and provide reference for other power grid.

1 Introduction

In order to realize the transformation of energy and power development mode and solve problems caused by the deterioration of the environment, the strong smart grid will accelerates the UHV construction and pays more attention to the coordinated development of power grids at different levels[1]. In order to ensure sufficient power supply reliability, the receiving-end power grid must be further strengthened and have more supporting power supply capacity. The electrical contact of different substations will become increasingly close and the receiving-end power grid will shows the features of substation-intensive and line-intensive. While promoting the power supply ability and the security and stability of power grid, the short-circuit current level of UHV substations and load-center substations grows rapidly, which brings security and stability risks to the power grid.Therefore, it is necessary to study the corresponding limiting measures to prevent the short-circuit current exceeding the standard.

Aiming at the problems of short-circuit current over-limited in the development of power grid, the scholars have made a deep research in the development of short-circuit current limiting schemes. Literatures [2-3] analysed the current-limiting measures from different aspects and showed their advantages and disadvantages, but this paper didn't use engineering examples to illustrate. Literature [4] proposed corresponding measures to solve the problems of short-circuit current over-limited of Pengcheng substation in Guangdong power grid and verified those measures from different aspects, which has certain guiding significance to the engineering practice, but this paper didn't touch on the UHV project. Literature [5] disclosed the short-circuit current over-limited mechanism of 500kV bus at Chang-sha UHV substation based on self-impedance analysis model of two-port network theory and compared the economic cost of different schemes, but those schemes only related to the installation of current limiting reactors. The short-circuit current over-limited influential factors would change with the period and structures of the grid. Therefore, the main factors should be the first to analyze for making the most effective current-limiting measures.

This paper focuses on the problems of three-phase short-circuit current over-limited of 500kV intensive receiving-end power grid in the early stage of UHV construction. Firstly, the main over-limited influential factors are analyzed by the self-impedance analysis model and equivalent calculation model. Then, the applicable current-limiting measures corresponding to different factors are put forward. Finally, the effectiveness of the above measures is verified by the example of solving the problems of short-circuit current over-limited of one actual power grid. The research route of this paper is reasonable, and the proposed measures are

[a] Xiao Hang ZHANG: zhangxiaohang206@163.com

simple and effective, which can provide reference to other power grid for making the current-limiting schemes in the early stage of UHV construction.

2 The analysis of three-phase short-circuit current over-limited influential factors

2.1 Over-limited influential factors

According to the periodic characteristics of the growth of short-circuit current in the power system development, the short-circuit current level of power girds with the highest two voltage levels is the highest and the consequences of short-circuit current over-limited are the most serious [6]. This chapter establishes the equivalent two port network model of 1000kV and 500kV two voltage levels to analyse the main influential factors of three-phase short-circuit current of 500kV bus of receiving-end power grid, and the impedance and power source in the model is the contact of each part. The self-impedance analysis model of short-circuit point is shown in Figure 1.

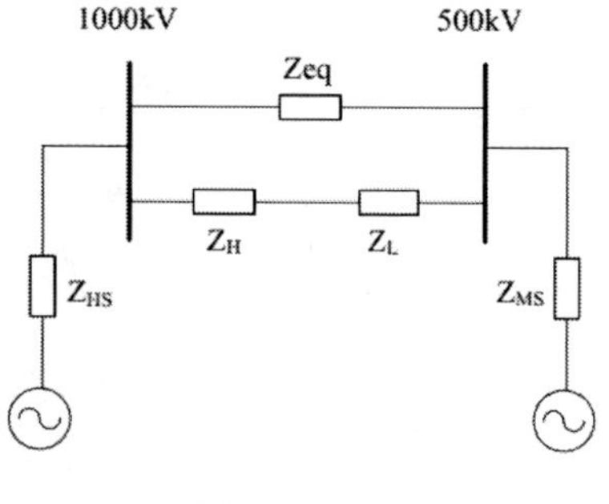

Figure 1. Positive sequence equivalent network of 500kV bus.

In Figure 1, Z_{HS} is the equivalent impedance of UHV bus to the ground. Z_{eq} is the equivalent contact impedance between the 1000kV power grid and the 500kV power grid. Z_H is the equivalent impedance of UHV transformer between the high voltage side and medium voltage side. Z_L is the equivalent contact impedance between the UHV substation and the fault point. Z_{MS} is the equivalent impedance of faulted bus of 500kV power grid to the ground.

From the above analysis model, the self-impedance of the 500kV short-circuit bus can be got in the formula (1).

$$Z_{KM}=\frac{Z_{TS}Z_{MS}}{Z_{TS}+Z_{MS}} \tag{1}$$

$$Z_{TS}=Z_{HS}+\frac{Z_{eq}(Z_H+Z_L)}{Z_{eq}+Z_H+Z_L} \tag{2}$$

In the formula (1) ~ (2), Z_{KM} is the self-impedance of the 500kV short-circuit bus, Z_{TS} is the transfer impedance of all the power supply on the UHV power grid to the fault point, and the smaller the value is, the greater the impact of the UHV on the fault point of 500kV power grid is.

In the early stage of UHV construction, the electrical distance between the UHV substations is relatively far. So, the value of Z_{eq} is relatively large, which shows that the UHV injects the short-circuit current to the 500kV power grid mainly though the UHV transformers. The more close the electrical distance from the UHV substation to the fault site is, the smaller the value of Z_L is, and the greater the impact of the UHV on the fault point is. In the intensive receiving-end power grid, Z_{MS} is the transfer impedance of all the power supply to the fault point in 500kV power grid, whose value is determined by the network density and the power supply quantity of the 500kV power grid. When the number of the substation tie-line and the power supply is large, the value of the Z_{MS} would be small and the short-circuit current level of the fault point would be high. Therefore, the three-phase short-circuit current level of the 500kV buses of UHV and its surrounding substations, load-centre substations and the power-concentrated area substations would be significantly higher than other 500kV buses. So, the three-phase short-circuit current over-limited influential factors of 500kV buses in the intensive receiving-end power grid can be summarized as follows: UHV construction, too much tie-lines in the load-centre substation and power-concentrated construction.

2.2 Influence of UHV power grid on the short-circuit current of 500kV receiving-end power grid

In the study of effect of UHV power grid on the three-phase short-circuit current of the 500kV power grid, the UHV power grid can be equivalent to a power supply which injects the short-circuit current to the 500kV power grid by the UHV transformers. The simplified system model is shown in the Figure 2.

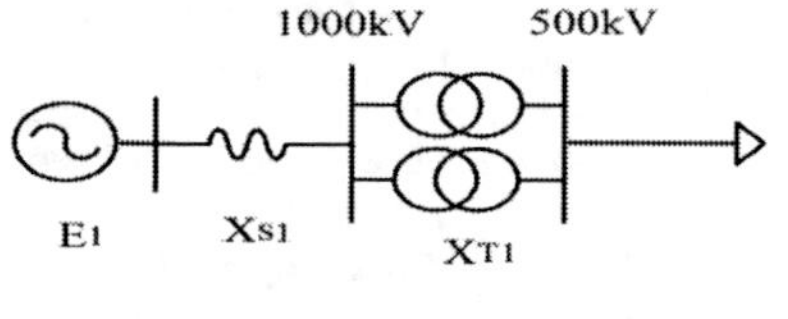

Figure 2. Simplified system model

When the three-phase short-circuit fault occurs in the 500kV bus, the short-circuit current of the fault point can be expressed in the formula (3).

$$I_{500}=\frac{1}{X_{S1}+X_{T1}} \tag{3}$$

Where,

$$X_{S1}=\frac{S_B}{\sqrt{3}\cdot I_{1000}\cdot U_B} \tag{4}$$

$$X_{T1}=U_{k1}\%\times\frac{S_B}{nS_{N1}} \tag{5}$$

In the formula (3) ~ (5), I_{500} is the short-circuit current of 500kV bus, E_I is the power electromotive force, X_{sI} is the equivalent impedance of UHV power grid, X_{TI} is the equivalent impedance of UHV transformers, I_{1000} is the short-circuit current of 1000kV bus, S_B is the reference system capacity, U_B is the reference system voltage, S_N is the rated transformer capacity, $U_{kI}\%$ is the transformer short-circuit voltage percentage, n is the number of the parallel UHV transformers.

The formula (4) ~ (5) show that, the change of I_{1000} and n would affect the value of short-circuit current of 500kV power grid. The value of I_{500} can be obtained under different I_{1000} and n by the formula (3) ~ (5), when S_B=100MVA, $U_{kI}\%$=0.21 and S_{NI}=3000MVA. The results are shown in figure 3.

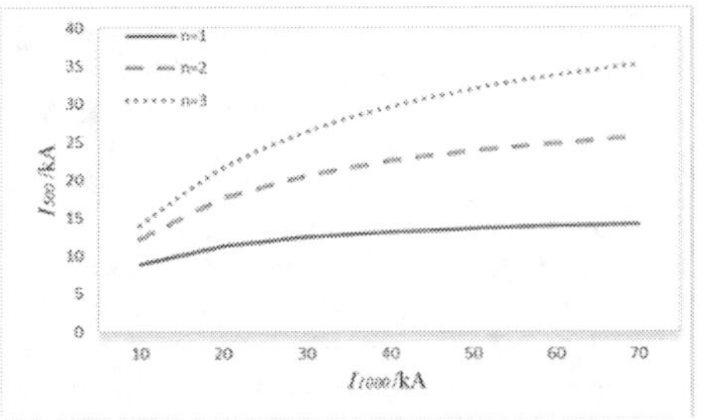

Figure 3. Three-phase short-circuit current of 500kV power grid

The Figure 3 shows that the UHV construction will have a great impact on the short-circuit current of 500kV power grid. When the short-circuit current level of UHV power grid reaches 30kA, the short-circuit current of 500kV power grid which is provided by UHV power grid will exceed 10kA, and it will be greater when the number of parallel UHV transformers increases. So, the reasonable arrangement of UHV transformer units will have great effect on limiting the short-circuit current level of 500kV power grid.

3 The short-circuit current limiting measures

The current-limiting measures based on the different mechanism have obvious advantages and disadvantages and have different scope of application. Under the specific structure of power grid and over-limited influential factors, the selection of the current-limiting measures with good applicability will be the best [7]. In view of the above three-phase short-circuit current over-limited influential factors, this chapter analyses the applicable and corresponding current-limiting measures.

3.1 Separated operation of 500kV bus at UHV substation

In the early stage of UHV construction, the UHV grid structure is relatively weak and the 500kV power grid does not have the ability for the partition. Therefore, other measures should be taken to control the short-circuit current level. The separated operation of 500kV bus at UHV substation is the available current-limiting measures. The way of the separated operation of 500kV bus at UHV substation is an economical and effective method to suppress the influence of UHV power grid on the short-circuit current of the 500kV power grid. This measure can greatly reduce the short-circuit current of the substation with the breakdown bus and can be operated easily. But to a certain extent, it reduces the power supply reliability of the substation and may cause the uneven distribution of the bus load [8-10].

After the separated operation of 500kV bus at UHV substation, the short-circuit current limiting effect of the substation in the near of the UHV substation will be influenced by the 500kV grid structure outside UHV station. The way of influence can be divided into two cases, one is that there is the loop network outside the UHV station (small loop) and the other is no loop network (large loop). Figure 4 and Figure 5 are schematic diagrams of the separated operation of 500kV bus at UHV substation under two cases.

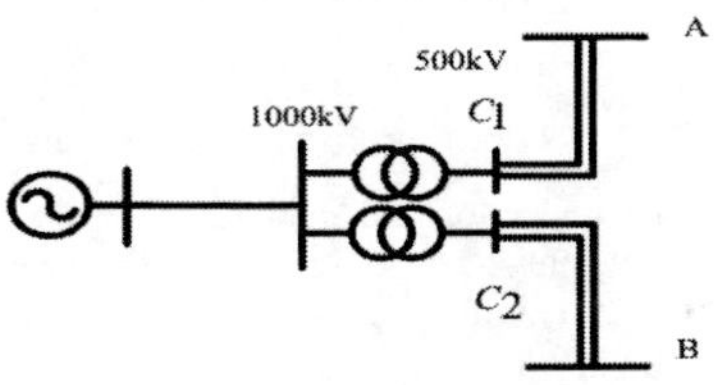

Figure 4. No loop network outside the UHV station.

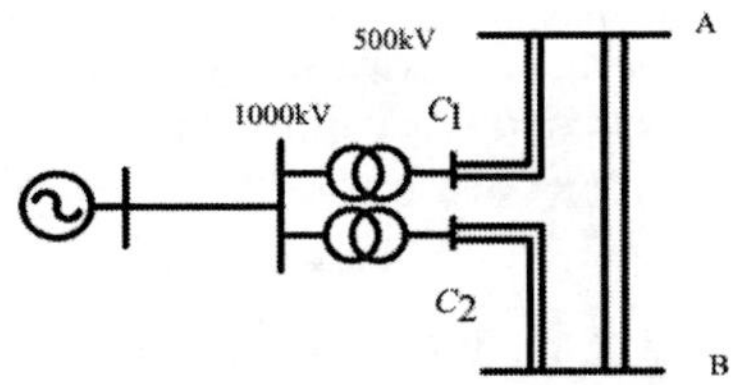

Figure 5. With loop network outside the UHV station.

Figure 4 shows that there is no 500kV loop network or there is a large 500kV loop network outside the UHV substation. The separated operation of 500kV bus at UHV substation will obviously increases the self-impedance value of the A node and B node to reduce the short-circuit current. From Figure 5 we can see that there is a small loop network outside the UHV substation. When the short-circuit fault occurs in the A node, the UHV power grid can inject the short-circuit current both by the transformer connected with the A node and the small loop after the separated operation of 500kV bus at UHV substation. The short-circuit current of the loop is mainly determined by the impedance of the loop network. Under the situation that the electrical distance of the loop network is short, the current-limiting effect of separated operation of 500kV bus for A substation is poor.

3.2 Lines out series or lapping outside the substation

At present, the transmission lines in the substation are generally 3/2 connection mode. In this mode, the lines in series can be connected only through the intermediate switch by turning off the switches on both sides. Thus, the transmission lines will don't connect to the substation, which can decrease the number of tie-lines in the load-centre substation and increase the self-impedance of bus to reduce the short-circuit current level. The schematic diagram of this measure is shown in Figure 6.

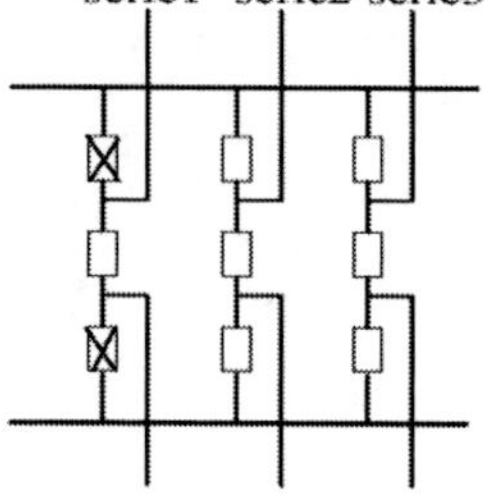

Figure 6. Schematic diagram of lines out series

Using 3/2 connection mode in substation to achieve the operation of line out series or lapping the lines outside the substation is one of the effective measures to restrain the excessive short-circuit current. The reliability of the power flow can be taken into account by the means of line out series or lapping lines outside the substation to reduce short-circuit current, but external electrical connection and the stable level of the substation are weakened. Therefore, it is recommended that the circuit breaker which are used for the line out series or lapping outside the substation should be in a hot standby state in order to be closed in time when the fault occurs, which can avoid serious accidents.

3.3 Decentralizing the power supplies

When the power supplies are connected to a certain area in clustering way, the short-circuit current level of the area is greatly increased. The greater the power density in the area is, the more serious the problem of short-circuit current over-limited is. In the beginning of power supply construction, the power supplies should be decentralized and connected to different trunk transmission lines based on the direction and scope of power transmission to make the network has better adaptability. Under the condition that the system reliability is not weakened, pulling open the electrical distance of power grid can both reduce the short-circuit current level and be prepared for the partition operation of 500kV power grid. The influence coefficient of the short-circuit current of power supply would vary with the power access point [11]. Therefore, in the beginning of power supply construction, the power access points should be chosen reasonably. Designers also should give full consideration to the influence of the power supply on the short-circuit current and reserve a certain space to the development of power grid.

4 Example analysis

4.1 The phenomenon of three-phase short-circuit current over-limited of 500kV buses in TQ power grid

TQ power grid is a typical urban receiving-end power grid with double loop network structure and is located at the intersection of two 1000kV UHV AC transmission channels. The schematic diagram of the TQ power grid structure is shown in Figure 7. The average electrical distance of adjacent substation of TQ power grid is only 15km and the power supplies of 500kV power grid are connected to the TQ eastern power grid in clustering way. TQ power grid shows obvious features of substation-intensive and line-intensive and the three-phase short-circuit current level is relatively high. Based on the operation planning data of TQ power grid in 2018, the three-phase short-circuit current of 500kV buses is calculated by the power system analysis software PSD-BPA, and the short-circuit current level of substations whose short-circuit current excess the standard is shown in Table 1.

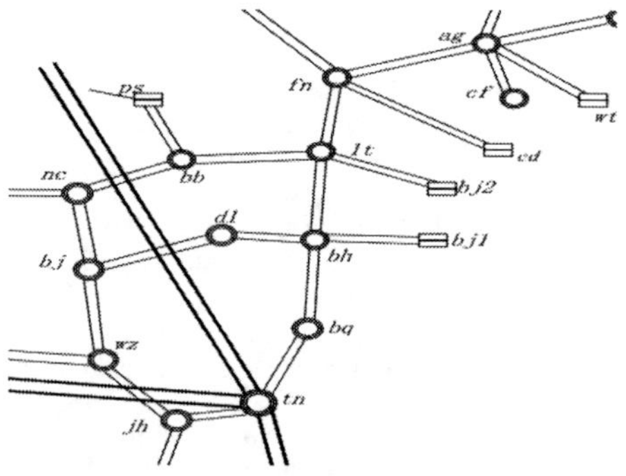

Figure 7. The structure of 500kV TQ power grid.

Table 1. Calculation results of three-phase short-circuit current of 500kV buses in TQ power grid in 2018

No.	Bus	Breaking current kA	Short-circuit current kA
1	*lt*	63	75.053
2	*tn*	63	70.806
3	*jh*	63	69.386
4	*fn*	63	68.443
5	*bh*	63	63.948

The Table 1 shows that the problems of three-phase short-circuit current over-limited of 500kV buses in TQ power grid are serious. The three-phase short-circuit current of *lt* substation has reached 75kA, which is far more than the rated current 63kA of circuit breaker. Besides, the three-phase short-circuit current of *tn* substation, *jh* substation and the *fn* substation is also high and all of them is more than 68kA. The short-circuit current over-limited substations mainly distribute near the UHV substation or in the power-concentrated area, which shows that the short-circuit current of TQ power grid is greatly influenced by UHV construction and power-intensive construction. From the above analysis, we can know that the short-circuit current over-limited influential factors of TQ power grid are consistent with the proposed over-limited influential factors of the intensive receiving-end power grid.

4.2 The schemes of current-limiting measures

The UHV construction and the power-concentrated construction are the main influential factors for the problems of three-phase short-circuit current over-limited of 500kV buses in TQ power grid. According to the current-limiting measures corresponding to the short-circuit current over-limited influential factors of intensive receiving-end power grid, the current-limiting schemes can be made of two measures.

(1) Measure One: separated operation of 500kV bus at *tn* UHV substation. The 500kV bus at *tn* UHV substation is separated into two parts, and one of them connects to *jh* substation and the other one connects to the *bq* substation. The schematic diagram of the structure is shown in Figure 8.

(2) Measure Two: grafting cd power plant units from *fn* substation to *cf* substation, lapping the line *fn-lt-bb* outside the *lt* substation. The schematic diagram of the structure is shown in Figure 9.

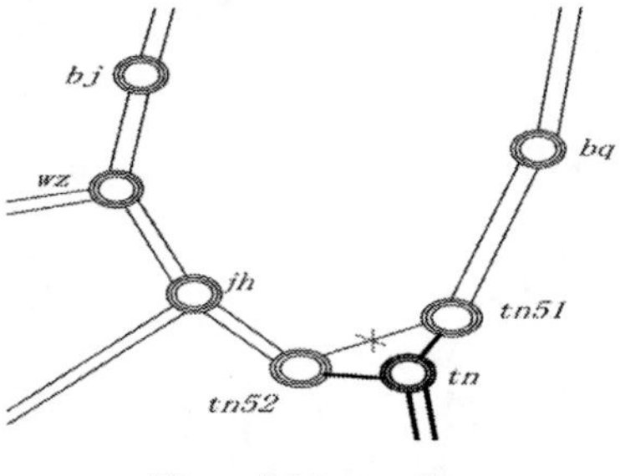

Figure 8. Measure One.

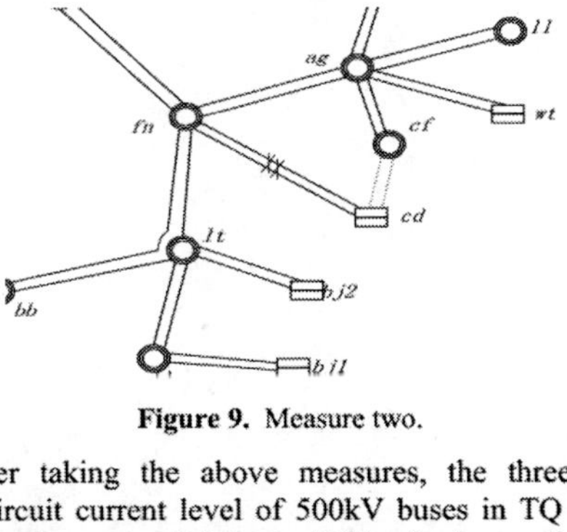

Figure 9. Measure two.

After taking the above measures, the three-phase short-circuit current level of 500kV buses in TQ power grid is shown in Table 2. Table 2 shows that the short-circuit current level of 500kV buses at different substations in TQ power grid decreases to less than the rated breaking current of circuit breaker. The short-circuit current of *tn* substation and the *jh* substation decreases more than 20kA, which shows that separating the 500kV bus of UHV substation can greatly weaken the impact of UHV power grid on the three-phase short-circuit current of 500kV power grid. The power density and the network density of eastern region of TQ power grid has been reduced after taking the second measure and the short-circuit current level of 500kV bus of *lt* substation, *fn* substation and *bh* substation decreases to less than the rated breaking current of circuit breaker.

Table 2. The three-phase short-circuit current of 500kV buses in TQ power grid after taking measures.

Bus	Initial short-circuit current	Short-circuit current after taking measures	Change in short-circuit current
	kA	kA	kA
Lt(lap joint)	72.16	38.280	33.88
lt	72.16	61.429	10.731
tn51	70.764	34.751	36.013
tn52	70.764	44.867	25.897
jh	69.335	48.806	20.529
fn	67.285	61.928	5.357

From the point of view of power flow distribution, there is certain uneven power distribution among the two transformers of *tn* UHV substation after taking the first measure. One of them is 1400MW, the other one is 900MW. But in the maintenance case of any one, the power of the other transformer is not overloaded. Besides, the power flow of TQ western power grid increases after taking the first measure. The single line power flow of *nc-bj* transmission line increases to 1300MW and the maintenance mode should be arranged reasonably to prevent the overload. The cd power plant mainly provides the power to the *cf* industrial zone after taking the second measure, which conforms the trend that the electricity consumption of *cf* industrial zone increases year by year. Lapping the line *fn-lt-bb* outside the *lt* substation does not affect the power flow transfers from the north power grid to the TQ western power grid and the power flow is reasonable.

To analyse the scheme from the aspect of security and stability, the power flow of all the single transmission line is below 1800MW and is not overloaded whether in normal operation mode or in maintenance mode after taking the measures, except the *nc-bj* transmission lines. The power flow of single *nc-bj* transmission line would reaches to 2300MW, which is a heavy load but not overloaded, when one of the two lines is broken. When the N-2 fault happens in the *ag* side of the *cf-ag* transmission lines, there is a need to shut down one unit in the cd power plant to maintain the stability of the system. When the N-2 fault happens in other transmission lines, the system can always keep the stable operation.

Seen from the above analysis, the problems of three-phase short-circuit current over-limited of 500kV buses in TQ power grid can be effectively solved.

Besides, the power flow distribution is reasonable and the power grid can ensure the safe and stable operation. Therefore, the proposed scheme can be used as a feasible solution to solve the problems of three-phase short-circuit current over-limited of 500kV buses in TQ power grid. This example shows that the phenomenon of three-phase short-circuit current over-limited of 500kV intensive receiving-end power grid will occur in the early stage of UHV construction, and short-circuit current over-limited problems of the substations near the UHV substation or in the power-intensive area will be more serious. The measure of separating the 500kV bus at UHV substation can effectively suppress the effect of UHV on the short-circuit current of 500kV power grid. Grafting the power plant units and lapping the lines outside the substation can effectively decrease the power density and the network structure density of power-intensive area to reduce the short-circuit current level.

5 Conclusions

In the early stage of UHV construction, the problems of three-phase short-circuit current over-limited of 500kV intensive power grid will be serious. The three-phase short-circuit current over-limited influential factors can be summarized as follows: UHV construction, too much tie-lines in the load-centre substation and power-concentrated construction. Aiming at these over-limited influential factors, the applicable current-limiting measures can be taken as follows: separating the 500kV bus at UHV substation, lapping the lines outside the substation and decentralizing the power supplies. The example of solving the problems of three-phase short-circuit current over-limited of TQ receiving-end power grid shows that the research route of this paper is reasonable, and the proposed measures are simple and effective, which can provide reference to other power grid for making the current-limiting schemes in the early stage of UHV construction.

References

[1] F. Han, F.L. Song, J.S. Luo, & L. Chang. Research on development of transmission power grid in 13~(th) five-year period. *Electric Power*. **48**(1): 11-14 (2015).

[2] Q.T. Ruan. Present situation of short circuit current control in shanghai power grid and countermeasures. *Power System Technology*. **29** (2): 78-83 (2005).

[3] G. Han, L. Han, & L. Wu. Application and development of methods on limiting power grid's short-circuit current. *Power System Protection & Control.* **38**(1): 141-144(2010).

[4] X.P. Yang, L. Li, Y.X. LI, etal. Running programs of limiting 500 kV short-circuit current in Guangdong power grid. *Automation of Electric Power System*. **33** (7): 104-107 (2009).

[5] G.D. Liao, X.T. Xie, Y.L. Hou, etal. Analysis on the Problems of Three-phase Short-circuit Current Over-limited of 500 kV Bus When UHV Connected to Hunan Power Grid. *High Voltage Engineering*. **41**(3):747-753 (2015).

[6] F. Wang, & Q.J. Li. Short-circuit current control for beijing-tianjin-tangshan grid. *North China Electric Power*. **4**: 17-19 (2007).

[7] Q.Z. Sun, Z.X. Cai, A.M. Li, etal. A short-circuit current over-limited mechanism of 500 kV power system and the adaptability of limiting measures. *Automation of Electric Power Systems*. **33**(21), 92-96 (2009).

[8] T. Luo, L.X. Liu, K. Wang, etal. The influence of UHV accesses to Tianjin 500 kV power grid on the short-circuit current and the limiting measures. *Electric Power Construction*. **36**(8): 79-83 (2015).

[9] J. Yuan, W.Y. Liu, M.Q. Dong, etal. Application of measures limiting short circuit currents in northwest china power grid. Power System Technology. **31**(10):42-45 (2007).

[10] H.Y. Ma. Research on method of limiting short-circuit current and partition principle for UHV interconnected power system. North China Electric Power University. pp: 21-24 (2014).

[11] J. Zhou, D. Jin, W.Z. Wang, etal. Analysis of influence of huge power supplies grid-connection modes on short-circuit current. *Advanced Technology of Electrical Engineering and Energy*. **31**(1): 48-51 (2012).

MATEC Web of Conferences **63**, 01040 (2016) DOI: 10.1051/matecconf/20166301040
MMME 2016

Optimized Design Method of the Prefabricated Hole for Flanging Forming of Square Tributary Junction on Spherical Shell

Zhong Lei WANG [1,a], Shi Ke HAO [1], Gang CHENG [1], Jin Jing XU [1]

[1]*School of Mechanical and Electrical Engineering, Shandong Jianzhu University, Jinan 250101, China.*

Abstract. The key problem for flanging forming of square tributary junction on spherical shell is to determine the shape and sizes of prefabricated hole. In order to investigate precise shape and sizes of the prefabricated hole, an optimized design model was established. The flanging forming process was simulated using the software of FE-simulation and the target function for optimized design was calculated based on results of the simulations. In order to increase the convergence rate of the optimized design, a mathematical model of prefabricated hole for flanging forming on spherical shell was established, the results were used as initial values for optimized design. To verify its accuracy, the optimized design was carried out on a typical flanging forming of square tributary junction on spherical shell. The results of the optimized design meet the technological requirements.

1 Introduction

With rapid development of economy and technology, especially the rapid development of urban pipe network, petrochemical industry, water conservancy project, oil and gas transportation, the large-scale pipe network is rapidly developed.The demands for for pipe joins, which is key component for connecting of large tube nets, is also increasing. For the process design of flanging of square tributary junction on spherical shell, the critical issue is to determine the shape and sizes of pre-machined hole precisely, for this an optimized design method is researched in this paper.

Luo Ya Tao[1]simplified the pre-machined hold as ellipse, and calculated the long and short axis of the ellipse through special positions, and further investigated the shape and sizes of the pre-machined ellipse.Liu Jian Shen et. [2-4] carried out experimental and numerical study on flanging on thick wall thickness tube. In order to obtain better quality part of the flanging forming, Chen Y.L[5] compared the two specific cases of the flanging forming process by using finite element software DEFORM.Gu Li[6]simplified the pre-machined hole as ellipse to optimize the parameters for flanging of T-shape tube with greater diameter, in which the long and short axis of the ellipse as optimized design variables , the neural network system as the optimized target function solver, and the genetic algorithm is used.

In the aforementioned researches the optimized design method based on optimized algorithm is of high degree of automation and reliability. Yet the accuracy of the optimized design method to determination of prefabricated hole is not high, the reason is that the prefabricated hole by simplifying is regarded as an ellipse, it could not describe the shape of the prefabricated hole precisely. In this paper an optimized model is established, in which the shape of the prefabricated hole is controlled with spline curves, and spline curves are controlled with points, so a high accuracy of prefabricated hole in the optimized design can be achieved.

[a] Corresponding author: wade_thunder@163.com

2 Establishment of Optimized Model

Optimization model including three aspects: the optimal objective function, optimization design variables and constraint conditions. Optimization problem of prefabricated hole of flanging forming on spherical shell can be described that choosing the size of appropriate prefabricated hole make the height of the flanging port meets the requirements of high value. Goal of optimization design is to make the height of the flanging port close to ideal flanging height, so we can put the minimum of the geometric average of the difference of the flanging port number of discrete point height and ideal height as the optimization objective function. Optimal design variables of the optimization model is the geometry of prefabricated hole of flanging forming. Since the prefabricated hole is a continuous space geometry curve, however the optimal design variables are limited data, so the continuous space geometry curve must be discrete. Some key nodes are used as the design variables of optimization design and the prefabricated hole is simplified as a spline curve which is determined by the key nodes. The prefabricated hole is a space curve, so the key points should be described by the three-dimensional coordinates. However, the prefabricated hole is on the spherical shell, so the Z coordinates can be calculated when its (X, Y) coordinates can be determined. From what has been discussed above, the optimal design variables is the (X, Y) coordinates of the key points of the hole.

In order to ensure the performance of the flanging process successfully and the parts of flanging forming can meet the technological requirements, the optimization process must meet the requirements of the flanging process. So it is the constraint conditions of the optimization model that there aren't defects, such as wrinkling and cracking, on the flanging forming workpiece. In order to meet the requirements of the following welding technology, the roughness of port must be guaranteed. So the maximum height difference of the port will be used as another constraint conditions of optimization model.

In summary, the optimization model is described as follow:

Design variables of the optimization model: $(x_1, y_1),(x_2, y_2)\Lambda\ (x_n, y_n)$

Target function of the optimization model:

$$F(x_1, y_1, x_2, y_2 \Lambda\ x_n, y_n) = \sqrt{((H_1 - H)^2 + (H_2 - H)^2 + \Lambda + (H_n - H)^2)/n} \quad (1)$$

Constraint conditions of the optimization model:

(1) cracks or wrinkles should not be aroused on the parts.

(2) $H_{max} - H_{min} \leq \Delta H$.

In which, $(x_1, y_1),(x_2, y_2)\Lambda\ (x_n, y_n)$ are coordinates of discrete points of the prefabricated hole, $(H_1\ \ H_2\ \ \Lambda\ \ H_n)$ are flanged height in various positions, H is the ideal height of workpiece, H_{max}, H_{min} are maximal and minimal height of flanged in all positions; ΔH is maximal allowed height tolerance, in this paper $\Delta H = 0.25mm$.

3 Initial Values of Optimized Variables

The convergence and convergence rate in optimized design are closely related with initial value of design variables, so an important work for optimized design is to ascertain the initial values for the optimized design variables. In order to determine the initial value of optimization variables, the mathematical model of prefabricated hole is established. In flange-forming process, the flowing in the thickness of the sheet in the forming region is more obvious than that in the radial direction. So the method of equal length is usually used to calculate the prefabricated hole of flange-forming.

In order to study the problem, the main coordinate system is set up. The axis of the spherical shell is the Z axis of coordinate system and the center of the spherical shell is the origin of coordinates. The straight lines which parallel to the flange square port side and through the origin line respectively is the X and Y axis. It assumes that the length of flanging hole sides are L and W and the radius of the round corners in the neutral layer is R3. It assumes that spherical shell radius in neutral layer radius is *R1*,flange corner in neutral layer radius is R2 and sheet thickness is t. If the forming part is cut down by the plane through the Z axis, we could get the schematic section as shown in figure 1.

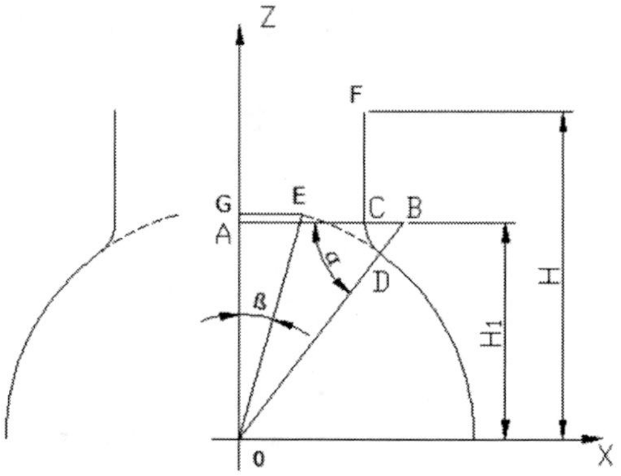

Figure 1. Schematic section of flanging forming workpiece.

From Figure1 ,we can see that H1 can be obtained by Eq2 using Pythagorean theorem.

$$H_1 = \sqrt{(\overline{OD} + \overline{DB})^2 - (\overline{AC} + \overline{CB})^2} \quad (2)$$

In which, $\overline{CB} = \overline{BD} = R_2, \overline{OD} = R_1$.

The geometrical relationship can be obtained from the Figure1.

$$\overline{CF} = H - H_1 \quad (3)$$

$$\tilde{CD} = R_2 \arccos \frac{\overline{AC} + R_2}{R_1 + R_2} \quad (4)$$

As shown is figure 1,the length of material before forming is the length of arc $\tilde{DE}$ and the length of material after forming is the sum of length of arc $\tilde{CD}$ and length of segment $\overline{CF}$.According to the method of equal length, Eq5 can be known.

$$\tilde{DE} = \overline{CF} + \tilde{CD} \quad (5)$$

$$\beta = \pi/2 - \tilde{DE}/R_1 - \alpha \quad (6)$$

$$\overline{EG} = R_2 \sin\beta \quad (7)$$

Obtained line segment $\overline{EG}$,the coordinates of point E in main coordinate system can be obtained.In order to calculate the length of segment $\overline{AC}$,we analyze the view of flanging forming workpiece in *XOY* plane, as shown in figure 3. The center coordinate of flanging forming square hole is ex in *X* axis and ey in *Y* axis. In the first quadrant, according to the geometric characteristics of port, the position of the cross-section is divided into three different regions, which are called latitude cross-sectional area, transition area and longitude cross-sectional area. Assuming that the angle between the cross section and the X axis is γ , the length of segment $\overline{AC}$ can be calculated could be known, as well as in other quadrants, the length of segment $\overline{AC}$ can be calculated, as shown in figue 2.

Let us assume that

$$\gamma_1 = \arctan\frac{W/2+e_y-R_3}{L/2+e_x}$$

$$\gamma_2 = \arctan\frac{W/2+e_y}{L/2+e_x-R_3}$$

When $\gamma < \gamma_1$

$$\overline{AC} = \frac{L/2+e_x}{\cos\gamma}$$

When $\gamma_1 \le \gamma < \gamma_2$

$$\overline{AC} = O_x \cos\gamma$$

$$O_x = L/2+e_x-R_3+\tan\gamma(W/2+e_y-R_3)$$
$$+\sqrt{(1+\tan^2\gamma)R_3^{\;2}-(\tan\gamma(L/2+e_x-R_3)-(W/2+e_y-R_3))}$$

When $\gamma \ge \gamma_2$

$$\overline{AC} = \frac{W/2+e_y}{\sin\gamma}$$

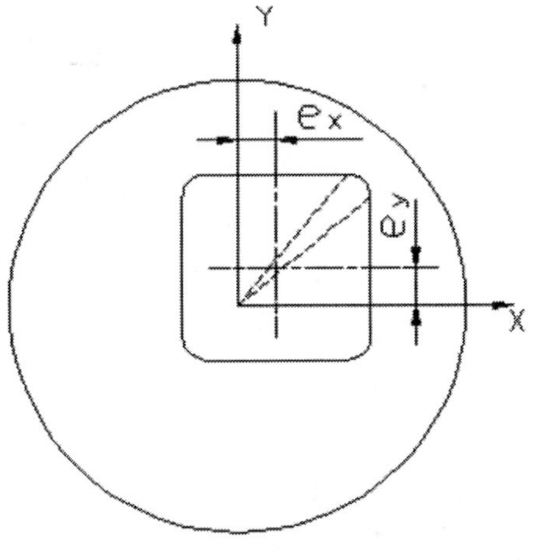

Figue 2. View of flanging forming part in *XOY* plane

4 Solution of Optimized Target Function

The target function of optimized model built up in this paper is the mean square difference between practical flanged height and ideal height, and it is related to the shape and sizes of the flanged part. Because of the complex flanging process, it is very difficult to find out the values with conventional method, for this reason FE-method is used to simulate the forming process, and using the results file of the simulation to find out the shape and sizes of the formed parts, further to find out the solution to target function, mean time to judge the constraints.

Taking a forming process as an example, in which the process parameters are as follows, the shell diameter is 120mm, width and length of hole is 60mm, square corner is 5mm, flange fillet is 5mm, height of flanging is 55mm and blank thickness is 2mm,the FE model is built. Through the simulation calculation, we can get all kinds of deformation parameters of the workpiece graphics, in which forming limit diagram as shown in Figure 3. In the optimization model, the objective function is related to by the height of the flanging port, which can be read out from the results of simulation calculation through independent development of the program.

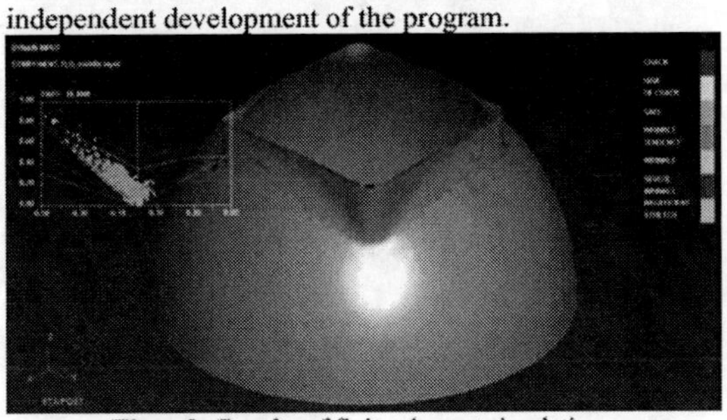

Figure 3. Results of finite element simulation

5 Optimized Design

According to the principle of length, linear material size of before and after flanging deformation is unchanged, so in the same direction, the flanging height changes is equal to the length change of radial length. Size difference of prefabricated hole and ideal prefabricated which can be calculated ideal prefabricated hole shape can be obtained through calculation difference of the flanging height and ideal flanging height. In order to obtain more accurate size of prefabricated hole, iterative method is applied in this paper.

Through the optimization calculation of flanging forming ,we obtain a workpiece such as shown in Figure 4. From the figure we can see port flanging is very smooth and there isn't any wrinkle and crack meeting the first constraint conditions of the optimization model. The average height of flanging forming is 55.01mm, the mean square difference is 0.004, the maximum value is 55.16mm, the minimum value is 54.78mm, maximum difference of height is 0.22mm meeting the second constraint conditions of the optimization model.

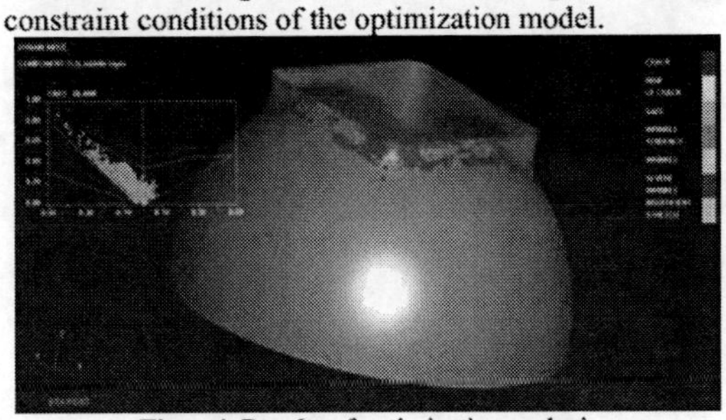

Figue 4. Results of optimization analysis

Summary

Study on the optimization design of the eccentric hole flanging forming prefabricated hole shape and size. We establish shell flanging forming prefabricated hole optimization design model, to solve the mathematical model value as the optimization design of initial value, to a limited simulation results as the objective function solution solver, on the shell side hole flanging forming prefabricated hole shape and size optimization design. Optimized design examples show that through two times of optimization design calculation, the mean square error is 0.004, the maximum height difference is 0.22mm, the optimal design is fast, and the design results meet the requirements of the process.

Acknowledgements

This work was supported by the Doctoral Foundation of Shandong Jianzhu University (XNBS1245).

References

[1] Y.T Luo. Rocket Propulsion,2,50(2009).
[2] J.S Liu, B Wang,J.H Tian. Plasticity Engineering. 12,74(2005).
[3]B Wang,J.S Liu. Journal of Taiyuan University of Technology. 3,161(2005).
[4]Y.P Zhang,B Wang,J.S Liu. Journal of Taiyuan University of Technology. 4, 437 (2006).
[5]S.L Heng,C.Y Lee,C.H Wub. Machine Tools & Manufacture, 47,168(2007).
[6]L Gu, J Zhang, L Zhu.Forging and Stamping Technology,3,71(2009).

Design of Electric Power Management System in Jilin Province based on SOA

Shou Ze REN[1], Yong Zhi WANG[1,a] and Tong WANG[1]

[1] *College of Instrumentation & Electrical Engineering, Jilin University, 130061 Changchun, Jilin China*

Abstract. Aiming at the problem of electric power data integration and sharing in Jilin Province, China, this paper proposed a method based on SOA which has high scalability, flexibility in strong power data and application integration. This approach establishes a web service-oriented system architecture, using SQL Server 2012 as its backend database and using C# as the programming language. And this approach has developed three kinds of terminal applications namely C/S(Client/Server), B/S(Browser /Server) and M/S(Mobile/Server). Client applications send all kinds of request to the electric power cloud computing centre, meanwhile web service components receive the requests and call corresponding functions to compute, and then return the results to the original app through the internet. The results show that system achieved the target we expected and successfully realized electric power information data integration, release and sharing.

1 Introduction

With the development of society and progress in technology, the demand for electricity in industry, agriculture, daily life increased sharply, the electric power industry is facing severe challenges. To dispose these problems, a variety of advanced technology has been applied to the power sector. The using of information and project management technology in the power industry have greatly improved the utilization of electric power resources and management efficiency and have reduced operating costs.

Lots of institutes developed different systems according to particular needs which have great difference in operating system, system model and data format, leading to different information exchange between systems, making it hard to realize information sharing and source exchange. Part of the network and application system become isolated systems, forming a large number of distributed heterogeneous "information isolated island".

The existence of "information isolated island" seriously blocks the effective sharing of information resources, and greatly limits the application of information resources in power system, and hinders the development of the electric power industry, resulting in a large number of repeated construction and waste of resources". The software system scale ,complex degree and operation flow changes deeply, so common system architecture is hard to realize fast dynamic adjustment and rapid optimization along with the changes in the market.

This paper proposes to build a power system for loosely coupled SOA-based collaborative design management platform, through coarse-grained services, loose coupling to shield the complex business logic, by witch to reduce the complexity of business processes through the organization of the professional division of labor, making the development of large software scale possible. Make various enterprise application systems combine into a unified platform for information sharing between the various isolated application systems to meet enterprise business needs across sectors by loosely coupled structure.

2 Electric Power Information Sharing Framework Based on Cloud Computing

Electric power information sharing framework is built based on cloud computing [1]. Electric power cloud center is implemented mainly by Service-Oriented Architecture (SOA)[2-3]. SOA is designed as development concept "software as a service, the service is the software"[4]. As a basic-unit programmer, Web service has good encapsulation, reusability, scalability, and maintenance and cross-platform operator in accordance with "data-service-application" organized by logic-levels. Meta data realizes opening directory in department data. Service composition and extension will leverage software sources scattered on internet in a form of web service in order to realize sharing capabilities. All in all, that is how SOA will create a loose coupling, scalable, managerial and self-service exchanging sources sharing environment of distributed spatial information

[a] Corresponding author: iamwangyongzhi@126.com

and complete construction of application system and maintenance upgrades in order to realize the public and mutual operation of spatial information among cities. In this case, we will regard web service as an object or component works on this web.

Recognized as a main operator of information sharing technology using web service, the SOA-based system builds a reusable source environment. System frame can be classified into database layer, data access layer, service process layer, network transmission layer and client application, as shown in figure 1.

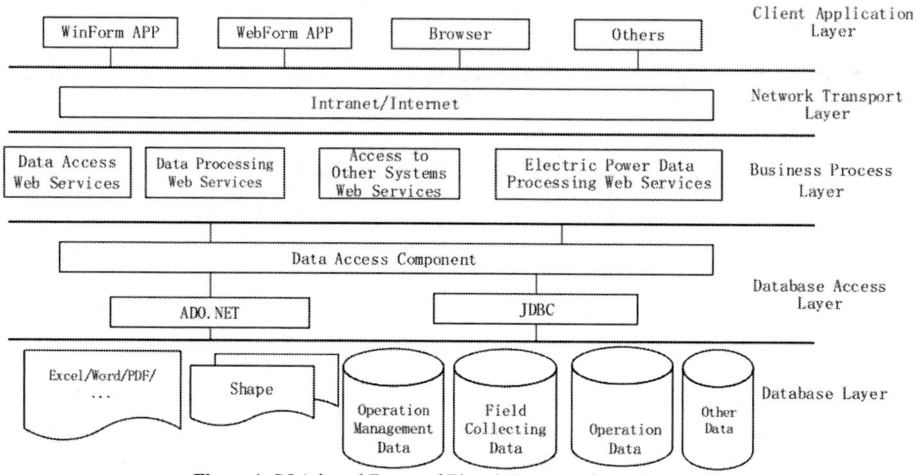

Figure 1. SOA-based Frame of Electric Power Information Sharing

(1)Database layer is the data source of electric power information management system. It is the core of the frame, including multi-media documents, Word, Excel, PDF, operation parameters of electric power, data of equipment monitoring collected by field bus system and stability data of power output voltage, the data of operation and management of power enterprise like electric price, sales of electric power, power consumption and client's information, etc.

(2)Database access layer is a service-processed provider for the business processing layer which accessing background database. It accesses to database and operates data by foreground as needed through ADO.NET and Java database connectivity.

(3)Service process layer consists of web service capable of querying data, which will change a request by client into web service's order. Service process layer can flexibly integrates with other system because of publicizing data access and analysis related to system through a loose coupling form of web service component.

(4)Network transmission layer refers to a channel to interaction between foreground clients and services. Requests by foreground clients transmitting under HTTP, TCP/IP protocols and format for data transmission is also XML.

(5)Client application layer uses based on shared data and the department of private data sources as information source. Supported by database, model database, network and sharing system, information sources, basic sharing data and private data will applied into trans-department and cross-region overall..

2.1 Distributed Data Storage

In order to effectively prevent redundant construction, information isolated island and the characteristics of electric power information resource heterogeneity, consider from the perspective of management and access efficiency, this system adopts the way combine the distributed storage and centralized storage to dispose the frequent base data in power engineering construction, make it centralized storage and management. These basic sharing data bridges between databases and this bridge will make different areas and departments have a spatial logic relation so as to achieve the goal of pooling electric power information sources.

Information with a lower shared frequency, also known as general shared data, is stored in the local database and provided to share in appropriate way. However, information used only by the department, also known as private local data, is stored in private local database.

2.2 Information Sharing Mechanism

The demand for sharing information can be divided into data-grade sharing and application grade sharing in accordance with granularity differences between systems. Data-grade sharing refers to data, also known as contents shared by system and shares a lower granularity. Data-grade sharing is appropriate to extract and send multitudinous data, which has nothing to do with controlled transaction logic and is a pure database level. Application grade sharing is improving frequency of business information system from application layer, however, functions for data exchange and general-purpose sharing related to controlled business logic shares a higher granularity.

 DOI: 10.1051/matecconf/20166301041

Based on two information-sharing levels, a core mechanism for all electric power shared information includes as followed:

(1)We can integrate and manage distributed data in order to realize the sharing and exchange of data-grade among cities based on open directory for electric power information sources with a core metadata.

(2)The discovery, configuration, interaction, assembling, operation and management of service under technical manual for web service realize the sharing of service logic and business logic.

(3)Combined by Meta data and web service, we can build and realize application systems and application-grade sharing.

3 Application Instance

Information sharing platform includes a number of subsystems. For example, information management system of electric power operation condition, electric power project information management system, electric power marketing formation management system and data analysis system for electric power equipment monitoring. We will see electric project information management system as an example and introduce it.

3.1 Functional Model for Electric Power Data Management System

Electric power data management system is an electric power project information publication platform which is built by electric power enterprises, an information management platform and a assistant decision-making platform from the leadership level. Leveraging electric power data management system will realize all information's sharing from the whole project and pool information source altogether. Well-equipped system optimized support system for information management and regulated measures for the management of business.

According to the demand for the use of productive management in electric enterprises, this system classified into four parts, namely, system management module, electric power production module, equipment management module and data management module. The functional block diagram of the system is shown in figure 2.

(1)The system management module lays foundation for ensuring the normal operation and also provides system log management sub-module, personal information management sub-module, system log sub-module and user privilege management sub-module.

(2)The electric power production module includes management on engineering plan, management on engineering schedule, management on the implementation of the information and management on engineering files.

(3)The equipment management module includes the maintenance of the equipment basic information, the maintenance of the equipment failure information, and the management of the information for equipment inspection, inquire of equipment information and the management of inventory information of the equipment.

(4)As a core module, the main functions of data management model are described as followed: adding data, browsing data, statistical data and analysis data etc.

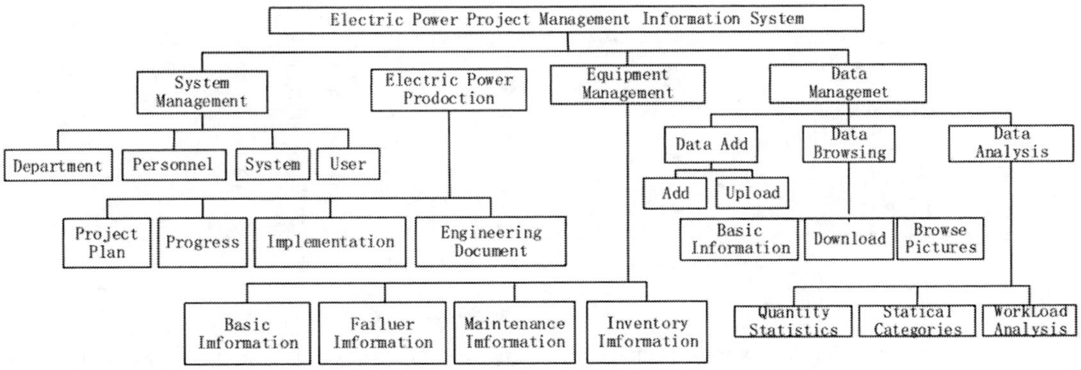

Figure 2. Functional Structure Diagram of System.

3.2 Data Mode

As a core of electric power data management system, the database is an entity of storing data related to electric power documents and lays foundations for all information sources. The design quality of database system influences directly the performance of the system. In order to improve the productivity and quality of development of the system, we ensure data independence and maintainability, shared data correctness and consistency, decreasing data redundancy, guarantee data security and reliability during database designing process in accordance with features of electric power document.

DOI: 10.1051/matecconf/20166301041

3.3 System Implementation

3.3.1 Implementation Technology

The MVC-based system adopts one-stop portal technology and technology concept. RIA-based (Interface Application Rich) technology to achieve the application of various types of client. The front-end UI communicates and interacts with data service through HTTP and SOAP protocol and service centre invoking other services and return results back to the foreground through HTTP and TCP protocol. The client processes and formats results that returned by invoking service. The results will be shown in table, word, diagram and so on.

Some documents need to be uploaded to server in the application of documents management system, which characterized by large files, size is up to 100MB and a large number of files. Digital signature and data encryption is required in the data security. HTTP-based transmission mode is our initial consideration. We found that the data being uploaded through HTTP protocol seems to meet the convenience of web programming after comparison. The speed of uploading documents is less than 1MB is a little bit faster than that through FTP protocol. But it's impossible to upload piles of documents and large files. If the documents are greatly larger than 1MB, the speed of FTP protocol is faster than that HTTP. The fact is that the larger the file is, the faster the speed of uploading is.

Now, FTP protocol adopts a more popular C/S mode that implements the documents transmissions which consists of FTP client program and server program. For example, remote server programs in Windows 2008 are installed on IIS information server and then we set up our environment. By setting up a FTP server, this setting specifies permission for operating documents in server by clients. He can only download documents from server if a client only have a read access; He can upload documents if a client have a write access; He can upload and download documents if a client have a read and write access.

3.3.2 System Realization and Effect

The interface of project information management which is shown in figure 5 is a core of documents sharing. A large number of documents relates to project. Browsing interface for picture information is shown in figure 3.

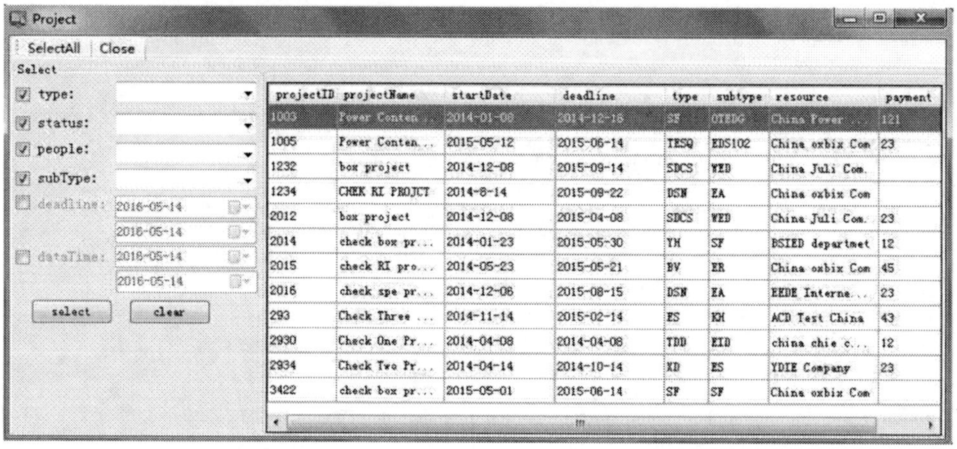

Figure 3. The Interface of Project Information Management.

4 Conclusion

Electric power project information management system can complete the fully sharing of project information, which will guarantee data consistency and real-time capability so as to revise discussion in time and complete the alternation of management. This purpose leads to a workflow management which is highly effective, provides effective support in the project of tomorrow and achieves the high efficiency of construction and management.

References

1. W. He,Y. Wang. Progress in geophysic,**29**,6,2886-2896(2014)
2. Y Wang, H Yang,L Lu. World Geology,**26**,4,447-452(2007)
3. Y Wang, H Pan, W He. Geological Bulletin,**28**,6,29-32(2009)
4. C. Hao,Y. Wang,et al. Applied Mechanics and Materials, (687-691),2952-2955(2014).
5. Y. Wang,Y. Zhang,J. Yi.,et al. Mathematical Problems In Engineering,2014,1-11(2014).
6. Y. Wang,Y. Zhang,F. Zhang,et al.Mathematical Problems In Engineering,2013,1-10(2013).

Garden class experiment connotation construction of practice training bases

Le Jing LIN

Ningbo City College of Vocational Technology.

Abstract : The quickening pace of construction of higher education and the deepening of teaching reform, higher vocational colleges have stepped up investment on construction of practice teaching facilities, experiment practice of improving the hardware environment, however, relative to the rapid growth of teaching in terms of hardware, practice teaching soft environment construction is slower, the connotation construction of practice training bases to ignore, practice teaching management mode as the soul and core of soft environment, is the most fundamental problems in the practice teaching, has brought to the attention of more and more scholars.

Key words : Landscape classes; Experimental training bases; The connotation of the construction

Strengthening the connotation construction of practice training bases, it is necessary to strengthen and standardize the laboratory management, straighten out the management procedure, to ensure implement management rules and regulations, to revitalize the existing training resources, improve the utilization efficiency of lab, optimize the laboratory construction and equipment configuration. Under current conditions, the characteristics of higher vocational colleges how to effectively to establish their own internal management mode of practice training bases, completes the connotation of the base construction, the rapid development of practice teaching system, to better meet the needs of the teaching of higher vocational colleges, has become a serious issue in front of higher vocational colleges. This paper mainly in view of the garden class experimental correlation analysis connotation construction of practice training bases.

1 The main content of landscape connotation the experimental practice base construction

1.1 The garden type campus experimental concept connotation of the practice base construction.

Botanical garden specialty is strong practical a discipline, students need to provide necessary experimental environment, the practice teaching to meet the colleges and universities, the training of specialists in the field of landscape gardening classes teaching task. Training base is the practice teaching system in order to strengthen the students theoretical knowledge jointly set up by the actual professional place, can provide students with practical training experiment and operation, is to improve the students' ability of operating one of the important places, including sites, laboratory equipment, teaching staff, teaching resources management, etc. Experimental practice link is to cultivate

Le Jing LIN: linlejing@nbcc.cn. **Fund project**: ningbo garden plant development key laboratory project a22008 (2014), a key project of ningbo city professional technology institute (JY1311)

students' professional quality as well as an important part of the actual operation ability, plays an important role in the professional education.

1.2Increase the funds investment, improve the practice teaching system.

Traditional laboratory due to insufficient funds, some problems such as lack of teaching equipment, cannot effectively carry out the teaching experiment; In the match the project curriculum integration in the training room, the traditional experimental teaching in the lab for the innovation and optimization, the more attention to the cooperation degree of project courses and teaching, to cultivate the students' dialectical thinking. Therefore, college should attach importance to the laboratory construction, increasing investment, perfect laboratory equipment, help to enhance the enthusiasm of students for learning. Schools to improve laboratory in various infrastructure construction, the construction of a comprehensive practice teaching system, and regularly to the laboratory equipment maintenance and update, ensure its role in the practice teaching, and regularly check for updates of laboratory equipment, ensure that the practice teaching work orderly, also need to adjust on the experimental equipment of spending.

1.3 Strengthen the human resource management, improve the overall service quality.

Professional education to teacher's requirements in addition to need to have professional quality, at the same time needs to have the practical application of the major of various quality. This requires teachers to the actual production and the enterprise, "practice" can not only came to lectures, but also to step down, is to achieve an important factor to improve students' comprehensive quality. School should attach great importance to the practice of backbone teachers and young teachers, education and vocational education should be strengthened in the course of implementation, vigorously introduce enterprises and technical personnel, improve the level of practice teaching. School attaches great importance to the cooperation between colleges and actively absorb the enterprise high quality education resources.

1.4 Perfect laboratory management system, strengthening laboratory management standardization.

Experimental training base management including equipment management, personnel management, curriculum management aspects of content, may adopt "combines unification" appropriate management style, set up perfect laboratory management system, reasonable distribution of managers at all levels, clear responsibilities and division of labor, strengthen the experimental base of normative management. Joint actual play laboratory effectively in the development of students' theoretical knowledge and the role of practice ability, helps to cultivate the practicing ability and specialized quality, strengthen the development of experiment teaching, in order to achieve the purpose of the experiment teaching, help to improve students' comprehensive ability, to adapt to the social development and landscape industry development demand for innovative talents.

2 Garden class experiment reform goals of the construction of the connotation of the practice base

2.1 Introduction of 7 s on-site management, optimizing experiment site management practice training bases.

7 S on-site management sort (Seiri), consolidation (Seiton), cleaning (Seiso), clean (Seikeetsu), quality (Shitsuke), security (safety), and speed/save (speed/saving), due to the first letter of the seven words for S, are referred to as "for 7 S. It is to point to carry out to sorting, reorganization, cleaning, clean, safe, save the content such as related activities, and to improve the quality as the main task. This method is mainly originated from Japan, and showed a

good application effect. 7 s on-site management to the botanical garden the experimental practice base site management, helps to cultivate the students' humanistic quality, and improve the efficiency of the experimental practice base site management, help to improve the quality of experimental base management.

2.2 The rectification laboratory setting, strengthening the management of resources and maximum realize resources sharing.

Because money is tight, can put in the laboratory construction funds is relatively small, and landscape architecture experiment is given priority to with agriculture production experiment. So, can consider to build a can accord with the actual situation of school public experimental platform. Through public experimental platform construction, to save the school bought infrastructure funds, and in the same layout and on the basis of unified planning, realize the combination of different professional requirements, will be effective for use in schools on the basis of the experimental teaching equipment purchase, and focus on key construction of professional school, optimizing the allocation of resources and sharing.

2.3The optimizing match the project curriculum integration training rooms and classification of traditional laboratory management, improve the efficiency of management.

Slam the door on the shortcomings in traditional classification of laboratory management, take with the project course to match the integration of laboratory management, and attaches great importance to the professional, practical, advanced, scientific, sharing principle, such as effective for production practice teaching, vocational training, research and development activities. According to the actual production process or in the production process to design practice, help to improve the effectiveness of practice teaching.

2.4 Strengthen the system construction and realize the scientific management.

Need attention in the construction of experimental teaching system of teaching, training, implementation of the scientific research and production, the need to build perfect equipment, advanced technology, characteristic, effective operation and production integration of modern practice teaching system, and modern enterprise cooperation, realize the resource sharing, and attaches great importance to the experimental practice base construction, according to the botanical garden the needs of students operation ability and the cultivation of strain capacity.

3. Landscape connotation the experimental practice base construction key problems to be solved

3.1How to build the garden type campus experimental connotation of the practice base construction system.

To enhance investment in experiment teaching, promote the botanical garden specialized teaching level, to promote the deepening of education reform, improve the comprehensive quality of students. Laboratory is an important part in modern school teaching infrastructure, and there are more and more attention to the importance of experimental teaching in school. Landscape gardening classes experiment bases most productive training project implementation, the site has particularity, experimental training project some use of means of production has a certain risk, some will use the toxic drugs, such as agricultural need more strict management. How to management, experiment practice base what management mode and so on all has become a problem to be solved. And now its input and output is out of practice training bases, people traditionally in general will simply "post professional ability" as the goal of practice teaching. Experimental teaching in order to make students in a real environment, increase the work experience, do joint theory to practice, but there was some difference of this approach compared with the actual, therefore, independent college can consider to cooperate with enterprises, let the students into the enterprise practice, help to improve the students' actual operation ability, and cultivate social needs of talents. Garden need to meet the engineering surveying and mapping class experiment practice base, landscape sketch operations such as training, can carry on the simulation under the guidance of the teacher project practice, and in the application of knowledge in project management, design and nursery, flower cultivation and management, and other activities.

3.2 How to realize the integration of match the project course training rooms and classification of traditional laboratory management.

School training base in higher vocational education is implement vocational skills training and vocational quality education for students in the best place, campus practice base created strong professional culture atmosphere, to improve higher vocational students' comprehensive professional quality and ability to adapt to society is of great significance, good practice base management system, practice base and can serve the teaching, so this topic choice school experiment connotation construction of practice training bases as the object of study has great practical significance. Traditional laboratory are less actual operation, teachers, students can't tracking guidance to all students, lack of systemic and purpose. Therefore, in addition to strengthen the practice teaching, can also through the simulation of the operation and the simulation experiment teaching to improve their strain ability, garden professional mainly in garden plants and maintenance, landscape nursery stock production, the botanical garden sketch, engineering construction and management, and other factors, therefore, to establish the virtual simulation platform has important practical significance. In virtual simulation experiment teaching, let students to simulate production activities to simulate the process of production and management of operation, help to improve the students' ability of operation.

3.3How to 7 s on-site management and experiment training room management to achieve docking.

At present, the domestic many managers and teachers in the management mechanism, the organizational structure and management system and other aspects are discussed in this paper and the research, some management concept has been done in the actual operation, and achieved good results. For the implementation of the new experiment bases management theory research, many teachers in the study of the practice teaching management analysis, combined with the actual situation of our school, puts forward the unified planning, unified construction, unified management, unified call, resource sharing, perfect function, make full use of the reform train of thought. 7 s on-site management attaches great importance to the efficiency, the principle of human nature and persistence, with the experiment training room management relative to meet can obtain good effect. This requires practice training management attaches great importance to the efficiency of management, help to promote the development of the experimental teaching; In addition, the establishment of expert team, establish the management direction and focus, make effective teaching management organization activities and assist teachers in teaching activities and equipment maintenance, and in strict accordance with the relevant procedures and requirements to carry out teaching activities, pay more attention to the specification of the experimental training rooms and management.

4 Conclusion

With the deepening of globalization, the traditional teaching mode is no longer adapt to the development of the society, under the background of rapid development of modern society, the experiment teaching of cultivating students' operating ability and innovation ability and teaching activities, improvement of teaching quality is an important part of . Laboratory construction as one of the school education quality evaluation standards, is an important performance of school teaching, scientific research and management, help to promote landscape engineering students the importance of theory union actual work.

References

[1] Zhang Yange, Mao Xiu yuan, yu-lan ding, etc. The nursery garden students in the construction of the experimental base for improving the employment competitiveness of research [J]. Journal of youth, 2014, 12 (8) : 176-177

Wide. [2] is high in garden "garden genetic breeding experimental teaching reform to explore" [J]. Journal of exam, 2012, 22 (63) : 19 to 20.

[3] wu Ming, basket. On strengthening practical teaching reform for advanced vocational garden technology discussion [J]. Chinese science and education innovation Tribune, 2012, 13 (18) : 29 and 30.

[4] Zhang Yunlu, li xiong, benji. Institute of Beijing forestry university garden Liang Xi experimental classes exploration and practice of talent training mode [J]. Chinese garden, 2015, 31 (5) : 23-26.

[5] jun-huo CAI, xu-ying wei, even aromatic green, etc. In garden "classroom, experiment and practice of the trinity" mode of teaching [J]. Journal of higher education BBS, 2012, 14 (9) : 39-41.

[6] Zhang Jinping, Pan Yuan wisdom. Agriculture and forestry colleges and universities the landscape architectural design mode of experimental teaching [J]. Journal of anhui agricultural science, 2012, 38 (28) : 16044-16045.

[7] car generation brother, have no opinion of eu. Gardening, landscape subject experimental teaching system, construction of exploration and practice [J]. Journal of laboratory science, 2013, 22 (6) : 3-5.

MATEC Web of Conferences **63**, 01043 (2016) DOI: 10.1051/matecconf/20166301043
MMME 2016

Research and Simulation of a flexible robotic fish tail fin propulsion system

Hong Yu Luo[1,a]

[1]*jilin teachers institute of engineering and technology, Changchun, China*

Abstract. This article uses a flexible crescent caudal fin tuna as the research object, sets up the robot fish physical model ,researches the propulsion and advancing speed of the model, discusses forward speed, sliding and swing amplitude, frequency and phase to the flexible tail fin propulsive performance, and uses MATLAB to simulate, motion simulation is consistent with the way to achieve the real movement of the fish.

Keywords: Flexible tail fin, Robotic fish, MATLAB simulation.

Introduction

Biologists found that in all ocean species, the efficiency of fish swimming mode is the highest. Through thousands of years of evolution, fishes have had an amazing swimming ability, high mobility, high hidden features. More and more Chinese and foreign scientists carried out research on ROBOFISH, trying to study the promoting mechanism of fish to implement a propeller which has movement equally efficient, high mobility, high hiding like fish imitation.

Based on previous studies, this article uses a flexible crescent caudal fin tuna as the study object,adds to the chord and spanwise deformation, and establishes a physical model ,researches and analysises the robot fish propeller thrust and speed of by MATLAB simulation, in order to get the motion parameters of thrust, efficiency and function.

1 Robotic fish tail fin model

Robot fish tail fin tuna model references the caudal fin shape (Figure 1). The shape of caudal fin tuna in the center is approximately NACA0015, in the end of the shape is approximately NACA0009. Therefore, this article uses NACA0015 airfoil in the middle of the caudal fin sectional robotic fish , uses at NAC0009 airfoil at the end , uses NACA0012 airfoil in the middle . Caudal fin dimensions shown in Figure 2, chord C = 35.22mm, exhibition length B = 130, sweep $\alpha = 47°$.

Fig. 1 Tuna caudal fin shape

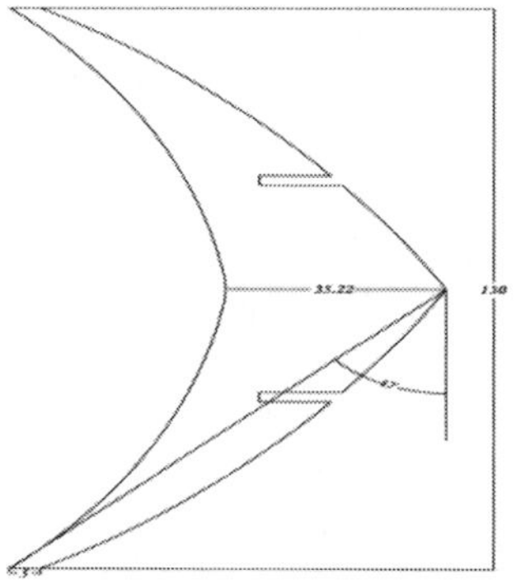

Fig. 2 Robotic fish fin size

1.1 Tangential deformation equation

Pengfei Liu proposed equations (1) tangential deformation equations shown in his article: δ_c is the string to the magnitude of the deformation, it is the product of tangential deformation factor δ_{c0} plus features of the chord length C_0. Namely; $\delta_c = \delta_{c0} C_0$; ε are 1.5,2,2.5,3,3.5,4 values respectively; Φ_c is a chord relative phase angle to traverse the caudal fin deformation, usually negative. Eq. (1) represents a chord deformation begins at the end of chord characteristic C_0, that is the Central point to the trailing edge deformation, while the leading edge to the midpoint of the deformation.

$$z = \delta_c (2x-1)^{\varepsilon} \sin(2\pi ft + \Phi_c) \quad x \geq \frac{1}{2}$$

(1)

On this basis, we can extend equations (1) to any point on the chord to the trailing edge deformation occurred, and the leading edge to the point is not deformed, the deformation equation of formula (2) below,in it $1<s<\infty$; $x \geq \frac{1}{s}$; $\varepsilon > 1$.

$$z = \delta_c \left[\frac{s}{s-1} x - \frac{1}{s-1} \right]^{\varepsilon} \sin(2\pi ft + \Phi_c)$$

(2)

[a] Corresponding author: dongnana860901@163.com

DOI: 10.1051/matecconf/20166301043

When s is larger,the deformation is closer to the leading edge; when ε is greater,the distortion is flatter near the leading edge , and the more intense near the trailing edge of the Transfiguration. When S=2, the equations (2) turns into the equations (1). Take $\delta_{c0} = 0.1$, in Figure 3s=2, respectively to take ε =1.5,2,2.5,3,3.5,4, when it is the maximum chord deformation; In figure 4 ε = 1.5, s is taken chordwise 2,3,4 maximum deformation.

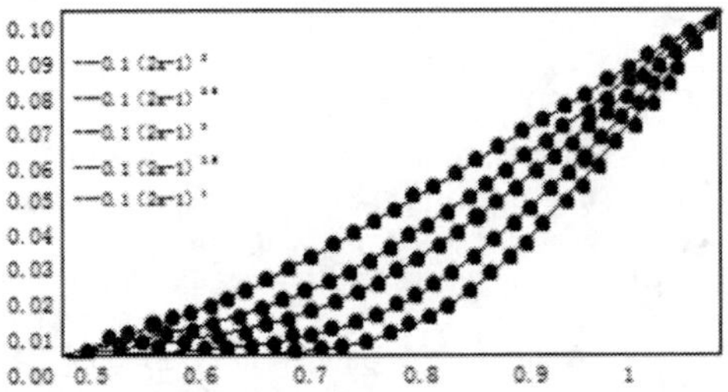

Fig. 3 Chordwise deflexion shapes of half chord length

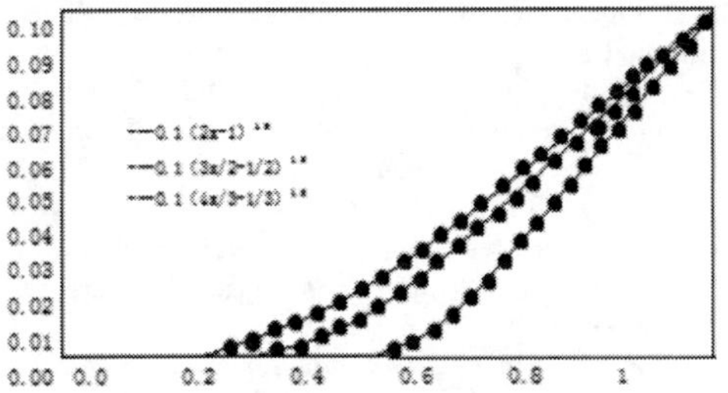

Fig. 4 Comparison of chorodwise deflexions

We assume that each interface of the caudal fin deform by the equations (2), leading edge is with no deformation , at the trailing edge it occurs maximum deformation . Also, because the deformation of the caudal fin, tail fin generates the relative speed between the bin ,which is determined by the time derivative of the equations (2). And after appropriate coordinate transformation, the speed should be the overall speed of the caudal fin at a point. In this way, calculating for each step time , the influence coefficient of each bin should be recalculated.

1.2 spanwise variable equation

Using the cantilever beam uniform deformation equations Pengfei Liu proposed based on evenly distributed load, such as equations (3), in it δ_s is spanwise deformation magnitude ,which is the product of the spanwise deformation factor δ_{s0} plus semi span l, that is $\delta_s = \delta_{s0} l$; ε should be greater than zero, and the greater the deformation near the strings more gentle, and more intense deformation near the wingtip; Φ_s is the phase angle of spanwise deformation to the tail fin traverse , usually negative. In equations (3) the number ±and the absolute value is produced by different about half the value of the symbol of the span y.

$$z = \delta_s \left[2(\frac{y}{l})^2 \mp \frac{4}{3}(\frac{y}{l})^3 + \frac{1}{3}(\frac{y}{l})^4 \right] \left| \frac{y}{l} \right|^{\varepsilon} \sin(2\pi f t + \Phi_s)$$

$$\varepsilon \geq 0 \qquad (3)$$

Suppose each tangential cross-sections of caudal deforms according to equation (3), string with no deformation, deformation of the tip is the largest. The relative velocity of spanwise deformation between the tail fin of each bin is produced by the equation (3) with respect to time to determine ,that is the total velocity of a point on the tail fin, and at each step time recalculating the influence coefficient.

After given to the chord of caudal fin and spanwise deformation equation, you can determine the location of any of the caudal fin in a time profile,set up the caudal fin is constant thickness after deformation, you can determine the shape of the deformed tail fin after deformation. In it , the tangential deformation parameters respectively are δ_{c0} = 0.1, s = 2, ε = 1.5 , $\Phi_c = -\pi/2$; spanwise deformation coefficient δ_{s0}= 0.1, ε = 0, $\Phi_s = -\pi/2$. Under normal circumstances, when the tail fin traverse zero, spanwise and chordwise deformats the largest, and their phases are lagging behind the traverse, so we take $\Phi_c = \Phi_s = -\pi/2$.

Tail fin motion parameters respectively are 为 $A_Z = C_0$, θ_0= 20° , f =0.5Hz , $\varphi_0 = \pi/2$. Tangential deformation parameters: δ_{c0} =0.1 , s=2 , ε =1.5, $\Phi_c = -\pi/2$; Spanwise deformation parameters: δ_{s0}=0.1, ε =0, $\Phi_s = -\pi/2$.

2 The flexible fins forward speed simulation

Because of the force of water, under normal circumstances the fin from root of the connection to the top of does not exhibit linear form, there will be a longitudinal bending. We assume that the water is evenly distributed load forces to establish a flexible fin geometry fluctuations. Take a point in the X axis , through point A parallel to do OYZ plane, the plane can be obtained online with flexible fins deflection surface, namely the deflection lines (Figure 5).

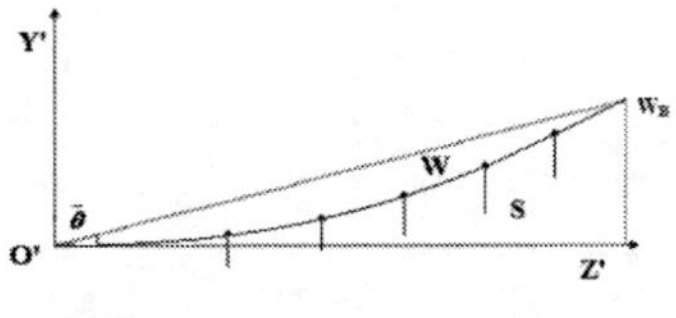

Fig.5 The deflection curve of section of fish fin

From the deformation formula of mechanical beam ,it is shown that the deflection line of flexible fins on OYZ plane can be represented by the formula (4), in it L is the length of the fin; E is the modulus of elasticity of the fins; I is the corresponding moment of inertia ; q is the fin surface uniformly distributed load to bear.

$$w = \frac{qz^2}{24EI}(z^2 - 4Lz + 6L^2)$$

(4)

Its end Deflection :

$$w_B = \frac{qL^4}{8EI}$$

(5)

Then q can be expressed as:

$$q = \frac{8EI}{L^4} w_B$$

(6)

So the fin deflection line can be expressed as:

$$w = \frac{w_B z^2}{3L^4}(z^2 - 4Lz + 6L^2)$$

(7)

w_B is the deflection of the curve in the ends of the fins and

$$w_B = |A\sin(kx)|$$

(8)

Under the flexible fins conditions,you can define nominal angle

$$\bar{\theta} = \arcsin\frac{w_B}{L}$$

(9)

In the case of a small deflection, it can be approximated that $z_0 = L\cos\bar{\theta}$.In figure 5 the area S is enveloped of fins curve, the dotted line and Z 'axis :

$$S = \int_0^{z_0} \frac{w_B z^2}{3L^4}(z^2 - 4Lz + 6L^2)dz$$

$$= \frac{w_B}{3L^4}\left(\frac{z^5}{5} - Lz^4 + 2z^3L^2\right)\Big|_0^{L\cos\bar{\theta}}$$

$$= \frac{w_B}{3}\left(\frac{\cos^5\bar{\theta}}{5}L - L\cos^4\bar{\theta} + 2L\cos^3\bar{\theta}\right)$$

(10)

The quality of water Fins enveloping is:

$$M_{water} = \frac{\rho}{3}\int_0^{\lambda}|A\sin(kx)|\left(\frac{\cos^5\bar{\theta}}{5}L - L\cos^4\bar{\theta} + 2L\cos^3\bar{\theta}\right)dx$$

$$= \frac{4}{15k}\rho AL\sin^2\left(\frac{k\lambda}{4}\right)(\cos^5\bar{\theta} - 5\cos^4\bar{\theta} + 10\cos^3\bar{\theta})$$

(11)

The flexible fins parameters can be set: a longitudinal length of the fin L = 0.2m; wavelength λ = 0.314m; fluctuation angular frequency $\omega = 2\pi$. Then you can substitute them into the formula (11), runs on MATLAB, obtain the results shown in Figure 6.

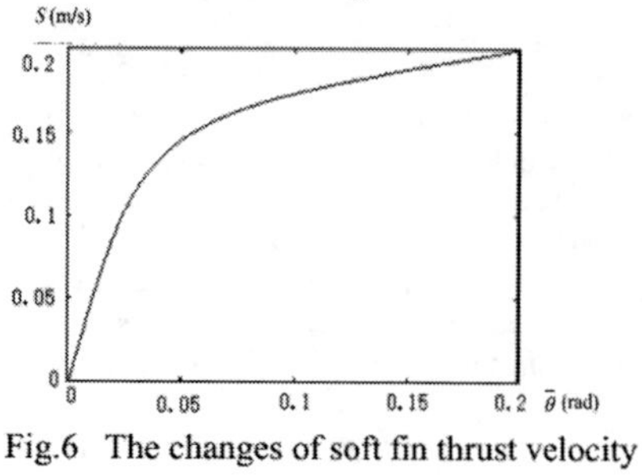

Fig.6 The changes of soft fin thrust velocity

3 Conclusions

(1) using the momentum theorem and over propeller theory, we can establish the relationship between the caudal fin propulsion and motion parameters, obtain the estimation method of caudal fin propulsion.

(2) Flexible crescent-shaped caudal fin propulsion mode is a novel way and can be used for micro and underwater diving underwater robots and other propulsion system.

References

[1] Fukuda T.Distributed type of actuators of shape memory alloy and its application to underwater mobile robotic mechanisms. In: Proceedings of the 1990 IEEE International Conference on Robotics and Automation, 1990,pp. 316-132.

[2] Zhang Xiaoqing,Wang Zhidong,Zhou Linhui.*Numerical simulation of three-dimensional oscillating caudal fin*[J].Journal of Jiangsu University of Science and Technology(Natural Science Edition,2006,22(1):1~10.

[3] Cheng Wei, Su Yu-min, Qin Zai-bai, Xu Yu-ru.*Flexible imitation crescent caudal fin tuna hydrodynamic analysis*[J].The Ocean Engineering,2004,22(4):1~7.

[4] Wang Sai.*Three joint robotic fish swimming free kinetic analysis of latent liter*.Xi'an University of Technology,2009.

[5] Pengfei Liu. *A time- domain panel method for oscillating propulsors with both chordwise and spanwise flexibility*[D].PhD Thesis. Memorial University of Newfoundland, Canada,1997.

MATEC Web of Conferences **63**, 02001 (2016) DOI: 10.1051/matecconf/20166302001
MMME 2016

Application of Numerical Simulation on the Manufacturing Process Qualification of Nuclear Power Plant Steam Generator Tube-sheet

Shao Jun ZHANG[1,a], Shu Hua LIANG[2], Wen Bin YANG[1], Zhao LIU[1] and Dong Hai ZHAO[1]

[1] *Equipment Supervision Research Centre, Suzhou Nuclear Power Research Institute Co. LTD, Suzhou 215004, China*
[2] *Manufacturing Process Qualification Centre, CGNPC, Suzhou 215004, China*

Abstract. Nuclear power plant key parts-steam generator (SG) tube sheet forging operate in harsh environment, so it has high requirements for internal quality. According to RCC-M specification requirements (take CPR1000 reactor type for example), the manufacturer should carry out workshop qualification and part process qualification before delivery product manufacturing. It is commonly used that centre compaction process and conical plate upsetting process regard as two kinds of large forging upsetting ways. The numerical analogue simulation has been carried out on nuclear power large tube sheet forgings using France Forge simulation software. The results show that: when the same pressure size, 3-direction compressive stress in the core of the billet always exist in the process of upsetting using centre pressing process, and the strain create in the central of the billet, but stagnant zone exist on the bottom and end plane; tensile stress in the core of the billet always exist in the process of upsetting using conical plate upsetting process, it will force the billet that contact with conical plate to deform, so eliminate stagnant zone, but effective deformation don't produce in the central of the billet. The conical plate continue press for conical plate upsetting process, tensile stress in the core of the billet disappear, and then gradually change into pressure stress, then tensile stress is generated at the bottom, finally tensile stress disappear and 3-direction compressive stress exist in the core of the billet.

1 Introduction

Steam Generator(SG) tube sheet forging for pressurized water reactor(PWR) nuclear power plant is a typical pancake forging, long-term operate in harsh environment with high temperature and high pressure. Thus it plate strength greatly weakened. Tube sheet will aggregate rather large internal stress, which will cause stress concentration on the crack tips, in particular, could have severity hydrogen damage in hydrogen-rich zone. Therefore, tube sheet forging must first ensure the purity of the molten steel in the manufacturing process. Secondly, it need adopt effective compaction method. In addition, it is difficult to compact in the centre of the tube sheet forgings and exceed the standard requirement easily in the ultrasonic testing [1-6].

As a key component of PWR nuclear power plant, SG tube sheet forging should follow the standard requirements of RCC-M. Before the formal part manufacturing, the manufacturer's workshop evaluation and part manufacturing process qualification are carried out according to the requirements of RCC-M M140, to verify the manufacturing factory technical ability and part whole/inner quality can meet the design and nuclear safety requirements [5].

At present, a big problem still present about tube sheet centre compaction. It is mostly that centre boss compaction and cone plate upsetting in the prior art. This paper only conducts analogue simulation and comparison about two compaction processes.

2 Introduce two compaction processes

2.1 Centre boss compaction process

Central boss compaction technology is first rotary pressure on the billet bottom end around, leaving a boss in the middle of the billet, the size of boss and billet into a certain proportion, then flat middle boss, on the end of the billet(top end) also performs the same operation. The process can form a similar effect-forging shell, thereby enhancing the compaction effect [7].

2.2 Conical plate upsetting process

The conical plate upsetting process first shallow recessing the bottom end of the billet with conical plate, conical plate pressure volume and billet height into a certain proportion, on the end of the billet also performs the same operation. The process can effectively improve the forging internal stress state, force hard deformation zone to change [8, 9].

3 Design model and set up boundary parameters

3.1 The establishment of model

Establish the billet and upper/lower mould model using 3D drawing software (figure 1), and save the file in a certain format, such as .stl, .may, .dou etc. Then import the file to forge software (France transvalor company) to compute.

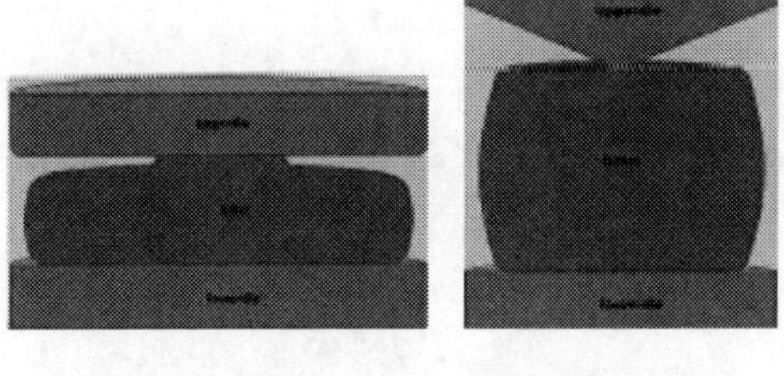

(a) center boss compaction (b)conical plate upsetting

Figure 1. 3D model building

[a] Corresponding author: zhangshaojun@cgnpc.com.cn

3.2 Set up boundary parameters

The materials 18MND5 is used for PWR nuclear power plant SG tube sheet [10], which comes from material database supplied by software company. Its main chemical composition is shown in table 1. The upper die and lower die are set to the rigid body in the whole simulation process. The whole model is divided into network structure consisted of many triangles and tetrahedron [11]. The main simulation parameters and boundary conditions are shown in table 2.

Table 1. 18MND5 chemical composition

Element	C_{max}	Mn	Si	P_{max}	S_{max}	Mo	Ni
Content%	0.20	1.15-1.60	0.10-0.30	0.012	0.012	0.45-0.55	0.50-0.80

Table 2. Simulation parameters and boundary conditions

billet					
Initial temperature/℃	Average mesh size/mm	Initial height/mm		Final height/mm	
		Center boss	Conical plate	Center boss	Conical plate
1250	20	1050	2200	850	2000
Upperdie/lowerdie					
Initial temperature/℃	Upperdie press/T	Upperdie depress rate/(mm • s-1)		stroke/mm	
				Center boss	Conica plate
25	15000	15		200	200

4 Results and analysis

4.1 Stress analysis

Figure 2 shows a longitudinal section variation process of stress by using the centre boss compaction technology. It can be seen that there has been greater compressive stress (up to 60.3MPa) in the centre boss upsetting process, and at the same time the heart of tube sheet billet always maintain the compressive stress, force is in good condition. The compressive stress extreme value appears in the upper die stroke 200mm, with a maximum of 16.1MPa in the central axis region.

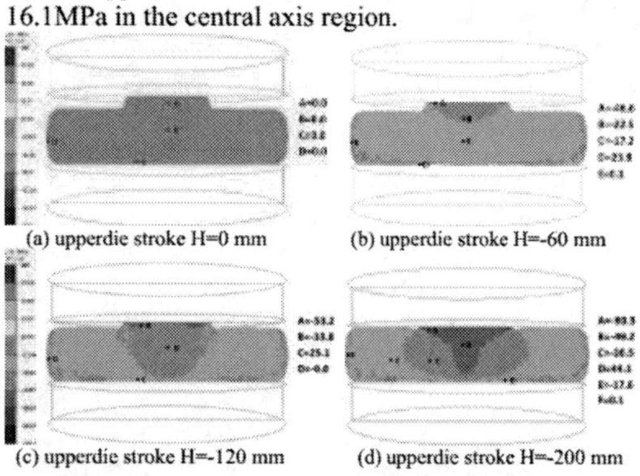

Figure 2. Stress analysis of center boss compaction process (Unit: MPa)

Figure 3 shows a longitudinal section variation process of stress by using the conical plate upsetting technology. From figure 3 can be seen, tube sheet billet internal stress is changing constant with the conical plate under pressure. The area near the conical plate produces compressive stress (region A), but there has been tensile stress near billet heart (C area of figure3 (b)), and the area is gradually moving down accompanied by a trend of expansion, eventually together with the tensile stress area of billet lower end face. The tensile stress extreme value appear in the conical plate pressure stroke 122mm under pressure process, the maximum tensile stress 16.1MPa, and the maximum tensile stress has always been located in the central axis of the billet at any time.

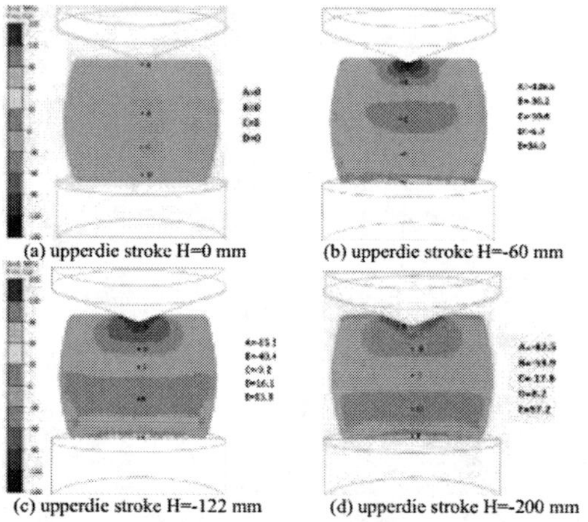

Figure 3. Stress analysis of concial plate upsetting process (Unit: MPa)

4.2 Equivalent strain analyses

Figure 4 show the equivalent strain change process of two upsetting compaction processes. As we can see, the range of the deformation of central boss process can be effectively delivered to the heart region of the billet under the same pressure stroke, but there has still stagnant zone contacted with upper and lower end plane, as shown in figure 4 (b). The range of the deformation of conical plate upsetting process mainly gathers near the top point of the conical plate, as shown in figure4 (d).

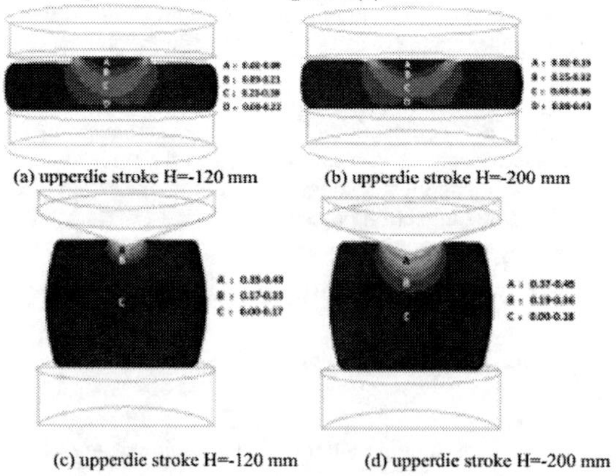

Figure 4. Billet equivalent strain analysis of two compaction processes

DOI: 10.1051/matecconf/20166302001

5 Optimization of conical plate upsetting process

Due to the billet core has been tension stress in upsetting process (pressure stroke 200mm, reduction rate of 9%), and the core has not produce effective deformation, so it need further optimize conical compaction process. When the conical plate continues under pressure (pressure stroke 298mm, reduction rate of 13.5%), billet heart tensile stress all disappear into compressive stress, but the edge zone of the billet produces tensile stress (minor stress value, about 18MPa), as shown in figure5(a). When under pressure reduction rate of 14.3%(pressure stroke 315mm), the inner of billet produces larger three direction compressive stress at the moment, the maximum tensile stress close to the core of billet is up to 100MPa, as shown in figure 5 (b). But when continue to press down to 14.4% (pressure stroke 317mm), and tensile stress appear in the edge of billet and near the outer circle, as shown in figure 5 (c). Continue to pressure reduction rate of 26.6% (pressure stroke 585mm), the tensile stress disappear, meanwhile the core produce larger three direction compressive stress which the value exceeds 100MPa, as shown in figure 5 (d).

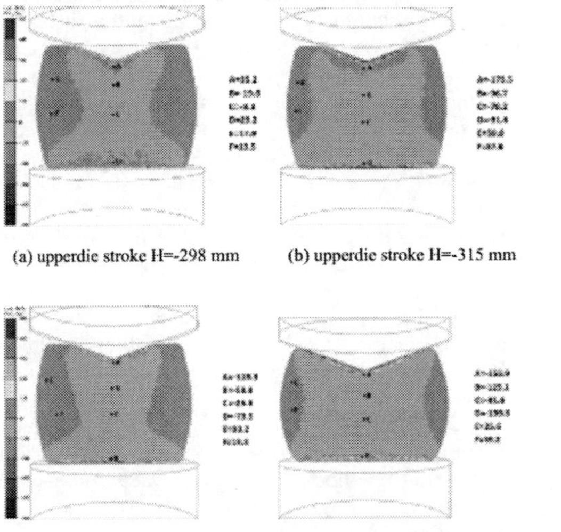

(a) upperdie stroke H=-298 mm (b) upperdie stroke H=-315 mm

(c) upperdie stroke H=-317 mm (d) upperdie stroke H=-585 mm

Figure 5. Billet stress analysis after conical plate optimization (Unit: MPa)

Figure 6 shows billet effective strain change process following billet under pressure stroke 298mm, 315mm, 317mm and 585mm.We can see that the deformation range of the billet constantly expand with conical plate under pressure, billet strain of conical top zone gradually increase and the effective deformation transfer to the core area of the billet. Combined with the analysis of billet deformation stress, the core zone of the billet accumulates certain tensile stress and the smaller effective strain when too small pressure stroke, so the effect of the billet core compaction is poor. When pressure amount is large, although the tensile stress of the billet core zone disappear and the larger effective strain, but it is not good for tube sheet forming and the final plate upsetting operation. Therefore, considering the actual production condition of the manufacturing plant, the factory should be comprehensive consideration combined with the stress and strain condition of the billet core and select the appropriate pressure stroke if the factory manufactures tube sheet using conical plate upsetting compaction technology.

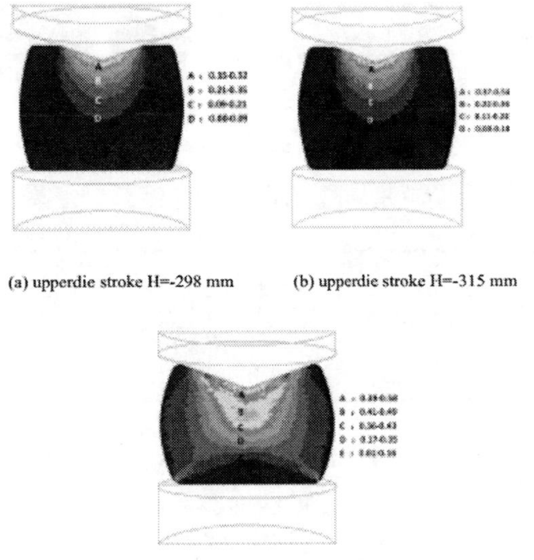

(a) upperdie stroke H=-298 mm (b) upperdie stroke H=-315 mm

(c) upperdie stroke H=-585 mm

Figure 6. Billet strain analysis of conical plate optimization

6 Actual result verification

The tube sheet forgings which two compaction processes produced are dissected and evaluated. The tube sheet is divided into 4 blocks according to the center symmetry, and is divided into 8 players parallel to the upper/lower end face, take off 4 layers to carry out mechanical properties test (as shown in figure 7). The tensile experiment is conducted using the above all samples and qualified test machine, and the test results are shown in table 3. It can be seen that Tube sheet produced by two kinds of compaction technology show different performance distribution rule. The tube sheet with central boss compaction technology has equal strength at the surface and the centre area, although there is a decreasing trend, but no significant weakening. The tube sheet with conical plate upsetting process has significant strength weaken from the surface to the core area, it is poor for tube sheet core zone compaction effect, do not get effective improvement. As a result, the actual product test results are consistent with the simulation results.

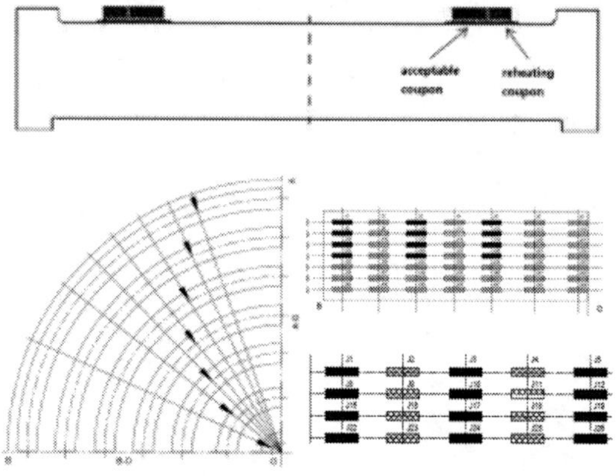

Figure 7. Sketch of tube sheet

Table 3. Test results of two compaction process

	Central boss		Conical plate	
	σ 0.2/MPa	σ b/MPa	σ 0.2/MPa	σ b/MPa
0(surface)	479/396	605/556	495/414	635/582
t/8(J5)	480/407	604/550	460/388	620/571
t/4(J12)	472/395	599/545	471/396	595/549
3t/8(J19)	477/402	603/560	443/382	587/540
t/2(J26)	465/389	593/552	444/370	589/545

7 Application of the simulation results

The simulation results of the tube sheet compaction can be used to guide the manufacturing process qualification, such as sampling plan, feasibility analysis of the forging process. It is found that the weak parts should increase the test items, sampling quantity, comprehensive anatomy inspection means and to verify overall quality of the tube sheet forgings. When conical plate upsetting process is used, the billet center has not produced effective deformation and tensile stress, so the region as a key test zone in the forming process. When central boss compaction process is used, there has still stagnant zone contacted with upper and lower end plane, add to low structure density, there can cause a certain influence on the mechanical properties, so this region as the focus of the test target.

8 Conclusions

(1) The results show that the billet core with central boss compaction technology has keep 3-D compressive stress under the upsetting process in the absolute pressure stroke, and heart area has certain deformation, but upper/lower have stagnant zone.

(2) The billet with conical plate upsetting technology has keep tensile stress under the upsetting process, and heart area has certain deformation, but upper/lower have stagnant zone. Although the billet has certain deformation with the conical top point area, stagnant deformation zone has been improved and eliminated, but there has not been effective deformation in the center region of the billet.

(3) It is need to optimize conical plate upsetting process. When the conical plate continues under pressure (reduction rate of 13.5%), billet heart tensile stress all disappear into compressive stress, but the edge zone of the billet produces tensile stress. When under pressure reduction rate of 14.3%, the inner of billet produces larger three direction compressive stress at the moment. But when continue to press down to 14.4%, and tensile stress appears in the edge of billet and near the outer circle. Continue to pressure reduction rate of 26.6%, the tensile stress disappear, meanwhile the core produce larger three direction compressive stress.

(4) The simulation analysis results can be used to guide the manufacturing process qualification, and can also test and verify weak areas targeted, can not only examine overall quality of tube sheet forgings and but also reduce the cost of testing.

References

1. Han J T, Zhang Y J, Zhao Z L, et al. The research of the upsetting process for controlling the inclusion defects in heavy disk forgings [J]. Heavy casting and forging, 2007, 1:1-5.
2. Zhang H, Liu D F. Scrap analysis about large forging of pipe plate [J]. Chinese journal of mechanical engineering, 2002, 38(1): 138-141.
3. Kan Y Q, Liang S H, Liu L, et al. Study on quality control of heat treatment for steam generator tube sheet forging [J]. Hot working technology, 2010, 39(4): 159-161.Luigi T.De Luca, Propulsion physics (EDP Sciences, Les Ulis, 2009)
4. Shanghai power equipment research institute. PWR nuclear island main equipment materials and welding [M]. Shanghai scientific and technological literature press.
5. Kan Y Q, Huang D P, Zhang S J, et al. Study meaning of manufacturing process qualification assessment method of nuclear island large-sized casting and forging parts [J]. Hot working technology, 2010, 39(1): 53-55.
6. Jin Z H, Guo W. Application of homogenization theory in study on tube-sheet [J]. Journal of Shenyang institute of chemical technology, 2007, 21(1):29-31.
7. Shi R G, Gao Q, Yan X P, et al. Manufacturing process of Large-scale heat exchanger SA765- II tube sheet [J]. Hot working technology, 2012, 41(19): 133-135.
8. Deng D M, Liu Z B, Ni L Y. Numerical simulation of the new technical for upsetting a cylinder between two cone plates [J]. Forging & Stamping Technology, 2005, 3:22-26.
9. Liu Z B, Ni L Y, Liu G H. New theory and technology of heavy forgings deformation [M]. China machine press, 2009.

DOI: 10.1051/matecconf/20166302001

10. Design and construction rules for mechanical components of PWR nuclear islands (RCC-M)[S].French association for design, construction and in-service inspection rules for nuclear island components (AFCEN):Edition 2000 & addendum June 2002.
11. Zhang S J, Liang S H, Liu Z . Numerical simulation and research of two kinds of compacting process for nuclear power large tube sheet forgings [J]. Hot working technology, 2013, 42(13):102-105.

MATEC Web of Conferences 63, 02002 (2016) DOI: 10.1051/matecconf/20166302002
MMME 2016

Conjoint utility analysis of technical maturity and project progress of construction project

Wei MA[1,]

[1]*Department of economics and management from Tianjin university of science and technology, Dagu South Road No. 1038, Hexi District, Tianjin, China*

Abstract. In this paper, taking construction project as the research object, the relationship between the project maturity index calculated by the construction project technical risks with different fine degree and the project progress index is studied, and the equilibrium relationship between the Party A's utility curve and the Party B's cost curve of using project maturity index and project progress index as the research variables is analyzed. The results show that, when the construction project technical risk division is more precise, the conjoint utility of the project's technical maturity index and the project progress is higher, and the project's Party A and Party B two sides are closer to the optimal equilibrium. This shows that the construction project technical risk must be finely divided, and managed and controlled respectively, which will help to improve the conjoint utility of the project Party A and Party B two sides..

1 Introduction

The complexity of construction project technology, the diversification of the management and the investment main body, the scale of the project, the multi degeneration of the project implementation environment have resulted in the construction project with the characteristics of large investment, many technical links, high risk, long cycle and so on. [1].These characteristics will cause the construction units to be inevitably faced with the solution of the inherent contradiction between the project technical risk and the duration of the project risk after the construction units have undertaken the construction task of the construction project [2]. This requires that after the signing of the construction contract, the construction side (i.e. construction project Part B) and construction units (i.e. construction project Part A) must refine calculations according to the contract schedule requirements and technical characteristics, try to avoid inherent contradictions between project duration risk and technology risk, and strive to reach the utility balanced state of project Part A and Part B two sides, find the meeting point to take into account the project time limit and technical requirements, strive to achieve the completion of the schedule and even ahead of schedule under the premise of ensuring the quality of construction projects, make the project Part A and Part B both to achieve the goal of maximizing utility. In this paper, the technology maturity is used to measure the technical risk, and the project schedule index is used to scale the project duration risk, and the utility curve and cost curve are used to describe the equilibrium utility analysis of the construction project Part A and Part B both sides.

2 TECHNICAL MATURITY INDEX

Domestic and foreign scholars have a different explanation for the concept of technology maturity degree, although there are differences in the expression of various interpretations, in essence, it refers to the estimation of the implementation capacity size of the technology to meet the project expected goal. The evaluation of technical maturity is to measure the stage or state of the results of the technical research and how much to achieve the expected goal of the project [4]. In this paper, the measurement method based on the technical risk of the construction project is adopted to measure the technology maturity, the greater the project technical risk is, the lower the technical maturity level of the project is. In the calculation of technical risks of the construction project, in order to explain the influence of technical risk division level on project technical maturity and project schedule, the construction project technical risk is divided into three levels in this paper. In Figure 1, earthwork engineering construction technology risk is taken as an example to explain the relationship between the three levels of technical risks.

[a] Corresponding author: maweilwtg@126.com

Table 1. Earthwork construction technology risk.

The first level technical risk	The second technical risk	The third technical risk
Earthwork construction technology risk	Supporting structure construction risk	Sheet pile engineering accident
		Fill embankment slope landslide
	Environmental risk caused by foundation pit dewatering	Site water
		Landslide
		Flow sand
		Piping effect
	Unreasonable foundation pit reinforcement risk	Foundation pit backfill soil subsidence
		Foundation trench backfill soil subsidence
	Foundation pit excavation risk	Excavation embankment slope landslide
		Foundation pit soaking
		Foundation trench soaking

In this paper, take a construction project example as the research object, and invite 18 industry experts to estimate and summarize the occurrence probability and influence degree of various level technical risk factors of the project. Calculate the weighted average number of each level technical risk to obtain that the first technical risk, the second level technical risk and the third technical risk of the project are 52.98, 112.41 and 581.76 separately. According to the construction project practical experience, experts give that the first level technical risk standard number of this general type project is 120, the second level technical risk standard number is 200, the third level technical risk standard number is 850.This paper defines that each level technology mature degree index = technical risk standard number / the project technical risk number, which is concluded that the technology maturity index divided by the first level technology risk is 2.27, the technical maturity index divided by the second level technology risk is 1.78, and the technical maturity index divided by the third level technology risk is 1.46.

3 Utility curve and cost curve

Project technical maturity index reflects the technical level of a construction project, and another important factor affecting the cost and quality of construction projects is the project progress, which reflects the comprehensive management level of the project. The project progress considered in this paper excludes the influence of technical risk factors, and the technical risk factors are reflected in the technical maturity index. The conjoint utility curve of a construction project is composed of two factors that one is the technical maturity index which reflects the project technical level, and another is the project progress index which reflects the project management level. According to the project practice experience, the industry experts give ten groups data of project schedule and technical risk level, which the project progress indexes are represented by the average number of the ten group project schedule data are divided by each project schedule number, and the project technology maturity indexes use the second level project technology risk classification as the standard, as shown in table 1. According to these data, the conjoint utility curve of the construction project Party A is fitted, as shown in figure 1. The X axis represents the project schedule index, the Y axis represents the project technical maturity index, and U represents the utility curve of the Party A. Any pair of combination on the utility curve is no difference for Party A.

Table 2. project progress index and technical maturity index.

	Pr .1	Pr .2	Pr .3	Pr .4	Pr .5	Pr .6	Pr .7	Pr .8	Pr .9	Pr. 10
progress index	1.78	1.40	1.27	1.09	0.80	0.83	0.70	1.06	0.87	0.62
technical maturity index	0.89	0.42	0.91	0.90	1.37	1.00	1.22	0.88	0.65	2.18

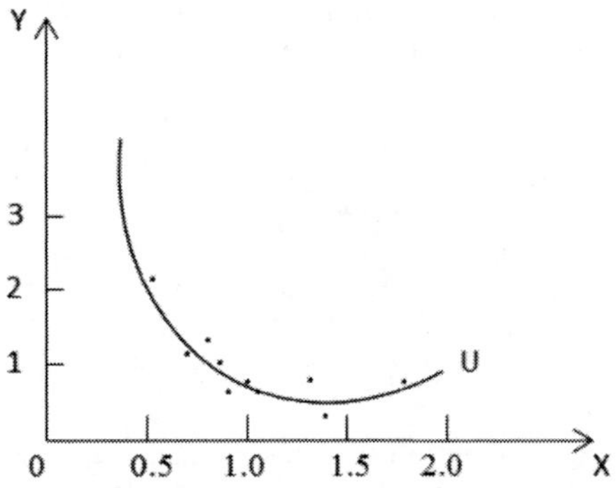

Figure 1. Party A conjoint utility curve.

The Party B's cost curve of the construction project can be represented by the $aX+bY+\mu = c$, where a represents the management cost ratio related to the progress in the project cost, b represents the technology cost ratio related to the technology maturity level in the project cost, μ represents the other controllable factors in project cost, and c represents the project cost. The Party B's cost curve of the construction project is shown in Figure 2. The slope and intercept of the cost curve are determined by the coefficient a, the coefficient b , the cost c and μ together. The triangle area composed by the costs line with the X axis and Y axis is the project Part B acceptable cost budget range, and the outside area is the part that cannot be realized in the project budget under the established technical maturity level and progress requirements.

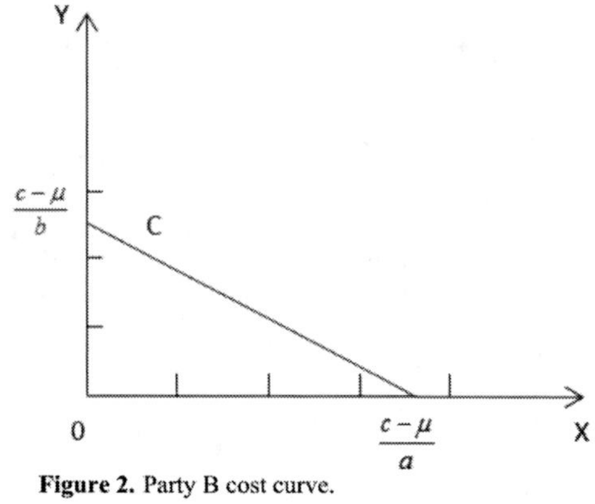

Figure 2. Party B cost curve.

4 Equilibrium

The Party A's utility curve indicates that any pair of combination of the technical maturity index and progress index on this curve is no difference to Party A, or it is impossible to be realized that speed up the project progress under the established technical maturity level. The upper right area of the curve indicates that the technology maturity is higher under the same project progress, the project scale is larger under the same technical maturity requirements. On the contrary, the curve of being more close to the origin means that the effectiveness of Party A is smaller.

Party B's cost curve represents the Party B's maximum cost of a construction project, and the slope of the cost curve will change, when the proportion of the cost related to the progress and the technical cost related to technical maturity happens to change. When the project cost is increased, the cost curve moves to the right. The cost curve of being more close to the origin indicates that the project cost is smaller.

In this paper, investigation of the construction project progress index is based on the same method as above. The average number of ten groups of project progress data divided by the project progress number obtains that the project progress index is 1.78, which is depicted in Part A and Part B equilibrium diagram, that is, the A point in Figure 3. The technical maturity index of the first level technical risk of the project calculated by the above is 2.27, which is depicted in the equilibrium graph of Part A and Part B, that is, the B point in Figure 3. The technical maturity index of the second level technology risk is 1.78, that is, the C point in figure 4. The technical maturity index of the third levels technical risk is 1.46, that is, the D point in figure 3.

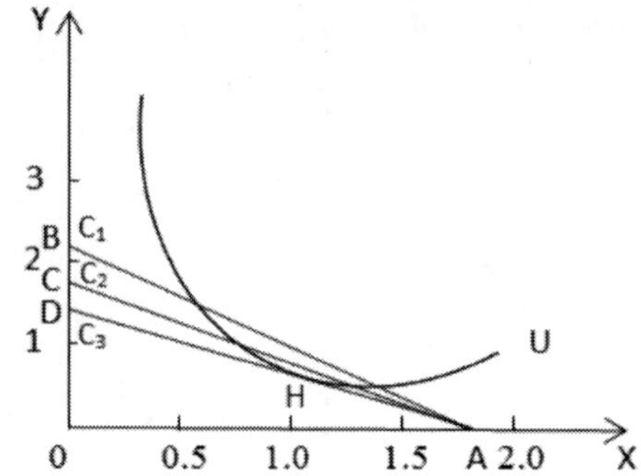

Figure 3. Part A and Part B two sides balance chart

In Figure 3 drawn according to the actual situation of the construction project, it can be seen that Party B cost curve AD and Party A utility curve U are tangent to the H point, AB and AC lines Intersect with utility curve U. It shows that in the case of the known Party A utility curve, the cost curve which is composed by the technical maturity index obtained according to the three level technology risk division and the utility curve of Party B can reach the optimal equilibrium state. The cost curve which is composed by the technology maturity index obtained according to the first level technical risk and the second level technology risk division and the Part B utility curve does not reach the optimal equilibrium. This shows that there is a better combination to make Part A and Part B two sides to achieve a balanced state. It is shown in figure 4.

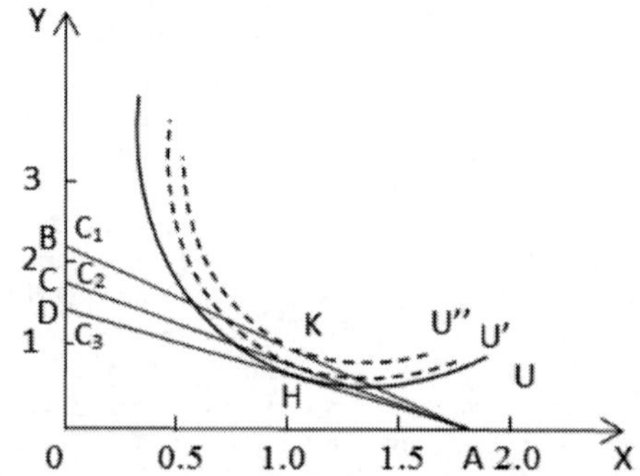

Figure 4. Adjustment of optimal equilibrium

In Figure 4, it can be seen that there are more efficient curves U 'and U " whose tangent lines are the cost lines AB and AC. On the cost curve AB which is composed by the technology maturity index obtained according to the first level technical risk division, when moves to the upper left corner from point A along the cost curve, it can be found that higher utility curve U 'and AB is tangent to the point K. The utility of U " curve is higher than U 'curve and higher than U curve, but it cannot be achieved under current project condition.

Because under the same technical level requirements, the U 'and U' utility curve require the project to have a larger scale, or under the same project scale, require a higher technical maturity level. All these are impossible to achieve under the premise of given the scale of the construction project, the construction period and the cost budget.

In the construction project of this paper research, there are only a set of optimal combination of the cost curve and the utility curve, that is, the cost curve AD of Party B and the utility curve U of Party A. This just shows that when the technology maturity index is measured according to the three levels of technology risk division, the combination of Part B's cost curve and Part A's utility curve of the project is the most economical and efficient. When the technology maturity index is measured according to the first level technology risk and the second level technology risk, due to the risk division being not precise enough, the measured technical maturity index is not accurate enough to reflect the actual technical level of the project, so that the combination with the progress index to describe the effectiveness of Party A and the cost of Party B cannot reach the optimal equilibrium state, resulting in the loss of economic efficiency. Under the premise of the measuring as accurately as possible of the project technology maturity, this paper tries to find the balance point between the technical maturity and the progress of the construction project through the utility curve and cost curve, and furtherly explain the intrinsic relationship between the technical risk and the progress time out.

5 Conclusion

In this paper, the relationship between the project maturity index calculated by the construction project technical risks with different fine degree and the project progress index is studied, and taking the concrete construction project as the research object, the equilibrium relationship between the Party A's utility curve and the Party B's cost curve of using project maturity index and project progress index as the research variables is analyzed. The results show that when the project technical risk is divided into the most precise three levels, the utility curve of the project Party A and the cost curve of the project Party B presents a balanced state. This indicates that when the construction project technical risk division is more precise, the measure of the project's technical maturity index is more accurate, and the project Party A and Party B two sides are closer to the optimal equilibrium which makes Party A's utility as high as possible while also making Party B's cost close to the minimum. This also explains the intrinsic relationship between technical risk and progress time out in a construction project from a new perspective, in order to provide a new solution for the future research on the technical risk and progress of the construction project.

References

[1]Forecasting Analysis and Control Measures of Risk Cost of Construction Project. Chinese and Overseas Architecture, 08(2009)

[2]Analysis on cost risk and cost control measures of construction project. City Architecture,15(2015)

[3]A Study of Influence of Technology Readiness Level on Spacecraft Development Schedule. Aerospace industry management , 18(2009)

[4]Research on the relationship between technology maturity and risk management. Aeronautic Standardization and Quality, **3** (2012)

Calculation and Simulation Study on Transient Stability of Power System Based on Matlab/Simulink

Xiu Feng SHI[1], Yan Rong YANG[1] and Jun Feng WANG[1]

[1]*School of Information Technology and Engineering Yuxi Normal University, Yuxi , Yunnan , 653100*

Abstract. The stability of the power system is destroyed, will cause a large number of users power outage, even cause the collapse of the whole system, extremely serious consequences. Based on the analysis in single machine infinite system as an example, when at the f point two phase ground fault occurs, the fault lines on either side of the circuit breaker tripping resection at the same time,respectively by two kinds of calculation and simulation methods of system transient stability analysis , the conclusion are consistent. and the simulation analysis is superior to calculation analysis .

Keywords. Matlab, power system, transient; stability, simulation

0. Introduction

The failure in stability of the power system would cause power outage to many users or even collapse of the whole system, thus bringing very serious consequences. Therefore, the study on the stability of the power system is of great significance in ensuring safe and reliable operation of the power system.

After the power system is subject to a large disturbance, the imbalance between mechanical torque and electromagnetic torque on the generator rotor leads to a change in the relative position between synchronous generator rotors, i.e. the current, voltage and electromagnetic power of the system are changed as the relative angle between generator potentials is changed. The study on the transient stability of the power system refers to researching whether parallel operated synchronous generators can still operate synchronously and the load is normal after the power system is subject to a large disturbance in an operation mode. The reasons for the large disturbance mainly include: short circuit fault; turning on or off main components of the system, such as generator, transformer and line; sudden change in load, such as addition of large users. The short-circuit fault is usually used to check the transient stability of the system as it is the most severe among them.

1. Calculation of transient stability of power system

Taking the diagram of single-machine infinite bus system shown in Fig. 1 as an example, the transient stability of the power system is calculated after the faulty line is cut off through simultaneous tripping of circuit breakers on both sides of the line in the case of two-phase ground fault at the point f.

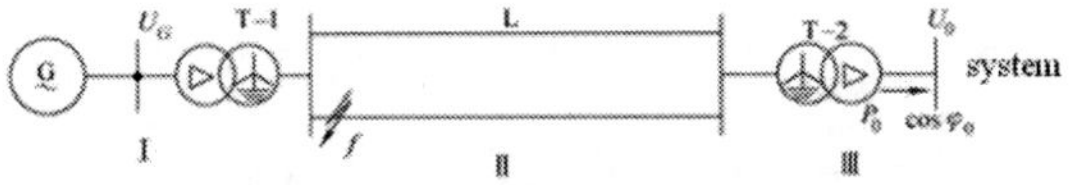

Fig. 1 Diagram of single-machine infinite bus system

The parameters of components in the single-machine infinite bus system shown in Fig. 1 are as follows[1]:

Parameters of the generator G:

$P_{GN}=300MW,\quad U_{GN}=10.5KV,\quad \cos=0.85,\quad X_d=X_q=2.36$

$X_d^{'}=0.32,\quad X_d^{''}=0.252,\quad X_q^{''}=0.243,\quad X_2=0.243,\quad T_j=6s$

Parameters of the transformer T-1: S_{TN1}=360MVA, U_{ST1}%=14%, K_{T1}=10.5/242;

Parameters of the transformer T-2: S_{TN2}=360MVA, U_{ST2}%=14%, K_{T2}=220/121;

Parameters of the transmission line L: U_N=220KV, l=200km, X_L=0.41Ω/km, X_{L0}=4X_L;

Operation: U_0=115KV, P_0=250MW, $\cos\varphi_0$=0.98.

1.1 Calculation of network parameters and operating parameters

S_B=250MVA, $U_{B\text{III}}$ = 115KV, $U_{BX}=U_X$ $U_{B\text{II}}$ = $U_{B\text{III}}$ × K_{T2}=115× 220/121=209KV,

$U_{B\text{I}}$ = $U_{B\text{II}}$ × K_{T1}=209× 10.5/242=9.07KV.

1.1.1 Per unit values are as follows after reduction of component parameters:

$$X_d=X_q=2.36\times\frac{250\times0.85}{300}\times(\frac{10.5}{9.07})^2=2.241$$

$$X_d^{'}=0.32\times\frac{250\times0.85}{300}\times(\frac{10.5}{9.07})^2=0.304$$

$$X_{T1}=0.14\times\frac{250}{360}\times(\frac{242}{209})^2=0.130$$

$$X_L=0.41\times200\times\frac{250}{209^2}=0.47$$

$$X_{L0}=4\times X_L=0.47\times4=1.88$$

$$X_{T2}=0.14\times\frac{250}{360}\times(\frac{220}{209})^2=0.108$$

$$X_2=0.23\times\frac{250\times0.85}{300}(\frac{242}{209})^2=0.218$$

$$T_J=6\times\frac{300}{0.85\times250}=8.46s$$

The equivalent circuit during the normal operation of the system is shown in Fig. 2a), and integrated impedances of the system are as follows:

$$X_{TL}=X_{T1}+\frac{1}{2}X_L+X_{T2}=0.130+\frac{1}{2}\times0.47+0.108=0.473$$

$$X_{d\Sigma}=X_{q\Sigma}=X_d+X_{TL}=2.241+0.473=2.714$$

$$X_{d\Sigma}^{'}=X_d^{'}+X_{TL}=0.304+0.473=0.777$$

1.1.2 Calculation results of per unit values of operating parameters are as follows:

$$P_0=\frac{P_0}{S_B}=\frac{250}{250}=1,\quad U_0=\frac{U_0}{U_{BX1}}=\frac{115}{115}=1,$$

$$Q_0=P_0\tan_0=P_0\tan(\cos^{1}0.98)=0.2$$

$$E_0=\sqrt{\left(U_0+\frac{Q_0X_{d\Sigma}^{'}}{U_0}\right)^2+\left(\frac{P_0X_{d\Sigma}^{'}}{U_0}\right)^2}=\sqrt{(1+0.2\times0.777)^2+(1\times0.777)^2}=1.392$$

$$\delta_0=\tan^{-1}\frac{\frac{P_0X_{d\Sigma}^{'}}{U_0}}{U_0+\frac{Q_0X_{d\Sigma}^{'}}{U_0}}=\tan^{-1}\frac{0.777}{1+0.2\times0.777}=33.92^0$$

1.2 Calculation of transferring impedance and power characteristic of the system

The equivalent circuit during the normal operation of the system, negative sequence and zero sequence equivalent networks in the case of two-phase short circuit at the point f, equivalent network in the case of short circuit and equivalent network after clearing of short circuit are shown in a), b), c), d) and e) in Fig. 2 respectively.

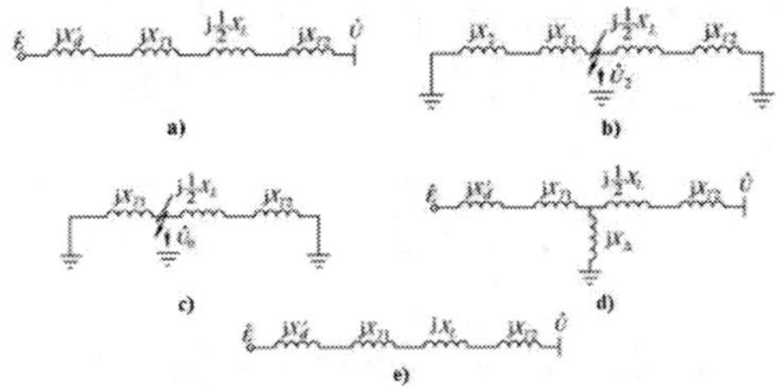

Fig. 2

a) Equivalent circuit during the normal operation b) Negative sequence network c) Zero sequence network

d) Equivalent network in the case of short circuit e) Equivalent network after clearing of short circuit

As seen in b) negative sequence network and c) zero sequence network in Fig. 2, the negative sequence and zero sequence equivalent reactance at the short circuit point are:

$$X_{2\Sigma}=\frac{(X_2+X_{T1})(\frac{1}{2}X_L+X_{T2})}{X_2+X_{T1}+\frac{1}{2}X_L+X_{T2}}$$

$$=\frac{(0.218+0.130)(\frac{1}{2}\times0.47+0.108)}{(0.218+0.130)+(\frac{1}{2}\times0.47+0.108)}$$

$$=0.173$$

$$\boldsymbol{X_{0\Sigma}=\frac{X_{T1}(\frac{1}{2}X_{L0}+X_{T2})}{X_{T1}+\frac{1}{2}X_{L0}+X_{T2}}=\frac{0.130\times(\frac{1}{2}\times1.88+0.108)}{0.130+\frac{1}{2}\times1.88+0.108}=0.116}$$

The additional reactance at the short circuit point in the case of two-phase grounding short circuit is:

$$\boldsymbol{X_\Delta=\frac{X_{0\Sigma}X_{2\Sigma}}{X_{0\Sigma}+X_{2\Sigma}}=\frac{0.173\times0.116}{0.173+0.116}=0.069}$$

From d) equivalent network in the case of short circuit in Fig. 2, the transferring impedance and power characteristic of the system are:

$$X_{II}=X_d'+X_{T1}+\frac{1}{2}X_L+X_{T2}+\frac{(X_d'+X_{T1})(\frac{1}{2}X_L+X_{T2})}{X_\Delta}$$

$$=0.434+0.343+\frac{0.434\times0.343}{0.069}$$

$$=2.93$$

$$P_{II}=\frac{E_0U_0}{X_{II}}\sin\delta=\frac{1.392\times1}{2.93}\sin\delta=0.48\sin\delta$$

From e) equivalent network after clearing of short circuit in Fig. 2, the transferring impedance and power characteristic of the system are:

$$\boldsymbol{X_{III}=X_d'+X_{T1}+X_L+X_{T2}=0.304+0.130+0.47+0.108=1.102}$$

$$\boldsymbol{P_{III}=\frac{E_0U_0}{X_{III}}\sin\delta=\frac{1.392\times1}{1.102}\sin\delta=1.38}$$

1.3 Calculation of limit clearing angle

The limit clearing angle δ_{cm} is calculated as follows through the equal area criterion [1]:

$$\boldsymbol{\delta_{cm}=\cos^{-1}\frac{P_0(\delta_h-\delta_0)+P_{IIIM}\cos\delta_h-P_{IIM}\cos\delta_0}{P_{IIIM}-P_{IIM}}}$$

$$\boldsymbol{=\cos^{-1}\frac{1\times\frac{\pi}{180^0}(133.6^0-33.92^0)+1.38\cos133.6^0-0.48\cos33.92^0}{1.38-0.48}}$$

$$\boldsymbol{=1.122}$$

Where, the critical angle δ_h is

$$\delta_h=180^0\ \ \sin^{-1}\frac{P_0}{P_{IIIM}}=180^0\ \ \sin^{-1}\frac{1}{1.38}=133.6^0$$

$\boldsymbol{\delta_{cm}}$ is converted to degrees: $\delta_{cm}=1.122\times\frac{180^0}{\pi}=64.3^0$

1.4Calculation of generator rocking curve $\boldsymbol{\delta-t}$

In the simple system above, the generator rocking curve $\boldsymbol{\delta-t}$ (i.e. motion equation of rotor) during the fault is

$$\begin{cases}\frac{d\delta}{dt}=(\omega-1)\omega_0\\ \frac{d\omega}{dt}=\frac{1}{T_j}(P_T-\frac{E'U}{X_{II}}\sin\delta)\end{cases}\qquad(1\text{-}1)$$

These are two first-order non-linear ordinary differential equations with known initial conditions, that is

$t=0;\omega=1;\delta=\delta_0=\sin^{-1}\frac{P_T}{P_{IM}}$.

After calculating the curve $\delta-t$ during the fault, the limit clearing time corresponding to the limit clearing angle can be found from the curve.

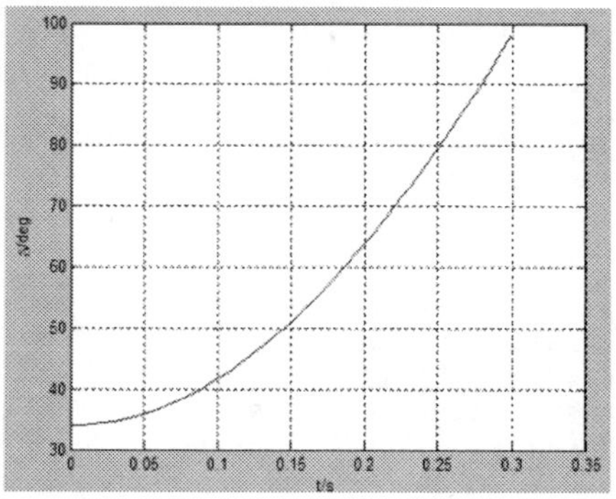

Fig. 3 Curve $\boldsymbol{\delta-t}$ during system fault

It is very difficult to conduct the analytical solution to two simple first-order non-linear ordinary differential equations (1-1). In this paper, the generator rocking curve $\delta-t$ during the fault is solved with Ronge-Kuta method in Matlab [2].] After running the program, the curve $\delta-t$ during the system fault is obtained, as shown in Fig. 3. The limit clearing time corresponding to the limit clearing angle $\delta_{cm}=64.3^0$ obtained from Fig. 3 (or output result) is 0.2s.

If the clearing time is known, the curve $\delta - t$ needs to be calculated to determine the stability of the system. When the fault clearing is calculated through the curve, another differential equation as shown below needs to be solved as changes in system parameters can lead to a change in the power characteristic of the generator

$$\begin{cases} \frac{d\delta}{dt} = (\omega - 1)\omega_0 \\ \frac{d\omega}{dt} = \frac{1}{T_J}(P_T - \frac{E'U}{X_{II}}\sin\delta) \end{cases} \quad (1\text{-}2)$$

Initial conditions for the equation 1-2 are t_c , where t_c is the given clearing time, δ_c and ω_c are δ and ω corresponding to t_c. δ_c and ω_c can be obtained from the curve $\delta - t$ and $\omega - t$ during the fault. The curve of δ and ω as a function of time can be obtained from the equation 1-2. Generally, the system is stable transiently if δ is always no more than 180^0 and the vibration amplitude value is becoming smaller and smaller in a few seconds of calculation.

It is very difficult to conduct the analytical solution to two simple first-order non-linear ordinary differential equations (1-2). The generator rocking curve $\delta - t$ after clearing of the system fault is solved with Ronge-Kuta method in Matlab. If the clearing time is 0.1s, the curve $\delta - t$ during the system fault obtained is shown in Fig. 4 after modifying system parameters and running the program. Fig. 4 (or output result) shows that reduction begins at 0.5455s and the

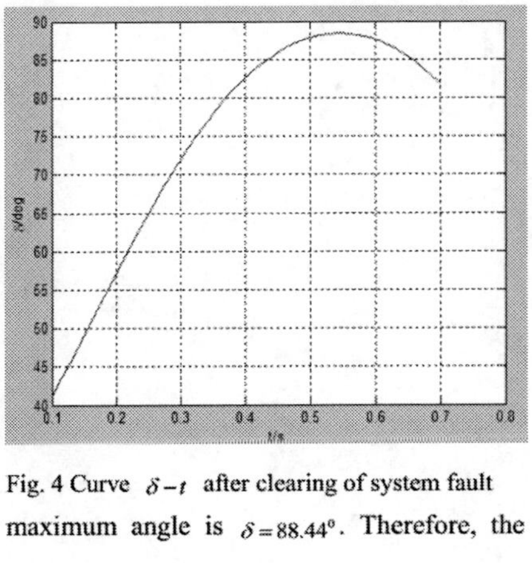

Fig. 4 Curve $\delta - t$ after clearing of system fault

maximum angle is $\delta = 88.44^0$. Therefore, the system is stable transiently.

2. Establishment of Simulation Model of Transient Stability

The simulation model [2], as shown in Fig. 5, is established to study the transient stability according to the single-machine infinite bus system in Fig. 1.

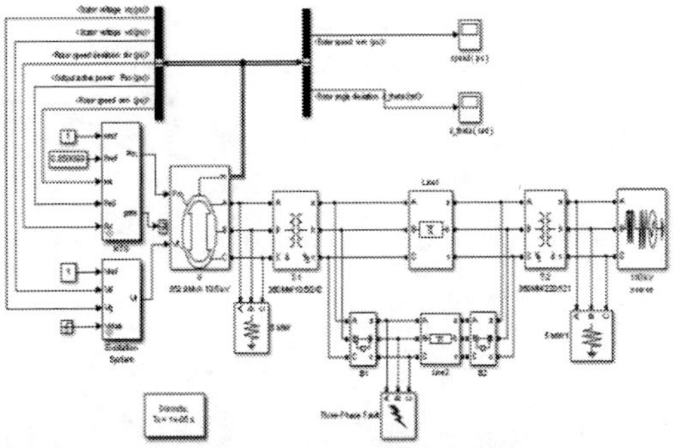

Fig. 5 simulation model of transient stability of power system

On the simulation diagram, parameters of the generator, transformer, transmission line, infinite bus system, circuit breaker B1 and B2 are set according to the given conditions, and then the generator is initialize d with Powergui module[3]. The generator bus is set to PV bus, the terminal voltage is set to 18KV, the output power is set to 300MV.A and then the system load flow is updated; the sampling time is set to $1\text{x}10^{-5}$s. Finally, simulation parameters are set, the discrete algorithm is chosen, the start time for simulation is set to 0, and the end time is set to 5s. In the fault point module, the system is set to have BC two-phase grounding short circuit at 0.05s and cut off the short circuited line at 0.1s after the fault[4].

3. Analysis on simulation result

The speed change curve, as shown in Fig. 6, of the generator is obtained through simulation according to the simulation model Fig. 5. The parameter settings of the circuit breaker module B1 and B2 are changed to cut off the line at 0.5s after the fault. The speed change curve, as shown in Fig. 7, of the generator is obtained through the second simulation.

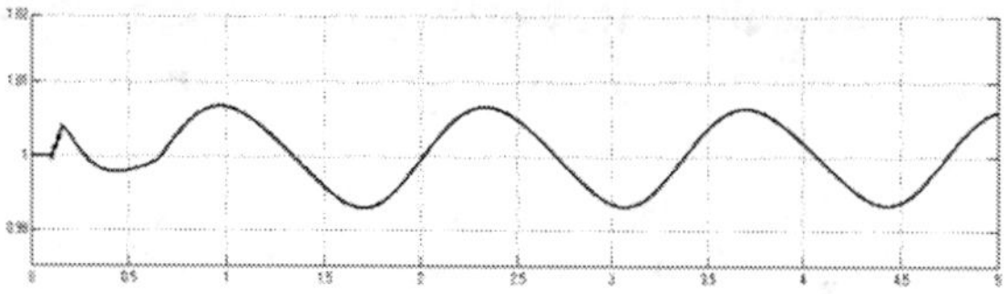

Fig. 6 Speed change curve of generator when cutting off the line at 0.1s after the fault

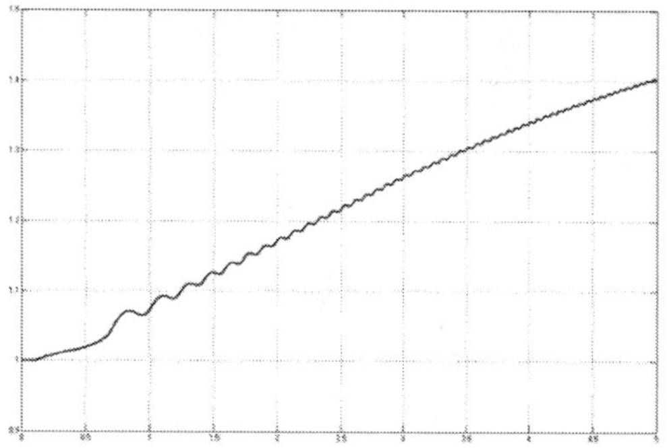

Fig. 7 Speed change curve of generator when cutting off the line at 0.5s after the fault

The simulation curve in Fig. 6 shows that the speed of the generator gradually decreases with the increase of the time (variation range: 0.99~1.01) and tends to become stable constants in the case of two-phase grounding short circuit fault at the point f after cutting off the faulty line after 0.1s (the clearing time is less than the limit clearing time), therefore the system is stable; the simulation curve in Fig. 7 shows that the speed of the generator gradually increases with the increase of the time in the case of two-phase grounding short circuit fault at the point f after cutting off the faulty line after 0.5s (the clearing time is greater than the limit clearing time), therefore the system is unstable

4. Conclusion

To sum up, the limit clearing angle obtained by calculating the transient stability of the power system is 64.3^0, and the limit clearing time obtained with Ronge-Kuta method in Matlab is 0.2s based on the limit clearing angle. If the clearing time is 0.1s (less than the limit clearing time), the maximum angle δ obtained with Ronge-Kuta method in Matlab is 88.44^0, which is less than 180^0, and the vibration amplitude value is becoming smaller and smaller, therefore the system is stable transiently.and for power system transient stability simulation analysis obtained the system is also stable transiently as the speed of the generator gradually decreases with the increase of the time after cutting off the faulty line at 0.1s after the fault. Therefore, the conclusions are consistent In comparison, the simulation analysis is superior to calculation, the reason is the complicated calculation can be avoided.

References

[1] Li Guangqi, Transient Analysis of Power System [M], Beijing: China Electric Power Press, 2007
[2] Yu Qun, MATLAB/Simulink Power System Modeling and Simulation [M], Beijing: China Machine Press. 05 (2011)
[3] Zhao Haibin, MATLAB Applications [M], Beijing: Tsinghua University Press, 05 (2012)
[4] Zhou Junjie, Matlab/Smiulink Example Explanation [M], Beijing: China WaterPower Press, 05 (2014)

Research on Optimization of Bearing Span of Main Reducer Gear System Based on the Transfer Matrix

Feng Yun HUANG, Li TIAN[a], Ding WEI and Huan Huan SUN

School of Mechanical and Electronic Engineering, Wuhan University of Technology, 430070, Wuhan, China

Abstract. The research object is the bearing span of the main reducer gear of the crossover vehicle for improving the performance of the Noise-Vibration-Harshness (NVH) on the rear axle. It could get the optimal bearing span by modal analysis of the driving gear shaft and the vibration test of the main reducer, based on the transfer matrix. It is verified that the vibration and transmission stability are greatly improved with this optimal bearing span by the dynamic simulations of different bearing spans in ADAMS. And it can provide reference for the structural optimization design of the main reducer and the research on the vibration and noise reduction.
Keywords. bearing span, transfer matrix, ADAMS, NVH

1 Introduction

The main reducer is one of the most important parts of the automobile transmission system, and its vibration and noise can directly affect the NVH performance of the car[1]. The bearing support which connects the gear shaft and the main reducer shell can affect the coupling vibration effect of the components in the main reducer. And its support stiffness can affect the meshing quality of the bevel gear pairs, thereby affecting the vibration response of the main reducer[2-3]. When the bearing preload is certain, bearing support span has a direct effect on support stiffness[4]. The bearing span is taken as the research object for achieving the purpose of improving the vibration response of the main reducer and the NVH performance of the crossover vehicle. The optimal bearing span under certain bearing preload is obtained by the modal analysis of driving gear shaft and the vibration test of main reducer, based on the transfer matrix. And the result is verified by the dynamic simulations in ADAMS.

2 The optimal bearing span based on the transfer matrix

Transfer matrix can describe the relationship between input and output of the linearized MIMO system (Multi Input Multi Output system). The input and the output are the end state vectors of each unit, including displacement, angular displacement, bending moment and shear stress. The basic idea of the transfer matrix method is to discretize the system into several feature units, and then the overall multi-body dynamics model is gotten by combining the each feature unit. The main reducer gear system is discretized into several feature units, and the transfer matrix of each unit is respectively established. Then the total transfer matrix of the system is obtained by combining the sub transfer matrices. The characteristic equation can be obtained with the free end as constraint conditions, then the system natural frequency under the different bearing span is gotten by calculating with relevant parameters[5-6]. The resonance frequency range which should be avoided can be determined by the modal analysis of the driving gear shaft and the vibration test of the main reducer, and the optimal bearing span is gotten.

2.1 System transfer matrix

The main reducer gear system is discretized in its section as shown in figure 1a. It contains 5 parts: driving gear, shaft-I, shaft-II, shaft-III and shaft-IV. The bearing is mounted on the shaft-II. The main reducer gear system is continued to be discretized into several feature units for establishing the transfer matrix model easily, as shown in figure 1b. Shaft-IV and shaft-III are set as a general shaft. The general shaft is composed of a non-mass elastic shaft and two centralized mass stations. Two bearing support are set as a non-mass elastic support station, and the acting point is in the center of the two support bearing. The shaft-II and bearings are simplified as a whole mass unit, and then it is divided into two parts: shaft-II-2 and shaft-II-1. The length of shaft-II-1 is $l/2$, and l is the bearing span. As the diameter of shaft-II-1 is larger than its length, shaft-II-2 and shaft-II-1 are set as the wheel stations. The shaft-I and the driving gear are integrally set as a shaft with an elastic support, and it can be equivalent to a non-mass elastic shaft and a rigid mass unit with an elastic support. The simplified mechanical models of the

[a]Corresponding author: 1422486766@qq.com

 DOI: 10.1051/matecconf/20166302004

feature units are shown in figure 2. The transfer matrix of each sub-unit can be obtained by the dynamic equations.

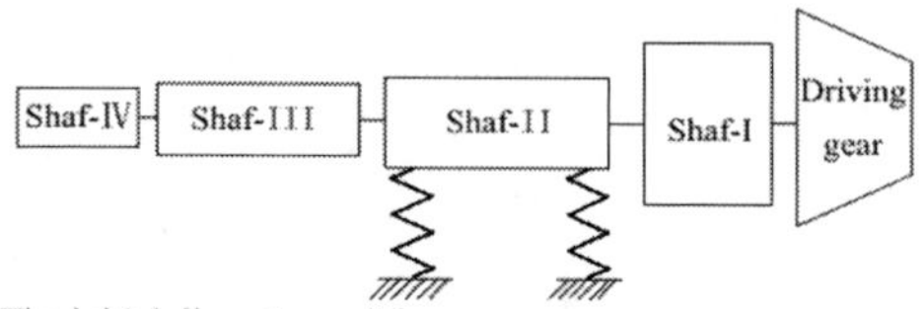

a. The initial discrete model

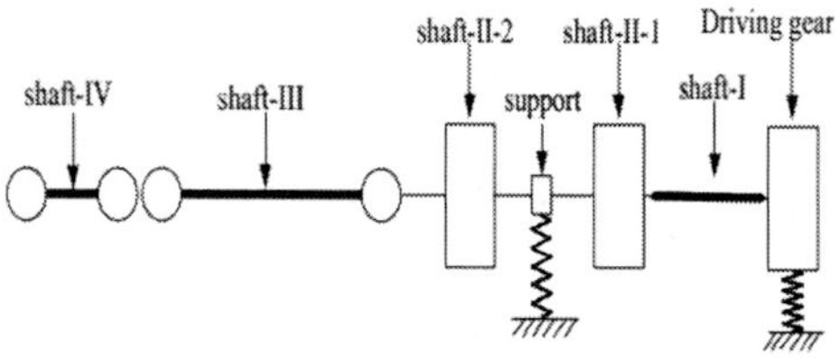

b. The final discrete model

Figure 1.The discrete model of the main reducer gear system

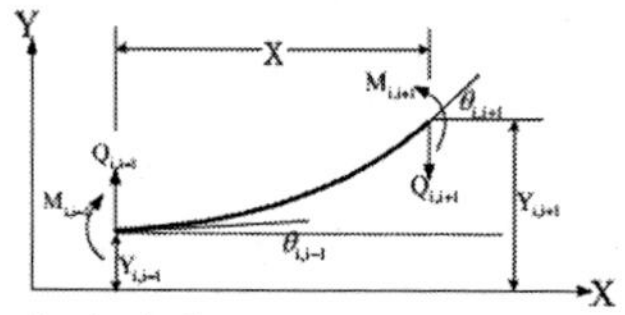

a. Non-mass elastic shaft

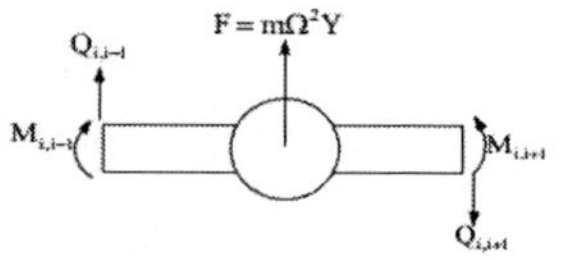

b. Centralized mass station

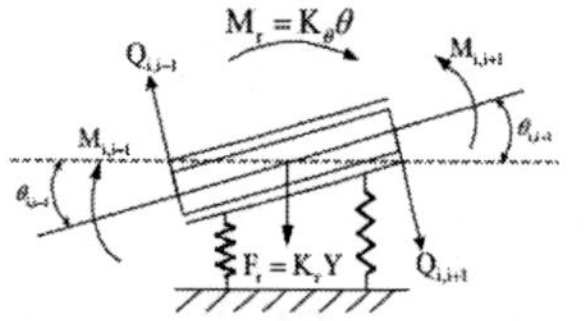

c. Non-mass elastic support station

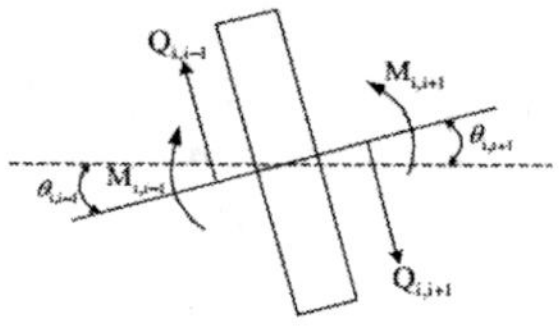

d. Wheel station

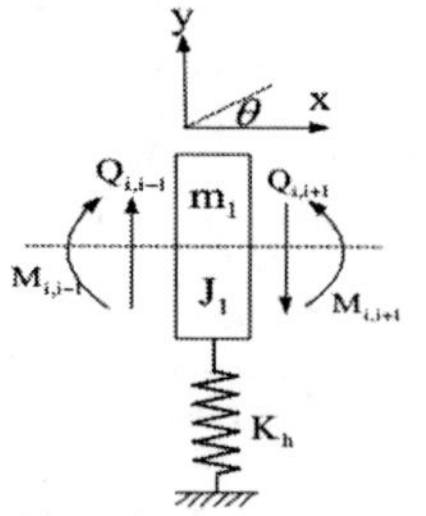

e. Rigid mass unit with an elastic support

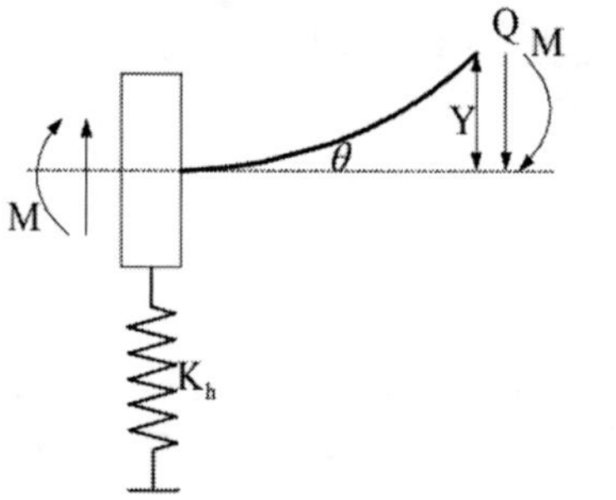

f. Shaft with an elastic support

Figure 2. The simplified mechanical models of feature units

2.1.1 General shaft

The general shaft is composed of a non-mass elastic shaft and two centralized mass stations. Figure 2a shows the definition of coordinate and the state vectors of the non-mass elastic shaft. The sections of shaft from $X_{i,i-1}$ to $X_{i,i+1}$ are called as i-1 section and i section. The end state vectors include: displacement (*Y*), angular displacement (*θ*), bending moment (*M*) and shear stress (*Q*). *I*is the section inertia, and *E* is the elastic modulus. The dynamic equations of the non-mass elastic shaft can be obtained by the mechanical model.

$$\begin{cases} Y_{i,i+1} = Y_{i,i-1} + x \cdot \theta_{i,i-1} + \frac{x^2}{2EI} M_{i,i-1} + \frac{x^3}{6EI} Q_{i,i-1} \\ \theta_{i,i+1} = \theta_{i,i-1} + \frac{x}{EI} M_{i,i-1} + \frac{x^2}{2EI} Q_{i,i-1} \\ M_{i,i+1} = M_{i,i-1} + l \cdot Q_{i,i-1} \\ Q_{i,i+1} = Q_{i,i-1} \end{cases} \quad (1)$$

Transforming the equation (1) into the matrix form, the coefficient matrix of the state vectors is the transfer matrix of the non-mass elastic shaft, denoted as T_{sh}.

$$T_{sh} = \begin{bmatrix} 1 & x & \frac{x^2}{2EI} & \frac{x^3}{6EI} \\ 0 & 1 & \frac{x}{EI} & \frac{x^2}{2EI} \\ 0 & 0 & 1 & x \\ 0 & 0 & 0 & 1 \end{bmatrix} \quad (2)$$

The dynamic equations of the centralized mass station can be obtained by the mechanical model shown in figure 2b.

$$\begin{cases} Y_{i,i+1} = Y_{i,i-1} \\ \theta_{i,i+1} = \theta_{i,i-1} \\ M_{i,i+1} = M_{i,i-1} \\ Q_{i,i+1} = Q_{i,i-1} + m\Omega^2 Y_{i,i-1} \end{cases} \quad (3)$$

Transforming the equation (3) into the matrix form, the coefficient matrix of the state vectors is the transfer matrix of the centralized mass station, denoted as T_m. *T* is the transfer matrix of the general shaft.

$$T_m = \begin{bmatrix} 1 & 0 & 0 & 0 \\ 0 & 1 & 0 & 0 \\ 0 & 0 & 1 & 0 \\ m\Omega^2 & 0 & 0 & 1 \end{bmatrix} \quad (4)$$

$$T = T_m T_{sh} T_m = \begin{bmatrix} \frac{m\Omega^2 x^3}{6EI}+1 & x & \frac{x^2}{2EI} & \frac{x^3}{6EI} \\ \frac{m\Omega^2 x^2}{2EI} & 1 & \frac{x}{EI} & \frac{x^2}{2EI} \\ \frac{m\Omega^2 x}{EI} & 0 & 1 & \frac{x}{EI} \\ m\Omega^2 + m\Omega^2\left(\frac{m\Omega^2 x^3}{6EI}+1\right) & m\Omega^2 x & \frac{m\Omega^2 x^2}{2EI} & \frac{m\Omega^2 x^3}{6EI}+1 \end{bmatrix} \quad (5)$$

The transfer matrix of the shaft-IV and the shaft-III can be gotten by putting their physical parameters into the equation (5). T_4 is the transfer matrix of the shaft-IV. T_3 is the transfer matrix of the shaft-III.

2.1.2 Non-mass elastic support station

The two bearing supports are simplified as an elastic supporting station, as shown in figure 2c. And it can be equated to a spring with radial stiffness and angular stiffness. K_r is the equivalent radial stiffness of two supporting bearings, K_θ is the equivalent angular stiffness, and l is the bearing span. The effect of rotational speed and load on the bearing stiffness can be not considered, thus the bearing stiffness is regarded as a constant. In the center of the two bearings, there will be a reaction force F_r and a reaction moment M_r.

$$\begin{cases} K_\theta = K_r \cdot l^2/4 \\ F_r = K_r \cdot Y \\ M_r = K_\theta \cdot \theta \end{cases} \quad (6)$$

The effect of damping of the rolling bearing will not bring large error to the calculation results, thus it can be ignored. The dynamic equations of the non-mass elastic support station can be obtained by the mechanical model shown in figure 2c.

$$\begin{cases} Y_{i,i+1} = Y_{i,i-1} \\ \theta_{i,i+1} = \theta_{i,i-1} \\ M_{i,i+1} = M_{i,i-1} + K_\theta \theta_{i,i-1} \\ Q_{i,i+1} = Q_{i,i-1} - K_r Y_{i,i-1} \end{cases} \quad (7)$$

Transforming the equation (7) into the matrix form, then its coefficient matrix of the state vectors is the transfer matrix of the non-mass elastic support station, denoted as T_b.

$$T_b = \begin{bmatrix} 1 & 0 & 0 & 0 \\ 0 & 1 & 0 & 0 \\ 0 & K_\theta & 1 & 0 \\ -K_r & 0 & 0 & 1 \end{bmatrix} = \begin{bmatrix} 1 & 0 & 0 & 0 \\ 0 & 1 & 0 & 0 \\ 0 & \frac{K_r l^2}{4} & 1 & 0 \\ -K_r & 0 & 0 & 1 \end{bmatrix} \quad (8)$$

2.1.3Wheel station

As shown in figure 2d, the wheel station is of mass and moment of inertia, and there will be the moment when it rotates with the angular velocity $\boldsymbol{\Omega}$. m is the mass of the wheel station. J is the moment of inertia of the wheel station. The dynamic equations of the wheel station can be obtained by its mechanical model.

$$\begin{cases} Y_{i,i+1} = Y_{i,i-1} \\ \theta_{i,i+1} = \theta_{i,i-1} \\ M_{i,i+1} = M_{i,i-1} + J\Omega^2 \theta_{i,i-1} \\ Q_{i,i+1} = Q_{i,i-1} + m\Omega^2 Y_{i,i-1} \end{cases} \quad (9)$$

Transforming the equation (9) into the matrix form, then its coefficient matrix of the state vectors is the transfer matrix of the wheel station, and the transfer matrix of the shaft-II-2 and shaft-II-1 are $T_{2\text{-}2}$ and $T_{2\text{-}1}$.

$$T = \begin{bmatrix} 1 & 0 & 0 & 0 \\ 0 & 1 & 0 & 0 \\ 0 & J\Omega^2 & 1 & 0 \\ m\Omega^2 & 0 & 0 & 1 \end{bmatrix} \quad (10)$$

$$T_{2-2} = \begin{bmatrix} 1 & 0 & 0 & 0 \\ 0 & 1 & 0 & 0 \\ 0 & J_{2-2}\Omega^2 & 1 & 0 \\ m_{2-2}\Omega^2 & 0 & 0 & 1 \end{bmatrix} \quad (11)$$

$$T_{2-1} = \begin{bmatrix} 1 & 0 & 0 & 0 \\ 0 & 1 & 0 & 0 \\ 0 & J_{2-1}\Omega^2 & 1 & 0 \\ m_{2-1}\Omega^2 & 0 & 0 & 1 \end{bmatrix} \quad (12)$$

2.1.4Shaft with an elastic support

The difference value between the diameter of shaft-I and the diameter of driving gear is small. The shaft-I and the driving gear can be integrally set as a shaft with an elastic support. And it can be equated to a non-mass elastic shaft and a rigid mass unit with an elastic support, as shown in figure 2f. The mechanical model of a rigid mass unit with an elastic support is shown in figure 2e. K_h is the meshing stiffness of the driving gear, m_1 is the overall mass of the shaft-I and the driving gear, and J_1 is the overall moment of inertia of the shaft-I and the driving gear. When the shaft-I and the driving gear rotate with the angular velocity $\boldsymbol{\Omega}$, there will be an inertia force and an inertia moment: $m_1\Omega^2 Y$ and $J_1\Omega^2\theta$. The dynamic equations of the rigid mass unit with an elastic support can be obtained by the mechanical model.

$$\begin{cases} Y_{i,i+1} = Y_{i,i-1} \\ \theta_{i,i+1} = \theta_{i,i-1} \\ M_{i,i+1} = M_{i,i-1} + J_1\Omega^2 \theta_{i,i-1} \\ Q_{i,i+1} = Q_{i,i-1} + m_1\Omega^2 Y_{i,i-1} - K_h Y_{i,i-1} \end{cases} \quad (13)$$

Transforming the equation (13) into the matrix form, then its coefficient matrix of the state vectors is the transfer matrix of the rigid mass unit with an elastic support, denoted as T_k.

$$T_k = \begin{bmatrix} 1 & 0 & 0 & 0 \\ 0 & 1 & 0 & 0 \\ 0 & J_1\Omega^2 & 1 & 0 \\ m_1\Omega^2 - K_h & 0 & 0 & 1 \end{bmatrix} \quad (14)$$

The transfer matrix of the non-mass elastic shaft has been derived, denoted as T_{sh}. x_1 is the overall length of the shaft-I and the driving gear. $I_1$is the section inertia of the shaft-I and the driving gear. T_g is the transfer matrix of the shaft with an elastic support.

$$T_g = T_{sh}T_k = \begin{bmatrix} (m_1\Omega^2 - K_h)\frac{x_1^3}{6EI_1}+1 & J_1\Omega^2\frac{x_1^2}{2EI_1}+x_1 & \frac{x_1^2}{2EI_1} & \frac{x_1^3}{6EI_1} \\ (m_1\Omega^2 - K_h)\frac{x_1^2}{2EI_1} & J_1\Omega^2\frac{x_1}{EI_1}+1 & \frac{x_1}{EI_1} & \frac{x_1^2}{2EI_1} \\ (m_1\Omega^2 - K_h)\frac{x_1}{EI_1} & J_1\Omega^2 & 1 & x_1 \\ m_1\Omega^2 - K_h & 0 & 0 & 1 \end{bmatrix} \quad (15)$$

2.1.5Transfer matrix of the system

According to the discrete relationship, all feature sub units are connected in series. It could get the total transfer matrix of the main reducer gear system by putting each transfer matrix together. T_N is the transfer matrix of the system, and it is a 4x4 matrix.

$$T_N = T_4T_3T_{2-2}T_bT_{2-1}T_g = \begin{bmatrix} u_{11} & u_{12} & u_{13} & u_{14} \\ u_{21} & u_{22} & u_{23} & u_{24} \\ u_{31} & u_{32} & u_{33} & u_{34} \\ u_{41} & u_{42} & u_{43} & u_{44} \end{bmatrix} \quad (16)$$

2.2 Calculation of natural frequency

According to the discrete model of the main reducer gear system, the initial cross section and the final section both are the free end constraints, thus M_I=0, Q_I=0, M_N=0 and Q_N=0. The relationship of the state vectors between the initial cross section and the final section of the main reducer gear system can be obtained.

$$\begin{bmatrix} Y \\ \theta \\ M \\ Q \end{bmatrix}_N = \begin{bmatrix} Y \\ \theta \\ 0 \\ 0 \end{bmatrix}_N = T_N \begin{bmatrix} Y \\ \theta \\ 0 \\ 0 \end{bmatrix}_1 \quad (17)$$

$$\begin{bmatrix} Y \\ \theta \\ 0 \\ 0 \end{bmatrix}_N = \begin{bmatrix} u_{11} & u_{12} \\ u_{21} & u_{22} \\ u_{31} & u_{32} \\ u_{41} & u_{42} \end{bmatrix} \begin{bmatrix} Y \\ \theta \end{bmatrix}_1 \quad (18)$$

Y_I and θ_I cannot both be 0 at the same time when the main reducer gear system is vibrating. Therefore, the equation (19) will exist based on the equation (18).

$$\Delta(\Omega^2) = \begin{vmatrix} u_{31} & u_{32} \\ u_{41} & u_{42} \end{vmatrix} = 0 \quad (19)$$

In this main reducer gear system, K_r=1023.65N/um, K_h=2.0436N/um, and the material of the gear shaft is 20CrMnTi. The other related parameters are shown in table 1. The relationship between the critical speed and bearing span can be obtained after calculating by the equation (19) with related parameters. The curve of the natural frequencies and the bearing span can be gotten because of $f=\Omega/60$, as shown in figure 3.

The calculation result shows that the first-order natural frequency of the main reducer gear system is mainly in the range of 2700~3500Hz. And the first-order natural frequency is under 3000Hz when the bearing span is in the range of 53~58mm.

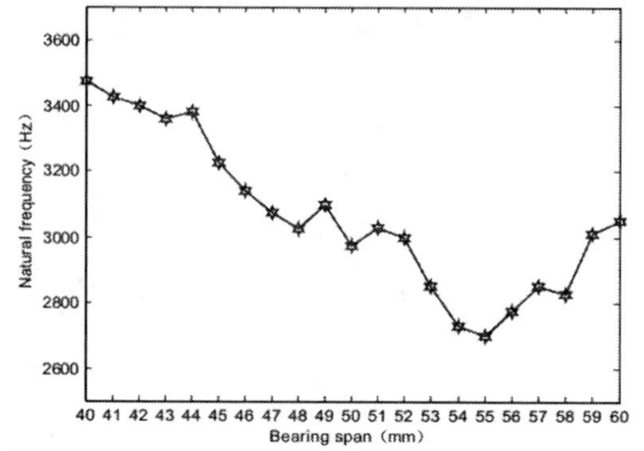

Figure 3. Natural frequencies - bearing span

2.3The optimal bearing span

The 3D model of the driving gear shaft was established in UG, based on the engineering drawing. The end faces of the bearings were constrained after the model imported into ANSYS. The modal analysis was performed by Block Lanczos method. The analysis result shows that the first-order natural frequency of the driving gear shaft is 3247.3Hz, and the two-order natural frequency is 3262.0Hz.

The natural frequency of the main reducer can be obtained by analyzing its vibration signal in the frequency domain [7]. The vibration test of the main reducer was completed by "the rear axle main reducer vibration performance testing system" which developed by the research group. The testing results of the three testing samples are shown in figure 4.

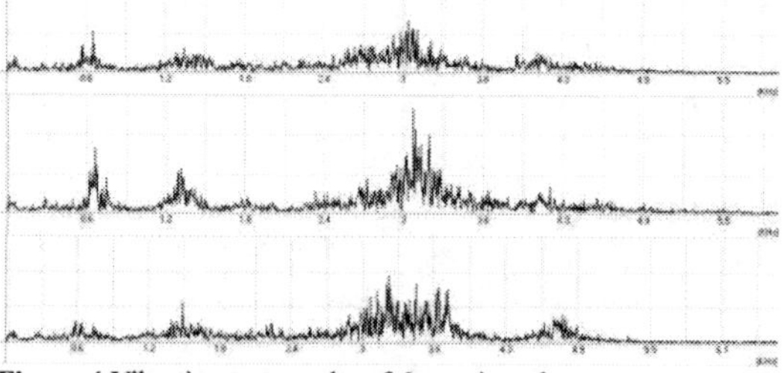

Figure 4. Vibration test results of the main reducers

The testing results show that the vibration of the main reducer is more obvious when the frequency in the range of 3000~3500Hz. That is the resonance frequency range of the main reducer. Thus the natural frequency in the range of 3000~3500Hz, especially around 3200Hz, should be avoided when designing the parts of the main reducer. And it could effectively reduce the vibration of the main reducer. It can be inferred that the better bearing span is in the range of 53~58mm by the curve shown in figure 3 which reflects the relationship between the natural frequencies and the bearing span. The first-order natural frequency is the most far away from 3000Hz when the bearing span is 55mm. Thus it can avoid resonance of the main reducer as far as possible when the bearing span is 55mm.

Table 1. The related parameters

	$x\ (10^{-3}\text{m})$	$m(10^{-3}\text{kg})$	$I(10^{-8}\text{m}^4)$	$J(10^{-6}\text{kg}\cdot\text{m}^2)$
Shaft-IV	25	49.6	0.5153	
Shaft-III	69	264.2	1.9175	
Shaft-II-2	57-*l*/2	254.562-2.233*l*		23196.96-203.48*l*
Shaft-II-1	*l*/2	2.233*l*		203.48*l*
Shaft-I & Driving gear	65	358.4	3.9761	4.032

3Dynamic simulations in ADAMS

The physical parameters and the constraints and loads were defined based on the actual situation after the 3D model of the main reducer gear system imported into ADAMS. The dynamic model of the main reducer gear system is shown in figure 5.

Figure 5. Dynamic model of the main reducer gear system

6 sets of bearing span models are prepared for the comparative simulation analysis, including 54mm, 55mm, 56mm, 57mm, 58mm and 59mm. At the same time, 6 sets of driving speed are set for the simulation, including 800r/min, 1500r/min, 2000r/min, 3000r/min, 4000r/min and 5000r/min. The angular acceleration and the contact force in the Y direction (vertical direction) of the driven gear are obtained by the simulation. The driven gear contact force in the Y direction can directly affect the vertical vibration when bevel gear pair meshing, thereby affecting the ride comfort of the car. The system is in the acceleration phase in the first 0.2 seconds. The simulation data of the acceleration phase should be excluded. The average values and difference values of the angular acceleration and the contact force in the Y direction at different driving speed are shown in figure 6 and figure 7.

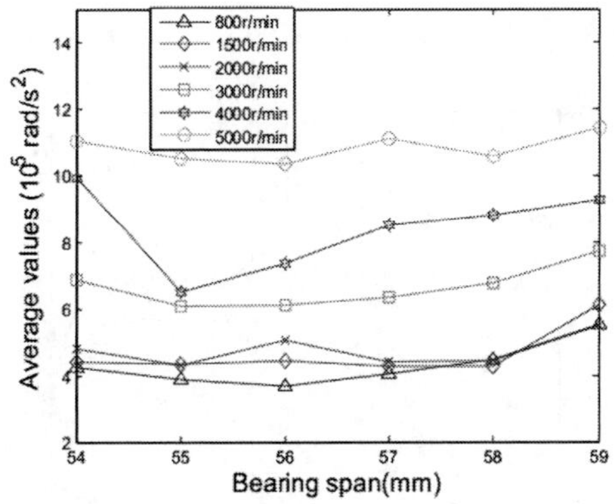

a. Average values of the angular acceleration

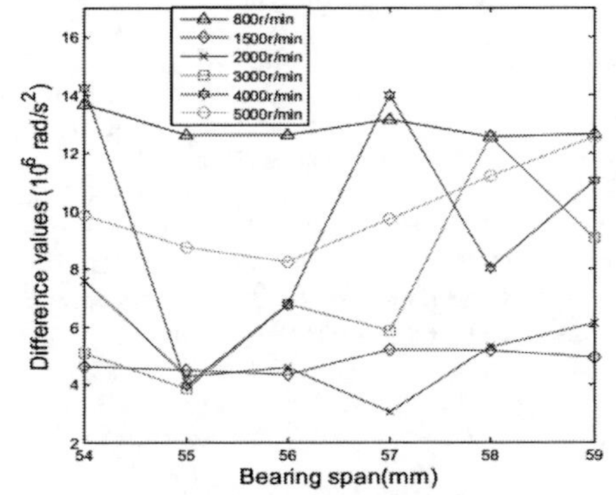

b. Difference values of the angular acceleration
Figure 6. Angular acceleration of the driven gear

The figure 6a shows that the average value of the angular acceleration of the driven gear closes to the minimum value when the bearing span is 55mm. The figure 6b shows that the difference values of the angular acceleration of the driven gear are fluctuating in different simulation driving speed, and the difference values is basically close to or in the minimum value when the bearing span is 55mm. From the above analysis, the vibration of the main reducer gear system is relatively stable when the bearing span is 55mm. The vibration and noise in the transmission process can be effectively reduced while the vibration fluctuation of the main reducer gear system is small.

DOI: 10.1051/matecconf/20166302004

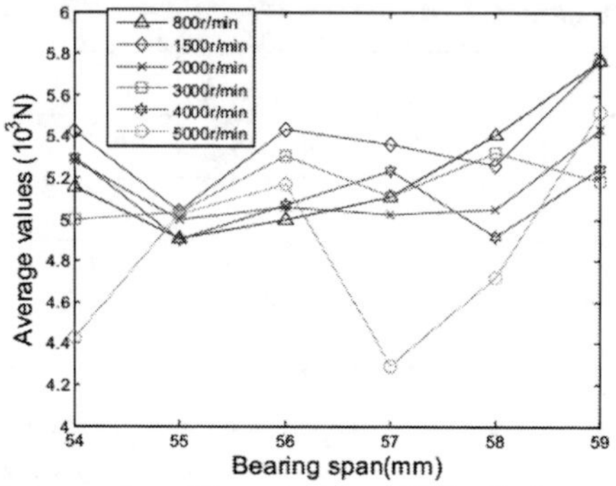

a. Average values of the contact force (Y direction)

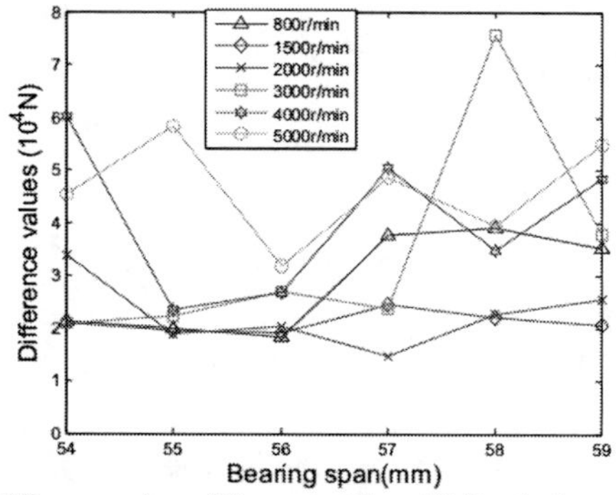

b. Difference values of the contact force (Y direction)
Figure 7. Contact force (Y direction) of the driven gear

The figure 7a shows that the average values of the contact force (Y direction) are relatively concentrated, and the average values are close to or in the minimum value when the bearing span is 55mm. The figure 7b shows that the difference values of the contact force (Y direction) are relatively small if the driving speed is not higher than 4000r/min when the bearing span is 55mm. It indicates that the impact of the transmission process is small, and the vibration is relatively stable.

Comprehensive analysis shows that the transmission of the main reducer gear system is relatively stable with small vibration fluctuation and the better vibration performance when the bearing span is 55mm.

4 Conclusions

The resonance of the main reducer can be avoided as far as possible when the bearing span is 55mm by the theoretical analysis and calculation, combined with the modal analysis of the driving gear shaft and the vibration test of the main reducer. The results of dynamic simulation shows that the angular acceleration and the contact force in Y direction of the driven gear are smaller when the bearing span is 55mm. This verifies the theoretical result. With this optimal bearing span, it can reduce the vibration of the main reducer effectively and improve the NVH performance of the crossover vehicle.

References

1. Shiwei Mao, Guangmin Li, Fuqiang Chen, Auto Engineer, 11, 40 (2011)
2. Jinli Xu, Shangguo Wan, Bing Xia, EMEIT-2012, 949, (2012)
3. Shaojun Han, Kang Liu, Xijin He, Journal of Highway and Transportation Research and Development, 11, 146 (2011)
4. Damin Pei, Pingfa Feng, DingwenYu, Machinery Design & Manufacture, 10, 44 (2005)
5. S. H. Gao, G. Meng, Archive of Applied Mechanics, , 10, 1473 (2011)
6. Shan Chai, Xianyue Gang, Fusheng Yao, Journal of University of Shanghai For Science and Technology, 01, 8 (2002)
7. Fengyun Huang, Huanhuan Sun, Hao Pan, WeiruZhang, Advanced Materials Research, 823, 417 (2013)

Simulation Research of Magnetically-coupled Resonant Wireless Power Transfer System with Single Intermediate Coil Resonator Based on S Parameters Using ANSYS

Liu Cheng, Peng Yan-qing[a], Zhang Ji, Jiang Lian-ming, Tang Jia-cheng

School of Electrical Engineering and Automation, Xiamen University of Technology, Xiamen 361024, China

Abstract. ANSYS can be a powerful tool to simulate the process of energy exchange in magnetically-coupled resonant wireless power transfer system. In this work, the MCR-WPT system with single intermediate coil resonator is simulated and researched based on scattering parameters using ANSYS Electromagnetics. The change rule of power transfer efficiency is reflected intuitively through the scattering parameters. A new method of calculating the coupling coefficient is proposed. A cascaded 2-port network model using scattering parameters is adopted to research the efficiency of transmission. By changing the relative position and the number of turns of the intermediate coil, we find some factors affecting the efficiency of transmission. Methods and principles of designing the MCR-WPT system with single intermediate coil resonator are obtained. And these methods have practical value with design and optimization of system efficiency.

1 Introduction

Wireless power transfer (WPT) has become a research interest in recent years. It was reported by Tesla a century ago[1], and offers the promise of cutting the last wire. It has been or will be used in costly, inconvenient, dangerous and even some occasions that cannot use wired transmission, such as implantable ventricular assistant devices[2] or for acquiring power from offshore wind driven generators and outer space solar power station[3].

In 2007, the Massachusetts Institute of Technology (MIT) has proposed a new scheme based on strongly coupled magnetic resonances, thus presenting a key breakthrough for mid-range wireless energy transmission[4]. Although the coupled mode theory can explain the process of energy exchange well[5], it can only apply to the conditions with small perturbations, and it is too hard for common electrical engineers to understand or use. So its application has been greatly restricted; on the other hand, measuring the parameters in the coupled mode theory is very difficult. In this work, we choose the ANSYS Electromagnetics, which is one of the most powerful tools to simulate the process of energy exchange in the MRC-WPT system as well as to calculate the parameters in it, to study the WPT system with single intermediate coil resonator. From an electrical engineering perspective, we derive curves for coupling coefficient versus the gap length between the sending coil and receiving coil using ANSYS Hfss and the curve fitting tool of MATLAB, thus compare it with the result calculated by Neumann formula. Based on the above work, we research the influence of single intermediate coil with different number of turns on transmission efficiency. Paper[6] has researched single intermediate coil resonator's impacts on transmission efficiency with different positions, but it ignores the geometrical factors of the intermediate coil. Paper[7] has researched multiple intermediate resonant coils' influence on transmission efficiency, but it uses the coupled mode theory that many electrical engineers are not familiar with. By introducing the efficiency of transmission, a cascaded 2-port network model using *S* parameters is adopted, which covers the principle of designing the single intermediate coil resonator and also avoids the blindness in designing the intermediate coil resonator.

2 Modeling and Inductance Analysis

2.1 Simulation Model of Non-intermediate System

As shown in Fig 1, the magnetic resonance coupling wireless power transfer system with non-intermediate resonant coil is composed of power supply, excitation coil, sending coil, receiving coil, load coil the load and for each coil there is one compensating capacitor installed in series. In this paper, only the efficiency of transferring energy from sending coil (labelled *Coil 1*) to receiving coil (labelled *Coil 2*) is considered, and for sake of simplicity, we suppose the load is a constant low resistive load which has no influence on the transmission efficiency analysis because each variable capacitor can tune the system to 2.8MHz that ensures all the coils are in resonance. The coil type in this analysis is planar spiral coil, and the

[a] Corresponding author: pyqxm@163.com

number of turns of both sending coil and receiving coil is 7. The material of the coil is set to copper. Then the 3D model of the two coils is built in ANSYS and Fig 2 shows the model. The skin effect and proximity effect can be ignored in calculating the inductance for the following three reasons: firstly, the frequency of the system is far lower than the very high frequency (VHF); secondly, the wire diameter is not very thick; thirdly, the wire section is a uniform circle[8].

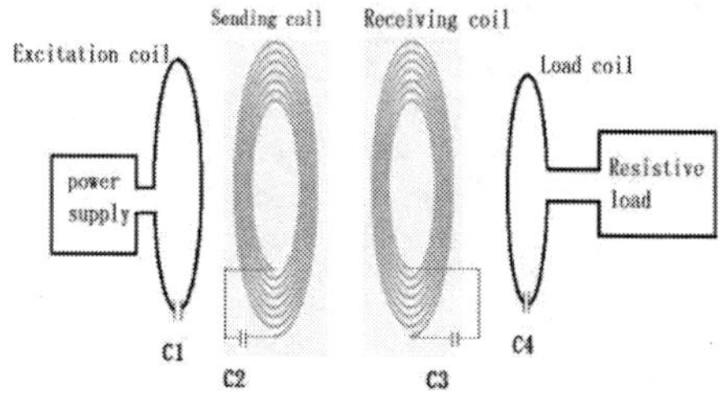

Fig 1. The MCR-WPT system with non-intermediate resonant coil.

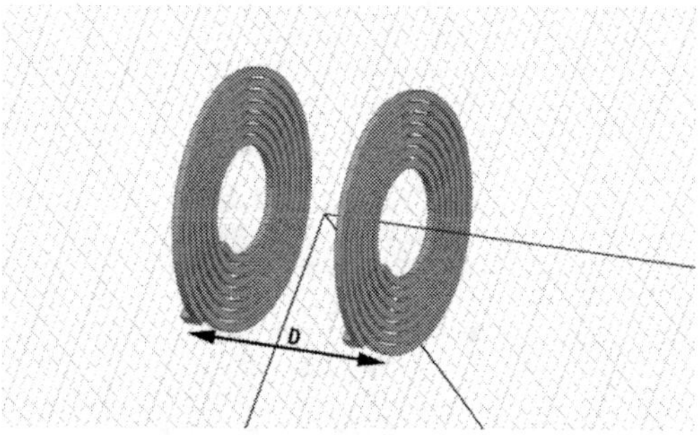

Fig 2. 3D model of sending coil and receiving coil.

2.2 Calculating the Inductance and the Capacitor Value

The inductance calculation of the coil is very important in the simulation, and the empirical formula for calculating the inductance of the planar spiral coil is given by equation (1) and the substitution is in equation (2), which indicates the relationship between the inductance value and its geometric size. The unit of the coil's size is inch and the unit of L (inductance) is μH. In order to make the meaning of the formulas more clear, all the coil sizes are labelled in Fig 3, which is the cutaway view of a common four-turn coil.

$$L = \frac{N^2 \bullet A^2}{30A - 11D_I} \quad (1)$$

$$A = \frac{D_I + N(W + S)}{2} \quad (2)$$

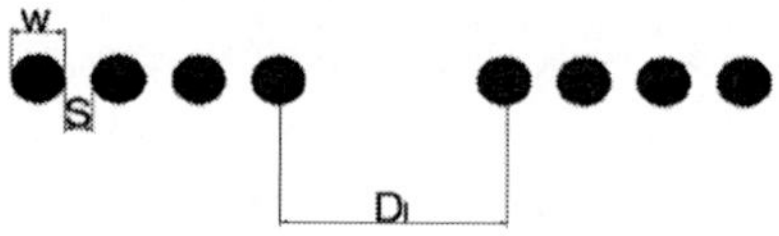

Fig 3. The cutaway view of a common four-turn coil.

In equation (1) and equation (2), W, S, D_I and N represent the wire diameter, the distance between two turns, inner diameter of the coil and the number of turns of the coil, respectively. In this paper, experimental parameters W=*11.8*, N=*7*, W=*0.6*, S=*0.31*, and the unit "inch" is for all the coil sizes. The theoretical result, as obtained from equation (1) and (2), and the simulation result, as obtained from ANSYS simulation, are shown and compared in Fig 4.

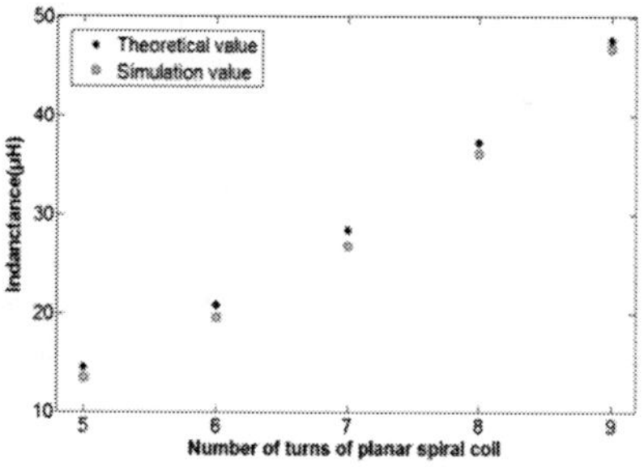

Fig 4. Results of the inductance calculated with two methods.

It can be seen from Fig.4 that the two results have only minor differences in calculating the inductance. But the method of simulation will be better than that using theoretical formula when the shape of the coil is irregular. The simulation values of the Coil 1 and Coil 2 are $27.3\mu H$ and $27.1\mu H$, respectively.

When the two coils are in resonance, the value of the compensating capacitor can be given by equation (3), where f represents the resonant frequency.

$$C = \frac{1}{4\pi^2 f^2 L} \quad (3)$$

The value of the capacitor for each coil 118.348 pF and 119.22 pF. And the characteristic impedance of each coil is set to 50 ohm for better impedance matching.

2.3 Calculating the Coupling Coefficient between Resonators

The scattering parameters are useful in analyzing the forward gain of the mid-range wireless systems.

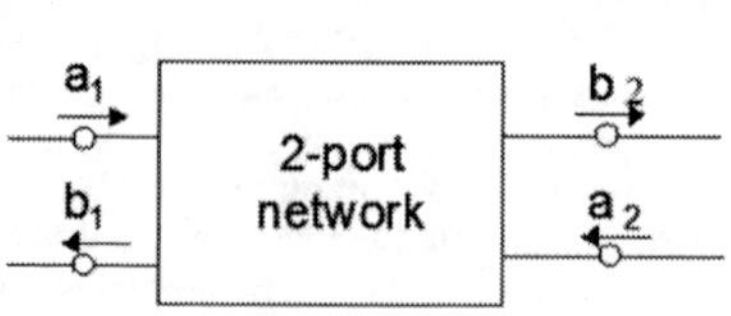

Fig 5. Sketch of a two-port network.

As shown in Fig 5, the relationship of the incident and reflected waves can be defined by equation (4),

$$\begin{bmatrix} b_1 \\ b_2 \end{bmatrix} = \begin{bmatrix} S_{11} & S_{12} \\ S_{21} & S_{22} \end{bmatrix} \begin{bmatrix} a_1 \\ a_2 \end{bmatrix} \tag{4}$$

where a_1 and a_2 are the incident power waves on Port-1 and Port-2, respectively; b_1 and b_2 are the reflected power waves on Port-1 and Port-2, respectively[9]. The forward transmission coefficient S_{21} is defined as

$$S_{21} = \frac{b_2}{b_1} = \frac{V_2^-}{V_1^+} \tag{5}$$

It can be seen from equation (5) that the S_{21} parameter can be an indicator for transmission performance of mid-range wireless power systems. ANSYS Hfss(High frequency structure simulator) can calculate the S parameters easily when the boundaries and excitations are arranged properly. Fig.6 and Fig.7 show the amplitude and phase curves of S_{21}, respectively. And the solid lines with different markers stand for the S_{21} parameters in different gap length between the two coils.

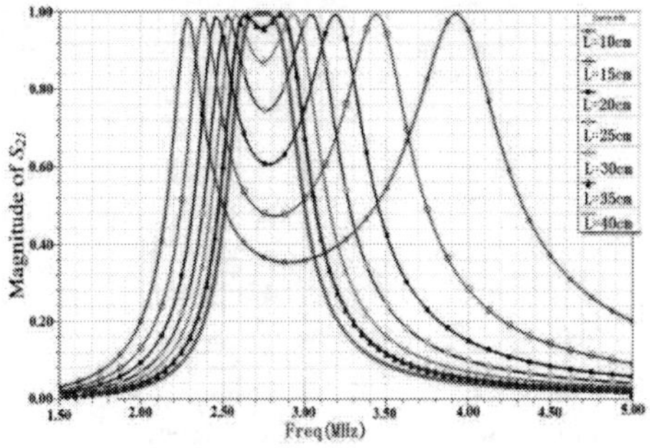

Fig 6. The amplitude curve of S_{21}.

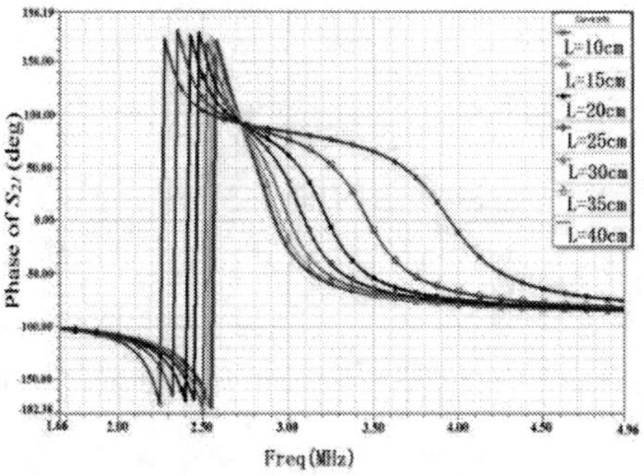

Fig 7. The phase curve of S_{21}.

It can be seen from Fig.6 that there are two resonant peaks when the gap between the two coils varying from 10 centimeters to 35 centimeters with the step 5 centimeters. This phenomenon is so called "frequency splitting" what has been explained in great detail in paper[10], and two coils are said to be over-coupled. And when the gap is 40 centimeters, the two resonant peaks combine with each other and the system is said to be critical coupled.

Paper[11] has proposed a formula to calculate the coupling coefficient between resonators based on S parameters when the two resonators are over coupled, and it is given by equation (6).

$$k = \sqrt{\frac{1}{2}\left[\left(\frac{f_2 - f_1}{f_0}\right)^2 + \left(\frac{f'_2 - f'_1}{f'_0}\right)^2\right]} \tag{6}$$

where f_1, f_2 *and* f_0 represent the frequency of the left peak, the right peak and the minimum point in Fig.6, respectively and f'_1, f'_2, *and* f'_0 represent the frequency of the ±180° jump point, the +90° point and the zero-crossing point in Fig.7, respectively. But with increasing distance, the two resonators will be critical coupled and even under coupled, thus the equation (6) will fail to finish the calculation. In this condition, a fitting expression like equation (7) is proposed. And a, b, c are deduced by the coupling coefficient ($D=0.35m,\ 0.3m,\ 0.25m,\ 0.2m$) obtained from equation (6) using the curve fitting tools of MATLAB. After the calculation, $a = 0.037, b = -1.341,\ and\ c = -0.035$ are obtained.

$$f(x) = ax^b + c \tag{7}$$

Another method of calculating coupling coefficient of the two resonators is to use Neumann formula (8) and equation (9).

$$M = \frac{\mu_0}{4\pi} \oint_{C_1} \oint_{C_2} \frac{dl_1 \bullet dl_2}{D} \tag{8}$$

$$k = \frac{M}{\sqrt{L_1 L_2}} \tag{9}$$

In equation (8), M and D represent the mutual inductance and the gap between the two coils, respectively. L1 and L2 represent the self-inductance of Coil 1 and Coil 2. Using equation (8) and equation (9) we can derive equation (10),

$$k = \frac{\mu_0 N_1 N_2}{4\pi \sqrt{L_1 L_2}} \int_0^{2\pi} \int_0^{2\pi} \frac{r^2 \cos(\theta_1 - \theta_2)}{\sqrt{2r^2 - 2r^2 \cos(\theta_1 - \theta_2) + D^2}} d\theta_1 d\theta_2 \tag{10}$$

where r is the average radius of the coils, and $r=0.242$m in this paper. Fig 8 shows and compares the results of coupling coefficient k versus gap length between the two coils using different methods.

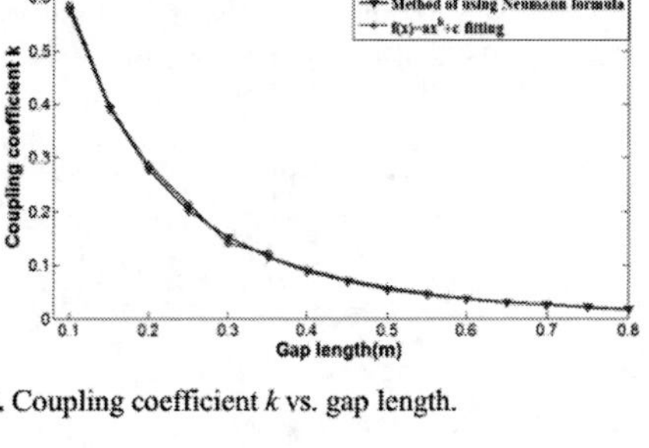

Fig 8. Coupling coefficient k vs. gap length.

The above simulation and calculation show that the results obtained from equation(6) is similar to that obtained from Neumann formula when the two coils are over coupled and the results obtained from the fitting expression(7) is similar to the one obtained from Neumann formula when the two coils are critical coupled or under coupled ($D \in [0.35m, 0.8m]$). Paper[12] uses the Method of Moments (MoM) to calculate coupling coefficient k whose results are the same as the one obtained from Neumann formula. All the above methods are valid, and our method is maneuverable and suitable for engineering applications when the system is critical coupled or under coupled while using Neumann formula or MoM needs lots of mathematical techniques.

3 Eficiency Analysis of WPT

According to the microwave theory, the relationship between S parameters and efficiency of transmission is given by equation (11).

$$\eta_{21} = |S_{21}|^2 \times 100[\%] \tag{11}$$

It can be seen from Fig.6 and equation (11) that efficiency of transmission will decrease rapidly when the gap length between the two coils increases. In order to solve this problem, wireless power transfer system with single intermediate coil resonator was proposed in paper[13].

When the gap length is 100 centimeters, the efficiency of transmission is lower than 8% and the coupling coefficient k_{12} is so tiny that can be ignored. And in this condition, one intermediate coil resonator (labelled Coil 3) is placed in the middle of two coils.

Efficiency of transmission of the system with three coils can be deduced from a cascaded 2-port network model, what is showed in Fig 9.

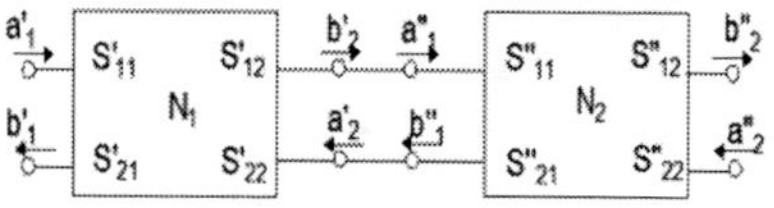

Fig 9. Sketch of a cascaded 2-port network.

T parameters of the cascaded 2-port network model can be given by equation (12),

$$T = T \cdot T' \tag{12}$$

where T is the parameters of the first 2-port network N_1 and T' is the parameters of the second one N_2. And the relationship of S parameters and T parameters can be given by equation (13), and equation (14).

$$\begin{bmatrix} S_{11} & S_{12} \\ S_{21} & S_{22} \end{bmatrix} = \frac{1}{T_{22}} \begin{bmatrix} T_{11} & T_{11}T_{22} - T_{12}T_{21} \\ 1 & -T_{21} \end{bmatrix} \tag{13}$$

$$\begin{bmatrix} T_{11} & T_{12} \\ T_{21} & T_{22} \end{bmatrix} = \frac{1}{S_{21}} \begin{bmatrix} S_{12}S_{21} - S_{11}S_{22} & S_{11} \\ -S_{21} & 1 \end{bmatrix} \tag{14}$$

3.1 Transverse Offset's Impact on Efficiency of Transmission

The position of intermediate coil resonator's transverse offset is shown in Fig 10. And all the three coils have the same number of turns 7.

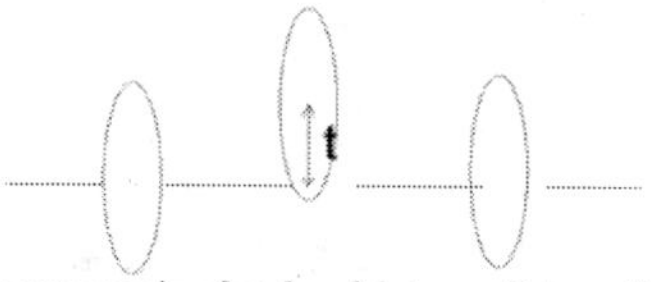

Fig 10. Diagrammatic sketch of intermediate coil's transverse offset

After the simulations and calculations, the relationship of maximum efficiency and intermediate coil resonator transverse offset can be seen in Fig 11.

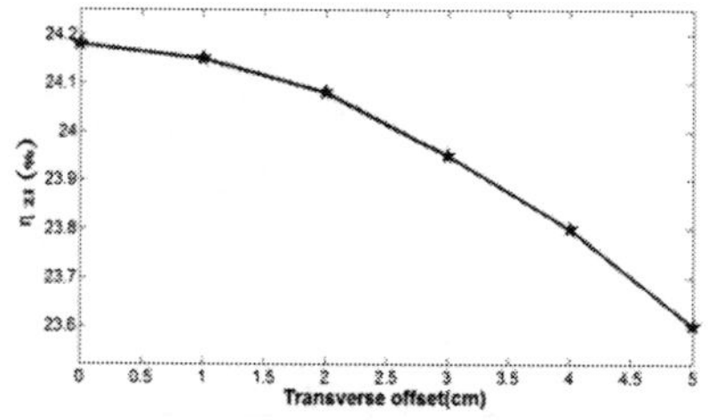

Fig 11. Maximum efficiency of transmission vs. transverse offset.

Fig 11 shows that small transverse offset of Coil 3 has minimal difference on the efficiency of transmission η_{21}. When designing the WPT system with single intermediate coil resonator, it's flexible for us to place the intermediate coil, instead of considering the small transverse offset of the intermediate coil.

3.2 Axial Offset's Impact on Efficiency of Transmission

The position of intermediate coil resonator's axial offset is shown in Fig 12. And all the three coils have the same number of turns 7.

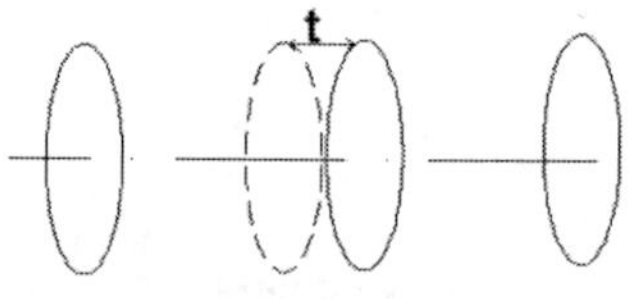

Fig 12. Diagrammatic sketch of intermediate coil's axial offset.

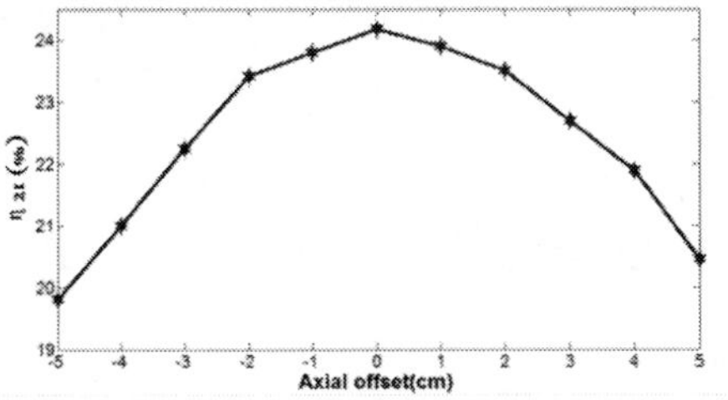

Fig 13. Maximum efficiency of transmission vs. axial offset.

Fig 13 shows the relationship of maximum efficiency and axial offset of *Coil* 3. Axial offset has significant effect on the efficiency of transmission. When designing the MRC-WPT system, we should place the intermediate coil resonator in the middle of the two coils to the greatest extent.

3.3 Efficiency of Intermediate Coil in Different Turns

In this simulation, the coil type of *Coil* 3 is the planar spiral coil and is also placed in the middle of *Coil* 1 and *Coil* 2. The number of turns varies from 5 to 9 whose inductance value has been calculated in section II.

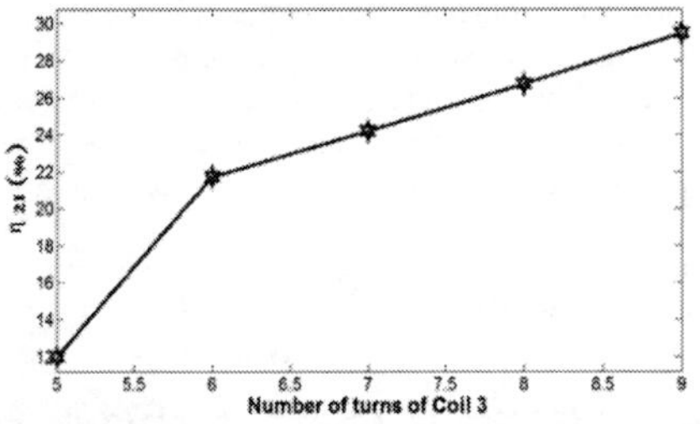

Fig 14. Maximum efficiency of transmission vs. number of turns of *Coil* 3.

With the increase in the number of turns, the inductance and radius of Coil 3 rise. As shown in Fig.14, higher efficiency of transmission can be obtained from increasing the number of turns of the intermediate coil. However, in the actual situation, radius of intermediate coil cannot be very large. And when the radius reach to some of extent, the resistance will be very large which can suppress the transmission efficiency [14].

4 CONCLUSION

The self-inductance of coils can be calculated accurately in ANSYS Electromagnetics. And the method of calculating coupling coefficient using scattering parameters is developed, and thus a fitting expression like $f(x)=ax^b+c$ is proposed to calculate the coupling coefficient. When the two coils are critical coupled or under coupled, this method is easy and useful in engineering application.

Based on the above work, magnetically-coupled resonant wireless power transfer system with single intermediate coil resonator is simulated and researched using scattering parameters. A cascaded 2-port network model is used to explain the above system. Simulations and calculations show that transmission efficiency of the system is relative to intermediate coil's axial offset while intermediate coil's small transverse offset has minimal difference on the efficiency of transmission. When designing the system, we can call for the powerful tool--ANSYS Electromagnetics to determine the proper number of turns and radius of intermediate coil.

ACKNOWLEDGMENT

The authors would like to thank Prof. Chen Li-an for useful communications and discussions. And the authors gratefully acknowledge the help of School of Electrical Engineering and Automation, Xiamen University of Technology for powerful computers. This paper is supported by Project of Xiamen Municipal Bureau of Science (Grant No. 3502Z20153016), Graduates' Science and Technology Innovation Project of Xiamen University of Technology (Grant No. YKJCX2015001), the High-level Personal Project of Xiamen University of Technology (Grant No. KJ14005R).

References

1. J.P. Barrett, *Electricity at the columbian exposition* (Madison: R R Donnelley, 1894)
2. Y. Zhou, X.J. Yu, Trans. of CES, **30**,103(2015)
3. N. Shinohara, S. Kawasaki, IEEE RWS, 13(2009)
4. A. Kurs, A. Karalis, R. Moffatt, J. D. Joannopoulos, P. Fisher, M. Soljačić, Sci., **317**, 83 (2007)
5. A. Karalis, J.D. Joannopoulos, M Soljačić, Ann. of Phys., **323**, 34 (2008)
6. B. LUO, M.T. SHENG, S.C. WU, M. ZHANG, *Proc. CSEE*, **33**, 171(2013)
7. Z.Q. LI, S.D. HUANG, M.S. YANG, X.F. YUAN, Trans. CES, **28**, 35(2013)
8. K. Tarloff, Caiyitelin. Inductance calculation manual (Beijing, China: China Machine Press, 1992)
9. S.Y.R. Hui, W. Zhong, C.K. Lee. IEEE Trans. on Power Electron., **29**, 4500(2014)
10. W.Q. Niu, W. Gu, J. X. Chu, Electron. Lett. , **48**, 723(2012)
11. S.E. ZHANG, C. MA, W.J. CUI. J. of NCEPU, **39**, 77(2012)
12. T. Imura, Y. Hori. IEEE Trans. on Ind. Electron., **58**, 4746(2011)
13. R.E. Hamam, A. Karalis, J.D. Joannopoulos. Ann. of Phys., **324**, 1783(2009)
14. Y. Zhai, Y. Sun, X. Dai, Proc. CSEE, **32**, 155(2012)

The Application Trends of Rapid Prototyping Manufacturing

Xiao Lin QIU, Yi Ren ZHOU[a], Zhi Qiang GUO, Jian Hui ZHANG

Nanchang Institute of Technology, No.901 Heroe Ave, Qingshanhu, Nanchang, Jiangxi, 330044.China

Abstract. The application of prototyping technology is progressing with each passing day, we studied the characteristics of laser stero lithography (LSL) selective laser sintering (SLS), three-dimensional printing (DP), fused deposition modeling (FDM), computer numerical control (CNC) and other rapid prototyping technologies. After discussed these five rapid prototyping technology materials, we presented the hotspot and direction of rapid prototyping technology and look forward to the development of its technique, the expansion of its field and the progress of its academic ideology.

Keywords. rapid prototyping, Three-dimensional model, laser, Crafts

1 Introduction

Rapid prototyping technology is the generic terms of group technologywhich produces sample pieces and spare parts in accordance with CAD model, developed in recent years. It is the important part of advanced manufacturing technologya and integrates CAD techniques, CNC technology,laser technology, materials technology and other modern scientific technological achievements. Different with traditional manufacturing methods, rapid prototyping parts based on CAD geometry model, adopt laser beam or other methods to form a solid material accumulation part by the software hierarchical and CNC forming discrete systems.See Figure 1. Due to its complex three-dimensional manufacturing transformed into a series of superimposed two-dimensional manufacturing, which can generate any complex parts without molds and tools,and greatly enhance theproduction efficiency and manufacturing flexibility.

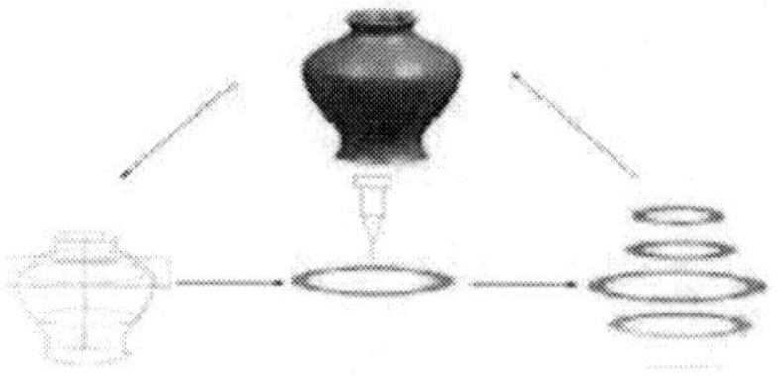

Figure 1 Solid Part generate schematic

(1)The advantages of rapid prototyping technology

Rapid prototyping as a design concept makes an important means of visualization, computer aided design mock-ups of parts can be machined in a very short periodof time, which can quickly design and processing capacity to assess the results. With rapid prototyping and direct digital manufacturing technology that enable cost reduction for the NC 1 / 3-1 / 5, cycles 1 / 5-1 / 10.

(2)Due to the rapid prototyping technology is a complex three-dimensional shape into a two-dimensional cross-section to solve, so it can manufacture high-precision parts arbitrarily complex shaped body, and without any tooling.

(3)Rapid prototyping as an important manufacturing technology, the use of appropriate materials, prototype can be used in subsequent production operations to obtain the final produc in subsequent production operations.

(4)Rapid prototyping can be used in mold manufacturing operation and obtain quickly and economically.

(5)Product manufacturing process is nearly independent of the complexity of the part it can achieve the free form fabrication, which substantially exceeds traditional manuf acturing methods.

2 The basic principle of rapid prototyping

Rapid prototyping process which based on the principle of accumulated material actually is manufacturing parts in a discrete layer by layer.To visualize this operation, it could image that the structure of a loaf of bread is a slice of bread fell on top of another piece of bread .There are many rapid prototyping process methods, but all methods are manufacturing layer by layer, the different is the manufacturing methods and materials only.

(1)Geometric Modeling

Get a description of the part CAD files in three-dimensional CAD design software (such as ProE \ UG \ SolidWorks \ SolidEdge, etc.), as three-dimensional parts 2 (a) in Fig. The general rapid prototyping supported file output format 5TL models that approximating the real surface, the so-called type of surface (Tessellation) treated with flat triangular facets approximation model surface. The advantage of this process is greatly simplified the data format of GAD model, thereby facilitating the subsequent slicing. Because of its relatively simple processing on the data, and regardless of the CAD system, it quickly developed into a standard of rapid prototyping manufacturing data between CAD

[a] Corresponding author: renyiz@126.com

systems and rapid prototyping machine. Each triangle patch represented with four data items is three vertex coordinates and normal vectors, and the entire CAD model is a collection of such a set of complexors.

When the three-dimensional CAD design software deal with the CAD model surface treatment type, general software system has output accuracy control parameters, by controlling this parameter can be reduced surface approximation error. Such as Pro / E software is selected by chord height value (eh-chord height) as a parameter precision approach, as shown as a sphere, given two value conversions. For a model, give range in software. Under normal circumstances this range can meet project requirements. However, if the value selected is too small, it should abnnegate processing time, storage space, moderately complex parts to be the number of megabytes or even tens of megabytes of storage space. And this data conversion process generates an error can not be avoided, such as the vertices of a triangle in the middle of another side of the triangle, triangle, etc. It will troubled the subsequent data processing, need further examination patch.

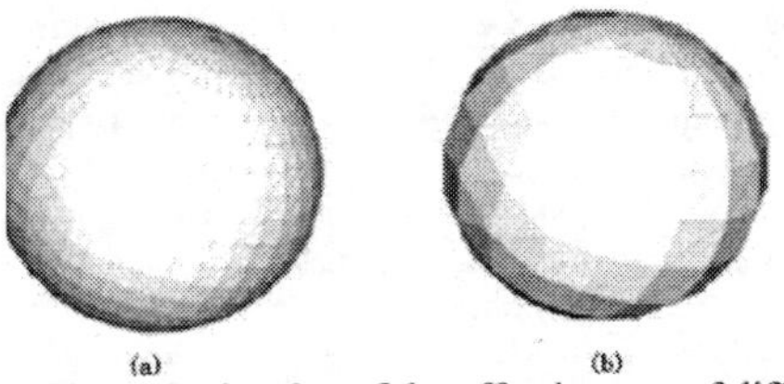

Figure 2 c h value of the effectiveness of different (a)c h=0.05 (b) c h=0.2

(2)Discrete processing three-dimensional model

Adopt specified blocking routine to layer a dedicated three-dimensional solid model (typically 5TL model), slicing is after selected production (bulk) direction, should discrete CAD model dimension get every piece of thin cross section contour and entity information. By making a cluster of parallel planes intersects the CAD model, the obtained thin line is the cross sectional contour information, and entity information through somecriterion to obtain. The distance between the parallel planes is layered thickness, i.e. accumulation of the layer thickness during molding. In this process, due to stratification, destroy the continuity of the surface of the slice direction CAD model inevitably lose some of the information model, resulting in part size and shape of errors .Slice thickness layer directly affects the surface roughness of the molding surface of the parts and the entire part of each layer obtained after slicing information is contour information of the upper and lower layers and physical information, and since the contour information is a plane CAD STL file (type of surface after the CAD model) model obtained intersection, so the outline is a series of sequential intersection after intersection even into the fold lin segments, so after stratification model profile has bee obtained by approximation, and contour information has been lost between the layers, thickness ofa large, multi lost information, resulting in an error of the molding surface in the molding process.

Currently, main rapid prototyping process for its classific ation as shown in 3.

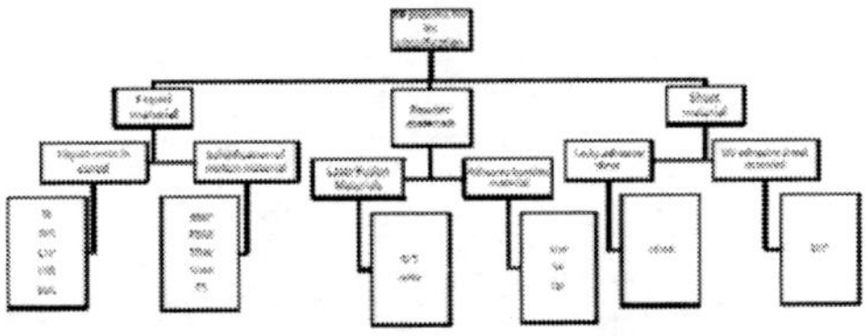

Figure 3 Current major rapid prototyping process for its classification

Despite the general steps of various rapid prototyping technologies are the same, but the method for producing aarticle different processes is different, here are several main types of RP processes.

(1) Stereo lithography

Laser stero lithography is a rapid prototyping manufacturing technique most widely used at present.it is actually earlier than fused deposition modeling. It used in the photocurable liquid photosensitive resin is cured (hardened) to a particular shape of the principle. Used thin photosensitive resin as raw materials, according to a predetermined part of each section of the outline layered ultraviolet laser under computer control for track-by-point scanning of the liquid resin, the resin thin layer of the scan area to produce photo-polymerization to form part layer section.

Forming at the beginning of the table in its highest position (depth a), at this level is higher than the thickness of a table, part cross-sectional profile of the first layer is scanned, so that the liquid photosensitive resin cured scanning area, forming part of the first a cross-section of the cured layer. Then drop a table thickness, so that the first surface of the cured resin before putting a new layer of liquid resin is then cured repeat the scan, at the same time the new cured layer firmly bonded to the previous layer, the process has been repeat to reach the height b. This time has resulted in a fixed wall thickness of the annular cylindrical parts. At this stage it can be noted from the dropped in a vertical direction. After reaching the height b, the movement range of the light beam in the x-y plane to produce a flange shape increasing in the front part of the molded part, generally should be added here is similar to FDM support [2]. When a certain thickness of the liquid is cured, the process is repeated to produce another cylindrical shaped section from a height of b to c. However, around the liquid resin remains flowable, because it is not within the range of the UV beam. Parts thus produced by the lower and upper layers.And when that portion of the liquid resin is not used in the manufacture of other parts can be reused or molding. It may be noted light curing is the same as FDM molding method requires a weak support material, the light curing method.Such a support is used in network structure. Components removed from the end of the table,remove the support structure, you can get three-dimensional parts.

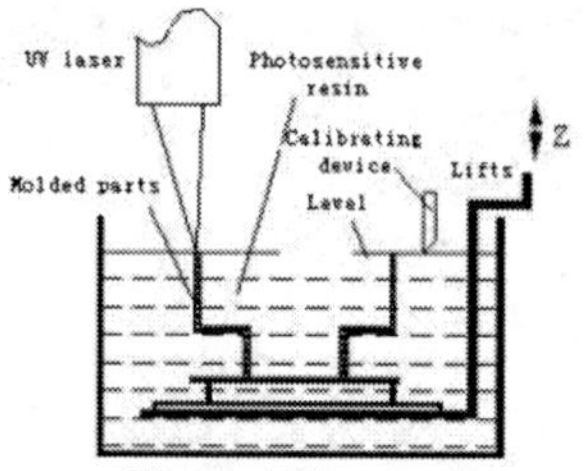

Figure 4 SL artwork

Laser stero lithography can achieve minimum tolerance depends on the extent of the laser focus is usually 0.0125mm (O.OOO5in). An inclined surface may also have a good surface quality. RF light curing method is first put into commercial application (Rapid Prototyping) technology. SL (light-curing) equipment accounts for about 70% RP total number of devices. SL (Stereo Lithography) technology advantages of high accuracy, precision control in general size 0.01 mm; market share, good surface quality, close to 100% utilization of raw materials, particularly complex shapes can be produced, extremely fine parts and equipment are high. The disadvantage is the need to support the design you can choose the type of material is limited, prone to warpage, materials more expensive. The process is suitable for prototyping and manufacturing complex of small pieces.

（2） selective laser sinering

(Selective Laser Sintering, referred to as SLS) is a non-metallic (metal or Normal) powder to selectively sinter process separate objects. The method uses CO2 laser as an energy source, currently used in the bottom of the process chamber is equipped with two cylinders:

1) In a powder supply barrel, its internal piston is gradually through a rolling mechanism to enhance the molding parts supply tube powder;

2) The other is part modeling.Its internal piston (table) is gradually reduced to local sintered partially formed.

Firstly, covered with a thin layer (l00 ~ 200 μ m) powder [3]on the table, under computer control the laser beam in accordance with a hierarchical outline part selectively sintered, so that the powder is solidified into a cross-sectional shape, after completion of a drop table layer thickness, the rolling bodies sintered powder spreading surface and then covered with a layer of powder layersintering. Unsintered powder remains loosely retained in theoriginal position, the supporting part being sintered, which secondary deformation limit, without designing a special supporting structure. This process is repeated untilthe entire three-dimensional model manufactured.

After sintering to remove all the excess powder, then grinding and drying process after obtaining the parts needed. Currently, mature technology materials for the wax and plastic powder, direct sintering process of metal powder or ceramic powder is research phase. It is possible to manufacture parts directly to engineering materials, with attractive prospects.

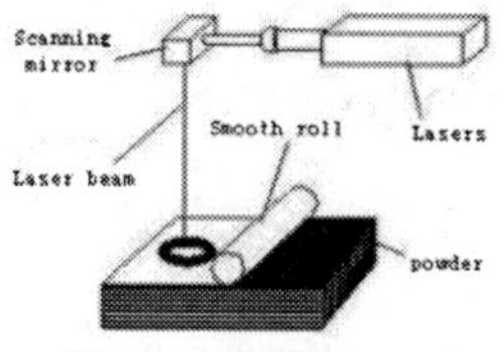

Figure 5 SLS artwork

Advantagesof SLS prototypes are good mechanical properties, high-strength; no need to design and build support; optional material kinds; utilization of raw materials close to 100%, the disadvantages are the prototype of a rough surface; prototypes porous, the need for post processing; high energy consumption; required before processing the material preheating 2h, 5 ~ lOh require cooling after molding, low productivity; molding process requires continuously purged with nitrogen, in order to nsure the safety of the sintering process, high cost; molding process produces toxic gases, certain environmental pollution. SLS process is particularly suitable for the production of functional test parts. Because it can take a variety of different components of the metal powder to sintered copper infiltration performother post-processing, and thus their prototypes may havefabricated metal parts with similar mechanical properties,it can be used directly to manufacture the metal mold. Because the process can be directly sintered wax, in contact with the investment casting process is particularly suitable for small quantities of small complex parts of the production.

（3） 3D Printing

In 1989, Emanuel M. Sachs, John S. Haggerty in Massachusetts Institute of Technology applied for a patent of three-dimensional printing technology, which has become one of the coe patents in this field in the future. Since then, the two researchers repeatedly to modify and improve the technology to get the formation of today's three-dimensional printing Rapid Prototyping (3DP) process. Figure 6 is a simplified schematic of a 3DP process.

3DP technology and selective laser sintering (SLS) have many similarities: both the powder material is selectively bond as a whole. But the biggest difference is that without 3DP the powder material is melted adhesive discharged through the nozzle but it sticking together. Its process usually is: a layer of adhesive is complete, forming cylinder (Fig.14) drops a distance (equal thickness),for the powder cylinder (Fig.18) rise for some height, introduced a number of powder and powder spreading roll onto the forming cylinder, paving and compaction. Head (Fig. 15) under computer control, the construction of a section of press molding data selectivelyspray adhesive construction levels. Dust roller (Fig.13) when the excess powder dusting powder collecting device is collected.

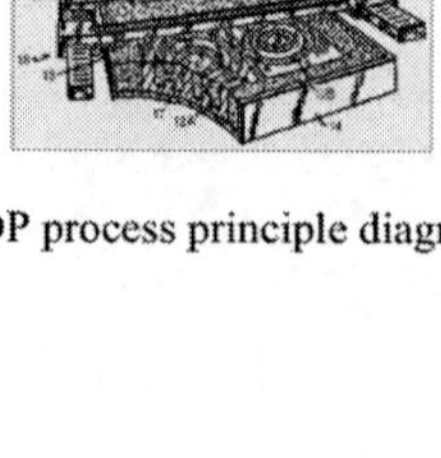

Figure 6 3DP process principle diagram

Feed powder and dusting and spray adhesives repeated, finalized a three-dimensional bonding powder to produce products.
Dimensional Spray adhesive characteristics: suitable for molding small pieces; the smooth surface of the workpiece is not enough, you need to scan the entire cross-section bonding, forming a long time; the use of multiple nozzles.
(4)Fused Deposition Modeling
As shown in Figure 7, during the process of fused deposition modeling (FDM), the gantry mechanical control head can be moved in two main direction of the table, the tablecan be moved up or down as needed. Thermoplastic or wax extruded from the heating fuse small mouth. The first layer is in accordance with a predetermined trajectory at a fixed rate on the fuse extruded foam matrixformed. Afterthe completion of the first layer, the thickness of a decline bench and began manufacturing superimposed layer. The key is to keep half FDM processflow just above the melting point of the molding material is usually higher than the melting point control in about 1 ℃.

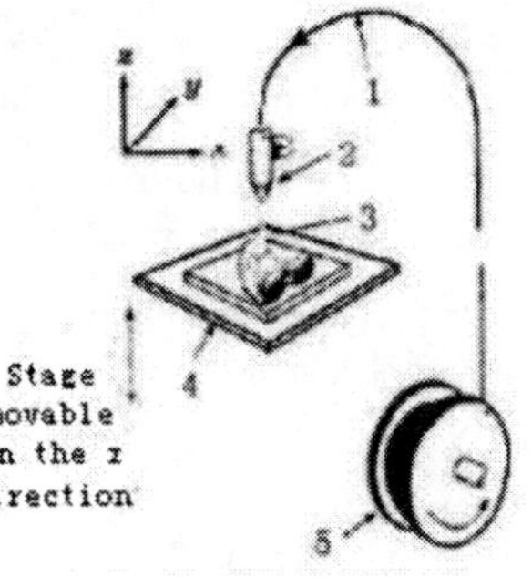

Figure 7 FDM Technology

1, Thermoplastic or wax fuse; 2, In the x-y plane can move FDM head; 3, Plastic models; 4, Not fixed base; 5, Provide a fuse
When FDM creating complex parts it must added supporttchnic. Such as the height of the figure7 (a), the next layer of fuse material will be spread without support space. The solution is making a single material support material out of the model, the support material may use low densityfuse, lower than strength material, after the completion of processing of parts it can be removed.
The diameter of the fuse depended on the thickness of themiddle of FDM machine.Usually from 0.5mm to 0.25mm(0.02in to from 0.01 in) This value represents the best of tolerance in the vertical direction can be achieved. In the x-y plane, as long as the characteristics of the fuse can be extruded, size precision can reach 0.025mm (0.001in).
The advantages of FDM is the high utilization, low-cost materials, the choice of the type of multi-material,processclean, simple, easy to operate and less impact on the environment. The disadvantage is low precision, easy to manufacture complex parts, poor surface quality,low molding efficiency, not suitable for the manufacture of large parts. The process is suitable for conceptual modeling products as well as its shape and function testing, moderately complex forming small, since methacrylic ABS material has good chemical stable, gamma sterilization may be particularly suitable for medical.

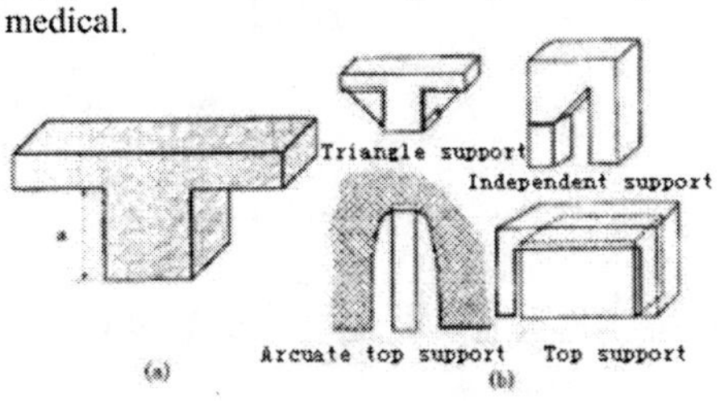

Figure 8 Rapid prototyping support structure diagram
(a)There is a need to highlight the cross-section of the support material parts;
(b)In rapid prototyping machines commonly used in the support structure
(5)Computer Numerical Control
CNC is an integrated computer-aided design (CAD), computer aided manufacturing (CAM),computer numeric acontrol (CNC) and other advanced technology, the three-dimensional data model on a computer constituted by a single piece of material excavated from cutting. After clamping the workpiece by a digital control system can control the machine according to different processes, and automatically selects the replacement tool automatically change the spindle speed, feed rate and the trajectory of the tool relative to the workpiece and other auxiliary functions, in order to complete the work of several surface working on a multi-step, the whole process is automatically controlled by the program, independent of the operator of human factorsl.
CNC machining method different from the above four RPprocess method, it is "subtractive" digital processing [6], is a modern digital manufacturing plant widely used way of machining, rapid prototyping service providers tend to CNC machining as a rapid prototyping model supplementary processing, CNC machining high speed, low cost, high precision, large processing size, features a wide range of material selection, to ensure the consistency of the material and surface quality and mechanical properties of production parts and materials may be hand model It reached a high level.

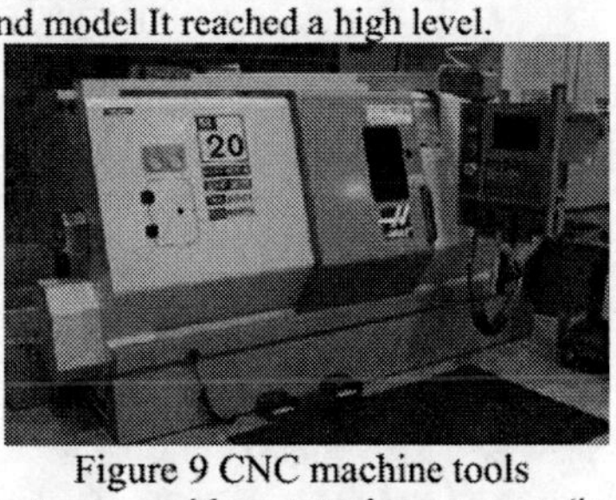

Figure 9 CNC machine tools
(6)Mainstream rapid prototyping process (including CNC) Comparison
SL, SLS, 3DP, FDM, CNC rapid prototyping several respective characteristics are shown in Table 1.

Table 1 Comparison of mainstream rapid prototyping process

	SL	SLS	3DP	FDM	CNC
Forming speed	Faster	Slow	Slow	Slow	Fast
Forming accuracy	Higher	Lower	Lower	Lower	Higher
manufacturing cost	Higher	Lower	low	Lower	Lower
Complexity	medium	complex	medium	medium	Complex (and associated multi-axis)
Parts size	Small pieces	Small pieces	Small pieces	Small pieces	The large
Commonly used materials	Thermoset ting photosensi tive resin	Paraffin wax, plastic, metal, ceramic powder, etc.	ABS plastic, which is also a mixture of aluminum and glass	Paraffin, nylon, ABS, and other low melting point metal	ABS, POM (race steel), acrylic, ylon, PC, PP, electric wood. Metal materials: aluminum, magnesium alloy, zinc alloy, copper, etc.

4 The direction of rapid prototyping technology development

Rapid prototyping manufacturing is a method based on the principle of accumulate material, it can be made arbitrarily complex three-dimensional structures. Currently the biggest problem of rapid prototyping development is the technical realization restricted by physical and chemical properties of the material. For example, in the molding material, the key is organic polymer materials; the directly formed of metal material is thehotspot of nearly 10 years research,the difficulty liesin how to improve the accuracy. On the molding material,thedirection of development is the study of ceramic materials and rapid prototyping composites. In the development of manufacturing equipment, 3D printer is hot at home andabroad in recent years, which as a peripheral application of computer. Rapid prototyping technology has a wide application prospect and great commercial value in industrial modeling, product innovation, arts and crafts and so on.

Rapid prototyping technology research focus and direction includes three aspects:

(1) The development of rapid prototyping technology itself. Such as 3D printing make the rapid prototyping into the information market; direct metal forming technology enables the structure and function parts can beproduced directly. Further development is rapid prototyping rapid prototyping technology of ceramic parts and composite materials.

(2) The expansion of rapid prototyping application fields.Such as rapid prototyping applications provide a quick technical support to the automotive manufacture field. Rapid prototyping applications in bioprosthetic tissue engineering, artificial prosthesis manufacture custom providea technical support to the prosthetic manufacture and dimensional scaffold. It is to further the creative design, aerospace manufacturers and the functional structure of devices in the field of development.

(3)The academic thought of rapid prototyping development.Rapid prototyping developed from the shape manufacture in the past to the material organizationand structure design. Try to achieve the organization from micro to the macro-controlled. For example, in the manufacture of composite materials, the ability to accomplish the design of composite material and the manufacture of structure shape. It can achieve the integrati of "design - Material - made" in broader terms.

References

[1] Hou Shu guang, ZHang Dong, HUANG Xiaoming, ZHAO Yongli. J Investigation of Micro-mechanical Response of Asphalt Mixtures by a Three-dimensional Discrete Element Model. Journal of Wuhan University of Technology(Materials Science Edition).2015 02

[2] Haiyan Liu,Xiaomei Bai,Dan Wei,Gengliang Yang. High-performance liquid chromatography separation of small molecules on a porous poly (trimethylol propane triacrylate-co-N-isopropylacrylamide-co-ethylene dimethacrylate) monolithic column [J]. Journal of Chromatography A2014

[3] Xu Zhifeng, Liang Pei, Yang Wei, Li Sisi, Cai Changchun.Effects of laser energy density on forming accuracy and tensile strength of selective laser sintering resin coated sands. China Foundry. 2014 03

[4] Yang Ju, Heping Xie, Zemin Zheng, Jinbo Lu, Lingtao Mao, Feng Gao, Ruidong Peng. Visualization of the complex structure and stress field inside rock by means of 3D printing technology. Chinese Science Bulletin. 2014 36

[5] S.Hwang, E.I.Reyes, N.S.Kim, K.S.Moon, R.C.Rumpf. Parameter study of Fused Deposition Modeling process on thermo-mechanical properties of the final 3D structures made by metal/polymer composite filaments. ICBAEE 2014. BEIJING, CHINA. 2014-05-22.

[6] Wang Linyun, Nuang Honghui, Rae W.West,Wang Dzhong. Intelligent manufacturing system of impeller for computer numerical control (CNC) programming based on KBE. Journal of Central South University. 2014 12

MATEC Web of Conferences **63**, 02007 (2016) DOI: 10.1051/matecconf/20166302007
MMME 2016

The design of Superpave compared with Marshall design method

Rui Nan JIANG[1], Bao Lin ZHU[1], Liang CHANG[2]
[1].*China academy of transportation sciences Beijing 100021*
[2].*Jiaoke Transport Consultants LTD 100019)*

Abstract: In order to compare and analyze Superpave and Marshall design method for asphalt mixture. Through the analysis about gradation selection, predicition asphalt content, the determination method of theoretical maximum density, the compaction methods of specimens, the determination method of optimum asphalt content and the test method of mixture performance of these two design methods. The following conclusions have been come to: Superpave design method is better than Marshall design method for asphalt mixture.
Keyword: Superpave; Marshall design method; gradatiom; compaction methods; optimum asphalt

1 FOREWORD

At present, the common asphalt mixture design method is Marshall (Marshell) method and the Superpave method. Marshall method first by Bruce Marshall (Brue Marshall) is put forward, in 1948 the U.S. army corps of engineers of this method is improved, and add some test performance test, standard development into the mix proportion design of asphalt mixture; Superpave method is 1987 ~ 1992, the United States SHRP plan of asphalt project research results, this system puts forward a new set of design based on performance related to traffic and climate on the basis of material selection and mixture design method.

The Marshall developed on the basis of experience and have some certain limitation at the field of method of sample shaping and test indexes. And the Superpave is not popularized in china because of the stiff price. There have some different between the grading choose, the method of sample shaping and OAC design. So most technician care for that whether have some point and relationship between two methods.

2 COMPARE WITH THE METHODS

(1) the grading choose

Marshall method is fit of the design of asphalt of continuity dense grading. In the past the grading choose usually use the mid-value that recommend by norm. The present norm was required to confirm the "engineering design gradation scope" by engineering practice, and according to the highway classification, climate and traffic condition to choose the coarse type (C) or thin type (F). Usually the gradation curve must continuousness, smooth and has large compactness. It is need to analyze whether form the skeleton structure .

Superpave gradation selection is through (area) to implement the control points and limits, locus of control is the requirement that aggregate gradation shall not exceed that of the specified range, respectively in nominal maximum size screen, medium sieve (2.36 mm) and minimum screen (0.75 mm). Its purpose is that to limit the amount of sand and provide enough VMA. Restricted area is located in the biggest density level wiring between the medium sieve and 0.3 mm sieve. There have same question of compaction if the asphalt beyond the restricted and lack of ability to resist permanent deformation during using [3] . And then, it is cause VMA too small to affect the asphalt usage when the gradation beyond the restricted by Superpave. Therefore, Superpave should make design gradation in between control points and avoid the limits when choose the gradating, and this grading are usually formed the skeleton structure.

(2) estimate the content of cementing agent

Marshall use this formula to estimate the asphalt usage:

$$p_a = \frac{p_{a1} \times \gamma_{sb1}}{\gamma_{sb}} \qquad (1)$$

among: p_a is the estimate asphalt-aggregate ratio,%; p_{a1} is normal proportion of asphalt in the same constructed project,%; γ_{sb} is the bulk volume relative density of aggregate; γ_{sbl} is the bulk volume relative density of aggregate of same constructed project.

This way can quickly to count the resolute. But there have some reasons to cannot confirm the estimate proportion of asphalt was very access the really one. The reasons are that there have much error because the data that from the constructed project is not fully reliable[4].

The formula about estimate the asphalt usage in the Superpave is :

$$p_{bi} = \frac{G_b \times (V_{be} + V_{ba})}{[G_b \times (V_{be} + V_{ba})] + W_s} \quad (2)$$

among： p_{bi} is the percent of cementing agent; W is the quality of aggregate, g; V_{be} is the volume of the cementing agent; V_{ba} is the volume of cementing agent of absorbent; G_b is density ratio of cementing agent;

$$V_{be} = 0.081 - 0.02931 \times [\ln(S_n)] \quad (3)$$

among: S_n is nominal maximum mesh size of grading of aggregate（inch）。

$$V_{ba} = \frac{p_s \times (1 - V_a)}{\left(\frac{p_b}{G_b} + \frac{p_s}{G_{se}}\right)} \times \left(\frac{1}{G_{sb}} - \frac{1}{G_{se}}\right) \quad (4)$$

among: p_s is the content of aggregate; p_b is the content of cementing agent; V_a is the volume of void; G_{sb} is density ratio of aggregate; G_{se} is the valid density of ratio of aggregate.

$$G_{se} = G_{sb} + C(G_{sa} - G_{sb}) \quad (5)$$

among: G_{sb} is the bulk volume density ratio of aggregate; G_{sa} is the surface density of aggregate ; C is the coefficient of hydroscopicity of aggregate.

Through the formula it is can be found that the superpave about the method of asphalt usage is conclude the volume based on the mixture. The Pb need a initial value is from the experience. And the coefficient C choose by hydroscopicity of aggregate. The more small the value, the more big the hydroscopicity, and common the rage is 0.5~0.8.

There have to setting parameter is Pb and C if measure the density of aggregate and cementing agent because the void ratio Va is 4% in the Superpave. The void ratio of Superpave is 4%, but marshall is a rage. Marshall will confirm a aim void ratio when count the asphalt-aggregate ratio, and the aim void ratio is confirm by experience.

(3) confirm the max theory density ratio

The Marshall using the vacuum test method for determining the biggest theory relative density for the modification of ordinary asphalt mixture, and for the modified asphalt and SMA mixture calculated using the following formula:

$$\gamma_{ti} = \frac{100 + p_{ai}}{\frac{100}{\gamma_{se}} + \frac{p_{ai}}{\gamma_b}} \text{或} \gamma_{ti} = \frac{100}{\frac{p_{si}}{\gamma_{se}} + \frac{p_{bi}}{\gamma_b}} \quad (6)$$

among： γ_{ti} is the max theory density of asphalt mixture relative to count the Pbi, dimensionless; p_{ai} is the asphalt-aggregate ratio of the asphalt mixture,%; p_{bi} is the asphalt usage, Pbi=Pai/(1+Pai)，%； p_{si} is the content of ore charge, Psi=100-Pbi,%； γ_{se} is the valid density ratio of ore charge, dimensionless; γ_b is the density ratio of asphalt, dimensionless.

among，the formula of valid density ratio of aggregate is：

$$\gamma_{se} = C \times \gamma_{sa} + (1 - C) \times \gamma_{sb} \quad (7)$$

This formula is same with the Superpave about valid density ratio of ore charge, and the different is coefficient C. the C of Superpave is the rage of 0.5 ~ 0.8, and conclude the hydroscopicity of aggregate, the Marshall is confirm by this formula.

$$C = 0.033w_x^2 - 0.2936w_x + 0.9339 \quad (8)$$

among : w_x is the hydroscopicity of mineral aggregate,% ;

$$w_x = \left(\frac{1}{\gamma_{sb}} - \frac{1}{\gamma_{sa}}\right) \times 100$$

compare with the Superpave, in the current specification, the C that asphalt absorptive of ore charge was regard as a constant. It is rational reflect the ability of adsorb asphalt of mineral aggregate and the adsorb condition of the open mouth of aggregate. The resolute is too access the really theory max density ratio.

Superpave completely adopt the vacuum field measurement method（AASHTO T209）. Form now on, this method was accepted by even more. But in the course of vacuumize the asphalt film may be destroyed and lead to the water into the void of aggregate. For the polymer modified asphalt, because of the biggest viscosity of asphalt, it is difficult to disperse the mixture and especially the fine aggregate was require disperse under the 6.4mm. It will make the measured value have much error. Therefore, it will use Marshall to count the mixture max theory density ratio of modified asphalt.

(4) the way of moding

Marshall compaction instrument is pure compaction which the hammer weight is usually constant and only increasing number of compaction work to increasing the compation. Although the price is cheap and simple structure, it can't accurately discriminate different traffic volume of asphalt mixture with the requirements of technical indicators. It is not suitable for large size asphalt mixture. The main reason is this way cannot simulate the situation that the road roller press the road. One thing is the compaction instrument of some particles will break easily thereby to change the grading of mixture. The other ting is this way cannot stimulate the press that road roller and load of car to the road.

Superpave shear compaction methods using rotational rubbing molding, it is sat the parameter of compaction through the traffic parameter. It is best stimulate the forming process of asphalt mixture, and the design process also consider the shote period aged of asphalt mixture. All of them make the Superpave to be more suitable, so Superpave design method mainly exist the following advantages:

①Because of Marshall stability have nothing to do with the correlate and asphalt mixture of road performance, it is have useless to improve the compactness of asphalt for increase the time to compaction. Rotating compaction method is not only close to actual pavement compaction process, but also the compaction effect is more close than the Marshall compaction method. Rotating compaction is super one way that have value to popularize.

②When it design the asphalt mixture with Superpave method, it is considering the aging of asphalt mixture, the design results are more in conformity with the actual.

③Superpave method have the clear purpose because the rotation of the compaction number apply with traffic.

④ Superpave method provision the compatibility with the different staged compaction by initial N, design N and max N. It is can adequate disclosure the character of compaction.

According to research, the speed of bulk volume density increased with rotating compaction times was slowly. And regardless of the times of rotating compaction, the bulk volume density by rotating compaction is higher than by Marshall at the same asphalt aggregate ratio. That means rotary compaction compaction can greatly improved than Marshall work[4].

(5) confirm the best asphalt-aggregate ratio

Marshall method using the compaction molding specimens, compaction number for 75 times, in order to forecast or experience of asphalt content is of value, molding specimens, 0.5% interval change test specimen density, calculate void fraction. It is calculate VMA and VFA, at the same time to test the Marshall stability and flow values of the specimens. And then draw the diagram of asphalt dosage and the indicators. To confirm the best asphalt-aggregate ratio by the range of the amount of asphalt which meet the various indexes. In the process of the optimum asphalt content, it is can't fully control the voidage at the range of reasonable after the pavement compaction because the voidage is not confirm. In addition, Marshall stability and not clear the relationship between road performance[6], therefore. In this sense, when it design the asphalt mixture , it is difficult to ensure optimum asphalt content accurately determined.

Superpave using rotation compaction to molding the mixture, and the rotation time is confirmed by the design traffic level. And then requires mixture should aged before the moding for the situation is more close the reality. It is demand to estimate the compaction of mixture, and separate provision the initial compaction times and the max compaction under the max compaction times. This aspect was dominance than Marshall because it is avoid produce the instability mixture and ensure the ability to resist the track.

The final dosage of asphalt of superpave is ensured by the 4% voidage when the rotating compaction. At the same time the VMA and VFA and rubber powder were fill the bill when the voidage was 4%. The initial and the final compaction degree meet the requirements.

Because of SGC compaction method and the compaction work more close to the site construction, no matter the way and the force of compaction is is quite different from the scene. It will lead to the Marshall method design of mixture ratio was to large and bulk volume to small. In field construction control, the degree of compaction often appear compaction of the overpressure phenomenon, to move forward a single step that the Marshall compaction merits is small and not close with the actual construction situation.

(6)the performance testing of mixture

The Marshall need to testing the performance on the base of the mix design according to the

DOI: 10.1051/matecconf/20166302007

standard for the highway and first-class highway. If the mixture is not satisfactory, it must reopen the design. Other grades highway asphalt mixture can be reference to perform. It is use the wheel tracking test to evaluation the high temperature stability of mixture (nominal maximum size is equal to or less than the mixture of 19 mm), and use the immersing Marshall test and freeze-thaw splitting test to evaluate the water stability of asphalt mixture, and use low-temperature bending test to evaluate asphalt mixture (nominal maximum size is equal to or less than the mixture of 19 mm) of the low temperature crack resistance.

Superpave have hear tester (SST) and indirect tensile test and instrument (IDT) to test and prediction the mixture performance. These two instruments were improve the forecast methods for performance test. But it was too expensive and the research are rarely. Superpave about water sensitivity of the test mixture, is not based on performance test, and there are two purposes: first, determine whether the combination of asphalt cement and aggregate is sensitive to water; Second, the effect of measuring antistripping agent. Superpave mixture water sensitivity of the test method is similar to the current specification of freeze-thaw splitting test[3].

Due to the shear test and indirect tensile test method and standard is perfecting, when the domestic scientific research units in the Superpave mixture design, they usually to test method to performance verification of the mixture by Marshall. Although it is not fully reflect the Superpave mixture excellent road performance, it yet be regarded as a kind of auxiliary Superpave mixture design method.

3 CONCLUSION

(1) when gradation selection, Superpave used the concept about restricted area and restricted, and the design of the grading was better than Marshall generally with skeleton structure.

(2) Superpave is more reasonable than experience of the Marshall method because it is based on the volume of the mixture properties for Forecast for cement content

(3) To determination the theoretical maximum density of mixture, the Marshall method use the vacuum test for ordinary asphalt mixture, and for modified asphalt mixture was used by calculation method. But Superpave are all made of the measurement method. It is due to the many shortcomings of measurement method for modified asphalt mixture, that use the Marshall to test the max theoretical density .

(4) Superpave conclude the traffic parameter for moding. At the same time to conclude the situation of pavement.Ttherefore the design of mixture density than the Marshall method design.

References

- [1]JTG D50-2006.Specifications for Design of Highway Asphalt Pavement[S].Beijing: China Communications Press, 2006.
- [2] JTG F40-2004.Technical Specification for Construction of Highway Asphalt Pavement[S].Beijing : China Communications Press, 2004.
- [3] Jiayu , Rognji-cao , Benjing-li compile.Superpave Fundamentals Reference Manual [M]. Beijing : China Communications Press, 2005.
- [4] Zhengxin. Study on Optimization Design of Asphalt Mixture Based on Superpave and Marshall Method [D].Xian: Chang'an University, 2008.
- [5]Qinsheng wang, Jinhong geng, Jianguo rong. Superpave method and Marshall method to determine the optimum asphalt content analysis [J].Journal of Shandong Jiaotong University, 2008, 16 (2) : 236-239.
- [6] Wenfeng li. Superpave volume design method and the mixture ratio design method of comparative analysis in our country [J].China Water Transport, 2009, 9(7): 233-235.
- [7] Fangfang zhang, Zhangjie, Hanguang, Yinxiang liu. Superpave and two types of asphalt mixture Marshall design method is discussed in this paper [J].Journal of China & Foreign Highway, 2008, 28 (6) : 236-239.

 DOI: 10.1051/matecconf/20166302008

Current Situation of Highway Daily Maintenance Management in Beijing

Feng Ming PAN , Bao Lin ZHU , Jia HUI, Rui Nan JIANG, Tong WANG

China Academy of Transportation Sciences, Beijing 100029, China

Abstract: Combined with the industry property, characteristics and management needs of highway maintenance project in Beijing, through investigation and research of the highway daily maintenance management, the present situation of highway repair and maintenance is expounded from the aspects such as the determination of maintenance funds and plan, maintenance management mode, and so on. Then in order to explore the new mechanisms for market management of highway maintenance, the advantages and disadvantages of highway maintenance and minor repairment of Beijing are analyzed.

Keywords: highway; maintenance management; maintenance and minor repairment; operating model

INTRODUCTION

In recent years, China's road construction has been developed rapidly as increasing investment in transportation infrastructure. By 2013, it has reached 6417km that road maintenance mileage of county level and above in Beijing. Rapid growth of highway construction obtains economic development and social benefits. The fast, safe and comfort features of road provide quality services for drivers and passengers, and earn widespread admiration. As society progresses, people's requirements for road is not only fast, safe, comfortable, but also green traffic environment of "smooth, clean, green and beautiful".

In order to improve the urban environment, this article synthetically considered factors of industry nature, characteristics and management needs of highway maintenance in Beijing, carried out road daily maintenance management and research, and explored new mechanisms of market management of road maintenance.

1 OVERVIEW OF BEIJING ROAD MINOR REPAIRMENT AND MAINTENANCE

By the end of 2012, Beijing's traffic capacity and accessibility has significantly increased. Total mileage of Beijing highways network has reached 21,492km; highways network density of the city has increased from 87km/100km2 by the end of 2005 to 131km/100km2. According to administration social estate, national expressway is 561 km, national highway is 754 km, city road is 2222 km, county road is 3838 km, country road is 8067 km, village road is 5563 km, and specified road is 487 km. According to technology classification, expressway is 922 km, one-level road is 1118 km, two-level is 3283 km, three-level road is 3641 km, four-level road is

 DOI: 10.1051/matecconf/20166302008

12335 km, and substandard road is 193 km. The proportion of secondary road and above grade in national and urban highway is 97.03%.

Beijing Municipal Roadway Administration Bureau takes charge of the maintenance management of the basic transportation facilities including highways, city roads, county road and etc in Beijing. Currently, the maintenance mileage of Beijing county road and above grade has reached 6417 km, including national expressway 628 km, urban road 1896 km, county road 3893 km.

The major and intermediate maintenance technology (including preventive maintenance) mileage of average annual program implementation accounts for no less than 17 percent of the total mileage of national, city, county trunk highway. The Pavement Performance Index (PQI value) of national, city, county trunk highway remains above 90; that is to say, the value remains above 90 per year, and the mileage of excellent road remained above 90%. The Pavement Performance Index (PQI value) of county road remains above 88; that is to say, the value remains above 88 per year, and the mileage of excellent road remained above 90%. The transformation rate of existing dangerous bridge was 100% between national, city trunk highway and county road, and the treatment rate of the newly found dangerous bridge in those years was 100%. The proportion of national, city trunk highway I, II bridge was 94.7%, and the proportion of county road I, II bridge was 93.7%.

1.1 Determination of maintenance funds and plan

(1) The allocation of highway maintenance and minor repairment funding is based on the structural features of each county's road network. According to the highway administration social estate, road width, traffic volume, date of construction, and geographical conditions (mountain/plain), after analyzing scientifically that the allocation of highway maintenance and minor repairment funds, highway maintenance and minor repairment funds have been allocated to 10 road sub-bureau by the administrative region considering local needs. The sub-bureaus make the secondary investment decomposition plan to allocate the funds to each route or road based on current road network service condition.

(2) Overhaul project investment plan is determined based on data collection, the road condition evaluation, and the decision optimization of fast road condition monitoring data. Generally the road sub-bureau of each county prepares and reports preliminary construction plans based on the evaluation results of the detection data analysis. Maintenance engineering department of each sub-bureau organizes experts to discuss and determine the plans. Then, according to the engineering project, each sub-bureau makes preliminary design based on the plans and reports to the roadway administration bureau for approval. The roadway administration bureau authorizes consulting firm to audit the project budget, and approve the preliminary engineering design and project budget. Planning management department of the bureau prepares annual investment plan in highway construction. The plan is audited by the transportation committee and reported to the municipal public finance for approval. After the municipal finance bureau for approval, the municipal roadway administration bureau let the branch and direct subordinate unit to execute.

1.2 Operation mode of maintenance management

(1) The major and intermediate maintenance project management has the full realization of market-oriented operation, open the way for the country to take a tender to select maintenance engineering construction enterprise and project management company. 10 suburban counties road branch as the main construction

DOI: 10.1051/matecconf/20166302008

unit project, according the Municipal Transportation Commission Road Council, approved the preliminary design of the project budget and tissue engineering project bidding. Under project implementation, each highway branch as the main manager, was responsible for project progress, quality, funds, and management of completion. With the supervision of the municipal roadway administration bureau, project payment was paied according the maintenance progress and test results of quality.

(2) It is adopted that mode of contracting out to manage highway maintenance and minor repairment. The contract is mainly awarded to a maintenance company in Beijing. In this mode, each sub-bureau signs contracts with the company's regional center. Those contracts clearly define maintenance operation standard, maintenance quality standard, maintenance operation frequency, and maintenance engineering quantity list etc. Projects of 100 million yuan and above take public bidding to determine the maintenance enterprises.

(3) In the maintenance quality supervision, the highway bureau directly commissioned with corresponding company of supervision qualification, which was responsible for the quality supervision of highway maintenance and minor repairment. Each highway bureau implemented process supervision for maintenance project, carried out compliance checks in accordance with the management approach. In general, the performance evaluation and assessment were carried out according to the monthly payment. Highway emergency management is a part of daily maintenance and management content, and is entrusted to undertake by conservation work units, while signing the contract.

The working relationships among highway branch, maintenance supervision, and maintenance companies are shown in Figure 1.

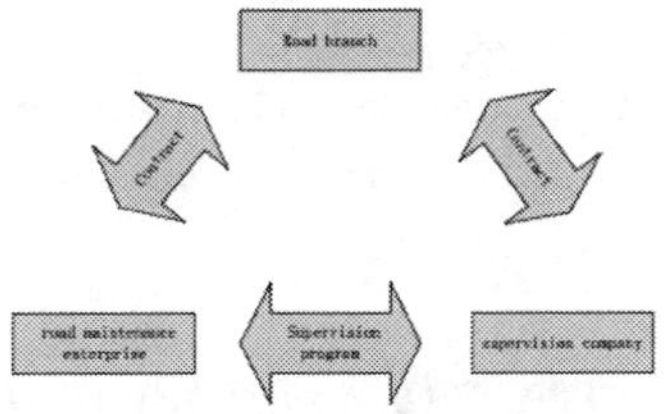

Figure 1 Highway maintenance management operation mode

2 ADVANTAGES OF BEIJING HIGHWAY MAINTENANCE AND MINOR REPAIRMENT

(1) Maintenance System

The strict rules of work are carried out, which makes division of responsibilities, work relations and work tasks clear and definite.

(2) Maintenance Management

Government departments have been streamlined. Then the maintenance vocational work with specifications and standards can be entrusted to the maintenance company and other social institutions to implement. The government improves management efficiency through supervision and restrictive contracts.

3 INADEQUACIES OF BEIJING HIGHWAY MAINTENANCE AND MINOR REPAIRMENT

3.1 Maintenance System

At present, for the maintenance system, it is separated of management and maintenance. However, effective competition mechanism has not been introduced in highway maintenance and minor repairmen because the maintenance enterprise in this market is monopolistic. This situation lead directly to lack of investment in road maintenance funds, inadequacies of technological innovation, and imperfections of enterprise management system. Therefore, it is affected that the improvement of the maintenance technology level and the innovation of management mechanism. So far, there is still a lack of effective market access mechanism, and competition can't really be introduced in the field of maintenance.

3.2 Operational Mechanism

Beijing traffic department implemented separation of administration from management in 2003. A maintenance company is an independent legal entity, that is to say, personnel, administration, finance and other aspects are full independence. This model is in line with the development trend of market economy. At present, maintenance and minor repairment has not been fully realized market-oriented management. There is a lot of work needs to be done to form conservation management mechanism in line with actual needs of this city.

3.3 New Demands of Urbanization Road

Beijing has entered a new stage of the development of the medium developed cities to the world city. The city has been expanding. The standard of living is constantly improving. With the rapid increases in the level of motorization, public's attention to the road service and demand is also getting higher and higher. The public has higher expectations of safe and smooth traffic so comprehensive service ability of highway infrastructures is facing a higher demand.

High-speed economic development of Beijing, promotes the city's management level, as well as the overall environmental requirements of the highway. This demands greater accountability from maintenance work since the highway and road-side facilities environment is an important part of the city. Virescence of separators, road shoulders and slopes is the first impression of the urban environment.

(2) Requirements of Improvement of Security Facilities

As urbanization processing, the road facilities need to be improved in accordance with the standards of urban roads. Street lamps, fences, traffic lights, warning signs and other safety facilities need to be upgraded.

(3) Requirements of Improvement of service quality

Integrity traffic, complete signs and marking, perfect auxiliary facilities, fine green and clean road are the keys to improve the quality of highway service. In addition to the daily maintenance work, the timely detection and treatment of emergencies on the highway is one of the important means to enhance the level of highway maintenance. Because of heavy traffic, large amount of ancillary facilities, and complex traffic composition, city highway needs to carry out maintenance according to the city standard, in order to guarantee the quality of service.

3.4 Funding

With the increasing public demand, the Beijing road maintenance management tasks become more difficult, and more and more money are demanded. After the fuel oil tax reformation, the ordinary highway construction and maintenance are facing huge financial pressure. During "Twelve-Five", the contradiction of maintenance funds shortfall becomes more and more obvious. Currently, maintenance funding problems are mainly embodied in the following aspects:

(1) Lack of maintenance cost accounting basis. In the current management system, maintenance quotas are not introduced, and maintenance cost accounting and maintenance costs to the financial application are baseless.

(2) Lack of the investment of maintenance science and technology. Due to the lack of investment in science and technology, which directly leads to the managing authorities and the maintenance units are lack of enthusiasm for the maintenance of new technologies, research and development of new material. Related research and development are seriously lagged behind,

which leads to lack of conservation technology content and low curing efficiency.

(3) Along with the quickening of the urbanization process, the new demand of highway maintenance is put forward. It will inevitably lead to increasing costs of the road cleaning, painting, paint, loss repair, replacement, personnel protection and other daily maintenance.

3.5 Maintenance Work

Since the separation between government and enterprises, highway maintenance in Beijing has a great leap and highway maintenance technical level was also rising. However, a survey shows that there aren't maintenance specifications in current contract, and the evaluation rules are not comprehensive, detailed, and uniform standard. These are not good for maintenance system standardization and refinement of management. The problems seriously affect the improvement of the level of highway maintenance. That mainly shown in following aspects:

(1) The detailed degree is insufficient. In the maintenance work, there is no specific requirement for flatness in pavement maintenance. That doesn't meet the idea of detailed management and detailed maintenance.

(2) The degree of standardization is insufficient. This mainly incarnates that the maintenance work site is informal, and the safety level of maintenance work is inadequate.

(3) The special mechanic is insufficient in maintenance operation. The main reasons for this problem are as following,

1. The maintenance workers' cultural level is low.

2. The investment in machinery and equipment is insufficient.

3. The special machinery and equipment are lack of development and application.

(4) The amount and pertinency of maintenance squad material reserves are insufficient. This increases the cost of material reserves, and reduces the efficiency of maintenance operations.

4 CONCLUSION

In recent years, highway maintenance in Beijing has developed enormously. The process of market-oriented operation and fine management has been pushing forward. However, there are some problems in practice. Combined with Beijing regional practical situation of development, the competitive market mechanism has been introduced appropriately. In order to satisfy the masses' growing transportation needs, and the new demands of highway maintenance management under the condition of urbanization, through tender document, management requirements have been paid much attention to, and transportation capability and level of transportation service have been tried hard to increase.

References

[1] Zhao Qin. Problems and solutions of road maintenance management [J]. Architectural Knowledge, 2013 (B12): 213.

[2] Jiang Bin. Brief Talk on Highway Maintenance Management [J]. China New Technologies and Products, 2013 (23): 146.

[3] Zhang Ruizhuo, Yang Jin. Current Situation and Development Direction of Expressway Maintenance Management [J]. East China Highway, 2013 (5), 3~5.

[4] Lin Wencheng, Song Yanda. The Research on Highway Maintenance Management [J]. China Science & Technology Panorama Magazine, 2013 (22): 140.

[5] Hu Fanjie, Ma Hongli. Discussion on Management of Administration mechanism and Improvement on Highway Maintenance. Journal of Henan Science and Technology, 2013 (11): 143.

MATEC Web of Conferences **63**, 02009 (2016) DOI: 10.1051/matecconf/20166302009
MMME 2016

Performance Comparison and Selection of Transformer Fluid

Yang LU [1], Shi Jia LIU [2]

Locomotive & Car Research Institute, China Academy of Railway Sciences, Beijing, China

Beijing Zongheng Electro-Mechanical Technology Development Co , Beijing, China

Abstract: Transformer fluid directly affects the working state of the components and the cooling efficiency of transformer. There are three kinds of transformer fluid used for electric locomotive, EMU and suburban rail vehicles: mineral oil, silicone liquid and synthetic ester based insulating oil. In this paper, the three kinds of oil are compared from the fire safety, environmental protection, reliability and low maintenance. It provides a strong basis for the selection of transformer fluid. By comprehensive analysis, synthetic ester based insulating oil can completely replace mineral oil and silicone liquid. With rail transport safety and environmental protection standards improving, synthetic ester based insulating oil will be the best choice for transformer.

1 Introduction

Traction transformer is an important equipment of electric locomotive and EMU, its cooling mode is by insulating oil. Transformer fluid directly affects the working state of the components and the cooling efficiency of transformer. Therefore, the choice of transformer fluid is essential.

Currently, there are three transformer fluid used in electric locomotive and EMU: mineral oil, silicone liquid and synthetic ester based insulating oil (synthetic ester). These three kinds of insulating oil can all meet the basic performance requirements of transformer fluid, but with the rapid development of rail transport technology, it puts forward higher requirements on the performance of transformer fluid. Meanwhile, safety and environmental protection must be considered in the selection of transformer fluid.

In the following article, three kinds of transformer fluid are compared from fire safety performance, environmental performance, reliability and low maintenance. The results can provide a strong basis for the selection of transformer fluid.

2 Fire Safety Performance

It is the most suitable transformer cooling method to use insulating oil (not other cooling medium) radiating heat, but this approach also determines that there is a fire hazard for transformer. In the field of rail transportation, the transformer fire is absolutely not allowed to happen. The fire will spread with terrifying speed, and result in a huge human and property losses. Fire safety performance of transformer fluid can be measured from the fire danger grade and ignition-resistant properties.

2.1 Fire Danger Grade Comparison

Fire performance comparison result of mineral oil, silicone liquid and synthetic ester is shown in Table 1.Among them, three indicators of mineral oil are the lowest, ignition point of silicone liquid is the highest, other indicators of silicone liquid are the same with synthetic ester.

 DOI: 10.1051/matecconf/20166302009

Table 1 Fire Performance Comparison Result

Parameter	Test Method	Mineral Oil	Silicone Liquid	Synthetic Ester
Flash Point	ISO 2719	150℃	260℃	260℃
Ignition Point	ISO 2512	170℃	>350℃	316℃
Fire Danger Grade	IEC 61039	O	K3	K3

2.2 Ignition-resistant Properties Comparison

Place the oxy-acetylene torch flame (over 2000 ° C) to the fluid surface in the metal pan, thermocouple is placed at the bottom of the dish to detect more distal surface temperature of the oil. Once the torch lit, began to record the oil temperature. Ignition-resistant properties comparison result of synthetic ester and mineral oil is shown in Figure 1. Through experiment observing, the surface of mineral oil began to burn in the first 3 minutes, there is not any change on the surface of synthetic ester in the first 3 minutes, even in the 70th minute no spark appears, shown in Figure 2.

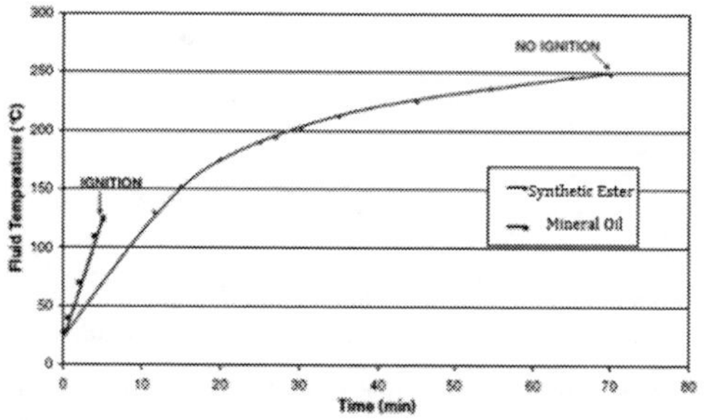

Fig 1 Ignition-resistant Properties Comparison Result

Mineral Oil 3min | Synthetic Ester 3min | Synthetic Ester 70min

Fig 2 Burning Test Phenomenon Comparison Result

In addition, synthetic ester passes through the following third-party fire certification:

- Underwriters Laboratory (UL® Classified) - EOUV Dielectric Medium Classification (UL fire certification)
- Factory Mutual (FM®) Global – FM Approved Less Flammable Transformer Fluid (FM fire certification)
- SGS BASEEFA, UK - ATEX Certification (Use authentication in potentially explosive atmospheres)

Thus, the higher fire safety performance of transformer fluid can greatly improve the safety performance of rail transportation equipment.

3 Environmental Performance

In the field of rail transportation, the selection of vehicle material is related to the survival of the human environment, transformer fluid is no exception. It could not be ignored how to deal with waste replacement fluid and unexpected leakage fluid. The main factors to judge the environmental performance of transformer fluid are biodegradability, the risk of water pollution and the influences on aquatic life.

3.1 Biodegradability Comparison

Biodegradation is a procedure that organic matter is decomposed and made harmless, which can be accepted by environment. Synthetic ester is non-toxic and easily biodegradable, its biodegradability has passed the standard test by laboratories officially recognized abroad. The biodegradation process of synthetic ester is shown in Figure 3.The degradation rate is 10% at the third day, 71% at 10th day, 89% at 28th day. This result meets the OECD biodegradable class requirement and the fully biodegradable requirement in IEC 61039 standard.

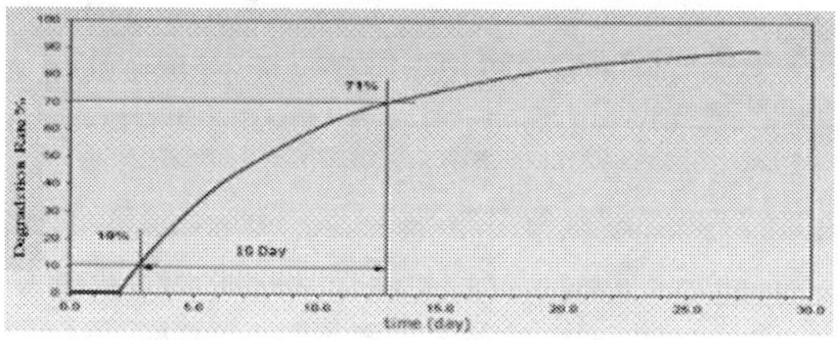

Fig 3 Biodegradation Process of Synthetic Ester

The biodegradation process of mineral oil and silicone liquid is shown in Figure 4. It is obvious that the degradation rates of both mineral oil and silicone liquid are less than 10% at the end of the test period 28 days, which makes a sharp contrast with synthetic ester.

DOI: 10.1051/matecconf/20166302009

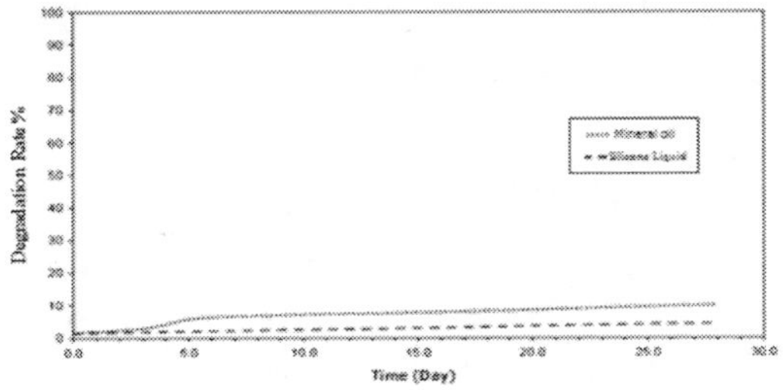

Fig 4 Biodegradation Process of Mineral Oil and Silicone Liquid

3.2 Risk of Water Pollution Comparison

Umwelt Bundes Amt（UBA） assess chemicals and divide levels based on biodegradability and the potential impact on aquatic organisms, synthetic ester is classified as non hazardous to water products. Mineral oil and silicone liquid are classified as hazardous to water classes, and the specific test parameters and reference limits are shown in Table 2. Mineral oil and silicone liquid will damage environment. It was confirmed that synthetic ester leaks into river will not cause harm to aquatic organisms at very high concentrations (1000 mg / l). Once mineral oil or silicone liquid occurs event of leakage it would cause serious environmental pollution, and this effect will not be eliminated in many years.

Table 2 Test Parameters and Reference Limits

Insulating oil	CAS number	UBA classification
mineral oil	manifold	1（hazardous to water）
silicone oil	63148-62-9	1（hazardous to water）
synthetic ester	68424-31-7	Nwg（non hazardous to water ）

In addition, synthetic ester also passed the following stringent international environmental certification:

- OECD Biodegradable Environmental Report
- UBA Environmental Report on non water pollution
- US government Water Environmental Report

Therefore, synthetic ester owns an obvious advantage of environmental performance, environmental pollution will not be worried.

4 Reliability

The main components of the transformer are immersed in transformer fluid. The reliability of transformer fluid directly impacts transformer normal work. It can not only slow down the deterioration rate of fluid ， but also reduce the failure rate of the components of the transformer by improving the reliability of transformer fluid. The reliability of transformer fluid is reflected by moisture resistance and lubrication property.

4.1 Moisture Resistance Comparison

Moisture resistance of transformer fluid is reflected by water content. With moisture content increasing, the performance of transformer oil will decline, mainly in the following points:

1）Increased water content will decrease dielectric strength.

The breakdown voltage and the water content of synthetic ester, mineral oil and silicone liquid at 20 ℃ is shown in Figure 5, Obviously, the water content of mineral oil and silicone liquid is very low, and the breakdown voltage decreases rapidly with the water content increasing. In contrast, the water content of synthetic ester is much larger than mineral oil and silicone liquid, even in the case of the water content over 600 ppm, synthetic ester also maintains a high breakdown voltage (more than 75kV).

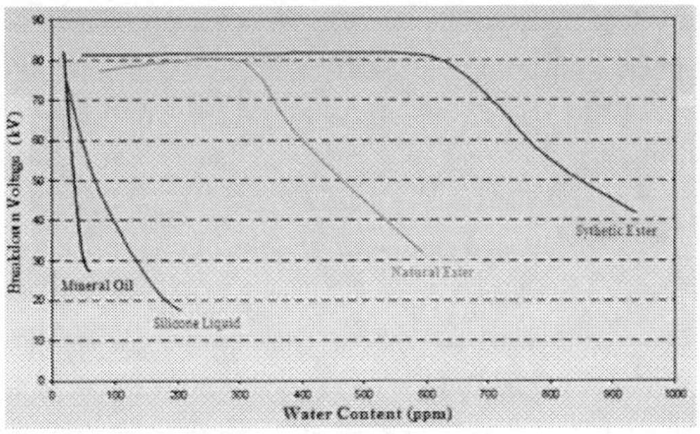

Fig 5 Breakdown Voltage and Water Content at 20 ℃

2) Increased water content will cause accelerated aging of the insulation paper.

Various studies show that the insulating paper aging rate has a direct relationship with the water content, the water content of the cellulose every 1% increase the life of insulating paper will be reduced by 10%. With the aging of cellulose, it will release water, which would aggravate the aging process. The good water absorption capacity of synthetic ester will reduce the water content of insulating paper, so it will slow down the speed of aging.

3) Increased water content will form bubbles easily.

Bubble will decline the electrical performance of insulating oil. According to IEC 60076-14 standard, the temperature of bubbles production is directly related to the water content of the cellulose. Under overload conditions, the temperature of iron core wrapped by insulation paper will rise. If the water content of insulating paper is 2.6%, the bubble generation temperature is 130 ° C. If the water content is 1.1%, the bubble generation temperature is 165 ° C. Since the synthetic ester can ensure insulation paper drier, the safety threshold is higher under overload condition.

4) Increased water content will increase the risk of releasing free water from the oil

In the transformer cooling process from working temperature to environment temperature, it is possible to release water from transformer fluid which depends on the water content saturation value. The value of synthetic ester is high, which means it is not easy to achieve its saturation limit and it is difficult to release free water.

4.2 Lubrication Property Comparison

Lubrication property is very important to transformer which works continuously and efficiently. If the transformer fluid does not reduce the friction of metal, it will cause the transformer premature failure. For example: transformer fluid pump requires effective lubrication to prevent the occurrence of breakdown. It can be got the lubricating effect of the different transformer fluid by Four Ball Wear experiment. The experiment result is shown in Figure 6, the wear value of synthetic ester is the minimum, which means the lubricating effect of synthetic ester is the best.

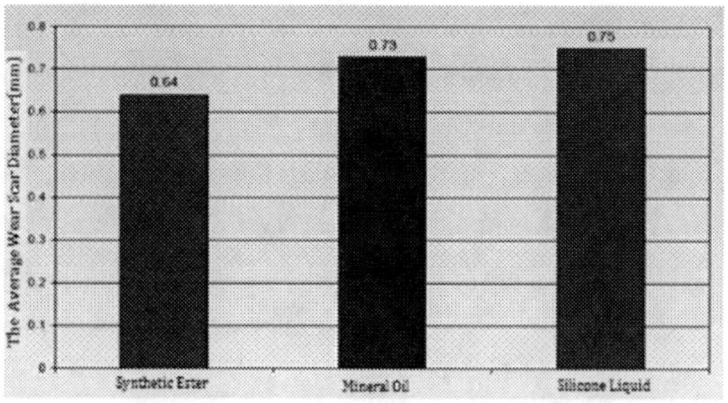

Fig 6 Result of Four Ball Wear Experiment

Synthetic ester shows an excellent performance in terms of moisture resistance and lubrication property. It can absorb more water, and guarantee the reliability of rail transportation equipment effectively.

5 Low Maintenance

The average life of the transformer is 30 years, the life and replacement cycle of transformer fluid is influenced by oxidation resistance and aging resistance. If transformer fluid and transformer have the same life, they can be replaced simultaneously.

5.1 Oxidation stability Comparison

Oxidation will accelerate the aging speed of transformer fluid, and cause deposition and degeneration of basic performance, high temperature will intensify the oxidation. The method to compare different oxidation stability of transformer fluid is the pressurized vessel oxidation test in ASTMD 2112 standard. The principle is to detect the oxygen consumption time to check the oxidation of oil. When the pressure drop is constant, the longer time-consuming means the stronger oxidation stability of transformer fluid. Three kinds of transformer fluid test result is shown in Figure 7. Oxygen consumption time of synthetic ester is the longest in the test, which means the oxidation stability of synthetic ester is high. However, mineral oil and silicone liquid are relatively poor.

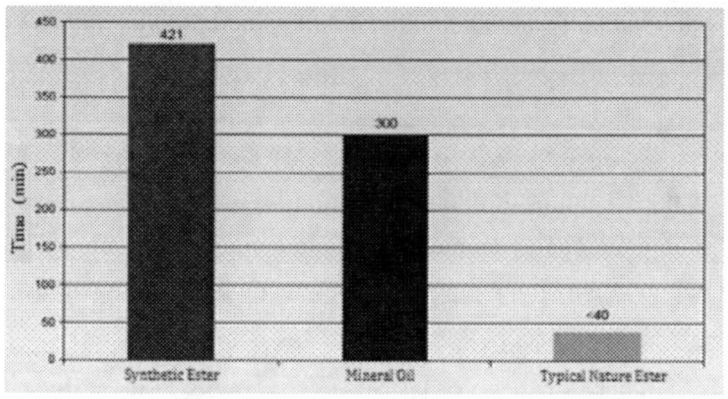

Fig 7 Three Kinds of Transformer Fluid Test Result

5.2 Aging Resistance Comparison

The acidity is a key indicator of transformer fluid aging. Transformer is experimented seal aging experiment which is made of common materials (such as copper) and using synthetic ester as transformer fluid. After more than a year of experiment period, the acidity of the synthetic ester is shown in Figure 8. Under the conditions of 125 ℃,

DOI: 10.1051/matecconf/20166302009

150 ℃ and 160 ℃ , the acidity of the synthetic ester are kept below prescribed limits in IEC 61203, which means the aging resistance of synthetic ester is very good. Meanwhile, viscosity, density and ignition parameters were monitored during the experiment period, and these parameters are not changed a lot. In addition, synthetic ester did not produce corrosive sulfur and other sediments during the experiment period. All these illustrate that the stability of synthetic ester is good, it may extend the transformer fluid replacement cycle.

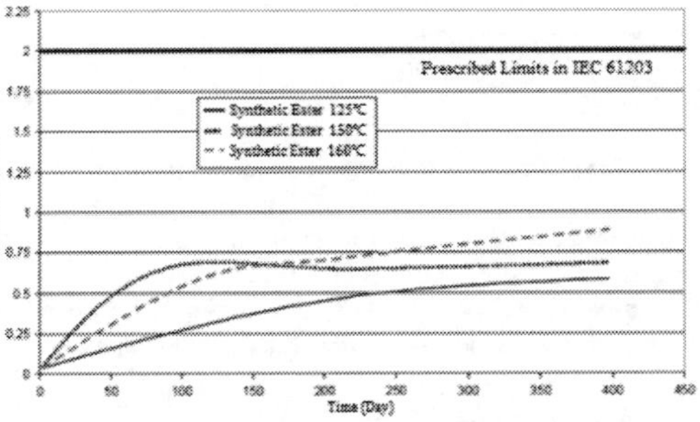

Fig 8 Acidity Change in Seal Aging Experiment

Therefore, oxidation resistance and aging resistance of synthetic esters are excellent, synthetic ester could improve the stability of the transformer, extend the maintenance cycle and reduce maintenance costs.

6 Conclusions

Synthetic ester is an excellent performance insulating oil, including fire safety, environmental protection and good moisture resistance. Compared with the other two transformer fluid, synthetic ester has a significant advantage to extend the life of transformer and reduce maintenance costs. Internationally, Siemens, ABB, Bombardier, Alstom and other well-known manufacturer of rail transportation equipment have been widely used synthetic ester, they have rich experience. Speed of 350 km Chinese standard EMU also chose synthetic ester as transformer fluid. After a comprehensive analysis, synthetic ester is the best choice as transformer fluid, and it could completely replace mineral oil and silicone liquid. As people pay more attention to safety, environmental protection, synthetic ester will be the preferred transformer fluid of electric locomotives, EMUs and suburban rail transport vehicles.

References

1. ZHANG Shuguang .CRH2 EMU [M] Beijing: China Railway Publishing House, 2008: 167.
2. ZHANG Shuguang .CRH5 EMU [M] Beijing: China Railway Publishing House, 2008: 248.
3. Liu Hong. High Speed Railway and EMU Traction Transformer Oil Analysis[J]. Petroleum Products Application Research, 2010 No.6.
4. ASTM D2112 - 01a (2007) Standard Test Method for Oxidation Stability of Inhibited Mineral Insulating Oil by Pressure Vessel[S].
5. IEC 60076-14. Liquid-immersed power transformers using high-temperature insulation materials[S].
6. IEC 61039. Classification of insulating liquids[S].
7. IEC 61203. Synthetic organic esters for electrical purposes; guide for maintenance of transformer esters in equipment[S].
8. ISO 2512. Furfural for industrial use; Determination of total carbonyl compounds; Volumetric method[S].
9. ISO 2719. Petroleum products and lubricants; determination of flash point; Pensky-Martens closed cup method[S].

The Application of the Taguchi Method in the Optimal Combination of the Parameters Design of the Spindle System

Di LIANG [1,a], Wei MENG [2]

[1] *School of Mechanical Engineering, Shenyang University, Shenyang, Liaoning Province, China*
[2] *School of Mechanical Engineering, Shenyang University, Shenyang, Liaoning Province, China*

Abstract. The paper combines with the application case in the field of parameter design of main spindle system. Firstly, the relevant parameters that affect the stability of the main spindle system transmission are determined, and then the Taguchi method and Minitab software are used to implement the orthogonal test for multiple factors. According to the experimental data, the analysis of signal to noise ratio and the analysis of variance are performed. Finally, the optimal combination of design parameters with different levels of spindle system is obtained. Case analysis shows that the Taguchi method has a significant effect on improving the transmission stability of the spindle system.

Keywords.the transmission stability of spindle system; Taguchi method; parameter design; Minitab; signal to noise ratio

1 Introduction

The Taguchi method was founded by Dr. Genichi Taguchi of Japan; its central content is regarded as Japan's "national treasure". The Taguchi method is widely used in automobile industry, mechanical manufacturing industry, electronic industry and other fields. In our country, colleges and the national defense science and technology industry have many theoretical researches on Taguchi method. Taguchi method is widely used in automobile manufacturing industry in the foreign countries, such as the U.S. automaker Ford and GM that apply the Taguchi method into practice, the cost of the two companies can be reduced, and the vehicle performance is improved, so Taguchi method gains remarkable achievements. In this paper, the thought of out-line quality control of Taguchi method is applied to the design stage of the spindle system.

2 Taguchi method

2.1 Basic thought

It is different from the traditional definition of quality; Dr. Genichi Taguchi defined the product quality as this [1]: The characteristic of avoiding causing losses to the society after the products get out of factories, the quality of products can be quantified by "quality loss". In order to improve the anti-interference ability of the product itself, Taguchi method can make the function and the property of products no insensitivity to the cause of deviation by adjusting the design parameters. In order to quantitatively describe the loss of product quality, Taguchi proposed the concept of "quality loss function [2]", and he used the signal-to-noise ratio to measure the stable level of designed parameters.

2.2 Parameter design

It means that the parameter design is carried out after the structure of the system is determined; the best combination of all parameters is found by using orthogonal experiment and the method of variance analysis. The so-called best combination is that the signal to noise ratio is introduced as evaluation indicator, the reliability, anti-interference, stability and uniform of products become stronger through the design of the stability.

2.3 Signal to noise ratio

In order to quantify the quality characteristics of the selected equipment, the concept of signal to noise ratio is introduced. Signal to noise ratios have three forms: The signal-to-noise ratio for the characteristic of nominal is best [3], the signal to noise ratio for the characteristic of smaller is best [4], the signal-to-noise ratio for the characteristic of larger is best.

[a] Corresponding author: sydxld@163.com

3 Instance verification

The paper researches on the problem which occurred when a new type CNC machine tool is processing parts in the MX Company, the transmission of the main spindle system is not stable, and the parts' defective rate is also increased.

3.1 Determine impact factors

The experts of the product design department, production operation department and equipment management department determined the main factors affecting the stability of the main spindle system by using the brainstorming method.

3.1.1 Controllable factors

a--bearing support rigidity (N/m): the increase of bearing support rigidity will enhance the stability of the spindle system.

b--spindle support type: The first support type is that the intermediate support bearings arrange in the place where the distance from the intermediate support bearings and the last support bearings is 35mm, the bearing span remains unchanged; the second support type is that one group of bearings selects individual bearing support, the last group of bearings selects three bearings support; the third support type is that there are three groups of bearings that arrange in the front, middle and back of the spindle, each group has two bearings.

c--bearing span (mm): The distance of adjacent two groups of bearings is called bearing span, this paper only considers the span of the first group of bearings and the last group of bearings.

d--pre tightening force (N): Using appropriate methods to make certain pre deformation between the bearing rolling body and the ferrules is called pre tightening force.

e--spindle inside diameter (mm): The inside diameter of high speed spindle is a ladder shape, changing the inside diameter spindle will affect the rigidity of the spindle.

The transmission stability of the spindle system belongs to the dynamic characteristic of the spindle system. For the convenience of the study, this paper gives a simplified schematic diagram of the high speed spindle system.

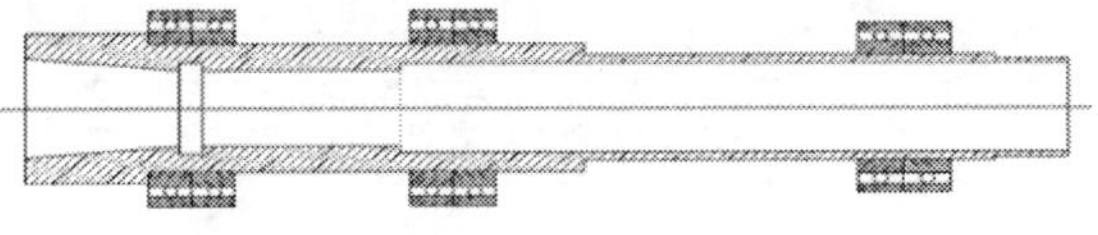

Figure 1. Simplified main spindle system

3.1.2 Noise factors

M- Spindle materials; N-The cutter point pressure fluctuations when spindles are processed by machine tool (MPa); Q - Cutting fluid type

3.2 Orthogonal test

3.2.1 Determine factors levels

Table 1. Different levels of controllable factors

factors	level1	level2	level3
a(N/m)	8×10^5	8×10^6	8×10^7
b	1	2	3
c(mm)	334	354	374
d(N)	80	150	280
e (mm)	diameter -1	diameter	diameter +1

Table 2. Different levels of noise factors

factors	level1	level2
M	Import	Domestic
N	60	80
Q	Q1	Q2

3.2.2 Make orthogonal table

Table 3. Outer table of orthogonal test

factors	levels			
M	1	1	2	2
N	1	2	1	2
Q	1	2	2	1
combine	S1	S2	S3	S4

Firstly, the orthogonal table is created by the Minitab software, then according to the 27 combinations of different levels in the orthogonal table, the production department performs the on-site experiments, produces CNC machine tools, observes the operation condition of the spindle system of the CNC machine tools, tests the qualified condition of the parts that are processed by CNC machine tools, and records qualified rate of the parts. Finally, the orthogonal test is performed by using Minitab software and the result of the test is shown as follows.

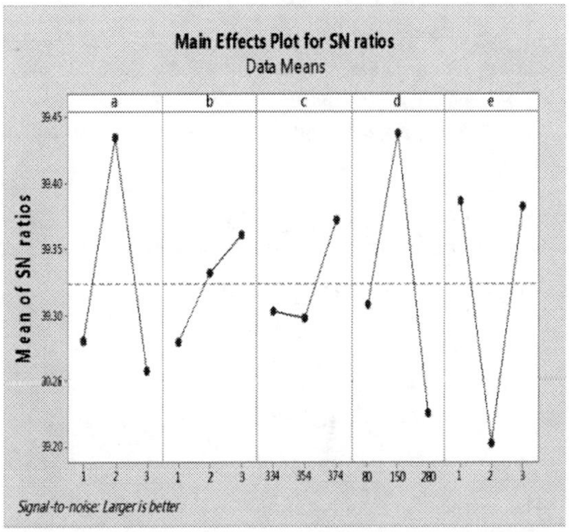

Figure 2. The main effect of each factor on signal to noise ratio.

↓	C1	C2	C3	C4	C5	C6	C7	C8	C9	C10	C11
	a	b	c	d	e	S1	S2	S3	S4	SN	Mean
1	1	1	334	80	1	90.80	97.44	86.77	90.17	39.1863	91.2950
2	1	1	334	80	2	87.44	85.80	89.00	90.97	38.9134	88.3025
3	1	1	334	80	3	97.75	95.44	91.10	93.47	39.4943	94.4400
4	1	2	354	150	1	96.43	95.44	94.13	93.47	39.5404	94.8675
5	1	2	354	150	2	96.67	97.42	86.46	89.68	39.2951	92.5575
6	1	2	354	150	3	90.61	92.32	93.14	92.55	39.2890	92.1550
7	1	3	374	280	1	88.68	92.46	96.12	93.21	39.3231	92.6175
8	1	3	374	280	2	90.72	90.77	94.11	90.17	39.2193	91.4425
9	1	3	374	280	3	90.35	89.35	95.44	92.35	39.2555	91.8725
10	2	1	354	280	1	97.75	91.79	91.13	91.79	39.3700	93.1150
11	2	1	354	280	2	94.65	92.75	89.79	86.68	39.1632	90.9675
12	2	1	354	280	3	95.48	89.77	91.75	90.75	39.2628	91.9375
13	2	2	374	80	1	92.35	96.44	97.35	95.35	39.5833	95.3725
14	2	2	374	80	2	90.75	91.72	93.31	92.15	39.2728	91.9825
15	2	2	374	80	3	95.35	96.52	93.75	95.20	39.5718	95.2050
16	2	3	334	150	1	96.76	98.59	95.35	97.35	39.7347	97.0125
17	2	3	334	150	2	92.35	93.35	89.88	91.75	39.2574	91.8325
18	2	3	334	150	3	97.38	98.21	96.89	94.20	39.7026	96.6700
19	3	1	374	150	1	90.48	88.68	95.79	97.79	39.3661	93.1850
20	3	1	374	150	2	93.72	87.77	95.01	95.15	39.3471	92.9125
21	3	1	374	150	3	97.79	89.38	98.02	89.68	39.4104	93.7175
22	3	2	334	280	1	97.79	87.52	91.50	88.95	39.1996	91.4400
23	3	2	334	280	2	90.75	90.35	91.13	92.06	39.1871	91.0725
24	3	2	334	280	3	92.13	89.38	87.35	89.76	39.0468	89.6550
25	3	3	354	80	1	92.47	88.68	90.36	92.65	39.1804	91.0400
26	3	3	354	80	2	85.94	95.59	93.50	89.16	39.1630	91.0475
27	3	3	354	80	3	95.44	87.75	95.01	96.35	39.4105	93.6375

Figure. 3 The orthogonal table created by Minitab

3.3 The analysis of signal to noise ratios

The figure.2 shows that the second level of bearing support rigidity is better, the third level of bearing support type is better, the third level of bearing span is better, the second level of pre tightening force is better, the first level of spindle inside diameter is better.

3.4 Variance analysis [5]

Here is the formula for the sum of squares of factors

$$SS=\frac{\text{N}}{l}\sum_{i=1}^{n}(\bar{y}_i-\bar{y})^2 \tag{1}$$

In the formula, N is the number of test; l is the value of factor level; $\bar{y}_i$ is the mean of the characteristic value corresponding to the 'i' level of the factor; $\bar{y}$ is the mean of the characteristic value.

The formula for calculating the degree of freedom of factor:

$$DOF=l-1 \tag{2}$$

The formula for calculating the variance of factor:

$$VAR=SS/DOF \tag{3}$$

This is the sum of squares of factor a.

$$SS_a=\frac{\text{N}}{l}\sum_{i=1}^{n}(\bar{y}_i-\bar{y})=\frac{27}{3}\begin{bmatrix}(92.17-92.64)^2+\\(93.79-92.64)^2\\+(91.97-92.64)^2\end{bmatrix}$$

$$=17.91$$

The degree of freedom of factor a:

$$DOF_a=l_a-1=2$$

The variance of factor a:

$$VAR_a=SS_a/DOF_a=8.954$$

The sum of squares of residuals of the factor a:

$SS_{error}=34.117$ (Derived from Minitab)

The degree of freedom of residual of the factor a:

$$DOF_{error}=n-1-\sum DOF=27-1-2\times5=16$$

The residual variance of factor a

$$VAR_{error}=SS_{error}/DOF_{error}=2.132$$

The variance ratio of factor a:

$$\text{F}=VAR_a/VAR_{error}=4.2$$

In the same way, other factors can be calculated and the variance analysis of the mean of the quality characteristic and the variance analysis of the signal to noise ratio are shown as follows

Table 4. Variance analysis of mean

source	df	Seq SS	Adj SS	AdjMS	F	P
a	2	17.91	17.91	8.95	4.20	0.034
b	2	3.006	3.006	1.503	0.70	0.509
c	2	3.419	3.419	1.709	0.80	0.466
d	2	24.37	24.37	12.18	5.71	0.013
e	2	22.71	22.71	11.36	5.32	0.017
error	16	34.12	34.12	2.132		

Table 5. Variance analysis of signal to noise ratio

source	df	Seq SS	Adj SS	AdjMS	F	P
a	2	0.170	0.170	0.085	4.61	0.026
b	2	0.031	0.031	0.015	0.83	0.454
c	2	0.032	0.032	0.016	0.85	0.445
d	2	0.207	0.207	0.104	5.61	0.014
e	2	0.201	0.201	0.100	5.43	0.016

error	16	0.295	0.295	0.018		

From the Table 4, we can see that the value of P is less than 0.05 for the factor *a*, *d* and *e*, so these factors are significant to the mean of the quality characteristic value. From the Table 5, we can see that the value of P is also less than 0.05 for the factor *a*, *d* and *e*, so these three factors are significant to the signal to noise ratio.

According to the variance analysis, we can know that the factors *a*, *d*, *e* have a significant effect on the signal-to-noise ratio and the mean of quality characteristic, so the three factors are important factors, the figure.2 shows that the level of factors *a*, *d*, and *e* should respectively select *a2*, *d2*, and *e1*. The effects of *b* and *c* on signal-to-noise ratio and the mean of the quality characteristic value are not very obvious, so they are minor factors, it indicated that the bearing support type and bearing span are not the main parameters that influence the dynamic characteristics of the spindle system, but it would impact on the flexibility of spindle if the span of the front and the back bearings is too big , so it is better for *c* to select the second level, and according to the signal-to-noise ratio response figure, factor *b* should select *b3*, finally the best combination of factor level is *a2*, *b3*, c2, *d2*, *e1*.

4 Conclusion

This paper has a in-depth study on the problem of unqualified parts that are made by the CNC machine tools of MX company, the research shows that the parameter design of the spindle system is not reasonable, this makes the spindle system transmission is not stable, and the anti vibration property is weak so that the accuracy of processing parts is not high and the qualified rate is low. In this paper, the main parameters affecting the dynamic characteristics of the spindle system and their levels are determined by using the brainstorming method, and the orthogonal experiment, the signal to noise ratio analysis, and variance analysis are carried on by using the Taguchi method and Minitab software, finally the optimal combination of design parameters of the spindle system is determined, then the production department does the production test with the optimal combination of parameters. The test result shows that the qualified rate of parts processed by machine tools is increased from 88.66% to 95.85% compared with before; production efficiency is also increased by 10%.

References

1. Z.J. Han, W.J. Zhang, Quality engineering–the quality management of out-line and on-line, Science Press, Beijing, 1991, pp. 33-37.
2. AI Wu, The second edition of quality management, edition, Jinan University press, Zhanjiang, 2004, pp. 20-25.
3. Genichi Taguchi, Development and design phase of the quality engineering, Japan specifications Association, 2011, pp. 11.
4. G.X. Wang, Z.J. Han, The relationship of quality loss and signal to noise ratio between the 'Larger is best' and 'Smaller is best', Mechanical science and technology, J. 19 (2000)236-238.
5. X.R. Shi, C.B. Yang, and Q.H. Hong, Optimal design of HPLC Chinese medicinal composition testing based on Taguchi and ANOVA methods, Strait pharmaceutical, J. 20(2008) 45-47.

MATEC Web of Conferences **63**, 02011 (2016) DOI: 10.1051/matecconf/20166302011
MMME 2016

Finite Element Analysis of the Effect on Edge Distance of the Tensile Bearing Capacity of Embedded Hanging Parts

Xian Hong MENG [1], Wei LIU [1,a]
1 School of Civil Engineering, Shenyang Jianzhu University

Abstract. In order to explore the trend of tensile bearing capacity of embedded hanging parts when change the edge distance. Based on the finite element analysis software ABAQUS, the four simulation model was established. The buried depth and strength of concrete remain unchanged, but the edge distance was gradient change. By the load - displacement curve of every model known, the greater the edge distance, the greater the bearing capacity. When the edge distance reaches 1.5 times buried depth, the effect of increasing edge distance for improving the bearing capacity will be impaired.

1 Overviews

The embedded hanging parts are special steel components, which are applied to the production, lifting and transportation of prefabricated structures. With the development of prefabricated structure, people gradually realize that the importance of embedded hanging parts of production, testing, application.

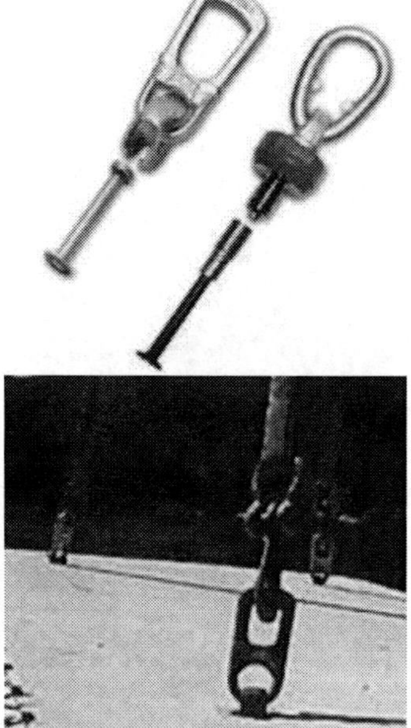

Figure 1 Embedded hanging parts

The traditional embedded hanging parts is made by the rings, Which is fixed at the special position before prefabricated parts are lifted. It will be the reliable connection between concrete and sling unless the strength of concrete reach its 70% of its design strength. People will cut the Exposed portion of the rings with angle grinder and grinding wheel, when the prefabricated parts are installed. But there is no doubt that the bar at the incisions will affect the beautiful of structure and the durability in lifetime. The rings is required that the burial depth over 30 times its diameter when it is used to lifting, in order to prevent it is pulled from the concrete. This way also waste a lot of steel to some extent. So, the ring is eliminated in the construction industry.

In order to meet the development of Prefabricated buildings, the requirements to embedded hanging parts will Increase year by year. The size of current embedded hanging parts have been optimized, and their height less than 100mm, diameter less than 30mm, decreasing the amount of the steel effectively. It's possible that the maximal lifting load can reach 50t with the high strength steel. Their exposed portions always are designed as ring, and the connection between the buried portion and sling can be finished by the threads. So, it makes sure the sling rotation, tilt freely in lifting, decreasing the extrusion and shear between embedded hanging parts and its surrounding concrete.

Embedded hanging parts stem from Europe and the United States, and relevant specifications have been established for guiding its engineering application in foreign, such as European standard*ETAG001 Annex A and Annex C*, American code*ACI318 Annex D*. The specifications define the failure mode, calculating capacity,parameters modification , but the control standards, loading system, unit conversion in foreign codes are different from them in home. Thus, the value

[a] Wei Liu:1280299968@qq.com

from foreign codes only has reference significance, which don't consider as the yardstick of measuring that whether embedded hanging parts are safe usage or not. In home, we have almost no relevant specification for embedded hanging parts, only the modification method of calculating Tensile bearing capacity of embedded hanging parts are mentioned by *Technical Specification for Post-installed Fastenings in Concrete Structures*, when it comes to define the Anchor bolt. In addition, a mass of relevant research define Four kinds of failure modes: anchor bolt steel fracture[1-3], Anchor bolt pulled up failure, Concrete vertebral failure by tensile, Concrete fracturing failure. However, the embedded hanging parts' transmit strength way, Stress analysis are different from Anchor bolts. So, the tensile bearing capacity of embedded hanging parts can't be calculated accurately according to the theoretical calculation method in Anchor bolt codes. It's similar for the way of embedded hanging parts, Anchor bolt and Planting bar between concrete, thus, the existing Anchor bolt and Planting bar research result can be learned when we research the effect of edge distance for embedded hanging parts. Wu Guifang[4] researched the four kinds of failure model mentioned in ACI318 Annex D, and come to a conclusion that it's no necessity to consider the side split of concrete, when the edge distance more than 4.8 times burial depth. The research of Yuan Tingpeng[5] indicated that the factors that effecting the tension capacity of Planting bar include Strength of concrete, burial depth, edge distance, spacing and so on, put forward the tension capacity empirical formula of Group of reinforcement, and involving the correction factor of every factors.

At present, the manufacturers of embedded hanging parts always provide the Nominal load to users, and it's also called safety load. Which is a safety value without considering the reduction effect of edge distance, burial depth and other factors. But it's common that embedded hanging parts have no enough safety edge distance in practical engineering application. The tension capacity must less than safety load in that case, and the reduction factor offered by engineers through the their experience. Besides, safety load always have no enough safety factor, Burying the hidden trouble for safety construction. From what has been discussed above, it is necessary to study the edge distance of embedded hanging parts.

From what has been discussed above, it is necessary to consider the effect of edge distance to tensile bearing capacity . But a great of time, manpower and material will be consumed at experimental stud, and there are many factors to affect the tensile bearing capacity of embedded hanging parts, which can't got from the experimental stud. However, the finite element analysis can solve the question by changing the factors, observing the analysis result easily. Besides, this way also can guide the following experiment, including the gradient change of factors, Size optimization, loading scheme and so on.

2 finite element analysis

2.1Model design

Based on the finite element analysis software ABAQUS, model is established, including the concrete and embedded hung part. There is no bar inside the concrete, and embedded hanging parts' material are Q345, Model size is shown in figure 2.

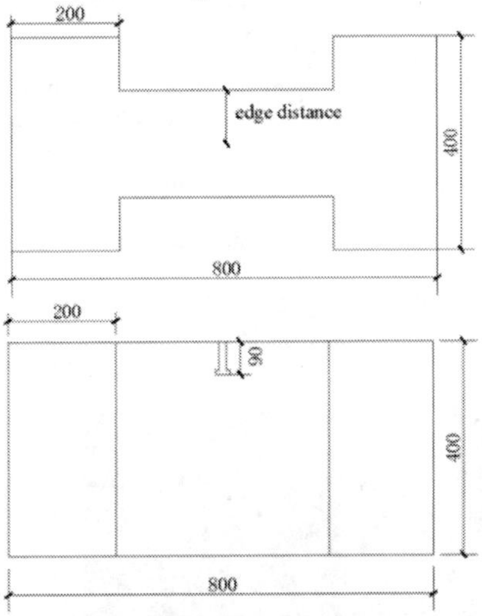

Figure 2 Model size

The model of embedded hung part in finite element analysis comes from a company product prototype. Its diameter is 12mm, effective depth is 90mm, safety load is 62.5kN. According to the provisions of Anchor bolt specification at home and abroad, it's no necessary to consider the reduction effect of edge distance, if the edge distance more than 1.5 times effective depth. Thus, the critical edge distance is defined as 135mm in this research. In order to explore the effect of edge distance ,author designs four models that their edge distance are gradient change. Main parameters are shown in Table 1.

Table 1. Main parameters

Model number	Y1	Y2	Y3	Y4
Edge distance（mm）	50	100	135	150

The Table 1 shows that, Y3's edge distance are critical, Y2's and Y1's edge progressively decreases. Y4 has no effect of edge distance.

2.2 Finite element simulation

2.2.1 Establishing the model

Establishing the model according to the model size before finished at previous step, the research focus on effcet of edge distance for tensile bearing capacity when concrete appear vertebral failure. So, we hold that there is no relative slip between embedded hanging parts and concrete, and the interface between them has enough cohesive force. Thus, the constraint between them is defined as embedded.

2.2.2 Define the material

DOI: 10.1051/matecconf/20166302011

The strength grade of concrete is C20, and the constitutive relation is concrete damaged plasticity model. Embedded hanging parts' material is Q345, established as ideal elastic-plastic model.

2.2.3 Load and Constraint

The load and constraint of models are shown in figure 3, two flanks of the concrete are on the surface Fixed constraint, and loading vertically at exposed end of embedded hanging parts. We can get the Ultimate load accurately by adjusting the load displacement time after time. I determine that concrete is poor, when Mises stress of any point reach 1.54MPa, the standard values of concrete ultimate tensile strength. I determine that embedded hung part is tensile failure , if the stress of embedded hanging parts over yield stress 345MPa before concrete is poor.

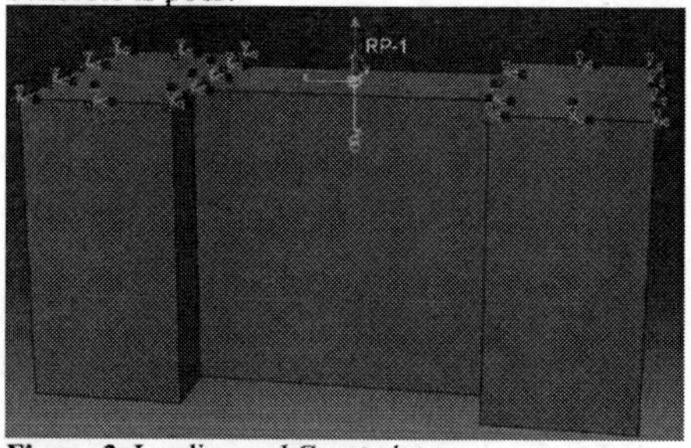

Figure 3. Loading and Constraint

2.3 Results analysis

2.3.1 Stress nephogram of model

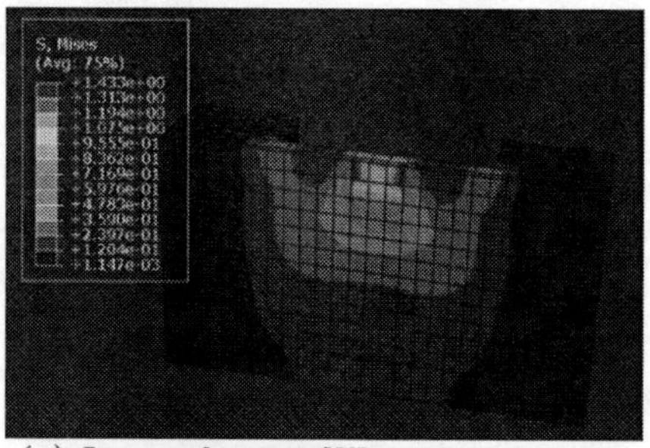

(a) Stress nephogram of YI

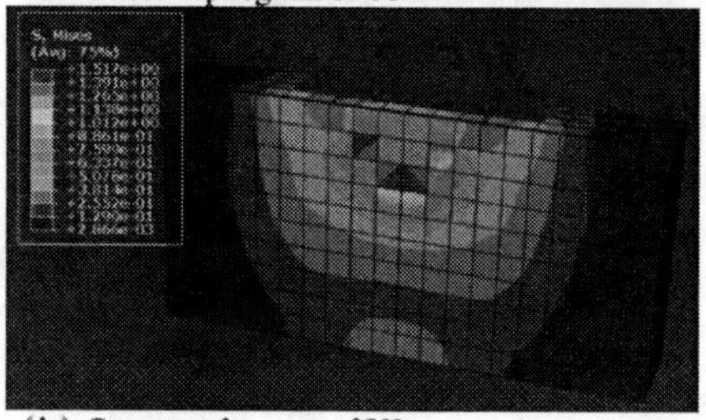

(b) Stress nephogram of Y2

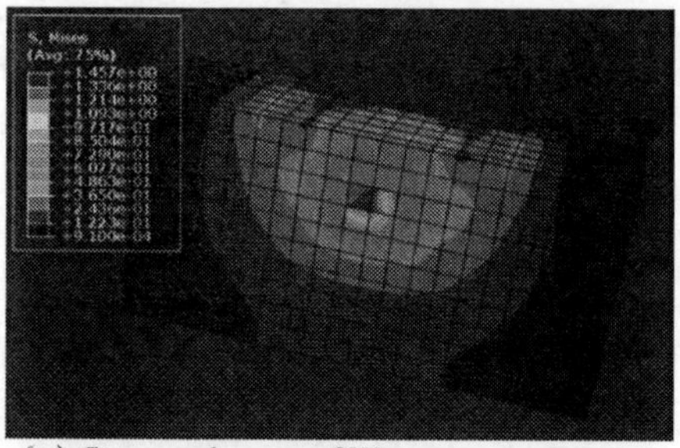

(c) Stress nephogram of Y3

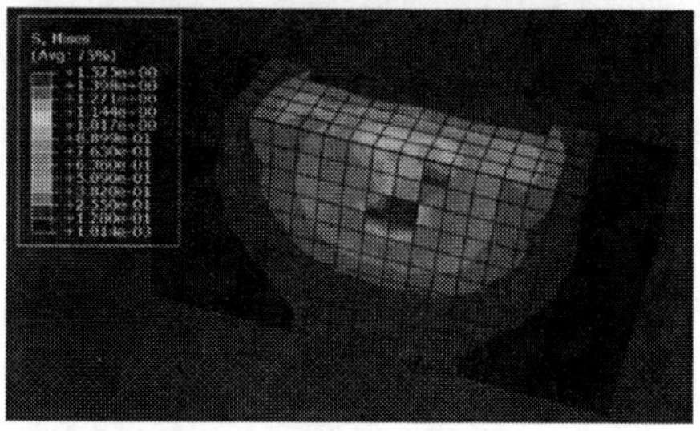

(d) Stress nephogram of Y4

Figure 4. Stress nephogram of models

As shown in figure 4, every models appear concrete vertebral failure result from the stress of concrete over ultimate tensile strength. Along the direction of burial depth, stress inside the concrete increase gradually, and the peak value appear at the bottom of embedded hanging parts. The stress inside the embedded hanging parts all less than yield stress, and maximal stress appear at the end of loading. Obviously, the damage radius of concrete vertebral body decrease gradually with the increasing of edge distance.

2.3.2 Capacity - edge distance curve

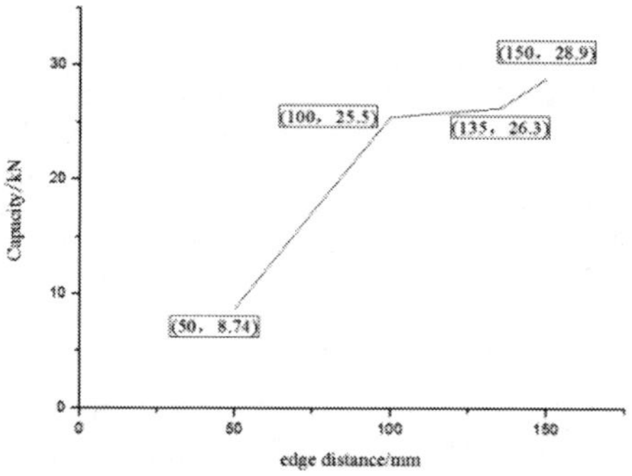

Figure 5. Capacity-edge distance curve

Table 2. Capacity summary

Model number	Y1	Y2	Y3	Y4	Safety load
Capacity (kN)	8.74	25.5	26.3	28.9	22.5

As shown in capacity-edge distance curve, the tensile bearing capacity of embedded hanging parts and the ultimate displacement increase gradually with the increasing of edge distance. The tensile bearing capacity increases slower when edge distance over critical value.

As shown in Table 2, the safety load offered by producer less than the capacity of critical edge distance, but it has no enough safety factor.

3 conclusion

(1) The safety load greater than capacity of critical edge distance, indicating it doesn't apply to guiding the lifting and transporting of prefabricated part ;

(2) The damage displacement of embedded hanging parts become smaller with decreasing the edge distance. And the greater the edge distance, the greater the capacity;

(3) The effect of edge distance to tensile bearing capacity of embedded hanging parts will be poor, when the edge distance more than critical value.

Acknowledgments

This work is financially supported by the Research Project of Urban and Rural Constriction system of Shen Yang (2011BAJ10B06)

References

1. Lin Zhi W, Dong Bin L, Fu Quan X. Earthquake Resistant Engineering and Retrofitting. J. 05, 101-107(2013)
2. Tie Zhu J, Fu Quan X, Lin Zhi W. Earthquake Resistant Engineering and Retrofitting. J. 05, 108-114(2013)
3. Lei S, Jie L, Zhou Dao L. Journal of Harbin Institute of Technology. J. 04,612-616(2010)
4. Gui Fang W, Building Structure. J.S1 ,436-438(2013)
5. Ting Peng Y, Zhou Dao L and Tao B. Structural Engineers. J. 02,97-100(2007)

The Finite Difference Method Of Discontinuous Constraint Slabs

J.C. WANG[1,a], J.H.ZHOU[1], M.X.PAN[1], D.Y.ZHANG[1]
[1] *Shenyang Jianzhu University, 110168, Shenyang, Liaoning, China*

Abstract: With the improvement of building standard, structural form becomes more and more complicated. The utility rate of special-shaped column, corner window and cantilever structure is higher and higher, which often makes reinforced concrete floor slabs in a discontinuous constraint state. To make the design of discontinuous constraint slabs more scientific, the author puts forward a kind of numerical method that can solve the problem of low calculation precision .On this basis, the numerical method is applied to solve the problem of rectangular slabs with two simply supported and the other with part of simply supported and part of clamped ,having a theoretical analysis of the result to verify that the calculation method is correct. The result shows :1. For discontinuous constraint slabs, the displacement difference is large and the average is 10%~20% when it comes to whether the method is adopted or not. 2. For discontinuous constraint slabs, higher precision is obtained by using the numerical method proposed in this paper. And it is necessary to consider the influence of boundary constraints when we design discontinuous Constraint slabs.
Keyword: Concrete slabs, Discontinuous constraint, Numerical method.

1 INTRODUCTION

The boundary conditions of floor can be divided into two categories. One is comprehensive boundary condition, the other is mixed boundary condition (referring to the edge of the rectangular slab with two or over two kinds of boundary conditions). In general, slabs with comprehesive boundary conditions are known as simple constraint slabs; Slabs with mixed boundary conditions are known as complex constraint slabs. Indeed,researches on simple constraint slabs are fruitful, whereas the complex constraint slabs are studied on starting late in engineering community[1~4].Eyre was the first to introduce a question with regard to free vibration problem of rectangular slabs with three simply supported boundaries and the other part of simply supported boundaries and part of built-in boundaries, and gave the relevant solutions.On this basis, WangGang, Qingxiang Wang and others also studied successively on the vibration and carrying capacity of discontinuous constraint slabs, but obtained few breakthrough results[5~7]. On the basis of the elastic slab theory, I will present numerical method for the complex constrained slabs to you combined with the finite difference method .Strict as it is,the method considers complex constraint situation,thus giving you more accurate calculation results than simplified processing method and providing technical support for the design of slabs.

2 BASIC EQUATION OF BENDING ABOUT THE COMPLICATED CONSTRAINT SLAB

For complex constraint elastic thin slabs made up of homogeneous isotropic materials,taking Kirchhoff assumption, the deflection of slabs $\omega(x,y)$ and the uniformly distributed load q_0 shall meet the following differential equation [1]:

$$\frac{\partial^4\omega}{\partial x^4}+2\frac{\partial^4\omega}{\partial x^2\partial y^2}+\frac{\partial^4\omega}{\partial y^4}=\frac{q_0}{D} \tag{1}$$

$D=Et^3/12(1-\mu^2)$ -The flexural rigidity of the slab, $E=$The elastic modulus of concrete; $t=$The thickness of the slab; $\mu=$The poisson's ratio of concrete.

The relationship between deflection and bending moment is as follows:

$$\begin{aligned} M_x&=-D\left(\frac{\partial^2\omega}{\partial x^2}+\mu\frac{\partial^2\omega}{\partial y^2}\right)\\ M_y&=-D\left(\frac{\partial^2\omega}{\partial y^2}+\mu\frac{\partial^2\omega}{\partial x^2}\right)\end{aligned} \tag{2}$$

WhereMx= the moment around y axis; My=the moment around x axis.

3 FORMULA OF FINITE DIFFERENCE METHOD ABOUT COMPLEX CONSTRAINT SLAB

Formula(1) is a four order partial differential equation, it is difficult to get its analytical solution. so the displacement of the slab is obtained through finite difference method. For example the node 0 in Figure 1.

[a] Corresponding author: wangjianchao005@163.com

DOI: 10.1051/matecconf/20166302012

$$\begin{cases}\left(\dfrac{\partial^4\omega}{\partial x^4}\right)_0=\dfrac{1}{h^4}[6\omega_0-4(\omega_1+\omega_3)+(\omega_9+\omega_{11})]\\ \left(\dfrac{\partial^4\omega}{\partial x^2\partial y^2}\right)_0=\dfrac{1}{h^4}\begin{bmatrix}4\omega_0-2(\omega_1+\omega_2+\omega_3+\omega_4)\\+(\omega_5+\omega_6+\omega_7+\omega_8)\end{bmatrix}\\ \left(\dfrac{\partial^4\omega}{\partial y^4}\right)_0=\dfrac{1}{h^4}[6\omega_0-4(\omega_2+\omega_4)+(\omega_{10}+\omega_{12})]\end{cases}\tag{3}$$

$$20\omega_0-8(\omega_1+\omega_2+\omega_3+\omega_4)+2(\omega_5+\omega_6+\omega_7+\omega_8)$$
$$+(\omega_9+\omega_{10}+\omega_{11}+\omega_{12})=\frac{q_0h^4}{D}.\tag{4}$$

ω_i = Nodal displacement in the slab; h = The length of the grid.

By the same token, the difference formula of moment 0 is as follows:

$$\begin{cases}(M_x)_0=\dfrac{D}{h^2}[2(1+\mu)\omega_0-(\omega_1+\omega_3)-\mu(\omega_2+\omega_4)]\\ (M_y)_0=\dfrac{D}{h^2}[2(1+\mu)\omega_0-(\omega_2+\omega_4)-\mu(\omega_1+\omega_3)]\end{cases}\tag{5}$$

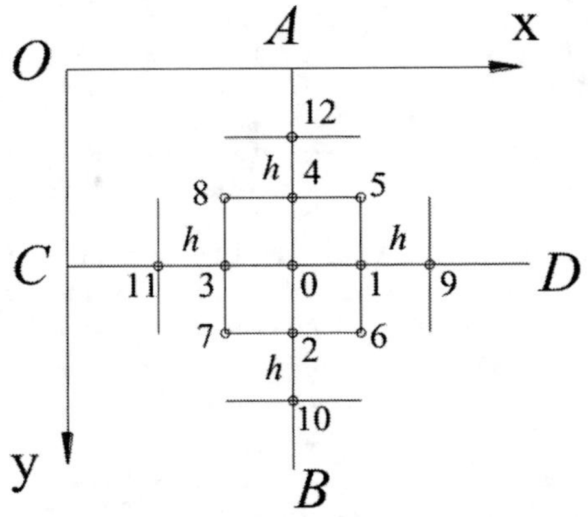

Figure 1. Schematic diagram of node.

4 BOUNDARY CONDITION PROCESSING ABOUT COMPLEX CONSTRAINT SLAB

For complex constraint slabs,there are at least a combination of two out of simply supported boundary , free boundary and built-in boundary in a straight edge with other straight edges simple or complex constraints. For example in Fig.2 ,this paper aims to present you with the thought of dealing with simple boundary and complex boundary.Assume that fig. 2 is a rectangular thin slab with boundary AB and AD clamped (simple boundary), the other for complex boundary (half clamped and half simply supported).

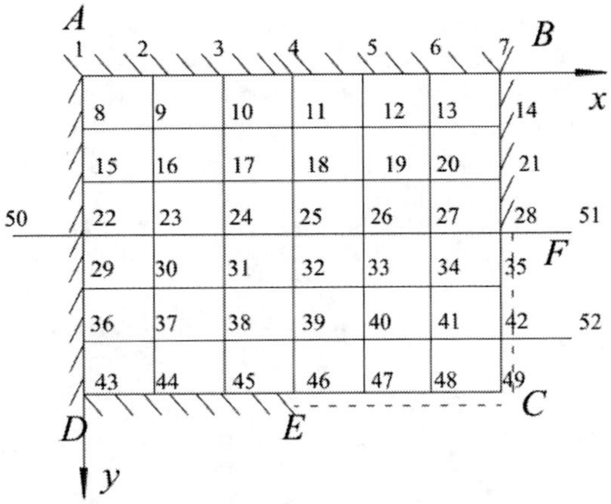

Figure 2. Boundary conditions of slabs.

Node 22: It is a built-in boundary, so the boundary displacement and first derivative of the displacement should be equal to zero,namely:

$$(\omega)_{22}=0,\qquad \left(\frac{\partial\omega}{\partial x}\right)_{22}=\frac{\omega_{23}-\omega_{50}}{2h}=0\tag{6}$$

Node 42: It is a simply supported constraint, so the boundary displacement and the second derivative of displacement should be equal to zero, namely:

$$(\omega)_{42}=0,\qquad \left(\frac{\partial^2\omega}{\partial x^2}\right)_{42}=\frac{\omega_{52}+\omega_{41}-2\omega_{42}}{h^2}=0\tag{7}$$

Node 28:In the intersection point with clamped and simply supported constraints, strictly speaking, the point neither belongs to clamped supported constraint nor belongs to the simply supported constraint. The point will be regarded as the combination of clamped and simply supported constraints in this article. According to the average of the two considerations, the specific calculation formula is as follows:

$$(\omega)_{28}=0,$$
$$\frac{1}{2}\left[\left(\frac{\partial^2\omega}{\partial x^2}\right)_{28}+\left(\frac{\partial\omega}{\partial x}\right)_{28}\right]\tag{8}$$
$$=[(2+h)\omega_{51}-(h-2)\omega_{27}-4\omega_{28}]=0.$$

If there is a free boundary, $\omega(x,y)$ of every point on the side should be regard as an unknown quantity.Equations of every point on the free boundary can be obtained through formula(8).

5 THE COMPARISON OF CALCULATION RESULTS ABOUT COMPLEX CONSTRAINT SLAB

In a high-rise residential building with Short shear wall structure, the corner window is set in the corner . Due to the need to set outtrigger ,the floor bove corner window will be a discontinuous constraint slab with a mix of clamped and simply supported . For a floor with supporting form shown in figure 2 , it can be solved through the above method . Suppose AB = AD = 3 m, for convenience the grid length are all h = 0.5 m and the grid nodal number are shown in figure 2.

Table 1 records z displacement about every point in the slab(positive direction displacement is said the same as the direction of q_0). $S_1=q_0h^4/D$,ω= $0.358q_0h^4/D$. when q_0 units off for kN/m^2, h units off for m, D units off for kN.m.The point displacement not listed in the table is zero.

Table 1. Nodal displacement of slabs with complex support.

n	Dis (S_1)	n	Dis (S_1)	n	Dis (S_1)	n	Dis (S_1)	n	Dis (S_1)
9	0.36	16	0.74	23	0.91	30	0.79	37	0.41
10	0.74	17	1.57	24	1.97	31	1.74	38	0.92
11	0.91	18	1.97	25	2.52	32	2.31	39	1.34
12	0.79	19	1.74	26	2.31	33	2.20	40	1.35
13	0.41	20	0.92	27	1.34	34	1.35	41	0.86

In order to obtain the difference of the results when complex constraints are imposed on the slab or not, so the author make boundary BC and CD in figure 2 respectively with fully clamped and simply supported constraint .

Table 2. Nodal displacement of slabs with BC, CD clamped constraint.

n	Dis (S_1)	n	Dis (S_1)	n	Dis (S_1)	n	Dis (S_1)	n	Dis (S_1)
9	0.34	16	0.68	23	0.81	30	0.68	37	0.34
10	0.68	17	1.39	24	1.66	31	1.39	38	0.68
11	0.81	18	1.66	25	2.00	32	1.66	39	0.81
12	0.68	19	1.34	26	1.66	33	1.39	40	0.68
13	0.34	20	0.68	27	0.81	34	0.68	41	0.34

Table 3. Nodal displacement of slabs with BC, CD simply supported constraint.

n	dis (S_1)	n	dis (S_1)	n	dis (S_1)	n	dis (S_1)	n	dis (S_1)
9	0.39	16	0.83	23	1.06	30	1.01	37	0.65
10	0.83	17	1.80	24	2.33	31	2.21	38	1.39
11	1.06	18	2.33	25	3.03	32	2.86	39	1.78
12	1.01	19	2.21	26	2.86	33	2.70	40	1.68
13	0.65	20	1.39	27	1.78	34	1.68	41	1.05

From table 1 to 3, you can see that the difference of displacement is large with the maximum near the midspan when applying to different constraint conditions .

Similarly, the moment corresponding to the above three kinds of boundary conditions are shown in table 4 to 6. the value for the moment in the table is about u=0.2. S_2= q_0h^3.

Table 4 . Node bending moment of slabs with complex support.

n	M_x (S_2)	M_y (S_2)	n	M_x (S_2)	M_y (S_2)	n	M_x (S_2)	M_y (S_2)
2	-0.72	-0.14	17	0.52	0.52	31	0.67	0.5
3	-1.48	-0.30	18	0.63	0.72	32	0.89	0.83
4	-1.82	-0.36	19	0.50	0.67	33	0.88	0.88
5	-1.58	-0.32	20	0.12	0.13	34	0.60	0.60
6	-0.82	-0.16	21	-0.37	-1.85	36	-0.16	-0.82
8	-0.14	-0.72	22	-0.36	-1.82	37	0.01	-0.10
9	-0.03	-0.03	23	0.26	-0.09	38	0.13	0.12
10	-0.05	0.20	24	0.72	0.63	39	0.45	0.48
11	-0.09	0.26	25	0.92	0.92	40	0.60	0.60
12	-0.10	0.24	26	0.83	0.89	41	0.45	0.45
13	-0.10	0.01	27	0.48	0.45	44	-0.82	-0.16
14	-0.16	-0.82	28	-0.11	-0.54	45	-1.85	-0.37
15	-0.30	-1.48	29	-0.32	-1.58	46	-0.54	-0.11
16	0.20	-0.05	30	0.24	-0.10	47	0.00	0.00

Table 5. Node bending moment of slabs with BC, CD clamped constraint.

n	M_x (S_2)	M_y (S_2)	n	M_x (S_2)	M_y (S_2)	n	M_x (S_2)	M_y (S_2)
2	-0.68	-0.14	18	0.63	0.65	33	0.53	0.53
3	-1.36	-0.27	19	0.53	0.53	34	0.21	0.01
4	-1.61	-0.32	20	0.21	0.01	35	-0.27	-1.36
5	-1.36	-0.27	21	-0.27	-1.36	36	-0.14	-0.68
6	-0.68	-0.14	22	-0.32	-1.61	37	0.00	0.00
8	-0.14	-0.68	23	0.24	0.00	38	0.01	0.21
9	0.00	0.00	24	0.65	0.63	39	0.00	0.24
10	0.01	0.21	25	0.79	0.79	40	0.01	0.21
11	0.00	0.24	26	0.65	0.63	41	0.00	0.00
12	0.01	0.21	27	0.24	0.00	42	-0.14	-0.68
13	0.00	0.00	28	-0.32	-1.61	44	-0.68	-0.14
14	-0.14	-0.68	29	-0.27	-1.36	45	-1.36	-0.27
15	-0.27	-1.36	30	0.21	0.01	46	-1.61	-0.32
16	0.21	0.01	31	0.53	0.53	47	-1.36	-0.27
17	0.53	0.53	32	0.63	0.65	48	-0.68	-0.14

Table 6. Nodal bending moment of slabs with BC, CD hinge constraint.

n	M_x (S_2)	M_y (S_2)	n	M_x (S_2)	M_y (S_2)	n	M_x (S_2)	M_y (S_2)
2	-0.77	-0.15	16	0.17	-0.11	30	0.17	-0.11
3	-1.65	-0.33	17	0.54	0.54	31	0.54	0.54
4	-2.12	-0.42	18	0.71	0.77	32	0.71	0.77
5	-2.02	-0.40	19	0.69	0.81	33	0.69	0.81
6	-1.29	-0.26	20	0.46	0.63	34	0.46	0.63
8	-0.15	-0.77	22	-0.42	-2.12	36	-0.42	-2.12
9	-0.07	-0.07	23	0.24	-0.16	37	0.24	-0.16
10	-0.11	0.17	24	0.77	0.71	38	0.77	0.71
11	-0.16	0.24	25	1.04	1.04	39	1.04	1.04
12	-0.13	0.27	26	1.00	1.07	40	1.00	1.07
13	-0.04	0.26	27	0.64	0.81	41	0.64	0.81
15	-0.33	-1.65	29	-0.40	-2.02	42	-0.40	-2.02

To obtain difference of the displacement value from every node under three constraints , we can see the figures below.Displacement diagram of node 22-28 are shown in Figure 3 and displacement diagram of node 3,10,17,24,31,38,45 are shown in Figure 4.

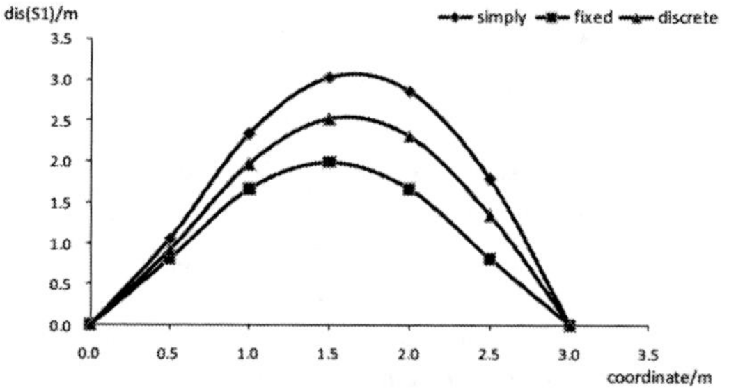

Figure 3. Displacement diagram of node 22-28.

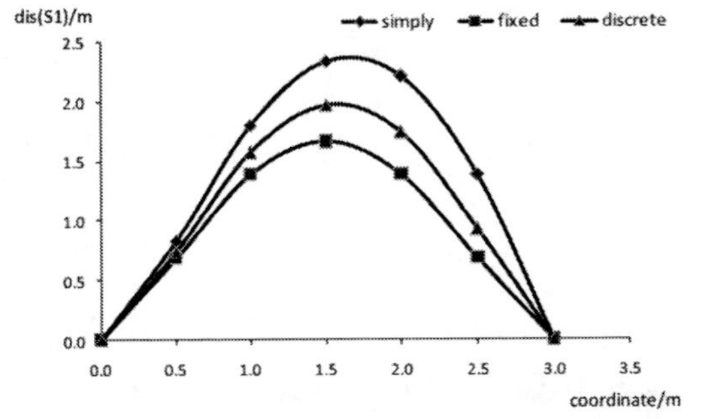

Figure 4. Displacement diagram of node 3,10,17,24,31,38,45.

As is shown in figure 5, My changes with X including node 22 to 28; As is shown in figure 6,Mx changes with y including node 3, 10, 17, 24, 31, 38. and they are all under three constraints.

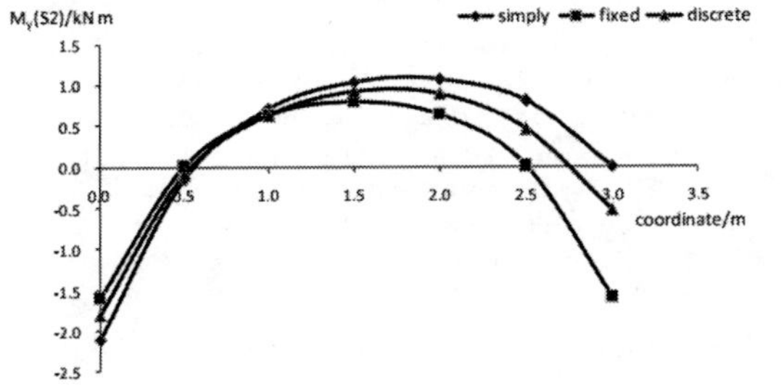

Figure 5. My diagram of node 22~28 .

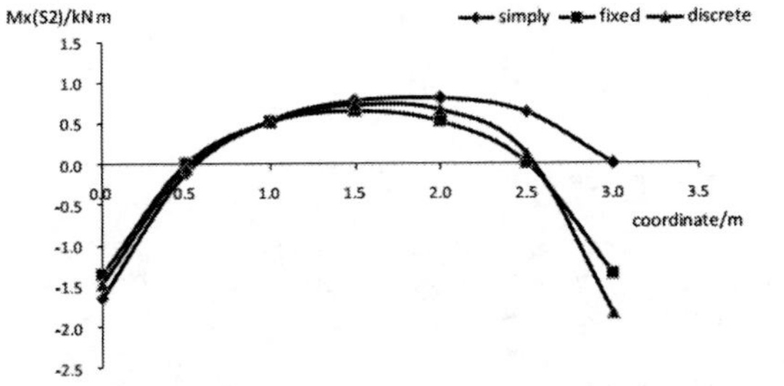

Figure 6. Mx diagram of node 3,10,17,24,31,38,45.

From figure 3 and figure 4, we can see that the difference of displacement is large when applying to different constraint conditions .We can learn that the difference near the midspan is largest and the difference at the lower right is greater than that at the upper left; From figure 5 and figure 6, you can see that the moment varies widely in the lower right, whears the difference in the cross is small. The result shows that the constraints have a significant impact on support moment, whears have little impact on mid-span moment.

6 CONCLUSION

1.The method proposed by the author is a good way to solve the problem such as displacement of complex constraint slabs and internal force calculation in the actual engineering.

2.When taking no account of problem of complex constraints, we will find the result of displacement and moment is different .And the result is neither unsafe or old-fashioned.As for this complex constraint slab in the actual project , the author recommend the proposed method in this paper.

ACKNOWLEDGEMENT

Project supported by Shenyang Jianzhu University (Grant NO.2014083)
Han Yu Project supported by Shenyang Jianzhu University(XKHY-19)

REFERENCES

1. American Concrete Institute & ACI Committee 318. 2008. Building Code Requirements for Structural Concrete and Commentary. Farmington Hills, MI: American Concrete Institute.
2. Eyre, J.R.2008.Surround stiffness to membrane action in concrete slabs.*Magazine of Concrete Research* 59(2):107-119.
3. Johnson, D. 2010. On the safety of the strip method for reinforced concrete slab design.*Computers and Structures* 79(11):2425-2430.
4. Muthu ,K .U. & Amarnath, K.. 2007. Load deflection behavior of partially restrained slab strips.*Engineering Structures* 29(5):663-674.
5. Taylor, S. E. & Mullin, B. 2009. Arching action in FRP reinforced concrete slabs. *Construction and Building Materials*20(2):71-81.
6. Wang Qingxiang & Wang Gang. 2010. Experiment Study on Compressive Membrane Action of Slab Strips Restrained by Shear Walls.*Key Engineering Materials*41(7):739-743.
7. Zheng, Y. & Robinson ,D. 2008. Analysis of compressive membrane action in concrete slabs. *Proceedings of the Institute of Civil Engineer*161(1):21-31.

MATEC Web of Conferences 63, 02013 (2016) DOI: 10.1051/matecconf/20166302013
MMME 2016

The Particle Number Emission Characteristics of the Diesel Engine with a Catalytic Diesel Particle Filter

Jia Qiang LI[1], Yan Yan WANG[1], Shu Chao HE[2], Huang Bin YAN[2] and Chao HE[1,a]

[1] *School of Mechanical and Traffic Engineering, Southwest Forestry University, Kunming 650224, China.*
[2] *Kunming Yunnei Power Co., LTD., Kunming 650224, China*

Abstract. Due to their adverse health effects and their abundance in urban areas, diesel exhaust ultrafine particles caused by the aftertreatment devices have been of great concern in the past years. An experiment of particles number emissions was carried out on a high-pressure, common rail diesel engine with catalytic diesel particle filter (CDPF) to investigate the impact of CDPF on the number emission characteristics of particles. The results indicated that the conversion rates of CDPF is over 97%. The size distributions of particles are bimodal lognormal distributions downstream CDPF at 1400 r/min and 2300 r/min. CDPF has a lower conversion rates on the nucleation mode particles. The geometric number mean diameters of particles downstream CDPF is smaller than that upstream CDPF.

1 Introduction

Particulate matter and nitrogen oxide (NOx) emissions from diesel engines are being recognized as the pollutants having adverse effects on the environment as well as on human health[1]. Many pathophysiological researches suggest that exposure to DEP, especially ultrafine particles, is associated with increased respiratory disease, cardiopulmonary mortality and other potential health effects [2]. Over the past few years, researches have begun to address the questions of the number concentration of DEP [3, 4] and regulations have been made to the number emissions of DEP in Euro V and VI emission standards. Among these after-treatment units, catalytic diesel particle filter (CDPF) has proved to be a very effective and promising technology for the abatement of particulate matter (PM) emissions from diesel engines. This technique not only reduces PM mass, but also provides good control over the number of ultrafine particles.

Maricq et al. studied the impact of DOC and sulfur content on the size distribution of DEP [5]. The results indicated that the nucleation mode particles would increase and the bimodal size distribution would appear only when the DOC and diesel with high sulfur were used together. Otherwise, the mono-modal distribution of the accumulation mode would occur. Biswas et al. investigated the size distributions , volatility, and agglomeration of particles from a heavy-duty diesel engine with DPF and SCR [6]. The advanced aftertreatment can decrease the mass of PM by 90%, but the number emissions of particles did not decrease on the stead modes and the nucleation mode particles increased. In this study, an experiment was carried out to investigate the number emission characteristics of particles emitted by the diesel engine with catalytic diesel particle filter (CDPF).

2 Experimental setup

2.1 Test devices

The test engine is a direct injection, high pressure common-rail, turbocharged diesel engine (WeiChai Power WP5, China), whose characteristics are shown in table 1. The engine with CDPF was tested on an engine test bench based on an AC dynamometer (Schenck HT350, Germany) with emission measurement system. An engine exhaust particle sizer (EEPS 3090, TSI, USA) was used for the investigation of particle matter (PM). To prevent the particle condensation and nucleation, two stages of ejector diluter (ED) upstream EEPS were used for sampling from raw exhaust stack (Burtscher, 2005). The fuel used in this study was locally available commercial low sulfur (50 ppm) diesel.

2.2 Measurement methods

The exhaust pollutant, temperature and pressure measurements were performed at the upstream and downstream of the after-treatment device. The results of emissions, fuel consumption and relevant engine parameters were recorded by the dynamometer control system. The stationary operation points (OPs), so called steps-tests were performed. The engine was operated at

[a] Corresponding author: hcsmile@aliyun.com

10%、20%, 30%, 40%, 50%, 60%, 70%, 80%, 90% and 100% of full load at seven engine speeds (1000, 1200, 1400, 1600, 1800, 2000 and 2300 r/min). All operation points were performed with a warm engine and for each esearch task always in the same sequence.

A portion of the exhaust gas emitted from the engine was extracted from the tailpipe and quickly diluted by two ejector dilutors (Dekati Ltd.) in series. The dilution air is preconditioned to remove the water and volatile organic compound (VOC). The dilution air and diluted exhaust in the first dilutor are heated to 200°C in order to prevent condensation. The second ejector dilutor is at room temperature to cool the sample prior to distribution into the EEPS. The dilution ratio during test procedure is around 64, which essentially inhibits all post-dilution particle dynamics such as coagulation and adsorption [7]. The dilution factor (DF) was found with the following expression:

$$DF = (C_e - C_a)/(C_t - C_a) \quad (1)$$

where C_e, C_t and C_a respectively denote the CO_2 concentration in the engine-out exhaust gas, in the dilution tunnel and in the air.

Table 1. Specification of the test engine.

Parameter	Feature/Size
Engine type	4-stroke, 4-cylinder, in-line
Bore×Stroke (mm)	108×10
Displacement (L)	4.76
Compression ratio	18
Fuel system	High pressure common rail
Max. Power (kW @ r/min)	132 (2300r/min)

3 Results and discussions

3.1. The total particle number emissions of DEP

Figure 1 show the total number emissions of DEP unstream and downstream CDPF. The total particle number emissions is about 2.8×10^6-1.8×10^8 $1/cm^3$ upstream CDPF. The emission peak of engine-out particles is at the full load of medium speeds. Because CDPF increased the backpressure of diesel engine, the total particle number upstream CDPF was higher than that upstream CDPF, especially at the heavy loads of medium speeds. The total particle number emissions downstream CDPF increased with the incresing engine speed, and is about 4.6×10^3-7.96×10^4 $1/cm^3$, which is reduced by 3 order of magnitude compared with that upstream CDPF. The emission peak of particles downstream CDPF is at the heavy loads of high speeds. CDPF has the high conversion rates on the particle number emission with the minimum value of 97% at high speeds and heavy loads. The conversion rates of CDPF can reach 99.9% at the heavy loads of medium speeds.

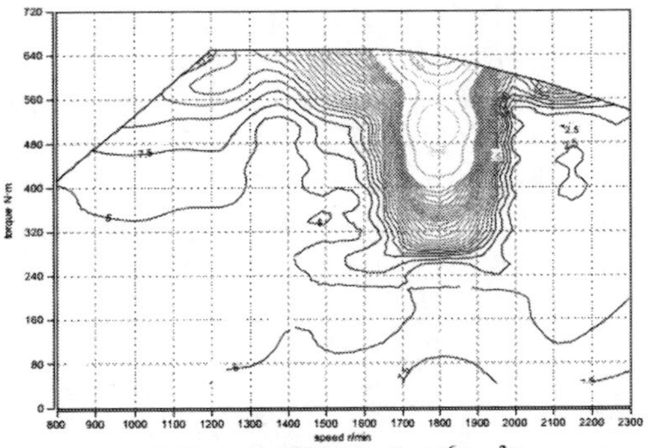

(a) Upstream CDPF （$\times10^6/cm^3$）

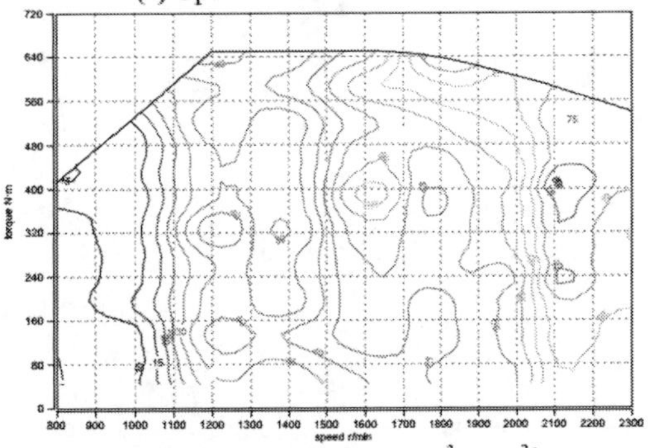

(b) Downstream CDPF （$\times10^3$ $1/cm^3$）

Figure 1. The total number emissions of DEP unstream and downstream CDPF.

3.2 The size distributions of DEP

Figure 2 showed the size distribution of DEP upstream and downstream CDPF and the conversion rates of CDPF at 1400 r/min. It can be seen that the size distribution of DEP upstream CDPF is mono-modal at 1400 r/min, and the peak concentration are at the diameter of 39-70nm at 10%-80% of full load, but 9-16nm at full load. The particle number concentration decreased firstly and then increased with the increasing engine loads. The maximum number concentration appeared at the full load, which is 5.8×10^7 $1/cm^3$, and the minimum number concentration is 5.8×10^6 $1/cm^3$ at the 60% of full load. Downstream CDPF, the size distribution is bimodal and the peaks are at the 7~10nm and 29~39nm. The conversion rates of CDPF keep over 99% except at the 20% of full load. CDPF has a lower reduction effect on the nucleation mode particles, that is 6nm~10nm.

DOI: 10.1051/matecconf/20166302013

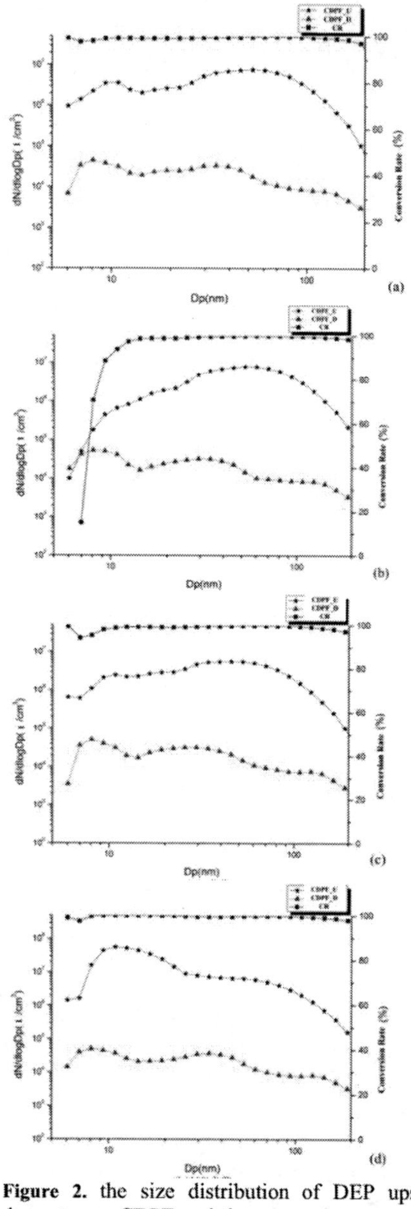

Figure 2. the size distribution of DEP upstream and downstream CDPF and the conversion rates of CDPF at 1400 r/min. engine load: (a) 10%, (b) 40%, (c) 80%, (d) 100%.CDPF_U: upstream CDPF, CDPF_D: downstream CDPF; CR:conversion rate

Figure 3 showed the size distribution of DEP upstream and downstream CDPF and the conversion rates of CDPF at 2300 r/min. At every load, the size distributions of particles are bimodal lognormal distributions upstream and downstream CDPF. The two peak concentrations upstream CDPF are at 6-9nm and 29-52nm, while at the diameter of 7-10nm and 29-39nm downstream CDPF. The conversion rate of CDPF decreased with the increasing particle diameter at the low load. The conversion rates of CDPF keep over 98%. CDPF has a lower reduction effect on the nucleation mode particles, that is 10nm~19nm.

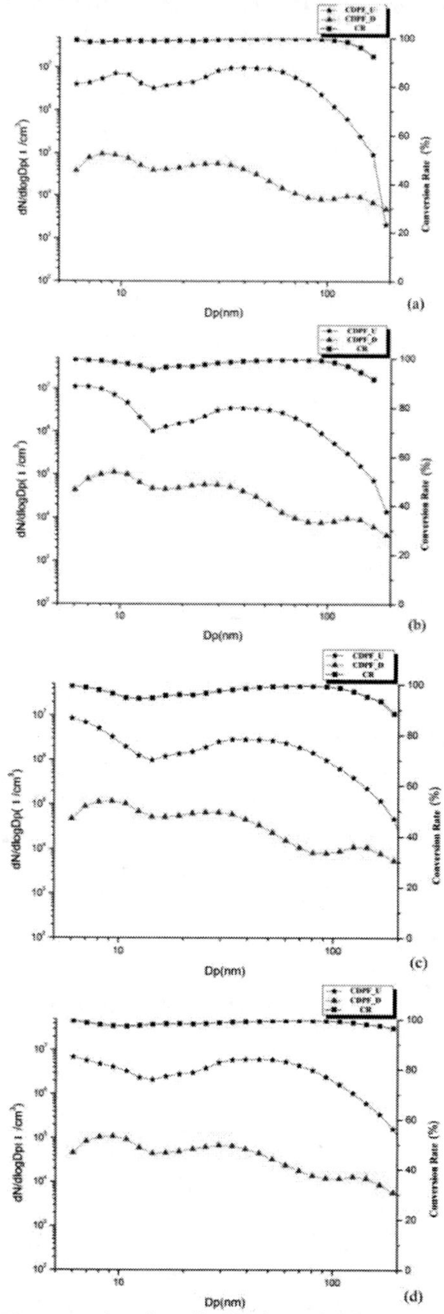

Figure 3. the size distribution of DEP upstream and downstream CDPF and the conversion rates of CDPF at 2300 r/min. engine load: (a) 10%, (b) 40%, (c) 80%, (d) 100%. CDPF_U: upstream CDPF, CDPF_D: downstream CDPF; CR:conversion rate

3.3 GMD of exhausted particles

Many literatures report that the experimentally measured size distribution of diesel exhausted particle can be fit to the lognormal form as following:

$$\frac{dN}{d\ln D_p} = \frac{N}{\sqrt{2\pi}\sigma}\exp[\frac{-(\ln D_p - \ln\mu)^2}{2\sigma^2}] \quad (2)$$

N is the number of particles, μ is the geometric number mean diameters (GMD) and σ is the corresponding geometric standard deviation (GSD).GMD and GSD are two important parameters regarding particle size. Table 2 gave the GMD and GSD of particles upstream and downstream CDPF. It can be seen that the GMD downstream CDPF is smaller than that upstream CDPF. The GMD of engine-out particles are about 38-58nm, and the GMD of particle downstream CDPF are about 17-30nm.

Table 2. The GMD and GSD of particles upstream and downstream CDPF.

Engine speed (r/min)	Engine load (%)	Upstream CDPF		Downstream CDPF	
		CMD (nm)	GSD	CMD (nm)	GSD
1400	20	42	1.94	29	2.53
	40	48	1.79	21	2.48
	60	40	1.88	21	2.43
	80	34	2.06	22	2.40
	100	16	1.83	22	2.43
2300	20	24	2.29	18	2.33
	40	15	2.41	17	2.26
	60	14	2.38	17	2.27
	80	18	2.56	18	2.29
	100	26	2.47	20	2.39

4 Conclusions

(1) The total particle number emissions of diesel engine is 2.8×10^6-1.8×10^8 1/cm^3 upstream CDPF and 4.6×10^3-7.96×10^4 1/cm^3 downstream CDPF. CDPF has the high conversion rates on the particle number emission with the minimum value of 97% at high speeds and heavy loads. The conversion rates of CDPF can reach 99.9% at the heavy loads of medium speeds.

(2) The size distributions of particles are bimodal lognormal distributions downstream CDPF at 1400 r/min and 2300 r/min. CDPF has a lower conversion rates on the nucleation mode particles.

(3) The geometric number mean diameters of particles downstream CDPF is smaller than that upstream CDPF. The GMD of engine-out particles are about 38-58nm, and the GMD of particle downstream CDPF are about 17-30nm.

Acknowledgements

This work was financially supported by the National Natural Science Foundation of China (NSFC, 51266015), Applied basic research project of Yunnan Province (2013FB052), Department of Education, Yunnan province (2013Z081) and Scientific Research Foundation of Southwest Forestry University (C14120).

References

1. J. S. Lighty, J. M.Veranth, A. F. Sarofim, "Combustion aerosols: factors governing their size and composition and implications to human health," J. Air. Waste. Manage, vol. **50**, pp.1565-1618, (2000)
2. C. A. Pope, D. W. Dockery, J. Schwartz, "Review of Epidemiological Evidence of Health-Effects of Particulate Air- Pollution," Inhal. Toxicol, vol. **7**, pp. 1-18, (1995)
3. M. M. Maricq, D. H. Podsiadlik, R. E Chase, "Size distributions of motor vehicle exhaust PM: a comparison between ELPI and SMPS measurements," Aerosol. Sci. Tech, vol. **33**, pp. 239-260, (2000)
4. D.B. Kittelson, W.F. Watts, J.P. Johnson, "On-road and laboratory evaluation of combustion aerosols-Part1: Summary of diesel engine results," J. Aerosol. Sci, vol. **37**, pp. 913-930 , 2006.
5. M. Maricq, R. Chase, N. Xu, The effects of the catalytic converter and fuel sulfur level on motor vehicle particulate matter emissions: Light duty diesel vehicles[J]. Envir Sci & Tech, **36** （*2*）: 283-289. (2002)
6. S. Biswas, S. Hu, V. Verma, Physical properties of particulate matter (PM) from late model heavy-duty diesel vehicles operating with advanced PM and NOx emission control technologies[J]. Atmo Envir, **42** （*22*）: 5622-5634. (2008)
7. I. Abdul-Khalek, D. Kittelson, F. Brear, "The Influence of Dilution Conditions on Diesel Exhaust Particle Size Distribution measurements," SAE Technical Paper Series, 1999-01-1142 , (1999)

MATEC Web of Conferences 63, 02014 (2016)
MMME 2016
DOI: 10.1051/matecconf/20166302014

On High Vacuum Densification of Soft Foundation Treatment

Mei Ling GUO[1,a]

[1]*Continuing Education College, Guangdong University of Science and Technology, 523083, Dongguan, China*

Abstract. The high vacuum densification, which inserted the vacuum tube to increase the drainage channel, set up the improved drainage system so that it can effectively reduce the underground water level, can be used for the reinforcement of the deep soft foundation. According to analyze the preparation and process of high vacuum densification, and summarize the control points of high vacuum densification, it renders great value to the constuction of soft foundation treatment.

1 Introduction

The high vacuum densification is a kind of soft foundation reinforcement method emerging in recent years. It is the development of the traditional dynamic compaction method: specifically, the main difference between the dynamic compaction method and the high vacuum densification method lies in the fact that the former does not take effective way of drainage so that it is only applicable for the reinforcement of backfill with a low water content and its applicable scope is thus limited.

As for Hubei Xiantao Station located in Jianghan Plain and featured by high underground water level, abundant water volume and thick silt layer, to adopt the dynamic compaction method will probably make the soft soil become the "spring soil", which thus cannot meet the design requirement; however, the high vacuum densification method has taken the effective way of drainage, inserted the vacuum tube to increase the drainage channel, set up the improved drainage system so that it can effectively reduce the underground water level; moreover, coupled with the variable energy densification, it can be used for the reinforcement of the deep soft foundation; therefore, both of its reinforcement depth and applicable scope are much larger than those of the traditional dynamic compaction method.

Xiantao Station of Hankou-Yichang Railway was located at Hubei. The Freight Terminal of Xiantao Station adopted the reinforcement way combined with high vacuum densification plus plastic drainage board: the plastic drainage board was a way to reduce the underground pore water over a long period of time; however, under the condition with no external load imposed, its dewatering effect would not be obvious so that only with the external load imposed could it effectively reinforce the foundation. Xiantao Station was featured by soft foundation soil, poor structure, high underground water level, abundant water volume and thick silt layer, upon which with the method combined with vacuum dewatering and dynamic consolidation technology adopted, the goals of reinforcement both for the shallow and deep layers could be achieved at the same time.

Its reinforcement principle lay in the fact that the "differential pressure", actually, the excess pore water pressure generated by the densification was called as the "positive pressure" while the pressure generated by the operation of the vacuum pump was called as the "negative pressure", manually produced in the soil layer was to rapidly dissipate the excess pore water pressure and drain the water in the soft soil layer quickly. The adoption of the high vacuum drainage made the densification effect increased greatly, which thus made the processed soil mass formed into a consolidated "hard crust" with a certain thickness: the existence of the "hard crust" led to the effective extension of the surface load, which reduced the uneven settlement generated by the uneven load.

2 Construction process

After the construction scheme had been determined, it is time to confirm the construction process for the high vacuum densification; the specific construction process was shown as follows.

[a] Corresponding author: hsp0121@126.com

DOI: 10.1051/matecconf/20166302014

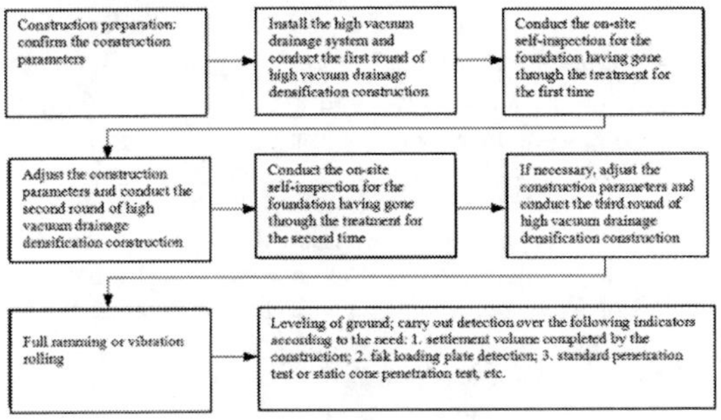

Figure 1. Construction process

3 Densification Preparation

3.1. Field clearing

Before the construction, the clearing, leveling and rolling were carried out first for the field to be treated so as to form a horizontal drainage slope larger than 1%; and then the clean sand with a thickness of 0.5m and the soil with a thickness of 1m would be filled as the working layer for the high vacuum densification.

3.2. Construction of the plastic drainage board

After the ground had been leveled, the workers carried out the construction for the plastic drainage board; before the construction, workers filled the coarse sand with a thickness of 0.2m as the underlying bed for the construction of the drainage board, upon which workers would implement the construction for the plastic drainage board. And then workers moved the machine into the construction site, put the drainage board into the drum and led the board into the pipeline through the pulley on the door frame; and then workers pulled out the drainage board from the pipeline, connected it with the pile tip, tightened it up, nestled it up to the pipeline shoe month and aimed it at the pile location. And then workers sank the pipeline into the design depth.

After the design depth had been reached, workers pulled out the pipeline. And then workers cut off the drainage board at the place 50cm above the sand bed and buried it into the sand bed towards the flowing water slope direction. During the process of driving, it was necessary to carry out the self-inspection board by board and to make the construction records appropriately.

Meanwhile, it was necessary to check the construction condition for each piece of board and only when it met the acceptance criteria could the machine be moved to drive the next piece; during the pulling of the pipeline, it was necessary to prevent the drainage board from being drawn out: if the drawn-out length was larger than 0.5m, it was necessary to carry out the supplementary driving at the adjacent board. During the construction for the plastic plug board, workers had plugged the board "one by one" according to the "reverse driving" to ensure that the mark of the plug point would not be substantially damaged or moved.

Moreover, the variety, specification and quality of the plastic drainage board shall meet the design requirements. It would be better that if the plastic drainage board driven into the foundation was an entire board; the board shall not be lengthened for application. The coil belt of the plastic drainage board stacked at the construction site shall be properly covered so as to prevent it from being aged due to the exposure to the air; during the driving of the plastic drainage board, it was necessary to pay attention to the control over the perpendicularity of the pipeline, the deviation of which shall be not more than ±1%; the steel pipeline shall not be bent and the permeable filter sleeve shall not be torn & polluted so as not to affect the drainage effect.

The holes generated by the construction of the drainage board shall be backfilled with the medium-coarse sand in a timely manner so as to prevent the falling of the soil block. After the construction of the drainage board had been completed, it was necessary to pave a medium-coarse sand bed with a thickness of 0.2m on it for coverage.

3.3. Determine the foundation coefficient

Before the formal densification, what workers shall do first was to determine the foundation coefficient. Workers had adopted the static penetrometer to carry out the further investigation over the field to be treated; as for the area with larger geological differences, workers would divide it into different sub-areas and adopt different construction technology parameters.

Through the static cone penetration test, workers found that the geological conditions for the section DK83+420-DK83+720 were basically the same; moreover, according to the mechanical equipment and labor force allocation, workers decided to divide the area into two construction sub-areas and to carry out the flow construction.

3.4. Dredged the excavated ditch

Workers deepened and dredged the excavated ditch between the freight yard and departure line to make it be the channel for vacuum drainage; meanwhile, workers dug the wells at the terminals of the ditch and drew the water by the pump to prevent the water accumulation in the ditch.

Moreover, workers dug the shockproof ditch between the mixing pile and the densification boundary to prevent the damage exerted by the densification on the pile body.

3.5. Set up control piles

Workers should set up control piles around the densification area; the control piles were 15 meters away from the densification boundary and one control pile was erected every 40 meters. Every densification construction sub-area would be equipped with one static cone penetration hole with a depth of 6-8m; before the treatment but after the dynamic compaction for each time, the test would be carried out. Two groups of hole pressure measuring points were set, the depth for both groups was 3m, 5m and 8m.

3.6. Plug the vacuum pipeline

After everything had been prepared, the key link was to plug the vacuum pipeline (according to the requirements of design, see the appended drawing); the vacuum construction adopted the 3m pipeline to majorly adjust the water content of the soil & sand and the 6m pipeline to majorly adjust the water content of the deep soil: specifically, the spacing between the vacuum pipelines was 3.5m*4.5m. The workers set up the peripheral high vacuum pipeline closed systems with spacing of 2m outside the construction area, which was 2m away from the construction boundary: specifically, the length of the deep sealing pipeline was 6-8m while that of the shallow one was 3-4m. The peripheral high vacuum pipeline closed systems could not be demolished until the foundation treatment within the area had been finished.

The flow construction was carried out within this construction area; according to the actual condition of the peripheral sealing pipeline was not set separately so that the vacuum pipeline within the construction area would be used as the peripheral sealing pipeline.

During the pipeline plugging, the high-pressure water gun was used to punch the hole to assist the manual plugging; meanwhile, the mechanical pipeline plugging could also be adopted to improve the work efficiency. The vacuum pipeline adopted the Φ 32 steel pipeline and the lower inlet opening was wrapped by two layers of nylon leakage membranes. The horizontal pipeline adopted the Φ 63PVC pipeline; the horizontal pipeline would be connected with the vacuum pipeline by the flexible rubber hose with the steel wire twined; meanwhile, the outside of the connection joint would be tightly sealed by the tape.

To prevent the damage of the filter screen, before the vacuum pipeline had been put into, it was necessary to check it carefully so as to ensure the intactness of the filter screen. After the completion of the pipeline plugging, it was necessary to connect the horizontal pipeline to the vacuum pump; each set of the unit shall be connected with 60-80 vacuum pipelines at the same time to form a vacuum system. It is easy to know that the principle of the high vacuum densification is: the vacuum would lead to the pressure and thus achieve the effect to treat the soft foundation.

After the completion over the layout of the vacuum pipeline, workers were to carry out the vacuum drainage and start up the vacuum system: specifically, during the operation of the vacuum system, the vacuum degree could reach 60KPa; during the construction period, the continuous water pumping would be conducted and the dual power supply would be prepared; specifically, the normal pumping rule would be "big then small, feculent then clear". 2-3 days before the pump starting, the pumping volume of each vacuum pump would be about 200 liters per hour while 4-7 days before the pump starting, the pumping volume of each vacuum pump would be about 50 liters per hour; during the construction process, through the measurement over the buried water level observation pipeline, when the water level dropped to the place 4 meters below the compaction surface, the densification construction shall be started.

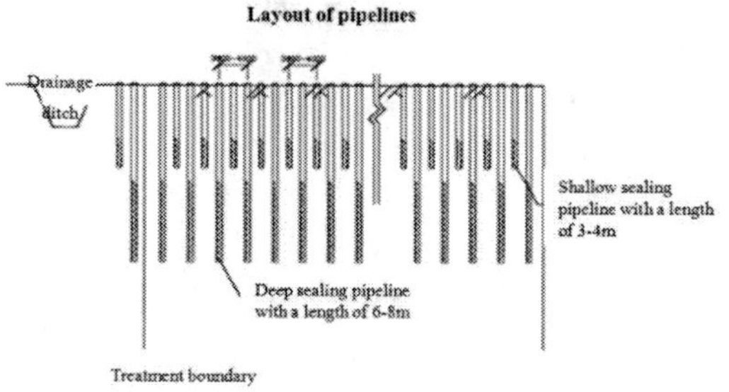

Figure 2. Plane schematic diagram for layout of vacuum pipelines

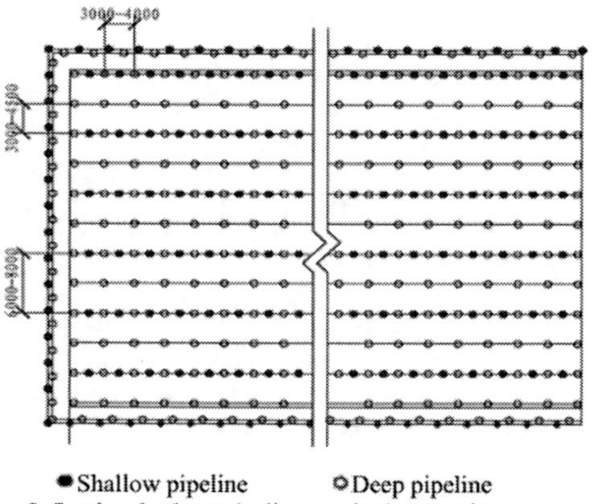

Figure 3. Sectional schematic diagram for layout of vacuum pipelines

4 Densification Process

After the above work had been prepared, workers can begin the first densification, which was distributed variable energy, according to the static cone penetration test and tracking tamping effect, and then adjusted densified energy according to the Ps-H curve. Before the formal densification construction, workers carried out single-point trial tamping within the field, to confirm the best tamping energy of the densification. Workers applied crawler crane of 50 tons for in-site densification, pounder with weight of 15 tons and diameter of 2.14 meters, with maximum lifting height of 14 meters.

After the in-site trial tamping (based on the last two tamping sinkages less than 5cm), workers confirmed that: point tamping energy was 2000KN*M (15 tons*10KN*13.3m), and the full tamping energy was 800KN*M. Effective reinforcement depth was: H=0.6 √200=8.6 meters (0.6 is geological factor, and takes 0.6 for sand layer; 200 is tamping energy, and its unit is tons meters.). The densification amount of each point was 4 and repeated 2 times, and full tamping was 1 time. The tamping points were distributed according to 3.5m*6m, while two times tamping points were arranged in the form of quincunx.

After the completion of first densification, workers carried out a static cone penetration test and compared it with results before tamping, in order to adjust the parameters of next construction step; workers used bulldozers to level the tamping holes formation in the densification; workers arranged square grid by 20m*20m to measure (level) settlement, and the measuring point location was consistent with that before tamping.

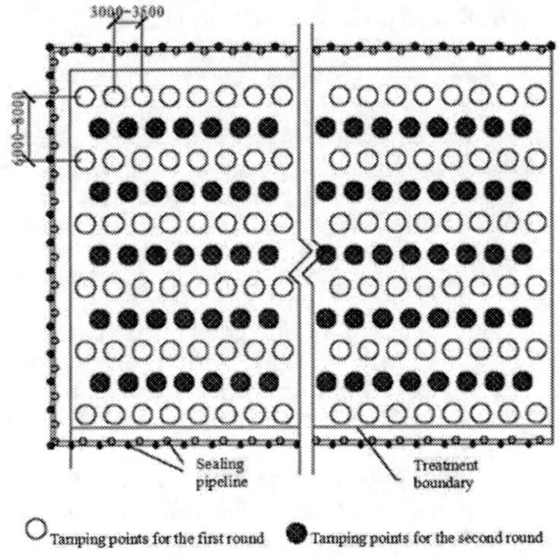

Figure 4. Tamping points during the densification process

After the completion of first densification, workers carried out a second high vacuum drainage, to continue adjusting the soil water content, in order to approach the best content for reinforced soil mass.

After the second high vacuum drainage, workers then began the second high vacuum densification. Through the embedded pore water pressure gauge, workers carried out the second one after the excess pore water pressure dissipated by 75% -85%, with tamping points arranged by 3.5m*6m, in the form of quincunx similar to those of the first time. Point tamping energy was 2000 KN*M, the densification amount of each point was 4, which was required to be same with the first densification.

Then workers had the third vacuum drainage which was the last one. After it, workers processed low energy full tamping; after excess pore water pressure dissipated by 75%-85%, and after the completion of the above two high vacuum drainage densifications, the foundation below the densification hole bottom can be effectively reinforced, and the bearing capacity meet the expected design requirements; the soil above hole bottom was relatively loose, so workers carried out a low energy full tamping in order to further increase bearing capacity and stiffness of the upper foundation. The densification amount of each point was 1.

So far, the whole construction process of high vacuum densification completed. The foundation after treatment reached its bearing capacity up to 125Kpa.

5 Conclusion

Control points of high vacuum densification include: confirmation of densification parameter, confirm densification energy, densification drop distance, densification points through trial tamping. Control of densification construction process: static cone penetration test, underground water level observation, densification settlement measurement, vacuum degree observation of vacuum pump.

References

1. Pan Quan-Xiang. Data Handbook of Safety Production and Civilization Construction for Building Installation Works. Beijing: China Building Industry Press, 1995.
2. Hong Sheng-Wei. Practical Handbook for Quality Engineer. Beijing: China Metrology Press, 2004.

Study on application and performance of OPPC under ice-phased condition

Yi AN [1,a]

[1]*Information and Telecommunication Company of Shanxi Electric Power Corporation, Taiyuan 031000, China*

Abstract. The optical fiber composite overhead line (OPPC) combines electricity transmission and information transmission, is used increasingly widely in the electric power system, further broadening the application area of our country's special cable in the meantime. In heavy ice-phased regions, the designing parameters and technical requirements should be of higher standards, or it will directly affect the safety and stability of electric communication system's operation. Therefore, OPPC under ice condition performance changes should cause enough attention. This article proposes simulation test of OPPC under ice-cladding condition, basing on which the mechanical properties and light transmission performance were calculated and analyzed. Then it comes to a conclusion that the ice-cladding has a variety of impact on OPPC in the stress and strain of fiber optic cable, optical transmission performance, residual RTS and other related elements. The test results show that the corresponding tension values under ice thickness would change and should be taken seriously into consideration when discussing the application of OPPC line under ice-phased condition.

1 Introduction

OPPC is developed in recent years gradually a kind of new type of electric power special fiber optic cable, and power transmission and double function of communication. And it has been widely used in related to OPGW. Compared with special ADSS optical cable, OPPC has many unique advantages, which make it in our country granted a broad application prospect in low voltage power distribution network. The emergence of OPPC and put into use, further broaden the application area of our country's special cable. Used special OPPC new technology has solved the power cable in electric corrosion, the disadvantages of theft, struck by lightning, provide low voltage intelligent power grid automation development in the future with optical fiber communication channel.

OPPC has been used in domestic voltage level of 220kV and below the line, to the reformation and the new line, it cannot erect OPGW, ADSS and ordinary optical cable lines, providing the effective solution. According to incomplete statistics, the country already has more than 4, 000 km of OPPC related voltage grade line application, in addition to the regular meteorological conditions and geographical conditions in application, especially at high altitudes, where the alpine areas and mountain areas, high temperature region were applied. But when it is applied in repeat ice should fully consider the ice to the influence of OPPC, otherwise the continuous low temperature sleet freezing disasters in the atmosphere, will cause transmission lines ice break down accident, seriously affecting the safe and stable operation of power grid China's vast territory. Meteorological condition difference is bigger, ice is inevitable, and thus the safety of the product and the safety of engineering application is particularly significant

Transmission line conductor ice coating side of the issue has caused the attention and research, but OPPC and conductor is different, which is formed in the element of optical fiber composite phase line of a new type of electric power special optical cable, mainly consisting of optical unit and wire unit. OPPC transmission performance by external force time unit is not affected.

Using laboratory simulation environment, a practical engineering application of OPPC ice simulation experiment was carried out, based on the certain condition of ice thickness OPPC performance of theoretical calculation and experimental study, analysis the influence of ice coating on properties of OPPC.

2 Simulation test description of ice-phase condition

Trials to select a voltage level of transmission line engineering construction of OPPC as sample, the structure of ground structure, diameter 17.50 mm, nominal tensile strength of 55.7 kN ice test ice thickness should be set according to the local weather condition. Table 1 shows the test sample application area of ice design weather conditions, climate conditions usually ice

[a] Corresponding author: 525492887@qq.com

thickness is generally 10 mm, considering the special situation, and design value should be adjusted.

Table 1. Ice design weather conditions.

No.	Meteorological parameters	Temperature(℃)	Wind speed(m/s)	Ice thickness(mm)
1	Lowest temperature	-38	3	0
2	Average temperature	-10	2	0
3	Fastest wind speed	-5	28	0
4	Ice-phased	-5	12	11
5	Highest temperature	38	0	0
6	The proportion of ice	0.892		

Table 2. Different ice thickness and tension comparison.

Parameters	Calculation methods			
Ice thickness(mm)	10	18	26	34
Ice weight(kg/m)	0.7652	1.9875	3.4750	5.8782
Tension(kN)	15.833	25.767	37.684	51.950

A lot of ice classification, according to the formation of ice, nature, appearance, etc. There are several classified methods, choose the cylindrical ice to facilitate the test. Ice density is 0.9 g/cm3, preliminary work for 16% RTS = 8.912 kN (RTS) = 55.7 kN o due to the actual line biggest span is about 269 m, 269 m to 30 m conversion laboratory will test file is apart from the corresponding ice conditions bligh. Calculate the 10 mm, 20 mm, 30 and 40 mm mm corresponding weight of ice and ice thickness by tension values, are shown in table 2 and table 3. As you can see, 10 mm and 20 mm when the thickness of ice cable tension by RTS 30.1% and 30.1% respectively RTS, the factory inspection qualified products can meet the test requirements, when the ice thickness of 30 mm, tension cable suffered 72.4% RTS, need a validation test.

Ice experiment plan: prepare OPPC samples of 100 m, exert tension span 30 m, on both ends of the concatenated fiber length 1 000 m; Calculation of 30 m span, different thickness ice weight (kg/m) are shown in table 3, exert tension 16% RTS 8.912 kN operation; Using simulated ice cover, points a few times evenly coated heavy, maintain after the completion of 96 h, during the test record of optical fiber attenuation, force value, the change of cable length, etc.

3 Simulation test results under 20 mm

3.1. Optical fiber attenuation test

Experiment using the ice line fiber 6.652 single-mode fiber, test using OTDR 1 550 nm wavelength window. 1550 nm wavelength in the process of test, the simulated ice about 72 h, fiber optic additional attenuation increases gradually, the surge in about 0.035 dB/km, which will affect the transmission performance of the fiber. 96 h after removing ice, additional attenuation fiber recovery to the initial state before the ice.

Test attenuation curve is shown in figure I, you can see, in the process of ice for a long time, the cable tension, make its produce strain, causes the fiber in the OPPC released in full, fiber began to stress, the additional damping increase.

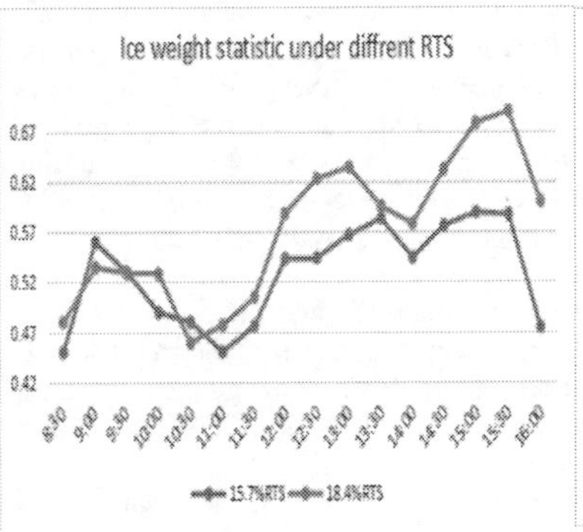

(a)

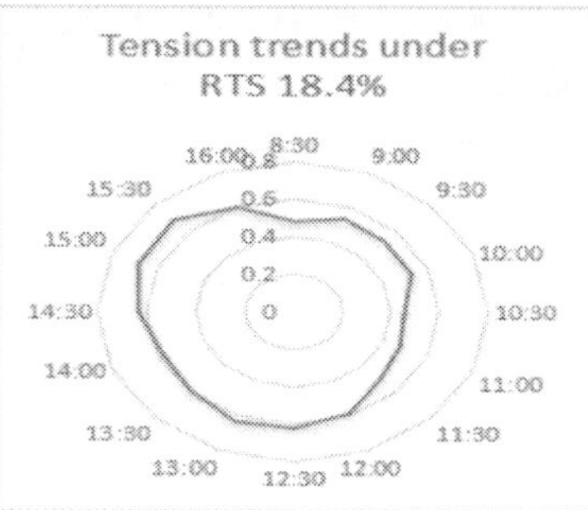

(b)

Figure 1. Ice weight statistics under different RTS and related tension analysis.

3.2 Stress strain test

Stress and strain test is used to determine its condition at a given load mechanical properties and fiber attenuation changes. Bligh for 30 mm thickness of ice conditions, in the case of 40% 40% RTS and RTS tensile force of divination for ice before and after the stress and strain test of the sample. Fiber tensile load and performance shall comply with and the provisions of the corresponding optical fiber length is not less than 100 m.

Stress and strain test is used to determine its condition at a given load mechanical properties and fiber attenuation changes. Bligh for 30 mm thickness of ice conditions, in the case of 40% 40% RTS and RTS tensile force of divination for ice before and after the stress and strain test of the sample. Fiber tensile load and performance shall comply with and the provisions of the corresponding optical fiber length is not less than 100 m.

(1) RTS tensile stress strain test

Acceptance criteria: 40% RTS tension, fiber optic strain, optical fiber no additional damping.

Test results: ice before the test curve as shown in figure II, after the ice test curve is shown in Figure III. Figure III and figure II contrast can be seen that the top 40% RTS ice force, fiber optic strain, optical fiber without additional attenuation; Ice after RTS pull down 40%, fiber optic strain, but there are additional two optical attenuation is greater than 0.03 dB, additional one optical attenuation is 0.05 dB, the ice has certain influence on the transmission performance of the fiber.

(2) RTS tensile stress and strain test

Acceptance criteria: RTS pulled in 60%, 0.25%, fiber optic strain optical fiber attenuation core 0.05, additional tension after cancel, no obvious residual additional fiber attenuation.

Test results: ice before the test curve as shown in figure 4, the ice after the test curve . Contrast, ice OPPC samples before fiber optic strain, fiber optic attenuation core 0.05, additional tension after cancel, no obvious residual additional fiber attenuation; Ice OPPC samples after fiber optic strain, with a single fiber 40% RTS tensile stress and strain test of the same root fiber, additional attenuation is beyond 0.05, after pulling force to cancel, no obvious residual additional fiber attenuation.

(3)30 mm ice OPPC cable length change test

Figure 6 to 30 mm thickness of ice conditions blight OPPC cable length change curve, fiber optic cable length before and after the ice by initial elongation of 28.50 m to 28.50 m, long after the ice to maintain phase line after a slight fluctuation change value tends to be stable, the length of 28.62 m and wood remains unchanged. 96 h after removing ice, sample length is 28.62 m.

3.3 After the ice tensile performance

After freeze disaster through the atmosphere, easy to pour tower bolt accidents, accident the main reason is that conductor ice coating thickness more than design ice thickness, ice produce longitudinal unbalanced than tower design load of experimental study on ice thickness exceeds design ice thickness of OPPC the effects of tensile performance.

After 30 mm thickness ice, tensile force test for OPPC tension RTS 95% and lasts 1 min. In the process of test, found that while maintaining time to 20, "pa" noise from the OPPC, tension not reduce at this time, still remained at 95% RTS until the end of the test, after dissection found cable outer single fracture, adjacent to the outer aluminum alloy wire, may cause broken stocks, as shown in figure III, fracture mono-filament largest number 6, at least one root. Through the shape and structure of the breaking point found for transverse tensile fracture. That ice thickness of 30 mm beyond the OPPC sample ice thickness design value, can lead to tensile performance drop. Easy to broken cable accidents.

4 Simulation test results under 40 mm

According to 40 mm thickness of ice conditions in table 3 groundwater calculation of tension and ice thickness data conversion, wooden article has carried on the laboratory span 40 mm ice simulation test, tension is 54.56 kN, OPPC suffered 98.0% RTS, approaching OPPC rated tensile strength values. Test using simulated ice points several times evenly coated. Covering all maintain after the completion of 96 h. Observe the change of the sample during the test.

(a)

(b)

Figure 2. Outer layer fracture of tested OPPC.

During the ice to maintain about 48 h, OPPC outer a mono-filament appear rupture, adjacent outer mono-filament fracture all at the same time, as shown in figure 2, fiber tensile strength is reduced, the experiment is over.

Because the OPPC samples during the ice test by heavy far beyond can withstand the design value of its own. Therefore mono-filament fracture. Illustrate the OPPC samples can't bear more than 40 mm and the thickness of the ice.

5 Conclusion

By stress strain test, the ice before the fiber attenuation curve is smooth and in the range of qualified, ice after 96 h, fiber optic attenuation increases obviously, the individual fiber beyond the standard requirements, ice that not only have an effect on fiber mechanical properties, especially the long fiber before and after the ice and strain, causes the fiber in the OPPC released in full, fiber began to stress, additional attenuation increases, resulting in ice impact on optical fiber attenuation.

Through the same sample in 30 mm and 40 mm compared the different thickness of ice conditions blight test, the sample during the end of the 40 mm ice test time is not the single fracture. Due to 40 mm ice simulation test of OPPC with 98.0% by tension RTS, approaching OPPC rated tensile strength value, therefore, the sample is hard to bear for a long time so the thickness of ice, illustrate the OPPC samples can't bear more than 40 mm and the thickness of the ice.

In addition, based on tensile performance test in tension 95% RTS mono-filament fracture phenomenon, maintaining time of test samples under 30 mm thickness of ice is just in the OPPC tensile performance of critical state, and 30 mm ice thickness corresponding tension OPPC suffered 72.4% RTS, therefore, considered in repeat ice OPPC under ice condition, accounting should be corresponding tension values under ice thickness, if the tension is greater than 70% RTS, should further through the test of OPPC tensile properties and stress and strain properties.

References

1. Jinesh Varia;Satia Munerze.OPPC fiber test and application study.Science and Electronic Equipment Press, (2009)
2. Mathew Perry.Textile pattern design based on changed-multi-rules of climate system. 2013 IEEE International Symposium on Instrument & Measurement, 29-33.
3. Oxford,RL.Optical fiber research:performance study under extreme condition .Newbury House /Harper Collins,NY, (2008)
4. Andrew D.Cohen.Electric optical communication transmision's path selection. EI Journal Assemble on 2009
5. Rod Ellis.Ice-phased altimation influences on OPPC linear parting.EI Journal Assemble on 2009.

MATEC Web of Conferences **63**, 02016 (2016) DOI: 10.1051/matecconf/20166302016
MMME 2016

Research on Dynamic Performance of Power Shift Clutch

Li WANG[1,2], Feng REN[3,a] and Ke Ke LI[4]

[1] *Zhuhai College of Jilin University, 519041Zhuhai, China*
[2] *School of Mechanical Science and Engineering, Jilin University, 130022Changchun, China*
[3] *Changchun Normal University, 130022Changchun, China*
[4] *China South Industries Group Corporation, 100089Beijing, China*

Abstract. Smooth shift and low shift shock are needed to improve power shift quality and comfort of construction vehicle. And higher dynamic performance of power shift clutch is needed. In this paper, the dynamic mathematical model and simulation model of the clutch engagement process were established, the factors affecting the shift quality were analyzed qualitatively and quantitatively. The simulation model of the transmission system was established based on the software AMESim, the corresponding vehicle tests were performed, and the shifting quality was improved by changing key parameters. Analysis results illustrate that the proposed mathematical model and simulation model are correct and effective and can be used to predict and evaluate construction vehicle shift-feel, which are the theoretical basis for analysis and research on the static and dynamic characteristics of the dynamic shifting process in the future.

1 Introduction

Shift quality is an important index to reflect the performance of the whole machine[1]. Most of the construction machineries are still using the hydraulic mechanical transmission nowadays. Because of the torque converter is arranged between the engine and gearbox, they can effectively reduce the load vibration and shift shock of the system. However, the construction machinery has more complicated working conditions such as frequent shift operation and bad working environment and etc. So the gear shifting shock greatly affects the performance and life of the transmission system, and greatly reduces the driving comfort.

Shift clutch is a key component to guarantee the good shift quality of vehicle. It uses a variable speed control valve to control the clutch engagement and disengagement to complete the shift process. If the engaging or disengaging time of the clutch is improper, it will lead to unsmooth shift. Premature lap will result in power interference, whereas late lap will produce power interruption[2]. By reasonable controlling the pressure characteristics of the oil that comes into the clutch cylinder through a variable speed control valve, it can effectively reduce shift shock in the process of shifting, reduce the friction of clutch disc, ensure dynamic performance in the process of shifting and get good shift quality. So it has important significance for improving shift quality by researching on the oil pressure characteristics, the binding time and the friction torque characteristics of clutch engagement[3].

2 Physical model of clutch engagement

Clutch shift process is mainly divided into three stages[4,5]:

(1) Rapid filling stage. The clutch cylinder is quickly filled with oil, and the clutch drive plate is driven to move under the action of the hydraulic pressure, and the clearance between the drive and driven plates is eliminated.

(2) Clutch sliding stage. With the pressure that entered into the clutch cylinder increasing, the torque transmitted by the clutch increases, and the rotation speeds of the clutch drive and driven plates are gradually uniform.

(3) Clutch fully integrated stage. In this stage, the clutch is fully integrated, the sliding wear is stopped, speeds of the clutch drive and driven plates are the same, and the transmission torque is stable.

In order to study the dynamic model of the clutch engagement, the following assumptions are made:① The clutch and oil channels have no leakage. ② The bearing and bearing seat are without flexibility, and the gear mesh elasticity and clearance are not considered.

The physical model of the clutch engagement during the shifting process is shown in Figure 1 [6]:

[a] Corresponding author: rfcust@126.com

DOI: 10.1051/matecconf/20166302016

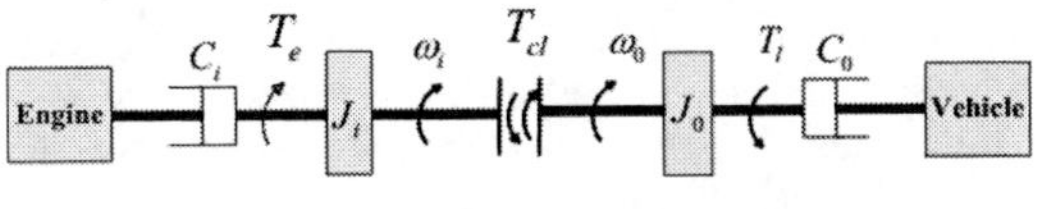

Figure 1. Schematic diagram of clutch engagement process

$$J_i \dot{\omega}_i = T_e - c_i\omega_i - T_{cl} \quad (1)$$

$$J_o \dot{\omega}_o = T_{cl} - c_o\omega_o - T_l \quad (2)$$

When the clutch is fully engaged, the model is shown in Figure 2:

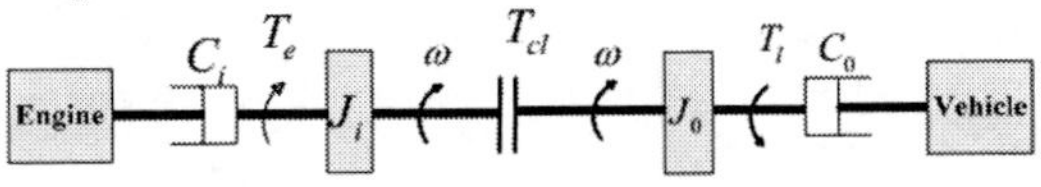

Figure 2. Schematic diagram after clutch engagement

$$(J_i + J_o)\dot{\omega} = T_e - (c_i + c_o)\omega - T_l \quad (3)$$

$$T_{cl} = \frac{T_e J_o + T_l J_i + (J_i c_o - J_o c_i)\omega}{J_i + J_o} \quad (4)$$

In the formula:

J_i ——Moment of inertia before the clutch drive plate;

J_0 ——Moment of inertia after the clutch driven plate;

ω_i ——Clutch drive plate speed;

ω_0 ——Clutch driven plate speed;

ω ——Speed when the drive and driven plates of the clutch are synchronized;

T_e ——Engine output torque;

T_{cl} ——Clutch disc Torque;

T_l ——Clutch output shaft torque which is translated from the road resistance;

C_i、C_0 ——Equivalent damping coefficients of the front-end and back-end of the clutch.

Formula (1) to (4) indicate that: when the clutch is in the process of sliding, T_{cl} is equal to the clutch friction torque. After the full engagement of the clutch, torque that the clutch can transmit is equal to the maximum value of T_{cl}.

3 Establishment of AMESim model and parameters setting

The Simulink MATLAB method, which is often used to do the dynamic simulation for the research of construction vehicle transmission system, is too dependent on the theoretical analysis in the modeling process. What's more, it usually ignores or equivalents some of the details of the factors. However, the transmission system involves the whole parameters of the vehicle, which leads to the calculation simulation require more details of the model. AMESim is a modelling and simulation platform that can be used in multidisciplinary fields for complex systems, through which the steady state and dynamic performance of any component or system can be studied. It can effectively improve the efficiency and accuracy of modelling and simulation analyzing.

In this paper, taking a certain type of loader as an example, the transmission system model was established with the AMESim software according to its working principle and actual structure, as shown in Figure 3. It contains some sub models such as the engine model, the torque converter model the tire model and etc.

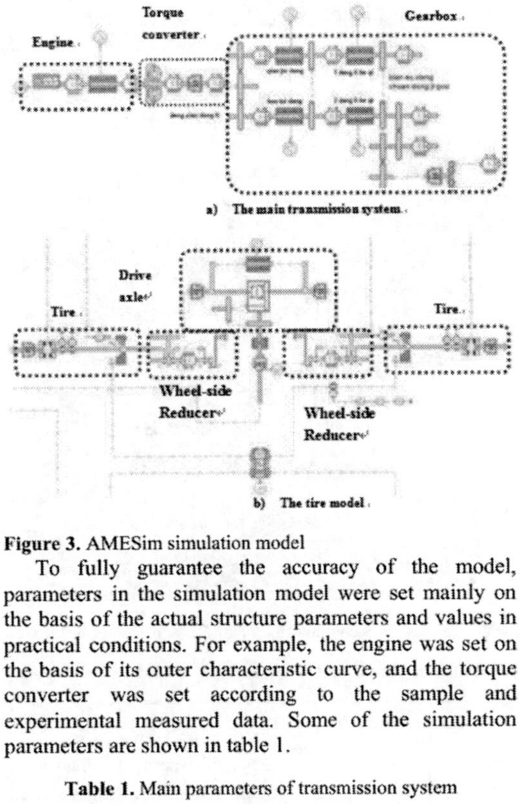

Figure 3. AMESim simulation model

To fully guarantee the accuracy of the model, parameters in the simulation model were set mainly on the basis of the actual structure parameters and values in practical conditions. For example, the engine was set on the basis of its outer characteristic curve, and the torque converter was set according to the sample and experimental measured data. Some of the simulation parameters are shown in table 1.

Table 1. Main parameters of transmission system

Effective friction radius of clutch	100mm
Gross mass	5600kg
Maximum longitudinal force	12400N
Maximum transmission torque of clutch	700Nm
Each moment of inertia	0.001Kgm^2
Speed ratio of each gear	2.4693 0.8784 2.5242 0.8980
Differential speed ratio	5.883
Tire inertia	0.5 Kgm^2
Tire radius	0.5m

DOI: 10.1051/matecconf/20166302016

Longitudinal slip ratio threshold	0.1
Road slope	0deg
Air density	1.226kg/m^3
Frontal area	4m^2
Resistance coefficient	0.3

4 Analysis of simulation and experiment results

The hydraulic pressure can be controlled easily. Reasonably controlling the hydraulic pressure change during the shifting process can effectively limit the torque disturbance on the output shaft of the gear box, and get good shift quality[7]. According to the research need, the experiment mainly measured the clutch pressures of the first gear, the second gear, the forward gear and the reverse gear, the inlet pressure of the variable speed control valve and the flow value. The arrangement of the measure points is shown in Figure 4.

Pressure sensor measurement range is 0~7MPa. Flow sensor measurement range is 100L/min. The 1~5V voltage output signals are converted to the 0~7MPa pressure output signals and 0~100L/min traffic signals.

Figure 4. Experimental arrangement of the transmission system

Table 2. Details of test points for the transmission system

No.	Signal type	Test point name	Range
P1	pressure	The inlet pressure of the variable speed control valve	7MPa
P2	pressure	The first gear clutch pressure	7MPa
P3	pressure	The second gear clutch pressure	7MPa
P4	pressure	The forward gear clutch pressure	7MPa
P5	pressure	The reverse gear clutch pressure	7MPa
Q1	flow	The inlet flow of the variable speed control valve	100L/min

5 Analysis results of simulation

This paper did a simulation research on a certain type of loader's shifting process. The simulation was set as follows: the system is in high speed no-load condition, start the car with the first gear, throttle maintains the maximum opening, shift to the second gear after travelling for about 10 seconds, keep travelling for about 10 seconds and then park and shut off the engine. The main parameters' characteristic curves of the prototype during the shifting process are shown in Figure 5.

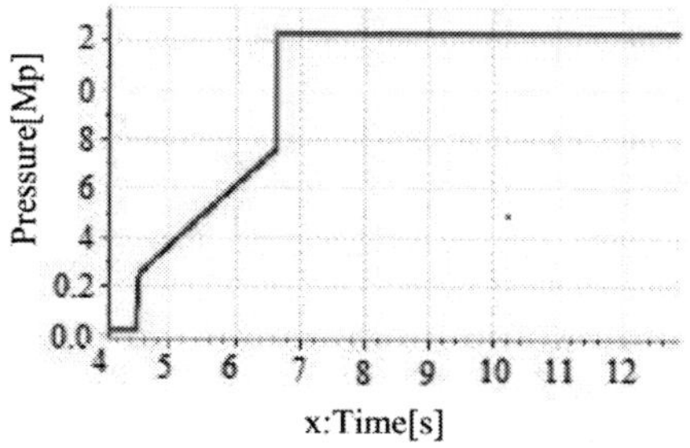

(a) Variation characteristics of hydraulic pressure

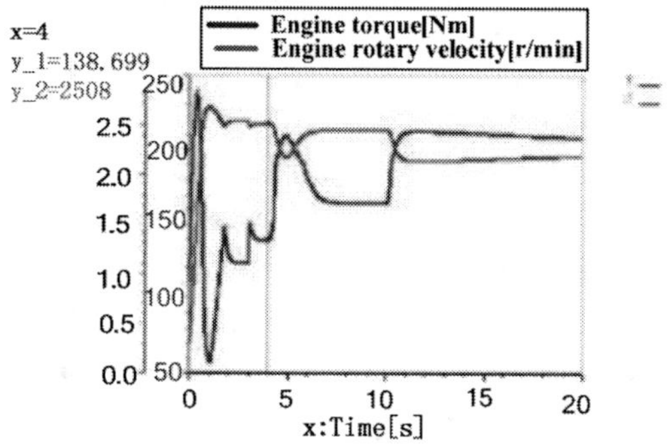

(b) Engine variation characteristics

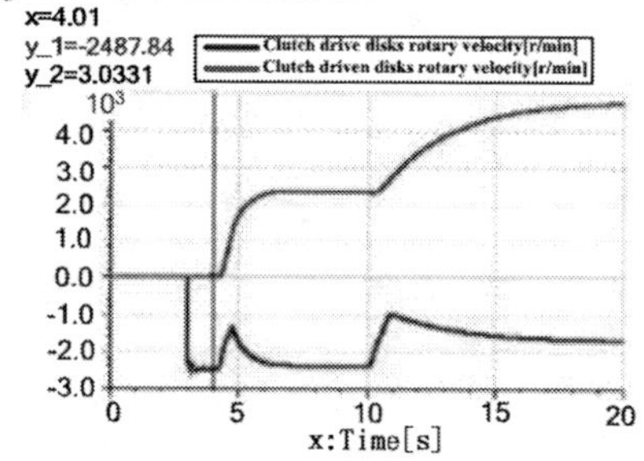

(c) Speed variation characteristics of the clutch drive and driven plates

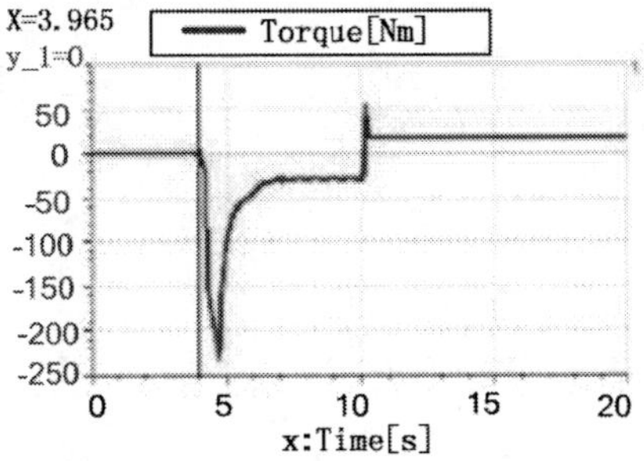

(d) Variation characteristics of the torque transmitted by clutch

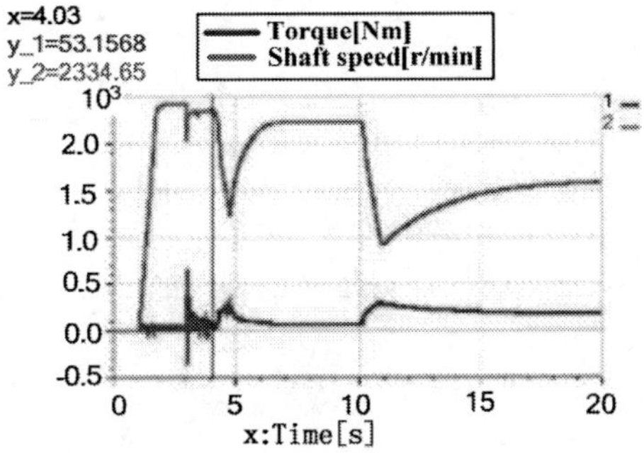

(e) Variation characteristics of torque converter

Figure 5. Simulation graph of parameters in shifting process

Simulation result indicates that: it begins to shift at the time of 4S and 10s. In the shifting process, the engine speed is reduced and the torque is increased. Figure (a) is the variation characteristics of hydraulic pressure that from the variable speed control valve acting on the clutch cylinder in process of shifting. Figure (b) is the speed and torque variation characteristics of the engine at this time. Figure (c) is the speed variation characteristics of the clutch drive and driven plates. Figure (d) is the variation characteristics of the friction torque transmitted by the clutch. Figure (e) is the speed and torque variation characteristics of torque converter. In this process, the clutch experiences the rapid filling stage, clutch slipping stage and fully integrated stage to complete the shifting process. In the rapid filling stage, the clutch drive and driven plates vary from the separation state to the beginning of contact state, the torque transmitted by the clutch increases rapidly, and the first shift shock is formed. The torque transmitted by the clutch is gradually increased with the increase of the control pressure. However, after the end of the sliding stage, the transmitted torque is quickly reduced to the torque value of vehicle acceleration, and another shift shock is formed.

6 Analysis of the vehicle tests results

Through simulation analysis, it is known that the main factors that affect the dynamic characteristics of the dynamic clutch are orifice diameter, preload and stiffness of the pressure regulating spring and energy storage spring. Therefore, during the experiment, the vehicle test was done by changing the damping hole diameter of shift buffer valve, the stiffness of pressure adjusting spring, the stiffness and preload of energy storage spring, and data mining equipment was used to collect data. Figure 6 to Figure 9 are the measured test curves.

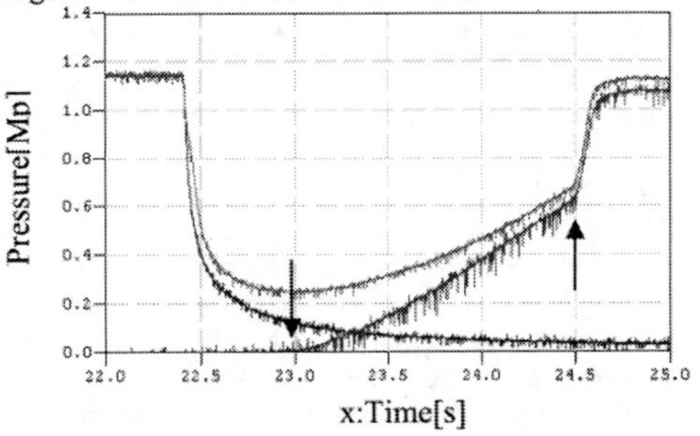

Figure 6. Experiment curves before improving the first gear

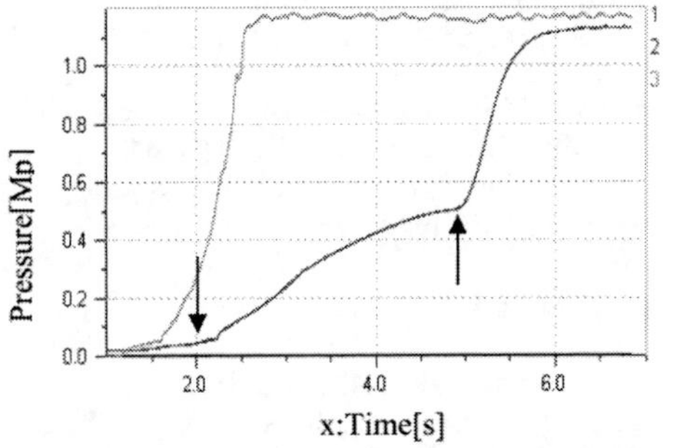

Figure 7. Experiment curves after improving the first gear

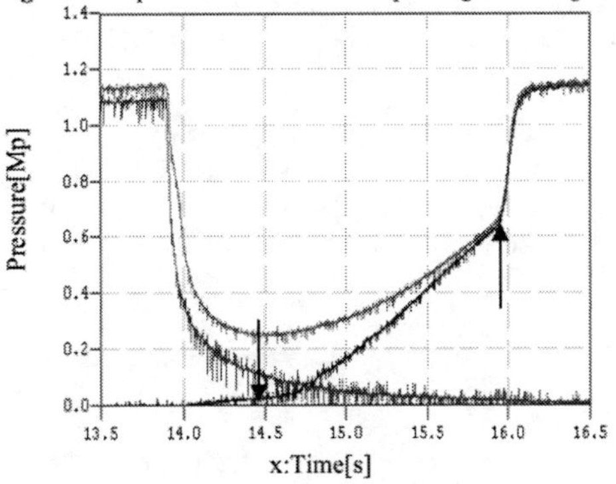

Figure 8. Experiment curves before improving the second gear

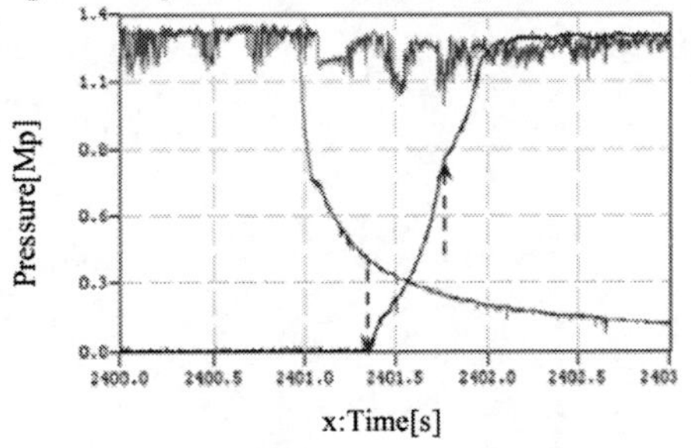

Figure 9. Experiment curves after improving the second gear

The experiment curves show that the buffer time after improving the first gear is changed to 3.5s, the pressure building process is more stable and the shift shock is effectively reduced. The buffer time after improving the second gear is changed to 0.5s, and the pressure of the forward gear is not affected, which effectively mitigates the "stall phenomenon"

7 Conclusions

(1) Through the analysis of the clutch engagement process in the transmission system, the simulation model of the transmission system was established using the AMESim simulation software. The correctness and validity of the simulation model were verified by the experiment of the vehicle, which provides a theoretical basis for the analysis and research on the static and dynamic characteristics of the dynamic shifting process in the future.

(2) The key parameters were tested using the vehicle experiment, and the shift quality of the vehicle was improved by changing the key parameters. This also indicates that the AMESim simulation can play an important role in designing and analysis of the vehicle transmission system.

References

1. Y. Zhang. JLU,**2012411037**(2015).
2. Zh.N. Ren, J.Y. Tian, Sh.X. Lan, P. Zhou. HPS, **03**(2014).
3. B. Ma. BIT, **20**,2(2000):188-192.
4. G. Cao, A.L. Ge, L. Zheng. CJME, **41**,12(2005):234-238.
5. X.H. Li, W. Ye, Zh. Liu, etc. JTJU: Natural Science,.**31,** 5 (2003):576-580.
6. D.X. Zhao, G.J. Cui, D.B. Li. JJSU: Natural Science Edition, **29,** 5 (2008):386-389.
7. Sh.J. Yang, Sh.H. Yuan, J.B. Hu. TCSAM, **11**, 11 (2005):39-41.

MATEC Web of Conferences **63**, 02017 (2016) DOI: 10.1051/matecconf/20166302017
MMME 2016

Relevant Applications of Tools in UG NX CAM Module

En Guang ZHANG[1,a], Li WANG[1]

[1] *Zhuhai College of Jilin University, 519041 Zhuhai, China*

Abstract. In this paper, the methods to establish the custom magazine have been proposed while the UG NX CAM module is employed to conduct the CNC milling programming. At the same time, how to set the parameters of the cutting tool has been also analyzed; in the creation of tools, the use of shank and holder as well as the use of handle check functions in NC machining has been introduced; meanwhile, the secondary rough machining of corner and the application of clean-cut features have been discussed.

1 Introduction

In recent years, the CNC machining technology has been greatly developed, which promotes the basic research on CNC tool materials and the development of new product[1]. The multifunctional complex tools and the modern efficient tools are the mainstream of tool development. The increasing development, the great changes of toll materials and the rapid development of coating technology have greatly promoted the tool development[2]. Currently, the high-speed cutting, the dry machining, the hard cutting and the composite cutting have become the important direction of development. The different cutting technologies use the different tools. For example, the ultra-fine grain carbide cutting tools have been developed to the nanoscale. At present, the main species have trungsten cobalt, (W,Ti)C-Co Cemented Carbide and WC TiC Tac Co. Other super-hard tool materials include the ceramics, CBN (Cubic Boron Nitride) and polycrystalline diamond, etc.

With the development of CNC machining technology, the tool types as well as the tool materials have obtained great progress. The types of tools are various. In CNC machining, the applications of tools have become increasingly important. For example, the magazine, the check functions of tools and the use of reference tools are defined in CAM software.

2 Creation of custom magazine in UG NX

2.1 UG NX software.

UG NX software is a CAD/CAE/CAM manufacturing information software, which has been widely applied in fields of automobiles, general machine, military industry, aerospace, consumer goods, high-tech products and ship design. At present, the latest version of NX software is NX10. The computer-aided design, the engineering analysis and the processing can be completed in the same software through employing the NX software without the format conversion, which can ensure the editable feature of model and greatly improve the secondary modification speed of NC program. Currently, the NX software has a wide range of users in Pearl River Delta and Yangtze River Delta.

2.2 CAM module of UG NX.

CAM (Computer Aided Manufacturing) refers to the process in which the computer is employed to control and operate the production equipment. The input information includes the process route and the process contents. The output information refers to the movement curve (cutter location file) and the CNC programs in the process of tooling.

CAM module of UG NX supports drilling, milling, turning and wire EDM as well as the three-axis and multi-axis milling. For the milling, the common processing methods include CAVITY MILLING, PLANAR MILLING, FACE MILLING, ZLEVEL PROFILE MILLING and FIXED_CONTOUR MILLING. The methods to control the tool path are extremely rich, which can meet the demands of any NC programming. The applicable scope and process of different operation types are shown in Table 1.

Table 1. The applicable scope and process of different operation types [3]

Operation type		Application scope	Applicable process
Drill	Drilling	Center hole	Hole rough drilling or finishing drilling
	Peck_Drilling	Deep hole	
Mill-Planar	Face_Milling	Model with plane bottom and vertical plane side wall	Face rough machining or finish machining
	Planar_Mill		MILL_ROUGH, finish machining
Mill-Contour	Cavity-Mill	Arbitrary shape model	mostly used for rough machining, and sometimes for

[a] Corresponding author: teamcenter@163.com

DOI: 10.1051/matecconf/20166302017

			the "steep"model of MILL_SEMI_FINISH and finishing machining
	Z level_Profile_ Steep	"Steep" model with a small angle between the Z axis	MILL_SEMI_FINISH、MILL_FINISH
	Fixed-Contour	Arbitrary shape model	MILL_SEMI_FINISH、MILL_FINISH

2.3 Creation of custom magazine in UG NX

Firstly, it is required to create an arbitrary file and "Manufacturing" module is entered from "Start". The processing environment can select "mill contour" initiation. Then, the processing module of UG is entered.

In "Resource Bar", the "operation navigator" can be found. It is required to select "tool view" and the subsequently created tools can be displayed in this view with the form of list. It is required to choose "creating tools" in "creation process". The subtype selects "Mill" and "OK" is clicked to enter "Milling Tool-5 Parameters" tool dialog box. In this dialog box, the geometry parameters of tools and the tool materials can be entered. And it is necessary to set up and generate the shank and holder. Then, it is needed to select "Export Tool to Library" to import the tools to the corresponding database.

It is required to execute the menu command "tool"----"Edit Machining Data Libraries" and input the parameters according to demands. Then, the cutting parameters of tools can be determined. These cutting parameters can be loaded when using this tool.

2.4 Use of shank and holder

When creating the tools, it is required to respectively define Tool, Shank and Holder. The tool is shown in Figure 1.

"Use of shank and holder" option helps avoid collision between handle and the workpiece and choose the cutting tool as short as possible during operation. If you want to use the handle to contain the tool path, please open this option.

"Use of shank and holder" is a common cutting parameter for "high milling", "surface contour milling"(according to the drive mode) and "cavity milling". During the "surface contour milling ", "Use of shank and holder" can be used to complete "Area milling" and "automatic cleaning".

The handle and holder in the tool definition dialog box are defined as a pile of stacked cylinders or tapered cylinders, as shown in figure 1. For the "milling area", "automatic cleaning root driving mode" and "high profile milling driving mode" of "fixed axis surface milling", this handle definition can be used to ensure that the tool path without collision. During this operation, these options must switch to "on" to identify the handle.

In the "surface contour milling" and "high milling", if a collision between handle and the workpiece was detected, the collision region would be saved as "two-dimensional workpiece" geometry during the operation. The "two dimensional workpiece" geometry can be displayed by in turn highlighting the operation, holding down the mouse button MB3 and selecting "workpiece —display two-dimensional". The geometry can also be displayed even when the other operation is performed. The geometry can be used as clipping geometry in subsequent operations in order to include cutting movement in the area where the left material with the collision between the handle and workpiece need to be removed.

In the "cavity milling", if the system detects a collision between the handle and the workpiece, the collision area will not be cut. In order to remove the non cutting area, all subsequent cavity milling operations must use the "Layer based IPW" option.

In Figure 2, the system uses the holder shape and the minimum clearance value to ensure the safe distance with geometry. The regions which can cause the collision will be excluded from the cutting area. Therefore, there is no toolpath collision when cutting materials. The materials which need to be removed should be updated after each completed cutting layer. It is required to maximum increase the available cutting area. At the same time, as the upper material has been removed, the handle has greater space at the bottom of workpiece. It must be necessary to use longer tools to cut the excluded (collision) area in subsequent operations.

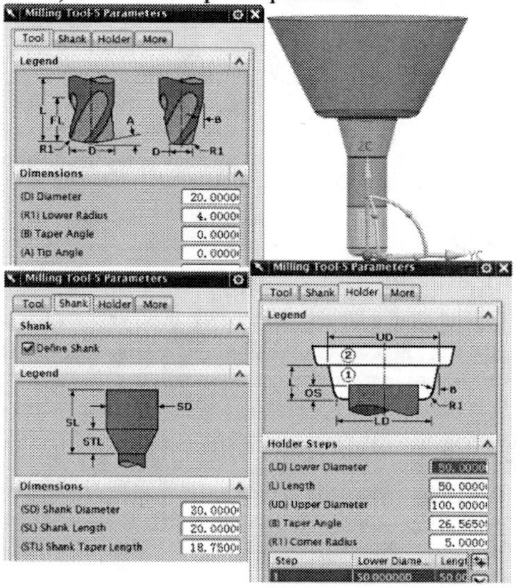

Figure 1. Defining Tool、Shank、Holder

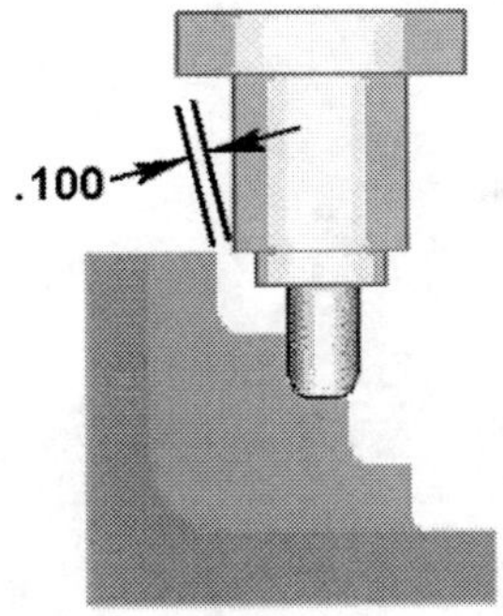

Figure 2. Use of shank detection

3 Reference tools are used for semi-finishing the corner

Many of the CAM modules in UG provide a "reference tool" option, such as cavity milling, high milling and fixed axis milling and etc.. Use the reference tool option to eliminate the tool movement in the place where there is no material. When creating a new process, if the using smaller tool refers to the larger one, only the none cutting material with the larger tool will be removed by the smaller one. The reference tool process can be placed before or after the process using the larger tool. It is often to do the rough cutting corners with smaller tool firstly as a process of reference tool before the process of contour milling with larger tool.

When you want to cut the remaining material in the corner missed by the last tool, you can use the "reference tool", as shown in figure 3. The remaining material may be that left over between the wall and the bottom due to the corner radius of the tool, or may be that left between the walls due to the diameter of the tool. This cutting is similar to the other "cavity milling" operations, however, it is limited to the corner area operation. The reference tool is usually used as a tool for rough machining of the area. The system calculates the remaining material for the specified reference tool, and then defines the cutting area for the current operation. Figure 4 shows the cavity milling path using automatic block as a blank. It has no reference tool.

Figure 3. Corner only

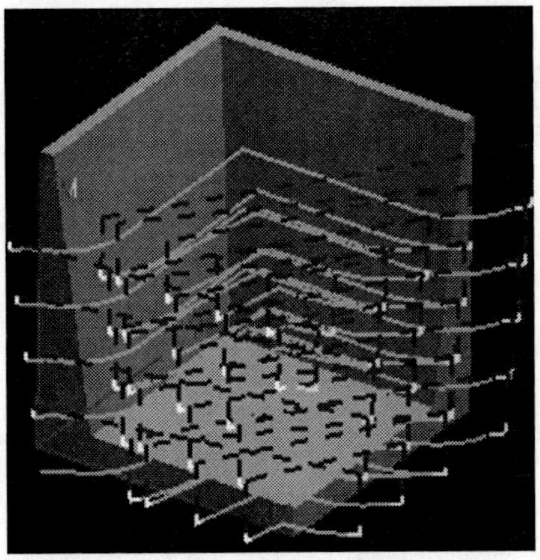

Figure 4. Tool path without reference tool

As shown in Figure 5, when the facing milling cutter with a diameter of 20 is roughly processed, the corner of workpiece has left more margin. At this time, the semi-finishing or finishing sidewall contour is not reasonable. It is required to select D8 tool for another processing through taking D20 tool as reference. D8 tool should be adopted to process the areas that D20 tool is not involved in. The specific toolpaths and the complete results are shown in Figure 6.

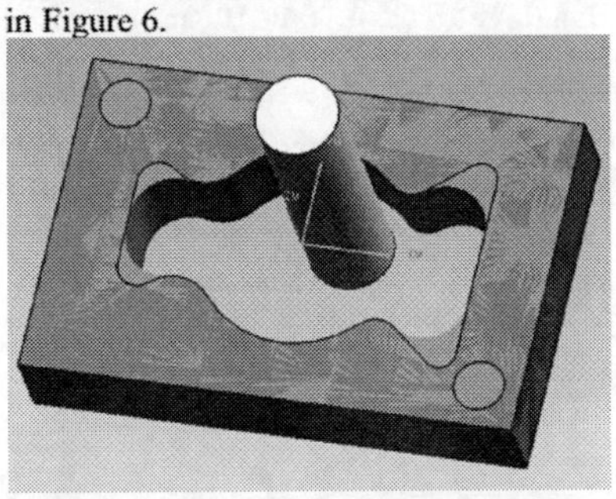

Figure 5. Facing milling cutter rough processing of D20

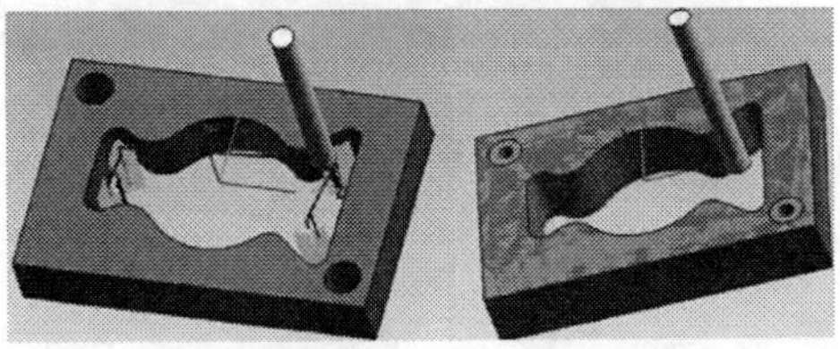

Figure 6. Tool paths and processing effects of D8 after using reference tool

4 Gouging cutting of reference tools

The gouging cutting is a cutter driving method in fixed axis milling surface profile. The gouging cutting method can be used in: high-speed machining; removing the excess materials in corner before finishing; removing the uncut materials left by larger ball facing mill. In gouging cutting, the reference tool can be employed for calculation. As shown in Figure 7, the diameter of roughing ball end mill is employed to specify the width

DOI: 10.1051/matecconf/20166302017

of finishing cutting areas. The software calculates the double-cut points according to the specified "diameter of reference tool". Then, these points are used to define the cutting areas of finishing process.

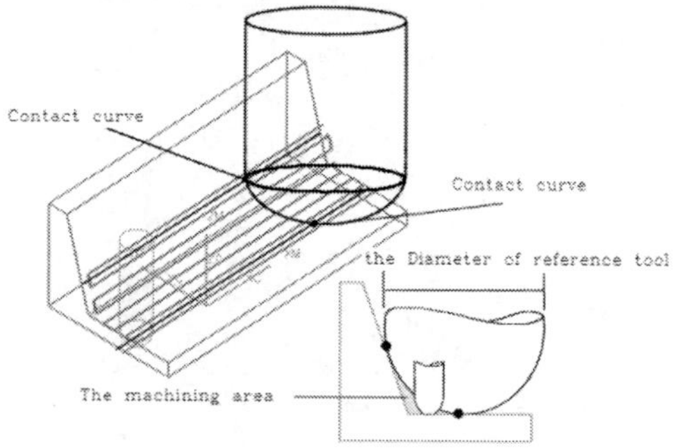

Figure 7. "Reference tools" are employed to conduct the gouging cutting

The part as shown in Figure 8 has been semi finished using ball end milling cutter with a diameter of 20 in previous process. In the example, a ball end mill with a diameter of 10 is used, which is suitable for the area that a ball end mill with a diameter of 20 is not suitable.

Figure 9 shows the tool path after machining using the ball end mill with a diameter of 10 and clear root driving using the ball end mill with a diameter of 20 as the reference tool. Figure 10 is the post processing effect diagram.

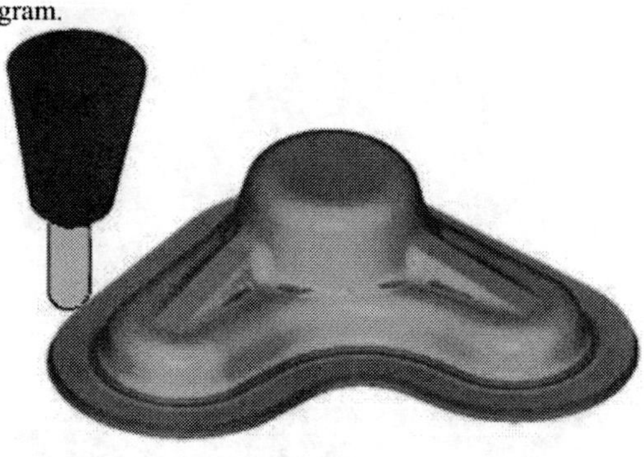

Figure 8. Part machined by the ball end mill with a diameter of 20

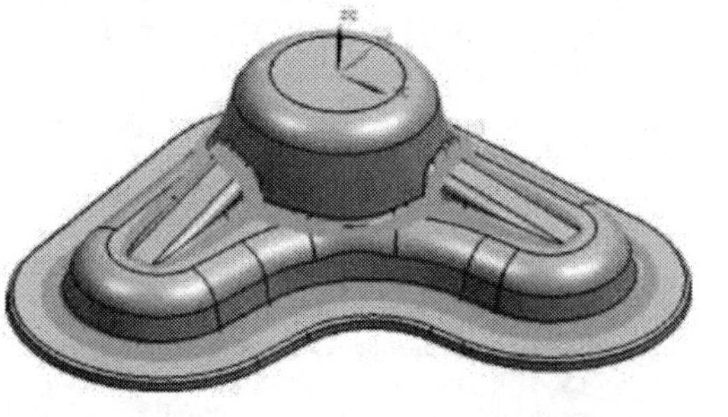

Figure 9. Tool path by the ball end mill with a diameter of 10 and the ball end mill with a diameter of 20 as the reference tool

Figure 10. The post processing effect diagram

5 Conclusions

The applications of tools in NX CAM module can greatly facilitate the programmer. UG NX can support the establishment of custom magazine and the size as well as the materials of tools can be set. At the same time, the extremely detailed process data can be provided, such as the cutting depth, the stepping, the surface speed and the cutting feed rate, etc. The option of "using shank and holder" helps to avoid the collision between shank and workpiece. In the operation, it is required to choose the shortest tool as possible. The handle shape and the minimum clearance value can ensure the safe distance with geometry. The regions which can cause the collision will be excluded from the cutting area. Therefore, there is no toolpath collision when cutting materials. The "reference tool" method can better process the corner and gouge, reduce the tool idling and improve the efficiency.

References

1. Y.X. Liang, H.B. Zhao. MW Metal Cutting, **4**(2001):8-10.
2. H.P. Zhang, Sh. Y. Meng, Y. Zh. Jia. MMAD, **23**,6(2008):20-23.
3. UG NX HELP Documentation.
4. S.T. Chen. Tool Engineering, **45**,7(2001):75-78.

Analysis of micro vibration in gas film of aerostatic guide way based on molecule collision theory

Shao Hua YANG[1] , Wei LONG[1,a] and Ling GONG [1]
[1]*Faculty of mechanical and electrical engineering, Kunming University of Science and Technology, Yunnan Kunming 650500, China*

Abstract. Micro vibration of the aerostatic guide way has a significant impact on its dynamic characteristics and stability, which limits the development of pneumatic component. High pressure gas molecules have been collided with the supporting surface and the internal surface of the throttle during the flow process. Variable impulse of the surfaces aside for the gas film are affected by the changes of impulse which formed irregular impact force in horizontal and vertical direction. Micro-vibration takes place based on the natural frequency of the system and its frequency doubling. In this paper, the vibration model was established to describe the dynamic characteristics of the gas film, and the formation mechanism of micro vibration in the film is defined. Through the simulation analysis and experimental comparison, formation mechanism of the micro vibration in the gas film is confirmed. It was proposed that the micro vibration of gas film can be produced no matter whether there is a gas chamber or not in the throttle. Under the same conditions, the micro vibration of the guide way with air chamber is greater than that without any chamber. The frequency points of the vibration peaks are almost the same, as well as the vibration pattern in the frequency domain.

0 INTRODUCTION

With the development of ultra precision motion platform, micro vibration has become a major problem which restricts the development of air flotation technology. Studying the mechanism of micro vibration is very important for improving the movement and positioning accuracy of air floatation platform[1-3]. Micro vibration is different from the "air hammer phenomenon" from the static pressure guide , It is mainly due to the gas film clearance between the film and the internal support platform of uneven force distribution has certain amplitude near the equilibrium position of the law on the deflection or vibration. Aoyamae et al. It is speculated that the micro vibration in gas bearing is caused by the flow of gas in the pressure chamber[4]:Chenxuedong et al. Through the simulation experiments, it is proved that the phenomenon of the air cavity of the small orifice throttle air pressure guide rail is related to the micro vibration, Gas flow in the gas film can also cause the micro vibration, but less than the micro vibration of the cyclone[5]; Zhang Ming et al. Think that the main reason of the micro vibration in the ultra precision working table system of the air bearing is that the turbulence of the gas flow in the gas flow is formed and the measures to restrain the vibration of the micro vibration [6].Dong Ju Chen et al by Reynolds equation derived row orifice form hydrostatic slideway of the gas film stiffness formula and Study on machining accuracy of workpiece caused by fluctuation of gas membrane effect[7].Ji Lujun et al. Use the linear perturbation method to establish the stability of the Reynolds equation and the Reynolds equation and the bearing motion equation to study the dynamic analysis model of the vertical impact caused by the static pressure rail fluctuations[8]. The amplitude of the micro vibration is typically between a few nanometers to tens of nanometers, and the frequency range from tens of hertz to several thousand hertz. For the cause of the vibration of the micro vibration, it is generally accepted that the gas flow pattern of the gas is caused by the flow of the gas, But there is no reasonable explanation for this kind of fluctuation, In this paper, the point of local impulse is changed from the velocity vector of gas molecule collision, and the micro vibration of the gas film is described by the simulation and experiment, This paper provides a new way of thinking for the fundamental suppression of gas film fluctuation.

1 Study on formation mechanism of gas film fluctuation

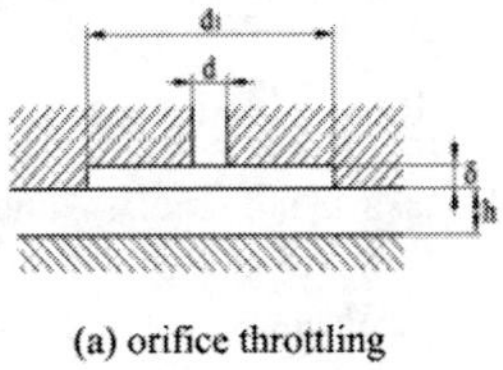

(a) orifice throttling

[a]Wei LONG: 827161564@qq.com

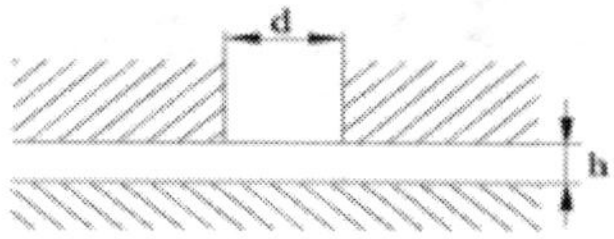

(b) torus throttling

Figure 1. common throttle.

Figure 1 (a) is shown as a small orifice throttling view, figure 1 (b) is a torus throttling view, In this paper, the characteristics of the gas flow in a complex throttle device with a gas chamber are described ,The torus throttle represented that no flow law of gas cavity internal gas throttling device

The orifice restrictor, external high-pressure gas into the gas chamber through the gas supply hole, gas velocity vector does not change in the high pressure jet, between the molecules by diffusion, the colliding molecules directly into the gas chamber, high pressure jet impact on the cavity at the bottom of the gas molecules From the beginning of the impact hypothesis, all the kinetic energy into pressure energy, The speed of the molecules that impact the bottom of the gas chamber is reduced, and the bottom temperature can be increased, which makes this part of the surface heat flow occurs at the instantaneous maximum, The support at the bottom of the thermal strain and cause fluctuation of gas film. The molecular impact area is A, then there is:

$$\frac{\rho Alv^2}{2} = \frac{Al\Delta p^2_{\max}}{2E} \quad (1)$$

In this formula: l is the equivalent gas film thickness of the gas chamber, $l = h + \delta$; E is a gas volume elastic modulus; c is the local speed of sound, $c = \sqrt{El\rho}$; $\Delta p_{\max}$ is the peak value of pressure; $\Delta p_{\max} = pv\sqrt{E/\rho} = \rho cv$, The energy absorbed by the elastic deformation is not considered in the derivation process, and the actual $\Delta p_{\max}$ is smaller than the calculated value.

The interaction between the gas molecules and the wall of the supporting surface is affected, Supposed that the σ_v part of the stream is completely diffuse, Then the rest of the $(1-\sigma_v)$ part is completely specular reflection. Specular reflection and diffuse reflection to the direction of movement of gas moleculars change. It is assumed that the molecules of the surface of the left surface are scattered by the Maxwell distribution, and the velocity distribution function is:

$$\begin{aligned} f_0 &= n(\frac{m}{2\pi kT})^{3/2} \exp(-\frac{m}{2kT}c'^2) \\ &= n(\frac{m}{2\pi kT})^{3/2} \exp[-\frac{m}{2\pi kT}(u'^2 + v'^2 + \omega'^2)] \end{aligned} \quad (2)$$

In this formula: n is the molecular number density, m is molecular weight, c' is the molecular heat movement speed, k is Boltzmann constant, T is the thermodynamic temperature in equilibrium state.

Since the H is micron level, Small part of the molecule after collision between molecules and the clearance of gas film inner surface collision fold, And then directly into the gas film thickness of h; Most molecules collide with each other, and the gas molecules collide with the surface of the gas, The velocity vector of the gas chamber is changed dramatically, and the impulse of the wall is changed rapidly, so that the wall is not balanced.

For torus throttling, The law of the movement of the gas molecules is similar to that of the molecular motion of the orifice, But because of the small hole in the wall, the pressure in the horizontal, vertical, and the velocity of the molecules is increased, and the pressure of the air floating block is not balanced in the three-dimensional direction. The main source of the micro vibration of the torus is the energy conversion of the high pressure jet molecules and the energy conversion and the imbalance of the gas molecules in the vertical direction after the impact of the support surface. The mechanism of the micro vibration of the static pressure guide rail is analyzed from the gas flow pattern, which is caused by the impact of the high pressure gas and the wall of the throttle.

The molecular interaction between molecules and the surface of the film space in the collapse of the collision, After entering the real meaning of the supporting lubrication region, the gas molecules in the pressure difference, the velocity vector in the horizontal component is much larger than the vertical direction, The change of molecular velocity vector is small, and the degree of micro vibration caused by this is very weak.

2 Simulation Analysis and experimental verification

2.1 simulation analysis

In this paper, the geometric parameters are used to enter the hole diameter $d = 0.15mm$ and the gas chamber diameter is $d = 1mm$, Torus throttle orifice floating block gas cavity height $\delta = 0.1mm$ floatation block and inlet diameter of $d = 0.15mm$, In the gas supply pressure of 0.3MPa and 0.6MPa, the air gap is 30 μm The results of CFD software simulation as shown in figure 2, With the increase of gas supply pressure, the inlet gas velocity of the throttle orifice and the torus is increased, the momentum of the vertical or the surface of the support is changed dramatically, which is characterized by the increase of the gas pressure. By the figure 2 (d) (c), the velocity vector is changed, the cyclone is formed in the gas chamber, with the increase of inlet pressure, the molecular velocity vector of the cyclone and the change of the surface impulse in the gas chamber. By (a) (c), (b) (d), it can be known that the presence of the gas chamber makes the molecular velocity of the gas chamber is larger than that of the same range, and the speed vector changes dramatically. Fig. 3 shows that the velocity of the gas molecules is greater than that of the gas flow in the center of the high pressure gas inlet. Because the simulation of CFD software is based on the assumption that the

DOI: 10.1051/matecconf/20166302018

simulation results of the laminar turbulence model are obtained, it is not be very accurate to simulate the motion trajectory of molecules with the collision avoidance, Figure 4 is the use of LAMMPS software to simulate the flow of molecules with the wall surface after the collision, the molecular velocity vector changes. It can be seen that the velocity vector of the molecules of the high speed movement changes sharply, resulting in the rapid change of the wall, which makes the stress of the wall is not balanced.

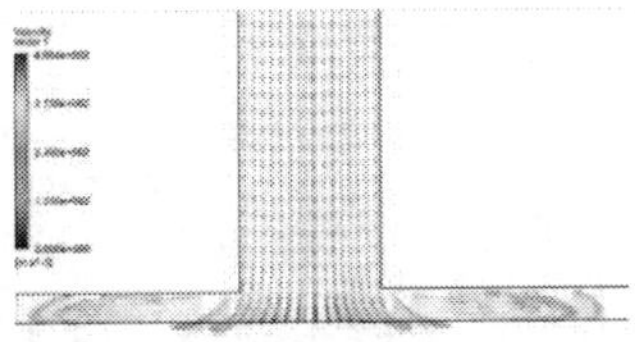

(a) P=0.3MPa torus throttling speed vector variation diagram

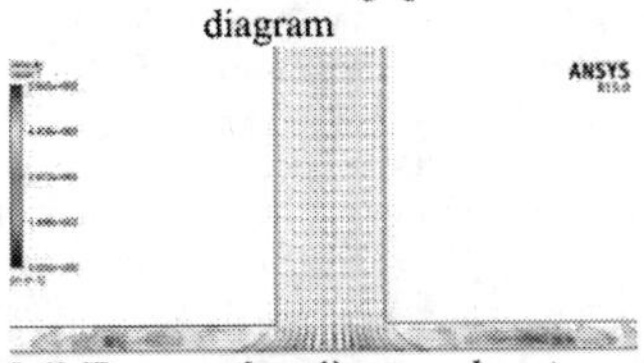

(b) P=0.6MPa torus throttling speed vector variation diagram

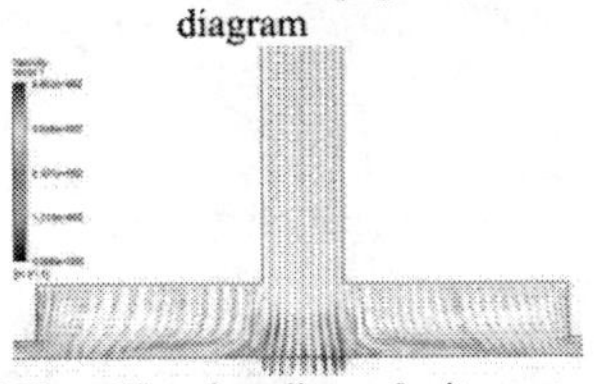

(c) P=0.3MPa orifice throttling velocity vector variation diagram

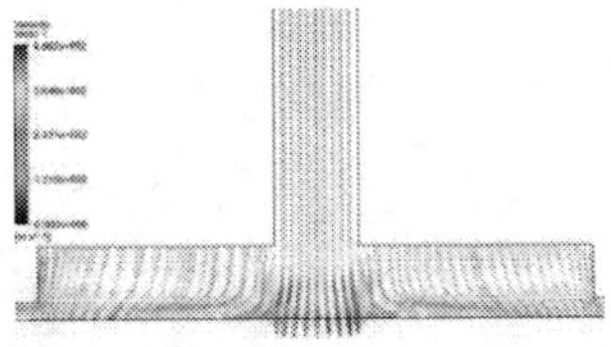

(d) P=0.6MPa orifice throttling velocity vector variation diagram

Fig 2. the variation of the internal velocity vector of two kinds of throttle under different inlet pressure.

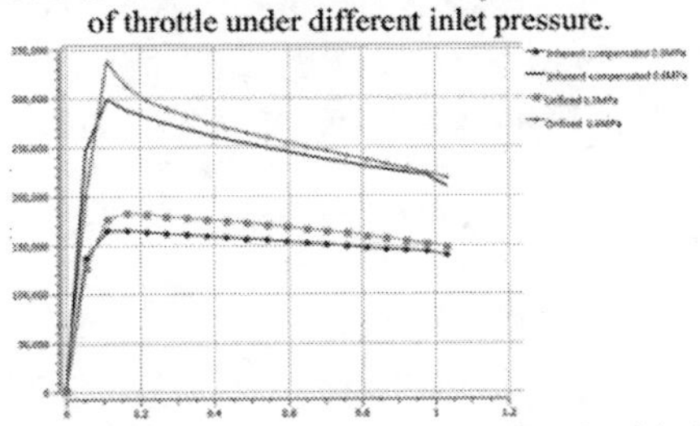

Fig 3 . gas molecular velocity diagram on the axis of the inlet center.

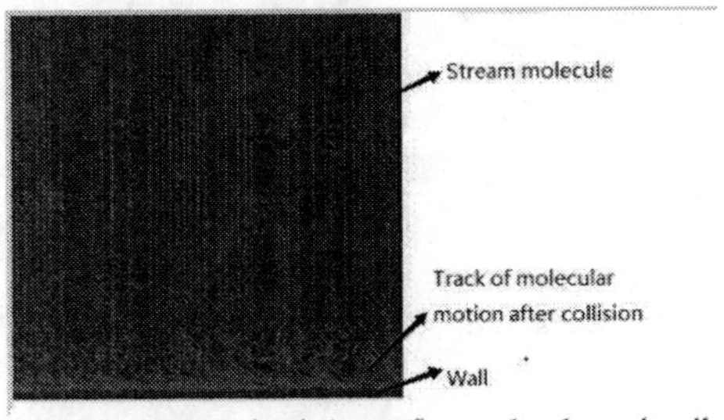

Fig 4. LAMMPS simulation to flow molecular and wall collision trajectories.

2.2 Experimental verification

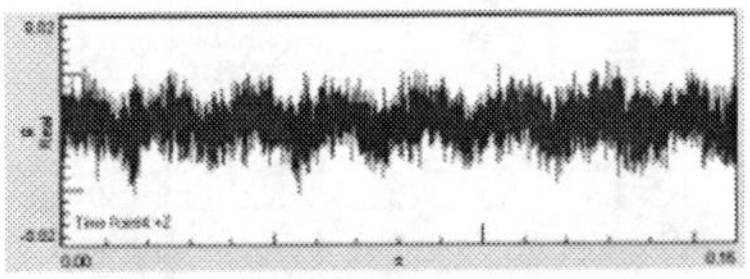

(a) P=0.3Mpa small hole throttle floating support signal

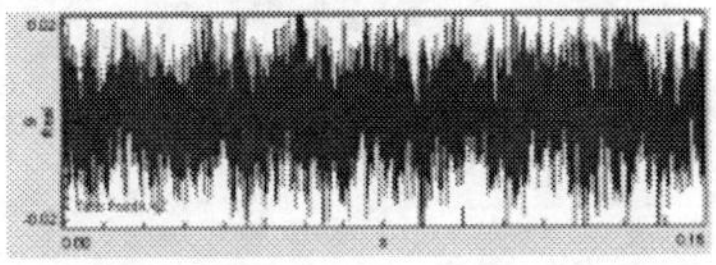

(b) P=0.6Mpa small hole throttle floating support signal

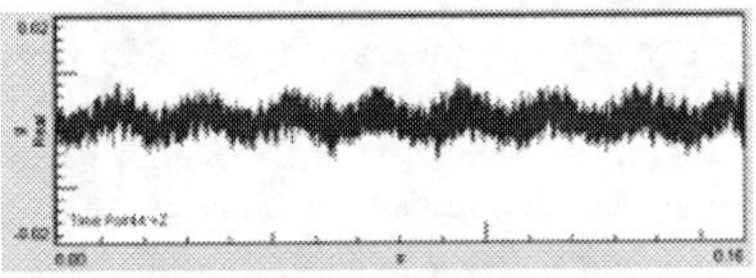

(c) P=0.3Mpa torus throttle floating support signal

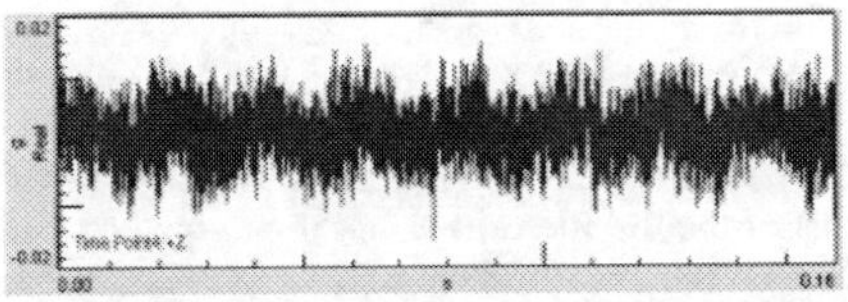

(d) P=0.6Mpa torus throttle floating support signal

Figure 5 .orifice and torus throttle floating support signal.

Figure 5 (a) (c) respectively as the gas supply pressure is 0.3MPa, gas film clearance h is about 30, the orifice and torus throttle floating support time domain signal. Figure 5 (b) (d) respectively as the gas supply pressure is 0.6MPa, gas film clearance h is about 50, orifice and torus throttle floating support time domain signal. Under the same air supply pressure and gas film clearance did not differ much, degree of micro vibration of throttle orifice is about 2 times the torus throttle micro vibrational degrees of, with the increase of the air pressure, small hole throttle cavity and the torus knots flow increased speed of gas molecules within the scope of the hole at the end of, fierce collision and velocity changes more frequently, rail surface by pressure variation has become even more unbalanced, the degree of micro vibration becomes large.

Figure 6 is a diagram of the orifice and the frequency domain signal torus throttle floating vibration of support, which is similar to the vibration between the two laws, The peak signal of vibration is 16KHZ, 25 KHZ, 32 KHZ, 38 KHZ, 45 KHZ, these frequencies are close to the natural frequency of air flotation system. In the low frequency range, the peak phase difference between the orifice and the torus is larger than that of the high frequency section. From the experimental results, we can know that there exists a micro vibration phenomenon in the throttle and the torus, but because of the existence of the hole wall, the velocity vector of the three-dimensional direction of the hole is increased, which makes the micro vibration of the orifice is greater than that of the torus.

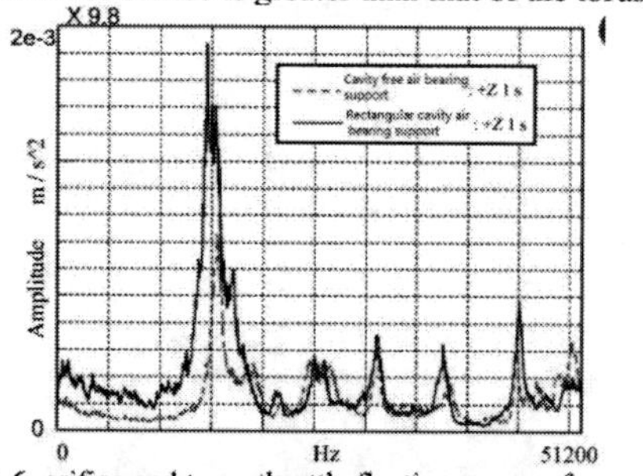

Figure 6. orifice and torus throttle floating support frequency signal.

3 Conclusions

(1) In this paper, the mechanism of micro vibration of static pressure guide rail is analyzed from the gas flow pattern, because of the impact of the high pressure gas and the wall and the supporting bottom surface of the throttle, the change of the velocity vector in the vicinity of the throttle is different, and the variation of the vector is different, which makes the interaction between the air floating block and the supporting surface is not balanced, resulting in the phenomenon of micro vibration.

(2) The micro vibration is generated by the air flotation support of the throttle orifice and the torus throttling, just because of the existence of gas cavity, it exacerbated by the complexity of the film inside the solid wall structure, and the influence of the molecular motion speed of air flow and change more, Performance of macro internal air cavity air film vibration is gas cavity is more obvious. The study also shows that although the micro vibration of the gas chamber is greater than that of the non gas cavity, but the same gas supply and load conditions, the vibration peak of the frequency points are basically the same; At low frequencies, there is a lot of vibration peak (nearly 2 times) of the throttle with a gas chamber.

(3) With the gas pressure increasing, the molecular velocity of the incident gas increases. The molecular and the support surface or the surface of the gas chamber, and the gas film internal gas molecular velocity vector changes dramatically, The component perpendicular to the support table or gas cavity surface changes rapidly, and the film inside the micro vibration severity.

Thanks

We show great acknowledgement of the experimental data from degree thesis of Dr. Ye Yixi" Study on ultra precision motion platform in support of the vibration characteristics of the floating ".

References

1. Zhu Yu, Wen,Sheng, Duan Guanhong . Research on ultra-Precision Stages of lithography. Lithography technology [J].2004.33(2).
2. Shinno H, Yoshioka H, Taniguchi K.A newly developed linear motor-driven aerostatic X-Y planar motion table system for nano-manchining . Annals of the CIRP,2007,56(1)
3. Verma Shobhit, Kim Won-Jong , Shakir Huzefa .Mu lti-axis maglev nanopositioner for precision manufac turing and manipulation applications. IEEE Transacti ons on Industry Applications,2005,41(5).
4. Aoyama T, Kakinuma Y, Kobayashi Y. Numerical and experimental analysis for the small vibration of aerostatic guideways. CIRP Annals-Manufacturing Technology[J], 2006,55(1).
5. [5] Chen Xuedong, He Xueming . The effect of the recess shape on performance analysis of the lubricated bearing in optical lithography, Tribology International[J], 2006,39(11).
6. Zhang Ming, Zhu Yu, Duan Guanhong . Micro-vibration of Ultra-precision Gas Bearing Linear Motion Stage and Its Elimination. Design and research [J].2005, (11).
7. Chen Dongju, Zhang Feihu , Fu Pengqiang . Identific ation Fluctuation of Aerostatic Guideway. Journal of Mecheanical Engineering[J].2010,46(21).
8. Ji Lujun , Zhang Ming, Hu Jinchun, Zhu Yu. Stabilit y to vertical impulse perturbation of plane externally pressrize gas bearings. Tsing Univ(Sci&Tech)[J].201 1,51(6).

The Parameters Selection of PSO Algorithm influencing On performance of Fault Diagnosis

Yan HE [1,a], Wei Jin MA [1] and Ji Ping ZHANG [1]

[1]*School of Mechanical Engineering and Power Engineer, North University of China, Taiyuan, Shanxi 030051, China*

Abstract. The particle swarm optimization (PSO) is an optimization algorithm based on intelligent optimization. Parameters selection of PSO will play an important role in performance and efficiency of the algorithm. In this paper, the performance of PSO is analyzed when the control parameters vary, including particle number, accelerate constant, inertia weight and maximum limited velocity. And then PSO with dynamic parameters has been applied on the neural network training for gearbox fault diagnosis, the results with different parameters of PSO are compared and analyzed. At last some suggestions for parameters selection are proposed to improve the performance of PSO.

1 Introduction

Particle swarm optimization (PSO) is an optimization algorithm which is based on swarm intelligent[1]. The PSO algorithm is initialized with a population of random candidate solution. Swarm intelligent produced by cooperation and competition of particles between population, guides optimization and search. PSO has much more profound intelligent background and could be performed easily and simply. Based on its advantages, the PSO has been applied in function optimization, artificial neural network training, fuzzy control system and some other fields at the present .

Parameters selection of PSO is the key influence on performance and efficiency of the algorithm. It is a complicated optimized problem that how to determine the optimum parameters for the optimal performance. There are no general methods to determine the optimum parameters, which are selected by user experience, because there are different parameter spaces between different parameters and relativity each other[2]. However, the regularity of different parameter which influences the performance of algorithm may be found. In this paper, the performance of PSO is analyzed, based on the control parameters variants including particle number m , accelerate constant c_1 and c_2 , inertia weight w and maximum limited velocity v_{max} . And then the effect of these parameters of PSO is applied in the neural network training for gearbox, the fault diagnosis results with different parameters of PSO are compared and analyzed. At last some suggestions for parameters selection are proposed. .

2 Particle Swarm Optimization Algorithm

Particle swarm optimization algorithm is a method which makes each particle in the population follow the current superior particle by the certain speed, searches problems' optimal solution in the solution space.

Particle flying speed is adjusted dynamically by individual and community's flight experience[3].The particle swarm algorithm formula as follows:

$$v_{id}(t+1) = wv_{id}(t) + c_1 r_1 (p_{id} - x_{id}(t)) + c_2 r_2 (p_{gd} - x_{id}(t)) \quad (1)$$

$$x_{id}(t+1) = x_{id}(t) + v_{id}(t+1) \quad (2)$$

where $x_{id}(t)$ is the current position of particle i , p_{id} is superior position of particle i , which searches until now, p_{gd} is the entire particle swarm' s superior position. w is positive constants, called inertia factor which is linearly reducing with iteration; c_1 and c_2 are non- negative constants, called cognitive learning rate. r_1 and r_2 are random numbers in range [0 1]; $v_{id} \in [-v_{max}, v_{max}]$; v_{max} is non-negative constant. The selected condition of stopping iteration is the max iteration number or the best position of particle searched till now meets for adaptive thresholds significance.

In formula (1), the velocity change of PSO includes three components: the momentum components, the cognitive components and the social components. It determines the performances of PSO how to balance

[a] Yan HE:zbhy91@163.com

three components. This evolution equation has been used as a standard PSO algorithm.

3 Parameters Selection

The PSO algorithm includes some tuning parameters, such as inertia weight w, particle number m, accelerate constant c_1 and c_2, maximum limited velocity v_{max}, the max iteration number T_{max} ,and computed precision ε .Of them, w , m , c_1 , c_2 and v_{max} are used as mainly controlling parameters ,at the same time T_{max} and ε are used as the conditions of stopping iteration.

3.1 Analysis of inertia weight selection

Inertia weight describes that the previous velocity influence on current velocity. Controlling its selection may tune the global and local search ability of PSO. If $w=0$ the velocity of particle depends on its current position pbest (the personal best) and gbest (the global best) . The velocity has no memory. If $w \neq 0$, the particle has the tendency of exploring new space, and the larger is the w ,the larger is the flying velocity of the particle, at the same time the particle will explore with longer steps. The smaller is the w, the smaller is the flying velocity of the particle, and then the particle will tend to dedicate local exploitation.

Presently, the inertia weight used more is linearly decreasing weight (simply called LDW), which is proposed by Shi, as following:

$$w = w_{max} - \frac{w_{max} - w_{min}}{T_{max}} t \tag{3}$$

Where w_{max} is the maximum inertia weight, and w_{min} is the minimum inertia weight, typically $w_{max}=0.9$, $w_{min}=0.4$; t is current iteration number; T_{max} is the maximum iteration number.

3.2 Analysis of accelerate constant selection

Accelerate constant c_1 and c_2 represent the particle stochastic acceleration weight toward the personal best (pbest) and the global best (gbest). Small accelerate constant may induce the particle wandering away in goal area; however, large accelerate constant may induce that the particle move quickly to the goal area, even fly away from it.If $c_1=c_2=0$, the particle will fly with current velocity till the border. At that moment, it is difficult for the particle to find the good solution, because it has to search in definite area. When $c_1=0$, the particle has no the cognitive ability, which is social-only, and the algorithm has the ability to get to the new searching area by particles' cooperation for each other. Its convergent speed is faster than standard algorithm. But encountering complicate problem, it is easily getting into local optima. When $c_2=0$, there isn't shared information between particles, which is cognition-only, so a swarm with particle number m equals to m single particle. The probability to get the optima is very small because individual has no interaction.

In previous researches, Kennedy and Eberhart describe that a relatively high value of the cognitive component, compared with the social component, will result in excessive wandering of the search space[1]. In contrast, a relatively high value of the social component may lead particles to rush prematurely toward local optimum. Moreover, they suggested setting either of the acceleration coefficients at 2 in order to make the mean of both stochastic factors in (1) unity. Later, Clerc deduced $c_1=c_2=2.05$, some other researchers thought that c_1 did not equal c_2, and reached a conclusion $c_1=2.8$ from experiments. In reality, these researches have been constricted to the application of part problems. So it had not popularized all fields. Suganthan tested a method of linearly decreasing both acceleration coefficients with time, but observed that fixed acceleration coefficients at 2 generate better solutions [4]. However, through empirical studies he suggested that the acceleration coefficients should not be equal to 2 all the time[5].

Generally, in population-based optimization methods, it is described to encourage the individuals to wander through the entire search space, without cluster around local optima, during the latter stages, it is very important to enhance convergence toward the global optima, to find the optimum solution efficiently.

3.3 Analysis of particle number of population selection

Shi and Eberhart found that PSO was not sensitive to particle number[6]. Their conclusion was based on the mean value of needed iteration under given precision. Zhang Li-ping has researched the mean values of all kinds of function with population variant, such as Sphere ,De Jong' s f4, Rosenbrock, Rastrigin ,Griewank Shaffer' sf6 Ackley[7]. The conclusion was that their effect on PSO was smaller, when particle numbers were more than 50; which were the same as the conclusion of Shi and Eberhart. However, when particle numbers were less than 50, they greatly influenced on the performance of PSO. From the computed complexity analysis, more particles require more function evaluation, thereby increase more computing time, and increase the reliability simultaneously.

3.4 Analysis of the max velocity selection

The velocity of particle is usually constricted to a range in order to decrease the probability of particle moving away from searching space in evolutionary process. v_{max} is the maximum limited velocity of a particle. If v_{max} is excessively large, it is thought that the velocity is not limited, then the particle may fly through the good solution. At that moment, the performance of PSO depends on inertia weight; in the contrast, If v_{max} is excessively small, the particle may not explore the space outer local best area. So it easily gets into local best. For these reasons, the excessively big or small velocity may lead the performance of algorithm decreasing.

4 The Application Of PSO Algorithm On Neural Network

4.1 Structure of neural network for gearbox fault diagnosis

In this diagnostic system, the gearbox is a researched object. Because gearbox is a very complicated transfer mechanism, its relations with faults and symptoms are not very definite and of nonlinear map, while BP neural network models have the properties such as good self-learning, self-adapting, associative memory and identification of nonlinear model, and particularly suitable for complicated pattern identification, so they are widely used in fault diagnosis of gear.

Here, the gearbox of a tractor is the researched object, and some feature parameters of frequency-domain are selected, based on gear fault. So gearbox fault diagnosis system is defined by three-layer BP neural network with topology architecture 15-31-3. The outputs of neural network are relative to fault types. There are 3 fault types, which are expressed no faults, root of tooth crack, and gear tooth collapsing.

4.2 The analysis of parameters selection of PSO influence on performance of fault diagnosis.

The simulation experiments were performed in neural network with standard PSO, when the main control parameters were tuned including particle number m and maximum limited velocity v_{max} .Then their effect on neural network training process and fault diagnosis results has been researched and analyzed.

4.2.1 The effect of different particle number

In simulation experiments, on the conditions including given $\varepsilon = 0.001$, $V_{max} = 1$ and $T_{max} = 2000$, particle numbers m was changed, , and 9 group samples of gearbox were inputted neural network to train it, the training error cures of network with different particle numbers have been showed in Figure.1(a) and (b).

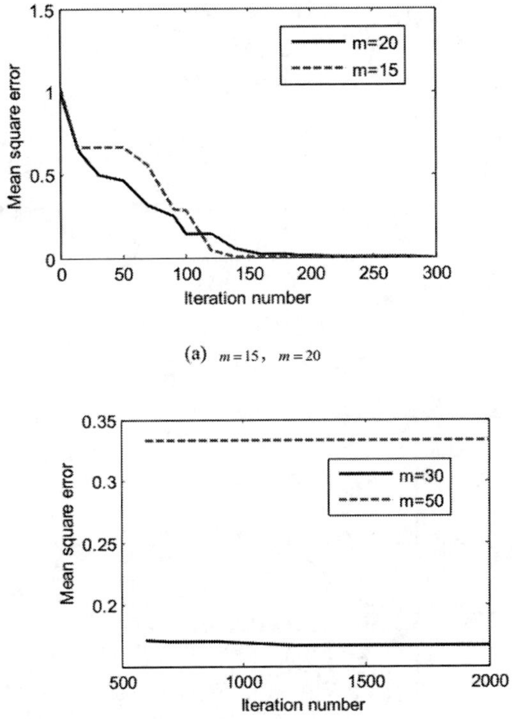

(a) $m = 15$, $m = 20$

(b) $m = 30$, $m = 50$

Figure 1. The training error cure of network with different m

When m is 15 and 20, the error cures convergent speedily, observed from Figure.1 (a); and iteration number being about 150, the mean square errors have reached specified precision 0.001. When m is 30 and 50, the error cures convergent slowly , observed from Fig.1(b); and iteration number being the maximum 2000, the individual mean square error is 0.1668 and 0.3333 , which has not reached specified precision 0.001 , and has got stuck in local best .

4.2.2 The effect of the max velocity

In previous researches, usually v_{max} was constant 1.In this study, the performance of PSO with changing v_{max} is observed. As formula (4) shows, v_{max} is changed with iteration number , and the PSO with dynamic v_{max} is called VPSO.

$$v_{max} = (e^{1-t/T_{max}})V_c \qquad (4)$$

Where v_{max} is the variable maximum which limited *velocity*, t is current iteration number, T_{max} is the maximum iteration number, and V_c is set velocity constant.

This algorithm was applied to simulation experiments in neural network with 15-31-3 topology architecture. Of them the other parameters were set as follow:

$m = 20$,fitness threshold $\varepsilon = 0.001$, $V_c = 1.5$, T_{max} =2000. 9 group samples of gearbox were inputted neural network to train it ,showed by Table1,ideal output type is set three types such as no faults (1 0 0),root of tooth crack (0 1 0)and gear tooth collapsing (0 0 1).The training error cures of network by PSO and VPSO have been showed in Figure 2. The error cure by VPSO convergent speedily observed from it. When iteration number reached 200, its error has satisfied a required accuracy target, which is fitting threshold $\varepsilon = 0.001$.

Table 1. Feather samples of Gearbox

Data number	Feature sample	Fault type
1	0.2286 0.1292 0.0072 0.1592 0.1335 0.0733 0.1159 0.0940 0.0522 0.1345 0.0090 0.1260 0.3619 0.0690 0.1828	1.
2	0.2090 0.0947 0.1393 0.1387 0.2558 0.0900 0.0771 0.0882 0.0393 0.1430 0.0126 0.1670 0.2450 0.0508 0.1328	1
3	0.0442 0.0880 0.1147 0.0563 0.3347 0.1150 0.1453 0.0429 0.1818 0.0378 0.0092 0.2251 0.1516 0.0858 0.0670	1
4	0.2603 0.1715 0.0702 0.2711 0.1491 0.1330 0.0968 0.1911 0.2545 0.0871 0.0060 0.1793 0.1002 0.0789 0.0909	2
5	0.3690 0.2222 0.0562 0.5157 0.1872 0.1614 0.1425 0.1506 0.1310 0.0500 0.0078 0.0348 0.0451 0.0707 0.0880	2
6	0.0359 0.1149 0.1230 0.5460 0.1977 0.1248 0.0624 0.0832 0.1640 0.1002 0.0059 0.1503 0.1837 0.1295 0.0700	2
7	0.1759 0.2347 0.1829 0.1811 0.2922 0.0655 0.0774 0.2273 0.2056 0.0925 0.0078 0.1852 0.3501 0.1680 0.2668	3
8	0.0724 0.1909 0.1340 0.2409 0.2842 0.0450 0.0824 0.1064 0.1909 0.1586 0.0116 0.1698 0.3644 0.2718 0.2494	3
9	0.2634 0.2258 0.1165 0.1154 0.1074 0.0657 0.0610 0.2623 0.2588 0.1155 0.0050 0.0978 0.1511 0.2273 0.3220	3

Note:Fault type: 1:no faults ,2:Root of tooth crack,3:Gear tooth collapsing

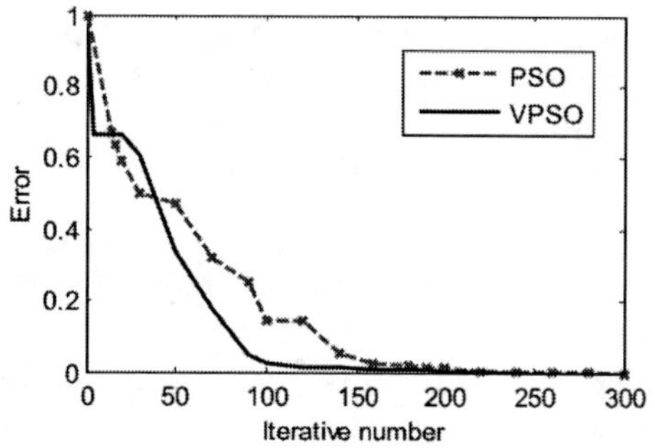

Figure 2. The training error cure of network based on VPSO and PSO

Figure 2 shows the training error cures based on VPSO and PSO. To the same group samples, two algorithms convergent quickly under 50 steps; but evolution steps being between 50 and 100, VPSO obviously convergent more quickly than PSO. It is indicated that, v_{max} linearly varying with iteration number, plays the role of accelerating convergent speed.

Two output mean error cures have been showed in Figure 3(a) and (b) ,when fault types of gearbox are root of tooth crack and gear tooth collapsing. The mean errors with VPSO are obviously smaller compared with PSO. It indicates that performance of algorithm is better, and the accuracy and efficient of fault diagnosis are obviously improved, when dynamic v_{max} is following with iteration number.

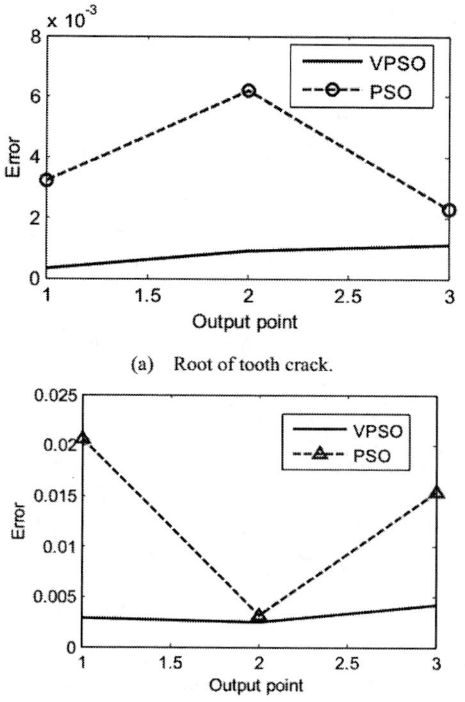

(a) Root of tooth crack.

(b) Gear tooth collapsing

Figure 3. The output mean error cure in different fault

5. Conclusion

In this paper the fault diagnosis results with different parameters of PSO have been compared and analyzed. The conclusion is as follows:

(1) As particle number $m < 100$, the effect of m variant on the training performance of networks and diagnostic accuracy, is obvious; As $m > 100$, the effect on them is not obvious. With particle number increasing,

the reliability of PSO algorithm improved, but more particles require more functional evaluation, so the required time of the algorithm increases, the efficient is poor.

Therefore, selection particle number should base on architecture of networks, and consider the reliability of the algorithm and the required time. Generally speaking, the particle number is 20 enough to usual problem, but the particle number may be selected 50 to some complicate problem.

(2) The performance of algorithm is better, and the accuracy and efficient of fault diagnosis are obviously improved, and the convergent speed is accelerating when the maximum velocity v_{max} is set a dynamic parameters following with iteration number. The v_{max} indirectly influences the global searching ability. Dynamic modification v_{max} may induce particle which always searches in rational space in order to prevent particle wandering through the searching space or toward the local optima.

Acknowledgment

The authors acknowledge the financial support of the National Nature Science Foundation of Shanxi Province contract/grant number: 2015011063.

References

1. J .Kennedy and R. C. Eberhart,Particle swarm optimization, *Proc. IEEE Int.Conf. Neural network*, pp.1942–1948,(1995)
2. Jian-chao Zeng, Jing Jie, Zhi-hua Cui, Particle swarm optimization algorithm, Sience publication, (2004)
3. Ioan Cristian Trelea, The particle swarm optimization algorithm: convergence analysis and parameter selection, Information Processing Letters, **85**, pp.317-325, (2003).
4. P.N. Suganthan, Particle swarm optimization with neighborhood operator, *Proc. IEEE Int.Conf. Evolution computation* ,**3**, pp.1958-1962,(1999),
5. Asange Ratnawwra, Saman K.Halgamuge, Harry C.Waston, Self-organizing hierarchical particle swarm optimizer with time-vary acceleration Coefficients, *IEEE Transaction on evolutionary computation*, **8**, pp.240-255, 3(2004)
6. Y.Shi, R.C.Ebethart, A modified particle swam optimizer, *Proc. IEEE International Conference on Evolutionary Computation*, pp. 69-73,(1998)
7. Li-ping Zhang, The Theorem and Practice upon the Particle Swarm Optimization Algorithm. Zhejiang University, Dissertation Submitted to Zhejiang University for the Degree of Doctor of Philosophy, (2005)

Kinematic and Dynamic Simulation Analysis of Hydraulic Excavator's Working Equipment based on ADAMS

Hong Yan YU[1], Ming Hai YUAN[1,2,a], Kun DENG[1] and Rong DONG[1]

[1]*College of Mechanical and Electrical Engineering, Hohai University, Changzhou 213022, China*

[2]*Guangxi Key Laboratory for Manufacturing System and Advanced Manufacturing Technology, Nanning, 530004, China*

Abstract. This paper establishes the 3D excavator model according to the actual size in UG firstly. Then based on the virtual simulation software ADAMS, the virtual prototype of the working device is built by adding interrelated constraints(kinematic pair and hydraulic cylinder driving function) and load secondly. This paper gets the main parameters of the excavator working scope and the pressure situation change curves of point of each hydraulic cylinder by making kinematic and dynamic simulation analysis of hydraulic excavator's working equipment at last. The conclusion providing design theory and improvement for the excavator's working device, which also play an important role in improving the level of China's excavator design, enhancing excavator's performance and promoting the rapid development of excavator industry.

Key words: Hydraulic excavator, Kinematic simulation, Dynamic simulation, ADAMS

1 Introduction

As the main mechanical equipment of earthwork excavation, hydraulic excavator is widely used in the construction of water conservancy and hydropower engineering, building engineering, municipal engineering and other engineering with the advantages of great digging force, stable transmission, simple manipulation and so on. And the excavator's working equipment plays an important part to make sure hydraulic excavator can complete all the functions. Under the actual working condition, the reliable performance and the working performance of the excavator is directly decided by the motion trajectory and the stress situation. Therefore, it is very important to design and study the working device of excavator. The traditional design method can not only cause the repetitive waste of manpower and material resources, but also result in some kinetic parameters are inconsistent with the design in manufactured products due to the dynamics of virtual model experiment was not carried out in time.

Scholars in and abroad did lots of research on the excavator's working equipment. For example, DiMaio S.P[1]ect. established a simulator with human-computer interaction interface. It simulated the interaction force between the excavator working device and the ground, and simulated the excavator's machine, hydraulic system successfully. Haga.M[2]ect. studied automatic mining control system, by controlling the horizontal motion of boom and bucket tooth to automatically control the mining depth. This method overcomes the problem that

[a] Corresponding author: ymhai@126.com

operators repeat to determine the mining depth, so that the mining efficiency is significantly improved. Imanishi, Etsujiro[3]ect. obtained some deformation characteristics of the excavator working device through dynamic simulation of working process, which also verify the superiority of the dynamics simulation analysis method in excavator structure design and machine performance analysis test. Xianghong Xu[4]ect. applied the grid computing technology to the finite element analysis, taking the excavator working device as an example, the optimization design was carried out. Caijun Xue[5]ect. carried on the static and dynamic analysis of the excavator working device by using the method of the finite element analysis, combined with the stress test, verified the validity of the finite element analysis results.

At present, the domestic and foreign research for excavator working equipment mainly concentrated in the finite element analysis, motion analysis and optimization of mining device, the optimization methods are mostly carried out by programming, and the research in the overall mining force optimization and operation stability is less, for the study in multi components and multi software of working device is relatively small too, this leads to the research data too independent, the lack of correlation, unable to thoroughly solve the problem of the engineering practice. For this reason, this paper is about the mechanical simulation analysis for the kinematics and dynamic of hydraulic excavator's working equipment.

2 The establishment of simulation model in ADAMS

The UG model can be imported into ADAMS, when a certain working environment which can simulate the real environment (including setting the unit, setting the gravity, setting the working grid, etc.) is given. The model is shown in Fig.1. After import, first to add constraints and driving function, such as adding fixed pair between the base and the ground, adding rotating pair, column pair and ball joint pair at each part of the connection, adding mobile pair between each hydraulic cylinder and its corresponding piston rod.second to define the driving function of the kinematic constraint as the step function (i.e. STEP function). After above, last to check the model with the model checking tool. Ensuring the model has 0 degrees of freedom and 0 redundant constraints[6].

Fig.1 After the introduction of ADAMS model

3 Kinematic simulation analysis

3.1 Motion parameters of Hydraulic cylinder working device

Backhoe device geometry position is determined by the lift arm hydraulic cylinder length L1, bucket rod hydraulic cylinder length L2 and bucket hydraulic cylinder length L3. Obviously, The anti shovel device will be a certain position when L1, L2, and L3 are fixed[7]. Table 1 is the movement parameters of the hydraulic cylinder for the anti shovel hydraulic cylinder.

DOI: 10.1051/matecconf/20166302020

Table 1. Movement parameters of HC working device

Symbol	Boom(mm)		Bucket rod(mm)		scraper bucket(mm)	
	MAX	MIN	MAX	MIN	MAX	MIN
Size	2100	1206	2240	1226	1840	1146
Max stroke	894		1014		694	
Diameter of HC	96		92		92	
Diameter of PR	90		80		90	

3.2 Hydraulic cylinder motion function and motion trajectory simulation

The Motion simulation of boom hydraulic cylinder, bucket rod hydraulic cylinder and bucket hydraulic cylinder are shown as follow:

3.2.1 Motion simulation of boom hydraulic cylinder

Condition: Adjust bucket rod cylinder in the full shrink state, motion function: step(time,0,0,5,400). Adjust the bucket cylinder in the full shrink state, motion function:step(time,5,0,10,452).

Movement function of boom hydraulic cylinder is step(time,10,0,20,-154)+step(time,20,0,40,154+740).

Transfer ADAMS/Post processor Modular, to replay simulation results and draw curves, Tracking the moving track of the mark points of the bucket tooth of excavator in Trace Marker, trajectory chart of the mark point is shown in fig.2.

3.2.2 Motion simulation of bucket rod hydraulic cylinder

Condition: Adjust the boom cylinder in the full shrink state,motion function: step(time,0,0,15,740).Adjusting the connecting point between the boom and the bucket rod,the connecting point between the bucket rod and bucket, the bucket tooth being a straight line,

Motion function: step(time,15,0,20,405).

Movement function of bucket hydraulic cylinder is step(time,20,0,35,-413)+step(time,35,0,50,413+400).

Transfer ADAMS/Post processor Modular,to replay simulation results and draw curves, Tracking the moving track of the mark points of the bucket tooth of excavator in Trace Marker, trajectory chart of the mark point is shown in fig.3.

Fig.2 Moving arm simulation

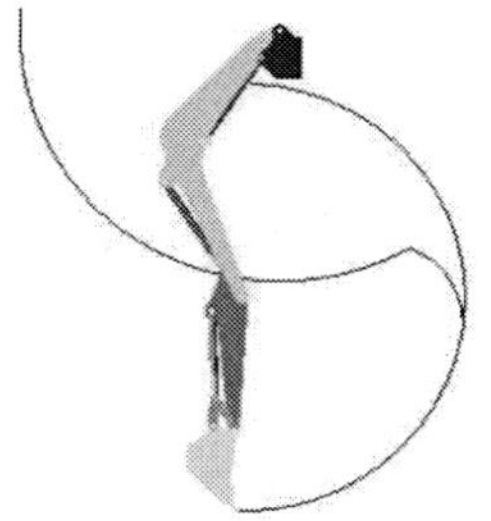

Fig.3 Bucket rod simulation

3.2.3 Motion simulation of bucket hydraulic cylinder

Excavator in mining only conducted by bucket hydraulic cylinder, the excavating trajectory is a circular arc, the arc line with the bucket and the bucket rod hinge point as the center, with the distance between the joint to the bucket tooth as the radius, and the wrap angle (i.e. the rotation angle of the bucket) and arc length is determined by the bucket hydraulic cylinder's stroke[8].

4 Dynamic simulation analysis

Dynamic Simulation is to study the role of external loads of the components and systems of internal power relations changes, for excavator, the main research is to study the variation force of bucket tooth and hinge point. The purpose is to ensure the effectiveness of the design[9].

4.1 Load Calculation in complex mining process

Soil resistance and friction are ignored. When the excavator working device is mining, the main external loads put on it are as follows.

4.1.1 Tangential and normal mining resistance

Generally speaking, the tangential and normal mining resistance can be considered acting on the top of bucket tooth approximately, and the directions are the tangential and normal direction of mining track. Their empirical formula are as follows:

$$W_1 = K_0 bh \quad (1)$$

$$W_2 = \psi W_1 \quad (2)$$

The k_0 is mining ratio resistance, it reflects the total resistance on bucket-type excavation devices when it is mining. The resistance exists in soil failure, loading soil and friction. k_0 's unit is N/cm^2 .Small excavators usually work in III level soil. According to look-up table, $k_0 = 15N/cm^2$ .b-Cutting width, it's unit is cm. According to the model, b=60cm. h-Cutting depth, it'sunit is cm. Generally, h=0.2b=12cm. Ψ-Coefficient of Mining resistance. Look-up table shows, Ψ=0.5.By substituting the above values into the formula(1), (2)and calculating, we get following data. W_1=10800N=10.8KN , W_2=0.5W_1=5.4KN.

4.1.2 Hoisting resistance

Hoisting resistance mainly refers to the gravity of the materials in the bucket when excavation finished. As usual, it is added to the centroid of bucket, and it's direction is always vertical downward.

Gravity formula: $$G = \rho Vg = \gamma V \quad (3)$$

V-Bucket capacity. According to the design value, V=0.28 m^3. ρ-Soil density. Density of III level soil is $1.8x10^4 kg/m^3$. On the basis of gravity set in ADAMS, g=9.8m/s^2. Then we can get the gravity value of the materials. It is about 5KN.

4.2 Add the load and drive in ADMAS

The direction of tangential and normal resistance correspond to the the tangent and normal direction of the bucket tooth. It shows that these there forces are in one direction[10].

By using common complex motion in digging as a work circle, we can determine the changes of force of each cylinder and hinge point when it works in the deepest position. Furthermore, we can get the kinetic parameters of excavation. Without considering the turn-around time, working cycle time including excavation time and unloading time. The excavation time usually depends on the speed of bucket digging. In order to facilitate the simulation, running speed values 0.5m/s, the digging speed values 0.75m/s, unloading time values 2s.

Based on the above analysis, we divide one cycle into three steps, and set the total cycle time as 8s:

(1) Bucket downward phase(0s-3s): We adjust working device's digging position to maximum digging radius which means that shrinking the bucket rod and boom cylinder. In this process, we assume there are no external loads applied.

(2) Digging phase(3s-5s): Bucket rod and bucket work together until the bucket filled with soil. It means the completion of the digging operation. In this process, the load is the tangential and normal digging resistance. They increase with bucket angle increases from the beginning. When digging into the deepest, digging resistance reaches the maximum, then decrease with the increase of the bucket angle.

(3) Hoisting and unloading phase(5s-8s): Hoisting phase, first we adjust the boom cylinder to the full shrinking state, at the same time adjust the bucket cylinder to ensure objects not scattered in the process of ascension in the bucket. then adjust bucket rod and bucket cylinder to the full shrinking state, and complete unloading,and complete unloading. In mining process, main load is the gravity of the material, the gravity of the material along with the increase of material, and boom lifting stage stays the same.Three load function curves are as shown in fig. 4.

DOI: 10.1051/matecconf/20166302020

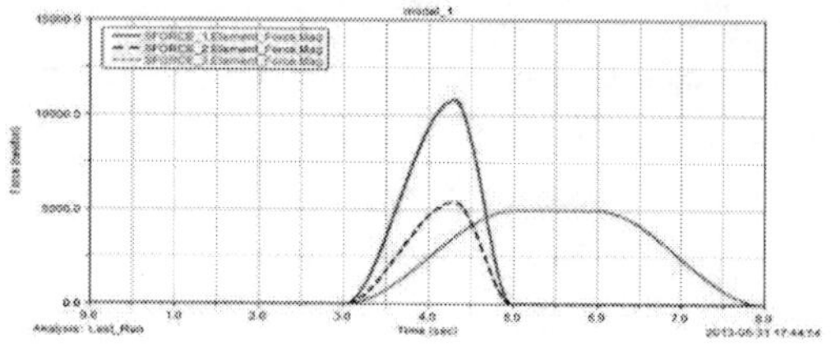

Fig.4 The change curve of the outside load

4.3 Dynamic simulation analysis

From motion simulation get the change of the connecting pointI between the boom and the bucket rod, the change of the connecting point K between the rocker rod and thebucket rod, the change of the connecting point M between the bucket rod and the bucket. the change of the connecting point N between the connecting rod and the bucket as shown in fig.5, The change of three hydraulic cylinder force as shown in fig.6: The figure shows that in the bucket down stage, each hinged point in 0-2.5 s basic is not affected by external forces, so the force variation is small, Lower movable arm due to mainly complete arm action, so the torque change largely as relative. In 2.5 s to 3.5 s mining process, the bucket tooth withstand from tangential and normal resistance, so the force changes in the rising trend. Mining process is mainly the bucket rod and the role of bucket hydraulic cylinder, relatively, hinge points I and M stress is larger; Mining end improve unloading phase 3.5 s - 6 s, which in ascending within 3.5 s to 4.5 s, movable arm plays a main role, but relative to the digging force is small lot, other hinge point more smoothly. In the unloading process of 5 s- 6 s, in order to be able to unload thoroughly, they are usually based on a great speed in a short period of time, this leads to the stress of the hinged points of ups and downs is bigger. Seeing from the diagram, there is an increase to reduce, the change of this force in the process is consistent with the actual working conditions. The result can be used as a reference, the subsequent design calculation and the change of three hydraulic cylinder force which can be taken as reference to the machine hydraulic joint simulation.

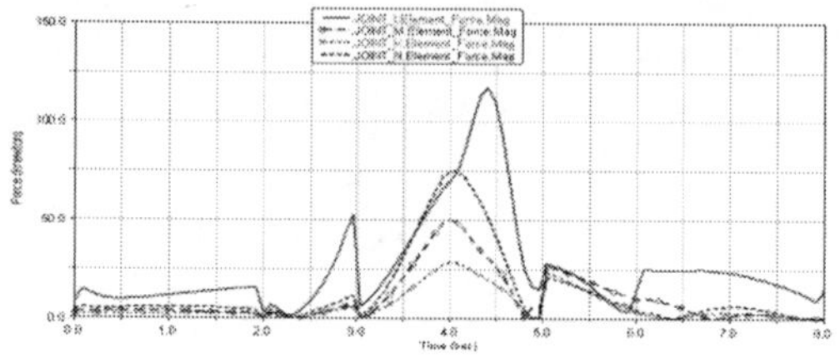

Fig. 5 Articulated point force curve

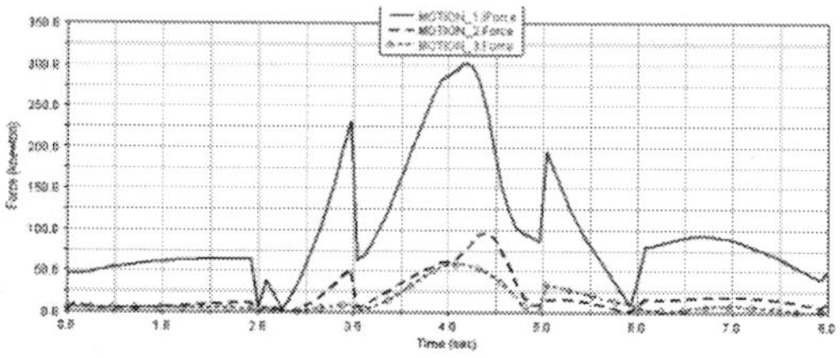

Fig. 6 Three hydraulic cylinder force change curves

5 Summary

In this paper, we analysis the light hydraulic excavator working device of the kinematics and dynamics simulation by using ADMAS software powerful analysis function. We have been obtained the scope of work and the main basic parameters of excavator boom, bucket rod, bucket and the whole machine by kinematics simulation; and obtained the articulated point and the force curve of hydraulic cylinder by dynamic simulation. we also got the curve of the movement process and model simulation analysis of the operation situation of excavator working device in this paper, the conclusion can be used as the theory basis of the design and optimization of the excavator working device.

Acknowledgments

This work was supported by the Applied Basic Research Programs of Changzhou under grant number CJ20140027, the Guangxi Key Laboratory for manufacturing system and advanced manufacturing technology under Grant number 13-051-09S09 and the graduate student innovation project of Changzhou campus of HoHai university under grant number XZX/14B001-05.

References

1. DiMaio S.P, Salcudean S.E, Reboulet C,Proc IEEE Int Conf Rob Autom **1**,52(1998)

2. Haga M, Hiroshi W, Fujishima K,Mechatronics **11**, 665(2001)

3.Imanishi E, Nanjo T, Hiroka E,Nippon Kikai Gakkai Ronbunshu C **69**,2336(2003)

4. Ye Li, Xiang Xu, Qin Qiu,Proc. World Congr. Intelligent Control Autom. WCICA **2**,6915(2006)

5. XUE Caijun, QIU Qingying, FENG Peien, ICED,772(2003)

6. Liu Jing, Pan Shuang-Xia, Feng Pei-En,J ZHEJIG UNIV **38**,1490(2004)

7. Weidong Zhou, Man. Auto. **7**,166(2010)

8. Zhenpei Qin,Junping Zhu, Agr. Mec. Research **4**, 46(2011)

9. Wei Huang,Qua. Tec. Supervision **5**,27(2008)

10. Deli Kong, Keli Zhao,*Hydraulic excavator*(Beijing, 2007)

MATEC Web of Conferences **63**, 02021 (2016) DOI: 10.1051/matecconf/20166302021
MMME 2016

Numerical simulation of the knotted nylon netting panel

Yuwei li[1,2,3,4,5,a], Lu Chen[1,2,3,4,5]

[1] *Shanghai Ocean University, College of Marine Sciences, Shanghai 201306, China*
[2] *The Open Foundation from Key Laboratory of Marine Fishery Equipment and Technology of Zhejiang, Zhoushan, Zhejiang, 316022, China*
[3] *Key Laboratory of Sustainable Exploitation of Oceanic Fisheries Resources (Shanghai Ocean University), Ministry of Education, Shanghai Ocean University, Ministry of Education, Shanghai, 201306, China*
[4] *National Engineering Research Center for Oceanic Fisheries, Shanghai, 201306, China*
[5] *Collaborative Innovation Center for Distant-water Fisheries, Shanghai, 201306, China*

Abstract. A piece of netting, consists of the 8 8 meshes, fixed on a square frame, was simulated and the tensions and their distribution, the positions of knots and netting shape were calculated by means of MATLAB in computer. The dynamic mathematic model was developed based on lumped mass method, the netting was treated as spring-mass system, the Runge-Kutta fifth-order and sixth-order method was used to solve the differential equations for every step, then the displacement and tension of each mass point were obtained. For verify this model, the tests have been carried out in a flume tank. The results of the numerical simulation fully agreed with the experiments.

1 Introduction

In fishing gear design and operation performance improvement, the drag, tension distribution and shape has directly affected capture efficiency and energy consumption. Netting panels are the basic composition of fishing gear system and aquaculture facilities such as trawl and purse seine, so the hydrodynamic performance of netting panel is the important content of the fishing gear performance study.

In the past decades, when the scientist studied the dynamic behaviour of netting gear operation process, they tried to use numerical method and model for predicting. So, AARSNES et al. considered the net cage consist of many discrete netting panels, and used the hydrodynamic coefficients from model test to calculate the shape and internal force[1].Theret built a three-dimensional numerical model to simulate netting gear, and successfully developed a software which can compute the shape and tension distribution of trawl in constant speed[2].Bessonneau and Marichal used rigid bar element to build the dynamical model for the netting panel of trawl. They combined gravity force, buoyancy force, hydrodynamic force and point force, applied finite difference method to disperse time and iterative method to model the shape and drag of the equilibrious netting panel in several trawl boundary conditions[3]. Based on the mapping coefficient method, Gignoux et al. used bar element and applied commercial software ABAQUS/AQUA to develop netting panel model[4], the results were in accordance with the results of Mannuzaa[5]. Priour et al. used two-dimensional triangular element based on the finite element method to model net shapes in order to build the trawl model. But the drawback of these methods was that the bars must be parallel to the net twines and the sides of the membrane element parallel to the diagonal of the net meshes, so the user was not free for the discretization of the net. The element can not describe the deformation outside of the membrane element. The twines only located and deformed along the side of the element, this great restricted the maximum size of the element and needed some additional elements to sufficiently describe the whole net construction[6-7].

Takagi et al. used the lumped mass method to develop the NALA software to predict the three-dimensional shape of netting[8-9]. Yuan used statics model to analyze the drag and deformation of plane netting, cylindrical netting, cone netting and frustum netting based on their boundary conditions. The primary study was compatible with the flume tank measurements. The three dimensional numerical netting model was established to calculate and estimate the shapes, inter-tensions and their distribution[10]. Lee and Zhao established the numerical model to describe the hydrodynamic behavior and deformation of net cage and fishing netting for optimization design[11-12]. Li et al. developed a numerical model based on lumped mass method to set up the motion equations for rectangle and diamond meshes. The Runge Kutta Verner fifth-order and sixth-order method was used to solve these equations. The results agreed well with the experimental data from the flume tank test. The deformation of the diamond mesh netting was larger than the rectangle mesh one. In square mesh netting, the highest tension distributed along the four diagonal lines. In diamond mesh netting, the tensions

[a] Corresponding author: ywli@shou.edu.cn

distribution along four sides were the highest,but inside the net were the lowest[13].

Depending on the lumped mass-spring method, Sun et al. established a mathematical model for simulating the dynamic behaviour of a fishing net. The model was solved by Newmark- β method which can take large time steps stably. Using Open GL library software, they developed a three-dimensional visualization model [14].

Song et al. and Cao et al. established the dynamic model of the tuna longline, and modelled the space shape distribution and tension distribution in different conditions [15-16]. Hosseini et al. refined the mass-spring method to consider the drag coefficient as a function the Reynolds number and the attack angle in order to build the numerical model for the purse seine gear and attain the sinking performance of different designs and netting materials[17]. Zhang et al. thought the inertial mass coefficient on knotless netting model used in tuna purse seine is the key factor for sinking performance [18].

These study mainly considered that the external condition affected the netting gear shape and the hydrodynamic performance, but the self-characteristics of netting gear affecting the netting gear shape and the hydrodynamic performance was less considered. There are many types of netting panels, such as knotted, knotless, diamond, square and hexagon. In actual operation, the diamond meshes are applied widely, and the hydrodynamic performance of netting panel is the important component of the netting gear study. So, we built the mathematical-mechanical model to simulate the space shape and tension distribution of the diamond meshes.

2 Materials and methods

2.1 Numerical Modelling of the knot nylon netting panel

In our study, the nylon knotted netting panel was shown in figure 1. A virtual mathematical mesh model with 8 by 8 meshes was simulated based on the lumped mass-spring method(figure 2). It was assumed that a net was composed of many lumped mass points interconnected by springs without mass. The lumped mass points were set at each knot. To simplify the model, each mesh knot was considered as a spherical shape where the fluid dynamical coefficients in all directions were constant. There was no external force existing on the spring, and buoyancy, gravity and hydrodynamic force (lift force and drag force) acting on the mass points. The layout of mass points were shown in figure 3.

2.2 Mathematics-mechanical Equations

It was assumed that the netting panel was in a state with spatial-temporally uniform current flow. The motion of point i can be expressed by the following equation based on the Newton second law:

$$M_i a = \vec{T} + \vec{F} + \vec{W} + \vec{B} \qquad (1)$$

Where a is the acceleration vector of point i; T is the tension force acting on point i; F is the hydrodynamic force including drag force and lift force; W is the gravity force; B is the buoyancy force, and M is the mass of point i including the added mass of i.

The mesh bar is looked as a spring, based on the Hooke's law, the tension T could be expressed as:

$$|T_{ij}| = \begin{cases} k\left(l_{ij} - l_{0ij}\right) & l_{ij} > l_{0ij} \\ 0 & l_{ij} \le l_{0ij} \end{cases} \qquad (2)$$

Where, k is the stiffness of the mesh bar. l_{0ij} and l_{ij} are the original length and deformed length between mass points i and j, respectively.

Figure 1. The knot plane net

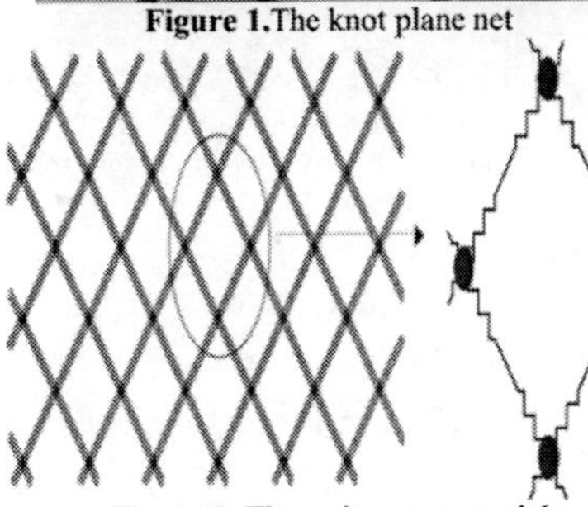

Figure 2. The spring-mass model

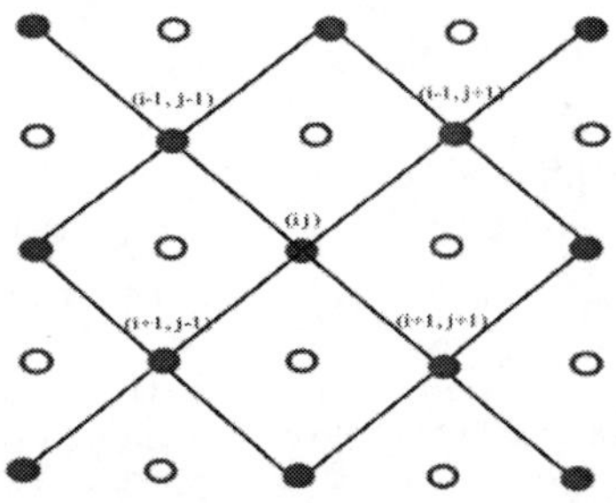

Figure 3. The layout of the quality point of plane net

As the mesh knot was considered as a spherical shape, so the hydrodynamic force can be expressed as the following equation

$$\mathbf{F}_s = \frac{1}{2} C_D \rho S_s \mathbf{V}^2 \qquad (3)$$

Where, $\mathbf{F}_s$ is the hydrodynamic force, C_D is the hydrodynamic coefficient, ρ is the water density, S_s is the projected area of mesh knot, **V** is the resultant speed vector.

Because the fluid dynamical coefficients of mesh bars in all directions are different. Then the drag force ($\mathbf{F}_d$), lift force ($\mathbf{F}_l$) can be expressed as:

$$\mathbf{F}_d = \frac{1}{2} C_d \rho S_b \mathbf{V}^2 \quad (4)$$

$$\mathbf{F}_l = \frac{1}{2} C_l \rho S_b \mathbf{V}^2 \quad (5)$$

Where, C_d and C_l are the lift force coefficient and drag force coefficient, S_b is the projected area of mesh bar.

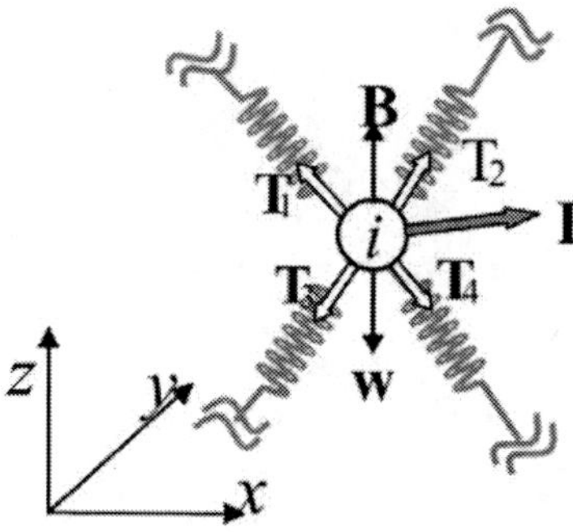

Figure 4. Schematic diagram of the model used for the calculations for mesh knots

So, the motion of the i-th lump mass point can be expressed as:

$$\begin{cases} M_i \ddot{x}_i = (T_x + F_{sx} + F_{dx} + F_{lx}) \\ M_i \ddot{y}_i = (T_y + F_{sy} + F_{dy} + F_{ly}) \\ M_i \ddot{z}_i = (T_z + F_{sz} + F_{dz} + F_{lz} + B_i - W_i) \end{cases} \quad (6)$$

So, the equation (6) can be changed into ordinary differential equations:

$$\begin{aligned} \ddot{x}_i &= f_2(x_i, y_i, z_i, \dot{x}_i, \dot{y}_i, \dot{z}_i, x_1, y_1, z_1, x_2, y_2, z_2; t) \\ \ddot{y}_i &= g_2(x_i, y_i, z_i, \dot{x}_i, \dot{y}_i, \dot{z}_i, x_1, y_1, z_1, x_2, y_2, z_2; t) \quad (7) \\ \ddot{z}_i &= h_2(x_i, y_i, z_i, \dot{x}_i, \dot{y}_i, \dot{z}_i, x_1, y_1, z_1, x_2, y_2, z_2; t) \end{aligned}$$

2.3 algorithm and simulation

The Runge-Kutta fifth-order and sixth-order method was used to solve the differential equations (7). The above algorithm was used to compile the program and to calculate the models in Matlab. These equations are solved numerically at all given points using a workstation HP 820. After that, the dynamics of the nylon knotted netting panel can be simulated.

3 The Results

After modeling and programming the nylon knotted netting panel using mechanical equations and implicit algorithm under Matlab, the motion process of the netting panel can be displayed (figure 5 and 6). The equilibrium state of netting panel agreed well to the observed data in flume tank test (figure 7). The tension distributions of the whole netting panel were shown in figure 8.

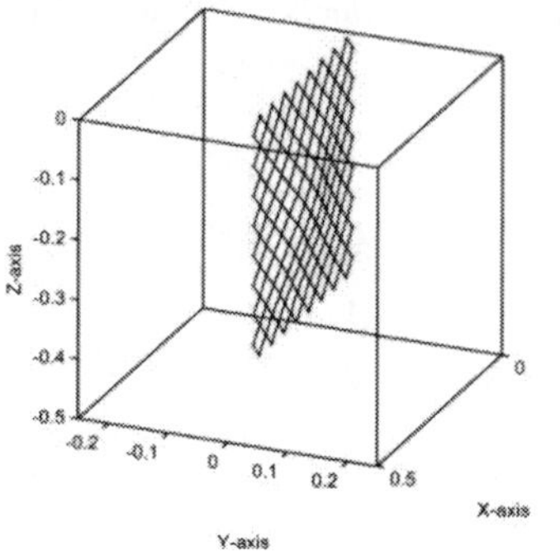

Figure 5. The initial position of the netting

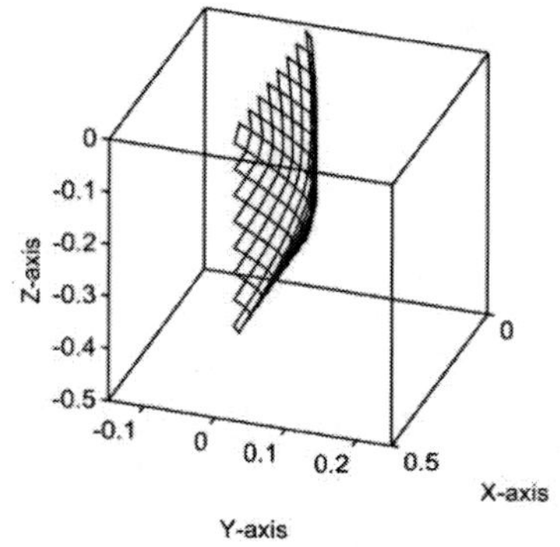

Figure 6. The equilibrium state of the netting

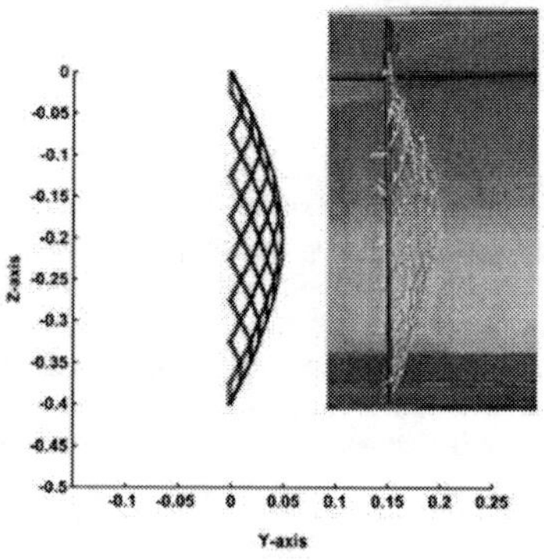

Figure 7. Comparison diagram

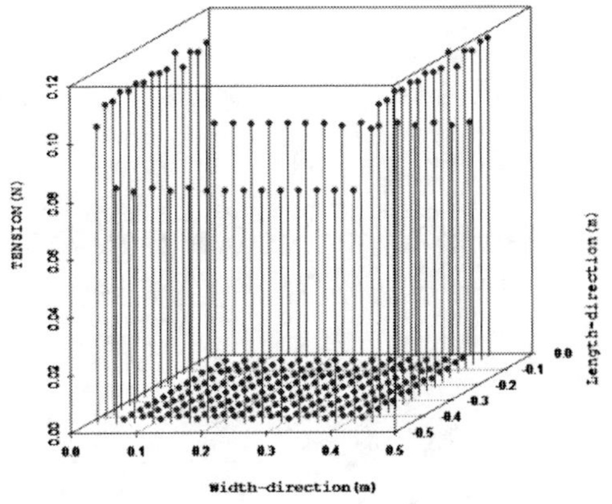

Figure 8. The Tension distribution of bars

Acknowledgment

This work was supported by: (1)the Open Foundation from Key Laboratory of Marine Fishery Equipment and Technology of Zhejiang(MFET201409); (2)Key Laboratory of Sustainable Exploitation of Oceanic Fisheries Resources(Shanghai Ocean University), Ministry of Education(A1-0209-15-0503-7); (3) Special funding for the development of science and technology of Shanghai Ocean University (A2-0209-15-200002);(4) The Youth Scholars of Shanghai Higher Education Institutions (A1-2035-14-0010-14);(5) Key technology in high efficient utilization of Chilean jack mackerel (*Trachurus murphyi*) resources (No. 2012AA092301).

References

1. J.V.Aarsnes, H.Rudi, G.Loland, Current forces on cage, net deflection. *In: Engineering for Offshore Fish Farming*, 137-152 (1990)
2. F. Theret, Etude de l'equilibre de surfaces reticules places dans in courant uniforme (application aux chalets), Ecole Centrale de Nantes: Universite de Nantes(1993)
3. J. S.Bessonneau, D.Marichal, Study of the dynamics of submerged supple nets (applications to trawls),Ocean Engineering, **25(7)**:563-583(1998).
4. M.Gignoux, H.Gignoux, R.H. Messier, Computational modeling for fin-fish aquaculture net pens,Oceanic Engineering International, **3(1)**:12-22(1999)
5. M.Matmuzza, Hydrodynamic forces on floating offshore aquaculture cages ,Orono :University of Maine(1995)
6. D.Priour,Calculation of net shapes by the finite element method with triangular elements, Communications in Numerical Methods, **15**:755-763, (1999)
7. D.Priour,Introduction of mesh resistance to opening in a triangular element for calculation of nets by the finite element method, Communications in Numerical Methods in Engineering, **17**:229-237 (2001)
8. K.Suzuki, T.Takagi, T.Shimizu, T.Hiraishi, K.Yamamoto,K.Nashimoto,Validity and visualization of a numerical model used to determine dynamic configurations of fishing nets, Fisheries Science, 69(4):695-705(2003)
9. T.Takagi,T.Shimizu, K.Suzuki, T.Hiraishi, K. Yamamoto,Performance of 'NaLA': a fishing net shape simulator,Fisheries Engineering, **40(2)**:125-134(2003)
10. J.T.Yuan, Development and Application of the 3D Numerical Netting Model,Shanghai: Shanghai Ocean University(2007)
11. C.W.Lee,Y.B.Kim,G.H.Lee,M.H.Lee,M.Y.Choe,M. K.Lee,K.Y.Koo,Dynamic simulation of a fish cage system subjected to currents and waves,Ocean Engineering, **35** :1521-1532.(2008)
12. Y.P.Zhao,Y.C.Li, G. H. Dong. Numerical simulation of the effects of structure size ratio and mesh type on three-dimensional deformation of the fishing-net gravity cage in current,Aquacultural Engineering, **36** :285-301, (2007)
13. Y.C.Li,Y.P.Zhao,F.K.Gui,B.Teng,Numerical simulation of the hydrodynamic behaviour of submerged plane nets in current,Ocean Engineering, 33: 2352-2368(2006)
14. X.F.Sun,Y.Yin, X.F.Zhang, Dynamic simulation and visualization of fishing nets based on a physical model, Journal of Dalian Fisheries University, **24(6)**:563-567(2009)
15. L.M.Song,J.Li,W.Y.Xu,X.F.Zhang,The dynamic simulation of the pelagic longline deployment, Fisheries Research, **167**:280-292(2015).
16. D.M.Cao,L.M.Song,J.Li,J.T.Yuan,Y.Q.Zhou, Determining the drag coefficient of a cylinder perpendicular to water flow by numerical simulation and field measurement, Ocean Engineering, **85**:93-99. (2014)
17. S.A.Hosseini,C.W.Lee,H.S.Kim,H.S.Kim,J.Lee,G.H. Lee,The sinking performance of the tuna purse seine gear with large-meshed panels using numerical method, Fisheries Science, **77**:503-520(2011)
18. X.F.Zhang,L.X.Xu,L.M.Song,J.Zhang,Y.W.Li,Effect s of inertial mass coefficient on knotless netting model used in tuna purse seine,Applied Mechanics and Materials, **2**:256-259,(2013)

Experimental Study on Tribological Properties of Laser Textured 45 Steel Surface

Zhi Peng LI[1,a], Jin He WU[1], Na WU[2] and Dong Xu LI[1]

[1]*school of mechanical and electrical engineering, Jiangsu Normal University, 221116 Xuzhou, China*
[2]*school of mechanical and electrical engineering, China University of Mining and Technology, 221116 Xuzhou, China*

Abstract. In order to study the influence of pits' size parameters on the tribological properties of textured friction pairs, using the Nd:YAG laser micro machining system and the "single pulse at the same point, interval more times" processing technics to process the pits on the surface of 45 steel. The dimension parameters of pits texture were obtained by orthogonal experimental design. The tribological experiment of GCr15 pin /45 steel disc was carried out by UMT-2 test machine. The surface morphology of the specimens was analyzed by using scanning electron microscopy. The experimental results show that the pits texture on the surface of 45 steel can effectively reduce the friction coefficient and the wear on the condition of oil-rich lubrication. The textured specimen with diameter 60μm, depth 6μm and surface density 10% has the lowest friction coefficient, and the friction coefficient is reduced by 21% compared with the smooth specimen. By analyzing the wear morphology on the surface of 45 steel, it is found that the surface of pits texture can obviously reduce the wear.

Keywords: Surface micro-texture; Orthogonal experimental design; Lubrication; Tribological properties

1 Introduction

The friction and lubrication performance have an important influence on the operation efficiency and service life of the mechanical system. These years, based on some research results, the surface with a certain texture and morphology has better wear resistance[1-5]. The texture with different sizes and shapes can be processed on the samples surface by using laser micro machining system. The processed textured samples were studied by friction and wear test, which show good effect of anti-friction and promote the lubrication. At the same time, it has been applied to the cylinder liner-piston ring friction pair of internal combustion engine[6-7]. Fu et al. [8] reported an experimental study on the tribological properties of the textured surfaces, they put forward that the pits texture have good lubrication performance. Meanwhile, when the pit surface density is 15% and the depth is 5μm, the friction coefficient of the texture sample is the lowest. Wan et al. [9] processed on the surface of GCr15 steel plate, and studied the effect of sliding speed on the lubrication state of textured surface, they put forward the lubrication state of surface texture changes from the mixed lubrication to the fluid lubrication at a lower speed. Moreover, the existence of micro porous texture can improve the bearing capacity of oil film and reduce the width of wear scar. Under the different surface densities, the effects of surface texture on the dry friction and wear properties were studied by using the ring-block line contact friction and wear test. The hardness increases

[a] Corresponding author: lizhipeng_19910122@163.com

and the wear rate decreases in the textured region. In a certain range, the wear rate decreases with the increase of the density of the textured surface[10]. At present, most of the research on the surface texture are the contrast experiments, which have not concluded the optimal surface texture profile and parameters. Therefore, the further research on tribology experiment are also needed.

In this paper, we used Nd:YAG laser micro machining system to process the pits on the surface of 45 steel, the dimension parameters of pits texture were obtained by orthogonal experimental design, and then carried out experiments on the UMT-2 test machine. The fundamental aim of this work is to investigate the effects of friction coefficient and wear morphology on the surface of 45 steel under different dimension parameters.

2 Experimental part

2.1 Laboratory equipment and materials

In this experiment, we used a Q switched diode-pumped Nd:YAG laser to process the pits on the surface of 45 steel. The material of the upper test piece is GCr15, the diameter is 6mm, the length is 20mm, and the hardness is HRC65. The material of the lower test piece is 45 steel, the diameter is 30mm, the thickness is 8mm, and the hardness is HRC50.

2.2 Orthogonal experimental design and the processing of pit textures

(1) Determine the experimental factors and levels: choose pit diameter, pit depth and pit surface density as the three factors of the experiment. The different values taken by each factor are the level of this factor, the factors and levels of the experiment are shown in Table 1.

Table 1. The graph of factors and levels.

Levels	*A* Pit diameter (μm)	*B* Pit depth (μm)	*C* Pit surface density (%)
1	40	4	5
2	60	6	15
3	90	8	10

(2) Choose the orthogonal table to design experimental plan: according to the factors and levels of the experiment, choose $L_9(3^4)$ orthogonal experiment table. In Table 2, the *A*, *B*, *C* and *D* respectively represent the pit diameter, the pit depth, the pit surface density and the empty column. Orthogonal experimental scheme is shown in Table 2.

Table 2. The orthogonal experimental form.

Experiment serial number	*A* Pit diameter (μm)	*B* Pit depth (μm)	*C* Pit surface density (%)	*D* Empty column	Experimental program
Untextured surface	0	0	0		
Sample 1	1 (40)	1 (4)	1 (5)	1	$A_1B_1C_1$
Sample 2	1 (40)	2 (6)	2 (15)	2	$A_1B_2C_2$
Sample 3	1 (40)	3 (8)	3 (10)	3	$A_1B_3C_3$
Sample 4	2 (60)	1 (4)	2 (15)	3	$A_2B_1C_2$
Sample 5	2 (60)	2 (6)	3 (10)	1	$A_2B_2C_3$
Sample 6	2 (60)	3 (8)	1 (5)	2	$A_2B_3C_1$
Sample 7	3 (90)	1 (4)	3 (10)	2	$A_3B_1C_3$
Sample 8	3 (90)	2 (6)	1 (5)	3	$A_3B_2C_1$
Sample 9	3 (90)	3 (8)	2 (15)	1	$A_3B_3C_2$

(3) Texture processing procedure: Firstly, the surface of 45 steel was grinded by using 160, 320, 500, 800, 1000, 1200 sandpapers, then the sample surface was polished by diamond polishing liquid with the diameter is 0.1μm, and the surface roughness of the sample is about 0.1μm after polishing. Secondly, use the "single pulse at the same point, interval more times" processing technics to process the pits on the surface of 45 steel. Figure 1 is the morphology of pit processed by laser. By using Keyence's high speed digital micro system can measure the diameter

of the pit is about 60μm, and the depth is about 4μm. As shown in Figure 1, the residue appeared around the texture is formed by the molten metal splashing and curdling in the process of machining. The adhesion between the specimen and the residue is small, it is easy to use sandpaper to wipe the residue. Thirdly, the 45 steel surface was lightly polished to remove some solid residues from the edge of the pits after the samples were processed, then use the ultrasonic cleaner to clean the surface of 45 steel. Finally, dry the 45 steel, and put it in the dry plastic box.

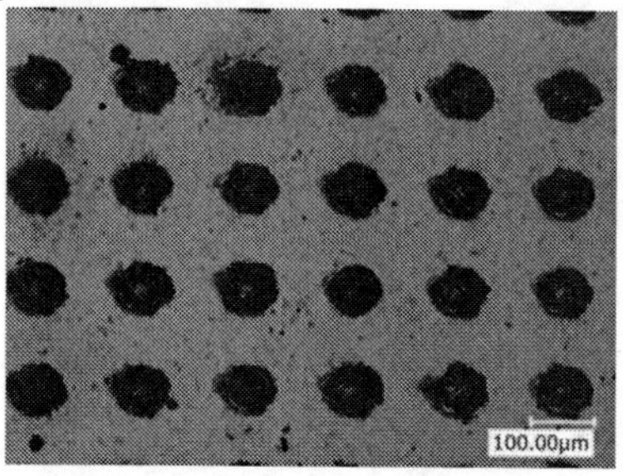

Figure 1. The surface texture were processed by laser with diameter 60μm and depth 4μm.

2.3 Tribology experiment

The UMT-2 testing machine is employed to conduct the tribological experiments, adopt the friction pair of GCr15 pin/45 steel disc. The upper test piece is still, the lower test piece is in a reciprocating motion and reciprocating stroke is 15mm. In the experiment, dripping about 0.2ml lubricating oil into the friction contact zone of pin and disc every 2min, the oil number is 5W-30. The experiment was carried out at 20℃ for 15min. After the end of the experiment, use the ultrasonic cleaner to clean the surface of 45 steel, then dry the 45 steel. The wear scar and surface morphology of the specimens were observed and analyzed by using scanning electron microscopy.

3 Results and discussion

The measured values of the average friction coefficient of the smooth surface and the nine groups of pits surface were obtained by the UMT-2 test machine. The measured values of the friction coefficient of orthogonal experiment are shown in Table 3.

3.1 Range analysis of specimens with different pit sizes

The range analysis method is used to deal with the results of orthogonal experiment, it can also determine the primary and secondary subsequence of influencing factors of pit diameter, pit depth and pit surface density. Table 4 is the range analysis of the experimental results of different pit diameters, different pit depths and different pit surface densities.

Table 3. The measured values of the friction coefficient of orthogonal experiment.

Experiment serial number	*A* Pit diameter (μm)	*B* Pit depth (μm)	*C* Pit surface density (%)	Friction coefficient
Untextured surface	0	0	0	0.143
Sample 1	40	4	5	0.142
Sample 2	40	6	15	0.134
Sample 3	40	8	10	0.194
Sample 4	60	4	15	0.143
Sample 5	60	6	10	0.113
Sample 6	60	8	5	0.148
Sample 7	90	4	10	0.126
Sample 8	90	6	5	0.140
Sample 9	90	8	15	0.119

Table 4. The range analysis of experimental results.

Experiment serial number	*A* Pit diameter (μm)	*B* Pit depth (μm)	*C* Pit surface density (%)	*D* Empty column	Friction coefficient

DOI: 10.1051/matecconf/20166302022

Sample 1	1 (40)	1 (4)	1 (5)	1	0.142
Sample 2	1 (40)	2 (6)	2 (15)	2	0.134
Sample 3	1 (40)	3 (8)	3 (10)	3	0.194
Sample 4	2 (60)	1 (4)	2 (15)	3	0.143
Sample 5	2 (60)	2 (6)	3 (10)	1	0.113
Sample 6	2 (60)	3 (8)	1 (5)	2	0.148
Sample 7	3 (90)	1 (4)	3 (10)	2	0.126
Sample 8	3 (90)	2 (6)	1 (5)	3	0.140
Sample 9	3 (90)	3 (8)	2 (15)	1	0.119
K_1	0.470	0.411	0.430		
K_2	0.404	0.387	0.396		
K_3	0.385	0.461	0.433		
$k_1=K_1/3$	0.157	0.137	0.143		
$k_2=K_2/3$	0.135	0.129	0.132		
$k_3=K_3/3$	0.128	0.154	0.144		
Range *R*	0.029	0.025	0.012		
Primary and secondary sequence	*A B C*				
Optimal scheme	$A_3B_2C_2$				

In Table 4, according to these three factors of the diameter , the depth and the density of the pits surface, the degree of the effect on friction coefficient should be: *A*(pit diameter)>*B*(pit depth)>*C*(pit surface density). The best solution should be $A_3B_2C_2$, namely the pits texture sample with the diameter is 90μm, the depth is 6μm and the surface density is 15%. The best experimental program hasn't been done, it can be found from the advantage of orthogonal table.

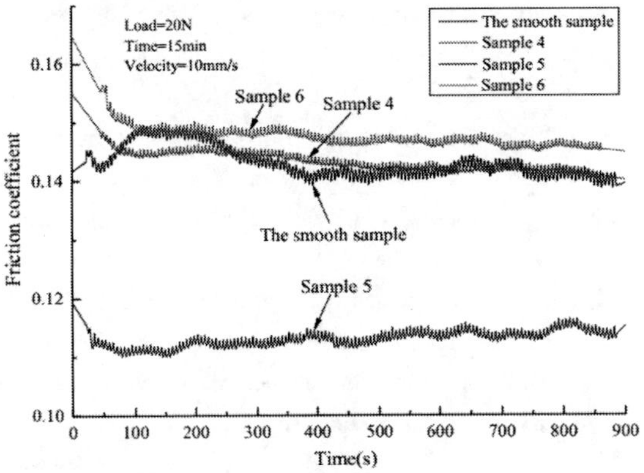

Figure 2. The friction coefficient of the pit diameter is 60μm and the smooth sample.

3.2 Tribological properties of specimens with different pit sizes

Friction coefficient is an important parameter to evaluate the friction characteristics of friction pairs. Through the experiment of pits samples with different size parameters, the friction coefficient of the pit diameter is 60μm and the smooth sample is shown in Figure 2. The friction coefficient of the pit depth is 6μm and the smooth sample is shown in Figure 3.

 DOI: 10.1051/matecconf/20166302022

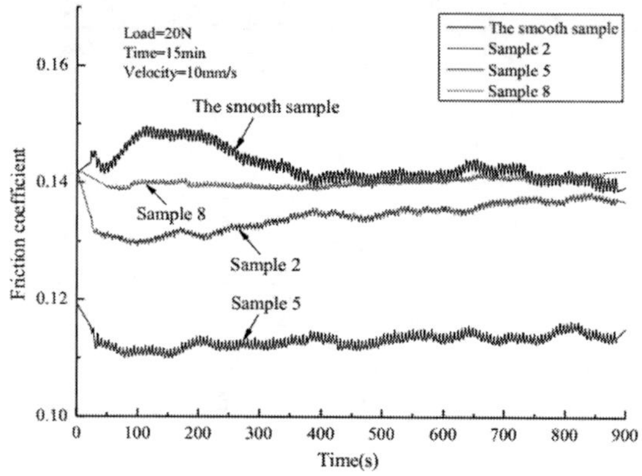

Figure 3. The friction coefficient of the pit depth is 6μm and the smooth sample.

From the Table 3, Figure 2 and Figure 3, we can find that the friction coefficient of sample 1, 2, 5, 7, 8 and 9 is lower than the smooth sample. The results show that the pits textured surface has the effect of reducing friction. However, the friction coefficient of sample 3 and sample 6 are higher than the smooth sample. This is because the diameter of sample 3 is small and the depth of sample 6 is deep, so that the film thickness becomes thin and the lubrication state of the friction pairs changes from fluid lubrication to boundary lubrication. The sample 5 with the diameter is 60μm, the depth is 6μm and the surface density is 10% has the lowest friction coefficient. It shows that the texture parameter of sample 5 is beneficial to increase the thickness of the oil film of the friction pair, it can also be better to store the wear debris and reduce the friction between the friction pairs.

3.3 Comparison of wear morphologies of specimens with different pit sizes

Figure 4(a) are the wear scars of smooth surface, which were observed by using scanning electron microscopy. Figure 4(b)~(f) are part of the wear scars between pits of different samples.

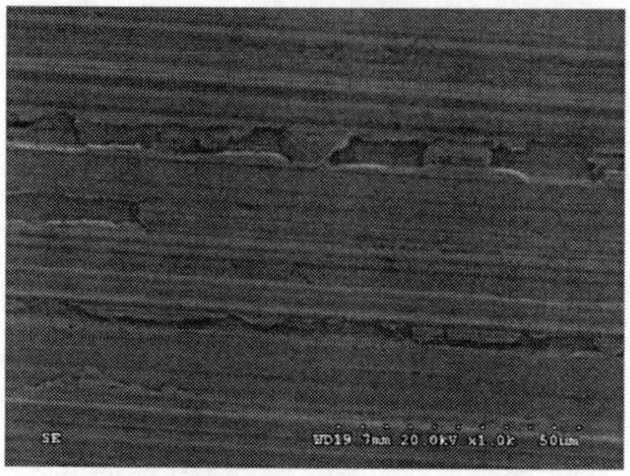

(a) the wear scars of smooth sample

(b) the wear scars of sample 1

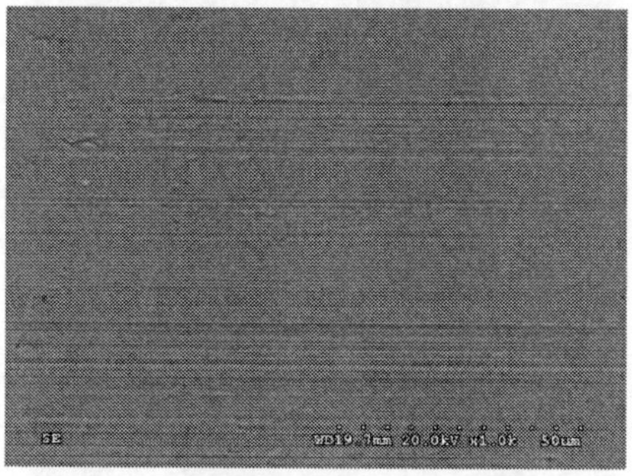

(c) the wear scars of sample 5

(d) the wear scars of sample 6

(e) the wear scars of sample 7

(f) the wear scars of sample 8

Figure 4. Microscopic photographs of the wear scars between pits of different samples.

It can be seen from Figure 4(a), the surface morphology of the smooth sample is relatively rough after wear, surface grinding marks are significant, surface scratches are more serious, the surface appears the obvious furrows, and because of fatigue and adhesive wear, the greater spalling were occurred locally. However, in Figure 4, among the micro images of the wear scars of sample 1, 5, 6, 7 and 8, it can be seen that the wear scars of the pits texture are relatively light compared with smooth sample. This shows that the pits texture can improve the wear resistance of the samples surface, which indicate the superiority of the textured surface. And in the samples of these textured surfaces, their wear traces are a little different. In Figure 4(c), the wear traces of the sample 5 are relatively small, it shows that the surface texture with this size can improve the lubrication on the surface of the friction pair.

4 Conclusion

By using the "single pulse at the same point, interval more times" processing technics to process the pits on the surface of 45 steel, and then the tribological experiment of GCr15 pin/45 steel disc was carried out by UMT-2 test machine. The following conclusions can be drawn:

(1) The pit texture can effectively reduce the friction coefficient of the contact surface. When the pit texture sample meets the conditions including that the diameter is 60μm, the depth is 6μm and surface density is 10%, the friction coefficient of friction pair is the lowest, and the biggest decline was 21% compared with the smooth sample.

(2) The surface scratches of the smooth sample are more serious than others, and the greater spalling were occurred locally because of fatigue and adhesive wear. The analysis of the wear morphology on the surface of 45 steel shows that the wear scars of the pits texture are relatively light compared with smooth sample. The surface with pits texture has a significant effect on wear resistance.

References

1. B. Bhushan, S.R. Ge, *Introduction to tribology* (China Machine Press, Bei Jing, 2007)
2. C.B. Ma, H. Zhu, J.Q. Li, J. China U. Min. Techno. **39**, 244 (2010)
3. Y.K. Zhou, H. Zhu, W. Tang, W.Q. Zhang, Lubrication Eng. **37**, 45 (2012)
4. J.H. Ji, Y.H. Fu, X.J. Hua, H. Fu, Z.Y. Kang, China Surface Eng. **27**, 107 (2014)
5. J. Gandra, H. Krohn, R.M. Miranda, P.Vilaca, L. Quintino, J.F. dos Santos, J. Mater. Processing Techno. **214**, 1062 (2014)
6. B.F. Yin, Z.T. Lu, S.J. Liu, Y.H. Fu, Y. Wang, J. Mech. Eng. **48**, 91 (2012)
7. B.F. Yin, Y.Q. Qian, Z.T. Lu, B.W. Wang, S. Sun, J. Xi'an Jiaotong Univ. **48**, 74 (2014)
8. Y.H. Fu, B.H. Tang, J.H. Ji, H.W. Zhang, X.J. Hua, Lubrication Eng. **38**, 19 (2013)
9. Y. Wan, J.L. Li, D.S. Xiong, J. Cent. South Univ. (Sci. Techno.) **46**, 4442 (2015)
10. B. Wang, Q.Y. Chang, Y. Qi, Lubrication Eng. **38**, 11 (2013)

MATEC Web of Conferences **63**, 02023 (2016) DOI: 10.1051/matecconf/20166302023
MMME 2016

3D modeling design and engineering analysis of automotive suspension beam

Zhi Lan JU[a], Fu Bao ZHANG , Guo Ran HUA ,Hua ZHANG

School of Mechanical Engineering, NanTong University, Nantong 226019, P.R.China

ABSTRACT: Automotive suspension is an important device for transmission and torque. The main parameters and dimensions of 40 tons of heavy duty truck spring suspension system are designed in the paper. According to the data, the 3D modeling and virtual assembly of the leaf spring suspension are carried out by using parametric design. Structural stress of spring suspension is analyzed which can provide a guide and basis for the design of the leaf spring suspension.
Keywords: Leaf Spring Suspension, Parametric Modeling, Finite Element Analysis

1 Introduction

Vehicle suspension system is a generic terms for all transmission gear and coupling arrangement between body-chassis units or frame construction and wheels or axle. The main function is to deliver the bearing force, lateral reaction force, vertical reaction force as well as the movements generated by those forces above to the body-chassis units or frame, which ensures the normal vehicle running [1]. The optimization design and static analysis of the suspension is one of the important contents of the vehicle movement. Currently for heavy duty automobile, the simple structure, low cost, mature and reliable technology of leaf spring suspension is the best choice, which is still worthy of study focus, in the design stage of the automotive suspension of stress field, temperature field and deformation analysis, and understanding automobile plate spring suspension thermal load state and comprehensive stress distribution situation and the suspension system improved, reducing the heat load, improve the thermal stress distribution, it is of great significance to improve the reliability.

Passive leaf-spring suspension has more advantages than active air suspension. The 40 tons of heavy truck suspension leaf-spring system was studied in this paper. Using parametric technology and finite element analysis to construct three-dimensional modeling and assemble suspension system virtually, according to the design requirements. Then the suspension system static analysis was carried out on the basis of the model.

2 The design of main structural parameter

2.1 suspension load

According to the leaf-spring placement, the designing of the leaf-spring is placed on the axle, the spring load distribution parameters as seen in the ***Tab .1.***

Tab.1. the arrangement of spring load

	Front suspension	Rear suspension
xial load mass	15 ton	25 ton
unsprung mass	1 ton	1.5 ton
Spring load	14 ton	23.5 ton

2.2 Determination of basic parameters of leaf-spring

In consideration of the overall layout of the car, the steel leaf-spring should be designed as long as possible. Because of the longer steel sheet can reduce the stiffness of the leaf-spring, and then can improve the ride comfort of the vehicle [2]. On the other hand, the long spring plate also can reduce the working stress, so it can improve the service life of the leaf spring. Comprehensive consideration of various factors, the length of the front suspension spring was calculated about 1570mm.

The total moment of inertia of the U type bolt can be cited to express as follow [3]:

$$J_0 = \frac{\delta(L-KS)^3 C}{48E} \qquad (1)$$

δ is the deflection coefficient and the deflection coefficient formula is calculated using ***Eq.(2).***

$$\delta = \frac{1.5}{1.04\times(1+\frac{n_1}{n_0})} \qquad (2)$$

Which n_1 is the number of overlapping slices with equal length, $n_1 = 2$ and n_2 is the valuation of the total number of springs, take $n_2 = 16$ [4], so $\delta = 1.357$ was calculated.

[a] **Corresponding author:ju.zl@ntu.edu.cn**

Where L is the spring stretched length, $L = 1570mm$. K is invalid length coefficient, generally $K = 0.5$, while s is the center of the U-shaped bolt distance, $s = 95mm$; C is the spring stiffness, $C=5.6\times10^5 N\cdot m^{-1}$, E is the elastic modulus, $60Si_2Mn$ is chose as the spring material, then $E = 210GPa$. Therefore J_0 can be obtained ,so the value is $2.66\times10^{-7}m^4$.

As seen from previous considerations, the depth of steel plate can be cited to express as ***Eq. (3).***

$$h \le \frac{8Jo}{Fw(L-kS)}[\sigma_c] \qquad (3)$$

σ_c is the material allowable bending stress, $\sigma_c = 450N\cdot mm^{-2}$, h can be obtained $\le 17.97mm$. So the spring thickness was selected 12mm.

The spring load is determined according to the 40 tons of heavy vehicle performance, followed by calculation of the basic parameters: static deflection, deflection, full of arc height and leaf spring sheet size parameters. It provides the necessary data and foundation for the three-dimensional modeling of the leaf spring. The dimension parameters of the rigid spring suspension system are shown in ***Tab.2.***

Tab.2. Size parameter of rigid spring suspension system

The Name of Part	Materials	Elastic Modulus *GPa*	Poisson's ratio μ
Leaf spring plate	$60Si_2Mn$	206	0.29
Leaf spring	Gray cast iron	145	0.25
U bolt	35	212	0.31
Upper cover plate	Gray cast iron	145	0.25
Spring pin	20	206	0.30
Spacer	35	212	0.31
ut	45	209	0.3
Washer	Q215	210	0.26

3 Simulation and analysis of vehicle suspension beam

3.1 Three Dimension model

Based on the design parameters presented in ***Table.2***, the three-dimensional model of rigid spring suspension system was established, as shown in ***Fig.1***.

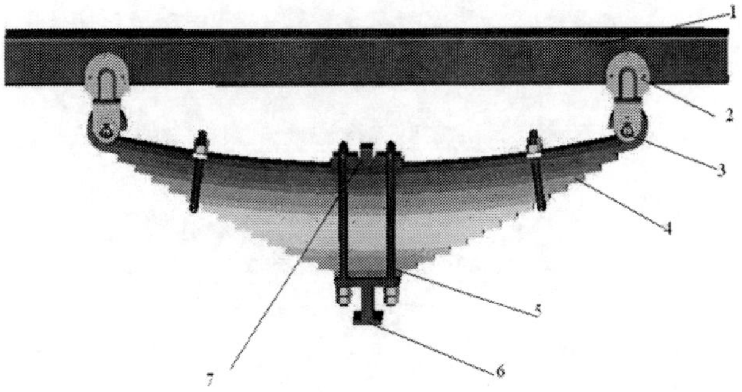

1-beam 2-bracket 3-eye axis 4-beam spring 5-U stud 6- the base of spring 7- Upper cover plate

Fig.1. the structure diagram of automobile suspension system

3.2 Engineering analysis of suspension beam

The performance of suspension system can be seen in ***Tab.3***

Tab.3. Material properties of suspension parts

The name of parameters	Selected or calculated values
Extended length	1570 mm
Static disturbance	625 mm
Dynamic disturbance	675 mm
Rigidity	5.6×10^5 Nm^{-1}
Moment of Inertia J_0	2.66×10^{-7} m^4
Thickness	12 mm
Width	120 mm
Number of tablets	16

3.3 FEM Analysis

The finite element modeling of leaf-spring suspension has been built in Para.2.1, then to solve it by loading in FEM software. The total deformation of suspension system results as shown in ***Fig.2*** and the equivalent elastic deformation are shown in ***Fig.3***, and so on.

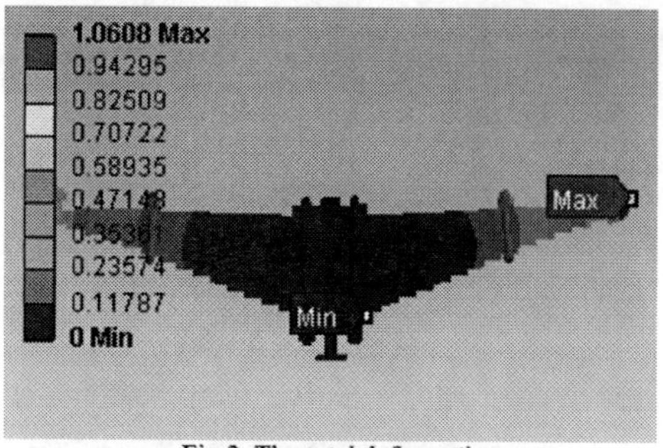

Fig.2. The total deformation

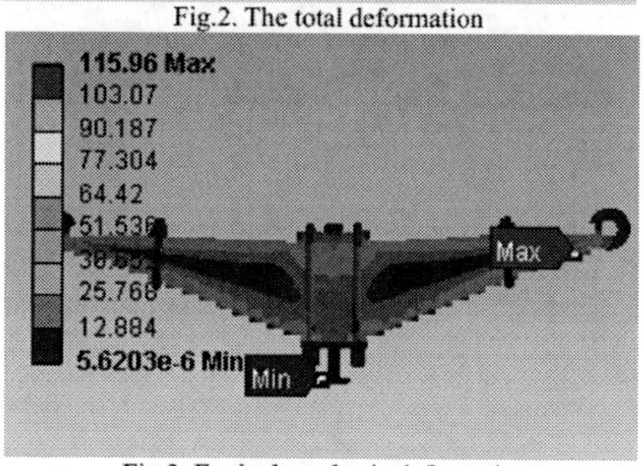

Fig.3. Equivalent elastic deformation

3. 4 The result analysis

The result shows that the maximum value of equivalent stress of the leaf spring suspension is $115.96\,MPa$. And the stress mainly concentrated on the vicinity of the leaf spring suspension at both ends of the mouse ear chickweed. The yield strength of 60Si2Mn spring steel was $800\,MPa$, and the tensile strength was 1000 MPa [5]. So the results of strain-stress of the vehicle suspension can meet the design requirements.

In consideration of the steel plate spring suspension, the Z_{max} load can be given by ***Eq. (4).***

$$Z_{max} = kG = 35000 \times 3 = 105000N \quad (4)$$

k and G are dynamic load coefficient and vehicle suspension spring static load, respectively. The safety factor K is also need to consider in addition to the safe of automobile driving process. The safety factor of the project is n=1.2. When the vehicle runs on rough road, the strength of the steel plate spring is satisfied with the design requirement that can be seen in ***Eq.5.***

$$\sigma_{max} = \sigma_1 \times k \times n = 115.96 \times 3 \times 1.2 = 417.46MPa$$
$$< [\sigma] = 800MPa \quad (5)$$

The maximum deformation value is 1.06mm. The zero of the maximum deformation zero of leaf spring suspension is appeared near the ends of the mouse ear chickweed, which can be seen from Fig.3. According to the finite element analysis, near both ends of the spring eye can be surface strengthened to improve the whole performance of the leaf-spring suspension system.

4 Conclusion

Based on the given parameter value, the suspension system was built through the parametric design software in this paper. Then the suspension is simulated and the stress analysis is obtained by FEM. The results show that the optimization was done correctly and effectively on improving the kinematic characteristics of the suspension system, which can provide the theoretical guidance for the following optimization design.

Acknowledgements

The authors express their deep gratitude to Natural Science Fund of China under Grant No.51205212, Natural Science Fund of Jiangsu Province of China under Grant No.BK2012233 and Nantong Municipal Science and Technology Bureau under Grant No.K2014037.

References

[1] Zhao Conglin.*Hebei University of Technology,* 2010.
[2] Xu Pingtao. *Changan University,* 2012.

[3]Wang Qinwu.*Kunming University of Science and Technology*, 2003

[4] Min peng, Zhao Jincheng. *Building Science.*29 (2013).
[5] Zhao Yuanyuan.Harbin Institute of Technology, 2009.

MATEC Web of Conferences **63**, 02024 (2016) DOI: 10.1051/matecconf/20166302024
MMME 2016

Influence of Stern Shaft Inclination on the Cooling Performance of Water-Lubricated Bearing

Li ZOU [a], Zheng Lin LIU
Wuhan University of Technology, Wuhan 430063, China

Abstract. The water film model of the marine water-lubricated stern bearing was established by FLUENT. The influence law of water flow rate on the cooling performance of water-lubricated bearing was studied in consideration of the stern shaft inclination. It will be helpful to improve the performance of marine water-lubricated stern bearing and both security and reliability of propulsion system. The simulation results show that the increase of cooling water flow ratein a certain range can effectively reduce bearing temperature.The bearing temperature rises sharply with thinning of water film thickness which is caused by the increase of inclination angle. Larger inclination angle can deteriorate the operating reliabilityof bearing.

1 Introduction

Many factors such as the cantilever action of propeller, shaft misalignment, bearing wear cancause the bending of marine stern shaft.The bending is either to increase the partial contact pressure of stern bearing or to result in the inclination of stern shaft, which will lead tothe boundary friction ofstern bearing, and the lubricating fluid cannot form hydrodynamic lubrication in this situation. What's more the increased friction factor will worsen the cooling performance of stern bearing.

The inclination of stern shaft both has the significant effect on the contact pressure of oil-lubricated and water-lubricatedbearings, especially a greater impact on the latter.Zhang et al. (2007) [1] analyzed the three-dimensional pressure distribution and temperature field of capillary-compensated water-lubricated hydrostatic journal bearings by solving the generalized Reynolds equation in the state of turbulent flow through the finite difference method, and the changes of the lubricant viscosity and density with temperature and pressure were taken into account. Dai (2012) [2] analyzed the water-lubricated rubber bearings by using finite volume method, and revealed the influence law of both bearing structure parameters and operation condition on bearing temperature. Cheng (2012) [3] conducted the research on the water-lubricated bearing with the finite element software ADINA, discussed the influence of bearing structure and operation conditions on lubrication performance.

These studies as mentioned above were conducted in ideal state without considering the stern shaftinclination. Actually, due to the lining material of water-lubricated bearing is usually non-metallic one whose elastic modulus is much lower than metal material (such as Babbitt metal), so the stern shaftinclination will has greater influence on the lining deformation and the cooling performance of water-lubricated bearing. The main objective of this articleis to conduct a research on the cooling performance of water-lubricated bearingconsidering stern shaft inclination. According to the structure of water-lubricated

[a] Li ZOU: 923309503@qq.com

bearing, a three-dimensional model of lubricating water film is established in GAMBIT. The influence of stern shaft inclination on the temperature distribution of water film is analyzed with finite volume method (FVM).

2. Model

2.1 Bearing structure

In this article,Thordon XL is selectedasthe lining material water-lubricated bearing. The material is of the superior resistance capacity such as impact, sediment and wear, but its limited operating temperature is 60 ℃ . The water-lubricated bearing consists of bush and lining. The axial groove is used to improve the cooling and tribological performances of the bearing. Bearing's structure is shown in Fig. 1, geometric parameters are listed in Table 1 and physical properties of the material in Table 2.

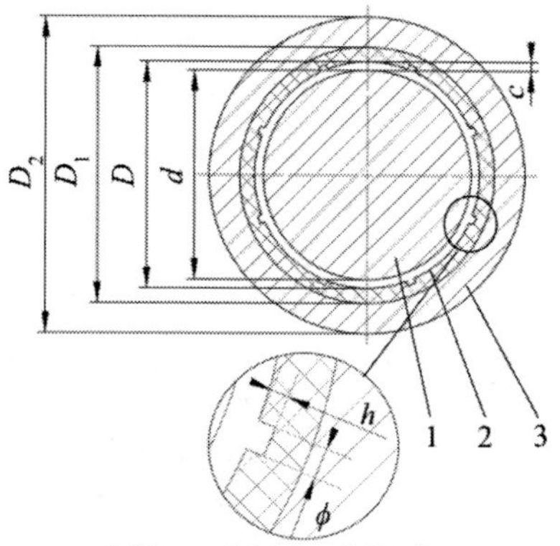

1-Sleeve 2-Lining 3-Bush

Figure 1. Section of shaft and bearing

Table 1. Structure parameters of bearing

Parameters(units)	Values
Shaft diameter(mm)	298.4
Lining inner diameter(mm)	300
Lining outer diameter(mm)	340
Bearing outer diameter(mm)	420
Bearing length(mm)	600
Groove depth(mm)	6
Bearing clearance(mm)	0.8
Groove width(°)	3

Table 2. Thermal properties of bearing materials

Part(material)	Thermal conductivity
Bush(Carbon steel)	48 $W\cdot(mK)^{-1}$
Lining(Thordon)	0.25 $W\cdot(mK)^{-1}$
Sleeve(Bronze)	400 $W\cdot(mK)^{-1}$

2.2 Calculation of bearing heat flux

The friction heat source of the bearing consistsofthe friction heat generated by lubricant's shear force and cooling water heat. The friction heat calculation formula of the bearing:

$$Q = \frac{2\pi n}{60} M \quad (1)$$

Where n ($r\cdot min^{-1}$) is the rotational speed of stern shaft, M ($N\cdot m$) is the frictional torque of stern bearing.

Stern bearing's frictional torque $M$is composed of two parts. One part is generated by the lubricant, the other part is generated bysupporting load which isunrelated to velocity. When the stern shaft is in the state of hydrodynamic lubrication, the frictional torque produced by the lubricant should be considered. However, in actual operation, due to the influence of various factors (such as the cantilever of the propeller's gravity, propeller hydrodynamic etc.), the bearing supporting load is fluctuated, which will lead to the partial contact between stern shaft and bearing. At this time, the frictional torque generated by bearing supporting load will be dominant while that oflubricant is negligible.Therefore, the calculation formula of friction torquecan be expressed as:

$$M = pDL\frac{d}{2}f \quad (2)$$

Where p(Pa)is the bearing supporting load, D (m)is the outer diameter of shaft, L (m)is the length of bearing, d (m)is the inner diameter of bearing, f is the equivalent friction coefficient derived from the relative experiment[4].

In this paper, the thermal conductivity of shaft sleeve [$400W\cdot(mK)^{-1}$] is much greater than that of lining[$0.25W\cdot(mK)^{-1}$], so that bearing friction heat can be totally applied on the outer surface of the shaft sleeve in the calculation. Considering energy loss in heat convection, the applied friction heat is actually 80% of totalheat[2].The bearing heat flux calculation formula:

$$q = \frac{kQ}{\pi dL} \quad (3)$$

Where $k$is the equivalent coefficient,$k = 0.8$.

Parameters of bearing heat flux are listed in Table 3.

DOI: 10.1051/matecconf/20166302024

Table 3. Parameters of friction heat calculation

Parameters(units)	Values
Bearing pressure(MPa)	0.4
Rotational speed($r \cdot min^{-1}$)	300
Eccentricity	0.9
Attitude angle(°)	10
Equivalentfriction coefficient	0.01
Bearing heat flux($W \cdot m^{-2}$)	3374

2.3 Governing equation

In this article, the governing equation used in FVM is:

$$\frac{\partial(\rho\varphi)}{\partial t}+\mathrm{div}(\rho v\varphi)=div(\Gamma_\varphi \mathrm{grad}(\varphi))+S_\varphi \quad (4)$$

Where $\varphi$is the variables, Γ_φ is the corresponding diffusion coefficient of φ, $S_\varphi$is thecorrespondinggeneralized source of φ, div(·) isthe divergence corresponding to a specific variable,grad(·) is thegradient corresponding to a specific variable.

2.4 CFD model, meshing and boundary definition

Under the conditions of both low rotational speed and heavy load, the water film thickness of water-lubricated bearing can reach micron level.Therefore, the water film temperature can be regarded as the same as one of bearing inner surface in this situation. So the cooling performance of the bearing canbe evaluated according to the water film temperature.

Beforethe establishment of CFD water film model, the eccentricity, deflection angle and the inclinational angle of the stern shaft are set.The related parameters are listed in Table 3.

To reduce the quantity of mesh cells and to improve calculation precision, computational convergence, the hexahedralstructured mesh is adopted in the model. And because the axial length of model is much large than that of the radial direction, the mesh at in latter direction is refined.The finite volume model is shown in Fig. 2.

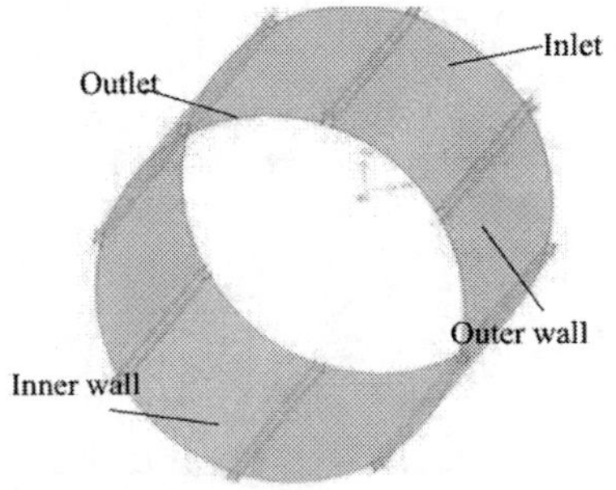

Figure2.CFD model of water film

In FLUENT, the boundary conditions of the model are set as follow:

(1)Inlet boundary is defined as velocity inlet;

(2)Outlet boundary is defined aspressure outlet;

(3) Outer wall boundary is defined as fix wall surface;

(4) Inner wall boundary is defined as rotating wall surface.

3 Result and discussion

China Classification Society (CCS)provides that the inclination angle of stern shaft shall not exceed 3.5×10^{-4}rad. In this article, the cooling performance of stern bearing is discussed when the inclinational angle is 0,1×10^{-4}, 2×10^{-4}, 3×10^{-4}, 4×10^{-4}rad respectively. The inclinational angle is shown in Fig. 3.

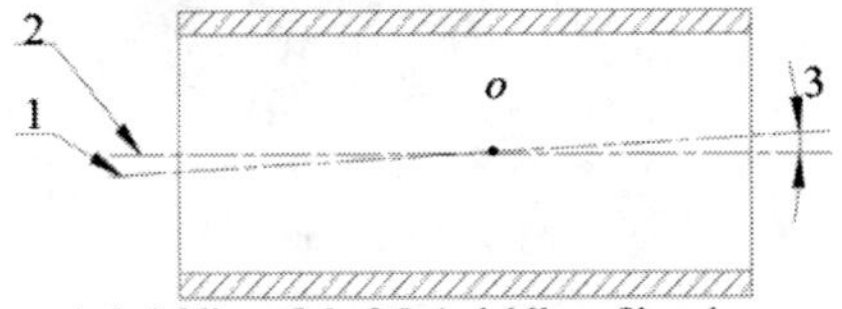

1-Axial line of shaft 2-Axial line of bearing
3-Inclination angle

Figure3. Inclination angle

Under the operating pressure 0.4MPa, rotational speed of shaft 300$r \cdot min^{-1}$, and bearing clearance 0.8mm, the temperature distributions at different water flow speeds(2, 4, 6, 8, 10 $m \cdot s^{-1}$) and inclinational angles (0, 1×10^{-4}, 2×10^{-4}, 3×10^{-4}, 4×10^{-4}rad) is analyzed, and the highest temperature and temperature distribution of water film are shown in Fig.4and Fig.5 respectively.

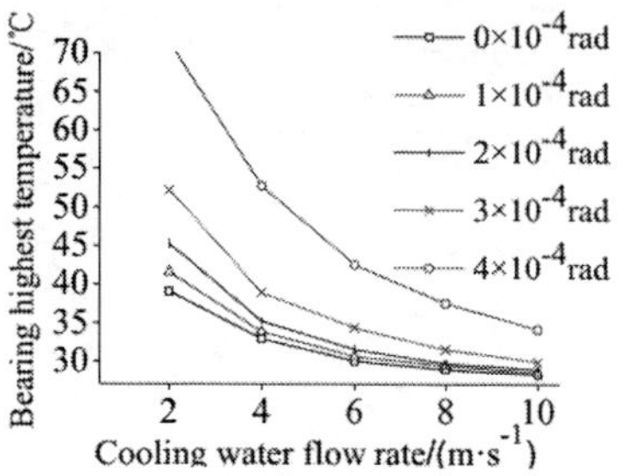

Figure4.Bearing highest temperature versus inclination angle for various flow rates

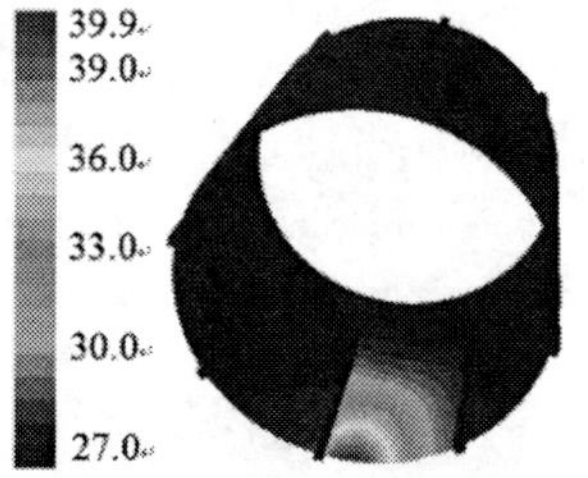

(a) Inclination angle is 0rad, Flow rate is $2m\cdot s^{-1}$

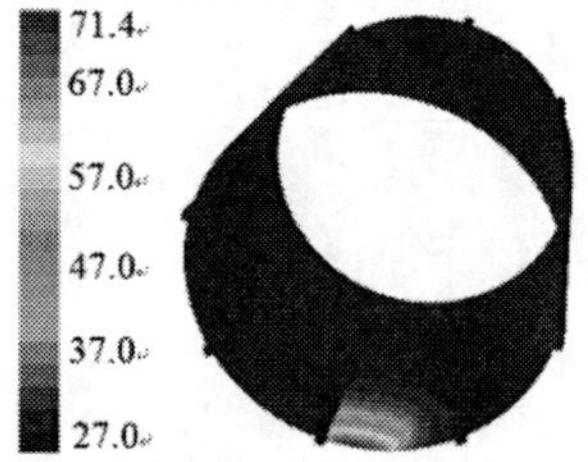

(b) Inclination angle is 4×10^{-4}rad, Flow rate is $2m\cdot s^{-1}$

Figure5.Bearing temperature distribution

It can be seen from Fig. 4, with the increase of cooling water flow rate, the highest temperatures of water film decline gradually. When the cooling water flow rate reaches $8m\cdot s^{-1}$, the water film temperature will be basically unchanged which indicates that thecooling water is sufficient to ensure the cooling effect at this time.The larger inclination angle is, the higher the water film temperature when the cooling water flow rate is constant. Contrasting the two curves whose inclination angles are 3×10^{-4}rad and 4×10^{-4}radseparately, it can be seen that when the cooling water flow rate is increased from $2m\cdot s^{-1}$ to $10m\cdot s^{-1}$, the temperature rising rates are 48.7%, 46.9%, 29.7%, 19.0%, 14.0%respectively.The results indicate that the flow rate is inversely proportional with the temperature rising rate.

It can be seen from Fig. 5, that the temperature at the bottom of water film is greatly changed and from the inlet to the outlet is gradually increased. The highest temperature region appears at the bottom of the outlet where the water film is the thinnest. The reason is that the thinner the water film is, the less the cooling water flow through the water film section and absorbed heat. And the more close to the outlet, the lower the cooling water flow rate is, the worse the cooling ability will be. As shown in Fig. 5(a) and 5(b), with the increase of inclination angle, the high temperature region of water film is closer to the outlet.

5 Conclusion

In this paper, the cooling performance of the water-lubricated bearing is studied with the finite volume method. The research conclusions as follows:

(1) The cooling water flow rate has a certain influence on the cooling performance of water-lubricated bearing.As the cooling water flow rate increases, the bearing temperature decreases.The minimum flow rate should be set to ensure the cooling effect.

(2) The inclination angle of stern shaft has a great influence on the cooling performance of water-lubricated bearing.As the inclination angle increases, the bearing temperature rises. Too large inclination angle will deteriorate the bearing working environment and result in excessive bearing temperature.

Acknowledgment

This work was financially supported by the project of Natural Science Foundation of China (No. 51379168), Key Natural Science Foundation of China (No. 51139005).

References

1. G. Zhang, X. Yuan, *Lubrication Engineering*, "A Theoretical Study on Three-dimensional Pressure Distributions and Temperature Field of Water Lubricated Hydrostatic Journal Bearings", No. **8**, 4-7(2006)
2. Y. Dai, *Wuhan University of Technology*,"Influence of Water-lubricated Rubber Stern Tube Bearings Cooling Lubrication Performance on FLUENT", (2012)
3. Y. Cheng, *Wuhan University of Technology*, "Research on Lubricating Properties of Water Lubricated Stern Tube Bearings Based on TFSI",

(2012)
4. Q. Yu, *Jiangsu Ship*, "Application of ThordonBearings on Ship", Vol. **17**, No. **5**, 26-30(2000)

Simulation Analysis of Spherical Mechanical Seal Property of Marine Stern Shaft

Xu Hui ZHOU [a] , Li ZOU
School of Energy and Power Engineering, Wuhan University of Technology, Wuhan 430063, China

Abstract. The finite element model of spherical mechanical seal wasestablished with ANSYS, and the influence of seawater pressure, shaft speed and other factors on the sealing performance was discussed. The studyresults show that local contact situation of the spherical mechanical seal is in the outside of the seal rings, and both maximum contact pressure and temperature appearat the same position.As seawater pressure and stern shaft rotary speed are increased, the contact pressure and temperature of the spherical seal surface are raised, and when the contact pressure of seal surface is 0, the spherical seal surface formstwo zones including contact one and clearance zone.The former is near the outside of the seal ring, the lateris close to the inside of one.These research results are of important theoretical significance and engineering application value for the development of new kinds of mechanical seals, and improvement of both safety and survivability of underwater vehicles.

1 Introduction

Stern shaft mechanical seal is an important device of underwater vehicle, whichcan affect seriously the safety and survival ability of underwater vehicles.

At present, domestic and abroad stern shaft seal of underwater vehicles adopts plane mechanical seal device, whose performance depends mainly on the materials of rotary andstator rings, and contact state of the friction pair. The material of rotary ringis hard alloy, while oneof stator ring is graphite carbon. Marine mechanical seal device is generally a kind of contact seal, and operates much in mixed friction state. British Deep Sea Seal L.t.d, USA SealLoLL.t.d, USA Durametallioc Co, Japanese Engle and some domestic companies did a lot of research work about mechanical seal, which made considerable progress. But because of the clearance between the stern shaft and stern bearing, plane mechanical seal elements appear radial runoutwhile working. The situation,such as wear of the stern bearing and sinks or bends of the stern shaft, is more serious; it results in overrunleakage, and affects seriously influence the reliability and safety of the propulsion systemof the vehicle.

Christophe Minet[1] points out, the mix lubricationof mechanical seal is complex, andrelative test research is less. In Andre' Parfait's study[2,3], he discussed theory calculation model,heat conduction, deformation, mixed lubrication region of seal rings, and so on. Theory analysis and calculation have applied on plane mechanical seal. Through a lot of experiments, E.Mayergot that the liquid film between two sealing surfaces of general mechanical seal rings is too thin, Newton fluid equations is a lack of effective application conditions, and sealing surfacesare in boundary lubrication state[4].In order to obtain the flow field, temperature fieldand main influence factorsof the mechanical seal, Dazhuan Wu[5] put forward that mechanical seal numerical analysis in high pressure, high temperature, and high speed should be carried out.

[a] Xu Hui ZHOU: 923309503@qq.com

GuopingYAN[6] gave the calculation expression of friction heatof sealing end face, deduced the friction coefficient of the end face in mixed friction state, analyzed the shortage of the calculation method of convective heat transfer coefficient, and gave a new calculation formula of inside and outside of rotary, stator ring's heat transfer coefficients. In the steady thermal-structure coupling analysis of mechanical seal rings, Zhu Xueming[7] made the direct coupling on the thermal structure field and simulationof the stress, temperature and deformation. Lu Sheng[8] carried out the research of dam surface mechanical seal, analyzed the influence levels of narrow seal structures on stress, temperature and deformation.

Overall, at present, there is no report about the theory analysis research on marine spherical mechanical seal. The difference between the spherical mechanical seal and the plane onelies in that the former structure takes spherical contact instead of plane contact, which can automatically adjust the seal surface contact status, enhance the tracing ability of the seal friction pair and make the spherical seal surface of bothrotary andstatorrings keep consistentlyin contact state. Therefore,spherical mechanical seal can be used to solve the above practical problems and to improvethe mechanical seal performances.

2 Spherical mechanical seal

The spherical mechanical seal is composed of a rotary ring, rotary ringseat, stator ring, stator ring seat, spring,spring seat and other componentsuch as stern shaft, fix ring, as shown in Figure 1.

The cemented carbiderotary ring is embedded in the rotary ringseat, and the rotary ringseat is fixed on the stern shaftand rotates with the shaft; the stator ring made of nonmetal material (FEROFORM) is fixed on the stator ringseat, and can move axially with the seat along stern shaft.FEROFORM is taken as the stator ringmaterial of the spherical mechanical seal instead of the commonly used graphite carbon, which can effectively improve both wear uniformity and wear resistanceof the contact surface.

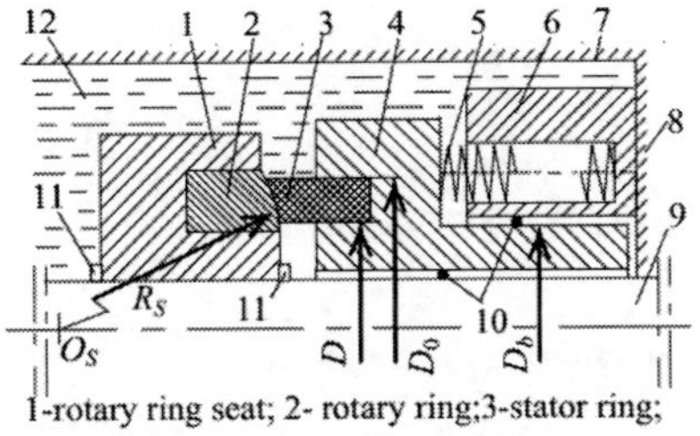

1-rotary ring seat; 2- rotary ring;3-stator ring;
4-stator ring seat; 5- spring;6- spring seat;
7- stern tube; 8- engine room 25℃; 9- stern shaft;
10- O-seal ring; 11- fix ring; 12-seawater25℃

Figure.1 Spherical mechanical seal drawing

The contact surface between rotary and stator rings is a spherical one that can plays a sealing role and has self-aligningfunction. The both spring force and pressure of sealed medium (seawater) can ensure the sphericalseal surfaceof bothrotary and the stator rings tobe consistentlycontacted so thatseawater is sealed outside of the spherical sealsurface, while air inside of one.

The ratio of areaof effective fluid pressure of the spherical mechanical seal to one ofsealing contact pressure is 0.634, which indicates that the seal is balance type. The width of spherical seal is 8 mm; the spherical contact area is 5275.2 mm^2.

The specific pressure of spring is 0.2 MPa.The key structure and material parameters of the spherical mechanical seal are listed in Table 1and Table 2.

Table 1 Key structure parameter of spherical seals

Seal type	Shaft diameter D /mm	Inside diameter of seal ring D_1 /mm	Outside diameter of seal ring D_0/mm	Slip diameter D_b /mm	Sealing spherical radius R_S/mm	Seal surface /(°)	
						θ_1	θ_2
Spherical seal	185	202	218	208	200	14.63	15.82

DOI: 10.1051/matecconf/20166302025

Table 2 Physical parameter of materials of seal rings and their seats

Item	Material	Modulus of elasticity *E*/MPa	Poisson's ratio μ	Coefficient of thermal conductivity λ/W/(m·K)	Linear expansion coefficient *a*/(m/℃)	Density ρ/(kg/m^3)	Maximum permissible temperature /℃	Friction Coefficient *f*
Stator ring	Feroform	300	0.48	0.5	70×10^{-6}	1.32×10^3	120	0.03
Rotary ring	bronze	11.5×10^4	0.32	63.8	17.8×10^{-6}	8.5×10^3		
Stator or rotary ring seat	C15	19.8×10^4	0.29	16.33	16.6×10^{-6}	7.9×10^3		

3 Finite element model

3.1 Thermal-structure model

The thermal-structure finite element model of the spherical mechanical sealis established with ANSYS software, which is a type of axial symmetrymodel as shown in Figure 2. The model includes 3069 elements, 3191 nodes in all.

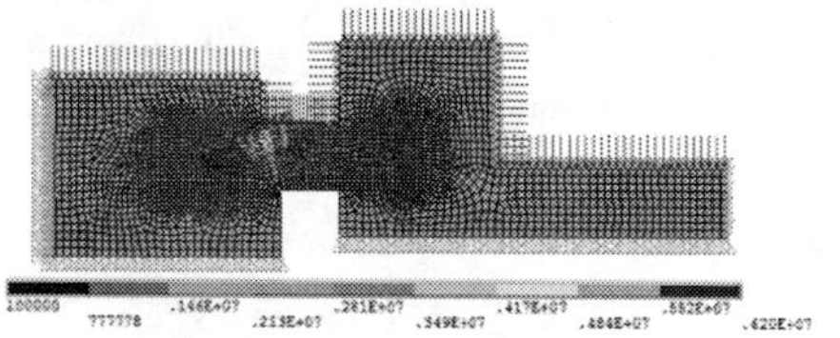

Figure 2. Finite element model of spherical seal

Assume that the spring pressure and external pressure of seawater are uniformly distributed, and the reaction force of seawater film between spherical sealsurfaces is in linear distribution; the fluid cooling of the sealing rings is dealt with according to convection heat transfer boundary; the friction heat between spherical seal surfaces is definedas heat flux boundary.

Seal contact surface is defined on the spherical seal surface of both rotary andstator rings. Contact unit is defined on the spherical seal surface of the stator ring by application of contact guide, and target unit is defined on the spherical seal surface of the rotary ring. Thus the automatic distribution of heat flux density can be realized and the corresponding node temperature of the spherical seal surfacesof both rotary and stator rings is guaranteed to be basically consistent.

3.2 Convective heat transfer coefficient

The convective heat transfer coefficient is expressed as follows:

$$h_m = \frac{\lambda_f}{d} Nu_f \quad (1)$$

Where *d*is external diameter of rotary ring or internal diameter of stator ring, m; λ_f is the thermal conductivity of the fluid, 2.63 W/(m·K) for air, 60.85 W/(m·K) for seawater; Nu_f is theNusseltnumber.

3.3 Boundary constraints

The boundary constraintof the finite element modelis shown in Figure 1. Constant temperature load 25℃ is applied on the regioncontacted with seawater of both rotary and stator ringseats; constant air temperature load 25 ℃ is applied on the region of both rotary andstator ringseatsfarfrom the sphericalseal surface.

The calculation formula of heat fluxdensity *q*of seal ring is shownas follow:

$$q = \frac{2\pi n}{60} f \cdot p_c(r) \cdot r \quad (2)$$

Where*q* is heat flux density, W/m^2; *f* is coefficient of friction; $p_c(r)$ is the contact pressure between the spherical seal surfaces of two rings, Pa; *r* is the radius(*X* direction) of seal ring, m;*n* is a rotary ring rotation speed, r·min^{-1}.

4 Performance analysis of seal ring

Theperformanceanalysis ofthe spherical mechanical seal ringsincludestemperature, deformation, contact pressure of the spherical mechanical sealand their influence factors.

4.1 Temperature of sealing surface

The highest temperature of the spherical mechanical seal is influencedbyboth seawater pressure and rotaryspeed on as shown in Table 3.

Table 3 The highest temperature of seal(℃)

Shaft rotary speed/(r/min)	Seawater pressure/ MPa					
	1	2	3	4	5	6
50	28.5	29.8	31.0	32.2	33.4	34.6
100	32.0	34.5	36.9	39.4	41.6	44.0
150	35.4	39.1	42.8	46.3	49.7	53.3
200	38.7	43.6	48.5	53.2	57.7	62.5
250	42.0	48.1	54.1	60.0	65.5	71.5
300	45.3	52.5	59.7	66.7	73.3	80.4

It can be seen inTable 3, that with the increase of seawater pressureor the rotary speed of stern shaft, thehighest temperature of the spherical seal surface is raised. The higher seawater pressure or rotary speedis,the more obvious temperatureincrease is.

4.3 Contact pressure and deformation of sealing surface

The elastic modulus of the stator ring (300MPa) ismuch lower than that of the rotary ring (11.5×10^4MPa) so thatthe deformation of the stator ring is larger.Therefore, the stator ring is only discussed.

When seawater pressure is 6 MPa and rotary speed of stern shaft is 50, 100, 150, 200, 250, 300 r·min^{-1}respectively, both contact pressure and axial deformation of the stator ring are shown in Figure 3 and Figure 4 separately.

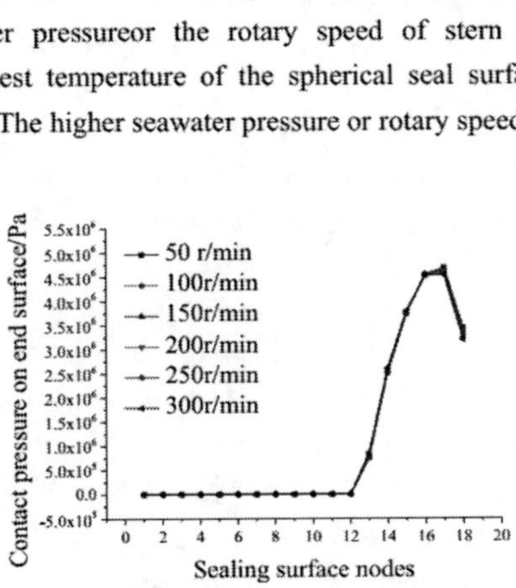

Figure 3. Contact pressure - rotary speed

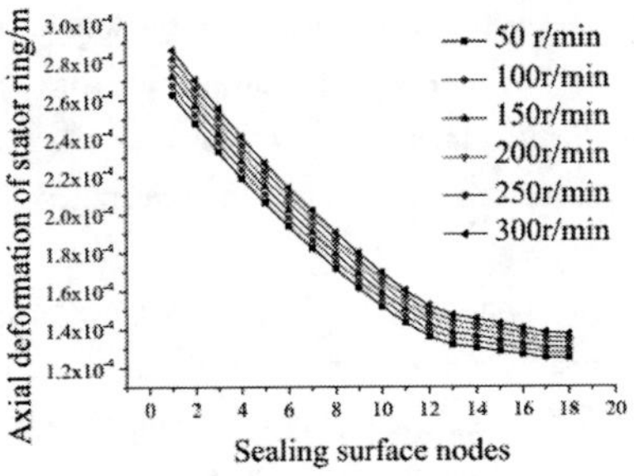

Figure 4. Axial deformation - rotary speed

In Figure 3, it can be seenthat the contact pressure of each spherical seal surface node is almost the samein different rotary speeds. The contact pressures are 0 in the region where the node number is less than 12, which indicates clearance exists in the region.

InFigure 4, it shows thatthe trend of axial deformations of the stator ring are the same along with node numberof the spherical seal surface in different rotary speeds,and the deformation decreases with the increase of the node number. When the node number is constant, the axial deformations of node areincreasedwith the enhancement of the rotary speed.

4.2 Discuss

Effects of seawater pressure on deformation and contact pressure should be seriously discussed.

When pressure of seawater 1~6 MPa and rotary speed is 300r/min, the axial deformation distribution of the stator ring is shown in Figure 5, and the contact pressure distribution in Figure6.

Figure 5 shows that under different seawaterpressure, the axial deformations of the stator ringis decreasedand tended to convergence with the increase of seal surface node number. The greater the pressure is, the more obviously the deformation is decreased.

Figure 6 illustrates that the contact pressure of some nodes near the inside is 0, and in certain node regionthe contact pressure is increased gradually with the increase of node number,but decreases correspondingly at the last two or threenodes.With the increase of seawater pressure, the contact pressure of each spherical surface node increasescorrespondingly,the number of nodes whose pressure is 0 is gradually decreased, and both maximum axial deformation and contact pressure of the stator ring are increased by almost 6 times when seawater pressure alters from 1~6 MPa.

When the contact pressure of seal surface is 0, it means

thespherical seal surface formstwo zones including contact one outside the rings and clearance zoneinside the rings.

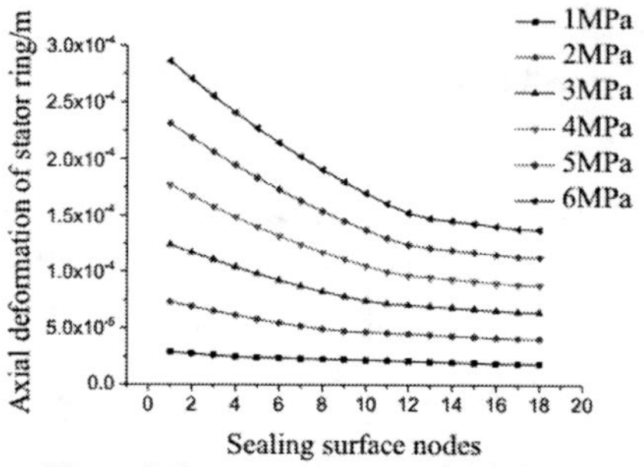

Figure 5. Seawater pressure - axial deformation

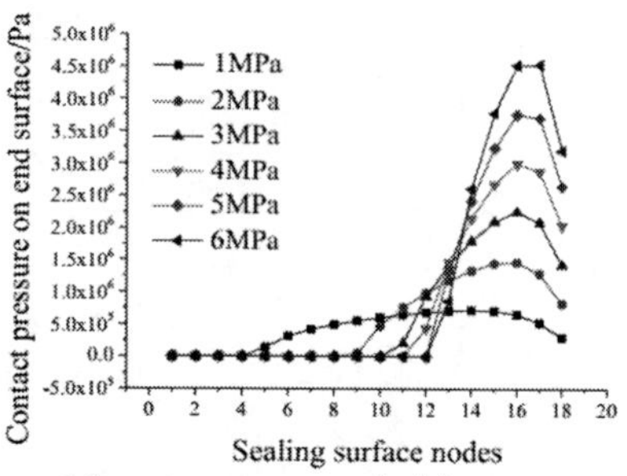

Figure 6. Seawater pressure - spherialcontact pressure

5 Conclusion

1) Axial deformation of spherical seal surface forms contact zone (outside the rings) and clearance zone(inside the rings). With the increase of seawater pressure and rotary speed, thedeformation of spherical seal surface increases, contact area is reduced and clearanceoneis increased.

2) As seawater pressure and rotary speed are raised, both contact pressure and temperature are increased. The maximum contact pressure and the higher temperature are near the outside of seal rings, but their valuesarein the allowed range of Feroformmaterial properties of stator ring.

Acknowledgements

It is aproject supported by the project of Key Natural Science Foundation of China (No.51139005), Natural Science Foundation of China (No.51379168).

References

1. C.Minet, A Deterministic Mixed Lubrication Model or Mechanical Seals, *Journal of Tribology*, **Vol. 133**, 1-13(2011)
2. A. P.Nyemeck, N.Brunetière, B.Tournerie. A Multiscale Approach to the Mixed Lubrication Regime: Application to Mechanical Seals, *TribolLett***47**, 417–429 (2012)
3. A. P. Nyemeck1, N.Brunetière, B.Tournerie, Parametric study of the behavior of a mechanical face seal operating in mixed and TEHD lubrication regimes,ESDA2012, July 2-4, (2012)
4. J. Sun, Wei Long, B. Gu, Development Course and Research Trend on the Mechanical Seal, *lubrication engineering*, **4**, 128-131(2004)
5. D. Wu, X. Jiang, S. Yang, Three Dimensional Coupling Analysis of Flow and Thermal Performance of a Mechanical Seal Journal of Thermal,*Science and Engineering Applications*, **Vol. 6**, 1-9(2014)
6. G.YAN, Z. LIU, X.ZHU,Numerical Analysis of the Thermal- field of Ship Stern- Shaft Mechanical Sealed Faces under the Variational Working Conditions,*Journal of Ship Mechanics.* **12**, 483-489(2008)
7. G.YAN, Z. LIU, X.ZHU,Numerical Analysis of the Thermal- field of Ship Stern- Shaft Mechanical Sealed Faces under the Variational Working Conditions,*Journal of Ship Mechanics.* **12**, 483-489(2008)
8. S. LU, Z. LIU, M. DAI, Performance analysis of dam sealing based on ANSYS,*Machinery Design & Manufacture*, **6**, 206-208(2011)

MATEC Web of Conferences **63**, 02026 (2016)
MMME 2016
DOI: 10.1051/matecconf/20166302026

The influence of damping layer on the modal of water-lubricated rubber bearing

Li HUANG[a], Yong JIN and Zheng Lin LIU
School of Power and Energy, Wuhan University of Technology, Wuhan 430063, China

Abstract. In order to find out the influence of damping layer on water-lubricated rubber bearing, element finite modal analysis in Workbench is used and three bearings with different damping layers are studied as well as an ordinary bearing without damping layer. The results show that the damping bearings have lower modal frequencies than the ordinary bearing, and the damping layer length and thickness can influence the deformation distribution of modal shape.

1 Introduction

Structural vibration can bring noise, reduce the life of components and even lead to failure of components in the field of ship, aviation, machinery and so on, thus always being the research hot spot. Damping layer is made of polymer material, and there is a phase difference between stress and strain in damping layer under the effect of vibration. This difference has to overcome great resistance which can reduce vibration of components by dissipating the vibration energy[1]. So, damping layer has been widely used to reduce structural vibration and has made distinct effect. For example, by applying it into the design of bellows, high-damping unreinforced U-shaped vibration reducing bellows was developed[2]; In the application of flat plate structure, damping layer was studied by scholars with variables to optimize beams with damping sandwich[3]. In these fields, the damping layer has made a contribution to the vibration reduction and noise reduction.

In recent years, water lubricated rubber bearings have been widely used in underwater vehicle[4,5]. Many scholars have used modal analysis to study the characteristics of water lubricated rubber bearings. Like Jin Yong[6,7] used simulation and experimental analysis to study the influence factors of modal properties of water lubricated rubber bearings, and found that the main influence of lower order modes is structure and material of liner and the main influence of higher order modes is structure and material of bush. Modal analysis can get the natural frequency and the relative deformation of the bearing under different natural frequencies, which has theoretical meanings and practical application values to the optimization of bearing structure and fault diagnosis.

In view of the good vibration performance of damping layer in other structure and the demand for reducing the vibration of water lubricated rubber bearings, this paper use finite element modal analysis to study the modal propertiesof water lubricated rubber bearings in Workbench. Here, influence of different damping layers on modal frequency and modal shape of bearings are taken into account.

[a]Li Huang :415540616@qq.com,Wuhan University of Technology

 DOI: 10.1051/matecconf/20166302026

2 Modal

Damping layer is added to the cylindrical surface of ordinary bearing, with an outer bush. The arrangement of damping layer in the water lubricated rubber bearing is shown in Figure 1. Three-dimension geometry modelsare built based on the water lubricated rubber bearing in Workbench. Bearing A is ordinary bearing, and bearing B2 and B3 are changed base on that of B3, to study the influence of different damping layers on the bearings.Material parameters of bearings are listed in table 1. Geometric parameters of bearings are listed in table 2. All the damping layers are symmetric arrangement. The difference between damping bearing B1 and any other damping layer bearing is listed in table 3.

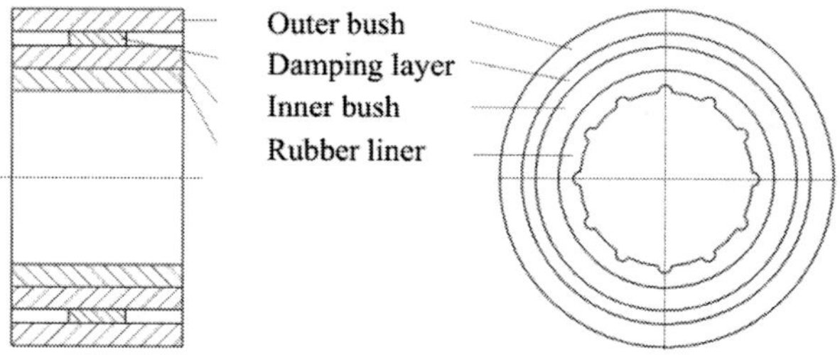

Fig.1 Profile of damping bearing

B1, B2, and B3 are damping bearings. The parameters of

Table1 Material parameters of bearings

Name	Material	Poisson ratio	Density/（kg/m3）	Young's modulus/（MPa）
Rubber liner	Nitrile rubber	0.47	1240	7.08
Inner/outer bush	Q235	0.288	7860	2.12e5
Damping layer	Nitrile rubber	0.47	1080	4.44

Table2 Geometric parameters of bearings

Type	Internal diameter	Thickness of liner	Thickness of inner bush	Thickness of damping layer	Length of damping layer	Thickness of outer bush	Length of bearing
Ordinary bearing A	150	20	20	\	\	\	150
Damping bearingB1	150	20	20	12	50	20	150

Table3 The difference between damping bearing B1 and any other damping layer bearing

Type	Particular	Thickness of damping layer/mm	Length of damping layer /mm	Young's modulus of damping layer/MPa
Damping bearing B1	Reference	12	50	4.44
Damping bearing B2	Change damping layer thickness	12	100	4.44
Damping bearing B3	Change damping layer length	14	50	4.44

3 Modal analysis

3.1 Finite element modal analysis

Modal analysis is mainly used to identify the vibrating characteristic of the framework.It has important significance on the dynamic response of the system, the generation and transmission of the dynamic load and the form of the system vibration, also it is the base for all other type of dynamics analysis. The Lanczos Block method is used to calculate the constraint mode of water lubricated bearings. According to the actual operation condition, the outer cylindrical surface of the outer bushing is fully restrained. Extract the modes of the first 20 orders of the bearings.

3.2 Modal frequency

According to the calculation modal frequencies of different bearings, we could draw Figure 2 and obtain the following results:

(1)The damping layer reduces natural frequencies of the bearing. It has a strong effect on the lower modes(sixth-order below) but a small effect on the highermodes(sixth-order above).

(2) The ordinary bearing has a larger arrangement of natural frequencies than the damping bearings.

(3) Influence of the damping layer length: natural

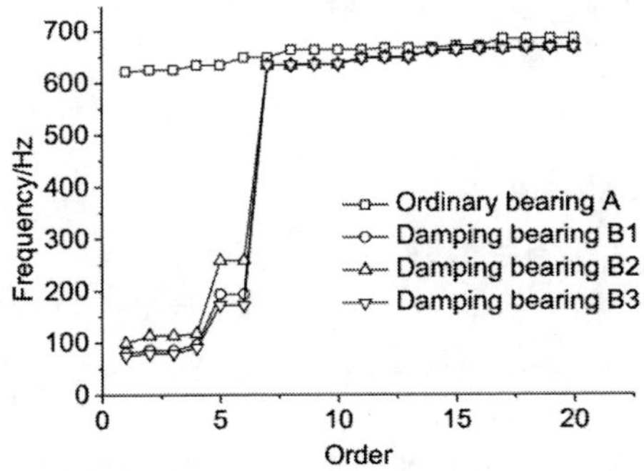

Fig.2 Influence of damping layer on modal frequency

frequencies of the damping layer increase with the increase of damping layer length, and the influence is significant on fifth and sixth order frequencies, and there is little influence on natural frequencies of sixth-order above.

(4) Influence of the damping layer thickness: natural frequencies of the bearing decrease with the increase of damping layer thickness, and the influence is significant on fifth and sixth order frequencies, and there is little influence on natural frequencies of sixth order above

3.3 Modal shape

Analysis different modal shapes of bearings with their corresponding modal frequencies, we could obtain the following results:

(1) The ordinary bearing has many orthogonal modes, like the first-order mode and the third-order mode, and the forth-order mode and the fifth-order mode, because there are many multiple roots in the modal solving equation.

(2) The damping bearings also has a lot of orthogonal modes, and its orthogonal modes correspond to the same order as the ordinary bearing.

(3) When two of the bearings have a close natural frequency, their corresponding modal shapes are similar. For example, as shown in Figure 3, around 634Hz, these

(a) Fourth-order modal shape of the ordinary bearing（634.28Hz）

(b) Seventh-order modal shape of the damping bearing B1（634.07 Hz）

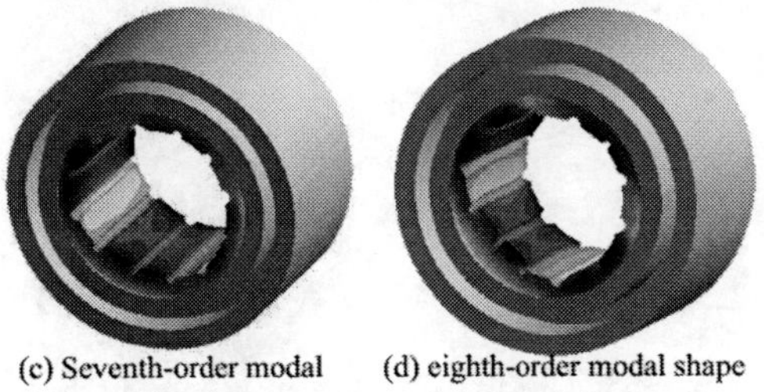

(c) Seventh-order modal shape of the damping bearing B2（634.06 Hz）

(d) eighth-order modal shape of the damping bearing B3（634.71 Hz）

Fig.3 Modal shapes of bearings relevant to close frequency

four bearings have a close natural frequency, so they have

a similar modal shape. We could find from every bearing in Figure 3 that the vibration is happened in the inner liner, and the respective maximum deformation occurs at the end of two adjacent strips, and between two couples of maximum deformation strips there is one strip of very small deformation.

(4) Compared with the ordinary bearing, the damping bearing has a larger deformation area. Figure 4 shows lower-order modal shapes of damping bearing B1. From Figure 4 we could find that the lower-order deformation of damping bearing B1 relates to its inner liner, inner bush and damping layer. However, the deformation of the ordinary bearing of the same order relates only to its inner bush, as shown in Figure 3(a).

(a) First-order modal shape of the damping bearing B1(80.943Hz)

(b) Fourth-order modal shape of the damping bearing B1(96.541Hz)

Fig.4 Modal shape of damping bearing B1

(5) The influence of damping layer length: compare bearing B1 with B2, we could find that the increase of damping layer length affects the fifth-order shape and sixth-order shape more than others. Figure 5 and Figure 6 show the modal shapes of fifth-order and sixth-order of bearing B1 and B2. These two bearings have different areas of deformation and the maximum deformation relates to a larger area in B1 than in B2.

(a) Fifth-order modal shape of the damping bearing B1（193.31Hz）

(b) Sixth-order modal shape of the damping bearing B1（193.31Hz）

Fig.5 Fifth-order and sixth-order modal shape of the damping bearing B1

(a) Fifth-order modal shape of the damping bearing B2(257.79Hz)

(b) Sixth-order modal shape of the damping bearing B2(257.79Hz)

Fig.6 Fifth-order and sixth-order modal shape of the damping bearing B2

(6) The influence of damping layer thickness: compare bearing B1 with B3, we could find that the increase of damping layer length affects the fifth-order shape and sixth-order shape more than others. Figure 7 shows the modal shapes of seventh-order of bearing B1 and B3. We can see that the inner bush of bearing B3 has a similar deformation with its inner liner, which is larger than that of bearing B1.

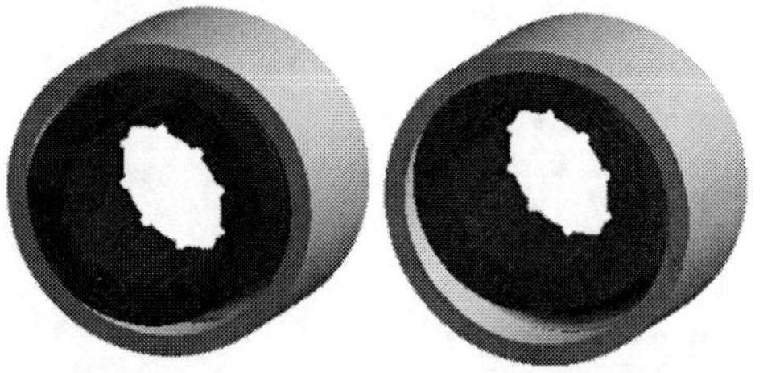

(a) Fifth-order modal shape of the damping bearing B3（172.18Hz）

(b) Sixnth-order modal shape of the damping bearing B3（172.18Hz）

Fig.7 Fifth-order and sixth-order modal shape of the damping bearing B3

4 Conclusions

The main conclusions are as follows:

(1) The damping bearing has lower natural frequencies than the ordinary bearing. When it comes to lower modes(six order below), the difference is significant, and the deformation area of damping bearing is larger than that of ordinary bearing.

(2) The damping layer length has a large influence on modal frequency, especially on fifth-order and sixth-order modal frequency, while the damping layer thickness has a small influence.

(3) The damping layer length and the damping layer thickness have larger effect on fifth-order and sixth-order shape than on other orders, and they can influence the distribution of relative deformation.

References

1. Huang Zhicheng, Qin Zhaoye, Chu Fulei. J.Vibra. Shock,**33**,105 (2014)
2. ZhongYuping, Chen Tongbiao. Pressure Vessel Technology,**22**, 21 (2005)
3. JinqiangLi , Yoshihiro Narita.,J. V. C.,**2**. 20（2014）
4. D. de Kraker, R. A. J. vanOstayen, D. J.Rixen,TribologyInternational,**403** (2006)
5. Zhou Huanhuan, Yu Qiang, Li Tianyun, Ship Engineering, **04**.25 (2015)
6. Jin Yong, TianYuzhong, Liu Zhenglin, Lubrication Engineering, **09**.10 (2011)
7. Jin Yong, TianYuzhong, Liu Zhenglin, Noise Vibra.Contrl. **32**.1 (2012)

MATEC Web of Conferences **63**, 02027 (2016) DOI: 10.1051/matecconf/20166302027
MMME 2016

Dynamics Research of rigid-flexible model of crank-rocker mechanism with Multi Clearance

Shao Jun BO[1,a], Tian Ci CHEN[1] and Qing Huai YE[2]
[1]*Institute of Mechanical Engineering Inner Mongolia University of Science and Technology,Baotou, Inner Mongolia, China*
[2]*Baotou Steel Group Design Institute,Baotou, Inner Mongolia, China*
Aerding Street on the 7th, Kundulun District,Baotou City, Inner Mongolia Province

Abstract. This paper takes the crank-rocker mechanism as the research object, purposed for the simulation of the motion pair with clearance utilizing the contact force model and the coulomb friction model, we build the virtual prototype of four-bar mechanism with clearance joints in the ADAMS. The article mainly contrasts the influence dynamics characteristics towards the mechanism multi-joint clearance between the rigid rocker and the flexible rocker . Considering the jointed arm' s flexibility, the four,five and six order modal of the rocker , generated by ANSYS as a neutral document , were imported into ADAMS for simulation.The result shows that Mechanism with four joint clearance increased the impact on the speed and the acceleration appeared greater fluctuation,the flexible rod has a buffer effect to the contact-impact forces,and the dynamic characteristics of fluctuations are improved.

Key Words: crank-rocker multi-joint clearance flexible body ADAMS dynamics characteristics

1 Introduction

With the high speed, overloading, the rapid development of precision mechanical and aerospace engineering, accurate prediction of mechanism motion demand is higher and higher. In most of the mechanical system, the mechanism of artifacts usually by kinematic pair connection. Due to error in the process of manufacture, assembly and mechanism motion produced by friction, wear, and the influence of such factors as movement exists between vice clearance. And the existence of the gap destroyed an ideal organization model, cause the collision of kinematic pair, cause the violent vibration of institutions, led to the decrease of the mechanical system stability and precision, and thus the efficiency and service life. In recent years, for mechanism with clearance, and there were many research of flexible components [1], but for considering four kinematic pair clearance and flexible connecting rod four-bar linkage study also is less. In this paper, based on two state model using coulomb friction model and collision model [2-4], and consider the connecting rod for the flexible body, including four crank rocker mechanism with clearance joints and soft model for the study of dynamics.

[a] Corresponding author: 15540180997@163.com

DOI: 10.1051/matecconf/20166302027

2 BUILT OF MODEL WITH MULTI-JOINT CLEARANCE

As the existence of the joint clearance, the center distance between the shaft pin and sleeve is no longer zero, and resulted collision and friction of shaft pin and sleeve. There are three kinds states of shaft pin: free state, collision, Continuous contact. Center distance is e, $r = R_i - R_j$,as shown in Fig.1 .

When e<r, shaft pin is in a free state.

When e>r, the shaft pin and sleeve generate collision.

When the shaft pin and the sleeve in collision, the contact force decompose into the normal collision force F_N and the tangential friction F_T .

In order to describe the rule of the collision between two objects, ADAMS software use the IMPACT function to solve the force of two objects at the time of the collision, and the motion rule of the two objects in the before collision and after collision. The function of the collision between two objects is equivalent to a nonlinear spring-damper model. As shown in Fig.2.

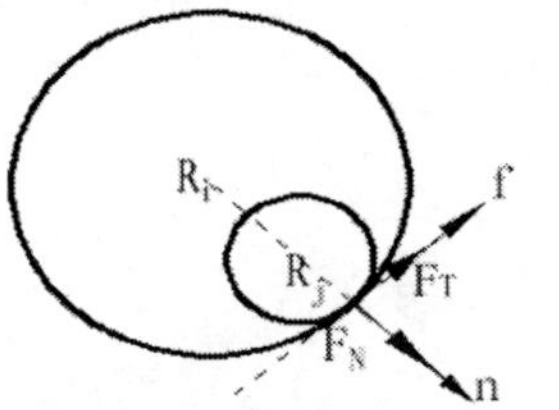

Fig 1. Normal and tangential contact force when the shaft pin and sleeve is in a collision.

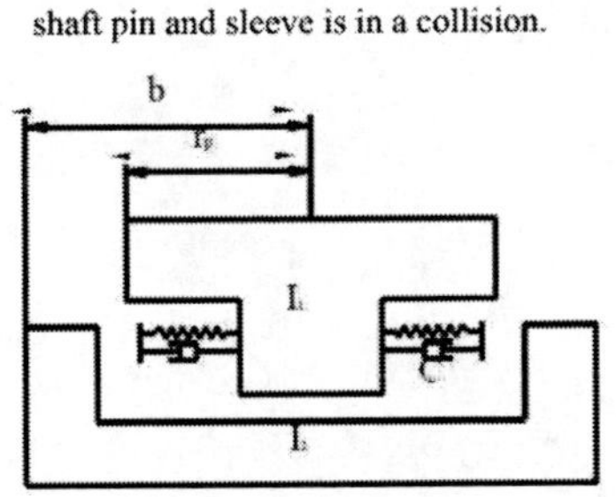

Fig 2. Figure4 Elastic damping model of impulse function.

The expression of the Impact function is as follows[5]:

$$F_N = \begin{cases} K\delta^n + Step(\delta, 0, 0, d_{max}, c_{max})\dfrac{d\delta}{dt}, & \delta > 0 \\ 0, & \delta < 0 \end{cases} \quad (1)$$

$$F_T = F_N \mu(v_N) \quad (2)$$

K is stiffness coefficient, the fomula is as follows:

$$K = \frac{4}{3\pi(h_i + h_j)} R^{\frac{1}{2}}, R = \frac{R_i R_j}{R_i + R_j}, h_k = \frac{1 - v_k^2}{\pi E_k}, k = i, j \,. \quad (3)$$

E_k, v_k, R_k represent Young's modulus, Poisson ratio and contact radius of the object.δ is the collision depth.Collision index n can reflect the level of material nonlinearity, normal metal materials takes the number of 1.5.In ADAMS, transient damping coefficient is the step function of collision depth, the value of c is determined by the Step function. Generally speaking, damping coefficient value take 0.1 ~ 1% of stiffness coefficient K[6], recommend d_{max} value is 0.01mm.Generally, friction coefficient data obtained from experiments, normal dynamic friction coefficient take 0.1, static friction coefficient take 0.3.

3 DYNAMICS SIMULATION OF PLANAR FOUR-BAR MECHANISM WITH MULTI-JOINT CLEARANCE

Fig.3 shows multi-joint clearance mechanism, put all parts material for steel.The four-bar mechanism parameters are as follows:

Crank=105×20×5(mm),Link=272×20×5(mm),

DOI: 10.1051/matecconf/20166302027

Rocker=175×20×5(mm),Frame=225×20×5(mm);k =2.53×10^{5}N/mm, Maximum damping coefficient c_{max} =253N×S/m; Young's modulus E=207Gpa, Poisson's ratio v_i =0.29,the rotation of the crank at a constant speed $\omega = 300r/\min$.In the crank and connecting rod, connecting rod and rocker, rocker arm and the frame, frame with motion pair with kinematic pair clearance between the crank, and then USES the ANSYS software component, the rocker to generate flexible body take flexible joystick four, five, six order modal system dynamics simulation, with a handle for the rigid rod when comparing the simulation results.

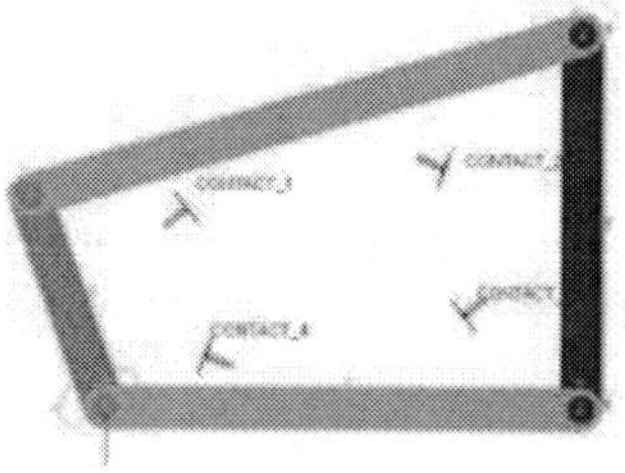

Figure3. Crank-rocker mechanism ADAMS.

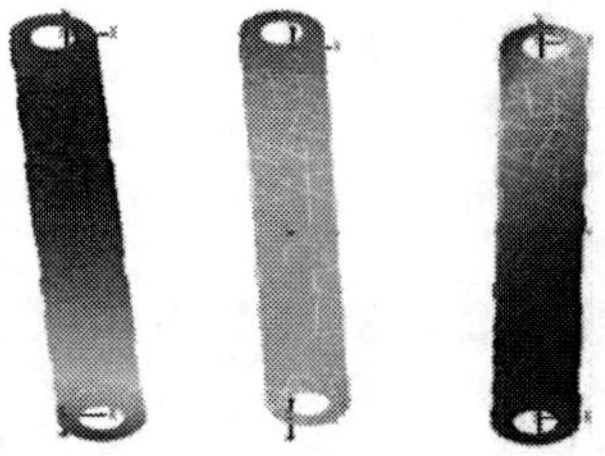

Figure4. Four,five and six modal of flexible rocker.

The simulation results and analysis. Figure 5 and figure 6 respectively a handle for the rigid and flexible connecting rod and rocker speed curve. Agencies began to work, two kinds of condition occurred on the connecting rod and a handle speed curve of a certain fluctuation, but the joystick is flexible, the speed is less volatile. At a later time, compared with the rigid rocker, flexible rocker rocker and connecting rod overall more smoothly, less fluctuations, shows that the flexible factors counter collision between shaft and shaft sleeve have buffer action.

Figure 7 and figure 8 respectively a handle for the rigid and flexible connecting rod and rocker speed curve. Rigid for analysis, rocker, connecting rod and rocker acceleration to the existence of the gap significantly in unstable state, shaft and shaft sleeve is in a state of collision and separation. When rocker for the flexible connecting rod and a handle fluctuates significantly decreases, and the acceleration amplitude greatly reduced, curve fluctuations more smoothly, also shows that the collision of flexible factor counter shaft and shaft sleeve have buffer action.

Figure 9 and 10 respectively a handle for the rigid and flexible component pin shaft and shaft sleeve contact between the collision force curve. Diagram curve clearly shows flexible joystick and shaft sleeve absorbs the shock, the collision has obvious cushioning, flexible factors to contain much clearance between the crank rocker mechanism motion state is very good, make body more close to ideal state, reduce the rigid adverse impacts on the state of motion artifacts.

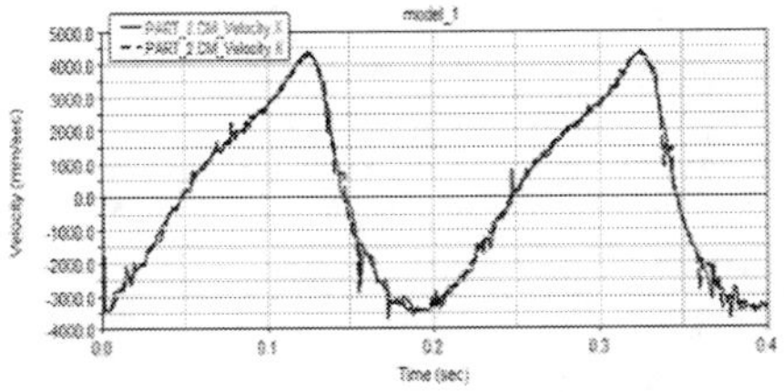

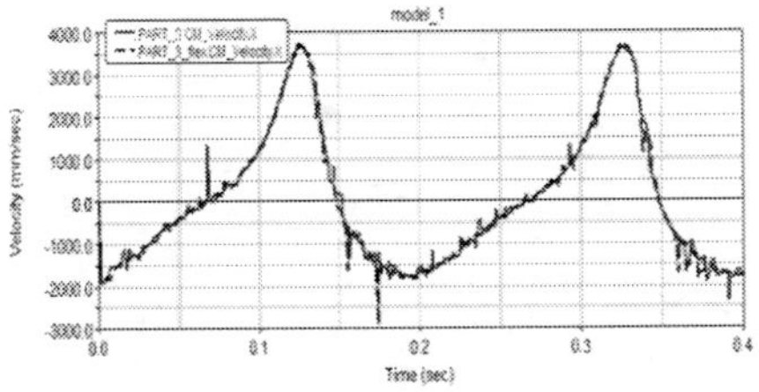

Figure5. Link X-axis speed in the case of the rigid and flexible rocker.

Figure6. Rocker X-axis speed in the case of the rigid and flexible rocker.

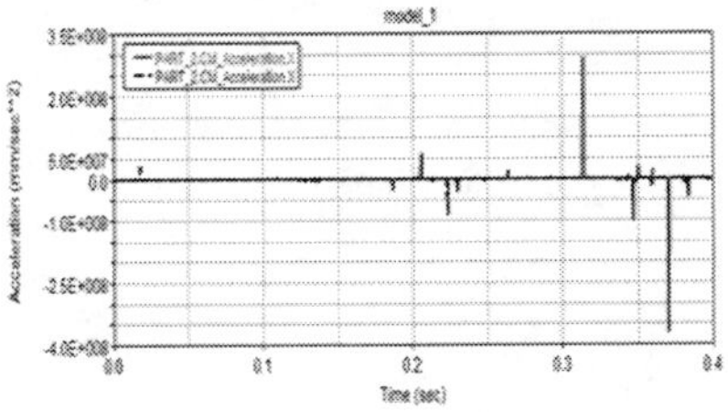

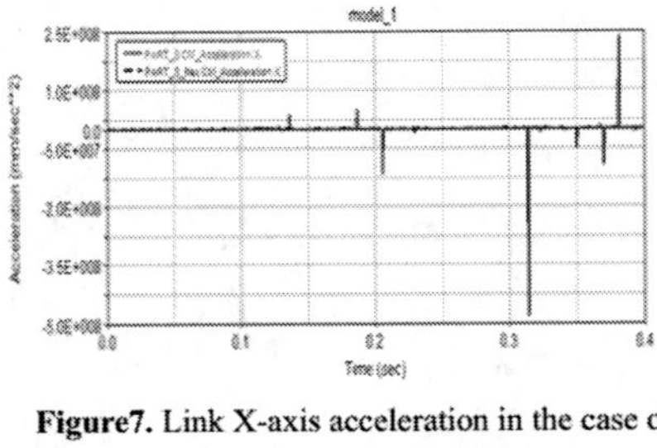

Figure7. Link X-axis acceleration in the case of the rigid and flexible rocker.

Figure8. Rocker X-axis acceleration in the case of the rigid and flexible rocker.

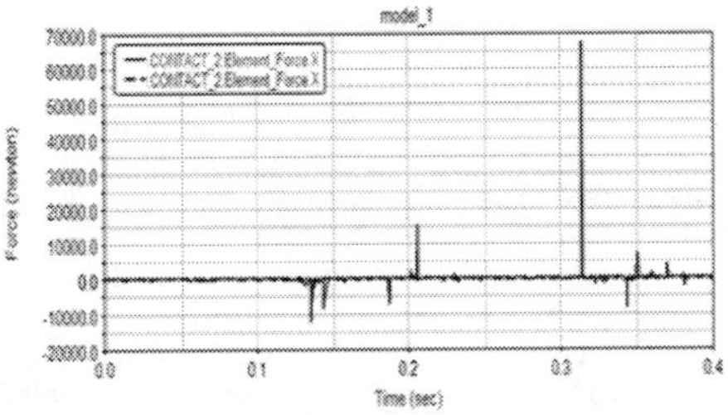

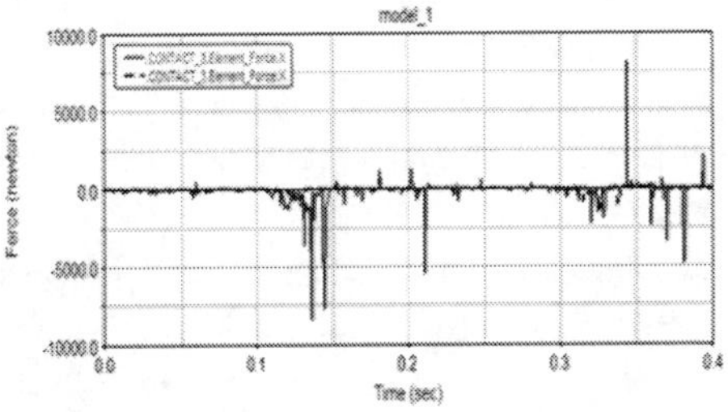

Figure9. X-axis contact force between shaft and sleeve of link and rocker in the case of the rigid and flexible rocker.

Figure10. X-axis contact force between shaft and sleeve of rocker and rack in the case of the rigid and flexible rocker.

4 Conclusions

In this paper, including many crank rocker mechanism with clearance joints as the research object, the introduction of flexible factor, rigid flexible rocker rocker and mainly analyzes the influence of the dynamic characteristics of a body. , according to the results of the existence of kinematic pair clearance more speed of center of mass of connecting rod and a handle, the acceleration and collision force of the pin shaft and shaft sleeve is especially significant, the influence of clearance on the dynamic stability of the crank rocker mechanism has a very adverse effect. When the rigid rocker into flexible rocker, connecting rod and a handle the velocity and acceleration of wave has improved to some extent, weaken the negative influence of clearance, relatively buffer intensity of the collision of pin shaft and shaft sleeve, the mechanism motion more tend to be ideal.

References

1.Xinglin Guo, Zikun Zhao: Dynamic Characteristics Analysis of the Crank-rocker Mechanisms with Joint Clearance[J], Mechanical Strength, 2010, 32(6):905~909 .

2.Dubinsky S, Freudenstein. Dynamic Analysis of Mechanical Systems with Clearance. Part2:Dynamic Response[J]. ASME, J. Eng. Ind, 1971, 93B(1):305-309.

3.Shaojun Bo,Ruibin Guo: Dynamic Characteristic Research of Planar Four-bar Mechanism with Joint Clearance[C],Advanced in Design Technology,ADME2012,2012.8,Taiyuan, china:912-916.

4.Bo Shaojun, Si Zhenjiu, Ye Qinghuai. Dynamics Simulation Research of Planar Four-bar Mechanism with Multi-joint Clearance[C]//Part1:Applied Mechanics and Materials,ICFMM2014.Guilin:539-542.

5.K.L.Johnson.ContactMechanics[M].Cambridge: Cam-bridge University Press,reprint,1987:92-98.

6.Imed Khemili, Lotfi Romdhane: Dynamic Analysis of Flexible Slider-crank Mechanism with Clearance[J]. European Journal of Mechanics A/Solids, 2008, 27(5):885-886.

MATEC Web of Conferences **63**, 02028 (2016) DOI: 10.1051/matecconf/20166302028
MMME 2016

Kinetic Analysis of Four-link Gantry Crane Hoisting System

Si Li FAN[1],Yu Feng SHU[1], Hai Ying ZHANG[1], Shuo OUYANG

[1]*DongGuan Polytechnic, DongGuan523808, China.*

Abstract: This paper establishes a flexible dynamics model of MQ2533 portal crane boom system. In the premise of ignoring control system, this paper analysis the mechanical dynamics of boom hoisting system. Simulation results show that: Dynamics response of the hoisting boom system have great fluctuation when excluding the control effects, this paper provide a theoretical basis for crane hoisting system to optimize control strategies, which committed to improve their performance and ensure safety, reliability and economy.

Keywords: Crane, Force, Displacement, Hoisting system.

1 Introduction

Gantry crane is a large construction machinery, mainly used for port, berth, open stockyards and outfitting site. It has good transport, big depth of spreader decentralization, better adapt to the temporary work and so on. we should full considerate its strength, rigidity and safety when design, but also to reduce the supplies and cost savings. Boom system is an actuator of portal crane, it is the design basis of electrical systems and other subsidiary organs. In order to simplify the calculation under the normal circumstances, we consider the mechanical components as rigid body[1], But when we need a high precision calculation or study the stress magnitude, distribution and deformation within the body, you need to regard it as a flexible member. Precise kinetic analysis can make sure the gantry crane safety and performance, minimize its weight、reduce noise and prevent the shock loads. Thus, kinetic analysis of four-link gantry crane hoisting system is particularly necessary[2-3].

ADAMS has unparalleled advantages in dynamics analysis of rigid body than the other software, but very weak in flexible mesh unit and other processing functions. Flex module provides bi-directional data exchange interface with other finite element software(such as ANSYS, NASTRAN etc). We can import the modal neutral file(*.mnf) of other finite element software into ADAMS through interface[4].

The main component of MQ2533 portal crane boom system comprises of nose, large rod, small rod, girder and etc. In order to understand its mechanical properties of displacement response of key nodes, stress, deflection, vibration frequency, we need to treat it as a flexible body; For the pulley, reel, small rod, balance beam and hoist and other ancillary components, we only need to know their motion parameters such as size, quality and moment inertia, thus regard them as rigid bodies[5-6]. MQ2533 portal crane is shown in Figure 1.

Figure 1. Four-link portal crane of MQ2533

This paper establish the flexible model through bushing which connected wire rope and many small cylindrical. The hinged position with revolute appeared in vice establishment of constraints. We regard a rigid sphere weighed 25e3Kg as loads, constrainting through the spherical jiont. We build a cuboid as ground wich force constrainted with the loads in order to simulate the lifting process. Rigid and flexible coupling model of boom system is shown in Figure 2.

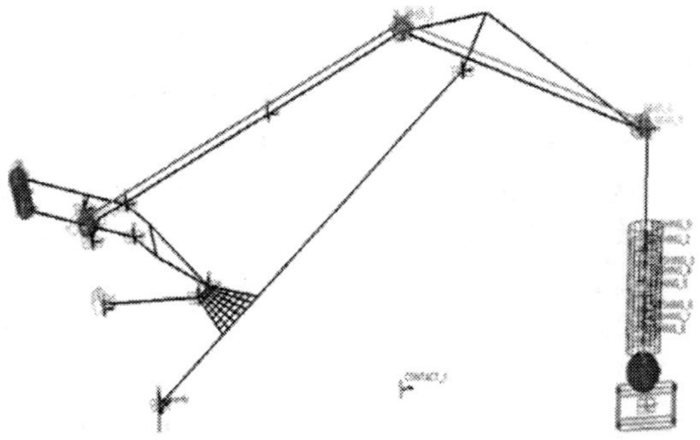

Figure 2. Rigid and flexible coupling model of boom system

2 Parameters Calculation

Table 1 lists the device types of power and drive system of MQ2533 gantry crane[7].

Table 1. Device types of power and drive system

Category	Model	Number
Motor	YZPF315L-8	2
Reducer	M3PSF80-40-124	2
Arrester	YWZ5-500 / 200	2
Nominal diameter of drum (m)	0.92	2
Wire rope	26NAT18X7SW+FC1670ZS/SZ	4

Seen from Tab.1 above, driving motor and transmission system both have two sets. motor is variable for frequencies and reducer magnification is 40, Then driving torque of drum to wire rope is 80η times of the single motor(η is equal to 0.9, which is the total efficiency of motor to drum[8]).

Nameplate data of the motor shows that ratio of starting torque and rated torque(1700N/m)is $\lambda_0 = 1.8$, ratio of the biggest torque and rated torque is $\lambda = 3$, The diving torque equation are as follows:

$$\begin{cases} T = 112531\omega + 203760, \omega \le 1.8107 \\ T = -1.03E10 + 1.06E10\omega - 2.73E9\omega^2, \text{other} \end{cases}$$

Drive system parameters are shown in Table 2:

Table 2. Drive system parameters

Denomination	Value
Equivalent inertia ofdrumJ_r(Kg·m^2)	7.168e4
Initial length of wireropel_0(m)	60
Rigidity coefficient ofwire rope E_c(N/m)	9.12e6
Damping coefficient ofwire rope C_c(N/m·s^{-1})	1.9e4

3 Simulation Results

We set parameters as follws in ADAMS: starting torque of equivalent drum is 1.8 times of the rated torque, which is equal to 203760N.m, then we can calculate the driving torque of equivalent drum acording to the equivalent above, and the terminal simulation time is 5s[9].

Figure3 to Figure8 show the front and rear displacement responses and stress changes of the elephant nose girder, load

DOI: 10.1051/matecconf/20166302028

torque characteristics of the conversion to motor is shown in Fig.ure9, Fig.ure10 depicts the tension value variation of wire rope along with time.

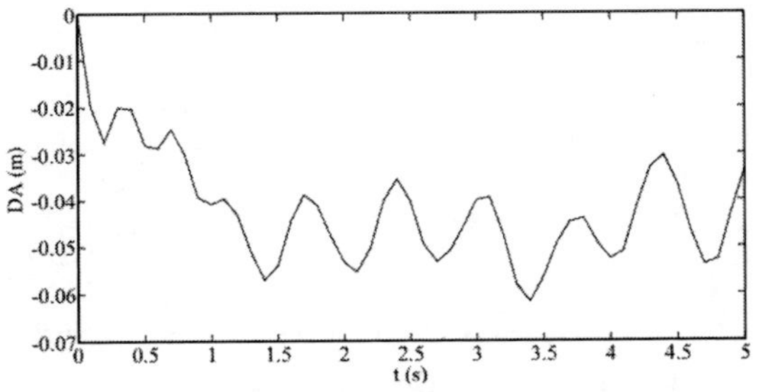

Figure 3. The front displacement response

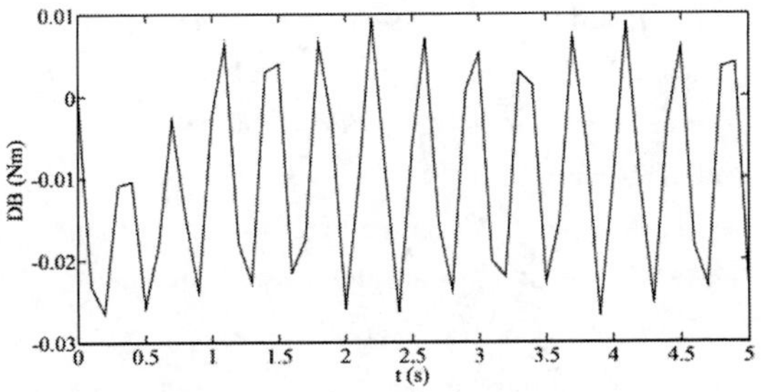

Figure 4. The rear displacement response

Seen from Fig.3, the front displacement of girder gradually decreases to -0.055m from static equilibrium position within the first 1.5 seconds, For the 3.5 seconds behind, the displacement changes in vicinity of -0.0466m, The biggest amplitude is -0.03m, and relative rate of change is 35.6%, which explains a large fluctuation.

Fig.4 shows the rear displacement of girder fluctuation both sides of the equilibrium position which is equal to -0.01m, and the biggest displacement exceed the static equilibrium position before lifting. Comparate Fig.3 and Fig.4 after lifting, we can see the front equilibrium value is 5.5 times the rear one, so the front stiffness of girder is worse than the rear.

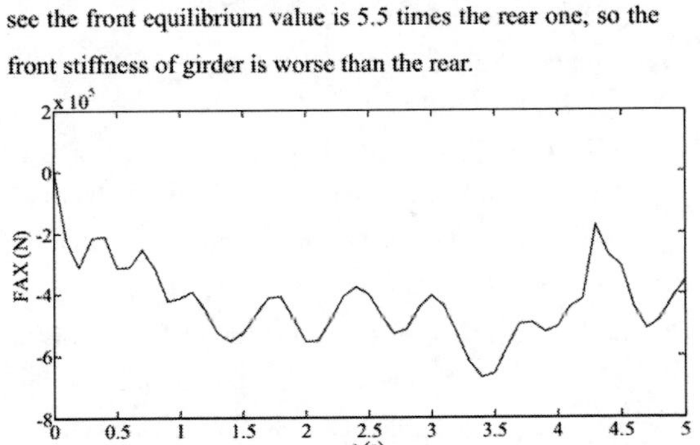

Figure 5. The front horizontal force

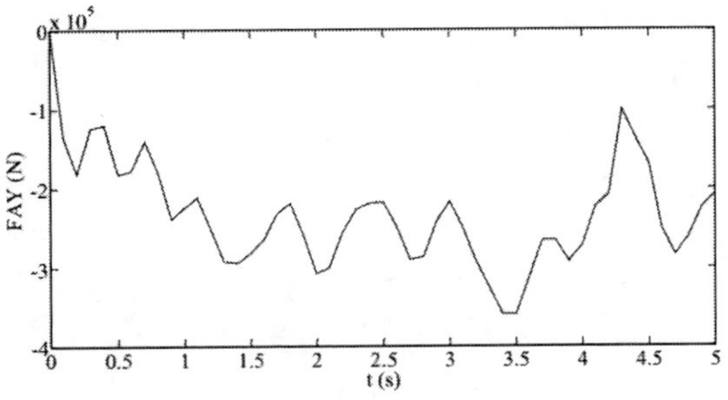

Figure 6. The front vertical force

Fig.5 shows that the front vertical force of elephant nose girder increase from 0 to 5×10^5 N within the first 1.5 seconds, then fluctuate in the vicinity of $4.68e^5$ N. The biggest amplitude is $2.03e^5$ N in the 3.5s later, and relative change rate is 56.6%.

Seen from Fig.6, the front vertical force of elephant nose girder gradually increase to $2.82e^5$N, then fluctuate in the vicinity of $2.54e^5$N, which lies in the static equilibrium position. The biggest amplitude is $1.53e^5$N in the 3.5s later, and relative change rate is 39.8%, indicating the vertical force has a big fluctuation.

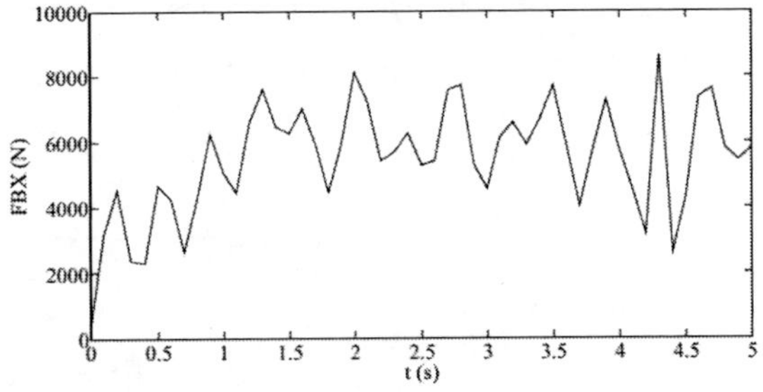

Figure 7. The rear horizontal force

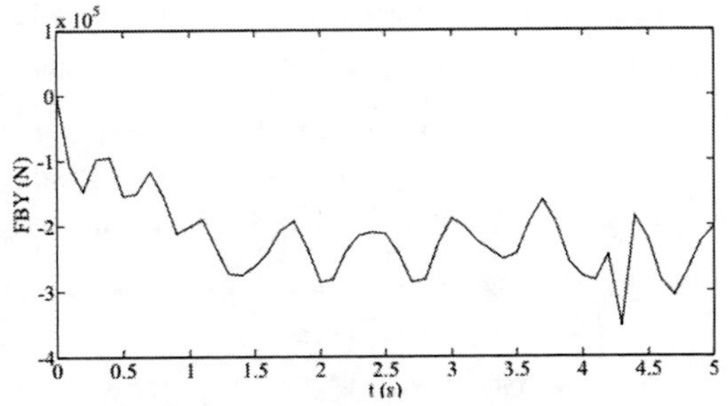

Figure 8. The rear vertical force

Seen from Figure 7 and Figure 8, the rear horizontal force has a significant fluctuation in the both sides of 5974 N. Nonetheless, it reduces two orders of magnitude than the load, but the magnitude vertical force remains the same order with load. The biggest amplitude of vertical force is $1.1343e^5$ N in 3.5s later, and relative change rate is 52.7% with respect to the average force 2.396e5 N.

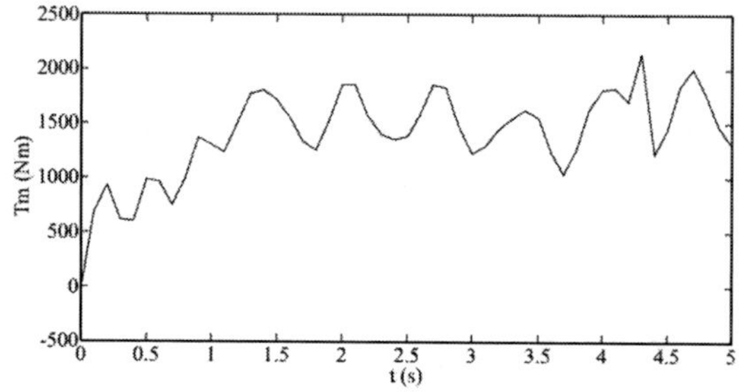

Figure 9. Load torque of motor

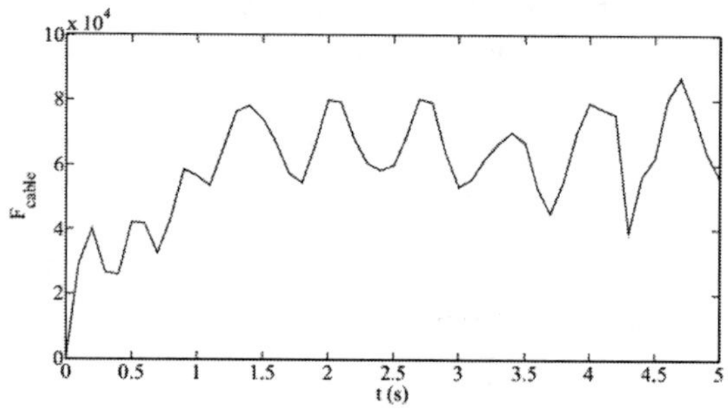

Figure 10. Rope force

We can only get the load torque of equivalent reel in simulation results, but can not directly obtain the input load torque of drive motor. The magnification formula: $T_m = T_r/80/$, (is the overall efficiency of motor to drum).

According to the simulation shown in Fig.9, load torque of motor increase to 1903 N·m erratically within the first 1.5s. The average value is 1727 N.m after 1.5s, the biggest change is 655 N·m, and relative change rate is 62.1% with respect to the average one.

MQ2533 crane has four ropes, according to the principle of total stiffness constant in ADAMS model, simplifing the four ropes as one. This paper regard a quarter of the measured force as one rope suffered.

Fig.10 demonstrate that rope force increase to 74200N within the first 1.5s, and it has a large fluctuation at the equilibrium position which is equal to 65756N, the biggest change is 26656N after 1.5s, accounting for 59.5% of the average one.

4 Summary

First of all, this paper uniformly convert the powertrain inertia into drum in accordance with the reasonable assumption, including the flexible vibration characteristics of wire rope and combining the mechanical properties of driving motor and Newton's law. Based on these, it determine the input torque equation of flexible dynamic model of boom system. Furthermore, establishing the flexible dynamics model of the boom system.

This paper analysis the mechanical dynamics of MQ2533 portal crane hoisting system ignoring the role of system control. Simulation results show that: Dynamics response of boom hoisting system has a great fluctuatation when excluding the effect of control system. Next, we will be able to optimize the hoisting system of portal crane considering the frequency vector control mode and improve its control strategy and performance, which can ensure its safety, reliability and economy.

References

1.Imanishi Etsujiro , Nanjo Takao, Kobayashi Takahiro . Dynamic simulation of wire rope with contact [J]. 2009, 23(4): 1083-1088.

2. BS EN 13001-2: 2004+A3, Crane saftey General desigen, Part 2: Load actions[S]. 2009.

3. Liming Li, Practical guide of finite element analysis, Tsinghua University Press. beijing, 2005.1, pp. 111-116.

4.Shuoouyang. Research on Dynamic Characteristics of For-bar Linkage Portal Crane Hoisting System[D]. Changsha, Central South University, 2013.5.

5.Zhong Liu. Dynamics Simulation Study of Hoisting Mechanism of Double-girder Overhead Travelling Crane[D]. Changsha, Central South University of Forestry and Technology, 2014.5.

DOI: 10.1051/matecconf/20166302028

6. Information on http://www.image.baidu.com.

7.Ghigliazza R M, Holmes P. On the dynamics of cranes, or spherical pendulum with moving supports[J]. International journal of non-linear mechanics. 2002,37(7);1211-1221.

8.Ellermann K, Kreuzer E. Nonlinear dynamics in the motion of floating cranes [J]. Multibody system dynamics. 2003, 9(4): 377-387.

9.QiuXiuyun, Zhao Tao etc.. Analysis on damping ratio of single factor and simulation of many factors optimize combination with suspension girder in ppr[J]. Journal of Xinjiang agricultural university, 2003,35(6): 2-3.

The affect distance analysis of tunnel blasting vibration on adjacent buildings

Hui Wang[1 a], Cai Xu[2]and Xuerui Yang[1,]

[1]Guizhou Construction Science Research & Desigh Institute Co. Ltd, Guiyang 550000, China

[2] Zunyi construction (group) coLTD ,Zunyi, 553000, China

Abstract: In the formation of the blasting seismic wave transmission is a complex mechanical process. Blasting seismic wave in different geological structure formation of the interface, diffraction, reflection, projection as the incident Angle is different, all kinds of waveform transformation, formation of different types, different amplitude, frequency and phase of various wave superimposition of random composite wave. Blasting seismic wave propagation distance (horizontal distance and height difference), and the performance of the explosive, explosive charge, charge structure, priming method, congestion state what international airport, the plane and direction, topography and geological conditions will affect the blasting vibration effect. In engineering by empirical formula to estimate main parameters of blasting seismic wave and the structure of the empirical formula is the result of the use of theoretical analysis, by blasting of similar rate to determine the parameters in the formula is made up of many engineering measured data from statistical analysis, or directly by the measured parameters of the blasting seismic wave is given. In this paper, through various points were set in the prison line large speed is the most value, using the mathematical statistics regression analysis method, attenuation coefficient is obtained, and then back to the formula of single ring allows maximum dose safety distance calculated..

1 Project Overview

A tunnel construction, the villagers reflect negatively with the vibration caused by blasting construction building, blasting excavation exposed rock is gray. Blasting area near a small building height is relatively high, most of the building elevation the blasting area is low, most of the houses have been built for brick structure since. 3 times of field vibration test were conducted on site.

2 The test is introduced

For scientific, fair and accurate judgment unit during blasting explosion scope of blasting vibration effect and the influence degree, must be based on the blasting site terrain and geological conditions, find out the blasting vibration wave attenuation law. I center analysis and judgement method is to use the blasting vibration is to actually test of blasting operation, by testing the ground points of different distance from the blasting vibration velocity, blasting vibration intensity analysis, obtained the attenuation law of explosion wave, to provide a scientific basis for reasonable determine a safe distance.

The implementation of blasting vibration test the chengdu zhongke measurement and control co., LTD. Of TC - 4850 type of blasting vibration. TC - 4850 type of blasting vibration is to speed speed sensor, with three sensors placed in the test point, blasting the instrument set to automatically trigger condition, when triggered, the instrument

[a]1638553132@qq.com

automatically start recording, the sensor by blasting vibration produced by automatic storage after analog signals to digital conversion. Through the instrument interface connected to the computer by special software for data analysis after processing vibration parameters is given.

3 The test results

According to the actual situation, considering the influence of different direction and geological conditions change, the three tests were conducted and the arrangement of measuring points are valid at the time of blasting, three test results are as follows .

Table 1 test results for the first time

result point	Channel X maximum (cm/s)	Half wave frequency	biggestsingle ring dose (kg)	distance (m)
1	0.06	36.70		296.27
2	0.07	25.89		255.94
3	0.04	32.13		383.06
4	0.04	40.00		431.82
5	0.07	16.81		371.00
	Channel Y			/
1	0.07	32.92		296.27
2	0.09	20.30		255.94
3	0.08	32.92	14.0	383.06
4	0.09	35.24		431.82
5	0.08	44.44		371.00
	Channel Z			/
1	0.14	60.15		296.27
2	0.01	23.32		255.94
3	0.09	36.36		383.06
4	0.08	55.17		431.82
5	0.10	51.95		371.00

Table 2 test results for the second time

result point	Channel X maximum (cm/s)	Half wave frequency Hz	biggestsingle ring dose (kg)	distance (m)
1	1.24	45.98		54.0
2	0.38	46.51		100.0
3	0.13	47.62		150.0
4	0.55	72.73.		24.0
5	0.55	33.06		51.0
6	0.89	22.60		75.0
	Channel Y			/
1	1.72	31.75		54.0
2	0.44	50.63		100.0
3	0.13	52.63	14.4	150.0
4	0.68	32.26		24.0
5	0.75	24.84		51.0
6	0.44	21.28		75.0
	Channel Z			/
1	2.67	55.56		54.0
2	0.67	60.61		100.0
3	0.21	51.95		150.0
4	0.67	64.52		24.0
5	0.77	68.97		51.0
6	0.93	42.11		75.0

Table 3 test results for the third time

result point	Channel X maximum (cm/s)	Half wave frequency Hz	biggestsingle ring dose (kg)	distance (m)
1	1.09	8.32	14.0	56.0
2	0.11	54.79		81.0

 DOI: 10.1051/matecconf/20166302029

3	0.19	250.00		106.0
4	0.13	63.49		131.0
5	0.16	60.61		156.0
6	0.38	59.70		181.0
	Channel Y			
1	0.62	61.54		56.0
2	0.13	60.61		81.0
3	0.58	85.11		106.0
4	0.11	54.05		131.0
5	0.25	95.24		156.0
6	0.35	60.61		181.0
	Channel Z			
1	1.11	65.57		56.0
2	0.21	50.63		81.0
3	0.34	266.67		106.0
4	0.16	74.07		131.0
5	0.22	58.82		156.0
6	0.53	60.61		181.0

4. Regression analysis to test results

By《safety regulations for blasting》(GB6722-2003) was 6.2.3

Type $$R = (\frac{K}{V})^{\frac{1}{\alpha}} \cdot Q^{\frac{1}{3}} \quad (1)$$

R —Blasting vibration safety allows the distance, the unit is（m）;

Q—Dynamite, simultaneous blasting for total dosage, time delay blasting is a maximum dosage, the unit is（kg）;

V —Allowed to defend the security of object location of particle velocity, the unit is cm per second （cm/s）;

K、a —Protection object and break point to calculate the terrain, geological conditions of related coefficient and the attenuation index, can according to the 《blasting safety regulations 》(GB6722-2003) [1] 43 selection table on page 5, or determined by field test. Blasting industry [2-5] many scholars in different topographic and geological conditions The size of the value and its change law, discussed a lot, in addition to a certain rule. But the author thinks that should be adopted by the field test of regression analysis to determine the two values. Will be in style

$$\rho = Q^{1/3} / R$$

Command $y = \ln V$, $x = \ln \rho$, $m = \ln K$, $n = \alpha$

（1） can be $y = m + nx$ （2）

Organizing, test data and draw curve as shown in figure 1: the figure in the linear regression curve, can be obtained, n=1.060，m=3.070。

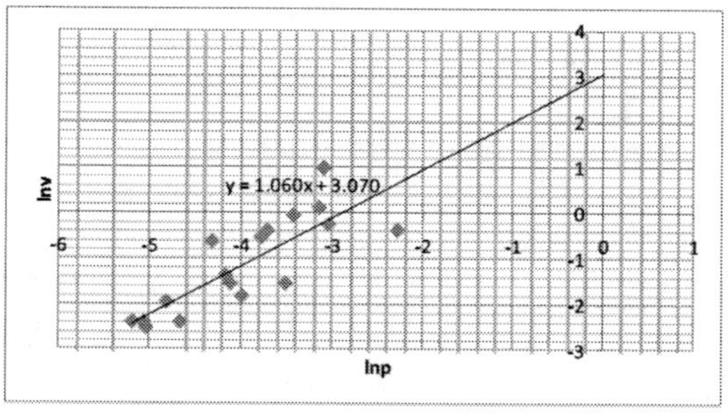

Figure 1. Test regression curve regression curve

To be on the safe side, n values, the curve move up (see f igure 2), so that you can include all points of m m value max imum, is obtained m=4.27。

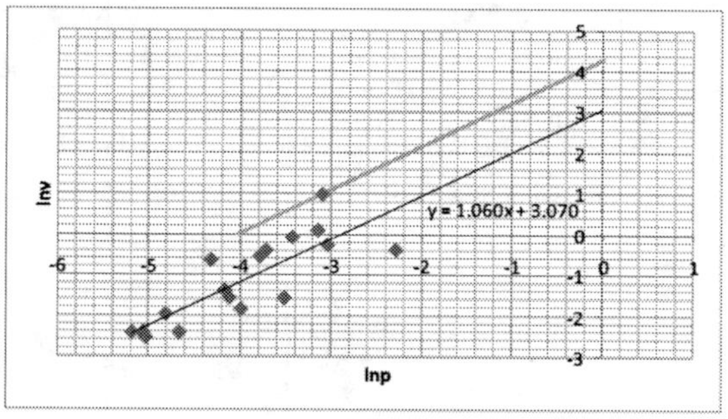

Figure 2 Test the envelope

Then,K=em=71.52,n=a=1.06.According to the The blasting safety procedurField test basic vibration frequency is greater

DOI: 10.1051/matecconf/20166302029

than 10HzBasic and local premises is generally brick, non seismic large block buildings, so take safety vibration velocity for 2.3cm/sThe distance formula (1) a

$$R=(\frac{71.52}{2.3})^{\frac{1}{1.06}}\cdot Q^{\frac{1}{3}} \quad (3)$$

5.Conclusion

1) Because of the influence of different geological conditions of blasting vibration, such as k、a coefficient should be determined by field test;

2) Proposed to ensure the safety of the housing m value is used in this paper, the testing method of envelope, make its can include a maximum of all m;

3) Should be carried out prior to the tunnel construction blasting vibration testing and related parameters, formulate reasonable blasting construction scheme, quality by holes or differential [J]. blasting,200825（3）; 96-99.

blasting method, control the largest single dose, control harmful blasting vibration effect.

References

[1] 《The blasting safety procedures》GB6722-2003 [S];

[2]Tao Lv.Hai bo Li, etc Spread the influence of the dielectric characteristics of blasting vibration attenuation law [J]Journal of disaster prevention and mitigation ofe ngineering.2008，28（3）：337-341；

[3]Yong quan.Liliang Wu.The three gorges underground powerhouse excavation blasting vibration attenuation law research[J] engineering blasting2009，15（2）：7-11；

[4]]Tao Lv.Yongqiang Shi.Cheng Huang.etc. The method of nonlinear regression to blasting vibration velocity attenuation formula parameters [J] Rock and soil mechanics, 2007，28（9）1871-1878；

[5] Xianlin liu. Chuanbo Zhou.Guosheng Zhang. Tunnel excavation blasting vibration monitoring and vibration velocity prediction analysis

MATEC Web of Conferences 63, 02030 (2016) DOI: 10.1051/matecconf/20166302030
MMME 2016

Tooling Design of Milling Compounding Machine

Wen Ju SHAN[1, a], Li Yu GUO[1]

[1] *Zhuhai College of Jilin University, 519041Zhuhai, China*

Abstract. Professional tooling design is one of the most important part of the Milling Compounding Machine design. This paper mainly introduces the fundamental process of the tooling Design of Milling Compounding Machine, which includes the choosing of tooling material, the development and the chosen principle of tooling structure, the chosen principle of the milling cutters, and the geometry design of the cutters etc. it has certain significance for the practice production discussed.

1 Introduction

Machining and tooling technology is the key element of advanced equipment manufacturing industry, which plays an important role in the revitalization of China's equipment manufacturing industry. We must make full use of the machining and tooling technology. Tooling Design is an important part of the design of modular machine tool. The tool design includes the chosen of tool material, the chosen of tool structure and the design of the whole tooling system.

Compared with general machine, the combination machine has certain limited using conditions for the tools, which makes the combination machine tool to adapt a variety of specific conditions, besides the general characteristics of ordinary machine tool. [1] The general machine tool requires a full use range, while the combination machine tool requires higher reliability, longer life time and convenience in adjustment and disassembly due to producing specific parts. The compounding machine has high production efficiency, the tools' running time for single workpiece is shorter, and supposing that the tooling changing and adjusting takes a long time, the whole machines' efficiency will be reduced. So in the process of machining, the requirements of compounding machine tool are as following:

1) The reasonable structure of the tooling to ensure the produce quality of workpiece

2) According to the characteristics of the workpiece, select proper material and suitable parameters of the cutting tool to ensure the durability of the tool

3) Convenient in disassembly and adjustment. The tool can be adjusted and changed in a relatively short time for the higher production efficiency.

2 Selection of milling cutters materials

The cutting performance of the tool is first determined by its material. The first in the design of the tool is to select the tool material. The tool material determines the performance of the tool. According to the specific type of tool, combined with the specific processing parts to select the appropriate tool material, and then according to the performance of the tool to determine the reasonable cutting parameters, so as to achieve the best cutting effect.

The material of the cutting tool plays a key role in the cutting performance. Modern cutting tool materials has a constant development, from tool steel, high speed steel, development to hard alloy, high performance ceramics, super hard materials etc. Heat resistant temperature has a higher improvement, from 600 degrees to more than 1200 degrees. The allowing cutting speed has more than 1000m / min, The cutting efficiency is improved by more than 100 times in the past less than 100 years.

In the metal cutting process, the cutting part of the tooling is in the condition of high temperature, high pressure, friction, shock and vibration and other harsh occasions. From the service life of the tool, the performance requirements of the tool material are mainly the wear resistance, strong toughness, and high temperature red - hardness and so on. Different types of tools and processing conditions requires different tool. Generally speaking, cutting tool material should have the following properties: enough hardness, enough strength and toughness, enough wear resistance and heat resistance, manufacturability and economy, etc...

With the development of materials science and engineering, to meet the needs of a variety of functions, advanced metal or non-metallic materials emerge in an endless stream, such as model materials in aerospace, nuclear materials, biological materials, nanometer materials, rare earth materials. [2] Facing a wide variety of engineering materials, to improve the productivity, reduce processing costs and reduce the consumption of

[a] Corresponding author:36486846@qq.com

resources, need to choose the right tool material for cutting.

Different tool materials or the same kind of tool processing different workpiece materials, cutting effect will often differs. Each tool material has its specific processing range, which can only adapt to a certain part of the workpiece material and a certain range of cutting parameters, in the other word, cutting tool material and workpiece material has the problem of reasonable match. The matching of cutting tool material and workpiece material mainly refers to the physical properties, mechanical properties and chemical properties of the two parts. According to the reasonable selection of cutting tool material, the ideal machining efficiency and low production cost can be obtained.

1) Match the physical properties of tool material and workpiece material. The physical characteristics of the cutting tool determine the workpiece material which is suitable for processing. For example, a poor thermal conductivity of workpiece needs a tool material of good thermal conductivity in order to facilitate the cutting heat spread quickly and reduce the temperature of the cutting area.

2) The mechanical properties of tool material and workpiece material are matched, mainly refers to the strength, toughness and hardness and other mechanical properties parameters.

3) The chemical properties of tool material and workpiece material are matched, mainly refers to the tool material and workpiece material chemical affinity, chemical reaction, diffusion and dissolution of chemical performance parameters. When tool material and workpiece material works in high temperature, chemical reaction occurs, chemical dissolution and diffusion elements between the tool and the workpiece. The tool occurs the oxidation wear and diffusion wear and chemical wear, especially in high speed cutting and the high temperature, the chemical wear is far more serious than the abrasive wear caused by mechanical wear.

At present, high speed steel, hard alloy, metal ceramics, ceramics, diamond and cubic boron nitride are widely used in cutting tool materials. With the development of metal cutting tool surface modification technology, coating tools used more and more widely used.

Tool surface coating technology is mainly through chemical or physical vapor deposition method, the tool surface to obtain a few microns to ten microns thick hard film. Hard coating has high hardness, good lubrication, excellent high temperature performance characteristics, so that cutting tool can achieve excellent comprehensive mechanical properties, a substantial increase in the cutting speed and the cutting dosage, high quality for the machined surface.

3 Selection of special milling cutter structure

Milling is a very common way of processing. Through the rotation of the Multiple cutting edge tool (cutting edge around its own axis of rotation of the main cutting edge movement), feed motion along the direction of the workpiece, thereby completing the metal cutting, forming the machined surface. Milling has the following characteristics: milling is a continuous cutting, the impact load is large, easy to produce vibration, the cutting thickness and cutting area in the cutting process is changing, and the cutting force is changed greatly. In milling, the total length of the cutting edge is longer, the milling area is relatively wide, the metal removal rate is high, and so the productivity is high. [3]

According to the whole structure, milling cutter can be divided into integral milling cutter, welding carbide milling cutter and Mechanical clamping indexable milling cutter.

1) The integral-type milling is the cutter tooth and cutterhead using the same material as a whole made of milling cutter.

The whole cylindrical cutter has spiral teeth on the surface and a cutter tooth on the end, while milling, mainly use spiral teeth on the surface for cutting. In the rough machining and other large margin occasions, the chip can be worn out of the main cutting edge to smooth chip removal. In machining process, if the blade wear or damage, integral type milling cutter need to replace the entire milling cutter, after installing the new cutter need to reset the tool. Replacing the tool takes a long time, which will affect the overall processing efficiency of the combined machine tool.

2) Welding hard alloy milling cutter generally refers to the cutter tooth welding on the cutter body.

The material of the cutter tooth is generally used for hard alloy or other good performance tool material, including welding carbide helical tooth end milling cutter, welding carbide helical tooth milling cutter, welding carbide helical corn milling cutter, etc. The advantage of welding carbide milling cutter is the simple structure, easy manufacture, low cost of the material. However, the precision of the milling cutter is affected by the welding technology, and after the tool wear, Need to replace the whole tool, Tool change time is long, the processing efficiency is low.

2) The machine clip indexable blade is fixed on the cutter body by a mechanical fixing method. General use perforated blade with clamp screw, Simple structure, large chip space.

The machine clip blade has good impact resistance, which is suitable for large cutting depth and fast feed, high production efficiency. The blade can be replaced, no need to regrinding and recoating, and the economic performance is better. In the cutting process, when a certain blade wears, the corresponding blade can be adjusted to continue processing, which greatly improves the use rate of the blade, reduces the time and the number of re adjustment. Also due to the blade is standardized in large quantities and centralized production and the size of each blade are the same, the tool is changed without resetting and can improve the productivity of machine tools.

4 Type of machine clip indexable blade

The machine clamp indexable blade mainly includes end mill, Face milling cutter and three face milling cutter.

The main cutting edge of the end milling cutter is in a cylindrical surface, and the cutting edge of the end face is auxiliary blade.

End mill is shown in Figure1

Figure 1. End mill

The difference between a face milling cutter and an end mill cutter is the diameter, an end mill with a diameter greater than 40 mm is called a face milling cutter.

Face milling cutter is shown in Figure2

Figure 2. Face milling cutter

The circumference of the three face milling cutter has a main cutting edge, and the two sides face also have an assistant cutting edge, which improves the cutting condition, improves the cutting efficiency, and improves the surface quality.

Three face milling cutter is shown in Figure3

Figure 3. Three face milling cutter

The pitch types of the indexable blade are divided into three types, unequal sparse pitch, close pitch and super dense tooth.

When the stability and power is limited in order to achieve the highest productivity, unequal pitch can be used to reduce the number of blades for a long time processing. The unequal tooth spacing is suitable for processing long chip metal, such as aluminum alloy and stainless steel. Unequal pitch is beneficial to the vibration reduction of milling. The number of blades can be increased by using super dense teeth in the steady working condition, and the optimal production effect can be obtained, the processing is more suitable for short chip materials and heat resistant materials. Super dense tooth milling cutter production efficiency is high, but also the most suitable for processing cast iron and other fine scrap metal, but it requires high machine power and clamping stability. The performance of close pitch is between the above two.

5 Selection of cutter blade

The indexable blade is a functional element which is composed of a variety of rotatable cutting tools. Milling performance depends on the performance of the selected blade. Therefore, select the right indexable blade is very important for the overall cutting performance of the tool.

Rough machining is best to use compacting blade, which can reduce the processing cost. Dimensional accuracy and the sharpness of the compacting blade are not as good as grinding blade, But the compacting blade's edge strength is better, The rough milling need high impact resistance and larger cutting depth and feed speed, A roll chip groove is in the front cutter face of the blade, which can reduce the cutting force, the friction between the workpiece and the chip and the power. [4]But compacting blade surface is poor in dimensional accuracy and not as compact as grinding blade, and the milling cutter tips are highly inconsistent. Because of the low production cost of the compacting blade, it is widely used in production.

The best choice for fine milling is grinding blade, it has a good size precision. The positioning accuracy of the cutting edge in milling is high, which can obtain higher machining precision and lower surface roughness value. In addition, the development trend of grinding blade for finishing is grinding volumes chip groove, forming a large is rake angle cutting edge, allowing the blade in the small feed, small depth of cutting quantity. Without sharp anterior horn of the hard alloy blade, when the small feed, small depth of processing capacity, the tip will be frict to workpiece reducing tool life.

6 Positioning and clamping structure of indexable blade

Positioning of indexable blade which is to determine the position of the blade on the cutter body, so that the blade can keep the original size and precision of the blade when converting the new cutting edge or the new turning blade. Therefore, first of all, it should ensure the reliable

DOI: 10.1051/matecconf/20166302030

positioning of the blade and the positioning accuracy. General should pay attention to the followings:

1) The positioning of the blade make sure that the cutting force should be in the range of the support surface of the blade, and it should be in accordance with the principle of positioning and clamping for cutting force.

2) Blade size positioning needs to meet the three point positioning principle.

The clamping way of the indexable blade is influenced by the shape of the blade, the size of the tool, and the situation of the use of the cutter.

Indexable inserts for clamping requirements of cutting force and clamping force of the force or the resultant force on the locating datum produced friction must to be greater than that caused by the centrifugal force generated in the cutting vibration and cutter the blade from the location of datum plane alternating force.

There are many forms of indexable inserts of the clamping mechanism, the mainly used is eccentric type, lever pin type, wedge type, pressure, pressure hole and vertical installation type etc.

7 Geometric angle of milling cutter

For milling cutters, it is very necessary to select the appropriate cutting angles for different materials and different shapes of the workpiece. The geometric angle of the milling cutter is shown in Fig4

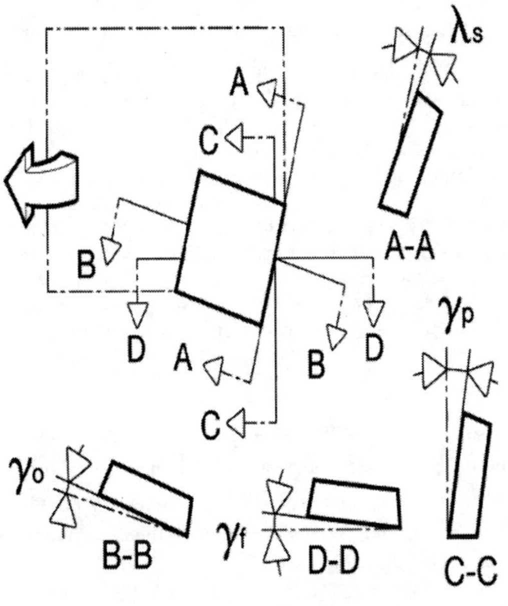

Figure 4. Geometric angle of milling cutter

Among them: γ_p is Axial rake angle,

γ_f is Radial rake angle,

γ_o is Rake angle

λ_s is Cutting edge inclination

Axial rake angle determines the direction of chip removal. Radial rake angle determines the cutting edge of the cutting edge.

When the Rake angle is positive, the discharge of the chip is better, but the tip of the cutting edge strength decreases. [5] Rake angle determines the degree of sharp cutting edges. Rake angle determine sharpness. Larger angle, better machinability, the chip is not fusible attached, When the negative angle is large, the cutting performance is poor, but the cutting edge of the cutting edge is high.

When axial rake angle and Radial rake angle are negative, it is suitable for milling cast iron and hard material. When axial rake angle and Radial rake angle are positive, it is suitable for milling the Aluminum Alloy. When axial rake angle is positive and radial rake angle is negative Suitable for milling of hardened steel.

The development of milling machine clip has gone through several stages

1) Plate blade. At this time, the cutter body is designed for axial negative rake angle.

2) Blade with back angle. The milling by shear processing instead of the original pushes processing.

3) Blade with back angle and the chip breaking groove is easy to chip breaking. The volume of the tool chip can be reduced, and the number of the blades is increased, and the chip flow direction can be easily controlled.

4) Blades with mounting holes. The cutter blade is clamped on the cutter body by screws; it is beneficial to increase the number of blades on the cutter body which requires high precision. The cutter body is usually processed after harden process.

5)The bellowing aspect should be considered in the selection of the blade: the blade shape, the main cutting edge angle, precision cutting blade, a blade fixed form, blade cutting edge length, thickness of the blade, blade cutting edge section shape, blade broken flutes etc.

References

1. D.P. Liang. WSTL, 68 (2008):243-245.
2. B. Bazrov, E. Z. Nasirov. RERV.27 (2012):95-97.
3. Dvorak, Paul. Machine Design (2005):115-116
4. Hj. Yang .EMT, 5(2011):8-10.
5. X. Zhang, Y. Zhang, DH .Liu .STI 14(2012):95-97.

MATEC Web of Conferences **63**, 02031 (2016) DOI: 10.1051/matecconf/20166302031
MMME 2016

Application and comparison of stability analysis of slope using circular arc method and strength reduction method

Bin Bin XU[1,a], Lin Bo XIE[1] and Wei SI[2]

[1] Tianjin Port Engineering Institution Ltd. of CCCC.; Key Lab. of Geotechnical Engineering of Tianjin; Key Lab. of Geotechnical Engineering, Ministry of Communication, Tianjin 300222, China
[2] Tianjin Port Engineering Institution Ltd. of CCCC.; Tianjin 300222, China

Abstract. In order to evaluate the accuracy of strength reduction method using FEM, the safety factors of the uniform clay slope and sand slope are investigated by Fellenius's method, Bishop's method using traditional limit equilibrium method and strength reduction method respectively. The limit equilibrium method is carried out based on the code for foundations in port engineering and the FE analysis is based on Plaxis3D. The results show that the safety coefficient obtained by strength reduction method of FEM is larger than that obtained by limit equilibrium method and the errors are 6.03% and 1.61% for clay slope and 8.54% and 3.99% for sand slope.

1 Research background

The side slope comes into being artificially or naturally and its destructional form is usually slide or collapse of the slide. The slope can be classified into natural slope and artificial slope by genesis, or soil slope and rock slope by composition, or high slope and general slope by height, or temporary slope and permanent slope. The reason of destruction can be attributed to two categories: the equilibrium of the stress state of the slope is disturbed by external actions; the strength of the slope soil is reduced by external influence. As the infrastructure construction such as irrigation works, transportation, mining develops, more and more engineers are related to the slope stability. How to evaluate the safety coefficient becomes the key point for the design, construction, controlling of the slope.

The limit equilibrium method is the main method for the stability analysis of the slope, which is based on the Coulomb's and Ranking's earth pressure. The characteristics of the limit equilibrium method only consider the static equilibrium condition and Mohr-coulomb's principle, namely analysing the equilibrium of the soil at destruction. During the analysis of the slope stability the plane strain condition is usually assumed and the sliding surface is regarded as artificial arc surface in early researches. The safety coefficient of the slope is defined by the ratio of anti-slide action to the slide action. In modern engineering, the concept of safety stock of the strength index is well admitted. Namely, the strength index of soil is reduced simultaneously until there is slide surface at the limit equilibrium state.

2 General methods for stability analysis

During the development of the slope research, lots of calculation methods of the safety coefficients have been proposed. In 1916, the Swedish method is proposed by Fellenius and Taylor[1], which is the earliest method for the slope stability analysis. In this method, the sliding surface is supposed to be arc and the positive slide force is on the tangent direction of the weight in the calculation. Supposing that there are many sliding surfaces and each sliding surface corresponds to one safety coefficient, the smallest safety coefficient is regarded as the final safety coefficient of the slope and it can be expressed as :

$$F_S^i = \frac{M_f^i}{M_i} \tag{1}$$

where F_S^i represents the safety coefficient of the ith sliding surface, M_f^i is the anti-sliding moment of the ith sliding surface to the upper sliding volume and M_i is the sliding surface of the upper sliding volume, and the final safety coefficient $F_S = min(F_S^i)$.

In 1955, Bishop[2] proposed another simplified method based on the Swedish method in which the arc-shape slide surface and the equilibrium of the moment are kept while the interaction between two strips is considered. The safety coefficient is represented by the ratio between the resistant shear strength and the actual shear strength on the sliding surface, which is called safety factor of shear stress :

[a] Corresponding author: xubinbin@tpei.com.cn

$$F_S = min(F_S^i) = min\left(\frac{\tau_f}{\tau}\right) = \min\left(\frac{c_f + \sigma \tan\varphi_f}{\tau}\right) \quad (2)$$

Here, *i* and *S* have the same physical meaning.

Besides, other shapes of the sliding surface instead of the arc surface are also used to compute the safety coefficients such as simplified Janbu method. Based on satisfaction of the equilibrium of the moment and the force, Morgenstern and Price[3] proposed a strict deduction of safety analysis assuming arbitrary shape of sliding surface.

Although there are a lot of definitions for the strength reduction method of FEM[4-8], the principle is almost the same. Ducan[9] pointed out that the safety coefficient can be defined as the reduction extent of the shear strength when the slope reaches the critical destroy state, which can be regarded as the strength reservation. The points of the strength reduction method is that the shear strength of the soil are reduced simultaneously assuming the external force keeps constant and then the finite element analysis is carried out to analyse the stress and strain until the destruction of the slope. Zhao and Zheng[10] believed that the safety coefficients obtained by Bishop method and strength reduction method possess the same physical meaning and the essential of two methods are also same.

3 Results by limit equilibrium method

In this chapter, the safety coefficient of the slope would be calculated by the Fellenius's method and Bishop's method which are the traditional limit equilibrium method. The slope is assumed to be composed by uniform soil and the detailed coordinates at each point are also shown in Fig. 1. The gradient of the slope is 1 :2 and the underground level is below the ground about 20m.

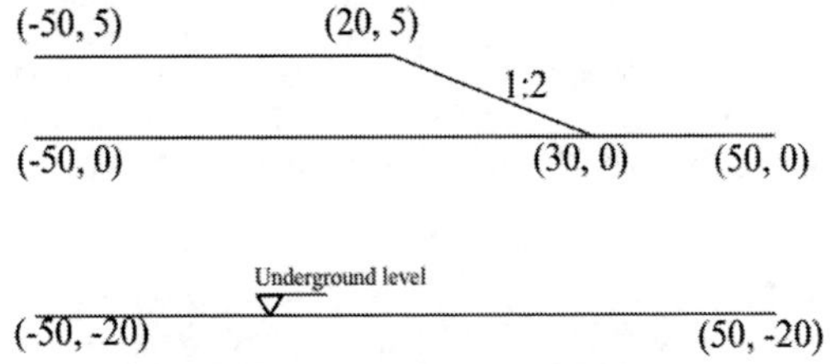

Figure 1. Sketch of the simple slope

The composition of the slope is regarded as clay or sand respectively. The soil parameters are listed in Table 1. As can be seen, the bulk densities of both clay and sand are the same. The cohesive force of clay is larger than that of sand while the frictional angle of clay is smaller than that of sand.

Table 1. Soil parameters of clay and sand

	Clay	Sand
Bulk density	18kN/m^3	18kN/m^3
Cohesive force	10kPa	1kPa
Frictional angle	20°	30°

The calculation is carried out by the computation system for port engineering which is developed by Tianjin port engineering institute. It is written by Fortran program based on Code for foundations in port engineering. It can deal with the settlement of the ground, vertical and horizontal consolidation, bearing capacity, slope stability. The calculation results are listed in Table 2 using each method for the uniform clay slope and sand slope. As can be seen, the safety coefficient of the clay is 1.84 and 1.92 determined by Fellenius's method and Bishop's method and the safety coefficient of the sand is 1.37 and 1.43 respectively. The safety coefficient obtained by Fellenious's method is smaller than that obtained by Bishop's method, which can be ascribed to that the interaction force between two slices is taken into consideration.

Table 2. Safety coefficient of the slope using limit equilibrium method

	Clay	Sand
Fellenius	1.84	1.37
Bishop	1.92	1.43

4 Results by strength reduction method

Although the strength reduction method has been widely used in the stability analysis of the slope, the final safety coefficient is always depended on the evaluation criterion of instability. Usually the numerical convergence is taken as the judge criterion, but the influence factors of the convergence of FEM are multiple and it is difficult to explain whether the divergency is attributed to instability or not. Now there are mainly three judge criterions for the instability of the slope: convergence of calculation, run-through of the plastic range and sudden change of characteristic position.

When using strength reduction method for the slope stability, the calculation convergence is usually used as the criterion[11-13]. If the convergent solution cannot be obtained under the designated convergence judgment, the slope loses its stability. For the run-through of the plastic range, some significant shear bands have been observed during the damage of the slope. Similar regulation can be forecasted. If there is run-through plastic deformation in the slope under certain reduction coefficient, such a coefficient can be regarded as the safety coefficient. For the sudden change of characteristic position, the treatment is building the relationship between the displacement of certain point and the reduction coefficient in the FE calculation. If there is a sudden change in the curve, the slope is in a critical state.

A FE code called Plaxis3D is used to carry out the strength reduction method. In Plaxis3D, a module called safety analysis can be used for the slope stability analysis. In the calculation, the reduction coefficient is gradually reduced and the total displacement at toe of the slope is also monitored for each reduction coefficient. When the relationship between the reduction coefficient and the

DOI: 10.1051/matecconf/20166302031

total displacement is plotted in the diagram, the safety coefficient is taken as the reduction coefficient if there is no variation even though the total displacement increases. The finite element mesh is shown in Fig. 2 and the variation of reduction coefficient of clay slope and sand slope is shown in Figs. 3 and 4. The four side surfaces are restricted on the nominal direction and bottom surface is restricted at both the horizontal and vertical direction. The hydraulic condition is the same as that in the limit equilibrium method.

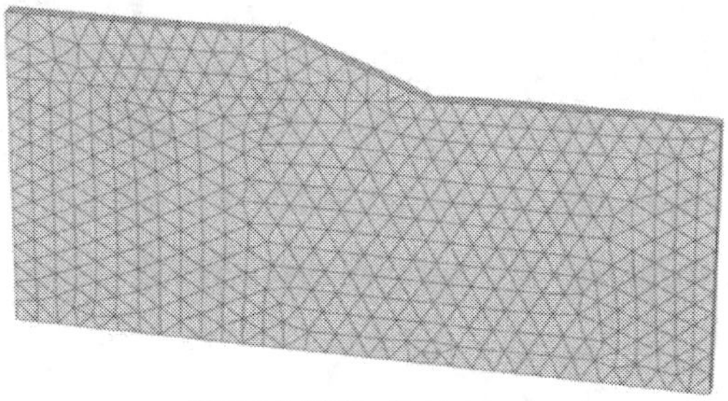

Figure 2. Finite element mesh

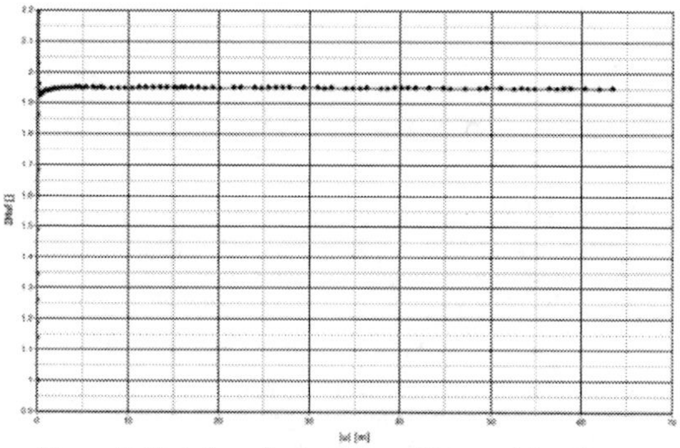

Figure 3. Variation of reduction coefficient of clay slope

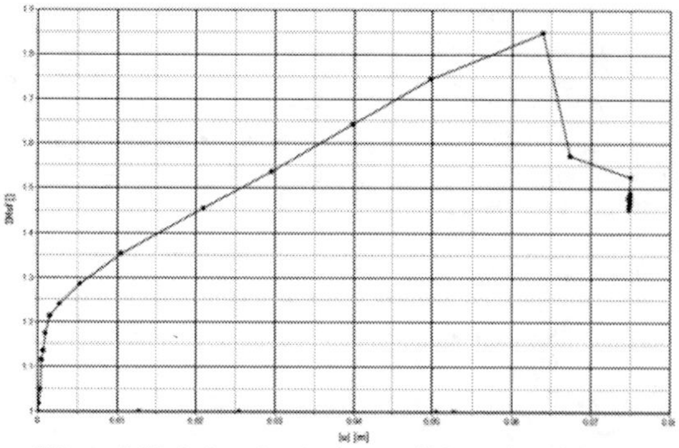

Figure 4. Variation of reduction coefficient of sand slope

As have been explained above, the method to determine the safety factor is to see whether the final reduction coefficient is constant or not as the total displacement of characteristic point increases. The variation of the reduction coefficient of the clay slope is shown in Fig. 3. As can be seen, the reduction factor firstly increases to as large as 2.2 and then decreases to 1.951. Therefore, the safety factor of the clay slope is 1.951. Similarly, the safety factor of the sand slope is shown in Fig. 4. The final reduction factor keeps 1.487. All in all, the reduction coefficients of clay slope and sand slope keep constant as the total displacement increases and the safety coefficients of the clay slope and sand slope obtained by strength reduction method are 1.951 and 1.487 respectively. Taking the safety coefficients obtained by traditional limit equilibrium method into consideration, the safety coefficients of clay and sand slopes are listed in Table 3. The safety coefficient obtained by strength reduction method of FEM is larger than that obtained by limit equilibrium method and the errors are 6.03% and 1.61% for clay slope and 8.54% and 3.99% for sand slope. There seems to be a fact that the clay slope is more suitable for strength reduction method because the errors are smaller.

Table 3 Safety coefficient of slope using different method

	Clay	Sand
Fellenius	1.84	1.37
Bishop	1.92	1.43
FEM	1.951	1.487
Error to Fellenius	6.03%	8.54%
Error to Bishop	1.61%	3.99%

In addition, Figs. 5 and 6 present the settlement of the ground and the slope. Initially, the ground stress is added to the finite element model and the settlement of the ground is zero, as shown in Fig. 4. Then the slope is regarded as an embankment to the ground and significant settlement occurs in the ground. The safety analysis of the slope is then carried out and the occurred settlement should be removed before the stability analysis. From the finite element analysis, we can see that the strength reduction analysis is easier to be understood because the entire process of the deformation can be monitored.

Figure 5. Initial state of the ground

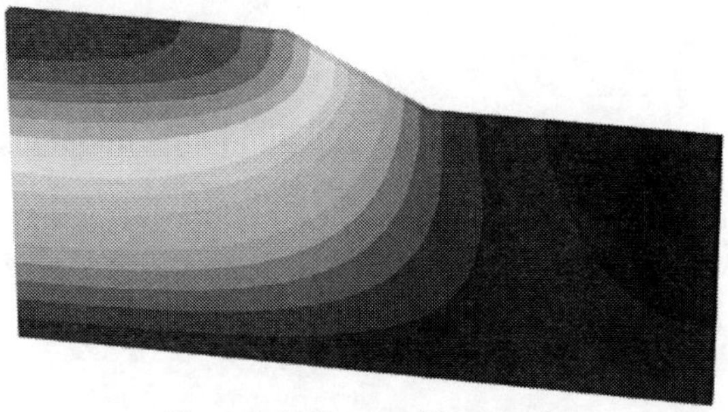

Figure 6. Settlement of the ground

After the safety analysis, the possible sliding surface can be shown by chosing the contour of total incremental displacement. The slide surfaces of the clay slope and the sand slope are shown in Figs. 7 and 8 respectively. The slide surface of the clay goes through the inside of the slope and the toe of the slope, which is in accordance with the traditional analysis method. While for the sand slope, the slide surface lies at the side surface of the slope. Actually, in Plaxis3D the plastic computation becomes very complex as the friction angle increases.

Figure 7. Slide surface of the clay slope

Figure 8. Slide surface of the sand slope

5 Conclusions

The stability of the slope is always the highlight of geotechnical engineering. Formerly, the limit equilibrium method is widely used due to its clear physical meaning and simple calculation. With the development of the finite element method, the strength reduction method is gradually recognized to determine the safety factor of complex ground. However, the accuracy of the strength reduction method is suspicious because the friction angle and cohesive force are reduced simultaneously. In this paper, the safety factors of one uniform clay slope and sand slope are firstly calculated by Fellinius's method and Bishop's method, which are then compared with the safety factor obtained from strength reduction method by FEM. The conclusions are as follows:

1) The safety coefficient of the clay slope is 1.84 and 1.92 determined by Fellenius's method and Bishop's method and the safety coefficient of the sand slope is 1.37 and 1.43 respectively.

2) The safety coefficient obtained by Fellenious's method is smaller than that obtained by Bishop's method, which can be ascribed to that the interaction force between two slices is taken into consideration.

3) The reduction factor firstly increases to as large as 2.2 and then decreases to 1.951 and the safety factor of the clay slope is 1.951.

4) The safety coefficient obtained by strength reduction method of FEM is larger than that obtained by limit equilibrium method and the errors are 6.03% and 1.61% for clay slope and 8.54% and 3.99% for sand slope.

5) There seems to be a fact that the clay slope is more suitable for strength reduction method because the errors are smaller.

References

1. W. Fellinius, Int. Cong. of Large Dams, Washington, **4** 445 (1936)
2. A.W. Bishop, Geotechnique, **5**, 7 (1955)
3. N.R. Morgenstern, Geotechnique, **15**, 11 (1965)
4. N. Janbu, Proc. of Eur. Conf. on Stability of Earth Slopes, Swenden, **3**, 43 (1954)
5. E. Spencer, Geotechnique, **17**, 1 (1967)
6. N.R. Morgenstern, V.E. Price, Geotechnique, **15**, 79 (1965)
7. Z. Chen, N.R. Morgenstern, Canadian Geotechnical Journal, **20**, 104 (1983)
8. I. Donald, Z.Y. Chen, Canadian Geotechnical Journal, **34**, 853 (1997)
9. J.M. Duncan, Journal of Geotech. Engineering, (1996)
10. S.Y. Zhao, Y.R. Zheng, China Geotech. Engineering, (2002)
11. K. Ugai, Soils and Foundations, **29**, 190 (1989)
12. D.V. Grifiths, P.A. Lane, Geotechnique, **49**, 387 (1999)
13. E.M. Dawson, W.H. Roth, Geotechnique, **49**, 835, (1999)

MATEC Web of Conferences 63, 02032 (2016) DOI: 10.1051/matecconf/20166302032
MMME 2016

Influence of strength parameters of soil on the slope stability

Bin Bin XU[1,a] and Wei SI[2]

[1] *Tianjin Port Engineering Institution Ltd. of CCCC.; Key Lab. of Geotechnical Engineering of Tianjin; Key Lab. of Geotechnical Engineering, Ministry of Communication, Tianjin 300222, China*
[2] *Tianjin Port Engineering Institution Ltd. of CCCC.; Tianjin 300222, China*

Abstract. The stability analysis of uniform clay and sand slope is carried out using traditional limit equilibrium method and strength reduction method. In order to evaluate the influence of strength parameters including friction angle and cohesive strength, the errors among Fellenius's method, Bishop's method and strength reduction method are calculated in detail. The results are: 1) the safety factor obtained from Fellenius's method is smaller than that obtained from Bishop's method; 2) the error between Bishop's method and strength reduction method usually firstly increases and then decrease as the friction angle/cohesive strength becomes larger.

1 Introduction

The stability analysis of the slope is always the highlight of geotechnical engineering including hydraulic engineering, transportation engineering, tunnel engineering etc. As the developing of the infrastructure, it is very necessary to deal with the slope engineering. The slope can be regarded as a volume with an inclining bottom surface. Under the action of self-weight of the slope and other external actions, there is a sliding tendency from upwards to downwards. If the active downward action surpasses the positive action of resistance, the destruction of the slope occurs.

During the development of the slope research, lots of calculation methods of the safety coefficients have been proposed. In 1916, the Swedish method is proposed by Fellenius and Taylor[1], which is the earliest method for the slope stability analysis. In this method, the sliding surface is supposed to be arc and the positive slide force is on the tangent direction of the weight in the calculation. In 1955, Bishop[2] proposed another simplified method based on the Swedish method in which the arc-shape slide surface and the equilibrium of the moment are kept while the interaction between two strips is considered. Besides, other shapes of the sliding surface instead of the arc surface are also used to compute the safety coefficients such as simplified Janbu method[3]. Based on satisfaction of the equilibrium of the moment and the force, Morgenstern and Price[4] proposed a strict deduction of safety analysis assuming arbitrary shape of sliding surface.

Built on the theory of the limit equilibrium method, it is impossible to consider the stress-strain relationship of the soil, the occurrence and development of the slide destroy and the influence of the local deformation on the stability. Actually, there is close relationship between the stability and the deformation. Before the occurrence of the entire destruction of the slope, significant deformation would also occur. The finite element method can deal with non-linear relationship between stress and strain and calculate the sliding surface without designating the shape in advance. When using the strength reduction method for the stability analysis the cohesive strength and the internal frictional angle are reduced simultaneously. However, the experimental results show that the sensitivity of the internal frictional angle is small and the cohesive strength is large, so it is unreasonable to treat them by the same method.

2 Calculation model

There are lots of influence factors of the slope stability, such as the range of computation model, the strength parameters including cohesive strength and inner friction angle. In order to evaluate the effect of strength parameters on the safety factor, a simple slope is used and the sketch is shown in Fig. 1. The gradient of the slope is 1 :2 and the underground level is below the ground about 20m. In the slope, the bulk density of the soil is $18kN/m^3$ and the soil of the slope is uniform.

[a] Corresponding author: xubinbin@tpei.com.cn

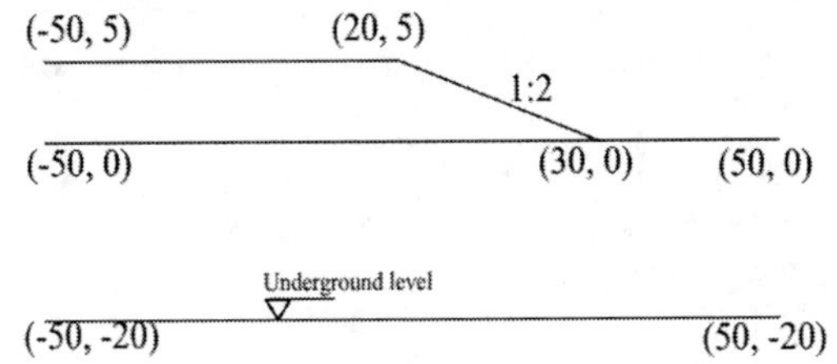

Figure 1. Sketch of the simple slope

The safety factor of the slope is determined by Fellenius's method, Bishop's method and strength reduction method repectively. Two kinds of slope including clay slope and sand slope are evaluted through changing one of the strength parameters while keeping another strength parameters constant. The calculation skeme is listed in Table 1. There are total 26 cases in the comparison.

Table 1. Calculation skemes

clay slope (c=10kPa) change φ						
5	10	15	20	25	30	35
clay slope (φ=20) change c						
2	4	6	8	10	12	14
sand slope (c=1kPa) change φ						
15	20	25	30	35	40	---
sand slope (φ=30) change c						
1	3	5	7	9	11	---

3 Calculation results

3.1 Influence of φ for clay slope

For the clay slope, in order to evaluate the influence of friction angle on the safety factor, the cohesive strength of the soil keeps constant and the friction angle changes from 5~35. Actually a soil with c=10 and φ=35 may not exist in the nature and here it is calculated as a extreme case. The obtained safety factors of clay slope are shown in Table 2. In the table, three safety factors obtained by Fellenius's method, Bishop's method and strength reduction method. The error between Fellenius's method and strength reduction method is marked as E1 and the error between Bishop's method and strength reduction method is marked as E2.

Table 2. Safety factor of clay slope under different φ

	clay slope (c=10kPa) change φ						
	5	10	15	20	25	30	35
FM	1.044	1.369	1.603	1.84	2.09	2.357	2.646
BM	1.144	1.412	1.667	1.92	2.186	2.471	2.777
SRM	1.045	1.364	1.655	1.95	2.249	2.559	2.894
E1%	0.1	-0.37	3.24	6.03	7.61	8.57	9.37
E2%	-8.65	-3.40	-0.72	1.61	2.88	3.56	4.21

p.s. FM, BM and SRM mean Fellenius's method, Bishop's method and strength reduction method; E1 and E2 mean error toward Fellenius's method and Bishop's method.

As can be seen, with increasing friction angle the error between Fellenius's method and strength reduction method increase gradually and the maximum error is as large as 9.37% when φ=35. While the the error between Bishop's method and strength reduction method decreases firstly and then increases as the friction angle grows. The maximum error is 8.65% when φ=35 and the minimum error is 0.72% when φ=15.

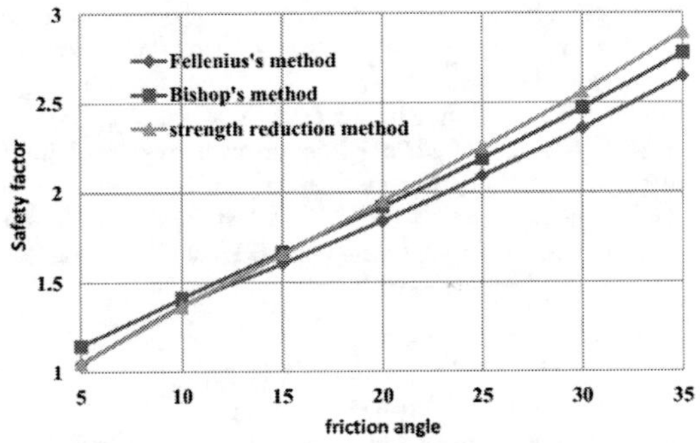

Figure 2. Safety factor v.s. friction angle (clay)

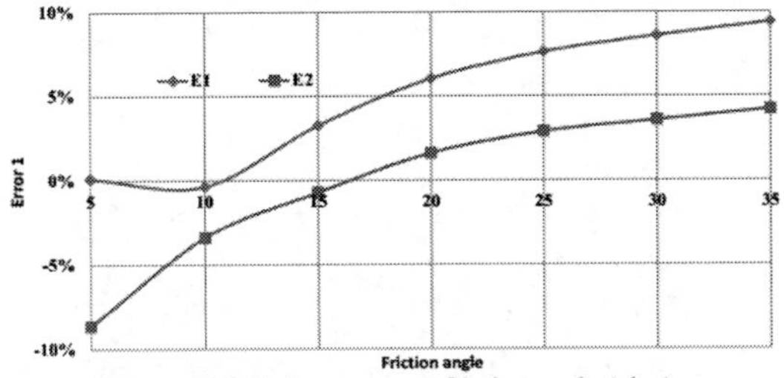

Figure 3. Relative error v.s. friction angle (clay)

The variation of safety factors and relative errors are shown in Figs 2 and 3. As the friction angle increases from 5 to 35, the safety factor obtained from strength reduction method becomes gradually larger than other two methods. There is a cross point between Bishop's method and strength reduction method.

3.2 Influence of c for clay slope

Also for the clay slope, in order to evaluate the influence of cohesive strength on the safety factor, the friction angle of the soil keeps constant and the cohesive strength changes from 2~14. The obtained safety factors of clay slope are shown in Table 3. In the table, three safety factors obtained by Fellenius's method, Bishop's method and strength reduction method. E1 and E2 have the same meaning.

Table 3. Safety factor of clay slope under different *c*

	clay slope (φ=20kPa) change *c*						
	2	4	6	8	10	12	14
FM	1.035	1.254	1.456	1.65	1.842	2.03	2.216
BM	1.084	1.318	1.526	1.73	1.92	2.11	2.303
SRM	1.137	1.38	1.584	1.77	1.951	2.11	2.282
E1%	9.86	10.05	8.79	7.45	5.92	4.19	2.98
E2%	4.89	4.7	3.8	2.84	1.61	0.0	-0.91

p.s. FM, BM and SRM mean Fellenius's method, Bishop's method and strength reduction method; E1 and E2 mean error toward Fellenius's method and Bishop's method.

As can be seen, when the cohesive strength increases the error between Fellenius's method and strength reduction method decreases gradually with small fluctuiation. The maiximum error is as large as 10.05% when c=4kPa. While for the error between Bishop's method and strength reduction method decreases firstly and then increases as the cohesive strength grows. The minimum error is 0.0% when c=12kPa and the maximum error is 4.89% when c=2kPa.

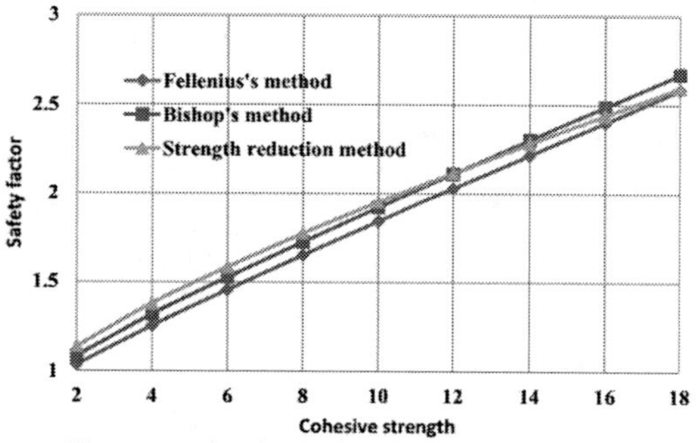

Figure 4. Safety factor v.s. cohesive strength (clay)

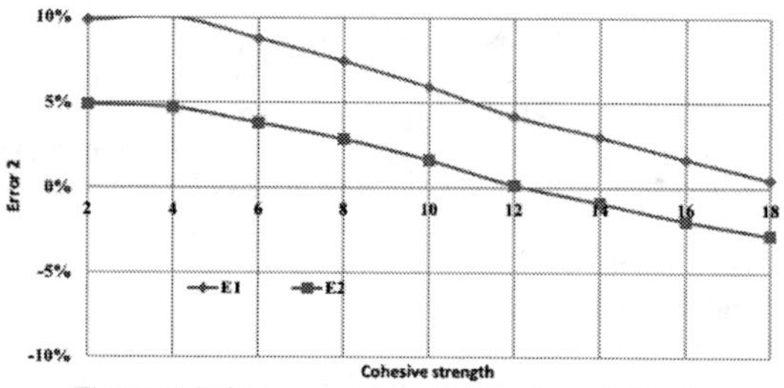

Figure 5. Relative error v.s. cohesive strength (clay)

Figs. 4 and 5 show the variation of safety factors and relative error when the cohesive strength increases. As can be seen, the safety factor obtained from strength reduction method is firstly larger than those obtained from other two methods and then becomes smaller gradually.

3.3 Influence of φ for sand slope

Similarly, for the sand slope, the influence of friction angle on the safety factor is investigated through changing the friction angle from 15 to 40. The obtained safety factors of clay slope are shown in Table 4. As the friction angle increases, the error between Fellenius's method and strength reduction method gradually decreases. The maximum and minimum error is 8.85% and 7.34% respectively. Also the error between Bishop's method and strength reduction method becomes smaller and the maximum error is 4.24% when φ=25.

Table 4. Safety factor of sand slope under different φ

	clay slope (c=1kPa) change φ					
	15	20	25	30	35	40
FM	0.706	0.913	1.18	1.37	1.63	1.92
BM	0.737	0.953	1.18	1.43	1.7	2.02
SRM	---	---	1.23	1.487	1.759	2.061
E1%	---	---	8.85	8.54	7.91	7.34
E2%	---	---	4.24	3.99	3.47	2.03

p.s. FM, BM and SRM mean Fellenius's method, Bishop's method and strength reduction method; E1 and E2 mean error toward Fellenius's method and Bishop's method.

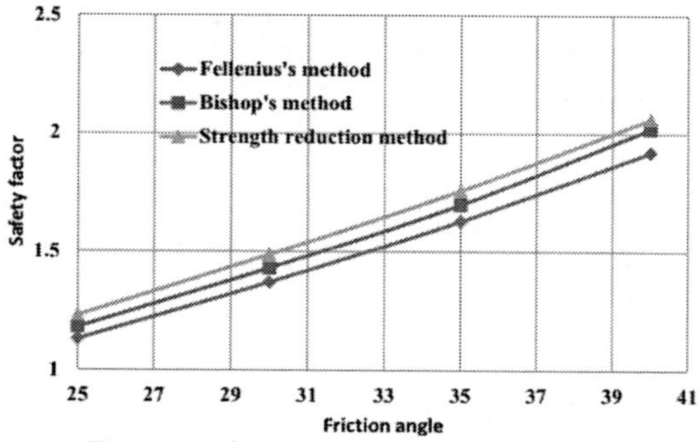

Figure 6. Safety factor v.s. friction angle (sand)

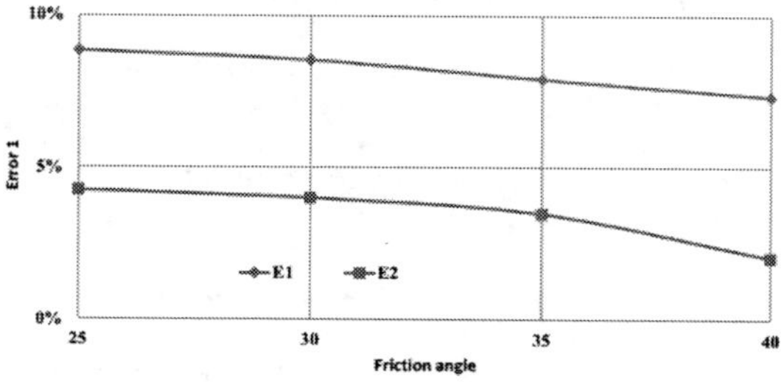

Figure 7. Relative error v.s. friction angle (sand)

Figs. 6 and 7 show the variation of safety factors and relative error when the friction angle increases. As can be seen, the safety factor obtained from strength reduction method is always larger than those obtained from Fellenius's method and Bishop's method.

3.4 Influence of *c* for sand slope

Also for the sand slope, the friction angle of the soil keeps constant and the cohesive strength changes from 1~11. The obtained safety factors of sand slope are shown in Table 5. In the table, three safety factors obtained by Fellenius's method, Bishop's method and strength reduction method. E1 and E2 have the same meaning.

As can be seen, when the cohesive strength increases the error between Fellenius's method and strength reduction method fluctuiates slightly and the maximum error is as large as 10.19% when c=3kPa. While for the error between Bishop's method and strength reduction method increases firstly and then decreases as the cohesive strength grows. The minimum error is 3.43% when c=11kPa and the maximum error is 5.0% when c=3kPa.

Table 5. Safety factor of sand slope under different c

clay slope (φ=30) change c						
	1	3	5	7	9	11
FM	1.367	1.62	1.847	2.056	2.26	2.454
BM	1.43	1.7	1.938	2.162	2.37	2.569
SRM	1.487	1.785	2.029	2.262	2.463	2.657
E1%	8.78	10.19	9.85	10.02	8.98	8.27
E2%	3.99	5.0	4.7	4.63	3.92	3.43

p.s. FM, BM and SRM mean Fellenius's method, Bishop's method and strength reduction method; E1 and E2 mean error toward Fellenius's method and Bishop's method.

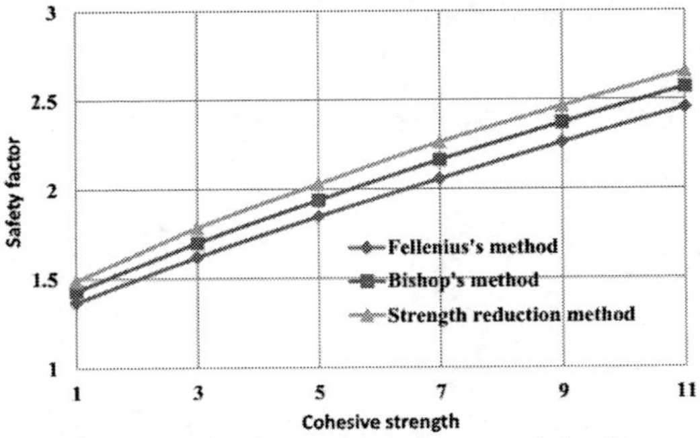

Figure 8. Safety factor v.s. cohesive strength (sand)

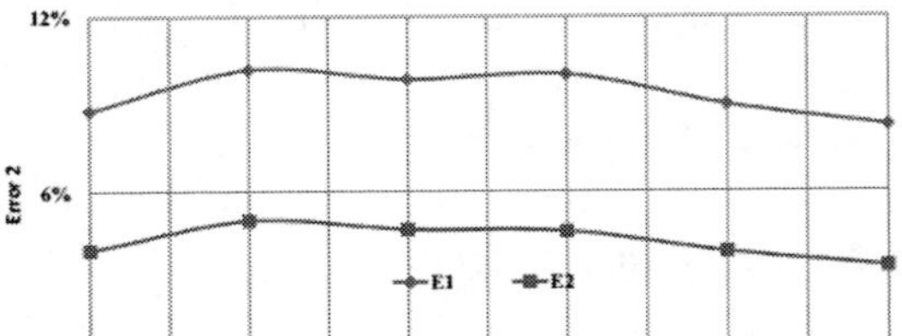

Figure 9. Relative error v.s. cohesive strength (sand)

Figs. 8 and 9 show the variation of safety factors and relative error when the cohesive strength increases. As can be seen, the safety factor obtained from strength reduction method is always larger than those obtained from Fellenius's method and Bishop's method.

4 Conclusions

The stability of the slope is influenced by the stiffness of the soil, friction angle, cohesive strength, underground water etc. It is generally believed that the strength parameters of soil including friction angle and cohesive angle have the greatest influence on the safety factor. In this paper, the safety factors uniform clay slope and sand slope are investigated through changing the friction angle and cohesive strength and the conclusions are as follows:

1) With increasing friction angle for the clay slope the error between Fellenius's method and strength reduction method increase gradually and the maximum error is as large as 9.37%. While the the error between Bishop's method and strength reduction method decreases firstly and then increases as the friction angle grows.

2) When the cohesive strength of the clay slope increases the error between Fellenius's method and strength reduction method decreases gradually with small fluctuiation. While for the error between Bishop's method and strength reduction method decreases firstly and then increases as the cohesive strength grows.

3) As the friction angle of the sand slope increases, the error between Fellenius's method and strength reduction method gradually decreases. The maximum and minimum error is 8.85% and 7.34% respectively. lso the error between Bishop's method and strength reduction method becomes smaller.

4) When the cohesive strength of the sand slope increases the error between Fellenius's method and strength reduction method fluctuiates slightly and the maximum error is as large as 10.19% when c=3kPa. While for the error between Bishop's method and strength reduction method increases firstly and then decreases as the cohesive strength grows.

References

1. W. Fellinius, Int. Cong. of Large Dams, Washington, **4** 445 (1936)
2. A.W. Bishop, Geotechnique, **5**, 7 (1955)
3. N. Janbu, Proc. of Eur. Conf. on Stability of Earth Slopes, Swenden, **3**, 43 (1954)
4. N.R. Morgenstern, Geotechnique, **15**, 11 (1965)

MATEC Web of Conferences **63**, 02033 (2016) DOI: 10.1051/matecconf/20166302033
MMME 2016

Analysis of the Influence Factors on the Efficiency of Marine Tail Shaft Transmission System

Jian WANG, Zheng Lin LIU, Li ZOU
Wuhan University of Technology, Wuhan 430063, China

Abstract. The transmission efficiency of the marine tail shaft was calculated with the finite element method and experimentvalidatedthe rationality and feasibility of finite element models. The research conclusions as follows: power consumption increases with the shaft length, shaft diameter and speed increase. The results can provide a theoretical support for further research andit is also helpful for engineering applications.

1 Introduction

When the ship is sailing on the ocean, itsshaft between the stern bearing and propeller is working in water, which is normally called the shaft in water [1]. There has always been scholars to question the power consumption of the shaft in water and then affect the ship's economy and efficiency. In particular, the shaft in water power consumption of the warship will further affect its operationalperformanceattract more attention from scholars and experts [2,3]. However, it is limited to directly measurethe shaft in water's power consumption in real environmental conditions when theshaft in water runs in the water [4]. So far there has been no relevant studies or reports about the shaft in water drive efficiency. In this paper, firstly, finite element model of the shaft in water is built in GAMBIT and its friction torque in water is calculatedvia FLUENT, and then the power consumed by the resistance is obtained. Our project team designed the relevant design experiments to test the friction torque produced by the shafting in water and then calculate the power consumption. The theoretical results and experimental results are comparatively analyzed to find out the factors affecting the efficiency of the shaft transmission in water.

2.Friction torque calculation with FEM

The power consumption caused by friction can be calculated by torque, so the friction torque should be calculated by FLUENT firstly.The power consumptioncan be calculated with:

$$P = \frac{T \bullet S}{9.55} \quad (1)$$

Where P(W) is the power consumption, S is the rotation speed of the shaft and T is the friction torque. The meaning is the same as in the following.

2.1 Finite elementmodel

Finite element model of shafting in water was established by the pretreatment software GAMBIT of FLUENT.First of all, establishing open water film axial segment model, and then placing the model used in the experiments of flow field area of the tank. This fluid region is relatively static.A rotating region on the outer cylindrical surface of the shaft is established, which is coaxial with the cylindrical surface. It is assumed that the flow field rotates around the Z axis, which is in a static state with the axial section.The calculation area adopts structured grid, and the water film model is divided into volume grid directly.The finite element model has three kinds of 240mm, 310 mm and 390mm in the diameter of the shaft section. Each diameter of the shaft section corresponds to two lengths of 1200 mm and 1500mm. For example, when the axial section of the shaft in water is 1200mm and the diameter of the shaft is 240 mm, the diameter of the rotating area is selected 360mm.Due to the large external flow field, the impact on the friction power consumption is small, the grid partition can be sparse, a total of 44350 units, the rotation of the basin and the axis of contact, the grid division of the higher density, a total of 43350 units.As shown in Figure 1. Thequality inspection of grid division meets the requirements and the finite element models of different axial lengths under different axial lengths are similar to that of example.

In the calculation of the friction powerconsumption of the shaft in the open water, the internal calculation area is only relative axial rotation, there is no mutual interference between other objects and axial segments. The boundary conditions of the inlet is velocity inlet, and outlet boundary pressure outlet and the external surface of the cylinder is set to the wall and the surface between the inner rotating basin and the outer cylinder is set to INTERFACE. When the finite element model is introduced into FLUENT, these two overlapping surfaces are connected together to form a pair of interface, which can realize the data exchange between the external static flow field and the inner rotating basin. At last, the axial section surface is set to the wall.

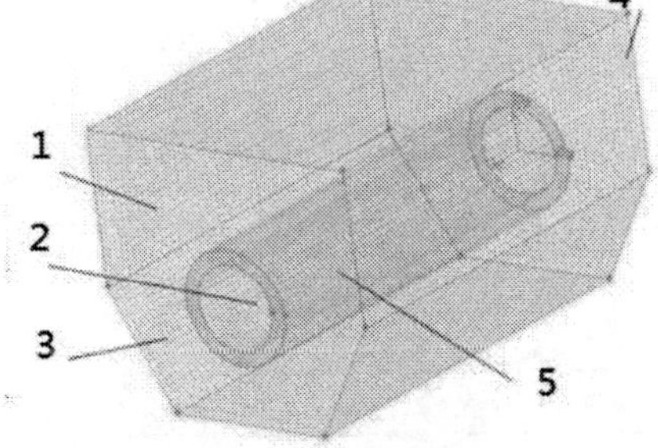

1-External flow field 2-Shaft 3-Inlet4-Outlet
5-Rotational flow field

Figure 1. Finite elementmodel

2.2 Determine the diameter of rotating basin

The external flow field can be limited to a certain range when calculating the rotational power consumption in the water, but the prerequisite is to determine the optimum diameter of the shaft section in the model.The external flow field can be limited to a certain range when the length and diameter of the shaft is rotated in the water. The friction torque values of different rotating basins can be obtained by the trial calculation, as shown in Table 1.

Table 1. Torquevalues under different rotating basin diameter

L (mm)	D (mm)	D_R (mm)	S(r/min)				
			50	80	110	150	200
			T(N·m)				
1200	240	360	0.4	1	1.6	2.7	4.5
		450	0.4	0.9	1.6	2.7	4.4
		600	0.4	1	1.6	2.8	4.5

Where, *L* is the length of the shaft in water, *D* is the diameter, $D_R$is the diameter ofrotating basin, *S* is the rotation speed of the shaft and *T* is the friction torque.

Obviously, there is no effect on the friction torque calculation of the rotating basin under different coaxial section diameter. According to the calculation formula of the power consumption and the friction torque, the power consumption is not related to the size of the rotating basin, so any diameter can be used in the FEM calculation.

2.3 Calculation results

The length of the shaft is 1500 and 1200 mm respectively. The theoretical friction torque and power consumption of the shaft in water at different rotational speeds (50r/min, 80r/min, 110r/min, 150r/min, 200r/min) are shown in Table 2.

3Friction torque measurement

3.1 Experimental device

According to the finite element model, we developed a ratio of 1:1 shaft in watering simulation experimental device which was installed on the comprehensive experimental platform for ship shafting.The friction torque of the shaft is measured in water and air respectively, and then the power consumption was calculated.The output end of the drive motor of the ship shafting integrated experiment platform is equipped with a torque meter to measure the output torque and power of the motor.Measuring the friction torque of the shaft in water, and then measuring the friction torque of the shaft exposed to air. Two experiments should be ensured that running, sealing, lubrication and other conditions unchanged. By comparing the results of two experiments, the power consumption to overcome the resistance of the shaft is obtained.

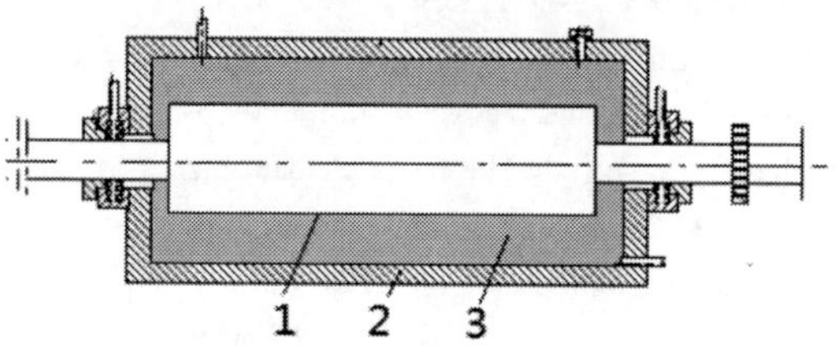

1-Shaft 2-Water tank 3-Water

Figure 2. Experimental device

3.2 Experimental Results

Power consumption under different conditions is shown in Table 3, which is the interpolationof the tail shaft power consumption in the water and in the air, that is, power consumed by the shaft to overcome the resistance of water.According to the experimental results, it is concluded that when the length and diameter of the shaft are constant, the power consumption of the shaft increases with the increase of the rotating speed.

Table2. Calculation results with FEM

L(mm)	D(mm)	S(r/min)				
		50	80	110	150	200
		P(W)				
1200	240	2.1	8.4	18	42.4	94
	310	5.2	19.3	46	107	230
	390	9.9	36.9	91	220	502
1500	240	2.6	8.4	20	50.3	111
	310	6.3	22.6	57	129	272
	390	13	41.9	115	267	607

Table3. Experimental results

L (mm)	D (mm)	S (r/min)				
		50	80	110	150	200
		P (W)				
1200	240	2.6	3.1	14.8	43.6	134.9
	310	9.5	36	55	91.2	230
	390	15.4	40	78	221	299.6
1500	240	5.5	30.4	70.6	229.4	281.4
	310	6.4	141.9	77	231.3	308.3
	390	20	46.2	85.3	228.4	390.7

DOI: 10.1051/matecconf/20166302033

4 Result and discussion

According to table 2, power consumption curvewith the change of the shaft diameter at the speed of 50r/min as shown in figure 3. With the shaft length increases, shaft in water power consumptionincreases; under the same shaft length and the same speed, with the increase of shaft diameter, power consumption increases too. According to table 2 and table 3, the theoretical calculation results and experimental results, the relative error remained at about 20% or less, in accordance with the requirements of engineering applications.

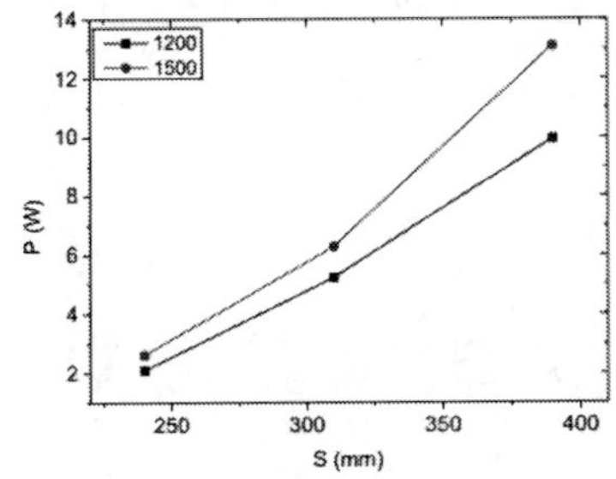

Figure 3. Power consumption varies with diameter

When the shaft length is 1500mm and the diameter is 390mm, the FEM calculation result is compared with the experimental results, as shown in Figure 4.Other comparison results are similar under different operating conditions, and here are no longer cited.

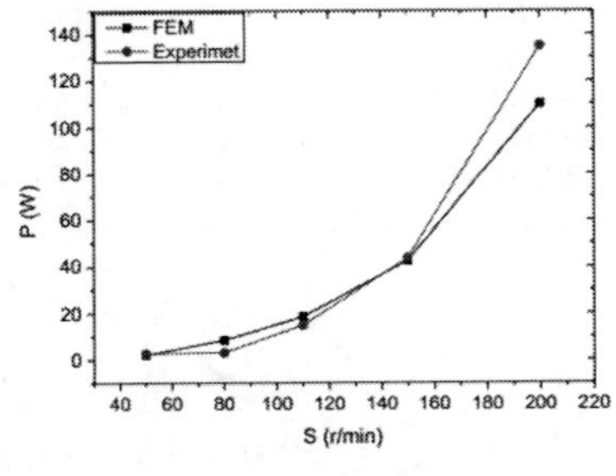

Figure 4. Comparison between experimentationand calculation

Easy to see the various experimental conditions calculated with the resultant power variation and trends basically consistent with the lower section of coaxial rotational speed increases, the corresponding increase in the power shaft section of the water by the rate will slow and fast.

5 Conclusion

The following conclusions can be drawn from this research: the experimental results verify the feasibility of the finite element model, and computational results by FLUENT also demonstrate the rationality of experiment. In finite element calculation, the author found that rotating area diameter size had no effect on the friction torque and power consumption, which can provide a theoretical reference for the other simulation. The power consumption of the shaft in water is increased with the increase of the length, diameter and rotational speed of the shaft.

Because of the gap between the experimental model and the actual ship, the experimental model is relatively small, the measured torque and power consumption are relatively small under different conditions. In the follow-up study the author will establish a larger experimental platform to get more experimental data to further explore the impact factor of the shaft in water drive efficiency and power consumption and describe them in a more specific numerical relationship.

Acknowledgment

This work was financially supported by the project of Natural Science Foundation of China (No. 51379168), Key Natural Science Foundation of China (No. 51139005).

References

1. Liang X X, Liu Z L, Liu J. , *Applied Mechanics & Materials*,"Analysis of Pre-Tightening Strength of Ship Tail Shaft Coupling Based on Workbench Software", 496-500:1028-1031(2014).

2. Yao Z S, Xue W P, Song H Q, et al. *Journal of Guangxi University*, "An experimental study on shift lining structure of frozen shaft in water-rich soft rock", (2014)

3. Zhang G C., *Ship Engineering*"Application of Energy-Feedback and Energy-saving Technology Based on SINAMICS in the Marine Shaft Generating", (2010)

4. Quan-Chao L I, Zhi-Heng L I., *Mechanical Engineer*"Research on Parametric Design for Marine Shaft" , (2013)

The Numerical Simulation Analysis of Hydro Forming of Hollow Crankshaft

Shi Gang WANG[1, 2,a], Shi Ming QU[2], and Shu Feng JIANG[2]

[1]*School of Electrical and Information Engineering, Guangxi University of Science and Technology, 545006 Liuzhou, China*
[2]*School of Mechatronics Engineering, Qiqihar University, 161006 Qiqihar, China*

Abstract. The hydro forming process of hollow crankshaft was numerically analyzed and simulated based on Dynaform. Then the influence to hydro forming process in different loading paths was studied and the match relations between optimum forming pressure and axial feeding were obtained. The forming result was ideal and the forming parts were qualified. Finally, to the same material, the change trend of pipe billet wall thickness under the same forming pressure and axial feeding was discussed.

1 Introduction

With the development of aerospace industry manufacturing, the space product is gradually moving towards lightweight and improving reliability. Crankshaft is one of the important parts of aircraft engine; it has a great influence on the overall performance of the aircraft engine whose performance directly affects the service life of the aircraft [1]. In hydro forming, hollow crankshaft is used to replace the original solid crankshaft in structure, which reduces the weight of materials and makes full use of the strength and stiffness of the material [2]. Tube hydro forming technology is an advanced plastic processing technology, which is developed to form variable cross-section hollow tubular components characterized by lightweight and integration. It is a perfect combination of lightweight structure and flexible processing technology; it is also of great significance to the development of manufacturing technology of automobile, aviation, aerospace, shipbuilding and other industries [3-4].

The traditional manufacturing process of the hollow variable cross section is first stamping into two half pieces then welding them to a whole, while the hydro forming is characterized in integral forming the hollow component whose section is variable along its axis [5-6]. Compared with stamping and welding process, hydro forming has the advantages of reducing quality, saving materials, reducing the number of parts and dies, reducing the cost of making dies, reducing the amount of machining and assembling, improving the strength and rigidity, reducing the production cost and so on [7].

In this paper, the aircraft engine hollow crankshaft was numerical simulated in finite element, the hollow crankshaft hydro forming factors that influence the performance were analyzed, the distribution of wall thickness under the different conditions of hollow crankshaft forming was obtained, which has important guiding significance to the actual production.

2 Part features and finite element model

Aircraft engine hollow crankshaft has complex geometrical shape of non symmetry, due to the large deformation of one side and no deformation in the other; it is difficult to keep the uniformity of wall thickness in the process of forming.

In this paper, PRO/E software was used to build the geometric model, the *.IGES file format was then converted to the former processor of DYNAFORM, a finite element analysis software. In hydro forming numerical simulation, the way of mesh was commonly used shell elements, but the four nodes BT shell element was used in meshing pipe. The numerical simulation model of hollow crankshaft, as shown in Figure 1, was including modelling, pipe billet, left and right punch, pipe billet material was selected stainless steel SS304.

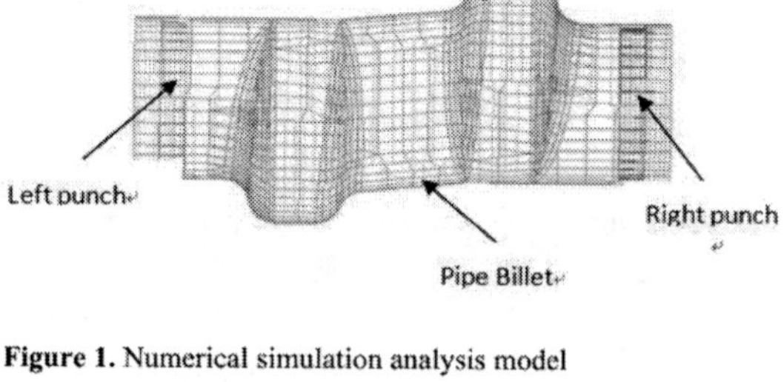

Figure 1. Numerical simulation analysis model

[a] Corresponding author: 446986768@qq.com

DOI: 10.1051/matecconf/20166302034

3 Simulation results analysis

3.1 The influence of different loading paths on the forming performance

The shape of hollow crankshaft parts were complex, the forming was difficult, in order to obtain forming part with good uniformity of wall thickness, reasonable match of the best forming pressure and axial feeding was required. The simulation scheme was shown in Figure 2, because it is easy to implement, the 2 fold linear loading paths and 3 trapezoidal loading path were used to simulate the actual forming, the maximum plastic pressure were 70 MPa, 80 MPa and 100 MPa. The forming limit diagram (FLD), the thickness distribution and the change trend of pipe billet wall thickness as shown in Figure 3- Figure 8.

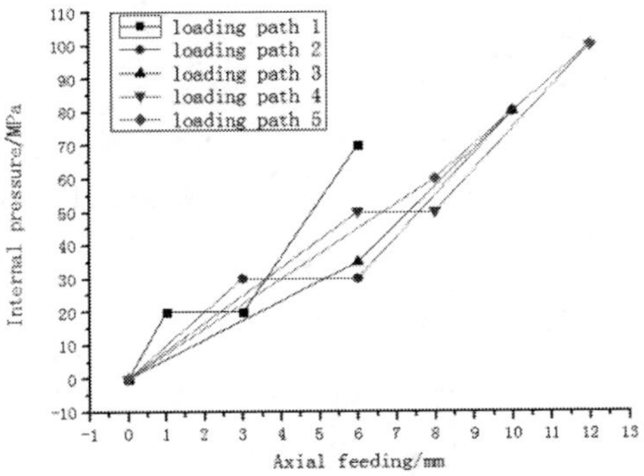

Figure 2. Five loading paths used in simulation

The internal pressure of loading path 1 was 70 MPa, the maximum axial feeding was 6 mm, and the minimum wall thickness after forming was 2.251 mm. But because the internal pressure was low and the forming of the curved part was not sufficient, the forming result is not obtained, the desired results have not got. The internal pressure and the axial feeding were the same in loading path 2 and the loading path 3, but the matching relation between the internal pressure and the axial feeding was different, and the forming results were ideal. As were illustrated in Figure 8 (The Trend of Wall Thickness Changing with Loading Paths) and Figure 4 (FLD and Thickness Distribution of Loading Path 2), the forming performance of the loading path 2 was optimal, the minimum wall thickness was 2.004 mm and the wall thickness reduction rate was 33%.

The inner pressure and the maximum feed rate of the loading path 4 and the loading path 5 are the same, but the matching relation between internal pressure and axial feeding is different. Loading paths 5 was better than loading paths 4 in formability, this was mainly due to the relationship between forming pressure and axial feeding in the early process of forming was not reasonable on loading paths 4. Too much pressure and lack of in-time fed-batch to metal curved part made the wall thicker and thicker.

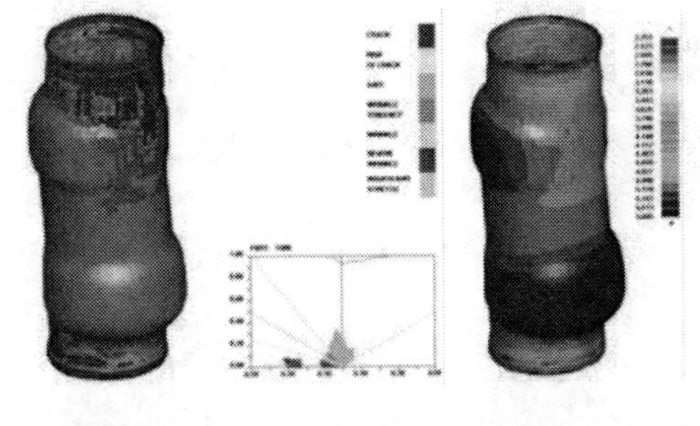

Figure 3. FLD and thickness distribution of loading path 1

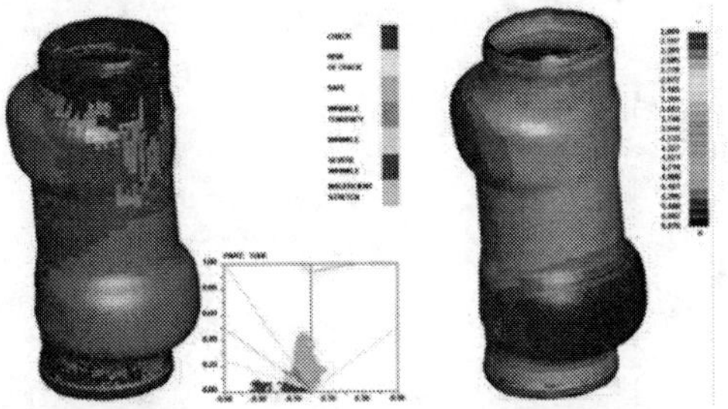

Figure 4. FLD and thickness distribution of loading path 2

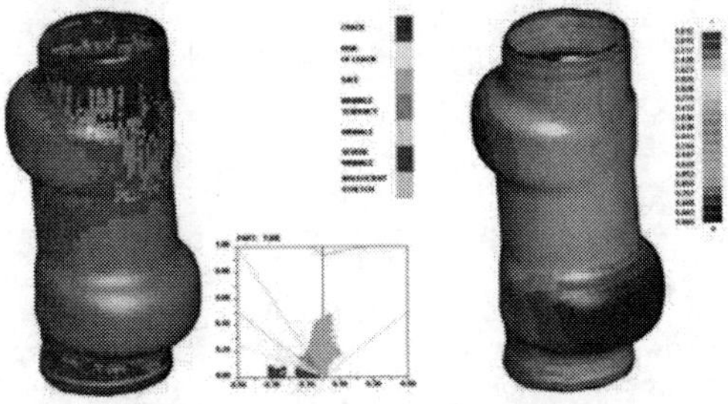

Figure 5. FLD and thickness distribution of loading path 3

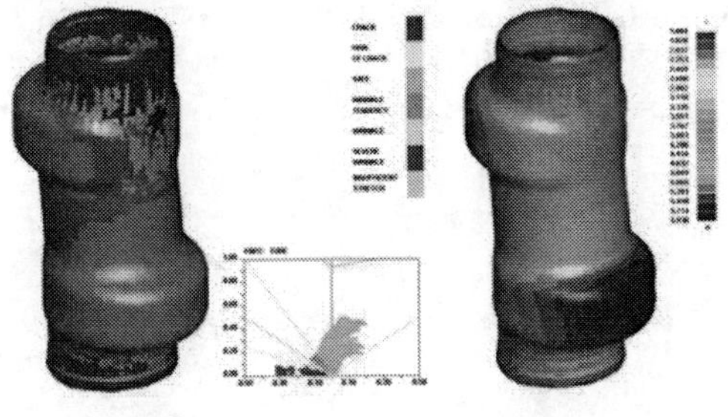

Figure 6. FLD and thickness distribution of loading path 4

 DOI: 10.1051/matecconf/20166302034

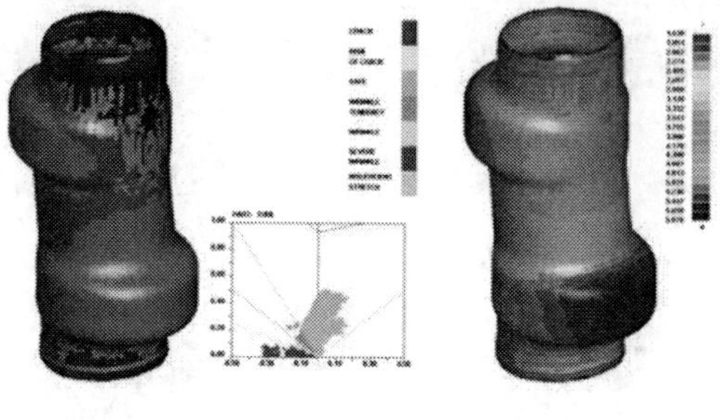

Figure 7. FLD and thickness distribution of loading path 5

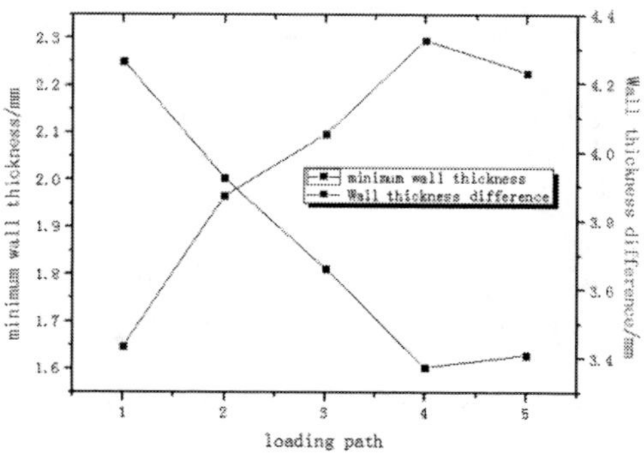

Figure 8. The trend of wall thickness changing with loading paths

3.2 The influence of pipe billet thickness on formability

The same material, though under the same inner pressure and axial feeding, had different forming properties of hollow crankshaft if the pipe billet thickness is different. In this paper, the thickness of tube is 2 mm, 2.5 mm, 3 mm, 3.5 mm and 4 mm, and the stainless steel SS304 was used as material. The forming results of pipe billet with different wall thickness were analyzed by the internal pressure and axial feeding in the loading path as shown in Figure 9.

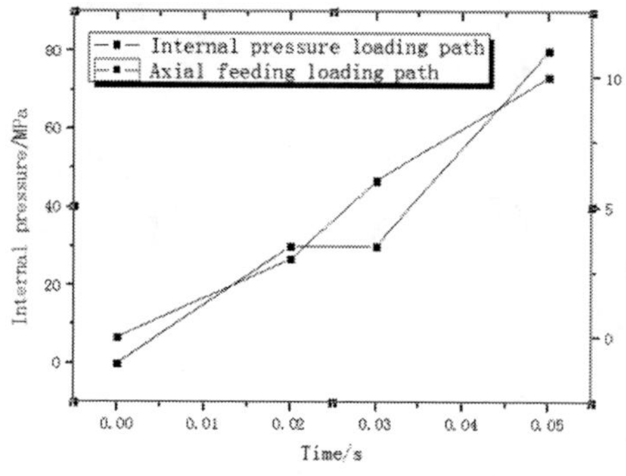

Figure 9. Pressure and axial feeding loading path

According to the parameters given in Figure 9, the wall thickness after forming changed with the thickness of the original pipe was shown in Figure 10. When the original thickness of the crankshaft is 2 mm and 2.5 mm, though the formed crankshaft was not in the state of rupture, the thickness of the crankshaft was too thin which caused the poor forming performance. When the original crankshaft thickness was 3 mm, the pipe was in the safe area, the wall thickness was uniform, and the forming performance was good. When the thickness of the original pipe was 3.5 mm and 4 mm, the pipe wrinkled and was not completely affixed to the mold, so the forming quality was poor. When the inner pressure was increased, the wrinkling of the pipe can be improved. This showed that at this time the internal pressure was not up to the distortion strength of the pipe, the pipe billet had the ability to continue plastic deformation.

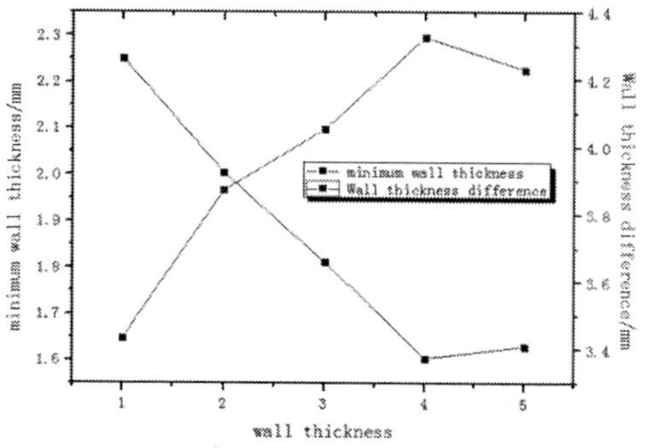

Figure 10. Wall thickness changed with the thickness of the original pipe

4 Conclusions

(1) The main influence factors of hollow crankshaft hydro forming are internal pressure and axial feeding. Through analyzing the forming results of five different loading paths, to get Optimum forming effect, the reasonable pressure in the hollow crankshaft should be 80 MPa, using trapezoidal loading path and the reasonable axial feeding should be 10 mm, using fold line loading paths. After the crankshaft hydro forming, the maximum wall thickness reduction ratio is in crank, the wall thickness increases gradually extending around, and the wall thickness distribution of the opposite is on the contrary. If the internal pressure and the axial feeding rate do not match, the internal pressure increases slowly while the axial feed rate is too fast which causes the pipe billet wrinkled. When the internal pressure is too low, the forming is not sufficient and the pipe billet could not be completely affixed to the mold. When the internal pressure rising rapidly and the axial feeding rate are slow, the pipe billet fracture appears due to excessive thickness reduction.

(2) Under the same hydro forming conditions, the thicker the pipe wall, the better the forming quality. But the pipe wall is thicker, the internal pressure is greater.

This work was supported by the Industry Project of Qiqihar City (Grant No. GYGG-201420) and Qiqihar University Graduate Innovation Research Project (Grant No. YJSCX2015-030X).

References

1. Sule Feng, Biyan Zhao, Yimin Luo, Gang Liu, Zhiqiang Luo. Manufacturing Technology 5 (2014)
2. Altan T, Jirathearanat S, Strano M. Proceedings of 2nd International Conference on Hydroforming, Fellbach, Germany (2001)
3. Bing Du, Changcai Zhao, Yijiang Liu, Guojiang Dong, Pipe hydro forming model. Chinese Journal of Mechanical Engineering 16 (2014)
4. He Yang, Luoxing Li, Qu Wang. Journal of Mechanical Engineering 12 (2010)
5. Shijian Yuan. Forging and Stamping Technology 1 (2004)
6. Junfeng Lin, Shijian Yuan. Journal of Plasticity Engineering 5 (2005)
7. Shijian Yuan, Zhubin He, Gang Liu, Xiaosong Wang, Han Cong. The Chinese Journal of Nonferrous Metals 10 (2011)

MATEC Web of Conferences **63**, 02035 (2016) DOI: 10.1051/matecconf/20166302035
MMME 2016

Discussion on Application of Intelligent Manufacturing Technology in SMT product assembly manufacturing system

De Jian ZHOU [1], Xiao Yong CHEN [2, 1,a]

[1]*School of Mechanical & Electrical Engineering, Guilin University of Electronic Technology, Guangxi Guilin 541004*
[2]*School of Mechanical Engineering, Xidian University, Shanxi Xi'an 710071*

Abstract. Intelligent manufacturing techniques has attracted much attention in SMT manufacturing system in recent years. A brief introduction to the basic idea of introducing intelligent manufacturing techniques to form SMT products intelligent manufacturing systems is described. Based on the analysis of SMT manufacturing system technical characteristics and problems, the basic composition and key technologies of SMT products intelligent manufacturing systems and technical system architecture are discussed.

1 INTRODUCTION

SMT (Surface Mount Technology) is the main technology of core component of modern electronics-PCB (Printed Circuit Board) circuit module assembly manufacturing. In SMT production line as the core component of SMT assembly manufacturing system (hereinafter referred to as SMT manufacturing system) is a typical modern electronic product manufacturing system, it's development influence and even determine the electronic product manufacturing ability. The research of SMT manufacturing system in our country started late. Such as SMT product assembly manufacturing process control technology, or assembly manufacturing system integration technology, has a large gap with foreign advanced technology. With modern electronics manufacturing automation and quality requirements for SMT manufacturing systems, some problems restricting SMT manufacturing technology are more and more highlighted. Based on the concept of intelligent manufacturing technology, combining with the characteristics of SMT products manufacturing, go on the research of SMT products intelligent manufacturing technology and intelligent manufacturing system, it is a breakthrough in SMT product manufacturing technology and it's manufacturing system development bottleneck.

2 SMT MANUFACTURING SYSTEM AND CHARACTERISTICS

2.1 SMT manufacturing system

SMT manufacturing system is mainly in the form of surface mount, including instrumentation mixed assembly forms such as PCB level electronic circuit assembly production system, the essence of surface mount circuit module (SMA) is workshop assembly manufacturing production systems for particular product. It has the characteristics of general assembly manufacturing system and production system, at the same time, the particularity of modern electronic products manufacturing system.

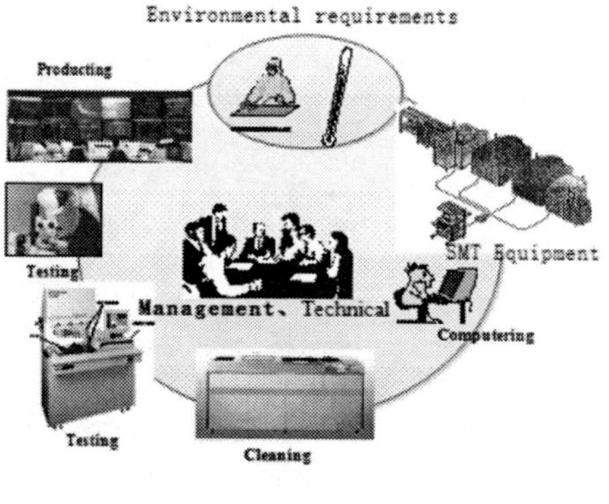

Figure 1. The basic composition of SMT

Its basic composition content is shown in figure 1 [1]. (1) People and the environment, system and equipment managers, operators, and the man-machine environment etc; (2) Computer hardware, software, main control computer and its peripheral equipment, operating systems, software, production management system software, etc; (3) Surface assembly production lines and equipment: solder paste printing machine, plastic welding (adhesive), SMC/SMD stick glue machine, reflow welding/curing furnace, etc. (4) Cleaning, testing equipment, cleaning machines, optical detector, needle bed type tester, flying probe tester, etc. (5) Auxiliary equipment: unload

[a] Xiao Yong CHEN:cxy160@gmail.com

machine, load machine, plate turnover machine, turning machine, transmission lines, SMC/SMD material library, and power-driven equipment, etc. (6) Others: SMT manufacturing system for surface mount and instrumentation mixed assembly product, its basic components in addition to the above content, the main content of the need to increase the wave crest welding equipment and automatic instrumentation equipment (or manual assembly line), etc.

There are many means of SMT manufacturing system, even correspond to the same type of SMT manufacturing system, according to the different requirements such as SMC/SMD type and quantity of assembly product, and the variety, quantity, quality and production efficiency of assembly product, also can have different equipment and system forms. With the change of forms, the biggest change of equipment adjustment is the core of system components, SMT production line, There are single - side assembly, double-sided assembly and other assembly methods, as well as single - line composition, two - line composition, etc.

2.2 The technical properties of SMT manufacturing system

The domestic using SMT manufacturing system have a certain level of computer integrated manufacturing and digital manufacturing technology foundation, especially as a core component of SMT production line, which has its relative independence in the system, is a self-contained computer integrated flexible manufacturing subsystem, already has had a high degree of digital manufacturing technology level. At the same time, compared with the general industrial product assembly manufacturing system, SMT manufacturing systems have some inherent characteristics, mainly in the following.

(1) System of complicated: Systems have a high degree of automation of SMT production line, and need to artificial auxiliary instrumentation or inspection, repair welding assembly line; In the production process is often embedded in the assembly line automatic up-down material and fully automated links such as automatic detection, and there was a need to artificial auxiliary antenatal preparation and semi-automatic link such as product testing at the end; The equipment of system both have high speed and high precision SMT machine automatic high-end CNC equipment, also have need of artificial auxiliary half automation equipment cleaning and inspection. System units of basic automation level, technology, and digital manufacturing with big difference, difficult to achieve overall computer integrated manufacturing system.

(2) Variety of products and their raw materials: SMT products consist of different specifications of PCB and electronic components, connectors, its breed type is quite rich. And the solder paste, flux used in assembly process have varieties of raw material types. Also, the raw materials dynamically change on different products and their technical requirements, most of them are with tiny body, complicated structure, various shapes, and electronic products material are easy to deform, oxidized and deteriorate, its storage, furnish and logistics management requirements are difficult.

(3) Factors influencing the product assembly quality complex: SMT has already developed to mount speed 150000 CPH (actual mount beat 0.024 seconds per point) of the high-speed assembly [2], device pin spacing of 0.3 mm limit spacing micro-package, solder joint density per square centimetre dozens or even hundreds of points of high density assembly, and lead-free welding technology stage. Many factors influence on the quality of product assembly and welding, assembly equipment running speed, accuracy and stability performance indicators, raw material quality status of electronic components, solder paste, flux and so on, as well as the production environment, temperature and humidity conditions, the stability of production power supply, resistance to electromagnetic interference will be a comprehensive impact on product assembly quality. The control process of product assembly quality is complex, and difficult to control in real time.

2.3 The main technical problems of the SMT manufacturing system

Due to the above characteristics, although the computer integrated automation level of SMT production lines has reached a considerable height up to now, there are some problems in the process of production, such as induction to the environment, lack of adaptive capacity, weak ability of product quality in real time measurement and control and low logistics control and management automation. For the whole of SMT manufacturing system, due to the influence of traits including the diversity of the system form and production mode, the diversity of the variety of product type and material, the complexity of process control parameters and information processing, its degree of automation and computer integrated level has obvious gap comparing to other advanced manufacturing system in the field of machinery manufacturing. The problems mainly displays in the following respects:

(1) Material management problems in SMT manufacturing system. The SMT products raw material types are rich. It is necessary to make automatic in material warehousing, sorting and distribution in order to make sure that it is in time, rapid, accurate and low cost. However, the model of the finding of product materials, storage and sorting are still behind. Most of them are performed by hand. Some problems such as high human cost, low efficiency, easy error, easily producing dull materials, large occupied space of stock, the waste of headroom, trouble of inventory, many customer stored materials and low customer satisfaction are universal in the process of material management.

(2) Quality control problems in the process of SMT production. The assembly of SMT products have high quality requirements, it mainly reflects in two aspects of the assembly precision and solder joint quality. In order to guarantee product quality, it needs to effectively adjust and control parameters of the assembly equipment in real time in the process of SMT production. In the current

control mode, in addition to the automatic adjustment function of optical detection counterpoint of SMT assembly device, setup and adjustment of critical process parameters including solder paste printing pressure and speed, welding temperature and PCB transmission speed mainly depends on the experience of technicians and technical level. Its control of the quality of the product process automatic control ability is weak. For assembly product quality problems occurring in the process of manufacturing, it basically relies on manual analysis to find the reasons but it can't real-time automatically detect the fault source and modify process parameters and it lacks scientific statistics analysis, optimization and the capacity of adaptive control of product quality.

(3) Perception and adaptive problems of SMT manufacturing environment and equipment working status. In the process of SMT production, some quality problems are caused due to some elements such as property changes of solder paste and glue etc. the oxidation of bonding pad, pins and other welding parts and dynamic changes of motion accuracy in key equipment's such as printing machine and SMT. These elements which belong to perception and the adaptive control problem of the manufacturing environment and working condition of equipment involved include air, humidity, temperature, power quality, and many other content. For the content, the current SMT manufacturing system also has certain standardization, standardization requirements but the monitoring of manual control which is based on the help of instrument, instrument testing. It cannot automatically perceive the dynamic changes of the manufacturing environment and working condition of equipment so that it cannot do the adaptive control and adjust accordingly.

(4) Comprehensive coordination problems in SMT manufacturing system. With the personalized development trend of electronic products, it requires rapid response ability, ability of cost adaption when producing many varieties and small batch production in SMT manufacturing system. To have these abilities, it involves the rapid response of the market, equipment and resources utilization, production process of high efficiency and high quality control, system management and comprehensive coordination, and many other factors, so it is a complicated system engineering. Current SMT manufacturing system in this ability is weak, the manufacturing process of the supply chain mode, load saturation, each manufacturing unit load balance scheduling and the coordinated ability with the market and so on still restrict the improvement of manufacturing efficiency ,so that it cannot furthest meet market and customers' needs of fast response, low cost manufacturing. Improving the comprehensive coordination of SMT manufacturing system and reducing the production cycle and the manufacturing cost are important tasks to SMT products manufacturing enterprises.

Based on the existing technology in SMT manufacturing, using the concept of intelligent manufacturing technology and its related technology to establish an efficient, reasonable, agile and resource sharing new mode of SMT intelligent manufacturing is an effective way of breaking through the problem and promoting SMT manufacturing ability and level.

3 SMT intelligent manufacturing system and its key technologies

3.1 SMT intelligent manufacturing system

Intelligent manufacturing is merged by manufacturing technology, digital technology, intelligent technology and a new generation of information technology. It is a manufacturing model for the whole lifecycle of products, which has the function of information awareness, optimizing decision and execution control and is aimed to manufacture products and service users efficiently, high-quality, cleanly and safely. Intelligent manufacturing includes intelligent manufacturing technology, intelligent manufacture equipment, intelligent manufacturing system and intelligent manufacturing service, etc. It derives from a variety of intelligent manufacturing products. Figure 2 provides a system of intelligent manufacturing technology and application [3].

SMT intelligent manufacturing system is an intelligent manufacturing system of electronic products, which is formed in the process of application in SMT intelligent manufacturing system, and designs SMT products as a particular object. It is a new kind of SMT product manufacturing system which is based on the basis of SMT intelligent manufacturing system, digital manufacturing technology and computer integrated manufacturing technology, etc. It is the result that the traditional electronic products manufacturing technology, the artificial intelligence science and the computer technology integrates with science, the network technology and the information management technology. Meanwhile it integrates into the production environment perception and self-adaption, intelligent production quality test and control, intelligent storage and allocating the material system and internal and external intelligent coordination and collaboration in the system.

SMT intelligent manufacturing system is the general term that finishes all production activities in the period of the whole lifecycle in SMT product digital manufacturing. It is a process of digital manufacturing, which is for the purpose of customer demand, by means of objective material means and uses intelligent manufacturing technology, and using an effective method which has the intelligent manufacturing technology characteristic to make products transform from conceptual design into final material products and put them on the market. It includes a series of interconnected activities within the whole lifecycle of products such as market research and forecasting, product design, process design, production and processing, quality assurance, production process management, marketing and after-sale. It has the common technology in a general intelligent manufacturing system. Its basic technical system and application can be seen in figure 2, and the main supporting technical system can be seen in figure 3.

DOI: 10.1051/matecconf/20166302035

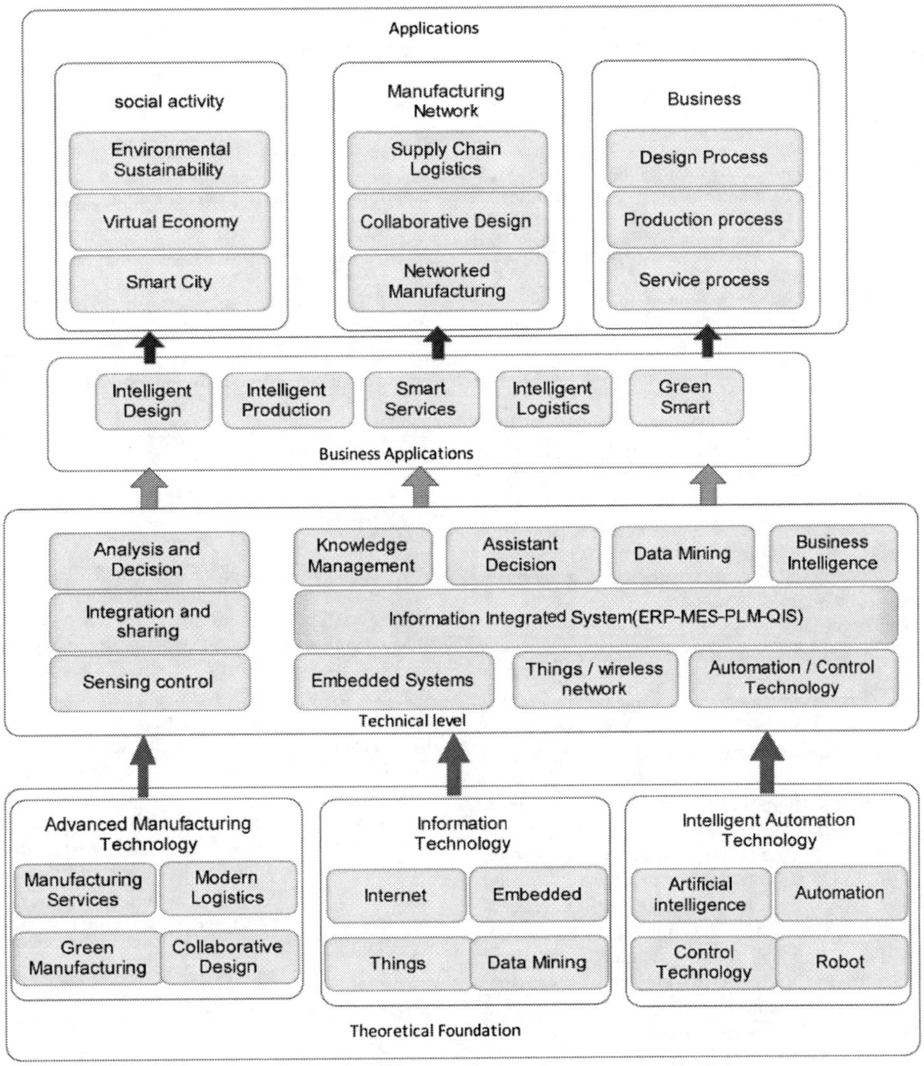

Figure 2. Intelligent Manufacturing System and Application of Technology

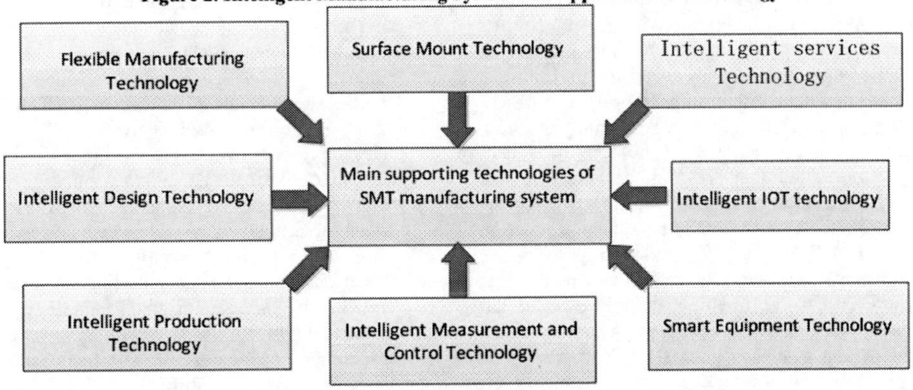

Figure 3. The Supporting technologies of **SMT Intelligent Manufacturing Systems**

3.2 The Key Techniques of SMT Intelligent Manufacturing System

The key technologies of SMT intelligent manufacturing systems required research priorities include the following aspects.

(1) Smart warehousing and distribution technology of SMT product material system. SMT Smart Material storage system is a kind Smart system with high integration of information technology and automation and control technology, and deeply optimizing on existing materials warehouse management mode of SMT industry. The intelligent material storage technology, SMT material access and sorting technology based on intelligent warehouse, stereoscopic warehouse and space storage technology required studied. It is based on network technology, radio frequency identification (RFID) and smart storage system, First of all, realize the intelligent storage of SMT product materials, then realize AutoFilter qualified materials according to SMT product BOM table with intelligent manipulator, after that study how to achieve SMT automatic load and unload materials by intelligent material handling device, and finally achieve market docking logistics management.

(2) SMT product assembly quality automatic detection and intelligent identification technology. In the SMT product manufacturing process, it generally relies on online test and complementally on manual inspection after critical process for assembling quality testing and statistical analysis. This traditional back-end process detection, measurement and control method of statistical analysis is difficult to form real-time quality feedback information, to achieve real-time quality control and adapt to modern electronic product variety, variable-batch, high reliability production requirements. The application of intelligent technology and SMT production process intelligence detection and real-time quality control technology based on optical or magnetic detection technology, so as to realized automatic real-time detection of product quality, intelligence analysis and production process data real-time control.

(3) IntelliSense and adaptive technology of SMT manufacturing environment. In the process of Micro-assembly and fine pitch assembly of the SMT product, which are sensitive to production environment, required to have the adaptive adjustment ability. IntelliSense and adaptive technology of SMT manufacturing environment means that the production environmental quality of SMT manufacturing system can be with real-time detection, analysis and automatic adjustment by use of sensor technology and intelligent control technology, for eliminating the dynamic changes of factors affecting the quality of products and ensuring the production environment stability in good condition, so as to ensure the stability of product quality. Research on this aspects, the first thing is to solve IntelliSense and self-adaption issue of climate change in manufacturing systems caused by external factors, and then gradual in-depth study the IntelliSense real-time and self-adaption problems on dynamic changes of the key equipment's system performance. Thereby, realize self-diagnosis and adaptive control in the manufacturing system's environment and production equipment.

(4) Research on Comprehensive information management system of SMT products intelligent manufacturing based on the Internet of things (IOT). Due to the lack of effective means information exchange means between management and automation of the production in traditional manufacturing workshop, poor control of the production process, some defects of efficiency and real-time in artificial data acquisition, so it has been difficult to meet the needs of intelligent information management systems. Therefore, means of data collection such as sensor, auto identification are necessary in enhancing the defects of advanced data capture tools and management systems should be applied in workshop to improve the disadvantage above, in order to achieve intelligent manufacturing informatization and modern management.

4 Conclusion

SMT manufacturing system is a complex manufacturing system with electronic product manufacturing characteristics, and there are many contents need special study in the process of technology advances towards system integration, digital manufacturing, intelligent manufacturing, which the investigative content discussed and the proposed in this article is just one part. There are many technical issues urgently to be solved for SMT Product Manufacturing. It needs the colleagues in the industry to work together. Meanwhile, the operation of intelligent manufacturing technology in the SMT manufacturing systems should also be scientific planning, step by step, gradually deepened based on the basic conditions of specific circumstances, integrated manufacturing and digital manufacture in corporation's existing manufacturing system.

References

1. Zhou Dejian. Electronics manufacturing electrical interconnect technology [M], Electronics Industry Publishing House, China, 2010.3 [A book reference in Chinese]
2. Zhang Qiang. SMT equipment manufacturing industry and the current situation facing the bottleneck analysis of China——NEPCON CHINA 2014 Exhibition Research Report. China Economic Net.http://intl.ce.cn/specials/zxgjzh/201405/19/t20140519_2835682.html [In Chinese]
3. Guo Wei. The development trend of intelligent manufacturing. http://articles.e-works.net.cn/amtoverview/article 108404.Htm
4. Shan Zhongde. The status and future development of Intelligent Manufacturing Technology. Advanced electrical interconnect technology seminar--China Machinery Manufacturing Technology Association Electronic Branch Conference Proceedings. 2014.10 [In Chinese]

MATEC Web of Conferences **63**, 02036 (2016) DOI: 10.1051/matecconf/20166302036
MMME 2016

The Research on Hidden Failures of Civil Aircraft

Lie Shu LIN , Shuang Xue FU [a]

Guangzhou Civil Aviation College, Guangzhou, Guangdong, 510403, China

Abstract: The hidden failure is often appear on civil aircraft, and it is a big risk for normal safe operation. The reason for this failure is the instability of the resistance of the electrical contact. The study on electrical contact is applied science of the reliability, the contact resistance is an important parameter of the reliability of the electrical contact system. In this paper, It also Produces the latest research results from factors that affect contact resistance and contact resistance microscopic model, and then it produces the method of the calculating and measuring engineering about the contact resistance. Lastly, it is point out the future direction of the electrical contacts.

Key words: Aircraft maintenance; Electrical contact; Contact resistance; Reliability.

1 INTRODUCTION

Hidden failures in the aircraft airborne equipment occur, are as the following common scenarios: (1) The operation of the airborne equipment in the air is not normal, while check everything is normal on the ground when power is on; (2) Equipment' operation is well when the power is on , and some time is over, failure occur suddenly, and then disappear; (3) the equipment occasionally did not work properly in the air;(4) In several successive days of flight, the equipment is working properly, and in a certain program instantaneous flight irregularities occur; (5) The equipment in a particular state is not normal randomness. Factors produced such failures are roughly divided into three categories:(a). Failure of the equipment itself;(b). improper influence of external conditions; (c) and / or using. Among them, the failure of the equipment itself is the main reason causing hidden failure of the aircraft, and its essence is the point of electrical connection in equipment , the characteristics of the contact resistance value is unstable.

Electrical connections are essential to aircraft systems and there are a lot of links, and electrical contact is the key of electrical connection [1]. Electrical connection can be divided into strong and weak categories .The main function of strong category is the transmission of energy, and electric arc and thermal effect are the main factors affecting the life and reliability of this connection; The main function of weak category is signal transmission, the environment is one of the main factors to the reliability and failure of this electrical connections[2]. Quality and level electrical connector had an important impact on the reliability of the equipment and systems. The electrical contact belong to multi- science. it use physical, mechanical, electrical materials, chemical, environmental concepts and theories in reliability issues of electrical , find electrical contact theories. Electrical contact is a common phenomenon in which aircraft systems exist .Between the elements of these systems, even between systems of the whole aircraft require electrical connections. If the contact point is in failure, it would affect the entire system reliable operation. This leads to issue about reliability, stability, lifetime of the associated electrical contact. While the contact resistance is a basic and important parameter of electrically connection. This parameter is closely related with the electrical contact material parameter, structure, manufacturing processes, the working environment .Thus, reliability issues caused by the contact resistance of the electrical connection is getting attention by researchers, and achieved some research results. This paper analyzes the affect of an unstable electrical contact resistance, and introduces its micro models for calculating. It has some significance about some guidance to trouble shoot/isolation of Failure in which aircraft [3].

[a] Corresponding author: 25272505@qq.com

2 THE CONTACT RESISTOR'S MAJOR FACTORS OF INSTABILITY

It is important to get stable contact resistance in applications, in order to ensure the reliability of the electrical contact. Factors Affecting the contact resistance value, and the main properties of the conductor material are as following: in the form of the contact, the contact pressure, temperature, corrosion, frequency, and etc.

2.1 Properties of Materials

The field for Metal material about electrically contacting is very wide. They are classified by the use, as copper, aluminum, and copper-aluminum alloy. They are used in a large current (ferroelectric) contacts in the field, while precious metals and their alloys are mostly used in small current (weak) contact application. Pure precious metal is mainly used as a coating material. Metallic material properties of constituting the electrical contact directly affect the volume of the contact resistance. these properties are the mainly resistivity of the material ρ、Material hardness H, etc. Shrink resistance volume of each conductive spot is proportional to ρ. The spots number on contact surface of the contact material are decided by material hardness . According to the definition of the hardness H of the contact material, Relationship between the small contact area of contact surface with the contact pressure F is as following :

$$F = \xi H \pi a^2 \quad (1)$$

Wherein, ξ is a constant, and less than 1; a is an radius of the small contact surface.

2 .2 Contact Form

An electrical contact form can be divided into three categories: point contact, line contact and surface contact. The number of contact points mainly affect a shrink resistor R_s . In general, the contact points (n) on the contact surface are at Maximum, the shrinkage resistance ($R_s = \rho / 2na$) will be minimum, in contrast, the contact points (n) on the contact surface are at minimum, the shrinkage resistance R_s will be maximum. The shrinkage resistance R_s volume on the line contact is in between the above two volume. The film resistors' volume is mainly affected by contact pressure at each point of the contact surface. If the pressure F is applied on the contacts , the number of contact points is n, then the pressure at each point of contact on, $F_1 = F / n$, thus, that each pressure on the point of contact is related with the contact form . In general, when the number of the contact points (n) of point contact is at minimum, and under the same contact pressure F is applied, so each contact point in point contact is in maximum pressure. this is the most likely to damage the contact surface of the contact film ,so that the resistance of film is reduced. In contrast, the number of the contact points (n) in surface contact is the maximum ,the ability to damage the surface film is small, the film resistance increases. The resistance volume on the line contact is between the point contact and surface contact. Moreover, the contact resistance is also related to the contact pressure. when the contact pressure is in small, the contact resistance volume of the contact surface is not sure smaller than a point or line contact.The data in Table 1 illustrates this point.

Tablel. The contact resistance of the contact form of relationship –(copper) [4]

Contact form	$R_j(u\Omega)$	
	F=9.807(N)	*F=980.7(N)*
Point contact	*230*	23 R_j
Line contact	330	15
Surface contact	*1900*	1

The vast majority of the pins and jacks installed on the aircraft are made of gold-plated for conduction and protection, their coating is worn after theyare frequently removed and installed , the contact resistance (R_j) will been changing . These will Cause circuits' failure. Aircraft manufacturers such as Boeing report : "SUNDSTRAND company show that: EICAS indicator frequently show "IDG TEMP" phenomenon on Beoing 757,767 display, they are proved that IDG nailing on the joint changed , and led to resulting coating worn [5] .

2.3 Contact pressure

We could see from Table 1, the contact pressure F has an important effect on the contact resistance. If the pressure is not enough , the resistance do not make a Significantly reduce only by increasing the contact surface dimensions When the contact pressure F increased enough, material's compression would be beyond the elastic deformation limits, and produced plastic deformation, then the contact surface would increase, the number of contact point would also increase,($R_s = \rho / 2na$) would reduced. At the same time, the contact pressure increased can crush the metal film surface, lead to the film resistance reduce. Therefore, increasing the contact pressure, the contact resistance can be reduce. Some parts installed in the joints are subjected to vibration, such as a aircraft wing or an engine joint section, they are frequently subject to vibration for a long time, resulting in looseness between the receptacles, the contact pressure changes, the contact resistance (R_j) will change. Therefore, when the current through the joints , They can produce a heat or even an arc, lead to wholly or partly pins burn, contact resistance increased further. For airborne precision instruments, their indication would result in a blank or swinging.

2.4 Temperature

As the contact point temperature raise, the resistivity of the metal will increase, while the hardness of the material will be reduce, so that the effective contact area will increase, and the number of the contact points will increase. The former made volume of resistance increase, the latter reduce. Both mutual compensation, the contact resistance variation is minima. However, if the current at the contact are more than the short-circuit current, the temperature would raise sharply, the contact surface would oxidize, the growth and thickening of the oxide film, the film resistance would increase, produce more heat and More serious consequences. In order to maintain a stable contact resistance, electrical contact with predetermined temperature is in very low for long time. Joints installed on aircraft in external environment (such as joints on the engine electrical components), due to the drastic changes in temperature, joints in the middle of the rubber pad are deformed , although the joint external connection is good, but no virtually and fully contact between the jacks or loosen, the system do not operate and display properly. Such as:CSD oil temperature indicator indicated a larger swing in China Southern Airlines B-2926 aircraft recently ,the reason is that the joint is un-completely contact or loosed in the middle of the rubber pad by sudden temperature changing lead to aging and deformation[6] .

2.5 Corrosion

The external environment around electrical contact also affects the electrical contact. Where in the contact corrosion can seriously affect the electrical contact resistance, and the reliability of the electrical contact. There are three types of the electrical contact corrosion or contamination : pore corrosion, creep corrosion and tarnishing. These three types affect the reliability of electrical contact. Pore Corrosion is a chemical or electrochemical reaction between air pollutants and base metal, it occurres through pores and other weaknesses of the coating surface. Insulating film and the corrosion products formed by the reaction can lead to unstable contact, accelerate friction, the contact resistance increase and so, eventually lead to the electrical contact failure. Copper such as other base metals and precious metals such as gold plating creep corrosion will react after bonding. The characteristics of the reaction process are obviously distribution matrix material in the surface of the gold plating creep corrosion products. Some trends in the metal sulfide generation exceed the oxides trend, the gaseous sulfur-containing substances around coating occurred chemical reaction on the surface of these metals, and produce diffusion phenomena, resulting in metal tarnish. The tarnish will affect electrical contact reliability.

Abbott, Antler and other researchers had extensively studied tarnish phenomenon about effect electrical connectors.

2.6 Signal frequency[7]

When the alternating current gone through the contact point, the current only gone through a conductor surface, due to the skin effect . it is more obviously at high frequencies. Skin depth δ could be calculated as:

$$\delta = \sqrt{\frac{\rho}{\pi f \mu_0}} \tag{2}$$

Wherein, ρ is the conductor resistivity; f is the frequency of the AC singnal; μ_0 is vacuum magnetic conductivity

Figure 1 showed the top contact between two spherical of the same configuration with conductor having the same radius of cylindrical.

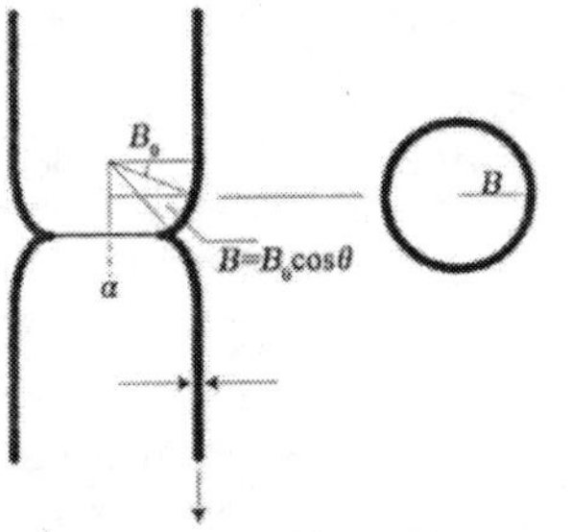

Figure 1. Skin Effect and contraction resistance of the substrate

Assuming skin effect, the effective current only gone through the thin shell in the defined depth of penetration, and assuming the thin shell's size smaller than the radius of the contact pattern, Marucci gave the formula for calculating the extra resistance due to the skin effect.

Wherein, ρ is the conductor resistivity; δ is the skin depth; a is a minimum radius of two conductors Correspondingly cylinders; B_0 is the maximum radius with contact cylinder.

3 ADVANCES CONTACT RESISTANCE MODEL

3.1 Holm electrical contact model [8]

Holm suggested that the total contact resistance should be a parallel value of all actual point of contact resistance (called self-resistance) and the resistance value caused by the interaction (called mutual resistance) in series, namely:

$$R_j = R_s + R_i = \frac{\rho}{2\sum_{i=1}^{n} a_i} + \frac{\rho}{2a} = \frac{\rho}{2na} + \frac{\rho}{2a} \tag{3}$$

In the formula, R_j is the total contact resistance, a is Holm radius which called mutual resistance or point set radius. Later, Mr. S .Timoshenko [9] estimated from prosperity by a dense point cloud, $R_j = \frac{16\rho}{3\pi^2 a}$, while $\frac{16\rho}{3\pi^2} = 0.5404 \approx \frac{1}{2}$ .The result is consistent with Holm equation.

3.2 Green Wood electrical contact model [10]

Green Wood gave the formula for calculating the contact resistance if that the current gone through the contact point between the circle model :

$$R_{G1} = \frac{\rho}{2\sum a_i} + \frac{\rho}{\pi}\sum\nolimits_{i \neq j}\sum \frac{a_i a_j}{d_{ij}} / \sum a_j^2 = R_{par} + R_{\text{int}} \tag{4}$$

Where, ρ is the resistivity, a_i is the i-th point of the radius, d_{ij} is distance between the i- point and the j-th

 DOI: 10.1051/matecconf/20166302036

point of circles, R_{par} is the parallel resistance of the dependent point of circle, and R_{int} is resistance of the each point between the interaction circles, If it is assumed that all circles are the same size, then the above equation can be approximated as:

$$R_{G2} = \frac{\rho}{2\sum a_i} + \frac{\rho}{\pi . n}\sum_i \sum_{i \neq j} \frac{1}{d_{ij}} \quad (5)$$

In 2001, Lionel Boyer [11] Promoted the use of the formula given by Greenwood through adding a shape factor S (N), such that the formula given by Greenwood could also be used to calculate the contact resistance to other conductive spots, Wherein the square form factor, $S(4) = (2 + \frac{1}{\sqrt{2}}) / \pi = 0.8617$,is very close to the 0.868 about Nakamura's. While, the triangular shape factor, $T(4) = 3(1 + \sqrt{3}) / 2\pi = 1.3$, is close R_{TN} ($R_{TN} = 1.27\rho / L$) in the reference, Nakamura[12].

3 .3 The Electrical contact for surface film

The electrical contact surface is covered with a poorly conductive material due to various factors, such as metal oxides, sulfides, dust, dirt, or sandwiched between the contact surfaces of the oil film, water film, etc. In general, the more oxide film cover with electrical contact surface. This oxide film is mostly semiconductor, and its resistivity is high. Because of the conductive film layer on the contact surface , while shrink resistance is advantages or disadvantages , depending on the resistivity between the film and the substrate The film can also change the hardness of the contact surface, and the deformation of the contact surface can change the contact resistance when the contact surface of film is thiner enough, and the film strength of the electric field will less than $10^8 \sim 10^9 v/m$,the resistivity of the tune of film from the Schrodinger equation:

$$\sigma = \frac{U}{J} \quad (6)$$

Wherein, U is the voltage between the contact surface; J is the current density flowing through the membrane. The electrons pass through the membrane resistance spot radius of a surface is encountered:

$$R_b = \frac{U}{I} = \frac{U}{J\pi a^2} = \frac{\sigma}{\pi a^2} \quad (7)$$

If there are n conductive spots, the mean radius of the inner contact surface is a_p , then the total resistance of the film is as following:

$$R_b = \frac{\sigma}{\pi a^2 a_p^2} \quad (8)$$

4 CONTACT RESISTANCE CALCULATING

Contact resistance consisted of two parts. The value of the total resistance is shrinkage resistance (R_S)plus resistance of the surface film ($R_b : R_j = R_s + R_b$).However, due to (n) the number of spots and the average radius are very difficult, so In general to calculate the contact resistance for engineering is as the following empirical formula:

$$R_j = \frac{k_j}{0.192F^m} \quad (9)$$

Where, R_j is contact resistance; F is the contact pressure; m is the actual index relating to the contact form, pressure and number of points of contact (proved, if the pressure is not too large, for point contact with $m = 0.5$;line contact with $m = 0.5 \sim 0.8$ (generally the $m = 0.7$);surface contact with $m = 1$; k_j is the relevant factor for the contact material, surface condition, etc. Whose value is determined by the experiment. It is shown in Table 2.

Table 2. Values of k_j for various contact materials

Contact materials	Surface condition	k_j
$A_g - A_g$	unoxidized	0.06
$Al - Cu$	unoxidized	0.98
$Cu - Cu$	unoxidized	0.08~0.14
$Al - brass$	unoxidized	1.90
$A_gC_dO(12) - A_gC_dO(12)$	unoxidized	0.17
$A_gC_dO(12) - A_gC_dO(12)$	oxidized	0.35

It must be note, there is significant limitations of this formula, we can not generalize the influence of various factors on the contact resistance. Contact resistance are frequently measured by measuring the contact voltage in application.

Figure 2 show a circuit diagram for measuring contact resistance. The R_j is contact resistance of electrical connector, V_m is electrical contact voltage measured, V_{in} is the supply voltage, the contact resistance can be calculated as:

$$R_j = \frac{V_m R}{V_{in} - V_m} \quad (10)$$

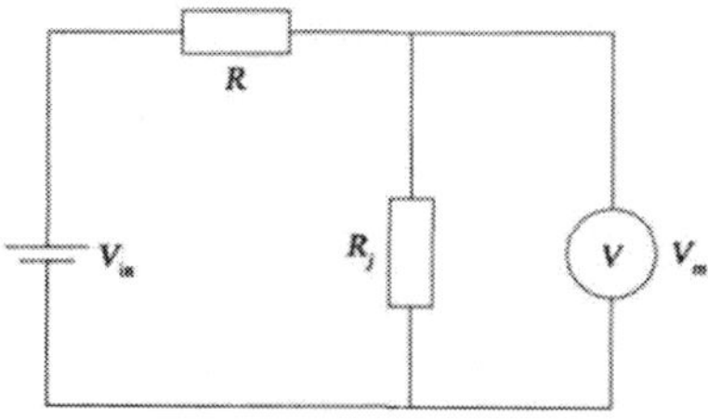

Figure 2. Showed a circuit diagram for measuring contact resistance

5 SUMMARY

There are a lots of high frequency signals used in the various aircraft systems, particularly in electronic systems. The study is not sufficient about the impact of the high frequency to the contact resistance, particularly in the case of multiple points of contact. As frequency increased, the contact resistance with only a parameter which is difficult to describe the good electrical contact of the model, then you need to study the electrical impedance model of the contact, including contact with capacitors and inductors. Although the research in this area has been reported, but it still need further studying.

Further, the contact nanoscale with the rapid development of nano-size electronic devices, especially with the microelectro mechanical system (MEMS) development, research issues related to the electrical contact became more urgent and more valuable.

ACKNOWLEDGMENTS

The authors gratefully acknowledge the support of training program of Guangdong province outstanding young teachers in colleges and universities (YQ2014178) .

References

1. Airframe and Powerplant Mechanics General Handbook US Department of Transportation. Printed by ASA Publication,INC
2. (Canada) Brown Novick, (Belarus) Kangqi Zi, (Russia) with Mishkin, Xu Liangjun translated electrical contact theory, application and technology Beijing: Mechanical Industry Press(2010).
3. Turbine engine aircraft structures and systems. Ordnance Industry Press, ISBN-7-80172-766-5
4. Guo Fengyi, Zhonghua. Electrical contact theory and its application technology , Beijing: Electric Power Press,(2008).
5. Aviation accidents and human factors . China Civil Aviation Press, ISBN7-80110-443-9 / V
6. Aviation education ISSN 1005-8176 / CN11-3411 / G4 (Total No. 59)
7. Malucci R D .High Frequency Considerations for Multi-Point Contact Interfaces .Proceeding of the FortyFifthIEEE Holm Conference on Electrical Contact s (2001) .
8. Holm R.Electrical Contacts. New York:Springer(1979).
9. Timoshen ko S, Goodier J N .Theory of Elasticity .New York: McGraw Hill (1951).
10. Greenwood J A .Constriction Resistance and the Real Area of Contact .Brit .J .App l.Phys , 17:1621 - 1632 (1966).
11. Lionel Boyer .Contact Resistance Calculations: Generalizations of Greenwood's Formula Including Interface Films .I EEE Transact ions on Components and Packing Technologies , 24(1):50-59 (2001).
12. Nakamura M, Minowa I .Computer Simulation for the Conductance of a Contact Interface.IEEE Transactionson Components , Hybrids and Manufacturing Technology, CHM T -9:150-155(1986) .

MATEC Web of Conferences 63, 02037 (2016) DOI: 10.1051/matecconf/20166302037
MMME 2016

Research on the Governance Mechanism of Aviation Complex Product Manufacturing Supply Chain Based on Dynamic Game Theory

Nan WANG [1,a], Bang Jun WANG [1] and Yu Jie YAN [1]

[1] *China Aero-Polytechnology Establishment, Quality Engineering Technology Center, 10028 Beijing, China*

Abstract. In the manufacturing process of aeronautical complex products, for the following problems: develop long production cycle, a large number of ancillary products, a plurality of participating units, etc, and the status of the frequency of quality problems, using single-phase static game model in the case of asymmetric information, study on Supplier Quality insufficient investment, which resulting in opportunistic behaviour reasons. And using KMRW dynamic game model, we quantitative analysis the mechanism and crucial role of reputation mechanisms to ensure aeronautical complex product manufacturing supply chain and effective operation.

1 Preface

With the rapid development of aviation equipment, aviation complex products is with a high degree of complexity, a large number of ancillary products, the participating units and more capital investment. This significantly increases the difficulty of aeronautical products quality management. Aviation complex products manufacturing, is not only within one enterprise, often also involve collaboration between several companies, and even extended to the global supply chain. As long as there is one company occurs problem in the supply chain, it is possible to cause the project schedule delays, cost increases, product performance degradation and even lead to the failure of the entire project. Thus, in the aviation complex manufacturing process, we should strengthen supply chain quality management to improve the overall performance of the supply chain. Existing empirical studies show that the effective use of quality management has a significant impact on the performance of overall supply chain and a single enterprise (Robinson and Malhotra, 2005; Flynn, 2005)[1-2].

Compared with traditional single enterprise quality management methods, supply chain quality management is focusing on the supply chain operations and structure management (Hsieh and Liu, 2010)[3]. The quality of the supply chain's product depends not only on the production capacity, level of quality of manufacturers and suppliers of, but also relies on its subjective factors. Therefore, some scholars research on the quality incentive mechanism of the supply chain from the perspective of sampling strategies, reward and punishment mechanism, profit distribution mechanism, payment and so on. Such as Starbird (2001)[4] in the process of the research of supply chain contract design quality, puts forward how to design the supply chain to punish and reward and check the strategy, and proposed the corresponding quality incentive mechanism; Chao et al. (2009)[5]in the study of supplier quality control problem, puts forward how to carry out quality improvement and the contract design of product buy-back; Zhu et al. (2011)[6] studied how to make the product quality control strategy in the two-stage supply chain, and established a quality control policy model, designed a quality control contract; Zhang et al. (2012)[7] analyzed how to control the real-time quality of the product bulk ordering in the supply chain. The above studies for the quality management of complex aerospace manufacturing supply chain provide an important reference, but they didn't discuss the important role of reputation in the supply chain quality management. Therefore, we use single-phase static game model to study the motivation of quality opportunistic behavior in the aviation complex product manufacturing supply chain. And we quantitative analysis the critical effects of reputation to ensure aviation complex product manufacturing supply chain effective operate by using dynamic game model.

2 Analysis of supply chain quality opportunism based on single stage static game

Opportunistic behavior means that damage the interests of others for oneself in the situation of information asymmetry. In the process of aviation complex products supply chain manufacturing, various actors are opportunistic behavior, opportunistic behavior largely affected the quality of the supply chain. By using the single phase static game model, we quantitative analysis

[a] Corresponding author: wn3322@163.com

the behavior of the constituent entities of the supply chain strategy. We suppose that in the supply chain, the action strategies of manufacturer A and Vendor B are “cooperative" or "opportunism". Manufacturer A action strategy "cooperative" is to provide reasonable purchase price for B, and ensure its profit level; "opportunistic" is in order to reduce its cost, and then give unreasonable price. Supplier B action strategy “cooperative” is to provide A quality qualified products; "opportunistic" refers to the shoddy. Benefits matrix of the single phase static game is shown in figure 1.

Table 1. Benefits matrix of the single phase static game.

		Supplier	
		Cooperative	Opportunistic
Manufacturer	Cooperative	U1，V1	U3，U2
	Opportunistic	U2，V3	U4，V4

U and V represent the revenue functions of actors A and B that under different strategy combination, and assuming U2 > U1 > U4 > U3 ,V2 > V1 > V4 > V3. The game can be seen as a typical prisoner's dilemma game. To meet the goal of maximizing their own interests, actors choose opportunistic behavior in driving short-term interests. Although the benefit that (cooperative, cooperative) brings to actors A and B to is more than (opportunistic, opportunistic), but in the end their strategic combination of rational choice is (opportunistic, opportunistic). This is a non-cooperative Nash equilibrium. Thus, in single-phase static game, the final result of two actors in the supply chain is: Manufacturers formulate unreasonable purchase price, suppliers provide inferior products, supply chain product quality problems worsened. As can be seen, in the supply chain collaboration, causes of opportunistic behavior are mainly the following two aspects:

(1) “Information asymmetry”. In the supply chain, quality information asymmetry will lead to two consequences: First, prior to the signing of the contract, Since the downstream buyers cannot determine the level of quality provided by suppliers upstream products, which tend to pay the same price to all types of providers. This will result in the interests of high-quality products suppliers have been violated and to change the type. Second, after the signing of the contract, Because of asymmetric information buyers cannot effectively monitor and supervise the quality of products offered by suppliers, thus the suppliers have an incentive to reduce the quality level of the product. It will further deteriorate the product quality of supply chain.

(2) “Unstable partnership”. Unstable partnership is another major cause of product quality defects in the supply chain. Due to the cooperation of upstream and downstream enterprises of the supply chain is a temporary alliance based on interest, it is lack of trust and the partnership is in a high degree of substitutability. These features result in the potential instability of cooperation in the supply chain, and partnership of members may collapse in the short term. In non-long-term expectations, members of the supply chain will choose to maximize short-term profit and ignore the overall interests that win-win situation brings.

3 Analysis of Reputation Governance Mechanism of Supply Chain Based on Dynamic Game

KMRW model is raised by Kreps, Wilson, Milgrom, Roberts four scholars in 1982. They noted that the role of reputation is to provide an implicit incentive to the participants who concerned the long-term interests. Thus, reputation can be a substitute for explicit contracts. This theory has been widely used in the field of management. Mailath and Samuelson(2001) put forward that the reputation as an asset also requires investment and maintenance. Schiff et al. (2002) emphasized the role of the flow of reputation in reducing information asymmetry. They noted that the widespread of reputation can improve the efficiency of operation of the market. In China, the reputation of the model is widely used in research of innovation collaboration. Related studies have shown that the establishment of the reputation mechanism will help alleviate the problems caused by asymmetric information.

As can be seen from the analysis of the second chapter, if the members of the supply chain pay too much attention to their own interests, do not establish mutual trust, and without enough expectations for the stable relations of cooperation, it will lead to the prisoner's dilemma and the loss of long-term interests. Reputation mechanism is to reduce the behavior uncertainty of members by providing information of the trust and willingness. It can enhance the effectiveness of the interaction between supply chain members. Members can use the "reputation" to accumulate the trust between members and reduce the transaction costs. The higher level of uncertainty in the external environment, the partners will increasingly concern about its reputation. The role of reputation will become more significant(Kreps and Wilson, 1982)[8]. In the trading between members, members who take opportunistic behavior can help obtain short-term benefits, but it can be not always succeed in the long run. The cost of short-term speculation may be greater. Obviously, the protective effect of cooperation by reputation is to prevent the impact of fraudulent acts of cooperation, and thus improving the interests of the members. To test this hypothesis, this paper introduces KMRW model in supply chain quality management and to explore the mechanism of reputation in the supply chain to ensure product quality.

3.1 Model assumptions

Hypothesis 1. To simplify the analysis, we assume that manufacturer A and supplier B to be the participants of the supply chain game. Manufacturer A has a good sense of cooperation and does not to take opportunistic behaviour. In the first phase, the manufacturer A will use cooperative strategy. The later stage strategy will be adjusted according to the supplier B’s last stage strategy.

Once B Select uncooperative, and then the manufacturer A will never cooperate with B.

Hypothesis 2. Supplier B has two types: cooperative and opportunistic. Uncooperative type B is also possible to disguise as cooperative in the pre-transaction process for the final stage. Supplier B's type is private information, under conditions of incomplete information, the manufacturer A does not know.

Hypothesis 3. For supplier B, the higher of occupation rate, the greater of the effectiveness. In the long process of repeated game, manufacturer A can observe the behaviour of B and consequent adjustment measures. So B's effectiveness obtained due to encroachment will decrement by the defensive measures of A.

Hypothesis 4. In the transaction process between manufacturer A and supplier B, the single stage utility function is:

$$U=-\frac{1}{2}V^2+\partial(V-V^s) \qquad (1)$$

V is the actual occupation rate of supplier B, $0\leqslant V\leqslant 1$; V^s is the expected occupation rate of the manufacturer A, it represents manufacturers A expected judgment for the supplier B's action, $0\le V^s\le 1$; α indicates the type of supplier B, When α =0, it represents supplier B is cooperative. When α =1,it represents supplier B is opportunistic.

Hypothesis 5. At the beginning of the game, P_0 indicates the priori probability that manufacturer A consider Supplier B is cooperative. 1- P_0 indicates the priori probability that manufacturer A consider Supplier B is opportunistic.

3.2 Model analysis

In the process of multi-stage supply chain dynamic game, q_t is assumed for the probability of the supplier B choose to cooperate in t stage. q'_t is the probability that manufacturer A consider supplier B will cooperate in t stage. If the manufacturer A don't observe the misappropriating behavior from supplier B in t stage, according to Bayes' rule, in $t + 1$ stage the posterior probability of B is considered of cooperative is:

$$P_{t+1}(\alpha=0|V_t=0)=\frac{P_t\times 1}{P_t\times 1+(1-P_t)*q'_t}\ge P_t \qquad (2)$$

By equation (2) can be obtained theorem 1.

Theorem 1: If the supplier B chooses to cooperate in t stage, then in the next stage the probability that manufacturer A consider supplier B is Cooperative will rise.

Suppose supplier B select uncooperative in t stage, it can get the formula (3):

$$P_{t+1}(\alpha=0|V_t=0)=\frac{P_t\times 0}{P_t\times 0+(1-P_t)*q_t}=0 \qquad (3)$$

By equation (3) can be obtained theorem 2.

Theorem 2: Once the supplier B select uncooperative, the manufacturer A will consider it is uncooperative model and won't cooperate with it. This is called the grim strategy, also known as the trigger strategy.

Suppose t stage is the final phase, then it is as a single game in t stage game. Suppliers B will not choose to cooperate, it's the optimal choice is $V_t=\partial=1$. And manufacturer A knows B' optimal choice. Therefore, the expectation for A to B is:

$$V_t^e=V_t\times(1-P_t)=1\times(1-P_t)=1-P_t \qquad (4)$$

In this case, the utility level of non-cooperative supplier B is:

$$U_t(1)=-\frac{1}{2}V_t^2+\alpha(V_t-V_t^e)=-\frac{1}{2}+(1-1+P_t)=P_t-\frac{1}{2} \qquad (5)$$

By equation (5) can be obtained: $\partial U_t/\partial P_t=1$,

And it can get Theorem 3,4.

Theorem 3: The utility of non-cooperative supplier B in final stage is an increasing function of the reputation that it develops in the transaction process with members. The non-cooperative B will disguise cooperation actively to improve the reputation in the supply chain.

Theorem 4: Manufacturer A more thinks B is cooperative, the bigger the P_t, and supplier B will encroach more utility in the final stages.

Now, we will study the non-cooperative supplier B's selection strategy in t-1 stage. Because t stage is the final phase, and the non-cooperative B is cooperative before the t-1 stage. Before t-1stage B cooperative behaviour ensures $P_{t-1}>0$. Therefore, the expectation for A to B is:

$$V_{t-1}^e=V_{t-1}^\theta\times(1-P_{t-1})(1-q_{t-1})=1\times(1-P_{t-1})(1-q'_{t-1}) \qquad (6)$$

$V_{t-1}^\theta=1$is the maximum cupation rate in t-1 stage. δ is the discount factor, reflecting the long-term patience of supply chain members. Here we only consider q_{t-1}=0,1,and to compare the utility of B in t-1stage by two strategic choice.

If the non-cooperative type B does not cooperate in t-1 stage, namely q_{t-1}=0, V_{t-1} =1, and P_t=0. In this case the total utility of B is:

$$U_{t-1}(1)+\delta U_t(1)=[-\frac{1}{2}+(1-V_{t-1}^e)]-\frac{1}{2}\delta=\frac{1}{2}-V_{t-1}^e-\frac{1}{2}\delta \qquad (7)$$

If the non-cooperative type B cooperate in t-1 stage, namely q_{t-1}=1, V_{t-1} =0. In this case the total utility of B is:

$$U_{t-1}(0)+\delta U_t(1)=-V_{t-1}^e+\delta(P_t-\frac{1}{2}) \qquad (8)$$

Therefore, if the following conditions are true, then the formula (8) is greater than the formula (7), namely

$$P_t\ge\frac{1}{2}\delta \qquad (9)$$

By the formula (9), we can know that the total efficiency if non-cooperative supplier B chooses cooperate in t-1 stage is greater than it doesn't. Therefore, if q_{t-1} constituting a non-cooperative equilibrium strategy , q'_{t-1} = 1, this means $P_t=P_{t-1}\ge\frac{1}{2}\delta$. Theorem 5 can be obtained.

Theorem 5: When the probability of A thought B is cooperative in t - 1 stage $P_t\ge\frac{1}{2}\delta$, non-cooperative B will continue to pretend cooperate.

In other words, if the supplier has more reputation, it will have more enthusiasm to maintain reputation.

Conversely, if the supplier due to take opportunistic behaviour and undermines the reputation, the opportunistic behaviour will occur earlier and more frequently. By using the KMRW model, it can be seen that reputation mechanism will constraint behaviour of opportunists in the conditions of incomplete information.

4 Conclusion

In the aviation complex product manufacturing supply chain, due to the formation of product quality has several structural features, such as information asymmetry, double marginal and unstable partnership, these problems make the supply chain quality management is more complex and difficult respect to internal quality control. Compared with traditional single enterprise quality management methods, supply chain quality management is focusing on the supply chain operations and structure management. Existing studies pay more attention to the interests of the settlement mechanism, and didn't consider the important role of the reputation. Therefore, we use the single-phase static game model in the case of asymmetric information, study on Supplier Quality insufficient investment, which resulting in opportunistic behaviour reasons. And using KMRW dynamic game model, we quantitative analysis the mechanism and crucial role of reputation mechanisms to ensure aeronautical complex product manufacturing supply chain and effective operation.

References

1. Robinson C J and Malhotra M K. Defining the concept of supply chain quality management and its relevance to academic and industrial practice[J]. International Journal of Production Economics, 2005, 96（3）: 315-337.
2. Flynn B B and Flynn E J. Synergies between supply chain management and quality management: emerging implications[J]. International Journal of Production Research, 2005, 43（16）: 3421-3436.
3. Hsieh C C, Liu Y T. Quality investment and inspection policy in a supplier-manufacturer supply chain[J]. European Journal of Operational Research, 2010,202(3):717-729.
4. Starbird S A. Penalties, rewards, and inspection: provisions for quality in supply chain contracts[J]. Journal of the Operational Research Society, 2001, 52(1): 109-115.
5. Chao G H, Iravani S M R, Savaskan R C. Quality improvement incentives and product recall cost sharing contracts[J]. Management Science, 2009, 55(7): 1122-1138.
6. Zhu L, You J. Moral hazard strategy and quality contract design in a two-echelon supply chain[J]. Journal of Systems Science and Systems Engineering, 2011, 20(1): 70-86.
7. Zhang S, Wang F, He D, et al. Real-time product quality control for batch processes based on stacked least-squares support vector regression models[J]. Computers & Chemical Engineering, 2012, 36: 217-226.
8. Kreps, D., P.Milgrom, Roberts and R.Wilson. Rational Cooperation in the finitely Repeated Prisoners Dilemma[J] . Journal of Economic theory, 1982 (27):245 -252.

MATEC Web of Conferences **63**, 02038 (2016) DOI: 10.1051/matecconf/20166302038
MMME 2016

A Design and Research on Protection of Architectural Heritage Based on Digital Technology

Shuang Qing HOU[1]
[1]*Yantai Nanshan University, Yantai, China*

Abstract: In the era of digital information, how to reasonably use new-tech and idea for the architectural heritage protection remains urgent. In terms of digital protection, 3D laser scanning technology can perfectly remedy the manual measurement limitation thus correctly obtaining the relevant data. Via analysis, we find that the architectural heritage bears in itself the parameterized characteristic. As Building Information Modeling (BIM) can help realize the parametric design, we proposed a feasible Architectural Heritage Information Model with the BIM "family" connecting architectural heritage and BIM. Finally, the parameter-induced graphic driving is achieved based on the relation between the two.

Keyword: Parameterization; Architectural Heritage; Information Modeling

1 APPLICATION OF DIGITAL PROTECTION TECHNOLOGY

1.1 *Three-dimensional laser scanning technology*

At present, for the protection of architectural heritage generally limited to repair and maintain the existing situation, and less for their own intrinsic architectural heritage law research in the understanding of how to design a building in the ancient heritage of not doing enough. Parametric design application ancient model, can create dynamic static measurement data to create a three-dimensional model based on French. Parametric design applied to the study of architectural heritage inherent law, the law can create complex and cumbersome architectural heritage members clearly expressed, so that the architectural heritage of the design concept and construction logic clearly demonstrated.

Three-dimensional laser scanning technology is a technology in the wake of the development of laser measurement technology formed, which is reflected by the object by measuring the time the laser pulses can be obtained from the information.

Due to the non-divergent laser, laser measurement precision is very high in kind, according to the reality learned that laser is used to measure the distance of Earth to the moon, can be accurate to centimeters or more. On the basis of the measurement by laser scanning on the physical distance, but also measure the height of the horizontal and vertical azimuth angle, so you can calculate the trigonometric method by the laser hit the horizontal distance between the vertical height of a point on the physical.

Can be calculated for each spot in the scanning system as defined in the three-dimensional spatial coordinates Cartesian coordinates based on the measured angle and distance. If a certain kind of continuous spatial sampling density scan measurement, overall measurement can be performed on the physical space relative positioning, which is the technical principles of three-dimensional laser scanning technology. Terrestrial laser scanning system consists of three-dimensional scanners, computer control and data storage, tripods and other accessories.

Development of digital technology to promote more accurate graphics technology equipment have been produced, a more accurate real digital preservation and archiving technology also will be born. And digital photography combined with laser scanning eye can capture subtle details, such as building surface cracks. By recording the laser light is emitted from the object to the reflected or refracted back in time, to calculate and processing software can understand the point of view of the object, as well as the location of the material, and finally a three-dimensional image.

Digital preservation is the use of advanced 2D and 3D scanning, digital photography, 3D modeling and image processing technology to achieve high-precision acquisition and preservation of heritage graphical structure and texture information. Aim is to establish relevant digital model in the computer, an information heritage preservation, restoration, research, development and utilization of visits and provide accurate digital material.

1.2 *Building information modeling technology*

The relevant information and data building information model is based on building projects as a basis for the model, building model were established, true information through digital information simulation building has. It has visualization, coordination, simulation, optimize and can plot of five characteristics.

BIM design process from the resource, behavior, delivered in three basic dimensions, given the implementation of corporate design standards specific methods and practices contents. BIM (Building Information Model) is not a simple digital information integration, but a digital information applications and can be used for digital approach design, construction and management. This approach supports the integrated management of the environment of construction projects, construction works can significantly improve the efficiency of its entire process, a significant reduction in risk.

For one building element, it itself carries a lot of information, how that information is combined with the model functions ancient building information models where the information into the system. BIM's information into the system is completely dependent on the model itself, the modeling process is the information input process. In ancient building information model, all information is recorded in the form of parameters or primitive type properties among members. The system comes with a conventional family model contains parameters related to the more comprehensive mechanism for designers demand entry, and for the family model to establish their own, it may be all the information needed to classify add member properties dialogue box. Making parameters added, the software offers a variety of parameter types and groupings to facilitate designers to model complex information classification entry, and each has its own parameter types and system functions, such as material type provides several default type, designers can be copied in a new material library or depending on the circumstances.

2 THE ESTABLISHMENT OF BIM BUILDING INFORMATION MODELING HERITAGE

2.1 *BIM advantage*

BIM drawing thinking has changed a lot, it means graphic information from a single line to carry a variety of engineering information into building components, it realized from the application point, line drawing building projection, architectural expression to use accurate data recording building information to explain the transition of the building. BIM is essentially equivalent to a database that records the various data complete information about all of the construction process, the physical geometry information including building components, cost and production management information, and so on.

Logical Construction built the logical and actual construction BIM model high degree of unity, compared to 3dmax and other three-dimensional modeling software, BIM models modified linkage great savings renderings companies repeatedly revised wasted human and material resources; the roof curve compared to the use of CAD drawing repeated Videos roof projection curve each facade also precise a lot of convenience; information and detailed data model parameter sharing applications enable orderly management of the entire document; reused family make the design work has become effective convenient etc.

In addition, the building information model in the protection of architectural heritage in archival work can also have a good application. Our legacy of rich and varied architectural heritage, imagine if all create a complete building information model for each legacy buildings and collect them together to create a heritage building information model repository that will enable the architectural heritage of great information Perfection and systematic. Architectural heritage information model repository will be built based on the database, which means that it is a good statistical work is completed within the data repository, so that we have a comprehensive and systematic understanding of Chinese architectural heritage. For example, when creating each building information model, we can record this building's construction, architectural form and style in which important parts of the project information, as an example of architectural heritage style under a specific category.

2.2 *"Family" establishment*

Architectural heritage information model for the protection of architectural heritage importance of archival work in that "family" establishment. "Family" for building components for, and when we were in-depth study of a construction element and classified in accordance with a certain condition, the summary of its shape structure rules, find out the universal significance of drive parameters established this family file type with common sense, it can be said that the model belonging to a particular style common characteristics of all of the components described or a member belonging to the construction of the building instance of this template form. Establishment of a "family" will enable us to recognize documents of a particular type of building elements from a macro point of view the whole picture, and more intuitive to show some form of classification of various components

The entire building information model contains two parts data and graphical information, BIM has a complete database mainly through the family to build. BIM can be said to be a combination of all races. All the family can be attributed to the broad categories into software systems. BIM software in the main categories of walls, columns, roofs, framework, doors and windows, each category has its own category with the characteristics, such as windows and doors can be inserted directly on the wall and open holes. Each category below, could create numerous family, family can be defined by a specific property information - the main control parameters to generate a complete building components.

For example, in a cylindrical system family, BIM software defines the column diameter, column height, material and other parameters to control the generation of a column family, and these parameters are part of the database, and can be viewed any time. Family can be modified on the basis of a model of the family. Group set up process is as follows: For the graphic topology relationships between specific member, controls the graphic to find several main constraint to use as the main drive parameters, ensure the same graphical topology relations on the basis of data obtained a new geometry as a follower point, or direct relationship between time constraints and the main constraint values and logical drive parameters to define the times, and driven by the main constraint point or the main constraints and constraints to jointly define a sub-family, in order to achieve graphical parameter control.

This column building elements form a more independent and relatively simple, easy to make an independent family directly into the project file. We know that the pillars of ancient buildings in general practice have collection points and the side of the foot, relatively fixed shape, if we find a good collection points and side foot control constraints, and define it as the main drive parameters to achieve the parameters of the control body then when we establish a good column family, this family model can be applied to many common projects, simply by changing the parameter value. Because in the same building, the type of beam member more, length, width, size of each member of the Higher slightly different but have similar shapes, and therefore these members can be classified into several production model family, then each family member made into the following types, which will greatly improve efficiency. Ancient building information model component elements of ancient building system mainly refers to large wood removed, the remaining part of the wall and roof, including pedestal and base class masonry, doors, windows, ceilings and other wooden parts. Similarly, models of ancient buildings can be part of the family member for accurate detailed performance.

3 PARAMETRIC DESIGN ARCHITECTURAL HERITAGE INFORMATION MODEL

3.1 *The concept of parametric design*

Parametric design model changes by means of a portion size or a few parts, auto-complete the relevant parts of the model changes to realize the size of the model-driven, in which geometric and topological information required is a computer-driven automatic extraction of. Parametric Design constraints drive is based on the relationship between the various elements of the development provided to achieve these constraints can be constructed in the building mode, it can be modulus relationship, it has certain features and rules. Architectural heritage as a carrier of ancient civilizations, today various studies feudal social, political, cultural, military, science and technology, and the people to provide valuable information. In the process of urbanization, the architectural heritage protection work there room for improvement. First, the mapping results of the architectural heritage is not intuitive, too can learn from drawings, architectural heritage in the process of mapping, the data is generally in the form of plans, elevations, sections that out, spatial relations and some architectural heritage construct more complex and requires a considerable number of cross-sectional and detail to fully articulate the building. Secondly, the internal laws of the architectural heritage of understanding is not deep enough. Since entering the information society, people use high-tech measurement technology to produce three-dimensional model, though give people more intuitive impression, but these models are mostly used for rendering or animation, the lack of project information model, the model is not the relationship between the various components of architectural heritage information.

3.2 *The method of parametric design*

Parametric design approach is the architectural heritage of the modulus, specifications and other information input model, no matter how the model changes, the inherent logic of the architectural heritage Construction will not change. Thus, the model can not only meet the needs of an intuitive visual, but also in the post-modification model, spatial location and size of the architectural heritage of each component member will change accordingly. This change is a computer automatically. Three-dimensional parametric model can automatically generate any portion of the flat, vertical, sectional view.

To create a three-dimensional architectural heritage parametric model, the first of its structural analysis, research which parameters can be compared with the full expression of architectural heritage features. The more parameters, the more the expression of the characteristics of the architectural heritage intact, but too many parameters also make less versatile, the resulting model is too large, the impact speed. In Tongwa for example, can record the complete four corners of the damaged portion accounted watt high proportion, as if it is a member of the filters need to be replaced, the unit price can also be used to replace the definition of statistical follow-up member of that repair costs to prepare. After recording the relevant data, you can build a residual bad member screening schedule according to specific conditions, for example Tongwa, it can be a complete screening of the damaged portion of the four corners of the tile high as more than a third of the member to be replaced member, statistics of these members as well as their cost-efficiency replacement required, which will need to be greatly improved before the cost of construction work on the budget. After recording the status quo after the model should be replicated archive repair design carried out in the presence information to copy this model, after the establishment of a model of the proposed repairs to the pre-construction and construction.

Since the status of the building information model is entirely based on the current situation of the building, including the practice of thatch backing layer and so on, so when making repairs required for the design model can be fully established on the basis of the model to modify the current situation, the need for replacement of residual bad member, simply select and change the type to. All changes in the status quo on the basis of the model should be based on a new stage, in order to facilitate viewing at any time to modify the operation of each phase. After the renovation is completed after the building information model, the software also can make use of BIM data exchange capabilities, import it into the structural analysis software or energy analysis software for simulation analysis through comparison of programs select the best solution.

4 SUMMARY

Architectural Heritage repair work is an important part of the architectural heritage protection, the architectural heritage information model used in this process, it will protect the architectural heritage of digital systems is significant. Architectural heritage information model based on the actual process of mapping a detailed record a variety of information architectural heritage itself, it can completely replace the traditional mapping drawings as design and construction data archive for future inspection carried out at the time of repair of architectural heritage. Component size and process data practices rely on information model architectural heritage recording, you can be a member of the residual bad repair or replacement; dimensions and materials in accordance with member data model records of the missing members can be made up with the original. In addition, it has been severely damaged and repaired urgently, on the verge of collapse ruined building information model can be applied to architectural heritage in the course of them were rehabilitated.

Through the above description is not difficult to find, the building information model can be applied to studies of ancient architecture. With the study of ancient ancient architecture building information model will greatly facilitate the research work. Building an accurate record of the information it can give historians provide rich and accurate research data; dimensional presentation contains details of its construction elements of buildings can be more intuitive to show the face of the building; it's powerful database functionality capable of building information very system statistics and classification; its component model can help family more clarity on the building in different ways to classify; to achieve its member based on the size of the bucket mouth parametric control helps clearer profound understanding of ancient buildings wooden structure to create a way and so on. In short, it parametric qualities enable it to study the ancient architecture provides a wide range of convenience and a variety of ideas.

Research on the integrated design strategy of green building

Yang WANG[1] ,Ping Qing ZHANG[2]

[1] *Yantai Nanshan University Longkou, Shandong, China*

[2] *Yantai Nanshan University Longkou, Shandong, China*

Abstract: With the increasingly serious environmental problems, energy saving and environmental protection of the green building began to have people's attention, and gradually become the development trend of future building, which came into being green building integrated design. How to apply the integrated design theory of green building to the concrete architectural design practice is a question that designers must seriously think about. This paper expounds the integrated design concept of green building and integration methods from the connotation of green building design, and puts forward the integrated design strategy in the green building design concept in theory and practical application, in order to provide reference to the designers or managers.

Keyword: Green building;Integration;Design strategy

At present, the concept of green building has gradually been known and accepted by people, and increasingly become a rookie in the construction industry. However, according to the specific conditions, the reasonable design of green building, green building has been a major problem faced by the builders. Integrated green building design is the green building design in a very important part, through the use of aesthetics, architecture and other related disciplines of knowledge, combined with the building of the actual characteristics of, with the vision of the part and the whole, of Ministry of inside and outside the building of the various elements of the intrinsic link of analysis and integration, harmony in order to achieve with nature and architecture between the three.

1 CONNOTATION OF GREEN BUILDING

The green building, as the name suggests, is shown the concept of green building, and the green concept, in fact, have a very broad meaning, which is not only a vertical greening, roof garden or green plants a lot of architectural concept, but based on and harmless to the environment and natural resources make full use of the concept of a sustainable, ecological, energy saving and environmental protection building. In fact, the green building, the use of materials and interior decoration and overall design has a very high requirements, on the one hand, the need to ensure indoor layout is reasonable at the same time, by reducing the selection of synthetic materials, and make full use of sunlight, thereby saving energy loss. On the other hand, emphasizes the implementation of human, architecture and nature in a harmonious development requirements in the use of artificial means or behavior, create a comfortable, good life, the use of the environment at the same time, it will be possible to nature damage to the lowest.

Generally speaking, green building not only includes temperature, solar lighting, air quality indoor environment, or in the green environment, natural ventilation and other external environment, have a high demand. Of course, on the green building, it is a fundamental way to realize the strategy of sustainable development of our country, the implementation of green building concept is extremely urgent.

2 CONNOTATION OF GREEN BUILDING INTEGRATION DESIGN

Between the integrated green building design is to achieve an important path of green building features, its position in the green building design, in ensuring the building body and the life at the same time, as far as possible according to the knowledge of related disciplines, from both the overall and local levels of, man, nature and construction of three harmony.

However, the design theory actually needs to carry on the guidance of certain design goals, in general, the main performance in target function, environmental objectives, and economic goals etc.

First, the green building integrated design and common building compared to higher, not only need to meet the basic functional requirements, but also should pay attention to environmental comfort considerations. In order to meet the requirements of this goal, building designers need to flexible use of physics, ergonomics and psychology knowledge.

Secondly, the integration of green building design is another important goal is the pursuit of environmental goals, in terms of energy use not only need to give priority to the use of renewable energy, but also should try to improve the utilization efficiency.

The end, the economic benefit of green building determines whether it can develop healthily and steadily, so it should be considered carefully in the process of green building design. That requires strict control of all aspects of the design, in the premise of ensuring the quality of the building, the maximum cost savings.

3 INTEGRATION METHOD OF GREEN BUILDING DESIGN

The main essence of integration is the scattered elements together, the different things of value of organic combine, make itself has no value or the value of the small things play a role in value, and ultimately the formation of value and efficiency as a whole. Green building integration has experienced many levels of integration, from disciplines to social systems integration, from small to large integration. Green building projects from the proposed to remove the whole life cycle should run through the concept of green. Including planning, design, construction, operation and management of the whole process of the completion of the whole process, the only way to be called a green building.

This paper only discusses the design phase of green building integrated design method. Integrated design is a design method and idea, to carry out a systematic research on construction can be by using the integrated design, through the relationship between the various elements of the building to achieve harmony between man and nature, through integrity, timing and association characteristics, this paper puts forward some corresponding solutions. Integrated design in the design concept, design technology, design and creative aspects of the new design concepts and design methods, to expand the scope of the architectural design.

3.1 *Manifestation of green building*

American architect Sim van der Ryan in his book "integration design," a Book of "integrated design" concept, gives the definition In 1981. He believes that the integration of design is in the construction process, according to the building environment, according to local conditions, the design and construction of housing, to reflect the harmonious unity of the building with nature, but also in his life.

Research, focusing from the point of view of the ecology of building with natural harmony and unity as the main research direction and the research goal, in the course of the study, he pays attention to the field of terrain and the local climate characteristics, can according to the terrain and geography climate design harmonious and comfortable building, in dealing with the problem, he is more from the ecological nature of thinking, find out problem-solving methods, such as how to deal with solid waste, construction waste, etc.

3.2 *Technology integration*

American architect Leonard R. Bachman mentioned in his book, "the integration of architecture and design elements of the system, a book, integration is a concept, is a way of thinking, architects are required to have this way of thinking, in the construction process to purpose and direction of architecture selection and combination, emphasize the complementary and technical design, he believes that the effective combination of the integration of technology and design, the integration will be subjective and sensory of harmony and unity, he will be the integration points for the integration of software and hardware integration of two parts.

3.3 *Method integration*

Germany's famous architect Thomas integrated design as a scientific design method to treat, he will be building research as an incentive, through the study of architectural design methods to find. Thomas believes that as a good architect needs to be curious about everything, architects should also have the scientist's thinking mode and on building full of imagination and creativity, and from the deep level of

DOI: 10.1051/matecconf/20166302039

construction of analysis and research, to find effective methods for integrated building.

4 PROBLEMS IN THE INTEGRATION DESIGN OF GREEN BUILDINGS

4.1 *By the limitation of traditional design patterns*

By the current situation, China's design company or design institute basically are the traditional design mode are adopted, the staff is divided obviously, different design personnel responsible for the area is relatively independent, generally speaking, mainly planning, architecture, structure, equipment, etc. professional personnel, each design personnel to carry out their duties is responsible for their own area of work. For the problems in the work of communication and less, that involves a long cycle. This traditional model because of its own shortcomings so that it can better adapt to the development of green building.

4.2 *Collaborative decision making*

Through the existing architectural design patterns of substance analysis and past experience reference, using collaborative design and concurrent design of green building for the integrated design, the for reconstruction of the design process, the life cycle of green building and other aspects of the coverage is wide and contains the content is more extensive.

The architect plays a central role in connecting the whole team. Planning, building, structure, equipment and other professional construction information communication must be clear. Is divided into two forms, one is according to the professional order delivery; one is according to the inverse professional order feedback. In the design of the whole green building in the process of the existence of the problem of transmission and feedback more timely, there are contradictions in time to coordinate and solve. For example, the choice of bricks: Architects according to the green building concept, to determine the type of specifications to other professional designers. This brick must also meet all kinds of requirements of other professional. Other professional designers if there are different views in a timely manner to the architect. Will consult, adjust and ultimately determine the best solution to meet the requirements of the parties.

More frequent exchanges between designers and various professional work is not only the total transfer of information, even under incomplete information can also be their own work is carried out by the relevant information, this greatly, the product life cycle, saving more time and improve the work efficiency.

5 INTEGRATED DESIGN STRATEGY OF GREEN BUILDING

5.1 *Integration of developer factors*

Project development is a kind of market behavior, so the economic benefit will directly affect the overall level of development of green building, the developers factor is the cost factor, but the fundamental factor is the input-output ratio, which is benefit for our evaluation of the economic benefits of green building, the main from the amount of funds used, results of these two aspects of the problem analysis of the use of funds less, good results, it can be judged as excellent economic benefits in assessment of green building cost, not only from the pre cast cost calculation, while also taking into account the late sales value and the total life cycle, in the current environmental problems are becoming more and more severe form, the market value of green building is to constantly improve, starting from the long-term perspective, the most obvious advantage for green building is reflected in the use of the process of building can be seen in many Reduce operation and maintenance costs, save a huge amount of money at the same time, the market value is generally high non green building.

5.2 *Integrated of operator factors*

China's "green building evaluation standard of land, energy and water saving, material saving, indoor environment and operation of these six items in half will produce benefits in the later operation, can be said that the biggest beneficiaries of the current evaluation system for green building operations, however, due to the development of green building technology is late, compared with the traditional technology factors still exist such problems as difficult maintenance, easy to damage and difficult to debug, maintenance cycle is short, the use is not convenient, the actual efficiency low so the latter part of the integration operators should also consider more green building technology has the advantages of easy maintenance, easy maintenance, easy operation, high use maintenance performance, should play the designer's creativity, innovation, based on the existing green on the improvement of Technology.

5.3 *Integrating of user factors*

China's "green building evaluation criteria in Chinese items of indoor environment is considered user factors and the specific provisions comprehensively consider the lighting, noise control, air quality and other factors, but on the current China's existing environmental problems is lack of consideration, integrating user factors to be considered in addition to the green building evaluation standard, basic requirements of indoor environmental requirements, should consider the healthy lifestyle, from the safety of drinking water, the actual air quality, waste disposal, environmental comfort aspects of further green design, green building become comfortable and healthy building, so as to enhance the market value

5.4 *Integrating and coordination balance*

Green building integrated design theory, the main consideration of the evaluation standard of green building, developers, operators, users factors factors factors of these four aspects, however, these four factors namely the ultimate interests of unity, even many contradictions and opposites, contradictions complex dialectical unity, the core task is to integrate the design theory in four big contradictions the factors to balance out a compromise best, let four factors to obtain a win-win situation which requires designers to gather four requirements in the entire design process, constant collision and modified design content, design, procurement, construction, commissioning, the use of five stages are needed to balance design

6 INTEGRATED DESIGN STRATEGY OF GREEN BUILDING

Amory. B. Lowens in the ideas from East and West fusion: sustainable development construction of overall design, this paper pointed out: green building contains very extensive, not only refers to the material side, including cultural, economic and spiritual needs. In the green building design, not only need to consider the natural lighting, solar power generation, sewage recycling and other factors, also need to involves the integration of the natural integration of the building and the surrounding environment in relation to man and nature, and many other aspects. With green building design specification GB /T50378 2006 requirements, the integrated application of green building design theory is summarized as follows: to meet the control; choose the general term; as far as possible the use of priority

6.1 *Sspace saving design*

Building itself is a commodity, and green building is a good product, to create economic benefits for their own at the same time improve people's quality of life, save the cost of production and living, improve economic efficiency. Green building design process, the project location should be in line with the principle of land, project location should be combined with specific urban and regional planning and meet the residential buildings per capita living index, and give full consideration to the rational development and utilization of the roof and the underground space.

6.2 *Energy saving, water saving, material saving design.*

Green building advocates a green, energy-saving, environmentally friendly architectural design. Therefore, we should pay attention to construction and the overall performance of traditional architecture compared to upgrade and fully pay attention to building the economic benefits, social benefits and environmental benefits of combining.

6.3 *Interior environment design*

Building good environment is the necessary conditions to ensure the normal life of the people, because reasonable light environment can not only reduce artificial lighting system energy consumption, but also effectively alleviate visual fatigue, and improve the comfort of people's lives.

In order to achieve the above purpose, now building design in green building design theory to guide the integration, on the one hand, in the use of solar light focusing on the shade, effective way to avoid direct sunlight glare phenomenon. On the other hand, when the lighting cannot meet the requirements of the building, the courtyard, the atrium for a simple processing to increase the amount of indoor lighting. At the same time, under the impetus of modern science and technology, some new technologies have been applied to the design of buildings, in order to achieve the purpose of improving the quantity of indoor lighting.

7 CONCLUSION

To sum up, have a certain understanding of the relevant aspects of our connotation of green building, green building and the importance of integrated design theory. From which we can deeply recognize, as a fully embodies the harmony between man and nature of green building and its specific construction up is not easy, in fact, whether it is in the integration of green building design theory or in the green building design theory of concrete operation, the need strictly in accordance with the internal and external environment and to a part and the whole, the whole and the part of visual concept, effectively achieve green goals. Anyway, we as the actual builders of green building should in the practical aspects of the work, and constantly strengthen their design knowledge, analyzing and summarizing the experience that will truly green building integrated design theory of utility to play to the maximum.

MATEC Web of Conferences **63**, 02040 (2016) DOI: 10.1051/matecconf/20166302040
MMME 2016

Research on green building design based on ecological concept

Ping Qing ZHANG[1],Yang WANG[2]

[1]*Yantai Nanshan University Longkou, Shandong, China*

[1]*Yantai Nanshan University Longkou, Shandong, China*

Abstract: At present, the protection of the ecological environment and the promotion of green building has been recognized and widely promoted.With the rapid development of the construction industry, Architecture design will inevitably require the resentation of its unique form and charm to reflect the ecological concept and ecological culture, because of the unique nature of the art and the particularity of the environment. To establish the ecological concept of green building design and vigorously develop the green green building has a complementary role to alleviate the pressure on resources,and to speed up the eco city planning design, and to realize the sustainable development of the city, and to protect the urban ecological environmental.

Keyword: Ecological concept; Green building; Sustainable development

"Green building" as a concept is more and more introduced into the urban planning and design, with the continuous deepening of China's reform and opening up. The idea of sustainable development based on ecological concept is also more and more recognized. In order to better for urban construction and development, we must also change the traditional way of building design, the concept of ecological, green building and ecological city construction as a new force of development to make our city more allow full of vigor and vitality.

1 ESTABLISH THE ECOLOGICAL VIEW OF THE SIMPLE GREEN BUILDING DESIGN

Plain green ecological architecture design concept is refers to the harmonious status of the urban ecological environment, an embodiment of the harmonious coexistence of man and nature values, it to the subject of participation and on the nature of interdependence as the premise, and the organic whole concept of body and urban environment object into the concept of ecological system, simple green building ecological design concept is a in contemporary ecological culture idea level grow out of and the concept of ecological city occurred resonance mode of thinking, reflecting the ecological culture of the city and green building design and the natural environment of symbiosis and mutual integration and sharing.

1.1 *Harmonious coexistence of green building and natural environment*

Green building should promote urban eco system, a virtuous circle angle as the starting point, fundamentally reduce resource and energy consumption, prevent and reduce waste and harmful substances on the urban environment pollution, realize the city sustainable development.

1.2 *Green building based on humanistic spirit*

Green building design as a kind of essence of life of art activities, not only can meet the aesthetic needs of urban people. It can guide the public culture, strengthen the green building and urban environment formed a sense of order, rhythm and rhyme, arouse people's life of beauty or ideal of life the beauty of feelings and yearning.

DOI: 10.1051/matecconf/20166302040

2 MANIFESTATION AND DESIGN METHOD OF GREEN BUILDING

Green building is defined on the basis for effective conservation of resources, to the best of our ability for people to provide a healthy living space, it is in the protection of the earth's ecological environment based on using renewable resources and environment construction of a building. In the sixties of the 20th century, the famous Italian architect Paul Soler Rui ever proposed the idea, with the people's attention to the environment, the idea has been recognized by the people and the implementation of, in the recent crisis in the earth environment, advocating and promoting the concept of green building has become an important topic of various countries, the mountain visible, green building is to maintain the ecological balance and the best choice for environmental protection. Green building is not just a green building, green building is actually in water, land, energy saving, material saving, environmental protection and other aspects, from design to operation are the implementation of more stringent energy efficiency targets, to obtain a high efficiency, low consumption, no waste, no pollution, ecological balance of the built environment, make people live more comfortable and healthy.

2.1 *Manifestation of green building*

Building energy saving. Green building must pay attention to "low carbon" benefits, should clean non polluting renewable energy more use to the building ventilation, lighting, heating and so on, in order to achieve energy saving and emission reduction.The construction of pollution prevention and control. The concept of pollution not only in production and daily life, but also in the city building construction to operation and maintenance process.

2.2 *Common design method of green building*

Make full use of natural conditions, such as climatic conditions, hydrological and geological conditions;The design of natural ventilation and natural lighting is adopted as far as possible;The use of renewable energy such as solar energy, wind energy and so on, in order to reduce building energy consumption;The importance of building materials and other resources recycling;And as much as possible the use of wood materials;To strengthen the building envelope insulation performance;The use of heat recovery design;The use of high performance glass design, lighting and shading system.

3 PRINCIPLES OF GREEN BUILDING DESIGN

3.1 *The principle of ecological protection*

Measure of the building is the primary criterion for green building is the on the ecological environment protection degree, whether to maximize the use of ecological resources, on the basis of this, follow the objective law of nature, and nature live in harmony. Such a building can get the masses of the masses, conform to the requirements of the development of the times.

3.2 *The principle of economy*

Green building construction projects, whether it is design, construction or use should follow the principle of economy, should pay attention to the price, in pay attention to efficiency and quality at the same time also should pay attention to the economic benefits, reasonable cost control is very important.

3.3 *The principle of healthy*

The quality of construction can bring good experience to living in households, more humanized design is conducive to the physical and mental health of residents. In the design of better new energy investment, reduce air pollution, making green construction, bring the inhabitant quality and healthy living environment.

3.4 *The principle of modernization*

Green building, as a new type of building, constantly adapt to social changes, the trend of social development is very necessary. Design should be in the appearance and performance, followed by the pace of modernization, designers should continue to improve their own aesthetic standards on the premise of independent innovation, to bring new vitality to the green building.

4 ECOLOGICAL DESIGN CONCEPT OF GREEN BUILDING

4.1 *Construction features*

In the process of green building, the location of the building site conditions have higher requirements, small and medium can be harmful to the health of the human body, such as toxic and hazardous substances, and to the ground temperature is suitable, groundwater to national verification standards. Secondly, the use of green building materials in the process must pass the relevant department for inspection, to ensure that no harm to human and ecological environment be put into use.

4.2 *Saving energy*

In the implementation of green building process, according to the climate conditions in the construction of "Bu and situation and layout, through natural ventilation design principle of air cooling system, the building to the effective use of the dominant wind direction in the summer, but also by building terrain conditions of the regeneration resources for the design of a solar heating facilities, power generation device and, in this way, we can effectively save energy and little on human health hazards.

4.3 *Saving water*

In architectural design, construction and building materials selection, introduces the theory considering the use and disposal of resources, reduce the use of resources on the basis, and strive to renewable resources. Secondly, through the investigation of water resources construction, making the referral of utilization of water resources, saving water resources effectively.

4.4 *Saving material*

Node dielectric construction characteristics of the climate and resources, and vigorously promote the use of energy-saving, safety, convenience of green building materials, as far as possible in the vicinity of the building from raw materials, in order to reduce the transport process on the environment caused by pollution and resource consumption.

5 GREEN BUILDING AND ECOLOGICAL ENVIRONMENT

5.1 *Green building is the need to maintain the ecological environment*

Ecological environment is good or not were since then affects the human physical and mental health, abide by the laws of nature, with the purpose of healthy, harmonious and sustainable development in short, to improve people's awareness of environmental and ecological, effective control of energy over consumption, encourage innovation and research and development of new technologies, so as to effectively use renewable resources, in the construction of the urban environment full of science for the people to solve and the environmental crisis problem

5.2 *Promote the significance of green building*

Ecological environment and each of us have Xiaoke segmentation, promote the effective implementation of the green building is countries to improve environmental measures, is the great project of human welfare. So, to promote green building, so that people in the concern of its process, to understand the effects of housing construction and consumption of material resources, as well as indoor air to the human body and environment, and in order to form the market demand for green building. With the gradual expansion of market demand, green building can be more widely used.

5.3 *The embodiment of ecological energy saving design in green building*

With the development of social economy, people's outlook on life, world outlook and values will change, they on green building standards will be changed. However, its basic principle is always the same. Is mainly reflected in the following aspects: as far as possible to reduce energy consumption, in order to maintain the ecological balance; try to according to the requirements of households make corresponding adjustment, to ensure the dominant position of the households in the architectural design; try to make human and nature as a whole.

6 GREEN BUILDING AND ECOLOGICAL CITY PLANNING

Green building and ecological city planning and design for the sustainable development of China's economy and the development of China's economy has a vital role in the development of. National planning and design of green building and ecological city has taken a high degree of attention. In the development requirement indicates we should promote green building and urban ecological planning and design work, in the later work more considering the quality of economic development, taking into account the relationship between economic development and ecological environment.

For the construction and urban planning supervision and regulation to be more stringent, the construction of the project is to reflect the concept of green building. In order to better conform to the times development to promote the development of their own ability, we have to do the planning and design of green building and ecological city, in the engineering of the overall design and concrete implementation process should strive to precision, as far as possible reduce carbon emissions and the environmental harmful emissions of ignorance.

National development plan clearly pointed out that the green building and eco city planning consistent strict

DOI: 10.1051/matecconf/20166302040

implementation of and in relation to the upper transfer, planning and use of stage strictly, only after the audit in order to make the project launched. After the end of the project to strict supervision and inspection, which requires parties and establish a sound strict inspection and supervision mechanism, ensure that there is no slip through the net. So for the requirements of an open business to be more stringent, check the test side cannot be relaxed. To promote the only way to green buildings and urban ecological planning and design of the real green environmental protection, rather than simple shouting slogans and pretend, really promote the plan implementation and scientific, systematic and rational, scale.

7 THE PROSPECT OF GREEN BUILDING

7.1 *Green building can be perceived by people*

At present, the development of green building in China has reached a bottleneck, the key is to popularize, let the people know what is the green building and green building, what are the benefits and so on. There are many popular innovative ways of green building:

First is to promote the development of people to cognition, monitoring and evaluation of mobile phone software, familiar with the green building, green building not only knowledge, but also to stimulate domestic demand and owner of such behavior.

Second is to promote the focus to the masses in the green building will bring the actual interests of party, such as energy saving economy. After calculation, the new green building will be able to recover the wood into 3-7 years, according to the calculation of life expectancy of 50 years of construction, the occupants and owners can enjoy 45 year average net profit period.

Third is the visual focus on the performance of green building in the design. With the development of IT technology, the visualization of green building design and comparable, imagine the future every day to open mobile phone, together with the bed or out of a green building energy saving, my water, rainwater, air quality is in the first of several similar buildings in the community to see a small screen, a what room for improvement. Research shows that the visibility is only due to saving energy and water saving, can enhance the degree of more than 15%.

Fourth is the property management of green building will become a huge emerging industry. This emerging industry focuses on the construction of renewable energy utilization and rainwater collection, water reuse, garbage reuse etc..

7.2 *Internet plus green building based on the integration of Internet and green building*

The first is the design of the internet. At present, China's building energy software introduction or the number of independent research and development complex, but the lack of the integration of cloud computing platform software. We should not only pay attention to integrate the use of cloud platform in the future, at the same time to Party and a large number of application data and network technology in the construction of new parts, green building materials, new materials, new technology and new management operation mode.

Second is new products, new parts, green building materials, new materials and new technology of internet. Through the Internet, new materials, new technology and new technology, designers can easily find a variety of climatic conditions in accordance with local or national standards, the new building materials has reached a new stage of the development of revolutionary. Almost every have emerged a variety of new building materials, many new building materials not only safety, corrosion resistance, heat insulation is very good, also can absorb harmful gases, even enough to release gas beneficial to people's health. These new materials can be quickly applied in the building through the internet.

Third is the Internet identity management. China city will be within two months after the launch of this session of the general assembly of green building assessment system identifies the network monitoring application consulting, and provide free software, realize identification application assessment management consulting supervision of online integration and convenient evaluation, can further reduce the cost of green building consulting.

Fourth is the construction of the internet. Agile production system similar to the invention of the Japanese Toyota Corporation (Just in time), the future construction of green building construction is like building a car that industrialization, the whole process is strictly regulated by the Internet, the components and parts manufacturers and logistics system, the construction site supervision, such as seamless connection, the whole system reaches zero inventory, low pollution, high quality and low cost, which is the green building construction must be the direction of development.

7.3 *Green building being more ecologically friendly, more humane*

Richard Smalley the Nobel prize winner made a list of the ten major challenges facing human in the next 50 years before his death, sorted according to the important degree, the first is energy, second water, third food, fourth is the environment, fifth is poor, sixth is terrorism, seventh war, eighth diseases, Ninth is education. The last is democracy and population. If the green building be more humane and more environmental protection, create Aquapcmic, circulation mode, can solve the above problems in whole or in part.

8 THE PROSPECT OF GREEN BUILDING

To sum up, with the increasing consumption of resources and energy, as an important cause of building energy consumption, green energy saving measures is imperative. Green building is the inevitable trend of the development of construction industry in our country, it takes Scientific Outlook on Development as a guide, adhere to the people-oriented strategy, the ecological concept of sustainable development and actively maintain ecological balance. Focus on the design of energy saving and environmental protection, in addition to the establishment of the quality of life and working environment for people at the same time, can effectively promote the development of green building.

MATEC Web of Conferences **63**, 02041 (2016) DOI: 10.1051/matecconf/20166302041
MMME 2016

Analysis of impact between the helmet and flying rock during earthquake

Lan Ying ZHAO, Gang SUN[a], Wei Hua LI, Xuan LIU and Jing ZHU

National Earthquake Response Support Service, Beijing100049,China

Abstract. The kinematics parameters like velocity of the flying rock during an earthquake and the collision between the helmet and the flying rock is presented. Since the seismic wave could be simplified as sinusoid individually in the horizontal direction and the vertical direction, the velocity of the flying rock is calculated separately with the acceleration and the frequency in the correspondingly orientation. Then, theoretical model of the collision process include velocity and the impact force is actualized based on the analysis of cases not only the impact between the rock and the helmet lonely but also the rock and the helmet well-worn on the rescuer's head.

1 Introduction

Except for causing a large number of people's lives and property losses, severe earthquakes also trigger a lot of geological disasters or secondary disasters such as the landslides, debris flows, barrier lakes, tsunami, flood, etc. For earthquake rescue missions, the rolling and flying rocks from the loose rock mass structure take serious threat to the lifeline engineering, vehicles and rescuers and affect the rescue mission. Meanwhile, severe earthquakes cause a large number of constructions collapsed and damaged thus unable to stay. Many concrete components of the building ruins explode outside and sometimes fall towards the earth which will hurt somebody or even cause somebody to die if one is just in the position[1,2].

The movement of rolling rocks during earthquakes had been an issue and many literatures studied it. The motion types were divided to several kinds like slipping, rolling, free-faller, projectile, etc and series equations were proposed in order to calculate track and speed of the rolling rocks[3,4]. Xiao Feng analyzed selected samples of perilous sector with potential rockfall and the farthest, the nearest distance of the rolling rocks [5]. Tang Hongmei proposed the reasons of rolling rock and obtained the kinematic trajectory equation using the collision theory, instantaneous friction role theory[6].Factors influenced the movement of the rolling rocks were analyzed and the mainly factor is sequential the slope and slope conditions, the shape of the rolling stones, and the rock quality according to their influence[7-9].The motion feature, the energy absorption way and the harm scope were analyzed taking Wenchuan earthquake as an example[10]. Based on the statistical analysis of the 399 collapsed stones profiles, the slope angle was respective 56 degrees, 39.6 degrees, 26 degrees to form different motion type of the rolling rock and good linear relationship existed not only between the maximum movement distance and slope but also between the spread of the rock and the slope[11].

There were less literatures about the impact force than the movement of the rolling rocks and mainly concentrated on engineering protection demand. The Hertz elastic collision theory and the impulse theorem was the base of the calculation and there were the Swiss formula, formula of Japan, Australian formula, roadbed code formula, Yang Qixin formula as typical algorithm [12-16]. Yuan Jinke compared the above algorithms and introduced his method to calculate the impact force [17].

For the kinematics and impact force of the flying rocks, there were few research and mainly about the engineering blasting. Chen Kang of the blasting flying rock based on the data of the Three Gorges Dam using the similarity theory[18]. Wu Chunping obtained the distance equation suitable for one same blasting project using the dimensional analysis method[19].In addition, the BP neural network model and the grey Elamn neural network were also applied to calculate the distance of flying rocks[20, 21]. However, there is little research about lying rocks caused by earthquakes. In this paper, the speed of the flying rock is analyzed with the acceleration and the frequency of the seismic wave in three orientations. Besides, the collision process between the rock and the earthquake rescuer helmet is presented and the calculate module of the collision is carried out.

[a] Corresponding author: sungis01@126.com

2 The speed of the flying rocks

2.1 Model and setting

Considering one rock flies out from the original position with the affection of the seismic wave in the horizontal direction and the vertical direction after an earthquake and it will get certain speed through the seismic wave. As known, seismic wave could be regarded as combination of a series of sinusoid, the affection of the seismic wave could be decompose as the combined action of the series of sinusoid. The wave records of the seismic station are divided to three parts as E-W, N-S, U-D and represent the seismic wave of east-west, north-south and vertical direction. We simplified each part of the seismic wave to be a sinusoid and influenced the rock on time t1, and the rock flied out with the affection as shown in Fig.1. The speed of the flying rock is the vector superposition of the speed in three directions, so we can calculate affect of each part and then synthesize the affects to obtain the final result. In order to analyze the affect, there are parameters setting as shown in Fig.2.

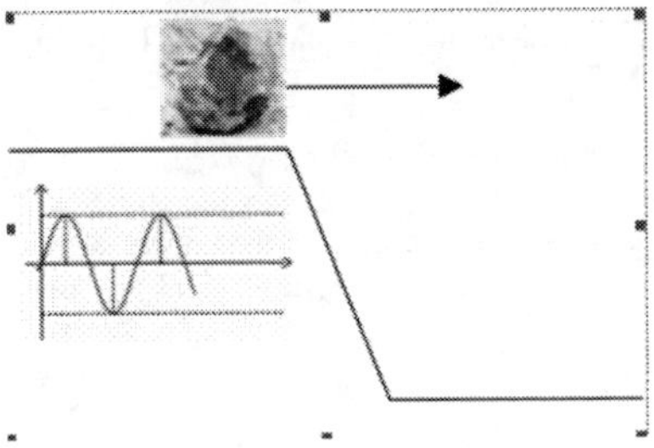

Figure1. The flying rock affected by the seismic wave

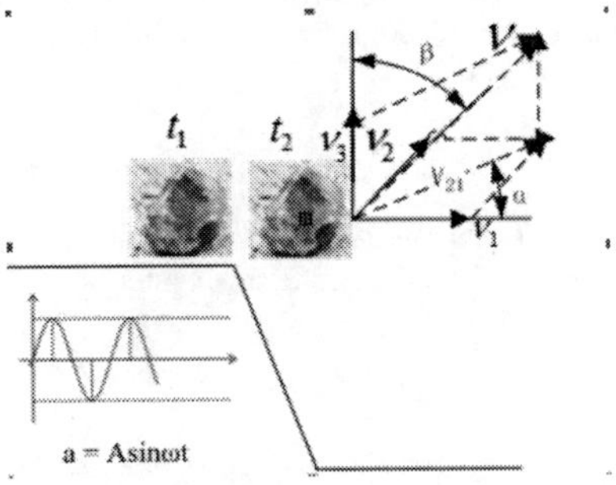

Figure 2. The parameters setting

t_1 - the time of seismic wave affect the rock

t_2 - the time of rock fly from the original position

v_1 - the speed in E-W direction

v_2 - the speed in N-S direction

v_3 - the speed in U-D direction

v_{21} - the vector superposition of v_1 and v_2

α - the intersection angle between v_1 and v_2

β - the intersection angle between v_{21} and v_3

m -the quality of flying rock

a = Asinωt-the expression of the seismic wave

2.2 The horizontal speed calculation of flying rock

The speed in E-W and N-S direction is similar and both belong to the horizontal speed and the E-W speed is used to expound the calculation here. With the influence of the E-W part of the seismic wave, the flying rock could get the starting speed:

$$v_1 = \int_{t_1}^{t} A_1 \sin \omega_1 t dt = A_1 \left(\frac{\cos \omega_1 t_1}{\omega_1} - \frac{\cos \omega_1 t}{\omega_1} \right) \quad (1)$$

and the maximum of v_1 is:

$$v_{1m} = \frac{2A_1}{\omega_1} \quad (2)$$

There will be air friction during the flying and we consider the air friction is opposite with the flying direction and it is a linear relation with them, the linear correlation coefficient k is:

$$f = -kv \quad (3)$$

During the flying of the rock, with the barriers of air friction,the speed of the rock is :

$$\frac{dv_1}{dt} = -\frac{kv_1}{m} \quad (4)$$

$$v_1 = c_1 e^{-\frac{kt}{m}} \quad (5)$$

and the constant

$$c_1 = \frac{2A_1}{\omega_1} e^{\frac{kt_2}{m}} \quad (6)$$

so the E-W speed :

$$v_1 = \frac{2A_1}{\omega_1} e^{\frac{kt_2}{m}} e^{-\frac{kt}{m}} \quad (7)$$

and similarly , the N-S speed

$$v_2 = \frac{2A_2}{\omega_2} e^{\frac{kt_2}{m}} e^{-\frac{kt}{m}} \quad (8)$$

Furthermore, there is the vector superposition of v_1 and v_2:

$$v_{21} = \sqrt{v_1^2 + v_2^2} = 2\sqrt{\left(\frac{A_1}{\omega_1}\right)^2 + \left(\frac{A_2}{\omega_2}\right)^2} e^{\frac{kt_2}{m}} e^{-\frac{kt}{m}} \quad (9)$$

DOI: 10.1051/matecconf/20166302041

$$\alpha = \arctan\frac{v_2}{v_1} = \arctan\frac{A_2\omega_1}{A_1\omega_2} \tag{10}$$

2.3 The vertical speed calculation of flying rock

The difference between the horizontal and vertical direction is the gravity. During the flying, the rock will barge up against both of the air friction and the gravity and there will be equation similar with the equation (4):

$$\frac{dv_3}{dt} = -\frac{kv_3}{m} - g \tag{11}$$

$$v_3 = \left(\frac{mg}{k} + \frac{2A_3}{\omega_3}\right)e^{\frac{kt_2}{m}}e^{-\frac{kt}{m}} - \frac{mg}{k} \tag{12}$$

Similarly, there is the vector superposition of v_{21} and v_3, we can get the speed of the flying rock:

$$v = \sqrt{{v_{21}}^2 + {v_3}^2} \tag{13}$$

$$\beta = \arctan\frac{v_{21}}{v_3} \tag{14}$$

3 Impact analysis

3.1 Type of collision

The collision happens in short time with big impulse when the flying rock impacts the helmet and there will be three types according to the displacement in the collision.

Firstly, if the rescue helmet is not fixed or worn with rescuer, it will obtain a movement after the collision and move to a certain direction with a velocity. This collision process can be simplified as the elastic collision that there will be no displacement and plastic deformation at the contact point.

Secondly, if the rescue helmet is fixed or well worn with the rescuer and the quality of the flying rock is not too big to affect the movement of rescuer head, which means there is no displacement or little displacement that can be ignored. This collision process can be simplified as the nonelastic collision with no displacement at the contact point.

Thirdly, if the rescue helmet is fixed or well worn with the rescuer and the quality of the flying rock is big enough to affect the movement of rescuer head, since the dimensions of flying rock is smaller than the rescuer height, the helmet will obtain angular rotation instead of little linear displacement after the collision. This collision process can be simplified as the nonelastic collision that there will be rotation movement at the contact point.

3.2 The elastic collision

There are assumptions as follow to analyze the elastic collision between the flying rock and the helmet:

1) The affection of the gravity force and the elastic force is ignored since they are fewer compared with the impact force.
2) Because the collision process is very short time, the flying rock and the helmet has no linear displacement during the collision and they will move with new velocity after the collision.
3) The flying rock and the helmet is regarded as one system and there is no outer energy or force influencing the system, so there is conversation of energy during the whole collision.

Based on the above assumption, the collision could be modeled as two rigid bodies collision as shown in Fig.3.

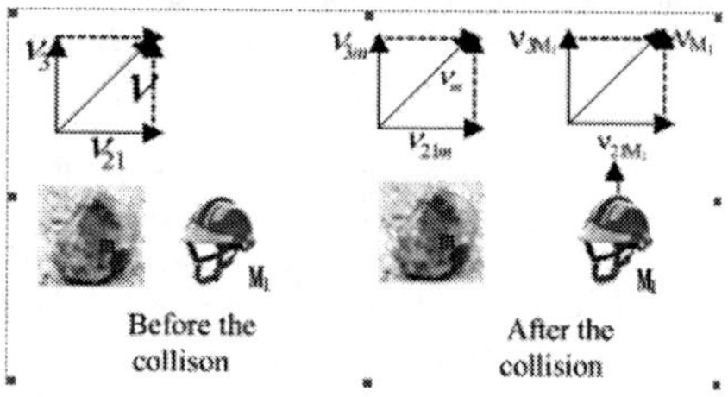

Figure 3. The model of the elastic collision between the rock and the helmet

M_1-the quality of rescue helmet
V_m - the speed of flying rock after the collision
V_{M1} - the speed of helmet after the collision

If there is coefficient of restitution as k_m, then

$$k_m = \frac{I_{M_1}}{I_m} = \frac{v_{M_1} - v_m}{v} \tag{15}$$

According to the momentum conservation theorem, there has equation about the velocity before and after the collision:

$$mv = mv_m + Mv_{M_1} \tag{16}$$

With the equations (15) and (16), we can obtain the velocity of the flying rock and the helmet:

$$v_m = v - (1 + k_m)\frac{m}{m + M_1}v \tag{17}$$

$$v_{M_1} = (1 + k_m)\frac{m}{m + M_1}v \tag{18}$$

Moreover, according to the impulse theorem, there will be:

$$mv_m - mv = \int_0^t Fdt \tag{19}$$

$$M_1 v_{M_1} = \int_0^t Fdt \tag{20}$$

With the equations (19) and (20), we can obtain the impact force during the collision:

$$F = \frac{M_1 v_{M_1}}{\Delta t} = \frac{m v_m - m v}{\Delta t} \quad (21)$$

The energy loss during the collision:

$$\Delta T = \frac{1}{2} m v_m^2 + \frac{1}{2} M_1 v_{M_1}^2 - \frac{1}{2} m v^2$$

$$= \frac{m M_1}{2(m + M_1)} (1 - k_m^2) v^2 \quad (22)$$

3.3The nonelastic collision with no displacement

In this type of collision, the flying rock, the helmet and the rescuer head are regarded as one system and the collision could be modeled as shown in Fig.4. Since there is no outer energy or force influencing the system, the conversation of energy is suitable during the whole collision and there is:

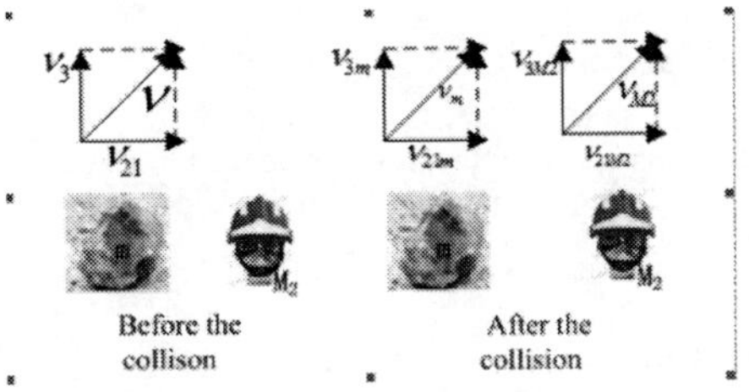

Figure 4. The model of the nonelastic collision with no displacement

$$\iint_s F\mu ds + \frac{LF^2}{2EA} + W_2 = \frac{1}{2} M_2 v_{M2}^2 + \frac{1}{2} \iiint_V e_{ij} X_{ij} dV \quad (23)$$

For the left side of the equation (23), the first part is the energy of the impact force.

The second part is the elastic deformation energy W_1:

$$W_1 = \frac{LF^2}{2EA}$$

L - the height of the head-helmet combination
E - the elastic modulus of the head-helmet combination
A - the sectional area of the contact point

The third part is the energy consumption of the vibration of the head-helmet combination W_2:

$$W_2 = k_{M_2} \int_0^L M_2 D(x) dx = k_{M_2} M_2 D^2$$

K_{M2} - the natural frequency of the head-helmet combination
D(x) - the vibration amplitude of the head-helmet combination
M_2 - the quality of the head-helmet combination

For the right side of the equation (23), the first part is the kinetic energy of the head-helmet combination W_3:

$$W_3 = \frac{1}{2} M_2 v_{M_2}^2$$

The second part is the elastic deformation energy of the head-helmet combination W_4:

$$W_4 = \frac{1}{2} \iiint_V e_{ij} X_{ij} dV$$

e_{ij} - the dynamic stress of the head-helmet combination
X_{ij} - the dynamic strain of the head-helmet combination

3.4The nonelastic collision with rotation movement

In this type of collision, the flying rock, the helmet and the rescuer are regarded as one system and the rescuer may have fixed axis rotation with the foot as the pivot axis and the rescuer height as the shaft length and the collision could be modeled as shown in Fig.5.

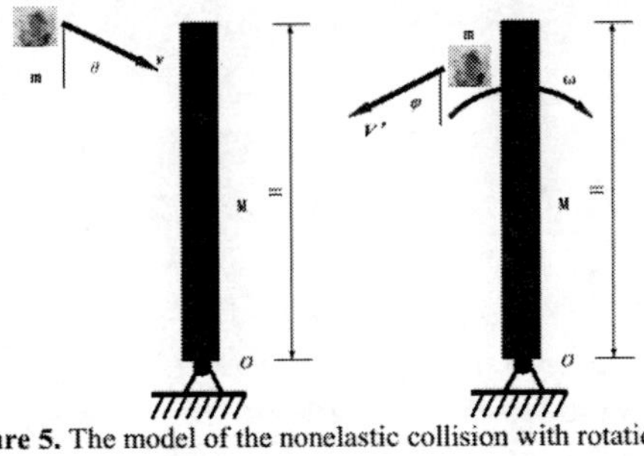

Figure 5. The model of the nonelastic collision with rotation movement

v _the velocity of the flying rock before the collision
v' _ the velocity of the flying rock after the collision
θ – angle between v and the vertical
ϕ – angle between v' and the vertical
M – the quality of rescuer and helmet
H – the height of rescuer
ω- the rotational angular velocity of rescuer

There is no outer energy or force influencing the system, so the system momentum is conserved and there is:

$$m v \sin\theta \bullet H = J_0 \omega + m v' \sin\phi \bullet H \quad (24)$$

For the left side of the equation (24), there is

$$J_0 = \frac{1}{3} M H^2$$

$$v' = \omega H$$

and the rotational angular velocity of rescuer is:

$$\omega = \frac{3m}{(3m+M)H}\frac{\sin\theta}{\sin\phi}v \tag{25}$$

According to the theorem of conservation of momentum, in the collision process there is:

$$mv'\sin\phi + M\omega H - mv\sin\theta = I_x \tag{26}$$

$$0 - mv'\cos\phi = I_y \tag{27}$$

Take the equation (25) into the equation (26) and (27), then

$$I_x = \left(\frac{3m}{3m+M} + \frac{3m}{3m+M}\frac{1}{\sin\phi}\right)mv\sin\theta \tag{28}$$

$$I_y = -\frac{3m}{3m+M}ctg\phi \bullet mv\sin\theta \tag{29}$$

If the impact time is signed as Δt, the impact force during the collision could be calculated as follow:

$$F_x = \frac{I_x}{\Delta t} \tag{30}$$

$$F_y = \frac{I_y}{\Delta t} \tag{31}$$

There is two special situation when θ is separately 0 and 90 degrees. When θ is 0 degree which means the flying rock has free-fall movement, there is no rotation and only displacement in the vertical direction like the elastic collision, the velocity of the rescuer and the flying rock will be:

$$v' = v - (1+k)\frac{m}{m+M}v \tag{32}$$

$$v_M = (1+k)\frac{m}{m+M}v \tag{33}$$

and the impact force will be:

$$F = \frac{Mv_M}{\Delta t} = \frac{mv_m - mv}{\Delta t} \tag{34}$$

When θ is 90 degrees which means the flying rock has horizontal movement and attack the side of rescuer, there is no displacement and only rotation, the rotational angular velocity and the impulse of rescuer will be:

$$\omega = \frac{3m}{(3m+M)H}v \tag{35}$$

$$I_x = \frac{6m}{3m+M}mv \tag{36}$$

$$I_y = 0 \tag{37}$$

and the impact force will be:

$$F = F_x = \frac{6m^2v}{(3m+M)\Delta t} \tag{38}$$

4 Conclusion

Focused on the flying rock caused by the earthquakes, the velocity in the horizontal and vertical orientation is calculated with the simplified sinusoid model of seismic wave and the speed could be obtained through the vector superposition of velocity in the different directions. When the flying rock shocks the helmet or the rescuer, three types of collision is introduced and with the analysis one by one, the kinematics parameters and the impact force is calculated in the paper using the theorem of momentum, impulse and angular momentum.

Acknowledgment

This work was supported by the Science for Earthquake Resilience of China Earthquake Administration (No.XH16045Y, No.XH16046).

References

Here are some examples:

1. X.P.Tian, J.Z.Sun, Q.Chen, L.P.Liu, A.P.Tian, J. Eng. Geol.**17**,3(2009).
2. Q.Xu, R.Q.Huang, J. Eng. Geol.**16**,6(2008).
3. Q.Lv, H.Y.Sun, S.K.Zhai, H.B.Wang, Y.Q.Shang, J. Nat. Disaster.**12**,2(2003).
4. H.Q.Yang,X.P.Zhou, Rock Soil Mech.**30**,11(2009).
5. X.Feng,J.Guo,J.Liu, Subgrade Eng.6(2013).
6. H.M.Tang, P.Y.Yi, J. Chongqing Jianzhu U. **25**,1 (2003).
7. R.Q.Huang, W.H.Liu, J. Rock Mech.& Eng. **28**,5 (2009).
8. R.Q.Huang, W.H.Liu, J.P.Zhou, X.J.Pei, J. Geo. E. **29**,9(2007).
9. X.J.Pei, R.Q.Huang, Z.Pei, X.J.Dong, J. Eng. Geol. **19**,4(2011).
10. B.L.Li, X.J.Pei, Q.Cheng, S.G.Li, Gansu Water Resources And Hydropower Technology .**46**,2(2010).
11. Q.Cheng, S.R.Su, Rock Soil Mech. **35**,3(2014).
12. S. Kawahara , T. Muro, J. Terramechanics. **43**,3(2006).
13. V. Labiouse, F. Descoeudres, S. Montani, STRUCT ENG INT.**6**,3(1996).
14. B. Pichler, C. Hellmich, and H. A. Mang, INT J IMPACT ENG.**31**,5(2005).
15. *JTG D30-2004, Specifications for Design of Highway Subgrades* (China Communication Press,2006).
16. Q.X.Yang,B.S.Guan, J Chin Railway Soc.,1(1996).

17. J.K.Yuan, R.Q.Huang, X.J.Pei, Rock Soil Mech. **35**,1(2014).
18. K.Cheng, C.S.Zhang, J. Rock Mech.& Eng. **19**,4 (2000).
19. C.P.Wu,L.S.Liu,J.L.Dou,G.Q.Zhang, Eng. Blasting, **18**,2(2012).
20. J.H.Chen,Y.Peng,S.L.Wu, Blasting,**32**,1(2015).
21. Q.Liu,G.Q.Zhang,C.P.Wu,T.J.Tao, Blasting,**30**,1 (2013).

Simulation Process Analysis of Rubber Shock Absorber for Machine Tool

Rong Xia CHAI[1] , Hua Ying WU[2,a] ,Chuan Wei ZHANG[1] and Wei GUO[1]
[1]*College of Mechanical Engineering, Xi'an University of Science and Technology, Xi'an 710054, PR China*
[2]*Department of Mechanical Engineering, Xi'an Jiaotong University, Xi'an 710049, PR China*

Abstract. The simulation on rubber shock absorber of machine tool was studied. The simple material model of rubber was obtained by through the finite element analysis software ABAQUS. The compression speed and the hardness of rubber material were considered to obtain the deformation law of rubber shock absorber. The location of fatigue were confirmed from the simulation results. The results shown that the fatigue position is distributed in the corner of shock absorber. The degree of deformation is increased with increasing of compress speed, and the hardness of rubber material is proportional to deformation.

1 Introduction

Guideway cover made is an important part of high-speed machine. The vibration of cover made impacts on the accuracy and stability of machine tool. So the rubber shock absorber has an important application in high speed machine[1-2].The deformation conditions is important for the life of shock absorber. Some research works on the rubber materials have been reported. Dynamic compression properties and constitutive model with strain rate effect of rubber material have been studied[3]. Large deformation finite element analysis of rubber isolator has been analyzed[4]. Effect of strain rate on mechanical properties of vulcanized rubber has been researched[5]. Kinetic friction characterizations of the tubular rubber seals has been studied[6]. Thermal stress development of liquid silicone rubber seal under temperature cycling has been analyzed[7]. With the advent of better simulation tools, engineers continue to aim at replicating the complex behaviour in the laboratory, followed by the methods of computer-based, such as finite element[8-9]. A shock absorber is be product by a material of rubber which is elastic type.

In this paper, the simulation of rubber shock absorber is studied. The location of fatigue has been found, and the effect of compression speed and rubber materials on deformation behaviour of rubber shock absorber has been discussed.

2 Experimental procedure

2.1Material property analysis

The deformation of the shock absorber belongs to the large elastic deformation. Therefore, there is a need to have the material uniaxial compression, biaxial compression, plane compression stress and strain parameters, the other did not provide the material of these performance parameters. At present, the performance parameters of the material with shore hardness of 88 and 72 are applied to the simulation. Stress and strain curves of uniaxial compression of materials of 72 and 88 were determined according to the data of the investigation. As many models are necessary to run an assembly performance FEA, there are elastic, elastic-plastic, orthotropic types and so on. In this paper, the numerical modelling of rubber requires non-liner elastic in FEA.

Selected shore hardness 72 of the material as the object, the material used to determine the model for the Ogden model. According to the data fitting, the basic experimental data of rubber mechanics were fitted with Ogden N=3,Fitting of rubber parameters for: μ_1= -201.161, σ_1=10.161,D_1=0; μ_2=114.144, σ_2=11.249,D_2=0; μ_3=95.885, σ_3=17.983, D_3=0. The curve of hardness 72 between fitted and experimental data can be shown in Fig.1.

[a] Corresponding author: wuhy@mail.xjtu.edu.cn

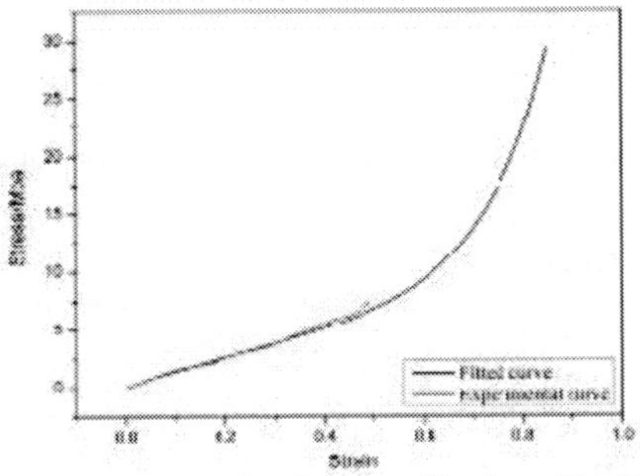

Fig.1. The curve of strain stress of hardness 72 rubber

2.2Modeling Building

According to the force of the shock absorber, Put the data into the ABAQUS software, and then a shock absorber in numerical simulation under load alternating. The model is simplified, and the model is shown in Figure 2:

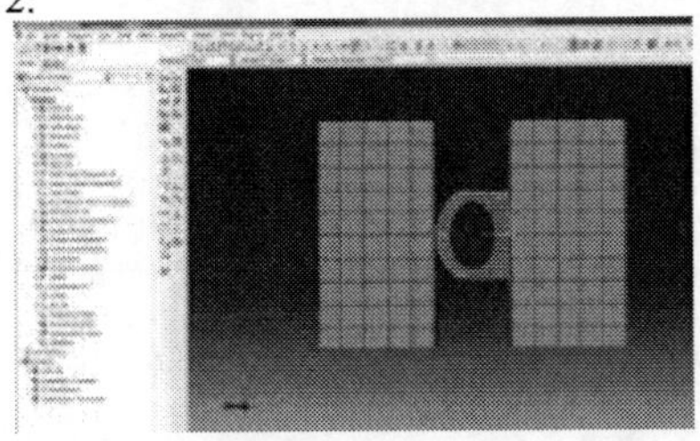

Fig. 2. Compression model of shock absorber

3 Result analysis

3.1Change trends of observation points of shock absorber

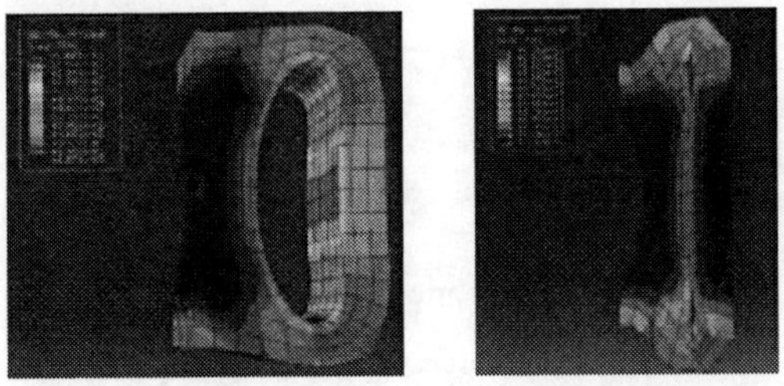

(a) Compression process (b) Compression end

Fig. 3. Elastic strain of the compression process of the shock absorber

Strain variation in the deformation process of the material is shown in Fig 3. According to compression process, material deformation distributions for some different can be seen in the upper part and the two sides of the cushion deformation is relatively large, the force analysis of shock absorber is obtained based on the deformation of the two sides. According to the compression of the shock absorber set two points as the observation point, its observation position as shown in figure 4.

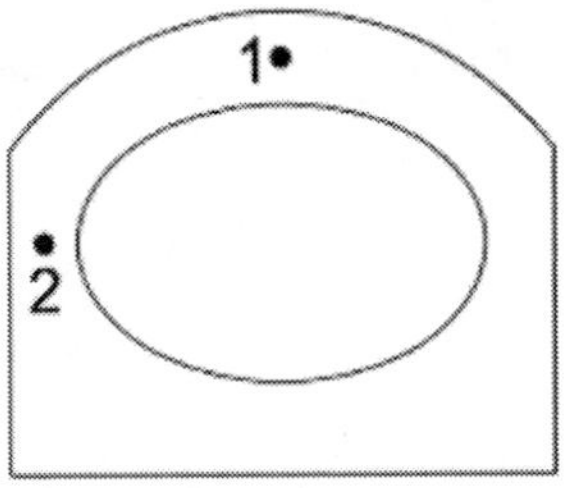

Fig. 4. Schematic diagram of the observation position

The changes of stress and strain at different compression positions are shown in Figure 5 and Figure 6, respectively. It can be seen that the stress and strain are first increased with the change of the compression stroke, and the performance is tensile stress. After reaching a vertex position, the rate of increase is slowed down. It is deeply related to material hardening caused by compress in deformation process.

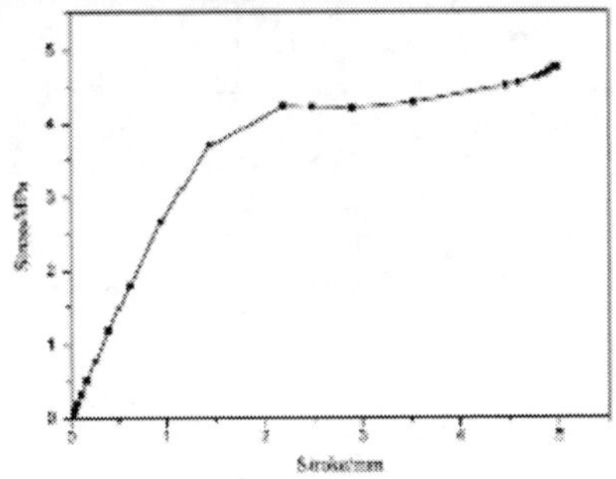

Fig. 5. Stress-stroke curve of point 1

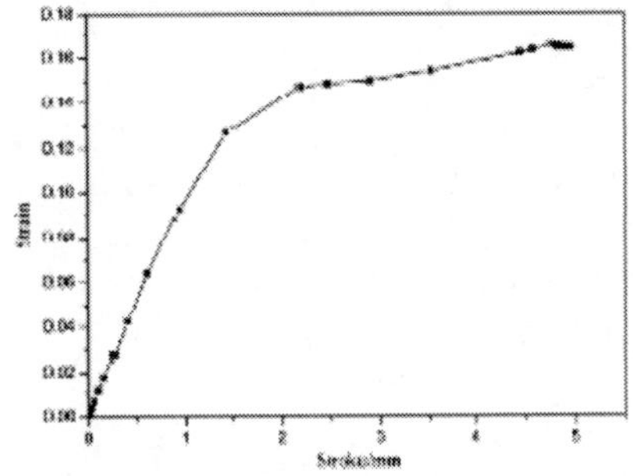

Fig. 6. Strain-stroke curve of point 1

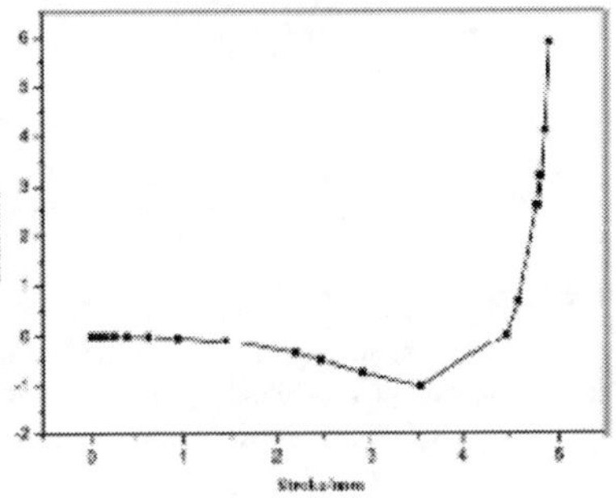

Fig. 7. Stress-stroke curve of point 2

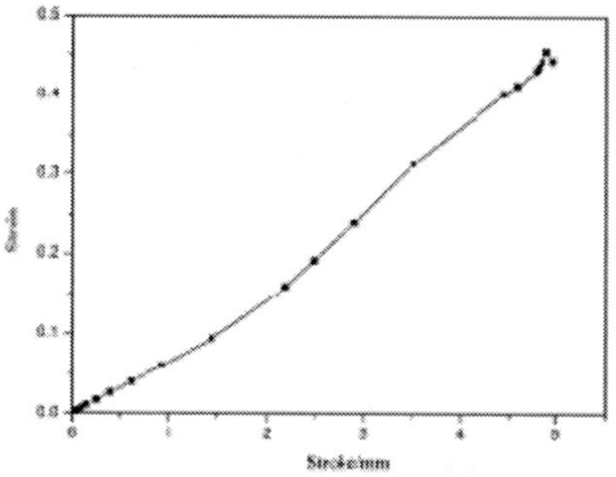

Fig. 8. Strain stroke curve of point 2

The changes of stress and strain at different compression positions are shown in Figure 7 and figure 8, respectively. It can be seen that the observation point 2 is first subjected to tensile stress in the compression process, With the increase of compression stroke, when the stroke is about 3.5mm, the maximum value of tensile stress is reached. As the compression stroke continues to increase, the observation point 2 gradually changed to compressive stress. And the compressive stress increases with the increase of the compression stroke at a faster acceleration. The strain has increased linearly with the effect of compression speed on the performance of shock absorber.

As shown in Figure 9 and 10 for the shock absorber speed of 60m/min and 120m/min, the shock absorber hardness is 88, the deformation of the observation point 2.

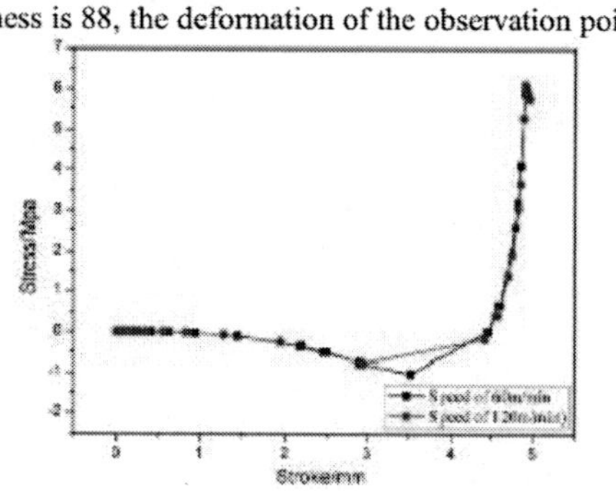

Fig. 9. Stress stroke curve under different compression speed

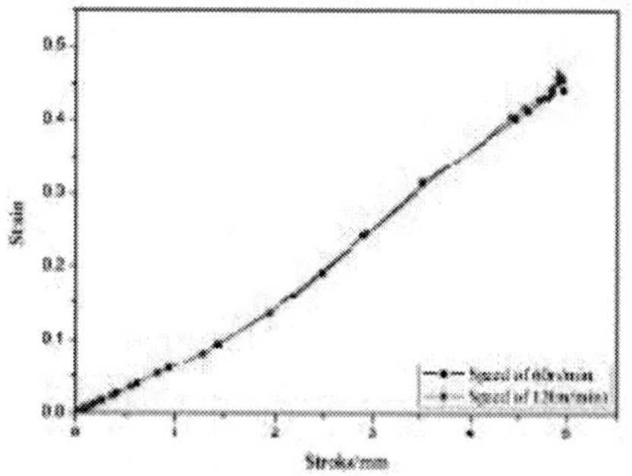

Fig. 10. Strain stroke curve under different compression speed

It can be seen from the figure above, the compression speed of the shock absorber is consistent with the change of the material, that is, the speed is different, the performance of the material is not affected by the change of the trend.

3.2 Effect of materials on the performance of shock absorber

The changes of stress, strain and compression stroke of materials with different hardness at the observation point 1 are shown in Figure 11 and 12.

From the figure 11 and 12, we can see that the hardness of the rubber material is consistent with the influence of the compression deformation of the shock absorber, the difference is that the higher the hardness, the greater the stress and strain in the deformation process.

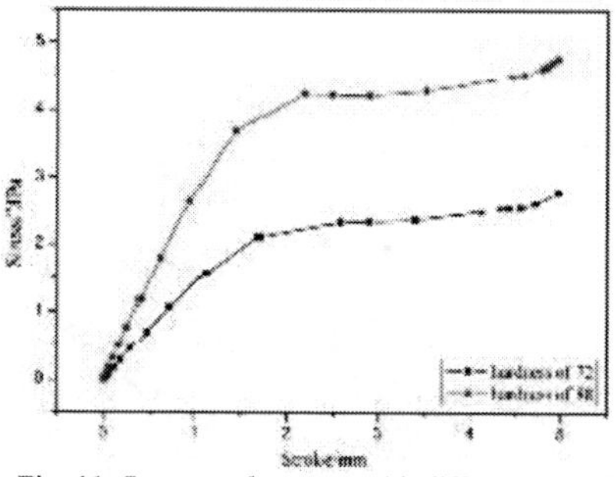

Fig. 11. Stress stroke curve with different material

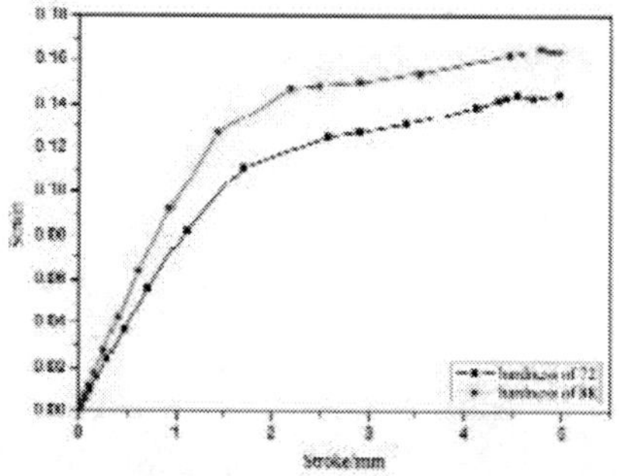

Fig.12. Strain stroke curve of shock absorber with different material

3.3 Effect of observation position

The different observation points at the speed of 120m/min under the compression process of the deformation for the material hardness of 88 as shown in figure 14 and 13.

DOI: 10.1051/matecconf/20166302042

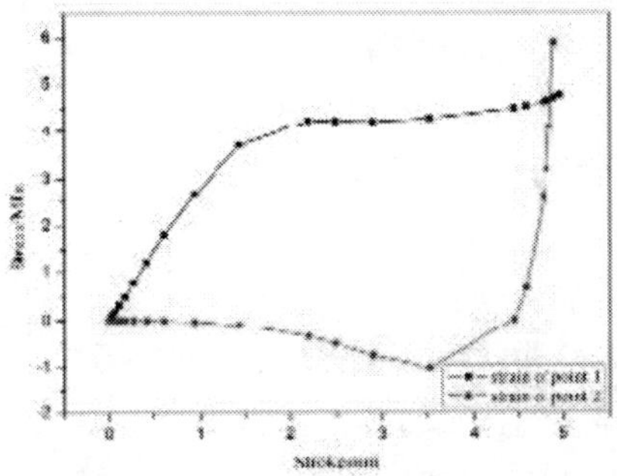

Fig. 13. Stress stroke curve of the shock absorber with different material

According to figure 14 and 13, it can be seen that Point 1 in the deformation process has been in the compressive stress, and the stress has been increasing, the point 2 is the first compressive stress. With the gradual increase of deformation, the point 2 is subjected to compressive stress. The stress of the point 2 should be greater than the stress of the point 1. It can be seen that the point 1 of the strain gradually increased to a certain amount, it showed a slow growth trend. The strain at point 2 appears to increase rapidly with acceleration.

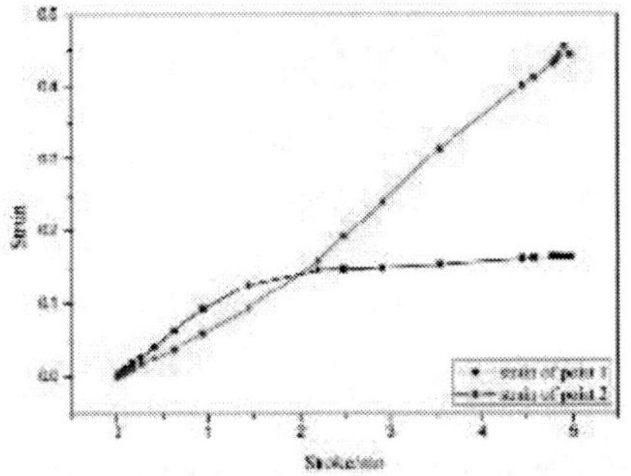

Fig. 14. Strain stroke curve of shock absorber with different material

4 Conclusions

(1) According to the simulation results, we can know that the speed of machine tool has no obvious impact on the deformation degree in the course of operation foe the shock absorber.

(2) In the simulation process, the risk point is in the edge position of the shock absorber. So the failure of the shock absorber can be determined according to the deformation of the risk point during the operation of machine tool.

(3)The same material, the hardness is not the same, for a higher hardness of material deformation resistance is also bigger.

Acknowledgement

The authors gratefully acknowledge the support from the Doctor Fund of the Major Project of High- end CNC Machine Tool and Basic Manufacturing Equipment of China(2014ZX04011021), Shanxi National Science Foundation (2015JM5257), Science Foundation for Education Department of Shanxi Provincial Government(14JK1462).

References

1. Bin Sun, Xiaoxu Bi. The structural analysis based on guideway cover made of high speed machine tool[J].Silicon valley,2013,124(4):43-44.
2. YANG Bing, GUO Wenjie, ZHANG Lu,etc., Manufacturing X—shield used in high speed horizontal machining center[J]. Manufacturing Technology & Machine Tool,2016,(1):129-131.
3. Pang baojun, Yang zhenqi,Wang liwen, etc., Dynamic compression properties and constitutive model with strain rate effect of rubber material[J]. Chinese journal of high pressure physics.2011,25(5):407-415.
4. Zhou zhenkai, Xu bing, Hu wenjun ,etc., Large deformation finite element analysis of rubber isolator[J]. Journal of vibration and shock.2013,32(3):171-175.
5. Wu changhe, Feng xiaowei, Ye pei,etc., Effect of strain rate on mechanical properties of vulcanized rubber[J]. Journal Of Functional Materials.2013,44(8):1098-1101.
6. Yuchao Ke, Xuefeng Yao, Heng Yang, etc., Kinetic friction characterizations of the tubular rubber seals[J]. Tribology International,2014,72:35-41.
7. Tong Cui, Y.J. Chao, J.W. Van Zee. Thermal stress development of liquid silicone rubber seal under temperature cycling[J]. Polymer Testing,2013,32: 1202-1208.
8. Dr Ben Chouchaoui. Rubber seal development via computer simulation[J]. Sealing Technology,2016,01:8-12.
9. Wu kaisong, Xu daping, Yan yongfa.etc., Data processing method of rubber testing based on incompressible large deformation analysis[J]. China Rubber Industry,2013,60:400-403.

DOI: 10.1051/matecconf/20166302043

The Design of Angle Auto Direction-adjusting System for Wind Tunnel Motion Mechanism

Xiao Tao Yu [1, a]

[1] *AVIC Shenyang Engine Design and Research institute, Shenyang, 110015, China*

Abstract. In this article a motion mechanism angle self direction-adjusting controlling system is designed. The main function of this system is to achieve the angle following control with the variation of inflow direction. This article describes the construction and working principle of the angle self direction-adjusting system, aiming to the controlling target the self-adapting fuzzy controller is designed, the controller is then simulated and analyzed after adding square wave signals. It is validated that the self-adapting fuzzy PID controller has a good dynamic, static and robust characteristics.

Keywords: angle control; self-adapting; fuzzy control; MATLAB simulation

Introduction

When the wind tunnel is built up, the performance of the wind tunnel required to be checked out through flow field verification. Flow field verification requires a set of equipment, by fixing probes and movable ones of certain requirements in order to test the parameters of flow stability, dynamic pressure field, flow deviation angle and pressure grads of model at each point on the flow section of testing segment. in the testing process it is required that the probe direction should right face to the inflow direction all the time, which means the motion mechanism need to have angle auto direction-adjusting function. To improve the flow field test efficiency, guarantee the testing accuracy, the motion mechanism angle self direction-adjusting control system is designed [1,2].

1 Motion mechanism system components

The motion mechanism is constructed by a DLK120 beeline anchor cell and two ELK40 single-slide ball screw drive positioning cells as well as 1D rotary platform and the accessories, the structure is shown in Fig1. There are 4 degrees of freedom for the motion mechanism, namely X, Y, Z and M axes. Each axis is driven by servomotor, where M axis is angle axis, by driving the M axis moving through motor it is able to achieve probe auto direction-adjusting.

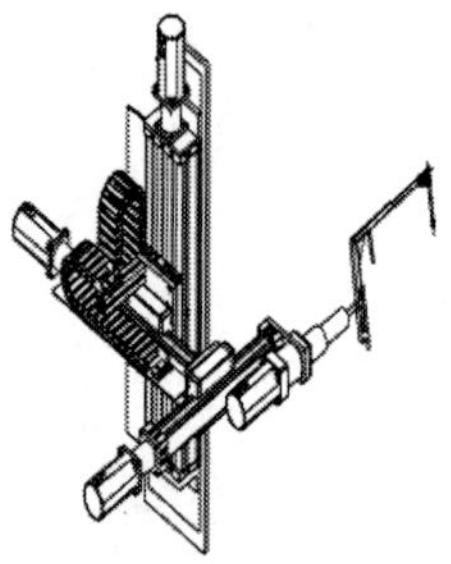

Fig.1. Self-adapting fuzzy PID controller

2 Working principle

The basic principle of the motion mechanism angle following system is to transform the pressure signal received by the controlled probes into standard signal through sensors. The signal is compared with real-time adjusting data collected by PLC, and then the deviation signal is sent into servo driver to operate reversible motor rotating in contrary direction; to control the overshoot of the system, Ka, the correction factor, is calculated out through the relative PLC measuring parameters to control the command impulse of the servo driver to make the probe taken by the servomotor rotate to the direction of smaller deviation voltage, in this way the probe is made to achieve self equilibrating in the flow direction.

3 The design and simulation for the controller of the angle control system

The self-adapting fuzzy PID controller structure is shown in Fig.2. The core of the self-adapting fuzzy PID controller is to apply PID controller to control system. The inputs of the fuzzy deduction is deviation and deviation varying rate ec, after fuzzy deduction three parameters of PID controller, namely

[a] Corresponding author: 505031090@qq.com

DOI: 10.1051/matecconf/20166302043

K_p, K_i, K_d are auto modified in the parameter modification stage to satisfy the system requirement. It can be seen that since the fuzzy deduction and parameter modification stages are added, even with different control system a good control result is able to be achieved by online setting modifying parameters, meanwhile by using this control method not only the controlled object has a good dynamical performance but also has a good static performance[3].

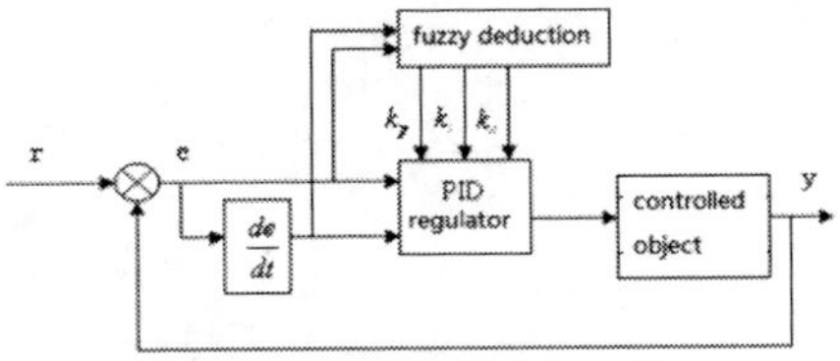

Fig.2. Self-adapting fuzzy PID controller

When using deviation and deviation varying rate to indicate a PID control, the representation is:

$$u(k)= K_p e(k)+ K_i \sum e(k)+ K_d ec(k) \qquad (1)$$

Where $u(k)$ is controller output, $e(k)$ is deviation, $ec(k)$ is deviation varying rate, K_p is scale coefficient, K_i is integral action coefficient, K_d is derivative action coefficient.

The calculation formulas that this system achieved self-adjusting PID parameters are as follow:

$$K_p = K_p' + \Delta K_p \qquad (2)$$

$$K_i = K_i' + \Delta K_i \qquad (3)$$

$$K_d = K_d' + \Delta K_d \qquad (4)$$

Where K_p', K_i', K_d' are PID initial values, $\Delta K_p, \Delta K_i, \Delta K_d$ are the fuzzy controller's outputs, K_p, K_i, K_d are the final output control parameters.

From the above we knew that self-adapting PID is to find the fuzzy relationships among three parameters K_p, K_i, K_d and e as well as ec, kept checking e and ec during the operation, based on the fuzzy principle to carry out online modification for the three parameters to satisfy different requirements for control parameters with different e and ec so that the controlled object had good dynamical and static characteristics. In the following sections design method for this system controller would be described particularly.

3.1 The fuzzification of precise value

The deviation e and deviation varying rate ec of this system are taken as inputs, K_p, K_i, K_d are as three outputs of the fuzzy controller. The domain of the deviation e and deviation varying rate ec is [-3,3], the aggregate in the domain is {NB,NM,NS,ZO,PS, PM,PB}, assume the basic domain of ΔK_p is {-3,3}, the aggregate in the domain is {PS PM PB}. Assume the basic domain of ΔK_i is {-0.1, 0.1}, the aggregate in the domain is {PS PM PB}. Assume the basic domain of K_d is {-1, 1}, the aggregate in the domain is {PS PM PB}. See Fig.3 to Fig.7.

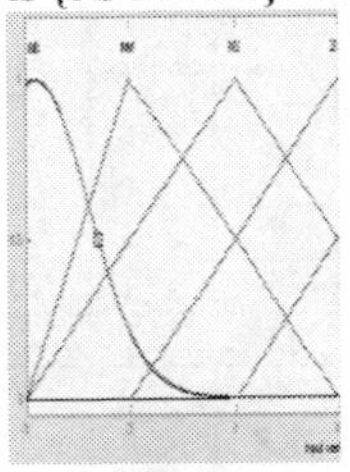

Fig.3 Membership grade of deviation e

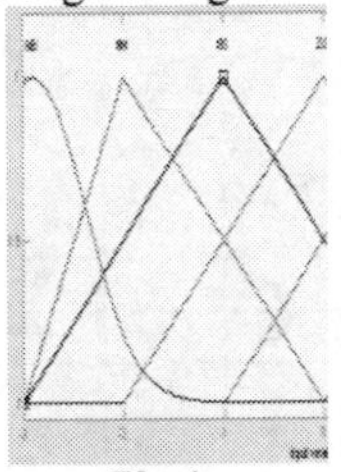

Fig.4 Membership grade of deviation ec

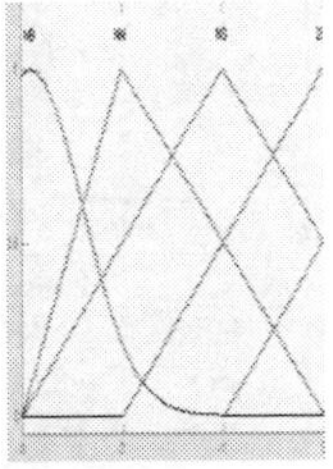

Fig.5 Membership grade of ΔK_p

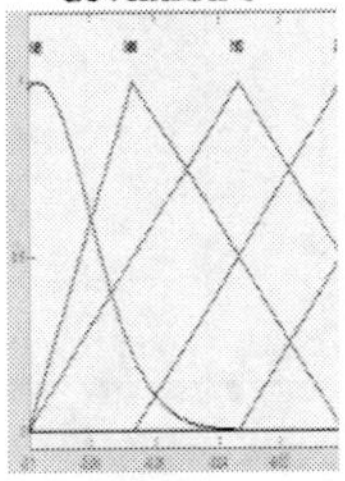

Fig.6 Membership grade of ΔK_i

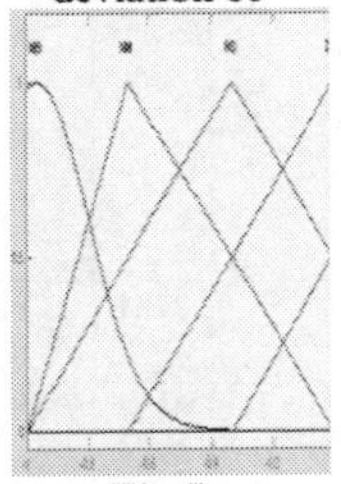

Fig.7 Membership grade of ΔK_d

3.2 The construction of control regulation

By a huge amount of experience gathering, there is a kind of non-linear relationship between PID parameter and input controller deviation e and deviation varying rate ec, the relationship can be described in fuzzy language. Deviation e and deviation varying rate ec have a following relationship with the three parameters of PID controller K_p, K_i, K_d. When $|e(t)|$ is larger, to get a faster system responding rate, a larger K_p shall be taken to reduce the system time constant and damping coefficient, too large one may lead to a system instability; to avoid the possibility of over-range controlling effect at the beginning of the system, a smaller K_d shall be taken to raise the system response. To avoid a bigger overshoot to take out integral action, take $K_i = 0$. When $|e(t)|$ is right a middle value, a smaller K_p shall be taken to get a slightly smaller overshoot; to assure the responding speed, a suitable K_d shall be taken, a larger K_i can be taken but not too much. When $|e(t)|$ is relatively small, to make sure a good stability, a larger K_p and

K_i shall be taken to avoid surge when system is at the equilibrium point, the value of K_d shall be proper[4,5]. See Table1 to Table3.

Table 1 The fuzzy regular table of ΔK_p

e \ ec (K_p)	NB	NM	NS	ZO	PS	PM	PB
NB	PB	PB	PM	PM	PS	ZO	ZO
NM	PB	PB	PM	PS	PS	ZO	NS
NS	PM	PM	PM	PS	ZO	NS	NS
ZO	PM	PM	PS	ZO	NS	NM	NM
PS	PS	PS	ZO	NS	NS	NM	NM
PM	PM	PS	ZO	NS	NM	NM	NB
PB	ZO	ZO	NM	NM	NM	NB	NB

Table2 Controlling regular table of ΔK_i

e \ ec (K_i)	NB	NM	NS	ZO	PS	PM	PB
NB	NB	NB	NM	NM	NS	ZO	ZO
NM	NB	NB	NM	NS	NS	ZO	PS
NS	NB	NM	NM	NS	NS	PS	PS
ZO	NM	NM	NS	NS	ZO	PS	PM
PS	NM	NS	ZO	PS	PS	PM	PM
PM	ZO	ZO	PS	PS	PM	PB	PB
PB	ZO	ZO	PS	PM	PB	PB	PB

Table 3 the controlling regular table for ΔK_d

e \ ec (K_d)	NB	NM	NS	ZO	PS	PM	PB
NB	PS	NB	NB	NB	NB	NM	PS
NM	PS	NS	NB	NM	NM	NS	ZO
NS	ZO	NS	NM	NM	NS	NS	ZO
ZO	ZO	NS	NS	NS	NS	NS	ZO
PS	ZO	ZO	ZO	ZO	ZO	ZO	ZO
PM	PB	NS	PS	PS	PS	PS	PB
PB	PB	PM	PM	PM	PS	PS	PB

Fuzzy regular table can be written in the form of conditional statement, for ΔK_d the 46 controlling regulars have following forms:

if e=NB and ec=NB then ΔK_d=PB

if e=NB and ec=NM then ΔK_d=PB

if e=NB and ec=NS then ΔK_d=PM...

In the same way there are also 49 fuzzy regulars for ΔK_i and ΔK_d respectively.

3.3 Software achievement

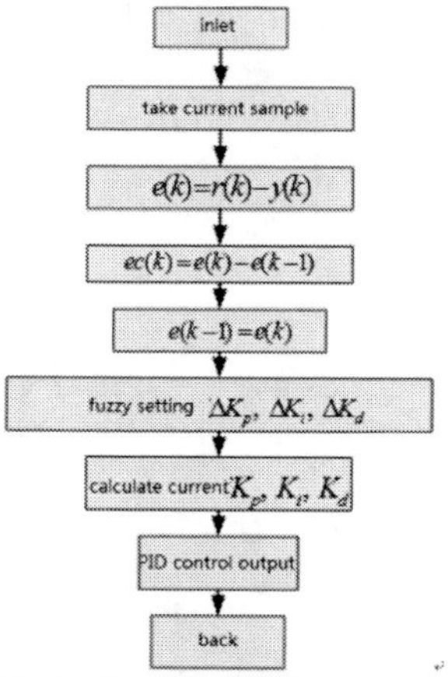

Fig.8. Self-adapting fuzzy PID control program flow chart

In the process of online operation, controlling system accomplish PID parameter online self-adjusting through result solving, table look-up and calculation for fuzzy logic rules. The working flow diagram is shown in Fig.8.

4 Simulation and analysis on self-adapting fuzzy PID control

Neglecting the system friction and load influence, utilized the control system simplified math model deduced in this article, then carry out system simulation research in the environment of Matlab/Simulink, with two conditions of non-self-adapting fuzzy PID control and applying self-adapting fuzzy PID control method, gave out the respond curves of stepped signal and sine following signal.

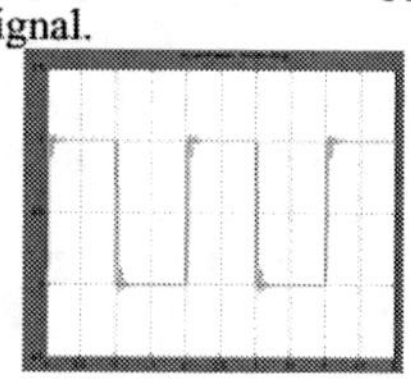

Fig.9. Square wave responding of the system

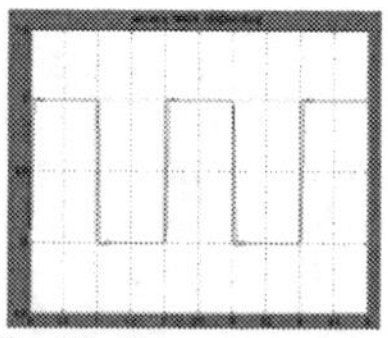

Fig.10. the square wave responding of self-adapting fuzzy PID control system

From Fig.9 it can be seen that the system without modification has a larger overshoot and long adjusting time, the static accuracy is unable to satisfy the technique requirements of control system. After adding up self-adapting fuzzy PID control strategy, as is shown in Fig.10, the system overshoot reduces and

DOI: 10.1051/matecconf/20166302043

static accuracy improves, which satisfies the need of system.

5 Conclusions

In this article base on the model of wind tunnel motion mechanism control system, the application of self-adapting fuzzy PID control arithmetic on motion mechanism control system is studied, the design method of self-adapting fuzzy PID control arithmetic is properly described, and the simulation model of self-adapting PID control arithmetic is built up in SIMULINK, the parameters of control arithmetic are debugged, and the simulation results are analyzed and compared. According to the analysis, self-adapting fuzzy PID controlling arithmetic is able to achieve a effective action on angle control system, and shows a better control effect and robust.

References

[1] Chen Taihong. Design of Moving and Measurement Shelf of Single Free Degree and its Control System for Wind Tunnel[J].Industrial Control Computer,2006,19(3) ： 1-2.

[2] Chen Taihong. Control System for Transfer Rack of Wind Tunnel[J]. Measurement &Control Technology, 2005,24(4):80-84.

[3] Song Fengmin. Study on electro-hydraulic servo fuzzy control system of hydraulic cylinder assembling specialized equipment. Master thesis: Shandong University of Technology, 2007.

[4] Zhang Pengcheng. Research on the Intelligent Control Algorithm of 6R Robot. Master thesis: South China University of Technology., 2011.

[5] Mao Bo. Study on self-learning fuzzy control strategy for electro-hydraulic servo system. Master thesis: Yanshan University, 2011.

Technical Design of Flexible Thin-Film Solar Heating Clothes with Switchable Output Power

Yu Xiao Zhao[1,a], Ya Li Li[1], & Li Min Shi[1]

[1] *The School of Fashion Art and Engineering, Beijing Institute of Fashion Technology, No. 2A Yinghua Road, Chaoyang District, Beijing, 100029, China*

Abstract. This research focuses on the research and development of thermal clothes through technical design, by adopting unique removable electronic equipment and applying carbon fiber material to thermal clothes against cold, so as to meet the requirements of active heating and passive warmth retention. Firstly, the specification of power supply system was determined in accordance with the requirements of power system, and the specification of charging system was determined according to the specification of power system. Then circuit system was designed and tested. Fianlly, the electronic device was configured on the clothes appropriately, so that it should be conforms to ergonomic principles, convenient and fast.

Keywords: Heating clothes, flexible thin-film solar cells, carbon fiber, wearable technology.

Introduction

With the social and economic development, people have increasingly high requirements for energy. Solar energy, as a new energy resource, has huge development potential. Solar heating clothes which can be used to replace traditional thermal clothes are a new type of self-heating clothes. On one hand, the clothes can eliminate the limitation of charging with power supply during outdoor activities. On the other hand, the application of the new material of carbon fiber makes the thermal clothing develop from passive warmth retention to active warmth retention. Thus, the clothing is lighter and more beautiful. At the same time, the unique electronic equipment and removable design facilitate the cleaning of clothes, to maximize the potential safety hazard, which is an innovative attempt to warm clothing.

Functional Requirement for Heating Clothes

Based on the heat exchange form between human body and external environment, firstly, it was confirmed that the applicable circumstance temperature for the clothes is 0℃. Secondly, it was determined that the thermal insulation of the clothes during heat exchange between human body and environment to reduce heat exchange to the greatest extent and maintain the balance between heat production and heat release of human body at a comfortable state. Thus, the passive thermal functional requirement for flexible thin-film solar heating clothes is reducing convection heat loss and increasing the resistance of internal and external thermal conduction of clothes. This requires the clothes having good warmth retention property for human body heat and good wind protective effect for cold air to reduce cold air invasion and cause air cooling effect. Then, it is required to supplement the inadequate heat by autonomic regulation of human body and the heat loss through various heat exchanges between human body and external environment. And electric heating units are required for active heating to reach thermal comfort. The functional requirements for specific active heating are as follows:

(1) The internal temperature of electric heating clothes should be maintained within the human body comfort temperature and can be manually adjusted in accordance with external environment and person's needs.

(2) Electric heating unit adopt portable rechargeable batteries. The batteries can be charged with the charger for direct charging indoor and it is inconvenient to charge outdoor. Thus, flexible solar cells can be used for charging.

Specification Setting and Electronic Device Selecting

Specification Setting

Firstly, the specification of power supply system was determined in accordance with the requirements of power system, namely load. Then the specification of charging system, namely solar cells was determined according to the specification of power system, namely lithium battery.

The specification of the load - heating unit is determined as 9W according to the applicable temperature of 0℃. The voltage of lithium battery is selected as lower than 12V according to the safety voltage which the body can bear (the voltage of single lithium battery is 3.7V, the voltage of lower than 12V includes 3.7V, $3.7\times2=7.4$V and $3.7\times3=11.1$V). Due to the heating quantity that human body can feel, the load current should reach up to 1A at least. In accordance with the formula 1, the selected battery specification is 7.4V at last. Then the specification of solar cells is determined in accordance with the charging principle of lithium battery.

[a] Corresponding author: yuxiao.z@163.com

I=P/U
(1)

In the formula: P-Power (W);
U-Supply voltage (V);
I- Current (A).

Lithium battery is the battery which has high requirements for charging voltage. Over charge or discharge will cause irreparable damage to battery and battery failure [1]. Explosion can be caused due to serious overcharge. Polymer lithium battery (PLB) is a new generation of lithium battery. Compared with the original Liquified lithium-ion battery (LIB), PLB is featured with thinness, arbitrary area and arbitrary shape. In addition, there is no liquid leakage, combustion explosion and other safety problems. At the same time, the specific energy of PLB is more than 50% of the current ordinary lithium battery. The capacity, charge and discharge characteristics, safety, operating temperature range, cycle life and environmental performance are improved greatly [2]. So the PLB is adopted in this research.

At present, the electrochemical window of the electrolyte of the anode and cathode material adopted by the PLB on the market is basically over 4.5V. That is to say, when charging voltage of battery reaches 4.5V, the battery is in a stale state. However, if the charging voltage of battery is too low (lower than 4.15V), the charge level of battery will be affected greatly. If the charging voltage is controlled at 4.13V, the charged electric quantity is about 60～80% of the battery capacity, which will greatly affect the battery performance [3]. To sum up, the charging voltage for single lithium battery shall be controlled within 4.2～4.5V. The lithium battery charging voltage of 7.4V obtained by two lithium batteries connected in series should be within 8.4～9.0V. That is to say, the charging voltage of solar cells shall be within the range mentioned above.

Electronic Device Selecting

Flexible Thin-film Solar Cells

The dimension of each solar cell is 190mm in length, 50mm in width and 1mm in thickness. And the single cell weights 20g. The optimum operating voltage is 1.5V and the optimum operating current is 300mA. The open circuit voltage is 1.5V and short-circuit current is 380mA, as shown in Fig.1.

Five solar cells can be selected for charging. Single measured charging voltage is 1.9V and the output voltage of the whole piece of solar panel composed by five solar cells is 9.5V. And the charging current is 250mA. As the output current of solar cells changes with illumination intensity, a diode should be connected in series between flexible solar cell and lithium battery. The diode shall be positive onset in reverse blocking state [4] to avoid the inverse current due to voltage and current dip, which is lower than the voltage of lithium battery, after sunlight is blocked or charging direction changes sharply. In addition, the diode has step-down function. After a diode is connected in series, the measured charging voltage is 9.0V and the charging current is 220mA, which basically meet the requirements of the charging voltage and current.

Fig. 1 Flexible thin-film amorphous silicon solar cell

Lithium Battery

Two specifications of the same voltage and different capacity are used. The battery capacity of the preferred outdoor solar charging equipment is 1300mAh and the charging time is within 5～6 hours. In addition, indoor charger for direct charging can also be used and the charging time is 1.3 hours. The battery capacity of the preferred indoor charger for direct charging is 2200mAh and the charging time is 2.2 hours (outdoor solar charging equipment can also be used but the charging time is relatively long for about 10 hours). The two batteries are provided with protective circuit with overcharge and over-discharge protection functions. The over-charged protection voltage is 8.4±0.05V and the over-discharged protection voltage is 6.0±0.05V. That is to say, if the voltage is more than 8.4±0.05V and lower than 6.0±0.05V after charging, the circuit will cut off automatically. And no-load removal will be delayed or charging will automatically release.

Lithium Battery Charger for Direct Charge

Input specification is U=AC100～240V, P=47～63Hz and I= 0.5A, output specification is U=DC8.4V and I= 1A. Charging mode is constant current first and then constant voltage. For charging, the charging light is red. For a full charge and on load, the charging light is green. This is because charger is equipped with guard plate. When battery voltage is about 8.4V and charging current is less than about 50mA, the protection circuit will cut off charging current.

Carbon Fiber Wire

The carbon fiber wire produced by Toray is adopted and the specification is 12k. Electrical resistivity R1=32Ω/m, as shown in Fig. 2.

First, the resistance and length of resistance wire, namely carbon fiber wire were obtained in accordance with the formula 2.

$$RT=U^2/P$$

In the formula: P——Power (W);

U——Supply voltage (V), $U=3.7\times2=7.4V$;

R_T——Resistance (Ω), $RT=7.4^2/9=6.08\Omega$.

Because the specification of the purchased carbon fiber resistance wire is 12k, that is to say, there are 12000 carbon fibers for each bunch of carbon fiber and the parallel connection with N=8 is adopted. The resistivity of the carbon fiber resistance wire per centimeter is $R2=R1/N=0.32/8=0.04\Omega$; the length of carbon fiber wire meets the requirement $L=R_T/R2=6.08/0.04=150cm$.

Temperature Controller

The single-chip micro-controller provided by Shenzhen Kanghualin Industry Company is adopted, as shown in Fig. 3, 2cm at the widest point and less than 5g in weight. The resin encapsulated type is featured with accurate control, low-voltage operation, less energy consumption and beautiful appearance. The controller can control output power through controlling the output current of lithium battery, so as to control the temperature of heating unit.

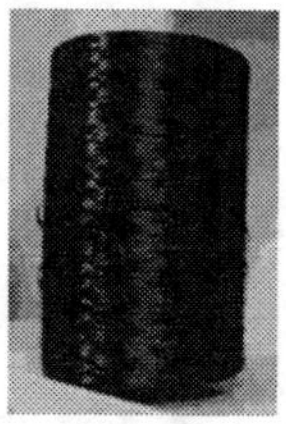

Fig. 2 Carbon fiber wire produced by Toray

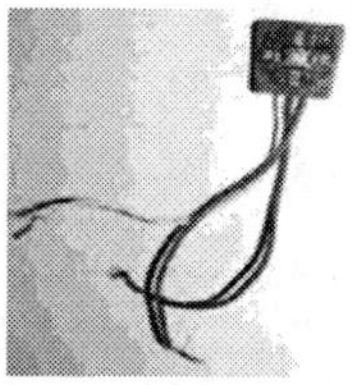

Fig. 3 Temperature controller

Work Principle of Circuit

Charging Circuit of Solar Cells

The charging circuit of solar cells is as shown in Fig. 4.

Solar Cell Charging Process

Flexible solar cell can generate current with sunlight and charge lithium battery by diode. After pressing the switch, when the lithium battery voltage is 8.4V displayed on the digital display module, the charging is complete (the charge cut-off voltage of 3.7V single lithium battery is 4.2V and the charge cut-off voltage of two lithium batteries is 8.4V). As the charging current of solar cells is small, which can be regarded as constant flow source low current charging, at this time, the voltage of lithium battery increases gradually? In general, when the voltage reaches up to 8.4V, the charging is completed.

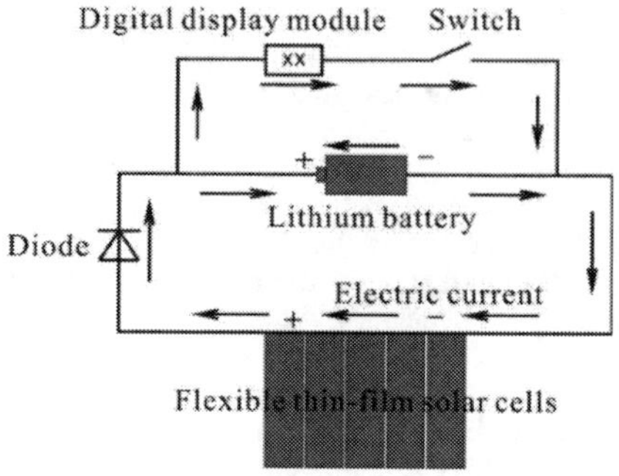

Fig. 4 Solar cell charging circuit diagram

Solar Cell Charging Test

When the lithium battery power is used until automatic over-discharge protection is started and the voltage decreases to 6V, the electric quantity is 0 at this time. After charging the lithium battery in the sun, the measured 1300mAh lithium battery charging curve is shown as in Fig. 5. From the figure, it can be seen that when the voltage is between 6V and 7.5V, the charging rate is fast. When the voltage is greater than 7.5V, the voltage increases slowly. When the voltage reaches 8.1V, the voltage is basically stable for 6 hours. At this time, the lithium battery is fully charged. Then the lithium battery can be removed for use.

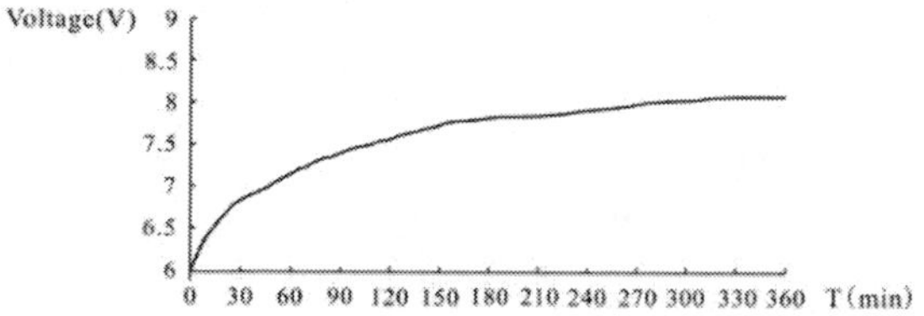

Fig. 5 Lithium battery charging voltage time profile

Temperature-Regulated Active Heating Circuit of Electric Heating Unit

Temperature-regulated active heating circuit of electric heating unit is as shown in Fig. 6.

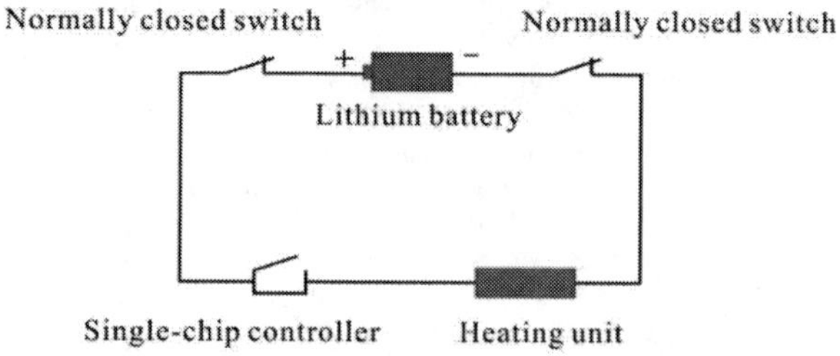

Fig. 6 Temperature-regulated active heating circuit diagram of electric heating unit

DOI: 10.1051/matecconf/20166302044

Working Process of Electric Heating Unit Active Heating

Long press the controller button for three seconds. And LED displays red flashing light. At this time, carbon fiber wire is at preheating condition. Then short press the button. The LED displays red freeze-frame. At this time, the power output is about 98%. Then press the button again. The LED displays white freeze-frame. At this time, the power output is about 75%. Press the button again. The LED displays blue freeze-frame. And the power output is about 60%. Then long press the button for 3 seconds, LDE indicator light is off without power output. Two 50℃ temperature switches are connected in parallel inside heater band. One switch is located at the chest position and the other switch is installed at the lumbar spine. When the circuit is normal working state, the bimetallic strip on switch is in free state, and the connector is closed. When the temperature reaches 50℃, the circuit can be cut off by any switch, with dual temperature protection, so as to avoid hazard due to excess temperature.

Lithium Battery Discharge Test Results

When 2300mAh lithium battery is fully charged, for red gear output, the measured service time is 160 minutes. For white gear output, the measured service time is 220 minutes. For blue gear output, the measured service time of lithium battery is 280 minutes. When 1300mAh lithium battery is fully charged, for red gear output, the measured service time is 1 hour. For white gear output, the measured service time is 80 minutes. For blue gear output, the measured service time is 100 minutes.

Overall Wiring Diagram inside Clothing

The design principle of the integration of the electronic device for flexible thin-film temperature-adjustable heating clothes and clothes is to configure the electronic device on the clothes appropriately on condition that the electronic device can be operated normally. The wire between devices should be reduced as far as possible. In addition, for electronic device operation, it shall be conforms to ergonomic principles, convenient and fast. For integration of electronic device and clothes, zipper, adhesive button and other design shall be adopted as far as possible. The used electronic equipment shall be easy to remove. The overall wiring diagram inside clothes is as shown in Fig. 7.

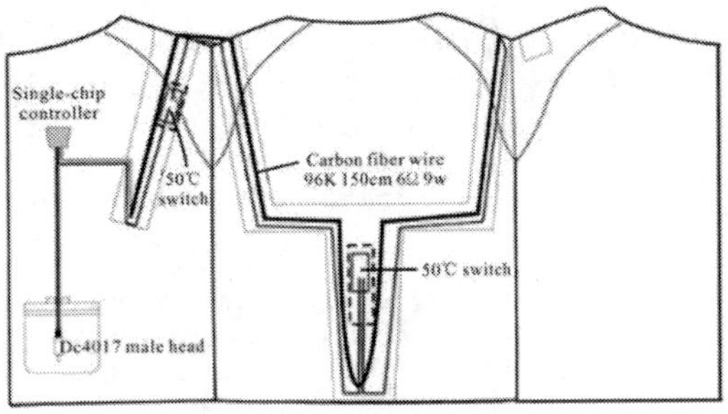

Fig. 7 Overall wiring diagram inside clothing

Conclusion

Based on the principle of reducing heat exchange as far as possible, this research determines the functional requirement of warm clothes which performs active heating on the basis of strengthening passive heat preservation to make up heat production and offset heat release in accordance with the heat exchange form between human body and external environment in cold environment, so as to determine the electric heating form of active heating.

The active heating of warm clothes is achieved as follows: in order to charge lithium battery through solar cells, with temperature-adjustable and heating electric heater bands, first, set the specification of electronic device and determine the specification of load, power supply unit and charging equipment successively. In accordance with the set specification of electronic device, select the mature electronic product on the market, design and test the whole circuit. The results show that the whole circuit reaches the requirements of technical design required for this research.

References

[1] C.H. Li, N. Wang, Analysis of Performance of Lithium-ion Battery, Inner Mongolia Science Technology & Economy. 2012(12) 71-74.

[2] B.Z. Qian, Development Situation and Prospects of Polymer Lithium-ion Battery, World Plastics. 2010(12) 44-47.

[3] Description of Polymer Lithium Battery Charging Voltage on http://www.hqew.com/tech/fangan

[4] X.C. Deng, C.Y. Lai, Role Recognization of Diode in the Circuit, Journal of Jinggangshan Normal College. 2002(S2)88-89.

Study on Damping Characteristics of Automobile Shock Absorber based on Fluid-Structure Interaction Simulation

Xin Wen Bi [1], Shao Cheng Song [1,a]

[1] *College of Information Technology and Media, Beihua University, 132013, Jilin City, China*

Abstract. The automobile's better performance of suspension system can improve good performance of power and mobility so that it has good driving operation, low vibration and low noise, comfortable driving environment. In this paper, the test to verify the simulation results of the damping characteristics by comparing the simulation external characteristics with the experimental external characteristic and analyzing the advantages and disadvantages of the simulation method.

Keywords: Shock absorber, Damping characteristics, Simulation.

Introduction

In the modern automobile industry, in addition to automotive safety, economy and power requirements in terms of comfort, reliability, durability and safety requirements are also increasing. Superior driving maneuverability, ride a good environment, low vibration and noise, it has become an important symbol of modern high-quality cars. The more superior performance suspension means the vehicle can make its full power and flexibility, its performance improved, therefore the suspension device becomes critical.

It ensures an elastic contact between the wheel and the car bearing system, and the ability to transfer loads, vibration damping, to ease the impact on the body and adjusting the height position of the moving body. Through the damper and damping effect of damping material on the suspension, the relative between the wheel and the vehicle body vibration damping off, therefore, the damper performance directly determines the performance of the suspension device. In recent years, electronically controlled variable damping shock absorber in foreign countries has been widely used. However, the current domestic and foreign design or shock absorber to complete a single design and the design of the task, with the rapid development of the automotive industry, it is necessary shock absorber designed to respond quickly to market needs, so that in a short period of completion time and cost savings. Shock Absorber performance quality of the car is critical, it directly affects the performance of dynamic resistance nits of vehicle dynamics.

The traditional design approach is usually damper according to the precision requirements and requirements of the vehicle, based on the accumulated experience to select the design parameters of the shock absorber, manufactured test sample, through experiments indeed whether the requirements of the design of the damper. This approach often been repeated to adjust the design parameters, the repeated manufacture of experimental samples, repeated experiments and adjustments, so that the design cost becomes expensive.

With the FEM (Finite Element Method, FEM) and computational fluid dynamics (Computational Fluid Dynamics, CFD) continues to develop, so reducing the shock absorber design and development of experimental dependence, reducing damper parameters different samples the cost of manufacturing and real vehicle test and shorten the development cycle, improve design efficiency damper, it becomes very meaningful.

Research status of automobile shock absorber

Domestic research with respect to foreign automobile shock absorber late start, but the beginning of the design methods are traditional rely on experiments to determine the design parameters. With the continuous development of computer technology as well as to foreign Shock Absorber continuous learning, Chinese scholars have gradually begun to use CAD/CAE software for shock absorber design and development. The basic process of this method is: the establishment of the actual structure of the damper according to its geometry and geometric models into analysis software simulation analysis to obtain the damping characteristics of the shock absorber, the damping characteristics according to the simulation analysis obtained it steering stability, ride comfort and ride comfort and other performance to improve it and optimize. The method of finite element software simulation of the shock absorber, shock absorber design and development of great significance. With the rapid development of China's auto industry, Chinese scholars study the deepening of shock absorber technology, to explore research can improve damper performance technology. To improve the performance of the shock absorber, it is necessary to make a study of the previous

[a] Corresponding author: Shaochengsong007@126.com

damper system analysis and summary for future research on this aspect of the work of scholars who have some enlightening. Mechanics (Computational Fluid Dynamics, CFD) and finite element analysis (Finite Element Analysis, FEA) and other methods to build and solve the hybrid model has become possible. Shock Absorber is a complex nonlinear systems dynamics model for the establishment of the shock absorber has been a concern of scholars. Foreign scholars on the shock absorber, mainly through its three modeling methods research: the physical parameters of the model, the model equivalent parameters and non-parametric models.

The basic theory of fluid - solid coupling

Through the analysis of the shock absorber above the work process, we can see this working process belongs to the category of fluid-structure interaction. Coupling, refers to the coupling between fluid and solid between each other, resulting in a series of phenomena. People in order to investigate the influence of the flow field, the deformation behavior of a variety of solids and solid-bit form convection properties produced interactive relationship between the two, thereby generating a new branch of mechanics - Fluid - solid coupling dynamics.

Coupling Dynamics is a solid stream involves both CFD (Computational Fluid Dynamics, CFD) and relates to solid mechanics (Computational Solid Mechanics, CSM) is a discipline, shown in Figure 2.2. Fluid-structure coupling Problem solving involves both solid, but also relates to fluid solver. Because taking their own characteristics, and solving these two aspects, it makes fluid-structure coupling problem solving results closer to the physical laws of the phenomenon itself. Since the fluid-structure coupling phenomenon exists widely in various fields, and therefore, fluid-structure interaction study has important significance. In recent years, with the continuous development of science and technology, using software simulation technique for solving fluid-structure interaction and design is becoming increasingly widespread in engineering.

CFD numerical simulation method

CFD (Computational Fluid Dynamics) is referred to as computational fluid dynamics. Computational fluid dynamics yes a branch of mechanics, numerical methods it and data visualization technology combine the relevant physical phenomena flow, heat transfer and other simulation analysis. The basic idea of CFD are: the original continuous physical field in time and spatial domain, a series of discrete points with variable set of values instead, and set up to reflect these variables of discrete points by playing certain principles and modalities relations between algebraic equations and then solving algebraic equations to obtain an approximation of field variables.

CFD technology is in conformity with the fluid control equations, numerical simulation of fluid flow. Fluid equations have the energy conservation equation, mass conservation equation, momentum conservation equation. CFD numerical simulation of flow distribution can be obtained for each parameter field and their circumstances change over time. CFD technology can compensate for the lack of theoretical analysis and experimental measurement methods.

Based on the working process of the shock absorber, coupled with a solid approach flow simulation analysis, that both the fluid analysis result to the analysis of solid structures, but also the results of the analysis of solid structures (such as displacement, velocity and acceleration) response to the fluid transfer analysis. In ANSYS WorkBench platform with ANSYS and ANSYS CFX FSI simulation joint general technical route. Among them, the establishment of three-dimensional solid model of this paper chose CATIA software, then geometry processing in ANSYS WorkBench geometry; meshing selection grid before choosing HyperMesh software, because of the simple structure of the solid portion, larger mesh can achieve accuracy , the fluid part of the complex structure, requiring a finish painting hexahedral grid to meet the accuracy requirements; set herein solid simulation model selection ANSYS WorkBench of Transient structural (ANSYS) module, the fluid domain calculations herein choice fluid Flow (CFX).

Simplified simulation system

This binocular damper hydraulic system consisting of a closed system, the internal fluid motion by the pull rod reciprocating motion caused. So the model should include the complete recovery of the damper valve, circulation valve, compensating valves and compression valve and mesh constraint set, driven by movement of the piston rod assembly valve stationary no-slip wall cavity, gas chamber (reservoir chamber above) and the gas chamber and the reservoir chamber of the gas-liquid interface, and so on. However, the establishment of a complete damper model brings two problems: First, they must use the grid to simulate the dynamic motion of the piston driven movement around the oil, which causes the system to simulate the mesh quality requirements will be higher setting parameters is complex, and most parameters are set empirically, there is no accurate theoretical basis, after setting the simulation is also very easy to make mistakes; second is to greatly increase the number of grid computing will consume a lot of time and resources, the performance requirements of the computer will be high. As can be seen from the shock absorber structure, the number of circulating valve and compensating valve little throttling effect at work with respect to the recovery valve and the compression valve is negligible. Therefore, only the recovery valve and compression were established flow field mesh valve.

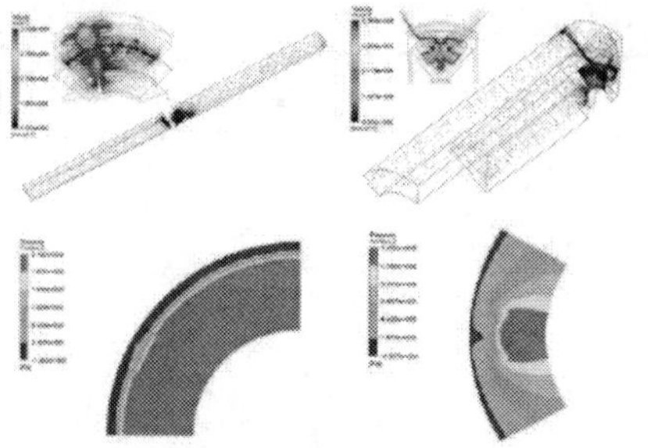

Fig. 1 Speed damping valve vector Fig. 2 Fluid-structure coupling surface pressure contours

Summary

Depending on performance shock absorbers, obtained by experiment specific performance parameters and conditions of this test piece, and the experiments of the speed curve and shown in FIG workers and simulation results were compared, the results can be seen by comparing the experimental results agree well with the simulation results to verify the feasibility of this simulation method, and to simulate errors and experimental results appear, error analysis.

Acknowledgement

1. Ministry of Education Chunhui projects: vehicle vibration damping characteristics, project number 104 900 150.
2. National Social Science Fund Project: integration and application of multi-modal intelligence crisis in the city emergency management (item number: 15BTQ062)

References

[1] Chen Jiarui, Jilin University, Department of Automotive Engineering car construction. China Communications Press, 2002.

[2] DIXON J. The shock absorber handbook [M]. John Wiley & Sons, 2007.

[3] Li Shimin, Lv Zhenhua car telescopic hydraulic dampers technology development of automotive technology, 2001, 8): 10-6.

[4] Lv Zhenhua, Wu Zhuoan, Li Shimin finite element simulation analysis of nonlinear characteristics of the throttle damper mechanical strength, 2004, 25 (6): 614-20.

[5] Studies valve throttle characteristic Li Xiaobo, Shen Sheng season, ROCKETS, et al restoration based on the shock absorber fluid-solid coupling analysis of mechanical manufacturing and automation, 2010, 40 (4): 20-2.

[6] He Ping Simulation Technology of Automotive Shock Absorber Dynamic characteristics; Beijing: Beijing Institute of Technology, 2010.

[7] Feng Xuemei, Liu Zuo people inside the hydraulic shock absorber characteristic finite element analysis of the behavior of the distribution of Wuhan University of Technology: Information & Management Engineering, 2002, 24 (5): 121-4.

[8] He Ping, Gu Liang, Xin Guoguo, et al annular damper valve large deflection precision analytical expression Beijing Institute of Technology, 2009, 6): 510-4.

[9] Lv Zhenhua, Li Shimin, head Liu Zhen, et al Experimental Research telescopic hydraulic dampers operating characteristics of Automotive Engineering, 2005, 27 (2): 203-8.

[10] Chen Yijie, Gu Liang, Yang Zhanhua, et al deformation of throttle slice large deflection damper mechanical design and manufacturing, 2008, 10): 90-2.

[11] Xu Zhongming, Zhang Yufeng, Yu Feng, et al variable damping shock absorber outer Simulation and Test of Chongqing University: Natural Science Edition, 2010, 33 (9): 29-34.

[12] The throttle plate and throttle opening week of the Great Wall research, Gu superimposed light absorber fluid machine, 2006, 34 (7): 19-23.

[13] Zhao Feng automobile shock absorber damping characteristics based on MATLAB environment simulation analysis; Huazhong University of Science and Technology, 2007.

MATEC Web of Conferences **63**, 02046 (2016)
MMME 2016

DOI: 10.1051/matecconf/20166302046

Time—window Complexity and its Application in the Fault Diagnosis of Bearing

Pei JIANG [1a], Liang GAO [1]

[1] *Post doctoral mobile station of mechanical engineering, Huazhong University of Science and Technology, 430033, Wuhan Hubei, China*

Abstract: Different from conventional spectral method, complexity analysis treaded with the signals' time domain structure feature. Before, due to the no stationary and the unevenness of state space of mechanical signals, the state information is likely to get a loss. Now, the time-window complexity is proposed to overcome certain limitations of complexity itself in some extent. It will help to extract the state features of mechanical systems in different states. The concept and algorithm of time-window complexity is introduced detailed. The way of applying the time-window complexity for fault diagnosis is discussed. The mechanical signals of ball bearing with slight flaw and that of four kinds of impact-rubbing states of a typical rotor are then studied. The results show that time-window complexity can reflect the early fault of ball bearing, and differentiate the four kinds of rotor impact-rubbing states properly, which provides another effective way for fault diagnosis of mechanical systems.

Key Words: Complexity; Rolling bearing; Rotor rub; Early fault diagnosis

1 Introduction

At present, with the development of nonlinear science, complex systems and complexity science have become cross disciplinary research, there are many applications in the stochastic resonance feature description [1], biological signal processing [2] and power electronic system fault prediction [3]. Complexity theory and nonlinear research is closely related, before, people are accustomed to linarite the nonlinear system, but with the development of science, people gradually realized that this simplification is difficult to maintain the essence of the system. Complexity concept is proposed in this background, and nonlinear is the source of the complex behavior.

Mechanical signals directly reflect the dynamic behavior of mechanical system operation. The mechanical signal analysis, feature extraction to reflect the running state of the system is the key in fault diagnosis of mechanical equipment. The classical spectrum analysis method plays an important role in fault diagnosis, such as FFT analysis, holographic spectrum analysis, and wavelet analysis. However, these methods are given. The results are mechanical image of signal characteristics, depends largely on the experience, especially the mechanical system itself has a very strong nonlinear, non-stationary, makes the diagnosis more difficult. In order to realize the intelligent monitoring and diagnosis of machine operating, need some simple and the quantitative index to evaluate the mechanical equipment running reliable.

To solve this problem, many methods are raised at this moment; complexity analysis is one of a kind. Complexity provides a method for the analysis of time domain structure of mechanical signal, some qualitative, fuzzy concept, can be given as a quantitative description index. Complexity analysis can effectively monitoring system status and sensitively reflect the system fault, make up the deficiency of traditional monitoring methods. In the field of fault diagnosis for mechanical equipment, people have begun to use complexity method to analysis the mechanical signal, and achieved certain results [4].

[a] Corresponding author: Jiangpei@vip.163.com

Due to non-uniform and non-stationary of signal's state space, will cause missing state information, in the extraction of signal characteristics, limits the application of complexity analysis [5]. The method of time window complexity can more effectively reflect the status of signal, to overcome the own limitations of complexity analysis, it will help to extract the characteristics of mechanical system at different fault distinguishing and early fault features. This paper focuses on the complexity and time window complexity concept, algorithm and its application in mechanical signal analysis. The time window complexity is used to study complexity of rolling bearing the early crack fault and rotor rub four running mechanical signals.

2 Complexity concepts and algorithms

In the study of complexity, there are many kinds of characterizations of the system complexity index, The Kolmogorov complexity, topological entropy, fractal dimension D and the formal language and automata are the famous represent [6,7]. And theoretical analysis and application results show that Kolmogorov complexity is very good at measuring the complexity.

Firstly, Kolmogorov defined the symbol sequence complexity concept in 1965 [8]. Kolmogorov complexity degree refers to the minimal number of bits required for a given "0,1" sequence with computer program, but this definition is difficult to establish a general algorithm. Until the Kasper Schuster [9], Lempel and Ziv [10] gave the algorithm of complexity. Lempel and Ziv complexity of a finite length sequence defined for with the length of the sequence increases the speed of new model.

The basic principle of the algorithm is as follows:

With the sequence $\{x_i\} = x_1, x_2, \text{L} \ , x_N$, this sequence is reconstructed:

$$s_i = \begin{cases} 1, & x_i \geq \bar{x} \\ 0, & x_i < \bar{x} \end{cases} \quad (1)$$

Among of those, $\bar{x} = (x_1 + x_2 + \text{L} \ + x_N)/N$

Based on the formula, the time series $\{x_i\}$ mapping sequence, $\{s_i\} = s_1, s_2, \text{L} \ , s_N$, which s_i=0 or 1. Kolmogorov think that the complexity of $\{s_i\}$ reflect the complexity of $\{x_i\}$. For the formation of the "0,1" sequence, Lempel and Ziv attributed the two kinds of operations: the first is "copy", namely a prefix of sequence produce a longer prefix; the second is "add", namely add symbols to the prefix had been generated, any length n sequences of symbols can be produced by the second operation from an empty string, but only the shortest description is the complexity of sequences.

First of all, from the empty string begin to generate. A prefix have been generated shilling, $S = (s_1, \text{L} \ s_r), r < N$. Assumed, $Q = s_{r+1}$ formed SQ string, $SQ\pi$ is the string after deleting the last character from the string SQ. If Q can be copied from a sub string from $SQ\pi$, put the character to be added later, known as "copy", if not belonging to, is said to "add" and "add" with a "·" before and after the separation. If $Q = s_{r+1}$ can be obtained from a sub string $SQ\pi$ by "copy", then continue to investigate whether $Q = s_{r+1}s_{r+2}$ obtained from a sub string $SQ\pi$ by "copy". If not, belong to "add", $S = (s_1, \text{L} \ s_r s_{r+1} s_{r+2} \bullet)$, so that in the back of $s_1 \text{L} \ s_r s_{r+1} s_{r+2}$ with "·". Next, all characters in front of"·" look as S. Repeat the above steps until generating the entire sequence.

For example: generate "0,1" sequence "0110".

(1) To generate a character by adding from the empty string denoted as $\rightarrow 0 \bullet$.

(2) $S = 0, Q = 1, SQ = 01, SQ\pi = 0$, Q not belongs to $SQ\pi$, noted as $\rightarrow 0 \bullet 1 \bullet$.

(3) $S = 01, Q = 1, SQ = 011, SQ\pi = 01$, Q belongs to $SQ\pi$, noted as $\rightarrow 0 \bullet 1 \bullet 1$.

(4) $S = 01, Q = 10, SQ = 0110, SQ\pi = 001$, Q not belongs to $SQ\pi$,记为$\rightarrow 0 \bullet 1 \bullet 10 \bullet$.

In this way, we get a string divided into segments by "·", the number of segments is defined as the complexity of C, so the complexity of sequence "0010" note as C(4)=3.

According to the Lempel and Ziv, almost all 0,1 random sequences, their complexity will tend to a value b(N):

$$b(N) = \lim_{N \to \infty} C(N) \equiv N / \log_2 N \quad (2)$$

Next, normalization, so that it becomes the relative complexity of $C_r(N)$.

DOI: 10.1051/matecconf/20166302046

$$C_r(N) = C(N)/b(N) \qquad (3)$$

We can find that relatively complexity $C_r(n)$ reflects similarity degree between a time series and random sequence, the $Cr(n)$ of completely random sequence will tends to 1, and periodic sequence will be 0, $Cr(N) \in [0,1]$ when series between the two cases. Thus, signal complexity is related with signal ordered and random behavior, relatively complexity $C_r(n)$ can be the characteristic parameters to represents the state of the system.

Complexity of system is very sensitive, when the machinery and equipment have early fault, complexity analysis of changes can reflect changes in the state of the system, and complexity is especially suitable for computer processing. In this paper, complexity analysis is used to detect the early crack of bearing inner ring and early touch rub of rotor.

3 Early crack fault detection of rolling bearing inner ring

Rolling pair local defects is the most typical failure mode of rolling bearing; especially it is most difficult to diagnose the inner ring rolling defects. This is because that the impact has a transmission link caused by raceway defects, the fault feature signal is transmitted to the base surface has been greatly weakened, flooded in background noise. It is not useful to diagnose the faults of rolling bearing by the characteristic frequency.

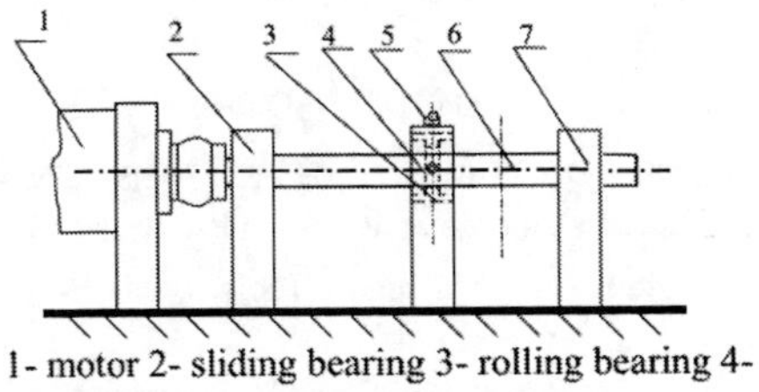

1- motor 2- sliding bearing 3- rolling bearing 4- horizontal direction (x) displacement sensor 5- vertical direction (y) displacement sensor 6- shaft 7- support

Figure 1. Experimental rotor-bearing rig

As shown in figure 1, the normal state and early crack failure state are all tested in the rotor bearing test.

In speed 3500r/min, the sampling frequency is 2.5 kHz; we get 10000 sampling points in horizontal and vertical direction. Then signal is processed by low-pass filtering, filter cutoff frequency is 1 kHz. Waveform and power spectrum in the state of fault and the inner crack at early stage are given respectively in the Figure 2.

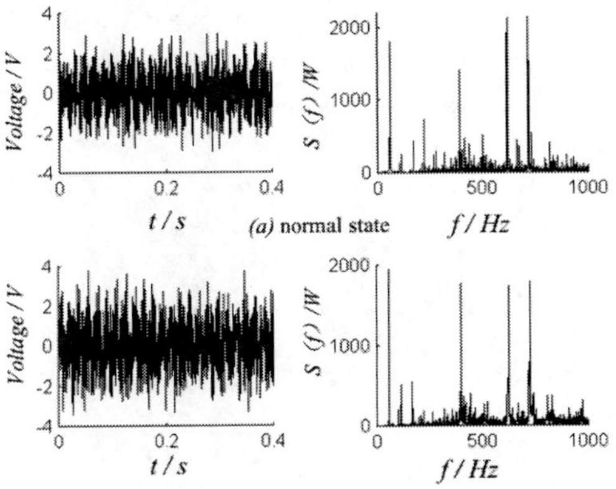

*(b) Early crack of inner ring*normal state

Figure 2. Waveform and spectrum of ball bearing vibration signals

From the above, we could not tell the difference between rolling bearing inner ring raceway early crack fault and no fault condition in the time domain and frequency domain. Following the complexity analysis is used to analyzed the 5 groups of sampled data, the length of each set of data for 2000 points, complexity are calculated in table 1.

Table 1. Relative complexity of ball bearing vibration signals

Bearing condition	Relative complexity C_r				
	First groups	Second groups	Third groups	Fourth groups	Fifth groups
failure-free	0.7406	0.7385	0.7621	0.7337	0.7428
Early crack	0.8574	0.8538	0.8577	0.8606	0.8581

The complexity of fault is more complex. Thus, it could be effective to extract the fault characteristic of rolling bearing inner ring raceway early crack fault detection by complexity analysis.

4 Early rub impact fault detection

Next, rotor rubs experiment is taken on the on test bench. Rotor frequent 3500r/min, axis horizontal and vertical displacement is measured by the measurement system using KEYENCE-LC2401 laser displacement sensor. Signal is also processed by low-pass filtering, filter cutoff frequency is 800Hz, and sampling frequency 5 KHz sampling number is 10000.

In the rub friction experiment, the signal is strong non-stationary between the rotor and the stator occurred early rub impact (the beginning touch). And, complex degree analysis can reflect the early rub impact fault of rotor. Waveform and the axis orbit of horizontal direction of signal are given without rubbing and early rub impact in Figure 3 (a) and (b).

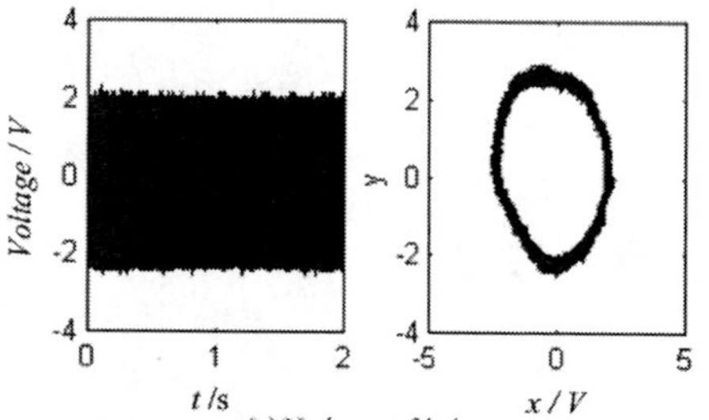

(a) No impact friction

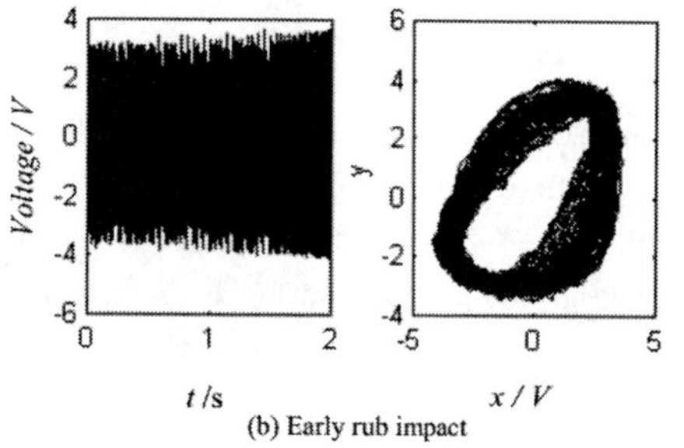

(b) Early rub impact

Figure 3. Waveform and axes track of rotor rub signals

From the figure, the shaft center trajectory becomes complicated and can qualitatively reflect the early rub impact conditions, but without a clear quantitative indicators. Following the relative complexity of multiple sets of data obtained by sampling analysis, each data length of 2000. Results are shown in Table 2.

From table 2, it can be seen that the signal of no rotor rub impact and early rub friction relative complexity is significantly different. The complexity of no rotor rub is smaller than early rub friction signal. So that, according to the relatively complex of signal, we can identify early detection of rub fault.

Table 2. Relative complexity of rotor rubs signals

Rub impact condition	Relative complexity C_r				
	First groups	Second groups	Third groups	Fourth groups	Fifth groups
failure-free	0.1020	0.1013	0.0987	0.1068	0.1043
Early crack	0.1325	0.1348	0.1311	0.1398	0.1360

5 conclusions

This paper introduces the concept and calculation method of time window complexity, and the bearing inner ring early crack fault signal and rotor rub fault signal of rolling are analyzed in this paper. The analysis results show that there is no obvious difference in the time domain and frequency domain between the mechanical signal of rolling bearing inner ring occurred early crack and free fault state, but the former time window complexity index is significantly higher than the latter; Another window complexity can overcome alone with complex index as characteristic parameters, can describe the strong characterizations of the running state of machine, which provide a method to extract fault features from mechanical signals, further provide a effective method to fault diagnosis.

References

1. P. Shi, X. Ding, and D. Han. Study on multi-frequency weak signal detection method based on stochastic resonance tuning by multi-scale noise, Measurement, 2014,47(1): 540-546
2. Moses O. Sokunbi, George G. Cameron, Trevor S. Ahearn, Alison D. Murray, Roger T. Staff. Fuzzy approximate entropy analysis of resting state fMRI signal complexity across the adult life span, Medical Engineering & Physics, 2015,37(11): 1082-1090
3. Liu Qing, Wang Zengping, Zheng Zhenhua, Chang Yanyan. On-unit Transient Protection Based on Signal Complexity for Series Compensated Transmission Lines, Proceedings of the CSEEN, 2010,30(13): 81-88
4. Xiaoli Li, Gaoxiang Ouyang, Zhenhu Liang. Complexity measure of motor current signals for tool flute breakage detection in end milling, International Journal of Machine Tools and Manufacture, 2008,48(3): 371-379
5. Long Fei, Zhang Daoxin, Fan Ling, Wu Xiaopei. Analysis of sleep staging by time-window complexity sequence of EEG. Journal of Anhui University Natural Science Edition, 2002,26(1): 56-61
6. Grassberger. Toward a quantitative theory of self-generated complexity. Int. J.Theroy.Phys.,1986,25: 907-936.
7. B.L.Hao. Symbolic dynamics and characterization of complexity. Physica D, 1991,51,161-176.
8. R. Wackerbauer, *et al* A comparative classification of complexity measures. Chaos, Solitons & Fractals,1994,

DOI: 10.1051/matecconf/20166302046

4(1)：133-173.

9. F.Kasper，H.G.Schuster. Easily calculable measure for the complexity of spatiotemporal patterns. Physical review A,1987,36(2)：842-848.
10. Abraham Lempel， Jacob Ziv. On the complexity of finite sequences. IEEE Transactions on information theory,1976，22(1): 75-81.

Effects of Mo substitution for Ni on the microstructure and property of V-based hydrogen storage alloys

Yanwei Tong[1,a] and Lin Deng[1]

[1]*College of Biology and Chemistry Engineering, Panzhihua University, Panzhihua 617000, China*

Abstract. In order to study the property the V-based hydrogen storage alloys, Ni in the alloys was partially substituted by Mo. The $V_2Ti_{0.5}Cr_{0.5}Ni_{1-x}Mo_x$ (x=0.02-0.08) hydrogen storage alloys were prepared by induction melting, and the effects of Mo content on the microstructure and electrochemical property were investigated systematically. The results show that the alloys mainly consist of a V-based solid solution phase with a BCC structure and a TiNi-based secondary phase. The electrochemical measurements indicate that with increasing the content of Mo, the maximum discharge capacity, the cycle stability and high rate discharge ability is increased first and then decreased.

1 Introduction

Hydrogen storage alloys used as negative electrode materials in Ni/MH secondary batteries are extensively studied in the past decades and a series of metal hydride electrode materials have been discovered, such as rare-earth-based AB_5-type alloys [1], AB_3-type rare-earth-based alloys [2], AB_2-type Laves phase alloys [3], Mg-based alloys [4], and V-based solid solution alloys [5,6]. As one kind of metal-based hydrogen storage alloys, V-based solid solution alloys have exhibited highly attractive properties for practical application because of their higher hydrogen storage capacity and lower hydrogen absorption/desorption temperature [7-9]. However, the V-based hydrogen storage alloy exhibited poor charge-discharge cycle stability because of the dissolution of the V constituent in the electrolyte. In order to improve the cycle stability of this type of alloy, the investigations of the element substitution and additive [10-12] were conducted. Some researchers have studied the influence of Mo additive in AB_5-type alloys [13,14], but the research of influence of Mo additive in V -based hydrogen storage alloys is seldom.

In this study, we prepared $V_2Ti_{0.5}Cr_{0.5}Ni_{1-x}Mo_x$ (x=0.02-0.08) alloys in which Mo was partially substitution for Ni, in order to improve the charge-discharge cycle stability of $V_2Ti_{0.5}Cr_{0.5}Ni$ alloy. Characteristics of the battery performance, such as discharge capacity, cycle stability and high-rate dischargeability were evaluated.

2 Experimental

The purity of raw materials is as follows: V>99.9wt.%, Ti>99.9wt.%, Cr>99.9wt.%, Ni>99.9wt.%, Mo>99.9wt.%, respectively. The alloys were arc-melted in the water-cooled copper crucible under argon atmosphere (purity, 99.999 wt.%). To ensure the homogeneity, the alloys were turned over and remelted for five times. The samples were mechanically crushed into fine powders of 200-320 mesh in mortar for X-ray diffraction (XRD) and electrochemical measurements analysis. The XRD data was collected by a step-scanning method using DX-2700 with CuK α .

The well-mixed alloy powder and carbonyl nickel powder in weight ratio of 1:4 was compacted onto an expanded nickel substrate by a 10 MPa to form a test electrode (about diameter of 10 mm, thickness of 1 mm) without binder.

The electrochemical property was performed on LAND battery testing instrument using a tri-electrode system consisting of a working electrode (MH electrode), a sintered $Ni(OH)_2$/NiOOH counter electrode with excessive capacity and a reference electrode (Hg/HgO, OH-). The electrolyte in the cell was 7 mol/L KOH aqueous solution. The electrode was charged for 6 h at a constant current density (100 mA/g) followed by 5 min resting time and then discharge at a constant current density (60 mA/g) to a cutoff potential of -0.6 V at 303 K.

3 Results and discussion

3.1. Alloy structure characteristic

[a] Corresponding author: 317435736@qq.com

The typical X-ray diffraction patterns of the $V_2Ti_{0.5}Cr_{0.5}Ni_{1-x}Mo_x$ (x=0.02-0.08) alloys are shown in Figure1. It can be seen that all of the alloys exhibit similar diffraction patterns, indicating that the phase constitutions of the alloys are a V-based solid solution phase with a BCC structure and a TiNi-based secondary phase. With increasing the content of Mo, the diffraction peak intensity of V-based solid solution phase increases.

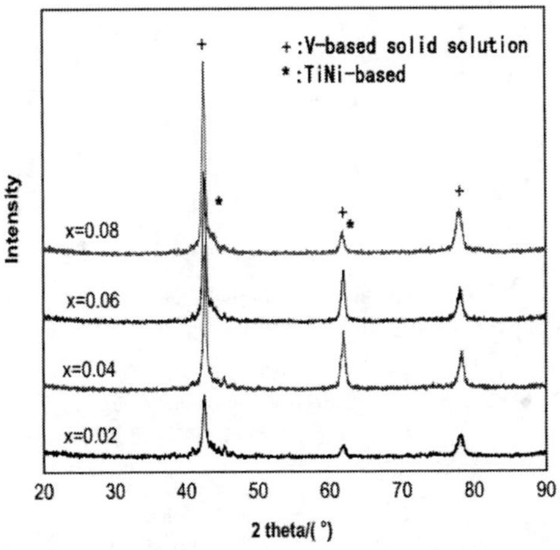

Figure 1. XRD pattern of the $V_2Ti_{0.5}Cr_{0.5}Ni_{1-x}Mo_x$ (x=0.02-0.08) alloys

The metallographs of $V_2Ti_{0.5}Cr_{0.5}Ni_{1-x}Mo_x$ (x=0.02-0.08) alloy electrodes are shown in Figure 2. It displays that all of the alloy samples are mainly composed of two distinct crystallographic phases: the 3-D interpenetrating phase is the TiNi-based phase, and the dendritic phase is the V-based solid solution phase. Akiba et al [15] pointed out that 3-D interpenetrating TiNi-phase could collect electrons and worked as a catalyst for the BCC phase which has a high absorption hydrogen capacity. The TiNi-based phase with 3-D network structure was very important for the electrochemical kinetics of the alloy electrodes.

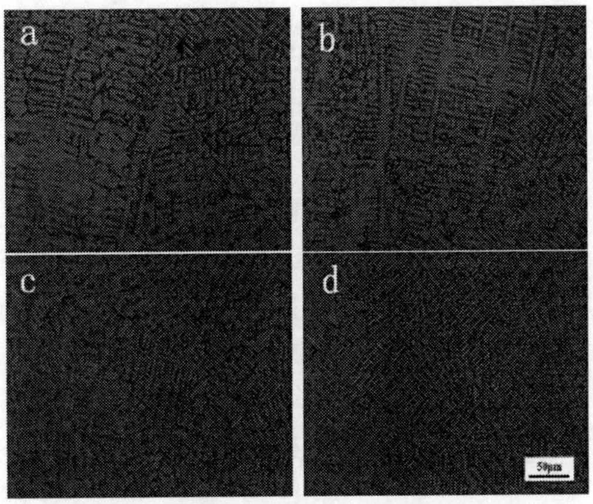

Figure 2. Metallographs of the $V_2Ti_{0.5}Cr_{0.5}Ni_{1-x}Mo_x$ (x=0.02-0.08) alloys. (a)x=0.02; (b)x=0.04; (c)x=0.06; (d) x=0.08

3.2 Discharge capacity and cycle stability

The curves of the discharge capacity versus the cycle number at 303 K for $V_2Ti_{0.5}Cr_{0.5}Ni_{1-x}Mo_x$ (x=0.02-0.08) alloy electrodes are shown in Figure 3. It can be seen that after the substitution of Mo for Ni, the activation property of the alloy electrodes is decreased. The maximum discharge capacity (C_{max}) and the rate of capacity retention S_{20} (S_{20} is defined as $S_{20}= C_{20}/C_{max} \times 100\%$, where C_{20} is the discharge capacity measured at the 20th cycle) of the alloy electrodes are summarized in Table 1. It can be seen that with increasing the content of Mo substitution for Ni, the maximum discharge capacity of the alloy electrodes increased from 362.5 mAh/g (x=0.02) to 385.7 mAh/g (x=0.06) and then decreased to 340.3 mAh/g. The rate of capacity retention S20 first increased from 80.4% to 85% and then decreased to 79.4%, which means that a proper component ratio of Mo substitution for Ni is benefit to the maximum discharge capacity and the cycle stability of the alloys. Maybe a proper component ratio of Mo substitution for Ni resists the dissolution of alloy element and increases the anti-corrosion of the alloy.

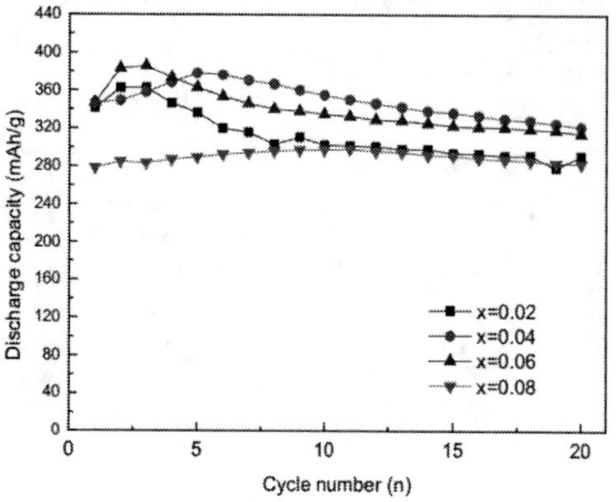

Figure 3. Discharge cycling performance of the $V_2Ti_{0.5}Cr_{0.5}Ni_{1-x}Mo_x$ (x=0.02-0.08) alloy electrodes

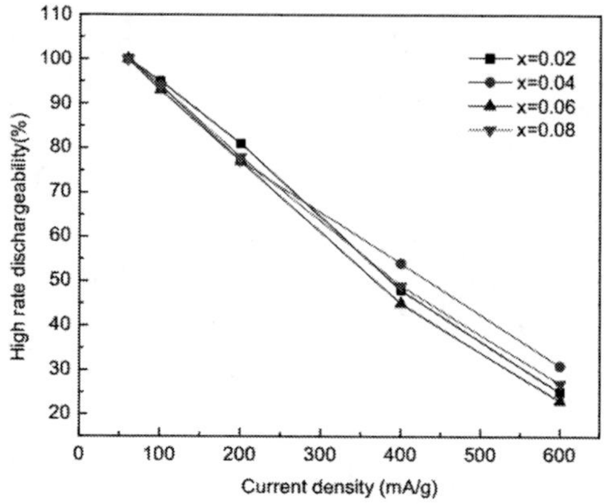

Figure 4. High rate discharge ability of the $V_2Ti_{0.5}Cr_{0.5}Ni_{1-x}Mo_x$ (x=0.02-0.08) alloy electrodes

3.3 High rate discharge ability

The high rate discharge ability (HRD) is measured at different discharge current densities (60 mA/g, 100 mA/g, 200 mA/g, 400 mA/g, 800 mA/g). The HRD is dendined as

$$HRD = \frac{C_n}{C_{\max}} \quad (1)$$

where C_{max} represents the maximum discharge capacity at a discharge current density of 60 mA/g, and C_n is the discharge capacity at a discharge current densities (n=60, 100, 200, 400, 800).

Figure 4. shows the high rate discharge ability (HRD) of the $V_2Ti_{0.5}Cr_{0.5}Ni_{1-x}Mo_x$ (x=0.02-0.08) alloy electrodes. The HRD is measured at different discharge current densities (60mA/g, 100mA/g, 200mA/g, 400mA/g, 600mA/g). It can be found that the HRD of the alloy electrodes decreases with increasing discharge current density, while with increasing the content of Mo substitution for Ni, the HRD first increases and then decreases. It is generally accepted that the high-rate dischargeability of a metal hydride electrode is mainly determined by the charge transfer process occurring at the metal electrolyte interface and/or the hydrogen diffusion process in the hydride bulk [16]. The substitution of Mo for Ni may improve the charge transfer and the hydrogen diffusion process.

Table 1. Electrochemical behavior of the $V_2Ti_{0.5}Cr_{0.5}Ni_{1-x}Mo_x$ (x=0.02~0.08) alloy electrodes

Samples	C_{max} (mAh/g)	S_{20} (%)	HRD (C_{400}/C_{60}) (%)	I_0 (mA/g)
x = 0.02	362.5	80.4	48	54
x = 0.04	378.0	85.0	54	68
x = 0.06	385.7	81.5	45	60
x = 0.08	340.3	79.4	49	38

The exchange current density (I_0) of $V_2Ti_{0.5}Cr_{0.5}Ni_{1-x}Mo_x$ (x=0.02-0.08) alloy electrodes is a measure of the forward and reverse electrode reaction rate at the equilibrium potential. It can be used to characterize electrocatalytic activity of the electrode surface. A high exchange current density means a high electrochemical reaction rate with good cyclic performance. The exchange current density may be calculated from the polarization resistance (R_p) using the simplified Butler-Volmer equation at low overpotential limit conditions as follows:

$$I_0 = \frac{RT}{FR_p} \quad (2)$$

Where R is the gas constant (J/(mol K)), T is the absolute temperature (K) and F is the Faraday constant (C/mol). The polarization resistance $1/R_p$ can be obtained from the linear polarization curves of the oxide electrode as shown in Figure 5. And the calculated current density is listed in Table 1, respectively.

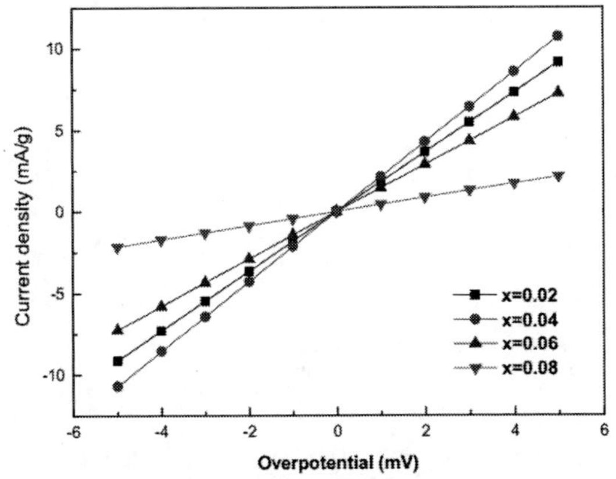

Figure 5. Linear polarization curves of $V_2Ti_{0.5}Cr_{0.5}Ni_{1-x}Mo_x$ (0.02-0.08) alloy electrodes

4 Conclusions

The structural and electrochemical properties of the $V_2Ti_{0.5}Cr_{0.5}Ni_{1-x}Mo_x$ (x=0.02-0.08) alloy electrodes have been systematically studied. All of the alloys are composed of a V-based solid solution phase with a BCC structure and a TiNi-based phase. With increasing the content of Mo substitution for Ni, the maximum discharge capacity (C_{max}) , the rate of capacity retention (*HRD*), the high-rate dischargeability and the exchange current density (I_0) of the alloy electrodes increase first and then decrease, and reach the best performance with x=0.04.

References

1. Liu YF, Zhang SS, Li R, Gao MX, Zhong K, Miao H, et al. Int J Hydrogen Energy **33**, 728 (2008)
2. Kohno T, Yoshida H, Kawashima F, Sakai I, Yamamoto M, Kanda M. J. Alloys Compd. **311**, 5 (2000)
3. Ovshinsky SR, Fetcenko MA, Ross J. Science **260**, 176 (1993)
4. Wang FX, Gao XP, Lu ZW, Ye SH, Qu JQ, Wu F, et al. J Alloys Compd. **370**, 326 (2004)
5. Tsukahara M, Kamiya T, Takahashi K, Kawabata A, Sskurai S, Shi J. J Electrochen Soc **147**, 2941 (2000)
6. Tong YW, Gao JC, Deng G, et al. Materials Letters **112**, 142(2013)
7. Tsukahara M, Takahashi K, Mishima T, Sakai T, Miyamura H, Kuriyama N. J Alloys Compd **224**, 162 (1995)
8. Song XP, Pei P, Zhang PL, Chen GL. J Alloys Compd **455**, 392 (2008)
9. Tong YW, Gao JC, Deng G, et al. Rare Metal Materials and Engineering **44**, 1052 (2015)
10. Parimala R, Ananth MV, Ramaprabhu S. Int J Hydrogen Energy **29**, 509 (2004)
11. Liu FJ, Suda S. J Alloys Compd. **231**, 666 (1995)
12. Wang YZ, Zhou MS. Journal of Rare Earths **30**, 146 (2012)
13. Yeh MT, Beibutian VM, Hsu SE. J Alloys Compd. **293**, 721 (1999)
14. Yang SQ, Han SM, Song JZ, Li Y. J Rare Earths **29**, 692 (2011)
15. Akiba E, Iba H. Intermetallics **6**, 467 (1998)
16. Li SC, Zhao MS, Wang LM, Liu Y, Wang YZ. Mater. Sci. Eng., B ; **150**, 168 (2008)

MATEC Web of Conferences **63**, 03002 (2016) DOI: 10.1051/matecconf/20166303002
MMME 2016

The innovative application studty on eco-packaging design and materials

Yong Min CUI[1], Xin Guang REN[2,a]

[1] *Rongcheng College,Harbin University of Science and Technology,Rongcheng;Shandong,China*
[2] *Rongcheng College,Harbin University of Science and Technology,Rongcheng;Shandong,China*

Abstract. The paper solves the increasingly deteriorate environmental problems by positively exploring how to utilize and develop eco-packaging design reasonably. The paper explores an effective method that combines eco-packaging and environmental protection materials, hoping to define the sustainable development road of packaging design. The paper is centered on the design application of eco-packaging and environmental protection materials, applies and analyzes the method to obtain innovative design requirements and development tendency of eco-packaging design by analyzing status and significance of eco-packaging design, combining with the development and main types of eco-packaging packaging materials, and based on the achievements acquired by eco-packaging and environmental protection materials. Meanwhile, the paper also reveals mutual dependence and mutual promotion of eco-packaging design and eco-packaging materials.

1 Status and Significance of Eco-packaging Design

1.1 Status of Eco-packaging Design

The generation of environmental design idea in the end of 1980s guided a new international design wave. Therefore, eco-packaging design has been generated correspondingly. Meanwhile, people also gradually realize that eco-packaging plays a positive role on natural environment and resources in their daily life. By constantly developing eco-packaging design, designers realize that packaging materials play a dominant role on the packaging design. Lots of successful eco-packaging designs have already concentrated on environmental protection and recoverability of packaging materials in the design concept. In the packaging design, innovative design techniques and new environmental technology are fully utilized, so as to satisfy packaging requirements and environmental requires of products.

Secondly, laws and regulations relating to eco-packaging materials have acquired greater development in recent years. The first sign (Der Grüne Punkt namely environmental pattern of product packaging) relating to eco-packaging in the world was generated in Germany in 1975. In later several decades, eco-packaging has been rapidly developed around the world until Environmental Control Board was officially established in ISO and eco-packaging standard was formulated officially in 1993. The standard was officially implemented in Jan. 1996 in the global. In recent decades of years, the developed countries have adopted measures successively and formulated laws and regulations about packaging. At present, packaging organizations around the world make efforts required by the ISO, research and develop new materials and new technology constantly.

In addition, with the social development in recent years, environmental consciousness from enterprises to designers, as well as the public has been greatly improved. Many new materials, new technologies and new packaging design idea have been utilized and acquired good performance. Designers adopted more environmental design ideas and techniques in a great number of product packaging designs, but the eco-packaging design is still encountered with further development space.

1.2 Significance of Eco-packaging Design

The eco-packaging occurrence indicates that designers attach importance to ecological environment and even embodies the urgent requirements of consumers for environmental protection. The product packaging design of environmental protection is related to future living environment of humans and has stronger practical significance. Reasonable utilization and development of eco-packaging design can save resources, protect environment, promote ecological sustainable development, and create a good life condition.

1.2.1 Realistic Significance on Developing Packaging Design

[a] Xin Guang Ren:renxinguang12@163.com

First of all, eco-packaging runs through the entire link of production, sales and utilization for products. The generation of eco-packaging promotes the further development of packaging design, points out the developmental direction of packaging design in the future, and develops a unique style in various schools of packaging design field. Secondly, eco-packaging design plays an important connecting role on the production and utilization process of eco-packaging. After a commodity is produced and completed, a designer designs it by blending in environmental idea in line with characteristics of the commodity and demands of producers. Then, the commodity is sold to consumers, and environmental idea is transmitted to consumers. As a result, eco-packaging design plays the guiding significance on the development and progress of packaging design. Thirdly, people have already had relatively mature environmental consumption behaviors from a radical environmental concept after entering into the 21st century and bring a new opportunity to develop eco-packaging, so that eco-packaging design trend with idealism gradually has realistic significance. With the constant packaging development, there is a definite requirement for eco-packaging design of products in the future. Obviously, it will present a brand-new page of packaging design.

1.2.2 Social Significance Generated by Eco-packaging

Eco-packaging is equipped with bright social significance. Commodities walk into thousands of households through complete environmental packaging. Thus, the environmental idea transferred by eco-packaging is also generalized with it. Consumers have already accepted propaganda of environmental idea, while getting commodities. Moreover, they become direct participators of environmental protection, improving consumers' consciousness of environmental protection. The packaging design of numerous products just pays attention to visual effects of product packaging, but neglect design of product functions and social property. It depends on excessive flaunty and wasteful packaging to attract consumers to some extent. However, such the excessive packaging design has already seriously threatened natural and ecological environment of human life. Facing to environmental pressure, people have already increasingly paid attention to green and environmental protection of product packaging. Under the promotion of green consumption, an increasing number of consumers select environmental green products. Adoption of eco-packaging products is easy to be accepted by consumers. Undoubtedly, there is a close relation between eco-packaging and consumers. As a result, the development of eco-packaging design definitely will guide the concept of consumers and promote constant progress of the society.

2 Characteristics of Eco-packaging Materials

Any kind of packaging design can't do without packaging materials. Only to fully know about eco-packaging materials and select environmental protection materials reasonably can it promote the constant development of eco-packaging design.

2.1 The Development of Eco-packaging Materials

Humans have a long history of packaging products with environmental protection materials. In the past, humans applied lots of packaging materials, which conformed to an environmental idea in our opinions. Moreover, some eco-packaging ways have already been used until now and they are deserved to be learned and referred by packaging designers. For example, Japanese natto packaging is a good traditional packaging design. It applies straws to regard as the natto packaging materials. Straws are necessary materials for natto fermentation. Moreover, sources of straws are extensive and reliable. After using straws, they can be abandoned in nature and are easy to be decomposed, making natto packaging is natural, kind, environmental and health.

Before the revolution of industrialization, lots of packaging materials are derived from nature directly and have smaller environmental pollution. However, all of these are based on consuming lots of natural resources. For example, woods and leather products are utilized in mass. From the perspective of a new environmental idea, it still needs to be further improved. With the constant development of social economy and technology, as well as constant innovation of technology, environmental protection materials also have new changes. Particularly, research and development of environmental protection materials in recent years have acquired outstanding progress in the improvement of traditional packaging materials.

2.2 Main Types of Eco-packaging Materials

With the development of industrialization and increasingly prominent environmental problems, the design idea of eco-packaging design also has a new change. People think that it is necessary to develop eco-packaging materials under the precondition of protecting natural resources. Moreover, eco-packaging materials in new period should be free of contamination, can be recycled, and should not depend on new requirements of natural resources excessively. Western countries apply biotechnology and genetic technology, etc., technical means, so eco-packaging materials develop rapidly. With the application of lots of new-type eco-packaging materials, eco-packaging design has been greatly changed from design ideas and techniques. Many design ideas and techniques can be performed by applying new materials. Eco-packaging materials have a profound influence on environmental design. Nowadays people mainly apply the following categories of eco-packaging materials:

2.2.1 Degradable Materials

Under the specific environment of causing performance loss, a material of which chemical structure is changed within the specific time is called as a degradable material. For example, after degradable materials are used, they are restored and degraded through the function of UV-light or microbial action and return to the natural environment in a non-poisonous form. Beets are used to make degradable materials in France, which are made into ecological packaging boxes, possess some advantageous characteristics of physiolysis, rapid decomposition and non-toxin property and won't cause pollution to the environment.

2.2.2 Recycling Materials

Recycling materials mainly refer to the packaging materials that can be recycling, used repeatedly and regenerated. For instance, glass is the repeated and regenerated packaging material. Glass packaging is equipped with some advantages, such as stronger visuality and easy to recycle, and has already become a main vessel of traditional packaging. After using, glass bottle can be used repeatedly by recycling disinfection system, having stronger environmental effects.

2.2.3 Upgrade of Traditional Packaging Materials

The upgrade of traditional packaging materials means that materials depend on modern technology to transform and update techniques on the basis of traditional packaging materials. Based on modern technology, China develops new-type packaging materials to replace traditional wooden packaging materials constantly. For example, new-type packaging materials are developed by virtue of wood flour, cotton and linen, bamboo, reeds, straws, wheat straws and corn bran, etc. This not only utilizes resources reasonably, but also protects natural resources.

One of main methods to update and replace traditional packaging materials is to expand varieties of packaging materials constantly and improve technical content of packaging materials. For example, strawboard made by processing straws not only is energy-efficient, chill-proof, heat-resisting and sound-proofing, as well as has good permeability and stronger shock resistance, but also performs excellently in waterproof and shock resistance and is convenient for packaging and transportation.

3 Innovative Design Application of Eco-packaging Design

In order to develop eco-packaging career constantly, designers must innovate design thought constantly, update design ideas continuously, and maintain high social responsibility, so as to innovate product packaging with a responsible attitude and method, transfer green and humane idea and spirit to consumers, and promote people's ideology on eco-packaging constantly.

3.1 Application of Simple Design Philosophy in Eco-packaging Design

Simple design needs designers to analyze and design required factors seriously, comprehensively and objectively on the basis of knowing about psychological life and economic ability of audiences and conducts reasonable and proper design on items from multiple aspects, including functions, materials and colors, so that designed packaging has proper product protective function, suitable materials, and reasonable costs. The essence of simple design is to make products reflect their value accurately, cater to consumption demands of consumers, but not decorate unnecessarily, so as to solve some problems generated by "excessive design". Simple design also respects and protects nature. There is a symbiotic relationship between humans and nature at the very beginning, but human activities destroy the relationship. Moreover, the contradiction and conflict between humans and nature are increasingly serious. Thus, we propose an idea of harmonious co-existence between humans and nature. Simple design exactly doesn't cause larger damage, but conforms to harmony between humans and nature and development of human society.

Designers should blend social responsibilities into the design process, but can't design for designing. Nowadays, packaging design presents the tendency of excessive packaging. Resources consumed by packaging also attach importance of humans. Though there is no definite definition about excessive packaging, designers should try to promote simple design.

3.2 Embodiment of Humanization Design in Eco-packaging Design

The concept of design humanization has already been proposed with the core of paying attention to various demands of humans. The ergonomics has already been mature. In terms of eco-packaging, humanization of eco-packaging design should pay more attention to pursuing for social value. In the pursuit for social value of contemporary people, it can't neglect the pursuit for environmental protection and sustainable development.

The significance of humanized eco-packaging design lies in concerning comprehensive social value of commodity packaging and satisfaction of human spirit and emotional demands, except that products should satisfy basic functions of substances. Because modern consumers pay more attention to packaging or even feel product quality by packaging, when they purchase commodities, such as cosmetics and gifts, visual effects of packaging will be associated with product quality in consumer opinions. Packaging satisfies spiritual needs of consumers to a large extent. Moreover, in the further present and bestowal, etc., links, packaging has a larger satisfaction on spiritual level. If eco-packaging can't satisfy demands in these aspects, it is hard to realize. Thus, designers should study consumption mentality of consumers for different products, study application of design techniques for human practical demands, and study different acceptance levels of different consumers for new ideas. The method of humanization design lies in studying human characteristics and demands of humans and combining with eco-packaging design. Meanwhile, eco-packaging design is the important perfection and improvement of humanization design.

3.3 Embodiment of Cultural Elements in Eco-packaging Design

It seems that product packaging is extremely important in modern international market. Packaging impacts quality image of products in human opinions deeply and becomes an important sign of brand material culture. In the design or product packaging, it generally should give consideration to materials, structure, modeling, colors and patterns, etc., elements Moreover, packaging can be provided with functional value and aesthetic value through skillful conception and scientific utilization of these elements, so that consumers feel brand individuality and cultural connotations from it. Packaging design is also a kind of cultural behavior of humans. Therefore, it is necessary to conduct an in-depth study on different cultural backgrounds in different areas, find out elements utilized by packaging design, blend environmental ideas, and finally realize it through design. In this way, it not only is easy to acquire obvious effects of propaganda, but also is easy to acquire cultural identification of consumers.

Lots of packaging materials are equipped with profound cultural connotations, especially for time-honored materials with profound cultural background. Many materials nowadays are still deserved to be referred by our designers. These materials are mainly derived from nature. Moreover, they are easy to acquire and regenerate. For example, zongzi packaging in Chinese tradition is the characteristic traditional packaging design. Steamed rice dumpling leaf is regarded as the appearance. The south of China generally utilizes bamboo leaf, while north of China applies reed leaf as the raw material of steamed rice dumpling leaf. All of them come from nature. The fragrance of steamed rice dumpling leaf not only makes zongzi lead a person to endless aftertastes. Moreover, steamed rice dumpling leaf is easy to be degraded after it is abandoned in nature. All of fully embody the "theory that man is an integral part of nature" in Chinese traditional culture.

To sum up, with the constant development of industrial technology, new-type environmental protection materials promote the development of eco-packaging. Therefore, people should reinforce the development of eco-packaging materials and its application in packaging design, while advocating eco-packaging design. Meanwhile, with the proposal of new environmental thought, the environmental climax with the center of protecting environment and saving resources has already become an inexorable trend of world eco-packaging revolution. With the constant improvement of living standard and environmental cognition, people don't satisfy with aesthetic functions of product packaging, but pay more attention to environmental protection property of product packaging. Undoubtedly, eco-packaging will bring a new lifestyle for commodities.

References

[1] Jie Chen, The Study on Packaging Design of Environmental Protection Materials [D], Yanbian University, 2013.5

[2] Tao Jiang and Lixia Wu, Development of Packaging Materials and Ecological Study [J], China Packaging, 2000

[3] Chunshan Wu, The Development of Green Packaging and Packaging Materials [J], Journal of China Packaging, 2000

[4] Jing He, Design Application of Eco-packing and Materials [D], Xi'an Academy of Fine Arts, 2011.3

[5] Yanfang Shen, Application and Development Study Based on Green Packaging Materials [D], Nanchang University, 2012.11

[6] Taichun Wang, The Role of Green Ecological Packaging Materials in Modern Packaging Design [J], the Academic World, 2015

[7] Jianxiang Peng, The Study on Packaging Design Method Based on Natural Materials [J], Northwest Fine Arts, 2015.(2)

Review on research and application of stainless steel reinforced concrete

Li GU[1,a], Xian Hong MENG[1]

[1]*Department of Civil Engineering,Shenyang Jianzhu University,Shenyang,China*

Abstract. For ordinary reinforced under corrosion environment corrosion problems and analysis the main factors affecting ordinary steel corrosion, proposed stainless steel bar rust and corrosion resistance advantages, introduce the related properties of the stainless steel reinforced, combined with the research status of domestic and international stainless steel bar, put forward the research and engineering application of stainless steel rebar for the related problems and direction, the prospects and development of stainless steel reinforced put forward suggestions.

1 Introduction

Currently most practical engineering, ordinary carbon reinforced concrete in a non-extreme environments can maintain good performance, the design life of the structure in the basic need for maintenance, when the bar is used in bridge engineering, marine engineering, underground engineering, high corrosion special environment, reinforced concrete often do not meet the design life, structural damage and destruction appeared, even failure. Corrosion of steel structure led to a serious degradation of performance, but also affect the safety and durability of the structure. Durability of reinforced concrete structures become an urgent problem.

Traditional reinforced anti-corrosion method (galvanized steel surface; use of high-strength concrete; concrete protective layer thickness increases) can not be avoided under a special environment of long-term corrosion of steel and structural performance degradation. The use of stainless steel instead of ordinary carbon steel with its excellent anti-corrosion resistance, machinability, high temperature impact resistance can be solved to a large extent from the steel corrosion problem.

Some foreign countries and has been used in the actual project for stainless steel has been more systematic design manual. In 2004, China in the "durability of concrete structures Design and Construction Guidelines" provisions, In particularly severe corrosive environment, to ensure that more than one hundred years of service life of the special important projects, can be used stainless steel. Our country is still a lack of systematic instruction for stainless steel, the stainless steel has caused tremendous obstacles in the development and application in china.

[a]Corresponding author: 799907264@qq.com

2 The reinforced concrete problems

2.1 reasons for corrosion of ordinary steel bars

There are many factors that cause the failure of the structure, the corrosion of steel bars is a key factor, and the main cause of the corrosion of the common steel bar is the accumulation of internal damage and the special environment.

Figure 1. The steel corrosion phenomena

Due to the concrete structure with cracks, in the marine environment of high corrosion environment, environment of water, chloride and salt and other substances through the concrete surface cracks or voids, towards the gradual spread of the concrete, reduce the concrete alkaline. At the same time, reinforced the original compact passivation film was damage caused by steel corrosion, corrosion products generated by the enable the reinforcement volume expansion, aggravate crack expansion, more corrosive substances invasion within the concrete, accelerate the steel corrosion, neutralization of concrete and the crack development.Since the bonding properties of rust matter, between steel and concrete fell, increasing the relative degree of slip, with the internal damage of development and accumulation, and ultimately lead to spalling of concrete steel corrosion.

In combination with the external conditions such as air, temperature and humidity, the acid and alkali salt through the concrete surface pore into the concrete inside and between the reinforcement of electrochemical corrosion, the corrosion damage is very strong, veryfast. Electrochemical corrosion usually occurs in high chloride high salt environment, such as: the marine environment, the use of de-icing salt deicing salt bridges, roads, high soil salinity environment and so on. External special environment allows ordinary carbon steel is difficult to resist corrosion, resulting in damage and destruction of the structure.

Corrosion of steel is the result of special external environment and internal damage to the role. External factors for the steel corrosion provides objective criteria such as: temperature and humidity, salinity content of the acid environment of the material is high, the role of structural fatigue load and the like. Internal factors aggravate the damage accumulation, for example: the concrete carbonation, crack expansion, corrosion of the accumulation of steel, such as volume expansion.

2.2 Methods to improve the durability of reinforced concrete structures

In the rational design and construction conditions, to prevent steel corrosion in order to ensure the safety of the structure, increasing the durability of the structure, thereby improving the service life of reinforced concrete. The main method of increasing the structural durability of steel and concrete can be from two aspects.

(1) in the high corrosion environment, the metal coating is commonly used in the measure, commonly used in the coating metal are: zinc, nickel, chromium, etc. For the purposes of ordinary steel, galvanized steel is anodic coating has excellent electrochemical protection; single nickel-plated steel is common for cathodic coating, and high porosity, corrosion resistance is poor; chrome-plated steel have good chemical stability, with a base, sulfide, nitric acid and other non-reactive.

(2) increasing the thickness of concrete protective layer or adding covering layer, but simply improving the thickness of concrete protective layer will not have a great improvement on the corrosion resistance of steel reinforcement, and the increase of thickness will increase the cost of the project. The use of high strength concrete is also commonly used to prevent steel corrosion method,with its characteristics of high compactness, reduce erosion damage to the reinforcement material.

Using galvanized steel, highstrength concrete,increasing the thickness of the protective layer measures are unable to complete requirements of reinforcement corrosion, while the use of corrosion-resistant reinforcement(such as: stainless steel reinforced) can effectively resist the corrosion factors of the environment,to increase the durability of the structure.

3 stainless steel research status at China and abroad

3.1 stainless steel performance

Stainless steel material has high chemical stability, especially in high corrosion environment, the superior corrosion resistance of stainless steel has become the focus of attention of researchers.

Stainless steel mainly containing chromium, nickel, molybdenum and other alloying elements, which vary depending on the chromium content in stainless steel, stainless steel and acid corrosion resistance will vary, chromium content in steel should be maintained at between 11% to 30%, only stainless steel with acid corrosion resistance, rather than the chromium content, the better, the chromium content in the usual sense of the stainless steel should be greater than 10.5%, equal to the carbon content of less than 1.2%.

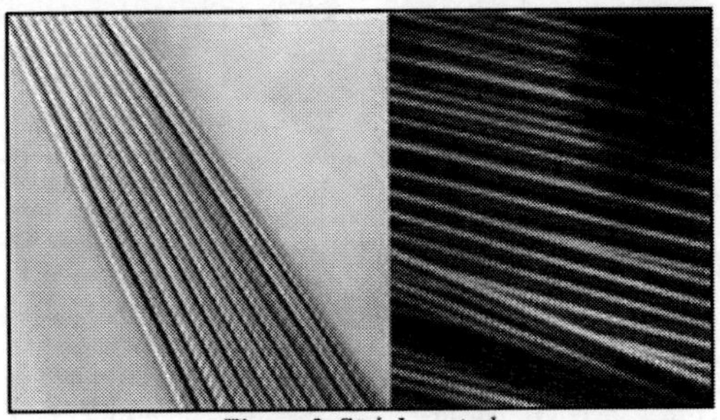

Figure 2. Stainless steel

Alloying elements nickel and molybdenum also improve its corrosion resistance, but play a major role is the chromium, and nickel element for strengthening inter-group if the microscopic crystal structure of stainless steel, to ensure its mechanical properties and mechanical properties.

In addition to these three elements, it will join the stainless steel, silicon, titanium, magnesium, niobium, aluminum and other elements, addition of these alloying elements can not only strengthen the stainless steel corrosion resistance to acids, reduce the carbon content, while improving to some extent stainless steel plasticity, weldability, toughness and processability, making it more suitable used in engineering practice. Holding chromium, nickel and other elements content showed a certain proportion, reduce the carbon content, the better the performance of the corrosion resistance of stainless steel.

Table 1. Mechanical properties of stainless steel (304,316,2205)

performance parameter	304	316	2205
elastic modulus(GPa)	193	193	190
chromium content(%)	18.0~20.0	16.0~18.5	21.0~24.0

yield limit(MPa)	≥205	≥205	≥450
tensile strength(MPa)	≥520	≥520	≥620
Elongation(%)	≥40	≥40	≥25

3.2 Status of foreign steel Stainless Steel

Stainless steel study abroad for relatively large, especially in the United States, Britain and other stainless steel has been included in the list of steel reinforced concrete, while in some areas of high corrosion and high durability requirements of the building project has been the practical application.

T.Zhongand Kim J. R. Rasmussen use three basic Ramberg - Osgood Parameters of ferritic stainless steel stress - strain curve, they found that the current based on Rasmussen (2003) and Wang (2014) stress-strain model is not suitable ferritic stainless steel or cold-formed hollow sections by cutting ferritic stainless steel tensile test specimens revised model, and the model for yield stress [250,550] MPa ferritic stainless steel has good predictive results.

Siebren J. DeJong, Patrick J.Heffernan, Colin MacDougall through a series of constant amplitude and periodicity overload test to study the MMFX and 316LN stainless steel in a corrosive environment, fatigue resistance, and found MMFX and 316LN stainless steel have a good super periodicity load corrosion fatigue performance, Determine the current steel fatigue design criteria can be used MMFX and 316LN stainless steel, features and functions of the two materials fatigue limit in corrosive environments periodic overload within 250Mpa range, while traditional low-alloy steel exhibited no fatigue limit .

M. Nakajima, M. Akita, Y. Uematsuc, K. Tokaji study of type 304 stainless steel fatigue performance problems in 3% NaCl solution and 15%, 30% and 60% tensile pre-strain samples effect, research shows under these conditions, the fatigue strength increases with prestrain increases, stainless steel fatigue strength under these conditions were: 280MPa, 320MPa and 320Mpa, in the role of the air sample of 15%, 30% and 60% pre-stretch after the strain fatigue strength are: 320MPa, 420MPa and 480Mpa; martensitic transformation with the prestrain increases; type 304 stainless steel corrosion fatigue behavior of the affected martensite.

Mesquita, Thiago J, Chauveau, Eric, Mantel, Marc, who studied in austenitic, ferritic, duplex stainless steel corrosion properties of Mo added Elements of stainless steel in an alkaline environment, identifying three stainless steel corrosion pitting and passivation characteristics of a bad environment.

Shayan. A, Xu. A study by chloride contamination of the concrete and stainless steel welding injury problems, has been reinforced stainless steel lap can effectively resist corrosion damage, the chlorine content in the appropriate environment, low-carbon by galvanic corrosion of steel is negligible, electrochemical properties were studied exposure to chloride contaminated concrete steel reinforcing bar. Further Shayan, A, Xu, A also studied restoration and stainless steel corrosion in concrete properties, the conclusion is: Stainless steel can effectively prevent the concrete structure due to key parts of chloride-induced corrosion damage.

Schiegg. Y, Voute. CH, Peter. H, Urlau. U study of ordinary carbon steel, duplex stainless steel (1.4462) and chromium steel (1.4003) for different chloride concentration, PH value and other test conditions, has been reinforced specimen corrosion test data presented stainless steel corrosion initiation and propagation in concrete structures conclusions.

3.3 research status of stainless steel in China

Since the domestic late start, so a lot of research on stainless steel still at the primary stage.

G.X.Zhang,Y.Zhangby stainless steel reinforced concrete slab and reinforced concrete slab of static test and fatigue test, it is concluded that the same conditions stainless steel reinforced concrete slabs than ordinary reinforced concrete plate ultimate bearing capacity increase 13.8%, stainless steel bar specimen failure strain is 5.31 times that of ordinary steel, cross deflection reduced 13.03%-30.27%, tensile reinforcement strain reduced 16.69%-46.76%, concrete strain decreased 8.9%-29.63%.

G.X. Zhang, Y.H. Xu, Z. Ding.They studied the flexural performance of the stainless steel reinforced concrete beams. The test results show that the tensile strength of the stainless steel reinforced than its yield strength; under the same conditions, the deflection and the maximum crack width of the stainless steel reinforced concrete beams are larger than that of the ordinary reinforced concrete beams; the calculation of the maximum crack width can be designed according to the current specification, and the deflection calculation is based on the American ACI code.

H.X. Yuan, Y.Q. Wang,Y.J. Shi.Their application in the building structure of stainless steel, the prospect is described, the stainless steel to improve the durability of the structure, reduce the corrosion rate of reinforcement and maintenance and reinforcement cost advantage.

G.J.Van den Berg, A.H. Liang analysis of ordinary steel corrosion reasons, the use of reinforced concrete and steel reinforced concrete comparative tests, indicating that the stainless steel good corrosion resistance and which has a yield limiting.

H.C. Zhu, J. Yao study the mechanical properties of stainless steel and stainless steel constitutive model drawn steel stress-strain relationship is nonlinear and anisotropic conclusion, through simulation and experimental results show that the use Quach model component designed to meet the requirements.

G.X. Zhang, F. Zhao, Z.H. Zhang, H.Y. Zhang, X.W. Zhou, D.J. Raoby the pseudo static experimental study on seismic performance of steel reinforced concrete beams. They made of stainless steel reinforced concrete beams under low cyclic loading, through elasticity, yield enhancement, strength degradation stage, and stainless steel reinforced concrete beam deformation is large still has a large bearing capacity reserves. Stainless steel ductility is good, full hysteresis curve, the total energy consumption and energy dissipation coefficient is 2.36-4.25 times that of ordinary reinforced concrete beams and 1.12-2.63 times. Capacity, yield strength, ductility, energy dissipation rate increased with the stirrup. Stainless steel reinforced concrete beams better seismic performance.

Table 2. Stainless steel reinforced concrete engineering

project name	country	Stainless steel models	position
Progreso Pier Bridge	Mexico	AISI304	
AbandoibataddBridge	Spain	2205	
York Millennium Bridge	Britain	2205	arched beams and sling
Sheikh Zayed Bridge	UAE	EN1.4462	lower portion ofbridge
Britain Hall	Britain	316L	
Herringbone lift bridge	Germany	0Cr17Ni12Mo2	
Auckland transmission tunnel	new Zealand	0Cr18Ni9	
Stonecutters Bridge	China	2205	
Hong Kong-Zhuhai-Macao Bridge	China	2205	

4 problems

Stainless steel with superior corrosion resistance can overcome an ordinary carbon steel corrosion problems in highly corrosive environments and durability and safety of structures have special requirements, which has a broad space for development. However, studies on stainless steel is still not perfect, which gives practical application to bring obstacles. Where the main problems are:

(1) The key factors that hinder the development of stainless steel is its higher initial cost of stainless steel per ton than ordinary steel about 6-10 times, but if considering the whole life cycle cost structure, stainless steel has a great advantage, ordinary carbon steel is usually the life of about 23 years, while the stainless steel does not require post-maintenance. Structures within their service period, maintenance expenses strengthening reinforced concrete structures is much higher than stainless steel.

In was built in the 1940s in Mexico Progreso Pier Bridge as an example, the use of 200 tonnes of AISI304 grade stainless steel, stainless steel exhibits good corrosion resistance and durability, but also during their service, almost no investment in maintenance costs, built in the 1980s, ordinary steel bridge in the marine environment under the long-term erosion occurred collapsed. To solve the high cost of the initial problem, you can take the stainless steel combination of ordinary carbon steel structural measures in key positions, and to improve the durability of the structure, the purpose of reducing ongoing maintenance costs.

(2) For high corrosive environmental conditions and long-term effects of fatigue loading structure (such as: sea-crossing bridges, offshore drilling platforms), the mechanical properties of stainless steel under fatigue cyclic loading conditions to determine the damage and failure mechanism and member attenuation of structural fatigue life by the effects of corrosion remains a key problem.

(3) damage evolution under study fatigue and corrosion both coupling concrete structure coupling analysis of the cumulative conditions under both structural damage, rather than a single consideration of fatigue or corrosion conditions on structural damage. Coupling effect on stainless steel reinforced concrete deterioration development.

(4) usually test difficult realization of component random amplitude fatigue loading, and many of the component functions of constant load or fatigue load. The actual load is difficult to determine the size and direction of research, amplitude fatigue loading has more practical significance to study the dynamic characteristics of stainless steel.

(5) Stainless steel and concrete bonding properties of the two materials, stainless steel with regard to corrosion environment fatigue failure mechanism of bond between steel and concrete, a clear test pieces in various stages of destruction anchoring force of the proposed bond stress relaxation law.

5 Conclusion

The corrosion resistance of stainless steel can have gradually displayed in engineering practice,for the development of marine engineering projects to improve durability under high corrosive environment structure has broad application prospects.

Meanwhile, Retrofitting existing buildings, structures replace key parts, extended service life of the structure, reduce maintenance costs structure has important practical significance. To solve the traditional steel corrosion under extreme environmental conditions is difficult to guarantee the life of structures and other issues of great reference value.Moreover,the country building a resource-saving and environment-friendly strategy has a positive role in promoting.Theoretical and experimental research carried stainless steel reinforced concrete, stainless steel rebar in the promotion of the use of engineering practice is very necessary.

Acknowledgements

This work is financially supported by National High-tech R&D Program (863 Program)(2015AA03A502-3).

References

1. G.J.VandenBerg,A.H.Liang.*Architecture Technology*.**J.31**.105-107(2000)
2. G.X. Zhang,M.M.Wu.*Journal of Foshan University (Natural Science)*.**J.24**.10-13(2006)
3. M.Nakajima,M. Akita,Y. Uematsu,K. Tokaji. *Procedia Engineering*.**J.2.**323-330(2010)
4. Siebren J. DeJong,Patrick J.*Materials In Civil Engineering*.**J.21**.1-9(2009)

MATEC Web of Conferences **63**, 03004 (2016) DOI: 10.1051/matecconf/20166303004
MMME 2016

Research on the anchorage measures of reinforced concrete beam and wall plane external connection joint

Xian Hong Meng[1], Lian Kai SHA[1,a]

[1]*School of Civil Engineering, Shenyang Jianzhu University*

Abstract. In the external connection of the beam and wall, the anchorage of the end of the longitudinal reinforcement plays an important role in the bearing capacity of the whole node.Relevant information at home and abroad were reviewed in this paper. In this connection the existing anchor improvement measures research progress and application were summarized,which includes the improving and characteristics of visible beam or beam head, double anchor plate and so on. Put forward the problems to be further studied in the future.

Introduction

Reinforced concrete beam and wall plane connection joints are not uncommon in frame shear wall structure system. In the last few years, a lot of work has been done at home and abroad and has achieved a lot of research and engineering application results for this node. Existing standard in our country on basis of research results also made improvements, such as for the setting of the concealed column width, buttress, along the beam axis direction of the shear wall are given specific recommendations.

But in this connection using 90° standard bend hook or other mechanical anchoring measures if frame beam longitudinal reinforcement' s diameter is large, or wall thick is small, usually encountered anchorage section of the horizontal projection length does not meet the requirements of the rules and standard of 0.4labE and this usually lead to brittle failure at the node. The so-called meet the anchorage requirements is that until the end of the steel yield, anchorage section of the strength is enough, sliding displacement of drawing end is small, beam crack is not malignant widening.The tensile stress of the steel bar is decreasing from the end of the drawing to the end of the steel bar, which requires that the anchorage reinforcement has enough horizontal section length, so that the reinforcement can yield before the large slip and crack. At the same time, under the same tension reinforcement effect, enough straight anchorage length will make critical reinforced section of shear force is small, effectively avoid reinforced pull off critical section concrete shear failure.Therefore, it is very important to study how to optimize the condition of reinforcement and reduce the length of the reinforcement. For this kind of situation, the present research has the following several aspects of the results.

1 Setting the transverse anchorage bars at the hook of the longitudinal reinforcement

This method is improved on the basis of the 90 degree bend hook anchoring at the end of the longitudinal reinforcement,by setting the transverse anchorage bars at the hook of the longitudinal reinforcement, the anchorage of beam longitudinal reinforcement in the wall is strengthened.

Zhihao Wang of Tsinghua University[1] has adopted this method to strengthen the anchoring, and the anchoring schematic diagram is shown in figure 1, The method has the advantages of simple connection operation and good economic performance. The final test results show that this method has a certain degree of improvement of the deformation capacity of the test piece, and it shows that the anchorage of the longitudinal reinforcement in the wall is effective. But it is lack of essential and quantitative analysis on the effect of welding transverse reinforcement in the bending hook of longitudinal bars.

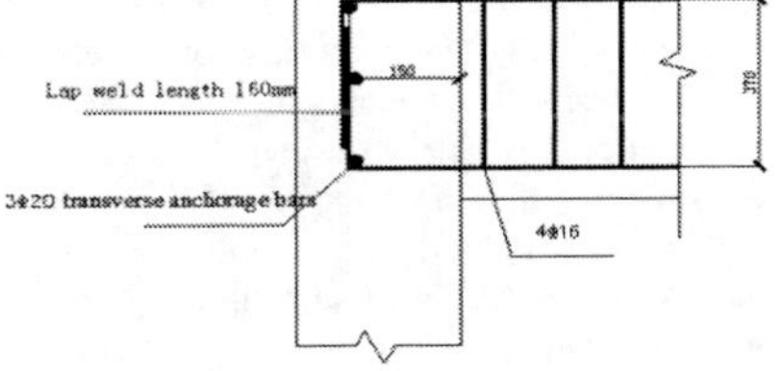

Figure 1. Anchorage details of beam longitudinal steel bars

[a] Lian Kai SHA:867310459@qq.com

DOI: 10.1051/matecconf/20166303004

2 double anchor plate

With anchor plates or nut at the end of longitudinal reinforcement of beam provide greater bearing surface to help share the pulling force of the anchor bar. As is shown in the figure2 and 3, the tension is provided by the steel bar and the anchor plate. The United States, Canada, South Korea, Japan and China has carried out to set up research and development of the anchor plate at the end part of the longitudinal reinforcement, summed up the anchoring plate reinforced than traditional bending reinforced with obvious mechanics and engineering application advantages. In our country, the existing standard has specific requirements on the bearing area and thickness of the anchor plate, but the difficulty still lies in the problem of the horizontal projection length of the anchorage segment.

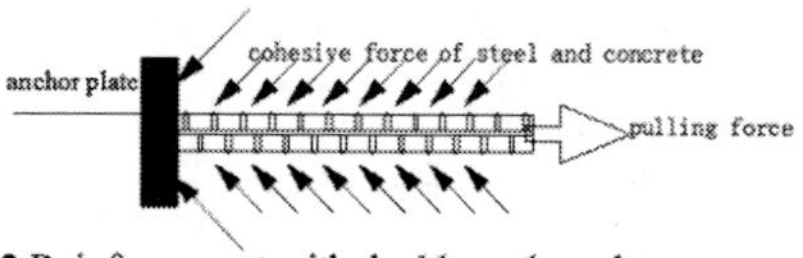

Figure 2. Reinforcement with double anchor plate

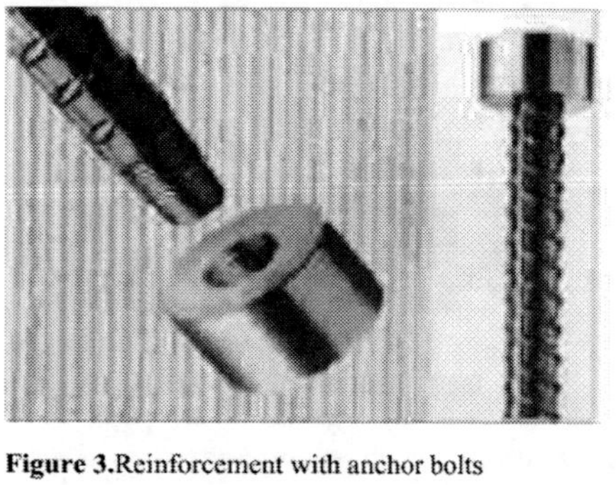

Figure 3. Reinforcement with anchor bolts

It is proposed by Qiaoling Gao of Chongqing University[2] that the double anchor plate of the tensile steel plate and the anti sliding plate at the end of the beam longitudinal reinforcement as shown in figure 4 and 5. And the comparison of the test pieces with the tensile anchor plate measure only at the end of the longitudinal reinforcement. And in the case of thin walls,it was compared with the specimen only with tensile anchor plate measures at the end of the longitudinal reinforcement. Test results show that only welding tensile anchor plate although can improve the node bearing capacity but the plate in reverse loading will easily be pushed from the wall, and the use of double anchor plate increases the critical section area of pull-off failure and reduces the critical section of the uneven shear, it also effective the control of the "pull" and "shirk" damage and effectively improve the bearing capacity and ductility planar of the beam-wall connections. And through further theoretical analysis, the failure mechanism of the force model of the beam-wall plane external connection node is proposed.
But it needs to be pointed out that although the double anchor plate measures can effectively improve the performance of the node, but the double anchor plate measures exist a lot of inconvenience in the actual construction, especially when the wall is thin.

Figure 4. Reinforcement with double anchor plate

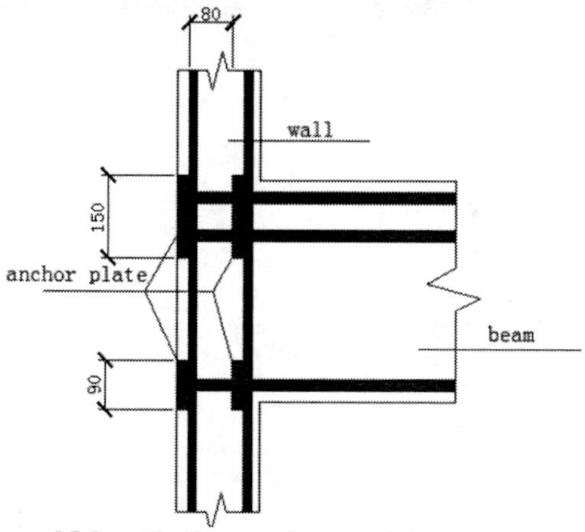

Figure 5.Schematic diagram of anchor plate

3 Visible beam and beam head

According to the specification that when the horizontal projection length of anchoring section does not meet the requirements, the floor beam can be put out of the wall to form a head beam, and the beam longitudinal reinforcement bent anchorage after extends into the beam head..And Zhiliang Li, Zujing Zhu of Chongqing University[3][4], completed 6 low cycle repeated load test of beam wall plane external connection node with the visible beam anchor measures and a preliminary approach to the design of the visible beam is presented. The two structures are shown in the figure below:

DOI: 10.1051/matecconf/20166303004

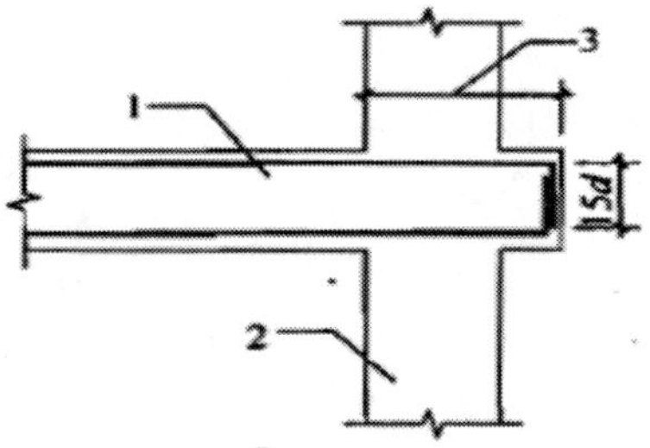

Figure 6. Schematic diagram of beam head

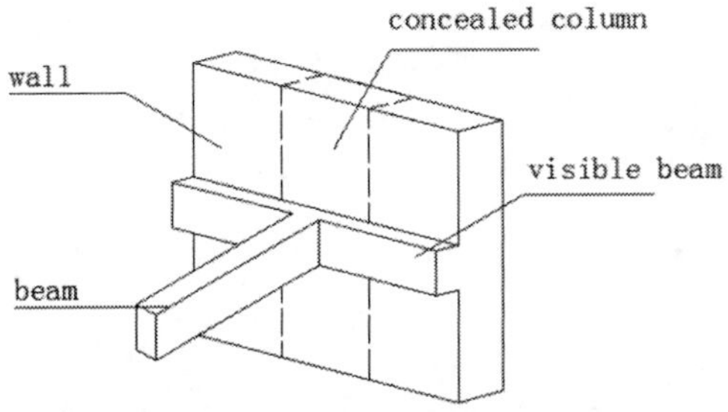

Figure 7. Schematic diagram of the visible beam structure

Figure 8. Schematic diagram of the visible beam

The two methods both can improve the anchorage performance, the bearing capacity and the ductility of the beam. But the common case is the frame core tube is connected with beam, the core tube is usually used as the elevator wells, and setting the beam head is not allowed. So there is a lot of limitations to set the beam head measures. And the setting of visible beam measures can effectively improve the bearing capacity and ductility of the node, And the deficiency is the process of construction is slightly complicated. the renovation will be needed in the late to avoid the lack of architectural aesthetics. But it is still a good choice to solve the problem of the lack of anchorage length by rational structure.

4 foreign related standards

4.1 ACI 318-02

The provisions of article 12.6 of the ACI318-02 code of the United States[5] on the mechanical anchoring of steel bars are as follows:

(1) Any mechanical device that can play the strength of a steel bar and does not damage the concrete can be used as anchorage;

(2) External force of reinforcement composed by mechanical anchoring force and cohesion force of reinforced steel;

(3) The full documentation of the mechanical anchoring device shall be provided in the contract issuing drawings, and shall be submitted to the relevant building authority.

4.2 ACI 352R-02

The relevant provisions of the United States ACI352R-02 [6]specification in section 4.5 on the use of steel bars with anchor plate in the joint shall be as follows:

(1) reinforced with anchor plate should meet the A970 ASTM specification;

(2) the anchor plate should be placed in the joint core concrete 50mm, the minimum anchorage length of ldt can not be less than 8db or 150mm;

4.3 CSA Standard CAN3-A23.3-94

The Canadian concrete structure design code 13.4.8 section CAN3-A23.3-94[7] allows the use of a strip of anchorage plate shear reinforcement, as follows:

(1)The reinforcement of the steel bar should be able to give full play to the yield strength of the whole reinforcement;

(2)The head area of the reinforced head is at least 10 times the area of the reinforcing steel, unless the test proves that it can be used with a smaller size;

(3)Total shear stress should not be more than $0.8\lambda\phi_c\sqrt{f_c'}$ after multiplied by the coefficient,the total value is greater than 1.33 times the traditional shear stirrup component allowed.

(4) The contribution of concrete should be not greater than 0.3 $\lambda\phi_c\sqrt{f_c'}$ after multiplied by the coefficient, the value is greater than 1.50 times the total shear force of the traditional reinforced concrete members.

(5)The increase of the shear capacity of concrete comes from the effect of the reinforcement provided by the reinforced bar.

5 Issues to be further studied

After the domestic vigorously advocate and a large number of scientific research institutions, enterprises and institutions continue to participate, the development of the exterior node of the beam-wall has made great progress and a lot of achievements. However, there are still many problems need to be further studied in the connection of the connection point with the hook or other mechanical anchoring measures: A lack of anchorage concrete bearing capacity calculation formula, so lack of detailed theoretical basis when using some other measures to strengthen anchorage. Looking for engineering practice in the domestic application of anchorage strengthening measures to form with independent intellectual property rights that applicable to domestic engineering products and technologies, is the important issue in the future we still have to face.

Acknowledgments

This work was financially supported by the Project of National Key Technology R&D Program in the12th Five Year Plan of China (2011BAJ10B04)

References

1. Wang Zhihao, Huang Caiguang, Qian Jiaru. Journal of Building Structures. J .12,33-39(2008)
2. Qiaoling Gao. Shanxi Architecture.J.11, 83-85(2011)
3. Zujing Zhu. Chongqing University.D.(2006)
4. Zhiliang Li,Chongqing University.D.(2009)
5. FarmingtonHills,Michigan,ACI318-02,American Concrete Institute.S.(2002)
6. FarmingtonHills,Michigan,ACI352R-02,American Concrete Institute.S.(2002)
7. Rexdale, Ontario,CSA Standard CAN3-A23.3-94,Canadian Standards Association.S.(1994)

MATEC Web of Conferences **63**, 03005 (2016) DOI: 10.1051/matecconf/20166303005
MMME 2016

Effect of desliming on the flotation of a complax copper ore from China

Yong CHEN[1,a], Yong Sheng SONG[1], Wen Juan Li[1] and Gui Ying ZHOU[1]

[1] *National Engineering Laboratory of Biohydrometallurgy, General Research Institute for Nonferrous Metals, No.2 Xinjiekouwai Street, Beijing, China*

Abstract. The complex copper oxide ore sample was taken from Deerni copper deposit, Qinghai Province of China. Batch flotation tests had been conducted to upgrade the copper concentrate by conventional amyl xanthate reagents under 73% -74 μm of the particle size; however, the unsatisfied results (Cu grade of 18.21% and recovery of 59.25%) were obtained. Also there are a large amount of slimes trapped in the concentrate and high-dosages consumption of reagents, for example, more than 2.5 kg/t sodium sulfide and 1.5 kg/t sodium silicate at only one-staged roughing. Based on the analysis of the sample, most of gangues are clay, feldspar and mica, which may easy to over-grind and deteriorate the flotation process either on reagents consumption or slurry fluidity or viscosity. A hydro-cyclone had been introduced to pre-concentrate the oxide ore by scrubbing the slime before flotation, which can result obviously in reducing the dosages of sodium sulfide from 2.5 Kg/t to 1 Kg/t, and raising the grade of Cu in the concentrate from 18.21% to 26.65% at the expense of about 1% recovery of Cu. In this paper, the effects on the recovery of Cu by the different dosages of flotation reagents with or without de-sliming were studied, with the objective of determining the different effects on the functions of flotation reagents by slime during flotation.

1 Introduction

The flotation of oxidized Cu minerals, is much more difficult than the flotation of corresponding sulfide minerals. Several factors may influence oxidized Cu minerals recovery during flotation, such assist dissemination and composition, particles size[1-3], flotation reagents, especially sodium sulfide, temperature of slurry effected on the behavior of collectors, especially fatty acid [4-5].

Sulfide compounds have long been used in oxide type base-metals mineral flotation practice for decreasing the level of metal species in solution and for modifying the surface of the mineral while still having a surface that will respond to flotation. Sodium sulfide is the most important reagents in oxidized Cu minerals flotation [6-8]. The surface characteristics of tenorite in sodium sulfide solutions have been investigated in detail by Herrera-Urbina et al. Sulfidization of the mineral surface significantly improves the flotation of tenorite with amyl xanthate as collector. The xanthate flotation of both sulfidized tenorite and malachite, however, is strongly affected by the sulfide dosage. Once the sulfide ion reports to the solution at the end of the conditioning period, mineral flotation ceases [5]. Both dispersion and depression were also found for sodium silicate [3, 6]. However, the only problem for the use of sodium silicate in oxidized Cu ore flotation is how to reduce its dosage in order to filter concentrate smoothly and fast.

Particle size has also been a subject of research for flotation of tenorite and malachite. The different sensitivities to the reagent concentration for fine (-400#) and coarse particles during phosphate ore flotation had been found by Santana et al. for carbonated and siliceous ores [9], slime has a very detrimental effecton the flotation of solids, which includes contaminating foams, raising the consumption of flotation reagents, reducing the speed of the process and causing an anarchic response in mineral separation.

Although the use of sodium sulfide and sodium silicate in copper oxide minerals flotation is well established, one of the main problems in phosphate ore in China is high-consumption of sodium sulfide, which is generally range from 5 to 8 kg/t, about 50% of total flotation agent cost [4-8], especially the mechanism between reagents and particles size and are not sufficiently understood. Singh et al. [9] indicated that "Spilt conditioning" can improve the flotation efficiency of slime (-37 micron) and coarse particles (-105+44 micron) due to hydrophobic aggregation between coarse particles and slimes from the Maton rock phosphate plant, India. Santana et al. [8] reported that the size of mineral particles plays a significant role in the sequence of events that lead to the flotation of a particle.

[a] Corresponding author: cy040311@163.com

The aim of this paper is to investigate effect on dosages of sodium sulfide and sodium silicate in flotation of copper-oxide by slime from a Deerni copper deposit ore in order to gain a better understanding of the behavior of different particle sizes in flotation performance when adding the flotation reagents.

2 Expertimental

2.1 Materials

The copper oxide ore was provided by Deerni copper deposit, Qinghai province of China, in which the main gangue were clay, feldspar and mica. Around 200 kg of representative ore samples was crushed to below 2 mm size with two-stage jaw crusher and one-stage roll crusher. The materials were then well mixed and divided into 1 kg samples for mineralogical and pre-concentration studies.

Table 1. List of the experimental conditions of flotation tests.

Experimental conditions	
Constants	
Cell type	3L XF-D
Feed ore concentration	30%(mass)
Feed size	70% -74μm
Flotation time (rougher)	10min
Impeller speed	1200rpm
Variables	
Dispersant	Sodium silicate
Sulfurizing reagent	Sodium Sulphide
Collector	Amyl xanthate
Frother	Pine camphor oil

Sodium sulphide was reagent-grade chemicals supplied by Kunming chemical reagent company. Sodium silicate was from metallurgical research centre of Kunming, which modulus was about 2. Amyl xanthate as collector was also from the same company.

2.2 Methods

2.2.1 De-sliming tests

1 kg sample was mixed with 600 ml tap water and ground to 75% -74μm in the ball mill, and then the pulp was diluted to 15% Solids and then fed to the hydro-cyclone (FX 25-PU Hydro-cyclone). The pressure was about 100kPa. The underflow from the hydro-cyclone was collected. A size analysis for the overflow from FX-25 PU by dry and wet-sieving had been conducted by a series of screens from 400 mesh to 200 mesh.

2.2.2 Bench flotation tests

Flotation tests were conducted in a 3.0L XF-D cell. In a typical ore sample flotation test, 1kg sample mixed with a certain amount of tap water was ground to 70% -74 μm for several minutes in the ball mill, then diluted to 3.0L cell XF-D which had been used to conduct roughing at 1200 rpm for a certain time.

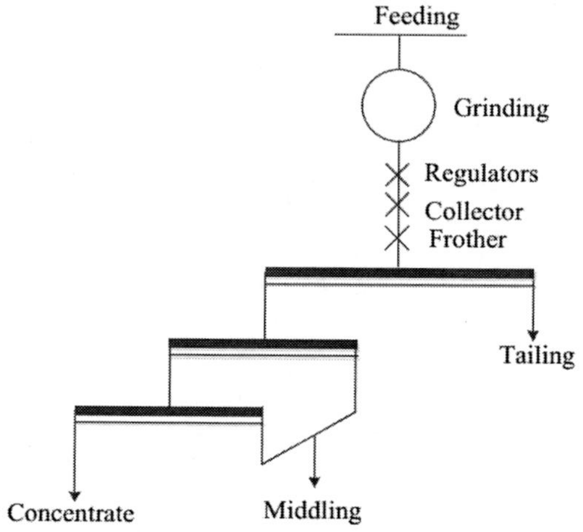

Figure 1. Flotation flowsheet

The flotation reagents had been added in the following order: depressants, collectors, before adjust the pH to a certain value by soda ash or hydrochloric acid as shown in Table 1, which presents the experimental conditions of bench flotation tests. In addition, the signs of Conc., Mid. and Tail. In the results from the flotation tests mean the concentrate, the total middlings included middling I, II and III and the total tailings involving the tailings from the flotation tests and the slime from hydro-cyclone if scrubbing was used before flotation tests.

3 Results and discussions

3.1. Size distribution

Figure 2 shows the distribution of the particles copper oxide in the different size fractions of the head-feed ore samples without grinding. As can be seen, most of copper oxide minerals distribute in the relatively coarse fractions, such as more than 80% of copper oxide minerals at the fraction of +74 μm, especially about 50% of copper oxide mineral sat the fraction of +210 μm, which indicated that it is necessary to grind the sample to ensure the reasonable liberation of value minerals before the bench flotation tests are conducted.

DOI: 10.1051/matecconf/20166303005

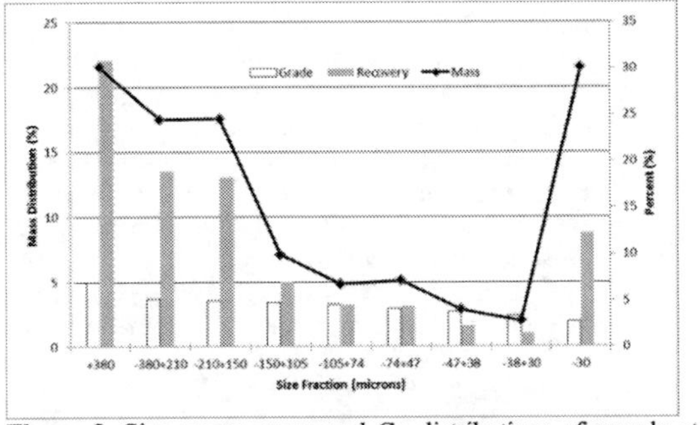

Figure 2. Size-assays, mass and Cu distributions of sample at different size fractions

Figure 3 presents the distribution of the copper oxide minerals at different size ranges at 200 mesh 73%, which indicated that a sharply increase in the mass and recovery either at the fraction of -30 μm or -74+30 μm. In the meanwhile, the low content of Cu in the fraction of -30μm may be possible to remove the slime by scrubbing effectively which can remove 30% slime from the head feed to flotation at a loss of about 13% of Cu in the slime.

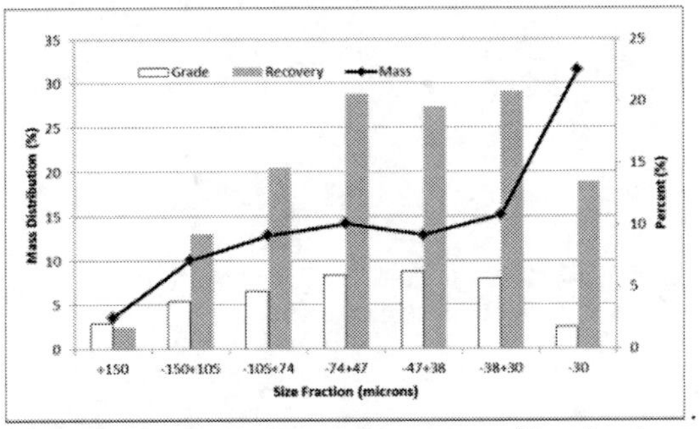

Figure 3. The distribution of the copper oxide minerals at different size ranges at 200 mesh 73%

3.2 De-sliming by hydro-cyclone

In order to remove the slime, de-sliming tests were carried out. Size analysis of the underflow and overflow from FX 25-PU Hydro-cyclone by dry and wet-sieving had been presented in figue 4. It was noted that only 28% of Cu and almost half of yields of feed distributed in the overflow. Also, -38μm size fraction is the main mass in the overflow while -105μm ~ +30μm size fraction is the main mass in the underflow. The de-sliming greatly reduces the fine mud, and create a good condition for subsequent flotation of underflow.

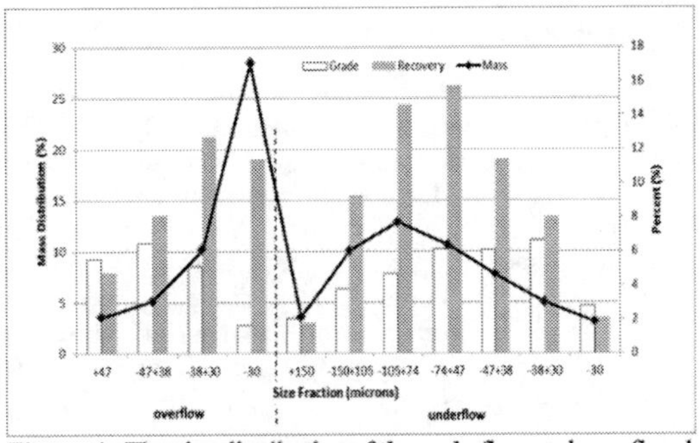

Figure 4. The size distribution of the underflow and overflow by de-sliming with FX 25-PU Hydro-cyclone.

3.3 Flotation

The flotation results without de-sliming was shown in Figure 5, which provides the grade and recovery of Cu in the concentrate, midding and tailing. As shown in Figue. 5, this illustrates the unsatisfied results (Cu grade of 18.21% and recovery of 59.25%) had been obtained. The better results (26.65% grade of Cu at 58.24% recovery) had been obtained by de-sliming with FX 25-PU Hydro-cyclone before flotation (show in Figure 6). De-sliming exhibited powerful effect on rising the grade of copper concentrate and little effect on the recovery of Cu. Therefore, it could be an efficient measure before flotation by hydro-cyclone.

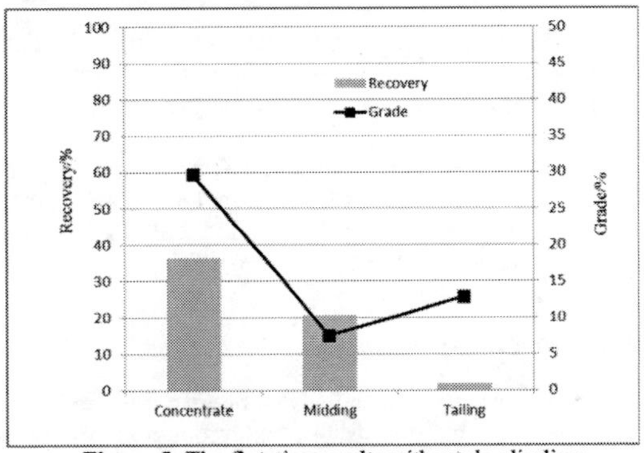

Figure 5. The flotation results without de-slimling

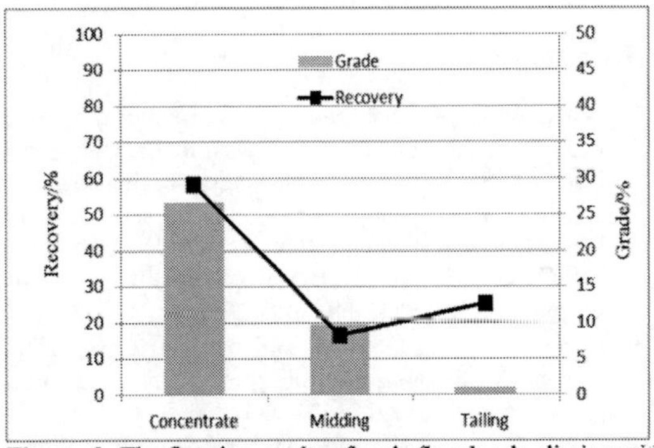

Figure 6. The flotation results of underflow by de-sliming with FX 25-PU Hydro-cyclone

3.4 Reagent consumption

The reagent consumption results with or without de-slimling before flotation was shown in figue 7. The flotation reagents consumption with de-sliming is much lower than without de-sliming, especially the amount of sodium sulphide and amyl xanthate. A hydro-cyclone had been introduced to pre-concentrate the oxide ore by scrubbing the slime before flotation reducing the dosages of sodium sulfide from 2.5 Kg/t to 1 Kg/t, which saving reagent consumption greatly and improving the beneficiation indexes.

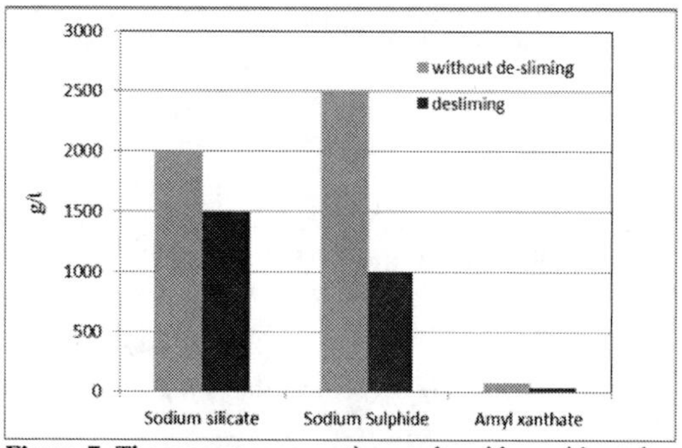

Figure 7. The reagent consumption results with or without de-slimling

4 Conclusions

(1)An increase of Cu in the concentrate from 18.21% to 26.65% at saving about 1/2th original dosages of sodium sulphide and amyl xanthate was obtained by scrubbing.

(2) The consumption of sodium sulphide may be very sensible to the amount of slime in the pulp, so may sodium silicate and collector, which all reduce the dosages after de-sliming.

Acknowledgments

This work was financially supported by High and New Technology Industrialization promotion programs of Qinghai province, China (No. 2015-GX-Q06A).

References

1. Dos Santos, M A, Santana, R C and Cappoin, F, 2010. Effect of ionic species on the performance of apatite flotation, Separation and Purification Technology, 76: 15-20)
2. Feng, D and Aldrich, C, 2004. Influence of operating parameters on the flotation of apatite, Minerals Engineering,17: 453-455)
3. Hernainz, F, Calero, M and Blazquez, G, 2004. Flotation of low-grade phosphate ore, Advanced powder technology, 15: 421-233.
4. F. De Lillo, F. Cecconi, G. Lacorata, A. Vulpiani, EPL, **84** (2008)
5. Houot, R, 1982. Beneficiation of phosphatic ores through flotation: review of industrial applications and potential development, International Journal of Mineral Processing, 9: 353-384
6. Santana, R C, Farnese, A C C and Fortes, M C B, 2008. Influence of particle size and reagent dosage on the performance of apatite flotation, Separation and Purification Technology, 64:8-15
7. Zhang, P, Yu, Y and Bogan, M, 1997. Challenging the "Crago" double float process II. Amine-fatty acid flotation of siliceous phosphates, Minerals Engineering, 10: 983-994
8. Yehia, A, Youseff, M A and Boulos, T R, 1999. Different alternatives for minimizing the collector consumption in phosphate fatty acid flotation, Minerals Engineering, 3: 273-278
9. Zheng, X and Smith, R W, 1997. Dolomite depressants in the flotation of apatite and cellophane from dolomite, Minerals Engineering, 10:537-545.
XXVI

MATEC Web of Conferences **63**, 03006 (2016) DOI: 10.1051/matecconf/20166303006
MMME 2016

Comparison of the design rules for the end of the prestressed concrete double tee component between China and the United States

Xian Hong MENG[1], Lu Xue ZHANG[1,a]

[1]*Department of Civil Engineering, Shenyang Jianzhu University, Shenyang,China*

Abstract. Precast prestressed concrete double tee is an economic bearing component which can be made into a large span, large coverage area .From the angle of the design for the end of the precast prestressed concrete double tee component, flexure and axial tension in extended end, direct shear, diagonal tension at re-entrant corner, diagonal tension in extended end and the connection of the end of precast prestressed double tee component in China and the United States are compared. The comparative studyprovide the direction for the future development of China in the double tee,including strengthening the end reinforcement and increasing end connections.

1 Introduction

Precast prestressed concrete double tee (hereinafter referred to as the double tee) is a pretensioned concrete element composed of two vertical precast beam and horizontal roof. Its cross section is a two "T" shaped, Precast prestressed concrete double tee is a prestressed reinforced concrete bearing structure, the high-strength prestressed reinforcement and high-strength lightweight concrete combined well, to improve the bearing capacity, reduce weight.

Double tee has simple shape with two high and narrow rib, a thin and widepanel, is a new type of environmentally friendly building materials is different with the traditional roof member, can be widely used in underground garage, industrial buildings, schools, hospitals and other large span building roof. Double T board itself has many advantages:

1.1 Strong bearing capacity, light weight

Double tee is beam plate combination, made of high-strength prestressed concrete and high-strength prestressed reinforcement. High strength prestressed reinforcement with high-strength lightweight concrete, not only can improve the bearing capacity of the component, reduce the weight, but also can effectively reduce the use of steel and concrete, reduce material costs, save cost, achieve economic applicable effect;

1.2 Simple appearance, convenient construction

Double tee has the cross section of double "T" type section, simple design, convenient construction, high construction speed, low material consumption and no large area of high formwork construction work,good economy, and can ensure the safety of construction process;

1.3 The structure of good performance

Compared with the steel structure, concrete frame, truss + roof panel, double tee structure has better comprehensive performance. The fire performance fire can reach grade two and above, good durability, good insulation properties, corrosion resistance, and rarely need maintenance. Compared with the traditional roof beam and roof panel system, double tee system has better overall performance, less welding, and has good aseismic performance. In addition, you can also save the indoor space.

Since Harry Edwards, the Americansdesign the first double tee with 2.4mwidth and 15m span in 1952, double tee as a kind of economical type of precast concrete element, develop rapidly in the construction of industrial plants and large span buildings. The United States as the origincountry of the double tee, is the most widely used, especially in the large parking garage as floor. In Japan, the UK, Canada and other countries have also promoted the double tee, double tee has further development and improvement, and lay a solid foundationfor the future popular. Chinese understanding and research on the double tee began in the 60's of last century, due to the production conditions and technical problems, the research on the double tee is mainly focused on the small span industrial plant. In the 70's of the last century, the main application of double tee was in electric power, oil field and other engineering, the maximum span of double tee was about 9m, it was flat, and then the standard design drawings SG432 was published. From the late 90's of last century so far, Daqing firstly transfer the study to large span industrial

[a]Corresponding author: 1316099859@qq.com

building roof, and developed a double slope shape double tee products. April 1, 2014,the first precast concrete parking building project: assembled parking building of China first automobile group, FAW Group Technology Center in passenger car project began hoisting construction, so that the popularization and development of the double tee achieve a leap in the progress, become a landmark in the history of the development of double t. At present, the double tee is mainly used in industrial plants and large span civil buildings in our country.

Figure 1. Double tee section in China

Figure 2. Double tee section in the United States

Figure 3. The application of double tee in China (a)

Figure 4. The application of double tee in China (b)

Figure 5. The application of double tee in the United States (a)

Figure 6. The application of double tee in the United State (b)

Compared to the United States, our country because of short development time, the limit of production technology, construction technology and production conditions, the study of double tee still belongs to the development of the state, still need to continue to learn. Based on few of domestic and foreign comparative study on design for the end of double tee component, the paper regard this as the starting point, the comparative study of the design for the end of double tee component, through the comparative study, find out the problems in double tee's design in China, and then improve the design of the double tee.

2 Technical Standards

In the 70's of the last century, China compiled the standard design atlas SG432. According to the local atlas and the engineering application experience, in 2006, the national building standard design drawings of the prestressed concrete double tee officially approved, a total of three, respectively 06SG432-1,09SG432-2,08SG432-3. So far, China has not yet issued special double tee product standards and technical regulations, in the engineering application person mainly refer to the national building standard design drawings.

The production and application of the double tee in the United States is mainly based on the PCI Design Handbook prepared by the Precast /Prestressed Concrete Institution. The design of the components in the PCI manual is mainly based on the concrete structure building code (318-05 ACI) and its provisions (318r-05 ACI). In addition, the load, section size, span and reinforcement amount are given in PCI manual, which provide the basis for the design and engineering application of precast and prestressed concrete components.

3 Contrast of the End of Component

3.1 Component-end bearing

In the United States, the end of double tee component often use the dapped-end , according to the PCI manual, calculate the dapped-end bearing .

Design of bearing areas that are recessed or dapped into the end of the component requires the investigation of several potential failure modes.

Dap reinforcement should be provided in all cases where any one or more of the following conditions occurs:

1. The depth of the recess exceeds 0.2 H or 8 in.

2. The width of the recess (l_p) exceeds 12 in.

3a.For components less than 8 in. wide, less than one-half of the main flexural reinforcement extends to the end of the component above the dap.

3b.For components 8 in. wide or more, less than one-third of the main flexural reinforcement extends to the end of the component above the dap.

These criteria indicate that only short, shallow recesses, having the minimum amount of main reinforcement or more extending into the nib above the dap, do not require all the dap reinforcement. Experience and some unpublished tests verify that, for short, shallow recesses, the hanger reinforcement A_{sh} and A'_{sh} is not necessary. However, in these cases, it is recommended that confinement reinforcement A_v and flexural reinforcement $A_{vf} + A_n$.

The potential failure modes are as follows:

1. Flexure (cantilever bending) and axial tension in the extended end. Provide flexural reinforcement A_f plus axial tension reinforcement A_n

2. Direct shear at the junction of the dap and the main body of the component. Provide shear-friction reinforcement composed of A_{vf} and A_h , plus axial tension reinforcement A_n .

3. Diagonal tension emanating from the reentrant corner. Provide shear reinforcement A_{sh} .

4. Diagonal tension in the extended end. Provide shear reinforcement composed of A_h and A_v .

5. Diagonal tension in the undapped portion. This is resisted by providing a full development length for A_s beyond the potential crack.

Each of these potential failure modes should be investigated separately. The reinforcement requirements are not cumulative, that is, A_s is the greater of that required by 1 or 2, not the sum. A_h is the greater of that required by 2 or 4, not the sum.

Detailed calculation can be seen in PCI manual.

China double tee end is flat-end, rarely involved in the end of the gap type. The position of double tee of the flat-end is generally at the top of the support member, the design calculation is relatively simple. For the calculation of the dapped-end, the standard GB50010-2010 of our country has not yet provided the corresponding calculation standards, and the lack of relevant design basis.

3.2 Component-end Connection

According to national building standard design drawings “Prestressed Concrete Double Tee", the end of double teecomponent is the whole section, the end of the rib beam mainly connect by welding. Specific connection mode can be seen in the Chinese Atlas “Prestressed Concrete Double Tee." And thedouble tee component-end connection style in the United States not only include Chinese connection style, but also put the binding bolt (on top of the end plate and welding combined to connect. Thus, compared with the component-end connection of double tee with the UnitedStates,component-end connection in China is relatively single, we can learn from the successful experience of the United States.

4 Conclusions

According to study on the difference from flexure and axial tension in extended end, direct shear, diagonal tension at re-entrant corner, diagonal tension in extended end and the connection of the end of precast prestressed double tee component in China and the United States, we can discover that the design for the end of precast prestressed concrete double tee component is still flawed,China national building standard design drawings"Prestressed Concrete Double Tee" should be revised to make it more perfect.

Acknowledgements

This work was financially supported by the Project of National Key Technology R&D Program in the12th Five Year Plan of China (2011BAJ10B04).

References

1. China Construction Science Research Institute.06SG432-1prestressed concrete double tee (slope board, width 2.4m) [S]. Beijing: China Planning Press, 2006
2. China Construction Science Research Institute.08SG432-3 prestressed concrete double tee (slope board, width 3.0m) [S]. Beijing: China Planning Press, 2008
3. China Construction Science Research Institute.09SG432-2 prestressed concrete double tee (slab, width 2.0m, 2.4m, 3.0m) [S]. Beijing: China Planning Press, 2009
4. MaoyuWang. Comparison of precast prestressed concrete double tee components between China and America [J]. Concrete and cement products, 2014, (8): 42-45
5. Precast/Prestressed Concrete Institute. PCI Design Handbook [M].7th Edition.Chicago,IL,2010.
6. QiongLiu. Research and application of precast concrete structures [J]. construction technology, 2014, (22): 09-14
7. Xiaofeng Wang, Yong Zhao.Application and technique of prestressed concrete double tee [J].The second session of the forum of industrialization of construction technology,2012,98-103
8. Wuhua Tang, Tao Qiu. Design analysis of prestressed concrete double tee [J]. Building structure, 2009,(39),58-59
9. Rui Pang, Shuting Liang. Study on precast concrete double tee floor system abroad[J].Industrial buildings 2011, 41(3), 12-126
10. Fenggang Jiang, Chengwu Chen. Application analysis of large span prestressed concrete double T slab construction[J].Journal of Qingdao Technological University. 2014, 35(6), 36-39

Attention to the production and construction of assembled concrete laminated plates

Xian Hong MENG[1], Wei GUO[1,a]

[1] *Department of Civil Engineering, Shenyang Jianzhu University, Shenyang, China*

Abstract. with the country of prefabricated construction vigorously promote, laminated concrete gradually into the line of sight of people, it not only has the factory production, convenient construction convenient, highlight the advantages of material saving, etc., this paper production and construction from the point of view of, the in the production process and Application of the discussion.

In the application of assembly construction, precast concrete composite plate has the advantages of saving formwork and supporting, quick and convenient construction, good economic benefits and so on. Due to its outstanding advantage countries began to vigorously promote the assembly of the type of construction, but because of our country in prefabricated construction area started late and some prefabricated components in production and construction exists many defects, this paper mainly from the perspective of the laminate floor of to explain it in the factory production process and problems that should be paid attention to in the construction site in the process.

1 production matters needing attention

In the production process should be considered during the construction of the floor and cast-in-place laminated layer as a whole, so in the production process of bottom surface should make bump not less than 4mm artificial roughness surface so that the base plate can and superimposed layer better placement as a whole.

In order to meet the building warm water pipeline, laminated need opening in some of the construction process often by on-site construction workers digging, inevitably cut inside the soleplate by reinforcement, thereby reducing the slabs. Therefore laminate for opening to reserved hole (hole reinforced temporarily cut off) in factory production, stack layer of concrete pouring time to set aside the holes, laminated arrival intensity after resection of the hole steel.

Floor openings

2 Transportation and stacking problem of Composite Plates

In order to prevent the laminate in transportation and stacking process failure, floor of the transportation, piling up process shall comply with the concrete structure engineering construction quality acceptance norms "GB50204-2002, the laminated prestressed concrete floor" GB/T16727-2007 provisions. In addition, the stack field shall be smooth and compact, stacking between the board and the ground should have a certain gap, and drainage facilities. Both ends of the board (board to end 200mm) and the cross position should be set when a pad pad wood, wood set cross sign in length 3.9m, two wooden cross flag set pad length 3.9m, pad wood should be on the alignment. Different plate number should be stacked, stacking height should not be more than 6 layers, stacking time should not exceed two months. Concrete strength to meet the design requirements of the rear to

[a] Corresponding author: 1979173001@qq.com

play and transport to firmly on a fulcrum lashing, beat to prevent movement or, in the concrete on the edge of the board or the rope at the point of contact should be protected by liner

Floor transport

3 hoisting problem

Laminated hoisting process is the construction of a important link, in order to ensure the laminate can be safely lifting to a reasonable position, production manufacturers should design embedded rings position, to lifting does not occur when the eccentricity. The design height of rings, ensure the crane hook and the rings may not be stuck or crooked. In addition, taking into account the moment and deflection of the hoisting process of floor should be generated during lifting dynamic coefficient of 1.5. And the floor should be lifting floor number, so that the floor can be accurately placed in proper position, to prevent the two handling events, the maximum saving manpower. The temporary support should be installed in the front of the base plate in the middle of the span and close to the supporting seat. When the axial cross 3.9m cross set in a support. When the axial cross 3.9m cross set two support, support the top surface should be strictly copy flat, in order to ensure the flatness of the bottom plate, the support of all levels should be arranged in a vertical line.

Floor hoisting

4 splice structure of the composite plywood

After floor hoisting in order to improve the laminated overall performance and shock resistance, formed in the bottom plate of rectangular holes interspersed with transverse reinforcement, the laminated stacked layers of cast-in-situ formed in two-way force, under normal circumstances plate and the plate gap mainly take two splicing way, a is in accordance with design drawings in the spaces placed transverse reinforcement, but this method at the construction site due to insecure reinforced easily lead to dislocation of reinforced, another it is directly on the floor laying the reinforcing mesh, the ways to save time and effort, and can improve floor carrying capacity.

Seam reinforcement

Steel mesh

In bottom plate and beam joint laminated prestressed

DOI: 10.1051/matecconf/20166303007

tendon anchorage to the beam, need a crowbar to laminated prestressed reinforcement force anchorage to the beam within, resulting in deformation of beams with stirrups, and also brings difficulty to the construction itself, this time can be combined with the construction practice, through the opening of the hoop method to improve the construction.

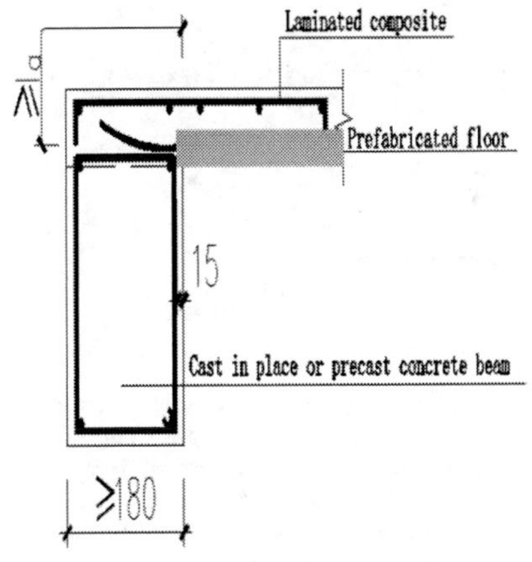

Plate beam Lap

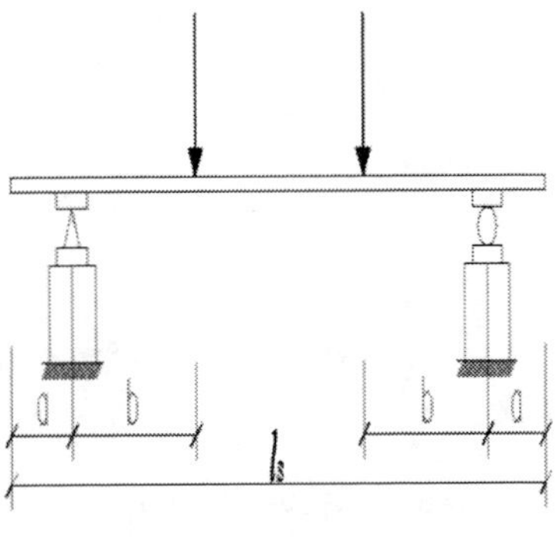

Loading sketch map

5 quality testing

The quality acceptance of the components shall conform to the relevant provisions of the national standard "code for acceptance of construction quality of concrete structures" (2011 Edition), GB50204-2002 (EDITION).

The floor should be according to "laminated with bottom slab of prestressed concrete GB/T16727-2007" request the appearance inspection and testing.

The compressive strength of the concrete cube in the floor should be 100% after the design of concrete strength grade.

When testing the crack resistance of the base plate, we can take two points. When the other loading method is adopted, the value and the test index should be converted according to the principle of equivalent load effect.

6 conclusions and Outlook

Precast ribbed composite not only has good integrity, two-way force, save supporting, highlight the advantages of convenient construction and obvious economic advantages, suitable for in housing construction vigorously promote the use of, the production and use of the prefabricated ribbed slabs should be the matters needing attention are reviewed, at present for ribbed precast laminated we still need to do the following research work.

To study the mechanical properties of prefabricated ribbed slab, to determine its mechanical properties, and give the calculation method, and finally form a complete design theory system.

To strengthen the construction of assembly type concrete construction work to ensure the smooth progress of the assembly and the quality of work.

Acknowledgements

This work is financially supported by Scientific research general project of Liao ning Provincial Department of Education(L2015310).

References

1. DBJT05-275 prestressed concrete laminated Atlas
2. GB50010-2010 design code for concrete structures
3. JCJ/T258-2011 Technical specification for prefabricated ribbed floor slab
4. Yao Yuan ,Zhang JingShu ,Liu Jun, Ni YongShong,Zhao Jun, Technical characteristics and economic comparison of precast slab with rib floor
5. Cui QingBiao , Prefabricated concrete structural member and construction matters needing attention
6. Pan Yanhua Study on mechanical behavior of PK prestressed concrete composite slab

Preparation and release kinetics of betulinic acid/ CS drug-loaded microspheres

Hao Ran Zhou[1], Xiao Jiao FANG[2], Jin LV[2], De Xin WANG[2]

[1]*The University Key Laboratory of Fine Chemical Industry of Heilongjiang Province ,Chemistry and Chemical Engineering, Qiqihar University 161006, China*
[2]*School of Material Science and Engineering, Harbin University of Science and Technology,Harbin 150080, China*

Abstract. Chitosan(CS) is the unique alkaline polysaccharide in nature, because of its perfect biocompatibility and degradability, it is widely used in medicine, soft release and control release. The betulinic acid is the derivative of betulin, it has many pharmacological activities, such as anti- inflammatory, antitumor, anti-malaria and anti-HIV. In this paper, the Span-80 was used as emulsifiers, the glutaraldehyde was used as crosslinker. With the method of orthogonal experiment, the preparation technology was optimized. The microspheres were characterized by SEM and its degree of crosslinking, drug-loading rate and encapsulation efficiency were tested at the same time.

Introduction

Chitosan(CS) is the unique alkaline polysaccharide in nature, because of its perfect biocompatibility and degradability, it is widely used in medicine, soft release and control release. When the CS is used for drug carrier, it can prolong the application time of the medicine, reduce the adverse effect, increase the stability of the medicine and increase the permeability of cell membrane, it is a widely studied pharmaceutical excipient. The betulinic acid is the derivative of betulin, at first, it is extracted from birch bark by physical methods, then, with the development of the study, many other chemical methods were found to synthesis the betulinic acid at a high rate. As a result, the betulin acid has become a widely source. It not only has many pharmacological activities, such as anti- inflammatory, antitumor, anti-malaria and anti-HIV, but also low toxicity and with high safety index, so it is widely studied in the area of antitumor and anti-HIV.

Because the microspheres particle size homogeneous degree will directly affect the loading of betulinic acid / CS microspheres, so we have to achieve a relatively uniform levels of the bland microspheres of CS. In this paper, the Span-80 and Twen-80 were used as emulsifiers, the formaldehyde and glutaraldehyde were used as crosslinker. With the method of orthogonal experiment, the preparation technology of betulinic acid / CS microspheres was optimized. Above all, we prepared the microspheres of CS with uniform particle size, then we prepared the microspheres of betulinic acid and CS. We characterized the microspheres with the use of FT-IR, SEM and the test of the degree of crosslinking, we tested the drug-loading rate and encapsulation efficiency at the same time.

We studied the factors which will have an influence on the microspheres, such as the dosage of emulsifier and crosslinker, the dosing ratio of betulinic acid and CS. Then, we determined the optimum technological conditions: the molecular weight

1 Preparation of the betulinic acid/CS microspheres

The CS was was added in a round flask and dissolved in acetic acid (the concentration is 2%), then put the flask in water bath and stirred until it mixed evenly, the betulinic acid was dissolved in chloroform. At the same time, the emulsification system was prepared: A certain amount of atolein and Span-80 were mixed in a round flask, the mixture was stirred evenly in water bath, too.

Emulsion reaction: 30mL of the CS-acetic acid mixture was taken out and added in the emulsification system with use of injector to control its flow rate, the temperature was 50° and the process lasted for about 30min.

Cross-linking reaction: After the emulsion reaction, a certain amount of glutaraldehyde was added, then the cross-linking started, keep the temperature of 50°C and it lasted for about 3h.

After cross-linking process, the mixture was poured out, laid up until the solution was layered, then poured out the supernatant, a certain amount of petroleum ether was added, stirred and laid up, then poured out the supernatant, repeated the process for two times.

Washed the mixture with a certain amount of isopropanol and vacuum filtration at the same time, the solid we obtained is the betulinic acid/CS microspheres.

DOI: 10.1051/matecconf/20166303008

The microspheres were laid on a clean watch glass and put it in an oven at the temperature of 50°C.

2 Test of the betulinic acid/CS microspheres Analysis of SEM

The relatively uniform particle size of betulinic acid/CS microspheres is quite important for drug-loading, encapsulation efficiency and slow-release effect. As a result, the best process condition was determined by orthogonal test. The SEM was used to test the change caused by modification, the result is as following:

(a) the SEM image before modification

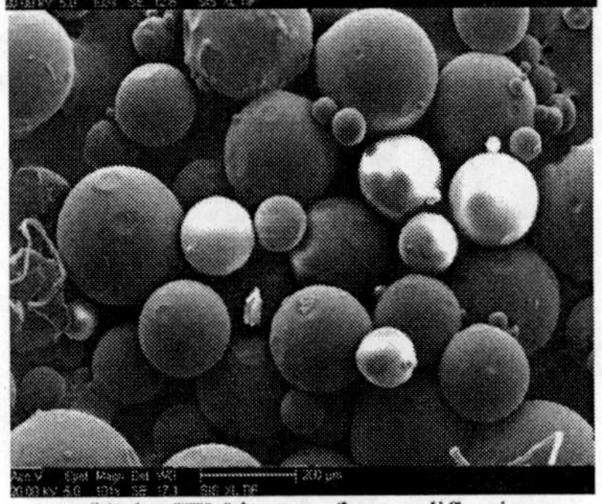

(b) the SEM image after modification

Fig.1 SEM images of the betulinic acid/CS microspheres

It can be concluded that the degree of particle size uniformity has been improved obviously, the particle size of (a) is ranged from 30 to 180μm, the particle sized distribution is so wide, but the uniformity of particle size after modification is improved, primarily fasten on 100-150μm.

3 Analysis of the degree of crosslinking

To determine the crosslinking degree of betulinic acid/CS microspheres[1-3], the drug was extracted with the use of extractor, and the solution used in this process is glacial acetic acid, its concentration is 0.5%, the extraction last for about 24h. The weight of the microspheres should be weighed before and after the extraction, the product was put in an oven and weighed it after dried. The degree of crosslinking could be calculated and the result is shown as Tab. 1:

Tab. 1 Effect of glutaraldehyde dosage on the crosslinked degree of microsphere

Dosage of glutaraldehyde (mL)	Mass before extraction(g)		Mass after extraction(g)	Degree of crosslinking (%)
	Filter paper	Microspheres		
1.0	0.314	0.250	0.180	72.0
1.5	0.309	0.272	0.199	73.1
2.0	0.314	0.242	0.182	75.2
2.5	0.314	0.250	0.192	76.8
3.0	0.310	0.282	0.224	79.4
3.5	0.309	0.250	0.195	78.0
4.0	0.315	0.235	0.171	72.7

It can known from the Tab. 1, with the increasing dosage of glutaraldehyde, the degree of crosslinking presents the trend of rise at first, but the trend reverse beyond a certain limit. When the dosage of glutaraldehyde is 3mL, its crosslinking degree can reaches 79.4%, but when the dosage exceeds it, the crosslinking degree will not rise as before. The reason is that the reaction of aldehyde group and amino group has been finished.

4 Analysis of encapsulation efficiency

In the process of prepare betulinic acid/CS microspheres, many factors can have influence on the encapsulation efficiency, as a result, the practical application would be influenced. In this paper, the molecular weight of CS is a million, the ratio of betulinic acid to CS is 1:4, the dosage of Span-80 is 3mL and the glutaraldehyde is 2mL, the calculation of encapsulation efficiency is as following:

$$\text{Drug-loading rate} = \frac{Wc}{Wm} \times 100\% \tag{1}$$

In the formula above, Wc is the mass of betulinic acid that has existed in the microspheres, g. Wm is the mass of betulinic acid/CS microspheres[4-6]. The mass of microspheres used here is 0.03g, so the drug-loading of microspheres is 8.3%.

In order to increase the drug-loading, the best process conditions was determined by orthogonal test as following:

Tab.2 Orthogonal analysis of chitosan microspheres

NO	Mass of microspheres(mg)	Mass of emulisifier (mL)	Molecular weigh of CS (ten thousands)	m(betulinin acid)/m(CS)	Dosage of glutaraldehyde (mL)	Drug-loading (%)
1	30	1	20	1:2	2	6.9
2	30	1	50	1:4	3	7.8
3	30	1	100	1:8	4	9.3
4	30	2	20	1:4	4	6.7
5	30	2	50	1:8	2	5.9
6	30	2	100	1:2	3	4.3
7	30	3	20	1:8	3	7.6
8	30	3	50	1:2	4	8.9
9	30	3	100	1:4	2	8.3

It can be known from the orthogonal test that, all of the dosage of emulsifiers and crosslinking, the ratio of betulinic acid to CS and the molecular weight of CS have influence on the drug-loading[7]. The effect of molecular weigh is the minimal and the ratio is the maximal. When the molecular weigh of CS is one million, the dosage of Span-80 is 1mL and glutaraldehyde is 4mL, the drug-loading can reach its maximum of 9.3%.

5 Analysis of sustained release

0.30g of betulinic acid/CS microspheres was taken out and dissolved in hydrochloric acid solution at the concentration of 0.1mol/L, constantly stirring for a while, put the mixture in a constant-volume of 50mL. Take out 2mL of the mixture with pipette at 1h、2h、4h、6h、8h、10h、12h、16h、20h and 24h, respectively, the corresponding volume of hydrochloric acid solution is added at the same time. Determined the absorbance of the mixture with UV-VISIBLE spectrophotometer at the wavelength of 224nm, then the cumulative release rate of betulinic acid can be calculated with following formula:

$$E_r = \frac{V_e \sum_{i}^{n-1} c_i + V_0 c_n}{m_1 D} \qquad (2)$$

In the formula above, V0, is the initial release of concentration, Ve is displaced volume of release medium, ci is the concentration of drug at the ih time, n is the time of displacement, m1 is the mass of microspheres, D is the drug-loading of microspheres[8,9]. 24h later, we can obtain the result of slow-release effect as following:

Tab.3 The relation between the cumulative release rate and time

Release time/h	Concentration of betulinic acid /g/L	Cumulative release rate /%
1	0.010405	20.8
2	0.020335	40.7
4	0.030054	60.1
6	0.037549	75.1
8	0.040256	80.5
10	0.043564	87.1
12	0.046026	92.1
16	0.048285	96.6
20	0.049520	99.0
24	0.049950	99.9

It can be known from the Tab.3 that, at the beginning of sustained-release, the betulinic acid in the microspheres was released at a high rate, the concentration of betulinic acid in solution increased rapidly, 4h later more than half of the drug has been released. When the microspheres released for about 16h, its cumulative release rate has passed 96%, so it can be affirmed that the drug has been released absolutely.

DOI: 10.1051/matecconf/20166303008

6 Conclusion

In the process of prepare the CS microspheres, the emulisifiers used here is Span-80, and the surface of the microspheres indicates that when the dosage of Span-80 is 1mL, the particle size uniformity is the best, it ranged from 30μm to 180μm.

Many factors could have an influence on drug-loading and encapsulation efficiency, and their order is the ratio of betulinic acid to CS is the maximal, the dosage of emulsifier and crosslinker take the second place, the molecular weigh of CS is the minimal.

Through the orthogonal test the optimum technological condition can be determined: the molecular weight of CS is one million, the dosage of Span-80 is 1mL and glutaraldehyde is 4mL, the reactive temperature is 60°C, the stirring rate is 300r/min. At this time, the maximum of drug-loading is 9.3% and the maximal encapsulation is 79.4%.

Acknowledgement

In this paper, the research was sponsored by the University Key Laboratory of Fine Chemical Industry of Heilongjiang Province from Qiqihaer University.

References

1. M. A. Berthol, S. K. Cremer, S. D. Kreuter. *Control Rel*, **39**, 17-25 (2006).
2. Y. Xing. The Application of Chitosan in Drug Control-released . J. **26**, 62-64 (2005).
3. T. D. Jiang, Chitosan, M, *Chemical Industry Press*, 1-4, (2007).
4. F. Li, M. Liu, K. Yao. *Biomaterials*. J. **23**, 343-347(2012).
5. H. Shi. Biomaterials, J, **23**, 4469-4471(2002).
6. C. Berkland, M. Kipper. *Journal of Controlled Release*. J. **94**, 129-141(2012).
7. P.G.Srinivasa, J.A.Ramseh, C.S.Kuma. *Journal of Food Engineering*. J. **63**, 79-85(2004).
8. M. L. Liu, S.D. Javalgill. *Adv Drug Deliver*. J. **51**, 81-96(2001).
9. X. B. Xiong, J.M. Li. *Sustained-Release Pellets*. J. **12**:705-720(2001).

MATEC Web of Conferences **63**, 03009 (2016) DOI: 10.1051/matecconf/20166303009
MMME 2016

Application and development of stainless steel reinforced concrete structure

Xian Hong MENG[1,a],Si Yuan ZHANG[1]

[1]*Department of Civil Engineering,Shenyang Jianzhu University,Shenyang ,China*

Abstract: Now reinforced concrete structure in our country develops very fast, and reinforced concrete structure has been widely applied by various buildings.But with the deepening of the research experts and scholars,theyfound in some areas where high corrosion of reinforced concrete structure with the increase of service time,the concrete cracks,and led to the internal steel bar corrosion conditions.In the face of these problems,the experts used stainless steel applied to the study of concrete.In this paper,the stainless stell reinforced concrete structure of the application and development status of made briefly.

1Research background

Since the reinforced concrete structure in the actual engineering application, its durability has become one of the major problems in the structure engineering, especially in the coastal areas and high corrosive areas, railway bridge and highway bridge, and so on serious erosion degree of the reinforced concrete structure. Due to construction in these areas is affected by the environment corrosion, lead to corrosion of rebar in reinforced concrete structures, with the extension of time, the corrosion of reinforcement is more and more serious, leading to the expansion of rebar in reinforced concrete, the concrete cover off, makes the steel skeleton is exposed to the air, it also accelerated the further corrosion of steel bar, not only appearance affected, and greatly reduce the bearing capacity and durability of the reinforced concrete, reduces the service life of structures. Reinforcement corrosion in reinforced concrete structures has caused the world very big economic loss [1].

In Europe, Japan and so on some countries of the rapid development of construction industry has been studied the stainless steel, and stainless steel bar in reinforced concrete structure with steel. Stainless steel reinforcement in the Marine environment and a series of easy corrosion area and service life of bridge in 75 ~ 100 years of engineering and the key node replaced as ordinary steel reinforced concrete with steel has received recognition of the industry, and has been widely used in large area.

2Introduction of stainless steel bars

Stainless steel is a kind of stable performance and superior corrosion resistance of metal materials. Using stainless steel instead of carbon steel can improve the durability of reinforced concrete structure, can largely improve the corrosion of the steel. Stainless steel is mainly for chromium and nickel alloy components, ithas a high chemical stability, can generate dense in oxidizing medium toughness of more than 11.7% when the content of chromium passivation membrane, can make the alloy electrode potential increased significantly, effectively to prevent further oxidation of the alloy, nickel in the chrome steel, can improve the corrosion resistance of alloy in the oxidation of the medium, when the chromium, nickel content must lower carbon content in steel, the corrosion resistant performance is good.

Stainless steel according to the microstructure of different can be divided into: austenite size steel, ferritic steel, martensite steel, two-way stainless steel and precipitation hardening stainless steel. According to the different chemical composition can be divided into: ferrite series, markov system. Normally, in the stainless steel reinforced concrete structure and the types

[a]Corresponding author: 1445827503@qq. com

of stainless steel is adopted in the bi-directional austenitic stainless steel and stainless steel reinforcement. The serial number of different countries on the stainless steel is different, as shown in table 1.1.

Table 1.1 stainless steel bars in the world (part) [2]

No.	Chinese number	Japanese number	The United States		EU	Australia	Taiwan
			ASTM	UNS			
1	12Cr17Mn6Ni5N	SUS201	201	S20100	1.4372	201-2	201
2	12Cr18Mn9Ni5N	SUS202	202	S20200	1.4373	-	202
3	12Cr17Ni7	SUS301	301	S30100	1.4319	301	301
4	06Cr19Ni10	SUS304	304	S30400	1.4301	304	304
5	022Cr19Ni10	SUS304L	304L	S30403	1.4306	304L	304L
6	06Cr19Ni10N	SUS304N1	304N	S30451	1.4315	304N1	304N1
7	06Cr19Ni9NbN	SUS304N2	XM21	S30452	-	304N2	304N2
8	022Cr19Ni10N	SUS304LN	304LN	S30453	-	304LN	304LN
9	06Cr17Ni12Mo2	SUS316	316	S31600	1.4401	316	316
10	06Cr17Ni12Mo2Ti	SUS316Ti	316Ti	S31635	1.4571	316Ti	316Ti
11	022Cr17Ni12Mo2	SUS316L	316L	S31603	1.4404	316L	316L
12	06Cr17Ni12Mo2N	SUS316N	316N	S31651	-	316N	316N
13	022Cr17Ni13Mo2N	SUS316LN	316LN	S31653	1.4429	316LN	316LN
	Austenitic, ferritic stainless steel (duplex stainless steel)						
14	0Cr26Ni5Mo2	SUS329J1	329	S32900	1.4477	329J1	329J1
15	00Cr18Ni5Mo3Si2	SUS329J3L	-	S31803	1.4462	329J3L	329J3L
	Ferritic stainless steel						
16	0Cr13Al	06Cr13Al	SUS405	405	S40500	STS405	1.4002
17	-	022Cr11Ti	SUH409	409	S40900	STS409	1.4512
18	00Cr12	022Cr12	SUS410L	-	-	STS410L	-
19	1Cr17	10Cr17	SUS430	430	S43000	STS430	1.4016

3Application of stainless steel reinforced concrete structure

The first high stainless steel reinforced concrete engineering is completed in 1941 ,it is Mexican Yucatan Progress pile foundation engineering of the port[3].The project usedthe model 304 stainless steel bar.Although during the period of construction cost more expensive compared with the ordinary steel, in the project is still not repaired, the corrosion in Marine environment completed still maintain good state of service.Up to now,inmarine corrosive environment still maintain good state of service, and built in 1963 near the project of ordinary concrete piles have been serious corrosion phenomena.It can prove that in the long run, the high stainless steel instead of stainless steel rebar corrosion area has obvious economic benefits and remarkable achievement. Table 1.2 is abroad some stainless steel reinforcement in practical engineering instance.

Table 1.2 examples of foreign stainless steel reinforced concrete structures[4]

The Bridges	Stainless steel type	Weight
The Brooklyn bridge in New York	2205	181吨
New Jersey Garden State Parkway the highway bridge	2205	165吨
Oregon Haynes Inlet Slough bridge	2205	400吨
Canada Church bridge	316LN	150吨
British Cradlewell underground passage	316	240吨
London's Guildhall Yard Eastproject	304	140吨
The uae Sheik zayed bridge	EN 1.4462	-

In our country, the AngChuan Zhou Bridgeis firstly made of stainless steel reinforcement project, the project adopts the stainless steel reinforced with a magnitude of 1.4462 types of stainless steel.Due to the bridge design to take into account the project cost and actual application effect, only the outermost layers of reinforcement in bridge components and prone to corrosion of bridge node adopts the stainless steel reinforced.ShenzhenWestern Corridor Bridge, the bridge length 3.2 km, in the design of the bridge, designers considered the construction cost and the actual application effect, adopted a diameter of 16 mm to 40 mm S31600 (JIS316) and S2205 two types of stainless steel as part of the bridge reinforcement, the rest for ordinary reinforced still adopted ordinarysteel, this would make it easier for Western Corridor Bridge not for large servicing within 120.Hong kong-Zhuhai-Macao Bridge is the most used stainless steel to reinforced concrete engineering amount of the project, the project costs the most expensive in the world, construction difficulty is the largest and longest span of the project.Hongkong-zhuhai-macao bridge using the duplex stainless steel rebar is made by shanxitaigang group independent research and development.Stainless steel is mainly used in the

DOI: 10.1051/matecconf/20166303009

bearing platform for part of the hongkong-zhuhai-macao bridge which is easy to rust, the bridge is about 50 kilometers long, the dosage of the stainless steel 3500 tons.

4The research status of stainless steel reinforced concrete structure

In recent years our country studies the mechanics properties of stainless steel bar:

In 2007, GuoxueZhang YongshengXu, in order to observe the specimen of crack width and deflection, the ultimate bearing capacity index, analyzing the mechanical properties of stainless steel reinforced concrete beam, design of the stainless steel reinforced concrete beam static load experiment. Experimental results demonstrate in the use of stainless steel reinforced concrete beam phase, the maximum crack width and the maximum deflection are within the scope of ordinary reinforced concrete design value. If using ACI American code of reinforced concrete beam deflection formula to calculate and compare the data and found stainless steel reinforced concrete beam specimen deflection and the whole loading process of specimen are closer to the actual situation of deflection.

In 2008, Wu miaomiao made 27 a study to test adhesive strength of pull-out tests to study the bond strength of reinforced concrete stainless steel. Test results show that, the stainless steel bond strength of reinforced concrete and ordinary reinforced concrete in the steel surface shape under the condition of identical is roughly same, so you can choose ordinary steel instead of stainless steel reinforcement design calculation anchorage length.

In 2010, Guo-xueZhang do the stainless steel reinforced concrete beam subjected to low-cyclic static test, they focus on the types of strength grade of concrete, steel bar, stirrup ratio and so on several factors. The results showed that the stainless steel reinforced concrete beam specimens of various performance including limit displacement, yield displacement and displacementductility coefficient is significantly higher than that of ordinary reinforced concrete specimen, such as limit displacement, yield displacement and displacement ductility coefficient.

In 2012, Chang-weiWang established the parameters formula of the stainless steel reinforced concrete column.This formula was established by using nonlinear program ensees simplifies the model of restoring force of the stainless steel reinforced concrete column and get, and explore the stainless steel reinforced concrete column under constant load under the model of hysteresis. Finally, through the contrast analysis, the original application of theoretical calculation and experimental result is close to the curve.

In 2013, Gary huangtry to do fatigue test, analysis the various performance betweenthe stainless steel reinforced concrete beam and ordinary reinforced concrete beams, including beam crack width and the development status, stress on cycle times, the relationship between stress and strain of steel bar and reinforced concrete, in order to study the various characteristics of stainless steel reinforced concrete beam.

Foreign scholars mainly study the stainless steel corrosion resistance:

Zelbel[5] did the reinforcement corrosion test, ordinary carbon steel reinforced concrete specimens respectively and is equipped with stainless steel specimen comparison study, through the concrete corrosion damage of the specimens tested the applicability of the stainless steel reinforcement in concrete structures. Experimental results show that the ordinary carbon steel concrete specimen crack due to reinforcement corrosion, but stainless steel concrete specimen without any crack. Damage reason is summarized as: volume of reinforcement corrosion in reinforced concrete, increasing the extrusion to produce ring tensile stress, when reach ultimate tensile strain of concrete fracture, and cracks, the internal corrosion of steel bar is more likely to be accelerated concrete destruction.

Cox and Flint[6] respectively put normal deformation of reinforced concrete specimen, solid stainless steel reinforced concrete specimen and stainless steel in seawater to analysis of observation, concluded that the corrosion resistance of stainless steel is very strong, the strength of the steel bar did not drop by, the surface of the steel is not corroded; Equipped with ordinary reinforced concrete cracks and steel bar intensity drops, steel surface corrosion is serious.

Nye and others put stainless steel reinforced concrete specimens in relatively normal atmospheric environment, the chloride ion corrosion resistance experiment was carried out. Using different kinds of stainless steel compared with ordinary reinforced concrete specimens, the purpose was to observe the damage of concrete block. Test conclusion: the common concrete cracks, the specimens reinforced corrosion occur; And stainless steel specimens without corrosion phenomenon, no cracks, also found in terms of corrosion resistance, corrosion resistance of austenitic stainless steel is better than other types of the corrosion resistance of stainless steel.

The domestic and international numerous stainless steel corrosion test results show that the

DOI: 10.1051/matecconf/20166303009

stainless steel reinforced under the serious environmental corrosion factors can maintain good physical properties and chemical properties, has a superior corrosion resistance.

5Conclusion

The corrosion problem of the ordinary reinforced concrete engineering is the important research direction of scholars, using stainless steel instead of common steel can effectively solve the problem of rust.Although in the project cost be more expensive than ordinary steel, is applied to some engineering structures or important buildings, can guarantee in many years, can need not big repair, so in the long run, it is very suitable.

Acknowledgements

This work is financially supported by National High-tech R&D Program (863 Program)(2015AA03A502-3).

References

1. ABDULLAH A, ALMUSALLAM.Effect of degree of corrosion on the properties of reinfarcaig steel bars, *Construction and Buildaig Material*.J.2007, 15(8): 361-68
2. GB/T 1220-2007, *The stainless steel rod*.S
3. Zhang Xiaojuan. Study on the corrosion inhibition effect of aluminum zinc layered double metal hydroxide and its derivatives on reinforced concrete. *Xiamen: Xiamen University*. D.2010.
4. Huang Jiawei. Experimental research on fatigue performance of [D]. steel reinforced concrete bridge girder.2013. *Guangzhou: Guangdong University of Technology*
5. E.Mirambell, E.Real. On the calculation of deflection in structural stainless steel beams: an experimental andnumericalinvestigation.JournalofConstructionalSteelResearch, J.2000,54(1):109-133.
6. Zin-Taek Park, Yoon-Seok Choi, Jung-Gu Kim and Lan hung. Development of a galvanic sensor system for detecting the corrosion damage of the steel embedded in concrete structure.*Cement and Concrete Research*, J.2005,35(9):1814-1819.

MATEC Web of Conferences **63**, 03010 (2016) DOI: 10.1051/matecconf/20166303010
MMME 2016

Effect of biochar on the greenhouse gas emissions from farmland and its physicochemical properties

Xiao Yi CHENG[1a], Yu LAN[1], Zun Qi LIU[1], Xiao Lin LIU[1], Wei MIAO[1], Tian Yi HE[1], Xu YANG[1], Jun MENG[1] and Wen Fu CHEN[1*]

[1]*Shenyang Agricultural University, Biochar Engineering Technology Research Center of Liaoning Province, Shenyang 10086, China*

Abstract. To investigate the effects of biochar on greenhouse gas (GHGs) emissions by static chamber-gas chromatograph method, a field experiment with three treatments was conducted: soil without fertilizer (CK), soil with conventional fertilizer (N) and soil with chemical fertilizer after biochar application (NB). The results showed that biochar reduced the N_2O emission significantly and increased the absorption of CH_4 to soil. Compared to N, the N_2O emission from NB decreased by 21.76%, and increased CH_4 absorption by 55.31%, but no significant difference of CO_2 emission was showed between NB and N. Meanwhile biochar increased soil pH, total nitrogen content and organic carbon content, and also decreased soil NH_4^+ -N content and bulk density. Biochar provided a new way to reduce GHGs from farmland and force air pollution of China's agriculture.

1 Introduction

Recently, The global warming has caused much attention around the world, and Promoting low carbon development mode has become consensus in the world[1]. Carbon dioxide (CO_2), methane (CH_4) and nitrous oxide (N_2O) are the main greenhouse gases (GHGs) forcing global warming in atmospheric[2]. Farmland is a main source of GHGs emission[3], and reducing GHGs from farmland and increasing carbon sequestration to terrestrial ecosystem play an important role in dealing with climate changes at present.

Biochar is pyrolysized from biomass under hypoxic conditions[4, 5]. Biochar could increase the storage of soil stability carbon[6] while decreasing GHGs emission[7] and improving crop productivity[8] when it applied to farmland soil. So it gradually becomes a kind of new way to increase carbon storage and reduce GHGs emissionsm [9]. It showed that after adding biochar to the soil of grassland and soybean field at rates of 20 $t \cdot hm^{-2}$ reduced the N_2O emission of soil by 80% and 50% respectively, the emission of CH_4 was completely restrain[10]. Zhang et al.[11] reported when applyed biochar at rate of 40 $t \cdot hm^{-2}$ into soil, CH_4 emission from paddy land increased by 34% and 41% respectively under the condition of applying and not applying nitrogenous fertilizer, and reduced N_2O emission by 40%-51% and 21%-28% respectively.

Yuan et al.[12] indicated that if weight ratio of biochar and soil was up to 20% or the more, CO_2 emissions of the soil could be reduced. Other researches showed that biochar application could led to an increase of CO_2 emission from SOC-rich soil[13]. When applied pasture waste biochar(10 $t \cdot hm^{-2}$) into soil, it showed increased CO_2 and N_2O emission than CK whereas net absorption of CH_4[14]. The researches on the effect of biochar on farmland GHGs emissions have not reached a consensus right now, and also the mechanism is not well understood. studies of that on maize field is even less, so this study is aimed to investigate the effect of biochar made from maize straw applied into soil on GHGs emissions from farmland and soil physical chemical properties. The objectives of this study were to provide theoretical basis for slowing the global climate change by promoting and application of biochar in agricultural.

2 Methods

2.1 Experiment site

The experiment was carried out at Shenyang agricultural university testing site in Shenyang, Liaoning province in 2013. The soil type is brown earth. The physicochemical properties of the soil were showed as Table 1.

Table 1. Basic physicochemical properties of soil and boichar

Physicochemical properties	Soil	Biochar
pH	7.9	9.2
Bulk density ($g \cdot cm^{-3}$)	1.31	—

[a] Corresponding author: cxy_beyond@126.com

DOI: 10.1051/matecconf/20166303010

Available N (mg·kg^{-1})	84.5	—
Available P (mg·kg^{-1})	15.9	—
Available K (mg·kg^{-1})	158.7	—
Total N (g·kg^{-1})	1.1	10.7
Total P (g·kg^{-1})	0.39	7.8
Total K (g·kg^{-1})	22.0	16.8
Organic C (g·kg^{-1})	11.3	679.0

2.2 Experiment material

He biochar used for test is produced by maize straw. Basic properties of biochar are shown in Tab.1 Maize (ZD 958) was selected as experimental material which was sowed on May 17 and gained on September 29 with density 3970 plants·hm^{-2} and conventional management during the crop growth.

2.3 Experiment design

Three experimental treatments marked as CK, N and NB. CK: soil without fertilizer, N: soil with conventional fertilizer, NB: soil with chemical fertilizer after biochar application. The biochar was applied at rates of 20 t·hm^{-2}. Fertilizer and biochar was applied before sowing. N, P and K Fertilizer were urea (46% content of N), potassium chloride (50% content of K_2O) and monocalcium phosphate (12% content of P_2O_5) respectively. Urea was applied at 160 kg N·hm^{-2} which was mixed with soil before dusting within the plot evenly. Also as basal fertilizers, monocalcium phosphate and potassium chloride were applied at 70 kg P_2O_5·hm^{-2} and 150 kg K_2O·hm^{-2} respectively to all plots. There were 15 experimental plots which was 60 m^2 consisted of 10 rows of maize, each 6 m in length with 60 cm between rows.

N_2O, CH_4 and CO_2 emission was monitored with a closed chamber method produced by transparent plexiglass which is 40 cm×40 cm×40 cm cube. There was a sunction hole on the top and a small fan and a thermometer. The closed chamber was pressed into the soil and covered with soil around to isolate inside of charmber from the outside air. The gas smple taking process was strictly controlled during 9 am-11 am. The sample were collected after 0 min, 15min, 30min and 45 min of monitor process respectively. Gas sample was simultaneously analyzed with a gas chromatograph (Agilent 7890A).

2.4 Measurement content and methods

N_2O emission was calculated by the method of [15] as following formula:

$$F=60\times[273/(273+T)]\times(P/P0)\times\rho\times V/A\times(dc/dt)\times 28/40$$

F: N_2O-N emission flux(μgN·m^{-2}·h^{-1}) P: the normal atmospheric pressure (mm Hg), P0: atmospheric pressure as a standard atmospheric pressure; Rho to 0 ℃ and the standard air pressure under the condition of CO_2, CH_4 and N_2O density (1.977 g·L^{-1}, 0.717 g·L^{-1}, 1.970 g·L^{-1}); V : sampling box volume (m^3); A : sampling box area over the soil(m^2); Dc/dt : the rate of the changes of gas concentration in the sampling box, T: the temperature of the interior of the sampling box (℃). The rate of gas concentration changed is calculated obtained by the three regressions.

Cumulative amounts of the gas emission is calculated as following formula:

$$M=\Sigma(F_{i+1}+F_i)/2\times(t_{i+1}-t_i)\times 24$$

M: cumulative amounts of gas (N_2O, CH_4 or CO_2), F: gas emission flux, i: number of times for sampling, t: the date for sampling (days after seeding)

The concentration of soil organic C, soil available nitrogen, soil available P and soil available K were measured in accordance with the method of Lu[16]. The soil total nitrogen and carbon content was measured by elemental analyzer (produced by Foss company, German). Statistical Analyses All data collected from this experiment were determined by means of one-way analysis of variance (ANOVA) (SPSS 20.0). In addition, the cartography tool is Microsoft Excel 2010.

3 Results and discussion

3.1 The influence of biochar on N_2O emission

As shown in the Figure 1, the N_2O emission of CK was stable and kept at a lower level during the whole growing season, which lay between 4.36-9.26 ug·m^{-2}·h^{-1}. During the first 10 days after nitrogen fertilization, the N_2O emission flux of N and NB increased rapidly and arrived a peak value by the 10th day, which were 65.54 ug·m^{-2}·h^{-1} and 52.01 ug·m^{-2}·h^{-1} respectively. After 10 days, the N_2O emission flux of N and NB dropped sharply so far as jointing stage(about 35 days after seeding), and then kept at a stable and low level. During the whole growing season, the means of emission flux of CK, N and NB were 5.98 ug·m^{-2}·h^{-1}, 16.97 ug·m^{-2}·h^{-1} and 13.06 ug·m^{-2}·h^{-1} respectively, and all the differences were significant. It showed that nitrogen fertilizer played a important role in N_2O emission, and biochar application could restrained the emission of N_2O efficiently. Cavigelli[17] showed that biochar improved aeration with decreased bulk density which depressed the denitrification in soil and decreased the N_2O emission. The another mechanism of leading to reduction in N_2O emissions in field was that biochar could potentially favor the activity of N_2O reductase from denitrifying microorganisms as soil pH was increased while enhancing the conversion of N_2O to N_2[18,19].

DOI: 10.1051/matecconf/20166303010

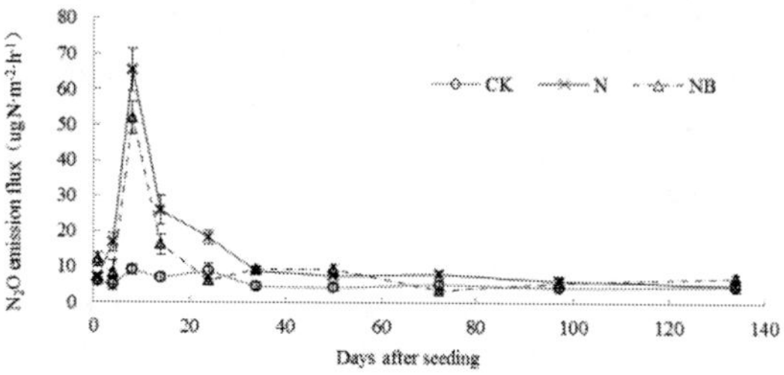

Figure 1. Emission flux of N_2O emission in soil

3.2 The influence of biochar on CH_4 emission

As shown in the Figure 2, soil CH_4 emission fluxes of three treatments all presented fluctuation which was sharp at beginning and then became mild, it was reflected the absorption and emission of CH_4 from soil. The changing trends of N and NB were basically the same during the 75 days after seeding, while CK was opposite. The variations of three treatments were consistent after 75th day. During the whole growing season, the average CH_4 emission flux of CK, N and NB were -0.0059 $mg \cdot m^{-2} \cdot h^{-1}$, -0.014 $mg \cdot m^{-2} \cdot h^{-1}$ and -0.019 $mg \cdot m^{-2} \cdot h^{-1}$ respectively, so CH_4 was absorbed by soil in three treatments, and the average value in NB was the highest.
It was reported that the mechanism of reducing CH4 emissions from the soil by way of biochar application might be due to the increasing soil permeability, which inhibited the activity of methanogenic bacteria or enhanced the activity of methane-oxidizing bacteria[11]. In this experiment, biochar decreased soil bulk density (Table.2) and increased soil porosity which might lead to the variations of methanogenic bacteria activity.

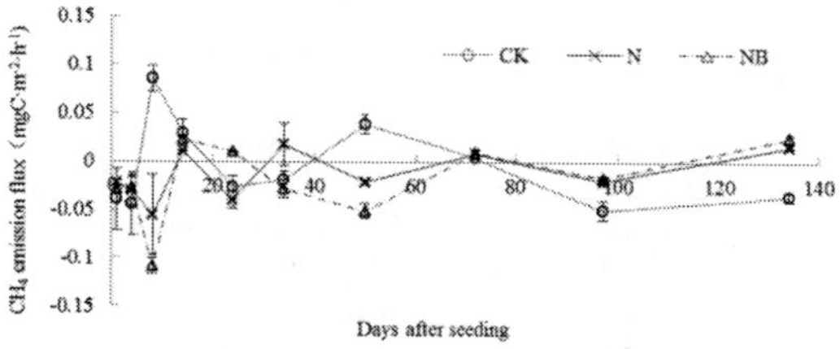

Figure 2. Emission flux of CH_4 emission in soil

3.3 The Influence of Biochar on CO_2 Emission

The CO_2 emissions of the three treatments during the whole growing season were shown in the Figure 3, and the CO_2 emission presented a distinct seasonal variation, which is increased firstly and then decreased. Two obvious peaks of CO_2 emission were found, one was after seeding and fertilizer application, the other one was in the grouting period, when where were abundant rainfall, the high temperature and sufficient soil moisture content, so the vigorous metabolism of root system promoted CO_2 emissions as a result. Nitrogen fertilizer application could promote the soil CO_2 emissions obviously, because the addition of nitrogen source contributed to the activities of soil respiration consequently. During the whole growing season, the average CO_2 emissions flux of CK, N and NB were 19.21 $mg \cdot m^{-2} \cdot h^{-1}$, 21.52 $mg \cdot m^{-2} \cdot h^{-1}$ and 24.29 $mg \cdot m^{-2} \cdot h^{-1}$ respectively.

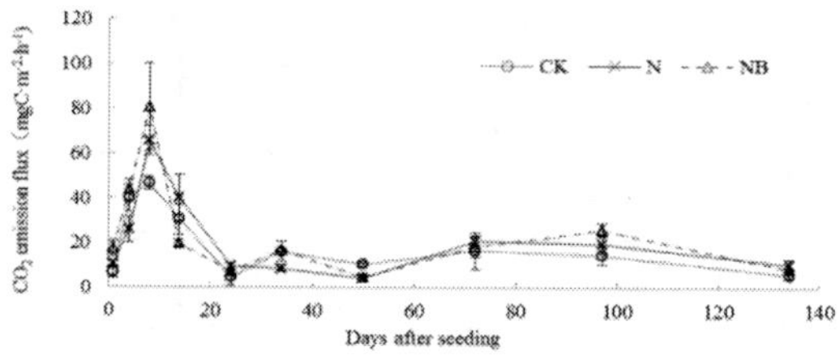

Figure 3. Emission flux of CO_2 emission in soil

3.4 The influence of biochar on physicochemical properties of soil

Incorporation of biochar into soil may provide benefits for the physical, chemical and biological properties of the soil and sustainable agriculture[20]. It was found that biochar had a great influence on soil physiochemical properties. Biochar application increased pH, contents of total N and total C in soil significantly, but reduced soil bulk density. The results showed that normal fertilization could increase the contents of total N, NH_4^+-N and NO_3^--N in soil significantly. Compared with N, NB significantly increased pH, content of total N and organic C of soil by 2.68%, 7.89% and 19.64%, respectively, while NB significantly decreased bulk density and NH_4^+-N content of soil by 9.80% and 26.17%.

Table 2. Effects of biochar on physical and chemical properties of soil

	CK	N	NB
pH	8.03±0.11b	8.18±0.07b	8.40±0.12a
Bulk density ($g.cm^{-3}$)	1.21±0.03a	1.21±0.02a	1.10±0.04b
Total N ($g.kg^{-1}$)	1.22±0.08c	1.52 ±0.01b	1.64±0.02a
Organic C ($g.kg^{-1}$)	18.74±0.87a	15.22±0.01b	18.85±0.54a
NH_4^+-N ($mg.kg^{-1}$)	1.68±0.30c	5.31±0.71a	3.92±0.39b
NO_3^--N ($mg.kg^{-1}$)	15.18±0.84b	21.69±3.66a	25.95±4.42a

None: Different small letter means signification at 0.05 level.

4 Conclusion

Biochar can improved the physicochemical properties of field soil and relieved the GHGs emission in a certain extent. In this study, total nitrogen, organic carbon contents and pH in the soil were increased by 7.89%, 19.64% and 2.68%, as well as, the bulk density was decreased by 9.8% under biochar application at 20 $t\cdot hm^{-2}$ compared to no biochar and N fertilization treatments. Therefore, biochar application into soil could improve microenvironment, porosity and fertilizer of the farmland soil whose physicochemical properties were modified. Meanwhile, the total N_2O emission was decreased by 21.76%, but the total CO_2 emission was unchanged with biochar application at 20 $t\cdot hm^{-2}$ compared to conventional fertilization treatment. The total CH_4 emission of all treatments are negative that showed the three treatments all had net absorption and the biochar application treatment had a maximum, more than 55.31% compared with conventional fertilization treatment.

Currently, To pursue high yield of crops, N_2O emission from farmland because of nitrogen fertilizer application made a great contribution to GHGs[3,21]. This study suggested N_2O emission from the farmland soil was decreased and the CH_4 uptake by farmland increased during the whole maize growing season, which achieve simultaneously high grain yield and low global warming potential intensity during maize production, and reaching a balance between the land-use and environmental pollution. It provided a new way to reduce greenhouse gas emissions from farmland and force air pollution of China's agriculture.

References

1. IPCC.Agriculture. In: climate change: mitigation, contribution of working group III to the fourth assessment report of the intergovernmental panel on climate change. Cambridge University Press. 499–532 (2007).
2. Mosier, A.R., Halvorson , A.D., Reule, C.A., Liu, X.J. Net global warming potential and greenhouse gas intensity in irrigated cropping systems in Northeastern Colorado. Environ Qual.35:1584–1598 (2006).
3. Smith, P., Martino, D., Cai, Z., Gwary, D., Janzen, H., Kumar, P., et al., Agriculture. In: Climate Change: Mitigation. Contribution of Working Group III to the Fourth Assessment Report of the Intergovernmental Panel on Climate Change, Cambridge University Press, Cambridge, United Kingdom and New York, NY, USA. 497–540(2007).
4. Lehmann J., A handful of carbon. Nature. 447:143–144(2007).
5. Chen, W.F., Zhang, W.M., Meng, J., et al., Researches on biochar application technology. Engineering Science, 46, 3324–3333(2013).
6. Fowles, M., Black carbon sequestration as an alternative to bioenergy. Biomass Bioenergy. 31, 426–432(2007).
7. Marris, E., Black is the new green. Nature. 442: 624–626(2006).
8. Major, J., Rondon, M., Molina, D., Riha, S.J., Lehmann, J., Maize yield and nutrition after 4 years of doing biochar application to a Colombian savanna oxisol. Plant Soil. 333, 117–128(2010).
9. Lehmann J., Gaunt J., Rondon, M., Biochar sequestration in terrestrial ecosystems-a review. Mitig Adapt Strat Glob Change. 11:403–427(2006).
10. Rondon, M., Ramirez, J.A., Lehmann, J., Charcoal additions reduce net emissions of greenhouse gases to the atmosphere. Proceedings of the 3rd USDA Symposium on Greenhouse Gases and Carbon Sequestration, Baltimore, USA. 208(2005).
11. Zhang, A.F., Cui, L.Q., Pan, G.X., et al., Effect of biochar amendment on yield and methane and nitrous oxide emissions from a rice paddy from Tai Lake plain, China. Agriculture, Ecosystems and Environment. 139, 469–475(2010).
12. Yuan, J.H., Xu, R.K., The amelioration effects of low temperature biochar generated from nine crop residues on an acidic Ultisol. Soil Use and Management. 27, 110–115(2011).
13. Kimetu, J.M., Lehmann, J., Stability and stabilization of biochar and green manure in soil with different organic carbon contents.Australian Journal Of Soil Research. 48, 577–58(2010).
14. Grace, P.R., Rowlings, D.K,, Kimber, S., et al., Effect of biochar amendment on the soil-atmosphere exchange of greenhouse gases from an intensive subtropical pasture in northern New South Wales, Australia. Plant and Soil. 345, 47–58(2011).
15. Liang, G.Q., Zhou, W., Xia, W.J., et al., Effect of optimized nitrogen application on N_2O emission from paddy field under wheat-rice rotation system. Plant Nutrition and Fertilizer Science. 16, 304–311(2010).
16. Lu, R.S., Soil agricultural chemical analysis method. Beijing: Chinese Agricultural Scientific Press(2000).
17. Cavigelli, M.A., Robertson, G.P., Role of denitrifier diversity in rates of nitrous oxide consumption in a terrestrial ecosystem. Soil Biol Biochem. 33, 297–310(2001).
18. Yanai, Y., Toyota, K., Okazaki, M., Effects of charcoal addition on N2O emissions from soil resulting from rewetting air-dried soil in short-term laboratory experiments.Soil Science and Plant Nutrition. 53, 181-188(2007).
19. Zwieten, V.L., Singh, B., Joseph, S., Kimber, S., Cowie, A., Chan, Y.K., Biochar and emissions of non-CO_2 greenhouse gases from soil. In: Lehmann, J., Joseph, S. (Eds.) Biochar for Environmental Management Science and Technology. Earth-scan Press, UK. 227–249(2009).
20. Lehmann, J., Joseph, S. (Eds.), Biochar for environmental management: science and technology. London, Sterling, VA: Earthscan(2009).

MATEC Web of Conferences **63**, 03011 (2016) DOI: 10.1051/matecconf/20166303011
MMME 2016

Study on diagonal hammer of three kind metals composite casting with block protecting handle

Hui Liu [1a], Shou fan Rong[2],Yan he Wu[1],Peng hui Yang[1],Xiu lan Duan[1],Yong chang Zhu[1]

School of Materials Science and Engineering, Jiamusi University,Jiamusi,154007,China

Abstract: Hammer crusher is widely used in cement, ceramic, mining and electricityother industries, hammer head is one of important parts in crusher, its abrasion performance directly affects the service life of hammer crusher and economic cost. According to bimetalcomposite hammer head is often appear "Hammer handle wear and tear","Composite difficult" and the set of casting alloy block hammer head set piece of "come off" phenomenon, which design a set piece type bimetal composite casting straight diagonal hammer, make full use of high chromium cast iron wear resistance and the toughness of low alloy steel.Design a protect block structure reasonably that solve the hammer when the actual production of composite interface is difficult to control and "the problem of hammer head wear"for use.Solve the problem of hammer head wear actively, achieve "energy saving, emission reduction and environmental protection".

Keywords: Wear and tear; Straight bevel hammer; Hammer handle wear and tear; Dissimilar metal composite

1 Introduction

Hammer crusher iswidely used to cement, metallurgy, mining, coal, building materials, transportation and other industries, mining of the ore is high speed rotating hammer broken into grinding material, the workload is very heavy.Hammer crusher is one of the shortest life and the biggest parts of consumption. And Hammer head basically has single material hammer head, cast in alloy block hammer head, doubleliquid metal composite hammer head. High manganese steel was made earlier crusher hammers and common one kind of material, but some broken material impact is not strong,the work hardening of high manganese steel can't display,high manganese steel wear fast, short service life and so on problems, and Single material of hammer head can't meet the needs of high toughness and high wear resistance. Cast in alloy block hammer head often set piece of "come off", Bimetal composite hammer head is often appear "Hammer handle wear and tear" and "Composite difficult" phenomenon.

According to the above problems, this paper designs a with a piece of straight bevel type hammer handle, with good abrasion resistance of high carbon high chromium steel manufacturing handle piece, hammer end work part adopts high wear resistance of high chromium cast iron, hammer handle part adopts good toughness of low alloy steel. Give full play to the material and cast-in superiority, solve the problem of hammer head production and application in [1].

2 Testing materials and methods

2.1 Determine the structure

This paper studies the with a piece of straight bevel type hammer handle, It contains the hammer side and with the hammer handle, hammer side is equipped with right angle trapezoid structure to protect the handle. To protect the handle (as shown in figure 1a)with good abrasion resistance of high carbon high chromium steel and cast in the hammer handle the wear parts.hammer handle (as shown in figure 1b) is good toughness of low alloy steel material, and the hammer end (as shown in figure 1c) is to use high wear resistance of high chromium cast iron material. The study of the hammer head of simple structure, not only solved the "handle" phenomenon, but also solves the caused by the external environment and the level of workers is not enough "composite difficulty" problem. And without affecting the original under the premise of use function, greatly extend the service life of the whole, reduce the maintenance cost and use cost of the enterprise. Straight bevel hammer structure as shown in figure 2 [2].

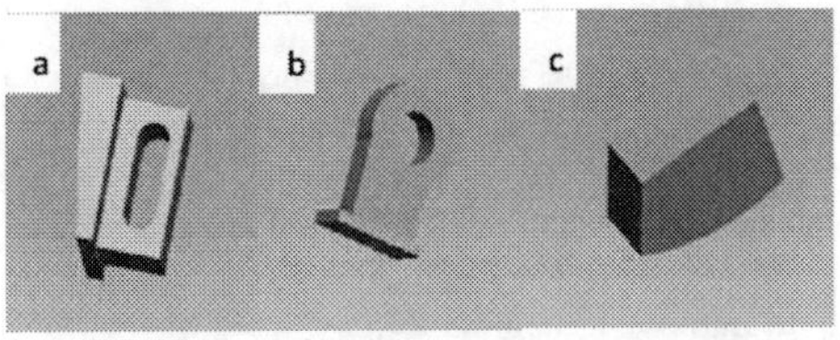

Figure 1protect the handle schematic block type straight diagonal hammer decomposition

a) protect the handle piece b) hammer handle c) side

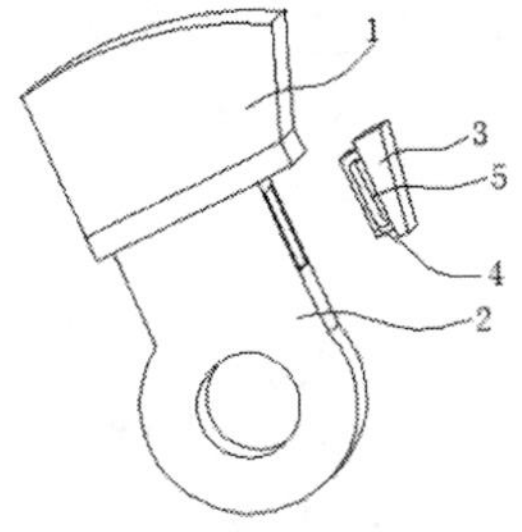

Figure 2 protect the handle type straight diagonal hammer piece of structure

1) Side 2) Hammer handle 3) Protect the handle piece 4) Jack 5) Waist type hole

2.2 Determine the chemical composition of composite materials

In order to make protect the handle piece, side, hammer handle to achieve their ideal abrasion resistance and toughness, in this article, use the high carbon high chromium steel armor handle with better abrasion resistance, and hammer end is high abrasion resistance of high chromium cast iron, the hammer handle is toughness good low alloy steel. According to the needs of the effect of alloy elements and the performance of the belt can be determined to protect the handle of the chemical composition of each place straight diagonal hammer as shown in table 1.

Table 1 chemical composition table (wt %)

Material	C	Cr	Mn	Mo	Si
High chromium cast iron	2.9 ~ 3.0	22 ~ 24	1.5 ~ 1.8	0.4	≤0.6
Low alloy steel	0.32 ~ 0.4	0.6 ~ 0.8	0.6 ~ 0.7	--	0.4
High carbon high chromium steel	1.0 ~ 1.2	8 ~ 10	1.2 ~ 1.5	0.2	0.4

2.3 Heat treatment process

This study chooses the wind quenching heat treatment process, the austenitizing temperature heated to 1020℃,then use the air-filled.Will tempering temperature heated to 230℃, and continue to air cooling.Research on heat treatment process to observe straight diagonal hammer's the law of the performance, thus ensuring a good comprehensive mechanical properties. Heat treatment process curve is shown in figure 3[3].

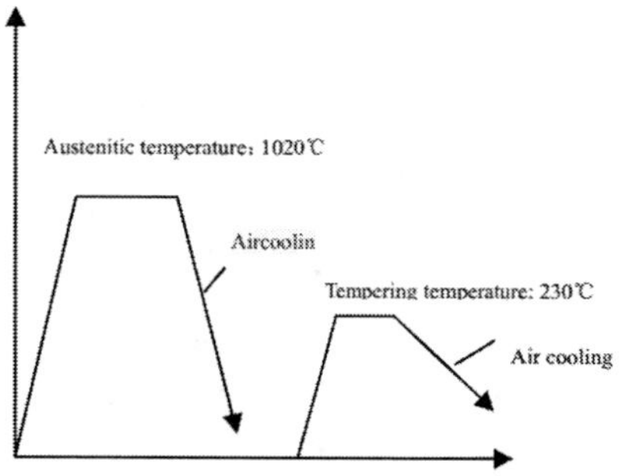

Figure 3 heat treatment process curve

2.4 Compound technology

First of all, protect the handle piece of the high carbon high chromium material heated to 300℃, put in the cavity and mould assembling, then poured into the low alloy steel liquid, At this point, the liquid steel temperature is about 1580 ℃, after a layer of solidified shell in low alloy steel, pouring into around 1550℃ high chromium cast iron liquid, through the heat capacity of high chromium cast iron melting low alloy steel solidification. Finally, to obtain the ideal metallurgical bonding[4].

3 Results of the study

3.1 Microstructure observation and analysis

a)

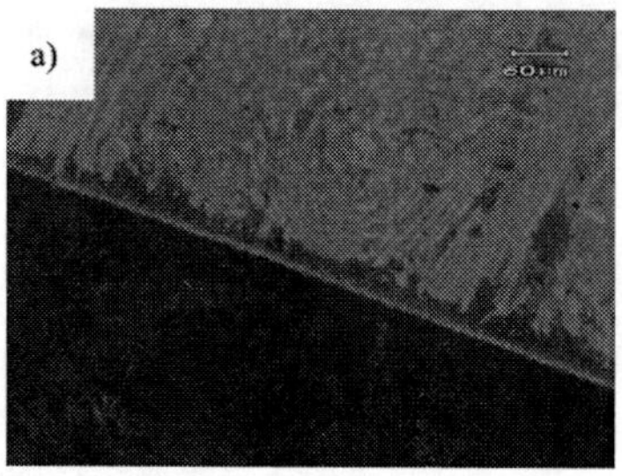

b)

Figure 4 three types metal composite microstructure

a) high chromium cast iron with low alloy steel

b) low alloy steel with high carbon high chromium steel

Through the observation of the microstructure, after heat treatment, low alloy steel group is composed of needle martensite and retained austenite. Figure in the dark is needle martensite, bright is residual austenite.

After heat treatment, high carbon high chromium steel constitution as the net eutectic carbide, distribution at the grain boundary and secondary carbide particles evenly distributed in bainite, martensite and retained austenite matrix. In A1 above a certain temperature, of the high carbon high chromium steel constitution convert into austenite, and in the matrix alloy elements diffuse more fully, more uniformity. After cooling, the eutectic carbide network disconnect, most constitution dissolve into matrix; Chromium in the form of secondary carbide precipitation. Therefore, in the matrix, a large number of distribution in the granular secondary carbides[5].

After heat treatment, the organization of high chromium cast iron (Fe, Cr)$_7$C$_3$ carbide, isolated distribution in continuous martensitic matrix. On the matrix, distribute the hexagonal rods and plate strip type M_7C_3 carbide, thus greatly reduced the continuity of matrix, reducing the degree of damage to the substrate, to improve toughness [6].

3.2 Composite interface bonding zone microstructure characteristics

DOI: 10.1051/matecconf/20166303011

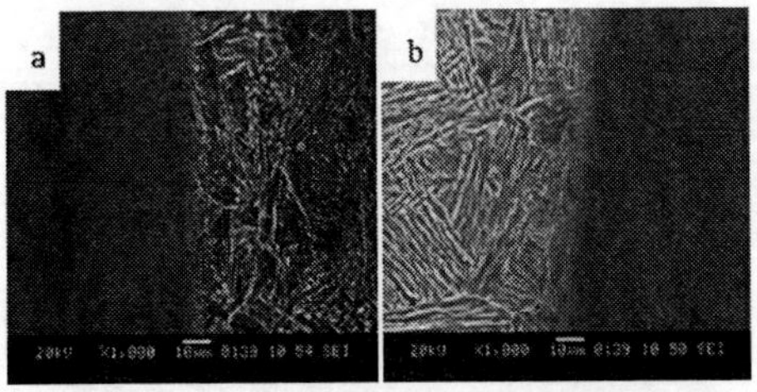

Figure 5three types metal composite interface organization characteristics

a) high chromium cast iron/low alloy steel low alloy steel

b) low alloy steel/high carbon high chromium steel

Figure 5(a) liquid-liquid composite icrostructure characteristics.Composite interface can be seen in the metallographic photos showing a small serrated, appear the phenomenon of melting and mutual penetration. And composite interface no shock phenomenon, compact structure, achieve ideal metallurgical bonding state [7].

Figure 5 (b) solid-liquid composite micr-ostructure observation, In metallographic figure, you can see the interface without micro crack, blowhole and inclusions, etc..Formed a similar liquid-liquid double metal composite of straight line.So in the combination of fluid of low alloy steel with high carbon high chromium steel interface to achieve complete metallurgical bonding state[8].

3.3 Microhardness

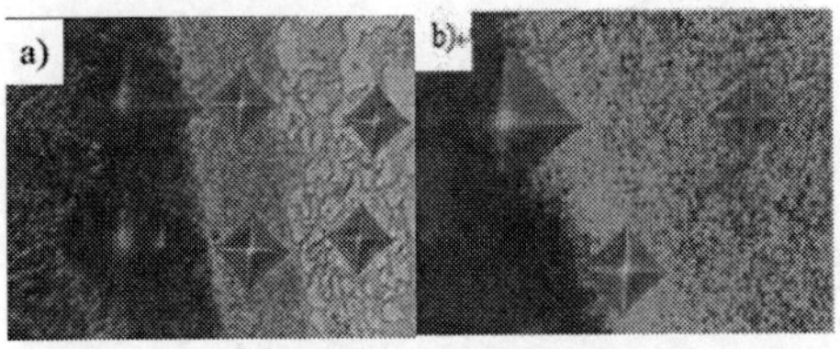

Figure 6 microhardness bimetallic interface bonding area

Through the analysis of the microhardness: If the interface has a good combination of area, the microhardness should gradient smooth transition, and measured the hardness of low alloy steel is about 350HV, solid interface combined with the hardness of about 510 HV, fluid interface in combination with the hardness of about 580 HV, the hardness of high carbon high chromium steel is about 750HV, the hardness of high chromium cast iron is 950HV.Obviously,the microhardness value is obvious gradient distribution.

4 Benefit analysis

In protecting the handle piece of crusher hammer straight Angle, the main technical indicators are the hardness of high chromium cast iron and low alloy steel, impact toughness and wear resistance and the impact toughness on bimetal hammer. After the wind quenching heat treatment, the hardness of high chromium cast iron is more than 62 HRC, impact toughness for 3~6 J/cm2; The hardness of low alloy steel is 180~250HB, impact toughness is greater than 70 J/cm2; To protect the handle piece of 55 HRC hardness above [9], impact toughness for 6~8 J/cm2.

Through installed test showed that protect the handle piece type straight diagonal hammer service life is 3 ~ 5 times of the high manganese steel, The cost of production is the single material of high manganese steel hammer 1.5 times, not only improves the production efficiency, increase the service life of hammer crusher hammer, and for the "energy conservation and emissions reduction, low carbon environmental protection" has made a great contribution. Extended life expectancy and the hammer head replacement cycle, reduce the working strength of workers, a working cycle will save the time for replacing hammer head many times, improve the production efficiency of crusher. To protect the handle piece type straight diagonal hammer schematic diagram and the real figure as shown in figure 7 [10].

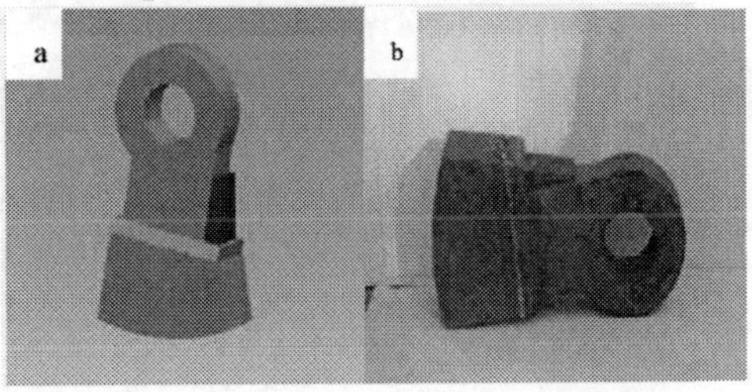

Figure 7 a) straight diagonal hammer handle piece of type b)Real figure

Conclusions

(1)This article can completely solve the set piece "come off" of casting in alloy block type hammer head, bimetal composite oblique hammer often appear "handle" and "complex difficulties" and so on.

(2)Using liquid-liquid composite casting echno-logy, the composite interface routine jagged state, can obtain complete metallurgical bonding of bime-tallic interface bonding area; Liquid-solid com-pound can achieve certain mechanical connec-tion and part of the metallurgical bonding.

(2)Through the wind quenching heat treatment process, under the conditions of equal protection of the handle to block the service life of straight bevel hammer, higher than the single material fierce steel hammer head high 3~5 times, abrasion resistance and toughness is improved obviously.

Acknowledgements

This work is by the national natural science foundation of China 51371090, jiamusi university entrepreneurial innovation project lzz2015-005, jiamusi university President fund xzyf2014-04 and xzyf2016-04.

References

1. Hanguang Fu.Metal casting ear-resistant materials research progress[J].China casting equipment and technology,2006.
2. Shoufan Rong,Zhang Yin, Jiwei Guo .Casting low djembe anti-wear ustenite steel excavator relieving material research[J].casting 2007.
3. Enze Liu,Shucheng Sun.Effects of heat treatmenton microstructure and perfor -mance of high chromium cast iron [J].The hot working processes,2008
4. Benkui Gong,Gu Hua,Yu Hai.Hammer crush -er hammer head[J].Sintered pellets,2006.
5. Dadong Shen,Baoshi Jin,Enlu Wang.Low alloy Mr Berlusconi ladle cast high chrom -ium cast iron hammer head devel -opment [J]. Journal of thermal proc -essing, 2003,
6. Wang Chao,Hongxiao Cheng.Xiangdong Wang.The high chromium cast iron, cast steel hammer head cast in [J].The casting equipment research,2004.
7. Zhang jun, zhenqing wu,Bowen Xiong.The development of the bimetal composite hammer and production [J].The hot working processes,2006.
8. Yaohui Liu,Haifeng Liu,Sirong Yu.Solid-liquid combined with double metal composite material interface research [J].Journal of mechanical engineering,2000
9. Bleck,Wolfgang,Kriangyut Phiu-on,Christo -ph.Hot Workability of as-Cast High Mang -anese-High Carbon Steels[J].Steel Research International, 2007,
10. Lixin Shi.Hammer crusher hammer head repair [J].Cement,2006.

MATEC Web of Conferences **63**, 03012 (2016) DOI: 10.1051/matecconf/20166303012
MMME 2016

Experimental study on Bond slip relationship of Steel sleeve

J.C. WANG[1,a], J.H. ZHOU[1], M.X. PAN[1], D.Y. ZHANG[1]
[1] *Shenyang Jianzhu University, 110168, Shenyang, Liaoning, China*

Abstract: The assembled concrete structure is usually connected with the steel bar through the sleeve。In order to realize the optimum design of sleeve structure and develop new sleeve, the author put forward a new test method, and made 18 specimens for uniaxial tensile test. Study the influence of the strength of grouting material, diameter of reinforcing steel bar on bond slip relationship between steel bar and grouting material and its failure characteristics by experimental research. This result suggests that with the increase of age, the bond strength between steel bar and grouting material is increasing gradually, the increase speed gradually slowed down after 20 days. For CT20H sleeve, with the increase of the diameter of reinforcement, the bond strength between reinforcement and grouting material increased gradually. Conclusion: In the optimization design of the sleeve, under the premise of ensuring the smooth injection of grouting material and be convenient for the positioning of the reinforced members, reducing the inner diameter of the sleeve can improve the bond strength between reinforcement and grouting material.

key word: Assembled concrete structure; sleeve; bond slip; Constitutive relation

1 INTRODUCTION

Steel sleeve grouting connection is made of high strength, rapid hardening inorganic slurry in the connection between reinforcement and special sleeve connecting pieces. In recent years, with the development of assembled concrete structure, Both the domestic and abroad did a series studies of grouting sleeve. Huang Hejun discussed the bonding mechanism of anchoring adhesive through tensile shear test of 20 metal sleeve bonding specimens, put forward the constitutive relation of anchoring adhesive slip. It can be used as a reference finite element analysis[1]. Wu Xiaobao did monotonic tension and one-way repeated tension test in 36 specimens, researched the influence of age (1, 4, 7, 28d) and steel type(HRB500 and HRB400) on the mechanical properties of the steel sleeve grouting connection[2]. Shu Ruibin obtained the bond slip constitutive relation of two interface slip, including the rubber band and the plastic tube by the anchorage experimental study on steel sleeve, through the equipartition method, they get the steel sleeve anchorage system of the rubber band interface and a rubber cylinder interface separate bond slip constitutive relationship[3]. Wang Yuxin choose smooth and ribbed reinforcement with different corrosion rate to make concrete center pull out specimens with same strength grade, he obtained the bond slip relationship curve between corroded steel bars and concrete under cyclic loading[4]. Zheng Shansuo made 20 trusses specimens pull-out test of bond slip behavior of solid web type steel reinforced concrete member, set up a formula about reinforced concrete bond strength and slip quantity[5]. Xu Feng proved that the bond slip relationship is independent of the path of the lateral stress field, then concluded that changes in lateral stress field under the bond slip curve is the superposition of numerous fixed lateral stress field under bond slip curve[6]. At the same time, foreign scholars have done some research on bond slip, Hunebum

[a] Corresponding author: wangjianchao005@163.com

DOI: 10.1051/matecconf/20166303012

Kocarried out the experiment of 54 members under cyclic loading and obtained the bond slip constitutive relationship of fiber reinforce d concrete[7]. Kankam C respectively tested the tensile members of the low carbon steel, cold rolled steel bars and hot rolled bars with a diameter of 25mm.After analyzed, he got the relationship between the bond strength, the stress of the steel bar and the slip[8]. Chen, G. And Baker, G studied the influence of bond slip on the crack spacing of concrete by simulation in 2003,and believed that the influence of bond slip on the crack spacing is very obvious[9]. Belleri、Andreaet al[10]put the grout sleeve for column and foundation connections in earthquake zones, and found that the grouting sleeve connection is similar to the traditional connection, which can ensure the ductility and energy dissipation capacity of the connection area.

So far, few people have studied the bond slip between the steel bar and the grouting material or the sleeve and grouting material. The reason was that the space in the slurry anchored bar connecting sleeve internal is too small to determine the displacement of steel relative grouting material and the sleeve relative grouting material by direct method. In this paper, we used test methods to give the bond slip relationship between the internal reinforcement grouting material in the sleeve by indirect method, and the bond slip relationship between the sleeve grouting material will be studied as a follow-up work.

2 SPECIMEN MAKING

This test grouting material comes from Beijing Jian Mao CGMJM-VI, 1d compression strength(35MPa), 20d compression strength(60MPa), 28d compression strength(85MPa); Sleeve is Beijing Jian Mao CT20H, diameter is 20mm; Special steel sleeve for reinforcement: The strength grade of reinforcement is HRB400, diameter are 16mm, 20mm and 22mm, the purpose is to determine how the steel bar diameter make the influence of reinforcement-grouting material bond slip relationship. It is made 18 specimens, 6 days of drawing test, respectively grout 1d、3d、7d、15d、20d and 28d,the purpose is to determine how the strength of grouting material make the influence of reinforcement-grouting material bond slip relationship. Test piece numbering rule "sleeve length - steel bar diameter".The specific grouping is shown in Table 1.

Table 1 Specimen grouping

Group number	Test piece number	Group number	Test piece number
1d	T160-D16	3d	T160-D16
	T160-D20		T160-D20
	T200-D22		T200-D22
7d	T160-D16	28d	T160-D16
	T160-D20		T160-D20
	T200-D22		T200-D22

The cross section size of the test block is 250mm*250mm, along the length of the reinforcement direction is consistent with the sleeve length, as shown in figure 1.

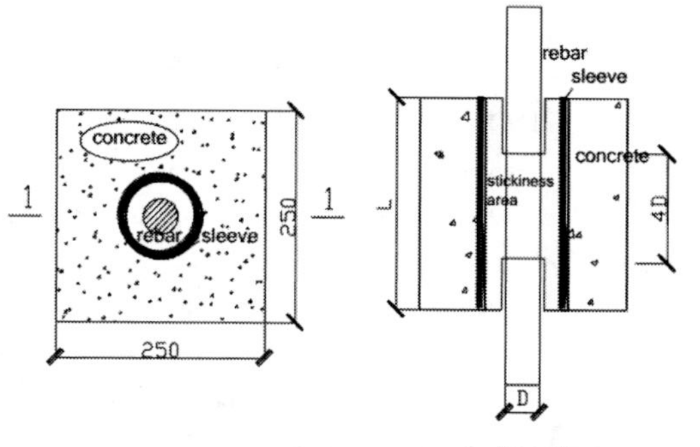

Fig. 1 Specimen shape and loading diagram

Sleeve length is 8d, d is the diameter of reinforcement, Intermediate 4d is the bond length between the reinforcement and the sleeve, the remaining 4d are distributed on both ends of the component. The advantage of this is that it can avoid the influence of the loaded end socket on the bond strength of the reinforcement-sleeve. In the specific production , according to the size of the test block to make wood mold, drill holes in the middle of the two cross cutting panels which parallel to each other. The diameter of a hole is the same as the external diameter of the sleeve, ensuring that the sleeve can be inserted into the mold smoothly from one side. The diameter of the other hole is equal to the diameter of the reinforcement , so that the steel bar can be passed through the mold smoothly. In order to facilitate grouting. at both ends of the sleeve, connected the groutig hole with outlet hole by pvc pipe.

In order to avoid the splitting failure of concrete, arrange a piece of steel mesh on both ends of the sleeves. For the convenience to fix free end displacement meter. Cut 50mm thread at one end of reinforcement , Wrap steel bar surface with PVC pipe at a appropriate position in the other end of the reinforcement, The gap between the PVC pipe and the reinforcement is filled with the rubber mud to prevent the grouting material from entering. In order to reduce the bond strength between the PVC pipe and the grouting material, the outside surface of the PVC pipe is brushed with a mold release agent. after fixed, insert the reinforcement from one side of the sleeve, arranged the bonded reinforcement in the middle of the sleeve, used the Rubber plug to fix reinforcement in the sleeve. At the end of all preparations, mix concrete in the field, Pour into the mold, vibrate it slightly, Water and conserve for 30 days, then prepare for grouting.

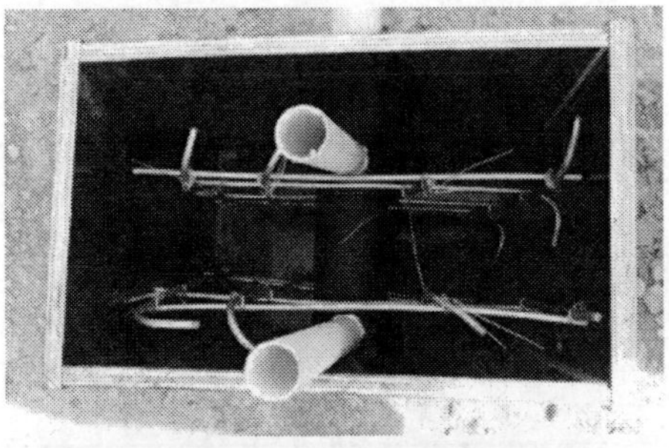

Fig. 2 Specimen fabrication diagram

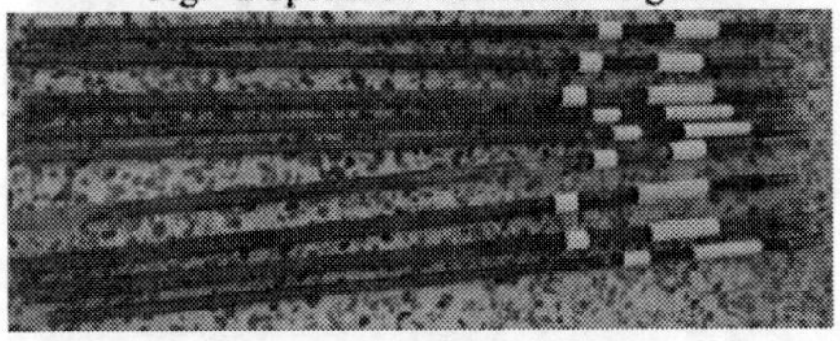

Fig. 3Rebar processing diagram

After the concrete strength is stable, start grouting. According to the requirement of grouting material product specification, water cement ratio should be controlled in 3:25. First, we weigh 10kg grouting material and put it into the plastic bucket, and then weighed 1.2kG water, add slowly to the plastic bucket. Stir and water at the same time, and ensure that the grouting material just like water. Finally, use the grouting lance to extract grouting material, grout from grouting hole, when the slurry flow out from grouting hole, the grouting is successful. At the same time, in order to get the strength of grouting material, make a batch of 40mm * 40MM * 160mm test blocks. carry out the first batch of drawing tests after grouting for 1d.

Fig. 4 Injection of grouting material

3 LOADING METHOD

In order to measure the displacement of the relative grouting material before the test. the hole of Loading plate on the test block surface should not be too large, the opening diameter should be less than or equal to the inner diameter of the sleeve. To avoid the relative displacement between sleeve and test block. we use 40t hydraulic jack and reflexive force principle design, The loading device is shown in Figure 5 and Figure 6.The size of the force is obtained by the sensor. The displacement of the relative grouting material of the reinforcement in the loading end is measured by the two displacement gauge at the end of the loading end. Displacement of the free end steel bar relative to the grouting material is used to measure the free end two displacement gauge. Loading system: 0→ maximum tension → 0, the load rate range is 0.5~1.0MPa/s.

Fig. 5 Loading device diagram

DOI: 10.1051/matecconf/20166303012

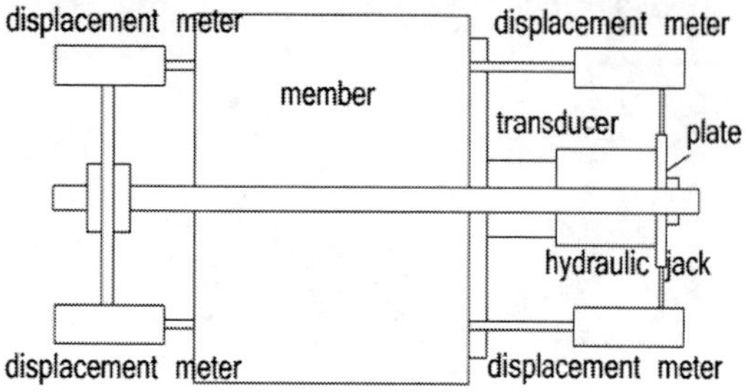

Fig. 6 Loading diagram

4 TEST CONTENT AND ULTIMATE LOAD

The test contents of the experiment are loaded end load, load displacement and free end displacement, compression strength of grouting material test block, rupture strength. The load in Table 2 is the ultimate load, Loading end displacement and free end displacement are displacement value of ultimate load. Rupture strength test block is three of each group, compression strength test block is six of each group.

Table 2 The test results

Days	number	Load(kN)	Loading end displacement(mm)		Free end displacement(mm)		Rupture strength(MPa)	compression strength(MPa)	
1d	T160-D16-I	45.29	3.0	3.1	0.5	0.6	3.1	32.6	33.5
	T160-D20-I	78.37	2.9	3.1	0.6	0.7	3.2	34.5	34.1
	T200-D22-I	100.38	3.1	3.3	0.6	0.8	3.4	35.5	34.5
3d	T160-D16-I	58.88	3.2	3.3	0.6	0.7	5.0	52.2	53.3
	T160-D20-I	101.89	2.9	3.0	0.7	0.5	5.3	54.1	54.7
	T200-D22-I	130.49	3.0	3.2	0.7	0.8	5.2	55.7	54.4
7d	T160-D16-I	72.46	3.3	3.2	0.6	0.6	5.8	61.6	64.2
	T160-D20-I	125.40	3.1	3.1	0.7	0.7	6.1	61.9	65.8
	T200-D22-I	160.61	3.1	3.2	0.7	0.8	6.6	64.0	63.6
15d	T160-D16-I	90.58	5.2	5.3	1.2	1.3	7.0	72.8	72.8
	T160-D20-I	156.79	5.4	5.5	1.3	1.3	6.9	74.8	76.1
	T200-D22-I	200.76	5.4	5.4	1.2	1.2	7.2	76.5	78.1
20d	T160-D16-I	108.70	5.3	5.3	1.3	1.3	8.0	82.8	82.8
	T160-D20-I	188.15	5.5	5.5	1.2	1.2	7.9	81.8	81.1
	T200-D22-I	240.91	5.4	5.4	1.4	1.3	8.2	80.5	82.1
28d	T160-D16-III	120.77	7.5	7.6	1.5	1.6	8.8	85.6	86.2
	T160-D20-III	209.05	7.5	7.5	1.7	1.7	8.7	86.9	85.8
	T200-D22-III	267.68	7.6	7.6	1.6	1.6	8.6	87.1	86.6

5 LOAD DISPLACEMENT CURVE

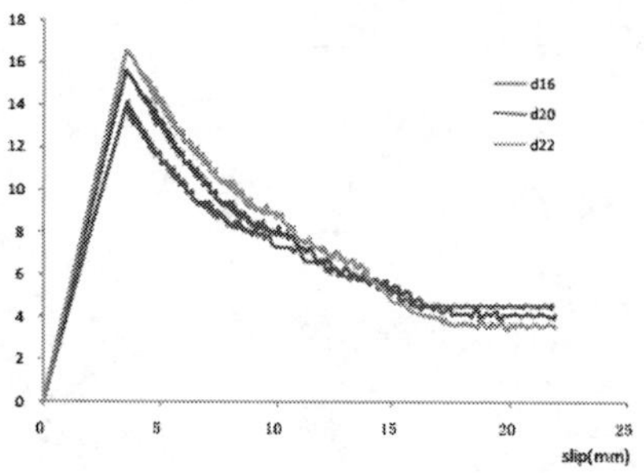

Fig. 7 The bond-slip curve of the first day

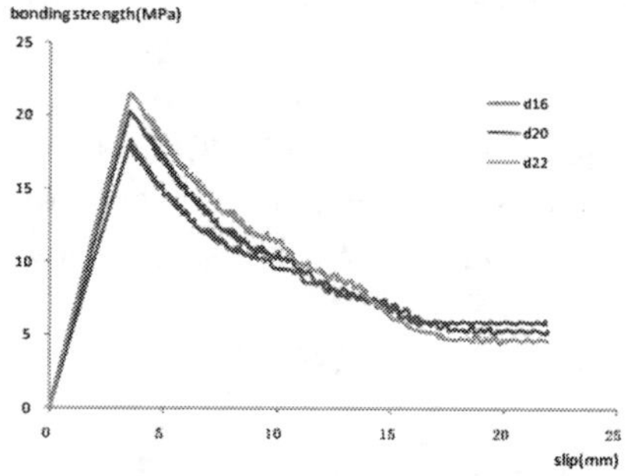

Fig. 8 The bond-slip curve of third day

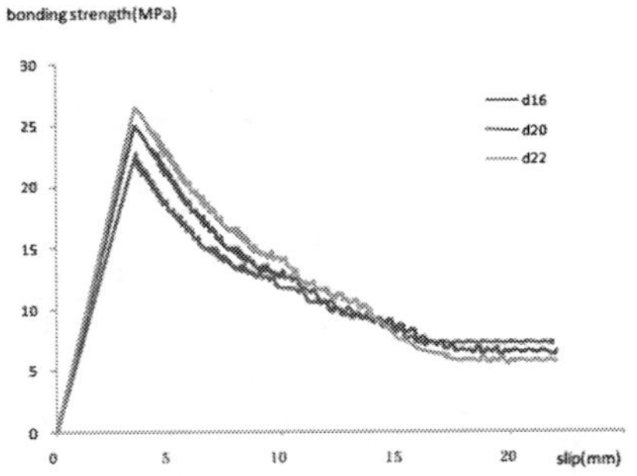

Fig. 9 The bond-slip curve of seventh day

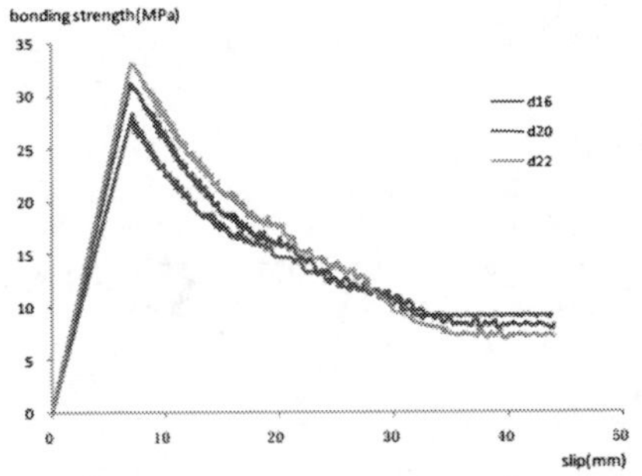

Fig. 10 The bond-slip curve of fifteenth day

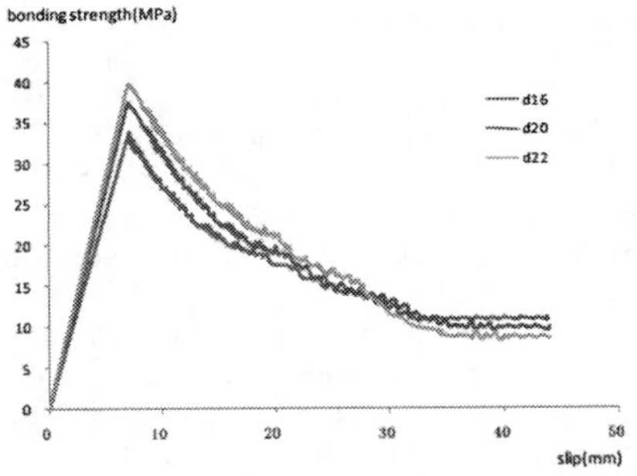

Fig. 11 The bond-slip curve of twentieth day

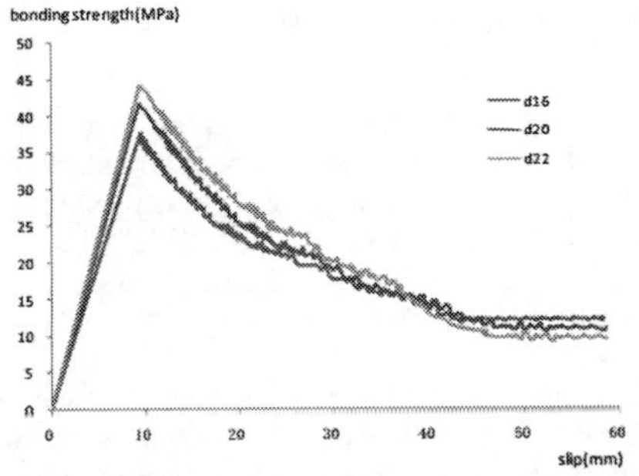

Fig. 12 The bond-slip curve of twenty-eighth day

6 CONCLUSIONS

1)with the increase of age, the bond strength between reinforcement and grouting material increased gradually, and the speed of 20d increased gradually.

2) for the CT20H sleeve, with the increase of the diameter of reinforcing steel bar, the bond strength between reinforcement and grouting material increases gradually. The reason is that the increase in the diameter of reinforcement will cause the increase of the mechanical bite force between reinforcement and grouting material.(The thickness of the grouting material between the steel bar and the sleeve is reduced, the height of the column formed after the cracking of the grouting material is reduced, and the compression strength is increased.)

3) It is possible to reduce the size of the inner diameter of the sleeve, but it is necessary to ensure the orientation of the reinforcing bar between the grouting material and the member during the construction.

REFERENCES

[1]Huang Hejun,Zhang Jianrong,Wu Jin,Song Lei. Testing study on tensile-shear resistance property of the bondof metal tube[J].(Sichuan Building S cience,2007,33(4):18-30)

[2]Wu Xiaobao, Lin Feng, Wang Tao. Experimental research on effects of grout age and types of steel bars on mechanical behavior of grout sleeve splicing for reinforcing bars[J]. (Building Structure, 2013,43 (14):77-82.)

[3]Shu Ruibin ,Zhang Jianrong ,Zhang Chun. FE analysis on load transfer mechanism of bonded rebars with steel sleeve considering slip of two interfaces. (2010)

[4]Wang Yuxin.Investigatian on bond-slip relationship between colloded steel bars and concrete under cyclic loading(.2014)

[5]Zheng Shansuo, Deng Guozhuan, Yang Yong,Yu Maohong,Zhang Junfeng. Experimnetal study of bond-slip performance between steel and concrete in src structures[J]. (Engineering Mechanics, 2003, 20(5):63-68)

[6]Xu Feng.Bond performance of reinforcement in concrete subjected tocompelx lateral pressure.2012

[7] Ko H,Sato Y. Bond stress - slip relationship between FRP sheet and concrete under cyclic load [J]. (Compos Constr, 2007,11(4), 419 - 426.)

[8] Kankam C .relationship of bond stress, steel stress, and slip in reinforced concrete[J]. (Struct Eng,1997,123(1), 79 - 85.)

[9] Chen G,Baker G.Influence of bond slip on crack spacing in numerical modeling of reinforced concrete[J]. (Struct Eng, 2003,129(11), 1514 - 1521.)

[10] Belleri, Andrea,Riva, Paolo. Seismic performance and retrofit of precast concrete grouted sleeve connections . Precast/Prestressed Concrete Institute, 2012, 97-109.

MATEC Web of Conferences **63**, 03013 (2016) DOI: 10.1051/matecconf/20166303013
MMME 2016

Study on Fluorescence Spectra of Thiamine and Riboflavin

Hui YANG [1, a], Xue XIAO [2], Xue Song ZHAO [2], Lan HU [1], Xian Feng XUE [1] and Jie Song YE [1]

[1] *New Star Application Technology Institute, Hefei, Anhui, 230031, China;*
[2] *Key Lab. of Environmental Optics & Technology, AIOFM, CAS, Hefei, Anhui, 230031, China*

Abstract: This paper presents the intrinsic fluorescence characteristics of vitamin B1 and vitamin B2 with 3D fluorescence Spectrophotometer. Three strong fluorescence areas of vitamin B2 locate at $\lambda ex/\lambda em$=270/525nm, 370/525nm and 450/525nm and one fluorescence areas of vitamin B1 locates at $\lambda ex/\lambda em$=370/460nm were found. The influence of pH of solution are also discussed, and with the PARAFAC algorithm, 9 vitamin B2 and vitamin B1 mixed solutions are successfully decomposed, and the emission profiles, excitation profiles, central wavelengths and the concentration of the two components were retrieved precisely through about 10 iteration times.

1 Introduction

The fluorescence of a folded protein or bio-aerosol is a mixture of the fluorescence from individual aromatic component and coenzyme. Riboflavin, known as vitamin B2, is an easily absorbed micronutrient with a key role in maintaining health in humans and animals. As such, vitamin B2 is required for a wide variety of cellular processes. Vitamin B2 plays a key role in energy metabolism, and is required for the metabolism of fats, ketone bodies, carbohydrates, and proteins [1]. Thiamine, known as vitamin B1, is one of 8 B vitamins, is used in many different body functions and deficiencies may have far reaching effects on the body, yet very little of this vitamin is stored in the body, and depletion of this vitamin can happen within 14 days. B vitamins are water-soluble and of great importance for the health of people, they can not be synthesized within the body, but be obtained from daily food, so the content measurement of B vitamins in the food is necessary for food security and research.

PARAFAC, a three way-decomposition method, has been found to be very useful in identifying the independent spectra of different types of fluorophores [2]. Compared to its predecessor, Principal Component Analysis (PCA) technique, PARAFAC provides both a quantitative and qualitative model of the data and separates the complex signal measured into its individual underlying fluorescent phenomena with specific excitation and emission spectra. It can track even small variations in EEM datasets by separating several independent groups of fluorophores from the overlapped components with a high resolution, so it is commonly used technique to monitor the mixed fluorescence EEMs. On the other hand, the weakness of PARAFAC model may include the assumption of the independence among the estimated components in the model, and potential inclusion of one or more poorly estimated components, which may significantly affect the spectra and scores of all other components [3].

Some previous studies reported that PARAFAC was applied to resolve the mixed and overlapped spectra of vitamin B1, B2 and B6 [4], accordingly, in this paper, discussed not only the fluorescence spectra of riboflavin and thiamine individually, but also the overlap of fluorescence spectra of vitamin B1 and vitamin B2 at the Ex/Em area with wavelength of 300~550/370~610nm, by means of Parallel factor analysis (PARAFAC) method.

2 Experimental Section

2.1 instruments and reagents

The Molecular ΣH_2O ultra pure water machine (Shanghai Molecular Co. Ltd) was used to generate the ultra purified water, UPW whose pH value is 5.4. The vitamin B2 mother liquid were compounded with Riboflavin from Amresco co.Ltd whose purity greater than or equal to 98%, and the vitamin B1 from Sigma co.Ltd with purity greater than or equal to 99%. All reagents and materials were weighed with Mettler Toledo precise electronic balance, and dissolved with Briton Robson Buffer with different pH values (1.95, 5.4, 5.66, 8.0 and 11.92).

[a] Hui YANG :sanpedroman@163.com

The solvents used in the experiments include 0.2mol/L Na_2HPO_4 buffer, 0.1mol/L $C_6H_8O_7 \cdot H_2O$ buffer, Na_2HPO_4 and $C_6H_8O_7 \cdot H_2O$ mixed buffer, 1% $K_3[Fe(CN)_6]$ buffer, 2mol/L NaOH buffer and 3mol/L H_2SO_4 buffer. 0.2mol/L Na_2HPO_4 buffer was compounded by putting 28.4g sodium phosphte dibasic anhydrous (Na_2HPO_4, Tianjin guangfu chemical research institute, China) into 1000mL UPW. 0.1mol/L $C_6H_8O_7 \cdot H_2O$ buffer was compounded by put 21.01g citric acid monohydrate($C_6H_8O_7 \cdot H_2O$, Sinopharm Chemical Reagent Co.Ltd) into 1000mL UPW. Na_2HPO_4 and $C_6H_8O_7 \cdot H_2O$ mixed buffer was compounded by mixing 0.2mol/L Na_2HPO_4 buffer and $C_6H_8O_7 \cdot H_2O$ buffer at the ration of 6:1; 1% $K_3[Fe(CN)_6]$ buffer was compounded by put 1g Potassium hexacyanoferrate buffer(Tianjin guangfu chemical research institute) into 100mL UPW, and was diluted by mixing with 3mL 2mol/L NaOH buffer to 10mL.

The vitamin B2 and B1 mother liquid concentration are 10mg/L respectively. The vitamin B2 and vitamin B1 mother reagent solutions were transferred through DragonLab whole disinfection manual single channel adjustable liquid shifter and dilute to working solutions of different concentrations. All reagents were of analytical grade, all solutions and put in amber glass bottles and stored in a refrigerator (4 ℃) because of the light sensitivity of vitamin B2.

3D fluorescence intensity measurements were carried out on an F-7000 FL spectrophotometer (Hitachi High-Technologies Corporation, Japan).

2.2 instrument settings and experiment procedure

500ul Briton Robson Buffers with different pH values and mother liquids of different volumes were injected into the 10ml test tubes, and diluted with purified water to form the working liquids and background liquids.

For the fluorescence EEM measurements of vitamin B2, the spectrophotometer excitation wavelength ranged from 200.0nm, to 550.0nm, emission wavelength ranged from 450.0nm to 650.0nm, scan speed was set at 12000nm/min with excitation and emission sampling interval of 10.0 nm, excitation and emission slit of 5.0nm, the PMT voltage was set at 700 V. Accordingly, for fluorescence EEM of thiamine, the excitation wavelength ranged from 300.0nm, to 400.0nm, emission wavelength ranged from 350.0nm to 550.0nm. All experiments were performed at room temperature at 25℃.

The 1st level and 2nd level Rayleigh scattering, Raman scattering and other background components within the fluorescence signals were corrected for the following analysis.

2.3 Multi-components discrimination using PARAFAC method

Based on the tri-linear decomposition theory, the parallel factor analysis(PARAFAC) method is a kind of mathematical model implemented through alternating least squares algorithm, which is widely applied to analyze three-dimensional or multi-dimensional data, to decompose N-dimensional data to the N load matrixes.

The measured fluorescence spectrum EEM data is a $I \times J \times K$ matrix, in which, I indicates the number of the samples, while J and K are the number of excitation wavelengths and emission wavelengths of samples respectively. Using Parallel Factor decomposition model, the fluorescence spectrum data matrix can be decomposed to score matrix A, load matrix B and C. The decomposition model can be represented as

$$x_{ijk} = \sum_{f=1}^{F} \alpha_{if} b_{jf} c_{kf} + \varepsilon_{ijk}, \quad i=1, 2, \ldots, I, \; j=1, 2, \ldots, J, \; k=1, 2, \ldots, K \qquad (1)$$

where, x_{ijk} is the fluorescence intensity of sample i at excitation wavelength j and emission wavelength k, F is the column number of load matrix, or the number of factors, ε_{ijk} is the residual element, α_{if}, b_{jf}, c_{kf} are the elements in load matrix A, B and C respectively. The algorithm will be aborted until convergence of the PARAFAC model, that is, the minimum loss function $f_{SSR} = \sum_{i=1}^{I} \sum_{j=1}^{J} \sum_{k=1}^{K} e^2{}_{ijk} < 10^{-6}$.

In this study, PARAFAC modeling was performed using the MATLAB 7.0 code. The appropriate number of components was determined primarily based on the three diagnostic tools including residual analysis, core consistency and visual inspection of spectral shapes of each component, which are widely used by other similar studies. The components extracted by PARAFAC represent groups of the organic components that exhibit similar fluorescence properties. The component scores indicate the relative concentration of the groups, not the actual concentration of a particular material/fluorophore. However, it is typically assumed that the scores are proportional to the concentrations of the different components [5, 6].

3 Fluorescence EEM Characteristics of Thiamine and Riboflavin

3.1 Intrinsic fluorescence EEM characteristics of riboflavin and thiamine

For riboflavin/vitamin B2, there are three strong fluorescence areas, whose center locate at λex/λem=270/525nm, 370/525nm and 450/525nm respectively, and the emission wavelength ranges from about 500nm to 600nm, as shown in figure 1. The fluorescence intensity excited by 270nm

DOI: 10.1051/matecconf/20166303013

excitation wavelength is much stronger than that by 370nm and 450nm, the ratio of fluorescence intensity is 1:0.41:0.25 approximately.

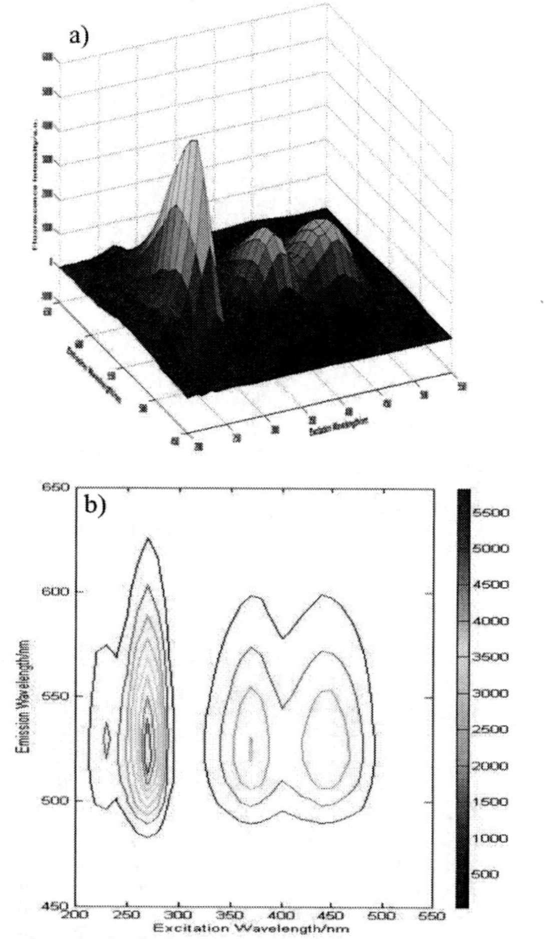

Figure 1. Fluorescence intensity distribution of riboflavin@200ug/L

Accordingly, the Thiamine/vitamin b1 has only one strong fluorescence areas, whose center locate at λex/λem=370/460nm, the excitation wavelength ranges from about 320nm to 400nm, and the emission wavelength ranges from about 370nm to 550nm, as shown in figure 2.

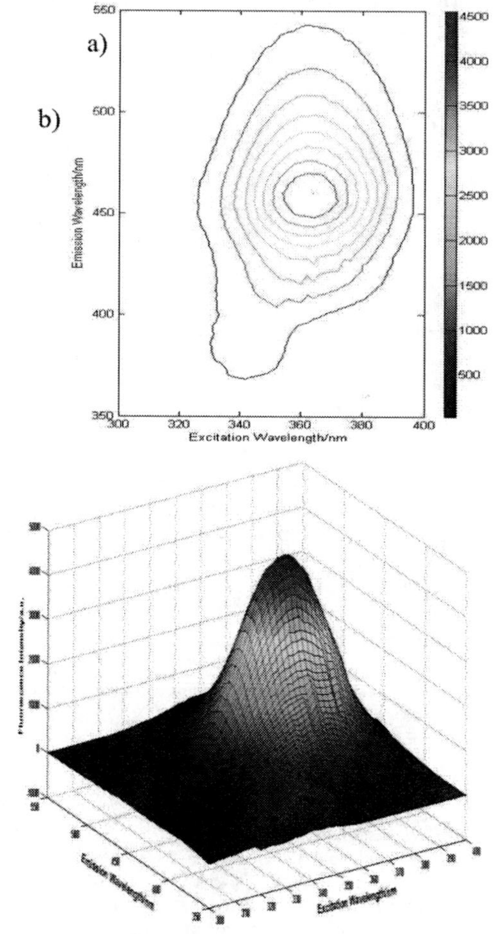

Figure 2. Fluorescence intensity distribution of thiamine@75ug/L

The fluorescence efficiency of riboflavin is about two times stronger than that of thiamine for the same solvent.

3.2 Affection of pH to the fluorescence emission intensity distribution of riboflavin

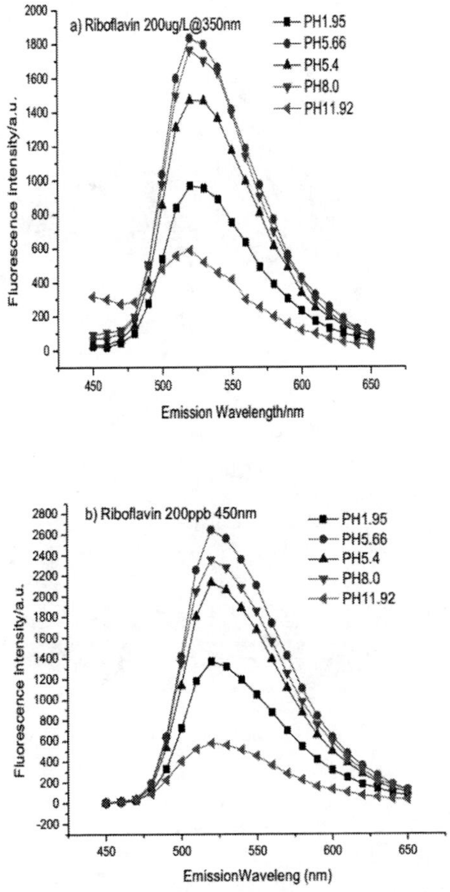

Figure 3. Fluorescence intensity distribution of riboflavin@350nm and 450nm affected by pH value of the resolvent.

When resolved in strong alkaline and strong acid solution, the structure of vb2 molecular is distorted, so the fluorescence intensity declines sharply compared to the weak alkaline and weak acid solution, and if resolved in weak alkaline and weak acid solution, the fluorescence intensity are much stronger(figure 3).

4 Analysis of Fluorescence Spectra of Riboflavin and Thiamine Mixed Solutions

4.1 Instrument settings

Firstly thiamine and riboflavin mother liquids of different concentrations were put into 10mL numbered test tubes, and then 1mL Na_2HPO_4 and $C_6H_8O_7{\cdot}H_2O$ mixed buffer and 0.22ml $K_3[Fe(CN)_6]$ buffer were added into the test tubes. The pHs of mixed solutions were then adjusted to neutral with added 3mol/L H_2SO_4 and UPW. Lastly, the mixed solutions were laid stationary for 10 minutes for the following measurements.

For the fluorescence EEM Matrixes decompose of the mixed solution of riboflavin and thiamine, the spectrophotometer excitation wavelength ranged from 300.0nm, to 550.0nm, emission wavelength ranged from 390.0nm to 610.0nm, scan speed was set at 12000nm/min with excitation and emission sampling interval of 5.0 nm, excitation and emission slit of 10.0nm, the PMT voltage was set at 700 V.

4.2 3D fluorescence spectrogram of thiamine and riboflavin mixed solutions

Fluorescence EEM intensity of thiamine and riboflavin mixed solutions of different concentrations (listed in table1) are shown in figure 4 (a) ~ (i). The two separated fluorescence EEM peaks at 370nm/460nm and 470nm/520nm are obviously when the thiamine's concentration is low at 10ug/mL, and with the concentration rising of thiamine from 10ug/mL to 25ug/mL and 50ug/mL, the two peaks merged together.

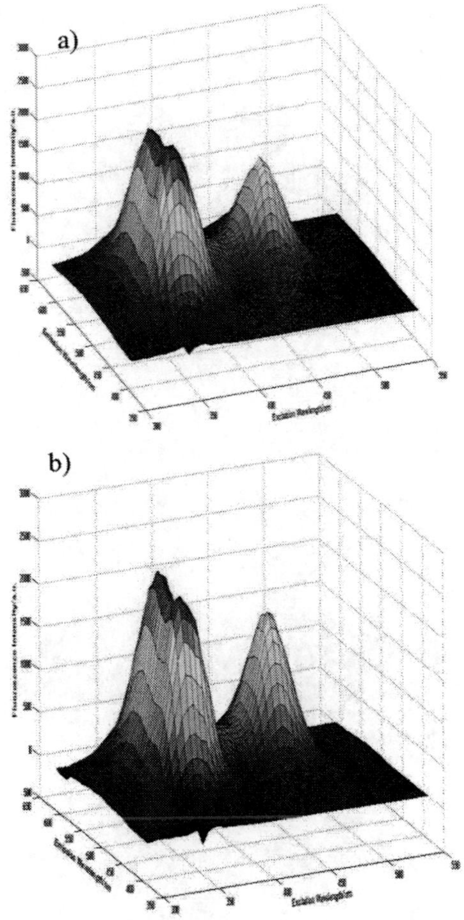

DOI: 10.1051/matecconf/20166303013

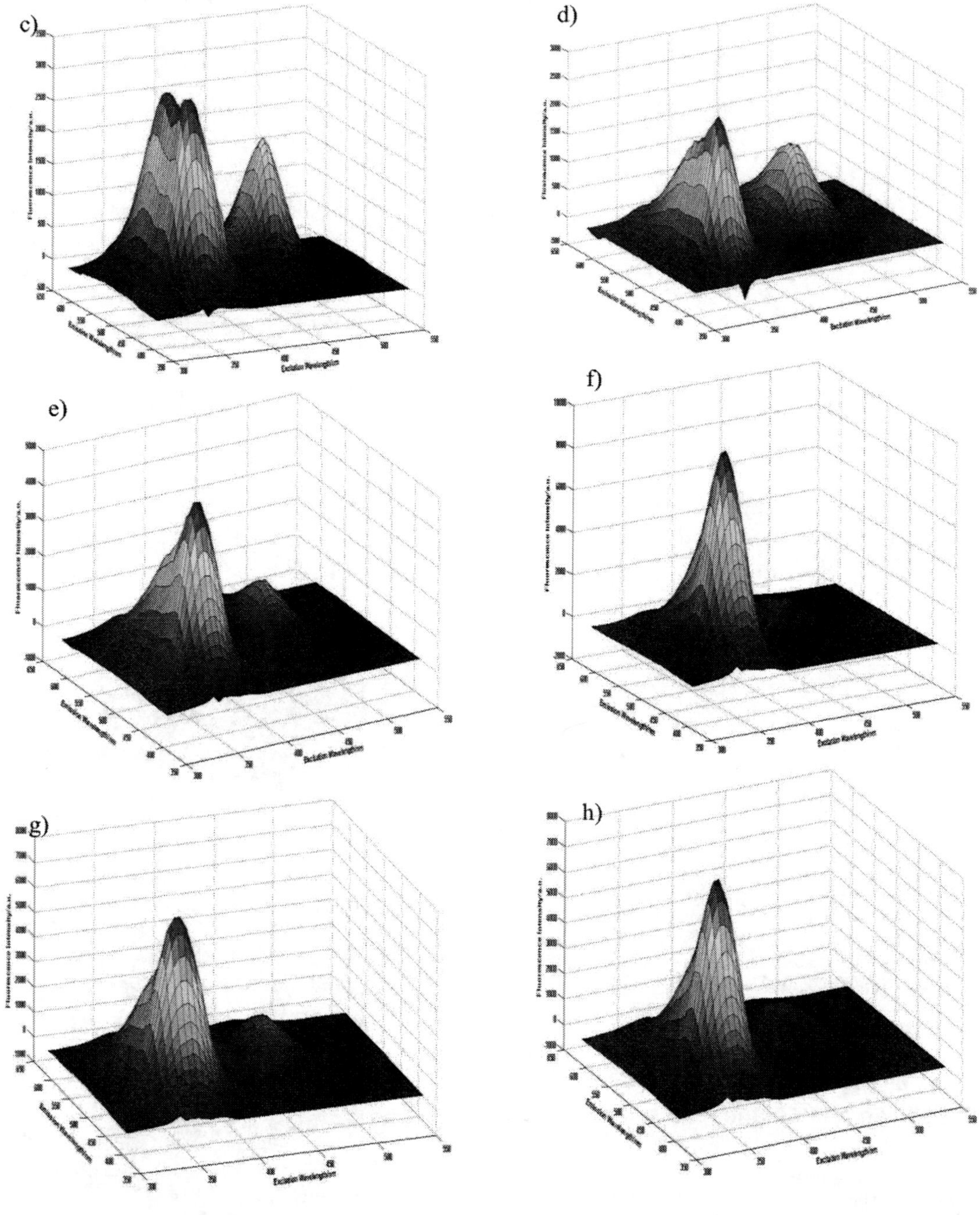

DOI: 10.1051/matecconf/20166303013

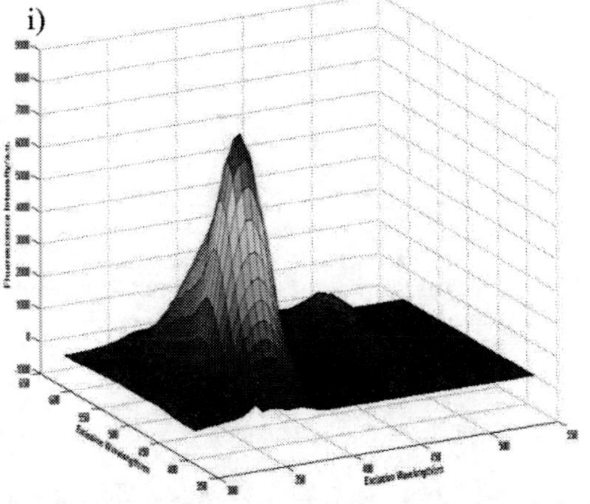

Figure 4. Fluorescence EEM intensity distribution of riboflavin and thiamine mixed solutions of different concentrations.

4.3 Results retrieved by PARAFAC algorithm

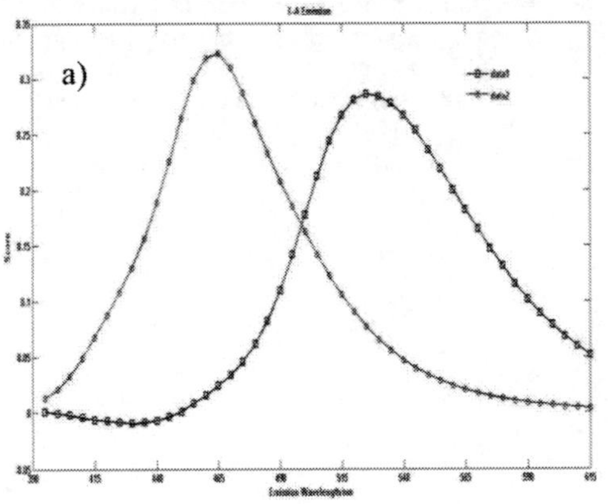

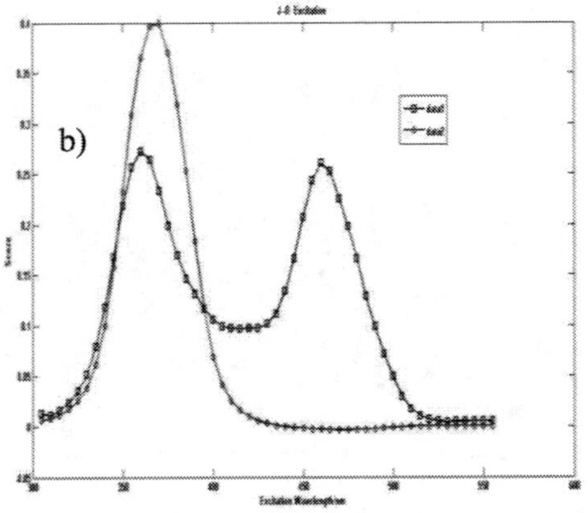

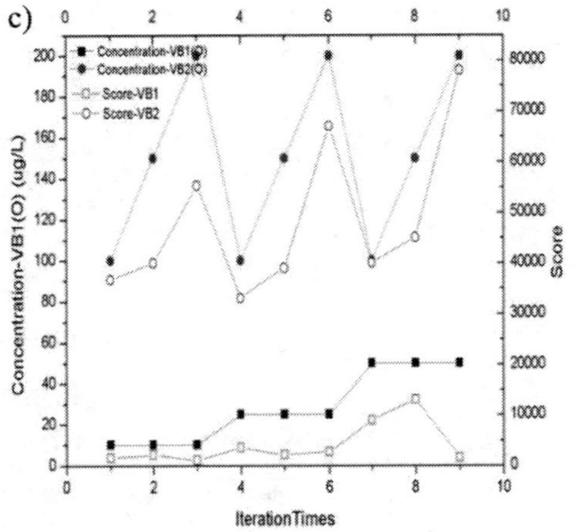

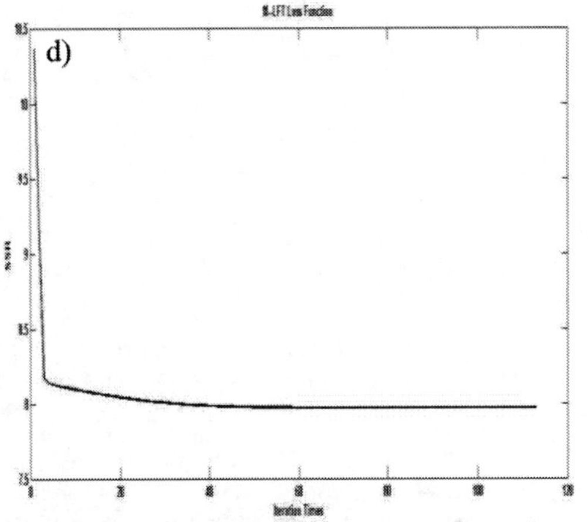

Figure 5. Excitation and emission spectral profile retrieved by PARAFAC algorithm

(a)Emission spectra, (b) Excitation spectra and (c) concentrations of original solutions and retrieved (d) SSR and iteration number

The emission spectra retrieved by PARAFAC are shown in figure 5 (a), the blue curve with block symbol (data1) indicates the retrieved emission profile of riboflavin and the green curve with circle symbol (sample2) indicates the retrieved emission profile of thiamine. From figure 5(a) and figure 1, figure 2, one can see that the retrieved emission profile and central wavelength of riboflavin and thiamine are coincident with their real emission profiles very well.

The excitation spectra retrieved by PARAFAC are show in figure 5 (b), the blue curve with block symbol (sample1) indicates the excitation profile of riboflavin and the green curve with circle symbol (sample2) indicates the excitation profile of thiamine. From figure 5 (b) and figure 1, it can be seen that the retrieved excitation profile and central wavelength of riboflavin are coincident with its real excitation profile, the two strong excitation wavelength at 370nm and 450nm are obvious. And for the retrieved excitation profile and central wavelength of thiamine, the single excitation wavelength peak is also clear and coincident with the real excitation wavelength profile.

The good linear correlations of original and retrieved concentrations of thiamine and riboflavin can be observed in figure 5(c). From figure 5 (d) one can see that SSR of PARAFAC logarithms decreases quickly and sharply at the beginning of the iteration times, the SSR is stable when iteration times≥5.

Table 1. Analytical concentrations of riboflavin and thiamine

Sample	Component			
	Thiamine		Riboflavin	
	Original (ug/L)	Retrieved score	Original(ug/L)	Retrieved score
a)	10	1500	100	36701
b)	10	2100	150	40010
c)	10	1000	200	55310
d)	25	3500	100	33090
e)	25	2100	150	39010
f)	25	2700	200	67001
g)	50	8900	100	40010
h)	50	13000	150	45010
i)	50	1520	200	78001

5 Conclusions

The fluorescence of a protein or bio-aerosol or bio-agent is a mixture of the fluorescence from individual aromatic residues and coenzyme. Using fluorescence Spectrophotometer, the intrinsic fluorescent characteristics of vitamin B1 and vitamin B2 are measured with solutions of different pH and discussed. Vitamin B2 and vitamin B1 mixed solutions are successfully decomposed and resolved by PARAFAC algorithm. The retrieved emission profiles, excitation profiles, central wavelengths and the concentration of the two components are coincident precisely with real emission profiles, excitation profiles, central wavelength of each component.

Acknowledgments

The work was supported by the National Natural Science Foundation of China named "Study on Technology of Ultraviolet Laser-induced Fluorescence LIDAR for Bioagent Remote Sensing" No. 41375026.

References

1. National Academy of Sciences, 2004. Sensor Systems for Biological Agent Attacks: Protecting Buildings and Military Bases, Committee on Materials and Manufacturing Processes for Advanced Sensors, Board on Manufacturing and Engineering Design, Division on Engineering and Physical Sciences, National Academies Press Washington URL: http://www.national-academies.edu/bmed [Accessed: Jan. 23, 2015].
2. Stedmon, C.A.; Markager, S.; Bro, R. Tracing dissolved organic matter in aquatic environments using a new approach to fluorescence spectroscopy. Mar. Chem., vol 82, pp 239-254, Aug. 2003
3. Murphy, K.R.; Hambly, A.; Singh, S.; Henderson, R.K.; Baker, A.; Stuetz, R.; Khan, S.J. Organic matter fluorescence in municipal water recycling schemes: Toward a unified PARAFAC model. Environ. Sci. Technol., vol 45, pp, 2909-2916. Mar. 2011
4. Stedmon, C.A.; Bro, R. Characterizing dissolved organic matter fluorescence with parallel factor analysis: A tutorial. Limnol. Oceanogr. Methods, vol 6, pp 572-579, Nov. 2008.
5. Ni Yong-nian, Cai Ying-jun, Simultaneous synchronous spectrofluorimetric determination of vitamin B1,B2 and B6 by PARAFAC, Spetroscopy and Spectral Analysis, vol 25, no. 10, pp 1641-1644, Oct. 2005
6. Li Weihua.; Sheng, G.aoping.; Liu, Xianwei.; Yu H.anqing. Characterizing the extracellular and intracellular fluorescent products of activated sludge in a sequencing batch reactor. Water Res, vol 42, pp, 3173-3181, Mar. 2008.

MATEC Web of Conferences **63**, 03014 (2016) DOI: 10.1051/matecconf/20166303014
MMME 2016

Development of Partial Tubular Flat Knitting Fabric Composite Preform

Wei Qing JIANG[1] Chun Xia CHEN[2]

[1]*Yan cheng institute of industry College， Yan cheng， Jiang su 224005*
[2]*Excellent Scientific and Technological Innovation Team of Jiangsu Province 2015*

Abstract: After building some structures of partial tubular flat knitting fabric composite preform, the influencing factor on tubular section was analyzed and the fabric was knitted selectively. The partial tubular flat knitting fabric composite preform were Knitted by changing different yarn, row number and two-sided partial tubular flat knitting fabric. Multilayer sheet would be got after hot pressing and it has big market prospects and good application value.
Key words: partial tubular; flat knitting fabric; composite preform; structure design; development
Top-notch Academic Programs Project of Jiangsu Higher Education Institutions 2015

1 Development of composite preform

In recent years, the thermoplastic composites material which is used textile preform as reinforcement develop rapidly, for the industrial use of textiles, the proportion is also increasing. Braided、woven and knitted fabric construction are often used as textile preform. among the Knitted fabric preforms ,warp knitted fabric preform's development is much maturer, because the weft knitted fabric preform has good extension flexibility and high production efficiency, so it has broad production prospects. At present, the research on the whole tubular structure of flat knitting fabric has been made at home and abroad. This knitting method achieve braiding cylindric structure on the knitting machine. The fabric is a whole tube. But the report of the structure of partial tubular flat knitting fabric as well as the fabric is few. For this paper the partial tubular flat knitting fabric composite preform structure is designed and trial-weaving. The purpose is to lay the foundation for the 3D knitting fabric production and industrialization promotion of the composite materials.

2 Structure design of partial tubular flat knitting fabric composite preform

The structure of partial tubular flat knitting fabric was designed, as shown in figure 1.

DOI: 10.1051/matecconf/20166303014

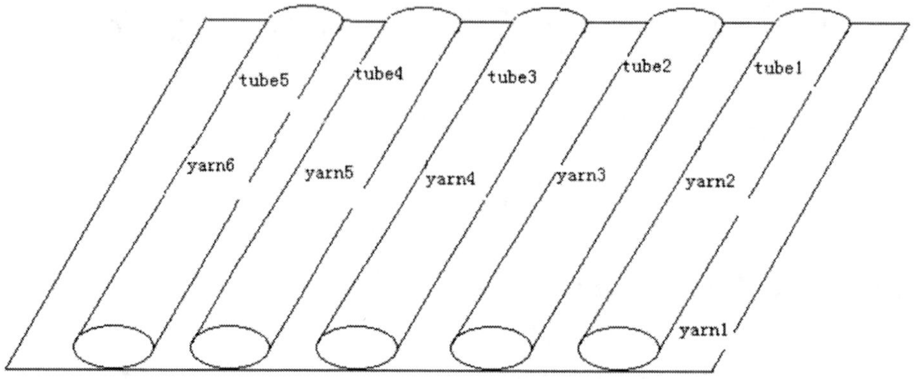

figure 1 Structure of partial tubular flat knitting fabric composite preform

In Figure 1, the yarn 1 is woven in the plane part of partial tubular flat knitting fabric composite preform. partial tubular structure 1-5 Knitted with yarn 2-6, The appearance presented by the tubular part mainly affected by yarn, fabric weave, row number etc.

For the Structure design of the partial tubular flat knitting fabric composite preform, the following changes can be choosed, Yarn 1-6 can choose the same or different colors and raw materials for weaving, every parts of tube 1-5 different can use different fabric structure.

3 The machine design process of partial tubular flat knitting fabric composite preform

partial tubular flat knitting fabric composite preform were knitted on a flat knitting machine. Taking organization cycle of 160 row circular as example, the knitting diagram is shown in Figure 2. When specific weaving, Organization cycle course number、yarn number that participate in knitting etc.can be designed according to the specific needs.

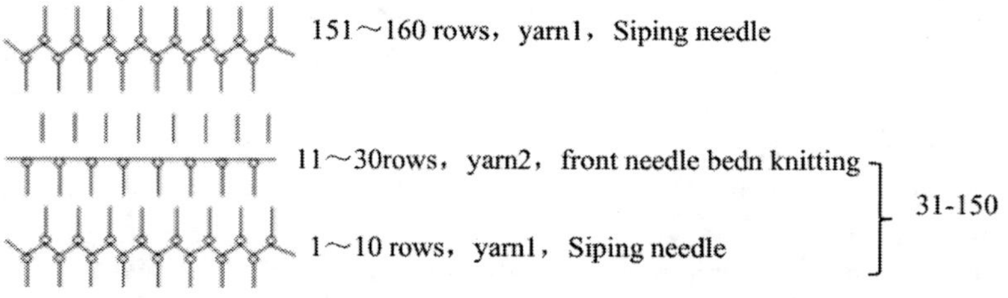

rows, repeats the process

Figure 2 knitting diagram of partial tubular flat knitting fabric composite preform

partial tubular flat knitting fabric composite preform knitting process is as follows

3.1knitting of one pipe structure

for the 1-50 longitudinal yarn, take it turns to knit10 rows by using the yarn 1 before and after needle bed, Siping needle。when knitting the 11-30 rows, close the knitting triangle of the back needle bed. change the Yarn nozzle 2, Knitting with yarn 2, Pay attention to give the tubular part pressure during the knitting process. In order to avoid the fabric floating on the fell affect knitting.the 30th rows, After opening the needle bed knitting triangle. Woven to form a partial tubular structure.

3.2multiple partial tubulars knitting

From the thirty-first row, according to figure 2 alternately repeating one pipe structure knitting, Successive formation of multiple partial tubulars structures, for the example of this paper, yarn2-yarn6 selected dfferent colors of yarn respectively, we can choose different colors, different raw materials, different thickness of the yarn according to the need of the Actual design。

4.Trial knitting of partial tubular flat knitting fabric composite preform

4.1 machine conditions

the design and trial knitting of the partial tubular flat knitting fabric composite preform can be knitting on hand-operated flat knitting machines or computerized flat knitting machines, The machine conditions are as follows.

4.1.1 hand-operated flat knitting machines

Machines: Nan tong sansi Technology，ongxing brand hand-operated flat knitting machines

Yarns: 28s/2、White, green, pink, red, blue, orange, mulberry /PP blended yarn

Machine No.: E9。

4.1.2computerized flat knitting machines

Machines: Jiangsu Xing Textile Machinery Co., Ltd. automatic computerized flat knitting machine

Machine No.: E3/5/7 Variable pitch

Yarns: 28s/2、White, green, pink, red, blue, orange, mulberry /PP blended yarn

Systems: Zhejiang Heng Qiang technology co.ltd, heng qiang plate making software system.

4.2 sample display

The pipe part of the partial tubular flat knitting fabric composite preform has stereoscopic sense, Presents a tubular appearance, Multilayer composite sheets can be formed at one time. The fabrich have strong integrity. the prepared composite materials have superior performance. The sample of partial tubular flat knitting fabric composite preform made on transverse machine is shown in Figure 3. The composite sheet can be hot pressed into three layers at one time, and the state diagram of the fabric is shown in Figure 4.

Figure 3 partial tubular flat knitting fabric composite preform

Figure 4 The state of partial tubular flat knitting fabric which was hot pressed into composite sheet

5. revised design of partial tubular flat knitting fabric composite preform

Braiding of partial tubular flat knitting fabric composite preform can be rivised by adjustingsome process parameters and the woven yarn.

5.1 Change the row number of the pipe part

During the development and design of the partial tubular flat knitting fabric composite preform，Change the row number of woven pipe part can change the size of the pipe part。for example，increase the row number of yarn 2-6，The braided fabric's pipe part will become long. And the shape of the "tube" becomes larger. It will need lesser pipes for the same length of composite sheet. As shown in Figure 5.

Small pipe structure, 5 tubes per 25cm

big pipe structure, 3 tubes per 25cm

Figure 5 Comparison of partial tubular flat knitting fabric with different pipes after hot pressing

Weave the tubular part by using different raw materials、styles of yarns

Figure 6 change the tubular part yarn

5.2 Change the yarn in the pipe part

Partial tubular flat knitting fabric has strong stereoscopic effect, and has the appearance of tubular structure. Using different yarn weaving, fabric surface presents different mutual spacing pipe which have different raw materials, different sizes, different styles of yarn. As shown in Figure 6, According to the different needs for different composite materials, we can choose the different yarns.

double-faced partial tubular flat knitting fabric's state after being hot-pressed

The state of the double side asymmetrical partial tubular flat knitting fabric after being hot pressed

Fig. 7 contrast diagram of different double-faced partial tubular flat knitting fabric after being hot pressed

5.3 Design of double-faced partial tubular flat knitting fabric composite preform

when design the partial tubular flat knitting fabric composite preform, knit one side tubular fabric, forming Sheet of three layers fabric after hot pressing。knit two side tubular fabric, forming Sheet of five layers fabric after hot pressing。As shown in Figure 7,

when knit double-faced partial tubular flat knitting fabric composite preform, the double-faced pipe structure can be designed symmetrical or asymmetrical.

6. Conclusions

Through the structure design 、 weaving technology develop and test weaving of partial tubular flat knitting fabric composite preform, The fabric's structure has strong stereoscopic sense, many kind of material and various style yarn can be chossed to knit, this [product has good market prospect and application value.

The partial tubular flat knitting fabric composite preform can be made into multilayer composite sheet one time , particular to the 5 layers, the composite materials have superior performance .

References

[1]TAO Li-jun,LIN Feng-ling,YIN Xiao-hai,LIU Xiao-hong,Preparation and Study on Weft Knitted Fabric Feinforced Composites [J].JOURNAL OF WUHAN UNIVERSITY OF SCIENCE AND ENGINERING,2010, 23 (4) : 4-7.

[2]Zhang Li,Ma Chongqi and Cheng Qingji, Research on Woven Method of a New Three-dimensional Textile Composite Perform[J].INDUSTRIAL TEXTILE, 2006 (12) : 5-9.

[3] CHEN Jie,QIN Xiao,WANG Jianming, Development of partial tubular flat knitted fabrics[J].Journal of Textile Research, 2013,9

DOI: 10.1051/matecconf/20166303015

Experimental study on fracture properties of Waste fiber recycled concrete

Jing Hai ZHOU [1,a], Tian Bei KANG [2], Feng Chi WANG [2], Wei DONG[2]

[1]Construction Energy Conservation Research Institute, Shenyang Jianzhu University, Shenyang, China, 110168
[2]School of Civil Engineering, Shenyang Jianzhu University, Shenyang, China, 110168

Abstract. In order to study the influence of the replacement ratio of recycled aggregate and the volume of waste fiber on the fracture performance of concrete, the three-point bending test is carried out on the waste fiber recycled concrete precast beams. Calculate the waste recycled fiber concrete fracture parameters according to the double-K fracture model. The results show that the fracture indexes of recycled fiber are related to the recycled aggregate replacement ratio and the volume content of waste fiber. The larger the replacement rate of recycled aggregate is, the smaller the fracture indexes of the specimens are. The volume content of waste fiber is 0.12%, which has the most significant effect on the fracture performance. The study shows that waste fiber can improve the fracture properties of recycled aggregate concrete.
Key words. recycled concrete; waste fiber; fracture criteria; fiber volume content

1 Introduction

As the environmental and resource issues have become increasingly prominent, the application of recycled concrete conforms to the development trend of green building. There are micro cracks in recycled concrete, and it has been found that the mechanical properties of recycled concrete are worse than ordinary concrete [1-3]. Fiber can improve the mechanical properties of recycled concrete. Therefore, in-depth study of fiber-recycled concrete has great significance for promoting the development of recycled concrete [4-6]. Cracks are an important factor affecting the quality and durability of concrete. How to control the generation and propagation of cracks effectively is the most important problem of fiber-recycled concrete.

The waste fibers can constrain the development of micro-cracks. Currently, the research on fiber recycled concrete is still in service behavior, basic mechanical function and durability [7-9]. In this paper, using three-point bending fracture tests, the study is carried out on the fracture properties of waste fiber recycled concrete with diffident replacement ratio of recycled aggregate and fiber volume.

2 Experiment

2.1 Materials

Ordinary Portland cement (P.O.42.5) with physical properties shown in Table 1 is used in this study. The fine aggregate is ordinary river sand with fineness modulus of 2.7, moisture content of 4.15% and apparent density of 2610kg/m³. There are two kinds of coarse aggregate, one for nature gravel and the other for recycled aggregate. Recycled aggregate from the abandoned concrete beams in laboratory of Shenyang Jianzhu University, the initial intensity is C40. After artificial crush and then through the jaw crusher, the recycled coarse aggregate is obtained and basic properties of recycled coarse aggregate indicators are shown in Table 2. Two kinds of coarse aggregate with gradation within 5~20mm, the maximum should be less than 25mm. The waste fibers are from waste polypropylene fiber carpet, which is split into length of 19mm.

[a] Corresponding author: jinghaiz@qq.com

Table 1 Physical Properties of Cement

surface area (m^3/kg)	Consistency (%)	setting time		stability	compressive strength (MPa)		flexural strength (MPa)	
		initial setting	final setting		3d	28d	3d	28d
348	25.0	2.3h	3.5h	eligible	25.9	43.6	5.9	9.1

Table2 Physical Properties of recycled aggregate

Index / project	apparent density (kg/m^3)	bulk density (kg/m^3)	crush index (%)	water absorption (%)	moisture content (%)	
					max	min
recycled aggregate	2461.0	1167.6	17	4.18	1.10	0.47
					0.83	

2.2 Test methods

Place a 3mm steel plate in the cross of the specimen, form 3mm preformed joint after maintenance for 24 hours, and then cure 28 days under natural conditions. The specimen size and breaking performance test are shown in Figure 1.

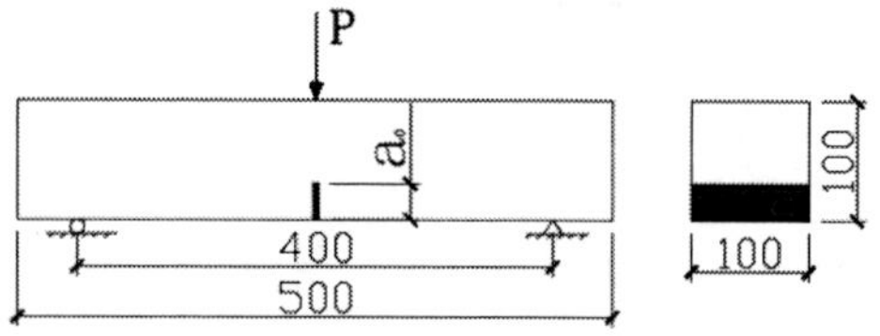

Figure 1. Schematic of breaking performance test

By using hydraulic jacks manually in the test, the load is applied with a loading speed of 0.02kN/s. The data are recorded in the DH3816N static strain test system. The main contents of this test are as follows: cross deflection of precast cracks, load-middle span displacement, load-crack mouth opening strain and load- crack tip opening strain. The recycled aggregate replacement rate and waste fiber volume fraction are the design variables. The specimen grouping is listed in Table3.

Table3 Specimen grouping

Series	Specimen size (mm)	Initial crack-depth ratio	Recycled aggregate replacement ratio (%)	Fiber length (mm)	Fibers content (%)
F1	500×100×100	0.3	0	19	0.12
F2	500×100×100	0.3	100	19	0.12
F3	500×100×100	0.3	50	0	0
F4	500×100×100	0.3	50	19	0.08
F5	500×100×100	0.3	50	19	0.12
F6	500×100×100	0.3	50	19	0.16

3 Results and discussion

3.1 Phenomena and results

Before waste fiber recycled concrete cracking, cracks expand slowly in the concrete, waste fiber plays a role in the junction in preventing the development of cracks. The plastic zone is formed in the connection of the waste fiber and the crack. In addition, when the load reaches the intensity of the crack strength, the specimens of waste fiber will produce a certain amount of load mutation, but fracture damage phenomenon has an obvious distinction with or without waste fiber. Waste fiber recycled concrete three-point bending beam after the load mutates, can continue to bear a greater load, and result in large deformation, until some or all of the waste fibers are pulled out of the test pieces to appear fracture damage of instability.

Fracture performance indicators of waste fiber recycled concrete are shown in Table 4. From the experimental results, the curve of load-crack mouth opening displacement ($CMOD_c$) and the load- crack tip opening displacement ($CTOD_c$) are obtained. Fracture parameters is calculated based on double - K fracture criterion proposed by Shilang Xu[10]. In the critical condition:

$$K_{IC}^{ini} = K_{IC}^{un} - K_{IC} \quad (1)$$

Where K_{IC}^{ini} is initial fracture toughness, K_{IC}^{un} is unstable fracture toughness, K_{IC} is stress intensity factor.

Table 4 Fracture performance indicators of waste fiber recycled concrete

Series	$CMOD_c$ (μm)	effective crack length a_c (mm)	$CTOD_c$ (μm)	K_{IC}^{ini} (MPa m$^{1/2}$)	K_{IC}^{un} (MPa m$^{1/2}$)	K_{IC} (MPa m$^{1/2}$)
F1	27.2	41.59	12.13	0.74	1.81	1.06
F2	25.8	41.64	11.52	0.43	1.35	0.92
F3	17.5	38.68	7.02	0.48	1.03	0.56
F4	24.1	40.67	10.44	0.36	1.40	1.04
F5	27.1	41.58	12.08	0.33	1.55	1.22
F6	24.5	43.76	11.58	0.37	1.34	0.97

3.2 curve of *P-CMOD$_c$*

The *P-CMOD$_c$* curve of waste fiber recycled concrete precast beam is shown in Figure 2. Figure 2 (a) is the influence of the replacement rate of recycled aggregate and figure 2 (b) is the influence of volume fraction of waste fiber on the curve.

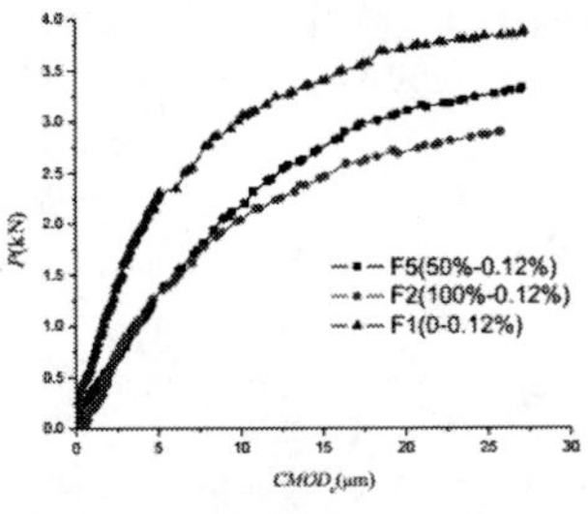

(a)the influence of the replacement rate of recycled aggregate

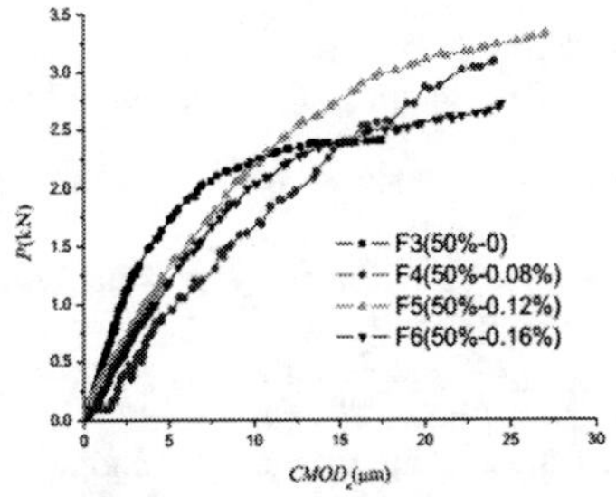

(b) the influence of volume fraction of waste fiber

Figure 2. circumstances of *P-CMOD$_c$*

As shown in Fig.2(a), the P-CMODc curve of natural concrete is above the other two curves. The smaller the replacement rate of recycled aggregate is, the greater the maximum crack load of recycled fiber recycled concrete is. The cracking load of recycled concrete specimens with replacement ratio of 50% and 100% is lower than that of the natural aggregate concrete by 29.6% and 34.2%. This shows that the greater the replacement rate of recycled aggregate is, the weaker is the ability of concrete to resist the development of micro and macro cracks growth and development.

It can be seen from Fig.2 (b), when the fiber length is 19mm, the curve of the test piece F5 is above the other curves. The average cracking load of waste fiber volume fraction of 0.08%, 0.12% and 0.16% is increased by 28.2%, 38.2% and 12.9%, respectively, compared with that of non - fiber specimens. The replacement rate of recycled aggregate is 50%, the fiber length is 19mm, the volume content is 0.12%, the maximum cracking load of recycled aggregate concrete.

The reasons for recycled concrete cracking load bigger than ordinary concrete are as the following three points. First of all, the mixing process is required to increase the amount of water in order to maintain the performance of recycled concrete; Secondly, there is a big difference between hardened cement mortar and new cement mortar. Due to the difference between hardened cement mortar and new cement mortar, there is a weak area in concrete. Finally, the initial damage and secondary injury cause a large number of micro cracks in recycled concrete. Waste fiber can improve the maximum cracking load of recycled concrete, and the optimal volume content of waste fiber is 0.12%.

3.3 Fracture toughness

The fracture index $CMOD_c$, $CTOD_c$ and K_{CI} decreases with the increase of recycled aggregate replacement rate, as illustrated in Fig.3.

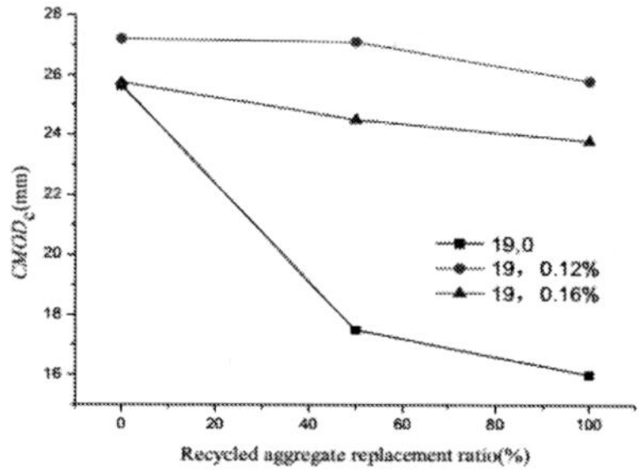

(a) $CMOD_c$

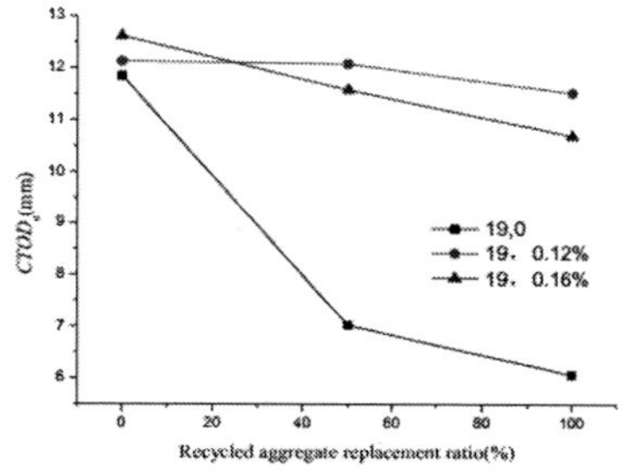

(b) $CTOD_c$

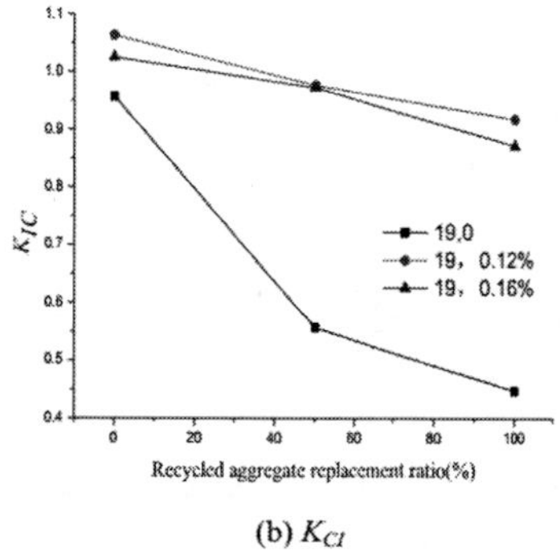

(b) K_{CI}

Figure 3. fracture parameters change with recycled aggregate replacement ratio

Compared with the replacement rate of 100% of the waste fiber concrete specimens, replacement ratio is 0 and 50% specimens of the $CMOD_c$ increases about 9.4%~60.3%, $CTOD_c$ increases about 15.8%~95.2% and the K_{CI} increases about 6.4%~15.9%. The greater the rate of recycled aggregate is, the worse the fracture properties of waste fiber recycled concrete is.

When the fiber length is 19mm and the volume content is 0.12%, the three indexes are decreased with the increase of the replacement rate of the recycled aggregate. The replacement rate is 100%, and the fracture index is the smallest. Recycled aggregate ratio of 0 and 50% the $CMOD_c$ upgrading from 5.4% to about 9.3%, the $CTOD_c$ lift about 4.9% ~ 5.3%, lifting the K_{CI} about 6.4% - 8.9%.

When the fiber length is 19mm and the volume content is 0.16%, the recycled aggregate ratio of 0 and 50% of the $CMOD_c$ increased by 5.1% ~ 8.2%, the $CTOD_c$ increase from 8.3% to 18.1% and the K_{CI} improve the 11.6%~17.7%.

With the increase of recycled aggregate replacement rate, the original defects increase and the maximum cracking load was decreased. The fracture and other mechanical properties of the recycled concrete are also deteriorated with the increase of the replacement rate of the recycled aggregate.

4 Conclusions

The following conclusions can be drawn from the results of the three-point bending fracture tests:

(1) According to the three-point bending fracture tests of waste fiber recycled concrete precast beam based on the study of double-K fracture criterion, waste fibers can significantly improve the fracture properties of recycled concrete.

(2) The fracture indexes of the specimen become smaller with the increase of the replacement ratio of recycled aggregate. The addition of recycled aggregates reduces the fracture properties of the specimens.

(3) The replacement rate of recycled aggregate is 50%, waste fiber length is 19mm, and volume content is 0.12%, the fracture performance of the specimen is the best, followed by volume content is 0.08%. When waste fiber volume content is 0.16%, the fiber is difficult to disperse in the concrete and forms an area of weakness, thus the fracture properties of the specimen decrease.

References

1. Qiong Liu, Jianzhuang Xiao, Zhihui Sun. Cem. Concr. Res.,**41**:1050-1057 (2011)
2. B Mas, A Cladera, TD Olmo, et al., Const. Bldg. Mater., **27**:612-622(2012)
3. Jian-zhuang Xiao, *Recycled concrete* (Building Industry Press of China, 2008)
4. Jinghai Zhou, ShuJun Bai, Chen Bian, Appl. Mech.Mater., **513-517**:16-19(2014)
5. Haitang Zhu, Danying Gao, Zhanqiao Wang, J.Bldg. Struct., **31**:41-46(2010)
6. Kejin Wang, Shah S P, Pariya P, ACI Mater. J., (6):458-464(2001)
7. Jinghai Zhou, Lin Cheng, Wei Dong, Appl. Mech. Mater., **387**:105-109 (2013)
8. A. Domingo-Cabo, C. Lazaro, F. Lopez-Gayarre. Const. Bldg. Mater. **23**:2545-2553(2009)
9. Jinghai Zhou, Chaobi Zhang, Xianhong Meng. Adv. Mater. Res., **430-432**:1446-1449(2012)
10. Shilang Xu, Guofan Zhao. CHN Civ. Eng. J., **25**:32-38(1992).

MATEC Web of Conferences **63**, 03016 (2016) DOI: 10.1051/matecconf/20166303016
MMME 2016

Study on Seismic Behavior of Recycled Concrete Energy-efficient Homes Structure Wall

Lan DONG[1,a]

[1]*Liaoning Water Conservancy Vocational College,Shenyang ,China*

Abstract. The main point is to study the seismic behavior of the lattice type recycled concrete energy saving wall under low-cyclic loading,to provide the basis for the seismic performance of application of recycled concrete lattice wall in energy-saving residential structure.Design two walls with the same structure measures, include Lattice type recycled concrete wall and natural concrete wall, they are tested under low-cycle repetitive loading, compared failure mode and seismic performance in different reinforcement conditions of side column. The bearing capacity and ductility of recycled aggregate concrete are better than natural aggregate concrete, The stiffness degradation curves and the skeleton curves of the walls are basically the same, both of them have better seismic energy dissipation capacity.Lattice type concrete wall is good at seismic performance, recycled aggregate concrete is good at plastic deformation ability, it is advantageous to seismic energy dissipation of wall, it can be applied in energy efficient residential structure wall.

0 Introduction

Lattice type concrete wall is made by cement polystyrene membrane shelljoining together to form internal vertically and horizontally run-through channel , in the channel concrete pouring molding and finally remove external shells.It sets light, bearing, seismic and environmental protection in a body, in line with the requirements for construction of green environmental protection.Concrete waste that is left after the construction dismantled is crushed to form a recycled aggregate, finally to form recycled aggregate concrete, which is called recycled aggregate concrete.

Lattice regenerative concrete wall is different to traditional solid abdominal wall. The abdominal wall of the wall is a lattice type cross section, in a certain extent save the amount of concrete wall, the inner wall of the recycled aggregate also can effectively solve the problem of waste concrete buried and disposal problems, and is conducive to the protection of the ecological environment.Since the 80's of last century, countries in the world have carried out the research of energy saving residence, mainly related to the wall insulation, green building materials, structural reform and other aspects of.

Zhou Jinghai et al. study on the lattice type concrete wall with thermal insulation, the ductility, stiffness degradation and energy dissipation capacity of the wall under different axial compression ratio and high width ratio were observed. Xiao Jianzhuangstudies on bending performance of recycled concrete beams and the confinement of steel tube concrete columns axial compression test, study different recycled aggregate replacement rate, the deflection of the beam, cracking moment, ultimate moment, bending index regeneration, recycled concrete columns under axial load - strain curve relationship and transverse deformation coefficient change trend. The results show that under the same conditions, the deflection and the crack of limit are greater than the natural aggregate concrete beams and steel tube confined cylindrical lateral variation coefficient changed little, the axial compression ultimate bearing capacity of the recycled aggregate replacement rate of increase is on a downward trend.

70s of last century, Austria invented called "Rastra structure system of wall, which is the prototype of the lattice type concrete wall.In 1996, the University of California in the United States to carry out the plane bending test, the results show that the wall has a good toughness.In 1999, ICBO announced that the system can be used as concrete lintel, bearing wall structure and other permanent members.With the extensive research of the system in European and American countries, the system has been applied to civil buildings, high-rise structures, railway noise walls and so on.

At present, the domestic and foreign about lattice type recycled concrete research rarely involves regeneration and natural wall contrast analysis, I regard this as the starting point, I study on recycled concrete lattice type

[a]Corresponding author: nddl2005@163.com

DOI: 10.1051/matecconf/20166303016

wall seismic performance which is made of recycled concrete, through under the same conditions of natural concrete wall compared to explore the lattice structure of regenerative concrete wall for energy saving residential structure in the energy dissipation capacity.

1 test profiles

1.1 specimen making

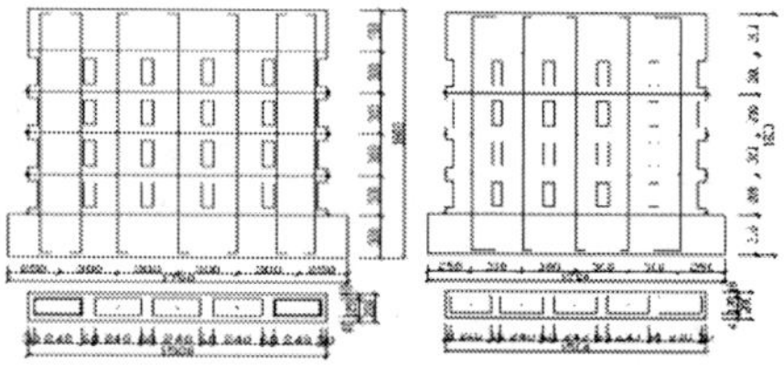

Figure 1. Side column strengthening and Side column is not strengthened

The test produced a total of 2 lattice structure type concrete wall, which recycled and natural concrete wall is a piece respectively for W1, W2. wall concrete strength for the C20, a top beam and a bottom beam concrete C30 and other parameters can be seen in Table 1.

Table 1 main parameters of test

No.	axial compression ratio	Side column stirrup	Horizontal reinforcem.	Side column reinforcement	Types of concrete
W 1	0.16	鉼 6@150	鉼 6@150	4 鉼 10	regenerate
W 2	0.16	鉼 6@150	鉼 6@150	1 鉼 6	natural

1.2 loading device and system

1.2.1The loading device is shown in Figure 2.

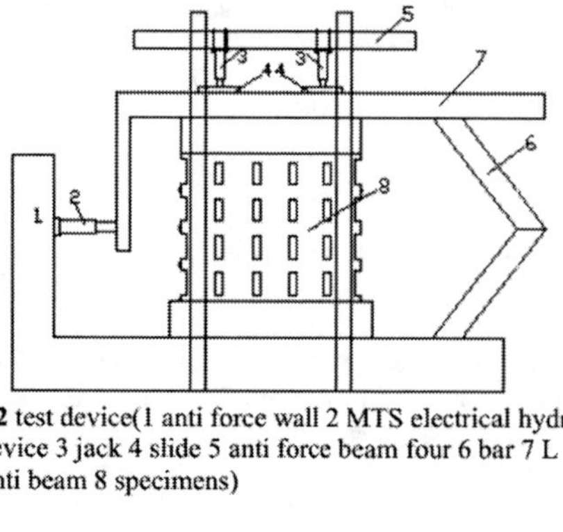

Figure 2 test device(1 anti force wall 2 MTS electrical hydraulic servo device 3 jack 4 slide 5 anti force beam four 6 bar 7 L shape anti beam 8 specimens)

1.2.2 loading system is shown in figure 3

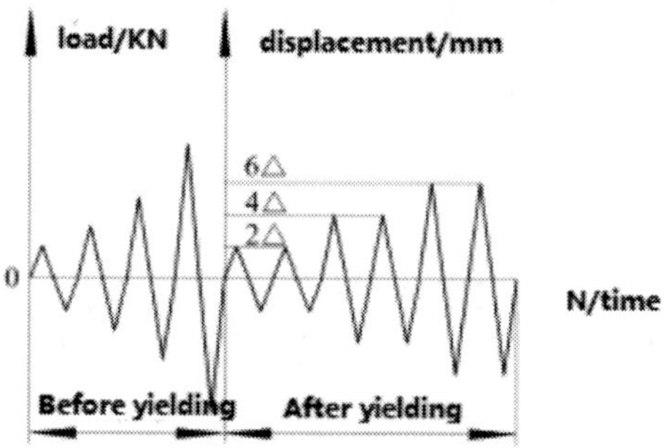

Figure 3.loading system

Provisions of the "rules for seismic method of buildings" (jgj101-96): first of all on the wall applied vertical load, and to maintain a constant, initial level load calculation load of 30%, to increase step by step, the range is 20KN.After yielding , change the displacement and control loading, record test pieces yielding displacement Δ, increase step by step 2 Δ, each load stage reciprocating 2 times.When the specimen deformation or bearing capacity limit down to 85%, as the damage, stop loading.

1.3 test parameters

The concrete performance in a large extent determines the wall seismic energy dissipation capacity. It is generally believed that the recycled aggregate concrete with many cracks. Therefore under reciprocating load inside the wall crack closure can produce large deformation, can improve the ductility of the wall, and to improve the seismic performance.By comparing the seismic energy dissipation capacity of concrete lattice wall with different performance under the same conditions, this paper studies the performance of recycled concrete in energy saving residential structure.

2 experimental phenomena

2.1 recycled concrete wall W1

Specimen W1 axis pressure ratio of 0.16, the initial load 110kN. When horizontal force increases to 250kN and wall body begins to local cracks appear, mainly concentrates in even the beam, as shown in Figure 4 (a) shows that test yield and yield displacement Δ for 3.54mm.Continue loading. The range is 2 Δ. With the increase of the displacement of the step by step, try a deformation increases, displacement control for the second delta, wall body incipient crack extends to the bottom beam, 4 Δ, at the bottom of the wall is inclined cracks significantly wider, and cross cutting; 6 Δ, even appears obvious X shape cracks, from Figure 4 (b) shows; 8 Δ, top beam in the lower part of the transverse through cracks, deformation increases sharply, as in Figure 4 (c); 10 Δ, concrete beam is crushed, is beginning to flake, as shown in Fig. 4 (d); 12 Δ, coupling beams of concrete at

DOI: 10.1051/matecconf/20166303016

the serious spalling, exposed steel bars, as Figure 4 (E) shows, bearing force is reduced to the limit bearing capacity of less than 85%, a announces destroyed.

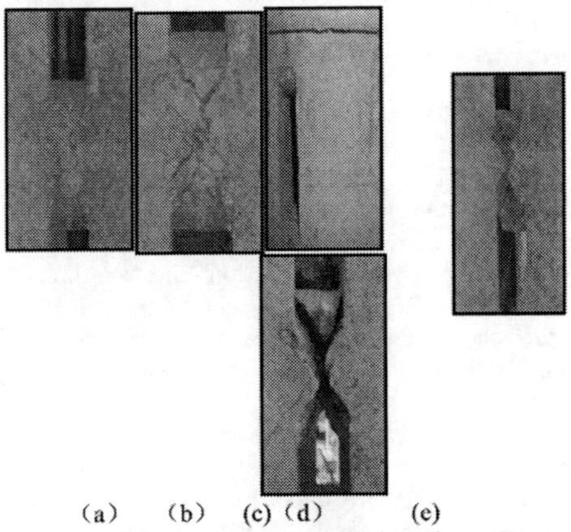

(a)　(b)　(c) (d)　(e)

Figure 4.failure process of specimen W1

2.2 natural concrete wall W2

Make the same axis pressure ratio and vertical load of W1 as W2, with the increasing of load level, W2 of the central beam area cracks appeared fine, as is shown in Figure 5 (a) , yield displacement Δ for 3.39mm.After the change of displacement loading displacement for 2 Δ, all beams have cracks, and gradually extends and shows a ring, as is shown in Figure 5 (b) ; 4Δ, even two ring cracks appears in the X type cracks, and to the outer ring extends out of the tall and slender slanting cracks, as is shown in Figure 5 (c) ; 6Δ, the top of the wall at the bottom of the wall are the vertical compression crack, wall body appears a large number of diagonal cracks, concrete flaking signs; 8Δ, the middle of the wall and the surrounding area have been crushed, spalling, as is shown in Figure 5 (d) that the specimen capacity decreases to the limit bearing capacity of 85%, and regards as failure , stop loading.

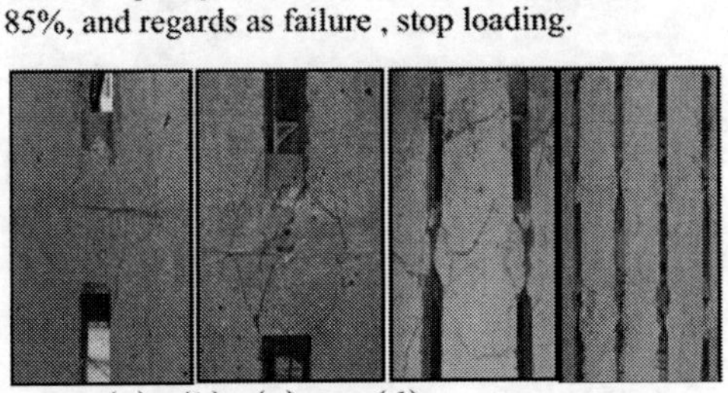

(a)　(b)　(c)　(d)

Figure 5.failure process of specimen W2

3 Results Analysis

3.1 hysteresis curve analysis

The hysteresis curve is the displacement-load curve of the specimen under cyclic loading, which can represent the bearing capacity, ductility and seismic energy dissipation capacity of the structure or component.Surrounded by the hysteresis loop area of the specimen under a cyclic load absorbs the energy, more full of hysteresis loop stronger of seismic energy dissipation ability.

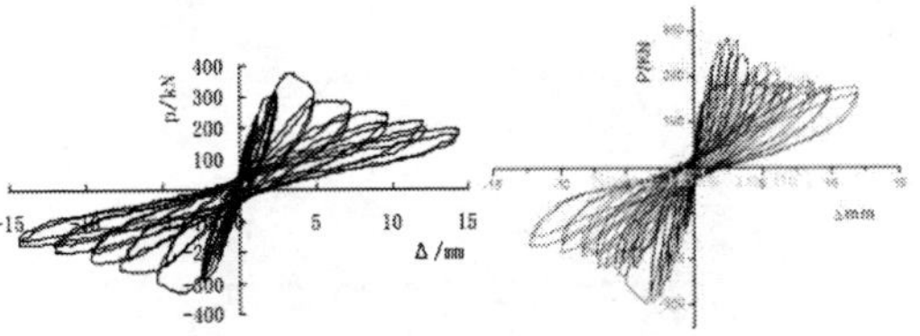

Figure 6. Hysteresis curve(W1 Hysteresis curve W2 Hysteresis curve)

Two specimens of hysteresis curves pinch the effect is obvious, indicating that lattice type concrete wall has better capability of plastic deformation, under cyclic loading it can absorb more energy, and transformed to deformation energy, to achieve energy consuming.Try a W1 of the hysteresis loop area is greater than that of W2, energy absorption also more, this is because recycled aggregate concrete internal large crack closure can produce a certain deformation and improve the ability of energy dissipation.

3.2 skeleton curve analysis

Each level lag back curve peak in turn connected curves become the skeleton curve, obviously characterizes structure or component of the cracking load and ultimate displacement of main characteristics, W1, W2 and skeleton curve as shown in Figure 7

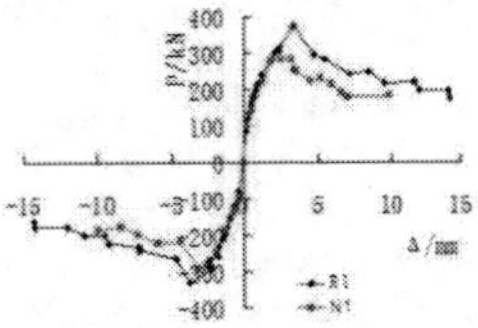

Figure 7.skeleton curve

The skeleton curve shows two bearing capacity of specimens decreasing slowly, it illustrates that the lattice type concrete wall has good ductility under cyclic loading.Comparison of W1 and W2 can be found: the skeleton curve of specimen by W1 skeleton curve of large area, that recycled aggregate concrete specimens of seismic energy dissipation capacity is better than natural concrete specimens, the bearing capacity of 375.6kN is greater than that of the natural concrete 286.7kN.

3.3 stiffness analysis

The stiffness degradation curve, as shown in figure 8.W1 specimen stiffness degradation is slightly less than W2,cause analysis:W1internal initial cracks are more than W2, so the lattice type recycled concrete wall reducing the same stiffness can produce larger displacement, can rely on increasing capacity of deformation to absorb more energy, is conducive to the seismic wall.

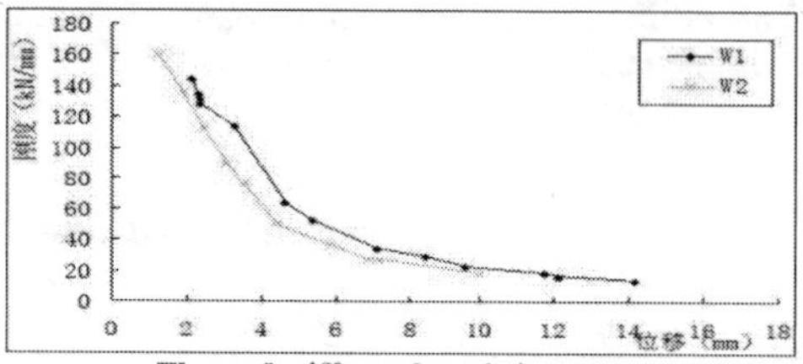

Figure 8.stiffness degradation curve

3.4 ductility analysis

The ductility coefficient is characterized by the parameters of structure or component plastic deformation ability, ductility coefficient, strain for curvature ductility coefficient, displacement ductility coefficient, the author calculates the displacement ductility coefficient, see table 2.

Table 2 ductility coefficient

Specimen	Load direction	Displacement			Ductility coefficient
		crack	yield	limit	
W1	Negative	0.94	3.79	5.37	1.42
	Positive	0.59	3.29	4.82	1.47
	mean value	0.77	3.54	5.10	1.45
W2	Negative	0.53	3.66	5.67	1.55
	Positive	0.56	3.12	3.64	1.17
	mean value	0.55	3.39	4.66	1.36

Table 2 shows the ductility coefficients of specimens W1 greater than W2, indicating the ability of plastic deformation is better than W2.Cause analysis: (1) closed cracks of recycled concrete, to a certain extent, improve the plastic deformation ability of specimens; (2) the study results show that the elastic modulus of recycled concrete is lower than that of natural concrete, so under the same stress condition, the deformation is greater than natural concrete specimens, indirectly increasing ductility.

3.5 energy consumption analysis

A specimen under cyclic loading, energy absorption and deformation energy ratio is called energy dissipation coefficient,is proportional to the hysteresis curve of the surrounding area.Specimens are subjected to cyclic loading and in elastic state, will produce some hysteresisdamping, the displacement is larger, the damping is greater, is conducive to the seismic energy component, equivalent viscous coefficient is the ratio of hysteresisdamping and 2 π.

Table 3 the energy dissipation coefficient and equivalent viscous coefficient

Sample number	W1	W2
E	0.389	0.401
h_e	0.059	0.063

Can be seen from table 3, try a W2 energy dissipation coefficient is 3.1% higher than W1, the equivalent viscous coefficient is 6.8% higher than that of W2, energy dissipation capacity of recycled concrete wall is slightly lower than that of ordinary concrete wall.

4 Conclusion

(1)Relative to the specimen of W2 , the energy dissipation coefficient and the equivalent viscous coefficient of W1 decreases, but the decrease slightly, the seismic system can be applied to the wall energy-saving residential structure;

(2)Under the same construction and reinforcement conditions, load bearing capacity and ductility of W1 are higher than those of W2, improve the bearing capacity of 31%, the ductility increases 7%, good capacity of energy dissipation;

(3)Two specimens of stiffness degradation are basically the same, are degraded slowly, illustrate the lattice type concrete wall under cyclic loading in plastic deformation capacity, good ductility, a re conducive to earthquake.

Acknowledgements

This work is financially supported byScientific research general project of Liaoning Provincial Department of Education(L2015310).

References

1. J.Z. Xiao,J. Yang, Y.J. Huang. *Journal of building structures*. **J,06**.92-98(2011).
2. X.F. Yin, L. Wang, S.F. Wang. Journal of Qingdao Technological University of China,**J.06**.34-38(2005)
3. W.L. Cao, J. Cheng, Y.B. Zhang et al. Journal of building structures,**J.36**.51-58(2015).
4. W.L. Cao, H. Ma, J.W. Zhang, et al. earthquake engineering and engineering vibration shear concrete shear wall,**J.04**.78-84(2015).
5. J.H. Zhou, J.W. Zhu, X.H.Meng.Journal of Shenyang Construction University,J.06.1095-1100(2010).

MATEC Web of Conferences 63, 03017 (2016)
MMME 2016 DOI: 10.1051/matecconf/20166303017

Experimental Study on Performance of SF_6+N_2 Mixed Gas Insulation

Gang LU[1,a], Zhen Xi SU[2], Ou XU[3], Liang ZHANG[1], Gong Yu LAN[1], Yang LIU[1], Hong Da ZHANG[1]

[1]*Heilongjiang Electric Power Science ResearchInstitude, Xiangjiang RoadNo.7,150030 Harbin,Heilongjiang Province, China*
[2]*Sulfur hexafluoride gas purification Laboratory of Analysis and Processing TechnologyofState Grid Corporation, Jinzhai Road No. 73, 230022 Hefei, Anhui Province, China*
[3]*China ENFI Engineering Corporation, Fuxing Road No. 12,100038 Beijing,China*

Abstract. This paper designed and constructed a 126kV GIS partial discharge test platform, performed frequency breakdown tests and lighting impulse tests at temperature of -30~20℃to SF_6/N_2 gas mixture. The present experiment studied frequency breakdown tests in different temperatue and slightly inhomogeneous electric field, obtained change in frequency breakdown voltage as a function of temperature; Analysed positive and negative lightning impluse test in gas mixture, in which SF_6 had a propotion of 30%~60%, in different temperature. Results indicated 50% probability breakdown voltage variation in positive lightning impluse tests, and in a low temperature environment, the breakdown voltage increased at negative lightning impulse, the polarity effects varied with condition of pure SF_6 gas. The experimental results showed implement of SF_6/N_2 mixed gas with 30% SF_6 supersedespure SF_6 gas, could effectively reduce the consume of SF_6, and remit greenhouse efect.

1 Introduction

High-voltage gasinsulated switchgear (GIS) is a key equipement in high-voltage transmission system which provides controlling and protectingfuncions under normal operation and fault conditions. It seals high-voltage components, such as circuit breakers, disconnectors, grounding switches, in a metal pipe, uses SF_6 as insulating medium[1]. SF_6 has a strong adsorption ability of electrons, the strong electronegative indicates its excellent properties of insulation and arc extinguishing[2,3], another feature of the SF_6 gas is having a high thermal conductivity[4]. Till now, SF_6 gas is the best insulation and arc extinguishing medium. Base on its excellent properties of insulation and arc extinguishing, there excists a widespread use of SF_6 in power industry including gas-insulated transformers (GIT), transformers, cables (GIC), line (GIL) and combination units (GIS). According to statistics, 80% of SF_6 gas is used in power industry, but during practical application of SF_6, there are also some problems, mainly in the following aspects[6]:

1)SF_6 gas is very serious greenhouse gas, which is about 25,000 times that of CO_2; the degradation of SF_6 under atmosphere condition is very slow, takes about 3200 years [7];

2)SF_6 gas is sensitive to non-uniform field, in engineering applications, tiny conductive particles and roughness of electrode surface cause the non-uniformity, result in reduction of SF_6 insulating properties[8];

3)In alpine region, SF_6 in GIS has a problem of liquefaction and density reduction;

4)Due to the normal operation of the circuit breaker or an internal short, occurrence of arc causes SF_6 decomposes, the reaction produces toxic gases for instance SF_4, SOF_2;

5)When the SF_6 gas contains water, condensation will produce in the insulation material surface, leads to significantly decrease of insulating properties, SF_4 and other decomposition products hydrolysis into HF and SO_2 and other acidic substance, internal members are corrosived, result in operation failure[9];

6) SF_6 gas is in higher cost, there exists no natural SF_6 gas, SF_6 gas in industry is synthetic so far[10].

When reducing the use and emissions of SF_6 gas, the selection of alternative gas of SF_6, 3 main indicatiors are considered: the electric strength (E/N)tbm, greenhouse index (GWP) and liquefaction temperature(℃). Since no gas could completely replace SF_6 gas, many studies tend to use a mixture of SF_6 and other gases. The SF_6 gas mixture can solve the liquefaction in alpine region, hybrid gas-insulated manner can also spare expensive SF_6 gas, and mixed gas is less sensitive to small defects in the electrode surface. All these properties show a good prospect. Therefore, research of insulating properties of SF_6 gas mixture under low temperature circumstance has great significance to development of grid and environmental protection.

2 Researching methods of insulation properties of SF_6/N_2 mixed gas

[a]Corresponding author: 13163417178@163.com

SF_6/N_2 mixed gas insulation theoretical foundation and main research methods includes: breakdown characteristics experiment, studies of discharge mechanism of electron avalanche stage, in which including Steady State Townsend (SST) and Pulse Townsend (PT), simulation methods, Boltzmann equation analysis and Monte Carlo method.

Breakdown experiment applies under certain experimental conditions: different electric field distribution and different voltage waveforms to directly obtain experimental electric strength of the insulating gas, but this method cannot access to the discharge mechanism of mixed insulating gas, it is difficult to propose effective method of inhibiting the breakdown. studies of discharge mechanism of electron avalanche stage, serve as methods complement to the experimental studies, could predict and control the gas insulation breakdown, theorically analyse the breakdown process. Simulation gas electron avalanche development process, mainly describe the relationship between collision cross-sectional and electron energy distribution in gas discharge. Boltzmann equation analysis is under the condition of known various cross sections, solving the microscopic parameters of the gas discharge. It obtains various cross sections and other appropriate data by experiment, by solving the Boltzmann equation, describes the gas electron avalanche development process with a series of elastic collision and inelastic collision, eventually derives electron energy distribution ionization coefficient, adhesion coefficient, drift velocity and diffusion coefficient and other parameters. The basic idea of Monte Carlo method is based on the probabilistic model on the basis of physical phenomena, and according to the unknown number of random variation, a random sample test is applied to calculate the approximate probability of occurrence of an event, uses the probabilistic models to obtain its estimated parameters. During the study of the mechanism of gas discharge, Monte Carlo method can simulate the behavior of particle dynamics and transport properties in a gas discharge process, simulate the occurence and development of electronic avalanche.

Currently discharge theories such as Townsend streamer discharge theory are incomplete. The breakdown of specific structure can not be accurately calculated. Engineering design and improvement of insulation structure often rely on experimental methods or variety of typical test data, discharge theory has important reference value of analysis and prediction of the insulating structure, and is helpful for analyse the problems of insulated structural. Combining the method of experimental and theoretical analysis to calculate the variety of insulation in different ratio of the mixed gas, supersede SF_6 gas with gas mixture of high electrical resistance performance, low greenhouse effect, low-temperature liquefied. Solves practical application problems of SF_6 alternative gases.

2.1 Frequencydielectric strength of SF_6 and SF_6/N_2 gas mixture

According to the experimental research literature and research institutions[2-4]. In frequency AC voltage, SF_6 gas gap electric strength (peak) can be represented by the following empirical formula:

$$Eb_{(SF6)}=65(10p)0.73(1)$$

SF_6 gap breakdown voltage can be approximatelyestimate:

$$Ub_{(SF6)}=65\eta d(10p)0.73 \quad (2)$$

Where: $Eb_{(SF6)}$as SF_6 gas gap frequency Dielectric strength, kV/cm; $Ub_{(SF6)}$as SF_6 gas gap frequency breakdown voltage, kV; Pas gas absolute pressure, MPa; ηas utilization factor of insulation, calculated, $\eta = 0.9$; $d$as the electrode spacing (cm).

For SF_6/N_2 gas mixture with SF_6 gas percentage of x%, $x \geq 10$, empirical formula of SF_6/N_2 gas mixture gap breakdown voltage present as:

$$Ub_{(SF6/N2)}= 65\eta d(10p)0.73x/100 \quad (3)$$

Where: $Ub_{(SF6/N2)}$as frequency gap breakdown voltage of SF_6/N_2 mixed gas,kV; $x$as the volume fraction of SF_6 gas in the mixed gas required $x \geq 10\%$.

2.2 Impactdielectric strength of SF_6 and SF_6/N_2 mixed gas

In the condition of operational negative impact voltage, the empirical formula of SF_6 gas gap $Eb_{(SF6)}$ (lower limit) is:

$$Eb_{(SF6)}=68(10p)0.73 \ (4)$$

By equation (2), (3), (4) Export gap electric strength (lower limit) $Eb_{(SF6/N2)}$ of SF_6/N_2 mixed gas is empirical formula (5) and gap breakdown voltage (lower limit) of SF_6/N_2 mixed gas (6)

$$E_{b(SF6/N2)}= 68(10p)0.73x0.18 \ (5)$$

$$Ub_{(SF6/N2)}=68\eta d\,(10p)0.73(x/100)0.18 \ (6)$$

Where: $Eb_{(SF6/N2)}$ gap impact resistance of SF_6/N_2 mixed gas, dielectric strength, kV/cm; $Ub_{(SF6/N2)}$ gap impulse voltage of SF_6/N_2 mixed gas, kV; $P$is gas absolute pressure, MPa;x is the volume fraction of SF_6 gas in the mixed gas required $x \geq 10\%$.

3Experimental study of power frequency withstand voltage and lightning impulse test of SF_6/N_2gas mixture under low temperature environment

3.1 Test Platform

Based on the structure and characteristics of switch-type device under condition of low temperature. Established a mixed gas test platform with 126kV GIS equipment carried out SF_6 gas mixture withstand voltage test at different temperatures. tested SF_6 gas mixture variety of mixing ratio, pressure and temperature and other factors along with frequency withstand voltage. 126kVGIS simulation test device is shown in Figure 3-1, there is an air chamber, including cables joints, can be set to a variety of types of electrodes, the discharge gap can be adjusted.

Figure 1.126kV GIS discharge test platform

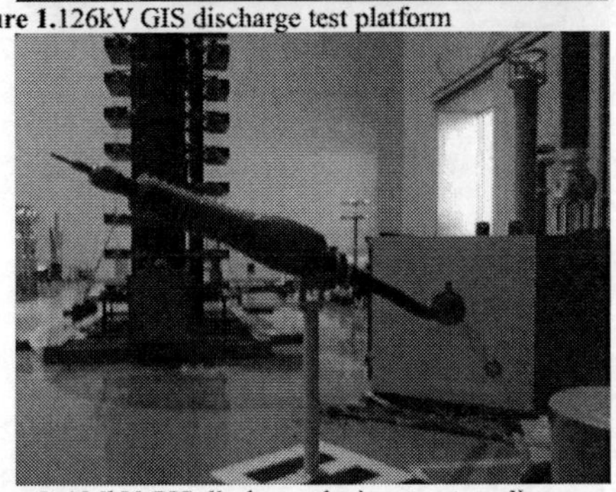

Figure 2. 126kV GIS discharge device structure diagram

3.2 Electrode size designand gap arrangement

According to experimental pre-discharge results of gaps in the air and SF_6 gas, set a bar-plate discharge gap in GIS simulation device. Objects are made of copper, plate electrode radius R = 21.5mm, the rod-shaped electrode radius R = 6mm, hemispherical electrode radius 21.5 mm, the discharge gap distance is adjustable between 1.0--15mm.

3.3 Temperature control

To realize temperature control, used low temperature test chamber, temperature, control range from -60℃ to room temperature, room size is 2500×1200×1500mm, temperature fluctuation ≤ ±0.5℃, temperature uniformity ≤±2℃. Cooling rate 0.7~1℃/min, heating rate of 2~3℃ / min.

3.4 Voltage Control

Frequency breakdown test used 70% step-up mode, i.e. First constantly increase voltage to 70% of estimated breakdown voltage, then increase voltage by 5% of starting voltage, at a speed of 4kV/s slowly rise until breakdown. Interval between 2 tests was 10 min, to ensure that the gas insulation strenthrecovered[11-17]. with a digital oscilloscope for recording and storing the voltage waveform, 50% of the breakdown voltage measured by lifting method to calculate the average of 50% probability breakdown voltage values.

Lightning impulse characteristic test, according to the requirements of GB/T16927.1 "High VoltageTestTechnology", test voltage started from 50% of estimated breakdown voltage, increased 5% in amplitude until the breakdown, detected full-wave lightning waves with the hatox of 1.2μs, and wave tail of 50μs. At the breakdown voltage, applied tests of both positive and negative full-wave, each ≥30 times the

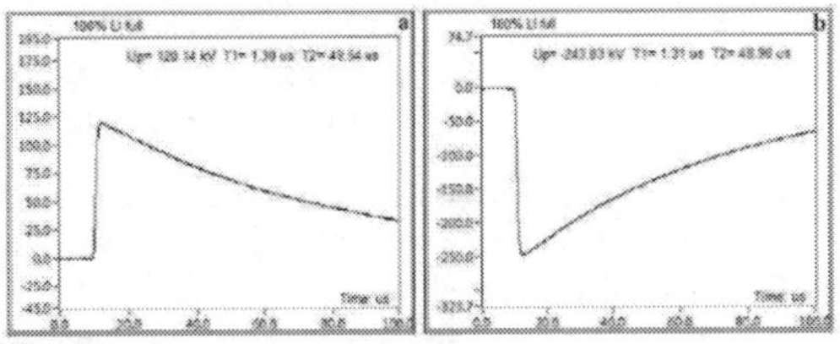

full-wave.

Figure 3. Standard lightning impulse voltage waveform

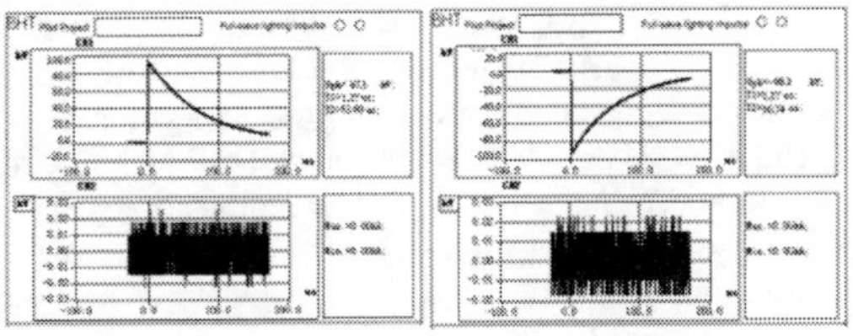

Figure 4. Positive and negative lightning impulse test waveform diagram

4 Test results and analysis

4.1 Impact of mixing ratio of SF_6 gas in the mixed gas to the frequency breakdown voltage

Test results of SF_6/N_2 mixed gas under different temperature conditions and slightly non-uniform electric field configuration showed that, in the temperature range of -30℃~20℃, the frequency breakdown voltage of mixed gas decline linearly as the temprature decreased. When the mixing ratio of SF_6 in the SF_6/N_2 mixed gas raised, its frequency breakdown voltage significantly increased.

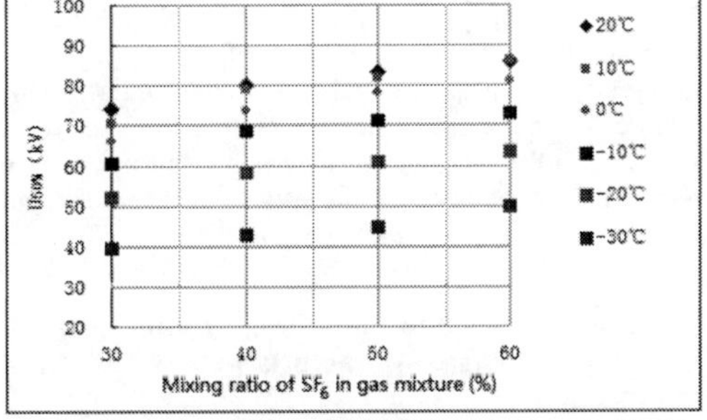

Figure 5. Variation of frequency breakdown voltage in SF_6/N_2 mixed gas at different temperatures under circumstance of slightly uneven frequency voltage

In the temperature range of -30℃ to -10℃, the frequency breakdown voltage of mixed gas increased as the the mixing ratio of SF_6 in the SF_6/N_2 mixed gas raised;

in the temperature range of 0℃ to 20℃, frequency breakdown voltage increase as the increase in the mixing ratio of SF_6 is not obvious, there existed a saturation trend. Frequency breakdown voltage and the mixing ratio of SF_6 showed positive correlation with saturation tendency in the normal temperature range, especially when the temperature is above 0℃, the saturation trend was particularly evident.

4.2 Impact of ambient temperature to the frequency breakdown voltage of mixed gas

The temprature of mixed gas decreased along with the ambient temperature, the frequency breakdown voltage declined. In the temperature range of -30℃ to -10℃, frequency breakdown voltage declined as the temperature decreased significantly; in the temperature range of -10℃ to 20℃, frequency breakdown voltage with decreasing temperature drop is not obvious; A positive correlation existed between frequency breakdown voltage and ambient temperature, in the normal temperature range trend slowed, especially at temperatures above 0 ℃ area, its decline is not obvious; In the temperature range of -30℃ to -10℃, mixed gas Power frequency withstand voltage greatly influenced by the ambient temperature decreased significantly.

4.3 Impact of mixing ratio of SF_6 gas in the mixed gas to 50% probability of lightning impulse breakdown voltage

The variation of 50% probability of lightning impulse breakdown voltage in mixed gas in different ambient temperature, the mixing ratio of SF_6 gas in the mixed gas varied within 30%~60%, showed in Figures 6 and 7, in the positive and negative lightning impulse, 50% probability of lightning impulse breakdown voltage of SF_6/N_2 mixed gas shows a downward trend at -30℃ to -10℃ range. As the mixing ratio of SF_6 gas in the mixed gas decreased, drop trend of 50% probability of lightning negative impulse breakdown voltage was not obvious, drop trend of 50% probability of lightning positive impulse breakdown voltage had a significant effect. When the mixing ratio of SF_6 gas was within 40%~60%, the impact of the mixing ratio of SF_6 gas to 50% probability of negative lightning impulse breakdown voltage was not clear. When the mixing ratio of SF_6 gas was under 40%, the impact of the mixing ratio of SF_6 gas to 50% probability of negative lightning impulse breakdown voltage was quite obvious.

4.4Impact of ambient temperature to the lightning impulse characteristics in the SF_6/N_2 gas mixture

Analysed Figures 6 and 7, As Ambient temperature went down, the 50% probability of lightning impulse breakdown voltage showed a decrease trend. The 50% probability of positive lightning impulse breakdown voltage decreased more distinct.

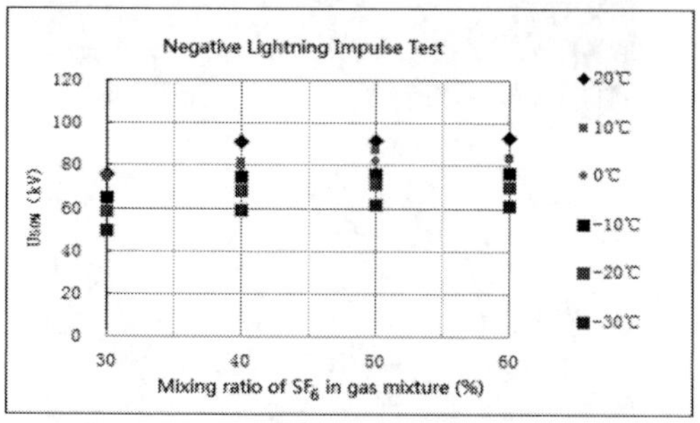

Figure 6. Variation of 50% probability of negative lightning impulse breakdown voltage in different mixing ratio of SF_6

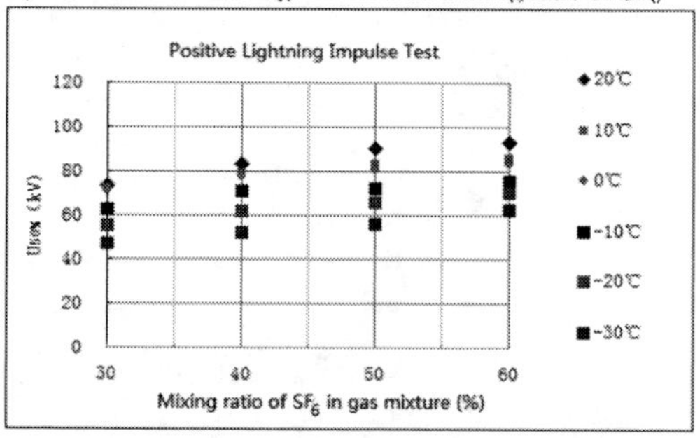

Figure 7. Variation of 50% probability of positive lightning impulse breakdown voltage in different mixing ratio of SF_6

In condition of same mixing ratio of SF_6, in the range of ambient temperature as -30~-10℃, positive impulse breakdown voltagewas lower than that in negative; in the range of ambient temperature was 0~20℃, negative impulse breakdown voltage was lower than that in positive. Under normal circumstances in pure SF_6 gas, since the test perfomed in a slightly nonuniform electric field, the breakdown voltage of negative polarity lightning impulse was sure to be lower than which of positive polarity lightning impulse at low temperature environment. Experimental results showed that the breakdown voltage of negative polarity lightning impulse was increased, its polar effect was different with pure SF_6 gas, polarity inversion occured at low temperature, the temperature range was about -30~-10℃.

5 Conclusion

This paper designed and constructed a 126kV GIS partial discharge test platform, performed frequency breakdown tests and lighting impulse tests at temperature of -30~20 ℃ to SF_6/N_2 gas mixture. By analysed the experimental results, received following conclusions:

In low temperature and slightly nonuniform electric field enviroment, the frequency breakdown voltage had a significant increase along with the augment of mixing ratio of SF_6, in the temperature range of -30℃ to -10℃; as in the temperature range of 0℃ to 20℃, the increase trend was unconspicuous. Frequency breakdown voltage and the mixing ratio of SF_6 showed positive correlation with saturation tendency in the normal temperature range, especially when the temperature was above 0℃, the saturation trend was particularly evident.

In the temperature range of -30℃ to -10℃, frequency breakdown voltage declined as the temperature decreases significantly; in the temperature range of -10℃ to 20℃, frequency breakdown voltage with decreasing temperature drop was not obvious. Mixed gas Power frequency withstand voltage greatly was influenced by the ambient temperature decreased significantly.

50% probability of breakdown voltage showed a decrease trend as the ambient temperature decreases under the positive or negative impulse. When the mixing ratio of $SF_6$was high, the decrease trend was obvious; when the mixing ratio of $SF_6$was low, the decrease trend was unconspicuous.

In the temperature range of -30℃ to -10℃, 50% probability of negative lightning impulse breakdown voltage decreased slowly, while probability of positive lightning impulse breakdown voltage decreased observably. In the condition of same mixing ratio of SF_6, polarity inversion occurs.

SF_6/N_2 mixed gas with 30% SF_6 supersede pure SF_6 gas was feasible in engineering applications. This replacement can effectively reduce the consume of SF_6, the greenhouse gas emissions and the insulating gas in GIS.

References

1. ChenQingguo,XiaoDengming,QiuYuchang Discharge characteristisc of SF_6/N_2 Gas Mixtures [J]Journal of Xi'an Jiaotong University2001，(04): 12-13
2. Christophorou LG,OlthoffJK,Van Brunt RJ.Sulfur Hexafluoride and Electric Power Industry[J].IEEE Electrical Insulation Magazine,1997，13(5):20-24
3. Hikita M,Ohtsuka S,kabe S,et a1.Insulation Characteristics of Gas Mixtures Including PerfluorocarbonGas[J].IEEE Transactions on Dielectrics and ElectricalInsulation,2008,15(4):1015-1022
4. Rigby M,Mfthle J,Miller B K et a1.History of Atmospheric SF,from 1973 to 2008[J].Atmospheric Chemistry and Physics,2010,10(21):10305-10320
5. Sato M.,Hoshina Y，Hanai M.，et a1. Dielectric Properties in High Pressure $C0_2$ and N_2 Gas using Coaxial CylindricM Electrode and on the Surface of Spacer[C],Transmission and Distribution Conference，2002,(2):1016-1019
6. J. S Rajan，K. Dwarakanath，N. Srinivasan，Performance of CycloaliPhatic Resin Spacers in SF_6-N_2 Gas Mixtures[J]，Proeeedings of 2001 International SymPosium on Eleetrical Insulating Materials，2001,19-22：375-378
7. Zhou J ， G.D.TheophiluS ， K.D.Srivastava Pre-breakdown Field Calculations for Charged Spacers in Compressed Gases under Impulse Voltages[C].Conference Record of the 1992 IEEE International Symposium on Electrical Insulation. Baltimore，1992:273-278
8. Moukengue Imano A,The Influence of the Dielectric Strength of theN_2/SF-Insulation by Conducting Particle on the Spacer Surface[J],Dielectrics and Electrical Insulation，2003,10(3):483-490
9. A.M.Imano,K.Feser,Flashover Behavior of Conducting Particle on the Spacer Surface in Compressed N_2,90 % N_2+10%SF_6 and SF_6 Under Lightning Impulse Stress[C]. Conference Record of the 2000 IEEE International Symposium on Electrical Insulation,Anaheim,2000:296-299
10. J.SundaraRajan , K.Dwarakanath,N.Srinivasan.Surface Flashover Strength of Different Insulating Materials in N_2-SF_6 Gas Mixtures Under Combined AC/DCVoltages[C].International Symposium on Electrical Insulating Materials,Himeji,2001:262.265
11. [20]B.B.Pati,K.C.Patra,M.S.Naidu.Variation of Surface Flashover Voltage in SF6 with Pressure[J],Dielectrics and Electrical Insulation,1994,l(6):1175-1179
12. J.Sundrar Rajan , K.Dwarakanath,S.N.Moorching.Investigations on Correlation Between Surface Flashover and Surface Roughness of Insulating Spacers[C].Conference on Electrical Insulation and Dielectric Phenomena.Atlanta,1 998:472-475
13. Jia zhijie,Liu Ping,Jianming,Yang Lin.Flashover Characteristics Along the EpoxyResin Insulator under DC Voltage in SF_6 Gas[C],Computer Distributed Control andIntelligent Environmental Monitoring,2012,338-342
14. S.Y.Li,K.D.Srivastava . Nanosecond Streak Photography of Discharges on SpacerSurface in Gases[J].IEEE Transactions on Dielectrics and Electrical Insulation,1995 , February.,Vol.2 No.1:114-120

 DOI: 10.1051/matecconf/20166303018

Research background, content and route of geological environment survey and evaluation for typical metal mines in Yunnan, China

Xian Feng CHENG [1,2a], Jun XU [2], Wu Fu QI [2], Qian Rui HUANG [2], Yi ZHENG [2], Ping HE [3]

[1] *China University of Geosciences, Beijing 100083, China;*
[2] *Yunnan Land and Resources Vocational College, Yunnan Kunming 652501, China;*
[3] *Kunming University, Yunnan Kunming 652501, China*

Abstract. Based on present condition analysis of survey and assessment of geological environment of mines in China, mining geo-environmental problems in Yunnan Province were discussed in this paper. Then, main content, critical technology and research approach of typical mines were given. The objective is to be helpful for technological means of survey and assessment of metal mines' geological environment.
Key words. Metal mines; Geological environment; Survey; assessment; Yunnan

1 Introduction

Mineral resources is the important material basis for the human survival and development.China is a great power with a long history of mining. However, the extensive development activity of mineral resources has caused many environmental geological problems in mines while protecting the national economy development needs and creating huge economic benefits. Some of the problems pose a threat to the mining area and the surrounding people's life and property safety and restrict the sustainable development of mining industry itself. The large-scale and high strength of mineral resources development has become the most active geological force and a "catalyst" to influence and change the geological environment of mining areas (figure 1). So, the environmental geological problems caused by the mineral resources exploitation need to be solved urgently. In the future, it is the inevitable choice to build resource-conserving, environmental friendly and green ecological mines.

As one of the main environmental geological research methods and approaches, mine geological environment survey and evaluation is six-in-one geological work integrating practicability, fundamentality, commonweal, perspectiveness (to formulate mine geological environment protection plans and prevention scheme), intersectionality (comprehensive application of multi-disciplinary) and systematicness (resource defacement, geologic hazards and environmental pollution). So, the work has been playing an increasing important role in the mine environment protection [4] .

In the paper, the work of mine geological environment survey and evaluation of the present situation in China was reviewed, and the mine environmental geological problems in Yunnan province was summarized. On the basis, it discussed the main contents, key technologies and research approaches of the typical metal mine geological environment investigation and evaluation in Yunnan. The purpose is to explore technical path of the typical metal mine geological environment survey and evaluation and provide reference for carrying out the next step work.

2. Current situation of mine geological environment survey and evaluation in China

It began to carry out the mine geological environment survey work systematically in the late 1990 s in China. Especially, a lot of work in mine environmental protection has been done since the foundation of the Ministry of Land and Resources, adhering to the principle of "protection in the development, development in the protection".

China Geological Survey deployed mine geological environment survey designedly in the great survey for the land and resources in 2000, marking the start of public-welfare and systematic mine geological environment survey in China. In 2002, China geological survey launched the national mine geological environment survey and evaluation program by the province as a unit, and it is the first time to know the real situation of geological environmental problems for all mines systematically and comprehensively. After five years of efforts, 113149 mines were investigated, involving 193 mine types. The survey found out the current situation of the mine geological environment and main environmental geological problems and the damages. It got A series of

[a] Corresponding author: 654145410@qq.com

achievements, such as "the mine geological environment investigation in China" by Jin-de Zhang, etc. (2009)[5] , "Survey and assessment on environmental geology problems of mine in northwest China" by You-ning Xu (2006) [6]and "Regional characteristics of mineral resources and mining environmental geological problems in Central south China" by Jin-hui Cai (2009)[7] .

The survey results was the basis to establish relevant laws, regulations and institution. In 2005, "the notice of the provincial planning and management work in mine environmental protection" was issued by the Ministry of Land and Resources the implementation. "Compiling code on the project of mining geo-environmental protection and integrated renovation" was issued in 2007, then the governance standard for mining industry was revised in 2011. The mine geological environment protection and management institution of recovery program was established in 2009. In April 2010, the national mine geological environment protection and treatment planning (2009-2015) was promulgated formally. In 2013, Xi'an center of China geological survey accomplished the first national comprehensive evaluation map of Chinese mine environment investigation and research-"the mine environmental geological map (1/4 million) and the specification of China".

By the end of 2010, all kinds of mines damaged land area of 3.868 million hectares cumulatively in China. Pumping and dewatering due to mining activities influenced groundwater aquifer of mining area about 4.96 square kilometers. Mine drainage added up to 69.41 billion tons in 2010. Mineral resources development produced a large amount of waste gas and dust and seriously affected the atmospheric environment around the mining area, of which most serious coal and sulfur chemical mines. The solid waste generated by mining activities added up to burden of 41.14 billion tons. According to the new data, the Chinese mining industry induced 21932 geological disasters by the end of 2011, caused a direct economic loss of about 33.53 billion yuan. Obviously, the mine geological environment protection and management work is still a long way in China.

The first round of the national mine geological environment survey and evaluation can't meet the needs of the national project planning and implementation, because of lack of time to carry out in accordance with the regional environmental geology survey specification, but the technical route-"previous information collection, primary to fill out a form investigation, combined with remote sensing investigation aided, field survey and verification of typical mine" [8].

One of the key problems was low accuracy in the investigation of mine geological environment status of "background"; the other was the lack of mine geological environment monitoring system, so the development of the mine environmental geological problems is not clear.Therefore, it is imperative to investigate major mines geological environment detailedly in China. Thus, 86 main mineral resources development zones (belts) (about 75 square kilometers) is carried out 1:50000 detailed investigation at present for current situation of geological environment, the main environmental geological problems and damage degree.

3 Profile of the mine environmental geological problems in Yunnan Province

3.1 The mine environmental geological problems

Known as the "kingdom of nonferrous metals", Yunnan province is rich in mineral resources.

9 kinds of mineral reserves rank first in China, such as lead, zinc, tin, phosphorus and so on. 12 kinds of mineral reserves are on the top three, such as platinum, germanium, diatomite, copper, antimony, nickel.

The potential economic value of proven reserves is about 3 trillion yuan, and mining industry is one of the five pillar industries in Yunnan province.

According to "the overall planning of the mineral resources in Yunnan province (2008-2015)", Yunnan will implement a number of mining development and major construction projects. But the planning also pointed out that one of the major problems existing in the development and utilization of the mine ecological environment is serious, with requirement of building a resource-conserving and environmental friendly mining industry.

Yunnan is one of provinces with the most complex and serious problems of the mine geological environment. According to 72 mining areas in 390 sampling evaluation on the quality of mine geological environment of mining enterprises, the results showed that the extremely serious accounted for 37.5%, the serious accounted for 44.4%, and the general accounted for 18.1%. Obviously, the mine geologic environment in Yunnan province was bad. Mine environmental status in whole province at present is deteriorating into worse stage, and the environmental situation is not optimistic. Summing up, the main problems in mines in Yunnan can be divided into 3 types and 15 sub-types, causing a direct economic loss of 307.3074 million yuan and 1203 deaths (table 1) [9]. It has become an important issue of the province's economic and social sustainable development and harmonious development of man and nature to to rationally utilize mineral resources and effectively protect the ecological and geological environment [10].

Table 1 Table of the main mine geological environment problems in Yunnan province

3.2 The important significance of geological environment survey and evaluation for typical metal mines

Mine geological environment survey and evaluation in Yunnan province synchronized with the national work, and part of the work is at the forefront. Since the 1980 s, some mining areas in Yunnan province have carried out geological environment investigation, such as " Yimen copper mine exploration and development influence of geological environment survey", "Jinshachang lead-zinc mine geological hazards survey in Yongshan Country",

DOI: 10.1051/matecconf/20166303018

etc. The objects of research in this period were mainly hydrogeology, engineering geology and geological hazards, while the mine geological environment survey and evaluation work basically focused on the individual points. Since 1997, the former Yunnan Bureau of Geology and Mineral Resources successively arranged two investigation projects, namely "Mine environmental geological investigation in Yuxi district " and "Mine environmental geology survey in Wenshan Prefecture" .The work adopted methods of "the surface general survey, major mines survey and key environmental indicators monitoring, semi-quantitative evaluation of mine geological environment quality". Some working experience were figured out.

As a component of the typical mine geological environment assessment projects in China, the systemic work in Yunnan began in 2002. According to the characteristic of the complex geological conditions and mining enterprises distributing widely, the investigation was on the basis of predecessors' information collection, filling out a form on the surface of census and key point investigation for the typical mines. In 2003, the mine geological environment survey and evaluation report of Yunnan province was formed, through the comprehensive analysis of the data and the present conditions assessment. It provided the basic data of protection of geological environment, mine environmental comprehensive renovation, mine ecosystem restoration and reconstruction, the implementation of the mine geological environment monitoring. Due to the limitation of conditions at that time, it lacked in-depth study of the inducing factors, formation mechanism of the mine geological environment problems and regional geological environment background on the sensitivity of the mining activities and restriction function. The mine geological environment quality evaluation is given priority to with expert evaluation method, so it is difficult to ensure objectivity of evaluation results and hard to meet the needs of the mining recovery treatment project design and implementation. Therefore, it is still necessary to select typical mines to carry out detailed geological environment survey and evaluation work.

4 Main content of geological environment survey and evaluation of representative metal mines in Yunnan

4.1 Geo-environment survey and sampling analysis of objective mines

In accordance with technical manuals and documents spirit of "Detailed rules for the implementation of the mine geological environment survey technical requirements in China (amended)", "The mine geological environment protection and management planning in China(2009-2015)" and so on, it will be carried out field investigation and remote sensing mine environmental status of Gejiu tin mine, Dongchuan copper mine, Lanping lead-zinc mine, Dahongshan iron-copper mine and Beiya gold mine in case study. Ore, soil, air, water and organism will be sampled to make pointed references for detection and analysis.

(1) Specific content of the survey:

1) To find out the basic information of objective mines basically;

2) To find out the background of objective mines geological environment roughly;

3) To find out the mine geological environment problem type, distribution, scale and damage degree in detail, including:

• Influence on and damage to land source and geomorphologic landscape by mine exploitation;

• Influence on and damage to water source, especially groundwater system by mine exploitation;

• Type, scale, loss and damage of mining geological hazards;

• Mine environmental pollution problems: pollution of soil and water, destruction of ecological resources by solid waste(spoil,tailings)stacking and waste water(pit water,beneficiation wastewater 、 dump leaching wastewater etc.)discharge;

4) To investigate measures and effect of preventing and controlling mine geological environment

• Land rehabilitation and environmental construction efficiency of mines

• Comprehensive treatment and efficiency of mine wastewater and waste residue pollution

• Mine geological disaster prevention and control efficiency

(2)Sampling and analysis

To sample ore, tail sand, soil, air, water and biological materials systematically in mining area, and to sample some representative specimen outside the mining area for comparing the samples as the background value. To select some key sample for full analysis. Soil, air and water would be analyzed according to the mine environmental assessment needs; For different indicators selected according to the characteristics of different mining special analysis.Sample analysis focuses on heavy metals and related to the concentration of organic pollutants.

4.2 Geological environment assessment of objective mines

On the basis of the geological environment investigation and sample analysis, to do special evaluation and comprehensive evaluation of the objective mines. To evaluate heavy metal pollution of mine using the total analysis method, chemical speciation, biological method and so on. To evaluate the geological environment in mines using weighted score index comprehensive evaluation method, fuzzy mathematics comprehensive evaluation method, grey situation judgment method, environmental pollution index method and other methods respectively, and then to compare their advantages and disadvantages. To elect some for targeted evaluation methods of different kinds of mines.

4.3 The basic characteristics,impact assessment, prevention and control measures of objective mines geological disasters

To summarize the essential features, damage characteristics and developmental rules on the basis of objective mines geological disasters survey. To select Dongchuan copper mine geological disasters carrying out researches of the environment impact assessment, dynamic changes and the prevention measures. To compare and analyze the forming conditions of the mine geological disasters according to the different types. To focus on the forming conditions of the mine debris flow.

Formation condition research of mine debris flow: to dissect Kafang debris flow in Gejiu mine area, Yinmingou and Heishangou debris flow in Dongchuan mine area, Nandagou debris flow in Lanping mine area in case study; according to the regional natural environment and geological conditions, to analyze and summarize typical mine debris flow gully watershed and channel characteristics using the method of remote sensing in combination with field investigation; according to field survey, to analyze and summarize typical mine debris flow gully sediment source condition, landform condition and water condition, etc., and combined with other aspects of judging the basic characteristics of the typical mine debris flow gully; according to the basic characteristics of the mine debris flow influence factors related to its starting, to establish the mine debris flow starting critical discrimination model; to analyze the starting mechanism of mine debris flow research finally .

4.4 Distribution characteristics, migration mechanism and control measures of heavy metals and organic pollutants in objective mines

On the basis of investigation and chemical pollution of heavy metals, to select 1-2 typical mines and find out distribution characteristics of heavy metals and organic pollutants in different environmental media (weathering profile typical ore deposits, mining pit, dump, tailings, surface water and groundwater, water sediment, sewage drainage sediment, river sediment, soil, air, etc.). to evaluate the pollution degree, the scope and the potential ecological harm of heavy metal in the soil of the different functional areas and water bodies using various methods.

5 Research strategy

It will be carried out investigation and evaluation of Gejiu tin mine, Dongchuan copper mine, Lanping lead-zinc mine, Dahongshan iron-copper mine and Beiya gold mine in case study. To collect the former data of objective mines adequately. On the basis of remote sensing interpretation results, to carried out the field investigation, verification and sampling, supplemented by discussion, questionnaire and interview, etc; to do 1:10000 ~ 1:50000 geological mapping in key locations. To analyze, summarize the survey data, and figure out the actual materials and present status. To evaluate geological environment of mines. To dissect 1-2 target mines, focusing on the study of environmental impact assessment, and environmental pollution repair exploratory research on key technologies(figure 2).

5.1 Data collection and analysis

① to collect and clear up the technical data, such as the evaluation index system and mine geological environment assessment related technical specifications; ② to collect the literature of science and technology domestic and overseas in recent 5 years; ③ to collect data of target mining natural geographical background, regional geological conditions, mineral resources and exploitation and utilization status, mining activities on the ecological environment influence, management information of mine geological environment restoration, the preliminary summary of the mine environmental geological problems.

5.2 Remote sensing

To deal with the image synthesis, integration, and enhance processing by getting different resolution of satellite remote sensing data and extract the mine geology environment information rapidly. To acquire the present situation of the mine geological environment, including quarry, tailings, dump, mine water, geological disasters (collapse, landslide, debris flow, ground subsidence, ground fissure, etc.), water and soil pollution, dust pollution, ecological destruction and so on. To make mine environmental geological problems of remote sensing interpretation map using visual interpretation, man-machine combination interpretation and computer automatic extraction according to the types and characteristics of remote sensing images.

5.3 Field investigation and verification

The contents include the basic situation of mine, occupation and destruction of land, land recovery governance, mining wastewater and effluent discharge, mine tailings solid waste, mine drainage effect of groundwater, the secondary geological disaster situation, etc. Adoption of 1/5 0000 topographic map as work base, combined with remote sensing interpretation chart. According to the above content item by item, to take photos of typical environment geological problems using a digital camera. If necessary, to attache sketches in accordance with the methods of field geological work. To verify remote sensing interpretation results by the chance of field investigation.

5.4 Questionnaire investigation, interview and discussion

The mine geological environment questionnaire form are distributed to the various mining enterprises, and filled in by the enterprise in accordance with the requirements for the questionnaire content one by one. Project professionals go into the first production line, and communicate with staff, and have an interview with the crowd near mines. Held by mining enterprises management and technical personnel to , to understand the status of mine geological environment and governance measures by mining enterprises management and technical personnel attending the symposium.

6 Epilog

Yunnan is a great province with mineral resources, biological resources and tourism resources, at the same time it is located in the upstream of six big rivers, namely Yangtze River, Zhujiang River, Lancangjiang River (Mekong River), Nujiang River (Salween River), Red River and Irrawaddy River. The benefit is not only in Yunnan province, but also producing good regional influence and international reputation by protecting the ecological environment of Yunnan.

As a prominent aspect of artificial environmental problems in Yunnan province, mine environmental geological problems should be solved urgently, and it has a long way for the protection and management work. It is an important basic, public welfare and social work to investigate and assess the mine geological environment. To make achievable mine environmental comprehensive renovation, mine ecosystem restoration and reconstruction scheme is beneficial to ease conflicts between tourism development and biological resources development and promote the sustainable, rapid and healthy development of the national economy.

Acknowledgments

Fund project: This study was supported by "Comprehensive utilization of resources and the evaluation of the mine environment survey of the typical metal mines in Yunnan province" (Grant No.[2013]1) and Department of Yunnan Education (Grant No.2015Y395).

References

1. Jun Wu. (1999). The mine environmental geological problems of Yunnan[J]. China geology, (4), 14-16.
2. Zhongli Tang et al. (2005). The mine geological environment problems and the countermeasures[J]. Journal of earth science and environment, 27 (2), 1-4.
3. You-ning Xu et al. (2011). Study on the difference of mine geo-environmental problem response to mineral resource exploitation :a case study about the mining areas in Tongguan and Daliuta of Shaanxi and Fuxin of Liaoning[J]. Journal of Earth Sciences and Environment, 33(1), 89-94.
4. You-ning Xu. (2008). Research status and prospects of mine geological environment survey[J]. Geological bulletin of China, 27 (8).
5. Jin-de Zhang, et al. (2009). The mine geological environment investigation in China [M]. Beijing: geology publishing house.
6. You-ning Xu. (2006) . Survey and assessment on environmental geology problems of mine in northwest China[M]. Beijing: geology publishing house.
7. Jin-hui CAI et al. (2009). Regional characteristics of mineral resources and mining environmental geological problems in Central south China [M]. Beijing: China university of geosciences press.
8. J D Zhang, D Q Zhang , L Tian.(2007). Methods of investigations and integrated assessments of the nationwide mine geo-environment[J]. Geological Bulletin of China, 26(2) : 136- 140
9. Yu Huang. (2007). Research on the mine environmental geological problems and countermeasures in Yunnan province [D]. Kunming: Kunming university of science and technology.
10. Jun Wu, et al. (2005). Mine environmental protection and renovation suggestions of Yunnan. China engineering science[J], 7, 397-400.

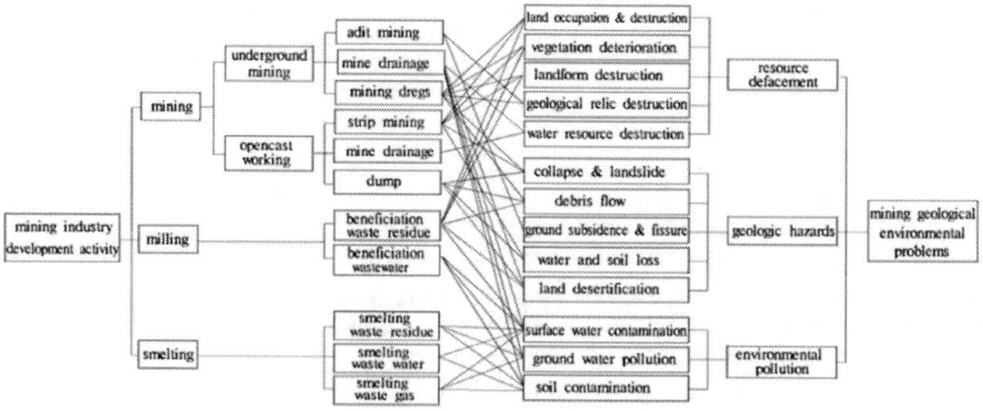

Figure 1. Schema of the relation between mine exploitation and environmental geological questions[1-3]

Table 1 Table of the main mine geological environment problems in Yunnan province

Types	number	land occupation and destruction(hm2)	economic losses (million yuan)	Deaths (person)
stope		30831.78		
waste pile		2994.31		
tailing pond		1128.22		
Surface water contamination	2000			
groundwater pollution	54		12	
soil contamination	56			
collapse	56	1038.84	8.06	135
landslide	171	1124.52	133.0228	594
debris flow	78	2491.8	69.6967	362
ground subsidence	169	1101.52	62.8733	30
ground fissure	179	1109.12	2.4476	
water inrush	39		11.077	30
springs unwatered	100			
total		42004.37	307.3074	1203

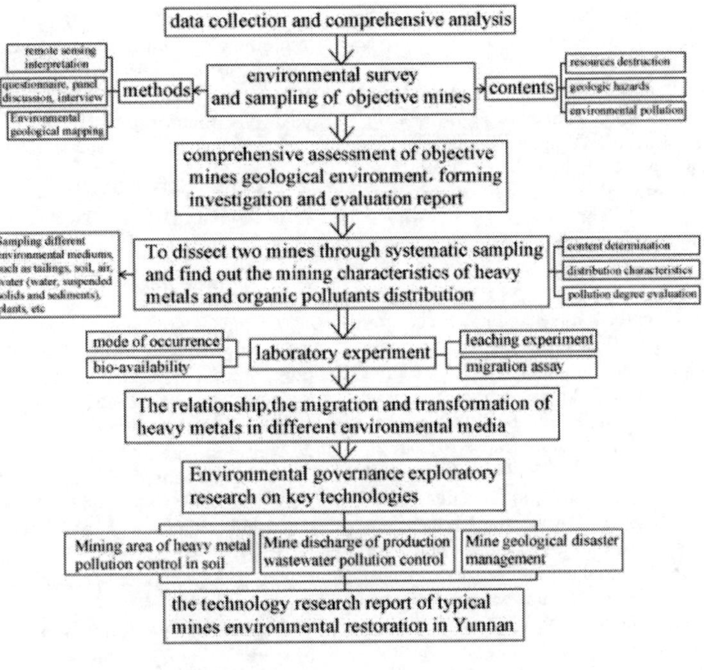

Figure 2. Technology Roadmapping of mine geological environment survey and assessment

 DOI: 10.1051/matecconf/20166303019

Research and Application on low molecular weight polymer fracturing fluid system

Xiao Chao WANG [a], Yuan Zhao JIA, Bin WANG, Peng WaNG, Zheng Dong XU, Gang Fei WU, Hui Juan LI
Huabei Oilfield Company Production Engineering Research Institute, 062552 He bei ,China

Abstract. In order to reduce the secondary damage of melon gum fracturing fluid for the low permeability alkali sensitivity reservoir, Optimized a kind of low molecular weight polymer fracturing fluid, and solve the technical problem of crosslinking time is not easy to control, fracturing fluid gel breaking difficult in low temperature. Through the study of the optimization of main additives, The fracturing fluid use low molecular highly active three yuan low molecular polymer as a thickener, which use of special synthetic process of polymerization with multiple active node. Its molecular weight is 200000 ~ 300000, water soluble speed is fast, site preparation easy, and thickening ability is strong. When the concentration is 0.2% ~ 0.6% in the fracturing fluid ,it can satisfy the requirement of fracturing operation. With organic zirconium and aldehyde compounds as crosslinking agent, The fracturing fluid have been achieved in good results in erlian uli of north China oilfield, ordos basin yishan slope of oil well and gas well fracturing.
Key words. Low permeability; Alkali sensitivity; Low molecular weight polymer; No residue; Fracturing fluid

1 Introduction

Wuliyasitai oilfield administrative is part of the Inner Mongolia Autonomous Region and there is 7km to the north of Xilin Gol League Dongwuzhumuqinqi. Its construction belongs to the erlian basin depression northeast side mani. North east to strip type distribution on the whole, Area of 3000 km2, Reservoir porosity is 8% ~ 10% on average, Permeability is 0.05 x 10-3 ~ 1.03 x 10-3 um2. Belongs to low porosity and low permeability reservoirs, At the same time have the feature such as water sensitivity and alkali sensitivity. Fracturing usually use gel fracturing fluid, But due to its basic crosslinking environment and the characteristics of large residue content. The damage of alkali sensitivity reservoir is bigger. Synthetic polymer has always been the research object of water-based fracturing fluid because of neutral, good solubility, no residue, etc. It is also the ideal substitutes of melon gel fracturing fluid, But due to the time of crosslinking polymer fracturing fluid is fast, not easy to control, not resistant to shear, Defects and poor heat resistance make the application very restricted. Experimental study of low molecular weight polymer fracturing fluid - hydrophobic low molecular weight polymer as the main association of the new residue-free fracturing fluid system, solve the above problems. Low molecular weight polymer fracturing fluid is the use of special synthetic polymerization process, it has plurality of nodes of low molecular weight active high active low-molecular polymer three yuan. Its molecular weight of 200,000 to 300,000, Only about 1/10 of a molecular weight of guar gum. It crosslinked under acidic conditions, and it has a strong tendency to self-polymerization association between molecules. In the high temperature polymer to form a similar spatial network structure by coupling side chain. While the rigidity characteristic side chain groups covalently effect multicore inside the bridging hydroxyl ion reduction, crosslinking increases steric hindrance, Gel formation with temperature, shear resistance and good features. In this paper, preferably a low, medium, high temperature and low molecular weight polymer fracturing fluid formulation series, And achieved good effect in Wuliyasitai Oilfield, Ordos Basin Yishan slope of oil and gas wells fracturing.

[a] Corresponding author: 155064851@qq.com

2 Low molecular weight polymer fracturing fluid laboratory study

2.1 Experimental part

Chemical reagents and materials including the preparation of fracturing fluid thickener SD-3, organic titanium cross-linking agent, ammonium persulfate breaker, cleanup additive BSA-101, anti-swelling clay stabilizer BYH-8. Natural core were taken from 53 wells, 29 wells too, yan 620 wells.

Test instruments including MARS rheometer, K12 automatic table interfacial tension instrument, XB220A electronic balance, 30 to 70 high-speed mixing device, CL10 high-speed centrifuge, BINDER heated blast drying box, High temperature and high pressure core flow meter.

Fracturing fluid formulation: Adding a certain amount of water in the 30-70 high-speed mixing vessel, adding a certain amount of polymer SD-3 under stirring, Preparation of the required experimental fracturing fluid.

Fracturing fluid temperature shear resistance properties, gel breaking performance, the content of the residue was determined and the extent of fracturing fluid core damage evaluation: The method according to the oil and gas industry standards SY / T 5107-2005 "water-based fracturing fluid performance evaluation methods" provisions.

Anti-expansion properties of fracturing fluid measurement: Based oil and gas industry standards SY/T5762-1995 method "Fracturing clay stabilizer performance measurement method" provisions.

2.2 Results and discussion

DOI: 10.1051/matecconf/20166303019

2.2.1 Determination of water solubility and thickening properties

Put 0.5% low molecular thickener SD - 3in 400 ml water, Stirring was continued at room temperature for 25 ℃, 300r / min Condition, With six speed rotary viscosimeter test sol viscosity of the liquid in different time. The results showed that: Stir the 15 min, Viscosity can be up to 91.1% of the final viscosity. Low molecular weight polymer water soluble speed, liquor easily at the scene.

Table 1. The relation between low molecular weight polymer fracturing fluid viscosity of sol solution and stirring time.

Time (min)	5	10	15	20	30
Viscosit (mpa•s)	19.3	26.5	33.8	35.6	37.1

When the room temperature is 25 ℃, With water make different concentrations of low molecular weight polymer fracturing fluid, Under the condition of the speed of 300 r/min stir 15 min. Standing swelling 1h, Use six speed rotating measuring sol viscosity fluid viscosity.The results showed that: Low molecular weight polymer to the thickening ability of water, The concentration of 0.2% ~ 0.6% of the fracturing fluid can satisfy the requirement of fracturing operation.

Table 2. The relationship of Low molecular weight polymer fracturing fluid viscosity of sol solution and the concentration.

Concentration (%)	0.2	0.3	0.4	0.5	0.6	1.0
Viscosit (mpa • s)	16.8	22.5	27.3	32.6	40.5	68.2

2.2.2 Crosslinking performance

According to the SD - 3 molecules contain -CONH2 and -COO- perssad, it can be crosslinked by aldehyde, dialdehyde, six methylene tetramine and polyvalent metal ion Al^{3+}, Cr^{3+}, Zr^{4+}, Ti^{4+}. But research shows that a kind of crosslinking agent to form gel performance is not ideal, optimizing the organic zirconium and aldehyde compounds as crosslinking agent. The pH value of base fluid has a big influence on the crosslinking fracturing fluid properties and heat resistance performance. When the pH of base fluid within the scope of 5.5 to 6.2, The fracturing fluid not only has a certain delay crosslinking, but also has good heat resistance performance. The experimental data are shown in table 3.

Table 3. The effect of PH value on the cross-linking properties and heat resistance of the fracturing fluid

Regulators (%)	pH	Crosslinking time (s)	AV105℃ (mpa • s)
0.5	6.5	15~25	357
0.6	6.2	30~40	320
0.7	6.0	40~50	305
0.8	5.7	60~80	284
0.9	5.5	90~120	243
1.0	5.2	250~320	115
1.1	5.0	Uncrosslinked	

2.2.3 Temperature resistance to shear performance

Low molecular weight polymer fracturing fluid has good heat resistance shear resistance, Continuous shear 90 min under 105 ℃、170 s-1, Viscosity retention value is still high (greater than 100 mPa • s), It shows that this fracturing fluid shear resistance of heat-resistant performance is good, it can meet fracturing created fracture and proppant requirements.The results shown in Figure 1.

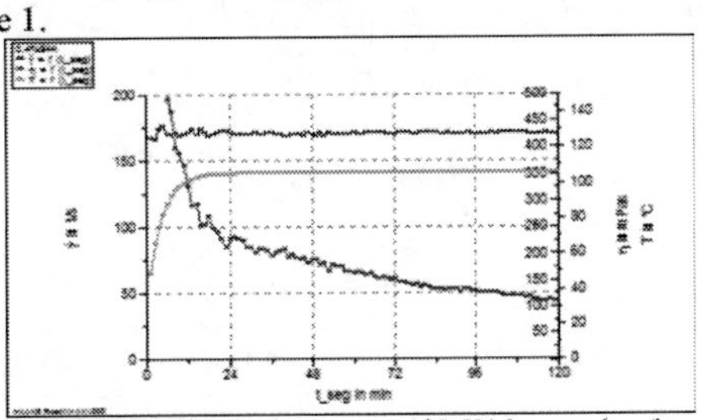

Figure 1. The rheological curve of 0.5% low molecular weight polymer fracturing fluid when 105 ℃,170 s-1

2.2.4 Gel breaking performance

The residue generated after gel breaking of gel fracturing fluid, it is difficult to discharge from the fracture, which will greatly reduce the fracture conductivity, and cause permanent damage to the formation. In order to reduce the formation fracturing and fracture conductivity damage, fracturing fluid gel breaking should as thoroughly as possible. And residue content in the gel breaking liquid should less as far as possible. Low molecular weight polymer fracturing fluid using oxidant ammonium persulfate gel water, Measured the gel breaking liquid viscosity of 0.5% low molecular weight polymer fracturing fluid were 3.5, 20.4 mPa • s. which use 0.03% ammonium persulfate solution respectively, 0.03% ammonium persulfate solution + reacted acid as gel breaker. Under the experimental conditions, 0.03% ammonium persulfate and residual acid (pH>4) can make 0.5% low molecular weight polymer fracturing fluid gel breaking completely. Gel breaking liquid viscosity meet the provisions of the viscosity is not more than 5 mPa • s requirement in SY/T 5674-2007 《General technical conditions for fracturing with plant gum》.

2.2.5 Performance Evaluation of core damage

Carried out a core damage experiment of tai 53 Wells and 29 well, yan 620 well with different fracturing fluid, The results showed that: Low molecular weight polymer fracturing fluid for core damage rate is less than 20%, Guar gum fracturing fluid for core damage greater than 40%, As shown in Table 4.

Table 4. Different fracturing fluid on the rate of core damage

Core Well No.	Fracturing fluid types	K_0/ 10^{-3} μm²	K_d/ 10^{-3} μm²	Damage rate / %
Tai 53	Hydroxypropyl guanidine gum	0.268	0.224	16.4
	Low molecular weight polymer	0.283	0.166	41.5
Tai 29	Hydroxypropyl guanidine gum	0.362	0.313	13.6
	Low molecular weight polymer	0.359	0.208	42.2
Yan 620	Hydroxypropyl guanidine gum	0.210	0.178	15.3
	Low molecular weight polymer	0.198	0.118	40.6

2.2.6 Comprehensive performance

The comprehensive performance of low molecular weight polymer fracturing fluid shown in table 5.

Table 5. Polymer fracturing fluid routine performance

Items	Conditions	Reference index（0.5% Melon glue）	Test results
Apparent	normal temperature	Without delamination, floc and floats	Without delaminatio n, floc and floats
ρ/（g/cm³）	normal temperature	1.00~1.02	1.01
PH	normal temperature	7~8	6~7
$AV_{Base\text{-}fluid}$ /mPa·s	normal temperature, 170s⁻¹, swelling 16h	63	48
$t_{Crosslinked}$/min	normal temperature	3~5	4
Gel breaking liquid surface tension /（mN/m）	normal temperature	≤32.0	30.0
Gel breaking liquid interfacial tension /（mN/m）	normal temperature	≤5.0	4.0
Residue content /mg/L	normal temperature	≤600	50
Demulsification rate /%	90℃	≥95.0	94.7

3 The field test and the analysis of the results

In erlian basin wulisitai oilfield tai 27 wellblock use gum fracturing hydraulic fracture 4 Wells, The early daily oil production is 40.3 t, the average single well oil production is 10.08 t. Using low molecular weight polymer fracturing hydraulic fracture 4 Wells, The early daily oil production is 48.9 t, the average single well oil production is 12.23 t. The oil production increased by 21.3% than the melon gum fracturing fluid. In yishan slope of ordos basin using low molecular weight polymer fracturing hydraulic fracture 4 Wells, output has increased by 2 times above. Shown in table 6. Using low molecular weight polymer fracturing hydraulic crack Y413 and Y620 gas Wells. Testing for production, Natural gas production were 1.2×10^4 m³ and 1.5×10^4m³, The gas production increased by 25% than using hydroxypropyl melon gum fracturing hydraulic. Shown in table 7.

Table 6. Yishan slope well fracturing construction

Well No.	Fracturing interval /m	Displacement /m³/min	Sand ratio /%	Sand proportion /m³	Before fracturing		After fracturing	
					Fluid producing /m³/d	Oil production /t/d	Fluid producing /m³/d	Oil production /t/d
P3155-3	508.6～525.2	1.8~2.2	26.3	20	2.46	0.81	6.75	2.96
P2171-7	726.8～737.9	1.9~2.1	27.6	16	1.69	0.59	5.86	2.85
P2181-5	761.5～769.5	1.5~1.8	28.1	12	1.36	0.35	4.89	1.56
P2834-4	738.6～741.8	2.1~2.3	26.6	19	2.85	0.68	7.16	2.81

Note：Construction pressures of P3155-3、P2171-7、P2181-5 and C2834-4 well is 18~21、19~25、17~22 and 23~26 MPa.

Table 7. Gas well fracturing operation table

Well No.	Perforation interval	Displacement /m³/min	Construction /MPa	Sand ratio /%	Sand proportion /m³	Infusion / m³	Flowback rate /%
Y413	3 402～3 408	3.2～3.6	41～47	24.6	35.2	289	81.9

	3 379～3 383	3.2～3.6	40～46	23.9	20.1	185	
Y620	3 211～3 215 3 222～3 225	3.3～3.5	43～49	24.5	35.0	293	81.3

4 Conclusion

（1）The SD-3 low molecular weight polymer thickener has good solubility in water and salt resistance, the thickening ability and the advantages of no residue, can form crosslinked gel under neutral acidic conditions.

（2）The low molecular weight polymer fracturing fluid system has characteristics of high viscosity, seam and carrying capacity is strong, good heat resistance and shear resistance, to support cracks and low formation damage, etc，able to meet the needs of low permeability, alkali-sensitive and water-sensitive reservoir fracturing.

（3）The fracturing fluid composition simple，easy to configure and easy application.

References

1. Zhou cheng-yu, Chen fu, Huang lei, et. Development of a high temperature polymer fracturing fluid shear [J]. Drilling and completion fluid，2008，25(1)：67-72.
2. Yang zhen-zhou, Chen mian, Xu yun, et. New synthetic polymer ultra-high temperature fracturing fluid system [J]. Drilling and completion fluid，2011，28(1)：49-5.
3. Lin bo, Liu tong-yi, Zhao zong-cong, et. Research and application of high-temperature fracturing fluid residue-free [J]. Drilling and completion fluid，2012，29(5)：70-73.

Influence of carbon nanofibers electric conductivity on infrared extinction capability

Tao SHEN[1,a], Wei Jun YANG, Bai He WANG, Min Li YAO

[1] *Xi'an Research Institute of High Technology, Shannxi, China*

Abstract. Using method of moment, the calculating model of electric current distributing and scattered electric field for carbon nanofibers is established in this paper. Based on it, the expression of absorption cross section, scattering cross section and extinction cross section are got. The relationship between infrared extinction cross section of carbon nanofiber and electrical conductivity is analyzed and calculated by using Matlab. The results show that the infrared extinction cross section increases with electrical conductivity increase and peak value appeares when wavelength is longer than the length of carbon nanofiber. The electrical conductivity on peak value is related to wavelength.

Keywords- carbon nanofibers; infrared extinction; absorption; scattering; extinction coefficient

1 INTRODUCTION

Smoke is a simple and effectual means of laser passive interference, and the infrared smoke materials should be taken on the better infrared extinction capability. Carbon nanofiber is a new kind of materials with good capability of electromagnetic wave absorption. It not only has the characteristic of common gas phase growth carbon fiber, but also has the characteristic of low density, high electrical conductivity, large specific surface and high specific intensity. In infrared band, the length and wavelength of carbon nanofiber is the same order. It can generate not only scattering attenuation, but also absorption attenuation. In modern warfare, carbon nanofiber has been a new kind of infrared smoke interference materials. To explore the applications of carbon nanofibers in passive interference by smoke screen, the integral equation of induced cueernt for carbon nanofibers is established based the electromagnetic field theory.

2 The integral equation for carbon nanofiber

To carbon nanofiber, the radius is much shorter than the length and incidence wavelength. The induced current inside carbon nanofiber is uniform distribution along the axisl direction and no component is in radial direction. According to the axial symmetry, the axisl direction current can replace the total induced current I. Suppose the length direction as Z axis, the scattered field produced by the axisl direction current in r point is following:

[a] Corresponding author: luckyshentao@163.com

$$E(r)=-jwu\int[1+\frac{1}{k^2}\nabla^2]I(z')\times G(r,z')dz' \quad (1)$$

$G(r,z')=\dfrac{\exp(-jk\,|r-z'|)}{4\pi\,|r-z'|}$ is Green function.

The carbon nanofiber produces the conduction current by field: $I_1=\pi a^2\sigma E_{nei}$, E_{nei} is the internal field, σ is carbon nanofiber conductivity; the dielectric property of the materials can also produce polarization current[1] $I_2=j\pi a^2\omega(\varepsilon-\varepsilon_0)E_{nei}$, ε is the dielectric constant, then the total induced current is:

$$I(z)=I_1+I_2=\pi a^2[\sigma+j\omega(\varepsilon-\varepsilon_0)]E_{nei} \quad (2)$$

The relationship between the induced current and the internal field of the carbon nanofiber is:

$$E_{nei}=\frac{I(z)}{\pi a^2[\sigma+j\omega(\varepsilon-\varepsilon_0)]} \quad (3)$$

The field and internal field of the materials is the same along the Z axis.The field is composed by incidence field Ei and scattered field Es and the boundary condition is

$$E_{nei}=E_{iz}+E_{sz}|\text{surface} \quad (4)$$

(1)and(3) is substituted in (4)

$$\tau gI(z)+j\omega\mu\int(1+\frac{1}{k^2}\frac{\partial^2}{\partial z^2})I(z')G(z,z')dz'=E_{iz} \quad (5)$$

$\tau=\dfrac{1}{\pi a^2[\sigma+j\omega(\varepsilon-\varepsilon_0)]}$ Equation(5) is the carbon nanofiber integral equation. The method of moment(MOM) is used to solve the integral equation.

DOI: 10.1051/matecconf/20166303020

First, dividing L length carbon nanofiber into homogeneous N segment, each segment length is $\Delta l = L / N$, the whole integrating range is composed by N sub-interval.

$$I = \sum_{n=1}^{N} I_n P_n \qquad n = 1,2,\mathrm{K}\ N \tag{6}$$

$$P_n = \begin{cases} 1 & inLn \\ 0 & others \end{cases}$$

(6) is substituted in (5)

$$\tau g \sum_{n=1}^{N} I_n P_n + j\omega\mu \sum_{n=1}^{N} I_n \times \int_{Ln} [1 + \frac{1}{k^2}\frac{\partial^2}{\partial z^2}] G(z, z') dz' = E_{iz} \tag{7}$$

Select point matching weight function $\omega_m = \delta(z - z_m)$ to multiply formula(7),

$$\tau g \sum_{n=1}^{N} I_n \delta_{mn} + j\omega\mu \sum_{n=1}^{N} I_n \times \int_{Ln} [1 + \frac{1}{k^2}\frac{\partial^2}{\partial z_m^2}] G(z_m, z') dz' = E_{iz}(z_m) \tag{8}$$

The induced current distribution inside the carbon nanofiber can be obtained by solving the matrix equation above.

3 carbon nanofiber extinction cross section

In far field condition, According to formula (1), the carbon nanofiber scattered field is

$$E_s(r) = \frac{j\omega\mu_0}{4\pi}\frac{\exp(-jkr)}{r}\sin\theta g\hat{\theta} \times \int_{-L/2}^{L/2} I(z')\exp(jkz'\cos\theta)dz' \tag{9}$$

(6) is substituted in (10)

$$E_s(r) = \hat{\theta}\frac{\eta_0}{4\pi}\frac{\exp(-jkr)}{r} tg\theta g \sum_{n=1}^{N} I_n\{\exp(jkn\Delta l\cos\theta) - \exp[jk(n-1)\Delta l\cos\theta]\} \tag{10}$$

$$\eta_0 = (\mu_0 / \varepsilon_0)^{1/2}$$

To unit amplitude incidence wave, the scattered field can be obtained from above and the carbon nanofiber scattering cross section is

$$\sigma_s = \frac{1}{I_0}\int I_s ds = 2\pi \int_0^{\pi} |E_s|^2 r^2 \sin\theta d\theta \tag{11}$$

the carbon nanofiber absorption cross section is

$$\sigma_a = \int_V k_0 \varepsilon_r'' |E_{nei}|^2 dV = \frac{\sigma\sqrt{\frac{\mu_0}{\varepsilon_0}}}{\pi a^2 |\sigma + j\omega(\varepsilon - \varepsilon_0)|^2}\sum_{i=1}^{N} |I_n|^2 \Delta l \tag{12}$$

extinction cross section is composed by absorption cross section and scattering cross section, that is

$$\sigma_e = \sigma_a + \sigma_s \tag{13}$$

4 Calculation and discussion

The code of carbon nanofiber induced current, scattered field and extinction cross section are programmed based on Matlab. The relationship between infrared extinction cross section of carbon nanofiber and electrical conductivity is analyzed and calculated. The calculation conditions are:

$\varepsilon_r = 10, a = 50nm, L = 5\mu m, N = 50$.

The calculation result is as followed Fig.1.

The results show that when the wavelength is 2μm, extinction cross section increases fast with electrical conductivity increase.

When $\sigma < 6\times10^5 \Omega^{-1} m^{-1}$, carbon nanofiber absorbs the incidence wave more than the scattered wave. When $\sigma > 6\times10^5 \Omega^{-1} m^{-1}$, extinction effect is mainly the scattering effect, absorbing effect is weak. When the wavelength is 5μm, the extinction cross section peak value appeares near the electrical conductivity $\sigma = 10^6 \Omega^{-1} m^{-1}$. The absorbing effect is tapered with electrical conductivity increasing, the scattering effect is enhanced and the extinction cross section reduces little. When the electrical conductivity is near $10^7 \Omega^{-1} m^{-1}$, the extinction cross section keeps constant. When incidence wavelength is 14μm and 20μm, the extinction cross section has peak value. It is $3\times10^6 \Omega^{-1} m^{-1}$ and $8\times10^5 \Omega^{-1} m^{-1}$ accordingly. To 14μm wavelength, when $\sigma < 7.8\times10^6 \Omega^{-1} m^{-1}$, the extinction effect mainly is absorbing effect. To 20μm wavelength, when $\sigma < 2\times10^7 \Omega^{-1} m^{-1}$, the extinction effect mainly is scattering effect. So, the longer the wavelength is, the larger the electrical conductivity span of absorbing extinction is.

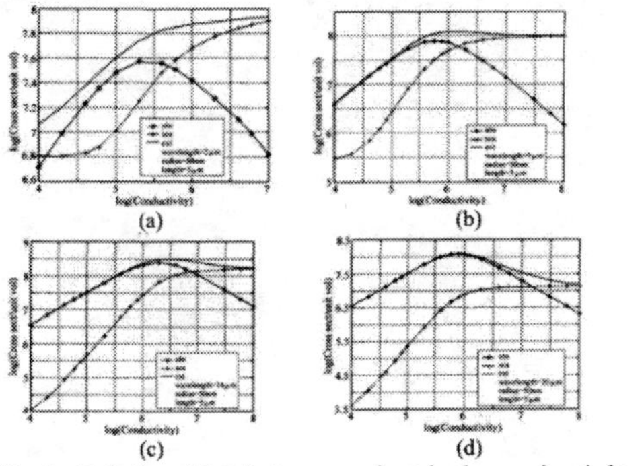

Fig.1 Relationship between electrical conductivity and extinction cross section

5 Conclusion

According to the electrical conduction and dielectric characters of carbon nanofiber, the electromagnetic field integral equation is built. The integral equation is solved and the expression of absorption cross section, scattering cross section and extinction cross section are got. The relationship between infrared extinction cross section of carbon nanofiber and electrical conductivity is analyzed and calculated based on Matlab. The results show that the infrared extinction cross section increases with electrical conductivity increase and peak value appeares when wavelength is longer than the length of carbon nanofiber. The electrical conductivity on peak value is related to wavelength. The analysis results can supply the theoretical basis for IR smoke materials in preparation and selection.

ACKNOWLEDGEMENTS

This work was in part supported by China Postdoctoral Science Foundation (2015M572698) and

Shaanxi Technology Committee Natural Science Basic Research Project(2014JM2-6108).

References

[1] DU Jin hong, SUN Chao, BAI Shuo, et al. Microwave electromagnetic characteristics of a microcoiled carbon fibers/paraff in wax composite in Ku band[J] . J Mater Res, 2002, 17(5) : 1232-1236.

[2] Lee B O, Woo W J, Park H S, et al. Infulence of aspect ratio and skin effect on EMI shielding of coating materials fabricated with carbon nanofiber/ PVDF[J]. J Mater Sci, 2002, 37: 1839.

[3] Xiao Chun Nie, Le Wei Li, Ning Yuan, et al. A fast analysis of electromagnetic scattering by arbitrarily shaped homogeneous dielectric object s[J] . Micro Opt Tech Lett, 2003, 38(1) : 30-35.

[4] Deitzel ,J M, Kleinmeyer ,J, Harris D, et al. The effect of processing variables on the morphology of electrospun nanofibers and textiles [J]. Polymer, 2011, 42(1): 261-272.

[5] Jun Kameoka, Reid Orth, Yanou Yang, et al. A scanning tip electrospinning source for deposition of oriented nanofibres[J]. Nanotechnology, 2003, 14(10): 1124-1129.

[6] Zhang Zhenhua, Peng Jingcui. The electronic structure and magnetic properties of chiral toroidal carbon nanotubes [J] . Act a Physica Sinica, 2011, 50(6) : 1150.

Synthesis and Properties of Nd-doped TiO_2 thin films by sol-gel dip-coating method

Jun WANG[a], Jia Le TANG, Yan Ling CHEN, Fang HE

Jiangxi key laboratory of surface engineering, Jiangxi Science and Technology Normal University, Nanchang, Jiangxi

Abstract TiO_2 thin films doped with neodymium(Nd) in the range from 0 at.% to 10at. % were prepared on glass substrates by sol-gel dip-coating method. Water contact angle on the Nd-doped TiO_2 films were measured by contact angle analyzer. The water contact angle measurement result showed that the hydrophily of the films improved when the Nd^{3+} concentration increased to 5 at.%. The photocatalytic activity of the undoped and Nd-doped TiO_2 thin films were investigated by the degradation of methylene blue under high-voltage mercury lamp. The results showed that 5 at.% Nd-doped TiO_2 thin film exhibited the highest photocatalytic efficiency.

1 Introduction

TiO_2 is a broad band gap (3.2 eV) semiconductor materials. TiO_2 have many excellent properties, such as non-toxicity, good chemical stability, good optical properties, photocatalyst, as well as biocompatibility(1-3). In recent years, the application of TiO_2 have been extend to dye sensitized solar, gas sensors, optical coatings, self-cleaning surface and biomedical materials(4-7).

One way to improve the properties of TiO_2 materials was coupling it with other semiconductors or metallic oxide. Fakhouri H. et al. (4) prepared TiO_2/TiN bilayer thin films. The photocatalytic properties were enhanced comparing signal layer TiO_2 films. Li C. et al.(8) found that the optical and photoelectrochemical properties were improved by using Ag_2O/TiO_2 double layer thin films. Another way to improve the properties of TiO_2 thin film was doping nonmetal or metal ions. Nonmetal (such as N(9),C(10)) and metal (such as Fe(5),Nb(3, 11, 12),Sm(13),Hg(14),Mn(15), et al.) ions doping TiO_2 thin films have been widely studied. The properties, especially optical and photocatalytic were improved.

Nd-doped TiO_2 films have been reported by many researchers. The photocatalytic properties were improved for Nd-doped TiO_2 thin film(16-19) and the mechanical properties were enhanced by Nd ions doping(20). However, the effect of Nd^{3+}-doping on the wettability and potocatalytic activity of TiO_2 thin films by sol-gel dip-coating method have not been elaborated.

Ion doping TiO_2 thin films can be prepared by CVD(1, 21), PVD(18, 22) and sol-gel(5) . Among them, sol-gel method have advantages such as ease of fabrication large area film,

[a] Corresponding author:wangjun460@qq.com

DOI: 10.1051/matecconf/20166303021

precise chemical composition control and reduced sintering temperature. In this paper, TiO_2 thin films doped with various Nd^{3+} ions (from 0 % to 10 at.%) were fabricated on glass substrates by sol-gel dip-coating. The effect of Nd^{3+} doping on the wettability and photocatalytic property of TiO_2 thin films were discussed in detail.

2 Experiment

The titania sols were prepared by mixed tetrabutyl titanate ($Ti(C_4H_9O)_4$, 98 wt.%), alcohol (CH_3CH_2OH, >99 wt.%) and concentrated nitric acid (HNO_3, 65~68 wt.%) with the mole concentration was 1:20:1.5. The detail process of preparing TiO_2 sols can be found elsewhere[23]. Neodymium was added to the TiO_2 sols in the form of neodymium oxide with five different mole ratio: 0 (undoped), 3, 5, 8, and 10 at.%. The undoped and Nd-doped sols were aged for 24h.

The absorbance spectra of the undoped and Nd-doped TiO_2 sols were shown in Figure1(a). Compared to the undoped TiO_2 sol, several absorbance peaks were found for Nd-doped TiO_2 sols. This may be ascribed to the 4f energy levels of neodymium ions. The absorption of photo can lead to electron transition between the energy levels, which displaying absorbance peaks. The intensity of the absorbance peak increased linear with Nd^{3+} concentration (shown in Figure 1(b)). It is consistent with Bouguer-Lambert-Beer law. The glasses substrates were cleaned in acetone, alcohol and distilled water respectively by supersonic purifier. The film deposition was carried out on glasses substrates by sol-gel dipping-withdrawing at 2000μm/s. The wet films were dried at 150°C for 30 mins and annealed at 500°C for 1h in air.

The wettability of the sol-gel films were tested by the water contact angle at room temperature. The photocatalytic activities of the undoped and Nd-doped TiO_2 thin films were evaluated by the degradation of methylene blue solution (2mg/L) under high-voltage mercury lamp. The photodegration were described by $(C_0\text{-}C)/C_0$, where C_0 and C are the initial absorbance and different reaction time absorbance at 654 nm.

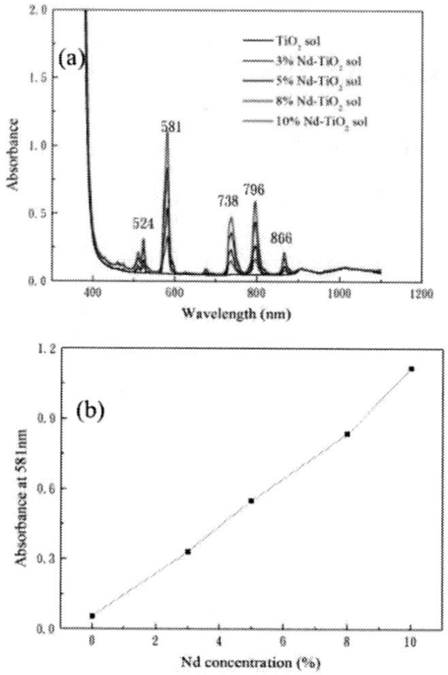

Figure 1 Absorbance spectra of undoped and Nd-doped TiO_2 sols (a) and the intensity of absorbance at 581 for different Nd^{3+} concentration (b)

3 Result and discussion

The wettability behavior of the films is important for photocatalytic activity, self-clean properties and biocompatibility. The water contact angle of uncoated, TiO_2 thin film coated and Nd-doped TiO_2 thin film coated glass substrates were measured. The contact angle images were exhibited in Figure 2. The contact angle of TiO_2 thin film (37.8°) was smaller than the uncoated glass substrates (53.4°), which indicating the TiO_2 thin film prepared by sol-gel method has hydrophily. The contact angle of 3 at.% Nd-doped TiO_2 thin film (36.7°) was similar to the undoped TiO_2 thin film. However, when the Nd^{3+} doping concentration exceeding 5

DOI: 10.1051/matecconf/20166303021

at. %, the contact angle decreased obviously (about 12 °-16 °). The improvement of the hydrophilic property for high content Nd^{3+} (>5 at.%) doping TiO_2 thin film may be ascribed to the varieties of the microstructure and chemical composition. The XRD results (will be published elsewhere) showed that the Nd^{3+} doping worsen the crystallization of TiO_2. The increasing of amorphous phase proportion and residual organics in Nd-doped TiO_2 thin film may enhance the hydrophilic property. Hydrophilic films can improve the self-clean and photocatalytic efficient. When Nd-doped TiO_2 films used as biomaterials, the improved hydrophilicity can promote the cell growth and proliferation, which is related to the biocompatibility[25].

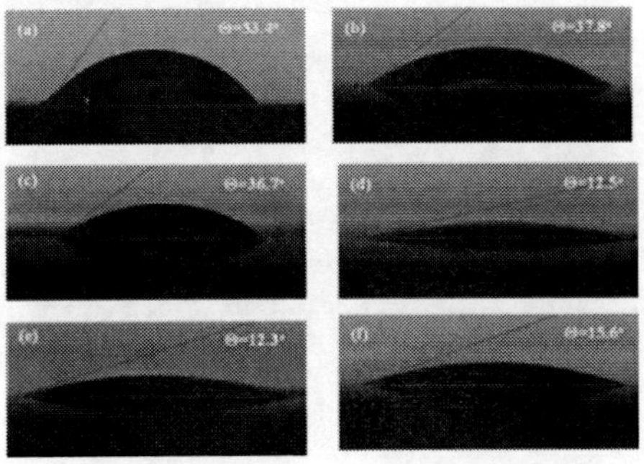

Figure 2 Water contact images of (a) glass substrate, (b) TiO_2 / glass, (c) 3 %Nd- TiO_2 / glass, (d) 5%Nd-TiO_2 / glass, (e)8%Nd- TiO_2 / glass and (f) 10 %Nd-TiO_2 / glass

The photocatalytic activity of TiO_2 thin films with different neodymium concentration was given in Figure 3. The degradation rate increase with the increasing of irradiation time. The effect of Nd^{3+} doping on the improvement of photocatalytic activity was obviously. As shown in Fig.6, the photocatalytic activity of 5 at.% Nd-doped TiO_2 films was higher than others under the same condition. Anatase TiO_2 and larger surface area are beneficial to photocatalytic activity. The increasing of Nd^{3+} ions doping decreased the grain size and reduced the proportion of anatase phase. In addition, good hydrophilicity also helps to improve the photocatalysis. The coordination of the reasons lead to 5 at.% Nd-doped TiO_2 thin film performed the best photocatalytic activity.

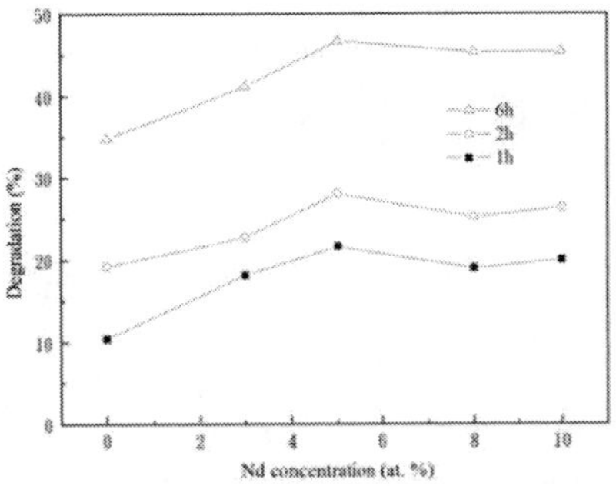

Figure 3 Photocatalytic activity of TiO_2 thin films with different neodymium concentration

4 Conclusion

Nd-doped TiO_2 films were successfully prepared by sol-gel dip-coating method. Wettability and photocatalysis of undoped and Nd-doped TiO_2 thin films were investigated. The concentration of Nd ions effect the wettability and photocatalysis . When the Nd^{3+} concentration exceeding 5 at.%, the hydrophilic of the films was improved obviously. The 5 at.% Nd-doped TiO_2 thin film performed the best photocatalytic activity under the same experiment condition.

Acknowledgements

This work was financially supported by the Natural Science Foundation of Jiangxi Province (20142BAB216016).

References

1. A. Sobczyk-Guzenda, S. Owczarek, H. Szymanowski, L. Volesky, B. Walkowiak, S. Miszczak, and M. Gazicki-Lipman, Ceram. Int. **41**, 7496 (2015).
2. M. Kopaczyńska, B. Sobieszczańska, A. Ulatowska-Jarża, I. Hołowacz, I. Buzalewicz, Ł.

Wasyluk, S. A. M. Tofail, D. Biały, M. Wawrzyńska, and H. Podbielska, Biocyb. and Biomed. Eng.**34**, 189 (2014).

3. A. V. Manole, M. Dobromir, M. Gîrtan, R. Mallet, G. Rusu, and D. Luca, Ceram. Int. **39**, 4771 (2013).

4. H. Fakhouri, F. Arefi-Khonsari, A. K. Jaiswal, and J. Pulpytel, Appl. Catal. A: Gen. **492**, 83 (2015).

5. H. Lin, T. Yang, C. Hsi, M. Wang, and K. Lee, Ceram. Int. **40**, 10633 (2014).

6. J. Zita, J. Krýsa, U. Černigoj, U. Lavrenčič-Štangar, J. Jirkovský, and J. Rathouský, Catal. Today. **161**, 29 (2011).

7. R. López and R. Gómez and S. Oros-Ruiz, Catal. Today. **166**, 159 (2011).

8. C. Li, J. H. Hsieh, J. C. Cheng, and C. C. Huang, Thin Solid Films, **570**, 436 (2014).

9. K. G. Grigorov, I. C. Oliveira, H. S. Maciel, M. Massi, M. S. Oliveira, J. Amorim, and C. A. Cunha, Surf. Sci. **605**, 775 (2011).

10. K. Rajendran and V. Senthil Kumar and K. Anitha Rani, Opt. – Int. J. for Lig. Electr. Opt.**125**, 1993 (2014).

11. D. Y. Lee, J. Park, Y. Kim, M. Lee, and N. Cho, Curr. Appl. Phys. **14**, 421 (2014).

12. G. Q. Wang, W. Lan, G. J. Han, Y. Wang, Q. Su, and X. Q. Liu, J. Alloys Compd. **509**, 4150 (2011).

13. D. de la Cruz, J. C. Arévalo, G. Torres, R. G. B. Margulis, C. Ornelas, and A. Aguilar-Elguézabal, Catal. Today. **166**, 152 (2011).

14. F. Abbas and R. Bensaha and H. Taroré, Comptes Rendus Chimie. **17**, 1176 (2014).

15. M. Z. Lin, H. Chen, W. F. Chen, A. Nakaruk, P. Koshy, and C. C. Sorrell, Int. J. Hydrogen Energy. **39**, 21500 (2014).

16. F. Miao, Z. Wang, B. Tao, J. Chu, and P. K. Chu, Electrochim. Acta. **112**, 32 (2013).

17. A. Bokare and M. Pai and A. A. Athawale, Sol. Energy. 91, 111 (2013).

18. M. Mazur, D. Wojcieszak, D. Kaczmarek, J. Domaradzki, G. Zatryb, J. Misiewicz, and J. Morgiel, Opt. Mater. **42**, 423 (2015).

19. D. SUN, K. WANG, Z. XU, and R. LI, Journal of Rare Earths. **33**, 491 (2015).

20. M. Mazur, J. Morgiel, D. Wojcieszak, D. Kaczmarek, and M. Kalisz, Surf. Coat. Technol. **270**, 57 (2015).

21. D. Pjević, T. Marinković, J. Savić, N. Bundaleski, M. Obradović, M. Milosavljević, and M. Kulik, Thin Solid Films. **593**, 612(2015).

22. Ş. Korkmaz, S. Pat, N. Ekem, M. Zafer Balbağ, and S. Temel, Vacuum. **86**, 1930 (2012).

Morphology tuning of CoMo catalysts: Effect of La and Fe as an additive

Pei YU, Ming KE[a], Qiang LIU , Xi Ming YAN

College of Science, China University of Petroleum-Beijing, 102249 Beijing, China

Abstract. Three catalysts CoMo/γ-Al_2O_3, CoMo-La/γ-Al_2O_3, CoMo-Fe/γ-Al_2O_3 with different morphologies were prepared for study the effect of introduction of La and Fe on morphology properties and hydrodesulfurization (HDS) activity. The morphology properties of catalysts were characterized by BET, XPS and HRTEM. In addition, the catalytic performances of catalysts were investigated. The results indicated that the introduction of La and Fe promoted the sulfidation of Mo species and made the average length and stacking layer numbers of MoS_2 slabs longer and higher. These effects could improve the activity of catalysts for HDS. The present results indicated that the catalyst modified by Fe had the best HDS activity with the optimal morphology among three catalysts.

1 Introduction

The production of clean gasoline has received considerable attention with the growing awareness of environmental protection. Hydrodesulfurization (HDS) as a key technique to produce clean gasoline is widely used in refineries. Therefore, numerous studies have focused on developing catalysts with high selective HDS performances [1, 2].

It is well known that the morphology properties of catalyst are closely related to their catalytic performances. Topsøe et al [3] reported that two kinds of CoMoS structure are formed during sulfuration process. Type Ⅰ CoMoS structure exhibits low HDS activity, which has strong interaction with support. Type Ⅱ CoMoS phase shows high HDS activity, which has weak interaction with support. So, more type Ⅱ CoMoS phase can be formed via weaking interaction of metal and support with maintaining metal dispersion [4]. According to the Brim-edge model [5], the MoS_2 phase exist brim and edge active sites. Both HDS and hydrogenation of olefin (HYDO) reaction are catalyzed by the brim sites, while only HDS reaction occurs at the edge sites. The adjustment of interaction of metal and support could control the morphology of metal sulfide with changing the ratio of brim sites to edge sites.

Many studies have focused on the nature of support which is one of major factors affecting the morphology of active metal phase. The supports incorporated with K, Mg, B, P and Zr has adjusted the morphology and dispersion of active metal phase, leading to the change of selective HDS performance [6, 7, 8, 9, 10].

2 Experimental

2.1 Materials

The supports modified by La and Fe were prepared by incipient wetness impregnation of γ-Al_2O_3 (specific surface area 315.4 m^2/g; volume of pores 0.55 ml/g; average pore diameter 6.96 nm). Then, the resulting supports dried at 100 °C for 6h and calcined at 550 °C for 4h. CoMo catalysts were prepared by co-impregnation of La/γ-Al_2O_3 and Fe/γ-Al_2O_3 using incipient wetness impregnation method, and then dried at 100 °C for 6h and calcined at 550 °C for 4h. The mass concentration of MoO_3 and CoO in each catalyst was 8% and 2%, La_2O_3 in CoMo-La/γ-Al_2O_3 was 1%, Fe_2O_3 in CoMo-Fe/γ-Al_2O_3 was 1%.

2.2 Methods

2.2.1 N_2 physisorption

N_2 adsorption-desorption experiments were performed on a Micromeritics ASAP 2400 equipment. The catalysts sample was evacuated at 300 °C under vacuum for 5 h.

2.2.2 HRTEM analysis

High-resolution transmission electron microscopy (HRTEM) images of the sulfided catalysts were obtained on a Philips Tecnai G_2 F_{20} Transmission Electron Microscope (Philips Co., Netherlands). The samples were

[a] Corresponding author: keming698@sohu.com

prepared by dropping the dispersed suspensions on carbon-coated copper grids.

2.2.3 XPS analysis

X-ray photoelectron spectroscopy (XPS) measurements of the sulfided catalysts were obtained using a Thermo Fisher K-Alpha instrument with monochromatic AlKα radiation. Before measurement, the freshly sulfided catalysts were put into cyclohexane to avoid contact with air.

2.2.4 Sulfidation conditions and catalysts evaluation

The catalytic of performance of catalysts were evaluated in a continuous flow fixed-bed micro-reactor. 5 ml samples were loaded in a stainless steel tube reactor with internal diameter of 10 mm. Before reaction, the catalysts were presulfided with 2 wt% CS_2 in petroleum ether at 310°C, 1.6 Mpa for 4 h. Then, the FCC gasoline was fed into reactor under the reaction conditions: 240°C, 1.6 MPa, liquid hourly volume space velocity (LHSV) of 4.0 h^{-1} and volumetric ratio of H_2 to oil of 300. The products were collected per 4 h, after a stabilization period of 12 h. The total sulfur content was determined by ANTEK-7000NS. The compositions of feed and products were analyzed by SP-3420A GC installed with HP-PONA capillary column.

3 Results and discussion

The textural properties of the catalysts are shown in Table 1. It shows that compared with CoMo/γ-Al_2O_3 catalysts, the surface area of CoMo-La/γ-Al_2O_3 have decreased and the mean pore diameter have increased significantly, which could attribute to the blockade of small pores by La_2O_3 particles [11]. And the CoMo-Fe/γ-Al_2O_3 has a little change on surface area and mean pore diameter, which indicate that the Fe_2O_3 particles were well dispersed on support. Fig.2 shows the pore size distribution of three catalysts. It can be seen that compared with CoMo/γ-Al_2O_3, the most probable diameter of the catalysts modified by La and Fe are little smaller.

Table1. Textural properties of the catalysts

Catalysts	S_g^a (m^2/g)	V_p^b (ml/g)	D_p^c (nm)
CoMo/γ-Al_2O_3	268.8	0.50	7.38
CoMo-La/γ-Al_2O_3	236.8	0.48	8.17
CoMo-Fe/γ-Al_2O_3	259.3	0.49	7.58

a BET surface area.
b Pore volume.
c Mean pore diameter

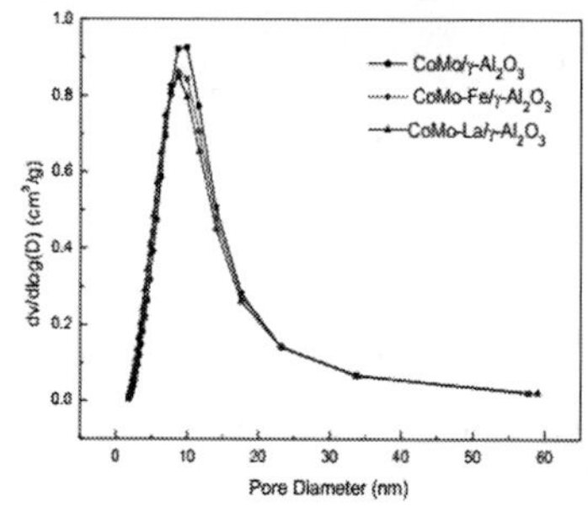

Figure1. Pore size distribution of catalysts

The Mo 3d XPS spectra and their deconvolutions of the three sulfided catalysts are listed in Fig.2. It shows that the spectra contained three Mo 3d doublets and the peak of S 2s about 225.9 eV [12]. According to literature [13], the binding energies at 232.3 eV and 235.4 eV were attributed to Mo^{6+}; the doublet with energies at 228.4 eV and 231.5 eV was related to Mo^{4+}; the last doublet with energies at 230.1 eV and 233.2 eV was attributed to Mo^{5+}. The sulfidation degree of Mo species, denoted as $Mo_{sulfidation}$, was calculated as $Mo_{sulfidation}=Mo^{4+}/(Mo^{4+}+Mo^{5+}+Mo^{6+})$ [14]. The results listed in Table 2 show that the $Mo_{sulfidation}$ of the three catalysts increasing as CoMo-Fe/γ-Al_2O_3>CoMo-La/γ-Al_2O_3Mo>CoMo/γ-Al_2O_3, demonstrating the promoting effect of the introduction of La and Ce. This can be concluded that the La or Fe incorporation weaken the interaction of active metal phase and support.

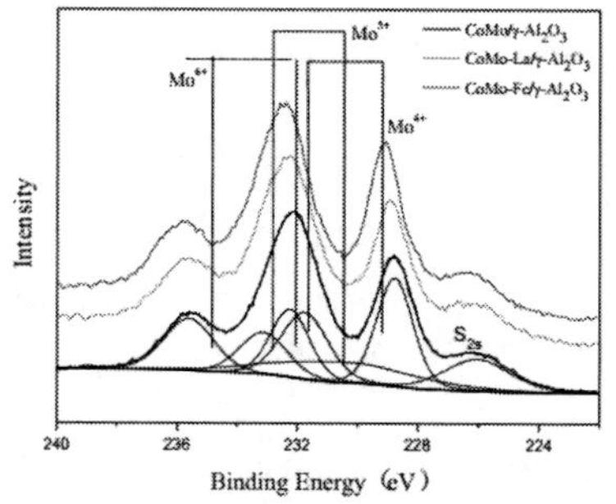

Figure2. Mo 3d XPS spectra of catalysts

Table2. Sulfidation degree of catalysts

Catalysts	$Mo_{sulfidation}$(%)
CoMo/γ-Al_2O_3	43.8
CoMo-La/γ-Al_2O_3	47.6
CoMo-Fe/γ-Al_2O_3	51.3

The representative HRTEM micrographs of three catalysts are provided in Fig.3. It could be seen that the MoS_2 slabs structure are well dispersed. The representative HRTEM micrographs of three catalysts are provided in Fig.3. It could be seen that the MoS_2 slabs

DOI: 10.1051/matecconf/20166303022

structure are well dispersed. Though the statistical analysis, about 400 MoS_2 slabs from 30 representative micrographs obtained from different part of each sulfided catalyst have been conducted to compare the average length ($\overline{L}$) and stacking layer numbers ($\overline{N}$).The average slab lengths ($\overline{L}$) and stacking layer numbers ($\overline{N}$) were calculated by the following formula[15],

$$\overline{L}(\overline{N}) = \frac{\sum_{i=1}^{n} x_i M_i}{\sum_{i=1}^{n} x_i} \quad (1)$$

where M_i is the slab length or stacking layer number of a MoS_2 unit, and x_i is the number of slabs or stacks in the determined range of length or stacking number.

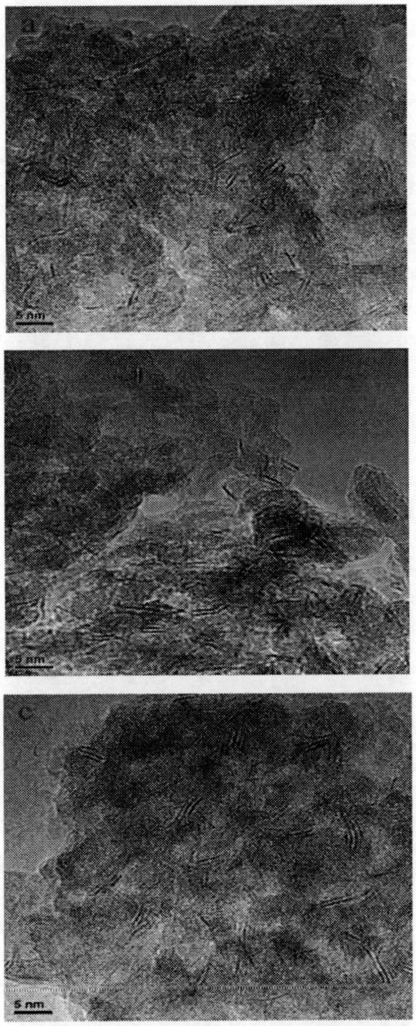

Figure3. HRTEM images of the sulfided catalysts. (a) $CoMo/\gamma\text{-}Al_2O_3$, (b) $CoMo\text{-}La/\gamma\text{-}Al_2O_3$, $CoMo\text{-}Fe/\gamma\text{-}Al_2O_3$

The statistical results are listed in Table 3. It is clear that the effects of La and Fe incorporation on $\overline{L}$ and $\overline{N}$. Compared with those in $CoMo/\gamma\text{-}Al_2O_3$, the MoS_2 slabs in $CoMo\text{-}La/\gamma\text{-}Al_2O_3$ have a longer average length and higher average stacking number. Moreover, $CoMo\text{-}Fe/\gamma\text{-}Al_2O_3$ has the longest average length and highest average stacking number of MoS_2 slabs. This indicated that the introduction of La or Fe weaken the interaction of metal and support, which could be consolidated with the results of the XPS characterization.

Table3. Average slab lengths and stacking layer numbers on the sulfided catalysts.

Catalysts	$\overline{L}$ (nm)	$\overline{N}$
$CoMo/\gamma\text{-}Al_2O_3$	2.34	2.01
$CoMo\text{-}La/\gamma\text{-}Al_2O_3$	2.49	2.21
$CoMo\text{-}Fe/\gamma\text{-}Al_2O_3$	2.58	2.34

The catalytic activities of three catalysts for HDS and HYDO were evaluated. The assessment results are given in Table 4. It is shown that both the $CoMo\text{-}La/\gamma\text{-}Al_2O_3$ and $CoMo\text{-}Fe/\gamma\text{-}Al_2O_3$ has the better HDS activity, but the HYDO activity of the former increased simultaneously. So, the $CoMo\text{-}Fe/\gamma\text{-}Al_2O_3$ catalyst has the best selective HDS activity. Contacted with the results of characterization, the $CoMo\text{-}Fe/\gamma\text{-}Al_2O_3$ catalyst has the maximum sulfidation degree of Mo species, longest average length and highest stacking layer numbers.

Table4. HDS and HYDO of the sulfided catalysts.

Catalysts	HDS%	HYDO%
$CoMo/\gamma\text{-}Al_2O_3$	94.3	16.1
$CoMo\text{-}La/\gamma\text{-}Al_2O_3$	95.4	16.8
$CoMo\text{-}Fe/\gamma\text{-}Al_2O_3$	96.7	16.2

4 Conclusions

The $CoMo/\gamma\text{-}Al_2O_3$ catalysts modified by La and Fe are prepared for selective HDS. These three catalysts haver been characterized by N_2 adsorption-desorption, XPS, HRTEM and tested with FCC gasoline. The sulfidation degree of Mo species was improved, with average slab lengths and stacking layer numbers of the MoS_2 slabs both increased. The $CoMo\text{-}Fe/\gamma\text{-}Al_2O_3$ catalyst has the best activity and optimal morphology among three catalysts.

References

1. S. Brunet, D. Mey, G. Pérot, C. Bouchy, F. Diehl. *Applied Catalysis A*: General, **278**(2005), 143-172.
2. E. Ito, J.A.R. van Veen. *Catalysis Today*, **116**(2006), 446-460.
3. H. Topsøe. *Applied Catalysis A*: *General*, **322**(2007), 3-8.

4. Y. Fan, G. Shi, H. Liu, X. Bao. *Applied Catalysis B: Environmental*, **91**(2009), 73-82
5. P.G. Moses, B. Hinnemann, H. Topsøe, J.K. Nørskov. *Journal of Catalysis*, **268**(2009), 201-208.
6. S. Damyanova, L. Petrov, P. Grange. *Applied Catalysis A: General*, **239**(2003), 241-252.
7. S. Shan, P. Yuan, W. Han, G. Shi, X. Bao. *Journal of Catalysis*, **330**(2015), 288-301
8. V. La Parola, G. Deganello, A.M. Venezia. *Applied Catalysis A: General*, **260**(2004), 237-247.
9. H. Li, M. Li, Y. Chu, F. Liu, H. Nie. *Fuel*, **116**(2014), 168-174.
10. J. Vakros, A. Lycourghiotis, G.A. Voyiatzis, A. Siokou, C. Kordulis. *Applied Catalysis B: Environmental*, **96**(2010), 496-507.
11. T.K.T. Ninh, D. Laurenti, E. Leclerc, M. Vrinat. *Applied Catalysis A: General*, **487**(2014), 210-218.
12. W. Chen, F. Maugé, J. van Gestel, H. Nie, D. Li, X. Long. *Journal of Catalysis*, **304**(2013), 47-62.
13. P.A. Nikulshin, D.I. Ishutenko, A.A. Mozhaev, K.I. Maslakov, A.A. Pimerzin. *Journal of Catalysis*, **312**(2014), 152-169.
14. E.J.M. Hensen, P.J. Kooyman, Y. van der Meer, A.M. van der Kraan, V.H.J. de Beer, J.A.R. van Veen, R.A. van Santen. *Journal of Catalysis*, **199**(2001), 224-235.
15. P.A. Nikulshin, V.A. Salnikov, A.V. Mozhaev, P.P. Minaev, V.M. Kogan, A.A. Pimerzin. *Journal of Catalysis*, **309**(2014), 386-396.

Research on mechanical properties and thermo-mechanical fatigue behaviours of Mg-12Gd-3Y-0.5Zr magnesium alloy at elevated temperatures

Xiao Ming YANG[1,a]

[1] *Jingchu University of Technology, JinMen City, HuBei province, china*

Abstract. In this paper, tensile and thermo-mechanical fatigue tests were carried out on Mg-12Gd-3Y-0.5Zr magnesium alloy at elevated temperatures. It shows that Mg-12Gd-3Y-0.5Zr magnesium alloy has good mehanical properties at high temperatures and thermo-mechanical fatigue mechanism is disscussed in this paper.

1 Introduction

Tensile strength and plasticity of magnesium alloys at room temperature are poor due to their close-packed hexagonal structure which leads to restriction on their application. Many measures have been adopted to improve their mechanical properties [1]. However, magnesium alloys are limited to some noncritical parts as a result of the restriction of fatigue strength and creep resistance at high temperatures [2]. With addition of rare earth elements, Mg-12Gd-3Y-0.5Zr alloy show high tensile strength and modified fatigue strength at elevated temperatures. In this paper, strengthening mechanism and fatigue mechanism of Mg-12Gd-3Y-0.5Zr magensium alloy were discussed.

2 Experimental

The ingots of Mg-12Gd-3Y-0.5Zr alloy(hereinafter referred to as GW123K) were rolled with a ratio of 25:1 at 400℃. The second phase whose constituent is approately $Mg_{24}(Gd,Y)_5$[3] was precipitated in microstructre in the process of extrusion.

Tensile specimens were cut by electric spark machining along the rolling direction with a gauge sections of 3mm×1.4mm×1mm^3. Tensile tests were carried out on an Instron 8801 testing machine equipped with a heater and temperature controlling system at room temperature, 130, 170, 200 and 235 °C and at a initial strain rate of $1\times10^{-3}s^{-1}$.

The fracture surface morphologies of the specimens were observed by scanning electron microscopy (SEM, Cambridge S360). TEM observation was carried out on TEM 2000FX II operating at 200 KV.

3 Experimental results and discussion

3.1 Tensile properties at different temperatures

Fig.1 shows optical micrograph of microstructure for GW123K. It can be seen that microstructure of GW123K is composed of α-Mg solid solution and Mg-riched precipitation which precipitaed along or inside grain. Grain size of as-rolled GW123K is about 10μm.

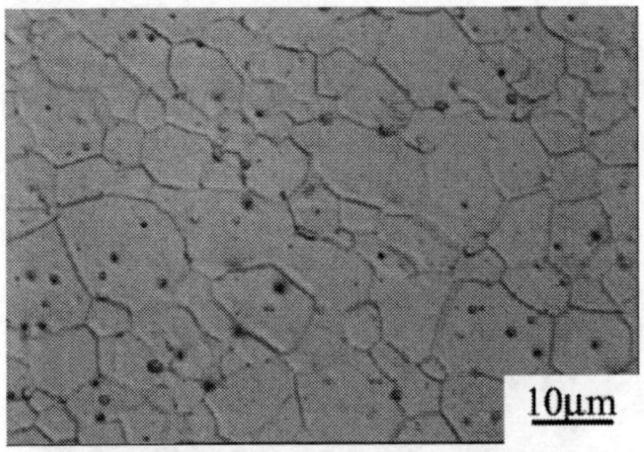

Figure 1. Optical micrograph of microstructure for GW123K.

Fig.2 and Fig.3 show temperature dependence of ultimate tensile strength and elongation-to-failure of GW123K, respectively. Tensile strength of GW123K at room temperature is about 405 Mpa which is well above other magnesium alloys. It is mainly caused by two reasons: on the one hand the grain size decreases to a little extent (Fig.1)after being rolled, on the other hand a large quantity of second precipitation phases impede

[a] Corresponding author: yang760507@163.com. This research was supported by university's scientific research project of JingChu University of Technology(No. 2013BK02)

grain boundary sliding. Futhermore, basal plane sliding can not be activated in microstructure of magnesium alloys at room temperature and twinning is main deformation mode which leads to their poor plasticity. Yttrium in microstructure of GW123K as an rare earth element addition plays a important role in grain refinement which results in improvement on strength and plasticity[4], and Gd-riched phases resulting from Gadolinium addition can effectively improve strength of GW123K by means of impeding grain boundary sliding[5]. Fig. 4 shows TEM micrograph of microstructures in as-rolled GW123k alloy after tensile test at 235 °C. Many slip bands can be observed from Fig. 4 that lead to a greatly decreasing strength and a greatly increasing plasticity for GW123K alloy at 235℃(as shown in Fig.2).

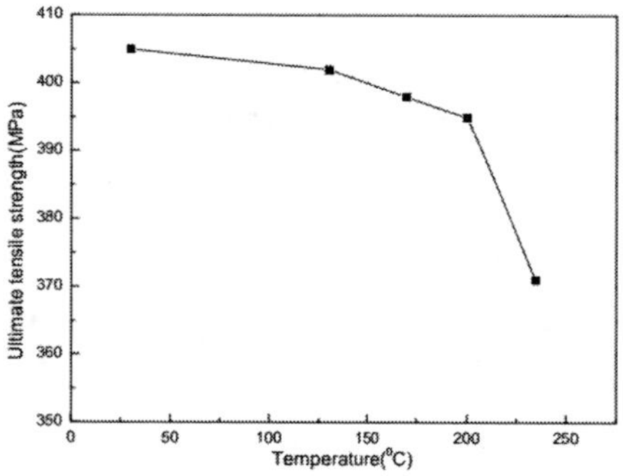

Figure 2. Temperature dependence of ultimate tensile strength of GW123K.

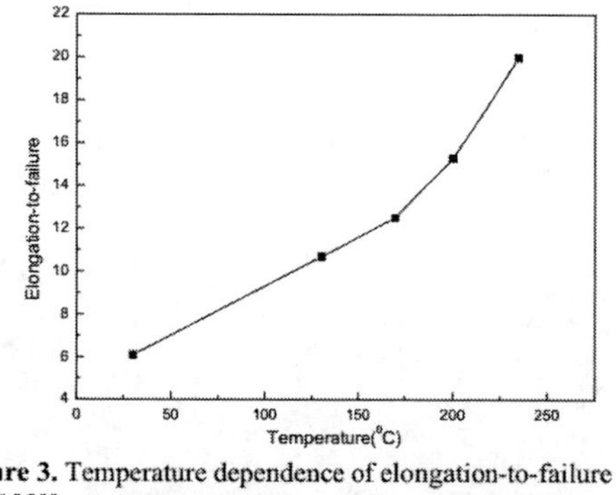

Figure 3. Temperature dependence of elongation-to-failure of GW123K.

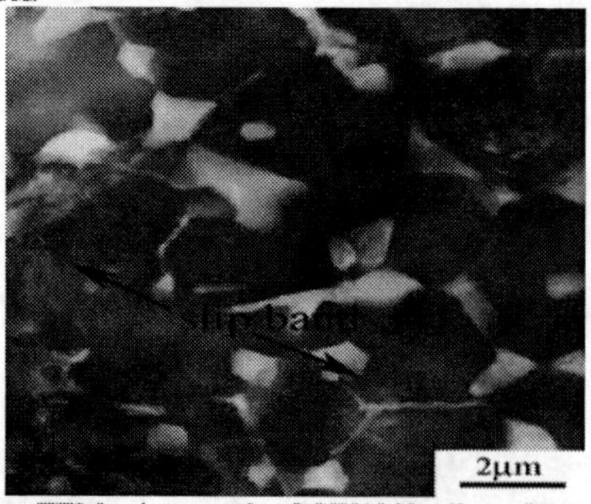

Figure 4. TEM micrograph of GW123k alloy after tensile test at 235 °C.

3.2 Thermo-mechanical fatigue behaviours

Fig.5 shows optical micrograph of GW123K tested by in-phase thermo-mechanical fatigue under stress amplitude 180Mpa at temperature cycle range from 150℃ to 250℃. It can be seen that average grain size of GW123K after thermo-mechanical fatigue test increases to 20μm, and grains are non-uniformly distributed. In the process of thermo-mechanical fatigue test, dynamic recrystallization led by temperature cycle and load loop results in formation of some grains with small size. However, the majority of grains in the microstructure become more and more coarse due to grain boundary diffusion and grain boundary migration at high temperatures. Coarse grains and non-uniform microstructure lead to descending of resistance to thermo-mechanical fatigue failure.

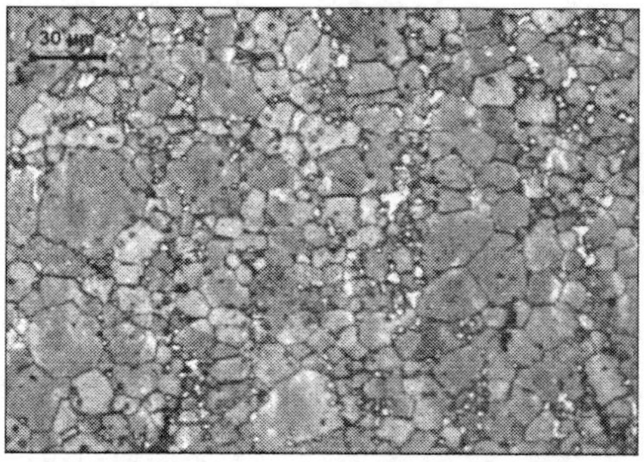

Figure 5. Optical micrograph of GW123K tested by in-phase thermo-mechanical fatigue under stress amplitude 180Mpa at temperature cycle range from 150℃ to 250℃.

Fig.6 shows hysteresis loops at 5 and100 cycles of GW123k alloy after in-phase thermo-mechanical fatigue test at the stress amplitude 180MPa. Hysteresis loop is charateristic of cyclic softening due to dynamic recrystallization at high temperatures.

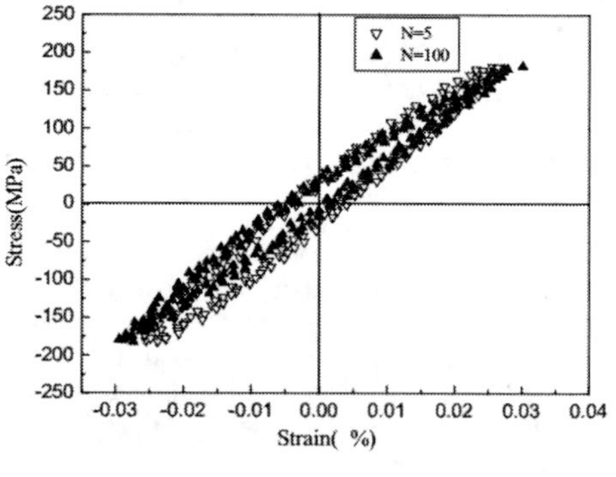

Figure 6. Hysteresis loops at 5 and100 cycles of GW123k alloy after in-phase thermo-mechanical fatigue test at the stress amplitude 180MPa.

Fig .7 shows SEM photographs of fatigue origin and analysis of fatigue crack initiate sites. It can be seen from Fig.7(a) that the overall fracture surface is very smooth and fatigue cracks initiate on the surface or subsurface. As shown in Fig.7(b) and (c), fatigue cracks initiate from oxide or slip bands led by cyclic slip. As a result of cyclic load under fatigue test, many slip bands become into being on the surface of specimens which will lead to stress concentration, at this moment fatigue cracks initiate from these slip bands. It can be concluded that fatigue crack initiating from slip bands at low testing temeratures is the main fatigue failure mechanism.With the increase of testing temperature, specimens are more prone to be oxidized, oxide inclusions will be main fatigue origins. In general, the main fatigue failure mechanism of GW123K under thermo-mechanical fatigue tests is oxide inclusion initiations, moreover, fatigue initiating from slip bands led by cyclic load plays a secondary role in fatigue failure.

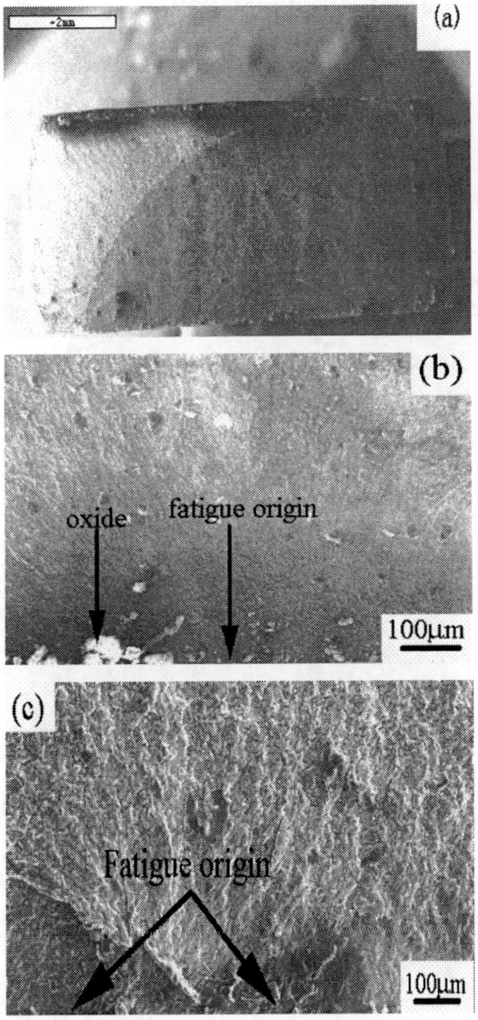

Figure 7. SEM photographs of overall fracture surface and analysis of fatigue crack initiate sites :(a)mophorlogy of thermo-mechanical fatigue test ; (b)fatigue crack initiate from oxide ; (c)fatigue crack initiate from slip bands.

Fig. 8 shows S-N fitting curves of GW123k alloy under thermo-mechanical fatigue test. It indicates that out-phase thermo-mechanical fatigue lifetime of GW123K is above that of in-phase thermo-mechanical fatigue lifetime. It is mainly because that fatigue specimens under out-phase thermo-mechanical fatigue test suffer from pressure stress at the highest temperature, and at the same time many micro-cracks caused by micro-voids(shown in Fig.9) were difficult to propagate due to crack closure effect led by pressure stress. Once fatigue cracks intiate from fatigue origins , they will propagate rapidly along micro-voids which result in decrease of thermo-mechanical fatigue lifetime[6]. In line with S-N fitting curves, the equations of in-phase (equation 1) and out-phase(equation 2) thermo-mechanical fatigue can be obtained and is shown as follows :

$$N_f = -13.23 + 4.52\times10^6 \exp(-\sigma_a/16.98) \quad (1)$$

$$N_f = -749 + 1.93\times10^6 \exp(-\sigma_a/22.76) \quad (2)$$

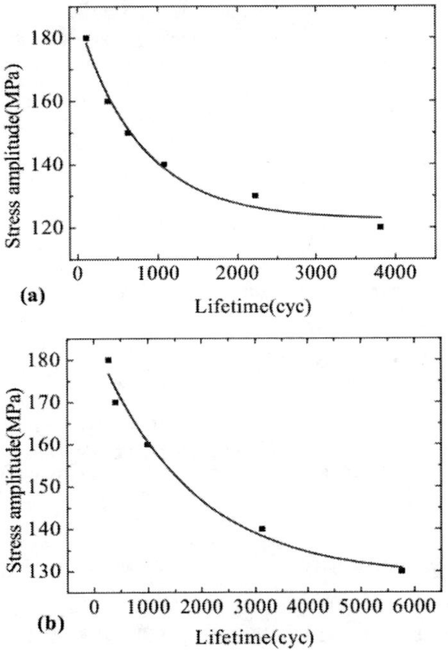

Figure 8. S-N fitting curves of GW123k alloy under thermo-mechanical fatigue test : (a)in-phase ;(b)out of phase.

DOI: 10.1051/matecconf/20166303023

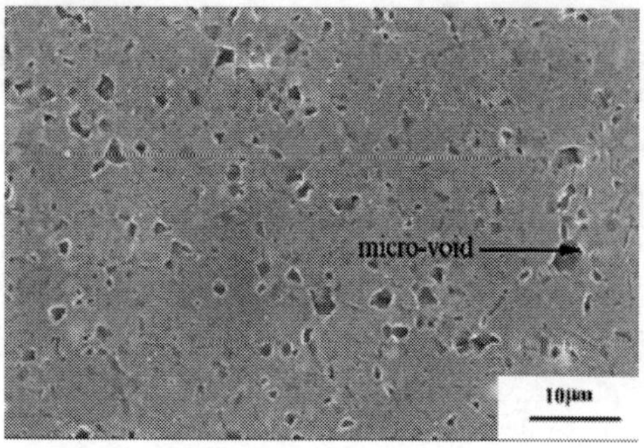

Figure 9. SEM images of GW123k alloys after thermo-mechanical fatigue test.

4 conclusions

(1) GW123K alloy has a good combination of ultimate tensile strength and plasticity at temperatures from ambient up to 200℃.

(2) The main strengthening mechanism of GW123K is precipitation strengthening. Tensile strength of GW123K decrease obviously at temperatue above 200℃ because of new slip systems are activated.

(3) Out-phase thermo-mechanical fatigue lifetime of GW123K is above that of in-phase thermo-mechanical fatigue lifetime due to crack closure effect.

(4) Micro-voids caused by cyclic load lead to decrease of thermo-mechanical fatigue lifetime.

Acknowledgement

This work was funded by scientific research project of Jingchu University of Technology (NO. 2013BK02). We would like to thank Yang Fan for kind assistance on SEM and TEM analysis.

Referencesw

1. T. Liu, W. Zhang, S.D. Wu, C.B. Jiang, S.X. Li, Y.B. Xu. Mater. Sci. Eng.A. **360**, 345 (2003)
2. A.A. Luo, Int.Mater.Rev. **49**, 13(2004)
3. Y. Gao, Q. D. Wang, J. H. Gu, Y. Zhao, Y. Tong. Mater. Sci. Eng. A. **459,** 117(2007)
4. Y. Gao, Q. D. Wang, J. H. Gu, Y. Zhao, Y. Tong, J. Kaneda. Journal of Rare Earths. **26,** 298(2008)
5. I. A. Anyanwu, S. Kamodo and Y. Kojima. Mater. Tran. **42(7),** 1212 (2001)
6. X. M. Yang, H.J Yang, F. Yang, S.M. Yin, W. Wang, S.X. Li, Q.D. Wang. J. Mater. Sci. Technol. **25(6),**731(2009)

MATEC Web of Conferences **63**, 03024 (2016) DOI: 10.1051/matecconf/20166303024
MMME 2016

Investigation on low cycle fatigue behaviours of Mg-12Gd-3Y-0.5Zr magnesium alloy

Xiao Ming YANG[1,a]

[1] *Jingchu University of Technology, JinMen City, HuBei province, china*

Abstract. In this paper, low cycle fatigue behaviours of Mg-12Gd-3Y-0.5Zr magnesium alloy under different strain amplitudes were investigated. It shows that Mg-12Gd-3Y-0.5Zr magnesium alloy has good cyclic stabilization. SEM fracture micrograph indicate that fatigue cracks initiate on the surface or subsurface. Low cycle fatigue mechanism of Mg-12Gd-3Y-0.5Zr magnesium alloy under different total strain amplitudes is disscussed in this paper.

1 Introduction

More and more importance is paid to magnesium alloys due to their high specific strength, good machinability and castability [1]. Many measures have been adopted to improve their mechanical properties [1]. However, magnesium alloys are limited to some noncritical parts as a result of the restriction of fatigue strength and creep resistance at high temperatures [2]. With addition of rare earth elements, Mg-12Gd-3Y-0.5Zr alloy show high tensile strength and modified fatigue strength at elevated temperatures.

In the present day, investigation on low cycle fatigue behaviours of magnesium alloys is only limited to AZ series and AM series. Begum investigated low cycle fatigue behaviours of as-rolled AZ31 magnesium alloy and the research indicates that it shows cyclic stabilization under low strain amplitudes and cyclic hardening under high strain amplitudes, respectively [3]. Park revealed that low cycle fatigue behaviours of AZ31 magnesium alloy with strong basal plane texture was predominated by the alternation of twinning and detwinning during each cycle [4].

In this paper, the cyclic strain response and fatigue machanism is disscussed.

2 Experimental

The ingots of Mg-12Gd-3Y-0.5Zr alloy(hereinafter referred to as GW123K) were rolled with a ratio of 25:1 at 400℃. The second phase whose constituent is approately $Mg_{24}(Gd,Y)_5$[5] was precipitated in microstructre in the process of extrusion.

Specimens for low cyclic fatigue tests were cut by electric spark machining along the rolling direction with a gauge sections of 5×4.5×14 mm^3. Low cycle fatigue testss controlled by total strain range from 1.5×10^{-3}～2×10^{-2} were carried out on an MTS 810 fatigue testing machine with an immanent frequncy of 0.5Hz.

The fracture surface morphologies of the specimens were observed by scanning electron microscopy (SEM, QUANTA600).

3 Experimental results and discussion

3.1 Microstructure

Fig.1(a) shows optical micrograph of microstructure for GW123K. It can be seen that microstructure of GW123K is composed of α-Mg solid solution and Mg-riched precipitation which precipitaed along or inside grain. The average grain size of as-rolled GW123K is about 10μm and many grains are characteristic of recrystallization. It can be seen from Fig.1(b) that there are a large quantity of uniformly distributed second precipitates in microstructure of GW123K alloy.

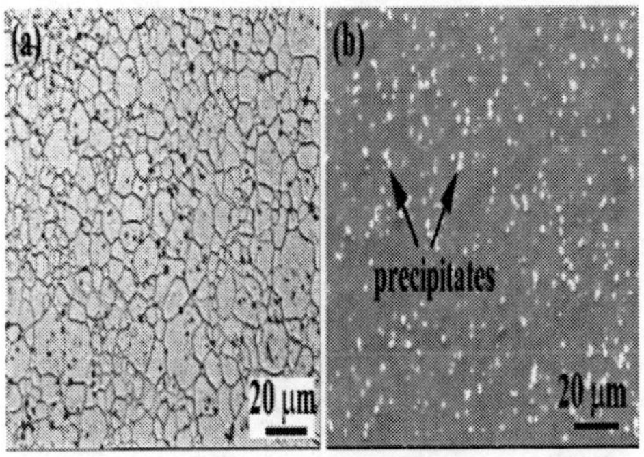

Figure 1. Microstructures of GW123k alloy, (a) OM and (b) SEM backscatter image where abundant fine precipitates can be observed.

[a] Corresponding author: yang760507@163.com. This research was supported by university's scientific research project of JingChu University of Technology(No. 2013BK02)

3.2 Cyclic stress response curve

Cyclic stress amplitude vs number of cycles for GW123K at different total strain. Amplitudes are shown in Fig.2. It indicates that GW123K maitain cyclic response stability which is different from cyclic hardening of other rolled magnesium alloys [6] and cyclic sofening of as-ECAPed magnesium alloys [7]. It is well known that both cyclic hardening and cyclic sofening depend on dislocation movement to a great extent. As a result of pinning effect of solid solution and precitation phase, dislocation density of rolled GW123K is not prone to propogate in the pocess of low cycle fatigue teses which impair cyclic hardening. Furthermore, initiation and propogation of fatigue cracks under cyclic loading will lead to loss of loaded area which causes a little cyclic sofening. In general, GW123K alloy shows cyclic response stablity as a combination of interaction of cyclic hardening and cyclic sofening `mention above.

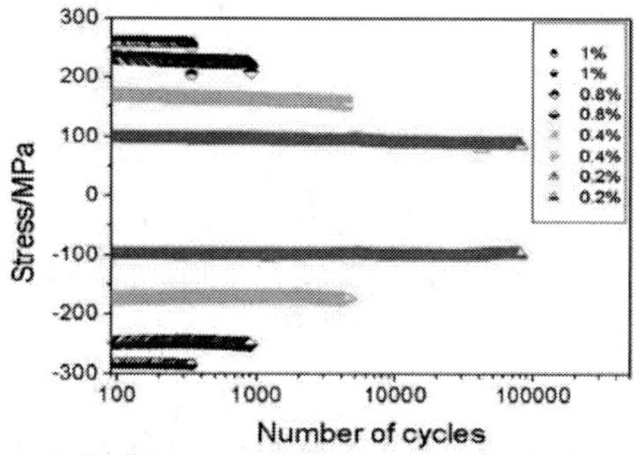

Figure 2. Cyclic stress amplitude vs number of cycles for GW123K at different total strain. amplitudes.

3.3 Hysteresis loop

Fig.3 shows hysteresis loops of GW123k alloy at the half-life cycle for different total strain amplitudes. As a result of its cyclic response stability, the shape of hysteresis loop remains unchanged. However, the shape of hysteresis loop is linear when strain amplitude is less than 0.2% which indicates that its cyclic deformation is mainly composed of plastic deformation.

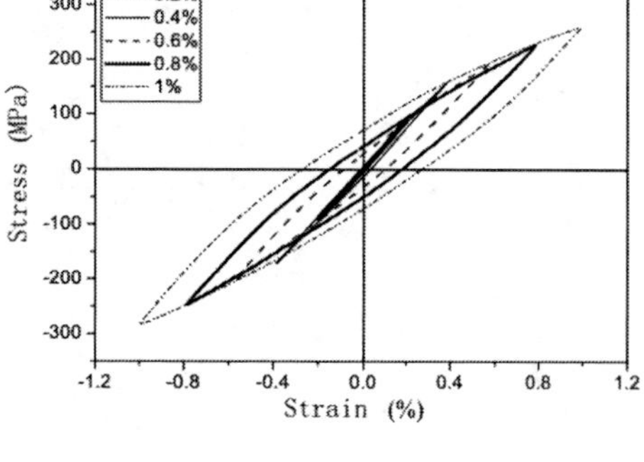

Figure 3. Hysteresis loops of GW123k alloy at the half-life cycle for different total strain amplitudes.

3.4 Strain-lifetime curves

Strain amplitude vs number of cycles to failure curves is shown in Fig.4. It can be seen from Fig.4(a) that both elastic strain and plastic strain are linear relationship with reversals to failure and all of them abide by Basquin and Coffin–Manson equations. Fig.4(b) indicates that low cycle fatigue properties of GW123K is lower than aluminium alloy but has the same conditon as that of other magnesium alloys. The prediction equation of low cycle fatigue of GW123K is as follows :

$$\frac{\Delta\varepsilon_e}{2} = 0.2076(2N_f)^{-0.218} \quad (1)$$

$$\frac{\Delta\varepsilon_p}{2} = 1.353(2N_f)^{-0.956} \quad (2)$$

$$\frac{\Delta\varepsilon_t}{2} = \frac{\Delta\varepsilon_e}{2} + \frac{\Delta\varepsilon_p}{2} = 0.2076(2N_f)^{-0.218} + 1.353(2N_f)^{-0.956} \quad (3)$$

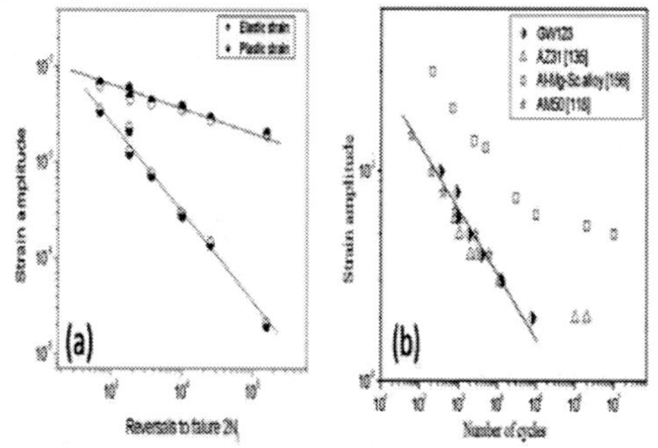

Figure 4. Strain amplitude vs number of cycles to failure curves for（a） as-rolled GW123k, (b) for different magnesium alloys and an aluminium alloy.

3.5 Analysis of low cycle fatigue mechanism

SEM images of overall fracture surfaces of specimens are shown in Fig.5. The fracture surface is very smooth and characteristic of fatigue failure. Fatigue cracks initiate from surface or subsurface and they initiate one or more faigue origins (shown as indicated by the arrows).

Fig.6 shows SEM images of fracture surfaces of two specimens and corresponding energy spectrum analysis of fatigue crack initiate sites. It can be seen that fatigue cracks initiate from Gd-riched intermetallic compounds with large size or large-sized inclusions(mainly composed of oxides of Mg and Gd). Stress concentration caused by intermetallic compounds and inclusions induces fatigue initiation which result in final fatigue failure of GW123K alloy under cyclic loading. A conclusion can be drawn that large-sized intermetallic compounds and inclusions impair fatigue property of GW123K alloy. As a result, fatigue cracks initiating from large-sized intermetallic compounds and inclusions is the dominated fatigue mechanism for GW123K under low cycle fatigue tests.

DOI: 10.1051/matecconf/20166303024

Figure 5. SEM images of overall fracture surfaces of specimens: (a) fatigue crack initiates from single origin, (b) fatigue crack initiates from two origins(fatigue origin shown as indicated by the arrows)

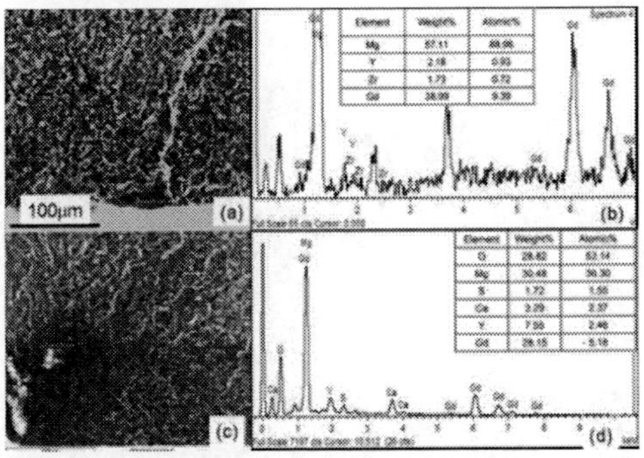

Figure 6. SEM images of fracture surfaces of two specimens and corresponding energy spectrum analysis of fatigue crack initiate sites (a) large second-phase, (b) composition of second-phase in (a), (c)oxide inclusion, (d) composition of oxide inclusion in (c).

Fig .7 shows fatigue fracture surface for specimen fatigued at strain amplitude of 0.2% . It can be seen that fatigue fracture can be divided into three obvious regions, that is to say, region Ⅰ of fatigue cracks initiation and stable propagation, region Ⅱ of unstable crack propagation and region Ⅲ of instaneous failure. region Ⅰ has typical feature of transcrystalline fracture in which the fracture surface is very complanate and mainly composed of cleavage planes (as shown in Fig.7(b)). A small amount of shallow dimples and cleavage planes can be observed in region Ⅱ(as shown in Fig.7(c)) and fracture surface in this region is coarse. Region Ⅲ(as shown in Fig.7(d)) has typical feature of a lot of shallow dimples and some second phase particle can be observed in the bottom of dimples which shows that GW123K has poor plasticity .

Fatigue fracture surface for specimen fatigued at strain amplitude of 0.5% is shown in Fig.8. It can be seen from frature morphology that there is a transitional region between region Ⅰ and region Ⅲ. Compared with fatigue fracture under low strain amplitude fagitue tests, volume fraction of clevage planes decreased to a great extent and fracture surface became more coarse than that of the former. There are a bit more dimples in transitional region but no large-sized transcrystalline clevage plane can be observed in this transitional region. The characteristic of region Ⅲ is similar to that of specimen fatigued at strain amplitude of 0.2%, and the differences between them are that the latter has smaller and fewer dimples.

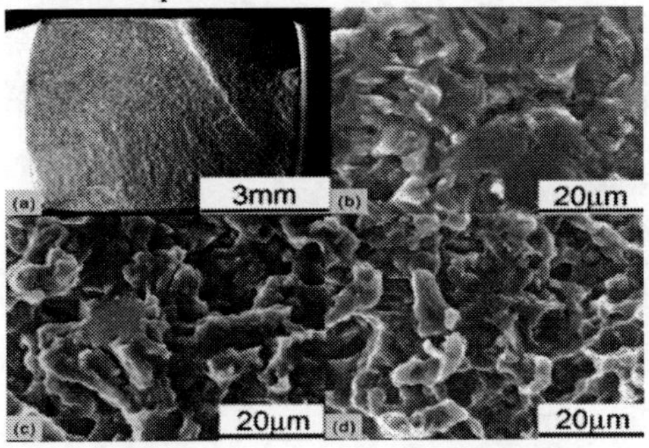

Figure 7. Fatigue fracture surface for specimen fatigued at strain amplitude of 0.2%: （a）overall fracture surface, (b) Region Ⅰ denotes fatigue crack initiation and steady propagation region; (b) Region Ⅱ denotes transition region of fatigue crack propagate from steady to unstable; (d) Region Ⅲ denotes fatigue crack unstable propagation and final failure.

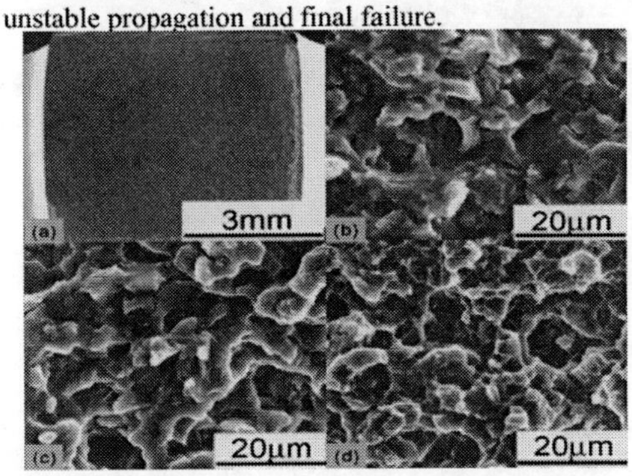

Figure 8. Fatigue fracture surface for specimen fatigued at strain amplitude of 0.5%: （a）overall fracture surface, (b) Region Ⅰ denotes fatigue crack initiation and steady propagation; (b) Region Ⅱ denotes transition region of fatigue crack propagate from steady to unstable; (d) Region Ⅲ denotes fatigue crack unstable propagation and final failure.

4 Conclusions

(1) GW123K alloy has charactersitic of cyclic stress response stability in the process of low cycle fatigue tests controlled by strain.

(2) Both elastic strain and plastic strain of GW123K abide by Basquin equation and Coffin–Manson equation, respectively. Low cycle fatigue property of GW123K is close to that of AZ31 and AM50 magnesium alloys.

(3) Fatigue cracks initiating from large-sized intermetallic compounds and inclusions is the dominated fatigue fracture mechanism.

(4) SEM observations on fatigue fracture surface show that it has three typical fatigue cracks propagation regions, and with the increase of strain amplitude it has smaller and fewer dimples.

AcknoWledgement

This work was funded by scientific research project of Jingchu University of Technology (NO. 2013BK02) and partly by subject groupings of green chemical industry and biological medicine. We would like to thank Yang Fan for kind assistance on SEM and TEM analysis.

References

1. W. J. Kim, S. W. Chung, C. S. Chung and D. Kum. Acta Mater. **49**, 3337 (2001)
2. A.A. Luo, Int.Mater.Rev. **49**, 13(2004
3. S. Begum, D. L. Chen, S. Xu, A.A. Luo. Inter. J. Fatigue. **31**, 726 (2009)
4. S. H. Park, S. G. Hong, B. H. Lee, W. Bang, C. S.Lee. Inter. J. Fatigue. **32**, 1835 (2010)
5. Y. Gao, Q. D. Wang, J. H. Gu, Y. Zhao, Y. Tong. Mater. Sci. Eng. A. **459**, 117(2007)
6. L. Wu, A.Jain, D.W. Brown, G.M. Stonica, S.R. Agnew, B. Clausen, D.E. Fielden, P.K Liaw. Acta Mater. **56**, 688(2008)
7. L.J. Chen, C.Y. Wang, W. Wu, G.M. Stonica, L. Wu, P.K. Liaw. Metall. Mater. Trans. A. **38**, 2235(2007)

MATEC Web of Conferences **63**, 03025 (2016) DOI: 10.1051/matecconf/20166303025
MMME 2016

Isothermal fatigue behaviours of Mg-12Gd-3Y-0.5Zr magnesium alloy at elevated temperatures

Xiao Ming YANG[1,a]

[1]*Jingchu University of Technology, JinMen City, HuBei province, china*

Abstract. Isothermal fatigue tests were carred out on Mg-12Gd-3Y-0.5Zr magnesium alloy and its heat-treated counterpart at different temperatures. The exprimental results shows that isothermal fatigue strength of two alloys decrease very slowly with increasing temperature up to 200℃. The ultimate tensile strength of heat-treated Mg-12Gd-3Y-0.5Zr is slight lower than that of as-rolled counterpart, however, the fatigue strength of heat-treated alloy is higher.

1 Introduction

Magnesium alloys are called green-engineering materials with great development potential because of their many outstanding advantages, such as low density, high specific strength and specific stiffness, good machinability and castability, etc.[1]. Numerous magnesium alloy products have been used in automobile, communication and aerospace industries [2]. However, there are some serious shortcomings, one of which is poor mechanical property at high temperature that limits the applications of magnesium alloys in the corresponding conditions. In recent years, many attempts have been made to improve the mechanical properties of magnesium alloys. Equal channel angular pressing (ECAP), as one effective method of severe plastic deformation (SPD), can modify the microstructure and texture of magnesium alloys and improve the tensile properties to some extent [3].

In this paper, the ultimate tensile strength and isothermal fatigue strength at different temperatures are reported. With added rare earth elements, the as-rolled and heat-treated Mg-12Gd-3Y-0.5Zr alloys show the high tensile strength and decent fatigue strength at elevated temperatures. The strengthening and the fatigue failure mechanisms are discussed in this paper.

2 Experimental

The ingots of Mg-12Gd-3Y-0.5Zr alloy were rolled with a ratio of 25:1 at 400 °C. For comparison, part of the as-rolled alloy was further heat-treated in a furnace at 415 °C for 16 hours and then cooling in air.

Tensile and isothermal fatigue specimens were cut by electric spark machining along the rolling direction with a gauge sections of $3mm \times 1.4mm \times 1mm^3$ and $20mm \times 4mm \times 3mm^3$ respectively. Tensile tests were carried out on an Instron 8801 testing machine equipped with a heater and temperature controlling system at room temperature, 130, 170, 200 and 235 °C and at a strain rate of $1 \times 10^{-3}s^{-1}$. Isothermal fatigue tests were carried out on an MTS810 testing machine at room temperature, 100, 200 and 235 °C and at a frequency of 20Hz under stress control. The stress wave form was triangular, and the load ratio R=0.1.

The fracture surface morphologies of the specimens were observed by a Cambridge S360 scanning electron microscope (SEM). For TEM observations, slices with a thickness of 0.6mm were cut from the section parallel to the rolling direction, mechanically thinned to about 50μm and cut into discs with diameters of 3mm. Thin foils were made from these discs with a twin-jet polisher with a current of 40μA using an electrolyte of 20% HNO_3 in methanol at 248 K. The thinned foils were observed using a TEM 2000FX II operating at 200 KV.

3 Experimental results and discussion

3.1 Tensile properties at different temperatures

Temperature dependence of ultimate tensile strengths of the as-rolled and heat-treated Mg-12Gd-3Y-0.5Zr alloys is shown in Fig. 1 for tests carried out at room temperature, 130, 170, 200 and 235 °C. Ultimate tensile strength of the as-rolled Mg-12Gd-3Y-0.5Zr reaches the highest at 405 MPa at room temperature. Ultimate tensile strength of the heat-treated Mg-12Gd-3Y-0.5Zr is a bit lower than that of the as-rolled alloy at corresponding

[a] Corresponding author: yang760507@163.com. This research was supported by university's scientific research project of JingChu University of Technology(No. 2013BK02)

temperatures. Fig. 1 indicates that the ultimate tensile strengths of the as-rolled and heat-treated alloys decrease very slowly with increasing temperatures up to 200°C which is desirable for high temperature applications obviously. However, the ultimate tensile strengths of both as-rolled and heat-treated alloys dropped apparently at 235 °C possibly due to new slip system of magnesium alloys activating at nearly 220 °C.

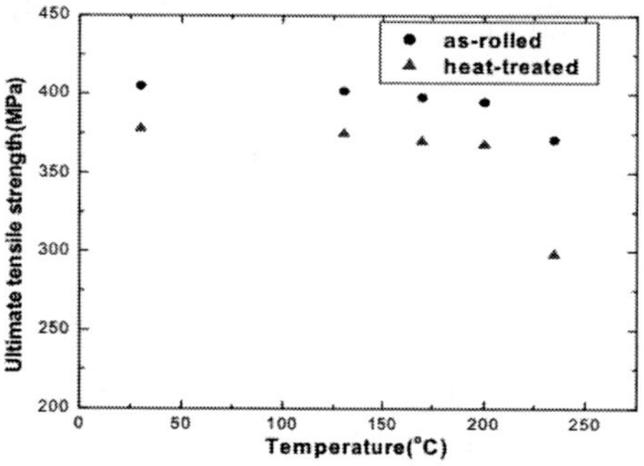

Figure 1. Temperature dependence of ultimate tensile strengths for the as-rolled and heat-treated Mg-12Gd-3Y-0.5Zr alloys.

3.2 Fatigue behaviors at different temperatures

Isothermal fatigue tests were performed at different temperatures under stress control. Figs.2a-c demonstrate the isothermal S-N curves of the as-rolled Mg-12Gd-3Y-0.5Zr at room temperature and 235 °C, and the heat-treated alloy at 235 °C, respectively. All tests were carried out on an MTS810 fatigue testing machine at a constant frequency of 20Hz. The isothermal fatigue strength at 5×10^6 cycles was obtained as shown in Fig.3a. The dependences of the ratio of fatigue strength over ultimate tensile strength on temperatures for the as-rolled and heat-treated Mg-12Gd-3Y-0.5Zr alloys are shown in Fig.3b. The ratio is within a range of about 0.5 to 0.2 when the temperature varies from ambient temperature to 200°C.

From Figs.2 and 3, we can see that although the heat-treated Mg-12Gd-3Y-0.5Zr has slight lower ultimate tensile strength than that of as-rolled counterpart, it has higher fatigue strength. It means that in the cyclic loading condition, the heat-treated alloy would be preferential to be used at the temperatures ranged from ambient up to 200 °C. However, when the temperature reaches 235 °C, both the as-rolled and heat-treated alloys would not be used due to their fatigue strengths decreasing drastically. It is believed that fatigue strength increases with increasing ultimate tensile strength and/or increasing elongation-to-failure[4,5]. Higher elongation-to-failure of heat-treated Mg-12Gd-3Y-0.5Zr offsets the negative effect of its lower ultimate tensile strength and finally the higher fatigue strength can be achieved at different temperatures except 235 °C.

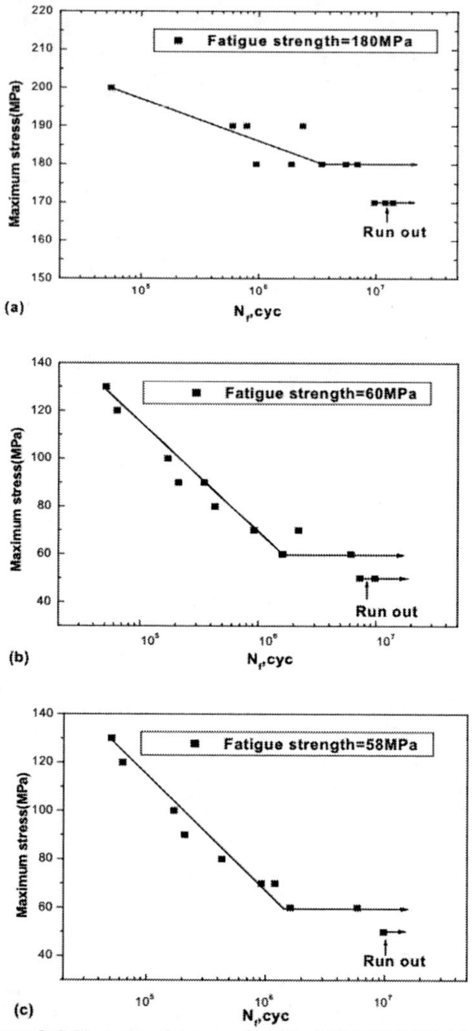

Figure 2. S-N curves of the as-rolled Mg-12Gd-3Y-0.5Zr (a) at room temperature, (b) at 235 °C, (c) S-N curve of the heat-treated Mg-12Gd-3Y-0.5Zr at 235 °C .

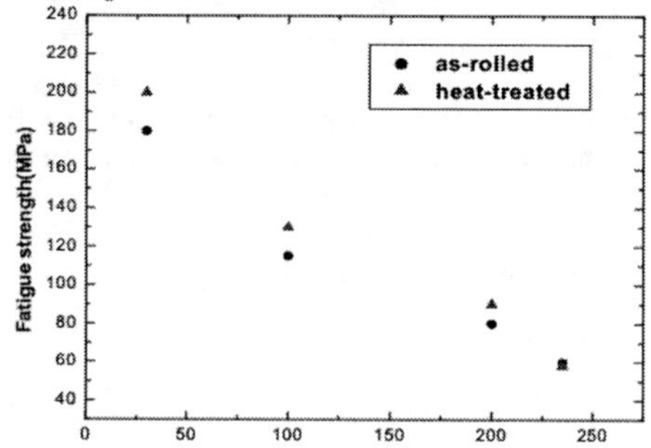

(a)

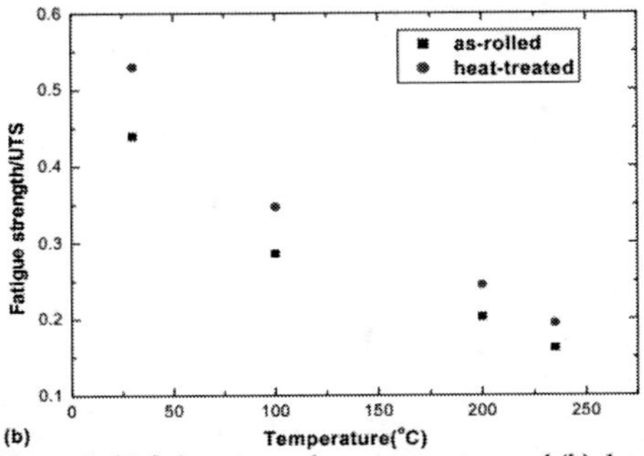

Figure 3. (a) fatigue strength vs temperature and (b) the ratio of fatigue strength over ultimate tensile strength vs temperature for the as-rolled and heat-treated Mg-12Gd-3Y-0.5Zr alloys.

3.3 Fatigue crack origins

Some of the fatigue crack initiation sites of the as-rolled and heat-treated Mg-12Gd-3Y-0.5Zr alloys tested under different stresses and at different temperatures are shown in Fig. 4. In the present experiment, it is noted that all of fatigue crack origins initiated at the sample surface no mater what stresses and temperatures applied. Ridges are obviously observed on the facture surface as shown in Fig. 4c which indicates that it has characteristic of multiple fatigue origins as marked by arrows with increasing temperatures. Similar to Fig. 4c, there are several fatigue crack origins in Fig. 4d due to high cyclic stress applied, however, the fatigue crack propagation zone is rather smooth for heat-treated alloy than that for as-rolled alloy observed in low magnification. The reason is partly due to more precipitate occurring in the heat-treated alloy which restricts the crack meandering.

In the extruded magnesium alloys without RE element additions, for example AZ31, usually the yield stress asymmetry between tension and compression caused by texture and twining exists. We found that twining and detwining [6] play key roles for crack initiation in the low cycle fatigue for AZ31. At the same time, the work in our group revealed that twining-involved fatigue damage process could also play a important role in the very high cycle fatigue regime for AZ31. However for Mg-12Gd-3Y-0.5Zr alloy the reasons for fatigue crack not be observed to initiate from twinning-detwining process may be as follows. In the present extruded alloys the texture is weak and no obvious yield asymmetry exists, in other words, twining and detwining during deformation are rare. Under isothermal fatigue test above 100°C, cyclic slip was the main form of plastic deformation and the cyclic twining and detwining would not take place substantially. During cyclic deformation the persistent slip leads to surface roughening which makes it possible for the crack to initiate at surface, particularly, at the position where the slip band impinges with the oxide inclusion or large second phase. At the same time environmental effect plays an important role in crack initiation for these two alloys tested at high temperatures. Oxides formed at the surface make the surface roughening and embrittlement, and the crack is prone to initiate around the oxides. It can be observed from Fig. 4b that there are some oxides near the crack initiation sites. As a result, cyclic slip combined with environmental effect may be the main crack initiation mechanism. RE element additions result in dispersion of second phases and fine grains which restrains the fatigue crack growth and hence improves fatigue strengths of the as-rolled and heat-treated Mg-12Gd-3Y-0.5Zr alloys.

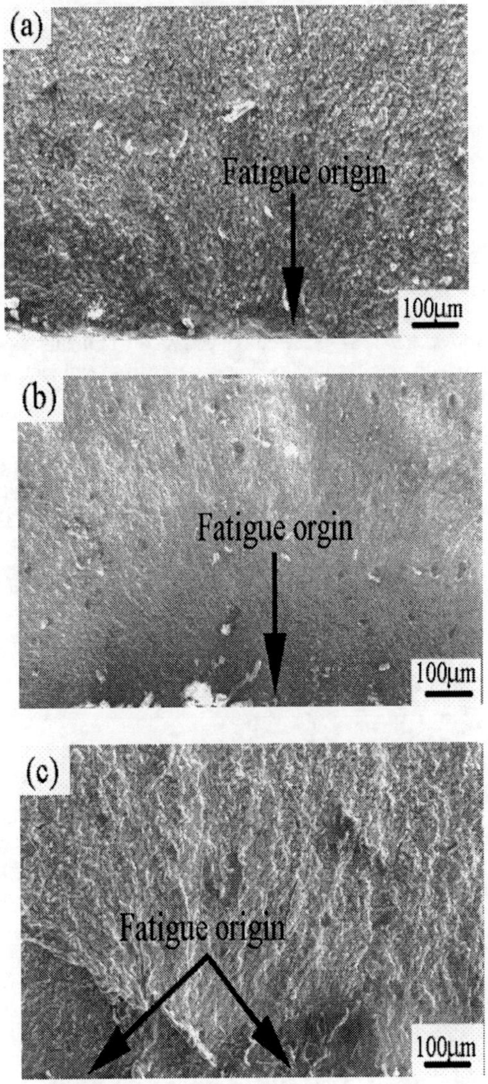

DOI: 10.1051/matecconf/20166303025

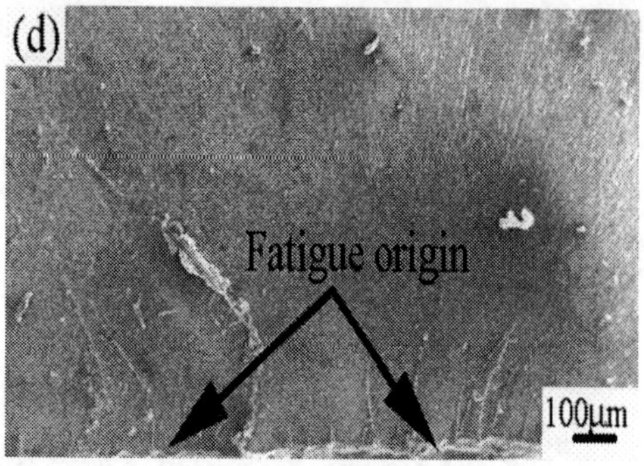

Figure 4. SEM photographs of fatigue origin: as-rolled Mg-12Gd-3Y-0.5Zr tested at (a)100℃,σ_a=110MPa, N_f=1.17×10^7cyc; (b) 200℃, σ_a=80MPa, N_f=2.42×10^6cyc; (c) 235℃, σ_a=50MPa, N_f=9.82×10^6cyc; heat-treated Mg-12Gd-3Y-0.5Zr tested at (d) 100℃, σ_a=130MPa, N_f=1.2×10^6 cyc.

4 Conclusions

(1) The as-rolled and heat-treated Mg-12Gd-3Y-0.5Zr alloys have a good combination of ultimate tensile strength and ductility at temperatures from ambient up to 200°C.

(2) The ultimate tensile strength of the heat-treated Mg-12Gd-3Y-0.5Zr alloy is slightly lower than that of the as-rolled counterpart, however, the fatigue strength of the heat-treated Mg-12Gd-3Y-0.5Zr alloy is higher than that of the as-rolled alloy.

(3) Fatigue crack initiates at the surface and has the characteristic of multiple fatigue origins with increasing temperatures and/or high cyclic stresses. Precipitates in the two alloys play essential roles in the tensile and fatigue tests.

Acknowledgement

This work was funded by scientific research project of Jingchu University of Technology (NO. 2013BK02). We would like to thank Yang Fan for kind assistance on SEM and TEM analysis.

References

1. W. J. Kim, S. W. Chung, C. S. Chung, D. Kum. Acta Mater. **49**, 3337(2001)
2. M.H. Idris: AFS Trans. **104**, 237（1996）
3. R.E. Brown: Light Metal Age. **55** (7-8), 72(1997)
4. Y. Yang and Y. B. Liu, Mater. Characterization. **59**, 567(2008)
5. X. M. Yang, H.J Yang, F. Yang, S.M. Yin, W. Wang, S.X. Li, Q.D. Wang. J. Mater. Sci. Technol. **25(6)**,731(2009)
6. S.M. Yin, F. Yang, X.M. Yang, S.D. Wu, S.X. Li and G.Y. Li: Mater. Sci. Eng. A. **494**, 397(2008)

MATEC Web of Conferences 63, 04001 (2016)
DOI: 10.1051/matecconf/20166304001
MMME 2016

Subsynchronous Oscillation Problem Research in the UHVDC System of a Regional Power Grid in China

Ying QU[1, a], Chao ZHANG [2], Hui Ping ZHENG[1], Min XUE[1] and Ying ZHANG [1]

[1] *State Grid Shanxi Electric Power Research Institute, Shanxi Province, 030001*
[2] *Shanxi Electric Power Cooperation of SGCC, Shanxi Province, 030001*

Abstract. Along with the grid structure being more and more complex and the rapid development of the HVDC system, studying the subsynchronous oscillation (SSO) problem on HVDC system has more engineering practice significance. The paper studies subsynchronous oscillations problem of generators near the ±800kV UHVDC converter station, and analyzes the subsynchronous oscillation possibilities through PSCAD/EMTDC simulation. At last, though the researched UHVDC thermal plants have none SSO risk but it needs other measures to make the relevant generators return on normal operation.
Keyword: UHVDC; subsynchronous oscillation (SSO); PSCAD/EMTDC simulation

1 Introduction

As we all know, though China is a big energy country, the energy resources and the development of productive forces have reverse distribution and serious imbalance. Long-distance large-scale electricity transmission becomes inevitable, HVDC technology has vast prospect in China and develops rapidly. Since 1989 Zhoushan HVDC project had been put into operation officially, China has become the first country in the world of largest DC transmission capacity, the highest voltage level, the fastest growing speed [1]. Subsynchronous oscillation belongs to the power system oscillation instability and it's caused by a special kind of electromechanical coupling effect. Subsynchronous oscillation problem caused by HVDC has become an important practical problem [2-6], so study of UHVDC transmission system's SSO question has great significance for power system stability.

2 The researched ±800kV UHVDC project in China

The researched ±800kV UHVDC project is the first UHVDC project which transfers Xinjiang province electric energy outwards and is also the first "wind, solar energy, thermal power "hybrid transmission engineering HVDC project[7]. The project's rated power is 8000MW, rated voltage is±800kV. And the converter station AC bus rated voltage is 500kV, which contacts the 750kV northwest regional power grid through transformers. It includes 8 thermal power units near the rectifier station. The researched UHVDC project was put into operation at the end of 2013, and the rectifier area has 3 power plants. Plant 1 has 2 same generators, Plant 2 has 2 same generators, Plant 3 has 4 same generators of which all generators' rated power is 660MWand their rated voltage is 22kV. The simplified model is shown in Figure 1.

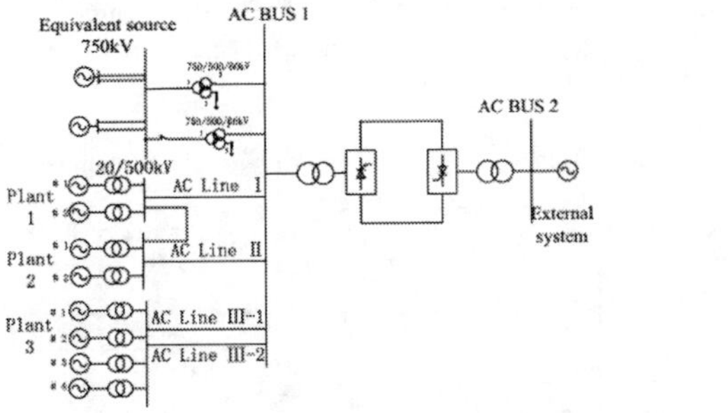

Figure 1. The Researched UHVDC Equivalent Network Diagram

According to the plan data, the electrical distance between thermal power units and the convertor station is relatively short. The negative electrical damping caused by HVDC system may result in SSO problem and seriously may lead to shaft fatigue and damage. The UHVDC has lager rated capacity and may have more serious SSO problem, so it is necessary to conduct study in-depth of the UHVDC's thermal power units.

[a] Ying QU: 865139180@qq.com

3 The time domain simulation analysis of steam turbines near the UHVDC system rectifier station

As shown in Figure 1, under normal operating conditions, the 750kV regional power grid and the rectifier station area units transfer electrical energy to the load center (the external equivalent source) together through the researched UHVDC project. When the UHVDC system is under worse conditions, the thermal units nearby is at island operation, which is isolated with the 750kV regional gird. Unit interaction factor method (UIF) is a screening method which is suitable only for the analysis of DC transmission system on SSO problem. It's always used to analyze simply whether the relevant unit has the SSO problem [8-9]. As the following equation shown:

$$UIF_i = \frac{S_{HVDC}}{S_i}\left(1-\frac{SC_i}{SC_{TOT}}\right)^2 \quad (1)$$

S_{HVDC} is the HVDC rated capacity, the unit is *MW*; S_i is the rated capacity of the researched generator *i*, the unit is *MW*; SC_i is the three-phase short-circuit capacity at the HVDC rectifier bus (its unit is *MVA*, and it doesn't conclude the effect of AC filter and the generator *i*); SC_{TOT} is also the three-phase short-circuit capacity at the HVDC rectifier bus, it concludes the effect of generator *i* , but does not contain the AC filter effect as well.

When UIF_i<0.1, we can think that the interaction is not obvious between researched units and the DC transmission system. The risk of subsynchronous oscillation in this unit is relatively small. On the contrary, when UIF_i>0.1, the researched unit has SSO risk, some other precise methods can be used to analysis later on. According to the UIF method, it can conclude the following Table1 and Table 2:

Table1. Each unit UIF coefficient with the 750kV equivalent source

Plant 1 Unit amount	Plant 2 Unit amount	Plant 3 Unit amount	Plant1 UIF	Plant 2 UIF	Plant 3 UIF
2	0	0	0.029	-	-
2	2	0	0.035	0.030	-
4	2	0	0.031	0.031	-
4	2	2	0.029	0.028	0.027

Table2. Each unit UIF coefficient without the 750kV equivalent source

Plant 1 Unit amount	Plant 2 Unit amount	Plant 3 Unit amount	Plant 1 UIF	Plant 2 UIF	Plant3 3 UIF
2	0	0	0.295	-	-
2	2	0	0.153	0.182	-
4	2	0	0.116	0.118	-
4	2	2	0.094	0.093	0.099

It can be concluded from the above two tables that when the UHVDC system has 750kV equivalent source , each unit's UIF coefficient at different conditions is less than 0.1which means no SSO problem; when the UHVDC system has none 750kV equivalent source, each unit's UIF coefficient of the first 3 conditions is more than 0.1,which indicates SSO problem probably. But the UIF method is only a rough way to estimate SSO, so it needs detailed time domain simulation.

3.1 PSCAD/EMTDC simulation with 750kV equivalent source

When the 750kV equivalent grid is associated with UHVDC transmission project, we can establish the simulation model in accordance with Figure 1 with the PSCAD/EMTDC software which is specifically used for the analysis of power system transient software. And the simulation conditions are set as Table 1. What's more, the fault type is A phase grounding; fault location is at the AC Line III-2; fault time is at the t=20s, and the whole fault time is 0.05s, and then the fault is removed.

The result founds that at each condition, the physical quantities of each unit would converge after the fault at last . It means that each unit has strong electrical damping at each shafting modal frequency, and the SSO probability is little. At each condition, the physical quantity of units in Plant 1(Figure 2. and Figure 3.) and Plant 2(Figure 4. and Figure 5.) converge easily. Note that each figure below is under 8 machines' condition.

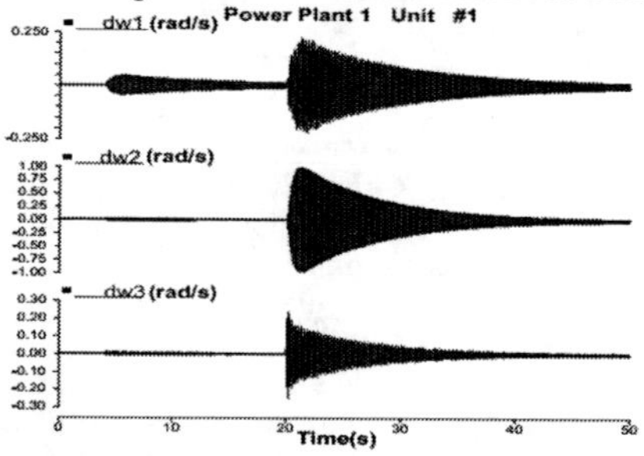

Figure 2. Power plant 1 -unit #1 each modal shaft speed deviation when having the 750kV equivalent source

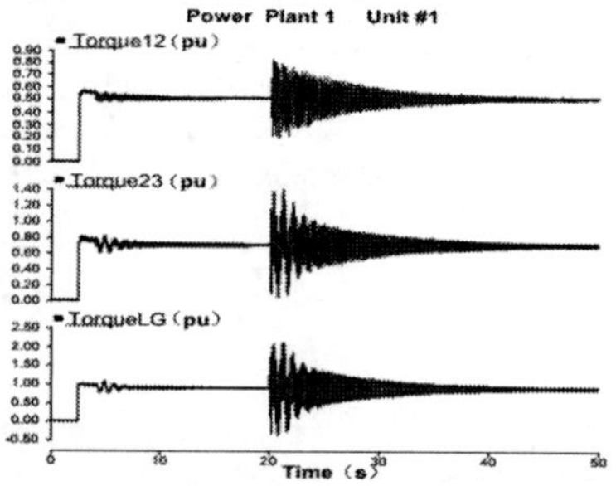

Figure 3. Power plant 1 -unit #1 torque change between each mass when having the 750kV equivalent source

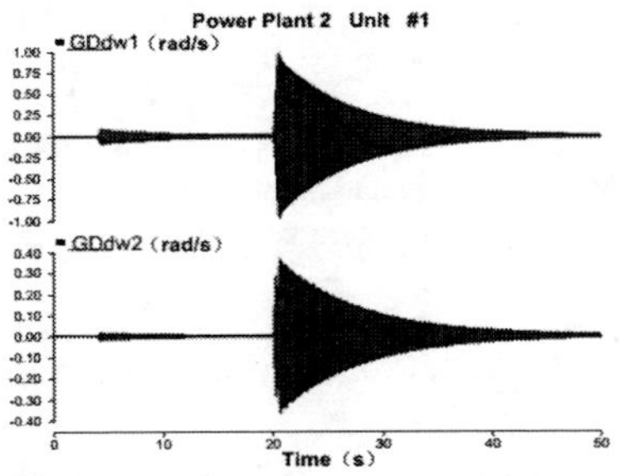

Figure 4. Power plant 2 -unit #1 each modal shaft speed deviation when having the 750kV equivalent source

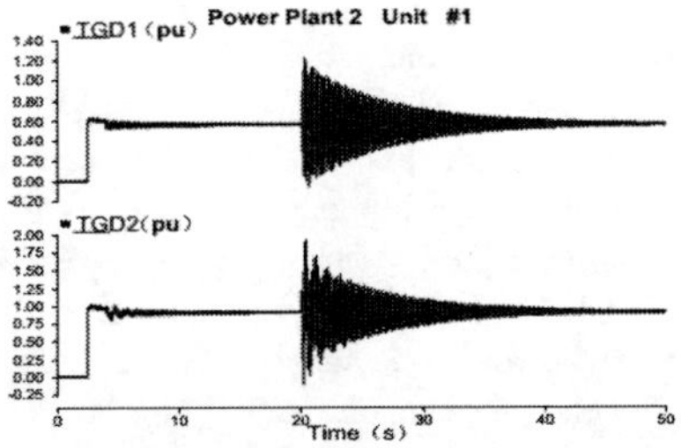

Figure 5. Power plant 2 -unit #1 torque change between each mass when having the 750kV equivalent source

But Plant 3's physical quantities have different trend, as shown in the following Figure 6~7.- unit #1 shaft speed deviation and the torsional torque(As units in the power plant have the same parameters, we choose the unit #1's figures under 8 machines' condition in the system for the example).

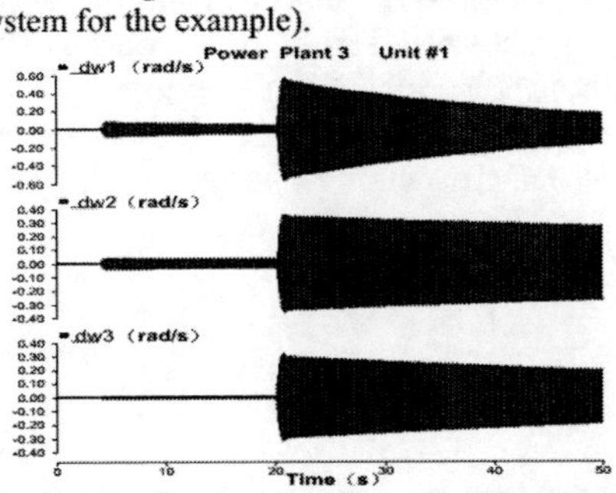

Figure 6. Power plant 3 -unit #1 each modal shaft speed deviation when having the 750kV equivalent source

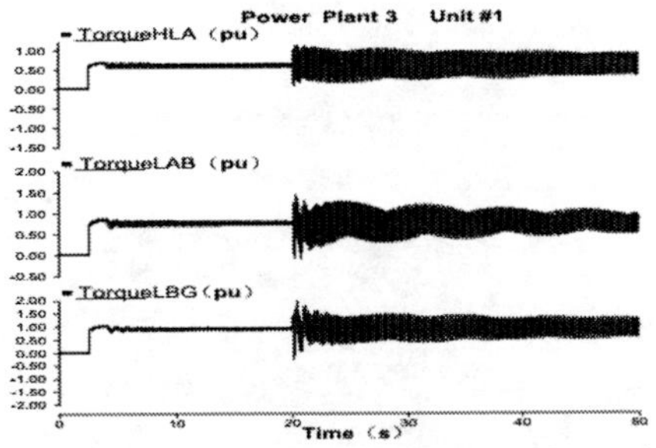

Figure 7. Power plant 3 -unit #1 torque change between each mass when having the 750kV equivalent source

The time domain simulation results show that, in the UHVDC system with 750kV equivalent source condition, whether the power plant 1, plant 2 or plant 3 has no torsional vibration divergence phenomenon, namely the system having none SSO danger which is consistent with the judge result using by UIF method. Though the power plant 3 has no subsynchronous oscillation phenomenon, but each modal component converges much more slowly than power plant 1 and power plant 2. And this phenomenon is also harmful for the whole power system.

3.2 PSCAD/EMTDC simulation without 750kV equivalent source

When the 750kV equivalent network is removed from the researched UHVDC system, the system's thermal power units near the converter station are at the islanding operation status. And the fault is same as that shown above. When the units' amount is 2, 4, 6, each generator's UIF is more than 0.1, which means that the generator and DC transmission system have closer electrical coupling, and the SSO problem is more obvious. When all plants units are put into operation, that's to say, the whole generators' number is 8, it can be found that, each unit UIF is less than 0.1, but close to 0.1. As the UIF method itself is a kind of rough way to estimate system SSO risk, so for this condition, it also needs PSCAD/EMTDC time domain simulation of SSO in detail.

Therefore, we can establish the UHVDC transmission system islanding model with the PSCAD software, and set up different simulation conditions in accordance with Table 2. Simulation results show that under various conditions, power plant 1, 2, 3 don't have SSO phenomenon. Then it can be found that physical quantities of power plant 1 and power plant 2's units also converge quickly under different conditions. Unit #1's shaft speed deviation of each modal frequency and the torsional torque between each mass in power plant 3 are shown in Figure 8 and Figure 9 below (take the 8 machines' condition for the example):

DOI: 10.1051/matecconf/20166304001

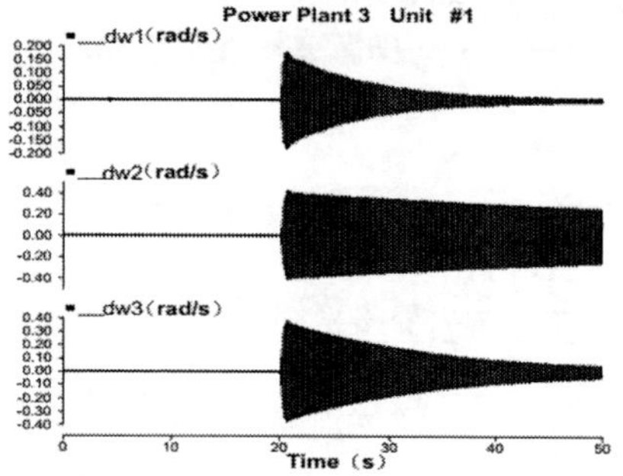

Figure 8. Power plant 3-unit #1 each modal shaft speed deviation at which not having the 750kV equivalent source

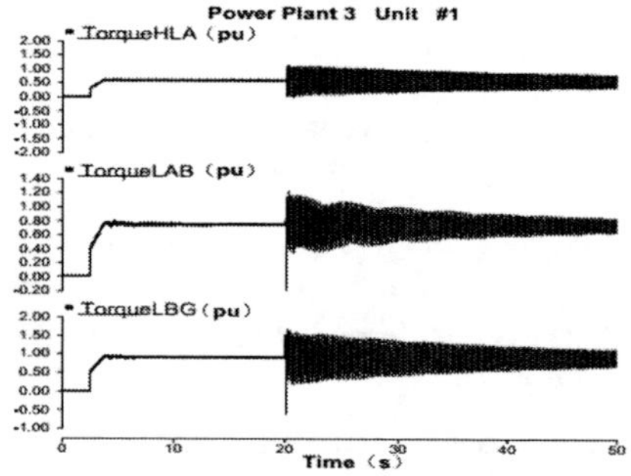

Figure 9. Power plant 3 -unit #1 torque change between each mass at which not having the 750kV equivalent source

Under the islanding operation, after the fault, although each physical quantity of power plant 3 units doesn't have the divergent phenomenon, the convergence speed is very slow especially the modal 2 speed deviation *dw2*, which is significantly lower than the other plants' generator. Power plant 3 should be taken as the important object in later SSO research.

4 Conclusions

From the PSCAD/EMTDC time domain simulation and analysis above, we can conclude that:

(1) The researched UHVDC thermal plants all have none SSO risk and none generator shafting torsional vibration divergence phenomenon whether the system have or not the 750kV equivalent source.

(2) Although power plant 3 don't have SSO phenomenon, but the modal components convergence speed is very slow.

(3) It's necessary to design subsynchronous damping controller (SSDC) for units in power plant 3 which can enhance the relevant system electrical damping and can make the generator return to normal operation as soon as possible.

References

1. Online Liu Zhenya. 2009. UHVDC Transmission Theory[M]. China Electric Power Press.
2. IEEE Subsynchronous Resonance Working Group. 1985, Terms, definitions and symbols for subsynchronous oscillations [J]. *IEEE Transactions on Power Apparatus and Systems,* **104**(6):1326-1334.
3. IEEE Committee Report. 1992, Reader's guide to subsynchronous resonance [J]. *IEEE Transactions on Power Systems*, **7**(1):150-157.
4. IEEE Subsynchronous Resonance Working Group. 1980, Proposed terms and definitions for subsynchronous oscillations [J]. *IEEE Transactions on Power Apparatus and Systems*, **99**(2):506-511.
5. IEEE Subsynchronous Resonance Working Group. 1977, First benchmark model for computer simulation of subsynchronous resonance[J]. *IEEE Transactions on Power Apparatus and Systems,* **96**(5):1565-1572.
6. IEEE Subsynchronous Resonance Working Group. 1985, Second benchmark model for computer simulation of subsynchronous resonance [J]. *IEEE Transactions on Power Apparatus and Systems,* **104**(5):1057-1066.
7. Hu Yanmei, Wu Junyong, Li Fang et c. Effects of ± 800kV Ha-Zheng UHVDC Control Mode in Henan Power Grid Voltage Stability [J]. *Power System Protection and Control*, **2013**(41):147-153.
8. IEC Publication 91923. 1993.Performance of High Voltage DC Systems, Part 3:Dynamic Contions[J].
9. Xu Zheng, Zhu Ruijin, Luo Huiqun. 1999,SSO Characteristic in East China Electric Power System[J]. *Power System Technology*, **23**(7):20-23.

MATEC Web of Conferences 63, 04002 (2016) DOI: 10.1051/matecconf/20166304002
MMME 2016

Research on the Safe Broadcasting of Television Program

Jin Bao SONG[1,a], Jin Hong SONG[2] and Jian Ping CHAI[1]

[1]*Information Engineering School, Communication University of China, Beijing, China*
[2]*Shandong Gold Mining Jiaojia Gold Mine (Laizhou) co.,LTD*

Abstract. The existing way of broadcasting and television monitoring has a lot of problems in China. On the basis of the signal technical indicators monitoring in the present broadcasting and television monitoring system, this paper further extends the function of the monitoring network in order to broaden the services of monitoring business and improve the effect and efficiency of monitoring work. The problem of identifying video content and channel in television and related electronic media is conquered at a low cost implementation way and the flexible technology mechanism. The coverage for video content and identification of the channel is expanded. The informative broadcast entries are generated after a series of video processing. The value of the numerous broadcast data is deeply excavated by using big data processing in order to realize a comprehensive, objective and accurate information monitoring for the safe broadcasting of television program.

1 Introduction

The existing way of broadcasting and television monitoring has a lot of problems in China. Firstly, the existing way of monitoring is the front-end monitoring finished before scrambling, rather than the user terminal monitoring [1]. Therefore it is not true effect of broadcasting monitoring in the network. Secondly, the existing way of monitoring is based on the video server. Sampling frame is no used. The monitoring cost is too expensive. In addition, the special line transmission is used in the present monitoring, which requirement for network transmission is very high [2]. So it only cover city level monitoring and don't cover the county and is not suitable for the county level monitoring. Thirdly, the existing way of monitoring is manual and passive without the information processing and analysis of the monitoring data. Therefore it cannot be automatic signal processing and data mining to realize active monitoring [3].

On the basis of the signal technical indicators monitoring in the present broadcasting and television monitoring system, this paper further extends the function of the monitoring network in order to broaden the services of monitoring business and improve the effect and efficiency of monitoring work. Firstly, the new equipment of sampling and processing video signals is researched which fits to be widely deployed at a lower cost in actual network. Secondly, the automatic or semi-automatic intelligence analysis software system is researched for mass program video signal. Thirdly, the watermark technology is researched for tracking certification of signal and program attributes in the actual broadcasting environment. The main purpose of this paper is the development of cheap monitoring hardware devices which can be widely deployed to the village, so the actual situation of the user terminal broadcasting can be monitored by the administration of radio, film and television. At the same time, the data mining and data statistical analysis of the monitoring data can make the monitoring data have a certain market value after deep processing of data information.

In order to achieve the above objectives, the main task of the paper are as follows. Firstly, the current central, provincial and prefecture level multistage program broadcasting management system and the cable, wireless and satellite broadcasting technology system are summarized and analysed. Using the cable television network as the breakthrough point, the technical architecture system suitable for identification of video content and television channels is studied and put forward. The technical architecture system must suit the actual situation needs of radio and television development and management in our country and has technical and economic feasibility of practical application promotion and deployment. Secondly, frame sampler with independent intellectual property right is developed which can recycle to sample, compress and store the frame of TV programs in the actual broadcasting network. Thirdly, the watermark recognition algorithm and image content recognition algorithm with independent intellectual property rights are developed for automatic or semi-automatic identification of the image/video, generating the broadcast channel information data and data mining in depth. The monitoring results and analysis, statistics and other information are output in a variety of forms such as reports and charts. Fourthly, according to the requirements of TV monitoring content and copyright

[a] Corresponding author: songjinbao@cuc.edu.cn

protection, considering the technical characteristics of the watermark, a feasible technical scheme is put forward. The watermark is overlaid to the video signal in the source end. After the transmission through the actual network, finally the watermark is reliable to extract from the video signal at the receiving end [4].

2 The architecture of television program safe broadcasting system

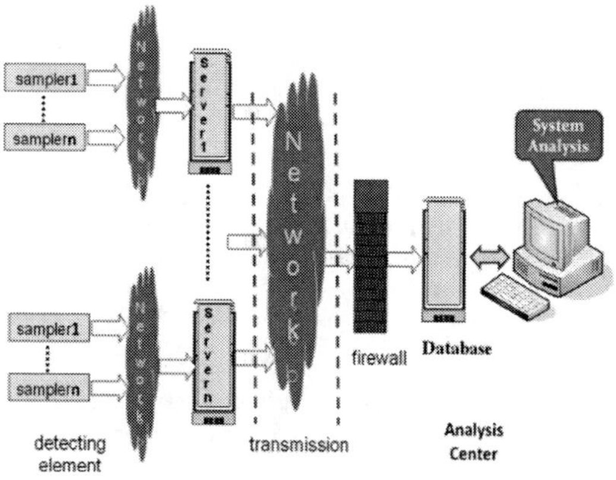

Figure 1. The architecture of television program safe broadcasting system.

The show program source is gained from detecting cities. There are two types of program, type Ⅰ superimposed by watermarking, type Ⅱ without any watermarking. The pictures of each shown program are sampled from cable television network using frame samplers. Then the data is stored in the disk, encrypted and transmitted back to analysis centre using the Internet at night when network isn't busy. All the information enters the data storage structure through the Internet according to frame dispatching mechanism in the analysis centre. For type Ⅰ, watermark is decoded. For type Ⅱ, it is treated with image matching retrieval. The formation of the actual radio and television broadcast programs data and audience rating data is produced by the two ways. Then all the information data entries processing and management system. In management system, through radio and television broadcast content analysis software will do data mining and statistical analysis of various content [5].

The concrete modules of television program safe broadcasting system are as follows:

- frame sampler module
- frame sampler dispatching module
- television content watermark superimposition and recognition module
- image matching and recognition module
- program background information data module
- storage structure, database module
- broadcasting and television public opinion analysis software modules

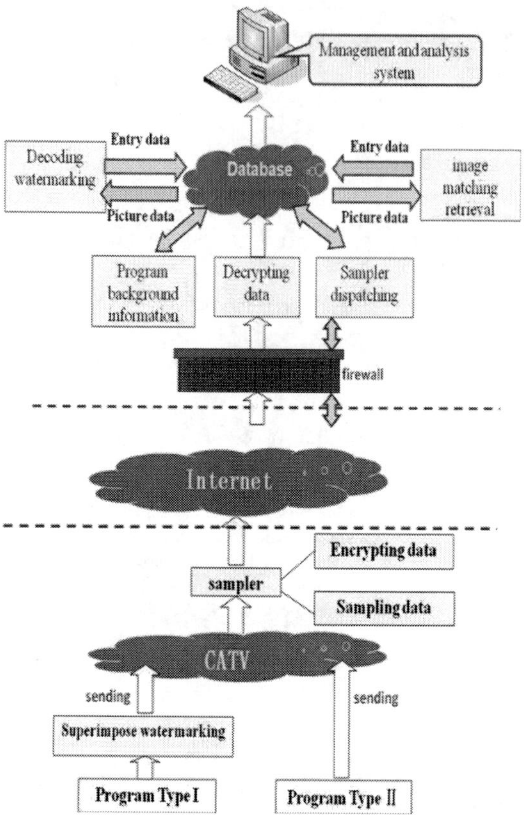

Figure 2. The concrete modules of television program safe broadcasting system.

The data collection layer is responsible for the original data collection at the bottom of the whole system. The original data refers to the key frames compressed and coded according to certain rules which are collected by the digital television sampler and the analog television sampler. Data collection layer is placed in cable TV network client. Daily work content of data collection layer is controlled and allocated by data scheduling layer.

The task of data transmission layer is to provide services for the communication between two nodes process. It includes the process of data scheduling layer sending instruction to data collection layer and the process of data collection layer sending data to the data storage layer, which are completed by the data transmission layer. Data transmission layer is deployed in the capital city of each province, directly connected with the provincial data collection layer.

The data processing layer is divided into data scheduling configuration system, data storage system and data analysis system. Data scheduling configuration system is task distribution and monitoring center. It is responsible for the output instruction of task allocation to the data collection layer, as well as the real-time condition monitoring of data collection nodes. It is also responsible for the monitor work of the system history data. The data scheduling configuration system is placed in the monitoring control center. Data storage system is placed in the monitoring center console. It is the data storage center of the whole system. It is responsible for dealing with a large number of key frame data. The frame data is classified and stored according to the corresponding rules. The data analysis system mainly complete the real-time monitoring and identification of broadcast content. It realizes automatically monitoring of TV advertising, program and news content. How to detect monitoring content is the key to the system. The data analysis system is placed at the top layer in the architecture.

Monitoring system as a whole and complex system, has wholeness, hierarchy and correlation characteristics. In order to make full use of the system's internal potential, ensure the stability and safety, at the same time expand the mining depth and scope of application and make the system not only meet the business requirements but also meet the need of development in the future, the following principles must be followed at the time of design.

1. The principle of safety.

Because the monitoring system has special position and role in the broadcasting and television system, high security is required by the system. The network security of the system is used to give full consideration to in this paper. The equipment used to make up the system takes effective measures to prevent the illegal invasion from the network. The corresponding authentication and certification are needed when all modules are visited. The key encryption can be configured to access module. In addition to the network security, fault inspection, alarm and processing mechanism, and the corresponding redundant backup mechanism are used to ensure the safe and reliable operation of the whole system.

2. The principle of reliability.

This paper not only considers advanced technology but also considers the aspects of the system structure, technical measures, equipment performance, system management and maintenance to ensure reliability and stability of the system operation. The fault inspection, alarm and processing mechanism are used to ensure that the data is not lost or damaged by accident situation. Load balancing is realized by using flexible task scheduling mechanism to prevent "bottleneck". When system failure occurs, seamless rearrangement is used in order to ensure the reliable stability of the whole system. The system design and equipment selection focus on stable and reliable products and solutions of large-scale use and long time running to ensure safe and reliable system operation.

3. The principle of extensibility.

The flexibility and scalability of the system are fully considered by this paper. Modular design is adopted to system hardware and software. With the increase of business types and growth of amount in the future, the system can be a smooth upgrade.

4. The principle of practicality.

The functions of the system in this paper must be based on the actual system and close to the operating habits and processes of the existing system. Although the requirements of the storage capacity is very big, but the image quality requirement of the playback is not high, which is good for long distance transmission.

The business model of the monitoring system is divided into two parts. One part is scheduling configuration business of the frame sampler on the client side in the network. The other part is data analysis and processing business of the sampled key frame. Scheduling configuration business of the frame sampler is completed by scheduling center and configuration center. With the aid of network transmission module, local real-time channel list can be obtained from the frame samplers deployed in the monitoring information network. The work task and work mode of each frame sampler are modified according to the specific tasks of monitoring center. Frame data analysis and processing business is completed through the task assignment components, combined with watermark identification, image matching automatic processing program and image artificial proofreading program. The analysis and processing of key frames data extracted by each frame sampler from each channel are completed in order to generate broadcasting entry information.

3 The key technology

The television program safe broadcasting system is intricate. We refine several key technologies as follows:

3.1. Analog/digital TV signal frame sampler

The mode of using frame sampling can reduce the amount of data. The existing radio and television broadcasting monitoring still do not use the way of key frames or I frame. This key technology provides a new way of thinking for the television program safe broadcasting monitoring.

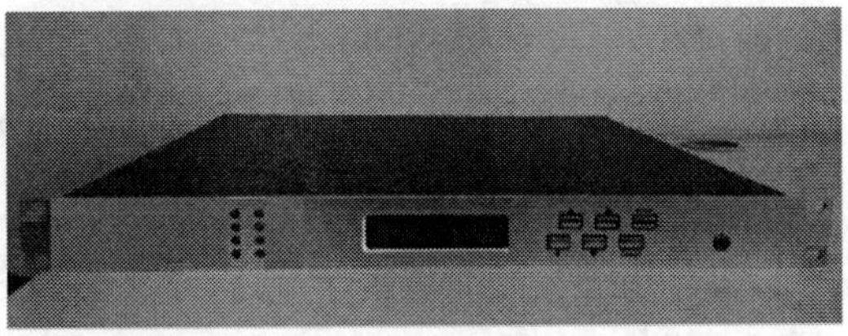

Figure 3. Analog TV signal frame sampler.

Analog frame sampler is carried out more developed version and reaches the following indicators: 1) a video key frame is extracted. Any extracted adjacent odd field or even field in any channel consist of a frame, which is compressed according to JPEG2000 and stored in separate files. 2) JPEG file is added city code, time code, channel number and the extraction time (accurate to millisecond) in addition. 3) Frames sampled per second is not less than 10. It can be classified according to the type of image content in full screen view (718 * 289 pixels) and station caption (230 * 74 pixels).

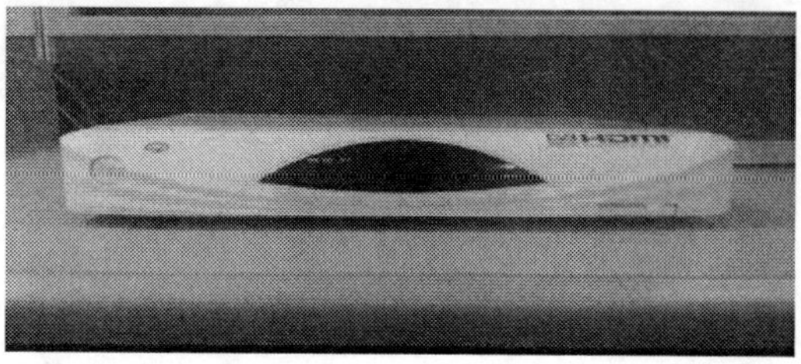

Figure 4. Digital TV signal frame sampler.

Digital frame sampler is carried out by a joint research group and reaches the following indicators: 1) a video I frame is extracted. The frame is compressed according to JPEG and stored in separate files. 2) JPEG file is added city code, time code, channel number and the extraction time (accurate to second) in addition. 3) Frames sampled per second is not less than one. The type of image content is only full screen view (720 * 576 pixels).

The functions of this key technology are as follows: Firstly, television signal frame sampler can be deployed to the bottom of the TV user terminal in an inexpensive way. Secondly, the amount of per frame data extracted by frame sampler is about 20 k bytes. The transmission bandwidth pressure of network is small by the way of sampling frame for television program safe broadcasting monitoring. The requirement of storage is also small.

3.2 Image watermark recognition

The vision image watermark identification process is as follows: Before the broadcasting, watermark is embedded into uncompressed video program with a watermark embedding device. The watermark includes information such as program copyright, etc. Watermark detection/extraction device tests I frame of broadcasting video programs, including the video programs with and without watermark, to identify the watermark and the content of the watermark. The identification results, namely the watermarking information is stored into the database. Watermark embedding device and watermark detection/extraction device must be matched. Several attacks which are one of the biggest impact on the watermark recognition rate and accuracy influence the actual use in television networks. The watermark system is relatively complex. We made a preliminary validation. For uncompressed program, watermark is superposed frame by frame. The superposition watermark includes embedding information for copyright. The program has no watermark visibility. The watermarking algorithm, the watermarking algorithm has compression resistance and can be blind detection. The watermarking algorithm has resistance to attacks such as stack, geometry transformation. It can enhance the information extraction speed and the recognition rate by using the image watermark identification.

3.3 Image matching and recognition

Image matching and recognition module mainly completes two tasks: station caption recognition and program content recognition. The feature extraction of original program frame samples is acquired including colour characteristic, shape characteristic and the texture characteristics, etc. The feature extraction of front-end broadcasting program frame samples is also needed. Using the image matching technology, the original program frame and broadcasting program frame sample feature matching samples. Whether the original program and the broadcasting program are the same is judged according to the matching result. The function of image matching and recognition is quick identification and to improve efficiency of information processing.

3.4 Data scheduling and processing

Data scheduling and processing system consists of three parts including the network server, the data centre server and database. Put in data analysis centre, extract entries information of all of the node distribution servers and extract JPEG image the hard disk from some frame sampler if necessary. It is to guarantee the effectiveness and reliability in the large amounts of data transmission in the limited bandwidth. According to the user access speed and trunk speed, we can make flexible configuration dispatching mechanism. The network server is a hub of data transmission in the whole project. The data flows are dispatched among the modules and the components by the network server as the role of a data centre station. The data centre processes large amount of data frames from the network server. The key information will be stored in the database. The frame data files are stored to disk at the same time. Data centre is responsible for saving and processing of the network server information and the frame extractor information at the same time. Database is responsible for the data itself and all kinds of information storage.

3.5 Program monitoring analysis

It is necessary to filter data in the database, including city, broadcasting television station, broadcasting time, broadcasting duration, program categories, keywords of program, program production company, broadcasting date. Graphics drawing pattern is set when statistical graphics are drawn after filtering data, which makes the final result of the statistical analysis more practical. Program monitoring analysis mainly includes statistical analysis based on program production companies, statistical analysis based on program type, etc.

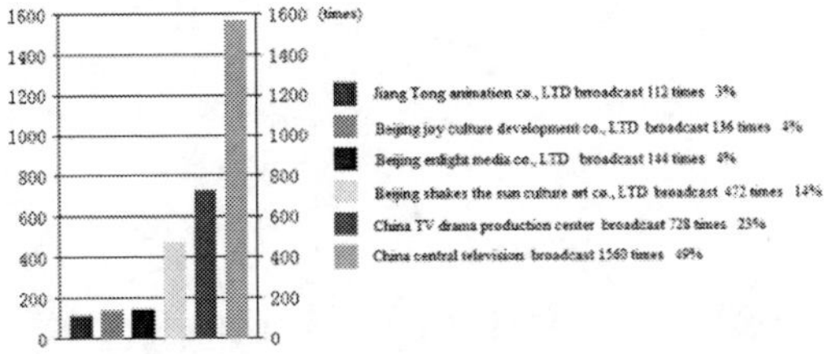

Figure 5. The ranking information of program production companies broadcast on CCTV-1.

4 Summary

The existing way of broadcasting and television monitoring has a lot of problems in China. On the basis of the signal technical indicators monitoring in the present broadcasting and television monitoring system, this paper further extends the function of the monitoring network in order to broaden the services of monitoring business and improve the effect and efficiency of monitoring work. The problem of identifying video content and channel in television and related electronic media is conquered at a low cost implementation way and the flexible technology mechanism. The coverage for video content and identification of the channel is expanded. The informative broadcast entries are generated after a series of video processing. The value of the numerous broadcast data is deeply excavated by using big data processing in order to realize a comprehensive, objective and accurate information monitoring for the safe broadcasting of television program.

References

1. Xu Xin, Zhang Zhicheng. Research on Application and analysis of Internet Mediated Public Sentiment[J]. Information Science, No 8,2008,1194-1201.
2. L.H. Wang. Public Opinion Overview: theory, method and realistic hot . Tianjin: tianjin academy of social science press, 2003.
3. G.LIU. in public opinion. The Function of Media Collection and Analysis of the Mechanism.Journal of Chinese reporters, 2005, (9) : 14-16.
4. Jinbao Song, Jinliang Wan, Xin Xin. An Internet TV Monitoring System Using Digital Watermarking Technology[J].The International Conference on Business Management and Electronic Information (BMEI2011)，2011,5,609-614.
5. Song Jinbao, Chai Jianping. An Analysis System of Broadcasting and Television Public Opinions[J]. Proceedings-International Conference on Management and Service Science(MASS2009),2009,9,923-927.

MATEC Web of Conferences **63**, 04003 (2016) DOI: 10.1051/matecconf/20166304003
MMME 2016

Accessibility Research of Highway Network System in Henan Province

Kai Guang ZHANG [1, a], Ming Ting BA [1,b] and Hong Ling MENG [1,c]

[1] *Institute of 3S Technology, Zhengzhou Normal University, Zhengzhou 450044, China*

Abstract. The accessibility index is designed by using the inverse of time distance and population weighted average value to research the city accessibilities in a region, with the definition of time threshold, which not only considers the transportation characteristics of each segment in the region, but also effectively reduces the digital effects from measurement on the analysis results, makes the analysis result could better reflect the reality at the same time. The results show that the Henan highway traffic network is relatively perfect, basically have been forming a high speed transportation system with the provincial capital, Zhengzhou, as the center. The city accessibility indexes in Henan province show the characteristics of normal distribution, the Accessibility differences are relatively obvious, the regional distribution of highway is very unbalanced. The accessibilities have high-high and low-low aggregation phenomenon, the Accessibility along the north to South and northwest to southeast direction becomes worse; the accessibility distribution and population distribution exist obvious departure.

1 Introdution

The accessibility means the interactions between the nodes in a traffic network and the traffic convenience from one node to others, which is an Important tools to evaluate the maturity of a regional traffic network structure [1-3], is firstly proposed by Dr. Hansen to study its impact on landuse in a region. In 1990, Professor Murayama studied the impact of the railway development on city accessibilities in Japan, and got some meaningful results[4]. Since the 1980s, Fractal theory, network theory, space syntax, shortest path, spatial topology were introduce in the study of city accessibility, which not only enriched the connotation of accessibility study, but also got some valuable research results. However these studies mainly focused on simple traffic network and used Euclidean distance measurement [5-14].With a complex highway traffic network gradually forming in China, city links gradually are becoming more diversity, national highway, fast-track lines, and more than one highway lines with different traffic characteristics constitute the main link between the cities in a region. The efficient highway traffic network closes the distance between cities, as well as gives a new connotation on the city accessibility research [15-18].

Henan province is the main body of Central Plains Economic Region and an important traffic hub in China. By the end of 2015, highway mileage is more than 6600 km forming a relatively well-developed highway transportation system with 9 national highway lines as the backbone, 22 local highway lines to complement. The developed highway has changed a lot in people's life style and greatly boosted the local economy development [19].

In the paper, based on Henan Province Ordinary traffic network, a new accessibility index is constructed by using the inverse of time distance and population-weighted average value, to evaluate the accessibilities between cities in the province and the sophistication of highway traffic network, in order to provide scientific reference for the region to further improve the network. Spatial data in the research comes from Henan traffic network database [19], the time measure bases on the empirical data of travelling through road junctions and segments between January to September in 2015, the shortest time distance between two points is calculated by NEDS algorithm[20].

2 City accessibility index

2.1 accessibility index

The city accessibility index in a highway traffic network is a degree of accessibility from one city to all adjacent cities, usually by using the weighted average distances deriving from population, regional GDP, passenger and freight traffic transport amount as the weighted, the optimum route number, the sum of shortest distance, they describe the numerical characteristics of city accessibility in different aspects[5-15]. However, for the complex highway traffic network as in China recently, the national highway traffic network is being intensified by the provinces (municipalities), two adjacent cities have

[a] Corresponding author: zzgis@126.com

 DOI: 10.1051/matecconf/20166304003

multiple highway lines and fast track lines to link, these descriptions hide some limitations.

Different highway line has different transportation characteristics with features complement each other, not a substitute for the function. Therefore, only using numerical characteristics of single line or the number of general lines to describe the correlation between cities is not comprehensive solution to analyze the whole city accessibility in a region. Highway traffic characteristics is an important attribute of highway traffic, mainly reflect on the comfortableness of driving and the average speed of traffic, the last could be is expressed as the time distance between the two points in time measurement.

There are various possibilities in direct connections and circuitous connections between two cities. In the study of city accessibility coming from highway traffic network, it is necessary to set a time threshold by only considering traffic routes within the threshold range. In the paper, the time threshold is set as the shortest travel time between cities on the ordinary traffic network(Correspond to highway traffic network), two routes on the highway traffic network are considered to be the same if their repetition rate is greater than 50%.

In this paper, the weighted inverse of time distance is used to carry on the city accessibility research. The accessibility from city i to city j is defined as the sum of the time inverse distance for all routes which fall in the time threshold range, and do not travel through the third city, denoted as A_{ij} .

$$A_{ij}=\sum\frac{1}{T_r} \qquad (1)$$

The larger A_{ij} means better accessibility form i to j . Here we should be paid more attention on $A_{ij}\neq A_{ji}$.

The city accessibility A_i within a region is defined as the weighted average of accessibility from all the other cities in the region to it,

$$A_i=\sum_{r=1}^{n}\frac{M_j A_{ji}}{M_j}, \qquad (2)$$

where, M_j is the weight of the city j indicating the influence degree of j on the accessibility of other cities in the region. Generally M_j used the residents population. The larger A_i means the better the accessibility of i is, and there are more attractiveness for the adjacent cities.

2.2 Probability distribution of accessibility

The probability distribution of accessibility describes the spatial distribution characteristics of accessibility in a region for all cities. The mean E indicates the average traffic capacity, the standard deviation σ and the range Z indicate the difference degree of traffic capacity in a region. The larger σ and Z are, the more significant the regional capacity difference is, the more departure between the distribution of highway traffic network and the distribution of population is in space.

$$E=\frac{1}{n}\sum_{r=1}^{n}A_r,\quad D=\frac{1}{n-1}\sum_{r=1}^{n}(A_r-E)^2,\quad Z=\max(A_i)-\min(A_i) \qquad (3)$$

2.3 Centrality correlation analysis of accessibility

The city accessibilities coming from highway traffic network in a region have some correlations with the location of cities and the weighted parameters. The accessibility of one city reflects its centrality degree in the region, and the convenience degree the people of adjacent cities traveling to it. The center of accessibility B_0 in a region is defined as the weighted average of city geographical coordinates, the center of population R_0 is defined as the weighted average of the number of residents population with its Accessibility as weight.

$$B_0=\left(\frac{A_i}{\sum A_i}x_i,\frac{A_i}{\sum A_i}y_i\right) \qquad (4)$$

$$R_0=\left(\frac{m_i}{\sum M_i}x_i,\frac{m_i}{\sum M_i}y_i\right) \qquad (5)$$

where (x_i,y_i) represents the geographical coordinates of the city i .

3 City accessibility analysis in Henan province

3.1 Time threshold in Henan province

The time threshold in the analysis of accessibility coming from highway traffic network is defined as the minimum time distance between two cities traveling in ordinary traffic network consisting of national ordinary trunk way (State Road), provincial ordinary trunk way (provincial road) (see Figure 1.), the calculation results are shown in Table1 (unit: minutes).

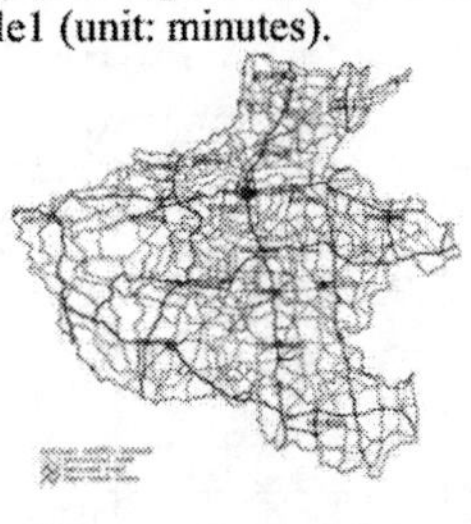

Figure 1 Ordinary traffic network

Figure 2 Highway traffic network

Figure 3. The distribution of city accessibility

Figure 4. The distribution of population

3.2 City accessibility in Henan province

The calculation of accessibility indexes from city j to i is based on highway traffic network (Figure 2.) by using formula(1), then calculate the shortest routes from city j to i and their traveling times which range in the time threshold, the results are shown in Table 2, then using formula (2), the number of residents population of j as the weight calculate the accessibility of city i, the results are shown in Table 3.

3.3 The spatial distribution of city accessibilities in Henan province

As shown in Figure 3., there is a local aggregation features in the spatial distribution of city accessibilities, presenting high-high aggregation in the central and the northern area, as well as low-low aggregation in the southern and the eastern area in the province. From the north to south and the northwest to southeast, it gradually deteriorated. Although the residents population weights are used to calculate the accessibility indexes, there is a significant deviation between the spatial distribution of accessibility and the spatial distribution of residents population (Figure 4.) , the most significant region is in the southern and the central area in the province.

Tablel. The time threshold in Henan province

	ZZ	KF	LY	PDS	AY	HB	XX	JZ	PY	XC	LH	SMX	NY	SQ	XY	ZK	ZMD	JY
ZZ		59	150	203			84	134		88						228		166
KF	59						106		232					140		220		
LY	150			168				134		221	283	175	329					91
PDS	203		168							121	163		199		318			
AY						48			150									
HB					48		71		166									
XX	84	106				71		128	199									
JZ	134		134				128											76
PY		232			150	166	199											
XC	88		221	121							66			237	226	133		
LH			283	163						66			256			102	70	
SM			175										375					
NY			329	199							256	375			215		225	
SQ		140								237						214		
XY				318						226			215			255	103	
ZK	228	220								133	102			214	255		156	
ZM											70		225		103	156		
JY	166		91					76										

Table 2. The accessibilities between cities in Henan province

	ZZ	KF	LY	PDS	AY	HB	XX	JZ	PY	XC	LH	SMX	NY	SQ	XY	ZK	ZMD	JY
ZZ		0.058	0.049	0.022			0.037	0.056		0.049						0.020		0.015
KF	0.058						0.030		0.014					0.038		0.022		
LY	0.049			0.034				0.029		0.018	0.008	0.020	0.004					0.042
PDS	0.022		0.034							0.035	0.017		0.023		0.007			
AY						0.040			0.034									
HB					0.040		0.031		0.024									
XX	0.037	0.030				0.031		0.033	0.013									
JZ	0.056		0.029				0.033											0.030
PY		0.014			0.034	0.024	0.013											
XC	0.049		0.018	0.035							0.027			0.018	0.007	0.047		
LH			0.008	0.017						0.027			0.010			0.030	0.022	
SM			0.020										0.005					
NY			0.004	0.023							0.010	0.005			0.009		0.008	
SQ		0.038								0.018						0.023		
XY				0.007						0.007			0.009			0.006	0.017	
ZK	0.020	0.022								0.047	0.030			0.023	0.006		0.010	
ZM											0.022		0.008		0.017	0.010		
JY	0.015		0.042					0.030										

Table 3. The accessibilities in Henna province

ZZ	KF	LY	PDS	AY	HB	XX	JZ	PY	XC	LH	SMX	NY	SQ	XY	ZK	ZMD	JY
0.038	0.035	0.025	0.023	0.036	0.033	0.031	0.041	0.021	0.030	0.018	0.011	0.010	0.026	0.009	0.020	0.012	0.027

3.4 Numerical characteristics of city accessibilities in Henan province

There are nine cities whose city accessibility are more than the mean $E = 0.0249$ and nine cities whose city accessibility are less than it. Jiaozuo is the most optimal accessibility city and Xinyang is the poorest accessibility city in the region, all of them are located in the border area of the province. The standard deviation and range respectively are $\sigma = 0.0104$ and $Z = 0.0371$ showing that the difference of city accessibility is relatively obvious, and the spatial distribution of highway is relatively unbalanced in the province.In Figure 5. the solid lines are the fitted curve of probability distribution for the standardized city accessibilities, and the dash curves are standardized normal distribution. The figure shows that the city accessibility indexes and the normal distribution fit very well, By hypothesis testing χ^2 and under the confidence of **0.10** , the probability distribution of city accessibility indexes meets the normal distribution.

3.5 Centrality correlation analysis of city accessibilities in Henan province

Accessibility center and the population distribution center of Henan province respectively are located in the northern part of Zhengzhou city and the northern part of Xuchang City (Figure 6., Figure 7.), there is 52 km between them. If the range of the accessibility and population are divided into 5 levels from large to small, and use the linear interpolation method ,we can get the spatial patterns of them as Figure 6., Figure 7.

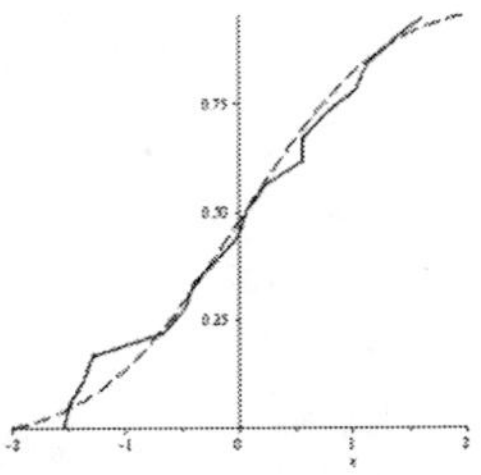

Figure 5. The probability distribution of urban accessibility

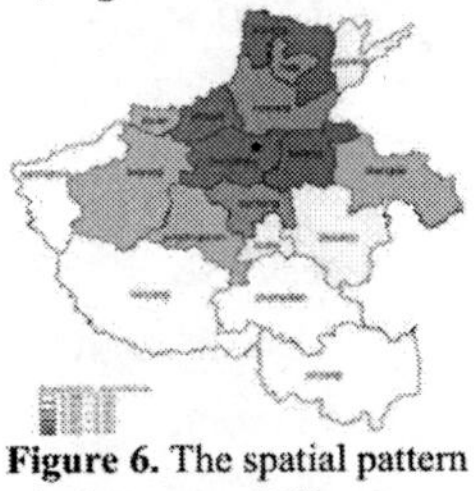

Figure 6. The spatial pattern of urban accessibility

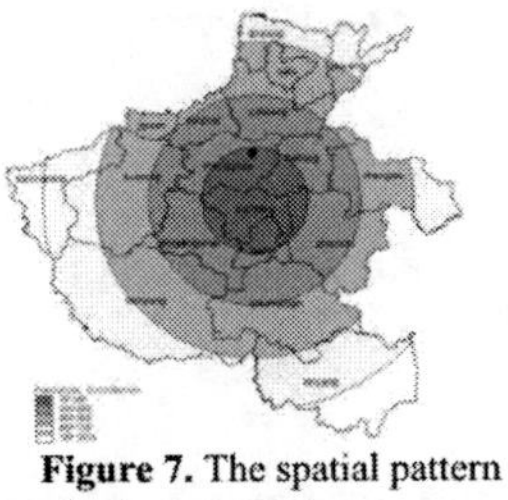

Figure 7. The spatial pattern of urban population

Contrasting Figure 6., Figure 7., It could be found that the central area of the population spatial distribution pattern fits with the central region of the accessibility spatial pattern in Zhengzhou and Kaifeng very well, but in other regions there are big differences. Unlike the population spatial pattern, the accessibility spatial pattern shows irregular distribution characteristics, the reason is that the city accessibility is not only affected by the condition of the highway traffic network, but also affected by the geographical position of the city.

4. Conclusion

(1) By using the inverse of time distance traveling through road junctions and segments and population weighted average value to research the city accessibilities in a region, with the definition of time threshold, which not only considers the characteristics of highway traffic network, but also effectively reduces the digital effects from measurement on the analysis results, makes the analysis result could better reflect the reality at the same time.

(2) There is a relatively perfect regional highway traffic network in Henan province basically forms a high speed transportation system with the capital, Zhengzhou, as the center. Compared with other domestic provinces (municipalities), the city accessibility is better.

(3) The city accessibility indexes in Henan province show the characteristics of normal distribution $N(0.0249, 0.0104)$, the Accessibility differences are relatively obvious, the regional distribution of highway is very unbalanced.

(4) The accessibilities have high-high and low-low aggregation phenomenon, the Accessibility along the north to South and northwest to southeast direction becomes worse;

(5)The accessibility spatial distribution and population spatial distribution exist obvious departure. the most obvious areas are in the south central part of Henan province.

References

1. W. G. Hansen, Journal of the American institute of planners, **25**, 73(1959).
2. L. Pinghua, L. Yuqi, Progress in geography, **24 (3)**, 69(2005).
3. R. Henrike, V. Amaya, Growth and Change, **43(4),** 667(2012).
4. L. Qinghua, W. Xinyan, Journal of Tibet University (Natural Science), **28 (1)**, 117(2013).
5. S. Jinghong, L. Yuqi, L. Xiaoji, Geographical Research, **31 (7)**, 1280(2012).
6. X. Jingge, Z. Tinggang, Journal of Geo-Information Science, **17 (1)**, 54(2015).
7. L. Baojie, G. Hehe, J. Yazhou, Human Geography, **6**, 76(2012).
8. X. Xu, C. Xiaoshu, Geographical Research, **26 (1)**: 180(2007).
9. L. Yating, Q. Yaoche, Y. Weiyang, Areal Research and Development, **29 (1)**,60(2010).
10. L. Hong, L. Xiaoyan, W. Chunguo, Areal Research and Development, **30 (1)**, 55(2011).
11. J. Awei, Progress in Geography, **31(12)**, 1591(2012).
12. C. Xiaoshu, X. Desheng, Y. XiaoPei, Acta Geographica Sinica, **60 (6)**, 903(2005).
13. J. Fengjun, W. Jiaoe, Acta Geographica Sinica, **59 (2)**, 29(2004).
14. W. We, C. Youhui, Acta Geographica Sinica, **61 (10)**, 1065(2006).
15. Y. Pei, highway, **10**: 141(2013).
16. Z. Li, L. Yuqi, Acta Geographica Sinica, **61 (12)**, 1235(2006).
17. Z. Bing, J. Fengjun , Economic geography, **26 (5)**, 776(2006).
18. Z. Kaiguang, S. Yu, B. Mingting, Areal Research and Development, **33 (3)**, 69(2014).
19. Henan Department of Transportation. Figure and row of Henan [EB / OL] (2015-10-10) [2015-10-10]. http: //road.hncd Gov.cn/page/txhn/txhn .jsp

20. Z. Kaiguang, M. Hongling, B. Mingting, Journal of Geomatics Science and Technology, **30 (1)**, 83(2013).

 DOI: 10.1051/matecconf/20166304004

Design of dual system teaching management system based on.Net Technology

Liang HU [1,a]

[1] *Jiangxi Science&Technology Normal University, Nanchang, Jiangxi, 330013, P.R. China*

Abstract. With college enrollment, the number of students continues to soar, significantly increasing the difficulty of teaching management. In the highly developed network of information technology today, how to achieve through modern information technology and efficient integration of network information resources, in order to achieve continuous improvement of the level of education, teaching and simplify management and reduce the human and financial investment, the university has become common Follow hot topic. Based on this, on a systematic and structural design of the .NET technology teaching management system based on a research point, the system through ADO.NET technology to access educational information and database retrieval, XML technology to achieve rights management, Finally, this system functions realization more thorough discussion. This system is used for the teaching management mode of German dual system localization, has a good effect

Key words: Dual system, XML, ADO.net, localization

1 Introduction

Today, the information age, based on the construction of information management system has become an important indicator of the level of teaching in Colleges and universities [1]. Teaching management system is the core of the construction of digital university. This kind of system can through modern information technology integration teaching resources, realize the sharing of resources, to provide a new and effective teaching platform for teachers and students, and optimize teaching resources investment, simplifying the task of teaching management. The following paper on a systematic and structural design of .NET technology-based education management system started to explore and implement this function of the system is analyzed.

2 Significance of Teaching Management System Design

Based on .net technology education management system is at the same time as the students and teachers to serve students in this system, through online technology, access to educational resources, courseware, learning and other auxiliary materials. Teachers are teaching in the system management module, the content of student information, teaching arrangements query and management, and through the campus network, publishing important information and teaching courseware, network instruction, etc. [2]. Thus, based on .NET technology while teaching management system can provide teachers and teaching management services, not only to provide network information management and teaching for teachers, but also be able to provide students with an efficient education resource sharing learning platform.

3 System design of teaching management

3.1 System Architecture Design

The learning management system access via ADO.NET technology and the transfer of educational information database to XML technology to achieve rights management, system design hierarchy, users can log in through a browser, the background provided by the WEB system and network database service support, the specific structure as shown below.

[a] Corresponding author: 596325696@qq.com

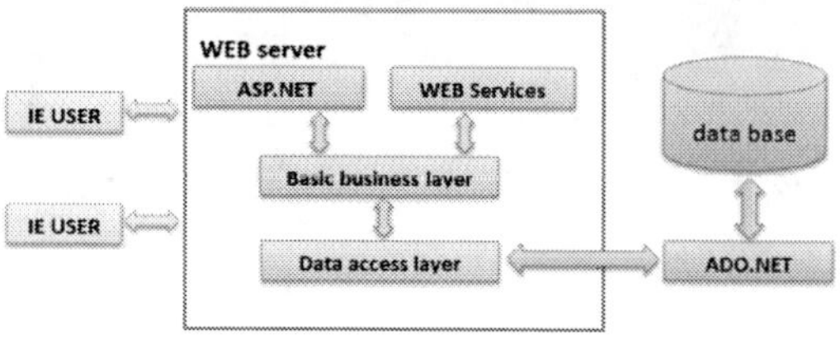

Figure 1. Schematic diagram of the structure of teaching management system based on.Net Technology

3.2 Function of the system design of each module

System also teachers and students as clients, according to the different needs of teachers and students teaching management, resource management module is divided into teaching, basic management module, user management module and switch modules and other teaching resources.

3.2.1 The function of education resource management module design

The main function modules include the preparation, revision, upload, review, revise teaching resources and so on.

Teaching Resources Upload: Teacher by logging in the client module to operate various types of teaching resources, storage and uploading.

Teaching resource sharing and download: teachers and students can be found teaching resources have been uploaded in the system, after registration, and free downloads.

Teaching Resources Audit: Before the formal uploading, you need to review the teaching staff responsible for evaluation, after passing the examination before officially uploaded.

Teaching Resources inquiry: teachers and students can be queried by entering the keyword, the system will enter keywords related resources with a display of all out, teachers and students can select from teaching resources they need to be viewed or downloaded.

Teaching Resources Delete: system administrator or upload resources educators can not meet the standards, or there is an error of teaching resources delete operation.

Teaching Resources Statistics and Analysis: The system of teaching resources Click to view the amount of downloads for automatic statistics, and ratings based on statistical results to help teachers and students selected the most valuable teaching resources.

3.2.2 The basic function design system management module

In this module can be serviced teaching management system, the system administrator and the system administrator privileges given to those who register before the landing module [3]. Key features include network configuration, fault repair, network information security, performance, maintenance, and other statistics and analysis. Also provided with extensible interface that enables the system to achieve system expansion and upgrade work through this interface.

3.2.3 Function design of user management module

Although to improve teaching quality, university teaching resources should share resources, but not all types of resources can be fully open, so as to achieve information security management, we need to set user permissions in this module. By the system management personnel, including teachers, students, auditors, unregistered persons (tourists) [4] set permissions. Such as registering students to teaching resources query, uploading, downloading and other operations, and teachers on the basis of students' rights, the courseware can be edited, modified, etc. operations. Tourists will only search and browse without downloading privileges.

3.2.4 Teaching Resources switching module functional design

Teaching Resources This module exchange is a resource sharing model, not a teaching resource management module campus resource sharing, but remote data exchange, sharing of resources between institutions at all levels. This module teaching resources in XML format for output, the system by means of information stored in the DOM XML format for access, then depending on the permissions of the user configure the menu, the process shown in Figure 2.

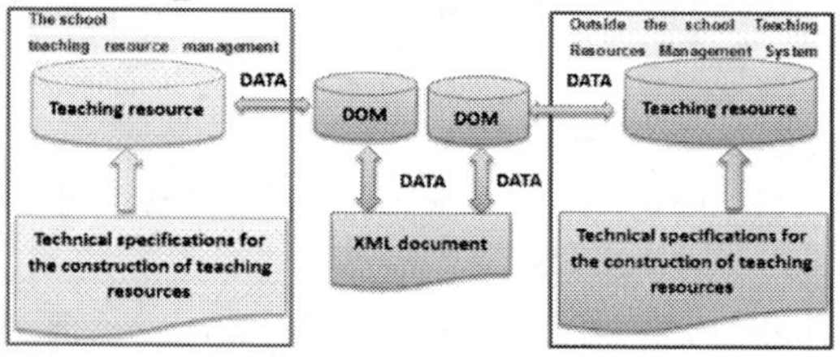

Figure 2. Data exchange process design of teaching resource exchange module

4 Function of the system to achieve

System is a hierarchical structure to achieve functional modules, divided into presentation logic, business logic, information layer.

4.1 Implementation of a logical layer

The presentation layer logic layer that is used to implement the interface between systems and the role of the user in the system, this logic layer mainly in the form of a dedicated web page rendering, customers and dialogue between programs in a dedicated web page, such as a user enter, log, etc., and feedback the corresponding results based on the user's actions.

4.2 The implementation of business logic layer

This logic layer is the core of the whole system, the whole system is responsible for data processing. System to design and deliver the convergence between the common database and be responsible for each database, and then developed based on the operational process design two kinds of business logic components of the system, are VB.net technology, namely basic business management and business components, responsible for basic component business is for users to achieve system operation and function of the client implementation [5]. Another component is responsible for managing the business, that system administrators use component management system management is mainly achieved through this component.

4.3 Implementation of information layer

System information layer is mainly responsible for the management and analysis of information and data, and analysis results back to the business logic layer, and then implemented by the business logic feedback function [6]. The system management module teaching resources, teaching resources through the switching modules are required to implement an information layer functions operate more modules to run in the database layer ADO.net information technology to connect and access the database to interact with the data transfer, as well as NET technology systems through greatly enhanced the familiar data access, and storage process more flexible.

5 Conclusion

To sum up, this paper firstly analyzes the design significance of the teaching management system based on.Net technology, and then discusses the overall structure design of the system, on this basis, respectively of educational resources management module, system management module, user management module and teach learning resource exchange module function design are discussed. Finally, the said logic layer, business logic layer and information layer of the concrete implementation scheme are discussed, hope that this research can provide some reference for the stakeholders.

This paper is the Achievements of Jiangxi Science and Technology Normal University Institute of Technical Education topics project that localization Practice Research of Germany dual system mode .

References

1. Li Jing, Zhang Jianhua. CD software and application of the design and research of [J]. ASP.NET technology in computer network teaching system based on 2014,20:295-296.
2. support, Fuchun rock, Xue Jiamei, Zhang Jingda, Guo Jiping, Ge Maosong, Liuyue, Qu Silong, Xiao Guang su. Based on. Net environment of the ordinary college undergraduate course teaching level evaluation management system design. Journal of Jiamusi University Journal (NATURAL SCIENCE EDITION), 2010,06: 894-896.
3. Yuan Hongjuan, Gu Sheng, Xu Yan. Based on XML and ASP. Net technology to achieve the teaching resource management system [J]. Xi'an University of Arts and Science (NATURAL SCIENCE EDITION) Journal. 2012,01:92-95.
4. Jia Wei Zhang, Longbin, Kan Jianxiong, Wang baijuan. Based on. Net technology class information management system design and implementation of wireless Internet technology [J]. And 2012,08:65-66.
5. Yang Chaohui, Wang Fang. And the implementation of [J]. technology and business vocational college educational management system design based on ASP.NET technology, 2013,13:144.
6. Zhong Wu Li, Chen Jingrui, summer Chunmei. Based on. Net technology of Baoshan college online school management system design and implementation of computer in Fujian [J]. And 2013,11:40-41.

Design of a WAV audio player based on K20

Yu XU, Chao Chao LI and Yong YANG[a]

College of Electric Information Engineering, Beihua University, 132021, Jilin, Jilin, China

Abstract. The designed player uses the Freescale Company's MK20DX128VLH7 as the core control ship, and its hardware platform is equipped with VS1003 audio decoder, OLED display interface, USB interface and SD card slot. The player uses the open source embedded real-time operating system μC/OS-II, Freescale USB Stack V4.1.1 and FATFS, and a graphical user interface is developed to improve the user experience based on CGUI. In general, the designed WAV audio player has a strong applicability and a good practical value.
Key words: WAV audio player; MK20DX128VLH7; μC/OS-II; FATFS; CGUI

1 Introduction

With the continuous innovation and progress of digital electronic technology, network technology, embedded technology and communication technology, the human life is undergoing a dramatic changing. The embedded system has become the most advanced technology in the present society, equipment of embedded system has exceed the PC in the application, and the application of embedded technology in NC intelligent equipment, industrial robots and automotive consumer electronics is also increasing [1]. A cutting system of WAV audio player based on embedded technology is designed in this paper, its function is reliable, the cost is low, the volume is small and the power consumption is low. There are many kinds of solutions to design audio player, but it still has a lot of room to improve in the overall power consumption, performance and sound quality. Therefore, the designed player is combined with the characteristics of the existing audio player and takes its essence to enhance the overall performance, and it has a certain practical significance [2].

2 The general design of the system

2.1 Principle analysis

This paper designs a WAV audio player system which uses MK20DX128VLH7 as the core controller. The main work of audio player is to read the audio file, the file decoding and information display [3]. Therefore, the overall design of an audio player can be divided into the following three processing stages:

a). accessing audio files, mainly including the transmission mode and the choice of storage media;

b). decoding audio files, the main decoding scheme is the choice of soft or hard to decode;

c). audio file information display, including the selection of the display, mainly to add more features and gorgeous interface to increase the user experience.

After a clear understanding of the different stages of the audio player design, we must complete the corresponding task at each stages, the next process mainly is the design of hardware circuit and software hierarchy.

2.2 Market demand analysis

The designed WAV audio player summarized the characteristics of the existing audio player and combined with the development trend of the audio player, it has the characteristics of compact and flexible, rich function, low overall design cost and so on. The specific functions are as follows:

a). USB download, support USB2.0 full speed;

b). SD card storage, support different types of SD card;

c). human-computer interaction and OLED display;

d). lyrics and audio information display;

e). support a variety of audio format, such as WAV, MP3 etc.

3 Hardware design

The WAV audio player uses MK20DX128VLH7 as the core control unit, it receives the audio files from the PC and then to decode audio files and plays them. The system integrates the keyboard input, OLED display, SD card storage, USB communication, power management and JTAG download function unit. According to the composition and function of the whole system, the principle block diagram of the system is designed as shown in figure 1.

[a]Yang Yong: 344861686@qq.com

 DOI: 10.1051/matecconf/20166304005

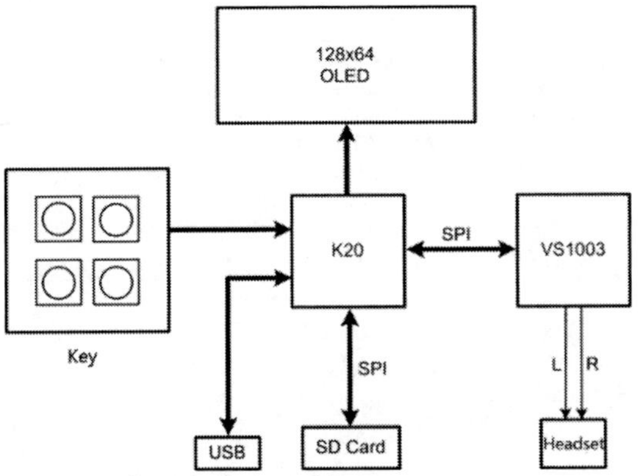

Figure 1. Schematic diagram of the system

The WAV audio player downloads WAV, MP3, WMA and other formats of digital audio files from the PC by USB interface, and then stores them on the SD card through the FATFS files system. The users can view and play the audio files on the SD card through the buttons and OLED display. When the player is playing the audio, the main control chip K20 reads the appropriate file content from the SD card and feds into the decoder VS1003 to decode and play through the SPI interface [4].

3.1. the Minimum system design

The peripheral circuit design of micro controller is shown in figure 2. S2 is a key switch for chip reset control; X1 is a crystal of 8 MHz which provides work clocks for MK20DX128VLH7; X2 is a crystal of 32.768 KHz which provides real-time clock for MK20DX128VLH7 (RTC); BT1 is a lithium ion battery of 3.7V to provide power for the system; J2 is output of serial port (UART) and is he output of debugging information of the system.

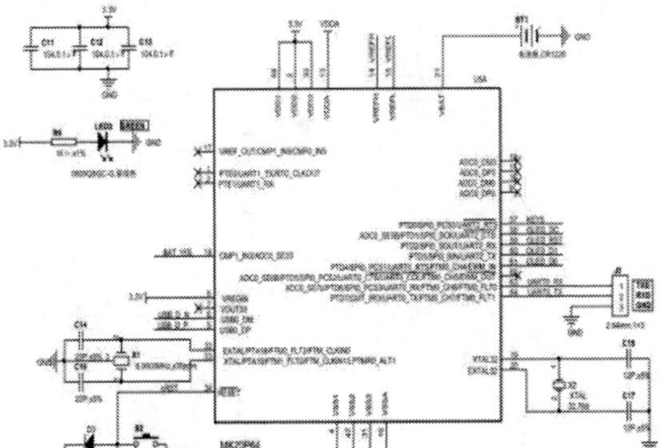

Figure 2. K20 minimum system

3.2 Voltage regulator circuit design

The WAV audio player uses 3.7V lithium ion as power source, and it can be charged through the USB interface. According to the desired voltage level, the whole system can be divided into three parts for 5V battery charging circuit, 3.3V power supply of system internal chip and 2.5V power supply of internal digital circuit VS1003.

The 5V battery power supply circuit uses power management integrated chip TP4056, it is a perfect function of a single lithium-ion battery constant current / constant voltage linear battery charger. The internal structure of the chip is designed with the PMOSFET structure, and the anti reverse charge circuit is designed, which needs no external isolation diode and simplifies the peripheral design. TP4056 charging voltage is limited to 4.2V, and the charging current can be adjusted by setting external resistors. If the charging current reaches the set float voltage value of 1 / 10, TP4056 will terminate charging automatically and it can be recovered when needed.

Because TP4056 has less number of external devices and it use the effective cooling package of SOP8, it is widely used in portable devices and variety of chargers. TP4056 charging circuit diagram is shown in figure 3, VCC is the 5V USB power input for internal circuit operation, BAT is connected to the positive side of the battery, which provides the current and 4.2V voltage to the lithium ion battery.

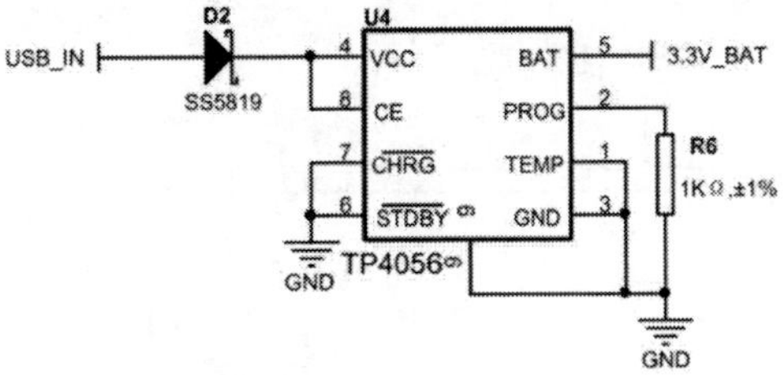

Figure 3. Lithium battery charging circuit

CAT6219-330TD is adopted in the 3.3V voltage regulator unit, it is a 500mA CMOS low dropout regulator and the initial output voltage accuracy is plus or minus 1.0%. And it possesses a fast response time during the fluctuations of the load current and line voltage. The stable operation can be done by a small ceramic capacitor which reduces the area and the cost of device greatly. CAT6219-330TD is becoming an ideal choice for equipment of 2.3V~5.5V battery power supplies because it's rapid startup characteristic, under-voltage lockout feature and low no-load current. The 3.3V voltage regulator circuit is shown in figure 4.

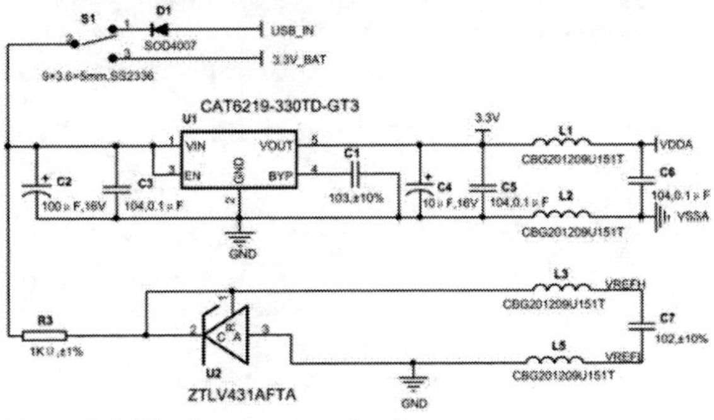

Figure 4. 3.3V voltage regulator circuit

 DOI: 10.1051/matecconf/20166304005

VIN is the voltage input which can choose the way of power supply mode for USB interface power supply or lithium ion batteries through the switch S1, VOUT is the stabilized voltage output which supplies for other IC in players system, and the ZTL431 is used to provide a 2.5V internal reference voltage source for MCU.

The 2.5V voltage regulator unit uses integrated chip SPX1117M3-2.5, it is a 800mA regulator of positive and ultra-low power consuming, and it has a very low static current. The chip of SPX1117M3-2.5 is very suitable for application in portable embedded devices and battery power supply equipment because its low differential pressure when full load is demand.

3.3 SD card interface circuit design

The WAV audio player's SD card storage circuit is shown as the figure.5. SD card module combines MCU through the SPI bus interface. CD/DATA3 is used as the selection line CS of SPI chip, CMD is used as the MOSI data output line in the SPI bus, DATA0 is used as the data input bus, CLK is the clock line, Card Detection is used to support hot plug of SD Card.

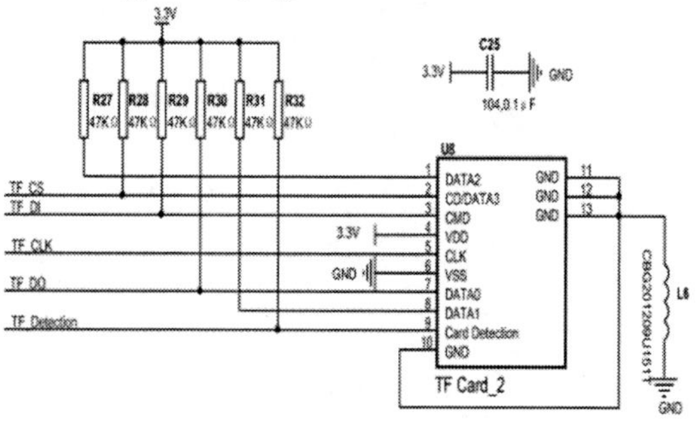

Figure 5. SD card interface circuit

3.4 The man-machine interface circuit design

The man-machine interface part is mainly composed by key control input, JTAG download, and OLED LCD display and so on. The key circuit design uses five buttons, as shown in figure 6. One button has integrated signals of multiple directions for the controller to identify. Nowadays, the handheld devices is more and more adopted the multidirectional integrated button, because it can reduce the circuit board space of multiple keys and easy to integrate.

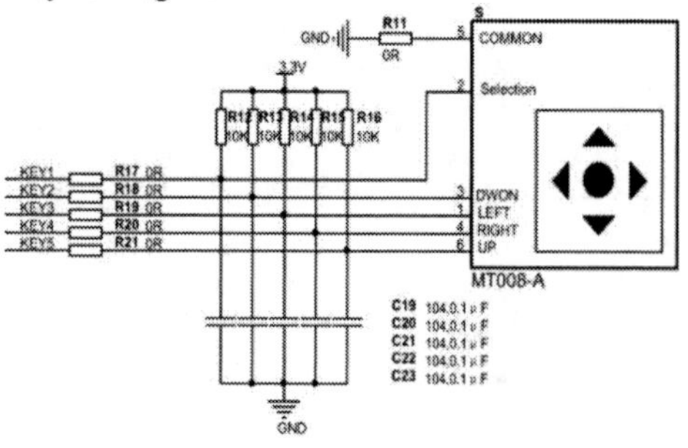

Figure 6. The key circuit

The OLED12864 is adopted in liquid crystal display. Its distinguish ability is 128 x 64 and its visible angle can reach 170 °. Its power consumption of the normal display is just 0.06W that far lower than TFT display, and it can support the communication mode of SPI and IIC. The circuit design of the OLED12864 is shown in figure 7. GND is power cathode, VCC is the positive, D0 is used as the CLK of serial input clock, D1 is used as the serial input of data, RST is the reset signal, and the DC port controls the input data or command.

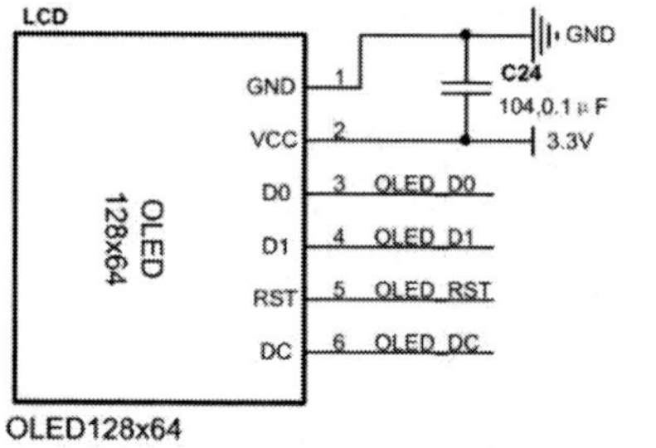

Figure 7. OLED display circuit

3.5 Audio playing circuit design

The data communication way of VS1003 and the core microcontroller MK20DX128VLH7 is SPI bus, circuit design is shown as figure 8. VS1003 receives audio file data and control commands of MK20DX128VLH7 through the serial data interface (SDI) and a serial command interface (SCI), and it confirms the interface whether is undering the state of transmission through the high or low level of XCS and XDCS. The main control function of VS1003 is the system initialization, play and pause, soft reset, volume control and the reading of audio playing time and so on, all of these functions are realized by write specific registers through the SCI. Between the two SCI instructions, the system need to judge whether the last data processing is completed through query the signal of DREQ [5].

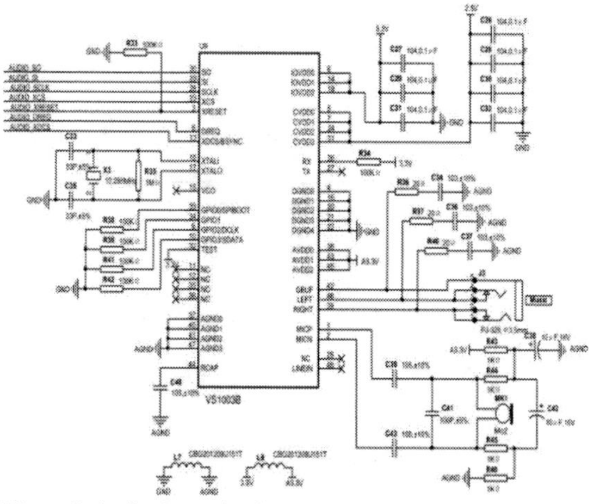

Figure 8. Audio player circuitt

4 Software design

The multi task operating system of μc/OS-II is adopted to manage all tasks of the player in this design, and the μCGUI graphics library is used to construct the graphical man-machine interface. The system's software is divided into four layers of driver layer, interlayer, application layer and system layer. The driver layer includes a bottom interface of the peripheral and board level peripheral devices. The intermediate layer is a transplanted code library of the file system, USB protocol stack and μ CGUI. The application layer is used to realize each function module of the system. The system layer uses μ c/OS-II to manage and schedule the task of the whole system.

Since the player uses the operating system, the overall task management of the system becomes very easy. The software procedure is shown as figure 9.

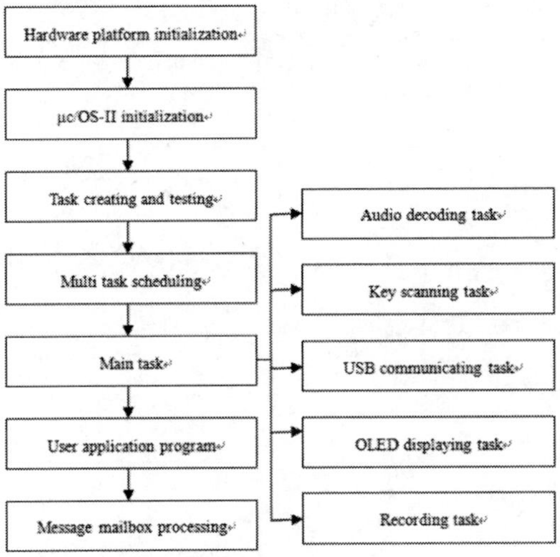

Figure 9. Main program flow chart

First of all, we need to initialize the system hardware platform, such as the basic peripheral registers and clock configuration. The next task is to initialize operating system and to write application task programs, and then starting the multi task operation and scheduling tasks after all the tasks are successfully created. Finally, the application layer drives programs indirectly in the management of the operating system through the middle layer, and the music playing and human-computer interaction is achieved ultimately.

4.1 Music playing program design

Music playing is the core of this system, the design of music player program should be important part of the software design. The function of music playing needs two procedures of reading music file and decoding broadcast. Because the size of an audio file is at least 512 bytes and reads it needs 1-2ms, while the audio decoder VS1003 data cache is only 32 bytes, the playing length is less than 1ms, so the data writing frequency of decoder is more frequent than the reading speed of SD card. There will be a pause if the scheme of single task reading and writing is used. The two tasks to read and decode respectively is designed in this paper, the program flow chart is shown in Fig. 10.

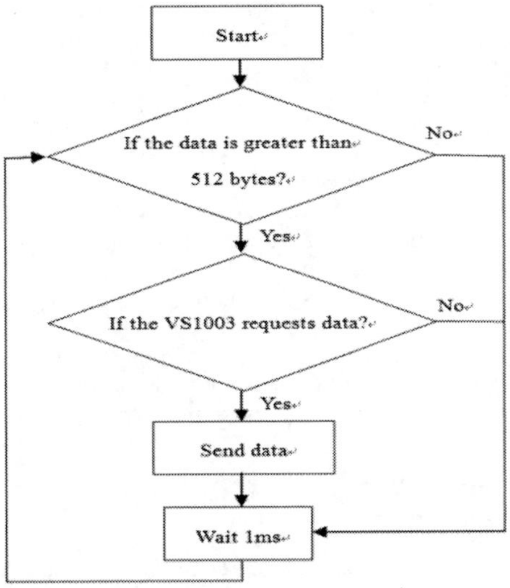

Figure 10. Music playing program flow chart

4.2 Key Program Design

In order to achieve more key functions, the key scanning program should obtain information of key codes, names and time when a button is pressed. The key structure has been defined to manage events of the key conveniently, and its definition is shown in following program.

```
typedef struct {
    INT8U    ucStatus;          /* 1, press; 0, release */
    INT8U    ucIsLongPress;     /* 1, long press; 0, short pres*/
    INT16U   usPressedTime;     /* Continuous time, ms*/
#if KEY_NAME_EN
    char     *pcName;           /*key name, Optional*/
#endif
    INT8U    ucKeyMask;         /* key code, equivalent to key value */
} KEY_DEF;
```

The key scanning program checks the status of buttons every 50ms as a separate system task. The program will wait for the button to loosen if it finds a button is pressed. If the button is pressed for a long time, the signal of long press will be sent to the system, or it will be the short. The key signal is sent to the system in the form of a μ COS system mailbox, the mailbox carries the instance pointer of the movable key, and the key processing program can get all the information of the active key through the pointer.

5 Performance test

The performance test of WAV audio player includes storage media reading and writing speed, USB transmission speed, audio performance index, key response speed and so on. The way of storage media reading and writing speed test is to copy a fixed size files into memory, and measuring the time of the whole process. A 1.3M file is tested in this design, the writing time is about 18 seconds to 20 seconds and the reading speed is about 2 times of writing. The actual test file is written and read successfully, and the contents of the file are correct. The method of key test is to measure the IO port waveform of the button by an oscilloscope, and then to calculate the interval time between the two interrupts signals. The actual test of the key response time is 100ms.

From the test, the overall design of the player has reached the expected design requirements, the extension mechanism and the decoding of different audio formats is realized, and the key input can be response correctly. After a long running test, the system runs stably and has good performance in the process of playing music.

6 Conclusions

A WAV audio player system based on K20 is designed in this paper. From the test results, the designed player meets the expected requirements. The designed audio playing system provides a complete solution to the design of an audio player. Due to the designed player has function of the modular expansion in the application software, the WAV audio player will have a high practical value in engineering applications.

References

1. Gao Meizhen. Based on embedded system design of MP3 player [J]. Hubei Normal University Journal (NATURAL SCIENCE EDITION), 2010, 04:92 - 94.
2. Rudy Susan F.The sounds of handheld audio players.[J].ORL - Head and Neck Nursing, 2007, 802:103-104.
3. Li Shiqi, Dong Haobin, Li Rongsheng. The SD card memory design [J]. control technology, FatFs file system based on 12:79-81. 2011.
4. Hong. Jia Ping VS1003 decoder MP3 player [J]. microcontroller and embedded system application, based on 11:54-56. 2010.
5. zoujie, Tang young year. Based on embedded system design of MP3 player [J]. China water transport (second half), 2013

MATEC Web of Conferences 63, 04006 (2016) DOI: 10.1051/matecconf/20166304006
MMME 2016

The improved relative entropy for face recognition

Qi Rong ZHANG [a], Ji Yan ZHANG and Yan WANG

Information Engineering, Guizhou University of Engineering Science, Bijie 551700, Guizhou , China

Abstract. The relative entropy is least sensitive to noise. In this paper, we propose the improved relative entropy for face recognition (IRE). The IRE method of recognition rate is far higher than the LDA, LPP method, by experimental results on CMU PIE face database and YALE B face database.

Keywords. Face recognition; Relative entropy; Improved relative entropy

1 Introduction

Through decades of development, face recognition has gradually get one of the most attract research domain in image analysis and computer vision, because a large number of applications. A large number of face recognition methods have been investigated such as Principal component analysis (PCA) [1], Linear discriminate analysis (LDA) [2], Independent component analysis (ICA) [3], Elastic bunch graph matching (EBGM) [4], Support Vector Machine (SVM) [5], Isometric feature mapping (ISOMAP) [6], Locally linear embedding (LLE) [7] [8], Laplacian eigenmap (LE) [9] [10], Locality Preserving Projections (LPP) [11]. Principal component analysis (PCA) [1], linear discriminant analysis (LDA) [2] and locality preserving projections (LPP) [11] are typical linear algorithms for face recognition. Sirovich and Kirby first applied PCA, and then the well-known Eigenface method is proposed. PCA and LDA intend to project the vectors which represent face images onto lower dimensional subspace. PCA does not work for the class information, so it's an unsupervised method. LDA can take full consideration of the class labels of the input data. So the LDA is should be a supervised method. In fact, LDA is to find the projection axes that when face image data points of different classes are projected as much as possible the far from, while face image data points of the same class are nearly as much as possible. LPP can construct a face subspace by preserving local information which is a recently appeared method. LPP is also an unsupervised method, but LPP is to maintain the human face image space of the local structure, while PCA is to maintain the global structure.

In statistical physics the most important concept is the Boltzmann's entropy [12]. This concept is introduced by Shannon into information theory, called Shannon's entropy [13]. Some scientific field research and application of the concept of entropy is a wide range of [14] [15] [16] [17] [18] [19] [20] [21]. Guang Deng found the relative entropy appears to be the least sensitive to noise through experiments [23]. However, Relative entropy has a drawback that it is potentially sensitive to noise in dark areas. Therefore, it is performing a pixel-wise multiplication for an image to improve the relative entropy of this shortcoming. In this paper, we will use the improvement of the relative entropy for face recognition.

The rest part this paper is organized as follows: Section 2 introduced the proposed the improved relative entropy for face recognition (IRE). Section 3 reports the results of the experiment on CMU PIE face databases. Finally the conclusion of this paper is given in Section 4.

2 The relative entropy

2.1 The relative entropy

2.1.1 Shannon' entropy

The Shannon' entropy can be defined as follows [13].

$$H = H(P) = -\sum_{n=1}^{N} p_n \log_2 p_n \quad (1)$$

$$p = \{p_n\} \quad (2)$$

p_n is the probability of occurrence of the i th outcome or state.

2.1.2 Relative entropy

The relative entropy is defined as follows [22]

$$D(Q \| P) = \sum_{n=0}^{N-1} q_n \log \frac{q_n}{p_n} \quad (3)$$

[a]Qi Rong ZHANG: zqr_hk@163.com

$$=\frac{1}{N}\sum_{n=0}^{N-1}(\log A-\log x_n) \quad (4)$$

$$=\log\frac{A}{G} \quad (5)$$

$$p_n=\frac{x_n}{S} \quad (6)$$

$$S=\sum_{n=0}^{N-1}x_n \quad (7)$$

$$q_n=\frac{1}{N} \quad (8)$$

$$A=\frac{S}{N} \quad (9)$$

$$G=\exp(\frac{1}{N}\sum_{n=0}^{N-1}\log x_n) \quad (10)$$

Where Q is probability model, P is also. A is the arithmetic mean. G is the geometric mean. So, Esq. (5) shows the relative entropy can be the ratio of the arithmetic mean to the geometric mean.

3 The improved relative entropy for face recognition (IRE)

The negative image is given as follows [23]

$$y=M-x \quad (11)$$

x is a pixel of the original image and y is a pixel of negative image.

We can compute a pixel-wise multiplication as follows [23]

$$Y=D_1\times D_2 \quad (12)$$

D_1 is the relative entropy of a pixel and D_2 is the relative entropy of a pixel .

The improved relative entropy for face recognition

Step.1. Input face images

Step.2. Compute the relative entropy of each pixel by Esq. (5), (9) and (10), using a neighboring (in this paper, the experiments using 3 × 3 neighborhood window).

Step.3. Computed by Esq. (11) and (12).

Step.4. Compute the Euclidean distance between of training face images and of testing face images.

$$d(Y_1-Y_2)=\|Y_1-Y_2\|_2 \quad (13)$$

4 Experiments

4.1 CMU PIE database

We checked the feasibility and effectiveness of IRE method on CMU PIE face database.

The CMU PIE face database includes 68 persons with 41,368 face images with varied pose, illumination and expression which were photographed by thirteen synchronized cameras and twenty-first flashes in different pose, illumination and expression. In this experiment, we used twenty-one face images of sixty-eight persons with illumination variations. Each image is resized to 92×112. We arbitrarily selected face image from CMU face database. Fig.1. displayed face sample images from the CMU PIE database.

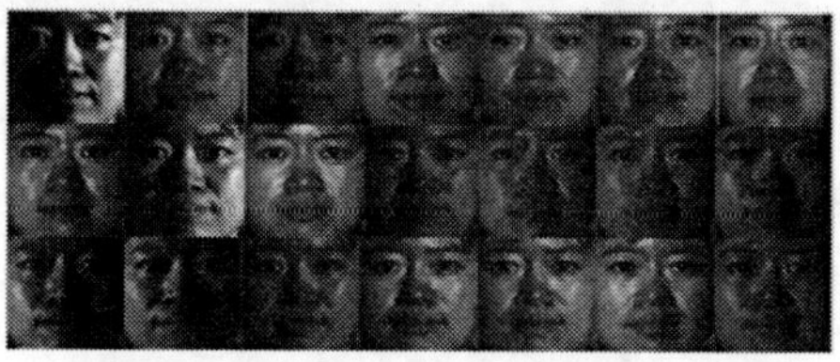

Fig.1 Illustration 21 images of one individual from CMU PIE face database.

Fig.2 Illustration 21 images of one individual from CMU PIE face database by the relative entropy.

Table 1 displayed the three kinds of methods the highest recognition rate on the CMU PIE face database.

Methods	10 Train	11 Train	12 Train
LDA	91.34	92.17	94.14
LPP	72.38	75.64	77.05
IRE	93.32	94.22	95.66

4.2 The YALE B database

On YALE B face database. We test positive images of 10 individuals who were shooting under 64 different lights. The above data set is divided into five subsets according to the direction of light: the best lighting conditions are Sub1, followed by Sub2, Sub5 worst.

We chose three images of each person as training samples Sub1, with the rest of the four samples for testing. Figure 3 displayed a sample image YALE B Sub1 face database.

Fig.3 Illustration 7 images of one individual from YALE B subset 1 face database.

Table 2 displayed the three kinds of methods the highest recognition rate on the YALE B sub1 face database.

Methods	3 Train
LDA	80.00
LPP	95.00
IRE	97.50

We chose each person's six Sub2 image as the training sample, with the rest of the six samples for testing. Figure 4 displayed a sample image YALE B Sub2 face database.

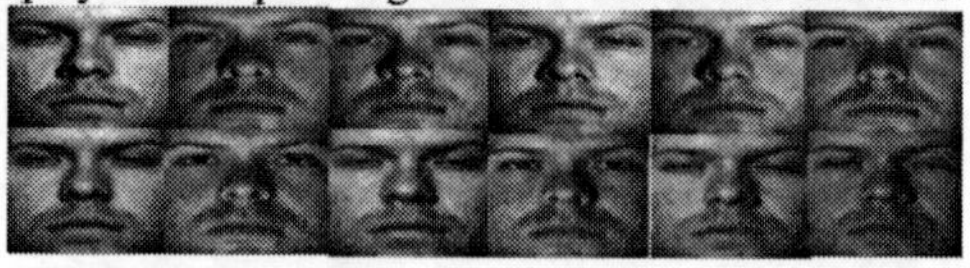

Fig.4 Illustration 12 images of one individual from YALE B subset 2 face database.

DOI: 10.1051/matecconf/20166304006

Table 3 displayed the three kinds of methods the highest recognition rate on the YALE B sub2 face database.

Methods	6 Train
LDA	81.67
LPP	91.67
IRE	96.67

We chose each person's six Sub3 image as the training sample, with the rest of the six samples for testing. Figure 5 displayed a sample image YALE B Sub3 face database.

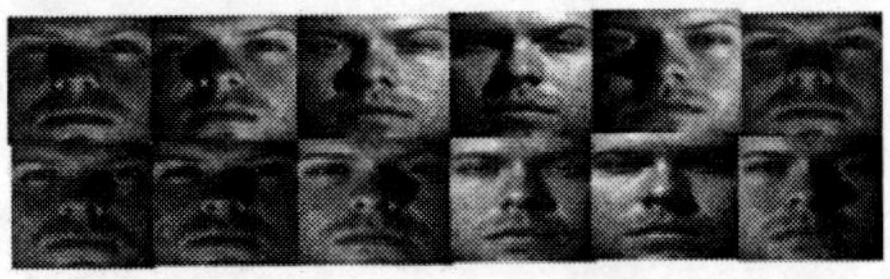

Fig.5 Illustration 12 images of one individual from YALE B subset 3face database.

Table 4 displayed the three kinds of methods the highest recognition rate on the YALE B sub3 face database.

Methods	6 Train
LDA	58.33
LPP	25.00
IRE	68.33

We chose each person's seven Sub4 image as the training sample, with the rest of seven the samples for testing. Figure 6 displayed a sample image YALE B Sub4 face database.

Fig.6 Illustration 14 images of one individual from YALE B subset 4 face database.

Table 5 displayed the three kinds of methods the highest recognition rate on the YALE B sub4 face database.

Methods	7 Train
LDA	38.57
LPP	22.86
IRE	75.71

5 Conclusions

In this paper, we proposed IRE method for face recognition. This method can make efficient recognition rate than LDA, LPP in CMU PIE face database and YALE B face database.

Acknowledgments

This work was supported by 2015 Annual Science and Technology Department of Guizhou Province, Bijie City Science and Technology Bureau, Guizhou University of Engineering Science Joint Science and Technology Fund Project, Cherry picking robot visionrecognition system research, Guizhou Branch of LH word [2015] No. 7599; the grants from 2014 Science and Technology Department of Guizhou Province, Bijie City Science and Technology Bureau, Bijie University of Science and Technology Joint Fund Program, Guizhou Branch of LH word [2014] No. 7536

References

1. M.Turk, A.P.Pentland, Eigenfaces for recognition, J. Cognitive Neurosci.3 (1) (1991)71-86.
2. P.N. Belhumeur, J. Hespanha, D.J. Kriegman, Eigenfaces vs. Fisherfaces: recognition using class specific linear projection, IEEE Trans. Pattern Anal. Mach. Intel. 20 (7) (1997) 711-720.
3. M. S. Bartlett, J. R. Movellan, T.J. Sejnowski, Face recognition by independent component analysis, IEEE Transactions on Neural Networks.13(6)(2002) 1450-1464.
4. L. Wiskott, J.M. Fellows, N. Kruger, C.V. Makburg, Face recognition by elastic bunch graph matching, IEEE Trans. Pattern Anal.Mach. Intell.19 (1997) 775–779.
5. G. Guo, S.Z. Li, K.L. Chan, Support vector machines for face recognition, Image and Vision Computing. 19 (9-10) (2001) 631–638.
6. J.B. Tenenbaum, V.de Silva, J.C. Langford, A global geometric framework for nonlinear dimensionality reduction, Science. 290 (5500) (2000) 2319-2323.
7. S.T. Roweis, L.K. Saul, Nonlinear dimensionality reduction by locally linear embedding, Science. 290 (5500) (2000) 2323-2326
8. L.K. Saul, S.T. Roweis, Think globally, fit locally: unsupervised learning of low dimensional manifolds, J.Mach. Learning. Res. (4) (2003) 119-155.
9. M. Belkin, P. Niyogi, Laplacian eigenmaps and spectral techniques for embedding and clustering, in: T.G. Dietterich, S. Becker, Z. Ghahramani(Eds.), Advances in Neural Information Processing Systems, vol.14, MIT Press, Cambridge, MA, USA, 2002, 585-591.
10. M. Belkin, P. Niyogi, Laplacian eigenmaps for dimensionality reduction and data representation, Neural Comput.15 (6) (2003)1373-1396.
11. X. He, S. Yan, Hu, P. Niyogi, H. Zhang, Face recognition using Laplacianfaces, IEEE Trans. Pattern Anal. Mach. Intell. 27 (3) (2005) 328-340.

12. C.H.Liu, Nonsymmetric entropy and maximum nonsymmetric entropy principle, Chaos, Solitons & Fractals. 40 (5) (2009) 2469-2474
13. C.E. Shannon, Bell Syst. Tech. J. 27 (1948) 379-623.
14. Szwast Z, Sieniutycz S, Shiner JS. Complexity principle of extremality in evolution of living organisms by information-theoretic entropy. Chaos, Solitons & Fractals.13 (9) (2002) 1871–1888.
15. Allegrini P, Grigolini P, Palatella L. Intermittency and scale-free networks: a dynamical model for human language complexity. Chaos, Solitons & Fractals. 20(1) (2004) 95–105.
16. Gontar V. Entropy principle of extremality as a driving force in the discrete dynamics of complex and living systems. Chaos, Solitons & Fractals .11 (1-3) (2000) 231–236.
17. El Naschie MS. Dimensions and Cantor spectra. Chaos, Solitons & Fractals.4 (11) (1994) 2121–2132.
18. El Naschie MS. On the topological ground state of E-infinity spacetime and the super string connection. Chaos, Solitons &Fractals.32 (2) (2007) 468–70.
19. El Naschie MS. Superstrings, entropy and the elementary particles content of the standard model. Chaos, Solitons & Fractals.29(1) (2006) 48-54
20. El Naschie MS. Internal Cantor distance and entropy of multidimensional Peano–Hilbert spaces. Chaos, Solitons & Fractals.1993; 3:361–8.
21. Tsallis C. Possible generalization of Boltzmann–Gibbs statistics. J Stat Phys.52 (1-2) (1988) 479–487.
22. D.J.C. MacKay, Information Theory, Inference and Learning Algorithms. Cambridge, U.K. Cambridge Univ.Press,2003
23. Guang Deng, An Entropy Interpretation of the Logarithmic Image Processing Model With Application to Contrast Enhancement, IEEE Transactions on Image Process.18 (5) (2009) 1135-1140.

Evaluation of Psychological Impact of Emergency Events Based on Bayesian Network

Xin WANG[1], Zhi Jie SONG[2,a], Rui SHI[2]

[1] *School of Economics and Management, and School of Science, Yanshan University Qinhuangdao, China*
[2] *School of Economics and Management, Yanshan University, Qinhuangdao, China;*

Abstract. Representation and inference of uncertain knowledge pose great challenge to the evaluation of the impact of emergency events on public psychology. With the advantages in solving issues on an uncertain basis, Bayesian network exhibits a superioty in effectively coping with the diversity, uncertainty and ambiguity of the evaluative information of emergency events. A model was established in this article to evaluate the impact caused by emergency events on the public's psychological health based on the Bayesian network. Moreover, the method of inference and decision-making was also presented, and an example analysis was performed. The result shows that the proposed model can improve the evaluation accuracy and the responsiveness of the managers when making decisions.
Keywords. Bayesian network; emergency events; public psychology; impact degree

1 Introduction

In recent years, the emergencies such as "9.11 Attacks" in the US in 2001, "SARS" outbreak in China in 2003 and Japan Tsunami in 2011 are reported all around the globe[1]. Emergency events bring threats to not only the human life and properties, but also the psychological trauma[1-2]. Many countries have attached great importance to emergency response management, with a release of a series of laws and regulations. The emergency response management is the subject of academic research in recent years[3]. The research and evaluation of the impact of emergency events on public's psychological health are complex and highly important[4]. It lays the basis for the prevention and control, rescue and handling of emergency events. The investigation in this respect also provides the basis for a reasonable decision making for the authorities. At present, some achievements have been made with respect to the evaluation of the impact degree of emergency events. The commonly used methods are fuzzy decision-making method, grey theory, neural network and genetic algorithm[5]. However, these methods can hardly solve the issue of the impact degree of the emergency events. That is, the inference based on uncertain information in combination with expert knowledge. Most of the existing studies concerning the impact of emergency events on the psychological health of the relevant parties are based on questionnaire survey or statistical analysis. Few have treated public psychology as a complex system comprising uncertain information.

Artificial intelligence can be applied to evaluate the impact of emergency events on public's psychological health, and inference is performed according to expert knowledge. Representation and inference of uncertain knowledge are two key processes in the evaluation of the psychological impact of emergency events. Bayesian network has found extensive applications as a framework of knowledge representation and probabilistic reasoning. The abilities to represent the uncertainty of and relevance between variables and to infer on an uncertain basis make the Bayesian network applicable to such evaluation [6]. Recently, the Bayesian network has been already employed in many fields, such as military decision-making, intelligent robot, pathological diagnosis, assessment of war situation, threat classification, fault diagnosis and the multi-attribute decision-making in uncertain environment [7-9]. A variety of adverse consequences resulting from emergency events were discussed in this article. The model for the evaluation of the impact of emergency events on public psychology was established based on the Bayesian network. The method of inference was expounded and its effectiveness was verified by example analysis.

2. Evaluation of the impact of emergency events on public psychology based on Bayesian network

To evaluate the impact of emergency events on public psychology based on Bayesian network, an evaluation model is first established. As the number of nodes

[a] Corresponding author: songzhj@ysu.edu.cn

increases in the Bayesian network, the establishment of model is a problem of combinatorial explosion. Reasonable principle and method have to be followed. The Bayesian network can be established through learning or artificially. Given the ample training samples, the learning is possible for establishing the network. Due to the small probability of emergency events, there will not be ample training data. Therefore, it is important to study the strategy and method for the establishment of Bayesian network. The evaluation of the impact of emergency events on public psychology based on Bayesian network consists of the following 6 steps.

2.1 Determining the contents of nodes

The Bayesian network is composed of nodes corresponding to different events. Hence it is necessary to identify the dimensions of adverse impact brought by emergency events on the psychology of the relevant parties. The adverse consequences of the emergency events can be multi-dimensional, including both direct and indirect impact. The time of post-disaster recovery and the situation of collaboration in post-disaster rescue have to be considered. Thus the psychological impact of emergency events is taken as the root node. The anxiety degree, sensitivity, degree of avoidance, coping style, time of post-disaster recovery and the collaboration in post-disaster rescue are taken as the nodes.

2.2 Determining the relationship between the nodes

After the determination of the contents of nodes, the casual relationship between the nodes is determined following a certain method. The evaluation of the psychological impact of emergency events is a highly complex issue. The impact analysis involves a large amount of uncertain knowledge, and the impact is either direct or indirect. The relevant evaluation work should be jointly undertaken by the experts and the decision-makers. The casual relationship between the contents of nodes and the nodes can be used to determine the structure of the Bayesian network. With the psychological impact of the emergency events as the root node and the different dimensions of the adverse consequences as the child nodes, the hierarchical structure of the impact of the emergency events is established, as shown in Fig. 1. The child nodes are subdivided to form more hierarchies of psychological impact of the emergency events. The hierarchical structure shown in Fig. 1 is considered as the topology of the Bayesian network for mapping the psychological impact.

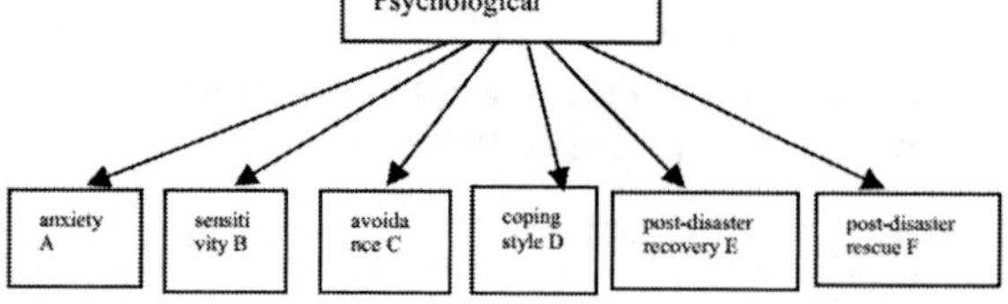

Figure 1. Hierarchical structure of psychological impact of emergency events

2.3 Classification of node status

In case of disastrous events, the decision makers usually have inaccurate and obscure information at hand. Moreover, due to the limitations of the perception and accumulated knowledge of the decision makers, the subjective judgment of the decision makers plays the decisive role in regard to the impact degree of the emergency events. Here the impact degree of the nodes is divided into different discrete statuses. For example, the psychological impact of the emergency events as the root node (H) is divided into 4 degrees: very low (H1), low (H2), high (H3) and very high (H4). Similarly, the child nodes are also divided into different discrete statuses. For example, the anxiety degree (A) is divided into 3 levels: low (A1), moderate (A2) and high (A3).

2.4 Construction of conditional probability matrix for child nodes

With the topology of the Bayesian network, the conditional probability distribution of the child nodes is determined. For data with continuous values, discretization is performed for the nodes. A large number of sample data are required for quantitative data, while for qualitative data, the inference is made based on expert knowledge. The determination of the conditional probability in the Bayesian network is a complex process, designed by expert experience or statistical tests. The row of the conditional probability matrix represents the status of the emergency event as the root node; the column represents the status of the child nodes. The elements in the matrix indicate the impact degree of the reasons on the results, which is estimated by expert knowledge or based on professional literatures and statistical techniques. For example, according to expert knowledge, the status of root nodes is divided into four levels, which are very low (H1), low (H2), high (H3) and very high (H4). The conditional probability of the anxiety degree (A) is expressed as

If the psychological impact of the emergency event is very low (H1), then the probability of the anxiety degree (A) being low (A1), moderate (A2) and high (A3) is, respectively. If the psychological impact of the emergency event is low (H2), then the probability of anxiety degree (A) being low (A1), moderate (A2) and high (A3) is, respectively. If the psychological impact of the emergency event is high (H3), the probability of the anxiety degree (A) being low (A1), moderate (A2) and high (A3) is 0.15, 0.30 and 0.55, respectively. If the psychological impact of the emergency event is very high (H4), then the probability of anxiety degree (A) being low (A1), moderate (A2) and high (A3) is 0.02, 0.03, 0.95, respectively. According to expert knowledge, the conditional probability matrix of the node 'anxiety degree' (A) with respect to the root node is expressed as follows. Similarly, the conditional probability matrices of the nodes 'sensitivity', 'degree of avoidance', 'coping style',

DOI: 10.1051/matecconf/20166304007

'time of post-disaster recovery' and 'collaboration in post-disaster rescue' with respect to root node are constructed.

$$M_A = \begin{array}{c} \\ H_1 \\ H_2 \\ H_3 \\ H_4 \end{array} \begin{array}{c} \begin{array}{ccc} A_1 & A_2 & A_3 \end{array} \\ \begin{bmatrix} 0.90 & 0.07 & 0.03 \\ 0.20 & 0.65 & 0.15 \\ 0.15 & 0.30 & 0.55 \\ 0.02 & 0.03 & 0.95 \end{bmatrix} \end{array} \quad (1)$$

In the same way, for the hierarchical structure of psychological impact of emergency events shown in Fig. 1, the conditional probability matrix MB, MC , MD , ME , MF of child nodes with respect to the root node is obtained.

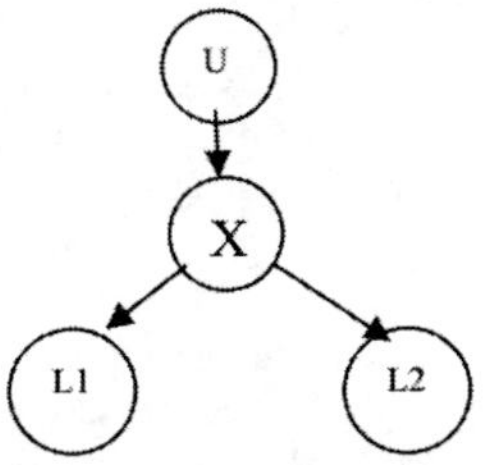

Figure 2. Tree-like Bayesian network structure

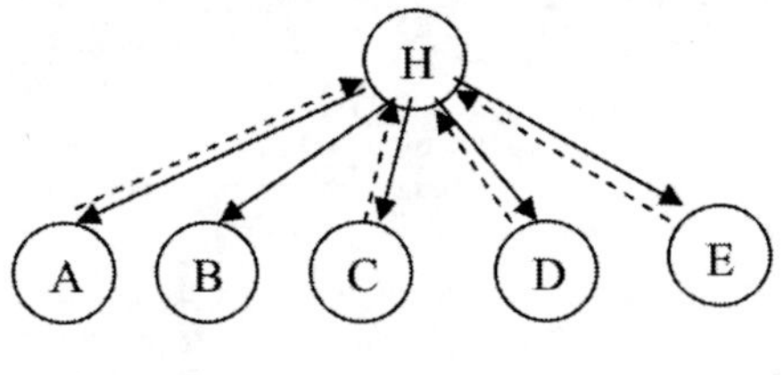

Figure 3. Persuasion propagation of influence degree assessment of emergent event

2.5 Algorithm of Bayesian inference

Bayesian inference is carried out based on priori information. With the conditional probability distribution of the leaf nodes being constant, the Bayesian network maintains the equilibrium. Once the status of the leaf nodes changes with the observation information, the status probability distribution will be updated for all the nodes in the entire network according to Peal algorithm. The simplified tree-like Bayesian network is used as the inference model. Every node of the tree-like Bayesian network only has one parent node, as shown in Fig. 2. The variables involved in the algorithm are listed below:

Bel(X): the status probability distribution of node reflects the occurrence probability of the event X under a certain scenario. It is a one-dimensional column vector, with the number of elements equal to that of the discrete values of node status.

$\lambda(X)$: the diagnostic information is obtained from the child node, and it supports the diagnosis. It is a one-dimensional column vector, with the number of elements equal to that of the discrete values of X .

$\pi(X)$: the casual information is obtained from the parent node, and it supports the forecast. It is a one-dimensional column vector, with the number of elements equal to that of the discrete values of parent nodes.

Using a single node as the center in the algorithm, $\lambda(X)$ is obtained from the child node, and $\pi(X)$ is obtained from the parent node. On this basis, $Bel(X)$, $\lambda(X)$ and $\pi(X)$ of the current nodes are calculated, and the adjacent nodes are triggered and updated.

These procedures are repeated until the posterior probability of all nodes is equal to the priori probability. At this time, the network reaches a new equilibrium. Based on the hierarchical structure of the psychological impact of the emergency events shown in Fig. 1, the calculation procedures are as follows:

(1) The priori information $\pi(X)$ of the root node H is determined. The confidence level $B(H)$ of the root node H is initialized. Let

$$Bel(H) = \pi(H) \quad (2)$$

(2) If the diagnostic information of a child node θ changes into λ_θ , then the diagnostic information of root node H will change accordingly:

$$\lambda_H = M_\theta \times \lambda_\theta \quad (3)$$

where θ represents a child node; M_θ is the conditional probability of a child node with respect to root node, in the form of (3). The element in conditional probability matrix represents the impact degree of root node (H) on the child node.

(1) The confidence level of the root node (H) (H) is upwards updated to:

$$Bel(H) = a \times (\lambda_H g \pi(H)) \quad (4)$$

where g is inner-product operator. That is, for two column vectors $\beta = (\beta_1, \beta_2, \text{K}, \beta_n)^T$ and $\gamma = (\gamma_1, \gamma_2, \text{K}, \gamma_n)^T$, the result of executing the operator is $\beta g\gamma = (\beta_1 \times \gamma_1, \beta_2 \times \gamma_2, \text{K}, \beta_n \times \gamma_n)^T$; a is the normalization factor, which is responsible for making the sum of confidence intervals of root node (H) under different status to be 1.

(2) The confidence interval of the child node (θ) is downwards updated. The updated confidence interval of the child node (θ) is expressed as

$$Bel(\theta) = a \times M_\theta^T \times Bel(H) \quad (5)$$

where M_θ^T is the transpose of conditional probability matrix M_θ ; a is the normalization factor.

The process of Bayesian inference in regard to the psychological impact of the emergency events is shown in Fig. 3. In Fig. 3, the bottom-up inference indicates the impact degree of the emergency event inferred from one dimension of the adverse consequence of the emergency events. The top-down inference indicates the influence of the changes of psychological impacts of emergency event on the adverse consequences. In this way, Bayesian inference can be employed for bottom-up diagnostic inference and top-down casual inference, until the posteriori probability is equal to priori probability for all nodes.

Calculation of the degree of psychological impact

Suppose $Q=\{H_1,K\ ,H_n\}$ is the set of status of root node (H), where is the fuzzy comment made by the decision makers on the psychological impact of the emergency event on public psychology. For example, the fuzzy comment given by the decision makers with respect to the psychological impact of the emergency event may be very low (H_1), low (H_2), high (H_3) and very high (H_4). $U(Q)=\{u(H),K\ ,u(H)\}$ is the set of utility values of each status in the comment set. $u(H_l)$ $(l=1,...,n)$ are the fuzzy utility values of status H_l. Its value range is $0\le u(H_l)\le 1$ .For the same set of fuzzy comments, different decision makers may assign different utility values.

The psychological impact of the emergency event is determined by the confidence level of root node status and the fuzzy utility value when the network reaches the equilibrium.

$$E(H)=\sum_{i=1}^{n}(u(H_i)\times Bel(H_i)) \qquad (6)$$

3. Example analysis

A stampede occurring during a large-scale activity was studied as an example. The topology of the Bayesian network built for the emergency event and the subsequent adverse events is shown in Fig. 1. For root node H , the set of fuzzy comment is Q ={very low (H_1), low (H_2), high (H_3) and very high (H_4 }. The corresponding set of fuzzy utility values is $U(Q)=\{0.25;0.50;0.75;1.0\}$. The set of anxiety degree (A) is {low (A_1), moderate (A_2), high A_3)}; the set of sensitivity (B) is {low (B_1), moderate (B_2), high (B_3). The set of the degree of avoidance (C) is {low (C_1), moderate (C_2), high (C_3)}; the set of coping style (D) is {positive (D_1), neutral (D_2), negative (D_3)}. The set of the time of post-disaster recovery (E) is {short (E_1), moderate (E_2) and long (E_3)}. The set of the status of collaboration in post-disaster rescue (F) is {positive (F_1), moderate (F_2) and negative (F_3)}. According to expert knowledge, the conditional probability matrix of child node (A) with respect to root node is expressed as formula (3). The conditional probability matrices of child nodes B , C , D , E and F with respect to root node are MB,MC,MD,ME and MF.

Next the impact degree of emergency events on public psychology is calculated based on priori information.

(1) At the early stage of the stampede, the psychological impact caused by this event cannot be judged at all. Thus it is assumed that the probabilities of the impact degree being very low (H_1), low (H_2), high (H_3) and very high (H_4) are equal. That is, the priori information in regard to the psychological influence is $\pi_1(H)=(0.25,0.25,0.25,0.25)^T$. At the early stage of the event, it is supposed by experts that the stampede has a considerably influence on the sensitivity, and its influence on the coping style is more likely to be of a moderate degree. However, there is no sufficient information on the child nodes such as degree of avoidance, time of post-disaster recovery and collaboration in post-disaster rescue. Thus the diagnostic information provided by the experts with respect to the child node is $\lambda_A=(0,0,1)^T$; $\lambda_B=(0.3,0.4,0.3)^T$; $\lambda_C=(0.4,0.3,0.3)^T$; $\lambda_D=(0.3,0.6,0.1)^T$; $\lambda_E=(0.3,0.4,0.3)^T$; $\lambda_F=(0.3,0.4,0.3)^T$.

Here the confidence level $B(H)$ of the root node (H) is initialized. The child nodes and the root nodes are updated under the given mode of information propagation. Hence the confidence level of the impact on the nodes at the early stage is $Bel(H)=(0.0051,0.2595,0.4861,0.2480)^T$. This indicates that the probability that the impact degree of the event is high is the largest. This is because the confidence level of the status (H_3) is 0.4861. The probability that the impact degree of the event is low and very high is equal, with the confidence interval being, respectively 0.2595 and 0.2480. The psychological impact of the emergency event determined by formula (8) is expressed as $E(H)=0.7431$.

(2) With the passage of time, the experts judge that the event has a great influence on the degree of avoidance based on the initial information $\pi_1(H)$. Moreover, it is judged that the influence on the time of post-disaster recovery and on the collaboration in post-disaster rescue is greater. Then the diagnostic information given by the experts on each child node is $\lambda_A=(0,0,1)^T$; $\lambda_B=(0.3,0.4,0.3)^T$; $\lambda_C=(0.4,0.3,0.3)^T$; $\lambda_D=(0,0,1)^T$; $\lambda_E=(0.1,0.5,0.4)^T$; $\lambda_F=(0.1,0.5,0.4)^T$. Hence the confidence level of the influence on the root node at the later stage of the event (H) is expressed as:

$$Bel(H)=(0.0000,0.0085,0.2210,0.7711)^T$$

It can be sure that the impact of the emergency event on public psychology is of high and very high degree. According to formula (8), the impact degree of the emergency event is calculated as $E(H)=0.9327$.

By using priori information, the Bayesian network can achieve a dynamic evaluation of the psychological impact of the emergency event. Thus the situation of the emergency event can be better evaluated to provide a basis for decision making.

(3) Priori information is highly important in the application of the Bayesian network. If the priori information is $\pi_2(H)=(0.10,0.80,0.10,0.00)^T$, the probability of the initial status of emergency event being very low (H_1), low (H_2), high (H_3) and very high (H_4) is 0.1, 0.8, 0.1, and 0.0, respectively. Thus the confidence interval of the psychological impact corresponding to situation (1) $\lambda_A=(0,0,1)^T$; $\lambda_B=(0.3,0.4,0.3)^T$; $\lambda_C=(0.4,0.3,0.3)^T$;

$\lambda_D = (0.3, 0.6, 0.1)^T$; $\lambda_E = (0.3, 0.4, 0.3)^T$; $\lambda_F = (0.3, 0.4, 0.3)^T$ is $Bel(H) = (0.0021, 0.8121, 0.1765, 0.0000)^T$. From formula (8), the impact degree of the emergency event is calculated as $E(H) = 0.5402$.

If the priori information is $\pi_3(H) = (0.10, 0.10, 0.80, 0.00)^T$, the probability of the initial status of emergency event being very low (H_1), low (H_2), high (H_3) and very high (H_4) is 0.1, 0.1, 0.8, and 0.0, respectively. Thus the confidence interval of the psychological impact corresponding to situation (1) $\lambda_A = (0,0,1)^T$; $\lambda_B = (0.3, 0.4, 0.3)^T$; $\lambda_C = (0.4, 0.3, 0.3)^T$; $\lambda_D = (0.3, 0.6, 0.1)^T$; $\lambda_E = (0.3, 0.4, 0.3)^T$; $\lambda_F = (0.3, 0.4, 0.3)^T$ is $Bel(H) = (0.0012, 0.0612, 0.9256, 0.0000)^T$. From formula (8), the impact degree of the emergency event is calculated as $E(H) = 0.7432$.

As seen from the evaluation results based on priori information $\pi_1(H)$, $\pi_2(H)$ and $\pi_3(H)$, the evaluation results differ for the same input information at the child nodes due to the differences in priori information. For the same input information at the child nodes $\lambda_A = (0,0,1)^T$; $\lambda_B = (0.3, 0.4, 0.3)^T$; $\lambda_C = (0.4, 0.3, 0.3)^T$; $\lambda_D = (0.3, 0.6, 0.1)^T$; $\lambda_E = (0.3, 0.4, 0.3)^T$; $\lambda_F = (0.3, 0.4, 0.3)^T$, the confidence interval of the root node corresponding to priori information $\pi_1(H)$, $\pi_2(H)$ and $\pi_3(H)$ is shown in Fig. 4.

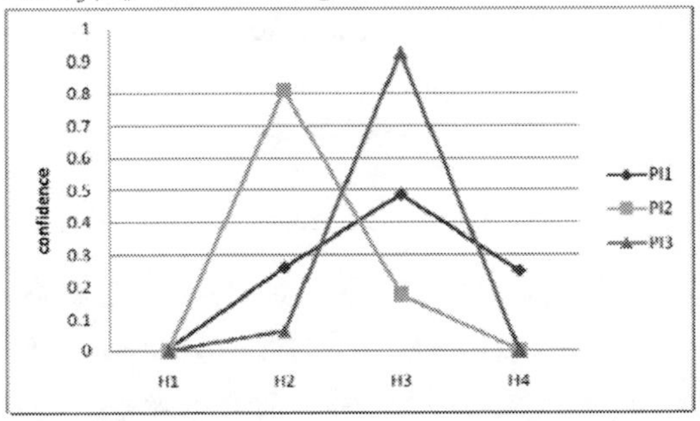

Figure 4. Confidence of different states corresponding to different priori information

It can be seen from Fig. 4 that using the priori information $\pi_1(H)$, the impact degree is mainly of low (H_2), high (H_3) and very high (H_4) level. The probabilities of being very low (H_2) and very high (H_4) are not significantly differed. However, under the priori information $\pi_2(H)$ and $\pi_3(H)$, there is an obvious difference in the probability of the impact degree, mainly due to the differences in priori information.

4. Summary

Bayesian network is a powerful graphical tool to represent the probability-based domain knowledge. With the ability of bidirectional inference, Bayesian network has shown prominent advantages in the evaluation of the impact of emergency events on public psychology. The representation and inference of uncertain knowledge are the main aspects in regard to the evaluation of the psychological impact of the emergency events. The model for the evaluation of the impact degree of emergency events on public psychology was proposed based on Bayesian network. This method not only integrates expert experience and knowledge, but also enables multi-stage inference by utilizing all available information. The model can accurately characterize the impact degree and satisfy the requirements for real-timeliness, dynamics and rapidity of evaluation. The execution of Bayesian network can enhance the degree of intelligence and the effectiveness of decision making during the emergency handling process.

Acknowledgment

This study was supported by a Grant from National Natural Science Foundation of China for Prof. Zhijie Song (71171175).

References

1. Wu,J.H,Ge,Y,Shi,Z.B,Duan,X.J,Wang,L,Sun,X.H.et al.Response inhibition in adolescent earthquake survivors with and without posttraumatic stress disorder:A combined behavioral and ERP study. Neuroscience Letters . 2010
2. Ma Q G,Shen Q,Xu Q,et al.Empathic responses to others''gains and losses:an electrophysiological investigation. Neuroimage . 2011
3. Lean Yu,Kin Keung Lai.A distance-based group decision-making methodology for multi-person multi-criteria emergency decision support. Decision Support . 2011
4. Undala Alam.From Disaster Response to Risk Management:Australia's National Drough Pol icy. Political Geograph . 2007
5. Lean Yu,Kin Keung Lai, A distance-based group decision-making methodology for multi-person multi-criteria emergency decision support, Decision Support . 2011
6. Bayraktarli Y.Y,Ulfkjaer J,Yazgan U,et al.On the application of Bayesian probabilistic networks for earthquakerisk management. Proceedings of the 9th International Conference on Structural Safety and Reliability . 2005
7. ALYSON G W,APARNA V H.Bayesian networks formultilevel system reliability. Reliability Engineering . 2007
8. Gregory F. Cooper,Edward Herskovits. A Bayesian method for the induction of probabilistic networks from data[J]. Machine Learning . 1992 (4)
9. Bonafede C E,Giudici P.Bayesian networks for enterprise risk assessment. Physica A:Statistical Mechanics and Its Applications . 2007.

Research on Magnetically Coupled Resonance Wireless Transmission Practice Teaching Platform

Yi Ren ZHOU[a], Xiao Lin QIU, Zhi Qiang GUO, Jian Hui ZHANG

Nanchang Institute of Technology, No.901 Heroe Ave, Qingshanhu, Nanchang, Jiangxi, 330044.China

Abstract. Magnetically coupled resonant wireless transmission of practice teaching platform is a high-tech system, and magnetically coupled resonant wireless power transmission experiment design and verification can be carried out on this platform by students. This thesis discusses the importance of efficient wireless energy transfer technology applied to distance magnetically coupled resonance technology in the practice of teaching from the practice teaching system of magnetically coupled resonant wireless transmission mechanism, the establishment and application of practice teaching system. It provides a system platform for students to establish wireless power transmission model and to apply the mutual inductance principle and coupling theory to derive the transmission efficiency for obtaining the key factors that affect the transmission efficiency such as the resonant frequency, the coil mutual inductance and the coil quality factor, etc., and to draw the conclusions of the above theoretical analysis verified by experiments

Keywords. Magnetically coupled,resonators, wireless transmission, ANSYS, practice teaching

1 Magnetically coupled resonance wireless transmission mechanism of practice teaching system

Magnetically coupled resonance photoelectric wireless energy transmission technology is the use of non-radiative magnetic coupled resonance principle for photoelectric distance energy transmission[1]. Research the use of magnetic fields as a medium of energy transfer by Maxwell equations. Two electromagnetic systems having the same natural frequency, at a certain distance, due to the electromagnetic coupling, resonant energy transfer can be performed between two systems. This transmission is unique in the use of the resonance principle, two coils for wireless energy transfer from resonance, and the impedance of the coil circuit is minimized, so most of the energy transmitted to the resonance direction.

1, Effect of magnetic coupling coil resonant frequency of the energy transmission efficiency

Maxwell's equations coupled with the use of model theory (Coupled Mode Theory, CMT) energy transmission coupling body in such a strong research. Essence coupling model theory is coupled by the following differential equation expressed[2].

$$\dot{a}_m(t) = (i\omega_m - \Gamma_m)a_m(t) + \sum_{n \neq m} ik_{mn}a_n(t) + F_m(t)$$

In the above formula, the subscript m, n represents different resonance body, variable am (t) to determine the energy contained in the resonator body m, is | am (t) | 2, ω m is the natural frequency of the resonator body Γ m system natural decay rate (absorption and radiation losses). Kmn resonator body indicates m, the coupling coefficient between h, Fm (t) represents the driving conditions.

Magnetically coupled resonant wireless energy transfer key is to make two resonant objects of the same frequency produce a very strong mutual coupling. In this way, compared with the electromagnetic induction method, use of the magnetic field is much weaker, but

[a] Corresponding author: renyiz@126.com

can achieve a more long-distance power transmission, the transmission shown in Figure 1.

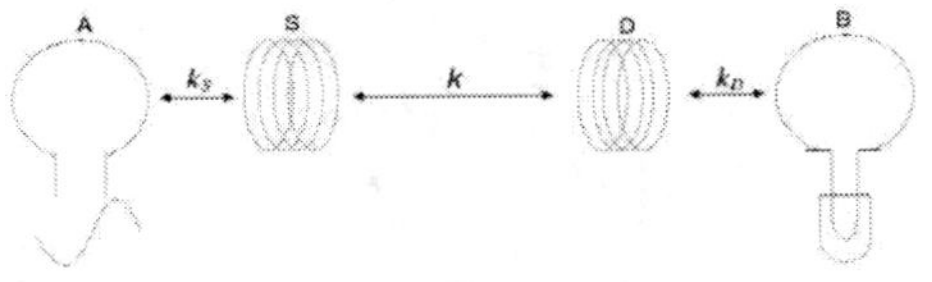

Figure 1 Magnetically Coupled Resonance Wireless Transmission Practice Teaching System Schematic

Schematic ks transmission distance S between the transmit coil and an oscillator coupled coils A, D and receiving coil coupled to resistive load coil B, A and S, D and B, kD is very small, and they are primarily between Proximity Coupling . Due to S and D of the resonant coil on the coil's inductance and distributed capacitance to achieve resonance matching focus, energy is coupled to the power transmitting coil by the oscillation circuit S A, S transmitting coil and a receiving coil D has the same resonant frequency since, in the coupling of the magnetic field under resonance, receiving coil loading coil B and D to achieve energy transfer by coupling, κ is magnetically coupled resonance distance.

The same frequency coil energy transfer

Resonant transmitting coil S and a resonant receiving coil D form a complete resonant system, the energy transfer between the two coils. Only the two resonant coils produced a strong resonance coupling in order to effectively transfer energy in the far distance. When no drive, using the coupled-mode theory (CMT) analysis of differential equations

$$\begin{cases} \dot{a}_s(t) = (i\omega_s - \Gamma_s)a_s(t) + ik_{SD}a_D(t) \\ \dot{a}_s(t) = (i\omega_D - \Gamma_D)a_D(t) + ik_{DS}a_S(t) \end{cases}$$

The above equation: as (t), aD (t) is the amplitude of the two coils midfield; ω s, ω D for their respective natural angular frequency of the two resonant coil, wherein, ω s = 2π fs ω D = 2π fD, fs, fD two a natural frequency of each resonance coil; Γ s, Γ D is inherent loss due to absorption and radiation caused; κ sD κ Ds is the coupling coefficient between the two resonant coils.

Coupling coefficient κ speed reflects energy transfer between the two resonant coils, the coupling coefficient between the two coils is larger, the energy from the transmitting coil to receiving coil used for shorter time spread, the faster the establishment of the two coils from a stable energy transmission channel, complete energy transfer. If the coupling coefficient κ and loss factor Γ appropriate values, you can complete the establishment of channels of energy transfer between the two coils before the energy attenuation finished, it can be said receiving coil energy loss due to various factors radiation emitted from a rate of less than coil energy transfer over the speed, it will have enough energy to supply the load back. If κ and Γ value is not appropriate, the transmission channel cannot be established, it cannot be transmitted energy.

Coupling coefficient κ distance between the two coils K and have a great relationship, the smaller the distance between the coils K, the greater the coupling coefficient κ . In WiTricity transmission system, the relationship showed the difficulty of energy transfer between the two coils, it is clear that the closer the distance between the two coils, the larger the coupling coefficient is, and the energy between two coils is easy to transport. Γ loss factor is directly related to the size of the coil energy decay rate, Γ larger own loss, loss of energy and radiation from the surrounding non-resonant objects between the greater by the coil, and therefore the system developed by WiTricity distance Γ ranges from K decision.

1) Coil frequency is not the same energy transfer

When WiTricity system natural frequency resonant coils two are not the same, namely ω s ≠ ω D, re-soluion equation:

$$\begin{cases} a_S(t) = e^{(x_S+x_D)t/2}\left[(x_S - x_D)\sin(\frac{dt}{2})\Big/d + \cos(\frac{dt}{2})\right] \\ a_D(t) = e^{(x_S+x_D)t/2}\left[2i\kappa\sin(\frac{dt}{2})\Big/d\right] \end{cases}$$

 DOI: 10.1051/matecconf/20166304008

$xs = i\omega s - \Gamma s$， $xs = i\omega D - \Gamma D$,

$$d = \sqrt{4k^2 - x_s^2 + 2x_s x_D - x_D^2}$$

Through the above mathematical model, to meet the strong coupling condition, the same natural resonant frequencies of the two coils with a different state, higher by experimental analysis of the energy-transfer efficiency which states the two resonant coils.

2, Analysis of the overall performance of the wireless transmission magnetically coupled resonance Practice Teaching System

Establishing mutual coupling model is magnetically coupled resonant wireless energy transfer, and ANSYS software tightly wound single coil excitation magnetic field simulation, and two closely wound coils magnetic field coupling between simulation, comparative analysis of the radius of the coil, the coil and the distance relationship receiving coil radius and the magnetic field coupling; transmission efficiency research and coil size, the relationship between the natural frequency, to obtain an effective method for improving transmission efficiency.

2 Magnetically coupled resonant wireless energy transfer system to establish teaching practice

According to the coupling resonator type photoelectric wireless energy transfer mechanism to transfer the device structure and circuit design, the transmission means comprises a DC power supply section, a signal generating section, a power amplification section, transmit coil section, the coil receiving portion, a rectifier circuit and a load seven parts, its transmission functional block diagram shown in Figure 2.

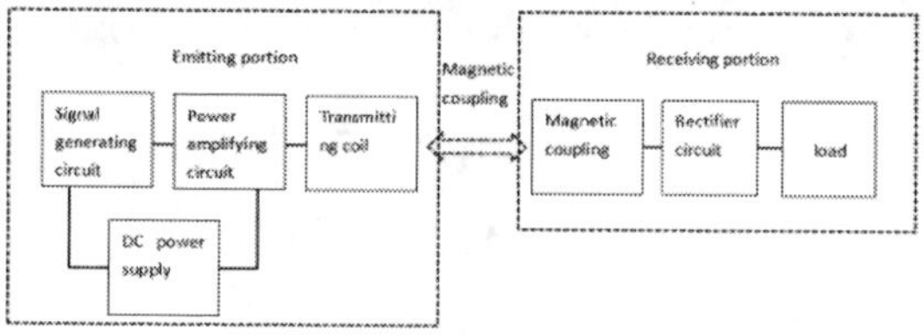

Figure 2 Transfer Schematic

Emitting portion signal generator to produce the desired frequency square wave signal by driving the power amplifier circuit and the load on the DC power supply to the square wave energy, the square wave generated non-radiative alternating magnetic field emitted by the coil there around. Receiving coil in the transmitter coil an alternating magnetic field generated by the resonance due to the coupling effect, the magnetic field receiving coil energy into electric field energy. Finally, after rectifying circuit load energy supply, complete energy transfer.

1、Using ANSYS software to analyze MAXWELL equations based on field

Using ANSYS software module electromagnetic simulation analysis of magnetic field coupling between the two coils and a single coil excitation; MAXWELL equations based on magnetic field analysis WiTricity transmission means and the coil coupling magnetic field simulation analysis; drive circuit design, power amplifier circuit and transmitting and receiving resonant circuit.

ANSYS is a kind of large general-purpose finite element analysis software[5], which can effectively block electromagnetic field analysis of a variety of devices. MAXWELL equations based on the magnetic field, by the scalar, vector potentials or boundary introduced MAXWELL equations and considering the electromagnetic nature of the relationship[6], developed equations suitable for finite element analysis is used to analyze a wide range of electromagnetic problems, such as inductors , capacitance, impedance, magnetic flux density, magnetic field strength, the electric field distribution, magnetic field lines, quality factor,

DOI: 10.1051/matecconf/20166304008

characteristic frequency, magnetic force and torque, motion effects and energy losses.

As ANSYS analysis of other types, for harmonic magnetic analysis, to establish the physical environment, to model district to impart properties, mesh, boundary conditions and loads increase, solution, and then observe the results using ANSYS finite element software for electromagnetic fields module for single coil magnetic field and magnetic field coupling the two coils were analyzed frequency of the excitation signal derived from the excitation signal size, number of turns, coil radius and effect of coupling the two coils of the magnetic field distribution, magnetic resonance wireless power coupling transmission designed to provide the basis. ANSYS software and magnetic field coupling the two tightly wound coils of a tightly wound coil excitation of a single simulation, comparative analysis of the relationship between the two coupling magnetic field from the coil and the coil radius, analyze the coupling efficiency and the efficiency of energy transfer between the magnetic field relationship calculate the distance energy transmission.

2. Magnetically coupled resonance wireless transmission system circuit practice teaching model

The entire transmission system, the key is to analyze the relationship between the coupling between the transmitter and receiver coils, and therefore will be sending and receiving circuit as the research object, coupled resonator circuit model shown in Figure 3.

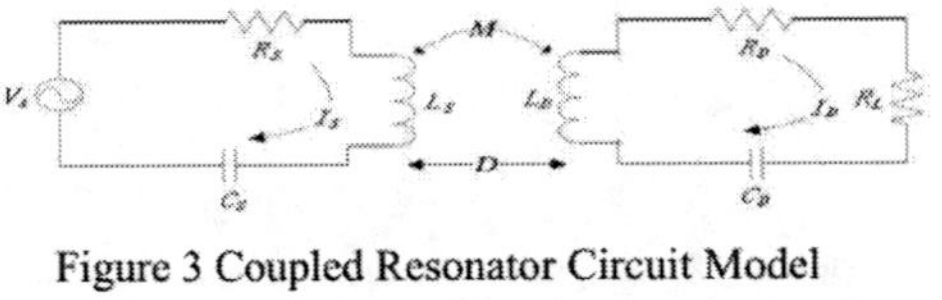

Figure 3 Coupled Resonator Circuit Model

Where l is the inductance of the high-frequency signal source, RS, RD are two resonant inductor parasitic resistance at high frequencies, CS, CD series resonant capacitor, RL is the load resistance, LS, LD are sending and receiving coils , M is the mutual inductance between the coils, D is the transmission distance.

3. Magnetically coupled resonance wireless transmission system set up practice teaching

Efficiency is the key to wireless power transmission systems electromagnetic emissions and electromagnetic receiving system. Magnetically coupled resonance wireless transmission practice teaching system configuration shown in Figure 4. System is mainly composed of two parts, the energy transmitter and energy receiver. Energy transmitter as a DC power input terminal, a high-frequency signal is amplified by the formation of a high-frequency excitation power source, so that the source is directly connected with the coil (the coil label 1) to resonate, and the source alternating field is formed around the coil. Transmit coil (coil label 2) alternating magnetic field induction source coil, and then with the formation of a resonance, a more powerful alternating magnetic field. Thus, the source of energy transmitted through the coil to the transmitter coil, and then passed out by the transmitter coil.

Energy receiving end comprises two coils, receiving coils respectively (label coil 3) and loading coil (coil label 4), wherein the load coil is directly connected with the electrical load. A receiving coil, the energy transfer of the transmit coil, and transmits it to the load coil. After loading coil connected energy conversion circuit, the high-frequency power conversion DC power supply directly behind load electricity use or storage.

The whole system is the most critical part of the high-frequency excitation and transmitter coil and the receiver coil design. Radio frequency excitation directly determine the entire frequency range of the energy transmission system, and the parameter transmitting coil and a receiving coil for energy transmission efficiency play a decisive role.

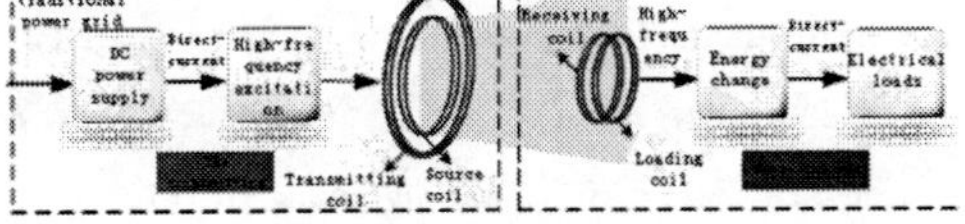

DOI: 10.1051/matecconf/20166304008

Figure 4 Magnetically Coupled Resonant Wireless Power Transmission Output Practical Teaching System Components

Magnetically coupled resonant wireless power transmission output circuit shown in practice teaching system 14a, b. inductively coupled power transfer technology to transmit and receive modules based wireless power transmission system was designed on the transmitting and receiving circuit structures. Send and receive circuit in FIG. 5a, b.

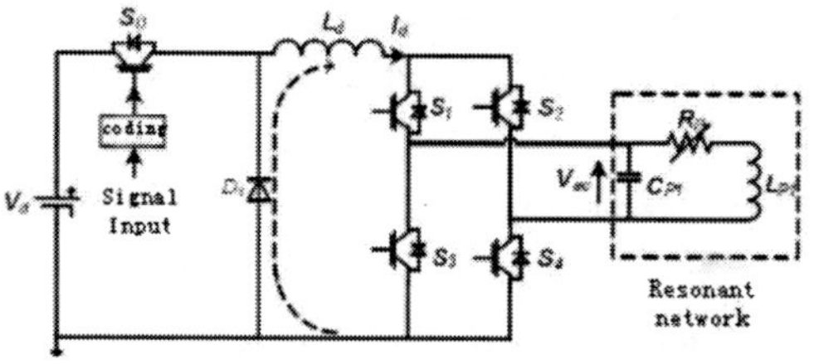

Figure 5a Transfer Schematic

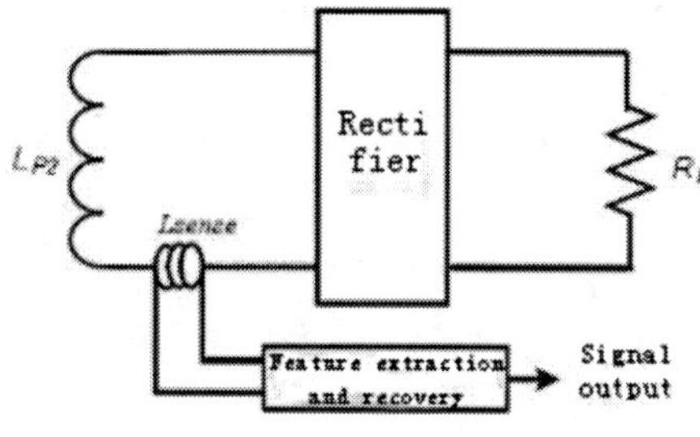

Figure 5b Receiving circuit

3.Magnetically coupled resonance wireless transmission Practice Teachi System

Magnetically coupled resonance wireless transmission system verifies practice teaching of modern power electronics technology theory, electromagnetic resonance technology to verify the effect of wireless power transmission;

Used to analyze a wide range of issues of electromagnetic fields, such as inductance, capacitance, impedance, magnetic flux density, magnetic field strength, the electric field distribution, magnetic field lines, quality factor, characteristic frequency, magnetic force and torque, motion effects and energy loss; using magnetically coupled resonance wireless circuit model of transmission through the system output power and efficiency of factor: the resonant frequency, distance, load, carry out a detailed analysis, the influence of magnetically coupled resonance wireless power transmission output power and efficiency of the impact factor.

Resolve conflicts magnetically coupled resonant wireless power transmission photoelectric transfer efficiency and distance cannot have both; technical issues to overcome transmission efficiency, transmission distance, safety and reliability aspects to provide a solution.

References:

[1]Schuder.J,Gold.J,Stephenson.H.AnInductively Coupled RF System for the Transmission of 1 kW of Power Through the Skin[J]. Biomedical Engineering,IEEE Transactions on, 1971, 18(4): 265-273.

[2] Lechner E,Schladover S.Th e Roadway Powered Electric Vehicle-An All-electric Hybrid System[C]. Washington: Proceedings of eighth international electric vehicle symposium, 1986.

[3] Jackson D K, Schultz A M, Leeb S B, et al. A multirate digital controller for a 1.5 kW electric vehicle battery charger[J]. Power Electronics, IEEE Trans on, 1997, 12(6): 1000-1006.

[4] D. M. Budgett, A. P. Hu, P. Si, W. T. Pallas, M. G. Donnelly, J. W. T. Broad, C. J. Barrett, S. Guild, and S. C. Malpas. Novel technology for the provision of power to implantable physiological devices. J Appl Physiol, 2007, 102: 1658-1663.

[5] A. Kurs, A. Karalis, R. Moffatt, J. D. Joannopoulos, P. Fisher, M. Soljačić. Wireless Power Transfer via Strongly Coupled Magnetic Resonances. Science, 2007, 317: 83-86.

[6]Esser A, Skudelny H. A new approach to power supplies for robots[J]. Industry Applications, IEEE Transactions on, 1991, 27(5): 872-875.

[7]Aristeidis,Karalis,J.D.Joannopoulos, MarinSoljac'"ic'. Efficient wireless non-radiative mid-range energy transfer[J]. Annals of Physics, 2008，3 (23): 34-48.
[8]Andre Kurs, AristeidisKaralis, RobertMoffatt, J. D.Joannopoulos, Peter Fisher, Marin Soljacic. Wireless Power Transfer via Strongly Coupled Magnetic Resonances[J]. Sciencexpress, 2007, 10 (1126): 1-4.
[9] Marin Soljacic. Wireless energy transfer can potentially recharge laptops cell phones without cords[R].San Francisco: Massachusetts Institute of Technology 2006.
[10] Elliott G, Boys J,Green A. Magnetically coupled systems for power transfer to electric vehicles[C]. Singapore: Proc Int Power Electronics and Drive Systems, 1995, 2: 797-801.

MATEC Web of Conferences **63**, 04009 (2016) DOI: 10.1051/matecconf/20166304009
MMME 2016

A Study on Optimal Resource Allocation of Taxis

Xiaopeng Lv[1,a], Xinyuan Ding[2] and Xian Chen[3]

[1] *Beijing Normal University, Zhuhai, Zhuhai Guangdong, China*

[2] *Beijing Normal University, Zhuhai, Zhuhai Guangdong, China*

[3] *Cardiff University, Cardiff, Wales, The UK*

Abstract. As taxis play an increasingly important role in urban traffic system, the research on the supply and demand of taxis and the design of an optimal model of taxis supply has an importantly practical significance. In this paper, we used the traffic bureau data of Guangzhou, and determined the rate of empty driving as the index to establish the model. Nonlinear programming method was uesd to establish the optimal model of taxis quantity to meet the maximum income and maximum passenger satisfaction. Finally we got the optimal number of taxis was 37537.59

Key words: Taxi, supply and demand, nonlinear programming.

1 Introduction

In recent years, the city taxi has become an important means of transport in the public transport system, it has some feature like convenient, time-saving, which get more and more young pro-Lai. Since the 1990s, under the relevant policy support, the taxi industry develop rapid, but in many medium-sized cities, due to lack of corresponding planning and urban taxi formal management, resulting in a high rate of empty driving and road jams. In addition, the approach of taxi services unhealthy and bad habits of commuters, and also a serious impact on city traffic. Although some cities taken some control measures, but most of them are lack of practical experience. Therefore, its significance that studying the supply and demand taxi and establish a model of optimal taxi supply and demand to alleviate traffic congestion and improve urban planning.

Guangzhou located in the southeast coastal areas of our country, adjacent to Hong Kong and Macao, the located advantage is obvious, since the reform and opening up, Guangzhou's economic developing rapidly, the service sector's proportion rapidly increasing in regional GDP, the corresponding urban development more mature, transportation usher the peak period of development. This paper will use Guangzhou taxi industry as the research object, by analyze data and establish optimization model, study the model of Guangzhou optimal taxi supply and demand in the two dimensions of space and time.

2 Problem situation

The modern economic growth was actually the change of economic structure, was main caused by the revolution of services industry which was represented by the transportation and communications. In the future economic development, transportation industry will become a pillar industry in the service industry.

BIAN Yang etc. (2006) consider the social and individual interests while building multi-objective function model, case analysis shows that the model can determine the taxis' split rate, and provide theoretical basis for transport policy decisions [1]. BIAN Yang etc (2007) to build the model of network equilibrium of city's taxi to determine a reasonable scale of development of the city Taxi, and performance behaviors objectively in the driver search off, by using passengers' waiting time to determine a reasonable scale of development of the city taxi[2]. LI Hong-yanetc (2007) establish the index system and its analytical method of taxis' travel characteristics, time's distribution of travel and the spatial distribution of travel. And using the example of Suzhou City, analyzes the characteristics of taxis' trip in each time, giving the quantitative basis for decision making for reference to taxi operators and urban traffic management [3]. ZHOU Jing etc (2000) obtained the proportion of the system Kongshi taxi cab and passenger under certain conditions in the macro point of view. From the microscopic point of view, using queuing theory analyzed the busy of site of taxi[4]. LUO Duan-gaoetc (2009) established the model of urban taxis' network's equilibrium under the influence of distributed demand, and designed the algorithm to solve

[a] Xiao peng Lv: 270464440@qq.com

the mode by using its' characterized. Its analysis results, can be determined reasonable scale development of urban taxi according to the passenger waiting time, and providing city taxi network planning and scientific basis of management[5]. MA Wu-bingetc(2014) proposed to use Hidden Markov theory to the model of trajectory of city taxi movement, constructed taxi motion model by analyze the actual data, predict the node position of probability distribution by calculating the model, and provide route suggestions ride support to different needs of users be basis on this model[6].

3 Index selection

Establish a rational analysis the taxi resource supply and demand matching degree in different indicators of time and space is important for the accuracy of establishment of a subsequent optimization model. we use empty rate as a measure of indicators for analysis. Empty rate is the empty of mileage's proportion of the total running. As empty rate higher, the effect is more enough for the vehicle, operating income decrease in the same mileage, operating costs increase, the formula is empty stroke / total stroke × 100%. Integrating the three factors that is the utilization of mileage, load factor of vehicle, taxi quantity of per thousand people, this paper define the new expression of empty rate, as follows:

$$Empty\ Rate = 1 - \frac{Total\ Stroke\ in\ Each\ Day}{n \times Attendance\ of\ Taxi \times Stroke\ of\ Each\ Taxi\ in\ a\ Day}$$

According redefined empty rate formula, the first calculation is the total mileage of the average daily passenger of molecule item, to consider the issue from a combination of the two dimensions of space and time, According to Sort of GDP, as well as the distance of District of Guangzhou City from the Guangzhou municipal government, the paper divided these two spaces in space. At the same time, in the time dimension, we define weekdays, weekends and holidays these three variables as a time period of our analysis.

4 Data analysis

District of Guangzhou City from 2015 GDP data, the largest GDP of total area is 3011.12 in Tianhe District, The total economy of Yuexiu District, and Huangpu District followed by it. The lowest total economic output is the 317.07 from Cong Hua District. According to the total amount of GDP of District of Guangzhou City, which divide Guangzhou City into the fourth ring.

Central District I: Yuexiu District, Tianhe District, Huangpu District, Central District II: Haizhu District, Baiyun District, Panyu District, Fringe District I: Liwan District, Huadu District, Nansha District, Fringe District II: Cong Hua District.

For total mileage of average daily passengerin different time and space, this paper according to the district population, employment and travel conditions of the Statistical Yearbook of Guangzhou, based on employment and GDP, calculate proportionally the number of trips and travel intensity,

Table 1. Travel situation of different area residents

	Central District		Fringe District		Total
	I	II	I	II	
number of people	30480 22	55466 41	314021 4	593415	123282 92
Number of trips	15652 50	28366	161259 4	304736	351094 6
Trip times	34664 849.4	33471 759	167634 04	275463 .2	851754 76.8
Trip intensity	11.372 9	6.0346	5.3383	0.4642	23.21

Table 2. Travel situation of residents in different periods

	population	Number of trips	Trip times	Trip intensity
workdays	6275101	6275101	20218187. 17	3.22197
weekend	12328292	1972527	3934574.3 92	0.31915
holiday	17167580	1687101 4.32	2950248.6 23	0.17185

We can saw clearly that the strength of travel of the central area is significantly higher than the strength of travel of the marginal zone from Table I Residents of different regions of Guangzhou travel case, but also the strength of travel of the central area 1 Guangzhou travel was significantly higher than the strength of central area 2, the travel intensity of Fringe 1 was significantly higher than Fringe 2. So the supply of taxi should be allocated by accordance with travel intensity level of each region. We can know Travel intensity of workdays of residents of Guangzhou is most in Table II Travel case of Guangzhou residents in different time, then, Saturday and Sunday, holiday travel intensity is lowest.

Structural modes of travel are related to characteristics of mode of transportation, urban form and land layout and policies of traffic management, so the travel structural mode of different urban transport system is different. Based on the number of population in each districts of Guangzhou, people travel intensity, the rate of taxi's use andrateof choosing taxi, daily mileage data and using GDP to calculated proportionally daily total passenger miles of District of Guangzhou and the total number of passenger miles in three time periods daily.

Table 3. Average daily passenger in different time and space

Central District I	259×10^5	workday	344×10^5
Central District II	137×10^5	weekend	341×10^5
Fringe District I	122×10^5	holiday	183×10^5
Fringe District II	106×10^5		

From table 3 Total miles of daily average passenger in different time and space, we can see, Total miles of daily average passenger of central area and workdays are greater than Total miles of daily average passenger of other time and space. Total miles of daily average passenger and positive correlation between economic, but also the impact is a two-way relationship, more total miles of daily average passenger in one time, the time create more in economic.

5 Establishment and solution of model

Then forecast the optimal number of taxis, if the fees of taxi and oil prices are fixed, the "optimal" of taxi should be reflected in these two following aspects: (1) for taxi ride passengers, the degree of satisfaction is the higher, the better; (2) for every taxi, the annual profit as the more, the better. By Analyzing the relationship between empty rate i, passengers' satisfaction S, every taxis' annual profit R, number of taxis in city n, to establish the optimization model. Symbols as follows:

Variable: i: empty drive rate; R: annual profit of taxi; n: the number of city taxi; pr: profit rate

Constant: co: average per taxi annual expenditure; re: industry annual total revenue; or: Taxi attendance; ad: average daily passenger total mileage; pd: average daily mileage per vehicle

For the time when the taxi number of "optimal" means rate of empty driving, passengers ride car easier, the satisfaction of passengers is also higher. Therefore, this article assumes that passenger satisfaction is uniquely determined by the rate of empty driving, and along with empty rate increases. For the mathematical model of empty rate iis:

$$i = 1 - \frac{\text{Total Stroke in Each Day}}{n \times \text{Attendance of Taxi} \times \text{Stroke of Each Taxi in a Day}}$$

Mathematical model of earnings R per taxi:

$$R = \frac{\text{Industry annual total revenue}}{n} - \text{Average annual expenditure per vehicle}$$

The annual average costs include of operating costs fixed expenses and fuel consumption on. According to the research analysis of supply relationship between domestic and foreign by China Taxi Association of Cities, rate of empty driving of Urban Taxi and traffic relationship between demand and supply has the following:

Table 4. Empty driving rate and the traffic supply and demand

empty drive rate	≤25%	≈30%	≥40%
state	Obviously in short supply	Basic saturation, balance of supply and demand	Obvious oversupply

Based on consideration of the state of supply and demand, which we determine the rate of empty driving range $0.3 \le i \le 0.4$ According to the report of taxi industry cost supervision and examination announced in 2015 by Guangzhou Municipal Price Bureau, can calculate the Guangzhou taxi companies profit margin is 13% at that time. In consideration of taxi companies and taxi drivers, we only determine the profits pr≥10% is passable. Based on the above analysis, establish optimization model:

$$\begin{aligned} & Max\ S;\ \text{R} \\ & s.t\ i = 1 - \frac{Ad}{n \times Or \times Pd} \\ & R = \frac{\text{Re}}{n} - Co \\ & S = S(i) \\ & \frac{dS}{dI} \ge 0 \\ & 0.3 \le i \le 0.4 \\ & \text{Pr} \ge 10\% \end{aligned} \tag{1}$$

As we can see from the above equation, passenger satisfaction S and taxis annual profit R are two conflicting requirements, so it has important practical significance to the optimal model while for allowing passenger satisfaction and annual earnings optimal at the same time. Model Solution:

1. Describe the satisfaction and its relationship of the rate of empty driving:

We use the number between 0 and 1 describe the passenger satisfaction, the highest satisfaction is 1, the lowest satisfaction is. At the same time, this paper argues that the passenger satisfaction can describe uniquely empty rat, and it is assumed that:

(1) When empty rate is more than 40%, the market presents the apparent oversupply state, the passengers can gets on the taxi easily, passenger satisfaction close to 1.

(2) When empty rate is 30%, the market is basically saturated, passengers can take the taxi, but need to wait a little time, the passenger satisfaction is 0.8;

(3) When empty rate is 25%, the market is in short supply, passengers is difficult to take a car, passenger satisfaction less than 0.6, with reduction of empty rate, passenger satisfaction index fell sharply.

In this paper, SPSS is used to fit the data, and the goodness of fit is determined as the three curve function. So, when the driving rate is $0 \le i \le 0.4$, the function relationship of rate of empty driving and passenger satisfaction:

$$S(i) = 5.714 \times 10^{-4} - 3.980i + 39.455i^2 - 58.133i^3 \tag{2}$$

2. The relationship between customer satisfaction and taxi number

According to the Bureau of transportation statistics, Guangzhou taxi-operating conditions are as follows:

Table 5. Guangzhou taxi status table

project	Number of taxi companies	Taxi number	Daily operating income	Monthly operating income	Mileage utilization
data	84	33 thousand	650 yuan/car	19500 yuan/car	73.79%

project	Operating mileage	Fuel cost	Vehicle fee	Driver income	Taxi attendance
data	400 mile/day	5328 yuan/ month	9300 yuan/ month	2436 yuan/ month	85%

According to the table data, the calculation of the empty drive ratio is:

$$i = 1 - \frac{132.1\times10^5}{n\times0.8\times400} = 1 - \frac{41250.2}{n} \quad (3)$$

The relationship between the number of taxis and the customer satisfaction of the formula(3) into formula(2) is as follows:

$$S(i) = 5.714\times10^{-4} - 3.9801\times(1-\frac{41250.2}{n}) + 39.455 \times(1-\frac{41250.2}{n})^2 - 58.133\times(1-\frac{41250.2}{n})^3 \quad (4)$$

3. The relationship between the annual profit of the taxi and the number of taxiIndustry annual total revenue: $650\times365\times3.3 = 782925 thousand\ yuan$

car fuel costs: $5328\times12 = 63.936 thousand\ yuan$

Contract cost per vehicle $9300\times12 = 111.6 thousand\ yuan$

We can get the profit R of a taxi for a year:

$$R(n) = \frac{78292.5}{n} - (11.16 + 6.3936) \quad (5)$$

4 the relationship between taxi profit and passenger satisfaction

Attention to the relational expression (4) between customer satisfaction and the number of taxis, s (n) is monotone decreasing by n; which is taxi relational expression (5) of the profit of taxi and the number, R (n) reduced with the increase of N, R and s are irreconcilable. In order to achieve the optimal value of R and S, the N in (5) is brought into (4) by R, and the relationship between S and R is obtained:

$$S(i) = 5.714\times10^{-4} - 3.9801\times(1-\frac{41250.2}{\frac{78292.5}{R+17.5536}}) + 39.455\times(1-\frac{41250.2}{\frac{78292.5}{R+17.5536}})^2 - 58.133 \times(1-\frac{41250.2}{\frac{78292.5}{R+17.5536}})^3 \quad (6)$$

5. Determination of the optimal number of taxis in Guangzhou City

The model of load rate requirement is $0.3 \le i \le 0.4$ by formula (2) can get the satisfaction S of should meet $0.788571 \le S \le 1.0$ And because the requirements of profit margin by model is $\Pr \ge 10\%$, according to basic hypothesis of annual expenditure per vehicle unchanged, annual profit $R \ge co\times10\% = 1.75536 yuan$. By formula (6), we can get the satisfaction of S easily should be consistent with $S \le 0.8856$. According to formula (4) to the relationship between the satisfaction and taxi, at this time, using *MATLAB* inverse solve the scope of the taxi number n is (because it is a cubic equation, so each calculation results are three numerical value, according to the number of actual number of 33 thousand taxi, we give away the actual value of the phase difference of another numerical)

$$37414.75 \le n \le 37660.43$$

That is to say, in the reference of Guangzhou city travel data, the number of city's taxis should best control in the range in 2015, the average value is:

$$n = 37537.59$$

6. The central region, the edge region and the time interval of the optimal number of taxi

Different time and space only change the number of taxi passenger, we still use this model, then d_x, e_x express regional center, edge area and the number of taxi passenger and turnover of different time intervals, as follows:

Table 6. The matching degree in different time and space

	Central District I	Central District II	Fringe District I	Fringe District II
workdays	1.495	0.793	0.702	0.061
weekend	0.148	0.0786	0.07	0.006
holiday	0.08	0.042	0.037	0.003

6 Conclusion

Urban taxi traffic is an important part of urban public transport.If the taxi number is too much, which results in a waste of public resources, and if there are too little taxis that will cause inconvenience to the public. These are different degrees of restricting the taxi industry healthy, stable and harmonious development.Finding the balance between the two is the focus of this paper.Through the reasonable analysis of the different time and space residents travel intensity, the optimization model is established to achieve the best matching degree of supply and demand, which provides a scientific, reasonable and feasible basis for the management department.

References

1. Y. Bian, W. Wang, J. Lu, J. Ma, Urb. T. T. M. S. R. P. Met, T. S. Eng. Inf. J. **1**, 95 (2006).
2. Y. Bian, W. Wang, J. Lu, A B. Mod. Urb. T. Ope. Net, J. Tra. Tran. Eng. J. **1**, 93 (2007).
3. Y. Li, Z. Yuan, H. Xie, S. Cao, X. Wu. Ana, T. Tri. C. Bas T. OD. Dat, Tra. Tran. S. Eng. Inf. J. **5**, 85 (2007).
4. J. Zhou, J. He, Z. Sheng. R, Ana. C. T. Ope. S, J. Man. Eng. J. **1**, 63 (2000)
5. G. Luo, F. Shi, Urb. T. Ope. B. Mod. C. Dem. Dis, J. Rai. Sci. Eng. J. **1**, 87 (2009).
6. W. Ma, M. Liu, H. Huang, S. Deng, C. Urb. T. Mov. P. Mod. Bas. His. L, J. Nat. Uni. Def. Tec. J. **3**, 129 (2014).

MATEC Web of Conferences **63**, 04010 (2016)
MMME 2016
DOI: 10.1051/matecconf/20166304010

The environmental impact of mining and its countermeasures

Jin LI[1],Tong Tong ZHANG[1],Wen YANG[1],Yu ZHANG[1]
China University Of Mining And Technology.BEI JING.China.100083

Abstract: Exploration of mineral resources had been part of the major means to promote economic development in China,but the devastating effect of mining on the environment is inevitable.Based on the analysis of factors of the environmental disasters caused by the mining methods,this paper systematically analyzes the influences of open pit mining on land resource,ecological system,geological and ecological environment,especially analyzes the effects on the environment and ecological system.Meanwhile,the paper points out that the mining scale,method and type are the main factors that affect the environment and ecology.At the end of this paper,some rational countermeasures, protection and control measures for environmental protection and governance are put forward on the basis of the research.

Key words: open-pit mining;environmental affect;inuence factors;countermeaflsures.

Mineral resources are an important material foundation of human survival and development, plays a role in the progress of the society.Till to 2010, there are more than 74590 metal non-metal mines in China, including 8032 underground mines and 67862 open-pit,provides a strong guarantee of the national economic development[1].Through China emphasized to avoid taking Western countries lay in the industrialization process of "pollution first, treatment later",but in fact we still on the same way,the environment is very grim. This paper focus on mining environmental problems, especially the open pit mining,such as the loss of resources,air pollution,damage of landscape and ecology.

1.The environmental impact of mining

1.1 The environmental impact of open pit mining

1.1.1 the destruction of land resources

The process of open-pit has a devastation impact on the land resources,such as overlying soil was dug and ground vegetation was destroyed.In China,most open-pit adopts external dump mode.The waste-dump soil covered 40~50% of total area of the open pit,at 1.5~2.5 times of mining land. Dump produced by each 1.0×104t coal occupation 0.04hm^2~0.33hm^2 of the land,an average of 0.16 hm^2.And covered 16300 hm^2 areas of land in 2010, occupied a large number of land resources[2]. Almost 1.4×106~ 2.0×106Hm^2 areas of land were destructive and occupied by directly mining, and increased 2×104 Hm^2 a year[3].Waste rock piles

DOI: 10.1051/matecconf/20166304010

and tailing piles also occupied a lot of land, according to the data show that produce 1t ore usually needs to strip 5~10t covered rock,while produce 1t copper can form on 400t waste rock and tailing[4].Construction plant,equipment placement,construction traffic divided into ground fire and outcrop fire.Outcrop is on the brink of soil,high temperatures can make the soil dehydration,dense and hard, make mineral particles molten,changes its composition and structure.A distance of soil made ground fire has a lower surface temperature,and lower degree of burnt soil.Burnt soil is easy to break and serious decline in fertility.Can be seen in the serious damage to the land of open-pit.

1.1.2 The destruction of the ecosystem

During in the open pit process,stripping the topsoil and leaving the subsoil and slag,add the weight of machinery and equipment,so that the soil becomes hard,compacted,extreme lack of moisture and organic nutrient,substantial areas of forest and grassland destruction.Noise,vibration, water damage generated during mining inevitably leads to animal migration and death,the impact area is five to 10 times of the men[5]. Range of biological activities in the vicinity of the mining area are reduced.That decline the function and stability of the ecosystem. All over the world,there are 593 kinds of birds,400 kinds of mammals,209 species of amphibians and reptiles and more than 2×104 species of higher plants on the verge of extinction[6]. Usually,this kind destroy of biodiversity is irreversible. Although try to recovery and reclamation of vegetation in the damage zone, the effect is not very obvious. Mainly metal is quarter,granite, granulite and chlorite schist, that has strong wind resistance capabilities but weak in the natural ecological restoration ability[7]. Compared to the developed countries, the rehabilitation rate low 50%,only 10% in China,the low rehabilitation rate expands the destruction region of vegetation[9].Effect of coal fire on the ecological environment cannot be ignored. On the one hand,the ground temperature rises changed the physical and chemical properties of surface soil, reducing soil moisture and organic matter content, leading to soil depletion,plant growth difficult,reduce the biomass production;on the other hand, the surface temperature directly lead a large area of vegetation to die,more seriously to ignite forests,pastures,causing network also takes up lots of land.Besides that,coal fire damage on land resources is also quite serious.widespread damage to vegetation.

1.1.3 Geological disasters

Stability of open pit slope is one of the foremost environmental geological problems of open pit [4]. Artificially shaped slope scale has been expanded with the increase of mining deep,seriously damaged the natural balance to the stress and lead to artificial slope deformation, damage and slippage.One waterworks of Shougang Group operation,located in Qian'an,Hebei Province,occurred 109 times landslides and slope deformation instability since its operation.

In the mining process,will inevitably damage to vegetation,water,mountains,soil.All of this lead to water retention capacity decreased in surfaces, coupled with the concentrated interference and change from solid waste in the mining of the original mining area, and the solid waste piled messy,loose soil,causing serious soil erosion,leading to further deterioration of the surrounding environment,finally evolution to debris flow ditch,causing landslides.Development is harmful to the vegetation surrounding the mine, also lead the land productivity gradual decline,surface in the state of sand, gravel and barren[9].Formed the desertification that gradually weakened from mining to outward.

Burning cavity,burnt rocks,combustion fracture and ground subsidence are the most obvious coalfield geological effect.Burning cavity is formed after the burning of shallow coal seam and the floating underground coal and coal pillar.It changes the balance of the seam roof and resulting in a large number of surface cracks and collapse pits.Grilled surrounding rock making the rocks,appeared to be different degrees of burns phenomenon, result the formation of burnt rocks.

1.1.4 The impact on the human living environment

①Water Resource

The waste-water pollution generated by open-pit

water is very obvious.Large amounts,of acid,heavy metal ions,various kinds of mineral processing reagents,suspended solids washing waste-water and domestic sewage are produced in the mining process. Features of waste water emission are large quantity, complex composition,long duration and wide range. Secondly,toxic and harmful substances of solid waste through the long term of oxidation,erosion and leaching flows into underground.These waste-water through the way of runoff or infiltration to pollution the environment of mining and the surrounding area.Opencast mining also pumping and drain of groundwater,lower the groundwater level,less the amount of water, aquifer of Quaternary will be drained in the open-pit mining process, seriously influence the use of water resources.

② Air Pollution

The main air pollution generated by open-pit mining include:gas generated by large open-pit mining operation of machine equipment,power equipment, transportation equipment.Float coal mining steps,coal under side slopes,waste dump and miscellaneous coal scattered on the ground during transportation of miscellaneous coal,after prolonged exposure to air,the phenomenon of spontaneous combustion will produce large amounts of harmful gases,there is a lot of smoke associated. Motor transport in large opencast coal mine is very common,especially more than 100 tons of large vehicles are favored by the mine.it generates a large amount of exhaust.The harmful gases like CH_4 and CO, CO_2, NO_x, H_2S etc.and other substances can lead to the destruction of the ozone layer,cause the greenhouse effect,but also the formation of acid rain,causeing secondary pollution,seriously affecting the production of human living environment.

Work of waste haulage along with a lot of dust generation and after the vehicle airborne dust effect is more obvious;Tailing surface like a desert,oxidation and wind erosion can make it become a source of periodicity dust storms,spring and winter is particularly serious.If the long-term without precipitation,settling dust blown by the wind and car driving again pollute the surrounding environment,a huge range of its spread.According to relevant information,dust can float to 10~12km outside and dust capacity reached about 300t / (h • m^{-2}),Grain yield decreased by 10% due to dust,and if people long-term inhalation of high concentrations of dust will affect their health and cause pneumoconiosis. [6]

③ Noise pollution

The sound generated by transport vehicles and large machinery are the main source of noise mine open-pit mining.The noise level can reach 90 dB (A) above,sometimes as high as 120 dB(A) This level often extends far beyond national standards.The noise environment not only can make listening mechanism of damaged,causeing insomnia, fatigue, dizziness, memory loss and other symptoms and affect sleep and rest,but also harm people's physical and mental health.[11]

④ The burning loss of coal resources and the destruction of the environment

Opencast coal mining and mining which due to open pit mine periphery generated vertical cracks, result in oxygen along the slit channel with its underlying coal seams through oxidation and lead to coal combustion(Figure 1).Burning coal goaf and ground subsidence become a threat to the safety of the residents of the district fire trap.Yining and Kerr base coal fire occurred,people suddenly collapsed due to burn empty beast into the fire case.So it brings about a series of environmental and disaster problems such as resource loss,surface subsidence,air pollution and so on.

Figure 1 Xinjiang Hami basin three Ling open-pitmines and surrounding landscape after coal combustion

1.2 The environmental impact of underground mining

1.1.1 the environmental impact of ground collapse

After mining,the surrounding mechanical balance is destroyed,as the mined-out area increasing,the surface subsidence appeares becase the surrounding rock is not enough to resist the overlying rock mass gravity.After the surface subsidence,the ground structure, farmland and irrigation facilities will be damaged.In addition, Bridges,railways,and transmission lines will be affected in different degree.Deformation,rupture, displacement,and caving sink occour to roof strata which followed by surface subsidence due to the change of sinking.In 2011 ,the total production of coal is 3.52 billion tons which 92 % came from underground mining.According to incomplete statistics, one million hectares of land has been damaged due to the national underground coal mining subsidence by the end of 2011 which still increases at a speed of 70000 hectares a year.Among all the ecological environment problems caused by coal mining,the problems caused by the eastern plain area coal base should be paid more attention such as the damage of human settlement, the change of rural landscape and the large amount of village relocation.According to incomplete statistics,the number of removal will reach 2.7334 million in five lianghuai coal base and Great Plains in West Shandong base which outnumbered the total more than 1.3 million immigrants of the three gorges. For Jinzhong mining area, 509 villages and 97802 hectares of land suffered varying degrees of damage such disastrous consequences which attribute to ground subsidence.

1.2.2 hydrogeological conditions

Influence of underground mining on hydrogeological conditions include both surface water and groundwater.Ground subsidence caused by massively extracting of mineral resources ruins the water and water systems,resulting in the fall of groundwater and increase of surface runoff. New underground mined-out area makes the drainage characteristics of groundwater change.Effects on runoff regulation of groundwater system is corrupted. Dynamic rules of surface runoff in the basin is not only controlled by atmospheric precipitation and groundwater storage function,but also by the mine drainage and drainage of aquifer dynamic static reserves.All the above factors makes the space-time distribution of water resources more unequal.(In addition,a large efflux of the mine wastewater not only seriously polluted surface water with in the basin, and possibly through osmosis contaminated groundwater,It is worrying.As for the influence on groundwater,To the exploitation,the inside of mine groundwater or threaten the safety of mine aquifer groundwater.must to be drainaged which changes the natural groundwater runoff and discharge conditions,thereby makes mine near of shallow layer groundwater was thinning dry. Near surface water turns to dry for drainage or artificial diversion.Huge waste of water resources lead to the decline of regional underground water level, deterioration of ecological environment,plant difficult growth,sand land of petrochemical.Coupled with the slow flow and invisibility of polluted groundwater, the contaminated area is difficult to determine.It is difficult to recover depending on natural purification of the aquifer itself alone.[15]

2.Influencing factors of the environment in mine exploitation

2.1 Mining scale

Mining scale and environmental damage have a certain relationship.[16]Improving the mine production will require more machinery and equipment,buildings and open spaces which will also generate waste solid water and gas,occupying surface and underground space larger as well. This will cause the reduction of production life required,far exceeding the environmental limits of self-regulatory,and more serious environmental pollution.

2.2 Mining methods

Open-pit mining needs lower-cost,conducive to mining of lower grade ore bodies,so as to improve the utilization of resources.On the other hand,open-pit mining severely damaged the surface landscape, seizing and destroying large areas of land resources, striggering geological disasters,seriously affecting living and production environment.Underground mining has relatively little influence on the surface landscape and environment but can cause surface subsidence,affect hydrogeological conditions and induce earthquake.

2.3 Mining types

In the process of exploitation,different types of mineral resources have environmental impacts are not the same.Combustible organic minerals will result in solid waste,waste rock dumps,as well as the emission of toxic and harmful substances,causing global warming or acid rain.Heavy metal pollution from mining of metal ores does great harm to humans and the environment.Building materials mining destroy mountains,vegetation and other natural landscapes directly.

2.4 others factors

Environmental impacts of mining is related with its surroundings.Levels of impact on the human environment different from the mining location, topography, and climate.Environmental damage close to the number of people affected,for example,A city with annual mine output of 50000 tons will receive more complaints than a remote deserted island with an annual output of 5 million tons when it comes to the environmental problems caused by mining.Local terrain and climate characteristics have an enormous impact on the mining environment.In the flat area,the river flows slowly,with less sediment in the water away.In contrast, at the high altitude wide, noise and dust air can spread quite far away.Hills rolling meant country matters,geomorphological features is very important.Located high in the top of the device is very conspicuous,noisy and dusty air will spread widely, installation of equipment in the Valley can only be seen in height at close range limited to a few;The surrounding hills have formed an effective barrier,so that noise and dust can not escape.The river at high altitude carries a large amount of sediment due to high flow velocity, and is deposited in a relatively flat low-lying area. And the precipitation, air temperature, humidity, wind and other climatic factors will strongly affect the diffusion of mine pollution to the surrounding environment. Therefore, the climate affects the intensity of pollution. The atmosphere plays a dominant role in the emission of waste gas, dust, noise and air, and the precipitation has a very important influence on the diffusion of the waste liquid.

3.Environmental protection measures and recommendations

3.1 Sound environmental laws, regulations

Although the our country has formulated the "environmental protection law","mineral resources law",the geological disaster prevention and control law"and other,but our country haven't formed a complete legal system, lead to long-term since the management of mine environment in our country is relatively weak,do not have a strong sense of the protection of the ecological environment,government and enterprises pay attention to the economic benefits, ignoring environmental protection,go a at the expense of the environment at the expense of economic development.First,sound and perfect our country's relevant legal system, in reference to foreign advanced environmental protection legal system based on, combined with the actual situation of our country's environmental protection, to develop suitable for the situation of our country's environmental protection system,which is currently carried out in our environmental protection an important guarantee. Secondly,intensify the propaganda of environmental protection laws and appealed to the people involved in environmental protection and supervision, improve the legal consciousness in the process of the exploitation of mineral resources, the resources according to the economic development of basic rules and national standardized mining development, can not be blind to predatory development;again,increase of environ -mental pollution violations punishment,to prevent local protectionism, from two aspects of economic and policy punish illegal behavior.

3.2 Reclamation and utilization of open-pit mining

Mining should take the method for mining and reclamation of all hands at the same time,inner dump open-pit mining method,We should take reclamation as an important link in the process of mining and take dragline method,back haul,back track freight.

Communicates with underground pits, reservoirs and ponds can be built,covering an area large enough to build artificial lakes and a water recreation area.For water depths shallower places,we can grow aquatic crops such as Lotus root and wild rice stem,for relatively flat water pits,through experimental studies, after transformation to the soil, planting suitable vegetation,Gradually formed a growing development way of combining farming,so that maximum benefit from the economic and environmental benefits .

For the destruction of land resources,biological reclamation of approaches you can take.Use of microbial fertilizer technology,aging and improvement of barren land,restore soil fertility;worked out a method of adopting"the mud into the sand,the sand into the mud"method,adjusting the proportion of sand and mud soil,thereby improving the tilth.For the protection of the ecological system,through the establishment of parks and protected areas to reduce the exploitation of mineral resources to habitat destruction,the maximum amount of formulation of different types of mineral resources development projects caused by the biological abundance index,makes in the actual production process in time to pay attention to protecting the development zone surrounding the biological diversity.For the prevention and controlling of geological disasters,due to mining formed a variety of slopes,in addition to restoration,we should evaluate the degree of slope stability and its influence on the surrounding environment,according to different stone revetment slope or slope.In order to prevent soil erosion can obstruct the solid waste;the dangerous slope,depending on the slope length divided by intercepting ditch and drainage ditch,drainage works and increase of vegetation coverage.

Sewage and waste water resource is the realization of opencast mine area to maximize the use and utilization of water resources in the most effective way, can build a small sewage treatment system can not only of the mining area of the sewage of purification treatment,but also can deal with domestic waste water from surrounding residents.To strengthen the rational design and management of drainage and irrigation system,to prevent the waste of water resources in the process of the development of mineral resources.In order to reduce the air pollution,the construction site should as far as possible the use of clean energy or liquefied petroleum gas as fuel;secondly,can also in the transport vehicle tarpaulin covered,try to reduce the loading and unloading height and the fallout promptly removed to prevent the spread of these wastes caused due to the action of wind,watering facilities in the construction area,transport vehicles for timely cleaning, concrete mixing as set in the outlet and other measures.

3.3 measures taken in underground mining

Of underground excavation in rock roadway and generated a lot of waste,not to the ground directly in the underground processing,can reduce the ground gangue heap,the elimination of pollution,can save the gangue underground,well on transportation costs. Gangue well technology is through a variety of technical means,gangue rock roadway excavated, transported to underground waste disposal field broken, through a pipe and the wind transport to the work surface backfill mined out area,do gangue underground. The gangue filling to fill both to eliminate the pollution of coal gangue surface emission,but also to reduce the surface and strata movement and subsidence area is reduced.

Perfect the management system of mine environment monitoring.Help and guidance in the local environmental monitoring station,routine of mine groundwater standard higher data monitoring,the establishment of groundwater environment database, provide detailed and reliable basic data for the development of enterprises and scientific research. Environment related departments establish regional groundwater monitoring network and compensation mechanism,the mainland area water environment real-time monitoring,in accordance with the "who develops protection,who destroyed who is restored, who pollution who governance"principle,urge the Mining Corporation to fulfill the groundwater environmental protection and restoration of the obligation,funds through multiple channels to raise, gradually complete the pit was closed after the groundwater environmental restoration,the simultaneous development of the mining area of economic development and environmental protection.

References

[1] Wang Yan, Cheng Hao, Zhang Yu, Wang Pengfei. Environmental protection and restoration of mining environment protection and restoration measures [J] Henan building materials, 2013 (4): 93-95

[2] Dai Liping. The influence of mining on ecological environment and ecological restoration in the coal mine area: a case study of [J]., 2010 (18): 109-110

[3] Ma Jing. Development and utilization of mineral resources and environmental protection [J] resource development and market.2003 (3): 151-153

[4] Li Tiefeng, Pan Mao. Environmental geology [M] higher education publishing, 2008.7

[5] Yu Lin, Wang Xueli. Discussion on the ecological protection in the environmental impact assessment of mine development [J] Xinjiang environmental protection, 2003, 25 (4):48-50.

[6] Gao Yaling, Yang Wei, Wang Shuo. The environmental effect of iron ore mining [J] Liaoning chemical industry, 2009, 38 (1): 62-65

[7] Jiang Zhen, Yun cable. Anshan Iron ore waste rock field ecological environment governance plan and technical method [J]. Jiangxi Journal of Agricultural Sciences, 2008, 20 (4): 102-104

[8] Heming Kui, Ding Hua. Strengthen the development of mineral resources on environmental protection and management of Sichuan [J]. environment, 2003, 22 (3): 86-90.

[9] Lin Chen. Effect of mining on geological environment and Control Countermeasures of [J]. technology and market, 2013, 8 (20): 180-181

[10] Tu Scott. Effect of coal mining on the environment and countermeasure research of [J]. mining research and development 2003, 23 (4): 8-10

[11] Wang Wen, Xiang you, Wang Guojun. Journal of Liaoning Technical University, 2002, (4), [J]. Journal of,, (in Chinese).

[12] Wang erde. China's coal mining destroyed 1 million hectares of coal base in the number of immigrants over the three Gorges [EB/OL]. http://finance.sina.com.cn/china/20121205/020613896525.shtml.

[13] song Hongjuan. Jinzhong coal mining impact on water environment evaluation and control measures of groundwater [J], 2009, 31 (2)

[14] Wang Zhihui. Analysis on the influence of coal mining on the environment and Countermeasures [J]. Jiangxi coal science and technology, 2012, 3:115-116

[15] Chen Xuejun, Bin Xiuling, the influence of mine development on environmental quality [J]. Journal of Guilin University of Technology, 1996, 1 (16): 83-86

MATEC Web of Conferences **63**, 04011 (2016) DOI: 10.1051/matecconf/20166304011
MMME 2016

Realization on the interactive remote video conference system based on multi-Agent

Yan ZHENG [1, a], Wen Ting LI [2], Li ZENG [3], Yu GE [1], Xiao Yan CAI[1], Xian Yu MENG [1]

[1] *Military Information Room, Wuhan Ordnance N.C.O Academy of PLA, Wuhan, 430075*

[2] *China Electric Power Research Institute, Wuhan, 430074*

[3] *Language Training room, Wuhan Ordnance N.C.O Academy of PLA, Wuhan, 430075*

ABSTRACT: To make people at different places participate in the same conference, speak and discuss freely, the interactive remote video conferencing system is designed and realized based on multi-Agent collaboration. FEC (forward error correction) and tree P2P technology are firstly used to build a live conference structure to transfer audio and video data; then the branch conference port can participate to speak and discuss through the application of becoming a interactive focus; the introduction of multi-Agent collaboration technology improve the system robustness. The experiments showed that, under normal network conditions, the system can support 350 branch conference node simultaneously to make live broadcasting. The audio and video quality is smooth. It can carry out large-scale remote video conference.

KEY WORDS: multi-Agent; interactive; forward error correction; Peer to Peer; video conference

0 Introduction

In daily work, we must participate in many large and small meetings and seminars, but because of a variety of reasons, such as in business trip, in different branch or sick; we cannot attend it in person. The above problem can be well solved by remote video conference system which is developed by computer network and multimedia technology. At the same time, this system can greatly improve the interactivity, through which the remote participants can discuss and communicate with each other through audio and video, letting all participants have face to face experience. In the course of design and implementation the system, the FEC forward error correction coding, P2P technology and multiple Agent theory are adopted, which greatly improved the quality of audio and video services and enhance the robustness, maintainability and adaptability of the system.

1 Related contents

FEC forward error correction

FEC error correction was used in the correction of bit stream. In this system, it is used to make group data error correction. It encodes the original data firstly to produce a certain amount of redundant data, and then the sender sends out the N groups. As long as we can receive any K groups of the N groups at the receiving end, the original data can be returned, which greatly reduce the probability of flowers screen caused by echo video when there is bad network packet loss.

Multi-Agent technology

Intelligent Agent, a computer program running in mobile environment, has highly autonomous ability, which can accept entrust of another entity like user, system or machine, and provide aid and service for it. Under the drive of the target it is also capable of initially carrying out the action including social contact and study, to sense and adopt the change of mobile environment, and cope it with properly response. As Agent has the autonomous ability, it itself can use its decision-making mechanism to select the strategy and make response without users' monitor and guidance. The sociality of

[a] Corresponding author: zy397037172@163.com

 DOI: 10.1051/matecconf/20166304011

Agent realizes the function that one Agent can communicate with another Agent, which can offer better service for users.

P2P Tree

A Multicast distribution tree is constructed in accordance to the conference situation, which is Audio and video dispatching strategy with push-pull combination based on tree structure. The branch conference node can link child node, distributing audio video data downward and setting the allowed maximum layers of tree by preset to make sure the postpone time of bottom node will not too big[1].

2 System functional structures

The structure diagram of the interactive remote video conferencing system based on multi-Agent is as follow (Fig.1).

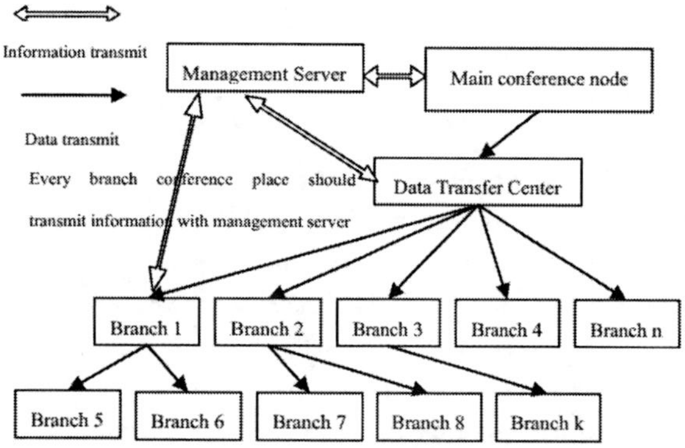

Fig.1 The structure diagram of the interactive remote video conferencing system based on multi-Agent

The system was divided into four modules: management Server, DTC（Data Transfer Center）, main conference node, branch conference node.

The scene of conference supports a main conference node and many branch conference nodes. After having applied interactive node, people can communicate, talk and discuss in branch conference place. You can join in the corresponding conference place, if you input Server IP, conference ID and local conference place' name and password in branch conference place port. The system can support more than one live branch conference branches at one time. The branch conference place which has spare bandwidth can be used as father node to send audio and video data to its child node of branch conference place port.

At the same time, the system has interactive function. Branch conference place can be applied as conference place interactive point, from which speech can be discussed and local audio and video data can be send out, and all conference place ports connected with this conference place port can receive the audio and video data. It guarantees that all branch conference places have opportunity to take part in the communication.

3 System realization

Interactive focus

Branch conference place can be applied as conference place interactive focus. After having approved by main conference place, it transmits the local audio and video data to DTC, and DTC transmits the data to main conference place and each branch conference place. In this way all receive port in the conference place can watch and listen the speech and discussion of the branch conference place. The system can support no more than five interactive focuses to take part in conversation and discussion simultaneously.

Applying branch conference place to be an interactive focus is realized as following (Table. 1).

Table. 1 branch conference node applying focus

```
void MainWindow::applyFocusFunction( )
{
    CreateApplyFocusMessage( );
     sendMessageToAgent( );
  }
```

Interactive focus transmitting local audio and video data to DTC is realized as following (Table. 2).

The multi-Agent

Information processing Agent, data processing Agent and new node processing Agent were set in the system. They coordinate with each other to finish the organization and control of the system. [3]

Every branch conference port requires sending information to management Server. As there are different

DOI: 10.1051/matecconf/20166304011

kinds of information, branch conference place port needs to engage in corresponding operation according to different information. In order to promote system processing capacity, an information processing Agent was set.

Table. 2 Interactive focus transmitting local audio and video data to DTC

```
//start transmit focus'audio and video data
void MainWindow::startSendFocusAVData(quint16 focusNum)
{
 AuAvHandle.initializeSendAV(9002, 9004);
  if ( focusNum == 1)
      {
AuAvHandle.startSendAudioToDTC
(DTCAddr.IPAddress.toLatin1( ).data( ),DTCAddr.F1AuPort);
AuAvHandle.startSendVideoToDTC
(DTCAddr.IPAddress.toLatin1( ).data( ),DTCAddr.F1ViPort);
AuAvHandle.stopForward(localDestAddr.F1AuPort, "127.0.0.1",
localAddr.F1AuPort);   // stop forward local, remove local echo
   return;  }
   else if ( focusNum == 2)
       {
AuAvHandle.startSendAudioToDTC
(DTCAddr.IPAddress.toLatin1( ).data( ),DTCAddr.F2AuPort);
AuAvHandle.startSendVideoToDTC
(DTCAddr.IPAddress.toLatin1( ).data( ),DTCAddr.F2ViPort);
AuAvHandle.stopForward(localDestAddr.F2AuPort, "127.0.0.1",
localAddr.F2AuPort);
   return;  }
}
```

DTC will transmit a great deal of data to branch conference places. Since the data have big volume, branch conference port can hardly process all of them in time. If the data processing Agent is set, it can receive data and decode it into audio and video format, which will be transmitted to matched port and played.

As to those new nodes which will add to the system, emphasis should place on considering that under which node it connect to receive data in high quality. The new added node Agent can finish this function.

The simple reaction Agent is used in the system. Its function procedure is as follow.

```
Function   SIMPLE-REFLEX-AGENT
(percept) returns   action
Static:   rules, a set of condition action rules
State ←—INTERPRET-INPUT(percept)
Rule ←— RULE-MATCH(state, rules)
Action ←—RULE-ACTION(rule)
Return   action
```

The simple reaction Agent is relatively simple, which should sense the current input state, and based on it, find matched rule of the state in predefined state set. The corresponding operation is finished according to this rule. Multi-Agent can work together to finish complicate task.

4 Key Technology

Forward error correction coding

By using the technology of FEC (forward error correction coding), receive port can use the received normal data package and redundancy data packet to decode the original audio and video data even though there are a few packet loss[5].

The technology of forward error correction coding is dramatically decrease the influence which is caused by network packet dropout for the quality of audio and video in system. When the rate of packet lost less than 20%, FEC can decode and repair it into normal audio and video to avoid flower screen.

Tree P2P

The Management Server is responsible for constructing, maintaining and upgrading multicast tree, which stores information of every node and supports node to find father node. The system will be more robust when some node leave or some node is failure. Typically, father node push data to their son nodes actively according to their data in NICE system[4].

After having acquired part of data DTC by using pull method as startup (pull), it transmits data package to its child node without requiring that the child node asks for data package (push). Pull model is of good robustness, which can work properly in highly disturbed P2P environment. Push model can effectively reduce cumulative delay observed at each node. Combing pull

DOI: 10.1051/matecconf/20166304011

with push model can deal with dynamic question of child node efficiently without producing high startup delay.

P2P technology can make good use of spare upstream broadband of every branch conference places' terminal to transmit video and audio data for new added node of branch conference place, which can partake DTC load volumes to increase conference places and guarantee the smoothness of audio and video.

5 Experiment and Results

The system had tested in computer lab, and the result showed that the testament is successful.

The system increased delay time when conference places increased (the lever number of P2P Multicast tree is within predicative value).It is shown in Fig. 2. Under normal network condition, delay time was controlled within 8s when branch conference place increased. The test result showed that the delay time was different under different father node.

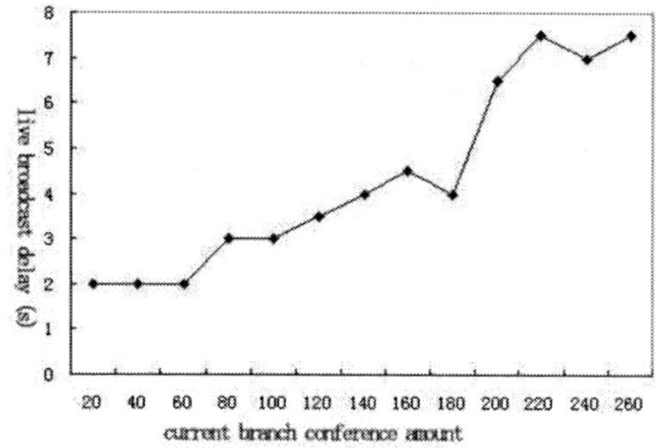

Fig.2 The relation between live broadcast delay and branch conference amount

When the current branch conference places added, the change of dropout rate is shown in Fig.3. Since FEC forward error correction used in the system could make code, no blurred screen was produced when the dropout rate was less than 20%. However, when the number of conference places were approaching to the limit that network card of Server can undertake, the dropout rate would be serious, and part of conference places could produce blurred screen.

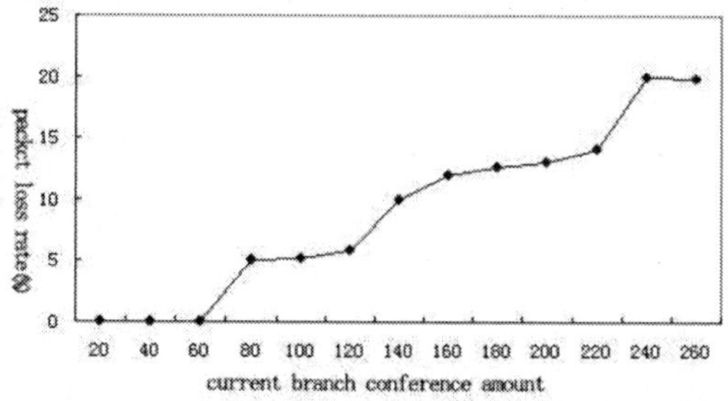

Fig.3 The relation between packet loss rate and branch conference amount

When the branch conference place increased, the change of the Server bandwidth occupancy rate is shown in Fig. 4. Since DTC server used kilomega broadband, it supported support 350 nodes simultaneously to make live broadcasting.

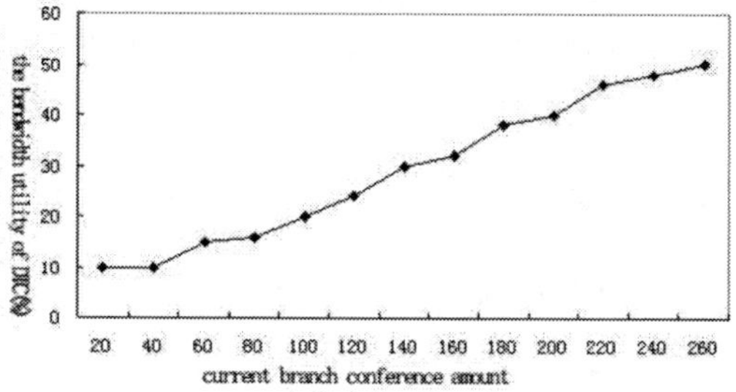

Fig.4 The relation between the bandwidth utility of DTC and branch conference amount

6 Conclusion

At present, the remote video and audio conference system is equipped in our academy to apply and test. The function and stability of this system has been widely recognized after having been applying it in several large-scale video and audio conferences. The imported Multi-Agent cooperation theory technology enhances the robustness and stability of the system dramatically. Meanwhile, FEC forward error correction coding and P2P tree are used to make sure that the system can transmit video and audio with good quality and liquidity. The applicability and maintainability of the system has been improved.

References

[1] Segarra J, Cholvi V. Convergence of periodic broadcasting and video-on-demand[J]. Computer Communications, 2007, 30 (5): 1136-1141.

[2] Cheng B, Stein L, Jin H, et al. GridCast: Improving peer sharing for P2P VoD[J]. ACM Transactions on Multimedia Computing, Communications, and Applications (TOMCCAP), 2008, 4 (4): 1-31.

[3] Cheng Feng, Study of Agent Reliability under Multi-Agent Cooperative Decision-Making[J]. Pattern Recognition And Artificial Intelligence;2008.10. Vol.21, No.5

[4] Zhong Yuzhuo et al. Media Streaming and Video Server[M]Beijing: Tsinghua University Press, 2003.

[5] Wu Zhigang, Bai Guangwei, Adaptive FEC based diffserv model for real-time media streaming[J]. Computer Engineering And Design;2009,30(6)

MATEC Web of Conferences **63**, 04012 (2016) DOI: 10.1051/matecconf/20166304012
MMME 2016

Development of Multifunctional Anti Aging Military Raincoat Fabric by Using Fitting Technique

Chun Xia CHEN

Yan cheng institute of industry College, Yan cheng, Jiang su 224005; Excellent Scientific and Technological Innovation Team of Jiangsu Province 2015; chx9941@126.com

Abstract: Use Woodland camouflage paint Oxford fabric as base cloth, first the light pressure processing was used on the fabric, and then use polyurethane which has high water vapor permeability performance and other various additives as face glue and Bottom glue, the fabric was compounded a thin macromolecule hydrophilic polymer PU film, the fabric was water allocation processed after stripping, At last the fabric will have high performance of the windproof, waterproof and moisture permeability. After testing, the performance index of the fabric are: the smoothness appearance of fabrics after cleaning≥grade4.0; hydrostatic pressure≥10000mmH_2O, after 20 times washing≥5000mmH_2O; The surface water repellency level≥grade4.0,after 20 times washing≥grade3.0; water vapor permeability≥3500g/(m^2·24h); Electrostatic charge density≥2.5uc/ m^2; humid air accelerated aging grade≥4.0.Test data shows that the designed functional fabric meet the technology requirement of the standards, meet the demand of the customer's use requirements.

Keywords: Fitting Technique; raincoat fabric;; design; properties; test;

Top-notch Academic Programs Project of Jiangsu Higher Education Institutions 2015

With the economic development and The continuous improvement of people's living standard, fabric and clothes continuously extended in practice , gradually formed a special fabric finishing technology. Especially the development of advanced science and technology, People proposed higher and higher、more and more requirements for textiles' performance. And the high waterproof and high moisture permeable fabrics were popular.

1 Experiment

1.1 Fabric specifications

Width:150cm Warp: 150D polyester filament (FDY)
Weft yarn: 150D polyester filament (FDY)
Warp and weft density:480 roots/10cm×430 roots /10cm;
Weight:150-160g/m^2
Fabric weave: 2/2twill

1.2 Choice of raw materials:

release paper
High moisture permeable polyurethane (solid content 30 ~ 55%)
solvent (two methyl amide DMF, butanone MEK , toluene TOL)
Desiccant, curing agent, promoting enzyme, color paste, etc.

1.3 Equipment

The production equipment of Multifunctional Anti Aging Military Raincoat Fabric is provided by jiang su nanwei Yueda, The equipment consists of the following parts: surface Glue table、oven、Bottom glue table、Laminating device and winding device

2 Technological process

base cloth → light pressure

face glue configuration → release paper gluing → drying → cooling → bottom side gluing → semi drying

→ fitting → cooling → winding → ripening → peeling → Water allocation processed → drying → end product

2.1 pretreatment process

The pretreatment process of base cloth are generally light pressure and water allocation, Considering the adhesion fastness between Oxford cloth and PU film , select the light pressure pretreatment on the base cloth, Considering the requirements of water resistance and moisture permeability index , water allocation finishing is processed after stripping.

2.2 transfer and fitting

This project uses the release paper transfer and fitting process. The process shown in figure 2-1:

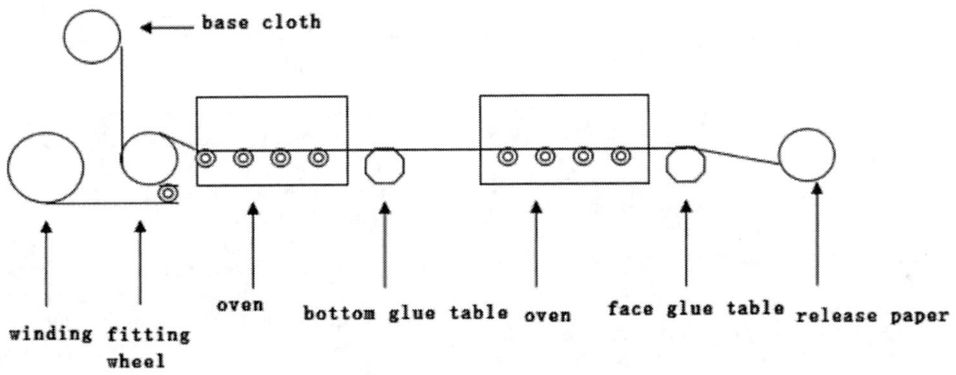

Figure 2-1 Transfer and Fitting process

2.3 Ripening process flow

In order to improve adhesion fastness. After transfer and fitting the fabrics are send to the ripening room, in the environment of 65℃ , ripen 48 hours.

2.4 Water allocation process flow

The specific process flow shown as below:

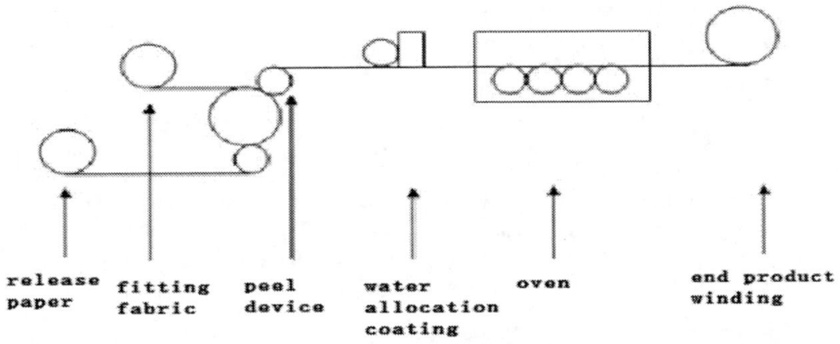

Figure 2-2 Water allocation process flow

3Process parameters and key technologies

3.1 Glue formula and viscosity control

The goal of this project is to produce high moisture and high waterproof raincoat fabric, The product is processed by using the single component Solvent type high moisture permeable polyurethane as the surface layer of the fabric. Because the composite base fabric is thin and light, Slip and anti blocking properties should be better for the single component polyurethane.

The tensile modulus of the standard skin coating should be about 8MPa when the elongation is 100%. According to the fabric specification the surface glue viscosity is 3000-4000cps,the bottom glue viscosity is

4000-5000cps， The glue formula is as follows:

Table 3-1 Glue formula

Reagent	Face glue	Bottom glue
PU resin	100（Solid content 30%）	100（Solid content 55%）
solvent	40--80	40-60
auxiliaries	1---2	
Bridging agent		6--8
Promoters		1--2

3.2 The gluing quantity and laminated pressure

The test shows that the surface layer thickness increase that is the increase of coating amount. Abrasion resistant and strength also increased. But the water vapor permeability decreased; when The Link layer thickness increased can increase the peeling strength, but effect the moisture permeability and handle. Coating thickness determines the PU coating agent amount, also an important factor which determine the product's cost. According to the fabric specifications, the coating thickness is 18-20μm。If the pressure is small, although the handle is good but it will result in no laminating or weakly laminated , If the pressure is too big, the fabric and substrate 's compression deformation will increase and the handle will become hard or stiff. For this product the laminated pressure is about 4.5Kg.

3.3 drying temperature, time and air volume

Solvent ratio, drying temperature, vehicle speed, air volume and so on are the important factors that affect the film forming properties of polyurethane, It is also an important factor affecting the waterproof and moisture permeability of fabric . If the film formation temperature is high, the solvent evaporation rate will be fas，And accelerate the movement of polymer molecular chain segment , also can raise the generating efficiency and improve the permeability. But when the speed is too high, the heating rate is too fast，Can cause the film surface temperature rise too fast，make the polyurethane film crack and generate air hole. Reduce the waterproof performance. This product use the stage heating mode drying when the temperature up to 80 ℃ ~ 120 ℃. Speed is also an important factor affecting the film formation and fitting. If the Speed is too fast, the film formation won't be good and it is easy to produce bubbles. At the same time, the glue viscosity is insufficient, the fastness is poor, feel hard. Generally the speed is slower controled, Film formation speed is 10~15m/min, fitting speed is 20~25m/min.

3.4 Ripening temperature and time determination

Since it is a semi dry lamination, So the curing effect of connecting layer depends on the ripening temperature and time, For this product the ripening temperature and time：65 ℃，48h. This condition can completely cure products, and enhance the peel strength.

4 Performance test

The Products were produced by Jiangsu Nanwei Yueda Fiber Technology Co.Ltd and were sent to CNTAC Testing Center-Suzhou Zhongke Textile Technology Service Co.Ltd. The Test standards and the corresponding index and test results are shown in the table 4-1.

Table 4-1 Testing Results

Item	Standard	Index	Result
Washing appearance	GB/T13769-2009	10 times washing≥4.0 grade	10 times washing: 4. 0 grade
Hydrostatic pressure	GB/T 4744-2013	≥10000mmH_2O, after20 times washing≥5000mmH_2O	14760, 5671
Surface water resistance	GB/T4745-2012	≥4.0 grade, after20 times washing≥3.0 grade	4–5
Moisture permeability	GB/T12704.1-2009	Moisture permeability rate≥3500g/(m^2·24h)	3990
Anti static	GB/T 12703-2009	≥2.5uC/m^2	2. 8
Ageing resistance	FZ/T75007-1995	≥4.0 grade	4–5 grade

Test data from table 4-1show hat the designed product's performance on Washing appearance、Hydrostatic pressure、Surface water resistance、Moisture permeability and anti static、Ageing resistance have all reached technical requirements of the standards, meet the requirements of the normal customers, fulfill the original intention of the design.

5 Conclusions

Multifunctional anti aging military raincoat Fabric, Use Woodland camouflage paint Oxford fabric as base cloth, The Multifunctional fabric is the preferred material for making the military raincoat, the mountaineering clothes and the ski clothes. The product has the performance of windproof、waterproof 、and high moisture permeability which ensure the product having competitive advantage in the market. The development of this product can improve the competitive advantage of the enterprise, It is of great significance to promote the product export and to improve the product value added.

References

[1]GB/T 13769-2009 Textiles-Test method for assessing the smoothness appearance of fabrics after cleaning[S].

[2]GB/T 4744-2013Textiles-Testing and evaluation for water resistance-Hydrostatic pressure method.

[3] GB/T4745-2012 Textiles-Testing and evaluation for water resistance-Spray test method.

[4]GB/12704-2009 Textiles-Test method for water-vapour transmission of fabrics.

[5]GB/T 12703-2009 Textile-Evaluation for electrostatic propertie-part 2:Electric charge desity.

[6]FZ/T75007-1995Testing Method for Accelerated Ageing of Coated Fabric under Hot and Damp Air Conditions

[7] WANG Chun-mei, LI Zhao-hui. Analysis of factors affecting performances of waterproof & moisture permeable laminated fabric[J].SHANGHAI TEXTILE SCIENCE& TECHNOLIGY, 2009, 37(7): 52-53.

MATEC Web of Conferences **63**, 04013 (2016) DOI: 10.1051/matecconf/20166304013
MMME 2016

Ageing of rice husk biochar along a freeze-thaw cycles

Wei MIAO [1a],Xiao Ying CHENG[1],Jun MENG[1] and Liang TANG[1*]

[1]*Shenyang Agricultural University, Biochar Engineering Technology Research Center of Liaoning Province, Shenyang 10086, China*

Abstract. In order to elucidate the aging possess of biochar, the experiment with treatment biochar with soil were performed. For accelerating aging process, freeing-thawing cycle were conducted to simulate the changing process of the physical and chemical properties of biochar and explore the roles of biochar in the changes of the soil nutrition. Aging treatment has a significant impact on the physical and chemical properties of biochar. The pH values, element composition, oxidation of the surface, absorption capacity of biochar changed in the ageing process. In the simulation test of freeze-thaw cycles, pH of biochar displayed upward trend after the first decline in 25 cycles, and then return to the initial level, indicating that the pH of biochar is not easy to change. Biochar surface oxidation occurs rapidly and significantly in the early, and then stabilized in the latter stage. Surface oxidation degree is higher than the whole. The results showed that surface oxide of biochar protect the internal particles from further oxidation. In the first 25 freeze-thaw cycles, adsorption capacity of biochar to hydroquinone changes from the 5.928 $mg.g^{-1}$ to 11.73 $mg.g^{-1}$, indicating that freeze-thaw cycle treatment increase adsorption capacity of biochar.

1 Introduction

Biochar is admitted as a valuable tool for carbon abatement increasingly, with pyrolysis–biochar systems potentially offering greater carbon-equivalent gain than bioenergy only [1]. biochar may offer additional benefits to soil fertility [2]. Biochar plays a positive role in improving crop yield and promoting low-carbon and sustainable development, due to good physical and chemical properties.

However, these analyses depend on assumptions of the absolute and relative stability of different biochar products in soil. The stable properties of biochar also possess some instability. Nguyen[3] indicated that with the increasing of soil temperature, the loss rate of biological carbon increased, more incubation experiment showed that biochar character changes in its physical and chemical properties under different conditions[4]. The chemical and biological oxidation constantly changing the functional groups and chemical properties when biochar applied into soil[5]. It may also changed adsorption ability of biochar[6]. Different soil environment interact with biochar, leading to the change of biochar properties is different, and then effect on soil. Cultivation experiments study shows that biochar adsorption increased, CEC decreased after aging[7], The physical and chemical properties of biochar are changed inordinately in - 22 ℃ to 70 ℃ cultivation experiments [8]. The pH, adsorption ability CEC, surface functional groups and other important physical and chemical properties will change in the process of biochar aging.

Biochar aging will inevitably affect the content of production function. In a lot of characteristic of biochar, pore structure, element distribution are important, is also the focus in the study of the aging process. Pore structure affected the specific surface area and pore size of biochar, and then affected surface adsorption capacity. Element distribution affects the functional groups and ion exchange capacity, further affects the pH with alkali metals, organic acids, etc.

Aging is a long-term process, conventional methods of field experiment is difficult to meet the needs of the research. So, In this paper, we use freezing and thawing cycle to accelerate the aging process, study on the change rule of physical and chemical properties of biochar, and then analyzes its reason. Explore biochar aging impact on soil nutrients, In order to use biochar reasonable and efficient and provide theoretical basis for a long time.

2 Materials and methods

2.1 Biochar sample

In this study, The rice husk were collected from paddy field in southland of Rice Research Institute, Shenyang Agricultural University, Shenyang, China. The rice husk were air-dried at room temperature then placed in ceramic crucibles each covered with a fitting lid, and pyrolyzed under oxygen-limited conditions in a muffle furnace. The pyrolysis temperature was raised to the values of 400 ℃ at a rate of approximately 20℃ min-

[a] Liang Tang , tl_biochar@126.com, weiwei_youl@126.com

DOI: 10.1051/matecconf/20166304013

1and held constant for 30min. After heating for 30min, the biochar samples were allowed to cool to room temperature. The rice husk biochar was used to study the ageing processes of biochar(Figure 1) .

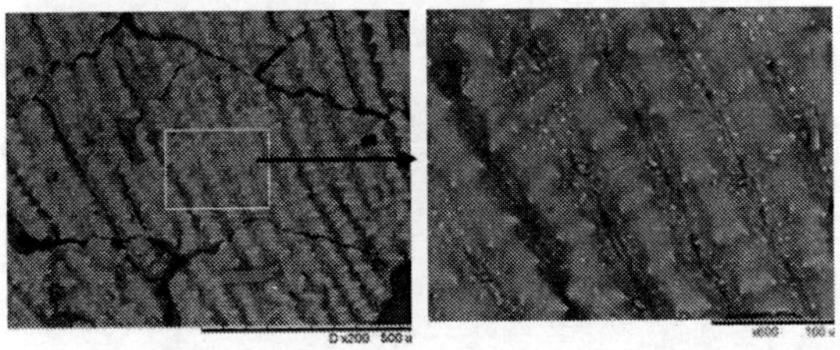

Figure 1. SEM of inside surface of husk biochar

2.2 Ageing experiment

Ageing of biochar was conducted in an aerobic incubation experiment along a freeze-thaw cycles at 30℃ /10d,-20 ℃ /10d,25 cycles total. Ten grams of biochar were placed in a PVC Pipe Scrap , which the hight is 60cm, diameter is 4.4cm and 4 mL of water were added to attain a moist environment, and then incubated in the dark.

2.3 Elemental and chemical analyses

Elemental analysis (C, S, N,) of the biochar samples was conducted using a Germany Elemental, The composition of C, N, S, and O of biochar samples were presented on a dry ash-free basis Energy Dispersive X-ray spectrometers (Bruker Nano GmbH).Morphological and surface elemental analyses were carried out by scanning electron microscopy (SEM) and Energy dispersive Spectroscopy (EDS) (Jeol JSM-6400).

The pH value of biochar was measured in 1:20 w/v ratio in either H_2O or 1 M KCl solution. The pH value in 1 M KCl presents the potential pH, since high ionic KCl solution could release exchangeable protons of biochar into solution. Each experiment was conducted in 3 duplicate

In the acid - base titrations, 0.500 g of the biochar samples (0.3-mm sieve) was placed in 50-mL Erlenmeyer flask. Then, 20 mL of deionized water was added to each bottle, and each of the bottles was stirred on a magnetic stirrer for 2 h at 25℃ The samples were then titrated with 0.1 M HCl at 25℃ to the end point at pH 2.0.

The amount of carbonates was determined by volumetric analysis of the CO_2 liberated through adding 4 M HCl solution to the biochar samples (Pansu and Gautheyrou, 2006). About 5 g of CaCO3 was dried at 104℃ and then 0.0000, 0.1000, 0.2000 and 0.3000 g of the dried CaCO3were each weighed as standard substances. Of the prepared biochar samples (0.15-mm sieve), 1.00 g was weighed to determine the carbonate content.

2.4 Statistical analysis

The experiment was set up in a randomized design with three replications. The data were analysed for significance of difference by analysis of variance using Data Processing System (DPS) (Tang and Feng 2006). Means with the same letters are not significantly different .

3 Results and discussion

3.1 Elemental and chemical analyses

The fresh biochar(0) showed a relatively high C concentration(47.985%). After 13 cycles, the C content of the biochar was very significantly lower than other cycles. With the ageing of biochar a decrease in N concentration was observed and the magnitude of these changes increased with the aging cycles, significantly for the last 2cycles In contrast, the concentrations of H did not display differences among biochar samples except the last cycles the H content significantly lower than 3 cycles.

With the ageing of biochar, An decrease in C/N ratio from 1-15 cycles and the last 2 cycles increased very significantly (Table 1).

Table 1Elemental composition of biochar along 25 freeze-thaw cycles (30℃/10d,-20℃/10d)

	C/%	N/%	H/%	C/N	C/H
0	47.985aA	1.278ABC	34.401abA	37.547cdBCD	1.395aA
1	46.627aAB	0.908bcBCD	37.542abA	51.435bB	1.246aA
2	44.642abAB	0.975abcABC	37.976aA	46.763bcBC	1.178aA
3	45.817abAB	1.157abABC	34.882abA	39.752cdBCD	1.332aA
5	45.235abAB	1.21aA	35.650abA	37.291cdBCD	1.278aA
7	47.281aAB	1.420aAB	34.315abA	33.580dCD	1.379aA
9	47.391aA	1.383abcABC	33.159abA	34.290dCD	1.436aA
11	46.287aAB	1.123aABC	33.095abA	41.185cdBCD	1.446aA
13	42.544bB	1.310aA	36.522abA	34.174dCD	1.169aA
15	44.673abAB	1.458dDE	33.10abA	30.652dD	1.355aA
20	46.438aAB	0.613dE	36.579abA	76.331aA	1.270aA
25	47.026aAB	0.554dE	31.667bA	85.408aA	1.494aA

The data indicate mean±SD Means followed by the same letter within a line are not significantly different at the 5% level

The O/C ratios of biochar surface increased rapidly in the first 9 cycles, then deposition till the 15 cycles. There were no significantly change in15-25 cycles. The O/C ratios of entire biochar were wavelike rises with cycle. In the first 9 cycles increases more slowly than surface biochar(Figure 2).

DOI: 10.1051/matecconf/20166304013

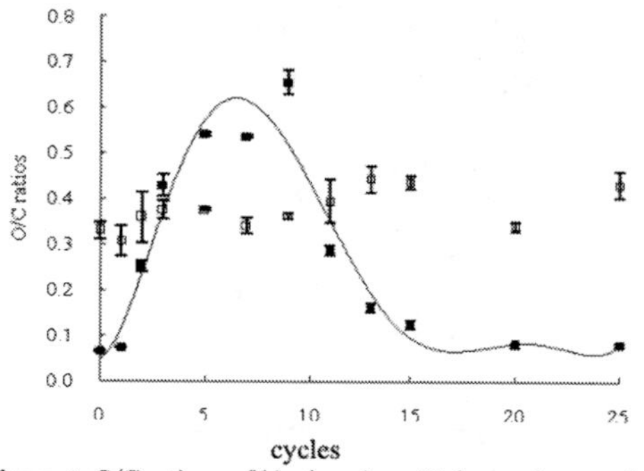

Figure 2. O/C rations of biochar along 25 freeze-thaw cycles
Note: Solid black squares were the observations of biochar surfaces(determined by EDS) and open squares those of entire biochar particle

The pH values of the biochar keep decreased in the first 11 cycles, particularly at 7, 9 and 11 cycles decreased significantly, then increased substantially from 13-20 cycles. No significantly change among 20,25and 0-3cycles.The lower pH values in 1M KCl solution compared to in H_2O were due to the release of exchangeable acidity in the KCl solution.

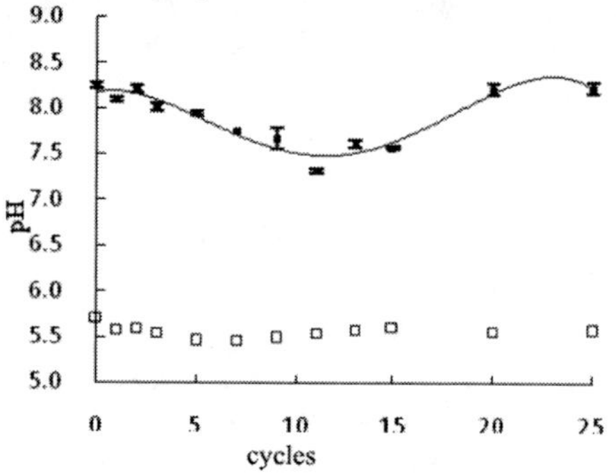

Figure 3. pH of husk biochar in 25cycles,
Note: Solid black squares were the normal pH with deionized water of biochar and open squares were the potential pH with 1M KCLof biochar

Biochar is alkaline, with high pH, Figure 3 show the pH changes of biochar in different Freeze-thaw cycles. In 25 cycle, the pH of biochar always higher than 7, the minimum is 7.33 (11 cycle), the highest is 8.25(0 cycle). The pH dropped significantly from the second cycle and then increased after the 13th cycle, appear frist falling and then rising. At the last 20, 25 period keep steady and there was no significant difference with original biochar pH. Although the pH value decline in the early treatment , the pH is above 7.33 in all the cycle and back to the original pH value at last. Indicated that biochar alkaline is relatively stable.

From Fig.3 we can see that the pH value between 5.45 - 5.71,PH value does not change significantly in 25 freeze-thaw cycle in1N KCL. There was a slight decline in early-stage,then rebound and maintain stability.

3.2 The influence of the freeze-thaw cycle treatment on carbonate content of biochar

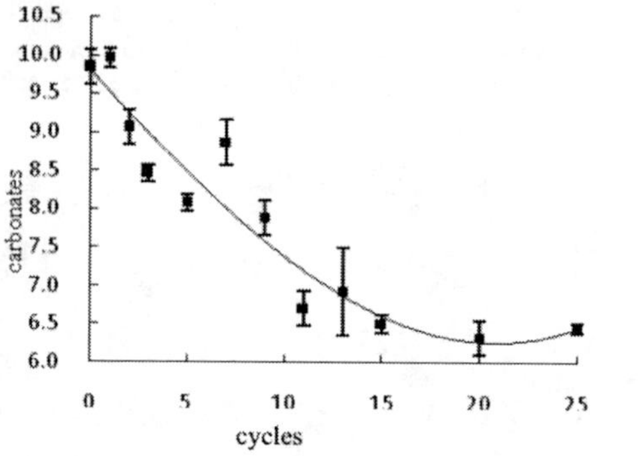

Figure 4. Carbonates content of biochar along 25 freeze-thaw cycles

Figure 4 is the condition of carbonate content changes of biochar. The carbonate content of biochar is low, Original biochar carbonate content is 9.835cmolkg^{-1},and then fall down .After freezing and thawing cycle treatment, carbonate content present down trend and keep stable after 11 cycles.

3.3 The influence of the freeze-thaw cycle treatment on Adsorption capacity of biochar

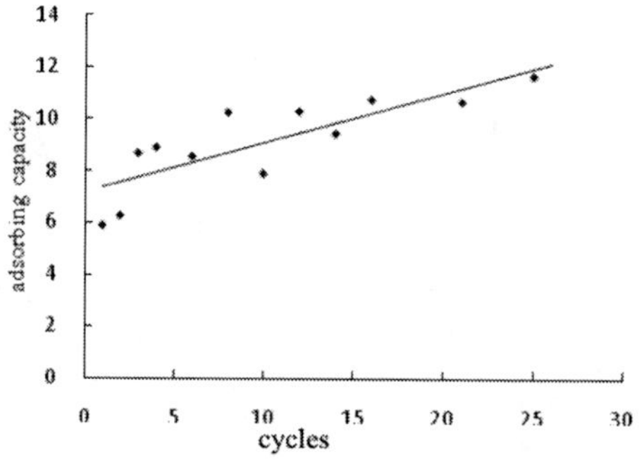

Figure 5. Adsorption capacity of biochar along 25 freeze-thaw cycles

The adsorption hydroquinone of biochar is showed in Figure 5, We can see that during the freeze-thaw cycle biochar adsorption ability slowly rising.This study use hydroquinone as adsorbent, because hydroquinone allelopathy to plant.

Biochar has rich pore structure, The increases of biochar for hydroquinone adsorption may be due to freezing and thawing cycle caused the change of pore structure.

4 Conclusion

Under different environmental conditions, the change of biaochar are different. In the nature soil, four seasons alternately, freeze-thaw cycle[9], Wind and rain erosion[10] can cause the physical and structures change of biochar. For example from larger particles to small particles, from a lower layer soil displacement to the deep soil.

In the aspect of chemical properties, the aromatic structure will increased when biochar stayin soi for a long time[11]. Elements will change and surface charge will also change.

The results show that biochar pH change in the freezing and thawing cycle, but in the end and return to the original. It proved that is not easy to change biochar pH. This study suggests that the pH changes associated with carbonate content and the change of functional groups. The oxidation on the surface of biochar more significant than inner during freezing and thawing cycle. Surface oxidation occurred earlier and bigger, The level of oxidation of overall biochar are less than surface oxidation. The freeze-thaw cycle of biochar effects of surface structure characteristics irregularly, shows that biochar surface structure characteristics affected by various factors. The freeze-thaw cycle can increase biochar adsorption ability of hydroquinonesignificantly.

study shows that biochar alkaline depends on the content of carbonate[12].But in this study the change of pH value during the process of freezing and thawing cycle is not completely consistent with the carbonate content. We speculate that pH value is not only related to the carbonate content and it also correlate with the carbonate content and the functional groups.

Carbonization process of biochar greatly increase its stability in nature,its the basis of the stability of structure. With the development of biochar industry, The character of the artificial preparation biochar is not set in stone, Different materials with different firing temperature and time produced properties of biochar are different. The stability of the under different environmental conditions are different. Studies have confirmed that biochar in the cultivation of the short time (6 months, 12 months) during22 ℃-70 ℃, the properties of the biochar are changed in different degrees [13]. This study have th same conclusion.

This study also concluded that the physical and chemical properties of biochar are changed in the process of freezing and thawing alternation. At the same time the study was done under the condition of indoor simulation With the actual situation may be different in nature. This study will provide theoretical basis for biochar research in the future.

References

1. Hammond J, Shackley S, Sohi SP, Brownsort P Prospective life cycle carbon abatement for pyrolysis biochar systems in the UK. Energy Policy,**39**, 2646－2655(2011).
2. Sohi SP, Krull E, Lopez-Capel E, Bol R, A review of biochar and its use andfunction in soil.Advances in Agronomy, **105**, 47–82 (2010).
3. Nguyen B T, Lehmann J, Kinyangi J.Long-term black carbon dynamics in cultivated soil. Biogeochemistry, **92**,163-176(2009).
4. Yunying Fang, Balwant Singh,Bhupinder Pal Singh Effect of temperature on biochar priming effects and its stability in soils. Soil Biology & Biochemistry,**80**,136-145 (2015).
5. Andrew Cross and Saran P. Sohia. Method for screening the relative long-term stability of biochar. GCB Bioenergy **5**, 215–220(2013).
6. Kgid. Kumari , Per Moldrup, Marcos Paradelo,Lis Wde Jonge. Phenanthrene Sorption on Biochar-Amended Soils: Application Rate, Aging, and Physicochemical Properties of Soil. Water Air Soil Pollut 225:2105(2014).
7. Tang Wei, Guo Yue, Wu Jing-gui, Huang Zhao-qin, Dai Jing-yu. Structural Changes of Aged Biochar and the Influence on Phenanthrene Adsorption. Environmental Science.**35**,2604-2611 (2014)
8. Chih Hsin Cheng, Johannes Lehmann. Ageing of black carbon along a temperature gradient.**75**,1021-1027(2009).
9. Carcaillet C.Are Holocene wood charcoal fragments stratified in alpine and subalpine soils? Evidence from the Alps based on AMS 14C dates, The Holocene, **11**,231–242(2001).
10. Skjemstad J O , Dalal R C, Janik L J,.Changes in chemical nature of soil organic carbon in Vertisols under wheat in southeastern Queensland. Australian Journal of Soil Research,**39**,343-359(2001).
11. Lehmann J, Gaunt J, Rondon M. .Biochar sequestration in terrestrial ecosystems a review. Mitig Adapt Strat for Glob Change,**11**,403-427 (2006).
12. Jin H Y, Ren K, Xu H Z.The forms of alkalis in the biochar produced from crop residues at different temperatures.**102**,34988-3497(2011).
13. Chih Hsin Cheng, Johannes Lehmann.Ageing of black carbon along a temperature gradient.**75**,1021-1027(2009).

MATEC Web of Conferences **63**, 04014 (2016) DOI: 10.1051/matecconf/20166304014
MMME 2016

The entire network topology display system of terminal communication access network

Yi AN [1, a]

[1]Information and Telecommunication Company of Shanxi Electric Power Corporation, Taiyuan 031000, China

Abstract. Now order terminal communication access network is network technology in Shanxi Province is diversiform, device type complex, lack of unified technical standard, the terminal communication access network management system of construction constitutes a great obstacle. Need to build a "unified communication interface and communication standard, unified communications network management" of the terminal communication access network cut in the integrated network management system, for the terminal communication access network of shanxi electric power company and subordinate power supply company comprehensive management standardization management, the integration of the auxiliary support.

1 Introduction

Smart grid development will focus on power generation, transmission, substation, power distribution and utilization, scheduling, seven links, such as communication information, and marketing automation, distribution automation is an important part of the current smart grid development, with the electric communication is the key to a successful decision with electrical automation system construction. Terminal communication access network has wide coverage, equipment type complex and numerous technical standard, the lack of unified technical standards and management norms, the construction of communication network management system constitutes a great obstacle.

Thus, according to the national grid company in terminal communication access network management system construction unified communication interface and communication standard, unified communications network management demand as the goal, the construction of a set of network communication access network terminal manufacturer based integrated network management system, implementation supported by communication comprehensive monitoring of computerized, networked, standardized terminal communication access network management system, to enhance data analysis ability and the sharing and communication management system construction; And focus to provide various manufacturers professional network management unified entrance, improving the efficiency and level of communication, communication management system for the company to provide technical support.

2 Project overall design and technical solutions

For the terminal communication access network equipment of large quantity and many kinds of Shanxi Province, the characteristics of wide distribution, EPON third-party network management system with electric network communication equipment manufacturers (EMS) or equipment on the basis of the integration of integrated function such as data acquisition, integrated monitoring module, for operation maintenance electrical communication with an intelligent management system. Provide a warning function of centralized monitoring system to achieve regional network management information, communication access network terminal manufacturers through production forward regional special VPN access to provincial company collection server, regional companies related certification by web access province system, data deployment in province centralized way, the third party network management system and provide the unified entrance, calls to various cities, each manufacturer of professional network management client, the user can of various system resources, configuration, topology, alarm query, and within the scope of the authority to allow for the configuration changes.

Third party for the construction of the network management system will make full use of modern communication and information technology, based on the development of digitization and automation system, intelligent power grid development needs as the core, constantly improve's ability to allocate with electrical communications resources at all levels, improve and

[a] Corresponding author: 525492887@qq.com

DOI: 10.1051/matecconf/20166304014

improve the carrying capacity of all kinds of communication business's ability to resist all kinds of natural disasters and the external force damage, electrical network construction in line with the strong smart grid with the integration of intelligent management system.

3 Software configuration

The system uses commercial relational database as the data source, through the uniform resource model, in the unity of the support across systems software framework based on the platform, according to the system function and the interaction between the relationship, formulate the corresponding software modules, the integrated acquisition platform, comprehensive monitoring, resource management, system management, auxiliary functions and main software module, realize the real-time monitoring terminal communication system, resource management, supporting the business analysis, and main application functions of support.

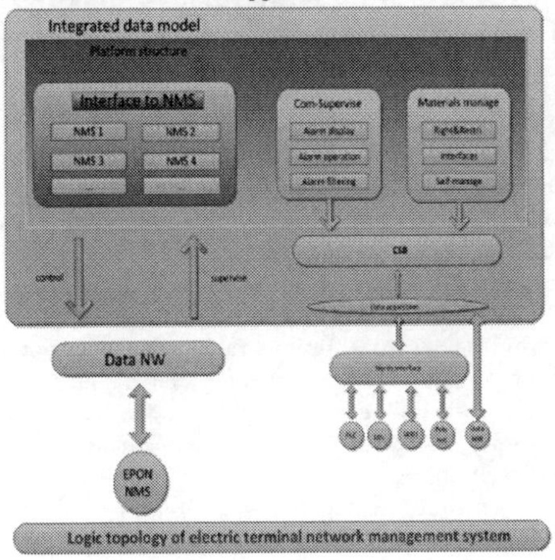

Figure 1. Integrated data model.

Monitored system: including terminal communication access network of EPON access system in different manufacturers, have a combination of electrical communication with PLC, industrial Ethernet, the expansion of the wireless network function.

Integration platform: based on multithreading and preparation of the acquisition agreement and realize the integration of collection and process monitoring, management functions, through standardized north interface protocol implementation of the access network system of data collection.

Communication comprehensive monitoring: with integration platform for interaction, obtained by the monitoring system of the alarm information and topology information, such as real-time data on the basis of the implementation of comprehensive monitoring communication system in the form of entire network topology.

System management: provide with electricity network intelligent management system since the system management, security management, and other basic functions.

Professional network and unified entrance: by each manufacturer professional network management in city bureau to provide local service and management system to provide the various cities, each manufacturer specialized network unified entrance, call professional network client, see the resources, such as topology, alarm information, and the permissions allow, within the scope of the configuration management.

4 Integration of entire network resources configuration management

To the north of the interface, protocol conversion function as the foundation, implementation for access to the factory for network management interface, realize the integrated network management system for the network management system configuration, alarm, performance, such as real-time data acquisition, and to build a unified data model, realize the terminal communication access network in the entire network topology display system for the entire network of different manufacturers and different technical standard network management network, the properties of the network elements of abstracting, realize the unity of the each type of resource storage format, at the same time through the unified data model will be cut in the manufacturer of different manufacturers, different technical standard configuration management functions of network management integration to the terminal communication access network in the entire network topology display system, realize the unified data model based on the entire network factory network configuration management functions and the integration of network resources management.

At present, China's main electric power communication transmission medium to optical fiber is given priority to, through their will power plants, substations and other electric power facilities and various communication site. Communication according to the classification of management level of power grid in China, can be divided into national backbone, provincial backbone, regional backbone level 3, at all levels with different optical communication network according to the function and demand rate level and different network topologies, effectively ensure the safe and stable operation of power system.

Sort is various, the main business is in the power system include: data services, such as power system protection and safety automation devices of power grid data, the electric power market supply and demand data, dispatching automation operation data and information management system and office automation system of information data, etc.; Voice services, such as grid internal conference calls and dispatching telephone, production management, etc.; Video business, such as video conference, industrial video; Multimedia services, such as E-mail, interactive videotex, the multimedia conferences, etc. These business requirement for time delay, bandwidth, reliability difference is bigger, cannot be treated in the same way when processing.

By adopting SDH and ATM as the bearing layer of the business, can realize the point-to-point communication service. However, if you want to realize the entire network data real-time interaction, must solve 1 point to multipoint and multipoint to multipoint communication problems.

5 The entire network topology display

Using the unified data model, data correlation and topological connection network elements such as technology, implement a topology based on IBM Ilog engine components, and extensions for the actual application requirements, with different ICONS show a variety of different network and topology, using dynamic layout, graphic topology automatically connected at the same time, form data serialization compression storage technologies such as show the diversification of network elements for multiple grouping (according to the area, sites, lines, and other forms of network and the regional organization or management requires grouping).

a. Integration assembling platform

Integration platform for multiple management system configuration data, alarm data and performance data acquisition design integrated process management tools, integrated platform for integration of multiple sets of EPON network, multiple acquisition agreement, through friendly interactive interface, the log processing methods such as; Control the acquisition modules (including the system configuration data, alarm data and performance data acquisition) working condition and working process, so as to realize the management and monitoring during the process of collecting all the system itself. In addition to the expansion of the system of support for a variety of other techniques. Integration platform for electric power communication network information collection can bring series of technical advantages.

At all levels are now running network of SDH network, telecommunication level Ethernet have compatible with the network function and characteristics of the SDH can gradually withdraw from operation and transition to the telecommunication level Ethernet, also can be used within the scope of the access layer is small to retain and realize SDH network and telecommunication level Ethernet network. Telecommunication level after the completion of Ethernet, SDH network, data network power system can be unified to the telecommunication level Ethernet transmission platform, reduce or even avoid the SDH network, data network points cause the waste of resources

b. Comprehensive supervision

Integrated network management through the management of machine and network management interface service for the data of the north, complete with the interaction of EPON system, so as to realize the access configuration data, alarm data, real-time monitoring of the performance data. At the same time to ensure the comprehensive network and access network system of loosely coupled, ensure the system security.

c. Alarm management

Alarm management is a comprehensive monitoring technology in the basic functions, including alarm filtering system, display, processing, analysis the whole process. Alarm management function module mainly summarized as: alarm processing, alarm operation, failure analysis, historical alarm query and statistics, several advanced features.

Management main detection parameters for different types of level alarm detection parameter setting management, main categories including emergency alarm level Settings, alarm filtering and warning, and other functions. The general alarm query, statistics, confirmation, remove, delete, storage, and other functions outside the system to provide a list of the alarm information display interface, real-time display of collecting the alarm; System can also according to the alarm information, the fault location to management of related objects, and show to the entire network topology.

6 Professional network management unified entrance

According to the unified data model for the system to provide various cities, each manufacturer specialized network client unified entrance, through unified entrance, a key to the user can log in the professional network management, resource view, topology, alarm, and the permissions allow, within the scope of the configuration management.

Sytem manegement

Provides the basis for integrated network management system management system system management functions, including user rights management, system monitoring and data backup recovery management, and other functions.

b. User rights management

User rights management is for users of the system and user rights management support functions, the system USES the tree structure to achieve the user based on the structure of the enterprise internal management system, realize the user's group management, managers can for new users of the system, delete, query and modify etc operation; On the user permission management system provides the powers of a matrix distribution management functions, to achieve user permissions rapid distribution management system.

d. Sytem supervision

Through uniform data model in storage manufacturers run data network management system, integrated network management also provides for EPON system and the third party server management system itself, the key software service and network environment monitoring, and in exceptional cases run warning trigger system itself. Monitoring objects including server, PC, its running state; the operating system, service process, such as database software running status.

e. System data backup and recovery

To ensure the system in the event of a failure in the first place to recover and use the system to provide the data backup and restore function. In the process of data backup recovery system will also support the classification of data backup, classification, delete, can according to user needs, in the case of ensure data integrity system implementation for system data backup and recovery quickly.

7 Conclusion

Through the integrated network management system through the north to help users to realize the interface and protocol conversion function more manufacturers focus on configuration management of network management system, and further use of unified data model achieved under the unified interface centralized monitoring of multivendor network management data, and on this basis using the data correlation and topological connection technology such as network elements implements access global network elements, topology and real-time warning information on the topology of the concentrated display and positioning, solves the user previously confined to the opposite of each subsystem is now monitoring, maintenance, enhances the management of operations staff and equipment repair efficiency; For the entire network transformation, equipment updating and open cut business support and implementation of auxiliary provides effective extension.

References

1. Flair Chouish.Basic software construction bulding.Didital and Information Press, (2007).
2. James Blunt.Real-time commmunication and information exchanges between software and hardware.2010 NECP International Symposium on computer science, 44-46.
3. QD.Armstrong. EPON transition analysis based on mordern communication techlology. Science Newly, London, (2010).
4. James Rogger.System re-construction on EPON fountain structure.2009 EI International Symposium on Software Development and Modeling, 112-114.
5. Stephen Chen. What doninates the future communication: on EPON technology. EI International Symposium on Software Development and Modeling, 23-26 (2009).

MATEC Web of Conferences **63**, 04015 (2016) DOI: 10.1051/matecconf/20166304015
MMME 2016

An Anti-jamming Algorithm for Attitude Measurement Based on Antenna Array

Jian Hua CUI[1,a], Nai Ping CHENG[1] and Pan Feng HE [1]

[1] *Department of Optoelectronic Equipment, Academy of Equipment, 101416 Huairou District Yanqi Town, Beijing*

Abstract. Aiming at the problems existing in the attitude measurement using carrier phase of navigation signal in the jamming environment, an antenna array for anti-jamming attitude measurement is constructed. The antenna array is constructed by three circle array with core. An anti-jamming measurement method based on antenna array is proposed. This method uses array antenna to estimate the direction of arrival jamming signal though MUSIC algorithm. Then the adaptive nulling algorithm based on the linear constrained minimum variance algorithm is used to eliminate the jamming and hold the navigation signal. Finally, the double difference method is used to measure the attitude. The whole attitude information is obtained by the attitude measurement of the two baselines. Simulation has been made for eliminating the jamming with adaptive nulling algorithm. Comparison of the attitude measurement accuracy is also made by simulation. The results show that the algorithm we proposed can significantly inhibit the jamming and improve the accuracy of attitude measurement.

1 Introduction

Attitude measurement though the carrier phase of the navigation satellite signal is an important part of the satellite navigation application. The attitude measurement system based on GPS has high precision, whose initialization time is relatively short. The system also has high stability. Compared with the traditional way of measuring attitude, it has its own advantages.

Attitude measurement system is of great significance in Aeronautics and Astronautics, and is widely used in various kinds of aircraft and ships. But because the accuracy of the position measurement depends on the carrier phase of the navigation signal, it needs to solve the integer ambiguity in the attitude calculation process, and the operation is larger. The measuring accuracy is affected by the baseline length of the measuring antenna too. Under the condition of modern war, the navigation signal is very vulnerable to various kinds of interference, so that the accuracy of attitude measurement algorithm is reduced. Even more, the attitude measurement algorithm is invalid.

At present, there are many applications and researches on the adaptive nulling antenna[1,2], but there are few researches on the application of the adaptive control algorithm in the attitude measurement. The anti-interference attitude measurement method is based on the antenna array to form an adaptive nulling antenna. The signal of each channel is weighted by amplitude and phase under certain constraint conditions, which forms a null in the direction of the interference signal. Then the synthesized signal is used for attitude measurement. At the same time, due to the difference of the attitude measurement antenna, we construct an antenna array which is suitable for anti-interference attitude measurement. The output of each measuring antenna is a signal after eliminating interference.

This method has certain effect for the attitude measurement based on carrier phase measurement, but there are also some problems. The reason is that in the amplitude and phase weighting process, the signal phase information of each array element received is destroyed. The carrier phase exists error. Through this way to measure the attitude can improve the accuracy in the interference environment.

As the anti-jamming attitude measurement algorithm and anti-jamming algorithms differ in the constraint conditions, so they should have different constraint criterion[3]. The constraint criterion used in anti-jamming algorithm is also not suitable for anti-jamming attitude determination algorithm. For this purpose, we choose the Linearly Constrained Minimum Variance (LCMV) as the optimization criterion of the nulling antenna. In this way, it can ensure the correct measurement of the carrier phase in the interference condition, and ensure the accurate measurement of the attitude.

2 Signal model

In order to realize anti-jamming attitude measurement on the carrier, the right antenna arrays are needed. There are many types of antenna arrays, such as linear array, planar array, uniform circular array, circular array with core, etc.

[a] Corresponding author: cjh_xilinx@163.com

DOI: 10.1051/matecconf/20166304015

Anti-jamming attitude measurement algorithm is realized based on a 7 element circular array measuring antenna with core. The middle element is the reference element. Studies have shown that in nulling antenna, the null can be generated effectively in the circular array with core, a flat pattern can be generated in the expected direction. These features meet the requirements of the array antenna, so we use a circular array with a core as the measuring antenna array. The antenna structure is shown in Fig. 1:

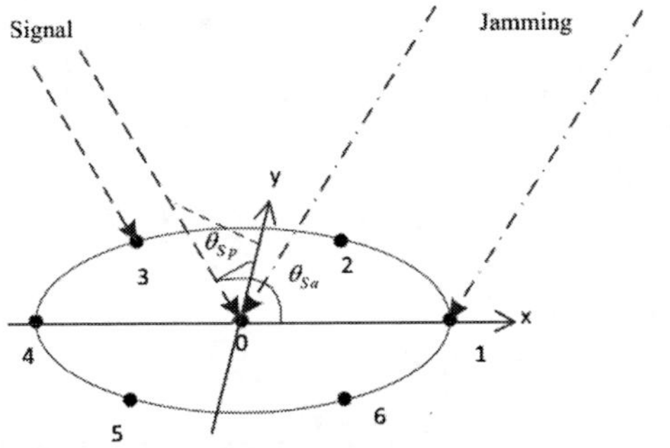

Figure 1. Formatting sections, subsections.

Element 0 is the reference, a circular array is formed by the element 1-6 which is uniformly distributed on the circumference. The X axis is established by the array element 4-0-1, the Y axis is perpendicular to the X axis in the antenna plane. The Z axis is perpendicular to the plane of the antenna. The vector coordinate system is established, and the position of the array element 0 is the coordinate origin. Each array element spacing is the base line length l_b, the signal carrier wave length is λ. In order to do not have the fuzzy degree, we take $l_b \le \lambda/2$.

For the 7 array elements antenna with a core, compared with the reference array element 0, the angle of the element i relative to the positive X axis is: $\theta_i = (i-1)\pi/3$. For a signal with an random angle (θ_p, θ_a), the phase delay of the array element i and the reference array element can be expressed as:

$$\Delta\theta_{Si} = 2\pi l_b \sin(\theta_{Sp})[\cos(\theta_{Sa} - \theta_i)]/\lambda \quad (1)$$

So the direction vector is:

$$D = (1, e^{\Delta\theta_{S1}}, e^{\Delta\theta_{S2}}, ..., e^{\Delta\theta_{S6}}) \quad (2)$$

where $\Delta\theta_{Si}$ is the phase delay.

3 Anti-jamming attitude measurement method based on array antenna

Our method for anti-jamming attitude measurement is composed of three parts, first of all need to direction of arrival estimation of signal to interference, arriving direction of the interference signals need to be estimated. After DOA (Direction Of Arrival) estimation, equations are established based on the geometric relationship between wave direction and antenna array. The optimal weight vector is solved according to the established equations and constraints. The array output signal after the interference suppression is calculated based on the weight vector and the antenna input. Finally, the attitude of the carrier is measured according to the output signal of the three antenna array. This is the whole process of anti-jamming attitude measurement.

3.1 DOA estimation of the jamming

In the processing of anti-jamming attitude measurement, in order to eliminate the interference, the DOA of the interference signal need to be estimated, interference is eliminated using the wave direction information to phase shift. In N element antenna arrays, N-1 interference can be eliminated. We use the traditional MUSIC method to measure the DOA of the interference. The interference direction is obtained by using the orthogonality of the interference signal and the noise signal in the characteristic vector. The specific process includes the following parts:

1 According to the input signal of each array element at different time $X_i = (x_0, x_1, ...x_7)_i$, the covariance matrix of the input signal of each array element of the antenna array is calculated:

$$\hat{R}_X = \frac{1}{N}\sum_{i=0}^{N-1} X_i X_i^H \quad (3)$$

2 After the establishment of covariance matrix, eigenvalues and eigenvectors are obtained by means of eigenvalue decomposition of $\hat{R}_X$:

$$\hat{R}_X V = V\Lambda \quad (4)$$

where $\Lambda = \mathrm{diag}\{\lambda_0, \lambda_1, ..., \lambda_{N-1}\}, \lambda_0 \ge \lambda_1 \ge ... \ge \lambda_{N-1}$ is the eigenvalues, and $V = [p_0, p_1, ..., p_{N-1}]$ is the eigenvectors matrix. The eigenvectors of the covariance matrix contains two parts, the signal subspace with larger values and the noise subspace with smaller values. Direction vector of the interfering signals belongs to the signal subspace, and is orthogonal to the noise subspace.

3 In the input covariance matrix, all the eigenvalues of the noise are relatively small. If the number is M, the corresponding interference number is K=N-M. The characteristic vector of the signal is orthogonal to the characteristic vector of the noise. Constructing the noise subspace by using the characteristic vector corresponding to the eigenvalues of the M:

$$V_n = [p_K, p_{K+1}, ..., p_{N-1}] \quad (5)$$

4 Structure MUSIC spectra for the incident signal:

$$P_{MUSIC} = \frac{1}{D^H V_n V_n^H D} \quad (6)$$

The characteristic vector of the interference signal and the noise is orthogonal, $D^H V_n V_n^H D$ should be zero. That is to say, at the peak of the MUSIC spectrum, the direction vector is corresponding to the direction of arrival. If at this time the corresponding peak of the incident angle is (θ_p, θ_a), the incident angle is the direction of the incoming wave.

3.2 The adaptive nulling based on antenna array

In the navigation system, the GPS navigation signal will be transmitted after the pseudo code spread spectrum, and the basic signal model received by the reference antenna in the receiving antenna is $S = sPe^{-j(\omega_c t + \varphi_{S0})}$, where P is the spreading pseudo code, s is the original signal, ω_c is the carrier angular velocity, φ_{s0} is the initial phase.

For the center reference array element of the array antenna, hypothesis in the process of signal propagation, the number of jamming is M-1. The jamming is $J_i = j_i e^{-j(\omega_c t+\varphi^0_{J_i})} (i = 1..M-1)$ [4], at the same time the environmental noise is N_0 ,the signal received by the central reference array is:

$$R_0 = (1 \quad 1 \quad \dots \quad 1)(S \quad J_1 \quad \dots \quad J_{M-1})^T + N_0 \qquad (7)$$

As shown in Fig.1, in the seven array element antenna array, azimuth is defined as the angle between the signal and the X axis, pitch angle is defined as the angle between the signal and the array antenna. The baseline length between each array element and the reference array is l_b , carrier wave length is λ . For arbitrary array elements, the incident azimuth angle of the signal is θ_{Sp} , the pitch angle is θ_{Sa} , the incident azimuth angle of the jamming is θ_{J_ip} , the pitch angle is θ_{J_ia} . If the angle between the array element k and the X axis is θ_k (k=1... 6), for any array element k, relative to the reference array, the carrier phase difference of the received signal is:

$$\Delta\theta_{S_k} = 2\pi l_b \sin(\theta_{Sp})[\cos(\theta_{Sa} - \theta_k)] / \lambda \qquad (8)$$

The carrier phase difference for jamming i received is:

$$\Delta\theta_{J_i^k} = 2\pi l_b \sin(\theta_{J_ip})[\cos(\theta_{J_ia} - \theta_k)] / \lambda \qquad (9)$$

Therefore, the signal model of any array element K received is:

$$\begin{aligned} R_k &= (e^{j\Delta\theta_{S_k}} \quad e^{j\Delta\theta_{J_1^k}} \quad \dots \quad e^{j\Delta\theta_{J_{M-1}^k}}) \\ &(S \quad J_1 \quad \dots \quad J_{M-1})^T + N_k \qquad (M = 1 \dots 6) \end{aligned} \qquad (10)$$

The model of the received signal of all the elements can be used to form the equation:

$$\begin{aligned} &(R_0 \quad R_1 \quad \dots \quad R_{M-1})^T = A(S \quad J_1 \quad \dots \quad J_{M-1})^T \\ &+ (N_0 \quad N_1 \quad \dots \quad N_M)^T \qquad (M = 6) \end{aligned} \qquad (11)$$

$$A = \begin{pmatrix} 1 & 1 & \dots & 1 \\ e^{j\Delta\theta_{S_1}} & e^{j\Delta\theta_{J_1^1}} & \dots & e^{j\Delta\theta_{J_{N-1}^1}} \\ \vdots & \vdots & \ddots & \vdots \\ e^{j\Delta\theta_{S_{N-1}}} & e^{j\Delta\theta_{J_1^{N-1}}} & \dots & e^{j\Delta\theta_{J_{N-1}^{N-1}}} \end{pmatrix}$$

Since our main application scenario is to measure the attitude, we need to find the signal's phase as far as possible without distortion. So in the interference environment, the linearly constrained to minimum variance(LCMV) can make signal output power minimized with the guaranteed gain in the signal direction[5], which can try to suppress the interference maximum, while maintaining signal produces less distortion.

According to the LCMV criterion, the optimal weights are generated:

$$\omega_{opt} = \hat{R}_X^{-1} A (A^H \hat{R}_X^{-1} A)^{-1} g \qquad (12)$$

where $g=(1,\dots,1,0,\dots,0)$, the number of '1' is the number of the signals, the number of '0' is the number of the jammings.

Under the influence of the optimal weight, the output after nulling is getting from:

$$Y = \omega X \qquad (13)$$

For the anti-jamming attitude determination, the output of each antenna is supposed to be the signal after the anti-jamming algorithm, because this is the only way to ensure that the distortion of the carrier phase is minimized. Therefore, in the application of attitude measurement, the three measurement antennas should be in the form of anti-jamming antenna array, the structure is shown as below:

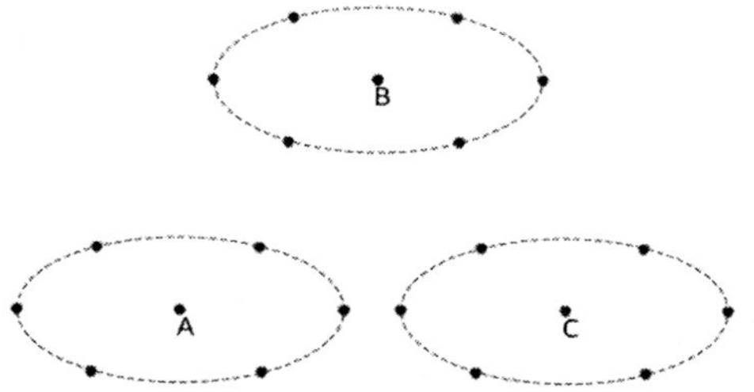

Figure 2 Anti-jamming attitude measurement antenna array

The antenna ABC are arranged in an equilateral triangle, each attitude measurement antenna is an array of 7 elements. The attitude is carried out by using the synthetic signal of each antenna array. The advantages of this array structure is that the attitude can be carried out in the interference condition, its disadvantage is the use of a large array of elements, but the carrier frequency of the navigation signal is higher, so the radius of each anti-jamming antenna is $r \le \lambda / 2 = 9.52$ cm , the structure of the antenna can be used in many application scenarios.

3.3 Double difference method for antenna attitude measurement

The attitude is carried out by means of the double difference method, the attitude information of the carrier is measured by the data obtained from the three position measuring antenna.

We observe measurement antenna's output of A and B received from satellite i, j. The difference between the measured value of the carrier signal of the two measuring antenna for the same satellite is called the single difference, the difference between the two single difference measurements of satellite i and j is called the double difference. As shown in fig. 3:

 DOI: 10.1051/matecconf/20166304015

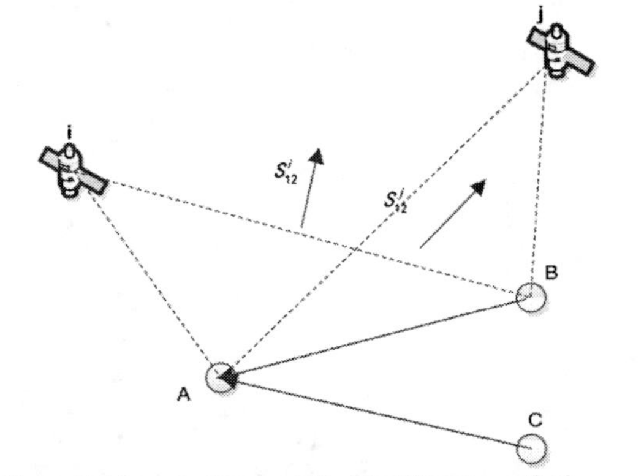

Figure 3 Schematic diagram of double difference method

The observation equation of the carrier phase for the double difference method is[6] (for the measurement antenna AB):

$$\begin{aligned}\Delta\Phi_{AB}^{ij} &= \frac{1}{\lambda}(d_{AB}^{i} + c\Delta\gamma_{m_{AB}^{i}}) - N_{AB}^{i} + \varepsilon_{AB}^{i} \\ &-\frac{1}{\lambda}(d_{AB}^{j} + c\Delta\gamma_{m_{AB}^{j}}) + N_{AB}^{j} - \varepsilon_{AB}^{j} \\ &= \frac{1}{\lambda}(d_{AB}^{ij} + c\Delta\gamma_{m_{AB}^{ij}}) - N_{AB}^{ij} + \varepsilon_{AB}^{ij}\end{aligned} \quad (14)$$

where λ is the signal wavelength, c is the speed of light, d_{AB}^{ij} is the double difference of the distance between the satellite and the antenna, $\Delta\gamma_{m_{AB}^{ij}}$ is the multipath error of the double difference, N_{AB}^{ij} is the double difference of the ambiguity, ε_{AB}^{ij} is the measurement error.

Combined with the antenna structure mentioned above, without considering the multipath error and measurement error, the double difference observation equation of carrier phase is:

$$\Delta\Phi_{AB}^{ij} = d_{AB}^{ij} / \lambda - N_{AB}^{ij} \quad (15)$$

According to the double difference carrier phase observation equation of multi satellites, the integer ambiguity can be solved by the LAMBDA method. The attitude of baseline AB and baseline AC is calculated by the least square method. The information of the two baseline is calculated to get the attitude information of the carrier.

4 Simulation results

Based on LCMV algorithm, the adaptive nulling algorithm of 7 array elements circle array with core is simulated[7]. In the simulation, the number of jamming is 1, the jamming direction is $(10\pi/9, \pi/18)$. Simulation result is shown as Fig.4.

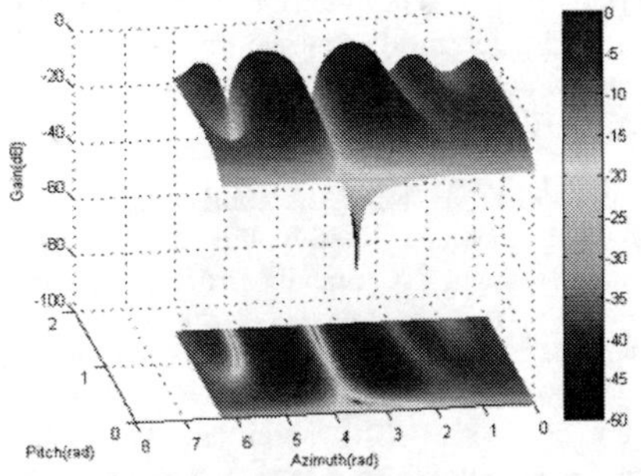

Figure 4 Pattern of adaptive nulling antennas

As can be seen in Fig.4, there is a nulling in the direction of interference whose depth is about -45dB. Strong jamming can be suppressed by the adaptive nulling antenna. The effect of the attitude measurement algorithm is ensured.

When the jamming exists, the received GPS signal is interfered by the same frequency jamming, and the simulation results are shown below.

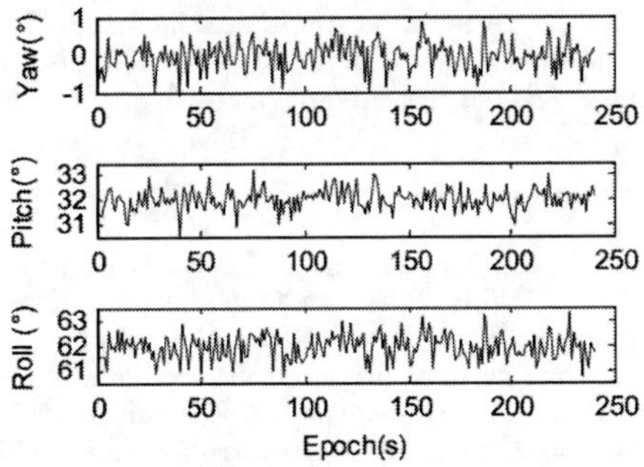

Figure 5 Attitude accuracy without anti-jamming algorithm

As can be seen from Fig.5, when there is no anti-jamming algorithm, attitude accuracy is reduced because of the presence of jamming. The accuracy is about 2 degrees.

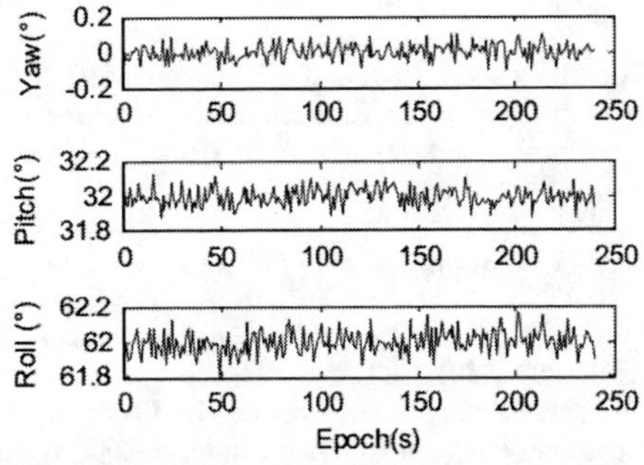

Figure 6 Attitude accuracy with anti-jamming algorithm

As we can be seen from Fig. 6, when using the nulling algorithm, the accuracy of the attitude measurement is about 0.4 degrees. This is much higher than that without the use of anti-jamming algorithm. and then verify the effectiveness of the attitude measurement algorithm using the zero modulation antenna.

The simulation results verify the effectiveness of the zero nulling antenna to improve the accuracy of the attitude measurement algorithm.

5 Conclusions

We propose a kind of anti-jamming attitude measurement method. Firstly, the antenna array is constructed, which is composed of 3 anti-jamming antenna array, and the antenna array is a 7 element antenna array. Before the attitude determination is carried out, the direction of the jamming signal should be estimated by the MUSIC algorithm, then the anti-jamming reception of the navigation signal is realized by using the nulling algorithm. Finally, the attitude measurement is realized by the double difference method.

The simulation results show that the anti-jamming attitude measurement method can realize attitude measurement with high accuracy in the presence of strong jamming. Compared with the attitude measurement method without anti-jamming algorithm, it has higher practical value.

The disadvantage of this method is that it needs more computation in the process. The constraint method is also should be studied for the carrier phase without distortion, which is a direction that is worth improving.

References

1. Zhang, Jian Jun and J. H. Xiang. Study on Effect of Channel Uncertainties on GPS Antenna Adaptive Algorithm. Radio Engineering (2010).
2. Raasakka J and Orejas M. Analysis of notch filtering methods for narrowband interference mitigation. Position, Location and Navigation Symposium Plans 2014, 1282-1292(2014).
3. Ardalan A and Rezvani M. An iterative method for attitude determination based on misaligned GNSS baselines. IEEE Transactions on Aerospace & Electronic Systems, 51:97-107(2015).
4. Li Qiang et al. A Robust Anti-Jamming Navigation Receiver with Antenna Array and GPS/SINS. Communications Letters IEEE 18.3:467-470(2014).
5. Wang Dawei, et al. Attitude aided space-time multi-beamformer anti-jamming approach for satellite navigation receiver.368-372(2014).
6. Henkel P and Iafrancesco M. Tightly coupled position and attitude determination with two low-cost GNSS receivers. Wireless Communications Systems, 2014 11th International Symposium on IEEE, 895-900(2014).
7. El Khatib, Alaa, K. Assaleh, and H. Mir. Space-Time Adaptive Processing Using Pattern Classification. IEEE Transactions on Signal Processing 63.3:766-779(2015).

Research of Partial Discharge Monitoring for Switchgear Cabinet Based on LSSVM

Shai Gen HAN [1,a], Shi Lei GUAN [1], Yan WU [1], Zhi Xin BO [1], Yan DING [1], Li An CHEN [2]

[1]*China Electric Power Research Institute, Beijing 100192, China*
[2]*Xiamen University of Technology, Xiamen 361024, China*

Abstract. Now, there are many influencing factors of partial discharge(PD) for switchgear cabinet, but very few methods used for switchgear cabinet monitoring. In this paper, Partial discharge of switchgear cabinet is identified based on Least Square SVM (LSSVM) used for comprehensive evaluation of switchgear cabinet. Partial discharge ultrasonic signal of time and frequency domain are gave to get two stable characteristic bands in MATLAB tool. It provides a theoretical basis for partial discharge monitoring to ensure proper and efficient operation of the switchgear cabinet.

Key words: Partial Discharge; Switchgear Cabinet; LSSVM; Ultrasonic Signal.

1 Introduction

The insulating materials in high voltage switchgear cabinet are vulnerable to the effects of chemical, mechanical, electrical, thermal and other factors. The defect of insulating materials can lead to partial discharge and prompt the cumulative charge and thus causes insulation failures, which accounts for more than 50% of electricity equipment failures. The insulation structures, materials and workmanship determine the reliability of electrical equipments' reliability. Papers[1-5] research the phenomenon of partial discharge, but the results are not complete, and the test is not very good. Papers[6-8] use ultrasonic method to test arc fault, but the dynamic performance and accuracy rate are difficult to meet the requirements, and the calculation algorithms are very complicated. Therefore, studying one new method which is suitable for high voltage switchgear discharge monitoring is particularly meaningful [1-3].

2 Factors affecting the partial discharge in switchgear

In the electric field, the gas gap in the vicinity of electrical conductors or in other regions is broken down, and the discharge phenomenon is called partial discharge. When partial discharge occurs, a variety of high-energy charged particles, such as electrons, positive and negative ions, etc., in the electric field strength of traction, these charged particles strike the walls of the air gap to the media, resulting in extremely high temperatures which causes the insulation material damage. At present, the mechanical, electrical faults high voltage switchgear are:

(1) Insulation malfunction. The main reason for this failure is that the internal insulation material is brokendown. This is the main fault of the switchgear.

(2) Refuse to move or malfunction failure. The main reasons for this failure are mechanical transmission, electrical control and auxiliary circuits. This is more fault occurred in the switchgear.

(3) Malfunction of breaking and closing. this is caused by that the breaking and closing force of the circuit breaker is insufficient, and gas leakage of arc extinguish chamber, what is the common fault of switchgear.

(4) Malfunctions caused by external forces.

3 LSSVM's application in detecting the partial discharge

With the research of support vector machine algorithm[4-9], it is found that the method of using standard support vector in partial discharge detection has many limitations, and many scholars have put forward the improved algorithm of support vector machine. Among them, least squares support vector machine algorithm, Bias support vector machine, Lagrange support vector machines, simplified support vector machines are widely used. When using the least square support vector machine to train samples, it only needs to solve a linear equation group, which greatly improves the computational speed

[a] Corresponding author: hanshaigen@epri.sgcc.com.cn

and training efficiency, and thus meet the requirements of system modeling and identification.

Least squares support vector machine(LSSVM) method converts solving nonlinear equations to a set of linear algebraic equations, and converts the uncertain solution into certain solution, and as well as converts the standard support vector machine constraints by the inequations into the equations, thus greatly reducing the training time. Experiments show that the generalization performance of LSSVM is higher than that of the standard SVM.

Partial discharge detection is a process of two classification, and according to the criteria of SVM conditions, the constraint conditions of LSSVM can be equivalent to:

$$y_i[w^T\phi(x_i)+b]=1-\xi_i,\quad i=1,2,3,4.....,N \quad (1)$$

The objective function of LSSVM is:

$$\begin{cases} \min J(w,\xi)=\frac{1}{2}w^Tw+\frac{c}{2}\sum_{i=1}^{N}\xi^2 \\ s.t.y_i[w^T\phi(x_i)+b]=1-\xi_i,\ i=1,2,3,4.....,N \end{cases} \quad (2)$$

$$\max Q(w,b,a,\xi)=J(w,\xi)-\sum_{i=1}^{N}a_i\{y_i[w^T\phi(x_i)+b]-1+\xi_i\} \quad (3)$$

The equation (3) is the dual Lagrange polynomials of equation (2), a_i is Lagrange multiplier. At the same time we calculate the partial derivation of w、b、a、ξ and then equation (4) is derived.

$$\begin{cases} \frac{\partial Q}{\partial w}=0\Rightarrow w=\sum_{i=1}^{N}a_iy_i\phi(x_i) \\ \frac{\partial Q}{\partial b}=0\Rightarrow \sum_{i=1}^{N}a_iy_i=0 \\ \frac{\partial Q}{\partial a}=0\Rightarrow y_i[w^T\phi(x_i)+b]-1+\xi_i\}=0 \\ \frac{\partial Q}{\partial \xi}=0\Rightarrow a_i=c\xi_i \end{cases} \quad (4)$$

where a_i and the training error ξ_i are linear proportional relation, and the equation can be obtained by being simplified as:

$$\sum_{i=1}^{N}\sum_{j=1}^{M}a_i[y_iy_j\phi^T(x_i)\phi(x_j)+\frac{\delta_{ij}}{c}]+by_j=1 \quad (5)$$

Let $\Omega_{ij}=y_iy_i\phi^T(x_i)\phi(x_j)+\frac{\delta_{ij}}{c}$, where $\delta_{ij}=\begin{cases}1 & i=j\\ 0 & i\neq j\end{cases}$. According to Mercer's theory we can get $\phi^T(x_i)\phi(x_j)=\psi(x_i,x_j)$, we can get linear equations equation (6) of a and b with the third equation in equations (4) and equation (5).

$$\begin{bmatrix}\Omega & y\\ y^T & 0\end{bmatrix}\begin{bmatrix}a\\ b\end{bmatrix}=\begin{bmatrix}\overset{r}{1}\\ 0\end{bmatrix} \quad (6)$$

In equation (6), Ω is a Semi-definite symmetric matrix whose diagonal elements are 1/C, and we can derive equation (7):

$$\begin{cases} a=\Omega^{-1}(1-yb) \\ b=(y^T\Omega^{-1}y)^{-1}y^T\Omega^{-1}\overset{r}{1} \end{cases} \quad (7)$$

Set up the two classification mode by the least squares method[10-13], and choose the best characteristic quantities and parameters based on the classification to improve the promotion and viability, which was divided into the following four steps:

(1) Data normalization. When partial discharge occurs, the ultrasonic frequencies is located in: (a) 20~40 kHz, (b) 80~140kHz. Through the signal conditioning circuit, the integrated value of these two frequencies of ultrasound is limited to 0-1.

(2) Establishment of SVM's sample database. By simulating six kinds of working conditions under the discharge model, to create the sample database, in order to achieve fully optimized results. supposing the threshold parameter is H, and the tag value is Y, when H ≥90pC, it is marked as partial discharge phenomena, and Y=1;when H<90pC, it is marked no partial discharge, Y=-1.

(3)Train and test samples. Training samples contain various partial discharge conditions, and we train a total of 150 sets of data samples, where 50 sets of data are used to test samples, 100 sets of data are used to train the samples. Table 1 shows part of the PD data.

Table 1. Part PD data

Sample number	Feature vector S1	Feature vector S2	Output
1	0.323457	0.143148	-1
2	0.321687	0.144001	-1
3	0.384538	0.174869	1
...	...	...	...
149	0.326145	0.156784	-1
150	0.381357	0.246034	1

(4) Select kernel function δ^2 and penalty factor c .In this work, grid optimization method is used. And we choose RBF function classification kernel function, and cross-validation is used to calculate the mean square error. The horizontal axis is set to the penalty factor c, and the vertical axis is set to kernel function σ2. After verification, the minimum MSE value set is derived, which are the optimal values. This strategy is fast, easy and has features of a high calculation precision. And using this method we can successfully get the best global

DOI: 10.1051/matecconf/20166304016

solution. After calculation and analysis, c =20, δ^2 =0.005 is derived.

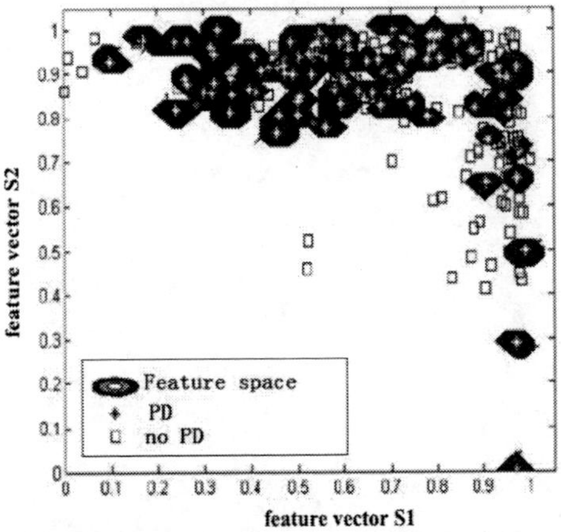

Figure 1. Recognition PD pattern rendering

As shown in Figure 1, "□" represents the samples that there are no PD existence, "*" represents the samples that there are PD existence. When there is no partial discharge the samples will be scattered outside the feature space, and samples that fall on the boundary are the feature vectors, and when there is no partial discharge the samples will fall within the feature space, the feature vector consists of the classification hyperplane, and samples without partial discharge can be apart from the sample. LSSVM classification model has a strong learning ability and generalization. This model can successfully identify partial discharge phenomenon.

4 Experimental platform for PD

Figure 2 is the sketch of experimental platform for PD. This experimental platform is consist of the transformer control box, water resistance, shielding network, metal plate frame, insulating rubber mat, step-up transformer and other necessary equipments. The transformer's capacity is 50kVA, The adjustment range is 0~400V, and the rated current is 125A, the transformer model is YDWB-50kVA/50kV, the voltmeter shows the setting voltage. Water resistance is used to buck and restrict the current, the gap adjusting wheel is composed of insulating rod, electrode, adjusting shaft, spinner wheel device. Control the discharge distance by adjusting the hand wheel through the rotary gap.

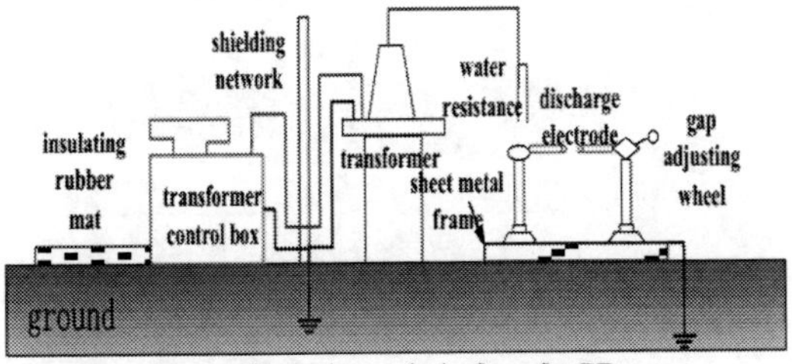

Figure 2. Sketch of experimental platform for PD

Use the experimental platform to simulate the six kinds of partial discharge, and each kind PD is tested for 8 times. DSP control system can accurately check the arc fault, and it can send warning signals and give trip command simultaneously. Actuator can do protected action, reliability, speed instantaneous trip, to meet the accuracy requirements.

MATLAB is used to analyze ultrasonic data processing. In this work, the frequency response range of ultrasonic sensor is 22kHz~220kHz, and the sampling frequency is 2MHz, which is in line with Shannon sampling theorem.

Fig.3 shows the time-domain characteristics of the six discharge models. The distance between electrodes of needle-stick model is 25mm, and the distance between electrodes of the other five situations is 60mm. Six discharge models are obtained by the z Butterworth bandpass filter in the frequency of 22kHz~220kHz and the mean filter is used to get the ultrasonic time-domain waveform.

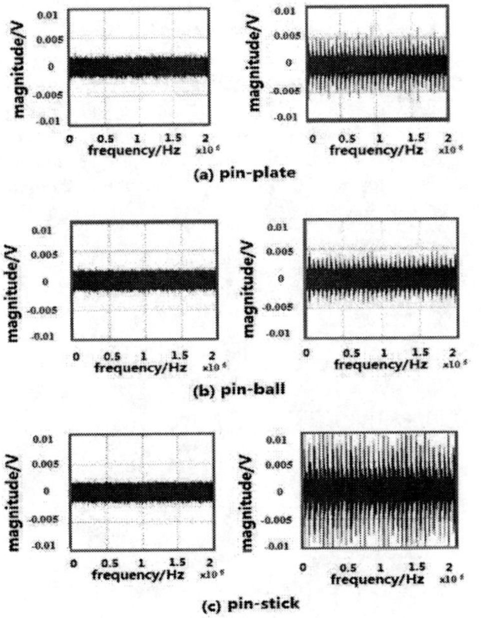

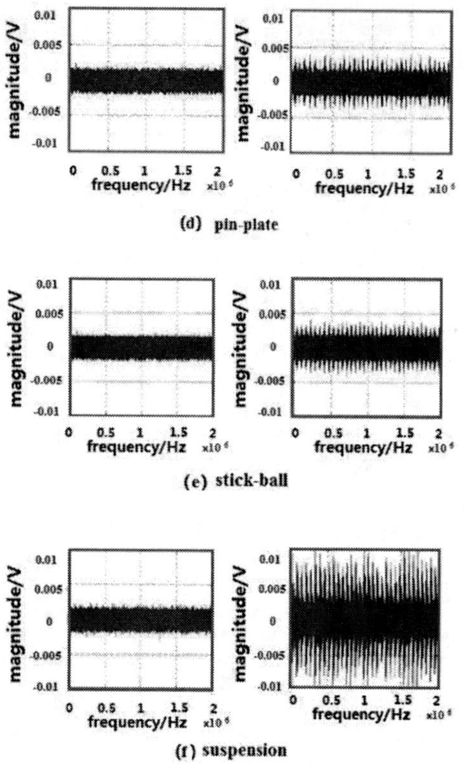

Figure 3. Six diagrams of discharge time-domain model

As shown in Figure 3, when there is no partial discharge, the ultrasonic wave signal is very smooth and has no obvious peaks. It's amplitude fluctuation is in ±2.8mV. When partial discharge occurs, the waveform has obvious burr and the ultrasonic signal's fluctuation increases to ±5.0mV. Among them, ultrasonic signal amplitude of needle-stick model and suspended model can even reach ±10.0mV, which is about three times larger than that of no PD occurring.

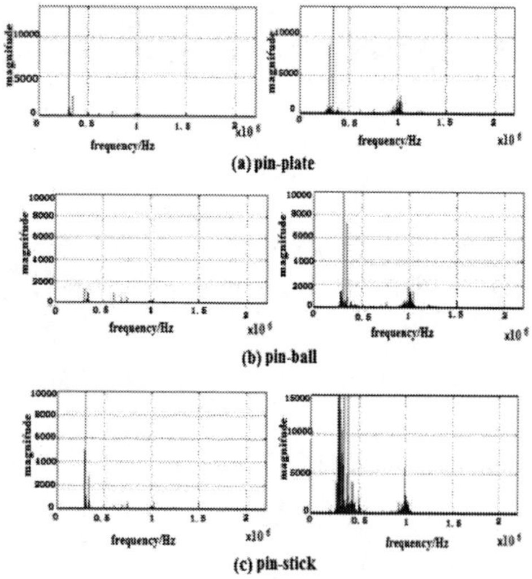

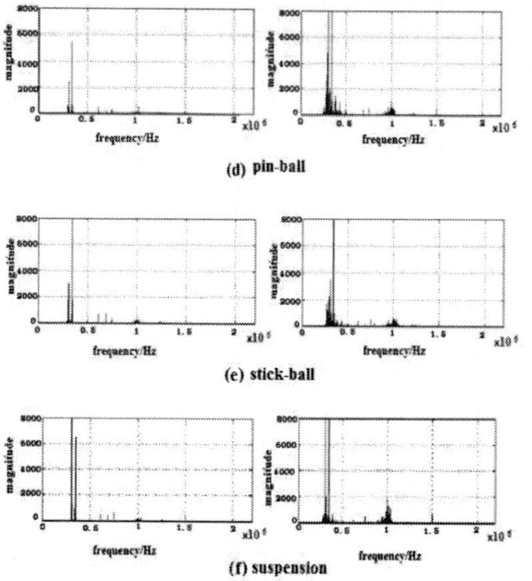

Figure 4. Six kinds of discharge model spectrums

As shown in Figure 4, when there is partial discharge, ultrasonic signal has been significantly enhanced, the signals have a very intensive frequency distribution. Its frequency energy band focused on the two frequency bands: (a)20~40kHz, (b)80k~140kHz. It is found that needle-stick model has the power spectrum value of the maximum discharge. In the absence of partial discharge ultrasonic frequency waveforms concentrated around 30kHz, and the frequency with relatively sparse distribution, the amplitude is relatively small. Since the electrode is relatively sharp. The needle-stick model suffers severe electric field distortion, resulting in a more intense partial discharge phenomena.

5 Conclusion

We use partial discharge experimental platform to analyze the partial discharge mechanism in switchgear cabinet. The strategy using LSSVM is proposed to detect partial discharge of switchgear according to causes and mechanism of partial discharge. Six kinds of PD are simulated, then we use MATLAB to do a great deal of experiments for different discharge models. According to time-domain waveforms and frequency-domain waveforms, the essence of the phenomenon of partial discharge is captured, and it provides an experimental basis for the characteristic values of partial discharge.

References

[1] D Hao-yin, T. Neng-ling, C. Xin-qi, G. Jian-zhong, F. Chun-ju. J. Electr. Power Sci. and Technol., **25**, 64(2010)
[2] W. Quan, C. Hong-song, ZH. Han-dong, Y. Da-wei. J. Gun Launch & Control, **3**, 59(2010)
[3] Z. Wen-qi, X. Li-yun, ZH. Li-lin. J. Electr. Power Sci. and Technol., **25**, 32(2010)
[4] H. Wei-tao.J. Electr. Power Sci. and Technol., **27**, 57(2012)
[5] M. Peiqing, L. Xiuwei, H. Yue, Sh. Gehao, J. Xiuchen. PEAM. IEEE,1(2012)
[6]R. Widodo; Z. Ming-Xiao; Suwarno; Zh.Guan-Jun. ICEEI, 181(2015)
[7] M. Kida, T. Kawashima, Y. Murakami, M. Nagao. CEIDP, 322(2015)
[8]T.S. Sidhu, M.S. Sachdev, G.S. Sagoo . Dev. in Power Syst. Prot., 157(2010)
[9] L. De-jun, SH. Wei, G. Zhi-qiang. High-voltage E.A., **45**, 99(2009)
[10] K. Siderakis, E. Pyrgioti, E. Thalassinakis, I. Vitellas. IEEE Trans. Dielectr. E.I. **18**, 122(2011)
[11] W. Zhang-jian, L. Cheng-rong, Q. Bo, H. Zhen, G. Bibo. Mod. Electr. Power, **27**, 31(2010)
[12] Z. Lei. Yunnan Electric Power, **39**, 49(2011)
[13] Z.Li-lin. SHH Electr. Power, **23**, 50(2011)

ADNP-CSMA Random Multiple Access protocol application with the function of monitoring in Ad Hoc network

Gang ZHAN[1],Yi Fan ZHAO[1],Hong Wei DING[1,a],Sheng Jie ZHOU[1] ,Long Jun LIU[1] and Zhi XU[1]

[1]*Institute of Information, Yunnan University, Kunming, Yunnan, 650504, China*

Abstract. In Ad Hoc networks,the net work of mobile nodes exchange information with their wireless transceiver equipment,the network throughput is in increased,compared to other such multiple hops network.Moreover along with the rapid development of modern information,communication business also will be increase.However,the access and adaptive of previous CSMA protocol are insufficient.According to these properties,this paper presents a kind of adaptive dual clock with monitoring function P-CSMA random multiple access protocol(ADNP-CSMA),and discusses two kinds of P-CSMA.ACK with monitoring function is introduced to maintain the stability of the whole system,and the introduction of dual clock mechanism reduces the channel of idle period.It calculate the system throughput expression through the method of average period,and the simulation results show that the system is constant in the case of high load throughput.

Keywords: Adaption; Dual Clock; The monitoring function; CSMA;Throughput

1 Introduction

With the rapid development of wireless technology,the dynamic nature of the network node draws pubic attention,thus it arises Ad Hoc network at the historic moment.It makes up the limited transmission distance with the help of technology for wireless devices through jumping forward,and it hosts networking communication freely through the moving mobile.It provides the successful solution for military communication,disaster relief and temporary communication[1].

MAC layer channel access protocol directly affects the performance of the whole system.How to use the channel effectively becomes the key issues of the whole network.We adopt a new kind of CSMA protocol to research this problem.[2]

There are three kinds of traditional CSMA, N-CSMA, 1-CSMA and P-CSMA.These three protocols have some shortcoming.P-CSMA is defined as pointing P probability to send in idle

DOI: 10.1051/matecconf/20166304017

time,and retreating in the probability of 1-P.When a packet arrives in busy time,there is two cases following:P-insist in idle period with non-insist in busy period CSMA and P-insist in idle period with 1-insist in busy period CSMA.Paper 2 discusses P-insist in idle period with 1-insist in busy period CSMA,and we call it 1P-CSMA.When the P equals to 1,it becomes 1-insist CSMA.We discuss the protocol in the circumstance of P-insist in idle period with non-insist in busy period CSMA,and we called it NP-CSMA.[3]

We need the data of each node,however,the nodes only send data in the access channel,and it can not know that the data is transmitted to the right recipient.If it adds a monitoring information section(ACK) at the end of each packet,then we called it P-CSMA protocol with the function of monitoring.After adding a monitoring information,the system will be able to know about the sending information group and channel.The monitoring information plays an oversight role,and this is an important improvement.It obtain a monitoring of the whole system with little throughput loss.[4]

The system uses discrete clock to control channel.If arriving a packet,it begins to send in next sequence.If arriving a packet in idle period,it also begins to send in next sequence of idle period,thus it wastes the channel utilization.If it uses continuous clock control in idle period.When it arrives a packet in a sequence of idle period,the packet do not have to wait for the beginning of the sequence to send,then sends immediately.Such control principle reduces the average length of idle period,and increases the average length of the transmission period.It is helpful to increase the throughput of the whole system.[5]

2 The analysis of the proposed protocol

2.1. The analysis of ACK mechanism

The channel of P-CSMA with ACK random multiple access protocol is shown in Figure 1.[6]

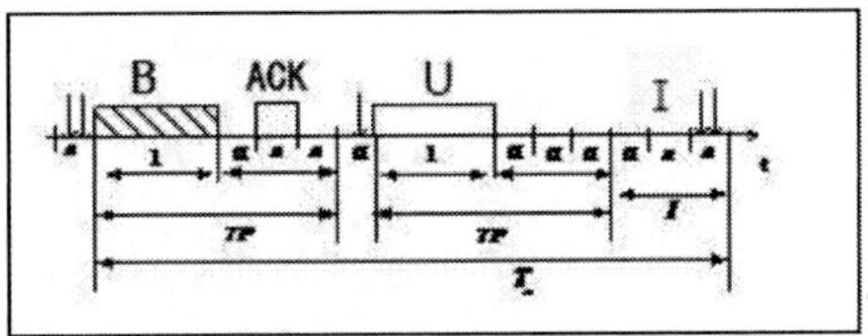

Figure.1 The channel of P-CSMA with ACK random multiple access protocol

The transfer cycle divided into three parts,the packet reached 1,transmission delay a,and the ACK with the function of monitoring

In this model,there will be three random events:

Event that information packets are sent successfully(U events).

Event that information packets collide with each other(the collision appears, B events).

Event that there are no information packets in the channel arrive,the channel is idle(I events).[7]

Supposing we put the B events and U events as BU events,so the channel appears on two event,I events and BU events(we called busy events).

Before the mathematical analysis of channel,we make the following assumptions:[8]

1.The channel access method is MAC control protocol of random multiple access,and there will be independent of process parameters for G poisson process;

2.The length of idle time sequence is a in the channel,and the length of sending packet time sequence is 1;

3.The channel is assumed to be the ideal state,and no noise channel;

4.The collision packet will be retransmitted at a certain moment,and the retransmitted packet has no effect on channel arrival process;

5.The length of ACK with the monitoring function is $\tau_A(\tau_A = a)$;

In a idle period,the probability of no reaching packet in the channel is:

$$q_0^0 = e^{-apG} \tag{1}$$

In a idle period,the probability of a reaching packet in the channel is:

$$q_0^1 = apGe^{-apG} \tag{2}$$

In a cycle,the probability of continuous i idle events is:

$$P(N_I = i) = (e^{-apG})^i \tag{3}$$

In a cycle,the probability of continuous j BU events is:

$$P(N_{BU} = j) = (1 - e^{-apG})^j \tag{4}$$

In a cycle,the probability of continuous i I events and j BU events is:

$$P(N_I = i, N_{BU} = j) = (e^{-apG})^i (1 - e^{-apG})^j \tag{5}$$

The averages length of i continuous j events in a cycle is:

$$E(I) = a\sum_{i=1}^{n}\sum_{j=1}^{n} ip(N_I = i, N_{BU} = j) = \frac{a}{1 - e^{-apG}} \tag{6}$$

The averages length of j continuous BU events in a cycle is:

$$E(BU) = (1 + 2a + \tau_A)\sum_{i=1}^{n}\sum_{j=1}^{n} jp(N_I = i, N_{BU} = j)$$

$$= \frac{1 + 2a + \tau_A}{e^{-apG}} \tag{7}$$

The averages length of information packet transmitted successfully is:

$$E(U) = \frac{q_0^1}{1 - q_0^0} = \frac{apGe^{-apG}}{1 - e^{-apG}} \tag{8}$$

The throughput of P-CSMA random multiple access protocol with ACK mechanism is:

$$S = \frac{E(U)}{E(I) + E(BU)} \tag{9}$$

$$= \frac{\frac{apGe^{-apG}}{1 - e^{-apG}}}{\frac{a}{1 - e^{-apG}} + \frac{(1 + 2a + \tau_A)}{e^{-apG}}}$$

$$= \frac{apG(e^{-apG})^2}{ae^{-apG} + (1 + 2a + \tau_A)(1 - e^{-apG})}$$

2.2. The analysis of dual clocks mechanism

When the system is idle,the system uses continuous-time control strategy.And if the system is busy,the system uses discrete-time control strategy.[9]

The period of idle time sequences obeys the geometric distribution of the mean:

$$E[N] = \frac{1}{1 - e^{-apG}} \tag{10}$$

the normalized probability of a reaching packet in a time sequence is:

$$P_{I_1} = \frac{apGe^{-apG}}{1 - e^{-apG}} \tag{11}$$

The normalized probability of two or more reaching packets in a time sequence is:

$$P_{I_2} = \frac{1 - apGe^{-apG} - e^{-apG}}{1 - e^{-apG}} \tag{12}$$

Because the reaching pockets in idle period will interrupt the idle period,so we can take the middle point of the mean.The averages length of I events is:

$$E(I) = (\frac{1}{1 - e^{-apG}} - 1)a + \frac{a^2 pGe^{-apG}}{2(1 - e^{-apG})} + \frac{(1 - apGe^{-apG} - e^{-apG})a}{2(1 - e^{-apG})}$$

$$= \frac{(1 + e^{-apG})a}{2(1 - e^{-apG})} \tag{13}$$

The throughput of the dual clocks P-insist CSMA random multiple access protocol with ACK mechanism is:

$$S = \frac{E(U)}{E(I) + E(BU)} \tag{14}$$

$$= \frac{\frac{apGe^{-apG}}{1 - e^{-apG}}}{\frac{a(1 + e^{-apG})}{2(1 - e^{-apG})} + \frac{1 + 2a + \tau_A}{e^{-apG}}}$$

$$= \frac{2apG\left(e^{-apG}\right)^2}{ae^{-apG}(1+e^{-apG})+2(1-e^{-apG})(1+2a+\tau_A)}$$

2.3. The analysis of the adaptive mechanism

The above is throughput expression.In order to realize the adaptive function,we need to adjust the size of the P according to the system load value,namely to P derivative:[8]

$$\frac{dS}{dP} = 0 \qquad (15)$$

Due to that getting for nonlinear equations,we obtain approximate solution by adopting the method of Newton type.

When the condition of light load $G \in (0,2]$, the probability P is 1. When the load increases slowly, probability P also corresponding reduces.We get the approximate solution of the probability P by adopting the piecewise idea to obtain the maximum throughput case. Supposing the relationship between P and G is:

$$P = \frac{1}{\beta \times G} \qquad (16)$$

P and G combination cases can be drawn from β value:

$$P|_{MAX} = \begin{cases} 1 & G \in (0,2] \\ \frac{1}{0.5G} & G \in (2,3] \\ \frac{1}{0.333G} & G \in (3,5] \\ \frac{1}{0.2G} & G \in (5,8.5] \\ \frac{1}{0.1G} & G \in (8.5,\infty) \end{cases}$$

3 Simulation results and analysis

With the simulation tool MATLAB R2014a, the simulation results are shown in Figure 2 to 5, in case a=0.1.

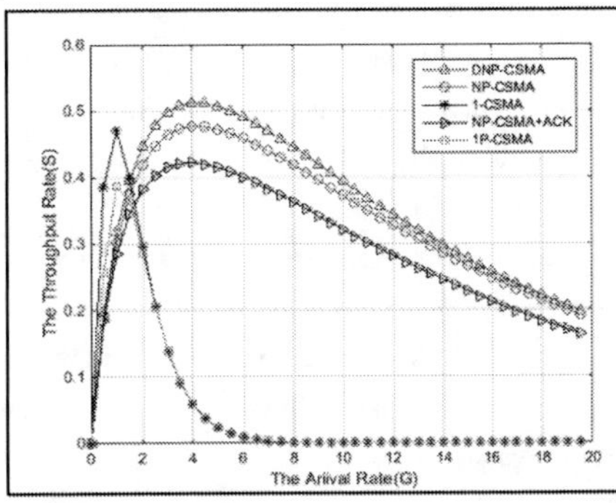

Figure.2 The comparison of several different CSMA

In the Fig.2,compared to DNP-CSMA, NP-CSMA, 1-CSMA,NP-CSMA with ACK, P-CSMA, when P=0.5,1P-CSMA reached the maximum throughput rate at 1.8 G,but NP-CSMA reached the maximum throughput rate at 4 G.Moreover NP-CSMA maximum throughput rate is bigger than 1P-CSMA.When G=6, 1P-CSMA throughput rate is 0,but NP-CSMA throughput rate would be 0 after 20 G.DNP-CSMA reduced idle time sequence,so the throughput rate increases a bit. NP-CSMA with ACK adds the monitoring fields,so the throughput rate decreases a bit.

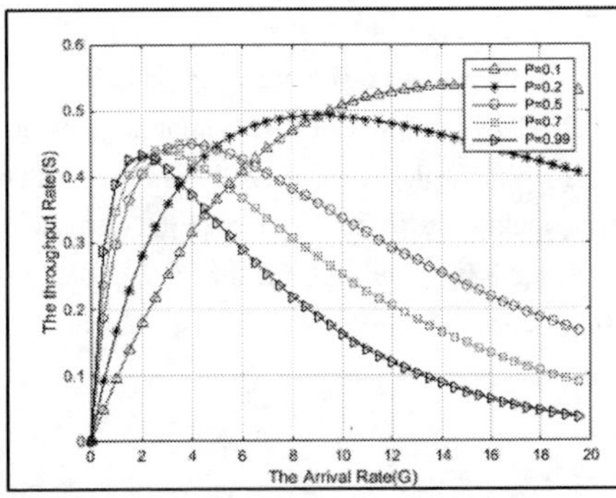

Figure.3 The throughput rate of DNP-CSMA with different P

In the Fig.3, it can be seen that DNP-CSMA in different P effects the throughput rate of the system.

With the increase of P,the probability of success that the sites send information packets in idle channel also grows,and the channel utilization increases, so the throughput rate of the system increases.However with the rapid of information packets arrival rate *G*, the system throughput rate also fells rapidly,finally tends to zero.

So we can improve performance of the system by controlling *P*. If we find the critical point, it will reach maximum throughput rate of the system.Then this reminds us of the adaptive mechanism.

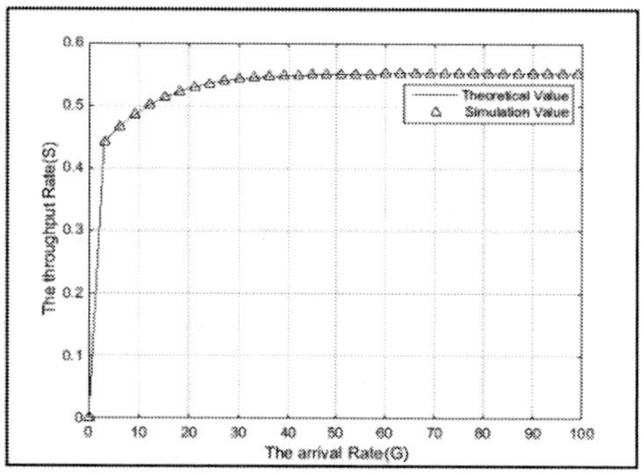

Figure.4 The throughput rate of ADNP-CSMA

In the Fig.4,after adding the adaptive mechanism, the system throughput rate is stable in arrival rate G reaching 30,and then with the increase of load, the system throughput is to maintain a constant value,so the system turns into a steady state.

In the adaptive system, the probability P in the whole process can be dynamic adjustment according to the circumstance of network load to make the optimal throughput rate.This mechanism can ensure that the throughput rate can still maintain large under the heavy load network.

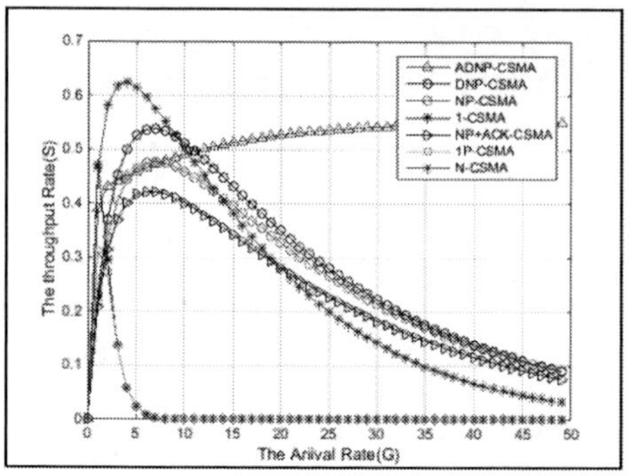

Figure.5 The comparison of several different CSMA when P=0.3

In the Fig.5,compared to ADNP-CSMA, DNP-CSMA, NP-CSMA,1-CSMA,NP-CSMA with ACK, P-CSMA,N-CSMA,the throughput rate of ADNP-CSMA is better than NP-CSMA's and 1P-CSMA's. For N-CSMA,although in case of light load, the throughput rate of N-CSMA is higher than ADNP-CSMA's, however, with the arrival rate increasing N-CSMA's throughput rate reduces, finally turns zero,and ANDP-CSMA's throughput rate is stabilize after arrival rate G is 30. Therefore, the new protocol for network has more advantages in heavy load.

4.Conclusions

In this paper, ADNP-CSMA protocol is proposed on the basis of traditional CSMA protocol, which distinguishes between NP-CSMA and 1P-CSMA. Moreover the Dual Clock mechanism and the function of monitoring mechanism are put forward, increasing the throughput rate and maintaining the stability of the system. After introducing the adaptive function, the system can dynamic adjustment to send probability P according to the weight of the load. It is ensure that the system can still maintain high throughput rate in case of heavy load,and it improves the performance of the wireless communication system. At the same time, the simulation experimental results show the consistency and the rationality of the theoretical analysis and

DOI: 10.1051/matecconf/20166304017

the simulation experiment. Compared to the throughput rate of different systems, ADNP-CSMA random multiple access protocol system has more advantages in the case of heavy load.

References

1. George, Y., and I. Bergel. "The Spectral Efficiency of Slotted CSMA Ad-Hoc Networks with Directional Antennas." IEEE Transactions on Wireless Communications 11.10(2012):3799-3809.
2. Dongfeng Zhao. Study on A New Method for Continuous-time Systems of Random Access Channel [J]. Journal of Electronics, 21(1): 37-41, 1999.
3. Dongfeng Zhao, Study on A New Method for the Slotted Access Channel [J], Journal of Electronics,Vol.19, pp.814-819, No.6,1997.
4. Hongwei Ding, Yingying Guo, Yifan Zhao, Shengjie Zhou, and Qianlin Liu. Research on the Multi-Channel Probability Detection CSMA Protocol with Sensor Monitoring Function. Sensor Lett. 13, 143-146 (2015).
5. Maomao Huang. The analysis of time-sharing CSMA wireless sensor network MAC control protocol analysis[D]. Kunming: Yunan University 2009
6. Dongfeng Zhao. Study on the Average Cycle Method for S-ALOHA Access Channel[J].Journal of Yunnan University, 1994, 16(2): pp133-138.
7. Liu Binbin. The Analysis of Multi-channel Random Multiple Access Wireless Communication Network Protocol based on Probability Detection [D]. Kunming: Yunnan University, 2006.
8. Shengjie Zhou et al., Research on the Discrete time Three-Dimensional Probability Csma Protocol In ad-hoc Network. International Journal of Recent Scientific Research Vol. 6, Issue, 5, pp.4257-4262, June, 2015.
9. Sheng Jie Zhou, Hong Wei Ding, Meng Yao Wang, et al. Multi-channel Dual Clocks three-dimensional probability Random Multiple Access protocol for Wireless Public Bus Networks based on RTS/CTS mechanism[J]. MATEC Web of Conferences, 2016, 44.

MATEC Web of Conferences 63, 04018 (2016) DOI: 10.1051/matecconf/20166304018
MMME 2016

Research on Improved Collaborative Filtering Recommendation Algorithm on MapReduce

Jie DONG[1,a], Yun QIN[1,2], Xue Yang SUN[1]and Li Ming DU[1]

[1]*Information and Control Engineering College, Shenyang Jianzhu university, Shenyang, 110168, China*

[2]*Liaohe Oilfield Company SAGD Project Department*

Abstract. Information overload is one of the most serious problems in big data environment, recommendation systems is a way to effectively mitigate the problem. In order to make use of rich user feedback and social networks information and to further improve the performance of the recommendation system ,This thesis makes a improvement on the user-based collaborative filtering algorithm by normalization method, Meanwhile the algorithm could be run on the MapReduce in the Hadoop platform. The experimental results show that the algorithm on Hadoop platform can effectively improve the accuracy of the data to recommend and computational efficiency, so as to improve the satisfaction of users.

1. Introduction

The rapid development of the Internet have brought about great changes in the world, which cause information overload[1-3]. Information overload has two sides .For information users, searching for what they need from the vast amounts of information accurately is becoming more difficult .For information manufacturers, making information they produce conspicuous in the vast amounts of information is also a problem need to be solved urgently. Recommendation system is based on the analysis of used behavior to meet users' need and interest information for the user in initiative.

In 2003 Deng Ailin put forward a collaborative filtering recommendation algorithm based on project score predicts [4] in order to solve the sharply fallen recommendation quality rating caused by extremely thin data .In 2004, Gao Fengrong adopt the method of clustering and classification respectively by divided sparse score matrix, reducing the scope of neighbor and the need to predict the number of resources, improving the efficiency and scalability of the collaborative filtering recommendation algorithm [5]

With the growing amounts of information, Research focusing on the recommendation system has changed from stand-alone mode to server cluster [6-8]

In 2011, Jing Jiang and others further divided collaborative filtering recommendation algorithm based on the project into three stages of the main computing [7].After the segmentation of each Map Reduce phase can be run in parallel in each node of the cluster. Meanwhile they put forward the data partitioning strategy based on Hadoop platform to maximize the local data storage, reducing the communication overhead between system and improving the recommendation efficiency of the recommendation system. In 2012, Juan Yang and others further used mongo as auxiliary storage database [8].

At present, collaborative filtering recommendation methods are mainly based on matrix decomposition [9][10]. However, the current method in scalability performance is poor, and unable to adapt to the massive data processing.

In this paper, aiming to the recommendation system's error range,. We analyzed deeply the traditional collaborative filtering recommendation algorithm based on user, run the improved the algorithmon on MapReduce in the platform of Hadoop ,which has made the effect more accuracy

2. The introduction of collaborative filtering recommendation algorithm

Through the ten year research, a variety of recommendation algorithm has been proposed and collaborative filtering recommendation algorithm is widely used.

Collaborative Filtering Recommendation algorithm usually uses the nearest neighbours technology and utilize the history of users' preference information to calculate the similarity between users' interest and use the nearest neighbours of the target users by items evaluation value to predict the target user preferences for certain products to recommend for target users. The largest advantage of collaborative filtering algorithm have no special requirements on recommend objects And it also can handle unstructured complex objects such as

[a] Corresponding author: dong_jie_2002@163.com

music and movies. At present, collaborative filtering has been used in e-commerce recommender systems by based on other users' evaluation to recommend for target users. Collaborative filtering recommendation algorithm are usually divided into two types: Item-based and user-based. The following will be introduced to these recommendation algorithm respectively.

2.1 User-based collaborative filtering recommendation algorithm

Collaborative Filtering Recommendation algorithm usually uses the nearest neighbors technology and utilize the history of users' preference information to calculate the similarity between users' interest and use the nearest neighbors of the target users by items evaluation value to predict the target user preferences for certain products to recommend for target users. Specific process is as follows:

Step1: Collection and storage, sorting out the history of user's behavior data such as the user's existing purchase records, browse list page, the collection behavior, attention, project ratings, etc. These data is the base for calculating the similarity of users.

Step2: Searching for collection of users similar to target users, the key to this step is calculating the similarity between two users.

In order to calculate user similarity, Cosine similarity calculation is often used to calculate the degree of similarity between two users. Assuming $\vec{i}$ and $\vec{j}$ represent n dimension vector of user *i* and *j* respectively, the similarity between the user *i* and *j* is defined as follows:

$$sim(i,j) = \cos(\vec{i},\vec{j}) = \frac{\vec{i}\cdot\vec{j}}{\|\vec{i}\|*\|\vec{j}\|} \quad (1)$$

In practice, the cosine similarity did not take into account user ratings subjective preference. cosine similarity is modified to solve the above problem. It subtracts the user average score for the item so as to reduce difference of the subjectivity. Assuming $I_{i,j}$ is scores set that user *i* and *j* give the items together, I_i and I_j are scores sets of users *i* and *j* respectively, $R_{i,c}$ or $R_{j,c}$ respectively stand for score of user *i* or user *j* on the item c, $\overline{R_i}$ and $\overline{R_j}$ are users *i* and *j* average' scores respectively, the user similarity between user *i* and user *j* is as follows:

$$sim(i,j) = \frac{\sum_{c\in I_{i,j}}(R_{i,c}-\overline{R_i})(R_{j,c}-\overline{R_j})}{\sqrt{\sum_{c\in I_i}(R_{i,c}-\overline{R_i})^2}\sqrt{\sum_{c\in I_j}(R_{j,c}-\overline{R_j})^2}} \quad (2)$$

The accuracy of the similarity calculation directly decides the recommendation algorithm quality.

Step3: By means of similarity calculation results, the most nearest neighbors of target user could be made of the top N users which similarity value are very similar to the target user similarity degree, at this time the value N can be set according to the need to manually. After that ,we can obtain recommend results for the target user .

Supposing N_u represent a nearest neighbors set to target user , $P_{u,c}$ represent a score of user u on the item c, calculating method as shown followed:

$$P_{u,c} = \overline{R_u} + \frac{\sum_{v\in N_u} sim(u,v)*(R_{v,c}-\overline{R_v})}{\sum_{v\in N_u}|sim(u,v)|} \quad (3)$$

2.2 Item-based collaborative filtering recommendation algorithm

Item-based collaborative filtering recommendation algorithm based on the similarity between items, finding the nearest neighbors set of items, according to the user preference information to predict item ratings, eventually object could be feedback to the user by those similar items which users like , but Item-based collaborative filtering recommendation algorithm does not use of the attributes contents of the item to calculate the similarity between items, but to calculate the similarity through the analysis of user behavior records.

3. Distributed Computing Framework Based on MapReduce

Under the circumstance of the big data, Hadoop is one of the basic infrastructure which make full use of the cluster ability of distributed computing. MapReduce is a programming framework which handle a large number of data based on the hadoop platform, and it is also a kind of application in large-scale parallel processing of information, a new distributed computing model. No matter how many hosts on Hadoop platform cluster, MapReduce technology can finish the tasks such as web searching, advertising pushing and rapid processing even for TB, PB, and even EB bytes of data on thousands of cluster nodes at the same time.

The core idea of MapReduce algorithm is dividing and then rule it, namely the operation of large-scale data sets, firstly distributed to various points node to complete together ,which is under the management of one master node, then getting an intermediate result. After that integrating the middle results of each node, and obtaining the final result . The entire MapReduce framework is composed of the Map and Reduce functions, run map stage first, then perform reduce stage. The Map is responsible for dividing a task into many subtasks and divided large-scale data into several smaller blocks, then transfer to a large number of cluster nodes. The output of map stages is the input of the Reduce phase. Reducing is responsible for integration and merge and then the child tasks processing results.

4. The Improvement to User-based Collaborative Filtering Recommendation Algorithm

4.1 The improvement of computing user similarity

The formula (1)-(3) show that the user similarity computing depends on the score of items given by users. In fact, some users are very casual, and all scores are higher. And some users are very strict with it, even if the quality of the goods is well; however the scores are also not high. The more the difference of evaluation standard is, the worse the accuracy of recommendation.

Although modified cosine similarity take into account above problem, but it still slightly rough. In this paper, the subjective normalization algorithm is introduced. By using the initial users-item evaluation scores, normalized scores of user-item are calculated in order to eliminate the error that caused by user different grading standard. The specific implementation process is shown below.

Step1: according to the initial scores, calculate each user's average score $\overline{r_u}$, calculation method is as follows:

$$\overline{r_u} = \frac{\sum_{i=1}^{n} r_{u,i}}{n} \tag{4}$$

In the formula, $r_{u,i}$ is the score given by the user u to the item i.

Step2: set user ratings interval by calculating averages of the minimum and maximum of all user scores as both ends of the interval, and calculation method is as follows:

$$[\frac{\sum_{j=1}^{m}\min(r_{u_j,i_1}, r_{u_j,i_2},\ldots\ldots, r_{u_m,i_n})}{m}, \frac{\sum_{j=1}^{m}\max(r_{u_j,i_1}, r_{u_j,i_2},\ldots\ldots, r_{u_m,i_n})}{m}] \tag{5}$$

$$b_{i1} = \min(r_{ui,j1}, r_{ui,j2}\ldots\ldots, r_{ui,jn}) \tag{6}$$

$$b_{i2} = \max(r_{ui,j1}, r_{ui,j2}\ldots\ldots, r_{ui,jn}) \tag{7}$$

$$k = \frac{\sum_{i=1}^{m} b_{i1} + \sum_{i=1}^{m} b_{i2}}{2m} \tag{8}$$

In the above formula, b_{i1} and b_{i2} stand for the minimum and maximum values by calculating each user the boundary value, k is average for all users interval.

Step3: recalculate score $r'_{u,i}$ that the user u give the item I in order to obtain a new user -item evaluation.

$$r'_{u,i} = \frac{k * r_{u,i}}{\overline{r_u}} \tag{9}$$

In formula, the normalized coefficient is $\frac{k}{\overline{r_u}}$.

By the formula (9), via the derivation transformation, the average score of each user has become interval average k, namely: $\overline{r'_{u1}} = \overline{r'_{u2}} = \ldots = \overline{r'_{um}} = k$

Using this normalization processing on the user's score, it can unify the standard of the evaluation. And it can effectively reduce impact which the different user rating scales on the similarity computations.

4.2 Improved recommendation algorithm for scoring prediction algorithm

When we calculate the user similarity by the traditional collaborative filtering algorithm, it can calculate on the item set that the user has scored together. If two users give the more same type of items, however the grade is less for the same item. They are similar to each other in theory, but according to the traditional similarity calculation formula, it will leads to the dissimilarity results between two users. Therefore, on the basis of the subjective normalization method, prediction score calculation method is further introduced. When we calculate the similarity between two users, we can increase user's common grading evaluation scores to reduce data sparse and improve the quality of recommendation.

The concrete implementation process of improved filtering recommendation algorithm is shown below.

Step1, according to user-item evaluation matrix A (m, n), we are able to calculate union sets of item evaluation among each users ,calculation method as follows:

$$I_{i\cup j} = I_i \cup I_j \tag{10}$$

Among the formula: I_i and I_j are sets that users i and j have evaluated scores respectively, $I_{i\cup j}$ is a union set between the user i and j.

Step2 using subjective normalization algorithm to construct normalized score matrix, calculation method is as follows:

$$A(m,n) \to A'(m,n)$$

Step3 according item set $I_{i\cup j}$ of users i and j after users-item evaluation matrix normalized calculation $A'(m,n)$ to calculate similarity, the calculation method of the improved formula(3) as follows.

$$sim(u_i,u_j) = \frac{\sum_{c\in I}(r'_{u_i,c}-k)(r'_{u_j,c}-k)}{\sqrt{\sum_{c\in I}(r'_{u_i,c}-k)^2\sum_{c\in I}(r'_{u_j,c}-k)^2}} \tag{11}$$

Among the formula: k is the normalization factor, I represents union set that the user i and user j have evaluated items commonly by the formula (3)

Step4 assumes the user u_a is the target user, c is for forecasting project, improve user's similarity calculation formula (4) to calculate the forecast scores of user u_a on item c, calculation method is as follows:

$$P_{u_a,c} = k + \frac{\sum_{u\in U} sim(u_a,u)(r_{u,c}-k)}{\sum_{u\in U} |sim(u_a,u)|} \tag{12}$$

5. The realization of the improved algorithm based on MapReduce

In the Collaborative filtering recommendation algorithm based on the user, first of all, according to the user-item evaluation matrix the degree of similarity among the projects are calculated. Then on the basis of similarity, calculating similarity between the user and the target users, predicting scores which target users do not have the evaluation score of the item, and obtaining the TOP-N value of highest score. Obviously these two steps is a serial process of order.

In the process of user similarity calculation, there is almost no coupling between them, so it can be parallel computing, which could assign a MapReduce job to implement. In the prediction of target users evaluation score ,it also can use MapReduce job parallel computing forecast rating of each item.

Thus, the essence of collaborative filtering recommendation algorithm based on the user for distributed computing is to calculate the similarity between users with a MapReduce task. And its result is another input of MapReduce task to calculate the target users forecast rating of the item. And the two tasks are executed serially.

Subjective score normalization algorithm is mainly responsible for the user-item rating matrix is converted to a unified coordinate system, the parallelization of subjective score normalization algorithm is divided into three steps. The first step is according to the user to solve the user-item rating matrix. The second step is according to the user-item rating matrix and solving the maximum , minimum and average score. The third step is based on the user-item rating matrix and maximum, minimum and the average score of solving normalized user-item rating matrix.

In the Improved User-based Collaborative Filtering Recommendation Algorithm, Each of three processes, which are user evaluation matrix normalized, users prediction score, calculate similarity .They can be realized in parallelization, these three parallelization process combined into a serialization process. Each intensive computing process segment with parallel ideas to speed up the whole operation process, and make the algorithm tolerated normal time to complete the calculation. Through the above steps, the parallelization of the recommendation algorithm has been realized, which reduce the running time of the recommendation algorithm and improve the efficiency of the recommendation system.

6. Test results

The experiment is needed so as to prove the effectiveness of the improved collaborative filtering recommendation algorithm based on Mapreduce, In the experiment, using the data of 6000 users to 1000000 scores of 4000 movies information and inputting the user ID and the recommended number, in the end the list of recommendation are displayed. For example to experiment , supposed the users NO.6 and the recommended list number is 20, the experimental results are shown in figure 1, in which the vertical axis shows recommended time used, the horizontal axis shows node number

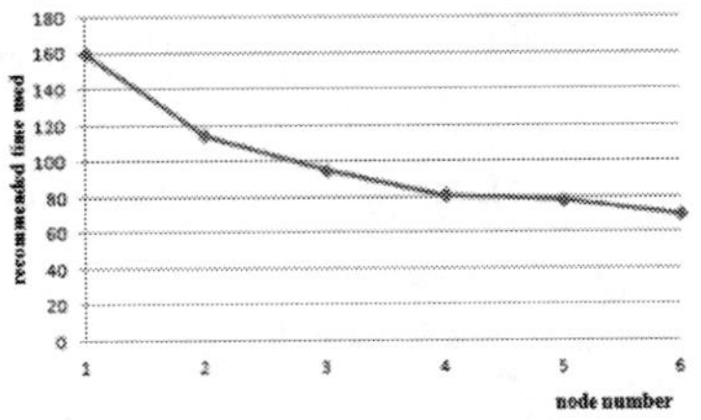

Fig1. Recommended performance comparison

Experimental results show that as the number of nodes increases unceasingly, Using the improved algorithm, it still get better recommendation results. recommended time used decrease, which can illustrate the performance of the system are on the increase and increase accuracy of the recommendation and the efficiency of recommendation, reduce the impact of the sparse data.

7. Conclusion

Aiming to the bottlenecks problem of traditional recommendation algorithm operations on a single machine performance, the improved collaborative filtering recommendation algorithm on MapReduce framework in the Hadoop distributed platform are put forward, Experimental results show that it increases recommending performance, which the computing tasks can be evenly distributed to each computing nodes in distributed parallel processing, and has the good expansibility.

Acknowledgement

In this paper, the research was sponsored by the Nature Science Foundation of Liaoning Province (Project No. 2014020068)

References

1. Yuanzhuo Wang, Xiaolong Jin. Journal of Computer, **36**,4(2013)
2. Jianguo Liu, Tao Zhou, Binghong Wang. Progress in natural science, **19**,7(2009)
3. Shahabi C,Chen Yishin. Distributed and Parallel Databases,**14**,9(2003)
4. Ailin Deng, Yangyong Zhu, Bole Shi. Journal of software. **14**,8(2003)
5. Fengrong Gao, Xiaoyong Du, Shan Wang. Microelectronics and computer. **21**,6 (2004)
6. Zhiyou Zhang. Laboratory research and exploration. **25**,3(2006)
7. Jing Jiang, Jie Lu, Guangquan Zhang, Guodong Long. . Services (SERVICES), 2011 IEEE World Congress . 8(2011)
8. Juan Yang, Han Du, Bin Wu, Xinxin Ge. CCIS, 2012 IEEE 2nd International Conference **1**,5(2012)

9. Ming-Sheng Shang, Linyuan Lü, Wei Zeng, Yi-Cheng Zhang, Tao Zhou. EPL(Europhysics Letters). **88**,3(2009)
10. E Haihong, Song Meina, Li Chuan，et al. Journal of Beijing University of Posts and Telecommunications.**37**,5(2014)
11. Ren Xiuchun, He Yaji. Electronic design engineering, 22, (2014)

Operation and management of Liaoning waste water treatment plants by STOAT Simulation

Yu Nan GAO[1,a], Xin Ze LIU[1], Rong Xin ZHANG[1], Jing Jing SHAN[1], Mei Qiao SUN[1],Zi Xiang ZONG[1],Chang Wei LI[1]and Xin Zhi WANG[1]

[1]*School of Municipal and Environmental Engineering, Shenyang Jianzhu University, Liaoning Shenyang 110168,China*

Abstract. Due to the irregular management, the effluent water of some waste water treatment plants (WWPT) of Liaoning province didn't meet the demand of the National Discharge Standard. Meanwhile, excessive of dosage and discharge of sludge made the results of operation costs increasing and environmental pollution during the processes. The use of mathematical models for the simulationof wastewater treatment processes has gained widespread acceptance as a tool to aid the design of newworks and the optimization of existing facilities. STOAT developed by the Water Research Center is the most widely used model for simulating the WWST processes. This paper invited the STOAT as simulation model for the design and optimization of Liaoning waste water treatment plant.

1 Introduction

With the development of industry, the northeast cities of Liaoning province has experienced an economic development at high speed. However, the population growth and industrial scale expansion of Liaoning province have led to an increased waste water discharge, which brought a lot of pressure to the current conventional waste water treatment plants.

At present,there are 160 waste water treatment plants distributed in different cities of Liaoning province, the total daily capacities are 8.467million tons theoretically,but actual treatment water capacities are only 4.67 million tons. Annual treatment water capacities are up to 23.62 billion tons which are amount to the total reservoir capacity of the Dahuofang reservoir.Shenyang is the capital city of Liaoning province, it has 34 waste water treatment plants (WWTP), the daily waste water treatment capacities are 2.508 million tons[1].

The "Discharge standard of pollutants for municipal wastewater treatment plant" (GB18918-2002) of the National Standard has revised in 2006, which has strengthened the effluent water quality of WWTP must meet the National Standard before discharged to the key drainage basins such as Liao River located in Liaoning province. Therefore the Environmental Management Departments have improved the requirements of wastewater treatment plants from large cities to the small towns[2].

Most of the wastewater treatment plants in Liaoning province mainly treat the domestic wastewater and the industrial wastewater.The treatment processes include biological methods, such as AAO, SBR, AO and other classical activated sludge process or the improved process of these classical methods[3].However, due to relaxed construction quality control and insufficient management, large number of small wastewater treatment plants were not meet the demands of National Standard. The running parameters and management methods were different from the original designs demand. Part of the wastewater treatment plant parameters were designed too conservative, which can not work well at full running time. Therefore, such wastewater treatment plants need to be simulated to optimize the running parameters.

Every month,the amount of the wastewater and water quality which need to be dealt are both changed.During the rainy season, the wastewater quantity of WWTP is usually larger than the dry season.Sometimes the amount of water needed to be handled during the rainy season is far greater than that of the design, which is beyond the processing capacity of the wastewater treatment plant.This phenomenon makes a mass of water enter wastewater treatment plant in a short period of time,reduction of hydraulic retention time in the biological reaction tank, a large amount of activated sludge lost ,total active species in the reaction pool decreased,the residual sludge quantity in the two settling basin suddenly increased. Therefore,the system is difficult to work normally.

In general, the small and medium-sized wastewater treatment plant is based on artificial experience to control the technological process,but many times artificial experience cannot meet the water treatment process

[a]Corresponding author: gaoyunan01@163.com; 873737660@qq.com

improvement and experience judgment error would reduce the capacity of wastewater treatment plant.A large number of wastewater above the standards were discharged into the river or cannot be recycled, it is easily to cause environmental pollution and economic loss.So the simulation software of wastewater treatment plant is very important in the rapid development of china.[4]

The online real-time monitoring system can provide real-time data for the waste water treatment plants conveniently. The system provided the real time data for the simulation models which can help the waste water treatment plants finding the optimal running parameters. By this system, the waste water treatment plant will reduce the energy consumption when running at full load and treat wastewater under the best condition, in this way, the discharge of WWTP could reach the demand of the National Standard.

2 Select simulation models

Now the mainstream simulation software of wastewater treatment plant are EFOR, GPS-X, STOAT, WEST, Biowin, etc.They are all widely used in developed countries. However, most of these simulation software are not free of charge, which will cost too much for some small wastewater treatment plants.So an economical and practical simulation software is needed to solve the existing problems for small WWTP of Liaoning province[5].

STOAT is the mainstream product of PLC WRc company which can simulate more than 40 kinds of process such as Inline Primary tank ,detention tank,SBR,Chemical P removal,IDEA,CSBR, trickling filter,Chlorination,Acid stripper humus tank,fluidised bed,Thickener,Chemical disinfection ,which basic covers a variety of processes of all the wastewater treatment plants in Liaoning.

STOAT is the only simulation program based on BOD and COD and has been used in many countries such as Britain, Europe, the United States and the Middle East.In order to help the industry better understanding and application of the wastewater treatment model,the British PLC WRc company in 2010 make the STOAT free to use in the world. Any enterprise or individual can go to the PLC WRc company website for free download or obtain STOAT simulation software.This is a good news for Liaoning wastewater treatment plants. For this reason, wastewater treatment plants do not have to consider the cost of the purchase of simulation software problems, also they do not need to solve the use of pirated software problems.

3Simulation of Liaoning wastewater treatment plant

3.1 Sanbaotunwastewater treatment plant

Sanbaotun wastewater treatment plant was conducted in Fushun, Liaoning, since 20 years ago to treat wastewater for surrounding cities. This plant served a population of 250000 P.E., the secondary treatment process of this plant used AAO process. And the capacity was 400,000m3/d. The STOAT simulation was used to build a new WWTP and optimize the running parameter. According to the original data of the water influent and the analysis of the online monitoring data to determine the modules used in the simulation framework and the models used in each module.The AAO process of this simulation wastewater treatment process was a multimode AAO technology, also it will be an improved process of traditional AAO.

Multimode is the multiple inlet and multiple return port, which can flexible control between AAO and inverted AAO on the basis of inlet and outlet requirements to achieve the processing requirement.The waste water passes through the thick grid, the fine grid and the micro grid,then adopting the method of subsection inlet makes the wastewater to be treated input in the anaerobic and anoxic zones respectively.Sanbaotun wastewater treatment plant is made up of four AAO reaction tanks,each tank with 161 meters long,102 meters wide and 6 meters effective depth,each anaerobic anoxic zone 19 meters long and 102 meters wide.In the front of the main reaction tank in AAO process is equipped with anaerobic anoxic zone which has the function of removing nitrogen and phosphorus.The back end of the reaction zone is dominated by the aeration zone which has two internal reflux pump to reflux mixed liquid in anoxic zone.Each AAO tank corresponds to two settling basin and each two settling basin's surface area is 1018 square meters and the effective pool depth is 4.5meters.

3.2STOAT model

The STOAT was used to simulate Sanbaotun waste water treatment plant processing. Because the waste water treatment plant was made up of four groups of AAO process, one of the AAO group was chosen for the STOAT simulation.The process was composed of waste water influent and flow divider to shunt waste water.Part of waste water have flowed into anaerobic tank and another part have flowed into anoxic tank.Waste water passes through anaerobic tank and anoxic tank to aerobic tank.After the aeration treatment,part of waste water have returned to anoxic tank and other parts have equally assigned tosecondary sedimentation tank 1 and secondary sedimentation tank 2. After treatment, water and sludge were discharged, part of the return activated sludge after the confluence back to tank Anaerobic

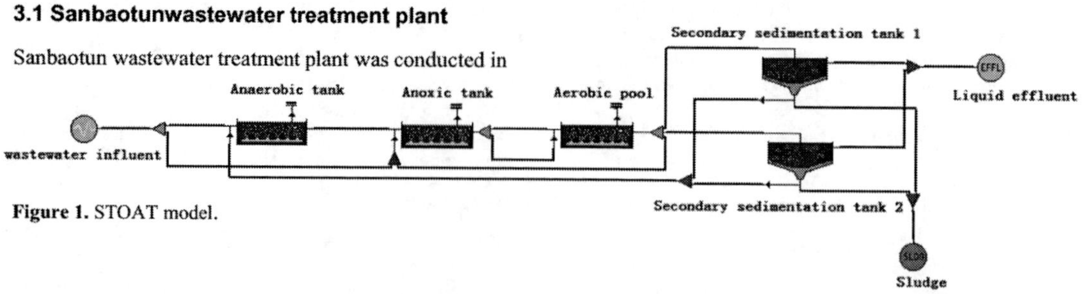

Figure 1. STOAT model.

3.3 Model comparison

ASAL1 : The WRc model comprises a number of differential equations written as a mass balance around a completely mixed reactor. Increasing the number of reactors in series increases the model's approach to plug flow. Twelve tanks in series provide an excellent approximation to a true plug-flow reactor. Each tank contains an identical number of components with a differential equation for each.The model recognised that utilisation of substrate can occur without this consumption being coupled to growth. Thus a more accurate description of the removal of substrate from a wastewater treatment system would include in the mass balance equation a Monod term for the conversion of substrate to new biomass and a Michaelis-Menten term for non-growth substrate removal.

ASAL2: Activated sludge model 1 assumes that all particulate BOD is solubilised immediately. This approximation is acceptable when the sewagewater retention time is long, greater than 4 hours. For shorter sewagewater retention times hydrolysis is incomplete and Model 1 is likely to over-predict the effluent BOD.This model has been evaluated at a UK sewagewater works on storm flows and found to predict the correct effluent BOD. The activated sludge unit at this works had a hydraulic retention time during the storm of less than one hour

ASAL3: This model is equivalent to Model 2 but with all the kinetic and stoichiometric parameters accessible to you and should be restricted to `expert' use especially when treating industrial effluents.

ASAL4: This model will be introduced in a later version of STOAT. It is equivalent to the IAWQ Activated Sludge Model #2 for biological phosphorus removal, modified to work with BOD rather than COD and to include the concept of nonviable bacteria having biological activity.

ASAL5: Activated sludge model 5 is the same as model 1 but includes a simple model of biological P-removal.

The basic model is described here as Model 1 and is intended for normal activated sludge modeling. This model ignores the breakdown of particulate BOD to soluble BOD, and this simplification can affect the predictions of detailed spatial oxygen demand within the aeration basin, and also the effluent BOD under storm conditions. Model 2 therefore includes the effects of solids hydrolysis to soluble products. Although

Model 2 is therefore more accurate than Model 1 it does require that you have some data on solids breakdown; where this is lacking Model 1 should be used. Model 3 gives the user full access to all the kinetic constants within the model. Model 5 is a simple biological phosphorus removal model, based on an extension of Model 1.

3.4Simulationg results

After inputting the data and selecting the model, the practical simulation result of waste water treatment plant will be achieved by adjusting operating parameters in STOAT simulation software according to the practical operating conditions of the wastewater plant. We input the real water quality into the simulation frame to obtain the simulated water quality.

Model parameters need to be revised according to the influent and effluent water quality indexes of aerobic tank, anoxic tank, erobic tank and secondary sedimentation tank. Each parameter adjustment to the reasonable numerical needs a large amount of data and the experiments to verify. Adjusting the parameters to achieve the perfect simulation of a wastewater treatment plant is a very difficult event. But it can simulate the effluent to the waste water treatment plant to the guidance of the proposed process modification.

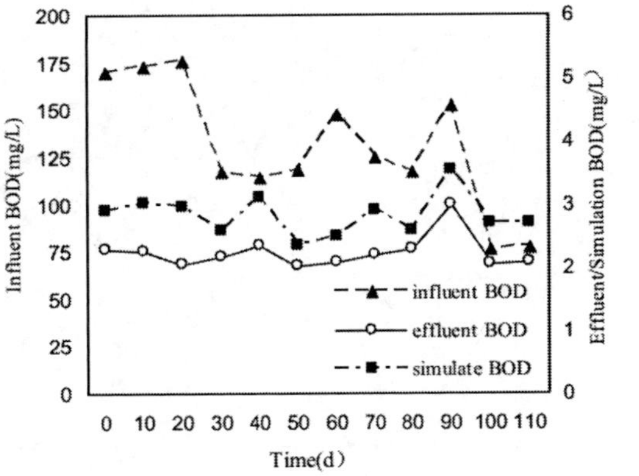

Figure 2. Influent / Effluent / Simulate BOD data

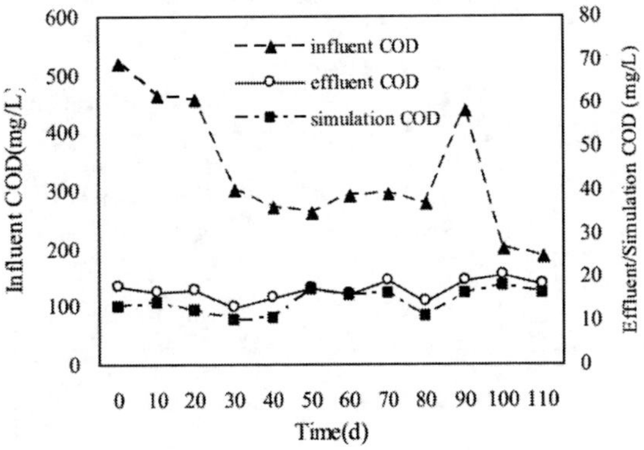

Figure 3. Influent / Effluent / Simulate COD data

DOI: 10.1051/matecconf/20166304019

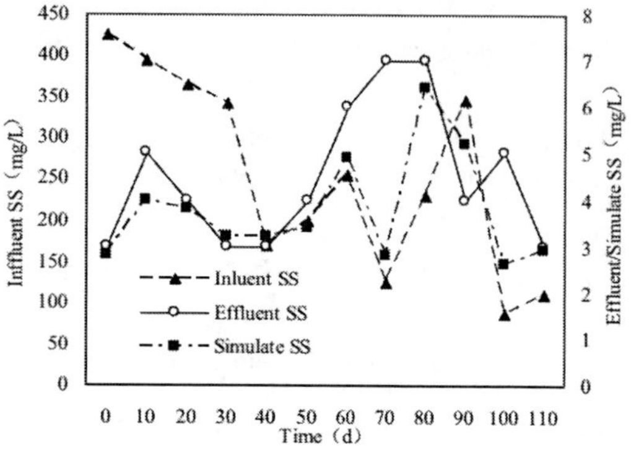

Figure 4. Influent / Effluent / Simulate SS data

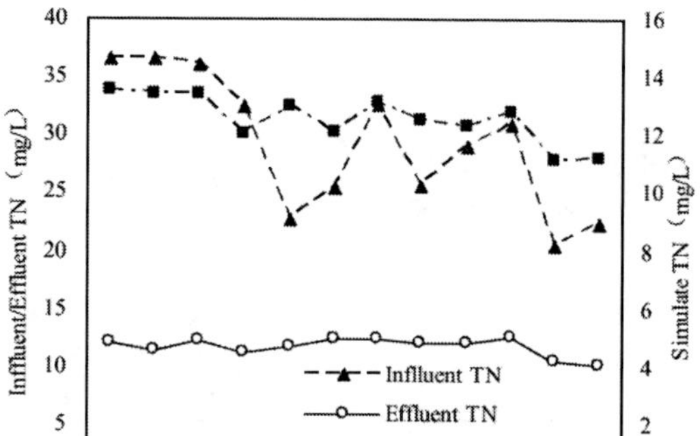

Figure 5. Influent / Effluent / Simulate TN data

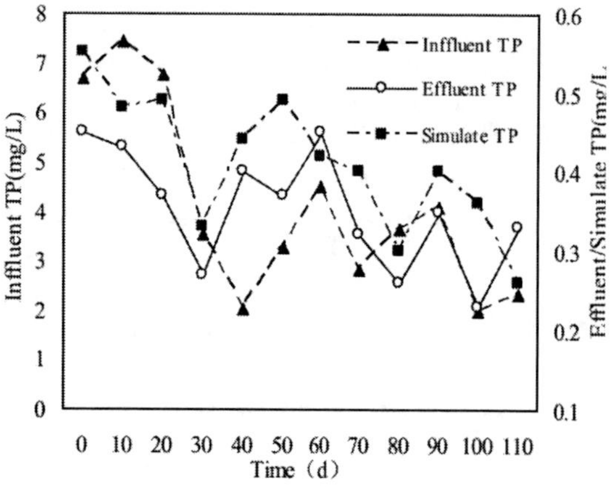

Figure 6. Influent / Effluent / Simulate TP data

4Conclusion

After comparing the real water quality index with the simulated water quality index, the quality of the waste water treatment plant can be simulated by the STOAT software. By this way, the simulation process of the wastewater treatment plant can be simulated by adjusting the reflux ratio and the amount of aeration according to the simulation software. The discharge of wastewater and the amount of aeration can be reduced as far as possible. The operation energy consumption of wastewater treatment plant will be reduced either, and the effluent water quality of wastewater treatment plant will meet the national standard.

Acknowledgments

This work was financially supported by National Science and Technology Subject (2012ZX07505002-002), National Natural Science Fund (51508342), Plan of Housing and Urban-Rural Development (2015-k7-019) and Plan of Talented Person Award of Liaoning Province (LJQ2015090).

References

1. Li Zhang.Analysis on the construction of urban sewagewater treatment plant in liaoning province[J]. Agriculture and technology,(2016)
2. YongChao Lu. The diagnosis and evaluation of the controltechnology of small towns wastewater pollution in Liaohe[D]. ShenYang Jianzhu University(2015)
3. Yang Yang. Study on Optimization and application of A2/O technology in urban wastewater treatment plant [D]. Harbin Institute of Technology, (2013)
4. Zou Ling,Study on Denitrification of Modified A2/O Process through the Small Pilot Plant and STOAT Simulation software during the rainy season [D],Chongqing University(2014)
5. Study on Denitrification of Modified A2/O Process through the Small Pilot Plant and STOAT Simulation Software during the Rainy Season[D],Chongqing University(2014)
6. Orhon D. and Artan N.Modelling of Activated Sludge systems.Technomic Publishing Co, Lancaster, USA(1994)
7. Smith M., Cooper P., McMurchie J., Stevenson D., Mann B., Stocker D., Bayes C. and Clark D. Nitrification trials at Dunnswood sewage treatment works and process modelling using WRc STOAT. J.CIWEM 12(1998)
8. Karaha-Gul, van Loosdrecht M C M and Orhon D.Modifying of Activated sludge model No.3. considering direct growth in primary substrate [J], Water Science and Technology, (2003)
9. Gujer W, Henze M, Mino T, etal. Activated Sludge Model No. 3 [J].Water Science and Technology, (1999)

MATEC Web of Conferences **63**, 04020 (2016) DOI: 10.1051/matecconf/20166304020
MMME 2016

The Application of MATLAB in Automatic Fingerprint Recognition System of Police

Hai Ping LI
College of Sciences, Hebei University of Science and Technology, Shijiazhuang 050018, China

Abstract. This paper studies on the automatic fingerprint recognition system, the former three fingerprints are extracted from the suspects, the fourth incomplete fingerprint is extracted in scene. Enhance the quality of fingerprint image through improved algorithm and image preprocessing method, at last matching the fingerprint of the three suspects to the fourth to find the criminal. All papers get the simulation results by matlab.

1 Introduction

Using fingerprint to confirm the identity can be track back to ancient China. At the beginning, fingerprint recognition technology widely applied on the contract, etc, now it began to develop in other areas. The 18th century, Europe's Mayer proposed each person's fingerprints are not the same in the world. In the early 19th century, scientific research found two important fingerprint characteristics: one is the fingerprint ridge lines are different in different finger, another is the pattern of the fingerprint ridge lines will not change all life. Those results were already used in fingerprint recognition of police. In the 20th century, as the computer can process images effectively, it was used to deal with the fingerprint analysis instead of the artificial processing traditionally. By this time, the fingerprint recognition technology reached a comprehensive development stage. At present, fingerprint recognition technology still faces many impediments, especially after the change into a 2 dimension images, some information will be missed, especially the fingerprints extracted is not very complete and clear, for example, lack of some parts, or have scars, twisted, stains, etc, will directly affect the efficiency of fingerprint recognition, which increases the difficulty of feature extraction.

In paper [1], the relative deformations are represented as an average deformation model based on minutiae locations and orientations using 2-D Thin Plate Splines (TPS). Paper[2] shows that the proposed algorithm is more robust and has better binarization performance . In paper [3-5], propounded a new fingerprints matching. Paper [6-7] show that how to enhance the fingerprint image and make it more convenient for the minutiae extraction. In paper [8-10], the thinning algorithm for fingerprint image is studied deeply in these articles. Paper [11-12] introduces a set of algorithms for extracting minutiae from fingerprint image. Paper [13-17] discuss preprocessing algorithm in fingerprint recognition.

[a] Corresponding author: lishuxue@126.com

The rest of this paper is organized as follows: In section 2, preprocessing of the fingerprint image. In section 3, feature extraction of the fingerprint. In section 4, elimination false of the fingerprints features. In section 5, matching of the fingerprints.

2 Preprocessing of the fingerprint image

The automatic fingerprint recognition system consists of acquisition, preprocessing, feature extraction and matching process fingerprint image, as shown in figure 1.

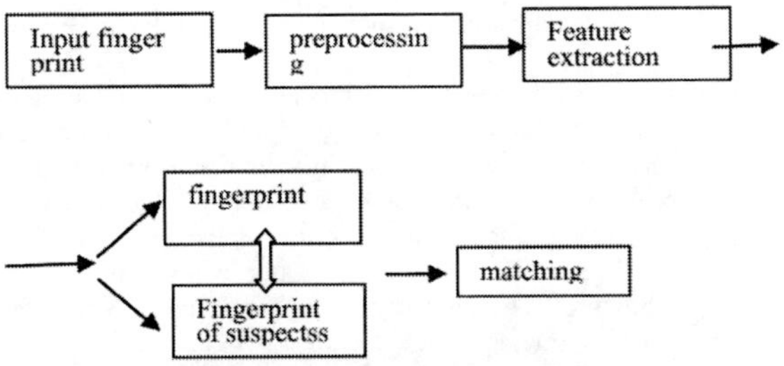

Figure 1. The automatic fingerprint recognition system.

The original fingerprint images are shown in figure2:

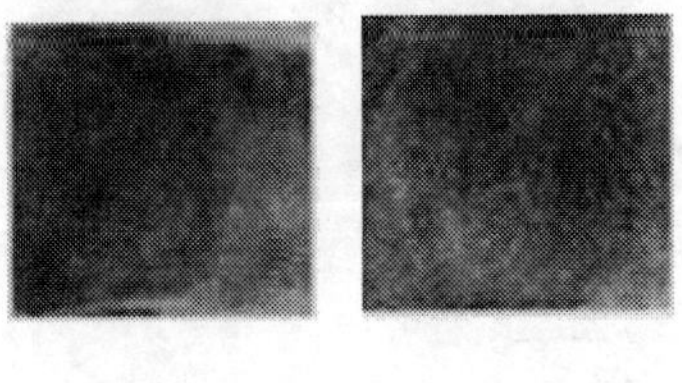

DOI: 10.1051/matecconf/20166304020

f1 f2

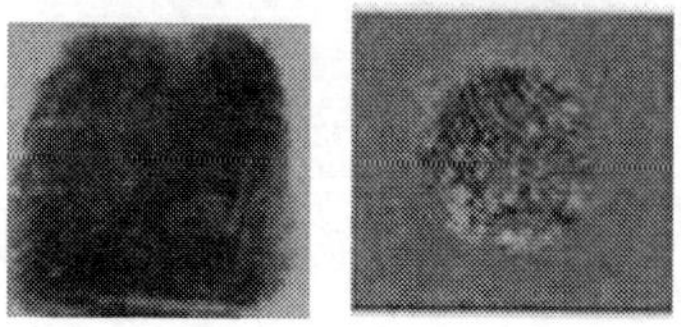

f3 f4

Figure 2. Original fingerprint image.

It is difficult in the extraction of fingerprint in fingerprint recognition system of police, so high demands in the preprocessing of the fingerprint system than in the company's fingerprint attendance system. Image preprocessing is the key in the fingerprint identification, it includes: grayscale transformation, binarization, filtration (noise reduction) and refinement.

(1) Grayscale transformation: using matlab's function rgb2gray, the results are shown in figure 3.

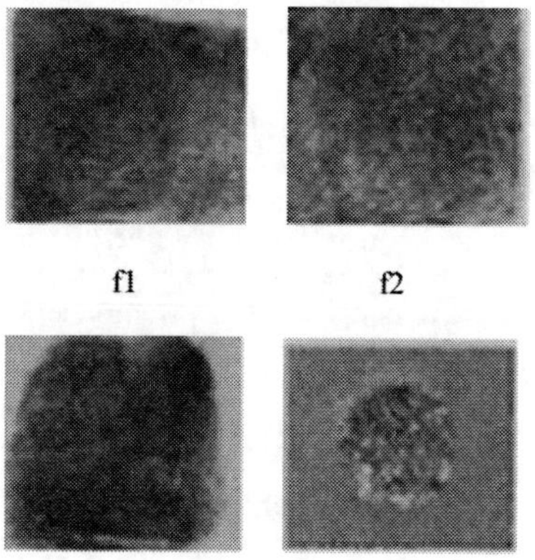

f1 f2

f3 f4

Figure 3. Grayscale of fingerprint image.

(2) The median filtering: median filtering is a pixel grayscale by the adjacent field, output the median pixels (here can also use the median filter). The greater median filtering window, the stronger filtering effected. But with the increase of the window, some of its details will be losed. The (4*4) median filtering is a nonlinear filtering, suitable the nose of granular type and can protect the edge of the image, the results are shown in figure 4.

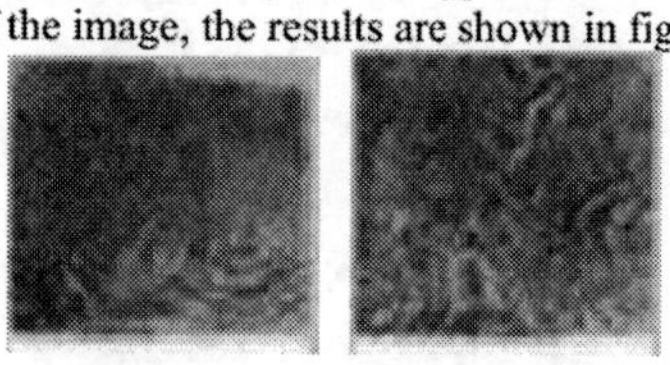

f1 f2

f3 f4

Figure 4. Denoising of fingerprint image.

(3) Binarization: fingerprint image binarization is to compare each pixel, if greater than the threshold value, let the value is 1; if less than the threshold, let the value is 0. After binarization, not only can reduce the amount of data greatly, and the make less interference of the process behind, to simplify the processing greatly, the results are shown in figure 5.

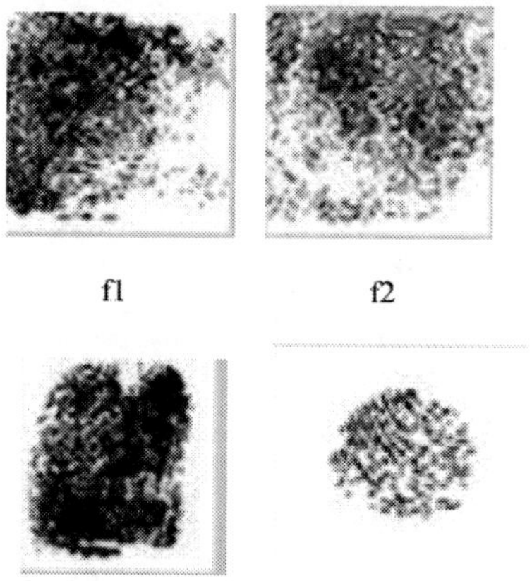

f1 f2

f3 f4

Figure 5. Binarization of fingerprint image.

(4) Thinning: fingerprint thinning is reduced the width of crest to a single pixel, get the skeleton of ridge line image process. This process further reduces the data quantity of fingerprint image, clear the form of the crest line, prepared for the feature extraction of the fingerprint. Using matlab function "bwmorph" carries on the thinning, the results are shown in figure 6:

s11= bwmorph(f13, 'skel',Inf); This statement is to extract the bones of the fingerprint image.

s1= bwmorph(s11, 'spur',8); This statement is to delete the burr extracted from fingerprint image skeleton.

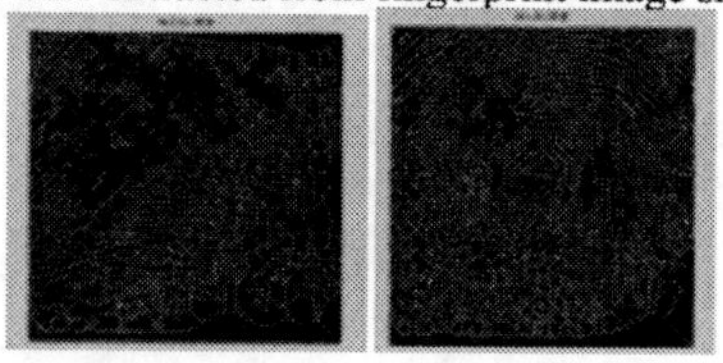

f1 f2

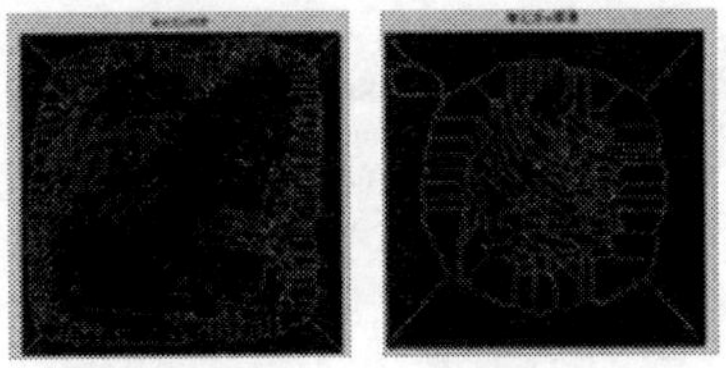

f3 f4

Figure 6. Thinning of fingerprint image.

3 Feature extraction of the fingerprint

Extract the feature points from the elaborated binary image to get the useable image, and then extract the feature point by using the improved image enhancement algorithm, at last matching the thinned fingerprint in the template.

Feature points have six types before the feature point extraction: endpoints, branch points, bifurcation, isolated points, ring and short grain. True and false of the feature points will affect the effect of matching directly, we often do post-processing after feature extraction to eliminate the false feature points, this is what we called the pseudo features. Since there are so many fingerprint characteristics, we only test the feature of bifurcation point and endpoint respectively in this simulation. Considering some information may be missed after the thinning, we use eight neibourhood algorithm which is the template matching method, but we has made great improvement in the process. The eight neibourhood method to extract feature points is shown in figure 7:

N4	N3	N2
N5	(x,y)	N1
N6	N7	N8

Figure 7. Eight neibour table.

After fingerprint image change from the gray image into binary image, take point (x, y) for example, judge it is effective pixels or not based on its around eight points in the neighborhood. If the "0", "1" change six times, then it is the bifurcation point; if change twice, it is endpoint. The main algorithm is described as follows:

```
n=p0+p1+p2+p3+p4+p5+p6+p7
if(n==7|n==5)
  t=t+1;
  x(t)=i;
  y(t)=j;
end
for i=1:t
  plot(y(i), x(i),);
end
```

4 Elimination false of the fingerprints feature

Elimination false is mainly to remove the fingerprint feature points which do not conform to the characteristics. The false features most in the edge of the fingerprint image, especially as we can see from f4 which is after refining and extraction of characteristic value that the edge of the fingerprint image is all pseudo features. According to these features, we put forward a new method of pseudo: first cut the edge of the image directly; then remove the feature points closed based on distance threshold value method. The main algorithm is described as follows:

```
for i=1: t-1
  for j=i+1
    f=sqrt((x(i)-x(j)^2+(y(i)-y(j)^2);
   if d< %Is a distance threshold
       typel(i)=0
     break
   else
    tpel(i)=1;
   end
  end
end
```

5 Matching of the fingerprints

Observe from f4: the most effective information concentrated in the lower right part of the fingerprint image after elimination false features of f4, thus interception the v4 fingerprint feature image using matlab, clockwise rotation 30 degrees respectively and then matching with the three fingerprints, the purpose is to improve the matching rate. The interception of information is shown as follows in Figure 8.

Figure 8. F4 intercept information graph.

By using the similarity of the two matrix, comparing matrix elements with the same position:

```
ht211=imcrop(ht2, [I, j, 9, 99]);
2=length(find(ht211-ht4==0);% two matrix similarity.
```

Stop when two matrix have similarity 99% and timing the used time by matching, when the program met "tic", automatically start timing by Matlab, when run to "toc" automatic computing the time between the last "tic", the matching result is shown in figure 9.

```
Command Window
Elapsed time is 15.148022 seconds.
Elapsed time is 14.989565 seconds.
match with f3
```

Figure 9. Matching.

6 Conclusion

This paper puts forward some different algorithms and Matlab simulations, simplifies the complexity of preprocessing, improves the accuracy of the feature extraction, strengthens the accuracy of matching.

The results show that the algorithm need long time to the mass of fingerprint recognition system, but have a much higher accuracy for the police. Though enhance the image by the algorithm, fingerprint minutiae can be extract accurately. The false feature points which caused by various reasons removed by different algorithms respectively. These algorithms not only do effect the precision of matching, but also improve the speed of fingerprint image matching recognition and performance.

References

1. Chen, Y.,Dass, S.,Ross, A.,&Jain, A. Fingerprint deformation models using minutiae locations and orientations.Appeared in Proc. of IEEE Workshop on Applications of Computer Vision (WACV), (Breckenridge, Colorado), 150-156.(2005)
2. Ji,C., Feng, W., Li, M., &Yang, J. Dynamic Threshold with Hole Padding Algorithm for Fingerprint Image BinariZation.Computer Simulation, 28(7), 258-261. (2011)
3. Mayur D Jain, Nalin Pradeep S, Prakash C and Balasubramanian Raman, Binary tree based linear time fingerprint matching.309-312.(2006)
4. Liu,Y.,Cao, J., Xu,Z.,Tian,Y.,Fu, T., &Wang,F., Improvement of image matching algorithm based on gray correlation.Journal of Applied Optics. 28(5), 536-540.(2007)
5. Zhao,J., &Wang, D.The implementation of a fingerprint matching algorithm. Computer Engineering and Applications,13, 66-69. (2005)
6. Wang,J.,Chen,S., &Sun,S., Realization of the fingerprint image enhancement based on matlab. Electronic Science and Technology. 6,37-39.(2009).
7. Wang,Y., &Su,C., Study of fingerprint image enhancement algorithm. science technology and Engineering,10(1),94-98.(2010)
8. Wang,Y.,Ning,X., &Yin,Y., Study on the fingerprint image thinning algorithm. Journal of Nanjing University (Natural Sciences), 39(4), 468-475.(2003)
9. Xiao,X., Wang,K., &Li,Z., mproved thinning method to fingerprint image. Computer Applications, 28(2), 466-468.(2008)
10. Yang,W., Guo, K., &Wei,Y., An efficient index thinning algorithm of fingerprint image based on eight neighborhood point. Journal of Sichuan University of Science& Engineering (Natural Science Edition), 21(2), 61-63.(2008)
11. Guo,J.,Wu,Q., &Shang,Q., Minutiae extraction of fingerprint image based on matlab. Computer Simulation,24(1),182-185.(2007)
12. Liao,K.,Zhang,X., Zhang,M.,&Pan X, Extraction of minutiae and significant feature from fingerprint image. Computer Applications, 28(9), 2312-2314.(2008).
13. Shang,Q., Wu, Q., &Guo,J. Preprocessing algorithm of fingerprint image based on matlab. Computer Simulation, 24(3), 215-218.(2007)
14. Tai,Y. Fingerprint processing algorithm based on matlab. Journal of Southwest University for nationalities (Natural Science Edition), 34(4), 836-838.(2008).
15. Wu,J.,Xu,C.,Yang,K.,&Zhang,G., Study of essential technology for fingerprint image Pre-processing.Computer Engineering and Applications, 44(3), 223-239.(2008)
16. Xue,J.,Wang,S., &Liu,Z., Adaptive pre-processing for fingerprint image. Computer Engineering and Design, 29(1), 157-159.(2008)
17. Ye,Q. Study of a fingerprint pre-processing and matching algorithm. Journal of North China University of Technology Beijing China, 24(3) , 6-10.(2012).
18. Wang Guwen, Cao Yu MATLAB6. 5 graphics image processing [M]. Beijing: national defence industry press, (2004
19. RuanQiuqi. Digital image processing [M]. Beijing: electronic industry press, (2007)

MATEC Web of Conferences **63**, 04021 (2016) DOI: 10.1051/matecconf/20166304021
MMME 2016

Exploration of the Modality of Artistic and Scientific Achievements

Zhuo JIN
Bozhou University, Bozhou, 236800, Anhui

Abstrect:Eyes are the windows to the soul. In all ages we are convinced that the most authentic and beautiful sights are in our eyes. This, however, is just like an indiscernible legal provision that confines the development of art all the time. Artistic vision may conflict with visual impression, so vision science for artistic creation is of the myriads of changes. Vision is in fact a reflection of a physical phenomenon, and "light" is the base for us to see the diverse world. Exploring vision theories and intentionally using them in artistic creation can be a good attempt. Vision advantages and characteristics will be effectively reflected when applying the vision theories to graphic design as well as to indoor and outdoor designs. This article will make an exploration of the vision science from several perspectives of artistic creation, with focusing on the relation between art and science.

Keywords: Artistic creation; Science of art; Achievement modality; Exploration and analysis

1. Science vs. Art

The reflection of "light" into eyes is the reason why human beings are visually sensible to surroundings. But, eyes will sometimes cause inauthentic feelings, and visual illusions may be generated in both dynamic and static observations, which, however, is the spring of art. Over a long period of observations, Leonardo da Vinci discovered and summarized the perspective principle. This discovery is a breakthrough to the superficial understanding of image-forming principle and has greatly raised the level of the beauty of paintings. Perspective is in essence a scientific physical phenomenon. An organic combination between perspective and art can not only be a great innovation for art but also bring new understandings about sculptures, paintings and even about beauty-appreciation. It is therefore necessary to combine science with art for creations.

1.1 Relation between Science and Art

1.1.1 Seemingly Related or Essentially Different

Some would think that science and art are two complete different fields and even that they have nothing in common for description. Science is subdivided into natural science and social science. Natural science includes mathematics, physics, chemistry, geography, astronomy and biology, while social science is mainly related to literature, history and philosophy, etc. For art, it is not logic and provable. It is an unrestrained creation that brims with fanciful things. From another angle, of course, art is related to science. For example: geometric knowledge is required in paintings; the image-forming principle in physics is applicable to photography, and; multiple sciences including mechanics, materials and aesthetics are useful for buildings. Different scholars have different views, and no consensus has been reached by far. However, art and science are developing all the same. As pointed out by Professor Jiang Xiaoyuan, more evidences are required for the argument of "science is related to art" or "science is not related to art". Although it is difficult to prove this argument, there are still evidences. In a word, science is seemingly related to art but they are essentially different.

1.1.2 Binary Opposition vs. Common Descent

The history of science is a system that reflects the objective laws of development and change of things including nature, society and thinking. It is the knowledge covering general laws, as obtained through logical deduction and full observation to describe and explain the objective world. Art contains different expressions of objects, images, voices and acts, in which, aesthetics is carried in creations by subjective factors such as imagination, skill and experience. In theory, therefore, people have often subjectively and aesthetically differentiated art from science, and deemed that they belong two unlinked aspects. For this reason, Caroline A. Jones and Peter Galison proposed the point view of binary opposition: intuition/analysis, induction/deduction, vision/logic, random/system, autonomy/collaboration, male/female and creation/discovery. However, this binary opposition has been soon challenged and questioned. From this point of view, the object of vision science shall be related to scientific images, with artistic paintings being excluded. But the fact is that scientists cannot plot a definite boundary to explain what is a scientific image and what is an artistic painting. Their difference is not that clear, which means that artists may collaborate closely with scientists in some fields. Seen from today, both science and art had enjoyed a rapid development during renaissance. Their common historical characteristics make up the indispensable historical background of vision science, which means the pursuit of rationality and wisdom. During renaissance, many artists improved their abilities of precisely painting objects through scientific observations and researches. Likewise, the artistic culture of scientists is also a favorable condition for doing scientific researches. The two are closely bound up that they cannot be clearly distinguished from each other.

1.2 Concept of Vision Science

In new fields about the history of science, it seems that deeper evidences are available to prove the relevance between science and art. Since 1980s, in-depth studies on the history of science have been performed in both the science circle and the art circle, as the vision culture has drawn increasing attention from the world. In 1990s, a new research branch, say, the history of vision science, appeared

The history of vision science involves many subjects, including science and its history, art and its history, history and historiography. In its true sense, the history of vision science is interdisciplinary research achievement. With changes in methods of researching topics, it will involve many other fields such as philosophy, art psychology, cognitive psychology, iconography, semiology, anthropology, linguistics, cultural study and even the history of religion. In the history of vision science, art and science have been historically combined along the direction of vision art.

2. Relation between Art and Vision Science

2.1 Photography vs. Vision Science

Photography has experienced a tortuous development of more than 170 years since it was born in 1839, and gradually, the commercial photography branched out. Nowadays, the art of commercial photography has permeated almost every facet of our lives. It attracts people's eyes by taking a candid photograph. Its fancy visual effect is not forgettable, which achieves the goal of promoting consumption and improving popularity. By advantages of unconstrained imaginary and insightful observations, commercial photographers have created a great number of photography works, providing the audience with a grand visual feast.

2.1.1 Vision Modality of Commercial Photography

Commercial photography is a part of photography. Its most prominent feature is that it is commercial. Unlike other photography types, commercial photography aims not only to capture an instant impression but also to present the whole product information to consumers. For different consumers, commercial photography would show product features from different perspectives. Therefore, it has been favored by more and more people. When taking a photograph, the photographer should first find out the connotative information of the commodity and stick to the commodity itself to design a satisfactory work by using photographic equipment. Then, under the technical level of the photographer, the satisfactory business photo can be finished. In recent years, people have spent a lot of energy on vision modality, which is related not only to research of "vision culture" but also to the extension of the culture of "visuality".

A uniqueness of vision modality presented by commercial photography is to compose a picture through visual elements. There is a variety of lines in nature, which dominates the composition of the picture and supports the framework of the structural image to extend from two-dimensional plane to three-dimensional space, so that deep meaning can be expressed from different vision perspectives and viewing distances. The center of a commercial picture is requisite to convey the visual sense, so as to achieve a clear and entire image expression as well as to deepen the sense of presence of the commodity, which provides the audience with a perfect visual enjoyment. In a commercial photography process, conflicts and combination of elements make the vision extending from one to another, and the visual image also becomes more smooth and natural. In addition, commodity features can be highlighted by concise words that match well with the theme. In this way, all desired elements are contained in the photography work. The purpose of a photography work is to highlight the theme with ignoring some unimportant episodes to enhance the inherent expressivity.

Visual image means the combination of all perceptible social formations in nature. It includes objective reflections of "shape" and "state". "Shape" represents the form of object presence, while "state" represents the internal development mode within a range that involves all spatial relationships and visual relationships. As the integration of " shape" and "state", modality put more highlights on existential state as well as on dynamic sense. By combining image with culture, vision modality is of multiple connotations such as thought, culture and aesthetics.

Unlike conventional photography, modern photography puts emphasis on psychology perception of the "vision" part of the visual image. It is equivalent to a psychological image perception. After being perceived by optic nerve cells, the information will generate a visual image in the brain. Vision can be either a simple perceptual mode or a cumulative feeling based on experience. Collocation of different shapes and colors gives people with different feelings. So we say, "Colors have temperature and emotion". Optical illusion" means the perception of the senses of taste, touch, pain and smell when observing an object under different light conditions. The texture, shape and state of an object can also bring us different feelings. These wonderful and colorful feelings come from our lives as well as from cultural deposit, making our subjectivity diverse and graceful.

2.1.2 Vision Language of Commercial Photography

To understand the commercial language under the background of art, we need to study the creative thoughts of the vision language from different angles, and extend both external and internal aspects of the language. In commercial photography, the application of new concepts, such as structuralism, Bauhaus and surrealism, has broadened the space for its development. Unique features of the vision language of commercial photography can be found by making analysis from different angles as well as by comparing this language with other classes of art.

All artistic expressions have their own language systems. Because of the relative short period of development, the language system of commercial photography still needs to be further improved. Language is a scientific system. Language develops at its own rhythms. Therefore, scientific methods are also required to study the system language of commercial language. During the evolution of the style of commercial photography' s vision language system, factors including color, light, shooting skill and lens have their own attributes. For focal length lens, there are normal lens, zoom lens, telephoto lens, wide-angle lens and micro-lens for selection. Different lenses have different attributes that are applicable to different objects and themes. The visual effect of the normal lens is the most similar to that of human eyes. Therefore, the images shot by normal lens are more real and natural. However, it is exact the too real sense that cannot produce a strong visual impact with special effect. The wide-angle lens can enlarge the viewing angle, broaden the depth of field and enhance the sense of perspective, so it can achieve a strong perspective sense at close-up shots to attract people' s attention. The telephoto lens that has small perspective for large formation of image is suitable for shooting, for example, natural landscapes at a long distance. The micro-lens is very useful for tiny objects such as insects and flowers. Commercial photography is just like the communication between people with different accent and personalities: some are frank and some are euphemistic. All the different languages composes unique images, which is exact the science aspect of the vision art.

2.2 Three-dimensional Paintings vs. Vision Science

In Cubist artwork, objects are analyzed, broken up and reassembled to express modality and emotion with the two-dimensional space plus the time dimension. It is a new attempt to creatively use the cubism for photography, in which, some new shooting principles and methods are proposed to create more unique artworks.

Cubism has three expressions--broken objects, planarizartion and blanking. Conventional image composition is based on the completeness and integrity of the object, while the cubism breaks the mode by decomposing the object into 2D pieces and then recomposing them into a broken object. In Cubist artwork, geometrical shapes are employed to organize the object. These geometrical element themselves do not have 3D illusions. Different lateral characteristics are arranged on a plane into a 2D flat geometrical shape, which, however, contains many information of the original subject. For the purpose of changing the conventional idea that puts the subject of the focus of the image, cubism blanks the subject or depicts several subjects to eliminate the optical center, which breaks the sense of wholeness of the image. Though this seems unique and even inconceivable, a different kind of style can be presented and applied even to buildings.

Cubism is people-oriented. But for architectural photography, modern architecture shall be the subject. Buildings are not suitable for cubism because they consist of only a few elements. But, if the top, the body and the bottom of a building are decomposed and recomposed again, a soft and gentle effect without strong visual impact can therefore be achieved, which can avoid the influence on the expression of the subject. Buildings are three-dimensional. If these three dimensions are expressed on a plane, then a cubism style can be attained with using special shooting and post-processing skills. If one desires to weaken the sense of the three-dimensional space, he shall take the shot under a light with diffuse reflection. For instance, in a cloudy day the shadow of a building would become blurry, which would generated the desired effect. It thus can be conclude that a 3D object can have the 2D sense in visual effect, in which, science is indispensable to achieve the effect we want.

2.3 Buildings vs. Vision Science

2.3.1 Style Design of Building Vision

Style design of building vision is an important aspect of building design. As a branch of design, style design of building vision includes: building structure design, color design, culture design and material design, etc. In practice, the style design focuses on tiny changes of point, line to plane. Those tiny changes are then integrated into a big one on the whole level. Concrete, as one of the essential materials for architectural projects, is also a good carrier for art. Its rough and pristine nature can be used for the "ruggedness" effect of the building. Besides, an intentional roughening during construction can better express the rhythm sense of the building. In vision art, "point" is the smallest unit. It has a centripetal force and is the most imprecise shape. A point is usually used to represents a location that has neither length nor width. Some parts are as abstract as points, so as to grasp the overall image. In the innovation of architectural style elements, inspirations consistently emerges, further to create products abiding by the characteristics of materials.

2.3.2 Super-vision Modality

The design research scope of super-vision modality mainly covers macroscopic super-vision modality, microscopic super-vision modality and other super-vision modalities. By making use of some modern technical apparatus, such as the astronomical telescope, spacecraft, optical microscope, electron microscope and others, all kinds of super-vision modalities could be observed clearly. The advanced science and super-vision modality are closely connected in terms of the application of architectural design. Nowadays, along with the rapid pace in science and technology, particularly the greater development during synthesizing and applying new material, it is no longer a problem with regard to satisfy any model in architectural design by means of applying Nanotechnology material. The development of super-vision modality becomes possible and gets mature gradually.

To sum up, the science and art are supplementing each other and inalienable to some extent. They are blended into an inter-discipline, which contains both the sensibility in art and the rigorousness in science. The artistic design and creation will open up new development direction and even the new development genre may emerge in future. At present, the artistic creation from photography, architecture and other aspects has obtained numerous achievements and innovation. The rise of various kinds of unique designs of advertising boards and buildings indicates the potential value about artistic creation in scientific research. After all the vision imaging and optical image-forming principle in physics are inseparable. During the photographic process, pictures with peculiar fairyland have been taken by utilizing different optical skills. Various kinds of artworks with peculiar shapes have been constructed with scientific material. Thus the artistic works with vivid and depth of field effect could be produced. In case of dividing the science into reason, dividing art into sensibility blindly, it will be very hard to break through and innovate. To explore the new mode for the co-existence of both will be bound to cause a great revolution between science and art. People' s most direct cognition towards the world is visual cognition, which brings more reality and control. However, usually the visual art is illusory or reaching the mysterious effect by taking advantage of illusion. Under the scientific guidance, the artistic creation will expand more dimensions for observing the world and bring new development direction for artistic creation. Behind the aesthetics there will always be visual scientific principle. Actually there are scientific reasons when we have discovered some phenomena and rules. In most cases, new inventions and creations are unexpected during unintentional creation. The art will open our fields of vision, which will not be confined into a vicious cycle and irresistible. While we are standing at the world boundary, all problems will be readily solved.

References

[1] Finite Element Analysis and Design Software of the Structural Space in Multi-layer and High-layer Building (Wall Element Model) User Manual and Technical Conditions.[j]. China Academy of Building Research PKPM CAD Engineering Department, October of 2008.

[2] Hu Wangshe, Dailei and Lintao. Indoor Modeling Elements---Floor Design and Originality.[j]. Chongqing Architecture, The 09 Phase in 2009.

[3] Translated by Li Zhicong and others. Teaching Materials from New York Institute of Photography in America, Second Volume. [j]. Wrote by New York Institute of Photography in America, China Photography Press, 2000.5.

[4] Rudolph Arnheim, Translated by Tengshou.

Visual Thinking: Intuitive Psychology in Aesthetics.[j]. Chengdu: Sichuan Publishing Group, People's Publishing House in Sichuan, 1998.

[5] Heinz Paetzold, Translated by Gao Jianping. Symbol Culture and City: Five Topics about Cultural Criticism Philosophy.[j]. Sichuan Publishing Group, People's Publishing House in Sichuan, 2008.

[6] Kent C Broome, Charles W Moore, Translated by Cheng Zhaohui. Body Memory

MATEC Web of Conferences **63**, 04022 (2016)
MMME 2016
DOI: 10.1051/matecconf/20166304022

The Influence of the Intangible Cultural Heritage About propagation medium in the Big data Era

Zhuo JIN
Bozhou University, Bozhou, 236800, Anhui

Abstrect: Film and television works have been attracting the attention of the modern people with their own characteristics. Image, vivid, intuitive, popular, the media uses its own advantage to spread out the information quickly, the transmission efficiency and the promotion is unmatched by other any medias, it also plays an important role for the dissemination of intangible cultural heritage. With our country pay more attention on protecting the intangible cultural heritage, digital media can use its own advantages to protect, promote, promote the intangible cultural heritage, to visual intangible cultural heritage, and present it to the public, in order to let the public understand what is intangible cultural heritage, take the initiative to protect the intangible cultural heritage. This paper discusses the influence of digital media on protecting intangible cultural heritage, with the power of digital media, we expect to protect it better, leave a valuable legacy for future generations.

Keywords: Big Data Era; Media; intangible cultural heritage; inheritance and protection

The present information era told the developed society that the "intangible cultural heritage" is submerged gradually and even has been withered away. Therefore, how to protect the "intangible cultural heritage" by utilizing the modern technological means becomes the issue that people explore. The film and television medium shall blend the "intangible cultural heritage" into the current background of social culture positively and present the "intangible cultural heritage" in front of people with the visible effect, making it no longer distant to people. By virtue of the unique intangible cultural heritage in our country, the film and television works could be the good ones with Chinese characteristics. Then by taking advantage of the transmission from the film and television media, the intangible cultural heritage could be carried forward. Next I will elaborate the influence to intangible cultural heritage by medium in the data era from several aspects.

1. Concept and Background of the Intangible Cultural Heritage

1.1 Concept of the Intangible Cultural Heritage

The intangible cultural heritage refers to all kinds of social practice, opinion expression, patterns of manifestation, knowledge, skills, as well as the tools, material objects, handiworks and cultural space related that are regarded as the intangible cultural heritage by communities, groups or individuals. Mostly it is manifested by relying on verbal convey, performance, customs and other forms, but with free and diversified modes. In order to accommodate the nature and social environment, people show their creativity according to the demand of group sense of identity, and thus the goal of living in harmony with the nature and social environment could be reached. Consequently the intangible cultural heritage is a kind of great social practice.

1.2 Characteristics of Intangible Cultural Heritage

Possessing its own unique charm, the intangible cultural heritage not only has shown the territoriality, national character and uniqueness, but also is characterized by group, inheritance and so on. All of these characteristics allow the intangible cultural heritage to become a kind of cultural form that is different from other culture.

1) Territoriality: due to the geographical environment, natural resources, living styles and thinking habits in every territory are widely different, the group in each territory has the unique living state, so the full-bodied territorial characteristics to intangible cultural heritage are produced, namely "the living atmosphere is different within hundred miles and the customs are different within thousand miles".
2) National character: being influenced by geographical environment, the costume, etiquettes, customs and others in different nations are greatly differed. All of these differences allow the different national culture to be divided obviously. In the meantime, these cultural symbols display the national differences in terms of thinking characteristics, world view and religious belief.
3) Uniqueness: it is established based on the unification of territory and nation. It makes the intangible cultural heritage express in the artistic and cultural form. The diverse cultural patterns, thinking modes, aesthetic tastes are revealed in line with the regional, time and national differences. Meanwhile, the inimitability and non-renewability in it have been shown.
4) Inheritance: the intangible cultural heritage relies on horizontal and longitudinal ways. The former refers to the propagation among the same generations, and the latter refers to the transmission from the last generation to the next generation. Not only the cultural genes, spirit, emotion and ways of production are kept, the skills, technologies and knowledge are saved and promoted under the unceasing repetition and circulation, and thus the pluralistic value systems are embodied.

1.3 Urgency to Protect the Intangible Cultural Heritage

In 2006, the State Council released the Notice in Regard to Publish the First Batch of National-level Intangible Cultural Heritage by the State Council, showing the importance paid to intangible cultural heritage by the country.

The main reason for protecting the intangible cultural heritage is that people have started to recognize its value and significance to the entire world. The world is diversified, so does the nation, all of which are concluded to the diversity of the culture. Each kind of culture is conveying the value, aesthetic and thinking mode of its own nation. Once a nation loses the peculiar nature in its culture, it has already been disappeared fundamentally. However, along with the fast-paced time, people's living space is shrinking gradually. Young people would rather accept new culture and they are indifferent to the traditional culture. The protection and inheritance of intangible cultural heritage has suffered extreme obstruction and is facing the extinction danger. The inheritance chain is in the process of breaking. Hence, it is imperative to protect the intangible cultural heritage. But the traditional ways of protection and inheritance spend vast manpower and material resources, leading some of the intangible cultural heritage lay aside. Under the big data era, to convert the intangible cultural inheritance into digit and to store in the computer by making use of multiple propagation media is not only fast, but also could save permanently. So it is worthy of exploration and research.

DOI: 10.1051/matecconf/20166304022

2. Propagation Characteristics to Intangible Cultural Heritage by Television Medium

2.1 Image and Intuition of the Expression Force

The television medium is oriented by visual effect and could bring super shocked visual impact by means of the vivid and bright images to audience. The artistic appeal from other traditional media, paper media and the media with single sound could not match with it. Particularly for its vividness, it allows the audience to be personally on the scene. For instance, the film and television work Ashima, taking love as the clue, it reveals the national customs and scenes in Yunnan, and meanwhile singing in antiphonal style, the Torch Festival and other customs are inserted into the story naturally. By the time of appreciating the story, the audience is able to understand the customs and humanity in Yunnan, rather than to be preached blindly. The dry and dull feeling could be avoided, and the effect of yielding twice the result with half the effort could be played under unconscious influence.

2.2 Extensiveness of Social Influence

Along with the growth in the living standard, the television, computer and other digital products are widely circulated among the public gradually, and appreciating movie and television play has become one of the entertainment ways among the young people. The visual culture plays subtle role in establishing the public cultural consciousness. Undoubtedly its advantage is self-evident. By the time of appreciating movie and television plays, people could recognize the huge value of the intangible cultural heritage. The public's protection and inheritance consciousness could be motivated, thus more and more people could participate in the corresponding work.

2.3 Value in Developing the Culture Industry

To protect the intangible cultural heritage shall be oriented by protection and reasonable development. Protection-oriented refers to strive for saving the intangible cultural heritage and preventing it from extinction. In the meantime, in order to promote the vitality of the intangible cultural heritage and further raise its value, and thus to protect and inherit in a better way, the development work shall be emphasized. In addition, after our country entered into WTO, the exchange among countries got close with each passing day, foreign countries had entered into the industry in every country. In case we could not be the first ones to develop the intangible cultural heritage in our country, other countries may race to develop. During developing the culture industry, the movie and television medium plays significant role. For example, after the film Liu Sanjie was broadcast, tourists from the world travelled to Guiyang, Yangshuo and other places for fame, development in local tourist industry had been driven and the industrialization of intangible cultural heritage had been promoted. Let's take another example, the first intangible cultural heritage film Love On Gallerv Bridge in our country, not only the "Beilu Drama" had been spread, the territorial and folk customs in Eastern Fujian region had been reflected, the protagonist's spirit in sticking and carrying forward culture had been delivered, making huge contribution for protecting and inheriting the intangible cultural heritage.

3. Current Status and Influence to the Protection of Intangible Cultural Heritage by Media and Medium

3.1 Positive Protection

3.1.1 Visualized Spread---Movie and Television Documentary

The development of digital technology under the big data has changed the ways in protecting and inheriting intangible cultural heritage. Meanwhile, by means of being packaged by fashionable media, the intangible cultural heritage could be kept the original lasting appeal and possessed fashionable appearance, which is greatly attractive to the young people. At the same time, the information receiver could receive and spread to the next audience via the Internet, the propagation scope could be enlarged constantly. The intangible cultural heritage just makes use of the timeliness of media and image to post message and makes adjustment according to the feedback. At this moment, many approaches could be taken as the carrier, e.g. E-mail, bbs, qq, MSN and other tools, whereas normally the recording forms are movie and documentary.

The protection of intangible cultural heritage by movie mainly lies in propagation, publication, popularization, promotion, record and storage. Due the national character and territoriality, the intangible cultural heritage is scattered in distant places and difficult to be collected. The characters pictures could be saved and recorded on certain basis. However, the static state is its boundedness. The movie could be revealed comprehensively by virtue of three-dimensional and dynamic state, voice, images and characters. Perhaps some customs or behaviors are hard to be comprehended by ordinary people, they are endowed with the new significance that is easy to be comprehended through the annotation and unscramble by the new generation. The movie has powerful cultural penetration force and it could convey the visual perception by crossing time and racial limitation. The vision has crossed different cultural levels and background gaps, making the public accept, probably there will be different levels of comprehension, but this is enough. The "liberal education" demands this effect. Pictures and vision are the global language that doesn't need translation. Movie could span the national boundaries and spread to further places, popularize to tens of thousands of families. The government and society are now endeavoring to build a favorable protective atmosphere. Besides movie, which is for watching, the protection of intangible cultural heritage has another function, namely data storage. It could store the relevant data in intangible cultural heritage safely and permanently, and also show the huge advantage.

The intangible cultural heritage is dynamic and developed. The departure of a folk artisan is likely to become a mark when a work of art vanishes. For this reason, the documentary is required to record all of the information in an authentic way. Compared to characters and pictures, to record an event without subjectivity, information loss or disguise is more visual. The documentary does not regard the characters as the intermediary agent. It takes the frames as the load completely and presents the original appearance in life, so that the information could not be lost. People's perception degree towards pictures is more sensitive than that towards characters. They seem to be personally on the scene, and their stronger sense of self-identity allows the acceptance degrees to be greatly promoted. Actually the originators to the diversified movies such as stories, cops and robbers, action, scientific fiction and other categories are documentary. Concerning to be accessed to natural materials, documentary is concluded to be the most objective and authentic one. Once it was referred to as "direct film", "non-fiction film", which has explained a principle in documentary intuitively, that is, the realistic intuitive recording. By virtue of direct image, the documentary has moved many people, and it particularly renders the intangible cultural heritage truly. The intangible cultural heritage has its primitive, simple and toughed places under the historical accumulation. It could reach good effects without deliberate embellishment or rendering. This kind of data recording is more reliable and important than movie. Just like there is no rehearsal in life, once some fragments are not recorded in time, the remedy methods are unavailable. Hence, documentary is the material that is hard to get, and it is also the way of protecting the intangible cultural heritage worthy of attention.

3.1.2 Stage-style Propagation

Foreigners from every country in the world are fond of the national culture to be conveyed in the way of artistic performance on the stage within the transitory period. During the mutual clash and exchange with other countries, a branch in Chinese intangible cultural heritage has blended itself into the world culture. It is natural that the medium under new media cannot do without the stage acted as the carrier. For instance, so far people could still remember the opening ceremony for Peking Olympic Games designed by Zhang Yimou in 2008, calligraphy, beating Fou, Chinese zither, dance, Tai Ji, Bejing Opera, Silk Road and other intangible cultural heritage had been displayed one by one. Meanwhile, by utilizing the modern elements like colored lantern and other digital devices, Zhang displayed the obsolete and classic intangible cultural heritage in a bright way. It made people's eyes brightened. The long scroll that was more and more colorful had been drawn. The freehand spirit in Chinese aesthetics had been revealed incisively and vividly. This example has proved well that the digital era does not mean the disappearance of the old intangible cultural heritage, better development under fusion should be the mainstream. Just because people have wrong thoughts all this time, the loss of intangible cultural heritage is caused at present.

3.1.3 Entertainment Propagation

Along with the popularization of fast-food culture, people regard entertainment culture as the vulgar and superficial box news. However, the truth is not just the case. Game, music, movie, television and other modes are a kind of entertainment, they have turned dry propagation from mouth to mouth into entertainment programs, and thus culture is spread unconsciously. For example, injecting intangible cultural heritage into game is a kind of popularization towards children and young people. During the game process, prop decoration, figure image, scene reappearance and other ways make the intangible cultural heritage no longer keep away from cities and crowd. There are even some games take creating the world by Pangu, removing mountain by Yugong as the background resources during the entire game, making children experience historical culture during entertainment.

3.2 Negative Influence and Existing Problems

3.2.1 Distortion

Nowadays, with the commercialization, some businesses weaken the "culture" in the name of intangible cultural heritage and enlarge the "artistry" infinitely. As a result, the visual feast has been brought to the audience and high box office has been gained, but the popular psychology has been misled. For those costume dramas that are indifferent to the etiquettes and norms in traditional culture and only paying attention to the plots are a good counter-example. On account of such films are short of historical and cultural accumulation, even they are popular within certain period, they cannot withstand the time experience and will be forgot by the audience quickly. Since the shocking goal cannot be reached, they could not carry forward the intangible cultural heritage.

3.2.2 Lacking Quality

Under the impact by digital era, the fast-pace rhythm has largely lowered the quality of many film and television works. A Dream in Red Mansions of version 87 took 3 years from selecting actors, teaching relevant knowledge to accomplishment, and its preliminary preparation work was longer. Although it could not match with the present make-up skill, definition, music configuration and so on, its public praise is far more than several works with the same name reproduced afterwards. This has sufficiently explained that the destiny of "paper plane" will never be long. In case the problem that only attaching importance to its superficial thing but without understanding the essence in intangible cultural heritage cannot be solved, the loss root will be seriously affected.

4. Propagation Mode and Planning to New Media and Medium under Digitalization

The digitized saving of the intangible cultural heritage under new media is a course that collects, saves, processes and propagates information. By utilizing three-dimensional image, three-dimensional model and three-dimensional animation, the intangible cultural heritage is recorded via taking the video, music and image as the carrier, meanwhile, archives is established. In this way, the intangible cultural heritage is protected systematically and comprehensively. In addition, the abundant cultural resources and concepts contained in it are dynamic and changing. Right now the mode of digitalized management is: traditional database does centralized management, shares the resources from other units and sets unit to manage as per requirements, all levels of cultural media enterprises, academic institutions and cultural departments build up multi-unit co-share bi-directional construction and propagation mode. The three-level service network system, namely the national center, provincial center and primary center has been established, the national center takes charge of putting the collecting resources, as well the resources

DOI: 10.1051/matecconf/20166304022

provided or collected by the province into the store, and also provides convenient and fast network resource service for each branch-center.

The technical support from new media allows the intangible cultural heritage to be inherited, but at the same time new opportunities and challenges are faced. A kind of measure that makes the intangible cultural heritage still develop in modern society is required, as far as the folk art is concerned, the digitalized intangible cultural heritage has abandoned the previous workshop mode and become the complete industry chain mode that enlarges scale gradually. As far as the development of folk art is concerned, this is a kind of favorable protecting way, which blends the intangible cultural heritage and new media, medium, and also it could make progress during saving and development. More development prospects will be got from creating old culture, and the breakthrough evolvement will be obtained by the cultural economy. Only by the propagation from all kinds of media and medium, the intangible cultural heritage could neither be vanished from this world nor become people's memory. The vivid depiction by image covers writing, portray, graving and other artistic means, making the intangible cultural heritage always surround people.

5. Conclusions

The society under big data era is a fast-paced and high amount of information age. However, people who live in this society forget the previous traditional culture gradually, and they even think it is the past that should be forgot. The history that has been selected essence and dropped dross still deserves to be existed. It is the sedimentation of 5000 years' historic culture in China and could not be discarded optionally. Because it represents the wisdom and glory of the persons of the same category in the past, it is our obligation to protect it. Under the perfect fusion between new science and technology, media and medium, after being blended the modern elements, the intangible cultural heritage will develop towards better direction. By the time of satisfying the popular psychology, the popularization towards intangible cultural heritage could be accomplished. It becomes the protecting responsibility for Chinese people. The intangible cultural heritage will be carried forward under the generation's common protection.

References

[1] He Xingliang. Protection of Intangible Cultural Heritage and Modernization of National Culture [j]. Yangyuan, He Xingliang. Research on National Costume and Cultural Heritage----China Ethnology Academy 2004 Annual Meeting Memoir. Kunming: Yunnan Press, 2005:360.

[2] Chen Jianxian. Cultural Innovation and Parent Reconstruction----Discussion of Functional Integration of the Intangible Cultural Heritage in Modern Society. [j]. Folk Culture Forum, 2006(04).

[3] Wangcan, Shizhimu. Function and Influence to Historical Propagation by Documentary.[j]. Film Literature, 2011(12).

[4] Gong Panpan, Enlightenment on Protecting and Developing "Intangible Cultural Heritage" by Utilizing Film and Television Means. [j]. Hubei Institute For Nationalities Journal: Philosophy and Social Sciences Version, 2010(06).

[5] Li Huailiang, New Media: Competition, Cooperation and Win-win.[j]. Communication University of China Press, 2009.

DOI: 10.1051/matecconf/20166304023

Coupling Model and Controller Design for Four-layer Register System

Jian LI[1,a], Shan Hui LIU[2] and Liu Jin CAI[1]

[1] *School of Mechatronics Engineering, Henan University of Science and Technology, No.48, Xiyuan Road, Luoyang, Henan, 471003, China*
[2] *School of Printing and Packaging Engineering, Xi'an University of Technology, No.5 South Jinhua Road Xi'an, Shanxi, 710048, China*

Abstract. A nonlinear and coupling model is established according to the multi-layer register system working principle in gravure printed electronic equipments, and then, the linear model of the register system is constructed based on one order Taylor formula. In order to improve the speed and accuracy of register control, a feedforward PID control algorithm is presented according to the linear model in multi-layer register system. This algorithm is unique in that it uses the PID to adjust the inputs of the register system, and feedforward compensation for the known disturbance, making the register control accuracy has been greatly improved. The simulation shows that the proposed register controller is able to realize a high precision control for the register system and endowed with better control performance than PID controller in register control.

1 Introduction

Just as the register process in traditional color printing, all parts of printed electronics can be precipitated by ink formation layer and layer correspondent to electronic materials through the method of gravure printing, and the quality of printed electronics is controlled by each layer of register accuracy. Register error, reflecting the direct index of register accuracy, comprises print-direction register and cross-direction register errors which dominated by the factors of each roller speed, web tension, web property, manufacturing and fixing errors. What's more, the cross-direction register is the key point of register system. Therefore, the thesis will discuss about it in terms of the cross-direction register, without the condition of explanation.

Kang designed his model based on the register error definition and feedforward PID controller with speed volatility of contiguous roller as the foregone interference[1,2]. But he ignore the influence of traction tension in register error. Although Li deduced detailed register model and devised feedforward controller in terms of disturbance of tension and speed, only with two layers of register system[3]. Yoshida established register error model based on web mass conservation and nonlinear control strategy, using Lyapunov's theory of stability[4]. Liu establish the relation among register error, roller speed and web tension, and design controller with method of active disturbance rejection control[5,6]. Chen put forward multi-layer register system controller strategy with the cooperation of expansion state observer and feedforward controller[7].

This thesis set up nonlinear coupling model for four-layer register system based on the working principle of cross-direction register errors and deduces the linear model of register system. On the basis of linear model, feedforward PID register controller is designed in terms of the disturbance of tension and speed. The simulation shows that the proposed register controller is endowed with better control performance than PID controller in register control.

2 Four-layer register system model and linearization

As shows in Figure 1, the register system of four-layer gravure press comprise printing unit, drying mechanism and detection section. The printing roller is compelled by servo motor and works under the model of speed. Where L^* is the nominal length in color cell, V_i is the linear speed of the roller, T_i is the tension of roller, and e_{ij} is register error between adjacent units i and j.

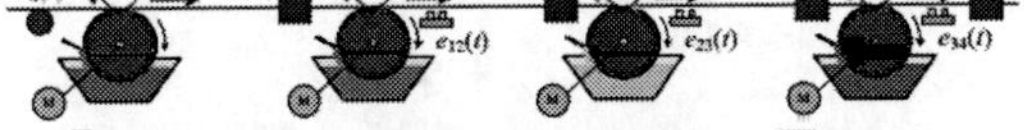

Figure 1. Structure of four-layer register system

Refer the model of double layer register system established in reference [5], the four-layer nonlinear coupling model for register system, using the relation of tension and strain, is established as Eq. 1. Where A is

[a] Corresponding author: ly-sword@163.com

cross-sectional area of web, E is elastic modulus, t_T is delay time of the web transmission, $V_2(t)$~$V_4(t)$ are system inputs, and $e_{12}(t)$~$e_{34}(t)$ are system outputs.

$$\left.\begin{aligned}
&AE\frac{de_{12}(t)}{dt}=T_0(t-t_T)V_1(t-t_T)-T_1(t)V_1(t)+\\
&\qquad AE\left(V_2(t)-V_1(t-t_T)\right)\\
&AE\frac{de_{23}(t)}{dt}=T_1(t-t_T)V_2(t-t_T)-T_2(t)V_2(t)+\\
&\qquad AE\left(V_3(t)-V_2(t-t_T)\right)\\
&AE\frac{de_{34}(t)}{dt}=T_2(t-t_T)V_3(t-t_T)-T_3(t)V_3(t)+\\
&\qquad AE\left(V_4(t)-V_3(t-t_T)\right)\\
&L^*\frac{dT_1(t)}{dt}=T_0(t)V_1(t)-T_1(t)V_2(t)+AE\left(V_2(t)-V_1(t)\right)\\
&L^*\frac{dT_2(t)}{dt}=T_1(t)V_2(t)-T_2(t)V_3(t)+AE\left(V_3(t)-V_2(t)\right)\\
&L^*\frac{dT_3(t)}{dt}=T_2(t)V_3(t)-T_3(t)V_4(t)+AE\left(V_4(t)-V_3(t)\right)
\end{aligned}\right\}\quad(1)$$

The four-layer nonlinear coupling model for register system is linearized with one order Taylor formula and change it with Laplace transform as Eq. 2.

$$\left.\begin{aligned}
&E_{12}(s)=G_A(s)V_2(s)+G_{D1}(s)V_1(s)+G_{E1}(s)T_0(s)\\
&E_{23}(s)=G_A(s)V_3(s)+G_B(s)V_2(s)+\\
&\qquad G_{D2}(s)V_1(s)+G_{E2}(s)T_0(s)\\
&E_{34}(s)=G_A(s)V_4(s)+G_B(s)V_3(s)+\\
&\qquad G_C(s)V_2(s)+G_{D3}(s)V_1(s)+G_{E3}(s)T_0(s)
\end{aligned}\right\}\quad(2)$$

System input shows as Eq. 3 in terms of the expression transfer function $G_A(s)$、$G_B(s)$、$G_C(s)$.

$$\left.\begin{aligned}
&G_A(s)=\frac{V^*}{L^*s+V^*}\\
&G_B(s)=\frac{L^*}{L^*s+V^*}\left(\frac{V^*}{L^*s+V^*}-e^{-t_Ts}\right)-\frac{T^*}{AEs}\\
&G_C(s)=\frac{L^*V^*}{(L^*s+V^*)^2}\left(\frac{V^*}{L^*s+V^*}-e^{-t_Ts}\right)
\end{aligned}\right\}\quad(3)$$

The transfer functions of the disturbance $V_1(t)$ to each error $G_{Di}(s)$ are shown in Eq. 4, and the transfer functions of the disturbance $T_0(t)$ to each error $G_{Ei}(s)$ are shown in Eq. 5.

$$\left.\begin{aligned}
&G_{D1}(s)=\frac{1}{s}\left(\frac{V^*}{L^*s+V^*}-e^{-t_Ts}-\frac{T^*}{AE}\right)\\
&G_{D2}(s)=\frac{V^*}{s(L^*s+V^*)}\left(\frac{V^*}{L^*s+V^*}-e^{-t_Ts}\right)\\
&G_{D3}(s)=\frac{(V^*)^2}{s(L^*s+V^*)^2}\left(\frac{V^*}{L^*s+V^*}-e^{-t_Ts}\right)
\end{aligned}\right\}\quad(4)$$

$$\left.\begin{aligned}
&G_{E1}(s)=-\frac{V^*}{AEs}\left(\frac{V^*}{L^*s+V^*}-e^{-t_Ts}\right)\\
&G_{E2}(s)=-\frac{(V^*)^2}{AEs(L^*s+V^*)}\left(\frac{V^*}{L^*s+V^*}-e^{-t_Ts}\right)\\
&G_{E3}(s)=-\frac{(V^*)^3}{AEs(L^*s+V^*)^2}\left(\frac{V^*}{L^*s+V^*}-e^{-t_Ts}\right)
\end{aligned}\right\}\quad(5)$$

From the Eq. 2, it could read that the influence of register error e_{12}、e_{23}、e_{34} are the traction tension T_0 and the speed of first roller V_1, besides the corresponding roller speed V_2、V_3、and V_4.

3 The design of feedforward PID controller

The control of register error is attained by adjusting the speed of after-layer roller in the design of feedforward PID controller. That is to say, adjusting V_2、V_3、V_4 to change e_{12} 、e_{23} 、e_{34} respectively. So the transfer function $G_A(s)$ stands for the property of register error system with other factors as system disturbance. Therefore, PID is designed in terms of transfer function $G_A(s)$, and feedforward controller compensate other systematical influencing elements. Because T_0、V_1、V_2、V_3 can be measured, feedforward controller can be designed by using T_0、V_1、V_2、V_3 based on the invariance principle. What's more, feedforward controller can eliminate the effect of T_0、V_1、V_2、V_3 in register error. The structure of feedforward controller reveals as the Figure 2.

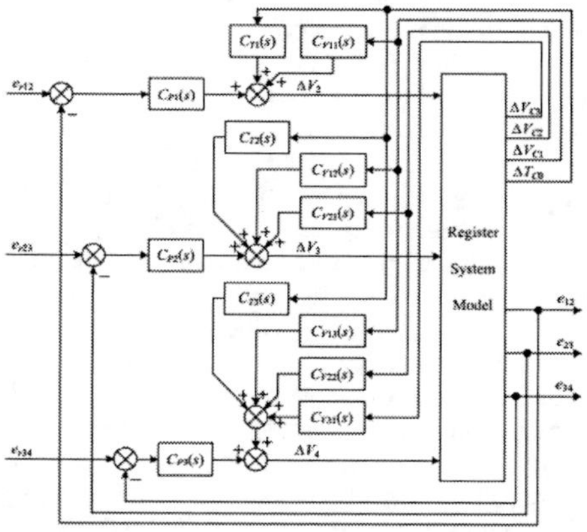

Figure 2. Block diagram of feedforward PID controller for four-layer register system

In Figure 2, Er_{12}、Er_{23}、Er_{34} are the reference inputs of register error, ΔV_{C1}、ΔV_{C2}、ΔV_{C3} are the variations of actual speed, and ΔT_{C0} is the tension of the actual traction unit. T_0 and V_1 are the disturbances for E_{12} and they have two feedforward controllers, $C_{T1}(s)$ and $C_{V11}(s)$ respectively. In a similar way, E_{23} has three feedforward controllers, $C_{T2}(s)$、$C_{V12}(s)$ and $C_{V21}(s)$ respectively. E_{34} Has four feedforward controllers, $C_{T3}(s)$ 、$C_{V13}(s)$ 、$C_{V22}(s)$ and $C_{V31}(s)$. The PID controller of each control

DOI: 10.1051/matecconf/20166304023

loop are $C_{P1}(s)$、$C_{P2}(s)$ and $C_{P3}(s)$. According to the Figure 2 and the Eq. 2, the expression of $E_{12}(s)$ can be achieved as Eq. 6.

$$\begin{aligned} E_{12}(s) = {} & \frac{G_A(s)C_{p1}(s)}{1+G_A(s)C_{p1}(s)}E_{r12}(s) + \\ & \frac{G_{D1}(s)+C_{V11}(s)G_A(s)}{1+G_A(s)C_{p1}(s)}V_{C1}(s) + \\ & \frac{G_{E1}(s)+C_{T1}(s)G_A(s)}{1+G_A(s)C_{p1}(s)}T_{C0}(s) \end{aligned} \tag{6}$$

Based on the invariance principle, feedforward controller is devised in terms of T_0 and V_1 in E_{12}.

$$\left.\begin{aligned} G_{E1}(s)+C_{T1}(s)G_A(s) = 0 \\ G_{D1}(s)+C_{V11}(s)G_A(s) = 0 \end{aligned}\right\} \tag{7}$$

According to Eq. 7, the feedforward controller $C_{T1}(s)$ and $C_{V11}(s)$ can be designed as Eq. 8.

$$\left.\begin{aligned} C_{T1}(s) &= -\frac{G_{E1}(s)}{G_A(s)} = \frac{V^*}{AEL^*s}\left(V^* - L^*s + V^*e^{-t_T s}\right) \\ C_{V11}(s) &= -\frac{G_{D1}(s)}{G_A(s)} \\ &= \frac{1}{L^*s}\left(\frac{T^*(L^*s+V^*)}{AE} + (L^*s+V^*)e^{-t_T s} - V^*\right) \end{aligned}\right\} \tag{8}$$

Similarly, the feedforward controller $C_{T2}(s)$、$C_{V12}(s)$ and $C_{V21}(s)$ can be designed as Eq. 9, and $C_{T3}(s)$、$C_{V13}(s)$、$C_{V22}(s)$ and $C_{V31}(s)$ can be designed as Eq. 10.

$$\left.\begin{aligned} C_{T2}(s) &= -\frac{G_{E2}(s)}{G_A(s)} \\ &= \frac{(V^*)^2}{AEL^*s}\times\left(\frac{V^*}{L^*s+V^*} - e^{-t_T s}\right) \\ C_{V12}(s) &= -\frac{G_{D2}(s)}{G_A(s)} \\ &= -\frac{V^*}{L^*s}\left(\frac{V^*}{L^*s+V^*} - e^{-t_T s}\right) \\ C_{V21}(s) &= -\frac{G_B(s)}{G_A(s)} \\ &= e^{-t_T s} - \frac{V^*}{L^*s+V^*} + \frac{T^*(L^*s+V^*)}{AEL^*s} \end{aligned}\right\} \tag{9}$$

$$\left.\begin{aligned} C_{T3}(s) &= -\frac{G_{E3}(s)}{G_A(s)} \\ &= \frac{(V^*)^3}{AEL^*s(L^*s+V^*)}\times\left(\frac{V^*}{L^*s+V^*} - e^{-t_T s}\right) \\ C_{V13}(s) &= -\frac{G_{D3}(s)}{G_A(s)} \\ &= -\frac{(V^*)^2}{L^*s\left(L^*s+V^*\right)}\left(\frac{V^*}{L^*s+V^*} - e^{-t_T s}\right) \\ C_{V22}(s) &= -\frac{G_C(s)}{G_A(s)} \\ &= -\frac{V^*}{L^*s+V^*}\left(\frac{V^*}{L^*s+V^*} - e^{-t_T s}\right) \\ C_{V31}(s) &= -\frac{G_B(s)}{G_A(s)} \\ &= e^{-t_T s} - \frac{V^*}{L^*s+V^*} + \frac{T^*(L^*s+V^*)}{AEL^*s} \end{aligned}\right\} \tag{10}$$

With this, Eq. 8, Eq. 9 and Eq. 10 constitute the feedforward controller of four-layer register system. Meanwhile, they and PID controller constitute the whole register controller, namely, feedforward PID controller.

4 The simulation

In order to verify the performance of the proposed feedforward PID register controller, simulation of register error with PID controller and feedforward PID controller are performed respectively.

4.1 Performance comparison under tension disturbance

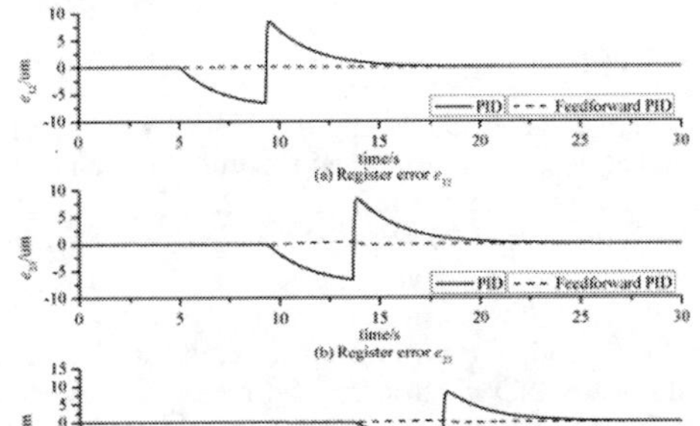

Figure 3. Simulation curve of register error with tension disturbance

The performance of PID controller and feedforward PID controller in simulation are shown in Figure 3 when the upstream tension T_0 has a step change from 100N to 120N after 5s stable operation of the system. Compared with the register error in PID control, the peak of register error controlled by feedforward PID controller is considerably decreased and there is no obvious time delay. It indicates that feedforward control can generate a control value in advance to compensate the register error caused by tension disturbance, which makes the performance of feedforward PID controller much better than that of traditional PID controller.

4.2 Performance comparison under speed disturbance

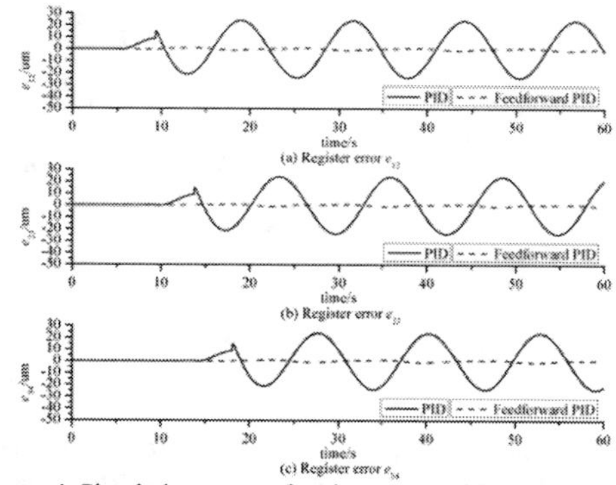

Figure 4. Simulation curve of register error with speed disturbance

In order to reflect the performance of different controllers in speed disturbance, simulation are performed with a man-made sinusoidal disturbance whose amplitude is 0.1r/min and frequency is 0.5rad/s after 5s stable operation of the system. As shown in Figure 4, the register error with the traditional PID control has a sinusoidal variation and the maximal register error is obviously larger than that with the proposed feedforward PID control. The results indicated that the proposed feedback controller can effectively reduce register error caused by the speed fluctuation.

5 Conclusion

The register error model and the control strategy of the sectional drive gravure printed electronic equipments are successfully studied in this paper. Based on the double layer register system model, a coupling mathematical model for the four-layer register error is derived, and linear model of register system is attained with one order Taylor formula. According to the established model, a feedforward PID controller is designed to alleviate the disturbance caused by the upstream tension and speed. In numerical simulation, the proposed feedforward PID controller was compared with traditional PID controller. The results showed that with the disturbances caused by the tension and speed, the performance of the proposed feedforward PID controller is much better than that of the traditional PID controller. The proposed feedforward controller can effectively alleviate the register errors caused by the disturbances from tension and speed.

Acknowledgements

This work was financially supported by the National Natural Science Foundation of China (51505376) and Initial Scientific Research Fund of Doctor in Henan University of Science and Technology.

References

1. Kang H K, Lee C W, Shin K H, Japanese Journal of Applied Physics, Vol.**50**, p. 6701 (2011)
2. Kang H, Lee C, Shin K, Control Engineering Practice, Vol.**21**, p. 645 (2013)
3. Li Jian, Mei Xuesong, TAO Tao, et al, Proceedings of the Institution of Mechanical Engineers, Part C, Journal of Mechanical Engineering Science, Vol.**226**, p. 626 (2012)
4. Yoshida T, Takagi S, Muto Y, et al, Electronics and Communications in Japan, Vol.**94**, p. 17 (2011)
5. Liu S, Mei X, Li J, et al, Mathematical Problems in Engineering, Vol.**2013**, p.10 (2013)
6. Liu S, Mei X, He K, et al, Control Theory and Applications in Chinese, Vol.**31**, p. 1574 (2014)
7. Chen Y, Chen Z, Deng Z, Control Theory and Applications in Chinese, Vol.**31**, p. 814 (2014)

MATEC Web of Conferences **63**, 04024 (2016)
MMME 2016
DOI: 10.1051/matecconf/20166304024

Relationship between urban heat island effect and land use in Taiyuan City,China

Shu Ting LI [1], Jia LI [1,a] and Ping DUAN [1]

[1] *School of tourism and geographical science,Yunnan normal university,Kunming,China*

Abstract: The land surface temperature is inversed by Landsat remote sensing images of 2000, 2009 and 2015 years in Taiyuan, China. The mono-window algorithm is used to remove the influence of the atmosphere. The land surface thermal radiation intensity is obtained by the mono-window algorithm. Then the land surface true temperature is converted by the land surface thermal radiation intensity. At the same time, the remote sensing images of Taiyuan city in three time periods are classified by supervised classification method. Finally, the relationship between different years of Taiyuan land surface temperature and land use change is analysed. The results show that Taiyuan city land surface temperature is positively correlated with land use. The land surface temperature is higher when the land is frequently used. Taiyuan city land surface temperature is negatively correlated with vegetation coverage. The land surface temperature is lower when the higher vegetation is covered in the area.

1 Introduction

Urban thermal environment influenced by the earth's surface physical properties and human social economic activities, it is the comprehensive summary of urban ecological environment and reflect [1]. In recent years, under the background of global warming and the high-speed urbanization process, urban thermal environment is considered to be one of the dominant elements of the city environment, it has a profound impact on urban public health, air quality, energy consumption, etc[2]. Due to the characteristics of urban heat reflects through the form of electromagnetic radiation, thus can be directly detected by means of remote sensing, it is also become the basic principle of thermal remote sensing detection of urban thermal environment[3]. There are many factors that can cause of urban heat island effect: human activities, the underlying ground structure, vegetation coverage, and so on. To study the changes of Taiyuan urban heat island effect in recent 15 years, combing with the land use classification. To better illustrate the relationship between human activities and the urban heat island effect, and gives some corresponding solutions.

2 The research area

Taiyuan is the capital of SHANXI Province. It is located in Taiyuan Basin, the northern section of the, central and southern plains, East by the TAIHANG Mountains, West LVLIANG, the overall high terrain north south, East and west across 114km, north-south span is about 107KM. The whole Taiyuan have GUJIAO, QINGXU, YANGQU, LOUFAN four counties (cities), and WANBAILIN area, YINGZE district, XINGHUALING district, JIANCAOPING district, JINYUAN district of six main area, the total area of 6959 km2 in the whole city area, the total population of 4.2565 million.(2012). The climate of Taiyuan city is the temperate zone monsoon climate, large temperature difference between day and night, ample sunshine [4].

3 METHODS

The calculation method of surface temperature inversion, the radiation brightness temperature is calculated and estimate the atmosphere on the effects of surface radiation brightness, using a single thermal infrared channel with Landsat. Thus the performance surface true temperature. Firstly, the image is calibrated by radiometric calibration and atmospheric correction. The method adopted is FLAASH's own ENVI5.1 atmospheric correction module, which can effectively remove the influence of the atmosphere on the earth surface albedo, and get the true surface reflectance. Mono-window algorithm is Tan Zhihao. (2004) according to the surface thermal radiation conduction equation, deduces data by Landsat TM/ETM + 6 band surface temperature inversion algorithm [5-6]. The formula for calculation is as follows:

$$T_s = [a(1-C-D)+(b(1-C-D)+C+D)T_6 - DT_a]/C \quad (1)$$

[a] Jia LI : keguigiser@163.com

In the formula, T_s is the surface temperature, a and b are constants, are respectively 67.355351 and 0.458606, $C=\varepsilon\tau$, $D=(1-\tau)[1+(1-\varepsilon)\tau]$, ε is the surface emissivity, T atmospheric transmittance.

The infrared brightness value received by the satellite sensor is composed of three parts: Atmospheric radiation brightness upwards, the energy of the earth's surface, which passes through the atmosphere, reaches the satellite sensor, atmospheric radiation reaches the ground down after the reflected energy. Announced on NASA's website (http://atmcorr.gsfc.nasa.gov), input image's time and center longitude and latitude, can get the atmospheric transmittance, atmospheric radiation brightness upward, atmospheric radiation brightness down three parameters. The radiative transfer equations is as follows:

$$L_\lambda=[\varepsilon B(T_S)+(1-\varepsilon)L\downarrow]\tau+L\uparrow \quad (2)$$

In the formula, ε is surface emissivity, T_s is the surface temperature, $B(T_s)$ as the blackbody radiation brightness, T for atmospheric transmittance. Blackbody temperature is T in the thermal infrared radiation intensity $B\,(T_S)$:

$$B(T_S)=[L_\lambda-L\uparrow-\tau(1-\tau)L\downarrow]/\tau\varepsilon \quad (3)$$

In the formula, T_s can be obtained by the formula of Planck's radiation law.

$$T_S=K_2/In(K_1/B(T_2)+1) \quad (4)$$

In the formula, for Landsat TM, K_1 = 607.76 W/(m^2*μm*sr), K_2 = 1206.56 k; for TIRS Band10, K_1 = 774.89 W/(m^2*μm*sr), K_2 = 1321.08 k.

The NDVI threshold method proposed by Sobrino is used to calculate the surface emissivity:

$$\varepsilon=0.004P_v+0.986 \quad (5)$$

In the formula, P_v is vegetation coverage, Calculating formula is:

$$P_v=[(NDVI-NDVI_{Soil})/(NDVI_{Veg}-NDVI_{Soil})] \quad (6)$$

In the formula, $NDVI$ is the normalized difference vegetation index, $NDVI_{Soil}$ is the $NDVI$ value of no vegetation or completely bare soil, $NDVI_{Veg}$ is the $NDVI$ value of completely covered by vegetation. Remote sensing image characterizes terrestrial object by brightness value and the pixel value and the spatial change. This is the physical basis of distinguishing different image features [7]. It is through the unique differences that can distinguish objects on the same image. Remote sensing image classification have supervised classification and unsupervised classification, it is the use of different spectral information and spatial information, and then each pixel is classified according to a certain rule.

Supervised classification is used in this paper, Select a certain number of training samples in remote sensing image, to carry out statistical or other information. After choosing the supervision of the support vector machine (SVM) classifier for classification image, SVM can automatically search for the support vector which has a large capacity for classification. Thus the separator constructed can maximize the distance between the class and class, and having a good classification accuracy. This article using the method of supervised classification to classify the land in Taiyuan City, in ARCGIS, the land use classification figure transformation, get vector boundary files, to the area of different land types. In ENVI software statistical analysis of different types of land surface temperature, get the average temperature of the different land types.

4 Conclusion analysis

4.1 Characteristics of urban heat island in Taiyuan

Taiyuan city administrative region plan as shown in figure 1, There are six municipalities directly under the jurisdiction of Taiyuan: JIANCAOPING district, WAN BAILIN district, JINYUAN district, XINGHUALING district, YINGZE district,Xi district. LOUFAN country and YANGQY country have a sparse population and natural vegetation coverage rate is relatively high, GUJIAO is Taiyuan energy heavy chemical industry base, QINGXU is main crop growth area in TAIYUAN [8].

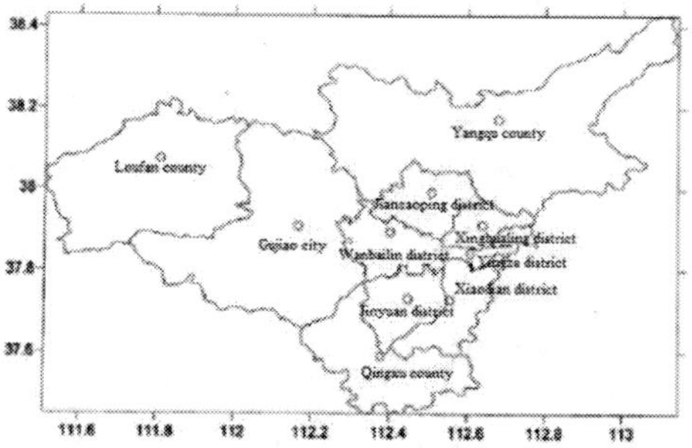

Figure 1. Taiyuan administrative chart

In this paper, the surface temperature of Taiyuan city is retrieved by remote sensing image of Taiyuan city in August 2009 and August 2000 in September 2015. The results are as follows:

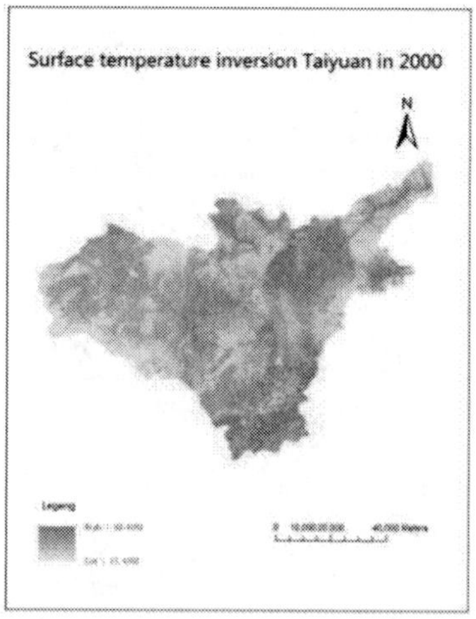

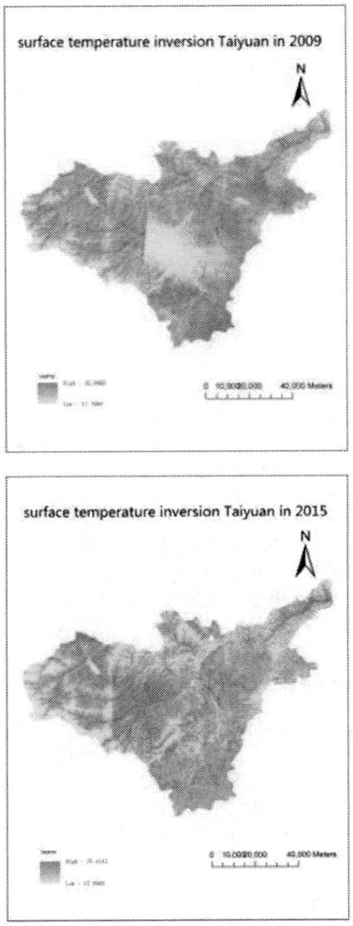

Figure 2. Surface temperature inversion 2000-2015 in Taiyuan

As seen in the figure, the surrounding temperature is higher in Taiyuan 2000, in 2009, due to the reasons of the image, it is not to be able to show it very well. But it can be seen that the temperature is higher in JIANCAOPING district. In 2000, QINGXU County's temperature has decreased. Taiyuan city high temperature region in 2009 has increased than 2000, heat island effect is more obvious. QINGXU county temperature decreased.

4.2 Analysis of urban heat island and land change in Taiyuan

There are several reasons for the emergence of the urban heat island: underlying surface structure, urban landscape, vegetation coverage, etc. It is one of the most closely related to human activities, through the comparison of land use planning in 2000, 2009 and 2015, explain that the emergence of the urban heat island has close relationship with human activity.

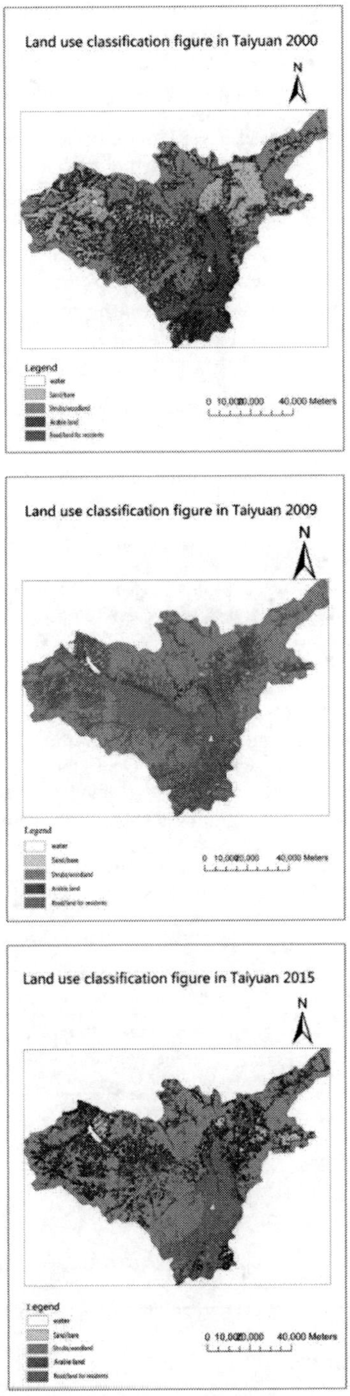

Figure 3. Land use classification 2000-2015 in Taiyuan

In addition to the image has a problem in 2009, It can be seen that Sand/bare land is more in 2000, in 2015, the area of the sand/bare land accounts for obvious less than in 2000. Impervious surface (road /residential land) in the

2000-2015 years is constant expansion trend. Combining with the land surface temperature inversion figure in 2000-2015, TAIYUAN city high temperature area is increasing, and has a tendency to expand outward. Combined with the above figure, found that most high temperature areas of TAIYUAN city are bare land/sand and the ridge. The following figure used the ARCGIS, it is the statistical analysis of the 2000 -2015 Taiyuan City Road / residential that different land type area changes in line chart:

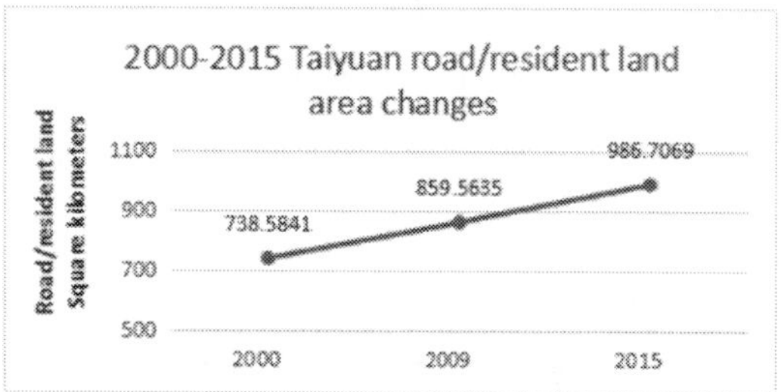

Figure 4 2000-2015 Taiyuan road/resident land are changes

Through by 2000 and 2015 remote sensing images of land use / cover types of surface temperature statistics analysis, we can see that Taiyuan City 2015 sand / bare area significantly reduced 23.8%, shrub / forest land occupied area ratio increased 18.85% road / residents with added 3.57%, it is indicates that the city still has expansion, and the average temperature of the road / resident is increased by 1.31 Centigrade. From 2000-2015, Taiyuan economy is developing fast, urban land area changed little,at the same time increased the vegetation construction, driven by economic development, urban heat island effect more obvious (because image reasons in 2009, there is no do statistical analysis)

Table 1. Surface average temperature of land cover type

Surface type	2000		2015	
	Proportion of area /%	**average temperature /℃**	**Proportion of area /%**	**average temperature /℃**
1sand/bare	37.32	36.62	13.52	34.69
2shrubs/woodland	45.69	26.04	64.81	29.44
3arable land	6.04	23.38	7.14	34.7
4residential land	10.61	32.74	14.18	34.05

5 conclusion

There are many factors cause urban heat island effect: artificial construction of urban land use planning, which will change the city underlying surface structure, thus will change the underlying surface properties; a large number of industrial emissions and vehicle emissions, these heat emitted in the air, will make the city's temperature increased. So urban heat island effect has a close relationship with human activities, is caused by human beings. Daniel believes that in the climate of the state, the urbanization and industrial pollution will cause downstream of the rainfall and snowfall increased[9].Sang Jianguo and other found that in the winter is also the heat effect of forcing effect, that urban heat island effect is easy to form a low pressure system for the center of the city as well as the direction of the city center pressure gradient force[10].It not only affects the environment of human life, but also has a great impact on human health.Urban heat island effect are also associated with vegetation coverage,research shows that the green coverage rate is higher, the lower the intensity of the heat island,when the coverage is more than 30%, heat island effect obviously weakened; coverage is more than 50%, the green space is extremely obvious to the reduction of heat island effect.In the city, planting more trees and the construction of the park, urban heat island effect will be eased. The wind also can reduce the quantity of heat. It will be in the urban planning design construction direction, and it is beneficial to the formation of the wind that will reduce heat in the city. The most important thing is the human activity, human reducing emissions of heat, it is also important measures that reduce heat island formation.

References

1. X.L. Zhang, S.W. Zhang, Y. Li, Y.F. Xie, W.H. Kuang, J. Resour Sci 30, 390(2008)
2. J. Zhang, X.P. Zhang, X.Y. Wang, J.M. Zhang,Y. Liu, Y. Xiao, Arid Land Geog 33, 51(2010)
3. G.L. Tian, Publishing House of Electronics Industry(2006)
4. Shanxi Statistical Bureau. *Shanxi statistical yearbook* (2012)
5. Z.Q. Lv, Y. Wen., Z. Sun, J. Acta Ecol Sin 19,1771(2010)
6. Z.H. Tan, W.J. Li, M.H. Zhang, J. RS for Land & Resour 23, 33(2002)
7. S.B. Deng., *Envi remote sensing image processing method (Second Edition)*
8. Y.H. Li, Z.C. Li, J.H. Zhou, C.P. Zhao, Journal of Arid Land Resour and Environ 12,173 (2013)
9. Rosenfeld. D, J. Sci 287,1793(2000)
10. J.G. Sang, W.J. Liu, J. Acta Meteoro Sinica 48, 459(1990)

Simulation of influence of disturbance on the release of endogenous phosphorus in sediment of HAOHE River of Nantong, China

Xiu LU [1], Jia LI [1,a] and Ping DUAN [1]
[1]*School of Tourism and Geographical Sciences of Yunnan Normal University Yunnan, China*

Abstract: There are less researches in the influence of disturbance of endogenous phosphorus release in sediment. The correlational analysis only takes the disturbance as one of the many factors in influencing the phosphorus release. It hasn't been able to make a deeper discussion from disturbance itself. In this paper, HAOHE River of Nantong, China is selected as study area. The whole process of disturbance is simulated in the laboratory. Disturbance intensity, time and depth is respectively simulated and analysed their effects on endogenous phosphorus release in sediment. The result shows that the external disturbance can significantly increase the release of endogenous phosphorus in sediment. From the time series of the experimental results, the release of phosphorus is not increased with the increasing time. The depth experiment has showed that the maximum phosphorus release is closest to the sediment and the closest to the water surface, while the phosphorus release in the position of middle depth is less.

1 Introduction

In the aspect of controlling the eutrophication of water body, the endogenous release of the sediment has been controversial. For the first time, foreign scholars put forward the endogenous release of water body in 1941. For a long time, the scientific research shows that N, P and other elements are the main factors that limit the growth of plankton, in which phosphorus the biggest factor leading to eutrophication among them. Nutrient elements, such as nitrogen and phosphorus, have important functions on primary productivity and nutrient level of water body. In recent years, the effects of sediment on the internal nutrient sources have been paid more and more attention in the control of exogenous phosphorus. At present, the release of endogenous phosphorus has been up to more than half a century. Studies show that the main factors affecting the release of endogenous phosphorus are environmental factors [1]. At home and abroad, scholars have carried out thorough study, most of the method is the release method. Various factors affect endogenous release, such as temperature, redox potential, PH, disturbance, microbial action, the ratio of iron and phosphorus in sediment and the condition of aquatic plants [2]. Li Dapeng [3] and others believe that the release of endogenous phosphorus has two kinds of static and dynamic release in some shallow water bodies. Relatively speaking, the release amount of the dynamic release is much more than the static one. So the dynamic release of endogenous phosphorus is more concerned. The factors which influence the large factors are the wind wave disturbance, the biological disturbance, the tidal current disturbance, the artificial disturbance and so on.

2 Materials and methods

2.1 General situation of study area

Nantong is known as "China's first modern city", with a long history and strong cultural heritage. It has a moat HAOHE river which has a thousand years of history, the spatial pattern of integrity and a very high historical value. But the HAOHE River also encountered many typical problems in the history. With the economic development, urban scale expansion and the increase intensity of the transformation of the old city in Nantong resulting in the residential and high-rise buildings continued warming. It is not only suffer from man-made destruction, industrial pollution caused by water black odor, the most serious time being the HAOHE River every day to bear nearly million tons of domestic sewage and industrial wastewater. Then the public likened HAOHE River being vividly "wash rice and vegetables in the 1950s, deterioration of water quality in the 1960s, fish and shrimp are dead in the 1970s, eighties wash dirty toilet cover cannot be washed neatly in the 1980s ."Later in order to change this situation, local government and the people conduct a series of reform measures. Until today, the HAOHE River has successfully promoted to the state 5A class tourist scenic spot. The HAOHE River line length of approximately 14.96 km, 1040 acres of water surface, the widest at 215 meters, the mighty waves, like the lake, the narrowest point is only 10 meters which like

[a] Jia LI: keguigiser@163.com

the drift ribbon. HAOHE River can be divided into the southeast, southwest and North River three sections. Nantong becomes a "water surrounds City, city surrounds water" city, which has a unique landscape. It was praised as "Oriental Venice" by some people. The HAOHE River water quality is gradually good, when the external pollution is under control, how the external disturbance will affect the HAOHE internal sediment pollution release? The author from August 2013 to April 2014 in the HAOHE River respectively in the vicinity of HAOHE River People's Park Bridge and North Bridge North HAOHE for sampling and testing.(As is shown in Figure1)

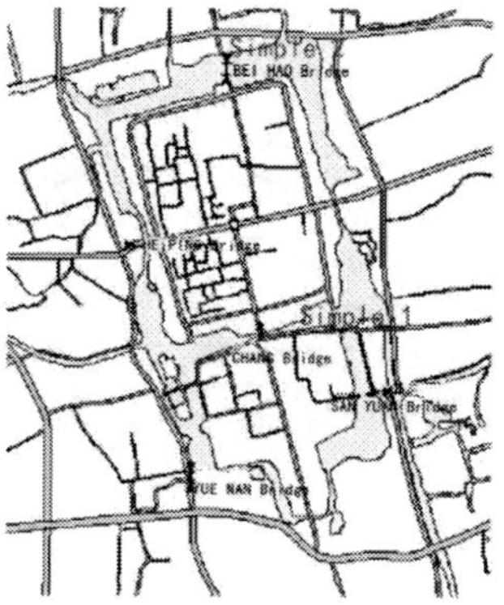

Figure1. Nantong HAOHE River [4]

Table1. List of sample points

Number	Geographical coordinates	Location
Sample point 1	32°00′56.19″ N 120°51′57.529″E	HAOHE River People's Park Bridge
Sample point 2	32°01′31.949″N 120°51′36.999″E	North Bridge North HAOHE

2.2 Reagent and instruments

Main materials and reagents: Nantong HAOHE River mud sediment and overlying water; sulfuric acid: 1:1; potassium peroxydisulfate: 50g / L solution; ascorbic acid: 100g / L solution (store in a brown bottle reagent); molybdate solution: dissolve 13g ammonium molybdate solution in 100 ml of water. Dissolve 0.35g antimony potassium tartrate in 100ml water. Metastasis after reagent slowly under constant agitation to ammonium molybdate solution adding 300ml sulfate (1:1), and tartaric acid potassium antimony solution and mixed evenly (store in brown bottles); phosphorus standard reserving solution: weigh 0.2197 ± 0.001g to 110 degrees centigrade and dried in a desiccator for two hours and put cold monopotassium phosphate. And then dissolve in water to 1000ml volumetric flask, add about 800ml water, add 5 ml of sulfuric acid (1:1), dilute with water to marked line and mixed evenly, the solution in the glass bottle can be stored for at least six months; phosphorus standard using liquid: 10.0 ml of phosphorus standard solution is transferred to a 250 ml volumetric flask and water dilution to marked line and mixed evenly (prepare the solution when use that day). Main equipment: high pressure steam disinfection device; centrifuge; spectrophotometer; constant speed blender.

2.3 Method

2.3.1 Analytical method

Potassium peroxydisulfate dissolution: the sample (25ml stoppered graduated tube with sample) was added 4ml of potassium peroxydisulfate, the stoppered graduated tube cover after plugged with a small piece of cloth and tight wire glass stopper, put placed in a large beaker was heated high-pressure steam sterilizer, until a pressure of 1.1kg / cm^2, corresponding to a temperature of 120 degrees centigrade. And maintain heating was stopped after 30min. After the pressure gauge is approaching zero, remove and let cool. Then dilute with water to the marked line. Chromogenic: respectively, each part was added 1ml ascorbic acid solution into the digestion solution and mix, 30s later adding 2ml molybdate solution and mix well. Spectrophotometry measure: 15min after standing at room temperature, optical path 30mm cuvette, at a wavelength of 700nm, with water as a reference, to measure the absorbance. After deducting the absorbance of the blank test, inquire the phosphorus content from the working curve.

2.3.2 Experimental method

2.3.2.1 Sample preprocessing

In Nantong HAOHE River sample points with Gravity Sampler collected sediment into the laboratory in the bucket is about 500g. And in the same place it above the acquisition of overlying water 50L. Bring back to the lab, the first disturbance strength and phosphorus of the relationship between a set of experimental samples to 40 degrees centigrade oven drying and grinding the standby. The sediment barrel needn't dry and grind. PH between 7-8 and, the temperature maintain at 10-14 degrees, the relative humidity between the stability of the 60-80. It has little change, and it also play a very good condition for considering the only one factor which is disturbance.

2.3.2.2 Sample handling

① Disturbance strength group: Take the drying and grinding the 40g samples into the volume of 1000ml cylindrical container, plus the cover water 800ml. A total of two groups of experiments: A group only on the overlying water, B group of sediment and the overlying water with a constant speed mixer for continuous disturbance 12h [5]. Change the disturbance intensity, that is, the performance of a certain gradient of speed, respectively is 100 r/min, 150 r/min, 200 r/min, 250 r/min, 300 r/min, 350 r/min, 400 r/min, and the interval is 50r/min. Each set of three groups of parallel samples, the error will be reduced to less than 0.05. ②Time series group: The weighed returned to samples of about 2cm in thickness, added to the overlying water to the bottom diameter is 16cm, 27 cm high cylinder shaped container, the depth of the water remained in the 25cm. A total of three groups of experiments: A group only on the water, the B group on the overlying water and sediment does not disturb the C group of sediment and overlying water and disturbance, the disturbance group disturbance intensity is set to 200r/min (because of its intensity of the maximum release amount). The time series are 1H, 3h, 6h, 12h, 24h (one day), 72h (three days), by simulating a natural wind. Each set of three groups of parallel samples, the error will be reduced to less than 0.05. ③Depth sequence group: The weighed returned to samples of about 2cm in thickness, adding overlying water to the bottom diameter is 16cm, high 27cm in cylindrical vessels. A total of three groups of experiments: A group only on the water, the B group on the water covered with sediment does not disturb the C group of sediment and overlying water and disturbance, the disturbance intensity of 200r/min, rotation time of 24h (due to its release of the maximum time). The depth of the water samples are: 10cm, 15cm, 20cm, 25cm. Each set of three groups of parallel samples, the error will be reduced to less than 0.05.

3 Result and analysis

3.1 Effects of disturbance intensity on phosphorus release from sediment

As is shown in Figure 2, a horizontal coordinate indicates the speed, and the vertical coordinate indicates that the phosphorus release concentration (released amount is equal to the later phosphorus content minus the original phosphorus content). (The standard curve is: R^2=0.9999 y=0.0204 x +0.0011) from the figure can be clearly known, at the speed of 200r/min, the maximum phosphorus release. After the release of phosphorus did not always increase, than 100 r/ min, emissions are not 400 r/ min, to 200 r/ min increases to a maximum and also decreases, and in the min and the emergence of a small peak, but did not affect the overall results, still less than 200 r/ min when the phosphorus release volume, that disturbance intensity and phosphorus release is not showed a positive correlation. It also verifies the experimental results of Cai Jingbo [5] et al. At low speed, the effect of mixing and exchange of sediment and water on phosphorus release from sediments is more important than that of pure dissolved oxygen. At the same time, the disturbance makes the phosphorus in the water diffusion in the water, and the results of the water exchange between the overlying water and the water gap will increase the pH value of the gap water, and promote more phosphorus from the solid phase into the gap water. In 200r/min, the maximum release amount is in a middle position, which is similar to the normal distribution.

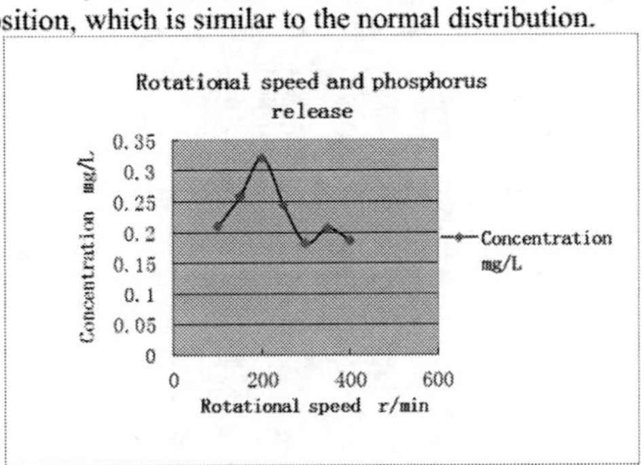

Figure2. Rotational speed and phosphorus release

3.2 Influence of disturbance time on the release of phosphorus from sediments

Two times respectively in HAOHE North Bridge, and HAOHE river People's Park Bridge of time series, the standard curve respectively: y=0.0234x+1.0E-4 R^2=0.9999; y=0.0202x-0.0031 R^2 = 0.9998. The curve is highly consistent. The results show that the release of phosphorus had a certain rule to follow in time, which was in the form of wave type. From Figure 3, figure 4, it is clear that the phosphorus in the sediments did not release to the aqueous solution. On the contrary, phosphorus in water solution back to the sediments and in time sequence in 72 hours to release at least, return to the sediment deposition in the most. That is to say, even if disturbed, it does not mean that the disturbance time more long, the release amount of the greater. It is released at 24 hours the most, that is, back to the sediment in the sediment at least. It can be seen that the phosphorus in the sediment did not show a "source" of the state, but rather the contrary, showing a "sink" of the state. And the first time series in December 2013 and January 2014 completed, and in March 2014 the results of the curve is quite consistent. It shows that the temperature for the experiment did not play a control and affect the conditions.

Preliminary can draw the following conclusions: in 24 hours, both the static release and dynamic release, the phosphorus release is the most. That is exactly the mud at least. And in 72 hours, whether it is static or dynamic, phosphorus release is at least and bottom mud in the most. Time series experiments results show that the phosphorus release amount and not with increase of time, the release quantity also increases, but after the release of the day release up to the amount of. Also is to say, if people for shallow lakes produced some disturbance, then at the same time the next day can phosphorus release is the largest, and relevant departments need to make a deal, reduce the nutrition level and then reduce the harm.

 DOI: 10.1051/matecconf/20166304025

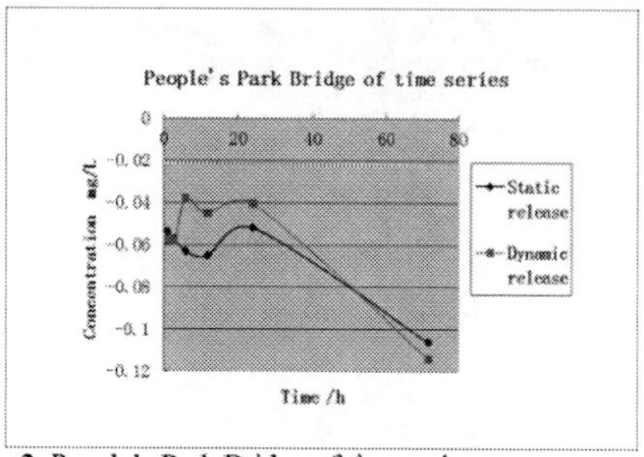

Figure3. People's Park Bridge of time series

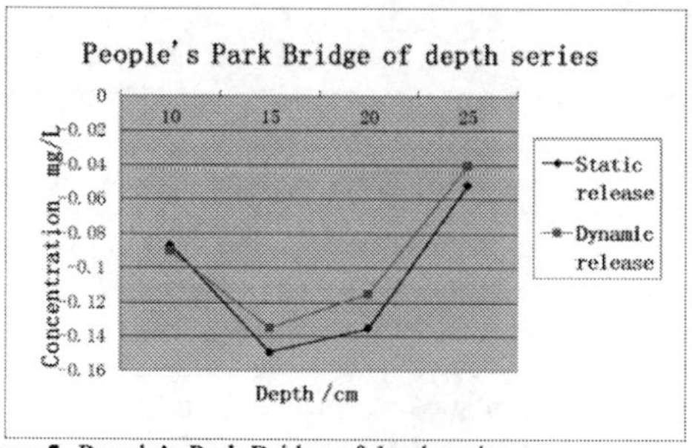

Figure5. People's Park Bridge of depth series

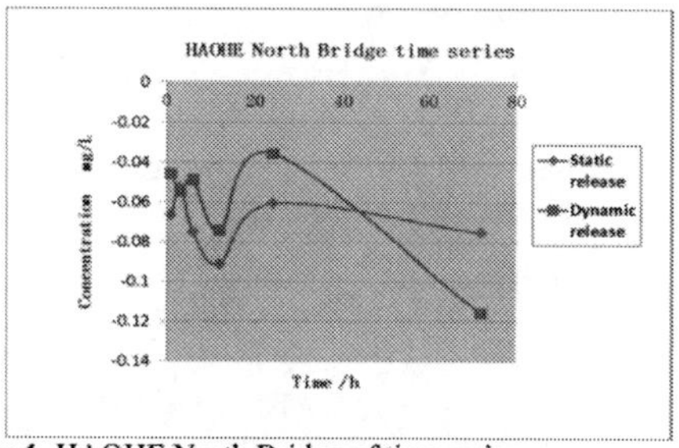

Figure4. HAOHE North Bridge of time series

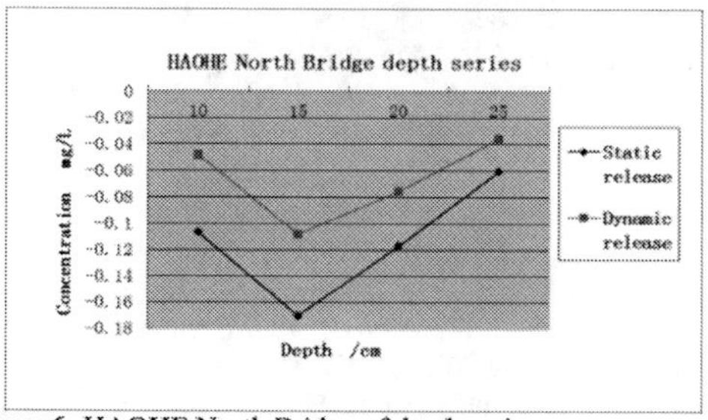

Figure6. HAOHE North Bridge of depth series

3.3 Effects of disturbance depth on the release of phosphorus from sediments

Two times respectively in HAOHE North Bridge, HAOHE River People's Park Bridge the depth sequence, the standard curve are y=0.0202x-0.0031 $R^2 = 0.9998$). By the following two pieces of test depth Figure 5 and figure 6 that: whether it is static or dynamic phosphorus do not release the, but return to the sediment, showing a "sink" state. Especially in 15cm performance is particularly evident, showing the shape of inverted triangle, breaking the conventional conclusion, is not closer to the sediment, phosphorus release quantity is more, in the 10 cm and 25 cm of the static and dynamic are in a state of high release. And the minimum value is at 15cm. Once emerging the disturbance, the most close to the bottom of the sediment and the closest to the water is the largest phosphorus release. And in the middle of the depth of the release of phosphorus is less, the more light is rich in nutrition. Perturbation due to simulate sediment thickness range covering the surface of the water with a certain distance on the surface of phosphorus release produced little effect. It can provide the references for the relevant pollution-control sections, also can increase the depth of the water to reduce endogenous pollution.

4 Conclusions

A. Experimental study shows that the external disturbance can significantly increase the release of endogenous phosphorus in sediment. The high speed disturbance results in the internal phosphorus release being not higher than the low velocity disturbance. Instead the distribution is approximately an approximate distribution of the intermediate disturbance velocity. In the static state of the water body, the internal phosphorus is from low concentration to high concentration of the transference process. And the disturbance is to speed up the transfer process, and also promote the release of phosphorus.

B. Overall, sometimes the role of the disturbance is not "promote" the release of endogenous sediment, that is, not "source", but a "sink" of the state. It owes to the disturbance of the water to strengthen the disturbed state. Make the phosphorus in a high speed scattered, suspended state, but makes it more conducive to the sediment for phosphorus adsorption. Or the perturbation increases the low gradient of dissolved oxygen, PH changes, so that the phosphorus is adsorbed by the sediment easily. From this view, the "sink" and "source" phenomenon exists at the same time, and also plays a role.

C. From the time series of the experimental results, the release of phosphorus is not increased with the increase in time. But after the release of the day release most. That is to say, if people for shallow lakes produced some disturbance, then in the second day of the same time phosphorus release is the most. The relevant departments to make a certain processing, reduce the nutrition level and at last reduce the harm.

D. Deep test, once appear the disturbance situation, the most close to the bottom of the sediment and the closest to the water for maximum phosphorus release. And in the middle of the depth of the phosphorus release is less and the nutrition is lighter. Because there is a certain distance between the mud depth and the water surface. Distance can result in a little effect on the phosphorus release from water surface. It can provide the basis for the relevant pollution control department. It also means that some relevant departments can increase the depth of the water to reduce endogenous pollution.

E. Follow-up researchers should sum up previous experience and put up new ideas. And constantly explore new knowledge, improve methods and strategies to establish a more complete, scientific research system or model. So as to we can find a more useful method to control the release of sediment phosphorus. And then control the eutrophication of the water body.

References

1. K. Kruusement, J.M. Punning, Proc. Estonian Acad. Sci. Biol. Ecol **49**, 163 (2000)
2. G.W. Zhu, B.Q. Qin, G. Gao, Chin. Sci. Bull **01**, 66 (2005)
3. D.P. Li, Y. Huang, Chin. Wt. Waste **04**, 17 (2013)
4. B. Yang, X.B. Yin, Jiangsu. EST **20**, 61 (2007)
5. J.B. Cai, X.F. Ding, H.Y. Peng, H.Q. Chang, X.E. Yang, P.M. Pu, J. Soil. Wt. Conserv **02**, 151 (2007)

Geochemical Characteristics and Geological Significance of the Tawangshan gold deposit in Dongfang City,Hainan Province

Ying XIONG [a], Zi Qiang ZHU and Guang Yin LU
School of Geoscience and Info-Physics, Central South University, Changsha 410083,China

Abstract: TheTawangshan gold deposit is located in the southwest of the Gezhen ductile shear zone in southwestern Hainan Island. Major, trace and rare earth element characteristics of the least altered granites and altered granite ores, as well as the oxygen and hydrogen isotopic features of the quartz, the sulfur and lead isotopic features of the pyrites, galena from this deposit were systematically studied. The altered granite ores content of $Fe_2O_3/(Fe_2O_3+FeO)$ ranges from 0.409 to 0.410,showing strong magma reducibility. REE characteristics of the granite ores are similar to those of the least altered granites, indicating a close relationship between the mineralization and granite. Hydrogen and oxygen analysis results show that the content of 18OSMOW ranges from 7.7‰ to 16.3‰, $\delta^{18}O_{H_2O}$ ranges from1.0‰ to 6.8‰, suggesting that the ore-forming fluid was the dual nature of the metamorphic and magmatic hydrothermal. The compositions of sulfur isotope and lead isotope imply that the metallogenic materials of the deposit derived from the deep magma mixing with the shell source material contamination. Ore deposit geological features reflect the obvious characteristics of magmatic hydrothermal and porphyritic granite is mineralized rocks, implying the deposit was formed in the post-collisional extensional tectonic setting.

1 Introdution

The Tawangshan gold deposit is located in the southeast of the Gezhen ductile shear zone, a small gold deposit newly discovered. After exploration, the deposit has submitted 1600kg gold resources. However, due to lack of theoretical research on mining gold mineralization characteristics, mineralization and other systems, further prospecting mining has been caused a lot of constraints. Based on the field of exploration and geochemical analysis, the metallogenic regularity and the genesis of mine have been studied, in order to provide theoretical guidance to further mining prospecting work, as well as enriching the theoretical knowledge Gezhen gold mineralization belt.

2 Regional geological setting

Tawangshan gold deposit is located in the western South China fold belt Wuzhishan fold system, Baisha depression belt and hold the plate crown area of the contact portion waterfront-Jinbo rift belt, south-eastern of Gezhen fault. The rock of the area is characterized by Kongliecun group, Tuolie group of Silurian System; Nanhao group, Qingxia group of Carboniferous System; E cha group of Permian System; Lu muwan group of Cretaceous System and Quaternary. Regional faults are

Corresponding authors: Xiong ying, E-mail:32973457@qq.com

DOI: 10.1051/matecconf/20166304026

well developed, there are Yalao fracture, Wangxia fracture, Yongwang fracture and Yan Woling fracture. The area magmatic is frequently activity, dominated by porphyritic biotite monzonite granite unit (porphyritic structure) →fine biotite monzonite granite (fine-grained granitic structure) → fine-grained biotite to life monzogranite unit (fine-grained granitic structure), composing of a sequence of structural evolution. The unit intruded into unit of Zhenghao and Permian E Cha group, and intruded by Bawangling super unit, Nam Young Ridge super intrusion. Zircon U-Pb isotopic age of the unit is 248±25Ma, whole rock Rb-Sr isochronal age is 245.9± 4.2Ma (Wang,1991).

3 Mining geology and mineralization characteristics

3.1 Geological characteristics

The strata exposed area is mainly Quaternary stratigraphy, Late Palaeozoic Permian strata. Quaternary strata are mainly distributed in the wash, by the alluvial and alluvial gravel, clay, and loam composition. Permian strata exposed in the south and southeast of the area, with wide 100-800m, length of 4km, tending to the north-west, dipping 50-65°.The rock of Permian strata is characterized by crystalline limestone, microcrystalline biolistic limestone and metamorphic siliceous sandstone.

The structural fault is dominated by NNE - NEE trend. The NNE-trending faults which strike 10 °, tend to the north-west, dip 70-80 °. This group faults cut through the Yanshanian biotite granite and Permian strata, with filling with broken quartz veins, lamprophyre vein and diorite vein. The NEE-trending group faults tend to northwest, dip 35 ° -55 °, followed by east-west fault was compressive echelon distribution, the local formation of gold-polymetallic ore body.

The area magmatism is frequently activity, dominated by Indosinian porphyritic granite and late Yanshanian biotite granite types. Indosinian porphyritic granite is explored in the north of the mining area, with a rock-like base. The late Yanshanian biotite granite is distributed in porphyritic granite batholith, with irregular rock stock. The west side of the rock strains of the long axis east to 1000m, minor 500m,and the eastern side semicircular rock strains, the major axis 700m, minor 500m.

3.2 Mineralization characteristics

According to the existing exploration project and expose fire assay analysis results, the Tawangshan gold deposit has delineated Ⅰ, Ⅱ,Ⅲ,Ⅳgold ore belt. The gold ores are always distributed in east-west faults, with no obvious boundaries between gold and rock .Mostly vein ore bodies form as lenticular, lenticular, along with a complex branching and tends to strike, pinch out the reproduction, the expansion and contraction of the local formation of barren skylights and other characteristics.

Ⅰ gold belt: with length of kilometres, width of meters to tens of meters, trending to east-west, this belt is composition of three ore bodies and two mineralized bodies, with lenticular shape. The gold ores has not been fully controlled, extending more than 200 meters.

Ⅱ gold belt: with length of several hundred meters, width of a few meters to tens of meters ,trending to east-west, this belt is a single ore body, composition of two ore bodies.

Ⅲ gold belt: with length of kilometers, width of meters to tens of meters , trending to east- west.

Ⅳ gold belt: as the smallest alteration zones ,this belt were circled a gold orebody ,with length of 50m, width of 3m, Au average grade of 1.2g/t .

4 Sample collection and analysis

The study samples were collected from the deep drill core. Major elements, trace elements and REE samples are mainly collected from unaltered and altered granite (ore) sample, and H-O isotope analysis samples are for the selection of the main mineralized quartz vein type ore of pure quartz minerals, but S-Pb isotopes analysis samples are for s the altered granite ore, the selection of pure mineral pyrite, galena minerals as a single sample.

Major elements, trace elements and REE elements are anglicised in ALS Laboratory (analysis results in Table 1). H-O isotope analysis was finished in Wuhan Institute of Geology and Mineral Resources isotope geochemistry laboratory, with MAT261 Mass Spectrometer. Sulfur and lead isotopes are analysis in State Key Laboratory of Geological Processes and Mineral Resources measured at China University of Geosciences (Wuhan).

DOI: 10.1051/matecconf/20166304026

4.1Granite of major, trace and REE characteristic

4.1.1 Major elements

The content of SiO_2 in unaltered granite ranges from 69.05 to 76.96wt.%.The total alkali content ranges from 7.75 to 9.40wt.%. The Al_2O_3 content ranges from 12.15 to 14.19wt.%.The K_2O content is higher than the Na_2O content. The rate of K_2O / Na_2O ranges from 1.24 to 3.35, with a potassium-rich feature,belonging to high-K calc-alkaline series. The MgO, TiO_2 and P_2O_5 content ranges from 0.06 to 0.88 wt.%, 0.04 to 0.43 wt.%,0.015 to 0.156 wt.%.The characteristic of the altered granites is dominated by rich in K, rich in alkali and poor in SiO_2 content. The rate of K_2O/Na_2O ranges from 1.02 to 1.66.The content of (K_2O+Na_2O) ranges from 6.99 to 8.59wt.%.The rate of $Fe_2O_3/(Fe_2O_3+FeO)$ ranges from 0.409 to 0.410, lower than China biotite granite average (0.44) and the world granite average (0.47), suggesting that the magma has strong reduction.

4.1.2 Ore-forming elements

The altered granite is rich in Au, Ag, Pb, Zn. The content of Pb, Zn and Au content was a significant positive correlation, suggesting that the gold mineralization is closely with lead-zinc mineralization. The Cu mineralization is not obvious. Either unaltered granites or altered granites ,the content of W, Mo, presently high-temperature ore-forming elements ,is very low, indicating that the mineralization temperature should be in medium-low temperature.

Table 1 Major, rare earth element and ore metal contents of the rocks and ores from the Tawangshan gold deposit

Sample name	unaltered porphyritic		unaltered biotite granite		altered porphyritic	
	ZK34-3-H1	ZK34-3-H4	ZK18-1-H1	ZK18-1-H2	ZK12-4-H6	ZK0-1-H7
SiO_2	76.96	69.05	72.56	74.42	57.96	51.23
TiO_2	0.10	0.43	0.25	0.04	0.44	0.50
Al_2O_3	12.15	13.82	14.19	13.55	15.56	15.60
Fe_2O_3	0.17	1.16	0.81	0.39	1.36	1.54
FeO	0.25	1.67	1.17	0.56	1.96	2.22
MnO	0.01	0.06	0.04	0.04	0.09	0.09
MgO	0.06	0.88	0.58	0.09	1.48	1.66
CaO	0.01	1.47	1.90	0.31	3.55	3.63
Na_2O	2.16	2.89	3.55	3.33	3.27	3.52
K_2O	7.24	4.86	4.43	5.63	3.72	3.61
P_2O_5	0.015	0.156	0.063	0.021	0.146	0.169
Total	99.13	96.45	99.54	98.38	89.54	83.77
Rb	367	218	178	194	397	335
Nb	11	11	8	9	18	16
La	39.5	44.8	27.9	29	78.4	36.8
Ce	85.8	84	53.8	53.9	149.5	69.2
Pr	8.95	9.18	5.79	5.87	15.35	7.24
Nd	31.4	31.2	21.4	21.6	52	25.5
Sm	6.21	5.55	3.91	3.83	8.82	4.48
Eu	0.44	0.93	0.93	0.79	1.38	0.88
Gd	5.28	4.12	2.55	2.41	6.56	2.71
Tb	1.13	0.56	0.34	0.33	0.93	0.4
Dy	7.19	2.76	1.87	1.93	5.26	2.21
Ho	1.52	0.48	0.35	0.33	0.94	0.43
Er	4.69	1.23	1.11	1	2.57	1.19
Tm	0.71	0.17	0.15	0.16	0.4	0.18
Yb	4.67	0.99	1	1.18	2.6	1.05
Lu	0.72	0.15	0.16	0.18	0.4	0.17
Y	37	13	9.9	10.4	29.3	11.3
ΣREE	198.21	186.12	121.26	122.51	325.11	152.44
LREE	172.3	175.66	113.73	114.99	305.45	144.1
HREE	25.91	10.46	7.53	7.52	19.66	8.34
LREE/HREE	6.65	16.79	15.1	15.29	15.54	17.28
LaN/SmN	4.11	5.21	4.61	4.89	5.74	5.3
GdN/YbN	0.94	3.44	2.11	1.69	2.09	2.14
δEu	0.23	0.57	0.84	0.74	0.53	0.71
δCe	1.07	0.96	0.98	0.96	0.99	0.98
Au	14	14	15	19	205	29
Ag	449	150	1971	1989	448	1049
Cu	11	6	16	22	15	1
Pb	57	57	77	54	133	45
Zn	59	81	103	77	84	45
Mo	2	1	19	1	1	0
W	6	0	1	0	0	0

4.1.3 REE elements

The content of rare earth elements in unaltered granite ranges from121.26×10^{-6} to 198.21×10^{-6}, distribution patterns rightist, LREE enrichment (Fig. 1a). The rate of LREE/HREE ranges from 6.65 to 16.79. The rate of (La/Yb) N ranges from 6.07 to 32.46. The rate of (Gd/Yb) N ranges from 0.94 to 3.44.The rate of (La/Sm) N ranges from 4.11 to 5.21. The δEu content ranges from 0.23 to 0.84. The δCe content ranges from 0.96 to 1.07.

The content of rare earth elements in altered granite ranges from 126.85×10^{-6} to 325.11×10^{-6}, distribution patterns rightist, LREE enrichment (Fig. 1b). The rate of

LREE / HREE ranges from 15.54 to 19.83. The rate of (La/Yb) N ranges from 21.63 to 25.22. The rate of (Gd/ Yb) N ranges from 1.94 to 2.21 .The rate of (La/Sm) N ranges from 5.30 to 6.62 .The δEu content ranges from 0.53 to 0.72. The δCe content ranges from 0.96 to 0.99.

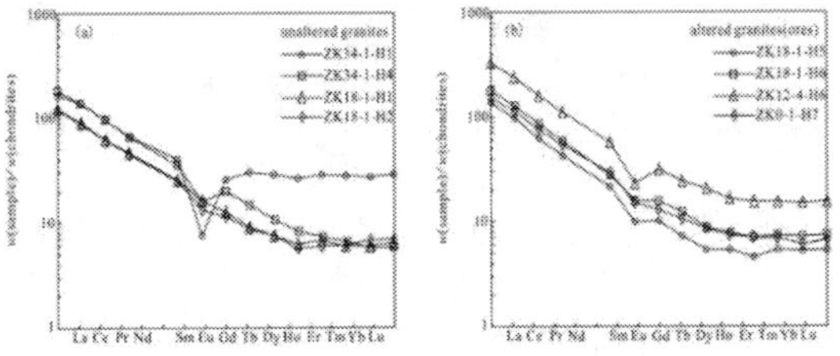

Fig.1 Chondrite normalized REE patterns of the granites

4.2 Isotopic Characteristics

4.2.1 H-O isotope

The hydrogen and oxygen isotopic date of the Tawangshan gold deposit are listed in Table 2. Theδ D of all samples range from -59.0‰-99.0‰ ,showing significant isotopic fractionation occurrence. The $\delta^{18}O_{SMOW}$ of all samples range from 7.7‰~16.3‰.The $\delta^{18}O_{H_2O}$ of all samples range from 1.0‰~6.8‰, with an average of 4.1‰。

Table 2 Oxygen and hydrogen isotopic date of the Tawangshan gold deposit

sample number	types ore	$\delta^{18}O_{SMOW}$/‰	δD_{SMOW}/‰	homogenization temp/℃	δO_{H_2O}/‰
Ta-7	alterd rock type gold ore	9.5	-96	289	2.2
Ta-8	alterd rock type gold ore	10.6	-75.8	306	3.9
Ta-9	alterd rock type gold ore	10.4	-87.1	296	3.3
Ta-10	alterd rock type gold ore	7.7	-59	357	2.6
Ta-11	alterd rock type gold ore	12.4	-99	315	6.0
Ta-12	alterd rock type gold ore	12.9	-91	283	5.3
Ta-13	quartz vein type gold ore	16.4	-64.3	240	6.8
Ta-14	quartz vein type gold ore	11.6	-77	219	1.0
Ta-15	quartz vein type gold ore	16.3	-70.1	240	6.7
Ta-16	quartz vein type gold ore	14.3	-59	219	3.6

4.2.2 S isotope

The sulfur isotope of pyrite, galena values are listed in Table 3.The $\delta^{34}S$ of all sample range from 6.3‰ to 9.1‰, with an average of 7.8‰ and range of 2.8‰.Sulfur isotope variation range is narrow, relatively homogeneous composition and exhibiting a relatively enriched in ^{34}S isotope characteristics.

Table 3 Sulfur isotope results from the Tawangshan gold deposit

sample number	types ore	sulfide	$\delta^{34}S$/‰
Ta-7	alterd rock type gold ore	pyrite	7.8
Ta-8	alterd rock type gold ore	pyrite	8.0
Ta-9	alterd rock type gold ore	pyrite	8.0
Ta-10	alterd rock type gold ore	pyrite	7.6
Ta-11	alterd rock type gold ore	pyrite	7.6
Ta-12	alterd rock type gold ore	galena	6.3
Ta-13	quartz vein type gold ore	galena	7.1
Ta-14	quartz vein type gold ore	galena	8.2
Ta-15	quartz vein type gold ore	pyrite	8.6
Ta-16	quartz vein type gold ore	pyrite	9.1

4.2.3 Pb isotope

The lead isotope of galena values are listed in Table 4.The rate of $^{206}Pb/^{204}Pb$ ranges from 18.09 to 18.39, with an average of 18.25.The rate of $^{207}Pb/^{204}Pb$ ranges from15.47 to 15.58，with an average of 15.53.The rate of $^{208}Pb/^{204}Pb$ ranges from 37.74 to 38.17, with an average of 37.91.The content of μ ranges from 9.24 to 9.43, with an average of 9.34. The content of ω ranges from 34.18 to 35.35, with an average of 34.56.

Table 4 Lead isotope results from the Tawangshan gold deposit

sample number	$^{206}Pb/^{204}Pb$	$^{207}Pb/^{204}Pb$	$^{208}Pb/^{204}Pb$	μ	ω	Δβ	Δγ
Ta-7	18.29	15.54	37.88	9.36	34.33	14.10	17.77
Ta-8	18.21	15.49	37.85	9.27	34.18	10.84	16.96
Ta-9	18.34	15.56	38.01	9.39	34.78	15.40	21.26
Ta-10	18.18	15.52	37.83	9.33	34.52	12.79	16.43
Ta-11	18.39	15.58	38.17	9.43	35.35	16.71	25.56
Ta-12	18.09	15.47	37.74	9.24	34.18	9.53	14.01

DOI: 10.1051/matecconf/20166304026

5 Discussion

5.1 Ore-forming fluid source

Projecting the isotope date (Table 2) onto hydrogen and oxygen isotopic composition diagram (Fig.2), showing that the almost sample of isotope date are fallen on the area near the magmatic water, and some sample is fallen metamorphic water area. This phenomenon indicates that ore-forming fluids have reaction with metamorphic rocks as it flowing from deep. As well as, the rain hydrothermal may be fallen in mineralization process, and may be very favourable. Therefore, the ore-forming fluid of the Tawangshan gold deposit has a characteristic with metamorphic and magmatic hydrothermal as a mixed fluid characteristic.

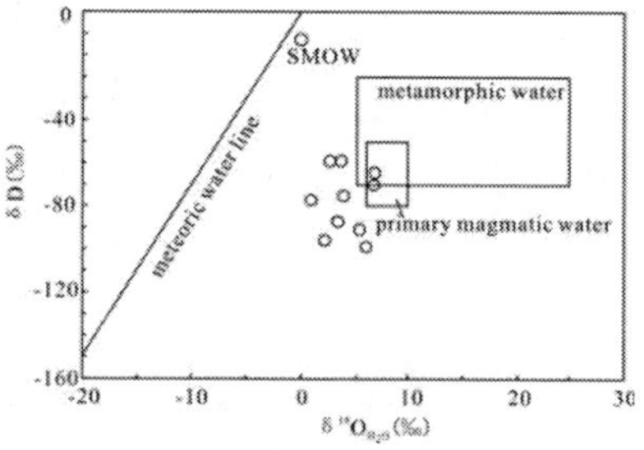

Fig.2 Plot of hydrogen and oxygen isotopes for gold ores of the Tawangshan gold deposit

5.2 Forming material source

5.2.1 S isotope tracer

Compared to sulfur isotope reservoir（Ohmoto and Rye，1979） (Fig. 3), the content of sulfur isotope in Tawangshan gold deposit is more than the content of sulfur isotope in magmatic sulfur ($\delta^{34}S = 0 \pm 5‰$), while lower than the content in seawater ($\delta^{34}S$ in Cambrian seawater limited of 15‰). Meanwhile, the value of $\delta^{34}S$ shows a high positive value, closely to oxidizing granitites（Coleman，1979）.It may be associated with fluid degassing. Thus we believe that the sulfur isotope in Tawangshan gold deposit is from magma sulfur mixing with the portion of the stratum sulfur.

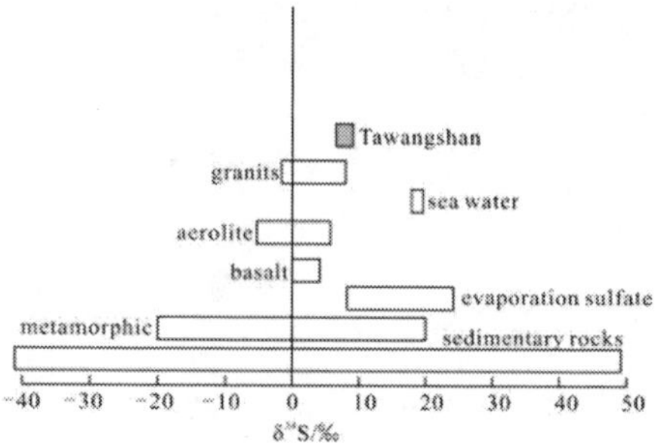

Fig.3 Sulfur isotope histograms for gold ores of the Tawangshan gold deposit

5.2.2 Pb isotope tracer

Project the lead isotope values of the Tawangshan gold deposit on the lead isotope evolution of tectonic setting diagram(Zartman,1981). In $^{207}Pb/^{204}Pb$-$^{206}Pb/^{204}Pb$ diagram (Fig.4a), the sample is located between the mantle and the orogenic belt. In $^{208}Pb/^{204}Pb$-$^{206}Pb/^{204}Pb$ diagram (Fig.4b), the sample is located between the orogenic belt and upper crust, reflecting that the lead is from the deep source material related to orogeny. IN the $\Delta\beta$-$\Delta\gamma$ diagram(Fig.5), the sample is located in the mantle and mixed upper crust and mantle-derived lead (closer to the magmatism).So we can think that the lead of the Tawangshan gold deposit is from deep mantle source, mixing some crust lead source.

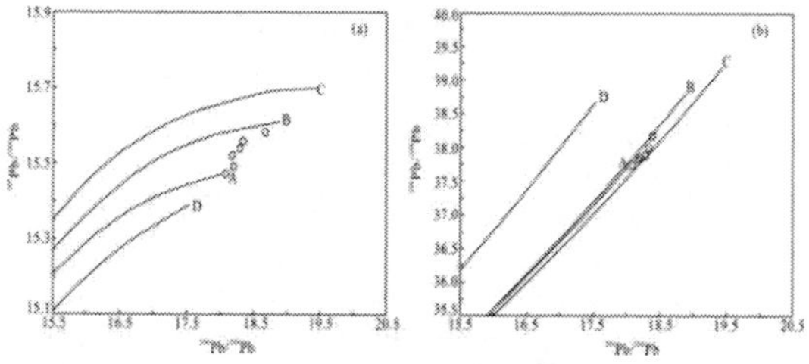

Fig.4 Diagram showing evolutionary tectonic setting of Pb isotope from the Tawangshan gold deposit

A-mantle; B- orogenic belt; C- upper crust; D- lower crust

5.3 Metallogenic tectonic setting

Generally, the ore-forming hydrothermal which the deposit related with granitites is from the richment of volatile components late of granitic evolution. And then the granitic magnatism should be earlier than the mineralization events. Therefore, granitic magmatism tectonic environment can reflects tectonic setting of mineralization. Projected the trace elements onto the tectonic discrimination diagram (Fig.7), the sample are mainly fallen in syn-collision granites area. The syn-collision granites include continent - continent collision and late the collision and continue to converge on the role (Han, 2007). Studies have shown that the collision organic belt magmatic hydrothermal deposits are formed in the collision extensional tectonic environment, not in the main stage of collision environment (Kerrich et al, 2000; Hou et al, 2011).This is consistent with Indosinian gold mineralization tectonic environment within Gezhen fault belt, which is in extensional tectonic environment in Indosinian(Ding and Peng,1993). Therefore, the tectonic environment of the Tawangshan gold deposit may be extensional tectonic environment after the collision.

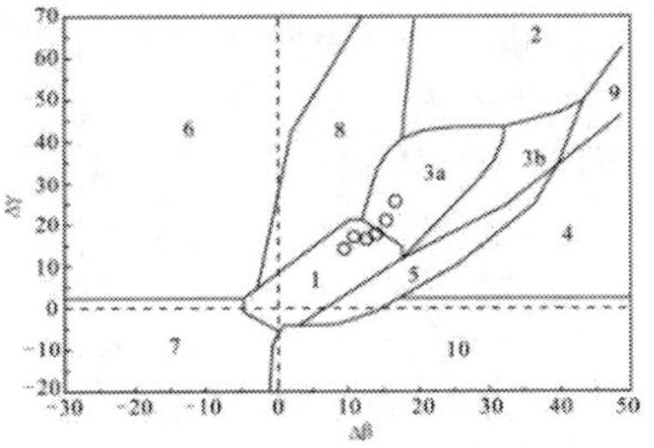

Fig.5 Δβ-Δγ diagram of genetic classification of ore lead isotopes from Tawangshan gold deposit

1-mantle-derived lead;2-upper crust –derived lead;3-mixed upper crust and mantle-derived lead（3a.magmatism；3b.sedimentation）；4-lead from chemical deposition；5-lead from submarine hydrothermalism；6-lead from medium-deep matamorphism ； 7-lower crust-derived lead from matamorphism；8-lead from orogenic belt；9-lead from upper crust of ancient shale；10-lead from retrogrande metamorphism

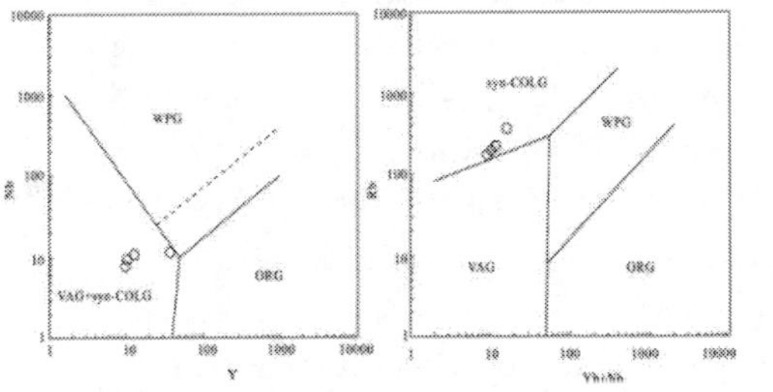

Fig.6 Diagrams of tectonic environment for granites of Tawangshan gold deposit

VAG-volcanic arc granites；ORG-oceanic ridge granites；WPG-within plate granites；syn-COLG-syn-collision granites

6 Conclusions

The ore body is mainly quartz veins type and altered rock type in the Tawangshan gold deposit, hosting in granite. The granite is rich in medium –low temperature ore-forming elements. The REE characteristics of REE in altered granite (ores) is similarly with the unaltered granite, showing the closely relation between mineralization with granite.

The ore-forming fluids is from magmatic hydrothermal and metamorphic hydrothermal, according to the H-O isotopic characteristics. The source of ore-forming elements is from deep magma, mixing with crust lead, based on the S, Pb isotopic characteristics.

The Tawangshan gold deposit is a magmatic hydrothermal deposits, formed after the collision extensional environment.

References

1 Li Xianhua, Zhou Hanwen, Ding Shijiang.Ocean ridge in metamorphic rocks of Hainan: ocean crust relic of Gut Tis?[J]. Chinese science bulletin, 45,1(2004).

2 Li Zhengxiang,LI Xianhua.Formation of the 1300 km-wide intracontinental orogen and postorogenic magmatic province in Mesozoic South China:A flat-slab subduction model[J].Geology,,35,2(2007).

3 Kerrich R, Goldfarb R, Garwin S. The characteristics, origins,and geodynamic settings of supergiant gold metallogenic provinces[J]. Science in China Series D: Earth Science, 43,1(2004).

4 Ohmoto H. Stable isotope geochemistry of ore deposits[J].Reviews in Mineralogy and Geochemistry,16,1(1986).

5 Wu Chuanjun, Xu Deru, Fu Yangrong. Research of metallogenic regularity and metallogenic mechanism of important types of gold deposits in Hainan island[J].Acta Mineralogica Sinica, 32,S1(2011).

6 Xia Bin,Cui Xuejun, Zhang Yanhua. Dynamic factors for the opening of south china sea and a numerical modeling discussion[J].Geotectonica et Metallogenia, 29,2(2005).

MATEC Web of Conferences **63**, 04027 (2016)
MMME 2016
DOI: 10.1051/matecconf/20166304027

Algorithm Robustness Analysis for the Choice of Optimal Time Delay of Phase Space Reconstruction Based on Singular Entropy Method

Zai Yu CHEN [1], Bao Heng YAO [2], Mei Na GUO [1]
[1]*China Offshore Oil Engineering Corporation, Ltd Tianjin, 300451*
[2]*Institute of Oceanology, Shanghai Jiao Tong University, Shanghai 200240*

Abstract. The method of Delays is commonly used in state space reconstruction of chaotic time series, and the quality of attractor reconstruction of phase space is sensitive to the choice of embedding parameters, i.e. embedding dimension and time delay. In this paper, singular entropy (SE) is applied to choose the optimal time delay in reconstructed dynamical system, and the comparisons of relevant results are made with the correlation function method by using singular value fraction (SVF). The robustness of singular entropy algorithm is further discussed, and it shows the advantage of the proposed method.

Keywords.singular entropy; phase space reconstruction; the optimal time delay; robustness

0 Introduction

The analysis of chaotic time series has become popular in many fields of science and engineering since the work of Packard [1] and Takens [2] who initially introduced the delay-coordinate embedding technique to reconstruct a smooth copy of the underlying dynamics. According to Takens' theorem, a scalar time series $\{x_i\}, i = 1, 2, \ldots, N$ can be embedded into a m-dimensional space to form multiple state-space vectors X_i if the embedding dimension m is equal or more than twice the topological n, i.e. $m \geq 2n+1$. The reconstructed trajectory X is given by

$$X = [X_1, X_2, \cdots, X_{N_m}]^T \qquad (1)$$

The reconstructed state of the system at each discrete time i is

$$X_i = (x_i, x_{i+L}, \ldots, x_{i+(m-1)L}) \qquad (2)$$

where m is the embedding dimension, L the reconstruction time delay defined as the integer multiple of sampling time τ_s, and N_m the number of vectors in state space which is given as $N_m = N-(m-1)L$.

In fact, the choice of embedding parameters is critical for the quality of attractor reconstruction due to the time series with noise and finite length. An excessively small value of L results in redundancies between successive delay coordinates with little information, while too large value of L may cause successive delay coordinates irrelevant so that the reconstruction attractors are no longer the representative of true dynamics [3].

Different methods have been proposed to choose the value of L based on the criteria of sequential correlation and dimensional expansion. Albano suggested using L as optimal time delay when the autocorrelation function $R_{xx}(L)$ first drops to 1/e of its initial value [4], and Fraser and Swinney proposed to choose the first minimum value of the mutual information function L as optimal time delay instead [5]. The geometrical concepts of *fill factor* [6], *wavering product* [7] and *average displacement* [8] are also applied to measure the expansion of attractors in the state space. Martinerie considered that it's more meaningful to estimate m and L simultaneously with a time window [9], and Kember proposed the recipe of correlation function termed *singular value fraction* (SVF) to choose proper time delay [10]. Such methods as high-order correlation [11], nonlinear correlation functions [12], and singular value spectrum entropy [13] are also proposed to overcome the deficiency of autocorrelation function.

In this paper, on the basis of global *singular value decomposition* (*SVD*), the concept of *singular entropy* (*SE*) [14] is adopted to determine the optimal time delay L. Section 1 presents the definition of singular entropy (*SE*) and its application in the description of information measure. In Section 2, *SE* is applied to choose the optimal time delay L of dynamical reconstruction, where the comparisons of relevant results with SVF method are concerned. Section 3 discusses the influences of the order of SE and the signal to noise rate (*SNR*) of time series on the choice of time delay. The corresponding conclusion is drawn in Section 4.

1 The definition of singular entropy

For a measured discrete time series $\{x_i\}$ that is normalized to have zero mean and unit variance, the SVD of reconstructed trajectory matrix (1) gives

$$X = S\Lambda U^T \qquad (3)$$

where S is the $N_m \times m$ matrix of eigenvectors of XX^T, U the $m \times m$ matrix of eigenvectors of $X^T X$,and Λ $m \times m$ diagonal matrix consisting of singular values σ_i, $\sigma_1 \geq \sigma_2 \geq \ldots \geq \sigma_m \geq 0$, i.e., $\Lambda = diag(\sigma_1, \sigma_2, \ldots, \sigma_m)$.

Rearranging equation (3)

$$(XU)^T XU = \Lambda^2 \qquad (4)$$

Thus, σ_i can be considered as a measure of power resident in the column vector of XU whose nonzero amounts represent different patterns of X and information measure of original signal. Λ is also related to the expansion of attractor, where symmetric spread with all equal σ_i is more preferable [10]. Instead of the description of singular value fraction function $f_{sv}(k)$ in references [10] for attractor expansion, in this paper singular entropy (*SE*) is adopted instead and defined as

$$E(k) = \sum_{i=1}^{k} \Delta E_i, \quad i = 1, 2, \cdots, m \qquad (5)$$

where k is the order of singular entropy ($k \leq m$), and ΔE_i the increment of singular entropy in order i which is given by

$$\Delta E_i = -p_i \log p_i \qquad (6)$$

p_i reflects the proportion of the *i*th singular value in the total singular value spectrum that contributes to the global pattern, which is defined as

$$p_i = \sigma_i^2 \Big/ \sum_{i=1}^{m} \sigma_i^2 \qquad (7)$$

Therefore, singular entropy can be considered as information measure of signals [14], and the example is given as follows.

The following of this paper is concerned with the application of SE for the choice of L, and the comparisons with SVF method for dynamical phase reconstruction is made as well.

2 Applications for the choice of L with Singular entropy (*SE*)

The choice of time delay of reconstructed dynamics with singular entropy involves in this section. Three equations of sinusoid and Rössler are employed. k is chosen to be 1 in the examples and whose influence on L will be further discussed in Section 4.

2.1 Choice of L for sinusoid

Fig.1 shows the curve of $E(1)$ versus L

DOI: 10.1051/matecconf/20166304027

for time series of $x(t)=\sqrt{2}\sin\pi t$ with sample time $\tau_s = 0.01$ s and embedding dimension $m = 2,3,4,5,7,9$ respectively. Larger value of $E(k)$ implies more information from the first k singular values of reconstructed dynamics, which is closely related to preferable expansion of attractors.

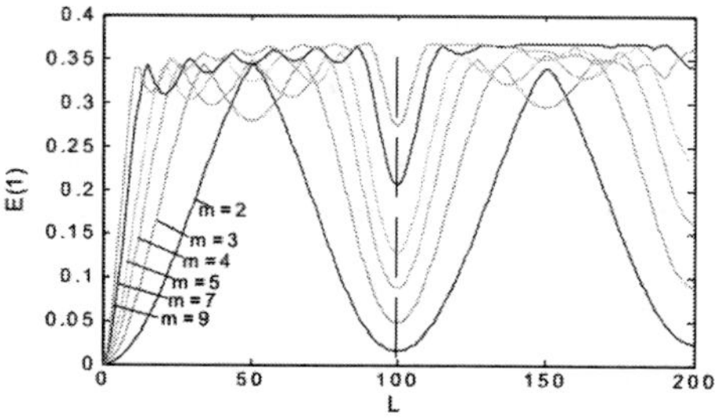

Fig.1 $E(1)$ versus L for a time series of $x(t)=\sqrt{2}\sin\pi t$ with $m = 2,3,4,5,7,9$

In fig.1, the moment when first maximum of $E(1)$ occurs is chosen to represent the optimal time delay denoted as L_m, which is closely related to the time of Δ_M when first minimum $E(1)$ occurs and given by

$$L_m \approx \Delta_M / m \qquad (8)$$

Optimal time window Δ_w is defined as

$$\Delta_w = (m-1)L_m \approx (m-1)\Delta_M / m \quad (9)$$

Similar conclusions can be found in reference [10], while the proposed method here applies singular entropy instead of using linear SVF function.

For comparisons, Fig.2 shows the plots of $fsv(1)$ versus L by using the method of SVF for the time series of $x(t)=\sqrt{2}\sin\pi t$ with $m = 2,3,4,5,7,9$ respectively.

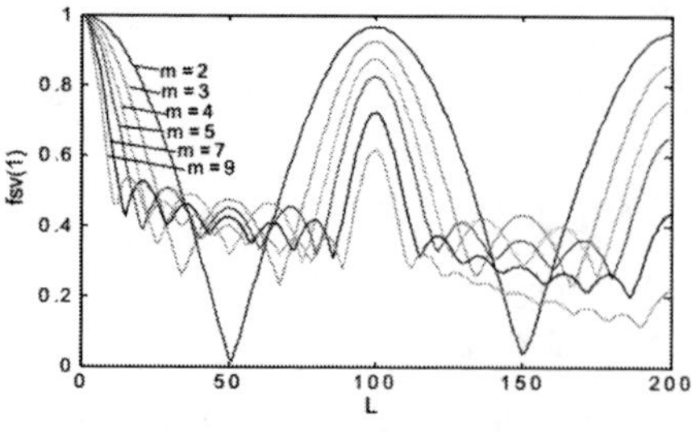

Fig.2 $fsv(1)$ versus L for a time series of $x(t)=\sqrt{2}\sin\pi t$ with $m = 2,3,4,5,7,9$

Table 1 shows the variation of L_m with different m for time series of $x(t)=\sqrt{2}\sin\pi t$, and the results are compared with SVF method as well.

Table 1. the variation of L_m with different m for time series of $x(t)=\sqrt{2}\sin\pi t$ and the comparisons with SVF method.

	L_m	Δ_M	Δ_M / m	Δ_w	m
51	2	51	100		50
68	3	34	100		33
78	4	26	100		25
SE 84	5	21	100		20
90	7	15	100		14
96	9	12	100		11
51	2	51	——		——
68	3	34	——		——
78	4	26	——		——
SVF 84	5	21	——		——
90	7	15	——		——
96	9	12	——		——

2.2 Choice of L for Rössler equations

Consider Rössler system described by the equations $\dot{x} = -y - z$, $\dot{y} = x + ay$, $\dot{z} = b + z(x - c)$, where $a = 0.15$, $b = 0.20$, $c = 10.0$. The differential equations are solved with numerical integration using a four-order Runge-Kutta algorithm. Initial values are chosen to be $x_0 = 10$, $y_0 = 0$, $z_0 = 0$ and the integral step is fixed at $\tau_s = \pi / 100$. The first 5000 data samples are discarded.

Fig.3 shows the curve of $E(1)$ versus L for Rössler equations using a time series of $x(t)$ with embedding dimension $m = 2,3,4,5,7,9$ respectively.

	L_m	Δ_M	Δ_M / m	Δ_w	m
SE	48	2	48	96	48
	64	3	32	96	32
	72	4	24	96	24
	76	5	19	96	19
	84	7	14	96	14
	88	9	11	96	11
SVF	48	2	48	——	——
	64	3	32	——	——
	72	4	24	——	——
	76	5	19	——	——
	84	7	14	——	——
	96	9	11	——	——

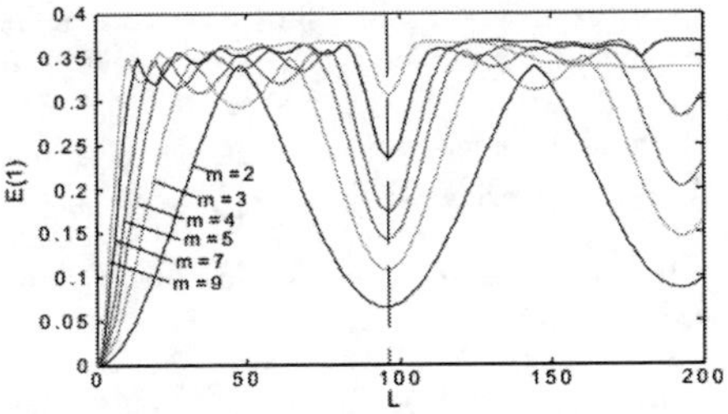

Fig.3 $E(1)$ versus L for Rössler equations using a time series of $x(t)$ with $m = 2,3,4,5,7,9$

Fig.4 shows the curve of $fsv(1)$ versus L for Rössler equations using a time series of $x(t)$ with the same parameters as those of Fig.5.

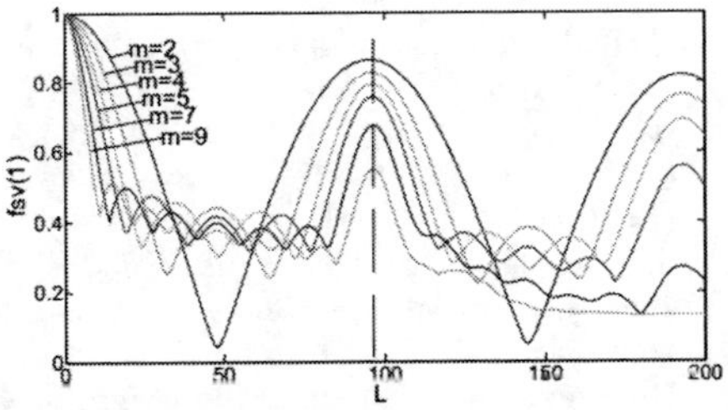

Fig.4 $fsv(1)$ versus L for Rössler equations using a time series of $x(t)$ with $m = 2,3,4,5,7,9$

Table 2 shows the variation of L_m with different m for Rössler equations using a time series of $x(t)$ and resultant comparisons as well.

Table 2 The variation of L_m with different m for time series of Rössler equations and the comparisons with SVF method

3 The robustness analysis of SE method

The robustness of SE method for the choice of L, including the effects of order k and the corresponding signal to noise rates (SNR) of time series, involves in this section.

Fig.5 shows the curves of $E(k)$ versus L for the same time series of (a) $x(t) = \sqrt{2}\sin \pi t$, and (b) Rössler equation respectively as mentioned in Section 2 with fixed $m = 7$ and the variation of k from 1 to 7.

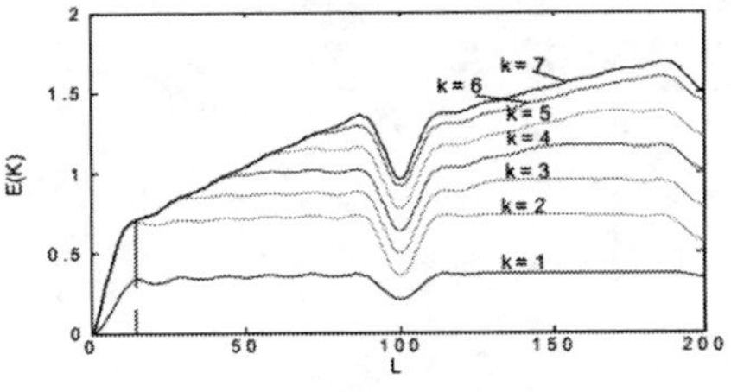

(a)

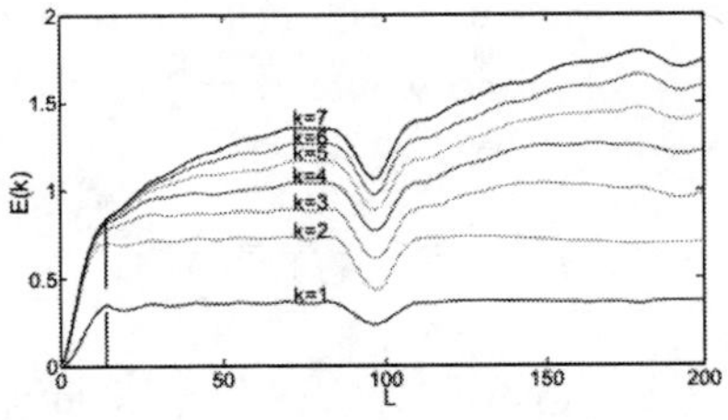

(b)

Fig.5 $E(k)$ versus L for time series of (a) $x(t) = \sqrt{2}\sin \pi t$ and (b) Rössler with $m = 7$ and $k = 1,2,3,4,5,6,7$

From Fig .5, the choice of larger k fails to reach satisfactory results, and therefore the smaller one is recommended instead.

To examine the sensitivity of SE method to

the SNR of time series, $m = 7$ and $k = 1$ are fixed. Fig.6 shows the plots of $E(1)$ versus L for clean, and noisy signals with SNR=20dB, 10dB, 6dB and 0dB respectively. For time series of (a) $x(t) = \sqrt{2}\sin \pi t$ and (b) Rössler equation, it is shown that the choice of optimal time delay is insensitive to the SNR of time series. It shows the advantage of the proposed method.

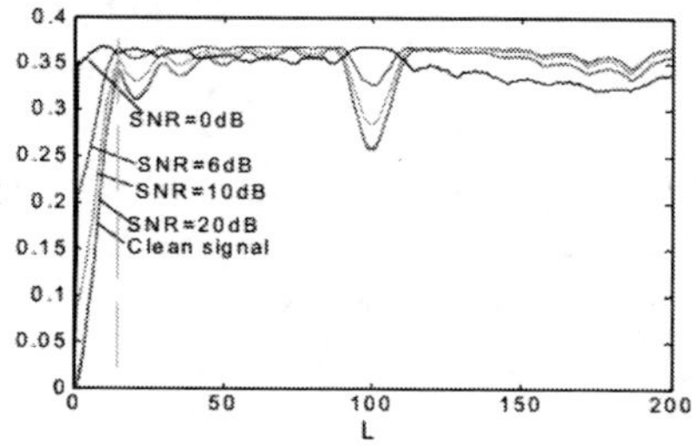

(a)

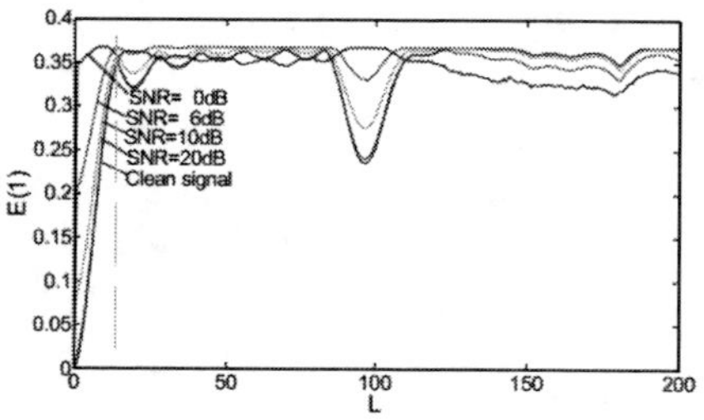

(b)

Fig.6 $E(1)$ versus L for clean and noisy signals with SNR=20dB, 10dB, 6dB and 0dB respectively of (a) $x(t) = \sqrt{2}\sin \pi t$ and (b) Rössler equation with $m = 7$

4 Conclusion

Singular entropy (*SE*) is applied to choose the optimal time delay in reconstructed dynamics. Algorithmic robustness for the choice of time delay of time series, including the effects of the order of *SE* and the corresponding signal to noise rates (*SNR*) ,is discussed. It's found that smaller order of *SE* is more preferable, and the choice of time delay is insensitive to the *SNR* of time series with appropriate order, and it shows the advantages of the proposed method.

Acknowledgements

Part of this work was supported by Natural Science Foundation of China, (Grant NO. 51279107).

References

[1] N. H. Packard, J. P. Crutchfield, J. D. Farmer and R .S. Shaw, Phys.Rev.Lett.45 (1980) 712.
[2] F. Takens, in: Dynamical Systems and Turbulence, Eds. D.A. Rand and L.S. Young (Springer, Berlin, 1981)
[3]M.Casdagli, S.Eubank, J.D.Farmer, J. Gibson, State space reconstruction in the presence of noise, Physica D 51(1991) 52
[4]A.M.Albano, J.Muench, C.Schwartz, Singular value decomposition and the Grassberger-Procaccia algorithm, Phys.Rev.A38 (1998) 3017-3026.
[5]A.M.Frazer, H.L.Swinney, Independent coordinates for strange attractors from mutual information. Phys.Rev.A 33(1986) 1134- 1140
[6]T.Buzug, G. Pfister, Optimal delay time and embedding dimension for delay-time coordinates by analysis of the global static and local dynamical behavior of strange attractors, Phys.Rev.A 45(1992) 7073
[7]W.Liebert, K.Pawelzik, H.G.Schuster. Optimal embeddings of chaotic attractors from topological considerations. Europhys.letter 14(1991) 521
[8] Michael T.Rosenstein, James J.Collins, Carlo J. De Luca, Reconstruction expansion as a geometry-based framework for choosing proper delay times, Physica D 73(1994) 82
[9] J.M.Martinerie, A.M.Albano, A.I.Mees and P.E.Rapp, Phys.Rev.A 45 (1992) 7058
[10]G.Kember, A.C.Fowler. A correlation function for choosing time delays in phase portrait reconstructions. Physics Letter A 179(1993)72~80
[11] A.M.Albano, A. Passamante and M.E. Farrell, Physica D 54(1991) 85.
[12]L.A.Aguire, A nonlinear correlation function for selecting the delay time in dynamical reconstruction, Physics Letters A 203(1995) 88-94
[13]YAO Baoheng. The Principle of optimum time lag choice based on singular value spectrum entropy, Journal of Shanghai Jiao Tong University, 2005,Vol 39,No.1,150-153
[14]YANG Wenxian, Jiang Jiesheng. Study on the singular entropy of mechanical signal, Chinese Journal of Mechanical Engineering, 2000,Vol 36: 9-1

MATEC Web of Conferences **63**, 04028 (2016) DOI: 10.1051/matecconf/20166304028
MMME 2016

Design and experimental studies of a novel circularly polarized antenna

Hai ZHANG [a], Wei Guo ZHANG and Yue He GE

College of Information Science and Engineering, Huaqiao University, Xiamen 361021, China

Abstract. A circularly polarized antenna with patches on the top is studied in this paper. The antenna is composed of two printed dipoles and four radiation patches which are loaded on top of the dipoles. The antenna is fed by microstrip balum in the middle of dielectric slab. And the balum is connected to a Wilkinson power divider which owns 90 degree phase shift. Numerical simulations of proposed antenna are performed using the 3D electromagnetic simulation software HFSS. The experimental studies of the antenna are also carried out. Measured results show that the parameter S11 is less than -10 dB from 1.1GHz to 1.8GHz while the 3dB axial ratio bandwidth is from 1.27 GHz to 1.62 GHz. The highest gain of 5dBi can be achieved during the operating frequencies.

1 Introduction

Circularly polarized (CP) micro-strip antenna has became one of the hottest research field of communication by its advantages, such as, small volume, light weight, simple construction, strong anti-interference capacity, tractable and so on. In recent years, the GPS and BD system have been used into military and civil communication fields extensively, while CP antennas are the important part of these systems. The frequency band of GPS contains L1 (1575.42 MHz), L2 (1227.6 MHz), L3 (1381.05 MHz), and E5a (1164~1189MHz), E5b (1189~1214 MHz) which belong to the European Galileo navigation system [1]. The frequency band of BD system contains L1 (1602 MHz), L2 (1246 MHz), B1 (1561 MHz), B2 (1207 MHz).With the further research of CP antennas, how to realize broad axial ratio has became a research hotspot. The methods of realizing CP antennas have trend towards diversify. Slotting a strip slot on the surface of patch antenna can realize circularly polarized [2-4]. We can obtain high gain using horn antenna, but its size is too big and it is difficult to process [5]. By using single feed patch to design CP antenna is easy to process, but its bandwidth was narrow [6]. By using circle structure to realize CP, its structure is simple. However, it also can't obtain ideal bandwidth [7, 8]. In addition, there are L strip antennas [9], dual-band CP antenna [10], and double cross patch antenna [11] .Some scholars have proposed using broadband CP antenna to cover many bands of GPS system [12]. Therefore the broadband CP antennas that can cover part of the communication bands of GPS system are designed in this paper.

The unit of CP antenna are designed by combining the two double polarization printed antenna elements, and this structure has good bandwidth and good isolation [13, 14] .There are four vertical radiation patches on the top of

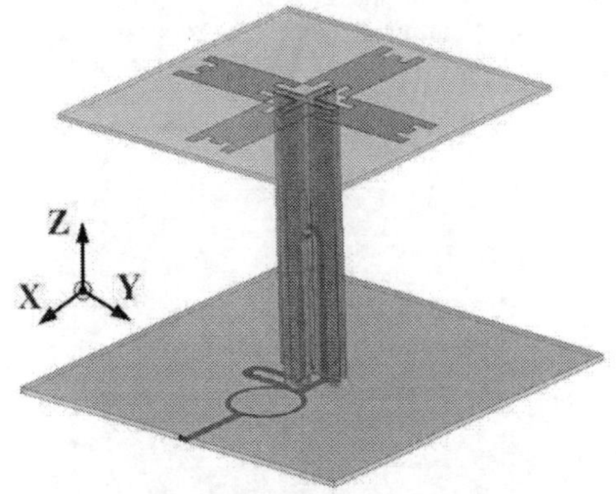

Figure 1. Simulation model of CP antenna

the antenna. And the antenna is fed by the one Wilkinson micro-strip power divider which has 90° port phase shift on the bottom of the antenna. The printed power divider has many advantages such as simple structure, easy processing, and so on, such it is always used as feed network to feed small antennas. In this paper, the power divider has been printed on the bottom of the baffle-board. It can feed the antenna more simply, and has no-increase of the antenna size, while it can realize the broadband circularly polarization.

[a] Hai ZHANG: xjtuzhanghai@163.com .

2 Simulation study of CP antenna

In this section, the simulation model of CP antenna is shown in figure 1. It can be found that the two linear polarizations which have constant amplitude and the phase difference of 90 ° can produce circular polarized wave. So in this paper, the halving feed network which has the phase difference of 90 ° was used to feed the unit of antenna. By emulating and optimizing process, the antenna can possess the wide-band characteristics and is also stable. Designing CP antenna on this basis can effectively expand the axis ratio bandwidth of CP antenna and can make the radiation pattern of the antenna stable. The structure of the antenna in this paper has been shown in figure 2(dual polarized printed elements).

This antenna unit is composed of two double-sided printing vibrators with 90 ° phase difference. In the actual design, the double-sided printing vibrator is composed of two dielectric-slabs added together, and it is fed by the micro strip Balun in the middle of it. The Balun feedings two vibrators have different height, and the slots were shaped on the bottom and top of two vibrators PCB. In this way, the two linear polarized vibrators can decussate together and realize the double polarization performance. There are four horizontal metal strips on the top. The antenna was designed using the 3D simulation software HFSS and the structure of the antenna is constructed as followed. The double-sided printing elements decussate together and the antenna is fed by the two J type Balums.

On the top of antenna, there is a dielectric-slab with the single sided printed metal strips. A cross bath that is used to package the antenna was slotted in the middle, and the top pastern is shorted with double-sided printed elements. The optimized size of the antenna is shown in table 1.

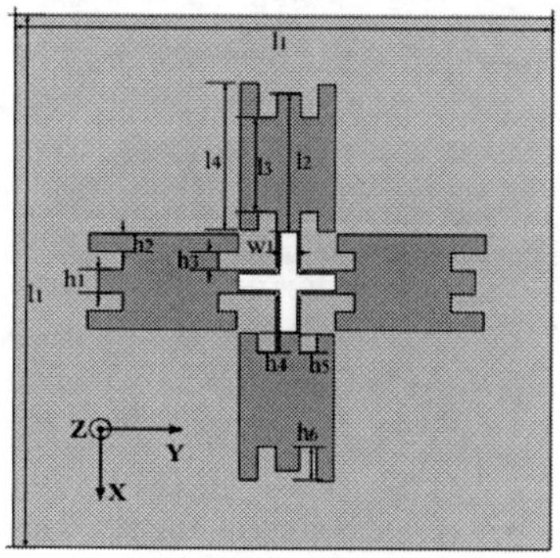

(a) The top loaded paches

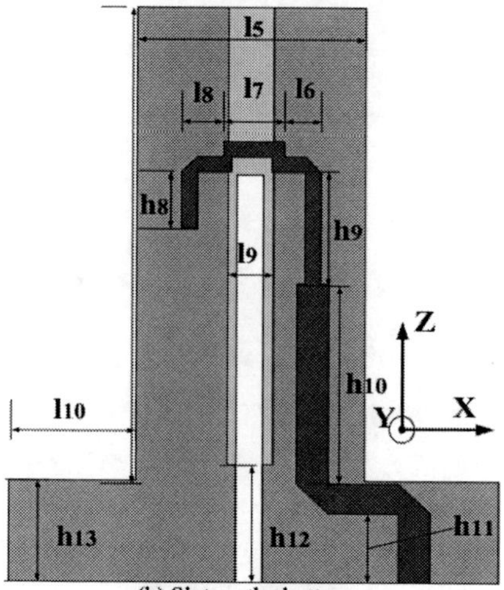

(b) Slot on the bottom

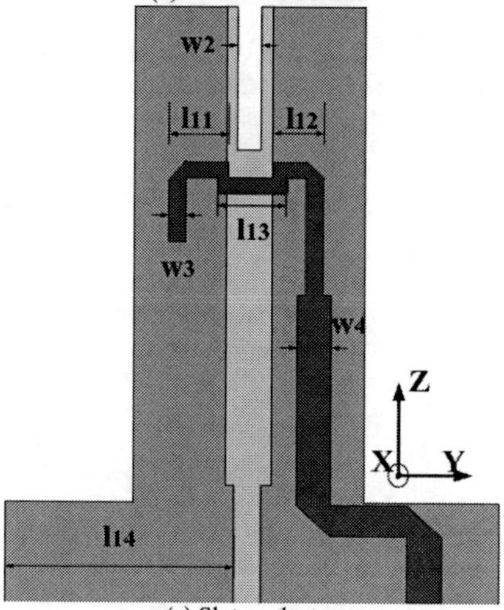

(c) Slot on the top

Figure 2. Structures of CP antenna with top patches

According to the design purpose, the antenna need to meet that the VSWR is less than 1.5 in 1.1~1.7 GHz and the isolation of the two ports need to be less than -20dB in the bandwidth. When the design of the antenna has reached the standard, we can design the antenna by adjusting the feeding port on this basis. By using the 3D simulation software HFSS to simulate this structure, the simulation results are shown in figure 3. It can be seen that the VSWR of the antenna is less than 1.5, and the bandwidth can cover 1.1~2.0 GHz. The isolation of the two ports is less than -30dB shown in figure 4.

Table 1. Parameters of the antenna. (unit: mm)

Parameters	Value	Parameters	Value
l1	110	h1	4.5
l2	31.5	h2	4

DOI: 10.1051/matecconf/20166304028

l3	20	h3	3
l4	38	h4	6
l5	25	h5	9
l6	2.2	h6	9
l7	7	h7	81.6
l8	3.6	h8	7
l9	5	h9	15
l10	17.5	h10	27
l11	4.6	w1	3.2
l12	3.2	w2	3.2
l13	7	w3	1.4

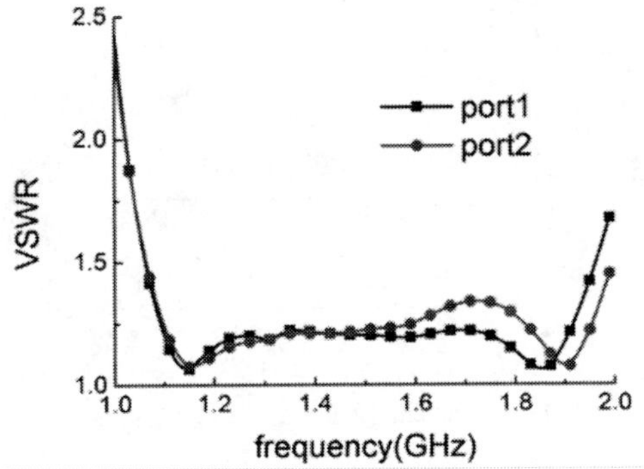

Figure 3. VSWR of antenna

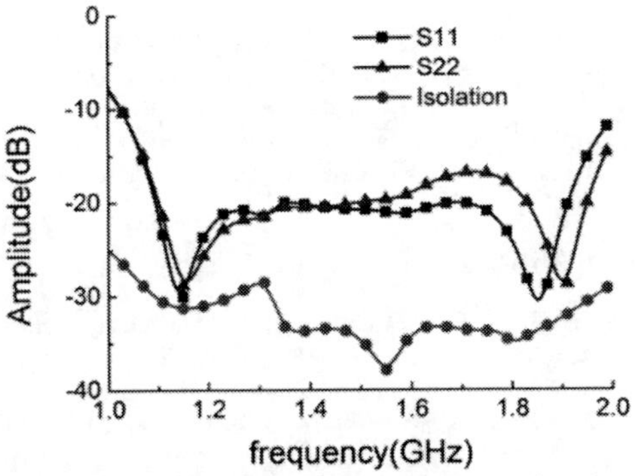

Figure 4. Reflection coefficient and isolation of antenna

3 Design of feeding network

It is a common way to use feeding network in CP antenna design. In the design, only one port act as the feeding port, and the signal reach the feeding port though the feeding network, which can reduce the design cost, simplify the structure, and the test and application will be easy. Considering about the bandwidth of the antenna in the context, we use a first order Wilkinson power divider to realize a circular polarization. We all know that the amplitude of the two feeding port of a CP antenna are equal as well as the phase difference between the two port is 90 degree. Therefore, a bisection of microstrip power divider has been used, which has a 90deg phase difference between the two feeding port.

4 Experimental study of CP antenna

The proposed CP antenna which has four patches on the top is shown in figure 5. It is machined by the substrate FR4 with thickness of 1mm. We use vector network analyzer to test the matching impedance, and the result is shown in figure 6.

The results of simulation and measurement are shown from figure 7 to figure 9, the output port and the input port of feeding networks is connected. As shown in figure14, during 1 GHz and 2 GHz, the simulated s11 is lower than -10dB, while the bandwidth of the measured result is from 1.1GHz to 1.8GHz. Figure 8 shows the bandwidth corresponding to the AR below 3dB. The simulation results are from 1.15 GHz to 1.68 GHz, while the measured bandwidth is from 1.27 GHz to 1.62 GHz, about 24%. Figure 9 showed the gain of the antenna.

The measurement was conducted in our microwave anechoic chamber. The antenna was fixed on the rotating platform, while one standard horn antenna was fixed one the other side. After processing the received signal we can achieve the measured results. The simulated results agree well with the measured results. The bandwidth of AR below 3dB is from 1.27 GHz to 1.62 GHz (about 24%), and the bandwidth of s11 below -10dB is from 1.1GHz to 1.8GHz. The peak gain is 5dBi.

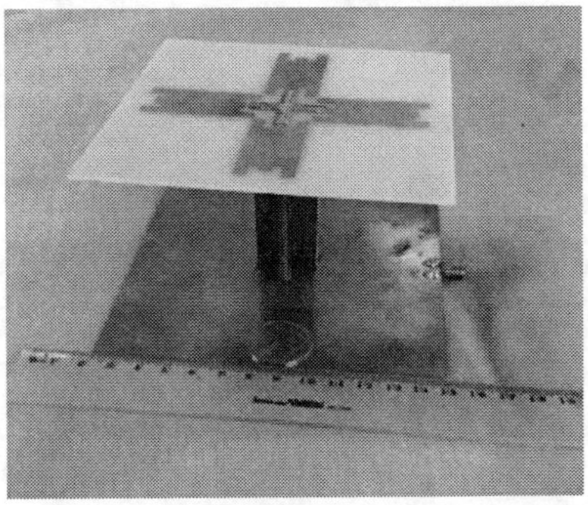

Figure 5. Photo of antenna

DOI: 10.1051/matecconf/20166304028

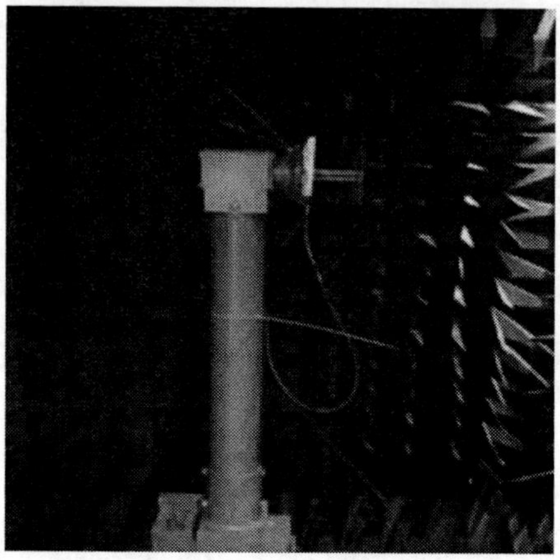

Figure 6. Photo of test equipment

5 Conclusions

The proposed CP antenna is made up of double sided printed dipoles, which is fed by a power divider network. Four radiation patches are loaded on top of the dipoles. The antenna is fed by a splitter located at the bottom of antenna. Both simulation and experimental studies are carried out, and the experimental results shown that the 3 dB axial ratio bandwidth of the antenna is from 1.27 GHz to 1.62 GHz (24%) and the highest gain can be up to 5 dBi within the operating frequency range. The antenna has good radiation performance.

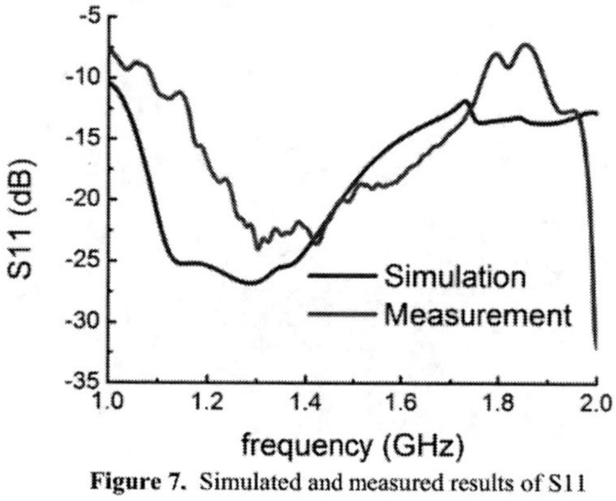

Figure 7. Simulated and measured results of S11

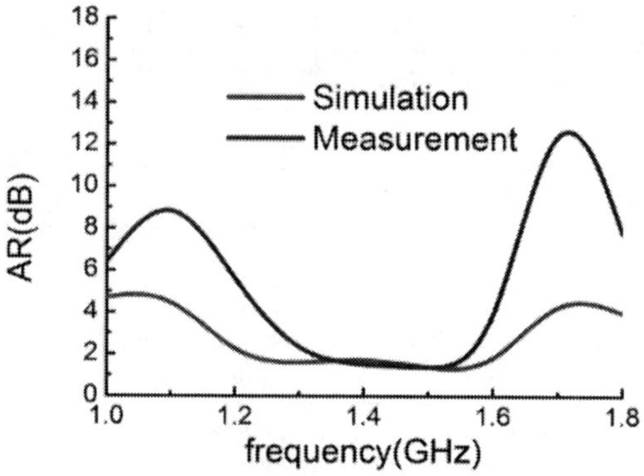

Figure 8. Simulated and measured results of AR

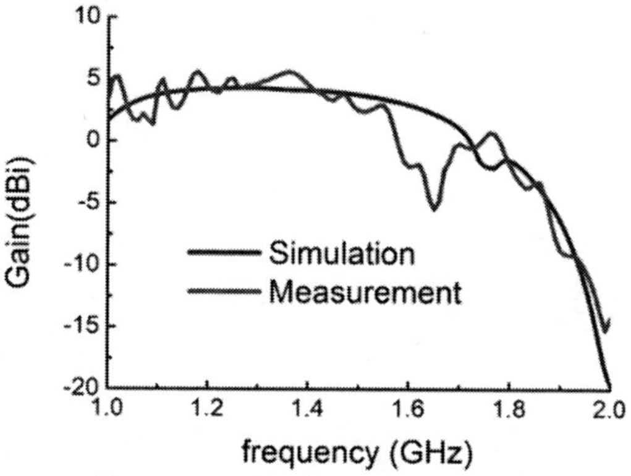

Figure 9. Simulated and measured results of gain

Acknowledgments

This work was supported by the Project of Education Department of Fujian Province, China (Grant No. JA14022), and the Undergraduate Education & Teaching Reform Project of Huaqiao University, China (Grant No.2013JGYB36), Xiamen, Fujian province, China.

References

1. ZHOU Y, CHEN C. Single-fed Circularly Polarized Antenna Element with Reduced Coupling for GPS Arrays [J]. Antennas and propagation, IEEE Transactions, pp.1469-1472, Vol.56 (5), May 2008.
2. TOMOYA I, DAISUKE K, KKUNIALI Y, KEIJI Y, AKIRA I, SHUGO F. Circularly Polarized One-Sided Directional Slot Antenna with Reflector Metal for 5.8-GHz DSRC Operations [J]. IEEE Antennas and Wireless Propagation Letters, 2014, 13:778-781.
3. FUJITA K, KANEMOTO D , YOSHITOMI K, YOSHIDA K, KANAYA H. UWB Circularly Polarized Planar Antenna on Flexible Substrate[J].IEEE MTT-S Internation,2013,1-3.
4. HANI A. Broad-Band and Circularly Polarized Space-Filling-Based Slot Antennas [J].IEEE Transaction Antennas Propagation, 2005, 53(6):1946-1950.

DOI: 10.1051/matecconf/20166304028

5. MA Xiao-liang, HUANG Cheng, PAN Wen-bo, ZHAO Bo, CUI Jian-hua, LUO Xiao-gang. A Dual Circularly Polarized Horn Antenna in Ku-Band Based on Chiral Metamaterial[J]. IEEE Transaction Antennas Propagation,2014,62(4);2307-2311.
6. ZHANG T, ZHANG Y. Single-Layer Wideband Circularly Polarized Patch Antennas for Q-Band Applications [J]. IEEE Transaction Antennas Propagation, 2015,63(10);409-414.
7. WANG LEI, GUO YONGXIN. Wideband modified L-probe fed circularly polarized patch antenna with conical beam radiation pattern [J]. Journal of Electromagnetic Waves and Applications, 2013, 27(27); 969-977.
8. LI Bo, LIAO Shao-wei, and XUE Quan . Omnidirectional Circularly Polarized Antenna Combining Monopole and Loop Radiators [J]. IEEE Antennas and Wireless Propagation Letters, 2013, 12(12):607-610.
9. ZHANG Xue-ying, LIN Shu, WANG Yu-die, LIU Sheng-ying, ZHANG Xin-yue, ZHANG Wen-bin. Broadband Circularly Polarized Antenna with a Ttype Fractal Boundary Wide-slot and a L-shaped Strip [J]. IEEE Microwave Conference Proceeding, 2011, 1-3.
10. ZHANG YQ, YANG L. Dual-band circularly polarized antenna with low observation angle performance enhanced for GNSS applications [J]. Journal of Electromagnetic Waves and Applications, 2014, 28(28); 1131-1141.
11. HE Ye-jun. A Wideband Circularly Polarized Cross-Dipole Antenna [J]. IEEE Antennas and Wireless Propagation Letters, 2014, 13(1):67-70.
12. RAUT S, PETOSA A. GPS Wideband Circularly Polarized Microstrip Antenna Array [J]. Antenna and Propagation, 2009 EuCAP, 2009, 2990-2993.
13. WANG Can, GE Yuehe. A Novel Design of Broadband Dual-polarized Printed Dipole Antenna for Base Stations [J]. Journal of Huaqiao university, 2014, 35(2):132-135.
14. WANG Can, GE Yue-he. A Novel Broadband Printed Dipole Antenna and Its Application for TD-LTE Communications [J]. International Journal of Antennas and Propagation, 2014, 960963, pp.1-7.

Research of Logistics Cost Based on Saving Algorithm: A Case of A Certain Logistics Company's Logistics Cost

Jing NIE[1,a], Wei Yuan ZENG[1] Hong Lei JING[1]

[1] *Department of Software Engineering, Xiamen Institute of Software Technology, Xiamen, 361024, China.*

Abstract: In recent years, logistics companies of different sizes has been considerably developed all over the country, with the intensified competitions. To stand out in the competitive market, a logistics company must improve its logistics situation, focus on the function of distribution center, reduce logistics costs for enhancing the impact of supply chain, and rapidly response to customer needs. Thus, how to better design or optimize the existing distribution route for reducing logistics costs becomes a difficult problem, and this article will do some researches on it.
Keywords: logistics company; distribution route; logistics costs; saving algorithm

1 Introduction

With the huge business opportunity and potential of modern logistics industry, it attracts China's attention more significantly. In 2009, the State Council issued the *Adjustment and Plan for Logistics Industry*, setting a milestone in the history of China's economic development. With the publication of this government document, the " logistics craze" occurred in domestic market: all kinds of new logistics companies were launched and the existing logistics companies accelerated the adjustments of their own long-term plan. Now in the international market, the share of logistics costs in GDP is supposed to symbolize the strength of a country. Although from 1991 to 2001 the share of China's logistics costs in GDP has dropped from 24% to 19.4%, by 4.6 percentage points, it is still unacceptable that from 2002 to 2009 that has fallen from 18.8% to 18.1%, by only 0.7 percentage points. However, the logistics costs of developed countries remain 10% in general, obviously indicating that China's logistics costs still have room for further reduction. According to saving algorithm, this article will combine with a certain case and put forward a countermeasure with actual value and operability to save logistics costs.

[a] Corresponding author: 56564995@QQ.COM

2 Problem Description and Modeling

2.1 Problem Description

Let goods be delivered from Distribution Center O to Transfer Station (Customer) v_i ($i=1,\cdots,m$): q_i ($i=1,\cdots,m$) is the volume of goods delivery from O to v_i ($i=1,\cdots,m$), and L_{ij} (i, j=1, ..., m) is the delivery distance between p_i and p_j; as delivery vehicles are classified into n categories by load (or capacity), there are x_i (i=1, ..., n) delivery vehicles with w_i, where $w_{i-1}<w_i$ and $w_i > \sum_{i=1}^{m} q_i$,i.e. the load (or capacity) of each delivery vehicle outweighs the total customer demands.

Thus,if without any limitations, goods should be delivered by a vehicle, transforming the issue into a shortest route problem.

In this article, the conditions of route optimization are defined to be that when a logistics distribution center delivers goods to multiple customers (or transfer stations), the locations of logistics center and customers are given; the distance among logistics center and customers (or transfer stations)and among different customers (or transfer stations) are given and fixed. Meanwhile, the traffic limit on transportation network is out of consideration; customer time demand is out of limitation; each customer demand for goods does not exceed the maximum load of delivery vehicles while the goods can be mixed in a same vehicle; each vehicle hasa fixed maximum load, so overload cannot be allowed; each vehicle is set out from logistics centerand back here after delivery task is completed.

2.2 Parameter Definition and Problem Model

V: Customer (or transfer station) set
O: Logistics distribution center
W: Capacity of delivery vehicle
q_i : Customer (or transfer station) demand
c_{ij} : Distance between customer (or transfer station) and another customer (or transfer station)
y_{ik} :1 represents that Vehicle k is driven from customer i to Customer j; 0 represents that Vehicle k is not driven from customer i to Customer j
x_{ijk} :1 represents that Customer i is served by Vehicle k; 0 represents that Customer i is not served by Vehicle k.
So mathematical modeling can be as follows:

$$\min \sum_{k=1}^{K} \sum_{i=1}^{V} \sum_{i \neq j, j=0}^{V} c_{ij} x_{ijk} \quad (1)$$

$$st. \sum_{i=0}^{V} q_i y_{ik} \leq W, \forall k \subseteq K \quad (2)$$

$$\sum_{k=1}^{K} y_{ik} = \begin{cases} k, & i = 0 \\ l, & i \subseteq V \end{cases} \quad (3)$$

$$\sum_{i \in V} x_{ijk} = y_{jk}, \forall j \in V, k \in K \quad (4)$$

$$\sum_{j \in V} x_{ijk} = y_{jk}, \forall i \in V, k \in K \quad (5)$$

$$x_{ijk} \in \{0,1\}, \forall i, j \in V, k \in K \quad (6)$$

$$y_{jk} \in \{0,1\}, \forall i \in V, k \in K \quad (7)$$

The meanings of equation(1) to equation(7) are explained as follows:
Equation(1): Find the minimum of route distance sum
Equation(2): Total volume of goods through each route does not exceed the capacity of delivery vehicles
Equation(3): Upper equation represents the number of vehicles in logistics distribution center; lower one represents that each demand site can only be served by a vehicle
Equation(4): Traffic conservation limit
Equation(5): Traffic conservation limit
Equation(6): Distribution site of Vehicle k
Equation(7): Route of Vehicle k

3 Algorithm Analysis

This article chooses C-W saving algorithm to optimize the distribution of a certain logistics company, due to the fact that in logistics distribution system route optimization has too many limits including vehicles' capacity limits, load limits, time limits, etc., while saving algorithm has the advantage of adding constraints flexibly.

3.1 Distribution Model

3.1.1 Point-to-point Transportation

Shortest route algorithm between two points is the most basic one in the category of logistics distribution. It determines the best route among multiple distribution sites by calculating the shortest route between two points. Given the actual constraints such as the one-way traffic system and delivery speed, as well as the road control on vehicle types, etc., the road network within distribution area can be taken as a Digraph G=(V, E) with weights, where the weight of each side is a non-negative real number. In addition, let the point of v be a starting point of distribution. Then calculate the shortest route from the starting point to other vertexes.

3.1.2 Single-loop Transportation - TSP

Single-loop transportation problem is to select an appropriate and closure-required route through all the nodes for a given node-set D in the route optimization. In transportation decision, it mainly aims at the route arrangement of a single vehicle for ensuring that the shortest distance is driven by the target vehicle through all the customers. TSP mathematical model is described as:

Based on the Digraph G=(V, E), where $V = \{v_1, v_2, \cdots, v_n\}$, $E = \{e_1, e_2, \cdots, e_m\}$

Unit cost among vertexes is $C = \{c_{ij} | i, j \in N, 1 \leq i, j \leq n\}$

Distance is $D = \{d_{ij} | i, j \in N, 1 \leq i, j \leq n\}$ (8)

Objective function is $\min \sum_{i=1}^{n} c_{i(i+1)} d_{i(i+1)}$ (9)

The traveling salesman sets out from and returns to the starting point, thus $v_{n+1} = v_1$.

3.1.3 Multi-loop Transportation - VRP

Multi-loop transportation is a common distribution issue in societyand an integration of single-loop transportation problems. It originates from TSP. VRP is generally defined as an appropriated route arrangement that makes vehicles pass a series of loading point and (or) unloading point orderly and achieve some certain purposes (such as the shortest distance, least cost, shortest time, least vehicles, etc.) under the constraints (such as the demands for goods, delivery volumes, delivery times, vehicles' capacity limits, distance limits and time limits, etc.).
VRP mathematical model is described as:

Based on the Digraph G=(V, E), the vertex set is $V = \{v_0, v_1, \cdots, v_n\}$, the set of two vertexes(or side set) is $E = \{(v_i, v_j) | v_i, v_j \in V, i \neq j\}$, where v_0 represents the

DOI: 10.1051/matecconf/20166304029

location of distribution center, and other nodes represent all the customers.

The basic condition of a typical VRP model is arranging N vehicles to deliver goods for L customers. Its purpose is to determine the number of vehicle M (M≤N)and to finish the arrangement of loop transportation with the minimum costs. Its target model can be represented as follows:

$$\text{Objective function: } \min Z = \sum_{m=1}^{M}\sum_{l=1}^{L} d_{lm}x_{lm} \quad (10)$$

Constraints:

$$\sum_{l=1}^{L} w_l x_{lm} \le b_m, (l = 1,2,\cdots,L; m = 1,2,\cdots,M) \quad (11)$$

$$\sum_{m=1}^{M} x_{lm} = L, (m = 1,2,\cdots,M)\ x_{lm} \in \{0,1\},$$

$$(l = 1,2,\cdots,L; m = 1,2,\cdots,M) \quad (12)$$

where b_m is the transport capacity of Vehicle m;L is the sum of customers;M is the sum of vehicles; w_l is the demand (by weight or volume) of Customer l.

3.2 Shortest Route Solution of Point-to-point Transportation - Dijkstra Algorithm

Dijkstra algorithm is also called dual labeling method. This method is to give the connecting Point v_j two labels l_j, k_j in the Digraph G=(V, E), where the first label l_j represents the shortest distance from the starting point v_s to v_j and the second one k_j represents the subscript of one adjacent point before v_j in this shortest route, further finding the shortest route.

3.3 Basic Principle of Saving Algorithm

Saving algorithm is put forward by Clarke and Wright in 1964, aiming at solving vehicle routing problem. Its basic principle is: Distribution Center O delivers goods to n distribution sites through one-node route connecting all the sites and distribution center. Therefore, the total cost of distribution transportation is:

$$z = \sum_{i=1}^{n} c_{oi} + \sum_{i=1}^{n} c_{io} \quad (13)$$

If Distribution Site i is connected with Site j and forms a route tothe distribution center, the saving cost will be:

$$\begin{aligned} s(i,j) &= c_{oi} + c_{io} + c_{oj} + c_{jo} - (c_{oi} + c_{ij} + c_{jo}) \\ &= c_{io} + c_{oj} - c_{ij} \end{aligned} \quad (14)$$

The more the value of s(i, j) is, the more costs will save when connecting Site i and Site j. In the practical route optimization, saving cost is sorted by the value of s(i, j) in descending order. This algorithm has an advantage of fast calculation speed. Especially in the small-scale distribution optimization, the calculation is completed fast by the close accuracy and best solution.

4 Calculation Example

As the distribution center, a certain logistics company delivers goods to 12 customers. The center has five vehicles with the load of 680 items per vehicle. Find the best solution for the shortest route.

This issue for VRP with single center and non-full loads has the following characteristics: the location of logistics distribution center is given and single; the locations and demands of transfer stations are given; the route and distance among transfer stations are given and determined; vehicles are only unloaded in transfer stations and return empty to the distribution center after distribution; the distribution center has only a sort of vehicles. Find the objective function: the minimum cost of vehicle delivery routes is reflected as the shortest distance in this solution.

According to the data provided by a certain logistics company, draw a route distribution from this company to each transfer station at last, as shown below:

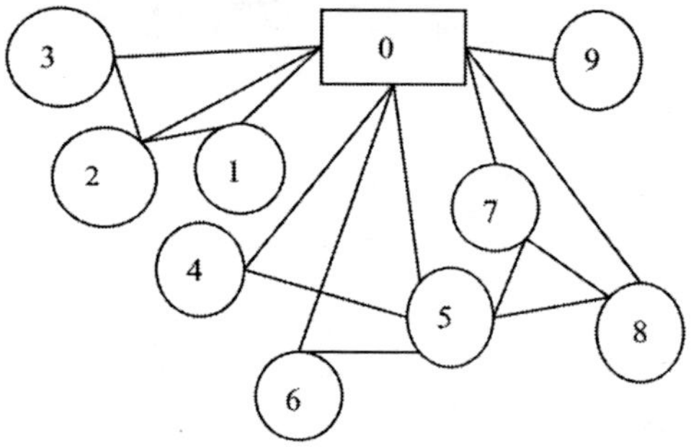

Figure 1. Route distribution from this company to each transfer station

4.1 Symbol Instructions of Transfer Stations

Table 1. Symbol Correspondences of Transfer Stations

P_1	P_2	P_3	P_4	P_5	P_6	P_7	P_8	P_9
1	2	3	4	5	6	7	8	9

Table 2. Demands of Transfer Stations (Unit: Item)

P_1	P_2	P_3	P_4	P_5	P_6	P_7	P_8	P_9
470	690	625	300	525	230	295	225	295

4.2 Solution Process.

First, calculate the distances among this company and transfer stations.

Second, calculate the shortest distances among this company and transfer stations.

Third, calculate the saving mileage s(i, j) among customers by saving algorithm, according to the shortest distances.

Forth, sort s(i, j) in descending order.

Fifth, find the initial solution.

Sixth, find the best solution (according to the savings in descending order).

According to the saving algorithm, we can find that only the route below can save the distance. The saving distance is as shown below:

Table 3. Route and Distance

Route / Distance	P_5, P_6	P_4, P_6	P_6, P_8	P_4, P_8	P_6, P_7	P_4, P_7	P_7, P_8
Corresponding Route	5-6	4-6	6-8	4-8	6-7	4-7	7-8
Mileage	2433	2438	2443	>2438	>2438	>2438	>2438

By the comparative analysis of the table above, we find that Route 5-6 can save the maximum mileage under the premise of equal vehicles.

According to the practical route, the saving mileages of Route 5-6 in practice is less than that in theory. Therefore, we should choose the route of the second maximum saving mileage, i.e. Route 4-6.

Theoretically, the goods delivery from logistics center to nine transfer stations respectively needs 10 vehicles once and 20 vehicles twice. According to the current situation of this company, goods are delivered to transfer stations twice per week basically. The working time of vehicles is from Tuesday to Saturday, i.e. five days per week. If goods are delivered to nine transfer stations twice in the five days, 18 transfer stations can be reached; and if five vehicles are on work in the five days, 25 places (transfer stations) can be reached. Therefore, the existing five vehicles with full load can achieve the annual freight volume by the route above, completing the task with the excess capacity of solving emergencies.

Compared with original route, this route saves costs as follows:

goods are delivered 2 per week and 8 times per month at most.

Monthly saving mileages (km): 218×8=1744

Monthly saving tolls (yuan): 25×8=160

Monthly saving oil costs (yuan): 1744×0.33×6.7=3856

（Logistics vehicle of this company costs 0.33L oil per kilometer, and oil price is calculated by 6.7 yuan/L.）

Annual saving oil costs (yuan): 3856×12=46272

Annual saving tolls (yuan): 160×12=1920

Total saving costs (equaling annual saving oil costs plus annual saving tolls): 46272+1920=48192.

Acknowledgments

Fund Project: the Project of Xiamen Science and Technology Program under Grant Nos. 3502Z20153006, 3502Z20133042.

References

1. Lv Xiongwei.Researches on post VRPTW based on genetic algorithm [J].Journal of Shandong University: Engineering Science, 2009, 44(6):46-50.
2. LIU Cheng.VRP with interval parameters and its improved C-W algorithm [J].Journal of Wuhan University of Technology, 2010, 32(2):182-185.
3. FAN Zhan, Liang Guolong.Design and expending of adaptive neighborhood searching algorithm for TSP [J].Computer Engineering and Applications, 2008, 44(12):71-74.
4. Zhang Xuezhi, Chen Gongyu.An improved saving method of the vehicle routing problem [J].Systems Engineering, 2008, 26(11):67-70.
5. Altinel 1,Oncan T.A new enhancement of the Clarke and Wright savings heuristic for the capacitated vehicle routing problem[J].The Journal of the

Operational Research Society,2005,56(8):954-961.

6. 2008International Conference on Wireless Communications.A Modified Max-min Ant System for Vehicle Routing Problems[C].Dalian[Zhao Gang, Luo Wenjuan,et al],2008:1-4.
7. Barbarosoglu G,Ozgur D.A tabu search algorithm for the vehicle routing problem[J].Computers&Operations Research,1999,26(3):255-270.
8. Remy Spliet, Guy Desaulniers.The discrete time window assignment vehicle routing problem [J].European Journal of Operational Research.2015.8:1-13.

A new method of on-line turned surface monitoring by digital image processing

Qingqun Mai[1,a], Yanming Quan[1], Peijie Liu[1] and Guo Ding[1]

[1]*School of Mechanical and Automotive Engineering, South China University of Technology, Guangzhou, China*

Abstract. A machined surface can be seen as a replica of the geometric outline of the cutting tool nose during the process of cylindrical turning, and the texture of the machined surface varies along with the tool wear. A new method of on-line turned surface monitoring is proposed in this paper by studying the grey value characteristics of the turned surface digital image. The regularity of the surface image is judged and analysed by using wavelet transform, fractal analysis, and discreteness analysis of the wavelength of the texture contour. The average wave contour, which can reflect processing state and the condition of the turned surface, is extracted from the regular texture image. Experimental results show that the turned surface condition can be monitored on-line effectively with the presented method.

1 Introduction

In automatic manufacturing, an appropriate on-line machined surface monitoring technology can effectively help to improve the production quality and efficiency. Traditionally, machined surface topography was observed by optical microscopy, scanning electron microscopy, and other methods off-line or by CCD cameras and other equipment on machine. The machined surface can be seen as a replica of the geometric outline of the cutting tool nose in cylindrical turning, and it varies along with the tool wear. Digital image processing (DIP) is an appropriate method to obtain information on the machined surface roughness and tool condition from the machined surface, which allows for on-line tool condition monitoring by combining a computer vision technique. [1]

In Ulf Persson's research, angular speckle-correlation was used in measuring surface roughness on machined surfaces, and a new technique to achieve increased repeatability by using an angle detection unit was presented[2]. Ghassan A. Al-Kindi et al. proposed a methodology to measure the surface roughness by using machine vision data. Two light reflection models (Intensity-Topography Compatible (ITC) and Light-Diffuse) were adopted and applied in this methodology[3]. S. Palani et al. presented a machine vision system, integrated with an artificial neural network, for prediction of surface roughness of end milled parts. A two dimensional Fourier transform was used to get the features of image texture in the spatial frequency domain. A back-propagation algorithm was developed, and used to train the artificial neural network. [4] Dutta, S et al. analyzed machined surface for detection of Surface roughness and tool wear, using an improvised grey level co-occurrence matrix (GLCM) technique with appropriate pixel pair spacing or offset parameter. The machined surface were imaged after cutting full length job. [5]BS Prasad et al. analyzed the surface topography image of machined surfaces during the progression of the tool wear. Their study indicated, when the tool is worn, the texture distribution strength in feed direction is weakened. With the increasing tool wear, the effect of randomly occurring deeper furrows along the feed direction on the machined surface is more apparent. [6] Narayanan M.R. et al. designed a image operator for illumination enhancement and noise filtering using evolutionary approach in the on line surface roughness measurement system in milling process. Relationship between the feature of the surface image and the actual surface roughness is obtained using regression analysis after image processing.[7] In theory, these studies paved the way for the development of an on-line and real-time machined surface-condition-monitoring system, which has not been implemented entirely yet.

Fractals have been used to analyse machined surface images for tool condition investigations. Kassim A.A et al. had investigated fractal analysis of the texture of machined surfaces. High directionality and self-affinity were dealt with by extracting the fractal features from machined surfaces images to extract tool wear information. And Hidden Markov Model is used to classify the various states of tool wear. [8] Kang M.C. et al. identified the relation between surface roughness and fractal dimension depending on coated tool wear using a generalized formula. The fractal dimension and the surface roughness showed similar tendencies depending on various cutting conditions. By obtaining images using light scattering method, the fractal dimension regularly

[a] Corresponding author: qingqmai@126.com

 DOI: 10.1051/matecconf/20166304030

increases as tool wear increases. [9] It is an effective method for on line monitoring of tool wear using fractal analysis of machined surfaces. However, the algorithms above were too complex to meet the requirement of on line real-time monitoring.

A computer vision system for on-line turned surface condition monitoring is presented in this paper. Images of the turned surface from a cylindrical turning process have been grabbed and analysed. To judge whether the turning process is on the rails, wavelet transform, fractal analysis, and statistics of discreteness are applied to analyse the regularity of the turned surface texture. Then, the average wave contour is extracted from the regular turned surface texture, which is proposed as a new method for processing state monitoring in this paper.

2 Methods and theory

2.1 Machining and computer vision systems

Dry turning experiments were conducted on CNC lathe. The workpiece material was stainless steel 1Cr18Ni9Ti, and the tool material was cemented carbide K20. The turned surfaces were imaged in the process as 250×190-pixel grey-level digital images using a CCD camera (with max a frame rate of 30 fps at a full resolution of 1624×1234-pixel), a lens (with magnification of 1.0×), and a ring light source. The hardware equipment of the computer vision system for capturing cylindrical turned surface images on-line is shown in Figure 1.

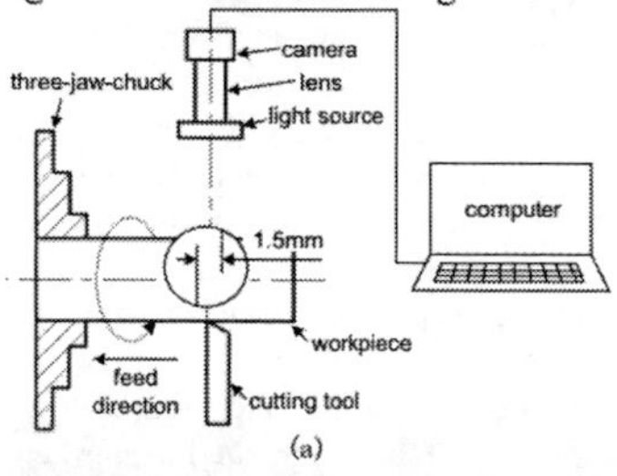

(a)

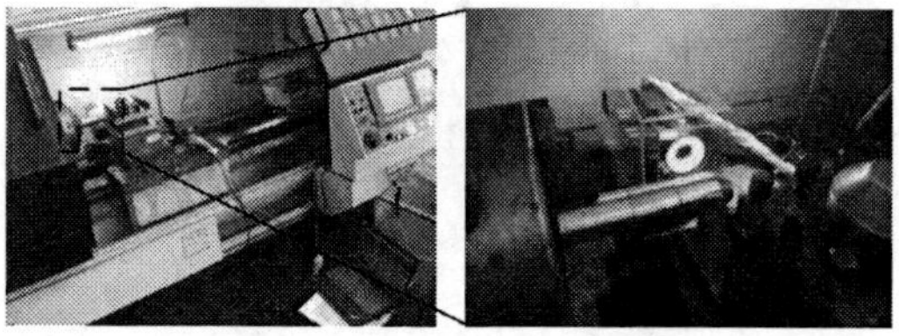

(b)

Figure 1 On-line monitoring system with computer vision for cylindrical turning

2.2 Image processing

2.2.1 Regularity of the turned surface texture

Figure 2(a) shows the image of the regular texture of the turned surface, and Figures 2(b) and (c) show images of the irregular texture of the turned surface. The significant adhesion of the turned surface is shown in Figure 2(b), and the vibratory texture is shown in Figure 2(c). As shown in Figure 2, the profiles of several rows in the turned surface image were extracted and compared. The waveform distribution of the different rows in the regular texture image was similar, but not in that of the irregular texture image.

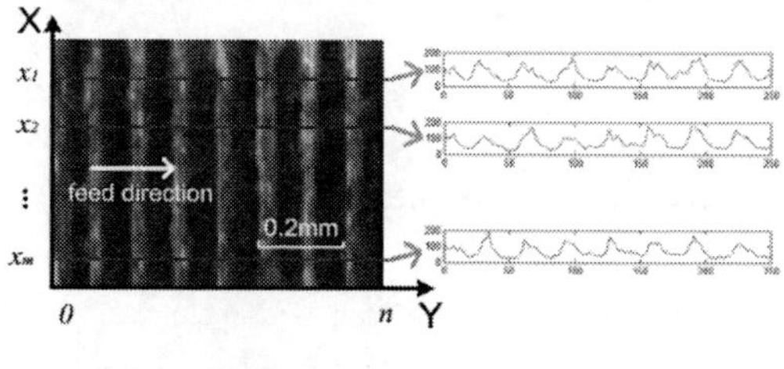

(a) Regular texture

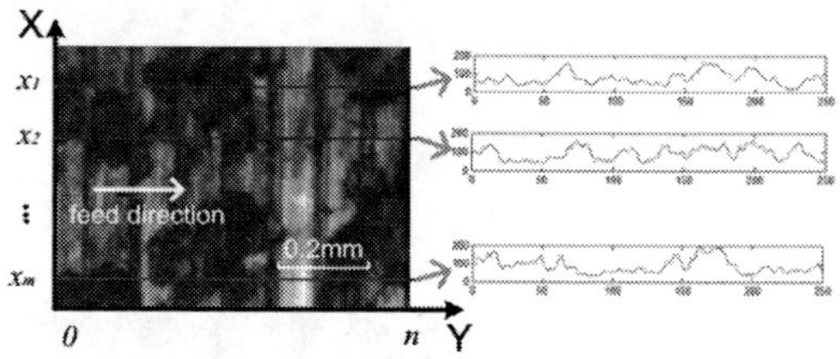

(b) Adhesive texture

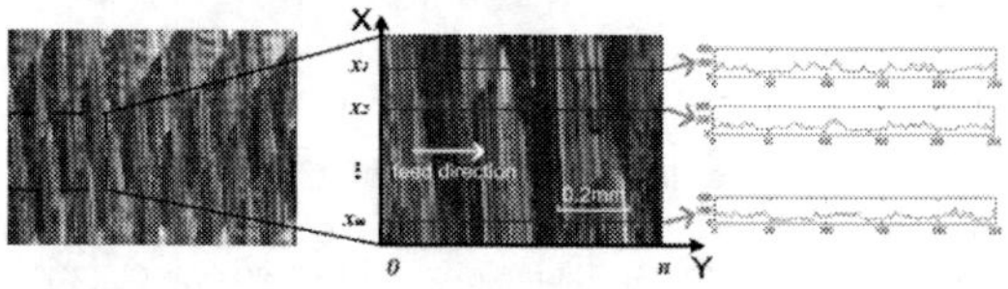

(c) Vibratory texture

Figure 2 Images of the turned surface

The regularity of the machined surface texture can be judged with the method of fractal analysis. In particular, for a cylindrical turned surface, because its texture shows parallel stripes, we can calculate the distance between two stripes as a wavelength and then analyse the discreteness of the whole wavelengths, to judge the regularity of the texture more quickly.

2.2.2 Fractal analysis of the machined surface image

Grey data were extracted from each row of the turned surface image (as shown in Figure 2) and then was calculated with wavelet transform based on the fractal theory to estimate the Hölder exponent. A series of Hölder exponents was acquired from every image. The difference between the maximum and the minimum Hölder exponent (*Ha*, Hölder amplitude) was taken as an indicator to judge the texture regularity.

The grey data vector extracted from the machined surface image was marked as x. Vector x was decomposed by wavelet transform after mirror-symmetric boundary extension. The boundary extension equation is shown as Equation 1.

$$X = \begin{cases} x_{n_m-i+1} & ,i < n_m - 1 \\ x_{-n_m+i+1} & ,n_m \leq i \leq n_m + n - 1 \\ x_{2n+n_m-i-1}, & i \geq n_m + n \end{cases} \quad (1)$$

where X denotes the data after boundary extension, n indicates the length of the original data x, and n_m is the size of the mirror. The expression of the Morlet wavelet is as Equation 2.

$$\Psi(t) = \pi^{-1/4} e^{i\omega_0 t} e^{-t^2/2} \quad (2)$$

where t is an independent variable, as Y in Figure 2, and ω_0 denotes a dimensionless frequency. After the Morlet wavelet decomposition of vector X (Equation 3), the wavelet transform coefficient matrix W and analysed scales s are obtained.

$$W(s) = \langle X, \Psi_s \rangle = \frac{1}{\sqrt{s}} \int_{-\infty}^{+\infty} x(t) \Psi_s^*(t) dt \quad (3)$$

Then, the Hölder exponent was calculated:

$$W_m = max_n\, W \quad (4)$$

$$M = \frac{1}{n} \sum_{i=1}^{n} |W_{mi}|^2 \quad (5)$$

$$log_2 M = b + a\, log_2 s \quad (6)$$

where coefficient a is the Hölder exponent.

After calculating each Hölder exponent $a1, a2, \ldots, am$ of each row $x_1, x_2, \ldots, x_m$, respectively, in the surface image, the texture regularity indicator Ha can be obtained through Equation 7.

$$Ha = max_m\, x_i - min_m\, x_i \quad (7)$$

In a regular machined surface image, the grey curves extracted from different rows are similar, just like the fractals. Therefore, Hölder exponents show less volatility in a regular texture image than in an irregular texture image. As a result, the texture regularity indicator Ha for a regular texture will be smaller than for an irregular texture.

2.2.3 *Discreteness calculation of turned surface wavelength*

The contour vector x was low-pass filtered and derived to obtain the maximum points. The distance between two adjacent maximum points is defined as a wavelength; thus, a series of wavelengths (wavelength vector λ) was obtained. The discrete coefficient ξ of elements in the vector λ can be calculated using Equation 8. The texture wavelength is considered to be more uniform when the discrete coefficient ξ is smaller. Therefore, the discrete coefficient ξ of a regular texture would be smaller than that of an irregular texture obviously.

$$\xi_\lambda = \frac{\sigma_\lambda}{\bar{\lambda}} = \left(\frac{1}{n_\lambda} \sum_{i=1}^{n_\lambda} (\lambda_i - \bar{\lambda})^2\right)^{\frac{1}{2}} / \bar{\lambda} \quad (8)$$

2.2.4 *The average wave contour extracted from the regular turned surface*

The regular turned surface can be seen as a collection of a large number of similar waveforms machined by the same tool nose. If the variation of tool wear during very short times can be ignored, then the waveforms machined during this period can be seen as repeated waveforms. With all repeats overlaid, the wave contour error caused by random noise will be effectively eliminated. As a result, a wave curve is obtained as the typical contour of the turned surface, which is seen as the replica of the tool nose in this moment.

3 Experiment and results analysis

3.1 Turned surface image captured on-line

The cutting parameters of the cylindrical turning of stainless steel were: Cutting speed of 100 m/min, feed of 0.1 mm/r, and cutting depth of 0.25 mm. Three cutting tools with the same material and size were used in this experiment. The geometric parameters of the turning tool were: Corner radius of 1 mm, cutting edge angle of 85° ,minor cutting edge angle of 13° , rake angle of 0° , clearance angle of 14° , and minor clearance angle of 5° . The tools had different degrees of wear: Tool I had light wear, tool II had serious wear in the tool nose, and tool III had obvious wear along the cutting nose and the main cutting edge. The images of the rake face contour are shown in Figure 3(a), and the tool nose morphology is shown in Figure 3(b). Cylindrical turning experiments with tools I -III were conducted, and the turned surfaces were imaged on-line, as shown in Figure 3(c).

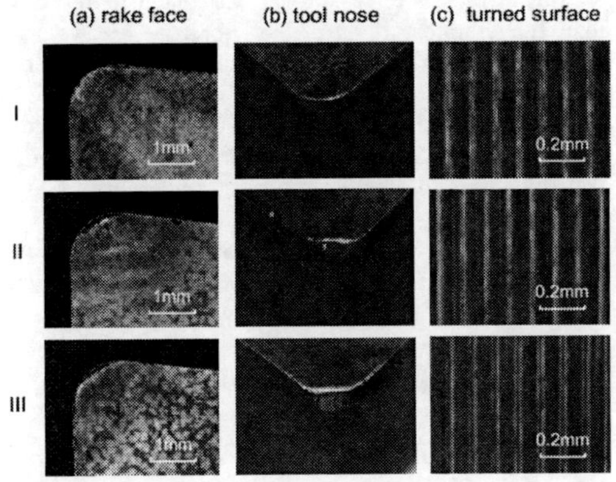

Figure 3 Images of tools and turned surfaces

3.2 Judgement of texture regularity

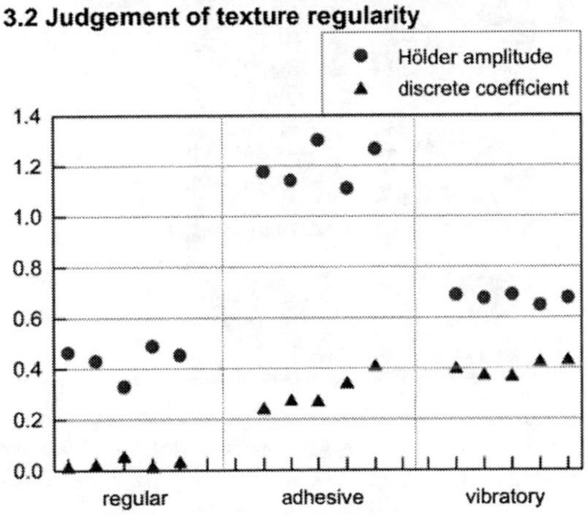

Figure 4 Judgement of image regularity

Adhesion and vibration can occur in the stainless steel machining process frequently. Adhesive texture and vibratory texture can have a serious impact on the integrity of the machined surface. The regular texture, adhesive texture, and vibratory texture were calculated for fractal and discreteness analyses. The calculated results are shown in Figure 4.

Two phenomena can be summarized from Figure 4. (1) With the illumination of the ring light, the Ha of the regular texture was smaller than that of the irregular texture, and the Ha of the adhesive texture was much greater than that of the others. (2) The discrete coefficient of the regular texture was close to 0, while the discrete coefficient of the adhesive texture was between 0.2 and 0.4, and the discrete coefficient of the vibratory texture was between 0.3 and 0.5.

Phenomenon (1) verified the principle described in Section 2.2.2. The contours extracted from the different rows from the regular machined surface image were similar, just like the fractals. However, this was not the case for the irregular machined surface image. Thus, the texture regularity indicator *Ha* of a regular texture would be smaller than that of an irregular texture. Because the texture of the adhesive surface was much more random than the vibratory texture, the vibratory texture had a greater *Ha* than the regular texture, and the adhesive texture had much a greater *Ha* than the others. The principle behind phenomenon (2) was described in Section 2.2.3. The regular texture had more uniform wavelengths and smaller discrete coefficients than the irregular texture.

To summarize, the difference in the Hölder exponent *Ha* and the discrete coefficient ξ between the regular texture and the irregular texture was obvious; thus, *Ha* and ξ can be used as indicators for the judgment of regularity of turned surface textures.

3.3 The average wave contour extracted from the turned surface image

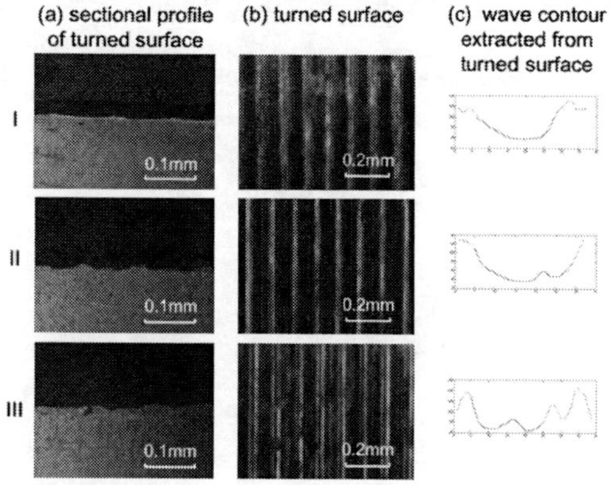

Figure 5 the average wave contour extracted from the turned surface

A small piece of sheet with a turned surface was cut down from the stainless steel workpiece with a wire-cutting method and then was buffed to view the sectional profile of the turned surface with an optical microscope, as shown in Figure 5(a). The turned surface images shown in Figure 5(b) were processed with the method described in Section 2.2.4 to extract the average wave contour (shown in Figure 5(c) in red after removing the interference). Comparing Figures 5(a) and 5(c), the average wave contour extracted from the turned surface conformed to the sectional profile of the turned surface, verifying the feasibility of extracting the average wave contour from a turned surface.

4 Conclusions

This paper concerns on-line turned surface monitoring in the cylindrical turning process. The main work and conclusions can be summarized as follows:

(1) An on-line turned surface monitoring system for cylindrical turning with computer vision and a digital image process was presented.

(2) The machined surface was analysed with fractals to judge the regularity of the surface texture and the operation of the whole machining system.

(3) For a cylindrical turning surface, the regularity of the texture can be judged simply from the wavelength discrete coefficient.

(4) The average wave contour extracted from the turned surface reflected the variety of the turned surface effectively.

(5) Turned surface condition monitoring was performed successfully with computer vision and digital image processing.

References

1. S. Dutta, S.K. Pal, S. Mukhopadhyay, R. Sen. CIRP Journal of Manufacturing Science and Technology 6, 212(2013)
2. Persson U., Journal of Materials Processing Technology, 180, 233(2006)
3. Ghassan A. Al-Kindi, Bijan Shirinzadeh. International Journal of Machine Tools & Manufacture 47, 697 (2007)
4. S. Palani, U. Natarajan. Int J Adv Manuf Technol 54,1033(2011)
5. Dutta, S., Datta, A., Chakladar, N.D., Pal, S.K., Mukhopadhyay, S., Sen, R., Precision Engineering, 36,458(2012).
6. Balla Srinivasa Prasad, M. M. M. Sarcar, B. Satish Ben. Int J Adv Manuf Technol, 55,1025(2011)
7. Narayanan M.R., Gowri S., Krishna M.M., in: Proceedings of the World Congress on Engineering 2007 (WCE, London, UK)
8. Kassim A.A., Mian Z., Mannan M.A., Machine Vision and Applications, 17, 327(2006)
9. Kang M.C., Kim J.S., Kim K.H., Surface and Coatings Technology, 193,259(2005)

A reasoning system for image analysis

Jin Sheng GAO[1,a], Chang Le ZHOU[2,3]

[1]*Department of philosophy, Xiamen University, Xiamen 361005, China*
[2]*Department of Cognitive Science and Technology, Xiamen University, Xiamen 361005, China*
[3]*Fujian Key Lab of the Brain-like Intelligent Systems, Xiamen 361005, China*

Abstract. For image analysis in computer, the traditional approach is extracting and transcoding features after image segmentation. However, in this paper, we present a different way to analyze image. We adopt spatial logic technology to establish a reasoning system with corresponding semantic model, and prove its soundness and completeness, and then realize the image analysis in formal way. And it can be applied in artificial intelligence. This is a new attempt and also a challenging approach.

1 Introduction

The emergence of the digital image analysis technology and its development is only a few decades. But it has deep into our life, such as, artificial intelligence [1], military [2], medical science [3] and other fields [4], and become our main production tools and modern life tools [5, 6]. By the current conventional methods, the image analysis based on computer is usually done with the following procedures: input high quality image, image segmentation, feature extraction, converting features into quantitative expression, image description, image understanding, and so on [7, 8]. However, in this paper, we don't take a usual measure to analyse image but spatial logic that is a special technique in topology and modal logic [9, 10]. We establish a corresponding semantic model and axiomatic system (notation: S_{IA}), and then realize the image analysis in formal reasoning method. It is not only a new attempt but also a challenging approach.

In order to express our new method systematically, the framework of this paper is roughly as follows: In section 2, we introduce a formal language to express some information which we need in the following system and computer can recognize. And in section 3, we present a semantic model to represent several key operators in following system. After the previous preparation, we establish a corresponding axiom system in section 4. In the end, we have some discussion in section 5.

2 Syntax

Firstly, we present the syntax of S_{IA}. By the syntax of S_{IA}, we can recognize which formula is the expression of S_{IA}. And our reasoning system just to describe what it can express.

Definition 1 (Language L_{IA}): Given a set of atomic proposition Atom ($p \in$ Atom), Boolean operators$\{\neg, \wedge, \vee, \rightarrow, \leftrightarrow\}$, spatial operators $\{\oplus_1, \oplus_2, \oplus_3, \oplus_4, \oplus_5, \oplus_6, \oplus_7, \oplus_8, \oplus_9\}$, arbitrary propositions $\{\varphi, \psi\}$ and punctuations {), , (}, then the language of S_{IA} (notation: L_{IA}) is defined by BNF as follows:

$$\varphi ::= p|\neg\varphi|\varphi \wedge \psi|\oplus_1\varphi|\oplus_2\varphi|\oplus_3\varphi|\oplus_4\varphi|\oplus_5\varphi|\oplus_6\varphi|\oplus_7\varphi|\oplus_8\varphi|\oplus_9\varphi$$

Where, the formula $\oplus_1\varphi$ means "φ is valid in the horizontal rightward", the formula $\oplus_2\varphi$stands for "φ is valid in the upper right", the formula $\oplus_3\varphi$ stands for "φ is valid in the lower right", the formula $\oplus_4\varphi$means "φ is valid in the overlapping point", the formula $\oplus_5\varphi$ stands for "φ is valid in the vertical upward", the formula $\oplus_6\varphi$ means "φ is valid vertical downward", $\oplus_7\varphi$ means "φ is valid in the horizontal leftward", the formulas $\oplus_8\varphi$ and $\oplus_9\varphi$ stand for "φ is valid in the upper left" and "φ is valid in the lower left" respectively.

In S_{IA} , we take operators of negation ($\neg$) and conjunction ($\wedge$) as basic Boolean connectives. And the other Boolean connectives can be abbreviated as usual, such as $\varphi \vee \psi := \neg(\varphi \wedge \psi)$, $\varphi \rightarrow \psi := \neg(\varphi \wedge \neg\psi)$ and $\varphi \leftrightarrow \psi := \neg(\varphi \wedge \neg\psi) \wedge \neg(\neg\varphi \wedge \psi)$. The binding force of Boolean connectives also as usual, i.e., negation ($\neg$) has a stronger binding force than conjunction ($\wedge$) or disjunction ($\vee$) or implication ($\rightarrow$) or bi-implication ($\leftrightarrow$). And spatial operators $\oplus_i (i \in [1,9])$ has the same binding force as negation ($\neg$). Therefore, if L_{IA} does not involve other symbols like $\oplus_i (i \in [1,9])$, then S_{IA} has nothing to do only to describe general proposition independent space location. The L_{IA} not only describes the relationship between image points, but also can express the proposition of relationship between any two points in the image.

[a]Corresponding author: Hi.Jackson@163.com

3Semantic

Definition 2 (The Frame $\mathcal{F}_{IA}$): Given an ordered pair $< X, Y, \prec_x, \prec_y >$, then $\mathcal{F}_{IA} =< X, Y, \prec_x, \prec_y >$ is a frame ofS_{IA}, if and only if, X is a nonempty set of horizontal points, Y is a nonempty set of vertical points, and $\prec_x, \prec_y$ are binary order relation on X and Y.

In theory, image domain can be infinite, but in reality, agent visual is blame, therefore the x axis and y axis value is bounded. For convenience, we take H_x and V_y to represent horizontal and vertical orientation respectively. Then, we have $X = [-H_x, +H_x]$ and $Y = [-V_y, +V_y]$. More specifically, for any x_1, $x_2 \in X$, $x_1 \prec_x x_2$ represents "x_1 is in the left of x_2". Similarly,y_1, $y_2 \in Y$, if$y_1 \prec_y y_2$ represents "y_1 is in the down of y_2".

Definition 3 (The Model $\mathcal{M}_{IA}$): Given a four tuples $< X, Y, \prec_x, \prec_y, V >$, then $< X, Y, \prec_x, \prec_y, V >$ is the model of S_{IA} (notation: $\mathcal{M}_{IA}$), if and only if, $< X, Y, \prec_x, \prec_y >$ is a frame of S_{IA} , V is an assignment function on $\mathcal{F}_{IA}$. And $(\mathcal{M}_{IA}, (x, y))$ is a notation of point model which means just to consider a point $(x, y)(x \in X, y \in Y)$ that belongs to model $\mathcal{M}_{IA}$.

Specifically, the point model is a major tool for measuring whether a formula $\varphi \in L_{IA}$ is satisfiable on point$s$ or not. Sometimes in order to convenient, the parentheses of point model can be omitted in the case of no confusion.

Definition 4 (Truth): Let $\varphi, \psi \in L_{IA}$ be arbitrary formulas, $(\mathcal{M}_{IA}, (x, y))$ be a point model ofS_{IA}, thenφ is true on (x, y) in $\mathcal{M}_{IA}$ (let $(\mathcal{M}_{IA}, (x, y)) \vDash \varphi$be a notation for it) can be recursively defined as follows:

(1) $(\mathcal{M}_{IA}, (x, y)) \vDash p$ iff$p \in V((x, y))$;

(2) $(\mathcal{M}_{IA}, (x, y)) \vDash \neg\varphi$ iff $(\mathcal{M}_{IA}, (x, y)) \nvDash \varphi$;

(3) $(\mathcal{M}_{IA}, (x, y)) \vDash \varphi \wedge \psi$ iff $(\mathcal{M}_{IA}, (x, y)) \vDash \varphi$ and $(\mathcal{M}_{IA}, (x, y)) \vDash \psi$;

(4) $(\mathcal{M}_{IA}, (x, y)) \vDash \oplus_1\varphi$ iff for all $x_1 \in X$, if$x \prec_x x_1$, then $(\mathcal{M}_{IA}, (x_1, y)) \vDash \varphi$;

(5) $(\mathcal{M}_{IA}, (x, y)) \vDash \oplus_2\varphi$ iff for all $x_1 \in X, y_1 \in Y$, if $x \prec_x x_1$ and$y \prec_y y_1$, then $(\mathcal{M}_{IA}, (x_1, y_1)) \vDash \varphi$;

(6) $(\mathcal{M}_{IA}, (x, y)) \vDash \oplus_3\varphi$ iff for all $x_1 \in X, y_1 \in Y$, if $x \prec_x x_1$ and $y_1 \prec_y y$, then $(\mathcal{M}_{IA}, (x_1, y_1)) \vDash \varphi$;

(7) $(\mathcal{M}_{IA}, (x, y)) \vDash \oplus_4\varphi$ iff $(\mathcal{M}_{IA}, (x, y)) \vDash \varphi$;

(8) $(\mathcal{M}_{IA}, (x, y)) \vDash \oplus_5\varphi$ iff for all $y_1 \in Y$, if $y \prec_y y_1$, then $(\mathcal{M}_{IA}, (x, y_1)) \vDash \varphi$;

(9) $(\mathcal{M}_{IA}, (x, y)) \vDash \oplus_6\varphi$ iff for all $y_1 \in Y$, if $y_1 \prec_y y$, then $(\mathcal{M}_{IA}, (x, y_1)) \vDash \varphi$;

(10) $(\mathcal{M}_{IA}, (x, y)) \vDash \oplus_7\varphi$ iff for all $x_1 \in X$, if $x_1 \prec_x x$, then $(\mathcal{M}_{IA}, (x_1, y)) \vDash \varphi$;

(11) $(\mathcal{M}_{IA}, (x, y)) \vDash \oplus_8\varphi$ iff for all $x_1 \in X, y_1 \in Y$, if $x_1 \prec_x x$ and $y \prec_y y_1$, then $(\mathcal{M}_{IA}, (x_1, y_1)) \vDash \varphi$;

(12) $(\mathcal{M}_{IA}, (x, y)) \vDash \oplus_9\varphi$ iff for all $x_1 \in X, y_1 \in Y$, if $x_1 \prec_x x$ and $y_1 \prec_y y$, then $(\mathcal{M}_{IA}, (x_1, y_1)) \vDash \varphi$.

For the definition 4(1), (2) and (3), their semantic definitions are the same as usual definitions in other formal systems. And for the others, in order to make more detail and visual, we will explain the semantic definitions of above with figures one by one. At the same time, in this process, it shows S_{IA} is how to characterize the image analysis by computer.

Firstly, we purposively extract some points in image (for instance, A, B, C, D, E, F, and G), such as figure 1. Secondly, given arbitrary real number $i \in [-H_x, +H_x]$ or $[-V_y, +V_y]$ (for example, $-e, -d, -c, -b, -a, 0,\ a, b, c, d, e$ and $-e \prec -d \prec -c \prec -b \prec -a \prec 0 \prec a \prec b \prec c \prec d \prec e (\prec \in \{\prec_x, \prec_y\})$), then we can draw the coordinates with real numbers on X axis and Y axis. Thirdly, the coordinates is placed on the image in the middle, and then it becomes the figure that A= (a, d), B= (e, d), C= $(-b, b)$, D= (c, b), E= $(-d, -b)$ and F=G= $(c, -b)$. Next, we analyze every semantic definition above in the following figures.

Figure 1. Extract the feature points

In order to highlight the relationship between point and point, the following coordinates omit the background image. For the definition 4(4), it means two points that are (x, y) and (x_1, y_1) in image have the relationship of horizontal rightward. And one other thing to note is the y in (x, y) is equal to $y_1$in (x_1, y_1). They can be depicted in the following figure 2. As we can see, $a \prec_x e$, $-b \prec_x c$, $-d \prec_x c$ and $y = y_1 = d$, $y = y_1 = b$, $y = y_1 = -b$, therefore B, D, F and G is respectively in the horizontal rightward of A, C and E. In other words, $\oplus_1\varphi$ is valid in point A= (a, d) if φ exist in point B= (e, d), $\oplus_1\varphi$ is valid in point C= $(-b, b)$ if φ exist in point D= (c, b), and $\oplus_1\varphi$ is valid in point E= $(-d, -b)$ if φ exist in points F and G= $(c, -b)$.

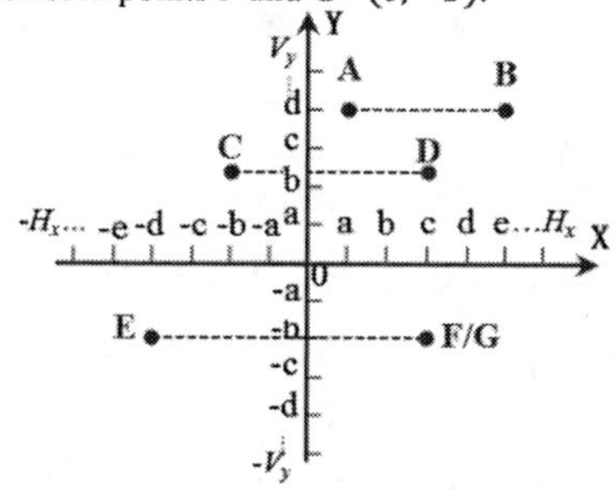

Figure 2. The relationships of horizontal rightward and horizontal leftward

For the definition 4(5), it means "φ is valid in the upper right of (x, y) ". As we can see in figure 3, A= (a, d) is in the upper right of C= $(-b, b)$ since $b \prec_y d$ and $-b \prec_x a$. Similarly, A= (a, d) , B= (e, d), C= $(-b, b)$ and D= (c, b) is in the upper right of E= $(-d, -b)$, and B= (e, d) is in the upper right of D= (c, b) and F/G= $(c, -b)$. Hence, $\oplus_2\varphi$ is valid in points C= $(-b, b)$, E= $(-d, -b)$, D= (c, b) and F/G= $(c, -b)$ if φ is existing in points A= (a, d) , B= (e, d) , C= $(-b, b)$, D= (c, b) and B= (e, d) respectively.

DOI: 10.1051/matecconf/20166304031

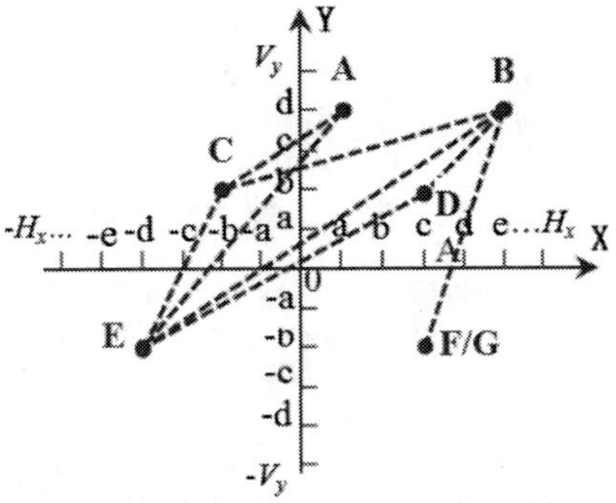

Figure 3. The relationships of upper right and lower left

For the definition 4(6), it means "φ is valid in the lower right of (x, y)". We show it in figure 4, D= (c, b) and F/G= $(c, -b)$ is in the lower right of A= (a, d) and C= $(-b, b)$ respectively for ($a \prec_x c$, $b \prec_y d$) and $(-b \prec_x c, -b \prec_y b)$. Similarly, Hence, $\oplus_3\varphi$ is valid in points A= (a, d) and C= $(-b, b)$ if φ is existing in points D= (c, b) and F/G= $(c, -b)$ respectively.

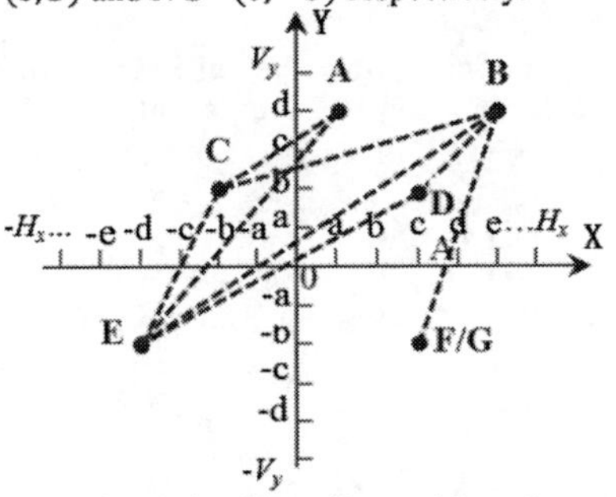

Figure 4. The relationships of lower right and upper left

For the definition 4(7), it means "φ is valid in (x_1, y_1) that is an overlapping point of (x, y) ". It can be represented in figure 5, since F=G= $(c, -b)$, $\oplus_4\varphi$ and φ are valid in points F and G= $(c, -b)$.

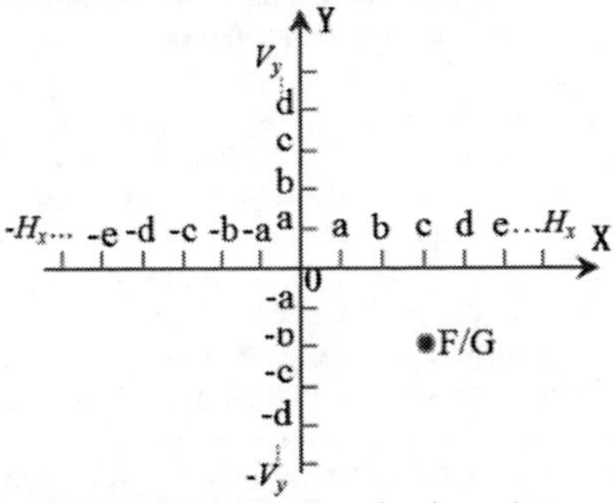

Figure 5. The relationship of overlapping points

For the definition 4(8), it means "φ is valid in the vertical upward of (x, y)". And as shown in figure 6, they are mirror image relationship for the definitions 4 (8) and (9). It is easy to find $\oplus_5\varphi$ is valid in D= (c, b) if $\varphi$is existing in F/G= $(c, -b)$, and similarly $\oplus_6\varphi$ is valid in F/G= $(c, -b)$ if φ is existing in D= (c, b).

For the definition 4(10), as shown in figure 2, $\oplus_7\varphi$ is a mirror image of $\oplus_1\varphi$, and it means "$\oplus_7\varphi$ is valid in (x, y) if φ is valid in the horizontal leftward of (x, y)". Similarly, $\oplus_8\varphi$ and $\oplus_9\varphi$ are mirrored to $\oplus_3\varphi$and $\oplus_2\varphi$.

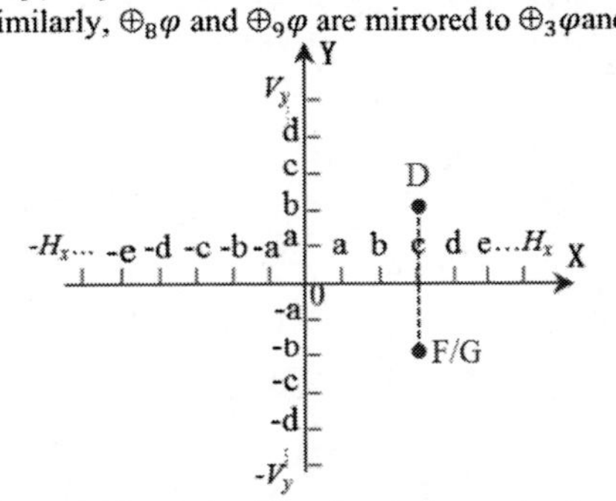

Figure 6. The relationships of vertical upward and vertical downward

4 The axiomatization of S_{IA} and its theorems

4.1Axiomatization

Definition 5 (Axiomatization): Let $p \in Atom$ be an atomic proposition, $\varphi, \psi \in L_{IA}$ be arbitrary formulas, and $i \in \{1,2,3,4,5,6,7,8,9\}$, then the axiomatization of S_{IA} consists of the following axiom models and inference rules:

(AM1)$\oplus_i(\varphi \rightarrow \psi) \rightarrow (\oplus_i\varphi \rightarrow \oplus_i\psi)$
(AM2)$\oplus_i\varphi \wedge \oplus_i\psi \rightarrow \oplus_i(\varphi \wedge \psi)$
(AM3)$\oplus_1\oplus_7\varphi \rightarrow \oplus_7\varphi$
(AM4)$\oplus_7\oplus_1\varphi \rightarrow \oplus_1\varphi$
(AM5)$\oplus_5\oplus_6\varphi \rightarrow \oplus_6\varphi$
(AM6)$\oplus_6\oplus_5\varphi \rightarrow \oplus_5\varphi$
(AM7)$\oplus_2\oplus_9\varphi \rightarrow \oplus_9\varphi$
(AM8)$\oplus_9\oplus_2\varphi \rightarrow \oplus_2\varphi$
(AM9)$\oplus_3\oplus_8\varphi \rightarrow \oplus_8\varphi$
(AM10)$\oplus_8\oplus_3\varphi \rightarrow \oplus_3\varphi$
(AM11)$\oplus_1\oplus_2\varphi \leftrightarrow \oplus_2\oplus_1\varphi$
(AM12)$\oplus_1\oplus_3\varphi \leftrightarrow \oplus_3\oplus_1\varphi$
(AM13)$\oplus_1\oplus_4\varphi \rightarrow \oplus_1\varphi$
(AM14)$\oplus_1\oplus_5\varphi \leftrightarrow \oplus_5\oplus_1\varphi$
(AM15)$\oplus_1\oplus_6\varphi \leftrightarrow \oplus_6\oplus_1\varphi$
(AM16)$\oplus_1\oplus_8\varphi \leftrightarrow \oplus_8\oplus_1\varphi$
(AM17)$\oplus_1\oplus_9\varphi \leftrightarrow \oplus_9\oplus_1\varphi$
(AM18)$\oplus_2\oplus_3\varphi \leftrightarrow \oplus_3\oplus_2\varphi$
(AM19)$\oplus_2\oplus_4\varphi \rightarrow \oplus_2\varphi$
(AM20)$\oplus_2\oplus_5\varphi \leftrightarrow \oplus_5\oplus_2\varphi$
(AM21)$\oplus_2\oplus_6\varphi \leftrightarrow \oplus_6\oplus_2\varphi$
(AM22)$\oplus_2\oplus_7\varphi \leftrightarrow \oplus_7\oplus_2\varphi$
(AM23)$\oplus_2\oplus_8\varphi \leftrightarrow \oplus_8\oplus_2\varphi$
(AM24)$\oplus_3\oplus_4\varphi \rightarrow \oplus_3\varphi$
(AM25)$\oplus_3\oplus_5\varphi \leftrightarrow \oplus_5\oplus_3\varphi$
(AM26)$\oplus_3\oplus_6\varphi \leftrightarrow \oplus_6\oplus_3\varphi$
(AM27)$\oplus_3\oplus_7\varphi \leftrightarrow \oplus_7\oplus_3\varphi$
(AM28)$\oplus_3\oplus_9\varphi \leftrightarrow \oplus_9\oplus_3\varphi$
(AM29)$\oplus_4\oplus_1\varphi \rightarrow \oplus_1\varphi$
(AM30)$\oplus_4\oplus_5\varphi \rightarrow \oplus_5\varphi$
(AM31)$\oplus_4\oplus_6\varphi \rightarrow \oplus_6\varphi$

(AM32)$\oplus_4\oplus_7\varphi\rightarrow\oplus_7\varphi$
(AM33)$\oplus_4\oplus_8\varphi\rightarrow\oplus_8\varphi$
(AM34)$\oplus_4\oplus_9\varphi\rightarrow\oplus_9\varphi$
(AM35)$\oplus_5\oplus_7\varphi\leftrightarrow\oplus_7\oplus_5\varphi$
(AM36)$\oplus_5\oplus_8\varphi\leftrightarrow\oplus_8\oplus_5\varphi$
(AM37)$\oplus_5\oplus_9\varphi\leftrightarrow\oplus_9\oplus_5\varphi$
(AM38)$\oplus_6\oplus_4\varphi\rightarrow\oplus_6\varphi$
(AM39)$\oplus_6\oplus_7\varphi\leftrightarrow\oplus_7\oplus_6\varphi$
(AM40)$\oplus_6\oplus_8\varphi\leftrightarrow\oplus_8\oplus_6\varphi$
(AM41)$\oplus_6\oplus_9\varphi\leftrightarrow\oplus_9\oplus_6\varphi$
(AM42)$\oplus_7\oplus_4\varphi\rightarrow\oplus_7\varphi$
(AM43)$\oplus_7\oplus_8\varphi\leftrightarrow\oplus_8\oplus_7\varphi$
(AM44)$\oplus_7\oplus_9\varphi\leftrightarrow\oplus_9\oplus_7\varphi$
(AM45)$\oplus_8\oplus_4\varphi\rightarrow\oplus_8\varphi$
(AM46)$\oplus_8\oplus_9\varphi\leftrightarrow\oplus_9\oplus_8\varphi$
(AM47)$\oplus_9\oplus_4\varphi\rightarrow\oplus_9\varphi$
(AM48)$\oplus_4\varphi\rightarrow\varphi$
(Rule1)If $\vdash\varphi$, then $\vdash\oplus_i\varphi$.
(Rule2)If $\vdash\varphi$ and $\vdash\varphi\rightarrow\psi$then $\vdash\psi$.

4.2Soundness and completeness

Theorem 1 (Soundness): Let $\varphi\in L_{IA}$ be arbitrary formula, if $\vdash\varphi$then $\vDash\varphi$.

Proof. It is easy to show all the axioms is valid on the frame, and (Rule1) and (Rule2) are closed under the frame. Thus, S_{IA} is soundness.

Definition 6 (Canonical Model): Let $<X^C,Y^C,<^C_x,<^C_y,V^C>$ be the canonical model of S_{IA} (notation: $\mathcal{M}^C_{IA}$), if and only if, (i) $X^C,Y^C=\{F^m|F^m$ is a maximal consistent formula set of $S_{IA}\}$, (ii) $F^m<^C_x F^{m'}$ iff $\oplus_i^- F^m\subseteq F^{m'}$, $F^m<^C_y F^{m'}$ iff $\oplus_i^- F^m\subseteq F^{m'}$, (iii) $V^C(p)=\{F^m\in X^C$and $Y^C|p\in F^m\}$.

Theorem 2: Let $Form(L_{IA})$ be the formula set of S_{IA}, $F\subseteq Form(L_{IA})$ be an arbitrary subset, $\varphi\in Form(L_{IA})$ be an arbitrary formula of S_{IA}, if F is consistent and $\neg\oplus_i\varphi\in F$, then $\oplus_i^- F\cup\{\neg\varphi\}$ is consistent.

Proof. By the structured induction on $\varphi\in Form(L_{IA})$, and the axioms and inference rules of S_{IA}, it can be proved.

Theorem 3: Let $Form(L_{IA})$ be the formula set of S_{IA}, $F\subseteq Form(L_{IA})$ be arbitrary subsets, $\varphi\in Form(L_{IA})$ be an arbitrary formula of S_{IA}, if $\neg\oplus_i\varphi\in F$, then there are $F'\subseteq Form(L_{IA})$ that is $F<F'$ ($<\in\{<_x,<_y\}$) and $\neg\varphi\in F'$.

*Proof.*Assume $\neg\oplus_i\varphi\in F$, then by "theorem 2", we have $\oplus_i^- F\cup\{\neg\varphi\}$ is consistent. And by the property theorem of consistent set, it can be expanded to a maximal consistent set F'. Thus $\neg\varphi\in F'$.

Theorem 4: Let $\mathcal{M}^C_{IA}$ be the canonical model of S_{IA}, $\psi\in Form(L_{IA})$ be an arbitrary formula of S_{IA}, then existing $(\mathcal{M}^C_{IA},F^m_x,F^m_y)\vDash\psi$ iff $\psi\in(F^m_x\cap F^m_y)$.

*Proof.*Consider the induction on ψ.

(i) Assume $\psi$is $p\in Form(L_{IA})$. By $\mathcal{M}^C_{IA}$, that $F^m_x,F^m_y\in V^C(p)$ iff $p\in(F^m_x\cap F^m_y)$.

(ii) Assume ψ is $\neg\varphi$. Then $(\mathcal{M}^C_{IA},F^m_x,F^m_y)\vDash\neg\varphi$, by "definition 4", iff $(\mathcal{M}^C_{IA},F^m_x,F^m_y)\nvDash\varphi$. By introduction hypothesis, iff $\varphi\notin(F^m_x\cap F^m_y)$. And by property theorem of consistent set, we have $\neg\varphi\in(F^m_x\cap F^m_y)$.

(iii) Assume $\psi$is $\varphi_1\wedge\varphi_2$. Then $(\mathcal{M}^C_{IA},F^m_x,F^m_y)\vDash\varphi_1\wedge\varphi_2$, by "definition 4", iff $(\mathcal{M}^C_{IA},F^m_x,F^m_y)\vDash\varphi_1$ and $(\mathcal{M}^C_{IA},F^m_x,F^m_y)\vDash\varphi_2$. By introduction hypothesis, iff $\varphi_1\in(F^m_x\cap F^m_y)$and $\varphi_2\in(F^m_x\cap F^m_y)$. And by property theorem of consistent set, we have $\varphi_1\wedge\varphi_2\in(F^m_x\cap F^m_y)$.

(iv) Assume $\psi$is $\oplus_1\varphi$. Suppose $\oplus_1\varphi\in(F^m_x\cap F^m_y)$. By "definition 6", for $F^{m'}_x\in X^C$and$F^{m'}_y\in Y^C$, if $F^m_x<^C_x F^{m'}_x$ and $F^m_y<^C_y F^{m'}_y$, then $\varphi\in(F^{m'}_x\cap F^{m'}_y)$. By induction hypothesis, for any $F^{m'}_x\in X^C$ and $F^{m'}_y\in Y^C$, if $F^m_x<^C_x F^{m'}_x$ and $F^m_y<^C_y F^{m'}_y$, then $(\mathcal{M}^C_{IA},F^{m'}_x,F^{m'}_y)\vDash\varphi$. And by "definition 4", $(\mathcal{M}^C_{IA},F^m_x,F^m_y)\vDash\oplus_1\varphi$. Suppose $\oplus_1\varphi\notin(F^m_x\cap F^m_y)$. By property theorem of consistent set, we have $\neg\oplus_1\varphi\in(F^m_x\cap F^m_y)$. And by "theorem 3", there are $F^{m'}_x\in X^C$ and $F^{m'}_y\in Y^C$, such that $\neg\varphi\in(F^{m'}_x\cap F^{m'}_y)$. By property theorem of consistent set, $\varphi\notin(F^{m'}_x\cap F^{m'}_y)$. And induction hypothesis, $(\mathcal{M}^C_{IA},F^{m'}_x,F^{m'}_y)\nvDash\varphi$. By "definition 4", $(\mathcal{M}^C_{IA},F^m_x,F^m_y)\nvDash\oplus_1\varphi$.

(v) Similarly, we can prove the other cases of ψ is $\oplus_i\varphi$ ($i\in\{2,3,4,5,6,7,8,9\}$), there is $(\mathcal{M}^C_{IA},F^m_x,F^m_y)\vDash\oplus_i\varphi$iff $\oplus_i\varphi\in(F^m_x\cap F^m_y)$.

Theorem 5(Completeness): Let $\varphi\in L_{IA}$ be arbitrary formula, if $\vDash\varphi$then $\vdash\varphi$.

Proof. Let $\mathcal{M}^C_{IA}=<X^C,Y^C,<^C_x,<^C_y,V^C>$ be the canonical model of S_{IA}, $F^m_x\in X^C$ and $F^m_y\in Y^C$ be any maximal consistent sets. Assume $\nvdash\varphi$. By property theorem of consistent set, we have $\varphi\notin(F^m_x\cap F^m_y)$. And by "theorem 4", $(\mathcal{M}^C_{IA},F^m_x,F^m_y)\nvDash\varphi$.

5Discussion

We have established a reasoning system S_{IA} that can be used in the computer image analysis. In this system, people just need to put the image under a computer with a camera, or through computer scanning, and then the computer can according to the coordinates of system analyzes the relationship between any pixels or compositions in this image. Moreover, it can be carried out a series of location analysis, so as to achieve the expected effect of image processing in computer. We just need to compile the system in the form of a language as a programming language that can be systematically reasoning. However, we must earnestly point out this system still have some shortcomings, for instance, it fails to directly exhibit what specific area have but through indirect and tedious iteration by different operators. To solve these defects, apparently, is our future work.

Acknowledgements

This work was supported in part by the National Natural Science Foundation of China under Grant Nos.61273338, and in part by the Major Program of National Social Science Foundation of China under Grant 11&ZD088.

References

1. S. S. Rautaray, A. Agrawal. Vision based hand gesture recognition for human computer interaction: a survey. Artificial Intelligence Review, 43(1), 1-54, (2015)
2. Y. Ahn, S. Lee. Hit Rate Prediction Algorithm for Laser Guided Bombs Using Image Processing. KIISE Transactions on Computing Practices, 21(3), 247-256, (2015)
3. X. Zhang, W. Liu, M. Dundar, S. Badve, S. Zhang. Towards large-scale histopathological image analysis: Hashing-based image retrieval.Medical Imaging, IEEE Transactions on, 34(2), 496-506, (2015)
4. B. Rieger, R. Nieuwenhuizen, S. Stallinga. Image Processing and Analysis for Single-Molecule Localization Microscopy: Computation for nanoscale imaging. Signal Processing Magazine, IEEE, 32(1), 49-57, (2015)
5. E. Boix, A. Ferrer, S. Fernandez, X. Sellart. Vehicle Guiding System Through Image Processing in Crash and Misuse Tests. In *Proceedings of SAE-China Congress 2014: Selected Papers* (Springer Berlin Heidelberg, 411-424, 2014)
6. E. Tabassi, P. Grother. Fingerprint image quality.In *Encyclopedia of Biometrics* (Springer US, 635-643, 2015).
7. J. C. Russ. *The Image Processing Handbook*. (CRC Press, 269-629, 2007)
8. M. Sonka, V. Hlavac, R. Boyle. *Image processing, analysis, and machine vision* (Cengage Learning,1-537, 2015)
9. M. Aiello, I. Pratt-Hartmann, J. van Benthem (Eds.). *Handbook of spatial logics* (Vol. 4)(Springer, 217-291, 2007)
10. P. Blackburn, M. De Rijke, Y. Venema. *Modal Logic* (Vol. 53) (Cambridge University Press, 50-426, 2001)

MATEC Web of Conferences 63, 04032 (2016) DOI: 10.1051/matecconf/20166304032
MMME 2016

The Analysis of Agent's Knowledge Based on Game Behavior

Jin Sheng GAO[1,a], Chang Le ZHOU[2,3]

[1]*Department of philosophy, Xiamen University, Xiamen 361005, China*
[2]*Department of Cognitive Science and Technology, Xiamen University, Xiamen 361005, China*
[3]*Fujian Key Lab of the Brain-like Intelligent Systems, Xiamen 361005, China*

Abstract. The process of game is actually the process of game behavior which is often accompanied by agent's epistemic process. And Knowledge is a key factor of epistemology. In the past, most of the time game theory confuses information with knowledge. This paper, based on game behavior, analyzes agent's knowledge differences from information. It also analyzes how agent reasons by hidden knowledge behind game behavior, as well as what kind of knowledge is involved in the process of reasoning, and what is the origin of the knowledge. This paper divides the agent's knowledge into two parts: common knowledge and private knowledge, which analyze the specific elements that are corresponding to them in games. In the end, it constructs a knowledge model of game agent, furthermore, by this model, the essence of knowledge held by agents in their game would be presented.

1 Introduction

In recent years,many logicians, economists and computer experts have been studying games using epistemic logic. Some researchers have studied the assignment games [1], some like to study games using logic techniques [2], such as the preference in strategy choice [3], the dynamic logic in extended games[4].As further research goes, the boundary between logic and games is getting increasingly blurring.In game behavior, according to the logical points of view, the reasoning involved in games can be divided into deductive reasoning and inductive reasoning.But both deductive reasoning and inductive reasoning will involve agent's cognitive behavior in the process of interaction with each other, i.e., the agent's choicesin strategy game are based on their common knowledge and background knowledge.In particular, common knowledge has a direct impact on the game agent's behavior.More topics involved are the relationships between common knowledge and agent's background knowledge, how they transform from each other, and the properties of knowledge in game behavior, which will be discussed and explored in this paper.

This paper consists of five sections.In section 2, we introduce the general expression of knowledge. In this section we define the concept of knowledge by its history and development. In section 3,it analyzes the types of agent's knowledge in game behavior. In section 4, we explain the process of knowledge formation and the structure of knowledge. In this section, we differentiate knowledge with data, information, and so on. In the end section, we discuss the results and of this paper and the future work.

2 The general expression of knowledge

Knowledge is a terminology that is frequently defined by many academic disciplines. There are a great of philosophers in epistemology had researched it.They have their own different ideas about knowledge. About the definition of the knowledge, they have been to debate endlessly. Here comes to a basic question: what is knowledge? For this, the most classic should be traced back to Plato directly. For example, his article "Theaetetus" is one of the most influential works about knowledge. In his book, he argues, knowledge is a true belief, which satisfies the following three conditions: The agent knows φ, when only if (i) φ is true, (ii) agent believes φ, (iii) agent hasdefinite evidences to believe φ.

Therefore, Plato defines knowledge as something that is justified, true, andbelieved. However, this traditional definition had been challenged byGettier. In"Is justified true belief knowledge?",Gettier puts forward some counterexamples on traditional definition. In his cases, even though these conditions are all satisfied, agent still doesn't know φ [5].In other words, these conditions actually cannot explain what knowledge is.Because we can hold a true belief that can be distinguished, but we believe that this belief is true just with some sort of accident. And then this belief is not as knowledge.

After Gatthier case, whether need to modify the initial conditions, or whether to introduce other conditions?In Epistemology, for example, some advocates have presented different ways based on Gattier case, and then

[a]Corresponding author: Hi.Jackson@163.com

there are forming fundamentalism,conheretism and contextualism, and so on[6].

Various suggestions have been proposed, but none of them got the consensus, and the question remains. In the current, the amendments of the knowledge definition or a new definition haven't arrived at a consensus. Obviously, Plato's ancient definition of knowledge has a profound impact on. Therefore, Plato's ancient definition of knowledge has been recognized by most people.

Based on the specific property of game behavior, and combine with the above analysis, we have the following definition about knowledge:

Definition 1(knowledge):Knowledge is a kind of cognitive stateincludingconnotative state (such as practical skills or expertise) and explicit state (such as disciplinary theory), which agent holds after through his experienceand education.

In fact, the factors of knowledge and the relations among them can be showed by figure 1 as follows:

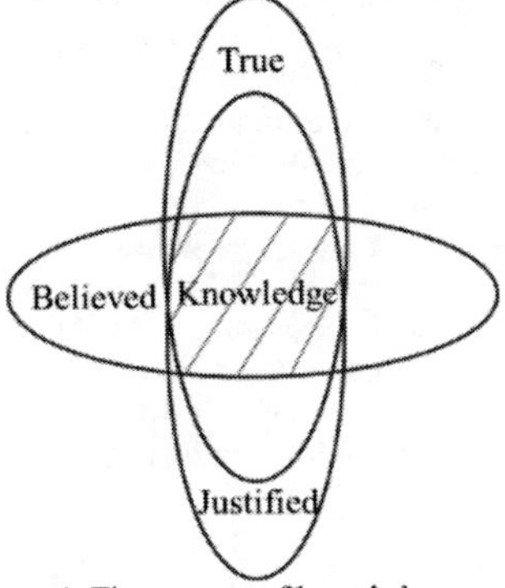

Figure 1. The concept of knowledge

3 Knowledge types of behavioral agent

Just as the above-mentioned definition of knowledge, game behavior involves agent cognitive, and agent cognitive involves external information, agent's own background, as well as the implied knowledge deduced by the agent through external information and its own background knowledge, *et al.* From the perspective of epistemology, knowledge has a variety of types; accordingly, for each specific discipline area, knowledge can be divided into corresponding types. For the game behavior, the knowledge of game agents can be divided into two categories: common knowledge and private knowledge according to these agents' situation.

3.1 Common knowledge of agents

The notation of common knowledge is first proposed by the philosopher, Lewis. He mentioned the general idea of common knowledge in his book "Conventions: A Philosophical Study" in 1969: It's a knowledge that everyone knows and everyone knows other knows and … recursively [7]. Though the common knowledge is first proposed by Lewis, Aumann firstly gave out the formal definition of common knowledge at 1976 [8]. This definition is aimed at game theory and can eliminate infinite recursions, such that people is accustomed to regard Aumann as the main contributor for the research of common knowledge.

Nowadays, there are two expressions of "Common Knowledge", the former introduce modal operators in epistemic logic into the definition, and the later one is a kind of Aumann's set structure expression.

3.1.1 Modal operator expression

To characterize the properties of "Common Knowledge", a corresponding formal system S_{CK} is usually constructed in epistemic logic. Similar to other epistemic logic systems, S_{CK} is also an extension of propositional logic. The extended parts include: a cognitive community N, modal operators$K_i (i = 1,2,3,\cdots,\mathrm{n})$, modal operator K_S ($S \subseteq N$), modal operatorE_N and the operator C, where $K_i\varphi$means "Agent $i$knows φ", E_N φmeans "Each member in community N knows", K_S φmeans "Some members in N know", $C\varphi$ means " φ is Common knowledge for everyone".In addition, following axioms are added to the classical logic system [9]:

$$K_S\varphi \equiv \vee_{i\in N} K_i\varphi \quad (1)$$

$$E_N\ \varphi \equiv \wedge_{i\in N} K_i\varphi\ (2)$$

$$C\ \varphi \equiv \wedge_{i=1}^{n} E^n\ \varphi (n = 1,2,3,\cdots)\ (3)$$

Ncan be infinite in theory, but the language of epistemic logic is usually finite.

Then, we usualy use Kripke model to define the above syntax structure, such thatwith semantic meanings.Given a cognitive model$< S, R_i, V >$, where

(i)S is a non-empty of cognitive states, $s_m \in S$;

(ii)$R_i$is an accessible relation between possible worlds, $(i = 1,2,3,\cdots,\mathrm{n})$;

(iii) V is a truth assignment for each proposition in possible worlds.

In this way, the semantics of modal operators can be determined: by definition, $K_i\varphi$is true at cognitive state s_0, if and only if $\varphi$is true in $s_i \in S$ where $s_0 R_i s_m$.

3.1.2 Aumman'sstructure expression

The Nobel economics laureate,Aumann, utilize set theory to define common knowledge in 1976. Let Sbe the set of all states and event E be the set of all subsets of S. For each member$i$ $(i = 1,2,3,\cdots,\mathrm{n})$, an information set is defined upon $S$and P_i. This information set represents the cognitive state of the member in this state. At state s_0, that the statement "member i knows" obtains a state from $P_i(s_m)$, but the obtained state is explicitly unknown. This fully embodies the essence of dynamic game under incomplete information. In such a dynamic game scene, agents can have several decision nodes in some information set in the game tree, but the agents can't tell which decision node they are located in, thus these decision nodes belong to the same information set. Now, we will define a knowledge function $K_i$by the following way [10]:

$$K_i\varphi \equiv \{s_m \in S |\ P_i(s_m) \subset \varphi\} \quad (4)$$

$K_i\varphi$is the set of all states that i knows the event φ can have. Similar to the above modal logic formulization, we can define a operator to represent "Everyone knowsφ":

$$E_i\varphi \equiv \cap_{i=1}^{N} K_i\varphi \quad (5)$$

Similarly, we can define the common knowledge in Aumann's way as follows:

$$C\varphi \equiv \cap_{n=1}^{\infty} E^n\varphi \quad (6)$$

The Aumman's structure expression can be transformed to Kripkemodel$< S, P_i >$, where the state set S is similar to cognitive states and the information set P_i defines the accessible relation between states.

3.1.3 Common knowledge in game behavior

AfterAumann's definition of common knowledge, game theorists always like to use something that similar to "someone knows what the other agent knows" as common knowledge. However, many game theorists are unwilling to define the form of common knowledge on the structure of game theory. There are some technical problems as it is necessary to adjust the previous research results in many ways. So they are used to stopping in the form of common knowledge. Because of the different research perspective, the common knowledge involves different elements. For the game behavior, the common knowledge is as follows:

(I) First of all, "the hypothesis of rational man" is the common knowledge. Rationality is as the most basic conditions for game occurrence and strategy selection. In game theory, behavioral agent is a rational person who will maximize the pursuit points of their own income. However, in epistemic logic, it is usually used to indicate the best response of the player who knows or believes certain knowledge. The best response, while, is a kind of preference logic, specifically to characterize the optimal reaction function of action logic. In this regard, it can be subdivided into "weak rationality" and "strong rationality":

Definition 2 (Weak Rationality, Strong Rationality): Weak rationality means behavioral agent is not stupid and will always choose the optimal strategy in his cognitive range to maximize his payoff, but the agent may not know the chosen strategy is strictly dominant. Strong rationality refers to the agent can find a complete measure in overall strategy, and select out a strictly dominant.

(II) Secondly, the structure of the game is a common knowledge. Basic elements of a game includes: the behavioral agent, the rules, the outcome, and the payoff [11]. The structural elements of the game must become common knowledge among the agents. For example, in the dynamic game of incomplete information, game players must recognize the game order between them.

(III) In addition, the characteristic elements of different games are common knowledge. Since the game can be divided into different types and each type of game has different characteristic elements, which become common knowledge among the agents and take play game behavior. For example, for the incomplete information game structure, although the game player i don't clear the opponent's type P_{-i}, player i must be clear his own type of belief P_i and the function $A_i(T_i)$ mapping types space T_i to action space A_i. All these knowledge must be the common knowledge between game players.

3.2 Private knowledge of agents

Game agent's knowledge is not just common knowledge, there is corresponding with the private knowledge. Specifically, private knowledge mainly embodied in the following aspects:

(I) Formulti-stage dynamic game with complete information and static game, the game agent cannot obtain the competitor's accurate action space. (II) The opponent type P_{-i}. Game agent just knows his type P_i, and reasons the probability of opponent type P_{-i}, but he cannot exactly know P_{-i}. In this case, game agent can only obtain Bayesian Nash equilibrium by the Bayesian formula. (III) For incomplete information dynamic game, at least one agent does not know the type.

If based on game agent's knowledge space, and its relationship with common knowledge, we can make the following analysis: game agent's knowledge space roughly consists of the factual propositions (notation: φ), public knowledge (notation:$E_G\varphi$), common knowledge (notation:$C_G\varphi$), and implicit knowledge (notation:$I_G\varphi$). The clear expression is as the following figure 2:

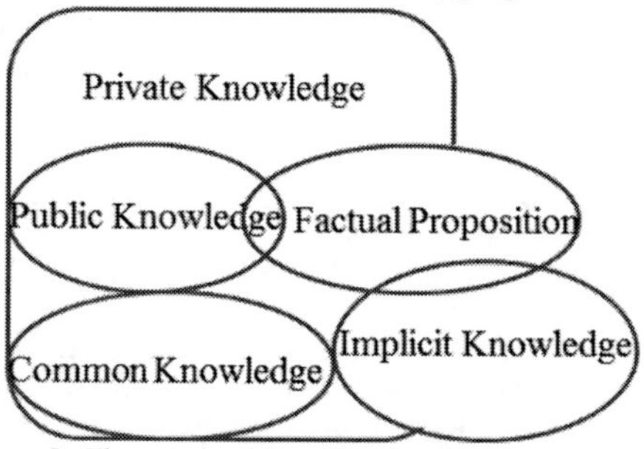

Figure 2. The relationship of private knowledge with other knowledges

4 Game behavioral agent's knowledge structure and its formation

As above, it is after Aumann's study about common knowledge that researcher begin to use common knowledge to analyze some hard problems. But in many cases, the research of game theory remains at information level. Information and knowledge are often lumped together or mixed up, and they don't distinguish. However, epistemic logic needs a clear cognitive concept. Therefore, it is necessary and significant to distinguish information and knowledge. The following retrospect can be the way of formation of knowledge.

4.1. The relationship among data, information and knowledge

There are different among date, information and knowledge, although we usually take information as knowledge.

4.1.1From data to information

Data, originating from the characterization of human, is generally believed to a sign of objective reality. Data is an unprocessed material which can show the objective existence of things and be perceived by agent, such as graphic symbol, number, letter, sound and so on [12, 13]. Before extracted, what showed up are messy symbols. They are original materials and carriers of information and knowledge. In the game behavior, data can be any symbol related to game-agent and external space between game-agents, such as space and time symbol of game behavior. They are disorganized discrete status distribution symbols without rules.

Although information has different interpretations in different fields,it must reflect essential attribute of information.It is clear that information comes from data.Data can be said to be a kind of objective existence. Data becomes information because it involves cognitive function of game-agent.As we all know,"manufacturer A, 2015, 3000" are data but not associated.Only when they are used to describe objective reality and their relationship and form ordered data flow,can data be called information.Information is not simply the stack of data,because the result of casual stack of data is likely to be a kind of data.So information is distinguished from data in the fact that information is ordered association of data and information has timeliness.For example,in incomplete information static game, manufacturer B gets data flow "manufacturer A, 2015, 3000"from outside world. Manufacturer B can process them into:(a) Single commodity cost of manufacturer B is 3000 in 2015,(b) Monthly output of manufacturer B is 3000 in 2015, and so on.

4.1.2 From information to knowledge

In game theory, information is a very important concept. Information economics, the most compelling part of modern economics, is based on game theory. According to the specific needs, game agent can get information of itself by extracting, marshaling and translating data with priori knowledge and cognitive function. Ultimately, only the information which conforms to "definition 1" can be the behavioral agent's knowledge. In other words,only the knowledge perceived and confirmed by a game agent can become the behavioral agent's knowledge.

4.1.3Fromprivate knowledge to common knowledge

In game behavior, the process from private knowledge to common knowledge of game agents involves game agent's belief update process by public announcements. The logic of public announcement is first proposed by logician Plaza in 1989 [14].He studied the impact of belief update on behavioral agent in detail. Moreover, a reasoning system for transforming from private knowledge to common knowledge can be referred to the literature [15].

4.2 knowledge structure of game agent

First, agent extracts the data from the external environment, and then transforms information to private knowledge. In the end,game agent adopts suitable strategy according to private knowledge. For this, we can construct a model figure 3 as follows:

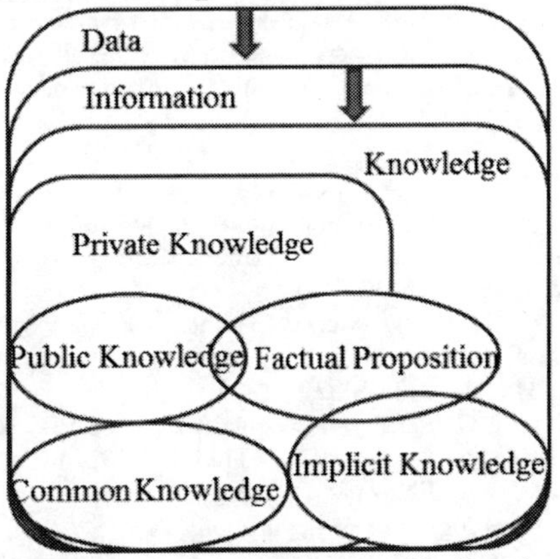

Figure 3. Agent's knowledge structure and its formation

In this model,we can clearly see the transform relationship and containment relationship among information,data,knowledge, individual knowledge and common knowledge.In the process of forming knowledge, incomplete information appeared when relevant part of information is not shared by all the agents.This often happens when some(or all)agents exclusively know their payoff function.

5 Conclusion

It can be found that the research about the knowledge of game agent is different from the analysis of knowledge in other fields.Because game behavior has the property of information economy,the analysis of game behavior inevitably contains the analysis tools of information economics.And then the research analyzes it gradually by epistemic logic. Colligating knowledge of different agents,reflecting the features of game behavior,this research is believed to be more scientific.On the basis of distinction of data,information,knowledge,common knowledge and private knowledge involved in game behavior,the transform process is analyzed and knowledge structure model of game agent is constructed eventually.This paper can see clearly the form of knowledge in game behavior, andobtain the essence of game agent'sknowledge. It's worth emphasizing that the analysis about the form of knowledge is a framework and cognitive game behavior needsfurther research.

Acknowledgements

This work was supported in part by the National Natural Science Foundation of China under Grant Nos.61273338, and in part by the Major Program of National Social Science Foundation of China under Grant 11&ZD088.

References

1. J. Hintikka,*Knowledge and Belief*(Cornell University Press, 1962)
2. H. H. Nax, B. S. Pradelski, Evolutionary dynamics and equitable core selection in assignment games. International Journal of Game Theory,44(4), 903-932(2015)
3. P. I. Cowling, S. Devlin, E. J. Powley, D. Whitehouse, J. Rollason, Player preference and style in a leading mobile card game.Computational Intelligence and AI in Games, IEEE Transactions on, 7(3), 233-242(2015)
4. J. van Benthem, Extensive Games as Process Models. Journal of logic, Language and Information, 11(3):289—313 (2002)
5. E. L.Gettier, Is justified true belief knowledge?. analysis, 23(6), 121-123(1963)
6. J.Dancy, E.Sosa,M. Steup,*A companion to epistemology*(John Wiley & Sons, 27-119, 2009)
7. D.Lewis, Conventions: *A Philosophical Study*(Harvard University Press, 1969)
8. R. J.Aumann, Agreeing to disagree. The annals of statistics, 1236-1239(1976)
9. J. Y.Halpern, Y.Moses, Knowledge and common knowledge in a distributed environment. Journal of the ACM (JACM), 37(3), 549-587(1990)
10. M. J.Osborne, A.Rubinstein, *A course in game theory* (MIT press, 67-84, 1994)
11. A. K.Dixit, S.Skeath,*Games of Strategy: Fourth International Student Edition*(WW Norton & Company, 2015)
12. M.Frické,Big data and its epistemology. Journal of the Association for Information Science and Technology, 66(4), 651-661(2015)
13. S.Yin, X.Li, H.Gao, O.Kaynak,Data-based techniques focused on modern industry: an overview. Industrial Electronics, IEEE Transactions on, 62(1), 657-667(2015)
14. J.Plaza, Logics of public communications. Synthese, 158(2), 165-179(2007).
15. J.Gao, C.Zhou, A reasoning system about knowledge acquisition in bi-agent interaction.*2016 Eighth International Conference on Advanced Computational Intelligence (ICACI)*. IEEE, 412-417 (2016)

MATEC Web of Conferences **63**, 04033 (2016) DOI: 10.1051/matecconf/20166304033
MMME 2016

Research on the thinking path of enterprise management in the era of big data

Ying GUO

Qilu Normal University, China Shangdong 250013

Abstract.With the acceleration of economic globalization, enterprise internationalization deepen and capital supply chain extension, Chinese enterprises will inevitably was involved in the economic tide to, the business was information and data of the high-frequency explosive growth. At the same time, the rapid expansion of the management network and the management functions of the cross redundancy is serious, so that the enterprise management, decision-making, execution are subjected to tremendous impact. Big data era has arrived, companies need new management ideas and solutions to cope with the challenges of the new situation. This article embarks from the enterprise management under the background of big data, according to three aspects: production, marketing, transportation, a full range analysis to in the era of big data for scientific and effective management planning, provide strong management support for the development of enterprises.

Key Word. Big data;Intelligence;data network;Management change

Introduction

Human society has changed from the industrial age to the material and energy to the era of knowledge economy based on science and technology. The economic driving force has been transferred from the capital management and labor force to the human resources, the cultural value and the innovation of the invention. With the fifth scientific and technological revolution, following the cloud space and Internet of things, the new era of change in the Internet Era - the era of big data is coming. Economic development trajectory has been towards the knowledge assets and virtual economy closer, the traditional tangible assets, fixed assets successor weakness. Rapid development momentum forcing enterprise management mode to update, must learn from the management perspective of application of information technology to support business analysis and decision-making, strengthening and solid business operation system, so in order to better adapt to market demand.

1 Impact on enterprise development in the era of big data

1.1 Modern deconstruction of big data

The improvement of scientific and technological level makes people's material and cultural life more abundant, and the popularity of mobile terminal and information network to enhance the adhesion of a huge volume of data. Data is more with the development of the Internet is a geometric multiplication, the traditional data processing methods have been difficult to control. International Data Corporation (IDC) is a digital universe on the topic research report shows: information data in 2003 the sum of only for 5EB (exabytes). Today, only 2011 statistics of world was created not to repeat the amount of data has already exceeded the 1.8ZB (zettabyte one hundred thousand million bytes), to produce 5EB of the data requires only two days. And by 2020 there will be at least 25.2ZB of the total amount of data, and 70 of the 2 parties almost. And will also follow the new Moore's Law (global data over the years based, double every two years) continued growth!

At the same time, the complexity of the data also increased dramatically. Data types (multi-source, heterogeneous, multi source heterogeneous, multi modal, multi class grammar writing, non coherent language, grammar hybrid, etc.) are diverse, and have a high degree of random combination. In such a complex data type structure, the value of the data is very low density (there is a lot of irrelevant information, interference information and debris incomplete information) knowledge of the "purification" is more difficult. And most of the data is real-time (data requires real-time acquisition, generation, storage and analysis) and the stage (data only in a particular time period, after which will be deleted). Into the "unable to in a short period of time using conventional analysis software or tools to grab the contents of the data analysis and processing of large data set", which is widespread in social information network under the "data" intuitive deconstruction.

1.2 Big data brought about by the era of change

With the development of social information, the mobile network is becoming more mature, large LAN architecture is becoming more and more perfect, the amount of data created by individuals is increasingly large, big data comes also brought a time change. Big data has the characteristics of 4V+1C, that is, speed (Velocity), quantity (Volume), value (Value), type (Variety) and complex (complexity) to complete the summary. The huge amount of data is its basic feature, but also the prominent feature. And in this era of big data, even the most simple data sharing will give artificial become great distress, a large amount of data processing in lower operational efficiency at the same time also makes the cost of infrastructure are high, system maintenance, repair increasingly difficult, data storage has become a problem.

So at the end of 2008, three from world famous universities of information in the field of senior scientists jointly organized "computing community Alliance (computing community consortium)": the worldwide impact huge white paper "big data calculation: Social Sciences, business of the revolutionary breakthrough", make the corporate executives to realize the data real value and meaning of the times is not the data itself, but data hidden behind the powerful use. May 2011, EMC company in the cloud computing encounter big data, World 2011 conference in the big data concept and technology trends. In October of the same year, Gartner will be the big data as the first 20 thousand and 120 strategic new technology. The advent of the era of big data, but also to provide a new big data processing technology. That collects the massive data information to carry on the analysis contrast, through the data mining to create its biggest value. Subsequently, led by the United States, the United Kingdom, Germany, Finland, China have added to the application of large data technology research, the era of big data changes brought about by sweeping the globe.

1.3 The prospect of social industry under the big data

Large data processing technology to the surface is very wide, from medical equipment, video monitoring equipment, instruments, internet intelligent equipment, non traditional IT equipment to mobile devices and Internet information deconstruction can join the big data processing methods. Enterprises can continuously generated from trading in huge data acquisition to trillion bytes as units of consumption and supply records, to crawl "purification" such as consumer demand for goods distribution, supplier distribution ratio, capital return period enterprise management plan the core of reference data.

At present, the big data has been in the retail, financial industry, manufacturing, logistics, telecommunications, medical services, led by a large number of industries in the start, resulting in a huge social value and appreciation of space. According to the professional market research firm IDC predicts that by 2010, big data technology and business services market will grow from $17 billion in 2015, has exceeded 100 billion, more than 50% years to achieve amazing growth rate. Had the highest growth rate of IT is less than 10% of the annual growth. Plus in recent years, domestic and foreign well-known enterprises, such as Taobao, mobile, Wal-Mart, Amazon, all have developed large data products and platforms, the use of big data technology has become a new generation of industry darling. Thus, big data in the enterprise development prospects.

2 Enterprise management model based on big data

Enterprise management mode with the development of information technology experienced a total of four stages of change, are based on a single stage of manual management, based on LAN online file system, based on Internet database system and based on mobile Internet data integration and analysis stage. Which promotes the management mode of the fundamental elements of constant change, is the increase in the amount of data and data types of differentiation and evolution. Now, push in the era of big data, and a data explosion, we need a new management model.

2.1 Intelligent production

Enterprise information system connected in factories and workshops placed sensing devices and the core control data through the integration of the physical information system (CPS), will be generated in the production of large data information through the cloud computing processing center for storage, analysis, and then draw the specific planning programs to know industrial production, which is the intelligent production management mode.

Further, is through the comprehensive production equipment data collection, access to the huge function of the production information, through the data capture and screening, for wireless communications will transmit data to the Internet, so as to realize the real-time monitoring, analysis, the Austrian data processing results feedback to the production chain, so that the entire production process for adaptive and self adjustment of intelligent network and management so as to achieve the optimization. And intelligent production, but also with the flexibility of human resources to speed up the analysis of search efficiency. For example, a Canadian named GoldCorp mining company to solve the problem of its RedLake mine veins positioning, to social media open the all the geological data in the area since 1948, in just a few weeks received a large number of Internet users in the positive feedback, and recommend users all 110 spots accurately found 80 multiple deposits, geological exploration of a large amount of money is saved.

2.2 Big data network marketing

Big data network marketing is the essence of the combination of marketing and the Internet, to build a cloud based platform for the marketing model. Including commodity geographic marketing, association rules mining marketing, user behavior analysis marketing, personalized marketing and big data analysis marketing. They are MapReduce parallel computing framework based on, through the collection of relevant data and with cloud platform for large data mining analysis, extract useful information integration, revised for the production of products, marketing strategy formulation, cargo personalized recommendation provide effective basic data support.

Thus in this mode, you can through the accumulation of user data, so as to compare the analysis of the customer's preferences and buying habits. And based on customer preferences to develop marketing strategy, which is the advantages and characteristics of large data network marketing. According to this, enterprises can better carry out the network marketing planning, to ensure the rational production of products, to protect the economic benefits of enterprises. For example, Taobao is based on large data analysis, to draw the purchase preferences of consumers, thus pushing a number of related products, thus greatly enhance the Taobao's trading volume and customer satisfaction.

2.3 Data logistics

Data of logistics is based on modern information data equipment (GPS, networking, mobile data terminal, GIS, cloud computing and Internet interaction, widely used in logistics transportation, loading and unloading, handling, storage, packaging, processing flow, distribution and after-sales service in the link. For enterprise products from production to sales, to carry out a full range of three-dimensional data collection, screening and analysis. Through the information network to carry out scientific scheduling, the whole process of supervision to reduce costs and potential risks, while improving the quality of logistics services.

Data logistics requires intelligent analysis, perception, optimization and real-time feedback. So big data processing cloud platform to build a key. Through the integration and reorganization of the multi-source data information, the server and data network is assimilated into a large data environment, and the data processing platform is constructed by using the matching model of intelligent terminal + data information. So as to solve the current customers, goods, courier and logistics management staff to see the current situation of information interaction and audit efficiency. In the end, the construction of the data logistics will greatly promote and accelerate the development and the rise of modern logistics business.

The management thinking and way of the enterprise should be amended with the change of the times. In today's era of big data should be closely linked to the data processing core technologyanalysis, implementation management of super flat to the data network of excessive. In order to allow enterprises to fully integrate into the era of change, and gradually optimize the enterprise system, and then to the international development.

3 Conclusion

The management of large data, is a new attempt in the new era, has great potential and space mining. Constructing a data management system, focused by a large number of resources, information sharing, management decision by the more broader participation, brainstorming, enhance decision-making efficiency and reduce the error probability. Also by processing the big data, cloud computing and related technology, "elastic" and "universal" and "building blocks" of the thought of technology has been popularized application, the system platform can provide ductility and durability of better quality service to customers. And in the expansion of enterprises to change the management and business module, give good technical support. And then realize the high quality, high efficiency and high intelligence of the new enterprise management mode.

References

[1] Feng Guo Xunhua. Zeng an army. Chen Guoqing Chen Yubo. Several frontier issues in business management under the background of big data [J]- Journal of management science 2013 (1)

[2] bell, a new management model based on large data processing [D] Hubei University of Technology -2015.06

[3] Jenny Tseng electricity business enterprise big data marketing application research [D], Guangdong University of Foreign Studies, -2015.06

[4]Hu Kaibao, moor-jankowski TaoQing. Corpus and translation studies: trends and problems -- - 2007 international academic seminar of corpus translation studies, and research review [J]. Journal of foreign languages (journal of

Shanghai international studies university), 2007 (05)
[5] nai-xing wei. John Sinclair linguistic heritage - on the thought and method of [J]. Journal of foreign languages (journal of Shanghai international studies university), 2007 (04)
[6] tian zhen xie. Translation studies after the "cultural turn" - the cultural turn of translation studies in comparative literature significance [J]. Journal of comparative literature in China. 2006 (3)

MATEC Web of Conferences **63**, 04034 (2016) DOI: 10.1051/matecconf/20166304034
MMME 2016

Research on influence factors of VSC's efficient operation

Jie Lu [1,a], Xin Fa Tang [2,b]

[1] *Research Center of Jiangxi Local Culture, Jiangxi Science & Technology Normal University, Nanchang, 330013, China.*

[2] *Tourism School, Jiangxi Science & Technology Normal University, Nanchang, China*

[a] *154416372@qq.com*, [b] *xinfatang@sina.com.cn*

Abstract: VSC is a league made up of some software firms of different advantages and resources aiming at common software developing tasks or relative services. the result of cooperation is based on it's cooperative efficiency. This article tries to find out and compute those primary influence factors of VSC' operation through mutiple statistical method and RST, and then some suggestions are given.

1 VSC' s connotation

VSC is abreviation for Virtual Software Corporation which is one kind of Virtual Enterprise. With the development of information technique, the hardware which software system relies on has changed much, at the same time, the scale, complexity and development cost of software system are increasing evidently. It's hard for one software firm to meet all these needs only depends its own resources and knowledge. As a creative organization form, VSC provides a fresh drain of thought to solve those problems mentioned above.

This paper considers that VSC is a league made up of several software firms of different advantages and resources aiming at a better job in common software development tasks or relative services.

2 Analysis on influence factors of VSC's efficient operation

Till now, little research has been done to find out pivotal influence factors in VSC's operation. This paper strives to reveal the factors by multiple statistics method and RST based on questionnaire survey.

2.1 Truth analysis on influence factors of VSC's operation

To make a thorouth study on influence factors of VSC operation, 3 different analytic methods are adopted, i.e., to gather data by questionnaire survey, factor analysis method is taken to find out those influence factors and RST is used to compute those factors' relative importance.

2.2 Questionaire devising and data gathering

Questionnaire survey was taken to get data which are needed in truth analysis, further analysis was done by one of multiple analysis methods——factor analysis method with SPSS12.0. in this survey, Sends out questionnaire 200, reclamation effective questionnaire 122, recovery pecent 61%, Invalid questionnaires (with incomplete answers) were eliminated, 103 valid questionnaires, the valid questionnaire return rate is 84.4%.

2.3 Reliability testing

Reliability is regarded as criterion to test the consistency and stability of gathered data. In this research, Cronbach's Alpha was adopted. Table 1 was gotten by SPSS:

Tab 1: the Result of Reliability Testing

Cronbach's Alpha	Cronbach's Alpha Based on Standardized Items	N of Items
.798	.804	29

Cronbach's Alpha is 0.798, Cronbach's Alpha Based on Standardized Items reaches 0.804, so we can believe that this questionnaire's reliability is satisfying and meets statistics analysis.

2.4 Factor analysis

From table 2, KMO is 0.760 which indicates these data are suitable for factor analysis, Bartlett's Test of Sphericity .000, less than 0.01, which shows that correlation matrix isn't identity matrix and suitable too.

Tab 2: the Result of KMO Measure of Sampling and Bartlett's Test of Sphericity

Kaiser-Meyer-Olkin Measure of Sampling Adequacy.		.760
Bartlett's Test of Sphericity	Approx. Chi-Square	3068.15 0
	df	406
	Sig.	.000

9 factors were extracted in principal component analysis as seen, totle variance explained reach 78.504%. After a rotation converged in 7 iterations, table 3 could be gotte

Tab 3: Rotated Component Matrix

	Component								
	1	2	3	4	5	6	7	8	9
C1	-.003	**.903**	.162	-.011	-.030	.052	.004	-.031	-.055
C2	-.105	.070	.325	.172	-.122	.134	-.144	.297	.134
C3	-.013	.075	**.708**	.033	.009	-.137	-.067	.146	.126
C4	.110	.197	.334	.134	-.087	**.630**	.248	.231	-.041
C5	.156	.229	.118	.063	-.150	**.587**	-.303	-.010	-.152
C6	.013	.094	.879	.057	.058	.132	-.038	-.128	-.202
C7	.103	-.117	.008	-.119	.034	-.016	-.187	**.922**	.064
C8	.043	.059	.160	.133	.167	-.007	-.191	.219	.054
C9	.052	.051	**.920**	.136	.085	.104	.062	-.005	-.057
C10	.024	.058	**.919**	.074	.079	.174	.067	.025	-.067
C11	-.034	**.961**	.052	.087	.041	.086	.002	-.001	-.056
C12	.160	.049	.081	.037	-.070	-.093	**.778**	-.030	.050
C13	-.047	**.934**	.014	.033	.020	.042	-.005	-.072	.005
C14	**.980**	-.051	-.022	-.002	.008	-.010	.025	.034	.027
C15	-.038	.094	.181	.305	.224	.441	**.728**	.144	.055
C16	**.944**	-.052	-.047	.002	.058	.035	.018	.019	.018
C17	-.038	-.010	.030	**.881**	.059	.076	-.025	.017	.030

C18	.021	.091	.071	**.918**	.033	.122	-.010	-.037	-.032
C19	-.016	-.023	.072	.453	.045	.103	.373	.487	-.037
C20	-.017	.117	.195	**.883**	.052	.041	-.017	-.025	-.009
C21	**.948**	.011	.028	-.044	.062	.065	.048	.016	.019
C22	**.943**	-.012	.019	.004	.026	.038	.027	.066	.009
C23	**.929**	-.024	.065	-.013	-.011	-.013	.010	-.046	.006
C24	-.042	-.004	-.021	.095	.282	**.804**	-.068	-.092	.108
C25	-.020	-.009	.032	.045	**.916**	-.001	-.112	-.013	-.088
C26	.126	.095	.124	.063	**.885**	.120	-.056	.002	-.074
C27	.032	-.005	.011	.042	**.932**	.032	.000	.031	.040
C28	-.046	**.952**	.070	.103	.056	.096	.010	.008	-.042
C29	.073	-.126	-.096	-.014	-.096	.017	.025	.052	**.817**

Extraction Method: Principal Component Analysis.
Rotation Method: Varimax with Kaiser Normalization.
a Rotation converged in 7 iterations.

Combined with table 4 and questionnaire, 7 factors were found out:

Factor 1: composed by C14、C21、C16、C22, named definity of software module; Factor 2: composed by C11、C28、C13、C1, named profit allocation; Factor 3: composed by C9、C10、C3, named cooperation attitude; Factor 4: composed by C18、C20、C17, named employee quality; Factor 5: composed by C27、C25、C26, named developing conditions; Factor 6: composed by C24、C4、C5, named communication ways; Factor 7: composed by C12、C15, named learning and creation; Factor 8: composed by C7, named developing standard; Factor 9: composed by C29, named management's support.

As is shown in the figure 1:

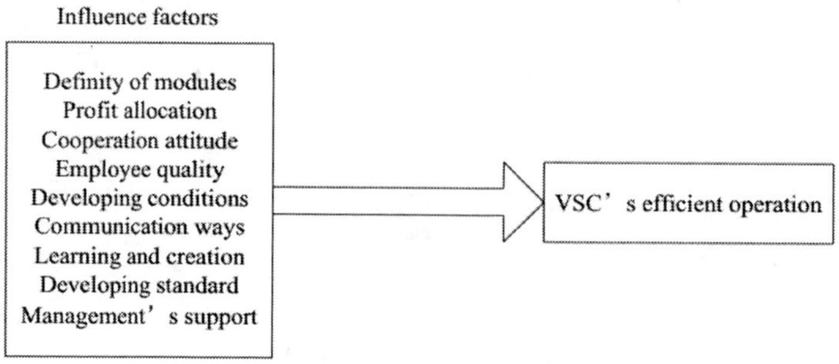

Figure 1: the primary influence factors of VSC' efficient operation

3 Analysis on importance of influence factors on RST

3.1 Train of analytic thought

(1) To compute relative importance of influence factors, formula is:

$$I(P)=\sum_{i=1}^{n}\frac{|X_i|}{|U|}\left[1-\frac{|X_{i.}|}{|U|}\right]=1-\frac{1}{|U|^2}\sum_{i=1}^{n}|X_i|^2$$

(2) Further analysis of each factor in terms of the

 DOI: 10.1051/matecconf/20166304034

result computed above.

3.2 Computing and analysis on importance influence factors of VSC' efficient operation

(1) Computing influence factors'importance

According to the train of analytic thought, RST was adopted to compute factors'importance. This information system of RST derived from the questionnaire, so there are 9 attributes in this information system and 103 records, a software program run in MATLAB6.5 was devised to do the job. The computing results are saved in an array IMP:

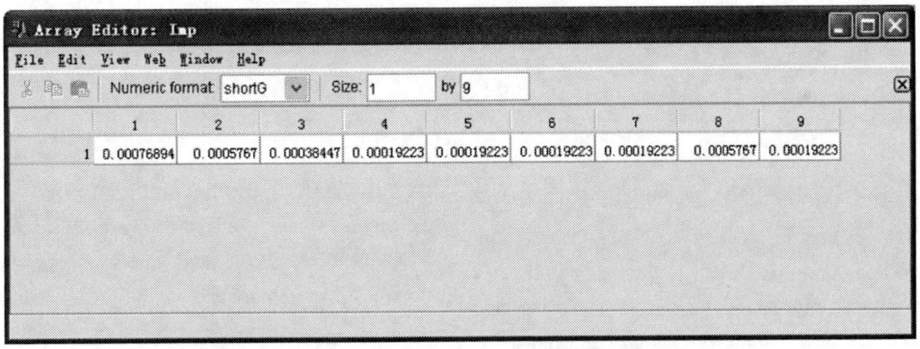

(Figure 2 influence factors's importance computed by program)

(2) Further analysis to results

①The result shows that cooperation attitude ranks third. Software developing is base on top-down decomposition layer modular architecture, different modules are integrated with stated interface and standards. Certain adjustment could be done well on the basis of favorable cooperation and communication among the developing teams. Employee belong to different teams or firms maybe never meet each other during the cooperative process, so completion of the task relies more on other connection ways (kinds of electronic ways).

② From figure 2, quality of employee, developing conditions, communication ways, the atmosphere of learning and creation and support from management would definitely bring some positive effect on VSC' operation. During the period, developing enviornment, management's support are crucial for smooth communication of firms or employee, at the same time, responsibility and attitude for work are also significant.

4 Conclusion

The motive of forming VSC is pursing a better completion of software developing task than any single software enterprise, for VSC can make use of all advantages of partners' firms. During cooperation process, lack of cooperative efficency may lead to failure of the task which isn't the original intention of VSC. This paper strives to find out the influence factors of VSCs' operation, so, necessary studies have been done by adopting factor analysis method and RST based on gathered data through questionnaire survey. At the end of this paper, some effective suggestions are put forth according to the analytic results.

References

[1] XuPei, ZhangRenJun, How to create effective project team[J]. Statistics and desicion, 2005 (3): 132~133

[2] KangZhuang, Study on Project managementb methods of Software developing [D] . Master's degree paper of University of International Business and Economics, 2006

MATEC Web of Conferences **63**, 04035 (2016) DOI: 10.1051/matecconf/20166304035
MMME 2016

Analysis on VSC' Efficient Operation

Jie Lu [1,a], Xin Fa Tang [2]

[1] *Research Center of Jiangxi Local Culture, Jiangxi Science & Technology Normal University, Nanchang, 330013, China*

[2] *Tourism School, Jiangxi Science & Technology Normal University, Nanchang, China*

Abstract: VSC is a league made up of some software firms of different advantages and resources aiming at common software developing tasks or relative services. the result of cooperation is based on it's cooperative efficiency. This article tries to find out the embodiment of VSC's efficient operation and those primary influence factors of VSC' operation.

Keywords: VSC, Efficient operation, Analysis, Suggestion

1 VSC' s connotation

VSC is abreviation for Virtual Software Corporation which is one kind of Virtual Enterprise. With the development of information technique, the hardware which software system relies on has changed much, at the same time, the scale, complexity and development cost of software system are increasing evidently. It's hard for one software firm to meet all these needs only depends its own resources and knowledge. As a creative organization form, VSC provides a fresh drain of thought to solve those problems mentioned above.

This paper considers that VSC is a league made up of several software firms of different advantages and resources aiming at a better job in common software development tasks or relative services.

2 VSC's efficient operation

During the process of cooperation, problems of sorts come forth because of the difference of destination, management's thought, organization mode and enterprises' culture of partners in VSC. The methods and attitudes been taken will definitely have direct influence on the cooperation efficiency. Communication, share and creation are of great importance so collision is unavoidable. Easily to understand, efficent cooperation relies on regular communication and connections among the partners in a VSC.

[a] Corresponding author: 154416372@qq.com

2.1 The meaning of VSC's efficient operation

To sum up, this article regards that efficient operation of VSC means favorable and smooth cooperation status present among partners in VSC by devising and establishing reasonable operation conditions or mechanism, each can solve kinds of problems in positive and serious attitude during cooperative process in order to promote and facilitate their software developing efficiency and effectiveness.

2.2 The embodiment of VSC's efficient operation

In VSC, to gain profit not only depends on separate firm but all, the cooperative efficiency has obvious importance on the quality, cost and market competition power of the software. According to the analysis made above. This paper holds the opinion that the cooperative efficiency is embodied as follows:

2.2.1 Partner firms are approbated and satisfied with the operation

Partner firms approbate with the aims of the collaboration and be pleased with the devised operation mechanism, respect each enterprise's culture etc, which are the foundation.

2.2.2 Customers' reasonable requirements can be converted into clear development demand

The pivotal job in software developing is to ascertain customers' reqirements, the significance of demand analysis is try to gain an exact and plain conception of functions and performance of the software system to be developed. In many software engineering books, all think that an excellent demand analysis should be integrated, correct, feasible, verifiable, but in fact, the analysis like

DOI: 10.1051/matecconf/20166304035

this is impossible. Actually, requirements of customers may need some changes during the developing process. So to keep close connections with customers is essential in developing a software system for some adjustments may be present, that these adjustments can be responded quickly is also one aspect of the representation of cooperation efficiency.

2.2.3 To keep close connections and to deal with kinds of problems or collision in time

As we know, software modules' developing, integration and testing need close communication among partners in VSC, problems and collision are inevitable, negative attitude may result in confict or failure of the project, obviously, to deal with kinds of problems or collision in time is indispensable.

2.2.4 Information resources sharing

To improve the efficiency of information resources sharing which helps to resopond rapidly to market and customers, it's good to improve efficiency that some effective measures are taken such as enrich communication ways etc.

2.2.5 VSC's competition power enhanced and software developing ability and efficency strengthened obviously

VSC is a new and powerful organization mode for it is a cooperative league based on complementary ability of partner' enterprises so its competition power is much greater than each one. By this kind of way, developing ablity and efficiency of each are improved.

3 Therotical analysis on influence factors of VSC's operation

As a new organization mode, VSC certainly receives many internal and external factors in restriction. External factors include economy, policy, market factors. Considering that these external factors can't be controlled by enterprises themselves and can't be changed in a comparatively short period, so they are left out of account in this article.

Combined with characteristics of software developing and VSC operation, this article tries to investigate those factors from 6 aspects as follows:

Characteristics of task includes knowledge of customers' demand, technique, task risk; firms strength includes culture of partners firms, organization structure, learning mechanism, cooperating skill, management, relative experience, technical ability, employee' responsibility; relativity quality includes communication, familiarity among partner firms; profit allocation includes profit allocation among partner firms, profit allocation among employees; developing process includes adjustment of demand, the quality of developing documents, information sharing effect, stabilization of cooperation.

3.1 Truth analysis on influence factors of VSC's operation

To make a thorouth study on influence factors of VSC operation, 3 different analytic methods are adopted, i.e., to gather data by questionnaire survey, factor analysis method is taken to find out those influence factors and RST is used to compute those factors' relative importance.

9 factors were extracted in principal component analysis, totle variance explained reach 78.504%. After a rotation converged in 7 iterations, 7 factors were found out:, As is shown in the figure 1:

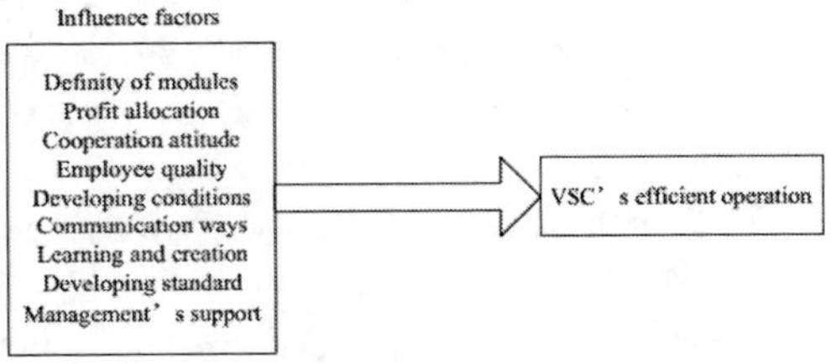

Figure 1: the primary influence factors of VSC' efficient operation

3.2 Further analysis to results

(1) It's easily to recognize that the most pivotal one among all these factors is definity of modules. In the beginning of the cooperative process, the analysis of customers' demand, the defintion to interface of software modules are of great significance. Construction of VSC is better completion of certain software developing tasks,that some uncertainty such as unreasonable disintegration occurs will directly influence the quality and efficiency of software, it would be a disaster if these problems are revealed later.

(2)Both partner firms' satisfaction to profit allocation and adherence to software development standards are essential. Easily to understand, partner firms are willing to join a VSC for they want to gain some profit. It's no doubt that cooperative efficiency would be suspicious if profit allocation isn't satisfying for these enterprises. So reasonable profit allocation mechanism should be devised in terms of characteristics of software developing.

Software developing should abide by thought of software engineering. Actually, so-called standard of software developing is a series of rules to improve the quality of software development, shorten the developing cycle, enhance software codes readability and useability. Partner firms must insist on these standards for which would guarantee software developing quality.

4 Suggestions to improve VSCs' operation efficiency

4.1 To attach importance to modules' decomposition and standards of modules' interface

Analysis to customers' demand is the base of software modules' decomposition architecture, in this phase, software systematic models are established after study on software's functions, performance which customers need. In order to make software architecture more reasonable, one of the most important things for software developing is to set down standards of every module's interface, by which can not only stability but efficiency of software developing can be improved greatly.

The cooperation of partners in a VSC is based on software decomposition layer modular architecture, so it's required that those stated standards and requirement be strictly complied with.

4.2 To establish communication mechanism among partners in a VSC

As far as VSC is concerned, different module's developing task is taken by different firms, discriminative communication strategy is required in terms of corresponding modules' characteristics and connections. Firms' credits and modules' relative importance should be considerd when module tasks' allocation.

4.3 To establish reasonable profit allocation mechanism

After field sampling research, in this article a creative viewpoint is put forward on profit allocation among firms in a VSC that 3 principles should be complied with: profit sharing principle, software modules' importance principle and firms' performance principle.

(1) Profit sharing principle

Profit a VSC gained should be shared by all partner firms, the sum of income of all firms in a VSC should be equal to the whole profit.

(2) Software module's importance principle

The income one firm earned must adapt to the importance of module the firm taken, that is, if the module is of greater importance, the corresponding firms should gain more.

(3) Firms' performance principle

By proper and approbatory appraisal, better performance means the firm's better earning. The performance evaluation should be done from 2 aspects, both developing effect and cooperative process.

5 Conclusion

The motive of forming VSC is pursing a better completion of software developing task than any single software enterprise, for VSC can make use of all advantages of partners' firms. During cooperation process, lack of cooperative efficency may lead to failure of the task which isn't the original intention of VSC. This paper strives to find out the influence factors of VSCs' operation, so, necessary studies have been done by adopting factor analysis method and RST based on gathered data through questionnaire survey. At the end of this paper, some effective suggestions are put forth according to the analytic results.

References

[1] XuPei, ZhangRenJun, How to create effective project team[J]. Statistics and desicion, 2005（3): 132~133

[2] KangZhuang, Study on Project managementb methods of Software developing [D] . Master's degree paper of University of International Business and Economics, 2006

[3] ZuoZhiJuan, Study onTeams efficiency evaluation structure[D] . Master's degree paper of Tianjin Normal University, 2004

MATEC Web of Conferences **63**, 04036 (2016)
MMME 2016
DOI: 10.1051/matecconf/20166304036

Function of One Regular Separable Relation Set Decided for the Minimal Covering in Multiple Valued Logic

Yu Zhen Liu[1,2] ,Min Liu[1,2],Xing Bao Liu[3] Xin Fen Zhang[4,a]

[1.] *Key Lab of Knowledge Processing and Networked Manufacturing, Xiangtan, Hunan, 411105, China.*

[2.] *Hunan University of Science and Technology,Xiangtan, Hunan, 411105, China.*

[3.] *School of Computer & Information Engineering, HunanUniversity of Commerce, Changsha, 410250, China.*

[4.] *School of Science ,Qilu University of Technology, Jinan 250353, China*

ABSTRACT: Multiple-valued logic is an important branch of the computer science and technology. Multiple-valued logic studies the theory, multiple-valued circuit & multiple-valued system, and the applications of multiple-valued logic included.In the theory of multiple-valued logic, one primary and important problem is the completeness of function sets, which can be solved depending on the decision for all the precomplete sets(also called maximal closed sets)of K-valued function sets noted by P_K^*, and another is the decision for Sheffer function, which can be totally solved by picking out all of the minimal covering of the precomplete sets.In the function structure theory of multi-logic, decision on Sheffer function is an important role. It contains structure and decision of full multi-logic and partial multi-logic. Its decision is closely related to decision of completeness of function which can be done by deciding the minimal covering of full multi-logic and partial-logic. By theory of completeness of partial multi-logic, we prove that function of one regular separable relation is not minimal covering of P_K^* under the condition of $m=2,\sigma=e$.

KEYWORD: Regular Separable Relation; Sheffer Function; Minimal Covering; Maximal Closed Set;

1 INTRODUCTION

Let $E_K=\{0,1,2,\Lambda,K-1\},K\geq 2$.Function $f(x_1,x_2,\Lambda,x_n,)$ is from E_K to E_K. Function $f(x_1,x_2,\Lambda,x_n,)$ is called complete K-valued function when if for all of $(\alpha_1,\alpha_2,\Lambda,\alpha_n)$ $f(x_1,x_2,\Lambda,x_n,)$ is a function. Otherwise $f(x_1,x_2,\Lambda,x_n,)$ is called non-complete K-valued function. All of the complete and non-complete K-valued function is called partial K-valued function. The set of all of the complete K-valued function is signed as P_K. The set of all of the non-complete K-valued function is signed as P_K^* [1-6].

Let $A\subseteq P_K^*,f(t_1,t_2,\Lambda,t_m)\in A,g_i(x_1,x_2,\Lambda,x_n)\in A$ or $g_i(x_1,x_2,\Lambda,x_n)$ is one of x_1,x_2,Λ,x_n ,here $i=1,2,\Lambda,m$. $f(g_1(x_1,\Lambda,x_n),\Lambda,g_m(x_1,\Lambda,x_n))$ is called one function which is composite from the function of A. The set of all of functions which is composite from the function of A, signed as A'. A is called one closed set if $A=A'$.

Let $A\subseteq P_K^*$,

$GEN(A)=A\cup A'\cup A''\cup\Lambda\cup A^{(i)}\cup\Lambda$, if $GEN(A)=P_K^*$, A is called a complete set.

For non-complete function $f(x_1,x_2,\Lambda,x_n,)$, if $f(\alpha_1,\alpha_2,\Lambda,\alpha_n)$ is not belong to E_K , we let $f(\alpha_1,\alpha_2,\Lambda,\alpha_n)=*$.

The sequence which includes m numbers of E_K called one m- sequence. m- sequence is signed as

[a]Corresponding author: 249781563@qq.com Tel.: +8618573206768,+8615910080626

$<\alpha_1,\alpha_2,\Lambda,\alpha_n>$. The set of m- sequence is signed as G_m.

$G_m=\overline{G}_m\cup\overline{G}_m^{\sigma_1}\cup\overline{G}_m^{\sigma_2}\cup\Lambda\cup\overline{G}_m^{\sigma_h}$ is called regular separable if it have these conditions below: existing one division of E_K:

$$D_i:E_K=A_1^i+A_2^i+\Lambda+A_m^i, i=1,2,\Lambda,n.$$

For all of the $<\alpha_1,\alpha_2,\Lambda,\alpha_m>\in\overline{G}_m$ there exists one division

$$D_i(1\le h\le n):\alpha_i\in A_i^{(h)}, i=1,2,\Lambda,m,$$

and also for all of $D_j,(1\le j\le h-1)$ there exiting $A_{j_1}^{(j)},\Lambda,A_{j_n}^{(j)},(1\le j_1,\Lambda,j_n\le m)$ have the condition of $\alpha_1,\alpha_2,\Lambda,\alpha_{r_1}\in A_{j_1}^{(j)},\Lambda,\alpha_{r_{n-1}+1},\alpha_{r_{n-1}+2},\Lambda,\alpha_{r_n}\in A_{n_1}^{(j)}$.

The set of all of regular separable function is called regular separable function set which is signed $S_{R,m}$.

One function $f(x_1,x_2,\Lambda,x_n)\in P_K^*$ is called Sheffer function if $GEN\{f(x_1,x_2,\Lambda,x_n)\}=P_K^*$. That is to say, $f(x_1,x_2,\Lambda,x_n)$ can produce all functions of P_K^*.

Let

$\Sigma=\{A_i\mid A_i\subset P_K^* \text{ and } A_i \text{ is maximal closed set}\},\Sigma'\subseteq\Sigma$

$S_{\Sigma'}=\{f\in P_K^*\mid \text{if there exists one } A_j',f\in A_j'\}$.

$\Sigma'(\subseteq\Sigma)$ is called the covering of Σ, if $S_{\Sigma'}=S_\Sigma$. For one set A(here A is maximal closed set), If there exists one function $f\in A$ and f is not belong to any maximal closed set except A, A is one of member of the minimal closed set.

Function $f(x_1,x_2,\Lambda,x_n)\in P_K^*$ is called protecting of G, if for any function

$g_i(x)\in G, i=1,2,\Lambda,n$,

$f(g_1(x),g_2(x),\Lambda,g_n(x))\in P_K^*$.

All of $f(x_1,x_2,\Lambda,x_n)\in P_K^*$ which is called protecting of G is signed as $T(G)$.

The relation of regular separable

$$G_2=G_2(\{1,2\})\cup\overline{G}_2$$

can be described by relation of $<a,b>$. It can be represented as one graph which has K vertices[1-6].

2 MAIN CONCLUSION

Theorem Under the condition of $\sigma=e$, if the basic graph of regular separable set

$$G_2=G_2(\{1,2\})\cup\overline{G}_2$$

is only one loop, then it is not the minimal closed set of P_K^*.

Proof Because of the Characteristics of loop, without loss of generality, we let vertices $0,1,2,\Lambda,k-1$ is in turn connected to each other. We will proof this loop can't be one strongly connected graph.

Let $<0,1>,<1,2>,\Lambda,<k-2,k-1>,<k-1,0>\in G_2$, According to the definition of regular separable set, let division $D_i:E_K=A_1^{(i)}+A_2^{(i)}, i=1,2,\Lambda,n$, for any $<a_1,a_2>\in\overline{G}_2$, there existing one division $D_h(1\le h\le n):a_i\in A_i^{(h)}, i=1,2$ and also for any $D_j(1\le j\le h-1)$, there exists

$$A_{j_j}^{(j)},1\le j_1\le 2$$

which satisfies $a_1,a_2\in A_i^{(h)}$.

Let $D_i:E_K=A_1^{(i)}+A_2^{(i)}, i=1,2,\Lambda,n$ satisfy the definition of regular separable set. For the sequence of binary regular separable relation $<0,1>\in G_2$, without loss of generality, let exist one division D_1 which satisfies $0\in A_1^{(1)},1\in A_2^{(2)}$.

Next we consider $<1,2>\in G_2$, it is clear that D_1 can't satisfy $1\in A_1^{(1)}$.

In the same way, we consider $<2,3>\in G_2$, it is clear that D_2 can't satisfy $1\in A_1^{(1)}$.

Now we consider $<k-1,0>\in G_2$, it is clear that D_{k-1} can't satisfy $k-1\in A_1^{(1)}$, So there also exists one division D_k which satisfies

$$k-1\in A_1^{(1)},0\in A_2^{(2)}.$$

But according to the definition of division, this gets contradiction: $0\in A_1^{(1)}$ and $0\in A_2^{(2)}$.

So we prove that this loop is not one strongly connected graph.

The following we will prove this theorem.

Because the basic graph of this relation graph is only one loop, this basic graph is connected. For any $f\in T(G_2)$, without loss of generality, let $f(0,0,\Lambda,0)=i, 0\le i\le K-1$. There must exist one path L from 0 to i, let $|L|$ be the length of L. we have possibilities as below.

1. if $|L|=0, f\in T(G_0)$,
2. if $|L|=1$, that is $f(0,0,\Lambda,0)=1$, that is to say $<0,1>\in G_2$. Next we will consider

$$<1,2>\in G_2, or <2,1>\in G_2.$$

Because the basic graph is only one loop, we have possibilities as below.

(1) $f(1,1,\Lambda,1)=1$,

(2) $f(1,1,\Lambda,1)=*$,

(3) $f(1,1,\Lambda,1)=2$,

For (1)(2),we can get $f\in T_{\{1\}}$, and $<2,1>\in G_2$. We get $f\in T(G_2)$ isn't the member of minimal closed set.

For (3), we can get $<1,2>\in G_2$. Next we consider $<2,3>\in G_2, or <3,2>\in G_2$. We also have possibilities as below.

(a) $f(2,2,\Lambda,2)=2$,

(b) $f(2,2,\Lambda,2)=*$,

(c) $f(2,2,\Lambda,2)=3$,

For (a)(b),we can get $f\in T_{\{2\}}$, and $<3,2>\in G_2$. We get $f\in T(G_2)$ isn't the member of minimal closed set.

For (c), we can get $<2,3>\in G_2$. Next we consider $<3,4>\in G_2, or <4,3>\in G_2$. We can do like above. We also have possibilities as below.

(e) $f(k-2,k-2,\Lambda,k-2)=k-2$,

(f) $f(k-2,k-2,\Lambda,k-2)=*$,

(g) $f(k-2,k-2,\Lambda,k-2)=k-1$,

For (e)(f),we can get $f\in T_{\{k-2\}}$, and $<k-1,k-2>\in G_2$. We get $f\in T(G_2)$ isn't the member of minimal closed set.

For (g), we can get $<k-2,k-1>\in G_2$. Because this basic graph is not strongly connected. We have $<0,k-1>\in G_2$. We have possibilities as below.

(h) $f(k-1,k-1,\Lambda,k-1)=k-1$,

(i) $f(k-1,k-1,\Lambda,k-1)=*$,

For (h)(i), we can get $f\in T_{\{k-1\}}$ .We get $f\in T(G_2)$ isn't the member of minimal closed set.

3. if $|L|>1$, without loss of generality, let $f(0,0,\Lambda,0)=i$, here $2\le i\le i-1$. Let $L_1=L$, we consider $f(1,1,\Lambda,1)$. We have possibilities as below.

(a_1) If $f(1,1,\Lambda,1)=i$, we consider the path L_2: from vertices 1 to vertices *i*. Here we get $|L_2|=|L_1|-1$,

(a_2) If $f(1,1,\Lambda,1)=i-1$, we consider the path L_2 : from vertices 1 to vertices *i*-1. Here we get $|L_2|=|L_1|-2$,

(a_3) If $f(1,1,\Lambda,1)=i+1$, we consider the path L_2 : from vertices 1 to vertices *i*-1. Here we get $|L_2|=|L_1|$,

(a_4) If $f(1,1,\Lambda,1)=*$, we have $f\in T_{\{1\}}$,

Now we consider path L_2 and we can get L_2 which have the same features like L_1. So we use the technology like above, we can get one of $f\in T_{\{1\}}$ or $f\in T_{\{2\}}$, or we can get L_3. If we do like above, we can have possibilities as below.

(b_1) $|L_{n+1}|=|L_n|-1$,

(b_2) $|L_{n+1}|=|L_n|-2$,

(b_3) This directional path is connected from 0 to $k-1$,

For (b_1) (b_2), because the path becomes smaller and smaller, $f\in T_E$.

For (b_3), because this loop can't be strongly connected graph, $<k-1,0>\notin G_2$, that is to say $<0,k-1>\in G_2$. And this path is one directional connected path from 0 to k-2. We can get $<k-2,k-1>\in G_2$. We can have possibilities as below.

(c_1) $f(k-1,k-1,\Lambda,k-1)=k-1$,

(c_2) $f(k-1,k-1,\Lambda,k-1)=*$,

For (c_1) (c_2) we can get the result $f\in T_E$.

According to the proof of above, we prove that if under the condition of $\sigma=e$, the basic graph of regular separable set $G_2=G_2(\{1,2\})\cup\overline{G_2}$ is only one loop, then it is not the minimal closed set of P_K^*.

ACKNOWLEDGMENT

This paper is supported by NSFC, under grant number: 61402167, 61300129, State key laboratory project, under grant number: SKLSDE-2014KF-08, Planned Science and Technology Project of Hunan Province of China, under grant number:2013FJ3002,
Fund of Hunan University of Science and Technology, under grant number: E51368, E51473, Fund of Scientific Project of Education Department of Hunan province, under Grant No. 14C0437, The educational reform project, under grant number: G31344.

REFERENCES

[1] Zhukai, Luo, Research on theory and application of multi-logic[M] (In Chinese), National University of Defense Technology, 2003.

[2] Liu R.R., The minimal covering of precomplete in partial four-valued logic, Proceedings of ICYCS'91, 1991.

[3] Liu R.R. Some results on the decision for Sheffer functions in partial K-valued logic, Multi. Val. Logic, 1996,1: 253-269.

[4] Liu R. R., Some results on the decision for Sheffer functions in partial K-valued logic (II), Proceedings of IMVL'98, IEEE computer Society Press,1998.

[5] Liu R.R., Some results on the minimal coverings of precomplete classes in partial K-valued logic functions, Proceedings of 2003 IEEE International Conference on Systems, Man & Cybernetics, 2003.

[6] Liu Y.Z., Liu X. B. etc., Decision on Function of One simple separable relation for the minimal covering of P_K^*. the 4th international conference on frontiers of manufacturing and design science (icfmd2013 Hong Kong China),2013.

MATEC Web of Conferences **63**, 04037 (2016)
DOI: 10.1051/matecconf/20166304037
MMME 2016

Fast Image Super-resolution with Sparse Coding

Zhi-chao Yuan [1,a], Ben-tu Li [1]

[1] *Shandong Vocational College of Science and Technology, Weifang, 261053, China;*

Abstract. In this paper, we introduce a novel fast image reconstruction method for super-resolution (SR) base on sparse coding. This method combine online dictionary learning and a fast sparse coding way, both of which can improve the efficiency of the reconstruction process and ensure the image visual quality. The new online optimization algorithm for dictionary learning based on stochastic approximations, which can drastically advance the learning speed, especially on millions of training samples. Meanwhile, we trained a neural network to speed up the reconstruction process, which based on iterative shrinkage-thresholding algorithm (ISTA), we called learned iterative shrinkage-thresholding algorithm (LISTA). It would produce best approximation sparse code with some fixed depth. We demonstrate that our approach can simultaneously improve the image fidelity and cost less computation.

Keywords: SRIR; sparse coding; super-resolution; fast image super-resolution.

0. Introduction

High-resolution image is valuable in many social areas. Such as reconstruct the high-resolution image from low-resolution image which take in Criminal Investigation, get high-accurate CT, MIR, Ultrasonic wave to help doctor diagnose disease. Additionally, this is also play an important role in HDTV, Video Rip and Military occasion.

The main contribution of this paper is combining the online dictionary learning and approximating sparse coefficient together to reconstruct the High-resolution image, so that the reconstruction rate is splendidly speedup. Learning the dictionary by online method process one element of the millions of training samples at a time, which based on stochastic approximations(Aharon & Elad, 2008). Instead of optimize the l1-norm optimization for accurate sparse coefficients, we employ a feed-forward neural network to construct approximated sparse coefficients(Gregor & LeCun, 2010) .

1. Fast image super-resolution

This section describe the two key component of SRIR—learn the over-complete dictionary and find sparse coefficient. Online dictionary learning proposed an iterative algorithm that solves eqn. by efficiently minimizing at each step a quadratic surrogate function of the empirical cost over the set of constraints (Mairal, Bach, Ponce, & Sapiro, 2009). We find the approximation sparse coefficient by train a non-linear, feed-forward neural network instead of find the accurate solution(Gregor & LeCun, 2010).

1.1 Online dictionary learning

According to we talked above, the dictionary learning can be get by optimize the energy function:

$$f_n(\tilde{D}) \triangleq \frac{1}{n}\sum_{i=1}^{n} l(\tilde{y}_i,\tilde{D}), \quad where \quad l(\tilde{y}_i,\tilde{D}) \triangleq \min_{\alpha \in R^k} \frac{1}{2}\|\tilde{y}_i - \tilde{D}\alpha_i\|_2^2 + \lambda\|\alpha_i\|_1 \qquad (1.1)$$

As point by Bettou and Bousquet(Bousquet & Bottou, 2007), maybe its not important in a perfect minimization of energy function $f_n(D)$, but interested in expected minimization cost:

$$f(\tilde{D}) \triangleq E_x\left[l(\tilde{y}_i,\tilde{D})\right] = \lim_{n\to\infty} f_n(\tilde{D}) \quad (1.2)$$

The outline of our algorithm is described in algorithm 1

Algorithm1 Online dictionary learning

Require: $x \in R^m$: $p(x)$ (random variable and an algorithm to draw i.i.d samples of p), $\lambda \in R$ (regularization parameter), $D_0 \in R^{m\times k}$ (initial dictionary). T (number of iterations)

1: $A_0 \leftarrow 0$, $B_0 \leftarrow 0$ (reset the "past" information); 2: For t=1 to T do

3: Draw xt from p(x). 4: Sparse coding: compute using LARS

$$\alpha_t = \arg\min_{\alpha\in R^k} \frac{1}{2}\|x_t - D_{t-1}\alpha\|_2^2 + \lambda\|\alpha\|_1 \quad (1.3)$$

5: $A_t \leftarrow A_{t-1} + \alpha_t\alpha_t^T$; 6: $B_t \leftarrow B_{t-1} + x_t\alpha_t^T$.

7: Compute Dt using Algorithm 2, with D_{t-1} as warm restart. So that

$$D_t \triangleq \arg\min_{D\in C} \frac{1}{t}\sum_{i=1}^{t}\frac{1}{2}\|x_i - D\alpha_i\|_2^2 + \lambda\|\alpha_i\|_1 = \arg\min_{D\in C}\frac{1}{t}\left(\frac{1}{2}Tr\left(D^TDA_t\right) - Tr\left(D^TB_t\right)\right) \quad (1.4)$$

8: End for; 9: Return DT (learned dictionary).

We supposed the training set is composed of i.i.d samples of a distribution p(x), every loop it draws one element from the training set. It alternatively solve α by classical sparse coding and D by minimizing the following function:

$$\hat{f}_t(D) @ \frac{1}{t}\sum_{i=1}^{t}\frac{1}{2}\|x_i - D\alpha_i\|_2^2 + \|\alpha_i\|_1 \quad (1.5)$$

The quadratic function $\bar{f}_t(D)$ is a surrogate. This function aggregates the previous information by the process of computing α . The convergence analysis has shown that $\hat{f}_t(D)$ and f_t convergence almost to the same limit(Mairal et al., 2009) . Because of $\hat{f}_t(D)$ is close to f_t , so it can be efficiently computed via D_{t-1} as warm restart.

Algorithm 2 Dictionary update

Require: $D = [d_1 \; d_2 \; K \; d_{k-1} \; d_k] \in R^{m\times k}$ (input dictionary),

$$A = [a_1 \; a_2 \; K \; a_k] \in R^{k\times k} = \sum_{i=1}^{t}\alpha_i\alpha_i^T \;, B = [b_1 \; b_2 \; K \; b_k] \in R^{m\times k} = \sum_{i=1}^{t} x_i\alpha_i^T \;,$$

1: Repeat;2: For j=1 to k do

3: Update the j-1 column to optimize for eqn. :

$$u_j \leftarrow \frac{1}{A_{jj}}(b_j - D\alpha_j) + d_j, \; d_j \leftarrow \frac{1}{\max(\|u_j\|_2, 1)}u_j \quad (1.6)$$

4: End for;5: Until convergence;6: Return D (update dictionary).

The algorithm 2 uses block-coordinate descent with warm restart to update the dictionary. The benefits of this method is parameter-free and does not require any learning rate tuning. It sequentially update D column by column when update the j-th column, when set the other ones fixed under constraint $d_j d_j^T \le 1$. Recent works shows that separable constraint could make the function convergence to a global optimum. Since we use Dt-1 to warm restart Dt , so the block-coordinate descent will more efficient.

DOI: 10.1051/matecconf/20166304037

1.2 Approximation sparse coefficient

In this section, we employ the same neural network structure in paper(Gregor & LeCun, 2010) to reconstruct the HR patches.

In paper(Gregor & LeCun, 2010) , the approach of train a neural network is inspired by the iterative shrinkage-thresholding algorithm (ISTA), The outline of ISTA is described in algorithm 3

Algorithm 3 ISTA

Require: L>largest eigenvalue of $W_d^T W_d$.

Initialize: Z=0.

$$1\text{:Repeat } Z = h_{(\alpha/L)}\left(Z - \frac{1}{L}W_d^T\left(W_d Z - X\right)\right) \quad (1.7)$$

2:Until convergence; 3:End function.

The elements of the eqn.are defined as:Filter matrix: $W_e = \frac{1}{L}D_y^T$.

Inhibition matrix: $S = I - \frac{1}{L}D_y^T D_y$

Shrinkage function: $h_\theta(z) = sign(z)(|z| - \theta)_+$.

Here, L is a constant, which is larger than the largest eigenvalue of $D_y^T D_y$, function $h_\theta(z)$ is a component-wise shrinkage function and θ is usually set as λ / L . In practice, because of this algorithm have to solve thousands of optimization and take tens of iterations to convergence, it' s too slow for practical applications. We use for reference paper (26), proposed a new idea which only need a fixed number of iteration T, they do not use fixed parameter in ISTA. The algorithm through minimize the following energy function to get the parameter W = (We,S, θ):

$$E(\mathrm{w}) = \frac{1}{N}\sum_{i=1}^{N} C(\mathrm{W}, \mathrm{y}_i), \text{with } C(\mathrm{W}, \mathrm{y}_i) = \frac{1}{2}\left\|z_i - f_e(W, y_i)\right\|_2^2 \quad (1.8)$$

Here, zi is the accurate sparse coefficient of yi which from traditional sparse method. And $f_e(W, y_i)$ is the transformation of eqn. .

Extensively, in our super-resolution algorithm we rewrite the energy function as:

$$C(\mathrm{W}, \mathrm{y}_i, \mathrm{x}_i) = \frac{1}{2}\left\|z_i - f_e(W, y_i)\right\|_2^2 + \frac{v}{2}\left\|D_x z_i - x_i\right\|_2^2 \quad (1.9)$$

This will work better to predict the HR image. The outline of LISTA is describe in algorithm 4 and algorithm 5.

Algorithm 4

Require : We, X.Initialize: B=WeX; Z(0)=h θ (B).Arguments are passed by reference.Variables Z(t), C(t) and B are saved for bprop.

1:For t=1 to T do; 2:C(t) = B+SZ(t-1);

3:Z(t) = h θ (C(t)); 4:End for;5:Return Z=Z(T)

Algorithm 5

Require: Z* (accurate sparse coefficient from traditional methods); Z(t), C(t), B (saved in fprop).

Initialize:

$\delta B = 0; \delta S = 0; \delta\theta = 0; \delta Z(\mathrm{T}) = (\mathrm{Z(T)} - Z^*)$

1:for t=T down to 1 do;

2: $\delta C(t) = h_\theta'(C(t)) \cdot \delta Z(t)$;

3: $\delta\theta = \delta\theta - sign(\mathrm{C}(t)) \cdot \delta Z(t)$

4: $\delta B = \delta B + \delta C(t)$;

5: $\delta S = \delta S + \delta C(t) Z(t-1)^T$;

6: $\delta(t-1) = S^T \delta C(t)$; 7:end for

8: $\delta B = \delta B + h_\theta'(B) \cdot \delta Z(0)$;

9: $\delta\theta - sign(\mathrm{B}) \cdot \mathrm{h}_\theta'(\mathrm{B}) \delta Z(0)$;

10: $\delta W_e - \delta B X^T ; \delta X = W_e \delta B$

2.Experiment results

In the following, we first compare the SRIR image with Bicubic interpolation method(Yang et al., 2010), which is a splendid approach. Then we discuss the efficiency with other splendid approaches.

2.1 Comparison with joint dictionary learning

We compare our algorithm with Bicubic interpolation in visual result and other index. In order to make the fair comparison, we adopt the same training set and initial D for both methods.

a) Visual result: fig. 1(a)-(d) compared the results of Bicubic interpolation method with our fast image super-resolution four test images. From the left to right are the original HR images, the reconstructed images by Bicubic interpolation method and the reconstructed images by fast super-resolution method. From the results, we can easily see that our recovery are closed to the original images, the edge is more sharper than the Bicubic recovery, such as the moustaches of the cat, the texture of the freckles of the girl, the texture of the leaves and the hair of lena.

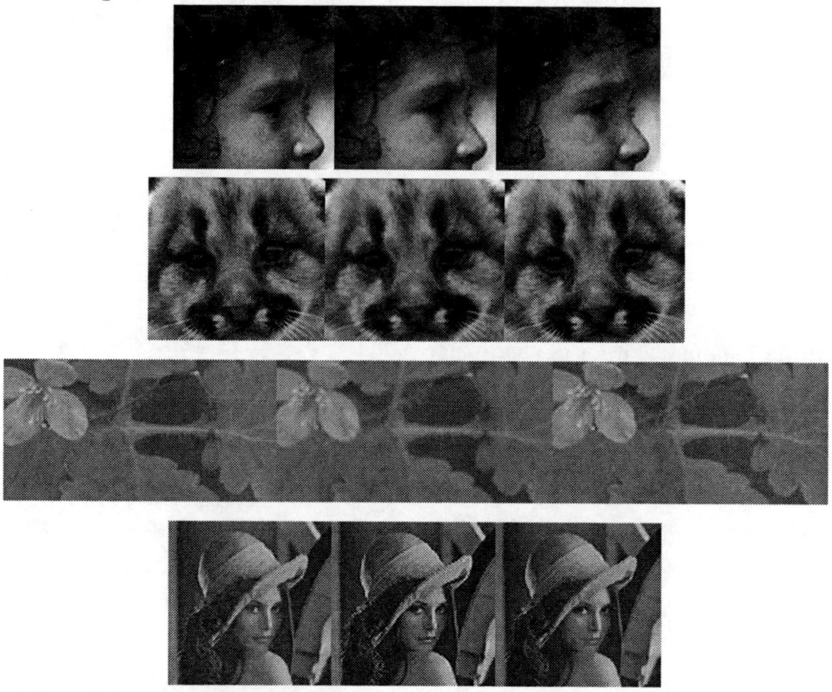

Fig. 1 from left to right: original image, Bicubic interpolation recovery, our recovery

b) Numerical result: now we compute the Peak signal-to-noise ratio (PSNR) of the fast super-resolution image. PSNR is most easily defined via the mean squared error (MSE), MSE is defined as:

$$MSE = \frac{1}{MN}\sum_{m=0}^{M-1}\sum_{n=0}^{N-1}[\mathrm{I(m,n)} - \mathrm{f(m,n)}]^2 \qquad (2.1)$$

Here, I and K are respectively represent the original image and our recovery. The PSNR is defined as:

$$PSNR = 10\log_{10}\left(\frac{MAX_I^2}{MSE}\right) = 20\log_{10}\left(\frac{MAX_I}{\sqrt{MSE}}\right) = 20\log_{10}\left(MAX_I\right) - 10\log_{10}\left(\sqrt{MSE}\right) \qquad (2.2)$$

Here, MAXI is the maximum possible pixel value of the image. When the pixels are represented using 8 bits per sample, this is 255. Typical values for the PSNR in lossy image and video compression are between 30 and 50 dB, where higher is betters.

images	Bicubic	Fast recovery	
		`T = 3	T=5
girl	31.699140db	33.324861db	33.336294db
cat	28.3597db	28.8539db	28.8669db
leave	37.197505db	38.060257db	38.060146db
lena	32.794678db	35.02261db	35.022206db

Table I the PSNR of Bicubic interpolation recovery and our recovery

Table 1 shows the PSNR of test images. From table I we can see that our recovery outperforms the Bicubic interpolation recovery well with high PSNR. When the iteration times is fixed as 3, the fast sparse coding is converging to optimal.

2.2 Algorithm efficiency

In this subsection, we evaluate the efficiency of our algorithm and other state-of-art algorithms (Glasner, Bagon, & Irani, 2009; Yang et al., 2010). All of the experiments are conducted in Matlab 8.1. Table II show the consumed times of different method.

iamge	mesuares	Bicubic	Yang at el.	Fast recovery
Girl	PSNR(db)	31.699140	33.34565	33.336294
254×256	Time(s)	0.012371	285.496759	34.6597
Cat	PSNR(db)	28.3597	28.8658	28.8669
326×298	Time(s)	0.017776	446.296511	48.6235
Leave	PSNR(db)	37.197505	38.0611	38.060146
328×170	Time(s)	0.012091	246.902859	41.3629
statuary	PSNR(db)	26.0253	26.3064	26.2923
256×162	Time(s)	0.007996	180.382172	23.26

Table II processing time of different SR method

From table II we can see that our algorithm speedup the reconstruction process significantly. In addition, we can ensure both of the PSNR and the cost of compute time, since our algorithm adopt the approximate sparse coefficients and train the online dictionary. From fig. 2 for "statuary" image, Yang' s algorithm has the higher PSNR than our algorithm, but both of two images almost have the same visual quality.

Fig. 2 top: original, Bicubic; bottom: Yang at el. Fast recovery.

3. Conclusions

In this paper, we propose a novel approach to reconstruct the high-resolution image based on online dictionary learning and approximate sparse coefficient.The preliminary experiment prove that our algorithm not only ensure the visual quality but faster than other state-of-art approaches. However, our algorithm is not have the optimal recovery than other SR approaches, so, our future research is to train accurate dictionary to represent the sparse coefficients.

References

[1] Aharon, M., & Elad, M. (2008). Sparse and Redundant Modeling of Image Content Using an Image-Signature-Dictionary. Siam Journal on Imaging Sciences, 1(3), 228-247.

[2] Beck, A., & Teboulle, M. (2009). A Fast Iterative Shrinkage-Thresholding Algorithm for Linear Inverse Problems. Siam Journal on Imaging Sciences, 2(1), 183-202.

[3] Bousquet, O., & Bottou, L. (2007). The Tradeoffs of Large Scale Learning. Paper presented at the Advances in Neural Information Processing Systems.
[4] Candes, E. J., Romberg, J. K., & Tao, T. (2006). Stable signal recovery from incomplete and inaccurate measurements. Communications on Pure And Applied Mathematics, 59(8), 1207-1223. doi: Doi 10.1002/Cpa.20124
[5] Daubechies, I., Defrise, M., & De Mol, C. (2004). An iterative thresholding algorithm for linear inverse problems with a sparsity constraint. Communications on Pure And Applied Mathematics, 57(11), 1413-1457. doi: Doi 10.1002/Cpa.20042

MATEC Web of Conferences 63, 04038 (2016)
MMME 2016
DOI: 10.1051/matecconf/20166304038

Routing Algorithm of Concentric Excursions for ITS Communication

Jianfeng Luo[1], Yuqiang Chen[1]

[1]Department of Computer Engineering, Dongguan Polytechnic, Dongguan Guangdong 523808, China

Abstract. In the modern society, intelligent transportation builds on the basic of real-time wireless communication with vast information. Only by adding and upgrading the high-costed equipment to keep the speed of communication with the development of the transportation is far from enough [1]. Thus, to make full use of the present equipment to achieve higher speed of information transmission is a practical and challenging project. The Routing Algorithm of this paper is taking the Traffic Management Center as the center. Starting from the center and covering all the traffic management station, we can get the best path program under multi-limits with the help of the combination Algorithm of Shortest Path Algorithms and Concentric Excursions. A lot of constraints has been taken into consideration so as to make the result preciser. The result of multiple experiments has testified the feasibility and efficiency of the algorithm even under the great information content.

Key words: Cumulative Conditions Queue, multi-limits Routing Algorithm, Concentric Excursions, value-Saving Transmission Queue, Shortest Path

1. Introduction

Traffic monitoring information grows with the rapid growth of the cars number. To do researches on the routing algorithm of information transmission to improve the speed is of great importance.[2] The present common wireless networks are Cellular System with complicated structure and Network Protocol which may cause problems such as louder noise, longer Transmission time, worse net jam etc.[3] The wide range of intelligent traffic management, a large amount of equipment, complicated routing paths and huge amount of information will force the system to upgrade continuously. Otherwise, the system has to keeping adding new equipment or other high-costed way to keep up with the need of the pace[4]. In order to get rid of the problems mentioned above, we have to figure out a simple-structured and time-saving routing solution which can cover all the connecting points as well work under the limitation of transmission path and data.

2. A Brief introduction of the intelligent transportation wireless network

Shown in Fig.1, the Traffic Management Center is responsible for the calculations of traffic control and coordination. The result will be sent back to the regional wireless receiving tower through wireless signal routing. The receiving tower with the relay function will send the information to the assigned server and then to the roadside equipment or to the mobile equipment fixed in the vehicle. Meanwhile, the traffic signal will be sent through the reverse way back to the Traffic Management Center [5].

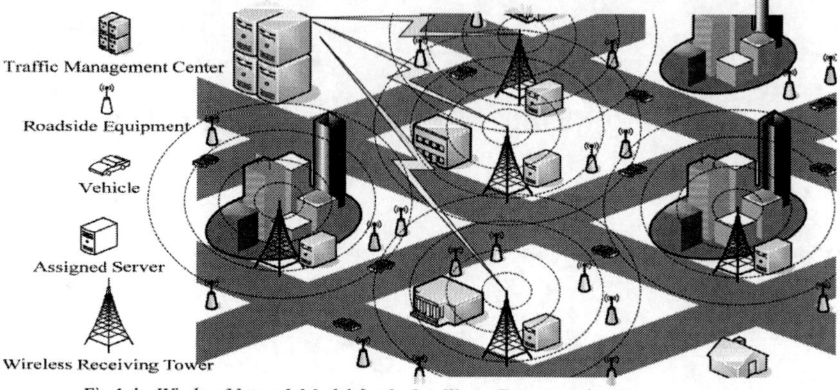

Fig.1 the Wireless Network Model for the Intelligent Transportation System

3. The Core idea of Excursion Algorithm

The Core idea of Excursion Algorithm is as follow: signal from Traffic Management Center covers as many as the receiving towers and collects as much as the local traffic information at the same time. What works importantly is to work out the shortest path to send the information back to the Traffic Management Center. Suppose that the information is fairly large and take up all the broadband resource, we can only exchange the information in the next tower. Under such situation, we have to prove that the Excursion Algorithm is the shortest Information transmission path among the receiving towers.

The following simple Model Diagram will show the wireless traffic network[6](Fig.2): C is the Traffic Management Center, Ti(i=1,2)is the receiving tower, Dj(j=1,2,3) is the Line-of-sight among the tower.

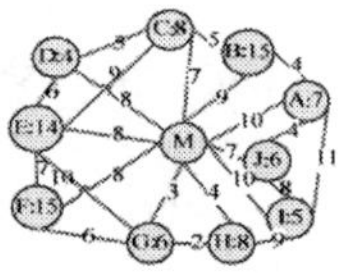
Fig.2 Model diagram

There are three information transmission paths:

Broadcast transmission. That is C->T1,C->T2,T1->C,T2->C, the total information transmission distance (TITD) is A=2*D1+2*D3;

Single path transmission. That is C->T1->T2->T1->C, the TITD is B=2*D1+2*D2;

Excursion Algorithm. That is C->T1->T2->C, the TITD is C=D1+D2+D3.

Now we have to prove: C<A and C<B.

Since A-C=2*D1+2*D3- (D1+D2+D3), the result is D1+D3-D2. According to the Triangle Theorem, the sum of any two sides of a triangle is great er than the third side, we can get that D1+D3-D2>0, that is A>C, hence, C<A.

Since B-C=2*D1+2*D2-(D1+D2+D3), the result is D1+D2-D3. According to the Triangle Theorem, D1+D2-D3>0, that is B>C, hence, C<B.

From the result mentioned above, we can get the point that the the Core idea of Excursion Algorithm is correct and achievable. More detailed analysis of the Algorithm will be explained in the following part. We take the value of A-C or B-C as the saved Transmission value, and TV(T1,T2) represents saved Transmission value from C(Traffic Management Center) passing through T1, T2 and back to C.

4.The Process of Excursion Algorithm

Taking Fig.3 (network diagram) as an example, M is the Traffic Management

 DOI: 10.1051/matecconf/20166304038

Center, from A-J are 10 receiving towers(10regions) , and the unit is MB for the collecting information content in each region. And information content is the nodal value in the fig while the bandwidth is 40MB. In order to avoid too much noise, the transmitting length is below 30 kilometers [7]. The edge weight is the line-of-sight (kilometer); the lines connect the receiving towers.

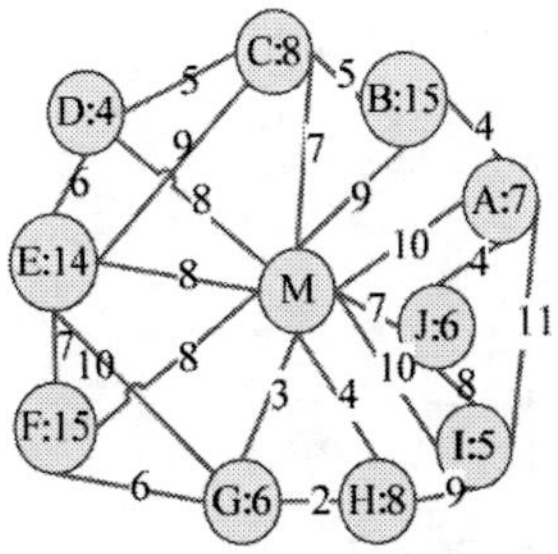

Fig.3 network diagram

4.1 Figure out the Shortest Path Matrix

Since in wireless transmission, speed is inversely proportional to distance [8], we need to find out the shortest path among receiving towers. Thus, Floyd Shortest Path Algorithm is adapted to put into use here to figure out the shortest path from M to other points as well as the shortest distance among the points[9]. Then, we will get the shortest path matrix [10]. Shown as in Fig. 4

	M	A	B	C	D	E	F	G	H	I
A	10	—								
B	9	4	—							
C	7	9	5	—						
D	8	14	10	5	—					
E	8	18	14	9	6	—				
F	8	18	17	15	13	7	—			
G	3	13	12	10	11	10	6	—		
H	4	14	13	11	12	12	8	2	—	
I	10	11	15	17	18	18	17	11	9	—
J	7	4	8	13	15	15	15	10	11	8

Fig.4 the shortest matrix for each point

	M	A
A	10	—
B	9	4

left

TV(A, B)=10+9-4=15

	M	A
A	10	—
B	9	15

right

Fig.5 the matrix transition diagram

4.2 The TV Matrix

From Line A, B (Fig. 5 Left), M->A, and the distance is 10; M->B, the distance is 9; A->B, the distance is 4. Based on the Core idea of Excursion Algorithm, TV (A, B) =15(Fig. 5 Right). Fig. 5 is the matrix transformation diagram of the first two lines. By examining it, TV (i,j)=D(i,0)+ D (j-1, 0)-D (i, j) can be used to calculate other TVs except M queue. In the matrix, i stand for the line, j stands for the queue. Thus, M is queue 0, A is line 0, TV (i,j) is the value of Line i and Queue j. In this way, we can get the Fig. 5.

	M	A	B	C	D	E	F	G	H	I
A	10	—								
B	9	15	—							
C	7	8	11	—						
D	8	4	7	10	—					
E	8	0	3	6	10	—				
F	8	0	0	0	3	9	—			
G	3	0	0	0	0	1	5	—		
H	4	0	0	0	0	0	4	5	—	
I	10	9	4	0	0	0	1	2	5	—
J	7	13	8	1	0	0	0	0	0	9

Fig.6 the TV matrix

A-B	15
A-J	13
B-C	11
C-D	10
D-E	10
A-I	9
E-F	9
I-J	9
A-C	8
B-J	8
B-D	7
C-E	6
F-G	5
G-H	5
H-I	5
...	...
...	...

Fig.7 the TV queue

From Fig. 6, the value represents the saved transmission value of the paths. Taking A-J as an example, the transmission distance can be saved up to 13 kilometers. However, with the constraints of 40M bandwidth and under 30 kilometers transmitting distance, not all the paths are the selection paths.

4.3 Path Selection Process.

The higher TV, the shorter distance. So we sort the information in Fig. 6 in descending order (Fig. 7) to get the TV queue. Dequeening it one by one, we will get the Conditional Additive Queue. As queueing, the value of the distance and the amount of the information will be added. Once the meeting the constraints of 40M bandwidth and 30 kilometers transmitting distance, the value ranged in the last place will be dequeened with the path deleted. At the same time, the queueing path will be marked on the network diagram (deleting all the paths in Fig.3 except the direct transmission from M). The marked path will form a Excursion Algorithm. The Path selection process is shown in Fig. 8.

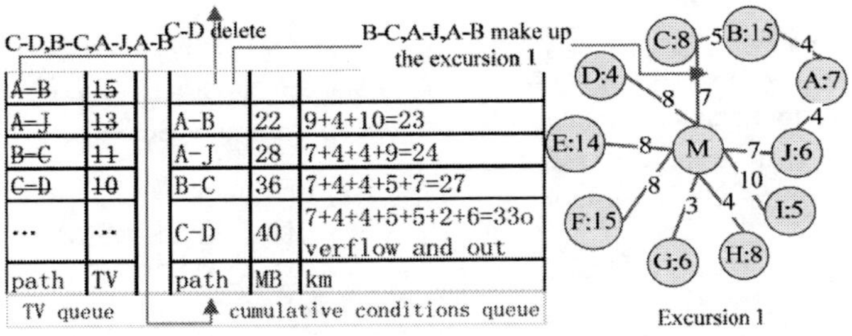

Fig.8 the process diagram about the path choice

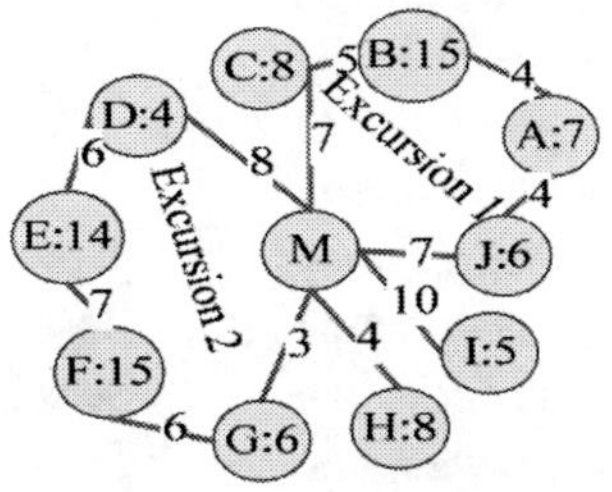

Fig. 9 the diagram about the second Excursion

Similar processes as the first Excursion, from D-E to G-H, the value are dequeened. Thus, A-I,I-J,A-C,B-J,B-D,C-E cannot be added back to the Excursion- value will not be marked on the network diagram. The process of D-E, E-F, F-G, G-H is the same as the one in Fig. 8. When G-H is adding to the Conditional Additive Queue, the distance is over 30 kilometers; it will form the second Excursion (Fig. 9). By repeating the procedure above, we will get the final Excursion Algorithm (Fig. 10).

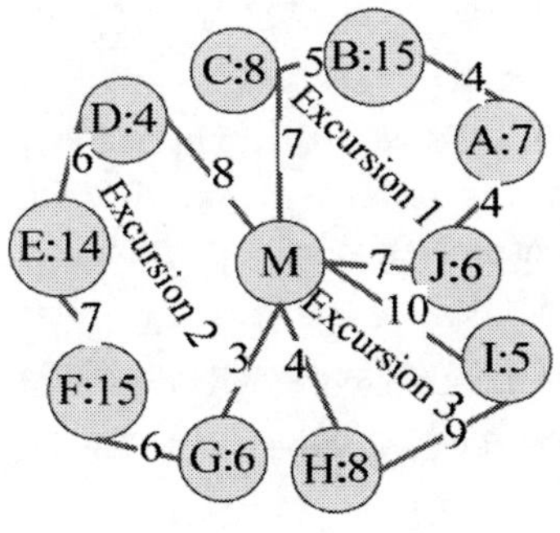

Fig.10 The final Routing Scheme use of Excursion

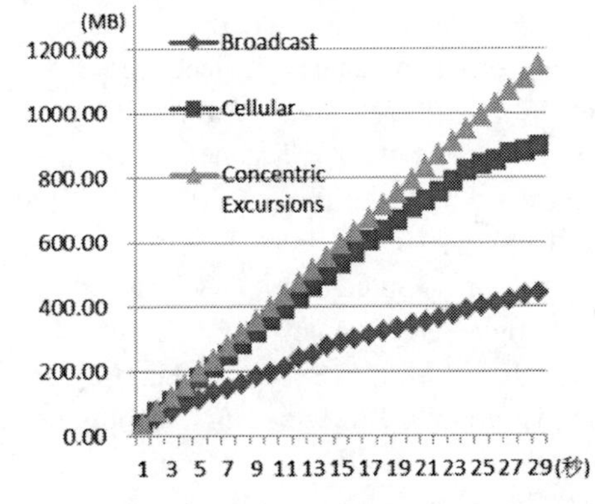

Fig.11 The throughput comparison chart

5. Performance Analysis

Through OPNET, we make a comparison on the Cellular System, simple Broadcast Method and Routing Excursion Algorithm. In the case of the same environment and parameters (the Node number and location in Fig.3), the throughput capacity is shown in

DOI: 10.1051/matecconf/20166304038

Fig.3 (The throughput comparison chart). From the chart, the basic linear of Routing Excursion Algorithm grows with the time, while the Cellular System tends to be lagging behind with the growth of time. And the Broadcast Method is quite slow in growing.

6. Conclusion

Routing Excursion Algorithm here is using the Traffic Management Center as the center to send out information to all the other regions as well receives the signals. It has to meet the demand of working under limiting condition as well to keep high efficiency. Besides routing method, there are still plenty factors may influence the efficiency of the net, such as the transmitting power of wireless devices, the relay capability of the receiving towers, the computational efficiency of the server and etc.. In this article, the author did not take them into consideration. However, these factors can be treated as numerical formation condition which can also be solved out with the method of this article. Sincerely hope that the readers can get help after reading this article.

Acknowledgments

Funds supported by 2014 Guangdong Province Outstanding Young Teachers Training program Funded Project (Grant No. 2015S02); 2013 Guangdong Province Outstanding Young Teachers Training program Funded Project (Grant No: Yq2013201); High Tech Industrialization Project of Guangdong Science and Technology Department of (Grant No: 2014A010103002); Special Funds for Demonstration Projects of Dongguan Polytechnic College (Grant No: ZXHQ2014d001).

References

[1] ZHAO Na, YUAN Jia-bin, XU Han. Survey on Intelligent Transportation System [J]. Computer Science, 2014,41(11): 7-11.

[2] Lu Huapu, Li Ruimin. Developing Trend of ITS and Strategy Suggestions[J]. JOURNAL OF ENGINEERING STUDIES, 2014, 06(1): 6-19.

[3] Wang Shundao, Song Rongfang. Joint optimization of sensing time and power allocation in small-cell networks [J]. Application Research of Computers, 2015,32(9),2766-2768.

[4] ZHOU Xian-wei, LIU Bin, QIN Bo-ping. Research on Routing Algorithm for Wireless Sensor Network [J]. Journal of Transduction Technology, 2006,19(2): 463-467.

[5] ZHANG Jian-kui,QIAN Meng-tao,QIAN Chuan-gen. Analysis and optimization of wireless network structure on the miniature type and medium-sized city [J]. Jiangsu Communication. 2011,(3): 58-60.

[6] JI X Y. Models and algorithm for stochastic shortest path problem. Applied Mathematics and Computation . 2005

[7] Luo Jianfeng Ji Shanshan Zheng Huijun. Design of Algorithm for Network Information Search in Information Integrated Platform of Cargo Tracking[J]. Science Mosaic, 2015,(2):51-53.

[8] LUO Kun WANG Jian-xin ZHAO Xiang-ning. Survey on Geographic Routing in Wireless Sensor Networks [J]. Computer Science, 2008,35(10): 28-32.

[9] WEI D C. An optimized Floyd Algorithm for the shortest path problem. Journal of Networks . 2010

[10] Da-chuan Wei. Implementation of route selection function based on improved Floyd algorithm. Proceedings of the 2010 WASE international conference on information engineering . 2010

Research and Application of WCF Technology in Data Acquisition of Ultra-high Speed Packaging Machine

Jie Qian[1,a], Li Ming Zhu[1], Zhen Xun Jin[1] and Huan Le[1]

[1] *China Tobacco Zhejiang Industrial Co. LTD., Hangzhou, China*

Abstract: By introducing WCF technology on data acquisition of ultra-high speed packaging machine, data acquisition system reads dates of machine in polling mode through the WCF client, which can achieve accurate data collection, and effectively isolate the data acquisition system and the machine control system. It enhances the security of data interaction between systems, but also reduces the coupling degree between systems.

Keywords: WCF, H1000, Data Acquisition, Ultra-high Speed Packaging Machine.

Introduction

In the Eleventh Five ex situ technological transformation, Hangzhou Cigarette Factory introduces ultra-high speed packaging machine, which is designed and manufactured in Italian GD Company. This is one of the most technologically advanced ultra-high speed cigarette packaging machine, with the production capacity of 1,000 packages per minute, which is one of the key models that the tobacco companies will introduce [1]. The data acquisition system of Hangzhou cigarette factory will acquire production data of packaging machine, which will used to analysis operating status, and also analysis input-output and raw materials consumption. The following paper will divided into four parts: technology analysis, communication interface protocol analysis, development and design, summarize.

Technology analysis

WCF is the short of Windows Communication Foundation, which is a communication technology framework to create and run a distributed system, it provides a service-oriented system to build secure, reliable, cross-platform distribution solutions [2]. The communication model structure of WCF is shown in Fig. 1. Address, binding and contract constitutes an endpoint. The endpoint client and the server is the basis for message exchange. The server public endpoint. When the endpoint match between client and server, both will be able to interact through messages [3]. WCF technology has been widely used in the field of distributed systems, business intelligence and data synchronization [4-6].

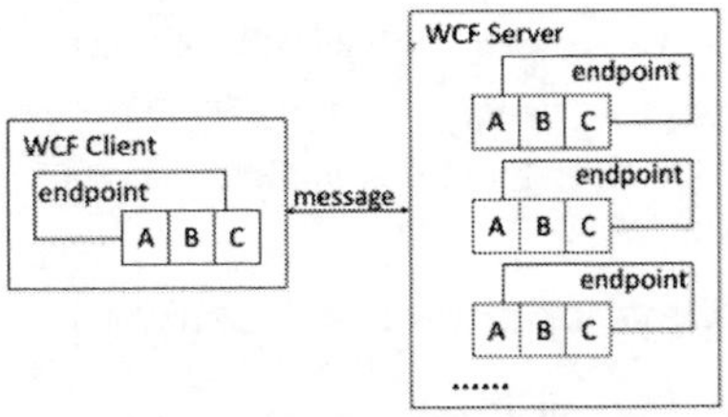

Fig. 1: communication model structure of WCF

GD Company uses new technology in their ultra-high speed packaging machine. They don't use MICROII system which is their own development electric control system, but use the

[a] Corresponding author: qianj@zjtobacco.com

DOI: 10.1051/matecconf/20166304039

production of Beckhoff which is much more popular recently. Corresponding HMI program is also replace by aHMI programs that is developed with .NET. HMI program communicates with data acquisition system through WCF.

Communication interface protocol analysis

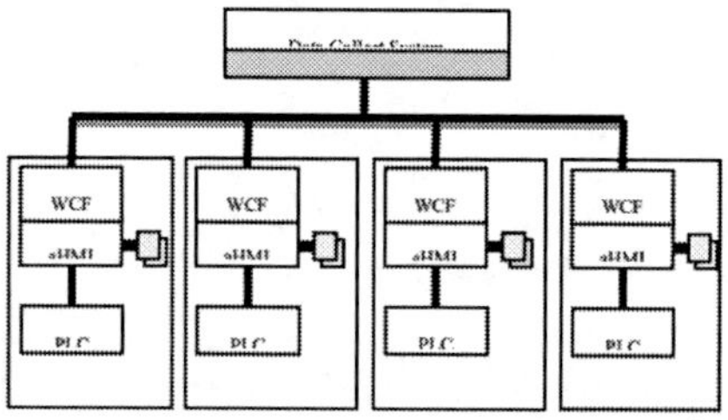

Fig. 2: the communication architecture

As shown in Fig. 2, the communication protocol is based on client-server architecture, while the aHMI program as WCF server, and data acquisition system as WCF client. We will use WSDualHttpBinding which is a secure and interoperable binding, and suits for communication in duplex service agreements or SOAP. Due to the use WSDualHttpBinding, we need to open two HTTP channels, one for service and one for the callback. Therefore, you need to configure two HTTP address, and the HTTP port number is configured as 8005 in GD.

Development and design

Ultra-high speed packaging machine consists of four parts which are H1000, W1000, C800BV and Material Automation, data acquisition systems via WCF client communicate with WCF services of four parts respectively. Interactive data between systems can be divided into three categories: machine information, statistical data and production grades. The details is shown in Table 1.

category	details
machine information	machine status, machine speed, machine user
statistical data	Machine rated speed, qualified products, excluding products, machine time (production, low speed, wait, internal stop and external stop, etc.), set production classes, material consumption, the date and time settings, machine failure statistics, history of mechanical failure
production grades	The setting of production grade

Table 1: details of the interactive data

After acquiring data from four aHMI server of each machine, Data acquisition system needs write data to system database in accordance with a standardized. System reads the data of machine in polling mode through WCF client. The details is shown in Fig. 4.

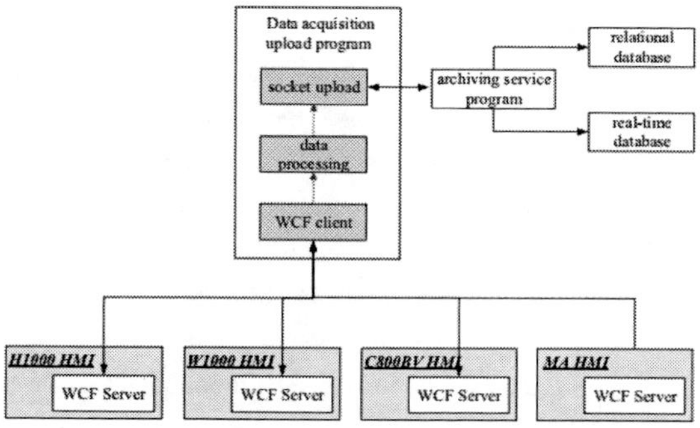

Figure 3: the details of data acquisition

Summarize

Through the application of WCF technology in ultra-high speed packaging machine, the system can successfully acquire machine information, statistics and production grades, etc. The control system only open binding address and service, the external system cannot interfere with internal control procedures of machine, this messaging model is also effective to reduce the degree of correlation between the two data exchange systems, to enhance the security between systems when data interaction, but also reduces the associated coupling between systems.

REFERENCES

[1] STATE TOBACCO MONOPOLY ADMINISTRATION, The policy of "12th Five-Year" technology and equipment in the tobacco industry[R].

[2] Jiang Jinnan, WCF Technical Analysis .Beijing: publishing house of electronics industry, 2009.

[3] Yan Shang, Huang Zhangchan. WCF: New Fundament of Communication Technology Based on Windows Platform [J]. Computer & Digital Engineering, 2008(4):86-89.

[4] Wu Qingshou. Research and Implementation on the Distributed System Model of WCF [J]. Jilin Normal University Journal(Natural Science Edition), 2012(3):61-64.

[5] Lin Xin, Zhou Yonggang, Yan Hua. Research and Application of Business Intelligence System Based on WCF [J]. Computer Technology and Development, 2010(7): 223-226.

[6] Zhang Qingfeng, Zhang Fengqin, Wang Lei. Design and Realization of Data Synchronization Model Based on WCF [J]. Journal of Shanxi Datong University(Natural Science Edition), 2013, 29(2): 9-11.

MATEC Web of Conferences **63**, 04040 (2016) DOI: 10.1051/matecconf/20166304040
MMME 2016

Research on the method of the Test and Evaluate（T&E）of Complex Simulation System

Li Liu[1,a], Xu Chang Li[1], Kun Wang[2], Xiao Min Zhang[2], Hua Xiao[2]
1.Air Force Engineering University, Xi'an China
2.Chinese Flight Test Establishment,Xi'an China

Abstract. With the development of science and evolution of society, the structure of a system becomes more and more complicated. There is an increasing need for the research, development and application of complex systems. Based on the brief introduction to the significance of T&E, the relationship between T&E and development status in the domestic and overseas, expounding the part of the establishment of the theory and methodology of T&E at the technical and methodological level, which offers references for the further development of the test and evaluation method and technology of T&E of complex systems.

Keywords- complex systems; Modeling and Simulation (M&S); Test and Evaluation (T&E)

I. INTRODUCTION

The simulation is the whole process to predict the future, aid for design, research and develop products and do scientific research with model operation. It is different from the theoretical and experimental research, but based on the scientific model operation and engineering activities. The establishment of a credible simulation system that meets the application target is the basic requirement of the user, detection and evaluation of operation data is necessary and a fundamental guarantee to meet this requirement. Therefore, the simulation must answer and prove the following three major issues seriously: ① From the source, if the model correctly to describe the actual system (process) features and characteristics or not? ② in operation, if the simulation effectively reflect data, character and behavior or not? ③ Analysis, if the simulation result really realize the application target and user requirement or not? These problems must be solved through the test and evaluation of complex simulation system.

II. THE SIGNIFICANCE AND ROLE OF COMPLEX SIMULATION SYSTEM T&E

T&E is the necessary work processes and technologies to ensure that its complex simulation system is special credibility or credibility, its significance and function is very important, the main reflect are the following several aspects:

①It can effectively improve the correctness of the simulation results; strengthen the confidence in application of the simulation system;

②It can reduce the risk of the building and application of simulation system, especially to avoid disastrous risk, to shorten the development cycle, improves the efficiency of development;

③It may reduce the total investment of the simulation system, and improve the effectiveness and efficiency;

④It is advantageous to the comprehensive analysis of the problems of the scientific research;

⑤It expands the scope of the simulation system, to improve the reusability of the simulation system;

⑥It can promote simulation system of total quality management; improve the management level and efficiency (reference to figure1).

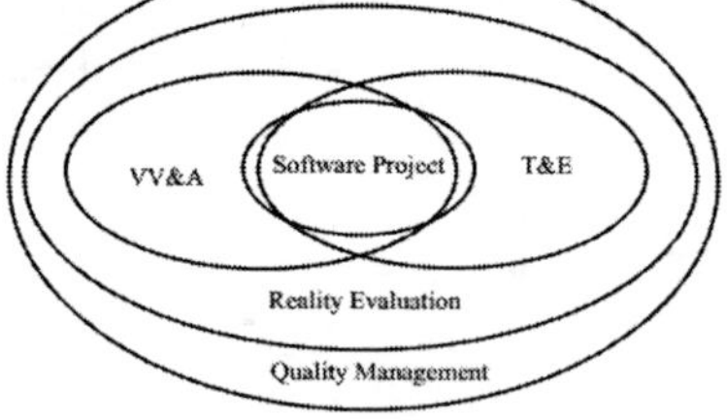

Fig1. The relationships among T&E, Reality Evaluation, VV&A and Quality Management of Complex Simulation System

[a] Corresponding author: ll200010@126.com

III. THE DEVELOPMENT TENDENCY IN CHINA AND ABROAD

Abroad, especially in the United States, the researches about complex simulation system T&E are treated seriously and always walk forward the whole world. In the last century the America defense systems Management College drawn up "T&E management guide". The modeling and simulation of the US department of defense office (DEMO) lists 76 kinds of software testing and system evaluation methods and 18 kinds of statistical techniques, which can be used for the simulation and verification of V&V. In recent years computer aided V&V and T&E tools have been developed, such as FVT, EE, Osman, Robin2002, VV&A Turbo tools etc. Currently T&E is of integration, automation, intelligence and commercial.

T&E has been development and application history in china earlier, especially in the field of aviation and space flight, in the 1980s the research application began. Many units (such as the Chinese flight test institute, 618, 623, etc.) have already adopted the advanced spectrum analyzer and programmable frequency characteristic to test instrument for real-time simulation system by running T&E orders. Recently, the intelligence and simulation in this field has become a hot spot.

IV. THEORY AND METHOD SYSTEM OF THE COMPLEX SIMULATION SYSTEM'S TEST AND EVALUATION

It is necessary to build corresponding theory and method system for any scientific research, and the complex simulation system's T&E is no exception. Research and perfecting theory and method system of the complex simulation system's T&E constantly is crucial to meet the new needs of the complex simulation system's T&E. Since Professor Osman Balci came up with the concept of VV&T, the technology and application of T&E have been developing constantly, and it has form a thorough theory and method system today. Author presents a high-level overview of the theory and method which is used by predecessors in the research of T&E, gives the framework of the theory and method system of the complex simulation system's T&E（as show in Figure2）.

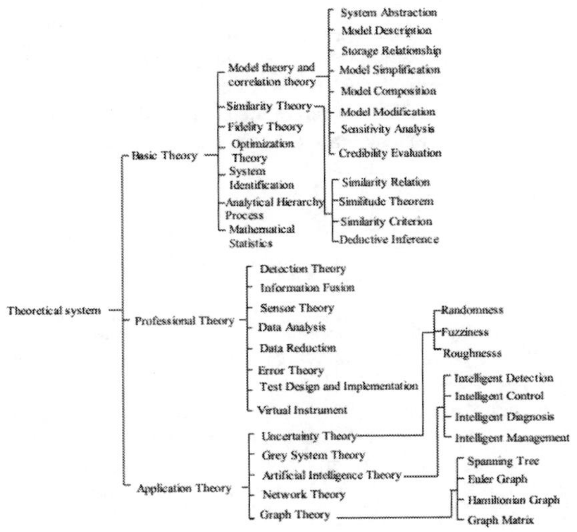

(a) Theoretical system

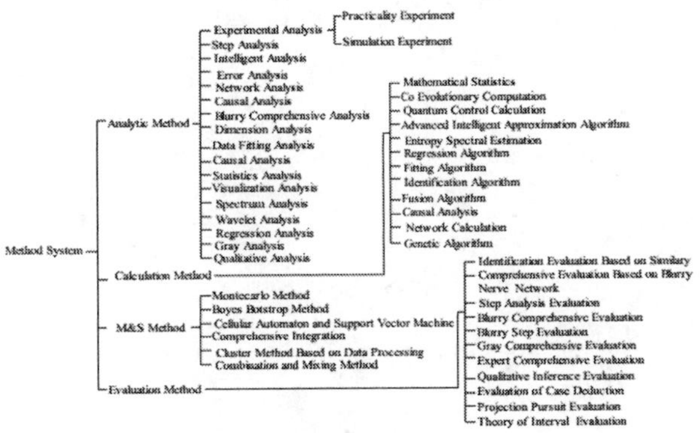

(b) Method System

Fig2 T&E theory and method system structure developed/moved by complex simulation system

At present, the T&E method for development /operation of the complex simulation system has been not a few, but for large complex simulation system T&E tend to work hard. As a result, it is very necessary to study the suitable T&E method of his kind of simulation system. At present, a T&E engineering method based on similarity theory and the identification technology has emerged, it is from the system and the similarity of concept model, using system identification techniques, can be directly by the similarity to determine the credibility of the complex simulation system. Practice has proved that this is a new credibility evaluation method with excellent operation. In addition, the identification evaluation method based on fuzzy neural network is a kind of combined modeling method, also suitable for large complex simulation system, and make the intelligent control of complex simulation system possible.

V. Conclusion

It is very difficult to monitor and evaluate the complex simulation system mainly because of the uncertainty (including randomness, fuzziness, and roughness, etc) in operation, serious nonlinear and the diversity of information and so on. For an efficient monitoring and evaluation, to provide effective and correct basis of the credibility, it is crucial to use artificial intelligence technology and computer aided analysis.

References

[1] Xingtang Liu, Li Liu. Simulation Science Technique and Engineering. Beijing: Sciences Publishing House, 2013.03

[2]Xingtang Liu, Bingcheng Liang. The Theory, Method & Technique for Complex System Modeling. Beijing : Sciences Publishing House,2011.04

[3]Xingtang Liu, Li Liu. Modern Modeling and Simulation Technology. Northwestern Polytechnical University Publishing House. 2011.03

[4]Jianjun Yang. Scientific Research Method Introduction. Beijing: National Defense Industry Press, 2006.01

[5]Henglin Wang, Jianguo Cao. Simulation System Design and Application. Sciences Publishing House., 2003.

MATEC Web of Conferences **63**, 04041 (2016)
DOI: 10.1051/matecconf/20166304041
MMME 2016

Risk Assessment on the Transition Program for Air Traffic Control Automation System Upgrade

Dong Bin Li [1,a], Yu Mei Liu[1] , Jian Wu Xu[1],Yuan Zhou[2] & Yu Song[2]

[1]*Civil Aviation Management Institute of China, 3rd Road East Huajiadi, Beijing, 100102, China*

[2]*East China Air Traffic Management Bureau, Hongqiao Airport, Shanghai, 200335, China*

Abstract. We analyzed the safety risks of the transition program for Air Traffic Control (ATC) automation system upgrade by using the event tree analysis method in this paper. We decomposed the occurrence progress of the three transition phase and built the event trees corresponding to the three stages, and then we determined the probability of success of each factor and calculated probability of success of the air traffic control automation system upgrade transition. In the conclusion, we illustrate the transition program safety risk according to the results.

Keywords: Air Traffic Control Automation System, Event Tree, Probability Calculation, Risk Assessment, System Upgrade Transition.

Introduction

Upgrading the air traffic control automation system can enhance the ATC operating environment and operational efficiency, and the upgrading process will inevitably lead to some risks, so risk assessment on the transition program is the key element to ensure the transition process safe.

Event tree analysis is a technique of risk analysis which deduces the possible consequences according to a chronological series of subsequent events from the initiating event. Event tree analysis is a temporal logical method of accident analysis, Which begins with an initiating event, analyzes step by step according to the development stage of the accident, where consequence of this event only takes one of the two diametrically opposed species status (success or failure, normal or fault, safety or hazard), progressively develops to the end until a system failure or an accident.

Air traffic control automation system upgrade transition program divides the upgrade transition into three stages, gradually develops until the final end of the transition, which is similar to the process of event tree, so this paper analyzes the safety risk of the three stages in the transition process using the event tree method.

The Transition Program

The transition program includes a total of seven parts: 1: General; 2: Organization and responsibilities; 3: Transition program for ATC operations; 4: Transition program for equipment support; 5: emergency response preplans; 6: organization and training of personnel; 7: annex. Wherein the third and fourth part is the core content of the transition program; The fifth part is to provide a preplan for the emergency during the program transition process, so as to make sure that transition go forward smoothly; The sixth part is to provide personnel preparation for the transition, to support on the relevant personnel to master the new system and make sure a smooth transition.

In accordance with the main content of the program, the transition process can be described by the diagram as follows: While the transition condition is ready, the transition will go through three stages and then a safe system transition is finished. Three stages are shown as Fig 1:

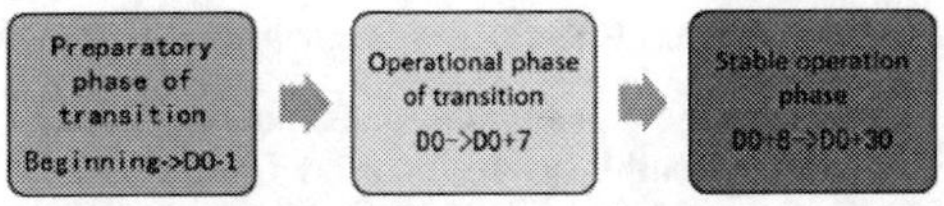

Fig 1. Transition phase diagram

Therein, D0 refers to the transition day, T0 refers to the transition time. D0-1 refers to the day before the transition day, D0+7 refers to seven days after the transition, D0+8 and D0+30 and so on.

Risk Assessment

Safety risk analysis of the transition program is to assess whether each stage can be successfully moving forward. In this part the three transitional phases (the preparatory phase of transition, the operational phase of transition, and the stable operation phase) are evaluated by using event tree method. The key elements those determine the probability of success are analyzed, finally the success probability is calculated based on the

[a] Corresponding author: lidongbin@camic.cn

 DOI: 10.1051/matecconf/20166304041

probability provided by the relevant technical personnel.

Preparatory Phase of Transition

There are three main elements those need to be considered in the preparatory phase of transition: 1. whether the operators have been familiar enough with the operation of the new system; 2. whether the new system has been proved to be acceptable in the technique; 3 whether the backup ATC automation system can be made sure to run independently during the transition period, because only backup system is in use during the upgrade process from the old version to the new version which will take about two hours; At that time air traffic controllers rely on this backup system to command, so to ensure the normal operation of the backup system is the key. According to the analysis above, the events tree for the preparatory phase of transition is illustrated in Fig 2.

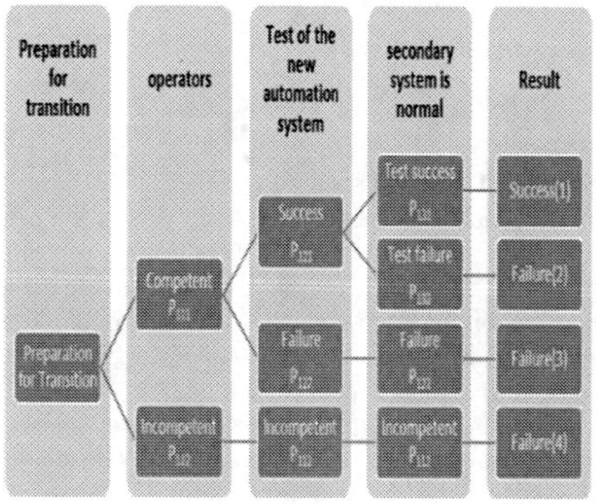

Fig 2. The Event Tree for the Preparatory Phase of Transition

Operational Phase of Transition

During the operational phase of transition, the secondary control system is used to control by the controllers, and the new air traffic control automation system synchronization is operating normally, if the above two conditions are satisfied, the new system will be switched to be used to command to air, but in more than two hours when the original secondary system is switched to be enabled as primary system, there is no other secondary system, so the system operating properly is very important. The remaining seats will be switched in if the operation is normal after the new system switched to be in use, if the phase of full load operation is normal, it will enter the next stage of stable operation. Based on the above analysis, we use event tree to illustrate the program of the operational phase of transition which is shown in Fig 3.

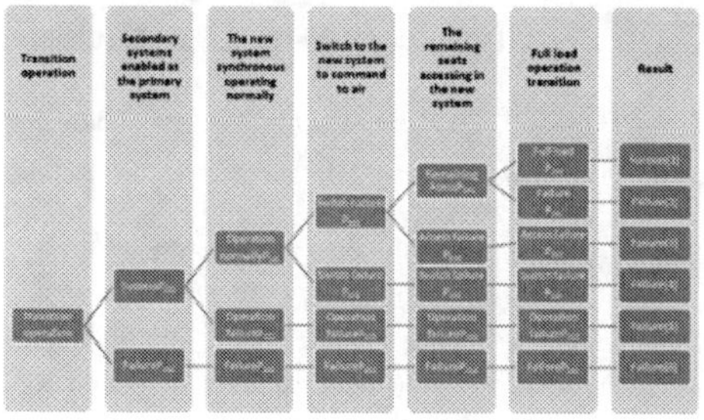

Fig 3. The Event Tree for the Operational Phase of Transition

Stable Operation Phase

The situation in the stable operation phase is much simpler comparing with last phase, the event tree with two main results is shown in Fig 4.

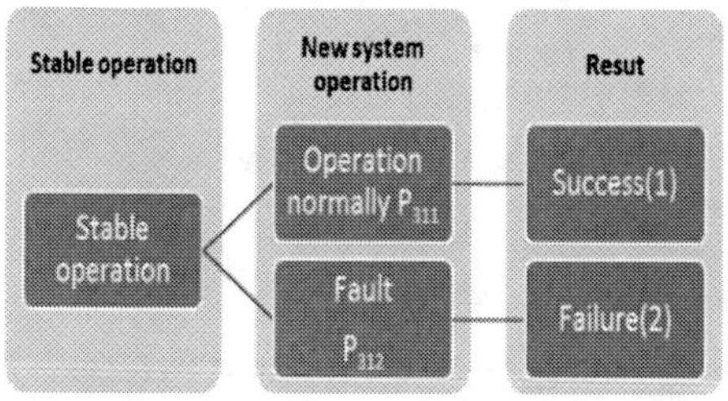

Fig 4. The Event Tree for the Stable Operation Phase

Risk Analysis

Finally, the probability of each step is calculated based on data and experience by the relevant departments. Because of a lack of data accumulation, the following data values is not precision, the probability of transition success can be increased by continuous improvement of the elements.

P_{111}=99%, P_{112}=1% ; P_{121}=90%, P_{122}=10% ; P_{131}=90%, P_{132}=10%;

Therefore, the probability of the successful completion of the preparatory phase of transition is: $P_1=P_{111}\times P_{121}\times P_{131}$=80%。

P_{211}=95%, P_{212}=5%; P_{221}=95%, P_{222}=5%; P_{231}=97%, P_{232}=3%; P_{241}=99%, P_{242}=1%; P_{251}=90% P_{252}=10%;

Therefore, the probability of successful completion of the operational phase of transition is : $P_2=P_{211}\times P_{221}\times P_{231}\times P_{241}\times P_{251}$=78%.

P_{311}=90%, P_{312}=10%;

Therefore, the probability of successful completion of the stable operation phase is: $P_3=P_{311}$=90%.

Assuming that the success probabilities of the preparatory phase of transition, the operational phase of transition and the stable operation phase are independent from each other, then the probability of the program to be a successful transition at present stage is:$P=P_1\times P_2\times P_3$=56%.

After the completion of the preparatory phase, P_1 will be 1, the probability P of the program to a successful transition on the condition that after completing is:$P=P_1 \times P_2 \times P_3=70\%$ (the preparatory phase of transition is completed).

While the transition proceeding, P_1 and P_2 are both value of 1 after the completion of the operational phase of transition, the probability P of the program to a successful transition on the condition that after completing the operational phase of transition is:$P=P_1 \times P_2 \times P_3=90\%$ (operational phase of the transition is completed).

while the new system upgrade is implemented successfully after the completion of the stable operation phase , the probability of the success of the transition program will reach the maximum value:$P=P_1 \times P_2 \times P_3=100\%$ (Stable operation phase is completed).

The data above can be used as reference values to evaluate the safety risk of the program, not the exact values. And this values indicate the probability of the success in completion of the new system with the program in this paper. If the transition to the new system fails, the program presents the detailed response measures and emergency response in case of failure in the three event trees, which is to say that this program can ensure not to affect the safe operation even if the transition to the new system cannot be completed.

Summary

We draw the conclusion that the early reference probability of the safety risk is 56% using event tree analysis method to study the program , although the probability is not high, it can be improved by increasing the stability of the new system, thereby improving the overall probability of success; as the transition proceeds, the probability of smooth transition after the completion of the preparatory phase of transition reaches 70%, the probability of smooth transition after completion of the operational phase of transition reaches 90%, the probability of smooth transition after completion of the stable operation phase reaches 100%. The program presents the detailed response measures for the transition failure, the program can ensure to return to the original operation and it will not affect the safe operation regardless of the transition is successful or not.

Acknowledgement

This research was financially supported by the National Science Foundation of China (Grant No.U1233130).

References

[1] Piero Baraldi, Enrico Zio, A Combined Monte Carlo and Possibilistic Approach to Uncertainty Propagation in Event Tree Analysis, Risk Analysis, Vol 28, No. 5, (2008) 1135–1489.

[2] Andrews, J.D., Dennett, S.J., Event-tree analysis using binary decision diagrams, Reliability, IEEE Transactions, Vol 49 , No. 2 , (2002) 230-238.

[3] Air Traffic Control Automation System, Information on http://indracompany.com.

[4] P. li, Summary of Air Traffic Control Automation System, Science Mosaic. No. 6 (2011):72-74

MATEC Web of Conferences **63**, 04042 (2016) DOI: 10.1051/matecconf/20166304042
MMME 2016

Research on Sports Tourism Image Design and Dissemination of Limiao Nationality

Jian Xiong[a]

Hainan Technology and Business College, Haikou, China 570203

Fund Project: Hainan philosophy and social sciences planning project (NO. HNSK（YB）15-114)

Abstract: This paper adopts the methods of documentary, field investigation, logic analysis etc., and takes Hainan Li Miao nationality cultural custom, traditional sports and experiential sports tourism as research content to discuss design and dissemination of Li Miao nationality sports tourism image and puts forward workable ideas so as to provide a theoretical reference for sports tourism resources development of international tourist island.

Key words: sports tourism, image design, national culture dissemination, tourism resources development

1. Introduction

Located in the most southern tip of China, Hainan has more than 1800 km long coastline and spring season all the year, with an average temperature of 22 ℃-25 ℃. Hainan has rich tourism resources of sea, sunshine, beach, forest, springs, tropical products, ethnic customs as well as the nation's marine resources which accounts for two-thirds of China. There are more than 30 nationalities in Hainan, among them the native nationalities are Li, Miao, Hui and Han. The weather in Hainan is tropical rainforest and monsoon climate, with abundant sunshine and rain, fresh air and beautiful environment, which is a holiday resort and has more than ten million tourists both at home and abroad every year. In 2012, Hainan attracted 33,203,700 domestic and foreign tourists, with the total revenue of 37.9 billion RMB and increased 17%. On January 4, 2010, the State Council promulgated "State Council's Opinions on Promoting Development of Hainan International Tourism Island", aiming to put forward general requirements of construction and development for Hainan, which contains strengthening construction of ecological civilization, enhancing the ability of sustainable development and initially building a world-class leisure and tourism resort island until 2020. The time Hainan international tourism island becomes national strategy, its sports tourism gets further development. Traditional sports activity has close relationship with Li Miao native history, custom, labor production and religious events. It has long history and rich cultural connotation. The development of Li Miao sports tourism resources can not only improve local cultural heritage but also promote regional economic development, thus it worth discussing on sports tourism image design and dissemination of Li Miao Nationality.

2. Design background of Limiao nationality sports tourism image

2.1 Historical and cultural customs of Limiao nationality

Li nationality is the first original inhabitants of Hainan. According to the exact historical records, there were Li nationality living in Hainan since pre-Qin. At that time, Li nationality belonged to a branch of Luo Yue people of hundred Yue nationalities. About five thousand to two thousand years ago, Lu Xu took the boat across the sea. Later scattered people came Hainan from Nanyang Island and integrated into Li nationality. In the Western Han Dynasty, they were called "Luo Yue," and was called "li", and "bullying" in Eastern Han Dynasty, and was called "Li" and "Liao" in Sui and Tang dynasty. The name of "Li" began in the end of Tang dynasty and was fixed in Song dynasty, and is still in use by now. During the time of Qin and Han, the Lingao people that branched by the Zhuang nationality also entered the Hainan Island. Then Han, Miao, Hui and other ethnic groups continued to arrive here. According to historical records, the Miao nationality were not active migration but be forced to brought here. They all lived in rural areas with hard conditions. Miao ancestors has kept their toughness so that they can survive in the poor mountains and create a unique culture, which enriches the connotation of the island culture.

Li nationality is a nation that pay much attention to ceremony. Li nationality etiquette included rich contents,

[a] Corresponding author: 419985168@qq.com

covering various aspects of production and life. Li has created live and wide range of topics and colorful oral literature, and has created more than 40 folk musical instruments with long history. Representatives of the eight traditional instruments include canoe drums, buzz wood, mouth bow, MPH blanket, beep da, mouth-Bai, nose flute and Zhuo-ba! Li nationality's songs use Fu, Bi and Xing with rich imagination, which uses language image, rigorous structure and artistic conception. Miao nationality is called "dancing nationality", the dance includes Lusheng dance, bench dance, Houer-gu dance, and the Lusheng dance is most common with high skills, ,which is deeply appreciated by domestic and foreign countries.

2.2 Costumes culture of Limiao nationality

Traditional dress of Li nationality has its own distinctive national style and characteristics. The main raw materials of Li nationality's colorful costumes comes from Hainan Island's cotton, linen and the like. Li's family women are with ingenuity, they can spin the wood cotton into thread, woven into cloth and dye in color and embroider with flowers. There are more than 160 species of Li's brocade fabric patterns, reflecting the Li's colorful life with rich connotations. For example, Li nationality's women in mountain area like using water deer, birds, and other animals and plants as a pattern; while women from plains love using fishes in the rive, shrimp and frogs in the pool and field egrets and other animals as brocade pattern material. "Hunting Picture" and "Wedding Picture" are representative works of Li nationality's . brocade. Most dress patterns are drawn from in a variety of vivid images in everyday life, which can identify race, branch and language. These image records are called "wearing epic" by the specialists.

2.3 Sports cultural traditional festivals of Limiao nationality

Hainan Island residents mainly include Han, Li, Miao, Hui residents, wherein Li and Miao are the island's indigenous people. There are 471, 500 people of Li nationality and 77,000 people of Miao nationality. They are located in Qiongzhong, Baoting,Baisha, Ledong, Changjiang, Lingshui, Sanya, Five Finger Mountain, Dongfang and other cities and counties. They are industrious and brave, can sing and dance. Living labor and the accumulation of wealth for thousands of years have formed a splendid traditional folk culture. Now the excellent traditional Li nationality's sports culture has been preserved include climb the perch to pass the tree, the dog to the slope, pull the tortoise, shooting, knee climbing to pick flowers, playing top, cow fighting, boat race with holding things, float match, climb the mountain to "grab the girls". These sports not only are with recreation and full of competition features; Tsien fluid double dance, Dachai dance and Dalu dance are sport projects that combines celebration, dance, fitness and entertainment. Li nationality always has the custom of dabbling leisure, and the main activities include water-sprinkling, water fights, throw water gage, tap water drum, pass water bridge, and diving etc.

March third of Love Festival is Li nationality's March Third Festival (On the 3rd of the 3rd lunar month), which is the greatest folk festivals in Hainan Province, is also a good day for Li nationality's youth. It is also know as Love Festival or Love day, and is called "Fu Nian Fu" in Li's language. Every festivals of the March Third Festival, Li and Miao residents get together in the beautiful Hainan Island with their beautiful songs and graceful dances. Men and women, the old and young of Li and Miao nationality are all dressed in festive costumes, and they merrily dance of the original ecological minority villages dance, and singing with Li and Miao's folk songs. "March Third" is the only festival to be held on behalf of the provincial government. It is stipulated by the Hainan Provincial People's Congress: "The 3rd day of the 3rd lunar month is Li and Miao's traditional festival" . In the traditional March third festival, people hold Limiao's traditional sports matches, intangible cultural heritage exhibition and cruise carnival with Limiao's features.

3. Development status of Limiao nationality sports tourism

3.1 Li miao nationality has rich tourism resources and great development potential

Since the founding of Hainan province, social economy gets very fast development and the number of residents and tourists increases a lot. Government provides policy support for Hainan tourism development especially during the construction of International tourist island. Every year, the increasing tourists from home and abroad lead to the development of catering, accommodation, transportation service industries, the GDP has been increased, the employment has been expanded and the regional economy has been promoted. Improving living standards makes people gradually transfer from pursuing material culture to spiritual culture, which leads them have more demand for sports fitness, entertainment and exploration etc. People's demands for sports tourism are becoming diversified. To integrate local ethnic culture with sports tourism is a new pursuit. Hainan has the characteristics of Li and Miao customs culture, clothing culture, sports of traditional festival, its has rich sports tourism resources and great development potential.

3.2 Lag of sports tourism image products development in Limiao nationality

Although Limiao area has relatively rich sports tourism resources, it lacks of representative products, professional image production institutions, and tourism products with local characteristics. Traditional sports with regional characteristics have not been promoted and exhibited, clothing culture failed to get publicity, sports tourism projects have no further development, besides, sports tourism products lack of unified and standardized design and production, they fails to form a new industrial system.

3.3 Sports tourism product design and dissemination has not been taken into account

Although Hainan international island has become a national strategy, with large increase in foreign and domestic tourists, improvement of Gross national product (GNP) and rapid growth in regional economy. The design and dissemination of national sports tourism image have not received enough attention, which makes sports tourism on a level of simple scenic spots and natural resources. Most of them are sightseeing-type tourism products, professional tourism products with rich local characteristics and recreation tourism have not developed at all, sports culture connotation of Limiao nationality has not been explored well, and the custom national traditional sports culture and modern sports do not become two organic synthesis.

4. Existing problems of sports tourism image design and dissemination of Limiao nationality

4.1 Loss of craft culture with rich national characteristics

Limiao plastic arts has distinctive national features and it is unique national culture. But Li miao plastic arts are in danger of dying out, the existing arts are traditional costume, tools for production and living, cotton and hemp products containing coats, dresses, headscarf, cotton cloth, hemp cloth, hemp quits and hemp bags etc. The knitting technique of Li nationality is also very famous, but there is less place to inherit it. Traditional costume culture's dying out will hinder development of sports tourism. National dancing needs national costume, which can attract tourists.

4.2 Imperfect management system, deficient in sports tourism products innovation

Enterprise management mechanism is not perfect and there is inadequate investigation for demand of tourism products. The scale of the production enterprises is less, with decentralized management, low level and insufficient overall design and planning of sports tourism. Most tourism products are less innovate, and the artwares that can represent national features are less, recreational programs are less and tourists' degree of participation is low and the visiting time is short.

4.3 Insufficient propaganda, inadequate input of funds and lag of infrastructure construction

Tourists know very little about Hainan sports tourism, which will undoubtedly hinder development of sports tourism. Insufficient implementation rules which are in accordance with actual development of Hainan can easily bring negative effects to environment. Beside, lag of supporting facility and low level of food, accommodations and transportation will bring inconvenience to tourists in sports tourism.

4.4 Comprehensive quality of sports tourism practitioners have not been improved

Since the beginning of the opening up policy, political and economic culture of Hainan have gotten rapid development excepting for tourism industry image. Brand image of international tourism island has not yet been formed, with the reasons of government's imperfect management system, input has fallen short of needs, and the low quality of practitioners. The comprehensive quality of practitioners represents the image of a region and will affect the number of tourists and economic development of the region.

5. Conclusions and Suggestions

5.1 Combine Limiao traditional sports tourism to promote sports tourism

Sports tourists have characteristics of general tourists containing property of participation and tourism as well as specificality. The culture of ethnic minorities, costume culture, food culture and traditional sports culture can be exhibited, inherited and spread through sports tourism. Future trends of tourism will pay special attention to participate nature and go back to nature. People hope to enjoy the beautiful nature scenery, to get a taste of regional ethnic culture, and especially to participate in regional minority culture and participation in local sports. Therefore, to establish some sports programmers containing recreation, ornament and local customs with regional resources in Hainan scenic spots will not only promote traditional sports and develop sports tourism but also establish the image brand of sports tourism.

5.2 Optimize the structure of traditional sports structure of Limiao and image design

Limiao has rich historical tradition culture, costume culture and traditional sports culture. People have created many ways to conquer nature combining with actual environment in daily living and production and have formed different styles of sport events. They have fully excavated important practical significance of traditional sports projects. Thus, we should study connotation of Limiao traditional culture and develop it into project with spread value, as well as to optimize the structure of sports project and make it suitable for mass sports participation, conducive to historical and cultural heritage and development. Thus, to increase minority sports competition and performance through improving sports facilities construction in tourism scenic spots and ethnic culture reserves and to make tourists involved in sports and enjoy ethnic sports will promote all-round development of minority sports tourism and lead to development of related industries to promote regional economy of related industries.

5.3 Strengthen the construction of web advertising and sports tourism facilities, enhance Limiao sports tourism brand image

Firstly, We can use internet propagation to help tourists know more local customs and distinctive culture, as well as to strengthen construction of sports tourism facilities so as to provide more sports fitness spots and recreation activity facilities. At the same time, we should use Limiao traditional festival to develop Limiao traditional sports, to push the combination of Limiao traditional sports and tourism and to create Limiao traditional sports band. Thirdly, we should design sports tourism sports with national features, optimize the structure and quality of sports tourism products, expand sales volume of national products and minority culture's influence, increase the diversity of products with national characteristics, to standardize price of scenic spots, to maintain a good image of international tourism island and to create Limiao sports tourism brand with the aim of developing sports tourism products of Limiao and enhance popularity and economic development drive of ethnic minority areas.

5.4 To strengthen construction of practitioner professional training and self-building, using traditional ethnic sports to shape regional image

Sports tourism contains natural resources and human resources which are full of sports elements, and it provides programs containing many services such as competition, fitness, relaxation, recreation and communication that are represented as tourism commodities. During sports tourism, it is the practitioners that makes contact with the tourists, thus the quality of practitioners will influence tourists' satisfaction to the scenic spots. Therefore, to strengthen construction of practitioners' professional training and self-building will shape regional image effectively. As the highest sports event of reflecting development of traditional minority national, minority nationality traditional sports meeting will promote minority sports significantly and contribute a lot to disseminate minority nationality traditional sports, thus it should be excavated, arranged and improved.

Acknowledgement

This research was financially supported by Hainan philosophy and social sciences planning project Foundation. (NO. HNSK（YB）15-114).

References

[1] The State Council. "The State Council's Opinions on Promoting Development of Hainan International Tourism Island". http://www.gov.cn/zwgk/2010-01/04/content_1502531.htm 2010.01.04

[2] Gao Yuanyuan, al. Research on Image Design and Communication Strategy of Sports Tourism Destination , J. Business Culture, 2010, 5.

[3] Min Jian. Sports Tourism and Its Definition, J. Journal of Wuhan Institute of P. E，2002（6）: 4-6

[4] Wang Guizhong. Research on Classification of Sports Tourism Resources and Its Renewability, J. Sports Science, 2003, 23(4):12-15

[5] Han Zheng, al. Investigation in Tropical Sports Tourism Resources of Hainan Li nationality, J. Social Science Journal of Hainan University: 2014,32(3):126-130

[6] Xia Minhui , al. Research on Development of Minority National Sports Tourism in Hainan, J. Natural Science Journal of Hainan Normal University:2010,23(2):224-228

[7] Ma Yifang，al. Research on Minority National Sports Tourism Image Design-Take Guangxi Nationality National Sports Tourism as An Example, J. Journal of Xi an Institute of Physical Education, 2004,21（4）, 27-30.

Image-Forming System Design of Dynamic Targets Based on Reflecting Mirror Splicing

Hong Mei Zhu[1,a], Yan Piao[1]
[1]*The College of Electronic Information Engineering, Changchun University of Science and Technology, Changchun, Jilin, China.*

Abstract. In this paper, the author designed an area-scan CCD reflecting mirror splicing image-forming system with area-scan CCD ICX415AL as its transducer module and this system can be used for tracking dynamic targets. By analyzing the theory of vignetting generating, the author made mathematic model of vignetting and confirmed the splicing and overlapping pixel number of the optical system. What's more, the sequential circuit and driving power circuit of ICX415ALwas designed and the correlated noise in video signals was strained with CDS technology. Therefore, the signal-to-noise ratio (SNR) of the system was elevated. With FPGA as its core controlling module, this system postponed the splicing image-forming system to a period during which a line of CCD data are read, thus the need of real-time tracking was completely met.

Keywords: reflecting mirror, area-scan CCD, dynamic targets, FPGA. .

Introduction

As for the tracking of dynamic targets, the image-forming system with big viewing field and high resolution can greatly increase the probability of target acquisition and recognition[1]. Therefore, most dynamic target tracking systems adopt area-scan CCD as their image-forming module. However, with the rising number of CCD pixels, frame frequency decreases inevitably, and the accurate tracking of dynamic targets fails. Thus, the method of splicing many minor area-scan CCDs can guarantee the high frame frequency in accurate tracking and at the same time meet the need of big viewing field and high resolution. At present, there are a lot of area-scan CCD splicing methods, such as mechanical splicing, multiple-lens splicing, prism splicing, reflecting mirror splicing and so on. Among them, CCD reflecting mirror splicing uses reflecting spectroscope to divide one image plane into two or more spaced planes. Then, many CCDs can be interlacedly placed on every plane and the pixels of the two interlaced neighboring CCD border are made to meet overlapping requirement. Thus, an equivalent big viewing-field detector is formed.

Formation of the mathematic model of vignetting

For Point A and Point B which form images in the infinite distance on reflecting surface CCD, when they pass through the optical system, they are supposed to converge at image points A' and B' on the reflecting surface. But, for Point A and Point B, with the existence of reflecting mirror on the optical path, part of their rays haven't been reflected by the reflecting mirror and instead they are projected directly behind the reflecting mirror. For Point A, this part of optical energy is lost and this can be reflected on the image by the fact that the digital value of this point is darker. When the position of the reflecting mirror and that of the plane on which CCD image plane is located are fixed, it can be known through geometrical relationship that for any point whose viewing angle is larger than B's angle ω, all their rays can be reflected on the above reflecting surface CCD through the reflecting mirror. Therefore, no energy is lost and all the points whose viewing angles are larger than Angle ω can be considered to have created no vignetting.

According to cosine theorem and the known conditions, we can get the values of Angles α and β as is shown in the following formula:

$$\cos\alpha = \sqrt{(2fd + \Delta aD + 2\Delta a^2)/(4f^2 + 4\Delta a^2 + 4\Delta aD + D^2)} \quad (1)$$

$$\cos\beta = \frac{\sqrt{(4f^2 - D^2 + 4\Delta a^2)/(4f^2 + D^2 - 4\Delta a^2)}}{\sqrt{4f^2 + 4\Delta a^2 + 4aD + D^2}} \quad (2)$$

Here D is the caliber of the optical system, f is the focal length of the optical system, d_0 is the distance from reflecting mirror to image plane, Δa is the distance from vignetting point on the image plane to the centre of viewing field.

The energy of a point light source distributes on an image plane as a round spot. For image-forming Point A' on the reflecting surface, the energy that a reflecting surface CCD receives should include all the energy of the entire light spot. However, the energy within Angle α has not been reflected. Therefore, the energy of Point A' can only be the difference between the energy of the light spot and that of the area within

[a]Corresponding author: custzhm@126.com

Angle α. As can be shown in Figure2,when the position of the optical system and that of the CCD image plane l are fixed, the size of the light spot created by the point light source is fixed. Suppose that the diameter of the round light spot is d, then the ratio of the light spot area of Angel α to that of Angle β is

$$\left\{\frac{d}{4}\arcsin[\frac{2\Delta a(l-d_0)}{d_0 d}]-\sqrt{\left(\frac{d}{4}\right)^2-\left(\frac{\Delta a(l-d_0)}{d_0}\right)^2}\cdot\frac{\Delta a(l-d_0)}{d_0}\right\} \quad (3)$$

Suppose E is the energy of a point in infinite distance, and E' is the energy of image-forming Point A', then

$$E':E=S_\beta:S_W \quad (4)$$

$$E'=E\left\{\frac{d}{4}\arcsin[\frac{2\Delta a(l-d_0)}{d_0 d}]-\sqrt{\left(\frac{d}{4}\right)^2-\left(\frac{\Delta a(l-d_0)}{d_0}\right)^2}\cdot\frac{\Delta a(l-d_0)}{d_0}\right\} \quad (5)$$

Formula (5) is the energy value that Point A' receives and Point A' is a random point within the vignetting area of reflecting surface CCD.

The establishment of mathematic model plays a guiding role in splicing optical system. The critical points B1and B2 of the vignetting on the two CCD image planes can be calculated through the mathematic model, as is shown in Figure 1.

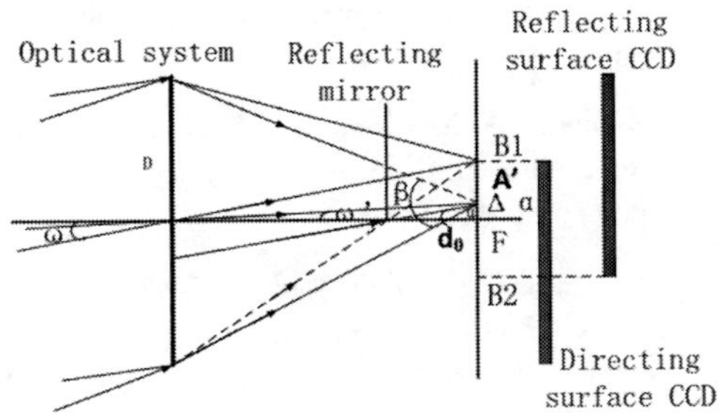

Figure 1: diagram of the production of vignetting of reflecting surface CCD

The reflecting surface CCD is placed on Point B2--the maximal critical point of the direct shooting surface vignetting, while the direct shooting surface CCD is located on Point B1-- the maximal critical point of the reflecting surface vignetting. When the optical energy of the shot target radiates respectively on the direct shooting and reflecting CCDs, a compensation can be made via adding the energy of corresponding pixels. According to calculation, if the optical system of reflecting mirrors is spliced in accordance with overlapping area, 324 pixels will be overlapped and these overlapped pixels will account for 1/5of the total horizontal pixels. After compensation, an integral radiometric calibration experiment of the system has to be carried out to eliminate the impact of overlapped noise resulted by adding on the image. Through radiometric calibration, not only compensation factor can be confirmed but also the impact of nonuniformity of pixels can be eliminated.

The composition of area-scan CCD image-forming system

The driving pulses for ICX415AL to produce, transfer and sense charges include vertical transfer clocks (ΦV1、ΦV2、ΦV3), horizontal transfer clocks (ΦH1、ΦH2), reset clock (ΦRG) and substrate clock (ΦSUB). Every driving pulse should meet exactly the requirement of phase time sequences in order to guarantee normal operation of CCD. Whenever a set of vertical transfer clockΦV1 、 ΦV2 、 ΦV3 time sequences are transmitted, the collected charges of every line of pixels of area-scan CCD will move a line to the horizontal shift register; Whenever a set of horizontal transfer clockΦH1、ΦH2 time sequences are transmitted, the charges in the horizontal shift register will transfer to the output terminal once. The transferred charges will be output to the output amplifier where these charges will be converted to the form of voltage and output from OS pins. The reset clock ΦRG clears away the remaining charges in the horizontal shift register while the substrate clock (ΦSUB) determines the adjusting time of CCD shutter. The more substrate clocks (ΦSUB) in a frame, the shorter the exposure time of CCD photosensitive areas.

According to the accuracy detecting result of optical system splicing, the error of straightness and coplanarity of the same horizontal lines is smaller than 1/3 of a pixel. So, every line of data are used as a caching unit. This method doesn't need external RAM chips and it takes up small storage space to use a line as a caching unit. ICX415AL only needs 782*10=7820bit storage space and FPGA on-chip RAM can meet this need. Meanwhile the caching time is very short. With ping-pong structure reading mode, the caching time is only the period during which a line of data are read by CCD. The advantages of using FPGA on-chip RAM as its caching memory lie in the facts that signal crosstalk caused by the usage of an external memory can be avoided and that the designing requirement for the circuit board is not very high. What's more, the real-time performance of image output is good and it only takes the time for reading a line.

The procedure of ping-pong caching structure is as follows: in the first buffering cycle, the first line of data are written respectively into RAM1 and RAM3; in the second buffering cycle, the data are stored into RAM2 and RAM4 through the switch of writing enabling signals and at the same time, the first line of data are read from RAM1 and RAM3; in the third buffering cycle, the third line of data are written into RAM1 and RAM3 and meanwhile, the second line of data are read from RAM2 and RAM4. In this cycling way, the seamless cache and processing of digital streams is fulfilled.

After multiplying the synthesized image data with corresponding adjusting factor of radiometric calibration, the adjusted final output image is obtained as is shown in Figure2 (c) and (d). The original image is

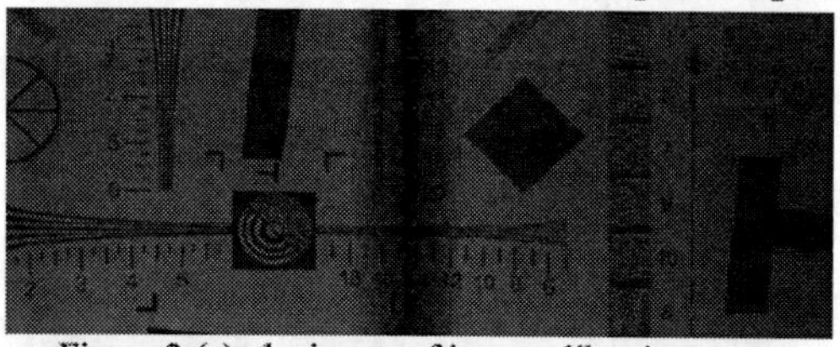

Figure 2-(a): the image of inner calibration

Figure2 -(c): the image of inner calibration after vignetting is eliminated

As for the effect of image restoration, apart from subjunctive evaluation on resolution with human-being's eyes, objective comparison with image quality evaluating factors can also be made. In this paper, the compensated image is evaluated with signal-to-noise ratio （SNR） of the image [6] and gray mean grads(GMG) with no reference evaluating index. The higher the values of SNR and GMG, the clearer the image and the better the image quality. After selecting the central region of the spliced image and making calculation, we can get the following results as are shown in Chart 1.

Chart 1: comparison of SNR

	Fig.1(a)	Fig.1(c)	Fig.1(b)	Fig.1(d)
SNR	12.79dB	27.34dB	13.23dB	28.67dB
GMG	4.542	7.462	4.911	7.636

It is obvious that when the image is processed with the above-mentioned vignetting processing method which is based on degradation model, its GMG is greatly elevated. The SNR of the image is elevated at least by 15dB. The grayscale of the processed image is in uniform distribution and this can be considered as ideal effect.

Summary

In this paper, the mathematic model of vignetting generating in reflecting mirror splicing optical system is established. With this model, the number of overlapping pixels of the viewing field centre in reflecting mirror splicing system is confirmed and the area-scan CCD image-forming system which can satisfy dynamic target tracking is designed. The compensation for two area-scan CCD data is made in accordance with the number of overlapping pixels. The vignetting in central

shown in Figure2 （a） and （b）. Through comparison, we can conclude that this imaging system can acquire clear images in both indoor roller imaging experiment and outfield imaging experiment and what's more, it can eliminate image degradation caused by vignetting on the premise of satisfying dynamic target tracking.

Figure 2-(b): the image of the outfield experiment

Figure 2-(d): the image of the outfield experiment after vignetting is eliminated

viewing field is eliminated on the premise of meeting real-time tracking need and the SNR of the image is increased at least by 15dB.

References

[1] Wang Qingyou. Application Technology of Image Transducer. [M], Beijing: the Electronic Industry Press, 2006: 343-414.

[2] Thomas W.Mccurnin. Signal Processing for Low Level、High Precision CCD Imaging[C]. Proc SPIE, 1991, 1448: 225-235.

[3] Li Guoning, Liu Yanyan, Jin Longxu. The Study of Area-Scan CCD Imaging System for Dynamic Target Tracking. Optical Precision Engineering [J], 2008, 3.. 558-564.

[4] Tong Shoufeng, Ran Jin, Hao Zhihang. The Study of the Technology for Reducing CCD Image Transducer Noise. Optical Precision Engineering [J], 2000,8(2): 140-146.

[5] Xu Xiuzhen, Li Zitian, Cao Lijun.CCD Noise Analysis and Disposing Technology. Infrared and Laser Engineering, 2004,33(4): 343-347.

[6] Tong Shoufeng, Ran Jin, Hao Zhihang. The Analysis of Several CDS Topology Circuits [J], Semiconductor Photoelectronics, 2000, 21(5): 358-362.

MATEC Web of Conferences **63**, 04044 (2016) DOI: 10.1051/matecconf/20166304044
MMME 2016

Multi-stages decision voice print recognition in telecommunication

Fa Jiang Ma [1], Lan Tian[1] & Xiao Shan Lu [1,a]

[1]*school of information Sci. and Eng., Shandong Univ., jinan, China*

Abstract. Using the cepstrum and pitch parameter, a VQ voice print recognition (i.e. speaker identification) method was realized and used in telecommunication. In this paper, we analyzed the characteristic of long-time LPCCEP and the pitch parameter for text-independent speaker identification. Considering the different contribution of each LPC cepstrum component, the weight was added to modify the distortion measurement. Based on the improved vector quantification (VQ), we constructed the speaker individual model and the system was training shorter, the response real time and the data memory smaller. A simple pitch measurement was introduced and applied in multi-stage decision to raise system robustness. The test results show that using about 10 seconds voice, in 60 speakers set, the correct identification rate is above 95%.

Keywords: Voice print recognition; Vector quantification(VQ); LPC cepstrum; Pitch.

Introduction

Recently, voice print recognition technology has caused widespread interests for its convenience, economic, and veracity. Particularly in identity automatic identification based on telecommunication, as one of the human biology characteristics, voice print is advantaged and irreplaceable.

Voice print recognition is realized when speakers provide the system with speech materials, considering whether speech materials is fixed or not, it is divided into two cases: text-dependent and text-independent. The later is more practical because of less limitation, and can prevent from voice bootleg in use. This paper introduces a rapid and accurate recognition method, which analyzes and processes readily, then tells the speaker form all set in high correct rate, which based on an arbitrary speech, when speech materials are unstable and irregular. We will discuss this method in the following aspects: feature parameter selection, recognition modeling, system structure, and test experiments. This technology has been applied to some business system,such as: SpeakerKey of ITT company, VoiceGuardian of Keyware company and so on[1]. Several problems should be solved in voice print recognition technology, such as, the unsteady recognition capacity because of voice drift[2], and the extracted information lacking of emotion factor[3].

Speech Feature Analysis

Feature parameters extraction is crucial in voice print recognition. Pitch is an obvious acoustic feature. The pitch periods of different speakers vary not so much, could be imitated easily, and change a lot over time. Therefore it has a poor effect to recognize, when using pitch parameters only.

Comparatively speaking, vocal tract features change a little over time, and can't be imitated easily. And it is the more important physiology difference of speakers' personality, and the spectrum envelope of speech is the acoustic performance of vocal tract. In comparison to pure unvoiced sound, voiced sound has a high frame loudness and high SNR, and contains more speakers' information. Applying linear prediction analysis technology, vocal tract and voice source feature can be separated effectively. Particularly, we analyze some typical vocal tract parameters, including LPC, PARCOR and LPC cepstrum coefficient. LPC cepstrum coefficient performs best in speaker recognition, and is easy to calculate. Through long-time average processing to LPC cepstrum coefficient , speech content information would be weakened, but vocal tract characteristic is almost invariant. Independent of vocal tract, pitch information can be applied to raise the system robustness.

Feature Parameter Extraction[4]

According to the above mentioned analysis, applying linear prediction analysis method, we extract long-time LPC cepstrum vector and pitch period, and the extraction algorithm is shown in the figure 1. In the minimum phase condition of all-pole vocal tract model, the solution formula of short-time LPC cepstrum can be deduced:

$$\hat{h}(1) = a_1 \quad \hat{h}(n) = a_n + \sum_{k=1}^{n}(1-\frac{k}{n})a_k\hat{h}(n-k) \quad 1 < n \le p$$

$$\hat{h}(n) = \sum_{k=1}^{n}(1-\frac{k}{n})a_k\hat{h}(n-k) \quad n > p \qquad c(n) = 1/2[\hat{h}(n)+\hat{h}(-n)] \tag{1}$$

[a]Corresponding author： 2281176576@qq.com

DOI: 10.1051/matecconf/20166304044

Where $a_k(k=1,2,...,p)$ is a predictive coefficient, and can be solved by Durbin recurrence method from speech signal $S(n)$, and p is the predictive order. $\hat{h}(n)$ is the LPC complex cepstrum, and $c(n)$ is the cepstrum. The lower part of the cepstrum decays rapidly as n increases, therefore a p-order cepstrum vector far less than frame width(N) could express the vocal tract characteristics of a voice frame, and P values 8 to 16 typically. Applying the adaptive endpoints detection method, to detect voiced sound, we average every 10 frames to consist a frame of long-time LPC cepstrum vector.

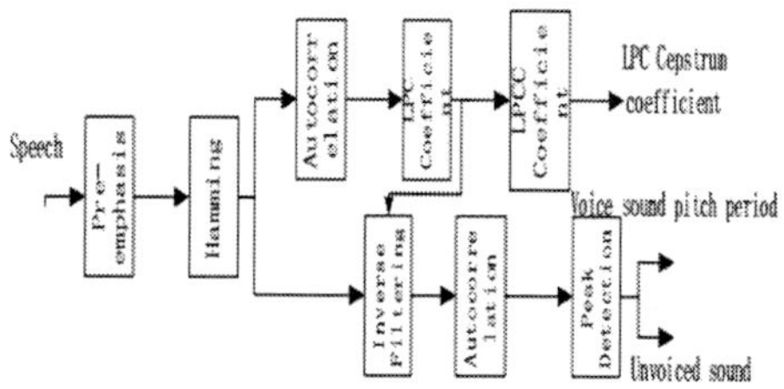

Fig. 1 The block graph of features extraction

In the quantitative analysis, when P values 16, the each component of 16-orders LPC cepstrum coefficient contributes differently, and 6-order to 16-order components contribute much more significantly and obviously. Using this phenomenon to modify the distortion measurement formula, the system response speed of speaker recognition would be raised.

Build Recognition Model on VQ[5]

In recent 20 years, Vector Quantification is a well-developed effective compression coding technology. Considering its great classification capacity, the statistical information of long time speech characteristic parameters would be quantified integrally to distinguish different speakers. Simultaneously, data would be compressed effectively , speech segmentation and time warping problem that hard to deal with would be avoided. VQ speaker identification system is shown in Fig.2.

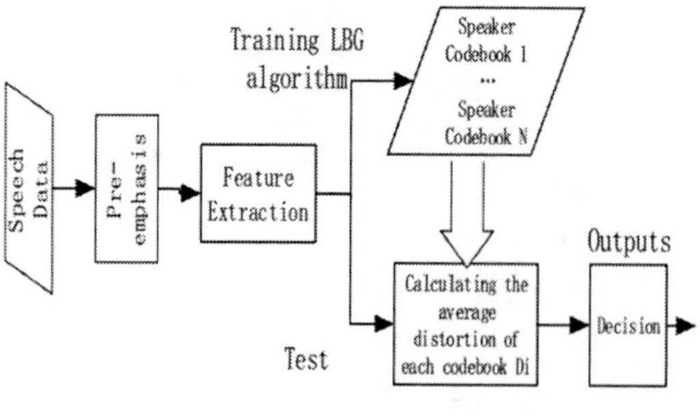

Fig.2 VQ speaker identification system

We regard the speaker personality characteristic as a information source, and model it using VQ technology. We use long time LPCCEP parameters as the information source characteristic of speakers and built N different code books corresponding to N speakers. Owing to the massive data and the uniform distribution of training vectors in feature space, the original code book is selected randomly to save time. Iteration steps are as follows:

1. Initialization: The number of given code word of the code book is L, distortion threshold $\varepsilon>0$, randomly select original code book $Y_L^{(0)}$, $X_k(k=0,1,...,K)$ is the training sequence, at first iterations $n=0$, average distortion $D_{-1}=\infty$;
2. According to the given code books $Y_L\{Y_i, i=1,2,...,L\}$, we use training sequences $X_k \in S_i$ make $d(X_k,Y_i)<d(X_k,Y)$ $(Y\epsilon Y_L)$, to get the optimum boundary $S_i^{(n)}$, then calculate the average distortion of the training sequence in this optimum zone boundary:

$$D^{(n)} = \frac{1}{K}\sum_{k=0}^{K-1}\min_{Y\in Y_M} d(X_k,Y)$$

(2)

3. If $\frac{D^{(n-1)}-D^{(n)}}{D^{(n)}} \le \varepsilon$, Y_L is the final code book and $S_i^{(n)}$ is the designed zone boundary, and the iteration is end. Otherwise, continue.
4. Calculating these L zones' centers of form, and using these L centers to make up the new code book $Y_L^{(n+1)}$ of the $(n+1)$ time iteration, $(n+1)$ replace n, and turn to step 2.

Considering the different contributions of each component of LPCCEP parameters, we should develop the average distortion measurement formula, that is, to weigh euclidean distance,

$$d(X,Y)=\sum_{i=1}^{p}\rho(i)(x_i-y_i)^2$$

(3)

Where p=feature dimension, $\rho^2(i)=1/\sigma^2(i)$, and $\sigma(i)$ is the variance mean of each dimension component. The improved distance formula reflects the recognition capacity of each component adequately, thus raises the system response speed. Test speech feature vector sequence $X_k(k=1,2,...,K)$ and all speakers' code books should calculate their average quantification distortion,

$$D_i = \frac{1}{K}\sum_{k=1}^{K}\min_{1\le j\le L} d(X_k,Y_j^i)$$

(4)

Then choose the smallest one or some among D_i as the result and make further judgment.

Multi-features and Multi-stages Decision Method

We adopt LPC cepstrum based on VQ to make the first decision in the identification process, then use Multi-stages decision strategy combining pitch characteristic to make further decision. In the first

DOI: 10.1051/matecconf/20166304044

decision, LPC cepstrum vectors of the test speech pick up M speakers which have the minimum distortions, $M<20\%*N$ generally. Then we propose the pitch distortion measurement formula,

$$D_i^P = \alpha_1|P - P_i| + \alpha_2|\sigma - \sigma_i| \quad 0 < \alpha_1, \alpha_2 < 1$$

(5)

Where P, σ are the pitch mean and variance of the test sample, $\alpha_1 \alpha_2$ value 1 usually. For M speakers' templates, combining this two features, the combined measurement formula is defined as the final identification result.

$$D_i^* = \beta * D_i + (1-\beta) * D_i^P \quad 0 < \beta < 1$$

(6)

If the distortion with the optimum code book is less than the specified threshold, then the corresponding speaker is the one to identify, otherwise the test speech is the sample outside the set.

Identification Experiments and Results

The original speech data was collected by telephone recording card or multimedia sound card, and the speakers consisted of 60 persons speaking different dialects, including 36 males and 24 females. All the samples are 5 to 30 seconds long, sampled in 4kHz with 16 bit resolution and the telephone speech bandwidth was within 4 kHz generally. The digital filter transmission function is $H(z)=1-0.96z^{-1}$ for pre-emphasis, then the speech was processed by Hamming window, and the window width is 30 ms. The pre-processing adopted adaptive multi threshold zero-crossing rate and short-time energy to detect steady voiced sound, to calculate LPC coefficient one frame by one frame, then to acquire pitch period and each order LPCCEP coefficient (p=16) as feature vectors. Using above multi-stages decision to test, the results were shown in Fig.3. The experiment results indicated that, test length, code book size L and false recognition rate had an obvious dependence relationship, that is, the system recognition rate is higher when test speech is longer or L is larger. When L values 64 and the speech length is 10 seconds, the recognition rate is above 95%.

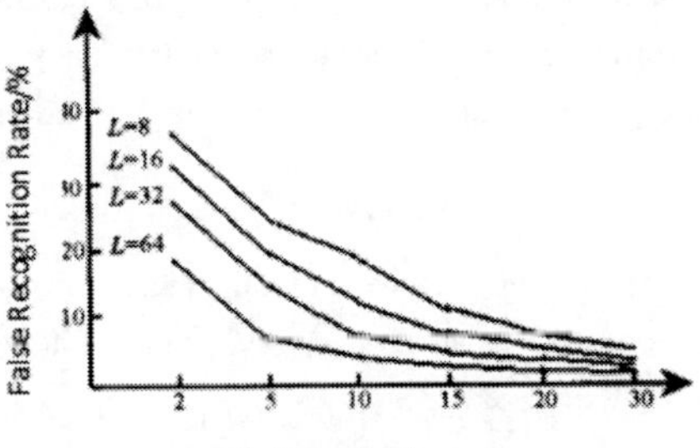

Fig.3 the results of test

Conclusion and Suggestion

In the speaker identification system based on VQ, adopting the pitch and cepstrum combined method, using weighting distortion measurement considering unbalance contribution of different components, therefore the founded voice print identification system is more complete than the cepstrum only system. In addition, this paper puts forward the decision strategy of voice tract characteristic identification firstly and then pitch feature recognition lastly, which is effective to avoid imitative testers to some extent, and it is also an important measure to raise the system robustness.

This method will be improved in the following aspect, when using LBG algorithm to cluster, the splitting method is time-consuming, and the random method is not precise enough. The system would be more practical if a precise and time-saving cluster algorithm could be used.

Acknowledgment

This paper, the research was sponsored by the National Nature Science Foundation of China (Nos.11474185, 61271453) and the Fundamental Cross-decipline Research Foundation of Shandong University(No. 2015JC029).

References

[1] J.P.Campbell. JR. Speaker Recognition A Tutorial, Proceedings of the IEEE, VOL.85, N0.9, September 1997.

[2] J.D.Woodward, Biometrics: Privacy's Foe or Privacy's Friend? Proceedings of IEEE, VOL.85, No.9, September 1997.

[3] ZHAO Li, JIANG Chun-hui, A Study on Emotional Feature Analysis and Recognition in Speech, ACTA ELECTRONICA SINICA, VOL.32, No.4, April 2004.

[4] Xingjun Yang, Huisheng Chi. Speech signal processing[M], Electronic Industry Press, Beijing, 1995.

[5] GISH H, SCHMIDTM. Text-independent speaker identification [J]. IEEE Signal Processing Magazine,1994,11(4):18-32.

MATEC Web of Conferences **63**, 04045 (2016) DOI: 10.1051/matecconf/20166304045
MMME 2016

The generation algorithm of arbitrary polygon animation based on dynamic correction

Ya Wei Hou[a], Yong Gang Li, Jin Biao Zhou, Jian Wei He and Yu Xin Zhang
China Satellite Maritime Tracking and Controlling Department, Jiangyin 214431, Jiangsu, China

Abstract. This paper, based on the key-frame polygon sequence, proposes a method that makes use of dynamic correction to develop continuous animation. Firstly we use quadratic Bezier curve to interpolate the corresponding sides vector of polygon sequence consecutive frame and realize the continuity of animation sequences. And then, according to Bezier curve characteristic, we conduct dynamic regulation to interpolation parameters and implement the changing smoothness. Meanwhile, we take use of Lagrange Multiplier Method to correct the polygon and close it. Finally, we provide the concrete algorithm flow and present numerical experiment results. The experiment results show that the algorithm acquires excellent effect.

Key words: Animation, Polygon,Vectors,Correction

Introduction

As the base of computer animation techniques, two-dimensional animation is a research highlights of current computer science. Two-dimensional animation can be implemented through boundary polygon deformation, so the main mathematical model of two-dimensional animation is polygon deformation techniques. That is to say, in some method to design medium growing change process, the initial object continuously changes to target object on the vision. It is also called 2D-Shape blending. Polygon deformation techniques is widely applied in many fields, such as computer graphics, virtual reality, industry simulation, visualization in scientific computing, bio-medical engineering and so on. And it has already been the important part of computer animation techniques. As an emerging research field, it is with important academic value and research value.

Fundamental assumption and Define

The deformation process of polygon, without considering the effect of adjacent cycle, need to meet following conditions.

(a) the track of feature points of polynomial(generally the peaks of polynomial) in change period is continuous, smooth and fairing;

(b) the length of polynomial sides is monotonous in change period.

In the process to implement continuous animation, in order to make the polygon continuously transit in different change periods, we need to weaken condition (b) according to requirements, so as to achieve the continuity of deformation track during different periods.

This paper, for the sake of convenience in researching, we give instructions for the following symbols. $A=[A_0,A_1,\cdots\cdots,A_n,\cdots\cdots]$, $T=[T_0,T_1,\cdots\cdots,T_n,\cdots\cdots]$, with

[a] Corresponding author: yaweizq@163.com

$A_k = \{a_{k,1}, a_{k,2}, \cdots\cdots, a_{k,n}\}$. $A_{k,i}$ means the peak i of the state polygon k, and record as $\Delta T_k := T_k - T_{k-1}$. $B_k(t)$ is continuous deformation sequence of $A_{k-1} \to A_k$. That is to say, the polygon continuous change sequence $(k = 1, 2, 3 \cdots)$ in interval $[T_{k-1}, T_k]$, in the form of edge vector is $B_k(t) = \{b_{k,1}(t), b_{k,2}(t), \cdots, b_{k,n}(t)\}$.

Second interpolation method of edge-vector

Define second interpolation vectors of $a_{k-1,s}$ and $a_{k,s}$

$$b_{k,s}(t) = (1-t)^2 a_{k-1,s} + 2t(1-t) p_{k,s} + t^2 a_{k,s} \quad t \in [0,1] \quad (1)$$

Among them:

$$p_{k,s} = (1 + \lambda_{k,s})(a_{k-1,s} + \theta_{k,s} a_{k,s}) / 2$$

$\lambda_{k,s} \in (-1, \infty), \theta_{k,s} \in (0, \infty)$

t is the mapping of $[T_{k-1}, T_k]$ in unit interval $[0,1]$. As Fig 1, the physical significance of second interpolation vector defined by (1) is that vector's terminal point track is quadratic Bezier curve of $p_{k,s} - a_{k-1,s}$ and $a_{k,s} - p_{k,s}$ as feature vectors. $p_{k,s} - a_{k-1,s}$ and $a_{k,s} - p_{k,s}$ form feature polygon of interpolation curve.

Vector $b_{k,s}(t)$ keeps good continuity and smooth in deformation period. But in initial deformation period, it is necessary to consider the monotony of vector length. For first-order continuously differentiable curve $b_{k,s}(t)$, it has following theorem:

Theorem 1: the sufficient and necessary condition for $|b_{k,s}(t)|$ as monotonous on [0, 1] is that $(b'_{k,s}(t), b_{k,s}(t))$ on [0,1] keeps same symbole.

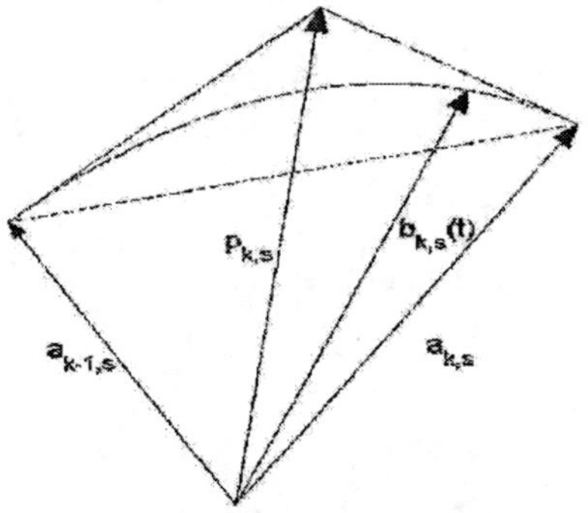

Figure 1 quadratic Bezier interpolation of edge-vector

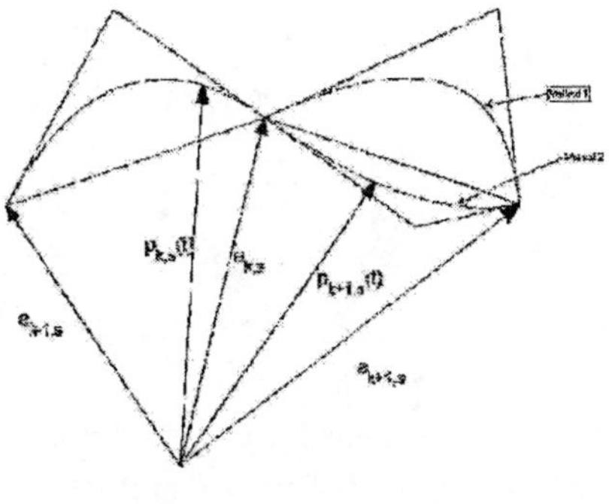

Figure 2 quadratic Bezier interpolation based on dynamic correction

Dynamic correction of interpolation parameter

The quadratic Bezier curve defined by the method mentioned above, keeps good characteristics. It only meet the continuation in adjacent periods,while the smooth and fairing cannot be realized. Therefor,in order to hold the polygon deformation smooth during animation restructuring procedure, every edge-vector need meet first-order continuity during whole deformation process.That is, $\frac{db_{k-1,s}(t)}{dT}|_{T=T_{k-1}} = \frac{db_{k,s}(t)}{dT}|_{T=T_{k-1}}$,according to Bezier curve characteristic, Bezier curve and its characteristic polynomial's first and last side respectively tangent to the initial and ending points,and the tangent vector is n times of edge-vector.That is:

$$\frac{2(a_{k-1,s} - p_{k-1})}{\Delta T_{k-1}} = \frac{2(p_k - a_{k-1,s})}{\Delta T_k} \quad (2)$$

It can get from above formula:

$$\lambda_{k,s} = \frac{\Delta T_{k-1}}{\Delta T_k}[2-(1+\lambda_{k-1,s})\theta_{k-1,s}]+1 ,$$

$$\theta_{k,s} = \frac{\Delta T_k(1+\lambda_{k-1,s})}{2\Delta T_{k-1} + \Delta T_k[2-(1+\lambda_{k-1,s})\theta_{k-1,s}]} .$$

The interpolation curve $\widehat{b_{k-1,s}(t)b_{k,s}(t)}$ got from above formula could meet deformation smooth as $T = T_k$. But because of randomness of time sequence, $p_{k,s}$ cannot be guaranteed to fall over the first quadrant of affine coordinate system built by $a_{k,s}, a_{k+1,s}$.So it is necessary to conduct limit to above formula:

$$\begin{cases} \lambda_{k,s} = \max(\frac{\Delta T_{k-1}}{\Delta T_k}[2-(1+\lambda_{k-1,s})\theta_{k-1,s}]+1,-1) \\ \theta_{k,s} = \max(\frac{\Delta T_k(1+\lambda_{k-1,s})}{2\Delta T_{k-1} + \Delta T_k[2-(1+\lambda_{k-1,s})\theta_{k-1,s}]},0) \end{cases} \quad (3)$$

As shown by Figure 3, the first method is the results by general second interpolation according to Guozhao Wang, and the second method is the results by correction on interpolation parameters according to this algorithm. By comparison, during polygon animation restructuring procedure,the method in this paper could make the transition much smooth and fairing in different periods.

Polygon closure correction

The polygon formed by above method cannot completely close.That is to say,it is not real sense of polygon. Therefore, we need to amendment the results. According to the characteristic of polygon,if polygon $B_k(t)$ meets the closed condition, it is necessary $\sum_{s=1}^{n} b_{k,s}(t) = 0$.

So, we could make correction on $b_{k,s}$ and add adjustment amount $\eta_{k,s}$ on it, and then as edge-vector of polygon $B_k(t)$, that is $b^*_{k,s} = b_{k,s} + \eta_{k,s}$

And then we could get:

$$\sum_{s=1}^{n}[b_{k,s}(t) + \eta_{k,s}] = 0 .$$

Define function

$$\varphi(\eta_{k,1}, \eta_{k,2}, \cdots, \eta_{k,n}) = \sum_{s=1}^{n}(b_{k,s}(t) + \eta_{k,s}) = 0 \quad (4)$$

Objective function is $f(\eta_{k,1}, \eta_{k,2}, \cdots, \eta_{k,n}) = \sum_{s=1}^{n} \frac{\eta^2_{k,s}}{|b_{k,s}(t)|^2}$ .It needs to confirm all $\eta_{k,s}$, and make f reach its minuteness under the constrain condition (4).Using Lagrange multiplicator method, take λ as multiplicator, and command $\Phi(\lambda, \eta_{k,1}, \eta_{k,2}, \cdots, \eta_{k,n}) = f + \lambda\varphi$.

From equation set

$$\begin{cases} \frac{\partial \Phi}{\partial \lambda} = 0 \\ \frac{\partial \Phi}{\partial \eta_{k,s}} = 0 \end{cases}$$

It can be got:

$$\eta_{k,s} = \frac{b^2_{k,s}(t)\sum_{i=1}^{n} b_{k,i}(t)}{\sum_{i=1}^{n} b^2_{k,i}(t)} \quad (5)$$

Therefore, the edge-vector of polygon on time t is

$$b^{*}_{k,s}(t)=b_{k,s}(t)+\frac{b^{2}_{k,s}(t)\sum_{i=1}^{n}b_{k,i}(t)}{\sum_{i=1}^{n}b^{2}_{k,i}(t)} 。 \qquad (6)$$

Thus we could calculate polygon $B_k(t)$.

Numerical experiment result

There are many experiments using the method mentioned above. As Figure 4, it is the deformation animation from Chinese character "凹"to a circular. The 1st,3rd and 8th frame is polygon key-frames of T=0,T=0.4,T=1.4. The other frames is polygon reconstructed by the method stated in this paper. We can see from the figure,for the implement of polygon sequence animation with time attributes, polygon gradual changing method based on dynamic correction can achieve the effect of smooth, fairing and natural transition.

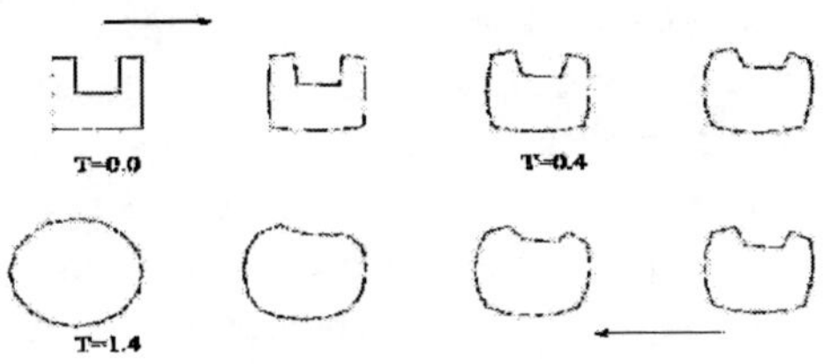

Figure 3 Deforming animation from Chinese character '凹' to circular

Conclusion

This paper proposes one method of polygon animation implement based on dynamic correction. Firstly, vectorize the metamorphosis polygon, and then carry out second interpolation to corresponding edge-vector, meanwhile in the interpolation process, keep first-order continuity of edge-vector and peak-point track, and to the greatest extent maintain the monotony of side.And then, by means of dynamic correction for interpolation parameters, implement continuous transition and fairing in different deformation periods.Finally,correct the deformation edge-vector which uses least square method to close the polygon. Numerical experiment results indicate that the method proposed in this paper implement the smooth and fairing of animation, and the effect is excellent.

References

[1] Zhao-guo Wang, Bao-gang Bai, Ruo-feng Tong, A vector blending approach for computer animation .J. Chinese Journal of Computers. 12 (1996). 881-886.
[2] LEE T Y, HUANG Pohua. Fast and intuitive metamorphosis of 3D polyhedral models using SMCC mesh merging scheme .J. IEEE Transition Visualization and Computer Graphics, 2003, 9 (1) : 85-98.
[3] The method of computer aided geometric design. Xi'an , Northwestern poly-technical university press,1986.
[4]Ya-Qing Xu, Bao-gang Bai, UNRBS as a new application of animation vector mixed method. J. Journal of Xinjiang Petroleum Institute 12(2000):69-74
[5] Chang-xu Dou ,Yu-mei Wang , Quadratic Interpolation Deformation Algorithm of Polygon Central Point Vector. J. Computer Engineering . 2010.36(16):189-191

MATEC Web of Conferences **63**, 04046 (2016) DOI: 10.1051/matecconf/20166304046
MMME 2016

Application of Multi-level Recursive Method for Ship-Sway prediction

Shang-cheng Hu, Yong-gang Li, Xiang-ming Li, Sheng-ping Li and Yi Wang

China Satellite Maritime Tracking and Controlling Department, Jiangyin 214431,China

Abstract. The ship-sway is a non-stationary time series while ship sailing in the sea , when using the traditional orthogonal polynomial fitting,index filtering or Kalman filtering to predict ship-sway, the prediction error may be large, because the model parameters are fixed and cannot be adjusted in real time according to the measured data. Multi-level recursive method treats the dynamic system as a time-varying parameters of the system, and can be more in line with the objective reality of rocking the ship. After analyzing the characteristics of the ship-sway, the predict model established by multi-level recursive has been built and test results showed that this model can improve the prediction accuracy of the ship-sway data, and has some practical value in the prediction of the ship-sway.

Key words: Multi-level recursive,Ship-sway prediction,EViews,Stationary time series.

1 Introduction

In the tasks of maritime spacecraft tracking and control, the prediction of the measure base directly influences the data accuracy of radar while tracking the target in the way of following number-leading calculated by computer. At present, the ship-sway is the most important factor in the measure base, and the common ship-sway prediction methods include statistics-predict, Kalman filter-predict, orthogonal polynomial fitting-predict and maximum likelihood estimation-predict [1]. In reality application, some weakness of these methods is found, just like needing too much prior condition, too difficult to build the accurate ship-sway model, and too big predict deviation etc. Using multi-level recursive method, we can treat the ship-sway data as a random dynamic time varying system, further, separate the system predict to two steps, predict the dynamic time varying parameters and system status based on ahead parameters. By this way, the dynamic parameters will update real-time while the ship-sway make a sudden change, and the system prediction can have a good adaptability.

2 Principle and Procedure of Multi-level Recursive

2.1 Principle of Multi-level Recursive

The Multi-level Recursive treats a dynamic system as a stationary time series composed by one or more dimensional. By analysis the external characteristic of the system, establish the connection between inputs and outputs. Under the general circumstance, all systems can be simplified to one model includes one or more one-dimensional model [2]. Here, we suppose the input of a stationary time series is:

$y(0)$、$y(1)$、$y(2)\cdots\cdots y(n)$

Inside the above expression, $y(k)$ is the observed value in the time of k, so the one-step prediction is:

$$\hat{y}(k+1|k)=a_1(k)y(k-1)+\cdots\cdots+a_n(k)y(k-n)+ b_0(k)u(k)+b_1(k)u(k-1)+\cdots\cdots+b_m(k)u(k-m)+e_k \quad \cdots\cdots(1)$$

Inside the above expression, $u(k)$ is weight coefficient, $a_1(k)\cdots\cdots a_n(k),b_0(k)\cdots\cdots b_m(k)$ are dynamic time varying parameters, n、m are the model points, $\hat{y}(k+1|k)$ is the one-step prediction.

Suppose:

$$\theta(k)=\left(a_1(k)\cdots\cdots a_n(k),b_0(k)\cdots\cdots b_m(k)\right)^T \quad \cdots\cdots(2)$$

And then $\{\theta(k)\}$ can be treated as a new stationary time series, just like predict $\hat{y}(k+1|k)$ above, we can have the second expression for $\{\theta(k)\}$:

$$\hat{\theta}(k+1|k)=c_0(k)\theta(k)+c_1(k)\theta(k-1)+\cdots\cdots+c_r(k)\theta(k-r) \quad \cdots\cdots(3)$$

Inside the above expression, c_0、$c_1\cdots\cdots c_r$ are dynamic time varying parameters, If the parameters are time varying, we can establish third, fourth or more further expression[3], else the whole expressions make up the predict progress.

2.2 Parameters Update

According to the expression (1) (2) and reference documentation [4], the common expression of a predict progress is:

$$y(k)=f(y_{k-1},U_k,\theta,k)+v(k) \quad \cdots\cdots(4)$$

In the real-time prediction of the ship-sway by the multi-level recursive method, the factor U_k can be ignored, only analyses the ship-sway data series. So, the estimate of parameters θ is:

$$\hat{\theta}(k)=\hat{\theta}(k-1)+\frac{\delta}{\|\varphi(k)\|^2}\varphi(k)\left[y(k)-\varphi^T(k)\hat{\theta}(k-1)\right] \quad \cdots\cdots(5)$$

Inside the above expression, $\varphi(k)=(y(k-1)$、$y(k-2)\cdots\cdots y(k-n))^T$, $\hat{\theta}(k)=\left(a_1(k)\cdots\cdots a_n(k)\right)^T$ are estimated values, δ is adjust factor.

According to reference documentation [5], generally speaking, the different initial values of the parameters will not influence the predict results. But in real application, the most suitable initial values can shorten the predict time. So, least square method is used to calculate the initial values of the parameters. With the initial values and the expression (5), the parameters series can be estimated and updated:

$$\hat{\theta}(1)、\hat{\theta}(2)、\hat{\theta}(3)\cdots\cdots\hat{\theta}(N)$$

Based on the known series $\hat{\theta}$, we can predict $\theta(N+l)$ by methods include AR model, circle variable model, or mean value approximation model. Combined with expressions (1) (4), the l steps prediction of y can be calculated.

2.3 Points of Multi-level Recursive Method

According to reference documentations [6][7][8], whatever the model points selected, the predict precision will satisfy requirement by parameters stochastic makeup, But just like the model parameters initial values, the most suitable model points can shorten the predict time. So, Final Prediction Error criterion is used to define the most suitable model points:

$$FPE(k)=\left(1+\frac{k+1}{n}\right)\times\left(1-\frac{k+1}{n}\right)^{-1}\delta_e^2 \quad \cdots\cdots(6)$$

Inside the above expression, k is number of the model independent parameters, n is the number of the samples, δ_e^2 is the error mean square deviation of the model.

3 Multi-level Recursive Model for Ship-sway

Using the method from reference documentation [9], we can simulate 1000 points ship-sway samples, after difference of first order, the new series is a stationary time series, and multi-level recursive model is built for predicting it. Here Eviews soft is used to get the model points and parameters initial values. Select Final Prediction Error criterion to have a unit-root test for the series and set the biggest lag order is 21 [10], the result shows figure 1.

Though the orders D(-1)、D(-14)、D(-15) have no conspicuousness, considering the randomness, the model points is defined 16. Using least squares method to estimate initial values of the parameters, the result shows figure 2.

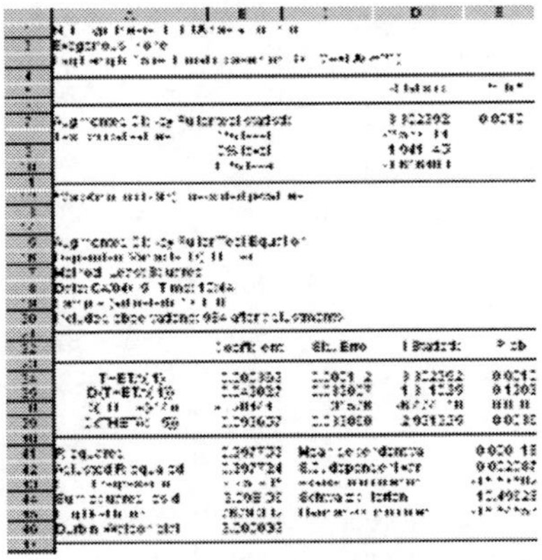

Fig1. Unit-Root Test Result

Dependent Variable: THETA
Method: Least Squares
Date: 04/04/15 Time: 15:04
Sample (adjusted): 17 1000
Included observations: 984 after adjustments
Convergence achieved after 8 iterations

	Coefficient	Std. Error	t-Statistic	Prob.
C	0.004155	0.008409	0.494117	0.6213
AR(1)	1.041339	0.032088	32.45168	0.0000
AR(2)	0.432263	0.046276	9.341001	0.0000
AR(3)	-0.226145	0.047922	-4.719032	0.0000
AR(4)	-0.019044	0.048371	-0.393713	0.6939
AR(5)	-0.014565	0.048316	-0.301459	0.7631
AR(6)	-0.377030	0.047940	-7.864660	0.0000
AR(7)	0.020329	0.049738	0.408716	0.6828
AR(8)	0.274969	0.049732	5.529053	0.0000
AR(9)	0.083236	0.049748	1.673152	0.0946
AR(10)	-0.044963	0.049897	-0.901129	0.3677
AR(11)	-0.200893	0.048403	-4.150433	0.0000
AR(12)	-0.073613	0.048839	-1.507273	0.1321
AR(13)	0.099657	0.048850	2.040076	0.0416
AR(14)	-0.200003	0.048357	-4.135992	0.0000
AR(15)	0.110696	0.046654	2.372685	0.0179
AR(16)	0.093391	0.032077	2.911458	0.0037

R-squared	0.999999	Mean dependent var	0.002779
Adjusted R-squared	0.999999	S.D. dependent var	0.066330
S.E. of regression	9.96E-05	Akaike info criterion	-15.57403
Sum squared resid	9.59E-06	Schwarz criterion	-15.48952
Log likelihood	7679.423	Hannan-Quinn criter.	-15.54189
F-statistic	46172992	Durbin-Watson stat	2.004930
Prob(F-statistic)	0.000000		

Fig2.Based on least squares method for estimating initial time-varying parameters

Thus, the initial values of the parameters $a(0) \sim a(16)$ are defined:

THETA=0.00415521684124+[AR(1)=1.04133890593,AR(2)=0.432262930578,AR(3)=-0.22614528621,AR(4)=-0.0190443017317,AR(5)=-0.0145654379696,AR(6)=-0.37703029319,AR(7)=0.0203289046708,AR(8)=0.274968931422,AR(9)=0.0832362795316,AR(10)=-0.0449634962796,AR(11)=-0.200892928773,AR(12)=-0.0736132949649,AR(13)=0.0996569990796,AR(14)=-0.200003294775,AR(15)=0.110696302707,AR(16)=0.0933908674749]。

And meanwhile, the predict model expression is:

$$\hat{y}(k+1|k) = a_{16}(k)y(k-1) + \cdots\cdots + a_1(k)y(k-16) + a_0(k) \qquad (7)$$

According to expression (5), estimate and update parameters a_i expressions are:

$$\begin{cases} \hat{a}_0(k) = \hat{a}_0(k-1) \\ \hat{a}_1(k) = \hat{a}_1(k-1) + \dfrac{1}{R(k)} y(k-16)V(k) \\ \cdots \\ \hat{a}_{16}(k) = \hat{a}_{16}(k-1) + \dfrac{1}{R(k)} y(k-1)V(k) \end{cases} \qquad (8)$$

Above expressions: $R(k) = \sum_{i=1}^{n} y^2(k-i)$ 、 $V(k) = y(k) - \sum_{i=1}^{n} \hat{a}_i(k-i)y(k-i)$

Predict Steps:

Step1: According to the expressions (8) and the initial values of $a(0) \sim a(16)$, using the before 30 points samples, each parameter will form a series;

Step2: For each parameter series, establish a AR(3) model, and the one step prediction of the parameter can be get;

MATEC Web of Conferences **63**, 04046 (2016) DOI: 10.1051/matecconf/20166304046
MMME 2016

Step3: With the predicted result of the parameters, the progress for predicting the ship-sway has been shown by the expression (7);

Step4: Compute the deviation between the samples and the prediction with the same time,figure them and the result shows figure 3.

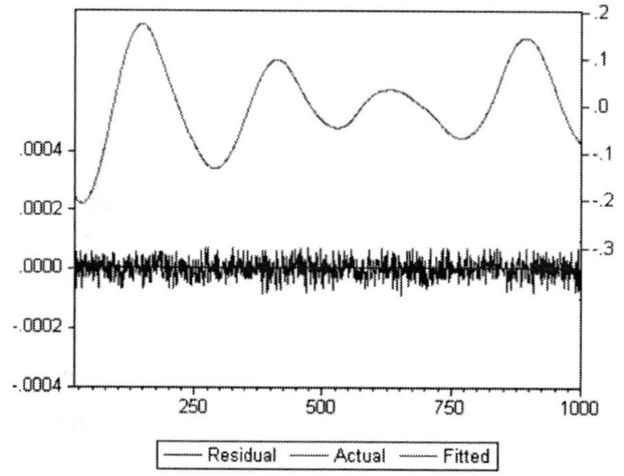

Fig3.Dynamic curve fitting step prediction and deviation results

Compute the maximum,mean and variance of the deviation,and compare with the results from reference documentation [10], the result shows table1.

Table1 Forecast accuracy deviation comparison

Method/Deviation	Maximum/Arcseconds	Mean/Arcseconds	Variance/Arcseconds
Index Filter	13. 899	3. 753	5. 512
Time series	6. 376	1. 662	2. 443
Multi-level Recursive	4. 352	1. 355	2. 459

For verify when varying parameters dynamically adjusting benefits, when the ship rolling simulation data, select the data points between 400-500 increases a trend item (endpoint regression), using the same method as described above and spreadsheet modeling, comparative prediction deviation maximum, mean and variance, the results shown in table 2.

Table2 Forecast accuracy deviation transition sequence comparison

Method/Deviation	Maximum/Arcseconds	Mean/Arcseconds	Variance/Arcseconds
Index Filter	24. 970	6. 501	10. 111
Time series	19. 017	4. 907	5. 337
Multi-level Recursive	12. 549	2. 014	3. 918

From the spreadsheet results, when the status of the observed series change suddenly,the multi-level recursive method advantage is more obvious, with a faster adaptability and higher prediction accuracy。

4 Conclusion

Compared with traditional fixed parameter prediction method, multi-level recursive method forecast by layer by layer when varying parameters, dynamic adjustment of the estimated parameters in the prediction process, the study can be closer to the actual situation, a more sensitive and state recognition ability to adjust the prediction accuracy has also been greatly improved, with high practical value and popularization meaning.

References

[1]Zhao Xi-ren.Status of Ship-Modeling Sport Very Short-term Forecasts.Ship Engineering,J.

2002,3

[2] Wang Yun-ge. Multi-level Recursive Forecasting Method and Its Improvement,Journal of Hebei Normal University,J.1998,22(4)

[3] Han Zhi-gang. The Progress of Theory and Application of Multi-level Recursive Method,

Control and Decision,J.2001,16(2)

[4] Han Zhi-gang. A New Method of Dynamic Prediction System,Acta Automatica Sinica,J.

1983,3

[5] Han Zhi-gang. Parameter Estimation Initial for Multi-level Recursive Method, Control and Decision,J.1986,1(1)

[6] Luo Qin-bin. The Application of Multi-progressive Method to weather Forecast,Harbin University Institute,J. 2003,10

[7] Wang Fu. Research and Design for Energy Management Information System Based on Multi-level Recursive and Grey Prediction Theory, Kunming University of Science and Technology,J. 2010,3

[8] Chen Yu-xiang. Prediction and Applications,Machinery Industry Press,BeiJing,1985, pp.98-135

[9] Liu Peng. Research on MATLAB based ocean wave and ship roll simulation models, ShanDong Science,J.2012,25(6)

[10] Zhang Zhong-hua. Boat Pose Data Processing Method for Space-tracking ship,National Defend Industry Press,BeiJing,2009, pp.219-293

The separability analysis of pitch parameters in speaker recognition

Rong Rong He[1, a], Lan Tian[1], Fa Jiang Ma[1]

[1]*School of Information sci. and Eng., Shandong Univ., Jinan 250100, China*

Abstract. Aseparability research of prosodic parameters in speaker recognition is studied.The space distribution of the prosodic parameters is drawn using the mean and variance of the speaker pitch parameters,through which the separability of the speaker is obvious.We do a speaker identification experiment with a small corpus,and the experiment results show that speaker identification based on the prosodic parameters is possible. The first-level decision using the prosodic parameters can reduce the candidate corpus effectively,demonstrating the practicability of the pitch parameters.

Keywords: speaker recognition; pitch parameters; feature space

Introduction

Speaker recognition is a branch of the traditional speech recognition and it has wide application prospects in telephone financial transactions, network security, multimedia and other fields. According to their final tasks , it can be divided into two kinds: speaker identification and speaker verification two kinds. According to the testing speech inputted, it can be divided into text-related and unrelated categories. The correct rate and speed of the speaker recognition system is an important indicator to evaluate a system, but there're contradictions between accuracy and efficiency in the general recognition system.

At present, most of the speaker recognition models of the system are based on the channel characteristic parameters with little regard to the sound source characteristics. However, the sound source characteristics (such as pitch) is an important information source of the physical characteristics of the speaker and their speaking habits, which should be given adequate attention. This paper presents a method for preliminary speaker identification which takes the features of pitch into accountWe plotted the diagram of the speaker's pitch feature space distribution according to the statistical properties of the mean and variance. It can be observed that the pitch figure region doesn't coincide for different speakers, and that pitch parameters in the feature space are separable, which provides a basis for the design of speaker recognition based on the pitch feature. he As we used this method for primary judgment it can effectively narrow the range of candidate set[1], if jointly use the typical speaker recognition method for careful judgment, we can ensure that the system can have good performance in accuracy and efficiency.

Analysis of the characteristic parameters of pitch

The extraction of the feature parameters is crucial for speaker recognition[2]. Currently, speaker recognition rely mainly on the acoustic characteristics of the lower level, among which the relatively effective features[3] are: short-term speech energy, pitch period, short-term spectrum, BPFG voice characteristics, the linear prediction coefficients, formant frequencies and bandwidths, the LPC cepstrums, etc., and some linear regression coefficients that reflect the dynamic changes of those characteristics. The fundamental frequency parameters (also known as pitch parameters) are the most important parameters of the speech signal, which describes the important features of the speech excitation source, and has been widely used in many areas. People's fundamental frequency range is generally 60 ~ 500 Hz, men mainly in the range of 60 ~ 220 Hz, while women and children in the range of 150 ~ 450Hz. Different people have different pitch distributions, in spite of the disadvantages of being easily imitated, it can be used as an auxiliary judgment in some specific occasions (such as studio monitors) for speaker recognition.

Pitch features are always mixed in the track characteristics and the speech signals of the speaker instead of existing independently in the actual speech signals. To accurately extract the signal, it is necessary to adopt appropriate detection algorithm. Currently, there is a variety of pitch detection method. [4], such as the central clipping autocorrelation method, average magnitude difference function method, method of parallel processing, the wavelet method and cepstrum method, etc. In analysis, we using the cepstrum method, its principle is as follows. The speech sound s (n) is inspired by the glottis pulse source, responded by track e(n) and filtered by v (n), namely: s (n) = e (n) * v (n). If the three corresponding cepstrums are s (n), e (n) and v (n), then: s (n) = e (n) + v (n). Because of the separation of e (n) and v (n) in the cepstrum signal, specifically the pitch information and track information are distributed in different time section, so simply adopting the cepstrum filtering method [5]can separate the glottis excitation signal e (n), and then get its circle value.

In the cepstrum analysis, speech widowing is also of great importance. We choose gradual widows as for the

[a] Corresponding author: herongrong1993@163.com

widows function, Hamming window is the usual choice. To predict the pitch cycle from the high point in the cepstrum waveform, in other words, to estimate the position of the peak in it, we can get corresponding incentive source pitch position, which is pitch cycle.

In actual pitch detection, going through a filter is necessary to filter caused by the corresponding high fluency of the sampling theorem and the interference 50 Hz power fluency. After framing widowing the pitch data, the endpoint detection is conducted to get rid of the silence segments or surd segments. Then we use cepstrum method to calculate the pitch cycle. Because the channel response may produce harmonic interference, such as 1, 2, 1, 3, 2, 3 times frequency errors, so generally pitch trajectory smooth processing are carried out to remove the "singularity" pitch trajectory. Smoothing process consists of 3 kinds, namely, median smoothing and linear smoothing and combination smoothing. Median smoothing can effectively remove the small amounts of the singular points and at the same time, do not destroy the staged transformation between the two pitch tracks in the surrounding area of pitch cycle. Although linear smoothing can improve the effect of smooth, meanwhile it will lead to staged fuzzy between two smooth sections. So these methods can be chosen according to their effects. Figure 1 is the design sketch of smoothing process in the pitch analysis.

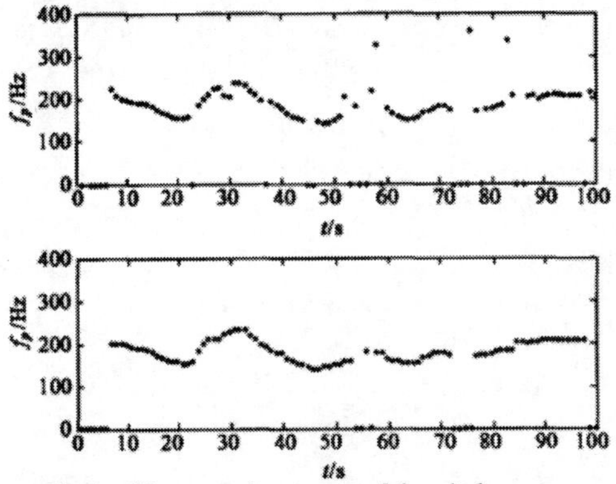

Fig.1 The smooth process of the pitch contour

Speaker pitch feature space analysis

We established a group of 11-persons voice library for analysis, six men and five women, the training voice and the testing voice of which are the natural speech flow which are not limited in the content. The speech flow is collected and recorded in the clean speech environment (the mute condition at laboratory) by telephone voice card, which is 120 seconds in length. The first 60 s are set for training, and the rest for test. The sampling frequency is 11 kHz, and it's in 16-bit-quantization storage.

According to the method above, we obtained the pitch period for each frame. After that, we smooth the track, and then calculated their long-time average, deriving the average of the pitch period.

Due to the speaker's individual physiological features and habits, the range of the pitch change is also different. In order to fully reflect the characteristics of the distribution of different speakers' pitch period, we consider adding another statistical property – the variance of pitch period, therefore the mean and variance constitutes the pitch feature space. Figure 2(a) is a collection of the mean and variance of pitch feature space distribution (when the window length is 40 ms, and the frame shift is 20ms) of the 11-person group; Figure 2(b) is characteristic spatial distribution when the frame-shift is not considered.When drawing the pitch characteristic spatial distribution diagrams, each speaker's voice materials are divided into six segments and the mean and variance of each segment are calculated respectively. Then we can obtain the corresponding six points in the feature space, and connect them into a block area with the largest square, which corresponds to the general distribution of its pitch.

It can be seen from the figure that the male speakers are distributed at the left bottom of the basic characteristics of space, while female speakers are distributed on the right of the feature space. This phenomenon is because the male voice tone is low with a relatively small range, while female voice is higher with a larger voice range. Theoretically, speaker identification can be implemented by mapping the mean and variance of the speaker's voice segments into the distribution diagram.

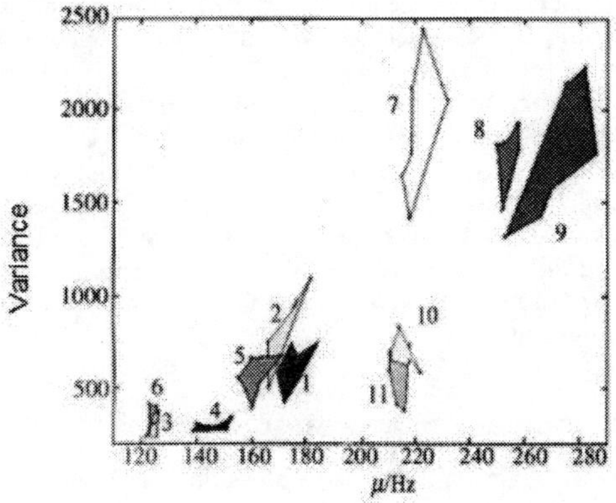

(a)

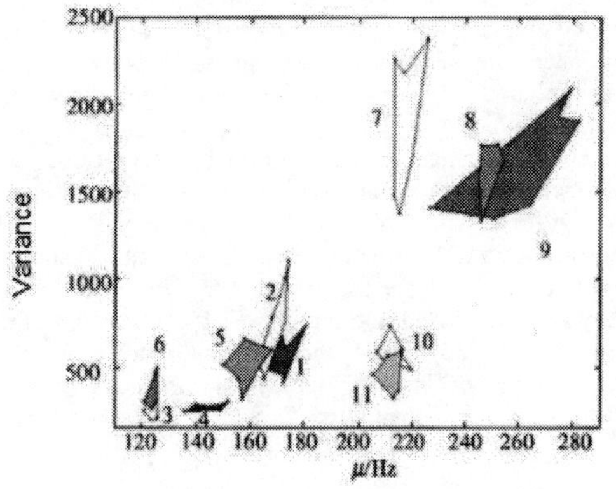

(b)

Fig.2 The space distribution of the pitch with the frame shift(a) and without the frame shift(b)

The experiment on identifying the speaker

Based on the foregoing analysis, We use the characteristic of pitch in the experiment on identifying the speaker from the text in set. The length of time of training words is assumed as 60s and that of identifying words is assumed as 10s. The models for reference of the speaker are the mean and variance in pitch period. Given that the vnit for measures of pitch period and variance is different and the distance measure can not be used directly. We put them together into the range from 0 to 1, and then we do distance measure. The way to match is to adopt the simple model, and contrasting the parameter of the speakers to identify to all the speakers'. The speaker who have minimal distance is the likely one. The outcome of experiment suggest that the percentage of identifying correctly is up to 78%.

Certainly, the accuracy of identifying speakers depend on the character of pitch is not high. But because the draw of character of sipn can be tested by specific hardware and is not relevant to common track character parametric modeling identification, and can not replace the contribution stage of system. When use speaker to identify the telephone channel, the parameter has strong anti-noiseness and robustness because of its low occupancy frequency range. We can do anticipation judgement before in the fixed text and variable text's identifying speaker. For a speaker identification system of 60 people, we adopt a pitch anticipation strategy to test. We broaden preliminary gauge threshold and take 30%-close speaker enter into candidate set, and then we use method of VQ and HMM to do the final judgment of speaker. On the condition that the identification rate of system not reduce, it can largely decrease the range of candidate, and the efficiency of identification are increased up to over 60%, the reliability of the system is also improved.

Conclusions and suggestion

By analysing the space distribution of the character of speaker's pitch and the experiment of identifying actual speaker, this paper confirms that the application value of character of pitch in experiment of identifying speaker, and also the applicating methods are given. The methods are not perfect. Because the character of pitch of speaker's character of space distribution is likely to influenced by the emotion, body condition and applicating environment. This pretreatment method is not fit for a variety of reasons not to cooperate with the speaker's words used specification(for example, imitate other's voice). For this, we should add other measures to avoid the unrecoverable error. On the occasion of appropriate control using occasions, to avoid error of initial judgment, sometime we must broaden the initial judgment's scope. As a result the efficiency of identification of system will drop. If the statistical features of pitch can consist more whole feature vector with other feature(like voice channel feature), the forecasting error can be avoid easily, and effect of recognition will be better.

Acknowledgement

In this paper, the research was sponsored by the Nature Science Foundation of China (Nos.11474185,61271453) and the Fundamental Cross-decipline Research Foundation of Shandong University(No.2015JC029).

References

[1] MOU Xiao-long,HU Qi-xiu,WU Wen-hu.Text-independent speaker identification system based on multiple strategies[J]. Journal of Tsinghua University(Sci&Tech),1997,37(3):16-19.

[2]Kong Y Y, Zeng F G. Temporal and spectral cues in Mandarin tone recognition. J. Acoust. Soc. Am., 2006; 120:2830-2840

[3]TIAN Lan,BAI Shu-zhong.Based onMulti-features and multistages decision speaker identification in telecommunication network[J].Journal of Shandong University(Engineering Science),2003,33(6):648-651.

[4]YI Ke-chu,TIANBin.Speech signal processing[M].Beijing:National Defense Industry Press.200.1-10.\

[5]ZHAO Li.Speech signal processing[M].Beijing:Mechanical Industry Press,2003.1-10.

Study of the bullwhip effect in Chinese coal supply chain under fuzzy environment

Xiao Hong Yun[1,2,a], Jin Suo Zhang[3]

[1]*School of management, Xi'an University of Science & Technology, Xi'an, 710054, China*

[2]*Research center for energy economic and management, Xi'an University of Science & Technology, Xi'an, 710054, China*

[3]*School of management, Yan'an University, Yan'an, 716000, China*

Abstract. Supply chain management is important for coal companies and organizations to improve their business and enhance competitiveness in the Chinese marketplace. The bullwhip effect problem of coal supply chain systems with all demands, lead times, and ordering quantities in an uncertain environment is addressed in this paper. To simulate the bullwhip effect, the Hong Fuzzy Time Series approach and Genetic Algorithm module are preferred as a superior forecasting model. And then a back propagation Neural Network module is added to defuzzify the output of the proposed model. So the bullwhip effect is calculated and analyzed here. The effectiveness and flexibility of proposed method is verified through simulation study.

Keywords: Supply chain management, bullwhip effect, coal production, uncertainty.

Introduction

China is now the largest energy consumer in the world. Providing sufficient coal supply to meet the requirements of a huge population with rising living standards will be a difficult task. The Chinese government is now facing severe challenges from a coal supply problem. With increasingly competitive coal market and high coal consumption, the coal supply chain has become critical and strategic to long-term development in China.

In coal supply chains, multiple independent mines are often connected to a shared rail network. A single rail operator often coordinates rail schedules with the mines, to schedule trains between the mines and the transport terminal in order to meet transport demand of each mine. A common phenomenon in the coal supply chain management that has been observed and justified is known as ''bullwhip effect''[1]. Lee, Padmanabhan, and Whang have identified five major causes for bullwhip effect in the supply chain. They are listed as demand forecasting, order batching, price fluctuations, supply shortages and non-zero lead-time[2]. According to Sterman, the bullwhip effect originates from the non-optimal solutions adopted by a supply chain participant while does not consider the supply system as a whole[3].

Among various causes of the bullwhip effect, forecasting methods are considered as one of the most important causes because the inventory system of a coal supply chain is directly affected by the forecasting method. Many scholars have tried to quantify and eliminate this effect. Graves quantifies the bullwhip effect for a supply chain in which the demand pattern follows an integrated moving average process[4]. Zhang investigates the impact of different forecasting methods on the bullwhip effect for a simple inventory system with a first-order autoregressive demand process[5]. Chen et al. show that increasing lead time enhances the bullwhip effect regardless of the forecasting methods employed[6]. The authors considered a supply chain as a business process which had to be redesigned assuming that the main cause of the bullwhip effect was the absence of coordination in the management of the supply chain. Following a systems analysis approach, Chen and Disney were able of reducing the bullwhip effect by controlling the order policy by using a proportional controller[7]. An improvement in the cost saving due to the reduction of order variance was obtained. Lee et al. measure the benefit of information sharing between a retailer and a manufacturer in a two-stage supply chain in which the demand pattern follows a first-order autoregressive process and an order-up-to policy with a minimum mean square error forecasting technique is used at the retailers[8].

As a matter of fact, for analyzing the bullwhip effect in the real world, few scholars consider it in a fuzzy environment. Some feasible methods have been proposed in supply chain management researches, and most of them focus on controlling the uncertainty causes of bullwhip effect. Pishvaee et al. proposed a robust optimization model and relative robust counterpart for handling the inherent uncertainty of input data in a closed-loop supply chain network design problem[9]. The widely used time series models for forecasting purpose especially ARIMA model is generally applicable to linear modeling but it hardly captures the non-linearity inherent in time series data.

In this paper, Fuzzy Theory is utilized as an analysis method to represent and interpret the uncertainty in the coal supply chain management. The authors try to study

[a] Corresponding author: yun_xiaohong@163.com

the effect of fuzziness, i.e., information uncertainty, on the bullwhip effect. Here, it is assumed that all demands, lead times and order quantities have fuzzy values, i.e., they are imprecise. Then, the bullwhip effect is simulated in such an environment. Firstly, to simulate the bullwhip effect in a fuzzy environment, the Hong Fuzzy Time Series approach and Genetic Algorithm module are preferred as a superior forecasting model. Next, a back propagation Neural Network module is added to defuzzify the output of the proposed model. In addition, the bullwhip effect is calculated and analyzed. Finally, we discussed the validation of the proposed solution approach.

Proposed Model and algorithm

Fuzzy time series model

Here we explains the time variant fuzzy time series model by Hong[10, 11]. Let ξ be the class of all fuzzy quantities defined on [0,∞] which are continuous non-increasing with the strict maximum 1 in the point 0, where its members are called shapes. Let N be a LR-fuzzy number whose membership function is defined as follows:

$$N(x)=\begin{cases} L((m-x)/a) & x\le m, a>0 \\ R((x-m)/b) & m>x, b>0 \end{cases}$$

(1)

where, $L, R \in \xi$. Symbolically, $N=(m,a,b)_{LR}$, respectively.

Let T_w denote the weakest t-norm defined by:

$$T_w(x,\ y)=\begin{cases} \min(x,\ y) & if \max(x,\ y)=1 \\ 0 & otherwise \end{cases}$$

(2)

Let $A_i=(m_i,\ L_i,\ R_i)$ be LR-fuzzy numbers, then weakest sum and minus showed by $+w$ and $-w$ are defined as:

$$A_1 +_w A_2=(m_1+m_2,\ \max\{L_1,\ R_2\},\ \max\{R_1,\ L_2\}) \quad (3)$$

$$A_1 -_w A_2=(m_1-m_2,\ \max\{L_1,\ R_2\},\ \max\{R_1,\ L_2\})$$

(4)

Let $F(t)$, $t\in T$ be a fuzzy time series with the observation $F(t)=(m(t),\ L(t),\ R(t))$ as fuzzy numbers. Then,

$$D^kF(t)=D^{k-1}F(t+1)-_w D^{k-1}F(t)$$

(5)

Where,

$$\begin{cases} D^kF(t)=(m_d^k(t),\ L_d^k(t),\ R_d^k(t)) \\ m_d^k(t)=m_d^{k-1}(t+1)-m_d^{k-1}(t) \\ L_d^k(t)=\max\{L_d^{k-1}(t+1),\ R_d^k(t)\} \\ R_d^k(t)=\max\{R_d^{k-1}(t+1),\ L_d^k(t)\} \end{cases}$$

(6)

$D^kF(t)$, which is a fuzzy time series, is called the k th order $(k=1,\ 2.\ 3,\ \Lambda)$ forward linguistic difference of $F(t)$ at t .

Then, we have:

$$F(t+1)=F(t)+_w D^1F(t-1)+_w \Lambda +_w D^kF(t-k)+_w \Lambda \quad (7)$$

If and only if for any $t\in T$, $m(t+1)=\sum_{i=0}^{\infty} m_d^i(t-i)$, where $m_d^0(t-0)=m(t)$, and

$$L(t+1)=\max\left\{\sup_{k\le t} L(k),\ \sup_{k\le t-1} R(k)\right\}$$

(8)

$$R(t+1)=\max\left\{\sup_{k\le t} R(k),\ \sup_{k\le t-1} L(k)\right\}$$

(9)

In a time variant model, the difference of the fuzzy time series at different times is determined by a window basis(k in formula (7)). The amount of a window basis determines how many of the previous data should be applied for the forecasting of a new value. The paper adopted a Genetic Algorithm (GA) to determine the value of k in each time t .

Genetic Algorithm procedure

The GA procedure begins with determination of the population of chromosomes[12]. One evaluates these structures and allocate reproductive opportunities in such a way that those chromosomes, which represent a better solution to the target problem, are given more chance to "reproduce" than those chromosomes, which are poorer solutions.

- ***Representation of the chromosomes.***

In order to encode each solution to a chromosome, one binary string is used for integer-valued variables. Population in our proposed algorithm incorporates 10 individuals, since this population size shows an appropriate performance.

- ***Initial population generation.***

As the initial population, we generate population size chromosomes. Each individual chromosome in the initial population is a candidate solution to the problem, so for forecasting $F(t+1)$, it must have a value between 1 and $(t-1)$.

- ***Fitness function.***

A commonly used transformation, which is applied in our proposed algorithm, is that of proportional fitness assignment. The individual fitness $F(x_i)$, of each individual is computed as the individual's raw performance, $f(x_i)$, relative to the whole population, i.e.,

$$F(x_i)=f(x_i)\Big/\sum_{i=1}^{N_{ind}} f(x_i)$$
(10)

where, N_{ind} is the population size and x_i is the phenotypic value of the individual i.

- ***Crossover.***

The single-point crossover is applied for this problem.

- ***Selection.***

A roulette wheel selection procedure has been applied for the selection operator in our algorithm.

- ***Mutation.***

Herr, the mutation probability is 0.001.

- ***Termination condition.***

In this algorithm method, we considered 200 generation to be termination condition.

Defuzzification procedure

This research adopted a back propagation Neural Network (NN) for the defuzzification. "Back-propagation" learning algorithm is an effective learning algorithm of multilayer perceptron and has directed attention to the information processing capability of artificial neural networks. This algorithm had been widely employed for various pattern classifications or inference problems expressed in terms of nonlinear functions[13].

- The k-layer perceptron is a model consisting of one output layer and $(k-1)$ hidden layers.
- The I/O relationship of NN is expressed by the equation $y=NN(x)$, where x and y are input and output matrices.
- The term k-layers $[u_0 \times u_1 \times \Lambda \times u_k]$ expresses the model size, where u_i is the number of neurons in the input layer, hidden layer(s), and output layer, and k is the number of layers.
- Each input layer and hidden layer has an extra unit of constant 1 in addition to those specified by upside. This unit has no connection to the lower-layer neurons and has connections to the upper-layer neurons.
- All neuron models in the neighboring layers are connected, but there should be neither connections within layers nor jump connections between layers.

Here, a 5-layer back-propagation NN has been implemented. There are 5 neurons in each layer, except the output layer.

Fuzzy variance and mean

At the end of time period t, after its demand $\widetilde{D}_{k,t}$ has been realized, the participant observes the inventory level, places an order of size $\widetilde{Y}_{k,t}$ to its supplier and receives this order at the beginning of time period $(t+\widetilde{l}_k+1)$, where $\widetilde{l}_k$ is the order lead time at stage k. Excess demand is backlogged[11]. We assume that the orders at any given stage become demands for the immediately upstream stage. Obviously we have $\widetilde{D}_{k+1,t}=\widetilde{Y}_{k,t}$.

Let $\widetilde{q}_t$ represent the ordered quantity in period t to be delivered in period $(t+l)$, the timing of the event and the conservation of flow imply that:

$$\widetilde{q}_{k,t}=\widetilde{Y}_{k,t}-\widetilde{Y}_{k,t-1}+\widetilde{D}_t$$
(11)

It should be noted that all demands and lead times are considered to be uncertain and are represented by fuzzy triangular numbers.

We assume that each participant of the supply chain applies the same inventory and order policy as follows:

$$\widetilde{D}_{k+1,t}=\widetilde{D}_{k,t}+\left(\widetilde{S}_{k,t}-\widetilde{S}_{k,t-1}\right)$$
(12)

where, $\widetilde{S}_{k,t}$ is the fuzzy value of "the order-up-to level" at stage k and period t.

$$\widetilde{S}_{k,t}=\widetilde{m}_{k,t}+z_k\sqrt{v_{k,t}}$$
(13)

$$\widetilde{m}_{k,t}=E\left(\sum_{i=1}^{l_k+1}\widetilde{D}_{k,t+i}\middle|\widetilde{D}_{k,t}\right)$$
(14)

$$v_{k,t}=\mathrm{var}\left(\sum_{i=1}^{l_k+1}\widetilde{D}_{k,t+i}\middle|\widetilde{D}_{k,t}\right)$$
(15)

$$z_k=\phi^{-1}\left(h_k/(p_k+h_k)\right)$$
(16)

In this case, all the demands are considered to be fuzzy sets. The fuzzy number $\sum_{i=1}^{l_k+1}\widetilde{D}_{k,t+i}\big|\widetilde{D}_{k,t}$ can be extracted by using extension principle and forecasting of fuzzy demand from time t to time $(t+\widetilde{l}_k+1)$, according to known demand at time t [14].

Simulate the bullwhipeffect

As described before, end customer's demand is generated by using an ARIMA process, and then these values are fuzzified into some symmetric triangular fuzzy numbers. These fuzzy values are considered as demands. The proposed approach is used for forecasting purposes. Since any lead time also has a fuzzy value, when computing the value of $\widetilde{m}_{k,t}$ by using extension principle and cutting the $\widetilde{l}_t$ in α level, the value of $\widetilde{m}_{k,t}$ can become a type-II fuzzy set. To overcome this problem, a defuzzification on the values of $\sum_{i=1}^{l_k+1} \widetilde{D}_{k,t+i} \left| \widetilde{D}_{k,t} \right.$ is implemented, by using the centroid method. Thus,

$$\text{Bullwhip effect} = \operatorname{var}(q_t)/\operatorname{var}(D) \quad (17)$$

Discussion

The characteristic ascending slightly in energy market shows China's energy supply being in shortage. Generally speaking, the energy consumers may keep the energy amount by increasing one and reducing the other in order to keep their utility steady under the situation of an adequate supply in energy market. To assess the performance of proposed uncertainty-dependent coal supply chain method and to gain further insights into the dynamical characteristics of supply chain systems with order placement lead time delays in an uncertain environment, especially the bullwhip phenomenon, we carry out simulations in this section. Fig.1 Comparison of supplier's order and distributor's demand

We adopt a simulation example of the bullwhip phenomenon between coal supplier's order and coal distributor's demand in Xi'an city to demonstrate the feasibility and practicability of the proposed model. The occurrence of the bullwhip is shown in Figs. 1. Results showed that bullwhip still exists in such an environment. Nevertheless, it is clear that the variance of demand is amplified in supply chain when demand is transferred from downstream to upstream stages. The model can also forecast the trends in fuzzy data.

The previous approaches for overcoming the bullwhip effect did not consider fuzziness of environment. Based on above simulation, we concluded that the model can forecast the trends in fuzzy data. Adoption of the forecasting model can reduce the bullwhip effect and help in improving the customer service level with decrease in total cost of the coal supply chain. The model can suggest the reasonable ordering policies. The results also show that the propose model is superior to the previous analytical methods in terms of discovering the best available ordering policies.

Conclusions

In this paper, we have studied the bullwhip effect problem in uncertain coal supply chain systems. We proposed a model to simulated bullwhip effect in a fuzzy environment, when all demand and lead times were considered to be fuzzy numbers. Taking into consideration the multiple sources of uncertainty in supply chain systems, we focused on uncertainties with respect to demand and production process. Simulation study indicates that order placement do play an important role in reducing supply chain stability and inducing the bullwhip effect phenomenon. The larger the uncertain range of order placement is, the larger the inventory deviation amplitude is and the longer the bullwhip phenomenon exists.

It is important to note that our model is open-ended. There are many avenues for future research. Extending the proposed model to more detailed study, such as intensive study of the relationship between different uncertainty sources in coal supply chain operation process, is currently under investigation.

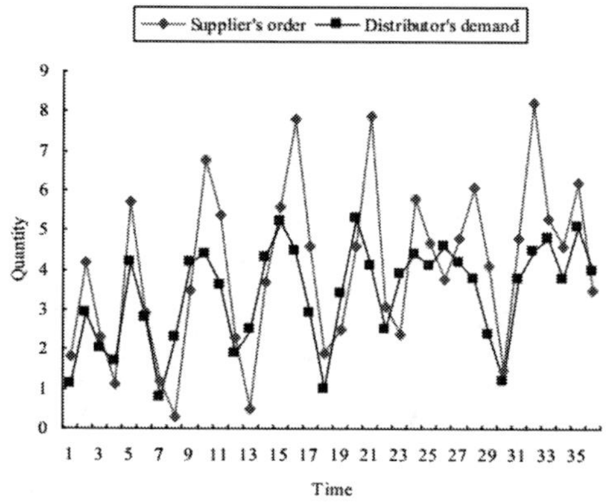

Acknowledgment

The research presented in this paper was supported by the Natural Science Foundation of China (NSFC) through Grant Nos. 71273206 and by the prosperity and development of philosophy and Social Sciences in Xi'an University of Science & Technology under the grant numbers: 2016SY04.

References

[1] Ouyang, Y., Daganzo, C. F., Characterization of the bullwhip effect in linear, time-invariant supply chains: Some formulae and tests. Management Science. 52 (2006)1544-1556.

[2] Lee, H. L., Padmanabhan, V., Whang, S., The Bullwhip effect in supply chains. Sloan Management Review Spring. 38 (1997)93-102.

[3] Sterman, J. D., Modeling managerial behavior: misperceptions of feedback in a dynamic decision

DOI: 10.1051/matecconf/20166304048

making experiment. Management Science. 35 (1989)321-339.

[4] Graves, S.C., A single-item inventory model for a non-stationary demand process. Manufacturing and Service Operations Management. 1 (1999)50-61.

[5] Zhang, X., The impact of forecasting methods on the bullwhip effect. International Journal of Production Economics. 88(2004)15-27.

[6] Chen, F., Drezner, Z., Ryan, J.K., Simchi-Levi, D., Quantifying the bullwhip effect in a simple supply chain. Management Science. 46(2000)436-443.

[7] Chen, Y.F., Disney, S.M., The order-up-to policy "sweet spot"— using proportional controllers to eliminate the bullwhip problem. In: EUROMA POMS Conference. 2(2003)551-560.

[8] Lee, H.L., So, K.C., Tang, C.S., The value of information sharing in a two-level supply chain. Management Science. 46(2000)626-643.

[9] M.S. Pishvaee, M. Rabbani, S.A. Torabi. A robust optimization approach to closed-loop supply chain network design under uncertainty, Appl. Math. Model. 35(2011)637-649.

[10] Hong, D. H., Some results on the addition of fuzzy intervals. Fuzzy Sets and Systems. 122(2001)349-352.

[11] Hong, D. H., A note on fuzzy time-series model. Fuzzy Sets and systems. 155(2005)309-316.

[12] M.H. Fazel Zarandi, M. Pourakbar, I.B. Turksen A, Fuzzy agent-based model for reduction of bullwhip effect in supply chain systems. Expert Systems with Applications. 34(2008)1680-1691.

[13] Takagi, H., Hayashi, I., NN-driven fuzzy reasoning. International Journal of Approximate Reasoning. 5(1991)191-211.

[14] Carlsson, C., Fuller, R., On possibilistic mean value and variance of fuzzy numbers. Fuzzy Sets and Systems. 122(2001b)315-326.

Exploration of low-carbon rural tourism in zhejiang: A Case Study of Tian Mu Village, Lin an County

Jian Qiang HU [1,2,a], Chu Yuan LOU [1] ,and Ling WANG [1]

[1]*School of Modern Service, Zhejiang Shuren University, Hangzhou, 310015, China;*

[2]*The Research Center of modern service industry of Zhejiang Province, Hangzhou, 310015, China*

Abstract: Along with the advancement of the construction of ecological civilization, low-carbon economy and low-carbon tourism are increasingly affecting the production and life style of human beings. Low carbon tourism is in the process of tourism development, through the use of low-carbon technologies, promote carbon absorbing mechanism and advocate low carbon tourism consumption to obtain higher tourism experience quality and greater tourism economic, social and environmental benefits of a sustainable development of tourism in new ways . Low carbon has become a recognized future development direction of the tourism industry. Closely linked to this paper, vigorously develop low carbon tourism industry requirements, to Lin an, Zhejiang Tian mu village, for example, through on-the-spot investigation, the problems existed in the village rural tourism low carbon analysis, and put forward the corresponding countermeasures, in order to rural tourism and low-carbon development in our province to find a feasible and with a demonstration of the significance of the road, and for other areas to carry out low carbon tourism experience.

Keywords: Low carbon Tourism; *Low carbon economy*; Rural tourism; Tian mu Village

1 Introduction

Rural tourism is based on the farmers' family management, with the characteristics of rural scenery and rural interest, in order to experience the main content of rural life as the main content of leisure travel [1]. The practice of rural tourism in the province, to speed up the beautiful countryside, the construction of ecological civilization, and strive to rural areas to build a "livable industry should swim better homes has played a positive role in promoting [2]. By the end of 2015, the Zhejiang province reported a total development of 3,246 rural tourism villages, 14,840 viable farm units, tourism reception of 183,608,400 visitors, and annual business income reached 1412 billion yuan. On the whole, however, from the current situation in our province, the various links and processes of rural tourism generally exist the problem of high carbon emissions, which directly affects the sustainable and healthy development of rural tourism [3]. Therefore, the low carbonization of rural tourism becomes a difficult problem in theory and practice.

2 The problems of the low carbonization of rural tourism

Based on the basic understanding of the dynamic development of the low carbon tourism education, the author selected Zhejiang province rural tourism model village ----Lin an Tian mu village as a case, through field research, personal visits the local tourists, tourism enterprises (mainly the farmhouse, the shop) operators, to further study the problem and Countermeasure of low carbon Tourism education.

Tian mu village is located in Tian mu Mountain National Nature Reserve in the South. Tian mu village has a beautiful environment, farmhouse economy and rich cultural connotation.

Tian mu village has won the "Zhejiang Province farmhouse characteristic model village", the first batch of "Hangzhou city rural tourism demonstration sites", "Hangzhou city comprehensive construction well-off demonstrative village", "Hangzhou Municipal farmhouse leisure tourism Village, Lin an County socialism new rural construction pacesetter Village, Lin an best leisure place" honor, and adopted the "Hangzhou Municipal eco village" acceptance.

In recent years, Tian mu Village with the guide of the local village committee, has done many beneficial

[a] Jian Qiang HU： hujianqiang2006@qq.com

attempts in building the "Happy farmhouse ", including construct a number of new tourist facilities, continue to strengthen the management of the local environment, such as road sclerosis, river dredging transformation and so on for constantly create a good environment for "Happy Farmhouse " development, while continuing to increase tourism product innovation, improve product efficiency. With the Tian mu village "Happy farmhouse" developed is getting better, the local village committee come to realize that a good ecological environment is an important prerequisite for sustainable development of "Happy farmhouse ", but also the biggest wealth of Tian mu village. Therefore, the village committee to actively explore the construction of "Happy farmhouse ", but it is also faced with many problems in the low carbonization of rural tourism.

2.1 Lack of enough attention to low carbonization of rural tourism

Due to lack the enough understanding to the construction of low-carbon tourism for long time, the local development of rural tourism construction is the pursuit of benefits, and there is a certain spontaneity and blindness, protection of the local ecological environment and resources are in a relatively subsidiary position. According to investigation, so far, Tian mu village has not organized a low-carbon tourism systemic science education activities, Tian mu village rural tourism practitioners' concept of "low carbonization of rural tourism" is not accurate cognition, whether the need to carry out low carbon tourism education also has fudged.

2.2 Willingness to carry out the low carbonization of rural tourism is not high

Based on the spot investigation found of Tian mu village, it is not very high for visitors and employees to accept the low carbonization of rural tourism. Most of tourists think that its to enjoy the natural scenery, travel beauty, rather than accept a the low carbonization of rural tourism to Tian mu village. To accept the low carbonization of rural tourism is bound to affect the tourist mood and the quality of tourism. It has following reason to Tian mu village practitioners unwilling to accept a low carbonization of rural tourism: Firstly, their " Happy Farmhouse " has done quite a low-carbon , there is no need to carry out the low carbonization of rural tourism ; The second is that the low- carbon or low carbonization of rural tourism reform needs to spend more equipment cost , the financial pressures leading to its unwilling to accept a low carbonization of rural tourism or simply training ; The third is considering a low carbonization of rural tourism of their Happy Farmhouse development can not play the role of economic efficiency , and therefore unwilling to accept low carbonization of rural tourism.

2.3 Lack of reasonable planning for low carbonization of rural tourism

Due to the lack of low carbonization of rural tourism enough attention, Tian mu village committee and the relevant departments are not reasonable planning for the development of low carbon Tourism education. In Tian mu village, part of the rural tourism, there are still rubbish dumped, low carbon tourism practitioners consciousness, imperfect carbon tourism facilities, tourism product development and irrational phenomena. By visiting the Tian mu village committee, the village committee has invested a number of power and money to strengthen the environment management, such as dredging rivers roads, the construction of seven waste transfer station, through various forms of low-carbon environmental advocacy However, due to the " Happy Farmhouse " practitioners cultural level is not high, inadequate attention to environmental protection, low-carbon tourism building did not achieve significant results. The main reason is the lack of low-carbon tourism-related construction personnel on how to conduct a low carbonization of rural tourism system still lacks theoretical understanding.

3 Countermeasures to solve low carbonization of rural tourism

The low-carbon tourism is not just a concept, but also a new way of life and travel behavior. It needs tourists, tourism planning and development departments, tourism enterprises, government and tourist -related management department to think and act. According to the development practical of Tian mu village "Happy Farmhouse "and the problem of low-carbon tourism, the author puts forward some countermeasures.

3.1 Strengthening the importance of low-carbon tourism promotion

In order to solve the weak low carbon consciousness of tourists and practitioners ,the lack of correct understanding of low carbon tourism and other issues, to develop low carbon " Happy Farmhouse " construction of promotional activities, through various forms including promotional pieces , posters, lectures and other forms to enhance " Happy Farmhouse " practitioners to build awareness of the importance of low-carbon tourism , establish a " Happy Farmhouse " building need to support the concept of low carbonization of rural tourism

3.2Strengthening the construction of low carbonization of rural tourism system

Carry out the construction of low carbonization of rural tourism, we must solve the problems of the willing of the low carbonization of rural tourism and how to carry out the deficiencies effectively, which requires to establish a long-term mechanism of low carbon tourism education system. The village committee and the relevant department of tourism should be established a low carbon tourism education system which in accordance with Tian mu village" Happy Farmhouse "of the actual mechanism, including low carbon tourism education planning, to determine the leader of low carbon tourism education, low carbon tourism education object, the form of low carbon tourism education, the content of low carbon tourism education, put forward the guarantee measures low carbon tourism education and evaluate the achievement of low carbon tourism education, etc. Meanwhile, Tian mu village also should strengthen the construction of low carbon tourism education service points, low carbonization of rural tourism service points as the carrier of low carbonization of rural tourism should be made clear its composition, function attributes, operational mechanism and management regulations, etc., so as to promote the normal development of carbon tourism education.

3.3 Strengthening rural tourism low-carbon tourism star standard construction

Through the rural tourism low-carbon tourism star standard construction, can greatly improve the Tian mu village rural tourism willingness to accept low carbonization of rural tourism practitioners, thereby promoting low carbonization of rural tourism. Rural tourism low-carbon tourism star means rural tourism indicators based on low-carbon assessment of the extent of its division, determining carbon levels. Specific indicators include low carbon content rural tourism level waste disposal, equipment operation and management degree, the extent of low-carbon tourism services, construction of low-carbon tourism products and tourism product is the situation with multiple aspects of educational and other low-carbon. Low-carbon tourism -star standard construction can take environmental protection and " Happy Farmhouse " stakeholders combine to greatly facilitate " Happy Farmhouse " low-carbon tourism practitioners willingness education .

4 Summary

To build a low carbonization of rural tourism is an important way to implement a low-carbon economy, is also to put the low carbon tourism concept into practice, and the effective way to guide practice [7]. At present, the low-carbon tourism construction of Tian mu village is still in exploring stage, if can from the perspective of the ideological level, to improve the " low carbon concept of Happy Farmhouse " practitioners through education , improve the relevant tourism services , and has the scientific guidance for " Happy Farmhouse " low carbon tourism facilities and product development . It will play an important role to improve the Tian mu village " Happy Farmhouse " level of development and the promotion of " Happy Farmhouse " sustainable development; At the same time, the practice of Tian mu village of low carbonization of rural tourism will be provide valuable experience and reference to other areas to develop low carbon rural tourism construction, so as to promote the development of low-carbon tourism. And low-carbon tourism will also bring a new era of development for rural tourism in the development of our country, let us concern, think and act get together.

Acknowledgements

This work is supported by the projects: "Zhejiang rural tourism low carbon demonstration esearch". (15JDNF02YB). Zhejiang province philosophy social sciences research base planning project.

References

[1] Hu Jianqiang, Shan Wenjun and Li Liang: Carbon Tourism Research Status and Prospects [J]. Zhejiang Shuren University, 2011.11(3):31 - 35

[2] Cai Meng: Theory and practice of low carbon Tourism -- a case in China, East China Normal University, 2012, http://www.cnki.com.cn/

[3] Wu Ying: Promote low-carbon tourism: a new opportunity for development agencies [J] economic and technological co-operation, 2010 (13):36- 37

[4] Feng Duo and Li Bin: The think of improving low-carbon economy of Happy Farmhouse Tourism – Dapingan Village Xiaying town Ji country TiLin ann city as an example of low carbon economy [J] Agricultural Environment and Development , 2010 (6) : 59 - 61

[5] Liu Wei and Yang Lei: Carbon Happy Farmhouse suburbs of Beijing Tourism Development Strategies [J]. Agricultural Sciences, 2012.51(20):4691 - 4696

[6]Zhou Jianfa: The current situation, problems and countermeasures of building a low-carbon rural areas[J]. Anhui Agricultural Sciences, 2012 (9):5640 - 5642.

[7] Al Bobby. Social research methods. Beijing: China Press.2009.

Design of College Students' employment management system based on WEB

Yu CAO [1,a]

[1] *Jiangxi Science&Technology Normal University, Nanchang, Jiangxi, 330013, P.R. China*

Abstract. Since at the end of last century, our country cancelled the allocation of national college students employment, college graduates employment pressure also with the increasing trend of the expansion of enrollment in Colleges and universities. At the same time, it also greatly increased difficulty in the college students employment management, under this background, the traditional manual management mode obviously can not meet the required at present. In the information technology increasingly perfect today, how it has become an important task of university reform through the digital management information system technology research and development of College Students' employment. In this paper, a management system based on Web Technology College Students Employment launches the analysis, aimed at through the information technology, to promote the management of information, efficient development, so as to provide an important reference for the employment decision.

Key words: Employment guidance; employment management; WEB; system design

1 Introduction

Based on the employment management system of modern information technology can timely query of college students employment whereabouts, unit of invite applications for a job related information, students' personal information and through this system to student's employment status were tracking and evaluation, and according to the evaluation results clear talent cultivation mode is there problems, to provide the direction of reform, so as to promote college students' employment ability, promote employment rate [1]. Following on a management system based on Web Technology College Students Employment launches the analysis, aimed at through this kind of management system, contributing to the management of information, efficient development, provide the basis for the employment decision, as follows.

2 The significance of the construction of College Students' employment management system based on WEB

University leadership needs to grasp the current employment trends and employment of students, which is also an important indicator of the strength of teaching in Colleges and universities [2]. In addition, many areas of education departments are required local colleges and universities on a regular basis to the employment situation statistics reported, including the signing of the number of posts, employment intentions, etc.. Therefore, it is necessary to take effective statistics and analysis means to ensure the timeliness and accuracy of employment. From the perspective of colleges and universities, the employment management system can make scientific statistics on the employment situation of students, and make clear the students' employment intention, so as to provide valuable reference for employment guidance. From the perspective of students, students can also be employed by the employment management system for the use of the unit to query, two-way selection [3], but also more conducive to helping students to carry out career planning.

3 Design of College Students' employment management system based on WEB

3.1 System structure design

Based on Web employment management system main interface and back office interface, which interface Is mainly the employment department staff on the current college students employment information management and related operations, the interface of the B / S structure, to ASP. Net on the web server site to interact. This design can fully realized by WWW browser to login the system, so as to realize the structure of the employment management based on the web, which greatly facilitate

[a] Corresponding author: 466167034@qq.com

the system [4]. The backstage interface plays to the information data statistics, the analysis function, in which the database is the core of the system. System database management system selects Microsoft SQL Server 2005, SQL Server Management Studio to create a database. A database and on the create including employment survey, user information, professional information, employment analysis a number of functional components. The SQL Server2005 a completely able to achieve web products, teachers can through the system access, edit, publish employment information, students can in this system of employment information release and checking operation. And the management and maintenance of the system is performed by the system administrator, as figure 1.

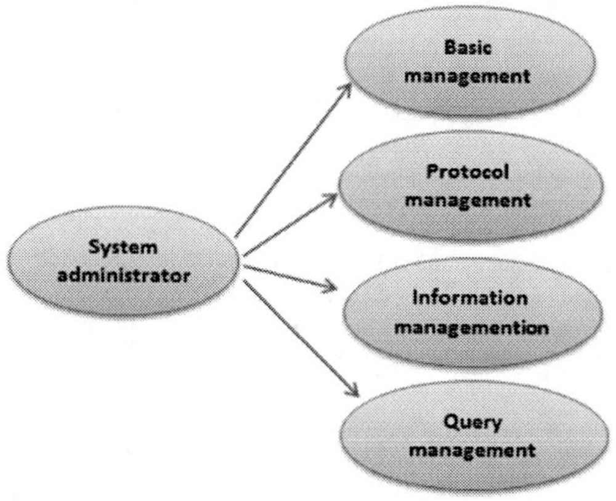

Figure 1. System administrator permissions diagram in the system functions

3.2 Functional module design of the system

System mainly includes the following function modules: user registration module, employment information management and employment guidance module and management module of the unit, the employment department management module, statistical module of employment, employment information query module, the employment agreement and policy management module and turnover and default function module etc.

3.2.1 User registration module

WEB based employment management system design for membership registration system, set by the system administrator privileges. There is no registered user that is a tourist, available information is limited. By the system administrators to different users of the rights to set [5], such as the students registered users can be information query, download, and employment guidance module in the interactive features. The teacher user can be registered on the basis of user rights, to achieve employment information editing, upload, modify, and other functions. User information can be released by the user, or according to the needs of the user, the system recommended by the system to meet the conditions of College students.

3.2.2 College students employment information management and employment guidance module

Each university student users in this module has a resume template, the student user login, you can edit your resume, you can download the resume template [6]. Students can also fill in their employment information management module in the employment information management module, the employment guidance module can be based on the students' employment intentions of screening results, to provide students with relevant employment information.

3.2.3 Unit management module

Employing units to provide relevant information to the University and registered in the system, after the audit, as a member of the employing unit. The employer to publishing in the module units have confidence, at the same time, the system also can according to unit needs and recommend to meet the requirements of the students resume, college employment personnel can also be according to the requirements of the enterprise employment, to unit of choose and employ persons recommended interview and interview results to release in the employing unit management module.

3.2.4 College employment department management module

Department of employment management in Colleges and universities is required to release the college students' employment related information in the system, employment sector managers not only from colleges and universities employment sector also includes various department of employment guidance teachers, in addition to the dissemination of relevant information, these personnel also responsible for employment management information system in the information audit and authenticity of the investigation.

3.2.5 Employment statistics module

This module is a module for the statistics and analysis of the employment situation, job position, salary and employment rate of college students. In addition, this module also has the print function, teacher users and system administrators can be directly retrieved and printed.

3.2.6 Employment information inquiry module

This module, whether teachers, students, students, or users, can be used to enter the key words, the search to the relevant information available to the public

3.2.7 Employment agreement and policy management module

Graduate students in the process of job search and recruitment, the university can check the employment agreement through this system module, so as to determine whether the student employment. In addition, the system will be based on the needs of the enterprise, the first time to inform students of the formal appointment of the money needed to do the preparation, the student after induction, the need to record in the system.

3.2.8 Turnover and default function module

College students in the employment process, there may be job hopping problem, there are some students after the signing of the entry agreement, may be due to the existence of one's own or other factors. Therefore in order to be employment management system more complete and need to set turnover and default module, in which recording their employment after the breach and turnover, and clearly is unit reason is their own reasons. Can also provide opportunities for students to a two-way choice, if an employer reason for the unit often students leaving the problem, based on for students to develop perfect employment planning principles, the need to remind students to avoid such enterprises signed.

4 Conclusion

To sum up, this paper first based on web of college students employment management system construction significance analysis. After the structure design of the system are discussed, finally, respectively, from the user registration module, employment information management and employment guidance module and management module of the unit, the employment department management module, statistical module of employment, employment information query module, the employment agreement and policy management module and turnover and default function module of system function design are analyzed and hope that this research can provide some reference for the stakeholders.

This paper is the Research Institute of Vocational Education in Jiangxi science and technology Normal University and the research on the effectiveness of the teaching effectiveness of Jiangxi Province Education Science Planning Project (NO:15YB084).

References

1. Xu Jingcai. Web based medical students' career planning of CAI multimedia courseware design [J]. Journal of electronic design engineering, 2015, 14:173-175.
2. Metcalf jv, li yu, Chen Lijun NiuLei. College student dormitory management system based on Web design and implementation of [J]. Computer knowledge and technology, 2014:8098-8100.
3. Shun, Liang Guicai Li Jiguo, Liu Ming. Based on Asp.net/.Net Web Service report of higher vocational college graduates employment information management system [J]. The design and implementation of enterprise and the development of science and technology, 2011 preceding: 19-21.
4. Lu Hongcai. Student employment management system based on WEB - in the design and implementation of industrial vocational and technical college as an example in gansu province [J]. Journal of nanyang institute of technology, 2011, 2011:24-27.
5. Li Yang ling. Based on field work and employment outside the WEB research and implementation of quality control management system [J]. Journal of information and computer (theory), cutflower production potentials of 2014:132-133.
6. Garrel Dr. Based on WEB3.0 technology vocational school graduates employment tracking system, the design and implementation of higher vocational and technical school in suzhou as an example [J]. Journal of digital technology and application, 2013 cutflower production potentials: + 119 116-117.

MATEC Web of Conferences 63, 05002 (2016) DOI: 10.1051/matecconf/20166305002
MMME 2016

Recommendation System of Program Based on REST Style

Jin Bao SONG[1,a], Jin Hong SONG[2], Jian Ping CHAI[1] and Ying LIN[1]

[1]*Information Engineering School, Communication University of China, Beijing, China*
[2]*Shandong Gold Mining Jiaojia Gold Mine (Laizhou) Co., Ltd*

Abstract. With the popularity of digital TV, TV programs have been on the increase no matter in both the number and species, which brought many choice to the users. Although the digital TV has increased largely in the selectivity, it has become a fussy process that users search for programs which they are interested in. So there is need to have an efficient program recommendation system to solve the problem that is "information overload" for users. It can not only help users to get the program which they require, but also bring convenience to people's life. The program recommendation system named MyView is planned and designed, aimed to providing an efficient information platform. The system also involves intelligent recommendation. The information guide will trigger the recommendation engine after users registering information, the engine will accord to the data in the guide information to make the personalized program recommendation. The system was deployed in the Tomcat and Apache integration servers on my localhost, so it also belongs to the Web application based on J2EE platform. AJAX is used that can achieve a good user experience to develop web presentation layer on MyView PC browser with flexible interface performance. The background of business services uses the hierarchical form. MyView System uses the CXF framework and Hibernate to equip controller and data persistence layer in the Spring container. The overall framework of the system uses the REST style, in order to extend the performance and function later. Background service layer with uniform interface, marked by the URI resource. At the same time, HTTP requestion is submitted by the AJAX to obtain services provided by resources. Finally, we can analyze and summary the features of MyView System.

1 Instruction

Along with a variety of video sites increasing in recent years, the way of people watching TV programs converts from the traditional TV to set-top box and then to the use of mobile phones and tablet computers to access the Internet to watch a variety of programs through different video sites without subjecting to the limitations of time and space. In order to gain more traffic and revenue, major video sites will provide more video resources and more personalized services as their competitive advantage. How to effectively integrate the user's own needs and existing video resources together becomes an exceptionally arduous task in the field of today's information technology services. Currently, Youku Tudou, iQIYI and Sohu video lead in the video program system, which basically complete the functional requirements including the user information sharing, personalized recommendation, and multi-display terminals. Especially, personalized recommendation is not only applied on the video site, but also even applied to the major online shopping platforms such as Taobao, Jingdong, etc. In order to allow users to have a better experience, and strive to be accurate and efficient on personalized recommendation [1].

Internet is based on HTTP protocol at the beginning. The HTML document is located through URL. Thus mutual communication is allowed between the various groups within the scope of the world. Web services are often used to solve problems between different interrelated platforms. But it seems to ignore the distributed programming platform features which support the Web success to improve the technology. The RESTful architecture allows Web service back up the essence of Web. With the rapid development of the mobile Internet and cloud computing today, REST style has dominated the field of Web development. More and more Web services begin to use design and implementation of REST style such as this Internet giant like Google, Amazon. The report notes that 95% of businesses are using the REST service from lightweight in Amazon bookstore. At the same time, the service system needs some individual business processes in order to attract users. A variety of complex services based on HTTP protocol survive on the Internet. In order to continue, these popular services not only attract basic groups of users by the unique function, but also promises the ability to open interfaces to other service systems. In order to achieve a personalized treatment pluggable module replacement, we need to decouple service

[a] Corresponding author: songjinbao@cuc.edu.cn

processing module and personalized processing module [2].

2 Development environment

2.1 Maven

In order to be able to build a large system to facilitate the release of project information and a shared JAR in multiple projects, the traditional architecture of Ant is not used when to the build cross-platform system, but the software project management tools –Maven is used. Maven is the software project management tools of the project construction management, reporting and documents by a short description information based on the Project Object Model (POM). When you use Maven, you use a well-defined project object model to describe your project and run the plug-in target to generate the site, manage JAR files according to their need.

2.2 Navicat

In order to make database operations easier, a visual database management tool Navicat is used. Navicat is a fast, reliable database management tools to simplify the management of database and reduce cost of management system. It is designed to meet the needs of developers. Navicat is based on intuitive graphical interface. So data resources can be created, organized, accessed and shared in a simple way.

2.3 Apache Http Server 2.0

Apache HTTP Server is open-source web server which can run on most computer operating systems. It can construct Web server with a good cross-platform and security. It also provides Web server management function. The professional Apache HTTP server can be used to handle static resources of our entire system.

2.4 Tomcat 7

Throughout the program recommendation system, a lightweight web application server Tomcat is used to deal with the occasions of small and medium systems and not many concurrent access users to develop and debug JSP program. Apache server is used to respond to HTML page access request. Tomcat is actually part of an extension of the Apache server. So when Tomcat is run solely, Tomcat and Apache actually run separately as a separate process. Dynamic resources are dealt with by Tomcat. Servlet and JSP support to improve the ability of the processing of requests. When Tomcat is configured correctly, Apache serves HTML pages, and Tomcat actually runs JSP pages and Servlet.

2.5 MongoDB

In the whole MyView system, the most popular MySQL is used as the relational database to store the platform basic information and data. It is worth mentioning that, a non-relational database MongoDB is also used mainly to store user authentication information and guidance information. The guidance information is very important for the start of the behind recommendation function. Mongo DB is attracted by much of the current IT professionals because of its flexible data storage and a good object-oriented thinking achievement. All data persistence operations needn't have to manually write SQL statements. You can directly call the method to save the user session state. The server achieves stateless, which can ensure the entire system uses a unified REST architectural style.

3 System technology selection

3.1. REST architectural style

REST-style Web services are primarily based on three basic Web technologies: XML markup language, HTTP application protocol and URI naming standard, also known as RESTfulweb services. RESTful web services and traditional services are somewhat different in interface design and use, loose coupling and expandability. But REST-style Web services have reached a simpler, more convenient and good performance [3].

REST refers to Representational State Transfer. When a well-designed Web composed by the network pages runs, the user selects the appropriate URL link is referred to state transfer. It refers to the state to advance in network applications and lead to a page, which is presented to users. The representation of REST refers to the next state in the transfer process, including the common HTML pages, XML pages, or certain data and services, these are the representations.

The most obvious REST several design criteria in Web applications are:

- All things on the network are abstracted to resource;
- Each resource has a unique resource identifier, which is the URL;
- Resource identifier will not change because of a variety of operations;
- Resource is operated through a common connector interface, which is the HTTP protocol PUT, DELETE, HEAD, GET, POST and other actions;
- All operations are stateless.

As it can be seen, URI is not only the resource name on the Internet, but also the address of the resources on Web. The statements involved in REST resources refers to resources can transfer between the client side and the server side described in the status of a specific moment [4].

3.2 RESTfulWeb service

RESTfulWeb service means to build Web services following the REST style. We mentioned earlier that the modern Web services are more and more away from the original Web development concept. The appearance of REST is to enable it to return to the resource-oriented

DOI: 10.1051/matecconf/20166305002

architecture. RESTfulWeb identification and location of resources are by URI. The operation of these resources are performed using HTTP methods (common HTTP methods DELETE, POST, GET, HEAD, and PUT). That is say, in RESTfulWeb, the place for scope information is URI path of the resource. The method information of data operation is placed in the HTTP method. From a technical point of view, RESTfulWeb service is related to basic URI and XML technology in addition to HTTP, known as three basic technologies of REST [5].

3.3 REST operational features

In the traditional Java web architecture, people often prefer to use MVC design pattern with Spring and Hibernate technology together development. The biggest feature MVC is which Model member to handle user requests and which View to display the returned data are determined by the Controller.

For REST, on the client and server assignment, it follows a client - server design specification. The service is the responsibility of offering resources and service methods of operating resources. The client only need to make page rendering in the front end according to resource data and status.

In order to enrich Web service dynamically personalized user experience, due to the characteristics of REST itself bounding the transfer of resources and data resources to the client and separated from the server separation, AjaxWeb with the most advantageous features is used in the front page rendering, which can interact with the server without needing to require a full refresh. So that it resides in the browser to execute application logic and state, to achieve partial refresh and to avoid the screen flicker.

4 System implementation.

As a RESTful Web project, MyView uses Maven build tool. Domain model is developed in stratified mode. Background service layer is abstracted to have a unified interface. Resource is identified by a URI. Service layer is divided into two layers DAO and Service is. DAO mainly located in the lower layer to package some common methods of operating entity Bean which waits for the call from the upper. Service is located in the upper layer, calls the DAO layer methods, and provides specific business logic as external manifestations of REST-style Web Service. Web front-end uses Ajax technology. The presentation layer uses Spring + MVC, using its flexible, expressive rich interface to get a good user experience.

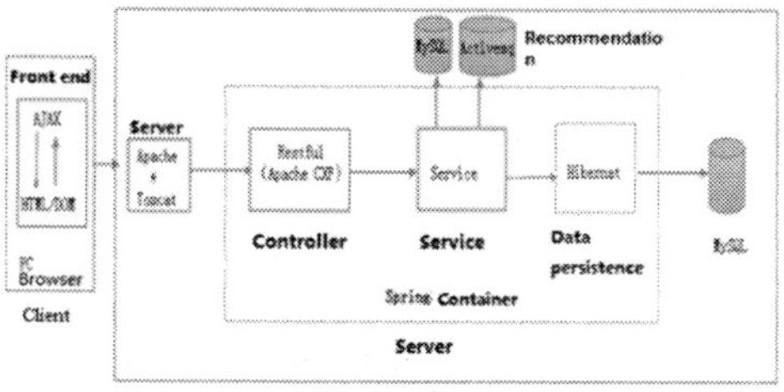

Figure 1. System structure diagram.

5 Recommendation system analysis.

After the birth of SlopeOne, the Internet publishers paid more and more attention to recommendation function as a mainstream business so as to bring more traffic. At present, the more extensive used personalized recommendation technology is primarily Collaborative Filtering Recommendation and Content -based Recommendation. In order to produce better recommendation, some mixture of different recommendation algorithms will also be used. The technology such as big data is used to optimize the algorithms. MyView mainly uses recommendation algorithm based on content to avoid the Matthew effect.

5.1 Classic recommendation algorithm

SlopeOne algorithm is essentially a kind of algorithm based on the project, but in conventional cognition is a kind of classic algorithms based on score predictions. SlopeOne makes predictions on the basis of the principle of linear regression. The predictor is in the form of f (x) = x + b. Free parameter b refers to the users' average deviation of two projects score. Based on the available parameters, for example, user B's score prediction of project j is represented as $r\_B_j = r\_B_i + (r\_A_j - r\_A_i)$.

	Item*i*		Item*j*		
	1		3		User*A*
	2		?		User*B*

Figure 2. Basic SlopeOne algorithm.

Slopeone mainly computes mean deviation {dev}_jk and the target project score {P(u)}_j.

$$dev_{jk} = \sum_{u_i \in I_{jk}} \frac{r_{ij} - r_{ik}}{card\left(I_{jk}\right)} \quad (1)$$

$$P(u)_j = \frac{\sum_{k \in R_j} (dev_{jk} + u_k)}{card(R_j)} \quad (2)$$

In the above basic formula, the mean deviation [dev] _jk is based on the benchmark of target project i_j with other project i_k. Score [P (u)] _j is used to predict the current active users u's score for the target project i_j. I_jk is collection for customer evaluation of project i_j, i_k. Card (S) stands for the number of elements in the collection S. u_j is user U's score for project i_j. R_j is project collection that is evaluated with project i_j at the same time. For a simple recommendation system, under the premise of not requiring the accuracy of recommendation, in order to avoid ignoring the number of the user's evaluation, a weighted SlopeOne algorithm is considered to use to improve the accuracy. In weighted algorithm, the weight of deviation is represented by the number of users that evaluate project i_j and project i_k at the same time. c_jk stands for the weight of deviation. c_jk = card (I_jk). The final prediction result is as follows:

$$P^w(u)_j = \frac{\sum_{k \in R_j} (dev_{jk} + u_k) c_{jk}}{\sum_{k \in R_j} c_{jk}} \quad (3)$$

5.2 Similarity calculation formula

In various kinds of similarity calculation methods with different benefits, the current common used methods are based on the vector. The calculation size stands for the similarity. In the MyView system, the size is used to calculate the similarity based on the Cosine Similarity.

Cosine Similarity is widely used in the document vector similarity calculation. Formula is shown as follows:

$$T(x, y) = \frac{xy}{\|x\|^2 \times \|y\|^2} = \frac{\sum x_i y_i}{\sqrt{\sum x_i^2}\sqrt{\sum y_i^2}} \quad (4)$$

T (x, y) function measures similarity by computing two vectors x, y angle Cosine. The similarity relation between vector and the change of internal elements are reflected. The calculated distance between the vectors is more to distinguish the differences from the direction. Cosine Similarity is used to distinguish user interest similarity and difference, as it is not sensitive to the absolute value [6].

5.3 Recommendation algorithm based on the content

In the process of recommendation, each user needs to create a user model. In order to be intuitive to recommend explanation and change the weights of different interest items according to the need, programs and users can be processed for domains. Firstly, the user preference information and programming model can be divided into n different areas and each domain corresponds to an interest item. It can be concluded that the user preference model.

User_Profile = {(k_1, w_1), (k_2, w_2),...(k_n w_n)}.(5)

Film_Model = {(l_1, v_1), (l_2, v_2),...(l_n, v_n)}. (6)

Then score between user and program can be calculated according to the Cosine similarity formula. When programs are recommended, system will sort the programs according to the score size in order to achieve the goal of recommending program. Sim (l, k) is for the similarity between the user and i domain of the program. After the cosine similarity formula, it is available:

$$sim(l_i, k_i) = \frac{\sum_{p=1}^{j} w_p' v_p'}{\sqrt{\sum_{p=1}^{j} w_p'^2 \bullet \sum_{p=1}^{j} v_p'^2}} \quad (7)$$

w_p' refers to the weight of the i domain the p word in user preference model. v_p' refers to the weight of the i domain the p word in program model. For the sake of accuracy, we can consider this kind of circumstance: without a word, then the weight of the word is 0.

For the film show, relative to the user, under the recommendation algorithm based on content, the score is Score_cb, then the calculation method of Score_cb represented by formula (8).

$$\text{Score}_{cb} = \frac{\sum_{i=1}^{n} sim(l_i, k_i) w_i v_i}{\sqrt{\sum_{i=1}^{n} (w_i v_i)^2}} \quad (8)$$

5.4 Lucene index mechanism

Our core algorithm module mainly realize the similarity algorithm based on content, by introducing the heat factor to get the recommended list. But in the core algorithm module, we need the data retrieval and filtering, using Lucene tools library to build indexes and create a query.

In recent years, Lucene has become a popular Java open source information retrieval tools libraries. It is often used to increase the index and search functions. Lucene provides simple and easy API to use index and search. A Lucene database is created in the database. The data unit is converted to inverted index through the database.

MyView establishes Lucene database in Navicat Lite. The data structure is usually stored in the file system as a set of index files. Lucene libraries join the search function to MyView recommendation engines to build indexes and create a query. Those programs that do not conform to the user interests will be screened out, greatly reducing the subsequent data calculation amount.

Figure 3. The main program for program data index.

6 Summary

With the popularity of digital TV, TV programs have been on the increase no matter in both the number and species, which brought many choice to the users. Although the digital TV has increased largely in the selectivity, it has become a fussy process that users search for programs which they are interested in. people gradually accept and rely on using the system such as video website to playback TV shows on demand, because the way of watching TV on television is accepted by the development of all kinds of the video website. So there is need to have an efficient program recommendation system to solve the problem that is "information overload" for users. It can not only help users to get the program which they require, but also bring convenience to people's life.

The program recommendation system named MyView is planned and designed, aimed to providing an efficient information platform. It mainly serves for audience and the users who publish video on Internet. Our users are called the Viewer in this system. Users can create your own personal resume list, and program list, searcing programs, publishing program and reviews, managing programs so on. The system also involves intelligent recommendation. The information guide will trigger the recommendation engine after users registering information, the engine will accord to the data in the guide information to make the personalized program recommendation.

References

1. J.B. Song, The characteristics and application of recommendation system, Journal of China digital TV, 46,2009, 10.
2. C. Chen, K. Y. Qi, J. B. Chen. Podcast show recommendation based on user clustering, Journal of computer applications and software, 2009 (01):7-10.
3. H.Tian, Employment system design and implementation based on REST style, Communication university of China, 2014.
4. J. Zhang, The research and implementation of Web services based on REST architecture, Beijing university of posts and telecommunications, 2011.
5. X. F. Cheng, The research and design of Web Services based on the REST architecture, Wuhan university of technology, 2010.
6. J. T. Ru, Christine. the position recommendation optimization research based on information retrieval, Communication university of China, 2014.

 DOI: 10.1051/matecconf/20166305003

Pattern-Based Coarse-Grained Component Modelling for Enterprise Application Software Developing

Hai Bo LI [a], Meng Xia LIANG and Ya Feng GAO

College of Computer Science&Technology, Huaqiao University,Xiamen,361021,China

Abstract. To improve the efficiency of ESA (Enterprise Application Software) development, component-based software development is a good solution. However, though those general software components, like JavaBean, COM/COM+ (Component Object Model), are maturely used to develop ESAs, it is still too inefficient to industrialize ESA development. This is because design of software component has a tightly correlation with the inner structure of an ESA and should not be independent from architecture of ESA. By analysing software patterns and characteristics, a business component (BC) modelling based on pattern is proposed first. Then a BC scheduling pattern and an executable workflow model are presented to show the principle by which BCs are organized and scheduled. The proposed method is a PSM (Platform-Specific Model) modelling which a part of MDA (Model-Driven Architecture) based framework. Finally, a case is showed that our method can provide theoretical and practical significance to the development software of ESA.

1 Introduction

Patterns of ESA (Enterprise Application Software) development have been evolved for several phases, from coupling code to data structure decades ago, and then separating code from database. Until now, object-oriented method is introduced to divide codes as many objects or components. At present, though SOA (Service-Oriented Architecture) presents more coarse-grained reuse mode than software component [1], it focuses on distributed applications [2]. To improve efficiency of building ESAs, the industrialized developing process still needs a method to construct coarse-grained component, even to be integrated into workflows.

The current research focuses on approaches based on MDD (Model-Driven Development). However, the MDD emphasizes the whole architecture of ESA functionality and behaviour, not the technology in which it will be implemented [3-5]. UML model can be considered as traditional software modelling for ESA developing [6], however, it lies in PIM level of MDA (Model-Driven Architecture) and can not be mapped into code directly. An object-oriented approach is needed for its advantageous modelling executable ESA, so as to attain a uniform overall approach to analysis, design, and implementation [7]. Though other object-oriented methods are provided for designing [8-13], executable component modelling is required for ESA developing nowadays.

In this paper, we propose a method of pattern-based coarse-grained component modelling for ESA developing (PCGCM). We first analyse the software patterns and characteristics of ESA, and then a BC (Business Component) scheduling pattern and an executable workflow model are presented in PCGCM. Next the principle is presented by which BCs are organized and scheduled by the execution workflow engine. The proposed method lies in PSM level in our MDA based framework-ICEMDA (Interoperable Configurable Enterprise Model Driven Architecture). Finally, a case is discussed using PCGCM.

The rest of the paper is organized as follows. In Section 2, the motivation of the paper is presented. In Sections 3, patterns and characteristics of component is analysed, and a case is discussed in Section 4. Finally, in Section 5, we present a summary.

2 Ideas of PCGCM

Relative to small-granularity components as elementary units in windows, such as those control objects, button, list and text box, etc., a big or huge-granularity component encapsulates many small-granularity components and process based on them to an integration entity. The big or huge-granularity component is considered as application (or application-like), and the process in it as micro process. This development pattern divides a large-scale software system into big or huge-granularity component and their integration. The big or huge-granularity components are called BC (Business Component), while the process between them is called macro process. From the point of view, ESA is a software system which composed of those applications and macro processes, and whose applications have same or similar

[a] Corresponding author: lihaibo@hqu.edu.cn

pattern. Namely, ESA represents the idea of application based on pattern.

3 Patterns and characteristics of BC

3.1 BC pattern analysis

ESAs show different patterns which are supported by different technologies and platforms. A comparison of different patterns is presented, as shown in Figure 1. Some pattern needs only one table, while another needs two or more table altogether. They all contain areas of query items and data presentations.

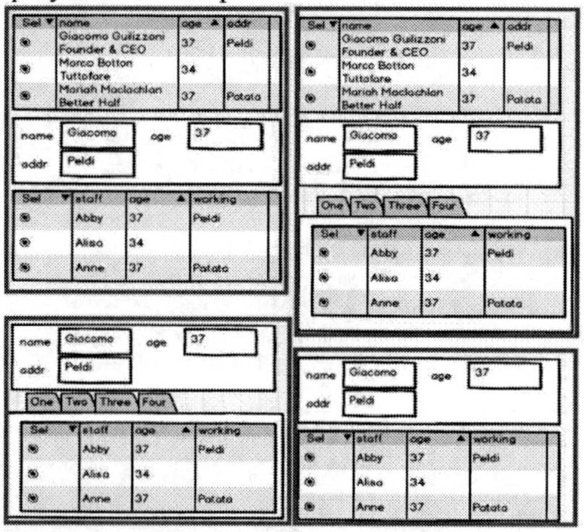

Figure 1. Several Traditional ESA Software Styles

In most ESAs, pattern is a kind of abstract from those common features which can be reused as template when developing. Pattern refers to application-oriented software pattern, whose characteristic is further divided into user-oriented (called user pattern) and system-oriented (implement pattern).

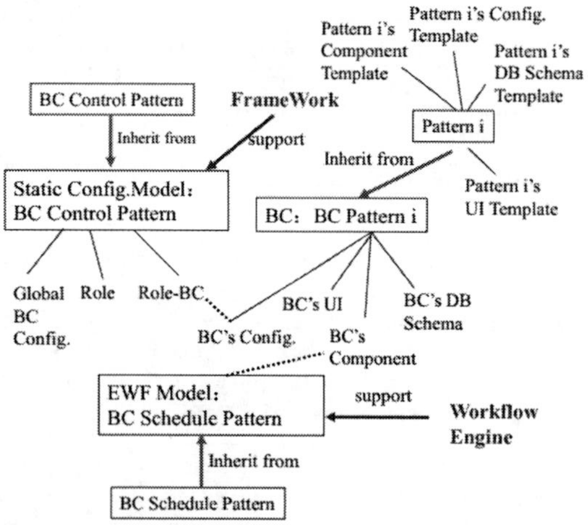

Figure 2. ESA Software Patterns

From view of user pattern, pattern represents interface style, operation style and control style. From view of implement pattern, pattern represents a common realization.

Relative to small-granularity component, BC is an independent application. BC pattern defines the style which inner structure of the applications should have. A BC application is composed of four parts, namely user interface, business logic, data pattern and configuration file, which describe BC's style and provide BC's template defining the common content of BC, as shown in figure 2.

As mentioned above, ESA is composed of more complete applications and macro processes, and participated by different roles. BC's control pattern defines how to control BC execution under system framework through access control. So the control pattern is also called framework pattern. BC's control pattern statically describes which BCs are used, which roles will participate and access associated with them. The framework determines which BC, and those small-granularity components such as button, data set and its attributes included in it, is executed by which role.

BC's scheduling pattern describes the execution sequence between each component, so to form macro process between BCs.

3.2 BC scheduling pattern

BC's scheduling pattern describes the execution sequences between each component, to form macro process between BCs.

In manufacture enterprise, for example, users can create an order from purchase requirement or purchase plan by clicking hyperlink to select them. The process involves three BCs, i.e. purchase order, purchase requirement and purchase plan. The process needs to be scheduled strictly.

First, users open the editing order interface, and then query, or skip the step. In any case, only those orders which are editing or have not summated to audit can be displayed.

Second, the purchase requirement and purchase plan interfaces opened by users only display those audited requirement and plan respectively. Those purchase requirement and plan which are edited or have not been audited can not be listed. Thus it can be seen that interaction between BCs is scheduled depending on state. Order audit is executed by a set of roles step by step. This audit process not only are constrained by audit rules, but also are controlled by states.

From the example, interaction between BCs have two aspects, data interaction and execution dependency (process). For the former, all the states of each BC are all determined. For example, when editing order, the selected purchase requirement must keep the state "audited". For the later, there exist several transfer patterns. For example, a state can only be transferred to its next state, depending on some conditions.

State transition produces event, and event triggers state transition depending on activity execution. For conveniently give expression to different state transition pattern, we adopt four workflow control pattern (WfMC)

DOI: 10.1051/matecconf/20166305003

[14]: sequential, and /or split and iterative. Under sequential routing, several activities are executed step by step. Under parallel routing, two or more activities are executed in parallel. Event triggers activity execution which must satisfy some constraint. This is ECA (Event Condition Action) pattern. An activity can produce many parallel events depending on some conditions or not, which is called event logic.

The items displayed on user interface are workflow task list (WfMC), which provides all tasks a user must do. For sequential state transition, an exact button can be determined. For example, button "audit" can be put on order edit interface. For other control patterns, next state(s) will depend on some conditions or not, so except button "next step", exact button can not be determined. An activity interface must own an exact state. State transition realizes BC interaction. This is workflow.

Some constraints may be satisfied before starting an activity when an event comes. For example, audit activity should satisfy such business rule: section chief audits orders if any value; if value>5000 and section chief has audited, then vice manager audits; if value>10000 and vice manager has audited, then manager audits. In the constraint, audit rule and audit state are described at the same time. Those business object attributes involved in the constraint and conditions used in event logic are called workflow relevant data. Because these constraint and conditions are used to control business logic, they are expressible in ESA. So workflow relevant data and constraint expression can be separated and converted to system configuration and modification.

3.3 Executable workflow model

Based on analysing above, executable workflow model should involve the four elements:

(1) State. State is used to control user interface, and to filter business object items as a part of query where expression.

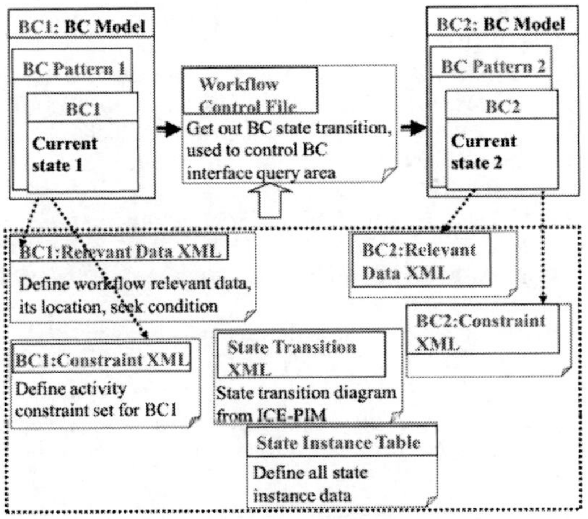

Figure 3. Executable Workflow Model

(2) Constraint condition. Constraint condition is constructed to executable SQL statements, associating with workflow relevant data in code. It is used to judge constraint of activity and condition to produce event.

(3) Workflow relevant data. Workflow relevant data refers to attributes which have association with constraint and condition expression used to control process.

(4) Workflow state instance table. State instance is used to track state transition. State instance is independent from framework and always stored in database.

Executable workflow model must configure these files at least, as shown Figure 3.

Workflow engine schedules process, as shown in Figure 4. When a user click a button in an interface, workflow engine computes the next state(s) according to workflow state instance and the current state, following the state transition diagram, then record the next state into workflow state instance table.

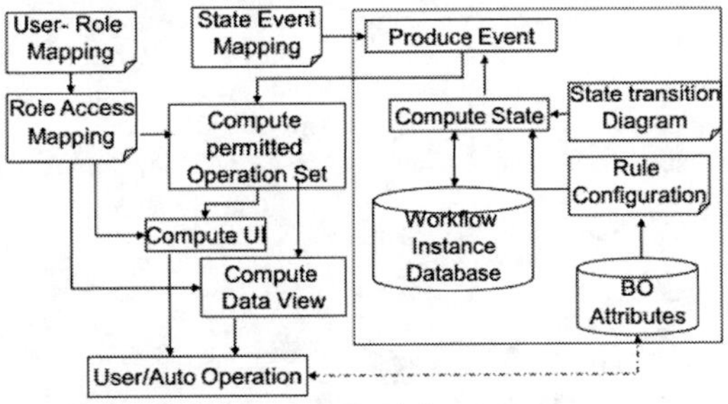

Figure 4. Workflow Engine Scheduling Process

The interaction between BCs involves two aspects: data interaction and execution dependency. Data interaction means that all user interfaces must have a determined state or state logic, which control business object items displayed. Execution dependency (process) means that when an interface is opened, and a business object instance is done, and next state needs to be computed. For example, an order is edited, and user clicks the button "audit" to request computing next state. Workflow engine receives the request, computes next state and updates the state instance table. After refreshing audit interface, the business object item will disappear.

PCGCM lies in the PSM level in our MDA-based framework ICEMDA, as shown in Figure 5.

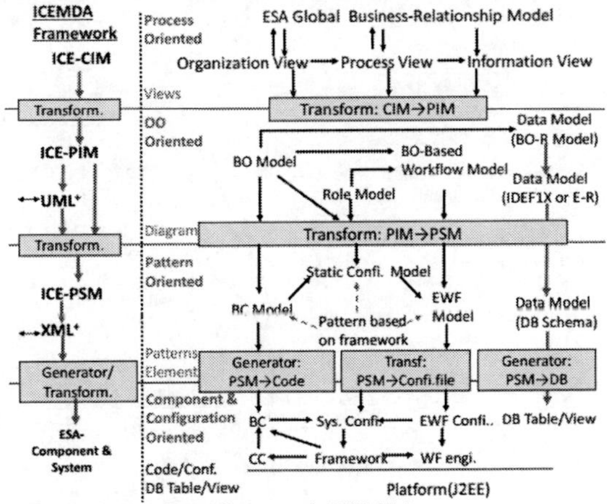

Figure 5. MDA-based framework ICEMDA

DOI: 10.1051/matecconf/20166305003

4 The case

The architecture of component based ESA can be seen in Figure 6. An execution of ESA can be divided into 3 stages. The first stage is constructing of component repository. In this stage, BC model which is descripted by PSM is transformed into execution component. Developers of components generate description files according to some standards, and store them into component repository all together.

The second stage is configuring of system. In this stage, some required components can be picked up from component repository, and then be initialized by some parameters.

The third stage is running of system. In this stage, besides J2EE application service, as basic executing platform, workflow engine and ESA framework are also needed. The workflow engine is used to access workflow configuration file, and the later is used to determine user roles. We have applied the modelling system and code generator toolkit to several ERP systems in Chinese enterprises. The efficiency of ESA developing is improved.

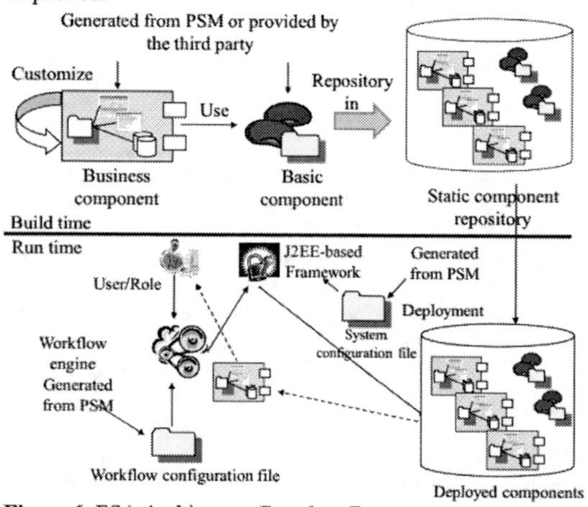

Figure 6. ESA Architecture Based on Component

Besides the patterns mentioned above, ESA developers can add their own patterns into to model, as shown in Figure 7.

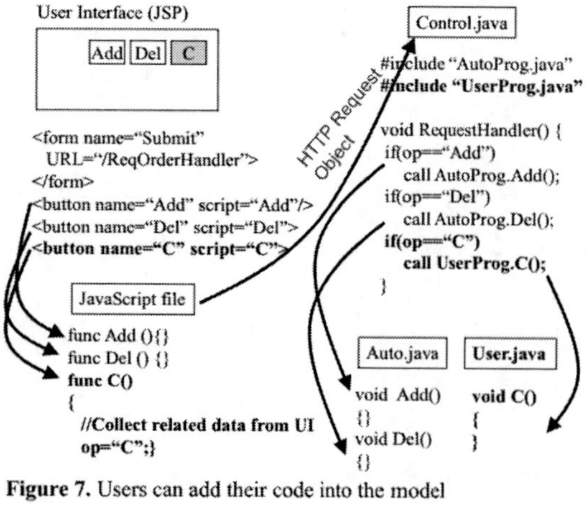

Figure 7. Users can add their code into the model

5 Conclusions

In this paper, a method of pattern-based coarse-grained component modelling for ESA developing, PCGCM is proposed to improve the efficiency of developing. The method presents a detail design for PSM in MDA-based framework. PCGCM can achieve this because it is object-oriented. Moreover, the principle by which BC is scheduled via an executable workflow model and its engine is represented. The proposed method has been applied to develop several ERP systems for Chinese manufacturing enterprises and has a good experience.

Acknowledgments

The authors are grateful to Prof. Dechen Zhan, Research Centre of Intelligent Computing for Enterprises & Service, School of Computer Science and Technology, Harbin Institute of Technology. This work was supported by Quanzhou Science and Technology Plan Projects (Nos. 2015Z125).

References

1. A. Khoshkbarforoushha, P. Jamshidi, M. F. Gholami, L. Wang, R. Ranjan. *IEEE Sys. J.*, **10**, 36, (2016)
2. H.B. Li, K.C.C. Chan, M.X. Liang and X.Y. Luo. *IEEE Trans.Indust. Inform.*, **12**,211(2016).
3. H. I-Ching, T. Der-Hong, H. Nien-Lin. *Comp. Stand. Interfaces*, **36**,648(2014)
4. Y. Rhazali, Y. Hadi and A. Mouloudi. *3rd Int. Conf. MODELSWARD*. 312(2015)
5. L. Zouhaier, Y. B. Hlaoui and L. J. B. Ayed. *IEEE 38th COMPSAC*. 535(2014)
6. Y. Rahmoune, A. Chaoui and E. Kerkouche.. *Procedia Comp. Sc.*, **56**,612(2015)
7. D. Ameller, X. Burgués, O. Collell, D. Costal, X. Franch and M. P. Papazoglou. *Inf. Softw. T.*, **62**, 42(2015)
8. G. Xie, F. Qian and X. Hei. *9th Int. Conf. CIS*. 688(2013)
9. D. Maplesden, E.Tempero, J. Hosking and J.C. Grundy. *IEEE Trans. Softw. Eng.*, 41, 691(2015)
10. S. Chandramohan, L. Ravikumar, G. Doolla, S. and S.A. Khaparde. *IEEE Trans. Power Syst.*, 30, 132(2015)
11. H. Fang, H. Zhu, J. Shi. *IEEE 16th Int. Conf. HASE*. 1(2015)
12. S. Al-Fedaghi. *IEEE 11th Int. Conf. ECTI-CON*. 1(2014)
13. A. Leva, D. Mastrandrea, M. Bonvini, A. V. Papadopoulos. *IEEE 7th Int. Conf. UCC*. 554(2014)
14. MC. *Workflow Management Coalition Specification: Terminology & Glossary. Document Number WFMC-TC-1011*. Brussels (1996)

Clone-based Data Index in Cloud Storage Systems

Jing HE[1,2], Yue WU[1] , Yang FU[2] and Wei ZHOU[2,a]

[1] *School of Computer Science and Engineering, University of Electronic Science and Technology of China, 611731 Chengdu, P.R.China*
[2] *School of Software, Yunnan University, 650091 Kunming, P.R.China*

Abstract. The storage systems have been challenged by the development of cloud computing. The traditional data index cannot satisfy the requirements of cloud computing because of the huge index volumes and quick response time. Meanwhile, because of the increasing size of data index and its dynamic characteristics, the previous ways, which rebuilding the index or fully backup the index before the data has changed, cannot satisfy the need of today's big data index. To solve these problems, we propose a double-layer index structure that overcomes the throughput limitation of single point server. Then, a clone based B+ tree structure is proposed to achieve high performance and adapt dynamic environment. The experimental results show that our clone-based solution has high efficiency.

1 Introduction

With rapid development of information technology, huge amount digital data brings great challenges to data storage and data retrieve. Recently, the cloud storage techniques were proposed to satisfy the huge storage requirements. These cloud storage systems can provide higher scalability and fault tolerance than existing relational database system, because they adopt distributed file systems to manage large scale datasets. Users share a "black-box" which known as Cloud that consists of a large number of interconnected data nodes. They can tailor the storage resources for their own purposes from the infinite amount of resources that Cloud systems can provide.

However, due to storage schema is changed in cloud environment, the traditional data index structures are unable to directly be transplanted to the current cloud storage systems. For example, the B+-tree has been successfully used in many disk file index systems, such as XFS[1] and JFS (Journal File System)[2]. Paper [3] proposed a distributed B+-tree structure to construct data index for cloud storage systems. Although this structure is not only easy to implement but also has high scalability and fault tolerance, it consume too much memory space with the amount of data increasing, besides, the enormous index is difficult to maintenance. Therefore it is not suitable to directly use traditional B+ tree in cloud environment.

Meanwhile, under the dynamic cloud environment, users can only see the current value of the data now. In practice, some specific data needs to be traced for its value history. For example, the products' price adjusting history is important to sales decision. It is a useful method that duplicated the different versions of the same records on the disk. But the versions' management is so complex. Even the disk space cost is unbearable for big data.

To address these problems, this paper proposes a double-layer index framework firstly. Then, a clone based B+ tree structure is designed, which can be dynamic modified during processing. Historical value of records can be traced and even to be modified according to specific requirements. Finally, the experimental results show that our clone-based solution has high performance.

2 Related works

The existing cloud storage systems mainly include Yahoo's PNUTS [4], Amazon's Dynamo [5], Facebook's Cassandra [6], Google's Bigtabe [7] and its open-source variant HBase [8], which are designed to carry out the large-scale distributed storage. They generally adopt pairs of key-value to organize data. It is efficiently when using the decided keyword to retrieve data in these systems, while it cannot satisfy users' time requirement when users send out non-key filed or multiple attributes query requests. Therefore, many research works have focused on construct data index for cloud storage systems.

Sai Wu et.al [9] proposed a double layer cloud index. In this framework, the cloud index is consists of two components: local index and global index. The local index is used to organize data on each individual data nodes, while the global index can locate data's storage location among the cluster.

Paper [10] presents an improved B+ tree index. This solution builds a local B+-tree index for each data node that only indexes data on that node. With an adaptive algorithm, a proportion of the local B+-tree nodes are

[a] Corresponding author: zwei@ynu.edu.cn

published to the BATON [11] overlay. It is efficiently for single attribute query.

An R tree and CAN (Content Addressable Network) based multi-dimensional index schema called RT-CAN is proposed in paper [12]. In RT-CAN, a CAN [13] overlay is construct on top of local R-tree index. Furthermore, a dynamic index nodes selection algorithm and a cost model are proposed for RT-CAN. This solution provides high performance for multi-attributes query.

Paper [14] proposed another efficiently double layers index that support multi-dimensional query. Its local index using the improved quadtree, and global index adopts Chord overlay. Based on quadtree's coding, this index structure makes local index to the global index node release strategy is more concise.

However, these index frameworks are difficultly be applied to existing cloud storage systems because they are mostly adopt a P2P overlay as their global index while current cloud storage systems are master-slave architecture. Moreover, above cloud index systems do not support dynamic changing tracing and specific historical records modifying. That means, although we have a mount of huge volumes storages space, we can only search data block through current index systems, but cannot access historical data block by index systems.

3. Double-layer index framework

In distributed storage systems, a large-scale data set is usually divided into many small data sheets called data shard through horizontal partitioning method. Then, these small data shards are distributed stored on many different data nodes with specific load balance algorithm. For improve the query performance, it is a simple method that built a global distributed data index for the whole data set. However, this kind of global distributed data index is difficult to maintenance, and it will take up a lot of memory space on each data node. Once appeared, double layer index framework has gained much attention in the field of cloud storage.

In this paper, a B+ tree based double layer index framework is showed below.

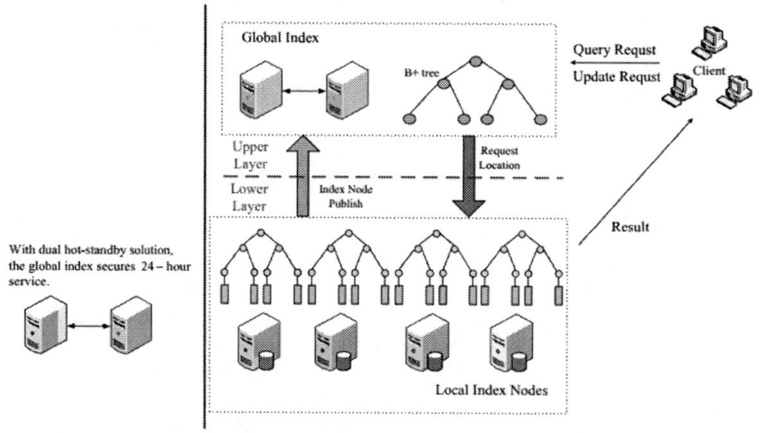

Figure 1. The B+ tree based double-layer index framework.

As shown in the Figure1, there are two layers (upper layer and lower layer). In the upper layer, it is a global index. That index store in a cloud server by the structure of B+ tree. This server stores the index nodes without the data. It uses the RPC (Remote Procedure Call) protocol to communicate between the cloud index server and local server. In the lower layer, the circles represent the index nodes of the local data server. The rounded rectangles represent final data blocks. Every local data server maintains a local index that is built by B+ tree.

When user proposed a query request, it is send to the global index server which performs index retrieve on global B+ tree to find the local data servers that may contain the query results. Then, the query request is redirected to the corresponding local data servers. Finally, each selected local data server start to retrieve the data on its own local indexes, and return the query results to the end user.

Compared with unitary key-value pair storage schema, our cloud index can help to cope with range queries. However, since the B + structure is adopted to the entire index, if a problem arises in any index node, such as, index information error and node information loss, the other nodes will be affected. The nearer the necrotic index node to the root node, the more data node will be affected. Therefore, the fast backup and recovery scheme of the index structure deployment is greatly important. It is necessary to study deeply in efficient clone-base B-tree in cloud environments.

4. Clone-based B+ tree Structure

Previous cloud index systems can only record current index information, which is just a one version index. However, due to huge storages, the cloud storage system can even store all data block update information. It requires that the cloud index system provide multi-version index function, which is used to trace historical records.

4.1. Background of clone

 DOI: 10.1051/matecconf/20166305004

Clone and snapshot are two important techniques which can be used to replicate the system in certain time. Compare with clone technique, the weakness of snapshot is that it cannot be modified. Any attempt to change the snapshot mirror is a kind of disaster, because the modification to the snapshot may cause the damage to the original data. Instead clone can solve this problem. Paper [15] proposed a basic clone model.

Let Tp is a B+ tree and Tq is a clone mirror of Tp, a tree-based structure's clone model should have following properties:

(1) Space usage rate: Tp and Tq should share the common node as much as possible.

(2) Time usage rate: creating Tq from Tp should take little time.

(3) Number of clones: Tp can be cloned many times.

(4) Multiple-clone: Tq is also can be cloned.

Typically, fully backup the whole tree is the easiest way to clone which can store a complete clone mirror for all over data at a certain time. However, this method is time inefficiency and the fully backup version will occupy much more storage spaces.

In this section, an improved clone model is designed to overcome above weaknesses. In which, an indicator named as reference count is added to every node for a B+ tree. Through check the value of this reference count, the index system determines a node is cited by how many index versions. Then, there is only partial index nodes should be replicated according to specific rules, while other nodes can be shared by different index versions.

4.2 Clone-based B+ tree Creation

An original B+ tree index is shown in figure 2(a). It is no double that each node is belong to one tree, so the reference count on each node is 1.

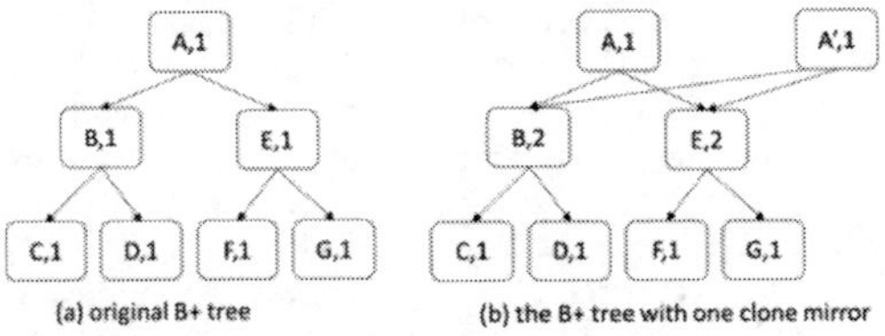

Figure 2. Create a clone-based B+ tree.

In figure 2(b), a clone tree *A'* is created from original B+ tree *A*. Firstly, there is only one index node that is the root node *A* needs to be replicated as *A'*. Then, the reference count of index node *B* and *E* is increasing 1 respectively. Finally, a newly cloned B+ tree is produced. Our B+ tree clone model meets all previous described properties. By the same clone method, the original B+ tree *A* can be cloned again, and the clone mirror *A'* also can be cloned. Due to few index nodes need to be duplicated, this clone model has constant time efficiency and space efficiency. With multiple-version index function, users can retrieve or update data from any index tree of them.

4.3 Clone-based B+ tree Maintenance

When data modification (inserting or deleting) happens to the index tree, in order to ensure the isolation between each tree, the key ideas are described below:

(1) If the reference count of the index node which would be modified is 1, it is indicated that this node is belong to only one tree. The modification to this node cannot affect other index trees, so the traditional B+ tree update process is suitable in this situation.

(2) Otherwise, if the reference count of the index node which would be modified is larger than one, it is indicated that this node is belong to more than one tree. In this case, this index node needs to be duplicated before updated.

A typical data updating procedure is show as figure 3.

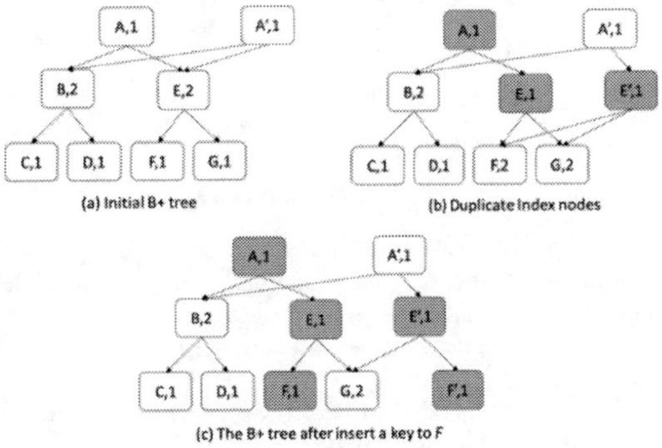

Figure 3. The procedure of data update

In figure 3(a), there are two B+ trees which are represented as original B+ tree *A* and its cloned B+ tree *A'*. Due to the index nodes *B* and *E* are belong to both index tree *A* and *A'*, their reference count are 2. Assume a new keyword will be inserted into index node *F* in B+ tree *A* now, the CSD (Clone Status Detect) function checks the reference count of all nodes in the path from root to *F* will be triggered. Firstly, since the reference count of index node *E* is 2, the node *E* is replicated as *E'* which cause the reference count of index node *F* and *G* increase 1 and the reference count of index node *E* decrease 1. Then, Root *A'* points to *E'* as show in figure 3(b). The node *F* is replicated as *F'* and so on. Finally, the node *F* is updated in B+ tree *A* by traditional B+ tree modification algorithm while the index node *F'* in clone tree *A'* keep the same. The node split in B+ tree *A* has no effect on clone tree *A'*.

Usually, data deleted from B+ tree would lead to index node merge operation. Therefore, an index node merge algorithm is considered in our clone model.

As can be seen from figure 4(a), there are two trees, one is the original tree *A*, the other is the clone tree *A'*. Assume a delete operation which remove some keywords from index node *F* triggered nodes merge. The detail description about nodes merge procedure is show in figure 4. In figure 4(b), on the path from node *A* to *F*, those nodes whose reference counts greater than 1 are copied. *F* and *G* merged into a new index node *H*. Then, because the original B+ tree is no longer balanced, it needs to be adjusted. The index node *B* and *E* merged into *I*. At last, the previous root node *A* is removed according to traditional B+ tree delete algorithm. In figure 4 (d), the original B + tree has completely changed while cloning B+ tree didn't has any effect. In our multi-version index system, the newly B+ tree *I* records current data information, at the same time, cloning B+ tree *A'* maintains historical information.

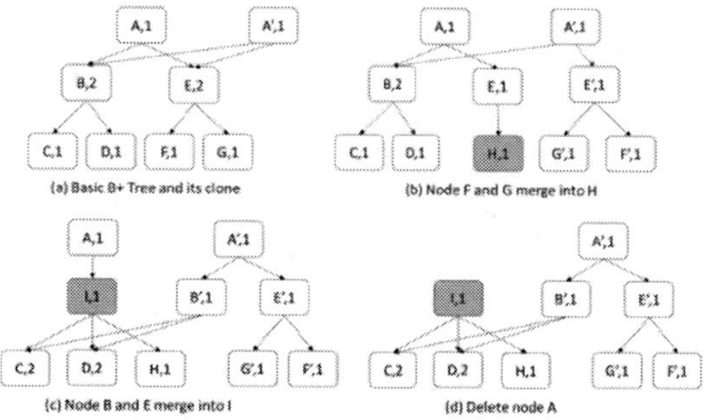

Figure 4. The procedure of index node merge

In order to save the storage space, cloning model should share the original node as much as possible. Therefore, the reference count of a node needs to be considered to determine whether it will be duplicated in our clone model. When data modification occurred, the CSD function is called to detect the value of a reference count. If a data update operation lead to a large number of index nodes adjustments, the efficiency of clone-base B+ tree maintenance algorithm will deteriorate for lots of CSD function recall. Fortunately, due to the branching factor of a B+ tree is usually big enough, the index adjustment is always small and local. The whole structure is stable. Let n represents the number of a B+ tree's leaf node, the time efficiency of our clone-base B+ tree maintenance algorithm is $O(\log_2 n)$.

As for the algorithm, the CSD function is independent encapsulation, we only add the CSD function to where the node needs to be modified. The pseudocode of CSD function is below:

Input

curRoot: the root node of the modifying index tree currently.

curNode: the index node is modified currently.

refNode: the brother of *curNode* that needs to be merged with curNode.

Algorithm 1 CSD (*curRoot, curNode, refNode*)

```
1: tmpNode=curRoot;
2: preNode=null;
3: tmpKey=curNode.entries.get(0).getKey();
4: WHILE tmpNode != curNode
5:    FOR tmpNode.entries.get(i=0).getKey() TO
              tmpNode.entries.get(i=entries.size()).getkey()
6:      IF tmpKey>=tmpNode.entries.get(i).getKey()
           && tmpKey<tmpNode. Entries.get(i+1).getKey()
        THEN
7:         preNode = tmpNode;
8:         tmpNode=tmpNode.children.get(i);
9:      IF tmpNode.refCount > 1 THEN
10:        tmpNode.refCount --;
11:        preNode.children.set(i, tmpNode.clone());
12:        cloneNode=preNode.children.get(i);
13         cloneNode.refCount = 1;
14:     IF tmpNode.children!=null THEN
15:       FOR TNode=tmpNode.children.get(0) TO
              tmpNode.children.get(tmpNode.children.size())
16:       TNode.refCount ++;
          ENDFOR
17:   IF refNode != null THEN
18:     i = preNode.children.indexof(refNode);
19:     refNode.refCount --;
20:     preNode.children.set(i, refNode.clone());
21:     preNode.children.get(i).refCount=1;
22:   IF refNode.children!=null THEN
23:     FOR TNode= refNode.children.get(0) TO
              refNode.children.get(refNode.children.size())
24:       TNode.refCount ++;
        ENDFOR
25:   ENDFOR
26: ENDWHILE
```

In algorithm1, there are three input parameters which are the root node of current index tree, the index node which will be modified, and the node that will be merged with *curNode*. Firstly, the current index tree is traversed from root node *curRoot* to the modifying node *curNode* (line 4-8). Within the traversing process, the previous node of the *curNode* is stored in *preNode* (line 7-8). Then, if the reference count of the scanned index node is larger than one, this index node is cloned (line 9-13). The reference count of the original node is minus by one. If this modified index node named as *tmpNode* is not a leaf node, the reference count of its all child nodes need to be updated (line 14-16). At last, if index node merge happened which means the *refNode* is not null, the process is from the bottom up while the nodes traverse above is top-down (line 17-24). For all children of this *refNode*, their reference count is plus by one (line 22-24).

Due to this function is trigger before any modification to the index node, the branching factor of the B+ tree is important to the efficiency of clone-based index maintenance. If the branching factor is low, the probability of this CSD function is called is high. On the opposite, the higher branching factor will make the probability lower.

5. Experimental evaluation

The test platform used in this paper is Intel Core Duo T9900 3.06GHz 2 core 2 threads CPU, 8GB memory, 64 bit windows8.1 operating system. JDK edition is 1.7.0_60, eclipse edition is 4.3.2. To ensure that the Java virtual machine has enough memory to use, the execution parameter -Xmx4000m -Xms3000m is added at run time. Specifically, 4GB available memory and 3GB available stack is allocated to Java virtual machine dynamically.

By this optimized parameters setting, there will be no stack overflow even if the amount of the data is too large or function recursive too many times.

Compared with fully backup the index tree, cloning can save huge amount of storage spaces. Meanwhile, clone an index node is so quickly, its time consume is almost can be neglected. However, due to there exists multi-version index trees for clone-based index structure at one time, index maintenance, such as inserting or removing a data, is turn to costly. In this section, many laboratories are built to evaluate the performance of data inserting and data deleting for clone-based B+ tree index.

Limited to the platform performance, each B+ tree index in our experiments only has one original tree and one clone tree. To guarantee the accuracy of experiment, a certain number of data are randomly generated and stored in a link list. Using this link list, it makes the order that each time inserting is the same.

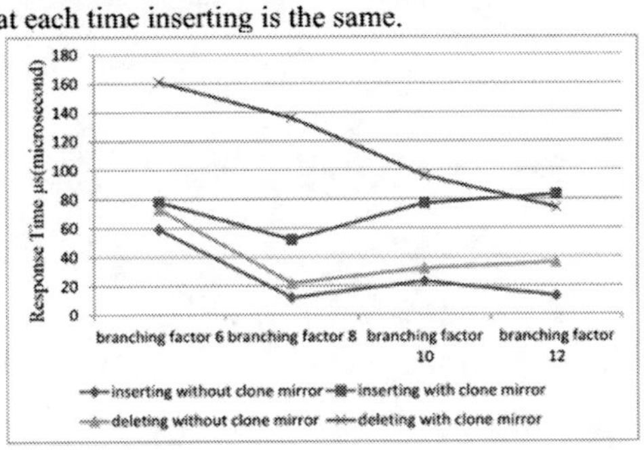

Figure 5. Time cost when inserting or deleting one thousands of data

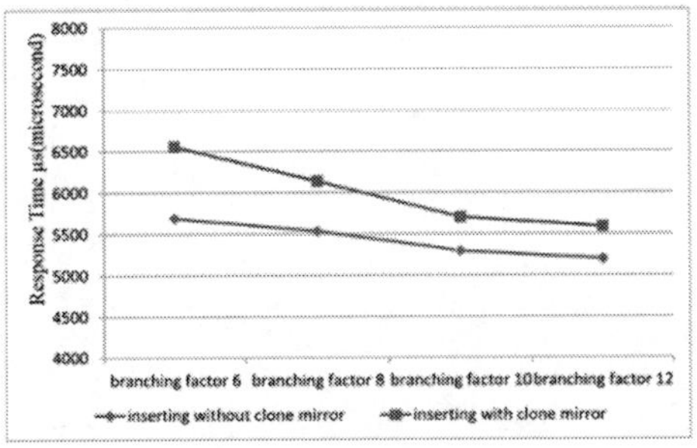

Figure 6. Time cost when inserting one million of data

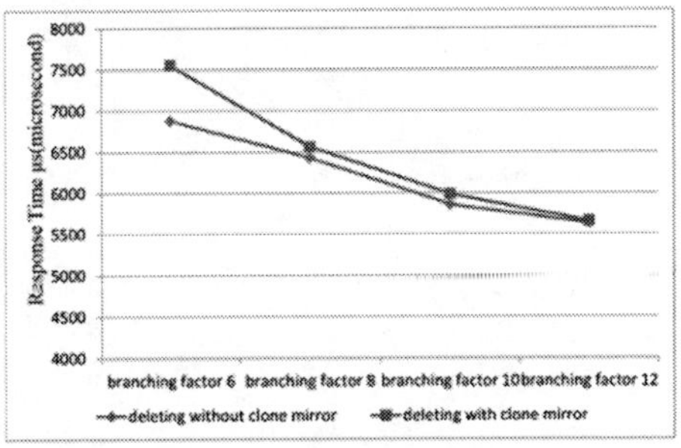

Figure 7. Time cost when deleting one million of data

Figure 5 shows that the addition of clone mirror makes the maintenance time cost much higher than original B+ tree structure. However, with the amount of data which inserting or deleting into the index tree increasing sharply, the extra time cost turns gently. In figure 6 and figure 7, when the amount of data that inserting or removing is one million, the operation response times which with clone mirror and without clone mirror have little difference. The loss of efficiency is acceptable. The reason is that the clone mirror is only duplicating a small part of index nodes which are modified. Under the big data environment, these duplicated index nodes is too small to the whole data set.

In addition, the above experiments verify the influence of the branching factor. In figure 7, when the branching factor is 6, due to deleting operation lead to some index nodes merge, then, the response time with clone mirror is much higher than the case without clone mirror because there is more CSD function call. Otherwise, when the branching factor is 12, the time cost is almost the same whenever with clone mirror or without clone mirror. The reason is the additional CSD function call turns to small.

6. Conclusion

This paper presents a double layer index structure based on B+ tree, this structure overcomes the throughput limitation triggered by single point server when index overload increase. Then a practical clone-based B+ tree model is designed. With clone mirror, our index tree is not a single version tree again, it is a multi-version tree. The different value of a record within a period of time can be traced through these different versions. Our clone model only backup modified nodes and share the other nodes in a B+ tree, therefore, this solution has advantages both in time complexity and space complexity. Finally, experimental results show that this clone-based B+ tree model has high performance and availability under dynamic cloud environment.

Acknowledgement

This work is supported by the National Natural Science Foundation of China (61363021); Science Research Fund of Yunnan Provincial Education Department (2014Y013);

References

1. Moshe Ba. *Linux File Systems*, 1nd ed., McGraw-Hill Companies, New York. (2001).
2. Vijayan Prabhakaran, Andrea C. Arpaci-Dusseau, and Remzi H. Arpaci-Dusseau. *Analysis and Evolution of Journaling File Systems*, USENIX Annual Technical Conference (USENIX Association) (2005).
3. Marcos K. Aguilera, Wojciech Golab, Mehul A. Shah. *A practical scalable distributed B-tree,*

Proceedings of the VLDB Endowment, **1(1)**, pp. 598-609. (2008).

4. Brian F. Cooper, Raghu Ramakrishnan, Utkarsh Srivastava Adam Silberstein, Philip Bohannon, Hans-Arno Jacobsen, Nick Puz, Daniel Weaver and Ramana Yerneni, *PNUTS: Yahoo!'s hosted data serving platform*, Proceedings of the PVLDB. Auckland, New Zealand, pp. 1277-1288. (2008).
5. Giuseppe DeCandia, Deniz Hastorun, Madan Jampani, Gunavardhan Kakulapati, Avinash Lakshman, Alex Pilchin, Swaminathan Sivasubramanian, Peter Vosshall and Werner Vogels, *Dynamo: amazon's highly available key-value store*, Proceedings of the 21st ACM Symposium on Operating Systems Principles. Stevenson, Washington, USA, pp. 205–220. (2007).
6. Avinash Lakshman, Prashant Malik. Cassandra: a decentralized structured storage system, ACM SIGOPS Operating Systems Review, **44(2)**, pp. 35-40. (2010).
7. Fay Chang, Jeffrey Dean, Sanjay Ghemawat, Wilson C. Hsieh, Deborah A. Wallach. Mike Burrows, Tushar Chandra, Andrew Fikes, Robert E. Gruber. Bigtable: A distributed storage system for structured data. ACM Transactions on Computer Systems, **26(2)**, pp. 1-26. (2008).
8. Craig Franke, Samuel Morin, Artem Chebotko, John Abraham, and Pearl Brazier. *Distributed semantic web data management in HBase and MySQL cluster*, IEEE International Conference on Cloud Computing. New York, pp. 105-112. (2011).
9. Sai Wu, Kun-Lung Wu. An indexing framework for efficient retrieval on the cloud, IEEE Data Engineering Bulletin. (2009).
10. Sai Wu, Dawei Jiang, Beng Chin Ooi and Kun-Lung Wu. *Efficient B-tree Based Indexing for Cloud Data* Processing, Proceedings of the VLDB Endowment, **3(1)**, pp.1207–1218. (2010).
11. H. V. Jagadish , Beng Chin Ooi , Quang Hieu Vu. *BATON: A Balanced Tree Structure for Peer-to-Peer Networks*, Proceedings of the 31st international conference on Very large data bases. Trondheim, Norway, pp. 661-672. (2005).
12. Jinbao Wang, Sai Wu, Hong Gao, Jianzhong Li, and Beng Chin Ooi., *Indexing multi-dimensional data in a cloud*, Proceedings of the 2010 ACM SIGMOD International Conference on Management of data. Indianapolis, Indiana, USA, pp. 591-602. (2010).
13. S. Ratnasamy, P. Francis, M. Handley, R. Karp, and S. Shenker. A scalable content-addressable network. ACM Sigcomm Computer Communication Review, **355(4)**:161-172. (2002).
14. LinLin Ding, Baiyou Qiao, Guoren Wang and Chen Chen, *An efficient quad-tree based index structure for cloud data management*, Proceedings of the 12th International Conference on Web-Age Information Management. Wuhan, China, pp. 238-250. (2010).
15. Rodeh O., B-trees, shadowing, and clones, ACM Transactions on Storage (TOS), **3(4)**, pp. 2. (2008).

Vision of Curriculum and teaching from Ecological Sustainability

Nai Zhu HUANG, Jia Ming ZHONG and Si DENG[a]

Xiangnan University,Chenzhou,Hunan,China, 423000

Abstract. Anthropocentrism of modern society leads to great damage to the environment. Mankind is faced with serious ecological crisis. In order to protect the environment, mankind must abandon political and religious differences and take the road to sustainable development. Modern education as delivering environmental information should contribute to protecting the environment. However, modern education based on utilitarian values can not cultivate global citizen possessing eco-consciousness. Reforming modern education model and advocating ecological education model is a reasonable choice in education reform.

Keywords: ecology, ecological sustainability, curriculum and teaching, vision

1 ECOLOGICAL CRISIS AND ECOLOGICAL SUSTAINABILITY

The word "ecology" is originated from Greek, which originally means "a dwelling place" or "a habitat" and now refers to the environment mankind lives in. At present, great damage is caused to global environment when mankind conquers the nature and satisfies self-demand. It is mainly shown as below: serious environmental pollution, reduction of rainforest and wetland, preventing disappearance of ozonosphere which makes mankind free from ultraviolet invasion and prevent global warming. Facing continual environment deterioration, the earth's ecological capacity in recovery, updating and rebalancing is badly insufficient. In this case, mankind will be faced with environmental disaster in the uncertain future. Facing global continual environment deterioration, far-sighted personages vigorously advocate environmental protection, advise policy makers of various countries to redefine the concept of "progress", criticize the hypocrisy and cruelty of market system with free competition and request the consideration of ecological sustainability in the process of making policies. The so-called ecological sustainability is to live in harmony between man and nature. In the process of development, mankind cannot abuse or squander natural resources and discharge plenty of waste, which leads to global environmental deterioration and near extinction of animals and plants. The idea of "syncretism between heaven and man" in ancient China reflects ecological sustainability. According to *The Book of Changes*, "The great man is he who is in harmony, in his attributes, with heaven and earth; in his brightness, with the sun and moon; in his orderly procedure, with the four seasons; and in his relation to what is fortunate and what is calamitous, in harmony with the spirit-like operations (of Providence). He may precede Heaven, and Heaven will not act in opposition to him; he may follow Heaven, but

[a] Si DENG:xnxytg@163.com

will act (only) as Heaven at the time would do. If Heaven will not act in opposition to him, how much less will men! How much less will the spirit-like operation (of Providence)!" [1]. Lao Zi also held that mankind was originated from nature and unified in nature, must survive under the condition of natural bestowal and can develop by following the nature law. The laws among heaven, earth and man are interlinked. That is, "Man models himself after Earth. Earth models itself after Heaven. Heaven models itself after Tao. Tao models itself after Nature." [2]. When the relationship between human society and nature is researched, it is proposed in modern western ecological philosophy that morality should be extended to animals, and the antiquated idea of "morality gap" between man and animal should be abandoned. French ideologist Schweitzer thought about the relationship between man and nature from the cultural perspective to form the ethics respecting life. He proposed the establishment of ethics showing reverence for life to get rid of cultural crisis. American ideologists Thoreau, Evans, Moore and Leopold et al emphasized that the earth was an organic life wholeness which forms the ecological ethics for natural wholeness. The ecological view in "Self-organizing Evolution of Nature" [4] by Erich Jantsch, a famous system philosopher in Austria, extremely widened people's vision and aroused people to protect the global environment. He held that the nature was organic, dynamic and growing and had the capacity of self-organizing evolution. Self-organization means that thing (system) is not externally forced but a dynamic process with ordered structure is spontaneously formed through the interaction among its internal constituent parts. Superficially speaking, the evolution process of modern western ecological philosophical thoughts is a direct response to the global ecological crisis. Deeply speaking, it is a revolution about way of thinking. It is a kind of introspection, criticism and transcendence for hypostatic and privative way of thinking by Newton and Ducal as well as world view and value of "anthropocentrism".

2 ECOLOGICAL CRISIS OF MODERN CURRICULUM AND TEACHING

The increasing ecological crisis will inevitably lead to human's quick destruction. How does mankind face currently serious ecological crisis? National governments must get across political, economic and religious disagreement to form a united alliance and formulate common environment protection strategy to promote ecologically sustainable development. As an important field to deliver environment information and cultivate global awareness, education should make contributions to environmental protection and ecologically sustainable development. Regrettably, modern curriculum and teaching field is based on utilitarianism value and dual opposition thinking and advocates the education model like large batch production in factory. This model is not beneficial to students' ecological awareness cultivation and healthy personality development. The poet Eliot applied prosaic tone to have profound disclosure: An individual requires more education not for wisdom but for further maintenance; a country requires more education to excel other countries; a hierarchy requires more education to excel other hierarchies or not to be excelled other hierarchies. On the one hand, education is related to technical effect. On the other hand, it is related to nationhood improvement...If education means more money, greater dominant authority, higher social status or at least a stable and decent job, then there will be few people receiving education [4]. Sloan, education professor of Columbia University, held that when the crisis human faces is not only natural environment destruction but also includes overuse of racialism and ultranationalism, endless spreading of science and technology in human's life, cultural diversity and significant global destruction. He warned that the world "is rapidly collapsing under the identical trend of modern way of thinking and its subsidiaries. In these subsidiaries, education system...forces children to enter infertile, greedy and vulgar adult cultural world in advance" [5]. The compulsory and control characteristics of modern education can be powerfully illustrated by using the Mississippi as a metaphor: Before white people came, Indians held great respect towards the Mississippi as a mother river. It was written in *Negro's Proclamation on Rivers* by Hughes: "I know rivers well: I know that the rivers are as old as the world. It is even older than the blood flowing in human's vessels. My soul is as long as the rivers." [6]. After the while people came, engineers applied modern technology to build a lot of flood prevention levees and overflowing flood ditches which surround the river and control its natural progress and flow. Thus, the flowing rivers have been controlled for several thousand years. The Mississippi is not a constituent part of natural living beings and biological laws any more but becomes a channel for human to pursue economic benefits and commercial development. The huge flood disaster in Middle West in 1993 was just a warning given by the nature to mankind. Even the most advanced technology including large-scale sluices and flood prevention levees cannot completely control the rivers. It shows how human's intervention with the rivers destroys the natural progress and threatens the whole global ecological system. It indicates by using the Mississippi as the metaphor for teachers and students that: if all the sites for student education activities are limited, controlled and polluted by those who want to conquer soul and spirit, it will be terrible for student development. Ironically, responsibility system and standardized examination implemented in the field of modern curriculum and teaching lead to the generation of curriculum schemes involving breaking away from environment, history, culture and personal interest. When learning these curriculum schemes, students cannot understand the knowledge. The consequences caused merely only mechanical memory are to determine the solution to complex problems, authoritative and rigid structure analysis. All of these control means have destroyed students' flowing capacities. Like the controlled river water, it has hindered the development of their creativity, discovering capacity and feedback capacity in perspective of self-reflection. Like natural water flow, students' learning should be rooted in the significant situation and special environment. It can be allowed to have free flow, exploration and creation. Then it can cultivate a person who possesses creativity, imagination and spirituality. There are classical characteristics about modern education concept including currently popular Tyler's Rationale, curriculum plan based on behaviorism, competitive external evaluation and dualism model. Teachers and students, significance and situation, subjective body and objective knowledge, body and soul, learning and environment are separated each other by transferring neutral value information. This is just the source for mankind to confront the ecological crisis. In fact, postmodern science is against separation between material and consciousness and among fact, significance and value, for such separation has destroys human spirit, and it is an important reason to result in human's current dilemma. As David · Kennedy said, "In modern west, children have become a victim which is considered to be separated from society and is researched and controlled by people to serve the preset social objective[7]. Modern education control leads to increasingly rebellious spirit or nonchalance from adolescents against teachers and school leaders. It can be interpreted as a query from the young. What is spent at school is only learning, serving a certain purpose or finalizing the definition about some people's fate. Deeply speaking, the adolescents are not clear about the real reason for the relevant authority to specify 9-12 years to study at school. To turn about increasingly serious ecological crisis and education crisis, school education's overstressing on obeying, control, order, level, competition, personal success and cultural environment must be changed, for this kind of cultural environment is the source to lead to children's dissimilation, isolation and being refused and advocates a postmodern education development concept based on ecological sustainability. The famous scholar Marley [8] used the following table to compare the education paradigm comparison during industrial and ecological periods, which is of great enlightenment for educators to recognize the concept of ecologically sustainable education,as Table 1.

Table 1. Education Paradigm Comparison during Industrial and Ecological Periods

Education paradigm during industrial period	Education paradigm during ecological period
Bureaucracy	Community
Control	Relation
Independence	Interdependence
Competition	Cooperation
Isolation, dissimilation	Attribution

Student is a tool.	Student is an objective.
Great scale, impersonalization	Small scale, personalization
Base on instructor's teaching	Service learning, team learning

3 CURRICULUM AND TEACHING VISION OF ECOLOGICAL SUSTAINABILITY

In modern education, education and teaching is carried out in a compulsory and control manner, which is increasingly mechanical, rigid and short of love. To reform the education based on knowledge theory is imperative, for this kind of education has enlarged the roles of knowledge, capacity, science and technology in social development and strengthened the importance of learning knowledge in school life, but it neglects the important value for students to experience life. This leads to an alienating state between education and people, and the students' diversity, difference and imagination will be strangled. This kind of education completely went against the concept of ecological sustainability as well as the education spirit and objective. To construct ecologically sustainable curriculum and teaching concept is a great reform initiative for mankind to protect the earth ecology and benefit the descendants.

3.1 Curriculum and teaching objective

Teaching is absolutely a "treasure chest" to obtain knowledge and possess curriculum knowledge such as concepts, facts and data. If it is believed that children can step off a right way in the stereotyped world only after they obtain those stuffs, it will be wishful thinking. David · Smith believed, "Those students who 'master' the contemporary norms most sufficiently are just the most short-sighted without insight in ideal, sentiment and social conscience" [9]. The objective of curriculum and teaching under the vision of ecological sustainability is to cultivate an integrated person. An integrated person is a life entity with a unity between body and mind, intellectuality and morality, rationality and irrationality. He can get rid of egoism control, link with the world with an open attitude and live in harmony with the world. The holism educator Ron · Miller thought that an integrated person included several essential elements [10]: intelligence, emotion, physique, communication, aesthetics and spirituality. Anyone is a complex system in which ability, potential and creative vitality are interrelated. To label students with reductionism (learning disability, hyperactivity and subversive) or classify them by scores and levels and discriminate them means neglecting the students' complexity. A student's learning problem might firstly lie in intelligence, but actually it relates a problem network: student's self-value, position in class, school and community, family expectation and a way of dissimilating individual's life and defining learning as an excessively narrow cultural strength. Holism calls for thorough self-criticism from public education institutions based on reductionism and technocracy, for anyone is not merely a "case of illness", "test score", "citizen" or "worker" but an integral and complex person. Children are not a "resource" utilized by the state and industry and must not be treated with cruel and utilitarian reductionism.

3.2 Curriculum and teaching content

Modern curriculum and teaching are originated from standards, government's order, great work collection or preset curriculum plans. This is wide apart from student's growth, learning, exploration of human life and living reality. Ecologically sustainable curriculum is not a preset plan brought by the teacher in classroom but experience from interaction among teacher, students and world. Alternatively, curriculum is generated from dialogue between teacher and students. The education experience created by teacher and students is the real curriculum. Ecologically sustainable curriculum particularly stresses on interconnectedness among knowledge, learning experience, international community, natural world and life and tries to eliminate the privative phenomenon in curriculum. The links among things in curriculum preparation are the ones between linear thinking and intuitive cognition, body and mind, knowledge area, individual and community, human and earth, ego and nature. As for ecological literacy, emotion quality and social responsibility cultivation proposed nowadays, ecologically sustainable curriculum goes against transforming them into "units" or "subject plans" as themes to instruct students, but advises teachers to bring ecological literacy, emotion quality and social responsibility to students by communicating with students. Ron · Miller clearly pointed out that, "If the teachers offering curriculum do not care about these things and request a school, a local education authority or a state government to increase our ideal objectives, it will not lead to significant and convertible learning for students"[11]. To illustrate the conversion quality of curriculum, holism educators cite "dance" as metaphor for curriculum, for dance is full of changes and passion. It shows that the curriculum must be open to all inherent possibilities of student development. We cannot change the world into what we imagine. This is the only way for curriculum to march towards mankind creation and liberation. Ecologically sustainable curriculum must combine the following problems into teaching [12]. Citizen activists and strict waste processing rules have exposed the danger caused by radioactive garbage, toxic substances and hazardous goods to human health and multiplication. In public and private territory, big groups are carrying out excessive deforestation in tropical rainforest and virgin forest in northern hemisphere at an amazing speed. The information about endangered species goes into the public consciousness, and excessive deforestation has been prohibited in some regions. The hazardous goods like pesticide and herbicide in food and water supply, and the adaptability of other alternative agricultural system has appeared. Direct and nonviolent behavior has become an acceptable and significant political protest means. Other feasible and non-patriarchal system forms in noology and other means in mainstream religious sects treat people as a keeper or an equivalent of the nature rather than a dominant player. These form and sects have been widely accepted. Ecological education and selection of other feasible lifestyles have significantly reduced energy consumption. To recycle the resources becomes popular in many schools and community. Resources should be supplemented, environment should be restored, and biodiversity should be supported by all industrial and commercial groups and particularly transnational groups. Reproductive health and family planning should be proper for all women. More and more people realize that the purpose can also be achieved by using simple method and avoiding unnecessary complex method. People are working hard to maintain and improve their sensitivity and understanding to food although food is sufficient for eaters. Understand all kinds of differences of ethics and culture, care about the status of the third and fourth worlds, try to avoid excessive differences among material living differences and avoid excessive waste in life beyond indispensable living standards. More and more people understand and select significant work, and they do not work only for life. People attempt to meet the main needs instead of desire and thus reduce the wealth. People struggle in conflict decisively and bravely, but their words and behaviors still remain nonviolent. More and more people begin to adopt vegetarianism and strict vegetarianism lifestyle. They are absolute or incomplete vegetarians.

3.3 Curriculum and teaching methods

Modern curriculum and teaching only stress on such teaching environments as desk, blackboard, textbook and semester. Teachers do not know well about local community, geography, history, economy and occupation and utilize them as a kind of education resource. They do not stress on the value of multiple experiences and avoid it on purpose. The purpose is to cancel human communication, reflect and think about all possibilities. Ecologically sustainable curriculum and teaching stress on interaction of classroom environment and experience and advocate reflective dialogue, autobiographical logs, non-rivalry debate, cooperative study and zetetic problems in classroom teaching. All of these are the key points of classroom experience. Teachers play a role as instructors and conductors, organize students to have learning experience for self-reflection and will not play a dominant and controlling role in the whole process. "A dancing circle" can be used as a metaphor for this kind of curriculum and teaching. The circle shape is soft and returned, and finally it returns to the original starting point. The circle keeps moving, so it is dynamic. It moves spirally inwards by using its internal energy and keeps reflecting. The circle moves spirally outwards and becomes a stuff which is called by chaotic and new science as partial model. These partial models keep moving annularly to finally a harmonious vortex. Therefore, it can be said that it is a heater for imagination no matter the annular movement is inward or outward. As for students' learning, we must get rid of the concept of "progress" in modern concept. This kind of progress means removing unknown, ascertained and controlled states. Thus, learning progress can be represented only by knowledge accumulation. It is better if the accumulation speed is quicker. However, this kind of learning cannot trigger the motivation and give us a better survival mode. The reason is that such learning is outward or otherwise inward. It only cares about how fast a learner can run but does not care about how far he can run. Ecologically sustainable learning advocates a kind of "cultivation" mode to get rid of learning dissimilation state, for cultivation learning means that instrumental learning is changed into spiritual learning view. Learning becomes a life source full of interest. Peter · Senge said, "Learning also includes spiritual right-about or operation. However, learning has lost its core significance in the present usage. In daily expression, learning has meat absorbing knowledge or obtaining information. This has a great distance with the real learning. The real learning refers to the core meaning about human being. Through learning, we can do what we fail to do, recognize the world again and the relationship between us and it, and expand the energy to create the future. " [13] The cultivation learning suggests balancing quickly successful learning and tedious and mechanical learning by slow learning, empathized learning and spiritual learning. On the one hand, the cultivation learning should be a kind of slow learning. That is, learning needs to be in harmony with natural pace. This can help experience the nature better and the world mystery, for the natural progress is slow and gradual. Only slow learning can make "precaution" and "deep cognition" appear. In teaching, teachers need to respect the natural progress in classroom instead of quickly changing from one topic to another, from one country to another and from one activity to another. On the other hand, cultivation learning calls for thoroughly utilizing such cognitive styles as emotion, body sensibility, intuition, empathy, care, sympathy, love and spiritual experience to recognize and experience the life. Park • Palmer held that such cognitive styles as "empathy, intuition, sympathy and belief" can "make us feel the subtle information in the universe. Its silence delivers our relevant information about limit, responsibility and potential. When we, as a whole self, learn about these relationships, we begin to enter a community in which we know each other. In this community, we will be changed as we change others. "[14]. Regardless of empathized learning or spiritual learning, it is a kind of "precaution" practice in the final analysis. It stresses the relationship between us and wider universe as well as feeling cultivation for human's devotion to life process. To realize cultivation learning can make children get rid of their sense of alienation with the world and self to achieve a state of everything at once and harmony of body and mind and experience the self's endless and short feeling or Maslow's "peak experience".

ACKNOWLEDGEMENTS

This study is supported by Hunan Education Science "the 12th Five-Year Plan" 2014 Approval Project: Multivariate Management Model Study of Local Colleges and Universities under Education Comprehensive Reform (XJK014BGD076).

REFERENCES

1. Zihong Xu, *Complete Interpretation of Classic of Change*(Guizhou People's Publishing House, Guizhou1991)
2. Dafu Feng, *Translation and Annotation of Laozi,* (Shanghai Classics Publishing House, Shanghai,1991)
3. Qi Zhan,*Cosmology of Self-organization,* (China Social Sciences Publishing House, Beijing,1992)
4. T · Eliot, compiled by Enzhong Wang,*Poetics Collection of Eliot*,(Int'l Culture Publishing House, 1989)
5. Sloan,Douglas,Foreworld, *In Ron Miller (ed.), The Renewal of Meaning in Education, 1,* (Brandon, VT: Hollistic Education Press,1993)
6. Hughes, Langston, The Negro speaks of rivers, In R. Ellmann & R.O,Clair (Eds.), The Norton anthology of modern poetry . New York: W.W. Norton,pp.634,(1973)
7. David · Geiffrey · Smith, translated by Yangsheng Guo, *Globalization & Postmodernism*,(Science and Education Press, Beijing,2001)
8. James Malley, Mitchell Beck, Delia Adorno. Building an Ecology for Non-Violence in Schools, International Journal of Reality Therapy, 1(2001)
9. David · Geiffrey · Smith,translated by Yangsheng Guo, *Globalization & Postmodernism*,(Science and Education Press, Beijing,2001)
10. Miller, R., Caring for New Life: Essays on Holistic Education. Brandon: Foundations for Educational Renewal, pp.23-24, (2000).
11. Miller, R., Make Connections to the World: Some Thoughts on Holistic Curriculum, Encounter: Education for Meaning and Social Justice,p p.33,**14**, 4 (2001)
12. Patrik · Slattery, Curriculum Development in the Postmodern Era; translated by Wenbin Xu et al, Guilin: Guangxi Normal University Press, 2007, 233.
13. Peter · Senge, The Fifth Discipline: The Art & Practice of the Learning Organization; translated by Jinlong Guo, Shanghai Sanlian Bookstore, 1998, 14.
14. Taylor, E. R., *Ecology, Spirituality, and Education: Curriculum for relational knowing*,(Peter Lang Publishing, New York,2002)

MATEC Web of Conferences **63**, 05006 (2016) DOI: 10.1051/matecconf/20166305006
MMME 2016

Comparison of Three retrieval accuracy of image retrieval technology

Bai Qing HE [1,a], Zi Min WANG [2]

[1] *lecturer, master, research direction: information processing, telecommunications management, unit: Nanchang Institute of Technology.*

[2] *lecturer, master, research direction: information processing, ideological and political management, unit: Nanchang Institute of Technology.*

Abstract. With the development of society, it began between the various sectors and industries increasingly high integration of information, and the image retrieval technology has played an increasingly important role in the current exchange of information. Image retrieval faster, higher accuracy, then the computer used in image retrieval technology is more advanced. By now several popular image retrieval technology introduction, pointed out that the principles of these types of image retrieval technology and difficulties, comparative study of the advantages and disadvantages of these three image retrieval technology, so that we can retrieve the image retrieval accuracy further understanding.

1 Introduction

In the field of visual information research, It has been greatly developed. International Organization for Standardization has developed the MPEG-7 GB [1] for this area, so that relevant research on an international scale with standardized criteria can exercise. With standardization in this field can make international researchers or better commonality between users, easy to exchange and research.

In the field of traditional visual information research for image retrieval is image-based text annotations. That is, when uploading pictures or video information, pictures or videos will be editorial comment, and then the user needs to retrieve text pictures or videos. This method is relatively simple, but there are difficult to solve several problems. First, you must carry out all the pictures or video text notes that in modern society is not very realistic, hundreds of thousands of pictures or videos are uploaded every day,it is unable to complete the task only rely on manpower to text notes. Secondly,Manual text annotation is very subjective.For the same picture or video, different people have different interpretations. This traditional method is only suitable for specific occasions particular search, it is not suitable for all occasions, the search accuracy is relatively low.Comparison and analysis of existing image retrieval technology, we can find that a lot of new search technology has very good advantages,these technologies are more efficient, accurate, and quickly retrieve what we're looking for. Here we briefly describe some of the practical application of the more mainstream image retrieval, retrieval accuracy comparative study between them.

2 color-based image retrieval

There are two main color-based image retrieval technology. The first technique is based on a picture of the main colors, it is the user through the palette depicted image colors to achieve image retrieval, such as a large blue sky picture main colors can be blue. The second technique is based on the legend, according to which electronic artwork provided by the user, characterized by color pattern analysis computer image retrieval system, and then retrieve the color characteristics similar images in the database. Wherein based on the legend of the technology is not suitable for ordinary users. The main colors in [2] based retrieval method can only describe a color picture information generally has a relatively large error. For example, the user is mainly blue, the computer may retrieve the blue sky or the sea. According to the actual situation of the picture, you can choose two different main colors. Than the picture depicts the desert under the blue sky, you can choose blue and yellow as a picture retrieved two main colors. So you can more accurately retrieve the user desired image.

The first main colors query mode is more commonly used. However, the user selects the main colors when it is based on the naked eye to judge, with the user's subjective color, all of this often with a certain randomness and fuzziness. So when the user selects a main color, it also determines a value, for example, select the red (R = 255, G = 0, B = 0) on the palette, then retrieve the image, the computer only searches for a red that has the value. But this color is only the user through the naked eye to determine the modulation of the main colors, with the real needs of the color still exist certain differences.In this case, the user may have missed out on

[a] Corresponding author: 86035262@qq.com、308809980 @qq.com

DOI: 10.1051/matecconf/20166305006

truly meet the requirements of image information in image retrieval process.

In fact, the range of red included is relatively broad, because the user's randomness and fuzziness, which requires a master tones expansion process. That is, when the user selects the red (R = 255, G = 0, B = 0), the computer system will take this color as the center, expanding the range of RGB, retrieve some other red within a certain range. So that you can visually identify colors and better linked to the theoretical value. So we built a HSV color retrieval model [2].

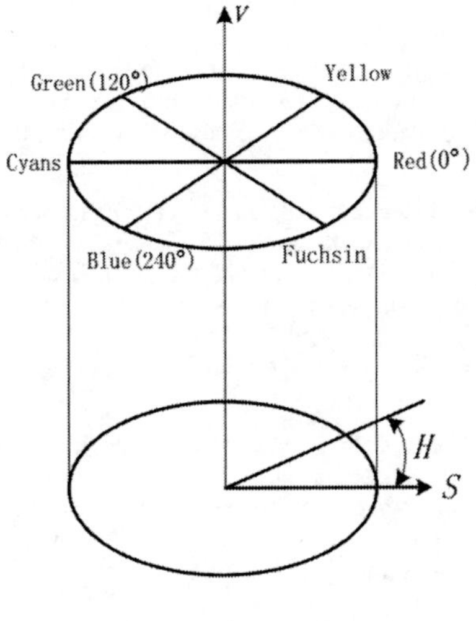

Figure 1 HSV color space schematic

HSV model refers to the color hue, saturation and brightness of the 3 properties, according to the 3 attributes to build a three-dimensional coordinate axes, where H is by color to identify the colors; saturation S is showing the depth of color, because also blue, light blue and dark blue also divided; the final V is light or dark. HSV values to obtain RGB values, the conversion process is as follows:Given the values of RGB color space (r, g, b), r, g, b∈ [0, 1, ···, 255], the conversion to HSV space, h, s, v value is calculated as follows:

$$\text{suppose} v' = \max(r,g,b), r', g', b' \text{ is defined as:}$$

$$r' = \frac{v'-r}{v'-\min(r,g,b)}, g' = \frac{v'-g}{v'-\min(r,g,b)}, b' = \frac{v'-b}{v'-\min(r,g,b)}$$

$$v = {}^{v'}\!/_{255}$$

$$s = \frac{v'-\min(r,g,b)}{v'}$$

$$h' = \begin{cases} (5+b'), r = \max(r,g,b) \,\&\, g = \min(r,g,b) \\ (1-g'), r = \max(r,g,b) \,\&\, g \neq \min(r,g,b) \\ (1+r'), g = \max(r,g,b) \,\&\, b = \min(r,g,b) \\ (3-b'), g = \max(r,g,b) \,\&\, b \neq \min(r,g,b) \\ (3+g'), b = \max(r,g,b) \,\&\, r = \min(r,g,b) \\ (5-r'), \text{other} \end{cases}$$

$$h = 60 \times h'$$

Here, $r,g,b \in [0\Lambda\ 255]$, $h \in [0\Lambda\ 360]$, $s \in [0\Lambda\ 1]$, $v \in [0\Lambda\ 1]$.

The HSV model we used that may represent a very large number of colors, if the histogram vector calculation, will be a complex process. In this case, if the first space proper quantization, then the histogram, will save many. HSV model Bar H space is divided into 8 parts, saturation is divided into 3 parts, the brightness is also divided into three points, then:

$$H = \begin{cases} 0 \; if \; h \in [316,20] \\ 1 \; if \; h \in [21,40] \\ 2 \; if \; h \in [41,75] \\ 3 \; if \; h \in [76,155] \\ 4 \; if \; h \in [156,190] \\ 5 \; if \; h \in [191,270] \\ 6 \; if \; h \in [271,295] \\ 7 \; if \; h \in [2961,315] \end{cases} \quad s = \begin{cases} 0 \; if \; \mathrm{s} \in [0,0,2] \\ 1 \; if \; \mathrm{s} \in [0,2,0,7] \\ 2 \; if \; \mathrm{s} \in [0,7,1] \end{cases} \quad v = \begin{cases} 0 \; if \; \mathrm{v} \in [0,0,2] \\ 1 \; if \; \mathrm{v} \in [0,2,0,7] \\ 2 \; if \; \mathrm{v} \in [0,7,1] \end{cases}$$

According to the above calculation, it can be thought of feature vectors:

l = H Q s Q v + SO v + V (1)

Wherein, Q s and Q v S and V components are the quantization levels, taking Q s = 3, Q v = 3. Thus (1) can be expressed as:

l = 9H + 3S + V (2)

Thus, H, S, V these components form a three-dimensional vector, and l know the range of [0, 1, ···, 71], the calculation to obtain a one-dimensional histogram 72 l handle (bin) of [4].

A hypothetical example of the image Q, which is a histogram of feature vectors H q (q 0, q 1, ···, q L), an image in the image database is S, which is a histogram of feature vectors H s (s 0, s 1, ···, s L), wherein, $0 \leqslant qi \leqslant 1$, $0 \leqslant si \leqslant 1$, a normalized scale value, L is the dimension of a one-dimensional histogram of the vector. then the histogram intersection algorithm [3] image similarity measure, the intersection of two histograms is defined as follows: According to HSV model it is more significant to distinguish between two different images. Color-based image retrieval technology with a certain degree of accuracy.

3 knowledge-based image retrieval

Knowledge-based image retrieval technology is based on artificial intelligence technology, through understanding and expression of the image, and image database of pre-built docking, the user needs to retrieve the comprehensive application of technology picture information. Premise of this technique is that the system must be able to recognize natural language and understand the meaning of the language, which can make a judgment.

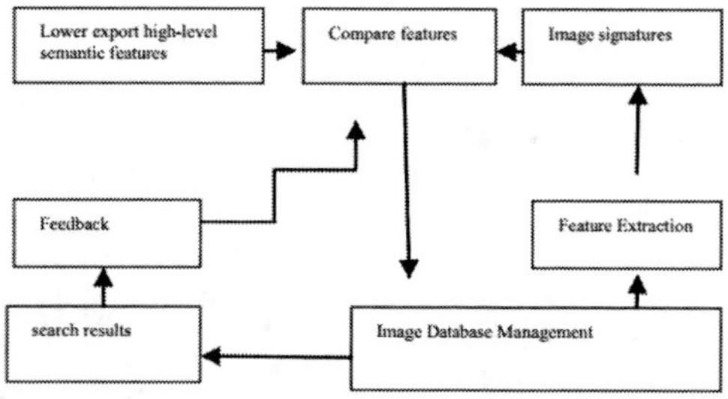

Figure 2 frames

Natural language [5] is the user input language query system. As we all know, natural language semantic structure and changeable, changeable, has the synonym. Let the system recognizes natural language, We need to build intelligent systems, and related language rules and syntax analysis introduced into the system to ensure the retrieval system is in understanding the basis of semantic retrieval, rather than simply matching text. Construction of the intelligent system, which we call the concept of Semantic Network, is highly portable and can be modified, to achieve user requirements for the system to accurately understand and implement the corresponding retrieval.

The construction of intelligent system needs the help of relevant experts. Development Engineer is responsible for building intelligent systems framework itself, but to be embedded in the language rules, we need to consult experts in related fields. First of all, experts in the field need to be grouped into specialized language knowledge of language rules, developed by the engineer in an appropriate form implant system to achieve a computer can preinstalled language database, retrieve the contents of the user to make appropriate judgments and simple reasoning, retrieve the information that users need.

Knowledge-based image retrieval technology in the actual production also has a wide range of applications, it also has a very important feature is the high artificial intelligence [6]. Retrieval system not only can be used to identify the natural language rules, but also through their own learning, the image retrieval process to retrieve the information stored in the database. Thus, in the process of image retrieval system, the system will be self-taught and self-storage mode, ready to improve their recognition rate. From this we can see that the core issue of knowledge-based retrieval technique is to establish contains a wealth of knowledge of the language database.

For now, the knowledge database with the subject-sidedness.For example, knowledge of modern information includes medical, electronic communications, mathematics, history and other aspects. In terms of the form of pictures stored to correspond to a very rich language information, the system has not been able to achieve in all areas of information for accurate identification and retrieval. To solve this problem, we need to work together various experts in the field.

Knowledge-based image retrieval technology to better understand the needs of users, in order to find a more accurate picture of the user needs to retrieve information. Therefore, this technique has higher retrieval accuracy, integrated knowledge-based image retrieval system should have the following characteristics:

1> The system is capable of doing the image NOTE: retrieval system according to the characteristics of the original image analysis knowledge base to derive the relevant semantic description, and then summed up with experts in the field of rule combine to form a new knowledge base.

2> The system is not only able to identify and retrieve relevant semantic input, but also according to semantics, for simple judgment and decision-making of the work content. This fully embodies the intelligent system.

3> system for the average user, it should have a simple user interface, which means that ordinary customers to retrieve the contents of the input, the system can make the search process according to the requirements, without the need for ordinary users with professional knowledge.

4 based on SVM relevance feedback image retrieval algorithm retrieval technology

First to introduce the principle of SVM [7], SVM principle refers to the object projected onto the high-dimensional space or build an optimal classification surface in the original space, so that two different samples can be distinguished separately. Constructed this classification is based on two different surface samples furthest from the hyperplane.

SVM principle need points note:

Assuming for the current set of samples (x i, y i) 1≤i≤N, x i ∈ R d,yi as the class symbols, ranging from (-1,1), linear functional form d-dimensional space corresponding to g (x) = w • x + b, and the corresponding function equation classification plane is w • x + b = 0. The g (x) is normalized to give xi are g (x) ≥1 solution, that is, the sample surface is a function of classification g (x) = 1, can be obtained classified operation interval 2 / ‖ w ‖. To solve the optimal classification surface is equivalent to minimize ‖ w ‖, the objective function is:

Using Lagrange multiplier method introduced multiplier A = (A 1, ···, AN), can be obtained w = ΣiA iyixi, xi is located spaced sample classification surface, this sample is called support vector [8], classification function:Of course, there is a situation when, this time on the need to introduce slack variables and the penalty factor, the objective function becomes:In the non-linear operation, SVM is also possible to convert the input space to a high-dimensional space, so that you can calculate solved in the new space. Which can become linear:Therefore, SVM FOUR built on the basis of the above statistics, in the actual application process also achieved good results.

Let's look at how to SVM algorithm associated negative feedback retrieval technologies. Relevance feedback process can be seen as two SVM classification model, and a specific search algorithm works as follows:

The first step: You can use the previous method for the user to retrieve a desired image.

Step two: user former N RT [9] images are labeled, can be related set of images I +, unrelated set of images IO,

Step three: Start Prepare the sample set of functions SVM algorithm (x i, y i),

Step four: the use of SVM algorithm to calculate the sample build classification surface.

Then added function sign (•),to obtain the maximum distance with the query image.

Step five: for every one picture I i, seek score (I i) = - f (x i)

Step Six: according to all the picture score, ascending order and returns the result.

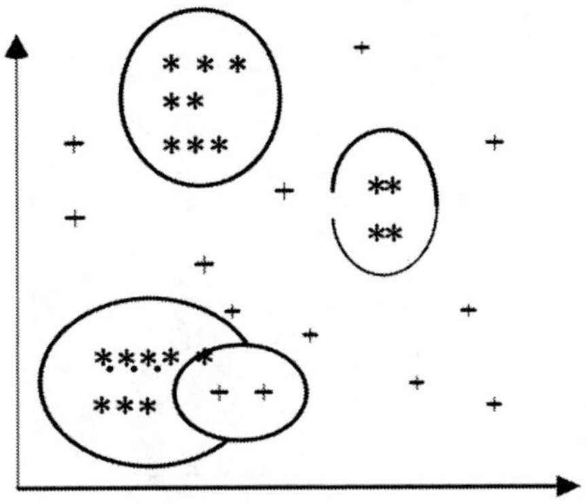

Figure 3 classification

Because users can get the image and the image characteristics similar to retrieve each time, but they can be classified surface [10] distinguish different image information, which can guarantee the retrieval of information is retrieved the next time the user can obtain the desired. It can be seen, with higher precision image retrieval based on SVM relevance feedback image retrieval algorithm retrieval technology.

5 Retrieval comparison summary

From the point of view of image retrieval accuracy, color-based image retrieval technology is relatively simple, from a technical level, relatively easy to implement, but the error is relatively large presence, the accuracy is relatively low; knowledge-based image retrieval technology need to be able to identify semantic intelligent system as a support, the system can identify the user relatively accurate information, has higher retrieval accuracy; based on SVM relevance feedback image retrieval retrieval algorithm is based on statistical evolved, the system can automatically retrieve the second, with a high retrieval accuracy.

Three image retrieval technology in applications complement its own advantages, widely used to meet the general user retrieval tasks, but in the face of emerging picture information,there is further room for improvement in retrieval accuracy, which requires researchers on the basis of prior art, painstaking research, the development of image retrieval technology to the next level with higher retrieval accuracy of image retrieval technology and promote.

References

[1] Xiao-Yu, Li sky. Matching algorithm based on local multi-scale spatial constraints [J]. Modern computer (Professional Edition) 2016 (03)

[2] Zhou Yi Fei Shumin, Shen Jie. Detect abnormal behavior based real feature point trajectory [J]. Industrial control computer. 2016 (02)
[3] Tao Tao, Zhang Yun. Logarithmic polar coordinate system scale-invariant feature detection and description of points [J]. Chinese Journal of Image and Graphics. 2015 (12)
[4] Feng Wang, Lanfen Lin, Min Tang. A new sketch-based 3D model retrieval approach by using global and local features [J]. Graphical Models. 2013
[5] Guang-Hai Liu, Jing-Yu Yang. Content-based image retrieval using color difference histogram [J]. Pattern Recognition. 2013 (1)
[6] Zhu Jie, Liu Bo, super wooden Japanese power grid, to the sword weighted representation method based on image characteristics of an object's color [J]. Journal of Nanjing University (Natural Science). 2016 (01)
[7] Huang Ningning, CONSERVATORY, Xiao Xuan Hong, Liang Fang Xi, Fan Min. Study Scene image similarity detection algorithm [J]. Science and Technology Outlook. 2015 (34)
[8] Lu waves multimodal media information retrieval research [D]. Northeastern University, 2013
[9] Qin Fei, Wang Ronggui, Kelvin Leung Hong, Zhang Dongmei, Li Xiang Better TLD-based targets key feature point tracking algorithm [J]. Computer Engineering and Applications. 2016 (04)
[10] Xu Chenghui, Liu Guihua, Liang Feng. Non-feature point Binocular Ranging Technique [J]. ITS Applications. 2013 (22)
[11] Zhao perseverance, Xia Yan. Disordered automatic image matching and recognition [J]. Journal of Computer Applications and Software]. 2014 (12)

MATEC Web of Conferences 63, 05007 (2016) DOI: 10.1051/matecconf/20166305007
MMME 2016

RFID Location Algorithm

Zi Min WANG[1,a], Bai Qing HE[2]

[1] *lecturer, master, research direction: information processing, ideological and political management, unit: Nanchang Institute of Technology.*

[2] *lecturer, master, research direction: information processing, telecommunications management, unit: Nanchang Institute of Technology.*

Abstract. With the development of social services, people's living standards improve further requirements, there is an urgent need for a way to adapt to the complex situation of the new positioning technology. In recent years, RFID technology have a wide range of applications in all aspects of life and production, such as logistics tracking, car alarm, security and other items. The use of RFID technology to locate, it is a new direction in the eyes of the various research institutions and scholars. RFID positioning technology system stability, the error is small and low-cost advantages of its location algorithm is the focus of this study.This article analyzes the layers of RFID technology targeting methods and algorithms. First, RFID common several basic methods are introduced; Secondly, higher accuracy to political network location method; Finally, LANDMARC algorithm will be described. Through this it can be seen that advanced and efficient algorithms play an important role in increasing RFID positioning accuracy aspects.Finally, the algorithm of RFID location technology are summarized, pointing out the deficiencies in the algorithm, and put forward a follow-up study of the requirements, the vision of a better future RFID positioning technology.

1 Introduction

With the rapid development of radio frequency identification (RFID) technology, RFID technology already have important applications in many fields. RFID technology originated during the beginning of World War II, the Allies to analyze and judge the enemy and friendly ships or aircraft using wireless technology. However, due to relatively high cost of the technology ,after the war it just used in military strategic studies for a long time. In the 1980s and 1990s, is widely used around the world began to chip and electronic technology, RFID only began to enter the field of civilian production, but has not been effectively developed and applied in our country. Until February 2006, China promulgated the "China RFID Technology White Paper", it means that China began to develop RFID technology. Now, RFID applications have been explored in a number of aspects of identification, security, anti-counterfeiting and so on.

In the aspects of positioning technology, the prior art are more mature US-developed GPS global positioning system, commonly used in Europe's Galileo global positioning system, Russia is good at Green NASDAQ Global Positioning System and China's own development of the Big Dipper global positioning technology, This positioning technology has a relatively high accuracy, commonly used in military combat and nautical terms. But these technologies have positioned a short board, when the target is in the room, then positioning is compromised. Because the indoor environment is more complex, and easy to interfere with the absorption of the satellite signal, resulting in greater positioning system can not locate or error. To solve this problem, we need to find ways to quickly prepare targeted indoor new technology, so the radio frequency technology run into people's vision to become a scientific research institutions and scholars sought after star.

Radio frequency identification technology is a non-contact technology, the main principle is through electromagnetic coupling or electromagnetic coupling to the propagation and reception of signals, even in relatively harsh environments, the system can run stable, signal communication's distance is also available may enter, can automatically identify the corresponding tags, can simultaneously identify multiple targets without human intervention. And it has a small, penetrating strong, convenient application features.

2 RFID technology summary

Radio frequency technology includes three aspects: the c ontrol chip, antenna tags and readers.

1, The control chip is the core component of the radio frequency system, equivalent to a host computer, controls the transmit electromagnetic signals, receiving, processing and judgment. For example, commonly used access control system, when a signal of the correct access cards, access control systems is to identify and determine the received signal, to open the action.

2, The antenna RF cards and tags essence of memory modules. Capable of storing identified items of information are transmitted RF card antenna related information. Depending on the article, the antenna can be made into a variety of label materials and shapes. It can be attached to the surface of an article to be implanted

[a] Corresponding author: 86035262qq.com、308809980qq.com

inside the article. According to the work principle,Tag antenna can be divided into active tags and passive tags. Internal affinity label containing a small battery to provide energy to provide power for the chip, which can be discriminated from the label is much greater than the passive; but the presence of short battery life, larger shortcomings, is not conducive to carry, in general for applications that require identification distance and special scene.

Internal battery is not passive tags provide energy, it is the use of electromagnetic coupling or electromagnetic coupling to an external reader generates a current work, although closer than active tags to identify short, but with better shape and more low plasticity prices are widely used in the actual production applications.

3, The reader is an important means to receive and identify electromagnetic signals.It emits an electromagnetic signal, when the electromagnetic signal coverage within the working area of the tag antenna, the receiving antenna is the automatic tag feedback signal, and then sends a signal to carry out comparative analysis and decoding chip, chip and then make the correct decoding and error analysis results decoding the corresponding action. The reader is to achieve recognition of information, which is the basis of RFID positioning technology.

3 the common base algorithm

In the field of RFID positioning technology, now commonly used positioning divided into three areas, namely: (1) to reach the positioning information, (2) positioning signal strength, (3) measuring the azimuth positioning. Let these three commonly used positioning method a brief introduction.

1, Positioning arrival information divided into two discussions:

(1) the arrival time of positioning.

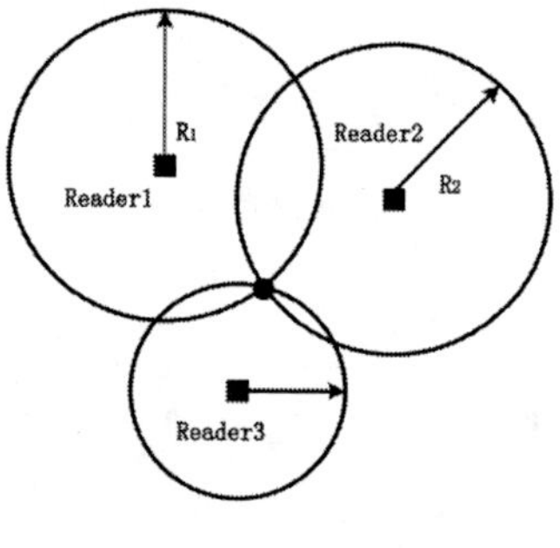

Figure 3-1 arrival time targeting method

The figure shows that there are three tag reader intersection is the sitting position. It is easy to locate a tag needs to know three readers to achieve, we know that the speed of propagation of electromagnetic waves in the air of C (3 * 108m / s), the coordinates of three readers also known (Xi, Yi), measured of electromagnetic waves from the target tag to the reader the propagation time t1, t2, t3, this way, they will know the distance of the target tag to the reader is Ri = C * ti, according to the mathematical method, can be assumed that the position coordinates of the target tag (X0, Y0), the following relation formula is as follows:

$$\begin{cases} (x_1 - x_0)^2 + (y_1 - y_0)^2 = R_1^2 \\ (x_2 - x_0)^2 + (y_2 - y_0)^2 = R_2^2 \\ (x_3 - x_0)^2 + (y_3 - y_0)^2 = R_3^2 \end{cases}$$

(2) Time Difference of Arrival Location

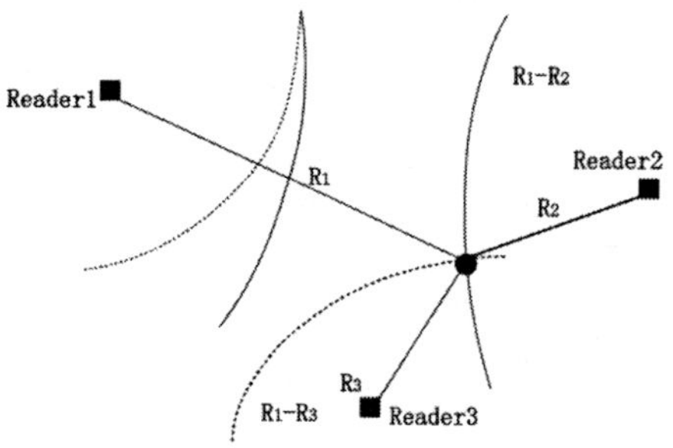

Figure 3-2 TDOA hyperbolic positioning method

3 or a known reader coordinates, as they are from the same distance from each part of the target tag, and electromagnetic wave propagation velocity is known, the time they received the target tag signal will be different. As shown, there is a time difference between the time t_{21}, through publicity R_{21} = C * t_{21}, the target tag certainly at the intersection of two readers, according to mathematical knowledge, the relationship can be written as follows:

$$\begin{cases} \sqrt{(x_0 - x_2)^2 - (y_0 - y_2)^2} - \sqrt{(x_0 - x_1)^2 - (y_0 - y_1)^2} = R_{21} \\ \sqrt{(x_0 - x_3)^2 - (y_0 - y_3)^2} - \sqrt{(x_0 - x_1)^2 - (y_0 - y_1)^2} = R_{31} \end{cases}$$

2, the signal strength of the positioning

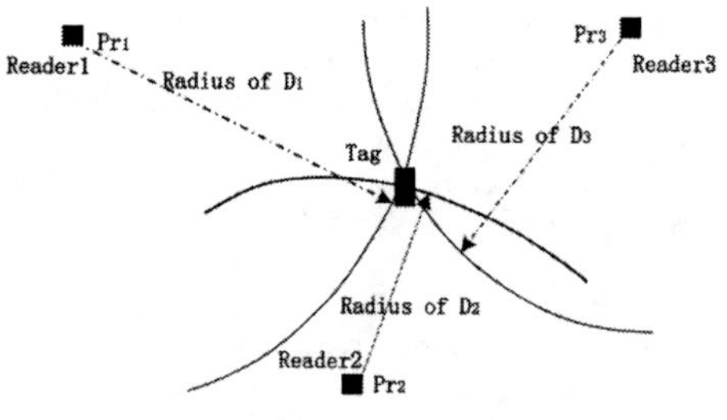

Figure 3-3 the signal strength of the positioning

In essence, is the energy of electromagnetic wave propagation, the spread will widen as the distance decreases, according to the theory of electromagnetic wave transmission, publicity can be listed:

$$P_{r_i} = \frac{P_t * G_t * G_{r_i} * \lambda^2}{4\pi * D_i^2}$$

As shown in FIG, P_t represents the transmission power of the target tag, the Pri reader represents the i-th power of the received signal shows the wavelength of electromagnetic waves, G_t, G_{r_i} said first target gain and the i-th label reader antenna, according to publicity to give the title to the first tag reader distance D_1, once on, you can also obtain the target tag to the other two readers distance D_2, D_3. The position of the target label is to the three readers radius circle intersection.

3, The direction of the measurement positioning.

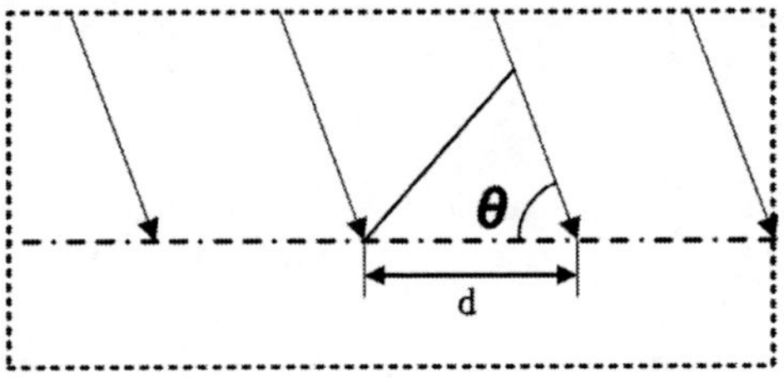

Figure 3-4 the received signal to determine the angle of arrival

Known coordinates of the three readers have been installed antenna array can be used to measure the tag to the reader emit electromagnetic wave angle, as shown, it can be listed according to the following formula known conditions:

$$\varphi = 2\pi * d * \cos\theta / \lambda$$

For different antenna arrays, we can measure the sign al has a different phase, so the angle can be calculated a tr ansmitted signal, can be calculated the same way three re ader receives the angle signal, thereby to obtain the locati on information of the target tag.

4 the topology of the network algorithm

In practice, often densely build RFID reader, when a label is placed where multiple readers will read the information and processing information, the process is a relatively complex process. We know RFID reader is a RS-232 serial port or Ethernet docking with the computer in order to transfer the collected information to a computer, make calculations and analysis by a computer.

The reader position is known, the calculation will be in the form of nodes, each node of the interconnected, it will form a network topology, as shown:

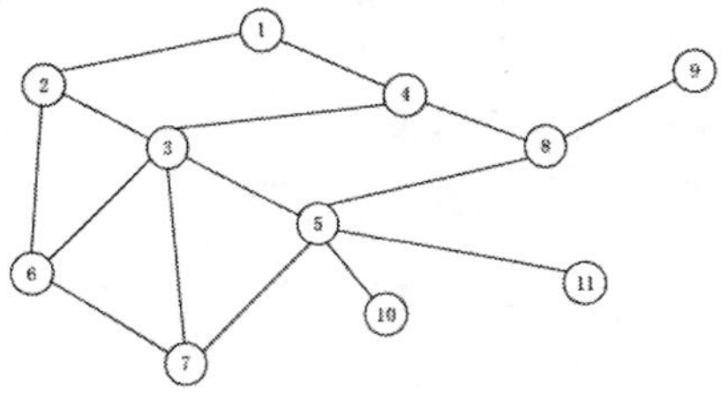

Figure 4-1 Reader topology

Read the entire topology map, each label and each reader has a unique identifier ID, so when the tagged item or person moves in the topology map, RFID reader will re-read the tag's location, It sends a signal to the computer in the background computing and storage, and finally all the information combinations reconstruction, reflecting the label walking paths. RFID reader topology can tell from the label very close, so the results will be measured more accurately. However, if the label in the topology map to move, space does not exist on two adjacent reader, then there will be a reader does not have access to information of the label, there is leakage reader to read, so the need to create new computer system to avoid this problem.

We can also locate to the original vector improve:

$$\langle TagID, T, \mathrm{Re}\,aderID, RSSI \rangle$$

Wherein TagID target represents a unique ID tag, T represents the tag reader signal reaches the moment and is detected, ReaderID represents the detected movement of the tag reader ID, RSSI indicates the received signal strength, to analyze in detail each of the following vector meanings.

(1) tag identification number TagID: This is a vector equation of the first variable, the label is a unique number within the chip, globally unique. During the movement of the tag, the reader will be real-time tracking of the received signal, consisting of a group targeting vector, then background computer systems start to realize the label positioning monitoring.

(2) Read time T: The vector equation is a second variable, which represents the time the label is detected. It is worth mentioning that the requirements of each reader clock to be synchronized, otherwise there is no meaning. When a label reader into a work area, just generate a targeting vector. But when the label reader returns back and forth movement of the edge of the area, it will produce multiple targeting vector, in this case is to be discussed separately, and do not here described.

(3) The reader identification number ReaderID: topology map reader corresponds to a different identification number, if they are to transmit signals to the computer, the computer system according to the signal reader sent a different identification number, by continuously reading the position, calculate the approximate location of the label.

(4)The signal strength RSSI: when the tag near the reader, to the signal is getting stronger, while the principle of the tag reader, the signal is in a slow decay process, by detecting signal strength, but also can help accurate positioning orientation coordinates of the tag.

Suppose a tag in the process of moving, there are two time information is submitted back to the computer syste m, according to the vector equation can be obtained:

$$\{\langle T_1,t_1,R_1,S_1\rangle,\langle T_2,t_2,R_2,S_2\rangle\}(t_1<t_2)$$

Both vectored means: at time t_1, the tag T_1 signal is to capture readers R_1, then the signal strength of S_1; Similarly, at time t_2, the signal label reader R_2 T_2 is captured, This signal strength S_2. The computer system will calculate a route that information. Formula $t_1<t_2$, so the second position from the first position in the past, so there is:

$$track = R_1 \rightarrow R_2$$

Easy to learn, improved topology calculation method is relatively simple, but the basic premise is to satisfy a condition that a label with an identification number must be the same. While many tags in the reader's mobile work area, the system must be based on the same identification number, filter out other labels, only the label to locate the target and monitoring.

5 LANDMARC Algorithm

Location technology based on signal strength, is the most commonly used method of LANDMARC, this algorithm makes use of different labels as a reference material, due to the different position of the reference material, the reader detects the signal strength is different, of LANDMARC way it is by comparing the different reference RSSI different labels, and then use the right to reuse the empirical formula to calculate the position coordinates of the label to be located.

LANDMARC nearest neighbor algorithm depending on the intensity of the signal, the reader will be divided into eight power level (1-8), and then assume a reader M, L reference tag, N a label to be located. This defines the signal strength of the matrix S: where Sij (i = 1,2 ... N, J = 1,2 ... M) represents the i-th reference values J-th tag reader to take.

$$S = \begin{bmatrix} S_{11} & S_{12} & K & S_{1M} \\ S_{21} & S_{22} & K & S_{2M} \\ K & K & K & K \\ S_{N1} & S_{N2} & K & S_{NM} \end{bmatrix}$$

Signal strength matrix:

$$T = \begin{bmatrix} T_{11} & T_{12} & K & T_{1M} \\ T_{21} & T_{22} & K & T_{2M} \\ K & K & K & K \\ T_{L1} & T_{L2} & K & T_{LM} \end{bmatrix}$$

Wherein, Tij for the first J-reader to get i-undetermined label values.

Refer to the label to be positioned with a tag associated matrix E:

$$E = \begin{bmatrix} E_{11} & E_{12} & K & E_{1M} \\ E_{21} & E_{22} & K & E_{2M} \\ K & K & K & K \\ E_{L1} & E_{L2} & K & E_{LM} \end{bmatrix}$$

among them

$$E=\sqrt{\sum_{k=1}^{M}(T_{ik}-S)^2} \qquad (\mathrm{i}=1,2,..,\mathrm{L},\mathrm{j}=1,2,..,\mathrm{N})$$

It represents the j-th reference label to be associated with the i-th degree positioning tag. If the smaller the value of their degree of relevance, the more that the closer the distance between the two labels.

The position coordinates of a known reference coordinate position of the label, the label is to be located:

$$(x,y)=\sum_{i=1}^{k} w_i(x_{i,}y_j)$$

$$w_i = \frac{\frac{1}{E_{pi}^2}}{\sum_j^k \frac{1}{E_{pj}^2}}$$

Wherein, W represents the weight of heavy reference frame, the closer the distance, the greater the weight, K represents the tag to be located away from the nearest reference coordinate tag number, for a label to be positioned P, that is, (E_{P1}, E_{P2}, .. ., E_{Pn}) selects the smallest K value corresponds to a reference label by calculating the value of W, the positioning error is:

$$e=\sqrt{(x-x_0)^2+(y-y_0)^2}$$

Wherein, (x_0 , y_0) is positioned to be the actual coordinates of the label, (X, Y) to calculate coordinates. Place a test tracking labels, 49 tag reference, four readers, can be obtained by calculating the difference between the maximum distance is 0.8M, the average error is 0.5M.

6 Summary

This paper describes several common RFID positioning technology algorithms and the other two advanced algorithms more accurate and efficient, overall grasp both tag and reader identification code, select the reader or the tag as a reference to establish the coordinates of each portion, It is obtained by positioning the label coordinate geometry mathematics calculations. But advanced algorithms can be more accurate, fast and stable to be calculated from the target location.

Of course, the existing RFID location there are still some technical problems to be studied and solved. For example,three-dimensional positioning technology,RFID technology and the fusion of things, would require further exploration. Under the new era of the new situation, RFID location technology is playing an increasingly important role for some breakthrough technology bottlenecks, can effectively promote the development of social production. Study more advanced RFID location algorithm is the key to break the technical bottlenecks.

References

[1] Li Fei high, Sun Remington. Analysis of RFID radio frequency identification technology [J]. Digital Technology and Applications. 2012 (05)

[2] Hae-Won Son, Hyuk Park, Kyong-Hee Lee.UHF RFID Reader Antenna With a Wide Beamwidth and High Return Loss. IEEE Transactions on Antennas and Propagation. 2012

[3] Yao-Lin Wang. RFID-based indoor localization of a movable object [D]. Nanjing University of Posts and Telecommunications 2015

[4] Tang Jun. Design and implementation of active RFID location system [D]. Southwest Jiaotong University 2014

[5] Yang Hang. Roadway precision personnel location system and data processing technology [D]. University of Electronic Science and Technology 2015

[6] Wang Xiaoyuan. RFID-based outdoor personnel positioning system research [D]. Lanzhou Jiaotong University, 2015

MATEC Web of Conferences **63**, 05008 (2016) DOI: 10.1051/matecconf/20166305008
MMME 2016

The wireless sensor network (WSN) triangle centroid localization algorithm based on RSSI

Chuan Wei ZHANG[1,a] ,Xin ZHAO[2]

[1]*College of Mechanical Engineering, Xi'an University of Science and Technology, Xi'an 710054 China*

[2]*College of Mechanical Engineering, Xi'an University of Science and Technology, Xi'an 710054 China*

Abstract: Node location is one of the key technologies in wireless sensor network. RSSI-based location is a hotspot in nowadays. For resolving biggish error in RSSI-based location, the paper presents a new method of location, RSSI-based triangle and centroid location, using triangle and centroid method to reduce the error of RSSI measurement. Simulation experiments prove that this algorithm can obviously improve the location accuracy compared to trilateration.

1 Instruction

Node location technology is an important supporting technology of wireless sensor network, the location information of sensor network monitoring activity is very important[1]. The sensor node positioning is based on a few known location of the node itself, to determine its own position according to some targeting mechanism.

Positioning technology become an important research direction and hotspot. Localization algorithms for wireless sensor networks can be divided into two categories[2]: distance based localization algorithm and range free localization algorithm. The distance based localization method needs to measure the absolute distance or orientation between the beacon nodes and unknown nodes, and use the actual distance between nodes to estimate the position of unknown nodes. The main methods are : TOA,TDOA,AOA and RSSI[3,4]. The localization algorithm don't relate to positioning distance need not measure the absolute distance between nodes or orientation. It uses the estimated distance between nodes and the network connectivity to compute the position of nodes. The main methods are centroid algorithm, APIT algorithm, DV-hop algorithm, DV-distance algorithm and so on. The RSSI positioning using the signal strength will be some loss during transmission, according to the theoretical and empirical model, this loss is estimated to be the distance between nodes. Sensors can satisfy the measurement of RSSI value itself , without additional hardware support, easy to implement. Because of the estimation of the theoretical and empirical models, RSSI has a large positioning error.

Based on RSSI technology, a new positioning algorithm combining RSSI measurement with triangle centroid algorithm is proposed. The algorithm uses triangle centroid algorithm to reduce the measurement error of RSSI.

2 Distance measurement based on RSSI

The distance measurement technology based on RSSI is used to measure the distance between nodes, which is based on the principle of the radio signal with the distance[5]. The relationship between RSSI and the transmission distance d is shown below[6]:

$$RSSI=-(10n\log 10d+A) \quad (1)$$

Among them:

n: Signal propagation constant, also called transmission coefficient

d: Distance between the transmitter and the receiving end; unit M

A:In one meter away from the launcher of signal strength

The accuracy of ranging is influenced by the n and A's actual value. A is an empirical parameter, can be worth to by measuring the distance from one meter of RSSI . N is used to describe the signal strength decrease with the increase of distance parameter. The size of N depends on the specific environment. In order to get the optimal values of n, all the beacon nodes be placed first then try different values of n_index to find the most suitable value of the specific environment of the n. According to equation (1), The distance d from the node to the other node of the radiation signal can be calculated, that is, the distance between the beacon node and the unknown node. In this paper, we use 3 beacon nodes A, B, C to locate the unknown nodes, so the distance phase should be at least three distance values d_1, d_2, d_3, as shown in figure 1.

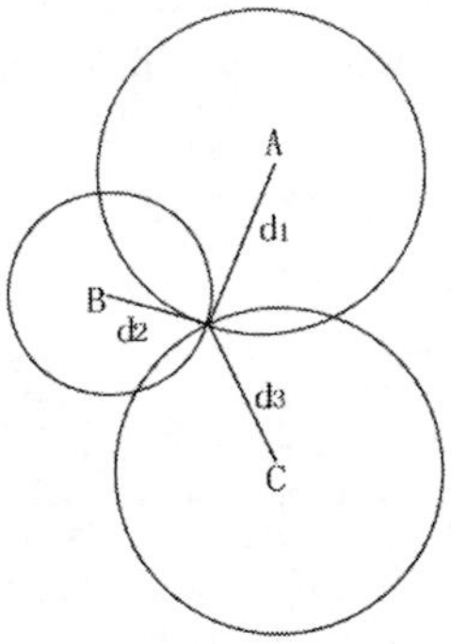

Figure 1. RSSI positioning of the ideal model of the distance

As shown in figure 1, in the ideal in the wireless sensor network (WSN), according to the beacon node and the distance between the unknown node model, can be calculated the coordinates of unknown nodes by three side measurements. However, due to the interference of the environment and the electromagnetic field signal of the irregular attenuation, lead to the beacon node to the unknown node RSSI value conversion value is greater than the distance from the reality of unknown node to beacon node distance real value, so distance model by RSSI conversion to the actual situation is shown in figure 2. The unknown node should be in respectively by A, B, C beacon as the center, d_1, d_2, d_3 is the radius of 3 circle intersection area.

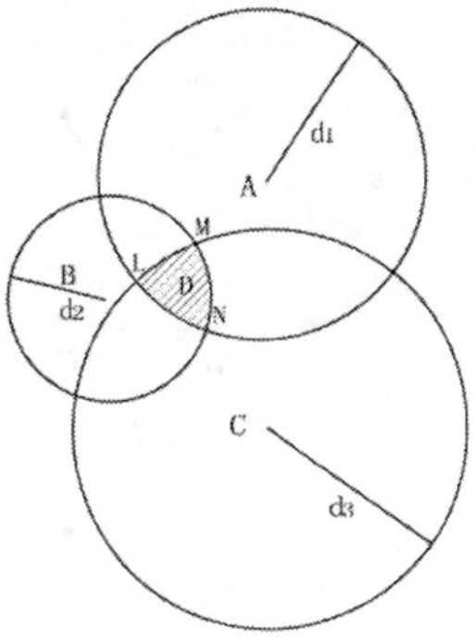

Figure 2. Actual distance model diagram of beacon nodes and unknown nodes

3 Localization algorithm correction

The positioning calculation is obtained from the d on the basis of the ranging stage. Position is obtained by the trilateral localization algorithm, and then make the positioning accuracy was improved by triangle centroid algorithm.

3.1The trilateral localization algorithm

Known of the beacon node location information, select appropriate coordinate origin, and establish a coordinate system. Assuming that beacon node coordinates for A(x_a,y_a), B(x_b,y_b), C(x_c,y_c).The unknown node coordinates for M (x_m,y_m). The beacon node as the center, the distance between the beacon nodes and unknown nodes as the radius of the circle equation is

$$\begin{aligned}\sqrt{(x_m-x_a)^2+(y_m-y_a)^2}&=r_a\\ \sqrt{(x_m-x_b)^2+(y_m-y_b)^2}&=r_b\\ \sqrt{(x_m-x_c)^2+(y_m-y_c)^2}&=r_c\end{aligned} \tag{2}$$

Formula (2) is the nonlinear equations obtained according to the formula of distance calculation. As shown in figure 2, circle A, B, C intersect at a point. Intersection point coordinates M (x_m, y_m) can be obtained by solving equations.

3.2Triangle localization algorithm

Because of the trilateral localization algorithm without considering the actual factors, so the error is bigger. The actual situation is shown in figure 2. Assuming that beacon node coordinates are A (x_a, y_a), B (x_b, y_b), C (x_c, y_c). The coordinates of the unknown nodes is D (x_d,y_d). Beacon node to the distance of the unknown node can be obtained by the RSSI ranging method, are d1, d2 and d3. Can get cross area by A, B, C as the center, d1, d2, d3 as the radius of the circle. The intersection point is M, L, N. Then can get equation:

$$\begin{aligned}\sqrt{(x_m-x_a)^2+(y_m-y_a)^2}&\le r_a\\ \sqrt{(x_m-x_b)^2+(y_m-y_b)^2}&=r_b\\ \sqrt{(x_m-x_c)^2+(y_m-y_c)^2}&=r_c\end{aligned} \tag{3}$$

Therefore, the coordinates of M can be obtained is (x_m,y_m).

$$\begin{aligned}\sqrt{(x_l-x_a)^2+(y_l-y_a)^2}&=r_a\\ \sqrt{(x_l-x_b)^2+(y_l-y_b)^2}&\le r_b\\ \sqrt{(x_l-x_c)^2+(y_l-y_c)^2}&=r_c\end{aligned} \tag{4}$$

Therefore, the coordinates of L can be obtained is (x_l,y_l).

$$\begin{aligned}\sqrt{(x_n-x_a)^2+(y_n-y_a)^2}&=r_a\\ \sqrt{(x_n-x_b)^2+(y_n-y_b)^2}&=r_b\,.\\ \sqrt{(x_n-x_c)^2+(y_n-y_c)^2}&\le r_c\end{aligned} \tag{5}$$

Therefore, the coordinates of N can be obtained is (x_n,y_n).

Then the coordinates of unknown nodes for D is

$$x_d=\frac{x_m+x_l+x_n}{3}\ \ y_d=\frac{y_m+y_l+y_n}{3}$$

The value of RSSI is affected greatly by the distance, the greater the distance, the greater the error. So weighting the results. The weighted coordinates is

$$x_d=\frac{\dfrac{x_m}{r_b+r_c}+\dfrac{x_l}{r_a+r_c}+\dfrac{x_n}{r_a+r_b}}{\dfrac{1}{r_b+r_c}+\dfrac{1}{r_a+r_c}+\dfrac{1}{r_a+r_b}}\ \ y_d=\frac{\dfrac{y_m}{r_b+r_c}+\dfrac{y_l}{r_a+r_c}+\dfrac{y_n}{r_a+r_b}}{\dfrac{1}{r_b+r_c}+\dfrac{1}{r_a+r_c}+\dfrac{1}{r_a+r_b}}$$

4 The process of triangle centroid algorithm based on RSSI

4.1Algorithm steps.

(1)Beacon nodes periodically broadcast information around, information including the node itself and coordinate ID. After the unknown node receives the information, take average of RSSI by the same beacon node.

(2) When the unknown node to collect a certain number of beacon nodes, no longer receive new messages. The unknown node based on RSSI from strong to weak to sort the beacon node, and establish distance map by the RSSI values with the unknown node to beacon node. Establish three sets, the beacon node set, distance between unknown node and beacon node set, location of beacon node set.

(3)Select larger RSSI value of several beacon nodes and positioning calculation itself. In turn according to formula of (3), (4), (5) calculate the three point coordinates. Finally by the centroid algorithm, it is concluded that the unknown node coordinates.

(4) Taking the average of the coordinates of unknown nodes for collection, get the unknown node coordinates.

4.2 Error

Error is defined as M. Assuming that the unknown node coordinate for (x_d, y_d), its true position is (x, y).Then,

$$M=\sqrt{(x_d-x)^2+(y_d-y)^2} \tag{6}$$

5 Simulation

This algorithm use ZigBee hardware platform as wireless communication platform, the ZigBee module as the unknown node and beacon node to complete the functions of wireless signal to send and receive[7]. Chip module using CC2431, and using matlab simulation tool to simulate triangle centroid algorithm, investigate the performance of the algorithm. Suppose there are 30 unknown nodes and 20 beacon nodes are evenly distributed in the square area of 50 m × 50 m. The simulation with trilateral localization algorithm and triangle centroid location algorithm, and randomly selected 7 groups of data. The results are shown in table 1. Table 1 shows that triangle centroid algorithm accuracy is higher than trilateral localization algorithm.

Table 1. Comparison of two algorithms error

algorithm	the actual coordinates /m	the calculated coordinates /m	error/m
trilateral localization algorithm	(8,12)	(9.56,14.88)	3.28
triangle centroid algorithm	(8,12)	(8.32,12.56)	0.64
trilateral localization algorithm	(15,19)	(13.77,20.33)	1.81
triangle centroid algorithm	(15,19)	(15.82,19.34)	0.89
trilateral localization algorithm	(25,20)	(23.22,18.45)	2.36
triangle centroid algorithm	(25,20)	(25.56,20.88)	1.04
trilateral localization algorithm	(40,38)	(41.55,39.12)	1.91
triangle centroid algorithm	(40,38)	(40.88,38.65)	1.09
trilateral localization algorithm	(27,30)	(28.66,31.32)	2.12
triangle centroid algorithm	(27,30)	(27.55,30.79)	0.96
trilateral localization algorithm	(20,12)	(21.33,13.12)	1.73
triangle centroid algorithm	(20,12)	(20.56,12.77)	0.95
trilateral localization algorithm	(10,5)	(11.23,6.34)	1.82
triangle centroid algorithm	(10,5)	(10.46,5.76)	0.89

6 Conclusion

The proposed method combines the RSSI method and the triangle centroid location algorithm, use RSSI value to measured distances d of unknown nodes and beacon node, according to the beacon node as the center of the circle, the distance d as the radius of three circle intersect to obtain the unknown node location of the triangle area, seeking the centroid and the location of the unknown node is obtained. Through the simulation experiment, compared to the algorithm and trilateration algorithm proved the superiority of the algorithm.

References

[1] Zhao Zhao, Chen Xiaohui. An improved location algorithm based on RSSI in WSN[J]. Chinese Journal of Sensors and Actuators, 2009(22) : 392-394.

[2] He T, Huang C D, Blum B M, et al. Range-Free Localization Schemes in Large Scale Sensor Networks[A]. In: Proc. of the 9th Annual Int'l Conf. on Mobile Computing and Networking[C].San Diego: ACM Press,2003:81-95.

[3] Arias J, Zuloaga A. An RSSI Based Ad Hoc Location Algorithm[J] . Microprocessors and Microsystems, 2004, 28(5):403-405.

[4] Paramvir Bahl, Venkata N Padmanabhan. An in-building RF-based user location and tracking system[C] // Proc. of the IEEE INFOCOM.IEEE,2000:775-784.

[5] Theodore S Rappaport. Wireless communications: principles and practice[M]. Prentice Hall PTR, 1996.

[6] Oka A, Lampe L. Distributed target tracking using signal strength measurements by a wireless sensor network[J]. IEEE Journal on Selected Areas in Communications,2010,28(7):1006-1015.

[7] Zhang J, Zhang L.Research on distance measurement based on RSSI of ZigBee[C]//Proc Computing, Communication, Control, and Manag-Ement. China:Shandong,2009:210-212.

MATEC Web of Conferences **63**, 05009 (2016) DOI: 10.1051/matecconf/20166305009
MMME 2016

Study on Resource Management Based on Cloud Computing

Fu Gui FANG [1,a]

[1] *Hexi University,734000 Gansu China*

Abstract. In view of that it is difficult to unify the resources management and the low system security in current cloud computing, the paper designs resource management system. First, it is the definition of cloud computing, architecture description of cloud services and cloud computing; followed by a detailed analysis of ant colony algorithm which is the core task scheduling in cloud computing; finally, it is the design framework and core code of resource management system based on cloud computing. This paper has a positive effect for big data staff and cloud computing researchers.

Key words. Cloud computing, Resource management, Task scheduling

1 Introduction

With the rapid development of information technology processes, network service requires large scale of resources and computing, which makes a separate computer cannot meet the needs of the current resource management. Cloud computing is the realization of commercialization network computing which is combined by parallel computing, network storage, load balancing, distributed computing and virtualization technology and the networks. As the characteristic of heterogeneity and distributivity, also the varied service types, when cloud computing manages resources, in order to meet the various needs of different users, it will have a huge scale of storage and calculation.

2 Cloud computing

2.1 Definition of Cloud Computing

In the era of information explosion today, it has become part of people's lives that we work in high-efficiency and low cost by use of internetwork communication for resources computing and storage. Along with the deepening of network applications, a lot of new concept and computational models such as on-demand computing, efficacy computing, grid computing which are proposed continuously and applied in the large-scale collaboration, cloud computing is presented in this case.

Since the introduction of cloud computing, experts and scholars at home and abroad proposed different definitions of cloud computing based on their own application scope and reality, but so far it does not have a clear definition. In this article, cloud computing is defined as a computing model which relies on the Internet, by increasing and using related network services and less management costs, gets dynamic, a lot of computing resources for users to use seamlessly.

2.2 Cloud Services

At present, cloud computing can provide various types of services, but it can be summed up as three types: Laas (infrastructure as a service), Paas (Platform as a Service) and SaaS (Software as a service). With underlying hardware, the structure of these three kinds of services is shown in Figure 1:

Figure 1 Structure of cloud services

2.2.1 LaaS

Related to the underlying hardware closely, the service of this mode provides storage, calculation and related infrastructure resources, its theoretical basis is virtual technology, and it will make the users feel no difference resource management when they use cloud computing.

2.2.2 Paas

This mode is a business model which provides development platform and server platform for developers,

[a] Corresponding author: 931773740@qq.com

by loading some configuration for operating system platform, it provides service interface for secondary development, by eliminating infrastructure work management, it brings great convenience for the users.

2.2.3 SaaS

This mode mainly aims at end users, which is specially designed for specific services of network, it reduces unnecessary infrastructure configuration for end-users.

2.3 System architecture of Cloud Computing

After years of development, Cloud computing has gradually formed relatively complete system architecture, its system architecture is shown in Figure 2:

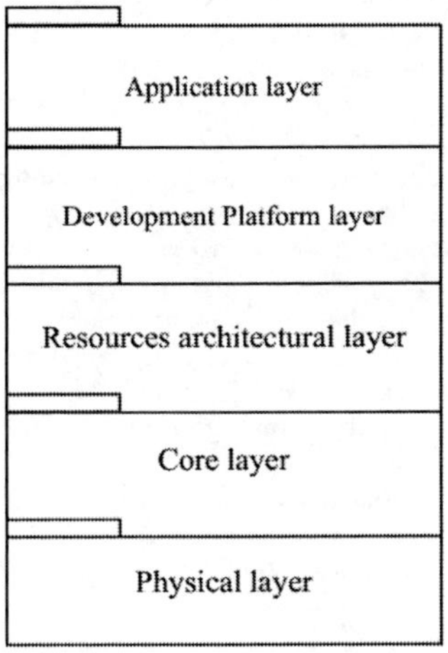

Figure 2. System architecture of cloud computing

As shown above, the entire system architecture is divided into five levels, of which the underlying physical layer is composed by the infrastructures distributing at different geographic locations; core layer manages physical resources, and provides abstract service of specific application for the upper layer; resource architecture layer ensures the reliable distributed communications and the construction of storage resources architecture system; development platform layer provides interface for specific application through API, which is the application platform of resource development; based on the development platform layer, application layer provides reliable quality service for the users.

3 Task scheduling of Cloud computing

There are many task scheduling algorithms in cloud computing, by comparison, ant colony algorithm has the advantages that it is easy to combine with other algorithms, so this article takes ant colony algorithm as the task scheduling algorithm of cloud computing.

3.1 Principle of ant colony algorithm

Ant colony algorithm was first proposed by Italian scholar Dorigo in 1992, it was a kind of bionic algorithm derived from nature. In nature, it was discovered that when ants look for food, they released a unique secretion along the way, which we called it pheromone, within a certain range, ant will decide its next route according to how much pheromones there are, when the longer the way was, the less pheromone was left, this will provide a reference to the ants back, when there were more pheromone left on a way, it showed that there were more ants had passed by, it had greater likelihood to find food on that way, and then it attracted different numbers of ants moving food along the best route.

Principle of ant colony algorithm is by use of ant groups transferring information in the path to achieve optimization. This algorithm has a lot of advantages including positive feedback mechanism, better robustness, self-organizing ability, good parallelism; but due to the short generating time, ant colony algorithm has some shortcomings in the application process, the main performance is slow convergence, long search time and ease of stagnant status.

3.2 Task scheduling process based on ant colony algorithm

3.2.1 Mathematical model of ant colony algorithm

Suppose that there are m ants to find n food nodes, between two food nodes the distance is dij, bi(t) is the number of ants located on Food node i when t time, then the total number of ants $m=\sum_{i=1}^{n} bi(t)$ wherein ants leaving 0 pheromone at the first time on the road, after some time of searching food, the pheromone in each path is expressed as formula 1:

$$\tau_{ij}(t+1)=(1-\rho)\tau_{ij}(t)+\Delta\tau_{ij}(t)$$

Formula 1

Where ρ is the evaporation coefficient of pheromone, Δ Γ is the increased volume of pheromone.

The probability of Ant k moving from node i to node j is represented by Equation 2:

$$p_{ij}^{k}=\begin{cases}\dfrac{[\tau_{ij}(t)]^{\alpha}\cdot[\eta_{ij}(t)]^{\beta}}{\sum_{s\in allowed_k}\tau_{is}(t)]^{\alpha}\cdot[\eta_{is}(t)]^{\beta}} & \text{if } j\in allowed_k\\ 0 & \text{otherwise}\end{cases}$$

Equation 2

The larger P is, the greater likelihood ant moves.

3.2.2 Flow of algorithm

Ants always tend to move on the way having more pheromone, by recording the nodes have been passed into the table T, after entire discovery process, the value of pheromone is updated, by repeated execution till the final termination conditions, the particular algorithm folw is as follows:

Step 1: Initialize the data, place m ants on n nodes randomly;

Step 2: Store initial node of all the ants into a specific table T;

Step 3: Determine whether the number of iterations has reached the set value, if it has reached, go to step 7, otherwise go to step 4;

Step 4: Calculate the probability of transfer by formula 2 above, and move to the corresponding node, modify the corresponding node value in table T;

Step 5: Check if table T meets the condition, if not skip to step 4, if it meets go to step 6;

Step 6: Update the value of the pheromone, and empty table T, plus 1 to iteration value;

Step 7: Get the best value, ends.

4 Resource Management System of Cloud Computing

4.1 Architecture Design of the System

According to the architecture of cloud computing and the analysis of ant colony algorithm, resource management system of cloud computing designed has different rights management for users and administrators. It includes checking and applying status of VPS (virtual machine) host, basic operation(switch, restart, etc.) to VPS (virtual machine) host, checking the task application and task execution schedule, and submit computing tasks. The framework of its system is shown in Figure 3:

4.2 system implementation

In the system, ant colony algorithm is the core, in the process of searching nodes, it make ants find all of the nodes, wherein the core code is as follows:

```
if (e.Node.Tag! = null)
  {
   try
   {
    SearchKey = "";
    string path = e.Node.Tag.ToString ();
    FileInfo file = new FileInfo (path);
    if ((file.Attributes & FileAttributes.Directory)! = 0)
    {
      TreeViewShow (e.Node);
    }
    else
    {
      ParameterizedThreadStart gr = new ParameterizedThreadStart (Get_FileText);
      Thread th_GetFileText2 = new Thread (gr);
      th_GetFileText2.Start (path);
     }
    }
    catch (Exception ex)
    {
    MessageBox.Show (ex.Message, "prompt", MessageBoxButtons.OK, MessageBoxIcon.Information);

    }
```

5. Summary

This paper researches the resource management of cloud computing, so the resource management of cloud computing is the main content. This paper introduces ant colony algorithm which is the major algorithm in cloud computing, and designs out the framework. Due to limited space, this paper does not offer the specific function description and database design of the system completely, readers who are interested can do further research and discussion on this basis.

References

1. Li Xiaolong, Li Ting ,Research on resource management of cloud computing, *Computer & Telecommunication*, 1,p.62-64(2010)
2. Yang Siwei, Gao Donghuai, Ning Yuwen, Research on network learning resource sharing based on Cloud Computing, *The Chinese Journal of ICT in Education*, 3,p.20-23(2012)
3. Aao li-bo, platform design of teaching resources sharing based on cloud Computing, *Information Technology*,10,p.46-48(2015)
4. Zhang Lei ,Cloud computing and information resource sharing management, *Electronic Technology & Software Engineering*,4,p.21(2014)
5. Fan Guisheng, Yu Huiqun,Cloud computing security model and management, *Microcomputer Applications*,1,p.1-3(2013)
6. Lv Yan,Research on human resource management information platform based on Cloud Computing, *Management & Technolosy of SME*,20,p.182-183(2015)
7. Jiang Zongpin, Luo Guofu, Liu Haidong,Design and implementation of logistics information platform based on Cloud Computing, *Logistics Technology*,13,p.256-258 (,2015)
8. Duan Zhongxiang, Construction of shared resource model of network platform based on Cloud Computing, *Computer Engineering & Software*,5,p.119-121(2013)
9. Fang Jinming,The decision making system of virtual resource scheduling in cloud computing, *Computer Measurement & Control*,12,p.2145-2148(2011)
10. Liu Guoying,Library resource management system in the era of cloud computing, *Journal of Baoshan University*,3,p.28-32(2015)

DOI: 10.1051/matecconf/20166305009

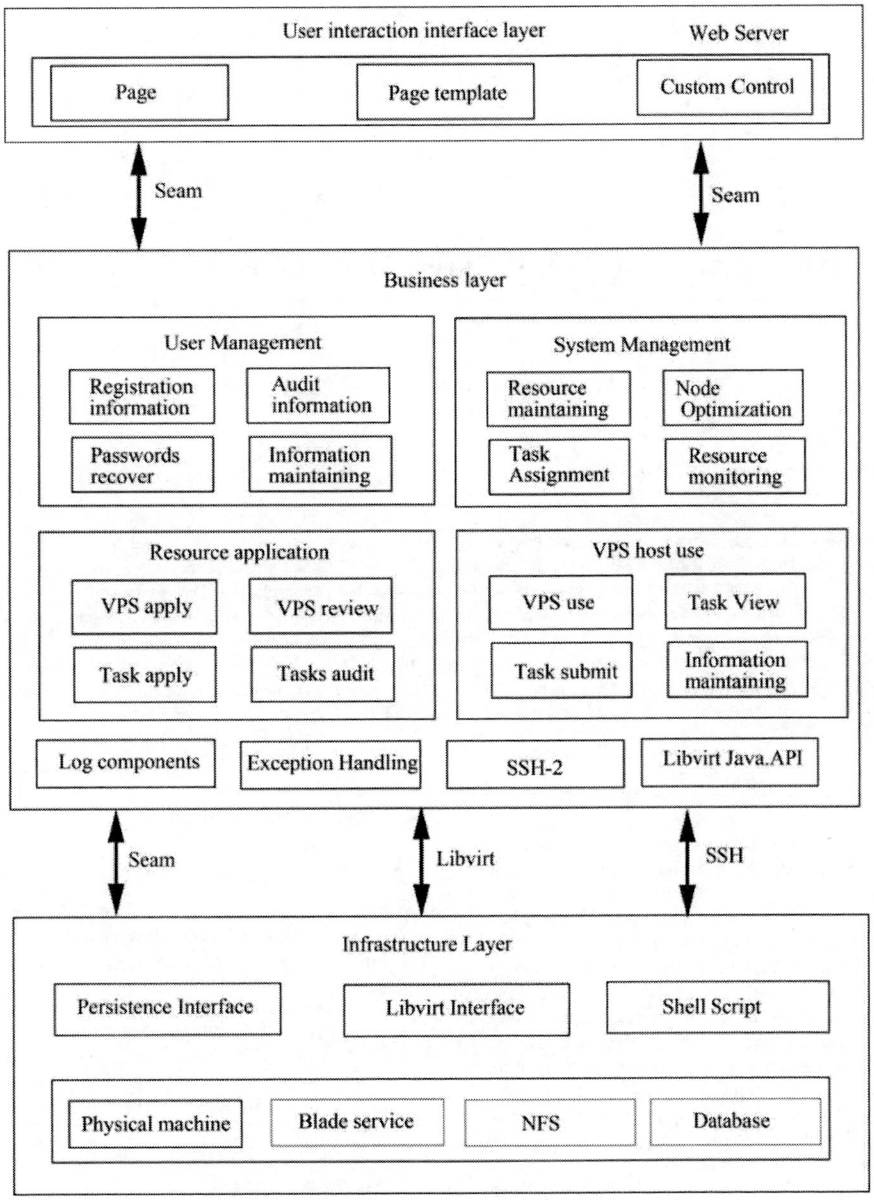

Figure 3. System Framework

DOI: 10.1051/matecconf/20166305010

Virtual Reality Analysis in Tennis Serve Technique Stability for Junior Masters

Chuan Jia DU[1,a], Ji He ZHOU[1] and Shuai WANG[1]

Chengdu sport university,Chengdu, Sichuan, China

Abstract: This study consists of virtual reality analysis of tennis serve technique was constructed based on two male professional players from 2016 Chengdu ITF Junior Masters: Casper RUUD and Miomir Kecmanovic. The purpose of the study is to find an effective way of building stability of serves for junior players. This study will provide some considerable data for coaches and players in improving the stability and quality of the serve technique. Results of the study show 5 main points of view: (1) RUUD has a higher stability on controlling the height and location of serve. (2) During the swing stage, RUUD' s "scratching back" position of racket head is deeper, but Kecmanovic' s consistency is better. (3) For the consistency of the height to hit the ball, RUUD has a better control. (4) The height of the ball toss is a key point to affect the quality and stability for serves. (5) Overall, RUUD has a higher serve consistency than Kecmanovic.

1 Introduction

Stability in tennis means when a player produces a technique repeatedly, he or she shows the consistency of stroke power, impact direction, speed, and angles of joints for the certain tennis technique. The tennis serve is one of the most important techniques that initiates each point in a match. Developing high speed and consistency of serve percentage will make every professional tennis player more successful. However, usually when players cannot produce a high quality serve it is because the great pressure from serve makes bodies hardly work with neuromuscular system ideally. The consistency of swing power, impact direction, speed and angles of joints is stable or not can make a huge potential impact on serves.

This paper is a study that based on the sophisticated 3-D motion analysis system and the virtual reality tech to analyze the two players, Casper RUUD and Miomir Kecmanovic, specifically. Based on the study of the serve technique for these two players to provide recommendations for coaches and professional junior players about improving and strengthening tennis serve technique.

2 Subjects and methods

2.1 Subjects

The study sample included 2 male tennis players from 2016 Chengdu ITF Junior Masters: Casper•RUUD and Miomi•Kecmanovic. Table 1 is the information of these two players.

Table 1. Basic information of the subjects

Name	Height(m)	DOB	Plays	Ranking
RUUD	1.83	12/22/1998	Right	2
Kecamanovic	1.81	08/31/1999	Right	8

2.2 Methods

2.2.1 Three-dimensional video analysis: During 2016 Chengdu ITF Junior Masters, matches were recorded simultaneously with two high-definition digital JVC GC-PX10AC video cameras, recording at 50 fbS. The first camera was located in the back of the doubles sideline, and the second camera was located in the back

[a] Corresponding author: 549828421@qq.com

of the base line. The angles of the two cameras were about 45° . (See Figure 1.) There were 10 first serves that be chosen randomly of each player, and then analyzed by the 3-D SignalTec software and Germany Dempster mannequin (16 links, 21 joints parameter). Due to the necessary of the study, the tennis ball and the racket head were also chosen as two parts of the test points. All the data were smoothed by the LPF method, and the cutting frequency was 8Hz.

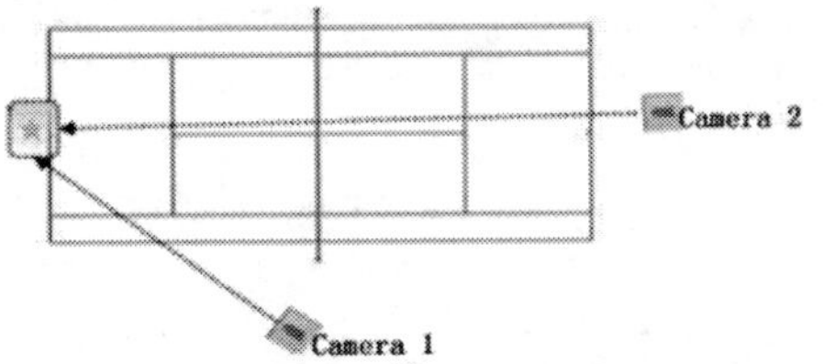

Figure 1. *Diagram of cameras position*

2.2.2 Three-dimensional virtual reality technology method: Based on the 3-D motion analysis, all the serve parameters of the two players were inputted into the virtual reality system, and analyzed by the 3-D animation motorically to analyze players' serves accurately.

2.2.3 Mathematical statistics: All the data were also analyzed by the spss19.0. The mean and standard deviation of the data can show the consistency and stability of the players.

3 Results

3.1 Stages of tennis serve and key parameters

According to the feet stance, the serve technique has two different ways: Foot-back and Foot-up. Casper RUUD used Foot-up in the games, and Miomir Kecmanovic used Foot-back. For the study, the serve technique is divided into two stages as below and each part has different key parameters about the stability.

1) Ball toss stage: the moment to toss the ball and the maximum angle of knees flexion (lowest body center of gravity). Key parameters: the distance between the hand and the ball (vertical distance from the start of ball toss to the ground), the height of ball toss (vertical distance from the highest point of ball toss to the ground), the maximum angle of the right and left knee flexion.

2) Swing stage: the moment that knee flexion to the maximum, and the feet of the player has touched the ground after impact. Key parameters: The distance between the ground and the racket head when the racket was dropped behind individual' s head, the height of hitting the ball, and the time to finish the contact point part.

3.2 Comparative analysis of the Ball Toss stage

Ball toss means the player is going to release the tennis ball to make the toss to the proper location and height. Meanwhile, the racket hand and the body will move inversely. The muscle chain of the lower limbs begin to function in order to facilitate the stretch-shortening cycle that will contribute to storage of elastic potential energy in the muscles that are acting within the upper limb and lower limb muscle chain (Ellenbecker, Roetert, Bailie, Davies, & Brown, 2002). In this stage, the location of the toss, the height of the toss, the maximum angle of knee flexion when the body center gravity is at the lowest, and etc. are all the important parameters of the quality for tennis serve technique. (See Table 2).

Table 2. Key Parameters of Ball Toss Phase

Name	Location (m)	Height (m)	Left Knee (°)	Right Knee (°)
RUUD	**1.833±0.040**	**3.215±0.061**	**114.7±4.5**	**119.2±3.2**
Kecmanovic	**1.786±0.083**	**3.382±0.120**	**95.0±4.8**	**126.5±7.3**

All the data show that the average heights of toss location of RUUD and Kecmanovic are 1.833m and 1.786m. The ratio of the location and their body height is 1.00 for RUUD, and 0.98 for Kecmanovic. According to the related research, the location of the toss should be the same height with the top of the player' s head, which means the ratio should be 1 (Hu & Wang, 2013). Therefore, RUUD' s toss location is better than Kecmanovic. Kecmanovic' s toss location is 2% lower than the standard location. Besides, from the standard deviation of the toss location, RUUD has a higher ratio stability of toss location than Kecmanovic.

The toss height is the vertical distance from the highest toss point to the ground. It is an important parameter to affect serve technique. The higher the height of the toss, the faster will the ball be when it goes through the contact point. On the contrary, if the toss height is too low, the transaction between the ball toss stage and the swing stage will not be smooth. In conclusion, if a player can have a high ratio stability of the toss height, then he or she will control serve rhythm better. The means and standard deviations of the toss height of RUUD and Kecmanovic are 3.215m, 3.382m, 0.061 and 0.120. This shows that RUUD has a better control of the toss height. Comparing to RUUD,

Kecmanovic has a lower toss location, which is 1.786m; but his toss height is 3.382m that is higher than RUUD's. Kecmanovic's toss height standard deviation is 0.120, which is also relatively large. The reason that Kecmanovic' s toss height stability is not as good as RUUD is because the force to the ball. When the location of the ball toss is too low, the player has to use a large force to the ball to make the impact. However, the large force will cause a large muzzle velocity, and at the same time the velocity works with the gravity, then the ball will uniform motion in a straight line (predict the force from the hand goes through the center of the ball). At that moment, it is hard for a player to control the force, so the height of toss could be too high and instable.

At the moment that the angle of knees flexion to the maximum, the average angles of RUUD' s and Kecmanovic' s left and right knee joints flexion are 114.7°, 119.2°; 95.0°, and 126.5°. According to related research, top tennis players' knee flexion of the opposite play hand is round 100° (Hu & Wang). RUUD and Kecmanovic are both right handed players, so based on the angles of their left knees flexion, Kecmanovic has a larger angle. However, the reason that there is a big difference between the two players is because their stances are different. Kecmanovic uses the Foot-back stance so the angle between his legs is too large to make a big difference between the angles of his knees flexion. While RUUD uses the Foot-up stance to keep his knees flexion similarly. Therefore, because the difference of their stances, it is not reasonable to compare their knees flexion. The current #1 ranking male tennis player, Novak Djokvoic, also uses the Foot-back stance to produce serves; and his minimum angle of knee flexion is 110.3° (Yang & Zhou, 2015). Therefore, Kecmanovic' s angle of knee flexion is too much to finish the following hitting process because the burden on the related muscle chain have a negative impact on the speed to stretch the extensor muscle group. Another top ranking player, Rafael Nadal, uses the Foot-up stance to serve, and the minimum angle of his left knee flexion is 109.3° (Hu & Wang). This tell us that Ruud' s knee flexion angle is not large enough to stretch the extensor muscle group to storage the potential elastic energy.

3.3 Comparative analysis of the Swing stage

The swing part of the tennis serve is a complicated process to decided the quality of serve technique. When a player is ready to swing and hit the ball, his or her feet will bend first and then stretch fast. The racket head will not be swung up immediately but to be dropped down like "scratching back." The chest and the shoulder will form a big radian angle. The range of the "scratching back" position not only depends on the inertial of racket head, but also on the control of the position. The form of the "scratching back" position will produce a swing like whipping. At the contact point, the racket head will extend more through the contact zone to make the forearm do an internal rotation and then follow through to finish the serve. The key parameters of the swing stage are the height from the racket head to the ground of the "scratching back" position (H_{re}), the height of the contact point (H_{cp}), and the time to finish the contact point (T_{cp}). (See Table 3).

Table 3. Comparison parameters in the swing stage

Name	Hre (m)	Hcp (m)	Tcp (s)
RUUD	1.068±0.072	2.780±0.025	0.11±0.02
Kecmanovic	1.093±0.060	2.821±0.041	0.12±0.02

The range of the "scratching back" can be noticed from the distance between the racket head and the ground. The shorter the distance, the obvious the position, and vice versa. The mean heights of RUUD and Kecmanovic are 1.068m, and 1.093m; and the standard deviations are 0.072 and 0.060. According to their mean and standard deviation, RUUD can form the position deeper, but Kecmanovic' s consistency is better.

After stretching the legs, the lower limbs are almost straight. At the moment that the two feet are off land, the racket head starts accelerating upward. The tennis ball will uniform linear motion from the highest point. When the ball falls to the contact point, the racket center should contact with the ball. The height of the contact point is the vertical distance from the racket to the ground. A proper height of the contact point will have a direct impact on hitting the ball. The average heights of the contact point of RUUD and Kecmanovic are 2.780m and 2.821m; and the standard deviations are 0.025 and 0.041. According to Yan and Wu' s study, the best ratio of the contact point height and a player' s height should be around 1.5 (Yan & Wu, 2000). This ratio is 1.51 of RUUD, and 1.55 of Kecmanovic. Therefore, RUUD' s height of contact point is better. Kecmanovic' s height of contact point is relatively high, so it will increase the instability to hit the ball. Besides, Hu and Wang also points out that the head between the highest ball toss point and the contact point should be around 0.5m (Hu & Wang). The mean head of RUUD is 0.435m, and of Kecmanovic is 0.561m. Both of them control the head within the effective head range. However, based on the free fall equation, the speed of the ball from Kecmanovic at the contact point is too fast, and this is a main reason that the standard deviation of his contact point height is relatively large.

The time to finish the contact point means the time that racket from the lowest point of the back to the contact point. It is an important parameter to evaluate the explosive power and motor coordination of swing stage. The mean time from the maximum knee flexion moment to contact point of RUUD and Kecmanovic is 0.11s and 0.12s. There is not difference on the standard deviation for the two players, which is 0.02. The lowest time that Nadal can finish is 0.066s (Hu & Wang). Therefore, RUUD and Kecmanovic have to increase their muscle explosive power to shorten the time to finish the contact point and increase the speed to impact.

4 Conclusion

1) The mean heights of ball toss location for RUUD and Kecmanovic are 1.833m and 1.786m. This shows that RUUD has a proper height of toss location, but Kecmanovic's toss location is relatively low. The standard deviations for RUUD and Kecmanovic are 0.040 and 0.083, which means RUUD has a higher ratio of stability of toss location.

2) The mean ball toss heights of RUUD and Kecmanovic are 3.215m and 3.382m, which means RUUD can control the height more ideally. The standard deviations are 0.061 and 0.120, so RUUD's stability of toss heights is also better.

3) The mean angles of the maximum knee flexion for RUUD and Kecmanovic are 114.7° and 95.0°. Therefore, the angle of Kecmanovic is too much, and the angle of RUUD is too small. The standard deviations are 4.5 and 4.8, so both of them have a relative high consistency on controlling the angles of knee flexion.

4) When RUUD and Kecmanovic are forming the "scratching back" position, the mean heights from racket head to the ground are 1.068m and 1.093m, and the standard deviations are 0.072 and 0.060. Therefore, RUUD can make the position deeper but Kecmanovic has a higher ratio stability of this position.

5) RUUD has a better control on the contact point height. The height of the ball toss is a key point to affect the stability of the height of the contact point.

6) There is no difference between RUUD and Kecmanovic on the time to finish the contact point. However, comparing to the top ranking player, Nadal, RUUD and Kecmanovic still spend too much time to impact in a high speed.

7) Overall, RUUD has a higher serve technique consistency.

References

1. David Whiteside, Bruce Clifford Elliott. *Coordination and variability in the elite female tennis serve*[J].Journal of Sports Sciences; 2015, **33** (7): 675-686.
2. Zhilei Wang,ShanfeiZhang. *Kinematic Analysis on the Serve Technique of lower limbs*[J]. Journal of ningbo university of technology. 2009,**21**(3):77-80.
3. Ellenbecker, T. S., Roetert, E.P., Bailie, D.S., Davies, G.J., & Brown, S.W.*Glenohumeral joint total rotation range of motion in elite tennis players and baseball pitchers.* Medicine and Science in Sports and Exercise; 2002, **34**(12): 2052-2056.
4. Chuanjia Du,Jihe Zhou,Shuai Wang. *Kinematic Analysis on the Serve Technique of Elite Tennis Player. Grigor • Dimitrov Based on 3D VR Technology*[J]. MATEC Web of conferences, 2016, **44**.
5. Botao Yan,Yanxi Wu. *Biomechanical analysis of the tennis serve*[J]. Journal of Sports science. 2000, **20** (4): 80-84.
6. Kai Hu,Yun Wang. *Serve action analysis of the Elite male tennis players*[J].Journal of Sports research and education: 104-106.

MATEC Web of Conferences **63**, 05011 (2016) DOI: 10.1051/matecconf/20166305011
MMME 2016

Research on Early dismantling head of numerical simulation and experiment

Fu Long ZHAO[1,a] , Xiao Li WANG[1] ,Wei WANG [2]

[1]*Guizhou construction science Research & Design Institute limited company of CSCEC, Guizhou Guiyang, 550006, China*
[2]*The third construction engineering co. LTD of China construction Forth engineering division CORP . LTD*

Abstract. The template support-system projects not only need to complete the range of quality optimization in one construction work, but also to make one of the sound security work. Installation and removal of template support-system are the largest amount of labor. To seek the template works rationalization and reduce the cost of the template works, it is an important way to save a lot of labor and reduce the cost of concrete structure. Therefore, how to improve the template engineering design and its production, installation, operation and removal efficacy effectively and make further reform of the template project for improving the quality of the project are of great significance in the construction of civilization.

1 Introduction

Early dismantling head of numerical is on the wooden and flute (keel) branch is on a wooden flute template, mean while the concrete pouring, conservation is 10 to 14 days.In order to remove all the templates and the support system, the strength of concrete should be more than 70%.The reasons of not earlier to remove the new pouring is that the early concrete strength is too low.It can not reach the predetermined strength yet under that condtion. However, the concrete structural engineering construction and acceptance are in accordance with the national standards (GB50204-2002) of China provides: Cast-in-place structure of the template and its support removing of the concrete strength should meet the design requirements and it shall meet the following requirements, such as board bottom mold, template span ≤ 2.0m, the concrete strength of 50% of the design strength before the form removal.

The concrete strength up to 75% of the design strength before the form removal, when beam arch shell span ≤ 8.0m, the concrete strength of 75% of the design strength, when the span is greater than 8.0m, the concrete strength up to 100% of the design strength before the form removal.

2 Early removal system

The key step of early removal template support system to achieve the early dismantling is to make the removal strength of the structure meet the requirements. If the form removal is premature then the concrete structure is easy to break. If the removal of the template is too late, and it is less than earlier demolition there will be other risks. To reinforced the concrete structure is the way to resolve the factors affecting problem. The strength of concrete has a lot of the main ingredients and thickness, including the spacing between the bar diameter and reinforced, etc. This paper use ANSYS to simulate the use of part of the early dismantling architecture and make the analysis of the stress and deflection of concrete and to reinforce ordinary concrete changes over times to gain the diameter of the reinforced ordinary concrete structure by that time. The analysis of the stress and deflection of the structure are in the most unfavorable position, calculated 10mm thick reinforced concrete floor when the early removal template support-system minimum form removal strength, and to simulate the cracking of the beam withstand construction loads. Design and make the manufacture of fittings welded to the material with Q235. φ48 pipe wall thickness 3.0 or 3.5mm steel pipe which is formed on top of the bar in the early removal system. The crossbar class, pole rod through and the connections connected is to form a structure of high dimensional with accuracy, durable, stable performance, There are flexible and convenient loading and unloading steel to support the frame.

2.1 Model analysis

2.1.1 Selection of Model unit type

Early dismantling head of numerical is to consider as the type of unit to process. Concrete and steel are divided into small units enough as each other, both the stiffness matrix can be solved separately, at the same time taking into account of the reinforcement is a thin material, it can be neglected its transverse shear strength usually, so the

[a] Corresponding author: 308369555@qq.com

 DOI: 10.1051/matecconf/20166305011

fixture can be treated as the line element processing.The bonding element can be inserted between the steel and concrete to simulate the bond between steel and concrete slip. Generally speaking reinforced concrete is in the presence of cracks, while the cracking will lead to steel and concrete deformation of the lack of coordination inevitably, that is to say, it should occur in the losing - efficiency slipping of bonding, so the application of such model is the most widely used. ANSYS can consider a separate model: concrete (with Solid65) + reinforced (Link unit or Pipe unit) the paper used unit-with Solid65 unit mainly: Solid65 unit with reinforced or non-reinforced three-dimensional solid model.

2.1.2 The most unfavourable conditions

This paper use the induction way to identify the affection of the structural safety performance under the most adverse conditions. If the structural is still safe and effective while the members is in the joint action. So we could summarize favorable and draw in the widespread use of the structure can be the safely using of conclusion relatively. Therefore it is the core to verify the early removal system board structure strength properties which is established by the most unfavorable model.

2.1.3 Finite Element Modeling

Concrete slabs. 5000mm × 5000mm × 100mm, with Solid65 unit, modular model, Concrete slab meshing automatic division plate width is divided into 50 equal portions, the plate thickness is divided into 2 equal facilities around the plate reinforced side constraints (Fig. 1~2), early dismantling stigma .100 mm × 150mm× 50mm with SOLID45 unit, in the end of the stigma applied reinforcement side constraints (equivalent to the rigid washer), weight loading .2.5 KPa = 2.5e-3MPa, the construction loads. 3 days (72 hours) construction load of 1.5kN /m² = 1.5e-3Mpa,to simulate the top-level conditions, 6 days construction loads for 2kN /m² = 2.0e-3MPa to simulation the top-level conditions. The stigma may not have the upper reinforced. Early removal template concrete slab crack width calculation for the design of concrete structures is given during the use phase, which does not apply during the construction period.

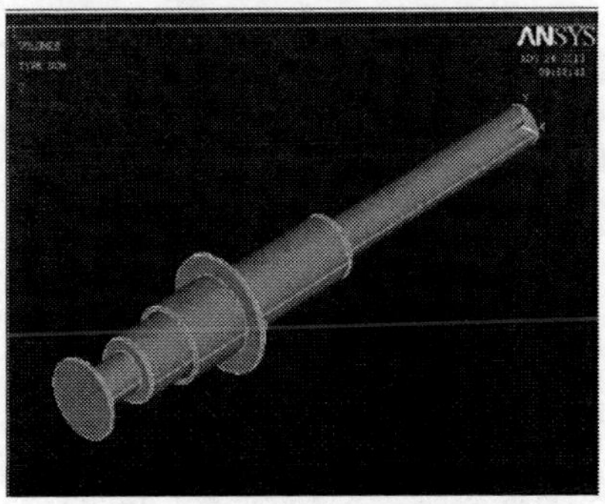

Fig. 1 Model meshing1

3 Finite element analysis of the bearing capacity

As previously mentioned, the stable capacity of attached template support frame is controlled by the stable of system. The computing stable theory is main of linear envalue buckling and nonlinear buckling.Thus, this paper is using the finite element method for the attached template support frame system capacity to analysis, and through the finite element software ANSYS to make the stability analysis of the attached template support frame.

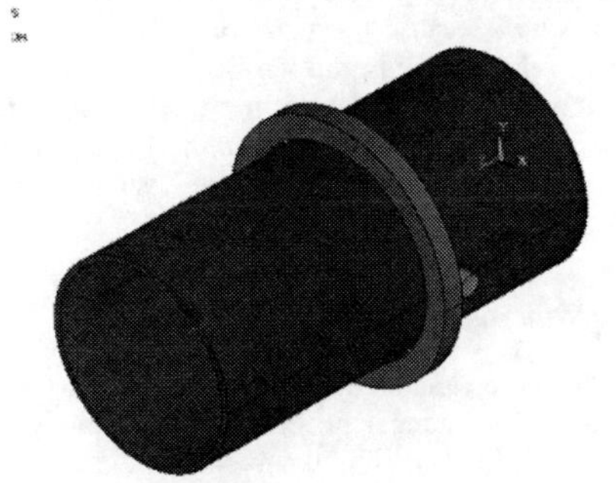

Fig. 2 Model meshing2

3.1 Captions/numbering

There are many reasons to cause cracking of concrete structures in the actual project and there are three main reasons: load crack, deformation, cracks and other physical or chemical changes cracks. The deformation cracks is caused by expansion or contraction of the concrete. Concrete is an unevenness brittle material, its tensile strength is low, when the contraction reaches a certain level, the shrinkage stress is greater than the tensile strength of concrete it will crack. The cracks are very different due to many reasons, sometimes in the design we will allocate a portion of the temperature tendons or set the expansion joints of reinforced concrete structures, the aim is to reduce this crack and its adverse effects .Captions should be typed in 9-point Times. They should be centred above the tables and flush left beneath the figures.

3.2 Positioning

Physical or chemical changes cracks are caused by the corrosion of carbide .When the carbonation depth of the steel is shown in the surface, the alkaline of concrete would be reduced and loss the protective effect of reinforced steel which rust in the concrete. Then make the concrete (mainly reinforced protective layer) be in tension around, when the tensile strain over concrete ultimate tensile strain value, it will cause the concrete cracked. In design, the measures taken in accordance with the design of concrete structures is under the minimum thickness of the protective layer structural members which is not to reach the steel surface and is in the normal period used. The two cracks before can be controlled with the design and construction techniques, for early dismantling template generated cracks load crack. The stigma resulting in structural members makes the stress state changed. But the measure is not suit for the board at the early dismantling era. The stigma may not have the upper reinforced. Early removal template concrete slab crack width calculation for the design of concrete structures is given during the use phase, which does not apply during the construction period. Concrete structure designs for specifications, the use of reinforced concrete structural members by cracks in the tension zone. Basically the homogeneous phase cracks is the concrete tension zone. Because the bond of concrete and steel is cracking, all rally withstand tensile reinforcement. Using finite element analysis can be an effective solution to the design concrete structures. Under the condition that do not consider concrete tension for the basic assumption to analyze the basic situation of the cracks.

3.3 illustrations

From the finite element analysis, the lower surface of the concrete slab is almost no cracks, so it can ignore the impact of the load crack in the middle of the construction, when the early dismantling stigma removed, the original early dismantling stigma at the crack part of concrete becomes under pressure then the cracks in the state with the growth of concrete shrinkage, creep, concrete slab it may be closed even healing, so the crack width is guaranteed.The results are shown in Fig. 3~6.

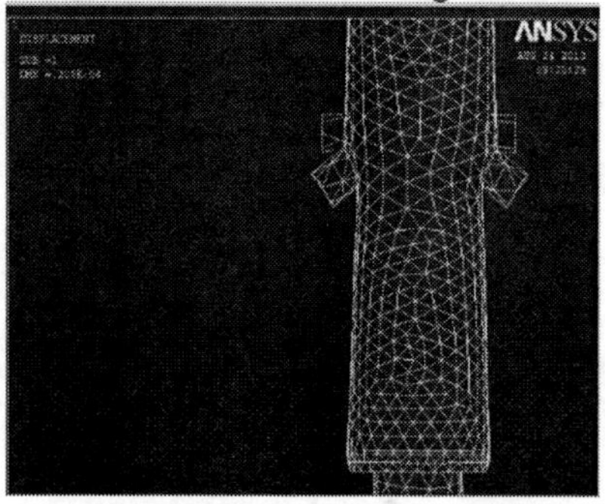

Fig. 3 Stress calculation results 1

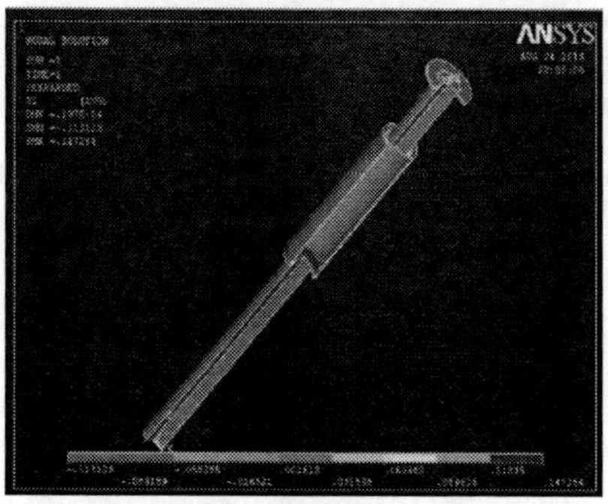

Fig. 4 Stress calculation results 2

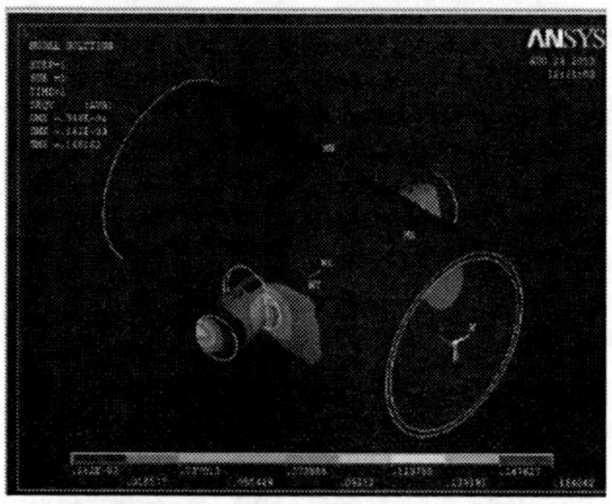

Fig. 5 Stress calculation results 3

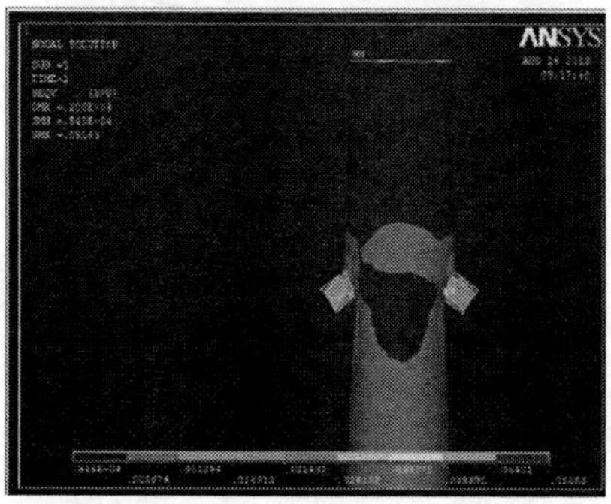

Fig. 6 Stress calculation results 4

In these Figs above, the deflection value is far less than the Design of Concrete Structures (GB 50010-2010) 3.4.3 provisions(China). At that time, the the deflection limits brackets values which apply to use deflection member of the early dismantling formwork construction technology on board deflection control is guaranteed.

At the same time, through the pole, adjustable push rod and the bar ,the composition of early removal template support system could be made.With outlet socket connections, it can form a structure size specification template support. Joint node of the support frame can withstand a certain moment,punching and torque. Plate bracket is to form self stability and spatial structure of the overall stability of the bracket. In the early dismantling system, the pole is as the basic force units through the crossbar composition bracket which is adjustable mandrel screw into hole to form a template early dismantling system. As mentioned above, with the early dismantling process principle, in fact, when the long-span is shortened to cross, the required form removal strength will reduce to 50% of the design strength of concrete, thus it can speed up the turnover rate of the template.

As previously mentioned, the stable capacity of attached template support frame is controlled by the stable of system. The computing stable theory is main of linear envalue buckling and nonlinear buckling.Thus, this paper is using the finite element method for the attached template support frame system capacity to analysis, and through the finite element software ANSYS to make the stability analysis of the attached template support frame.

4 Summary

The paper stars with model of early removal template support system and through ANSYS to make the model analysis. Then through the induction to find the worst case model, and in accordance with the model of ANSYS calculations to verify the structural panels tensile stress, compressive stress, cracks, deflection which meet the engineering requirements.So the destruction does not occur, so the attached early removal template support system is reliable.

References

1. Bangash,M Y H,Concrete and Concrete Structure:Numerical Modeling and Applications. Elsevier Science Publishers Ltd,1989.
2. Huyse L,Hemmaty Y,Vandewallc L,Finite Element Modeling of Fiber Reinforced Concrete Beams,Proceeding of the ANSYS Conference,Vol.2,Pittsburgh,Pennsylvania,1994.
3. Hemmay Y.Modeling of Shear Force Transferred Between Cracks inReinforced and Fiber Reinforced and Fiber Reinforced Concrete Structures, Proceedings of the ANSYS
4. S-Timoshenko.Theory of plates and shells,1959.
5. Isenberg J. Finite Element Analysis of Reinforced Concrete Structures II [J],ASCE,1993.
6. Farming P. Nonlinear Models of Reinforced and Post-tensioned Concrete Beams. Electronic Journal of Structural Engineering, 2001（2）:111-119.
7. Erduran E,Yakut A, Drift Based Damage Functions for Reinforced Concrete Columns. Computers and Structures,2004,82（2-3:121-130.
8. Barzegar,Maddipudi, Srinivas,Three-Dimensional Modeling of Concrete Structures.1:Plain Concrete Structures.1:Plain Concrete, Journal of Structural Engineering,1997,123（10:1339-1346.

The Application of Computer Technology in Art Design

Xin TAO
He University, Shenyang, 110163, Liaoning

Abstrect: After entering the twenty-first Century, computer technology has been greatly developed, applied in various fields gradually, and gained remarkable achievement. Computer technology not only is a project of great importance in China, but also cannot be ignored in the world. It has already attracted attention of numerous people, gradually integrated in daily life and closely related with it. Now how to apply it into more fields has been the main focus of our nation and also the topic we care about all the time. It is also an important project in our country to apply computer technology in art design in order to higher the art design quality. The purpose of this paper is to introduce people problems of the application of computer technology in art design, and find ways to promote cooperation and common development of them.

Keywords: art design; computer technology; application

Art design is a highly comprehensive subject, involving society, culture, economy, market, science and technology and so on. Besides, it is a strong independence of artistic discipline, which includes production and design of arts and crafts, environmental design, graphic design, and multi-media design. In the final analysis, art design is the embodiment of the designer's own comprehensive quality.

Computer technology has wide-ranging contents, which can be roughly divided into computer systems technology, computer devices technology, computer components technology and computer assembly technology. Computer technology is one of the most applied technologies in China and is of great significance for the future development of our country.

1.Computer Technology and Art Design

Computer technology is obviously comprehensive, closely relating to electronic engineering, applied physics, mechanical engineering, mathematics and other disciplines. With the changing times, computer technology is developing rapidly and now has been used in various fields.

1. Systems technology

Computer technology itself is a complete system, whose systems technology is mainly divided into four aspects: system structure technology, system management technology, system maintenance technology and system application technology.

1.1 System structure technology

The system structure technology can improve the problem-solving efficiency and promote reasonable and orderly cost performance. The combination of these four subjects including system structure technology, computer hardware, solid body, and software covers electrical engineering, microelectronic engineering, and theoretical computer science and so on.

1.2 System management technology

System management technology is completed through the operation. The main function of the operating system is the application of computer software and hardware, thereby enhancing the throughput capacity and problem-solving efficiency. It is easy to operate and conducive to improving the system reliability and reducing the computing cost.

1.3 System maintenance technology

System maintenance technology can automatically repair and diagnose the computer system. Auto-diagnosis is about examination and non-manual diagnosis for the system function. Functional examination procedure can make a comprehensive tidy-up for the micro-function of all the components within the computer system, and compare the test results to determine whether the components are normal by strict data, graphics or action retry test.

1.4 System application technology

The system application technology is the most widely used, which is most closely associated with programming automation and software engineering. Programming automation is the automated design of the computer program, which is the necessary condition for the computer to receive welcome. It also changes the early situation in which the computer relied on manual work to write programs for instructions so that much time and effort was wasted, errors were frequent and the reading and the mode changes were very difficult.

2. Art design

The art design itself has aesthetic and utilitarian function. Art design is a combination of material and spiritual function in the process of social development, and also the inevitable product of the social development. Art design is featured by independence, comprehensiveness, thought, mobility, serviceability, science and rationality.

2.1 Independence and comprehensiveness

The research content and the service object of art design are different from that of other traditional artistic disciplines. Art design involves many aspects such as society, culture, science, market and so on and its aesthetic function changes with these aspects. Art derives from life and art reflects life.

2.2 Thought and mobility

Art design is continuously developing in practice and is finally crated through the creation and practice. Whatever design it is, the first thing to do is to comprehend the concept of design, namely cultural background, geographical, historical and humanistic knowledge related to design.

2.3 Serviceability

Art design is to create a new way of life in order to achieve the goal of serving the society and allowing people to get more value. The ultimate goal of art design is simplicity, comfort, and efficiency, and to create better products, to change people's way of life, and to promote the development process.

2.4 Science and rationality

The means of art design is reasonable and orderly within the rigorous and scientific spirit level. Art design is not conducted blindly, but in a planned and purposeful way. It sums up different views and analyzes the pooled data so as to maximize the value of art design and benefit the people.

3. The Value Embodiment of Computer Technology Application in Art Design

The application of computer technology can play a significant role in the art design, thereby promoting the overall development of art design and giving full play to its true level. Such a success is achieved because the computer technology itself has a very high value and proper use of it can push art design forward.

3.1 Features of computer technology

DOI: 10.1051/matecconf/20166305012

Nowadays, all fields are inseparable from the application of computer technology, and its status becomes increasingly important. This phenomenon is mainly due to the features of computer technology itself. Application of computer technology in art design has the following features:

(1) The application of computer technology in art design will speed up the development process of art and design. The introduction of computer technology into the art design changes the manual activities such as graphic design, animation design, three-dimensional design, etc., which not only saves time and reduces the loss of labor, but also greatly improves the work efficiency of the art design.

(2) The application of computer technology prevents a lot of trouble for the previous manual art design and simplifies art and design. Most of the art design work is completed by inkjet machines, which is different from the previous complicated process and reduces the burden. By using computer technology, art design can be conducted in a quick and simple manner.

3.2 Advantages of computer technology

Computer technology has great advantages in art design. It makes art design more efficient, faster and more accurate and is a major innovation in the art design. Advantages of computer technology are embodied in the following aspects:

(1) It makes teaching more flexible and lively. The spirit of innovation in the art design is the main goal of cultivating students. The introduction of computer technology can effectively improve students' spirit of innovation, thus enhancing the level of art design. By using computer technology, the contents of art design can be more visually presented in front of students, thereby arousing students' interest. Computer technology can also make class full of more information. The teacher can impart knowledge to students by operating a computer, which will strengthen the interaction between the teacher and students, get rid of the traditional boring teaching methods, and provide a broader space for students' art creation. Different from the attending mode in the past, students' creative ability therefore is improved.

(2) It improves students' learning enthusiasm. The traditional teaching mode is passive learning for students, lack of initiative and interaction with the teacher in time, resulting in poor learning efficiency. The application of computer technology can make full use of multimedia technology, create new teaching mode, arouse students' enthusiasm, make the class more lively and flexible with sound effect, bring unlimited freshness to students, encourage students to learn further, and provide a pleasant environment for them to study art design.

(3) It improves the teaching efficiency of art design. When there was no application of computer technology in the art design, the main teaching tools include blackboard, teaching materials, chalk, etc. Using these teaching tools is time-consuming as well as inefficient. The application of computer technology just changes the way of teaching. The teacher imparts knowledge to students by simply moving the mouse. Besides, the traditional teaching method will be restricted by the pigment, paper and other factors, while these restrictions no longer exist by using computer technology. Drawing artistic work by using a computer can improve the efficiency and the teaching quality.

(4) It is conducive to promoting the further reform of China's modern education. The application of computer technology not only plays a good role in the domestic art design teaching, but also builds a bridge between Chinese art design and outstanding international art design through opening this new field. The strong function contained in computer technology greatly improves the teaching quality and level of art design, allows students to experience to see the international art design works, and draws lessons from foreign outstanding achievements so as to improve their own abilities. Students can also download learning materials through the computer technology, widen and deepen knowledge, enhance learning initiative and promote the reform of China's art design teaching.

3.3 Significance of computer technology

Computer technology plays an important role in the art design. The application of computer technology into the art design is historically inevitable and to meet the needs of the times.

Computer technology and art design is the perfect combination of science and art. Computer technology can bring art designers more convenience, simplify the complex process, provide a new form of artistic expression for the designers, and create a channel between the art designer's conception and drawing works, so that the designers can present pictures conceived in the mind by drawing works, which further improves designers' working efficiency.

The rich expression contained in computer technology promotes the technological innovation in art design. Art designers use advanced computer technology to create new works. The application of computer technology provides a virtual environment for art designers, allowing them to produce works in such environment including design of color, light and some other effects, and finally create works with higher value after many attempts and efforts.

The application of the computer technology will make art design get better and better towards a more brilliant future with its value fully put to good use.

4. Computer technology application and development in art design

Computer technology has been applied in the Art and Design generally and effectively. It is a great undertaking for our country, which can promote the development of the art field.

4.1 The application of computer technology in art design

The application of computer technology in Art and Design falls into three aspects: 1. highly qualified teachers; 2. a series of changes of students based on new teaching methods; 3. art design diversity.

(1)Highly qualified teachers

As China values education very much, to develop high quality education is key to cultivate talents of our country. With the advent of the 21st century, science and technology has been greatly developed so that computer technology will inevitably be part of education reform as it is irreplaceable to our education. Teachers are leaders of our education, as their responsibility is to guide students study effectively, and strive for the construction of our country. As for the development of art design, an excellent teaching team is indispensable. To introduce computer technology, teachers need to use their own methods which allow students accept and understand it, to analyze systematically the teaching materials and characteristics of each student, and to impart them knowledge with effective ways. At the same time, teachers should also have rich professional knowledge, improve themselves and guide students to learn art design.

(2) A series of changes of students based on new teaching methods

Since the computer technology has been introduced, the traditional teaching mode cannot be used any more. With the change of teaching form, students should also make preparation for it as the computer technology will bring them more enlightenment and inspiration. Computer technology enables students to enjoy larger scope of activities, to look up information they need, to create works of art, and to solve problems left by traditional teaching methods. Students can search for new knowledge and information by themselves, improve themselves and enhance their comprehensive quality by using computer technology.

(3) Art design diversity

Computer technology has laid good foundation for Art design diversity as it promotes the development of Art design in each aspect. It makes the tool of art design brush into mouse instead of a pen. And compared with a pen, a mouse is more flexible, easier to control, time-saving and effort-saving, and has better control over color debugging. Moreover, computer technology has also brought new thought and education methods for art design, and precious experience both for teachers and students, who can learn more advanced things through simple and acceptable methods. Computer technology provides students a good learning environment by creating them a virtual environment to show art design vividly.

4.2The application of computer technology in industrial design practice

Computer technology becomes more and more mature day after day, which enables the art design to enrich and develop greatly. The art field has been improved in certain degree, for example, computer technology nowadays is playing an important role in such areas like advertising design and production, industrial design, flash effects design and production, web pages design and editing and so on.

For example, computer technology can be applied in industrial design. In traditional industrial design, market research should be conducted firstly before making complete system design scheme. And then draw a manual sketch, which is a 3D sketch. After finishing, the sketch would be checked and modified repeatedly. Until the final confirmation is made, can industrial product renderings, three views be drawn, and models are made. There were too much cost of labors, materials and money in the traditional industrial design, as its complex designation wasted time and could not make sure the correctness. However, with the computer technology, the market research can be done through Word and Excel and can create diagrams later, which is much more accurate. At the same time, we can also integrate data available by Access to generate a database so that they can be found easily. Then we can create relative diagrams by the three-dimension software in the computer. It is easy and convenient. Personnel of industrial design can use computer technology to save time of designation and lower the cost. The application of such computer technology can not only promote industrial design development by realizing its goals, but also benefit our national economy.

Conclusions

In conclusion, the application of computer technology not only leads the art to a new level, but also benefits computer technology research in China a lot. As computer technology is the product of the development of the times, we should adapt ourselves to this trend, creating and developing it continuously in China and expanding its application to more fields in order to advance our science and technology, enhance our economic and technological strength, improve our national status and enhance our comprehensive national power.

References

[1]Chen Haolei, Zou Xiangjun, Chen Yan, Liu Tianhu. The Latest Development and Prospect of Virtual and Real Technology[J]. China science and technology papers online, 2011(1):1-5.

[2]Liu Qinglai, On the Role of Computer Technology in the Stage Art[J]. Drama Collection, 2014,20(2）: 44-45.

[3] Tang Xuewen. Art design Must Combine with Market Closely[J]. Economist.2008,(07).

[4]Wei Wei. On the Curriculum Reform of Art design Education in Colleges and Universities in the era of quality education[J]. Science and Technology Innovation Herald.2010,(12).

The Exploration of Maker Teaching Method in the Teaching of Multimedia Courseware Making

Ying ZHOU

LinYi University, ShanDong, China

Abstract. Being the embodiment and general way of Maker education in classroom teaching, the Maker teaching method is based on such views as "learning means to create new things through intelligence", and characterized by creative, designable, handing ability and sharing. The general process consists of several steps, including putting forward innovative project according the situation or questions, designing the completing path, completing the project through study and practice, organizing people to share, then final perfection. Based on teaching cases, here I will make a study and exploration on the practice of Maker Teaching Method in Maker Education.

Key Words.Maker education; Maker teaching method; the teaching of multimedia courseware making

Introduction

In the context that new technology and internet are developing, now we have entered the era of internet education, which is people-oriented and innovation-emphasized. The Ministry of Education points out in Several Problems on Cultivation of Innovative Talents that, talents training should adapt to the development of economic society, such orientations should be emphasized, and we must try our best to explore a new way of cultivating innovative talents from the close integration of education and economic, social demand. The cultivation of innovative talents is not only the mission of school, but also the common expectation of society, employers are attaching greater importance to student ' s ability to operate practically, which reflects the urgent social demand for innovative talents. As the employer of preschool teachers, the kindergarten realizes that, the modernization of educational measures is an important condition for preschool education reform, while the renovation of educational concepts is the key to reform. Preschool teachers in new era should cultivate their innovative consciousness, take good advantage of information technology, apply it into preschool education, and enjoy the information technology superiority we get from practice.

1. Analysis on Problems Existed in Making of Multimedia Courseware

Normal college has opened an important course, namely The Making of Multimedia Courseware, particularly for students majored in preschool education. The course is set to firstly adapt preschool teachers to the demand of computer-aided instruction, help preschool teachers to design multimedia courseware that will accord with the teaching demand of kindergarten and the law of students' learning, so as to increase the effectiveness of teaching. Secondly, the aim is to cultivate innovative talents in information technology so as to meet the demand of kindergarten. While because of imperfect training conditions and preschool teachers' incomplete knowledge system, which has resulted that the teachers' curriculum design is far from reality, as a result, the development of innovative capacity has been restrained.

The issues highlighted in traditional teaching practice are mainly reflected in the following aspects.

1.1 Learning the tool software in isolation, failing to accomplish the integration

The teaching of multimedia courseware making focus mainly on the learning of various kind of material handling software, too much importance has been attached to the applying learning of tool software, while this will easily lead students into wrong direction, they tend to neglect the value meaning of learning is to apply tools to practice, to solve real problems comprehensively. However, the participation of various kind of software is essential in the process of preschool courseware making, the loss of comprehensive ability in operating software usually cause students' failure in achieving good ideas. Although some students may show much more creativity in operating certain tool software, unfortunately, their creativity has not been given full play to.

1.2 Failing to protrude student subjectivity, lacking ample practical space and creativity room

The teaching methods that are commonly adopted in school curriculums, such as demonstration teaching method, task-based approach, cannot arouse students' interest in learning or enthusiasm. Such teaching methods will limit students' learning to a certain space or a certain period of time, will impede their taking good advantage of current tools to gain much more resources from internet, also, it fails to be the embodiment of student subjectivity or the student-centered teaching ideas.

1.3 Teacher' s role as traditional knowledge imparter still existing

According to fixed curriculum criteria, referring to the same teaching plan, teachers impart fixed knowledge to students in class. Once students raised questions beyond textbooks, teachers even cannot give proper directions or answers.

1.4 Failing to make innovative courseware or attract interest of young children

Lacking necessary teaching experiences, students majored in preschool education cannot take the need and characteristics of kindergarten teaching activity into full consideration, they are making the courseware for the courseware's sake. Such courseware fails to take creativity and practicality into account, fails to combine fun and practicality skillfully, certainly it is hard to arouse students' enthusiasm to take part in.

2. Maker education in Classroom Teaching, the Maker Teaching Method

The flourish of global Maker movement has offered a new opportunity for the innovation and reform of education [1]. Being a new education mode, Maker education absorbs information technology, hold the teaching concept of "open and innovative, study and experience", adopts "learning from creation" as major learning method, and aims to cultivate innovative talents in various fields. The core ideas behind Maker education are "learning by doing" "pleasure education" "metasynthetic wisdom" "constructionism". [2] Maker education, being a kind of problem-based education, usually point directly to the resolution of problems, lead students to detect realistic problems, lead students to seek creative solutions and turn it into reality through hard work.

The Maker teaching method is the embodiment and general way of Maker education in classroom teaching, it holds such teaching view as "learning means to create new things through intelligence", starts with students' interests, and creates interactive and cooperative space for students' creation. [3] However, it should be clear that greater importance should be attached to cultivate the innovative consciousness of students.

According to Professor Jiahou Li(Shanghai Normal University), in his lecture, The Changes of Time and the Rise of Maker Teaching Method , he put forward five major steps of Maker teaching method in teaching practices , namely Idea, Design, Maker, Share and Evaluate.

3. The Application of Maker Teaching Method

In the teaching of multimedia courseware making, the Maker teaching method can be applied not only to a single lesson, but also to a unit or even to the whole volume of textbook. During the practice, the applied class hours is determined by the difficulty of innovative project and the cognitive level of students. Now let's take the example of project named The making of demonstrating courseware, to talk about the application of Maker teaching method in class.

3.1 The class teaching design about The making of demonstrating courseware

Here I choose the The making of demonstrating courseware as the teaching topic. Demonstrating courseware is made on the base of teaching goals, and is made from the proper organization of teaching contents. It can be used widely in the story time and cognitive activities of kindergarten, and help to enhance the cognitive ability of young children enormously.

3.1.1 Teaching goal

1) To learn about the basic knowledge of multimedia software like PowerPoint, Authorware or Flash, and the operation skills of one of them.

2) Be able to make the demonstrating courseware by using one of above software.

3) Cultivate creative consciousness and ability with multimedia software.

3.1.2 Teaching analysis

1). Analysis on teaching objects

Senior students majored in preschool education in normal colleges has furthered their thinking development to a new high, they are independent, creative, can think logically to some extent, and have clear planning capacity and clear-defined objectives in learning; with the development of their autonomy, they can choose the learning contents independently, remarkable progress in their thinking activities can be seen; besides, with strong study consciousness, students can get to learn contents they are interested rather actively.

2). Analysis on learning contents

For Maker education, supported by rich teaching materials from internet, the contents of it comes from the actual need of kindergarten teaching, choice will be given to students, so they can determine the contents and topics they are interested in. Instead of telling students how to use educational resources, the Maker education is aimed to motivate students to take good advantage of educational resources, to do what they want to do, they are interested in, or things they already have some vision. As long as it will motivate students to act, to learn by doing, then step by step, turn the idea into reality.

3.2 The analysis on the teaching implementation of The making of demonstrating courseware

I choose to implement the class teaching design of The making of demonstrating courseware, in the meanwhile, I refine and complete it constantly. After the class, students prefer to utilize the internet resource to make the kindergarten courseware, so as to enjoy the true fun of learning.

3.2.1 Idea

To cultivate employability, students had better take up internship at kindergarten frequently. During the course of internship, students can have direct access to the real kindergarten teaching activities. Through investigating the kindergarten they interned, students can get a profound knowledge about when and where the courseware is needed, what kind of courseware is needed, and furthermore to pay more attention to the use of demonstrating courseware. Thus students can detect the merits and faults of other kindergarten teachers' courseware, which in turn will motivate students to thake active part in the courseware making.

3.2.2 Design

Students had better be divided into different groups, then through intercommunication, different groups share their ideas, determine proper topics after discussion, furthermore, to make overall plan and design, take the material handling software needed during the process into account, and develop the rough steps of learning software and making courseware.

3.2.3 Maker

This stage consists of two steps. The first step is to reserve skills. Students can study relative tool software with reference to the teaching resource recommended by teachers, or master the basic operations of tool software through internet resource independently. In this way, students will gain the initiative in study. The second step is to implement students' ideas. How to monitor students' learning process is the most difficulty in this step, it demands teachers to observe students carefully and make corresponding record. More time and space is needed in Maker learning, so the completion status of task and the making status of works are determined mainly by students' self-control and their interest in the topic, this is also the theme of Maker teaching method.

3.2.4 Share

Share and perfection are values of Maker teaching method, the Maker teaching method is a course that people are sharing and perfecting continuously, are studying, summarizing and practicing continuously. The sharing process also has its general steps: first of all, why should I share or why not; secondly, how to share, and what about the result; thirdly, the problems I encountered and the solutions I came up with; fourth, what do you think I should do.

After students completed their courseware, kindergarten teachers should be the first to give suggestions. Then, teachers can check students regularly in class teaching, give them opportunities to stimulate teaching on the platform. Only through practical exercises can students understand whether their courseware is suitable for the class, or if there have any things that can be improved. Students can also upload their courseware to network, or share it to the twitter, circle of friends, video space and so on. This way, the creative works would spread though sharing.

3.2.5 Evaluate

The evaluations of Maker teaching, which focus not only on the learning results of students , but also on the learning process of them, are made by specialists, scholars, kindergarten teachers, the learner himself (herself), his classmates and teachers.

3.3 Summary

After completing the teaching design of The making of demonstrating courseware, I interviewed several students that have participated in the Maker teaching randomly. Majority of students have high enthusiasm to participate in, and show increase in liking of the course after their participation in the whole teaching process, some students even generate ideas of making courseware that is suitable for the kindergarten teaching. Unlike other curriculums, in the class of Maker teaching, students take up much of the class time and can make full use of free time, their creative passion has been motivated. Furthermore, educational resource from network also provides choice in learning materials for them; besides, the class has given students right to choose the topics they are interested in. Through the feedback of teaching effects, I find that the Maker teaching method brings about many positive effects on class teaching, for instance, it contributes to construct new study atmosphere, to increase students' enthusiasm in study and creation.

4. Summary

Being part of educational reform and innovation, Maker education has began to bring changes to the traditional teaching and learning. Maker education encourages creativity and innovation, it guide students to utilize new digital technology to design, to make, to share and to study without the limitations of time and space. Maker education will also improve and optimize itself continuously, and play its specific role in promoting students' comprehensive ability, which is based on the communication skill, collaborative ability, problem-solving ability and innovative capacity [4].

As the course to cultivate the creative ability of students majored in preschool education, the teaching of multimedia courseware making should keep up with the trend, change the traditional concept of isolated operating skills teaching on tool software. Teachers should equip themselves with abundant, multi-disciplinary knowledge, so as to support students' continuous innovation consciousness and their increasing learning enthusiasm. Meanwhile, school should be able to supply students with enough learning resource and fertile learning environment, to support students' increasingly high innovative ability and practical ability, so as to ensure that more innovative talents suitable for the social development will be provided to the employers.

References

1.Maker,[EB/OL].http://zh.wikipedia.org/wiki/%E5%88%9B%E5%AE%A2, 2014-10-24.

2.Zhu Zhiting, Maker Education, Information Technology- Enabled Excising Ba of Innovation [J]. China Educational Technology, 2015(1):14-21.

3. Li Jiahou, The Change of Time and the Rise of Maker Teaching Method [EB/OL].http://byite.cn/7/viewspace-5109.html,2014-10-21.

4. Blue Book on Development of Basic Education Informatization in Shanghai in 2014, The Maker Style Learning Influence Development of Basic Education[EB/OL].http://www.shanghai.gov.cn/shanghai/node2314/node2315/node4411/u21ai881314.html, 2014-09-24.

MATEC Web of Conferences 63, 05014 (2016) DOI: 10.1051/matecconf/20166305014
MMME 2016

ACO-AFSA Algorithm in Function Optimization and Its Application

Shi Gang WANG[1, 2, a], Yu Chao JIANG[2], Feng Juan WANG[2] and Shu Feng JIANG[2]

[1]*School of Electrical and Information Engineering, Guangxi University of Science and Technology, 545006 Liuzhou, China*
[2]*School of Mechatronics Engineering, Qiqihar University, 161006 Qiqihar, China*

Abstract. ACO and AFSA as two kinds of new intelligent bionic algorithm, has better robustness, and the design method is simple and easy to understand. At present, in many engineering fields, they have been a pivotal position, but there are some defects in solving function optimization problems. Because of the lack of pheromone in the initial stage of ACO, the speed of the solution is relatively slow, and AFSA has the advantages of strong global convergence ability and fast speed. In this paper, based on the advantages and disadvantages of ACO and AFSA, two kinds of algorithms are integrated, and the advantage is enhanced, a better optimization algorithm is put forward: ACO-AFSA. The initial solution and improved state transition probabilities are obtained by AFSA, and the basic ACO is improved, the congestion degree of AFSA is introduced in the algorithm, makes most of ants' initial optimization will not be randomly selected at a strong crowding restrictions. Experimental results show that the new algorithm can get the optimal solution with less number of iterations, which greatly saves computation time, and has higher accuracy and better convergence performance. Engineering, computing the shortest distance between the surface and plane of the problem has also been good results, the experimental results are satisfactory.

1 Introduction

Optimization of the 21st century, most of the combination is computer and mathematical theory, when it comes to mathematics, most of the issues will relate to the function. In many fields, many problems are solved by mathematical modelling. This requires professional staff to optimize the function, so that the results can achieve the best, so as to find out the optimal solution to the problem. The optimization is to "take the essence, to its dregs". The optimization of the function problem is as follows: Given a function $F(x)$, the domain is D and range is R. The problem of global maximum (minimum) of a function is to find a point $X_{\min}$ ($X_{\max}$) in the domain, that makes this function value $Y_{\min}$ ($Y_{\max}$).

The traditional optimization algorithm of the function itself requires more things. For example, the function can be satisfied with the derivative, also can differentiate, etc. In recent years, many scholars have been inspired by the behaviour of many kinds of creatures in nature, and the bionic optimization algorithm is applied to function optimization. Zhao Hui [1] uses distributed artificial bee colony algorithm for solving function optimization problems, makes the outer layer sub-bee colony can quick search. Zhang Yuli [2] uses the GSO algorithm to solve the problem of function optimization, fixed step size and motion mechanism, increase the self discovery mechanism of firefly by using the value of smaller decision range. Chen Liang, Lu Houqing [3] uses the cuckoo algorithm to achieve self adaptive adjustment, improve the performance of the continuous function optimization problem. Ye Wen [4] use the technique of memory operator, restrain operator and recombination operator to simulate the optimization problem of multi peak function, the optimization effect is obvious. Zhang Xinming [5] improves its local search capabilities, more effective find the global optimal solution for the low dimensional complex functions. But the single bionic optimization algorithm is more complex, and still has many shortcomings.

Therefore, in many bionic algorithms, select the ant colony algorithm and artificial fish swarm algorithm, according to their respective advantages and disadvantages, two kinds of algorithms are integrated, the advantages of the disadvantages are complementary, and the advantage of the algorithm is enhanced. The authors propose a better optimization algorithm ACO-AFSA. The initial solution and improved state transition probabilities are obtained by AFSA, and the basic ACO is improved, the congestion degree of AFSA is introduced in the algorithm, in the initial optimization, the majority of ants are not randomly selected by the strong crowding degree. Experimental results show that the new algorithm can get the optimal solution with less number of iterations, which greatly saves computation time, and has higher accuracy and better convergence performance. Engineering, computing the shortest distance between the surface and plane of the problem has also been good results, the experimental results are satisfactory.

[a] Corresponding author: 446986768@qq.com

2 Ant colony algorithm and artificial fish swarm algorithm

2.1 Ant colony algorithm

Colony is also called ant colony optimization, the earliest proposed is by the Italy Dorgio and its research team in early 1990s [6,7]. Since then, the algorithm has been concerned by many scholars and experts, and has been systematically studied by many people. The basic model of ant colony algorithm is as follows:

Ants from position i to j, at the beginning, the path node can be chosen for A, $A=\{0,1,\ldots\ldots, n-1\}$. Initially, the ants number are m; the information inspire factor is α, relative importance of the trajectory of representation, reflecting the ants in motion the process of accumulation of pheromones in ants sports role, α is larger, the greater the probability description ants choose this path back; inspired desired factor is β, indicating the relative importance of visibility, reflecting the ants during exercise in inspiration factor α importance in the whole path; pheromone evaporation coefficient is ρ, is a constant, represents the weight; the number of iterations for N_c, and have a *tabu* list. Initialization set ϕ, select the starting point S and the destination point F, let the ants start walking at the starting point S. Ants from i to j of the probability are $p_{ij}^k(t)=\dfrac{\tau_{ij}^k(t)\eta_{ij}^\beta(t)}{\sum\tau_{ij}^\alpha(t)\eta_{ij}^\beta(t)}$ $(j\in c)$, Otherwise 0.

Among them, $\eta_{ij}(t)=\dfrac{1}{d_{ij}}$. In equation, d_{ij} represents ant distance from i to j, τ_{ij} expresses the pheromone, every time the ant k is transferred, the node j is added to the *tabu* list until all the ants in the loop reach the destination, each ant is calculated by taking the path length at the finish line, saved. Update pheromone is $\tau_{ij}(t+1)=\rho\tau_{ij}(t)+\sum_{k=1}^{m}\Delta\tau_{ij}^k(t,t+1)$; in formula when an ant k from i through j, $\Delta\tau_{ij}^k(t)=\dfrac{Q}{l_k}$ Otherwise 0.

2.2 Artificial fish swarm algorithm

In 2002, Dr. Li Xiaolei proposed a new bionic optimization algorithm by simulating four kinds of behavior of fish in water——Artificial fish swarm algorithm [8,9]. The algorithm provides a top-down design idea. Starting from the individual, shoal of fish was observed, the local optimization of the individual in the fish stocks reached the final global optimization. The basic idea of artificial fish swarm algorithm is:

In a space, there are N artificial fish, they form a group, and the state of one of the artificial fish is $X_i=(x_{i1},x_{i2},\mathrm{L}\ ,x_{iD}),i=1,2\mathrm{L}\ ,N$. The food concentration of artificial fish is used as the fitness value of the objective function, a single fish state as optimization variables, the adaptation value can be obtained by the optimization variable, that is $Y=f(X)$. The distance of artificial fish X_i and X_j is $d=|X_i-X_j|$. Congestion factor is δ, it indicates the degree of congestion near. Congestion formula is $q_{ij}=\dfrac{2\tau_{ij}(t)}{\sum_{i\neq j}\tau_{ij}(t)}$. Artificial fish can feel the range is visual, according to the behavior of other fish in the perception range, the direction of movement is determined. The step size of the artificial fish was step, and the number of the feeding process was try-number.

3 Fusion of ant colony algorithm and artificial fish swarm algorithm

The parameters of artificial fish swarm algorithm are more, but the influence of the number of artificial fish and the congestion factor on the performance of the algorithm is not obvious. The global search ability of artificial fish swarm algorithm is very strong, and the search speed is fast, but when it encounters a flat search range, search speed will become slow, the search ability will be weakened and easy to fall into local optimum in the flat area, which can not get the real optimal solution. Ant colony algorithm has better robustness, parallelism, and many other advantages. But the basic ant colony algorithm has less pheromonen in the early period of the initial period, resulting in the slow speed of searching for the search optimization, according to their respective advantages and disadvantages, two kinds of algorithms are integrated, the advantages of the disadvantages are complementary, and the advantage of the algorithm is enhanced. The author propose a better optimization algorithm ACO-AFSA. The steps of the algorithm are as follows:

Step 1: Ant colony initialization settings.

Step 2: The artificial fish swarm algorithm is used to initialize the individual optimal state.

Step 3: The direction of the movement of ants from position i to position j is determined by the concentration of pheromone.

Step 4: The artificial fish swarm algorithm is introduced to calculate the congestion degree of the current ant on the path. If it is not crowded, the ants continue to walk on the path. Otherwise, select a path to walk in the feasible range.

Step 5: Updated information element τ_{ij}.

Step 6: Do not change the initialization settings, repeat the Step 3 - Step 5, when all ants choose the same path as the path to find the optimal solution.

Step 7: Output results.

Ant colony algorithm and artificial fish swarm algorithm are swarm intelligence algorithm, when the function is optimized, the crowding degree of the fish swarm algorithm is put into the ant colony algorithm, if the number of ants in a path beyond the allow crowded range, even if this path pheromone concentration is high, ants do not choose this path to walk, but in the vicinity to elect a path walk. Such an approach can avoid the local optimal solution in the initial stage of the ant colony algorithm, and then improve the algorithm's ability to find the optimal results and the accuracy of the results. The detailed flow chart of the ant colony algorithm is shown in Figure 1.

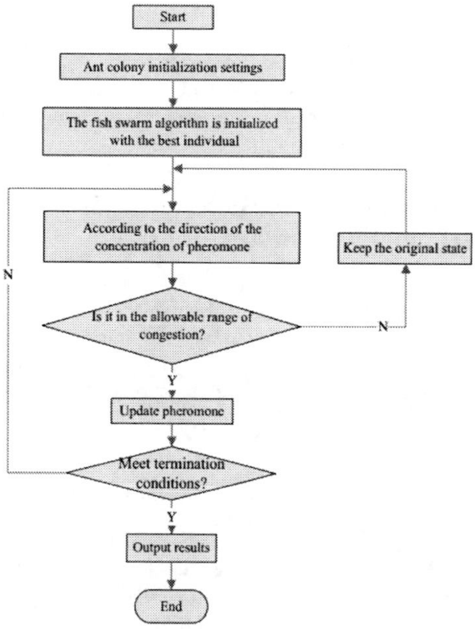

Figure 1. Flow chart of ACO-AFSA

4 Numerical experiments

In order to further prove the combination of the two algorithms can make the effect better, the result is better, in this paper, the author uses the standard test function in the literature [10]. Because of the limited time, only four functions are tested. Four functions are:

(1) $F_1(x)=4+4.5x_1-4x_2+x_1^2+2x_2^2-2x_1x_2+x_1^4-2x_1^2x_2 \quad (-100<x_1,\ x_2<100)$

(2) $F_2(x)=\{\sum_{i=1}^{5} i\cos[(i+1)x_1+i]\}\cdot\{\sum_{i=1}^{5} i\cos[(i+1)x_2+i]\}+0.5[(x_1+1.4251)^2]+[(x_2+0.8003)^2]$
$(-100<x_1,x_2<100)$

(3) $F_3(x)=0.5+\dfrac{\sin^2\sqrt{x_1^2+x_2^2}-0.5}{\left[1+0.001(x_1^2+x_2^2)\right]^2}$
$(-100<x_1,\ x_2<100)$

(4) $F_4(x)=4x_1^2-2.1x_1^4+\frac{1}{3}x_1^6+x_1x_2-4x_2^2+4x_2^4$
$(-100<x_1,\ x_2<100)$

Use Windows 7 operating system, Matlab 7.0 simulation platform for testing. Set the parameters as follows: α =1.0, β =5.0, ρ =0.55, the number of ants is 60, congestion factor is δ =0.62. The algorithm runs 30 times continuously, the global minimum and the number of iterations are used as the standard. The results of the tests are shown in Table 1.

Table 1. Comparison of simulation results

F(x)	Basic algorithm		ACO-AFSA	
	Iteration number	Optimal value	Iteration number	Optimal value
(1)	615	-0.513408173	15	-0.513400629
(2)	415	-186.73089128	107	-186.73088436
(3)	205	$8.2734178861\times10^{-8}$	31	$8.8757427369\times10^{-8}$
(4)	199	-1.0316281257	91	-1.0316290327

5 Examples of application

The shortest distance of free surface to plane is a common problem in geometric modeling. The traditional algorithms are often dependent on the form of the surface and the plane, and the robustness is not strong. In this paper, the fusion algorithm is applied to the analysis of the shortest path in the surface to the plane, and can find the optimal solution, and the calculation process is very simple. Take the surface $f_1=4x^2-3xy+4y^2-4z$ and the plane $f_2=x-y+4z-1$ as an example. Find the shortest distance coordinate and the shortest distance. The shortest distance is 0.2571. Table 2 is the shortest distance between the coordinate points. Figure 2 is an initial set of paths. Figure 3 is the global optimal solution for the path group approximation.

Table 2. The shortest distance points coordinate

x	y	z	x_0	y_0	z_0
-1/11	1/11	1/44	-0.091	0.091	0.023

6 Conclusion

The two basic algorithms have shortcomings in solving function optimization problems. Fusion algorithm in solving this problem is very easy to make up the shortcoming of the single algorithm stagnation. Through numerical experiments, we can see that the convergence and convergence speed of the algorithm can be improved by the fusion algorithm, so that the optimal value can be obtained in the case of less iteration number, and the convergence of the algorithm is still 100%. In addition, the fusion algorithm has some effect on the shortest path analysis of engineering application surface to the plane, and can find the optimal solution, and the calculation process is very simple.

In this paper, the advantages of the two basic algorithms are enhanced, make up two basic algorithms of their respective shortcomings. For function optimization problems, the convergence speed and accuracy of the algorithm is better than the basic algorithm, and in engineering application, the experimental results are satisfactory.

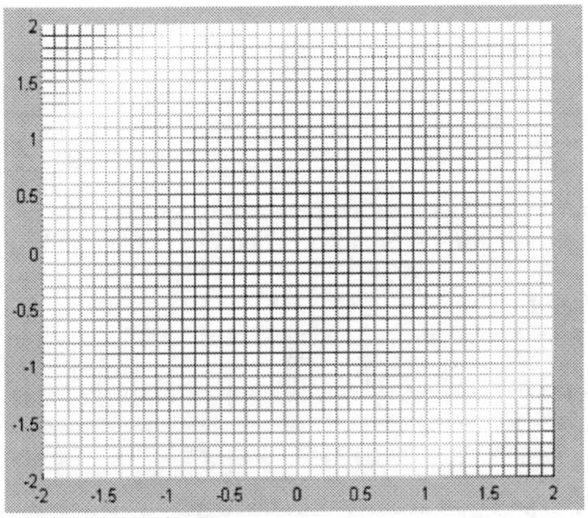

Figure 2. Initial set path group

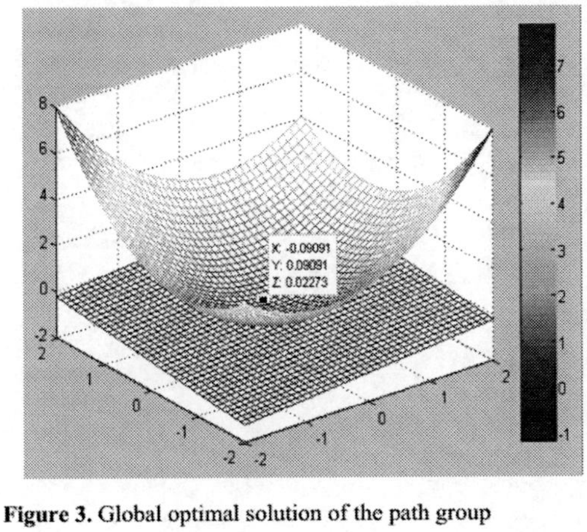

Figure 3. Global optimal solution of the path group

This work was supported by the Industry Project of Qiqihar City (Grant No. GYGG-201420) and Qiqihar University Graduate Innovation Research Project (Grant No. YJSCX2015-ZD15).

References

1. Zhaohui, Li Mudong, Weng Xingwei. Control and Decision **30**, 7 (2015)
2. Zhang Yuli, Ma Xiaoping. Chinese science and Technology Theory **10**, 8 (2015)
3. Chen Liang, Hou Qing Lu. Journal of the PLA University of Science and Technology **16**, 3 (2015)
4. Ye Wen, Zhong Hui Ouyang, Zhu Aihong, Hongda fan. Systems Engineering and Electronics **35**, 5 (2010)
5. Zhang Xinming, Feng Mengqing, Yin Xinxin, Duan Xintao. Mini Micro Computer Systems 35,12 (2014)
6. Dorigo M, Maniezzo V, Colorni A. Systems, Man, and Cybernetics, Part B: Cybernetics, IEEE Transactions on 26, 1 (1996)
7. Dorigo M, Gambardella L M. Evolutionary Computation, IEEE Transactions on, **1**, 1 (1997)
8. Li Xiaolei, Shao Zhijiang, Qian Jixin. System Engineering Theory and Practice 11 (2002)
9. Li Xiaolei, Lu Fei, Tian Guohui et al. Journal of Shandong University **34**, 5 (2004)
10. Li Hong, Jiao Yongchang, Zhang Li. Computer Engineering and Application **42**, 16 (2006)

Research of The Deeper Neural Networks

You Rong XIAO[1], Qing Xiu WU[1,a], Shu Qing LI[1] and Jun OU[1]

[1] *Hainan College of Software Technology, QiongHai, Hainan, 571400, China*

Abstract. Neural networks (NNs) have powerful computational abilities and could be used in a variety of applications; however, training these networks is still a difficult problem. With different network structures, many neural models have been constructed. In this report, a deeper neural networks (DNNs) architecture is proposed. The training algorithm of deeper neural network insides searching the global optimal point in the actual error surface. Before the training algorithm is designed, the error surface of the deeper neural network is analyzed from simple to complicated, and the features of the error surface is obtained. Based on these characters, the initialization method and training algorithm of DNNs is designed. For the initialization, a block-uniform design method is proposed which separates the error surface into some blocks and finds the optimal block using the uniform design method. For the training algorithm, the improved gradient-descent method is proposed which adds a penalty term into the cost function of the old gradient descent method. This algorithm makes the network have a great approximating ability and keeps the network state stable. All of these improve the practicality of the neural network.

1 Introduction

Neural networks are based on Multilayer Perceptron. Hubel and Wiesel [1] have researched the cat's visual cortex. It is known that the complex cells belong to visual cortex. The cells are apt to catch the little sub-regions of the visual object. The captured field is named with receptive field. In spatial filed, the characters of the sub-regions are correlated with the whole visual object. Also, the cells is divided into two cells: the simple cells and the complex cells. The former is sensitive to the edge of the receptive field; the latter covers the most of receptive fields. The receptive field cells produce the strongest response to the exact position of the visual object, but are unresponsive to the outside of receptive field.

2 Related work

2.1 Neural networks

The function can be represented in the form of a network diagram as shown in Figure 1. The process of evaluating can then be interpreted as followed.

As illustrated in Figure 1, input layer is consisting of vector $x=[x_0......x_n]^T$. Layer n computes an output vector a^n using the output a^{n-1} of the previous layer, with parameters b^n (a vector of offsets) and θ^n (a matrix of weights). The parameters θ^n as weights and the parameters b^n as biases. The bias parameter allows for any fixed offset in the data. The quantities a^n are known as activations. The formula is

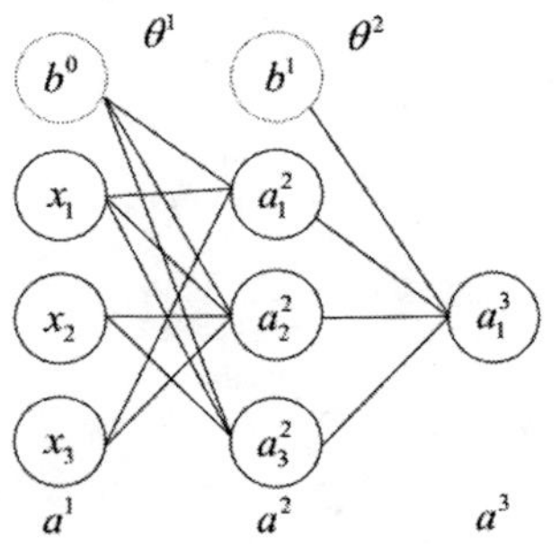

Figure1. Network diagram for neural network.

$$a^n = \text{sigmoid}(b^n + \theta^n a^{n-1}). \quad (1)$$

The logistic sigmoid function defined by

$$\sigma(a) = \frac{1}{1+\exp(-a)}. \quad (2)$$

It can be replaced by tanh(a)=2$\sigma(a)$-1, and so a general linear combination of logistic sigmoid function is equivalent to a general linear combination of 'tanh' functions. The activation function is used to emulate the frequency of the impulses sent to another neuron (or group of neurons) or sampled by the external environment. The sigmoid function $\sigma(a)$ has a real-valued input (the signal's final sum) and squashes it to range from value=0 to value=1. Linear functions are that the dependent variable (y) changes with independent

[a] Corresponding author: xhogh@hotmail.com

variable (x); otherwise, nonlinear functions mean that the dependent variable (y) is not a corresponding fixed change with the variable (x).

2.2 Convolutional neural networks

Convolutional neural networks contain convolutional layers, pooling layers, fully-connected layers. The features of one part are densely correlated with the other part in the input image. Learned from one part of the image [2], the features of one part of the image can applied to the other parts of the image. For example, there is a 5×5 input image. 3×3 patches represent the learned features which have been sampled randomly from the image. 3×3 patches slide over the large image. The new feature image can be produced.

The first value of the final matrix is computed as followed: [4]=1×1+1×0+1×1+0×0+1×1+1×0+0×1+0×0+ 1×1. The other values can be calculated by the same method. Then, the different feature activation value will be obtained in the image.

Pooling layers are usually used after the convolutional layers. The layers simplify the number of neurons. It can obtain the condensed feature map from the convolutional layer. As a concrete example, max-pooling always is exploited. A pooling unit outputs the maximum activation in 2×2 input region. Obviously, 14×14 neurons outputs from the pooling layers while 28×28 neurons as input from the convolutional layers. The convolutional layers always produce many feature maps. Max-pooling is applied to the feature maps individually [3].

Fully-connected layers are the final layer of connections in the network. In one word, the layers connects every neuron from the max-pooled layers to every one of the output neurons.

Given N×N neuron layer, (N-k+1) × (N-k+1) neurons will output from the convolutional layer with a filter of k×k size. Suppose that a standard two-layer network of the convolutional net. The weight parameters in the first layer of the network are shared between the various outputs. The first layer of the network can be thought as performing a nonlinear feature extraction, and the sharing of features between the different outputs can save on computation and can also lead to improved generalization.

2.3 Deeper neural networks

In this report, the architectures of deeper neural networks will be designed. Its structure splits the two hidden layers. One of the hidden layers is a feed-back hidden layer and the other one is a self-organizing hidden layer. With many 2×2 blocks, the feed-back hidden layer gives the network a weak stability and wealthy dynamics. The self-organizing hidden layer can transform the information from different blocks, improve the approximating ability and organize the different dynamic subsystems to work. Considering the features of the deeper neural networks, the dynamic subsystem of the network is defined in this report. The network adjusts the structure by adding the subsystems. Additionally, this report also proposed another self-organizing method to adjust the structure by activating the different subsystems when the network is used to approximate different system. These make the network can adjust its structure depending on the task.

Gradient descent is local-optimization algorithm. Function G(x) decreases fastest if goes along the negative gradient direction. By the method, G(x) obtains the min or max value. A modified gradient-based learning algorithm is proposed for the two- layer interval perceptron. In the modified learning algorithm, for the radius of each interval weight, the function of the absolute value in the original learning algorithm is replaced by a quadratic term. This modified learning algorithm, compared with original learning algorithm, can prevent the weights oscillation during the learning procedure. The monotonicity of the error function and convergence for the modified learning algorithm are proved [4].

A deeper neural networks model (DNNs) is proposed, which could recognize images from multiple problem fields. Using a neural network based dimension reduction technique, low-dimensional embedding for arbitrary reorder features can be learned; through a multi-layer network, these feature embedding are integrated with word embedding features into a linear ordering reorder models. According to the truth that DNNs captures image features level-by-level, a special method is used to combine DNNs of multiple fields. The bottom level of DNNs are shared to capture the low-level features in data space, while the high level neurons are still detached to capture the high-level features of the specific fields. Besides capturing low-level features, shared neurons also lead input data into correct high level neurons, which will finally make the reorganization in specific fields. This cross-field DNNs is trying to simulate human visual system, which has bionic meanings. Meanwhile, cross-field unlabeled sample sets could be used to train this model due to the shared neurons, which could improve the recognition performance in current internet environment with large scale cross-field unlabeled images. Because of the detached neurons, paralleled computing is possible, which could reduce the time cost of training. The DNNs is designed and realized with MATLAB library. Finally it is tested in MNIST data set of handwriting characters and COIL data set of object images. Experimental shows that the DNNs is more efficient than the traditional neural network and convolutional neural networks. In a deep neural network, the supervised training of bottom neurons could be replaced by unsupervised training. According to this, some improvements are applied on DNNs training algorithm, which made the paralleled computing of DNNs easier. The DNNs model's parallelization can be realized by (Compute Unified Device Architecture) CUDA framework.

2.4 Model test

The deeper neural networks (DNNs) can be constructed by MATLAB or C++. A graphics processing unit (GPU) can accelerate MATLAB function. MATLAB Executable (MEX) is intended to directly use C/C++ and FORTRAN codes within the MATLAB environment to accomplish higher executing speed and avoid application bottlenecks. CMEX is called for the C/C++ MEX, and C-MEX is only for the purpose of deploying a GPU device. Since c-mex requires building C/C++ executable and CUDA requires hardware-specific (NVIDIA GPU) codes, extra installation steps is needed in addition to a standard MATLAB installation. Table 1 shows the image recognition result using the DNNs built.

Table 1. The image recognition result using the DNNs.

	Times(s)	Error rate	Epoch
DDNS	2.4295	0.068	1
NNS	0.8518	0.2367	5

From the above compared data, the running time of DNNs is longer than normal neural networks. But Full-batch train mean squared error on training set of the former is better than the latter. In short, the DNNs run well than the normal neural network.

2.5 Applications of model

2.5.1 Online course quality evaluation

Based on neural networks model, the thesis puts forward the "Resource Strategies", "Teaching Strategies" for improving online course perceived service quality with the calculation and analysis of the index system. The evaluation process is shown in Figure 2.

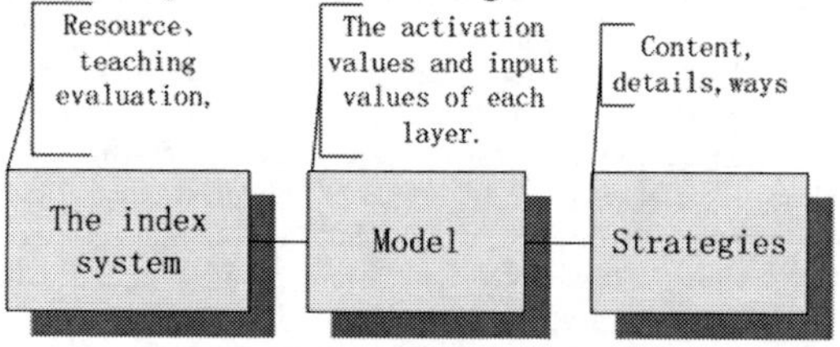

Figure 2. Online Course Quality Evaluation.

The index system includes the evaluation of learning resource (content resources, positioning, easy or difficult), learning process (the actual operation, communication and interaction), and personnel. It can analyze the rules and trend of the learners visiting the website, then find out the problems in the learing course, and provides some strategies. The strategies can strengthen the quality of the content resources, improve the user operation,and provide clearly detailed study guide to enhance interaction. It can modify and provide a reliable basis for making the adjustment curriculum of the network, in order to improve the credibility of the learners.

2.5.2 Water and soil data analysis

Water and soil data can be inputted the model by the sensors. The model can put forward water diagnosis and soil conditions. In the conclusions, water quality belongs to which kind of water. The proportion of salt and water can be obtained in the soil tested. The process is shown in Figure 3.

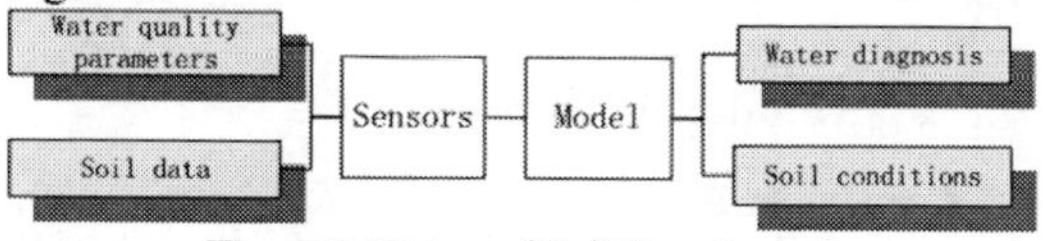

Figure 3. Water and Soil Data Analysis.

Water parameters include chemical indicators: the first kind is organic matter and the other is inorganic index.The former includes biochemical oxygen demand(BOD), chemical oxygen demand (COD), total organic carbon (TOC), and total oxygen content (TOD). The latter includes plant nutrient elements that is N element and P element of wastewater, PH which indicates the PH of the water samples, and heavy metal. Heavy metals mainly refer to the elements of mercury, cadmium, lead, chromium, nickel, and arsenic and other biological toxic elements. Soil parameters include soil moisture, temperature, PH, and salt content. According to the data, the soil tested will be irrigated or carrying out soil fertilization. The measures would enhance the soil properties.

3 Conclusion

Based on the study of neural network, a deeper neural network architecture has been designed. The method can be applied to image auto-annotation [5][6]. It can be an efficient method to solve semantic gap problems. Image auto-annotation can decomposed into two steps, that is, the basic image auto-annotation based on the relationship between image and tag, and the annotation enhanced based on the mutual information of the tags. The basic image auto-annotation is viewed as a multi-labelled problem. Therefore the prior knowledge of the tags, the dependent relationship of the tags is used to improve the annotation result. The experiments prove that the network has a great approximating ability and a weakly stable condition. It can efficiently solve the image recognition and labeling.

Acknowledgment

This work is supported by the Natural Sciences Foundation of Hainan Province under Grant No. 20156240, No.614238, No.20161009 and No.613167. The work is also supported by the Scientific Research Project of Hainan Province Colleges under Grant No.Hnky2015-79. *Corresponding author at: Department of Network Engineering, Hainan College of Software Technology; QiongHai. Hainan Province, 571400, P.R.China; E-mail address: xhogh@hotmail.com (Qx Wu).

References

1. Hubel, D. and Wiesel, T. (1968). Receptive fields and functional architecture of monkey striate cortex. Journal of Physiology (London), 195, 215–243.

2. http://ufldl.stanford.edu/tutorial/supervised/FeatureExtractionUsingConvolution/.
3. H Lee, R. Grosse, R. Ranganath, and A.Y. Ng, "Convolutional deep belief networks for scalable unsupervise learning of hierarchical representations.", ICML 2009.
4. H Lee, C. Ekanadham, and A.Y. Ng., Sparse deep belief net model for visual area V2, in Advances in Neural Information Processing Systems (NIPS) 20, 2008.
5. Y.Bengio, X. Glorot, Understanding the difficulty of training deep feedforward neuralnetworks, AISTATS 2010.
6. Socher, R., Huval, B., Bhat, B., Manning, C. D., and Ng,A. Y. Convolutional-recursive deep learning for 3d object classification. In Advances in Neural Information Processing Systems (NIPS). 2012.

Application of Gray Neural Network Combined Model in Transformer Top-oil Temperature Forecasting

Ya Yuan YI[a], Ran LI

School of Electrical Engineering, South West Jiaotong University, Chengdu 610031, China

Abstract. In order to forecast the transformer top-oil temperature accurately, quickly and efficiently, the paper proposes a prediction model of the transformer top-oil temperature based on Grey Neural Network. Such factors impacting transformer top-oil temperature as load current, ambient, active power and reactive power are comprehensively considered in the prediction of top-oil temperature, and the combined forecasting model which verified the applicability, accuracy and feasibility is established. The example shows that the forecasting results by the proposed method are better than those by BPNN method and the generalization ability of BPNN is improved. The proposed method possesses the following good properties: high precision of forecasting, fast convergence, nice commonality and its average relative error is within the range of 1%.

1 Introduction

Transformers are the important equipments in the process of power generation and power supply. Having a very high position in the power grid, transformers are the key equipments to ensure network security, reliability, economic operation of power system and people's production and living electricity.

In recent years, with the rapid growth of electricity demand, high-speed construction and investment of power grid drive the market demand of power transmission and transformation equipments. 《Action Plan of Distribution Network Construction and Reconstruction (2015-2020) 》 points out that the investment of distribution network construction will be not less than 2 trillion yuan in 2015-2020. There will be not less than 1.7 trillion yuan in a cumulative investment during the period of "the 13th five-year plan". Therefore, the upgrade and optimization of transformers are the new development trends of future. The high efficiency and energy saving, and the high overload capacity of transformers will be widely used. However, the transformer top-oil temperature is a key factor that affects load capacity, manufacturing cost and service life of the transformer. The transformer top-oil temperature overrun will bring huge damage to the transformer and affect the normal operation of the power grid seriously. There are many strict requirements on transformer top-oil temperature in the specifications. The actual transformer is commonly alarming in 70℃~75℃, and tripping operation in 80℃~85℃ and the highest temperature is 95℃. When the cooling system fails to remove all the coolers, the forced oil cooling transformer and the strong oil water cooling transformer will be allowed to operate 20min under the rated load. It is allowed up to 75℃ when the oil surface temperature has not reached 75℃. But the longest running time will be not more than 1h after the removal of the cooler. Thus, the accurate prediction of transformer top-oil temperature has a very important significance on our understanding of the transformer operation and preventing the significant thermal accident.

At present, there are many methods to predict the top-oil temperature, such as the indirect measuring method[1-3], the direct calculating method[4-9], the intelligent prediction method[10-12] and so on. Qing He[13] of Arizona State University established top-oil temperature prediction model by using three different methods: static neural networks, temporal processing networks and recurrent networks. Considering the influencing factors of load current and ambient temperature, the forecast results were compared with those of regression models. The feasibility of prediction model was verified while the prediction accuracy needed to be improved. Chen Weigen[14] of Chongqing University proposed the generalized neural network prediction model which using the environmental temperature, top-oil temperature, bottom temperature, dead end temperature and lower dead temperature as input layer neurons. Experiments were carried out under different loads and different oil flow velocities, and the results show that the prediction method was accurate. Due to the defects of the neural network itself, the model lacked the actual physical meaning. Kalman filtering algorithm was used by Wang Yongqiang[15] to predict the top-oil temperature. Constructing the top-oil temperature state equation and measurement equation based on Kalman filter, the forecasting method was reliable and accurate by the simulation. Because of the hysteresis of the model, the

[a] Corresponding author:319306111@qq.com

forecast results had deviation when the load and the environment temperature changed fast.

The paper establishes a prediction model of transformer top-oil temperature based on Grey Neural Network which considering the transformer load current, environment temperature, active power, reactive power and the top-oil temperature of the previous time. The prediction method is validated by the data collected from a power supply company in Chengdu. The results show that the prediction model of transformer top-oil temperature is accuracy and superior to BP neural network model. The proposed model improves the top-oil temperature prediction accuracy and can be used to instruct the transformer load operation mode and enhances the safety and reliability of the operation of the power grid.

2 Basic theory of grey neural network

Considering the different factors influencing transformer top-oil temperature, the top-oil temperature can be regarded as a complex process with large randomness. The grey theory and neural network theory are combined, which can improve the prediction accuracy of the top-oil temperature. The combined model has a good capability of nonlinear mapping and can weaken the randomness of the raw data.

2.1 Grey theory

Grey theory which is a new method to study the problem of less data, poor information and uncertainty was proposed by Professor Deng Julong in 1982.

It is a "small sample", "poor information", uncertain system which known to some information and unknown to part of the information as the research object. The proposed model achieves the correct description and effective monitoring of system's function and disciplinarian through the understanding of part known information.

The models of the grey system are called grey models, which are differential equations based on the original data sequence. Assuming that there is a sequence of raw data $x^{(0)}$:

$$x^{(0)} = (x_t^{(0)} \mid t = 1,2,\cdots,n) = (x_1^{(0)}, x_2^{(0)}, \cdots, x_n^{(0)}) \quad (1)$$

$x^{(0)}$ makes a accumulation to get a new data sequence $x^{(1)}$. The item t of $x^{(1)}$ is the sum of the previous t terms of $x^{(0)}$ and has the form:

$$x^{(1)} = (x_t^{(1)} \mid t = 1,2,\cdots,n) \quad (2)$$

Where

$$x_n^1 = \sum_{t=1}^{n} x_t^{(0)} .$$

According to the new data sequence $x^{(1)}$, the whitening equation will be established and is given by:

$$\frac{dx^{(1)}}{dt} + ax^{(1)} = u \quad (3)$$

Solving (3) we get,

$$x_t^{*(1)} = (x_1^{(0)} - u/a)e^{-a(t-1)} + u/a \quad (4)$$

$x_t^{*(1)}$ is the estimate of $x_t^{(1)}$. Made a regressive, $x_t^{*(1)}$ gets the predictive value $x_t^{*(0)}$ of $x^{(0)}$ and is given by:

$$x_t^{*(0)} = x_t^{*(1)} - x_{t-1}^{*(1)} \quad t = 2,3,\cdots n \quad (5)$$

2.2 Gray neural network

According to the analysis of grey theory, $x_t^{(0)}$ (t=0,1,2,⋯,N-1) makes a accumulation to get a new data sequence $x_t^{(1)}$ (t=0,1,2,⋯,N-1) which is in accordance with the law of some kind of functions. Described by a continuous function or differential equation, the new data sequence can be predicted by data fitting.

Through the above analysis, the differential equation of the grey neural network can be expressed as follows

$$\begin{cases} \dfrac{dy_1}{dt} + ay_1 = m \\ m = b_1y_2 + b_2y_3 + \cdots + b_{n-1}y_n \end{cases} \quad (6)$$

In the formula, $y(t)$ represents the $x_t^{(1)}$ above; y_i(i=2, ⋯, n) is a input parameter for the system; y_1 is the output parameter for the system; a, b_j(j=1, 2, ⋯, n) are the coefficients of the differential equation.

The time response equations of (6) are

$$\begin{cases} z(t) = (y_1(0) - \frac{b_1}{a}y_2(t) - \frac{b_2}{a}y_3(t) - \cdots - \frac{b_{n-1}}{a}y_n(t))e^{-at} \\ \quad + n \\ n = \frac{b_1}{a}y_2(t) + \frac{b_2}{a}y_3(t) + \cdots + \frac{b_{n-1}}{a}y_n(t) \end{cases} \quad (7)$$

Mapping to an extended BP neural network, the transformed formula (7) will be the gray neural network which has n input parameters and 1 output parameter. The topology of the network is shown in Figure 1.

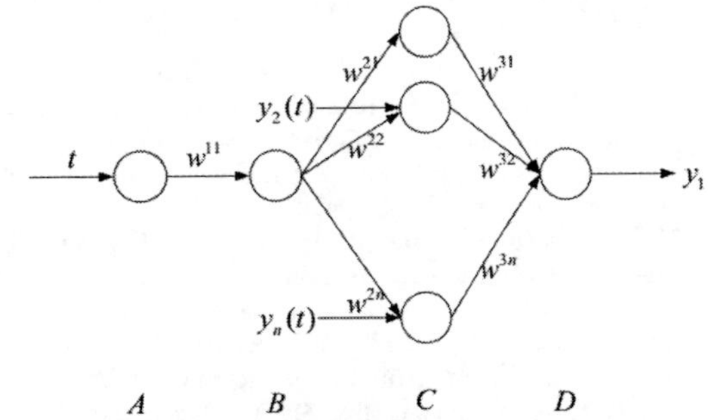

Figure 1 The topology of the network

In the figure 1, t is the serial number of the input parameters; y_i(i=2, ⋯, n) is the input parameter of the network; w^{lk}(l=2,3;k=1,2, ⋯ ,n) is the weight of the network; y_1 is the prediction value of the network; A、B、C、D are respectively represented four layers of gray neural network.

2.3 Gray neural network learning process

Through the above analysis, the gray neural network learning process is shown in Figure 2.

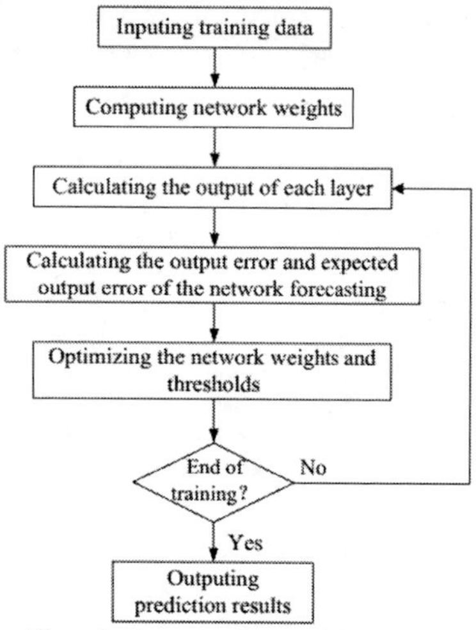

Figure 2 Grey neural network learning process

As shown in the figure: firstly, the network structure and network parameters are initialized according to the characteristic of training data. Then, the weights of network are calculated by the difination and the outputs of A, B, C, D four layers are calculated separately. We obtain the errors between the network forecast output and the expected output. Finally, we get forecast results by optimizing the weights and thresholds according to errors until the end of the training,.

3 The establishment of grey neural network model of top-oil temperature

3.1 The input and output parameters of the network

The transformer top-oil temperature often depends on three parameters: the first category is environmental parameters, including transformer temperature, humidity, wind speed, solar radiation, etc; the second category is characteristic parameters, mainly refers to transformer winding structure, cooling methods, etc; the third kind is running parameters, such as: load current, active power, reactive power, etc. It is crucial to choose the reasonable parameters for the accurate calculation of transformer top-oil temperature. This paper determines the final characteristic parameters from two aspects: the one is there have a great relevance between the parameters and transformer top-oil temperature; the other one is the network training effect achieves a higher technical indicator by entering the parameters of sample data.

According to the requirements of the two aspects above, the paper determines the environment temperature, load current, active power, reactive power and the top-oil temperature of the previous time as the input parameters of the network.

3.2 Data acquisition

Selecting the actual operation data of a transformer substation (the model is the SSZ11-50kV/110 power transformer) of Chengdu in August 2, 2015, we collect the data as a training sample every 5min from 00:00 to 23:55. Measuring the temperature of the transformer using optical fiber temperature sensors, the substation provides the historical data of the corresponding environmental temperature, load current, active power, reactive power. In order to meet the requirements of the input of the network, the sample data are normalized and are distributed in the [0.2, 0.8] range. As shown in Figure 3-6.

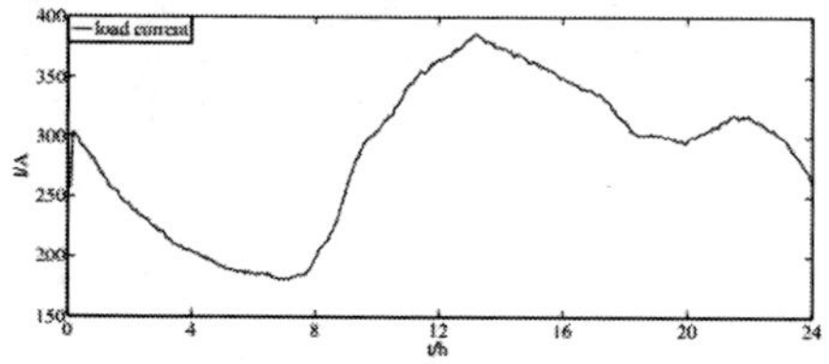

Figure 3 Variation of load current of transformer

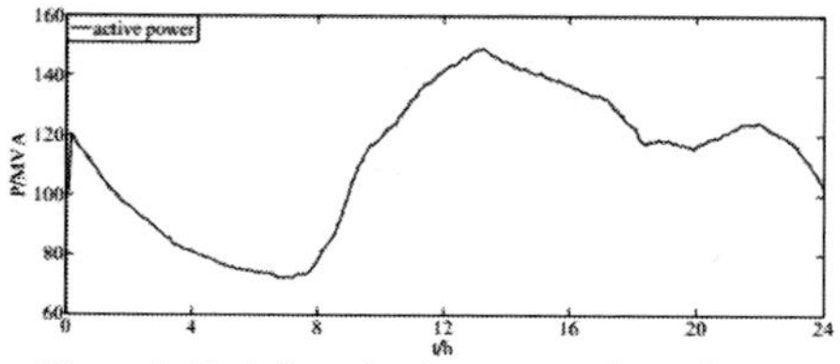

Figure 4 Variation of active power of transformer

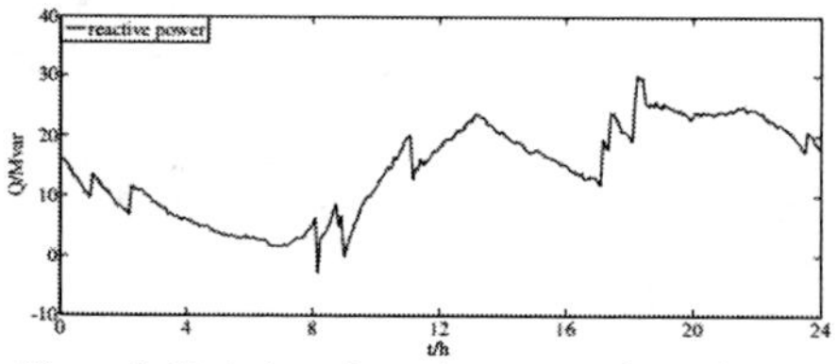

Figure 5 Variation of reactive power of transformer

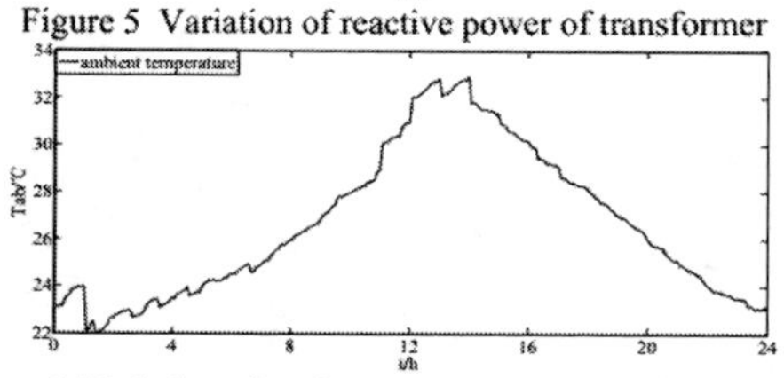

Figure 6 Variation of environment temperature of transformer

3.3 Training data and test data

In this paper, the transformer is working under the condition of natural oil circulation. Under the rated load of 24 hours of continuous work, a total of 288 sets of the sample data are measured. The training set and test set are classified according to the 5:1. The 240 groups are finally determined as the training data, and the 48 groups are used as the test data.

3.4 Network structure design

The inputs are the environment temperature, load current, active power, reactive power and the top-oil temperature of the previous time. The output is the transformer top-oil temperature. So the structure of the gray neural network is 1-1-5-1.

3.5 Network training and forecasting

The forecasting process of transformer top-oil temperature includes three steps based on grey neural network: the construction of grey neural network, the training of grey neural network and the prediction of grey neural network.

First, the impact factors of the transformer top-oil temperature are analyzed, the input and output parameters of the network are determined.

Then, the 240 sets of historical data consisting of a 5×240 matrix including the environment temperature, load current, active power, reactive power and the top-oil temperature of the previous time are entered. The output is 240 history data of transformer top-oil temperature expressed as a 1×240 matrix and the grey neural network begins to train.

Finally, we get the desired gray neural network at the end of the training and predict transformer top-oil temperature according to the trained network.

4 Forecast result and analysis

Table 1 shows the measured value, BP neural network prediction and grey neural network prediction of 48 groups of top-oil temperature for the test transformer. In order to compare the data visually, the three sets of data are represented in the graph as shown in Figure 7. As can be seen in figure 7, the dynamic changes of two kinds of the prediction models can reflect the changes of top-oil temperature. While compared with the BP network forecasting model, the proposed forecasting model has the better fitting degree and higher accuracy. This shows that the grey neural network model is superior to the BP neural network model in this experiment.

Table 1 Transformer top-oil temperature measured value and BPNN predictive value and grey network predictive value

Time[min]	Measured value [/℃]	BPNN predictive value [/℃]	Grey network predictive value [/℃]
25	67.88	66.81	67.99
55	67.50	65.50	66.95
85	67.13	67.35	67.45
115	65.44	64.38	65.56
145	63.75	64.08	63.67
175	62.44	63.31	62.22
205	61.50	61.96	61.85
235	59.63	58.65	59.73
265	57.94	58.94	58.07
295	56.81	57.04	57.01
325	56.06	56.41	56.56
355	54.94	55.42	55.04
385	54.94	53.94	55.14
415	54.00	54.47	54.38
445	53.81	53.96	53.78
475	53.81	54.81	53.61
505	54.19	53.93	54.10
535	55.69	53.78	55.09
565	57.94	56.78	58.09
595	59.81	59.68	59.71
625	61.50	61.35	61.31
655	64.13	63.18	63.05
685	65.81	65.60	65.94
715	67.50	66.46	66.84
745	69.56	69.23	68.98
775	71.25	71.47	71.34
805	71.81	71.02	71.18
835	73.13	72.36	72.38
865	74.25	74.57	74.42
895	74.44	74.37	74.33
925	74.81	75.77	74.74
955	74.81	74.88	74.90
985	75.19	76.14	75.29
1015	74.81	75.24	75.20
1045	73.88	74.29	74.33
1075	72.56	72.88	72.90
1105	71.81	71.72	71.99
1135	70.31	71.63	71.15
1165	69.38	69.02	69.07
1195	68.06	68.53	68.73
1225	66.94	67.49	66.74
1255	66.19	66.77	65.73
1285	66.19	66.03	66.00
1315	66.19	68.19	66.49
1345	65.44	64.50	65.21
1375	64.88	65.09	65.07
1405	64.50	64.67	64.68
1435	63.75	64.46	63.81

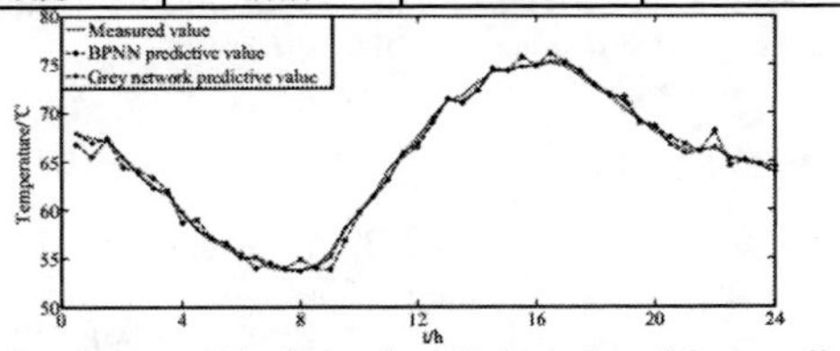

Figure 7 Measured value and prediction value of the top-oil temperature

In order to analyze deeply, a comparative analysis of the relative errors between the BP network model and the grey network model are given by figure 8 and table 2. From figure 8, we can see that the relative error of the gray network prediction model is near the zero point and the fluctuation is small. It shows that the relative error is small and stable. As is listed in Table 2, the minimum relative error, average relative error and the maximum relative error of the grey forecasting model are less than those of BP network prediction model. The maximum relative error of the grey forecasting model is 1.68%, which is quite acceptable in engineering.

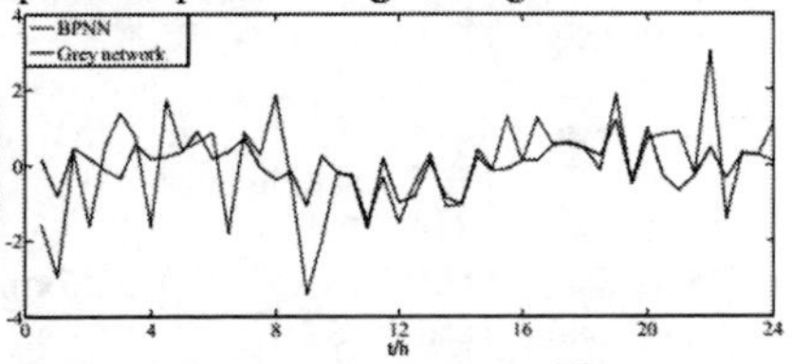

Figure 8 The relative errors between top-oil temperature prediction value and the measured value

Table 2 The relative errors in two model

Model	δ_{min}	$\bar{\delta}$	δ_{max}
BPNN(%)	0.10	1.03	3.44
Grey network(%)	0.05	0.47	1.68

In the actual operation of the transformer, the environment is complex. In the prediction of top-oil temperature, if we need to consider the wind speed, solar radiation, altitude and other factors, and we can directly change the input data of the grey neural network prediction model. For the test transformer, we select the load current, environment temperature, active power, reactive power and the top-oil temperature of the previous time as input parameters.

5 Conclusion

The accurate prediction of transformer top-oil temperature is of great significance for the safe operation of power system. Considering the factors affect the transformer top-oil temperature, the paper selects the top oil temperature, load current, active power, reactive power and the top-oil temperature of the previous time as the characteristic parameters, and establishes the prediction model of top-oil temperature based on Grey Neural Network. The applicability, accuracy and feasibility of the model are verified. The example result shows that the prediction model has a fast convergence, good versatility, high precision of the forecasting model which indicats that this is a kind of an effective prediction method of transformer top-oil temperature.

References

1 Wu Jianjun, Li Xiyuan, Li Sihua, et al. Research on transformer winding fibre optic temperature monitoring system and practical application[J]. Transformer, **50(11)**:47-50(2013).

2 YANG Honglei, XIONG XIN, LIANG Shibin, et al. Research of depth adjustable FBG oil temperature sensor for transformer[J]. Transducer and Microsystem Technologies, **33(2)**:47-49(2014).

3 HONG Dan, WEI Guangyuan, QIAN Zheng, et al. Research on application of fiber brag grating sensor in hot-spot temperature measurement for transformer winding[J]. Electrical Measurement & Instrumentation, **51(21)**:47-51(2014).

4 IEEE guide for loading mineral-oil-immersed transformers[J]. IEEE guide for loading mineral-oil-immersed transformers, (1996).

5 Lesieutre B C, Hagman W H, Kirtley J L. An improved transformer top oil temperature model for use in an on-line monitoring and diagnostic system[J]. IEEE Transactions on Power Delivery, **1(12)**:249-256(1997).

6 Tylavsky D J, He Q, Si J, et al. Transformer top-oil temperature modeling and simulation[J]. IEEE Transactions on Industry Applications, **5(36)**:1219-1225(2000).

7 CHEN Weigen, SU Xiaoping, CHEN Xi, et al. Influence Factor Analysis and Improvement of the Thermal Model for Predicting Transformer Top Oil Temperature[J]. High Voltage Engineering, **37(6)**:1329-1335(2011).

8 WEI Bengang, HUANG Hua, FU Chenchao, et al. Algorithm for Transformer Top-Oil Temperature and Winding Hot-Point Temperature Based on Modified Thermal Circuit Model[J]. East China Electric Power, **44(3)**:444-447(2012).

9 CHEN Jinming, WU Yi， ZHU Haibing, et al. Study and application of Transformer Top-oil predicting[J]. Electrotechnical Application, **22(33)**:89-93(2015).

10 XIONG Hao, CHEN Weigen, DU Lin, et al. Study on prediction of Top-oil Temperature for Power Transformer Based on T-S Model[J]. Proceedings of the CSEE, **27(30)**:15-19(2007).

11 CHEN Weigen, TENG Li, LIU Jun, et al. Transformer winding hot-spot temperature prediction model of support vector machine optimized by genetic algorithm[J]. Transformer of China electrotechnical society, **29(1)**:44-51(2014).

12 LU Fangcheng, MA Lun, WANG Liu, et al. Calculation of Winding Hot Spot Temperature of Oil Immersed Transformer by the Discontinuous Galerkin Method[J]. High Voltage Apparatus, **51(3)**:28-34(2015).

13 He Q, Si J N, Tylavsky D J. Prediction of top-oil temperature for transformers using neural networks[J]. IEEE Transactions on Power Delivery, **4(15)**:1205-1211(2000).

14 CHEN Weigen, XI Hongjuan, Su Xiaoping, et al. Application of Generalized Regression Neural Network to Transformer Winding Hot Spot Temperature Forecasting[J]. High Voltage Engineering, **38(1)**:16-21(2012).

15 WANG Yongqiang, YUE Guoliang, HE Jie, et al. Study on Prediction of Top Oil Temperature for Power Transformer Based on Kalman Filter Algorithm[J]. High Voltage Apparatus, **50(8)**:74-79(2014).

MATEC Web of Conferences **63**, 05017 (2016) DOI: 10.1051/matecconf/20166305017
MMME 2016

Design and Implementation of Cherry Picking Robot Vision Recognition System Based on Face Recognition

Qi Rong ZHANG [a], Ru Hui HU and Ting Zhong YU

Information Engineering, Guizhou University of Engineering Science, Bijie 551700, Guizhou, China

Abstract. This paper analyzes the necessity of cherry picking robot vision recognition system. It introduces the system function and the system structure, and provides the system operation environment, the development platform and the database design. This paper designs and implements the cherry picking robot vision recognition system, based on C# technology, using MATLAB, using face recognition technology.

Keywords. face recognition; Cherry picking robot; Vision recognition system

1 Introduction

Along with the development of agriculture in our country, in order to reduce the cost and improve the productivity, the requirement of the agricultural mechanization is more and more high. Cherry's price is higher, all over the country in the development of cherry cultivation. Cherry planting difficulty is not great, but its picking cost of artificial is more, which is a reason for its high cost, so it is imperative to use the robot picking cherries. But if the price of cherry robot is high, it is bound to cause its promotion difficulties and the value of the use of reduced. This system uses C# and MATLAB mixed programming technology, which can realize the basic function of cherry picking robot vision recognition system. In addition, the price of computer hardware is reduced, which makes it possible to extend the computational intelligence to the practical robot. So this system will use face recognition technology to replace the general system using password function, bring convenience to users. It prevent forget passwords and password to remember, at the same time, to prevent theft cherry picking robot. No face recognition successful login, cherry picking robots will not be able to work.

2 Face recognition technology

Face recognition is new technology in the development of recent decades, which has great practical value. Face recognition is based on multiple disciplines on the basis of a comprehensive technology. With the development of The Times, people is higher and higher requirements for protection of privacy, therefore, it produced a lot of way to protect privacy, such as using the id, password and smart card, etc. But these methods are not convenient, easy loss, easy to forget, theft, which can not adapt to the need for the development of the society. So the researchers began to target the biological recognition technology. This technology not only has the uniqueness of the individual, but these features are usually more stable. The biometric technologies mainly include: face recognition, iris recognition, fingerprint recognition, voice recognition, etc. Compared with other biometric, facial recognition has the friendly, convenient and not easy to forge, natural advantages, at the same time face data collection is also relatively simple. The cost is not high, not psychological barriers for being collected. Face recognition system is not high cost, more and more people to study it [1-4]. It has been widely used in many fields, such as the entrance of Beijing Olympic Games, such as in Xinjiang highway, of all the passengers were identity cards and the actual face matching for face recognition, to find wanted terrorist criminals, such as the Hong Kong customs using face recognition technology for identification.

3 System design

3.1. System architecture

The cherry picking robot visual recognition system based on face recognition has four function modules. They are face recognition login module, image acquisition module, cherry recognition module, binocular positioning module.

[a] Qi Rong ZHANG: zqr_hk@163.com

The overall structure of the system is shown in the following figure.

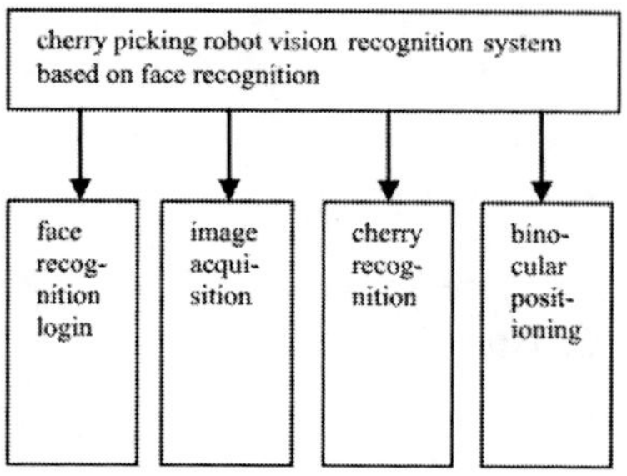

Figure 1. The overall structure of the system

As follows, we illustrate the functions of each subsystem.

Users can through the face recognition login module, after the camera to take pictures to identify the successful to login system. The module includes access to the image, image preprocessing, face detection, face recognition, face comparison of 5 sub functions. User face database is established in the function, because the users of the system is not much, and shooting is according to the positive, so recognition speed is fast and recognize faces time is short, accord with the actual needs. But the difficulty is only a test image, large difficulties of identification. Then we enter the image acquisition module, the system through the module collected in the actual picking environment of the cherry image, and then the image is stored in the module. Then the most important module of the system is cherry recognition module. In this module, eight sub function operations are performed on the cherry image, which are median filtering, color recognition, fixed threshold segmentation, morphological opening operation, morphological closing operation, Canny operator and circle Hough transform, picking point acquisition. Through a series of operations of the module, the cherry of the cherry image was identified by the picking point, so the cherry picking robot can pick cherry by the robot hand. The function of the module is sequential, one by one to carry out the operation, which can not be any choice of a sub module to perform. You must complete all the sub function, cherry recognition function to complete. This is different from general software system, which can choose subsystem operation according to the user. Binocular positioning module is through the left and right camera to get the internal calibration parameters on both sides, and then gets the external calibration parameters. The 3D coordinates of picking point can be obtained by the internal and external calibration parameters.

3.2 Development environment

Cherry picking robot vision recognition system based on face recognition selects C# as the system development language and Microsoft Visual Studio 2010 for online system environment, using MATLAB image processing, with SQL Server 2010 as the system database on the Server. The development process and the debugging of the whole system are conducted on Microsoft's Windows 7 operating system. Cherry picking robot visual recognition system based on face recognition uses SQL Server 2010 to build the database server. The database involves the main data table including the users table, cherry collection images table, cherry circle Hough transform table, cherry picking point table and cherry three dimensional picking point table. Class diagram of data relationship between them is shown in table 1.

Table 1. List of main data tables

The serial number	The name of the table	explain
1	Users	users table
2	CollectionImages	cherry collection images table
3	CircleHoughTransform	cherry circle Hough transform table
4	PickingPoint	cherry picking point table
5	ThreeDimensionalPickingPoint	cherry three dimensional picking point table.

The following is a list of the main data tables.

(1) The user table: this table is used to store the information of the user.

(2) Cherry collection images: the table used to store the actual scene of the camera to capture the image of the cherry.

(3) Cherry circle Hough transform table: the table is used to store the image cherry the circle Hough transform of image recognition module.

(4) Cherry picking point table: this table is used to store the image after the circle Hough transform.

(5) Cherry three dimensional picking point table: this table is used to store the cherry image, which is determined by the three-dimensional binocular calibration picking point.

3.3 The key technology of system implementation

The database stores a large amount of data, so design of the database directly affects the realization of the system. The database of the cherry picking robot vision recognition system is a very important part of the system, and a large number of data are stored in the database. Cherry picking robot vision recognition system includes a number of functional modules, so the following is part of the code.

(1) Part of the database connection code is as follows: [5]
//The database connection

```
string
dat="server=.;database="+textBox1.Text.Trim()+";u
id=sa;pwd=";
connn=new SqlConnection(dat);//Create a
SqlConnection object
connn.Open(); // Open a connection
if(connn,State==ConnectionState.Open)// Determine
the current connection status
{
 MessageBox.Show( "Connect successfully" ); //
Pop-up prompts
//Close the connection
String dat = " " ;// Define a string variable
Connn.Close();//Use Close to close the connection
If(connn.State==ConnetionState.Closed) // To
determine whether the current connection closed
{
dat= "Database has been successfully closed \n" ; //
If it has been closed, then pop-up tips
}
```

(2) Face recognition part of the MATLAB code

```
t_all=zeros(nr*nc,tp*ts); % The initial training
sample set is 0
for num_tp=1:tp
res=[];
for num_ts=1:ts
num_sequ=sequ(num_ts);
img_name=strcat('C:\Documents and
Settings\Administrator
\ModifyFERET\ModifyFERET\','s',int2str(num_tp),'\',int
2str(num_sequ),'.jpg');
img=imread(img_name); % Read the face image
img=double(img)./256;
img=imresize(img, [12 12]); %Cut out the image of
12 x 12
mid=(num_tp-1)*ts+num_ts;
res=reshape(img,size(img,1)*size(img,2),1);
t_all(:,mid)= res; %Read the images are put into the
training sample set
end
end
```

(3) Median filter MATLAB code

```
Img=imread('img.tif');
Noi=imnoise(Img,'salt & pepper',0.02); % Add salt,
pepper, noise density of 0.02
KK1=medfilt2(Noi);            % Median filtering is
performed in a 3 x 3 neighborhood window.
KK2=medfilt2(Noi,[5 5]);     % Median filtering is
performed in 5 * 5 neighborhood windows
```

4 Conclusion

The system can complete the visual identification of the cherry in the cherry orchard, so cherry picking robot can run smoothly. With the maturity of face recognition technology, the continuous improvement of computer hardware performance and the continuous decline in hardware prices, it is possible to use face recognition technology in the cherry picking robot vision recognition system. Cherry recognition module is the most critical modules in this system. This system can also apply to other fruit and vegetable picking robot visual recognition, which is for fruits and vegetables with a large difference in color and background color, and shape close to the circular shape.

Acknowledgments

This work was supported by 2015 Annual Science and Technology Department of Guizhou Province, Bijie City Science and Technology Bureau, Guizhou University of Engineering Science Joint Science and Technology Fund Project, Cherry picking robot vision system research, Guizhou Branch of LH word [2015] No. 7599; the grants from 2013 Annual Science and Technology Department of Guizhou Province, Bijie City Science and Technology Bureau, Bijie University Joint Fund Scheme under Project No. Guizhou Branch of J word LKB [2013] No.14.

References

1. S. Dongmei, Q. Zhengding. Review of biometric identification technology [J]. Electronic journal，2001,12A（12）：1744-1748
2. T. Jie, Y. Xing. Theory and application of biometric identification technology [M]. Beijing: electronic industry press, 2005
3. Qi Rong Zhang, Zhongshi HE, The Two-dimensional Bidirectional Inverse Laplacian Margin Maximum Criterion for Face Recognition, Journal of Computational Information Systems, 2010, 6：1101-1109.
4. Qi Rong Zhang, Zhongshi HE, Dual-tree Quaternion Wavelets for Face Recognition, Journal of Computational Information Systems, 2010, 6：1151-1160
5. Tomorrow Technology. C # from entry to the master (3rd edition) [M]. Beijing: Tsinghua University press.2012. 9.

MATEC Web of Conferences **63**, 05018 (2016) DOI: 10.1051/matecconf/20166305018
MMME 2016

Research on DTX Technology and Power Consumption Performance of Mobile Communication Terminal

Shui Zhen XIE [1,a]

[1] *School of electronic engineering , Jiujiang University, Jiujiang 330025, China*

Abstract. In order to reduce the power consumption of GSM and TD-SCDMA mobile communication system terminal, the paper starts with DTX (Discontinuous Transmission) technology of GSM and TD-SCDMA systems, offers a detailed analysis of the DTX's function in optimizing power consumption of GSM and TD-SCDMA mobile communication system terminal and reducing system interference, and verifies DTX's positive role in reducing the power consumption of the mobile terminal by experiment.

Key Words. Mobile terminal, green energy and energy conservation, power consumption performance, DTX

1 Submitting the Manuscript

Green energy, energy conservation and environmental protection has become a social consensus, awareness of energy conservation and emission reduction has been firmly established, as well as universal access to smart mobile phones and fierce increase of communications traffic, energy conservation of the communications industry allows of no delay. The mobile phones especially smart phones are closely related to people's daily work, life and entertainment, of which the standby time, power consumption, and recycling of used batteries, reflects the concept of green communications directly or indirectly. When people use mobile phones especially smart phones to experience with a range of voice, data, video and other multimedia services, the phone power consumption will be very fast, there will be insufficient of power supply of mobile phone batteries, resulting in frequent phone charging, which will consume more energy. If the performance of mobile phone is poor, frequent use and charging of mobile phones may make power battery hot, which affects the user's experience, even human health. In short, energy saving of mobile phone battery is the key protection to green communication, is the core of a low-carbon economy. Phone energy saving technologies, energy saving program has important practical significance for the construction of green communications and satisfaction improvement of mobile users.

2 Main Factors Affecting Power Consumption Performance of Mobile Terminal

The internal factors of mobile terminal, including: (1) battery performance of the mobile terminal; (2) power consumption of wireless transceiver (RF chipset); (3)power consumption and process characteristics of baseband chip; 4) peripheral devices power consumption of mobile terminal, such as keyboard backlight, speakers, screen displays; (5) power controls and power management scheme of the mobile terminal, etc.

External factors related to network environment: (1) the mode of mobile terminal, connected mode (ie, work status) has more power consumption than the idle mode (ie, standby); (2) movement of the mobile terminal, more frequent the mobile terminal moving costs more power consumption; (3) the quality of wireless network; (4) wireless network environment; (5) protocol stack control; (6) the set of network parameter values, such as DRX parameters, DTX parameters, and various of registration timers. DRX, DTX determines the interval that mobile terminal wakes to monitor the paging channel message in the idle state, the long interval means that the mobile terminal wakes up less, which will extend the standby time of the terminal. These parameters have obvious effect for the standby power consumption of mobile terminal.

The enhancement and improvement of mobile terminal power consumption performance is not only closely related to the user experience, the quality of mobile terminal, but also closely related to the green

[a] Corresponding author: xieshuizhen@126.com

communication concept. In general, according to the user's daily habits, the application state of the mobile terminal can be divided into the work state and the idle state, the corresponding technical indicators are operating current and standby current which measure overall power consumption performance of the mobile terminal. The power consumption of mobile terminal is relatively less in idle state, while it is more in the call state. For example, 6 TD Smart phone (TD smart phones), the power consumption in different states is shown in Figure 1. It can be seen in Figure 1, in the call state, TD terminal average power consumption is 140 ~ 160mA; in the idle state, TD terminal power consumption is at an average of 5 ~ 7mA. It is visible that power consumption in call state is much more than idle state. Thus, users are more concerned about how to reduce the operating current of the mobile terminal. There are a lot of factors affecting the power losses and operating current of mobile terminal, such as DTX period, re-election threshold, signal coverage strength, paging cycle, neighbor cell number, the paper researches DTX technology.

3 DTX Technical Parameters and Introduction

3.1 Introduction of DTX Technical Features

When users use mobile phone to call via mobile communication network, typically the two sides take turns to speak, that is, speech time of one side accounts for about 50% of the total communication time, the other time is silence period and listen to the other one talk, there is only background noise without voice signal during the silence. If all background noise information is transmitted to the mobile communication network, it will cost great waste of network system resources, and also increase interference within the network system. In order to reduce the total air emission levels in mobile communication system, improve mobile communication system efficiency and reduce the power consumption of wireless transmitter (especially mobile station), we use Voice Activity Detection (VAD) and Discontinuous transmission / comfortable noise generation (DTX / CNG) technology, discontinuous transmission / comfort noise generation (DTX / CNG) technology can encode in 13Kb / s during a call, encode with 500b / s and send comfort noise in the standstill period, so that the average rate is lowered down. If it is a waste of channel resources that using the same bit rate to encode the voice signal, the discontinuous transmission / comfort noise generation (DTX / CNG) technology, can control encoders with different bit rate to transmit data in voice stage and background noise stage, which will reduce power consumption of wireless transmitter and a mobile station, while also reduce the total power of the wireless channel, so as to improve the channel utilization and reduce interference.

3.2 The main control parameters

DTXD: This parameter specifies whether the mobile station use DTX functionality in the downlink of the cell site. The application selection is functional for voice and non-transparent data, and functional for all TCH of all non-BCCH carrier within the cell. This parameter should be set on each cell.

DTXU: This parameter specifies whether the mobile station use DTX functionality in the uplink of the cell site. This parameter should be set in each cell.

DTXU = 1: shall

DTXU = 2: shall not

DTXU = 0: may shall

3.3 Special adjustable parameters

DTXFUL: Define the number of measuring reports which selecting full set measurements requires.

4 Test Method of GSM/TD-SCDMA Terminal Power Consumption

This study tests in the current network environment, the test model is shown in Figure 2. The advantage of existing network environment is that the test environment itself is the actual working environment of the mobile terminal, test results of power consumption can directly reflect the power consumption performance of mobile terminal in use. The disadvantage is that it is difficult to control the wireless network system parameters, test results of power consumption are related to specific network parameters and wireless network environment. In this study, the test device comprises a plurality of GSM / TD-SCDMA dual-mode mobile terminal and a synthetical terminal tester SP6010, etc. In order to simulate terminal power performance more accurately in various conditions, we used the following 3 network environment for testing: moving state, fixed good signal quality, fixed poor signal quality. Test indicators take average current of DTX Enable before and after in call state as the main basis for judgment.

5 Test Results and Analysis of DTX Settings Affecting Terminal Power Consumption Performance

Respectively in the TD-SCDMA, GSM mode, use test instrumentation (terminal tester SP6010) DTX Enable function before and after, test the average current that GSM / TD-SCDMA dual-mode mobile terminal initiates a voice call, wherein, for TD-SCDMA mode, when DTX Enable, SBGP = 40ms (SBGP value is cycle of no data transmitted). Test results are respectively shown in Figure 3 and Figure 4.

The test results of the experiment above show that after enabling DTX the GSM mobile terminal can save almost half of the power, namely the use of DTX reduces the power consumption of 2G terminals significantly,

DOI: 10.1051/matecconf/20166305018

optimizes user's experience feelings, and does not adversely affect mobile communication system performance. For TD-SCDMA mobile terminal, it has a similar conclusion, if the network enables DTX function and set reasonable data transmission interval (ie DTX cycle SBGP), it so can save nearly half of the power after enabling DTX, and will be a great help for terminal power saving and reducing the system interference.

6 Summary

Improving the performance of terminal power consumption has important practical significance for improving users' satisfaction and enhancing users' experience feelings. Through the practice research of the project, we can see the introduction of DTX technology to mobile communication network system can significantly reduce battery consumption of mobile terminals, reduce wireless system interference, application of DTX can greatly reduce the power consumption of the mobile station. At the same time, through continuously optimizing and improving DTX technology, it can improve the flexibility and controllability in various scenarios, which has a very important practical significance for strengthening the green channel idea of mobile communication system and enhancing the user experience.

Acknowledgment

Note: This paper is the research achievement of school level topic 'DTX technology application in mobile communication network' of Jiu Jiang school, topic number: 2014KJYB004.

References

1. Li Jiandong. Mobile Communications, Xi'an University of Electronic Science and Technology Press, 2006
2. Wang Shiliang. Research on power consumption performance of 3G mobile communication terminal, Mobile Communications, 2005.12.25,37-38
3. Liu Dongdong. Research on technology enhancing mobile voice quality (MOS), Master's degree thesis of Beijing University of Posts, 2012.9
4. Zhou Dejun. Discontinuous transmission technology in voice communication, Communication Technology, 2001(9), 46-48
5. Li Fangwei, Zhang Yaqing. Green energy technology of TD-SCDMA terminal, Communication Technology of Guangdong province, 2010.3
6. Dong Yanjie, Zhou Meng. Power consumption test methods and test results analysis of TD-SCDMA terminal, Telecommunications engineering technology and standardization, 2008.10

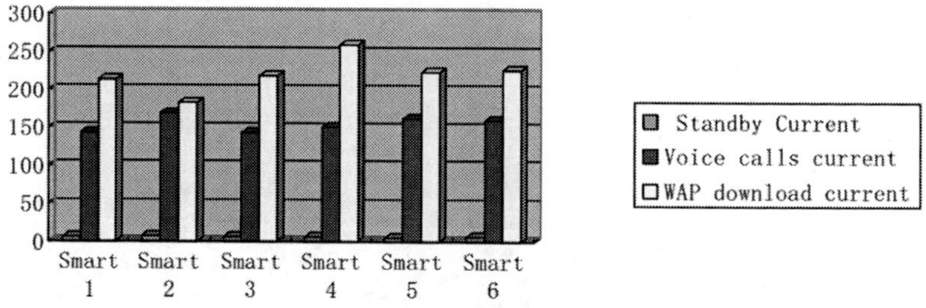

Figure 1. Test sample of TD terminal operating current and standby current (Current unit: mA)

 DOI: 10.1051/matecconf/20166305018

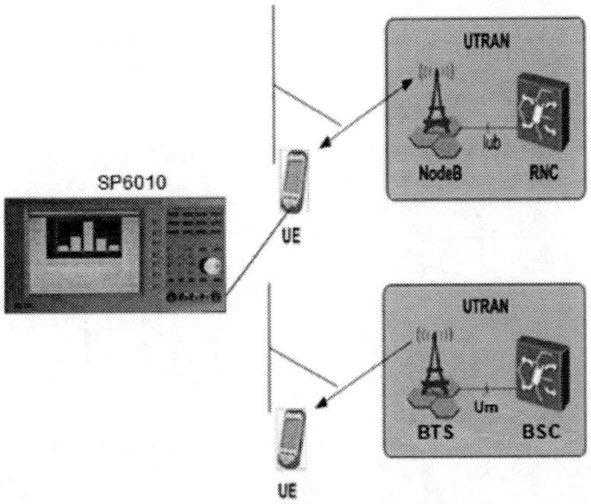

Figure 2. Test model of terminal power consumption

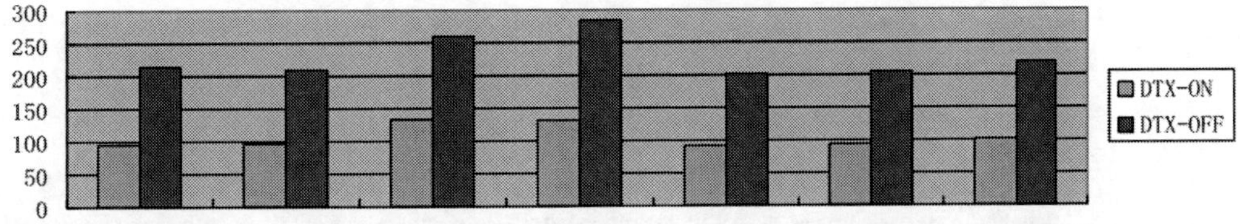

DTX-ON	95. 654	97. 327	133. 986	132. 497	92. 723	93. 759	101. 597
DTX-OFF	213. 425	208. 346	260. 598	284. 115	202. 187	204. 835	220. 314

Figure 3. The impact of DTX Enable before and after to GSM terminal power consumption (current unit mA)

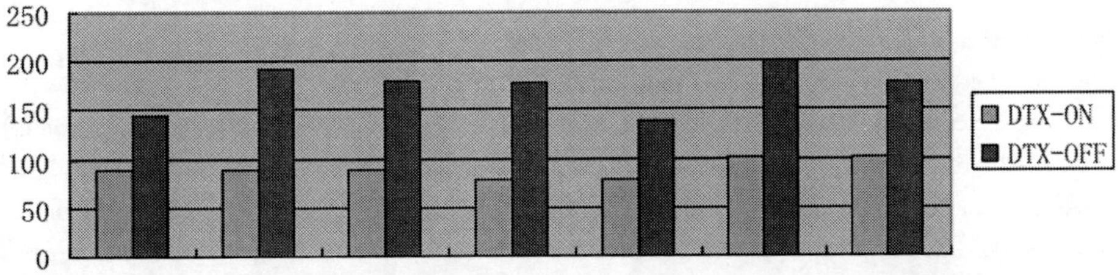

DTX-ON	88. 634	87. 879	88. 438	79. 335	78. 427	101. 164	100. 936
DTX-OFF	145. 254	191. 283	180. 534	176. 748	139. 218	201. 377	178. 583

Figure 4. The impact of DTX Enable before and after to TD terminal power consumption (current unit mA)

MATEC Web of Conferences **63**, 05019 (2016) DOI: 10.1051/matecconf/20166305019
MMME 2016

Research on the application of hybrid cloud desktop design method

Hong WANG, Ya Jing XUE

College of Electronic and Electrical Engineering,Shanghai engineering technology university,Shanghai, China

Abstract: Through the study of hybrid cloud desktop , proposed technical solutions based on hybrid cloud desktop, On the basis of this, a detailed research and design of the hybrid cloud desktop is carried out, including the architecture of a hybrid cloud desktop solution and the PNS Desktop Virtualization (IDV), PNS Server Virtualization (VDI) and PNS QuickCloud cloud storage system integration, to achieve the centralized management and control of the client operating system, convenient and fast to meet the individual needs, centralized data management, the realization of the hybrid cloud desktop computer management, and improve the management level of the computer, improves the efficiency,it is virtual desktop changes to traditional PC desktop.

1 Current status

With the advance of information construction, The popularity of computer use Every year a large number of computers increase, installation and debugging software wasted a lot of manpower and material resources, it reduces efficiency, as a professional IT management, they hope to have a, the terminal is no need of maintenance and management, System and software installed on a server ,it become managers of the dream scenario.

In daily work, computer viruses, data loss, application software error, system failure, some management problems have become the norm, these repetitive work invested a lot of manpower and occupied all the time. it has been unable to let managers again sufficient time to improve the construction of information.

At present, the mobile office will be a new office mode. In the process of daily office, mobile office demand is also increasing, office personnel in addition to the unit can handle the affairs and they can also in a local area network (LAN) of any piece of equipment, access to the unit of public resources, greatly improving work efficiency.

2 Design objectives and features

There are thousands of computers in schools, in view of the existing PC, they use IDV architecture to achieve a unified management of the virtual desktop, unified deployment, they improve management efficiency and reduce the failure rate. Virtual desktop to provide users with the current use of the PC as a variety of functions, they smooth use of the required software, smooth browsing web pages, and so on.

In view of the demand, we adopt the IDV virtual desktop and existing VDI, virtual desktop ,we can meet the needs of the existing management and teaching, the virtual desktop data of VDI andI IDV can be through a virtual storage to achieve interoperability.

IDV features: scalability, reliability, management

We use functional modularity. When the number of terminals or application requirements change, it can be easily extended by increasing the server, without the need

DOI: 10.1051/matecconf/20166305019

to redesign the entire program. In this way, in order to maintain business continuity, not because the single point of failure affects the overall structure, resulting in the system is not available. The system has active monitoring service and alarm mechanism, which supports load sharing or fault switching mechanism, and realizes seamless transfer of load.

Through the management end can for the whole platform for unified management , including servers, virtual machines, virtual machine operating system, application software and network, etc.. The console can monitor the running state of the whole platform, including the server, the terminal and the network.

3 Hybrid cloud virtual desktop Technology Architecture

3.1 Program architecture

Hybrid cloud desktop virtualization topology includes three parts: VDI virtual desktop, IDV virtual desktop, virtual storage, it is a traditional PC desktop to the transformation of the virtual desktop

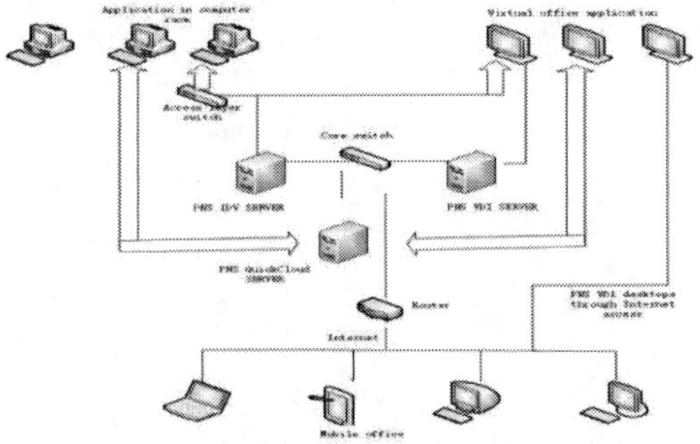

Figure 1. Program architecture

Working principle :

In the local area network, using the PNS desktop virtualization IDV server to push all kinds of operating system, software and services. In computer applications, have higher request for the client to use graphics performance, using common PC as the client, using PNS unique high performance local caching technology, it can fully play the PC local hardware performance, in order to meet the performance requirements.However, the client cannot access through the external network.

In need of remote access through the Internet to personal desktop, you can use PNS server virtualization VDI to create multiple virtual machine, virtual machine operating system unified by the PNS desktop virtualization push.. At this point, customers can use the PNS virtual client access to personal desktop through the Internet and telecommute.

Customer data are stored in the PNS cloud storage server.

Through the integration of PNS Desktop Virtualization (IDV), PNS Server Virtualization (VDI) and PNS cloud storage system, you can achieve:

- The client operating system centralized control
- Quick and easy to satisfy personalized needs
- Personal data of the centralized storage management .
- Internet remote office needs and remote data access requirements
- The local virtual desktop application of high performance

3.2 Host characteristics

1. Cloud desktop image (virtual desktop) system in the form of a sector is stored in operating system image
2. Desktop operating system in the form of data stream from the master mirror according to the need to transfer and run on the client terminal
3. With PNS HPLC technology will operating system cache in terminal local storage medium
4. The PNS OS over IP data link transmission protocol, in the network transmission is not affected by viruses and other cyber attacks and modifying the infection
5. Using TCP/IP based protocol, network transmission is not limited by network equipment features
6. The PNS Boot protocol and compatible with INTEL PXE protocol, they can unconditional across the gateway, routing, VLAN and VPN

3.3 Storage cloud system

DOI: 10.1051/matecconf/20166305019

Figure 2. storage cloud system

QuickCloud PNS storage cloud system supports PNS High Performance Cloud desktop (virtual desktop) infrastructure, anytime, anywhere you can easily store data into the cloud. Cloud storage system to support multiple terminal (computers, mobile phones, ipad, etc.) unified storage, Make full use of existing idle resources, resource integration, unified management. With increasing capacity, performance and access speed of the linear expansion, providing similar drive mode of operation, in line with the day to day operating habits, physical storage group to realize the support power resume, FTP / SSH can direct transmission of large files, it can high and stable operation of security, Extensible security supervision mechanism, applicable toward secrecy demands a high level, extensible differential backup mechanism, ensure data security.

3.4 PNS IDV virtual desktop

PNS IDV virtual desktop uses is server network push mode, On the server side to install various user application system image, all the client uses the network boot mode and by server push operating system and service.

PNS desktop virtualization technology principles:

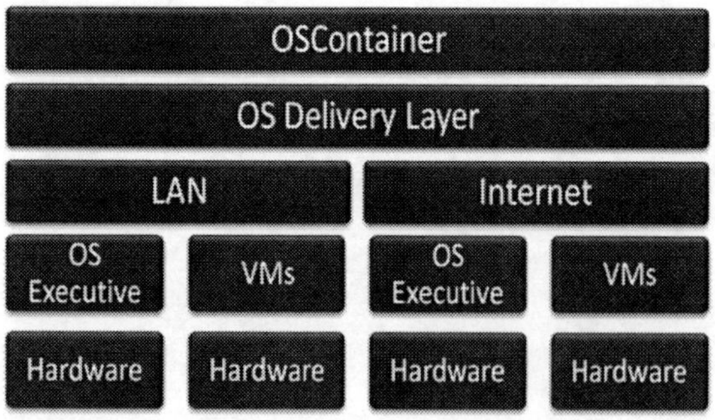

Figure 3. PNS desktop virtualization technology rinciples

In the top is the operating system container (OSContainer), in the container it can accommodate multiple operating system package, the operating system package by operating system push (OS delivery layer) system between the client and the server, network media can make local area network (LAN) or the Internet (Internet), operating system pushed to the client, by the operating system execution (OS) executive or virtual machine (VMS) execution, thereby driving the computer hardware.

IDV desktop virtualization features:

- Support load balancing
- can be flexibly expanded to support the virtual desktop in future
- Support Windows XP SP3/Windows 7/linux
- can remote management

3.5 PNS VDI virtual desktop

PNS server virtualization using VDI mode, you can create and use multiple virtual machines at the same time, Through the Internet remote access personal desktop in order to realize the remote office. This pattern and the combined desktop virtualization, all virtual client through desktop virtualization server push operating system and service, it has achieved a more centralized management.

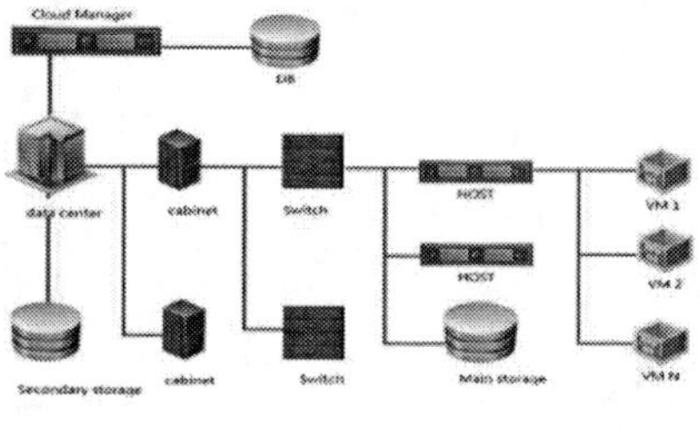

Figure 4. PNS VDI virtual desktop

VDI desktop virtualization features:

according to different application scenarios, the corresponding hardware configuration server, and unified management by the virtualization platform.

The centralized storage scheme, to ensure high availability and load balancing.

We need to consider in late stage development, hardware and software are able to support smooth expansion.

DOI: 10.1051/matecconf/20166305019

3.6 PNS cloud storage system

Personal data are centrally stored by the PNS, you can access a variety of terminal equipment through the LAN or Internet .

Virtual storage characteristics:

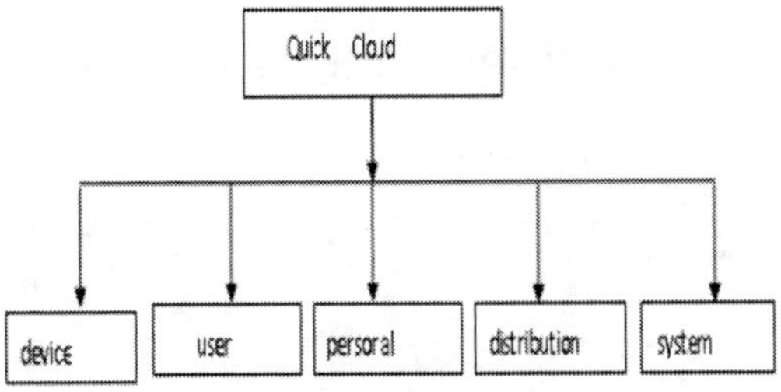

Figure 5. PNS cloud storage system

- can browse data by PC, ipad, mobile phone
- The virtual memory can be mapped into the letter in the use of virtual desktop;

4 Summary

For schools, computing center is a school for the students practice, is representative of the image of the school, building based on cloud desktop virtualization is a great revolution to the traditional desktop management, For the future schools has important significance for the development of computer technology.

References

1. Wang Qingbo, why virtualization and cloud computing [M].28-39. Electronic Industry Press.
2. stone Yirong, Duan Yong. Cloud computing in the field of application of Telecom Telecom I T science, 2009,25 (9).
3. Zhu Jinzhi wisdom of the cloud computing [M].68-108. Electronics Industry Press.
4. Liu Peng. Cloud computing [M].47-89. electronics industry publishing Press.
5. Wang Feng, Jiang Feng, Li Zhaoyang. Analysis of the virtual desktop and key technology. Telecom technology, 2011,1 (8).
6. Tao Caixia, Chen Kang. Cloud computing in telecom support system in the field of application and analysis, telecommunication science, 2010, 26 (10) (deadline: 2011-05-27) based on Virtualization cloud desktop technology program research and design of communication hot)

MATEC Web of Conferences **63**, 05020 (2016) DOI: 10.1051/matecconf/20166305020
MMME 2016

VANET Routing Replay Attack Detection Research Based on SVM

Qing Gang FAN[1,a], Li WANG[2], Yan Ning CAI[1], Yong Qiang LI [1] and Jing CHEN [1]

[1] *Computer Staff Room, Xi'an Research Institute of High Technology, 710025 Xi'an, China*
[2] *Library, Xi'an Research Institute of High Technology, 710025 Xi'an, China*

Abstract. In the process of establishing routing in VANET, because of the interference of routing request (RREQ) and routing response (RREP), the nodes in the network are busy with looking for routing or establishing routing. The impact on the network performance is extremely serious, such as increasing a lot of network overhead, consuming valuable bandwidth resources. In this paper, the influence of routing replay attack in VANET is studied. Four typical characteristic are extracted by "cross layer" selecting feature vector. The feasibility of VANET routing replay attack IDS based on SVM is verified through simulation experiment.
Keywords. SVM, VANET, Routing Replay Attack

1 ROUTING REPLAY ATTACK

In the communication process based on AODV routing protocol of VANET, establishing routing between nodes is on-demand type. If the source node needs to send a message to the destination node, firstly, the source node will query their own routing table, if the path to the destination node is not exist in the table, source node will be broadcast a route request packet RREQ in the network. Adjacent nodes will query their own routing information table after receiving the RREQ, if the path to the destination node is exist in the table, the RREP message will be sent back directly, passing by relay node to the source node. If not, continued to send RREQ to adjacent nodes, until find the path. The forward routing is established after source node received the RREP. This is the normal communication process based on AODV routing protocol.

In general, a relay node will forward RREP in order to set up the forward routing, when there are malicious nodes in network, malicious nodes would deliberately to interference of communication process, such as not forwarding RREP, makes the forward routing cannot be established. Or not forward RREQ, making the source node can't find the destination node. As shown in figure 1, B is the source node, P is the destination node, B will be broadcast a RREQ message to find the path of the P, in the process of normal communication, relay node, such as C, D, will forward RREQ, and continue to backward forward RREP after find the destination node P. In exceptional circumstances, such as D is a malicious node, after receiving RREP, D does not forward RREP to C, but drops the RREP and give an error, result in C and B can,t receive RREP of P, unable to form traffic route to the destination node P, forming a routing replay attack.

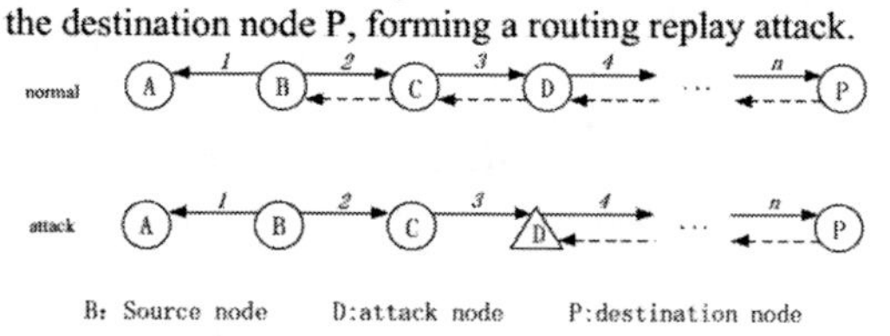

Figure 1. routing replay attack

2 ROUTING REPLAY ATTACK DETECTION METHOD

2.1 Routing replay attack Scenario simulation

This linear model is used to set up routing response team attack scenario in this paper, the specific scene parameter settings are as follows:

Table 1. Routing replay attack Scenario parameter settings

Parameter settings	value
Channel type	wireless channel
Routing protocol	AODV
Interface queue type	Queue/DropTail/PriQueue
Maximum length interface queues	50
Distance between the nodes	70m
Effective communication distance	75m

[a] Qing Gang FAN: fangang3232@126.com

Network interface type	Phy/WirelessPhy
Type of antenna	Antenna/OmniAntenna
Number of vehicles	30
vehicle speed	18m/s
Communications link	3→12 15→20
Frequency of packets	4packets/s
Malicious nodes	Node 10
the size of scene	4000m*500m
simulation time	1000s
Flow type	TCP
CBR packet size	512 bytes
MAC layer protocol type	IEEE 802.11
Wireless propagation mode	Two Ray Ground

By software NS-2, the corresponding simulations with tcl script is run in the normal and routing reply attack scenario, and the corresponding trace files are got. The attacked trace file as shown in figure 2.

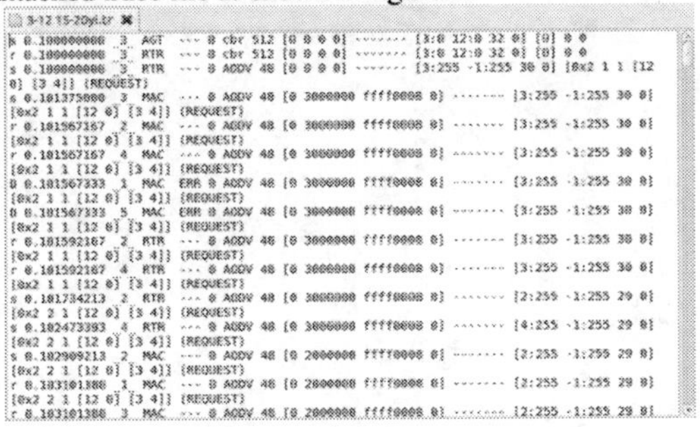
```
3-12 15-20yi.tr
s 0.100000000 _3_ AGT --- 0 cbr 512 [0 0 0 0] ------- [3:0 12:0 32 0] [0] 0 0
r 0.100000000 _3_ RTR --- 0 cbr 512 [0 0 0 0] ------- [3:0 12:0 32 0] [0] 0 0
s 0.100000000 _3_ RTR --- 0 AODV 48 [0 0 0 0] ------- [3:255 -1:255 30 0] [0x2 1 1 [12 0] [3 4]] (REQUEST)
s 0.101375000 _3_ MAC --- 0 AODV 48 [0 3000000 ffff0008 0] ------- [3:255 -1:255 30 0] [0x2 1 1 [12 0] [3 4]] (REQUEST)
r 0.101567167 _2_ MAC --- 0 AODV 48 [0 3000000 ffff0008 0] ------- [3:255 -1:255 30 0] [0x2 1 1 [12 0] [3 4]] (REQUEST)
r 0.101567167 _4_ MAC --- 0 AODV 48 [0 3000000 ffff0008 0] ------- [3:255 -1:255 30 0] [0x2 1 1 [12 0] [3 4]] (REQUEST)
D 0.101567333 _1_ MAC ERR 0 AODV 48 [0 3000000 ffff0008 0] ------- [3:255 -1:255 30 0] [0x2 1 1 [12 0] [3 4]] (REQUEST)
D 0.101567333 _5_ MAC ERR 0 AODV 48 [0 3000000 ffff0008 0] ------- [3:255 -1:255 30 0] [0x2 1 1 [12 0] [3 4]] (REQUEST)
r 0.101592167 _2_ RTR --- 0 AODV 48 [0 3000000 ffff0008 0] ------- [3:255 -1:255 30 0] [0x2 1 1 [12 0] [3 4]] (REQUEST)
r 0.101592167 _4_ RTR --- 0 AODV 48 [0 3000000 ffff0008 0] ------- [3:255 -1:255 30 0] [0x2 1 1 [12 0] [3 4]] (REQUEST)
s 0.101734213 _2_ RTR --- 0 AODV 48 [0 3000000 ffff0008 0] ------- [2:255 -1:255 29 0] [0x2 2 1 [12 0] [3 4]] (REQUEST)
s 0.102473393 _4_ RTR --- 0 AODV 48 [0 3000000 ffff0008 0] ------- [4:255 -1:255 29 0] [0x2 2 1 [12 0] [3 4]] (REQUEST)
s 0.102909213 _2_ MAC --- 0 AODV 48 [0 2000000 ffff0008 0] ------- [2:255 -1:255 29 0] [0x2 2 1 [12 0] [3 4]] (REQUEST)
r 0.103101380 _1_ MAC --- 0 AODV 48 [0 2000000 ffff0008 0] ------- [2:255 -1:255 29 0] [0x2 2 1 [12 0] [3 4]] (REQUEST)
r 0.103101380 _3_ MAC --- 0 AODV 48 [0 2000000 ffff0008 0] ------- [2:255 -1:255 29 0]
```

Figure 2. the attacked trace file

2.2 Routing replay attack extracting feature vector

Because the routing request packet and data packets are affected in routing replay attack, the method of cross-layer statistical feature vectors are using in this paper. The send and receive packet statistics and the packets send and receive situation of related to node communication in VANET MAC layer and routing layer are selected as routing replay attack feature vectors. The characteristics of the specific record as shown in table 2.

Table 2. routing replay attack feature

Feature name	Feature description
RREQ_s	Generate RREQ and sends it out
RREQ_r	Receive RREQ
RREQ_d	Drop RREQ
RREQ_f	Forward RREQ
RREP_s	Generate RREP and sends it out
RREP_r	Receive RREP
RREP_d	Drop RREP
RREP_f	Forward RREP
CBR_s	Generate CBR and sends it out
CBR_r	Receive CBR
CBR_d	Drop CBR
CBR_f	Forward CBR

These feature attributes record represents in a transceiver packet types, VANET communication packets record. Such as:

s 2.000000000 _0_ RTR --- 0 AODV 48 [0 0 0 0] ------- [0:255 -1:255 30 0] [0x2 1 1 [19 0] [0 4]] (REQUEST)
r 2.151811992 _7_ RTR --- 0 AODV 44 [0 6000000 8 0] ------- [19:255 0:255 17 5] [0x4 14 [19 4] 10.000000] (REPLY)
f 2.151811992 _5_ RTR --- 0 AODV 44 [0 6000000 8 0] ------- [19:255 0:255 16 4] [0x4 15 [19 4] 10.000000] (REPLY)

The first record as RREQ_s type, show node 0 at the time of 2.0s, broadcast a RREQ packet of node 0 as source node, node 1 as the destination node.

The second record as RREP_r type, show node 7 at the time of 2.151811992s, receive a RREP packet of node 0 as source node, node 19 as the destination node.

The third record as RREP_f type, show node 5 at the time of 2.151811992s, forword a RREP packet of node 0 as source node, node 19 as the destination node.

Reading data from the detailed records, each node in any period of packet types and transceiver can be counted, then formed the available SVM feature vector. For example, by NS - 2 a piece of data simulation to generate the following records:

r 2.016590249 _3_ RTR --- 0 AODV 48 [0 4000000 ffff0008 0] ------- [4:255 -1:255 26 0] [0x2 5 1 [19 0] [0 4]] (REQUEST)
r 2.016590249 _5_ RTR --- 0 AODV 48 [0 4000000 ffff0008 0] ------- [4:255 -1:255 26 0] [0x2 5 1 [19 0] [0 4]] (REQUEST)
s 2.019681040 _5_ RTR --- 0 AODV 48 [0 4000000 ffff0008 0] ------- [5:255 -1:255 25 0] [0x2 6 1 [19 0] [0 4]] (REQUEST)
s 2.021656040 _5_ MAC --- 0 AODV 48 [0 5000000 ffff0008 0] ------- [5:255 -1:255 25 0] [0x2 6 1 [19 0] [0 4]] (REQUEST)
r 2.021848206 _4_ MAC --- 0 AODV 48 [0 5000000 ffff0008 0] ------- [5:255 -1:255 25 0] [0x2 6 1 [19 0] [0 4]] (REQUEST)
r 2.021848206 _6_ MAC --- 0 AODV 48 [0 5000000 ffff0008 0] ------- [5:255 -1:255 25 0] [0x2 6 1 [19 0] [0 4]] (REQUEST)
r 2.021873206 _4_ RTR --- 0 AODV 48 [0 5000000 ffff0008 0] ------- [5:255 -1:255 25 0] [0x2 6 1 [19 0] [0 4]] (REQUEST)
r 2.021873206 _6_ RTR --- 0 AODV 48 [0 5000000 ffff0008 0] ------- [5:255 -1:255 25 0] [0x2 6 1 [19 0] [0 4]] (REQUEST)
s 2.023514660 _6_ RTR --- 0 AODV 48 [0 5000000 ffff0008 0] ------- [6:255 -1:255 24 0] [0x2 7 1 [19 0] [0 4]] (REQUEST)
s 2.024139660 _6_ MAC --- 0 AODV 48 [0 6000000 ffff0008 0] ------- [6:255 -1:255 24 0] [0x2 7 1 [19 0] [0 4]] (REQUEST)

According to the detailed statistics records, RREQ_r number of node 3 is 1, RREQ_r number of node 4 is 1, RREQ_r number and RREQ_s number of node 5 and node 6 are 1.Note that in the routing request packet number statistics, if only considering the routing layer data communication, the data records are identified as the fourth field in the "MAC" records do not conform to the requirements, identified as "RTR" on behalf of the routing layer data record is the record of this article needs. So, the RREQ_s number of node 5 is 1 instead of 2, by the same token, the RREQ_r number of node 4 and node 6 is 1 rather than 2.

In order to effectively detect the routing replay attack, the "Cross-layer" method is used to choose feature vector. According to different layer, the refinement of feature vector as shown in table 3. the MAC layer of CBR package delivery number, routing layer receives the number of CBR packages, the routing request packet routing layer number of RREQ and RREP response are chosen as feature vector of detecting routing replay attack, taking a unit time interval eigenvector statistics.

Table 3. routing replay attack feature format

feature	CBR_MAC_s	CBR_RTR_r	RREQ	RREP
format	1：n_1	2：n_2	3：n_3	4：n_4

Table 3 represents a period of time of a node is in the light of different packet types. In order to reduce the amount of calculation and improve the efficiency of detection, when choosing the characteristics of the raw data vector, not all feature vectors are counted, only some representative feature vectors are did for SVM training.

After the routing reply attack feature vectors ready, the awk program should be write according to the requirement. Detailed traversal raw data, each node can be got normal state and routing reply attack state eigenvector count.

Using awk programs on normal simulation data and simulation data of routing reply attack feature vector extraction, 2001 samples from positive and negative feature vector are got. As shown in figure 3, it is statistical feature vector data samples every 0.5 s under the routing replay attack from node 15.

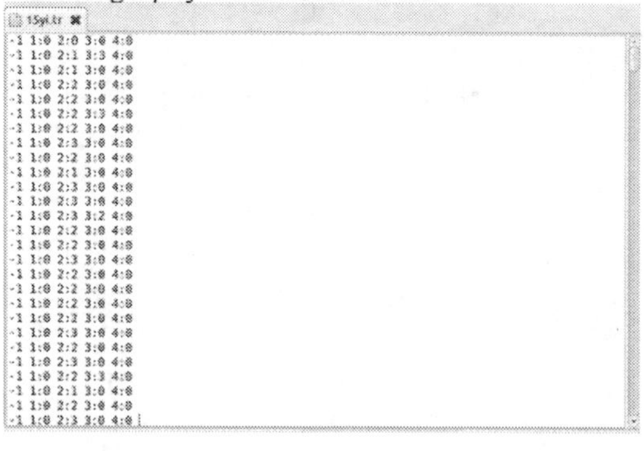

Figure 3. routing replay attack feature vector samples

3 THE SVM TRAINING PARAMETER OPTIMIZATION

Using libSVM software to training samples parameter optimization. First of all, the extracted feature vector samples can be divided into two parts, some are used for parameter optimization and the SVM training, the other part is used as the SVM classifier classification effect of a test. Training parameter optimization process and results shown as figure 4 and figure 5, the final optimal parameters: $C = 0.5, \gamma = 0.0078125, \sigma = 128$, rate of cross validation is 99.95%.

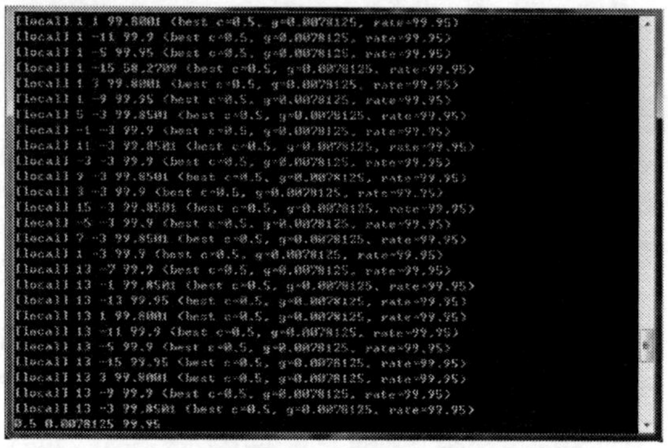

Figure 4. routing replay attack parameter optimization

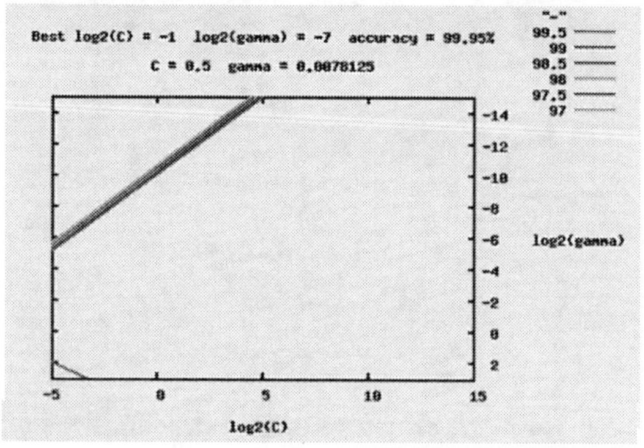

Figure 5. routing replay attack optimization result

4 Intrusion detection classifier training results and analysis

The optimal parameters and the positive and negative feature vector are input to SVM training, the routing replay attack intrusion detection classifier model as shown in figure 6.

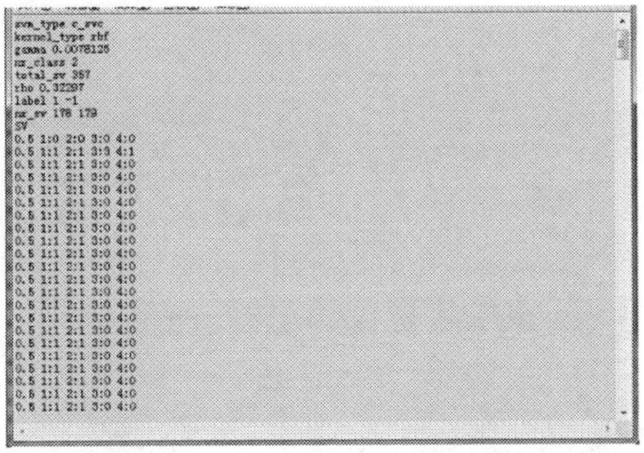

Figure 6. intrusion detection classifier model

Reoccupy has trained SVM classifier model forecast test sample, the prediction results are obtained as shown in figure 7.

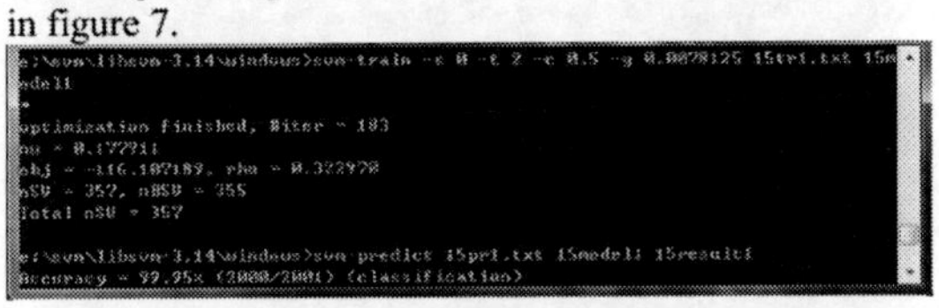

Figure 7. prediction results

From the result, Number of iterations of the intrusion detection classifier training algorithm is 183. The offset item in the discriminant function is 0.322970. The value of objective function is -116.107189. The number of support vector is 357, and the positive number is 178, the negative number is 179. The prediction accuracy reach 99.95%.

The simulation results of routing replay attack intrusion detection show that, on the choice according to the characteristics of the routing replay attack vector, the selected four features for routing replay attack detection is feasible, the establishment of VANET routing replay attack IDS based on SVM can obtain better testing result.

References

1. E.R. Gilles, B. Martine, P. Samuel, Q. Alejandro. Computer Communications, 44: 1-13 , 2014
2. G.J. Yan , O. Stephan, C.W. Michele. Computer Communications, 12: 2883-2897, 2008
3. T.W. Chim, S.M. Yiu. IEEE Transactions on Computers, 2: 510-524, 2014
4. J.A. Martinez, D. Vigueras, F.J. Ros, P.M. Ruiz. Communications and Networks, 2: 122-131, 2013
5. L.Y. Yeh, C.C. Yang, J.G. Chang, Y.L. Taai. Journal of Network and Computer Applications, 1: 284-292, 2013
6. M.M. Adankon, M. Cheriet. The Journal of the Pattern Recognition Society, 12: 3264-3270, 2009
7. G. Ratsch, S. Mika, B. Scholkopf, K.R. Muller. IEEE Transactions on Pattern Analysis and Machine Intelligence, 9: 1184-1199., 2002
8. W. Ju, H.D. Cheng. New Mathematics and Natural Computation, 1: 27, 2013
9. Y.C. Xiao, H.G. Wang, L. Zhang, W.L. Xu. Knowledge-Based Systems, 59: 75-84, 2014
10. R.S. Hamid, E. Mahdieh. Water Resources Management, 7: 2623-2644, 2013
11. A. Hamieh, O.J. Ben, L. Mokdad. Global Telecommunications Conference, 1-5, 2009
12. P.M. Nishant, K.V. Sunil, D.D. Jayesh. International Journal of Engineering Trends and Technology, 5: 2151-2157, 2013
13. D. Thierry, M. Nathalie, I. Sergio, H. Thomas. Mobile Information Systems ,4: 329-359, 2011
14. Y.G. Zhang, W.K. Lee, Y. Huang. Wireless Networks, 5: 545-556, 2003

MATEC Web of Conferences **63**, 05021 (2016) DOI: 10.1051/matecconf/20166305021
MMME 2016

An improved AODV routing protocol based on tower structure

Yong Qiang LI [1,a], Qing Gang FAN [1], Li WANG [2], Zhong WANG [1], Jing CHEN [1]

[1] *Computer Staff Room, Xi'an Research Institute of High Technology, 710025 Xi'an, China*

[2]*Library, Xi'an Research Institute of High Technology, 710025 Xi'an, China*

Abstract. The paper proposed a new routing protocol(IAODV) based on tower structure in the Ad Hoc network for the problem which Location Routing Protocol need hardware and Complex algorithm. By the simulation, The complexity of the new routing protocol is reduced without reducing the performance of the network.

1 INTRODUCTION

There have a category called routing based on geographic location Among many routing protocols, which use the sensor node location information to guide the discovery of routing, maintenance, and data forwarding, realize the directional transmission of information, avoid blind flooding routing probe packets, and reduce control overhead caused by routing protocols; Using the node location information to construct the network topology, easy for network management, realize the global optimization of network, has good network scalability and robustness. GPSR protocol is a kind of typical location-based Routing protocol, it forwards the greed of grouping and Stateless detour around. NADV is an improved routing protocol based on geographic location, by estimating the link cost improvement over a period of time greedy forwarding mode. EPD algorithm by using expectation transmission time ETT and geometric Distance criterion to construct EPD as the next hop node selection criterion.

These protocols are usually on the flooding mechanism of simple improvements and have higher routing efficiency. Its limitations are typically require positioning device and the antenna and other hardware to support, or need to be calculated according to the situation of peripheral node's own location coordinates, and also need to calculate the launch point of view, the complexity of the algorithm is quite high. This paper proposes a tower type network topology structure, the structure can reflect its location information by node address, and put forward an improved method of AODV protocol. Depending on the address of the neighbor node and the distance from itself, The new net nodes determine its own address. In the broadcast, It can do not rely on the location of the location of the directional broadcast information, to avoid the whole network broadcast!

[a] Corresponding author: lyq200381@163.com

2 PROTOCOL DESCRIPTION

2.1 The topological tower structure

According to the communication relationship between nodes, we divide the wireless ad-hoc network topology. The general three kinds of topology are a star network, mesh network and cluster tree nets, as shown in figure 1. The star network structure is simple, convenient organization; The expansion of the mesh network performance is the best, suitable for all kinds of complicated network environment; Cluster Tree network is a special case of the mesh network, support Cluster - Tree routing algorithm, there is no route discovery process, and the nodes do not need to maintain the routing table. Type topology structure, this paper proposes a tower structure of the node addresses can be mapped in the hierarchy, which effectively support the routing protocol based on location information.

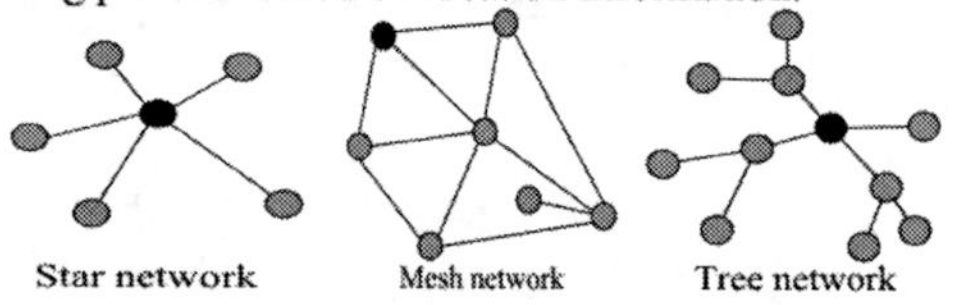

Figure1. three kinds of network topology

Definition 1: tower structure topology. Using specific address allocation algorithm, the wireless ad-hoc network node will be organized into a hierarchy of pagoda form of node address to map out what is the layer information. The characteristics of the structure is that node location and its level are one-to-one relationship, the more to the lower network can hold the more the number of nodes.

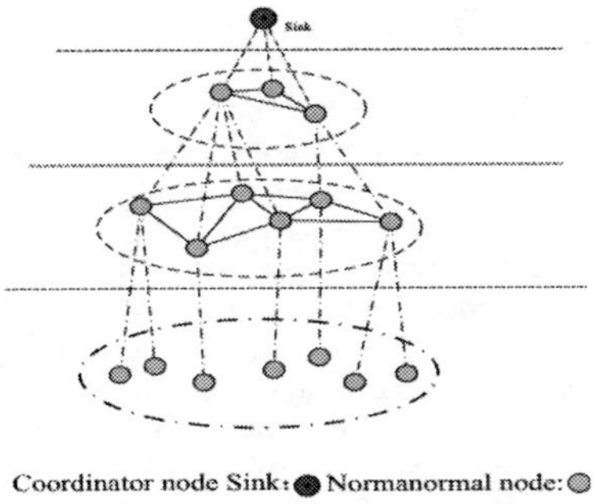

Figure 2. tower structure topology

2.2 Distributed dynamic address assignment algorithm based on four point method

There have a 16-bit network address and 64 - bit IEEE extension in mobile ad-hoc network node , which is 64 - bit IEEE extension address equipment factory set of unique identification number, 16-bit network address assigned by the use of address allocation algorithm.

Distributed dynamic address assignment algorithm based on four points method (hereafter referred to "four points method"), for construct the tower structure topology network , a distributed address assignment algorithm is put forward. Algorithm has low complexity, there is no conflict, address assignment and uniform, the communication overhead is small, time delay is little, realized the address mapping relationship between node position and meet the requirements of increasing number along with the network address radius.

The four method divides the N (N) address ADDr () into the segment address (dN) and the layer node address (N) in two parts, as in table 1. The layer segment address is prefix code, and the node address in the segment is dynamically allocated by the upper layer node. Network internal in addition to the tower base node, and the rest of the nodes have the ability to allocate nodes in the next level node address, the address pool Ac calculated by its own address, the formula 1.

$$Ac=(ADDr(A)*22 ,(ADDr(A)*22+4)) \qquad (1)$$

Tower cross-section shape by the shape of the spire decision method in addition to the network only the coordinator node set another address auxiliary nodes constitute spire layer, the address 0 and 1, respectively, the possession of a 32 address address pool. HM said tower height of Hm<=6, FN said node n the floor height.

Table 1. Division of the address of the four division method

	Segment address (dn)	Address pool (N)
The tower tip	0x0 (coordinator)	0~11111
	0x1(auxiliary ode)	100000~111111
The sixth layer odes	0000000011	$0 \cdot 2^{8}\ 1$
The fifth layer odes	00000011	$0\sim2^{10} - 1$
The fourth layer nodes	000011	$0\sim2^{12} - 1$
The third layer nodes	0011	$0\sim2^{14} - 1$
The second node	11	2^{16}
Tower base	without	without

The basic idea is as follows:

(1)At the beginning, there are only the coordinator node and secondary nodes in a network. There have a fixed address and pool, as see table 1.

(2)A node which is not configured wants to join the network, A request at the top of its adjacent node B for distribution within A free layer node address, the address B in the pool is A free address assigned to A and recorded, A computing their height FA = FB - 1,

ADDr (A) = FA * 22 (FA - 1) + (A).

(3)Nodes can be normal to take off the net. When a node is normal to take off the net, its address was returned to the parent node, and send the information to their child nodes to select the parent node. When the nodes to take off the net, The hole problems of address will arise. To deal with this kind of situation , it can only wait for the node synchronization.

(4)node synchronization. Auxiliary node synchronization service depend on beacon messages by the coordinator node and address radio Beacon information .Message to the coordinator node and address secondary node as the center, the edge of the ripple effect to network began to spread, if a node is as a parent, it should be in every HELLO_INTERVAL (namely the HELLO message cycle) to check whether they have sent a radio message (in a RREQ packet, or forwarding data needle). If inspection failed to send, then the nodes in a Beacon news must be forwarded to the upper nodes. Once the node as the parent node address change, it must immediately notify its nodes, modify child node address needs to be modified immediately. HELLO_INTERVAL cycle is used to control response to the mobile node sensitivity.

According to the algorithm constructs the tower type and distribution of basic ideas of topology network structure diagram as shown in figure 3, in the node location information and address into one-to-one correspondence relationship, each layer holds the number of nodes from the top down approximation into geometric sequence. Address of f1 is 0000000011000001, its address pool Ac is [00000100,00000011] 1, S1 as the fifth layer of segment address (dS1) = 00000011, from formula one assigned to layer inside the node address (S1) for 00000100, S1 node address ADDr (S1) = 0000001100000100.

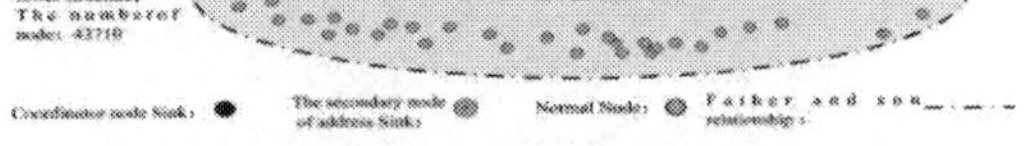

Figure 3. tower robot structure

DOI: 10.1051/matecconf/20166305021

2.3 Improved AODV routing protocol based on the structure of Tower (AODV in Tower)

This paper puts forward the improved AODV routing protocol based on the tower structure, on the basis of tower structure under the node address and level there is a one-to-one relationship between the characteristics of the existing AODV routing protocol RREQ packet forwarding mechanism and routing maintenance query mechanism and link mechanism is improved, the improved protocol can without a peripheral global position service equipment, do not use directional antenna under the condition of implement roughly broadcast-oriented, greatly reduce the consumption of the radio news of network resources.

2.3.1 Forwarding mechanism of RREQ

Protocol fully use the existing AODV routing request message of frame format, the production of its frame , the maintenance of serial number, and the establishment of reverse routing techniques such as completely the same, the RREQ forwarding mechanism was improved, which inherits the advantages of the original AODV routing protocol on the basis of further performance improvement.

The intermediate node Received RREQ packet, according to the destination address of the packet network segment address of a source node and network segment address, judge to the forwarding of a message or discard processing. If receive message is processed directly discarded; Address forwarding node and destination node address provided, the forward; Forwarding node address between the source node network segment level and the destination node address segment level were forwarded, or discarded. See the formula 2.

$$\text{Self node of RREQ packet processing} = \begin{cases} \text{farward} & (dS)<(dSelf)<(dD),\ (dS)>(dSelf)>(dD),\ (dD)=(dSelf) \\ \text{discard} & \text{other} \end{cases} \quad (2)$$

As shown in figure 4, a layer of the source node S request and destination node D build new routing, the RREQ request message will only be 2 parts of the node forwarding, also in three layers is only part of the node forwarding, all nodes forward in the fifth floor. Can be seen from the figure RREQ news is basically a S node center band radio, in the direction of the D nodes effectively implements the network RREQ request message transmission control and reduce the consumption of network resources.

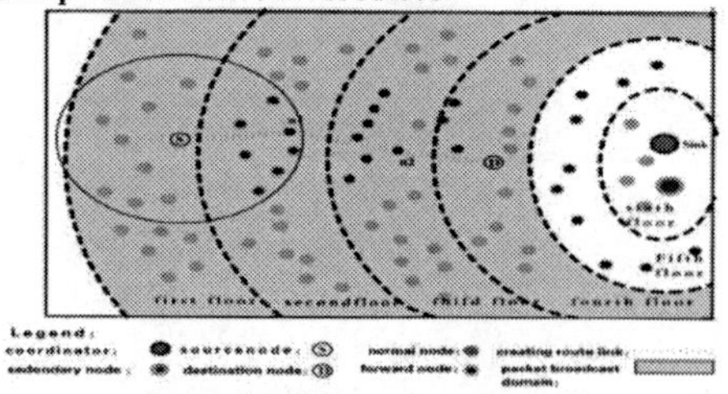

figure 4 RREQ message forward

2.3.2 routing query mechanism

RREQ forwarding mechanism proposed in this last section of route query methods usually can always find an available route, but in some special cases, if can't find available routing, the agreement will forward relaxing constraints, to expand the search scope. After the first route search, modify the RREQ message in the Time domain. Intermediate nodes will be disposed of in accordance with the conditions of the looser the forwarding of a message, by adding a middle node address and the source node layer equals the address forwarding, such as the formula 2.

The RREQ message transmission range will be expanded to all layer between the source node and destination node as shown in figure 5 although the news spread an enlarged than before, but still than simple AODV protocol has better performance.

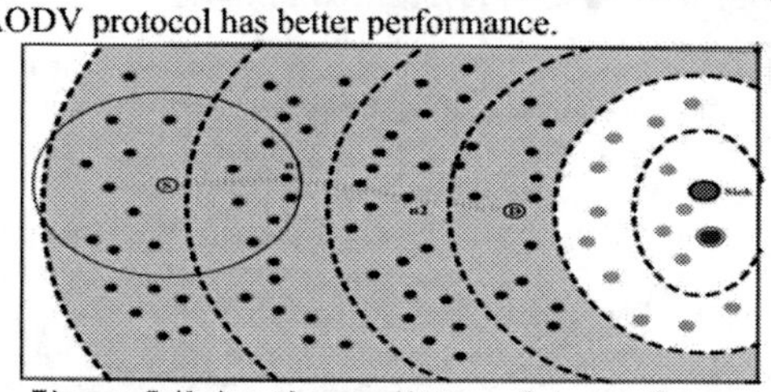

Figure 5 Under the condition of loose forward

3 SIMULATION

The specific parameters of the simulation environment such as the following Table1. The improved AODV(IAODV) protocol's performance is analyzed using NS2 simulator.

Table 1 The results of simulation analysis

application layer	**CBR**
Transport Laye	UDP
Routing Layer	IADOV 、AODV
MAC Layer	802.11DCF
Wireless transmission model	shadowing mode β =2.4
Wireless transmission frequency	2.4GH
Number of nodes	20, 40, 60, 80, 100, 120, 140, 160, 180, 200
Scene size	500m×500m
simulation time	100 s.

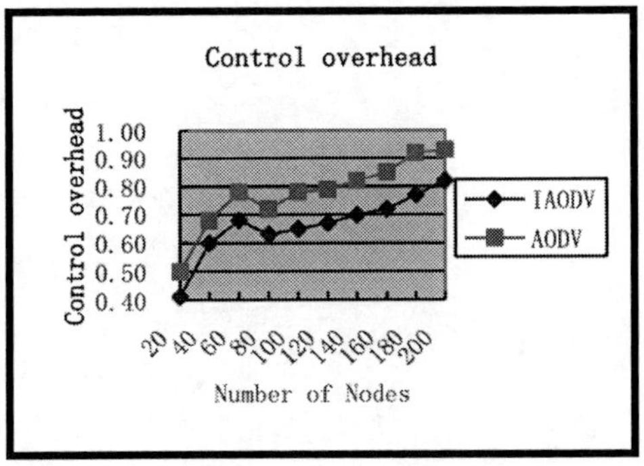

Figure6 The comparison of packet control overhead

As you can see from figure 6, with the increase of the number of nodes, the packet control load of proposed routing protocol is low then the packet control load of AODV routing protocol.

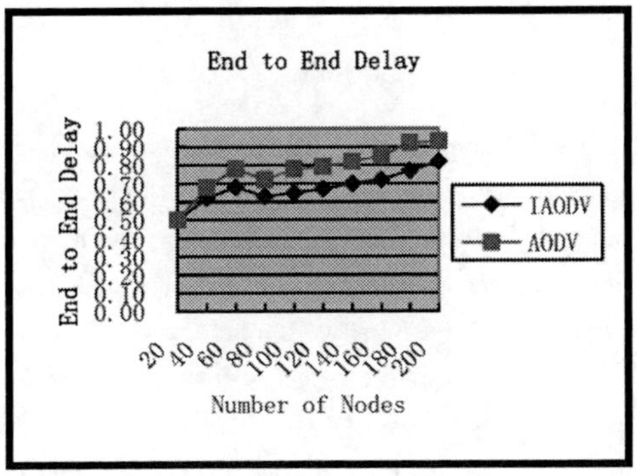

Figure 7 end-to-end delay

Figure7 said IAODV routing protocol of the average end-to-end delay less than AODV, because the cost of control of improved routing protocol and the cost of compute are little.

4 RESULT

Advantages: The network delay and the overhead of control are little, The node does not need to be stored long routing table saves storage space, realize the whole sky broadcast-oriented offline data, improve the channel utilization ratio and reduces the network consumption.

Faults: the route that was fond may not be the shortest routing, the performance is not the most stable routing and time delay is not the smallest routing. Network address is waste, and does not support the concentrated distribution of a large number of nodes and request node distribution is relatively uniform.

References

1. Meschenmoser R ,IP Address Assignment in Mobile Ad Hoc Network [J]. SAE2000 World Congress, USA, 2009.

2. Ghulam Bhatti and G. Yue, A Structured Addressing Scheme for Wireless Multi-Hop Networks. MERL Technology Lab,2006

3. Schmitz R, Leiggener A, Festag A. Analysis of path charac- teristics and transport protocol design in vehicular Ad Hoc networks [J]. Proc of the 63rd IEEE Semiannual Vehicular Technology Conf on VTC-Spring. Melbourne, Australia, 2012.

4.Yih-Chun Hu,(2003), Enabling Secure High-Performance Wireless Ad Hoc Networking, PhD Thesis, Carnegie Mellon University (CMU)

5. Rudack M, Meincke M, Lott M. On the dynamics of Ad Hoc networks for inter vehicle communication (IVC) [J]. Proceedings of the ICWN 2002, Las Vegas, USA, 2002.

6.Ilyas M.,(2003), The Handbook Of Wireless Ad Hoc Network, CRC

7.Sanjay Kumar Dhurandher,Sudip Misra, Mohammad S. Obaidat. EEAODR:An energy-efficient ad hoc on-demand routing protocol for mobile ad hoc networks. INTERNATIONAL JOURNALOF COMMUNICATION SYSTEMS. 2009(22):789–817

MATEC Web of Conferences **63**, 05022 (2016) DOI: 10.1051/matecconf/20166305022
MMME 2016

Research on Routing Protocol of Mobile Ad Hoc Network

Ming Fu TUO, Min CHE

Air Force Engineering University,Xi'An, China

Abstract: Routing protocol is an important content of mobile Ad Hoc. The typical mobile Ad Hoc network routing protocols is introduced. The advantages and disadvantages of these routing protocols are analyzed. A hybrid routing protocol is put forward based on this.

1. Introduction

Mobile Ad Hoc network is usually composed of a set of wireless mobile nodes. It doesn't depend on existing fixed communication network infrastructure. It does not have any central entity. It is a type of self-organization and self-healing network. Nodes in Mobile Ad Hoc network can move quickly and the positions of the network node are rapidly changing. Each node in the network are both routers and hosts. They communicate by sending and receiving data packet[1-3].

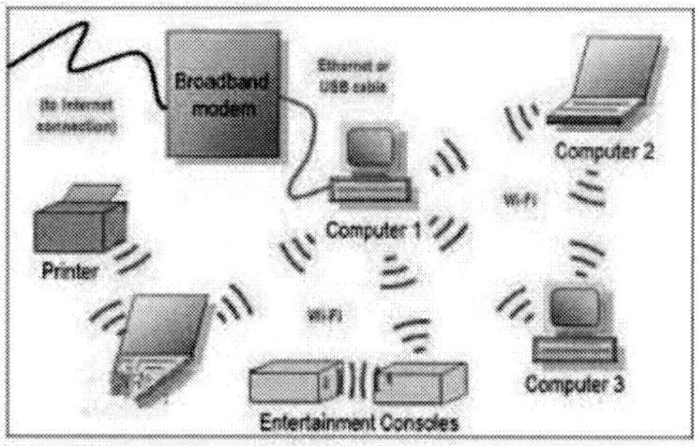

Figure 1. Mobile Ad Hoc network.

2. The characteristics of mobile Ad Hoc

The main characteristics of mobile Ad Hoc are as follows:

(1)The network topology structure changes frequently. There is no fixed in Ad Hoc communications facilities and central management equipment, network node can randomly move in any direction at any speed, combined with wireless transmitters sending power change, the influence of the environment and the factors such as mutual interference between signals, can cause the dynamic change of network topology[4].

(2) Limited resources. Energy for mobile host in Ad Hoc working energy is limited. The network performance will degrade as the energy decreases. On the other hand, there limited bandwidth in mobile Ad Hoc network, and the signals may bring interference. The available transmission bandwidth of mobile host will be far less than the maximum theoretical bandwidth.

(3) Multiple hops communications. Two network nodes will not be in the same network because of the limitation of available resources. This problem can be solved by multiple hops between the source host and the destination host[5].

(4) Low security. Ad Hoc inter-nodes communicates by the wireless channel. The transmission of information is very vulnerable to listen and retransmission, tampered with, forge attack, if a routing protocol once suffered the malicious attacks, the self-organizing networks would be unable to work normally[6].

3. Unicast routing protocol

There are three communication ways in Ad hoc networks, that is, unicast, multicast broadcast.

DOI: 10.1051/matecconf/20166305022

The mobile Ad Hoc network routing protocols can be divided into table driven routing protocol and on-demand routing protocols according to the routing driven mode.

Table driven routing protocol, also called a priori type routing protocols, each node in the network to maintain one or more tables to store other nodes in the routing information, through radio and update information to maintain routing tables and consistency between the change of network topology. Once the source node to send a message, can be gained through the look-up table immediately arrived at the destination node routing. Typical table driven routing protocol includes: DSDV (destination - sequenced short - vector routing), CGSR (cluster head gateway switch r outing) and WRP (wireless routing protocol), etc.

On-demand routing protocols, also called reactive routing protocols, it is a kind of when need to send the data to find the routing routing algorithm. In this kind of routing protocol, the node does not need to maintain the routing information accurately and in time, when the destination node to send it a message, only the source node in the network routing lookup process, find the corresponding routing. Compared with the prior type routing protocols, reactive routing protocol overhead is small, but the datagram transmission delay is larger. Commonly used reactive routing protocol AODV (Adhoc on - demand vector routing), DSR (dynamic source routing), TORA (temporally ordered routing algorithm), etc.

Logic structure according to the network, it can be divided into the flat structure routing protocols and hierarchical structure of two kinds of routing protocols.

In the flat structure routing protocols, each communication nodes are equal. Common agreement include DSDV, WRP, FSR and AODV, and so on. When network scale, the flat structure routing protocol performance decline rapidly.

The more widespread use of hierarchical Routing protocols are mainly CGSR (Cluster head Gateway Switch Routing), ZRP and LEACH agreement. CGSR routing protocol is based on DSDV prior type, in CGSR network model, the network was divided into overlapping groups, each group by group of first selection algorithm to select a group first. Each node holds a group of members of the table and routing table, the former record of each node in the network group of the first and periodic broadcast updates; The latter group for each save a table item, record the next node to the group first. In two or more group of the first communication within the scope of the node as a gateway node, group of the first must be through the gateway to realize.

ZRP protocol use based on ideas, but do not produce group of the first group of use should be first type routing protocols, group of on-demand routing between the arms. LEACH agreement also use group and gateway to mutual information, and use dynamic way to group of the first election

All three of these protocols has itself advantages and disadvantages, CGSR protocol response speed, but expensive. ZRP combines prior type and the advantages of on-demand routing, but slow route setup between group, and establish a route using flood, large energy consumption. LEACH agreement through the rotation group of the first random election, balanced relay communication business, but the node with the same probability in the agreement first competition group, factors such as the lack of energy, and must ensure that all nodes are able to direct communication with a gateway node. Another CGSR and LEACH group information must be used between the group first and gateway transmission, increased the group of the first, and the gateway load at the same time, the routing is not necessarily optimal.

4. Broadcast routing protocol

4.1 Reverse path forwarding broadcast protocol

Reverse path forwarding a bit is both efficiency and simplicity. Reverse path forwarding agreement, node after receiving the broadcast packets to check whether the packet is the latest package had never received. If so, that the package is the path to the source to the shortest path of the node, the node will receive a packet to all nodes except to forward; Otherwise don't do any processing, discards the packet. Agreement on the basis of was the first to arrive packets often corresponds to the shortest path.

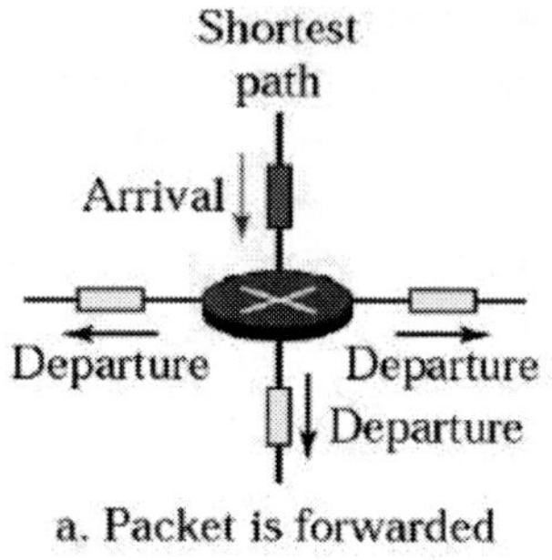

Figure 2 Reverse path forwarding broadcast protocol a

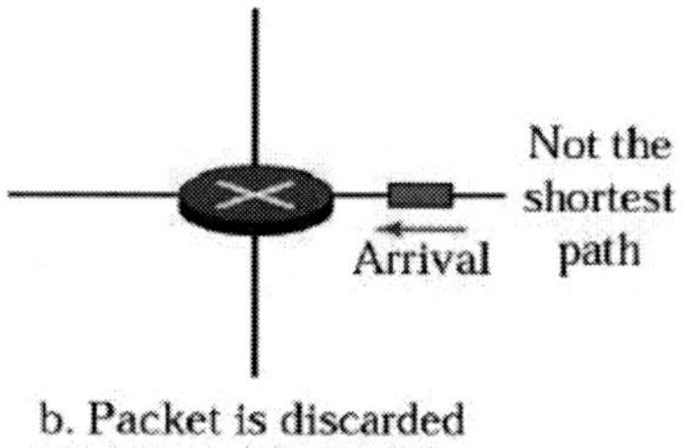

Figure 3 Reverse path forwarding broadcast protocol b

4.2 Support the tree broadcast protocol

Support all node tree agreement save a spanning tree, it is a subset of the subnet, it contains all of the nodes, but does not contain any ring. Every node knows to link its which lines belong to the spanning tree. When a packet arrived at nodes, it will forward it to all other spanning tree except arrival path path, this method can save the bandwidth, and generates a copy of the group is the least. The downside to this method is that each node must save a spanning tree.

5.Conclusions

Mobile Ad Hoc network has self-organizability, multiple hops and characteristics on the dynamic change of network topology, the special of the network make the traditional cable network technology can not be directly used in Ad Hoc network, which requires for Ad Hoc network many new protocols and technologies, such as channel access technology, security technology, routing technology, etc. This article mainly focus on the in Ad Hoc network routing technology.

References

1.Sung-Ju Lee,Matio Geria. Split Multipath Routing with Maximally Disjoint Paths in Ad Hoc Networks. IEEE 2011,10 (6):3201-3205.

2.Mahesh K.Marina, Samir R.Das.On-demand Multipath Distance Vector Routing in Ad Hoc Networks. Network Protocols,2011.Ninth International Conference.2001.

3.Stephen Mueller, Rose P .Tsang, Dipak Ghosal. Multipath Routing in Mobile Ad Hoc N etworks:Issues andC hallenges.2014,

4.Prasant Mohapatra, Jian Li, ChaoG ui." QoSin M obileA dH ocN etworks".2012,

5.Mingfu Tuo,Xingshe Zhou.etc.A method for Cyber-Physical System behavior modeling and safety verification based on extended hybrid system description language,ICEICE 2016.

6.Das S.K.,Mukherjee A , BandyopadhyayS , PaulK ." Improving Quality-of-Service in Ad Hoc Wireless Networks with Adaptive Multi-path Routing".IEEE Global Telecommunications Conference. 2015.

MATEC Web of Conferences **63**, 05023 (2016) DOI: 10.1051/matecconf/20166305023
MMME 2016

Server Program Analysis Based on HTTP Protocol

Min CHE[1] Ming Fu TUO[1]
[1]College of Science, Air Force Engineering University 710051 Xi'an, China

Abstract—The overwhelming majority of Web developments are built based on Web application of HTTP protocol. To analyze the server program based on the HTTP protocol and the process of interaction between the browser and server, a simple HTTP-based server program is taken as an example to describe how to use the theories of multi-threading and asynchronous operation, and the core section of HTTP-based server program is gradually resolved with Winsock programming. This method can achieve the purpose to avoid blocking the calling thread, and improve the responsiveness of a server program.

Keywords- HTTP; HTTP protocol; Server; WinSock

1 Introduction

Hypertext Transfer Protocol (HTTP) is the protocol of web's application layer. It is the web's core issue. HTTP can be realized in the web client's program or server's program. The programs running on different end systems of client and server can exchange with each other through exchanging HTTP message. Web Page is also called document composed of various objects. Object is a document which is only single URL addressable. It might be an HTML file, or JPG image, or GIF image, or JAVA applets, or voice clip, etc. Usually, most web pages consist of one basic HIML file and a number of the referenced objects. HTTP protocol defines how to request a web page from the client (web browser) and how to transfer the web pages to client from web server.

2 HTTP PROTOCOL

HTTP is a protocol of application layer basing on stateless, request and response model. Usually it is based on TCP connections. The vast majority of web developers are the web applications on the basis of HTTP protocol. A lightweight web server is accomplished via the program in this study [1].

2.1. HTTP request consists of three parts namely consumer request, line interest header, and request body.

Request line begins with a method symbol separated by a space, then followed by the Request-URI and the version of HTTP. The form is defined as follows: Method Request-URI HTTP-Version CRLF. Method represents request method; Request-URI is a uniform resource identifier; HTTP-Version indicates the request version of HTTP protocol; CRLF represents carriage return and line feed. Except the CRLF at the end, CRLF cannot be separated into CR or LF characters in other part [2].

2.2. HTTP response consists of three components namely the status line,message header,and response body .

The format of status line is: HTTP-Version Status-Code Reason-Phrase CRLF. HTTP-Version indicates the version of HTTP protocol used in web server; Status-Code indicates the status code of a response sent back from web server; Reason-Phrase indicates the text description of status code. The status code is made up of three digits. The first digit defines the categories and there are five possible values:

1xx: Indication Information It indicates that the request has been received and processing was continued;

2xx: Success It indicates that the request has been successfully received, understood, and accepted;

3xx: Redirection It indicates that further operating is needed to complete the request;

4xx: Client Error It indicates that the request has a syntax error or cannot be achieved;

5xx: Server Error It indicates that the server failed to achieve a legitimate request

Common status code, status description, explanation:[3]

200 OK //successful client requests

400 Bad Request//Syntax error in the client request. It cannot be understood by the server

HTTP message is composed of requests from the client to the server and responses from server to client. Request and response messages consist of start line (For request message, the start line is the request line; For response message, the start line is the status line), message header (optional), blank lines (only CRLF line), and message body (optional).

DOI: 10.1051/matecconf/20166305023

3 HTTPSVR PROGRAM ANALYSIS

HTTPSVR is a simple web server with a graphical interface, you can specify the port and the main directory of web page.You can generate web accessing log. It is constituted by an endless loop: Receive the request from client, parse and process the request according to the HTTP protocol, then sent the response back to the client.

3.1. HTTPSVR main work flow chart

HTTPSVR is a powerful software with rich graphical interface. The main workflow is shown in Figure 1.

Firstly ClistenSocket: OnAccept (int nError-Code) function generates CRequestSocket class. When monitoring the connection requests, Accept function creates a new socket pRequest and returns to the handle of sentence. AsyncSelect function monitors two events of FD_READ and FD_ CLOSE in port 80. When an incoming event is FD_READ, the receiver is prepared and OnReceive ( ) function is triggered. When an incoming event is FD_WRIT, OnSend( )function will be triggered and the data is sent.

Void Crequest Socket::On Receive (int nError Code) function of ReqSock.cpp will be executed when data arrive. The code int nBytes = Receive (m_buf, GetData (), m_buf, GetSize()) store the data packets uploaded from tcp layer in the m_buf buffers of request and response.

Next, use the switch statement to process three kinds of request status (REQ_REQUEST, REQ_HEADER, and REQ_BODY). When requested state of m_reqStatus == REQ_DONE, call Star-tResponse ( ) to began constructing the response message.

When the response message constructing with Start Response ( ) is finished, Async Select (FD_WRITE | FD_CLOSE) is used. The void Crequest Socket :: OnSend (int nError Code) function is called to send the data in buffer m_buf [4].

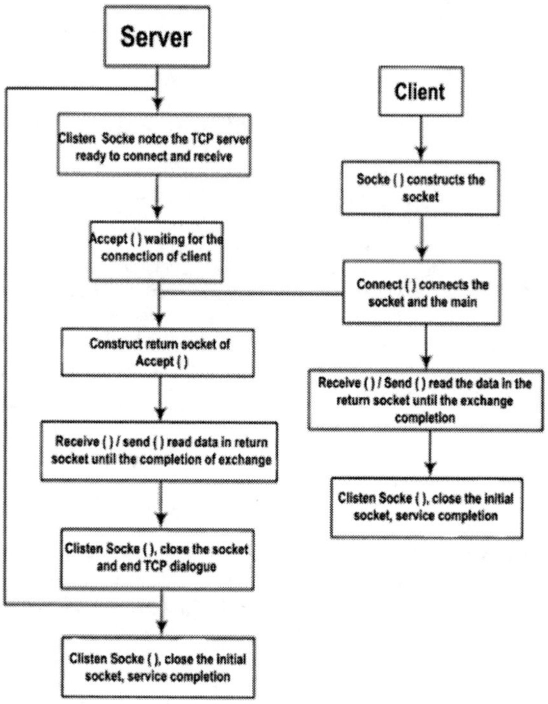

Figure 1. The main workflow of HTTPSVR

3.2. Asynchronous non-blocking CAsyncSocke

Create () function of CasyncSocket In addition to creating a Socket, WSAAsyncSelect ( ) is used to associate this socket with the window object so that the object can handle the events (message) from the socket. However, after CSocketWnd receives the event of Socket, it just calls back the virtual functions such as Casync Socket:: On Receive () Casync Socket :: On Send (), CasyncSocket :: OnAccept (), Casync Socket :: On Connect (), etc. So CAsyncSocket derived class only needs to add code or send code in these virtual functions. When using CasyncSocket, all networks IO are asynchronous operation if you use Create to create socket by default. The network data transfer needs the following functions: OnAccept, OnClose, OnConnect, OnOut-OfBandData, OnReceive, OnSen [5].

3.3. HTTPSVR KEY CATEGORY ANALYSIS

Void CListenSocket:: OnAccept (int nErrorCode) Create a new CRequestSocket and monitor port 80. Return to a socket when requests arrive. Through AsyncSelect (FD_READ | FD_CLOSE), the void CRequestSocket :: OnReceive (int nErrorCode) function is called to receive data.

Void CrequestSocket :: OnReceive (int nError-Code) All major features of httpsvr are achieved including accepting request from web browser, analyzing the request packet according to http protocol, constructing the response message, and sending the processed data packets back to the browser. Receive (m_buf, GetData ( ), m_buf, GetSize ( )) function receives the request packet from browser. Switch statement processes it depending on the state. When the browse's request status m_reqStatus is REQ_REQUEST, process with a while loop. Process Line ( ) function parses the each line of received packets and then initialize m_pRequest class. After the swich statement is finished, StartResponse ( ) is called according to the parsing results of m_pRequest to construct the response packet. Finally, the function AsyncSelect (FD_WRITE | FD_CLOSE) recalls the function OnSend to sent.

void CRequestSocket :: OnSend (int nErrorCode) call int nBytes = Send (m_buf. GetData ( ), m_cbOut) function to send the data in the cache and process the possible errors by closing or re-transmitting.

void CRequestSocket :: ProcessLine (void) conducts processing according to different states of m_reqStatus to complete the initialization of CRequest * m_pRequest.

BOOL CRequestSocket :: StartResponse (void) [6] This function is used to construct the response message. It saves the constructed response packet in the cach of m_buf and sends m_buf using void CRequestSocket :: OnSend (int nErrorCode). It can construct different response packet according to different Method (GET, HEAD, POST). For example, when the Method is GET, it indicates that the browser needs to browse the web page. The FindTarget (strFile) is started to search in the default directory. If it exists, the packet is open and the http protocol requiring head is constructed using StuffHeading ( ). Then using StartTargetStuff, the contents of opened web page is put into m_buf and waiting to be sent. After the construction is completed, return to BOOL CrequestSocket::StartRes ponse (void) and then call AsyncSelect (FD_WRITE | FD_CLOSE) function. This function can call function of void CrequestSocket :: OnSend (int nErrorCode) and send the data packs in m_buf catch.

BOOL CRequestSocket :: FindTarget (CString & strFile) Because the requests sent from browser are usually the network address format, so the function changes the network address format files into windows path format system. For example, GET/default.html is converted c:\ WebPages \ default. Html (httpsvr default root directory c\WebPages) after processed by this function.

void CRequestSocket :: StartTargetStuff (void). This function reads the content of a web page required by client and transfer in into m_buf for further sending.

UINT CGIThread (LPVOID pvParam) [7] The function can run in the thread model, call the corresponding GCI process, and output the results to a temporary file. After the thread model process is completed, the temporary files will be sent to the browser by BOOL CRequestSocket :: StartResponse (void).

4 HTTPSRV PROGRAM EEXCUTION PROCESS

4.1. Run Function

Program is started to be executed by the _tWinMain function of APPMODULE.CPP file in MFC. The function of return AfxWinMain is executed.

4.2. Link Server

A worker thread pThread is created by the int AFXAPI AfxWinMain function in WINMAIN.CPP file and initialize htt psvr. The BOOL CHttpSvrApp :: InitInstance method in HttpSvr.cpp file is called to initialize the HTTP server and then run the main function of thread. Finally, int CWinThread :: Run function is run in THRD-CORE.CPP file. The loop of server begins to run.

Within loops, the void CLis-tenSocket OnAccept method in Listen.cpp is called to generate CRequestSocket, which is set for monitoring port 8080. The port is set as 8080, the address of web service file is directed to the address of root. Now, the result is displayed as shown in Figure 2.

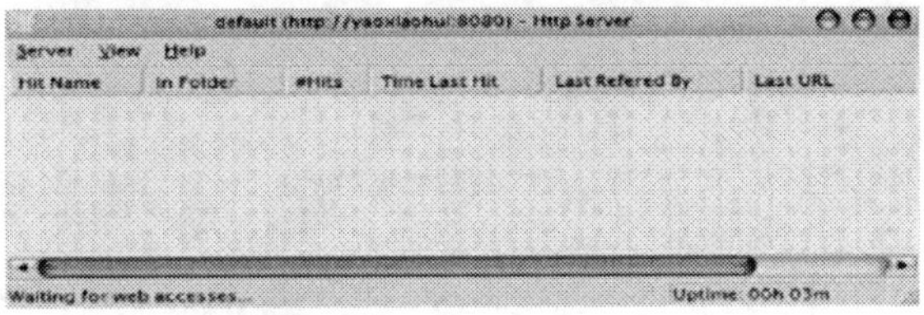

Figure 2. Link the server.

4.3. Requesting Client

In ReqSock.cpp file, different response-processing is conducted for the received data packets according to the difference of response status m_reqStatused. When the browser sends the first data packet, the response status m_reqStatus is set to the REQ_REQUEST. Then, each row of REQ_QUEST packet is processed. The m_pRequest is initialized using ProcessLine method according to http protocol. After the above operation is completed, the StartResponse method is called to construct the response message. The function of AsyncSelect is run. Thereafter, void CRequestSocket :: OnSend method is called to send the response message in m_buf cache to the client.

When you click submit, the return page is shown.

5 CONCLUSION & DISCUSSION

The primary function of a web server is to store, process and deliver web pages to clients. The communication between client and server takes place using HTTP. Pages delivered are most frequently HTML documents, which may include images, style sheets and scripts in addition to text content.

A user agent, commonly a web browser or web crawler, initiates communication by making a request for a specific resource using HTTP and the server responds with the content of that resource or an error message if unable to do so. The resource is typically a real file on the server's secondary storage, but this is not necessarily the case and depends on how the web server is implemented.

While the primary function is to serve content, a full implementation of HTTP also includes ways of receiving content from clients. This feature is used for submitting web forms, including uploading of files.

Many generic web servers also support server-side scripting using Active Server Pages (ASP), PHP, or other scripting languages. This means that the behavior of the web server can be scripted in separate files, while the actual server software remains unchanged. Usually, this function is used to create HTML documents dynamically ("on-the-fly") as opposed to returning static documents. The former is primarily used for retrieving and/or modifying information from databases. The latter is typically much faster and more easily cached but cannot deliver dynamic content.

Web servers are not always used for serving the World Wide Web. They can also be found embedded in devices such as printers, routers, webcams and serving only a local network. The web server may then be used as a part of a system for monitoring and/or administering the device in question. This usually means that no additional software has to be installed on the client computer, since only a web browser is required (which now is included with most operating systems).

In 1989 Tim Berners-Lee proposed a new project to his employer CERN, with the goal of easing the exchange of information between scientists by using a hypertext system. The project resulted in Berners-Lee writing two programs in 1990:

- A browser called World Wide Web.
- The world's first web server, later known as CERN httpd, which ran on NeXTSTEP

Between 1991 and 1994, the simplicity and effectiveness of early technologies used to surf and exchange data through the World Wide Web helped to port them to many different operating systems and spread their use among scientific organizations and universities, and then to industry.

In 1994 Tim Berners-Lee decided to constitute the World Wide Web Consortium (W3C) to regulate the further development of the many technologies involved (HTTP, HTML, etc.) through a standardization process.

In current study, by analyzing the HTTP server program, the authors will be familiar with HTTP service workflow and further understand the Socket, and then can create asynchronous Socket methods. In this article, source code of HTTP server with graphical interface is analyzed and debugged. The main functions of program, process running, the various kinds of function are introduced. Of course, this program is just a lightweight server whose function is relatively simple. However, after familiar with the basic principles of server creation basing on HTTP protocol, the authors will construct server with rich functions basing on this.

The HTTP is an application protocol for distributed, collaborative, hypermedia information systems [8]. HTTP is the foundation of data communication for the World Wide Web.

HTTP functions as a request-response protocol in the client-server computing model. A web browser, for example, may be the client and an application running on a computer hosting a web site may be the server. The client submits an HTTP request message to the server. The server, which provides resources such as HTML files and other content,. The response contains completion status information about the request and may also contain requested content in its message body.

A web browser is an example of a user agent (UA). Other types of user agent include the indexing software used by search providers (web crawlers), voice browsers, mobile apps, and other software that accesses, consumes, or displays web content.

HTTP is designed to permit intermediate network elements to improve or enable communications between clients and servers. High-traffic websites often benefit from web cache servers that deliver content on behalf of upstream servers to improve response time. Web browsers cache previously accessed web resources and reuses them when possible to reduce network traffic. HTTP proxy servers at private network boundaries can facilitate communication for clients without a globally routable address, by relaying messages with external servers.

HTTP uses Internet Media Types (formerly referred to as MIME Content-Types) to provide open and extensible data typing and type negotiation. For mail applications, where there is no type negotiation between sender and receiver, it's reasonable to put strict limits on the set of allowed media types. With HTTP, where the sender and recipient can communicate directly, applications are allowed more freedom in the use of non-registered types.

When the client sends a transaction to the server, headers are attached that conform to standard Internet e-mail specications (RFC 822). Most client requests expect an answer either in plain text or HTML. When the HTTP Server transmits information back to the client, it includes a MIME-like (Multipart Internet Mail Extension) header to inform the client what kind of data follows the header. Translation then depends on the client possessing the appropriate utility (image viewer, movie player, etc.) corresponding to that data type.

HTTP is an application layer protocol designed within the framework of the Internet Protocol Suite. Its definition presumes an underlying and reliable transport layer protocol, and Transmission Control Protocol (TCP) is commonly used. However HTTP can use unreliable protocols such as the User Datagram Protocol (UDP), for example in Simple Service Discovery Protocol (SSDP).

HTTP resources are identified and located on the network by Uniform Resource Identifiers (URIs)—or, more specifically, Uniform Resource Locators (URLs)—using the http or https URI schemes. URIs and hyperlinks in Hypertext Markup Language (HTML) documents form webs of inter-linked hypertext documents.

HTTP/1.1 is a revision of the original HTTP (HTTP/1.0). In HTTP/1.0 a separate connection to the same server is made for every resource request. HTTP/1.1 can reuse a connection multiple times to download images, scripts, style sheets, etc after the page has been delivered. HTTP/1.1 communications therefore experience less latency as the establishment of TCP connections presents considerable overhead.

The term Hyper Text was coined by Ted Nelson in 1965 in the Xanadu Project, which was in turn inspired by Vannevar Bush's vision (1930's) of the microfilm-based information retrieval and management "memex" system described in his essay As We May Think (1945). Tim Berners-Lee and his team are credited with inventing the original HTTP along with HTML and the associated technology for a web server and a text-based web browser. Berners-Lee first proposed the "World Wide Web" project in 1989 — now known as the World Wide Web. The first version of the protocol had only one method, namely GET, which would request a page from a server. The response from the server was always an HTML page [9].

The first documented version of HTTP was HTTP V0.9 (1991). Dave Raggett led the HTTP Working Group (HTTP WG) in 1995 and wanted to expand the protocol with extended operations, extended negotiation, richer meta-information, tied with a security protocol which became more efficient by adding additional methods and header fields. RFC 1945 officially introduced and recognized HTTP V1.0 in 1996 [10].

The HTTP WG planned to publish new standards in December 1995 and the support for pre-standard HTTP/1.1 based on the then developing RFC 2068 (called HTTP-NG) was rapidly adopted by the major browser developers in early 1996. By March 1996, pre-standard HTTP/1.1 was supported in Arena, Netscape 2.0, Netscape Navigator Gold 2.01, Mosaic 2.7, and in Internet Explorer 2.0. End-user adoption of the new browsers was rapid. In March 1996, one web hosting company reported that over 40% of browsers in use on the Internet were HTTP 1.1 compliant. That same web hosting company reported that by June 1996, 65% of all browsers accessing their servers were HTTP/1.1 compliant [14]. The HTTP/1.1 standard as defined in RFC 2068 was officially released in January 1997. Improvements and updates to the HTTP/1.1 standard were released under RFC 2616 in June 1999.

In 2007, the HTTPbis Working Group was formed, in part, to revise and clarify the HTTP/1.1 spec. In June 2014, the WG released an updated six-part specification obsolescing RFC 2616.

An HTTP session is a sequence of network request-response transactions. An HTTP client initiates a request by establishing a Transmission Control Protocol (TCP) connection to a particular port on a server (typically port 80). An HTTP server listening on that port waits for a client's request message. Upon receiving the request, the server sends back a status line, such as "HTTP/1.1 200 OK", and a message of its own. The body of this message is typically the requested resource, although an error message or other information may also be returned .

HTTP defines methods to indicate the desired action to be performed on the identified resource. What this resource represents, whether pre-existing data or data that is generated dynamically, depends on the implementation of the server. Often, the resource corresponds to a file or the output of an executable residing on the server. The HTTP/1.0 specification defined the GET, POST and HEADS methods and the HTTP/1.1 specification and added 5 new methods: OPTIONS, PUT, DELETE, TRACE and CONNECT. By being specified in these documents their semantics are well known and can be depended upon. Any client can use any method and the server can be configured to support any combination of methods. If a method is unknown to an intermediate it will be treated as an unsafe and non-idempotent method. There is no limit to the number of methods that can be defined and this allows for future methods to be specified without breaking existing infrastructure. For example, WebDAV defined 7 new methods and RFC 5789 specified the PATCH method .

In conclusion, this study reviews the concepts, history, development, and future direction of web server and HTTP. The server program based on HTTP protocol and the process of interaction between the browser and server were analyzed to show how to use the theories of multi-threading and asynchronous operation. The core section of HTTP-based server program is gradually resolved with Winsock programming. This method can achieve the purpose to avoid blocking the calling thread and improve the responsiveness of a server program.

References

1. Ye Q. "Hypertext Transfer Protocol HTTP1.0". Information Development & Economy,2004; (8):66-68
2. Xiao G. "HTTP protocol". Technical Analysis Jiangxi Communication Technology, 2001; (1):39-41.
3. Xu J., Wang T. "Analysis of HTTP protocol 1.1". Journal of Southwestern Normal University (Science edition), 2004; (2):47-48.
4. Tanenbaum AS. "Internet". Tsinghua University Press, Beijing, 2004, pp. 150–153.
5. Li L, Li C. "The computer netwotk technique". National Defense Industrial Press, Beijing, 2004, pp. 250–266.

6. Wang Y, Zhang Y. "Windows network and communication program design". People's Posts and Telecommunications Press. Beijing, 2006, pp. 340–345.
7. Zhang H, Liu Y. "Research on HTTP based on socket". Journal of Qiqihar University, 2004; (2):70-72.
8. Roy T, James G, Jeffrey CM, Henrik FN, Larry M, Paul JL, Lee B. Hypertext Transfer Protocol -- HTTP/1.1. IETF. RFC 2616, 1999 Jun.
9. Berners-Lee T. "The Original HTTP as defined in 1991". World Wide Web Consortium. 2010 July.
10. Raggett D, Berners-Lee T. "Hypertext Transfer Protocol Working Group". World Wide Web Consortium. 2010 September.
11. Simon S. "Progress on HTTP-NG". World Wide Web Consortium. 2010 June.

MATEC Web of Conferences 63, 05024 (2016) DOI: 10.1051/matecconf/20166305024
MMME 2016

Based on the "Internet plus" era of enterprise management innovation path exploration

Jian LI

Qilu Normal University, China Shangdong 250013

Abstract. The world has entered the information age, all kinds of information technologies such as cloud technology, big data technology are in rapid development, and the "Internet plus" appeared. The main purpose of Internet plus "is to provide an opportunity for the further development of the enterprise, the enterprise technology, business and other aspects of factors combine. For enterprises, grasp the "Internet plus" the impact of the market economy will undoubtedly pave the way for the future development of enterprises. This paper will be on the innovation path of the enterprise management Internet plus "era tied you study, hope to be able to put forward some opinions and suggestions.

Keyword. Internet plus;Business management;Innovation exploration

Introduction

"Internet" was made by the chairman of the board of directors of the Tencent and CEO Ma Huateng. In the eyes of many people, "Internet plus" may have a certain degree of current business impact, on the contrary, the promotion of traditional industrial production technology and management mode of the continuous upgrading from some aspects. The so-called "Internet plus" actually refers to a new form of market economy, it will be the Internet and social economy to create with reasonable configuration and integration of [1] production factors and resources of the enterprise to take advantage of the internet. This will greatly improve the cohesion of enterprise, but also in the Internet and all kinds of advanced research and production technology integration, the formation of a broader "Internet plus" form of economic development. At present, the application of Internet plus "has been more and more widely, so enterprises should take advantage of this opportunity, the management mode of the corresponding change, to adapt to the future there will be a wave of industry changes.

1 "Internet plus" for enterprises

"The birth of Internet plus" can be said to cause a tremendous impact on the enterprise, because it makes the operation mode and management mode of enterprises have changed greatly. First of all, in the traditional business activities, although the service object should occupy a high position, but due to various factors, it is difficult to achieve. Customer service evaluation can not play its due role, it is difficult for customers to communicate fully, which makes the enterprise services and customer needs are out of touch. "Internet plus" is a better solution to this problem. Such as the catering and entertainment industry have many in the use of "beauty", the use of every customer will use mobile phone as a medium, collecting customer feedback on the service enterprise, also between the customer can also communicate with each other to give service unit a fair evaluation. So that enterprises can not only products as the main competitive point, if not change can not be in the current fierce competition in the market to occupy a place. Second, because the Internet has a strong openness, business management framework can also produce a certain change. The traditional mode of enterprise management is to take from the top to the bottom of the pyramid management framework. In this way, the decision-making right of enterprises into the hands of a few people, some grass-roots workers unable to put forward valuable suggestions by combining their own experience and the development of the enterprise. The application of "Internet plus" will change the status, the customer service personnel from the nearest base for the enterprise development planning, bring the whole enterprise reform.

2 "The opportunities and challenges Internet plus" for enterprise development

"Internet plus" can be said to bring great convenience for the development of various enterprises at present, enterprises can rely on the Internet for their potential for development, expand the scope of business, improve brand competitiveness. But this also means that enterprises must be based on the relevant characteristics of the Internet to reform management mode, if you can not follow the direction of market economic development, it is bound to be left behind by other companies.

2.1 "Internet plus" has brought opportunities for the development of enterprises

The emergence of the Internet makes consumption more diversified, people at home can browse to the hundreds of millions of different kinds of commodity information, can also according to the evaluation of other consumers of a certain brand has a more intuitive understanding. Not only that, the use of the Internet can reduce the cost of physical stores, greatly reducing the cost of work. With the continuous development of e-commerce, the market operation mode Internet plus "increasingly mature, this also leads to the traditional market share continued to decline, all kinds of products, business model, business channels and marketing methods will produce some changes. The current world is in the information age, people for computers, electronic equipment, communications and other technology needs more . Although the "Internet plus" is the basis of market business model change, but to some extent, change the model of market economy indirectly promoted the upgrading of products of all kinds of information and use. It can be said that the use of electronic information technology and business marketing model has a good interaction between the transformation. Although the current our country market economy mode is not fundamentally change, but we can still rely on it in promoting the development of various traditional enterprises, network and economy closely integrated to form a new industrial chain. Still in the current food and beverage industry, for example, different customers demand for services is not the same, but in the past it is difficult to get effective information before the official consumer spending. But now it is different, customers can make full use of the network has been a large number of customers have been consumed by the evaluation to understand the characteristics of its services, and further determine whether it can be accepted. If you are not satisfied with the service, but also through the Internet for complaints and evaluation of enterprises play a good role in promoting.

2.2 "Internet plus" for enterprise development challenges

First of all, the current enterprise management concept is not completely. Our country's enterprise management work is in the reform and opening up after the official. Just a few decades of time is not enough to make our country's enterprise management has the perfect idea, so that our country's management theory is only from abroad for reference. But "Internet plus" requirement is the enterprise to customer service as the business center of the enterprise, the traditional business management theory "to the interests of the enterprise as the center" on the contrary. If you can not solve this problem, it is difficult to form a match with the "Internet plus" management style, efforts will be cast to the wind.

Secondly, the current enterprise management structure and management mode is more obsolete. In order to be able to Surety Company centralization, most of the management structure of the enterprise is Pyramid, the management of the enterprise only by a few people at the top of Pyramid. But the real exposure to corporate customers, to understand the needs of customers are often at the grassroots level of Pyramid staff, which may lead to the management and business development needs of the actual need to be separated from each other. It can be said that the traditional enterprise management structure is difficult to deal with the current market environment, the information transmission speed is too slow will become the biggest drawbacks of enterprise development. Not only that, the enterprise of our country because of the influence of the planned economy is difficult to carry on the thorough reform and the innovation, thus receives the certain degree the restraint. For example, in the past, companies only need to send advertising can be better to complete the work of enterprise product marketing. But now is twenty-first Century, the simple use of advertising marketing is far from enough. If the customer can not get enough information from the marketing, they will not choose to consume. Therefore, most of the current enterprise management structure and management model is not able to meet the market development needs of the information age.

Finally, the enterprise talent management work is not attached importance to. For enterprise management, how to recruit high-quality talent and retain high-quality personnel is the most important. In the past business management, the more excellent people tend to be more rigorous treatment, which is the current "people-oriented" management approach. Even in a short period of time to allow employees to create wealth, but also lose the loyalty of employees. Therefore, to take appropriate measures for human resources management is particularly important, only pay attention to the communication between management and employees, to establish a bridge of mutual attention, in order to improve the working efficiency of the staff.

3 "Enterprise management innovation path Internet plus" era

Through the enterprise management of the "Internet plus" era we can see, to carry on the innovation of enterprise management mainly need to proceed through the following aspects.

3.1 Innovative enterprise management concept

Enterprise management concept is the basis of enterprise management, it can be said that all the work of the enterprise management will be around the idea of the management of enterprises to carry out. In the "Internet plus" era, customers for the enterprise means more and more important. Whether the enterprise can have a strong and comprehensive competitiveness in the current market environment depended on not only in the traditional enterprise capital strength, power and other technical factors, customers to create the effect of relatively more important. This will require the enterprises "to their own interests as center" management philosophy to "in order to meet the actual needs of customers" management philosophy, and in work pay attention to the connection between the client and the, construction of active cooperation, openness and win-win cooperation environment. This can allow enterprises to develop the dominant direction of the customer, to achieve the effective development of the market environment.

3.2 The innovation of management structure and management mode

Restrict the further development of enterprise management structure greatly on the shape of the Pyramid enterprises, the news spread slowly makes it difficult to respond to market changes. However, in the "Internet plus" era, which is one of the most fatal flaw. Therefore, enterprises must change the management structure, reduce the management of no great importance, so that between the grass-roots workers and decision-making and communication more efficient. This can be said to be a Pyramid shaped flat management structure, not only can accelerate the speed of decision-making, but also can eliminate the influence of working pressure on the enthusiasm of the staff of the class. In addition, enterprises also need to innovate management model, the Internet technology is applied to them, the use of new management tools to carry out the production, marketing and transportation and other aspects of the enterprise. This not only can improve the level of business, to speed up the pace of customer demand response, but also to improve customer satisfaction, have a stronger market competitiveness.

3.3 Innovative talent management mechanism

According to the previous analysis we can see, talent for enterprise development has a crucial role, but past enterprise talent management mechanism will only compress to a great extent on the staff of the enthusiasm, greatly affected the efficiency of the work of the various departments of the enterprise. Such a talent management mechanism is undoubtedly a failure, because it can not stimulate the enterprise employees to devote to the work of the work, to create more wealth for the enterprise value. And this management mechanism was too strong, so that employees tend to have psychological conflict. Therefore, the management mode of enterprise development to flexibility is necessary. The so-called flexible management is to take the "people-oriented" concept of talent management, by all the production activities of enterprises by the staff to complete all must, therefore the staff should be the foundation of enterprise development. Therefore, the enterprise in the management of staff need to consider to staff the dignity and get sufficient attention, only in this way to let the staff have relatively strong sense of belonging, to enhance the cohesion of all the staff inside the enterprise.

4 Epilogue

With the application of information technology in people's lives more and more widely, the Internet for the impact of the major enterprises in the market is also growing. For the current enterprises, "Internet plus" era can be said that if the enterprise can There are both advantages and disadvantages., innovation of enterprise management under the current market situation, so they can allow enterprises to further development; but if the enterprise still complacent, followed the old management mode and management concept, a large wave that will be difficult to survive the market changes.

References

Feng Yanwei [1]. Green logistics problems and countermeasures in the development of research in China [J]. Journal of commercial economy, 2013, (9) : 242-245.

[2] weeks. Green of green logistics management, the role of environmentalmanagement system in modern green logistics analysis [J]. Journal of transportation enterprise management, 2013, 10 (3) : 105-108

[3] gao jie analysis of modern green logistics management strategy [J]. China's foreign trade business, 2013, 12 (8) : 156-159

MATEC Web of Conferences **63**, 05025 (2016) DOI: 10.1051/matecconf/20166305025
MMME 2016

Computer teaching process optimization strategy analysis of thinking ability

Liang LUO
ChongQing City Management College，China Chongqing401331

Abstract. As is known to all, computer is a college student in a university course, one of the basic course in the process of education for college students which lay a theoretical foundation for the next professional learning. At the same time, in recent years, countries and universities attach great importance to and focus on computer teaching for young college students, the purpose is to improve students' thinking ability, eventually to promote college students' ability to use computational thinking to solve and analyze the problems of daily life. Therefore, this article on how to the calculation of optimization in the process of computer teaching college students thinking ability on further discussion and analysis, and then explore the strategies and methods, so as to promote the computer teaching in the process of the cultivation of thinking ability and optimize the computer.

Key Word. Computational thinking; Computer teaching; Ability training; Implementation strategy

1 Basic computer thinking is introduced

1.1 Basic connotation

When it comes to computing thinking, academic scholars and experts are given a variety of views and opinions. In 2006, for example: the head of the department of computer science at Carnegie Mellon university weeks to really (Jeannette m. Wing) professor put forward its views, can be roughly summed up as the computer is a computer technology to solve problems through thinking, at the same time with the computer deeper levels of related thinking activity, the essence of which is the abstraction and automation. This view shows and puts forward the computer thinking is a must have for the masses, the essence of thinking mode, it is indispensable to People's Daily life and study. All in all, for a variety of different views and opinions, though there are differences, but take its generality, namely computer thinking is nothing but a kind of scientific and practical thinking, in People's Daily study and work plays an important role.

1.2 Current situation of the development of

As is known to all, computer courses generally opened in college students of the first semester of school, to improve college students' knowledge and understanding of computer basic knowledge and theory, so as to lay a foundation for the professional course learning. On the one hand, the development of the computer teaching to a certain extent, improve the college students a basic understanding of computer technology; But, on the other hand, college students' understanding of computer technology in the foundation stage, only failed to fundamentally understand the working principle of the computer technology, let alone the cultivation of thinking ability and optimize the calculation. Aiming at the existing problems and the insufficiency of computational thinking ability, can be roughly classified into the following several aspects.

1.2.1 Heavy practical,Theory of light

Combined with students' daily learning, can learn the current college students in the learning process exists the phenomenon of "practical, light theory". That is to say, in the process of computer courses, there are only takes the computer technology, ignore the phenomenon of computer theory. To this end, the school internal education workers should pay attention to the teaching of computer theory and computer thinking, so as to improve and optimize the calculation of students thinking ability.

1.2.2 Unconscious thinking teaching

Because the computer thinking in China's development time is shorter, enough attention in the internal teaching in colleges and universities, on the other hand is due to the essence of the computer technology in order to solve the problem, so in general, within the general computer educators in colleges and universities in the unconscious of computer teaching. Therefore, for this problem, the computer education in colleges and universities practitioners should focus on awareness of computer course teaching, so as to improve computing the existence and the importance of thinking ability.

1.2.3 Not from strategic level to realize computer thinking on the importance of the cultivation of talents

At this stage for computer thinking there are incomplete and lack of the understanding of the place, just think it is a part of computer theory, but not from the overall and strategic level to realize calculation thinking instead of other aspects below. Future, therefore, in the process of optimization calculation thinking ability, want to notice from the overall and strategic level general calculation combined with computer teaching of thinking, so as to promote the computer teaching strategy.

2 The need for analysis

From the previous illustration and analysis can know, in the internal to develop computer education in colleges and universities, on the one hand is to college students in the future to better learn professional class foundation, on the other hand through the computer teaching, and improve college students' thinking ability, enhancing the comprehensive quality of ascension, the finally adapt to the development of today's information age. In particular, on the one hand, improve the students' computer skills, help it to better solve the problems and deal with; , on the other hand, improve the students' judgment ability in information, namely in the current rapid development of computer technology, while the use of science and technology brought great convenience to our daily life, but the technology is a double-edged sword, it will also bring challenges to our life, therefore, the cultivation of thinking ability by calculation and optimization, can fundamentally improve the cognitive ability of information technology, to form his own judgment. Finally, calculation of thinking and ability can promote college students on their cognitive ability. Due to the thinking mode of thinking is a kind of scientific calculation, so college students after the formation of computational thinking can improve on their cognitive abilities.

3 Computational thinking ability training strategy and approach

To sum up, at this stage in the process of enhancing the teaching of computer calculation of the cultivation of thinking ability, is very necessary. Therefore, the author will for optimization strategy and way of thinking ability, can be summarized as the following respects.

3.1 The teaching mode and teaching method reform

In terms of the traditional computer teaching mode and method, its basic in computing thinking ability is introduced. Specifically, under the traditional teaching mode and methods, calculation of thinking ability is mainly intangible cover with the other abilities, such as: innovation ability, application ability, eventually led to the thinking ability is the unconscious. Therefore, in order to improve the present situation, colleges and universities and teaching patterns and methods to reform. Mainly embodied in: on the one hand, calculates the thinking ability of the unconscious teaching into teaching consciousness, namely the clear will calculate the cultivation of thinking ability is put forward to teaching, so as to calculate the existence and importance of thinking ability; , on the other hand, change teaching mode, the practical technology to be teaching mode by the previous teaching mode into teaching model of cultivating the ability of thinking and. In addition, still should strengthen the reform of teaching method, which USES the category and professional teaching method. In other words, the computer teaching of the college students according to the professional category for professional education, to better promote students' thinking ability.

3.2 Expand the extracurricular teaching, encourage students to participate in science class competition

First, by expanding the extracurricular teaching, on the one hand, can enrich student's study life, to a certain extent, make up for the deficiency of the classroom teaching; On the other hand, expand the extracurricular education, can widen students' view. Second, the discipline competition is the present college students show their professional skills and talents of the important carrier, through the discipline competition, to some extent, stimulate students' initiative and enthusiasm, eventually to improve the calculation for the cultivation of thinking and ability. At the same time, through the discipline of competition can promote study and communication between college students, help to calculate the cultivation of thinking ability.

3.3 Insist on computational thinking as the core concept, improve the basic teaching curriculum system

In recent years, the state attaches great importance to the construction of the college computer teaching course, at the same time, the ministry of education will also be set for the college computer curriculum reform project. In addition, some colleges and universities to calculate thinking as the core, improve the basic teaching course system, to improve and optimize the effect of college students' thinking and ability, mainly reflects in: on the one hand, some colleges and universities with its actual situation and the student's basic qualities, compiled computer teaching material has its own characteristics, to a certain extent, it reflected the color according to their aptitude; Education, on the other hand, practitioners, also want to combine their own years of experience in education, through their understanding of the present college students thinking ability level, to actively explore suitable for college students' thinking ability training calculation and optimization of ways and methods, to enhance and optimize the computing thinking ability in the process of computer teaching. Anyhow, in any country or university itself, it should pay attention to the optimization of computer basic teaching curriculum system, to some extent, this is the key link of improving the computational thinking ability of university students and part.

3.4We will improve the mechanism of teaching evaluation

As is known to all, the way of teaching evaluation is mainly the test, in order to achieve the purpose of evaluating students' mastery of knowledge level. In terms of the current examination system of the computer courses, the mainly through "written examination + machine try" mode for appraisal, can to some extent this evaluation to students mastering knowledge of computer theory and practice, but not the assessment to the students thinking ability. Therefore, colleges and universities should be in the present teaching evaluation mechanism, on the basis of the perfect for computing the examination link of thinking ability. Based on the improvement of the teaching evaluation system, to achieve the optimization calculation in computer teaching thinking ability, eventually to promote college students calculate their thinking skills.

Conclusion

Society played a great role in promoting. Computer technology in the calculation, on the other hand, the cultivation of thinking ability and improve requirements in the process of computer teaching should pay attention to the optimization calculation for college students thinking, so that every college students comprehensive quality accords with the demands of the development of human society. For this, big to small countries and universities, to each one education practitioners, to pay close attention to the quality education of college students, fundamentally improve college students' thinking ability calculation, prompting its computing thinking ability perfect fusion to social development and individual life.

References

[1] when yao, soubo ni, xue-dong zhang. The teaching of computer basic computational thinking training methods of research and discussion [J]. Journal of agricultural network information, 2015 (03) : 108-110.

[2] druid, lan-shun nie. Computer curriculum reform the basic idea of computational thinking and university [J]. Journal of China university teaching, 2013 (02) : 56-60.

[3] wang zhiqiang, fang-fang liu. Based on the computational thinking of reform of computer basic course study [J]. Journal of China university teaching, 2013 (6) : 59-60 + 36.

[4] druid, wang hao. University computer courses teaching content system for computing thinking [J]. Journal of China university teaching, 2014 (7) : 59-66.

[5] Zhu Minghua Zhao Mingwei, each, hong-fei Lin. Discusses the computational thinking ability training in computer basic teaching [J]. Journal of China university teaching, 2012 (3) : 33-35.

TongQi [6] guo-xing peng, liu, Zhu Yanhui, Chen Fangqin. Around computational thinking in university computer foundation course teaching content optimization [J]. Journal of business and the development of science and technology, 2012 (01) : 62-64.

[7] Gong Pei once, Yang. Computational thinking training in university computer foundation teaching [J]. Journal of China university teaching, 2012 (05) : 51-54.

[8] ZhanTao Deng Lei, zhi-gang liao. University computer foundation teaching for computing thinking culture [J]. Journal of computer education, 2014 (05) : 1-4.

[9] YiMingJing, unpaid. Computational thinking ability training as the core of university computer [J]. Journal of curriculum reform of computer education, 2014 (05) : 5-9.

MATEC Web of Conferences **63**, 05026 (2016) DOI: 10.1051/matecconf/20166305026
MMME 2016

A Survey of Key Technology of Network Public Opinion Analysis

Su Ying Li[a]
Department of Literature and History at Party school of the C.P.C. Shandong Province Committee, Shandong ,China

Abstract: The internet has become an important base for internet users to make comments because of its interactivity and fast dissemination. The outbreak of internet public opinion has become a major risk for network information security. Domestic and foreign researchers had carried out extensive and in-depth study on public opinion. Fruitful results have achieved in the basic theory research and emergency handling and other aspects of public opinion. But research on the public opinion in China is still in the initial stage, the key technology of the public opinion analysis is still as a starting point for in-depth study and discussion.

Keywords: network public opinion; information collection; information extraction; public opinion analysis

Introduction

After newspapers, radio, television, the internet has become the fourth media. It has become the main medium of information dissemination and main carrier of social public opinion reflected [1]. Internet public opinion refers to the internet users' subjective reflection of the various phenomena and problems in society, and the public's comments and views with a certain tendency propagated through the network [2]. Because of the internet's virtuality, arbitrariness, and rapid propagation etc., false information, reactionary remarks, malicious speculation and other negative things are developing rapidly when the normal public dissemination of information. It disrupts the social order and harms the network information security. If we let this phenomenon go unchecked, negative public opinion will form a larger threat to public safety. In order to maintain social stability and prevent danger, monitoring online public opinion and early warning have become increasingly important.

With the emphasis on the network public opinion, the analysis about public opinion has become a hot research. Internationally, the United States' TDT [3] (Topic Detection and Tracking) system is the most well-known network public opinion analysis system. TDT had developed a number of related algorithms about unknown topics to identify and existing topics to track. In 2005, the British software company developed a public opinion analysis software about emotion. The software is able to analyze online news stories and judge their emotion is positive, negative or neutral. Domestic public opinion research network units and agencies have Chinese Academy of Sciences, Academy of Social Sciences and some colleges and universities. Liu Yi wrote "Introduction to network public opinion research," this literature conducted in-depth discussions about the network of public opinion from theory and practice [4]. Pan xin etc.put forward new public opinion propagation model based on social network analysis[5]. Zhang Yu carried out fruitful exploration about actual forum, blog, website with news comment function and proposed a comprehensive variety of algorithms for network public opinion[6]. Jiang Fan [7] etc. conducted research about the forum network, and set up a theme discovery system. Zhou Yadong [8] analyzed the demand for network public opinion and delimited the hotspot of public opinion. Li hongtao [9] proposed gray evaluation method for network public opinion. Panchong Xia [10] completed the network public opinion evolutionary trend forecast with gray theory.

This paper analyzes the main technology in online public opinion. To collect the information rapidly and

[a] Corresponding author: lisuyign@sina.com

accurately, and achieve the extraction about acquired information and denoising, and ultimately finish accurate judgment the trends of public opinion and forecast analysis of public opinion is the core of the work in the network public opinion analysis and early warning.

A public opinion information collection technology

Because of diversity and complexity of public opinion, public opinion information collection network has a certain degree of difficulty. The traditional method of information gathering public opinion is finished manually, it has large limitations and it is inefficient. Collecting information on the internet depends on web crawler technology. It mainly collects web pages through making use of the relationship between the pages. Research directions about crawler technology are introduced as follows.

(1)Information collection based on the entire web. It can expand from some seeds URL to the entire web information collection and finish the entire web collection. It is dominant in practical applications currently.

(2)Incremental web information collection. It only collects the pages which has changed or generated newly when the page is refreshed, but it can not collect pages with no change.

(3)Web information collection based on the topics. Collection is finished through selective search for those pages associated with the theme predefined.

(4)Personalized information collection. It can collect information meeting the different needs of users by means of user interaction.

(5)Information collection based on agent. Intelligent agent system refers to a computer system, It can be flexible activities independently in the environment like humans and it can perceive changes in the user's interest and adjust acquisition strategy independently .

The results of different information collection methods are different in dealing with the same query asked to return, and the quality is uneven. This is mainly due to its own advantages and disadvantages. It was found in actual collection process that a method alone is not very satisfactory. It will miss a lot of qualified page. So we usually use two or more kinds of ways to complete the acquisition of information..

Public opinion information extraction technology

Web content will generally contain navigation, title, text, related links and advertising information. Web information extraction is to extract the web page content from the source file and identify the information related with the themes [11]. Extraction techniques commonly used are the following:

(1) the extraction technology based on the feature of web pages

Web page layout uses HTML tags generally. It has a hierarchical structure between the data with each other and is arranged in chronological order. To use structure characteristics of the html page layout, or to use a regular expression technology can get relatively pure text messages. The technology can remove additional noise information of web pages, such as page ads, links to other pages, descriptive information pages and so. It can effectively make the information extracted more accurate and pure.

(2) information extraction based on natural language processing

The technology completes processing by taking the entire web page as a text document. Therefore it is more suitable when the relatively large of information needs extraction. This method is the application of the traditional language processing technology, including parsing, syntactic analysis, semantic analysis and identification. The extraction process also includes the use of some extraction rules.

(3) information extraction technology based on ontology

Ontology is a formal concept. And it is used to describe the relationship of the related fields in order to provide for a common understanding of knowledge and realize shared knowledge ultimately. The approach based on ontology utilizes information to describe the data itself and it can establish an ontology library and use the specified extraction rule to complete information extraction. The advantage of this method is that does not depend on the pages' structure. As long as the body is large enough, it can extract various information in this field.

(4) information extraction technology based on Hidden Markov Model

Hidden Markov model is a finite state automaton and it is an important means of signal processing. Currently it has been quite successful applications in speech recognition, behavior recognition and other fields. In recent years, this model is widely applied to information extraction, and there is now a very significant effect. It is possible to deal with new data robustly.

Public opinion analysis and prediction

Public opinion analysis is the core of the public opinion analysis system. At present clustering and classification methods in data mining are used to solve this kind of problem. The ultimate goal is to carry out public opinion analysis to predict the future trend of public opinion. Internet public opinion prediction is to study how to use the massive alarm data to predict future network of public opinion, that is, to predict the future by analyzing the past. Public opinion analysis and forecasting technology mainly consists of the following tasks.

(1) identify hot topics and sensitive topics

Mass information from the internet contains a large number of topics. According to topic areas they can be divided into: military, sports, entertainment, technology, society and so on. According to the importance of the topics, they can be divided into general topics and hot, sensitive topic. Currently, mainly found in the web-based text mining on the topic, and then measure the relevant reports on the topic of the total amount, the growth rate, fluctuation level, there are users of the topic replies intensive, emotional and other dimensions tendency to identify certain period of hot topics. At the same time, we determine whether it belongs to the sensitive topic through keyword matching as well as further semantic analysis.

(2) the topic tendentious analysis

Topics propensity analysis mainly refers to the emotional orientation analysis. Further extension may be subdivided into political orientation analysis, product preference analysis AND interest preference analysis. Its main purpose is to analysis published articles, blog, reply and microblog, semantic analysis on the network. It can determine its emotional tendencies, such as commendatory, derogatory or neutral.

(3) the trend forecasting network public opinion analysis

Currently, the main focus of analysis to predict, there are two aspects, one is affected by the degree of concern of public opinion, mainly to predict trends reports its numbers and number of replies, the other is the degree of complexity of contradictory public opinion, mainly on trends to predict their emotional tendency. There are now widely used algorithm BP neural network model based on the integration of autoregressive moving average model (ARIMA), decision tree model and gray theory.

Conclusion

This article focuses on the key technology of internet public opinion analysis and introduces technical points and the corresponding research status. It can lay the foundation for further study of public opinion. When network provides a convenient way to send message, virtualization and concealment of the network also become a major social stability risk on the political and cultural security. Therefore, strengthening the monitoring of public opinion information and timely grasping public opinion trends are important.

Acknowledgments

This work is supported by the National Natural Science Foundation of China(Grant No.61402271), and Shandong Social Science Planning Project(15CXWJ08 and 14CZXJ14), and the Natural Science Foundation of Shandong Province (Grant No.ZR2015JL023 and Grant No.ZR2015FL025),and Key R & D program in Shandong Province(Grant No. 2015GGX101012).

References

[1]Dai Yuan, Cheng learned Flag practical overview of the key technologies for the Internet public opinion analysis [J]. Computer Research and Development, 2008: 62-65.

[2]Liu Zeguang network public opinion analysis[D]. Northeastern University Master Thesis, 2013.

[3]Leek T, Schwartz R, Sista S. Probalistic approaches to topic detection and tracking [M].New York:. Springer US, 2002.

[4]Liu Yi Public Opinion Research Network Introduction [M]. Tianjin: Tianjin People's Publishing House, 2007.

[5]Pan new Denggui Shi Tong Bin public opinion propagation model based on social network Construction and analysis [J]. Operations Research and Management, 2011: 176-179.

[6]Wu Yu. entertainment network public opinion analysis[D].Chengdu: University of Electronic Science and Technology, 2011.

[7]Jiang Fan, high Junbo, Zhang Min and other .BBS themes found in the design and realization of the prototype system [J]. Computer Engineering and Applications. 2005: 155-157.

[8]Zhou Yadong, Sun Qin East, Guan et Traffic content words related to the extraction of the hot topics of the network [J].2007 Xi'an Jiaotong University: 1142-1145.

[9]Lee Yun Tao, Liu Yan, Liu Yi Grey alarm assessment of network public opinion [J]. Information] .2011: 24-27.

[10]Pan Chongxia stage evolution of network public opinion analysis [J].Computer and Modernization, 2011 (10): 179-182.

[11]parties pay to build a network of public opinion research to quantify Situation Analysis Application [J]. Journal of Information, 2014 (33): 47-51.

MATEC Web of Conferences 63, 05027 (2016) DOI: 10.1051/matecconf/20166305027
MMME 2016

The Application of Floyd Algorithm for Returning Path Algorithm with Constraints of Time and Limit of Load

Jian Feng Luo[1,a] , Yu Qiang Chen[1]
[1]*Department of Computer Engineering, Dongguan Polytechnic, Dongguan Guangdong 523808, China*

Abstract. Under ideal condition, there are lots of Shortest Path Algorithms such as Dijkstra, Floyd, Johnson algorithms. However, putting these algorithms into use is not as we expected. The reasons cause the problem may be in the Returning Problem mentioned in the following paper. Shortest Path Algorithms only focus on the shortest path between the points, without paying attention to figure out the direct solution from the Starting Point to the Passing Points and finally to the Starting Point. A Shortest Path Algorithm applied in actual practice will be introduced in here, then another case will be introduced and analyzed to explain the following Algorithm: Under the Constraints of time and limit of Loading capacity, figure out the Shortest Path Algorithm through the Starting Point to the Passing Points and back to the Starting Point. At last , the algorithm is of feasibility and practicability.

Keywords: Shortest Path Algorithm, Returning Path, Single Source Point and Multi-sources Point, Time Window Constraints, Load Limit, Weighted Node

1. Introduction

There are many cases indicate that an effective algorithm to calculate the best route from the starting point, covering the nodal points and finally returning to the starting point under the restrictions on time and loading value is in great need. Here are some common cases in real life:

Case 1: Goods delivered by unmanned aerial vehicles (UAV). With the restrictions on battery energy and cargo weight, the normal flying time of a UVA to pick up goods and deliver the goods back to the starting point is about 30 minutes. Hence, an energy-saving route is needed.

Case 2: Self-driving tour to an unfamiliar city. In an unfamiliar city, the tourist will be found of a time-saving route to to more tourist attractions and come back to the hotel easier to have a rest and restart in better condition for the next day. So an easier plan and time-saving route is very practical.

Case 3: School buses pick up and send back students. The number of school buses is fixed . Thus, to make maximum of the school buses in a time and energy saving route is in need.

Since the need of shortest path algorithm is so important, the purpose is to figure out the algorithm that can be put into use with high efficient. The author applies the most commonly used shortest path algorithms into real practice, gets the shortest path matrix. With matrix, the most ideal rerouting path is chosen to put into the schematization to return when the time and loading value surpasses the designated value. The same method is applied to arrange the rest of the nodal points.

2. The choices on the Shortest Path Algorithms

The most famous and commonly accepted Shortest Path Algorithms are Dijkstra, Floyd, and Johnson Algorithms. A brief analysis is necessary to find out a better one. It is as followed:

[a] Corresponding author: rocrocroc@126.com

2.1 Dijkstra Shortest Path Algorithm is based on the path from the starting point to nodal point.

The time complex rate is O(n2) [1]. However, the returning path not only contains this path, but also other passing points and the distances among the these passing points. Hence, only applying the Dijkstra Shortest Path Algorithm is not enough to solve the complex real problems. Someone may think that if repeatedly apply the Dijkstra Shortest Path Algorithm to the passing points; then the shortest paths among the passing points will be acquired [2]. In real practice, this may cause the repeating calculation of the same nodal points which may make the time complex rate to O(n3) may lead to the instability of the passing points. Taking the UAV as an example, the passing point to fetch the goods may differ from the next time, and this will lead to the changes in the passing points. Thus, to get the shortest path, the computer has to spend a lot of time to calculate the path again.

2.2 Floyd Shortest Path Algorithm is based on the shortest path algorithm of multiple starting points whose time complex rate is O(n3) [3].

In spite of the relevant high time complex rate, this algorithm is easy to figure out the shortest path of all the nodal points, without recalculating when the passing points are changed[4]. What it needs to do is to find out the shortest path among the path matrix. Thus, the Floyd Shortest Path Algorithm is very practical, and it will be applied in the algorithm of this paper.

2.3 Johnson Shortest Path Algorithm is also based on multiple starting points.

The time complex rate of the sparse matrix path is O(VElgV). Theoretical, Johnson Shortest Path Algorithm is regarded better than the other two shortest path algorithms[5]. If we apply the Ford and Dijkstra Algorithms, Authorization means the traveling time and the loading value on the nodal points.In practice, most of the path graphs are not sparse matrix path[6] , let alone its algorithm which is using the re-authorization with proper influences on the time and loading program. Therefore, the Johnson Shortest Path Algorithm is not suitable for the calculation of this paper[7].

From all the analysis listed above, Floyd Shortest Path Algorithm is applied in this paper.

3. The Detour Algorithm

From the starting point, figure out the shortest path to cover the other passing points and finally get back to the starting points. To prove that this path is the shortest, a simple model diagram[8] (Fig.1) can show us. Since the path of the diagram has adopted the shortest path algorithm to filter the shortest path, the paths between the nodal points are the shortest[9]. S is Starting Point, Pi (i=1,2,3) are the nodal points , Lk (k=1,2,3,4,5) are the shortest weights of the nodal points. Three path routes:

A: Starts from S, cover Pi, then get back from the Pi(S↔P1, S↔P2, S↔P3);

B: Starts from S, cover all the Pi, then get back from the same path(S->P1->P2->P3->P2-P1->S);

C: Detour Algorithm. Starts from S, cover all the Pi, then gets back directly the LAST P(S->P1->P2->P3->S).

Detailed analysis can be shown as followed:

Route A: A=2*L1+2*L4+2*L5, Route B: B=2*L1+2*L2+2*L3, Route C: C=L1+L2+L3+L4 .

Now we need to prove that: C<A and C<B.

A-C=2*L1+2*L4+2*L5-(L1+L2+L3+L4) =L1+L4+2*L5-L2-L3, according to the Triangle Theorem, we can get: L4+L5>L3 and L1+L5>L2, so L1+L4+2*L5>L2+L3, and so A-C>0,the result is C<A.

B-C=2*L1+2*L2+2*L3-(L1+L2+L3+L4)=L1+L2+L3-L4, suppose that L2=0, then, L1,L3,L4will form a triangle, according to the Triangle Theorem, L1+L3>L4,however in real life, L2>0, thus L1+L2+L3-L4>0,we can get B-C>0, so C<B.

From this simple diagram, Detour Algorithm is shorter than the other two Algorithms. In the following paper, we take the value of A-C or B-C as the Path Value of the Detour Algorithm, then PV[10](P1,P2) means that starts from S and bypasses P1,P2 and gets back to S.

DOI: 10.1051/matecconf/20166305027

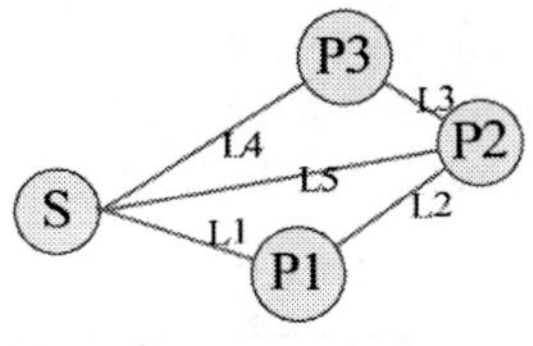

Fig.1 Simple model diagram

4.The Returning Path Algorithm

To make it easier to understand, we take the school bus picking up the students as example:

All together, 10 points to picking up the students, and the number of the students are shown in Fig. 2(network diagram) as the nodal points. There are two kinds of school buses, they are 20seats and 40 seats and share the same driving time--within 30 minutes. The Edge Weight in the diagram means the time (minute). We put the time as the value of the distances. Details are as followed:

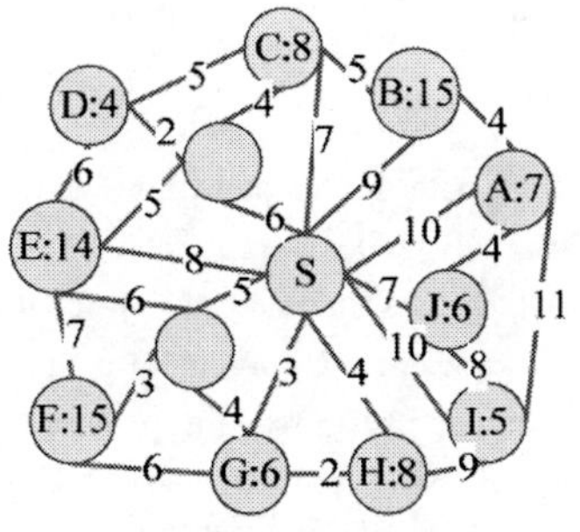

Fig.2 network diagram

	S	A	B	C	D	E	F	G	H	I
A	10	—								
B	9	4	—							
C	7	9	5	—						
D	8	14	10	5	—					
E	8	18	14	9	6	—				
F	8	18	17	15	13	7	—			
G	3	13	12	10	11	10	6	—		
H	4	14	13	11	12	12	8	2	—	
I	10	11	15	17	18	18	17	11	9	—
J	7	4	8	13	15	15	15	10	11	8

Fig.3 the shortest matrix

4.1 Figure out the shortest matrix

Floyd Shortest Path Algorithm is applied here to get the shortest matrix shown in Fig.3 to show the distance from S to the students picking up points.

4.2 Figure out the PV matrix.

According to the Detour Algorithm, the fist two lines of shortest matrix should be noticed firstly. As in Fig. 4, the Edge Weight from S to A is 10, to B is 9, while the Edge Weight from A to B is 4. Thus, the PV(A,B)=15. Fig.4 shows the detailed matrix transformation of the first two lines.

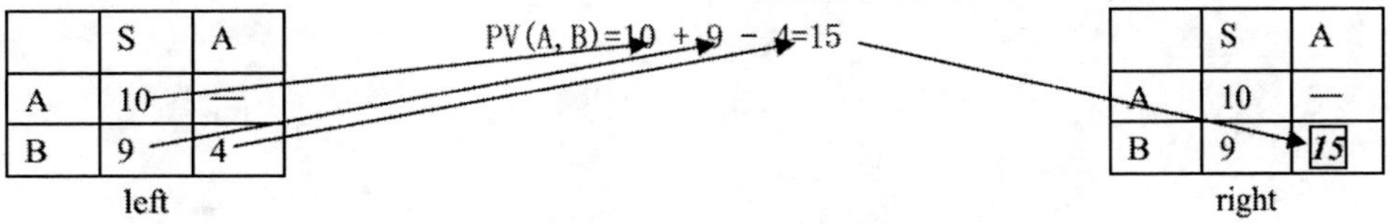

Fig.4 the matrix transition diagram

Employing the same method, the first three lines of the shortest matrix will be shown in Fig. 5.

The Edge Weight from S to A is 10, to C is 7; the Edge Weight from A to C is 9, thus the PV(A, C)=8.

Fig.5 shows the detailed matrix transformation of the first three lines.

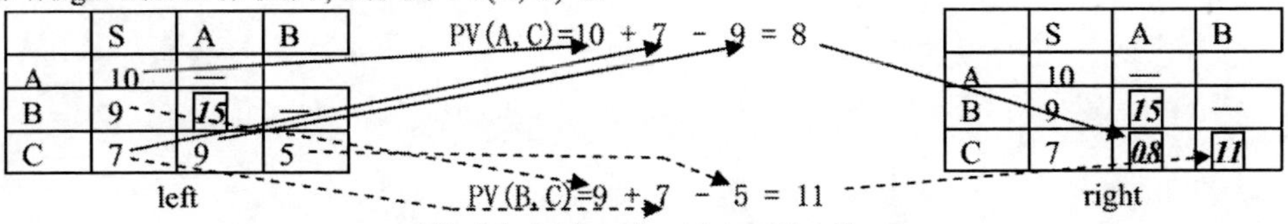

Fig.5 the 1-3 line of matrix transition diagram

 DOI: 10.1051/matecconf/20166305027

We can deduce the rest from this method: add one more line, get the PV from this point to the other points, then replace the value of the shortest matrix. When it comes to the 10th line (replacing from line A to line J), the result will be shown in Fig.6 (PV matrix).

	S	A	B	C	D	E	F	G	H	I
A	10	—								
B	9	15	—							
C	7	8	11	—						
D	8	4	7	10	—					
E	8	0	3	6	10	—				
F	8	0	0	0	3	9	—			
G	3	0	0	0	0	1	5	—		
H	4	0	0	0	0	0	4	5	—	
I	10	9	4	0	0	0	1	2	5	—
J	7	13	8	1	0	0	0	0	0	9

Fig.6 the PV matrix

ID	path	PV	ID	Path	PV	ID	path	PV
1	A-B	15	9	A-C	8	17	B-I	4
2	A-J	13	10	B-J	8	18	F-H	4
3	B-C	11	11	B-D	7	19	B-E	3
4	C-D	10	12	C-E	6	20	D-F	3
5	D-E	10	13	F-G	5	21	G-I	2
6	A-I	9	14	G-H	5	22	C-J	1
7	E-F	9	15	H-I	5	23	E-G	1
8	I-J	9	16	A-D	4	24	F-I	1

Table.1 PV Value ranking table

From the Fig. 6, except for column S, if other value stands for the shortest path for the school bus covering the nodal points, then the Contribution Value raised. The school bus does not need to pass all the nodal points. There are two kinds of school buses, they are 20seats and 40 seats which share the same driving time--within 30 minutes. The following two situations:

A: the school bus is fully occupied, and it needs to take them back to school.

B: the driving time has reached 30minutes, and it needs go back to the school.

4.3 The choices for the path and the arrangement of the buses

The PV and the path information from Fig.6, arrange the PV in descending order and get Table 1.

If choose A-B, A-J, B-C as the paths, the contribution value of the efficiency is the utmost. If make a detour,,choose these paths first. If we choose C-D, the driving time will be over 30 minutes, thus, the bus need to drive back to the school. In Fig. 7(picking up students Route 1), the route for picking up the students. Starting from S (hides the other paths) , Route 1 is the chosen path and a 40-seat school bus is needed. The number of the students is 36 and the driving time is 27 minutes.

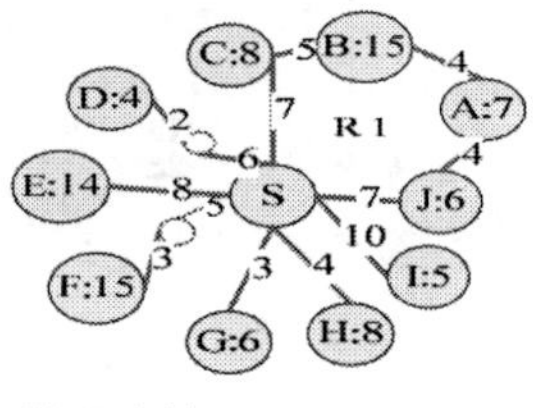

Fig.7 picking up students R 1

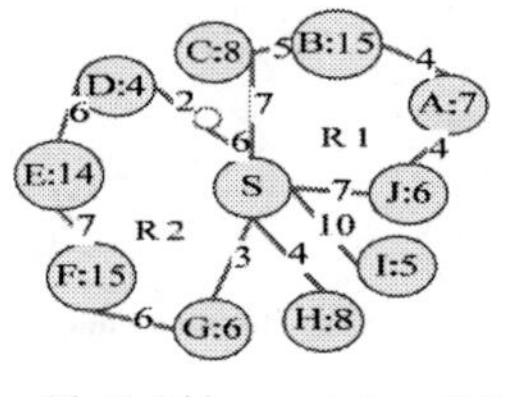

Fig.8 picking up students R 2

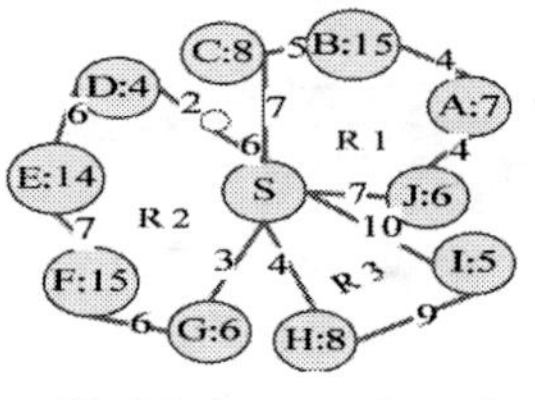

Fig.9 the best returning path

The details are as followed:

Since adopting Route 1, skip C-D and pay attention to D-E, A-I, and E-F. As Route 1 is not connected with other route, then A-I is excluded. Thus it can connect with D-E, E-F. Closely exam I-J,A-C,B-J,B-D,C-E in the Fig.7, they share the same situation as A-I. They can be excluded.

If we connect F-G, Route 2 is the chosen path. Why not connect G-H? Because the driving time for Route 2 (S-D-E-F-G) is 30 minutes, if we connect H, the time will be more than 30 minutes. In this situation, a 40-seat school bus drive within 30 minutes to pick up 39 students (Fig. 8).

Connect H-I, and form Route 3. A 20-seat school bus drive 23 minutes to pick up 13 students.

So far, the detour of the route has accomplished. Fig.9 is the best returning route and we need not to make reference to Table 1(PV Table). From Fig.9, we can get the conclusion that: the school needs two 40-seat and one 20-seat school buses. And the 40-seat school buses will take Route 1 and 2; while the 20-seat school bus will take Route 3. If the school begins the lessons at 8 o'clock, in ideal condition, the school bus can start out at 7:20, and get back at 7:50.

5. Summary

After testing in Baidu, Tencent and Goethe map, the result shows that: if without the demands as time limits, loading limits, returning path and more choose-able nodal points, Baidu Map has more advantages than Tencent map and Goethe map. Since Baidu Map can cover more view spots, but it can not fulfill the need of other demands. The first author of this paper has adopted this algorithm in the software system and got funds from the Guangdong Province Outstanding Young Teachers Training program Funded Project (Grant No. 2015S02 - Jianfeng Luo). And Special Funds for Demonstration Projects of Dongguan Polytechnic College (Grant No. ZXHQ2014d001 - Jianfeng Luo). The project ZXHQ2014d001 cooperates with the Dongguan Hu Yitong Freight Ltd and gets funds from this company. The software whose Copyright Registration Number is 2014SR146664 has been put into use of the company and received high evaluation. The authors sincerely hope that this paper will be beneficial. Please contact with QQ:251048547 for further cooperation.

Acknowledgments

Funds supported by 2014 Guangdong Province Outstanding Young Teachers Training program Funded Project (Grant No. 2015S02); 2013 Guangdong Province Outstanding Young Teachers Training program Funded Project (Grant No: Yq2013201); High Tech Industrialization Project of Guangdong Science and Technology Department of (Grant No: 2014A010103002); Special Funds for Demonstration Projects of Dongguan Polytechnic College (Grant No: ZXHQ2014d001).

References

[1] WANG Zhan-hong, SUN Ming-ming, YAO Yao. Analysis and Improvement of Dijkstra Algorithm [J]. Journal of Hubei University of Education,2008,25(8),12-14.

[2] ZHENG Sifa, CAO Jiandong, LIAN Xiaomin. Sector Dijkstra algorithm for shortest routes between customers in complex road networks[J]. Journal of Tsinghua University(Science and Technology),2009,49(11): 1834-1837.

[3] ZHENG Hai-hong. Analysis and Comparison of Common Shortest Path Algorithms[J]. Journal of Anhui Vocational College of Electronios & Information Technology, 2013,12(4):31-33.

[4] Luo Jianfeng Ji Shanshan Zheng Huijun. Design of Algorithm for Network Information Search in Information Integrated Platform of Cargo Tracking[J]. Science Mosaic, 2015,(2):51-53.

[5] WEI DC. An optimized Floyd Algorithm for the shortest path problem. Journal of Networks 2010.

[6] Lu Feng,Zhou Chenghu,Wan Qing. An optimum vehicular path algorithm for traffic network based on hierarchical spatial reasoning[J]. Geo-spatial Information Science . 2000 (4).

[7] U. Pape. Implementation and efficiency of Moore-algorithms for the shortest route problem [J]. Mathematical Programming. 1974 (1).

[8] JI X Y. Models and algorithm for stochastic shortest path problem. Applied Mathematics and Computation . 2005.

[9] Boris V. Cherkassky,Andrew V. Goldberg,Tomasz Radzik. Shortest paths algorithms: Theory and experimental evaluation[J]. Mathematical Programming . 1996 (2).

DOI: 10.1051/matecconf/20166305027

[10] Da-chuan Wei. Implementation of route selection function based on improved Floyd algorithm. Proceedings of the2010WASE international conference on information engineering . 2010.

MATEC Web of Conferences **63**, 05028 (2016) DOI: 10.1051/matecconf/20166305028
MMME 2016

Load forecasting method considering temperature effect for distribution network

Xiao Fang Meng [1], Shu Xuan Zhang [1], Jun Wang [1,a], Shi Hong Wu [1]

[1] *The College of Information and Electric Engineering, Shenyang Agricultural University, Shenyang 110866, China;*

Abstract. To improve the accuracy of load forecasting, the temperature factor was introduced into the load forecasting in this paper. This paper analyzed the characteristics of power load variation, and researched the rule of the load with the temperature change. Based on the linear regression analysis, the mathematical model of load forecasting was presented with considering the temperature effect, and the steps of load forecasting were given. Used MATLAB, the temperature regression coefficient was calculated. Using the load forecasting model, the full-day load forecasting and time-sharing load forecasting were carried out. By comparing and analyzing the forecast error, the results showed that the error of time-sharing load forecasting method was small in this paper. The forecasting method is an effective method to improve the accuracy of load forecasting.

Keywords: Load forecasting, Temperature, Multiple regression analysis, Time-sharing forecasting.

1 Introduction

The load level and structure of the power network has greatly changed [1], which has increased the difficulty of load forecasting. There are more factors that affect the load forecasting [2]. One of the more important factors is temperature. It's particularly vital to study the relationship between temperature and load [3]. Temperature has impacted on the prediction results largely [4].

When the weather is cold or hot, there will be a large number of heating or cooling loads put into operation, which will result in a substantial change in power consumption [5]. Therefore, the temperature has been become a common understanding factor in the short-term load forecasting. It is necessary to analyze the relationship between the load and the temperature [6]. There are many short-term forecasting methods, such as extrapolation method, regression analysis method, grey theory, neural network, etc. [2, 4]. Because of the advantages of the regression prediction, the method has been widely used in short-term load forecasting.

Using multi-variant regression method, this paper proposed the of distribution network load forecasting method considering the influence of temperature on load forecasting.

2 Analysis of temperature factors in load forecasting

Table 1 Peak load and temperature data in summer 2001

Date	Daily maximum load [MW]	Daily mean temperature [℃]
July 10th	2439	32.6
July 11th	2473	32.0
July 12th	2036	26.8
July 13th	2097	28.5
July 14th	1799	26.0
July 15th	1723	26.4
July 16th	1929	26.7

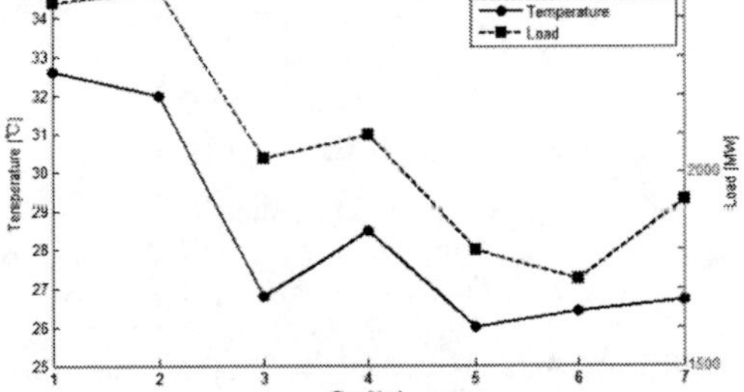

Fig. 1 Curve of load changes with temperature

[a]Corresponding author: 842638598@qq.com

Using the daily maximum load as an example, this paper researched temperature factor. Based on the data of power consumption and temperature in summer of 2001 [7], this paper listed the typical summer daily peak load and the daily average temperature in Table 1. According to the table 1, the temperature-load change characteristic curve was shown in the figure 1. In the figure 1, the trend of daily power was very similar with the average temperature.

3 Mathematical model of load forecasting

It is assumed that the load is random variable y. The variable y is related to the control variable x. The variable x has n components x_i, which were $x_1, x_2, x_3, \ldots, x_n$, ($n$>1). Supposed that there is only linear relationship between x and y. The two-variable linear regression model is as follows:

$$y = a_1 x_1 + a_2 x_2 + a_3 \tag{1}$$

Where y is load（MW）, and x_1 is time（h）, and x_2 is temperature（℃）, and **a** is regression coefficient, $\mathbf{a} = \begin{bmatrix} a_1 & a_2 & a_3 \end{bmatrix}^{\mathrm{T}}$.

For the sake of the regression coefficient **a**, assuming that the total data number is m, the i-th data is d_i. As follows:

$$d_i = \begin{bmatrix} y_i & x_{i1} & x_{i2} \end{bmatrix}, \quad i = 1, 2, \cdots, m \tag{2}$$

Further, the d_i is put into formula (1), and written as a matrix form:

$$\begin{bmatrix} y_1 \\ y_2 \\ \vdots \\ y_m \end{bmatrix} = \begin{bmatrix} x_{11} & x_{12} & 1 \\ x_{21} & x_{23} & 1 \\ \vdots & \vdots & \vdots \\ x_{m1} & x_{m2} & 1 \end{bmatrix} \begin{bmatrix} a_1 \\ a_2 \\ a_3 \end{bmatrix} \tag{3}$$

Where $\mathbf{Y} = \begin{bmatrix} y_1 \\ y_2 \\ \vdots \\ y_m \end{bmatrix}$, $\mathbf{X} = \begin{bmatrix} x_{11} & x_{12} & 1 \\ x_{21} & x_{23} & 1 \\ \vdots & \vdots & \vdots \\ x_{m1} & x_{m2} & 1 \end{bmatrix}$, and the formula (3) can be written as:

$$\mathbf{Y} = \mathbf{X}\mathbf{a} \tag{4}$$

In order to get the regression coefficient a, according to the least square method [8], the type (4) is multiplied by $\mathbf{X}^{\mathrm{T}}$, and there is:

$$\mathbf{X}^{\mathrm{T}}\mathbf{Y} = \mathbf{X}^{\mathrm{T}}\mathbf{X}\mathbf{a} \tag{5}$$

Where $\mathbf{X}^{\mathrm{T}}\mathbf{X}$ is m×m symmetric positive definite matrices. So, the regression coefficient **a** can be obtained as follows:

$$\mathbf{a} = \left(\mathbf{X}^{\mathrm{T}}\mathbf{X}\right)^{-1}\mathbf{X}^{\mathrm{T}}\mathbf{Y} \tag{6}$$

4 The steps of load forecasting

For the load forecasting, the first step is to collect and analyze data. Then the content of the forecasting is determined, and the load forecasting model can be established. Next step is to calculate regression coefficient and the predictive value. The load forecasting steps are shown in figure 2.

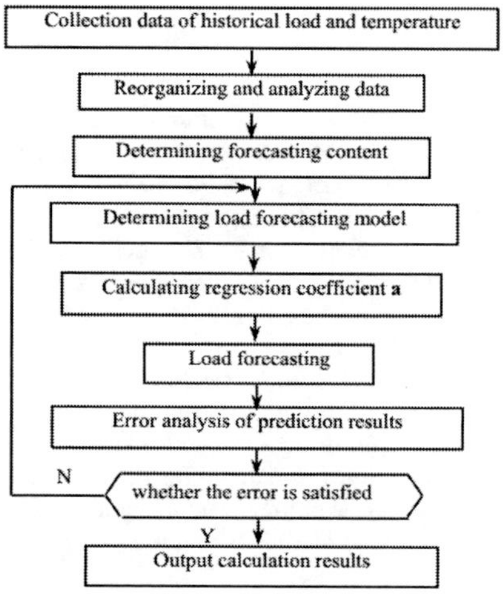

Fig. 2 The steps of load forecasting

5 Case analysis

Collect data and determine predictive content The January 28 of 2002 was selected as the forecast date [9]. Establishing mathematical model based on MATLAB, 24 point load value was forecasted on the forecast date.

All day load forecasting and error analysis According to the data and formula (6), the regression coefficient **a** was equal to

$[0.7345 \quad 0.2502 \quad 65.6889]^{\mathrm{T}}$.Based on the formula (1), the load forecasting model of the all day was established on Monday. The forecasting model was:

$$y = 0.7345x_1 + 0.2502x_2 + 65.6889 \tag{7}$$

According to the formula (7), the predicted result was compared with the actual load as shown in Table 2, and the average relative error was 5.23%, and the maximum relative error was 16.28%.

Table 2 Analysis of forecast load and actual load

Time	Forecasting load [MW]	Actual load [MW]	Absolute error[MW]	relative error	Time	Forecasting load [MW]	Actual load [MW]	Absolute error[MW]	relative error
0:00	61.1853	64.10725	2.9219	4.56%	12:00	71.2503	70.68528	0.5650	0.80%
1:00	61.6696	62.39206	0.7225	1.16%	13:00	72.2350	69.86855	2.3665	3.39%
2:00	62.1539	62.32896	0.1751	0.28%	14:00	73.2197	74.26696	1.0473	1.41%
3:00	63.1386	60.7861	2.3525	3.87%	15:00	74.2044	74.25833	0.0539	0.07%
4:00	64.1233	64.23827	0.1150	0.18%	16:00	75.1891	75.68637	0.4973	0.66%
5:00	65.1080	67.55168	2.4437	3.62%	17:00	76.1738	84.10879	7.9350	9.43%
6:00	65.5923	70.55515	4.9628	7.03%	18:00	77.1585	86.70985	9.5513	11.02%
7:00	66.5770	78.02583	11.4488	14.67%	19:00	77.6428	81.05289	3.4101	4.21%
8:00	67.5617	73.76723	6.2055	8.41%	20:00	78.1271	83.56409	5.4370	6.51%
9:00	68.5464	73.32247	4.7761	6.51%	21:00	78.1110	79.84229	1.7313	2.17%
10:00	69.5311	75.80108	6.2700	8.27%	22:00	78.5953	75.42288	3.1724	4.21%
11:00	70.5158	75.69858	5.1828	6.85%	23:00	78.8294	67.79508	11.0343	16.28%

Time-sharing load forecasting and error analysis

To reduce the error, using the time-sharing load forecasting, the 24 models were established respectively. The mathematical model is as follows:

$$y_i = a_{i1}x_1 + a_{i2}x_2 + a_{i3} \tag{8}$$

Where i is time (h), and y_i is the load of the time i (MW), $\mathbf{a} = [a_{i1} \quad a_{i2} \quad a_{i3}]^{\mathrm{T}}$ is the regression coefficient of the time i。

Table 3 Analysis of forecast load and actual load

Time	Forecasting load [MW]	Actual load [MW]	Absolute error[MW]	relative error	Time	Forecasting load [MW]	Actual load [MW]	Absolute error[MW]	relative error
0:00	59.2476	64.10725	4.8596	7.58%	12:00	70.9898	70.68528	0.3045	0.43%
1:00	58.1464	62.39206	4.2457	6.80%	13:00	67.6811	69.86855	2.1875	3.13%
2:00	58.3339	62.32896	3.9950	6.41%	14:00	70.1820	74.26696	4.0850	5.50%
3:00	58.3395	60.7861	2.4466	4.02%	15:00	71.0760	74.25833	3.1824	4.29%
4:00	59.9535	64.23827	4.2848	6.67%	16:00	72.3135	75.68637	3.3729	4.46%
5:00	64.1283	67.55168	3.4234	5.07%	17:00	87.0174	84.10879	2.9087	3.46%
6:00	70.2550	70.55515	0.3001	0.43%	18:00	83.7786	86.70985	2.9313	3.38%
7:00	75.9125	78.02583	2.1133	2.71%	19:00	79.4430	81.05289	1.6099	1.99%
8:00	70.3329	73.76723	3.4343	4.66%	20:00	77.6185	83.56409	5.9455	7.11%
9:00	69.0813	73.32247	4.2412	5.78%	21:00	76.2866	79.84229	3.5557	4.45%
10:00	71.4448	75.80108	4.3563	5.75%	22:00	69.8029	75.42288	5.6200	7.45%
11:00	74.1321	75.69858	1.5665	2.07%	23:00	63.7088	67.79508	4.0862	6.03%

The results predicted compared with the actual load as shown in table 3. To compare the accuracy of the whole day load forecasting and the time-sharing load forecasting, the two load forecasting waveforms were output into Figure 3. And the comparison of relative error was shown in Figure 4.

According to the table 3 and Figure 4, the average relative error of time-sharing forecast was 4.56%, the maximum relative error was 7.58%. The time-sharing forecast load values were similar with the actual values. So, the time-sharing method can meet the need for short-term load forecasting.

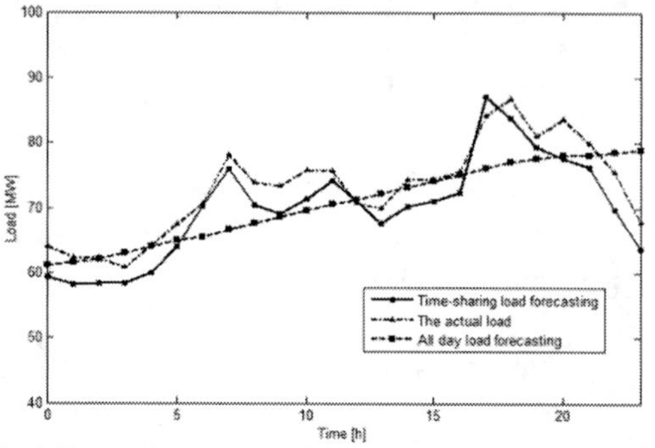

Fig. 3 Comparison curve of the forecast load and actual load

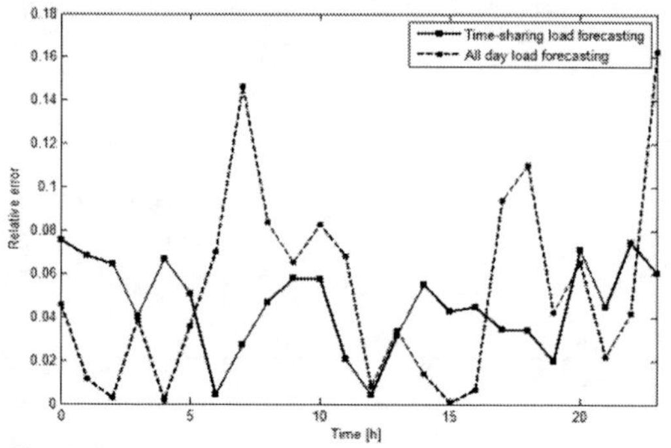

Fig. 4 Comparison curve of the forecasting relative error

6 Conclusion

(1) Temperature has become an important factor that affects the accuracy of load forecasting, and the effect of temperature on short-term load is most significant.

(2) Considering temperature effect, the linear regression forecasting method can meet the demand of short-term load forecasting of power system.

(3) Because the impact of temperature on the load is not exactly the same in different area, it is necessary to choose the forecast model to according to the specific circumstances of the region, which may avoid to increased load forecasting error.

Acknowledgement

This work was financially supported by the national science and technology support branch project (2012BAJ26B00), and the natural science foundation of Liaoning Province (2014027001).

References

[1] Liu Jidong, Han Xueshan, Chu Chengbo, et al, Cooling load of summer grid considering non-meteorological factors, Electric Power Automation Equipment, 2 (2013) 40-46.

[2] Piao Zailin, Meng Xiaofang, Distribution network planning, first ed., China Power Press, Beijing, 2015.

[3] GUAN Yi-bin, CHEN Zhe, LI Qiang, et al, Research on Influences Which Meteorological Factors Have on Jiangsu Summer Peak Load, Jiangsu Electrical Engineering, 6 (2012) 47-50 .

[4] Ren Youming, Ma Lixin, Qu Nana, Study on the load forecasting system of intelligent meteorological information processing, Control & Automation, 12(2011) 79-81.

[5] Leng Beixue, Application of support vector machine in power system short-term load forecasting. Sichuan: Southwest Jiaotong University, (2011) 29-30.

[6] Jun-Min Cha and Bon-Hui Ku, A Study on the Summer and Winter Load Forecasting by using the Characteristics of Temperature Changes in Korean Power System, Journal of International Council on Electrical Engineering，4 (2014)293-296.

[7] Kang Chongqing, Xia Qing, Liu Mei, Power system load forecasting, first ed. , China Power Press, Beijing, 2007, pp.71-72.

[8] Chen Baolin, Optimization theory and methods, Tsinghua University press, Beijing, 2005.

[9] Dong Zehong, Linear regression short-term load forecasting, Beijing: North China Electric Power University, (2012)16-24.

Data Visualization of the Advertising Exchange Flow Statements Generation System on Demand Side Platform Based on Highcharts

Jin Bao Song[1,a], Jin Hong Song[2], Jian Ping Chai[3] and Yue Yang[4]

[1]*Information Engineering School, Communication University of China, Beijing, China*

[2]*Shandong Gold Mining Jiaojia Gold Mine (Laizhou) Co., Ltd, Shandong, China*

[3]*Faculty of Science and Techonology, Communication University of China, Beijing, China*

[4]*Information Engineering School, Communication University of China, Beijing, China*

Abstract. Research shows that the information human obtained about the external world are more than 80% by visual access channel, especially in the background of today's era of the big data. The value of data visualization is particularly important. The flow statements generation system for advertising exchange on Demand-Side Platform based on Real Time Bidding is mainly to solve the large volume of data of flow observation statements and the inefficiency of artificial sorting. It sets RTX Single Sign On and Reverse Login to certificate user's identity through JAVA development, screening flow data according to the platform and region, and invoke Highcharts to generate advertisement exchange flow observation report automatically. In this paper, the data visualization and charts showing are regarded as the final goal, through the introduction of the Spring MVC framework, Eclipse development environment, Highcharts charts generation, JAVA programming language, MySQL database and some other related knowledge. It is to show readers the whole development ideas clearly of the flow statements generation system for advertising exchange on Demand-Side Platform, at the same time the visualized charts by running the whole program codes will be shown.

Keywords: Data Visualization, Flow Statements of Advertising Exchange, Java , Highcharts.

Introduction

The flow statements generation system for advertising exchange based on RTB is to develop a practical Internet interactive system to solve the complex and low efficiency problems of manual sorting data flow statements. The secondary development is done based on Tencent office software RTX in order to increase the function of the single sign-on (SSO) and reverse login and to make the office automation a great extent display. At the same time the database is connected, the classified and filtered advertising exchange flow data is shown on the Web page in the form of lists and charts. The function of the chart print and preservation is realized. The advertisers can monitor the change of advertising exchange flow more clearly and conveniently [1].

Real Time Bidding (RTB) is a kind of bidding technology to evaluate and bid for each user showing behavior on hundreds of millions of websites using third-party technology. Traditional Internet advertising ecosystem usually only has up to three sides including advertisers, advertising agents and the Internet media. However, in Real Time Bidding (RTB) advertising exchange mode, the whole advertising chain has four main bodies including advertisers, the advertising Demand Side Platform (DSP), advertising exchange Platform (Ad Exchange) and the Internet media. Advertisers put own advertising demand on DSP platform. The Internet media puts own advertising flow resources into the advertising exchange platform. DSP completes bidding to buy through docking with the advertising exchange platform technology. When the users access a web site, the Sell Side Platform (SSP) sends the users access signal to advertising exchange platform. Then the detail advertising information will be sent to DSP through the analysis matching of Decision Making Process (DMP). DSP will be bidding. The advertising position will be obtained by the highest price provider. The corresponding advertisement will be shown to target users. The advertising delivery will finish. The advertising real time bidding process flow chart is as shown in Fig. 1.

[a]Corresponding author： songjinbao@cuc.edu.cn

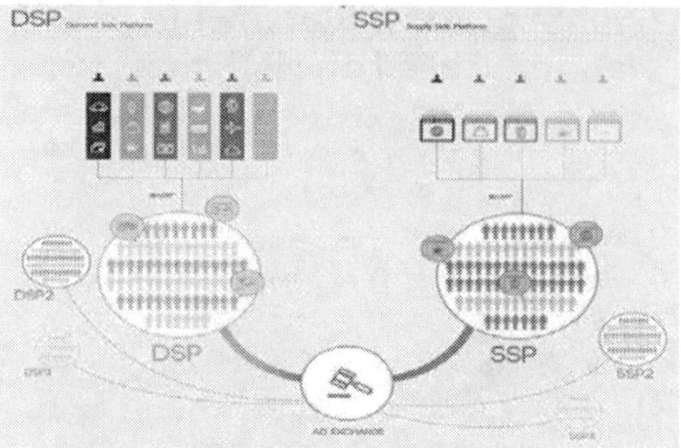

Figure 1. The advertising real time bidding process flow chart

There is the new change in the Internet advertising industry since 2011. Real time bidding (RTB) technology appears in the field of display advertising. The traditional artificial negotiation and flow package pattern slowly are reversed. The previous media advertising has been unable to meet the more and more refined requirements of advertisers. By the end of 2012, the domestic RTB advertising proportion in the whole display advertising is still very small. Today, more than 40 domestic online advertising DSPs are born, such as Easy Media, IPINYOU, etc. On the media side, Tencent, Baidu, Alibaba launch advertising exchange platform. DSP services for the advertisers or agency, in the form of RTB, purchases a variety of media resources (including Ad Network, SSP, Ad Exchange, etc). It is believed that real time bidding (RTB) advertising patterns will be roundly transplanted to advertising market in China soon.

In order to allow advertisers to understand the practical effects of the advertising, the demand side platform (DSP) need for real-time monitoring of advertising exchange flow. The data flow information obtained from advertising exchange flow monitoring tool is stored in DSP internal database. It is necessary for advertisers and DSP internal operating personnel to check and monitor advertising data flow regularly. By checking the flow percentage chart, click rate and the click unit price comparison chart, and participate rate and success rate in the bidding comparison chart of each platform, the development stage of advertising exchange flow can be seen clearly and the corresponding measures can be put forward in time. The required data for chart show was artificially and manually selected from the database ever before. Then a chart is made to view by using Excel. However, the rapid development of real time bidding business also has increasing advertising exchange flow observation data. And it is more and more difficult for flow monitoring and data analysis. The manual sorting flow data reports is very big workload. There are the shortcomings of low efficiency and time-consuming. Thus it becomes a pressing problem that DSP can realize the automatic generation of advertising exchange flow observation report.

Aiming at the three mainstream advertising exchange platform, this paper mainly researches the pie chart of flow occupancy rate for different platform generated by using the Highcharts, participate rate and success rate in the bidding each platform, and the comparison chart of the advertising effect and the flow utilization of each platform. The data visualization of related aspects is also researched [2].

The flow statements generation system for advertising exchange based on RTB mainly regards the data chart show as the core. It is the secondary development to office automation (OA) system RTX in order to increase the function of the single sign-on (SSO) and reverse login. Then the database is connected. The advertising exchange flow data is classified and filtered. The chart show is achieved together by invoking the Highcharts. The problem of low efficiency of manual sorting and complexity of advertising exchange flow data is resolved. The shown chart is not only beautiful and easy but also can realize data visualization. It makes the enterprise internal operational staff more convenient to analysis and sort data at the same time. The rise and decline of advertising exchange flow data and effectiveness of the market can be seen more intuitively.

Spring MVC

The flow data filtering and statements show include two parts of data statement and data graph display. The two parts are completed in the Spring MVC framework. The statement display is connected to MySQL database. The required data tables are queried for the advertisers or development operation personnel. Through the JAVA programming, the required data is displayed on the JSP page in the form of table. And data graph display part is mainly developed by using JAVA language and calling a graph library Highcharts written by pure JavaScript. Highcharts is simple and convenient to add interactive charts in the Web site or Web application. Highcharts interface is aesthetic. Due to the use of JavaScript code, so Highcharts don't need plug-in to run like Flash and JAVA. And Highcharts runs fast. HighCharts had a good compatibility. It will be able to support most of the current browsers. In the development process, by connecting to databases, the flow data can be filtered for each platform and can be made into needed chart data items through the operation. Then data can be converted to the JSON data format. The data can be gotten by using AJAX. The pie chart, rectangle and line - rectangular mixed chart are automatically generated by Highcharts [3].

Spring is a layered Java SE/EE application one-stop lightweight open source framework [4]. Spring's biggest feature is as a kind of container providing class. Spring is one of the factory pattern evolution version. The core in Spring is the BeanFactory. The object instantiation of the programmer is concentrated in the Spring configuration file [5]. On the suggestion of adopting programming to an interface, the highly coupling between the programs is realized by Spring. Two key technologies of Spring are inverse of control (IOC) and aspect oriented programming (AOP) [6].

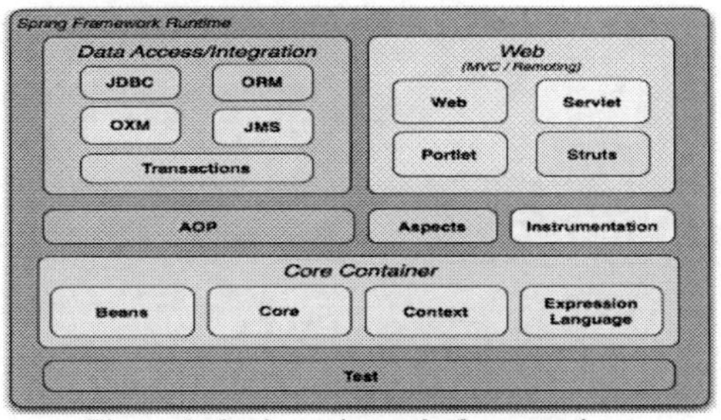

Figure 2. Spring schematic framework

Module Demand Analysis and Function Design

The Internet advertising exchange industry is continuously flourishing today. This new exchange form of the advertising real time bidding gradually win the favor of many advertisers and advertising exchange platform. More and more advertisers choose to place their own advertising demand on the demand side platform (DSP). At the same time, the Internet media advertising and other media resources will be provided on demand side platform (DSP). When the user browses the mainstream web site which has established cooperation with the demand side platform, the Sell Side Platform (SSP) sends the users access signal to advertising exchange platform. The advertising exchange platform launch real time bidding for different size of advertising position. The highest bidding price is success. Advertising can be placed. The whole process of advertising exchange is not only fast but also can advertise according to the crowd orientation. It has particularly good effects. But it also can produce a new problem at the same time. The increasing advertising exchange flow makes the advertisers become very complex for flow monitoring. So the demand side platform will need to build a platform that can automatically call flow data in the database and automatically generate flow analysis chart, which is advantageous for the advertisers to access and monitor.

The flow statements generation system for advertising exchange based on RTB includes three modules: the module of user register and login, the module of statement display and the module of chart display. The module of chart display includes three charts: the flow percentage chart of each platform as shown in Fig. 3, click rate and the click unit price comparison chart of each platform as shown in Fig. 4, and participate rate and success rate in the bidding comparison chart of each platform as shown in Fig. 5.

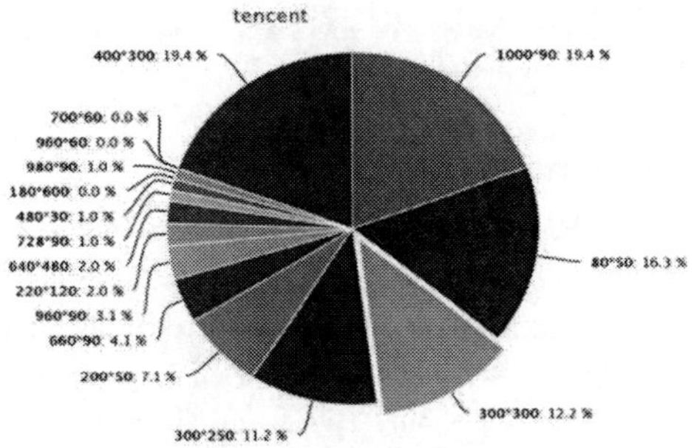

Figure 3. The flow percentage chart of each platform

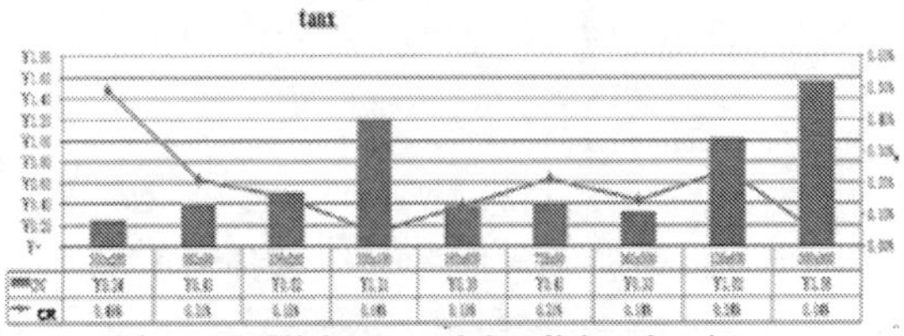

Figure 4. Click rate and the click unit price comparison chart of each platform

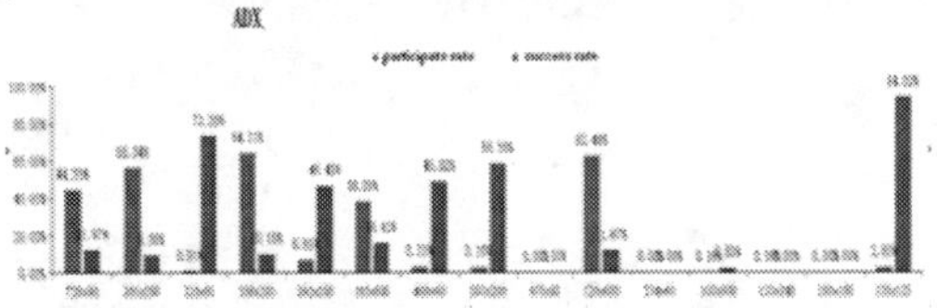

Figure 5. Participate rate and success rate in the bidding comparison chart of each platform

Summary

The flow statements generation system for advertising exchange based on RTB is mainly completed in the Spring MVC framework, using the Eclipse development platform, the MySQL database knowledge, the JAVA language development technology and JSP technology, etc. Research shows that the information human obtained about the external world are more than 80% by visual access channel, especially in the background of today's era of the big data. The value of data visualization is particularly important. The flow statements generation system for advertising exchange on Demand-Side Platform based on Real Time Bidding is mainly to solve the large volume of data of flow observation statements and the inefficiency of artificial sorting. It sets RTX Single Sign On and Reverse Login to certificate user's identity through JAVA development, screening flow data according to the platform and region, and invoke Highcharts to generate advertisement exchange flow observation report automatically. The flow statements generation system for advertising exchange based on RTB solves the problem of low efficiency of manual sorting and complexity of advertising exchange flow data and has good practical applications. The data visualization based on Highcharts also has a broad development prospect.

References

[1] Z. B. Liu, Java WEB core framework. Beijing: Posts & Telecom Press, 2013, pp. 123–135.

[2] R. Johnson and J. Hoeller, Expert one-on-one J2EE Development without EJB, 2015.

[3] J. Kuan, Learning Highcharts, Packt Publishing Limited, 2013.

[4] Johnson R and Hoeller J,The Spring Framework Reference Documentation, 2012, http: / / static. springframework.org/ spring/ docs/ 2. 5. x/ reference/.

[5] Alur D， Crupi J and Malks D, Core J2EE Patterns：Best Practices and Design Strategies, American：Prentice Hall，2013.

[6] M. Fowler, Patterns of Enterprise Application Architecture, 2011.

Force rebalance control for a MEMS gyroscope using ascending frequency drive and generalized PI control

Chun-Hua HE[1,2], Qin-Wen HUANG[2], Qian-Cheng ZHAO[1], Zhen-Chuan YANG[1] , Da-Cheng ZHANG[1] and Gui-Zhen YAN[1]

[1]*National Key Laboratory of Science and Technology on Micro/Nano Fabrication, Institute of Microelectronics, Peking University, Beijing 100871, P. R. China*

[2]*Science and Technology on Reliability Physics and Application of Electronic Component Laboratory, No.5 Electronics Research Institute of the Ministry of Industry and Information Technology, Guangzhou, 510610, Guangdong, P. R. China*

Abstract. This paper has proposed a novel force rebalance control method for a MEMS gyroscope using ascending frequency drive and generalized PI control. Theoretical analyses of ascending frequency drive and force rebalance control methods are illustrated in detail. Experimental results demonstrate that the electrical anti-resonant peaks are located at the frequency responses in the RFD system, which seriously deteriorates the original response characteristics. However, they are eliminated in the AFD system, and the electrical coupling signal is also suppressed. Besides, as for the force rebalance control system, the phase margins approximate to 60deg, gain margins are larger than 13dB, and sensitivity margins are smaller than 3.2dB, which validates the control system is stable and robust. The bandwidth of the force rebalance control system is measured to be about 103.2Hz, which accords with the simulation result. The bias instability and angle random walk are evaluated to be 1.65deg/h and 0.06deg/√h, respectively, which achieves the tactical level.

Keywords: MEMS gyroscope, mode-matching, force rebalance control, ascending frequency drive, electrical coupling suppressing.

Introduction

Electrical coupling is one of the major error sources for a MEMS gyroscope, which is induced by the direct coupling from the drive combs to the sense combs through the parasitic capacitances [1-2]. It will bring about the non-ideal anti-resonance at the amplitude-frequency responses of drive mode and sense mode, which deteriorates the performance and affects the reliability of the closed loop system. Therefore, it is significant to suppress the electrical coupling signals by modifying the control and readout circuit. Electromechanical amplitude modulation (EAM) [3] method can modulate the actual vibration signal by exerting a high-frequency carrier (usually several MHz) to the public mass. Thus, coupling signals can be separated and eliminated from the useful sense signal easily. However, a demodulation process should be added to restore the useful information, which will complicate the control system. Furthermore, the circuit has to handle the demodulation in high frequency domain, which increases the hardware cost, power consumption and high frequency noise.

By introducing a drive modulation process, ascending frequency drive (AFD) method [4-5] is proved effective to suppress the electrical coupling for the drive mode of a MEMS gyroscope. Compared with EAM, it can simplify the readout circuit since it needs no extra demodulation process. In addition, compared to resonant frequency drive (RFD) method, the noise in AFD system can be further suppressed. Hence, in this paper, AFD method will be applied to the force rebalance control system for sense mode of a MEMS gyroscope. Nowadays, force rebalance control for the sense mode is widely adopted in high-performance gyroscopes, since it can adjust the bandwidth, improve the nonlinearity, extend the measure range [6-9]. As for EAM

DOI: 10.1051/matecconf/20166305030

method, a lot of researches about the force rebalance control have been fulfilled and reported in our previous work [8-9]. In this work, novel theoretical analysis of force rebalance control with AFD method will be conducted in detail.

Theoretical analysis

Electrical coupling suppressing

The simplified schematic of a Z-axis doubly decoupled MEMS tuning fork gyroscope is shown in Fig. 1 (a). It mainly includes drive combs, drive-sensing combs, sense combs, force feedback combs, and mode-matching combs. The key parameters of the tested gyroscope at room temperature are listed in Table 1. The resonant frequency of the sense mode ω_s will equal the resonant frequency of the drive mode ω_d after mode-matching control. Fig. 1 (b) shows that there are some parasitic capacitances between the drive combs and sense combs since they are close to each other in the layout.

Table 1. Key parameters of the tested gyroscope at room temperature.

Parameter	Value
Drive mode resonant frequency f_d	9451.5Hz
Drive mode quality factor Q_d	11385
Sense mode resonant frequency f_s	9481.1Hz
Sense mode quality factor Q_s	2300

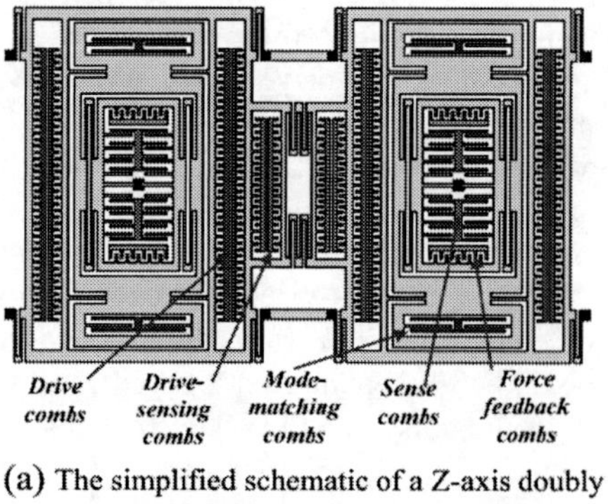

(a) The simplified schematic of a Z-axis doubly decoupled MEMS tuning fork gyroscope

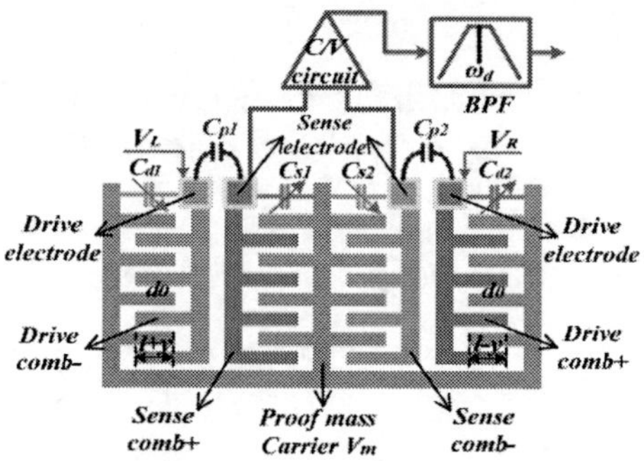

(b) Electrical coupling analysis between the drive combs and sense combs of a MEMS gyroscope

Fig. 1. Electrical coupling analysis of a MEMS gyroscope.

In order to fulfill the differential detection, a *DC* carrier signal V_m is exerted to the proof mass. C_{d1} and C_{d2} are the differential drive capacitances, while C_{s1} and C_{s2} are the differential sense capacitances. C_{p1} and C_{p2} stand for the differential parasitic capacitances between the drive combs and sense combs. V_L and V_R represent the two differential drive signals. Owing to the existence of parasitic capacitances, drive signals can be coupled to the sense combs, which will disturb the original detection signal and deteriorate the closed loop system's stability and performance. Fig. 2 and Fig. 3 depict the schematics of the readout circuit and AFD circuit for the sense mode of a MEMS gyroscope, respectively. In the readout circuit, the output of differential amplifier can be derived as:

$$V_{out} = \frac{k_g[V_m(C_{s2}-C_{s1})+(V_R C_{p2}-V_L C_{p1})]}{C_f} \quad (1)$$

Where, k_g is the gain of the differential amplifier, and C_f is the feedback capacitance of the charge amplifier. In the RFD circuit, the frequencies of V_L and V_R equals to those of C_{s1} and C_{s2}, thus it is hard to distinguish $V_m(C_{s2}-C_{s1})$ from $(V_R C_{p2}-V_L C_{p1})$. Due to the influence of coupling signal, there will be a non-ideal anti-resonance appeared at the amplitude-frequency response [4-5]. However, in the AFD circuit, the frequency of V_L and V_R is far larger than that of C_{s1} and C_{s2}, thus the coupling signal $(V_R C_{p2}-V_L C_{p1})$ can be eliminated by a suitable band-pass-filter (BPF).

Therefore, after filtered by BPF, y_{out} can be described as:

$$y_{out} = \frac{k_g V_m (C_{s2} - C_{s1})}{C_f} = \frac{2k_g V_m N_s \varepsilon h}{C_f d_0} y \quad (2)$$

Where, y is the displacement of the sense mode, ε is the permittivity, N_s is the number of sense capacitances, h is the thickness, and d_0 is the gap. From equation (2), the gain k_{cv} of the readout circuit can be obtained as:

$$k_{cv} = \frac{2k_g V_m N_s \varepsilon h}{C_f d_0} \quad (3)$$

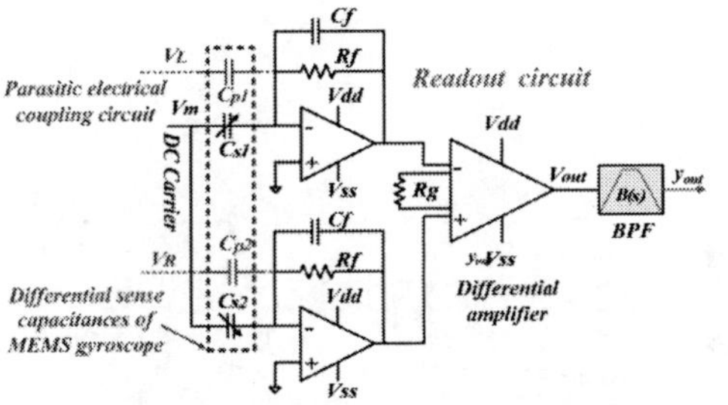

Fig. 2. The readout circuit for the sense mode of a MEMS gyroscope.

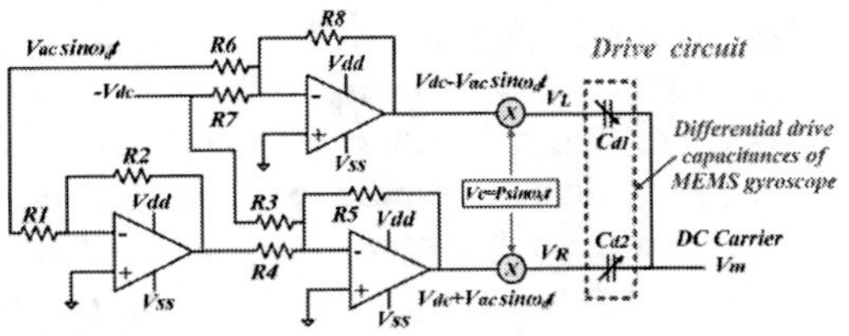

Fig. 3. AFD circuit for the sense mode of a MEMS gyroscope.

From Fig. 3, V_L and V_R can be yielded as:

$$V_R = (V_{dc} + V_{ac}\sin(\omega_d t))P\sin(\omega_c t) \quad (4)$$

$$V_L = (V_{dc} - V_{ac}\sin(\omega_d t))P\sin(\omega_c t) \quad (5)$$

Where, V_{dc} and $V_{ac}sin(\omega_d t)$ are the DC and AC drive signals. $Psin(\omega_c t)$ is the high frequency carrier, and ω_c ? ω_d . Take the noises n_1 and n_2 into account, the drive signals V_1 and V_2 of the differential drive capacitances can be derived as:

$$V_2 = V_R - V_m + n_2 = (V_{dc} + V_{ac}\sin(\omega_d t))P\sin(\omega_c t) - V_m + n_2 \quad (6)$$

$$V_1 = V_L - V_m + n_1 = (V_{dc} - V_{ac}\sin(\omega_d t))P\sin(\omega_c t) - V_m + n_1 \quad (7)$$

From Fig. 1 (b), the energies of the differential drive capacitances can be expressed as:

$$E_2 = \frac{NV_2^2 \varepsilon h(l - y)}{d_0} \quad (8)$$

$$E_1 = \frac{NV_1^2 \varepsilon h(l + y)}{d_0} \quad (9)$$

Where, N is the number of drive capacitances, and l is the overlapped length. Then, the total drive force can be derived as:

$$F_{tot} = \frac{-\partial E_1}{\partial y} + \frac{-\partial E_2}{\partial y} = \frac{N\varepsilon h(V_2^2 - V_1^2)}{d_0} = \frac{N\varepsilon h}{d_0}[2P^2 V_{dc} V_{ac}\sin(\omega_d t) + (n_2 + n_1 - 2V_m)(n_2 - n_1) \quad (10)$$

$$-2P^2 V_{ac} V_{dc}\sin(\omega_d t)\cos(2\omega_c t) + 2(n_2 - n_1)V_{dc}P\sin(\omega_c t) + 2(n_2 + n_1 - 2V_m)V_{ac}P\sin(\omega_d t)\sin(\omega_c t)]$$

Since the frequency response of a gyroscope can be regarded as a BPF, the lower and higher frequency components will be suppressed, and the real useful force can be yielded as:

$$F_{useful} = \frac{2N\varepsilon h P^2 V_{dc}}{d_0} V_{ac}\sin(\omega_d t) \quad (11)$$

Equation (11) indicates that the noises can be suppressed, and there is no need to demodulate the readout signal y_{out} since the frequency of y is ω_d, rather than $\omega_c \pm \omega_d$. Hence, the gain k_{vf} of the drive circuit can be expressed as:

$$k_{vf} = \frac{2N\varepsilon h P^2 V_{dc}}{d_0} \quad (12)$$

Force rebalance control

AFD method is effective to suppress the coupling signal, which is detailedly illustrated above. k_{cv} and k_{vf} of the AFD circuit are yielded in equations (3) and (12). Thus the force rebalance control system for the sense mode can be introduced as Fig. 4. Where, $-\Omega(t)$ and $q(t)$ are input angular rate and quadrature coupling signal, respectively. F_c, F_q and F_b represent Coriolis force, quadrature force, and feedback force, respectively. m_pA_v, k_{qf} and k_{vf} are corresponding force coefficients of them. m_s and m_p represent the sense mass and proof mass, respectively. A_v stands for the amplitude of the drive velocity. k_{dc} is the conversion coefficient from displacement to capacitance, and θ is the demodulation phase. The transfer functions of the sense mode $G_s(s)$, BPF $B(s)$, low-pass-filter (LPF) $L(s)$, and generalized-PI (G-PI) controller $P(s)$ are defined as follows:

$$G_s(s)=\frac{k_{dc}/m_s}{s^2+s\omega_s/Q_s+\omega_s^2} \quad (13)$$

$$B(s)=\frac{2\xi_b\omega_b s}{s^2+2\xi_b\omega_b s+\omega_b^2} \quad (14)$$

$$L(s)=\frac{\omega_l^2}{s^2+2\xi_l\omega_l s+\omega_l^2} \quad (15)$$

$$P(s)=\frac{k_p(s+\omega_2)}{(s+\omega_1)(s+\omega_3)} \quad (16)$$

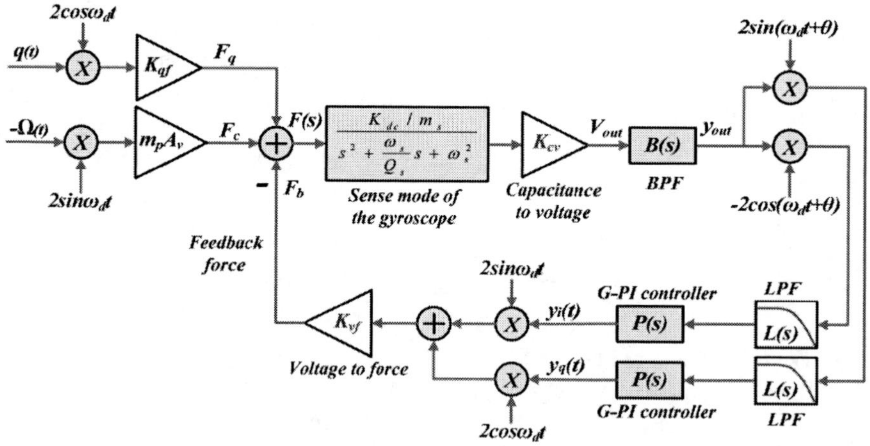

Fig. 4. Force rebalance control system with AFD method for the sense mode of a MEMS gyroscope.

Where, ξ_b and ω_b are the damping ratio and centre frequency of BPF, ξ_l and ω_l are the damping ratio and cut-off frequency of LPF, k_p, ω_1, ω_2 and ω_3 are the key parameters of G-PI controller. There are two control loops shown in Fig. 4, one is utilized to rebalance the Coriolis force, while another is used to rebalance the quadrature force. As for this force control system, the transfer function of the open loop system $D(s)$ can be deducted easily based on Fig. 4:

$$D(s)=\frac{F_b(s)}{F(s)}=2jk_{cv}k_{vf}[e^{-j\theta}L(j\omega\text{-}j\omega_d)P(j\omega\text{-}j\omega_d)\text{-}e^{j\theta}L(j\omega+j\omega_d)P(j\omega+j\omega_d)]B(j\omega)G_s(j\omega) \quad (17)$$

Where, complex frequency s is equal to $j\omega$. Thus the transfer function of the closed loop system $T(s)$ and the sensitivity function $S(s)$ can be yielded as:

$$T(s)=\frac{D(s)}{1+D(s)} \quad (18)$$

$$S(s)=\frac{1}{1+D(s)} \quad (19)$$

DOI: 10.1051/matecconf/20166305030

Experimental test

Swept frequency test

The simplified schematic of a Z-axis doubly decoupled MEMS tuning fork gyroscope is shown in Fig. 5. It mainly includes an analog circuit and a digital circuit. Swept frequency tests of the sense mode are conducted in the RFD system and AFD system, respectively. Experimental results demonstrate that the electrical anti-resonant peaks are located at the amplitude-frequency and phase-frequency responses in the RFD system, which seriously deteriorates the original response characteristics, as shown in Fig. 6 (a). Due to the existence of electrical anti-resonant peaks, it is very hard to design a suitable G-PI controller to guarantee the reliability and robustness of the force rebalance control system. Therefore, it is urgent to suppress the electrical coupling. On the other hand, in Fig. 6 (b), there are no anti-resonant peaks appeared in the AFD system, and the electrical coupling signal is eliminated, which indicates the theoretical analysis is accurate.

Fig. 5. The simplified schematic of a Z-axis doubly decoupled MEMS tuning fork gyroscope.

Force rebalance control test

After the theoretical deduction above, equations (3) and (12)-(19) can be utilized to design the closed loop system. Simultaneously, simulations based on these equations are accomplished with Matlab software, as shown in Fig. 7 and Fig. 8. In Fig. 7, owing to narrow-band force rebalance control, there are two crossover frequencies appearing in the frequency-amplitude response of *D(s)*. Thus, similar to our previous analysis [9], there are two corresponding phase margins (PM_l and PM_r), gain margins (GM_l and GM_r), cut-off frequencies (f_{tl} and f_{tr}) and sensitivity margins (M_{sl} and M_{sr}), as marked in the figure. The key stability indices of the feedback control system obtained by Matlab simulation are listed in Table 2. According to classical linear control theory, the force rebalance control system is stable and robust enough. Besides, the bandwidth (BW_{lr}) of *T(s)* is calculated to be 103.18Hz, which means the force outside this narrow-band will never be balanced.

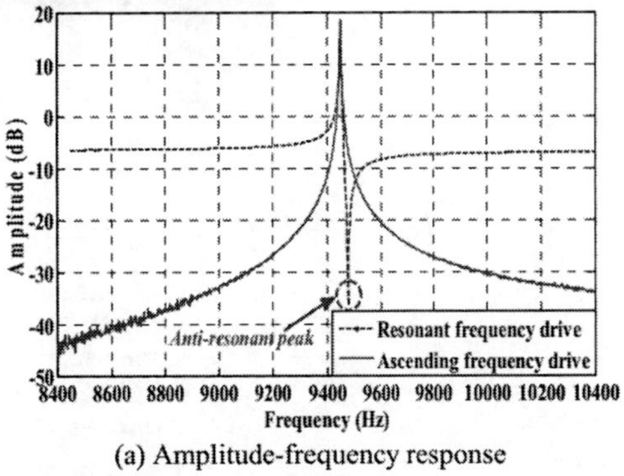

(a) Amplitude-frequency response

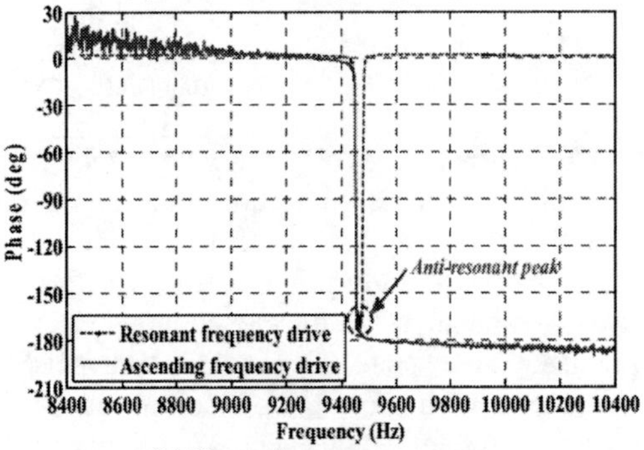

(b) Phase-frequency response

Fig. 6. Swept frequency tests of the sense mode in RFD and AFD methods.

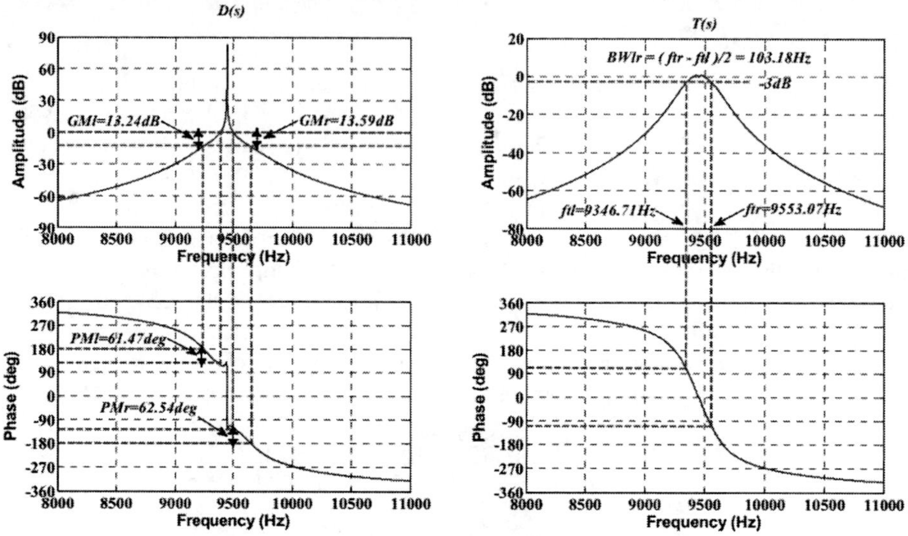

Fig. 7. The frequency responses of the open loop system *D(s)* and closed loop system *T(s)*.

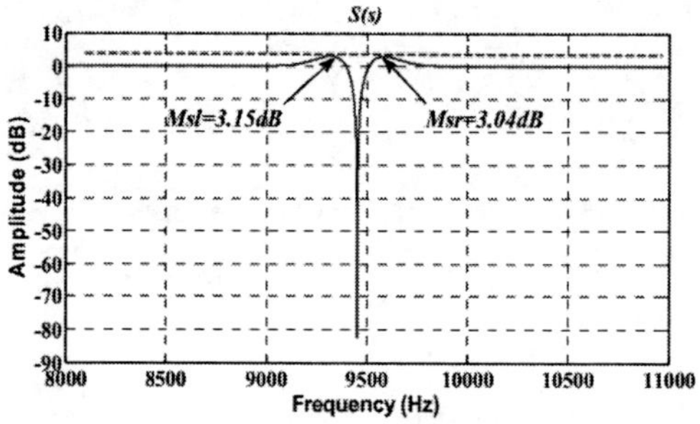

Fig. 8. The frequency response of the sensitivity function *S(s)*.

Table 2. Key stability indices of the feedback control system obtained by Matlab simulation.

Type	Key stability indices	
Phase margin PM (deg)	PM_l=61.47	PM_r=62.54
Gain margin GM (dB)	GM_l=13.24	GM_r=13.59
Sensitivity margin Ms (dB)	M_{sl} =3.15	M_{sr}=3.04
Cut-off frequency f_t (Hz)	f_{tl}=9346.71	f_{tr}=9553.07
Bandwidth BW (Hz)	BW_{lr} =(f_{tr} - f_{tl})/2=103.18Hz	

The bandwidth (BW) of the force rebalance control system is measured to be about 103.2Hz, which approximates to the simulation result, as shown in Fig. 9. It further verifies the theoretical analysis of the closed loop system is precise. The scale factor of the MEMS gyroscope with closed loop controlled sense mode is 65.9mV/deg/s with nonlinearity of 0.03%. Fig. 10 illustrates that the bias instability and angle random walk are evaluated to be 1.65deg/h and 0.06deg/√h, respectively, which reaches the tactical level.

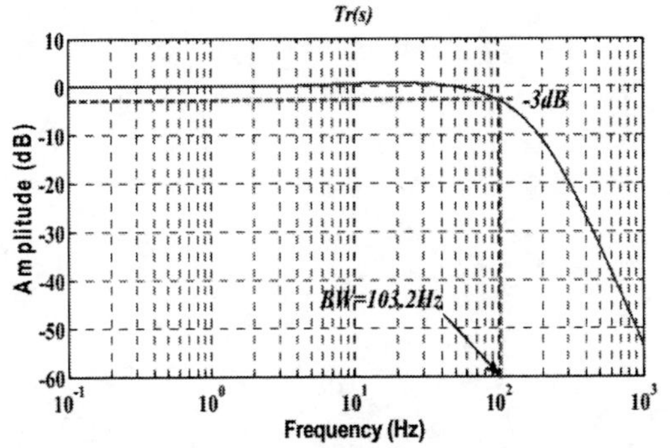

Fig. 9. Bandwidth test results of the closed loop control system.

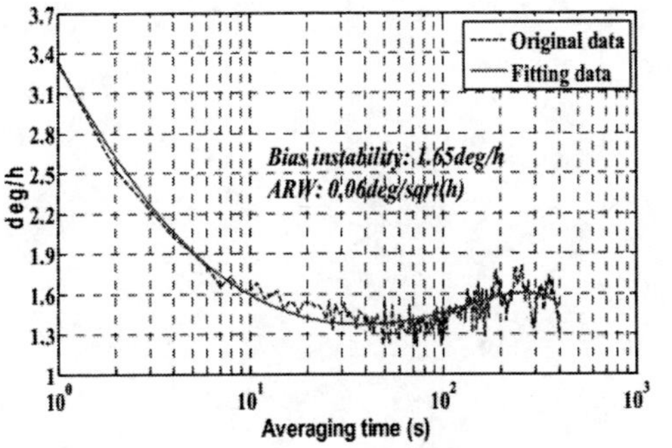

Fig. 10. Allan variance graph of bias drift of the MEMS gyroscope with closed loop controlled sense mode.

Summary

In this paper, a novel force rebalance control method has been presented for a MEMS gyroscope using ascending frequency drive and generalized PI control. Theoretical analyses of ascending frequency drive and force rebalance control methods are illustrated in detail. Experimental results demonstrate that the electrical anti-resonant peaks are located at the amplitude-frequency and phase-frequency responses in the RFD system, which seriously deteriorates the original response characteristics. However, they are eliminated in the AFD system, and the electrical coupling signal is also suppressed. On the other hand, the frequency responses of the force rebalance control system indicate that the phase margins approximate to 60deg, gain margins are larger than 13dB, and sensitivity margins are smaller than 3.2dB, which validates the control system is stable and robust. The bandwidth of the force rebalance control system is measured to be about 103.2Hz, which verifies the theoretical analysis is accurate. The bias instability and angle random walk are evaluated to be 1.65deg/h and 0.06deg/√h, respectively, which reaches the tactical level.

Acknowledgement

This work is partially supported by the National Natural Science Foundation of China under Grant No. 61434003.

References

[1] Lee, H. Ko, et al., "Non-ideal behavior of a driving resonator loop in a vibratory capacitive microgyroscope", Microelectronics J., vol. 39, 2008, pp. 1-6.

[2] M.S. Weinberg and A. Kourepenis, "Error sources in in-plane silicon tuning-fork MEMS gyroscopes", J. Microelectromech. Syst., vol. 15, 2006, pp. 479-491.

[3] Acar, A.M. Shkel, "Structurally decoupled micromachined gyroscopes with post-release capacitance enhancement", J. Micromech. Microeng., 2005, 15 1092-1101.

[4] J. Cui, Z. Y. Guo, Z.C. Yang, et al., "Electrical coupling suppressing for a microgyroscope using ascending frequency drive with 2-DOF PID controller", Proc. Transducers 2011, pp. 2002-2005.

[5] J. Cui, Z.Y. Guo, Z.C. Yang, et al., "Electrical coupling suppression and transient response improvement for a microgyroscope using ascending frequency drive with a 2-DOF PID controller", J. Micromech. Microeng., 2011, 21 095020.

[6] S. Sonmezoglu, S.E. Alper, T. Akin, "An Automatically Mode-Matched MEMS Gyroscope With Wide and Tunable Bandwidth", J. Microelectromech. Syst., Vol. 23, No. 2, 2014, pp.284-297.

[7] R. Mirjalili, H. Wen, D. E. Serrano, and F. Ayazi, "Substrate-decoupled silicon disk resonant having degenerate gyroscopic

modes with Qin excess of 1-million", Proc. TRANSDUCERS 2015, pp. 15-18.

[8] C.H. He, Q.C. Zhao, Y.X. Liu, et al., "Closed loop control design for the sense mode of micromachined vibratory gyroscopes", Sci. China Tech. Sci. 2013, Vol. 56, pp. 1112-1118.

[9] C.H. He, D.C. Liu, Q.C. Zhao, et al., "A novel narrow-band force rebalance control method for the sense mode of MEMS vibratory gyroscopes", Measurement, vol. 62, 2015, pp. 197-204.

MATEC Web of Conferences 63, 05031 (2016) DOI: 10.1051/matecconf/20166305031
MMME 2016

The New Approach to the Synthesis of Single-Channel Consistent Estimates of the Time Signal Parameters

Oleg V. CHERNOYAROV[1,2], Yury A. KUTOYANTS[1,3] and Boris I. SHAKHTARIN[4]

[1]*National Research University "MPEI", 14, Krasnokazarmennaya st., 111250 Moscow, Russia*

[2]*International Laboratory of Statistics of Stochastic Processes and Quantitative Finance of National Research Tomsk State University, 36, Lenin avenue, 634050 Tomsk, Russia*

[3]*Universite du Maine, Avenue Olivier Messiaen, 72085 Le Mans, France*

[4]*Bauman Moscow State University 5, Baumanskaya 2-ya st., 105005 Moscow, Russia*

Abstract. Based on the introduced modification of a maximum likelihood method we suggest the new technique for synthesis of single-channel quasi-optimal processing algorithms of signals with multiple unknown parameters. We illustrate the proposed approach by the estimation of the appearance time and duration of a rectangular video pulse. We show that unlike technically complex multichannel maximum likelihood estimates, the single-channel quasi-optimal estimates of time signal parameters are consistent for the implementation. We also find the analytical expressions for characteristics of passed estimates which allow us to assess the reasonableness of the application of this or that processing algorithm in each specific case.

Keywords: rectangular pulse, maximum likelihood method, unknown appearance time and duration, joint estimation, single-channel reception, characteristics of estimates.

Introduction

The estimation problem of the time parameters of pulse signals observed against noise is actual for many applications of radio- and hydrolocations, radio communications, seismology etc. [1,2]. In [3] the measurer of the appearance time and duration of a rectangular video pulse, which is synthesized using a maximum likelihood method, is presented. However, its technical implementation is possible as a multichannel device only and the estimates obtained are not consistent with a finite number of measurer channels. In work [4] the more general problem is considered of reception of the signal with stepwise changes at the M random points in time. But, in order to find the estimates of these points even in the simplest case when $M = 2$, it is necessary to solve rather complex nonlinear stochastic equations. In [5] the two-channel estimation algorithms of the moments of appearance and disappearance of a rectangular pulse are synthesized based on the maximum likelihood and Bayesian approaches. However, the domain of their applicability is limited to the tasks of localization of an information signal (when the prior interval of possible values of the moment of pulse appearance and disappearance does not exceed two minimum possible pulse durations).

Below we show that using the modification of a maximum likelihood method the very simple and easily feasible estimation algorithm of the appearance time and duration of a rectangular video pulse cab be obtained, without restrictions to the domain of like values of unknown parameters.

The synthesis of the estimation algorithm

So, let the additive mix

$$x(t) = s(t, \lambda_0, \tau_0) + n(t) \qquad (1)$$

is passed to the receiver input during time slice $[0, T]$. Here

$$s(t, \lambda_0, \tau_0) = aI\left(\frac{t - \lambda_0}{\tau_0}\right),$$
$$I(x) = \begin{cases} 1, & |x| \le 1/2, \\ 0, & |x| > 1/2 \end{cases} \qquad (2)$$

is the useful signal, which is a rectangle video pulse with amplitude a, appearance time λ_0 and duration τ_0, and $n(t)$ is Gaussian white noise with one-sided spectral density N_0.

On observable realization (1) it is necessary to estimate the parameters λ_0, τ_0 possessing the values from appropriate prior intervals $[\Lambda_1, \Lambda_2]$, $[T_1, T_2]$. And we presuppose that for Λ_1, Λ_2 the following condition is satisfied: $0 \le \Lambda_1 - T_2/2 < \Lambda_2 + T_2/2 \le T$, that is pulse (2) is always located within observation interval $[0, T]$.

According to [6] we write down the logarithm of the functional of likelihood ration (FLR) as function of current values λ, τ of the unknown parameters λ_0, τ_0 in the form of

$$L(\lambda, \tau) = \frac{2a}{N_0} \int_{\lambda - \tau/2}^{\lambda + \tau/2} x(t)\,dt - \frac{a^2 \tau}{N_0}. \qquad (3)$$

When maximizing Eq. (3) on variables λ, τ, we obtain the maximum likelihood estimates (MLEs) of the pulse (2) appearance time and duration defined as [3]

$$(\lambda_m, \tau_m) = \underset{\lambda \in [\Lambda_1, \Lambda_2], \tau \in [T_1, T_2]}{\arg\sup} L(\lambda, \tau). \qquad (4)$$

As can be seen, the measurer (4) is multichannel device and the number of channels should be infinite for its exact implementation that is not likely to take place in practice.

Let us consider the possibility to obtain the technically simpler and practically realized estimates of parameters λ_0, τ_0. For this purpose we study the behavior of the logarithm of FLR (3) as function of variable λ at the fixed value $\tau = \tau^*$. Following [6], we present the functional $L(\lambda, \tau^*)$ as sum of signal $S(\lambda) = \langle L(\lambda, \tau^*) \rangle$ and noise $N(\lambda) = L(\lambda, \tau^*) - \langle L(\lambda, \tau^*) \rangle$ functions:

$$L(\lambda, \tau^*) = S(\lambda) + N(\lambda). \qquad (5)$$

Here $\langle \ \rangle$ designate the averaging operation on all realizations of the observable data (1) at the fixed values λ_0, τ_0.

From Eq. (3) for the signal function we get

$$S(\lambda) = 2a^2 C(\lambda - \lambda_0)/N_0 - a^2 \tau^*/N_0,$$

where

$$C(\lambda) = \begin{cases} \tau_0 - \max(0, \tau_0 - \tau^*), & |\lambda| \le |\tau_0 - \tau^*|/2, \\ (\tau_0 - \tau^*)/2 - |\lambda|, & |\tau_0 - \tau^*|/2 < |\lambda| \le (\tau_0 + \tau^*)/2, \\ 0, & |\lambda| > (\tau_0 + \tau^*)/2. \end{cases}$$

Noise function $N(\lambda)$ is centered Gaussian random process with correlation function of the kind of

$$\langle N(\lambda_1)N(\lambda_2)\rangle = \left(2a^2/N_0\right)\max\left[0, \tau^* - |\lambda_2 - \lambda_1|\right].$$

As is widely known [6], the logarithm of FLR (3) as well as any biunique instantaneous its transformation is the sufficient statistics with all the information on unknown parameters λ_0, τ_0 which can be extracted from the realization of the observable data (1). As such transformation we apply the derivation (on a variable λ) to Eq. (5). Then we have

$$M(\lambda) = L'(\lambda, \tau^*) = \frac{2a^2}{N_0}\left[I\left(\frac{\lambda - \theta_{01}}{\tau_{min}}\right) - I\left(\frac{\lambda - \theta_{02}}{\tau_{min}}\right)\right] + \xi(\lambda). \quad (6)$$

(6)

Here $\theta_{01} = \lambda_0 - \tau_{max}/2$, $\theta_{02} = \lambda_0 + \tau_{max}/2$, $\tau_{max} = \max(\tau_0, \tau^*)$, $\tau_{min} = \min(\tau_0, \tau^*)$, $\xi(\lambda)$ is Gaussian white noise with one-sided spectral density $N_\xi = 8a^2/N_0$.

From Eq. (6) follows that when

$$\tau^* \le T_1 \quad (7)$$

the estimation problem of the appearance time and duration of the pulse (2) can be reduced to the estimation problem of the appearance times of two heteropolar rectangular video pulses by duration τ^* everyone. Indeed, by forming the estimates θ_{1q}, θ_{2q} of the parameters θ_{01}, θ_{02} based on Eqs. (6), for the estimates λ_q, τ_q of the parameters λ_0, τ_0 we get

$$\lambda_q = \left(\theta_{2q} + \theta_{1q}\right)/2, \quad \tau_q = \theta_{2q} - \theta_{1q}. \quad (8)$$

The estimates θ_{1q}, θ_{2q} can be found by one of possible optimal ways, for example, by means of the correlation reception [6]. Then we have

$$\theta_{1q} = \underset{\theta_1 \in [\Lambda_1 - \tau^*/2, \Lambda_2 - \tau^*/2]}{\arg\sup} M_1(\theta_1), \quad \theta_{2q} = \underset{\theta_2 \in [\Lambda_1 + \tau^*/2, \Lambda_2 + \tau^*/2]}{\arg\inf} M_1(\theta_2), \quad M_1(\theta) = \int_{\theta - \tau^*/2}^{\theta + \tau^*/2} M(\lambda)\, d\lambda. \quad (9)$$

At the same time, obviously, the following condition should be satisfied

$$\tau^* \le \theta_{2q} - \theta_{1q} \le T_2. \quad (10)$$

In order to ensure the fulfillment of inequalities (10) for any sizes of prior domains $[\Lambda_1, \Lambda_2]$, $[T_1, T_2]$ and signal-to-noise ratios (SNRs), instead of algorithm (9) one of its possible modification can be used, for example,

$$\theta_{1q} = \underset{\theta_1 \in [\Lambda_1 - \tau^*/2, \Lambda_2 - \tau^*/2]}{\arg\sup} M_1(\theta_1), \quad \theta_{2q} = \underset{\theta_2 \in [\theta_{1q} + T_1, \min(\theta_{1q} + T_2, \Lambda_2 + T_1 - \tau^*/2)]}{\arg\inf} M_1(\theta_2) \quad (11)$$

or

$$\theta_{1q} = \underset{\theta_1 \in [\max(\Lambda_1 - T_1 + \tau^*/2, \theta_{2q} - T_2), \theta_{2q} - \tau^*]}{\arg\sup} M_1(\theta_1), \quad \theta_{2q} = \underset{\theta_2 \in [\Lambda_1 + \tau^*/2, \Lambda_2 + \tau^*/2]}{\arg\inf} M_1(\theta_2).$$

The expressions (3), (6), (9), (11) determine the organization of the quasi-optimal measurer of the appearance time and duration of a signal (2). The block diagram of one of possible measurer implementations is shown in Fig. 1 where are the designations: 1 is the switch that is open for time $[0,T]$, 2 is an integrator, 3 is a delay line for the period τ^*, 4 is the substracter, 5 is the differentiator, 6 and 10 is the retrievers of the location of the input signal greatest maximum and minimum (extremators), respectively, 7 is the summator, 8 is the half-gain amplifier (halver), 9 is the switch that is open for time $\left[\theta_{1q}+T_1, \min\left(\theta_{1q}+T_2, \Lambda_2+T_1-\tau^*/2\right)\right]$.

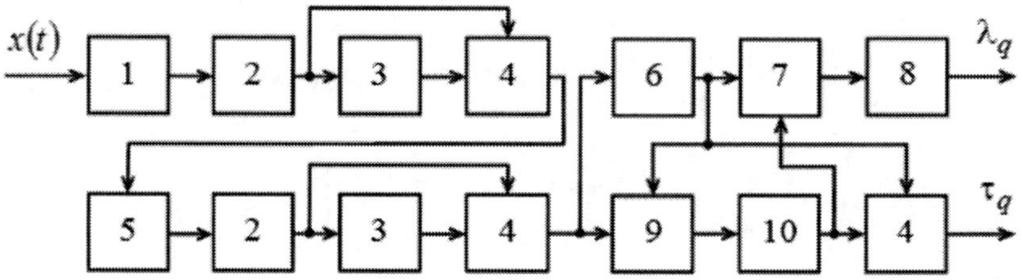

Figure 1. The quasi-optimal measurer of the appearance time and duration of a rectangular pulse.

As shown in Fig. 1, the introduced measurer is only two-channel and technically simpler than known analogues. And the closer τ^* to τ_0, this measurer tends to optimal.

The characteristics of the passed estimates

Let us find the characteristics of the estimates (8). We presuppose that SNR z^2 for algorithm (11) is great enough, i.e. the following condition is satisfied

$$z^2 = S_I^2(\theta_{01})/\langle N_I^2(\theta_{01})\rangle = S_I^2(\theta_{02})/\langle N_I^2(\theta_{02})\rangle = a^2\tau^*/N_0 = z_0^2\Delta_\tau/2 >> 1. \tag{12}$$

Here

$$S_I(\theta) = \langle M_I(\theta)\rangle = \left(2a^2/N_0\right)\left[\max\left(0, \tau^* - |\theta-\theta_{01}|\right) - \max\left(0, \tau^* - |\theta-\theta_{02}|\right)\right]$$

is the signal component, $\langle N_I^2(\theta)\rangle = 4a^2\tau^*/N_0$ is the dispersion of the noise component $N_I(\theta) = M_I(\theta) - \langle M_I(\theta)\rangle$ of the decision statistics (9), $z_0^2 = 2a^2\tau_0/N_0$ is SNR for the optimal receiver of a signal (2) with prior known duration, $\Delta_\tau = \tau^*/\tau_0$ is the relative pulse duration deviation.

According to [7] with increased z^2, the estimates θ_{1q}, θ_{2q} (11) converge to true values θ_{01}, θ_{02} in mean square. So, if the condition (12) is satisfied, then the estimates θ_{1q}, θ_{2q} are located in small neighborhoods of points θ_{01}, θ_{02}, respectively, with the probability close to 1. In addition, the estimates (11) are statistically independent, as integrals of white noise on nonoverlapping intervals. Therefore, in the conditions of high posterior accuracy the conditional biases $b_0\left(\lambda_q|\lambda_0,\tau_0\right)$, $b_0\left(\tau_q|\lambda_0,\tau_0\right)$ and variances $V_0\left(\lambda_q|\lambda_0,\tau_0\right)$, $V_0\left(\tau_q|\lambda_0,\tau_0\right)$ of the estimates λ_q, τ_q (8) can be expressed through the conditional biases $b_0\left(\theta_{1q}|\theta_{01}\right)$, $b_0\left(\theta_{2q}|\theta_{02}\right)$ and variances $V_0\left(\theta_{1q}|\theta_{01}\right)$, $V_0\left(\theta_{2q}|\theta_{02}\right)$ of the estimates θ_{1q}, θ_{2q} as follows:

$$b_0\left(\lambda_q|\lambda_0,\tau_0\right) = \left[b_0\left(\theta_{1q}|\theta_{01}\right) + b_0\left(\theta_{2q}|\theta_{02}\right)\right]/2, \quad b_0\left(\tau_q|\lambda_0,\tau_0\right) = b_0\left(\theta_{2q}|\theta_{01}\right) - b_0\left(\theta_{1q}|\theta_{02}\right),$$

$$V_0(\lambda_q|\lambda_0,\tau_0)=\left[V_0(\theta_{1q}|\theta_{01})+V_0(\theta_{2q}|\theta_{02})\right]/4,\quad V_0(\tau_q|\lambda_0,\tau_0)=V_0(\theta_{2q}|\theta_{02})-V_0(\theta_{1q}|\theta_{01}). \quad (13)$$

The characteristics of the estimate of the appearance time of a rectangular video pulse were studied in a number of works [7-9, etc.].

Using these results for conditional biases and variances of the estimates θ_{1q}, θ_{2q} we have

$$b_0(\theta_{1q}|\theta_{01})=b_0(\theta_{2q}|\theta_{02})=0,\quad V_0(\theta_{1q}|\theta_{01})=V_0(\theta_{2q}|\theta_{02})=13\tau_0^2/2z^4=26\tau_0^2/z_0^4\Delta_\tau^2. \quad (14)$$

From Eqs. (13), (14) follows that at high values of SNR the estimates λ_q, τ_q (8) are conditionally unbiased and possess the conditional variances

$$V_0(\lambda_q|\lambda_0,\tau_0)=13\tau_0^2/z_0^4\Delta_\tau^2,\qquad V_0(\tau_q|\lambda_0,\tau_0)=52\tau_0^2/z_0^4\Delta_\tau^2. \quad (15)$$

With fixed values of the signal (2) amplitude and spectral density of white noise the variances (15) is the less when the relative detuning Δ_τ is close to 1. Thus, the value τ^* is expedient for choosing as maximum possible, i.e. taking into account Eq. (7) as $\tau^*=T_1$. If we take $\Delta_\tau=1$ ($\tau^*=\tau_0$) in Eq. (15), then we obtain the expressions for conditional variances of MLEs (4) of the appearance time and duration of a rectangular video pulse without considering the hardware implementation of a measurer (when the number of channels is infinitely large) [3]. We shall note, that the variances of the combined estimates (15) are twice the variances of corresponding separate estimates [8,10], i.e. there is a nonlinear statistical dependence between estimates λ_q and τ_q [3].

For unconditional variances of the estimates (8) we have

$$V_0(\lambda_q)=\int_{\Lambda_1}^{\Lambda_2}\int_{T_1}^{T_2}V_0(\lambda_q|\lambda_0,\tau_0)w_{pr}(\lambda_0,\tau_0)d\lambda_0 d\tau_0=\int_{T_1}^{T_2}V_0(\lambda_q|\lambda_0,\tau_0)w_{pr}(\tau_0)d\tau_0,$$

$$V_0(\tau_q)=\int_{\Lambda_1}^{\Lambda_2}\int_{T_1}^{T_2}V_0(\tau_q|\lambda_0,\tau_0)w_{pr}(\lambda_0,\tau_0)d\lambda_0 d\tau_0=\int_{T_1}^{T_2}V_0(\tau_q|\lambda_0,\tau_0)w_{pr}(\tau_0)d\tau_0. \quad (16)$$

Here it is designated: $w_{pr}(\lambda_0,\tau_0)$ – joint probability density of the parameters λ_0, τ_0, $w_{pr}(\tau_0)$ – prior probability density of the parameter τ_0 and it is also taken into account that the variances (15) do not depend on λ_0.

If the number of channels N is finite, then similarly to [11] the quality of the estimate λ_{mN} of the appearance time of a signal (2) can be characterized its limit (maximum) conditional $V_0(\lambda_{mN}|\lambda_0)$ or unconditional $V_0(\lambda_{mN})$ variance, as well as the quality of the estimate τ_{mN} of the duration – the conditional $V_0(\tau_{mN}|\tau_0)$ or unconditional $V_0(\tau_{mN})$ variance of the estimate for multichannel reception. We introduce the following designations: $\tau_i=T_1+(i-1/2)\Delta$, $i=\overline{1,N}$ is the duration, to which i-th channel is tuned, $\Delta=(T_2-T_1)/N$ is the distance between the two adjacent channels, $\delta=(T_2-T_1)/(T_2+T_1)$ is the maximum possible absolute deviation of the signal duration τ_0 (2) from the average duration $\tau_c=(T_2+T_1)/2$, $z_c^2=2a^2\tau_c/N_0$ is SNR for a pulse (2), the duration of which is equal to average duration τ_c, $w(\tau_m|\tau_0)$ is the conditional probability density of MLE τ_m (4) in the measurer with the continuous change of current value of the estimated parameter. Then, in terms of the results [11,12], we have

$$V_0(\lambda_{mN}|\lambda_0)=V_0(\lambda_{mN})=\frac{\delta^2}{8N^2}+\exp\left(\frac{z_c^2\delta}{N}\right)\left[1-\Phi\left(z_c\sqrt{\frac{2\delta}{N}}\right)\right]\times$$
$$\times\left(\frac{13}{z_c^4}-\frac{13\delta}{z_c^2N}+\frac{8\delta^2}{N}-\frac{4z_c^2\delta^3}{3N^3}\right)+\sqrt{\frac{\delta}{\pi N}}\left(\frac{2z_c\delta^2}{3N^2}-\frac{13\delta}{3zN}+\frac{13}{z_c^3}\right), \quad (17)$$

$$V_N(\tau_{mN}|\tau_0)=\sum_{i=-(\nu-1)}^{N-\nu}(i\Delta)^2\int_{T_1+(\nu+i-1)\Delta}^{T_1+(\nu+i)\Delta}w(\tau_m|\tau_0)d\tau_m\,, \quad (18)$$

$$V_N(\tau_{mN})=\sum_{i=-(N-1)}^{N-1}(i\Delta)^2\sum_{\nu=1}^{N}\int_{T_1+(\nu-1)\Delta}^{T_1+\nu\Delta}w_{pr}(\tau_0)\int_{T_1+(\nu+i-1)\Delta}^{T_1+(\nu+i)\Delta}w(\tau_m|\tau_0)d\tau_m d\tau_0\,. \quad (19)$$

In Eq. (18) the number of the channel, nearest to τ_0, is meant by ν.

The probability density $w(\tau_m|\tau_0)$ can be written down through the joint probability density $w(\lambda_m,\tau_m|\lambda_0,\tau_0)$ of the estimates (4) found in [3] as

$$w_2(\lambda_m,\tau_m|\lambda_0,\tau_0)=\frac{w_1[\lambda_m-\lambda_0-(\tau_m-\tau_0)/2]w_1[\lambda_m-\lambda_0+(\tau_m-\tau_0)/2]}{\int_{\Lambda_1}^{\Lambda_2}\int_{T_1}^{T_2}w_1[\lambda_m-\lambda_0-(\tau_m-\tau_0)/2]w_1[\lambda_m-\lambda_0+(\tau_m-\tau_0)/2]d\lambda_m\,d\tau_m},$$

where

$$w_2(\lambda_m,\tau_m|\lambda_0,\tau_0)=\frac{w_1[\lambda_m-\lambda_0-(\tau_m-\tau_0)/2]w_1[\lambda_m-\lambda_0+(\tau_m-\tau_0)/2]}{\int_{\Lambda_1}^{\Lambda_2}\int_{T_1}^{T_2}w_1[\lambda_m-\lambda_0-(\tau_m-\tau_0)/2]w_1[\lambda_m-\lambda_0+(\tau_m-\tau_0)/2]d\lambda_m\,d\tau_m},$$

$$w_1(x)=(z_0^2/2\tau_0)w(z_0^2x/2\tau_0),\qquad w(x)=\Phi\left(\sqrt{|x|/2}\right)-1+3\exp(2|x|)\left[1-\Phi\left(3\sqrt{|x|/2}\right)\right].$$

It should be noted that unlike the quasi-optimal estimates (8), the MLEs of the appearance time and duration (4) of a signal (2) are not consistent with a finite number of channels N.

The expressions (15)-(19) allow us to assess the reasonableness of the application of this or that estimation algorithm of time parameters of a rectangular video pulse depending on the available prior information and the required accuracy and simplicity of hardware measurer implementation.

Conclusion

In order to jointly estimate the several parameters of information signals the technique based on additional transformation of the logarithm of the functional of likelihood ratio can be used in some cases. The purpose of the specified transformation is to divide a regular component of decision statistics into the components depending on one parameter, to carry out a separate estimation of parameters and to find the values of required parameters by the found separate estimates. Quasi-optimal measurers obtained in that way are technically simpler in comparison with optimal (maximum likelihood) analogues and can provide higher performance in actual conditions. The synthesis of two-channel estimation algorithm of the appearance time and duration of a rectangular video pulse represented in the work confirms the efficiency and usefulness of the introduced approach in practical applications.

Acknowledgement

This research was financially supported by the Russian Science Foundation (research project No. 14-49-00079).

References

[1] A.G. Tartakovskii, Detection of signals with random moments of appearance and disappearance, Problems of Information Transmission, 2 (1988) 115-124.

[2] V.G. Repin, Detection of a signal with unknown moments of appearance and disappearance, Problems of Information Transmission, 1 (1991) 49-58.

[3] A.P. Trifonov, V.K. Buteyko, Joint estimation of two parameters of an analog signal against a white-noise background, Soviet Journal of Communications Technology & Electronics, 12 (1990) 25-32.

[4] A.A. Mal'tsev, A.M. Silaev, Optimum estimates of the times of random irregular variations in signal parameters, Soviet journal of communications technology & electronics, 13 (1989) 116-124.

[5] A.P. Trifonov, Yu.E. Korchagin, Optimal reception of a rectangular pulse with unknown times of appearance and disappearance, Izvestiya Vysshikh Uchebnykh Zavedenij. Radiofizika, 3 (2000) 271-282.

[6] E.I. Kulikov, A.P. Trifonov, Estimation of Signal Parameters Against Hindrances (in Russian), Sovetskoe Radio, Moscow, 1978.

[7] A.P. Trifonov, Yu.S. Shinakov, Joint Discrimination of Signals and Estimation of Their Parameters against Background (in Russian), Radio I Svyaz', Moscow, 1986.

[8] A.S. Terent'ev, Probability distribution of the absolute maximum time position at the output of matched filter. Radiotekhnika I Elektronika, 4 (1968), 652-657.

[9] O.V. Chernoyarov, Sai Si Thu Min, A.V. Salnikova, B.I. Shakhtarin, A.A. Artemenko, Application of the local Markov approximation method for the analysis of information processes processing algorithms with unknown discontinuous parameters, Applied Mathematical Sciences, 90 (2014) 4469-4496.

[10] A.P. Trifonov, Yu.E. Korchagin, Receiving a signal with unknown duration, Izvestiya Vysshikh Uchebnykh Zavedenij. Radiofizika, 7 (2002) 625-637.

[11] O.V. Chernoyarov, M. Vaculik, A.E. Rozanov, A new calculation technique of characteristics of multichannel measurer, Science Journal of Circuits, Systems and Signal Processing, 3 (2013) 78-84.

[12] O.V. Chernoyarov, Sai Si Thu Min, Yu.A. Guseva, B.I. Shakhtarin, A.A. Artemenko, Application of the local Markov approximation method for the analysis of information processes processing algorithms with unknown discontinuous parameters under violation of the consistency property of their estimates, Applied Mathematical Sciences, 126 (2014) 6267-6294.

Design and Developmental Research on the VV&A of Complex Simulation System

Li Liu[a], Yan Bin Li, Da Wu, Lin Zhang, Ning Wang
Air Force Engineering University, Xi'an China

Abstract. The Verification, Validation and Accreditation(VV&A) of a complex simulation system is a complex systems engineering. Based on the brief introduction to the concept of VV&A, this paper puts forward its design principles, approaches and basic contents, expounds the typical developing process and predicts its up-to-date technology developing trend of complex simulation system.

Keywords- Complex Systems; Modeling and Simulation(M&S); Verification, Validation and Accreditation(VV&A)

I. VV&A CONCEPTION

Complex simulation system reliability can be judged by verification, certificated by validation, so that the complex simulation system can serve a specific purpose, this process is called verification, validation and accreditation, which is referred to as VV&A.

Verification is to determine whether or not the simulation system model and its simulation realization accurately have expressed the iterative process which the developer needs. It focuses on the inspection of the simulation modeling process, namely whether the code and logic are correct and whether the expected functions have been realized.

Validation is to confirm the authenticity of simulation model, a thorough inspection of mathematic and logic models. It uses the data from real world and creditable source comparing with the output and behaviors of simulation system to confirm whether it is vivid or not. Therefore, the fundamental methods of validation are experiment, observation and comparison, the most obvious purpose is to provide a set of credibility evidence of models or simulation for special applications.

Accreditation is to address the issue of whether simulation system is suitable for a particular application and whether to accept it or not by the government or authority. The process is based on the results of VV&A, and compared with an acceptable standard.

Verification involves the question of whether or not develop the simulation system correctly, validation involves the question of whether or not develop a correct system model, accreditation involves the issue of whether can use the simulation system. The relationship between them is shown in Figure1.

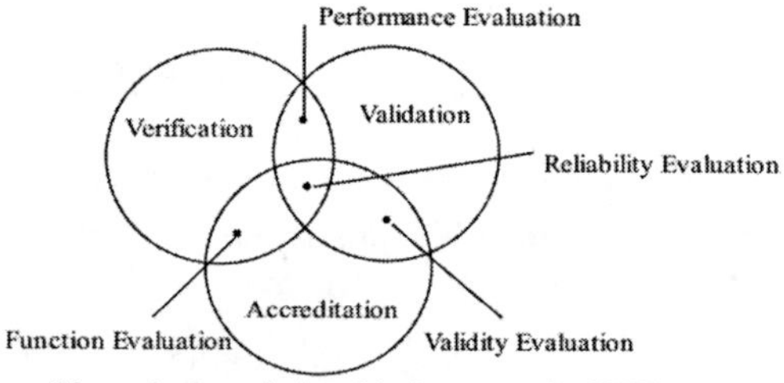

Figure1 the relationship between the VV&A

II. THE PRINCIPLES, STEPS AND BASIC CONTENTS OF VV&A SCHEME DESIGN

The result of complex simulation system VV&A highly depends on the established schemes. The establishment process of scheme must have the principles as follows:

① The VV&A scheme should cover the total lifecycle of simulation system;

② Considering comprehensively the influential factor of VV&A comprehensively;

③ The VV&A scheme must be feasible;

④ Some VV&A work which has large influence on the creditability should be put in the first position to enhance the effectiveness under limited input;

[a] Corresponding author: ll200010@126.com

DOI: 10.1051/matecconf/20166305032

⑤ Comprehending sufficiently the importance of VV&A work, then establishing the most effective VV&A scheme according to the reality;

⑥ Modifying the scheme according to the specific requirement based on the standard VV&A process model.

The VV&A scheme design steps as follows:

Collecting information → Analyzing the risk → Confirming the task of VV&A → Modifying VV&A activities → Designing VV&A scheme in detail → Reviewing VV&A scheme → Amending VV&A scheme → Submitting the document of VV&A plan

The basic contents of VV&A plan include two parts: V&V and validation.

To V&V:①The purpose and description of application ②The background of simulation system ③The definition and references ④The correlative people and their duty of V&V⑤The expected use of simulation system⑥The information source of VV&A implement ⑦The description of V&V complete activities projected ⑧The tool⑨Budget⑩The V&V report; ⑪The appendix;

To the validation:①The purpose②The background of simulation system③The definition and references④The correlative people and their duty of V&V⑤The schedule, milestone and resource⑥The expected use of simulation system⑦The information source of VV&A implement⑧The acceptable standard ⑨The validation methods.

III. THE TYPICAL DEVELOPING PROCESS OF VV&A

A. *The lifecycle VV&A developing process based on DIS*

The lifecycle developing process of DIS is as follows:

Planning drills and defining requirement → Designing, developing and testing drills → Executing drills → Analyzing and evaluating drills → Returning simulation results. The model frame is shown in Figure2.

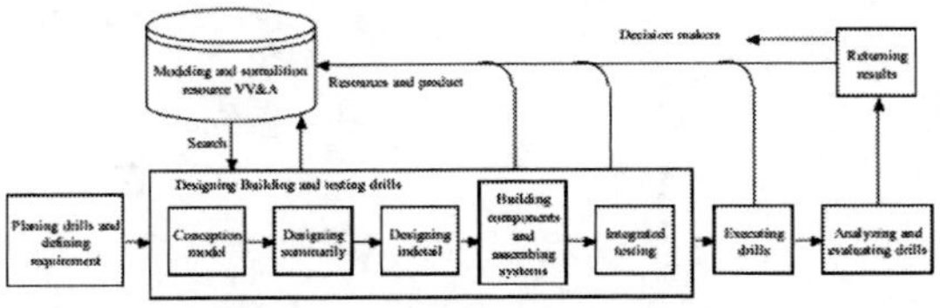

Figure2 Developing process model of DIS lifecycle VV&A

Drill means the execution of a simulation system in its development process. The given parameters, character data, initial conditions, working personnel and external systems represent special and ordinary thoughts.

B. *The lifecycle VV&A developing process of simulation system based on HLA*

The total lifecycle VV&A developing process can be concluded as six phases or steps: ①Defining federation targets; ②Developing federation conception models; ③Designing federation; ④Developing federation; ⑤Integrating and testing federation; ⑥Executing federation and analyzing results.

C. *The lifecycle VV&A developing process of mixed simulation system based on DIS/HLA*

DIS/HLA simulation system is a mixed and complicated simulation system. The core problem is the process of mutual information transfer and exchange. The model of developing process is shown in Figure3.

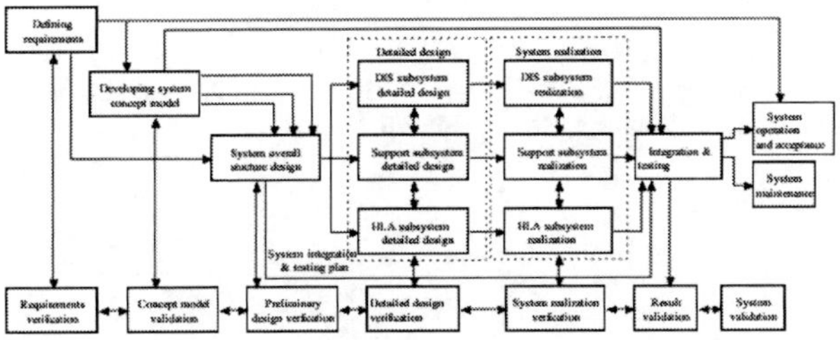

Figure3 VV&A development process modeling of DIS/HLA simulation system in life cycle

IV. THE DEVELOPING TREND OF VV&A TECHNOLOGY

Now, the technology and application of VV&A has made a great progress at home and abroad. It presents some new trend as follows:

① With the development of information technology, the emphasis of VV&A research has turned to the simulation system VV&A from model verification. At present, VV&A technology is progressing toward practicability, automation, standardization and integration to replace the partial and scattered research. There has been the "expanding VV&A technology".

②The US DoD puts the effective and efficient VV&A work in the scheme of M&S in 2007 while gives it priority.

③The GM-VV which is being developed by the NATO will become the current standard of VV&A.

④The western developed countries focus on the verification techniques of "black box" simulation. They launched the "M&S international test and operation process", including the data dependency, data flow analysis and dynamic VV&A technology (Alpha testing, operational testing, regression testing, statistical technology, comparing of graphics, visualization analysis and sensitivity analysis etc.). Meanwhile, they pay attention to the technology checked. In this field, the MCA has been the research hotspot.

REFERENCES

[1] Xingtang Liu, Li Liu. Simulation Science Technique and Engineering. Beijing: Sciences Publishing House, 2013.03
[2] Xingtang Liu, Bingcheng Liang. The Theory, Method & Technique for Complex System Modeling. Beijing：Sciences Publishing House,2011.04
[3] Xingtang Liu, Li Liu. Modern Modeling and Simulation Technology. Northwestern Polytechnical University Publishing House. 2011.03
[4] Henglin Wang, Jianguo Cao. Simulation System Design and Application. Sciences Publishing House., 2003.

Study on the Curvature Reducing Method of Non-linear Regression Model

Jin-mei Wu , Xiao-dong Ling, Ya-wei Hou, Yu-xin Zhang, Yi Wang

China Satellite Maritime Tracking and Controlling Department, Jiangyin 214413, Jiangsu, China

Abstract. The method to reduce the non-linear strength (curvature)of non-linear regression model was studied in this paper. Firstly, the reference point of the non-linear strength was analyzed. Based on the definition of curvature cubic matrix, a computing method of curvature cubic matrix was proposed based on the Cholesky disassembling. Then the common ways to reduce the non-linear strength was also discussed. Pointed at some common non-linear models in real engineering applications, such as non-linear models used for multiple-measurement and mutual-calibration of different instruments, or non-linear models prior information, a new least square method with weight was given, which can evidently reduce the curvature of these multi-structure non-linear regression models, therefore evidently reduce the non-linear strength. Finally, the Numerical simulation results were given to validate the effectiveness and feasibility of this weighted least square method. The method to reduce the non-linear strength (curvature) of non-linear regression model was studied in this paper. Firstly, the reference point of the non-linear strength was analyzed. Based on the definition of curvature cubic matrix, a computing method of curvature cubic matrix was proposed based on the Cholesky disassembling. Then the common ways to reduce the non-linear strength was also discussed. Pointed at some common non-linear models in real engineering applications, such as non-linear models used for multiple-measurement and mutual-calibration of different instruments, or non-linear models with prior informations, a new least square method with weight was given, which can evidently reduce the curvature of these multi-structure non-linear regression models, therefore evidently reduce the non-linear strength. Finally, the Numerical simulation results ware given to validated the effectiveness and feasibility of this weighted least square method.

Keywords: Non-linear regression, curvature, cholesky disassembling, least square with weight.

Introduction

Because of the complexity of non-linear function, currently the parameter estimate of non-linear regression mainly solved by linear approximating. The precision of the estimate result rests with the approximate level with linear model .That is the. Beale put forward the measurement of non-linear strength firstly in [1]. Bates and Watts defined inherent curvature and parameter effect curvature in 1980[2]. Their definition applied successfully. Document [3] put forward the concept of curvature cubic matrix. We are investigating how to reduce curvature of non-linear model.

Definition

Non-linear regression usually can be expressed as $Y=f(X,\beta)+\varepsilon$, in which measurement data $Y=(y_1,y_2,\text{L}\ ,y_n)^\tau$,unknown parameter $\beta=(\beta_1,\text{L}\ ,\beta_p)^\tau$, $f(X,\beta)$ is the non-linear function with design matrix X and unknown parameter β . Measurement error $\varepsilon=(\varepsilon_1,\varepsilon_2,\text{L}\ ,\varepsilon_n)^\tau$ obeys $\varepsilon\sim(0,\sigma^2 I_{n\times n})$. The definition is as follows: The inherent curvature and parameter effect curvature cubic matrixes of non-linear model $Y=f(X,\beta)+\varepsilon$ are

$$I_{(n-p)\times p\times p}=\left[N^\tau_{(n-p)\times p}\right]\left[U_{n\times p\times p}\right],$$
$$P_{p\times p\times p}=\left[Q^\tau_{p\times n}\right]\left[U_{n\times p\times p}\right],$$
$$Y_{sij}=\sum_{t=1}^{n}A_{st}X_{tij},Y_{\text{m}\times\text{p}\times\text{q}}=\left[A_{m\times n}\right]\left[X_{n\times p\times q}\right]. \quad (1)$$

The QR disassembling of Matrix V is the most important step. We can get form Cholesky disassembling theory that if $G=(g_{ij})_{p\times p}$, $G=LDL^\tau$,as a result that

$$\begin{cases} d_j=g_{jj}-\sum_{r=1}^{j-i}c_{jr}^2 d_r^{-1}, \\ c_{ij}=g_{ij}-\sum_{r=1}^{j-1}c_{jr}c_{ir}d_r^{-1},(j=1,\text{L}\ ,p,i=j+1,\text{L}\ ,p), \\ l_{ij}=\dfrac{c_{ij}}{d_j}. \end{cases} \quad (2)$$

From this easily get L and D. Because V is symmetry positive definite matrix, disassembling it as Cholesky disassembling theory, $V^\tau V=LDL^\tau=\left(L\sqrt{D}\right)\left(L\sqrt{D}\right)^\tau$, given $R=\left(L\sqrt{D}\right)^\tau$, $Q=V\left[\left(L\sqrt{D}\right)^\tau\right]^{-1}$,that is QR disassembling of V .Curvature cubic matrix is the standard of non-linear strength. An important direction is reducing the non-linear strength of the model or reducing the curvature. Sum up the methods of reducing the curvature. We get three methods, Parameter varying , increasing sampling, least square method with weight. Document [3] pointed out that inherent curvature is the inherent characteristic of the model. It will remain unchanged if parameter varies. Parameter effect curvature relies parameter. In practice, different parameter varying should be chose suitably according to concrete problem at different models. Because parameter varying has not simple and explicit calculation method, many learned man tried different ways to reduce curvature. Document [4] put forward the method of increasing sampling and proofed that if provided enough sampling data, the non-linear model can approach the linear model adequately. In practice, the prior information $\forall\hat{\beta}\in\Theta$ can be defined and then model curvature will be reduced on creasing sampling. When the non-linear model approaches linear model adequately, the least square iteration method can be used to calculate $\hat{\beta}$.

Be aimed at non-linear multi-structure, least square estimation with weight can reduce curvature and raise precision. Set up $\rho_i\in[0,1],\sum_{i=1}^{m}\rho_i=1.$ Non linear multi-structure model

$$Y(\rho)=F_\rho(X,\beta)+\varepsilon(\rho),\ \varepsilon(\rho)=\left(0,diag(\underbrace{\rho_1\sigma_1^2,\cdots,\rho_1\sigma_1^2}_{n_1},\underbrace{\rho_2\sigma_2^2,\cdots,\rho_2\sigma_2^2}_{n_2},\cdots,\underbrace{\rho_m\sigma_m^2,\cdots,\rho_m\sigma_m^2}_{n_m})\right). \quad (3)$$

The discussion about the existence and superiority of the optimal weighting can be referred to document [5].

Example

Grow model is the most wide applied model in agriculture、biological、economy、chemical. The harvest of the crops y increases with x, but suffered from environment and weather, so we built Gompertz model: $Y_1=\exp(\beta_1-\beta_2\beta_3^x)+\varepsilon_1$, in which β_i are the influencing factor we should think about. Because there are function relation between harvest and density of crops, so we can build another Holliday model. $Y_2=(\beta_1+\beta_2+\beta_3x^2)^{-1}+\varepsilon_2$. Unite the two models. We can get the non-linear multi-structure model. Set up $\varepsilon_1\sim N(0,0.01^2),\varepsilon_2\sim N(0,0.01^2)$, $\beta=(\beta_1,\beta_2,\beta_3)^\tau=(8,1,2)^\tau$, x=1，2，…，100，and might as well z= 1，2，…，100.

Simulation the measurement data $\{Y_{1i}\}_{i=1}^{100}$ and $\{Y_{2i}\}_{i=1}^{100}$ .We simulated 50 times and then statistic the data. Lead weight into model. Set up weights $\rho_1=\rho,\rho_2=1-\rho$, iteration initial value (8.5 , 1.5, 1.7) and step length 0.1. Iterate with G-N formula. The computing result is as follows.

Table 1.computed results

	Single structure model Y_1	Single structure model Y_2	Multi structure model with Traditional treatment	Multi structure model with Optimal weight
ρ	0	1	0.5	0.2
$\hat{\beta}(\rho)$	7.6494 1.2331 1.4045	8.0387 1.0298 1.9793	7.9964 0.9973 2.0018	7.9989 0.9992 2.0006

$MSE(\hat{\beta}(\rho))$	0.17726	0.00093	$2.3775E\text{-}5$	$2.4438E-7$

Table 2. curvature cubic matrix

	Single structure model Y_1			Single structure model Y_2		
ρ	0			1		
Parameter effect curvature $P_{3\times3\times3}$ cubic matrix	19.1346	-3.0727	0.6540	-0.0025	-0.0000	0.0000
	-3.0727	10.1343	-3.4131	-0.0000	-0.0025	1.9827
	0.6540	-3.4131	4.5696	0.0000	1.9827	-319.3462
	-3.0727	10.1343	-3.4131	-0.0000	-0.0025	0.0000
	10.1343	22.6259	2.8524	-0.0025	-0.0179	0.7496
	-3.4131	2.8524	21.9232	0.0000	0.7496	-95.2534
	0.6540	-3.4131	4.5696	0.0000	0.0000	-0.0025
	-3.4131	2.8524	21.9232	-0.0000	-0.0020	-0.0457
	4.5696	21.9232	54.9739	-0.0025	-0.0457	-0.9965
	Multi structure model with Traditional treatment			Multi structure model with Optimal weight		
ρ	0.5			0.2		
Parameter effect curvature $P_{3\times3\times3}$ cubic matrix	-0.0055	-0.0000	0.0000	-0.0035	-0.0000	0.0000
	-0.0000	-0.0055	4.4324	-0.0000	-0.0035	2.8038
	0.0000	4.4324	-713.7058	0.0000	2.8038	-451.5646
	-0.0000	-0.0055	-0.0000	-0.0000	-0.0035	-0.0000
	-0.0055	-0.0400	1.6757	-0.0035	-0.0253	1.0600
	-0.0000	1.6757	-212.8816	-0.0000	1.0600	-134.6911
	0.0000	-0.0000	-0.0055	0.0000	-0.0000	-0.0035
	-0.0000	-0.0046	-0.1021	-0.0000	-0.0029	-0.0646
	-0.0055	-0.1021	-2.2277	-0.0035	-0.0646	-1.4091

Conclusion

The curvature of Gompertz model is larger than the one of Holliday model. The curvature of Holliday model is very small. It explains that the Holliday model is close to linear model. The estimate result also displayed that Holliday model is more accurate than Gompertz model. The curvature of multi structure is nearly zero. It shows that multi-structure model can use all information fully to get precise least-square estimate. The method with optimal weight can get litter curvature than traditional treatment. The least square with weight can get more precise result.

References

[1] E. M. L. Beale, Confidence regions in nonlinear situation, J.R. Statist .Soc. B 22, 1960, 41-88.

[2] D.M. Bates, D.G. Watts, Relative curvature measures of nonlinearity, J. R. Statist, Soc. B 42, 1980, 1-25.

[3] Wei. Bo-cheng, Non-linear regression analysis in modern times, Southeast university press, Nanjing, 1989.

[4] Wu Yi, Yi Dongyun, Reducing curvature of a class of nonlinear regression models, Science

in china(Series E:Technological Sciences), 2000(2).

[5] Wu Jinmei, Ling Xiaodong, Li Yonggang. Least square regression model with weight for non-linear multi-structure, Statistics and decision, 2010(24):11-14.

[6] Cheng Xiru, Wang Songgui, Regression analysis at modern times, Anhui education press,1987.

[7] Qian Jin, Wu Jinmei, Ling Xiaodong. Computing method of least square regression model with weight, Statistics and decision,2007(9).

[8] Yi Wu， Zhu Jubo. Trajectory data processing fusion algorithm, Science in China(Series E:Technological Sciences),1998(6).

[9] Zhou Haiyin, Xie Meihua, Fusion algorithm for error satechment based on measure data of aero-craft, Journal of Ballistics ,Vol .13 No.1 March 2001.

MATEC Web of Conferences **63**, 05034 (2016) DOI: 10.1051/matecconf/20166305034
MMME 2016

Real-time Compressing Algorithm based on Outer-trajectory Measurement Data

Jin Mei Wu[a], Xiao Dong Ling, Shang Cheng Hu, Zhen Ping Wang
China Satellite Maritime Tracking and Controlling Department, Jiangyin 214413, Jiangsu, China

Abstract. Since huge sample datum has to be compressed properly in pre-processing to be sent out, a good compression algorithm will evidently improve the precision of the data-processing. In this paper, a compressing algorithm was studied based on polynomial fitting method. During the process of the real-time trajectory data compression, datasets were successively accumulated according to compression ratio. To apply all the information in the dataset, a series of orthogonal polynomial basis were applied to fitting the function, the least square estimation method was used to filter noise, and the estimated values of the position and the speed from differentiation of object datum in the dataset were sent out as compressed datum. And to get the best filter parameters, the mathematical expression of the error expectations and variances were studied. The compressing principle was given by considering the truncation error and random error simultaneously, which showed that, the best filter was the one by 21-point 3-order polynomial for position data compressing, while for speed data the filter by 41-point 2-order polynomial was better. The theoretical analysis and the simulation results were also provided to prove the effectiveness of this algorithm in data-compression and noise filtering.

Keywords: Outer-trajectory measurement data ; Polynomial fitting; Compressing; Noise filtering

Introduction

Nowadays every kind of outer-measurement equipment collects data in some fixed sampling frequency. But in data processing, sampling frequency is often amplified. It is obligatory to compress data in preprocessing. Traditional method is giving up the rest points. Obviously this method wastes a large number of data and easily leads into new error. Frequently-used method averages all points, but is still unable to approach the real value. Whereas polynomial fitting is widely applied in time domain filtering and signal detection. In this paper, we study how the polynomial fitting can be presented into real-time trajectory data compression.

[a]Corresponding author：wjm_83@aliyun.com

Definition

In real-time processing, a dataset accumulates and then the estimated value of position data in centre point on three coordinates and speed data from differentiation are outputting. Repeat to complete the compression from huge data to one point. In every dataset, a series of orthogonal polynomial basis were presented and the smoother formula of output point in every dataset with the least square filtering was applied to use every data fully and filter the measure noise.

Suppose sampling interval is h; sampling time is $t_i = ih$ in every dataset, $i = 1,\cdots,n$; flying goal motion function is $f(t)$, sampling equipment measurement value is y_i at t_i, so $f(t_i)+\varepsilon_i$, in which ε_i is random error. On the basis of Weierstrass No1 approximation theorem, $f(t)$ can be fitted by some m-order polynomial, such as

$$y_i = \alpha_0 P_0(t_i) + \alpha_1 P_1(t_i) + \cdots + \alpha_m P_m(t_i) + \varepsilon_i \quad (1)$$

Begin with uncomplicated algebra polynomial $\{1, t, t^2, \cdots, t^m\}$, we construct orthogonal polynomial foundation. If P is the assemblage of all polynomial, any $p,q \in \mathrm{P}$, define the scalar product $\langle p,q \rangle = \sum_{i=1}^{n} p(t_i) q(t_i)$, if $\langle p,q \rangle = 0$, p,q are orthogonal. By Schmidt orthogonalization method, we can get :

$$\begin{aligned}
p_0 &= 1 \\
p_1 &= t - \frac{n+1}{2}h \\
p_2 &= t^2 - (n+1)ht - \frac{n^2+3n+2}{6}h^2 = (t - \frac{n+1}{2}h)^2 + \frac{1-n^2}{12}h^2 \\
p_3 &= t^3 - \frac{3(n+1)}{2}ht^2 - \frac{6n^2+15n+11}{10}h^2 t - \frac{(n^2+5n+6)(n+1)}{20}h^3 \\
&= (t - \frac{n+1}{2}h)^3 + \frac{7-3n^2}{20}h^2 (t - \frac{n+1}{2}h)
\end{aligned} \quad (2)$$

If

$Y = (y_1, y_2, \cdots, y_n)^\tau$,

$$X = \begin{pmatrix} p_0(t_1) & \cdots & p_m(t_1) \\ \vdots & \ddots & \vdots \\ p_0(t_n) & \cdots & p_m(t_n) \end{pmatrix}_{n\times(m+1)},$$

$\alpha = (\alpha_0, \alpha_1, \cdots, \alpha_m)^\tau$, consequently $Y = X\alpha$, and the least-square estimate $\hat{\alpha} = (X^\tau X)^{-1} X^\tau Y$, where $X^\tau X = diag(S_0, S_1, \cdots, S_m)$, $S_j = \langle p_j, p_j \rangle$, get that:

$$\begin{aligned}
S_0 &= n \\
S_1 &= \frac{n(n^2-1)h^2}{12} \\
S_2 &= \frac{n(n^2-1)(n^2-4)h^4}{180} \\
S_3 &= \frac{n(n^2-1)(n^2-4)(n^2-9)h^6}{2800}
\end{aligned} \quad (3)$$

So we get that

$$\hat{\alpha} = \left(\sum_{i=1}^{n} p_0(t_i) y_i S_0^{-1}, \sum_{i=1}^{n} p_1(t_i) y_i S_1^{-1}, \cdots, \sum_{i=1}^{n} p_m(t_i) y_i S_m^{-1} \right) \quad (4)$$

Smoother formula of position data and speed data at output point are:

$$\hat{f}(t) = \sum_{j=0}^{m} \hat{\alpha}_j p_j(t) = \sum_{j=0}^{m} \left(\sum_{i=1}^{n} p_j(t_i) y_i S_j^{-1} \right) p_j(t) = \sum_{i=1}^{n} \left(\sum_{j=0}^{m} p_j(t_i) S_j^{-1} p_j(t) \right) y_i \quad (5)$$

$$\hat{f}'(t) = \sum_{j=0}^{m} \left(\sum_{i=1}^{n} p_j(t_i) y_i S_j^{-1} \right) p_j'(t) = \sum_{i=1}^{n} \left(\sum_{j=0}^{m} p_j(t_i) S_j^{-1} p_j'(t) \right) y_i \quad (6)$$

Example

We choose theoretical value and mixed noise to produced outer-measurement data. In Matlab simulation environment , we choose component product in X axis coordinate, then mixed a set of white noise with 0 mean value 10 variance, and select data of 250 seconds. Compare with four methods: choosing second points, averaging, 3order 21 point polynomial fitting, 3order 41 point polynomial fitting. Results showed as pictures.

 DOI: 10.1051/matecconf/20166305034

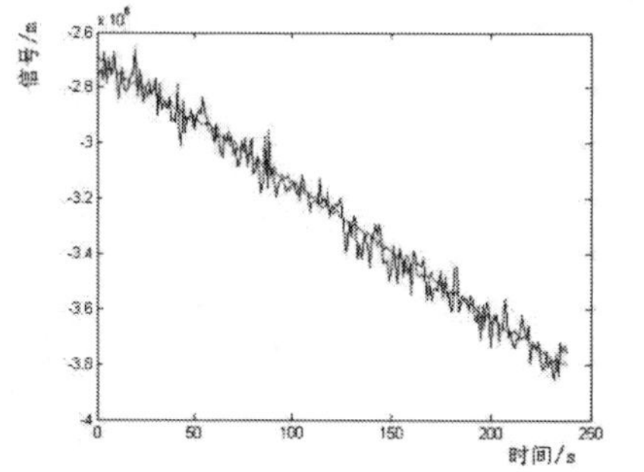

Fig. 1 Comparison of simulation data and compressed signals fitted by 21-point 3-order polynomial

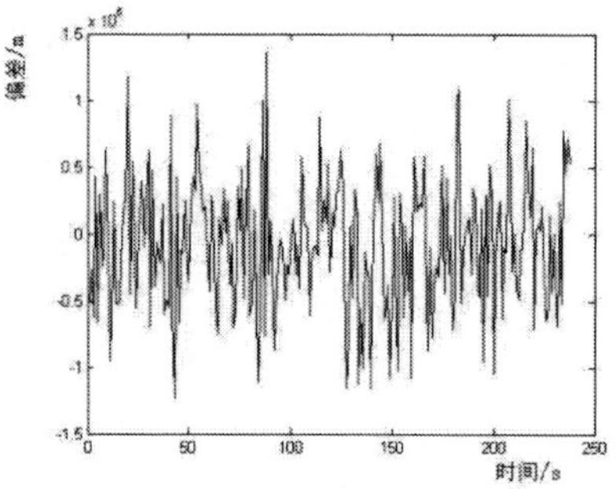

Fig. 2 Error of output data every second

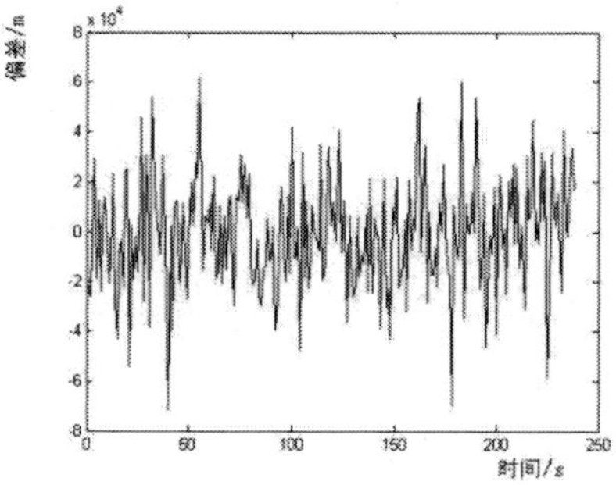

Fig. 3 Error of average method

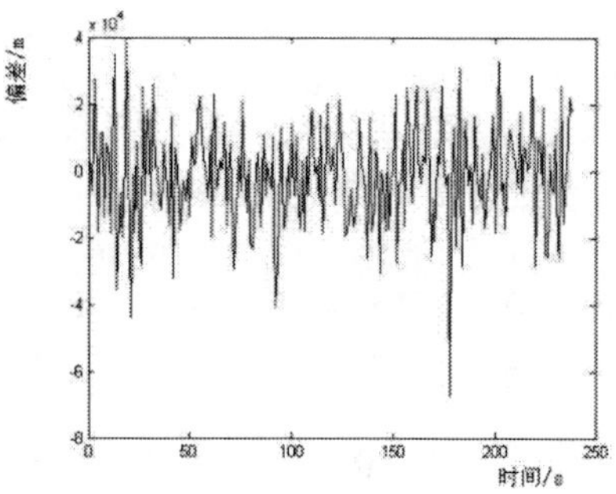

Fig. 4 Error of data fitted by 41-point 3-order polynomial

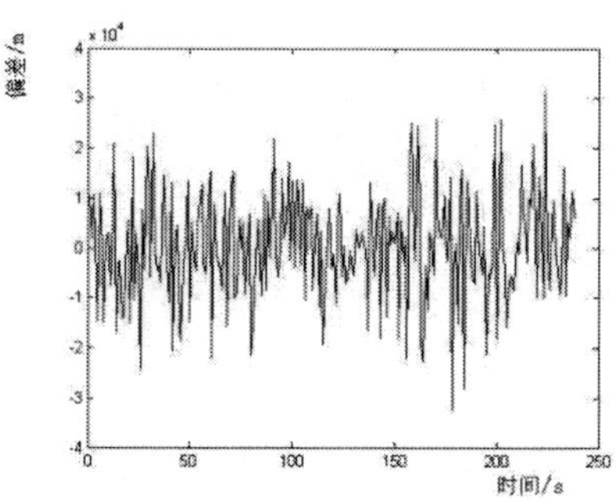

Fig. 5 Error of data fitted by 21-point 3-order polynomial

We can perceive that 3order21point polynomial fitting compression method reflect data characteristic, the error with real value obviously is smaller than methods of choosing second points and averaging.

Conclusion

Using polynomial fitting to compress outer-measurement data is uncomplicated, practical and can satisfy the requirement of real-time outer-measurement preprocessing. As the simulation result shows, the novel algorithm is a promised compression and noise filtering method which can reflect real curve characteristic in higher precision.

References

[1] Zhenjun An, Min Wang, Shaolin HU, et al, Tracking Data Collection and Processing Technologies , Journal of Spacecraft TT&C Technology, 2005,24(2):56-60.

[2] Lisheng Liu, Data Processing of Outer-Trajectory Measurement, Beijing:National Defense Industry Press,2002:151-170.

[3] Zhengming Wang, Haiying Zhou. Several Calculating Problems of the Processing of Trajectory Tracking Data, Journal of National University of Defense Technology, 2001, 23(6):42-47.

[4] Bo Yu, Wei Jian, Jianxun Chen, et al, A Method Based on Orthogonal Polynomial Partitioned Imitation for Image Compression Coding , Microcomputer Application, 2007,28(12):1316-1320.

[5] Zuoping Qin, Puyin Liu. Non-central Differentially Moving Average With Polynomial, Journal of Astronautics, 1997, 18(2):43-49.

MATEC Web of Conferences **63**, 05035 (2016) DOI: 10.1051/matecconf/20166305035
MMME 2016

Research on Evaluation Index System of Air to Ground Missile Testing Quality

Wei Chen[1,a] Xue Gang Zeng [1] Yong Qiang Ren [1] Si Yong Zhao [2]
[1] *Air Force Logistics College, Xuzhou Jiangsu ,221000, China*
[2] *95605 Troops, Chongqing ,402361, China*

Abstract: According to the environmental factors which were optimized, with the reference of opinions from expert, the initial establishment of the influence factors of air to ground missile quality evaluation. Using principal component analysis and fuzzy prioritization method to develop a preliminary index system of quantitative indicators and indicators were screened, air to ground missile quality evaluation index system is established, getting comprehensive, system reflect missile quality characteristics from many aspects.

Keywords: detection quality; air to ground missile; evaluation index system

Introduction

In the context of new military changes in the world, air to ground missile can play its powerful combat effectiveness is crucial to the future high technology war in the informationization of the rational use. Therefore, in order to enable the missile to fully play its role in the process of use, has very important significance to establish the quality evaluation index system is evaluated. The platform of the missile test parameters, proposed must comply with certain principles, establish[1][2]. For missile detection parameters using appropriate methods, the quality and effect of missile combat training as the research object, to calculate the weights of the various indicators based on fuzzy AHP method, analysis of the weight percentage of each index in the influencing factor the determination of the indexes at all levels, build missile quality evaluation index system, provides a theoretical basis for the analysis of the quality of the missile.

1 The Method of The Evaluation Index System for The Quality Evaluation of Air to Ground Missile

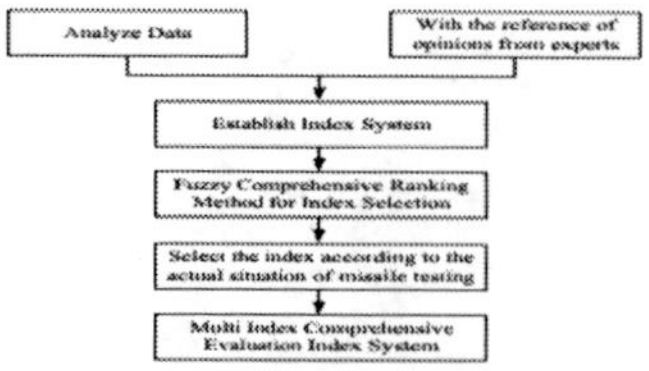

Fig1. the method of establishing the multi index system

According to the performance parameters of air to ground missile, tentatively the quality evaluation index system of a certain type of air to ground missile. Considering the missile indicators may be the existence of a large number, information duplication, the phenomenon of mutual interference by fuzzy analytic hierarchy process (AHP) screening indicators, the establishment of air to ground missile detection quality evaluation index system as Fig.1 follows.

Using the missile impact the effectiveness of many factors, the establishment of comprehensive evaluation index system of missile quality, to take a one-time all details are difficult to, so the use of system theory of decomposition and coordination principle, namely scientific system level principle, the problem can be decomposed into multiple levels, each level and a plurality of components, from coarse to fine from the outside to the inside, from the global to local to a gradual in-depth analysis, the effect of Missile Quality Evaluation of many factors structured, hierarchical, so as to establish hierarchical analysis model, as shown in Figure 2. The model by the evaluation object a, level

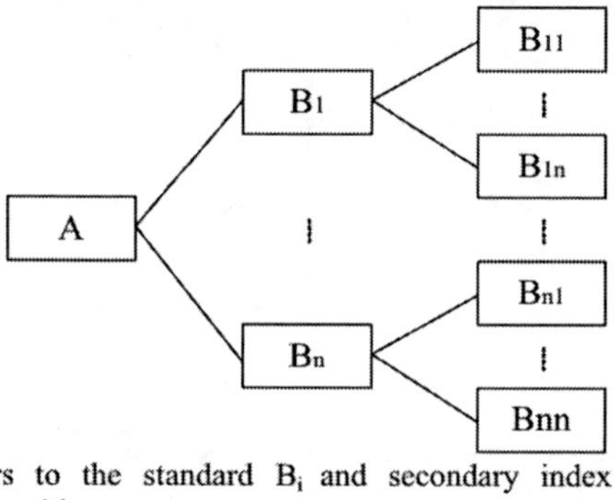

refers to the standard B_i and secondary index b_{ij} composition.

Fig. 2 hierarchical analysis model

[a]Corresponding author: chenweijiangyiwa@126.com

According to this principle, we tentatively index system mainly includes the level indicators and secondary indicators, the television homing seeker system, control and navigation systems, image / instruction transmission system and power system four system as the first grade indexes.

Because the four indicators can only reflect the missile general technical performance, by their reflect missile quality is too general, so, should according to the technical requirements of each detection parameters of a certain type of air to ground missile, missile quality evaluation indicators at the second level is established. Second level indicators to establish such as:

Guidance performance of main line of sight angle speed precision, stable platform isolation, heading and pitch open-loop optical axis drift, photoelastic shaft angle accuracy, roll stability range, roll angle zero bit accuracy, heading and pitch search, tracking range, maximum search speed and tracking precision of the secondary indicators.

Control and navigation performance mainly in the heading angle and angular accuracy, position accuracy, angular velocity range, sensitivity, angular velocity, angular velocity is zero, air data system is tight, air data system height accuracy, radio altimeter altimetry precision, radio altimeter sensitivity, maximum output torque, steering gear band, zero position of the steering gear, steering sensitivity, the maximum rudder angle, no-load angular velocities of the secondary indicators.

Image / instruction transmission performance is mainly transmitted by the image transmitting device, the command receiver sensitivity, the command receiving antenna gain, the image transmit antenna gain, image transmission system transmission delay and so on two level indicators.

Dynamic performance of the engine thrust, fuel consumption rate, servo valve current and fuel flow, exhaust temperature, the output voltage of the generator, lube oil consumption and secondary indicators. Initial formulation of the index is more, may overlap information exists, to be screened. Fuzzy analytic hierarchy process (AHP) is based on a certain number of data samples, using mathematical method to calculate and analyze, pick less index to reflect the connotation of selected indicators of the upper indexes, suitable for screening of quantitative indicators. According to the characteristics of this method, combined with the selected indexes do not have quantitative properties of, the performance of the steering, control and navigation performance, image / instruction transmission performance, dynamic performance and other qualitative indicators indicators at the second level using fuzzy analytic hierarchy process (AHP) screening.

2 The Principle of The Evaluation Index System for The Quality Evaluation of Air to Ground Missile

Air to ground missile detection quality evaluation index should be able to reflect the influence of air to ground missile application, selection of various factors. Selection index is appropriate, directly affect the final evaluation result is reasonable and reliable. To establish an air to ground missile comprehensive evaluation index system of decision-making, intends to set evaluation index should follow the following principles:

(1) the aim is to point out the purpose is the starting point, the index system should cover the basic content of evaluation purposes, can reflect all the information of the object.

(2) indicators should be independent, each index should be clear, relatively independent, the same level of indicators should try not to overlap each other; the index system to be distinct.

(3) indicators should be representative and feasible. It can reflect the characteristics of the evaluation objects, and is in line with the objective reality, easy to operate, easy to evaluate and compare.

3 Multi Index Comprehensive Evaluation Index System

The air to ground missile performance parameters and technical indicators as the basis, reference of air to ground missile expert opinion survey, the initial establishment of the influence factors of air to ground missile quality evaluation. On the guidance performance, control and navigation performance, image / instruction transmission performance, dynamic performance of the four first level indicators screened secondary indexes, the establishment of air to ground missile quality evaluation index system as shown in Figure 3.

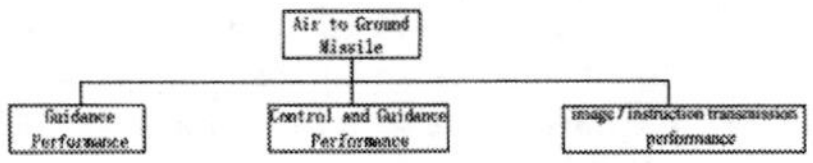

Fig. 3 index system of Missile Quality Evaluation

The first level indicators of missile development are as follows:

(1) guidance performance: the performance of the seeker is reliable.

(2) control and navigation performance: control and navigation system is able to successfully complete the task.

(3) image / instruction transmission performance: the image / instruction transmission system communication is smooth, reliable.

(4): the dynamic performance of the engine thrust, fuel consumption rate is normal.

Analysis of the relevant test parameters, the two indicators of the missile analysis are as follows:

Guidance performance of the two level indicators include LOS angle speed precision, stable platform isolation, heading, pitching axis drift, light missile axis angle accuracy and other indicators. The accuracy of line of sight is higher, and the precision of seeker is higher, which can improve the hit precision of missile;

The stability platform isolation is good, which can improve the hit precision of the missile;

The technical indexes for the drift of the optical axis of the pitching open loop are the key indexes;

The precision technology index of the angle precision of the optical elastic axis is a general index.

Control and navigation system, two indicators, including the heading angle and attitude angle accuracy, angular velocity range, zero angular velocity, position precision, air data system is tight, air data system height accuracy, radio altimeter altimetry precision, radio altimeter sensitivity index. The heading angle accuracy is high, which is helpful for the missile to fly in the planned route and improve the hitting accuracy of the missile.

The position accuracy is high, it is helpful for the missile to fly in the planned route, and improve the hit precision of the missile.

The technical index of the angular velocity range is a general index.

The technical index of the zero position of the angular velocity is a general index.

The air data system has good air tightness, which is helpful for the missile to fly in the planned route and improve the hit precision of the missile.

Air data system measurement accuracy is high, there is help to missile according to a predetermined plan flight routes, improve the missile hit precision. The index as the key indicators.

Radio altimeter altimetry, help to missile according to a predetermined plan flight routes, improve the missile hit precision. The index as the key indicators.

The radio altimeter has high sensitivity, which is helpful for the missile to fly in the planned route and improve the hit precision of the missile. The index is the key index.

The image / instruction transmission system includes two levels: transmission power of the image transmitting device, transmission delay of the image transmission system, and the sensitivity of the instruction receiver.

The increase of transmission power of image transmitting device is beneficial to the missile hitting target and improving the hit precision. This index is the key index.

The technical index of the delay of image transmission system is the general index. The higher sensitivity of the command receiver is beneficial to improve the hit precision of the missile, which is the key index.

Summary

Through the air to ground missile related data analysis and expert scoring, tentatively missile quality evaluation index system, taking into account the initial formulation of the index system of many indexes, each other also exist the phenomenon of information overlap, of quantitative indicators and qualitative indicators were screened by using the fuzzy analytic hierarchy process (AHP), the missile quality evaluation index system was established, as a basis for quality evaluation of missile.

References

[1]Melissa Dan. Models for Threat Assessment in Networks. [D]. Dissertation of California State University, 2006, 9

[2]N. Okello, G. Thorns. Threat Assessment Using Bayesian Networks[C]. Proceedings of the Sixth International Conference of Information Fusion, 2003, 2: 1102~1109

MATEC Web of Conferences **63**, 05036 (2016) DOI: 10.1051/matecconf/20166305036
MMME 2016

Research on The Operational Effectiveness of Air to Ground Missile Based on The Detection Parameters

Xue Gang Zeng [1] Wei Chen [1,a] Yong Qiang Ren [1] Si Yong Zhao [2]

1.Air Force Logistics College, Xuzhou Jiangsu ,221000, China;

2.95605 Troops, Chongqing ,402361, China

Abstract: Air to ground missile is the first weapons to attack, and due to its operational effectiveness and the quality of the air to ground missile is particularly important. This paper through the air to ground missile tactical performance indexes, data detection analysis, extracted factors from air to ground missile optimization assessment which should be considered, as formulated in the air to ground missile optimization evaluation index based.

Keywords: detection parameters; air to ground missile; operational effectiveness

Introduction

Nowadays the detection quality of the missile is only qualified or unqualified, with the missile shooting at the target before the selection, and need to make reasonable analysis of shooting results, we need to analyze each part, especially the detection of the parameters of the key components[1][2]. But at home the analysis of the missile has not been generally accepted by the evaluation index of effective and authoritative ranking method. This has resulted in the need for the selection of missile combat environment when there is no basis phenomenon. To achieve the effect of security, have a detailed study of various performance parameters of the missile, analyzed the objective quality of the missile and get to scientific election bomb.

1 Missile Test Parameters

The combat effectiveness of air to ground missile depends on the function of each component of missile normal play, and these components at run time, affected with a variety of external factors, various factors for the following detailed analysis of the impact of air to ground missile, as the development of air to ground guided bomb optimization evaluation index on the basis.

1.1Seeker System Related Parameters

TV seeker mainly have three key performance indicators: line of sight angle speed precision, stable platform isolation, optical axis open-loop drift. Other indicators for general index. Seeker angular speed of the line of sight of high precision, stable platform isolation degree and axis open-loop small drift, improve the missile accuracy. Seeker of typical buildings) detection, identification, tracking performance of the main influence factors have visibility, illumination and other environmental factors.

1.2 Control and Navigation System Parameters

Control and navigation system has eight key performance indicators: the integrated navigation system of heading angle / attitude angle precision navigation system navigation accuracy, air tightness of air data system, height accuracy of air data system, the sensitivity of the radio altimeter, radio altimeter altimetry precision actuator, the maximum output torque and servo band. Other indicators for general index. Under the same conditions, the integrated navigation system of heading angle / attitude angle with high accuracy, high precision of integrated navigation system, air data system has good air tightness, high precision, high accuracy of radio altimeter altimetry, contribute to the missile at a predetermined flight route planning, improve the hitting accuracy of the missile. The sensitivity of the radio altimeter the high elevation range is larger, the range of indicators can meet the requirements.

1.3 Image / Instruction Transmission System Related Parameters

Image / instruction transmission system are two key performance indicators: image transmission device can transmit power, command receiving device sensitivity. Other indicators for general index. Under the same conditions, image transmission device to transmit power, command receiving device has the advantages of high sensitivity, to improve the effective distance of image / instruction transmission system, the system margin and anti-jamming performance.

[a]Corresponding author: chenweijiangyiwa@126.com

DOI: 10.1051/matecconf/20166305036

1.4 Power System Related Parameters

Dynamical systems, mainly in two key performance indicators: engine thrust, fuel consumption rate. Other indicators for general function index. Engine thrust directly affect the engine can for the missile provides enough power, so that the stability of missile flight, engine fuel consumption affects missile can meet the shooting process requirements.

2 Typical Targets and Environmental Conditions

TV seeker for visible light imaging system, and the outside world is closely related to weather conditions, only visibility cannot determine the guided seeker role distance guide seeker distance but also consider the target size, the contrast of target and background, outside light conditions (illumination), flight course.

1)Typical targets: ground command and communication center, bridge, technical weapon positions, (semi) permanent underground works, port terminals, transportation hubs, and ships parked.

2) Environmental requirements: during the day, sunny, cloudy or cloudy. There are clouds, cloud base height is higher than 1500m.

3) Flight course requirements: in the task planning in terminal trajectory seeker boot point and target point connection direction as far as possible fairing or side fairing using seeker in fairing or side fairing) conditions, the goal is usually white, target and background contrast can be significantly improved, is conducive to the pilots of the target lock and seeker of target tracking; secondly is the backlight seeker, try to avoid in the backlight using seeker, because in the backlight and backlight conditions, the goal is usually black, will reduce the target and background contrast, also in the search easily in the Fig.1 formed on the spot, not suitable for use, Fig.1as follows:

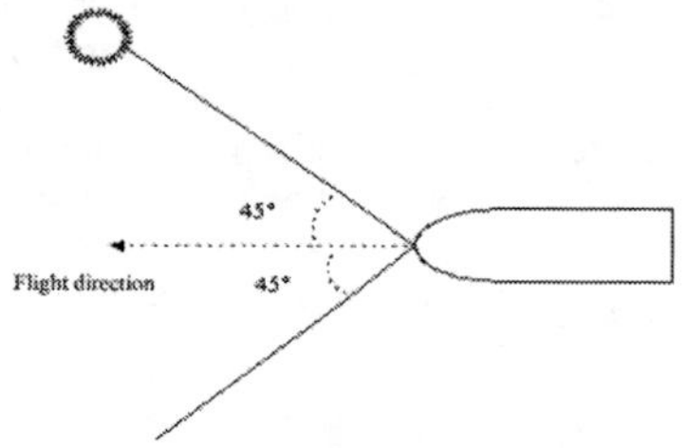

Fig.1 requirements of missile flight direction

4) Contrast requirements: TV seeker by gray television imaging system, background images of the target is generally dark gray, the target image and background image has certain difference, this is TV guide the inherent characteristics of homing guidance missile.

5) Air to ground missile image and command transmission system using microwave wireless communication of TV missile image and catching and command transmission. According to the image / instruction transmission system anti-interference ability and may encounter interference pattern to determine the missile using can avoid unnecessary losses in the war, and achieve optimal missile using effect.

3 Usage scenario

Missile penetration ability is an important factor to determine its effectiveness. The penetration ability of the missile is mainly reflected in the following aspects:

1) Formation

Ensure 8 aircraft with a certain missile in the same airspace in the same time to implement the attack:

(1) 8 plane formation (composed of two machine 4 wedge team): elevation 1 vertical interval of 1 of our knowledge of our knowledge;; the horizontal interval of 5 of our knowledge;

(2) 4 wedge: interval 400m; height 50m.

2) Routes planning

Each aircraft can be planned 4 goals, each target corresponds to 3 routes, the target and route choice can be automatically loaded, you can manually load through the pilot in the Fig.2 .

3) Lateral planning

To avoid the threat of lateral route planning. If the missile with lateral route planning, an acute angle between the adjacent segment of should is less than or equal to 25 degrees, and different leg lengths on the corner is limited, therefore, planning, should pay attention to try to plan smaller leg corner. If there is a war zone information map, according to the actual situation, choose reasonable emission area to avoid the threat.

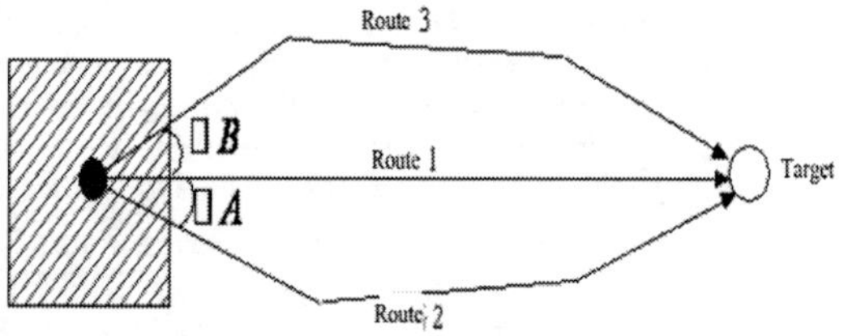

Fig.2 routes planning

4) Zooming

Missile terminal can achieve leap or not jumped two ballistic. Select terminal trajectory for jumps and requirements at the end waypoint height and the height of the target is less than equal to 350m and effective range is less than or equal to 110km. for (semi) permanent underground fortifications, airport runway, bridge deck planar targets and aircraft shelters, fixed radar bunkers target, it is recommended by terminal "zooming" effective attack.

5) Low altitude penetration Certain air to ground missile minimum flight altitude of up to 50m, against the established operational environment of the typical target.

4 Summary

This paper searches the detection parameters of the system composed of a certain type of air to ground missiles each. Analysis, the influences of the parameters, for each system to make a detailed and objective evaluation. According to the type of air to ground missile detection parameters, the type of air to ground missile quality and combat training effect as the research target, for the analysis of the quality of the air to ground missile provides the theoretical basis.

References

[1] M. Robert , L. Craig , S. Jeffrey. Software Framework in Support of Dynamic Situation Assessment and Predictive Capabilities for JSB-RD. Enabling Technologies for Simulation Science VIII, Proceedings of SPIE, 2004, 5423: 468~478

[2] S. Greenhill. Multiple Hypotheses Situation Assessment [C]. Proceedings of the Sixth International Conference of Information Fusion, 2003, 2: 972~978

[3] Melissa Dan. Models for Threat Assessment in Networks. [D]. Dissertation of California State University, 2006, 9

[4] N. Okello, G. Thorns. Threat Assessment Using Bayesian Networks[C]. Proceedings of the Sixth International Conference of Information Fusion, 2003, 2: 1102~1109

Representations for the Drazin inverses of block matrices

Li Guo[1,2], Jie Wang[2], Wenming Du[2]

[1] *Department of Mathematics, Southeast University, Nanjing, china*
[2] *College of Mathematics, Beihua University, Jilin, china*

Abstract. In this paper, we investigate representations of the Drazin inverse of a 2×2 block matrix. The Drazin inverse of a matrix is very important in various applied mathematical fields like machinery and automation, singular differential equations . We give a explicit representation of the Drazin inverse of a block matrix.

Keywords: Block matrix; Drazin inverse; Index

1. Introduction

Let $C^{n\times n}$ denote the set of $n\times n$ complex matrices. The smallest nonnegative integer k such that $rank(A^{k+1})=rank(A^k)$, denoted by $ind(A)$, is called the index of A. The Drazin inverse is the unique matrix $A^d\in C^{n\times n}$ satisfying

$$AA^d = A^d A,\ A^d AA^d = A^d,\ A^{k+1}A^d = A^k,$$

where $k=ind(A)$. We denote by $A^\pi = I - AA^d$. For $A\in C^{n\times n}$ we define $A^0 = I_n$, the identity of order n, even if $A=0$.

The Drazin inverse of a matrix is very important in various applied mathematical fields like machinery and automation, singular differential equations, singular difference equations, Markov chains, iterative methods and so on[1,2]. We introduce briefly an application of the Drazin inverse of a block matrix[2].

In this paper, representations for the Drazin inverse of 2×2 block matrix M , under the following conditions that $BD=0,\ D^\pi CB=0,\ D^\pi CAB=0,\ D^\pi CA^2=0$ and so several results are generalized.

Lemma 1.1 [3]Let $A\in C^{m\times n}$ and $B\in C^{n\times m}$, then $(AB)^d = A[(BA)^2]^d B$.

Lemma 1.2 Let $A\in C^{m\times m}$, then $(A^2)^d=(A^2)^2$.

Lemma 1.3 (see [5],) Let $M=\begin{pmatrix} A & B \\ C & D\end{pmatrix}$ be a 2×2 block complex matrix, where $A\in C^{m\times m}$ and $D\in C^{n\times n}$ with $ind(A)=r,\ ind(D)=s$, if $BC=0,\ BD=0$ then

$$M^d=\begin{pmatrix} A^d & (A^d)^2 B \\ \sum_0 & D^d+\sum_1 B\end{pmatrix},$$

and $k=0,1$,

$$\sum_k=(D^d)^2\sum_{i=0}^{r-1}(D^d)^{i+k}CA^iA^\pi+D^\pi\sum_{i=0}^{s-1}D^iC(A^d)^{i+k}(A^d)^2-\sum_{i=0}^{k}(D^d)^{i+1}C(A^d)^{k-i+1}.$$

with $ind(M)\le ind(A)+ind(D)+1$.

Lemma 1.4 (see [5],) Let $P,Q\in C^{n\times n}$, where $ind(P)=r,\ ind(Q)=s$, if $PQP=0,\ PQ^2=0$ then

$$(P+Q)^d = Q^\pi \sum_{i=0}^{s-1} Q^i (P^D)^{i+1} + \sum_{i=0}^{r-1} (Q^D)^{i+1} (P^i) P^\pi + Q^\pi \sum_{i=0}^{s-1} Q^i (P^D)^{i+2} Q$$
$$+ \sum_{i=0}^{r-2} (Q^D)^{i+3} (P^{i+1}) P^\pi Q Q^D P^D Q - (Q^D) P P^D Q.$$

2. Main Results

Theorem 2.1 Let $M = \begin{pmatrix} A & B \\ C & D \end{pmatrix}$ be a 2×2 block complex matrix, where $A \in C^{m\times m}$ and $D \in C^{n\times n}$ with $ind(A)=r$, $ind(D)=s$, if

$$BD = 0,\ D^\pi CB = 0,\ D^\pi CAB = 0,\ D^\pi CA^2 = 0,$$

then

$$M^D = \begin{pmatrix} A^D + (A^D)^3 BC + (A^D)^4 BCA & (A^D)^2 B + (A^D)^4 \mathrm{BCB} \\ \psi_0 + \psi_0 (A^D)^2 BC + D^D \psi_1 BC + \Delta DD^D C & D^D + \psi_1 B + \Delta BD^\pi \end{pmatrix}, \quad (2.1)$$

where

$$\Delta = \psi_0 (A^D)^3 BC + D^D \psi_0 (A^D)^2 BC + \psi_0 D^D \psi_1 BC,$$
$$\psi_k = \sum_{i=0}^{r-1} (D^D)^{i+k+2} CA^i A^\pi - \sum_{i=0}^{k} (D^D)^{i+1} C (A^D)^{k-i+1},\ k = 0,1.$$

Proof. Let $M = P + Q$, where

$$P = \begin{pmatrix} 0 & BDD^D \\ D^\pi C & 0 \end{pmatrix},\ Q = \begin{pmatrix} A & BD^\pi \\ DD^D C & D \end{pmatrix}.$$

By $BD = 0$, $D^\pi CB = 0$, then

$$PQP = \begin{pmatrix} 0 & BDD^\pi \\ D^\pi C & 0 \end{pmatrix} \begin{pmatrix} A & BD^\pi \\ DD^D C & D \end{pmatrix} \begin{pmatrix} 0 & BDD^D \\ D^\pi C & 0 \end{pmatrix}$$
$$= \begin{pmatrix} BDD^D \mathrm{C} & BD^2 D^D \\ D^\pi CA & D^\pi CBD^\pi \end{pmatrix} \begin{pmatrix} 0 & BDD^D \\ D^\pi C & 0 \end{pmatrix}$$
$$= 0.$$

Since $D^\pi CB = 0$, $D^\pi CAB = 0$, $D^\pi CA^2 = 0$, we have

$$PQ^2 = \begin{pmatrix} 0 & BDD^D \\ D^\pi C & 0 \end{pmatrix} \begin{pmatrix} A & BD^\pi \\ DD^D C & D \end{pmatrix} \begin{pmatrix} A & BD^\pi \\ D^D DC & D \end{pmatrix}$$
$$= \begin{pmatrix} BDD^D \mathrm{C} & BD^2 D^D \\ D^\pi CA & D^\pi CBD^\pi \end{pmatrix} \begin{pmatrix} A & BD^\pi \\ DD^D C & D \end{pmatrix}$$
$$= \begin{pmatrix} BDD^D CA + BD^2 D^D C & BDD^D CBD^D + BD^3 D^D \\ D^\pi CA^2 & D^\pi CABD^\pi + D^\pi CBD^\pi \mathrm{D} \end{pmatrix}$$
$$= 0.$$

By $P^2 = 0$, we get that $P^D = 0$, $P^\pi = I$. Noting that $BD = 0$, by Lemma 1.3, we have

$$Q^D = \begin{pmatrix} A^D & (A^D)^2 B \\ \psi_0 & D^D + \psi_1 B \end{pmatrix},$$

where

$$\psi_k = \sum_{i=0}^{r-1} \left(D^D\right)^{i+k+2} CA^i A^\pi - \sum_{i=0}^{k} \left(D^D\right)^{i+1} C\left(A^D\right)^{k-i+1}, \quad k = 0,1.$$

Since $PQP = 0$, $PQ^2 = 0$, By Lemma 1.4, we get

$$(P+Q)^D = Q^\pi \sum_{n=0}^{k-1} Q^i \left(P^D\right)^{i+1} + \sum_{i=0}^{m-1} \left(Q^D\right)^{i+1} P^i P^\pi + Q^\pi \sum_{i=0}^{k-1} Q^i \left(P^D\right)^{i+2} Q$$
$$+ \sum_{i=0}^{m-2} \left(Q^D\right)^{i+3} P^{i+1} P^\pi Q - Q^D P^D Q - \left(Q^D\right)^2 P P^D Q,$$

where $k = ind(Q) \le r+s+1$, $m = ind(P) = 2$, ,

By $P^D = 0$, $P^\pi = I$, then

$$(P+Q)^D = \sum_{i=0}^{m-1} \left(Q^D\right)^{i+1} P^i + Q^\pi \sum_{i=0}^{m-2} \left(Q^D\right)^{i+3} P^{i+1} Q$$
$$= Q^D + \left(Q^D\right)^2 P + \left(Q^D\right)^3 PQ \quad ,$$

$$\left(Q^D\right)^2 P = \begin{pmatrix} A^D & \left(A^D\right)^2 B \\ \psi_0 & D^D + \psi_1 \mathrm{B} \end{pmatrix} \begin{pmatrix} \left(A^D\right)^2 BD^\pi C & 0 \\ \psi_1 BD^\pi C & 0 \end{pmatrix}$$
$$= \begin{pmatrix} \left(A^D\right)^3 BC + \left(A^D\right)^2 \psi_1 BC & 0 \\ \psi_0 \left(A^D\right)^2 BC + D^D \psi_1 BC + \psi_1 B \psi_1 BC & 0 \end{pmatrix}.$$

By $BD = 0$, $BD^D = 0$, $B\psi_1 = 0$, so we have

$$\left(Q^D\right)^2 P = \begin{pmatrix} \left(A^D\right)^3 BC & 0 \\ \psi_0 \left(A^D\right)^2 BC + D^D \psi_1 BC & 0 \end{pmatrix}.$$

By $B\psi_0 = 0$, $B\psi_1 = 0$, $BD^D = 0$, we get

$$\left(Q^D\right)^3 P = \begin{pmatrix} A^D & \left(A^D\right)^2 B \\ \psi_0 & D^D + \psi_1 \mathrm{B} \end{pmatrix} \begin{pmatrix} \left(A^D\right)^3 BC & 0 \\ \psi_0 \left(A^D\right)^2 BC + D^D \psi_1 BC & 0 \end{pmatrix}$$
$$= \begin{pmatrix} \left(A^D\right)^4 BC & 0 \\ \psi_0 \left(A^D\right)^3 BC + D^D \psi_0 \left(A^D\right)^2 BC + \psi_0 D^D \psi_1 BC & 0 \end{pmatrix},$$

and

$$\left(Q^D\right)^3 PQ = \begin{pmatrix} \left(A^D\right)^4 BC & 0 \\ \psi_0 \left(A^D\right)^3 BC + D^D \psi_0 \left(A^D\right)^2 BC + \psi_0 D^D \psi_1 BC & 0 \end{pmatrix} \begin{pmatrix} A & BD^\pi \\ DD^D C & \mathrm{D} \end{pmatrix}$$
$$= \begin{pmatrix} \left(A^D\right)^4 BCA & \left(A^D\right)^4 BCB \\ \Delta DD^D \mathrm{C} & \Delta \mathrm{BD}^\pi \end{pmatrix}.$$

where

$$\Delta = \psi_0 \left(A^D\right)^3 BC + D^D \psi_0 \left(A^D\right)^2 BC + \psi_0 D^D \psi_1 BC \, .$$

So we have

$$
\begin{aligned}
(P+Q)^D &= Q^D + \left(Q^D\right)^2 P + \left(Q^D\right)^3 PQ \\
&= \begin{pmatrix} A^D + \left(A^D\right)^3 BC + \left(A^D\right)^4 BCA & \left(A^D\right)^2 B + \left(A^D\right)^4 BCB \\ \psi_0 + \psi_0\left(A^D\right)^2 BC + D^D \psi_1 BC + \Delta D D^D C & D^D + \psi_1 B + \Delta B D^\pi \end{pmatrix}.
\end{aligned}
$$

This result generalizes the result under the condition $BD = 0$, $D^\pi CB = 0$, $D^\pi CA = 0$ in [4]

References

[1] Campbell S L, Meyer C D, Jr. Generalized Inverses of Linear Transformations. London: Pitman Press, 1979.

[2] Campbell S L. The Drazin inverse and systems of second order linear differential equations. Linear Multili -near Algebra, Vol. 14 (1983), 195-198.

[3] D. S. Djordjevic, Y. Wei,, Additive results for the generalized Drazin inverse, J. Aust. Math. Soc. Vol. 73 (2002), 115 -126,.

[4] Dopazo E, Matrinez-Serrano M F. Further results on the representation of the Drazin inverse of a 2×2 block matrix. Linear Algebra Appl, Vol. 43 (2010), 1896-1904.

[5] Li Guo, Xiankun Du. Representations for the Drazin inverses of 2×2 block matrices. Applied Mathematics and Computation, Vol. 217 (2010), 2833-2842.

AUTHOR INDEX

AUTHOR INDEX

AUTHOR INDEX

AUTHOR INDEX

AUTHOR INDEX